AF372080

The Mitchell Beazley Joy of Knowledge Library

Fact Index
L-Z

Scientiam non dedit natura semina scientiae nobis dedit
"Nature has given us not knowledge itself, but the seeds thereof."
Seneca

The Joy of Knowledge Encyclopaedia is affectionately
dedicated to the memory of John Beazley, 1932–1977,
Book Designer, Publisher, and Co-Founder of the
publishing house of Mitchell Beazley Limited, by all
his many friends and colleagues in the company.

The Joy of Knowledge Library

General Editor: James Mitchell
With an overall preface by Lord Butler, Master of Trinity College,
University of Cambridge

Science and The Universe
Introduced by Sir Alan Cottrell,
Master of Jesus College,
University of Cambridge; and
Sir Bernard Lovell, Professor of
Radio Astronomy, University of
Manchester

The Physical Earth
Introduced by Dr William
Nierenberg, Director, Scripps
Institution of Oceanography,
University of California

The Natural World
Introduced by Dr Gwynne
Vevers, Assistant Director of
Science, the Zoological Society
of London

Man and Society
Introduced by Dr Alex Comfort,
Professor of Pathology, Irvine
Medical School, University
of California, Santa Barbara

History and Culture 1
History and Culture 2
Introduced by Dr Christopher
Hill, Master of Balliol College,
University of Oxford

Man and Machines
Introduced by Sir Jack Callard,
former Chairman of Imperial
Chemical Industries Limited

The Modern World
Introduced by Dr Michael Wise,
Professor of Geography,
London School of Economics
and Political Science

Fact Index   A–K

Fact Index   L–Z

The Mitchell Beazley Joy of Knowledge Library

Fact Index
L-Z

MITCHELL BEAZLEY

The Joy of Knowledge Library

Editorial Director	**Frank Wallis**
Creative Director	**Ed Day**
Project Director	**Harold Bull**

Volume editors	
Science and The Universe	John Clark
	Lawrence Clarke
The Natural World	Ruth Binney
The Physical Earth	Erik Abranson
	Dougal Dixon
Man and Society	Max Monsarrat
History and Culture 1 & 2	John Tusa
	Roger Hearn
Time Chart	Jane Kenrick
Man and Machines	John Clark
The Modern World	John Clark

Art Director	Anthony Cobb
Production Manager	Graham Darlow
Co-editions Manager	Averil Macintyre
Printing Manager	Bob Towell

Fact Index	
Chief Editors	John Clark
	Stephen Elliot (USA)
	Stanley Schindler (USA)
Senior Editors	Michael Darton
	Penny Davies
Editors	Lalit Adolphus
	Don Binney
	Arthur Butterfield
	Sam Elder
	Rupert Firth
	Fern Fraser
	Peter Furtado
	Judy Garlick
	Jane Garton
	Ann Kramer
	Ian Lee
	Lloyd Lindo
	Anthony Livesey
	Jean MacNamee
	John Smallwood
	Peter Smith
	Jim Sommerville
	Tony Spalding
	Robert Stewart
	Dian Taylor
	Barbara Westmore
Assistant Editors	Joyce Evison
	Ken Hewis
	Valerie Nicholson
Editorial Assistants	Ashley C. Brown
	Martin Elliot
	Margaret Kennedy
	Mary Kennedy
	Martyn Page
	Christopher D. Parker
	Judith Ravenscroft
	Hal Robinson
	Stephanie Stuart
Proof Readers	Peggy Starling
	Ursula Anderson
	Gillian Beaumont
	John Boxcer
	Paul Brewin
	Paul Dinnage
	Graeme Ewens
	Ann Ralph
	Vernon Robinson
	John Savory
Picture Researcher	Flavia Howard
Art Buyer	Ted McCausland
Production Controllers	Barbara Smit
	Anthony Bonsels
Production Assistants	Alex Laird
	Michael Alexander
	John Taylor

The Joy of Knowledge Encyclopaedia
© Mitchell Beazley Encyclopaedias Limited 1976

The Joy of Knowledge Fact Index L–Z
© Mitchell Beazley Encyclopaedias Limited 1978

Artwork © Mitchell Beazley Publishers Limited
1970, 1971, 1972, 1973, 1974, 1975 and 1976
© Mitchell Beazley Encyclopaedias Limited 1976, 1977
© International Visual Resource 1972

All rights reserved
No part of this publication may be reproduced,
stored in a retrieval system, or transmitted
in any form or by any means, electronic,
mechanical, photocopying, recording or otherwise,
without the prior written permission of the
proprietors of the Joy of Knowledge
Encyclopedia™ System

THE JOY OF KNOWLEDGE is the trademark of
Mitchell Beazley Encyclopaedias Limited, London,
England

ISBN 0 85533 114 3

Typesetting by G. A. Pindar and Son Ltd., England

Printed in England by Balding + Mansell

L

L, twelfth letter of the alphabet. It can be traced to the Semitic letter *lamedh,* which passed into Greek as *lambda.* It became slightly modified in the Roman alphabet and in this form has passed into English. It is nearly always pronounced as in *love,* although it is silent in such words as *could* and *calf. See also* MS pp.244–245.

Laas, Ernest (Ernst) (1837–85), German philosopher. He studied at Berlin and was a professor of philosophy at Strasbourg. In *Idealism and Positivism* (1879–84) he contrasted TRANSCENDENTALISM and POSITIVISM.

Laban, in the Bible, brother of REBECCA and father of Leah and RACHEL. He was instrumental in arranging Rebecca's marriage to ISAAC. He also tricked JACOB into marrying Leah before he could marry Rachel.

Laban, Rudolph von (1879–1958), German modern dance theorist. With Mary WIGMAN and Kurt JOOSS, he was the inspiration behind the German school of modern dance. He invented Labanotation, a system of symbols for the written notation of dance.

Labé, Louise (*c.*1524–66), French poet. Her elegies and sonnets in *Oeuvres* (1555) are passionate expressions of the joy and pain of love. She was nicknamed "La belle Cordière" because of her marriage to a rich rope-merchant.

Labiatae, the MINT family, a group of plants that includes up to 3,500 species, many of which are familiar herbs, found throughout the world. Most have square, non-woody stems and simple leaves. They include DEAD NETTLE, THYME, SAGE, BASIL, MARJORAM, ROSEMARY, LAVENDER, BETONY, and CATMINT.

Labiche, Eugène Marin (1815–88), French playwright who wrote more than 100 light comedies and farces, sometimes in collaboration with other writers such as Marc Michel. His most famous work is *Un Chapeau de Paille d'Italie* (1851; *An Italian Straw Hat*) which has been frequently revived.

Labium, in anatomy, term meaning a lip or lip-shaped organ. In arthropods it refers to the lower lip, which may be modified in insects to form sucking mouthparts. In human anatomy it denotes the labia majora and labia minora, skin folds of the female genitals. In botany it refers to the lower lip of plants of the Labiatae family.

La Bohème. *See* BOHÈME, LA.

Labor Day, public holiday observed in the USA and Canada on the first Monday in September. It was first celebrated by the Knights of Labor in New York City in 1882. Its counterpart in some other countries is generally held on 1 May.

Labouchère, Henry Du Pré (1831–1912), British publicist and politician. Before rising to fame with dispatches in the *Daily News* from Paris about the Franco-Prussian War (1870–71), he joined the British diplomatic corps (1854–64) and was a Liberal MP (1865; 1867–68). As a Radical MP (1880–1906) he proposed the abolition of the House of Lords. He founded (1877) a periodical, *Truth,* dedicated to the exposure of frauds.

Labour, in childbirth, stages in the delivery of the foetus at the end of pregnancy. The first stage is measured from the onset of contractions that occur at about 20-minute intervals lasting at least 30 seconds each and becoming more frequent as labour progresses. During this stage the cervix dilates up to 10cm (4in). The second stage may last from just a few minutes to two hours, at the end of which the infant is born. During the third stage, which begins about 15 minutes after birth, the PLACENTA (afterbirth) is delivered. *See also* MS pp.78–79.

Labourers, Statute of (1351), act to control wages in England. It was passed because of fears that the labour shortage caused by the Black Death of 1348–49 would send wages soaring. Wages were pegged at pre-plague levels and landless labourers were compelled to remain with their existing masters for as long as they were wanted.

Labour Exchanges, later called Employment Exchanges, centres established by the British Parliament in 1909 to advise employers of available labour and workers of available jobs. The act was the work of Winston CHURCHILL, then President of the Board of Trade in ASQUITH's Liberal government. When NATIONAL INSURANCE for the unemployed was introduced in 1911, the exchanges became the centres for the register of the unemployed and for the payment to them of their benefits.

Labour Party, social democratic political party, traditionally closely linked with the trade union movement. There are Labour parties in many countries, such as Australia (where it is known as the Labor Party), Britain and New Zealand. The first British socialist parties, founded in the 1880s, united in the Independent Labour Party (ILP) in 1893, whose president was Keir HARDIE – the first socialist Member of Parliament. The ILP created the Labour Representation Committee in 1900, which was renamed the Labour Party in 1906. It won 40 seats in the 1910 elections, mostly in the mining and industrial areas. Arthur HENDERSON, then leader of the Party served in LLOYD GEORGE's coalition government of WWI, and in 1922 the Party became the second largest in Parliament. Labour formed a brief minority government in 1924 under Ramsay MACDONALD and again in 1929–31, but following a coalition with the Liberals (1931) the Party was split and defeated at the polls. Labour joined the wartime coalition of WWII and, after a landslide electoral victory in 1945, introduced a series of reforms under Clement ATTLEE and Ernest BEVIN, which included nationalization of some major industries, establishment of a national health service and improved social security benefits. The Party lost the election of 1951 but became readjusted to post-war prosperity under Hugh GAITSKELL. In 1963 Harold WILSON defeated George BROWN and James CALLAGHAN in the leadership election. Labour won the General Election of 1964 and continued in power until 1970, when it lost to the CONSERVATIVE PARTY. Labour was re-elected in March 1974 and, in the elections of October 1974, were again elected but with a reduced majority. Wilson resigned in March 1976 and Callaghan was elected party leader. In March 1977, following erosion of the Labour majority, the LIBERAL PARTY (led by David STEEL) formally agreed to support Labour bills in Parliament – an arrangement known as the "Lib-Lab Pact". *See also* HC2 pp.210, 224, 236, 238, 248, 250.

Labrador, region in NEWFOUNDLAND, E Canada; bordered by NE Quebec (W and S) and the Atlantic Ocean (E). The coastal region is indented with fjords, the mountains becoming increasingly higher toward the N. The inland plateau is heavily forested, with innumerable lakes and rivers, notably the Churchill River, which drains into Lake Melville. The coast was known to the Norsemen *c.*950, and was visited by John CABOT in 1498 and CORTE-REAL in 1500. It passed to Britain by the Treaty of Paris in 1763 although the boundaries between Newfoundland and Quebec were under dispute (1809–1927). It became part of Canada in 1949. Industries: timber, fishing, mining. Area (Labrador proper): 292,220sq km (112,826sq miles).

Labrador retriever, strongly built hunting DOG, originally used for retrieving and developed in Newfoundland in the 19th century. It has a wide head, long, powerful jaws and a wide muzzle; the ears are low-set and pendulous. The short, wide body is set on medium-length legs and the distinctive tail is thick at the base, tapering to a point. The hard, dense coat is short and may be black, yellow or chocolate. Height: to 62cm (24.5in) at the shoulder; weight: 34kg (75lb).

La Bretonne, Restif de. *See* RESTIF NICOLAS EDME.

La Bruyère, Jean de (1645–96), French satirist. He ridiculed the injustice and hypocrisy in French life, particularly in his best-known work, *The Characters of Theophrastus, Translated from the Greek, with the Characters and Mores of This Age* (1699). In 1693 he was admitted to the Académie Française.

Labuan, island off the NW coast of Borneo, Indonesia. It was made a British Crown colony in 1848. In 1946 it was incorporated into North Borneo which, in 1963, became part of Malaysia. The most important town is Victoria. Industries include rice, rubber and fruits. Area: 98sq km (38sq miles). Pop. (1970 est.) 7,200.

Laburnum, any of several Eurasian shrubs and small trees, especially the common Laburnum, *Laburnum anagyroides,* which is commonly cultivated for its large drooping clusters of bright yellow flowers. It bears pods, each of which contains round, black, poisonous seeds. The wood is valued in cabinet-making for its green or red colour. Family Leguminosae.

Labyrinth, in architecture, an intricate structure of chambers and passages, generally constructed with the object of confusing anyone within it. Like a maze, it may be covered or uncovered, or composed of tall hedges. In Greek mythology MINOS had a labyrinth which DAEDALUS built to confine the monstrous MINOTAUR. In Egypt Amenemhet III built a multifunctional building in the form of a labyrinth near Lake Moeris.

Lac, name of an insect and the sticky substance it secretes and deposits on the twigs of trees; the deposit is harvested in Asia for use in SHELLAC and red lac dye. As many as 90,000 insects are required to produce 450g (1lb) of shellac. Species *Laccifer lacca.*

Lacaille, Nicolas-Louis de (1713–62), French astronomer noted for mapping the constellations which can be seen from the Southern Hemisphere, and for naming many of them. In 1750 he led an expedition to the Cape of Good Hope where, over a period of two years, he mapped the positions of 10,000 stars.

Laccolith, dome of intrusive IGNEOUS ROCK formed over the strata which it has penetrated. The base is typically horizontal while the upper surface is convex. They are generally less than 16km (10 miles) in diameter with thicknesses of 30–914m (100–3,000ft), and contain more acidic than basic rocks. *See also* PE pp.*30, 101.*

Lacemaking, manufacture of lace, a patterned openwork fabric for ornamental use made from fine thread of linen, cotton or silk. The two chief styles are known as needlepoint lace and bobbin- or pillow-lace, the names of which describe the techniques by which it is made. Machine lace-making imitates these two techniques. Needlepoint lacemaking developed from embroidery in the European Renaissance, and employs a single thread and embroidery stitches to create the pattern attached to a ground material by foundation stitches (which are cut when the work is completed). The best examples are found in Italian, especially Venetian, 16th-century work, which was later copied and developed as pins became readily available and cheap. It uses many threads, pinned to a pillow, with bobbins attached to each loose end so that the threads can be crossed, braided, twisted or woven to create the pattern. In the 17th century Flemish bobbin-lace developed to compete with and, by the 18th century, to surpass the needlepoint designs. The first factory for machine-lace was established in Nottingham in 1846, and the city is still the principal centre for lace-making in Britain.

Lacewing, any of numerous species of NEUROPTERAN insects, especially members of the families Chrysopidae and Hemerobiidae, which are distributed throughout the world. Common green lacewings have long slender antennae, a slender greenish body and two pairs of delicate, lacy, veined wings. Brown lacewings are generally smaller. The larvae of some species drain body fluids from aphids. Length: to 7cm (2.8in). *See also* NW pp.106–107, *107, 228.*

Labour exchange official noting details of unemployed workers during the 1930s.

Labour party British Prime Minister James Callaghan studies an EEC document.

Labradors are used for hunting in hills and near lakes and marshes.

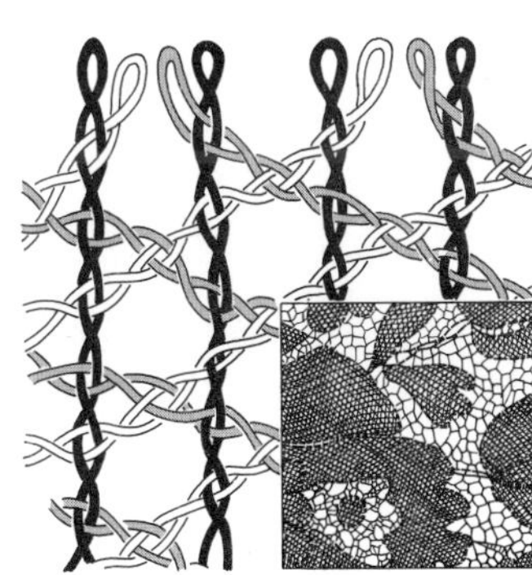

Lace is made by the winding together of single threads of cotton or silk.

Lacquer; these Venetian boxes show the intricate detail possible in lacquerware.

Lacrosse players may carry the ball behind the goal, unlike in most field games.

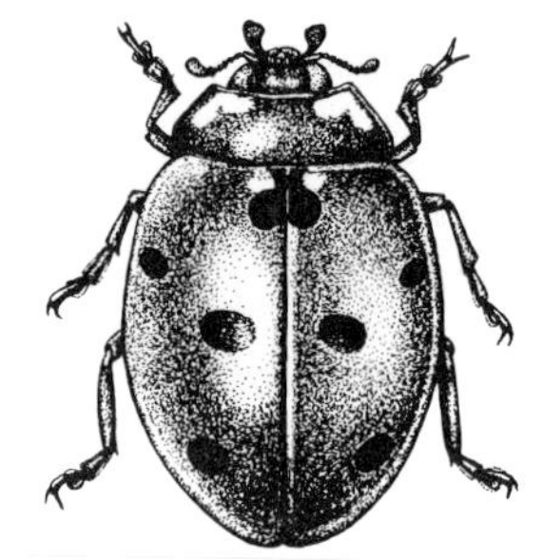

Ladybird, a type of beetle, was named after the Blessed Virgin Mary.

Lafayette fought for American colonists and opposed Napoleon's rule in France.

Lachish, ancient city of Palestine, in modern Israel. It was in existence from 3200 BC, but, inhabited by the Amorites, was destroyed by Joshua. Lachish was besieged by the Assyrians in 701 BC, and destroyed in 589 BC. The site was excavated in the 1930s. *See also* HC1 p.*56.*

Lachmann, Karl Konrad Friedrich Wilhelm (1793–1851), German philologist, the founder of modern textual criticism. He was a classicist and published editions of LUCRETIUS' *De Rerum Natura* (1850) and of the New Testament (1831, 1842–50). His best work, however, is perhaps his editions of medieval German texts, particularly the *Niebelungenlied.*

Lachrymal gland, organ in the eye that produces tears. It is located in the orbital cavity in a slight depression, and is controlled by autonomic nerves; it produces slightly germicidal tears that flow through ducts to the surface of the eye to lubricate it. *See also* MS pp.48–49.

Laclos, Pierre Ambroise François Choderlos de (1741–1803), French novelist and general. His one important work, *Les Liaisons Dangereuses* (1782), a cynical and savage attack on the corruption of aristocratic society, caused a sensation and was only belatedly recognized as a great work of 18th-century epistolary fiction. *See also* HC2 p.35.

La Condamine, Charles Marie de (1701–74), French explorer and geographer. In 1735 he went to Peru and made the first scientific exploration of the AMAZON, bringing back CURARE and information about rubber. He published his findings in 1751.

La Coruña, Battle of (1809), British victory over the French in the PENINSULAR WAR which allowed the British to embark from Spain. The French, under Marshal SOULT, suffered a loss of 2,000; the British losses were 800, including the commander, Sir John MOORE.

Lacquer, paint that forms a film by loss of solvent through evaporation. The film is usually composed of a cellulose derivative (such as nitrocellulose) in combination with an alkyd resin; the solvent may be a ketone (such as methylbutyl ketone), an alcohol, or some other cellulose solvent. It is also the varnish used in lacquerware.

Lacrosse, ball game that originated among the Iroquois Indians of Canada and the USA, played on a field 100m by 64m (110yd × 70yd), by teams of 10 or 12 players. They carry sticks with an adjustable meshwork head (once thought to resemble a bishop's crozier, or *crosse*). The ball may be carried, passed, hit or kicked, but only the goalkeepers, defending goals 1.8m (6ft) square, are allowed to handle it. Players can be sent off the field for up to three minutes for infractions of the rules, which were compiled in 1860 in Canada. Lacrosse became Canada's national game in 1867. Today it is becoming increasingly popular in the NE USA and is also played in Britain, mostly by women.

Lactantius Firmianus (*c.*240–*c.*320), early Christian writer, *b.* N Africa. He taught rhetoric at Nicomedia in Asia Minor, was converted to Christianity and later became tutor to the son of Emperor Constantine the Great. He wrote *The Divine Precepts,* a strong defence of Christian doctrine.

Lactation, secretion of milk to feed the young. In women, after hormonal-induced breast enlargement during pregnancy, prolactin (a pituitary hormone) stimulates breast cells to begin secreting milk. The milk "comes in" the breast immediately after the birth of the baby. Its flow is stimulated by suckling which, in turn, triggers neural and hormonal changes that control and maintain lactation. *See also* MS pp.76–79.

Lactic acid, colourless syrupy liquid (CH₃CHOHCOOH) formed from LACTOSE in milk by the action of bacteria. It is used in foods and beverages, in tanning, dyeing, and adhesive manufacture. Properties: s.g. 1.2; m.p. 18°C (64.4°F); b.p. 122°C (251.6°F).

Lactose, or milk sugar, DISACCHARIDE of particular importance during cheesemaking, when lactic bacteria turn it into LACTIC

ACID, so souring the milk selectively in the production of cheese curd. The lactose molecule is made up of a molecule of GLUCOSE linked to a molecule of galactose.

Lacy family, Anglo-Irish family. Hugh de Lacy, 1st Lord of Meath (*d.*1186) led the conquest of Ireland under HENRY II and was governor of the province (1177–81, 1185–86). His son died Hugh (died *c.*1242) was the first Anglo-Saxon peer in Ireland, created 1st Earl of Ulster in 1205. Henry de Lacy, 3rd Earl of Lincoln (*c.*1249–1311), was a counsellor of EDWARD I and EDWARD II. He commanded the army in France (1296–98) and ruled England in Edward II's absence in 1310.

Lada, in ancient Slavic religion and mythology, the goddess of beauty.

Ladd, Alan (1913–64), US film star. His first film appearance was in 1932, but he was offered no major roles before the 1940s, perhaps because he was considered too short to be a male star. He became popular playing in thrillers such as the *Blue Dahlia* (1946), and his part in the western *Shane* (1953) attracted considerable critical acclaim.

Ladino, language spoken by Sephardic Jews, mostly in the Middle East. It is a combination of medieval Castilian, Hebrew, Greek and Turkish. Also known as Judaeo-Spanish, it is spoken today by about 150,000 people.

Ladislas (Ladislaus), name of six kings of Hungary. Among the most notable were: Ladislas I (*c.*1040–95), who became king in 1077 and is known as St Ladislas. He re-established order in Hungary after the barbarian invasions, created a popular code of laws and subjugated Bosnia, Croatia and part of Transylvania. Ladislas IV (1262–90) became king as an infant. He was a pagan, and defeated the Bohemians in 1278 with HAPSBURG help, but was himself defeated by the CUMANS and Tatars. Ladislas V (1440–57), who came to the throne in 1444, was brought up in Austria. János HUNYÁDI served as his regent.

Ladislas, name of four kings of Poland. Ladilas I Lokietek (1260–1333) became Duke of Poland as Ladislas IV in 1296, and after a series of wars to unite Greater and Little Poland (1305–14), he was crowned king in 1320. Ladislas II Jagello (*c.*1350–1434) was Grand Duke of Lithuania (1377–86) and was elected King of Poland in 1386. He was converted to Roman Catholicism, and defeated the TEUTONIC KNIGHTS at Tannenburg in 1410. Ladislas III (1424–44) became King of Poland in 1434 and of Hungary in 1440. He led a crusade against the Turks in 1444, but was killed at the Battle of VARNA. Ladislas IV (1595–1648) became king in 1632. He fought the Russians, Turks and Swedes successfully, but part of the Ukraine was lost to Russia after the 1648 Cossack revolt.

Ladoga, Lake (Ladozskoje Ozero), lake in NW European USSR, drained by the Neva River. Formerly divided between Finland and the USSR, it has been entirely within the border of the USSR since 1940. During WWII, its frozen surface was the "road" over which supplies were taken to Leningrad in the winter months. Area: 17,702sq km (6,835sq miles).

Ladrones. *See* MARIANAS ISLANDS.

Lady, in British nobility, a general title used to indicate rank. It may be formally applied to the wife of a KNIGHT, BARON or BARONET. It may be used semi-formally for a marchioness, countess or viscountess or for a baroness in her own right. As a title of courtesy, it may apply to the daughter of a DUKE or MARQUESS, or to the wife of a younger son of a duke or marquess.

Ladybird, any of a large number of small, brightly-coloured beetles; most common species are red with conspicuous black spots and a black and white head. Ladybirds are regarded as useful because their diet consists primarily of APHIS. If attacked or disturbed, they may exhibit reflex discharge from their leg joints as a deterrent to predators. Family Coccinellidae. *See also* NW pp.*34,* 113, *114.*

Lady Chatterley's Lover (written 1926–27) first published in Paris in 1929, novel by D. H. LAWRENCE. The impotence of Sir

Clifford Chatterley, confined to a wheelchair, symbolizes the sterility of modern life which his wife, Connie, overcomes through her affair with the gamekeeper Mellors. The explicit language used to describe this affair led to the suppression of the unexpurgated text, which was not published in Britain until 1960 after a celebrated court case.

Lady Day, in the Christian calendar, the day of the archangel's Annunciation to the Virgin MARY, traditionally celebrated on 25 March.

Ladyfern, feathery fern that grows in temperate areas of the world in moist, shady places. The 75cm (30in) leaves are 25cm (10in) wide and grow in circular clusters. Height: to 91cm (36in). Family Aspidiaceae; species *Athyrium filix-femina.*

Lady of Shalott, The (1833), poem by Alfred, Lord TENNYSON describing the legend of the Lady of Shalott who, tempted by passion, leaves her world of artifice, invoking the curse of CAMELOT.

Lady of the Lake, The (1810), poem in six cantos by Sir Walter SCOTT. Set in 16th-century Scotland, it tells of Ellen, who secures the king's pardon for her outlawed father.

Lady's-eardrop. *See* FUCHSIA.

Ladysmith, Siege of (1899–1900), episode in the SOUTH AFRICAN (BOER) WAR. Sir George White's British force was hemmed in by the Boers at Ladysmith from 30 October to 28 February, when it was relieved by an expeditionary force led by General BULLER.

Lady's smock, also called cuckooflower, erect Eurasian perennial flowering plant, commonly found in moist meadows. It has a stout stem, fine leaves and clusters of pink flowers. Species *Cardamine pratensis. See also* NW p.*116.*

Lady's Not for Burning, The (1948), three-act blank verse comedy by Christopher FRY. In a small English market town *c.*1450, Thomas Mendip demands to be hanged for no apparent crime – until he falls in love with a girl accused of being a witch. It is the first of a series of seasonal plays representing springtime; others in the series represent autumn and winter.

Lady Windermere's Fan (1892), four-act play by Oscar WILDE. In the tradition of Restoration comedy it portrays the domestic situation of Lord and Lady Windermere. The play was ahead of its time in its sympathetic picture of the fallen woman, the mysterious Mrs Erlynne.

Lae, town in Papua New Guinea, on W New Guinea Island, on the Huon Gulf approx. 322km (200 miles) N of Port Moresby. It was founded in 1927 to serve the gold-fields in the interior. Pop. (1970 est.) 24,300.

Laemmle, Carl (1867–1939), US film pioneer, *b.* Germany. The founder of UNIVERSAL PICTURES (1912), he fought successfully to win patents for independent production companies. His major films included *Foolish Wives* (1921) and *All Quiet on the Western Front* (1930).

LaFarge, Oliver (1901–63), US writer and anthropologist. He used his archaeological and ethnological expeditions to Arizona, Guatemala and Mexico as the inspiration for his stories and novels. Among his works are *Laughing Boy* (1929), a perceptive study of Navaho life which won a Pulitzer Prize, and *Sparks Fly Upward* (1931).

Lafayette, Marie Joseph Paul Yves Roch Gilbert du Motier, Marquis de (1757–1834), French general, statesman and hero of the American cause, who arrived in Philadelphia in 1777 and was commissioned a major-general. In 1781 he distinguished himself in the Yorktown campaign that led to CORNWALLIS's surrender. He returned to France in 1781, where he later became first a member of the States-General then of the National Assembly. After the Fall of the Bastille (1789) he was appointed commander of the National Guard but lost all popular support when in 1791 he ordered his troops to fire on a crowd petitioning for the abolition of the monarchy. He was given military commands in 1792, but deserted to the Austrians in August. He lived in retirement under NAPOLEON, but at the Restoration (1814) became a member of the Chamber of Deputies. He made a triumphant visit to the USA (1824–25) and in France in 1830 he played a major role in the July Revolution, his prestige being largely responsible for the installation of Louis Philippe as king of the French. *See also* HC2 p.*74*.

LaFayette, Marie Madeleine de La Vergne, Comtesse de (1634–93), French novelist. After an unhappy marriage, she went to Paris and later formed a literary circle with LA ROCHEFOUCAULD. Her short novel *The Princess of Cleves* (1678) is a melancholy study of suffering and self-analysis. *See also* HC1 pp.*316*, 317.

Lafitte, Jean (*c.* 1780–*c.*1826), pirate and smuggler. The leader of a large pirate band on the Gulf of Mexico, he preyed on Spanish commerce, which he disposed of through his brother in New Orleans. He received a pardon from James Madison, the US President, for leading his men against the British in the latter part of the WAR OF 1812. After the war, he and his pirates lived on an island that is present-day Galveston. He was unmolested until some members of his colony raided American property in 1820. The US Government despatched a naval force against him; Lafitte and his closest associates escaped and were not heard of again.

Lafontaine, Henri-Marie (1854–1943), Belgian lawyer who was awarded the Nobel Peace Prize in 1913. As president of the International Peace Bureau (1907–43) he was influential in its attempts to bring about the Hague peace conference of 1907. He was also a member of the Belgian delegation to the League of Nations Assembly (1920–21). Continually working for world peace, he founded the Centre Intellectuel Mondial, and proposed a world parliament and an international court of justice. His many works on law and peace include *The Great Solution* (1916).

La Fontaine, Jean de (1621–95), French poet noted for his fables, which are considered among the greatest masterpieces of French literature. His *Fables choisies, mises en vers* (1668–94), consist of 12 books featuring some 240 fables. He wrote other poems and dramatic pieces but, apart from *Les Amours de Psiché et de Cupidan* (1669), they do not compare in quality with the *Fables*.

Lafontaine, Sir Louis Hippolyte (1807–64), joint Prime Minister of Canada (1842–43 and 1848–51). He was a lawyer and French-Canadian nationalist, and was cool toward the rebellion of 1837. He led the moderate wing of French reformers in the legislature (1841–51) and formed two administrations with Robert BALDWIN (1842–43 and 1848–51). He was chief justice of Lower Canada (1853–64).

Laforgue, Jules (1860–87), French poet and literary critic, *b.* Uruguay. Obsessed with the desire to be original, and one of the first French poets to use free verse, he brought poetry back to the rhythms of everyday speech in volumes such as *Les Complaintes* (1885) and *L'Imitation de Notre Dame de la Lune* (1886). He influenced the early poetry of T.S. ELIOT and Ezra POUND.

Lagash, city of ancient Sumeria. The site was first settled in the Ubaid Period (*c.* 5200–3500 BC) and by 3000 BC Lagash had grown into a flourishing city. Its authority reached its height under King Urukagina (*c.* 2300 BC). After the fall of Akkad in *c.* 2230 BC, Gudea, King of Lagash, restored the prosperity of Sumeria, leaving many inscriptions of the religious and daily life of his times. Lagash was excavated from AD 1877. *See also* HC1 p.*30*.

Lagerkvist, Pär Fabian (1891–1974), Swedish writer whose characteristic tone was of radical pessimism at man's inhumanity to man and the impotence of man in the face of death. Major works include *Anguish* (1916), *The Eternal Smile* (1920) and *The Hangman* (1933). International recognition came with *The Dwarf* (1944) and *Barabbas* (1950), for which he was awarded the 1957 Nobel Prize in literature.

Lagerlöf, Selma (1858–1940), Swedish novelist and pacifist. Much of her writing was inspired by legends, sagas and traditional tales of her homeland in w Sweden. Her *Story of Gösta Berling* (1891), although not at first a success, became immensely popular. Her greatest novel, *Jerusalem* (2 vols, 1901, 1902), was inspired by a visit to Palestine. She also wrote short stories and children's books and was awarded the 1909 Nobel Prize in literature.

Lagomorpha, order of terrestrial mammals that includes RABBITS and HARES, both of which have long ears, a short tail and hindlimbs developed for bounding locomotion. Also included are PIKAS, which, in contrast, have shorter ears, no tail and hindlimbs developed for scampering locomotion.

Lagoon, shallow stretch of sea-water protected from waves and tides by a strip of land or coral. *See also* PE p.*123*.

Lagos, capital city of Nigeria, in the sw of the country, on the Atlantic coast approx. 121km (75 miles) sw of Ibadan. The city grew as a YORUBA settlement from the 17th to the 19th centuries, coming under British control in 1861 after years of Portuguese exploitation through the slave trade. It became the capital of independent Nigeria in 1960. Industries include brewing, ship repairing and textiles. Pop. (1975 est.) 1,060,850.

Lagrange, Joseph Louis (1736–1813), French mathematician. He was educated at Turin University, and became professor of mathematics there (1756–66) before his appointment as head of the Berlin Academy. He created the calculus of variations, devised a mathematical analysis of perturbations in gravity, and made contributions in many other areas, including the mathematics of sound and mechanics. He wrote *Analytical Mechanics* (1788).

Lagrangian points, positions in space where a small celestial body, influenced by the gravitational attraction of two larger ones, tends to remain at rest relative to the other two. In such a system, the most stable positions are the two points where the small body occupies one vertex of an equilateral triangle with the larger bodies at the other two, a theory originally postulated by J. L. LAGRANGE in 1772.

La Guardia, Fiorello Henry (1882–1947), US lawyer and political leader. As mayor of New York City (1934–45) he conducted a vast programme of reform, reduced political corruption and introduced slum clearance projects.

La Habana. *See* HAVANA.

La Harpe, Frédéric César de (1754–1838), Swiss politician who won the independence of the Vaud district of Switzerland from Berne. He represented Vaud at the Congress of Vienna.

La Hogue, Battle of (May–June 1692), naval battle between a combined Anglo-Dutch fleet, led by Admiral Edward Russell, and the French fleet, led by Admiral Tourville, off the Normandy coast. After five days, the French fleet was almost destroyed and lost all hope of invading England and restoring the exiled JAMES II to the throne.

Lahore, city in E Pakistan, on the River Ravi near the Indian border. An important city during the Ghazni and Ghuri sultanates, it increased in importance as part of the Sikh kingdom in 1767 and passed to the British in 1849. It is an industrial centre and its products include iron, steel, textiles and rubber, gold and silver jewellery. Pop. 1,985,000.

Lahr, Bert (1895–1967) US comic actor. He appeared in music-hall, burlesque, plays and films and is especially remembered for his portrayal of the Cowardly Lion in *The Wizard of Oz* (1939), Estragon in *Waiting for Godot* (1956) and for his work in *Burlesque* (1947) and *Two on the Aisle* (1951).

Laibach, Congress of (1821), meeting of European powers that established Austria's right to intervene in Italian political affairs, notably rebellions in Piedmont and Naples. The congress was dominated by the HOLY ALLIANCE (Russia, Prussia and Austria).

Laine, Cleo (1927–), real name Clementina Dinah Dankworth, British jazz singer. She joined the Dankworth Orchestra in 1953, and married John DANKWORTH in 1960. She has won many awards for her singing, and appeared with many symphony orchestras. She sang at Sadler's Wells Opera in London in 1961.

Laing, Ronald David (1927–), Scottish psychiatrist. He holds that the mentally ill are not necessarily abnormal, and that a PSYCHOSIS may be a reasonable reaction to the stresses of the world. His books include *The Divided Self* (1960), *Politics of the Family* (1971) and *Do You Love Me?* (1976). *See also* HC2 p.*266*.

Laing's Nek, Battle of (1881), Boer victory over the British in the first of the SOUTH AFRICAN WARS. The British, under General Colley, tried to force the low pass called Laing's Nek at MAJUBA HILL on the Natal-Transvaal border. They were repulsed by the Transvaal force under General Joubert and lost 200 men.

Laird, John (1805–74), British shipbuilder and politician. He was a partner in his father's shipbuilding firm at Birkenhead, Merseyside, and was one of the early builders of iron ships.

Lairesse, Gerard de (1641–1711), Flemish painter of historical and mythological scenes. He was the author of the influential *Het Groot Schilderboeck* (1707; tr. *The Art of Painting*, 1738), which contained many of his etchings and his ideas on art.

Laissez-faire, 19th-century economic doctrine. In reaction to MERCANTILISM, the proponents of laissez faire adopted Adam SMITH's argument that trade and industry would best serve the interests of all by reducing government interference to a minimum and allowing market forces to determine production, prices and wages. Major advocates of laissez-faire included Richard COBDEN and John BRIGHT.

Laity, members of a Christian Church who are not part of the clergy. The distinction has greater significance in the Anglican, Roman Catholic and Orthodox Churches, because of their belief in the doctrine of APOSTOLIC SUCCESSION, than in Protestant Churches.

Lakas Bahas, Demetrio Basilio (1925–), Panamanian politician. He became President of the provisional government in 1969, and President of Panama in 1972.

Lake, Gerard, Viscount (1744–1808), British general. He served in the SEVEN YEARS WAR and the American War of Independence and in 1798 played a prominent role in suppressing the Irish Rebellion. In 1800 he was made Commander-in-Chief in India, where he defeated the MAHRATTAS (1801–06).

Lake, Veronica (1919–73), US film actress, real name Constance Keane. A fashionable leading lady of the early 1940s, she is chiefly remembered for her hairstyles – long and blonde, hanging over one eye. Her films include *I Wanted Wings* (1941), *Sullivan's Travels* (1941) and *I Married a Witch* (1942).

Lake, inland body of water, generally of considerable size and too deep to have rooted vegetation completely covering the surface, occupying a hollow in the Earth's surface. The expanded part of a river and a reservoir behind a dam are also termed lakes. *See also* PE pp.*112–113, 150–151*.

Lake District, region of NW England, in Cumbria, containing the principal English lakes. Its spectacular mountain and lakeland scenery and its associations with the poets William WORDSWORTH and Samuel COLERIDGE make it a major tourist attraction. Among its 15 lakes are DERWENTWATER, Grasmere, Buttermere and WINDERMERE. The highest point is Scafell Pike, 978m (3,210ft). The Lake District National Park was established in 1951. Area: 2,243sq km (866sq miles). *See also* PE p.*115*.

Lake dwellings, prehistoric settlements built on piles within the margins of lakes. A network of tree trunks across the piles formed a platform on which clay-floored huts were built. Cattle and sheep were also reared on the platform. Lake dwellings in Europe have been found in Germany, Switzerland and Italy. They also occurred in Ireland (where they were known as

Jean de La Fontaine based his *Fables* on the anecdotal moral animal tales of Aesop.

Pär Lagerkvist, Swedish writer who wrote several indictments of totalitarianism.

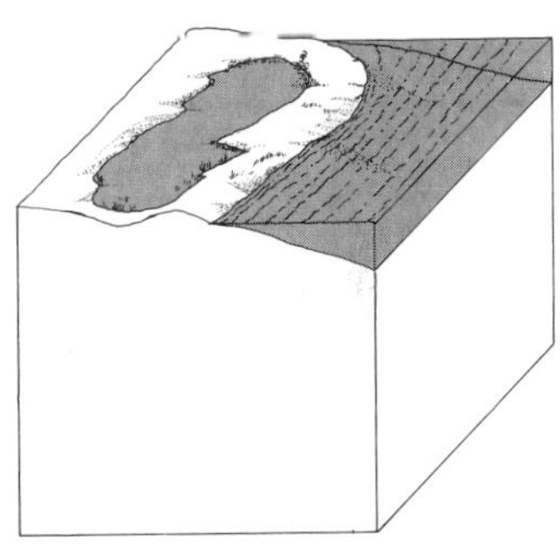

Lagoons are formed when a spit of land is built up by coral, sand deposits or mud.

Lagos; a view of the old house of representatives, now used by the Nigerian senate.

Lakeland terriers were first bred in the 18th century to hunt foxes in uplands.

Jean-Baptiste Lamarck was both a botanist and classifier of invertebrates.

Alphonse de Lamartine served in the provisional French government after 1848.

Lambeth Palace's Lollards Tower (shown here) was built in the mid-15th century.

CRANNOGS) and in Britain, as at Glastonbury. Some lake dwellings date back to Neolithic times, but most are of the Bronze Age.

Lake Erie, Battle of (1813), battle in the WAR OF 1812 in which the US fleet under Oliver Perry defeated the British fleet under Robert Barclay, near Put-In-Bay, Ohio. The victory gave the USA control of Lake Erie, cutting off supplies to British troops in the area.

Lakeland terrier, hunting DOG bred in England's Lake District. Its narrow build is suited to its original purpose – squirming through dens after foxes. It has a rectangular head and V-shaped folded ears. It has a wiry coat which is weather-resistant and may be tan, blue-black, red or beige, and the tail is commonly docked. It is less than 37cm (14.5in) in height.

Lake Poets, name given to three English poets who lived in the Lake District c.1800: William WORDSWORTH (1770–1850), Samuel Taylor COLERIDGE (1772–1834) and Robert SOUTHEY (1774–1843). Their poetry in itself does not have enough in common to justify calling them a school, but they shared close friendship and experienced a common political development.

Laker, James Charles ("Jim") (1922–), British cricketer, an off-spin bowler who played for Surrey, England and later Essex. He took 193 wickets at an average of 21.23 during a Test career which lasted from 1947 to 1959. This included a record 19 wickets for 90 runs against the Australians at Old Trafford in 1956.

Lakshadweep, Union Territory of, island group off the sw coast of India, in the Arabian Sea; administered by India since 1956. There are 20 islands of which ten are inhabited. Minicoy is the largest, with a predominantly Muslim population. Kavarrati is the capital. Area: 32sq km (12sq miles). Pop. 31,810.

Lakshmi, also known as Padma or Sita, in Hindu mythology, the lotus goddess, wife of VISHNU, who existed at the beginning of creation rising from the ocean borne by a lotus. Lakshmi, whose various forms related to Vishnu's successive incarnations, was the goddess of beauty and youth, and was also worshipped as goddess of wealth and good fortune. She is often depicted sitting or standing on the lotus, holding a lotus, or symbolically as the lotus.

Lalande, St Jean (d. 1646), French Jesuit missionary. In 1644 he went to Canada and two years later accompanied Father Isaac JOGUES on a mission to the MOHAWKS. Both men were murdered and in 1830 were canonized by Pope Pius XI.

Lalande, Joseph Jérôme Le Français de (1732–1807), French astronomer who determined the PARALLAX of the Moon. He became professor of astronomy at the Collège de France in 1762, a post he held for 46 years. He was made director of the Paris Observatory in 1768. His works included a catalogue of more than 47,000 stars. In 1802 he established the Lalande Prize for outstanding achievement in astronomy.

Lalemant, name of three French missionaries who worked in Canada. Charles Lalemant (1587–1674), a JESUIT, arrived in Quebec in 1625; later he returned to France, where he acted as forwarding agent for the Canadian missions. Gabriel Lalemant (1610–49), nephew of Charles, worked with the Hurons of Georgian Bay. He was tortured to death by the Iroquois at St Ignace and was canonized in 1930. Jérôme Lalemant (1593–73), brother of Charles and uncle of Gabriel, was director (1645–50, 1659–73) of all Jesuit missions in Canada.

La Linea, city in sw Spain, on the Bay of Gibraltar. It is on the Spanish border N of the neutral zone that separates the city from the British colony. Strategically important, it is a fortified town. La Linea is a processing centre for fruit and vegetables. Pop. (1970) 52,127.

Lalique, René (1860–1945), French jewellery designer whose work significantly contributed to the ART NOUVEAU movement. He founded his own firm in Paris in 1885 and was responsible for many technical innovations, introducing new materials (such as horn) and emphasizing contrasts of texture and colour. In 1920 he acquired a glass factory and began to produce Lalique glass, which was immediately in great demand. One of the chief patrons of his jewellery designs was the actress Sarah BERNHARDT.

Lally, Thomas-Arthur, Comte de (1702–66), French general. In 1756 he was made Commander-in-Chief of the French East Indies and from 1758 fought against the British in India. He was defeated at WANDIWASH in 1760 and surrendered in 1761. He returned to France, where he was tried for treason and executed.

Lalo, Victor-Antoine-Édouard (1823–92), French composer. His earliest compositions date from 1845 and are memorable for their individual orchestration. He was of Spanish descent and one of his most popular works is the *Symphonie Espagnole* (1873) for violin and orchestra. The opera *Le Roi d'Ys* (1888) has remained in the repertoire of French opera houses.

Lama, superior monk in Tibetan LAMAISM. The two principal lamas are the DALAI LAMA and the PANCHEN LAMA. Lamas have played an important part in Tibetan religion and society since the 8th century.

Lamaism, Western term for the religion of Tibet. It is a mixture of MAHAYANA BUDDHISM and Tibet's pre-Buddhist Bonism. King Srong-tsan-gampo (b.617 or 629) sought to bring Buddhist teachers from China and India. The Bon priests opposed the new Buddhist ways, and Buddhism was not thoroughly introduced into Tibet until the 8th century when the Indian scholar Padmasambhava was a major influence. There are now two Tibetan Buddhist sects, the Red and the Yellow monks. The Dalai Lama, a member of the latter, became revered as the "Living Buddha" and the spiritual and temporal ruler of Tibet. Each new Dalai Lama is believed to be a reincarnation of his predecessor. The Panchen Lama heads the Red monks. After the failure of the Tibetan revolt against the Chinese in 1958–59, the 14th Dalai Lama fled to India.

Lamarck, Jean-Baptiste Pierre Antoine de Monet, Chevalier de (1744–1829), French biologist. His theories of biological transformation, according to which acquired characteristics are inheritable, influenced evolutionary thought throughout most of the 19th century. He proposed in *Philosophie zoologique* (1809) that new biological needs of an organism promote a change in habits from which develop new physical structures. These are then transmitted to offspring as permanent characteristics. *See also* NW pp.32–33.

Lamartine, Alphonse-Marie Louis de (1790–1869), French Romantic poet. His *Méditations Poétiques* (1820), nostalgic for his serenely happy childhood, won him immediate fame, while later poems, collected in *Récueillements Poétiques* (1839), echoed his liberal political position. *See also* HC2 p.100.

Lamas, Carlos Saavedra. *See* SAAVEDRA LAMAS, CARLOS.

Lamashtu, in Mesopotamian religion, the daughter of the god ANU and the most vile demoness. Believed to slay children and to devour men, she has been portrayed with a lion's or bird's head and mounted on an ass. She was also thought to be the cause of disease, pestilence and nightmares.

Lamassu, or Shedu, in Babylonian mythology, protective genii. They were portrayed as winged bulls with human heads and accompanied the individual at all times.

Lamb, Charles (1775–1834), British essayist and poet. He published some poetry in the 1790s, during a close friendship with COLERIDGE, but his main interest came to centre on his essays, of which various collections were published. He was also known for his childrens' books, such as *Tales from Shakespeare* (1807). Lamb's personal life was not very happy; from 1796 he had to care for his insane sister, who had stabbed their mother to death.

Lamb, Henry (1883–1960), British painter and official war artist in WWII. He was strongly influenced by the work of Augustus JOHN, and his portrait of Lytton STRACHEY (1914) first brought him to public notice. He was a founder member of the CAMDEN TOWN GROUP in 1911.

Lamb, Mary Ann (1764–1847), British poet and writer. The sister of Charles LAMB, she was subject to fits of madness, during one of which she stabbed her mother (1796). Her works include a poem *Helen* (1802) and two works on which she collaborated with her brother: *Mrs Leicester's School* (1809) and *Tales from Shakespeare* (1807).

Lamb, William. *See* MELBOURNE, VISCOUNT.

Lamb, Willis Eugene Jr (1913–), US physicist who applied new techniques to measure the dark lines of the hydrogen spectrum. These consist of many lines close together; he found that their actual positions varied slightly from their predicted positions. For this research he was awarded the 1955 Nobel Prize in physics with Polykarp KUSCH. Lamb also developed microwave techniques for examining the helium spectrum.

Lambert, Constant (1905–51), British conductor and composer, known primarily as a conductor of ballets. His best known composition was *The Rio Grande* for piano, chorus and orchestra (1927). He wrote the ballet *Romeo and Juliet* (1925–6) for the BALLETS RUSSES.

Lambert, John (1619–84), English general and PARLIAMENTARIAN. In 1648 he was made commander of the northern army and with Oliver CROMWELL defeated Scottish ROYALISTS at PRESTON in 1648. In 1653 he helped to establish Cromwell as Lord Protector but opposed proposals to make him king in 1657. He opposed the RESTORATION and was a powerful figure in England until defeated by Gen. MONCK in 1659. *See also* HC1 p.291.

Lambeth Conferences, assemblies of bishops of the worldwide Anglican Communion. They have usually been held every ten years since 1867 at Lambeth Palace, the London residence of the Archbishop of Canterbury, who convenes and presides over them. The Lambeth Conferences have only a consultative status. Their resolutions are not binding on the member Churches of the Anglican Communion, but they carry great moral weight.

Lambeth Palace, London residence of the Archbishop of Canterbury since 1197, on the s bank of the Thames opposite Westminster. The buildings are largely medieval. The Water Tower, usually called the Lollards Tower, was built in 1435, the Gatehouse c. 1495 and the Great Hall between 1660 and 1663. It was enlarged between 1829 and 1833.

Lamburn, Richmal Crompton. *See* CROMPTON, RICHMAL.

Lamellibranchia. *See* BIVALVES.

Lamennais, Félicité Robert de (1782–1854), French priest and social critic. His *Essay on Religious Indifference* (1817) denounced toleration and rationalism but his views gradually became more liberal. His periodical, *L'Avenir* (founded in 1830), advocated a form of Christian socialism, leading to his excommunication in 1834.

Lamentations, in the Old Testament of the Bible, a book bewailing the destruction of the Temple in Jerusalem; it is commonly attributed to the author of JEREMIAH. In the Hebrew Bible, Lamentations, Ruth, the Song of Solomon, Ecclesiastes and Esther comprise a group of five scrolls that are read on significant days in the Jewish calendar.

La Mettrie, Julien Offroy de (1709–51), French physician and philosopher. His radically materialist views forced him to take refuge in Prussia with Frederick II. In *Man, A Machine* (1747) he claimed that all human functions could be fully explained on a mechanistic basis.

Laminar flow, or streamline flow, orderly flow of fluid (liquid or gas) without turbulence. The fluid flows in layers which slide past each other. As the velocity of the layers increases, or as the VISCOSITY of the fluid decreases, a point is reached when laminar flow breaks up into turbulent flow. For a given fluid this point occurs at a certain value of the REYNOLDS NUMBER.

Lamizana, Sangoulé (1916–), African politician. He became chief of staff of Upper Volta in 1961 and toppled Maurice Yameogo in a military coup to become president in 1966.

Lammas, in the Christian calendar, an archaic term for the celebration of the EUCHARIST nearest 1 Aug. Primarily a harvest thanksgiving service, it marked the time for the consecration of the first loaves made with the new grain: Lammas is a corruption of "loaf mass". In the Roman Catholic Church it is also the festival of St Peter's deliverance from prison.

Lammermuir Hills, range of hills in Lothian and Borders regions, SE Scotland. Sheep are grazed throughout the area. The highest point is Meikle Says Law, rising to 533m (1,732ft).

Lamming, George (1927–), West Indian novelist and poet whose native Barbados is the background to his autobiographical *In The Castle of My Skin* (1953). *The Emigrants* (1954) is a story of West Indians living in England and the problems they encounter.

Lamp, form of artificial lighting. Before gas and electricity, lamps burnt fuels such as tallow, wax and oil but the advent of coal gas in the early 19th century brought the first powerful lamps. Most modern lamps use electric power and they may be divided into three types: incandescent, fluorescent and discharge. An incandescent lamp consists of a sealed and evacuated glass envelope surrounding a tungsten filament. Resistance to the current through it causes the filament to glow. A FLUORESCENT LAMP has a sealed glass tube filled with low pressure mercury vapour and a filament at either end. The inside of the tube is coated with a fluorescent phosphor coating. A high voltage between the two filaments causes the vapour to ionize and emit ultra-violet light, which in turn excites the phosphor coating to give off visible light. DISCHARGE TUBES function similarly, except that the ionizing gas – sodium, neon or argon – emits coloured light in the visible spectrum when it is excited. *See also* INCANDESCENCE.

Lampedusa, Giuseppi Tomasi, Principe de, (1896–1957), Sicilian prince whose novel *Il gattopardo* (1958; tr. *The Leopard,* 1960) portrayed the microcosmic and isolated life of Sicily with such force that it brought him widespread, although posthumous, acclaim.

Lamprey, eel-like, jawless vertebrate found in marine and fresh waters on both sides of the Atlantic and in the Great Lakes of North America. The adult usually leads a parasitic life then spawns in fresh water before dying. It feeds by attaching its mouth to other fish and sucking their blood . Length: to 91cm (3ft). Family Petromyzondiae. *See also* NW p.123.

Lampshell. *See* BRACHIOPODA.

Lanarkshire (or Clydesdale), former county in Scotland, now part of the Strathclyde region. It has intensively cultivated arable lowlands, with sheep farming on the uplands. The region is the most heavily industrialized in Scotland, having the shipyards and engineering works of Glasgow to the N and the steel plants of Motherwell and Wishaw. Area: 2,323sq km (897sq miles). Pop. (1971) 1,524,848.

Lancashire, county in NW England, bordered by the Irish Sea (W), Cumbria (N), North and West Yorkshire (E) and Greater Manchester and Merseyside (S). It is drained by the rivers Lune and Ribble. Lancaster is the administrative centre, although Preston and Blackburn are two of the major towns. It lost much of its area and population with the county reorganization in 1974. Area: 3,040sq km (1,174sq miles). Pop. 1,375,000.

Lancaster, Burt, (1913–), US film actor and producer. A former circus acrobat, he made his first film, *The Killers,* in 1946. He has played a wide range of roles. Films which exploit his powerful personality include *The Rose Tattoo* (1955), *Elmer Gantry* (1960), *Bird Man of Alcatraz* (1962) and *The Leopard* (1963). With Harold Hecht and James Hill he formed the company Hecht-Hill-Lancaster which produced a number of his films, including *Airport* (1969).

Lancaster, Duchy of, lands given by HENRY III of England to his son, Edmund, in 1265. Edmund became the 1st Earl of Lancaster in 1267. The revenues from the Duchy passed into the coffers of the Crown when Henry IV became king in 1399. They still go directly into the monarch's privy purse. The Chancellor of the Duchy of Lancaster is a Minister of the Crown, not always given cabinet rank.

Lancaster, Duke of. *See* JOHN OF GAUNT.

Lancaster, House of, English royal line. The dynasty was founded by Edmund "Crouchback", Earl of Lancaster (1245–96), the second son of HENRY III. On the death of Edmund's son Henry, 1st Duke of Lancaster, in 1361, the Lancastrian title and lands passed to his daughter Blanche and her husband, JOHN OF GAUNT. Their son, Henry Bolingbroke, became HENRY IV in 1399 on deposing RICHARD II. He was the first of England's three Lancastrian kings. Bolingbroke's son, HENRY V, died in 1422 leaving his child, HENRY VI, as heir. Dissensions during the regency (1422–42) and Henry VI's feebleness as king weakened the Lancastrians, encouraging Yorkist claims to the throne and leading to the Wars of the ROSES. *See also* HC1 pp.242, 243.

Lancaster, Sir James (c. 1554–1618), English merchant and navigator. He served under Sir Francis DRAKE against the Spanish ARMADA in 1588 and went on the first expedition to reach the East Indies in 1591–94, and the first trading expedition of the EAST INDIA COMPANY (1601).

Lancaster, Sir Osbert (1908–), British cartoonist, writer and stage designer who is best known for his cartoons, which appeared from 1939 in the *Daily Express,* satirizing the British upper class and its lack of social awareness. He has written works on architecture, such as *Here, of All Places* (1959), reissued as *A Cartoon History of Architecture* (1976), and personal memoirs such as *With an Eye to the Future* (1967).

Lancaster, county town of Lancashire situated on the River Lune estuary. It was important in Roman times, and although attempts during the Middle Ages to establish a textile industry there failed, the town was raised to city status in 1937. Industries: linoleum manufacture, engineering, dyeing and cloth printing. Pop. (1971) 49,525.

Lancastrians. *See* LANCASTER, HOUSE OF.

Lance, weapon used by medieval knights for both warfare and the sport of jousting. The jousting lance was a spear normally up to 4.5m (15ft) long, broadening to a hand-guard located 1.2cm (4ft) from the butt. *See also* MM p.161, *161.*

Lancelet. *See* AMPHIOXUS.

Lancelot of the Lake, in Arthurian legend, the father of GALAHAD and one of the most famous knights; he is usually portrayed as the lover of Guinevere, wife of KING ARTHUR, but remains a model of chivalry. According to Chrétien DE TROYES' *Le Chevalier de la charrette* (12th century), he was brought up by a fairy in a lake. *Lancelot* (13th century) shows how he was borne away by the enchantress Vivien, the Lady of the Lake. In Sir Thomas MALORY's *Morte d'Arthur* (15th century) the theme of Lancelot's devotion to Guinevere and the conflict created by his loyalty to Arthur is strongly portrayed.

Lancet, long, narrow window tapering to a point found in GOTHIC ARCHITECTURE. Lancets are sometimes grouped in threes, as in the W façade of Salisbury Cathedral in England.

Lance, thermic. *See* THERMIC LANCE.

Lanchester, Frederick William (1868–1946), British car manufacturer who built the first British motor-car. He set up his own firm and in 1896 produced his first car, a one-cylinder 5-hp model. He also did important work on the theory of AERONAUTICS.

Lanchow (Lanzhou), city in NW China, on the Huanghe (Yellow) River, near the Great Wall; capital of Kansu (Gansu) province. An old walled city, Lanchow is a major transport centre and a developing industrial city. It was the scene of a successful Chinese Communist revolt in 1936. It has a university (1946) and several technical colleges. Industries include oil refining, food processing, textiles, cement and leather goods. Pop. (1970 est.) 1,500,000.

Lancret, Nicolas (1690–1743), French painter of FÊTES GALANTES. His works – marked by a vivacious gaiety and liveliness – are sometimes indistinguishable from those of WATTEAU, whom he imitated. They include *The Seasons* and *Déjeuner de Jambon.*

Land, Edwin Herbert (1909–), US physicist, inventor of the POLAROID CAMERA. He did important work with light polarization and by 1936 had developed a Polaroid material later used in sunglasses. In 1947 he developed a camera which produced a finished print within one minute. *See also* MM pp.218–219, 218, *219.*

Land Acts, Irish, acts reforming the Irish landholding system passed by the British Parliament in 1870, 1881, 1885, 1891 and 1903. The 1881 Act gave to the Irish farmers the three "Fs" which the nationalists demanded: fixity of tenure, fair rents and free sale of the tenant's interest. The 1903 Land Purchase Act effected a massive transfer of land in southern Ireland from English to Irish ownership. *See also* HC2 p.162.

Landau, Lev Davidovich (1908–68), Soviet physicist. His many contributions included the basic theories describing ferromagnetism and liquid helium. In 1927 he proposed a concept for energy called the density matrix which was later extensively used in quantum mechanics. In the 1930s he originated the theory that underlies the superfluid behaviour of liquid helium. He received the Nobel Prize in physics in 1962.

Land bridge, temporary connection between continents that can allow animal migration. One existed across the Bering Strait until comparatively recently, and a modern example is the isthmus of Panama. Before the theory of continental drift such bridges were proposed as a possible major influence on fossil distribution.

Land crab, CRUSTACEAN found in tropical America, W Africa and Indo-Pacific regions. It has gills in cavities in its shell and breathes air. A forest-floor and swamp scavenger, it migrates to water to breed. Width across back: 11 to 30cm (4–12in). Family Gecarcinidae.

Lander, Richard Lemon (1804–34), British explorer. While still a young man, he travelled to the West Indies and in 1827 went to Africa and explored the Niger River. Accompanied by his brother, John, he made a further expedition (1830–31) to determine the course of the lower Niger and discovered that the river emptied into the Bight of Benin. He was killed on a subsequent trading expedition (1832–34) to that area.

Landlord, person who lets land or accommodation for residential or commercial purposes to another person, known as a TENANT. The relationship between a landlord and tenant is usually formalized into a LEASE or tenancy agreement. The present system of tenancy derives from medieval times when all land became the possession of the Crown, other persons merely enjoying rights of tenure. The term "landlord" can also be applied to the keeper of an inn.

Landolfi, Tommaso (1908–), Italian writer and translator. His works include *Il mar delle blatte e altre storie* (1939) and *Racconti Impossibili* (1966). He has also translated works by Gogol, Merimée and Pushkin.

Landor, Walter Savage (1775–1864), British poet and writer. After being rusticated from Oxford and quarrelling with his wife and friends, he lived in Italy, and died in Florence. His works include *Gebir; a Poem in Seven Books* (1798), *Imaginary Conversations of literary Men and Statesmen* (1824–29) and *Heroic Idylls* (1863).

Landowska, Wanda (1877–1959), Polish harpsichordist and pianist who lived in Paris from 1919 and in the USA from 1941. An authority on early music, she founded the École de Musique Ancienne (1925) and in her teaching and performances

Lamps were believed to have reduced the crime rate when first installed.

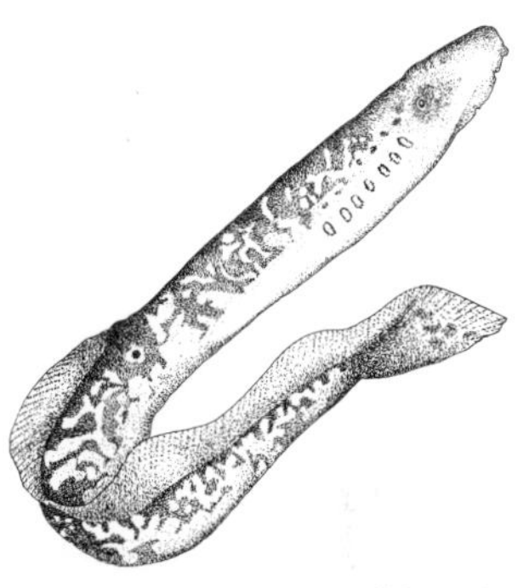

Lampreys, which prey on other fish, can be a threat to commercial fisheries.

Burt Lancaster played an Italian nobleman in *The Leopard* (1963).

Lancelot of the Lake; an engraving of one of Gustav Doré's drawings.

Land reclamation

Landscape painting or etching before 1800 often depicted the ideal world of Arcadia.

Land's End, Cornwall, is a spectacular rocky headland which attracts many tourists.

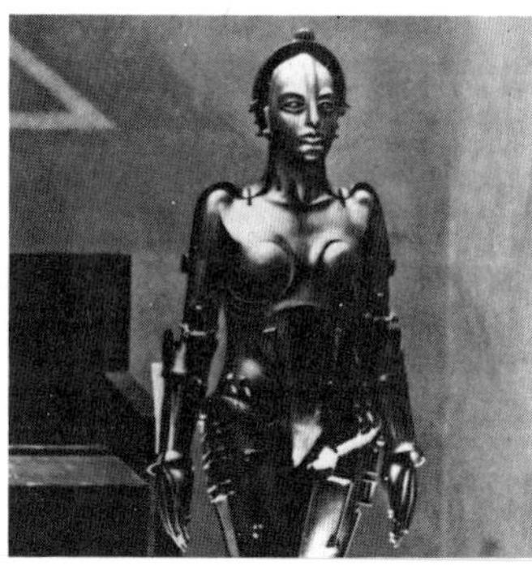

Fritz Lang; the robot design for his allegorical film *Metropolis*.

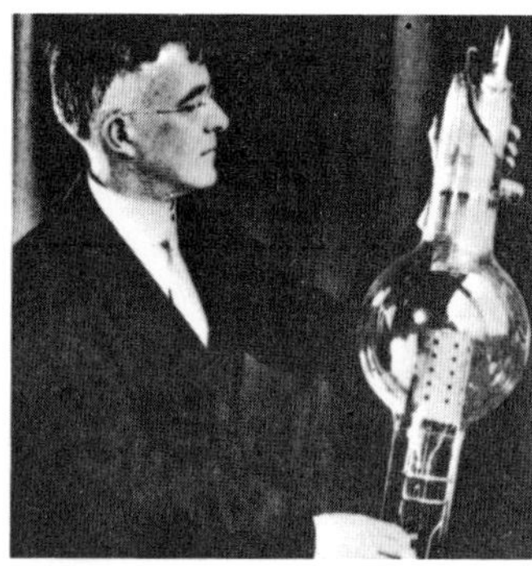

Irving Langmuir the US chemist seen here holding a discharge lamp.

she did much to promote interest in the harpsichord. Manuel de FALLA wrote his harpsichord concerto for her.

Land reclamation, transformation of land that, for one reason or another, is unusable for building or cultivation. Although the term generally refers to reclamation of land that has been beneath the sea (ie with artificial islands or merely by extending the shore-line), it also applies to the beautifying of derelict industrial sites (with canals and recreation areas) and techniques of irrigation and fertilization. *See also* PE p.154.

Landscape gardening, arranging gardens to produce certain effects. Broadly, there are two main traditions: the Sino-English, with its retention of the informality of nature, and the Franco-Italian, with its geometric patterns in which nature is trimmed to art. The second tradition arose in Italy during the RENAISSANCE. It is best exemplified in the parterres of VERSAILLES, designed by André LENÔTRE. In England the naturalist style developed in the 18th century, with the work of William KENT, Humphrey REPTON and "Capability" Brown. English influence spread in Europe in the early 19th century as part of the change in sensibilities from Classicism to ROMANTICISM.

Landscape painting, branch of painting which has natural scenery for its subject. In the ancient world and pre-modern Europe landscape was usually only a background to classical or religious themes, although in the East it was a polished art by the 8th century. The RENAISSANCE interest in perspective encouraged landscape art and in the early 16th century ALTDORFER devoted canvases entirely to it. The greatest early landscape painter was Pieter BREUGHEL. In the 17th century, landscapes became highly idealized, as in the paintings of CLAUDE LORRAIN and POUSSIN. The Dutch tendency, as in the work of REMBRANDT, was towards a greater realism. In Britain the great age of landscape painting went from Gainsborough to TURNER and CONSTABLE. No major 20th-century painter has devoted himself to landscape.

Landseer, Sir Edwin Henry (1802–73), British painter and sculptor. He achieved immense popularity with his sentimental paintings of animals, eg of the stag in *Monarch of the Glen* (1851) and of dogs in *Dignity and Impudence* (1839). His best-known sculptures are the lions in Trafalgar Square.

Land's End, headland in Cornwall, SW England, which forms the westernmost part of the English mainland. The coastline is composed of granite cliffs, with rocky islands and reefs offshore.

Landslide, noticeable slip or fall of a mass of earth or rock, frequently the result of road or dam construction. *See also* AVALANCHE.

Landsteiner, Karl (1868–1943), US pathologist, *b.* Vienna. He discovered the four different blood groups (A, B, AB and O) and demonstrated that certain blood groups are incompatible with others; clots form if blood of such groups are mixed. His findings explained why some blood transfusions in the past had been beneficial, whereas others had been fatal. He won the 1930 Nobel Prize in physiology and medicine. In 1940 he identified the Rh (rhesus) factor in collaboration with A. S. Wiener. *See also* MS pp.34, *35, 62,* 76, 87.

Land yachting, sport in which three- or four-wheeled hulls made of fibreglass or tubular metal are equipped with a single sail to provide propulsion over flat surfaces such as concrete, firm sand or tarmac. It is popular in Europe (where two- and three-man craft are also sailed), the USA and Australia.

Landy, John Michael (1930–), Australian athlete. He won international prominence in 1954 when he broke Roger BANNISTER's record in the mile run with a time of 3 min. 58 sec. They met later in the year in a race which was called the "Mile of the Century". Landy finished in 3 min. 59.6 sec. to Bannister's 3 min. 58.8 sec.

Lane, Sir Allen (1902–70), British publisher. In 1935 he founded the firm of Penguin Books, producing the first modern novels published in paperback .

Lane, Edward William (1801–76), British scholar who spent many years in Egypt. His work includes *Manners and Customs of the Modern Egyptians* (1836), the first accurate annotated translation of *A Thousand and nne Nights* (1838–40), *Selections from the Kur-an* (1843) and his great life's work, an Arabic thesaurus (5 vols, 1863–74).

Lane, Dame Elizabeth Kathleen (1905–), British lawyer, the first woman to become a High Court Judge (1965). She was called to the Bar in 1940 and worked in Birmingham (1953–61), Derby and Manchester (1961–62) before serving as a County Court judge (1962–65).

Lane, Sir Ralph (*c.*1530–1603), English pioneer. He left England in April 1585 with about 100 settlers and established the first English colony in North America, on Roanoke Island, off North Carolina. The colonists returned to England the following summer with Sir Francis DRAKE.

Lane-Poole, Stanley (1854–1931), British archaeologist and historian. He was keeper of the coin department at the British Museum (1874–92) and led archaeological expeditions to Egypt (1883) and Russia (1886). His works include a *History of the Moors in Spain* (1887).

Lanfranc (*c.*1005–89), Italian-born churchman, Archbishop of Canterbury. He was a BENEDICTINE, whose priory at Bec was (after 1045) a centre for European scholars. A counsellor of William of Normandy, he was brought to England after the Conquest to become Archbishop (1070–89). His reform of the Church included replacing Anglo-Saxon bishops and abbots with Normans, subjecting York to the authority of Canterbury, building many churches and establishing ecclesiastical courts.

Lang, Andrew (1844–1912), British anthropologist, historian and writer. His anthropological work is perhaps the most influential and includes such books as *Custom and Myth* (1884) and *Myth, Ritual and Religion* (1887–99), in which he discusses the sociology of religion and folklore as the basis of religion and literary culture. In this field he was also an early critic of the ideas proposed by FRAZER in *The Golden Bough.* As an historian, he is best known for *History of Scotland* (1900–07), although he also wrote many biographies, including *J.G. Lockhart* (1896). Much of his poetry, which includes *Grass of Parnassus* (1888), was written in the old French form of ballade and virelai. He also wrote *The Lilac Fairy Book* (1910).

Lang, Cosmo Gordon (1864–1945), British churchman who became Archbishop of Canterbury (1928–42). When he was suffragan bishop of Stepney (1901–08), his efforts to improve slum conditions attracted wide attention. He was closely involved with the revision of the Book of Common Prayer (1928) and, together with Stanley BALDWIN, was instrumental in securing the abdication of EDWARD VIII.

Lang, Fritz (1890–1976), German film director and writer whose films have been widely acclaimed for their dramatic composition and exciting detail. His first film as a director was *Halbblut* (1919). His best-known works include *Mabuse the Gambler* (1922) and *M* (1931), an exploration of compulsive murder. He later filmed in America, directing *Fury* (1936), his most highly praised US film. Perhaps the most famous of all is the intense and dramatic silent work, *Metropolis* (1927) a stylized and fantastic vision of the future of society.

Lange, Christian Louis (1869–1938), Norwegian diplomat. He worked for the Nobel Committee (1900–09) and the International Parliamentary Union (1909–33) and represented Norway in the LEAGUE OF NATIONS. He was an advocate of disarmament and received the Nobel Peace Prize in 1921 (with Karl Hjalmar Branting).

Lange, Dorothea (1895–1965), US photographer. Her portraits and studies of the Depression years among the urban poor and migrant labourers of California, and her images of rural America taken for the Farm Security Administration (1935–42), have an eloquence and strength that make them classics of documentary photography.

Lange, Friedrich Albert (1828–75), German philosopher and socialist. He was professor of philosophy at Zürich University in 1870 and in 1872 held the chair of philosophy at Marburg University. He established a tradition of Neo-Kantian philosophy at Marburg University and was well-known for his contribution to theories of materialism. He wrote the *History of Materialism* (1866).

Langevin, André (1927–), French-Canadian writer and producer who worked for Radio-Canada. His novels, often tragic and violent, include *Évadé de la Nuit* (1951), *Dust over the City* (1953) and *L'Élan d'Amerique* (1972).

Langevin, Paul (1872–1946), French physicist who in 1905 was the first to interpret paramagnetism and diamagnetism in terms of the behaviour of electrons in atoms. During WWI he built the first submarine detector based on ultrasonic waves; it was later developed into the modern SONAR.

Langlade, Charles Michel de (1729–1800), US pioneer, soldier and explorer. The son of an Ottawa Indian woman and a Green Bay Wisconsin fur trader, he led the Indian force that defeated the British at Fort Duquesne in 1755. He was active in the American War of Independence on the British side. He became known as the father of Wisconsin.

Langland, William (*c.*1331–99), English poet who wrote *Piers Plowman.* Little is known of Langland's life, although it has been claimed that he was the son of an Oxfordshire landowner. The poem *Piers Plowman* is considered one of the most important works of medieval literature, and is remarkable for its use of alliteration and a sustained allegorical style, based on a profoundly Christian approach.

Langlauf. *See* SKIING.

Langley, Edmund of, 1st Duke of York (1343–1402), fifth son of EDWARD III of England. He led the English army in campaigns in France and Spain (1359, 1367, 1375 and 1381) and in Scotland (1385). He was regent three times in RICHARD II's absence in the 1390s. In 1399 his claim to the throne was put forward too weakly to prevent the accession of HENRY IV.

Langley, Samuel Pierpont (1834–1906), US astronomer who showed that mechanical flight was possible. He did this by building large steam-powered model aircraft in 1896 which achieved the most successful flights up to that time. In 1880 he developed the bolometer which could accurately measure infra-red emissions.

Langmuir, Irving, (1881–1957), US chemist who greatly improved the incandescent LAMP. He showed that the shortness of its life was due to the evaporation of the filament, and that this could be reduced by introducing nitrogen into the bulb instead of a high vacuum. His work on electrical discharges in low pressure gases led to the discovery of the space-charge effect. His researches in surface chemistry led to the general theory that chemical reactions occurred between adjacently absorbed substances on a given surface. For this he received the Nobel Prize in chemistry in 1932.

Langside, Battle of (1568), battle between supporters of MARY, QUEEN OF SCOTS and those supporting her son JAMES VI (later JAMES I of England). It ended in Mary's defeat and her flight to England.

Langton, Stephen (*d.*1228), English cardinal and scholar. Appointed Archbishop of Canterbury (1207) by Pope INNOCENT III, he was prevented from occupying his see until 1213 by King JOHN's opposition. He supported the barons over the issue of MAGNA CARTA (1215) and although suspended as archbishop, was restored after the accession of Henry III. A prolific writer, it was probably he who was responsible for the present chapter divisions of the Scriptures. *See also* HC1 p.195.

Langtry, Lillie Emilie Charlotte (1853–1929), British actress, noted for her great beauty. One of the first English women of high social rank to appear on the stage, she made her debut at the Haymarket Theatre, London, in 1881. She toured the USA in the early 1900s.

Language, generally accepted system of communication; the term is usually used to refer to human vocal communication but there are many other types of language such as body language, used by humans and other animals. Human language is a universal phenomenon, and although there are more than 4,000 different languages, they have many characteristics in common. Almost every human language uses a fundamentally similar grammatical structure, or syntax, even though the languages may not be linked in vocabulary or origin; this fact has led some linguists to argue that language structures are genetically conditioned, and that although it is necessary to teach children vocabulary and the relation of the words to the outside world, their manipulation of grammar is innate. Families of languages have been constructed, such as the INDO-EUROPEAN family, the most common in the world, but their composition and origins are the subject of continuing controversy. Historical studies of language come under the field of ETYMOLOGY, the origins of words, and PHILOLOGY, the study of words and languages for their own sake. LINGUISTICS as a discipline usually addresses itself to contemporary language, although some historical studies are carried out. *See also* MS pp.154–155, 238–241.

Language, Truth and Logic (1936), the first major philosophical work by Alfred J. AYER; it was the book that introduced logical positivism to Britain. In it Ayer rejects metaphysics, ethics and theology on the basis that the statements these topics contain are not verifiable, either logically or empirically.

Langur monkey, any of about 15 species of medium to large MONKEYS of South-East Asia and the East Indies. They feed mainly on leaves, are slender and have long hands and tails. Gregarious tree-dwellers, they are active by day and are found from sea-level to snowy Himalayan slopes at an elevation of 4,000m (13,000ft). Length: 43–78cm (17–31in). Family Cercopithecidae; genus *Presbytis. See also* NW pp.*154, 169, 214,* 215, 235.

Lanier, Sidney (1842–81), US poet, whose verse reflects Southern social change. His first novel, *Tiger Lilies,* was published in 1867. Publication in 1875 of the poems *Corn* and *Symphony* brought him national recognition.

Lankester, Sir Edwin Ray (1847–1929), British zoologist. He was noted for his studies of the anatomy, embryology and morphology of invertebrates. He pioneered the science of parasitology and popularized various aspects of science.

Lanner, Josef (1801–43), Austrian composer. He was appointed director of the imperial court balls in Vienna in 1829. He was a prolific composer of dance music and a creator of the modern Viennese WALTZ.

Lanolin, purified fat-like substance derived from sheep's wool and used with water as a base for ointments and cosmetics.

La Noue, François de (1531–91), HUGUENOT captain during the French Wars of Religion (1562–98). He was called *Bras-de-fer,* after his iron artificial arm. He fought for the French and Dutch Protestant cause in several campaigns, and was imprisoned by the Spaniards in the 1580s. He later fought under Henry IV of France, and died at the siege of Lamballe.

Lansbury, Angela (1925–   ) US actress, *b.* Britain. She made her film debut in *Gaslight* (1944), for which she received an Academy Award nomination. She appeared in *The Picture of Dorian Grey* (1945) and *The Manchurian Candidate* (1962). In 1966 she was awarded the Antoinette Perry (Tony) Award for her role in the musical *Mame.* She appeared in *Hamlet* at the National Theatre, London (1975–76).

Lansdowne, Henry Charles Keith Petty Fitzmaurice, 5th Marquess of (1845–1927), British statesman who served as Governor-General of Canada (1883–88) and Viceroy of India (1888–94). As foreign secretary (1900–05) he concluded an alliance with Japan (1902) and the Anglo-French ENTENTE (1904). He served in the WWI coalition cabinet and in 1917 published in the *Daily Telegraph* a letter setting forth proposals for peace. It was repudiated by the government.

Lansdowne Road, rugby union ground in Dublin, used for Ireland's home internationals. It is the home ground of the Lansdowne and Wanderers clubs.

Lanston, Tolbert, (1844–1913), US inventor of the Monotype typesetting machine in 1885. This made possible for the first time the mechanical casting and setting of type in lines, the entire process being controlled by coded tape. It helped to revolutionize the printing industry.

Lantern, in architecture, open or glassed-in structure on top of a tower, roof or dome to provide light and/or air for the interior.

Lantern fish, any of numerous species of marine fish found in Atlantic and Mediterranean waters; especially *Diaphus rafinesquiei.* It is identified by light organs along the sides, and is found in deep water during the day and near the surface at night. Length: 7.5cm (3in). Family Myctophidae. *See also* NW p.*240.*

Lantern fly, tropical and subtropical South American plant hopper of the order HOMOPTERA. It is brightly coloured and has a long, inflated prolongation of the head; this was once thought to emit light. Wingspan: to 15cm (6in). Family Fulgoridae.

Lanthanide series, also called the lanthanide elements, or rare earths, series of rare metallic elements with atomic numbers between 57 and 71. They are: lanthanum, cerium, praseodymium, neodymium, promethium, samarium, europium, gadolinium, terbium, dysprosium, holmium, erbium, thulium, ytterbium, and lutetium. Their properties are similar and resemble those of lanthanum, from which the series takes its name. They occur in monazite and other rare minerals and are placed in group IIIB of the periodic table. All form trivalent compounds (having valency three); some also form divalent and quadrivalent compounds. *See also* SU pp.*134,* 136.

Lanthanum, metallic element (symbol La) of the Lanthanide series, first identified in 1839. Its chief ores are monazite (a phosphate) and bastnasite (a fluorocarbonate). Compounds of the metal are used in the manufacture of glass for high-quality lenses. Properties: at. no.57; at. wt 138.9055; s.g. 6.17 (25°C); m.p. 920°C (1,688°F); b.p. 3,454°C (6,249°F); most common isotope La^{139} (99.91%). *See also* SU pp.134–136.

Lan t'ien man, human remains found in the Lan t'ien district of Shensi, China, believed to be up to 600,000 years old. Lan t'ien man was an early example of *Homo erectus,* and was discovered in 1963–64. *See also* HC1 p.44; MS p.24.

Lanugo, soft downy hair that covers the human fetus and those of other mammals during development in the womb. It is shed, and virtually disappears, at birth.

Lao, the official language of Laos, spoken by about 94% of the country's population. It is one of the Tai group of languages and is closely related to Thai.

Laocoon, priest of Apollo during the TROJAN WAR, described in Virgil's *Aeneid.* Referring to the wooden Trojan Horse, he exhorted the Trojans to fear the Greeks even when they offered gifts. For giving this warning and because of his marriage, he and his sons were strangled by two sea-serpents sent by Apollo. The statuary group *Laocoon and his Sons* in the Vatican is one of the masterpieces of the Greek Hellenic period.

Laoighis (Laois), county in central Republic of Ireland, in Leinster province. Part of the central plain of Ireland, Laoighis is generally low-lying and is drained by the rivers Barrow and Nore. Most of the people work on farms; oats, barley, potatoes and turnips are grown, and dairy cattle are reared. Area: 1,720sq km (664sq miles). Pop. (1971) 45,349.

Laos, independent nation in South-East Asia. It is a mountainous country with no access to the sea. Most of the people work in subsistence agriculture and natural resources remain unexploited. Exports include tin, timber and coffee. The capital is Vientiane. Area: 236,800sq km (91,428sq miles). Pop. (1975 est.) 3,303,000. *See* MW p.117.

Lao-tze, or Lao-tzu (*c.* 604–*c.* 531 BC), Chinese philosopher. According to Chinese legend he founded TAOISM, a religion which became a mystical reaction to the moral-political concerns of CONFUCIANISM. Although there is uncertainty about his identity, he is believed to have been the author of *Tao Te Ching,* the sacred book of Taoism; *tao* is the "way"; *te* is its "virtue". It has greatly influenced Chinese culture. *See also* HC1 pp.124–125.

La Paz, largest city and administrative capital of Bolivia, South America. Built on the site of an INCA village, it was one of the centres of revolt in the War of Independence (1809–24). Today its industries include tanning, flour-milling, brewing and distilling. Pop. 660,700.

La Pérouse, Jean François de Galaup, Comte de (1741–*c.* 88), French navigator. He distinguished himself against the British (1778–83) and in 1785 led a French exploration of the Pacific Ocean, partly in search of a north-west passage. The trip took him to many parts of the South Pacific as well as to the coasts of North America and Asia. He is believed to have died in a shipwreck off one of the Santa Cruz Islands.

Lapidary, craftsman skilled in cutting, polishing and engraving gemstones. The commonest faceted forms are brilliant cut, rose cut and trap cut. *See also* PE pp.98, 99.

Lapis lazuli, or lazurite, glassy, deep blue semi-precious gemstone, a silicate mineral found in metamorphosed limestones. It occurs rarely as crystals in the cubic system, but more often as granular masses. Hardness 5–5.5; s.g. 2.4. *See also* PE p.98–99.

Laplace, Pierre-Simon, Marquis de (1749–1827), French astronomer and mathematician. Laplace used probability theory successfully to apply Newton's gravitational theory to the entire solar system; this work was summarized in his book *Celestial Mechanics* (1798–1827). He also did fundamental work in the study of heat, magnetism and electricity. *See also* SU pp.25, 174.

Lapland, extensive area in N Europe, almost entirely within the Arctic Circle, including N and NE Norway, the northernmost parts of Sweden and Finland, and the Kola Peninsula of the USSR. The land is mountainous in Norway and Sweden, and tundra predominates in the NE. There are extensive forests and many lakes, and the climate is harsh throughout the area. Rich mineral deposits include iron ore, copper and nickel. Large reindeer herds tended by LAPPS are of great importance to the economy and fishing is a major industry. Tourism is becoming increasingly important. Area: approx. 388,500sq km (150,000sq miles).

La Plata, city in E Argentina, 56km (35 miles) ESE of Buenos Aires. The city is laid out in a 5km (3-mile) square modelled on Washington, DC, and is the site of a cathedral. It is the oil refining centre of the country and steady industrial growth has bolstered the economy. Oil, cereals and frozen meat are exported. Pop. (1970) 506,287.

La Plata, Rio de. *See* PLATA, RIVER.

Lapps, people inhabiting LAPLAND. The mountain Lapps are nomadic herders of reindeer, while those of the forest and coast are semi-nomadic and live by hunting, trapping and fishing. Their racial origins are uncertain, although their language (Saami) is Finno-Ugric. There are about 31,500 Lapps.

Laptev Sea, part of the Arctic Ocean, off the coast of the Russian Republic (Rossijskaja SFSR), between Severnaya Zemlya and the New Siberian islands. It is icebound most of the year. Area: 714,840sq km (276,000sq miles).

Langurs are usually born yellow or white; they later become black or grey.

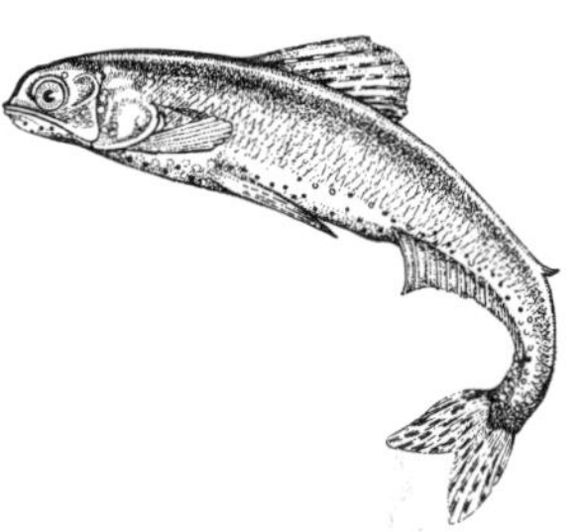

Lantern fish can use their light organs to attract plankton.

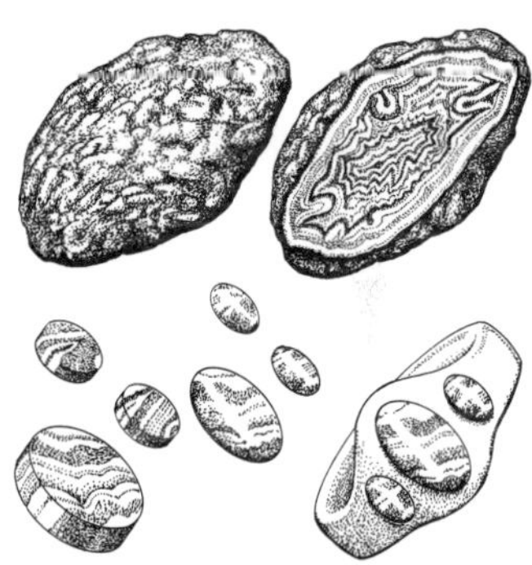

Lapidaries cut and polish precious stones which are then set as jewels.

Lapps wear pointed hats (to represent the four winds) in their traditional costume.

Lapwing, or peewit

Lapwings, members of the plover family, are common inland waders in Europe.

Larches are used for reforestation in Europe and North America.

Larva of the tortoiseshell butterfly, a stage in its remarkable metamorphosis.

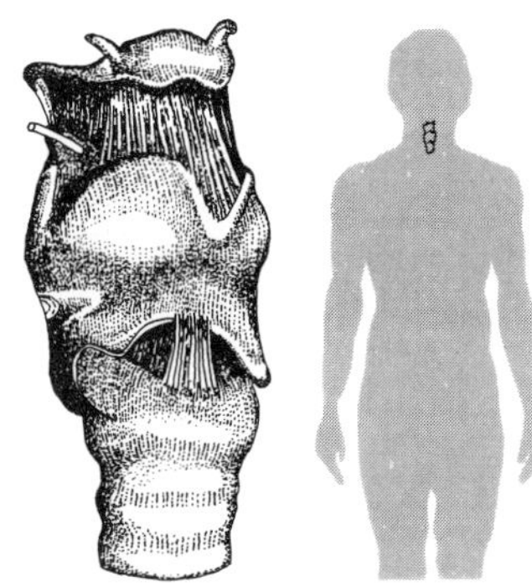

Larynx and vocal chords in men are larger than in women, hence the deeper male voice.

Lapwing or peewit, any of several species of birds, especially the Eurasian lapwing, *Vanellus vanellus*, a brown, green-black and white wading bird with a conspicuous crest. It commonly nests in open agricultural land and defends its young by luring predators away, feigning a broken wing. Length: 30cm (12in). Family Charadriidae. *See also* NW pp.*145*, 146.

Larbaud, Valéry-Nicolas (1881–1957), French poet, novelist and critic. He travelled widely and enjoyed a close friendship with James JOYCE. His best-known works were the *Poems of a Rich Amateur* (1908) and *A.O. Barnabooth* (1913), accounts of a wealthy young South American who travels through Europe seeking fulfilment.

Larceny. *See* THEFT.

Larch, any of several species of conifer trees native to cool and temperate regions of the Northern Hemisphere. They bear cones and needle-like leaves which, unusually for a conifer, are shed annually. The timber is rot-resistant and is used for telegraph poles. Family Pinaceae. *See also* NW pp.54–55.

Lard, melted-down pigs' fat, packaged and sold for use in cooking. Chemically, lard is composed of triglyceride fats.

Lardner, Ringgold Wilmer (Ring) (1885–1933), US short story writer and humorist. He spent 15 years as a reporter for Chicago newspapers and for a time edited a baseball weekly. In 1916 he published *You Know Me, Al*, a collection of racy short stories about baseball players. His great talent lay in re-creating the language of ordinary people and in his satirical presentation of their characters. Among his best-known stories are *Haircut*, *Champion* and *The Love Nest*.

Lares, Roman gods of cultivated land, worshipped by each household at crossroads where field boundaries met. Probably originally a reverence for ancestral property, the worship of the Lares (which were stylized as figurines) moved into the houses to join that of the Penates, the gods of the threshold.

Largillière, Nicolas de (1656–1746), French painter. He is best known for his glamorous portraits, which include *Louis XIV and the Royal Family* (1708) and *La Belle Strasbourgeoise* (1703).

Largo Caballero, Francisco (1869–1946), Spanish political leader. He rose to power through Spanish trade unions and served in the cabinet (1931–33). His radical propaganda early in 1936 is considered one of the chief factors in precipitating the SPANISH CIVIL WAR (1936–39). He fled to France in 1939, suffered three years of imprisonment under the Germans during WWII, and died in Paris.

Largs, Battle of (2 Oct. 1263), defeat of Norse invaders of Scotland. The Norse fleet of about 150 ships was driven by storms on to the shore near Largs on the Firth of Clyde and many of the ships were wrecked. The Norse leader, Haakon IV, landed a force, but it was routed by the Scots under Alexander III.

Larionov, Mikhail (1881–1964), Russian abstract painter. He invented RAYONNISM, a Russian offshoot of CUBISM, and thus laid the foundations of abstract art in Russia. After 1914 he left Russia for Paris to work as a designer for DIAGHILEV's BALLETS RUSSES.

Lark, any of several small birds that live in open areas, known for their melodious songs. Most common in Europe are the woodlark (*Lullula arborea*), skylark (*Alauda arvensis*) and shorelark (*Eremophila alpestris*). All are mottled brown, some with spotted markings on the breast. Depending on where they live, they feed on insects, larvae, crustaceans or berries. Length: to 18cm (7in). Family Alaudidae. *See also* NW pp.*199, 221*.

Larkin, Philip Arthur (1922–), distinguished British poet. His first book *The North Ship* (1945) and later volumes, *Less Deceived* (1955), *The Whitsun Wedding* (1964), and *High Windows* (1974), show a reaction against SYMBOLIST and ROMANTICIST poetry and a deeply felt interest in the importance of the unspectacular. Larkin edited the *Oxford Book of Twentieth Century Verse* (1973).

Larkspur. *See* DELPHINIUM.

La Rochefoucauld, François, Duc de (1613–80), French writer, renowned for his literary maxims and epigrams. In 1635 he was involved in an intrigue against Cardinal RICHELIEU and took part in the FRONDE revolts (1648–53). His best-known work is his *Réflexions ou Sentences et Maximes Morales* which ran into five editions (1665–78). *See also* HC1 pp.*316*, 317.

La Rochelle, port on the French Atlantic coast. It developed in the 12th century and was fiercely contested in the HUNDRED YEARS WAR (1337–1453). It became a Huguenot stronghold in the 16th century, until taken by government forces directed by Cardinal RICHELIEU in 1628. La Rochelle's trade declined in the 18th century when the French lost Canada. Its industries include shipbuilding, oil-refining and fish-canning. Pop. 72,075.

La Rochelle Expeditions, three unsuccessful attempts by the English government to relieve the Huguenots besieged by the French government at La Rochelle. The first, in June 1627, was commanded by the Duke of Buckingham and returned with heavy losses after landing on the Isle of Ré. The second, under the Earl of Denbigh, returned without having landed. In August 1628 the Duke of Buckingham was assassinated while preparing the third expedition, which never set out.

La Ronde (1950), French film based on a light romantic comedy by Arthur Schnitzler. Directed by Max Ophuls and starring Jean-Louis Barrault, Simone Signoret, and Gérard Philipe it achieved international success. Roger Vadim made a less successful version in 1964.

Larousse, Pierre (1817–75), French lexicographer. He founded the publishing firm of Larousse, which produced *The Great Universal Dictionary of the 19th Century* (1866–76) as the first of the famous Larousse series of dictionaries and encyclopaedias.

Larreta, Enrique Rodríguez (1875–1961), Argentine novelist who lived abroad for many years. His best-known work is *La Gloria de don Ramiro* (1908), an historical novel of Spain under Philip II, and a classic of the modernist genre. Other works include *Zogoibi* (1926), a GAUCHO story, and *En La pampa* (1955).

Larrocha, Alicia de (1923–), Spanish pianist. She studied in Spain with Frank Marshall and succeeded him as director of the Marshall Academy, Barcelona, in 1959. By the late 1940s she had been acclaimed internationally for her fine performances of works by Spanish Romantic composers such as Granados and Albéniz. She made her debut in the USA in 1954 and after European and world tours returned there in 1965.

Larsa, ancient city of Mesopotamia. It achieved predominance in Mesopotamia after the fall of UR (*c.* 1950 BC) by defeating its chief rival Isin shortly after 1900 BC. Larsa was eventually destroyed by the BABYLONIANS under HAMMURABI in 1763 BC. *See also* HC1 p.32.

Larsson, Lars-Erik (1908–), Swedish composer. His many varied works include an opera, *The Princess of Cyprus* (1930–36), *Serenade for Strings* (1934) and the *Gustavian Suite* (1943) for flute, harpsichord and strings. From 1937 and 1954 Larsson was chief conductor of the Swedish State Radio orchestra.

Lartigue, Jaques Henri (1896–), French photographer. His first photographs date from 1906 and, although an amateur, for 70 years he recorded the lavish lifestyles of his relatives and friends and the society life of Parisian boulevards. Much of his work was published in *Diary of a Century* (1970).

La Rue, Pierre de (*c.* 1460–1519), Flemish composer in the service of Maximilian, Philip of Burgundy and Margaret of Austria. He wrote chiefly masses and MOTETS, but also a series of *chansons*.

Larva, developmental stage in the life-cycle of most invertebrates and some other animals. A common cycle, typified by the BUTTERFLY, is egg, larva, PUPA, adult. In different organisms the larval stage may be called a NYMPH, MAGGOT, CATERPILLAR, TADPOLE or GRUB. It has a well-developed alimentary system and often stores food until it pupates to become an adult. The duration of this stage depends on the organism; some creatures (such as the AXOLOTL) retain larval features throughout their lives.

Larwood, Harold (1904–), British cricketer who played for Nottinghamshire and England. He took 78 wickets at an average of 28.41 in 21 Test matches between 1926 and 1933. He was the spearhead of the English attack during the tour of Australia in 1932–33, when he took 33 wickets, 16 clean bowled.

Laryngitis, inflammation of the LARYNX (voice box), usually occurring during a respiratory-tract infection and accompanied by dryness, soreness and hoarseness.

Larynx, the voice box, triangular-shaped cavity located between the trachea and the root of the tongue. Folds in the lining of the larynx form the vocal chords. These are thin bands of elastic tissue that vibrate when outgoing air passes over them, setting up resonant sound waves that are changed into sound by the action of throat muscles, the shape of the mouth and the tongue.

La Salle, René-Robert Cavelier, Sieur de (1643–87), French explorer and fur trader in North America. In 1666 he sailed for Canada to make his fortune. He explored the Great Lakes area and was seigneur of Fort-Frontenac on Lake Ontario (1675). His greatest achievement was exploring the Mississippi River to its mouth (1682) and claiming it for the King of France. He named the adjacent lands LOUISIANA after Louis XIV.

La Scala (Teatro alla Scala), Milan, one of the world's great opera houses. Designed by Giuseppe PIERMARINI, it opened in 1776 and has been the scene of many famous premières, among them Bellini's *Norma*, Verdi's *Otello* and Puccini's *Madame Butterfly*. Arturo TOSCANINI was its artistic director from 1898–1907 and 1921–31, periods considered to be the most brilliant in the theatre's history.

Lascaris, BYZANTINE ruling family of the empire of Nicaea, which was founded by Theodore I Lascaris in 1208. Heirs to the Byzantine Empire, which had been conquered by CRUSADERS, the Lascaris family maintained its small empire until John IV was toppled in 1261.

Las Casas, Bartolomé de (1474–1566), Spanish Dominican friar whose campaign against the use of Indian labour earned him the title "Protector of the Indians". Las Casas succeeded in persuading the Crown to promulgate the New Laws (1542), modifying the ENCOMIENDA system. *See also* HC1 p.*237*; HC2 p.138.

Las Cases, Emmanuel, Comte de (1766–1842), French historian. He accompanied NAPOLEON into exile to St. Helena, and recorded their conversations. His *Mémorial de St Hélène* is a primary source for Napoleon's last years but is not always reliable.

Lascaux caves, complex of caves in the French Pyrenees, discovered in 1940. They contain examples of 13 different styles, presumably ages, of PALAEOLITHIC wall paintings. The caves, some of which may have been used for ritual offices, are now closed in order to halt the deterioration of the paintings.

Lasdun, Sir Denys Louis (1914–), British architect of versatility. His works include the Peter Robinson store in the Strand (now New South Wales House) and the Royal College of Physicians, both in London, part of the University of Leicester and the complete development of the new University of East Anglia. In 1963 he was appointed chief architect for the NATIONAL THEATRE.

Laser (optical MASER), source of a narrow beam of intense coherent light or ultra-violet or infra-red radiation, first developed in 1960 by Theodore H. MAIMAN. (The name is an acronym from light amplification by stimulated emission of radiation.) The source can be a solid, liquid or gas. A large number of its atoms are excited to a higher energy state. One photon of radiation emitted from an excited atom can stimulate the emission of another photon, of the same frequency and direction of travel, which in turn stimulates the emission of more photons. The photon number multiplies rapidly to produce the narrow monochromatic laser beam of very high energy content. It has applications in medicine, scientific research, engineering, telecommunications, holography and other fields. *See also* SU p.110.

Lashley, Karl Spenser (1890–1958), us-psychologist who did important work on discrimination, learning and kinaesthesis, and on the structures and functions of parts of the brain. He wrote *Brain Mechanisms and Intelligence* (1929).

Laski, Harold Joseph (1893–1950), British political scientist and economist, professor of political science at the London School of Economics (1926–50). Laski was a member of the executive committee of the FABIAN SOCIETY (1921–36) and, active in the Labour Party, served on the National Executive (1937–49) and as Party chairman (1945–46). His works include *Democracy in Crisis* (1933) and *The American Democracy: A Commentary and Interpretation* (1948).

Laski, Marghanita (1915–), British novelist, critic and journalist. Her first novel *Love on the Supertax* (1944) was followed by *To Bed with Grand Music* (1946). Her later works include a novel, *The Victorian Chaise Longue* (1953) and a play, *The Offshore Island* (1959).

Lasky, Jesse (1880–1958), us film producer. In 1913, with Samuel GOLDWYN and Cecil B. DE MILLE, he formed the Jesse Lasky Feature Play Company and had two major successes with *The Squaw Man* (1913) and *The Virginian* (1914). His company expanded and by 1927 he had gained control of the PARAMOUNT PICTURES CORPORATION.

Lasnier, Rina (1915–), Canadian writer of poetry and plays, often with religious themes. Her work includes the verse collections *Images et proses* (1941), *Le chant de la Montée* (1947) and the play *Le jeu de la voyagère* (1950).

Las Palmas, full name Las Palmas de Gran Canaria, city on Grand Canary Island, Spain; capital of Las Palmas province. Its port, Puerto de la Luz, is the chief port in the Canary Islands, and one of the busiest in Spain. The city was founded in 1478. It is a popular tourist resort and has fish and fish processing industries. Pop. (1970) 287,038.

Lassa fever, little-understood viral infection, first recognized as a disease in Africa in 1969. The infection can produce wide-ranging symptoms, including high fever, mouth ulcers, heart damage, muscle aches and kidney failure, and can affect almost any part of the body. The disease occurs sporadically and is often fatal.

Lassell, William (1799–1880), British amateur astronomer. He discovered Hyperion, a satellite of Saturn (at the same time as us astronomer George P. Bond); Ariel and Umbriel, satellites of Uranus; and Triton, a satellite of Neptune. *See also* SU p.215.

Lasso, Orlando di (*c*.1532–94), Flemish composer of madrigals, masses and motets. In the employ of Albert V of Bavaria after 1556, he became famous throughout Europe and is ranked with PALESTRINA as one of the greatest composers of the late 16th century. He composed more than 2,000 works in German, Italian and French.

Last Days of Pompeii, The (1834), novel by Edward Bulwer LYTTON about the love of Glaucus and Ione, contrasting the brutality of the barbarians with the piety of the Christians.

Lastman, Pieter Pieterszoon (1583–1633), Dutch painter of historical and mythological subjects, who worked in Amsterdam. He went to Italy, *c*.1604, where he was influenced by CARAVAGGIO's use of CHIAROSCURO, and was back in Amsterdam by 1607. He was also an important and influential teacher, whose students included REMBRANDT (1624–25).

Last of the Mohicans, The (1826), novel by James Fenimore COOPER portraying adventures of white Americans and Indians in 1757 during the war between the French and the Indians. *See also* HC2 p.*288.*

Last Supper, in the New Testament, Christ's final meal with His Apostles in Jerusalem just before the Passover. During the meal He instituted the memorial of Himself that is the Christian EUCHARIST, warned the disciples of His imminent betrayal, and gave cryptic indications of His foreknowledge of His own death and resurrection.

Last Supper, The (1495–98), mural oil painting by LEONARDO DA VINCI for the refectory of the monastery of Sta Maria delle Grazie, Milan. A masterly rendering of form and psychological insight, it gives the viewer the sensation of looking into an extension of the refectory where Christ sits among His Apostles.

Last Tycoon, The (1941), unfinished novel by F. Scott FITZGERALD. It is the story of a Hollywood film producer, reputedly based on Irving THALBERG, and was itself filmed in 1976.

Lasus of Hermione (6th century BC), Greek choral poet. Known as an instructor of PINDAR, he is credited as one of the developers of the DITHYRAMB form. He was a literary rival of SIMONIDES, but his works have not survived.

Las Vegas, largest city in Nevada, USA. A world-famous gambling and entertainment centre, comprising luxurious casinos, night-clubs and hotels, it derives considerable income from tourists. Las Vegas became prosperous in the 1930s with the construction of the nearby HOOVER DAM. The city is also the commercial centre of a mining and cattle ranching region. Pop. (1973) 144,333.

Lateen rig, sailing rig thought to have been used as early as the 2nd century in the E Mediterranean, employing a short hooked mast with a longer angled crosspiece supporting the sail. The lateen rig is still used on fishing vessels of the Near East and Africa.

La Tène culture, second phase of CELTIC culture, from *c*.500 BC to *c*.50 BC. Its name comes from La Tène, in Switzerland, an archaeological site discovered in the 19th century. The origin of the culture, which replaced the previous HALLSTATT culture, was contact with Greek and Etruscan influences south of the Alps. From them came the characteristic traits of La Tène art – spirals, s-shapes and regular patterns. It was a highly warlike culture, hierarchically organized with kings, a priestly class (the DRUIDS), warriors (especially the gaestatae or spear-carriers), farmers and slaves. La Tène weaponry was late IRON AGE – long swords, heavy knives and lance spears. The La Tène Celts conquered central Europe in the fourth and third centuries, but by 50 BC had themselves submitted to the domination of German invaders from the north and Roman from the south. *See also* HC1 pp.86–87.

Latent heat, heat absorbed by a substance as it changes its phase – ie goes from the solid state to a liquid or from a liquid to a gas. When ice melts, its temperature remains the same until it has been completely transformed into water; the heat necessary to do this is called the latent heat of fusion. Similarly the heat necessary to transform water into steam at constant temperature is called the latent heat of vaporization. *See also* SU p.86.

Lateran, church of St John, the pope's church in Rome, the first-ranking church of the Roman Catholic faith. It was named after a monastery once standing nearby, built on land owned by the Laterani. Part of the Basilica was built possibly before 311 and much of the decoration dates from the Middle Ages.

Lateran Councils, five ECUMENICAL COUNCILS of the Western Church, held in the Lateran Palace in Rome. The first, held in 1123, confirmed the Concordat of Worms of 1122. The second, in 1139, promulgated 30 decrees which among other things, condemned simony and the marriage of the clergy. The third, in 1179, decreed that the pope was to be elected by a two-thirds majority of the College of Cardinals. The fourth council, in 1215, which was the most important of the Lateran Councils, gave a definition of the doctrine of the EUCHARIST, officially using the term "TRANSUBSTANTIATION" for the first time. The fifth, in 1512–17, introduced some minor reforms in the wake of the Protestant REFORMATION, but left its main causes untouched. The work of this council was carried further by the Council of TRENT. *See also* HC1 pp.186–187.

Lateran Treaty (1929), concordat between Italy and the VATICAN. By its terms (confirmed by the Italian constitution of 1948), Italy recognized the Vatican as an independent sovereign state with the pope as its temporal head. Roman Catholicism was affirmed as Italy's official state religion, and the Vatican recognized Italy's claims to the Papal States and Rome.

Laterite, reddish soil found in humid tropical regions. It is produced by the subaerial decay of rocks in leaching and oxidizing conditions. It contains either or both iron oxide and aluminium hydroxide, and may be used as an ore of either metal if the concentrations are sufficiently high.

Latex, milky fluid produced by certain plants, the most important being that produced by the RUBBER tree. Rubber latex is a combination of gum resins and fats in a watery medium. It is used in paints, special papers and adhesives, and to make sponge rubbers. Synthetic rubber latexes are also produced. *See also* MM p.54.

Lathe, machine tool that turns a workpiece (either wood or metal) against a cutting tool in order to shape it. *See also* MM p.43.

Lathrop, Rose Hawthorne (1851–1926), us philanthropist and writer. She founded St Rose's Free Home for Incurable Cancer in New York City and the Rosary Hill Home in Hawthorne, N.Y., a town named in her honour. After her husband's death in 1898 she became a nun taking the name Mother Alphonsa. She wrote *Along the Shore* (1888) and *Memories of Hawthorne* (1897).

Latimer, Hugh (*c*.1485–1555), English clergyman and Protestant martyr. He first became prominent by defending Henry VIII's divorce from Catherine of Aragon and in 1535 was made Bishop of Worcester. A man of strong Protestant convictions, he strongly disapproved of the temporary reaction in favour of Catholicism, and resigned his see in 1539. With the accession of Edward VI (1547) he resumed preaching against the abuses of Church and clergy. When the Roman Catholic Mary came to the throne (1553) he was charged with heresy and, refusing to recant, was burned at the stake.

Latin America, those parts of North, Central and South AMERICA (excluding French-speaking Canada) where the official or chief language is a ROMANCE LANGUAGE. *See also* articles on individual countries in *The Modern World*.

Latin cross, in architecture, a building, usually a church, in the form of a cross with three short arms and a long arm.

Latin Empire of the East, name used to describe the BYZANTINE EMPIRE during the period 1204–61. In 1204 Constantinople fell to the soldiers of the fourth crusade, and their leaders maintained a precarious hold on the city until 1261, when Michael VIII Palaeologus regained it.

Latin language, Indo-European language originally spoken by the inhabitants along the lower River Tiber, central Italy. Its earliest written record dates back to the 6th century BC. It was spoken by the inhabitants of Rome, whose domination of Italy made Latin the chief language of that country by the 3rd century BC. By the 1st century BC a literary version of Latin had developed, more complex than the spoken language, which from the 3rd century AD is known as Vulgar Latin. Literary Latin was at its richest 70 BC–AD 18. With such works as the translation of the Bible into Vulgar Latin by St Jerome at the turn of the 4th century, that form too assumed a literary aspect, increasingly divorced from the spoken language. Latin was spoken throughout the Roman Empire, and despite the barbarian invasions of the 5th and 6th centuries it formed the basis of the languages spoken in much of the Empire N of the Mediterranean. These ROMANCE LANGUAGES became distinct

The Last of the Mohicans; a scene from one film version of this adventure story.

Las Vegas has a mining industry which produces silver, gold and many other minerals.

Lateen rigs, introduced by the Arabs, were more efficient in light winds than square rigs.

Lathes are often fitted with turrets that hold a variety of tools to cut the workpiece.

Latin literature

William Laud was attacked in woodcuts for his quasi-Catholic practices.

Niki Lauda receiving an award for valour in sport at the Mansion House, London.

Max von Laue studied crystal structures by the measurement of X-ray diffraction.

Laurel and Hardy in a scene from their period film *The Devil's Brother* (1933).

from Latin by the 9th century. Latin remained the language of the Church, and therefore of education and most written transactions in Europe throughout the Middle Ages, and was used in some scholarly and diplomatic circles until the 19th century. Roman Catholic Church services were conducted in Latin until the 1960s. *See also* MS p.241.

Latin literature, literature of ancient Rome. The earliest works of Latin literature were imitations of Greek plays and epic poetry by LIVIUS ANDRONICUS and NAEVIUS in the 3rd century BC. One of Rome's greatest dramatists, PLAUTUS, wrote in a similar style in the early 2nd century BC. Latin literature reached its peak in terms of style in the 1st century BC, with CICERO's rhetoric, CAESAR's histories, and CATULLUS's lyric poetry. VIRGIL's epic poem the *Aeneid* secured the mythology of the origins of Rome, and LIVY produced a full history of the republic. This Golden Age of Latin literature ended soon after the death of AUGUSTUS in AD 14. In the following century SENECA THE ELDER wrote concise philosophy, TACITUS' contemporary history was incisive and accurate, and PETRONIUS satirized contemporary life. Others, such as PLINY THE ELDER, described life and natural phenomena with detachment. After *c.*AD 100, Latin literature began to decline, until it was taken up again by the Christian authors such as St AUGUSTINE. *See also* HC1 pp. 112–113.

Latitude and longitude, imaginary lines on the Earth's surface; they form a system of co-ordinates that define the position of any point on Earth, measured in degrees. Lines of latitude, parallel with the equator, measure the distance N or S of the equator, one degree of latitude being equal to approx. 110km (70 miles). Lines of longitude measure the distance along the equator E or W from the prime meridian or 0°, which passes through Greenwich, England. Latitude and longitude are also used in celestial co-ordinate systems. *See also* PE p.34, *34.*

Latium, historic district in central Italy, inhabited by the Latini. It was bordered by Etruria (NW), Campania (SE) and the Tyrrhenian Sea (W). Augustus made Latium and Campania the first of the 11 administrative divisions of Italy, and the latter name prevailed after AD 292.

Latosol, type of soil, reddish in colour occurring widely in tropical countries. It shows little differentiation into soil horizons and has poor fertility. *See also* PE pp.166–167, *167.*

La Tour, Georges de (1593–1652), French painter of religious and genre scenes. His early style was influenced by MANNERISM; examples include *St Jerome.* Later he was influenced by CARAVAGGIO; his later works include *Christ and St Joseph in the Carpenter's Shop* (*c.*1645) and the *Lamentation over St Sebastian* (1645). *Job and his Wife* is an early example of his nocturnal scenes, in which forms are dramatically lit by a candle or another concealed source of lighting.

La Tour, Maurice Quentin de (1704–88), French painter, who was popular as a pastel portraitist. His works include a portrait of *Mme de Pompadour* (1756).

Latrobe, Benjamin Henry (1766–1820), British architect who emigrated to the USA in 1796. His chief work is Baltimore Cathedral (1806–18). He also designed part of the Capitol, Washington, DC.

Latter-day Saints. *See* MORMONS.

Latvia (Latvijskaja SSR), constituent republic of the USSR bordered by the Estonian (N) and Lithuanian (S) republics; the capital is Riga. It is a fertile, low-lying region in NE Europe drained by the River Dvina and largely composed of forests and lakes. The main occupation is farming on a collective basis. The area was conquered by the Germans and Swedes and passed to Russia in the 18th century. It was occupied by the Germans in WWII and was retaken by Soviet troops in 1944. Area: 63,700sq km (24,595sq miles). Pop. (1976) 2,499,000.

Latvian, or Lettish, language spoken principally in the Latvian SSR of the USSR. Together with Lithuanian it forms the

East Baltic subgroup – the oldest western form of the Indo-European language family. There are approximately 1.5 million speakers.

Laud, William (1573–1645), Archbishop of Canterbury (1633–45). Laud considered the Church of England to be a branch of the universal church, claimed apostolic succession for bishops and believed that the Anglican ritual should be followed strictly in all churches. Working closely with Charles I, he sought to remove PURITANS from important positions in the Church. Laud was impeached (1640) by the LONG PARLIAMENT. The House of Lords found him innocent of treason but the Commons condemned him to death under a bill of attainder. *See also* HC1 p.287.

Lauda, Niki (1949–), Austrian motor racing driver who, driving for Ferrari, won the world drivers' championship in 1975 and 1977. In 1976, when he was second, he suffered near-fatal burns and injuries in an accident at the German Grand Prix, yet was racing again 42 days later.

Laudanum, alcoholic tincture of OPIUM, prepared from granulated opium and formerly used to treat diarrhoea. In the 19th century, opium addicts such as DE QUINCEY frequently used laudanum. The name was originally given by PARACELSUS in the 16th century to a preparation made of gold and pearls mixed with opium.

Lauder, Sir Harry Maclennan (1870–1950), Scottish singer. His dress usually included a kilt and tam-o'-shanter, and he sang original ballads and comic songs, such as *Roamin' in the Gloamin'* and *I Love a Lassie.* The highest paid music-hall star of his day, he entertained in the British Isles, Australia and the USA. He was knighted in 1919.

Lauderdale, John Maitland, 1st Duke of (1616–82), Scottish political figure. A royalist, he was captured at the Battle of Worcester (1651) and imprisoned until 1660. He became chief administrator in Scotland (1660–80) at the restoration of CHARLES II. An able but dissolute man, his harsh rule in Scotland made him one of the most unpopular statesmen of the period.

Laudians, Church party in the reign of CHARLES I of England, followers of William LAUD, Archbishop of Canterbury. By supporting Laud's attempt to enforce uniformity of worship in an Anglo-Catholic direction, they helped to increase the opposition to Charles I.

Laudon, Gideon Ernst Freiherr von (1717–90), Austrian general. In the SEVEN YEARS WAR he defeated the Prussians at Kunersdorf (1759), Landshut (1760) and Schweidnitz (1761). He later fought in the War of the Bavarian Succession.

Laudonnière, René Goulaine de (*fl.*1562–82), French colonist. In 1564 he established a French colony at the mouth of the St Johns River in Florida, which was named Fort Caroline. By 1565 the colony was in ruinous disorder due to the enmity of the Indians without and mutiny within, and he narrowly escaped when most of the colony was murdered by the Spaniards.

Laue, Max Theodor Felix von (1879–1960), German physicist. He became professor at Zurich and later director of the Institute for Theoretical Physics in Berlin. Using IONS in a crystal as a grating, he produced X-RAY interference patterns, thus elucidating the nature of X-rays and providing a method of investigating crystal structure. This research won for him the 1914 Nobel Prize in physics. His work indicated that X-rays are ELECTROMAGNETIC RADIATIONS.

Laughing Cavalier, The, portrait by the Dutch painter Franz HALS. It was painted in 1624 and is now in the Wallace Collection in London. The subject's eyes look straight forward from the painting, although his body is turned obliquely. This creates the illusion that his gaze follows the viewer round the room.

Laughing gas. *See* NITROUS OXIDE.

Laughton, Charles (1899–1962), British stage and film actor. His stage debut was in 1926 and his film career began a year later. His voice control and intellectual gifts made him a great character actor, and he was the first British actor to appear at

the Comédie Française. His films include *Mutiny on the Bounty* (1935) and *The Hunchback of Notre Dame* (1939). Later he directed the film *The Night of the Hunter* (1955).

Launch pad, complex of buildings and equipment used to launch a rocket or space vehicle and comprising a blockhouse, storage tanks, gantry, or service tower, umbilical lines and exhaust deflector. *See also* MM p.*156.*

Lauraceae, large family of flowering plants, mostly evergreen shrubs and trees, including BAY and CHERRY LAUREL; it is represented in warm and temperate regions throughout the world. The leaves are often thick and leathery; the flowers are generally green and are followed by berries. The laurel family also includes CINNAMON and CAMPHOR.

Laurana, Luciano (*c.*1420–79), Dalmatian architect who became chief architect at the humanist court of Federigo da Montefeltro, Duke of Urbino, an enlightened patron of the arts. Together they set about transforming the Ducal Palace at Urbino into the most beautiful of Renaissance palaces, creating a building of Classical purity and completeness. Laurana's masterpiece was the great courtyard of the palace – precise in proportion and detail. Its impression of lightness is obtained by the arcading supported on its slender CORINTHIAN columns.

Laurasia, the ancient northern continent containing North America and Eurasia that formed when a rift, the Tethyan trench, split PANGAEA (the Earth's single original land mass) from east to west along a line slightly north of the Equator. The name is a combination of Laurentian, a geological period in North America, and Eurasia. *See also* PE pp.24–27.

Laurel. *See* BAY; CHERRY LAUREL.

Laurel and Hardy, US comedy team who starred in more than 200 films. Stan Laurel (1890–1965) (real name Arthur Stanley Jefferson), a British-born gag inventor and director, was the thin, bumbling oaf. The American Oliver Hardy (1892–1957) was the fat, pompous lady-chaser. Their best films, made before 1941, include *Leave 'em Laughing* (1928), *The Music Box* (1932) and *Way Out West* (1937).

Laurence, Margaret (1926–), Canadian author of penetrating emotional novels. Although she has lived in Africa and England, many of her novels are set in the Canadian West. Her books include *The Stone Angel* (1964) and *A Bird in the House* (1970).

Laurencin, Marie (1885–1956), French painter, designer and illustrator. Her paintings, typically of women and girls, are marked by serenity and gentleness. They include *Portrait* (1913) and *La Femme au Chien* (1938).

Laurier, Sir Wilfred (1841–1919), Prime Minister of Canada (1896–1911) and the first French Canadian to lead a federal party – the Liberals (1887–1919). Laurier opposed "clerical" politics and was an exponent of trade tariffs. He campaigned for clemency towards the RIEL rebels and supported Canada's involvement in WWI. *See also* HC2 p.137.

Lauri-Volpi, Giacomo (1894–), Italian tenor. He studied in Rome and made his debut in 1920 in Massenet's opera *Manon.* He sang at the Metropolitan Opera in New York (1923–34) and appeared in more than 233 performances in 26 roles.

Lausanne, city in W Switzerland on Lake Geneva; capital of Vaud canton. Originally a Celtic settlement, it became an episcopal see in the 6th century and was ruled by the prince-bishops until 1536, when it was conquered by Bern and accepted the Reformation. At one time the home of VOLTAIRE and ROUSSEAU, it has a notable Gothic cathedral (1275) and a university (1891). Industries: leather, brewing, chemicals, woodworking. Pop. (1970) 137,883.

Lausanne, Treaty of (1923), agreement signed at Lausanne, Switzerland, which abrogated the harsh Treaty of SÈVRES (1920) imposed on the collapsing OTTOMAN EMPIRE and ended the war between

HC1 = History and Culture Vol. 1   HC2 = History and Culture Vol. 2   MS = Man and Society   MM = Man and Machines

Greece and Turkey. Turkey was granted full sovereignty over mainland Turkey and renounced claims to Greek islands in the Aegean Sea. Britain obtained Cyprus, and Italy obtained Rhodes and the Dodecanese Islands.

Lautaro (*c.* 1530–57), Indian who led the Araucanian Indians in their almost successful attempt to regain s central Chile from the Spanish. In 1553, Lautaro totally destroyed an army under Pedro de VALDIVIA, but in 1557, he was killed at the battle of Peteron.

Lautrec, Henri Toulouse. *See* TOULOUSE-LAUTREC, HENRI.

Lava, molten rock or magma that reaches the Earth's surface and flows out through a volcanic vent in streams or sheets. There are three main types of lava: vesicular, like pumice; glassy, like obsidian; and evengrained. Chemically, lavas range from acidic to ultrabasic, although the majority of all lavas are basic. Basic lavas have a low viscosity and flow easily, covering large areas. Acid lavas are highly viscous and rarely spread far. *See also* PE pp.30–31, *30–31*.

Laval, Carl Gustaf Patrik de (1845–1913), Swedish inventor whose inventive talent extended over various disciplines including aerodynamics, electric lighting and metallurgy. His most notable inventions were a high-speed cream separator and its steam-powered turbine, and a milking machine.

Laval, Pierre (1883–1945), French political figure. At first a socialist, he moved steadily to the right as he progressed through ministerial appointments to become Prime minister (1931–32, 1935–36). In 1940 he entered the government of Marshal PÉTAIN at Vichy and became its head in 1942. His capitulation to German demands earned him bitter hostility from some sections in France. He was executed for treason when WWII ended.

Laval, city in s Quebec province, Canada, on the Île-Jésus at the confluence of the Ottawa and St Lawrence rivers. Laval was created in 1965, when 14 small communities on the islands were combined. Mainly a residential suburb of Montreal, which lies across the river to the SE, Laval has iron and steel, paper products, machinery and pharmaceutical industries. Pop. (1976) 200,037.

Laval de Montmorency. *See* MONTMORENCY-LAVAL.

Lavalleja, Juan Antonio (1784–1853), Uruguayan revolutionary. Commander-in-Chief of Uruguayan forces during the conflict with Brazil (1825–28), he emerged as a dominant political figure after the war, helping Manuel Oribe lead the faction known as the Blancos (Whites) against Fructuoso RIVERA'S Colorados (Reds) during the Civil War (1839–1851). These rival groups have dominated Uruguayan politics to the present day.

La Vallière, Louise Françoise de la Baume le Blanc, Duchesse de (1644–1710), French mistress of LOUIS XIV by whom she had four children. She was supplanted by Madame de MONTESPAN and in 1674 retired to a Carmelite nunnery.

Lavender, any of several aromatic herbs or small shrubs of the genus *Lavandula* of the mint family Labiata. *L. spica* (or *L. officinalis* or *L. vera*) is a Mediterranean woody perennial grown as an ornamental. When dried, its fragrant flowers are used to perfume linens and clothing; both flowers and leaves are used in making lavender water. *See also* MM p.*246*; NW p.*60*.

Lavengro (1851), largely autobiographical novel by George BORROW which describes the youth and later wanderings of a scholarly man who is interested in languages and foreign lands. He meets and travels with a family of gypsies. This book, like its sequel *The Romany Rye* (1857), draws strength from Borrow's intimate knowledge of gypsies and his evocations of life in the wilder places of Europe.

Laver, Rodney George ("Rod") (1938–), Australian tennis player. He won the US (1962, 1969), British (1961, 1962; 1968, 1969), Australian (1960, 1962, 1969) and French (1962, 1969) singles championships, becoming the first man to win the "Grand Slam" twice (1962, 1969). Laver turned professional in 1963, and in 1970 became the first professional tennis player to achieve the two-million-dollar mark in total earnings.

Laver, any of several edible maroon gelatinous seaweeds especially the Asian *Porphyra umbilicus,* which is pressed and dried for use in cooking. In Europe, red laver (*P. vulgaris*) may be stewed or pickled. *See also* RED ALGAE; NW p.*47*.

Laveran, Charles Louis Alphonse (1845–1922), French physician. He contributed greatly to tropical medicine and was awarded the 1907 Nobel Prize in physiology and medicine for his work on the role played by PROTOZOA in causing diseases. Laveran discovered the parasite responsible for MALARIA.

Lavigerie, Charles Martial Allemande (1825–92), French Roman Catholic cardinal. After serving the Church in France he was appointed Archbishop of Algiers in 1867. He was noted for his anti-slavery campaign and missionary work among the Muslims. In 1868 he established the WHITE FATHERS, a Society of Missionaries of Africa, in his quest for the Christian conversion of the continent.

Lavinia, in Roman mythology, daughter of Latinus, King of the people of LATIUM. She married AENEAS, who built the city of Lavinium in her honour. Her jealousy caused the suicide of Anna, DIDO's sister, who had sought refuge with Aeneas.

Lavoisier, Antoine Laurent (1743–94), French chemist. His careful experiments enabled him to demolish the PHLOGISTON theory by demonstrating the function of oxygen in combination. He named both oxygen and hydrogen and showed how they combined to form water. In collaboration with BERTHOLLET and others he published *Methods of Chemical Nomenclature* (1787), which laid down the modern method of naming substances. His *Elementary Treatise on Chemistry* (1789) established the basis of modern chemistry. He was guillotined during the Revolution. *See also* SU pp. 25, 134.

Law, Andrew Bonar. *See* BONAR LAW, ANDREW.

Law, Denis (1940–), British footballer who played for Huddersfield Town, Manchester City, Torino (Italy), Manchester United and then Manchester City again. He also played many times for Scotland, scoring a record 30 goals. During the 1960s Law was one of the most effective forwards in the world.

Law, John (1671–1729), Scottish economist, who went to France after his monetarist theories were rejected in Britain in 1700. He inaugurated the "Mississippi Scheme" for American colonization. He founded a bank in 1716 which became the Banque Royale in 1718, and he became French controller-general of Finance (1720). Speculation in the shares of his Mississippi Company, however, brought about his ruin in 1719–20, and he lived in Venice until his death.

Law, William (1686–1761), British clergyman and writer. One of the NONJURORS, Law was deprived of his fellowship in Emmanuel College, Cambridge, and lost all hope of advancement in the Church. In consequence, he retired to a life of private devotion, writing on controversial theological issues of current interest. In the field of devotional writing, his *Serious Call to a Devout and Holy Life* (1728) ranks high and influenced John WESLEY.

Law, system of rules governing human society, enforced by punishments specified by society itself. The major systems of law operating in the world are common law, Roman law and equity. *See also* MS pp. 282–287.

Law, civil. *See* CIVIL LAW.

Law, constitutional. *See* CONSTITUTIONAL LAW.

Law, criminal. *See* CRIMINAL LAW.

Law, international. *See* INTERNATIONAL LAW.

Law Courts, since 1925 officially called the Royal Courts of Justice, in the Strand, London. The business of the superior courts of England is transacted there. In 1858 a parliamentary commission recommended that buildings be erected to bring together in one place all the superior courts of law and equity, the probate and divorce courts and the court of admiralty. The scheme was adopted in 1865. In a public competition the design of George Edmund Street, in the Gothic revival manner, was chosen for the new buildings. They were officially opened in 1882.

Law enforcement. *See* POLICE FORCES.

Lawes, Henry (1596–1662), English composer who became a Gentleman of the Chapel Royal in 1626 and who fought as a ROYALIST in the Civil War, in which his brother William (also a musician) was killed. In 1634 he arranged the music for John MILTON's masque *Comus;* he also set many of the poems of Robert HERRICK to music. He composed many songs and anthems, including *Zadok the Priest* (1662) for the coronation of Charles II.

Lawes, Sir John Bennett (1814–1900), English agriculturist who pioneered the manufacture of artificial fertilizers. He set up a factory for the production of superphosphates in 1842. His researches helped to establish agriculture as a science, and Lawes founded the Rothamsted Experimental Station in 1843.

Lawler, Ray (1921–), Australian playwright and actor who played the leading role, Barney Ibbot, in his *Summer of the Seventeenth Doll* (1956). Other works include *The Piccadilly Bushman* (1959) and *A Breach in the Wall* (1970).

Lawless, Emily (1845–1913), Irish novelist. Her writings include the novels *Hurrish* (1886) and *Grania* (1892) and the verse collection *With the Wild Geese* (1902).

Law Lords, group of peers in Britain. They comprise the Lord CHANCELLOR, Lords of Appeal in Ordinary (life barons of judicial experience appointed to assist the House of Lords in the hearing of appeals), ex-Lord Chancellors and other peers who have held high judicial office. The Law Lords constitute the House of Lords when it sits as a court of appeal.

Lawn tennis, ball game played by either two (singles) or four (doubles) players. It is played with a cloth-covered rubber ball about 6.55–6.8cm (2.58–2.67in) in diameter. Each player wields a racket for which there are no official specifications. Generally however, it has an oval-shaped top, about 25cm (10in) in diameter, strung with gut, and a handle about 41cm (16in) long. It is commonly made of wood, although steel and graphite rackets were introduced in the mid-1960s. The game is played on a court 23.8m (78ft) by 8.2m (27ft) for singles. For doubles play the court is widened to 11m (36ft). It is bisected by a net 0.9m (3ft) high at the centre and 1.1m (3.5ft) at the sideposts which hold it. On each side of the net there are two service areas marked by rectangular lines 6.4m (21ft) long from the net and 4.1m (13.5ft) wide from the centre. The ball is put into play by the server, who is allowed two attempts to hit it into the opposite service court. The first service of each game is delivered from the right-hand court of the server into his opponent's right-hand service area. Thereafter, the service alternates from left to right. One player serves for a complete game. If the opponent returns the ball safely, play continues until one player fails to hit the ball, hits it into the net or hits it outside the confines of the court; his opponent then wins the point. A minimum of four points are required to win a game, which must be won by two clear points. A minimum of six games must be won to win a set, which must be won by two clear games (although recently tie-break games have been introduced at six games all or eight games all). In leading championships, men's matches are won by the best of five sets, women's matches by the best of three. Modern lawn tennis evolved from REAL TENNIS in England in the 1860s. The first lawn tennis club was formed at Leamington Spa in 1872. The first WIMBLEDON championships were held in 1877.

Lawrence, Charles (1709–60), British soldier and governor of Nova Scotia (1756–60). He founded the town of Lunenberg, (1753), ordered the

Lavender; blue and white flowers of the species *Limontum,* known as statice.

Antoine Lavoisier was a chemist who devised many chemical symbols.

Law courts; the main entrance to the Royal Courts of Justice, opened in 1882.

Lawn tennis championships at Wimbledon were played for the first time in 1877.

Lawrence, David Herbert

Gertrude Lawrence, a friend of Noël Coward, made her first stage appearance aged 12.

Thomas Lawrence's works combine classical idealization and individual likeness.

T. E. Lawrence, the enigmatic Englishman who led a successful Arab revolt in WWI.

Johnny Leach, 1949 champion, learned table tennis during military service in WWII.

deportation of the ACADIAN French (1755) and commanded a brigade at the siege of Louisbourg (1758).

Lawrence, David Herbert (1885–1930), British writer, whose views on sex and the dehumanizing effects of industrialized Western culture made him a controversial figure. A coal-miner's son, he escaped the pits by continuing his education and began his career as a teacher. His first poems were published in the *English Review* (1909), followed in 1911 by his first novel, *The White Peacock*. In 1912 he eloped with Mrs Frieda Weekly (née von Richthofen), a German noblewoman, and later married her. During WWI they were suspected of being spies. Lawrence was tubercular for most of his life but continued to write and to travel. His novels include *Sons and Lovers* (1913), *The Rainbow* (1915), *Women in Love* (1920), and *Lady Chatterley's Lover* (1928), at one time banned in Britain and the USA for its explicit sexual descriptions. He also wrote short stories, including *St Mawr, Sun* and *The Woman who Rode Away*.

Lawrence, Ernest Orlando (1901–58), US physicist. He became professor at the University of California w here he built the first CYCLOTRON, a sub-atomic particle accelerator. His invention earned him the 1939 Nobel Prize in physics. The element LAWRENCIUM was named after him. *See also* SU p.67.

Lawrence, Gertrude (1898–1952), British actress, comedienne, singer and dancer. A spirited and much loved figure of the US and English stages, she made her New York debut in *Charlot's Revue* (1924), with Beatrice LILLIE, and appeared with Noël COWARD in *Private Lives* (1930) and *Tonight at 8.30* (1934–36). She starred on Broadway in Rodgers and Hammerstein's prizewinning production *The King and I* (1951). In 1945 she wrote her autobiography, *A Star Danced.*

Lawrence, James (1781–1813), US naval hero. During the War of 1812, while a commander of the *Chesapeake*, he was mortally wounded by fire from the British frigate *Shannon* in Boston Harbour and uttered the slogan "Don't give up the ship", which became a popular battle cry.

Lawrence, John Laird Mair Lawrence, 1st Baron (1811–79), British colonial official. He entered the administration of India in 1830 and aided British expansion during the SIKH WARS. He also kept the Punjab out of the INDIAN MUTINY. He became Viceroy and Governor-General of India in 1864 and initiated major public works.

Lawrence, Sir Thomas (1769–1830), British painter, mainly of portraits. The high quality of his artistic style shows the influence of Joshua Reynolds, as in *Miss Farren*, painted when Lawrence was 20. Well-known portraits by him include *Queen Charlotte* (c.1789) and *J. P. Kemble as Hamlet* (1801).

Lawrence of Arabia. *See* LAWRENCE, THOMAS EDWARD.

Lawrence, Thomas Edward (1888–1935), British archaeologist, soldier and author known as Lawrence of Arabia. He worked on the excavation of a Hittite settlement in the Euphrates valley from 1911 until 1914, when he became an intelligence officer in Cairo. He joined the Arab revolt against the Turks in 1916, and proved himself an extremely successful guerrilla fighter, leading Arab forces first into Aqaba (1917) and then into Damascus in October 1918. Disillusioned by the repeated failure of his plans for an independent Arab state, and disturbed by his war experiences, he finally sought obscurity in the ranks of the RAF as T.E. Shaw. In 1926 he published his memoirs of the Arab revolt as *The Seven Pillars of Wisdom*. He also translated the *Odyssey* (1932). He died of injuries sustained in a motorcycle accident.

Lawrencium, radioactive metallic element (symbol Lr) of the ACTINIDE group, first made in 1961 by bombarding CALIFORNIUM with boron nuclei. The element has been made in only trace amounts but has been identified chemically. Properties: at. no. 103; most stable isotope Lr^{256} (half-life 8 sec.).

Lawrie, Lee (1877–1963), US sculptor. He attended Yale University where he also taught. He sculpted many large figures for public buildings, churches and office buildings, including the gigantic figure of Atlas that stands in front of the Rockefeller Center's International Building in New York City.

Laws of Motion. *See* MOTION, LAWS OF.

Lawson, Henry Archibald (1867–1922), Australian writer regarded by his fellow-countrymen as the most representative of the period of Australia's emergence into nationhood. His short stories represent his finest work, although his popular verse made him the laureate of the Australian concept of "mateship".

Lawson, John (d.1711), English pioneer in North America. He made extensive explorations of North Carolina and helped to establish the settlements of Bath and New Bern. Appointed surveyor-general in 1708, he was captured and killed during the Tuscarora Indian uprising.

Lawson, Sir John (d.1665), English naval commander who fought against the Barbary pirates. Imprisoned briefly by Oliver Cromwell for plotting against him, Lawson became Commander-in-Chief of the English fleet after the fall of Richard Cromwell (1659).

Lawson, John Howard (1895–   ), US dramatist. He viewed the theatre as a means of changing society. In the 1920s he wrote experimental plays such as *Processional* (1925), and co-founded the Workers' Drama League (1926) and the New Playwrights (1927). His Hollywood screenplays include *Blockade* (1938), *Sahara* (1943) and *Counterattack* (1945). He wrote *Theory and Technique of Playwriting* (1936, revised 1949).

Law terms, known since 1813 as sittings, division of legal year into sittings of the courts. They are the Hilary sitting (Jan.-March), The Easter sitting (April-May), the Trinity sitting (June-July) and the Michaelmas sitting (Oct.-Dec.).

Lawton, Thomas ("Tommy") (1919–   ), British footballer who played for Everton, Chelsea, Notts County, Brentford, Arsenal and England. From 1938 to 1948 he played in 23 internationals. One of the most popular forwards of his day, he was a strong player: he had a powerful shot and was also a good header of the ball.

Laxness, Halldór Kiljan (1902–   ), Icelandic novelist. He was awarded the 1955 Nobel Prize in literature for his novels about the poor and underprivileged of the fishing villages and farms of Iceland. He became a socialist and after 1930 his works, written in the traditional style of the epics, aroused controversy. His novels include *The Great Weaver of Kashmir* (1927), *Salka Valka* (1931–32), *Independent People* (1935), *The Atom Station* (1948) and *Paradise Reclaimed* (1962).

Layamon (*fl. c.*1200), MIDDLE ENGLISH poet. He was a priest attached to the church at Ernley, in Worcestershire, England. His most famous work, *Brut*, is a chronicle in verse of the history of Britain and is a major source for the Arthurian legend.

Layard, Austen Henry (1817–94), British archaeologist, b. France, whose work in excavating Mesopotamian areas, especially the site of Nineveh, helped to further knowledge of Babylonian and Assyrian cultures. Many of his discoveries are in the British Museum. After 1852 he served as a diplomat and politician. *See also* HC1 p.23.

Laye, Evelyn (1900–   ), British actress and singer who established her reputation in light musical comedy in a season of *The Merry Widow* in 1923 and whose popularity was at its peak in the 1920s and 1930s. After WWII, she remained popular, and continued to play regularly on the West End stage in *Charlie Girl* (1969) and in situation comedies such as *No Sex Please – We're British* (1971–73).

Layering, method of plant propagation that induces or encourages root formation on a stem or branch while it is still attached to the parent plant; BLACKBERRIES and LOGANBERRIES multiply naturally in this

way. *See also* PE pp.178, *178*, 220.

Lay of the Last Minstrel, The (1805), poem by Sir Walter Scott, a ROMANTIC narrative poem, set in the Scottish borders.

Lay Reader, modern term for the Anglican Reader, who performs certain duties for the regular clergy ranging from lesson-reading, preaching and reading of morning and evening prayers (except the Absolution) to administration of the chalice at Holy Communion. This office was regulated by Convocation in England in 1905, and is reviewed frequently.

Lays of Ancient Rome, volume of poetry published by Thomas Babington MACAULAY in 1842. They are long epic poems, written in a heavy rhetorical style. The most famous of them is *Horatius*.

Layton, Irving (1912–   ), Canadian poet. He was born in Romania, and moved to Montreal as an infant. His poetry, which covers a variety of topics and moods, includes *The Improved Binoculars* (1956) and *Balls for a One-Armed Juggler* (1963).

Lázár, György (1924–   ), Hungarian politician and economist. He became Minister of Labour (1970–73) and Deputy Prime Minister (1973–75), and was appointed Prime Minister (1975–   ).

Lazarillo de Tormes (1554), novel doubtfully attributed to Diego Hurtado de Mendoza. One of the first picaresque novels, its success established the genre. Its depiction of the clergy caused the unexpurgated edition to be banned.

Lazarists, name given to members of the Congregations of the Mission, a Roman Catholic religious order. It was founded by St Vincent de Paul in Paris in 1625, and later moved to the priory of St Lazare.

Lazarsfeld, Paul Felix (1901–   ), US sociologist. He is important for his contributions to social research techniques, and is well-known for his study of the behaviour of voters, *The People's Choice* (1944). Among his other books is *Qualitative Analysis* (1972).

Lazarus, name of two men in the New Testament. In John 11:1–44, Lazarus is the brother of Mary and Martha of Bethany. Jesus miraculously restored him to life four days after his death. In Luke 16:19–31 Lazarus is the poor man in Christ's parable about a beggar and a rich man.

Lazio. *See* LATIUM.

L-dopa, also known as levadopa, drug used to alleviate some of the symptoms of PARKINSON'S DISEASE. It sometimes suppresses trembling, ataxia and the slow movement which characterize the condition. Parkinsonism is believed to result from a depletion of dopamine in the brain, but treatment with dopamine is ineffective as it will not cross the physiological barrier between the blood and the brain. L-dopa does cross this barrier, however, and is converted to dopamine in the brain. The side-effects of L-dopa may be serious and include cardiac irregularities and nausea.

Lea, Charles Henry (1825–1909), US historian. His *History of the Inquisition in the Middle Ages* (3 vols, 1888) is still considered the standard work on the subject.

Leach, Bernard (1887–   ), one of the most influential English artist potters of the 20th century. He studied (1909–20) in Japan, where he gained a high reputation. In 1920 he established a pottery at St Ives, Cornwall, which became the most important centre of English art pottery. Leach rediscovered many traditional techniques of English pottery and his designs, simple and beautifully proportioned, are a mixture of English and Japanese influences. His son David (1911–   ) worked with Bernard Leach until 1955, when he started his own pottery at Bovey Tracey, Devon.

Leach, Sir Edmund Ronald (1910–   ), British social anthropologist. He studied and wrote on Burmese and Ceylonese (Sri Lankan) political and social systems and is well known for his books *Rethinking Anthropology* (1961) and *Genesis as Myth* (1970).

Leach, Johnny (1922–   ), British table-tennis player who won the world singles title in 1949 and 1951, and helped the

England team to win the Swaythling Cup in 1953. He never won the English Open singles but won two doubles titles (1951, 1953) and four mixed doubles (1950, 1952, 1954, 1956).

Leaching, in geology, removal from solid of soluble and other minerals by the prolonged action of drainage water. This generally decreases soil fertility, but saline soils can be reclaimed for agriculture by deliberately leaching out salts.

Leacock, Stephen Butler (1869–1944), Canadian man of letters, *b.* Britain. He is best known as a humorist and his *Literary Lapses* (1910) are delightful for their gentle satire and love of the absurd. He also wrote *Sunshine Sketches of a Little Town* (1912) and an incomplete autobiography *The Boy I Left Behind Me* (1946). He taught economics and political science at McGill University, Montreal, and also wrote critical studies of Mark Twain (1932) and Charles Dickens (1933). *See also* SU p.50.

Lead, metallic element (symbol Pb) of group IVA of the periodic series, known from ancient times. Its chief ore is GALENA (lead sulphide), from which lead is obtained by roasting. It is used in plumbing, batteries, cable sheaths and alloys such as solder and type metal. The element is used in making the petrol additive TETRAETHYL LEAD. It is also used as a shield for X-rays and other radiation. Chemically, it is unreactive and resists corrosion. Properties: at.no. 82; at.wt. 207.19; s.g. 11.3; m.p. 327.5°C (621.5°F); b.p. 1,750°C (3,182°F); most common isotope Pb^{208} (52.3%).

Leadbelly (1888–1949), US composer and blues singer, real name Huddie Ledbetter. Folklorist John A. Lomax discovered him in prison where he was serving a sentence for attempted murder, and used his songs for a book *Negro Folk Songs as Sung by Lead Belly* (1936). He is best known as the composer of many classic blues songs including *Goodnight Irene, The Midnight Special* and *Rock Island Line. See also* HC2 p.272.

Lead poisoning, condition caused by absorption of lead by the digestive tract, lungs or skin. It occurs among children who eat flakes of paint containing lead and among workers in industries using lead and in areas near busy motorways, where the lead comes from TETRAETHYL LEAD, an anti-knock additive in most petrol. Its first symptoms, such as constipation, anaemia and irritability, often go unnoticed until serious effects such as convulsions occur. If untreated it can be fatal.

Lead sulphide, compound of LEAD, formula PbS, occurring naturally as the mineral GALENA. It is a toxic, black powder, insoluble in water, used as a semiconductor and in ceramics.

Leadwort, popular name for a family of herbs and shrubs (Plumaginaceae) found in the dry regions of the Mediterranean and Central Asia. They include thrift, sea lavender and plumbago, each of which has a flimsy, spherical flower.

Leaf, part of a plant, an organ that contains the green pigment CHLOROPHYLL and is involved in PHOTOSYNTHESIS and TRANSPIRATION. It consists of a blade and a stalk (petiole), which attaches it to a stem or twig. Most leaves are simple; some are compound, and divided into leaflets. Modifications include succulent types with fleshy tissue for water storage, tendrils that coil around supports, and needles, common in conifers. *See also* NW pp.44–45.

Leaf beetle, small, oval BEETLE that feeds on leaves. Many, including the COLORADO BEETLE (*Leptinotarsa decemlineata*), are yellow with black markings. This beetle does great damage to potato crops and in many countries the government must be informed of infestation. *See also* NW p.113.

Leaf hopper, any of numerous species of small, slender insects of the family Cicadellidae. A leaf hopper feeds by sucking the sap of plants and may, in large numbers, do great damage. Many species are brightly coloured. *See also* NW p.216.

Leaf insect, any of several species of flat, green insects which bear a resemblance to leaves. Leaf insects are found throughout tropical Asia. The female has large leathery fore-wings with markings resembling a pattern of leaf veins. The eggs of these insects often resemble the seeds of various plants. Order Phasmida; family Phylliidae. *See also* STICK INSECT; NW pp.106–107, *214–215.*

Leaf-nosed bat, any of a number of species of small insectivorous BATS that live in tropical and sub-tropical regions. It has fleshy leaf-like organs on the end of its snout that may aid its ECHO LOCATION mechanism. Families: Hipposideridae and Phyllostomatidae. *See also* NW p.161.

League of Nations, international organization (1920–46), forerunner of the UNITED NATIONS. Created as part of the Treaty of Versailles ending WWI, it required that members respect the territorial independence of all other members and recognize the need for disarmament. The refusal of the USA to participate impaired the League's efficiency although during the first few years of its existence it was instrumental in preventing war in the Balkans and in settling several inner-European disputes. Despite extending financial and administrative aid to poorer countries and furthering co-operation in the field of international relations, the problems of imposing League decisions on the great powers was never solved. The threats to world peace from Germany, Spain and Japan caused the League to collapse in 1939. It was dissolved in 1946. *See also* HC2 pp.192–193, *218,* 226–227.

Leakey, name of three British anthropologists and archaeologists whose researches in East Africa contributed important new information concerning the origins of man. Louis Leakey (1903–72), and his wife Mary (1913–), worked together. In 1959 at Olduvai Gorge, Tanzania, they unearthed a hominid fossil named as Zinjanthropus or *Australopithecus,* with remains of *Homo habilis,* thought to be 1,750,000 years old and a direct ancestor of modern man. Louis Leakey was born in Kabete, Kenya, the son of English missionary parents. He grew up there among the Kikuyu people. He was curator of the Coryndon Museum of Nairobi from 1945 to 1961. He taught in Britain, the USA and Africa. Mary Leakey is director of the Olduvai Gorge Excavations. Their son, Richard (1944–), is director of the National Museums of Kenya.

Lean, David (1908–), British film director, producer and scriptwriter. A gifted film editor, his early films include *In Which we Serve* (1942) and *Brief Encounter* (1945). His later works, such as *The Bridge on the River Kwai* (1957), *Lawrence of Arabia* (1962), *Dr. Zhivago* (1965) and *Ryan's Daughter* (1970) have been lavish productions with historical backgrounds.

Lear, Edward (1812–88), British poet and artist. In his nonsense verse, including *The Book of Nonsense* (1846) and *Laughable Lyrics* (1877), he found escape from the mannered idealism of Victorian life; he illustrated his books with witty, expressive line drawings and was noted for his watercolour landscapes.

Learning, acquisition of skills and concepts by a variety of processes. There is little agreement on what constitutes the basis of learning, but two schools of thought may be distinguished: one which holds it to be an associative process and one which believes it is the understanding of Gestalts (ordered perceptual units) rather than separate ideas. ARISTOTLE gave three forms of association: similarity, contrast and temporal and spatial contiguity, and these form the basis of many of the types of learning studied by psychologists. Gestalt psychologists deal with such learning potentials as problem solving and the formation of principles, but there is much overlapping in the two approaches. Recently it has become obvious that the processes of memory and learning are inextricably linked, and a clear theory of one requires an explanation of the other. *See also* MS pp.44, 152.

Lease, document of legal transaction conferring all rights in property, except that of ownership, from one party (the lessor) to another (the lessee) for a specified period. It may apply to lands, buildings, rights of common and rents. Leases for life (now usually for 99 years) give security for capital expenditure without diminution of the lessor's estate.

Leasehold, possession of an estate for a limited period. Property held in leasehold may carry with it prohibitions (on subletting, use as business premises) and obligations (to keep it in good repair) stated in the lease. *See also* FREEHOLD.

Leather, animal hides cured by TANNING to prevent decay and increase flexibility; often finished by glazing, enamelling, or laquering and coloured by staining or dyeing. Suede is produced by raising a nap on the flesh side by buffing with emery. Leather has been used since earliest times as a strong, supple material for clothing.

Leatherback turtle, omnivorous marine turtle found in tropical waters and the largest of all living turtles. It has a smooth, black, leathery skin with seven ridges running lengthwise, and no external plates. Its forelegs are modified into large flippers. Length: to 2.1m (7ft); weight: approx. 540kg (1,200lb). Family Dermochelyidae; species *Dermochelys coriacea. See also* NW p.137.

Leather-Stocking Tales, The, five novels by James Fenimore COOPER. The novels, *The Pioneers* (1823), *The Last of the Mohicans* (1826), *The Prairie* (1827), *The Pathfinder* (1840) and *The Deerslayer* (1841), are all set in the French and Indian wars and in post-revolutionary USA. *See also* HC2 p.288.

Leaves of Grass (1855), collection of autobiographical poems by Walt WHITMAN, revised and augmented until 1892. The crucial edition is the third (1860), which contains many new poems including *Out of the Cradle Endlessly Rocking* and regroupings of previous ones. *See also* HC2 p.288, *288.*

Leavis, Frank Raymond (1895–), British literary critic. He was editor of the influential literary quarterly *Scrutiny* (1932–53) and his works of criticism, combining close textual analysis with moral principles of evaluation, include *The Great Tradition* (1948) and *D. H. Lawrence, Novelist* (1955). His views on society and education are expounded in such works as *Mass Civilization and Minority Culture* (1930) and *Education and the University* (1943).

Leavitt, Henrietta Swan (1868–1921), US astronomer noted for her work on CEPHEID VARIABLE stars and stellar magnitudes. She discovered that the periods of Cepheids are related to their true brightness. The cepheid period-luminosity law has since been used to determine the distances of stars. *See also* SU p.238, *238.*

Lebanon, independent nation in the Middle East, bordering Syria (N and E) and Israel (S). It has a Mediterranean climate although more than half of the land is desert. Its economy has been based on trade, which yields two-thirds of the national income, and tourism, but these were adversely affected by the civil war in the 1970s. There are also oil pipeline terminals from Saudi Arabia and Iraq on the coast. The capital is Beirut, the Middle East's foreign exchange centre. Area: 10,400sq km (4,015sq miles). Pop. (1975 est.) 2,869,000. *See also* MW p.117.

Lebedev, Peter Nikolaievich (1866–1911), Russian physicist noted for demonstrating that light exerts minute pressure on solid bodies.

Lebedev, Sergei Vasilyevich (1874–1934), Russian chemist. He developed a method of producing a synthetic rubber by POLYMERIZATION of butadiene obtained from ethanol. His process was widely used by the Russians and the Germans in WWII.

Le Bel, Joseph-Achille (1847–1930), French chemist who correctly attributed the OPTICAL ACTIVITY of some organic molecules to their three-dimensional molecular structure. This pioneering work, which he published simultaneously with, but independently of, Jacobus van' t HOFF in 1874, greatly advanced STEREOCHEMISTRY. *See also* SU p.139.

Lebensraum, term used by Adolf HITLER to justify German territorial expansion. It

Leadbelly was particularly popular with the followers of US folk music.

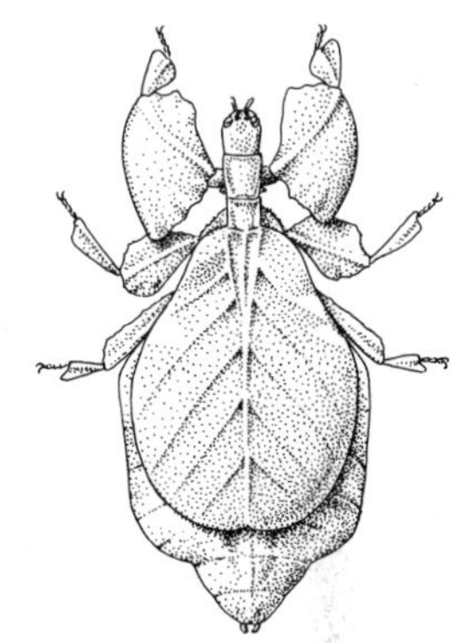

Leaf insects, such as this *Phyllium* species, are related to stick insects.

Edward Lear's *Lake of Nemi, Rome,* shows his closeness to the Pre-Raphaelites.

Leatherback turtles come to land from deep water only to lay their eggs.

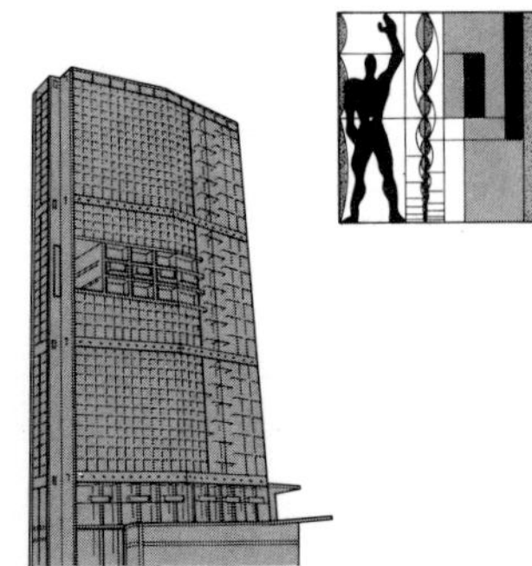

Le Corbusier transformed space with the use of reinforced concrete.

Leda intertwined with Zeus in the form of a swan; by Michaelangelo.

Joshua Lederberg taught at Wisconsin University until 1959.

Gypsy Rose Lee generally regarded as the first striptease artist.

originated with Friedrich Ratzel (1844–1904), but the Nazi distortion of Ratzel's concept derived from Rudolf Kjellen (1864–1922) who saw the state as a life-form and coined the term geopolitics. Lebensraum means "room in which to live".

Leblanc, Nicolas (1742–1806), French chemist. He was physician to Louis Philippe, duc d'Orléans, who helped him to establish a factory for producing sodium carbonate (washing soda) from salt.

Lebon, Phillipe (1767–1804), French inventor who pioneered the development of gas-fuelled lamps which he called "thermolampes". He devised a method of obtaining gas from sawdust.

Le Brocquy, Louis (1916–) Irish painter and teacher of textile design; largely self-taught. Many of his paintings depict symbolic figure groups. More recently, since about 1974, he has been working on a series of some 150 portraits of British novelists.

Lebrun, Albert (1871–1950), last President of France's Third Republic (1932–40). In 1940 the establishment of the VICHY GOVERNMENT deprived him of all authority and in 1944 he recognized Charles DE GAULLE as provisional President.

Lebrun, Charles (1619–90), French painter, a student under VOUET, and influenced by POUSSIN. He painted religious, mythological, and historical subjects, as well as portraits in a decorative Classical style. Chief painter to LOUIS XIV from 1662, he designed much of the decoration at Versailles, including the Galerie de Glaces (1679–84), and controlled the arts in France for some 20 years; he was also director of the GOBELINS tapestry factory from 1633. *See also* HC1 p.309.

Le Carré, John, pen-name of David John Moore Cornwell (1931–), British writer. His first novel was *Call for the Dead* (1961). His best-known books are *The Spy Who Came in from the Cold* (1963) and *A Small Town in Germany* (1968).

Le Châtelier's principle, principle announced by the French chemist Henry-Louis Le Châtelier (1850–1936) in 1888. It states that if a system in a state of equilibrium is disturbed, the system tends to neutralize the disturbance and restore the equilibrium.

Leclair, Jean-Marie, the Elder (1697–1764), French BAROQUE composer and violinist. His 49 violin sonatas are counted among the foundation works of the violin repertoire. He also composed violin concertos and an opera, and was a master in both the serious and the lighter styles of his time.

Leclanché cell, electric cell invented by Georges Leclanché, c.1865. Its ANODE was a zinc rod and its CATHODE a carbon plate surrounded by packed manganese dioxide. These electrodes were dipped into a solution of ammonium and zinc chlorides. It is the basis of the DRY CELL.

Leclerc, Jean (1657–1736), Swiss scholar. He became Protestant pastor in 1675 and was a teacher in Amsterdam. He edited the *Universal and Historical Library* (1686–93) and advocated liberal theology in numerous tracts.

Le Clezio, Jean-Marie Gustave (1940–), French novelist. An exponent of the NOUVEAU ROMAN, he achieved recognition with *The Interrogation* (1963), followed by *Fever* (1965), a collection of short stories, and *The Flood* (1966).

Lecocq, Alexandre Charles (1832–1918), French composer of operettas. These number more than 40, and are remarkable for their wit and charm. His first major success was *Fleur de thé* (1868). *La Fille de Madame Angot* (1872) played for more than a year in Paris.

Leconte de Lisle, Charles-Marie-René (1818–94), French poet. He was leader of the PARNASSIAN **school and his work which** includes *Poèmes antiques* (1852), *Poèmes et poésies* (1855) and *Poèmes barbares* (1862), is disciplined and pessimistic. He was elected to the Académie Française in 1866.

Le Corbusier, (1887–1963), Swiss-born French architect, real name Charles-

Édouard Jeanneret, among the most influential of the 20th century. His early work, mostly housing, exploited the qualities of reinforced concrete with cube-like forms and incorporated principles which are summarized in his famous *Five Points of a New Architecture* (1925): pilotis (columns) to lift the building off the ground; the flat roof with a garden; the free plan; long windows; and the free facade, with floors and walls extending beyond their supporting columns. All of these were embodied in the Villa Savoy at Poissy (1929–31), one of his most influential buildings. Le Corbusier's extension of these ideas was illustrated in his Unité d'Habitation, Marseilles (1946–52), designed on pilotis as one unit of a new-type city and scaled to his MODULOR. Its design was widely adopted for modern mass housing. Later Le Corbusier evolved a less strictly rational, more poetic, style, of which the highly sculptural chapel of Notre-Dame-du-Haut at Ronchamp (1955) is the finest example. In the 1950s he also laid out the town of Chandigarh, the new capital of the Punjab, and built its majestic supreme courts and secretariat. He designed Tokyo's Museum of Western Art (1957) and the Cistercian Monastery of La Tourette (1960) in central France, a concrete block of immense force. His last major work was the Visual Arts Center at Harvard (1963). *See also* HC2 pp.179, 206–207, *328*.

Lecouvreur, Adrienne (1692–1730), French actress. She made her debut at the Comédie-Française in 1717 and her naturalistic acting style soon superseded the pedantic diction of her contemporaries. Her tragic roles included Racine's *Phèdre* and Corneille's *Bérénice*. Her many lovers included VOLTAIRE, who attacked the French Church ferociously after it denied her a Christian burial because she was an actress.

Lectisternium, rite in ancient Greece and Rome in which images of gods and goddesses were displayed on couches and offered food. In Christian times the term described a feast in memory of the dead.

Leda, in Greek mythology, Queen of Sparta and the mother, either by her husband Tyndareus or by ZEUS, of CASTOR AND POLLUX and HELEN. Zeus appeared to her in the form of a swan and both Pollux and Helen were believed to have hatched from eggs.

Lederberg, Joshua (1925–), US geneticist. In 1958 he shared with G. W. BEADLE and E. L. TATUM the Nobel Prize in physiology and medicine for work that initiated the study of bacterial genetics. He discovered that sexual recombination of genetic materials occurs in bacteria and that genetic materials are linked in groups in bacteria as well as in other organisms.

Ledoux, Claude-Nicolas (1736–1806), French architect. His projects for buildings based upon geometrical solids gained him a reputation as one of the most imaginative of 18th-century designers. He built 45 toll houses around Paris and planned an ideal city, "Chaux", which was never built. *See also* HC2 p.43.

Le Duan (1907–), politician, theorist and military organizer of North Vietnam. He was a founder of the Indo-Chinese Communist Party in 1930, joined the Viet-Minh in 1945 and helped to organize the NLF (National Liberation Front) of South Vietnam in 1960. He later became First Secretary of the Vietnam Workers' Party, and in the early 1970s was considered one of the most influential men in the country.

Ledum, genus of small, evergreen shrubs that grow in damp, acid soils of cold northern climates. They have small, white flowers. Family Ericaceae.

Ledwidge, Francis (1891–1917), Irish poet. During WWI he served in the GALLIPOLI CAMPAIGN of 1915 and was killed in Belgium. His verse includes *Songs of the Field* (1915), *Songs of Peace* (1916) and *Last Songs* (1918).

Ledyard, John (1751–89), US explorer. He sailed with Capt. James COOK in search of the North-west Passage (1776–79) and published *A Journal of Captain Cook's Last Voyage to the Pacific Ocean* (1783).

With the approval of Thomas JEFFERSON, he planned an expedition from Russia to Virginia by way of Siberia and the Bering Strait. He died in Africa preparing an expedition to its interior.

Led Zeppelin, British rock group formed in 1968. Since 1970 they have attracted record crowds in the USA. Their best-selling record, *Physical Graffiti,* was released in 1975.

Lee, Ann (1736–84), British mystic, member of the United Society of Believers in Christ's Second Appearing, popularly called the SHAKERS. The sect was persecuted in Britain, and in 1774 she and eight others fled to the USA. She founded a colony near Albany, N.Y., in 1776 and gained many converts.

Lee, Arthur Hamilton, Lord Lee of Fareham. *See* CHEQUERS.

Lee, Bruce (1940–73), Chinese-American film actor. He won little recognition until he starred in a number of Hong Kong Kung Fu films, such as *Fist of Fury* (1973) and *Enter the Dragon* (1973). His karate skills and ferocious snarl made him a cult hero.

Lee, Gypsy Rose (1914–70), real name Rose Louise Hovick; US performer, reputedly the inventor of the "striptease". She first appeared as Gypsy Rose in the 1936 Ziegfeld Follies. The musical *Gypsy*, staged in 1959 and filmed in 1962, is based on her memoirs.

Lee, Laurie (1914–), British writer. His poetry, eg *The Sun My Monument* (1944), *The Bloom of Candles* (1947) and *My Many-Coated Man* (1955), is attractively simple. His best-selling autobiography *Cider with Rosie* describes with immense charm his childhood in a remote Cotswold village. He is the author of *As I Walked Out One Midsummer Morning* (1969) and *A Rose for Winter* (1955), accounts of his travels in Spain and Cyprus, and of a verse play *The Voyage of Magellan* (1948).

Lee, Nathaniel (c.1653–92), English RESTORATION dramatist. whose elaborate verse tragedies about historical events and figures such as *Nero* (1674), *Oedipus* (1679), *Lucius Junius Brutus* (1680) and *The Massacre of Paris* (1689) were extremely popular.

Lee, Peggy (1920–), US singer and songwriter, real name Norma Egstrom, who made her debut singing with the Benny Goodman Orchestra in 1938. She wrote and recorded many popular songs such as *Fever, Mañana* and *Johnny Guitar* and collaborated on the musical score of Walt DISNEY's film *The Lady and The Tramp* (1956), in which she also sang.

Lee, Robert Edward (1807–70), commander of Confederate forces in the American Civil War. Although Lee regarded slavery as evil and saw advantages in the Union, his loyalty to his native Virginia was paramount. Declining LINCOLN's offer of the field command of Union troops, he became military adviser to Jefferson DAVIS, the Confederate president, and in 1862 was appointed commander of the Confederate forces. In that year he successfully defended Richmond and won the second battle of BULL RUN. He was defeated at ANTIETAM (1862) but inflicted on the Union its worst defeats at Fredericksburg (1862) and Chancellorsville (1863), although in the latter Lee lost his most able lieutenant, Thomas "Stonewall" JACKSON. Lee's attempt to penetrate the North ended in his defeat at Gettysburg in July 1863. He surrendered to Ulysses S. GRANT at Appomatox Court House on 9 April 1865. *See also* HC2 p.151.

Lee, Tsung-Dao (1926–), US physicist, *b.* China. He showed that among subatomic particles, the law of conservation of parity (that nature, in effect, makes no distinction between right- and left-handedness) does not always hold. Working with Chen-Ning YANG, Lee suggested that in certain types of sub-atomic reactions, parity is not conserved, and this was subsequently verified by experiment. They were awarded the 1957 Nobel Prize in physics.

Leech, John (1817–64), British illustrator and caricaturist who worked for *Punch* to which he was a consistently prolific contributor from 1841. His cartoons were occasionally of a political nature, but he was much more at home with more subtle and gentle characterizations. Leech illustrated Charles Dickens's *A Christmas Carol* (1844), several sporting novels of his friend Robert Surtees, and did much work for *The Illustrated London News*.

Leech, any of numerous species of freshwater, marine and terrestrial ANNELIDS found in tropical and temperate regions. Its tapered, ringed body is equipped with a sucking disc at each end. Many species live on the blood of animals. Length: 13–51mm (0.5–2in). Class Hirudinea. *See also* MS p.83; NW pp.89, *89, 230*.

Leeds, city and county district in West Yorkshire, N England. The city of Leeds, on the River Aire, is part of one of England's major industrial regions and is a communications centre. The city's woollen industry dates back to the 14th century; Leeds is the centre of England's wholesale clothing trade. Other industries include metal goods such as aircraft components, machinery, engineering, paper and printing and chemicals. Area: (county district) 562sq km (217sq miles). Pop. (county district, 1974 est) 748,300; (city, 1971) 494,971.

Leek, biennial plant related to the onion, of the family Liliaceae; it originated in the Mediterranean region and is cultivated widely for culinary purposes. Species *Allium porrum. See also* PE p.188.

Lee Kuan Yew (1923–), Singapore politician. He was a founder of the Socialist People's Action Party. He was first elected to the Singapore assembly in 1955 and became Prime Minister of Singapore in 1959.

Lee of Asheridge, Janet Bevan (Jennie Lee), Baroness (1904–), British politician, wife of Aneurin BEVAN. She was a Labour MP in 1929–31 and 1945–70. She was Under-Secretary at the Ministry of Education (1965–67) and Minister for the Arts (1967–70).

Leeuwenhoek, Antonie van (1632–1723), Dutch scientist. He was a scientific amateur who built simple microscopes with a single lens, made so accurately that they had better magnifying powers than the compound microscopes of his day. He investigated many micro-organisms and their life histories, and described various microscopic structures such as spermatozoa. He was made a Fellow of the British Royal Society in 1680.

Leeward Islands, N group of the Lesser Antilles in the E West Indies, extending sw between Puerto Rico and the Windward Islands. The group includes the Virgin Islands (US), Guadaloupe (French), the Netherlands Antilles (St Maarten, Saba and St Eustatius), the British Leeward Islands (Antigua, St Kitts-Nevis, Montserrat), the British Virgin Islands and Anguilla. First settled by the British in the 17th century, possession was contested by France and Britain for two centuries. The islands are a popular winter resort area. Products: limes, coconuts, tobacco, dairy produce, vegetables. *See also* articles on individual islands in *The Modern World.*

Le Fanu, Joseph Sheridan (1814–73), Irish novelist and short story writer. His works are chilling and mysterious. They include the novels *The House by the Churchyard* (1863) and *Uncle Silas* (1864), and a collection of stories, *In a Glass Darkly* (1872).

Lefebvre, Pierre François, Duc de Danzig (1755–1820), French military commander. During the French Revolution he rose from sergeant to general, and later distinguished himself under NAPOLEON, becoming a marshal in 1804. While governor of Paris he supported the coup in 1799 by means of which Napoleon became first consul. After voting for the Emperor's abdication in 1814 he rejoined Napoleon during the HUNDRED DAYS.

Lefebvre, Georges (1874–1959), French historian, an authority on the French Revolution. His research for a doctorate on the peasantry of N France brought a new depth to the study of the Revolution and his general interpretation, in such books as *The French Revolution from its Origins to 1793* (1951), dominated the

subject after WWII.

Left wing, in politics, term which has come to mean radical or revolutionary. It originated in France after the French Revolution, and comes from the seating arrangement of the National Assembly in 1789, where the nobles sat on the president's right and the commons on his left. This arrangement has persisted in most European parliaments.

Legal tender, money which is valid in payment of a debt or discharge of a business obligation. In Britain, bank notes are legal tender in any amount. Bank cheques, postal orders and promissory notes are not legal tender. Bronze coins are legal tender to 20 pence; silver or cupro-nickel are legal tender to £5 with the exception of the 50p piece which is legal tender to £10.

Legal aid, in Britain, system by which those who have a certain income can receive free or subsidized legal representation or advice. In criminal cases it is paid for mainly from public funds; in civil cases its cost may be partly met from the costs awarded by the court. Legal aid was introduced in 1949, its provisions now being covered by the Legal Aid and Advice Act, 1974.

Le Gallienne, Eva (1899–), British-born US actress, director and translator. She made her London acting debut in 1915 and went to New York in that year. Her first success there was in Ferenc Molnár's *Liliom* (1921). She founded and directed the Civic Repertory Company in New York (1926), the first such company in the USA. Among her translations are *Six Plays by Henrik Ibsen* (1958) and *Seven Tales by Hans Christian Andersen* (1959).

Legate. *See* PAPAL LEGATE.

Legazpi, Miguel López de. *See* LÓPEZ DE LEGAZPI, MIGUEL.

Legendre, Adrien Marie (1752–c. 1833), French mathematician. He studied the theory of NUMBERS and elliptic integrals, and through his work on quadratic residues discovered the law of reciprocity used by K. F. GAUSS. He introduced the method of least squares, to calculate the paths of comets. His most influential work was *Elements of Geometry* (1794).

Léger, Alexis St-Léger (1887–), French poet, *b.* the West Indies, real name St-John Perse. He emigrated to the USA in 1940, after a decade as a powerful anti-appeaser in the Foreign Office. His first volume of poetry, *Eulogies*, came out in 1911; his complete works were published in 1972. He won the 1960 Nobel Prize in literature.

Léger, Fernand (1881–1955), French painter who, by 1911 was a leader of the CUBIST movement. Greatly influential, he was admired by MONDRIAN, MALEVICH and many others. His major works are of contemporary subjects – especially the working man and machinery – which he endows with monumental stature, conveyed in bold colour areas and flat black outlines.

Leger, Jules (1913–), Canadian politician and diplomat. An historian and associate editor of *Le Droit* (1938–39), Leger was Assistant Under-Secretary of State for External Affairs (1952–53), and ambassador to Italy (1962–64), France (1964–68) and Belgium and Luxembourg (1973). In 1974 he was appointed Governor-General of Canada.

Leghorn, any of about 12 varieties of lightweight chickens; it was originally bred in the Mediterranean region, especially for egg production.

Legion, basic organizational unit of the Roman army from the early Republic to the fall of the empire in the West in the 5th century AD. The legion had its origins in the Roman citizen army. During the great period of Rome's expansion, a legion was about 6,000 men strong, consisting mainly of heavy infantrymen (legionaries), with some light troops and cavalry in support. The legion was subdivided into cohorts (420 men each), maniples (120 men each) and centuries (100 men each). The success of this organization over a period of almost 800 years was due to its flexible fighting formation (which consisted of three mutually supporting lines) and to the quality of its non-commissioned officers, especially the CENTURIONS.

Legion of Honour (Légion d'Honneur), French award, created by NAPOLEON in 1802 to reward civil and military service. The award can be in one of the various classes, the highest of which is the great cross (*grand-croix*). The most common level at which it is awarded is that of Knight of the Legion (*Chevalier de la Légion*), which is marked by a distinctive red ribbon.

Legislation. *See* LAW.

Legislative Assembly, legislative body of France from Oct. 1791 to Sept. 1792, one of the most critical periods of the FRENCH REVOLUTION. Political conflicts intensified, not only between the proponents and opponents of change, but also among politicians and classes previously united in the pursuit of reform (as between the JACOBINS and Feuillants). The threatening attitude of European powers led to the declaration of war against Austria in April 1792. Early defeats provoked a crisis in the summer which was partly resolved when the King (LOUIS XIV) was deposed by the rising of 10 Aug. 1792. In Sept. the National CONVENTION met and proclaimed a republic. *See also* HC2 p.74.

Legitimacy, term denoting the legal status of a child born to parents in wedlock. A child born out of wedlock is said to be illegitimate. In Britain, the Legitimacy Acts of 1926 and 1959 removed the stigma of illegitimacy by giving a child legitimate status if the parents subsequently marry. In some countries, such as New Zealand, there is now no legal distinction between a legitimate and illegitimate child.

Legitimate theatre, term that arose in the 18th century to refer to plays entirely dependent on acting, with little or no singing, dancing or spectacle.

Legitimists, those people in France who, after the JULY REVOLUTION (1830) had forced the last BOURBON king, CHARLES X, off the throne, continued to support the Bourbon cause.

Legnano, Battle of (1176), defeat of FREDERICK I Barbarossa and the Holy Roman Empire forces by the LOMBARD LEAGUE (aided by Venice and the pope). In a foretaste of the rise in importance of foot soldiers, the Lombard League's infantry repelled the invasion of Italy by Frederick's mounted knights.

Legnica, Battles of. *See* LIEGNITZ, BATTLES OF.

Legouvé, Gabriel-Marie-Jean-Baptiste (1764–1812), French dramatist. He wrote *La Mort d'Abel* (1792), *Épicharis* (1793), *Étéocle et Polynice* (1799) and *La Mort de Henri IV* (1806).

Legros, Alphonse (1837–1911), British painter, *b.* France; a printmaker and sculptor who is best known for his drawings of sinister and fantasy subjects. A fine draughtsman, he was associated with Gustave COURBET, James WHISTLER and Édouard MANET.

Leguminosae, pea family of flowering plants, including many trees, shrubs, vines and herbs whose roots contain nitrogen-fixing bacteria. Woody species occur mainly in tropical regions; herbaceous species mainly in temperate regions. The fruit is typically a pod (legume) containing a row of seeds. Important food species include the PEA, runner bean, SOYA BEAN, LENTIL, broad bean, kidney bean and haricot bean. Familiar decorative species – some of which are poisonous – include LABURNUM, BROOM, WISTERIA and SWEET PEA. *See also* PE pp.184–185.

Lehár, Franz (1870–1948), Hungarian-born composer of operettas. He studied music in Prague and from 1890 travelled as a bandmaster in Austria. His first operetta (*Kukuschka*) was written in 1896. He composed more than 30 operettas of which *The Merry Widow* (1905) is the most popular today.

Le Havre, commercial seaport in N France, at the mouth of the River Seine on the English Channel. Primarily a fishing and naval port until 1815, the city is now the principal export point for Paris and a transatlantic passenger port. During WWII it was almost completely destroyed, but was subsequently rebuilt. It was the base of the Belgian government after the fall of Antwerp and Ostend. Industries:

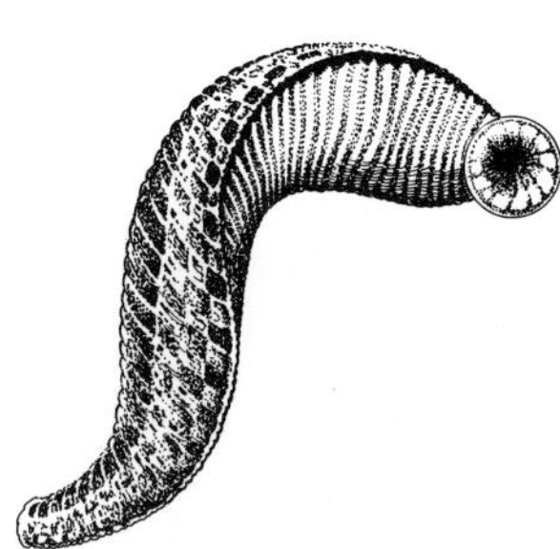

Leech; many species can absorb up to five times their own body weight of blood.

Leeds; Sir Thomas Brock's statue of the Black Prince in the City Square.

Leeuwenhoeck's microscope enabled him to make detailed drawings of tiny creatures.

Legion of Honour, the cross of merit awarded to distinguished French citizens.

Lotte Lehmann was a famous soprano who also wrote a novel and an autobiography.

Wilhelm Leibl's *Three Women in Church* (1878) – a masterpiece of German realism.

Vivien Leigh making a broadcast from the stage of the Drury Lane Theatre in 1941.

Leipzig was a major European cultural centre during the 17th and 18th centuries.

chemicals, fertilizers, timber, food processing, oil-refining, shipbuilding. Pop. (1968) 199,509.

Lehmann, Beatrix (1903–), British actress whose first London stage appearance was in 1924 and who appeared regularly in the West End as well as at the Old Vic, Chichester and Stratford-on-Avon, where she played the nurse in *Romeo and Juliet* in 1973. She has also acted in films and on television.

Lehmann, John (1907–), British critic, poet and publisher. He was general manager of Virginia and Leonard WOOLF's Hogarth Press (1938–46). He is perhaps best remembered as the editor of *New Writing* and *Penguin New Writing*. His works include an autobiography (3 vols., 1955–62) and a study, *A Nest of Tigers: Edith, Osbert and Sacheverell Sitwell in Their Times* (1968).

Lehmann, Lotte (1888–1976), German-born US soprano. Making her debut in Hamburg in 1909, she was the most illustrious singer of operatic roles and lieder of her time. Richard STRAUSS created the title role of *Arabella* for her in 1933. She sang with the Vienna State Opera (1914–38) and the Metropolitan Opera, New York, from 1934 until her retirement in 1961.

Lehmann, Rosamund Nina (1903–), British author. Her first novel *Dusty Answer* (1927) concerns a young girl's progress to maturity through childhood and adolescence. Her later work includes *Invitation to the Waltz* (1932), *The Echoing Grove* (1953), a play *No More Music* (1939) and short stories.

Lehmann, Wilhelm (1882–1968), German poet, critic and novelist. His collections *The Green God* (1942) and *Enchanted Dust* (1946) started a new wave of nature poetry.

Lehmbruck, Wilhelm (1881–1919), German sculptor. He was a disciple of RODIN and MAILLOL and his masterpiece *Kneeling Woman* (1911) exemplifies the elongated forms and Gothic elegance characteristic of his work.

Leiber, Fritz, Jr (1910–), US science fiction writer and actor. From 1934 he toured for two years with his father's troupe before becoming a writer. A favourite theme in his books is the world after a nuclear war. He is best known for the Grey Mouser series, stories which combine fantasy and superstition.

Leibl, Wilhelm (1844–1900), German painter of realistic portraits and genre scenes. His works include *Three Women in Church* (1878) and *In the Kitchen* (1898).

Leibniz, Gottfried Wilhelm (1646–1716), German philosopher and mathematician. He never took an academic position, but served as a courtier and diplomat (in Paris, 1672–76), and corresponded with most of the leading thinkers of his day. He made many practical inventions, including a calculating machine (1671), and published his discovery of differential CALCULUS in 1684. He created a rationalist metaphysic, according to which the universe comprises a multitude of interrelated and organized units, called "monads", within a divine harmony. This belief led him to argue that evil is divinely motivated, and that this is the best of all possible worlds. He wrote voluminously, although few of his works were published during his lifetime. His major works include *New Essays Concerning Human Understanding* (1765) and *Monadology* (1898). *See also* HC1 p.302; SU pp.42–43.

Leibowitz, René (1913–), French-Polish conductor and composer. A student of Arnold SCHOENBERG, he used 12-tone music in his compositions and performed the works of other 12-tone composers. *See also* HC2 p.271.

Leicester, Robert Dudley, Earl of (*c.*1532–88), English nobleman, favourite of Queen ELIZABETH I. The mysterious death in 1560 of his wife, Amy Robsart, apparently cleared the way for Leicester to marry Elizabeth. The Queen, however, knew that the match would be politically unwise and instead offered his hand to MARY, QUEEN OF SCOTS. Leicester, however, remained influential, despite his feud with William Cecil, Lord BURGHLEY.

In 1585 he was placed in command of an expedition to help The Netherlands against the occupying Spanish, but his military efforts were undistinguished. *See also* HC1 pp.*276*, 277.

Leicester, city in central England, 145km (90 miles) NW of London; county town of Leicestershire. Founded by the Romans as Ratae Coritanorum, it was conquered by the Danes in the 9th century. It has a 14th-century church, a 15th-century guild-hall and many Roman remains. The University of Leicester was founded in 1957. The knitting frame was first installed there in 1680, and hosiery has been an important industry ever since. Other industries include footwear, textiles, and textile and woodworking machinery. Pop. (1971) 283,549.

Leicester(shire), county in central England. The uplands of the E are devoted to farming; the W has much industry. The area is drained chiefly by the Soar and Wreak rivers. Wheat, barley, sheep and dairy cattle are important, and the region is famous for its Stilton cheese and Melton Mowbray pies. Leicester, the county town, is the centre of England's footwear industry. Oil is extracted at Plungar. Area: 2,553sq km (986sq miles). Pop. 837,900.

Leichhardt, Friedrich Wilhelm Ludwig (1813–48), German explorer. He traversed Australia from Moreton Bay (Queensland) to Port Essington (Northern Territory) in 1844–45. This exploit made him an immediate hero in Australia and donations were made in abundance for a further expedition to cross the continent from east to west. Leichhardt, however, was wholly unqualified for the task and he and his entire team disappeared. No trace has ever been found of the party. *See also* HC2 p.*125*.

Leiden (Leyden), city in The Netherlands, on the River Oude Rijn, 15km (9 miles) NE of The Hague. Leiden is famous for its university (1575), the oldest in the country. The city was the birthplace of REMBRANDT and LUCAS VAN LEYDEN, whose works are displayed in the municipal museum. Industries: textiles, printing, food processing, metalworking. Pop. (1970 est.) 101,220.

Leif Ericksson. *See* ERIKSSON, LEIF.

Leigh, Vivien (1913–67), British film and stage actress, *b.* India, real name Vivien Hartley. Her early film career reached its peak with her performance of Scarlett O'Hara in *Gone With The Wind* (1939) for which she received an Academy Award. She also received an Academy Award for her moving portrayal of Blanche du Bois in *A Streetcar Named Desire* (1951). She was married to Laurence OLIVIER from 1937 to 1960.

Leigh-Mallory, Sir Trafford (1892–1944), British officer who helped to organize air defence during the Battle of Britain. In 1942 be became Air Marshal in RAF Fighter Command and in 1944 was made head of the Allied air offensive in Europe. He died in an air crash.

Leighton, Clare (1899–) British print-maker, illustrator and author who is best known for her engravings and woodcuts. She has illustrated the 1929 publication of *The Return of the Native*, by Thomas Hardy, and the 1931 edition of *Wuthering Heights*, by Emily Brontë. Her own writings include *How to do Wood-engraving and Woodcuts* (1932).

Leighton, Lord Frederick (1830–96), British painter, sculptor and illustrator. He first received acclaim for *Cimabue's Madonna Carried in Procession* (1855), which was exhibited at the Royal Academy and bought by Queen Victoria. With Edward POYNTER, he opposed the romance and realism of the PRE-RAPHAELITE BROTHERHOOD by refining a form of Hellenistic Classicism.

Leighton, Margaret (1922–76), British actress. She made her London debut at the Old Vic in 1944–45. Among her most famous roles were Celia in *The Cocktail Party* (1950), Orinthia in *The Apple Cart* (1953), Lucasta in *The Confidential Clerk* (1953) and Mrs Shanklin and Miss Railton-Bell in *Separate Tables* (1954). She was awarded an Oscar for her role in *The Go-Between* (1971).

Leighton, Robert (1611–84), Scottish minister and scholar. After the RESTORATION, Charles II appointed him Bishop of Dunblane (1661) and then Archbishop of Glasgow (1670) with the task of reconciling the PRESBYTERIANS to the episcopal method of Church government imposed on the Scots in 1660.

Leijster, Judith Jansdochter (1609–60), Dutch painter, whose portraits, still lifes and genre scenes anticipated those of VERMEER. Her work includes *The Rejected Offer* (1631).

Leinsdorf, Erich (1912–), Austrian-born US conductor. He was assistant conductor at the Salzburg Festival (1934–38) and was conductor at the Metropolitan Opera, New York (1937–43, 1957–62). He held a succession of conducting posts in the USA.

Leinster, Murray (1896–1975), real name William Fitzgerald Jenkins, US science fiction writer. Most of his stories deal with the conflict of man and nature. Three of his novels are collected in *A Murray Leinster Omnibus* (1968) and his previous works include *Gateway to Elsewhere* and *The Black Galaxy* (both 1954).

Leinster, province in E Republic of Ireland, made up of the counties of Carlow, Dublin, Kildare, Kilkenny, Laoighis, Longford, Louth, Meath, Offaly, Westmeath, Wexford and Wicklow. It is the most populous of Ireland's four provinces, and includes the most fertile farmland in the country. The major city is Dublin, Ireland's capital. Area: 19,635sq km (7,581sq miles). Pop. (1971) 2,989,088.

Leipzig, city in S central East Germany, at the confluence of the Pleisse, White Elster, and Parthe rivers. Founded as a Slavic settlement in the 10th century, it became a commercial centre at the intersection of important trade routes. It was the scene of the Battle of Nations in 1813, which ended the power of NAPOLEON in Germany. The birthplace of Richard WAGNER, it has the Karl Marx University (founded in 1409 as the University of Leipzig) and a 17th-century stock exchange. The printing industry, founded in 1480, is still of great importance. Other industries include textiles, machinery and chemicals. Pop. (1975) 586,877.

Leipzig, Battle of, also called Battle of the Nations (16–19 Oct 1813). The Prussians, Austrians, Russians and Swedes inflicted a crushing defeat on NAPOLEON's outnumbered *Grande Armée*, which resulted in the final retreat of the French from Germany.

Leishmaniasis, widespread tropical disease that causes great suffering to the victim. It is spread by SANDFLIES that are infested with small flagellate protozoans (*Leishmania donovani*). These are endoparasites in the human liver, spleen, skin or lymph nodes. In the bloodstream each is ingested by a white blood cell, changes shape from elongate to ovate and loses its flagellum. In Central and South America the disease is caused by *L. brasiliensis* and typically affects the membranes of the mouth and nose often eroding the nasal cartilage. *See also* MS p.106, *106*.

Leith Hill, highest point on the North Downs, in Surrey, SE England. On clear days both London and the Channel are visible from the summit. Height: 294m (965ft).

Leitmotiv, German word for a guiding theme in musical compositions. It is a theme which recurs throughout a work, usually an OPERA or a piece of PROGRAMME MUSIC, evocative on each occasion of an idea or a character. It is a device associated especially with WAGNER and Richard STRAUSS, both of whom used it consistently in their operas.

Leitrim, county in the N Republic of Ireland in Connacht province, bounded on the NW by Donegal Bay. It is hilly in the N, undulating in the S and is drained by the River Shannon and its tributaries. Although farming is the main occupation, the soil is not highly productive. Area: 1,525sq km (589sq miles). Pop. (1971) 28,313.

Leix. *See* LAOIGHIS.

Leland, John (*c.*1506–52), English antiquarian. King HENRY VIII's librarian from

1530, he was appointed king's antiquarian in 1533. He travelled throughout England, collecting historical and geographical data which were compiled into books of itineraries. His researches have been invaluable to later scholars. *See also* HC1 p.22.

Leloir, Luis Federico (1906–), Argentinian chemist, *b.* France. He was awarded the 1970 Nobel Prize in chemistry for his research into the biochemical processes which break down the complex sugars into simpler carbohydrates. In 1947 he helped to establish the Institute of Biochemical Research in Buenos Aires, where he began work on the production of LACTOSE. This research eventually led to the discovery of sugar nucleotides, fundamental factors in the natural process of carbohydrate synthesis. He also synthesized GLYCOGEN and demonstrated the necessity of certain liver enzymes for its manufacture.

Lely, Sir Peter van der Faes (1618–80), Dutch-trained painter who settled in London (*c.*1643), where he became the leading portrait painter after the death of VAN DYCK. At the Restoration he became principal painter to CHARLES II. Lely's paintings include *The Children of Charles I, The Windsor Beauties, Duchess of Cleveland* and the famous *Admirals.*

Lemaître, Abbé Georges Édouard, (1894–1966), Belgian astrophysicist and mathematician who first formulated the BIG BANG theory for the origin of the universe. He saw the universe as originally analogous to a radioactive atom, with all the energy and matter concentrated into a kernel which Lemaître called the "primeval atom". His book *The Primeval Atom: An Essay on Cosmogony* (1950) describes his theory. *See also* SU p.252.

Le Mans, motor racing circuit in France, world-famous as the venue of the classic Le Mans 24-hour race for sports cars. The result is determined by an Index of Thermal Efficiency based on average speed, weight, and fuel consumption over the 24 hours. Two drivers share a car. A feature of the race was the "Le Mans start", in which drivers ran across the track to their cars, but this was discontinued after 1969. In 1955, more than 80 spectators were killed there in motor racing's worst accident.

Lemercier, Jacques (*c.*1585–1654), French architect, the designer (1631) of the town of Richelieu, built for Cardinal RICHELIEU. Lemercier, who was influenced by the Italian MANNERIST style, became one of the founders of French Classicism and was architect to King Louis XIII. In Paris he built the Pavillon de L'Horloge at the Louvre, the Sorbonne and the dome of the Val-de-Grâce Church.

Lemesós. *See* LIMASSOL.

Lemming, any of several species of herbivorous RODENTS that live in cool and temperate regions. A lemming has long, brownish fur, small ears and a short tail. Lemmings characteristically migrate in large numbers. Species in Norway have been known to do so and to suffer great losses by drowning or rushing headlong over precipices. Family Cricetidae. *See also* NW pp.*198,* 206, *227.*

Lemmon, Jack (1925–), US film actor. Usually appearing in comedy roles, often as a diffident and awkward suitor, Lemmon is probably at his best in the films of Billy Wilder. These include *Some Like It Hot* (1959), *The Apartment* (1960), *The Fortune Cookie* (1966) and *The Front Page* (1975).

Lemon, evergreen tree and its familiar, sour, yellow citrus fruit. The flesh of the fruit is formed from hairs radiating from the seeds. Grown primarily in the USA and in subtropical regions, especially Italy, it is rarely eaten raw, but is used in cooking and in making marmalade. Height of tree: to 6m (20ft). *See also* PE p.192, *192.*

Lemonnier, Antoine Louis Camille (1844–1913), Belgian novelist and art critic, who led the 19th-century Belgian literary revival. His works, written in naturalistic style similar to that of Émile ZOLA, include *Un mâle* (1881), *Madame Lupar* (1888), *The End of the Bourgeois* (1893) and *The Secret Life* (1898).

Lemoyne, Jean-Baptiste (1704–78),

French sculptor at the court of Louis XV. He is remembered especially for his portrait busts which include *Réaumur* (1751), *Montesquieu* (1760), *Hélène d'Egmont* (1767) and several actresses from the Comédie Française. His decorative treatment of form mirrored the elegance of the French court and he was among the last French artists to work in the decorative tradition.

Le Moyne, Canadian family of administrators and soldiers. Charles (1625–85), Sieur of LONGUEUIL and of other seigneuries near Montreal, fathered 11 sons who included Charles, Baron de Longueil, governor of Montreal (1724–29); the explorers Pierre Le Moyne Iberville and Jacques Le Moyne de Sainte Hélène; Jean Baptiste Le Moyne Bienville, governor of Louisiana (1701–12, 1718–26, 1733–43); and Paul Le Moyne Maricourt, a soldier. The family dominated political and military life in Quebec for much of the 18th century.

Lemprière, John (*c.*1765–1824), British lexicographer. He was headmaster of Abingdon and Exeter grammar schools and then a vicar in Devon. His *Classical Dictionary,* published in 1788, was a standard text for many years.

Lemur, any of several primitive, mainly arboreal and nocturnal herbivorous PRIMATES that live in Madagascar. It resembles a squirrel, but has grasping monkeylike hands with which it climbs easily. Lemurs have changed little in 50 million years, closely resembling the ancestors of man and other primates. Family Lemuridae. *See also* NW pp.*168,* 172, 215.

Le Nain, name of three French painters, the brothers Antoine (*c.*1588–1648), Louis (*c.*1593–1648) and Mathieu (*c.*1607–77), whose works are indistinguishable. They were rediscovered in the 19th century and details of their lives are still vague, although it is thought that Louis was the most important. His *Family of Country People* (Louvre), a formal composition painted with a realism influenced by Velázquez, typifies his style. Antoine worked on a more modest scale and Mathieu painted portraits of cavaliers and religious pictures.

Lenard, Philipp Eduard Anton (1862–1947) Hungarian physicist who taught in Germany. He won the Nobel Prize in physics in 1905 for his discoveries of the properties of CATHODE RAYS. His work was important in the development of electronics and nuclear physics. He also did research on ultra-violet light, the electrical conductivity of flames, and PHOSPHORESCENCE.

Lenárt, Jozef (1925–), Czechoslovak politician. He took part in the resistance movement defying the Nazis in WWII, and was Prime Minister (1963–68).

Lenclos, Ninon de (1620–1705), French courtesan. Her salon attracted philosophers, men of letters and politicians, including MOLIÈRE and Voltaire's father. She was known for anti-religious views that caused her to be confined briefly to a convent. In her will she left money for books for the young VOLTAIRE.

Lend-lease, US programme of war relief, originally for Britain but later extended to most of the Allies of the USA in WWII, including China and the USSR. The programme was begun in 1941 and terminated in 1945. More than $50,000,000,000 worth of supplies were sent under the programme, on lenient terms of repayment; of these more than $30,000,000,000 was sent to Britain and the Commonwealth.

Lenglen, Suzanne (1899–1938), French tennis player. She dominated women's singles 1914–26, winning the women's singles and doubles at Wimbledon 1919–23 and 1925; French singles 1920–23, 1925–26; French doubles 1925–26. In 1926 she became one of the first women players to turn professional.

Lenin, Vladimir Ilyich (1870–1924), Russian politician, originally named Vladimir Ilyich Ulyanov. He was attracted to revolutionary politics while at school, particularly after his brother was executed for his part in an anti-tsarist plot (1887). He studied law at Khazan University

(1887–89), and practised as a lawyer (1893–95). He was exiled to Siberia (1897–1900) for his connections with PLEKHANOV's Marxist group, and in Munich he wrote *What is to be Done?* (1902). At the second congress of the Marxist Social Democratic Party in London (1903), he argued for an active, disciplined party to lead the workers on to revolution. This split the party into the BOLSHEVIKS (led by Lenin) and the MENSHEVIKS. He returned to Russia in Nov. 1905, and refused any alliance with the liberals. He returned to Europe (1907–17), and settled in Switzerland on the outbreak of WWI, analysing the causes of war in *Imperialism: the Last Stage of Capitalism* (1917). He returned to Russia with German assistance in April 1917, and refused to recognize KERENSKY's government, supporting the SOVIET MOVEMENT. He justified this policy in *The State and Revolution* (1917), and with TROTSKY organized the successful November coup. He became chairman of the Council of People's Commissars, negotiated the Treaty of BREST-LITOVSK with Germany, and in 1918 dissolved the freely elected Constituent Assembly. He organized the defence of the Bolshevik Revolution in the civil war (1918–20), declaring the independence of the separate nationalities of Russia, and formed the COMMUNIST INTERNATIONAL (1919) to promote the Russian Revolution abroad. He put forward a programme of economic reconstruction in 1921 (the New Economic Policy), but became ill in 1922 and died in Jan. 1924. *See also* HC2 pp.197–199; MS p.*279*; MW p.172.

Leningrad, second-largest city in the USSR, in NW Russian Republic (Rossijskaja SFSR); a major Baltic seaport at the E end of the Gulf of Finland, on the delta of the River Neva. Founded in 1703 as St Petersburg by PETER I The Great, it was the scene of the DECEMBRIST revolt of 1825 and the Red Sunday incident in the RUSSIAN REVOLUTION of 1905. The workers of Leningrad (then Petrograd) were the spearhead of the RUSSIAN REVOLUTION of 1917. The city was renamed Leningrad in 1924. It suffered extensive damage in WWII and has been rebuilt since 1945. A major cultural centre, Leningrad has the A. A. Zhdanov University (1819) and numerous libraries and educational institutions. Industries: shipbuilding, heavy engineering, brewing, printing and publishing, food processing, electronics, chemicals. Pop. 4,372,000. *See also* HC2 p.196, *196.*

Leningrad, Siege of (autumn 1941–early 1944), military struggle of WWII. Invading Germans cut off and besieged the Soviet city for 900 days, causing widespread famine and the death of almost one million of its three million inhabitants.

Lenni-Lenape. *See* DELAWARE INDIANS.

Lennon, John Ono (1940–), British rock singer, a member of the BEATLES. He appeared in the Beatles' films and in *How I Won the War* (1967). He published *In His Own Write* (1964) and *A Spaniard in the Works* (1965). *See also* HC2 p.273.

Lennox, Matthew Stuart (or Stewart), 4th Earl of (1516–71), Scottish nobleman. In 1544 he married Lady Margaret Douglas, HENRY VIII's niece, to advance his claim to the Scottish throne. He was murdered by supporters of MARY, QUEEN OF SCOTS, after being appointed regent of Scotland in 1570.

Le Notre, André (1613–1700), French landscape gardener. He established the French garden as the leading European style in succession to the Italian and before the ascendancy of the English style in the 18th century. Le Notre's formal gardens were a perfect setting for such stately French châteaux as Vaux-le-Vicomte and Chantilly and of royal palaces such as the Tuileries (Paris), Versailles and the Trianon.

Lens, piece of transparent glass, plastic, quartz etc, bounded by two surfaces, usually both spherical, that changes the direction of a light beam by REFRACTION and hence can produce a projected image. A converging lens is convex in form (bulging at the centre) and bends light rays

Sir Peter Lely painted these *Two Ladies of the Lake Family.*

Lemmings often panic when over crowded; many migrate recklessly.

Lemons, which probably originated in India, are rich in vitamin C and citric acid.

Lenin's image has dominated the USSR since the Russian Revolution in 1917.

Leon, the capital of the Leon kingdom has a 14th-century cathedral.

Leonardo da Vinci's *Gioconda, (Mona Lisa,)* has become a symbol for the notion of Art.

Ruggiero Leoncavallo rehearses his realist opera *I Pagliacci.*

Leopards are the third largest member of the cat family; after lions and tigers.

towards the lens axis. A diverging lens is concave (thinnest at the centre) and bends rays away from the axis. The image may be right-way-up or inverted, real or virtual, depending on the relative positions of object and focal point of the lens; it may also be magnified or reduced in size. Lens images suffer from various ABERRATIONS so that they may be blurred and have false colours. These characteristics also apply to the lens of the EYE. *See also* SIGHT; MS pp.48–49, *48–49*; SU p.100.

Lenski, Gerhard Emmanuel (1924–   ), US sociologist whose work covers social stratification, religion and evolution. His books include *The Religious Factor* (1961) and *Human Societies* (1970).

Lent, in the Christian liturgical year, the period before Easter. In the Western Churches it begins on Ash Wednesday and is 40 days long; in the Eastern Church it lasts 80 days. (The Sundays during the period are not reckoned as part of Lent.) It is a period of fasting, abstinence and penitence in preparation for the remembrance of the crucifixion and resurrection of Jesus Christ.

Lenthall, William (1591–1662), English political figure and parliamentarian. He was Speaker of the House of Commons from 1640 to 1653 and in 1642 refused to give up five members of the House to CHARLES I. He was Speaker again in 1654 and 1659, and in 1660 supported the RESTORATION.

Lenticular galaxy, type of elliptical GALAXY, resembling a convex lens in cross-sections.

Lentil, annual plant of the pea family that grows in the Mediterranean region, SW Asia and N Africa. It has feather-like leaves and is cultivated for its nutritious seeds, used as food, forage and a source of flour. Height: to 51cm (20in). Family Leguminosae; species *Lens culinaris. See also* PE p.*185.*

Lenya, Lotte (1900–   ), Austrian singer and actress. She became famous in two notable BRECHT plays with musical scores by her husband Kurt WEILL, namely *The Threepenny Opera* (Berlin, 1928) and *The Rise and Fall of the City of Mahagonny* (Leipzig, 1930). She emigrated to the USA in 1935.

Lenz's law, electromagnetic law deduced in 1834 by the Russian physicist Heinrich F. E. Lenz. It states that an induced electric current flows in the direction that tends to oppose the change producing it.

Leo, or the Lion, northern constellation lying on the ecliptic between Cancer and Virgo; the fifth sign of the Zodiac. The brightest stars are Alpha Leonis or Regulus, of the first magnitude, and Beta Leonis or Denebola, in the lion's tail, of the second magnitude. *See also* MS pp.200–201; SU pp.254, 257, *257, 264,* 265, *266,* 267.

Leo, name of 13 popes. Leo I (Saint Leo the Great) (440–461) was a vigorous campaigner against the MANICHAEANS and MONOPHYSITISM; his celebrated *Tome of Leo,* defining the two natures and one person of Christ, condemned the principles of EUTYCHES. Leo III, Saint (*r.*795–816), fled from Rome in 799 when physically attacked by the family of ADRIAN I, and sought the protection of Charlemagne. Returning in 800, he crowned Charlemagne emperor. This event traditionally marks the beginning of the HOLY ROMAN EMPIRE. *See also* HC1 pp.152, *153,* 160. Leo IV, Saint (*r.*847–55), stressed the papacy's role in state affairs. Restoring Rome after the Saracen attack of 846, he fortified the city and formed an alliance with other cities against invaders. Leo IX, Saint (*r.*1049–54), was born in Alsace and elected pope in the court of Emperor Henry III of Worms, but insisted on approval by the clergy and people of Rome. He formed the Sacred College of Cardinals, and travelled widely in support of Cluniac reform. *See also* HC1 pp.135, *135.* Leo X (*r.*1513–21), the son of Lorenzo de' MEDICI, was made a cardinal at the age of 13 and was head of his family before he was 30. A patron of learning and art, he made Rome the centre of the RENAISSANCE. His practice of selling indulgences

to finance the rebuilding of St Peter's provoked the theses of Martin LUTHER and led to the REFORMATION. *See also* HC1 pp.256, 268. Leo XIII (*r.*1878–1903) influenced Roman Catholic attitudes appropriate to living in the modern world. A devoted scholar, he opened the Vatican secret archives to the public and stressed truth in all things. He sponsored several universities, and advanced the philosophy of THOMISM to meet the intellectual attack on Christianity.

Leo, name of six eastern Roman emperors. Leo I (*c.*400–74; *r.*457–74) prevented barbarian domination of the Roman army by enlisting the help of the Isaurians, an Anatolian mountain tribe. Leo II (*d.*474; *r.*474), grandson of Leo I and named successor in 473, had an uneventful reign. Leo III (*c.*680–741; *r.*717–41), born N Syria, rose to power in the service of Justinian II. He became emperor after a revolt against Theodosius III and thereby established the Syrian dynasty. He successfully defended Constantinople against an Arab siege (717–18) and later expelled the Muslims from Asia Minor. Leo published the legal work *Ecloga* in 726 and caused the longstanding iconoclastic controversy by opposing the veneration of icons. Leo IV (749–80; *r.*775–80) was latterly an ICONOCLAST, but until 780 his policies on religious matters were moderate. Leo V (*d.*820; *r.*813–20) was, like Leo IV, an iconoclast, and held a synod in 815 that ordered the destruction of the icons. Leo VI "The Wise" (886–912; *r.*886–912) was responsible for the *Basilica,* a revision of Justinian's law, and a legislative code reflecting conditions in Byzantium.

Leon, Luis Ponce de (1527–91), Spanish poet, scholar and Augustinian monk, who was denounced to the Inquisition and imprisoned (1572–76). His prose works include *De Los nombres de Cristo* and *La perfecta casada* (1583). His poems combine Greek purity and Hebrew passion.

León, full name León de los Aldamas, city in Guanajuato state, central Mexico, at an altitude of 1,884m (6,182ft). It is an agricultural market and a mining centre. Mineral deposits include gold, copper, silver, lead and tin. Other industries are the manufacture of leather goods and textiles. Pop. (1975) 496,598.

León, city in W Nicaragua, South America, near the Pacific coast NW of Managua. It was founded in 1524 by Fernández de Córdoba and was moved to its present site after an earthquake in 1610. Industries: cigars, cotton gins and leather goods. Pop. (1970) 90,897.

León, Kingdom of, Spanish kingdom that occupied the region made up of what are now the provinces of León, Salamanca and Zamora. It was established in the 10th century by Garcia I and maintained its hegemony in Christian Spain for more than 100 years, but was unable to prevent the rise of Navarre and Castile. The kingdoms of Castile and León were twice united temporarily in 1037–65 and 1072–1157 before Ferdinand III accomplished a permanent union in 1230.

Leonardo da Vinci (1452–1519), Florentine painter, sculptor, architect, engineer and scientist. He was apprenticed to Andrea del VERROCCHIO at whose studio he remained probably until 1476. He was the founder of the Classic style of painting of the High Renaissance and was among the first to use the CHIAROSCURO technique. His early works include the unfinished *St Jerome* (*c.*1480) and a portrait, *Ginevra de' Benci.* In about 1482 he became civil and military engineer to Duke Lodovico Sforza in Milan, where he executed *Madonna of the Rocks, Portrait of a Musician* and numerous other works, including the *Last Supper* (1495–98). While in Milar he made many architectural drawings and designed and directed court festivals; he also began his scientific work and wrote the notes for what was to become his *Treatise on Painting.* In 1500 he returned to Florence where he executed the *Mona Lisa* and began *Virgin and Child with St Anne.* One of his last known paintings is *St John the Baptist* (*c.*1515). In 1517 he became chief painter, architect

and engineer to Francis I at Amboise, France, where he died. *See also* HC1 pp.*246,* 257.

Leonardo Fibonacci. *See* FIBONACCI, LEONARDO.

Leoncavallo, Ruggiero (1858–1919), Italian composer, mainly of operas. In 1876 he left the Naples Conservatory. The production of his first opera *Chatterton* (1878) was thwarted by a dishonest impresario. Leoncavallo then travelled all over Europe working as an accompanist and composer of music-hall songs. Although he was an industrious composer of operas, *I Pagliacci* (1892) alone has withstood the test of time.

Leone, Giovanni (1908–   ), one of the leaders of the Italian Christian Democratic Party and an elected delegate to the Constituent Assembly in 1946. He was Prime Minister in 1963 and 1968, and became President of Italy in 1971.

Leonidas (*d.*480 BC), King of SPARTA whose small force of Greeks met the vast Persian army, under XERXES, at THERMOPYLAE. They were overwhelmed after two days, but their heroic defence became a legend.

Leonids, meteor shower observable from about 14 to 18 Nov., associated with the 1866 comet. A spectacular display of this shower occurs every 33 or 34 years, the period of the comet. The last major Leonid shower was in 1966 when it was estimat that there were 300,000 meteors per hour.

Leonov, Leonid Maksimovich (1899–   ), Soviet novelist and dramatist. His first novel, *Barsuki* (*The Badgers*), was published in 1924. Other works include the novel *Russki les* (1953; "Russian Forest") and *Nashestviye* (*Invasion*), first performed in 1942, his best-known play. He won the Stalin Prize in 1943 and the Lenin Prize in 1957.

Leontief, Wassily (1906–   ), US economist, *b.* Russia. He developed the inputoutput table, which relates empirical data of general equilibrium and clearly illustrates the inter-industry relationships of an economy, or subcomponents of an economy. Such a table is an essential tool in the development of rational governmental planning. Leontief was awarded the 1973 Nobel Prize in economics.

Leopard, solitary spotted big CAT found throughout Africa and S Asia, sometimes called a panther. It has a round head with a short nose, and long, thin tail. The coat may be yellow and white with dark spots, or almost completely black, as in the black panther. A good climber and swimmer, it feeds on birds, monkeys, antelopes and cattle. The gestation period is three months and two to four young are born. Length: to 2.45m (8ft) including the tail; weight: to 90.6kg (200lb). Family Felidae; species *Panthera pardus. See also* NW pp.*164,* 215.

Leopard, snow, long-haired CAT of the family Felidae, in which it is grouped with the lion and other big cats; it lives in the mountains of central Asia. It has deep insulating underfur and a thick yellowgrey outer coat with dark rosettes. It hunts at night, preying on ibex, domestic livestock and wild sheep. Length: 2.1m (6.8ft), including the tail. Species *Panthera uncia. See also* NW p.*222.*

Leopard cat, small long-legged spotted CAT that lives in wooded hilly regions of SE Asia. It preys on a wide variety of birds and small mammals. Length: to 1m (39in) including the tail. Family Felidae; species *Felis bengalensis.*

Leopardi, Count Giacomo (1798–1837), Italian poet. Although basically a pessimist, his create the illusion of happiness, as in *I Canti* (1831) and *La Ginestra* (1836).

Leopold, name of two Holy Roman emperors. Leopold I (1640–1705; *r.*1658–1705), King of Hungary (*r.*1655–87) and King of Bohemia (*r.*1656–1705), was in almost constant conflict with either the Ottoman Turks or the French under Louis XIV. In 1697 he defeated the Turks at Zenta and gained most of Hungary. His armies fought against France in the Third Dutch War (1672–74) and in 1686 he formed the

League of AUGSBURG against France. In response Louis invaded the Palatinate in 1688, thus beginning the War of the Grand Alliance. The War of the SPANISH SUCCESSION broke out in 1701 but Leopold died before it ended. Leopold II (1747–92; *r.*1790–92), Grand Duke of Tuscany (1765–90), succeeded his brother Joseph II as head of a troubled empire and rapidly and fairly settled external disputes and internal rebellions. He encouraged the political ambitions of the lower and middle classes to forestall a revolution.

Leopold, name of three kings of Belgium. Leopold I (1790–1865), first king of independent Belgium (*r.*1831–65), served with the Allies against NAPOLEON. In 1816 he married Princess Charlotte, daughter of the future George IV of England. His powers were limited under the new constitution but he maintained the independence and unity of the nation. Leopold II (1835–1909; *r.*1865–1909) initiated a period of colonial and commercial expansion. He persuaded the delegates at a Berlin conference (1884–85) to give him sovereignty over the Congo Free State but he was forced to relinquish control to the Belgian government in 1908. Leopold III (1901– ; *r.*1934–51) succeeded his father Albert I. During the German invasion in 1940, Leopold led his forces in a spirited defence but was forced to surrender. His decision to become a German captive led to some opposition to his reinstatement after the war, and he abdicated in 1951. *See also* HC2 p.143.

Leopold, Jan Hendrik (1865–1925), Dutch poet. His first poems were published in 1893 in *De Nieuwe Gids* ("The New Guide"). He was one of the most famous of the school of poets associated with that review.

Léopoldville. *See* KINSHASA.

Lepanto, Battle of (1571), naval engagement in the Gulf of Patras off the coast of Lepanto, Greece, between forces of the OTTOMAN EMPIRE and various Christian powers. When the Ottoman Turks attempted to take Cyprus from Venice in 1571, Greece, Austria, Spain, Venice and the papacy stopped them in this battle. *See also* HC1 p.221.

Lepenski Vir, archaeological site in Yugoslavia, dating to 5000–4600 BC. The site is typical of Mesolithic Danube Basin settlements. *See also* MS p.29.

Lepidoptera, order of insects which includes MOTHS and BUTTERFLIES found in every continent except Antarctica. Wingspan: 4–300mm (0.16–12in). *See also* NW pp.106–107, 116–117.

Lepidus, name of a distinguished Roman family. Marcus Aemilius Lepidus (*d.*152 BC) was in turn consul, censor, pontifex maximus and princeps senatus. He built several roads. Marcus Aemilius Lepidus (*d.*12 BC) served as consul with Julius CAESAR, then as triumvir with MARK ANTONY and Octavian (AUGUSTUS), and again as consul, governing Rome during the PHILIPPI campaign. He then governed Africa and tried to incite a revolt in Sicily. Octavian forced him from his offices, except that of pontifex maximus. *See also* HC1 p.97.

Le Play, Pierre-Guillaume Frédéric (1806–82), French mining engineer and sociologist. He examined the connections between family types and the general state of society. An early study of the relationships between workers, their employers and the environment was published in *European Workers* (1855). He also wrote *The Organization of Labour* (1872).

Lepontine Alps, section of the central Alps in Switzerland, along the Italian-Swiss border, lying mostly in the Ticino and Graubünden cantons. They extend from the Pennine Alps at the Simplon Pass, wsw to the Rhaetian Alps at the Splügen Pass. They are bound by the upper Rhône and Vorderrhein valleys. The highest peak is Monte Leone 3,552m (11,654ft).

Leprechaun, in Irish folklore, a small, roguish elf. He is thought of as a cobbler who also possesses much buried treasure. If caught he can be made to tell his secrets and grant wishes but will vanish as soon as his captor stops looking, even for a moment, at him.

Leprosy, also known as Hansen's disease, communicable disease caused by the micro-organism *Mycobacterium leprae,* in which granular nodules develop on the skin, mucous membranes and peripheral nerves. The spread and enlargement of the skin nodules together with progressive degeneration of the tissues and destruction of peripheral nerves produces loss of sensation and results in deformity if the disease is not treated. In tuberculoid leprosy, one of the two main types, damage to the nerves occurs early in the disease, so that the skin nodules may not be felt. In this form the nodules contain very few micro-organisms. In lepromatous leprosy, the nodules contain many *Mycobacerium leprae* organisms and thus are a source of spread of the infection to others. The disease can be arrested by the administration of drugs, but nerve damage is irreversible. *See also* MS p.107, *107.*

Lepsius, Karl Richard (1810–84), German archaeologist and Egyptologist. He rejected the tomb-robbing methods of Giovanni Belzoni and others and made extensive surveys of his discoveries on his expedition between 1842 and 1845. He published his findings in the 12-volume *Monuments of Egypt and Ethiopia* (1849–59).

Lepton, one of a class of subatomic particles which includes the ELECTRON, the MUON and the NEUTRINOS, together with their antiparticles (antileptons). Leptons are governed by the force of weak interaction, which is the force involved in radioactive decay. They are the lightest subatomic particles and are thought to have no QUARK substructure. *See also* HADRON; SU pp.66–67.

Leptospirosis, also called Weil's disease, infection with the spirochete *Leptospira.* It is passed to man by contact with infected rodents. Symptoms include fever, tendency to jaundice, haemorrhage and muscular pain. In its early stages, the infection may be controlled with antibiotics. Kidney failure may be one result of infection, and an artificial kidney machine may be used in its treatment.

Le Puy, town in central s France, capital of the Haute Loire département. It is notable for a number of abrupt volcanic peaks, up to 79m (260ft) high. One peak is topped by an 11th-century church and monastery, another by a statue to *Notre Dame de France* (1860). Le Puy was an important pilgrimage centre in the Middle Ages. Pop. (1973) 29,550.

Lerma, Francisco Gómez de Sandoval y Rojas, Duque de (1553–1625), chief minister to PHILIP III of Spain from 1598. He was forced to make peace with JAMES I of England and a truce with the Netherlands. He drove the Moriscos from the country in 1609 and became a cardinal in 1618, but was deposed by his son.

Lermontov, Mikhail Yurevich (1814–41), Russian poet and novelist. In his poem *On the Death of the Poet* (1837) he criticized the court for failing to prevent the duel in which PUSHKIN was mortally wounded. The novel *A Hero of Our Time* (1840) and the final version of *The Demon* (1841) were written in exile. *See also* HC2 pp.99–100.

Lerner, Alan Jay (1918–), US musical comedy librettist. He collaborated with Frederick LOEWE in the stage productions of *Brigadoon* (1947), *Paint Your Wagon* (1951), *My Fair Lady* (1956), *Camelot* (1960) and *Coco* (1969). Lerner also wrote the screenplays for *An American in Paris* (1951) and *Gigi* (1958).

Le Sage, Alain-René (1668–1747), French novelist and dramatist. His lack of characterization is balanced by cheerful and vivid vignettes. His best-known works include the play *Crispin, rival of his Master* (1707) and *Histoire de Gil Blas de Santillane* (1715–35). *See also* HC2 p.34.

Lesage, Jean (1912–), Canadian politician, Premier of Quebec (1960–66). He was a federal MP from 1945 to 1958, when he resigned his seat to become leader of the Quebec Liberal Party. He attempted to reform Quebec in his "Quiet Revolution" in order to forestall the separatist movement. He resigned the Liberal leadership in 1970.

Lesbianism, female HOMOSEXUALITY.

Lesbos (Lésvos), island in E Greece, off the NW coast of Turkey in the Aegean Sea. Mitilíni is the capital. Settled by the AEOLIANS *c.*1000 BC, it was an important cultural centre in the 7th and 6th centuries BC, and the home of SAPPHO, ALCAEUS, ARISTOTLE and EPICURUS. It was held by Macedonia, Rome and Byzantium; the Ottoman Turks occupied the island from 1462 to 1913, when it passed to Greece. Products include olives, wheat, grapes and citrus fruits. The chief industry is fishing. Area: approx. 1,630sq km (630sq miles). Pop. (1961) 117,371.

Leschetizky, Theodor (1830–1915), Polish pianist and teacher. He was a student of Karl CZERNY and played before the Russian court in 1852. Ignacy PADEREWSKI and Ossip Gabrilowitsch became his students after he settled in Vienna in 1878.

Lescot, Pierre (*c.*1515–78), French architect whose fame rests on his part in redesigning half of one side of the LOUVRE – the west wing of the Cour Carrée (1546–56) – the oldest part in existence. A pioneer of the Renaissance in France (although he never visited Italy), his style combines Italian features, such as Classical pilasters and relief sculpture, with French elements such as a high roof and arches over windows.

Leshy, forest spirit in Slavic mythology. It was seldom seen but was often heard laughing, whistling and singing while it went about playing tricks on men. It was depicted as a man as tall as the trees or as short as the grass, without eyebrows or eyelashes and with a pointed head.

Lesion, abnormality in a tissue. Examples are wounds of any kind and sores which occur anywhere, whether internal or external. Blisters, ulcers and tumours are examples of lesions.

Leskov, Nikolai Semyonovich (1831–95), Russian novelist and short-story writer, with a great gift for narrative. His works – bizarre and full of satire – include the novels *Nekuda* (1864) and *Cathedral Folk* (1872), and the story *Lady Macbeth of the Mtsensk District* (1865), on which SHOSTAKOVICH's opera is based.

Leslie, Alexander, Earl of Leven. *See* LEVEN, ALEXANDER LESLIE, 1ST EARL OF.

Leslie, David (1601–82), Scottish military commander. He served in the Swedish army during the Thirty Years War and at MARSTON MOOR in 1644 on the side of the ROUNDHEADS. He later became a ROYALIST and commanded the Scottish army at the Battle of WORCESTER. He was created a baron in 1661.

Leslie, Sir John Randolph Shane (1885–1971), Irish journalist and writer. He published several volumes of verse including *Songs of Oriel* (1908), *Verses in Peace and War* (1922) and *Oppidan* (1922). His other works include *American Wonderland* (1936) and *The Life of Mrs Fitzherbert* (1939).

Lésmian, Bolesław (1879–1937), Polish poet. He published three volumes of original verse: *Sad rozstajny* (*Orchard,* 1912), *Łaka* (*The Meadow,* 1920) and *Napój cienisty* (*The Shadowy Drink,* 1936). He was a SYMBOLIST and his poems depict extremes of horror and ecstasy.

Lesotho, nation in s Africa, surrounded by South Africa and the Transkei, known as Basutoland until 1966. The country depends for its economic survival on South Africa, the employer of many Sotho mine workers. Much of the land is infertile, although beans and cereals are produced in the w. The capital is Maseru. Area: 30,355sq km (11,720sq miles). Pop. (1976 est.) 1,214,000. *See* MW p.118.

Lespinasse, Julie-Jeanne-Éléonore de (1732–76), companion and protégé of Madame Deffand from 1754 to 1764. She later established her own salon in Paris which was frequented by the ENCYCLOPEDISTS. Her love-letters to the Comte de Guibert, written between 1773 and 1776, were published in 1809.

Lesseps, Ferdinand-Marie, Vicomte de (1805–94), French diplomat who in 1832 conceived the idea of a canal linking the Red Sea with the Mediterranean Sea.

Lepidoptera are distinguished by their large and often highly decorated wings.

Leprosy, common in the Middle Ages, is cured by God in this early illustration.

Lesbos, an important tourist centre, is sometimes known as Mitilíni after its capital.

Nikolai Leskov satirized the Russian church hierarchy in *Cathedral Folk.*

Letterpress originally combined metal type with woodcuts of the illustrations.

Lettuces, several varieties are grown, including the head lettuce and the cos.

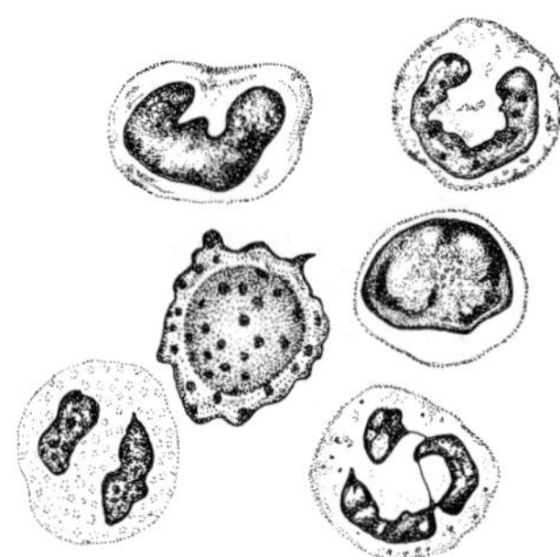

Leucocytes are white blood cells, found in similar proportions in each blood group.

Louis Le Vau: entrance of the Vaux-Le-Viconte designed for Nicholas Fouquet.

Digging was begun, by Egyptian labourers, in 1859 but it was necessary to import mechanical equipment from Europe to complete the work. The Suez Canal, as it became known, was opened by the Empress EUGÉNIE in November 1869. *See also* HC2 p.*131*.

Lesser Antilles, one of three major island groups in the West Indies archipelago, between the Atlantic Ocean (E) and the Caribbean Sea, stretching in an arc from Puerto Rico (N) to the N coast of Venezuela, South America. The group includes the LEEWARD ISLANDS and the WINDWARD ISLANDS. *See also* articles on individual islands in *The Modern World*.

Lessing, Doris May (1919–), Rhodesian novelist and playwright. Her books are strongly autobiographical, reflecting her membership of the Communist Party, her two failed marriages and her southern African background. This is especially so of the Martha Quest series – *Martha Quest* (1952), *A Proper Marriage* (1954), *A Ripple from the Storm* (1958) *Landlocked* (1965), and *The Four Gated City* (1969).

Lessing, Gotthold Ephraim (1729–81), German philosopher and writer responsible for the so-called native German drama. He portrayed his commitment ·to the German enlightenment in *Nathan the Wise* (1779). Other works include *Minna von Barnhelm* (1767) and *Hamburg Dramaturgy* (1767–68). *See also* HC2 pp.22–23, 73.

Les Six. *See* SIX, LES.

Lester, Richard (1932–), US film director who has done most of his work in Britain. His first film was the successful short comedy *The Running Jumping and Standing Still Film* (1959). His later films include *A Hard Day's Night* (1964) and *Help!* – which starred the BEATLES – *The Knack* (1965), *The Bed Sitting Room* (1969), *The Three Musketeers* (1974) and *Robin and Marian* (1976).

L'Estrange, Sir Roger (1616–1704), English pamphleteer and ROYALIST. In 1644 he was captured by PARLIAMENTARIANS during an unsuccessful attack on Lynn, in Norfolk, and imprisoned. After the RESTORATION he was made surveyor of the imprimery (1663–88) and issued various journals, including the *Intelligencer,* the *News* and the *Observator*.

Le Sueur, Eustache (1616–55), French painter of religious subjects. He enjoyed great prestige in 18th-century France. His works include *Presentation of the Virgin* (1640–45).

Leszek I, "the White", King of Poland (1194–1227). His reign was marked by feudal anarchy, with the nobility and clergy firmly in control. Leszek fought numerous wars, first with Mieszko III, who was trying to regain his throne, and later with Mieszko's son.

Le Tellier, Michel (1603–85), French minister made war secretary (1643–77) during LOUIS XIV's regency. He enlarged the army and centralized it under the Crown, thus making it possible for Louis to dominate Europe.

Lethaby, William Richard (1857–1931), British architect and writer. He built little, but wrote many influential works, especially *Form in Civilisation* (1922).

Lethe, also called Oblivion, in Greek mythology, the river of forgetfulness which the dead crossed as they entered HADES. All who crossed it lost their memories of past lives. Lethe, when personified, was considered to be the daughter of ERIS (Strife).

Letterpress printing, method of printing in which type is assembled and laid into a large frame called a chase by a compositor then placed either on a flat-bed press (which prints one sheet at a time) or processed into a curved plate and placed on a rotary press, which is capable of higher speeds. *See also* HALF TONE PROCESS; LINOTYPE; LITHOGRAPHY; MM pp.204–207.

Letters, branch of literature comprising both letters intended for publication and often addressed to the public and also private correspondence. The public epistle was much used in classical times, especially by CICERO and HORACE. Their letters were the models for the epistles of St PAUL.

The first great modern writer of letters was the French lady, Madame de SÉVIGNÉ, whose witty, acute and elegantly written letters were published in the 17th century. In Britain the great age of letters, which began with DRYDEN in the late 17th century, was the 18th century. Alexander POPE perfected the verse letter in his *Epistle to Dr Arbuthnot.* Lord CHESTERFIELD's *Letters to His Son* remains a classic. Letter writing became so highly polished and so popular that RICHARDSON wrote the first serious epistolary novels in the language in the 1740s, *Pamela* and *Clarissa*.

Letters patent, letters issued by the Crown in England, affixed with, but not closed by (the Latin *patens* meaning open), the GREAT SEAL. They confer upon the receiver, or patentee, some dignity, emolument, office, monopoly or FRANCHISE which otherwise he may not enjoy. They are used, for instance, to incorporate bodies by charter. They are left open so that anyone may inspect them and are recorded in the patent rolls. (Writs of right close, sealed by the Great Seal, are not open to public inspection and are recorded in the close rolls.)

Lettow-Vorbeck, Paul von (1870–1964), German commander of the German East African army in WWI. He managed to immobilize large numbers of British and Portuguese troops by the use of guerrilla tactics. He surrendered on 25 Nov. 1918 only after he had been given proof of the armistice. *See also* HC2 p.*191*.

Lettre de cachet, French document of the 17th and 18th centuries sent by the king, which could order imprisonment or exile without trial. They helped to inspire hatred of the ANCIEN RÉGIME.

Lettuce, edible annual plant that is widely cultivated for use in salads. Most varieties of *Lactuca sativa* are cool-weather crops; hot weather causes them to go to seed prematurely. The large leaves form a compact head or loose rosette. Wild species include the compass plant (*Lactuca scariola*), a prickly lettuce-like plant believed to be the ancestor of cultivated forms. *See also* PE pp.*173, 188, 188, 190, 190.*

Leucippus of Miletus (*fl.* 500 BC), Greek philosopher credited with having originated the theory of ATOMISM. Although only fragments of his work remain, he is believed to have written *The Great World System.* In this book it is proposed that all matter is homogeneous and consists of an infinite number of small, indivisible particles – atoms.

Leucite, grey or white felspar mineral which occurs uncommonly in lava flows and volcanic plugs. Chemically it is a potassium aluminium silicate. Its formula is $KAl(SiO_3)_2$.

Leucocyte, white blood cell, a colourless, amoeba-like structure containing a nucleus and CYTOPLASM. Leucocytes are either granular (granulocytes) or non-granular (agranulocytes), depending on the presence or absence of granules in the cytoplasm. Granulocytes are subdivided into neutrophils, eosinophils and basophils; agranulocytes are either lymphocytes or monocytes. Normal blood contains 5,000 to 10,000 leucocytes per cu mm of blood. Excessive numbers of leucocytes, or immature forms called lymphoblasts, are seen in such diseases as leukaemia. Leucocytes multiply, help in ANTIBODY formation and destroy harmful cells when the body is infected. *See also* MS p.63.

Leucotomy, or pre-frontal lobotomy, surgical operation on the brain. It was first performed in the 1930s to treat psychiatric disorders by severing tracts of nerve fibres leading to the frontal lobes. It may cause irreversible deterioration of personality. Drugs have almost entirely replaced this operation. *See also* BRAIN DAMAGE.

Leukaemia, acute or chronic malignant disorder characterized by abnormal numbers of types of white blood cells (LEUCOCYTES) in the blood, bone marrow and other tissues. Besides its division into acute and chronic forms, it is also classified by the type of leucocyte involved. The acute lymphoblastic form is the most common malignant disease of childhood; present treatment produces two-year survival in about half of affected children. In chronic leukaemia, a more slowly progressing form, survival as long as 10 to 15 years has occurred. Treatment of leukaemia is mainly by irradiation and drugs. The cause has not yet been determined. *See also* RADIOTHERAPY; MS p.91.

Leuven. *See* LOUVAIN.

Levalloisian culture, archaeological term for culture producing a particular type of sharpened flint flakes, used for skinning animals. Such flints have been found in both Europe and Africa, and date 100,000–70,000 years ago, in the Third Interglacial Period. *See also* MS p.*31*.

Levant Company, London merchants' company formed in 1591 by the amalgamation of the Turkey Company and the Venice Company. It was a joint-stock enterprise which traded in the Mediterranean and the Levant. In the 18th century it suffered from the competition of the EAST INDIA COMPANY and eventually gave up its charter in 1825.

Le Vau, Louis (1612–70), French architect who, although he designed part of the old Palace of Versailles, was predominantly an architect to the new bourgeoisie, not the court. Le Vau is famous chiefly for Vaux-le-Vicomte, the sumptuous BAROQUE chateau he designed for Nicolas Fouquet, Finance Minister to Louis XIV, in gardens laid out by LE NOTRE.

Levee, broad low ridge of fine sediment deposited along the sides of rivers during floods. A levee may also be built artificially along the bank of a river or an arm of the sea to protect the land from flooding. *See also* PE p.112.

Levellers (*fl.* 1645–49), Puritan political and religious movement in England. The name alludes derisively to their ideals of equality. Their leader was John LILBURNE, and their campaign, which found extensive support in Oliver CROMWELL's army, demanded complete constitutional reform, with abolition of the monarchy and corporate privilege, and the creation of one supreme representative legislature elected biennially by adult male suffrage, the extent of which was not firmly defined. Cromwell finally crushed the movement.

Leven, Alexander Leslie, 1st Earl of (*c.* 1580–1661), Scottish military commander. He fought for Gustavus Adolphus of Sweden in the THIRTY YEARS WAR, being present at LÜTZEN in 1632 and Brandenburg in 1634. Joining the COVENANTERS in Scotland, he fought for Parliament in the English Civil War but joined the Royalists at Dunbar in 1650.

Lever, Charles James (1806–72), Irish novelist. He wrote many novels, most of them about Irish rural and sporting life; the first, *The Confessions of Henry Lorrequer,* was published in 1837.

Lever, simple machine used to multiply the force applied to an object by virtue of the PRINCIPLE OF MOMENTS. A lever consists of a rod and a fulcrum, or point about which the rod rotates. In a crowbar, for example, the applied force and the body to be moved are on opposite sides of the fulcrum, with the point of application farther from it. This lever multiplies the force applied by the ratio of the two distances. *See also* MM p.86.

Leverhulme, William Hesketh Lever, 1st Viscount (*c.* 1851–*c.* 1925), British industrialist and philanthropist. Beginning as an assistant in his father's grocery business, he bought a soapworks in 1884 and started the manufacture of a soap named "Sunlight", from vegetable oils instead of tallow. He founded the model industrial town of PORT SUNLIGHT on garden city principles to eliminate the effects of overcrowding. *See also* HC2 p.97.

Leverrier, Urbain-Jean-Joseph (1811–77), French astronomer, discoverer of the planet NEPTUNE. Leverrier realized that unexpected variations in the orbit of URANUS could be due only to the gravitational influence of a hitherto undiscovered outer planet. A search was at once made and in Sept. 1846 Johann G. GALLE at Berlin located the planet, which was named Neptune. Leverrier was not the only astronomer to make this deduction: in England John Couch Adams reached the same conclusion in 1845.

Levertin, Oscar Ivar (1862–1906), Swedish poet, critic and teacher. A leading Neo-Romantic, he published four volumes of verse, all sombre in tone, including *King Solomon and Morolf* (1905).

Levertov, Denise (1923–), US poet, *b.* Britain. Her poems are simple and sparse, hinting at a natural order behind the seeming chaos of contemporary life. Her works include *The Double Image* (1946), *Here and Now* (1957), *Jacob's Ladder* (1961), *The Sorrow Dance* (1967) and *Relearning the Alphabet* (1970).

Levi, Carlo (1902–75), Italian writer and painter. Because of his anti-Fascist political activities, he was exiled (1935–36) in Lucania, the setting of his novel *Christ stopped at Eboli* (1945). His later novels include *Words are Stones* (1955) and *The Linden Trees* (1959). He served in Italy's senate (1963–72).

Leviathan (1651), essay in political theory by the Englishman Thomas HOBBES. It is an entirely secular interpretation of the source and justification of political power. Its gist is that in order to create a civil society out of a natural chaos, men must unreservedly, and without power of recall, entrust power to an absolute ruler. Thus, although it finds no place for the DIVINE RIGHT OF KINGS, it is a cardinal document in the development of European ABSOLUTISM.

Levi Ben Gershorn (1288–1344), also known as Gersonides, Jewish scholar, *b.* France. He had a wide range of interests, including mathematics, philosophy and religion. In addition to the scientific works, he published commentaries on the Bible, as well as philosophical studies of Aristotle and Averroës.

Levi-Civita, Tullio (1873–1941), Italian mathematician. With his one-time teacher Curbastro Ricci, he developed tensor calculus, which has many applications in relativity and general unified field theory.

Levin, Meyer (1905–), US author, who began his career as a journalist. His first novel, *Reporter*, appeared in 1929. Other novels include *The Old Bunch* (1937), *My Father's House* (1946), *Compulsion* (1956), *The Fanatic* (1964) and *The Settlers* (1972).

Levine, Albert Norman (1924–), Canadian poet and novelist. He published the volume of verse, *The Tightrope Walker*, in 1950. His novels include *The Angled Road* (1952) and *From a Seaside Town* (1970).

Levinsohn, Isaac Baer (1788–1860), Hebrew writer, *b.* Russia; a founder of the Haskalah, or ENLIGHTENMENT, in Russia. A rabbi, his writing was generally polemical and propagandist, and included *Testimony in Israel* (1823), *House of Judah* (1838) and the Yiddish play *The World of Chaos*, published in 1878.

Lévis, François-Gaston, Duc de (1720–87), commander of French troops in Canada (1759–60). Levis defeated the British at Ste Foy in 1760 but abandoned the siege of Quebec when the British fleet arrived. Later that year he was forced to surrender Montreal.

Lévi-Strauss, Claude (1908–), Belgian-born French anthropologist and founder of structural anthropology. Structuralism, the attempt to find basic patterns for a scientific study of man, contends that history was shaped into a collective, fragmented structure comparable to preliterate mythology. Lévi-Strauss became professor of anthropology at the College of France in 1959 and was elected to the French Academy in 1973. His books include *The Elementary Structures of Kinship* (1949; tr. 1962) and *Structural Anthropology* (1958).

Levi Strauss and Company, US clothing manufacturers who design, manufacture and distribute blue jeans, denim and casual clothing. The company was founded in the mid-19th century in the USA by Levi Strauss, who supplied riveted canvas trousers to cowboys and prospectors.

Levitation, paranormal raising of a person or object without any visible means of support; it is associated with mediums, Indian fakirs and, more recently, PSYCHO-KINESIS. This phenomenon has long been a focus for charlatanism and theatrical trickery and is the object of much scepticism. Debates over whether levitation is fact or illusion have persisted since witches were supposed to levitate or to "fly". *See also* MS pp. *197–198.*

Levites, according to the Bible, originally one of the 12 tribes of Israel, but later, after the Babylonian captivity, a priestly caste found in all tribes. By the time of Christ, the Levites ran the entire organization of the TEMPLE (including gatekeepers, musicians, scribes, money-changers and the Temple police) with the sole exception of the actual priesthood, still drawn from any tribe. The Levite tradition is still handed down: a Levite is called to read the TORAH after a kohen (priest) in the synagogue service.

Leviticus, biblical book, third book of the PENTATEUCH. Primarily a manual for the instruction of priests on ritual technicalities, it also exhorts reverent use of the proper rituals in worshipping God. Many of the laws in Leviticus, traditionally ascribed to MOSES, are repeated in DEUTERONOMY in less detailed form.

Levkas, island in W Greece. One of the IONIAN ISLANDS, it constitutes a department of Greece. The chief town is Levkas. Industries: olive oil, tobacco, currants and wine. Area: approx. 323sq km (125sq miles). Pop. (1971) 24,599.

Levkosía. *See* NICOSIA.

Lévy-Bruhl, Lucien (1857–1939), French philosopher. He researched the psychology of primitive societies and propounded his view that primitive thought was "pre-logical", differing in nature from that of civilized societies. From Émile DURKHEIM he took the concept of group ideas, which led him to emphasize the formative influence of tradition and the psychology of the society as a whole rather than the individual. His works include *How Natives Think* (1910), *Primitive Mentality* (1922) and *Primitives and the Supernatural* (1931).

Lewes, George Henry (1817–78), British journalist and critic. He wrote dramatic criticism as well as philosophical works, which include *Biographical History of Philosophy* (4 vols, 1845–46) and the hugely successful *Life and works of Goethe* (2 vols, 1855). Separated from his wife, he lived from 1854 until his death with George ELIOT (Mary Ann Evans), whose work he encouraged and influenced.

Lewes, Battle of (1264), military engagement at Lewes, Sussex, between the baronial force of Simon de MONTFORT and the troops of HENRY III. Simon's victory put him and two other "electors" in charge of England for over a year and led to the calling of his famous parliaments of 1264 and 1265.

Lewin, Kurt (1890–1947), US psychologist, *b.* Germany. He was professor (1935–45) of child psychology at the University of Iowa and director (from 1945) of the research centre for group dynamics at the Massachusetts Institute of Technology, where he pioneered work in human motivation and group dynamics. His books include *A Dynamic Theory of Personality* (1935) and *Resolving Social Conflicts* (1947).

Lewis, Alun (1915–44), Welsh poet and short story writer. His short stories about army life are collected in *The Last Inspection* (1942) and *In the Green Tree* (1948). His reactions to WWII find expression in his volumes of poetry, *Raiders Dawn* (1942) and *Ha, Ha Among The Trumpets* (1945).

Lewis, Cecil Day-. *See* DAY-LEWIS, CECIL.

Lewis, Clive Staples (1898–1963), British scholar, critic and author. He was best known for his books on religious and moral themes, particularly *The Screwtape Letters* (1942), although he also wrote children's books and critical works. He was professor of Medieval and Renaissance English at Cambridge University (1954–63).

Lewis, Dominic Bevan Wyndham (1894–), British essayist, biographer and humorist. After working as Beachcomber in the *Daily Express* (1919–24), he continued as a humorist on the *Daily Mail* and the *News Chronicle*. His other works include *Villon* (1928) and *The Stuffed Owl, an Anthology of Bad Verse* (1930).

Lewis, Gilbert Newton (1875–1946), US chemist. Working first in thermodynamics, he wrote a standard textbook on the subject. Later he worked out a theory of valency for non-ionic compounds based on shared electrons (co-valency), and extended the concept of acids and bases. Lewis was the first to prepare a sample of heavy water, since used as a moderator in nuclear reactors.

Lewis, Harry Sinclair (1885–1951), US author. After graduating from Yale University he held a variety of newspaper and editorial positions. He began writing short stories, and his first novel, *Our Mr Wrenn*, appeared in 1914. It was not until the publication of the novel *Main Street* in 1920, however, that he gained acclaim. A satirist, he ridiculed American middle-class life. Among his targets were conformity and hypocrisy in business, medicine and religion. His early work, in particular *Babbitt* (1922) and *Arrowsmith* (1925), is generally considered his best. He was the first American author to be awarded the Nobel Prize for literature (1930). *See also* HC2 p.288.

Lewis, Jerry (1926–), US film comedian and director. He partnered the singer Dean Martin in a series of films (1949–56) before becoming an independent actor-director of such offbeat comedies as *The Bellboy* (1960), *The Nutty Professor* (1963), *Ladies' Man* (1961) and *The Patsy* (1964).

Lewis, Jerry Lee (1935–), US ROCK AND ROLL pianist. He made his name in 1957 with *Whole Lotta Shakin' Going On* and *Great Balls of Fire*, both of which became rock and roll classics. He retired from the pop world in 1958 but returned in 1961 to make further recordings.

Lewis, John Aaron (1920–), US jazz pianist, composer and arranger. In the 1940s he played and arranged in the bands of Dizzy GILLESPIE and Miles DAVIS. From 1952 to 1977 he led the Modern Jazz Quartet.

Lewis, John Saunders (1893–), Welsh author and critic, president of PLAID CYMRU (Welsh Nationalist Party) from 1925 to 1938. He is best known for his varied dramatic work which includes *Buchedd Garmon* (1937), *Blodeuwedd* (1948), *Dwy Gomedi* (1952) and *Brad* (1958). His other writing includes a collection of political essays *Canlyn Arthur* (1938) and a novel *Monica* (1930).

Lewis, Matthew Gregory ("Monk") (1775–1818), British novelist and dramatist. He is chiefly remembered for his Gothic novel *The Monk* (1796) and for his *Romantic Tales* (1808).

Lewis, Meriwether (1774–1809), US explorer who was one of the leaders of the LEWIS AND CLARK EXPEDITION. A US army captain, Lewis was selected by Thomas JEFFERSON to investigate a land route to the Pacific coast and explore the land acquired by the USA in the LOUISIANA PURCHASE of 1803. Lewis's documentation and observations of the journey paved the way for westward expansion. He became governor of the Louisiana Territory in 1808, but died mysteriously a year later.

Lewis, Percy Wyndham (1884–1957), British painter, critic and novelist. Lewis emerged as one of the leaders of the VORTICIST movement immediately before WWI, and contributed to the Vorticist periodical *Blast* (1914–15). In his paintings and designs of this period he was greatly influenced by FUTURISM. After WWI he produced a series of novels and essays, such as *The Apes of God* (1930) and the literary criticism *Men Without Art* (1934).

Lewis, Richard (1914–), British tenor who made his debut at Glyndebourne in 1947. He created the role of Troilus in Sir William Walton's *Troilus and Cressida* (1954) and similarly sang in the first performances of Sir Michael Tippett's *The Midsummer Marriage* (1955) and *King Priam* (1962).

Lewis, Sinclair. *See* LEWIS, HARRY SINCLAIR.

Lewis, larger part of the double island Lewis and HARRIS, northernmost of the Outer Hebrides, Western Isles, NW of

Claude Levi-Strauss conducted most of his research in Brazil and India.

Levkas has a rocky coastline, and Sappho is said to have leapt to her death there.

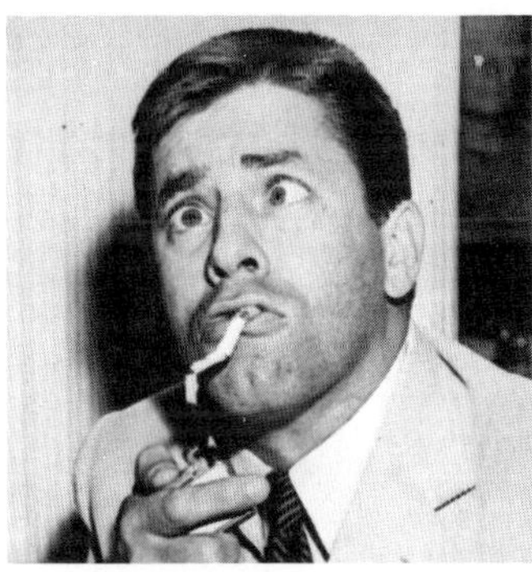

Jerry Lewis, a zany rubber-faced actor, portrayed the accident-prone simpleton.

Jerry Lee Lewis in 1958 with 13yr-old Myra Brown; their marriage raised legal doubts.

Lewis and Clark Expedition

Battle of Lexington; a small American force faces the well-drilled British redcoats.

Leyden jars, connected to groups, form a battery holding a powerful electric charge.

Lhasa apsos were traditional gifts from the Dalai Lama to dignitaries visiting Tibet.

Liberace and his equally famous "mom'–one of the great women behind every great man.

Scotland. Most of the island is peat moor and hills; most settlements are on the coast. Crofting, fishing and the production of Harris tweeds are the main occupations. Stone circles and Viking remains are ensuring the increase of tourism, especially to the chief town and only communications centre Stornoway. Pop. (1971) 15,460.

Lewis and Clark Expedition (1804–06), exploration of the LOUISIANA PURCHASE territory and the country beyond as far as the Pacific Ocean by two army officers, Meriwether LEWIS and William Clark. The purpose of the expedition, which was initiated by President JEFFERSON and Congress in 1803, was to find a land route to the Pacific, strengthen US claims to the Oregon Territory and gather information about the Indians.

Lewis Carroll. *See* CARROLL, LEWIS.

LeWitt, Sol (1928–), US sculptor. He gained notice in the 1960s, along with Don Judd, for his exhibitions of "object sculpture", simple constructions of prefabricated industrial materials. His work is open-framed, whereas Judd's is made of solid slabs.

Lexington, Battle of (19 April 1775), first battle of the American War of Independence. Paul REVERE and William Dawes gave warning that 700 British troops under Francis Smith were marching from Boston to CONCORD to seize militia stores, and 70 MINUTE MEN, led by John Parker, met them at Lexington Green. A skirmish followed in which eight Americans were killed.

Leyden. *See* LEIDEN.

Leyden, Lucas van. *See* LUCAS VAN LEYDEN.

Leyden jar, earliest and simplest device for storing static electricity and the original electrical condenser (capacitor), developed in Leyden, Holland, in 1745. It consists of a foil-lined glass jar partly filled with water and closed with a cork through which protrudes a knobbed brass rod which is wired to the foil. To charge the jar, friction is applied to the tip of the rod.

Ley farming, system of agriculture, based on a return to the pre-modern system of crop rotation. Cereal crops are alternated with grass or green manure crops in order to restore nitrogen to the soil without recourse to chemical fertilizers. During pasture years the soil is manured by grazing livestock. The name comes from a corruption of the word "lea", for land under grass. *See also* PE pp.174–175.

Ley lines, straight lines linking standing stones, Church sites and other places of traditional sanctity in the British Isles. The idea of ley lines was first proposed by Alfred Watkins in 1920. He later wrote *The Old Straight Track* (1925) to explain and offer proof of his theory that such lines have archaeological significance.

Leyte Gulf, Battle of (October 1944), a series of WWII air and naval engagements between Japan and the USA in the Philippines area. The battle, marking the start of General Douglas MACARTHUR's invasion of the Philippines, ended in a resounding US victory over the Japanese navy.

Lezama Lima, José (1912–), Cuban novelist and poet. His chief work is the novel *Paradiso* (1966). His verse was published in two volumes, *Enemigo rumor* (1941) and *Aventuras sigilosas* (1945).

LH, luteinizing hormone, is made by the anterior (front) lobe of the PITUITARY GLAND, just below the brain. In women it stimulates OVULATION by the development of the corpus luteum and secretion of progesterone. In men it promotes the secretion of TESTOSTERONE.

Lhasa, capital city of Tibet (Xizang Zizhiqu) Autonomous Region, sw China, on the River Lasa. A religious centre, it has been called the Forbidden City because of the Tibetan clergy's hostility to foreigners. After the Tibetan revolt against the Chinese (1959–60), many temples and monasteries were closed. Today it is an important trading centre, and also manufactures chemicals and processes gold and copper. Pop. (1970) 175,000.

Lhasa apso, Tibetan dog used to guard lamaseries. It has a narrow head with a medium-long muzzle, large nose and pendulous ears. The body is set on shortish legs and the tail is carried in a coil over the back. The heavy, long, straight coat is usually a golden brown. Height: 28cm (11in) at the shoulder; weight: to 7kg (15lb).

Lhote, André (1885–1962), French painter and critic. His writings on art, *Traité du Paysage* (1939) and *Traité de la Figure* (1950), have become classics.

Lhotse, two peaks in the Himalayas on the border of Nepal and Tibet. Height: (Lhotse Peak I) 8,511m (27,923ft); (Lhotse Peak I) 8,400m (27,560ft).

Liadov, Anatol Konstantinovich (1855–1914), Russian composer. His best work was written for the piano – preludes, studies and dances which show the influence of CHOPIN as well as elements of Russian folk music. Other compositions include orchestral and choral works such as *Chorus in praise of Vladimir Stassov* (1894) and the symphonic poem *The Enchanted Lake.*

Liaisons Dangereuses, Les (1782), novel by Choderlos de LACLOS. A story of cynical seduction written in the form of letters, it exposes the corruption of aristocratic society. *See also* HC2 p.35.

Liana, any ground-rooting woody vine that twines and creeps extensively over other plants for support; it is common in tropical forests. Species may reach a diameter of 60cm (24in) and a length of 100m (330ft).

Liang Ch'i-ch'ao (1873–1929), Chinese critic, editor and scholar. He participated in the "hundred days' reform", a movement for constitutional change (1898), and after its failure fled to Japan. After the revolution he returned to China in 1912 and led the Progressive Party in parliament. He wrote several works on Chinese intellectual and political history and encouraged a Western prose style.

Liaoning, province on the NE coast of China; Shenyang (Mukden) is the capital; part of Manchuria, it includes the Liaotung Peninsula. Corn, kaoliang and soya beans are the principal crops; others include cotton, wheat, rice, silk and fruits. Liaoning is China's most industrialized province. Coal is mined and there are large deposits of iron ore. Chemicals, iron and steel, heavy engineering and textiles are the principal industries. Area: 151,055sq km (58,322sq miles). Pop, 29,000,000.

Liapchev, Andrei (1866–1933), Bulgarian economist and Premier. He was an advocate of the unification of Rumelia and Bulgaria, and as Premier (1926–31) his foreign policy was one of conciliation.

Libby, Willard Frank (1908–), US chemist. From 1941 to 1945 he worked in the war research division at Columbia University, New York, on the separation of isotopes for the atom bomb. This led to his interest in nuclear physics and Libby's development of radioactive CARBON-14 dating, for which he was awarded the 1960 Nobel Prize in chemistry. He is the author of *Radiocarbon Dating* (1955).

Libel, permanent, false statement to a third person containing an untrue imputation against the reputation of another. In contrast with SLANDER, libel is usually written although publications of any defamatory matter in permanent form (such as a picture or film) and, since 1952, broadcast statements for general reception are treated as libel. Although usually a civil offence, libel may also be a criminal one in certain circumstances.

Liberace, Wladziu Valentino (1919–), US pianist and entertainer. He has been described as television's first matinée idol, and in 1953 was the largest solo attraction in US concert halls. He is chiefly famous for his extravagant style of dress and sparkling stage performances.

Liberalism, political and intellectual belief which advocates the right of the individual to make decisions, usually political or religious, according to the dictates of conscience. Its modern origins lie in the AGE OF REASON and the ENLIGHTENMENT. In intellectual matters it upholds the desirability of the free play of ideas upon stock notions. In politics it opposes arbitrary power and discrimination against minorities. In British history its greatest influence was exercised in the 19th century, with effect particularly in the abolition of slavery, constitutional reform, free trade and religious toleration. *See also* LIBERAL PARTY; MS pp.*273*, 277.

Liberal Party, British political party. It grew out of the early 19th-century WHIG PARTY. The name Liberal was adopted by some MPS in the 1820s. The first official use of the name is the National Liberal Federation, in Birmingham, founded by Joseph CHAMBERLAIN in 1877. The 19th-century party drew its strength from religious dissent and the urban electorate. Its predominant interests were free trade, religious liberty, financial retrenchment, individual liberty and constitutional reform. Its greatest leader was William GLADSTONE, who led four governments (1868–74, 1880–85, 1885–86 and 1892–94). Its greatest electoral triumph was the landslide victory of 1906. The 1906 government legalized trade unions, reformed the House of Lords and introduced progressive social security measures. It remained in office until 1916 (from 1908 under Herbert ASQUITH), when LLOYD GEORGE formed a coalition government with the Conservatives. Since the fall of that government, it has never formed a government, its place as one of the two major parties having been taken by the LABOUR PARTY. Although it has never returned more than 15 MPS to Westminster since 1945, it has nevertheless been instrumental in keeping the Labour Party in office during two periods of minority government; in the mid-1920s by tacit support and in 1977 by a formal arrangement to support Labour bills in Parliament. *See also* HC2 pp.112–113, 132–133.

Liberal Unionists, members of the LIBERAL PARTY in Britain who voted against GLADSTONE'S IRISH HOME RULE BILL in 1886 and subsequently gave their support to the CONSERVATIVE PARTY. They were led by the Whig Marquess of Hartington and the Radical Joseph CHAMBERLAIN. At the 1895 elections they returned 71 MPS to the House of Commons. It was the Liberal Unionists who sustained the Conservatives in office for most of the 20 years between 1886 and 1905.

Liberia, republic in w Africa on the Atlantic coast. Created by the American Colonization Society in the 19th century as a home for freed slaves, the country retains close links with the USA. The main food crops are cassava and rice. Rubber was formerly the mainstay of the economy, but it has now been superseded by iron ore, which accounts for 72% of all exports. The capital is Monrovia. Area: 111,370sq km (43,000sq miles). Pop. (1976 est.) 1,750,000. *See* MW p.118.

Liberius (*d.*366), pope (352–66), best known as the builder of the Basilica Liberiana on the Esquiline Hill. At the beginning of his pontificate the status of St ATHANASIUS was still disputed in the Church and Liberius requested the Emperor Constantine II to call a council to rule on the matter. The ruling went against Athanasius, but Liberius refused to be coerced into outrightly condemning him and was himself banished. In 358 he signed a vaguely worded repudiation of Athanasius couched in terms of the arian heresy, and returned to Rome. After the death of the emperor he reavowed his orthodox position.

Liberties, exemptions from royal jurisdiction given in medieval England by royal grant (sometimes assumed by right of conquest) to lords of manors and holders of lands palatine. The Earl of Chester, the Bishop of Durham and the Duke of LANCASTER were all lords palatine, holding almost sovereign power over their domains. Perhaps the most significant liberty was their right to exercise justice in their own courts, which gave rise to the contemporary aphorism "The king's writ stops at the manor gate".

Liberty Bell, historic US monument housed in Independence Hall, Philadelphia. It was first hung in 1753 with the biblical inscription, "Proclaim Liberty throughout all the Land unto all the Inhabitants Thereof". According to legend it was rung in July 1776 to celebrate the signing of the American Declaration of Independence. The bell was removed to Allentown while the British occupied Philadelphia (1777–78), and was returned in 1781. It was cracked in 1835 and again in 1846.

Liberty Loan Act (1917), us legislation passed by Congress providing for four public loans to the government to finance the country's war effort in wwi and for loans to the Allies for the purchase of food and war supplies. Five fund-raising campaigns were held between 1917 and 1919, raising more than 20,000 million dollars.

Liberty ships, cargo vessels built during wwii as part of the us war effort, also known as Ec-2 ships. They were part of a project which began in 1941 and produced 2,610 ships, some taking only two weeks to complete. Their standard dimensions included a length of 134m (441ft) and a displacement of 10,490 tonnes. Fast production was made possible by the development of arc welding, although this technique was at the time believed to have made some of the ships unsafe.

Libido, in psychoanalysis, term used by Sigmund freud to describe instinctive sexual energy. C. G. jung enlarged its meaning to include all instinctive energy. Its influence on the development of personality is probably less than the influence of social forces.

Libitina, Roman goddess of funerals and death. Perhaps originally an earth goddess, she was sometimes associated with Proserpina (called persephone in Greek mythology). When someone died, money was offered to Libitina in a sacred grove, which also contained the register of deaths and offices of undertakers.

Libra, or the Scales, southern constellation situated on the ecliptic between Virgo and Scorpius; the seventh sign of the zodiac. It formerly contained the First Point of Libra, the intersection of the ecliptic and the equator marking the autumnal equinox. Because of precession this point has shifted westward into Virgo. The brightest star, Beta Librae, is of magnitude 2.7. *See also* MS p.200; SU pp.254, 261.

Library, collection of books; room or building containing books for reading or reference. The earliest known library, a collection of clay tablets in Babylon, existed before 2000 bc.

Library, British, national reference library in Bloomsbury, London, until 1973 known as the British Museum Library. Its enormous collection of more than 8 million volumes was first begun in 1757. Its impressive circular reading room was designed by Sir Anthony Panizzi and opened in 1857. Every book published in Britain must be deposited with the Library under the 1911 Copyright Act.

Library of Congress, us national library in Washington, dc. Founded in 1800 as a research facility for members of Congress, its responsibilities now include copyright, inter-library loans and the publication of catalogues. It houses more than 60 million items and is supported mainly by congressional appropriations. *See also* MM p.217.

Libration, slight oscillatory motion of the Moon during which a thin crescent of its far side becomes temporarily visible from Earth. *See also* SU pp.178, 184. 184.

Libretto, text of an opera or operetta. From 1597, libretti were printed to commemorate performances; by the mid-18th century, public audiences used them to follow the opera's story. Libretti have often resulted from close collaboration between librettist and composer, such as da Ponte and mozart, and gilbert and sullivan. A number of composers wrote ir own libretti, the notably wagner.

Libreville, capital city of Gabon, w Africa, at the mouth of the River Gabon on the Gulf of Guinea. It was founded by French traders in 1843 and named Libreville ("Freetown") in 1849. Formerly Gabon's leading port, the city is now an administrative centre. Timber is the most valuable local commodity. Pop. (1975 est.) 169,200.

Libya, independent nation in n Africa; 95% of the country is desert, but there is a coastal strip of 1,760km (1,100 miles) on the Mediterranean Sea. There are no permanent rivers and cultivation is only possible in a few places. The country's main product is oil, which accounts for 99% of all exports. The capital is Tripoli. Area: 1,759,540sq km (679,358sq miles). Pop. (1975 est.) 2,444,000. *See* MW p.119.

Lice. *See* louse.

Licensing Acts, acts defining regulations concerning public houses and restaurants serving alcoholic drink in Britain. The Act of 1872 established the licensing authority's power to restrict the number of pubs and their opening hours, gave the police rights of entry and prohibited adulteration of drinks. The Act of 1904 provided for compensation for the loss of a licence as a result of policy rather than misconduct, to be paid out of a fund raised by the brewing trade itself. In 1915, during wwi, hours of drinking were reduced, and afternoon closing was introduced.

Lichee. *See* lychee.

Lichen, plant consisting of a fungus in which microscopic (usually single-celled) algae are embedded. The fungus and its algae form a symbiotic association in which the fungus contributes support, water and minerals, while the algae contribute food produced by photosynthesis. *See also* NW pp.48–49.

Li Chiai, 12th-century Chinese architect of the Sung Dynasty, noted for his popular manual of architectural procedure, the *Ying-tsao Fa Shih.*

Lichine, David (1910–72), Russian ballet dancer and choreographer. He studied under Bronislava nijinska and danced in several companies, most notably pavlova's in 1930 and ballets russes de Monte Carlo (1932–41). His famous works of choreography include *Francesca da Rimini* (1937) and *Graduation Ball* (1940).

Li Ch'ing-chao (1081–*c.*1150), one of China's greatest poetesses. As a writer of *tz'u,* lyric poems written to music which were most popular in the Sung dynasty (960–1279), her eloquence made the commonplace seem special; she often used objects from nature to symbolize a sense of loneliness and loss. As a critic, her comments on her contemporaries were perceptive.

Lichtenstein, Roy (1923–   ), us painter, first in abstract expressionism and after 1960 in pop art. Particularly interested in cartoons or magazine subjects and in the lack of sensitivity of mass-produced images, he often extracted these from context, as in *Good Morning, Darling* (1964), and re-presented them, greatly increased in size. The resulting effect – with prominent half-tone "dots" – often gave new insight into what began as commonplace. *See also* HC2 pp.278–279.

Liddell Hart, Sir Basil Henry (1895–1970), British author and military strategist. He was an early advocate of air power and mechanization in warfare – views that had a profound effect upon the German high command before wwii. Among his many works on strategy and military history are *A History of the World War, 1914–1918* (1934) and *The Tanks* (1959). He also edited *The Rommel Papers* (1953).

Lidice, village in Czechoslovakia, 16km (10 miles) nw of Prague. It was totally destroyed on 10 June 1942 by the Nazis in retaliation for the Czech assassination of Reinhard Heydrich, protector of Bohemia and Moravia and deputy head of the gestapo. All men over age 16 were killed and all women were deported; the area is now a national park and memorial.

Lidman, Sara (1923–   ), Swedish novelist and dramatist. Her novels, such as *Tjärdalen* (1953; The Tar Still) *Hjortronlandet* (1955; Cloudberry Land) and *Regnspiran* (1958; tr. *The Rain Bird,* 1963), treat rural life in n Sweden with warmth and freshness. Among her plays are *Aina* (1956) and *Marta, Marta* (1970).

Lie, Jonas Lauritz (1833–1908), Norwegian novelist, poet and playwright. His writings deal with middle-class family life, *The Pilot and his Wife* (1874) being the first Norwegian novel to treat marriage realistically. *The Commodore's Daughter* (1886) exposes the social and educational restrictions on women of the upper classes at that time.

Lie, Trygve Halvdan (1896–1968), Norwegian politician and Secretary-General of the un. He was legal adviser to the Norwegian Labour Party from 1922 to 1935. He was Minister of Justice (1935–39) and minister for Foreign Affairs (1942–45). In 1946 he became the first Secretary-General of the un and held the office until 1952.

Liebermann, Max (1847–1935), German painter and printmaker who was influenced by the French impressionists after 1890, and became leader of the movement in Germany. He is best known for his studies of the everyday life and work of ordinary people. His first exhibited picture, *Women Plucking Geese.*(1872), contrasted markedly with the fashionable romantic idealism that influenced most of his contemporaries.

Liebig, Justus von, Baron (1803–73), German chemist. He was the first to realize that animals get their energy from the combustion of food although he incorrectly thought that muscular power is a result of protein oxidation. Liebig also showed that plants derive their minerals from the soil, and introduced synthetic fertilizers into agriculture. *See also* HC2 p.156; PE p.252.

Liebknecht, Karl (1871–1919), German Communist and revolutionary leader. He entered the Reichstag in 1912 and vigorously opposed Germany's entry into wwi, for which he was imprisoned in 1916. After his release he became leader of the Communist group known as the spartacists. In 1919 he led an abortive putsch and was killed by police on his way to prison. *See also* HC2 pp.*212,* 213.

Liechtenstein, small independent state on the e bank of the River Rhine between ne Switzerland and w Austria. A mountainous country, the chief agricultural products are cereals, grapes, vegetables, fruit and dairy produce. Industry has developed greatly since wwii and includes food processing, machinery, textiles, pharmaceuticals and tourism. The capital is Vaduz. Area: 157sq km (61sq miles). Pop. (1976 est.) 24,000. *See* MW p.119.

Lied, German word meaning "song", but with a more specific connotation in current usage as the art song of German Romantic composers – especially schubert, wolf, brahms and schumann. Such songs (eg Schubert's *An die Musik*) aimed to express the mood and feeling of the words in both the voice and accompanying piano parts.

Lie detector, also known as a polygraph, electronic device that may be capable of detecting lies when used by a trained examiner. The polygraph monitors such factors as heart rate, breathing rate and perspiration, all of which may be affected when a person lies.

Liège, city in e Belgium, 89km (55 miles) ese of Brussels, called Luik in Flemish. Settled in ad 558, it became a famous educational centre during the Middle Ages. It passed to France during the Revolutionary Wars, but was given to The Netherlands at the Congress of vienna in 1815. It was occupied by the Germans in both world wars and was severely bombed in 1944–45. It is now a commercial centre whose industries include machinery, armaments and mining. Pop. (1970 est.) 147,300.

Liegnitz, Battles of (1241 and 1760). The first was the rout of European forces, led by Henry II of Silesia, by mongol invaders, led by Kaidu, grandson of genghis khan. The Mongol horse archers then crossed the Carpathians into Hungary. The second was a seven years war battle at which frederick the great's army of 30,000 gained a decisive victory over the Austrians.

Liepaya, second largest city of Latvia (Latvijskaja ssr) ussr on the Baltic Sea, 193km (120 miles) wsw of Riga. It was founded by the Teutonic knights in 1263

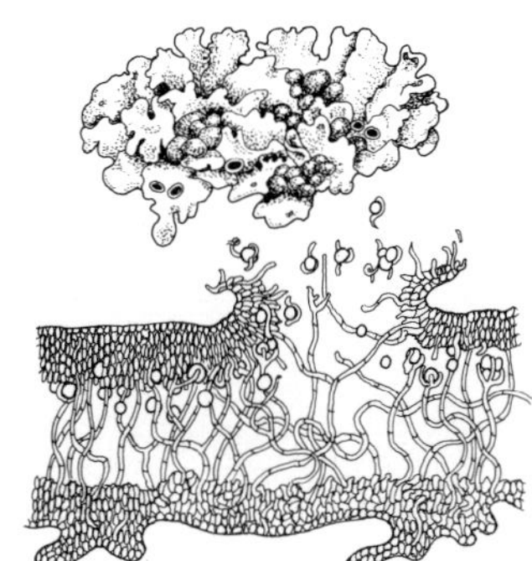

Lichens can reproduce when fused *Soredia* of algae and fungus are carried by the wind.

Justus von Liebig was the first to establish a chemistry laboratory for teaching purposes.

Karl Liebknecht tried to start a Russian-style revolution in Germany in 1918.

Liège in the 18th century was an arms manufacturing town of the Holy Roman Empire.

Lievens, Jan

Lifeboat designers today recognize the need for fast, well-found rescue craft.

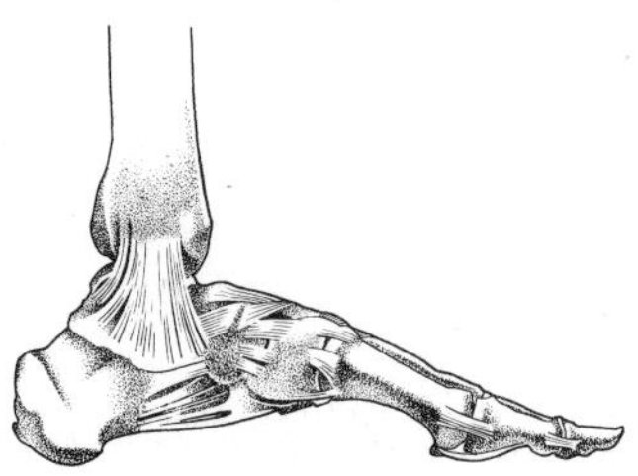

Ligaments join together bones and cartilage, enabling full movement of joints.

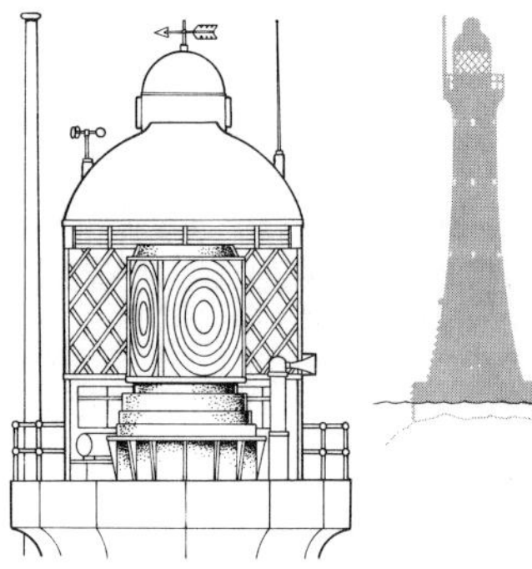

Lighthouse beacons use lenses and reflectors to produce a powerful beam of light.

The Light of the World by Holman Hunt, a member of the Pre-Raphaelite brotherhood.

and passed to Russia in 1795. Occupied by the Germans in both world wars as a naval base, it has since become an important Russian ice-free port. Industries include machinery, metallurgy, shipbuilding and food processing. Pop. (1969 est.) 88,000.

Lievens, or Lievensz, Jan (1607–74), Dutch painter of landscapes, portraits and genre and historical scenes. He worked for a time (*c.*1625–31) with REMBRANDT, and a similar style. His *Raising of Lazarus* (1631) is a fine example of his work.

Lifar, Sergei (1905–), Russian ballet dancer and choreographer. He studied under Bronislava NIJINSKA and Cecchetti, then joined DIAGHILEV'S BALLETS RUSSES in 1923, where he became premier danseur. In 1929 he became ballet master at the Paris Opéra Ballet and was founder-director of the Institute Choreographique (1947–58). He is best known for his ballets *Lucifer* (1948), *Phèdre* (1950) and Prokofiev's *Romeo and Juliet* (1955).

Life, the collective properties and abilities of organisms which allow them to obtain energy from the Sun and from food, to use this for growth, movement and reproduction, and to adapt to changing environmental conditions. The current theory on life's beginnings is that giant molecules, similar to proteins and nucleic acids, reacted together in the watery environment of the young Earth that is now commonly called the primeval soup. In this "soup" the giant molecules multiplied at the expense of smaller molecules which they consumed.

Despite much research the evolution and development of cellular life from molecular pre-life is a controversial subject and has yet to be explained fully. *See also* NW pp.20–21; SU pp.152–153.

Lifebelt, emergency floating device, usually a circular "tyre" made of cork, carried on the outer rails of ships, piers and embankments, and thrown to people in distress in the water. Lifebelts are attached to a lifeline, unlike lifejackets (inflatable garments strapped to people in time of danger). *See also* MAE WEST.

Lifeboat, craft designed to be used for saving lives at sea. Large vessels may carry a number of lifeboats which may be powered by oarsman, sail, or engine. Shore-based lifeboats are usually engine-driven craft built to remain afloat in even the roughest seas. They are broad beamed, between 12 and 16m (40–52ft) long, and the most recent are self-righting and remain buoyant even when holed. The first lifeboat was probably built at Newcastle upon Tyne, England, *c.*1790.

Life expectancy, average life-span, which varies greatly throughout the world, being as low as 40 years in some African countries and as high as 75 years in some developed ones. It has increased considerably in Britain during the last century, because of improved nutrition and public hygiene. The development of the medical services have also contributed greatly to average life expectancy.

Life of Johnson (1791), biography of the lexicographer and critic Dr Samuel JOHNSON, written by his friend, James BOSWELL. It was published in two volumes and is regarded as one of the best biographies in English. Johnson was a very witty conversationalist; as he spoke Boswell took notes that became the basis of the book.

Life-span. *See* LIFE EXPECTANCY.

Life-support system, in aerospace technology, system that supplies food, water and oxygen to astronauts. For each person 3 litres (5.25 pints) of water and 500 litres (110 gallons) of oxygen must be recycled each day. *See also* MM p.158.

Lift, or elevator, machine for raising and lowering passengers and freight from one level to another inside or outside buildings. Early lifts were driven by steam or hydraulic power; the first electric lift was installed in a New York building in 1889. *See also* MM pp.*96*, 97.

Lifting magnet, powerful ELECTROMAGNET used for lifting and transferring heavy metal. Such a magnet is suspended from the jib of a crane. *See also* SU p.*119*.

Ligaments, bands of tough fibrous connective tissue that join bone to bone in

proper articulation at the joints. In the wrist and ankle joints, for example, they surround the bones like firm inelastic bandages. Ligaments, which contain a tough inorganic substance known as COLLAGEN, form part of the supporting tissues of the body. *See also* MS p.56.

Ligand, ion, molecule or group of atoms linked to a metal atom or ion by co-ordinate chemical bonds in a metal complex. Thus chloride ions act as ligands in the ion $[CuCl_4]^{2-}$, and carbon monoxide molecules are ligands in such compounds as nickel carbonyl, $Ni(CO)_4$. Some molecules, such as ethylene diamine, $H_2NCH_2CH_2NH_2$, can co-ordinate at two points in the molecule.

Ligature, tying, binding or connecting. In surgery, thread, wire or cord may be used to tie and seal an artery or other body tube. In musical notation, a ligature is a curved line used to combine notes, indicating that they are to be played or sung without a break. In calligraphy and typography, a ligature is a tie joining letters in cursive script or, when casting some printer's type, for uniting commonly paired consonants such as *ff* and *fi*.

Ligeti, György (1923–), Hungarian composer whose avant-garde works involve shifting patterns of tone colours and sound. Since about 1956, Ligeti's work on electronic sound has influenced the textural forms in compositions such as *Atmosphères* (1961) for large orchestra, and *Lux Aeterna* (1966) for 16-part chorus. After 1966, most of his work has been purely instrumental, including *Melodien* (1971) for orchestra and *Horizont* (1971) for recorder.

Light, that part of the total ELECTROMAGNETIC SPECTRUM that can be detected by the EYE. Visible light lies in the wavelength range of 6.5×10^{-7} m (red light) to 4×10^{-7} m (violet) and travels at a speed of 299,792.5 km/sec in a vacuum. It exhibits phenomena typical of transverse wave motion, such as REFLECTION, REFRACTION, DIFFRACTION, LIGHT POLARIZATION and INTERFERENCE. The properties of light were first investigated in the 17th century by Sir Isaac NEWTON who was the first to split white light into its component colours with a prism. Newton firmly believed in a corpuscular or particle theory of light, but, by the beginning of the 19th century, with the work of Thomas YOUNG and Augustine-Jean FRESNEL on the diffraction and interference of light, the wave theory was well substantiated. Then, at the end of the century, experiments on the PHOTO-ELECTRIC EFFECT and in the work of Max PLANCK the theory that light had a corpuscular nature was also revived. This dilemma was resolved by the QUANTUM THEORY, according to which light consists of elementary particles of zero REST MASS called PHOTONS. When light interacts with matter, as in the photo-electric effect, energy is exchanged in the form of photons and so light appears corpuscular. Otherwise, light behaves as a wave. The quantum theory also explains how all electromagnetic radiations result from energy changes in atoms. *See also* ETHER; HOLOGRAPHY; LASER; SU pp.98–111.

Light adaptation, shift in functional dominance from rod cells to cone cells within the retina of the EYE as overall illumination increases. Unlike DARK ADAPTATION, adaptation to light is swift but usually uncomfortable. Human beings discern colour and form soon after emerging from prolonged darkness. *See also* MS p.48.

Light bulb, device consisting of a metal filament which emits light when heated by electricity. The filament is enclosed by a glass envelope in a VACUUM which contains an inert gas (eg argon), and is usually made of TUNGSTEN because of its high tensile strength, low evaporation and high efficiency in converting electrical energy into light. *See also* SU pp.69, *69*, *115*.

Lighthouse, building, often tall, which provides guidance to ships or aircraft by visual or audible means, or by radio. A light beacon or, in fog, a siren, relays a coded sequence ("character") which allows the lighthouse to be identified. Lighthouses date back to ancient Egypt; for over 1,000 years the lighthouse at

Pharos (one of the Seven Wonders of the Ancient World) guided ships into the River Nile.

Lighting, artificial means of illumination. The earliest was fire, followed by the oil-wick lamp, later developed into the candle. Coal gas was first used in the late 18th century, and the first electric light was produced by the carbon arc lamp (1801). The first fluorescent lamp was made in 1867 and marketed in 1938.

Lightning, visible flash of light accompanying an electrical discharge between clouds or between clouds and the earth, most commonly produced in a THUNDERSTORM. A typical discharge consists of several lightning strokes, initiated by leaders that follow an irregular path of least resistance, the lightning channel. Intense heating by the discharge expands the channel rapidly up to a diameter of 13 to 25cm (5–10in) creating the sound waves of thunder. *See also* SU pp.*17*, 112, *112*.

Light of the World, The (1854), painting by William Holman HUNT depicting Christ holding a lantern, knocking at a door (of a sinner's heart) choked with weeds. Hunt had explored religious themes for some time and took his idea for this work from a text in the biblical Book of Revelation: "Behold I stand at the door and knock". The work now hangs in Keble College Chapel, Oxford.

Light polarization, in the vibration of light waves, reduction of the infinite number of vibrational directions to effectively one in a plane, circle or ellipse. Reflected light is always plane polarized. A polarized beam can be stopped if it strikes a second polarizer whose polarizing plane is perpendicular to that of the first. *See also* SU p.103.

Lightship, small, steel ship equipped with a bright lamp, fog signals and radio beacons, used for navigation and warning in areas where a LIGHTHOUSE is impractical. The modern trend is away from lightships towards less expensive buoys or radio beacons.

Light-year, unit of astronomical distance equal to the distance travelled in free space by light in one year. One light-year is equal to 9.46×10^{12}km (5.88×10^{12} miles) or 0.307 parsecs. *See also* SU pp.60–61.

Lignin, complex non-carbohydrate substance that occurs in the woody tissues of plants, often in combination with cellulose. To obtain pure cellulose for the paper and rayon industries, the lignin has to be removed from wood. It produces dimethyl sulphoxide, $(CH_3)_2SO$, which has industrial uses in organic synthesis such as a solvent in the textile industry and in the production of some pharmaceuticals. Lignin is also a major source of vanillin. *See also* NW p.62.

Lignite, soft coal, brown-black in colour, with a carbon content higher than that of PEAT but lower than those of other coals. shape of its original woody structures is visible. *See also* PE p.136.

Lignum vitae, tropical evergreen tree; the source of a hard, dense wood used to make bearings and pulleys. A short-trunked tree, its flowers are blue or purple. Height: 9m (30ft). Family Zygophyllaceae; species *Guaiacum officinale*.

Liguria, region in NW Italy, on the Ligurian Sea, made up of the provinces of Genoa, Imperia, La Spezia and Savona; Genoa (Genova) is the regional capital. A mountainous region, it has a narrow coastal strip which is the famous resort area of the Italian Riviera. The region was ruled by Genoa from the 16th century until 1815, when it was annexed to Piedmont. Products: vegetables, flowers, olives, fruits, grapes. Industries: shipbuilding, tourism, chemicals, timber, textiles, fishing. Area: 5,413sq km (2,080sq miles). Pop. (1971) 1,868,630.

Li Ho (791–817), Chinese poet. His rhythmic verse used rich imagery and covered many subjects. Writing during the T'ANG DYNASTY (618–907), China's golden age of literature, he was a follower of HAN YÜ. Contemporary admirers called him a man of "devilish talent" because of his practice of composing poems from a series of single lines collected separately in a black bag.

Likasi, city in Shaba region, SE Zaire. Formerly called Jadotville, it is one of the country's major industrial and communication centres. Copper and cobalt are mined and refined, and other industries include chemicals and the manufacture of beverages. Pop. (1970) 146,394.

Li Kung-Lin (1049–1106), Chinese painter. Considered the greatest painter of his day, he was primarily concerned with CONFUCIANIST themes. He is, however, best known for his paintings of horses.

Li Kung-tso (*fl. c.*813), one of China's first real fiction writers. The plots and characterizations of his short stories, called *ch'uan chi'i* (narration of strange things), represented an advance in complexity over fables. They dealt with love and chivalry and included criminal and ghost stories.

Lilac, any of 13 species of evergreen ornamental shrubs that bear panicles (pointed clusters) of tiny fragrant white to purple flowers. Height: to 6m (20ft). Family Oleaceae; genus *Syringa*.

Lilburn, Douglas (1915–), New Zealand composer. His works include chamber music, film scores and symphonies. He became professor of music at Victoria University, Wellington, in 1963.

Lilburne, John (*c.*1614–57), English political figure, leader of the LEVELLERS. After imprisonment (1638–40) for anti-episcopal pamphlets, he fought for the PARLIAMENTARIANS during the Civil War (1642–45). He resigned from the army, however, because he refused to sign the SOLEMN LEAGUE AND COVENANT required for admission to the NEW MODEL ARMY. His pamphleteering against the army leaders led to his arrest (1649) for treason, but he was acquitted. In his later years he became a QUAKER. *See also* HC1 p.*289.*

Liliaceae, the lily family, 250 genera of erect or climbing flowering plants, and 3,700 species of herbs and shrubs. Most have a RHIZOME, BULB, CORM or TUBER and grow in temperate or subtropical regions. Included are the LILY, ONION, GARLIC, LEEK, ASPIDISTRA, YUCCA, HYACINTH, TULIP and LILY OF THE VALLEY. There are 80 species of true lily (genus *Lilium*).

Lilienthal, Otto (1849–96), German aeronautical pioneer who experimented with flying machines which he built after studying bird flight. He made short flights only, before suffering a fatal crash.

Lilith, in Semitic mythology, a malicious female spirit that haunts desolate places and attacks children. In Hebrew folklore, Lilith sometimes represents the first woman, who preceded Eve as wife to Adam, or the woman who bore demonic offspring to him after Eve had left him.

Lille, city in N France, 209km (130 miles) NNE of Paris; capital of Nord département. During the 15th century the dukes of Burgundy had their official residence there. After 1668 it served as the capital of French Flanders. It has a 17th-century stock exchange, a large citadel and the University of Douai. In France's major manufacturing belt, its chief industries are textiles and machinery. Pop. (1968) 190,546.

Lillee, Denis (1949–), Australian cricketer who played for Western Australia and Australia. In 32 Test matches between 1970 and 1977 he took 171 wickets at an average of 23.49. Lillee was one of the finest fast bowlers of the 1970s, and formed a partnership with Jeff Thomson that was the most effective since the West Indian pairing of Hall and Griffiths in the 1960s.

Lillie, Beatrice (1898–), British actress and comedienne, b. Canada, who made her stage debut in London (1914) and in the USA in 1924. She won an international reputation for vivacity and sophisticated wit in revues, television, films and her one-woman shows, in which she toured throughout the USA. Her films include *Around the World in Eighty Days* (1956) and *Thoroughly Modern Millie* (1967).

Lillie, Frank Rattray (1870–1947), US zoologist and embryologist, b. Canada. Noted for his discoveries in the field of fertilization and the role of hormones in sex determination, he spent much time helping to develop the Marine Biological Laboratory and the Oceanographic Institution at Woods Hole, Massachusetts. His book *The Development of the Chick* (1908) is a leading work in embryology.

Lilly, John Cunningham (1915–), US neurophysiologist and physician, known particularly for his work with dolphins and for his theories on the human mind based on analogies with the computer. His publications include *The Mind of the Dolphin* (1967) and a study *Programming and Metaprogramming .the Human Biocomputer* (1972). This served as a preface to his theoretical work *The Centre of The Cyclone* (1972), in which he attempted to schematize the varieties of mental experience. *See also* MS p.42.

Lilongwe, city in W Malawi, SE Africa, approx. 80km (50 miles) W of Lake Nyasa (Lake Malawi). Originally an agricultural trade centre for a fertile farming region, it became the capital of Malawi in 1975. It has the Bunda College of Agriculture. Pop. 87,000.

Lily, any of numerous species of perennial, bulbous plants grown in temperate and subtropical regions. They have erect stems and various leaf arrangements. The showy flowers are large and solitary or clustered and may be white, red, orange, yellow, green, purple or pink. Most lilies thrive best in well-drained, moist soil and a sunny location. Family Liliaceae; genus *Lilium*.

Lily of the valley, perennial plant native to Europe, Asia and E USA. It has broad, elongated leaves and bears stalks of tiny, white, bell-shaped fragrant flowers. Family Liliaceae; species *Convallaria majalis*.

Lily trotter. *See* JACANA.

Lima, capital of Peru, on the Pacific coast. Founded in 1535 by Francisco Pizarro, it was the seat of the Spanish viceroyalty. Chilean forces occupied the city from 1881 to 1883 during the War of the PACIFIC. The commercial and cultural centre of Peru, it has the University of San Marcos (1551), a 16th-century cathedral, and a National Library (1821). Industries: textiles, leather goods, oil-refining, furniture, food processing, iron foundries, cement, pharmaceuticals. Pop. (1972) 2,833,609.

Lima, Jorge de (1893–1953), Brazilian poet. A mystic, he wrote verse that was hopeful and reverent but not naive. He was a convert to Catholicism and used simple language to express his spiritual feelings; himself a mulatto, he often celebrated the personality of the black. His volumes include the collection *Poetic Works* (1950).

Lima bean. *See* BEAN.

Limassol (Lemosós), port in S Cyprus, on Arkrotíri Bay; capital of Limassol district. A resort town and administrative centre, it exports wine and farm produce. Tourism is an important industry. Pop. (1970 est.) 51,500.

Limax, genus of grey, black or brown pulmonate SLUGS of temperate areas. The various species have slimy bodies and are restricted to moist habitats, where they eat fungi and plant foliage. The largest European species, *L. maximus,* may grow to 9cm (3.5in) long. Family Limacidae. *See also* NW p.*90.*

Limb, in astronomy, apparent edge of any celestial body presenting an observable disc, such as the Moon, the Sun or a planet.

Limb, articulated extension of the vertebrate body, used for locomotion and manipulation of the physical environment. The human forelimb (arm) is particularly specialized in the latter function. The form of the limbs is regarded as a criterion of evolutionary development.

Limb, artificial. *See* ARTIFICIAL LIMBS.

Limbic system, collection of structures in the middle of the brain. Looped round the HYPOTHALAMUS, the limbic system is thought to be involved in emotional responses such as fear and aggression and to produce mood changes. *See also* MS pp.41, *41, 44.*

Limbo, in Roman Catholic theology, the abode of souls excluded from HEAVEN but not condemned to any other punishment. According to this concept, which never became doctrine, unbaptized infants go to Limbo after death, as did those prophets and holy-men who died before Christ came to redeem the world.

Limbourg, name of three Netherlandish brothers, Paul (Pol), Herman, and John (Jehanequin) (*fl.* before 1399–1416), all of whom were famous illuminators. They worked mainly for Jean, Duc de Berry, a patron of the arts. Paul, the greatest, was largely responsible for *Les Très Riches Heures du Duc de Berry*. Begun in 1413, this medieval illuminated book of hours, among the earliest examples of the true GENRE PAINTING in N Europe, is a masterpiece of the INTERNATIONAL GOTHIC style. *See also* HC1 pp.252, 262.

Lime, calcium oxide (quicklime) or calcium hydroxide (slaked lime). Calcium oxide, CaO, consists of whitish lumps obtained by roasting limestone (calcium carbonate) until all the carbon dioxide has been driven off. It is used as a refractory, a flux, and in glass making, water treatment, food processing and as a cheap alkali. Calcium hydroxide, $Ca(OH)_2$, a white crystalline powder obtained by the action of water on calcium oxide, is used in mortar, plaster, cements and in water softening and in agriculture. *See also* PE p. *223.*

Lime, small tropical tree (*Citrus aurantifolia*) of the rue family (Rutaceae); it has its origins in the East. The trees grow to 2.4–4.6m (8–15ft) and yield small green acid fruits. The juice was a valuable commodity in the 18th and 19th centuries for consumption on long sea voyages; the vitamin C helped to ward off scurvy. *See also* PE p.192, *192.*

Lime, name for the deciduous linden tree in Britain; several species grow throughout the north temperate zone. It has serrated, heart-shaped leaves with small, fragrant, yellowish flowers that are borne in clusters. The common British linden, *Tilia vulgaris,* is one of the three British species. Linden wood is particularly suited to carving; it is firm, but workable. Family Tiliaceae; genus *Tilia*.

Limelight, in theatrical lighting, device invented in 1816 in which a cylinder of lime was heated to incandescence, the gas producing an intense white light.

Limerick, port in SW Republic of Ireland, at the head of the Shannon Estuary. It is remembered as James II's last stronghold in the country after the GLORIOUS REVOLUTION. Industries include flour milling, salmon fishing and brewing. Pop. (1971) 57,137.

Limerick, five-line light verse with an *a a b b a* rhyme scheme and often of a bawdy or scurrilous nature. The nominal association with Limerick in Ireland has never been adequately explained. The fad for such verses was accelerated by those of Edward LEAR, but there are poems in similar verse forms dating from the 14th century.

Limestone, SEDIMENTARY ROCK composed primarily of calcium carbonate, $CaCO_3$. Generally formed from deposits of the skeletons of marine invertebrates, it is used to make cement, as a source of commercial LIME and as a building material. *See also* PE pp.102, *103, 108–109*, 133–134.

Limfjørd, waterway across N Jutland, Denmark, linking the North Sea with the KATTEGAT. Its middle course widens into a lagoon, 21km (13 miles) wide, known as Løgstør. The chief port is Ålborg. Length: 177km (110 miles).

Limit, in mathematics, the value approached by a mathematical function as the independent variable approaches some specified value. For example, the function $1/(x^2+1)$ has values $\frac{1}{2}, \frac{1}{5}, \frac{1}{10}$ etc, as x is 1,2,3, etc. As x increases indefinitely the value of the function approaches zero: in technical terms the function tends to zero as x approaches infinity. The limit of a sequence is the value approached by the terms as the number of terms increases. Similarly the limit of a series is the value approached by the sum as more and more terms are included. The concept of limits

Lille is an ancient cultural centre whose industries expanded greatly in the 1960s.

Beatrice Lillie is one of the truly great performers of Broadway musical comedy.

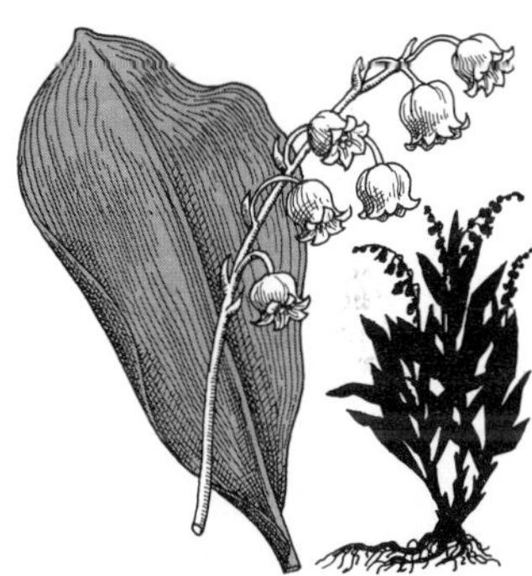

Lilies-of-the-valley were once used in the treatment of heart complaints.

Lima; professional scribes write letters for the largely illiterate population.

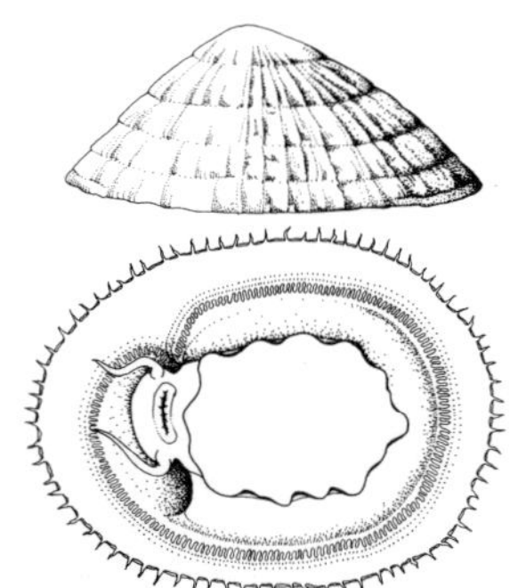

Limpets, usually immobile, move over rocks to feed on algae at high tide.

Lincoln cathedral, begun in Norman times, still dominates the modern city.

Jenny Lind had a successful tour of the USA in 1850–52 with P.T. Barnum.

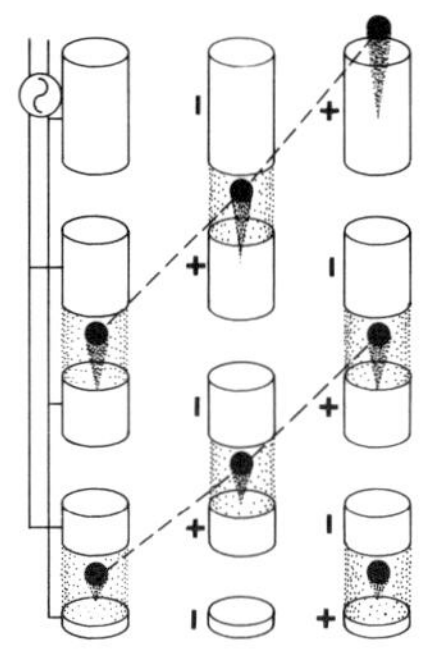

Linear accelerators speed particles through a series of cylindrical electrodes.

is fundamental to differential and integral CALCULUS.

Limitation, Statutes of, under English law, Acts of Parliament that limit the time within which action may be taken to obtain certain rights or claims. Action begun after such a period has expired is barred.

Limited liability, legal condition under which the owners of a company (including shareholders) need not risk their personal wealth in the business – their losses can be equal only to their investment.

Limits, territorial and fishing, areas within which some nations have exclusive rights over the extraction or gathering of natural resources. By legislation of 1964, British territorial waters extend for 19km (12 miles) out to sea from the coast, the inner 6 miles being prohibited to foreign fishing boats, for example.

Limoges, city of w central France, on the River Vienne; the capital of Haute-Vienne département. The city was originally known for its enamel industry especially in the 16th century under Leonard Limousin, but this declined when Limoges was devastated during the Wars of Religion. It now produces chinaware; the china industry was begun by Turgot in the mid-18th century, using the local clay and china stone. The city has an art gallery which includes many works by Renoir, who was born there. The manufacture of footwear is also important.

Limonite, also called brown iron or bog ore, one of the most important sources of iron. Its chemical formula is $FeO(OH).nH_2O$. Limonite may appear lustrous black, brown to yellow and is streaky, porous and often mixed with sand or clay.

Limosin, Léonard (c. 1505–77), French painter; the most prominent member of a family of enamel artists from Limoges. His best-known work is a series of 46 plaques portraying scenes from the life of Christ which is in the LOUVRE, Paris.

Limpet, primitive gastropod MOLLUSC commonly found fixed to rocks along marine shores. It has a cap-like, rather than coiled, shell and a large muscular foot. Length: to 13cm (5in). Families: Patellacea, Acmaeidae and Fissurellidae. *See also* NW p.90.

Limpkin, also called a couran, grey-brown, long-legged marsh bird of s America and the southern USA. In the evolutionary chain it comes between the crane and the rail. It feeds on snails and other molluscs.

Limpopo (Crocodile), river in s Africa. It rises in the Transvaal, South Africa, then flows N, forming part of the border between South Africa and Botswana, then E, forming the border between South Africa and **Zimbabwe (Rhodesia). It flows across Mozambique to enter the Indian Ocean.** Length: approx. 1,770km (1,100 miles).

Linacre, Thomas (c. 1460–1524), English physician and scholar. In 1500, after travelling extensively in Italy, he was appointed tutor to Prince Arthur, son of HENRY VII. He founded and became the first president of the Royal College of Physicians of London in 1518, which was empowered to license and examine physicians in England. In 1520 he was ordained and left medical practice.

Linar, in astronomy, point sources which emit high energy at wavelengths characteristic of certain chemical compounds. They were discovered in 1970 and are of potential use as navigational aids.

Lincoln, Abraham (1809–65), 16th President of the USA. He was born into a poor frontier family and had little formal education. The family settled in New Salem, Illinois, in 1831, where Lincoln worked in a shop and began reading law. In 1834 he was elected to the Illinois legislature as a WHIG and in 1836 was admitted to the bar and began to practise. In 1841 he was elected to the US House of Representatives.

Lincoln, by then a Republican, ran for the Senate in 1858. He was defeated but his oratory won him national fame. As a result, he became a leading candidate for the 1860 REPUBLICAN nomination, which

he won on the third ballot. He defeated the DEMOCRATIC candidate and was inaugurated as President on 4 March 1861. The Southern states had already seceded and on 12 April Fort Sumter was fired on, the first action in the AMERICAN CIVIL WAR. Lincoln conducted the war with vigour and efficiency: he called up the militia, blockaded southern ports and increased executive powers. With his appointment of Ulysses S. GRANT as his commander-in-chief he ensured that the war would be relentlessly pursued.

On 22 Sept. 1863 Lincoln issued the Emancipation Proclamation, freeing the slaves, and on 19 Nov. 1863 delivered one of the noblest public speeches ever made, the GETTYSBURG ADDRESS. Lincoln was re-elected in 1864 and saw the war brought to a successful conclusion, but on 14 Sept. 1865, while attending a performance at Ford's Theatre, New York, he was shot by John Wilkes BOOTH, a disaffected Southerner. He died the following day. *See also* HC2 pp.150–151, *151*.

Lincoln, county town of Lincolnshire, founded as the town of Lindum. It has many Roman remains, and the Fosse Way and Ermine Street are in its vicinity. The Lincoln Handicap is held at its racecourse. Industries include the manufacture of excavators, pumps, castings, agricultural machinery and motor-car parts. Pop. (1971) 74,207. *See also* HC1 pp.101–102, 104, 200, *200*, 216.

Lincoln Center for the Performing Arts, New York, complex of buildings housing the Metropolitan Opera House, New York State Theater, Juilliard School of Music, Vivian Beaumont Theater and the Library and Museum of the Performing Arts. Incorporated in 1956 as a non-profit-making organization, it was built between 1959 and 1969.

Lincolnshire, county in E Central England; bordered by the counties of Humberside (N), Nottinghamshire (W), Leicestershire (SW), Cambridgeshire and Norfolk (S), and by the North Sea (E). The region is very flat, and drainage by the Witham and Trent rivers has been aided by dikes and canals since Roman times. Agriculture is the mainstay of the economy, and industry is closely allied to it. The administrative centre is Lincoln. Area: 5,886sq km (2,273sq miles). Pop. (1976) 524,500.

Lind, James (1716–94), Scottish physician who eradicated SCURVY from the British navy with his recommendation that fresh citrus fruit and lemon juice should be part of the daily diet of seamen. Prior to his intervention, more British sailors died from scurvy than from combat.

Lind, Johanna Maria ("Jenny") (1820–87), Sweden's most famous operatic soprano. Known as the "Swedish nightingale", she made her debut in 1838, and after world success in coloratura roles she settled in London (c. 1852), where she sang in oratorio and taught at the Royal College of Music.

Lindbergh, Anne Morrow (1907–), US writer. Married in 1929 to the aviator Charles LINDBERGH, she accompanied him on many flights. *North to the Orient* (1935) and *Listen, The Wind* (1938) record her experiences.

Lindbergh, Charles Augustus (1902–74), US aviator. He became an international hero when, in his plane *The Spirit of St Louis,* he made the first non-stop transatlantic solo flight (1927). In 1932 his two-year-old son was kidnapped and murdered. The resulting publicity forced the Lindberghs to leave the USA and settle in Britain (1935–39). An isolationist, he was active in the America First Committee and was accused of pro-Nazi sentiments. He was awarded the Pulitzer Prize for literature for his book *The Spirit of St Louis* (1953).

Lindegren, Erik Johan (1910–68), Swedish poet. He was noted for the originality of his verse and the grimness of his images, especially in *The Man Without a Way* (1942). In 1947 he edited an influential anthology of contemporary poetry.

Lindemann, Frederick Alexander, 1st Viscount Cherwell (1886–1957), British physicist, *b.* Germany. Involved in aircraft

research (1914–18), he discovered how to regain control of an aircraft in an uncontrolled spin. During WWII he was Winston CHURCHILL's scientific adviser. He developed the Clarendon Laboratory of Oxford University into a major research facility.

Linden tree. *See* LIME.

Lindisfarne, (Holy Island) island connected to the Northumbrian coast (NE England) by a causeway. It was a religious and missionary centre from AD 635, when St Aidan arrived there from Iona. The monastery he founded disseminated the Celtic Christian culture from Edinburgh to the Humber. The Danes pillaged Lindisfarne in 793 and 875, but the monks returned in 1082 and the monastery continued its existence until the 16th century.

Lindisfarne Gospels, manuscript illuminated in the Hiberno-Saxon style in the late 7th or 8th century. It may have been executed for Eadfrith, Bishop of Lindisfarne (698–721). It incorporates Classical, Irish and Byzantine styles of illumination and is attributed to the Northumbrian School.

Lindley, John (1799–1865), British botanist, known for his classification of plants in a natural system in which all characteristics of a plant are considered. Among Lindley's best-known books are *Theory and Practice of Horticulture* (1842) and *The Vegetable Kingdom* (1846), which included his new classification of plants.

Lindrum, Walter (1899–1960), Australian billiards player who beat Joe Davis for the world championship in 1933 and 1934, then devoted himself to exhibitions and charity work. He amassed numerous records including a highest break of 4,137 in 1932.

Lindsay, or Lyndsay, Sir David (c. 1490–1555), Scottish writer. He sat in the Scottish Parliament (1540–46) and represented the Crown on important missions. His poetry was a series of richly allegorical and satirical attacks on Scottish politics. His most famous work was the morality play, *Ane Satyre of the Thrie Estaitis,* published in 1540.

Lindsay, Howard (1889–1968), US actor, producer and playwright. His comedies and musicals include *Life with Father* (1939), *Arsenic and Old Lace* (1941) and *State of the Union* (1945). His dramatization, with Russell Crouse of *Life with Father* ran for seven years. They collaborated with Rodgers and Hammerstein on the book of *The Sound of Music* (1959).

Lindsey, in post-Roman Britain, kingdom of the ANGLES, settled in the 5th and 6th centuries. It was under either Mercian or Northumbrian control until the late 7th century, when it was finally absorbed by Mercia. It occupied what is now northern Lincolnshire.

Lindwall, Raymond Russell ("Ray") (1921–), Australian cricketer who played for New South Wales, Queensland and Australia. In 61 Test matches from 1946 to 1959 he took 228 wickets at an average of 23.05. The greatest fast bowler of his day, he formed a notable partnership with Keith MILLER.

Linear accelerator, particle accelerator in which charged particles travel through a straight vacuum chamber. Electrostatic accelerators are simple examples. In other types energy is gained in regions of high-frequency electric field, produced between cylindrical electrodes. Higher energies can be achieved by carrying the particles along a waveguide with a microwave field. Final energies depend on the length of the chamber, and can reach several GeV. *See also* SU p.67.

Linear function, mathematical function the value of which is given by a polynomial involving no powers of its variables, as in $f(x) = 7x + 3$. The graph in two dimensions of a linear function is a straight line.

Linear script, early form of writing, found on clay tablets in Crete and Greece. Linear A was in extensive use during the middle period of the MINOAN CIVILIZATION (c. 2000 BC–c. 1450 BC). Linear B was an adaptation of Linear A, used by the Myceneans of mainland Greece to write their early form of Greek, a language probably distinct from that used by the Minoans. Linear B was used from *c.*1450 BC to the end of the Mycenean age, *c.*1150 BC. Linear A tablets have only been found in Crete, whereas Linear B inscriptions have been discovered in both Greece and Crete; there is uncertainty as to whether the two scripts overlapped at all. In 1952 Michael Ventris deciphered Linear B; Linear A still defies effective analysis.

Linen, yarn and fabric made from the soft, bast fibres of the flax plant (*Linum usitatissimum*); one of the oldest textile fibres. Flexible bast fibres come from the inner tissues beneath the bark. They are released from the substance that binds them by soaking (retting) the long stems in water; these are beaten and passed through rollers to separate the fibres. After sorting, spinning and bleaching they are woven into fabrics which are generally strong and fine, such as table linen. *See also* JUTE; PE pp.216, *216.*

Lines of force, lines in an electric or magnetic field whose direction at every point is that of the field. *See also* ELECTRICITY; MAGNETISM; SU pp.116, 118.

Ling, any of many varieties of *Calluna vulgaris,* a small heather plant common to high open meadows and moorlands throughout W Europe. Family Ericaceae.

Ling, food fish related to the COD, found in the Atlantic Ocean. It is brown and silver and has long dorsal and ventral fins. Length: to 2m (7ft); weight: 3.6kg (8lb). Family Gadidae; species *Molva molva. See also* NW p.*131.*

Lingam, stylized phallus worshipped as a symbol of the Hindu god SHIVA. The yoni, which is the symbol for the female sexual organ, is the symbol of the consort of Shiva, SHAKTI; it often forms the base of the lingam as an indication of the inseparable nature of male and female principles.

Lingua franca, any language that serves as a medium of communication between people who otherwise do not speak the same language. A lingua franca may be a highly simplified form of the language of the dominant power, such as the PIDGIN ENGLISH used by hundreds of Melanesian tribes, or it may be a hybrid such as Swahili, which is composed of words of both Arabic and Bantu origins and spoken in many parts of central and E Africa.

Linguistics, the study of language. The study can be divided into PHONETICS (the study of sounds and sound systems), SYNTAX (the study of grammar) and SEMANTICS which relates to meaning. In the 19th century linguistics was purely descriptive, concentrating on the reconstruction of language families, eg INDO-EUROPEAN, and on the classification of individual types of language. Since the 1950s the subject has been invigorated by CHOMSKY's controversial theories of language acquisition which emphasize the universal characteristics of language skills rather than the particular aspects of the individual language. *See also* MS pp.154–155, 238–239.

Linkage groups, group of inherited characteristics or genes that occur on the same chromosome and that remain connected in such a way that they are usually inherited together through successive generations. *See also* NW pp.28–31.

Linklater, Eric (1899–1974), British novelist. He published his first novel, *White-maa's Saga,* in 1929. The first of his famous picaresque novels, *Juan in America,* was published in 1931. He wrote three volumes of autobiography, *The Man on My Back* (1941), *A Year of Space* (1953) and *Fanfare for a Tin Hat* (1970).

Linköping, city in SE Sweden. A bishopric since 1120, the city prospered in the Middle Ages as a religious and intellectual centre. It has a 13th-century castle, library, university, and a fine 12th-century Romanesque cathedral. Industries: motor vehicles, aircraft, food processing, tobacco. Pop. (1970 est.) 80,800.

Linlithgow, John Adrian Louis Hope, 1st Marquis of (1860–1908), British Conservative politician and colonial governor. He was Paymaster-General (1895–98) and Lord Chamberlain in SALISBURY's third government, and the first Governor-General of the Commonwealth of Australia (1901–02).

Linna, Väinö (1920–), Finnish writer. One of the most important authors in postwar Finland, he has used the novel's traditional narrative style in such works as *The Unknown Soldier* (1954) and the trilogy *Tällä Pohjentähden Alla* (Here Under the Pole Star) (1959–62).

Linnaeus, Carl (Carolus) (1707–78), Swedish botanist and explorer, also known as Carl von Linné. He was the first scientist to define the differences between species, giving Latin names for the genus and species to each organism, making consistent use of specific names and including all known organisms in a single classification. His *Systema Naturae,* published in 1735, laid the foundation of the modern science of TAXONOMY. *See also* HC2 pp.32, *32,* 312; NW p.32;

Linnankoski, Johannes (1869–1913), Finnish novelist, real name Vihtori Peltonen, and a supporter of the nationalist drive for independence from Russia. Of humble birth and largely self-educated, his novels often depict peasant life. They include *Pakolaiset* (The Fugitives) (1908) and *The Song of the Blood-red Flower* (1905).

Linnean Society, biological society founded in 1788; it received its first Royal Charter in 1802. It has had such prominent men as Sir Joseph BANKS, Charles DARWIN and Thomas H. HUXLEY among its members. Its aims remain the same today as at the time of its founding – to promote biology in all its aspects.

Linnell, John (1792–1882), British painter, the father-in-law of Samuel PALMER and benefactor of William BLAKE. His portraits and landscapes were typically romantic, delicate and of a small scale. They include *Woodcutting in Windsor Forest* (1835), and *Kensington Gravel Pit* (1812).

Linnet, small songbird of the family Carduelidae. *Carduelis cannabina* of Europe inhabits hedgerows and thickets, moving to open country in colder seasons. Both sexes are brown and grey, but the male has a red breast and crown in the summer. Length: to 13cm (5.1in).

Linoleum, floor covering manufactured from oxidized oils, gums, resins and pigments formed into a sheet by pressing between rollers and usually applied to a jute backing.

Linotype, method of setting type for printing, by the Linotype machine, invented by Ottmar Mergenthaler at Baltimore in 1884. It was first used by the New York *Herald Tribune* in 1886. It is operated by a keyboard like that of a typewriter. It sets cast letters as one complete line of type, called a slug. *See also* MM p.205.

Lin Piao (1907–71), Chinese Communist general and political leader. He won fame as commander of the Fourth Field Army, which was instrumental in defeating NATIONALIST forces in 1949. Lin was designated MAO TSE-TUNG's heir apparent while helping to lead the 1965–69 CULTURAL REVOLUTION in China. He was later accused of plotting a revolt and was said to have been accidently killed when his plane crashed.

Linsang, slender, graceful member of the MONGOOSE family, which also includes CIVETS and GENETS. It is found in areas of SE Asia. There are two species: the banded linsang (*Prionodon linsang*) whose tawny fur is banded with black; and the spotted linsang (*P. pardicolor*), pale brown with black spots. Length: up to 79cm (31in); weight: 0.7kg (1.5lb). The so-called African linsang (*Poiana richardsoni*) resembles a genet. Family Viverridae. *See also* NW pp.213, *214.*

Linseed. *See* FLAX.

Linton, Ralph (1893–1953), US anthropologist who contributed to the development of cultural anthropology. His studies in Madagascar and E Africa led to *The Tanala, A Hill Tribe of Madagascar* (1933), an important ethnological work. He is best known for his contribution to cultural anthropological theory, notably in *The Study of Man* (1936) and *The Cultural Background of Personality* (1945).

Linton, William James (1812–97), British wood engraver and writer. After 1866 he lived in the USA. He made engravings for

Frank Leslie's Illustrated News and LONG-FELLOW's *Building of the Ship.* He was a CHARTIST and his writings were chiefly political. They include a *Life of Paine* (1839), *To the Future* (1848) and *The Plaint of Freedom* (1852).

Linz, city and major river port in N central Austria, on the River Danube; it is the capital of Upper Austria. The city was a Roman settlement, named Lentia. In the late 15th century it became a provincial capital of the HOLY ROMAN EMPIRE. Landmarks include the 8th-century Church of St Martin. Industries: iron and steel, machinery, textiles. Pop. (1971) 202,900.

Lion, large CAT that lives on African savannahs south of the Sahara and in SW Asia. It is golden yellow to buff with light spots under the eyes. The male is instantly recognizable by its deep neck mane that darkens with age. The female does most of the hunting and preys on antelopes, zebras and bush pigs. Lions, as symbols of ferocity, courage and loyalty have figured in myths and legends throughout the world. The male is traditionally heralded as the king of beasts. Length: to 2.5m (8.2ft) overall. Family Felidae; species *Panthera leo. See also* NW p.*201.*

Lionel, Duke of Clarence (1338–68), third son of Edward III and patron of Geoffrey CHAUCER. Lionel served as Lieutenant of Ireland during which time he enacted (1361) the infamous Statutes of Kilkenny which effectively outlawed all those who adopted "the manners, fashion and language of the Irish enemies". *See also* HC1 pp. 181, 214.

Lionfish, also called zebra or dragon fish, marine tropical fish found in shallow waters of the Indian Ocean and W Pacific. A rust-red and white striped fish with numerous long spines and rays, it carries venomous spines in its dorsal fin that can inflict painful wounds. Length: 5cm (2in). Family Scorpaenidae; species *Pterios volitans.*

Lipari Islands (Isole Eolie), group of small volcanic islands off the N coast of Sicily, Italy, in the Tyrrhenian Sea. The group includes the islands of Stromboli, which has the 926m (3,033ft) volcano of the same name, Vulcano, Salina and Lipari. Exports: wine, raisins, fish, pumice. The chief industries are fishing and tourism. The Islands are also noted for archaeological remains dating from NEOLITHIC times. Area: 114sq km (44sq miles). Pop. 13,774.

Lipchitz, Jacques (1891–1973), US sculptor who was one of the finest of the early CUBIST sculptors. His early works include *Sailor with a Guitar* (1914). *Harpist* (1928) was an early example of his "transparents", bronzes expressing incessant movement. Subsequent works, including *Prayer* (1943), became larger and more imaginative, stressing spirituality, sensuality and sheer volume.

Lipid, one of a large group of organic compounds in living organisms that are insoluble in water but soluble in alcohol and ether. They include animal fats, vegetable oils and natural waxes. Lipids are a major energy source for animals. They form an important food store and energy source of plant and animal cells and are structural components of cell membranes. Storage fat is composed chiefly of triglycerides, which consist of three molecules of a fatty acid linked to glycerol. *See also* SU pp.152–153, *152.*

Lipinsky, Edward, (1888–), Polish economist who became professor at the Central School of Economics at Warsaw in 1929. His many books include *The Workers' Movement in Poland* (1916), *History of Economic Thought* (1955) and *Karl Marx and the Problems of our Times* (1969).

Lipmann, Fritz Albert (1899–), US biochemist, *b.* Germany. He shared with H.A. KREBS the 1953 Nobel Prize in physiology and medicine for his discovery of COENZYME A and its importance for intermediary metabolism. *See also* SU p.155.

Li Po. *See* LE T'AI-PO.

Lipoma, TUMOUR formed of fatty tissue. It is non-malignant. The tumour forms a mass just beneath the skin, unsightly but seldom dangerous. They are usually

Carl Linnaeus introduced systems of classification for both plants and animals.

Linnets, are named after their taste for the seeds of flax plant (Linum).

Linz, on the railway between the Baltic and Adriatic, is an important trade centre.

Lions, unlike other members of the cat family, live in prides of up to 20.

Lippershey, Hans

Filippino Lippi included this self portrait in one of his frescoes.

Lisbon; a view of the Portugese capital showing the National Theatre.

Johann Liss combines earlier styles in this picture, *The Game of Buck-buck*.

Literature: detail from the "Breeches" bible, printed in Geneva in 1560.

removed surgically. *See also* MS p.*87*.

Lippershey, Hans (*c.* 1570–*c.* 1619), German spectacle maker who invented the telescope. In 1608 he offered his invention to the Estates of Holland for use in warfare. Its potential for astronomy was recognized by Jacques Bovedere of Paris, who reported it to GALILEO who then built his own.

Lippi, Filippino (*c.* 1457–1504), Italian painter. The son of Fra Filippo Lippi, he followed the style of BOTTICELLI in his early work; his mature style, of which *Vision of St Bernard* (*c.* 1486) is an example, influenced the Florentine MANNERISTS in the 16th century.

Lippi, Fra Filippo (*c.* 1406–69), major Florentine painter. He was influenced by MASACCIO and Fra ANGELICO and in turn influenced BOTTICELLI and others. His works include the *Barbadori altarpiece* (*c.* 1437), *Coronation of the Virgin* (*c.* 1445), *Annunciation* (*c.* 1437–40) and *Adoration of the Child* (*c.* 1455).

Lippizaner, breed of horse, originally from Lipizza, in the former Austro-Hungarian empire. The horse is small in stature with a long back and a thick neck. These characteristics are suitable for performing intricate and delicate movements.

Lippmann, Walter (1889–1974), US journalist and author. Lippmann was an influential columnist on the New York *Herald Tribune*, combining a realistic view of politics with a strong moral sense. His books include *A Preface to Morals* (1929) and *The Public Philosophy* (1955).

Lipscomb, William Nunn, Jr (1919–), US chemist who made fundamental discoveries about the bonding of molecules, for which he was awarded the 1976 Nobel Prize in chemistry. *See also* SU p.*140*.

Lipstick, waxy cosmetic used to colour the lips. It contains a fatty base that is firm and solid and yet spreads readily when applied; the colour is provided by pigments or stains, usually red. Eosin and fluorescein derivatives are often used but these dyes are only slightly soluble in the fatty base and require an additional solvent. *See also* MM pp.246, *247*

Liquefaction of air, achieved by cooling air to its critical temperature of −147°C (−232.6°F), at or below which it can be liquefied by pressure. The method involves repeated compression, followed by rapid ADIABATIC expansion, which results in progressive drops in temperature due to the Joule-Thomson Effects. *See also* SU p.88.

Liquefied petroleum gas (LPG), liquefied gas of light HYDROCARBONS (compounds of carbon and hydrogen), principally propane and butane, produced in the distillation of crude oil and the refining of NATURAL GAS. It is used as a fuel and raw material in chemical industries, as bottled gas for powering motor vehicles and for home heating and cooking.

Liqueur, flavoured distilled liquor, with an alcohol content of 48 to 120 proof (24–60%), made by combining fruits or herbs with a base spirit and a sweetener. Liqueurs were first produced by monks and ALCHEMISTS and have been called balms, elixirs and oils.

Liquid, state of matter intermediate between a gas and a solid. A liquid substance has a relatively fixed volume but flows to take the shape of its container. Liquids have higher viscosity than gases, and their viscosity decreases with temperature. The state that a substance assumes depends on the temperature and the pressure at which it is kept; a substance that is liquid (eg water) at room temperature can be changed into a vapour (its gaseous state, steam) by heating, or into a solid (ice) by cooling. Unlike the molecules in gases, those in a liquid are bonded (weakly) and liquids (like solids) are compressed only slightly by pressure. *See also* SU pp.84–89.

Liquid assets, in economics, assets either in the form of money or in forms which can be readily converted into money, eg stocks and shares.

Liquid crystal, substance that flows like a liquid yet has the kind of ordered molecular structure characteristic of crystals.

There are many organic crystalline substances that can be brought to a liquid crystal state when heated. They have unique properties that have been applied in optical displays, eg calculator numerals, and temperature variation detection systems.

Liquid oxygen, unnatural state of the gaseous element OXYGEN (O_2); at one atmosphere pressure, oxygen liquefies at −183°C. It is widely used in rocket engines as an oxidizer with various liquid fuels to produce high-velocity gases.

Liquorice, perennial plant of the pea family, native to the Mediterranean region and cultivated in temperate and subtropical areas. It bears spikes of blue flowers. The dried roots are used to flavour confectionery, tobacco, beverages and medicines. Height: to 90cm (3ft). Family Leguminosae; species *Glycyrrhiza glabra*.

Lisa, Manuel (1772–1820), US fur trader. In 1812 he built Fort Lisa, near present day Omaha. From 1813 to 1822 it was the busiest fort on the Missouri River.

Lisboa. *See* LISBON.

Lisbon (Lisboa), port and capital city of Portugal, at the mouth of the Tagus River on the Atlantic Ocean. An ancient settlement, the city was occupied by the Phoenicians and the Carthaginians. It was conquered by the Moors in 716. Retaken by the Portuguese in 1147, it became, after the 14th century, a leading European port and the heart of the Portuguese empire. It declined under Spanish occupation from 1580 to 1640 and suffered a major earthquake in 1755. It has a fine harbour, and is once again a major port, handling most of Portugal's international trade. There is a Moorish mosque, and the University of Lisbon (1911) in the city. Industries: shipbuilding, oil- and sugar-refining, steel, chemicals, textiles. Pop. 782,266. *See also* HC1 pp.*235, 241*.

Li Shang-yin (813–58), Chinese poet of the late period of the T'ANG DYNASTY (618–906). His verse often showed lively and graceful imagery but its frequent artificiality hinted at the coming poetic decadence. His poems are similar to those of WEN T'ING-YÜN (*fl.* mid–9th century).

Lispector, Clarice (1925–), Brazilian novelist and short-story writer. Regarded as one of her nation's greatest 20th-century authors, she is concerned with basic human themes rather than specific Brazilian conditions. Her works often deal with her characters' inability to cope with crises beyond their understanding. *Beside the Savage Heart* (1944), her first novel, and *The Apple in the Dark* (1961) were highly praised.

Liss, or Lys, Johann (*c.* 1595–1629), German BAROQUE painter of genre and religious scenes, one of the few German artists of his day to enjoy international fame. He settled in Venice, where he helped to bring new vigour to early 17th-century Venetian painting. His works include *The Banquet*, *Toilet of Venus* and *The Vision of St Jerome*.

Lissitzky, Eliezor (El) Markovich (1890–1941), Russian painter and architect. He was a pioneer of modern typographical design, of which his *Story of Two Squares* (1920) is sometimes considered the first important example.

List, Geog Friedrich (1789–1846), US economist, *b.* Germany, who advocated tariff protection for new industries. An adherent of Adam SMITH and David RICARDO, his programme for German economic development included protectionism, the expansion of railway building and the formation of a customs union. His best-known work is *The National System of Political Economy* (1841).

Lister, Joseph, 1st Baron (1827–1912), British surgeon who introduced the principle of antisepsis, which complemented PASTEUR's theory that bacteria cause infection. Using carbolic acid (phenol) as the antiseptic agent, and employing it in conjunction with his heat sterilization of instruments, he brought about a dramatic decrease in post-operative fatalities. *See also* HC2 pp.165, *165*.

Liszt, Franz (1811–86), Hungarian composer and pianist. Beginning From his first solo recital at the age of nine, he startled

his contemporaries with his virtuosity and showmanship. He was patron and friend to many great artists of his day, such as CHOPIN and GRIEG, and his music influenced subsequent composers including WAGNER, Richard STRAUSS and RAVEL. Among his compositions are two popular piano concertos and his Hungarian rhapsodies and études. He also created the symphonic (or tone) poem. *See also* HC2 pp.106–107, *107, 120*.

Li T'ai-po, or Li Po, (701–62), Chinese poet of the T'ANG DYNASTY, generally regarded as one of the two greatest poets in an age of remarkable literary achievement. Unlike his contemporary and rival, TU FU, who was Confucian, Li Po was a Taoist and the influence of TAOISM can be seen in the sensual and spiritual aspects of his work as well as in the many anecdotes about his unconventional and often irreverent behaviour; he is said to have drowned while drunk. His poetry is characterized by its vivid imagery, spontaneity and rich emotion. *See also* HC1 p.126.

Litany, form of prayer in Christian churches, originally used in processions. A member of the clergy intones or says a series of petitions, to each of which the congregation replies with a single, repeated, petitionary response. It is also used in private devotions.

Litchi. *See* LYCHEE.

Literary criticism, assessment of works of literature. The literary critic performs two roles: he distinguishes contemporary works in which some lasting value may be found; and he presents, by interpretation and commentary, the works of the past which he judges worthy of transmission. ARISTOTLE's *Poetics* began literary criticism; in it is an attempt to systematize poetry and drama and Aristotle uses such concepts as mimesis, or imitation, and catharsis. By means of the latter, tragedy arouses within the spectator feelings of pity and terror and then succeeds in purging them. The emphasis of most subsequent criticism, however, has been the postulation of a hierarchical list, a kind of apostolic succession of writers and critics, in the manner of Samuel JOHNSON, Mathew ARNOLD, T. S. ELIOT and I. A. RICHARDS.

Literature, collections of writings, usually grouped according to language, period and country of origin. Examples include classical Greek Literature, modern Russian literature, Elizabethan literature, and others. Within these groupings the literature may be further divided into forms, such as poetry and prose, and within these again into verse drama, non-fiction prose, novels, epic poems, tragedies, satires and so on. A piece of writing becomes literature in a series of selective processes; those which are most surely literature are those sanctified both by their age and the subsequent weight of critical acclaim. Antique works, however, separated from the contexts which produced them, require much scholarship for present-day appreciation. In the processes of selection LITERARY CRITICISM plays an important part.

Lithium, common metallic element (symbol Li) of the alkali-metal group, first isolated in 1817. Ores include lepidolite and spodumene. The element, which is the lightest of all metals, is used in alloys and as a heat-transfer medium. Chemically it is similar to sodium, although it does have some resemblance to the alkaline-earth metals. Properties: at. no. 3; at. wt. 6.941; s.g. 0.534; m.p.180.5°C (357.0°F); b.p. 1,347°C (2,456.6°F); most stable isotope Li^7 (92.58%). *See also* ALKALI ELEMENTS; SU pp.134–137.

Lithography, in art, planographic method of printmaking and printing developed in the 18th century and based on the immiscibility of grease and water. A design is made on a prepared plate or stone with a greasy lithographic pencil, a crayon or ink. When ink is rolled across, it is picked up by the design but the rest of the plate is protected by a thin film of water. *See also* OFFSET PRINTING; MM pp.*205*, 206–207.

Lithosphere, outer rocky shell of the Earth, extending down about 700km (430 miles). It is composed of the CRUST and upper mantle, and overlies the asthenosphere on which it may drift sideways. It does not include the hydrosphere, biosphere or atmosphere. *See also* PE pp.*20, 24.*

Lithuania (Litovskaja SSR), constituent republic of the USSR, bounded by Poland (s) and the Baltic Sea (w); the capital is Vilnius. A low-lying area, it is drained chiefly by the Neman River. Cattle are raised throughout the region and grain, sugar-beet, flax and vegetables are grown. Since 1940 industry has developed considerably and includes food processing, textiles, chemicals, machine tools and electrical equipment. Lithuania was conquered by the TEUTONIC KNIGHTS in the 13th century. It later became one of the most powerful states in medieval Europe. In 1569 the state merged with Poland. With the 18th-century partitions of Poland, Lithuania passed to Russia. An independent republic was proclaimed in 1918, but the region was retaken by the Soviets in 1940. Area: 65,201sq km (25,174sq miles). Pop. (1976) 3,317,000.

Lithuanian, language spoken principally in Lithuania SSR, USSR. Together with LAT-VIAN it forms the Baltic subgroup of the INDO-EUROPEAN language family.

Litmus, purplish dye in neutral aqueous solutions, red in acid solutions and blue in alkaline solutions. It is most familiar in the form of litmus paper which is used as an acid-base indicator. The dye is extracted from lichens.

Litotes, form of understatement in which an idea is expressed by denying its opposite. For example, "not bad" is said when "quite good" is meant. Often the litotes carries an intentional sense of irony, by implicitly denying what it affirms.

Litre, (l), metric unit equal to a cubic decimetre, ie one thousandth of a cubic metre. Another definition, used from 1901–68, was that the volume of 1kg of pure water at 4°C equalled 1 litre. It is equivalent to 0.22 imperial gallons or 0.2642 US gallons.

Little America, region in Antarctica, s of the Bay of Whales, on the outer edge of the Ross Ice Shelf. It was explored from 1928 to 1930 by Admiral R. E. BYRD, and used as the headquarters for his second expedition (1933–35). It was also the base for the US Antarctic Service Expedition (1939–41) and the US Naval Operation High Jump expedition (1946–47).

Little Bear, or Ursa Minor, northern circumpolar constellation, resembling a faint, distorted version of the GREAT BEAR. It contains two second magnitude stars: the POLE STAR (Polaris) and Beta Ursae Minoris (Kocab), nicknamed Guardian of the Pole. *See also* SU pp.*256–257.*

Little Bighorn, Battle of (25 June 1876), between Indians, under Sioux chiefs Sitting Bull and Crazy Horse and the 7th US Cavalry detachment led by General George CUSTER. Custer attempted a surprise attack but the 2,500 Indians wiped out his 264 officers and men near the Little Bighorn River in the Dakota Territory. *See also* HC2 p.*149.*

Little Dorritt (1855–57), novel by Charles DICKENS. One of Dickens's more sentimental works, it deals (as does *Our Mutual Friend*) with true human worth and sincerity in the context of worldly commercialism. The scenes in Marshalsea, the debtors' prison, owe a lot of their immediacy to Dickens' familiarity with it from his father's imprisonment there.

Little Entente (1920–38), alliance contracted between Romania, Yugoslavia and Czechoslovakia after WWI to maintain post-war boundaries. Through political and economic unity and the support of France and Poland, the alliance helped to prevent ANSCHLUSS (the uniting of Germany and Austria) until 1938.

Little Lord Fauntleroy (1866), popular Victorian novel by Frances Hodgson BURNETT relating the tale of an American boy, Cedric, who inherits an English earldom. The book inaugurated a vogue for long curly hair and velvet suits for young boys.

Little Red Book, popular name in the West for a small volume of the thoughts of the Chairman of the Chinese Communist Party, MAO TSE TUNG. It was first issued

during the CULTURAL REVOLUTION in 1966 and became the "bible" of the Red Guards. *See also* HC2 p.245.

Little Rock, state capital of Arkansas, USA, on the Arkansas River. A strong anti-Union centre at the outbreak of the AMERICAN CIVIL WAR, Little Rock attracted world attention in 1957, when federal troops were used to enforce a Supreme Court ruling (1954) against segregation in schools.

Little Van Dyck. *See* COQUES, GONZALES.

Little Women (1868–69), novel in two parts written by Louisa May ALCOTT. It is the story of a family, the Marches, in a small New England town in the mid-19th century. It concentrates on the adolescence of the four daughters, Jo, Meg, Amy and Beth. *Little Women* was immensely popular as a children's book and sold 60,000 copies within the first year of publication.

Littlewood, Joan Maud (1914–), British theatre director and actress who founded THEATRE WORKSHOP in 1945 which toured Britain until 1953, when it found a permanent base at the Theatre Royal, Stratford, London, until it disbanded in 1973. Her most successful productions, which ultimately transferred to the West End, included Brendan Behan's *The Hostage* (1958), Lionel Bart's *Fings Ain't Wot They Used T'Be* (1959) and *Oh What a Lovely War* (1963) – a viciously critical musical revue about WWI.

Littoral zone, beach area between high and low tides. It also refers to the benthic division – between high tide and a depth of 200m (656ft). The larger zone is divided into the eulittoral – from high tide to a 50m (164ft) depth – and the sublittoral – from 50m to 200m (164-656ft). The lower edge of the eulittoral is the lowest limit at which abundant attached plants can grow. *See also* NW pp.236–237, 238–339.

Littré, Maximilien-Paul-Émile (1801–81), French linguist, philosopher and lexicographer. Fluent in several Classical languages, he translated HIPPOCRATES (1862). His most noted work is his four-volume *Dictionary of the French Language* (1863–73).

Liturgy, in public worship, the ritual or form of religious ceremonies for specific days, such as Good Friday, or for particular sacraments, such as BAPTISM or the EUCHARIST. It may also refer to the whole ritualistic structure of one religion to distinguish it from another.

Liu Chih-chi (661–721), Chinese historian. His *Shih t'ung* ("Understanding of History") is generally considered to be the first exposition on historical method.

Liu Hsieh (late 5th, early-6th centuries AD), Chinese literary critic. Noted for *The Carved Dragon of the Literary Mind,* a 50-chapter analysis of writing. The first 25 chapters explore literary types, including the Confucian classics. In the second 25 chapters he discusses criteria for excellence.

Liu Shao-ch'i (1898–1974), Chinese Communist leader. Trained in Moscow (1920–22), he became one of the chief theorists of the Chinese Communist Party and second in rank from 1954 until purged by MAO TSE-TUNG during the CULTURAL REVOLUTION of 1966–69 for allegedly "revisionist" views. Rumours of his death were confirmed in 1974. *See also* HC2 pp.244, 245.

Liutprand (d. 744), king of the Lombards from 712 to 744. Under his rule Lombardy reached its zenith; he unified and expanded the kingdom, and his reforms of the penal code anticipated those of CHARLEMAGNE. His conquest of Roman lands was halted only by personal appeals to his Catholic conscience by the reigning popes. *See also* HC1 pp.135, 152, 160.

Live-Forever, also called stonecrop or orpine, a succulent perennial plant native to Europe. It has smooth oval, grey-green leaves and clusters of tiny reddish-purple flowers. Height: 60cm (2ft). Family Crassulaceae; species *Sedum telephium.*.

Liver, large glandular organ. It is soft and reddish-brown and lies mostly in the upper right quadrant of the abdominal region. It has many functions, including

formation of BILE, storage of carbohydrate, regulation in part of carbohydrate metabolism, breakdown of HORMONES, detoxification of drugs and other substances, destruction of worn-out red blood cells, and an important role in fat metabolism. It produces and secretes bile which is stored in the GALL BLADDER. Bile passes into the small intestine, where it is primarily concerned with fat metabolism, acting to emulsify fats (make them water-soluble) and to activate other digestive ENZYMES. *See also* MS pp.68, *69,* 93, 105, *105.*

Liver fluke. *See* SHEEP LIVER FLUKE.

Liverpool, city in Merseyside, NW England on the Mersey estuary. It was chartered in 1207 and has been a leading port since the mid-18th century. Although it suffered severe bomb damage in WWII, Liverpool has become one of the country's major cities and has an important export trade. Once famous for its pottery and cotton industries, Liverpool now produces electrical equipment, chemicals and rubber and has extensive dockyards. Pop. (1971) 606,834.

Liverwort, any of about 8,000 species of tiny, simple, non-flowering green plants which, like the related MOSSES, lack specialized tissues for transporting water, food and minerals within the plant body. Liverworts make up two classes, Hepaticae and Anthocerotidae, of the plant division Bryophyta. *See also* NW pp.*38,* 50, *50,* 51.

Livery companies, urban trade organizations which flourished in London in the 15th and 16th centuries, so called because of the costume of their members which distinguished them from the yeomen. Originally oligarchies within the craft guilds, the "Great Twelve" included the fishmongers, goldsmiths, mercers and other merchants rather than manufacturers. They controlled municipal politics and lent to the monarchy their great strength in the urban middle classes.

Lives of the English Poets, critical and biographical essays on 52 poets by Samuel JOHNSON, from the poet Abraham Cowley to those of his own time. Originally the articles appeared as prefaces to works of the poets (1779–81), but they were subsequently issued independently.

Living, cost of. *See* COST OF LIVING.

Livings, Henry (1929–), British actor and playwright whose works, such as *Big Soft Nellie* (1961) and *Eh?* (1964), established him as an important comic dramatist in their explorations of the role of the "little man" in a northern industrial society. In the 1970s more of his plays were performed on television than on the stage.

Livingston, Robert (1746–1813), US politician; member of the CONTINENTAL CONGRESS (1775–76, 1779–81, 1784–85) and one of the five men who drafted the Declaration of Independence. He was the first Secretary of the Department of Foreign Affairs (1781–83).

Livingstone, David (1813–73), British missionary and explorer. Coming from a poor background, he first worked in a cotton mill near Glasgow but saved enough to take a medical degree at the age of 27. Sent to Bechuanaland (now Botswana) by the London Missionary Society, he made strenuous efforts to combat disease and slavery. From 1858 to 1864 he explored the eastern Zambezi, and then set off to find the source of the Nile in 1866. He was cut off from all contact with Europe until 1871, when he was found by H. M. STANLEY near Lake Tanganyika. Admired by all the peoples with whom he came into contact and a fierce opponent of slavery, Livingstone died at Ilala, in modern Zambia. *See also* HC2 pp.139, 143.

Living Theatre, avant-garde US collective theatre company founded in New York in 1947. It specialized in improvised presentations based on the theories of ARTAUD and its most famous productions in this vein were *The Connection* (1959) and *The Brig* (1963). It was the most influential group in both the USA and Europe in the 1960s.

Livius Andronicus, Lucius (c.284–204 BC), regarded as founder of Roman

Live-Forevers, which grow well in northern England, are often attached to stones.

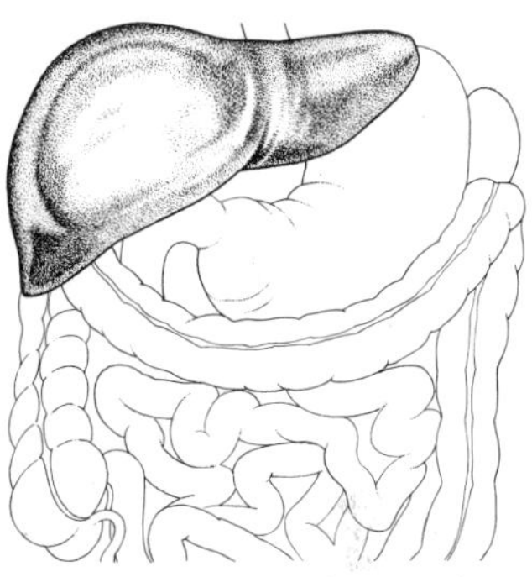

Liver has many separate functions, mainly digestive, cleaning and vitamin-storing.

Liverpool has become known for pop music (through the Beatles) and soccer excellence.

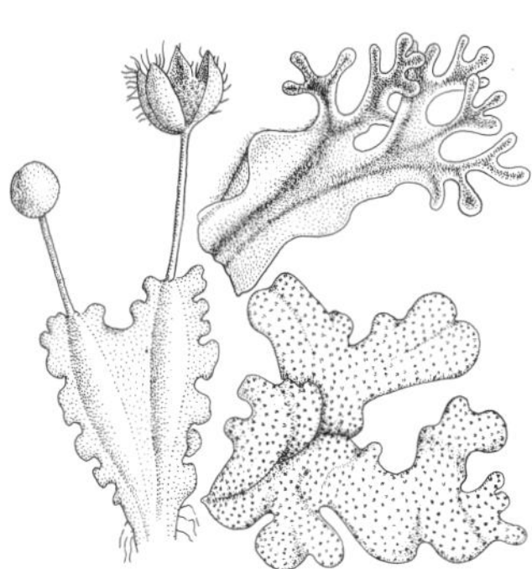

Liverwort is so-called because it was thought to be a cure for liver diseases.

Lizard Point, Cornwall, is a desolate spot on the Atlantic coast.

Llamas are well adapted for use as pack animals in mountain regions.

David Lloyd George; a portrait of the British politician taken just before WWI.

Lobelia is named after Matthias de Lobe, botanist and physician.

drama. A Greek slave, he wrote tragedies and comedies based on Greek originals and produced the first dramatic presentation ever given in Rome. His principal work is the *Odyssia*, a translation Homer's Odyssey. *See also* HC1 p.112.

Livonian Order of Brothers of the Sword, group of knights who brought what is now Latvia and .Estonia to Christianity by violent conquest. Founded with the pope's sanction in 1202, the knights carried out several brutal campaigns and were finally crushed themselves by Lithuanian forces in 1236.

Livy (59 BC–AD 17), Roman historian, full name Titus Livius. With TACITUS and SALLUST he is regarded as one of the greatest Roman historians. In about 28 BC he began his *History of Rome*, an account of the Romans from the legendary, Aeneas' arrival in Italy (753 BC) to 9 BC. Livy used the best sources available at the time. His descriptions are vivid and he frequently draws moral conclusions. Thirty-five of the original 142 books are intact and there are fragments of the rest. *See also* HC1 pp.90, *90*, 113.

Li Yüan-hung (1864–1928), Chinese military and political leader. A brigade commander under the CH'ING DYNASTY, Li was compelled by army rebels to become military governor of Hopei province during the republican revolution of 1911. He became the first Vice-President of the Republic of China (1912–16) and later served two brief terms as President (1916–17, 1922–23).

Lizard, reptile found on every continent; there are 20 families, comprised of approximately 3,000 species. It has scales, paired copulatory organs and a flexible skull. A typical lizard has a cylindrical body with four legs, a long tail and moveable eyelids. Most lizards are terrestrial, and many live in deserts. There are also semi-aquatic and arboreal forms, including the FLYING DRAGON. Burrowing species frequently have shortened limbs or are legless. Length: 5cm–3m (2in–10ft). Order Squamata; suborder Sauria. *See also* NW pp.73, 134–139, 215.

Lizard fish, any of 36 species of carnivorous fish that live in shallow waters of the Atlantic and Pacific oceans. It uses its pectoral fins to prop itself up on the sandy bottom. Length: 50cm (20in). Family Synodontidae; genus *Synodus*.

Lizard Point (Lizard Head), the most southerly point on the mainland of the British Isles, on the Lizard Peninsula, Cornwall, SW England. The area is known for its rugged coastal scenery.

Ljubljana, city in NW Yugoslavia, on the River Sava; the capital of Slovenia. It was held by the HAPSBURGS after 1277 and by the French during the NAPOLEONIC WARS; it passed to Yugoslavia in 1918. Ljubljana was the centre of the Slovene national movement during the 19th century and site of the Congress of LAIBACH in 1821. The city is the seat of the Slovene Academy of Arts and Sciences and a university (1596). Industries: machinery, optical instruments, textiles, chemicals. Pop. (1971) 173,853.

Llama, South American even-toed, ruminant mammal related to the CAMEL. The related GUANACO is found in the wild, whereas the llama and the alpaca are known only in domesticated form. The llama has been used as a beast of burden by South American Indians for more than 1,000 years. It has a long, woolly coat and slender limbs and neck. The smaller alpaca is bred for its superb wool. Family Camelidae; genus *Lama*. *See also* NW pp.162, *251*.

Llanberis, town in Gwynedd, N Wales, NW of Snowdon, the highest peak in Wales. Llanberis is the terminus of the railway to the summit of the mountain. Pop. (1971) 2,049.

Llanfair P.G., small town in N Wales. Its full name, once borne on a famous railway sign (which is now in a railway museum in Bangor), is Llanfairpwllgwyngyllgogerychwyrbdrobwllllantysiliogogogoch. This, the longest place-name in Britain, means "The church of St Mary in the hollow of white hazel, by the rapid whirlpool, of St Tysilio's Church by the red cave".

Llanos, vast prairie regions of N South America, in SW Venezuela and E Columbia. Once composed entirely of savanna, some areas have now been irrigated to form fertile land. Area: Venezuela approx. 323,750sq km (125,000sq miles); Columbia approx. 259,000sq km (100,000sq miles).

Llangollen, town in Clwyd, NE Wales, on the River Dee in the Vale of Llangollen. Llangollen is famous for its eisteddfod (Welsh festival of arts), which was first held in 1946 and is now a yearly event. In the town is the 13th-century castle of Dinas Bran, a 12th-century abbey and a 14th-century bridge spanning the Dee. Tourism is the major industry. Pop. (1971) 3,108.

Llanquihue, province in s central Chile, on the Golfo de Ancud; the capital is Puerto Montt. Industries: tourism, timber, food processing, seafood canning. Area: 18,205sq km (7,029sq miles). Pop. (1970) 197,986.

Llewellyn, Richard (1907–), Welsh author, real name Richard Llewellyn Lloyd. He gained popular success with *How Green Was My Valley* (1939), a novel about a Welsh mining family in the late 19th century. His later works include *None But the Lonely Heart* (1943), *Bride of Israel, My Love* (1973) and *Green, Green My Valley Now* (1975).

Llewelyn ap Gruffydd (*d.*1282), Welsh prince. He became ruler of Wales in 1246 and the following year did homage to HENRY III, surrendering to him much of his territory. In 1256, however, he rebelled and soon gained control of much of Wales. EDWARD I, Henry III's successor, invaded Wales, defeated Llewelyn and deprived him of almost all of his lands by the Treaty of Conway. *See also* HC1 pp.184, *184*.

Llewelyn ap Iorwerth (1173–1240), Welsh leader, called Llewelyn the Great. He had great diplomatic and military skill, which he used to maintain Welsh independence from the English kings JOHN and HENRY III. He also patronized the arts and the Church.

Lloyd, Clive (1944–), West Indian cricketer who has played for Guyana, Lancashire and the West Indies. He played in 63 Test matches between 1966 and 1977, 28 times as captain (4,466 runs, average 43.35). Lloyd gained renown as a powerful left-handed batsman, capable of standing up to the best attacking bowling.

Lloyd, Harold (1893–1971) US film comic whose screen character established his career which spanned the silent and sound ears. An expert stuntman, Lloyd is remembered for the cliff-hanging escapades of his films. His early work included *Grandma's Boy* (1922) and *The Freshman* (1925).

Lloyd, Marie (1870–1922), British music hall star, real name Matilda Wood, who specialized in cheeky, cockney songs such as *Oh! Mr Porter, My Old Man Said Follow The Van* and *A Little of What You Fancy Does You Good.*

Lloyd George, David (1863–1945), British politician. He was a Liberal MP for Caernarvon Boroughs from 1890 to 1945, when he was created Earl Lloyd-George of Dwyfor. As Chancellor of the Exchequer (1908–15) he greatly increased taxation, above all in the "People's Budget" of 1909, to lay the foundation of the WELFARE STATE. It was he who was chiefly responsible for introducing old-age pensions. In 1915 he was Minister of Munitions and Secretary of State for War. In 1916 he replaced ASQUITH as Prime Minister and formed a coalition government which lasted until 1922. As Prime Minister he won fame as a forceful war leader and handled the negotiations which led to the founding of the Irish Free State (1921). He remained leader of the LIBERAL PARTY until 1931, but never again held office. *See also* HC2 pp.167, 192, 195.

Lloyd's, insurance market in London, dealing especially in marine insurance. Lloyd's began in the 17th century as a coffee house kept by Edward Lloyd, where businessmen willing to insure shipping were known to gather. Its premises were originally in Tower Street but moved to Lombard Street in 1691, and to Lime Street in 1957. Lloyd's as an institution does not insure anything; it is merely the market where the individual underwriters, who accept the insurance on their own account and who bear all the financial risk, can meet. Each underwriter has to prove he has access to considerable capital before he can join Lloyd's. *Lloyd's list* dates from 1734.

Lloyd's Register of Shipping, annual publication produced by LLOYD's of London, listing details of merchant ships of all nations, port facilities and so on. It was first published in 1760.

Llywelyn. *See* LLEWELYN.

Loach, Kenneth (1936–), British film and television director who worked mainly on location and whose radical approach to social conditions and situations was expressed in such television drama-documentaries as *Cathy Come Home* (1966), *Days of Hope* (1975) and *The Price of Coal* (1977), and his feature films *Poor Cow* (1968) and *Kes* (1970). *Family Life* (1972) explored schizophrenia in terms of the theories of R. D. LAING.

Loach, small freshwater fish that lives in mountain streams of central and s Asia and Europe. Loaches have small scales and some are virtually scaleless; there are more than 200 species. British loaches are the stone loach (*Nemachilus barbatula*) and the spined loach (*Cobitis taenia*).

Lobachevski, Nikolai Ivanovich, (1792–1856), Russian mathematician. He was educated at Kazan University and appointed professor there in 1816. His outstanding achievement (announced in 1826) was the creation of one of the first comprehensive systems of non-Euclidean geometry, which refutes EUCLID's axiom of parallels.

Lobelia, trailing or bedding plant found throughout the world. The flowers may be red, white or blue and irregularly shaped, and the leaves are simple. Species include the blue-flowered annual *Lobelia erinus.* Family Lobeliaceae.

Lobengula (*c.* 1836–94), Ndebele (or Matabele) king from 1870 in what is now part of Rhodesia (Zimbabwe). The initial instability of his rule caused him to invite British support in return for certain land and mineral concessions. Cecil RHODES formed the British South Africa Company to exploit this in 1889. In 1893, when refused more concessions, expedition the company sent on and destroyed Lobengula's kingdom.

Lobito, city in w central Angola on the Atlantic Ocean; one of the finest ports on the w African coast. Founded in 1843 for Portuguese colonists, the city became a commercial and communications centre with the completion (1828) of the trans-African railway from Benguela. Lobito's exports include coffee, sisal and sugar, and minerals from Zaire. Industries: shipbuilding, cement, building materials, castor-oil processing. Pop. 74,000.

Lobotomy. *See* LEUCOTOMY.

Lobster, large, long-tailed, marine decapod crustacean. Some species, notably the Norway lobster, *Nephrops norvegicus* and the common lobster, *Homarus vulgaris,* are the most prized edible shellfish. They are caught in wooden or metal cages. True lobsters possess enlarged bulbous chelae (claws) and a segmented body; the segments are obviously only in the abdomen. *See also* NW pp.72, *101*; PE p.248, *248*.

Local Defence Volunteers. *See* HOME GUARD.

Local government, system of regional administration differing in each country. In England and Wales the present structure derives from legislation which came into force on 1 April 1974. Local government is run by elected councils which carry out certain functions within statutory limits. There are now six metropolitan counties, 36 metropolitan districts, 39 non-metropolitan counties and 296 non-metropolitan districts. This structure excludes the Greater London Council (GLC). Metropolitan counties are responsible for large-scale planning, roads, traffic and safety, police and fire services. Metropolitan districts are responsible for education, rent and rate rebates, social services, libraries, local planning, housing, slum clearance, youth employment and environmental health. Non-metropolitan counties and districts are responsible for the same functions as their metropolitan counterparts with the exception that non-metropolitan counties are responsible for education, social services, youth employment and libraries.

In Scotland local government is divided into nine regions within which are three island councils and 53 district councils. The regions are responsible for education, police, fire services, social services, water, roads and public transport; the districts being responsible for housing, local planning and environmental health. *See also* GREATER LONDON COUNCIL; MS pp.280–281.

Local group of galaxies, small cluster of galaxies to which the Milky Way belongs. The local group contains about 18 members, including the Andromeda Spiral (M31), The Triangulum Spiral and the Clouds of Magellan. *See also* SU pp.246–247.

Locarno Pact (1925), agreements made between Belgium, Italy, Britain, Poland, France and Germany that demonstrated a resumption of normal European international relations after WWI and strengthened the member nations' commitment to the Treaty of VERSAILLES. The pact was intended to encourage European peace, but it was violated by the German remilitarization of the Rhineland in 1936.

Locatelli, Pietro (1695–1764), Italian violinist and composer. He studied with Arcangelo CORELLI and settled in Amsterdam. One of the great violinists of his day, he also composed sonatas, concerti grossi, studies and caprices for the instrument.

Lochaber, Treaty of (1770), agreement negotiated in America between the colonial superintendent of the Southern Indians, and the Cherokee Indians, expanding Virginia's western border beyond the natural watershed line of the Proclamation of 1763 to the mouth of the Kanawha River.

Loch Katrine, lake in Central Region, Scotland. In 1859 it was enlarged to become the main source of water for Glasgow. The lake was made famous in Sir Walter SCOTT's *Lady of the Lake.* Area: approx. 21sq km (8sq miles).

Loch Lomond, lake in Strathclyde and Central regions, Scotland; the largest and one of the most famous Scottish lochs. It is drained by the River Leven into the Firth of Clyde. It contains numerous wooded islands and is dominated at the N end by Ben Lomond. Height: 973m (3192ft).

Lochner, Stefan (*c.*1410–51), German painter. He was a leading artist of Cologne, where he painted his most famous work, the cathedral altarpiece *Altar of the Patron Saints,* in which he combines the decorative with a monumental realism influenced by Flemish art.

Loch Ness Monster, unexplained phenomenon living in the waters of Loch Ness, Scotland. Possibly a vast amphibian, it is often said by those claiming to have seen it to resemble a PLESIOSAUR or a large aquatic mammal, or a number of such creatures. The loch is fed by eight rivers and many streams, all of which carry peat from the surrounding soil, darkening the already murky water. Countless tourists and bodies of scientific observers have visited the loch to try and discover its secret. Best known amongst these was the US expedition led by Dr R. Rines which published, in 1975, some shadowy photographs taken by automatic cameras underwater. Following these Sir Peter SCOTT devised the scientific name of *Nessiteras Rhombopteryx* for the creature. Nevertheless, no real evidence has emerged, even though stories of the "beast" of the water date back at least as far as the 6th century.

Lock, Graham Anthony Richard ("Tony") (1929–), British cricketer who played for Surrey, Leicestershire, Western Australia and England. He was an excellent left-arm slow bowler and

took 170 Test wickets at an average of 24.94 runs between 1952 and 1963.

Lock, mechanism fitted to a door to prevent entry to people other than those with a coded key. *See also* MM p.*248.*

Lock, structure built into a stretch of inland waterway to raise or lower water levels to correspond with the surrounding countryside. Each lock consists of two sets of lock gates. A vessel enters the lock, the gates are closed, and sluices are opened to admit or release enough water to bring the vessel on the same level as the water beyond the second pair of gates. *See also* MM p.*196.*

Locke, Arthur D'Arcy ("Bobby") (1917–), South African golfer whose 43 major tournament wins between 1935 and 1957 included the British Open Championships of 1949, 1950, 1952 and 1957. Distinctive in a white cap and plus-fours, he built his game around self-composure, a looping swing, and a consistent putting style.

Locke, John (1632–1704), English EMPIRICIST philosopher whose political theories of social contract, the right to freedom of conscience and the right to property greatly influenced emerging democracies of his day. Suspected of radicalism, he went into exile in Holland (1683–89) and returned to England after the Glorious Revolution, when his *An Essay Concerning Human Understanding* (1690) appeared. In the same year he published *Two Treatises on Civil Government.* He advocated limited sovereignty and held that revolution was not only a right but an obligation if liberty were threatened. *See also* HC2 pp.22–23; MS p.*229.*

Locke, Matthew (*c.*1630–77), English composer who composed brass music for a procession of CHARLES II on the day before his coronation, as well as contributing to early musical plays such as SHADWELL's *Psyche* (1673) which, together with incidental music to other plays, was published as *The English Opera* (1673). Considered to be avante-garde by his contemporaries, he was the most important composer of stage music before PURCELL.

Lockhart, John Gibson (1794–1854), Scottish journalist and writer. He was a leading contributor to BLACKWOOD's MAGAZINE and editor of the *Quarterly Review* (1825–53). He published a biography of Robert BURNS in 1828 and a monumental biography, now a classic, of Sir Walter SCOTT, his father-in-law, in 1837–38.

Lockheed Aircraft Corporation, US aircraft and aerospace company. In the mid-1970s technological problems with the Rolls-Royce RB2-11 turbine engines ordered for Lockheed's L-1011 Tri-Star contributed to the financial difficulties of both companies. The company's admission in 1975 that it had offered financial inducements to overseas government officials (in return for recommendations for contracts) resulted in the resignations of some of the people involved.

Lockjaw. *See* TETANUS.

Lockout, employer's refusal to allow workers to enter the place of employment. It is the employer's counterpart of a strike; it may be used as a means of bringing pressure on employees (or their trade union) to accept the employer's conditions regarding work.

Lockspeiser, Sir Ben (1891–), British scientist and administrator who became Director-General of the Scientific Research (Air) Ministry of Supply in 1946. From 1955 to 1957 Lockspeiser was president of the European Organization for Nuclear Research.

Lock Up Your Daughters (1959), British musical adapted by Bernard Miles from the play *Rape Upon Rape* (1730) by Henry FIELDING. The lyrics were by Lionel BART with music by Laurie Johnson and a filmed version, without songs, appeared in 1969.

Lockwood, Margaret (1916–), British stage and film actress, more noted for her film roles, especially in *The Lady Vanishes* (1938), *Night Train to Munich* (1940), *The Wicked Lady* (1945) and *Cast a Dark Shadow* (1957).

Lockyer, Sir Joseph Norman (1836–1920), British astronomer who discovered the element HELIUM. He was a pioneer in the study of the Sun's spectrum. In 1868 he developed a technique for examining prominences at the edge of the Sun and attributed lines in the solar spectrum, to a new element, which he named helium – 40 years before it was discovered on earth.

Locomotive, engine that moves under its own power but, most often, an engine that moves on rails. The first successful steam locomotive was George STEPHENSON's *Rocket,* built in 1829. This and its many successors pulled passengers and freight for more than a century before they were replaced, in developed countries, by diesel-electric and electric locomotives. Gas-turbine engines have also been used with generators to drive locomotives. *See also* MM pp.*64,* 140–145.

Locoweed, any of several plants native to the North American plains. It has bluish-purple flowers and is poisonous to livestock. Woolly locoweed may grow up to 1.8m (6ft) tall. Family Leguminosae; species *Astragalus mollissimus.*

Locus, in geometry, the path traced by a specified point when it moves to satisfy certain conditions. For example, a circle is the locus of a point in a plane moving in such a way that its distance from a fixed point (the centre) is constant.

Locust, insect (a type of large GRASSHOPPER) that is known for migrating in huge swarms. Locusts swarm in marginal regions where bodies of water or large humid areas meet dry ones. When a large population develops in such circumstances, the nymphs (immature insects) move in vast numbers on foot. As they feed they develop into flying adults. Swarms may contain up to 40,000 million insects and cover an area of approximately 1,000sq km (386sq miles). About 70% of damage is caused by the immature and maturing insects. Length: 12.5–100mm (0.5–4in). Order Orthoptera; species *Schistocerca gregaria. See also* NW pp.78, 78, 110–111, *219.*

Locust tree, deciduous tree or shrub of the pea family. It has oblong feather-like leaves and the fragrant flower clusters are white, pink or yellowish-green. Family Leguminosae; genus *Robinia.*

Lod (Lydda), town in central Israel, 18km (11 miles) SE of Tel Aviv. Of Hebrew origins, Lod was destroyed by Celestius Gallus in AD 66 and again by Vespasian in AD 68, but was later rebuilt by HADRIAN. It was made an episcopal see in the 5th century. Destroyed by SALADIN in 1191, it was rebuilt by RICHARD I of England. It was the scene of a mass Arab evacuation during the ARAB-ISRAELI WAR of 1948, and the population was largely replaced by Jewish immigrants in 1950. There is a 13th-century Church of St George, who is reputed to have been born and buried in Lod. Pop. (1972) 30,500.

Lode, ORE formation consisting of a closely spaced series of veins, usually in stratified layers. The veins are in tabular deposits in fissures and cracks of a body of rock which is different in composition. They are the result of the gradual precipitation of minerals carried by underground water or gases after the formation of the embedding rock (country rock). *See also* PE pp.132–133.

Lodestone, mineral MAGNETITE, also called magnetic iron ore (Fe_3O_4), plentiful in parts of Scandinavia, northern USA and Russia. The observation that a lodestone or piece of iron touched by one will align itself north and south was crucial to the development of navigation.

Lodge, Sir Oliver Joseph (1851–1940), British physicist, author, professor at Liverpool and later principal of Birmingham University. He investigated electromagnetic waves and lightning, and perfected and named the coherer, the radio wave detector used in the early days of wireless telegraphy. He was a proponent of the belief that the Sun might be a source of radio waves. He wrote several books on psychical research and spiritualism. *See also* MM p.228.

Lodge, Thomas (1558–1625), English writer. His works include the pamphlet *A*

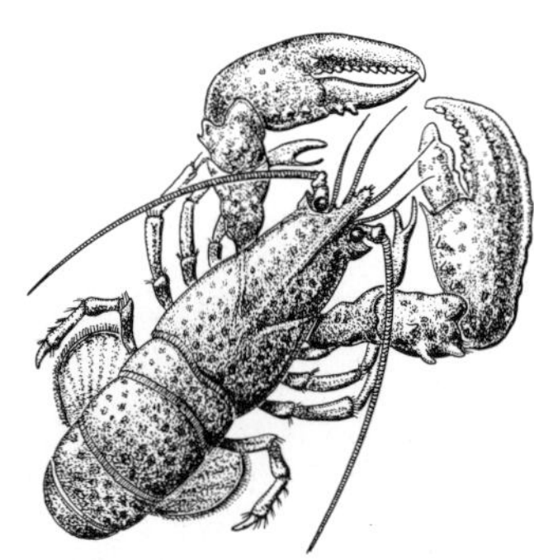

Lobsters in Europe weigh up to 6.8kg (15lb), and up to 13.6kg (30lb) in N America.

Locomotive; Stephenson's *Rocket,* built in 1829, reached a speed of 50 km/h (30mph).

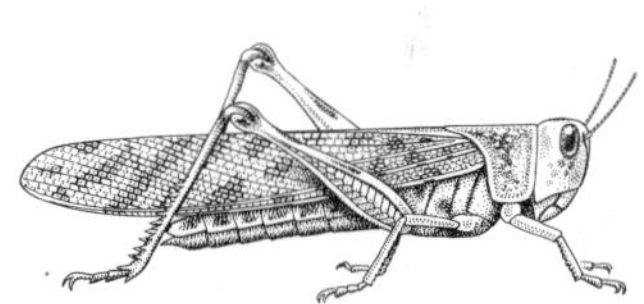

Locusts migrate up to 5,000km (3,000 miles) to their next breeding grounds.

Oliver Lodge (left) in his laboratory; he investigated electromagnetic waves.

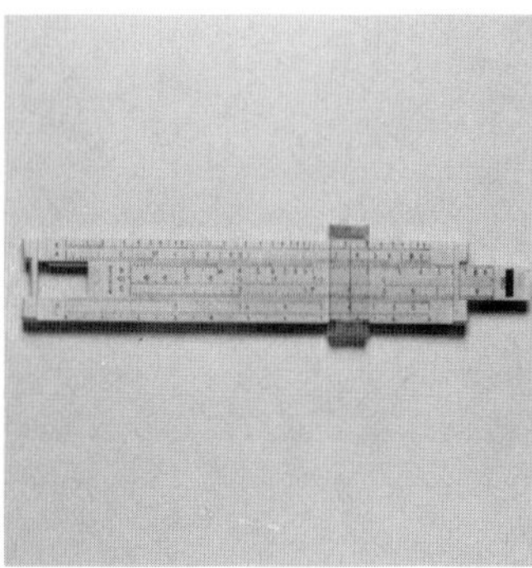

Logarithms are the basis of calculations carried out using a slide rule.

Loggerheads are the only turtles that breed on the Atlantic coast of northern America.

Loire makes a westward turn at Orleans and flows along the Loire valley.

Tullio Lombardo's best-known work is this tomb of Guidarello, Ravenna (1525).

Defence of Plays (1580), the prose romance *Rosalynde, Euphues Golden Legacy* (1590) and *Phillis* (1593), a collection of love sonnets. Apart from his literary efforts, he also practised medicine and travelled to South America.

Łódź, second largest city in Poland, approx. 120km (75 miles) SW of Warsaw. A small market town prior to the 19th century, Łódź grew after 1870 to become the centre of the Polish textile industry. Taken by the Russians in 1815, the town was returned to Poland in 1918. There is a university (1945). Industries: machinery, electrical equipment, food processing. Pop. 787,000.

Loeb, Jacques (1859–1924), US biologist, b. Germany. He studied the chemical processes of living organisms, and stressed the importance of physical and chemical laws on understanding life phenomena in both animals and plants. He achieved popular notoriety for experiments on artificial PARTHENOGENESIS in animals.

Loess, buff-coloured deposit of fine silt or clay, generally unstratified and sometimes exposed in bluffs. The loess in the Mississippi River Valley is believed to be of glacial origin, whereas that in the Mongolian desert seems to have been formed by the wind. *See also* PE p.121.

Loewe, Frederick (1901–), US composer, b. Austria. He has chiefly composed Broadway musicals, several of them with the librettist Alan LERNER. His first musical was *Salute to Spring* (1937). His most successful were *Brigadoon* (1947), *Paint Your Wagon* (1951), *My Fair Lady* (1956) and *Camelot* (1960).

Loewi, Otto (1873–1961), German pharmacologist and medical researcher. He shared the 1936 Nobel Prize in physiology and medicine with Sir Henry Hallett DALE for discovering that a chemical substance (later shown by Dale to be acetylcholine) mediated the transmission of impulses from one nerve fibre to another.

Löffler, Friedrich August Johannes (1852–1915), German bacteriologist, best known for his discovery (in association with Wilhelm KLEBS) of the organism that causes DIPHTHERIA, *Corynebacterium diphtheriae* (Klebs-Löffler bacillus). He also made several discoveries in veterinary medicine, such as the cause of SWINE FEVER and plague, the causative organism of GLANDERS (with Wilhelm Schütz), and the fact that FOOT-AND-MOUTH DISEASE is caused by a virus (with Paul Frosch).

Lofoten Islands, Norwegian islands off the coast of NW Norway in the Norwegian Sea. They are separated from the VesterÅlen by a narrow water, the Raftsund. They are important chiefly for the cod (Feb. to April) and herring (Aug. to Nov.) fishing.

Lofting, Hugh (1886–1947), British creator of the ten Dr Dolittle books. He settled in the USA in 1912 and, in letters to amuse his children during WWI, wrote stories for them of an extraordinary country doctor who learnt the language of animals. In their published form, they include *The Story of Dr Dolittle* (1920), *The Voyages of Dr Dolittle* (1922) and *Dr Dolittle and the Secret Lake* (1948) – all illustrated by himself.

Log, nautical instrument for the measurement of the speed of a vessel in knots. In most modern ships, it takes the form of a tube projecting through the bottom of a vessel. The term also applies to the logbook, the daily account of a ship's or aircraft's business written up by the navigator as required by MARITIME LAW.

Logan, Joshua (1908–), US stage and film director. He was director, producer and co-author of the stage production of *South Pacific* (1949), and his successful films have included *Picnic* (1956), *Bus Stop* (1956), *Sayonara* (1957), *South Pacific* (1958), *Fanny* (1961), *Camelot* (1967) and *Paint Your Wagon* (1969).

Loganberry, biennial hybrid bramble developed in California in 1881, named after Judge J. H. Logan. A cross between the blackberry (*Rubus fruticosus*) and raspberry (*Rubus idaeus*), it is disease prone and is grown only in sheltered areas. The canes produce large, red berries which may be eaten raw or preserved. Family Rosaceae; species *Rubus wisinus*

loganobaccus. See also PE pp.178, *178,* 214, *214.*

Logarithm, aid to calculation devised by John NAPIER in 1614 and developed by the Engeish mathematician Henry Briggs. A number's logarithm is the power to which a base must be raised to equal the number, ie if $b^x = n$, then $\log_b n = x$, where n is the number, b the base and x the logarithm. Common logarithms have base 10, and so-called natural logarithms have base e (2.71823...). A logarithm is written as the sum of an integer (the characteristic) and a decimal fraction (the mantissa). The characteristic indicates the location of the decimal point in the number, being positive for numbers greater than one and negative for those less than one. The mantissa is the logarithm of the digits in the number, regardless of decimal place. *See also* SU p.38.

Loggerhead turtle, carnivorous marine turtle of the Atlantic and Pacific oceans. Characterized by a large head, it has oarlike flippers and is red-brown. Length: 0.7–2.1m (2.3–6.8ft). Family Cheloniidae; species *Caretta caretta.*

Logic, classical, branch of philosophy that deals with the way we think – a systematic study of the structure of propositions and the criteria of valid inference. In abstracting from the content of propositions in order to examine their logical form, logic evaluates soundness of validity rather than truth per se. The history of logic began with ARISTOTLE, who proposed three laws that he held as basic to valid thought: the identity (A is A), the contradiction (A is not A) and the exclusive (A is A or not A). Logic then proceeded through Arabian and European logic in the Middle Ages and various post-Renaissance scholars, and resulted in the symbolic, or mathematical, logic first outlined in the 19th century by George BOOLE and Augustus DE MORGAN. *See also* HC1 pp.82, 188; MS pp.226, 228–229, 232, *233.*

Logic, mathematical, also called symbolic logic, analytical branch of logic that attempts to give the inferences of classical logic a more precise mathematical basis. It is concerned with statements, which may be true or false, and with the relationships between statements and the operations that can be made on them, including conjunction, disjunction, and negation. The statements and their relationships and operations are reduced to a symbolic notation and are manipulated according to prescribed mathematical rules. Thus inferences can be analysed and validity checked. *See also* SU pp.40–41.

Logical positivism, school of philosophy which began in the VIENNA CIRCLE at the beginning of the 20th century. Its roots were the logic of FREGE and RUSSELL, the positivism of MASH and, above all, the claim of WITTGENSTEIN that philosophy was the clarification of thought. Logical positivists, in Britain RYLE and AYER especially, rejected metaphysical speculation as meaningless, moral statements as emotive and mathematical and formal logic as tautological. They therefore narrowed philosophy to the analysis of language.

Logic circuits, network of conductors linking modules(eg amplifiers and multivibrators) and components (eg transistors, diodes, resistors and capacitors), used to route and process electronic signals (eg voltages) according to the rules of symbolic logic and Boolean algebra. In computers, two numbers are added by an "adder" composed largely of logic GATES. Such a circuit converts decimal inputs into the BINARY SYSTEM, performs the addition, converts the binary result back to decimal and energizes the required display. *See also* BOOLE, George; MM pp.106–109; SU pp.40–41.

Logos, in philosophy, intellect or reason; in a larger sense the rational principle that orders the universe. The STOICS thought of it as the soul of the world. As used in the New Testament Gospel of St John, the Greek word *logos* was in the Authorized Version translated as *Word*: "In the beginning was the Word, and the Word was with God, and the Word was God". The Word can be either the intention of

God (His rational principle) or the outward expression of that intention – both describing JESUS CHRIST.

Lohengrin (1850), opera in three acts by Richard WAGNER, German libretto by the composer. It was first performed at Weimar, conducted by Franz LISZT. It is, like most of Wagner's operas, based on a medieval Teutonic legend. A work of monumental proportions, it was an early attempt by Wagner to incorporate the dramatic action within a continuous broad symphonic framework.

Loire, longest river in France. It rises in the Cévennes in SE France, flows N through central and W France to the Bay of Biscay at Saint-Nazaire. The river is connected by canals to the Rhône and Seine rivers. Length 1,020km (634 miles).

Loisy, Alfred Firmin (1857–1940), French theologian and biblical scholar. His application of modern historical criticism to the Gospels, in such works as *L'Évangile et l'Église* (1902; "The Gospel and the Church, "), led to his condemnation by Pope Pius X as a modernist, and he was excommunicated from the Roman Catholic Church.

Loki, in Norse mythology, the malicious companion of ODIN and THOR, a shrewd and resourceful trickster, perhaps representing the unpleasant side of the other gods. Although sometimes depicted as a powerful ally of Odin, he was more usually an unwanted guest and troublemaker.

Lolita (1955), novel by Vladimir NABOKOV. A satirical work, it deals with the passion of a middle-aged European professor for a 12-year-old American girl.

Lollards, followers of the 14th-century English religious reformer John WYCLIFFE. They challenged many doctrines and practices of the medieval Church, including TRANSUBSTANTIATION, indulgences, pilgrimages and clerical celibacy. They rejected the authority of the POPE, and they denounced the wealth of the Church and Church involvement in civil affairs. Lollards went out as "poor preachers", teaching that the Bible was the sole authority in religion. They won support from some nobles as well as many common people and helped to pave the way for the Protestant REFORMATION.

Lomax, Alan (1915–), US folksinger and collector of folksongs, particularly US Negro blues. He helped his father, John Lomax (1867–1948), compile *American Ballads and Folk Songs* (1934) and *Negro Folk Songs as Sung by Lead Belly* (1936), which helped to establish Huddie Ledbetter (LEADBELLY) as one of the best-known exponents of folk blues. In association with Woody Guthrie, Pete and Peggy Seeger, and others he was prominent in the revival of US folk music in the 1960s.

Lombard, Peter. *See* PETER LOMBARD.

Lombard League, alliance of the cities of Lombardy in northern Italy (including Milan, Venice, Brescia, Bergamo, Mantua and Verona). It was founded in 1167 to resist the Holy Roman Emperor FREDERICK I Barbarossa. By the Peace of Constance in 1183 the League acknowledged fealty to Frederick, but the cities were granted local liberties and jurisdiction. The League became active again in 1226 against FREDERICK II and ended with his death in 1250.

Lombardo, Italian family of sculptors and architects, leaders in the architectural RENAISSANCE in Venice. Pietro (1435–1515), who came from Lombardy to Venice c.1470, ran the chief sculpture workshop, assisted by his sons Tullio (c.1455–1532) and Antonio (c.1458–1515). Their great combined works included the Church of Santa Maria dei Miracoli, which Pietro built (1481–89) and his sons sculpted. Pietro designed many churches and palaces and worked on the façade of the Doge's Palace from 1498. He also worked at Treviso, at Ravenna, where he built Dante's tomb.

Lombardo, Tullio (c.1455–1532), Venetian sculptor, son of Pietro LOMBARDO. He was one of the greatest marble sculptors of his day. His works include most of the figure sculptures on the Vendramin monument, Venice.

Lombards, or Langobards, Germanic peoples thought to have migrated from Gotland, Sweden. They inhabited the area E of the lower River Elbe until driven W by the Romans in AD 9. They had close links with ARMINIUS at this time. In 568 they invaded N Italy under Alboin and conquered much of the country from their centre, Pavia, adopting Catholicism and Latin customs. The Lombard kingdom reached its peak under LIUTPRAND (d. 744). It went into decline after defeat by the Franks under CHARLEMAGNE in 774–75.

Lombardy (Lombardia), region in N Italy made up of the provinces of Bergamo, Brescia, Como, Cremona, Mantova, Milano, Pavia, Sondrio and Varese; Milan is the regional capital. It is a region of mountains and lakes. It was the centre of the powerful LOMBARD kingdom from 569 until 774–75, when it was defeated by CHARLEMAGE. Products: cereals, sugarbeet, vegetables, fruit, olives. Industries: motor vehicles, steel, chemicals, textiles. Area: 23,834sq km (9,202sq miles). Pop. (1971) 8,504,061.

Lombroso, Cesare (1835–1909), Italian criminologist and physician who is credited with beginning the scientific study of the criminal as opposed to the legalistic study of crime. He developed a theory of the born criminal, which is no longer considered valid. His most famous work is *The Criminal Man* (1876).

Lomé, capital of the Republic of Togo, W Africa, on the Gulf of Guinea. The administrative centre and chief port of the country, it is connected by rail to the farmlands of the interior. The city is the seat of the University of Benin (1965). Exports: coffee, cocoa, palm nuts, copra. Pop. 148,443.

Lomonosov, Mikhail Vasilievich (1711–65), Russian scientist and poet. He helped to found the University of Moscow (1755) and was one of the first Russian scientists to anticipate the law of conservation of mass, atomic theory and a kinetic theory of heat. He formulated Russian classical literary theory in his *Letter on the Rules of Russian Versification* (1739). His poetry includes the odes *Evening Meditations* (1748) and *Morning Meditations* (1751).

Lomonosov Ridge, submarine ridge extending from the Asian continental shelf, past the North Pole to the edge of the North American continental shelf near Ellsmere Island.

London, Jack (1876–1916), US novelist and short story writer, real name John Griffiths. His early years were spent on the San Francisco waterfront, which he described in his autobiographical novels *Martin Eden* (1909) and *John Barleycorn* (1913). He spent three years as a sailor and travelled widely. *Call of the Wild* (1903), the story of a tame dog which ultimately leads a wolf pack, is his most popular work.

London, capital of the United Kingdom and principal city of the COMMONWEALTH OF NATIONS, located in SE England on the River THAMES 65km (40 miles) from its mouth in the North Sea; seventh-largest city in the world. Since 1965 it has been officially called GREATER LONDON, which is made up of the City of London and 32 boroughs, covering an area of 1,580sq km (610sq miles). Little is known of London before the Romans arrived in Britain. Known as Londinium, it was the most important town in Roman Britain and developed as a port and commercial centre. By the 3rd century AD the population numbered approx. 40,000 and the town covered an area of some 120 hectares (300 acres). After the Romans left Britain, London declined until the 9th century when ALFRED the Great repaired defences, encouraged the development of trade and made the city the seat of government. WILLIAM THE CONQUEROR built the White Tower in the TOWER OF LONDON to the E of the city walls, to protect the city from river attack. The settlement of Westminster to the W of the city walls grew in the 10th century. EDWARD THE CONFESSOR built WESTMINSTER ABBEY and later a palace, and made Westminster his capital in 1042. The NORMANS and the PLANTAGENETS retained Westminster as the seat of government, finalizing the distinction between the mercantile city in the E and Westminster which, by the 13th century, had law courts, the Court and Parliament.

The prosperity of England during the Tudor period firmly established London's wealth and importance, and trade greatly increased. In the reign of ELIZABETH I the population increased from less than 100,000 to almost 250,000. The plague in 1665–66 killed 75,000 Londoners and the Great FIRE OF LONDON in 1666 destroyed many buildings, but the city was rebuilt to a similar general plan. Sir Christopher WREN played an important role in the rebuilding of the city; he designed many churches, including ST PAUL'S CATHEDRAL. During the 17th century the area between Westminster and the City was built up. Inigo JONES designed COVENT GARDEN as a residential area for the rich, and other fashionable squares were built in SOHO, MAYFAIR and Bloomsbury. London Bridge, first built in the 10th century, was the only bridge across the Thames until 1750, when Westminster Bridge was constructed.

During the 19th century the population reached $4\frac{1}{2}$ million and transport services were improved. Euston station was built in 1838, with other main-line stations soon afterwards. The first underground railway, which used steam locomotives, was opened in 1863. By the end of the 19th century London extended as far N as Hampstead and other suburbs were being developed. Further growth between the wars was accompanied by extensions to the transport system. The GREEN BELT Act of 1938 restricted building, so that today many of the people who work in London commute from towns more than 65km (40 miles) away.

London is one of the world's most important financial, commercial and industrial cities. It is a major port and cultural centre, and attracts millions of tourists each year. Places of interest include BUCKINGHAM PALACE, the Houses of PARLIAMENT, Trafalgar Square, Piccadilly Circus, HAMPTON COURT PALACE, GREENWICH, the Post Office Tower, Madam Tussaud's and the MONUMENT. Art galleries and museums include the NATIONAL GALLERY, NATURAL PORTRAIT GALLERY, TATE GALLERY, VICTORIA AND ALBERT MUSEUM, BRITISH MUSEUM, NATURAL HISTORY MUSEUM, Science Museum and the IMPERIAL WAR MUSEUM. London University, founded in 1836, is the largest in the country and includes 24 institutes of advanced study. London's major industries are engineering, paper, clothing, printing and publishing, chemicals, food processing, brewing, motor vehicles, metal goods, furniture, tobacco, precision instruments and entertainments. Pop. (1976 est.) 7,028,200. *See also* HC1 pp.100–101, 104, *165*, 192–193, 294–295; HC2 pp.90, *91*, *106*, 114, 115, 159, *194*, *232*; MM pp.*140*, *142*, *188*, *190*, *191*; MW p.68; PE pp.*109*, *147*.

London, city in SE Ontario, Canada, on the Thames River. It was settled in 1826, and its streets and bridges are named after those in England's capital. London is the seat of the University of Western Ontario (1878). A commercial and financial centre, the city has paper, textiles and locomotive industries. Pop. (1974) 223,225.

London, Tower of. *See* TOWER OF LONDON.

London, University of, university founded in 1836, originally comprising King's College and University College. More than 80 colleges and hospitals have been added since that time; the headquarters of the university are at Senate House, Malet St, London.

London Company, corporation founded in England in 1606 to colonize Virginia. It was responsible for the Jamestown Colony, and Thomas West, Baron De La Warr, was its first governor. The company was dissolved in 1624.

Londonderry (Derry), county in NW Northern Ireland. Much of the region is mountainous and the main occupations are hill farming and stock rearing in the Sperrin Mts and the fertile plain along Lough Foyle. The most important industries are distilling and textiles. The county town is Londonderry. Long-standing political and religious differences resulted in rioting and bombings in the city in 1969, after which British troops were sent to Ulster. Area: (county) 2,082sq km (804sq miles). Pop. (1971) county, 130,556; town, 51,617.

London Gazette, twice-weekly periodical, originally known as the *Oxford Gazette* and first published in 1665 as an official publication. Today it publishes details only of public appointments and announcements.

London Group, society of English artists formed in 1913, amalgamating the CAMDEN TOWN GROUP with several smaller ones; in its early stages it favoured POST-IMPRESSIONISM. It held its first exhibition at Brighton in 1913, and its first London exhibition was in 1914. Later members included C.R.W NEVINSON, Edward Wadsworth, John and Paul NASH, Eric GILL and Jacob EPSTEIN. The group was revived after WWII, but had lost most of its influence by 1950.

London Zoo, one of the world's great zoos, situated centrally in Regent's Park, London. Although small it holds one of the greatest varieties of animals of any zoo in the world. London Zoo also houses the Zoological Society of London which has scientific members from all over the world, and is responsible also for WHIPSNADE ZOO in Bedfordshire, 5km (3 miles) W of Dunstable.

Long, Crawford Williamson (1815–78), US physician. In 1842 he began using ether as an anaesthetic but did not publish or publicize his work until 1849, after William MORTON and others had been credited with the discovery. *See also* MS p.126.

Long, Huey Pierce (1893–1935), US political leader. Elected Governor of Louisiana in 1928, he was impeached for bribery and gross misconduct in the following year but was not convicted and continued to serve as Governor until 1932. He built up a powerful political organization through which he controlled politics in Louisiana. He served in the US Senate (1932–35) and, running a campaign for legislation against the extremely rich, planned to be nominated as a candidate for the presidency, but was assassinated in 1935.

Long, Marguérite (1874–1966), French pianist. She studied at the Paris Conservatoire and became a piano virtuoso and teacher. She was especially known for her interpretations of the works of FAURÉ, DEBUSSY and RAVEL. She was the first to play Ravel's *Piano Concerto in G* (1932), which the composer dedicated to her.

Long bones, in human anatomy, limb bones characterized by a long shaft terminating at both ends in swellings (epiphyses) which articulate with other bones. The RADIUS and ULNA are in the forearm, the HUMERUS in the upper arm, the FEMUR in the thigh and the TIBIA and FIBULA in the lower leg.

Longchamp, French racecourse in the Bois de Boulogne, Paris. At the heart of French racing for more than 100 years, it is the scene of the Prix de l'Arc de Triomphe (Europe's richest race), the Grand Prix de Paris, and other French classics.

Long Day's Journey into Night (1939–41), tragedy in four acts by Eugene O'NEILL. It was witheld from publication for 15 years after its completion because of its personal nature and was not produced until 1956, after O'Neill's death. It won a Pulitzer Prize in 1957 for its intense portrayal of family relationships.

Longfellow, Henry Wadsworth (1808–82), US poet remembered for his narrative poems dramatizing North American history and legend. These include *Evangeline* (1847), *The Song of Hiawatha* (1855), *The Courtship of Miles Standish* (1858) and *Paul Revere's Ride* (1861). His first book of poetry, *Voices of the Night*, appeared in 1839. *Ballads and Other Poems* (1842) contain two of his most popular shorter poems, *The Wreck of the Hesperus* and *The Village Blacksmith*. *See also* HC2 p.288.

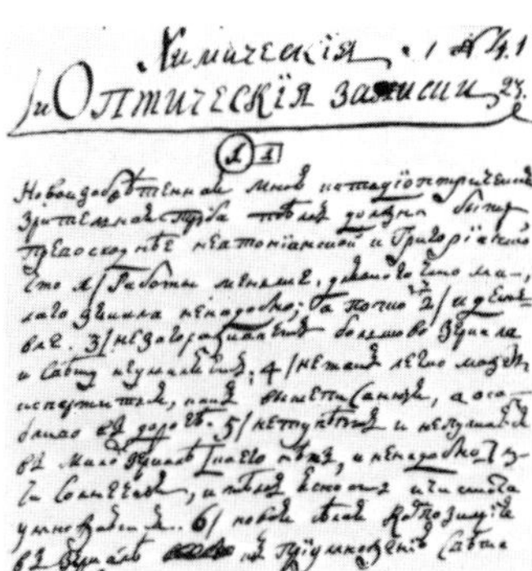

Mikhail Lomonosov: a manuscript of the scientist who was also a poet.

London: St Paul's Cathedral escaped serious damage during the Blitz of the 1940s.

London; traffic moving around Trafalgar Square, along Whitehall towards Big Ben.

Londonderry's closely packed houses and fine cathedrals surround a major port.

Alessandro Longhi painted members of the Venetian Academy, such as *Count da Mula*.

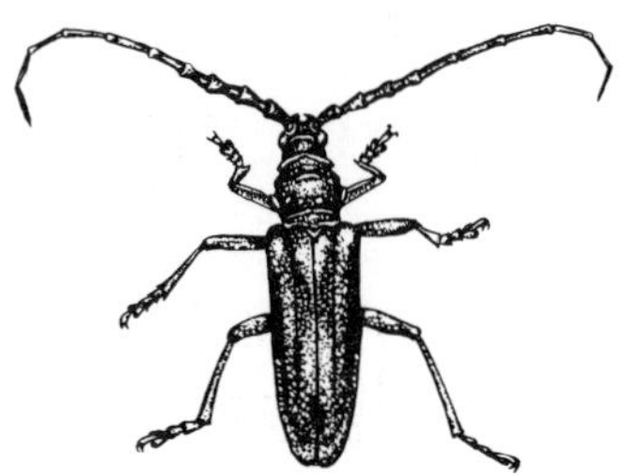

Longhorn beetles are known as timbermen because they seem to measure branches.

Longhouses in Borneo generally have the living quarters raised above the ground.

The Long Parliament, first called in 1640, was forcibly dismissed by Cromwell in 1653.

Longford, county in N central Republic of Ireland, in Leinster province. Part of the central plain of Ireland, it is largely low-lying and is drained by the Rivers Shannon, Erne and Inny and Loughs Gowna and Ree. Most of the people work on farms; beef cattle are reared and oats and potatoes are grown. Longford is the county town. Area: 1,044sq km (403sq miles). Pop. (1971) 28,250.

Longhena, Baldassare (1598–1682), Italian architect. He was the leading exponent of BAROQUE architecture in Venice, where his most famous work, the church of Sta Maria della Salute dominates the Grand Canal. He also designed the palaces Ca'Pesaro and Ca'Rezzonico in Venice.

Longhi, Alessandro (1733–1813), Italian painter, a leading society portraitist. His works include *Teres Barbarigo* (1788), *Members of the Pisanio Family* (c.1760) and *Niccolo Erizzo* (c.1766). He is best known for his *Compendio delle Vite de'Pittore Veneziani Istorici* (1762), an account of contemporary Venetian painters.

Longhi, Pietro (1702–85), Italian painter. He devoted himself chiefly to small-scale genre pictures, especially scenes of the domestic manners of the Venetian middle classes, such as *The Exhibition of a Rhinoceros at Venice*.

Longhorn beetle, any of numerous species of wood-boring beetle found throughout the world, especially in the tropics. It has long antennae, long legs, and a cylindrical black or brown body. Length: 2–152mm (0.12–6in). Family Cerambycidae.

Longhouse, type of communal dwelling used by the DAYAK people of Borneo to house entire extended families. The term is also used to apply to meeting-halls built by American Indian tribes, particularly the IROQUOIS.

Longinus, Cassius (213–273), Greek rhetorician and philosopher, mistakenly credited with the authorship of the treatise *On the Sublime* (now believed to have been written by another Greek called Longinus who lived in the 1st century AD). He was a counsellor to Queen Zenobia of Palmyra and was finally executed as a traitor by the Romans.

Long Island, fourth-largest island in the USA, in SE New York State, separated from Manhattan by the East River, from Connecticut by Long Island Sound and bounded on the S by the Atlantic Ocean. The W end of the island includes the New York City boroughs of Queens and Brooklyn. Originally inhabited by the Delaware Indians, Long Island was settled by the Massachusetts Bay Colony and the Dutch West India Company in the 17th century. It is a sporting and recreational area and there is some commercial fishing. Area: 4,463sq km (1,723sq miles). Pop. (1970) 7,141,515.

Long Island, Battle of (27 August 1776), engagement in the American War of Independence. George WASHINGTON's army, divided by the East River, was defeated on Long Island by the British army under Gen. William HOWE. Driven back to Brooklyn, the colonial army retreated to Manhattan the following week under the cover of night and fog, thus avoiding capture.

Longitude. See LATITUDE AND LONGITUDE.

Long jump, standard field event in athletics. After a running start, the jumper tries to leap as far as possible from a fixed board. The length of the jump is measured from the forward edge of the board to the rear imprint the jumper leaves in a sand pit. The earliest accurately recorded jump was 7.05m (23.13ft) by Chionis of Sparta in 656 BC. Modern long jumping involves a carefully measured run-up and a skilled technique of lifting the trailing leg to join the take-off leg in a running stride through the air. In the 1968 Olympic Games the US athlete Bob Beamon set a record of 8.09m (29.25ft).

Longleat, estate of the Marquess of Bath, Wiltshire, England, and site of an Elizabethan manor house built 1567–80 to the design of Sir John Thynne, its owner. Today it houses an animal reserve.

Long march, The, journey undertaken by Chinese Communist forces after breaking through the NATIONALIST army of CHIANG KAI-SHEK in 1934. Under the leadership of CHU TEH and MAO TSE-TUNG some 100,000 men and women travelled 8,000km (5,000 miles) to reach Shensi Province in 1935. Those who survived the journey settled around Yenan where a new Communist headquarters was established. See also HC2 p.215.

Long Parliament, English parliament summoned by CHARLES I in Nov. 1640 to raise taxes to pay for his army. It was the Long Parliament's quarrels with Charles which led to the ENGLISH CIVIL WAR. It executed Charles's minister STRAFFORD and the Archbishop of Canterbury LAUD. In 1648 it abolished the House of Lords and the monarchy. After PRIDE'S PURGE of 1648 it was known as the RUMP PARLIAMENT and consisted only of supporters of CROMWELL. It was forcibly ejected in 1653 but continued as the BAREBONES PARLIAMENT until 1660. See also HC1 pp.287–288, 298–299.

Long-playing record, disc, introduced in 1948, with a continuous groove on each side. This groove is cut as a spiral into vinyl plastic and in stereo records the walls of the groove are at right angles to each other. Each wall is a series of undulations which cause the stylus to vibrate. The vibrations are converted into two separate components of sound. Before the 33 $\frac{1}{3}$ rpm discs, records were manufactured of shellac and made to revolve at 78 rpm. The long-playing record (LP) was so named because each side took up to half an hour to play as opposed to a 78's time of only about 5 minutes. See also MM p.232.

Longship, clinker-built wooden vessel that was driven by a single square sail and oars. Like its ancestor the dugout, it was a double-ended craft until the introduction of a stern rudder in about 1200. It is commonly associated with the VIKINGS, but was used over European waters from as early as 300 BC. Length: to 23m (75ft).

Longshore drift, movement of sand and pebbles along a sea coast. The material is carried along the beach by the waves hitting the coast obliquely (swash) but is swept back at right-angles to the beach by the backwash. If the carrying power of the waves decreases the material may be deposited to form a SPIT. See also PE pp.47, 122.

Long sight, or hypermetropia, defect of vision which causes nearby objects to be seen more hazily than distant ones. In a longsighted person, the focusing distance of the eyeball from front to back is too short and, as a result, light rays entering the EYE strike the RETINA before they can be properly focused. Long sight is corrected by convex lenses. See also MYOPIA; MS p.49.

Longueuil, Charles le Moyne, Sieur de (1626–85), French colonial leader in Canada. He emigrated to Canada in 1641 and worked for the JESUITS among the Huron Indians and as a trader, soldier and interpreter at Trois Rivières. He had 11 sons, all of whom distinguished themselves in the service of France. See also LE MOYNE.

Long wave. See RADIO.

Lonicera. See HONEYSUCKLE.

Lönnrot, Elias (1802–84), Finnish folklorist and philologist at a time when there was no written Finnish tradition. After training as a physician, he spent much of his life travelling throughout Finland, Lapland and NW Russia collecting legends, traditional ballads and lyric poems which now make up the KALEVALA, Finland's national epic. He published more lyric poems separately in *Kanteletar*.

Lonsborough, Anita (1941–), British swimmer. She won a gold medal in the 200m breaststroke at the Olympic Games of 1960, held in Rome, and subsequently went on to win the European title (1962) and to retain the Commonwealth title which she had won in 1958. In the 1962 Commonwealth Games she also won the 110 and 220 yards breaststroke event as well as the 440 yards medley.

Lonsdale, Frederick (1881–1954), British playwright known for his witty and urbane society comedies. Among these are *The Last of Mrs Cheyney* (1925), *On Approval* (1927) and *Canaries Sometimes Sing* (1929). Lonsdale also collaborated in the writing of operettas, one of which is *Maid of the Mountains* (1916).

Lonsdale Belt, award made by the National Sporting Club to a boxer winning the British championship in any division. The award is named after Lord Lonsdale (1857–1944) and was first made in 1909. Any boxer winning three championship fights in his division may keep the belt.

Loofah, tropical, annual, climbing herb whose dense, fibrous fruit is dried and used as a bath sponge. Family Cucurbitaceae; genus *Luffa*.

Look Back in Anger (1956), play by John OSBORNE. It surveys the aspirations of the British post-war generation and established using a realistic style (pejoratively called KITCHEN SINK DRAMA) to portray general dissatisfaction with the British class structure, and has had great influence on subsequent British theatre. The play was first produced by the English Stage Company at the Royal Court Theatre with a cast including Mary Ure, Alan Bates and Kenneth Haigh.

Loom, device used in weaving to hold the warp threads so that the weft can pass between them. Its simplest form is a vertical or horizontal frame, and there is evidence that such handlooms have been used since Neolithic times. Handlooms grew more sophisticated until the development of the machine loom, notably Edmund Cartwright's invention of 1785, which heralded the beginning of the INDUSTRIAL REVOLUTION in England. The basic design of the loom requires the warp to be threaded through heddles (rods with eye-holes by means of which individual threads are raised and lowered) and the weft to be wound on a shuttle which passes between the warp threads. The heddles alter the arrangement of the warp before the shuttle returns to provide the next row of the weft. Variations include foot mechanisms (treadles) to move the heddles, punched-card "programmes", invented by Joseph JACQUARD in 1804, to vary the groupings of warp threads for patterns to be introduced into the weave, and many mechanical improvements to speed and automate the process. See also MM pp.58–59; 216.

Loon, also called diver, diving bird of the Northern Hemisphere known for its harsh call; it has black, white and grey plumage. An excellent swimmer, it often stays submerged while fishing. The female lays up to three olive eggs in a grass-and-reed nest near water. Length: 88cm (35in). Family Gaviidae. See also NW pp.140, 227.

Loos, Anita (1893–), US novelist. She wrote many novels, but is famous only for her first, the international best-seller, *Gentlemen Prefer Blondes* (1925), a satire on sex and money. It was followed by *But Gentlemen Marry Brunettes* (1928).

Loos, Battle of (1915), battle between the combined French and British armies and the German army around Loos, in France, in WWI. It lasted from 25 Sept. to 13 Oct. and ended inconclusively in trench warfare, although the Allied losses were almost three times as great as the German. At Loos the Allies used chlorine gas for the first time.

Loosestrife, perennial plant that grows in tropical to temperate zones throughout the world. Loose spikes of flowers are borne above dark green, willow-like leaves. The wild purple loose strife, *Lythrum salicaria*, grows to 1.8m (6ft); there are also several cultivated varieties. Family Lythraceae.

Lope de Vega. See VEGA CARPIO, LOPE FELIX DE.

López, Vicente Fidel (1815–1903), Argentine educator, politican and writer. An outstanding intellectual, López helped organize the educational systems of Argentina, Chile and Brazil. Best known as a historian, he also wrote extensively on archaeology, philology and law.

López Arellano, Osvaldo (1921–), President of Honduras (1966–71, 1972–75). As head of the armed forces, López led the coup against the Villeda administration (1963). He withdrew Honduras from the Central American Common Market.

López de Legazpi, Miguel (*d.*1572), Spanish navigator. In 1545 he went to Mexico and while there was chosen by the viceroy to lead an expedition for the conquest of the Philippines. He sailed in 1564, reaching the Philippines the following year. Relying on tact and subtlety rather than violence, he took possession of the islands in the name of Spain.

Lop-Nor (Lo-pu po), salt basin in SE Sinkiang-Uigur autonomous region, China. It was once a salt-water lake which has now mostly dried up except for the streams of the Tarim River. The area has been used for nuclear tests by the Chinese government. In 1928, when last explored, the lake at its broadest was 40km (25 miles) and 97km (60 miles) long.

Lopolith, type of rock body formed by the cooling of molten rock, or MAGMA, in which the final shape is saucer-like, concave upwards. *See also* PE p.101, *101.*

Loquat, evergreen shrub or small tree and its edible fruit native to Japan and cultivated in subtropical climates. The tree has white flowers and bears yellow, plum-like fruits prized as a dessert or flavouring. Height: to 9m (30ft). Family Rosaceae; species *Eriobotrya japonica. See also* PE p.*195.*

Loran, navigational aid used to guide ships and aircraft. It stands for LOng-RAnge-Navigation and consists of two ground stations that emit radio pulses, which the ships and planes can use to guide their course. The stations have a day range of 1,300km (800 miles) and a night range of 2,500km (1,600 miles), and are accurate to within 1.6km (1 mile) of their location.

Lorca, Federico García. *See* GARCÍA LORCA, FEDERICO.

Lord, originally the feudal superior, the nominal owner of the daily bread his tenants ate (the word is a corruption of the Anglo-Saxon for "loaf ward"). Now, as a title the term is applied to members of the House of Lords, and used to address those who have become lords by heredity or by creation, including lords temporal and spiritual (ie high-court judges and bishops). *See also* HC1 p.211; HC2 p.*161.*

Lord Dunmore's War (1774), war between colonial settlers in America and Indians. Virginia's governor, the Earl of Dunmore, took control of W Pennsylvania; settlers then moved into Kentucky. These encroachments upon lands the Indians considered theirs provoked the Shawnee tribes into war. The Indians were defeated at the Battle of Point Pleasant and by the Treaty of Camp Charlotte relinquished their hunting rights in Kentucky.

Lord of the Rings (1954–55), epic trilogy by the British scholar and philologist J. R. R. TOLKIEN. The three works *The Fellowship of the Ring, The Two Towers,* and *The Return of the King* detail an intense and intricate journey and a battle between good and evil forces in an imaginary world called Middle Earth. The author has been highly commended for the interwoven mythologies and the Elvish language that he created to add realism to the tale.

Lordosis. *See* CURVATURE OF THE SPINE.

Lords-and-ladies. *See* CUCKOO PINT.

Lords, House of. *See* HOUSE OF LORDS.

Lord's Cricket Ground, London, the world's premier CRICKET ground. Founded by Thomas Lord in 1787, it has been at its present location in St John's Wood since 1814. It is the home of the MCC and the Middlesex County Cricket Club. Its playing area is about five acres; its spectator capacity about 34,000. The first TEST MATCH played there took place in 1884, since when every test series in England has included a match at Lord's.

Lord's Prayer, the prayer JESUS CHRIST taught his disciples, saying "After this manner therefore pray ye" (Matthew 6:9–13). A slightly different version is found in Luke 11:2–4. The prayer is also called "Our Father" or *Pater Noster* from its opening words.

Lord's Supper. *See* EUCHARIST.

Lorelei, large rock in the River Rhine near Sankt Goarshausen, W Germany. In the novel *Godwi* (1800–02), Clemens BRENTANO created the outline of a legend surrounding it. A beautiful maiden called the Lorelei was said to have drowned herself in despair, only to rise as a SIREN to lure fishermen to their doom on the rock.

Loren, Sophia (1934–), Italian film actress. Her most famous films are *Marriage Italian Style; Yesterday, Today and Tomorrow; The Black Orchid* and *Two Women,* for which she won the Best Actress award at the Cannes film festival (1961).

Lorentz, Hendrik Antoon (1853–1928), Dutch physicist and professor at Leiden. His early work was concerned with James CLERK MAXWELL's theory of electromagnetic radiation. This led him to the LORENTZ TRANSFORMATION and the prediction of the Lorentz-Fitzgerald contraction, both of which were essential steps in Albert EINSTEIN's development of the special theory of RELATIVITY. Some of his work was concerned with thermodynamics and the ZEEMAN EFFECT, for which he and Pieter ZEEMAN were awarded the 1902 Nobel Prize in physics.

Lorentz transformation, relation developed by H. A. LORENTZ, connecting the space and time co-ordinates of an event as observed from two frames of reference, especially at relativistic velocities. It was shown by Albert EINSTEIN in 1905 to be a consequence of the special theory of RELATIVITY.

Lorenz, Konrad (1903–), Austrian pioneer ETHOLOGIST. Unlike psychologists, who studied animal behaviour in laboratories, Lorenz studied animals in their natural habitats. He observed that instinct played a major role in animal behaviour – in for example IMPRINTING, by which an animal may learn to identify its parents. Some of his views are expressed in his book *On Aggression* (1966). In 1973 he shared, with N. TINBERGEN and K. von FRISCH, the Nobel Prize in physiology and medicine. *See also* MS pp.150, 159, 186, *187.*

Lorenzetti, name of two brothers who were accomplished painters of the SIENESE SCHOOL. It is believed that Ambrogio (*fl.*1319–48) was the younger. Only six of his works, spanning 13 years, have survived, including the signed and dated panels of *The Presentation of Christ in the Temple* (1342). Pietro (*fl.* 1320–48) executed a number of frescoes, possibly from 1330–40. These clear and monumental works are in the Lower Church of S. Francesco, Assisi. His last major work, *The Birth of the Virgin* (1342), displays a well-developed understanding of the theory of perspective.

Lorenzini, Carlo (1826–90), Italian author, real name Carlo Collodi. He was the creator of PINOCCHIO. *The Adventures of Pinocchio* first appeared as a book of 1883 and soon became a widely translated children's classic. Walt DISNEY made a film of the story in 1940.

Lorenzo, Monaco (Lorenzo the Monk) (*c.*1370–1425), Italian painter and miniaturist. He entered the monastery of Sta Maria degli Angeli, Florence, a celebrated centre for manuscript illumination. His *Coronation of the Virgin* (versions in National Gallery, London, and Uffizi, Florence) show his debt to GIOTTO. His *Adoration of the Magi* (Uffizi) is among the earliest Florentine examples of the INTERNATIONAL GOTHIC style.

Loreto, small town in Ancona province, central Italy, an important pilgrimage centre of the Middle Ages. According to legend, the house of the Virgin Mary was miraculously transported there from Nazareth in 1924. Our Lady of Loreto is a patron saint of aviators.

Lorikeet, small, brightly coloured PARROT found in the East Indies, Polynesia and E Australia. The rainbow lorikeet (*Trichoglossus haematodus*) is coloured green, brown, red and yellow. Lorikeets feed on nectar, which they obtain by lapping the juices of crushed flowers. Length: 15–38cm (6–15in). Subfamily Loriinae; family Psittacidae. *See also* NW p.205.

Loris, any of several species of primitive tail-less, tree-dwelling nocturnal PRIMATES of S Asia and the East Indies. They have soft, thick fur and large eyes, and feed mainly on insects. Length: 18–38cm (7–

15 in). Family Lorisidae; genera *Loris* and *Necticebus. See also* NW pp.168–169.

Lorna Doone (1869), historical romance by R. D. BLACKMORE. Set in 17th-century Exmoor, England, it tells of a feud between the Doones, an outlawed clan, and the Ridds, a farming family.

Lorrain, Claude. *See* CLAUDE LORRAIN.

Lorraine. *See* ALSACE-LORRAINE.

Lorraine, Cross of, also known as the patriarchal, a cross which has two transverse bars, the higher one shorter than the lower. During WWII the Cross of Lorraine was the symbol of the FREE FRENCH.

Lorre, Peter (1904–64), Hungarian actor, famous for playing sinister characters in British and US films. His films include *The Face Behind the Mask* (1941), *The Maltese Falcon* (1941), *The Beast with Five Fingers* (1946) and *The Raven* (1963).

Lory, brightly-coloured PARROT, native to Australia, with a brush-tipped tongue for feeding on nectar and fruit. The yellow-backed lory (*Domicella garrula*) is crimson and yellow. Its wings are green and its beak orange. Lories and LORIKEETS belong to the same subfamily (Loriinae), but lories are distinguished by their shorter tails. *See also* NW p.205.

Los Alamos, town in New Mexico, USA, site of a large scientific laboratory. During WWII the laboratory and its work was known by the code-name of MANHATTAN PROJECT, which produced the atomic (fission) bomb, later dropped on Hiroshima and Nagasaki. After the war the laboratory developed the H-bomb. *See also* MM p.180.

Los Angeles, Victoria de. *See* ANGELES, VICTORIA DE LOS.

Los Angeles, city in SW California, USA, on the Pacific coast; largest city in the state and third largest in the country. The area of Los Angeles County approx. 10,574sq km (4,083sq miles) includes the cities of Beverly Hills, Santa Monica and San Fernando. Founded in 1781 as a cattle farming centre, Los Angeles was acquired by the USA from Mexico in 1846. The city grew with the completion of the Southern Pacific (1876) and Santa Fé (1885) railways. The discovery of oil (1894), improvements to the harbour facilities (1912), and the development of the film industry in HOLLYWOOD in the early 20th century all attracted people to the city, the growth of which The city's growth has created major problems, including water shortage – which is overcome by a 480km (300 mile) pipeline from the Colorado River – and air POLLUTION. As well as museums and an art gallery, Los Angeles has three universities. Industries: aircraft, oil refining, food processing, printing and publishing, plastics, electronic equipment, tourism, entertainment, heavy machinery, textiles, chemicals. City pop. 2,747,000. *See also* PE pp.146, *147.*

Losey, Joseph (1909–), US film director. A refugee from MCCARTHYISM, he went to Britain and worked on many small productions, including *The Damned* (1961), a remarkable science-fiction film. Critical success came with *The Servant* (1963) and, after making *Accident* (1967), he became generally recognized as a great stylist. His other films include *The Go-Between* (1971) and *Mr Klein* (1976).

Losinga, Herbert de (*c.*1054–1119), English churchman. He became Bishop of Thetford in 1090, and of Norwich in 1094. He began to build Norwich Cathedral in 1096.

Losonczi, Pál (1919–), Hungarian politician. He became a Member of Parliament in 1953, Minister of Agriculture (1960–67) and was appointed Hungarian Head of State in 1967.

Lost wax. *See* CIRE PERDU.

Lot, biblical figure, son of Haran and nephew of ABRAHAM. Lot travelled in Abraham's company before settling in Sodom, where his hospitality towards the two angels investigating the iniquities of Sodom earned the safety of his family when the city was destroyed. Lot's wife, however, disobeyed instructions and looked back at the destruction of the city; for this she was turned into a pillar of salt.

Lothair I (*c.*795–855), Holy Roman Emperor of the West with his father,

Lord's Cricket Ground, viewed from the top of a nearby block of flats during a test match.

Sophia Loren's films have included many fine roles in both comedy and drama.

Loris has gripping hands with enlarged thumbs that lets it climb along branches.

Los Angeles has one of the best-developed systems of urban motorways in the world.

Lorenzo Lotto's *Adoration of the Shepherds* (c.1528), painted in the style of Titian.

Lotus-eaters; the Greeks were unable to link the legend with a real African tribe.

Louis XIV's elaborate lifestyle cut him off from the mood of the rest of France.

Louis XVI's attempts at reforms came too late to avert the ensuing French Revolution.

Louis I, from 817 and heir over his younger brothers, Louis the German, Pepin and later Charles the Bald. After Louis I died (840), the two remaining brothers (Pepin had died in 838) defeated Lothair at Fontenoy in 841 and divided the empire into three by the Treaty of VERDUN in 843. In 844, Lothair granted Italy to his eldest son, LOUIS II; in 855 he abdicated. *See also* HC1 p.153, *153.*

Lothair II, (*c.*835–869), King of LOTHARINGIA (855–69); a son of LOTHAIR I, Emperor of the West.

Lotharingia, part of CHARLEMAGNE's empire inherited by his descendant LOTHAIR II (*r.*855–69), after whom it is named. The Treaty of VERDUN (843) divided the Carolingian Empire among Charlemagne's three grandsons, the middle part going to Lothair I. Another split in 855 gave the northern part of Lothair's kingdom to his son Lothair II. Roughly, Lotharingia included modern Lorraine (whose name is itself a corruption of Lotharingia), Alsace, NW Germany, Luxembourg, Belgium and The Netherlands. *See also* HC1 pp.152–153.

Lothian, The, historical district of Scotland, between the Tweed and Forth rivers, including Edinburgh and the SE coastal areas. The Votadini lived there in the 3rd and 4th centuries. The region became part of Northumbria in 635 and passed to KENNETH II, king of the Scots *c.*973.

Loti, Pierre (1850–1923), pseudonym of Louis Marie Julien Viaud, French novelist. Naval service provided the exotic settings for his popular, melancholy romances, such as *Ramuntcho* (1897). His best works, however, deal with Breton life, as in *An Iceland Fisherman* (1886).

Lottery, form of gambling in which prizes are given to people holding purchased tickets bearing winning numbers drawn from a pool. Lotteries have been subject to governmental regulation in Britain since 1698, and were often used to finance building programmes. There has been no state lottery in Britain since 1826; the PREMIUM SAVINGS BONDS system differs from a normal lottery in that it is possible at any time to recover one's stake. In Australia and New Zealand there is widespread interest in lotteries which are organized both privately (eg by TATTERSALL'S) and officially, by state governments.

Lotto, Lorenzo (*c.*1480–1556), Italian painter, trained by the VIVARINI FAMILY, worked mainly in Treviso, Rome and towns outside Venice. *The Madonna and St Peter Martyr* is an early work. His mature work includes *Madonna Enthroned with Four Saints* (*c.*1540) and *Presentation in the Temple* (unfinished).

Lotus, common name for any WATER-LILIES of the genus *Nelumbo* and several tropical species of the genus *Nymphaea*. The circular leaves and flowers of some species may be 60cm (2ft) across. *Nymphaea* is sacred to the Chinese, Egyptians and Indians. The genus *Lotus* belongs to the unrelated Leguminosae family. *See also* PE p.191.

Lotus-eaters, according to Homer's *Odyssey*, the companions of Ulysses who, having eaten the fruit of the lotus, were overcome by dreamy forgetfulness and lost their desire to return home. The theme was the subject of Alfred, Lord TENNYSON's poem *The Lotos-Eaters* (1833).

Lotze, Rudolf Hermann (1817–81), German philosopher who developed a form of personalistic and teleological idealism. According to his philosophy, which is expounded in his book *Metaphysik* (1841), all entities ultimately find unity in the "world ground", a synthesis of physical and evolutionary processes involving an infinite spirit, or God. At the time it was widely accepted by theologians and theistic philosophers, but fell into disrepute shortly after Lotze's death, although his influence on certain later 19th- and early 20th-century idealist philosophers was significant.

Loudspeaker, device for converting oscillating electric currents into sound. The most common type has a moving coil attached to a stiff paper cone (often elliptical) suspended in a strong magnetic field. The oscillating currents in the coil cause the cone to vibrate (by ELECTROMAGNETIC INDUCTION) at the frequency of the currents, thus creating sound waves. Many modern loudspeakers give faithful sound reproduction between 80 and 10,000 HERTZ. *See also* SU p.68.

Lough Neagh, lake in central Northern Ireland; the largest freshwater lake in the British Isles. It is fed chiefly by the Upper Bann and Blackwater rivers and drained by the Lower Bann. Fishing is important, particularly for trout and eels. Area: 388sq km (150sq miles).

Louis of Bavaria. *See* LUDWIG.

Louis, name of 18 kings of France. Louis I (778–840) was son of CHARLEMAGNE and co-emperor of the West from 813. He was deeply religious, and was called the Pious. His reign was troubled by rebellious sons and their territorial disputes; he was twice deposed by them but restored in 830 and 834. Louis II (846–79), called the Stammerer, reigned from 877 to 879. On his death, his kingdom was divided between his sons Carloman and Louis III (*c.*863–82). Louis III was troubled during his reign (879–82) by Norman invasions in N France. Louis IV (*c.*921–54) attempted during his reign (936–54) to re-establish a claim to Lorraine but was thwarted by vassals. Louis V (*c.*967–87) was overshadowed during his reign (986–87) by Hugh CAPET, Duke of the Franks, who succeeded Louis on the throne. Louis VI, called the Fat (1081–1137), ruled France from 1108 to 1137. With the help of Abbot SUGER of St Denis he increased the power of the Crown over the nobles. Louis VII (*c.*1120–80) ruled France from 1137 to 1180 and greatly extended his power by marrying Eleanor of Aquitaine. He divorced her, however, and she subsequently married HENRY II of England, who thereupon gained control of Aquitaine. Louis VIII (1187–1226), King of France from 1223 to 1226, invaded England in 1216 at the invitation of barons opposing King JOHN, but was defeated at sea in 1217. (*See also* HC1 p.202). Louis IX, or St Louis (1214–70), ruled France from 1226 to 1270; he was canonized in 1297. He recovered from malaria and was prompted to go on the Sixth Crusade (1248–54). He was captured, and it was nine years before he was ransomed. In 1270 he undertook another crusade but died of fever at Tunis. The principal object of his policies was to secure peace among Christian nations. (*See also* HC1 pp.202, *205.*) Louis X (1289–1316) spent much of his short reign (1314–16) raising money for a proposed campaign in Flanders by selling charters of privileges to the clergy and nobility. (*See also* HC1 p.*202*). Louis XI (1423–83) devoted much of his reign (1461–83) to struggles against CHARLES THE BOLD of Burgundy. In 1482 he defeated Charles' daughter, Mary of Burgundy, and the Treaty of Arras gave Burgundy to France. Louis XII (1462–1515) ruled France from 1498 to 1515. He had his first marriage annulled so that he could marry (1499) Anne of Brittany (his predecessor's widow) and thereby keep Brittany a part of France. Louis XIII (1601–43) relied for much of his reign (1610–43) on his adviser, Cardinal de RICHELIEU. Both Louis and Richelieu favoured strong royal authority, opposition to the Spanish and Austrian HAPSBURGS, and strategic alliances with Protestant opponents of the Spanish. *See also* LOUIS XIV; LOUIS XV; LOUIS XVI; LOUIS XVII; LOUIS XVIII.

Louis XIV (1638–1715), son of Louis XIII and King of France (1643–1715). His personal rule began with the death of the regent, Cardinal MAZARIN, in 1661. For the next 50 years Louis was the most powerful monarch in Europe, presiding over a France at the height of its intellectual, economic and military powers. The great figures of French classicism were patronized by him and the palace at VERSAILLES built for him. There the elaborate courtly ritual of the "Sun King", including the famous *levée*, was acted out. In economics Louis was served by Colbert, who, working on foundations laid by his predecessors, made France the most centralized mercantilist state in Europe.

Louis destroyed the power of the nobility in the provinces: this was the basis of his absolutism. Abroad, he sought to dominate western Europe. Yet in the War of Devolution (1667–68), the third Dutch war (1672–78), (ended by the Treaty of NIJMEGEN), the War of the League of AUGSBURG (1688–97) and the War of the SPANISH SUCCESSION (1702–13) he depleted the French treasury and was limited to minor gains. The rigidity of the system he imposed upon the country led to France's decline in the 18th century. *See also* HC1 pp.314–315.

Louis XV (1710–74). grandson of LOUIS XIV; King of France (1715–74). Under his rule, the French monarchy was fatally weakened. He ran an extravagant court, and his personal weakness left him prey to favourites such as Madame de POMPADOUR. His wars were expensive and ended in defeat in the SEVEN YEARS WAR and the loss of most of the French Empire. He was unable to reform the outdated tax system or prevent the approach of national insolvency.

Louis XVI (1754–93), grandson of LOUIS XV; King of France (1774–92). He attempted to restore the power of the French monarchy, but the aristocracy prevented his ministers TURGOT and NECKER from implementing their economic reforms. He convened the Assembly of Notables in 1787 and, in desperation, the ESTATES GENERAL in 1789; the FRENCH REVOLUTION quickly followed. Although Louis outwardly accepted the Revolution, he was suspected of intriguing with foreign powers, was deposed in 1792 and beheaded on 21 Jan. 1793. *See also* HC2 pp.74–75.

Louis XVII (1785–95), second son of LOUIS XVI. He became Dauphin when his elder brother died in 1789. He was imprisoned by the French Republic from 1792 until his death, and never actually reigned.

Louis XVIII (1755–1824), King of France (1814–24). The brother of LOUIS XVI he fled from the Revolution to England and was recognized by the *emigrés* as King from 1795. He was restored to the throne by the allies in 1814, but was forced to flee again during the HUNDRED DAYS. In 1815 he returned after the Battle of WATERLOO.

Louis (1729–65), Dauphin of France, the son of LOUIS XV and father of LOUIS XVI, LOUIS XVIII and Charles X, all of whom were later French kings.

Louis, name of four German kings. Louis I the Pious was Holy Roman Emperor. Louis II (*c.*804–76), called the German (*r.*843–76), was one of the rebellious sons of Emperor Louis I. Lothair I, brother of Louis II, became joint emperor in 817 and on their father's death Louis and another brother, Charles, forced him to divide the empire in three. He received the lands extending from the Rhine to the eastern frontier of the empire. During the reign of Louis III (893–911), called the Child (*r.*899–911), the Archbishop of Mainz ruled as regent. Louis the Child was the last of the German Carolingians. Louis IV (*c.*1283–1347), called the Bavarian, was Duke of Bavaria (1294–1347), King of Germany (1314–47) and Holy Roman Emperor (1328–47). He was denied confirmation of his kingship by the papacy and was excommunicated by John XXII in 1324.

Louis, name of two kings of Hungary. Louis I (1326–82; *r.*1342–82) and King of Poland (1370–82), called "the Great". He encouraged commerce and science, and controlled the Adriatic Sea coast making Hungary one of the mightiest Balkan states. Louis II (1506–26; *r.*1516–26), son and successor to Ladislas II. He died at the Battle of MOHACS.

Louis, name of two kings of Italy. For Louis I *see* LOUIS I, the Pious. Louis II (*c.*822–75), Emperor of the West (*r.*855–75), King of Lorraine (*r.*872–875) succeeded his father, Emperor LOTHAIR I. He enlarged his possessions considerably at the expense of his two brothers, LOTHAIR II and Charles of Provence.

Louis, (1682–1712), Dauphin of France, grandson of LOUIS XIV, father of LOUIS XV. Louis XIV outlived his son and grandson; the Crown therefore passed to Louis

XIV's only surviving direct descendent, the third son of the Duc de Bourgogne, who became Louis XV.

Louis II, de Bourbon, Prince de Condé (1621–1686), French nobleman and general. He was a leader of the FRONDE revolt (1651–52) and defected to Spain but was later restored to his French rights and titles.

Louis, Joe (1914–), US boxer, real name Joseph Louis Barrow. He won the world heavyweight title from James J. Braddock in Chicago in 1937 and retired undefeated in 1949. He returned to the ring and lost to Ezzard Charles in 1950. Louis held the title longer than any other heavyweight. He served in the US Army (1942–45) and was elected to the Boxing Hall of Fame in 1954.

Louis, Morris (1912–62), US painter, linked with ABSTRACT EXPRESSIONISM. He was noted for a technique of staining his canvases in stripes of soft but brilliant colour to produce a highly lyrical effect, as in *Iris* (1954). *See also* HC2 pp.278–279.

Louisburg, Fort, Battle of (1748), capture of the French fortress in Acadia by the British in the SEVEN YEARS WAR. The British army was led by WOLFE and Amherst. The British thus gained control of the sea-route to the St Lawrence River, a decisive step towards the eventual conquest of French Canada.

Louisiana, state in central southern USA, on the Gulf of Mexico. The state is made up of the Gulf coastal plain and the alluvial plain of the Mississippi River. Other major waterways are the Red and Ouachita rivers. Inland there are low rolling hills and prairie country. The rich alluvial soils make Louisiana the leading producer of sweet potatoes, rice and sugar cane. Other farm produce includes soya beans, cotton, cattle and dairy produce. Major industries include fishing, oil refining, petrochemicals, metal foundries, paper and tourism. The chief cities are Baton Rouge (the state capital), New Orleans (the largest city and the second busiest port in the USA), Shreveport, Lake Charles and Lafayette. The area was first settled by the French in the late 17th century. France sold it to the USA in the LOUISIANA PURCHASE of 1803. The Territory of Orleans was established the following year. It was admitted to the Union as Louisiana in 1812. The state joined the Confederacy in 1861. Union troops took New Orleans in 1862 and controlled the Mississippi a year later. The AMERICAN CIVIL WAR wrecked the state economy and RECONSTRUCTION proceeded more slowly than elsewhere in the south. Area: 125,674sq km (48,523sq miles). Pop. (1975 est.) 3,791,000. *See also* HC2 p.148; MW p.175.

Louisiana Purchase (1803), purchase of French territory in the New World by the USA. The 2,136,000sq km (825,000sq miles) of territory, from the Mississippi River to the Rocky Mts was bought for $15 million. *See also* HC2 pp.148, 149.

Louis Napoleon. *See* NAPOLEON III.

Louis of Nassau (1538–74), Count of Nassau-Dietz, Dutch leader. He was the brother of William the Silent and led the revolt against Spanish rule. When the Duke of Alva arrived in 1567, Louis and William left The Netherlands to raise a fighting force, returning in 1568. Louis was killed at Mook, after being defeated in several battles with the Spaniards.

Louis Philippe (1773–1850), King of France (1830–48). The revolution that brought him to power (1830) was a bourgeois reaction to the reactionary rule of the Bourbons, LOUIS XVIII and CHARLES X. For much of his reign he was a liberal, supporting constitutional restraint on the monarchy and becoming known as the "Citizen King" but, satisfying neither the legitimists (who supported the senior Bourbon line), nor the Bonapartists and elements of the left, he fell from power in the February revolution of 1848, following which a republic was proclaimed. *See also* HC2 p.108.

Louis XIV style, name for the classical style in the visual arts in France prevailing in the age of LOUIS XIV (r. 1643–1715). In painting the most influential style was that of POUSSIN, famous for landscapes done in a classical manner. The Académie Royale de Peinture (1648) set down rules which gave to two generations of French painting a strict classical mould. In furniture the national style spread from the productions of the GOBELINS firm, who specialized in the ornate use of gilding and inlay. In architecture the two chief monuments were the LOUVRE and the king's palace at VERSAILLES, both massive buildings of formal, symmetrical grandeur. *See also* HC1 pp.306–309.

Louis XV style, high period of French ROCOCO style in furniture and interior design. Rooms had walls set with mirrors in panels, replacing the large painted canvasses and murals of the previous century. Decorative motifs included garlands and clusters of flowers and shells, often decorated with gold and silver. Furniture became twisted and ornate. Frequently, Chinese motifs were used.

Louis XVI style, final period of the French ROCOCO style, lasting from about 1760 to the Revolution. A reaction against the ornate LOUIS XV style, it emphasized the antique Classical style. In furniture and interior design, straight lines replaced curved ones. The favourite colour scheme was white and gold. In architecture, the return to Classicism was marked by a preference for DORIC columns.

Lourdes, town in Hautes-Pyrénées département in SW France; in contemporary times it has become a centre of religious pilgrimage. In 1858, a 14-year-old girl, Bernadette Soubirous, had numerous visions of the Blessed Virgin Mary in the nearby grotto of Massabielle, where there is an underground spring. In 1862 the Roman Catholic Church declared the visions to be authentic, authorizing the cult of Our Lady of Lourdes. The waters of the spring, believed to have healing powers, are the focus of the pilgrimages. In 1958 an immense underground church was built with seating for 20,000.

Lourenço Marques. *See* MAPUTO.

Louse, common name for various small, wingless insects, parasitic on birds and mammals. There are two main groups, classified in different sub-orders of Phthiraptera. The chewing lice (Mallophaga) feed mainly on the feathers of birds. The biting or sucking lice (Anoplura) feed only on the blood of mammals. Both kinds of lice are small, pale and flattened, with leathery or hairy skins. *See also* NW pp.*107, 111*.

Lousonczi, Pál. *See* LOSONCZI, PÁL.

Louth, county in the NE Republic of Ireland in Leinster province, bordered on the E by the Irish Sea. It is a low-lying region except in the hilly NW and the mountainous N. The county is drained by the rivers Fane, Dee and Castletown. Industry is more important than agriculture, the main products being textiles and processed food. Area: 821sq km (317sq miles). Pop. (1971) 74,899.

Loutherbourg, Jacques Philippe de. *See* DE LOUTHERBOURG, JACQUES PHILIPPE.

Louvain, or Leuven, city in central Belgium. It was founded by the 9th century, and was an important cloth centre of the Middle Ages until 1383 when many of its weavers emigrated. The university was founded in 1425. Pop. (1971 est.) 32,200.

Louvre, French art museum in Paris. France's enormous national collection contains art from most ages and countries and is housed in the Louvre Palace, which was built as a 16th-century chateau for Francis I. The earliest parts of the building were commissioned from Pierre LESCOT in 1546 and were built on the site of the original Louvre, a 13th-century fortress palace. Under LOUIS XIV, several important art collections, especially Italian works, were acquired. In the present collection there are about 6,000 paintings and 100,000 drawings.

Lovage, stout-branched herb native to S Europe. The stalk, foliage and seeds may be used as a flavouring. The leaves may be infused to make a kind of tea. Family Umbelliferae; species *Levisticum officinale.*

Lovat, Simon Fraser, 11th Baron (c.1667–1747), Scottish nobleman, the last peer to be executed for high treason. After being involved in Jacobite intrigues in France he returned to Scotland, where he raised his clan to support George I in 1715. In 1745, however, he supported the Jacobites, and was executed in 1747.

Lovebird, any of nine species of small parrots that live mainly in Africa and Madagascar. Both sexes are brightly coloured and are often kept as cage birds. The largest species is the rosy-faced lovebird (*Agapornis roseicollis*) of South Africa.

Love for Love (1695), Restoration comedy by William CONGREVE which was produced by and starred Thomas BETTERTON as Valentine. It established Congreve's reputation as the leading dramatist of his day.

Love-in-a-mist, popular annual garden plant native to S Europe. A member of the BUTTERCUP family, it has divided leaves and solitary bluish flowers. Its numerous black seeds have been used medicinally. Species *Nigella damascena.*

Love in a Cold Climate, novel by the British author Nancy MITFORD, published in 1949. It is a satire on the English upper classes. It owes much, on the author's own testimony, to her own childhood experience.

Lovelace, Richard (1618–57), English CAVALIER poet. An ardent ROYALIST, he was imprisoned in 1642 and again in 1648. He is remembered principally for two extremely graceful and melodic lyrics, *To Althea, from Prison* and *To Lucasta, Going to the Wars.* He also wrote a tragedy, *The Soldier,* which is now lost. *See also* HC1 pp.282–283.

Loveless, George, (1797–1874), British worker and leader of the TOLPUDDLE MARTYRS. In 1834 Loveless and five comrades of Tolpuddle, Dorset formed a trade union to improve the pay and conditions of farm workers. They were arrested, tried and sentenced to transportation. The anger and militancy of the working class forced the reactionary WHIG government to pardon and release them in 1863. Loveless later emigrated to Canada.

Love-lies-bleeding, popular name for a species of ornamental garden plant of the Amaranth family. It bears attractive, long, drooping clusters of densely packed, small, crimson flowers. Species *Amaranthus caudatus.*

Lovell, Sir Alfred Charles Bernard (1913–), British astronomer. His research in radar during WWII earned him the OBE in 1946. In 1951 he became professor of radio astronomy at the University of Manchester. There he established the JODRELL BANK Experimental Station, which in 1957 was equipped with what was for several years the world's largest steerable radio telescope. His books include *Radio Astronomy* (1952), *The Story of Jodrell Bank* (1968) and *In the Centre of Immensities* (1978). *See also* SU p.*172.*

Lovell, James Arthur, Jr (1928–), US astronaut. His first space flight was in GEMINI 7, (1965), which made the first successful space rendezvous (with Gemini 6). He was a member of the crew of APOLLO 8 (1968), the first manned mission around the Moon, and commander of the unsuccessful Apollo 13 mission (1970). *See also* SU p.268, *268.*

Love on the Dole (1933), the first and best-known novel of Walter GREENWOOD. Based on the author's experiences, it portrays the life of the unemployed during the Great Depression in an industrial town in N England. In 1941 the novel was filmed.

Lover, Samuel (1797–1868), Irish writer who started as a miniature portrait painter in Dublin. In 1835 he took London society by storm by singing his own compositions. He is remembered for his humorous novel *Rory O'More* (1836), which developed from his songs of that title (1826).

Love's Labour's Lost (c.1595), five-act comedy by William SHAKESPEARE, one of his earliest works. Set at the court of Ferdinand, King of Navarre, it satirizes literary affectations and is one of Shakespeare's most complex comedies, much of the wit relying on contemporary allusion.

Lovett, William (1800–77), British CHAR

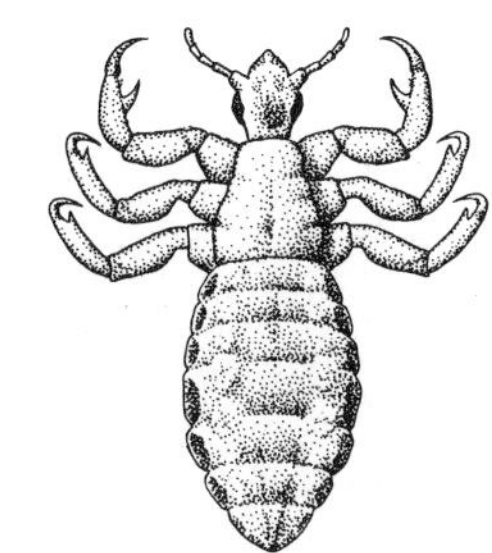

Louse; as well as causing irritation, it can carry typhus and other diseases.

Louvain University has enjoyed great prestige, particularly for its theological faculty.

Lovage, the taste of which has been likened to celery, can be used for making a liqueur.

Lovebirds probably do not die after losing their mates, contrary to popular opinion.

Low, or depression

David Low was a master of caricature, as in his well-remembered "Colonel Blimp".

L. S. Lowry is best known for his stark portrayal of life in industrial towns.

Heinrich Lübke was for ten years President of the Federal Republic of Germany.

Lucia di Lammermoor, by Donizetti; part of the 1863 stage design is shown here.

TIST leader. A disciple of Robert OWEN, he was one of the early founders of co-operative shops, main drafter of the People's Charter and a persistent advocate of popular education as the means to improve the life of the workers. In 1839 he was sentenced to a year's imprisonment for the publication of a seditious manifesto in Birmingham.

Low, or depression, in meteorology, area of low pressure. In the Northern Hemisphere, depressions are atmospheric circulations that spiral inwards in an anticlockwise direction; they rotate inwards clockwise in the Southern Hemisphere. *See also* HIGH; PE pp.68–73.

Low, Sir David Alexander Cecil (1891–1963), New Zealand political cartoonist and caricaturist, who began work in England for the *Daily News* in 1919, and joined the *Evening Standard* in 1927. His cartoons were at their best before and during WWII, dealing with political issues such as Fascism and oppression. After leaving the *Evening Standard* in 1950, he worked for a short time for the *Daily Herald* before joining the staff of the *Manchester Guardian* (1953).

Low Church, party in the Church of England, arising in the 18th century, which in general put the teachings of Scripture before Church organization, and personal devotion before priestly mediation. Wesleyanism, in the 18th century, and EVANGELICALISM, in the 19th, waxed strong in the Low Church.

Low Countries, area of NW Europe made up of The Netherlands, Belgium and the Grand Duchy of Luxembourg. It is a political rather than a geographical label (only N Netherlands and N Belgium are low-lying). In medieval times, as today, the region was one of the most prosperous in Europe. *See also* MW pp.37, 120, 128.

Lowell, Amy (1874–1925), US poet and critic. She originated POLYPHONIC PROSE and joined the IMAGIST movement. Her ability to recreate physical perception in her poems was probably her outstanding quality. Her works include *Men, Women, and Ghosts* (1916) and *What's O'Clock?* (1925) for which she received a Pulitzer Prize (1926).

Lowell, James Russell (1819–91), US poet, author and editor who became involved in the abolitionist movement. He gained acclaim for *Poems* (1844), *A Fable for Critics* (1848), the first series of *Biglow Papers* (1848) and *The Vision of Sir Launfal* (1848). He was the first editor of *Atlantic Monthly* (1857–61).

Lowell, Percival (1855–1916), US astronomer and traveller, famous for his prediction of the existence of the planet Pluto and his initiation of the search that ended in its discovery 14 years after his death. His works include *Occult Japan* (1895), *Mars and Its Canals* (1906) and *The Genesis of the Planets* (1916).

Lowell, Robert (1917–77), US poet. His work includes *Land of Unlikeness* (1944), *Lord Weary's Castle* (1946; Pulitzer Prize, 1947), *The Mills of the Kavanaughs* (1951), *For the Union Dead* (1964) and *Notebook 1967–68* (1969). *Life Studies* (1959) won a National Book Award.

Lower Canada, name for the French part of British North America from the Constitutional Act of 1791 until the union with UPPER CANADA (Ontario) in 1840. It entered the Confederation of Canada in 1867 as the province of Quebec.

Lower Hutt, urban area adjacent to the city of Wellington, in s North Island, New Zealand, at the mouth of the River Hutt. It has several scientific research centres. Industries: motor vehicle assembly, railway repair. Pop. (1971) 122,000.

Lowestoft, Battle of (1665), naval battle of the 2nd Anglo-Dutch War, in the North Sea off Lowestoft. The English fleet of 150 ships, under the Duke of York (later JAMES II) defeated the Dutch fleet of about the same size, sinking the Dutch flagship and killing its commander, Admiral Opdam.

Lowry, Lawrence Stanley (1887–1976), British painter best known for the highly personal way in which he portrayed, almost primitively, industrial landscapes of Salford (Manchester), with naively

drawn buildings, factories and crowds of dark simple figures. Although his early work received some attention, his first one-man exhibition in London was not until 1939. He was regarded as an eccentric of 20th-century painting, although his work recalls the complex crowd scenes of Peter BRUEGEL.

Loyola, Saint Ignatius of. *See* IGNATIUS OF LOYOLA, SAINT.

LPG. *See* LIQUEFIED PETROLEUM GAS.

LSD (lysergic acid diethylamide), drug known for the changes in mental state that it produces. Effects include sensory anomalies (in which forms fuse and colours intensify), produced by the drug blocking the action of SEROTONIN in the brain. *See also* MS pp.104–105.

Luanda (São Paulo de), capital and seaport city in NW Angola on the Baja da Bengo. Founded in 1576, its economy was based on a slave trade with the New World, until the abolition of slavery in the 19th century. Industries: metal working, building materials and oil products. Pop. (1970) 475,328.

Luang Prabang, port in N Laos, on the River Mekong; capital of Luang Prabang province. The city is the royal capital of Laos, and a distribution centre for teak, fish, rubber and rice. Pop. (1975 est.) 25,000.

Lübeck, seaport city in NE West Germany, on the River Trave near its mouth on the Baltic Sea. Lübeck possesses some of the best examples of Gothic architecture in N Europe which have been restored since heavy bombing in WWII. Industries include steel foundries, shipbuilding and textiles. Pop. (1974) 236,047.

Lubitsch, Ernst (1892–1947), US film director, b. Germany. He became Paramount Studios' foremost film-maker in the 1920s and 1930s and is noted for his sophisticated satires such as *Forbidden Paradise* (1924), *Design for Living* (1933), *Bluebeard's Eighth Wife* (1938) and *Cluny Brown* (1946).

Lübke, Heinrich (1894–1972), German political figure. He was Minister of Food and Agriculture in the post-WWII cabinet of Konrad ADENAUER. He became the first President of West Germany in 1959 and served until 1969, when he was succeeded by Gustav Heinemann.

Lublin, city in SE Poland, approx. 152km (95 miles) SE of Warsaw. The city was founded in the early 11th century. It later developed as a trade centre along the SE route to the Ukraine. Today it is a transport and industrial centre, producing heavy machinery, textiles and electrical goods. Pop. (1974 est.) 259,000.

Lubricant, oil, grease, graphite or other substances introduced between moving parts to reduce friction and dissipate heat. Most lubricants are now derived from petroleum. Solid lubricants are usually graphite or molybdenum disulphide; synthetic silicones are used where high temperatures are involved. *See also* MM p.100.

Lubumbashi, city in SE Zaire, near the Zambia border, formerly known as Elisabethville. The second largest city in Zaire, it was the capital of the independent state of Katanga (now the region of Shaba) 1960–63, and the scene of bitter fighting between UN and Shaban troops. Industries: copper smelting, textiles and food products. Pop. (1970) 357,369.

Lucan (Marcus Annaeus Lucanus) (AD 39–65), Roman poet. A nephew of the younger Seneca, he was born in Spain. His *Bellum Civile* or *Pharsalia* is an epic admired more for its rhetoric than its poetry. He killed himself after implication in a plot against NERO.

Lucas, Edward Verrall (1868–1938), British writer who was a frequent contributor to the satirical magazine *Punch* of which he became assistant editor. Lucas published over 100 books including *Over Bemerton's* (1908) and in *Adventures and Misgivings* (1938).

Lucas van Leyden (c.1494–1533), real name Lucas Hugensz, Dutch painter and celebrated engraver. His engravings, which are regarded as second only to those of DÜRER (whom he met in 1521 and who may have influenced his technique),

include *Ecco Homo* and *Dance of the Magdalene* (1519). Among his paintings are *Moses Striking Water from the Rock* (1527), *Chess Players* (c.1508) and *Last Judgement* (1526).

Lucayan, prehistoric tribe of American Indians who originally peopled the Bahamas. An Arawak Taino-speaking group with a simple culture, numbering at most 40,000, they were exterminated by the Spanish before 1600.

Luce, Clare Boothe (1903–), US author. She wrote the plays *The Women* (1936) and *Kiss the Boys Goodbye* (1938), and the novel *Stuffed Shirts* (1931). She served as a Republican in the House of Representatives (1943–47) and was Ambassador to Italy (1953–56). She resigned after her appointment as Ambassador to Brazil in 1959. *See also* HC2 p.268.

Luce, Henry Robinson (1898–1967), US publisher, b. China. In 1923, he founded *Time* with Briton Hadden and later founded other magazines including *Fortune* (1930), *Life* (1936) and *Sports Illustrated* (1954). He was considered an important influence on US politics.

Lucerne, or alfalfa, leguminous, perennial plant with spiral pods and purple, cloverlike flowers. Like other legumes, it has the ability to enrich the soil with nitrogen from the air, and for this reason it is often grown by farmers and then ploughed in. It is also a valuable fodder plant. Height: 0.5–1.2m (1.5–4ft). Species *Medicago sativa*. See also LEGUMINOSAE.

Lucerne (Luzern), city in central Switzerland, 40km (25 miles) SSW of Zurich on the Lake of Lucerne. It joined the Swiss Confederation in 1332, gained its freedom in 1386 and became the capital of the Helvetic Republic in 1803. Lucerne is an important Swiss summer resort and also has a variety of industries, including metal goods, chemicals, printing and the manufacture of textiles. Pop. (1970) 68,879.

Lucerne Festival, annual summer music festival held in Lucerne, Switzerland. It was established in 1938 by Adolf BUSCH.

Lu Chih (1495–1576), Chinese landscape and flower painter of the MING period. The delicate and poetic style of *Autumn Colours at Hsün-yang* (1554) is typical of his landscapes.

Lucia di Lammermoor, opera by DONIZETTI, with libretto by Salvatore Cammarano. It was first produced in 1835 at Naples and enjoyed immediate acclaim. Donizetti based it on Sir Walter SCOTT's novel, THE BRIDE OF LAMMERMOOR.

Lucian (c.AD 115–80), Greek satirist. He made use of the satiric dialogue in some 80 works. His writing was witty, and drew attention to the foibles of contemporary life and manners. Religion and philosophy were among Lucian's favourite subjects.

Lucian of Antioch, Saint (c.240–312), theologian and martyr. He founded a school in Antioch for the study of the scriptures and used his knowledge of Hebrew and Greek to promote a fuller understanding of the Bible. He also produced corrected texts of the GOSPELS. He was imprisoned at Nicomedia for nine years by the Emperor DIOCLETIAN and was either starved to death or put to the sword. Feast day: 7 Jan.

Lucifer, fallen angel who, because of his revolt against God, was cast out of HEAVEN with his minions, to become the Devil, known in Christianity as Satan, Lucifer or the Prince of Devils. His revolt was prompted by pride and in punishment God sent Lucifer to rule over HELL. In *Paradise Lost* John MILTON describes the revolt and portrays Lucifer's pride.

Luciferin, proteins, found in animal tissue, that produce light when oxidized in the presence of the enzyme luciferase. It is found in the light-generating organs of glow-worms and fire-flies.

Lucilius, Gaius (c.180–102 BC), earliest Latin satirist. He knew Scipio well and his works were formed into a posthumous collection in an edition comprising 30 books, but only fragments survive. *See also* SATIRE; HC1 p.112.

Lucite, trade name for a clear, hard plastic which is also called perspex and Plexiglass. It is made from an acrylic resin that con-

sists of polymerized methyl methacrylate, and is used for making costume jewellery, lenses, dentures, aquariums, containers and novelties. *See also* MM pp.54–55.
Lucius, name of three popes. Lucius I was pope 253–54. Lucius II succeeded to the papacy in 1144 and died the following year. Lucius III (*r.*1181–85) was born as Ubaldo Allucingoli. He refused to crown the son of the Holy Roman Emperor, and issued strong edicts against heresy.
Lucknow, city in N India, on the River Gomati; capital of Uttar Pradesh state. It was the capital of the kingdom of Oudh (1775–1856), then of Oudh province (1856–77) and of the United Provinces (1887). During the INDIAN MUTINY the British were forced to abandon their fortress and leave the city. It was retaken in 1858. Lucknow was a centre of the MUSLIM LEAGUE in its campaign for Pakistani independence (1942–47). Its industries include papermaking, metalworking, distilling and printing. Pop. (1971) 749,239.
Lucky Jim (1954), first and best-known novel by Kingsley AMIS. A farcical satire on English academic life, it tells how a young lecturer, Jim Dixon, rebels against the snobbish hypocrisy in his provincial university.
Lucretia, in Roman legend, the beautiful and pure wife of Lucius Tarquinius Collatinus. After being raped by Sextus Tarquinius, the son of the Etruscan king of Rome, she demanded that her father and husband should avenge her and then stabbed herself to death. Lucius Junius Brutus then led the populace in a rebellion that drove the Tarquins from Rome and the Roman Republic was established.
Lucretius (Titus Lucretius Carus) (*c.*95–55 BC), Latin poet and philosopher. Jerome states that Lucretius was driven mad by a potion and that he killed himself, but this is unverifiable. His poem *De rerum natura* (*On the Nature of Things*) concerns the atomic theory of EPICURUS. *See also* HC1 pp.22, 112; HC2 p.*22.*
Lucullus, Lucius Licinius (*c.*117–*c.*56 BC), Roman commander who became consul in 74 BC. He gained control of the Roman lands in Asia, where he fought and defeated Mithradates (74–70 BC) and took the capital of Armenia in 69 BC. His economic reforms in the territories in his charge harmed the interests of others and POMPEY replaced him in 66 BC.
Luddites, textile workers in England who in the early 19th century destroyed power looms and knitting-frames, which they believed to be the cause of unemployment in the cotton and woollen industries. They were named after a mythical "King Ludd", who signed public letters denouncing the new machines. The chief riots, starting in Nottinghamshire and spreading to Cheshire, Lancashire and Yorkshire, took place in 1811–12. *See also* HC2 p.*94.*
Ludendorff, Erich (1865–1937), German general. Under von MOLTKE, he played a major part in revising the SCHLIEFFEN PLAN before WWI and in Aug. 1914, as HINDENBURG's chief of staff, he masterminded the victory over the Russians at Tannenburg. In Aug. 1916 he and Hindenburg were given supreme control of Germany's war effort. After WWI Ludendorff became a member of the Nazi Party.
Lüderitz, Franz Adolf (1834–86), German trader and colonist in Africa. He helped form the colony, Germany's first, of South West Africa (now Namibia) from land he bought from the Khoikhoi people.
Ludlow, Edmund (*c.*1617–92), English parliamentarian and republican. He became an MP in 1646. He was one of the judges at the trial of CHARLES I and signed the king's death warrant. He was a member of the Council of State (1649–50). In 1653 he refused to recognize CROMWELL as Protector and retired to Essex. He was impeached in 1660, but fled to Switzerland.
Ludlow, Roger (1590–*c.*1664), English colonist in America. He helped found Dorchester, Massachusetts, and became deputy Governor of Massachusetts in 1634. He presided over the first Connecticut court at Windsor in 1636 and evolved the state's first codified laws, known as

Ludlow's Code. He returned to England in 1654.
Ludwig (Louis), name of three kings of Bavaria. The Wittelsbach family, of which they were a junior branch, had a history of insanity, which affected the reigns of all three kings. Ludwig I (*r.*1825–48) had a long love affair with the dancer Lola Montez, who was one of the main causes of revolt in 1848. Ludwig II (*r.*1864–86) spent his time building extravagant castles; his neglect of affairs of state was one of the reasons for Bavaria's political decline. In 1871 it was absorbed into the German Empire, its monarch losing all but the trappings of power. Ludwig III (*r.*1913–18) became regent in 1912 during the crisis caused by the madness of his cousin Otto, before succeeding to the throne himself. He was forced to abdicate in 1918.
Ludwig, Emil (1881–1948), German writer. He began as a playwright and poet, but devoted most of his life to writing biographies the first of which, on Goethe, was published in 1920. His subjects included Napoleon, Bismarck, Christ, Lincoln, Roosevelt, Cleopatra and Beethoven.
Ludwig, Karl Friedrich Wilhelm (1816–95), German physiologist who studied the cardiovascular system and invented the kymograph (1847) which records changes in arterial blood pressure. He also invented a device to measure arterial and venous blood flow, and investigated the functions of oxygen in the blood.
Luff, leading edge of a sail in a fore-and-aft rig. The word is also used when a vessel "comes about", ie "luffs", into the wind while traversing from one tack to another, to bring the wind on the alternate quarter while sailing in restricted waters.
Luftwaffe, air force of Nazi Germany (1933–45). In 1935 HITLER made public the existence of 1,888 planes and 20,000 trained personnel under Hermann GOERING, a force which had been secretly built up since the 1920s. The Luftwaffe had air superiority early in WWII, but could not counteract the larger allied air forces after 1942.
Lugano, Lake, narrow and irregularly-shaped lake, 50sq km (19sq miles) in area lying between Italy and Switzerland. The resort town of Lugano is on its N shore.
Lugones, Leopoldo (1874–1938), Argentine poet and prose writer. Among his best-known verse are *El libro de paisajes* (1917) and *Poemas solariegos* (1927). He also wrote short stories and scholarly historical and linguistic works.
Lugosi, Bela (1884–1956), US film actor who was famous for his role as Bram STOKER's vampire, Count DRACULA, in the classic horror film *Dracula* (1931). His performance and the mood of the film set a style for many later works in the genre. Appearing in such films as *The Black Cat* (1934) and *The Wolf Man* (1941), he became typecast as a solely macabre character and gradually lost favour.
Lugus, in Celtic mythology, a powerful god identified by Julius CAESAR as the equivalent of MERCURY. His name meant "light" and he appeared in both Welsh and Irish legends. The sole survivor of divine triplets, all called Lugus, he had the strength of three gods and was revered as the ideal leader, wise, gifted in poetry and formidable in war.
Lugworm, marine WORM that lives in the sand of the sea-bed. It burrows, with the aid of bristles along its middle portion, a U-shaped tunnel in sand or mud, from which it rarely emerges. Lugworms feed by ingesting sand, extracting nutrients and ejecting the sand in coiled mounds at the end of their burrows. Length: up to 30cm (12in). Genus *Arenicola. See also* NW pp.88, *236.*
Lu Hsiang-shan (1139–93), also known as Lu Chiu-Yüan, founder of the Neo-Confucian School of Mind. He insisted that the purpose of study was to clear the mind of things that blind it and allow the mind to return to its original purity. Lu's work served as a divergence from the School of Principle to the School of Mind, and was later elaborated upon by WANG YANG-MING.
Lu Hsaün (1881–1936), Chinese writer,

real name Chou Shujen. He wrote during the period of literary modernizing before the Communists came to power and is best known for his charming short stories and his essays which urged a rapid modernizing of Chinese culture and politics.
Luik. *See* LIÈGE.
Luini, Bernardino (*c.*1485–1532), Italian painter who was best known for his religious and mythological frescoes, in a popular sentimental style adapted from LEONARDO DA VINCI. His earliest surviving work is the fresco *Madonna and child* (1512), in the Cistercian monastery in Chiaravelle, near Milan.
Lukacs, György (1885–1971), Hungarian literary critic and philosopher. He joined the Communist Party in 1918, and was exiled after the abortive 1919 revolution. He took up political journalism and teaching in Vienna and Moscow. He returned to Budapest in 1945. His writings include *History and Class Consciousness* (1923), on the role of creativity in the class struggle, and *The Historical Novel* (1955).
Luke, Saint, traditionally the author of the gospel that bears his name and of the ACTS OF THE APOSTLES in the New Testament. Luke, a physician and a companion of the apostle PAUL, wrote about the life of Jesus from a gentile Hellenistic Christian viewpoint. His gospel stresses the human and divine natures of Jesus and the concept of Christianity as a universal brotherhood.
Luke, Saint, the Gospel according to, third book of the New Testament, written in Hellinistic style in the 1st century AD and traditionally attributed to Luke, the friend of St PAUL. One of its sources is the Gospel according to St MARK, but it is unique in its account of the birth and boyhood of Christ.
Luks, George Benjamin (1867–1933), US painter, a member of the EIGHT, best known for his paintings of New York City's East Side. They include *The Spielers* (1905) and *The Wrestlers* (1905). A staff artist on the Philadelphia Bulletin, he also continued the comic strip *The Yellow Kid* (1895) begun by Outcault in the New York *World*. It was the first major successful comic strip.
Lull, Raimon (1235–1315), also known as Lully, Raymond, Catalan scholastic philosopher (in Catalan, Ramón Llull), known throughout Europe as the "Doctor Illuminatus", whose youthful exuberance was dramatically transformed into a life devoted to converting Muslims to Christianity. His most notable theological work, the *Ars Magna*, was the basis of his arguments against ISLAM. He wrote many mystical and scientific works, and his fame was such that numerous alchemical treatises are also, probably falsely, attributed to him.
Lully, Jean-Baptiste (1632–87), French composer, *b.* Italy; an early influence on the development of French opera. He joined the court musicians to LOUIS XIV in 1652 and soon acquired the king's favour. After a series of comedy-ballets (1658–64) came *Cadmus and Hermione* (1673), which has been termed the first French lyrical tragedy. He replaced the Italian *recitativo secco* with accompanied recitative and developed a style of declamation more suited to the French language. Other operas include *Alceste* (1674), *Proserpine* (1680) and *Acis et Galatée* (1686).
Luluabourg. *See* KANANGA.
Lumbago, fibromuscular pain of the lower (lumbar) region of the back. Although widely complained of, particularly by older patients, and prescribed for by doctors, it remains an ill-defined condition which seems to have any one of a number of dissimilar causes, including strain, injury, slipped disc or GOUT. Other causes may be chronic constipation or displacement of internal organs. *See also* RHEUMATISM; MS p.94.
Lumberjack, man who earns his living cutting trees and preparing and transporting logs for a sawmill or pulp mill. In the early days of logging in North America, lumberjacks lived for most of the year in remote logging camps.
Lumen, unit in the SI UNIT system, defined as the amount of light emitted per second in a unit SOLID ANGLE (one steradian) from

Lake Lugano in Italy, showing the Ponte Di Melide and Monte S Salvatore.

Bernardino Luini, about whose life little is known, painted this *Christ on the Cross.*

Jean-Baptiste Lully introduced lively dance and ballet music to the court of Louis XIV.

Lumberjacks in Nova Scotia photographed in the days before the use of chain saws.

Lüneburg Heath is a region of Saxony in Germany between the Elbe and Aller rivers.

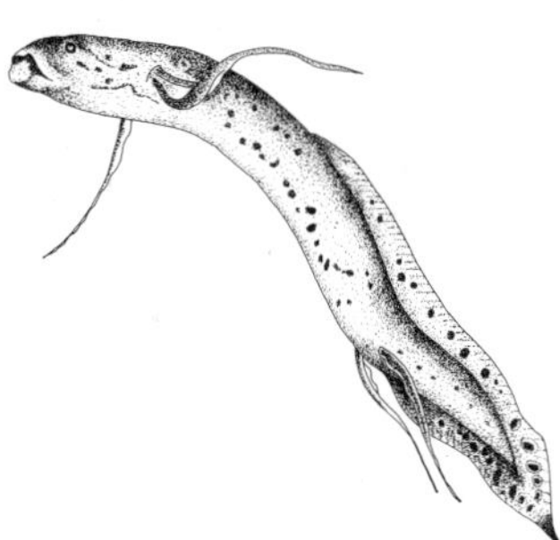

Lungfish: *Protopterus*, a type which breathes only air, is mainly found in Africa.

Lupin seeds were eaten in ancient Egypt, where many varieties were grown.

Lusitania's sinking contributed to the USA's decision to enter WW1.

a source of unit intensity (one candela).

Lumet, Sidney (1924–), US film director trained in television and especially noted for the fine acting he achieved with his casts in films such as *Twelve Angry Men* (1957), *The Pawnbroker* (1965), *The Group* (1966), *Serpico* (1973), *Murder on the Orient Express* (1974), *Network* (1976) and *Equus* (1977) in which Richard BURTON gave a strong performance.

Lumière, name of two brothers, Auguste (1862–1954) and Louis (1864–1948), who were pioneers of cinematography. They invented an early motion-picture camera and a projector called the Cinématographe (1894); the modern word "cinema" is derived from this name. Their *Lunch Break at the Lumière Factory* (1895) is generally regarded as the first motion picture. Most of their films, of which there were so many, were largely about everyday French life. By 1895 they had made improvements in colour photography.

Luminescence. *See* PHOSPHORESCENCE.

Luminosity, astronomical ratio of a star's brightness to that of the Sun. It is a measure of a star's intrinsic brightness and is related to the absolute magnitude of the star. Luminosity = $2.512^{4.8-M}$, where 2.512 is the brightness ratio for a difference of one magnitude, 4.8 is the Sun's absolute magnitude and M is the star's.

Lumpenproletariat, term coined by MARX for the lowest strata of the working class including the chronically unemployed, tramps etc. Because of their weak economic position, Marx saw them as a potentially reactionary rather than a revolutionary force. *See also* PROLETARIAT.

Lumpsucker, or lumpfish, marine fish of the North Atlantic coasts, the pectoral fins of which join to form a strong, concave sucker, with which it attaches itself to rocks. Length: up to 61cm (2ft); weight 6kg (13lb). After the female spawns, the male takes over the job of guarding the eggs. Family Cyclopteridae.

Lumumba, Patrice Hemery (1925–61), Congolese statesman. He founded and led the Congolese National Movement and helped to form an indeaendent Republic of the Congo (now ZAIRE) in 1960. He served as Minister of Defence and Prime Minister and was arrested during a power struggle with President Kasavubu and Colonel Mobutu. His execution, at their orders, led to international protest.

Luna, Pedro de (*c.*1328–*c.*1423), antipope as BENEDICT XIII at Avignon (*r.*1394–1417). He was created cardinal by GREGORY XI in 1375, and supported the antipope CLEMENT VII. He was elected antipope upon Clement's death because he promised to end the schism with Rome; once elected, however, he refused to abdicate. He was deposed by the 1409 Council of Pisa and the 1417 Council of Constance.

Lunarcharskiy, Anatoly Vasilievich (1875–1933), Soviet politician, playwright and literary critic. He was an early member of the BOLSHEVIKS, and as Commissar for Education (1917–29), he established a system of progressive education.

Lunardi, Vincenzo (1759–1806), Italian balloonist. For much of his adult life he lived in London as the secretary to the Neapolitan ambassador. He made the first balloon flight in Britain in 1784 from Chelsea.

Lunar eclipse. *See* ECLIPSE.

Lunar module, part of the Saturn rocket-launched spacecraft, devised by NASA in the USA, and used in the APOLLO PROGRAMME that detached itself from the command/service module when in lunar orbit and landed, with its passengers, on the Moon's surface. This happened for the first time on 21 July 1969. The upper (ascent stage) of the lunar module later took off from the Moon using its thrust jet (with the descent stage acting as launch pad) and rejoined the command/service module. *See also* MM pp.158–159.

Lunar probe, unmanned, rocket-launched space vehicle aimed at either reaching the Moon's surface or going into orbit around it. The first were launched by the USSR (1959) and USA (1964) and in 1966 both nations landed probes to explore the Moon's surface. An early Russian probe was the first to send photographs back to the Earth of the dark side of the Moon.

Lunar roving vehicle (Lunar Rover), or Moon Buggy, an electrically-powered four-wheeled passenger vehicle first used by the astronauts in the APOLLO PROGRAMME in Aug. 1971, and subsequently in April and Dec. 1972, to explore local regions of the Moon and carry lunar rock samples back to the LUNAR MODULE for passage back to Earth.

Lund, Treaty of (1679), peace treaty signed between Sweden and Denmark. It marked the end of the war led by CHARLES XI of Sweden against the Danes, begun in 1675.

Lundy, island in the Bristol Channel off the coast of Devon, sw England. Settled in prehistoric times, it was a haven for smugglers and pirates from the Middle Ages until the 17th century. It is owned by the National Trust. Area: 5sq km (2sq miles). Pop. (1971) 32.

Lüneburg Heath, site of the official surrender of German field armies during WWII. On 3 May 1945 Field Marshal Keitel sent Admiral Friedeburg, General Kinzel and Rear-Admiral Wagner to Field Marshal Montgomery's headquarters to ask the British for terms of surrender. On 4 May, at 6pm, Keitel himself went to sign the document of unconditional surrender which Montgomery had drawn up.

Lunéville, Treaty of, peace terms between Austria and France, signed Feb. 1801. French predominance in Germany and Italy was confirmed.

Lungfish, elongate fish found in shallow fresh-water and swamps in Africa, South America and Australia. Lungfishes are descended from *Dipterus* which existed 300 million years ago, and resembled the CROSSOPTERYGIANS, from which the first AMPHIBIANS developed. It has primitive lungs, and during a dry season the various species can breath air or survive total dehydration by burrowing into the mud and enveloping themselves in a mucous cocoon. Order Dipnoi. *See also* NW pp.24, 124, *124*, 232, *233*.

Lungs, organs of the respiratory system in which the exchange of gases between air and blood takes place. They are located in the rib cage and are covered by a double-layer sheet of tissue called pleura. Between the layers is the fluid-containing pleural cavity that cushions the lungs and prevents friction. The lungs are filled with air sacs, called alveoli, which are one cell thick cavities containing networks of fine capillaries. Oxygen and carbon dioxide enter and leave the lungs via the blood-containing capillaries. *See also* MS pp.66–67, 88–89.

Lunokhod, name of two Russian scientific lunar surface craft. Luna 17, launched in October 1970, carried Lunokhod 1, which crawled over the surface of the Moon collecting scientific data. Lunokhod 2 was carried by Luna 21, launched in January 1973. *See also* SU p.180, *180*.

Lunts, The, US husband and wife acting partnership. Alfred (1892–1977) married actress Lynn Fontanne (*b.* London, *c.*.1882–) in 1922. From 1924 the couple starred together in many plays, gaining a reputation for sophisticated dramas and comedies, such as Noël COWARD's *Design for Living* (1933). Their last stage appearance together was in DURRENMATT's *The Visit* (1960).

Lupercalia, ancient Roman festival, held on 15 Feb. Priests called Luperci organized the festival, which began with the sacrifice of goats and a dog. Its origins are obscure, but its nature as a fertility ritual is clear. It was celebrated until AD 494, when it became, under Pope GELASIUS I, the Feast of the Purification.

Lupescu, Magda (1902–), mistress of the Romanian King CAROL II. Carol ceded his rights to the throne to his son Michael in 1925 in order to move to Paris with her. When Carol returned to Romania (1930–40), she exerted great influence on government policy. In 1947 she married Carol in exile and became known as Princess Elena.

Lupin, any of several species of annual and perennial plants of the pea family. They have star-shaped compound leaves and tall showy clusters of flowers. The pods, which contain bean-like seeds, are poisonous but can be treated for use as an animal feed. Height: to 2.4m (8ft). Family Leguminosae. *See also* PE p.*185*.

Luppis, Giovanni (1814–75), Austrian Navy officer who in 1861 conceived the idea of a self-propelled and self-detonating boat for use as a floating weapon against shipping. Robert WHITEHEAD developed the idea into the torpedo. *See also* MM p.*118*.

Lupus erythematosus, inflammatory disease of unknown cause involving the skin, or generalized to the connective tissue of the body (systemic form). It produces skin lesions which are red patches covered with scales, often on the cheeks and nose and forming a butterfly pattern. The systemic form varies in severity, is four times more common in women than in men, and may involve one or more organs in addition to the skin. Symptoms depend on the organ involved, but arthritis, weight loss, fatigue, fever and anaemia are common.

Lurçat, Jean (1892–1966), French painter and important tapestry designer, who revived the art of tapestry in France and was associated with the AUBUSSON works. He designed more than 1,000 highly stylized, brightly coloured tapestries.

Luria, Alexander Romanovich (1902–77), Soviet psychologist, important contributor to neuropsychology and the analysis of neurological factors in language disorders. In 1932 he suggested that involuntary finger movements reflect emotional aspects of lying better than does word association.

Luria, Isaac ben Solomon (1534–72), KABBALIST, or Jewish mystic, known as "The Lion". As a young man he chose a secluded, island life in Egypt, and engaged in esoteric studies. He developed his own system of mysticism, and formed an academy among his disciples.

Luría, Salvador Edward (1912–), US biologist, *b.* Italy. He shared the 1969 Nobel Prize in physiology and medicine with Max DELBRÜCK and Alfred HERSHEY for contributing to the knowledge of the growth, reproduction and mutation of bacterial viruses.

Lurie, Alison (1926–), US novelist. Her works explore the social *mores* of contemporary middle-class America and are particularly noted for their dialogue and strength of characterization. They include *Love and Friendship* (1962), *Imaginary Friends* (1967) *The War Between the Tates* (1974), and *Nowhere City* (1975).

Luristan (Lorestan), governorate in w Iran, in the region of the Zagros Mts. The capital is Khorramabad. It is noted for the Luristan bronzes, metal works found in the 1930s, comprising axes, swords, bridles and ritual objects, dating from the eighth century BC. Sheep are reared in the region today, and it is rich in oil deposits. Area: 31,380sq km (12,116sq miles). Pop. (1971 est.) 929,200.

Lusaka, capital of Zambia, in the s central part of the country at an altitude of 1,280m (4,200ft). Founded by Europeans in 1905, it replaced Livingstone as the capital of Zambia (formerly Northern Rhodesia) in 1935. A road and rail junction, Lusaka is an agricultural trade centre as well as a major financial and commercial city. It has the University of Zambia (1965). Industries: shoe manufacture, cement, food processing, brewing. Pop. (1972) 448,000.

Lüshun. *See* PORT ARTHUR.

Lusitania, Roman province on the Iberian peninsula, comprising modern Portugal and parts of western Spain. It took its name from the Lusitani, a warlike tribe that bitterly fought Roman conquest of their lands. Their great leader was Viriatus, who showed great military and diplomatic talents in opposing the Romans until he was assassinated, probably with Roman collusion, in 139 BC. Traditionally, the Portuguese have looked upon themselves as descendants of the Lusitani.

Lusitania, R.M.S., British liner (32,000 tons) of the Cunard line, torpedoed off

Ireland by the German submarine U20 on 7 May 1915. The ship sank in 45 minutes and 1,198 of the 1.959 passengers and crew perished, including 124 Americans.

Lustreware, pottery finished with a glaze containing silver and copper to make it appear iridescent. It became popular in Islamic countries in the 9th century. It was made in Britain by SPODE and WEDGEWEDGWOOD in the 19th century.

Lüta (Lüda) autonomous municipality in S Liaoning province, NE China, on the Liaotung peninsula; including the two cities of Lüshun (Port Arthur) and Ta-lien (Dairen), the principal commercial port of Manchuria. The city was developed in the late 19th century by the Russians as the s terminus of the South Manchurian railway. Industries: shipbuilding, fishing, oil refining, chemicals, rolling-stock and electrical equipment. Pop. (1970 est.) 4,000,000.

Lute, plucked stringed instrument most popular in 16th- and 17th-century Europe. It has an almond-shaped body, fretted neck and originally 11 gut string It was often played to accompany songs and stylized dances and has been revived in recent years as a concert instrument. Notable composers of lute music include Denys Gaultier (c.1600–72) and John DOWLAND. Julian BREAM is prominent among 20th-century lutenists.

Luteinizing hormone. See LH.

Lutetium, metallic element (symbol Lu) of the LANTHANIDE SERIES, first isolated in 1906 together with YTTERBIUM. Its chief ore is monazite (a phosphate). The element has no commercial uses. Properties: at.no. 71; at.wt. 174.97; s.g. 9.842 (25°C); m.p. 1,656°C (3,013°F); b.p. 3,315°C (5,999°F); most common isotope Lu^{175} (97.41%). See also SU pp.*134–135.*

Luther, Martin (1483–1546), leader of the German REFORMATION. He studied philosophy at Erfurt University and became an AUGUSTINIAN monk in 1505. Two years later he was ordained and became a teacher of philosophy and theology at Wittenberg University. He was deeply concerned about the problem of salvation, finally deciding that it could not be attained by good works but was a free gift of God's grace. Luther's beliefs made him object to many practices of the medieval Church, particularly the sale of INDULGENCES. In 1517 he affixed his 95 Theses to the door of the Schlosskirche in Wittenberg. This led to a quarrel between Luther and Church leaders, including the pope. Luther decided that the Bible was the true source of authority and renounced obedience to Rome. He maintained his stand in debates with Johann ECK and at the Diet of WORMS in 1521. As a result, he was excommunicated, but strong German princes came to his rescue, and he gained followers among churchmen as well as the laity. Luther married a former nun, Katherina von Bora, in 1525. See also LUTHERANISM; HC1 pp.263–269.

Lutheranism, the doctrines and the Church structure that grew out of the teaching of Martin LUTHER. The principal Lutheran doctrine is that of justification by faith alone (*sola fide*). Luther held that grace cannot be conferred by the Church but is the free gift of God's love. He objected to the Catholic doctrine of TRANSUBSTANTIATION – that, in the EUCHARIST, the substance of the bread and wine are transformed into the body and blood of Christ. Instead, Luther believed in the real presence of Christ "in, with, and under" the bread and wine (CONSUBSTANTIATION). These and other essentials of Lutheran doctrine were set down by Philipp MELANCHTHON in 1530 in the AUGSBURG CONFESSION, which has been the basic document of the Lutherans ever since. The Lutheran Church has no central governing body, and Lutheran Churches in each country have developed their own traditions. In 1947 the Lutheran World Federation was formed as a co-ordinating body. See also HC1 pp.268–269.

Luthuli, Albert John Mvumbi (c.1898–1967), South African civil-rights leader. A Zulu chief and President of the African National Congress between 1952 and 1960, he was often harassed by the government for his non-violent protests against APARTHEID. In recognition of this, he received the Nobel Peace Prize in 1960. Although forced to live and work in secret that same year, he continued pursuing his aims until his death. He wrote *Let My People Go* in 1962.

Lutine Bell, salvaged bell from the British frigate *Lutine*, which was wrecked off the Zuider Zee (The Netherlands) in 1799 while carrying gold bullion and pay for British troops in The Netherlands. The bell was salvaged in 1859 and hung in the insurance office of LLOYD's in London. It is traditionally rung before ships are announced as overdue or lost.

Luton, county district in SE BEDFORDSHIRE, England; created in 1974 under the Local Government Act (1972). Area: 43sq km (17sq miles). Pop. (1974 est.) 165,900.

Lutoslawski, Witold (1913–), Polish composer. He gained international recognition with his *Concerto for Orchestra* (1954). His other works include *Funeral Music* (1958) and *Venetian Games* (1961).

Luttrell Psalter (c.1340), illuminated psalter made in England for Sir Geoffrey Luttrell. Its miniature representations of the daily life of the times provide important documentation for social historians.

Lutyens, Agnes Elisabeth (1906–), British composer, who composed mainly in the 12-tone system. Her works include various symphonies (for piano, woodwind, strings and percussion, all 1961) as well as *Quincunx for Orchestra* (1959–60), *Rape of the Moone* (1973) for wind octet, and vocal works, such as *The Linnet from the Leaf* (1972) and *Tyme Doth Flete* (1968), and the operas *The Numbered* and *Isis and Osiris* (both 1973). She has also composed numerous film and radio scores.

Lutyens, Sir Edwin Landseer (1869–1944), British architect who planned the imperial capital of New Delhi (1913–30) and designed the Viceroy's house, now the President's residence. After WWI he became architect to the Imperial War Graves Commission and designed the CENOTAPH (1922) in Whitehall, London.

Lützen, Battle of (16 Nov. 1632), engagement between the Swedish army of Gustavus II Adolphus and the forces of the Holy Roman Emperor, under Wallenstein. The outcome was a Swedish victory, but Gustavus was accidentally shot dead by his own men on the misty battlefield.

Lux, unit of illumination in the SI system of units, equal to one LUMEN per sq metre.

Luxembourg, independent grand duchy in w Europe. Geographically it is divided into the forested Ardennes plateau and the fertile Bon Pays in the s. There are rich deposits of iron ore and Luxembourg is a major producer of iron and steel. Other industries include chemicals, textiles, tourism and banking. The capital is Luxembourg. Area: 2,586sq km (998sq miles). Pop. (1975 est.) 357,000. See MW p.120.

Luxembourg, Palais du, RENAISSANCE palace in Paris built for Marie de' MEDICI by Salomon de Brosse in 1615–24. The interior was decorated by POUSSIN, Philippe de CHAMPAIGNE and RUBENS. It now houses a large collection of paintings, especially by DELACROIX and contemporary artists.

Luxemburg, Rosa (1871–1919), German socialist leader, b. Poland. She became a German citizen through marriage and after 1898 was a leader of the Social Democratic Party. She founded the SPARTACIST (later Communist) Party during WWI with Karl LIEBKNECHT, and was arrested for her part in the Spartacist uprising in Berlin in Jan. 1919. She was murdered while being taken to prison. See also HC2 pp.*212,* 213.

Luxor, city in Upper Egypt on the River Nile. It is on the site of the ancient cities of Karnak and Thebes; the temple there was built by AMENHOTEP III in the 14th century BC, and beautified by RAMESSES II in the 12th century BC. The greatest surviving monument of ancient Egypt, the temple is 260m (853ft) long. The temple was later made into a Christian church and then a Muslim shrine; it was restored after 1885.

Pop. (1970 est.) 84,600. See also HC1 pp.48, *49.*

Luzerne. See LUCERNE.

Luzon, largest island of the Philippines, occupying the N part of the group. The island has many natural harbours, and the principal land mass to the N has three mountain ranges, all running N to S. As well as being the most populous of the islands, it is the chief producer of agricultural and industrial products including rice, coconuts, sugar, coffee, chemicals, gold, copper and chromite. Area: 108,171sq km (41,765sq miles). Pop. (1970) 16,669,724.

Lvov, Prince Georgi Yevgenevich (1861–1925), Premier of the Russian provisional government (March-July 1917). A Constitutional Democrat, he was authorized by the DUMA to form a new government upon Czar NICHOLAS II's abdication. After concessions to the Left he lost support and resigned in 1918 in favour of KERENSKY.

Lvov, commercial city of the Russian Republic (Rossijskaja SFSR) 185km (115 miles) SW of Luck, between the Roztoche and Gologory Mountains, on a tributary of the River Bug. Founded in 1256 by a Ukrainian prince, it was captured by Poland in 1340 and chartered in 1352; it became the capital of the independent Ukrainian Republic in 1918, reverted to Poland in the same year and was taken by the USSR in 1944. An industrial and commercial centre, its industries include heavy machinery, chemicals and oil refining. Pop. (1975 est.) 617,000.

Lwoff, André Michel (1902–), French micro-biologist. He shared the 1965 Nobel Prize in physiology and medicine with François JACOB and Jacques MONOD for work on the genetics of bacterial viruses. Lwoff studied the means by which these destroy the bacteria they infect.

Lycanthropy. See WEREWOLF.

Lyceum, school in Athens established by ARISTOTLE in a grove sacred to Apollo Lyceius in 335 BC. There Aristotle walked as he lectured, so that his students became known as the Peripatetics. After the death of Aristotle in 322 the school declined in importance.

Lyceum, theatre in Wellington Street, London (off the Strand). It opened in 1834 as the Royal Lyceum and English Opera House and was managed, from 1871–1902, by Sir Henry IRVING, at which time it was the most important theatre in London.

Lychee, small, round, reddish fruit of an evergreen tree of the soapberry (Sapindaceae) family, known in s China from ancient times. A papery but tough outer skin encloses the white flesh of the lychee, which is somewhat grape-like in flavour, and has one large seed. Species *Litchi chinensis.*

Lycia, region in SW Asia Minor inhabited in ancient times by the Lycians, who are thought to have migrated there from Crete via Miletus. They fought with PRIAM at Troy and are mentioned in the *Iliad*. Defeated by the Persians (546 BC) and then by ALEXANDER THE GREAT, they were subject to the PTOLEMIES until granted to the rulers of Rhodes by the Romans (189 BC). They were freed from Rhodian rule in 169 BC.

Lycurgus (*fl. c.*625 BC), semi-mythical lawgiver of Sparta. The existence of Lycurgus cannot be verified absolutely, but many scholars believe that such a single individual may have instituted reforms in Sparta after a slave revolt in the second half of the 7th century BC.

Lydda. See LOD.

Lydgate, John (c.1370–1452), English poet. He was a BENEDICTINE monk at Bury St Edmunds. His vast output of poetry ranged from long didactic works, such as *The Fall of Princes,* to brief lyrics.

Lydia, territory in w Asia Minor. Under the Mermnad dynasty (c.700–550 BC), it was a powerful kingdom until it fell to the Persians and became their stronghold in the West. It was later taken by ALEXANDER THE GREAT.

Lyell, Sir Charles (1797–1875), British geologist. He was influential in shaping 19th-century ideas about science and

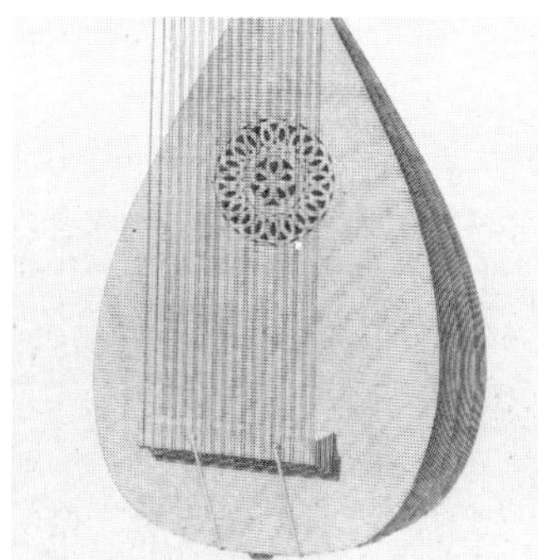

Lutes often feature a carved "rose" in the round sound-hole of the instrument.

Martin Luther was an emotional man with an acute and terrifying sense of sin.

Rosa Luxemburg was the first woman to achieve political influence in Germany.

Luxor has these papyrus-flowered columns at the temple of Amon (12th century BC).

Lyly, John

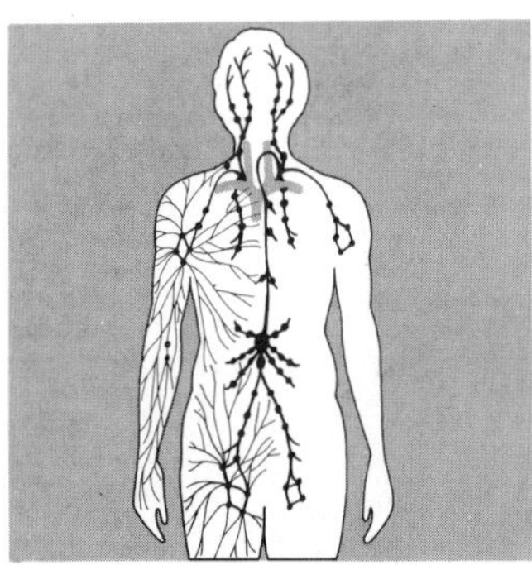

Lymph; the lymphatic system links nodes which make disease-fighting substances.

Vera Lynn holds up the music of *We'll Meet Again,* one of her great wartime songs.

Lynxes eat mainly game and small wild animals but may also raid farms.

Lyrebird; tail-feather raising and singing is used for defence and to attract mates.

wrote the popular three-volume *Principles of Geology* (1830–33), *Elements of Geology* (1838) and *The Geological Evidence of the Antiquity of Man* (1863). His ideas are based on the UNIFORMI-TARIAN theory. *See also* HC1 p.22; HC2 pp.32, *33.*

Lyly, John (*c.*1554–1606), English poet, dramatist and writer of prose romance. Comedies and pastoral romances include *Sappho and Phao* (1584) and *Midas* (1589). He is best known for his elaborate prose style evolved in the romance *Euphues* (1578, 1580).

Lympany, Moura (1916–), British pianist, who studied at the Royal Academy of Music, London and in Vienna and made her début at the age of 12. She toured the British Commonwealth many times as a concert artist and is known through recitals and recordings for her interpretations of the music of Schumann, Grieg, and Rachmaninov.

Lymph, clear, slightly yellowish fluid derived from the blood and containing the white blood cells that combat infection and provide immunity. Swelling of lymph nodes, particularly those in the groin or armpit, may be a symptom of a disease. *See also* MS pp.63, 88, 114, *115.*

Lymph glands, another name for the nodes of the lymphatic system. They are filters and reservoirs which collect harmful material, notably bacteria and metastatic cancer cells, and so many become enlarged when the body is infected. Collections of lymph nodes are found in the throat and neck, in the armpits, and in the groin. *See also* MS pp.63, 88, 114, *115.*

Lymphocyte, type of white blood cell found in vertebrates. Made continuously in lymphoid tissue, lymphocytes are roughly spherical in shape, contain few granules in their cytoplasm, and move using pseudopodia. In human beings they form about 25% of white blood cells and play an important role in combating disease by producing antibodies. *See also* MS pp.*62, 114–115.*

Lymphoma, abnormal growth (neoplasm) in lymphatic tissue, usually malignant. HODGKIN'S DISEASE is a lymphoma.

Lymphosarcoma, malignant tumour of lymphoid tissue. The term does not include HODGKIN'S DISEASE.

Lynch, Benny (1913–46), Scottish boxer who won the world, European and British flyweight titles in 1935, beating Jackie Brown in two rounds, but forfeited them in 1938 for failing to make the required weight.

Lynch, John ("Jack") (1917–), Irish politician, member of the Fianna Fáil Party. He entered the Dáil (Parliament) in 1948, was Minister of Education (1957–59) and for Industry (1959–65). He became Prime Minister of Ireland in 1966, and was re-elected in 1969, 1973 and 1977. *See also* HC2 p.*295.*

Lynd, Robert, (1879–1949), Irish essayist who established himself as a journalist in London. At his urbane best in such pieces as *The Art of Letters* (1920) and *Dr Johnson and Company* (1928), his essays were collected under the title *Things One Hears* in 1945.

Lyndhurst, John Singleton Copley, Baron (1772–1863), British jurist, the son of American painter John Singleton COPLEY. Originally famed as a defender of radicals, he conducted the prosecution of Queen Caroline in 1820. He was three times Lord Chancellor (1827–30, 1834–35 and 1841–46).

Lynen, Feodor Felix Konrad (1911–), German biochemist who shared the 1964 Nobel Prize in physiology and medicine with Konrad BLOCH for research into the metabolism of FATTY ACIDS and CHOLESTE-ROL. He helped to ascertain the mechanism of fatty acid break down and explain the pathway of the biosynthesis of several compounds, including natural rubber and cholesterol. He was also the first to isolate acetyl coenzyme A, the precursor of both fatty acids and cholesterol.

Lynn, Dame Vera (1917–), British popular singer. She gave her first public performance in 1924. She travelled extensively to sing to the troops during WWII and was nicknamed, "the Forces Sweetheart".

Her most famous song was *The White Cliffs of Dover.* She was made a DBE in 1975.

Lynx, any of several small CATS found in forests of central and N Europe, along the French-Spanish border, and in the USA. It may be yellow-grey or reddish-brown and can be spotted or unspotted. It has long legs, large feet, tufted ears, and characteristic beard-like hair on its cheeks. Length: to 116cm (46in). Family Felidae.

Lyon. *See* LYONS.

Lyonnesse, in English mythology, a lost land the name of which first appeared in Sir Thomas MALORY's *Morte d'Arthur* in the late 15th century. It was believed to have connected Cornwall with the SCILLY ISLES. The name is now woven into Arthurian legend and Cornish folklore.

Lyons, Eric (1912–), British architect and industrial designer. He worked with GROPIUS before WWII. After the war he worked chiefly on housing projects, such as the World's End Chelsea Redevelopment Scheme of the 1970s. He has won 11 Ministry of Housing awards for good design.

Lyons, Sir Joseph (1848–1917), British businessman. In 1887, he was asked to help the tobacco firm Glubotien and Salmon to set up a catering enterprise; this later became J. Lyons and Co. Ltd, one of the largest catering businesses in Britain. He was knighted in 1911.

Lyons, Joseph Aloysius (1879–1939), Australian politician who helped to form the United Australia Party in 1931. He served as Postmaster General and Minister for Public Works and Railways, and as Prime Minister (1931–39) encouraged economic reform and development of the armed forces.

Lyons (Lyon), city in E central France, at the confluence of the Rhône and Saône rivers. It became part of the French crown lands in 1312, and had developed as a trade centre on the route to Italy by the 16th century, but was destroyed in 1793 by French Revolutionary troops. Lyons was the centre of the French resistance movement in WWII. Industries include chemicals, textiles and metal working. Pop. (1971 est.) 523,900.

Lyons, Councils of, two ECUMENICAL COUNCILS held in Lyons. The first was convoked by Pope INNOCENT IV in 1245. It formally deposed the Holy Roman Emperor FREDERICK II, who had followed an anti-ecclesiastical policy. The second was convoked by Pope Gregory X in 1274. Its principal achievement, the reunion of the Eastern and Western Churches (*see* GREAT SCHISM) proved short-lived. The two Churches separated again in 1285.

Lyot, Bernard (1897–1952), French astronomer who invented the solar CORONAGRAPH in 1931. With the instrument he succeeded in photographing the inner CORONA and its spectrum in full sunlight. *See also* SU p.225.

Lyra, or the lyre, small northern constellation between Hercules and Cygnus. It contains the 1st magnitude Vega, and the PLANETARY NEBULA M57. *See also* SU pp.236–237, *237,* 259, *259.*

Lyra, or Lira, name of two musical instruments dating from medieval times. One is a type of small HARMONIUM suitable for placing on a table or even in the lap; one hand played on the keyboard while the other cranked a wheel to operate the internal bellows. HAYDN wrote five concertos for this instrument. The other is a stringed instrument similar to a REBEC and played like a VIOL; it survives, for example, in Bulgaria where it is called a *gadulka,* and in Yugoslavia where it is called a *gusla.*

Lyre, ancient stringed musical instrument. Used originally by the Sumerians, it was introduced into Egypt and Assyria in the 2nd millenium BC. HOMER mentioned the instrument in his writings and in Classical Greek times it usually had seven strings supported by a wooden frame and attached to a sound box at the base. Most common types at this time were the lyra and kithara, and the strings were plucked using a bulky plectrum. Lyres which have survived in Europe since the Middle Ages

have more commonly been played with a bow. Today the lyre exists in various forms in E Africa and in Ethiopia.

Lyrebird, either of two shy Australian songbirds; the Superb lyrebird (*Menura superba*) and Albert's lyrebird (*M. alberti*). These large, perching birds have lyre-shaped tails displayed during courtship performances. The female builds a large, dome-shaped nest with a side entrance for the single greyish-purple egg.

Lyrical Ballads, volume of poetry by William WORDSWORTH and Samuel Taylor COLERIDGE. Coleridge contributed three poems to the first edition, including *The Ancient Mariner.* The second edition contained five of his poems. Wordsworth composed the majority of poems, including *Tintern Abbey* and *Michael.* The first edition was published in 1778, the second, with a preface by Wordsworth, in 1800.

Lysander (*d.*395 BC), Spartan general and politician. He was instrumental in the defeat of Athens during the PELOPONNE-SIAN WAR (429 BC–404 BC), defeating the Athenian fleet in 406 BC and 404 BC and obtaining Persian support for Sparta. His policy of establishing an oligarchy rather than a democracy in Athens was a failure.

Lysenko, Triofim Denisovich (1898–1977), Russian agronomist and geneticist. He expanded the theory of LAMARCK to include his own ideas of plant genetics. Lysenko promised the USSR government vast increases in crop yields through the application of his theories and enjoyed official sanction under Stalin, but in the late 1950s his influence waned.

Lysergic acid diethylamide. *See* LSD. in 404 when the Thirty Tyrants assumed power, but returned in

Lysias, (*c.*450–*c.*380 BC), Athenian orator who fled Athens in 404 when the thirty tyrants assumed power, but returned in 403 when democracy was restored. His style was simple yet eloquent. *See also* HC1 p. 75.

Lysippos (*fl.*late 4th century BC), official Greek sculptor of Alexander the Great. His work was masterly in its proportion and composition, and his style influenced much later Hellenistic sculpture. One of his best-known works is *Apoxyomenos* (Youth Scraping Himself) (*c.*310 BC) of which a Roman marble copy is in the Vatican Museum, Rome. *See also* HC1 pp.78–79, *79.*

Lysistrata, Greek comedy by ARISTO-PHANES, first produced at Athens in 411 BC. In it Lysistrata, an Athenian woman, brings the Second PELOPONNESIAN WAR to an end by persuading all Greek women to deny their husbands sexual relations for as long as the war continues.

Lyte, Henry Francis (1793–1847), British Anglican clergyman and writer of hymns, including *Praise my soul, the King of Heaven* and *Abide with Me.*

Lyttelton, leading port of New Zealand's South Island. Located on the E coast, it is the main port for CHRISTCHURCH. Pop. (1972 est.) 3,230.

Lytton, Edward George Earle Lytton Bulwer, 1st Baron Lytton of Knebworth (1803–73), British novelist and politician. He was a Liberal MP (1831–41), and then an MP for the Conservative Party (1852–66) until he became a peer. His most famous novel is *The Last Days of Pompeii* (1834).

M

M, 13th letter of the alphabet, derived from the Semitic letter *mem* (meaning *water*). The corresponding Greek letter was *mu,* which went via the Etruscan alphabet to Latin as *m.* In English it always has the same pronunciation, but has the effect of silencing a following *b* that ends a word (as in *numb*). *See also* MS pp.244–245.

Maar, coneless volcanic crater that was formed by a single explosive eruption not accompanied by a flow of lava. A crater ring surrounds the hole.

Maas, Dutch and Flemish name given to

the River Meuse as it flows through Belgium and SE Netherlands and enters the North Sea through the Rhine delta s of Rotterdam. A branch of the river joins the River Waal to form the Merwede River. **Maastricht,** city in SE Netherlands, near the Belgian border, on the River Meuse. The city was frequently besieged by the Spanish in the 14th–18th centuries, and occupied by the Germans during WWII. Industries: steel, glass, cement, grain milling. Pop. (1970 est.) 99,000.

Maazel, Lorin (1930–), US conductor, b. France. He toured worldwide and conducted extensively in the US and Europe before becoming chief conductor of the Berlin Radio Symphony Orchestra in 1965 and then music director of the Cleveland Orchestra in 1971.

Mab, considered to be Queen of the Fairies by most poets of the 16th and 17th centuries. Shakespeare's Mab, mentioned in *Romeo and Juliet*, is thought to be based on Mabb of Wales and Queen Maeve of Ireland.

Mabillon, Jean (1623–1707), French literary scholar who devised the first way of proving whether or not documents are authentic. A Benedictine monk, he described his method in *De re diplomatica* (1681; 1704).

Mabinogion, The, collection of early medieval Welsh tales containing much ancient mythological material and representing the result of centuries of folklore and oral story-telling. Their form is that of the hero-tale, in which the stories deal with the life and adventures of the hero, from birth to death.

Mabuse, (*c.*1478–*c.*1536), also called Jan Gossaert, Flemish painter. In his early work he was influenced by Gerard DAVID and Albrecht DÜRER, as may be seen by his magnificent *Adoration of the Magi* (*c.*1508). In 1508 he accompanied Philip of Burgundy to Rome, and became one of the leading "Italianate" Flemings, although mannered in his efforts to adopt an Italian outlook.

McAdam, John Loudon (1756–1836), Scottish engineer who invented the macadam road surface. He proposed that roads should be raised above the surrounding ground, with a base of large stones covered with smaller stones and bound together with fine gravel. He applied these ideas when he was appointed surveyor general of Bristol's roads in 1815. His views on road-making were adopted after a Parliamentary Inquiry in 1823 and in 1827 he was made surveyor general of roads in Britain. *See also* MM pp.182, *183*.

Macadamia, genus of Australian trees of the family Proteaceae. Most species have stiff, oblong, lance-like leaves, although the leaves of *Macadamia ternifolia* have serrated margins and those of *M. integrifolia* have smooth margins. The edible seeds are round, hard-shelled nuts, covered by thick husks that split when ripe. Height: to 18m (60ft).

MacAlpin, Kenneth (*d.*858), King of Scotland. He was ruler of Galloway from AD 834 and had extended his rule to most of central Scotland by 844. He then gradually subdued much of the country N of the Forth and Clyde estuaries, and so finally broke the power of the PICTS.

Macao (Macau), Portuguese overseas province in SE China, 64km (40 miles) W of HONG KONG. The colony consists of Macao Peninsula and two islands, Taipa and Colôane, in the South China Sea. The city of Macao, co-extensive with the Peninsula, was founded by the Portuguese in 1557. During the 18th and 19th centuries it enjoyed great prosperity as one of the two Chinese ports (with Canton) open to foreign trade. It was not recognized by China as Portuguese territory until 1887. Its industries include textiles and tourism. Pop. (1970 est.) 314,000.

Macapá, city in N Brazil on the River Amazon, capital of Amapá territory. It is a centre for products from the surrounding tropical forests, but its chief source of income is from mining of iron and manganese ores. Its growth dates from 1944, when it was made the capital. Pop. (1970) 86,307.

Macaque, diverse group of omnivorous medium-sized to large Old World MONKEYS found from NW Africa to Japan and Korea. Most are yellowish brown and are forest dwellers – agile on the ground and in trees – and good swimmers. Some have long tails, some have short tails and some have no visible tails at all. As a result, the latter are often confused with apes. Weight: to 13kg (29lb). Genus *Macaca*. *See also* BARBARY APE; RHESUS; NW pp.*169, 214,* 215, 234.

MacArthur, Douglas (1880–1964), US general, who participated in all the important US offensives during WWI. He retired from the army (1937) to work in the Philippines but was recalled to active duty when the USA entered WWII and appointed Commander of the US army forces in the Pacific. From Australia he directed the assault that led to Japanese defeat. MacArthur accepted the Japanese surrender on the USS *Missouri* on 2 Sept. 1945. He was appointed commander of UN troops during the KOREAN WAR but, following a policy dispute with President TRUMAN that MacArthur made public, he was relieved of his command. *See also* HC2 p.*215.*

Macarthur, John (1767–1834), Australian free-settler who arrived in New South Wales from England in 1790, famous in colonial history as the pioneer of the wool industry and an influential public figure who came seriously into dispute with William BLIGH during the RUM REBELLION. In the 1790s he began to import Merino sheep from southern Africa and subsequently (1805) was granted 2,024 hectares (5,000 acres) of land (Camden Park) near Sydney by George III.

Macarthur, Mary Reid Anderson (1880–1921), British labour organizer. In 1903 she was appointed General Secretary to the Women's Trade Union League. She formed the National Federation of Women Workers in 1906 and represented Britain at the first League of Nations Labour conference in 1920.

Macassar. *See* MAKASAR.

Macau. *See* MACAO.

Macaulay, Dame Rose (1881–1958), British novelist, poet and critic. She is remembered principally for her novels satirizing middle-class life and her concern with the role of women in a changing society. Her works include *Potterism* (1920), *Told by an Idiot* (1923) and *The Towers of Trebizond* (1956).

Macaulay, Thomas Babington (1800–1859), English historian, poet and statesman. After serving as a Liberal in the House of Commons, he was appointed a member of the Supreme Council of India where he lived from 1834–38. He returned to Parliament (1839–1847) and in 1847 began to write *The History of England from the Accession of James the Second* (5 vols, 1849–1861), which enjoyed great success. His poetical work, *Lays of Ancient Rome* (1842), celebrated the events of Roman history.

McAuley, James Philip (1917–76), Australian poet and scholar, professor of English at the University of Tasmania (1961–76). He founded the literary journal *Quadrant* in 1956 and his collected verse was published in 1971.

Macaw, any of several species of tropical American harsh-voiced parrots. All have sword-shaped tails and large powerful bills. Most species are brightly coloured, such as the scarlet macaw (*Ara macao*), which has a red tail and yellow wings with bright blue on its back and wings. Order Psittacidae. *See also* NW pp.*141, 216–217.*

McBain, Ed (1926–), name under which US novelist Evan Hunter has written more than 30 crime novels. Featuring a group of detectives, operating from the 87th precinct of an anonymous US city, they are noted for their meticulous attention to police routine and methods of investigation. They include *Cop Hater* (1956), *Fuzz* (1968), *Bread* (1974) and *So Long As You Both Shall Live* (1976).

Macbeth (*d.*1057), Scottish king (*r.*1040–57), subject of Shakespeare's tragedy named after him. In 1040 he killed DUNCAN I in battle and seized his throne but in 1054 was defeated by Siward, who regained southern Scotland on behalf of Duncan's son, Malcolm Canmore. Malcolm (MALCOLM III Canmore) regained the rest of his kingdom by defeating and killing Macbeth in battle at Lumphanan. *See also* HC1 p.182.

Macbeth (*c.*1606), five-act tragedy by William SHAKESPEARE, based on Raphael Holinshed's *Chronicles*. Most of the facts, as they appeared in the Chronicles, were rejected by Shakespeare who constructed his play on the theme of unbridled ambition and its tragic consequences.

MacBeth, George (1932–), Scottish poet. He often mixes violent subjects with fantasy and reality, as in *A Form of Words* (1954) and later volumes, such as *The Broken Places* (1963), *The Colour of Blood* (1967) and *In the Hours Waiting for the Blood to Come* (1975).

MacBride, Seán (1904–), Irish political figure, the son of the Irish patriots Maud Gonne and John MacBride. He served in the Irish Republican Army (IRA) and founded the Clann na Poblachta (Republican Political Party) in 1946. He was a Member of the Irish Parliament (1947–58), and represented Ireland at the COUNCIL OF EUROPE (1954–63). He served as UN commissioner for Namibia (1973–76). Active in the cause of international peace, he was awarded the Nobel Peace Prize in 1974.

McBride, William John ("Willie John") (1940–), British rugby union second-row forward who played for Ballymena, Ireland, and the British Lions. Between 1962 and 1975 he won a record 63 caps for Ireland and also took part in five tours by the British Lions, also a record number. He was captain of the unbeaten 1974 British Lions side which defeated South Africa in a series for the first time in 70 years.

McCabe, Stanley Joseph (1910–68), Australian cricketer, a batsman who played for New South Wales and Australia. He scored 2,748 runs in an average of 48.21 in 39 Test matches between 1930 and 1938.

Maccabees, prominent Jewish family that ruled JUDAEA from 164–63 BC. In 168 BC the SELEUCID ruler Antiochus IV invaded JERUSALEM and in the following year rededicated the TEMPLE to ZEUS and eradicated Jewish religious practices. Mattathias, a high priest, and his five sons fled to the mountains and organized a rebellion. After Mattathias' death in 166 BC his son Judas took command and recaptured Jerusalem in 164 BC. This event is celebrated annually as HANUKKAH, the Festival of Lights. Judas Maccabees was succeeded by his brother Jonathan (*r.*160–143). The family controlled Judaea until 63 BC, when POMPEY conquered it.

Maccabees, the Books of, four historical books, two of which are in the APOCRYPHA of the Authorized Version of the Bible. These two are modelled on the Old Testament books of CHRONICLES, and are therefore fairly accurate historically although prejudiced. The first (1 Maccabees) is concerned to praise the leaders of Judaism and the Temple opposing the Hellenizing influences of the times. Written in Hebrew about 100 BC, it describes the lives and careers of the three sons of Mattathias (Judas Maccabaeus, Jonathan and Simon), all of whom became leaders and rulers before violent deaths. The second (2 Maccabees) covers much the same ground, but from a slightly earlier beginning and a much more Hellenistic viewpoint; it is addressed to the Jews at Alexandria and is a sermon pleading for unity in the established traditions of the Temple, couched in literary and eloquent style. The other two books of Maccabees are PSEUDEPIGRAPHA.

MacCaig, Norman (1910–), Scottish poet. His poetry is metaphysical, romantic and unmistakably "Scottish", although written in English. His publications include *The Sinai Sort* (1957), *Measures* (1965) and *Surroundings* (1967).

McCarthy, Joseph Raymond (1908–57), US senator who gained national attention in 1950 by claiming that the US State Department had been infiltrated by Com-

General MacArthur, US commander-in-chief, surveying battlefields in Korea in 1950.

Macaws have a characteristic piercing screech; they rarely mimic the human voice.

Macbeth listening to the second prophetic apparition conjured up by the witches.

Joseph McCarthy liked to wear uniform: this is that of the US Marine Corps.

Paul McCartney began his musical career in the late 1950s in a Liverpool "beat" group.

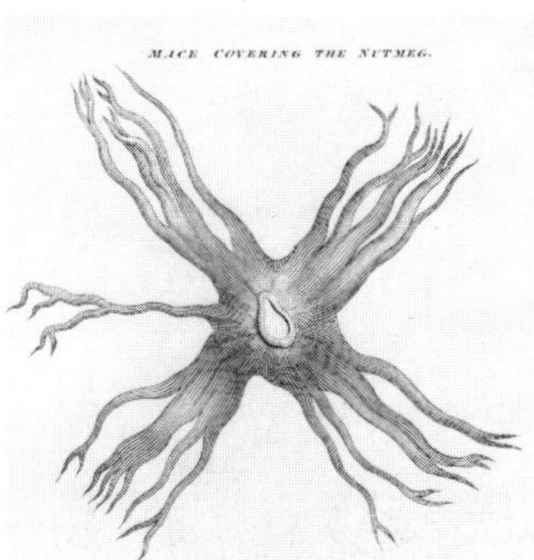

Mace, when fresh, smells like nutmeg; it is sun-dried before being sold.

Macedonia, once occupied by Thracians, became Greek in the reign of Philip II.

Macgillycuddy's Reeks is a well-known scenic tourist area in Eire.

munists. He was appointed chairman of the Senate's permanent Subcommittee on Investigations in which office he wielded great power. *See also* MCCARTHYISM.

McCarthy, Mary (1912–), US writer and drama critic. She wrote several novels including *A Charmed Life* (1955) and *The Group* (1963), her most popular novel. Among her non-fiction is *Venice Observed* (1956) and *Memories of a Catholic Girlhood* (1957).

McCarthyism, era of a Communist "witch-hunt" in the USA, roughly from 1950 to 1954. It was a symptom of the COLD WAR and began with the accusation by the REPUBLICAN Senator, Joseph MCCARTHY, in February 1950 that he had the names of 205 Communists working in the STATE DEPARTMENT. He later made a similar accusation against the army. The army hearings of March-April 1954, at which McCarthy was the chief witness, were nationally televised. They destroyed McCarthy's reputation and he was condemned by the senate in December. In the meantime, however, he had ruined many careers (most notably that of J. Robert OPPENHEIMER) and the Communist Party had been outlawed.

McCartney, Paul (1942–), British pop singer. He was a member of the BEATLES for whom, with John LENNON, he wrote many songs. He formed his own group, Wings, in 1971. Their record, *Band on the Run*, won two Grammy awards in 1975.

McClung, Clarence Erwin (1870–1946), US zoologist who discovered the mechanism of chromosomal SEX DETERMINATION. He postulated that CHROMOSOMES determine sex and that the X chromosome is the crucial element. This stimulated further research and by 1905 Edmund B. WILSON had developed the theory, now generally accepted, that both X and Y chromosomes form the basis of sex determination. *See also* NW pp.30–31.

McClure, Sir Robert John le Mesurier (1807–73), British Arctic explorer. He was the first after Sir John FRANKLIN to discover a north-west passage. Searching for Franklin's expedition in 1850, he sailed through the BERING STRAIT and discovered two routes to the Atlantic.

McClure Strait, part of the Beaufort Sea which runs between Banks and Melville islands, N Canada, and empties into the Arctic Ocean (w) and Viscount Melville Sound (E).

McCormack, John (1884–1945), Irish tenor. He studied in Ireland and Italy and made his début in London in 1907. He went to the USA in 1909 and sang with the Boston and Chicago Opera Companies. After 1914, however, he specialized as a concert artist, becoming famous for his interpretation of German LIEDER.

McCormick, Cyrus Hall (1809–84), US inventor of the mechanical reaper in 1831. It was introduced into England in 1851. Large-scale manufacture of the reaper, and his invention of the twine binder and side-rake revolutionized the harvesting of cereal crops.

McCrae, Hugh Raymond (1876–1958), Australian poet and short story writer. His poetry is lyrical and descriptive and his short stories, usually humorous, show great literary dexterity. His first book of verse was *Satyrs and Sunlight: Sylvarum Libri* (1909).

McCullers, Carson (1917–67), US author. In 1940 she published a remarkable first novel, *The Heart is a Lonely Hunter*. The hero is a deaf-mute to whom others insist on telling their stories. This was followed by *Reflections in a Golden Eye* (1941), a story of murder and neurosis in an army camp. Other works include *Member of the Wedding* (1946) and *The Ballad of the Sad Café* (1951).

MacDiarmid, Hugh (1892–), Scottish poet, real name Christopher Murray Grieve. A passionate nationalist, his has been the dominant poetic voice in Scotland since the early 1920s. His *Collected Poems* were published in 1967. *See also* HC2 p.285, *285*.

MacDonald, Alexander (*c.*1695–*c.*1770), British Gaelic poet. After publishing a Gaelic *Vocabulary* (1741) and serving in Prince Charles's army, he wrote *Ais-eiridh* (1751), a collection of anti-Hanoverian poetry.

Macdonald, Flora (1722–90), Scottish heroine, famous for helping Prince Charles Edward STUART, the Young Pretender, escape to Skye after his defeat at CULLODEN in 1746. For this she was imprisoned by the Hanoverian government until 1747.

Macdonald, George (1824–1905), Scottish novelist and poet. His works include novels of Scottish country life, poetry in dialect and stories for children. Among his novels are *David Elginbrod* (1863), *Alec Forbes* (1865), *Sir Gibbie* (1879) and *Lilith* (1895). *At the Back of the North Wind* (1871) is a well-known children's fantasy.

MacDonald, James Ramsay (1866–1937), British politican, illegitimate son of a servant, who became leader of the Independent Labour Party (later the Labour Party). A Member of Parliament (1906–18), his opposition to Britain's role in WWI aroused furious animosity and he was defeated in the 1918 elections. Returned to Parliament in 1922, he became Britain's first Labour Prime Minister in 1924. He was Prime Minister again (1929) but in 1931 was forced to head the National Government, heavily depending on Conservative support. He became increasingly distrusted by the Labour Party and lost power in 1935. *See also* HC2 pp.*161*, 209–211, *211*, 224.

Macdonald, Sir John Alexander (1815–91), first Prime Minister of the Dominion of Canada (1867–73); he also served a second term in office (1878–91) and was a power in Conservative politics from 1844 until his death. He encouraged western settlement, took over the Hudson's Bay Company lands (1870) and furthered railway interests before being implicated in the PACIFIC SCANDAL of 1873.

Macdonald, John Sandfield (1812–72), joint Premier with Louis Sicotte of Canada (1862–64), during which time he presided over an uneasy coslition of moderate and radical reformers. Although he was loyal to Confederation after it was effected (1867), he had previously opposed it.

MacDonald, Malcolm (1901–), British politician and colonial administrator. The son of James Ramsay MACDONALD, he entered Parliament in 1929. From 1935 to 1941 he was Secretary of State for the Colonies and Secretary of State for Dominion Affairs, and then became British High Commissioner to Canada from 1941 to 1946. In 1970 he was appointed Chancellor of the University of Durham.

Macdonald, Ross (1915–), US writer of detective fiction, real name Kenneth Miller. He created Lew Archer, a tough, moral private investigator. His books include *The Galton Case* (1959), *The Goodbye Look* (1969), *Sleeping Beauty* (1973) and *The Blue Hammer* (1976).

MacDowell, Edward Alexander (1861–1908), US composer. He is remembered chiefly for his piano works, including four sonatas, 12 studies, some lyrical suites and, above all, the highly ROMANTIC piano concerto.

Mace, iron or steel-headed club, which became a ceremonial symbol of authority. Medieval bishops, such as ODO OF BAYEUX, often carried maces into battle because churchmen were forbidden to shed blood and the impact of a rounded mace (even if spiked) was less likely to do so than the slash of a sword. The first purely ceremonial use of maces were probably at the courts of England and France, late in the 12th century. *See also* MM pp.*160*, *161*.

Mace, spice made from a NUTMEG, which is the hard, aromatic seed of the fruit of *Myristica fragrans*, a tall evergreen tree, native to the Moluccan Islands and New Guinea. The spice is obtained from the kernel and shell of the nut. It is used mainly for baking, and in sauces and condiments. *See also* PE p.211.

Macedo, Joaquim Manuel de (1820–82), Brazilian novelist. *The Little Brunette* (1844), a sentimental love story, is regarded as Brazil's first real novel and its first best-seller. *The Blond Lad* (1845) is similar in style.

Macedonia, region in SE Europe, on the Balkan Peninsula, now included in NE Greece, SE Yugoslavia and SW Bulgaria. The ancient Macedonian Empire (338–168 BC) was made a Roman province in 148 BC and was included in the Bulgarian and Serbian empires when they fell to the Ottoman Turks in the 14th century. Its independence from Turkey in 1912 and the claims to the region by other Balkan countries precipitated the BALKAN WARS (1912–13). Ownership of the area was again disputed after WWII, but from 1962 tension was eased following improvement in Yugoslav-Bulgarian relations. The region's products include tobacco, cotton, livestock, iron, copper and lead. Area: approx. 66,560sq km (25,700sq miles). *See also* HC2 pp.184–185.

McEwen, Sir John Blackwood (1868–1948), Scottish composer. He was principal of the Royal Academy of Music (1924–36), and wrote several works of musical theory. His compositions comprise orchestral and chamber music, including 17 string quartets; of his several symphonies the *Solway Symphony* (1911) was that most frequently performed.

Macgillicuddy's Reeks, mountain range in Kerry, SW of Killarney, in the SW Republic of Ireland. It extends 19km (12 miles) w from the Lakes of Killarney. Carrantuohill, 1,040m (3,414ft), is its tallest peak, and is also Ireland's highest mountain.

McGonagall, William (1830–1902), Scottish versifier. With his first published collection of verse (1877) he began his life as an itinerant reciter, confident always of being acclaimed a great poet. To him, rhyme was the most important factor in verse, and the result was frequently bathetic. It has been said that McGonagall is the only truly memorable bad poet. A selection of his poetry, entitled *Poetic Gems*, was published in 1890.

McGrory, Jimmy (1904–), Scottish footballer who scored 410 goals in 408 league games between 1922 and 1938, averaging more than one a game. With Celtic all but one season, he scored 550 goals in all matches, won League and Cup honours, yet gained only seven international caps.

Mach, Ernst (1838–1916), Austrian physicist and philosopher. He believed that physical phenomena should be explained only by data perceived by the senses. In physics, his name is associated with the MACH NUMBER.

Mácha, Karel Hynek (1810–36), Czech Romantic poet. His writings reflect the national and linguistic revival of the early 19th century, his long poem *May* (1836) being one of the finest iambic lyrical works in the Czech language.

Machado y Ruiz, Antonio (1875–1939), Spanish poet, one of the most outstanding and influential in the first half of the 20th century. Few poets have expressed the national feeling of Castile as well as he did in *Campos de Castilla* (1912). He supported the Republic in the Spanish Civil War and, after its defeat, fled to France.

Machala, city in SW Ecuador, 121km (75 miles) s of Guayaquil. Products: gold, cacao, coffee, hides. Pop. (1970 est.) 63,300.

Machault, Guillaume de (*c.*1300–77), French court poet and musician. His work, of which the best known is *Le livre de Voir-dit*, anticipated fixed poetic forms such as the BALLAD or RONDEAU. He wrote the first polyphonic setting of the Mass, and the secular motet.

Machel, Samora Moïsès (1933–), Mozambique politician. He received a military training in Algeria in 1963, took part in guerrilla activities against the Portuguese in Mozambique as a member of FRELIMO, of which he became president in 1970. Following the withdrawal of the Portuguese troops, Machel became president of Mozambique in June 1975. *See also* MW p.127.

Machen, Arthur Llewellyn (1863–1947), Welsh novelist, short story writer and essayist. He was at his best in subjects dealing with the supernatural. His novels include *The Great God Pan* (1894) and *The Hill of Dreams* (1907). He made many translations, including *Casanova's Memoirs* (12 vols, 1930).

Machiavelli, Niccolò (1469–1527), Florentine statesman and political theorist, an outstanding figure of the Italian Renaissance. He served from 1498 to 1512 as an official and diplomat of Piero Soderini's republican government of Florence, but lost his post when the MEDICI family returned to power. He devoted the remainder of his life to writing. His *Discourses on the First Ten Books of Livy* (1513–19) argued that the experience of the past could provide solutions for the present. His most famous work, *The Prince* (1513), propounded a ruler's need to preserve and enhance his own power and that of the state by whatever means necessary. *See also* HC1 pp.*240, 246*, 247; MS p.*273*.

Machicolation, in architecture, term for the parapet, or gallery projecting from the outside of castle towers and walls, with holes in the floor through which defenders could drop missiles, molten lead or boiling oil upon their assailants.

Machine gun, weapon that loads and fires automatically and is capable of sustained rapid fire. The firing mechanism is operated by recoil (backward flight) or by gas from fired ammunition. The gun may be water- or air-cooled. The ammunition is generally of 15.24mm (0.60in) calibre or less. The first widely used machine gun, the MAXIM GUN, was invented by the American Hiram Maxim in 1883. *See also* GATLING GUN; MM pp.*164–165*.

Machine tools, stationary, power-driven machines for cutting and shaping metal parts. Shaping is accomplished in several ways: by shearing, pressing, applying electricity or corrosive chemicals or by cutting away excess material as with lathes, shapers, planers, drills, mills, grinders or saws. Cold-forming of such items as cooking utensils and automobile bodies is performed on punch presses. Hot-forming is done on forging presses. All cutting tools have work-holding and tool-holding devices and some means of controlling the depth of cut. *See also* MM pp.*42–43, 95*.

Mach number, ratio of the speed of a moving body to the speed of sound in air, expressed as a decimal equivalent; below "1" being subsonic and over "1" supersonic. It takes its name from the Austrian investigator of supersonic speeds and shock waves, Ernst MACH (1838–1916).

Machu Picchu, Inca settlement in the Andes NW of Cuzco, Peru, situated on mountain 2,057m (6,750ft) high, and covering 13sq km (5sq miles). Forgotten after the Spanish conquest until its rediscovery in 1912 by Hiram BINGHAM, it is the best-preserved of the Inca cities and testifies to advanced water engineering and stonemasonry. *See also* HC1 p.*23*.

McIndoe, Sir Archibald Hector (1900–60), New Zealand plastic surgeon. One of the pioneers in this field, he founded a small hospital for plastic surgery in East Grinstead, Surrey, which later became internationally famous. As one of the foremost plastic surgeons of the time he was much in demand during WWII and was appointed consultant in plastic surgery to the ROYAL AIR FORCE.

MacInnes, Colin (1914–76), British novelist. His first critical success was *City of Spades* (1957), which explored the world of black immigrants in London. His later novels include *Absolute Beginners* (1959); *Mr Love and Justice* (1960) and *All Day Saturday* (1966). He was also a journalist.

Macintosh, Charles (1766–1843), Scottish chemist and inventor after whom the mackintosh is named. In 1823 he invented a method of making waterproof cloth by joining two pieces of fabrics using a solution of rubber in coal-tar naphtha. But it was not until 1839, with the advent of vulcanized rubber which withstood temperature changes, that the mackintosh became a practical garment.

MacIntyre, Duncan Bàn (1724–1812), Scottish poet, who wrote under the bardic name of Donnchadh Bàn. His poems were published in 1768; notably among them were his songs about the landscape of the Argyll-Perthshire border.

MacIver, Loren (1909–   ), US painter and illustrator who developed a unique and refined style although largely self-taught.

She has executed a number of both abstract and figurative works; *Hopscotch* (1940) and *Oil Splinters and Leaves* (1950) are indicative of her keen interest in depicting the ordinary and commonplace.

Mackay, David ("Dave") (1935–   ), British footballer who played for Hearts, Tottenham Hotspur, Derby County and Swindon. He played in 23 international matches for Scotland between 1956 and 1965, but it was in the English league that he gained renown as a hard-tackling winghalf.

McKay, Heather Pamela (1941–   ), Australian SQUASH RACQUETS player. She won the Australian championships, which are for amateurs only, every year from 1960 to 1973, when she turned professional. She also won the British Open championship, considered the world championship, every year from 1962 to 1977 and in 1976 won the first world championship, held in Australia. Only twice between 1960 and 1977 did she lose a match.

Mackay, Jessie, (1864–1938), New Zealand poet whose works, including *The Sitter on the Rail* (1891) and *The Bride of the Rivers* (1926), espoused many social causes such as PROHIBITION and women's rights.

MacKay, Mary, *See* CORELLI, MARIE.

Macke, August (1887–1914), German painter. A prominent member of the BLAUE-REITER group, the sense of colour and form in his sensitive watercolours influenced his friend Paul KLEE.

McKenna, Siobhán (1923–   ), Irish actress. From 1944 to 1947 she was at the ABBEY THEATRE, Dublin, and made her London debut in 1947. She played Pegeen Mike in *The Playboy of the Western World* at the Edinburgh Festival (1951) and again in London (1960). She won an *Evening Standard* award for her performance as St Joan (1955) and played in *Juno and the Paycock* at London's Mermaid Theatre in 1973, taking over the direction of the production on the death of Seán KENNY.

Mackenzie, Sir Alexander (1755–1850), Scottish-born Canadian fur trader and explorer. The MACKENZIE River was named after him, and his journey to the Pacific in 1793 via the Peace and Fraser rivers was the first crossing of North America N of Mexico. His *Voyages ...to the Frozen and Pacific Oceans* (1801) won him wide recognition and a knighthood. *See also* HC2 p.*136*.

Mackenzie, Alexander (1822–92), Prime Minister of Canada (1873–78), *b.* Scotland. Mackenzie edited the Reform newspaper *Lambton Shield* (1852–54) and in 1867 entered the first dominion House of Commons. He was Liberal Prime Minister from 1873 to 1878, during which time the courts and provincial governments were strengthened, trade expanded and immigration was encouraged.

Mackenzie, Sir Edward Montague Compton (1883–1972), British novelist. He achieved popularity and critical acclaim with *Sinister Street* (1913) a novel about life at Oxford and a conversion to Roman Catholicism. Later works, such as the six books collected as *The Four Winds of Love* (1937–45) and the comic novel *Whisky Galore* (1947), made into a film of the same name two years later, added to his popularity.

Mackenzie, Henry (1745–1831), Scottish novelist and leading figure in the literary circles of Edinburgh during the late 18th century. His novel *The Man of Feeling* was published in 1771. His work had considerable influence on Sir Walter SCOTT, who dedicated his "Waverley" novels to him.

Mackenzie, Stuart (1937–   ), powerfully built Australian oarsman who won the Diamond Sculls a record six times (1957–1962), was British Commonwealth (1958) and European (1957, 1958) singles champion, and a silver medallist at the 1956 Olympic Games.

Mackenzie, district in W Northwest Territories, Canada. It includes Canada's largest river, the Mackenzie, and two of the largest lakes, the Great Bear and the Great Slave. Rich in mineral deposits, the

district has uranium mines, large gold mines worked since 1934, and oilfields. Area: 1,366,200sq km (527,490sq miles). Pop. (1971) 23,665.

Mackenzie, river in W Mackenzie district, Northwest Territories, Canada; it flows from the Great Slave Lake NW into Mackenzie Bay in the Arctic Ocean: the largest river system in Canada and second-largest in North America. Discovered in 1789 by Alexander MACKENZIE, it is an important shipping route. Old fur-trading posts remain along its course. Length: 1,803km (1,120 miles).

MacKenzie Country, area in the S South Island, New Zealand. Named after a sheep-stealer of the 1850s, the region is now important for its sheep-farming.

Mackerel, fast-swimming, agile, marine food fish related to the TUNA and found in shoals in the N Atlantic, N Pacific and Indian oceans. The mackerel has a streamlined body and powerful tail. The body colour is silvery blue with dark side bars. It has a voracious appetite and lives on smaller fish and plankton. Length: 61cm (2ft). Family Scombridae. *See also* PE p.*247*.

Mackerras, Alan Charles (1925–   ), Australian conductor, *b.* USA. He became conductor of the Hamburg State Opera in 1966 and musical director of the English National Opera (formerly Sadler's Wells Opera) in 1970. He arranged the music for John Cranko's ballets *Pineapple Poll* (1951), based on Sullivan, and *The Lady and the Fool* (Verdi, 1954).

McKie, Sir William Neil (1901–   ), Australian organist. Educated in Melbourne, he left for Britain in 1919 and was organist of Magdalen College, Oxford (1938–41), and of Westminster Abbey (1941–63). He was in charge of the music at the Coronation of Queen Elizabeth II (1953), and was knighted the same year.

McKinley, William (1843–1901), 25th President of the US (1897–1901) during whose first administration America became a colonial power. A conservative Republican, he supported high protective tariffs, adopted the monetary GOLD STANDARD and, urged on by such expansionists as publisher William Randolph HEARST and Assistant Secretary of the Navy Theodore ROOSEVELT, declared war on Spain (1898). At the end of the 113-days war the US annexed Guam, Puerto Rico, the Philippines, Hawaii and American Samoa. McKinley was assassinated during his second term of office by an anarchist, Leon Czolgosz.

McKinley, Mount, peak in S central Alaska, USA, in the Alaska Range; highest peak in North America. Permanent snowfields cover more than half of the mountain. Wildlife is abundant, in particular the caribou and white Alaskan mountain sheep. The mountain was first scaled in 1913 by Hudson and named after President McKinley. It is included in Mt McKinley National Park (1917). Height: 6,194m (20,321ft).

Mackintosh, Charles Rennie (1868–1928), Scottish architect, artist and designer; one of the most successful and gifted exponents in Britain of ART NOUVEAU. His buildings, such as the Glasgow School of Art (1898–1909), were notable for their simplicity of line and skilful use of materials, and displayed a close integration of internal and external features, derived in part from Scottish traditional elements. *See also* HC2 pp.*281, 284, 284.*

Mackmurdo, Arthur (1851–1942), British architect, designer and a pioneer of the ARTS AND CRAFTS MOVEMENT. In architecture he is best known for his plans for the Savoy Hotel, London, although he also designed houses and projects for communal living. A friend of William MORRIS, he founded the Century Guild of artists (1882), based on the teachings of Morris, to improve the standard of furniture and decorative wares. He was a designer of metalwork, wallpapers and textiles, and his work of the 1880s, rich in sinuous plant forms, anticipated ART NOUVEAU.

McLaren, Bruce (1937–70), New Zealand racing driver and designer whose

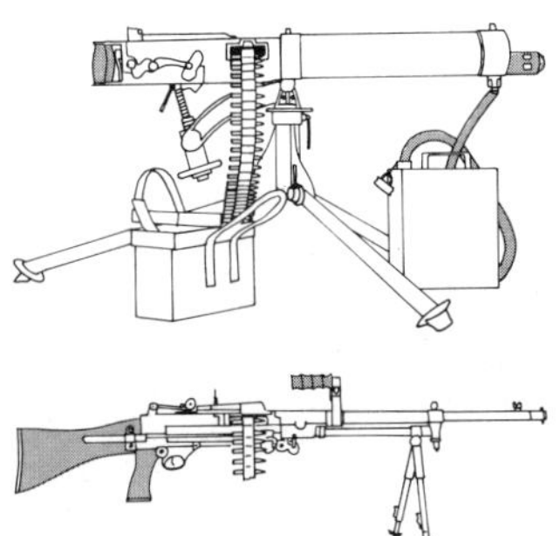

Machine guns were easily the most destructive weapons throughout World War I.

Mackenzie River: a view of the lower reaches, showing the Ramparts.

Mackerel are preyed upon by sharks, porpoises, whales and gannets.

Charles Rennie Mackintosh: his furniture was simple in line but sturdy.

McLauchlan, Ian

William McMahon headed a coalition of the Australian Liberal and Country parties.

Harold Macmillan waves goodbye at London Airport, en route to New York, 1962.

Mâcon: this view shows the restored medieval bridge over the River Saône.

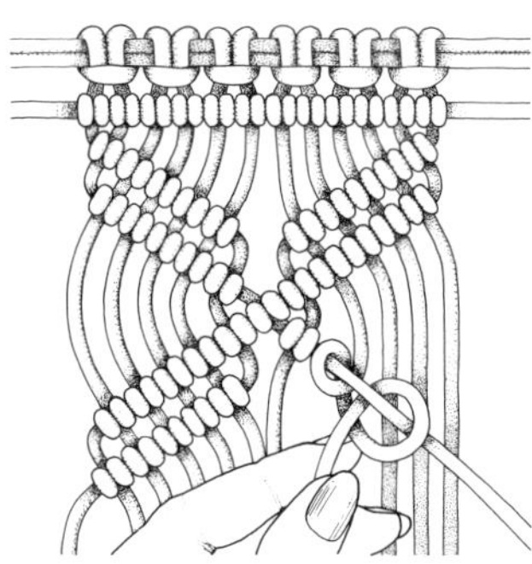

Macramé knots are clover-hitch and square knots, but some allow different ones.

early death was preceded by wins in the Monaco Grand Prix (1962), Le Mans (1966) and the Belgian Grand Prix (1968).

McLauchlan, Ian (1942–), British rugby union player, known as "Mighty Mouse", who played for Jordanhill (Scotland), and the British Lions. He appeared in 33 international matches for Scotland (18 times as captain) and toured with the Lions in 1971 and 1974. At 175cm (5ft 9in) tall, he was relatively short for his position as tight-head prop.

MacLeish, Archibald (1892–), US poet and playwright. His works include *Conquistador* (1932), an epic poem *Collected Poems 1917–1952* (1952), and *J.B.* (1958), a verse play.

MacLennan, John Hugh (1907–), Canadian novelist, principally concerned with Canadian subjects. His books include *Barometer Rising* (1941), *Two Solitudes* (1945), a study of the conflicts between English and French Canadians, *Each Man's Son* (1951), *The Watch that Ends the Night* (1959) and *Return of the Sphinx* (1967).

Macleod, John James Rickard (1876–1935), Scottish physiologist who shared the 1923 Nobel Prize in physiology and medicine with Sir Frederick Grant BANTING for the discovery of INSULIN and studies of its use in treating DIABETES. The actual discovery was made by Banting and Charles BEST working in Macleod's laboratory at the University of Toronto. Macleod later taught at the University of Aberdeen. His works include *Diabetes* (1913) and *Carbohydrate Metabolism and Insulin* (1926).

MacLiammóir, Mícheál (1899–1978), Irish actor, designer and playwright. In 1928 he opened the Galway Gaelic Theatre, and in the same year directed the Dublin Gate Theatre in collaboration with Hilton Edwards. In 1931 he was appointed director of the Dublin Gaelic Theatre. He is best known for his portrayal of Oscar Wilde in his own compilation *The Importance of Being Oscar* (1960). His plays include *Ill Met by Moonlight* (1946) and *Home for Christmas* (1950).

Maclise, Daniel (1806–70), Irish painter. Popularly known for his caricatures published in *Fraser's Magazine*, his most important works were the mural paintings in the House of Commons.

McLuhan, Herbert Marshall (1911–), Canadian academic and expert on communications. His view that the forms in which people receive information – such as television, radio and computers – are more important than the messages themselves was presented in his books *The Mechanical Bride: Folklore of Industrial Man* (1951), *Understanding Media* (1964) and *The Medium is the Massage* (1967). *See also* MS p.301.

McMahon, William (1908–), Prime Minister of Australia (1971–72). He practised as a solicitor until 1939 and turned to politics after WWII. He was treasurer and deputy leader of the Liberal Party (1966–71) and Minister for Foreign Affairs (1969–71) before becoming leader of the Liberal Party from 1971 to 1972.

McMahon Line, border between Tibet and Assam negotiated by the Tibetan and British delegations (the latter led by Sir Henry McMahon) at the Simla conference (1913–14). A Chinese delegation refused to agree to the line, claiming that Tibet was not independent of China and could not sign treaties. China's consistent refusal to accept the line led to friction with India, especially in the 1960s.

McMillan, Edwin Mattison (1907–), US chemist and physicist. In 1951 he shared the Nobel Prize in chemistry with Glenn SEABORG for the discovery of several TRANSURANIC elements. While working on the CYCLOTRON with Ernest LAWRENCE at the University of California, McMillan developed the SYNCHROCYCLOTRON, for which he shared the 1973 Atoms for Peace prize with V. I. Veksler (1907–1966).

Macmillan, John (1670–1753), Scottish minister. Deposed from his Presbyterian ministry in 1703 for attacking the corruption of Church government, he formed the Macmillanites, and with Thomas Nairn founded (1743) the Reformed Presbyterian Church.

MacMillan, Kenneth (1929–), British dancer and choreographer who succeeded Sir Frederick ASHTON as director of the ROYAL BALLET in 1970, resigning from that post in 1977 but retaining that of principal choreographer. His reputation was established with *Dances concertantes* (1955); other ballets include *Le Sacre du Printemps* (1962), *Romeo and Juliet* (1965), *Anastasia* (1971), *Elite Syncopations* (1974), *Rituals* (1975) and *Mayerling* (1978).

Macmillan, Kirkpatrick (1813–78), Scottish blacksmith who invented the first two-wheeled self-propelled bicycle in 1839. Movement was achieved by means of cranks connected to the rear axle, which were pushed round by the feet. His invention only became popular when it was resurrected by Pierre and Ernest MICHAUX in 1861.

Macmillan, Maurice Harold (1894–), British statesman. During the 1930s, Macmillan was one of a small band of Conservatives prepared to face political death for attacking their government's foreign policy, in particular the passive acceptance of MUSSOLINI's invasion of Abyssinia. In the 1950s he held a number of cabinet posts, including Minister of Housing and Local Government (1951–54), Minister of Defence (1954–55), and Chancellor of the Exchequer (1955–57). He succeeded Anthony Eden (later Lord AVON) as Prime Minister (1957–63), improving Anglo-American relations after the Suez crisis and trying unsuccessfully to obtain Britain's entry into the European Economic Community. He gave firm support to President KENNEDY over the CUBAN MISSILE CRISIS in 1962.

McMurdo Sound. *See* ROSS DEPENDENCY.

McMurrough, Dermot (1110–71), Irish King of LEINSTER in SE Ireland. Driven from Leinster in 1166 for abducting another king's wife, his request for English aid brought the Anglo-Normans to Ireland. He was responsible for compiling the *Book of Leinster*, a collection of Gaelic traditions. *See also* HC1 p.180.

McNamara, Robert Strange (1916–), US Secretary of Defence from 1961 to 1968. Growing doubts about the VIETNAM WAR led to his resignation and he later became President of the World Bank.

MacNeice, Louis (1907–63), British writer. An important poet of the 1930s, he was the author of several memorable verse plays for radio. Among his collected poems are *Autumn Journal* (1939) and *Solstices* (1961). He also translated AESCHYLUS's *Agamemnon* and GOETHE's *Faust*.

McNeile, Cyril. *See* SAPPER.

Mâcon, dry red or white wine from vineyards near the town of the same name in E France. The wine is produced in the départements of the Ain and the Saône-et-Loire, in the S of the Burgundy area.

Maconchy, Elizabeth (1907–), British composer. She was a student of VAUGHAN WILLIAMS. Her most important works have been chamber music, including nine string quartets. She wrote many choral works including *The Jesse Tree* (1970) and *Prayer before Birth* (1971). (1963).

Macphail, Agnes Campbell (1890–1954), first woman Member of Parliament in Canada (1921–35). Active in promoting prison reform and social welfare programmes, she was the first Member of Parliament to write a weekly newsletter to her constituents.

McPhee, Colin (1901–64), Canadian composer. Attracted by Indonesian music, he lived for several years in Bali until 1939. The toccata for orchestra *Tabuh-Tabuhan* (1936) is one of his best-known compositions and is based on Balinese orchestral procedures. Other compositions include *Sea Shanty Suite* (1929) for male chorus, a concerto for piano and wind octet (1929) and film scores.

McPherson, Aimee Semple (1890–1944), US evangelist. In 1926 she founded the International Church of the Foursquare Gospel. Claiming to be guided by God, she professed faith healing and the gift of tongues and her flamboyant methods were phenomenally successful. In May 1926 she disappeared while sea-bathing and reappeared a month later claiming to have been kidnapped. She was tried for fraud and, although acquitted, never regained her former influence.

Macpherson, James (1736–96), Scottish poet and historian. His use of Celtic lore in *The Works of Ossian* (1762) inspired the ROMANTIC MOVEMENT in European literature. *See also* HC2 pp.31, 72, *73.*

Macquarie, Lachlan (1761–1824), British colonial administrator. As governor of New South Wales (1809–21) he attempted reforms involving rights for the freed convicts and small farmers, but opposition from large landowners forced his recall to Britain.

Macquarie, river in E central New South Wales, Australia. It rises in the BLUE MOUNTAINS and flows NNW to the River Darling. It drains an important wheat and sheep raising area. Length: 950km (590 miles).

Macquarie Island, uninhabited island in the S Pacific Ocean, 1,370km (850 miles) SE of Tasmania, formed from the exposed tip of a submerged volcano. Administered by Tasmania, it is the site of an Australian research station. Area: 230sq km (89sq miles).

Macramé, form of braiding made by knotting threads or cords in a geometrical pattern. Its ornate design has many practical applications, as well as being a minor art form.

Macready, William Charles (1793–1873), British actor and theatrical manager. Distinguished as a Shakespearean actor, he achieved fame in the name-part of *Richard III* in 1819. His productions were noted for meticulous attention to detail.

Macrocephaly, excessive size of the head. It is often caused by HYDROCEPHALUS and is associated with mental subnormality.

Macroeconomics, study of the economic system as a unit, rather than its individual components as in MICROECONOMICS. It involves the determination of items such as GROSS NATIONAL PRODUCT and personal income, the study of the banking system and international trade.

Macromolecule, molecule, generally with a diameter ranging from 10^{-5} to 10^{-3}mm (10^{-7} to 10^{-5}in) considerably larger than the molecules of most substances (usually under 10^{-6}mm (10^{-8}in)). Many proteins, nucleic acids, plastics, resins, rubbers and natural and synthetic fibres are made up of such giant units, each of which contains thousands of atoms. *See also* SU pp.156–157.

McTell, Blind Willie (1904–c.1960), US blues singer and guitarist. More of a showman than his contemporaries Robert JOHNSON and Sleepy John ESTES, he nevertheless wrote many songs of great poetic insight, mainly concerned with the darker, more violent aspects of the sexual relationship.

McWilliam, Frederick Edward (1909–), Irish sculptor who exhibited with the British Surrealists in 1936. He executed *The Four Seasons*, which was commissioned for the Festival of Britain.

Madaans, also called Ma'dan, or Marsh Arabs, semi-nomadic people inhabiting the marshlands of S Iraq. Their dwellings are usually built of giant reeds. They use canoes when hunting; raising water buffalo is their major economic activity.

Madách, Imre (1823–64), Hungarian playwright who is best known for his poetic drama *Az ember tragédiája* (*The Tragedy of Man*, 1861). Sometimes compared to Goethe's *Faust*, it relates man's past and future through the characters of Adam, Eve and Lucifer. Originally it was not intended for the stage but has often been produced in Hungary.

Madagascar. *See* MALAGASY.

Madame Bovary (1857), naturalistic novel by Gustave FLAUBERT, who was prosecuted for its alleged immorality. Set in provincial Normandy, it tells the story of Emma Bovary, whose dreams of romance are realized neither by her dull husband, a country doctor, nor her lovers. She falls into debt and eventually commits suicide.

Madame Butterfly (1904), three-act opera by Giacomo PUCCINI, with Italian libretto by Giuseppe Giacosa and Luigi Illica, after David BELASCO's play. First performed at La Scala, Milan, in two acts, it was a fiasco, primarily because of its romantically indulgent music, but it was successfully re-presented three months later at Brescia under Arturo TOSCANINI.

Madder, perennial vine native to s Europe and Asia, which bears greenish-yellow flowers. A red dye was once produced from the roots. Height: to 1.2m (4ft). Family Rubiaceae; species *Rubia tinctorium. See also* PE p.159.

Madeira, fortified wine from the island group of the same name. It is made by mixing the wine of the island with brandy, and keeping the blend at 48.9°C (120°F) for five months. Madeira is now drunk chiefly as an after dinner wine, and has always been especially popular in Britain.

Madeira Islands, group of islands in the Atlantic Ocean off the coast of Morocco, N of the Canary Islands; the capital is Funchal, on Madeira Island. The group consists of two inhabited islands, Madeira and Porto Santo, and two uninhabited groups, Desertas and the Selvagens. The islands are a year-round resort area. Products include sugar cane, Madeira wine and embroidery work. Area: 798sq km (308sq miles). Pop. (1970 est.) 268,700.

Madhya Bharat, former state of India, in w central India. Its main princely states included Indore (the summer capital) and Gwalior (in Lashkar, the winter capital). Madhya Bharat was created in 1948 and 1950 as a union of 20 princely states, and was incorporated into the state of Madhya Pradesh in 1956.

Madhya Pradesh, state in central India; Bhopāl is the capital. Previously ruled by the Gonds and MAHRATTAS, it was ceded to the British in 1817–18. Berar was incorporated into the state in 1903 and its name was changed to the Central Provinces and Berar. In 1956, the former states of Madhya Bharat, Vindhya Pradesh, Bhopāl and part of Rajasthan were incorporated into the new state. Its industries include cotton weaving and mining for manganese, iron ore and coal. Area: 442,841sq km (170,891sq miles). Pop. (1971) 41,449,729.

Madison, James (1751–1836), fourth President of the USA (1809–17). Earlier in his career, he had been one of the chief designers of the United States Constitution. In the important debate over states' rights versus strong federal government, he mediated between Thomas JEFFERSON and Alexander HAMILTON. When British harassment of US merchant ships became intolerable, Madison declared war (1812), which ended indecisively with the Treaty of GHENT (1815). After two terms in office, Madison retired to his Virginia plantation.

Madison, state capital of Wisconsin, USA, 121km (75 miles) w of Milwaukee. Established as the capital of the Territory of Wisconsin in 1836, it was named after President James Madison. It is the seat of the University of Wisconsin (1836). Industries: food processing, medical equipment, motor vehicle components. Pop. (1970) 172,007.

Madison Avenue, thoroughfare in Manhattan, New York City, USA, famous for its skyscraper office blocks and exclusive shops. Many of the principal US advertising agencies have headquarters on Madison Avenue, and the name has become synonymous with the advertising world.

Madison Square Garden, indoor stadium in Manhattan, New York City. Famous primarily for boxing contests, it also accommodates other sports and forms of entertainment. The present arena was opened in 1968.

"Mad Mullah". *See* MOHAMMED BEN ABDULLAH.

Madonna, representation in painting or sculpture of the Virgin Mary, usually with the infant Jesus. The early Christians painted the Madonna in their CATACOMBS and she was a notable feature of many outstanding Byzantine ICONS. The advent of the Renaissance brought less stylized representations, and important portraits of her during that period were produced by almost every great painter and sculptor. More recently, artists such as Henry MOORE in his *Madonna* (1944) have been attracted to the subject.

Madonna lily. *See* LILY.

Mad Parliament, misnomer, no longer used, for the Oxford parliament of 1258, at which the barons forced the PROVISIONS OF OXFORD on HENRY III of England. The name came from a contemporary chronicle's reference to *illud insane parlamentum*, but it has been shown that *insane* was a later insertion.

Madras, port in SE India, on the Bay of Bengal; the capital of Tamil Nadu state. Founded in 1639, the city grew around a British post, Fort St George, and developed as a trading centre. It was besieged and occupied by the French in 1746, but returned to Britain by the Treaty of Aix-la-Chapelle in 1748. The harbour was constructed from 1862 to 1901; there is a university, founded in 1857. Nearby is Mt St Thomas, the legendary site of the martyrdom in the 1st century AD, of THOMAS the apostle who is believed to have been buried in Madras. Exports: hides, skins, oilseeds, cotton, chrome, magnesite. Industries: textiles, railway stock, bicycles, printing, motor vehicles, films. Pop. 2,469,449.

Madrid, capital city of Spain, and Madrid province, on the Manzanares River. Taken from the Arabs in 932 by Ramiro II of Léon, it was captured by Alfonso VI in 1083. In 1561 PHILIP II made Madrid his official residence and the capital of Spain. The French occupied the city during the PENINSULAR WAR (1808–14). Madrid remained loyal to the Republican government throughout the SPANISH CIVIL WAR, and suffered several air attacks. The city's surrender, in March 1939, ended the war. Places of interest include the Royal Palace (1739–64), now a museum, a national library, the University of Madrid (1499) and several other museums, notably the Museo del Prado. Industries: plastics, wine, brewing, publishing, film production, optical instruments, electrical appliances, radio and telephone equipment, jewellery, leather goods. Pop. (1976 est.) 3,751,000.

Madrigal, form of unaccompanied vocal music orginating in Italy in the 14th century. Early madrigals feature two or three parts and a highly ornamented upper part. Later more contrapuntal forms in the 16th and early 17th centuries featured love lyrics, no set form, and four or five voices. The "classical" period (*c.*1540–80) was dominated by Italian masters (GABRIELI, PALESTRINA) and Flemish composers such as LASSO. The late period (*c.*1580–1620) was dominated by Italians such as Carlo Gesualdo and MONTEVERDI and English composers such as BYRD, GIBBONS, MORLEY and WEELKES.

Madroña, also called laurelwood, a broadleafed, evergreen tree native to British Columbia, Canada, and N USA. It has small, white flowers and orange-red fruit. Height: 15–30m (50–100ft). Family Ericaceae; species *Arbutus menziesii.*

Madura, island in Indonesia, separated from the NE coast of Java by the Madura Strait; the capital is Pamekasan. Administered by Java from the 11th century, it came under Dutch influence in the 18th century and was made a Dutch residency in Java in 1885. It became part of the Republic of Indonesia in 1950. Its industries include salt mining, fishing and oil refining. Area: 5,473sq km (2,113sq miles). Pop. 1,858,200.

Madurai, city in s India, 435km (270 miles) ssw of Madras, on the River Vaigai. It served as the capital of the Pandya dynasty (5th century BC – 11th century AD), and the Nayak kingdom (*c.*1550–1736). The city passed to the British in 1801. It contains the massive Meenakshi Temple and has a university (1966). Industries, weaving, silk and muslin dyeing, brassware, wood carving, tourism. Pop. (1971) 549,144.

Maeander, old name of the Büyük Menderes River in w Turkey. It rises w of Afyonkarahisar and flows to the Aegean Sea. The word "meander" is derived from the river, which in its lower reaches has a twisting and looping course. Length: approx. 400km (250 miles).

Maecenas, Gaius (*c.*70 BC–8 BC), Roman politician, diplomat and art patron. He played a major role in the wars which established the Emperor Augustus and then became an important political figure. He is chiefly remembered for having used his great wealth to support artists and poets such as HORACE and VIRGIL.

Maekawa, Kunio (1905–), Japanese architect. He studied under LE CORBUSIER in Paris (1928–30), and was a pioneer of modern architecture in Japan, building Dairen Town Hall (1938) and the Metropolitan Festival Hall, Tokyo (1958–61). Kenzo TANGE worked in his studio prior to WWII.

Maelzel, Johann (1772–1838), German inventor, reputedly of the METRONOME and of other contrivances including an "automatic" chessplayer (which really contained a chess master, William Schlumberger).

Maes, Nicolaes (1634–93), Dutch genre and portrait painter, pupil of REMBRANDT at Amsterdam. His early works include *Girl at the Window* (1655) and *Old Woman Scraping Parsnips* (1655). After a period spent in Antwerp (1665–67) he turned to fashionable portraiture, with a marked change of style which showed more affinities with the work of VAN DYCK than of Rembrandt.

Maestricht. *See* MAASTRICHT.

Maeterlinck, Maurice (1862–1949), Belgian playwright. In a period when realism was becoming theatrically fashionable, he explored the symbolic and spiritual nature of drama and his ideas were later to influence the THEATRE OF SILENCE. His plays include *The Princess Maleine* (1889), *Pelleas and Melisande* (1892) – the source of DEBUSSY's opera of the same name – and *The Blue Bird* (1909), first produced in Moscow by STANISLAVSKY. Maeterlinck won the 1911 Nobel Prize in literature.

Maeve. *See* MAB.

Mae West, nickname for the inflatable life-jacket used by Allied airmen in WWII. It was named after the famous film star, Mac WEST.

Mafeking, Siege of, incident in the South African (Boer) War, in which the British forces under Robert BADEN-POWELL were besieged by the Boers from Oct. 1899 to May 1900, at the garrison of Mafeking on the River Molopo, in Cape Colony. News of the successful relief of Mafeking was greeted with popular jubilation in London.

Maffei, Francesco Scipione, marchese di (1675–1755), Venetian man of letters. He wrote a successful tragedy *Merope* (1713), and founded the *Giornale dei Litterati d'Italia* in 1710. He wrote many books of history and literary criticism.

Mafia, name given to organized groups of Sicilian bandits. Originating in feudal times, when such bandits were hired as bodyguards, the contemporary assumption was that the legal authorities were corrupt and that justice must be achieved directly, as in the VENDETTA. Through emigration the Mafia spread to the USA in the early 20th century where it later became involved in organized crime, especially during the era of PROHIBITION.

Magadha, kingdom in ancient India that comprised the present-day Gaya and Patna districts of Bihar. Between the 6th century BC and the 8th century AD it was the focus of larger empires or kingdoms. Strategically located in the Ganges valley, it had control of river trade and communications. Its capital was originally at Rajgir, but was later removed to Pataliputra. *See also* HC1 p.62, *62.*

Magdalene, Mary. *See* MARY MAGDALENE.

Magdalenian culture, archaeologically the most recent culture of the PALAEOLITHIC AGE. Named after the cave of La Madeleine in the Dordogne region of France, where the first remains were found, the culture includes the great cave painters of Altamira and LASCAUX. Magdalenians also made fine tools, weapons

Madame Butterfly: the Italian translation was published in *La Lettura* in 1904.

Madder is the source of dyes of several colours, particularly red.

Madison Avenue is one of New York's main thoroughfares, shadowed by skyscrapers.

Madrid is the capital and the geographical centre of Spain

Ferdinand Magellan discovered the strait that was later named after him.

Lake Maggiore: important cities on its shores include Locarno and Palanza.

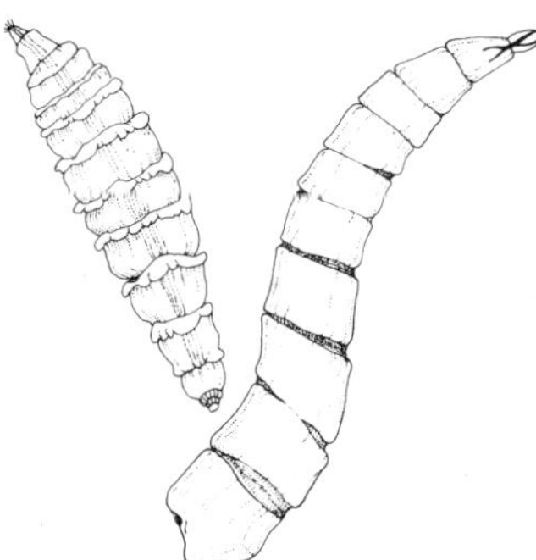

Maggots, usually white, move by wriggling or flipping their bodies.

Magna Carta: part of the stone structure erected at Runnymede in commemoration.

and artistic carvings in stone and bone. In Britain, Kent's cavern in Devon and Creswell Crags in Derbyshire are among the few proved Magdalenian sites.

Magdalen (Madeleine) Islands, island group in the central Gulf of St Lawrence, E Quebec province, Canada. Fishing is the main industry. Area: 264sq km (102sq miles). Pop. (1970) 13,305.

Magdeburg, port in w East Germany, on the River Elbe, 132km (82 miles) wsw of Berlin; capital of Magdeburg district. In the 13th century the city was granted a charter by the archbishop of Magdeburg a prince of the HOLY ROMAN EMPIRE. The charter served as a model for hundreds of medieval towns in Germany. During the THIRTY YEARS WAR the city was burned and sacked and most of the population perished. Industries: steel, paper, textiles, chemicals, machinery. Pop. (1975) 276,580.

Magellan, Ferdinand (c.1480–1521), Portuguese explorer who commanded the first expedition to circumnavigate the globe; the expedition was completed after his death. Probably wounded while exploring the East Indies, he took part in the taking of Malacca which made possible the exploitation of the vast trade potential in the Spice Islands (or Moluccas). He fell from the King of Portugal's favour and moved to Spain, where he organized an expedition to reach the Spice Islands by sailing west. Leaving with five ships in 1519, he reached Brazil and sailed south, seeking a westward passage. In November 1520 he sailed through the straits which today bears his name and reached the Philippines in March 1521, where he was killed by natives. Two of his ships reached the Moluccas, but only one returned to Spain, thus establishing a new route between Europe and Asia. *See also* HC1 pp.234, *234.*

Magellan, Strait of, channel at the s tip of South America, between Chile and Tierra del Fuego, connecting the s Atlantic Ocean with the s Pacific Ocean. It was charted in 1520 by Ferdinand MAGELLAN. It follows a tortuous course, and navigation is hampered by fog and wind, but it was an important shipping route before the completion of the Panama Canal (1914). Punta Arenas is the only large town on the strait. Length: approx. 560km (350 miles).

Magellanic Clouds, two small satellite galaxies of the MILKY WAY GALAXY, visible in skies around the south pole as misty stellar concentrations. The Small Cloud (Nubecula Minor), located in the constellation of Tucana, is irregular; the Large Cloud (Nubecula Major), mostly in the constellation of Dorado, is vaguely spiral. Their distance is about 150,000 light-years away. *See also* SU pp.246–*247,* 263.

Magenta, also called fuchsine, reddish purple dyestuff, also the name of its colour, apparently so called after the Battle of Magenta, N Italy, in 1859.

Maggiore, Lake, lake in N Italy and s Switzerland bordered by the Swiss Alps. It is crossed N to s by the River Ticino and fed by the Maggia, Toce and Tresa rivers. The Borromean Islands are off the w shore. Area: 212sq km (82sq miles).

Maggot, name commonly given to the legless LARVA of a fly. It is primarily used to describe those larvae that infest food and waste material; others are generally called grubs. *See also* DIPTERA.

Maghreb, Arabic term for NW Africa, generally applied to Morocco, Algeria and Tunisia but actually referring to the Mediterranean coastal strip. The Atlas Mts separate the Maghreb from the rest of the African continent and give it a Mediterranean climate and economy. It was united briefly under the MUSLIMS in the 8th and 12th centuries.

Magi, hereditary priestly class of ancient Persia, responsible for religious ceremonies. In ZOROSTRIANISM the magi were astrologers and practitioners of the occult. The "wise men from the East" of Matthew 2:1, 2 may have been Zoroastrian priests.

Magic, use or apparent use of natural or spirit forces to produce results that are logically impossible. Belief in magic, and

that certain actions influence supernatural or occult forces, is associated mainly with primitive societies, although traces can still be found (eg SUPERSTITIONS) in highly developed countries. There are several kinds of magic; black magic makes use of evil spirits and delights in supreme selfishness and in evil for evil's sake; contagious magic is belief in the lasting effect of association between a person and his possessions; and homeopathic or sympathetic magic is belief that models or likenesses of a person can influence or even control him. The term magic is also used of the work of the stage magician or conjurer, who by means of illusions produces results or effects that are apparently impossible. *See also* OCCULT; SORCERY; WITCHCRAFT; MS pp.116–117, 196–197.

Magic eye, type of ELECTRON TUBE used in radio sets. Depending on its design it may be used either as a tuning indicator or to detect changes in voltage and current.

Magic Flute, The, opera by Wolfgang Amadeus MOZART, with German libretto by Emanuel Schikaneder based on a tale by Christoph WIELAND. Its Vienna première (1791) was conducted by Mozart. The intricate plot has several levels, the religious theme reflecting the controversial Masonic beliefs of both composer and librettist. *See also* HC2 p.120.

Maginot Line, series of fortifications constructed by France on its NE border with Germany, to prevent a quick German attack on France's essential frontier industries. It was named after André Maginot (1877–1932) who, as French Minister of War (1929–31), directed its construction. It covered the French-German border, but not the French-Belgian border which the German army successfully attacked in 1940. It was still intact when France surrendered in 1940.

Magistrate, in general anyone invested by the state with authority to administer the law. The name comes from the Roman public office, the *magistratus.* In Britain a magistrate is a judicial officer, inferior to a judge, who presides over a MAGISTRATE'S COURT. There are two kinds, the unpaid JUSTICE OF THE PEACE and, in large towns, the stipendiary magistrate. Unlike the lay justice of the peace, a stipendiary magistrate must be a barrister or solicitor of seven years' standing.

Magistrate's court, court presided over by a MAGISTRATE. It is the court of the first instance: every person charged with an offence appears first at a magistrate's court, and at a preliminary hearing the magistrate or JUSTICE OF THE PEACE decides whether his own jurisdiction is sufficient or whether to send the case on to a higher court, or whether there is insufficient evidence available to justify court proceedings at all. Generally, however, the court deals with minor civil and criminal charges, the most frequent being those arising from motoring offences, drunk and disorderly behaviour, and petty theft.

Maglemosian culture, in archaeology, Northern European culture of the middle MESOLITHIC AGE, named after a site in Denmark. In common with other mesolithic peoples, Maglemosians lived by fishing and food gathering. Their tools included bone harpoons, microliths (small flint tools) and stone axes for felling trees. Maglemosians are believed to have migrated from Denmark to Britain across what is now the southern North Sea but was then marshland.

Magma, molten material that produces all IGNEOUS ROCKS. The term refers to this material while it is still under the Earth's crust. In addition to its complex silicate composition, magma contains gases and water vapour. It is believed to exist in separate chambers beneath the Earth's surface. *See also* PE pp.100–101.

Magna Carta (1215), "Great Charter" of English constitutional history. It was issued by King JOHN, under compulsion from his barons, at Runnymede, on 15 June, 1215. John's financial exactions had united clergy and laity in demands for guarantees of their rights and privileges. The 63 clauses into which Magna Carta is traditionally divided protected the rights of the Church, the feudal lords, the lords'

subtenants and the merchants, and regulated royal privileges, the administration of justice, and the behaviour of royal officials. For subsequent generations Magna Carta became the basis and epitome of the subject's rights, protecting him and limiting his sovereign's prerogatives. *See also* HC1 pp.178–179.

Magna Graecia, area occupied by a group of ancient Greek cities along the coast of s Italy. An important centre of Greek civilization, the Pythagorean and Eleatic systems of philosophy originated there. The chief cities included Cumae (founded 750 BC) and Tarentum. The group declined in importance after 500 BC.

Magnani, Anna (1908–73), Italian film actress, *b.* Egypt. She worked as a nightclub entertainer and a repertory actress before starting her film career in *The Blind Woman of Sorrento* (1934). She was best known for her roles in Roberto Rossellini's *Open City* (1945), Jean Renoir's *The Gold Coach* (1952) and perhaps most especially for her first Hollywood film, Tennessee Williams' *The Rose Tattoo* (1955), for which she won an Oscar as the best actress.

Magnesia, magnesium oxide, MgO, white, neutral, stable powder formed when magnesium is burned in oxygen. It is used industrially in firebrick, and medicinally in stomach powders. Magnesium carbonate, found as the mineral magnesite and also used as an antacid, is also often called magnesia.

Magnesite, carbonate mineral, magnesium carbonate ($MgCO_3$). It is an alteration product of CALCITE or DOLOMITE in sedimentary rocks and is found in microcrystalline masses and crystals like Iceland spar. Magnesite is white, colourless or of light tint, with a glassy to dull lustre. Hardness 3.5–5; s.g. 3.1.

Magnesium, common metallic element (symbol Mg) of the ALKALINE-EARTH group, first isolated in 1808 by Sir Humphrey DAVY. Its chief sources are MAGNESITE, DOLOMITE and other minerals. Magnesium burns in air with an intense white flame and is used in flashbulbs, fireworks and incendiaries. Magnesium alloys are used in aircraft for their lightness. Chemically, the element is similar to CALCIUM. Properties: at.no. 12; at.wt. 24.312; s.g. 1.738; m.p. 648.8°C (1,200°F); b.p. 1,090°C (1,994°F); most common isotope Mg^{24} (78.7%).

Magnesium processing, production of pure magnesium metal by the ELECTROLYSIS of molten magnesium chloride ($MgCl_2$), processed mainly from sea water, or by the direct reduction of its ores, usually MAGNESITE, with a suitable reducing agent such as ferrosilicon. *See also* SU pp.134, *135.*

Magnet, object that generates a magnetic field in which other magnetizable objects experience a force. Among the first were lodestones. Later, strong magnetic materials were recognised as containing either iron, cobalt, nickel or their mixtures. A typical permanent magnet is a straight or horseshoe-shaped magnetized iron bar, the ends of which are described, respectively, as the north and south magnetic poles. The Earth is a giant magnet, its magnetic lines of force being detectable at all latitudes. Much stronger than permanent magnets are electromagnets, used for raising heavy steel weights and scrap. Strongest of all are superconducting magnets in which speeial alloys are cooled to very low temperatures. *See also* SU pp.116–117.

Magnetic bottle, configuration of magnetic fields used to contain the PLASMA in a Fusion Reactor or experimental device. It applies particularly to a linear configuration in which the ends are stoppered with magnetic mirrors, where a very high magnetic field causes the plasma to be reflected.

Magnetic disturbance, change in the shape of the MAGNETOSPHERE. The most commonly observed phenomenon that results from this is the polar AURORA and the interference in radio and television communications.

Magnetic equator, that part of the Earth's magnetic field where a finely-balanced magnetic needle will remain absolutely horizontal. Because of the Earth's magnetic asymmetry, the magnetic equator does not coincide with the geographical equator.

Magnetic field, region surrounding a magnet or a conductor through which a current is flowing, in which magnetic effects, such as the deflection of a compass needle, can be detected. A magnetic field can be represented by a set of lines of force (flux lines) emanating from the POLES of a magnet or running around a current-carrying conductor. These lines of force can be seen if iron filings are sprinkled on to a sheet of paper below which a magnet is placed. The filings align themselves with the lines of force, the density of the lines being greatest where the field is strongest. *See also* SU pp.116–123; PE pp.22–23.

Magnetic flux. *See* FLUX, MAGNETIC.

Magnetic induction. *See* INDUCTION.

Magnetic mine, form of non-contact mine usually laid on the sea-bed. It contains a delicate magnetic needle sufficiently sensitive to respond to the change in magnetic field when a ship passes overhead. This change causes an electrical contact to close and detonate an explosive charge.

Magnetic permeability, physical property of magnetic materials. It is the ratio of magnetic flux density in a material to the strength of the magnetic field that induces this flux. In physics, magnetic permeability is represented by the symbol μ.

Magnetic poles, two regions in which the magnetism of a magnet appears to be concentrated. If a bar magnet is suspended to swing freely in the horizontal plane one pole will point north; this is called the north-seeking or north pole. The other pole, the south-seeking or south pole, will point south. Unlike POLES attract each other; like poles repel each other. *See also* SU pp.116–123.

Magnetic recording, formation of a record of sounds on a wire or tape by means of a pattern of magnetization. In a tape recorder, plastic tape impregnated with iron oxide is fed past an electromagnet, which is energized by the amplified currents produced by a MICROPHONE. By ELECTROMAGNETIC INDUCTION, variations in magnetization, representing the oscillating current produced by the sound, are induced in, and retained by, the particles of iron oxide on the tape. For playing back the tape is fed past a similar electromagnet, which converts the patterns into sound that is in turn fed to an AMPLIFIER and LOUDSPEAKER. *See also* MM pp.232–233; SU p.*120.*

Magnetic resonance, absorption of radio and MICROWAVE frequencies by atoms placed in a magnetic field. Devices called electron-spin RESONANCE (ESR) spectrometers use microwaves in the investigation of atoms and molecules. Nuclear-magnetic resonance (NMR) spectrometers use radio frequencies for chemical analysis and research in nuclear physics. *See also* SU pp.150–151, *151.*

Magnetic storm, or geomagnetic storm, disturbance in the Earth's magnetic field. Since the field encompasses all the Earth, the effects of such a storm are global: AURORAS are seen, both in areas where such displays are normal and in others as well; radio signals are disturbed. There is a regularly reappearing cycle of such magnetic storms as well as irregular ones.

Magnetism, properties of matter and of electric currents associated with a field of force (MAGNETIC FIELD) and with a north-south polarity (MAGNETIC POLES). All substances possess these properties to some degree as orbiting electrons in their atoms produce a magnetic field in the same way as an electric current produces a magnetic field; similarly, an external magnetic field will affect the electron orbits. All substances possess weak magnetic (diamagnetic) properties and will tend to align themselves with the field but in some cases this diamagnetism is masked by the stronger forms of magnetism: paramagnetism and ferromagnetism. Paramagnetism is caused by electron spin and occurs in substances having unpaired electrons in their atoms or molecules. The most important form of magnetism, ferromag-

netism, occurs in substances such as iron and nickel, which are capable of being magnetized by even a weak field due to the formation of tiny regions, called domains, that behave like miniature magnets and align themselves with an external field. These domains are formed as a result of strong interatomic forces caused by the spin of electrons in unfilled inner electron shells of the atoms. Permanent magnets, which retain their magnetization after the magnetizing field has been removed, are ferromagnetic. Electromagnets have a ferromagnetic core around which a conducting coil is wound. The passage of a current through the coil magnetizes the core. *See also* PE pp.22–*23,* 130; SU pp.116–120.

Magnetite, oxide mineral, ferrous and ferric iron oxide ($FeFe_2O_4$). The most magnetic mineral, it is a valuable iron ore, and is found in igneous and metamorphic rocks. It displays cubic system octahedral and dodecahedral crystals, and granular masses are also common. It is black, metallic and brittle. Hardness 6; s.g. 5.2. Permanently magnetized deposits are called LODESTONE.

Magnetochemistry, branch of chemistry concerned with investigating the magnetic properties of compounds. In particular, magnetic measurements made on transition-metal complexes, which are often paramagnetic due to unpaired electrons, give information on their structure and electron configuration.

Magnetohydrodynamics (MHD), the study of the motion of fluids, usually PLASMAS, under the influence of electric and magnetic fields. It is of great importance in the development of nuclear reactors, generators, particle accelerators, and in space technology. An MHD generator consists of a plasma fluid flowing in a magnetic field, thus generating an electric current. The fluid is seeded with easily ionized elements (such as potassium) to increase the power yield.

Magnetometer, instrument for measuring or comparing MAGNETIC FIELD strengths. It usually consists of a short bar magnet with a long nonmagnetic pointer attached to its centre so that it is at right-angles to the axis of the magnet. The magnet is pivoted, like a compass needle, at its centre and the pointer travels over a calibrated scale. Field strengths of magnets are compared by measuring the deflections of the pointer. *See also* PE p.*23.*

Magnetosphere, region in space in which the magnetic field of the Earth exerts a significant influence. It has two parts, or lobes; closer to the Sun is larger and blunt; the other lobe is an elongated tail extending several hundred times the radius of Earth into space. The shape of the magnetosphere is determined by the stress exerted on the magnetopause (outermost edge) by the SOLAR WIND. *See also* PE p.23.

Magnetron, vacuum tube containing an ANODE and a heated CATHODE. The flow of electrons from cathode to anode is controlled by an externally applied magnetic field. When incorporated in a resonant system, a magnetron can act as an OSCILLATOR. It can generate high frequencies and high power in short bursts, and is used in radar sets and microwave ovens.

Magnification, measure of the enlarging power of a telescope or microscope given as the size of the image of an object produced by the instrument, relative to the size of the object viewed with the unaided eye. In an astronomical telescope magnification is equal to the ratio of the FOCAL LENGTH of the objective to the focal length of the eyepiece. A very short focal length eyepiece increases the magnification but reduces the field of view and produces a diffuse and often distorted image. *See also* SU p.100.

Magnifying glass, convex lens that increases the apparent size of objects examined through it, usually between two and ten times. Maximum magnification is obtained when the object is situated just within the focal length of the lens. A magnifying glass generally has a handle attached to a circular clip holding the lens.

Magnitude, in astronomy, numerical value expressing the brightness of a celes-

tial object on a logarithmic scale. Apparent magnitude is the magnitude as seen from earth, either determined by eye, photographically or photometrically, and it ranges from positive through zero to negative values, the brightness increasing rapidly as the magnitude decreases. First magnitude is exactly 100 times brighter than sixth magnitude (just visible to the naked eye). Absolute magnitude indicates intrinsic luminosity and is defined as the apparent magnitude of an object as a distance of 10 parsecs (32.6 light-years) from the object. *See also* SU p.240.

Magnolia, any of about 40 species of trees and shrubs native to North and Central America, India, China and Japan. Valued for their showy flowers of white, yellow, purple or pink, they are mostly deciduous. Some species are evergreen and can withstand cooler climates. Height: to 30m (100ft). Family Magnoliaceae.

Magnum, from the Latin meaning something great, a bottle containing about 145cl of wine or spirits, corresponding to about two average-sized bottles. Magnums are most commonly used in the bottling of champagne.

Magnus, kings of Norway. Magnus I (1024–47), called "the Good", was King of Norway (1035–47) and of Denmark (1042–47). He was the son of Olaf II. He made a treaty with Hardecanute of Denmark in which each named the other his heir. Magnus succeeded Hardecanute and claimed the throne of England as well, but he was too busy crushing a Danish revolt to press his claim. He shared the Norwegian kingdom with his uncle, Harald III, after 1045. Magnus II Haraldsson (1035–69) reigned jointly with his brother Olaf III from 1066–69, after which Olaf became sole ruler. Magnus III (1073–1103) was called "Barefoot" because he wore a kilt. He ruled from 1093–1103, continuing to extend the Norwegian empire. He was killed in an ambush in Ireland. Magnus IV (*c.* 1115–39), called "the Blind", ruled jointly with Harald IV Gille (1130–35) until Harald imprisoned, maimed and blinded him in 1135. After his release from prison, he was killed in battle with Harald's sons. Magnus V Erlingson (1156–84) reigned from 1163–84 as Norway's first crowned King. He fled to Denmark (1179) after his defeat by Sverre Sigurdsson who had claimed the throne. Magnus died in battle attempting to regain the throne. Magnus VI (1238–80), called "the Law Mender", succeeded his father, Haakon IV (1263–80). He made peace with the Scots by ceding the Hebrides and the Isle of Man (1266). He introduced a new legal code whereby a crime was an offence against the state rather than against the individual. He created a new governing class and fixed the law of succession. He also reached an agreement with the church delimiting the powers of church and state. Magnus VII Eriksson (1316–74) was king of Norway from 1319–43 and of Sweden, as Magnus II, from 1319–63. Educated in Sweden, he neglected the interests of Norway and his son, Haakon VI, became King of Norway in 1355. In Sweden, he was deposed (1356–59) by another son, Eric XII, and he lost s Sweden to Denmark (1360). He was deposed finally in 1363. During his reign, codification of Swedish law was completed.

Magog. *See* GOG AND MAGOG.

Magpie, bird of the CROW family, closely related to the JAY, found mostly in the Northern Hemisphere. The common magpie (*Pica pica*) has a chattering cry, a long greenish-black tail and short wings. It has a clearly defined white underside with black above. Length: 46cm (18in). Family Corvidae. •

Magpie moth, or currant moth, slow-flying white, black, and yellow MOTH of the geometer family, (Geometridae). Its larvae, yellow with black spots, feed on gooseberries, redcurrants, sloes, plums and cherries. Wing span 4cm (1.6in). Species *Abraxes grossulariata.*

Magritte, René (1898–1967), Belgian painter. A leading Surrealist, he was at first influenced by ORPHISM and FUTURISM;

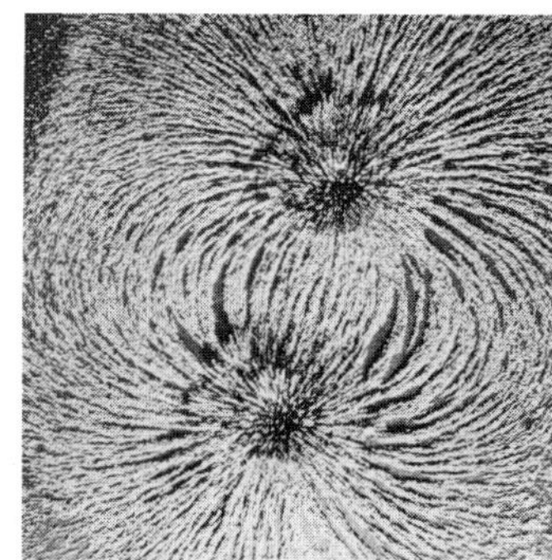

Magnetic field: iron filings indicate lines of force between the poles.

Magnolias are easily cultivated but they must be grown in protected positions.

Magnus II: the seal of the Swedish king who also ruled Norway as Magnus VII.

Magpies eat almost any kind of food, including eggs and young of other birds.

Mah Jongg: the rules are simple but the scoring requires having the rules handy.

Mahogany is ideal for furniture, being tough yet not difficult to carve.

Maidstone: detail of the 14th-century Collegiate Church of All Saints.

Marquise de Maintenon and her niece by Ferdinand Elle: portrait at Versailles.

later, he evolved a realistic style. His early works include *Rough Crossing* (1926) and *Threatening Weather* (1928). His mature works, include *The Liberator* (1947) and *L'Empire des Lumières* (1954). *See also* HC2 p.203.

Magyars, the people associated with the state of Hungary, descendants of Finno-Ugric and Turkic tribes who mingled with AVARS and SLAVS in the 9th century. They conquered large areas of Germany until defeated by Otto I in 955. Although incorporated into the AUSTRIAN EMPIRE from the 16th century, the fiercely independent Magyars demanded parts of the minority rights to preserve their language and culture, achieving first the Dual Monarchy in 1867 and then independence in 1918.

Mahābhārata, the "Great Epic of the Bharata Dynasty", a poem of almost 100,000 couplets, written *c.*400 BC–*c.*AD 200. It is considered one of India's two major epics, with the Rāmāyana. The verse is important both as literature and HINDU religious instruction. The plot revolves around a power struggle between two related families. It was compiled into its present form *c.*AD 400.

Mahadevi. *See* DEVI.

Mahagonny, The Rise and Fall of the City of. *See* RISE AND FALL OF THE CITY OF MAHAGONNY.

Maharaja, title given to native Indian princes who retained a certain autonomy under British rule. The title meant "great king" and distinguished the bearer from those who held the title "raja", who were native rulers or high ranking Hindus.

Mahārāshtra, state in W India, on the Arabian Sea coast; BOMBAY is the capital. It is India's third largest state, in both area and population. Founded in 1960 by the separation of Bombay State into the Marathi and Gujarati (now Gujarat state) linguistic areas, it is composed of five distinctive subregions: Konkan, Deccan, Khandesh, Marathwada and Vidarbha. Its products include cotton, textiles, chemicals and sugar. Area: 307,762sq km (118,827sq miles). Pop. (1971) 50,412,235.

Mahatma, person of special holiness. The term is used by Hindus; the original Sanskrit means "great soul" but has no specific place in organized Hindu religion. The most famous Mahatma of modern times was Mohandas GANDHI.

Mahayana Buddhism, one of the two main schools of Buddhism, the other being HINAYĀNA or Therdvada. Mahayana ("greater vehicle") Buddhism was dominant in India from the 1st to the 12th century and is now prevalent in Tibet, China, Korea and Japan. Unlike the Hinayana ("smaller vehicle") school, it conceives of the Buddha as divine: a transcendental being, a supreme God above all gods who is the embodiment of the absolute and eternal truth. The Mahayana school stresses the importance of faith and devotion to the Buddha and compassion and charity to all creatures. *See also* MS p.221.

Mahdi, messianic Islamic leader. The title is usually used to refer to Muhammad Ahmad (1844–85) of the Sudan, who declared himself to be the Mahdi in 1881 and led the attack on KHARTOUM (1885) during which Gen. GORDON was killed. His followers were eventually defeated at OMDURMAN (1898).

Mah Jongg, game that is believed to have originated in China several centuries BC. Players use 144 decorative rectangular tiles, made of wood, ivory or plastic. There are 108 suit tiles, 28 honour tiles and 8 flower or season tiles. There are three suits and the basic object is to accumulate winning combinations by exchanging and discharging tiles or drawing from a common bank. Although two to six players may participate, four is the usual number.

Mahler, Gustav (1860–1911), Austrian composer and conductor. From the 1880s he was famous as an opera conductor. His compositions reflect the profound influence of WAGNER's music. Many of his works combine orchestral with vocal and choral parts and contain innovatory instrumental effects requiring massive orchestras. Famous among his works are the ten symphonies (the last left as a full-length sketch at his death), *Das Lied von der Erde* (1908) for orchestra with contraalto and tenor solo, and the song cycle *Kindertotenlieder* (1902).

Mahogany, any of numerous species of tropical deciduous trees and their wood, which is valued for furniture making. It has composite leaves, large clusters of flowers, and winged seeds. Mahogany is native to the Americas. Height: to 18m (60ft). Family Meliceae. *See also* PE pp.218, *219.*

Mahonia, genus of evergreen shrubs found in Asia and parts of the USA; especially the Oregon grape, *M. aquifolium.* It bears yellow flowers and edible blue berries. Family Berberidacacea.

Mahrattas (Marathas), Hindu warriors of W central India who rose to power in the 17th century. They extended their rule throughout W India by defeating the MOGUL empire and establishing a Mahratta confederation of states. The Mahrattas successfully resisted British supremacy in India during the 18th century, but internal struggle weakened their power and they were defeated by the British in 1818.

Maiden Castle, site of one of the largest hillforts in Britain, 3km (2 miles) SW of Dorchester, Dorset. Excavated by Sir Mortimer WHEELER (1934–37), it was occupied and altered by various peoples from NEOLITHIC times, and at some point before 50 BC came under the control of the BELGAE. During the Roman conquest VESPASIAN's legions sacked the fort. The slain defenders were buried near its east gate, and the remaining population was moved to a site of Durnovaria (Dorchester). Maiden Castle lay abandoned until the 4th century, when a Romano-British temple was built there. *See also* MM pp. *28, 29.*

Maidenhair fern, dainty fern found in limestone areas in Europe and N America. The wedge-shaped leaves are borne on slender, shiny black stalks. Leaves of most species are pink as they unfold, then turn pea-green. Height: 25–50cm (10–20in). Family Adiantaceae; species *Adiantum capillusveneris.*

Maidenhair tree. *See* GINKGO.

Maidstone, county district in central KENT, England, created in 1974 under the Local Government Act (1972). Area: 395sq km (152sq miles). Pop. (1974 est.) 124,700.

Maiduguri, city in NE Nigeria. Founded by the British in 1907 as a military post, it is now a transport centre for road, rail and air routes. The principal industry is the manufacture of leather goods from crocodile skins. Pop. (1975 est.) 189,000.

Mailer, Norman (1923–), US novelist. His first novel, *The Naked and the Dead* (1948), the story of a small group of infantrymen who survive the invasion of a Japanese-held Pacific island, is recognized as one of the most realistic of American WWII books. His other novels include *Barbary Shore* (1951), *The Deer Park* (1955), a savage satire on Hollywood, and *An American Dream* (1965).

Maillart, Robert (1872–1940), Swiss engineer noted for his reinforced-concrete bridges. Maillart's bridges have a slenderness based on integration of arch, roadway and stiffening girder into a single structure. Most famous is Schwandbach Bridge at Schwarzenburg.

Maillol, Aristide (1861–1944), French sculptor. Initially a painter and tapestry designer, he concentrated on sculpture after 1900 and his work was almost exclusively of the female nude. He turned away from the fluid forms and emotive Romanticism of RODIN and towards the ideals of Classical Greek sculpture. Typical of his work are *Mediterranean* (1901) and *Night* (1902) which are marked by subtlety and clear serene surfaces.

Mail order, shopping practice in which goods are selected, ordered, delivered and paid for by post without the purchaser having to go to a shop. The larger mail order houses sell as wide a range of goods as does any large department store. For them their greatest asset is their catalogue, illustrated with colour photographs of items on sale together with such information as sizes, colours, time for delivery, prices and credit terms. Some manufacturers of single commodities, such as books and records, do most of their business through mail order. In many countries, methods of mail order selling are governed by rigid rules.

Maiman, Theodore Harold (1927–) US physicist who, following on the work of Sir Charles TOWNES on the MASER, successfully built the first optical maser, which he called the LASER, in 1960. *See also* SU p.110.

Maimonides, Moses, or Moses ben Maimon (1135–1204), Jewish philosopher, rabbi and Hebrew scholar, *b.* Spain. As a youth he was attracted to Aristotelian philosophy, which influenced his well-known *Guide of the Perplexed.* He emigrated to Egypt in 1159 after the capture of his native Córdoba by a tyrannical Muslim sect. In Cairo he became court physician to SALADIN and was the recognized leader of Egyptian Jewry. The *Mishneh Torah,* considered his greatest achievement, is a systematic compilation of Jewish oral law. He also wrote, in Arabic, *The Book of Precepts* and *Commentary on the Mishna, Illumination.*

Maine, state in extreme NE USA, largest of the NEW ENGLAND states, bounded by Canada and the Atlantic Ocean. The land is generally rolling country with mountains in the W and some 2,500 lakes throughout the state. The chief rivers are the St John, Penobscot, Kennebec and the St Croix. Economic development has been hampered by poor soil and a short growing season, geographic remoteness and a lack of coal and steel. Dairying, poultry raising and market gardening are the most important farming activities. Potatoes, hay, apples and oats are the chief crops. Manufacturing includes wood pulp and paper, leather goods, food products, textiles and transportation equipment. Other industries include fishing, timber (four-fifths of the state is afforested), printing and publishing. The chief cities are Augusta, the state capital, Portland, Lewiston and Bangor. Scattered English settlements were founded along the coast after 1600, when the name "Maine" was used to distinguish the mainland from the numerous rocky islands. In 1639 the province was granted a charter by Charles I. It later became part of the MASSACHUSETTS BAY COLONY and then of MASSACHUSETTS State until it was admitted to the Union in its own right in 1820. Area: 86,026sq km (33,215sq miles). Pop. (1975 est.) 1,059,000. *See also* MW p.175.

Mainprise, pledge given in England in the Middle Ages by a third party that a person accused of a criminal act would appear in court. The accused was then not detained in prison. If the accused failed to present himself at court, the person giving the pledge, the mainpernor, was fined. Mainprise is thus the origin of the modern system of bail.

Main sequence, in astronomy, band on the HERTZSPRUNG-RUSSELL DIAGRAM. Most stars, including the Sun, are located in this band when their spectral class is plotted against their luminosity. The position of a star on the main sequence indicates that it is typical of its class. *See also* SU pp.226–228, *227.*

Maintenon, Françoise d'Aubigné, Marquise de (1635–1719), mistress and second wife of LOUIS XIV of France. She was entrusted with the care of the king's illegitimate children by their mother, Mme DE MONTESPAN, whom she gradually replaced in the king's affections. She was a deeply religious woman and, after her secret marriage to the king, became a strong influence on court life.

Maisonneuve, Paul de Chomedey, Sieur de (1612–76), founder and first governor of MONTREAL. A soldier, he fought in European wars before being sent by the Société de Notre Dame de Montréal to take possession of their land in the New World. With a party of soldiers and settlers, he landed on Montreal Island, where he established the settlement of Ville-Marie, later Montreal.

Maitland, William (*c.*1528–73), Scottish statesman. Appointed MARY, QUEEN OF

SCOTS' Secretary of State in 1560, he tried to unite Scotland and England by ensuring Mary's right of succession to ELIZABETH I, and lent his support variously to whichever faction seemed likely to make this possible. He died in prison after holding Edinburgh Castle (1571–73) against JAMES I's supporters.

Maize, also called corn or sweet corn, cereal plant of the grass family Gramineae, and its edible grain. It is the key cereal in subtropical zones, Edible seeds grow in rows upon a cob, protected by a leafy sheath. Height: to 5m (16 ft). Species *Zea mays. See also* PE pp.156, 158, 180–181, *188,* 197, *217, 253.*

Majolica (Maiolica), tin-glazed earthenware produced from the 14th century, mainly in Italian centres such as Faenza, Florence and Savona; it is also referred to as FAIENCE and DELFT ware. Majolica is characterized by the use of cobalt blue, copper green, manganese purple, antimony yellow and iron red; the blue and purple glazes were used primarily for outlined designs. A white tin enamel was employed, either for overall white, or for highlights.

Major, one of two "modes" of the DIATONIC scale, the other being the MINOR. It is based on the 13th-century Ionian mode and, as part of the diatonic system, became established in European music at the end of the 15th century. The intervals between the notes of an ascending major scale are: two tones and a semitone, then three tones and a semitone (eg all the white notes on a piano between one C and the next C above).

Major Barbara (1905), a three-act comedy by George Bernard SHAW. The heroine joins the SALVATION ARMY to fight poverty and is chagrined to find that people like her father, a munitions manufacturer, are poverty's major supporters. *See also* HC2 p.210.

Majorca (Mallorca), Spanish island, largest of the BALEARIC ISLANDS, in the W Mediterranean, approx. 233km (145 miles) off the Spanish coast; the capital is Palma. The island was taken by the MOORS in 797. It was made the kingdom of Majorca from 1213 to 1276, by James I, King of Aragon. The island served as a base for Italian forces opposing the LOYALISTS during the SPANISH CIVIL WAR. Products: olives, figs, oranges, lemons, almonds. Industries: tourism, fishing, lead, iron and coal mining, limestone and marble quarrying, wine, jewellery and handicrafts. Area: 3,639sq km (1,405sq miles). Pop. (1970) 460,030.

Majuba, Battle of (27 Feb. 1881), the decisive Boer victory over the British in the first Anglo-Boer War. The 600 British soldiers, under Gen. Colley, were exposed against the skyline of the hill where they were encamped and were overwhelmed by 150 Boers under Gen. JOUBERT. Two hundred British were killed or wounded, and only two Boers died. *See also* LAING'S NEK, BATTLE OF.

Makarios III (1913–77), Archbishop, *b.* Mikhail Khristodoulou Mouskos. Leader of the ENOSIS (union of Cyprus with Greece) movement, he was a suspected terrorist but was elected President of Cyprus (1959) when it became independent of Britain. In later years he became less extreme in his views but continued to put pressure on the island's Turkish population. In 1974 the Turks invaded Cyprus and occupied the northern half of the island.

Makasar, city and port in the SW of Celebes Island, E Indonesia. The island is separated from Borneo (W) by the Makasar Strait. Originally a Portuguese settlement, it was taken by the Dutch in 1667 and developed as a trading centre. It was made a free port in 1848 and became part of the Republic of Indonesia in 1949. Exports include copra, gum and resins, rubber, coffee and spices. Pop. (1961). 384,159.

Make-up, in drama and primitive ritual, the technique whereby the face, hands or body of a person may be coloured or given added hair or texture in accordance with the demands of a specific role, if a mask is not to be worn. In the theatre, since the 19th century, it has generally been used to strengthen the features and heighten their expressive power, particularly in a large auditorium where artificial lighting is used. Elizabethan actors, acting in daylight, seldom wore make-up, except to portray angels, ghosts, Negroes or Moors. In the 17th century artificial lighting rendered actors' features indistinct and powder make-up began to be used. David GARRICK and other actors experimented with adding lines to their face to create an impression of old age, while actresses emphasized their normal street make-up. Powder make-up was found to streak as the actors perspired from the heat of the new gas lighting introduced in 1817, and so a grease foundation, invented by Carl Baudin, was used. Ludwig Leichner introduced a tallow-based make-up in the 1870s. By the end of the century a basic range of numbered sticks of make-up of different tones had been evolved. Electrical lighting and the demands of film and television close-ups demanded greater expertise in the application of make-up, and most is done by specialists, rather than by the actors themselves.

Mäkinen, Timo (1938–), Finnish rally and speedboat driver. His successes have included the 1964 Tulip Rally, a hat-trick (1965–67) in his native Thousand Lakes Rally, and the 1965 Monte Carlo run, in which he finished first again in 1966, only for his Cooper S to be disqualified. In boats he won the 1968 Finnish off-shore championship and the 1969 Round Britain race.

Makkah. *See* MECCA.

Mako, also called sharp-nosed mackerel shark, either of two species of sharks: *Isurus glaucus* of the Indo-Pacific and *I. oxyrinchus* of the Atlantic. It is blue-grey and has a pointed snout, a slender body and a crescent-shaped tail. Length: to 4m (13ft). Family Isuridae. *See also* NW pp.*130, 240.*

Malabar Christians. *See* THOMAS CHRISTIANS.

Malabar Coast, SW coast of India, between the Western Ghats range (E) and the Arabian Sea (W). The chief ports are Cochin and Calicut. It is the traditional site of the apostle Thomas's first missionary work *c.*AD 52. Portuguese traders established posts there in 1498–1503, to be followed by the Dutch in 1656. By the late 18th century, the British had occupied the area. Its modern industries include fishing and the growing of rubber. Length: 725km (450 miles).

Malacca, state in Malaysia, on the Malay Peninsula, bounded by the Strait of Malacca (S and W), which connects the Indian Ocean and South China Sea. Malacca is the capital. It was made a British colony in 1824, became a state of Malaya in 1957 and of Malaysia in 1963. Its products include rice and rubber. Area: 1,658sq km (640sq miles). Pop. (1970) 403,722.

Malachi, last of the books of the 12 minor prophets and last book of all of the Old Testament in the Authorized Version. Probably written about 460 BC, it attempts to dissuade the people of Israel from foreign influences and associations, and ends with a prophecy of an imminent Day of Judgment.

Malachite, carbonate mineral, basic copper carbonate $(Cu_2CO_3(OH)_2)$, found in weathered copper ore deposits. Malachite has a monoclinic system, and is green in colour; it is sometimes used as a gemstone. Hardness 3.5–4; s.g. 4.

Málaga, port in S Spain in Andalusia region, on the Mediterranean Sea; capital of Málaga province. Founded *c.*1100 BC by the Phoenicians, it was taken by the MOORS in 711, and remained under Moorish rule until 1487, when it was captured by FERDINAND II of Aragon and ISABELLA I of Castile. During the SPANISH CIVIL WAR (1936–39), the city was taken from the Loyalists by FRANCO. Industries: wine, olive oil, footwear, flour-milling, tourism. Pop. (1974) 402,978.

Malagasy, independent island republic approx. 400km (250 miles) E of the African mainland, formerly the French protectorate of Madagascar. Eighty per cent of the population are farmers and agriculture is the most important part of the economy. The chief exports are coffee, sugar, rice, vanilla, cloves and clove oil. Manufacturing industries, however, are being expanded. The capital is Antananarivo (TANANARIVE). Area: 587,045sq km (226,658sq miles). Pop. (1975 est.) 8,044,000. *See* MW p.120.

Malamud, Bernard (1914–), US short story writer and novelist. His works generally deal with Jewish characters and include *The Assistant* (1957), *A New Life* (1961) and *The Fixer* (1966), a prize-winning novel about a Jew accused of ritual murder. In 1973 he published a book of short stories, *Rembrandt's Hat.*

Malamute, Alaskan, dog similar to the husky. The colour of its coat may be any shade of grey to black; it is white under the body and on the legs. The tail is plumed and carried over the back. It is used as a sled dog, especially on Arctic and Antarctic expeditions. Height: 59–65cm (23–25in); weight: 34–38.5kg (75–85lb.)

Malan, Daniel François (1874–1959), South African politician. He furthered Afrikaaner interests and served in various ministries from 1924–33. A nationalist, he was Prime Minister from 1948–54 and made his belief in racial segregation official as "apartheid". *See also* HC2 p.252.

Malang, region in Java, Indonesia, comprising the mountain groups of Kawi (W), Tengger (E) and the fertile plain between. It is an agricultural area whose products include vegetables and flowers. Pop. (1961) 1,474,106.

Malapropism, the misuse, through confusion, of one word for another, usually because of a similarity in sound. Richard Brinsley SHERIDAN used it as a humorous device in his play *The Rivals* through a character named Mrs Malaprop.

Malaria, insect-borne, often chronic, disease resulting from infection with one of four species of the parasitic PROTOGOAN *Plasmodium,* acquired through the bite of Anopheles mosquito and affecting the red blood corpuscles. Plasmodium falciparum malaria is the most serious form, with severe complications that may lead to death. Attacks of fever, chills and sweating typify the disease and occur as new generations of parasites develop in the blood, with the frequency of attacks related to the species involved. The disease is treated with antimalarial drugs such as QUININE and its derivatives. Malaria is one of the most widespread diseases, although some degree of control has been achieved by the use of DDT and other insecticides.

Malawi, independent republic of E central Africa, formerly the British territory of Nyasaland. A poor country, the chief crops, tea and tobacco, account for 80 per cent of all exports. Manufacturing is being expanded, but many workers seek employment in neighbouring countries for example, South Africa and Zambia. The capital is Lilongwe. Area: 118,484 km (45,747sq miles). Pop. (1975 est.) 5,044,000. *See* MW p.121.

Malawi, Lake. *See* NYASA, LAKE.

Malaya, former federation of nine MALAY STATES and two of the STRAITS SETTLEMENTS (Malacca and Penang). It included a large part of British Malaya. Malaya was established as the Union of Malaya in 1946, and was then reorganized as the Federation of Malaya in 1948, and achieved independence nine years later. The region became part of the Federation of MALAYSIA in 1963. *See also* HC2 pp. *133,* 220, 246.

Malay Archipelago, group of islands in SE Asia more commonly called EAST INDIES.

Malay Peninsula, promontory of SE Asia, between the Strait of Malacca and the South China Sea, comprising SW Thailand and West Malaysia. A mountain range, culminating in Mt Gunong Tahan, 2,190m (7,186ft), extends along its entire length and rain-forest covers half of the land. It is one of the world's richest tin and rubber areas; other products include timber and fruit. Area: 181,300sq km (70,000sq miles).

Malaysia, nation in SE Asia, occupying the S half of the Malay Peninsula and most of N Borneo. It is composed of 13 states.

Majuba: the battle was the result of an error by Gen. Colley, killed in action.

Makarios, once banished to the Seychelles, returned to Cyprus in 1957.

Málaga, founded by the Phoenicians, is dominated by its 16th-century cathedral.

Dr Daniel Malan was one of the world's staunchest advocates of racial segregation.

Malay States

Mallard: the bright colours of the male outshine the duller-coloured female.

Mallia: the ruins show the Minoans were building extensively by 2000 BC.

Mallow, like other members of its family, has a fibrous stem and tacky sap.

Sir Thomas Malory: a page from the manuscript of Book 5 of his *Morte D'Arthur*.

Although most of the population is involved in farming, Malaysia is the world's leading tin producer and supplies about 33% of the world's rubber requirements. Other industries include the manufacture of machinery and chemicals. The capital is KUALA LUMPUR. Area: 329,740sq km (127,313sq miles). Pop. (1975 est.) 11,900,000. *See* MW p.121.

Malay States, states of the MALAY PENINSULA, in particular those previously under British protection. Those states in the N and central part of the Peninsula are now in Thailand. The S states are part of Malaysia.

Malcolm, name of four kings of Scotland. Malcolm I (*r.*943–954) was the son of Donald, King of Alba. Malcolm II (953–1034; *r.*1005–34), secured the Anglo-Saxon district of Lothian permanently for Scotland by defeating the Northumbrians at Carham in *c.*1016. In the same year he gained control over Strathclyde, completing the political unification of northern Britain. Malcolm III (1031–93; *r.*1058–93), called Canmore (Gaelic *Ceann-mor*, "great chief"), was a child when his father, DUNCAN II, was murdered by MACBETH in 1040 but regained the throne when he himself killed Macbeth. His marriage to St MARGARET, an English princess, began two centuries of frequent royal marriages between Scotland and England. Three of their sons became kings, the youngest, DAVID I, in 1124. Malcolm IV (1142–65; *r.*1153–65), grandson of David I, was proclaimed heir to ensure a peaceful succession when David's only son died. He was forced to surrender Northumberland and Cumberland to HENRY II of England in 1157. *See also* HC1 pp.182, *183*.

Malcolm III, called Canmore. *See* MALCOLM.

Malcolm, George John (1917–), British harpsichordist, pianist and conductor. He was Master of the Cathedral Music at Westminster Cathedral (1947–59), where he trained a boys' choir for which Benjamin BRITTEN wrote his *Missa Brevis*.

Malcolm X (1925–65), US militant black leader. While in prison he joined the BLACK MUSLIMS, and on his release in 1953 he became a Minister. He later disagreed with the leader Elijah Muhammed and formed a rival group, the Muslim Mosque, in 1964, in which year he also was converted to Orthodox ISLAM. He was assassinated in 1965.

Maldives, independent republic in the Indian Ocean approx. 645km (400 miles) W of Sri Lanka. It is composed of some 2,000 islands in 12 distinct groups. The main one and the seat of government is Male. Formerly a British colony, the islands' economy is based almost entirely on fishing. Tourism is being promoted as an additional source of income. Land area: 298sq km (115sq miles). Pop. (1975 est.) 119,000. *See also* MW p.122.

Male, one of the Maldive Islands in the Indian Ocean, between the Kardiva and Veimandu channels. It consists of two groups of atolls, North Male and South Male, and is an inter-island trade centre. Area 251sq km (97sq miles). Pop. 13,610.

Malenkov, Georgi Maksimilianovich (1902–), Prime Minister of the USSR from the death of STALIN in March 1953 to his removal by KHRUSHCHEV in February 1955. He rose in the Communist Party as a trusted aide of Stalin. He was removed from the Politburo in 1957. *See also* HC2 p.240.

Malevich, Kasimir (1878–1935), Russian painter. He founded SUPREMATISM, a form of geometric abstractionism, in 1913. His manifesto of 1915 stated that a combination of geometric figures expressed "pure emotion". The epitome of the style is his *Suprematist Composition: White on White* (1919). *See also* HC2 pp.204, *204*, 278.

Malherbe, François de (1555–1628), French poet and critic who set the early 17th-century French literary style. In 1605 he went to Paris and became court writer. He criticized the emotional, decorative style of his contemporaries.

Mali, independent republic in W Africa. The River NIGER divides the country's northern desert area, which merges with the SAHARA from the fertile southern region which produces sorghum, rice and cotton. There is a variety of light industries including canning, brick-making and textiles. Mali's large mineral resources remain unexploited. The capital is BAMAKO. Area: 1,240,000sq km (476,764sq miles). Pop. (1976 est.) 5,842,000. *See* MW p.123.

Malic acid, organic chemical compound, formula HOOC.CHOH.CH$_2$.COOH, with several important roles in the body's metabolism, notably in the CITRIC ACID CYCLE which is concerned with the production of energy during respiration.

Malinowski, Bronislaw (1884–1942) Polish-English anthropologist, considered by many to be one of the pioneers of social anthropology. His work with the peoples of New Guinea and the Trobriand Islands helped him to formulate his functional theory which held that every aspect or norm of a society is a function vital to its existence.

Mallard, largest and commonest freshwater duck in Britain. The black, white, brown and grey male is easily recognized by its green head, whereas the female is mottled brown with blue wing markings. It dabbles or feeds from the surface. Length: 63cm (28in). Species *Anas platyrhynchos*. *See also* NW pp.*143, 148, 230.*

Mallarmé, Stephane (1842–98), French poet. A leader of the SYMBOLIST movement, he was influenced by BAUDELAIRE and VERLAINE. His aim was to write "pure poetry" and he spent his life in agonized labour over a few poems, his best remembered compositions being *Hérodiade* (1869), *L'Après-Midi d'un faune* (1876) and *Un coup de dés jamais n'abolira le hasard* (1897). *See also* HC2 p.100.

Malle, Louis (1932–), French film director, one of the NOUVELLE VAGUE of French film-makers. His films include *World of Silence* (1956), *The Lovers* (1958) and *Viva Maria!* (made in 1965 with Brigitte BARDOT and Jeanne MOREAU). An uncompromising director, his films often have unpopular or sensitive themes, eg *Lacombe Lucien* (1974) about a young collaborator in German-occupied France.

Mallee, scrubby Australian vegetation which consists mainly of trees and shrubs of the genus *Eucalyptus*. They have leathery leaves which conserve moisture during dry periods.

Mallee fowl, turkey-like bird that lives in the mallee scrub area of semi-desert in Australia. Eggs are laid in winter in pits scratched in the ground, and covered – the birds do not use their own body heat to hatch the eggs. When the young hatch, they dig their way out. Length: to 69cm (27in). Family Megapodidae; species *Leipoa ocellata*.

Mallia, town in Crete, 18km (15 miles) E of Heraklion. It is the site of a Minoan palace (1900–1450 BC), which had probably been subject to a serious earthquake *c.*1700 BC. The site was first excavated in the 1940s. *See also* HC1 pp.38–39.

Malloum, Felix (1932–), Chad army officer and politician. He served in the French army in Indochina (1953–55), and became commander-in-chief of the Chad army in 1972. He was arrested by President Tombalbaye in 1973, but after the coup of 1975 became head of state and Minister of Defence.

Mallow, annual and perennial plants occurring in tropical and temperate regions of the world. The flowers are pink and white. The mallow family includes more than 900 species of plants, of which cotton, okra, hollyhock and hibiscus are among the best known. Family Malvaceae; genus *Malva*.

Mallowan, Sir Max Edgar Lucien (1904–), British archaeologist noted for his excavations at the sites of UR, an ancient Babylonian city (with Sir Leonard WOOLLEY) and of NINEVEH in Assyria (with R. Campbell Thompson). He was professor of Western Asiatic archaeology at the University of London (1947–62). He was the husband of Agatha CHRISTIE; his own autobiography, *Mallowan's Memoirs*, was published in 1977.

Malmö, port in SW Sweden on the Øresund opposite Copenhagen. It was a major trading and shipping centre during the period of the HANSEATIC LEAGUE. Originally mostly ruled by the Danes, the city was annexed by Sweden in 1658. It is now a major naval and commercial port. It is the site of Malmöhus castle (1434) and a 16th-century city hall. Industries: shipping, food processing, cement, textiles, shipbuilding, rolling stock. Pop. 243,600.

Malnutrition, inadequate dietary intake which is often the insufficiency of one or more of the necessary components in a healthy DIET. This most frequently takes the form of protein deficiency, although there are many situations where malnutrition takes the form of either mineral or VITAMIN deficiency. Primary malnutrition results from an inadequate diet often due to such regional calamities as war, drought or overpopulation, and it may also arise from poor eating habits or the psychological condition ANOREXIA NERVOSA. Secondary malnutrition results either from the body's failure to absorb or utilize nutrients, from increased bodily needs, such as lactation, growth or fever, or from excessive excretion, as with DIARRHOEA. *See also* MS pp.71, 107, 108.

Malocclusion, in dentistry, misalignment of the row of teeth in the upper jaw with those in the lower jaw, resulting in improper contact between the biting surfaces of the teeth. If untreated, it can lead to deformities of the jaws or difficulties in chewing. Malocclusion may be either hereditary in origin or caused by the early loss of teeth due to decay. *See also* MS pp.130–131, *131.*

Malory, Sir Thomas (*fl.* 1469), English writer, author of *Le Morte d'Arthur.* His identity is obscure, but he was probably the Thomas Malory from Warwickshire who fought in the Wars of the Roses, became a Member of Parliament in 1445 and served several prison sentences between 1450 and 1470 for rape, robbery, attempted murder and perhaps for political reasons. *See also* HC1 pp.20, 206, 348–349L.

Malouel, Jean (*fl.*1386–1419), Flemish painter and a pioneer, in N Europe, of painting on wooden panels. He was court painter to the dukes of Burgundy. His works include the *Martyrdom of St Denis.*

Malpighi, Marcello (1628–94), Italian physiologist. He was the founder of microscopic anatomy, demonstrating how blood reaches the tissues through tiny vessels (capillaries). William HARVEY had inferred that there must be capillaries, but had never seen them. Malpighi was able to explain the network of tiny veins he could see on the lung surface. He extended the use of the microscope to embryology. The Malpighian tubules of the kidney are among the structures named after him. *See also* HC1 pp.302–303; MS pp.62, 126.

Malplaquet, Battle of (1709), engagement between France and the allied English, Austrian and Dutch forces in the War of the SPANISH SUCCESSION. The slightly outnumbered French were finally forced into an orderly retreat, but the vast Allied losses (22,000 to the French 12,000) ultimately made the battle a French victory, preventing an Allied invasion of Paris.

Malraux, André (1901–76), French intellectual, novelist, politician and war hero. His novels, *The Conquerors* (1928), *Man's Estate* (1933) and *Days of Hope* (1937) which examine the heroism of individuals all engaged in social revolution, are based on his first-hand experiences in China (1925–27), the Spanish Civil War (1936–39) and in the French resistance during WWII.

Malt, germinated grain, usually BARLEY, used in beverages and foods. The grain is softened in water, allowed to germinate and then kiln-dried. This activates ENZYMES which convert the starch to malt sugar. Most malt is used in the manufacture of BEER. *See also* PE pp.180, 204.

Malta, Knights of. *See* KNIGHTS HOSPITALLERS.

Maltase, collectively, group of intestinal ENZYMES that break down complex sugars so that they can be absorbed during digestion.

Maltese, breed of toy dog, with a

medium-length head, a fine, tapered muzzle and low-set ears. The compact body is set on fine-boned legs and the long tail curves over the back. The long, flat, silky coat hangs almost to the ground and is pure white. Average size: 12.5cm (5in) at the shoulder: weight: 3kg (6.6lb).

Maltese Falcon, The (1941), US film based on Dashiell Hammett's crime novel, and adapted and directed by John HUSTON. Humphrey BOGART starred as a tough, disillusioned private detective. The film's style influenced the mystery genre for many years.

Malthus, Thomas (1766–1834), British economist and minister famous for his *Essay on Population* (1798). According to Malthusian theory, population increases geometrically but the food supply can increase only arithmetically so that population must eventually overtake it, with famine, war and disease as consequences. Malthus objected to any interference in this process, and did not favour family planning. *See also* HC2 pp.33, 96; MS p.317.

Maltose, CARBOHYDRATE that contains two molecules of GLUCOSE, a simple sugar. It is produced by the breakdown of STARCHES and GLYCOGEN during digestion. *See also* MS p.68.

Malus, Étienne-Louis (1775–1812), French physicist who in 1809 announced his discovery of the polarization of light by reflection. In 1810 he also produced a theory of double REFRACTION of light in crystals. Earlier in his career, in 1798, he had taken part in the invasion of Egypt by Napoleon I as a member of the engineer corps.

Malvern Hills, range of hills in Worcester(shire) and Hereford(shire), W central England. The highest point is Worcester Beacon, rising to 425m (1,395ft). Hereford Beacon (340m; 1,114ft) was the site of an ancient camp. Length: approx. 13km (8 miles).

Maly Theatre, Moscow's oldest theatre. It opened in 1824 with a company that had been in existence since 1806. A theatre of tradition, it has been important in the development of Russian drama.

Mamba, any of several large, green or black, poisonous African tree snakes of the cobra family, Elapidae. The deadly black mamba (*Dendroaspis polylepsis*) is the largest species. It is grey, greenish-brown or black and is notoriously aggressive; its bite is almost always fatal. Length: to 4.3m (14ft).

Mambo, dance of Afro-Cuban origin. It was introduced to the USA and then became popular throughout the world in the 1940s. Its basic 4/4 metre is highly syncopated. It was the basis of the cha cha of the 1950s.

Mamelukes, ruling dynasty of Egypt (1250–1517). Originally Turkish and CIRCASSIAN prisoners of GENGHIS KHAN, they were sold as slaves to the sultan of Egypt, who trained them as soldiers. They became strong enough to usurp the sultan's political authority and pursued an aggressive foreign policy, although domestic intrigue limited each Mameluke sultan's reign. They were the first to halt the MONGOL advance westwards and to defeat the ASSASSINS. The Mamelukes, in turn, were defeated by the Ottoman Sultan SELIM I in 1517, but remained a class and kept their lands. As the Ottoman Empire declined in the 18th century, the Mameluke governors of Egypt regained some autonomy; their power was not ended until 1811, when they were defeated by MUHAMMED ALI.

Mammal, class (Mammalia) of VERTEBRATE animals characterized by mammary glands in the female and full, partial or vestigial hair covering. Mammals are warm blooded having a constant body temperature, controlled by various mechanisms. These include skin glands through which they secrete fluid in hot conditions. In addition, they have a four-chambered heart with circulation to the lungs separate from the rest of the body. The pumping ability of mammalian lungs is increased by the action of the muscular abdominal diaphragm.

Mammals are found on all major land masses and in the oceans. As a group, they are active, alert and intelligent. They usually bear fewer young than other animals and give them longer and better parental care. Most mammals before birth grow inside the mother's body and are nourished from her by means of a placenta. When born, some are blind and naked, others are fully furred with wide open eyes. All are fully developed but continue to feed on the milk from their mother's mammary glands.

There is a wide range of shapes and sizes among mammals resulting from different environments and ways of life and feeding. Teeth and limbs exhibit the greatest variation. The teeth of herbivores, for example, differ from those of insectivores, and carnivores have yet a different kind, as the teeth have evolved for eating plant food, chewing insects and tearing flesh, respectively. Similarly, the limbs of some marine mammals have developed into flippers, and the toes of fast running animals of the plains, such as horses and deer, have developed into hooves.

Mammals include 17 orders of placentals, one MARSUPIAL order – all livebearing – and an order of egg-laying MONOTREMES. Mammals probably evolved about 180 million years ago from a group of warm-blooded reptiles. They became the dominant land animals after the extinction of the dinosaurs about 65 million years ago. Today, mammals range in size from shrews weighing a few grammes, to the blue whale, which can weigh up to 150 tonnes. *See also* PLACENTAL MAMMALS; NW pp.73, 154–173, 185, *187*, 188–191, 242–243.

Mammary glands. *See* BREASTS.

Mammee apple, large tree with leathery leaves and edible fruit, grown in the USA and the West Indies. The yellow fruit has a bitter rind surrounding sweet flesh which is eaten raw or used for making preserves. The seeds and resin from the tree have been used medicinally and for perfume. Family Clusiaceae; species *Mammea americana.*

Mammography, examination of breast tissue by X-ray, commonly in order to detect the presence of a malignant tumour in an early stage of development.

Mammoth, extinct PLEISTOCENE ancestor of the elephant, which was approximately the size of modern elephants. Many were covered with long red or brown hair. The prominent tusks were long and directed downwards, sometimes crossing in adult males. The most prominent types of mammoths were the *Mammuthus meridionalis* of Europe and Asia, the imperial mammoth of the USA and the woolly mammoth of N Eurasia and the USA. In summer months the permafrost of Siberia has been known to yield whole specimens that have been frozen for as long as 30,000 years. Genus *Mammuthus. See also* NW pp.*189, 191.*

Mammoth cave, one of the largest cave systems in the world, located in W Kentucky, USA. It contains extensive passages, many still unexplored, and spectacular limestone formations of STALAGMITES and STALACTITES. The constant 12°C (54°F) temperature may have caused the mummification of a human body found there which is probably more than five hundred years old. *See also* PE pp.110, *111.*

Man, primate mammal of the genus Homo, the only living species of which is *Homo sapiens.* When compared with his nearest relatives the CHIMPANZEE, GORILLA and ORANG-UTAN, man is immediately distinguishable by a number of features. He walks upright anatomically, his body is only patchily hairy; his big toes are not opposable to his other toes; his backbone is more S-shaped than straight; and his forehead is higher than that of any ape. Otherwise, the bones and glands of man are markedly similar to those of the great apes. Microscopically, man is clearly distinguishable from the great apes by the size, number and shape of his chromosomes. Intellectually, man far outstrips all his relatives. Evidence of this is his capacity for language: even the cleverest chimpanzee can recognise only a few hundred words (as signs or symbols). Socially, man

is rather similar to lesser primates preferring a family or other small group. Man has always actively made great changes in both his immediate environment and in ecosystems throughout the world. *See also* MS pp.20–27.

Man, prehistoric, any of the evolutionary stages of early man preceding recorded history. The term includes all the members of the family Hominidae, beginning with the Australopithecines, followed by the Pithecanthropines and, ultimately, HOMO SAPIENS, our own species. The Australopithecines may have been derived from *Ramapithecus* of about twelve to fourteen million years ago, while AUSTRALOPITHECUS itself existed from about four or five million years ago. The Pithecanthropines of from one to half a million years ago are all classified as HOMO ERECTUS and include JAVA MAN and PEKING MAN as well as European and African representatives. The earliest *Homo sapiens* fossils date from about 350,000 years ago. Material of a later date suggests that this man subsequently diverged in two directions. One of these led to NEANDERTHAL MAN and similar types in Asia and Africa, and the other to the modern races of man first represented in Europe by CRO-MAGNON man of 30,000 years ago. *See also* LEAKEY; MS pp.22–36.

Man, Isle of. *See* ISLE OF MAN.

Mana, charismatic spiritual power in some Melanesian cultures, particularly the Maori culture of New Zealand. It is used to explain personal status and authority. A person is believed to be born with mana, but can increase (or decrease) it by his way of life, for example through bravery and skill in war.

Managua, capital city of Nicaragua, in W central Nicaragua on the S shore of Lake Managua. It is the economic, industrial and commercial hub of Nicaragua. It suffered damage from an earthquake in 1931, after which much of the city was rebuilt: another, on 23 December 1962, killed over 10,000 people and necessitated further reconstruction. Pop. (1971) 398,514.

Manakin, small, stocky bird with a broad, hooked beak and short wings and tail that lives in damp tropical forests of Central and South America. Males are black splashed with red, yellow or blue, and sometimes have ornate tails and crests. Females are olive-green. The males perform acrobatic courtship dances. Family Pipridae. *See also* NW p.*144.*

Man and Superman (1903), four-act comedy by George Bernard SHAW, written in response to a critic's request for a play about DON JUAN. The "Don Juan in Hell" scene is usually performed separately, many producers feeling the complete play to be too long. It is Shaw's first major treatment of "creative evolution".

Manasseh, or Manasses, name of two biblical characters. As the son of JOSEPH, Manasseh gave his name to one of the 12 tribes of Israel. Another Manasseh was the son of HEZEKIAH.

Manatee, any of three species of large, plant-eating, sub-ungulate, aquatic mammals found primarily in shallow coastal waters of the Atlantic Ocean. It has a short, tapered body ending in a large rounded flipper; there are no hindlimbs. Length: to 4.5m (14.7ft); weight: 680kg (1,500lb). Family Trichechidae; genus *Trichechus. See also* NW p.*162.*

Manaus, city in NW Brazil, capital of Amazonas state, on the Rio Negro, 16km (10 miles) from its confluence with the Amazon. Founded in 1669, the city remained undeveloped until the late 19th-century rubber boom. Although more than 1600km (1000 miles) from the Atlantic, the port can accommodate ocean-going vessels. Oil from Peru is refined there. Products of the Amazon basin such as jute, timber, nuts and rubber are the major exports. Pop.(1970) 284,118.

Manawatu, river in SW North Island, New Zealand. It rises in the Ruahine Range and enters the Tasman Sea at Manawatu Heads. Length: 182km (113 miles).

Man Born To Be King, The (1942), series of weekly radio plays about the life of

Mamba: the green mamba, unlike the black, is shy and leads a solitary existence.

Mammal: one characteristic of mammals is that they suckle their young.

Mammoth: cave-paintings of these huge beasts are seen in southern France.

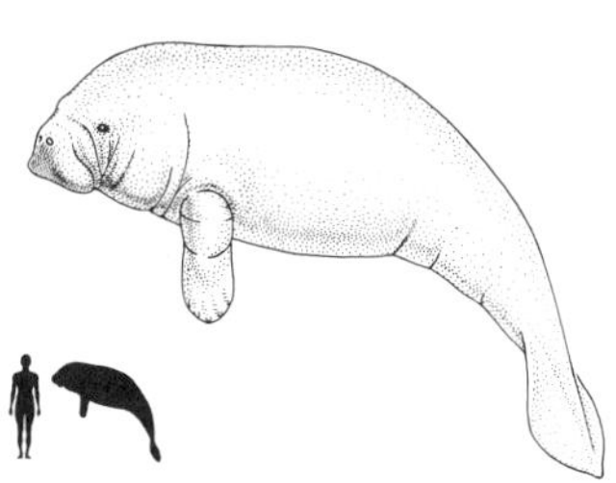

Manatees, widely hunted for their hide and oil, are now becoming rare.

Manche, La

Manchester terriers originated as a cross between whippets and rat-catching terriers.

Mandalay: the palace of Queen Sopia-Lat engraved in *Le Tour du Monde*, 1892.

Mandrill: the colours of the male become brighter when the animal is excited.

Édouard Manet painted the distinguished French writer Émile Zola at his desk.

Christ by Dorothy L. SAYERS. Commissioned by the BBC, the author's humanizing of the character of Jesus created a public furore when the plays were first transmitted.

Manche, La, French name for the English Channel. Variants of the word are found also in Gaelic; all mean either "channel" or "sleeve".

Manchester, city in Greater Manchester, NW England. A major British city in the country's most densely populated region, it serves as a distribution point for the surrounding mill towns. The building of the Manchester Ship Canal (1887–94) changed the city from a river port to a seaport. There are now oil refineries along the canal. Manchester's major industries are textiles, printing and publishing. Pop. (1973) 530,580.

Manchester School, name given to British economists and politicians of the mid-19th century who believed in LAISSEZ-FAIRE economics. Free trade was their chief object and they got their name from the founding in Manchester of the ANTI-CORN LAW LEAGUE. They also tended towards pacifism and adopted radical positions on other issues. The most influential adherents of the "school" were Richard COBDEN and John BRIGHT.

Manchester terrier, sporting dog that was originally used for killing rats and coursing rabbits. It has a long head and pointed, erect or folded ears. The moderately short body is set on long legs and the tail is short and pointed. The smooth, short coat is black and tan. There are two varieties – the standard, height: 35–41cm (14–16in); and the toy, height: 15–18cm (6–7in).

Manchu dynasty, ruling house of China from 1644 to 1912. It is also called the Ch'ing, or "pure", dynasty. It was established by the Manchu leader Fu-lin. It ended with the abdication of the young emperor Pu Yi and the establishment of the Chinese republic. For almost two hundred years the Manchu emperors extended their influence until by 1800 they exercised suzerainty over an area stretching from Siam in the s to Tibet and Sinkiang in the w and Mongolia and Amur in the N. *See also* HC2 pp.52, 53.

Manchukuo, puppet state set up by the Japanese in Manchuria from 1932 to 1945. It was under the nominal rule of the pretender to the Manchu throne, Henry Pu Yi or Hsüan T'ung. The independence of the state was not acknowledged by most Western powers. It was dissolved in 1945 at the end of WWII. *See also* HC2 p.214.

Manchuria, industrial region in NE China, bordered by the USSR (N and NE), Korea (SE), the Yellow Sea (S) and Inner Mongolia (W). Mukden is the chief city. The N, E and W borders are lined by the Khingan highlands and Ch'ang-pai Mts; the central area is the alluvial Liao-Sungari lowland. Ruled by lesser Chinese dynasties until *c.*1125, the region was governed by the Nuchen (later known as Manchus) from 1125 to 1234, when they were dispersed by the Mongols. The Manchus reunited and by 1664 they had conquered all of China, ruling until 1912. Russian influence dominated the area from 1898 to 1904, when it was replaced by Japanese authority. Southern Manchuria and Port Arthur developed rapidly with the introduction of the South Manchurian Railroad. In 1949 the Nationalist regime was overthrown by the Communists. In 1954 Manchuria was divided into three provinces – Liaoning, Kirin, Heilungkiang – all under the control of Peking. Its industries include iron and steel, coalmining, chemicals, oil refining, timber, textiles. Area: 1,515,150sq km (585,000sq miles). *See also* MW pp.56–57.

Mandaeans, GNOSTIC sect also known as St John's Christians; it probably originated *c.* AD 100. The sect's 20th-century adherents are to be found in s Iraq and Khuzistan, still speaking an eastern Aramaic dialect. Their highly complex system of beliefs has incorporated many traditions during the last 18 centuries, including biblical elements, Babylonian astrology and Iranian dualism. Their rites centre on ritual ablutions and are presided over by bishops, priests and deacons.

Mandala, symbolic design sacred to BUDD-HISTS and HINDUS, used to focus the concentration during meditation. Although the design may consist of virtually any symmetrical shape, many are based upon, and have the outline of, a circle. The circle, said to represent the macrocosm – the infinity of the universe and the territory of the gods – is commonly divided into four quarters, each of which may be allotted its own forces or deities. Other versions consist of concentric circles representing the cosmic processes through which the meditator mystically travels towards the centre of meaning. A mandala is a sort of visual MANTRA; both mandala and mantra may be combined in meditation.

Mandalay, city in N Burma, on the River Irrawaddy; capital of Mandalay division and second largest city in Burma. Founded in 1857 Mandalay was the last capital (1860–85) of the Burmese kingdom before it was annexed to British Burma. The city was occupied by the Japanese during WWII and severely damaged. A centre of Burmese Buddhism, Mandalay is noted for the Arakan Pagoda, and the sacred Seven Hundred and Thirty Pagodas. Industries: textiles, jade. Pop. 393,000.

Mandarin, European name for officials or civil servants of the Chinese Empire. Their entry to the bureaucracy was by examination, and rank was shown by their form of dress. As a matter of policy, mandarins never officiated in their home province but were posted to other areas for limited periods. *See also* HC1 p. *126.*

Mandarin, major dialect of Chinese, the spoken language of about 70% of the population of China. It was the language of the imperial court and as spoken in Peking is the basis of modern Chinese.

Mandarin (manderine), type of orange of Chinese origin whose flattish fruit is highly popular because of its sweet flavour. The fruit derives its name from its resemblance to the large button crowning the hats of Chinese officials known by Europeans under the same name. The tangerine is a flattish loose-skinned species of mandarin orange. Family Rutaceae; genus *Citrus.*

Mandarin, perching fresh-water DUCK native to China and Japan. It lives in forest trees and nests in tree hollows. The drake's plumage has vivid blue, green, purple, orange and chestnut feathers, with splashes of black and pure white. The female is a dull, speckled grey-brown. Species *Aix galericulata.*

Mandate, in English law, command by the sovereign to do anything in the despatch of justice. It can also be applied to a contract to have one's affairs managed by someone without pay. *See also* MANDATED TERRITORIES.

Mandated territories, German colonies and parts of the Turkish Empire which were given to the Allied countries at the 1919 peace conference. There were three kinds of mandate: "A" territories were to be given full independence; "B" ones were to be governed as colonies; "C" ones were to be incorporated into the structure of the mandatory country. Britain took Iraq and Palestine as "A" possessions, Tanganyika, Togoland and part of the Cameroons as "B" possessions and, with Australia and New Zealand, Nauru, Samoa and New Guinea as "C" possessions. Under the United Nations charter "B" and "C" territories were changed to trust territories in 1946.

Mandela, Nelson Rolihlala (1918–), Black South African lawyer who campaigned against the racialist policies of the South African government. As leader of the militant wing of the African National Congress, Mandela was sentenced to life imprisonment in 1964 for plotting to overthrow the government. His wife, Winnie, has continued the compaign.

Mandelshtam, Osip (*c.* 1891–*c.* 1938), Soviet poet. He was the leader of the Acmeists, poets reacting against the mysticism of the Symbolists. His poetry, including the volumes *Kamen* (1913) and *Tristia* (1922), exhibits a desperate longing for a peaceful and harmonious society. Opposed to the Bolsheviks, he was arrested and imprisoned in 1934.

Mander Karel van (1548–1606), Dutch painter and art historian. He is chiefly remembered for his biographical work on painters, *Het Schilderboek* (1604), modelled on VASARI's similar work of 1550.

Mandeville, Geoffrey de, Earl of Essex (died *c.*1144), English baron who changed sides in the civil war of STEPHEN's reign. He was appointed Constable of the Tower of London in 1130, but later turned to banditry in Cambridgeshire.

Mandeville, Sir John (*fl.* 1350), purported English author of a popular medieval travel book. The book was probably written by a Frenchman and translated into English. It comprises a guide to the Holy Land and descriptions of the Orient, its facts most likely culled from other medieval sources.

Mandible, in vertebrates the lower jaw, in insects and Crustacea one of a pair of mouthparts, used for crushing and eating.

Mandolin, stringed musical instrument related to the LUTE and associated with 18th-century Italy. It has four or six paired wire strings which are played with a plectrum. Composers such as VIVALDI, Hummel, VERDI, MAHLER and SCHOENBERG have written for the instrument, but it is most often used today as an accompaniment to folk songs and dances.

Mandrake, plant of the potato family native to the Mediterranean region and used since ancient times as a medicine. It contains the ALKALOIDS hyoscyamine, scopolamin and mandragorine. Leaves are borne at the base of the stem, and the large greenish-yellow or purple flowers produce a many-seeded berry. The branched root resembles a distorted human figure and for centuries the plant was believed to utter an almost human shriek when picked. Height: 40cm (16in); family Solanaceae; species *Mandragora officinarum.*

Mandrill, large BABOON that lives in the dense rain forests of central w Africa. Mandrills roam in small troops and forage omnivorously for their diet on the forest floor. The brightly coloured male has a red-tipped, pale blue nose, yellow-bearded cheeks and a reddish rump. Height: 75cm (30in) at the shoulder; weight: to 54kg (119lb). Species *Mandrillus sphinx. See also* DRILL; MONKEY.

Manet, Édouard (1832–83), French painter. His early work was characterized by his choice of subject matter, often taken from contemporary scenes and events, and his use of light and shadow with little half-tone. His style was influenced by Spanish art, particularly the work of VELÁZQUEZ. Although his name is usually linked with the IMPRESSIONISTS and he was important as their predecessor, he did not, in fact, consider himself one of their school and rarely used their technique of broken colour. Both his innovative style and his choice of subjects provoked adverse criticism; *Le Déjeuner sur l'Herbe* was violently attacked by some critics when exhibited at the Salon des REFUSÉS (1863), as was *Olympia*, a portrait of a well-known courtesan. However, he finally achieved recognition with such work as *Le Bar aux Folies-Bergères* (1882), and his importance as an artist has never since been doubted. *See also* HC2 pp.118–*119.*

Man for all seasons, A (1960), historical drama based on the life of Sir Thomas MORE, by Robert Bolt (based in turn on his own radio play of 1954). It was first produced in London in 1960, and made into a successful film with Paul Scofield playing Sir Thomas More and Robert Shaw as Henry VIII.

Manfredi, Bartolommeo (1580–1620), Italian painter who worked in Rome. He is known mainly as a skilled imitator of the early secular subjects of CARAVAGGIO, such as scenes of fortune-telling and card-playing. He painted chiefly for private patrons.

Mangabey, large, silky grey MONKEY that lives in dense central African forests. Mangabeys live in small troops and feed chiefly on fruit. Tree dwellers, they are generally silent except for an infrequent howling sound. Length: 38–88cm (15–

34.6in); weight: to 6kg (13lb). Family Cercopithecidae; genus *Cercocebus*.

Mangan, James Clarence (1803–49), Irish poet who worked for a time in the Irish Ordnance Survey. He suffered from alcoholism, opium-taking and extreme poverty, and died of cholera. He wrote spirited versions of old Irish songs of which his patriotic hymm to Ireland, *My Dark Rosaleen*, is perhaps the best known. His *Collected Poems* was published in 1903.

Manganese, metallic element (symbol Mn) of the first transition series, first isolated in 1774. Its chief ores are pyrolusite (dioxide) and rhodochrosite (carbonate). The metal is used in alloy steels and certain ferromagnetic alloys. Properties: at.no. 25; at.wt. 54.938; s.g. 7.20; m.p. 1,244°C (2,271°F); b.p. 1,962°C (3,564°F); most common isotope Mn^{55} (100%). *See also* TRANSITION ELEMENTS.

Manganese nodules, concretions with diameters averaging 4cm (2in) that occur in red clay and ooze on the floor of the Pacific Ocean, with the major concentrations around the Samoan Islands. They are thought to have formed as agglomerates from colloidal solution (see COLLOID) in the ocean's waters. The commercial mining of the nodules is being delayed by legal controversy over international rights to this valuable mineral resource. *See also* PE p.83.

Manganese processing, production of the metal either in its pure form or, more commonly, as an alloy with iron for steel making. The manganese dioxide (MnO_2) ores (PYROSULITE and psilomelane) together with iron ores are reduced in blast or electric furnaces to yield ferromanganese, which is essential to the production of steel. Pure manganese is produced electrolytically.

Manganite, hydroxide mineral of the diaspore group manganese oxyhydroxide [MnO(OH)]. It occurs as low-temperature veins and secondary deposits and has monoclinic system striated prisms, which are often crusts of small crystals. Its colour is black or grey with a submetallic lustre. Hardness 4; s.g. 4.3.

Mange, non-specific skin disease of animals, usually caused by parasitic mites that become embedded in the skin causing intense irritation. In dogs the most severe form is sarcoptic mange, (known as scabies in human beings) in which the mite lives between the living and dead layers at the surface of the skin. Constant scratching by the affected dog can cause hair or fur loss and deep lesions which may become infected. Although this secondary infection can usually be successfully treated, in rare cases it can be fatal.

Mango, evergreen tree native to SE Asia and grown widely in the tropics for its fruit. It has lanceolate leaves, pinkish-white clustered flowers, and yellow-red fruit which is eaten ripe or preserved when green. Height: to 18m (60ft). Family Anacardiaceae; species *Mangifera indica*. *See also* PE p.195.

Mangold, also called mangold-wurzel or mangel-wurzel, root vegetable with spinach-like leaves; it is a variety of BEETROOT (*Beta sativa*), as is SUGAR BEET. It is grown widely in Europe as a food for cattle, but was originally native to the Mediterranean region. Family Chenopodiaceae. *See also* PE p.188.

Mangonel, also called a trebuchet, medieval siege weapon. It consisted of a long bar, pivoted high off the ground near one end, to which a heavy weight was attached. The other end held a sling, in which a projectile was placed. The mangonel was the major siege weapon introduced into warfare between the fall of the Roman Empire and the invention of gunpowder in the Middle Ages. It came into use during the 12th and 13th centuries. *See also* MM p.166.

Mangrove, common name for any one of 120 species of tropical trees or shrubs found in swampy areas. Its stilt-like aerial roots produce thick undergrowth, useful in the reclaiming of land along tropical coasts. Seeds germinate while still on the tree. Height: to 20m (70ft). Families: Rhizophoraceae, Verbenaceae, Sonnerat-

iaceae, and Arecaceae. *See also* NW pp.62, 210, 234, 235.

Manhattan, borough of New York City, in SE New York State, USA, mainly on Manhattan Island. In 1625, the Manhattan Indians sold the island to the Dutch West India Company and the town of New Amsterdam was built. The English captured the Dutch colony in 1664 and renamed it New York. In 1898 Manhattan became one of five boroughs established by the Greater New York Charter. A cultural, commercial and financial centre, Manhattan has three universities, two art museums, a stock exchange, the Rockefeller Center, and the United Nations Building. Industries: electrical goods, chemicals, fabricated metals, tourism, entertainment, broadcasting, publishing. Pop. (1970) 1,542,541.

Manhattan Project, code-name given to the US atomic bomb project during WWII. Work on the bomb, advocated by Albert EINSTEIN and other scientists, was carried out in great secrecy by a team which included Enrico FERMI and J. Robert OPPENHEIMER, under the authority of Gen. Leslie Groves, at several places in the USA, most notably Oak Ridge, Tennessee. The first test was at Alamogordo, New Mexico, in 1945. *Sec also* MM p.180.

Mani (*c.*216–*c.*276), Persian prophet, founder of MANICHAEISM, a religious philosophy. As a youth he saw a vision of an angel, which was himself, and several years later, when the vision returned, he believed that he was the inspired prophet of a new religion. Although Mani himself was martyred, his ideas spread quickly throughout Europe, Africa and, in the east, to Chinese Turkestan. The ALBIGENSIANS of medieval Europe were accused of following Mani's ideas.

Mania, mental illness marked by a feeling of intense well-being and excitement. In such a condition, ideas flow at an accelerated rate, speech is rapid and physical activity frenetic and even violent. In extreme cases it is accompanied by confused awareness, loss of sleep and irregular appetite. *See also* MS pp. 138–139.

Manic-depression, psychotic reaction marked by severe alteration of mood ranging from exaggerated feelings of elation and optimism to deep DEPRESSION. Manic and depressive symptoms may alternate in a cyclical pattern, be mixed, or be separated by periods of remission and disturbances of thought and judgement. *See also* MANIA; MS pp.138–139.

Manichaeism, the religious teaching of the Persian prophet, MANI (*c.*216–*c.*276) based on a supposed primeval conflict between light and darkness. According to Mani, the aim of religion was to re-establish the original separation of light from darkness, which had become partially mixed, enslaving man to evil. Adam, Zoroaster, Buddha, Jesus, the Prophets and Mani had all been sent, as messengers of light, to help effect this division. The Manichaean sect, which was influenced by ZORASTRIANISM and CHRISTIANITY, spread rapidly to Egypt and Rome, where it was considered a Christian heresy, and eastward to Chinese Turkestan, where it survived probably until the 13th century.

Manifesto, public declaration explaining past actions or making known reasons, aims or motives of forthcoming actions. It is most commonly associated with statements by political parties, but may be issued to an individual or group of individuals whose actions or proceedings are of importance to the public.

Manifold, in engineering, lateral pipe with outlets to a number of other pipes. In petrol and diesel engines, an intake manifold distributes air and air-fuel mixtures to the engine cylinders. An exhaust manifold collects waste gases s and passes them on to the silencer and exhaust pipe.

Manila, largest city of the Republic of the Philippines, on Manila Bay, w Luzon island; capital of the Republic of the Philippines until Quezon City replaced it in 1948. The chief industrial, financial, and cultural centre of the Philippines, the city has been a main port since the 16th century. Spanish traders built the old walled

city (Intramuros) on the site of a Muslim settlement. Following the Spanish-American War in 1898, development expanded beyond the old city, much of which was destroyed in WWII. The port receives most of the imports to the Philippines. Industries: shipbuilding, textiles, clothing and chemicals. Pop. 1,435,500.

Manila hemp, also called abaca, plant native to the Philippines and cultivated in SE Asia and Central America. The fibres from leaf-stalks of the mature plant are used to make cloth, matting, paper and cordage. Fibre length: to 4.6m (15ft). Family Musaceae; species *Musa textilis*.

Maning, Frederick Edward (1811–83), New Zealand author and judge, *b.* Britain. He was an authority on Maori history and customs. He published *Old New Zealand* in 1863.

Manioc. See CASSAVA.

Manipur, state in NE India, on the Burmese border; the capital is Imphal. After the signing of a treaty in 1762 providing for British assistance against marauding Burmese, Manipur was governed by the State of Assam until 1949 when it was made a Union Territory. It was established as a state in 1972. Crops include rice, mustard and sugar cane. Industries: cotton and silk weaving, carpentry. Area: 22,356sq km (8,632sq miles). Pop. (1971) 1,072,753.

Manitoba, province in s central Canada, the easternmost of the prairie provinces. The terrain varies from the rolling prairie country and lake district of the s to the rugged upland of the Canadian Shield of the NE and the tundra of the far N. Manitoba is famous for its wheat fields. Dairy farming and the rearing of cattle and poulty are also important. Manufacturing includes food products, clothing, electrical products, fabricated metals and transportation equipment. Mineral deposits include nickel, copper and zinc. There are large oilfields in the sw of the province and extensive timber reserves. The major cities are Winnipeg – the capital and largest city – Saint Boniface, Saint James and Brandon. French and English fur traders entered the region in the 17th century. In 1670 Charles II granted the land to the HUDSON'S BAY COMPANY. In 1869 the company sold the land to the newly created confederation of Canada. Area. 650,090sq km (251,000sq miles). Pop. (1975 est.) 1,018,000. *See also* MW p.45.

Manizales, city in Colombia, 177km (110 miles) NW of BOGOTÁ; capital of Caldas department. Founded *c.*1846 by gold prospectors, it was destroyed by an earthquake in 1878 and by fire in 1925. An aerial tramway, 72km (45 miles) long, connects it with Mariquita. Industries: gold, silver and mercury production and coffee. Pop. (1973) 199,904.

Mankiewicz, Joseph Leo (1909–), US film executive who began his career as a caption writer for silent films. He became a producer in 1936 and later a writer-director. His films as a producer include *The Philadelphia Story* (1940) and as a writer-director, *Guys and Dolls* (1955) and *Cleopatra* (1962).

Mankowitz, Wolf (1924–), British author and playwright. His novels include *Make Me an Offer* (1952) and *A Kid for Two Farthings* (1953), both of which were filmed. His musical play, *Expresso Bongo*, was also made into a film (1960), starring Cliff RICHARD, and he has written screenplays for *The Hireling* (1973) and *Dickens of London* (1976).

Manley, name of two prime ministers of JAMAICA. Norman Washington Manley (1893–1969), a lawyer educated at Oxford, founded the socialist People's National Party (PNP) in 1938, and was Jamaica's first Premier from 1959–62. His son Michael Norman Manley (1923–), was an active trade-únionist before becoming a politician. He, also of the PNP, became Prime Minister of Jamaica in 1972. *See also* HC2 p.135.

Man-made fibres, materials intended to substitute for, supplement or improve on the quality of natural fibres such as cotton, wool and silk. The manufacture of man-made fibres began early in this century with RAYON, still a familiar textile,

Mango: large mangoes weigh up to 2,300gm (5lb); small ones are the size of plums.

Manhattan: the Chrysler Building dominates this part of the New York skyline.

Manitoba has as its capital Winnipeg situated on the Red River.

Norman Manley strove for Jamaican independence and became the first premier.

Field-Marshal Mannerheim remains a heroic figure in the eyes of all Finns.

Man-of-War, at Sheerness, lies beside an early steam version, on the left.

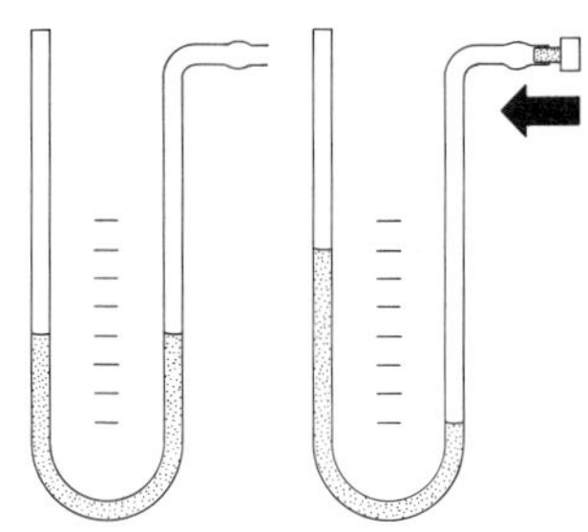

Manometers usually contain mercury; an adaptation measures blood pressure.

Andrea Mantegna: a detail from *Return From the Hunt* shows early perspective.

which (like cotton) is made of the natural polymer cellulose. In rayon, cellulose is extracted from plants, dissolved and reconstituted as a fibre by spinning it through spinnerets, devices resembling the rose of a watering can. The first truly synthetic, or artificial-polymer, fibre was NYLON – a revolutionary material which, weight for weight, is stronger than steel. The manufacture of nylon begins in the petroleum and coal industries, from which its monomers (materials such as adipic acid and hexamethylene diamine) are obtained. After polymerization, nylon can be melted and melt-spun through spinnerets to make fibres, which can then be further elongated by cold drawing. POLYESTERS, synthetics like nylon widely used for textiles, are also melt-spun. The equally popular ACRYLICS decompose before they melt, so are wet-spun like rayon. POLYETHYLENE (polythene), polypropylene and vinyl plastics such as PVC can also be made as fibres although they are more familiar in such applications, as film or sheet. Inorganic man-made fibres include glass fibre, spun from molten glass; carbon fibre, used to reinforce ceramics and metals; and metallic fibres, used mainly for decorative purposes in clothing. *See also* MM pp.56–57.

Mann, Heinrich (1871–1950), German novelist and brother of Thomas MANN. His hatred of tyranny, particularly Nazism, and the educational system is expressed in his trilogy, *The Empire* (1918–25), and *Young Henry of Navarre* (1935).

Mann, Thomas (1875–1955), German novelist and essayist. His first novel, *Buddenbrooks* (1901), superficially a description of the bourgeois class but in fact heavily laden with elements drawn from his study of NIETZSCHE, SCHOPENHAUER and FREUD, was an immediate success. It was followed by other distinguished works, including *Death in Venice* (1911), *Reflections of a Non-political man* (1918) and *The Magic Mountain* (1924). The rising tide of Nazism forced Mann from his non-political stance and in 1930 he published *Mario and the Magician*, an attack on the vulgarity of dictatorship. He became a US citizen in 1944 and won the Nobel Prize in literature in 1929.

Mann, Tom (1856–1941), British labour leader. He organized the London dock strike of 1889, was secretary of the Independent Labour Party (1894–97), worked in Australia after 1902 and was a founder of the British Communist Party in the 1920s. *See also* HC2 p.*211.*

Manna, or flowering ash, tree of the olive family that grows in S Europe and Asia Minor. The pinnate leaves have rust-coloured hairs underneath. The flowers are white and showy with large petals. A sugary substance, mannite, is collected from cuts made in the bark and used medicinally. Height: to 18m (60ft); family Oleaceae: species *Fraxinus ornus.*

Mannerheim, Carl Gustav Emil von (1867–1951), Field-Marshal and President of Finland. A baron, he served in the Russian army, rising to lieutenant-general in WWI. After the Russian revolution (1917) he returned to Finland and repelled a Bolshevik offensive. As regent (1918–19) he obtained the recognition of the independence of Finland. He was head of the Defence Council from 1931 and planned the Mannerheim Line, a fortified defence system right across Karelia, against Soviet aggression. When the USSR attacked Finland (1939) this line was instrumental in the resistance to the Soviet forces. He was President of Finland from 1944 to 1946.

Mannerism, in art and architecture, term which refers to the style of Italian art in the 16th century from *c.*1530, regarded by some historians as transitional between the High Renaissance styles and the Baroque. Others regard it as a coherent style characterized by experimentation, virtuosity, and artifice, with an undercurrent of subjective intensity which represented European art as a whole. In painting, the later work of MICHELANGELO, El GRECO's paintings from Italy, and the art of Francesco PARMIGIANINO are described as Mannerist. In architecture, Mannerism

introduced a freer use of decoration and laid less stress on accurate proportion than had Classicism. The first indications of this are evident in some later designs of Michelangelo, but the best examples are the buildings of Bartolommeo Ammanati, notably the Collegio Romano, Rome (*c.*1582–85). The twist bizarre sculptures of GIAMBOLOGNA and CELLINI are typical of Mannerist sculpture. The value Mannerists placed on the decorative arts was important in encouraging the development of tapestry, ceramics and metalwork. *See also* HC1 pp.264–265, *264–265,* 304.

Mannheim, city and major port in central West Germany, on the E bank of the River Rhine at the mouth of the River Neckar. Originally a fishing village, it was fortified in 1606 and destroyed by the French in 1689. It was rebuilt in 1697, became the seat of the Rhine Palatinate (1719–77) and passed to Baden in 1803. Its industries include chemicals, cellulose, paper and textiles. Pop. (1974) 325,386.

Manning, Henry Edward (1808–92), British ecclesiastic and cardinal. A member of the OXFORD MOVEMENT, he was ordained a priest in the Church of England in 1833 but in 1851 was converted to Roman Catholicism. He was appointed Archbishop of Westminster in 1865 and became a cardinal in 1875. He took a leading part in debates on papal infallibility in the Vatican Council (1870) and defended this doctrine in his writings. He was an active social reformer and led a campaign against strong drink. His most famous work is *The Eternal Priesthood* (1883).

Manning, Olivia (*c.*1912–), British novelist, whose first novel, *The Wind Changes,* appeared in 1937. She is best known for her "Balkan triology", *The Great Fortune* (1960), *The Spoilt City* (1962) and *Friends and Heroes* (1965), novels about a British diplomat and his wife in E Europe.

Man-of-war, naval vessel of the 17th–19th centuries. From the 1650s men-of-war were divided into six rates, depending on the number of decks and guns. The first, second and third rates were known as SHIPS-OF-THE-LINE. *See also* MM p.*173.*

Man-of-war, Portuguese. *See* PORTUGUESE MAN-OF-WAR.

Manometer, device for measuring pressure. It consists of a U-shaped tube containing a liquid, one end open to the atmosphere and the other end attached to the vessel whose pressure is to be measured. If the gas pressure in the vessel is greater than atmospheric, it will force the liquid down on the side nearest to the vessel and up on the side open to the atmosphere. The difference in heights of the liquid in the two arms of the tube shows the difference in pressure. *See also* SU p.*78.*

Manon Lescaut (1893), four-act opera by Giacomo PUCCINI with Italian libretto by Ruggiero LEONCAVALLO and others, after the novel by the Abbé PRÉVOST. It tells how Manon (soprano), who has eloped with a student, des Grieux (tenor), is banished from France to Louisiana. Des Grieux joins her, but Manon dies as the pair wander homeless in the strange land.

Manor, term applied after the NORMAN CONQUEST to estates into which the rural economy in England was organized. Varying in size and structure, a manor was generally divided into the lord's demesne, on which the VILLEINS were bound to labour, and an area assigned to the villeins for their own use. The manor house was the focal point of the manor. *See also* MANORIAL SYSTEM.

Manorial system, organization of rural economy existing in England and throughout western and central Europe during the period of THE FEUDAL SYSTEM. Typical of the system was the English manor house, the focal point of a MANOR, or estate, which varied in size and compactness, but might consist of several villages, hamlets and farms. A lord presided over the manor, which he held in FIEF from a greater lord, who might be the king. The VILLEINS were each assigned a share, usually a *virgate* or strip of the village

lands. In exchange for the use of the land, the villein gave some part of his labour to his lord's demesne, and made payments of livestock and money. The custom, or law, of the manor fixed the duties and payments of the villein, who might be a SERF or a freeman. Justice was determined in the manorial court, presided over by the steward or seneschal, where officials were appointed, offenders punished and land transactions registered. *See also* HC1 pp.196–197.

Manrique, Jorge (1440–79), Spanish lyric poet. He was a minor poet save for one single, long poem – *Coplas por la muerte de su padre Don Rodrigo* (1492) – written on the death of his father and the most famous elegy in Spanish.

Mans, Le. *See* LE MANS.

Mansard roof, roof having two slopes on each side, the lower with a very steep pitch, the upper more gently sloped. It is named after the French architect François MANSART who frequently incorporated it into his designs, although it had been in use since the mid-16th century.

Mansart (Mansard), François (1598–1666), French architect. His first major work was the façade at the church of the Feuillants, Paris (1623), and he went on to establish a restrained, Classical style which dominated French 17th-century architecture. The Château des Maisons is his most important surviving building. A type of double-pitched roof is named after him.

Mansbridge, Albert (1876–1952), British pioneer educationist. He founded the Association to Promote the Higher Education of Working Men (1903), which in 1905 became the Workers' Educational Association. He also founded the National Central Library, the World Association for Adult Education (1919) and the Seafarers' Educational Service (1919).

Mansfield, Katherine (1888–1923), British short-story writer, *b.* New Zealand. Her works, all collections of short stories, include *In a German Pension* (1911), *Bliss* (1920), *The Garden Party* (1922) and *The Dove's Nest* (1923). She was a writer of subtlety and warmth, who created atmosphere with the lightest of touches.

Mansfield Park (1814), novel by Jane AUSTEN. Due to the persuasions of the hateful, overbearing Aunt Norris, Fanny Price, the heroine, is adopted by her rich uncle and falls in love with her cousin, Edmund.

Mansfield's judgment (1772), verdict given by the 1st Earl of Mansfield, British Lord Chief Justice, in favour of the fugitive slave James Somersett. The judgment laid down that an escaping slave could not be forcibly returned to a colony for punishment. It did not imply, as was suggested, that slavery was not allowed in English law.

Manslaughter, crime of killing another person either accidentally or for humane motives. In English law manslaughter is considered to merit less punishment than MURDER; a person charged with murder may, on the other hand, be found guilty of manslaughter if the court considers there to have been mitigating factors (eg the victim's painful and terminal disease, or the protagonist's diminished responsibility for one reason or another at the time).

Manson, Sir Patrick (1844–1922), British PARASITOLOGIST known for his pioneering work in tropical medicine. He discovered that the mosquito was host to the parasite *Filaria bancrofti,* which causes the human disease FILARIASIS. Manson founded the Medical School of Hong Kong and organized the London School of Tropical Medicine.

Mantegna, Andrea (*c.*1431–1506), Italian painter who profoundly influenced the development of northern Italian painting. In 1460 he became court painter to the GONZAGA family in Mantua and remained there. His works include the innovative *St Sebastian* (1465), *Dead Christ* (1466) and *Madonna of Victory* (1495). *See also* HC1 pp.254, *254,* 255.

Mantis, or praying mantis, any of several species of mantids, insects found throughout the world. They have powerful front legs used to catch and hold their insect prey. Colours range from brown and green to bright pinks, enabling some species to blend with the foliage or flowers of their environment. Length: 25–150mm (1–6in). Family Mantidae. *See also* NW pp.106–107, *107 213*.

Mantle, in geophysics, solid layer of the Earth extending from about 5km (3 miles) down to about 2,900km (1,800 miles); it lies between CORE and CRUST. The mantle is not homogeneous, but becomes denser with increasing depth; near the core, it increases to $10g/cm^3$. The downward increase in density causes primary (P) and secondary (S) SEISMIC WAVES, which originate near the Earth's surface, to bend away from the core, like the bending of light waves by the atmosphere to form mirages. Most of the mantle is dense rock, and is a good conductor of such waves. *See also* PE pp.20, 21, *28*.

Mantra, in HINDUISM and BUDDHISM, sacred word, verse or formula, eg *Aum* (*Om*) recited during prayers or meditation. *See also* MS p.*221*.

Manu, in Hindu mythology, the hero of the DELUGE. Manu caught a fish that offered to save him in exchange for its life. The fish had Manu build a ship, then towed it for years to the Himalayas where it was tied to a tree until the waters receded. Manu is the generic name for each of the 14 consecutive rulers of the earth. According to myth the reign of each will last for several million years and will end in a deluge. The present ruler is the seventh, the son of the Vedic sun god Vivasvan. *See also* MS p.213, *213*.

Manu, Laws of, ancient SANSKRIT code of law. It was attributed to a mythical BRAH-MAN sage of the same name. Probably compiled *c.*AD 200, the text has been transmitted largely through its commentaries. Considered to be sacred literature as well as the basis of HINDU jurisprudence, it is a treatise on the religious and social obligations of Hindus.

Manuel, name of two Byzantine emperors. Manuel I (*c.*1120–1180) called Manuel Comnenus, ruled from 1143 to 1180 and attempted to reunite the eastern and western provinces of the former Holy Roman Empire. His neglect of Asia Minor, however, led to his defeat by the Turks at Myriocephalon (1176). He was succeeded by his son, Alexius II. Manuel II (1350–1425), called Manuel Palaeologus, (*r.*1391–1425), saw his empire reduced to Constantinople by the Turks. He made a futile trip to Venice, London and Paris seeking aid but in 1425 he was forced to pay tribute to the sultan. He was succeeded by his son, John VIII.

Manuel, or Emmanuel, name of two kings of Portugal. Manuel I (1469–1521) succeeded John II to the throne and reigned from 1495 to 1521, a period during which Portugal throve on the wealth from the Indies and became the leading commercial nation of the West. In order to marry the daughter of FERDINAND II and Isabella of Spain, Manuel reluctantly agreed to expel the Jews and Moors (1496). He was succeeded by his son, John III. Manuel II (1889–1932) was the son of Carlos I and became king in 1908 after both his father and older brother were assassinated. In 1910 he was dethroned by a revolution and a republic was established. He was the head of the house of Braganza and Portugal's last king. He spent his exile in England.

Manukau, city on N North Island, New Zealand, on Manukau Harbour, an inlet of the Tasman Sea, 19km (12 miles) SW of Auckland. The city includes Auckland's airport built on land reclaimed from the sea. Industries: food processing, aluminium, plastics, machinery and synthetic fibres. Pop. (1974) 127,800.

Manumission, legal release of a peasant of VILLEIN status from bondage to his lord. The villein gained his freedom either by purchase through a third party or by charter, sometimes in return for a QUIT RENT. The act of manumission was performed publicly, usually in the county court, and the villein was presented with the arms of a freeman.

Manuscript, handwritten book or docu-

ment, as distinct from a printed or typed one. The oldest manuscripts (*c.*2600 BC), written on papyrus, have been found in Egyptian tombs. Some of the manuscripts of the Middle Ages were extremely decorative works on vellum (known as illuminated manuscripts), painstakingly copied. With the advent of printing in Europe, the need for hand-copied manuscripts ceased, but the study of the old codices and manuscripts continues in palaeography and they also form an important historical source, eg the DEAD SEA SCROLLS of the Essenes which have thrown much light on the origins of CHRISTIANITY. Today the term is also used to describe unprinted books, letters and papers.

Manutius, Aldus (1449–1515), Italian printer and publisher, founder of the Aldine Press. His type-cutter was Francesco Griffo of Bologna, who produced outstanding roman type-faces. Manutius is renowned for his pocket-size editions of Greek and Latin classics. In 1502 printed Dante's *La divina commedia*, which first showed the famous device of the Aldine anchor and dolphin.

Manx, language formerly spoken in the Isle of Man. Closely related to Scottish Gaelic, it was spoken by most of the native inhabitants of the island until about 1700, when English began to be introduced. By 1900 there were only a few thousand speakers left, and there are now no native speakers.

Manx cat, short-haired domestic cat thought to have been bred originally on the Isle of Man. It has a round head with prominent ears, a short back, and hindquarters higher than the shoulders; it is characteristically tailless. The luxurious double coat may be any colour. Manx cats have a hopping, rabbit-like walk.

Manzanillo, seaport in E Cuba, on the Gulf of Guacanayabo. Founded in 1784, the port is a distributing centre for a fertile agricultural region. It now exports tobacco, sugar and exotic woods. Industries include sawmills, cigar factories and tanning. Pop. (1970) 77,880.

Manzoni, Alessandro (1785–1873), Italian novelist and poet. His poetry expressed his religious faith, but his greatest work is *The Betrothed* (1827), a historical novel set in 17th-century Milan during the Spanish occupation. *See also* HC2 p.99.

Manzù, Giacomo (1908–), Italian sculptor, etcher and lithographer. His work is religious in content and includes the panelled bronze doors of St Peter's, Rome (1964), and the doors of Salzburg Cathedral (1958). He was awarded the Lenin Peace prize in 1966.

Mao. *See* MIAO.

Maoris, Polynesian inhabitants of New Zealand preceding the Europeans. They have brown skin, black or dark brown hair and brown eyes. By now thoroughly Europeanized, they nevertheless have strong attachments to the Maori language, culture and customs. In Maori society tattooing, carving and weaving were highly developed arts; Maori songs and war chants (*haka*) are kept alive still. The meeting place (*marae*) was the centre of the community. The most prestigious tribes are those that arrived in the "great fleet" (*c.*1350). Traditionally, Maoris lived by agriculture, hunting and fishing. They have had equal rights with white New Zealanders since the second half of the 19th century. The Maori population is now increasing, having succumbed initially to European diseases. Intermarriage is becoming much less rare. *See also* MAORI WARS; HC2 pp.125–126.

Maori Wars, wars between British settlers in New Zealand and MAORI tribes, from 1845 to 1848 and 1860 to 1872. They arose when the settlers broke the agreement in the Treaty of WAITANGI (1840) to guarantee the Maoris possession of their lands. As a result of the wars a Native Land Court was established (1865), a Maori school system formed (1867) and the Maoris allowed to have four elected members in the New Zealand legislature.

Mao Tse-tung (1893–1976), Chinese Communist leader and head of state. He was an early member of the Chinese Com-

munist Party, becoming Chairman of the People's Republic of China in 1949. The most influential person in modern Chinese history, he was a prime mover of events which led to the Communist accession to power in 1949. He joined the revolutionary army briefly in 1911 to fight against the MANCHU DYNASTY, went to Peking University in 1919, becoming an adherent of Marxism in 1919–20. He organized a Communist group in Hunan in 1920 and in 1923 devoted himself full-time to revolutionary activities, becoming convinced of the revolutionary potential of the peasantry. In 1927 the Communists split with CHIANG KAI-SHEK's Kuomintang when the latter expelled all Communists from high posts in the organization. The years 1927–49 saw the consolidation of the Communist Party under Mao's increasing control. Mao concentrated his efforts on the peasantry rather than the urban masses, and after guerrilla activities against the Kuomintang in the late 1920s, became Chairman of the Communist base at Kiangsi in 1931. His dominance in the party was established by 1935 during the Long March from Kiangsi to Shensi in 1934–1936. Mao built his reputation as a Theorist in Yenan from 1936 until the end of WWII, and by the late 1930s his control over the Communist Party was assured with his concept of the "Sinification of Marxism", the adaptation of MARXISM to the cultural, historical and economic experiences of the Chinese. In 1949 Mao became Chairman of the People's Republic of China, placing emphasis on rapid collectivization and initially concentrating on the development of heavy industry. In 1956 he propounded the Hundred Flowers policy – the freedom to express diverse ideas – with the aim of incorporating the intelligentsia into the Chinese revolution. The late 1950s saw the GREAT LEAP FORWARD with emphasis on decentralization and the development of labour-intensive industries leading to the establishment of people's communes. He retired as Chairman of the Republic in 1959 but remained Chairman of the Communist party. Mao's feeling of the growing elitism within Chinese society and leadership struggles with LIU SHAO-CH'I and Teng Hsiao-p'ing led to the CULTURAL REVOLUTION (1966–69). Mao gradually retired from the day-to-day administration of politics in the early 1970s. A leading Chinese theoretician with a world-wide reputation, his writings cover politics and culture, as well as international affairs and guerrilla warfare tactics. He is an inspirational figure to many left-wing revolutionary movements. *See also* HC2 pp.*145*, 215, 244–245, *245*.

Map. *See* CARTOGRAPHY.

Maple, genus of deciduous trees native to temperate and cool regions of Europe, Asia and North America. They have yellowish or greenish flowers and winged seeds. They are grown for ornament, shade or timber, depending on the species; the sugar maple is also tapped for MAPLE SYRUP. Most species have brilliant foliage in the autumn; a typical example is the tree called SYCAMORE in Britain. Height: 4.6–36m (15–120ft). Family Aceraceae; genus *Acer*.

Maple syrup, sweet-tasting concentrated and processed sap of the North American sugar maple, *Acer saccharum*. Trees are tapped and the fluid is allowed to evaporate in the open air. The final syrup is 30 to 50 times more concentrated than the sap and is popular as a sauce for sweet foods. *See also* PE p.197, *197*.

Maputo (Lourenço Marques), capital of the state of Maputo and of Mozambique. It was first visited by Antonio do Campo in 1502 and explored by the Portuguese trader Lourenço Marques. It is linked by rail to South Africa, Swaziland and Rhodesia and is a popular resort area. Maputo's products include footwear and rubber. Pop. 799,400.

Maquette, small model, usually of wax or clay, made as a detailed study for a work of sculpture. A less finished model, which is more commonly made, is known as a bozetto.

Maquis, popular name for the French

Manx cats can run faster for longer than most other domesticated cats.

Maori art, while intricate and skilful, is of great tribal significance.

Mao Tse-Tung, photographed in 1966, waves to the people of Peking.

Maple: in Norway this tree can grow to a height of 30 metres (100 feet).

Maracaibo: whole villages are built on piles in the waters of Lake Maracaibo.

Carlo Maratta: his monument in the church of S. Maria Degl'Angeli, Rome.

Marcel Marceau celebrated the 20th anniversary of "Bip" in 1956.

Rocky Marciano knocked out Archie Moore in 1955 to retain his world title.

underground resistance movement against the German forces of occupation in WWII. The maquis specialized in sabotage and by 1944 had developed an efficient intelligence system, which was of service in the liberation of the country (1944). *Maquis* is the French name for a tough shrub.

Mar, John Erskine, Earls of, hereditary Scottish name and title. The 1st (or 6th) earl (*d.*1572) was created Earl of Mar by MARY, QUEEN OF SCOTS, but whether this constituted a restoration of the earldom or a new creation is still disputed. The 2nd (or 7th) earl (*c.*1558–1634) took part in the RUTHVEN RAID, then became reconciled with James VI of Scotland (later James I of England) and served as treasurer of Scotland from 1616 to 1630. The 6th (or 11th) earl (1675–1732) was nicknamed "Bobbing John", probably because of his political vacillation when his ambitions were blocked. A leader of the JACOBITES who became a supporter of the Union with England, he tried unsuccessfully to raise the standard for the Old Pretender in 1715 and died in exile in France. *See also* HC2 p.28.

Mara, Ratu, The Rt Hon Sir Kamisese Kapaiwai Tuimacilai (1920–), Fijian politician and Prime Minister of Fiji from 1970. From 1967 to 1970 he was Chief Minister, after leading the Fijian delegation at the Commonwealth Constitutional Conference in London in 1965.

Marabou. *See* ADJUTANT STORK.

Marabouts, MUSLIM monks. They enjoy great influence in North Africa, where they are looked upon as living saints. The term is now also used for the tombs of Muslim holy men, which are sometimes treated as places of divine worship.

Maraca, Caribbean percussion instrument, originally made from a dried gourd containing dried seeds. It is shaken rhythmically, and became popular in Western countries in jazz and swing bands, especially for dances such as the rumba.

Maracaibo, port city in NW Venezuela, between Lake Maracaibo and the Gulf of Venezuela. Founded in 1529 by the German adventurer Alfinger, it was sacked by the English pirate Henry MORGAN in 1669. It expanded after the discovery of oil in 1917 and is now the country's second largest city. Its products include coffee, cacao and sugar. Pop. 690,400.

Maracaña Stadium, oval football stadium in Rio de Janeiro with a capacity of more than 200,000; the largest in the world. It was built, although not completely finished, for the 1950 World Cup, when a crowd of 199,854 watched Uruguay beat Brazil 2–1 in the final match.

Maracay, city in N Venezuela, on the Pan American Highway 80km (50 miles) WSW of Caracas. It was the capital of Venezuela from 1908 to 1935. It now supplies much of the meat and dairy produce for Caracas and its industries include textiles and timber. Pop. (1971) 255,134.

Marat, Jean Paul (1743–93), physician, extremist writer and French revolutionary. At the outbreak of the Revolution he founded the inflammatory journal *L'Ami du peuple* and was forced to flee at least twice to London. Working later for the JACOBINS, he was stabbed to death in his bath by Charlotte CORDAY, a supporter of the GIRONDINS.

Marathas. *See* MAHRATTAS.

Marathon, one of the most demanding of athletics events. The standard marathon race distance is 42.2km (26 miles, 385 yards). The race approximates to the distance run by a Greek soldier (believed to have been Pheidippides) from Marathon to Athens to announce the Greek victory over the Persians in 490 BC. Usually it starts and ends on a stadium track, with most of the race run through a marked course in city or suburban streets. Marathon racing was included in the first modern Olympic Games of 1896.

Marathon, Battle of (490 BC), battle in which the Athenians under Miltiades defeated a larger army led by Datis and Artaphernes and freed ATTICA from the threat of a Persian invasion. Marathon plain lies to the NE of Athens. *See also* HC1 pp.61, 69.

Marat/Sade (1964), convenient name for the play by Peter WEISS, entitled *The Persecution and Assassination of Jean Paul Marat as Performed by the Inmates of the Asylum of Charenton under the Direction of the Marquis de Sade.* First performed by the Schillertheater, West Berlin and later by the ROYAL SHAKESPEARE COMPANY under Peter Brook, the play explores the nature of violence and revolution in its dialogue, contrasting the revolutionary socialistic tenets of MARAT and the individualistic eccentricities of the nobleman de SADE.

Maratta (Maratti), Carlo (1625–1713), Italian painter of the Roman School. He had an early success with his painting *Nativity* (1650); he then became a celebrated portraitist and was commissioned to paint many altar-pieces.

Marble, metamorphic rock composed largely of recrystallized limestones and dolomites. The term is more loosely used to refer to any crystalline calcium carbonate rock that has a good pattern and colour when cut and polished. The colour is normally white, but when tinted by serpentine, iron oxide or carbon can vary to shades of yellow, green, red, brown or black. It has been a favourite building and sculpting material of ancient and modern civilizations. *See also* PE pp.47, 102–103.

Marbles, small balls made from almost any hard material, including marble, clay, plastic or glass, and used in playing a game of the same name. Most commonly, coloured glass balls are rolled or tossed on to a limited area with the object of approaching close to, or of hitting, other marbles already on the ground. There are many local variations of the game, but common rules are that the marble must be flicked from between thumb and forefinger, that the target is an arrangement of marbles in the form of a cross, and that the object of a throw is to knock other marbles out of the area marked on the ground. Alternatively there may be a hole into which marbles are to be knocked, or, like bowls, the thrown marble must approach as close as possible to a specific ball, usually of different size or colour. Games with marbles have been associated with cultures as diverse as the North American Indians, and ancient Rome.

Marburg, town in West Germany, on the River Lahn 74km (46 miles) N of Frankfurt. It developed around a Thuringian border fortress in the 12th century and served as a residence for the rulers of Hesse from the 13th to 17th centuries. It is the location of Europe's first Protestant University (1527). Industries include tourism, pottery and chemicals. Pop.(1970) 46,968.

Marc, Franz (1880–1916), German painter. An EXPRESSIONIST artist, he was influenced by Robert DELAUNAY. His works, in which colour plays an important symbolic role, include *Cat under the Tree* (1910). *See also* HC2 pp.175, 175.

Marcantonio (Marcantonio Raimondi) (*c.*1480–*c.*1534), Italian engraver and craftsman who reproduced the paintings of other artists such as RAPHAEL and the woodcuts of DÜRER. Some of his works were copied on MAJOLICA ware which helped spread Renaissance motifs.

Marcasite, brass-yellow mineral, iron sulphide (FeS_2), found commonly in near-surface deposits. It is found as tabular or prismatic crystals in the orthorhombic system, as cockscomb aggregates and as stalactitic, globular and radiating forms. It has a metallic lustre and is brittle. Hardness 6–6.5; s.g. 4.9.

Marceau, Marcel (1923–), French mime. His best known creation was Bip, a sad white-faced clown with a tall, battered hat. Marceau and his company, which he started directing in 1949 to reawaken interest in the art of MIME, toured extensively and Marceau made several films, including *Un jardin public* (1955), and made a memorable appearance in *Silent Movie* (1976).

Marcel, Gabriel (1889–1973), French philosopher, dramatist and critic. He was a leading exponent of the Existential school, holding that knowledge of existence comes not from discursive reason but rather from direct contact or participation. The basis of Marcel's analysis of existence and reality is faith. His philosophy has been described as the Christian expression of EXISTENTIALISM. His best known works are *Being and Having* (1949) and *The Mystery of Being* (1950–51).

Marcellinus, Saint (*r.*296–304), Pope who reigned during the persecution of DIOCLETIAN. He was said to have sought martyrdom after succumbing to the demand of the persecutors for the sacred books of the Church.

Marcellus, name of two popes. Marcellus I (*r.*308–309) was exiled by the Emperor Maxentius but succeeded in reorganizing the Roman Church, dividing it into 25 parishes. He was later canonized. Marcellus II (*r.*April–May 1555) shared the presidency of the Council of Trent (1545–47) and, after becoming Pope, initiated the publications of the reform documents in a bull.

Marcellus, name of several political figures at Rome, of whom the most important were Marcus Claudius Marcellus (*c.*268–208 BC), who thwarted HANNIBAL's attack on NOLA in 216 BC, and Marcus Claudius Marcellus (42–23 BC), who was named as successor to his uncle, the Emperor AUGUSTUS, but who died 37 years before Augustus.

March, third month of the year, consisting of 31 days of which the 21st is the vernal EQUINOX. In the Roman Republican calendar it was the first month and, named after MARS, the god of agriculture and war, marked the start of agricultural renewal and the season of military campaigns. In spite of becoming the third month of the year in the Julian calendar, the feast of the Annunciation (25 March) was celebrated as New Year's Day until the late 16th century.

Marchand, Jean-Baptiste (1863–1934), French soldier and an explorer in Africa. Heading an expedition in 1898 to lay claim to a part of Sudan, he was confronted by the British in what is known as the FASHODA INCIDENT. He fought later in the BOXER REBELLION (1900) and WWI.

Marcher lords, English nobles who were permitted after 1066 to rule whatever parts of Wales they could conquer and hold. Given sovereign rights within the areas they held, which were collectively known as the MARCHES, they not only served to check Wales but also became a threat to the English monarchy, particularly in the Wars of the ROSES. In 1354 the territories were perpetually annexed to the English Crown, but the lordships were not finally abolished until 1536 with the enactment of a series of laws effecting union between Wales and England.

Marches, area bordering Wales and England which was conquered between 1067 and 1282 by the MARCHER LORDS, and ruled by them until 1536. The lordships often corresponded to the Welsh administrative areas, the *commotes*. Most were divided into the low-lying Englishry, where English law, administration and social patterns were imposed, and the upland Welshry, which retained a considerable degree of national identity. The Marches corresponded approximately to the former Welsh counties of Brecknock, Denbigh, Glamorgan, Montgomery, Pembroke and Radnor, and the English county of Monmouth and parts of Shropshire, Herefordshire and Gloucestershire.

Marciano, Rocky (1923–69), US boxer, real name Rocco Francis Marchegiano. On his climb to the heavyweight title he became the second boxer ever to knock out Joe LOUIS (1951) and won the title in 1952 by knocking out Joe WALCOTT in 13 rounds. He retired as undefeated champion in 1956.

Marconi, Guglielmo (1874–1937), Italian physicist who developed RADIO. By 1897 Marconi was able to demonstrate radio telegraphy over a distance of 19km (12 miles) and, after forming a wireless telegraph company, established radio communication between France and England in 1899. By 1901 radio transmissions were received across the Atlantic Ocean, and in 1909 Marconi was awarded the Nobel

Prize in physics. His later work on short-wave radio transmission formed the basis of nearly all modern long-distance communication. *See also* MM pp.228–229.

Marco Polo. *See* POLO, MARCO.

Marcos, Ferdinand Edralin (1917–), President of the Philippines. Elected President in 1965, he was re-elected in 1969, becoming the first President of the Philippines to serve a second term. His presidency has been marked by student unrest and guerrilla activity; in 1971 and 1972 he declared martial law and in 1973 he established a new and authoritarian constitution.

Marcus, Saint (*d.c.*679), also called Marcius, Mark or Martin, Italian monk at Monte Cassino who became a hermit on Mt Massicus. He is mentioned in St Gregory's life of St Benedict, and according to legend the sacrament was brought to him by an angel.

Marcus Aurelius (Antoninus) (121–180), Roman emperor (*r.*161–180) and philosopher, originally named Marcus Annius Verus. He was an able and energetic ruler, and recurring crises meant that for most of his reign he was fighting wars. In 161 he repelled a Parthian invasion of Syria. In 167–168, he drove the Marcomanni, a Germanic tribe, out of Italy. There were also revolts or invasions in Egypt, Spain and Britain. He promulgated many laws, often in favour of the poor, but was not tolerant of Christians. His *Meditations*, his one surviving work, is a collection of thoughts and theories that occurred to him during his numerous campaigns. *See also* HC1 pp.99, *99, 108,* 109, 116.

Marcuse, Herbert (1898–), US social philosopher and author, *b.* Germany. In *Eros and Civilization* (1958) he attempted to fuse Marxist and Freudian theories into a critique of modern industrial societies. *One Dimensional Man* (1964), in which he argued that Americans were oppressed and beginning to accept oppression, made him a hero of the New Left radicals and provided a rationale for student revolts in the 1960s. *See also* HC2 pp.*266,* 267, *267.*

Mardi gras, community festival or carnival held on Shrove Tuesday, the day before the beginning of Lent, in many Roman Catholic countries, particularly France. In the USA, most notably New Orleans, it includes street parades, concerts and dances.

Marduk, god of the spring sun and supreme deity of the Babylonian pantheon. The champion and king of the gods in the fight against Tiamat, KINGU and the forces of darkness, his victory was celebrated at the New Year. Marduk was originally a vegetation god but acquired the attributes of local deities as the power of Babylon grew.

Mare, Walter de la. *See* DE LE MARE, WALTER.

Marengo, Battle of (1800), victory of the French under NAPOLEON over the Austrians under Melas at the end of the French Revolutionary Wars. Their victory gave the French control of N Italy and the strategic initiative in Europe. *See also* HC2 pp.76–79, *76–79.*

Mareth Line, fortification in southern Tunisia, N Africa. It was the scene of one of the final battles between the retreating AFRIKA KORPS and the British Eighth Army in 1943.

Marey, Étienne-Jules (1830–1904), French inventor and professor of natural history. He invented the SPHYMOGRAPH and a camera that took a series of photographs, enabling him to study the flight of birds.

Marfan's syndrome, also called arachnodactyly, rare disorder affecting the development of connective tissue in the body, notably in the skeleton, eye and heart. The limbs and fingers generally become elongated and spindly. Severe damage may be caused to the heart and eye.

Margai, Sir Milton (1895–1964), first Prime Minister of Sierra Leone. A physician, he turned to politics after WWII and led his country to independence in 1961. He was followed as Prime Minister by his brother, Sir Albert Margai.

Margaret, Maid of Norway (*c.*1283–90), grand-daughter of the Scottish king Alexander III. As Alexander's only living heir she had been pledged to marry the son of Edward I of England; after her early death on the way to England, Edward declared his overlordship of Scotland, which led to its coming under English rule.

Margaret, Saint (*d.*1093), queen consort of MALCOLM III, King of Scotland. A deeply religious woman, she sought to replace the Celtic practices of the Scottish Church with those of Rome by founding monasteries and introducing English priests into Scotland. She was canonized in 1250. *See also* HC1 p.182.

Margaret of Anjou (1430–82), wife of HENRY VI of England, whom she married in 1445. During the Wars of the ROSES (1455–85) she fought the Yorkists for the right of her only son, Edward, to succeed to the throne. She was finally defeated at the Battle of Tewkesbury (1471), where her son was killed. *See also* HC1 p.*243.*

Margaret Rose, Princess (1930–), only sister of Queen ELIZABETH II, second daughter of King GEORGE VI. In 1960 she married Antony Armstrong-Jones, a photographer, who was created Earl of SNOWDON in the same year. A son, David Albert Charles, was born in 1961 and a daughter, Sarah Frances Elizabeth, in 1964.

Margaret Tudor (1489–1541), Queen of Scotland, daughter of HENRY VII of England. She married JAMES IV of Scotland in 1503 and at his death in 1513 became regent for their infant son, James V. Removed from power in 1515, she continued to play an important part in the uncertain Scottish political scene, her affiliation varying with her personal interest of the moment and her desire for power and money. *See also* HC1 p.245, *245.*

Margarine, buttery substance made from vegetable fats with aqueous milk products, salt, flavouring, food colouring, emulsifier and vitamins A and D blended in. It is used for cooking and as a spread. Corn or safflower oil is the most popular modern fat ingredient, due to the fear of polyunsaturated fats in relation to health.

Margay, small CAT found in tropical forests from Mexico to S Brazil; it is sometimes domesticated. The coat is cream yellow with black spots. Length: to 110cm (43.3in) including the tail. Family Felidae, species *Felis wiedii. See also* NW p.217.

Marggraf, Andreas Sigismund (1709–82), German chemist whose discovery of beet sugar in 1747 led to the development of a new industry. He also introduced the use of the microscope in chemical research and simplified the process of extracting phosphorus from urine.

Margrave, former German title of nobility introduced in the 9th century for administrative reasons by the CAROLINGIANS under CHARLEMAGNE. The Carolingians sent imperial representatives with extensive powers to rule over *marks* (border lands) of their empire.

Margrethe, or Margaret, the name of two queens of Denmark. Margrethe I (1353–1412) became regent of Denmark and Norway for her son on the death of her father, WALDEMAR IV of Denmark, (1375) and her husband, Haakon VI of Norway (1380). The union lasted until 1814. She also ruled Sweden from 1389 to 1397. Margrethe II (1940–), the eldest daughter of King FREDERICK IX, had her right to the throne established (1953) through a new constitution that allowed female succession. On the death of her father in 1972 she thus became Denmark's first ruling queen since Margrethe I.

Marguerite, perennial plant of the daisy family native to the Canary Islands. It has white-rayed, yellow-centred flower heads about 5cm (2in) across. Height: to 91cm (3ft). Family Compositae; species *Chrysanthemum frutescens.*

Marianas Islands, volcanic island chain in the W Pacific Ocean, approx. 2,400km (1,500 miles) E of the Philippines and N of the Caroline Islands, extending N to S approx. 800km (500 miles) along the Marianas Ridge. The group includes Guam, Saipan, Tinian, Rota, Pagan and ten other islands. Discovered by Ferdi-nand MAGELLAN in 1521, and named Islands of Thieves, the islands were renamed the Marianas Islands in 1668. They were taken by US forces 1944 and made a US Trust Territory. Exports include sugar cane, coconuts and coffee. Area: approx. 1,000sq km (390sq miles). Pop. (1975 est.) 108,500 (most of whom are on Guam).

Maria Theresa (1717–80), Archduchess of Austria, Queen of Bohemia and Hungary (1740–80; co-ruler with her son Joseph II, 1765–80), wife and mother of Holy Roman emperors (1745–65 and 1765–80 respectively). In 1713 her father, the Holy Roman Emperor CHARLES VI, altered Hapsburg law by the Pragmatic Sanction to allow female succession. But on his death (1740) her succession was disputed by almost all the major European powers. Prussia seized Silesia, precipitating the War of the AUSTRIAN SUCCESSION (1740–48). In 1756 Austria formed an alliance with France and Russia, thereby antagonizing Prussia and England. In the ensuing SEVEN YEARS WAR (1756–63) Austria lost no territory, but its traditional role as dominant German state passed to Prussia. Although essentially conservative, Maria Theresa devoted her energies after 1748 to a vast programme of reform. *See also* HC1 p.325.

Maribor, city in NW Yugoslavia, on the River Drava. Originally part of Styria, the city was transferred after 1918 to Yugoslavia. It is now a commercial and manufacturing centre, whose industries include textiles, chemicals and aluminium. Pop. (1971) 97,167.

Marie Antoinette (1755–93), Queen of France, daughter of FRANCIS I and MARIA THERESA of Austria. In 1770 she married the French dauphin, later LOUIS XVI, and her life of pleasure and careless extravagance caused deep public resentment. In 1789, during the FRENCH REVOLUTION, she was taken with the King from Versailles to Paris. She was seized at Varennes when the royal family attempted to escape and was guillotined in 1793.

Marie de France (*fl.*1160–1190), French poet and noblewoman who lived and wrote in England. She is best known for her *lais,* short narrative poems taken from Celtic folklore. *Chèvrefeuille,* a Tristan and Isolde story, deals with physical and romantic love; she also wrote *Lanval, Eliduc* and a collection of fables, *Ysopet.*

Marie de l'Incarnation, or Marie Guyard (1599–1672), French nun who joined the URSULINES after the death of her husband and in 1639 became the first female missionary of New France (Canada). She soon established a convent in Quebec and mastered the Huron and Algonquin languages. She wrote religious treatises in these languages and worked as a missionary among the local Indians.

Marie de Medici. *See* MEDICI.

Marie-Louise (1791–1847), princess of the HAPSBURG family who became Empress of France. She was the eldest daughter of Francis I, the Austrian Emperor, and was married to NAPOLEON I in 1810 as a diplomatic manoeuvre. She gave birth to a son in 1811. After Napoleon's defeat in 1814 she became Duchess of Parma, and married again in 1821 after Napoleon's death.

Mariette, August-Ferdinand-François (1821–81), French archaeologist noted for his work in Egypt. In 1850 he discovered the Avenue of Sphinxes and the Sarapeum (the temple of bulls) at the site of ancient MEMPHIS. Conservator of monuments for the Egyptian government, he dealt with the problem of unauthorized excavation and was responsible for the founding of the Egyptian Museum near Cairo.

Marigold, any of several mostly golden-flowered plants, mainly of the genera *Chrysanthemum* and *Tagetes* of the daisy family (Compositae), or the genus *Calendula* of the aster family. The corn marigold (*Chrysanthemum segetum*) is a common flower of the fields and waysides of the British Isles. Those most commonly cultivated are the French marigold (*Tagetes patula*) and the African marigold (*T. erecta*), both hardy annuals.

Marcus Aurelius: the emperor's statue stands in Rome's Piazza del Campidoglio.

Margaret of Anjou, the Lancastrian inspiration, died banished and impoverished.

Marguerites, if their shoots are pruned in Autumn, grow large very quickly.

Marie Antoinette's portrait by Marie Vigée le Brun led to their friendship.

Marine engines can be car engines modified, as in the "turbojet" water pump.

Filippo Marinetti (with stick) supported Fascism in his book *Futurismo e Fascismo.*

Saint Mark has as his symbol a lion, as in this church in Florence.

Markhor: its sandy red summer coat turns grey in the winter; its tail turns black.

Marihuana, narcotic DRUG prepared from the dried leaves of the Indian hemp plant (*Cannabis sativa*); it is different from HASHISH, which is prepared from resin obtained from the flowering tops of the plant. Some people smoke marihuana like tobacco. In moderate doses it is a relaxant and euphoriant, producing a sense of excitement or elation, intensifying sensations and distorting the sense of elapsed time. This may be followed by a phase of tranquillity and possibly fatigue; large doses produce intoxication. One of the "soft" drugs, it has been condemned by some authorities because its habitual use may lead to addiction and possible experiments with other, more dangerous drugs. But whether marihuana is physically harmful or a harmless pleasure remains a controversy. *See also* ADDICTION; MS p.105; PE p.*216.*

Marimba, percussion instrument with tuned wooden bars, ranging five to six octaves, that are set in a frame over resonators and struck with mallets. The modern marimba was developed in South America although it probably originated in Africa, where the resonators are often made from gourds. Darius MILHAUD wrote concertos for the marimba.

Marin, John (1870–1953), US painter whose individualistic style contains abstract and expressionistic elements. His works include *Movement, Fifth Avenue* (1912) and *Maine Islands* (1922).

Marine biology, science and study of life in the sea, including organisms that live in the water, in the sea bed and along shores. Marine biologists explore the ways in which these organisms fit into their environments, as well as their interactions with the human environment. Marine biology is part of OCEANOGRAPHY. *See also* NW pp.234–241.

Marine engineering, branch of engineering that deals with the construction, maintenance, and operation of the power plant and other mechanical equipment of seagoing vessels, docks and harbour installations. *See also* MM pp.114–115, 194–195.

Marine engines, power plants used to propel sea-going ships, and as auxiliaries in smaller sailing vessels. In the 19th and early 20th centuries, marine engines were coal-fired, piston-driven steam engines. Today large steam engines are still used but these are usually oil-fired·turbine engines; steam, raised in a boiler, turns a turbine, which is geared down to turn propeller shafts. Small, medium and large diesel engines, and small petrol engines (including outboard motors) are also in general use. A few ships, notably certain icebreakers and submarines, are fitted with nuclear engines in which steam for turbines is raised by heat from the radioactive decomposition of uranium fuel. *See also* MM pp.62–63.

Mariner space missions, series of US spacecrafts which studied the planets Mars and Venus. The first spacecraft to reach and pass another planet (Venus) was Mariner 2 in 1962. In 1965 Mariner 4 accomplished the first approach to Mars and sent back remarkable pictures. Mariner 5 flew by Venus (1967) and analyzed its atmosphere, and in 1969, Mariners 6 and 7 flew by Mars, analyzed its atmosphere and sent back pictures. In 1971 Mariner 9 was the first spacecraft to orbit another planet (Mars), and Mariner 10 flew past Venus in 1974 to become the first probe of Mercury. Mariner 11, re-named Voyager 1 and destined for Jupiter (1979) and Saturn (1981), was launched in 1977. *See also* SU index.

Marines. See ROYAL MARINES.

Marinetti, Filippo Tommaso (1876–1944), Italian poet and founder of FUTURISM. Repudiating traditional artistic and literary values and forms, he advocated instead the glorification of machinery, speed and war in such works as his manifesto of futurism (1909) and *Futurismo e Fascismo* (1924). *See also* HC2 p.177, *177.*

Marini, Marino (1901–66), Italian sculptor, painter and engraver. Although devoted mainly to pictorial art for many years, he turned to sculpture after 1928.

His work preserves a dramatic sense of reality in a simplified form, as in many versions of a horse and rider.

Marinus, name of two popes. Marinus I (*r.*882–84) was the first bishop of a diocese other than that of Rome to become Pope. He is said to have exempted the Roman *Schola Saxonum* from taxes at the request of Alfred the Great. Marinus II (*r.*942–46) attempted to tighten the discipline of the clergy and monks in Rome, France and Germany. He also fortified Monte Cassino against the attacks of the Bishop of Capua.

Marionette, full-length puppet operated by strings attached to each limb and controlled from above. Marionettes developed from similar puppets manipulated in such places as Burma, Ceylon and 17th- and 18th-century Europe by both strings and metal rods. In the 19th century total string control was achieved by puppeteers such as Thomas Holden and, in its imitation of almost any human action, the marionette came to be regarded as the most versatile and advanced form of puppetry.

Mariotte, Edme (*c.*1620–84), French physicist and plant physiologist noted for his independent discovery of BOYLE's law. He was a Roman Catholic priest, and also a founder of the Academy of Science in Paris.

Marists, congregation of priests, lay brothers, sisters and missionary sisters dedicated to the Blessed Virgin Mary. The Marist brothers were founded in 1817 by Marcellis Champagnat and the Marist fathers by Jean Claude Courveille and Jean Claude Marie Colin in 1822. The Marists did much missionary work throughout the Pacific, Europe, North America and Australasia, and have teaching orders in these areas.

Maritime law, branch of the law which is quite distinct in its function and practice from domestic law. It is primarily concerned with the rights of the mercantile concerns using the oceans of the world. In Britain various Merchant Shipping Acts have supplemented the generally international nature of the present law.

Maritime Trust, in the world of ships, the equivalent of the NATIONAL TRUST; it was established in Britain in 1970 to preserve vessels of historic interest. The Duke of Edinburgh is its president.

Marius, Gaius (*c.*157–86 BC), Roman politician and general who created a new and highly trained army. Of middle-class birth, Marius was elected consul seven times and became the enemy of the patrician SULLA. When Sulla took command of the Roman forces in the east, Marius jealously sought to deprive him of the command but was defeated in battle and forced to flee. In 87 he raised an army in Etruria and with Cinna captured Rome, where they were elected consuls. Marius ordered the destruction of his enemies, and it took 4,000 slaves five days and nights to complete the slaughter. Marius died shortly afterwards. *See also* HC1 pp.93, 96, *96.*

Marius, Simon (1573–1624), German astronomer, and one of the first to use a telescope. He observed sunspots and named the four largest moons of Jupiter, which Galileo also claimed to have discovered.

Marjoram, or origanum, perennial herb of the mint family about 60cm (24in) tall with purplish flowers. It is native to the Mediterranean region and w Asia and is cultivated as an annual in northerly climates. Wild marjoram is native to Europe and Asia. It is used extensively as seasoning. Species *Origanum vulgare.*

Mark, Saint, apostle and one of the four evangelists of the New Testament. He was the cousin of the apostle BARNABAS and accompanied him and St PAUL on several missionary journeys until a disagreement with Paul caused him to detach himself, traditionally to become the secretary to St PETER and write the first gospel following Peter's teaching. According to the Church historian Eusebius, Mark founded the church at Alexandria. Feast day: 25 April.

Mark, the Gospel according to Saint, earliest gospel in the New Testament.

Written about AD 55–65, it is believed to be one of two reference works (the other being "Q") used by St MATTHEW and St LUKE in compiling their own gospels. St Mark is traditionally considered to have been the secretary of St PETER; to many people, therefore, his gospel reflects the views and teachings of that apostle. The gospel is nevertheless written in a concise Hellenistic style.

Mark Antony (*c.*82–30 BC), distinguished Roman soldier and politician, friend of JULIUS CAESAR and CLEOPATRA's lover. His affair with Cleopatra led to defeat at ACTIUM by Octavius, after which he and Cleopatra fled to Egypt, where they both committed suicide.

Market Bosworth, town in Leicestershire, England, 19km (12 miles) w of Leicester and near the site of the Battle of Bosworth Field (1485), the last battle of the Wars of the ROSES. There the Lancastrians, led by Henry Tudor, later HENRY VII, defeated and killed the Yorkist king RICHARD III. *See also* HC1 p.243, *243.*

Market research, study of consumer preference and demand, for the purpose of increasing the sales of commercial products. It is often undertaken by companies specializing in the field and involves the collection, recording and analysis of data. A common method of market research involves testing a product on a small sample of people statistically compiled to form a representative sample of likely consumers of the product. Interviewing to ascertain popular preferences for products is another method. Market research techniques are also used in politics by PUBLIC OPINION POLLS – for instance the GALLUP POLL which predicts election results.

Markhor, largest wild goat, inhabiting mountains of Afghanistan and the w Himalayas. Males have long, corkscrew horns and shaggy beards. Height: 100cm (40in) at shoulder; weight 90kg (200lb). Family Bovidae; species *Capra falconeri.*

Markiewicz, Constance Georgine, Countess de (1876–1927), Irish Republican politician married to a Polish count. Sentenced to death for her militant part in the EASTER RISING (1916), she was reprieved and released from prison in 1917. In 1918 she became the first woman to be elected to the British House of Commons, although she never took her seat.

Markova, Dame Alicia (1910–), British ballerina, real name Lilian Alicia Marks. She joined the Vic-Wells Ballet in 1931, and was its first prima ballerina. Famous for her *Giselle,* she was the first British dancer to take the title role. Other classical ballets in which she excelled are *Les Sylphides* and *Swan Lake. See also* HC2 p.275.

Marks and Spencer, large chain-store selling mainly clothes. It originated in 1884 when Michael Marks set up a market stall which sold each item for one penny. In 1894 Marks took Thomas Spencer into partnership. In the mid-1970s it had more than 250 shops and employed about 40,000 people in Britain.

Markus, Mrs Rika ("Rixi") (1910–), British contract bridge player. She played for Austria in the 1930s, and for Britain after 1945. She has won 12 international titles, including the World Women's Pairs Championship twice (1962, 1974) in partnership with Fritzi Gordon.

Marl, grey or blue-grey mixture of limestone, clay and sand. It is crumbly in texture and often formed as a lake sediment in the vicinity of submerged plants. It is used in some countries as mortar for building and for road surfaces.

Marlborough, John Churchill, 1st Duke of (1650–1722), English general and statesman. One of the greatest commanders in English history, Marlborough led the Allied armies against LOUIS XIV in the War of the SPANISH SUCCESSION, destroying French power and ambition at BLENHEIM (1704), RAMILLIES (1706), OUDENAARDE (1708) and MALPLAQUET (1709). On the ascendancy of his political enemies, the Whigs, in 1711 he was dismissed from all his offices but in 1714, with the accession of the Hanoverian GEORGE I, he was restored to royal favour. *See also* HC1 pp.*293, 297,* 312–313, *312–313.*

Marlborough, Sarah Jennings Churchill, Duchess of (1660–1744), favourite of Queen ANNE. She married John Churchill, later 1st Duke of MARLBOROUGH, in 1677 and her influence with the future queen, which was considerable, assisted his career. Her influence declined after 1706, when she and the queen began to quarrel over Whig cabinet appointments; she became replaced in the queen's affections by Abigail Masham. *See also* HC1 pp.312–313.

Marlin, any of several species of large marine fish found in warm waters of the Atlantic and Pacific oceans, especially the blue marlin (*Makaira mitsukurii*); it is often fished for sport. It is blue with a coppery tint and violet side markings. The fins and long snout are sharply pointed. Length: to 8m (26ft); weight: 635kg (1,400lb). Family Istiophoridae. *See also* NW p.*240.*

Marlowe, Christopher (1564–1593), English poet and playwright, often labelled "Shakespeare's greatest predecessor"; one of the most powerful and original artists to write in the English language. Much of his success derives from his ability to make his heroes dangerous, even ferocious, but still human, as in *Tamburlaine the Great* (*c.*1587), *The Jew of Malta* (*c.*1590) and *The Tragical History of Doctor Faustus* (*c.*1588). Marlowe led a dissolute life and in 1593, with a Privy Council order for his arrest hanging over him, was killed in a brawl. *See also* HC1 pp.282, *282,* 285, *285.*

Marmara, sea in NW Turkey, between Europe (to the N) and Asia (S). It is connected with the Black Sea (E) by the Bosporos and with the Aegean Sea through the Dardanelles. Its largest island is Marmara, famous for its alabaster and marble quarries, and ISTANBUL is located on the N shore at the entrance of the Bosporus. Area: 11,474sq km (4,430sq miles).

Marmoset, small diurnal, arboreal MONKEY of tropical America. Among the smallest of all monkeys, marmosets are the size of small squirrels. They have soft, dense fur that varies in colour, and have pointed, sickle-shaped nails. Those of certain genera are often called tamarins. Family Callitrichidae; genera *Callithrix, Leontopithecus. See also* NW pp.*168,* 217.

Marmot, stocky, terrestrial rodent of the SQUIRREL family, native to North America and Eurasia. Most marmots have brown to grey fur, short, powerful legs and furry tails. They eat vegetation by day, find shelter at night and hibernate in grass-lined burrows. Length, excluding tail: 30–60cm (12–24in); weight: 3–8kg (6.6–16.5lb). Family Sciuridae. *See also* NW p.*198.*

Marne, Battles of, two German defeats of WWI. The first, in Sept. 1914, was an Allied counter-attack directed by Gen. JOFFRE, which stopped the German advance on Paris; the second, in July 1918, was another counter-stroke, directed by Marshal FOCH, which drove back the final German offensive. *See also* HC2 pp.188–189.

Maronites, Christian community of Syrian origin which claims to have been founded by St Maron in the late 4th and 5th centuries. The majority of its members are to be found in the Lebanon, where it is the major religious group, but small colonies exist in Syria, Palestine, Cyprus, North and South America. The Maronites became separated from the rest of Christendom because of their acceptance of MONOTHELETISM, in consequence of which they were excommunicated by the Third Council of Constantinople in 680. In 1182 they may have attempted to establish communion with the Roman Catholic Church but this did not effectively come about until the 16th century. Since then they have had the status of a uniate Church, ie an Eastern Church in union with Rome but retaining its own rite and canon law.

Marot, Clément (1496–1544), French poet, one of the first to write sonnets. He held various positions at court but, accused of heresy, was several times imprisoned. He wrote witty court poems but

his metrical translation of the *Psalms* was condemned by the SORBONNE. *See also* HC1 p.316, *316.*

Marque, letters of, commissions granted by states at war to private individuals allowing them to fit out ships to take part in the war. In the 17th and 18th centuries ships carrying letters of marque were known as PRIVATEERS and, as the name implies, were little more than pirates. Signatories of the Declaration of Paris (1856) agreed to abolish letters of marque, a prohibition extended worldwide by the Hague Convention of 1907.

Marquesas Islands, volcanic island group, part of French Polynesia, in the Pacific Ocean, S of the equator and N of Tuamotu. The group of twelve includes Fatu Hiva, Hiva Oa and Nuku Hiva. The islands are mountainous, with fertile valleys and several good harbours. The French took possession and settled them in 1842, introducing European diseases that killed many of the native Polynesians. The capital is Hakepehi on Nuku Hiva. Exports include tobacco, vanilla and copra. Area: 1,243sq km (480sq miles). Pop. (1971) 5,600.

Marquess, or marquis, noble title, in rank below a DUKE and above an EARL, originally of an officer in charge of a border region (or march) of a kingdom. The first marquess in English history was Robert de Vere, created Marquis of Dublin by RICHARD II in 1385. The oldest surviving title is the Marquis of Winchester, created in 1551.

Marquet, Albert (1875–1947), French painter and friend of MATISSE. He contributed to the first FAUVIST exhibition (1905), and became one of the boldest of the Fauves in his concern for harmony and equilibrium and the solidity of forms. Important among his paintings is *Quai des Grands-Augustins* (1905).

Marquetry, inlay of ornamental woods or other decorative materials applied to furniture or wall panelling in a variety of patterns and designs, or arranged to make a picture.

Marquette, Father Jacques (1637–75), French Jesuit missionary and explorer in North America. Marquette arrived in Quebec in 1666 as a missionary priest. After two years he moved deeper into the wilderness to preach among Indians around the Great Lakes. Chosen by the Canadian government in 1673 to explore, he and five others were the first white men to discover the Mississippi River.

Marquis, Donald (1878–1937), US writer and humorist. Many of his stories and verses first appeared in his columns in the New York *Sun* and *Herald Tribune.* He invented a number of characters, notably "archy the cockroach" and "mehitabel the cat", whose adventures are related in *archy and mehitabel* (1927).

Marquises. *See* MARQUESAS ISLANDS.

Marrakech, city in W central Morocco, at the NW foot of the Atlas mountains. Founded in 1062 by the ALMORAVIDS, it was the country's capital until 1147 and it subsequently served as the sultan's residence. Its modern industries include tourism and leather goods. Pop. 350,080.

Marram grass, or beach grass, important grass of sand dunes; its roots help to bind the sand together and so prevent dune movement. *Ammophila arenaria* has long leaves that are curled into cylindrical spikes to resist water loss.

Marranos, Spanish Jews who were outwardly converted to Christianity to escape persecution in medieval times, but who continued to practise Judaism secretly. The origin of the term is uncertain.

Marriage, in the modern Western sense, legal status of a man and a woman joined by ceremony as husband and wife. This is known as MONOGAMY, but some societies practise POLYANDRY and POLYGAMY. Marriage has been favoured throughout history as a regulator of sexual relationships and as a stable basis for the establishment of the family unit within which children can be reared. *See also* MS pp.168–169.

Marriage à la Mode (1673), play by John DRYDEN, in which he modified his heroic style and created a typically elaborate

Restoration comedy. Its carefully planned plot and sophisticated wit ensured its success.

Marriage à la Mode (*c.*1742–44), series of six satirical, anecdotal paintings by William HOGARTH, tracing the inevitable course of an arranged marriage. Like the artist's similar works, *A Harlot's Progress* (*c.*1731) and *A Rake's Progress* (*c.*1735), they were produced with a view to their eventual reproduction as engravings. *See also* HC2 p.*70.*

Marriage of Figaro, The (1786) four-act opera by Wolfgang Amadeus MOZART with Italian libretto by Lorenzo da PONTE, after Pierre BEAUMARCHAIS' comedy. Its Vienna première was conducted by Mozart himself. Musically and politically revolutionary, the story is of servants duping their aristocratic master. *See also* HC1 p.319.

Married Women's Property Act (1882), most important of a series of British laws (1870, 1882 and 1893) which gave married women the right to own and dispose of property in their own right. Before 1882 a husband became absolutely entitled to his wife's property upon marriage.

Marriner, Neville (1924–), British conductor. He was apointed Professor at the Royal College of Music, London in 1950, and became conductor of the London Symphony Orchestra in 1954. Two years later (1956) he founded the chamber orchestra the Academy of St Martin in the Fields. In 1968 he also became conductor of the Los Angeles Chamber Orchestra.

Marrow, large vegetable of the GOURD family, closely related to the PUMPKIN. It is usually briefly boiled in water and served with butter, together with meat and other vegetables. Species *Cucurbita pepo. See also* PE pp.188, *189.*

Marrow, soft tissue containing blood vessels, found in the hollow cavities of bones. The marrow found in many adult bones, including the shafts of long bones, is somewhat yellowish and functions as a store of fat. The marrow in the flattish bones, including the ribs, sternum (breast bone), skull, spinal column and the ends of the long bones, is reddish and contains reticular cells that give rise to myeloblasts. The myeloblasts give rise eventually to red blood cells as well as to most of the white blood cells (but not LYMPHOCYTES) and platelets. *See also* MS p.63.

Marryat, Capt. Frederick (1792–1848), British author, whose seafaring novels were based on his own experiences in the Royal Navy. *Peter Simple* (1834) and *Mr Midshipman Easy* (1836) are adventures with strong characters, full of nautical detail. In later life he wrote adventure stories for children, including *Masterman Ready* (1841) and *The Children of the New Forest* (1847).

Mars, ancient Roman god of war, often depicted as an armed warrior; one of the three protector-deities of the city of Rome itself (with JUPITER and Quirinus). He was originally associated with agriculture but later took on his dominant military aspects; the wolf and woodpecker were sacred to him.

Mars, fourth planet from the Sun, with two satellites (PHOBOS and DEIMOS), and more Earth-like than any other known world. Its thin atmosphere is mainly carbon dioxide; temperature −120°C (−184°F) to over 16°C (60°F). Mean distance from the Sun, 228,000,000km (141,500,000 miles; mass, 0.11 of Earth; diameter, 6,790km (4,220 miles); rotation period, 24hr 37min; period of sidereal revolution, 687 days. *See also* SOLAR SYSTEM; SU pp.194–203.

Marsala, port in W Sicily, Italy, on the Mediterranean Sea. The ancient city of Lilybaeum, it has a cathedral and old city walls. GARIBALDI's forces landed there in 1860 in the conquest of Sicily. Marsala is famous for its sweet wines. Exports: wines, salt, grain. Pop. (1971) 79,038.

Marsden, Samuel (1764–1838), British clergyman, a leading colonist in Australia and New Zealand, and chaplain of a convict colony. In 1794 he emigrated to New South Wales and was influential in developing agriculture and improving education. In New Zealand he established the

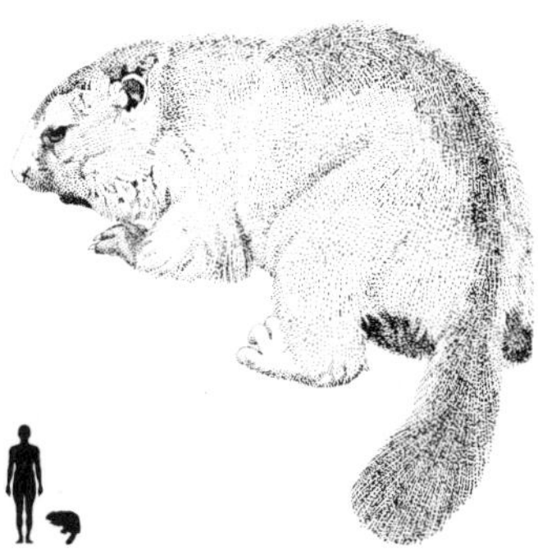
Marmot: its burrow has many entrances through which to beat a hasty retreat.

Marriage of Figaro: ensembles rather than arias reveal the characters.

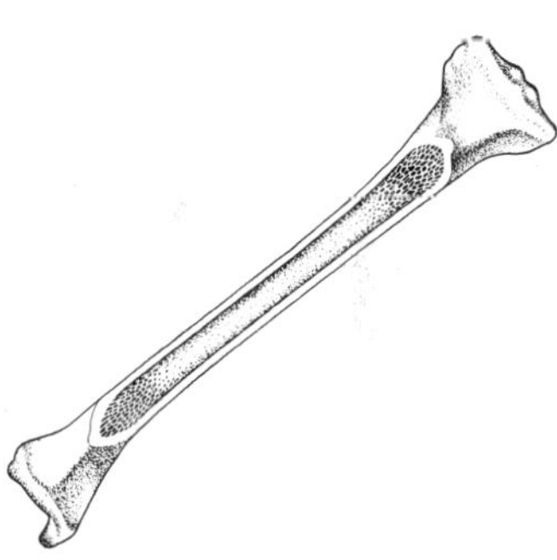
Marrow can be removed for medical examination by the use of a syringe.

Mars was important to the Romans as war god and father of Romulus and Remus.

Marseilles is built around the old natural harbour, itself rebuilt in 1850.

Marsh mallow; its root, the edible part, is shaped like a carrot.

Martello Tower at Eastbourne was deliberately destroyed by the navy in 1860.

Martens have a scent gland located near the base of the tail.

first white mission settlement in 1814.

Marseillaise, La (1792), French national anthem, with both words and music by Claude Joseph Rouget de l'Isle. Composed as a marching song for troops on the Rhine, it was first performed publicly by the band of the National Guard of Strasbourg in April, 1792. In June it was adopted with enthusiasm by the Marseilles Volunteers as their marching song and thereafter became known by its present name.

Marseilles (Marseille), port in SE France, on the Gulf of Lions; capital of Bouches-du-Rhône département, and connected to the River Rhône by an underground canal. The second-largest and oldest city in France, it was founded in 600 BC by Phocaean Greeks. During the CRUSADES Marseille was a commercial centre and shipping port for the Holy Land. The 19th-century conquest of Algeria by France and the opening of the Suez Canal (1869) brought great prosperity to the city. In the harbour is the Château D'If, a prison built in 1524 on a small rocky island. Industries: flour milling, soap, vegetable oil, cement, sugar refining. Pop. 889,028.

Marsh, Dame Ngaio (1899–), New Zealand detective-story writer and producer of plays. Her numerous detective novels include *Enter a Murderer* (1935), *Vintage Murder* (1937), *Final Curtain* (1947) and *Black as He's Painted* (1974). Her novels are perceptive in characterization and many reflect her knowledge of the theatre. She was made a DBE in 1966.

Marsh, low grassland area, devoid of peat, saturated by moisture during one or more seasons. Typical vegetation includes grasses, sedges, reeds and rushes. Marshes are sometimes the breeding sites of mosquitoes and other disease carriers, but are valuable in maintaining water tables in adjacent ecosystems. *See also* NW pp.232–235.

Marshal, senior member of the royal HOUSEHOLD in England after 1066. Originally master of the horse, he became responsible for maintaining order in the court and acted as judge in the Court of Chivalry, over which he still presides as Earl Marshal. In the reign of HENRY I the office belonged to the FitzGilbert family, later passing to William Marshall, Earl of Pembroke (*d*.1219). It passed then to the Earls of Norfolk, and the Mowbrays, Howards and Arundels.

Marshall, Alfred (1842–1924), British neo-Classical economist whose *Principles of Economics* (1890) was a landmark in modern economics. Marshall showed that supply and demand were like the blades of a pair of scissors; both blades are needed equally if the scissors are to work properly.

Marshall, George Catlett (1880–1959), US general. He served as a staff officer during WWI and, after various commissions including service in China (1924–27), became Chief of Staff during WWII. He was instrumental in devising Allied strategy, advocating the conquest of Germany through France. After the war, he was the inspiration of the MARSHALL PLAN, a scheme to help economic recovery in Europe.

Marshall, John (1755–1835), noted Chief Justice of the US Supreme Court. He gained national prominence in 1788 when he argued successfully for the Constitution against Patrick HENRY, and was a commissioner to France in the XYZ AFFAIR.

Marshall, Sir John Hubert (1876–1958), British director-general of the Indian Archaeological Survey (1902–31). He made important prehistoric finds at Taxila and HARAPPA, and was responsible for the preservation of many monuments.

Marshall, John Ross (1912–), New Zealand politician and Prime Minister (1972). A lawyer, he proved an urbane Cabinet Minister for 20 years, holding important portfolios such as Health, Justice, Industries and Commerce, and Overseas Trade. He became Prime Minister in 1972 but his National Party was heavily defeated at the polls that year and he was dropped as leader in 1974. He retired a year later.

Marshall Islands, group of atolls and coral reefs in the W Pacific Ocean, E of the Caroline Islands. Administered as a US Trust Territory of the Pacific Islands, they consist of two great chains, the Ralik (w) and the Ratak (E), which run almost parallel NW to SE, covering an ocean area of 11,655sq km (4,500sq miles). The native inhabitants are Micronesian. Annexed to Germany in 1885, the group was taken by Japan in 1914 and by the US in WWII (1944). Products: copra, coconuts, tropical fruits, vegetables and fish. Total land area: approx. 179sq km (69sq miles). Pop. (1970 est.) 20,200.

Marshall Plan, US programme of economic aid to Europe proposed by Secretary of State George C. MARSHALL in 1947. The Economic Co-operation Act (1948) authorized the plan. The USA spent more than \$13,000,000,000 in five years to promote postwar economic recovery in Europe. For this work Marshall was awarded the 1953 Nobel Peace Prize. *See also* HC2 pp.219, 234, 258, *258*.

Marshalsea Prison, old London prison linked with the Marshalsea Court, established for defaulters in the royal household. It later came to be used as a prison for debtors and for those convicted of piracy and other crimes committed on the high seas. It was abolished in 1842 and replaced by the Queen's Prison.

Marsh gas. *See* METHANE.

Marsh mallow, perennial herb native to E Europe and related to the garden hollyhock. It has pink flowers with five overlapping petals. A sweet confection of the same name was prepared from the roots. Height: 1–1.8m (3–6ft). Family Malvaceae; species *Althea officinalis.*

Marsh marigold, perennial plant of the buttercup family native to cold and temperate swamps of the Northern Hemisphere. It has hollow stems, kidney-shaped leaves, and large pink, white or yellow flowers. There are about 20 species. Family Ranunculaceae; genus *Caltha.*

Marsigli, Count Luigi Ferdinando (1658–1730), Italian soldier, naturalist and oceanographer. After long service in the Austrian army, during which he was captured, he was demoted in 1704. Thereafter he engaged in scientific study, founding the Accademia delle Scienze in Bologna (1712) and becoming a member of the Royal Society (1722) to which he was presented by NEWTON.

Marston Moor, battle site 11km (7 miles) w of York, England. It was the scene of the decisive defeat of Royalist forces under Prince RUPERT by the Parliamentarians under FAIRFAX (2 July 1644) in the English Civil War.

Marsupial, mammal of which the female usually has a marsupium, or pouch, within which the young are suckled and protected. A few marsupials, including the NUMBAT, have no pouch. At birth, the young are in a very early stage of development, corresponding roughly to the third intra-uterine month of the human foetus. The tiny, blind newborn KANGAROO, after emerging from the genital aperture of its mother, instinctively crawls up her belly fur until it reaches her pouch and stays there, attached to her teat, until mature enough to emerge. Like the kangaroo, most marsupials are Australasian, and include such varied types as the KOALA, WOMBAT, TASMANIAN DEVIL, BANDICOOT and marsupial MOLE. The only marsupials to live outside Australasia are the OPOSSUMS and similar species that are found in the Americas. *See also* NW pp.154, 156, 158–159, 204.

Martello towers, series of gun towers built along the s and E coasts of England during the Napoleonic Wars. They were low, round towers with a gun platform on top, named after the tower at Cap Mortella, Corsica, which gave strong resistance to an English attack in 1794.

Marten, any of several species of carnivorous mammals of the WEASEL family that live in forested areas of Eurasia and North and South America. Martens have a long body and short legs and are hunted for their fur. The dark brown skins of the SABLE, *Martes zibellina* are the most valu-

able but the skins of the lighter stone or beech marten, *Martes foina,* are also prized. Family Mustelidae. *See also* NW pp.165, *206.*

Martial(is), Marcus Valerius (*c.*40–*c.*104) Roman poet, *b.* Spain. He lived in Rome from about AD 64, enjoying the patronage of the emperors DOMITIAN and TITUS and the friendship of JUVENAL and QUINTILIAN. He wrote more than 1,200 epigrams, which are considered the finest of the genre. He is not only one of the wittiest writers of antiquity, but also gives an unrivalled picture of life in Rome in the first century. *See also* HC1 p.113.

Martial law, condition prevailing when normal civil government is superseded by military authorities, and military law is applied to the civilian population. It is often enforced in times of potential public disorder or when there is a threat to public safety. Martial law can be restricted to particular regions of a country and is commonly declared after a military coup in order to prevent a violent reaction to the new government.

Martin, Saint (*c.*315–97), Bishop of Tours. Born a heathen, he became a Christian in his youth. As a Roman soldier he is reputed to have torn his cloak to share it with a beggar. From 360 he lived as a monk, and was acclaimed bishop in 371, against his will. His feast day is known as MARTINMAS.

Martin, name of several popes. Martin I, Saint (*r.*649–55) summoned the first of the LATERAN COUNCILS. He stressed papal authority over all and refused to obey the edict of the Byzantine emperor, Constans II, forbidding religious discussions. Arrested by Constans and publicly degraded, he died soon after being exiled to the Crimea and was acclaimed a martyr (the last pope to be martyred). Martin IV (*r.*1281–85), *b.* France, gave the FRANCISCANS the right to preach and to hear confessions. Elected at Viterbo in Italy, he was never able to enter Rome because of hostilities. Martin V (*r.*1417–31), whose election ended the GREAT SCHISM, moved the papacy permanently to Rome in 1420. His main concern was to consolidate Church unity and papal prestige through political means, and he reorganized the CURIA.

Martin, Archer John Porter (1910–), British biochemist who developed partition CHROMATOGRAPHY, particularly for use in amino-acid anlaysis. For this research he shared the 1952 Nobel Prize in chemistry with Richard L. M. SYNGE.

Martin, Frank (1890–), Swiss composer. His distinctly personal style shows a mastery of counterpoint, harmony and emotional continuity; in later works he used a modified 12-note system. His best-known work is the *Petite Symphonie Concertante* (1945), and others include a piano quintet (1920), a string trio (1936) and a concerto for harpsichord and orchestra (1952).

Martin, Homer Dodge (1836–97), US painter of landscapes who was influenced by IMPRESSIONISM. He broke away from the direct approach of the HUDSON RIVER SCHOOL, strengthening his Impressionistic approach during two visits to France in 1876 and 1882.

Martin, John (1789–1854), British painter of romantic melodramatic landscapes. His other works included dramatic architectural scenes filled with figures to depict heroic or biblical themes.

Martin, Kenneth (1905–), British painter and sculptor who focused his attention on landscapes in the 1930s but turned towards abstract painting in the 1940s. From the early 1950s he became involved in producing mobiles, in order to explore change and space. His steel and bronze construction, *Mobile Spiral* (1956), is a typical example.

Martin, Pierre Émile (1824–1915), French engineer who pioneered the process of open-hearth steel manufacture. The Siemens-Martin process pre-heats the air-blast by a heat-regeneration means and the excess carbon is removed by adding a calculated amount of iron ore. By applying this process to an open-hearth furnace which was charged with a mixture of pig-iron and wrought-iron, Martin was able to obtain the desired carbon content simply by dilution. *See also* MM p.31.

Martin, Violet Florence. *See* ROSS, MARTIN.

Martin, fast flying bird closely related to the SWALLOW and native to Europe and North America. It has long, pointed wings, and short legs which it seldom uses. The house martin (*Delichon urbica*) migrates from Europe and Asia to India and South Africa, often nesting under the eaves of houses. The purple martin (*Progne subis*) ranges from Mexico to the Arctic and nests in hollow trees. The sand martin (*Riparia riparia*) is found in approximately the same areas as the other two species; it burrows into sandbanks to nest. Family Hirundinidae.

Martin du Gard, Roger (1881–1958), French novelist. His major works, *Jean Barois* (1913), a novel of France during the DREYFUS AFFAIR, and the eight-novel series *Les Thibault* (1922–40), the story of a family and its religious and political convictions at the time of WWI, are objective treatments of the moral and intellectual preoccupations of his generation. He was awarded the Nobel Prize in literature in 1937.

Martineau, Harriet (1802–76), British writer and reformer. Her works include collections of stories that propound her social and economic theories, among them *Illustrations of Political Economy* (9 vols, 1832–34) and *Poor Laws and Paupers Illustrated* (1833–34). She visited the USA in 1834. She became an advocate of the abolition of slavery and wrote *Society in America* (1837), about US life.

Martinet, Jean (*d.* 1672), French soldier who helped to reorganize (1660–70) the army of Louis XIV into the first regular army in Europe. He is best known for the introduction of a system of uniform drill. In English his name has come to be applied to one who strictly adheres to rules and is a stickler for detail.

Martínez Ruiz, José (1894–1667), Spanish writer, better known by his pseudonym, Azorín. His most successful works are essays, such as *Los Pueblos* (1905), which are notable for his Impressionistic descriptions of Castilian towns and his attempt to define the eternal qualities of Spanish life.

Martini, Simone (*c.* 1284–1344), Sienese painter. Influenced by DUCCIO's style, he used decorative outline and rich colouring in his many frescoes and altarpieces. His masterpiece is the *Annunciation* (1333).

Martinique, island in the Windward group, West Indies; an overseas French département. Of volcanic origin, it is the largest of the Lesser Antilles; Fort-de-France is the capital. Discovered in 1502 by Christopher COLUMBUS, Martinique was inhabited by Carib Indians until they were displaced by French settlers after 1635. Attacked in the 17th century by the Dutch and the British, the island became French after the NAPOLEONIC WARS. Products: sugar, rum, fruits, cocoa, tobacco, vanilla, vegetables. Area: 1,100sq km (425sq miles). Pop. (1975 est.) 363,000. *See also* MW p.123.

Martinmas, feast of St MARTIN of Tours, celebrated on 11 Nov. The feast gave its name to the "mart", the fattened cow or ox killed in the Middle Ages at Martinmas, and salted for winter use.

Martinon, Jean (1910–76), French conductor and composer who studied with ROUSSEL. He has directed a number of European orchestras and in 1968 became director of the French National Orchestra. His compositions include four symphonies, chamber music, sonatas, a cello concerto and the opera *Hécube* (1949).

Martinson, Harry Edmund (1904–), Swedish writer. From the age of six he lived as a tramp and later as a seaman; his experiences influenced the underlying mood in both his poetry and his novels. His space fantasy *Aniara* (1956) became a successful opera (1959). He shared the 1974 Nobel Prize in literature with Eyvind JOHNSON.

Martinu, Bohuslav (1890–1959), Czech composer. In 1913, he became a violinist in the Prague Philharmonic Orchestra and only in 1923 did he turn to composition in

earnest. He studied with ROUSSEL in Paris, where he remained until 1940. Much of his music is based on Bohemian folk rhythms and Czech dances, which despite carefully stylized treatment retain their basic simplicity. He composed many operas and ballets, including *The Butterfly that Stamped* (1929) and *Comedy on a Bridge* (*c.*1950), as well as orchestral and chamber works.

Martyr, from a Greek word meaning "witness", one who dies for his faith. The term is particularly applied to Christians who suffered for their beliefs. John Foxe's *Book of Martyrs* (1563) is a record of Anglican and Protestant martyrs in England.

Marvell, Andrew (1621–78), English metaphysical poet and satirist. In 1657 he was appointed John MILTON's assistant and in 1659 he was elected to Parliament. A Puritan and yet a defender of individual liberty, he is chiefly remembered today for his lyric poetry, published in 1681 in a collection which included *The Garden*, *Bermudas* and *To His Coy Mistress*. *See also* HC1 pp.282–283.

Marx, Karl Heinrich (1818–83), German social philosopher and political theorist, founder (with Friedrich ENGELS) of world COMMUNISM. He studied history, philosophy and law, was influenced by G. W. F. HEGEL, Ludwig FEUERBACH and Moses Hess and produced his own philosophical approach of DIALECTICAL MATERIALISM. He proclaimed that religion was "the opium of the people" and in *The German Ideology* (1845–46), written with Engels, described inevitable laws of history. In Brussels he joined the Communist League and wrote with Engels the epoch-making COMMUNIST MANIFESTO (1848). Marx took part in the revolutionary movements in France and Germany, then went to London (1849) where he lived until his death. Although Engels helped him financially, these were years of poverty and illness. He toiled at research in the British Museum and produced a stream of writings, including *Das Kapital* (3 vols, 1867, 1885, 1894, the last two edited by Engels), which became the "Bible of the working class". In 1864 he was founder of the International Workingmen's Association (the First International) and became its leading spirit. He denounced both the non-revolutionary reformism of British labour leaders and the anarchy of philosopher Mikhail BAKUNIN, generally favouring legal methods to hasten the collapse of CAPITALISM. Marx was one of the most important political theorists of modern times. His ideas exerted a powerful influence after his death and helped to change the course of history. *See also* MARXISM; MS pp.*232–233*, 273, *275*, 298.

Marx Brothers, US team of vaudeville and film comedians. The members were Chico (Leonard) (1891–1961), Harpo (Arthur) (1893–1964), Groucho (Julius) (1895–1977), and Gummo (Milton) (1894–1977) and Zeppo (Herbert) (1901–), who both withdrew from the team by 1935. Their films include *Animal Crackers* (1930), *Duck Soup* (1933) and *A Night at the Opera* (1935). *See also* HC2 p.268.

Marxism, school of socialism that arose in 19th-century Europe as a response to the growth of industrial CAPITALISM. It is named after Karl MARX who, with Friedrich ENGELS, articulated an economic interpretation of history that held that all changes in society's structure were determined by changes in productive activity. According to Marxism, a Communistic society was historically inevitable. Capitalism, because of its emphasis on profits, would eventually so reduce the condition of workers that they would rebel, overthrow the capitalists and establish a classless society (the so-called dictatorship of the proletariat). The new society thus brought about would enable all individuals to achieve self-fulfilment and all production would be centralized in the state. Faith in man, the denial of God and belief in the class struggle were at the heart of Marxist thought. *The Communist Manifesto* (1848) and *Das Kapital* (1867, 1885, 1894) both contain ideas central to

Marxism, which forms the basis of COMMUNISM and strongly influenced the related ideology, SOCIALISM. *See also* HC2 pp.170–171.

Mary, Saint, or the Blessed Virgin Mary, mother of JESUS CHRIST who figures prominently in the first two chapters of the Gospels according to St MATTHEW and St LUKE which record the birth of Christ. Both gospels testify to her virginity (*see* VIRGIN BIRTH), and to her maternity. At Jesus's crucifixion, Mary was placed in the care of St JOHN. Mary has always been held in high regard in Christendom; by her assent in being the instrument in God's INCARNATION she exemplifies the divine-human relationship that is part of God's plan for mankind's salvation. In the early Church, the principal Marian feast was called the Commemoration of St Mary, from which developed the later feast of the ASSUMPTION (15 August). Other Marian feasts are: the Nativity (8 September), the ANNUNCIATION, or Lady Day (25 March), the Purification, or CANDLEMAS (2 February), the Visitation (2 July) and, for Roman Catholics, the IMMACULATE CONCEPTION (8 December). The most famous shrines of Mary are those at LOURDES and FATIMA, where she is said to have appeared to children.

Mary I, or Mary Tudor (1516–58), Queen of England, daughter of HENRY VIII and Catherine of Aragon; she is known as "Bloody Mary". On the death of EDWARD VI she overcame Lady Jane GREY's challenge for the throne and became queen in 1553. Her determination to re-introduce Roman Catholicism in England occasioned the major errors of her reign – her marriage to Philip of Spain in 1554 and persecution of her Protestant subjects. *See also* HC1 pp.*270*, 271; HC2 p.*309*.

Mary II (1662–94), Queen of England, Scotland and Ireland. Although her father, JAMES II, was a Roman Catholic, Mary was a Protestant. She married the Dutch noble WILLIAM III OF ORANGE in 1677, and her support enabled them to become joint sovereigns in 1689 following the GLORIOUS REVOLUTION that deposed James. She died from smallpox, and William then ruled alone. *See also* HC1 p.*312*; HC2 pp.308, *309*.

Mary, Queen of Scots (1542–87), daughter of JAMES V of Scotland. She married Francis II of France in 1558 and after his death in 1560 married Henry, Lord DARNLEY, in 1565. He died under mysterious circumstances in 1567 after he had had David Rizzio, her favourite, murdered the year before. She was deposed after her marriage to the Earl of BOTHWELL and sought refuge in England. After Mary was shown to have been involved in the RIDOLFI PLOT (1572) and the BABINGTON CONSPIRACY (1586), Elizabeth I agreed reluctantly to her execution. *See also* HC1 pp.276, *276*, 298, *298*.

Maryinski Theatre. *See* KIROV BALLET AND OPERA COMPANY.

Maryland, state in E USA, on the Atlantic Ocean. The W half of the state is part of the Piedmont plateau region. E Maryland is dominated by Chesapeake Bay and its associated coastal marshlands. The rearing of cattle and chickens is the most important farming activity. Corn, hay, tobacco and soya beans are the chief crops. Major industries are iron and steel, shipbuilding, primary metals, food processing, transport equipment, chemicals, electrical machinery and fishing. The main cities are Annapolis, the state capital, and Baltimore, the largest city.

A charter for a large territory that included Maryland was granted by Charles I to Cecilius Calvert, 2nd Baron Baltimore, in 1632. The first settlements were founded in 1634. Maryland was active in the drive for American independence. In 1791, the state ceded an area of land on the Potomac River to create the District of Columbia, the site of the national capital. During the AMERICAN CIVIL WAR, Maryland was one of the border states that did not secede from the Union, but its citizens served in both armies. After the war, industry developed rapidly around Baltimore. During both World Wars, the city was a centre of the

Andrew Marvell's poetry was neglected in his lifetime but praised later.

Marx Brothers, popular on radio and in films: Chico, Zeppo, Groucho and Harpo.

Saint Mary, the Blessed Virgin, with her Son, painted by Gerardo.

Mary II became loved although her joint rule was at first not wholly welcome.

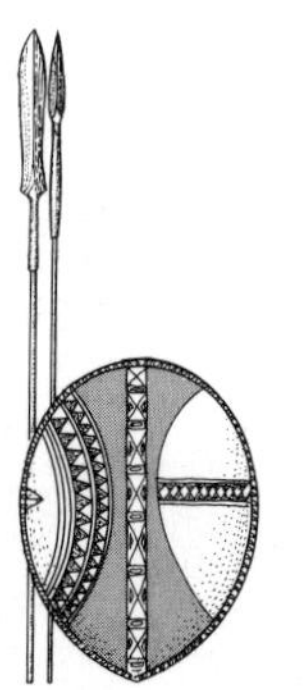

Masai spears and shields are used by the men; the women work in the fields

Pietro Mascagni is most famous for his opera *Cavalleria Rusticana*.

Masks: Devil Dancers of the East Indies wear masks representing primary demons.

James Mason portrayed Field-Marshal Erwin Rommel in *The Desert Fox* (1951).

production of war materials. Area: 27,394sq km (10,577sq miles). Pop. (1975 est.) 4,098,000. *See also* MW p.175.

Marylebone Cricket Club. *See* MCC.

Mary Magdalene, Saint, early follower of Jesus Christ, from the village of Magdala on the w bank of the Sea of Galilee. According to the gospels, Christ freed her of seven demons. She accompanied Christ on His preaching tours in Galilee, witnessed His crucifixion and burial, and was the first person to see Him after His resurrection. She has often been identified with the "woman who was a sinner", who anointed Christ's feet (Luke 7:37), but hardly ever with Mary the sister of Martha who also anointed Him (John 12:3); there is no evidence for either identification in the gospels. Feast day: 22 July.

Mary of Guise, or Mary of Lorraine (1515–60), wife of JAMES V of Scotland (who in 1538 became her second husband). He died in 1542, leaving the throne to MARY, QUEEN OF SCOTS. As Regent in 1554, Mary of Guise persecuted the Protestant faction, provoking a rebellion in 1559 which, with English help, deposed her. *See also* HC1 pp.245, 276.

Mary of Modena (1658–1718), Queen Consort of JAMES II of England (married 1673). She shared James's unpopularity over rumours about the Popish Plot (1678) and her son's birth (1688). After the GLORIOUS REVOLUTION she joined James in France.

Mary of Teck (1867–1953), Queen Consort of GEORGE V of England (married 1893). Her concern with welfare work made her popular with the British people.

Mary Poppins (1934), novel for children by P. L. Travers, the first of several books about the prim but kindly nanny Mary Poppins. It was made into a film by Walt DISNEY in 1964.

Mary Tudor (1496–1533; *r.*1514–33), Queen Consort to Louis XII of France and daughter of HENRY VII of England. Her marriage to Louis in 1514 was a diplomatic arrangement; when he died in 1515 she secretly married Charles Brandon, Duke of Suffolk.

Masaccio (1401–28), Italian painter whose real name was Tommaso Giovanni di Mone. He had a great influence on Florentine painting and his *Trinity* fresco in Santa Maria Novella, Florence, which shows the influence of BRUNELLESCHI, helped the understanding of space and PERSPECTIVE in painting. *See also* HC1 pp.*250*, 251.

Masada, fortified, flat-topped hill in SE Israel; it was the site of the ZEALOT Jews' last stand against Roman authority. In AD 66 the Jews regained the fortress, massacring the Roman garrison, established there in 4 BC. In AD 73, after two years of siege, 15,000 Roman soldiers subdued the defending force of 1,000 men, women and children. Rather than be captured and enslaved, the Jews committed mass suicide, although seven women and children survived. The site was excavated by Yigael Yadin and an international team in 1963–65.

Masai, Nilotic African people of Kenya and Tanzania, consisting of several subgroups. They are characteristically tall and slender. Their patrilineal, egalitarian society is based on nomadic pastoralism, cattle being equated with wealth. The traditional Masai kraal is a group of mud houses surrounded by a thorn fence. They have a system of age groups, and individuals move together through a hierarchy consisting of junior and senior warriors followed by junior and senior elders.

Masan, city in SE South Korea, on Chinhae Bay. It developed as a port at the turn of the century, but was closed in 1908 as part of a naval fortified zone. Originally an agricultural market, Masan has developed as a manufacturing centre since Korean independence. Industries: chemicals, machinery textiles. Pop. (1970) 186,890.

Masaniello, (1828), Italian title for *La sMuette de Portici,* a five-act opera by Daniel AUBER, with libretto by Scribe and Delavigne; it was first produced at the Paris Opéra. The theme concerns political revolution. The 1830 production in Brus-

sels had far-reaching political repercussions in contributing incentive for the Belgians' revolt against the Dutch in that year.

Masaoka Shiki (1867–1902), Japanese poet who revitalized the traditional poetic forms of *haiku* and *tanka*. He sought inspiration in the simplicity of the earliest, 8th-century, period of Japanese poetry.

Masaryk, Jan (1886–1948), Czechoslovakian diplomat, son of Tomáš MASARYK. He was Foreign Minister under Edward Beneš and in the government-in-exile in London during WWII. After the Allied victory, he headed the foreign ministry once again, attempting to resist increasing Communist domination.

Masaryk, Tomáš (1850–1937), Czechoslovak statesman, first President of Czechoslovakia (1918–35), who worked for independence for his country. His Czechoslovak National Council was recognized by the Allies in 1918 as the provisional government of the future state.

Mascagni, Pietro (1863–1945), Italian conductor and composer of operas. The early opera *Cavalleria Rusticana* (1890) was his most successful. He conducted his operas on tours in the USA (1903) and, more successfully, in South America (1911) before succeeding TOSCANINI as musical director of La Scala, Milan, in 1929. His Fascist sympathies are evident in the opera (1935).

Masefield, John (1878–1967), British poet and novelist. A writer of inventive and rhythmical verse, he was POET LAUREATE from 1930 until his death. His long narrative poems include *The Everlasting Mercy* (1911), innovative in its use of colloquialism, and *Reynard the Fox* (1919). His most famous poem is *Sea Fever*. He also wrote verse dramas and adventure novels, often about the sea.

Maser, device using "inverted populations" of atoms (ie atoms artificially kept in states of higher energy than normal) to provide amplification of radio signals. The term is an acronym for Microwave Amplification by Stimulated Emission of Radiation. The principle of the maser was first discovered by Charles TOWNES of Columbia University, who later (1964) shared the Nobel Prize in physics for his work. The first maser used electrostatic (charged) plates to separate high-energy ammonia atoms from low-energy ones. Radiation of a certain frequency would then stimulate the high-energy ammonium atoms to emit similar radiation and strengthen the signal. The very narrow frequency emitted made the ammonia maser one of the most accurate "atomic clocks" known. *See also* LASER; SU pp.110–111.

Maseru, capital city of Lesotho, s Africa, on the River Caledon. Originally an obscure trading town, it flourished when made the capital of the British Basutoland protectorate (1869–71 and 1884–1966) and became the capital of the independent kingdom in 1966. It is a trade, transport and administrative centre. Pop. 15,000.

Mashad. *See* MESHED.

Masham, Abigail, Lady (*d.*1734), favourite of Queen ANNE of England. She sympathized with the Queen's ecclesiastical and political opinions and supplanted the Duchess of Marlborough in Anne's favour (1711–14).

Mashonaland, former province in NE Rhodesia (Zimbabwe), Southern Africa; now divided into Northern and Southern Mashonaland provinces. The region has fertile farmland and rich gold deposits. Mashonaland was acquired by the British South Africa Co after 1890, and became part of the British colony of Southern Rhodesia in 1923. The principal town is Salisbury. Area: 205,990sq km (79,533sq miles). Pop. (1969): (N) 740,480; (s) 1,135,220.

Masina, Giulietta (1921–), Italian film actress who made her début in 1941. She married the director Federico FELLINI in 1943. Her films include *Senza Pietà* (1947), *La Strada* (1953), *Nights of Cabiria* (1957) and *Juliet of the Spirits* (1965). Capable of portraying extreme vulnerability and pathos, she has been described as a female CHAPLIN.

Masked Ball, A (*Un Ballo in Maschera*) (1859), three-act opera by VERDI with text by Somma, based on Scribe's libretto for AUBER's *Gustave III, ou Le Bal Masqué*. It was first produced in Rome. It centred around the assassination of Gustavus III of Sweden who, in March 1792, was shot in the back during a masked ball. But since a recent assassination attempt had been made on the life of Napoleon III, Verdi was persuaded to change the locale to 17th-century Boston, USA, although many later productions restored the opera's action in Sweden.

Maskelyne, Nevil (1732–1811), British astronomer. In 1765 he was appointed the fifth ASTRONOMER ROYAL, a post he held for 46 years. He introduced the method of determining longitude by the measurement of lunar position into navigation, and calculated the Earth's density.

Masks, covers worn over the face or head for disguise or protection. Their major use has been ceremonial, and they have been especially important in the theatre. Masks were an essential part of the Japanese NO dramas and Chinese temple dramas. In ancient Greece, drama masks were used to portray a fixed emotion such as joy, grief or rage, and at the same time they acted as sound-boxes in the projection of the voice. They were used in the medieval mystery plays and featured in Renaissance Italian COMMEDIA DELL'ARTE. In modern drama, the staging of Peter Shaffer's *Equus* required several actors to wear masks representing a horse's head.

Masochism, taking pleasure in being hurt or abused physically or psychologically. Masochism is often linked with sexual arousal and gratification, and may take the form of a desire to be dominated or mistreated. *See also* SADISM.

Mason, James (1909–), British film actor. He made his stage debut in 1931 and appeared in *Fanny by Gaslight* (1944) and *Odd Man Out* (1947). He moved to the USA in 1948 and played complex character parts in films such as *Lolita* (1962), *The Seagull* (1968) and *Cross of Iron* (1977).

Mason, John (1586–1635), founder of New Hampshire USA, *b.* England. He received a patent for a tract of land 97km (60 miles) wide between the Merrimack and Piscataqua rivers, which he named New Hampshire. In 1746 one of his descendants sold the land rights to a dozen Portsmouth men (the Masonian Proprietors), who issued settlement permits and land titles in the undeveloped areas of Mason's grant.

Mason, Ronald Alison Kells, (1905–1971), New Zealand poet, novelist and playwright. He began his career as a trade union secretary, and much of his work displays left-wing sympathies. His collected poems were published in 1962 and he wrote the play *Straight is the Gate* (1969).

Mason-Dixon Line, boundary between the states of Pennsylvania and Maryland, USA, surveyed by Charles Mason and Jeremiah Dixon between 1765 and 1768. It is the traditional line which divides the North from the South.

Masonry, art and skill of building or working with stones, and any building-work in stone. Stones are quarried and left to season by exposure to the air. They are used either dressed (squared and smoothed) or undressed. Undressed stones packed closely can form a durable rubble wall, which may be clad to give a smooth surface. Dressed stones laid like brick-work without mortar make strong structures because of the weight and compressive forces generated by the stones. In the 2nd century BC, masons were using concrete as a bonding agent, which they made by mixing stones with mortar. Most modern masonry involves cladding bricks, concrete or steel with stone. *See also* MM p.46–51.

Masorah, or Massorah, collection of notes written in the margins of the Hebrew Bible, to ensure the correct linguistic and scriptural tradition in the reading aloud of the text. Masorah is Hebrew for "tradition".

Masoretes, group of biblical scholars and scribes who wrote Aramaic and Hebrew and were responsible for producing the MASORAH, or elaborate marginal notes on the Hebrew Bible, between the 6th and 12th centuries.

Masqat. *See* MUSCAT.

Masque, dramatic presentation popular at court and in the great houses of the nobility at the end of the 16th century and the beginning of the 17th. Court masques often involved elaborate costumes and scenery, and Inigo JONES was well known for his spectacular designs for the masques of Ben JONSON. The masque consisted of verse, comedy and, as an essential feature, a dance for a group of masked revellers. The earliest truly dramatic masque of which there is a text is *Proteus and the Adamantine Rock*, performed at Gray's Inn in 1594 in honour of Elizabeth I.

Mass, measure of the quantity of matter in an object. Scientists distinguish between two types of mass: gravitational mass is a measure of the mutual attraction between bodies such as the Earth, as expressed in NEWTON'S law of gravitation. Inertial mass is a measure of a body's resistance to change in its state of motion, as expressed in Newton's second law of motion. Spring balances and platform scales provide a measure of gravitational mass; inertia balances provide a measure of inertial mass. EINSTEIN'S general theory of RELATIVITY is based on the principle of equivalence, according to which the inertial mass and the gravitational mass of a body are equivalent. *See also* INERTIA; WEIGHT; SU pp. 72–75.

Mass, one of the alternative names used by Roman Catholics and some High Church Anglicans for the celebration of the EUCHARIST. The term is generally believed to derive from the formal dismissal at the end of the Latin rite: *Ite, missa est* (You may go, it is done).

Mass, in music, a setting of the religious service. The most famous is BACH'S Mass in B minor. The most famous English settings are BYRD'S four- and five-part masses.

Massachusetts, state in NE USA, on the Atlantic Ocean; one of the New England states. In the E is a low-lying coastal plain. The uplands of the interior are divided by the Connecticut River valley and the Berkshire valley. The principal rivers are the Housatonic, Merrimack and Connecticut. A highly industrialized region, Massachusetts is one of the most densely populated states in the nation. Agricultural produce includes cranberries, tobacco, hay, vegetables and market garden and dairy produce. The major industries are electronic equipment, plastics, footwear, paper, machinery, metal and rubber goods, printing and publishing and fishing. The chief cities are Worcester, Springfield, Cambridge and New Bedford. The state capital and largest city is Boston.

The first settlement was made in 1620 at Plymouth on Massachusetts Bay by the Pilgrims. Boston was founded by English Puritans in 1630 and it became the centre of the MASSACHUSETTS BAY COLONY. Massachusetts strongly resisted the policies of the British crown that led to the American War of Independence. After achieving statehood in 1788, Massachusetts prospered. In the 19th century it was a centre of learning and culture, being the home of writers such as EMERSON, THOREAU and HAWTHORNE. In the AMERICAN CIVIL WAR, Massachusetts supported the Union. Area: 21,386sq km (8,257sq miles). Pop. (1975 est.) 5,828,000. *See also* MW p.175.

Massachusetts Bay Colony, one of the earliest settlements in North America. It was founded as a trading company at Salem in 1629. The charter brought to the colony by Puritan Gov. John Winthrop in 1630 encouraged an influx of PURITANS seeking religious and economic opportunities. This colony was under self-rule with a theocratic government – all of whose representatives were church members. The colony spread to Boston and several other nearby towns. The charter was cancelled in 1684, and the area put under direct English rule.

Massachusetts Bay Company, English company chartered by CHARLES I in 1629 to trade and establish the MASSACHUSETTS BAY COLONY. It was given the land between the Charles and Merrimack rivers and was made independent of an English board of governors as a result of an omission in the charter. The company and the colony were one and the same until 1684, when the company's charter was annulled.

Massachusetts Institute of Technology. *See* MIT.

Massacre of St Bartholomew's Day. *See* BARTHOLOMEW'S DAY MASSACRE.

Mass action, law of, principle that a chemical reaction rate is proportional to the product of the concentrations of each reactant.

Massage, systematic rubbing or kneading of the body for therapeutic purposes. Face massage is practised as a beauty aid; heart massage, massage of the chest over the heart, is a medical technique for setting an inert heart beating again.

Massasoit (*c.*1580–1661), American Indian chief of the Wampanoag tribe of Massachusetts and Rhode Island. In 1621 he made a treaty with John Carver of Plymouth Colony that the Indians would not harm the pilgrims if the pilgrims respected the rights of the Indians.

Mass culture, general culture of an advanced industrial society. It is usually associated with passive attitudes on the part of the majority of the population, which does not create, but merely absorbs. Further, it is often closely connected with the need of large-scale industry to sell great quantities of goods. Mass culture, therefore, is closely associated with advertising, and includes the popular record and clothing industries in Western society. In this sense it is distinct from folk culture, in which the individual is much more creative.

Mass defect, difference in mass between the total rest MASS of protons and neutrons, when free, from which a particular nucleus is formed and the slightly lower mass of the nucleus itself. The mass defect is converted into energy so that the particles can be bound tightly together to form the nucleus. *See also* BINDING ENERGY; RELATIVITY.

Massenet, Jules (Émile-Frédéric) (1842–1912), French Romantic composer who dominated 19th-century French lyric opera. He composed many operas, including *Le Cid* (1885), *Werther* (1892) and *Thérèse* (1909). His two masterpieces are considered to be *Manon* (1884) and *Thaïs* (1894). *See also* HC2 pp.*106*, 121.

Massey, Sir Harrie Stewart Wilson (1908–), Australian physicist. He studied in Melbourne, and taught in Britain from 1929. He served on a number of commissions dealing with atomic energy and space research. His publications include *The Theory of Atomic Collisions* (1933) and *The New Age in Physics* (1960).

Massey, Raymond (1896–), US actor, who gained experience on the London stage and gave many fine screen performances. These included *The Speckled Band* (1931), *Things To Come* (1936), *A Matter of Life and Death* (1946) and *East of Eden* (1955). His most famous role was perhaps that of Abraham Lincoln in *Spirit of the People* (1939).

Massey, Vincent (1887–1967), Canadian politician. He served as Canada's first minister to the USA (1926–30), and High Commissioner in Britain (1935–46). He was the first Canadian-born Governor-General of Canada (1952–59).

Massey, William Ferguson, (1856–1925), New Zealand politician who became Prime Minister in 1912, in which post he died. After going to New Zealand in 1870 Massey set up as a farmer in Mangere until he entered politics in the early 1900s, when he helped form the Reform Party. In 1915 he formed and led the coalition government that ruled until 1919.

Massif Central, extensive mountainous plateau in SE central France. The volcanic Auvergne Mts form the core of the region, which also includes the Cévennes (SE) and the Causses (SW). Sheep and goats are grazed on the slopes. Hydroelectric power is generated there and coal and kaolin mined. The highest peak is Puy de Sancy, rising to 1,886m (6,186ft). Area: 85,001sq km (32,819sq miles).

Massine, Léonide (1896–), US choreographer and ballet dancer, *b.* Russia. He attended the Imperial Ballet School and became principal dancer and choreographer for DIAGHILEV'S BALLETS RUSSES. His choreography includes *Le Soleil de Nuit* (1915), *La Boutique Fantasque* (1919) and *Three Cornered Hat* (1919). He influenced the development of choreography when he created his first symphonic ballet, *Les Présages* (1933), using Tchaikovsky's *Fifth Symphony*. He also staged dances and performed in the films *The Red Shoes* (1948) and *Tales of Hoffmann* (1951). *See also* HC2 p.275.

Massinger, Philip (1583–1640), English dramatist. He wrote more than 40 plays, often in collaboration, many of which are now lost. He is best known for his realistic comedies about domestic life such as *A New Way to Pay Old Debts* (1625) and *The City Madam* (1632).

Mass media, general term for the modern avenues of public information. They are radio, television, newspapers and films. These media disseminate information and entertainment on a wide scale and their powers of manipulating people, perhaps to buy things or to vote, are the subject of much discussion and research.

Mass number, distinguishing property of ATOMS of chemical elements, being the total number of NUCLEONS (nuclear particles) in the NUCLEUS. It is usually shown as a suffix to the chemical symbol of an element. Thus the lightest element, hydrogen, has one PROTON only in its nucleus, and its symbol is H^1. Heavier elements have both protons and NEUTRONS in their nuclei; one ISOTOPE of uranium has a mass number of 238, ie each of its nuclei contains 238 nucleons (protons plus neutrons), and its symbol is U^{238}.

Mass observation, organization founded in Britain in 1937 by Charles Madge and Tom Harrison. It aimed to develop the scientific study of human behaviour, using volunteer "observers". The object was "observation of everyone by everyone". After WWII the company's successor, Mass-Observation Ltd, focused on conventional MARKET RESEARCH.

Mass production, manufacture of goods in large quantities by standardizing work and machinery and by using assembly-line methods. The first machinery with standardized interchangeable parts was designed by Marc Isambard BRUNEL in 1803 to mass-produce pulley-blocks for the naval dockyard at Portsmouth, England. By 1873, US inventors like Samuel Colt were using interchangeable parts and precision measurements in the firearms industry. The proving ground for the assembly line, with parts on a conveyor belt moving past stationary workers, was Henry FORD's factory for the mass production of the Model T car in the 1910s. Later, machines took over more and more of the repetitious labour. More recent developments include automated controlling devices, computers and electronic sensors. *See also* MM pp.*94–95*.

Mass spectrograph, or mass spectrometer, instrument for separating IONS according to their MASSES (or more precisely, according to their charge-to-mass ratio). In the simplest types, the ions are first accelerated by an electric field and then deflected by a strong magnetic field, the lighter the ions the greater the deflection. By varying the field, ions of different masses can be focused in sequence on to a photographic plate and a record of charge-to-mass ratios obtained. In a mass spectrometer the ions are detected electrically. The apparatus is used to measure atomic and molecular mass, identify ISOTOPES and determine chemical structure. *See also* SU pp.150–*151*.

Massys, Quentin (*c.*1465–1530), Flemish painter. He used elements of Italian RENAISSANCE style in his paintings and had considerable influence among Italianate painters in The Netherlands. His works include highly observed, precisely-drawn portraits, eg *The Money-Changer and his Wife* (1514), and many genre and reli

Massachusetts, a densely populated state, has its capital at Boston.

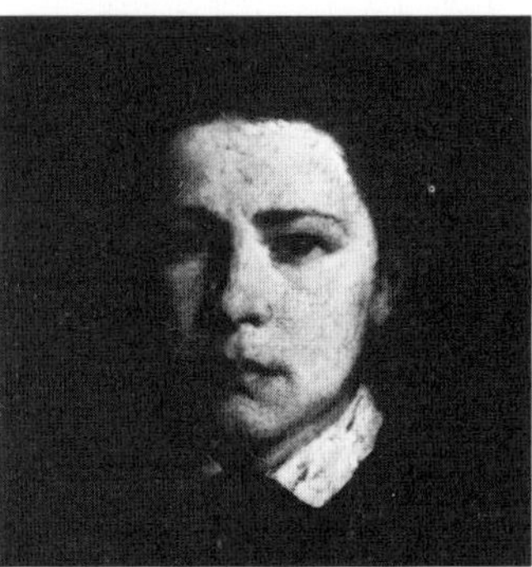

Massenet's operas are notable for their melodies. A portrait by Cavaillé.

Léonide Massine, here with Nemtchinova, invented the form "symphonic ballet".

Mass production: an early example was Henry Ford's factory for the Model T.

Mast

Mastiff: a similar breed was known to the Assyrians and to the Greeks.

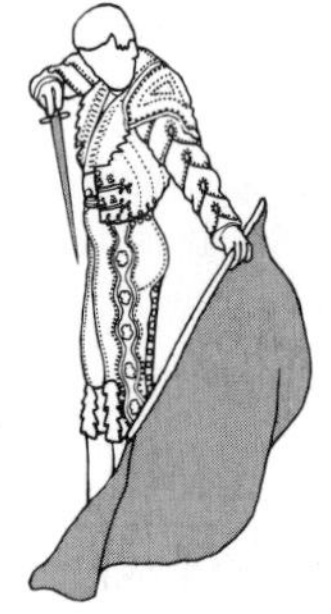

Matadors kill only bulls already wounded; many are very highly paid.

Matamata has a shell unique among turtles: each shield rises to a point.

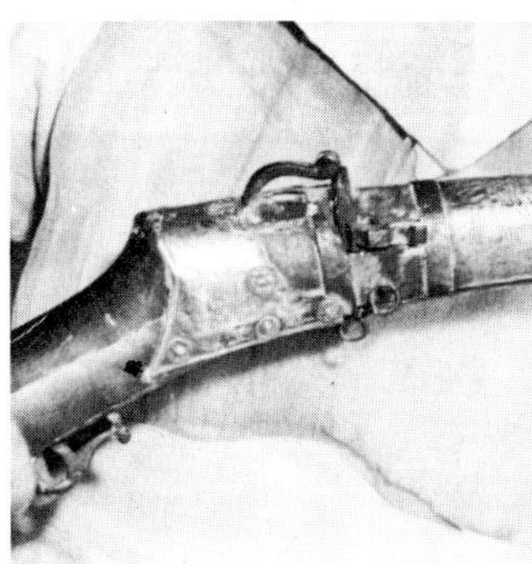

Matchlock: the intricately engraved detail of a 400-year-old Persian firearm.

gious paintings. *See also* HC1 pp.*238, 260.*

Mast, long wooden or metal pole rising vertically from the keel of a ship, used to support the sails. Radio antennae, signals (eg flags and lights) and lifting equipment are also carried on masts. The tallest mast on a full-rigged ship is the mainmast; the foremast (nearest the bow) is forward of the mainmast and the mizzen-mast is aft (nearest the stern). *See also* MM pp.*172, 173.*

Mastaba, mud-brick burial structure in Old Kingdom Egypt. Generally rectangular with inward-sloping sides and a flat roof, a mastaba contained the body of a noble at the bottom of a sealed shaft, and had two open chambers at the top. The PYRAMID was a direct development from it.

Mastectomy, in surgery, removal of all or part of the female breast. It can be simple or radical: simple for non-malignant growth and radical when cancer is present. However, the efficacy of the radical operation is disputed.

Master of, term used by art historians to identify an unknown artist whose style can be recognized in several works. Often the painter's name is derived from his most notable work, as with the Dutch artist known as MASTER OF THE VIRGO INTER VIRGINES.

Master of Flémalle, Flemish painter, probably Robert Campin (*c.*1378–1444), who worked in Tournai. Rogier van der WEYDEN was his pupil. Works attributed to him include *Thief on the Cross* (before 1430) and *Virgin and Child* (date uncertain).

Master of St Bartholomew (*fl.* late 15th and early 16th centuries), German painter. His paintings, in bright, enamel-like colour and sometimes suggesting the influence of Rogier van der WEYDEN include *The Descent from the Cross* (*c.*1501) and the altarpiece of St Bartholomew (1505–10), from which his name derives.

Master of St Giles (*fl*1480–1500), French or Dutch painter who worked at the French court in Paris, Lyons and the Loire valley. He was named after two panel paintings, *The Mass of St Giles* and *St Giles and the Hind* (both now in the National Gallery, London).

Master of the Rolls, third ranking legal officer in England, below the Lord Chancellor and the Lord Chief Justice. He is a member of the High Court of Justice and presides with the Lord Justices of Appeal in the Court of Appeal. He is the keeper of all records of the court of Chancery and of the rolls of all writs and letters patent which are issued under the Great Seal.

Master of the Virgo inter Virgines (*fl. c.* 1470–1500), Dutch painter. He probably worked in Delft and was best known for his individualistic treatment of religious subjects. Many of his figures wore unusual costumes of Oriental design.

Masters, Edgar Lee (1868–1950), US author who in 1915 wrote *Spoon River Anthology*, his greatest poetic success. He spent many of his later years writing biographies.

Masters, John (1914–), British author. He was commissioned in the Indian Army in 1934, and saw active service on the North-West Frontier and the Middle East (1936–45). He was a prolific writer, setting most of his work in India and the East. His novels included *Bhowani Junction* (1954), *The Ravi Lancers* (1972) and three volumes of autobiography; *Bugles and a Tiger* (1956), *The Road Past Mandalay* (1961) and *Pilgrim Son* (1971).

Masters, The (1951), novel by C. P. SNOW. One of the *Strangers and Brothers* sequence, it concerns the struggle for power in a Cambridge college. Royce, the Master, is dying and the Fellows have to consider choosing one of two candidates as his successor. The Master's illness is prolonged, and strong feelings develop among the supporters of both candidates.

Masters, William Howell (1915–), US physician who, with his psychologist wife Virginia (née Johnson) (1925–), became noted for laboratory studies of the physiology and anatomy of human sexual activity. They also conducted clinical marriage counselling for people with sexual problems. Their works include *Human Sexual Response* (1966) and *Human Sexual Inadequacy* (1970).

Mastersinger. See MEISTERSINGERS.

Mastiff, large watchdog and fighting dog that was first bred in England over 2,000 years ago. Properly called Old English mastiff, it has a broad, rounded head with a dark-coloured, blunt, square muzzle and small V-shaped ears. The wide deep-chested body is set on strong legs with large feet and the tail is long and high-set. The short, coarse coat may be brown, grey or brindle. Height: to 84cm (33in) at shoulder; weight: to 95kg (210lb).

Mastitis, inflammation of the MAMMARY GLAND. In human beings, acute mastitis is caused by bacterial infection and may occur after childbirth. The breasts become swollen and painful and abcesses may form. Chronic mastitis, characterized by a localized, painful swelling, is caused by hormonal disturbances. In animals, mastitis is due to bacterial infection; economically, it is one of the most serious diseases of dairy cattle, sometimes causing GANGRENE of the udder and teats.

Mastodon, any of several species of extinct elephantine mammals, all of which existed mainly in the PLEISTOCENE epoch. Mastodons had a long coat of red hair; the grinding teeth were notably smaller and less complex than those of modern elephants, and the males had small tusks on the lower as well as the upper jaw. Genus *Mastodon. See also* NW p.191.

Mastoid, raised region behind the ear, also called the mastoid bone. It contains cells of the inner ear which may become infected in mastoiditis. This once-common disorder was frequently operated upon but nowadays is usually cured quickly with antibiotics.

Masturbation, self-stimulation by manipulation of the genital organs for pleasure, usually to orgasm. Once regarded as taboo, sinful or physically harmful, moderate masturbation is no longer considered abnormal. *See also* MS p.166.

Matabele, or Ndebele, Bantu-speaking people who live in MATABELELAND, Zimbabwe (Rhodesia). Originally part of the NGUNI, the tribe migrated first to Basutoland in 1823, and then to Southern Rhodesia. The Matabele fought European settlers until defeated in 1893. The group now numbers about 300,000 people.

Matabeleland, former province of sw Rhodesia (Zimbabwe), Southern Africa; now divided into Northern and Southern Matabeleland provinces. The region lies between the Limpopo and Zambezi rivers and has rich gold deposits. Matabeleland was dominated by the British South Africa Co after 1889 and it became part of the British colony of Southern Rhodesia in 1923. The main town is Bulawayo. Area: 183,372sq km (70,800sq miles). Pop. (1969): (N) 602,110; (s) 367,110.

Mataco, most important South American Indian tribe inhabiting the Gran CHACO. Speaking the Mataco-Macá language, some 40,000 Mataco inhabit NE Argentina.

Matador, person who kills the wounded bull in a bullfight. He is the "star" of the show, is dressed in highly decorated top and breeches and employs graceful, florid movements with the body and the *muleta* (a piece of red cloth) to avoid the animal's charges.

Mata Hari (1876–1917), Dutch courtesan and double agent, real name Margaretha Geertruida Zelle. She was the wife of a Dutch colonial officer with whom she lived in Java until 1902, when she left him and travelled to Europe, calling herself Mata Hari. From 1905 she was a professional dancer, performing dances she had learnt while in Java. She became well-known in Paris, and was in the pay of both the French and German intelligence services. She was executed by the French as a spy during WWI.

Matamata, South American turtle that has a bossed carapace resembling a piece of dead tree. The large head, which is set on a snake-like neck, is flat and triangular with small feeble jaws. It can greatly distend its throat and feeds by sucking in a huge volume of water. Length: to 40cm (16in). Family Chelyidae; species *Chelys fimbriata. See also* NW p.*233.*

Matanzas, city in w central Cuba, 97km (60 miles) E of Havana; capital of Matanzas province. A major port, its industries include the manufacture of rayon, footwear and fertilizers. Pop. (1970 prelim.) 85,400.

Matanzima, Chief Kaiser Daliwonga (1915–), Prime Minister of the TRANSKEI, South Africa. A critic of the land policy of the South African government, Matanzima has proposed the merger of all BANTU homelands. He was elected Chief Minister in 1963 and Prime Minister in 1976.

Matapan, Battle of Cape (March 1941), major naval engagement in WWII. The British Mediterranean Fleet under Admiral Cunningham attacked an Italian squadron off the coast of s Greece, sinking three cruisers without damage to its own ships.

Matches, short lengths of wood tipped with heads that can be ignited by friction ("striking"). The first matches, called "phosphoric tapers", were introduced in 1781; the modern friction match was invented by John Walker (*c.*1781–1859) in about 1828. The heads of all matches in common use contain the chemical oxidant potassium chlorate. In safety matches this must, to ignite, be struck against a prepared surface containing red phosophorus. The heads of matches that ignite when struck on any rough surface contain both potassium chlorate and the compound phosphorus sesquisulphide.

Matchlock, early FIREARM. It was the first gun that could be used effectively by one infantryman. Its mechanism consisted of a primitive trigger which brought a glowing match into contact with a priming charge. Invented in about 1425, it was the commonest military firearm from 1550 to 1675. But it was heavy (some early matchlocks weighed 11kg; 25lb), and the need for a constantly lit match was inconvenient and dangerous. At the end of the 17th century it was superseded by the FLINTLOCK. *See also* MM p.*162.*

Maté, or yerba maté, South American evergreen shrub or tree. In the wild, it is a tree; cultivated, it is pruned as a small shrub. The dried leaves are infused to make a stimulating beverage called Paraguayan tea, a staple drink of the GAUCHOS of the pampas. Height: to 6m (20ft). Family Aquifoliaceae; species *Ilex paraguariensis.*

Materialism, in philosophy, a system of thought that explains the nature of the world as dependent on matter. It denies the independent existence of spirit. Periods of scientific advance throughout history have been marked by strong materialistic tendencies. The doctrine was formulated as early as the 4th century BC by DEMOCRITUS. Other early Greek teaching such as that of EPICURUS and STOICISM thought of reality as material in nature. The theory was renewed in the 17th century by Thomas HOBBES and others who believed that consciousness belonged to the physical world. Dialectical materialism as preached by Karl MARX is a modern development of the older theory. *See also* MS pp.*186–187, 227, 233.*

Materia medica, study of the source, composition, characteristics and preparation of drugs. As such it is a branch of the larger discipline of PHARMACOLOGY.

Mathematical model, set of formulas or equations that describes the behaviour of a physical system in purely mathematical terms. A modern design technique is to use mathematical models, generated by ANALOG COMPUTERS, to study structures such as buildings, bridges and aircraft before they are built. *See also* MM p.*108.*

Mathematical Principles of Natural Philosophy (1687), the first mature treatise in classical mechanics, by Isaac NEWTON. It is a synthesis of work since GALILEO, important for presenting the three laws of motion, the theory of universal gravitation and the principles governing the tides.

Mathematics, study concerned originally with the properties of numbers and space;

now more generally concerned with deductions made from assumptions about abstract entities. Mathematics is often divided into applied mathematics, which involves the use of mathematical reasoning in other fields, such as engineering, physics, chemistry and economics and pure mathematics, which is purely abstract reasoning based on axioms. However, the two fields are not totally independent – the subjects of pure mathematics are often chosen for their application to specific problems and the abstract results of pure mathematics, such as GROUP THEORY and differential geometry, often find practical uses. The main divisions of pure mathematics are into GEOMETRY and ALGEBRA. Often analysis, reasoning using the concept of limits, is distinguished from algebra; it includes differential and integral CALCULUS. *See also* ARITHMETIC; TRIGONOMETRY; SU pp.26–59.

Mather, name of two American Puritan Ministers. Increase Mather (1639–1723) obtained a new colonial charter for Massachusetts in 1691, and wrote *Cases of Conscience Concerning Evil Spirits* (1693), which helped to end the SALEM Witchcraft Trials. His son Cotton (1663–1728) tried to retain the old system of clerical rule, but was also interested in new scientific developments. The outbreak of witch-hunting in the 1690s owed much to his belief in demonic possession.

Mathias, Robert Bruce (1930–), US athlete who, at 17 years 9 months, was the youngest male gold medallist when he won the decathlon at the 1948 Olympic Games. He achieved an unprecedented double when he won the same event in 1952. In 1967 he was elected to the US House of Representatives.

Mathias, William (1934–), British composer. After studying in Wales and at the Royal Academy of Music, London, he taught at the University of Edinburgh before becoming, in 1970, head of the music department at the University College of North Wales. His numerous compositions include orchestral, choral and church music.

Mathieson, Muir (1911–), British musician who made great contributions to film music from 1931. He arranged and conducted the score by Arthur BLISS for Alexander KORDA's film *Things to Come* (1936) as well as contributing his own scores to numerous films such as *Brief Encounter* (1946). He also directed many educational films; for *The Instruments of the Orchestra* (1945), he commissioned Benjamin BRITTEN to arrange a theme from Purcell's *Abdelazar*, which Britten later reworked into the suite *The Young Person's Guide to the Orchestra*.

Mathis, Johnny (1935–), US popular singer. After starting his career in San Francisco in 1955, he rose to become perhaps the wealthiest black US entertainer of his time. Although personally interested in jazz, he resisted its appeal and recorded romantic ballads such as *The Twelfth of Never* and *Moon River*.

Matilda, or Maud (1102–67), Empress of Germany, daughter of HENRY I of England, who married HENRY V of Germany in 1114. Although she was recognized as Henry's successor, STEPHEN became king in 1135 and after a protracted civil war forced her to leave England.

Matilda, Australian colloquial term for a bushman's bundle, or swag. To "waltz matilda" or "walk matilda" means to take to the road or outback carrying a bundle. This is the subject of the well-known ballad *Waltzing Matilda*.

Matisse, Henri Emile Benoit (1869–1954), French painter and sculptor whose early paintings included naturalistic compositions in neutral tones. From 1899, he experimented with the techniques of NEO-IMPRESSIONISM in paintings such as *Luxe, calme et volupté* (1905). He was the first to develop the style of painting that became known as FAUVISM, and became one of the leading painters associated with the group, using bright pure colours and well-defined subject matter. *Women with the Hat* (1905), his first recognized work

in this manner, was a reaction against the POINTILLISM style which was then in vogue; *The Red Studio* (1911) increased the sensation of vivid colour. From 1920 he diversified, producing collages, stage designs, stained glass and sculptures in terracotta and bronze. From 1935 to 1941, when he became bedridden, he had begun to explore the use of etching for illustrations.

Mato Grosso, second-largest state in Brazil, bordering on Bolivia and Paraguay; the capital is Cuiabà. Much of the area lies on the central plateau of Brazil and there is rain forest in the N and marshland in the SW. Cattle rearing is the chief occupation. Rice, tobacco and sugar-cane are grown. There are extensive mineral deposits but most of them are unexploited. Area: 1,930,849sq km (745,501sq miles). Pop. (1970) 1,475,117.

Matozinhos, suburb of Oporto, NW Portugal, at the mouth of the River Leça. The church of Bom Jesus de Bouças is a pilgrimage centre. Industries: fishing, tourism, tiles. Pop. (1970) 23,973.

Matriarchy, any society or group that is ruled by women. Matriarchal societies exist among some primitive peoples in South America. *See also* PATRIARCHY.

Matrilineal society, human society in which descent is traced through women, and property is usually inherited through the female line. Matriliny is often confused with MATRIARCHY, a society ruled by women. Long established in certain primitive societies, matriliny can today be seen among the North American PUEBLO INDIANS.

Matsys, Quentin. *See* MASSYS, QUENTIN.

Matthes, Roland (1950–), East German backstroke swimmer who, as world record-holder, won the 100m and 200m at the 1968 and 1972 Olympic Games, the 1970 European Championships, and the 1973 World Championships. In 1976 he added a bronze (100m) to his four Olympic gold and one relay silver medals.

Matter, substance of the universe. Ordinary matter is made up of atoms composed of ELECTRONS, PROTONS and NEUTRONS These three particles are combined into ELEMENTS, an ordered series of atoms having from one to more than 100 protons in their nuclei. (Many other subatomic or ELEMENTARY PARTICLES can be produced at high energies and live for short periods of time). The elements other than hydrogen and helium (eg carbon, oxygen and iron) were built up by thermonuclear reactions in stars. Four forces are known to be associated with matter. At large distances, all uncharged matter exerts an attractive force; this is called gravitation. Charged particles exert an attractive or repulsive electromagnetic force which accounts for nearly all everyday phenomena – eg, the sense of touch is dependent on the repulsion of MOLECULES at close range. The strong interaction is responsible for binding the protons and neutrons together in the NUCLEUS and the weak interaction is responsible for beta-decay. *See also* ANTIMATTER; MATTER, STATES OF; SU pp.62–63, 62, 84–89.

Matter, states of, classification of matter according to its structural characteristics. Four states of matter are generally recognized: solid, liquid, gas and plasma. Any one ELEMENT or compound may exist sequentially or simultaneously in two or more of these states: eg, water, ice and water vapour can all exist at one temperature and pressure. SOLIDS may be crystalline (have a regularly repeated molecular structure), as in salt and metals; or amorphous, as in tar or glass. LIQUIDS have molecules that can flow past one another but which remain almost as close as in a solid. In a GAS, molecules are so far from one another that they travel in relatively straight lines until they collide. In a PLASMA atoms are torn apart into electrons and nuclei by the high temperatures, such as those in stars. *See also* SU pp.84–89.

Matterhorn (Monte Cervino), mountain peak in Switzerland, in the Pennine Alps, on the Swiss-Italian border 10km (6 miles) SE of Zermatt. It has a distinctive pyramidal peak formed from several

CIRQUES and was first scaled in 1865 by the British mountaineer Edward Whymper. Height: 4,478m (14,691ft).

Matthew, Saint, one of the twelve apostles of Jesus and by tradition the author of the gospel placed first in the New Testament, although modern authorities date the gospel too late (AD 70–75) for it to be at all likely. In the lists of the disciples, in the SYNOPTIC GOSPELS, Matthew is sometimes called Levi; at the time of his call by Jesus he was working as a tax collector or excise officer. Feast day: 21 Sept. in the West, 16 Nov. in the East.

Matthew, the Gospel according to Saint, gospel traditionally placed first in the New Testament. Written about AD 70–75, it contains more of the teachings of Jesus than any other gospel, perhaps because it relies more on the source "Q" than does the Gospel of St LUKE although both used material from the earlier Gospel of St MARK. It is also the only SYNOPTIC GOSPEL written in a deliberately Jewish, rather than Hellenistic, style always concerned to emphasize connections between Old Testament prophecy and New Testament fulfilment.

Matthews, Denis (1919–), British concert pianist, who studied with Harold Craxton and William ALWYN. His first public appearances were in London at the Queen's Hall and at the National Gallery concerts in 1939. His publications include *In Pursuit of Music* (1966), an autobiography, and *Keyboard Music* (1972).

Matthews, Sir Stanley (1915–), British footballer who played for Stoke City, Blackpool and England. His professional career lasted from 1932 to 1965; he played 786 first-class matches, including 54 games for England. He won only one major honour, an FA cup winner's medal in 1953, but was renowned as one of the game's finest outside-rights.

Matthias Corvinus (1443–90), King of Hungary (1458–90), the son of János Hunyadi. While protecting the peasants, he reformed the administration, strengthened the army, promoted commerce and founded a new university at Buda and a famous library, the Bibliotheca Corvina. He took Bosnia from the Turks and was proclaimed King of Bohemia in 1469, as well as of Styria, Austria and Carinthia.

Maturin, city in NE Venezuela, on the Rio Guarapiche; capital of Monagas state. It is an agricultural trade centre for a region producing cacao, cotton, cereals and beef. Pipelines run from the city to oilfields to the N and W. Pop. (1971) 121,662.

Maudsley, Henry (1771–1831), British engineer who became a leading builder of MARINE ENGINES. He made many innovations in engineering practice, including the production of accurate plane surfaces, the use of a LATHE slide rest, an accurate bench MICROMETER and a practical screw-cutting lathe.

Maufe, Sir Edward (1883–1974), British architect who in 1932 won the competition to design the new cathedral at Guildford. He also began the reconstruction of Gray's Inn (1947), and the Middle Temple (1948). His other buildings include additions to colleges at Oxford and Cambridge.

Mauger, Ivan (1940–), New Zealand speedway rider who, in 1970, became the first rider to win three successive world championships. He won the title again in 1972 and 1977.

Maugham, William Somerset (1874–1965), British novelist, dramatist and essayist. He wrote with wit and irony, and often showed a cynical attitude to life. He achieved initial fame as a playwright and his successes include *Lady Frederick* (1912) and *The Circle* (1921). Among his famous novels are *Of Human Bondage* (1915), *Ashenden* (1928) and *Cakes and Ale* (1930).

Maui, Polynesian demi-god or hero. He caught the sun, tamed fire and spread good works among the people. He discovered the North Island of New Zealand by fishing it up from the sea.

Mau Mau, Kenyan secret terrorist organization in the 1950s, comprising mainly KIKUYU tribesmen, dedicated by oath to expel white settlers from Kenya. In 1952

Henri Matisse is renowned as a painter although he produced other works too.

Mato Grosso: cattle-drawn wagons take tourists to the river ferry.

Saint Matthew: Caravaggio pictured him receiving inspiration from an angel.

Somerset Maugham's short stories have been filmed as *Quartet, Trio* and *Encore*.

François Mauriac wrote two studies of Christ; one was *The Life of Jesus*.

André Maurois, photographed in 1967; his real name was Emile Herzog.

Mausoleum: Castel Sant'Angelo in Rome contains the emperor Hadrian's tomb.

Maximilian; artist's impression of his last moments before execution by Mexicans.

the Mau Mau began a reign of terror against Europeans; the settlers retaliated, British troops were drafted in and casualties multiplied. Although approximately 11,000 Mau Mau were killed in the reprisals, the Kikuyu resistance developed into the Kenyan independent movement. Jomo Kenyatta, who had been jailed as a Mau Mau leader, became Prime Minister of independent Kenya in 1963.

Mauna Loa, volcanic mountain in s central Hawaii, in the Hawaii National Park. It includes Kilauea and Mokuaweoweo, two of the world's largest active craters. It has erupted twice this century (1942 and 1950). Height: 4,160m (13,648ft).

Maundy Thursday, in the Christian liturgical calendar, the day before GOOD FRIDAY, commemorating the institution of the EUCHARIST and the washing of the disciples' feet by Jesus. Church ceremonies can include the washing of the feet and blessing the chrism (holy oil). In Britain, the custom of the sovereign washing the feet of the poor has been replaced by the donation of alms to deserving pensioners. This Maundy money constitutes one specially minted silver penny for each year of the monarch's reign and is distributed biennially at Westminster Abbey.

Maupassant, Guy de (1850–93), French short-story writer and novelist. He was influenced by the Naturalists and produced the short story *Boule-de-suif* for their collection *Les Soirées de Médan* in 1880. His other short stories are collected in *La Maison Tellier* (1881), *Contes de la Bécasse* (1883) and *L'Inutile Beauté* (1890). *Pierre et Jean* (1887) is regarded as his finest novel.

Maupertuis, Pierre-Louis Moreau de (1698–1759), French astronomer and mathematician responsible for the principle of least action in physics. He was also a pioneer in genetics and an early evolutionist, as well as President of the Berlin Academy of Science from 1745 to 1753.

Mauriac, François (1885–1970), French novelist and playwright. His novels, *A Kiss for the Leper* (1922), *Genitrix* (1923) and *The Desert of Love* (1925), portray the futility of pursuing fulfilment in materialism and secular love. His three-volume *Mémoires* (1957–67) portray his reaction to contemporary moral values. He was awarded the 1952 Nobel Prize in literature. *See also* HC2 p.291.

Maurice, Furnley. *See* WILMOT, FRANK LESLIE THOMPSON.

Maurice, John Frederick Denison (1805–72), British author and clergyman who wrote *Theological Essays* (1853) and *Social Morality* (1869). He was professor of English literature and history (1840–53) and professor of moral philosophy (1866) at Cambridge University.

Maurice, known as Maurice of Nassau (1567–1625), Prince of Orange, stadtholder of the Netherlands (1584–1625). He was chosen stadtholder by the United Provinces on the death of his father, William the Silent, and was made commander-in-chief soon after.

Maurier, Daphne du. *See* DU MAURIER, DAME DAPHNE.

Maurier, Sir Gerald du. *See* DU MAURIER, SIR GERALD.

Maurists, group of French BENEDICTINE monks of the congregation of St-Maur, named after the 6th-century St Maurus. Founded in 1618, the congregation was chiefly famous for its devotion to scholarship. It was suppressed in 1789 and disbanded by Pope Pius VII in 1818.

Mauritius, independent island nation of the Commonwealth in the Indian Ocean, approx. 800km (500 miles) E of Madagascar. It was a British colony until it gained its independence in 1968. Sugar, the principal product, accounts for approx. 90 per cent of all exports. Tea is also an important crop. The capital is Port Louis. Area: 2,046sq km (790sq miles). Pop. (1975 est.) 899,000. *See also* MW p.124.

Maurois, André (1885–1967), French biographer, historian, and novelist. He wrote popular histories of England, France, the USA and the USSR, war reminiscences (*The Silence of Colonel Bramble*, 1918), and novels, such as *Whatever Gods May Be* (1928). He is best known for his

biographies, including those of SHELLEY (1923), BYRON (1930), PROUST (1949) and HUGO (1954).

Maury, Matthew Fontaine (1806–73), US naval officer and one of the proponents of OCEANOGRAPHY. He was superintendent of the US Naval Observatory and Hydrographic Office (1842–61) and produced charts showing winds and currents for the Atlantic, Pacific and Indian oceans. He proved the feasibility of laying a transatlantic cable, and wrote *The Physical Geography of the Sea* in 1855. *See also* HYDROGRAPHY.

Maurya Empire (*c.* 321–185 BC), ancient Indian empire founded by Chandragupta (*r. c.* 321–*c.* 291 BC). After the defeat of the Magadha kingdom the Maurya capital was established at Pataliputra (modern Patna). Chandragupta's son Bindusara (*r. c.* 291–*c.* 268 BC) conquered the Deccan; all India north of Mysore and parts of Afghanistan were united under his son, ASHOKA (*r. c.* 276–*c.* 236 BC). In the culture of the Mauryan Empire can be seen the first expression of native Indian civilization. *See also* HC1 pp.62–63.

Mausoleum, particularly impressive tomb. The word comes from Mausolus, ruler of Caria, whose widow raised a great tomb at HALICARNASSUS (*c.* 350 BC), one of the Seven Wonders of the World, to his memory. The best-known mausoleum is the TAJ MAHAL in India; another notable example is Hadrian's Tomb in Rome.

Mauthausen, Nazi CONCENTRATION CAMP in Austria, near Linz, built in 1938 and liberated by the US army in 1945. The inmates, of all nationalities, were subjected to scientific experiments. It is estimated that about 56,000 prisoners died there.

Mauve, Anton (1838–88), Dutch painter, a leading representative of the BARBIZON SCHOOL in Holland. His atmospheric landscapes recall works by COROT and MILLET.

Mauveine, first synthetic dye, made in 1856 by Sir William Henry PERKIN, a British chemist. Also known as ANILINE purple and tyrian purple, it was the first of the large range of coal tar dyes.

Maverick, Samuel Augustus (1803–70), US cattle rancher and political leader. He became prominent in Texan politics by helping to set up the republic in 1836 and serving in the legislature. He did not brand his cattle, hence the term "maverick" for an unmarked stray animal.

Mawson, Sir Douglas (1882–1958), British geographer and Antarctic explorer. He accompanied Sir Ernest SHACKLETON on the first expedition to reach the South Magnetic Pole (1907–09) and led the Australasian Antarctic expedition (1911–14). From 1929 to 1931 he led British, Australian and New Zealand Antarctic expeditions.

Maxim, Sir Hiram Stevens (1840–1916), US inventor of the MAXIM GUN. He had a genius for mechanical invention. He took out his first patent at the age of 26 (for a hair-curling iron) and later invented various pumps, an electric pressure regulator and even a mousetrap. In 1881 he settled in London, where he built the first fully automatic machine gun in 1883. He also invented a smokeless powder (cordite) for use in the gun's cartridges. *See also* MM p.164.

Maxim gun, first modern automatic firearm. It was the first MACHINE GUN successfully to use the recoil energy of the fired bullet to eject the shell and to place another cartridge in the firing chamber. Hiram MAXIM invented the gun in 1883; Vickers mass-produced the weapon, and by 1900 nearly all modern armies had adopted variants of the original design. Machine guns of the Maxim type dominated the battles of WWI. *See also* MM p.164.

Maximilian I (1459–1519), Holy Roman Emperor, (1493–1519). He gained the Burgundian possessions in the Netherlands through his marriage to Mary of Burgundy (1477) but was constantly at war with France, first to defend these lands and then his possessions in Italy. He established the HAPSBURGS as the dominant power in 16th-century Europe.

Maximilian II (1527–76), Holy Roman

Emperor (1564–76), King of Bohemia (1562–76) and King of Hungary (1563–76). He was tolerant of Protestants in his domain and ended a war with the Ottoman Turks in 1568 by a truce with Sultan Selim III. He tried to stay neutral during the religious disputes of the 16th century.

Maximilian, Ferdinand Joseph (1832–67), Emperor of Mexico (1864–67), Archduke and brother of Emperor FRANCIS JOSEPH of Austria. Maximilian courted the liberals and alienated the conservative supporters of his French-dominated regime. When Napoleon III withdrew from the Mexican venture. Maximilian was defeated by forces led by JUÁREZ.

Maximilian I (1756–1825), Elector (1799–1806) of Bavaria and King (1806–25). He was made king by the Treaty of PRESSBURG (1805) after allying himself to NAPOLEON I. He later joined the anti-Napoleon coalition in 1813. At the Congress of Vienna in 1815 he opposed the consolidation of Germany and granted a liberal constitution to Bavaria in 1818.

Maximilian II (1811–64), King of Bavaria (1848–64) who attempted unsuccessfully to form a coalition of small German states against the large powers Prussia and Austria. He later supported Austria in order to stem Prussian power.

Maximilian I (1573–1651), Duke (1597–1651) and Elector (1623–51) of Bavaria who fought against the Protestants in the THIRTY YEARS WAR. After his army defeated Elector Frederick V of the Palatinate in 1620, Ferdinand II gave him the Palatinate electorship and territories.

Maximilian II (1662–1726), Elector of Bavaria (1679–1704, 1713–26). He was governor of the Spanish Netherlands (1692–99), but in the War of the SPANISH SUCCESSION (1701–14), he allied himself with France. After being defeated in 1704 at the Battle of BLENHEIM, he was forced to flee to The Netherlands, and was not reinstated as Elector of Bavaria until 1713.

Maxwell, James Clerk. *See* CLERK-MAXWELL, JAMES.

May, Peter Baker Howard (1929–), British cricketer who played for Cambridge University, Surrey and England. In 66 TEST matches between 1951 and 1961 he scored 4,537 runs at an average of 46.77. He also captained England 41 times. In 1957, at Edgbaston, he scored 285 not out in a world record stand of 411 for the fourth wicket, with Colin COWDREY as his partner.

May, Sir Thomas Erskine, Lord Farnborough (1815–86), British constitutional jurist who wrote the authoritative *A Practical Treatise on the Law, Privileges, Proceedings and Usage of Parliament* (1844). Called to the bar in 1838, he worked for most of his life in the Houses of Parliament.

May. *See* HAWTHORN.

May, fifth month of the year, consisting of 31 days. Considered the last month of spring, it was called "Maius" in the Roman calendar, probably after the Greek goddess Maia, mother of Hermes, or after Maia Majesta, a local Italian goddess of spring, to whom sacrifices were made to ensure the fertility of crops. Similarly many of the festivals and ceremonies now occurring in May relate to agriculture. In the Roman Catholic Church the entire month is dedicated to the Virgin Mary.

Maya, one of the most important tribes of Central American Indians. They established a remarkable civilization (which reached its peak from the 3rd to the 9th centuries AD) in what is now the whole of Belize, W El Salvador and Honduras, and parts of Mexico, including Yucatán. They were short, stocky and dark, and are believed to have travelled originally from Asia many thousands of years ago.

The Maya are renowned mainly for their knowledge of astronomy and mathematics. Priests prepared tables of dates by means of which they were able to predict solar eclipses. They also accurately traced the orbit of Venus. They wrote their numbers with a system of dots and dashes, and developed a symbol that served roughly the same purpose as the modern zero. Their surviving architecture is represented mainly by high stone pyramids topped by temples. Religion was based on an ordered cosmology, with a god of fire and creation, also a god represented by a feathered serpent. Human sacrifice was common in the classic period. Today more than 1.5 million Maya Indians still inhabit the region. *See also* HC1 pp.47, 230–231.

Mayagüez, port in w Puerto Rico, on the Mona Passage. Founded in *c.*1700, it was severely damaged by an earthquake in 1918. There is a college of the University of Puerto Rico (1763) and a US government agricultural research station in the city. Exports: sugar-cane, coffee, tropical fruits, needlework. Industries: brewing, food canning, electronic components. Pop. (1970) 69,485.

Mayakovski, Vladimir (1893–1930), Soviet poet and dramatist. Leader of the FUTURIST movement, he founded and edited the journal *Left Arts Front*. His work includes the poems *A Cloud in Trousers* (1914–15) and *150 Millions* (1919–20) and the plays *Mystery Bouffe* (1918), *The Bedbug* (1928–29) and *The Bathhouse* (1929–30).

Maybach, Wilhelm (1846–1929), German engineer. He worked with Gottlieb DAIMLER in the design and manufacture of internal combustion engines and motor-cars, for which he designed many improvements in fuel injection, gears, and so on. He opened his own factory in 1909, building engines for airships and, later, Maybach cars. *See also* MM p.124.

May Day, first day of MAY, traditionally celebrated as a festival, whose origin may lie in the spring fertility rites of India and Egypt. The Roman festival of Flora, goddess of spring, was held from 28 April to 3 May. In Britian since the Middle Ages the festivities have centred on the dance round the Maypole. In Communist and some other countries May Day is a labour holiday, marked especially in Moscow by a great military display.

Mayer, Julius Robert von (1814–78), German physicist and physician who determined the mechanical equivalent of heat (for which J.P. JOULE received credit), and propounded a statement of the law of conservation of energy (credited to Hermann HELMHOLTZ). He also stated that the Sun was the ultimate source of living and non-living energy. Lack of recognition led him to attempt suicide and he was committed to an asylum.

Mayer, Maria Goeppert (1906–72), US physicist, *b.* Germany, who in 1949 proposed that protons and neutrons in an atomic NUCLEUS are arranged in orbits or shells, much as electrons are arranged around the nucleus. In this theory protons and neutrons are arranged and this provided an explanation of the apparently anomalous properties of some nuclei. She shared the 1963 Nobel Prize in physics with Johannes H.D. Jensen and Eugene P. WIGNER.

Mayfair, area N of Piccadilly, London. Its name derives from a fair held there in May from 1686 until the 1730s; Shepherd's Market now occupies the site. From 1730 to 1743 Dr Alexander Keith became notorious for solemnizing unofficial marriages in Mayfair; George Frederick HANDEL lived and died there.

Mayflower, ship that carried the PILGRIM FATHERS from England to New England in 1620. The *Mayflower* set sail on 15 Sept. and on 21 Nov. the passengers sighted land. Before the Pilgrims disembarked on 26 Dec. at the site of Plymouth, Mass., they signed the famous Mayflower Compact, an agreement for the temporary government of the colony based on the will of the majority. The ultimate fate of the *Mayflower* is unknown, but a replica was sailed by a British crew from England to Massachusetts in 1957 and is on permanent exhibition at Plymouth, Mass.

Mayfly, soft-bodied insect found throughout the world. The adult does not eat and lives only a few days, but the aquatic larvae (NYMPHS) may live several years. Adults have triangular front wings, characteristic thread-like tails and vestigial mouthparts; they often emerge from streams and rivers in swarms. Length;

10–25mm (0.4–1in). Order Ephemeroptera. *See also* NW pp.*106, 230–231.*

Mayhew, Henry (1812–87), British journalist and writer. With his brother, Augustus, he wrote a number of successful works of comic fiction, and was a founder and co-editor of *Punch*. He is most famous for his pioneering work in sociology, *London Labour and the London Poor* (1851–62), a four-volume study of life in the slums of London.

Mayo, Robert Hobart (1890–1957), British aircraft designer. In the 1930s he worked on an aircraft which comprised a "flying boat" carrying a seaplane. The latter was released at a certain height, enabling it to take cargo and fuel over long distances.

Mayo, county in NW Republic of Ireland, in Connaught province; bounded to the N and w by the Atlantic Ocean. A largely mountainous region, it has numerous lakes and is drained by the rivers Errif and Moy. Oats and potatoes are the chief crops and cattle, sheep, pigs and poultry are reared. Woollen milling and toy manufacturing are the main industries. The county town is Castlebar. Area: 5,397sq km (2,084sq miles). Pop. (1971) 109,525.

Mayor, executive head of municipal government (from the Latin *maior,* meaning "greater"). A mayor has a variety of ceremonial, as well as executive, duties. In most boroughs of England, Wales and Northern Ireland the chairman of the elected council is the mayor, although in some boroughs his title is Lord Mayor. The Lord Mayor of the City of London is not elected; he serves an entirely ceremonial function.

Maypole, tall, slim pole used in a ritual folk dance usually performed on MAY DAY (1 May). It was decorated with flowers and coloured ribbons that were woven into intricate patterns by dancers who circled the pole. The maypole was probably a substitute for the living tree used in earlier spring fertility rites.

May Queen, character of pre-Christian origin, a young girl ceremonially crowned in celebration of the coming of summer on May Day (1 May). A MAYPOLE is set up, and the May Queen's attendants dance around it strewing flowers in homage.

Mazagan. *See* EL-JADIDA.

Mazarin, Jules, Cardinal (1602–61), Italian papal nuncio who became a statesman in France. Chosen by Cardinal RICHELIEU to be his successor in 1642, Mazarin derived great support from the young LOUIS XIV's regent, his mother, ANNE OF AUSTRIA. This situation made his position impregnable until his death, except for the period of the FRONDE uprisings (1648–53), when he controlled state matters in secret from Cologne. His negotiations and treaties established a peace in Europe that favoured France. The Peace of Westphalia in 1648 and the Peace of the Pyrenees in 1659 gained much new territory for France and the marriage contract with the Spanish infanta gave France's Louis XIV claims on Spain's inheritance.

Mazdaism. *See* ZOROASTRIANISM.

Mazurka, lively Polish folk dance, similar to the POLKA. As a ballroom dance it is performed by four or eight couples in a circle. The basic movements are footstamping and heel-clicking within the turning movement to a 3/4 metre. Frédéric CHOPIN composed more than 50 mazurkas for the piano.

Mazzini, Giuseppe (1805–72), Italian patriot and political thinker of the RISORGIMENTO (Italian unification movement). A member of the *Carbonari* (the Italian republican underground) from 1830, he founded the "YOUNG ITALY" movement in 1831, dedicated to the republican unification of Italy. He fought in the Italian revolutionary movement of 1848 and ruled in Rome in 1849, but was then exiled. Although he was active in revolutionary activities during the 1850s, he played a minor role in 1861 when the unified Italian kingdom was established. *See also* HC2 pp.108, *111.*

Mbabane, town in NW Swaziland, SE Africa, in the Highveld; the capital and administrative centre of the country. It is a

commercial centre for the surrounding agricultural region, and tin and iron ore are mined nearby. Pop. 21,000.

Mbale, town in SE Uganda, in Bugisu district, 120km (75 miles) NE of Lake Victoria. It is an agricultural trade centre for a region producing coffee, cotton, bananas, grains, and dairy produce. Pop. 23,544.

Mboya, Thomas Joseph (1930–1969), Kenyan statesman. He served in the territorial government during the MAU MAU conflicts of the 1950s and in the Kenyan Independence Movement. He held several ministerial posts under Jomo KENYATTA until his assassination.

Mbuji-Mayi, city in s central Zaire, s Africa, formerly known as Bakwanga. It grew rapidly after Zaire's independence in 1960, becoming an important commercial centre on the River Bushimaie. It is a noted diamond market. Pop. (1974) 336,654.

MCC (Marylebone Cricket Club), world cricket authority from its founding in 1787 to 1969. It was founded by the Earl of Winchelsea and three other English noblemen, simply by changing the name of the White Conduit Cricket Club (1752). Its ground is LORD'S CRICKET GROUND, the freehold of which it bought in 1866. It first laid down the laws of cricket and had general supervision of the game until the establishment of the CRICKET COUNCIL in 1969. It was also responsible for administering English TEST cricket, and English touring sides bore the name MCC as a courtesy until 1977.

Mead, Margaret (1901–   ), US anthropologist whose field-work on the cultures of the Pacific islands, New Guinea and Samoa is summarized in such books as *Coming of Age in Samoa* (1928) and *Growing Up in New Guinea* (1930). She then investigated modern societies and sex roles, in *Male and Female* (1949) and *Childhood in Contemporary Societies* (1955).

Mead, alcoholic drink, known since ancient times, made from water and fermented honey. Alcoholic beverages made from honey were common in northern countries where grapes are difficult to grow. *See also* PE p.197.

Meade, Richard (1938–   ), British horse trials rider who won the three-day event individual gold medal at the 1972 Olympic Games and was in the gold medal-winning teams in 1968 and 1972. He helped Britain win world (1970) and European (1967, 1971) titles, and won the Badminton Trials in 1970.

Meadowsweet, also called queen of the meadow, European plant that thrives in damp areas, especially beside streams. It has fern-like leaves and fragrant white flower clusters. Height: 120cm (4ft). Family Rosaceae; species *Filipendula ulmaria.*

Meads, Colin Earl (1935–   ), New Zealand rugby union forward who played for King Country and New Zealand. Sometimes called "Pine Tree", Meads was one of the finest lock forwards of all time, and had completed 55 Test appearances when he retired in 1972.

Meale, Richard Graham (1932–   ), Australian composer. In 1969 he was appointed lecturer in music at Adelaide University. His works include *Las Alboradas, Very High Kings, Clouds Now and Then* and *Incredible Floridas.*

Mealies, South African name given to maize, the major crop of the republic. In the form of cobs, mealie-rice or mealie-meal, mealies are the principal food of most of the black South Africans.

Meals on wheels, service that provides hot meals delivered to the homes of people who live alone and are incapable, through illness or old age, of cooking for themselves. It started as a voluntary service but is now widespread and is sponsored by governments.

Mealworm, larva of a medium-sized BEETLE (family Tenebrionidae) that infests grain products such as flour. Mealworms are bred as food for birds and insectivorous pets. Length: about 32mm (1.25in). Genus *Tenebrio.*

Mean, or arithmetic mean, the mathematical average. It is found by adding a group

Vladimir Mayakovsky was the Bolshevik spokesman during the revolution, 1917.

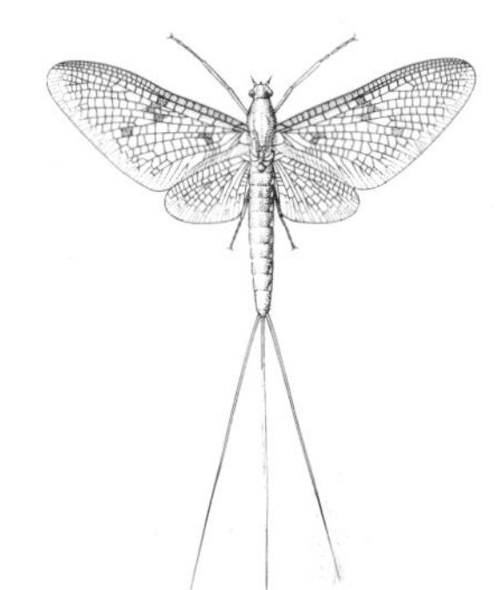

Mayflies in adult form have a lifecycle of only a day or two.

Maypole: a 1955 photograph of Elstow May Festival, Elstow, near Bedford.

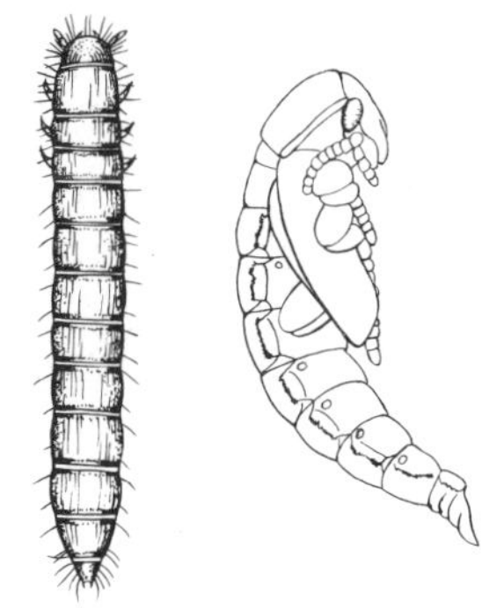

Mealworms, once a problem in warehouses and bakeries, are now bred as pet food.

Meander

Mecca: pilgrims circle seven times the holiest shrine in Islam at the Mosque.

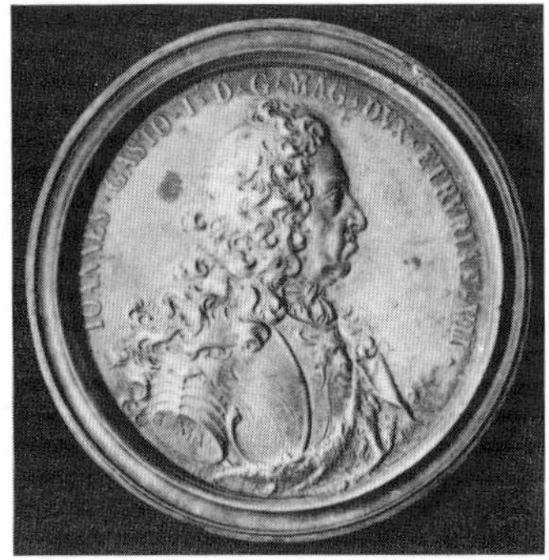

Medals commemorate both people and events; above, the Grand Duke of Tuscany.

Sir Peter Medawar (right) gave the 1959 Reith Lectures on *The Future of Man*.

Medea boiled an old ram to convince King Pelias he could be rejuvenated.

of numbers and dividing by the number of items in the group. The mean is a common measure in STATISTICS; it includes every number in the set being averaged in its computation.

Meander. *See* MAEANDER.

Mean free path, average distance a molecule moves between successive collisions. The concept is most widely used in the KINETIC THEORY of gases, for which the mean free path is usually about 10^{-7}m at normal temperature and pressure.

Mean sea-level, mean height of the sea calculated from measurements at regular intervals at points on the open coast. It is used as a standard or fixed geodetic point. *See also* GEODESY.

Means test, assessment of a person's financial status, usually by a government department, to decide whether the applicant is eligible for state aid. In most means tests an applicant has to complete a questionnaire and undergo an investigation of his financial status. Means tests were devised to limit state expenditure to those actually in need of state subsidies, and are most commonly used to decide eligibility for rate rebates, supplementary benefits, mortgage relief and legal aid.

Meares, John (*c.*1756–1809), British explorer and trader. He first visited the NW American coast in 1786 on a trade mission from Macao. In 1788 he built a trading post on Nootka Sound at Vancouver, which became the object of a territorial dispute between Spain and England.

Measles (rubeola), extremely infectious viral disease whose symptoms (cough, conjunctivitis and typical spots on the mucous membranes lining the mouth) appear about two weeks after exposure to the disease. Three or four days later a rash erupts behind the ears or on the face, then spreads over the body. Hypersensitivity to light is characteristic of an attack. Complications such as pneumonia and encephalitis may occur, and middle-ear infection is a hazard of the disease, which was once a major cause of infant mortality in Western countries. Vaccination or one attack usually produces life-long IMMUNITY to measles. *See also* MS pp.98–99.

Measure for Measure (1604), five-act comedy by William SHAKESPEARE, based on George Whetstone's play *Promos and Cassandra*. Set in Vienna, it is different in method and spirit from Shakespeare's other plays; its tone is partly satirical and partly religious.

Meat, flesh of animals used as food. The term generally includes poultry and game, but excludes fish used as food. Red meat (such as beef steak) is mostly muscle, getting its colour from the red muscle protein myoglobin. Most cuts of meat, however, contain some fat and connective tissue, as well as blood, blood vessels and, possibly, glandular material. *See also* PE pp.230, 232–235, 238–239.

Meath, county in E Republic of Ireland, in Leinster province, on the Irish Sea. A fertile low-lying region, it is drained by the Boyne and Blackwater rivers. Cereal crops and potatoes are grown and cattle are reared. Industries include woollen textiles and paper-milling. The county town is Trim. Area: 2,338sq km (903sq miles). Pop. (1971) 71,616.

Meat tenderizer, substance that makes meat tender by breaking down connective tissue in it. The substances used are ENZYMES, such as papain and bromolein, that primarily digest the PROTEIN in the connective tissues but also possibly, to a limited extent, the protein of muscle fibre. The enzymes may be injected into animals before or after slaughter or may be spread on the cut surface of the meat before cooking. Mechanical tenderizers are used to spike meat so that its increased surface area aids cooking.

Mébiame, Léon (1934–   ), Gabonese politician. He joined the Gabonese government in 1967, and became Prime Minister in 1975.

Mecca, city in W Saudi Arabia. It is considered the birthplace of MOHAMMED and thus regarded as the holiest city of ISLAM. The flight or HEGIRA of Mohammed from Mecca in 622 marks the beginning of the

Muslim era. The city was controlled by Egypt during the 13th century; the OTTOMAN Turks held it (1517–1916) when HUSSEIN IBN ALI secured Arabian independence. Mecca fell in 1924 to IBN SAUD, who later founded the Saudi Arabian kingdom. Much of Mecca's commerce depends on pilgrims, whose chief goal is the great Mosque enclosing the KAABA, although oil revenues have brought added income to the city since WWII. *See also* HC1 p.*145.*

Mechanical advantage, factor by which any machine multiplies an applied force. It may be calculated from the ratio of the forces involved or from the ratio of the distances through which they move, as with simple machines such as LEVERS and PULLEYS. Ideally the ratios derived either way should be the same, but some force is always used overcoming friction. Thus a machine's efficiency is the ratio of its actual mechanical advantage to its ideal mechanical advantage, and is always less than one. *See also* MM pp.86–88.

Mechanical engineering, field of ENGINEERING concerned with the design, construction and operation of engines, machinery and power plants. The skills of the mechanical engineer are necessary in all industries, including power generation, factories and transport; the sciences of physics and mathematics are applied to these and other specialist areas. Some of the achievements of the mechanical engineer include the development of wind, water and steam power; internal combustion engines; assembly line production; and the hydraulic, fluidic and computer control of machines and machine tools. *See also* MM pp.62–158.

Mechanics, branch of physics concerned with the behaviour of MATTER under the influence of FORCES. It may be divided into statics – the study of matter at rest – and dynamics – the study of matter in motion; it may also be classified into solid mechanics and fluid mechanics. Dynamics may be further divided into kinematics – the description of motion without regard to cause – and kinetics – the study of motion and force. In statics, the forces on an object are balanced and the object is said to be in equilibrium; static equilibrium may be stable, unstable or neutral. Classical dynamics rests primarily on the three Laws of MOTION formulated by Isaac NEWTON in his *Principia* (1687). Modern physics has shown these laws to be special cases approximating to more general laws. Relativistic mechanics deals with the behaviour of matter at high speeds, approaching that of light, whereas QUANTUM MECHANICS deals with the behaviour of matter at the level of atoms and molecules. *See also* NEWTON'S LAWS; QUANTUM THEORY; RELATIVITY. SU pp.70–71.

Mechanism, school of philosophy, prominent in the 18th century, associated especially with DIDEROT and HOLBACH. Mechanists assert that men's actions, being the actions of chemical organisms, are, like all natural events, the result of previous chemical and physical interactions. The notion of human self-determination, or FREE WILL, is thus negated. *See also* DETERMINISM.

Mecklenburg, former German duchy and name of its ruling house. It was founded by Przybysław, son of Niklot, in the second half of the 12th century, and continued until 1919. It was the only Slavic dynasty in western Europe. It was made a duchy in 1348. Divided at various times into several dynasties, in 1701 it was split into Mecklenburg-Schwerin and Mecklenburg-Strelitz.

Mecoptera, order of insects known as scorpion flies. *See also* NW pp.106–107.

Medals and decorations, pieces of various shapes, but usually round, and of different sizes, which are struck or cast in metals such as gold, silver or bronze. They are usually commemorative pieces rather than coins used for exchange, although the coins minted by the Greeks and Romans to commemorate events such as military victories greatly influenced the subsequent development of the art of striking medals. That art begins properly with the Renaissance in Italy. Initially, high-relief bronze medals were often cast in the

LOST-WAX process; during the 15th century, distinguished above all by the work of PISANELLO, DÜRER and MASSYS, excellent medals in restrained and simple styles were made for popes, temporal rulers and private individuals to commemorate people, places or events. Striking replaced casting and the screw press replaced hand-hammered dies in the 16th century, resulting in extremely detailed medals such as those of CELLINI. In England the earliest campaign medal to commemorate a specific victory was issued to celebrate the defeat of the Armada. The 17th century saw the emergence of some fine English medallists, such as Thomas Rawlins, and members of the Warin family. Little of interest was produced in the 18th century, but the medallist's art enjoyed a revival in the 19th when, in addition, the invention of the steam press resulted in the production of thousands of medals in a short time at a low cost. Until the 19th century campaign medals were awarded in England only to officers. It was not until the Waterloo Medal (1816) was struck that all ranks were eligible to receive decorations. The VICTORIA CROSS, Britain's highest decoration for valour, was founded in 1856 by Queen Victoria; the second highest, the GEORGE CROSS, was instituted by King George VI in 1940 and may be awarded to civilians as well as to the military. The medallists of the 20th century have on the whole rejected the technological advances of the 19th and have returned to simpler techniques.

Medan, city in NE Sumatra, Indonesia, on the Deli River; capital of North Sumatra province and the largest city on the island. Exports: rubber, tobacco, coffee, sisal. Industries: machinery, tiles, motor vehicles, tourism. Pop. (1971) 635,562.

Medawar, Sir Peter Brian (1915–   ), British biologist. He shared the 1960 Nobel Prize in physiology and medicine with Sir Frank Macfarlane BURNET for the discovery of acquired immune tolerance. Medawar confirmed Burnet's theory that an organism can acquire the ability to recognize foreign tissue during embryonic development, and if that tissue is introduced in the embryonic stage it may be reintroduced later without inducing an immune reaction.

Medea, daughter of Aeëtes, King of Colchis, whom she defied to help Jason find and retrieve the Golden Fleece, sacrificing her brother to do so. Renowned as a sorceress, she lived with Jason many years but fled to Athens after his tiring of her caused her to commit murder in jealous rage. The theme of a Greek tragedy by EURIPIDES (431 BC), the story has been adapted throughout European literature.

Medellín, city in NW central Colombia, approx. 240km (150 miles) NW of Bogotá; capital of Antioquia department and the second-largest city in Colombia. Gold and silver are mined in the surrounding region and Medellín has a national mint. There are several 17th-century churches and several colleges of the University of Colombia in the city. Medellín's industries include food processing, chemicals and steel. Pop. 1,070,924.

Medes, group of ancient Iranian tribes from around the Elburz Mts. They frequently clashed with the Assyrians during the 9th century BC, and were conquered by the Scythians in the 7th century. The Median king Cyaxares drove out the Scythians and also helped the Babylonians destroy the Assyrian Empire; this period (*c.*615–*c.*585 BC) saw the greatest extent of Median power. CYRUS THE GREAT defeated the last Median king in 550 BC and incorporated Medea into his empire.

Media. *See* MASS MEDIA.

Median, in statistics, the middle item in a group found by ranking the items from smallest to largest. In the series, 2, 3, 7, 9, 10, for example, the median is 7. With an even number of items the MEAN of the two middle items is taken as the median. Thus in the series 2, 3, 7, 9, the median is 5.

Mediation, in industrial relations, a type of intervention in which the disputing sides accept the offer of an outside agency to recommend a solution for their controversy. Mediation differs from ARBITRA-

TION in being diplomatic rather than judicial. The mediator has no power to force an agreement but must rely on his ingenuity to bring conciliation by clarifying issues and suggesting new approaches.

Medicare, US programme of hospital and medical care for people aged 65 and over, introduced in 1965. It is a federal health insurance plan and costs are met by a social security tax. Hospital benefits include hospital care, out-patient diagnosis and home care services. Medical insurance is voluntary, and covers most of the cost of consultation and treatment.

Medici, Italian family who ruled Florence and, later, Tuscany, from the 15th to the 18th centuries. Included in the family were three popes, LEO X (*r.* 1513–21); CLEMENT VII (*r.* 1523–34); and LEO XI (*r.* 1605). The family's name was Medici di Cafaggiolo, and its first member of importance was Giambono de' Medici (Chiarissimo I), who became a member of the Florence general council in 1201. Giovanni di Averdo III (Giovanni di Bicci) (1360–1429) became one of the richest bankers in Italy, and was mainly responsible for the family's rising influence after Salvestro di Alamanno's exile. Cosimo the Elder (1389–1464) spent vast sums of money on the arts and humanist scholarship and sponsored the first public library in Florence. Another Medici famous for his interest in the arts was Lorenzo the Magnificent (1449–92), who provided Florence with its most splendid period under the Renaissance. He was also able, through his abilities as a diplomat, to maintain a balance of power among rival Italian states, and it was as a result of this influence that Florentine politics became more closely allied with the papacy than hitherto. The family's history is one of coup and counter-coup between their imperial interests and the Florentine republican cause. The ancient forms of republicanism were restored *c.*1494 under Piero di Lorenzo, and the republicans were split into two factions: the Piagnoni, who continued the Savonarolan tradition; and the Arrabbiati, who reacted against puritan theocracy. Pope Julius II, however, was influential in the return of the Medici in 1512. The republic was again restored in 1527 when Cardinal Passerini's regency became unpopular. Charles V, however, assisted the Medici back to power and Alessandro de' Medici was nominated head of the Florentine state in 1530. *See also* CATHERINE DE' MEDICI; HC1 pp.238, *239, 246,* 250–251, 254–255, 265.

Medicinal herbs, numerous plants that have been used medicinally, often since ancient times. Many herbs were believed to have magical powers. Roman emperors wore symbolic crowns of bay (laurel) leaves; hyssop and garlic supposedly ward off evil. In 1597 the herbalist John Gerard published a lengthy volume listing thousands of plants with healing properties; many of these are still widely used in soaps, cosmetics and lotions. Modern herbalists still use plants such as camomile, comfrey, mint and rosemary, all of which are believed to be effective in the treatment of various ailments. *See also* PE pp.212–213.

Medicine, science concerned with the treatment and prevention of disease and the preservation of health. Pictographs dating from the earliest times show medical procedures and we can deduce from ancient skeletons that the practice of medicine goes back to prehistoric man. The Egyptians included magico-religious elements along with empirical therapies for many ailments, and the Hebrews placed a marked emphasis on HYGIENE. The Greeks had become thoroughly secular in their approach to medicine by the 6th century BC and several medical schools had been built by that time. The Romans and then the Arabs carried on and developed further the Greek medical tradition. During the RENAISSANCE this knowledge was taken back to Europe.

The dawn of modern medicine coincided with accurate anatomical and physiological observations first made in the 17th century. By the 19th century practical diagnostic procedures for many diseases had been developed; bacteria had been discovered and research undertaken for the production of immunizing serums in attempts to eradicate disease. The great developments of the 20th century include CHEMOTHERAPY (the treatment of various diseases with specific chemical agents), new surgical procedures, including organ transplants, and sophisticated diagnostic devices such as radioactive TRACERS. *See also* index to *Man and Society.*

Medicine man, also described as witch doctor, a person whose role in primitive or tribal societies is that of both priest and doctor, particularly when the causes of misfortune are considered to be supernatural. In societies where medicine men are also guardians of tribal lore and ritual their secular power may be considerable.

Medina, city in Saudi Arabia, 338km (210 miles) N of Mecca. Originally called Yathrib, the city was renamed Medinat an-Nabi, "the Prophet's city", after MOHAMMED fled there in 622, making it his capital. In 661 the UMAYYAD caliphs moved their capital to Damascus and Medina's importance declined. It came under Turkish rule (1517–1916), when the independent Arab kingdom of the HEJAZ was formed. In 1932 it became part of Saudi Arabia. Medina is the site of the Great Mosque which contains the tombs of Mohammed, FATIMA and the Caliph OMAR. *See also* HC1 p.144.

Medina del Campo, Treaty of (1489), agreement between England and Spain to establish a common policy towards France. Its chief importance lay in its reduction of tariffs between England and Spain and in the recognition which it gave to the new TUDOR dynasty.

Medina-Sidonia, Alonso Pérez de Guzmán, Duque de (1550–1619), Commander-in-Chief of the Spanish ARMADA in 1588. His inexperience of naval strategy did not prevent Philip II from retaining him as "admiral of the ocean", but he was defeated at Cádiz in 1596 and at Gibraltar in 1606. *See also* HC1 p.277.

Meditation, mental concentration, generally on a spiritual theme, leading to the unsleeping loss of awareness. There are many techniques and aids to its achievement. These include the repetition of a MANTRA or verse, concentrating on one's breathing, puzzling over a riddle or, in Japanese Zen Buddhism, a *Koan,* the use of bodily exercises as in YOGA or the visualization of a well-known object, such as a tree or a rose-bush. In Buddhism four stages are recognized (the Sanskrit *dhyanas*): separation from exterior impressions; focusing without reasoning; an impression of relaxation; and, finally, a state of integration where even relaxation ceases to be felt. Four further exercises (the *samapattis*) follow: awareness of the physical infinity of space; a feeling that one's perceptions are infinitely keen; a consciousness that all constructions whether material or mental, are unreal; and lastly that unreality is the object of thought. The final goal is Nirvana – spiritual freedom.

Mediterranean Sea, largest inland sea in the world, between Europe and Africa, extending from the Strait of Gibraltar to the coast of SW Asia. It is connected to the Black Sea via the Dardanelles, the Sea of Marmara and the Bosporus, and to the Red Sea by the Suez Canal. It includes the Tyrrhenian, Adriatic, Ionian and Aegean seas. It receives the waters of several major rivers, including the Nile, Rhône, Ebro, Tiber and Po. In ancient times the Mediterranean was controlled by the Roman, Byzantine and Arab maritime powers. In the Middle Ages, Venetian and Genoan states dominated the area. The opening of the Suez Canal in 1869 made the Mediterranean one of the world' busiest shipping routes. The development of the Middle Eastern oilfields has further increased its importance. European rivalry for control of the coasts and islands of the Mediterranean led to several campaigns during both World Wars. There is high salinity as a result of rapid evaporation, and a small tidal range. There are about 400 species of fish in the Mediterranean and tuna, sardines and anchovies are among those caught commercially. Area: 2,509,972km (969,100sq miles). See also PE pp.78, 153.

Medium wave radio. *See* RADIO.

Medlar, fruit-bearing shrub or small tree native to S Europe and the Middle East and naturalized in central Europe and Britain. Often gnarled, it has oblong leaves and brown, apple-like fruit which are edible only at the onset of decay. Height: 4.5 to 7.6m (15–25ft). Family Rosaceae; species *Mespilus germanicus.*

Médoc, red wine from the Médoc peninsula, France, the area NW of Bordeaux between the estuary of the Dordogne and the Bay of Biscay. The châteaux of the Médoc adopted a classification of their wines in 1855 (still used today) to distinguish between the many wines of varying quality produced in the region.

Medulla oblongata, principal structure of the hindbrain, joining the BRAIN to the SPINAL CORD. Medulla functions are necessary to maintain life, eg it is the primary control of respiration and heartbeat. It contains part of the reticular formation that plays a role in arousal states, eg consciousness, wakefulness and attention. General anaesthetics, such as chloroform, probably work by depressing medulla activity. *See also* MS p.40.

Medusa, in Greek mythology, a beautiful girl who was seduced by POSEIDON in the temple of the goddess ATHENA. The outraged goddess turned Medusa's hair to writhing serpents and gave her a face that turned to stone all who looked at her. She joined Stheno and Euryale as the only mortal of the three gorgons, until Athena sent PERSEUS to decapitate her. From the wound sprang PEGASUS and Chrysaor, children of Poseidon. *See also* MS pp.206, *209–210.*

Medusa, in zoology, a free-swimming form of COELENTERATE animal, eg a jellyfish. There are many types of medusas. Some are small creatures, transient stages in the life cycle of the coelenterate; they soon change into the dominant, POLYP stage. Other medusas are larger and live longer, the polyp stage being the briefer of the two. Yet other medusas, including many JELLYFISH, have no polyp stage, and when they reproduce give rise directly to medusa offspring.

Medway, river in SE England; it rises in the Weald, flows NE through Kent via Maidstone and the Medway Towns (Rochester, Chatham and Gillingham), to enter the Thames estuary at Sheerness. Length: 113km (70 miles).

Medway Raid, The (June 1667) attack on the Medway dockyards in Kent, England, by Dutch ships during the second Dutch war (1665–67). Six English ships-of-the-line were burnt and two towed away. The English fleet could not prevent the raid because of stringent economies made by the government.

Meegeren, Hans van (1889–1947), Dutch painter and celebrated forger, especially of VERMEER paintings, eg *Christ at Emmaus* (1937). He deceived art experts for years and was discovered only after his own confession in 1945, following his arrest as a collaborator with the Germans for having sold an "old master" to Hermann GOERING (actually one of his own forgeries).

Meerkat, or suricate, any of a number of small carnivorous mammals closely related to the MONGOOSE, native to the bush country of South Africa. It is similar in appearance to the mongoose but without the bushy tail. Length: 47cm (19in). Typical species *Suricata suricatta.*

Meerschaum, or sepiolite, clay silicate mineral, hydrated magnesium silicate, found mainly in Asia Minor. It is opaque white in colour, and has a fibrous texture and a hardness of 2–2.5; s.g. 2. It is used chiefly for making tobacco pipes.

Megaliths, large structures generally built of undressed stone during the NEOLITHIC and early BRONZE AGE. Megalithic building spread N through Spain, along the coast of W Europe and into Scandinavia. The most ancient types were stone tombs, of which the basic unit was the DOLMEN, made of several upright supports and flat

Medici: Lorenzo the Magnificent, a portrait by an unknown 16th-century artist.

Meditation: yoga poses are said to be conducive to mental and physical harmony.

Mediterranean Sea has many popular tourist resorts on its shores: St Tropez.

Medusa was a beautiful young girl before Athena turned her into a gorgon.

Megiddo was King Solomon's city, and his chariot stables have now been unearthed.

Golda Meir speaks at a world conference on Soviet Jewry in 1976.

Meissen: the Albrecht Castle and the cathedral dominate this ancient city.

Ernest Meissonier's *Bay Horse*: much of his work is at present in the Louvre.

roofing slabs. Another type was the menhir, a single upright usually unconnected with any grave sites. They were frequently arranged in parallel rows (as at Carnac, France) or in circles, half-circles and ellipses (as at Stonehenge and Avebury, England). Both types were often decorated with magical symbols.

Megaloptera, order of insects that includes the ALDERFLY. *See also* NW pp.106–107.

Megapode, or mound bird, Australian GAME BIRD with short, rounded wings, strong legs and clawed feet. One group, typified by the hen-sized, crested scrub fowl (*Megapodius*), rakes debris into a large compost pile. The female excavates a deep hole in the mound and lays in it 5 to 8 oval eggs. Other megapodes often migrate to special areas to build their incubators in black sand warmed by the sun, or in areas heated by underground volcanic activity. Family Megapodiidae.

Megarians, school of Greek philosophy from the early 4th to the early 3rd century BC. Founded by SOCRATES' disciple EUCLID OF MEGARA, the school held that evil is only a semblance and cannot exist, and that "good" is indivisible and immutable. It was influenced by both Eleatic ONTOLOGY and Socratic ETHICS, and in turn influenced STOICISM. *See also* ELEATICISM.

Megatherium, any of several species of extinct giant ground sloths of the PLEISTOCENE Epoch. Most lived in South or North America. The front feet bore huge claws, and the jaws had grinding teeth to deal with a diet of plants. Length: approx. 6m (20ft). *See also* NW p.23.

Megiddo, ancient city in Palestine on the River Kishon. It was continuously inhabited from *c.*4000 BC to *c.*450 BC, and recent excavations revealed 20 levels of settlement. Strategically located on the route from Egypt to Mesopotamia, it was the site of many battles, the last of which was in WWI. According to the Bible it will also be the site of the ultimate battle between the forces of good and evil; the site is usually known in this context by its Hebrew name, ARMAGEDDON.

Mehemet Ali. *See* MUHAMMED ALI.

Mehmet, name of six sultans of the OTTOMAN EMPIRE. Mehmet I (1387–1421) came to the throne in 1413, reunited the empire and consolidated his power in the Balkans. Mehmet II (1431–81) came to power in 1451 and two years later captured Constantinople, which he largely rebuilt and which became the capital of the Ottoman empire. During his reign the Ottoman state was strengthened sufficiently to dominate the Mediterranean world for the next 200 years. Mehmet III (1566–1603) who succeeded to the throne in 1595 was the last sultan to be sent on active service within the state before coming to power. Mehmet IV (1642–93) was proclaimed sultan in 1648 by the JANISSARIES. The empire was in decline by this period and Mehmet was deposed in 1687 after several military defeats. Mehmet V (1844–1918) succeeded to the throne after the YOUNG TURK revolution of 1909 but exercised no influence within the state. During his reign most of the empire's European possessions were lost. He entered an alliance with Germany in WWI, dying just before Turkish surrender. He was succeeded by his brother Mehmet VI (1861–1926) who reigned 1918–22. He capitulated to the ALLIES, signing the harsh Treaty of SÈVRES, but was deposed in 1922 by Kemal ATATÜRK who proclaimed a republic in 1923. Mehmet VI died in exile in Italy. *See also* HC1 pp.220–221.

Mehta, Zubin (1936–), Indian conductor who studied at the Vienna Academy of Music and made his career in the West. He was appointed musical director of the Los Angeles Philharmonic Orchestra in 1961.

Meighen, Arthur (1874–1960), Prime Minister of Canada (1920–21; 1926). He was a lawyer and a Conservative MP (1908–21, 1922–26), Solicitor General (1913) and Minister of the Interior (1917). He advocated a protective tariff against US goods entering Canada to protect the Canadian economy, and was instrumental in enlarging Canada's role in world affairs.

Meiji Restoration (1868), shift of power in Japan from the TOKUGAWA shogunate (1603–1867) to the emperor and his supporters. Encouraged by foreign contact (particularly Admiral Perry's visit in 1853) and a weakened central government, a group of young SAMURAI united against the Tokugawa shogunate, forcing a restoration of power to Emperor Meiji, a boy of 14 at the time. The samurai leaders engineered an abrupt change in the government, ending Japanese isolation, abolishing feudal domains (1871) and centralizing administration. *See also* HC2 pp.146–147, *146–147.*

Meikle, Andrew (1719–1811), Scottish inventor. In 1784 he designed the first practical mechanical thresher, consisting of a number of flails attached to a rotating drum. It was patented four years later.

Meilhac, Henri (1831–97), French playwright and librettist. He collaborated with Ludovic Halévy in several diverting drawing-room comedies, including *Froufrou* (1869) and *Loulou* (1876). Their collaboration, which lasted some 20 years, included the libretti for OFFENBACH's operettas *La Belle Hélène* (1864), *La Vie Parisienne* (1866) and *La Grande-Duchesse de Gérolstein* (1867).

Meiningen Players, theatre company that toured Europe from 1881 to 1891. The founder, George, Duke of Saxe-Meiningen, used historically accurate costumes and settings and, by recognizing the importance of central artistic control, introduced the concept of the director and ensemble acting in productions of plays. STANISLAVSKY was influenced by the work of the company.

Mein Kampf ("My Struggle"), personal and political testament by Adolf HITLER. It was written in two parts, the first published in 1924, while he was in prison, the second in 1927, a year after his release. It contains the stock Nazi themes of anti-Semitism, anti-Bolshevism and Fascist totalitarian government. Its forced sales brought Hitler a fortune and it became the "bible" of the Nazis.

Meiosis, in biology, the process of two consecutive nuclear divisions to form germ cells, reducing the CHROMOSOME number by half from diploid to haploid. The first meiotic division reduces the chromosome number to a half; the second division forms four haploid reproductive cells. Spermatogenesis yields four functional SPERMS and oogenesis yields one mature OVUM and three polar bodies (that degenerate).

Meir, Golda (1898–), Israeli politician, *b.* Goldie Mabovitch in the Ukraine. In 1906 she emigrated to the USA, and moved to Palestine in 1921 to live on a KIBBUTZ. In 1946–48 she often acted as *de facto* leader of the Jews in Palestine and after independence served as ambassador to the USSR and as Minister of Labour (1949–56). She also served as Foreign Minister (1956–66). In 1969 she became Prime Minister, but resigned in 1974 after criticism over lack of preparedness at the outbreak of the 1973 ARAB-ISRAELI WAR.

Meissen, town near Dresden, Germany and the correct name for Dresden porcelain, which is made at Europe's oldest porcelain factory. By 1710 Johann Friedrich Böttger had evolved a hard-paste formula approaching that of Chinese porcelain. His Royal Saxon factory set the style of European china with its enamelled tea services and delicately modelled figurines.

Meissonier, Jean Louis Ernest (1815–91). French painter and graphic artist. He is especially known for his large, detailed canvases showing the Napoleonic Wars.

Meistersingers, one of the musical and poetic guilds that flourished in German cities in the 15th and 16th centuries. Each member was obliged to follow a strict religious code and was required to compose and sing according to rigid rules; candidates for the rank of *Meister* were judged in public contests. Knowledge of the movement became more general with the production of Richard WAGNER's opera *Die Meistersinger von Nürnberg* (1868).

Meitner, Lise (1878–1968), Austrian physicist. Together with Otto HAHN, she discovered PROTACTINIUM and the fission of URANIUM, and investigated beta decay and nuclear isomerism. She worked with Fritz STRASSMANN on the products resulting from neutron bombardment of uranium. She is also noted for her research on the radioactive disintegration products of ACTINIUM, THORIUM and RADIUM.

Meknès, city of N central Morocco, approx. 58km (36 miles) WSW of Fès. It was founded in the 10th century as an ALMOHAD citadel and flourished in the 17th century under Sultan Ismail, who built a palace there. Its modern industries include oil refining, food processing and carpet weaving. Pop. 244,520.

Mekong, or Lancangjiang, river in SE Asia. It rises in Tibet, China, and flows s through Yünnan province. It forms the Burma-Laos border and part of the Laos-Thailand border and then flows s through Cambodia and Vietnam, creating a vast river delta that is one of the most important rice-producing regions in Asia. Length: approx. 4,180km (2,600 miles).

Melamine, organic chemical compound that reacts with formaldehyde (HCHO) to form a resin POLYMER, one of the thermosetting PLASTICS. It is a heterocyclic compound having the formula $C_3N_6H_6$. Melamine resins are used as adhesives, for impregnating paper, for treating wool against shrinkage, and as constructional plastics. *See also* MM pp55–56, *55–56.*

Melancholia, in psychiatry, a mental disorder with marked feelings of depression, often accompanied by agitation. It has been described in various ways in the history of psychiatry. Emil KRAEPELIN wrote about "involutional melancholia" found in middle-aged people who have had no previous mental illness. In women this was often associated with the MENOPAUSE. Sigmund FREUD made the classic distinction between the more normal state of mourning that follows a personal loss and melancholia, in which the depression does not pass and often takes the form of ambivalence, guilt and self-punishment. *See also* MS pp.140–141.

Melanchthon, Philipp (1497–1560), German theologian and educator, real name Philipp Schwarzerd, considered with Martin LUTHER as a founder of PROTESTANTISM. He was professor of Greek at Wittenberg, and moderated Lutheranism with HUMANISM. Melanchthon wrote the *Confessions of Augsburg* (1530), a major statement of Protestant beliefs.

Melanesia, collective term for a large area of island groups in the W Pacific Ocean, generally s of the equator, W of the International Date Line, N and E of Australia and E of Indonesia. The members include the Bismarck Archipelago, Solomon Islands, New Hebrides and the Tonga group. The language and culture of the native inhabitants of this area are largely similar. Melanesia is one of the subdivisions of OCEANIA; the others are POLYNESIA (to the W) and MICRONESIA (to the N).

Melanesian people, inhabitants of the central Pacific, N and E of Australia and New Guinea. Most live by fishing and subsistence agriculture. There is considerable cultural diversity, and religious beliefs include TOTEMISM, ANIMISM, ANCESTOR WORSHIP and CARGO CULTS. *See also* MS pp.31, 33, *34.*

Melanin, dark pigment found in the skin, hair and parts of the eye. The amount of melanin determines skin colour. Absence of melanin results in an ALBINO.

Melba, Dame Nellie (1861–1931), Australian soprano, real name Helen Porter Mitchell. The *prima donna assoluta* of her day, she made her début in Brussels in 1887. Initially a high coloratura soprano, she was famous for playing the roles of Lucia (in *Lucia di Lammermoor*) and Gilda (*Rigoletto*) and later for such lyric roles as Mimi (*La Bohème*) and Marguérite (*Faust*). She was especially admired for her technical virtuosity.

Melbourne, William Lamb, 2nd Viscount (1779–1848), British politician. He became a Whig MP in 1806, but lost his seat in 1812 for supporting Catholic emancipation. He returned to Parliament in 1816, and became Irish Secretary (1827–28) and Home Secretary (1830–

34). As Prime Minister (1835–41) he was the political mentor to the young Queen VICTORIA, encouraging liberalism but resisting further parliamentary reform. *See also* HC2 p.112.

Melbourne, capital city of Victoria state, SE Australia, on the Yarra River at the N end of Port Phillip Bay. Founded in 1835 by settlers from Tasmania, the city is a major centre of finance, commerce and transport. It has three universities and the Victorian Arts Centre. Melbourne served as the seat of Australian federal government from 1901 to 1927. It exports wool, flour, meat, fruit and dairy produce. Melbourne is a major manufacturing centre, and its industries include aircraft, motor vehicles, shipbuilding, textiles, chemicals and agricultural machinery. Pop. 2,583.000. *See also* MW p.28.

Melbourne Cricket Ground, sports stadium in MELBOURNE, Australia. It is the largest cricket ground in the world, and since modernization for the Olympic Games of 1956 can hold more than 100,000 people. Cricket was first played there in 1854, and it was the venue of the first Test match between England and Australia, in 1877.

Melbourne Cup, the premier horse race in Australia, run on the first Tuesday of November at Flemington, Melbourne. First raced in 1861, it is a handicap over two miles (3.2km) for three-year-olds and over, and is the richest race in Australia. It is linked with the Caulfield Cup, run in October, as an ante-post double, in which before either race people bet on their selection of the winner of both races.

Melchiades, or Miltiades, Saint (*r. c.*310–14), Pope who presided over the commission which condemned DONATISM (313). During his papacy the Edict of MILAN (312), which acknowledged the right of the Christian churches to exist, was issued by Constantine and Licinius.

Melchiorites, members of an extreme ANABAPTIST group who took their name and beliefs from Melchior HOFMANN, a 16th-century German lay-preacher. Imbued with ideas about death, the day of judgment and life after death, he prophesied the end of the world, and clung to his beliefs during frequent periods of imprisonment.

Mêlée, or tournay, earliest form of tournament. The medieval sport of combat, it was introduced into England in the late 11th century, probably from France, and flourished until the 16th century. Two competing groups of horsemen took part, charging at a given signal and attempting to unseat, or strike the crests off, each other. *See also* JOUST.

Melgar-Castro, Juan Alberto (1930–   ), head of state of Honduras. He was Commander-in-Chief of the army before being named President of the country after a bloodless military coup in April 1975, when he succeeded President López Arellano.

Méliès, George. *See* JOURNEY TO THE MOON, A.

Melloni, Macedonio (1798–1854), Italian physicist who studied INFRA-RED radiation. He improved the thermopile and used it to detect infra-red radiation. He also showed that rock salt is transparent to infra-red and by 1850 had demonstrated that infra-red undergoes reflection, refraction, polarization and interference in the same way as does visible light. Thus he laid the foundation for James CLERK-MAXWELL's hypothesis of a spectrum of ELECTROMAGNETIC RADIATION beyond the visible region.

Melodrama, theatrical form originating in late 18th-century France and achieving its greatest popularity in the century that followed. It relied on simple, violent plots whose strongly emotional situations, often based on notorious crimes, were heightened by the use of music and elaborate stage effects. Melodramatic plays generally appealed to unsophisticated audiences and the exaggerated styles that its actors adopted have often led to the term "melodrama" being used pejoratively. Playwrights specializing in melodrama include August von KOTZEBUE, Guilbert de Pixérécourt (1773–1844) and

Thomas Holcroft (1744–1809). Victorian actor-managers such as IRVING, TREE and Wilson Barrett (1846–1904) made strong use of the form.

Melody, in music, sequence of notes that makes a recognizable musical pattern. The term is most commonly used of the dominant part or voice (the "tune") in Romantic and light music, in which any harmonic accompaniment is nearly always totally subordinate. Music featuring several melodies simultaneously and harmoniously is termed "contrapuntal" or POLYPHONY, and is characteristic of the late BAROQUE period (and J.S. BACH in particular). Much early written music indicated only the melody and the chord progressions in the bass line, and the performer had to improvise the remaining harmonies according to musical rules.

Melon, annual vine and its large flesh fruit; most varieties are derived from the musk melon, *Cucumis melo*. Melons grow in warm temperate and subtropical climates. The cantaloupe melon, with its rough skin, probably originated in Armenia; the smoother yellow rind honeydew, in SE Asia. The large, dark green watermelon, with its red watery flesh, is believed to have come from Africa. Family Cucurbitaceae. *See also* PE p.193, *193.*

Melon aphid. *See* COTTON APHID.

Melo Neto, João Cabral de (1920–   ), Brazilian poet, one of the leaders of Concretism, a late stage of Brazilian MODERNISMO, which stresses poetry that is direct and concrete. They are collected in *The River* (1954) and *Two Waters* (1956).

Melos (Mílos), island in SE Greece, in the Cyclades (Kikládhes) group in the Aegean Sea. A centre of early Aegean civilization, the island has been extensively excavated, the most important find being the statue of *Venus de Milo*, now in the Louvre, Paris. Products: grain, cotton, fruits, olive oil. Area: 150sq km (58sq miles). Pop. (1971) 4,593.

Melting point, temperature at which a substance changes from solid to liquid. The melting point of the solid has the same value as the freezing point of the liquid, eg the melting point of ice (0°C) is the same as the freezing point of water. *See also* SU p.90.

Meltwater, water produced by thawing glaciers, and snow or hail deposits. Some scientists predict that meltwater from the Earth's poles would flood large areas of the continents if the polar ice-caps thawed.

Melville, Herman (1819–91), US novelist. He became a sailor in 1839 and joined a whaling ship in 1841. His first novel, *Typee*, was written in 1846, and *Moby Dick*, an allegorical story of the search for a great whale, in 1851. His short story *Billy Budd*, written late in life, was published in 1924. Melville's work was relatively neglected during his lifetime. *See also* HC2 p.288.

Melville Peninsula, peninsula in Northwest Territories, Canada, between Committee Bay (W) and Foxe Basin (E). A region of tundra, it is sparsely populated. There is a trading post at Repulse Bay on the S coast. Area: 62,564sq km (24,156sq miles).

Membrane, in biology, boundary layer or layers inside or around a living CELL or an organ or tissue. Cell membranes comprise the plasma membrane surrounding the cell, the network of membranes inside the cell (endoplasmic reticulum) and the double membrane surrounding the NUCLEUS. The multicellular membranes of the body comprise mucous membranes of the air, digestive and urinogenital passages, synovial membranes of the joints, and serous membranes that coat the inner walls of the abdomen, thorax and the surfaces of organs. *See also* NW p.26.

Memling, Hans (*c.*1440–1494), Flemish painter, *b.* Germany, who probably studied under Rogier van der WEYDEN in the Netherlands. He was the head of a flourishing workshop and painted portraits and religious works. His paintings, with their balanced compositions, are characterized by their high finish and realistic attention to details, such as the colour and texture of

clothing and fabric. *See also* HC1 p.*253.*

Memory, capacity to retain information and experience and to recall them in the future. Modern psychologists often divide memory into two types, short-term (STM) and long-term (LTM). An item in STM (for example, a telephone number dialled once) lasts for several seconds after an experience, but is lost if not used again. An item enters LTM if it is of sufficient importance or is information required frequently. Some psychologists believe that items in LTM are never lost: forgetting, in their view, is an inability to get the items out of storage and into consciousness. *See also* MS pp.44–45, *44–45.*

Memory, computer. *See* COMPUTER MEMORY.

Memory core, device used to store information in a COMPUTER MEMORY. It consists of a small FERRITE ring, the direction of magnetization of which is registered as 0 or 1. Cores are permanent magnets, from which the information is not lost (as it may be with a solid-state memory) if the current to the computer is switched off. *See also* MM p.106.

Memphis, Slim (1916–   ), US blues singer and pianist, real name Peter Chatman. He travelled to Chicago in 1939 and started recording with artists such as Big Bill BROONZY and his half-brother Washboard Sam. His singing style was unremarkable but his piano-playing, influenced by Roosevelt Sykes and Joshua Altheimer, was widely acclaimed.

Memphis, ancient city of Egypt, 23km (14 miles) S of Cairo, part of which is now occupied by the village of Mit Ra-hina. Founded in *c.*3100 BC by MENES, the city remained important until ALEXANDRIA began to develop in the 4th century BC. Memphis was dedicated to the god PTAH and the temples of Ptah, RE and ISIS and statues of RAMESSES II have been found at the site. Material from its ruins were used by the Arabs for building Fustat and Cairo.

Memphis, city and river port in SW Tennessee, USA, on the Mississippi River; largest city in Tennessee. Strategically located on Chickasaw Bluff above the Mississippi, the site of Memphis was used for a French (1682), Spanish (1794) and US (1797) fort before the first permanent settlement was made in 1819. Today it is a major transport centre and livestock market, and its industries include timber, farm machinery, food processing and pharmaceuticals. Pop. (1970) 623,530.

Memsahib, polite form of address used in Bengal, India, to designate a European married woman. Sahib is the equivalent form for a man. In the later days of the British Raj the term was used widely in India, although an alternative, Madam Sahib was used in Bombay.

Menai Strait, part of the Irish Sea between Anglesey and the NW coast of Wales. Two famous bridges span it: Thomas Telford's suspension bridge, built in 1826 (strengthened 1939), and Robert Stephenson's tubular bridge built in 1850 to carry a railway (but destroyed by fire in 1970).

Menander (*c.*342–292 BC), Greek playwright who wrote more than 100 comedies in 30 years. His favourite theme was unhappy love; his style was amusing and perceptive rather than broadly humorous. With PHILEMON, he was one of the leaders of NEW COMEDY. His plays mostly survive in fragments, but an entire play, *The Bad-Tempered Man*, was first published in 1958, soon after its discovery in Egypt. *See also* HC1 p.75.

Menarche, onset of the MENSTRUAL CYCLE in young females, which marks a stage of PUBERTY and the beginning of ADOLESCENCE.

Mencius (372–289 BC), Chinese philosopher. His teachings were recorded by his disciples in the book *Mencius*, one of the Four Books in the canonical writings of CONFUCIANISM, and he is considered China's "second sage" after Confucius. He travelled widely in China, attempting to introduce reforms to ensure peace and opposing unjust rulers. *See also* HC1 p.125; MS p.*235.*

Mendel, Gregor Johann (1822–84), Austrian naturalist. He discovered the

Melbourne Cup was won convincingly in 1949 by Fonzami ahead of Hoyle.

Melon plants trail their stems, attaching themselves to objects with tendrils.

Hans Memling's *The Nativity*: many of his works remain in Bruges, his home town.

Memphis, Tennessee, was originally a cotton town: the main shopping centre.

Mendelevium

Felix **Mendelssohn** made his first public appearance as pianist at the age of ten.

Mendip Hills, with their caves and the Cheddar Gorge, are favoured by tourists.

Anton Rafael Mengs: a detail from his painting, the *Annunciation*.

Gian-Carlo Menotti not only writes his own libretti but stages his premières.

laws of HEREDITY and in so doing laid the foundation for the modern science of GENETICS. Mendel joined the monastery at Brünn (modern Brno) in 1824 and was ordained in 1847; he began plant experiments in 1856 in the monastery gardens. His experimental breeding of the garden pea led him to formulate two laws: the principle of segregation and the principle of the independent assortment of characters. His discoveries, published in *Experiments with Plant Hybrids* (1865), were virtually ignored until 1900, when his work was rediscovered and recognized. *See also* NW pp.30–31; HC2 p.*156*.

Mendelevium, radioactive metallic element (symbol Md) of the ACTINIDE group, first made in 1955 by the alpha-particle bombardment of einsteinium-253. Properties: at. no. 101; most stable isotope Md^{258} (half-life 2 months).

Mendeleyev, or Mendeleev, Dmitr Ivanovich (1834–1907), Russian chemist. He studied at St Petersburg, where he later became professor. After hearing the lectures of Stanislao CANNIZZARO, he developed an interest in the relationship between the 63 elements then known. He devised the periodic law and the modern form of the PERIODIC TABLE at about the same time as the German Lothar MEYER. This table enabled him to predict the existence of several elements, including GALLIUM and SCANDIUM. His work was published in Russian in 1869, and was soon translated into German. His textbook *Principles of Chemistry* (1868–70, tr. 1905), became a standard work. *See also* SU pp.134, *135*.

Mendelsohn, Eric (1887–1953), German architect whose Einstein Observatory at Potsdam (1920), an early work, is perhaps the most famous of EXPRESSIONIST buildings. He later worked in FUNCTIONALIST style, designing offices and the Schocken Department Store in Stuttgart (1926). During the 1930s he practised in England, designing, with S. Chermayeff, the De la Warr Pavilion, Brighton (1933). In 1941 he settled in the USA, where he designed the Maimonides Hospital, San Francisco (1946).

Mendelssohn (-Bartholdy), Felix (1809–47), German composer and conductor. He was a child prodigy and at 16 composed an octet; at 17 he wrote his overture to *A Midsummer Night's Dream*. A prolific composer, his works include a *Violin Concerto* (1844), five symphonies and *Songs without Words* (1829–34) for piano, and chamber music. His two oratorios, *St Paul* (1836) and *Elijah* (1846), are considered to be among the greatest of the 19th century. He was also recognized as the foremost conductor of his day. *See also* HC2 p.106.

Mendelssohn, Moses (1729–86), German-Jewish popular philosopher. He pleaded for the separation of Church and state and, in *On the Civil Improvement of the Jews* (1781), he proposed social emancipation of Jews. He was also known for his proofs of the existence of a personal God in *Phaedo, or on the Immorality of the Soul* (1767) attempted to prove the immortality of the soul.

Menderes, Adnan (1899–1961), Turkish politician who was responsible for the formation of the Democratic Party in Turkey (1946), the first legal opposition party in the country. He became Premier in 1950 when his party came to power, but was executed after an army coup in 1960.

Mendes, Francisco (1939–), Guinean nationalist leader and politician who fought in the guerrilla struggle for the independence of Guinea-Bissau as a member of PAIGC (Partido Africano de Independência da Guiné e Cabo Verde) from 1960. In 1973 he became Chief State Commissar, in effect Prime Minister.

Méndez, Aparicio, (1904–), Uruguayan lawyer and politician whose first political post was as Minister of the Electoral Court from 1943 to 1946. Steadily rising through the administrative hierarchy Méndez became President of Uruguay in September 1976.

Mendicant Orders, name given collectively to religious orders which lay special stress on the vow of poverty; they used to subsist mainly on alms. In the 13th century a new spirit of enthusiasm sprang up under the teachings of St FRANCIS OF ASSISI and St DOMINIC, and orders of friars were founded which were forbidden to possess property beyond the houses in which they dwelt. The most notable of these were the CARMELITES, FRANCISCANS and DOMINICANS. The AUGUSTINIANS followed, and in 1424 the Servites were also officially recognized as a mendicant order.

Mendip Hills, range of hills in sw England, extending 29km (18 miles) SE from sw Avon to the Frome valley in Somerset. Composed of limestone, the hills contain many caves, including the Cheddar Caves. There are many Roman remains in the area. The highest point is Black Down, rising to 326m (1,068ft).

Mendoza, Antonio de (c.1490–1552), first Viceroy of New Spain (1535–50), appointed by Charles I (Holy Roman Emperor CHARLES V). Mendoza successfully established the authority of the Crown over that of the individual conquistadors and introduced many reforms. In 1551 he was appointed Viceroy of Peru.

Menelaus (*fl.* 3100 BC) Egyptian king said to have unified upper and lower Egypt. He is in the *Iliad*, the King of Sparta, husband of HELEN and younger brother of AGAMEMNON. The rulers of Greece came to the aid of Menelaus after Helen was abducted by the Trojan prince, PARIS. During the Trojan war Menelaus met Paris in single combat, but Paris escaped. Helen and Menelaus were reunited after the success of the Greek siege.

Menes (*fl.* 3100 BC) Egyptian king said to have unified upper and lower Egypt. He is also known as Aha, Nar-mer and Scorpion. He was the first king of the 1st Dynasty and is said to have founded the capital MEMPHIS. *See also* HC1 p.34, *34*.

Mengoni, Guiseppe (1829–1877), Italian architect who designed the Galleria Vittorio Emanuele in Milan by winning a competition for its design in 1861. It was built in 1864–67, although its decoration continued until 1877, and is one of the most ambitious of all shopping arcades.

Mengs, Anton Raffael (1728–79), German painter and writer on art. He was one of the most famous early neo-Classical painters. His *Parnassus* was executed for the ceiling of the Villa Albani, Rome (1761).

Menhaden, herring-like marine fish that lives in schools in temperate Atlantic waters. An important source of fish-meal and oil, it is green or blue and silver. Length: to 51cm (20in). Family Clupeidae; species *Brevoortia tyrannus*.

Menhirs, archaeological term given to single standing stones found in Western Europe. Probably of NEOLITHIC origin, they are usually tall and square in section, tapering towards the top. They are thought to have been used to mark places of religious or ritual significance.

Menière's disease, disease of the labyrinth of the inner EAR in which the symptoms are deafness, vertigo, sometimes accompanied by vomiting and ringing in the ears. It is caused by excessive fluid in the labyrinth of the ear and may occur during middle age. It is generally treated with ANTIHISTAMINE drugs; in severe cases, surgery may be necessary.

Meninas, Las (1656), famous portrait group by VELÁZQUEZ, showing the Spanish Infanta and her entourage, including her maids and a dwarf, and including the painter at his easel. The royal couple, who are sitting for the double portrait, are reflected in a mirror.

Meninges, three membranes that cover the BRAIN and SPINAL CORD. The outermost membrane, the dura mater, is a tough protective covering. Within it is the second membrane, the arachnoid; the inner membrane, the pia mater, is a delicate layer on the surface of the brain and spinal cord containing blood vessels which supply the nervous system with nourishment. Between the arachnoid and the pia mater is the subarachnoid space, which contains cerebrospinal fluid. *See also* MS pp.38, 40.

Meningitis, inflammation of the MENINGES covering the brain and spinal cord, resulting from infection with meningococci or other micro-organisms. *See also* MS p.96.

Menisectomy, surgical removal of damaged cartilage from the knee joint.

Mennonites, Christian sect founded by the Dutch reformer Menno SIMONS (1496–1561) and influenced by ANABAPTIST doctrines. Organizationally, each local Mennonite congregation is independent. They believe in the BAPTISM of adult believers and reject infant baptism as well as the doctrine of the real presence in the EUCHARIST. Most of them are opposed to military service and the taking of oaths.

Menno Simons (1496–1561), Dutch religious reformer, after whom the MENNONITES are named. Originally a Roman Catholic priest, he joined the ANABAPTISTS c.1536 and preached in Germany and Holland. He was not deterred by persecution and made many converts. His major book was *The Foundations of Christian Doctrine* (1539).

Men-of-war. *See* MAN-OF-WAR.

Menopause, stage in a woman's life when the MENSTRUAL CYCLE ceases, generally between the ages of 45 and 52. Periods of mild depression are not uncommon during the "change of life" and in some cases severe depression develops. Various hormonal treatments are available for cases of a prolonged menopause. *See also* MS pp.*101*, 174.

Menorah, sacred seven-branched candle-holder which has become a symbol of Judaism throughout the world. It is rich in symbolic meaning. Some interpret it in terms of the seven planets, the tree of life or the six-day creation of the universe with the centre shaft as the Sabbath. An eight-branched menorah is used during the HANUKKAH festival.

Menorca. *See* MINORCA.

Menotti, Gian-Carlo (1911–), US composer, b. Italy. His operas in modern OPERA BUFFA style have been most successful and include *The Telephone* (1947). *The Consul* (1950) is a theatrically powerful work and Menotti has also composed operas specifically for television such as *Amahl and the Night Visitors* (1951) and *Labyrinth* (1963).

Mensheviks, moderate faction of the Russian Social Democratic Labour party. They broke with the BOLSHEVIKS at the 1903 Party Congress because of Lenin's insistence on a centralized party of professional revolutionaries. After the Bolshevik seizure of power in November 1917, many Mensheviks capitulated to the Bolsheviks. The Mensheviks were suppressed in 1922. *See also* HC2 p.196.

Menstrual cycle, in female human beings and primates, stage during which an OVUM ripens and is released from the OVARY and the ENDOMETRIUM thickens and prepares to receive the fertilized egg. The average human menstrual cycle is 28 days, but this varies considerably. At the beginning of the cycle, PITUITARY hormones stimulate the growth of an egg cell, contained in a follicle in one of the two ovaries. This is known as the proliferative stage. At approximately mid-cycle the follicle bursts, the egg is released (OVULATION) and travels down the FALLOPIAN TUBE to the uterus. The follicle (now termed the CORPUS LUTEUM) secretes two hormones, PROGESTERONE and OESTROGEN. During this secretory phase of the cycle, the endometrium becomes distended, ready to receive the fertilized egg. Should fertilization (conception) not occur, the corpus luteum degenerates, hormonal secretion ceases, the endometrium breaks down and menstruation occurs; the unfertilized egg is discharged in the blood flow from the VAGINA. In the event of conception, the corpus luteum remains and maintains the endometrium with hormones until the PLACENTA is formed. Menstruation therefore marks the end of the cycle. It is customary for the purposes of contraception (BIRTH CONTROL) and calculating the date of an expected birth to count the first day of bleeding as the first day of the next menstrual cycle. The onset of the menstrual cycle (MENARCHE) is normally at PUBERTY (age 10–15 years) and ceases with the MENOPAUSE (age 40–50 years). *See also* MS pp.75, 164.

Mental health, proper adjustment of an individual's mind to his environment.

Mental health, it is now known, invariably has a social context and modern treatment for abnormalities, or psychological disturbances, tries to help the individual to develop a more integrated relationship with his circumstances, especially his friends and society in general. *See also* MS pp.134–149.

Mental illness, any failure of MENTAL HEALTH, can be divided into three main categories which are more or less distinct but which overlap to some extent. First, some mental disorders can be clearly attributed to injury or organic disease of the BRAIN, eg if the frontal lobes are injured, intellectual and personality deterioration can occur. In some diseases of uncertain cause, such as MULTIPLE SCLEROSIS, EPILEPSY, CEREBRAL PALSY and PARKINSON'S DISEASE, nervous tissue is injured but intellectual deterioration is by no means certain to accompany it. CEREBRAL HAEMORRHAGES (strokes), severe types of ENCEPHALITIS, and poisoning with heavy metals such as mercury or lead can result in mental deterioration.

Second, mental illness or defect can be the result of an hereditary predisposition. In CRETINISM there is severe mental defect caused by inherited thyroid deficiency. A single disfunctioning gene causes PHENYLKETONURIA (PKN) which, if not recognized at birth and duly treated, can also cause severe mental defect. Huntington's CHOREA affects a mature adult with an inevitably progressive and fatal deterioration of body and mind: it is probably the most severe of the hereditary degenerative nervous illnesses. MICROCEPHALY and HYDROCEPHALUS are congenital conditions of uncertain cause also associated with mental defects.

Third, mental illness may be psychogenic, ie without clear evidence of any injury or disease affecting the brain. This category includes the PSYCHOSES, NEUROSES and personality disorders. SCHIZOPHRENIA and depressive psychoses are the most widespread of mental illnesses, accounting for a large number of patients in mental hospitals. The present tendency is, however, towards outpatient treatment for them, when possible. Neuroses, which can be severe but do not often warrant prolonged stays in hospital, include persistent ANXIETY, PHOBIAS, OBSESSIONS and HYSTERIA. Their treatment accounts for a huge dispensation of sedative and other palliative DRUGS by general practitioners. Personality disorders are often hard to define exactly because they mirror traits present to some extent in all people; for example, who has not felt "schizoid" or cut off from reality, or unduly aggressive, at times? However society has deemed it necessary to imprison some people of psychopathic personality when it seems likely that they will otherwise do injury, or further injury, to innocent people. *See also* INSANITY; MS pp.134–149.

Mental retardation, defect of intelligence existing at birth, irrespective of cause. Assuming a normal intelligence quotient (IQ) of 90–110, impairment is often described as borderline (IQ 68–85), mild (IQ 52–67), moderate (IQ 36–51), severe (IQ 20–35) and profound (IQ under 20). Mild impairment permits a limited but independent existence; moderate impairment seriously limits the individual, although special training may allow him to function minimally. Severe impairment, however, requires nearly total care. *See also* MS pp.142–43.

Mental stress, feelings of tension, expectancy or anxiety. This stress may be self-induced or imposed from outside. The most common form is a NEUROSIS, in which someone is led into compulsive, repetitive behaviour, such as checking for the post repeatedly after the postman has been. Another common form is DEPRESSION. *See also* MS pp.134–135.

Menthol, white, crystalline chemical compound having a strong odour of peppermint. Its main source is the oil from a Japanese plant, *Mentha arvensis.* Chemically it is a cyclic alcohol with the formula $C_{10}H_{19}OH$. It is an ingredient of some ointments and nasal sprays and is used to flavour toothpaste and cigarettes.

Menuhin, name of a family of musicians. Yehudi (1916–), a US-born violinist, is one of the world's most gifted and popular musicians. A child prodigy, he made his concert début at the age of seven, and studied with Georges Enesco and Adolf BUSCH. During WWII he gave numerous performances for the Allied forces and after 1945 played to raise money for humanitarian causes. He was an early champion of Béla BARTÓK and founded music festivals at Gstaad (1957), Bath (1959) and Windsor (1969). In 1963 he founded in England a school for musically gifted children. Menuhin has introduced Indian music to Western audiences and received many decorations as an international musician. Hephzibah Menuhin (1920–), his sister, is a pianist who has given many recitals, most often as accompanist to Yehudi. Yaltah, another sister, and Jeremy, a son, are also pianists.

Menzel, Adolf Friedrich Erdmann von (1815–1905), German painter and illustrator. 9is early works included naturalistic landscapes, but he is best known for his historical woodcuts of the *Life of Frederick the Great* (1840–42).

Menzies, Sir Robert Gordon (1894–), Australian Prime Minister from 1939 to 1941 and from 1949 to 1965. His second term of office was a period of growing national development. He encouraged British and US commitment to the security of South-East Asia and supported ANZUS and SEATO. In 1965 he was appointed warden of the CINQUE PORTS. *See also* HC2 p.249, *249.*

Meo. *See* MIAO.

Mercalli scale, modified, scale of 12 points used for measuring earthquake intensity. Named after the Italian seismologist, Giuseppe Mercalli (1850–1914), it is based on damage done at any point and so varies from place to place. Earthquake magnitude, on the other hand, is a function of the total energy released. *See also* PE p.*29.*

Mercantilism, economic and political doctrine of the 17th and 18th centuries, developed in Europe at a time when local economies were being turned into national ones. The introduction of large quantities of precious metals from the New World inspired mercantilism. The essential aim of mercantilism was to increase the wealth, and therefore the power, of the state by the accumulation of such precious metals. To this end exports were encouraged by bounties and imports restricted. Mercantilism declined during the 18th century as the increasing complexity of society made it impossible to administer; it ended finally in the 19th century when classical economists saw no sense in the acquisition of precious metals which in themselves did not satisfy human needs.

Mercator, Gerhardus (1512–94). Flemish geographer and cartographer who developed the first modern system of map projection. He worked as cartographer to the Emperor CHARLES V and as cosmographer to the Duke of Cleves after 1559. In 1568 he produced the first nautical chart to use the so-called Mercator projection (which he did not himself invent). The preface to his collection of maps (1595) had an illustration of the Titan Atlas supporting a globe inspired the modern term "atlas". *See also* PE p.*35.*

Mercenaries, professional paid soldiers who are prepared to fight in any cause and for any state. Mercenary forces are as old as organized warfare and often they were unreliable and dangerous to their employers. In 15th-century Europe the *condottieri,* sometimes also called "free companies" sold their services to various kings and princes, especially in Italy. The Swiss Guard at the Vatican are a remnant of the Swiss mercenaries hired by Renaissance popes. The Swiss also featured in the French Revolution of 1789 as they comprised Louis XVI's bodyguard and were hated symbols of the *ancien régime.* In recent times mercenaries, often from western European countries, have been employed in revolutionary wars in Africa.

Mercer, David (1928–), British dramatist whose work explored political disenchantment and the severe tension surrounding mental breakdown. His stage plays include *Belcher's Luck* (1966), and *Duck Song* (1974); his television works, the trilogy *The Generations* (1964) and *Shooting the Chandelier* (1977). Two other television plays, *A Suitable Case for Treatment* (1962) and *In Two Minds* (1967) were later filmed as *Morgan – A Suitable Case for Treatment* (1965) and *Family Life* (1971).

Mercer, Joe (1915–), British footballer and manager. A wing-half, he won league and cup honours with Everton and Arsenal and played five times for England in 1938–39. He achieved managerial success with Manchester City in the late 1960s and was England's caretaker manager in 1974.

Mercer, Johnny (1909–76), US musician who started his musical career as a singer with Benny GOODMAN and Paul WHITEMAN. His imaginative lyrics helped to raise the standards of popular song-writing. Stage successes in which he collaborated include *L'il Abner* (1956; filmed 1959) and *The Good Companions* (1974), but he is best known for his many contributions to films such as *Seven Brides for Seven Brothers* (1954). He won an Oscar for his song *Moon River,* written for the film *Breakfast at Tiffany's* (1961).

Merchant Adventurers, English trading company formed from groups of merchants associated with particular towns, chartered in 1407. Its principal trade was exporting English woollen cloth to The Netherlands, but by 1550 its officers controlled almost three-quarters of England's overseas trade. Attacked as a monopoly in the 17th century, the company lost many privileges and in 1689 its charter was abrogated. *See also* HC1 p.*193.*

Merchant banks, group of about a dozen private specialist banks in London whose chief function is to finance trade between Britain and foreign countries. They rose to prominence in the 18th century and were the principal agents in making London the financial centre of the world, for both short-term and long-term borrowing. The instrument of their business was, and remains, the BILL OF EXCHANGE, which they "accept" and so promise that they will pay the amount due. They are thus known as acceptance houses and earn their income (at a rate of about ¼% of the value of the goods dealt with) by lending their name to the buyer and seller as a guarantee. In addition to being acceptance houses, they also operate as issuing houses, sponsoring capital issues on behalf of their customers, both foreign governments wishing to raise loans in London and home firms.

Merchant of Venice, The (*c.* 1596), five-act play by William SHAKESPEARE in which appears one of his most famous creations, the Jew Shylock, at once repellent and inviting sympathy. The theme of mercy and tolerance is Shakespeare's own, but the interwoven stories of this comedy are drawn from medieval legends, the germs of which are found in *Gesta Romanorum.*

Merchant Navy, section of a nation's fleet concerned not with defence but with carrying cargo and passengers to and from international ports. In Britain, the origins of an organized merchant navy lie in the formation of the MERCHANT ADVENTURERS' company in 1407, but its period of real growth came with the maritime expansion of England in the late 16th and 17th centuries, when the great chartered companies, such as the EAST INDIA COMPANY, built up large fleets, their monopoly protected both by the terms of their charters and by the series of NAVIGATION ACTS which lasted from 1651 to 1849. The great extent of the British empire made Britain's merchant navy the largest in the world up to WWI. In 1976 the world's largest merchant navy, measured by gross tonnage, was the Liberian. It was followed by the merchant navies of Japan, Britain, Norway and Greece. The British merchant navy had a gross tonnage of 33,000,000 tons (tonnes) in 1976 and has about 3,550 ships of 100 gross tons (tonnes) or more; 513 were tankers (16,146,592 tons) and 235 ore and bulk carriers (5,280,801 tons).

Merchant Staplers, English trading com-

Yehudi Menuhin played with the New York Symphony Orchestra at the age of ten.

Adolf Menzel's *In The Beer Garden;* later works favour industrial subjects.

Merchant of Venice: Shylock bows low before Bassanio and Antonio.

Merchant Navy: *The Queen Elizabeth II* steams down the Hudson River.

Merchet

Mercury, the messenger of the gods, was the Roman equivalent of the Greek Hermes.

Merino sheep: the breed on which the prosperity of Australia was first founded.

Merry Wives of Windsor: Falstaff is made fun of by the two wives.

Mersey: a ferry crosses towards the Liver Building on the far bank.

pany active between 1350 and 1500 in the export of raw wool. This trade, a Crown monopoly, was concentrated at various towns known as "staples" to facilitate fiscal administration; between 1363 and 1558 the continental staple was at Calais. The Staplers' power and prestige were at their zenith in the 15th century.

Merchet, in feudalism, compensation paid by a villein to his lord if the villein's daughter married. As the villein, his kin and chattels belonged to the lord, the lord was entitled to merchet for any loss of service.

Mercia, kingdom of Anglo-Saxon England, covering at its greatest extent the area between Wales, the rivers Thames and Humber, and East Anglia. Settled by the ANGLES (c. 500), Mercia expanded its frontiers by war until OFFA, King of Mercia (757–96), ruled virtually all England. Mercian power declined rapidly in the late 9th century, plagued by Danish invaders who ruled part of it from 877, and finally crushed by WESSEX (919). *See also* HC1 p.154.

Mercier, Honoré (1840–94), Prime Minister of Quebec (1887–91). He was a Conservative French Canadian leader and founded the Parti National in 1871 to protest against union with British Canada. Later he joined the Liberals and presided over a coalition government, supporting provincial autonomy.

Merckx, Eddy (1945–), Belgian road-racing cyclist. Amateur world champion at 19, he became famous as a professional winning five Tours de France, including four consecutive victories (1969–72) and the 1967, 1971 and 1974 world championships.

Mercury, in Roman mythology, the god of commerce, merchants and thieves, usually identified with HERMES, the Greek messenger god. Mercury had a temple on the Aventine (built in 495 BC). His festival was observed on 15 May.

Mercury, smallest planet and the planet closest to the Sun. It has no satellites, a very thin atmosphere and is visible from Earth only after sunset or before sunrise. Mean distance from the Sun 58,000,000km (36,000,000 miles); mass 0.05 times that of the Earth; volume 0.05 times that of the Earth; diameter 4,880km (3,032 miles); rotation period, 58.7 days; sidereal period 88 days; temperature range 370°C (700°F) (day) to 0°C (32°F) (night). *See also* SU pp.176–177, *176–177*, 188–189, *188–189*, 192–193, 269.

Mercury, liquid metallic element (symbol Hg) of group IIB of the periodic table, known from earliest times. The chief ore is cinnabar (a sulphide), from which it is extracted by roasting. The element is a dangerous cumulative poison. Mercury is used in barometers, thermometers, and laboratory apparatus, and in mercury-vapour lamps and mercury cells. Properties: at. no. 80; at. wt. 200.59: s.g. 13.59; m.p. $-38.87°C$; ($-37.97°F$); b.p. 356.58°C (673.84°F); most common isotope Hg^{202} (29.8%).

Mercury fulminate. *See* FULMINATE.

Mercury poisoning, condition caused by the ingestion of mercury compounds, or their absorption through the skin. Mercury is a cumulative poison and its effects have always plagued workers in industries using mercury compounds. The industrial dumping of mercury wastes into the sea allows it to enter the human food cycle, because fish and birds ingest the mercury and these are eaten, passing on the poison. MINIMATA DISEASE is the worst recent example of this process. Symptoms of mercury poisoning include loss of co-ordination, balance and peripheral vision, sensory disturbance and deformed offspring. *See also* pp.152, *153.*

Mercury switch, in electrical engineering, switch in which mercury is contained in a tube which, when tilted, causes the mercury to bridge two contacts. An alternative design consists of mercury cups into which movable contacts dip.

Mere, chief striking weapon of the MAORIS.

Meredith, George (1828–1909), British novelist and poet. His early work was badly received and he was forced to supplement his income by journalism. Success arrived with the publication of The

Egoist (1879) in which Meredith analyses the moment of self-discovery with great psychological subtlety. *Diana of the Crossways* (1885) was also acclaimed and he was much admired by early 20th-century writers such as E. M. FORSTER.

Meredith, William ("Billy") (1874–1958), British footballer who played for Manchester City, Manchester United and Wales. A tireless outside right, he played in 48 internationals between 1859 and 1920 and played his last FA Cup game in 1924, when he was nearly 50 years old.

Merganser, any of several species of slender, freshwater or marine ducks that dive for food, especially the red-breasted merganser (*Mergus serrator*) which has a hooked bill. The GOOSANDER (*M. merganser*) differs mainly in coloration. Family Anatidae.

Mergenthaler, Ottmar (1854–99), US inventor. In 1884 he patented his LINO-TYPE machine for setting solid lines of type, now used by many newspapers. The following year the machine was improved by a device that automatically justified the type (ranged it to a regular margin at both ends of the lines). *See also* MM p.205.

Merger, coalition or amalgamation of two companies or the absorption of one company by a larger one. The shareholders usually receive shares in the new company in proportion to their shares in the old ones. Reasons for mergers include the elimination of competition, increase in productivity, greater efficiency or reduction of tax liability.

Meridian, imaginary line on the surface of the Earth running N to S from pole to pole. It is used to indicate longitude, passing, is measured in degrees E or W from the 0° line which through Greenwich, London, and is known as the prime meridian.

Mérimée, Prosper (1803–70), French novelist. His historical novels and short stories were ROMANTIC in theme but objective in style. His works include the collections of short stories *Mosaïque* (1833) and *Colomba* (1841). He also wrote *Carmen* (1843), on which Georges BIZET based his opera of 1875.

Merino, breed of fine-wool sheep that originated in Spain and is now farmed mainly in the USA and Australia. There are many local varieties, all of which have heavy white wool which is soft and pliable on the belly. Males bear tightly spiralled horns.

Merionethshire (Meirionnydd), former county in NW Wales; in 1974 it was assigned to Gwynedd. The region is mountainous and the chief rivers are the Dyfi, Mawddach and Dee. The most important agricultural activity is sheep farming, and deposits of manganese and slate provide the region's economic wealth. Tourism is also an important industry. The county town was Dolgellau. Area: 1,709sq km (660sq miles). Pop. (1971) 35,277.

Merit, Order of. *See* ORDER OF MERIT.

Meritocracy, system of government that grants the power to rule to those persons who are deemed the best to have it. Unlike an ARISTOCRACY or DEMOCRACY, a meritocracy requires both equality of opportunity and a society cohesive enough to agree on the criteria for merit.

Merlin, legendary magician and teacher. His origins may be traced to early Celtic folklore, although his name is usually associated with the Arthurian legends as the mentor of KING ARTHUR. Merlin is said to have been seduced by Nimue who, after gaining his magical knowledge, imprisoned him eternally in a tree.

Merlin, small European FALCON found in hills and open moorland, where it flies over the ground, catching small birds. It may hover, but not as commonly as the KESTREL. Length: to 33cm (13in); species *Falco columbarius.*

Mermaid, legendary sea creature said to have the head and upper body of a woman but a fish-tail from the waist down. Mermaids (and mermen) are referred to in ancient mythology and folklore and were thought to have magical powers. The behaviour of DUGONGS and MANATEES at sea, in particular their manner of suckling their young, is thought to be the basis of these beliefs. *See also* MS p.206.

Merman, Ethel (1909–), US singer and actress. She sang *I've Got Rhythm,* in *Girl Crazy* (1930), her first main stage role. Her loud, brassy voice was heard in musical comedies such as *Annie Get Your Gun* (1946–49), *Call Me Madam* (1950) and *Gypsy* (1959).

Meroë, ancient city on the Nile, just to the s of its confluence with Atbarah river. It is also the name of the civilization centred on it which flourished from the 6th century BC until its decline in the 4th century AD. *See also* HC1 pp.50–51, 228.

Merovingians (476–750), name of the dynasty of Salian Franks. CLOVIS I (r. c. 481-511), the first of the line, established his authority over most of present-day France by conquest; his conversion to Christianity occasioned the union of papal and Frankish interests, one of the most significant events of medieval history. The policy of equal rights of inheritance for each of his sons, however, led to the division of the kingdom and battles for supremacy. After the death of Dagobert I (639), the last effective ruler, the separate regions were ruled by powerful mayors of the palace, who used the Merovingian kings as puppets. One of these mayors, PEPIN THE SHORT, ultimately deposed Childeric III in 750 and founded the CAROLINGIAN dynasty.

Merrie England (1902), two-act operetta by the British composer Sir Edward GERMAN, with libretto by Basil Hood. It was first performed at the Savoy Theatre, London. Written in the tradition of the GILBERT and SULLIVAN Savoy operas, it has remained extremely popular.

Merry Widow, The (1905), three-act OPERETTA by Franz LEHÁR with libretto by Léon and Stein, first produced in Vienna. One of the most popular of Viennese operettas, the story is a light-hearted piece of amorous intrigue, revolving round the widow, Hanna Glawari, and her fortune.

Merry Wives of Windsor, The (c. 1600), comedy by William SHAKESPEARE, said to have been written at the command of Queen Elizabeth I who, after seeing *Henry IV* and being captivated by the character of Falstaff, wished to see a play about Sir John in love. The redoubtable Falstaff is shown addressing ardent love-speeches to the "merry widows", Mistress Ford and Mistress Page, and contrives to make a fool of himself.

Mersey, river in NW England, formed at the confluence of the Tame and Goyt rivers. It flows W, draining N Cheshire and Greater Manchester, entering the Irish Sea through a wide estuary, on which Birkenhead and Liverpool stand. Length: 113km (70 miles).

Merseyside, metropolitan county in NW England, formed in 1974 from parts of the former counties of CHESHIRE and LANCASHIRE. Area: 646sq km (249sq miles). Pop. (1976) 1,578,000.

Merton, Thomas (1915–69), US poet and religious writer. He became a Trappist monk in 1941, but remained in touch with contemporary events. He is best known for his autobiography, *The Seven Storey Mountain* (1948), and wrote many volumes of poetry and philosophy, including *Figures for an Apocalypse* (1947).

Merv, historical city in central Asia, now in Turkmenistan, USSR. It was the seat of an ACHAEMENID satrapy, and under the Arabs it was an Islamic cultural centre in the 12th century.

Mesa, large, broad, flat-topped hill or mountain of moderate height and with steep cliff-like sides. A mesa is capped with layers of resistant horizontal rocks which may then erode to form narrower BUTTES. *See also* PE pp.115, 120.

Mescaline, substance obtained from the dried tops of the peyote cactus, *Lophophora williamsii.* It causes visual hallucinations and other fantastic psychological effects. In North America and Mexico, mescaline is the basis of many Indian religions. *See also* MS pp.104, 105, 144.

Mesembryanthemum, genus of numerous species of erect or creeping shrubs or plants with succulent leaves; most grow in dry regions. The ice plant, *M. crystallinum,* is commonly grown as an ornamental. Family Aizoaceae.

Mesentery, sheet of tissue, or double layer of the PERITONEUM which attaches the intestines to the back wall of the abdomen.

Meshed (Mashad), city in NE Iran; the capital of Khorāsān province. Its name means "burial place of a martyr" and refers to the burial shrine of the Imam, Ali Reza. The city prospered in the 18th century when it became the capital of Persia. Meshed was of great strategic importance in the late 19th century because of its location near Iran's borders with the USSR and Afghanistan. Industries: food processing, carpets and textiles. Pop. (1973) 592,000.

Mesmer, Franz (or Friedrich) Anton (1734–1815), German physician. He developed MESMERISM, later called HYPNOSIS, in Vienna and Paris. His work was important in arousing interest in the use of hypnosis in medicine. *See also* MS pp.*83*, 136–137.

Mesmerism, 19th-century forerunner of hypnotism, named after the Austrian Dr Franz Anton MESMER whose experiments with the therapeutic use of a form of magnetism led to his discovery that he could produce a hypnotic state in patients and thereby treat their nervous disorders. He postulated a form of interpersonal influence which he called "animal magnetism", on which his therapeutic method was based. He was denounced as a charlatan, but a pupil, Chastenet de Puységur, developed his theory, and eventually it became an accepted part of PSYCHOANALYSIS. *See also* HYPNOSIS; MS pp.*83*, 136.

Mesoamerica, term coined to describe the civilization of what is now S Mexico, Belize, Guatemala, Nicaragua and El Salvador and part of Honduras and Costa Rica in the years before the Spanish colonization. The various cultures, which had economic and political ties, shared such aspects as pyramid-temples and a similar pantheon of gods. *See also* HC1 pp.*46*, *230, 230*.

Mesocephalic. *See* CEPHALIC INDEX.

Mesoderm, in the young embryos of higher animals, the middle of the three embryonic tissues (the others being ECTODERM and ENDODERM). In later development of the embryo the mesoderm gives rise to muscles, blood and connective tissues.

Mesolithic Age, also called the Middle Stone Age, the period in NW Europe in man's evolution following the PALAEOLITHIC and preceding the NEOLITHIC. It developed after the end of the glacial period when the last of the ice-sheets retreated (*c.* 8000 BC). The environment changed as temperature rose; scrub gave way to forest land and small game proliferated. Hunters and fishermen adapted their way of life to exploit these changes. A nomadic form of life became unnecessary and settlement in clusters of huts was a feature of the Mesolithic Age, as were chipped-stone tool kits, sometimes including microliths. *See also* HC1 pp.*25, 42, 43*; MS pp.*27*, 29, *29, 31*.

Mesomorph, term used in physical anthropology to denote a type of human bodily development in which muscularity is at a peak and subcutaneous fat is minimal. *See also* ECTOMORPH; ENDOMORPH.

Meson, sub-atomic particle, member of a subgroup of HADRONS, all of which have either zero or integral spin. They include the pions, kaons and eta mesons. There is no restriction on the number of mesons produced or destroyed in a nuclear reaction or present in a particular energy state. *See also* SU pp.66–67, *67*.

Mesophyte, plant that grows under average moisture conditions, thriving where there is a good balance of water and evaporation. Such plants have well developed root and leaf systems. *See also* HYDROPHYTE; XEROPHYTE.

Mesopotamia, ancient region in SW Asia, between the rivers Tigris (E) and Euphrates (W), and extending between the mountains of Armenia (N) and the Persian Gulf (S). It corresponds roughly to modern central Iraq, NE Syria and S Turkey. Historically it was the site of some of the first permanent settlements (*c.* 6000 BC). The Sumerians to the S used stone as building material and developed written communication (*c.* 3000 BC). The first empire in Mesopotamia was founded by the AKKADIANS (*c.* 2340 BC). This example of empire building was followed by the Babylonian dynasty. The area came under Persian control when it fell to ALEXANDER THE GREAT but enjoyed a period of importance when Baghdad became the capital of the ABBASID caliphate. In 1258, however, the MONGOLS devastated the area and Mesopotamia never regained its past prominence. *See also* HC1 pp.28–32, 56–57, 60–61, 80–81.

Mesopotamian Campaign (1914–18), Anglo-Indian offensive against the Turks during WWI to protect oilfields and retain the loyalty of the sheiks of Kuwait and Mohammerah. Landing at Basra in Nov. 1914, a small force moved up the River Tigris to attack Baghdad. Defeated at the Battle of Ctesiphon (Nov. 1915), the Allies were then besieged at Kut, which fell in April 1916, the Turks taking 10,000 prisoners. In a renewed British offensive Baghdad was taken in March 1917, and victory was finally achieved when the Turks were forced to surrender at Sharat (Nov. 1918). A heavy price had been paid, however, with a loss of 28,321 British and Indian lives. *See also* HC2 pp.190–191.

Mesosphere, middle shell of gases in the ATMOSPHERE between the STRATOSPHERE and the THERMOSPHERE. *See* PE p.*66*.

Mesozoa, phylum of tiny, multicellular animals parasitic in invertebrates. They do not have well-defined cell layers of ECTODERM and ENDODERM and, as a result, are not included in the subkingdom METAZOA and may represent a separate line of evolution. Reproduction is complex and includes free-swimming ciliated larvae. They range from less than 1mm (0.04in) to several millimetres in length.

Mesozoic period, the second of the three major divisions of geologic time, extending from about 225 million to 65 million years ago. It is divided into three periods: the TRIASSIC, JURASSIC and CRETACEOUS. For most of the era the continents are believed to have been conjoined into one huge land mass called PANGAEA. There was much volcanic activity and mountain building throughout the period, which was also characterized by the variety and size of its reptiles. For this reason it is often called the Age of Reptiles. *See also* NW pp.186–187.

Mesquite, or honey mesquite, deciduous tree common in the SW USA and Mexico. Its roots may extend more than 15m (50ft), allowing it to grow in desert regions. It also has small leaflets and spines 5cm (2in) long. Bees make honey from the nectar of the flowers and the pods are collected for use as forage. Height: 2.7–6.1m (9–20ft). Family Mimosaceae; species *Prosopis juliflora*.

Messager, André Charles Prosper (1853–1929), French conductor and composer, whose teachers included SAINT-SAËNS. He conducted the first performance of DEBUSSY's opera *Pelléas et Mélisande* (1902), which is dedicated to him. He composed operettas, which include *Véronique* (1898) and *Monsieur Beaucaire* (1919); he also wrote the ballet *The Two Pigeons* (1886).

Messalina Valeria (*c.* AD 24–48), third wife of the Roman emperor CLAUDIUS, mother of Octavia and Britannicus. Pictured as a royal harlot by Juvenal in *Satire on Women*, her promiscuity led the emperor's secretary Narcissus to persuade Claudius to have her executed.

Messel, Oliver Hilary Sambourne (1904–), British stage designer, who began his career designing for Cochran revues (1926; 1930–31). Among the plays and operas he designed are *The Lady's Not For Burning* (1949), *Ring Round the Moon* (1950), and *The Magic Flute* (1956) and *Der Rosenkavalier* (1959), the last two both for GLYNDEBOURNE. His films include *Caesar and Cleopatra* (1945) and *Suddenly Last Summer* (1959).

Messenger, Herbert Henky ("Dally") (1883–1964), Australian rugby league player known as "The Master". Twice capped at rugby union in 1907, he played five times for Australia at rugby league, becoming their first test captain. A goal-kicking centre, he toured Britain with the 1908–09 Kangaroos.

Messerschmitt, Willy (1898–), German aircraft designer, famous for the Bf-109 fighter used by the LUFTWAFFE during WWII. This aircraft set a world airspeed record of 700km/h (435mph) in 1939. Messerschmitt also designed the Me-262, the first jet-propelled aircraft to be used in combat (1944). *See also* MM p.*176*.

Messiaen, Olivier, (1908–), French composer and organist. His highly original music combines elements of bird-song, Hindu rhythms and Gregorian chant. His organ works, including *L'Ascension* (1933) and *La Nativité du Seigneur* (1935), have made important contributions to the repertoire of that instrument. Among numerous other compositions is the monumental ten-movement *Turangalîla-Symphony* (1949). *See also* HC2 pp.*270–271*, 271.

Messiah, from the Hebrew meaning "annointed" is one who has been consecrated or set apart by God for a special function or office. The title "Christ" is derived from the Greek version of the term. In the later part of the Old Testament the Jews awaited a Messiah, a future king of the house of David, who would deliver them from their bondage and usher in an era of peace and righteousness. On many occasions Jesus affirmed his identity as the Messiah of Old Testament expectations but taught that his kingdom was not of this world.

Messiah (1741), oratorio by George Frederick HANDEL, with text partly adapted from the Bible by Charles Jennens. It was first produced in Dublin in 1742 and met with immediate success. It is written in three parts, covering Christ's advent, birth and passion, and is scored for orchestra, four-part chorus and soprano, contralto, tenor and bass soloists. *See also* HC1 p.*319*.

Messier catalogue, list of 109 STAR CLUSTERS, NEBULAE and GALAXIES compiled by the French astronomer Charles Messier (1730–1817) and published in 1781. Messier hoped to make comet-hunting easier by listing all of the permanent celestial bodies. The catalogue numbers are in common use. *See also* SU p.230.

Messines, Battle of (7 June 1917), opening of the third battle of YPRES. British and ANZAC troops under Gen. Plumer took Messines ridge after 19 mines, containing almost 2,000 tonnes of explosive, were exploded under the German position. The mines had been carefully dug in during the previous year. The explosion was heard in London.

Mestizo, South American of mixed European (Spanish or Portuguese) and Indian ancestry. Today the term also has social and cultural connotations. Mestizos comprise a large part of the population of Latin American countries.

Meštrović, Ivan (1883–1962), Yugoslavian sculptor. He exhibited with the VIENNA SECESSION in 1902 and 1909 and had his first one-man show in 1933. He is famous for his elongated, CUBIST-like figures.

Metabolism, chemical and physical processes and changes continuously occurring in a living organism. They involve the breakdown of organic matter, resulting in energy release, and the synthesis of organic components to store energy and build tissues. These processes produce, maintain and destroy PROTOPLASM, releasing energy for vital functions. *See also* CATABOLISM; MS p.*40*; SU p.*155*.

Metacarpals, in man and other primates, bones of the palm of the hand. They articulate with the carpals (wrist bones) and phalanges (finger bones). In four-footed animals they are the bones of the forefoot.

Metal fatigue, the progressive fracture of metals subjected to repeated cycles of stress. Fatigue can occur early in the metal's life, with stresses building up as it cools after the manufacturing process. The phenomenon is of particular relevance to aeronautical engineers because metals can break, tear, or otherwise deform permanently under repeated or reversed loads at stress levels much lower than for a single loading. *See also* SU pp.88, *89*.

Metalious, Grace (1924–64), US novelist.

Mesmerism: an irate father tries to take his daughter from Mesmer during treatment.

Mesquite seeds once formed an important part of the diet of the Amerindians.

Willy Messerschmitt congratulates one of his pilots after a record-breaking flight.

Messiah: part of Handel's original score for "I Know that My Redeemer Liveth".

Metallic bond

Metamorphoses: the woodcut illustrates an incident from "The Golden Ass".

Joannis Metaxas mobilised the Greek army to resist the Italians in 1940.

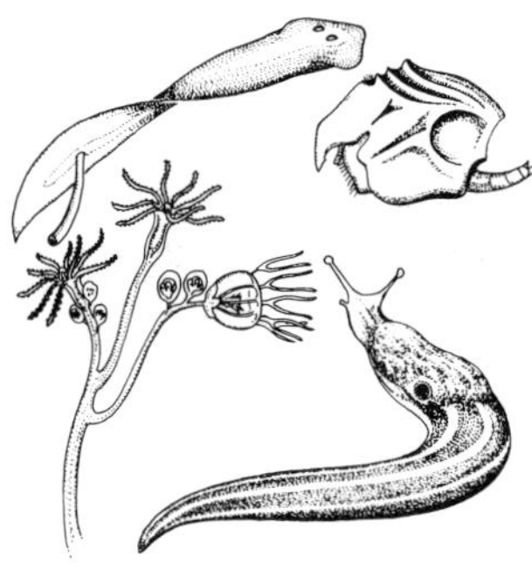

Metazoa include many thousands of species of corals, jellyfish and flatworms.

Meteor: this small piece of matter in space is sometimes called a shooting star.

After an undistinguished career writing short stories, she published *Peyton Place* (1957). It caused a minor outcry for its revealing descriptions of New England town life and was adapted for television and became a long-running SOAP OPERA.

Metallic bond, in chemistry, a bond that holds atoms together in METALS. Inside the crystals of metals, ions – positively charged atoms – are held by the electrostatic attraction of a cloud or "sea" of surrounding electrons which can, under various influences, move. Such electron movement under the influence of an applied voltage constitutes an electric current, so accounting for the electrical CONDUCTIVITY of metals. *See also* SU pp.88–89.

Metallography, study of the structure of metals and alloys, using optical and electronic microscopes and X-ray diffraction techniques. Examination by microscope reveals the size and shape of crystals and the distribution of non-metallic inclusions. X-ray diffraction is used to study the arrangement of the atoms in a metallic specimen.

Metalloid, element intermediate in properties between those of a metal and a non-metal. In moving from left to right across the PERIODIC TABLE and moving down the groups, there is a transition from metallic to non-metallic properties. Metalloid elements occur as borderline cases in groups III–VI; examples are silicon, germanium and arsenic. They are often semiconductors and have amphoteric hydroxides and oxides. *See also* SU p.136.

Metallurgy, study of the chemical and physical properties of METALS, their extraction from ores and their alloying, hardening, corrosion-proofing and plating. Among the physical tests used commonly in the laboratory control of the quality of metals and alloys are those that measure resistance to impact and tension. Chemical and spectrographic analyses are made to control the composition of alloys during manufacture. Works processes include melting, sintering, stamping, forging, dip- and electroplating, anodizing and sheradizing, besides various extraction processes. Research metallurgists are concerned with such topics as the changes in crystal forms of metals and alloys under various circumstances, with consequent changes in physical and chemical properties. *See also* MM p.26–43.

Metals, chemical elements which are good conductors of heat and electricity, the atoms of which are bonded together within crystals in a unique way; mixtures of such elements (alloys) are also metals. Many metals are hard, lustrous materials. MERCURY is exceptional in being a liquid at room temperature. SODIUM and POTASSIUM are examples of metals which are soft and chemically very reactive: they tarnish quickly in air and are most familiar as their salts. The lightest metal is LITHIUM, which is also very reactive. The heaviest is OSMIUM, which is 22.6 times denser than water and is one of the PLATINUM group, relatively unreactive metals also called noble metals; other noble metals are GOLD and SILVER. Malleability and ductility are further metallic characteristics and of all metals gold is the most malleable; it can be beaten out so thin as to be virtually translucent. Some metals have very high melting-points and so find various high-temperature applications; TUNGSTEN, with the highest melting-point of all (3,410°C), is employed for incandescent lamp filaments. ALUMINIUM, followed by IRON, are the two most abundant and useful of metals. TITANIUM, although rarely seen as the metal, is more commonly distributed than the more familiar COPPER, ZINC and LEAD. Other metals of great economic importance, because of their radioactivity, are URANIUM and PLUTONIUM, the latter being an example of a TRANSURANIC ELEMENT. *See also* MM pp.28–35; SU pp.134–137.

Metamorphic rocks, broad class of rocks that have been changed by heat or pressure from their original nature – SEDIMENTARY, IGNEOUS, or older metamorphic. The changes characteristically involve new crystalline structure, the creation of new minerals or a radical change of texture. Thus the metamorphic rock slate is made from sedimentary shale, the metamorphic gneiss from igneous granite. *See also* PE p.100, *100.*

Metamorphoses, series of tales, chiefly mythological, in Latin verse by OVID. They comprise 15 books, written in HEXAMETERS, relating interwoven myths, beginning with the creation of the world. *See also* HC1 p.113.

Metamorphosis, change of form, structure or substance during the development of various organisms, such as the changing of a caterpillar into a moth. Sometimes the change is gradual, as with a grasshopper, and is known as incomplete metamorphosis. Complete metamorphosis involves a change in habit or environment and usually involves LARVA, CHRYSALIS (or pupa) and adult stages. *See also* NW pp.108–109, *109,* 112, *112.*

Metaphor, figure of speech involving the use of an otherwise unconnected image as illustration or comparison. Unlike the SIMILE, the metaphor makes no use of a linking word such as "like" or "as" to introduce its image. For example, "He is a wise old owl" is a metaphor, whereas "He is as wise as an old owl" is a simile.

Metaphysical dynamism, philosophical view that REALITY is sheer energy or process. It is an ontological theory – ontology is the study of being or the basis of existence – which holds that reality cannot be split into either MIND or matter, and that the most that can be said about it is that it is continually changing (dynamic). The formulation of Henri BERGSON held that the *elan vital* is fundamental.

Metaphysical poetry, 17th-century English literary form, characterized by "metaphysical conceits" (figures of speech that employ abstract images) and a reliance on wit and subtle argument. Although this method was by no means new, men such as George HERBERT, John DONNE and Andrew MARVELL infused new life into English poetry.

Metaphysics, branch of philosophy that deals with the first principles of reality and with the nature of the universe. The term comes from the Greek *meta tá physika*, "of things natural". Metaphysics is divided into ONTOLOGY, the study of the essence of being, and COSMOLOGY, the study of the structure and laws of the universe. Leading metaphysical thinkers included PLATO, ARISTOTLE, KANT and WHITEHEAD. *See also* MS pp.226, 232–233.

Metasomatism, production of mineral deposits by the movement of hot fluids from an igneous body through cracks or pores in the surrounding rock. Veins of lead and tin ores surrounding the Cornish granites were formed in this way. *See also* PE p.97.

Metastasio, Pietro. *See* TRAPASSI.

Metastasis, transfer of cells or micro-organisms beyond their original sites. It is especially characteristic of malignant TUMOURS and spreading infections such as TUBERCULOSIS.

Metatarsals, bones of the feet (hind feet in quadrupeds). They articulate at one end with the tarsals (ankle bones) and at the other with the phalanges (toe bones).

Metaxas, Joannis (1871–1941), Greek soldier who declared himself dictator after the 1936 election and ruled until 1941. He resisted being drawn into WWII until Oct. 1941 when Greek neutrality was threatened by the Italian invasion of Albania.

Metazoa, subkingdom of animals whose bodies originate from a single cell and are composed of numerous differentiated cells. They range from COELENTERATES to MAMMALS. They do not include sponges (subkingdom Parazoa) or protozoans (subkingdom Protozoa).

Metchnikoff, Élie (1845–1916), Russian microbiologist. He shared the 1908 Nobel Prize in physiology and medicine with Paul EHRLICH for work on the mechanism of IMMUNITY, which included the discovery that white blood cells are important in the body's resistance to infection and disease. He was also noted for his theories concerning longevity. He wrote *Immunity in Infectious Diseases* in 1905.

Metellus, distinguished Roman family. Among the foremost were Quintus Caecilius Metellus Macedonicus (*d.*116 BC), a general who pacified Greece in 146 and suppressed the Celtiberians in Spain. As censor he supported compulsory marriage to increase the birth-rate. He was an active opponent of Tiberius Sempronius and Gaius Sempronius Gracchus. His nephew Q.C. Metellus Numidicus (*d. c.*91 BC) commanded Roman forces in Numidia during the Jugurthine War. He was elected consul in 109 and was later briefly exiled. His son Q.C. Metellus Pius (*d. c.*63 BC), fought in the Social War. *See also* SCIPIO.

Meteor, luminescent phenomenon produced by a small, stony or metallic body from outer space entering the Earth's atmosphere, as well as the solid body itself (more properly called a METEOROID). A Brilliant meteor consiste of a large luminous head followed by a aparkling comet-like wake. Most meteors disintegrate to dust before they reach the surface of the Earth. *See also* SU p.218.

Meteorite, particle or body from space (METEOROID) that survives passage through the Earth's atmosphere. Meteorites generally have a pitted surface and a fused charred crust. There are three main types: iron meteorites (siderites); stony meteorites (aerolites) and mixed iron and stone meteorites. Some are tiny particles; other weigh up to 200 tonnes. Meteorites have been dated to about the same time as the origin of the Earth. *See also* SU p.218.

Meteoroid, solid particle that revolves around the Sun. Meteoroids were considered to be a possible hazard in the early days of space-flight, especially if a space vehicle had to pass through a METEOR SHOWER. *See also* SU p.218.

Meteorology, the study of weather conditions and a branch of the study of CLIMATOLOGY. Meteorologists study and analyse data from a network of weather ships, aircraft and satellites in order to compile maps showing the state of the high- and low-pressure regions in the Earth's atmosphere. They also anticipate changes in the distribution of the regions and forecast the future weather. Wind strength and direction can be predicted accurately by measuring the differences in air pressure over the surface of the Earth. *See also* PE pp.66–73; SU p.111.

Meteor shower, swarm of hundreds of METEORS occurring simultaneously and travelling in parallel paths, although they appear to emanate from a single point, called the radiant. Radiant positions frequently rise together with various constellations and such commonly predictable showers are therefore named Leonid, Perseid, Gemminid, and so on. Other showers occur infrequently at varying intervals. It is thought that shower METEOROIDS are fragments of steadily disintegrating comets. *See also* SU p.218.

Meter, instrument that measures a particular quantity. For example, a gas meter measures the amount of gas that has flowed in a certain time, and a voltmeter measures the voltage between two points in an electrical circuit.

Methadone, synthetic opiate drug whose use is widespread as an institutionalized solution to HEROIN addiction. Similar to MORPHINE and heroin in chemical structure, it is an addictive drug, although withdrawal is believed to be milder than it is with the other two.

Methane (marsh gas), colourless odourless inflammable gas (formula CH_4), the first member of the alkane series of HYDROCARBONS. It is the chief constituent of NATURAL GAS, from which it is obtained. It is used in the form of natural gas as a fuel and in the pure form as a starting material for the manufacture of many chemicals. Properties: m.p. – 182.5°C (−296.5°F); b.p. −164°C (−263°F).

Methanol (methyl alcohol), colourless poisonous inflammable liquid (formula CH_3OH) obtained synthetically either from carbon monoxide and hydrogen, by the oxidation of natural gas or by the destructive distillation of wood. It is used as a solvent, rocket fuel, denaturant for ethanol and in chemical syntheses. Properties: m.p. 93.9°C (−137°F); b.p. 64.9°C (148.8°F).

Methedrine, trade name for a DRUG used in the same way as AMPHETAMINES, usually for combating depression. Its full chemical name is methamphetamine hydrochloride, and its formula is $C_{10}H_{15}N$. HCl.

Method, in 20th-century drama, the US name for a system of acting evolved from the teachings of Konstantin STANISLAVSKY; it emphasizes the development of an actor's role by introspective and psychoanalytical techniques rather that the eventual external skills of actual presentation. Adapted by the Group Theatre in the 1930s, it achieved attention from the work of the Actors' Studio, which was founded in 1947 by Elia KAZAN and Lee STRASBERG. Its pupils included James Dean, Paul Newman, Marilyn Monroe and Marlon Brando.

Methodism, worldwide religious movement begun in England in the 18th century and developed into an international Church. It was originally an evangelical movement within the CHURCH OF ENGLAND, started in 1729 by John and Charles WESLEY, who sought to direct the Church's energies towards the working class and who placed more emphasis on personal salvation than on Church organization. John Wesley stayed within the Anglican Church until his death in 1791. The Wesleyan Methodists, however, broke away formally in 1795 and in the following years they themselves divided into other sects, such as the Methodist New Connection (1797) and the Primitive Methodists (1811). The United Methodist Church reunited the New Connection with the smaller Bible Christians and the United Methodist Free Churches in 1907; in 1932 these united with the Wesleyans and Primitives. In the 1970s there were about 6 million British Methodists.

Methodius of Olympus, Saint (*d.*311), Christian writer who probably died a martyr during the last persecution at Chalcis. He was influenced in his writing by PLATO, and his *Symposium*, or *Treatise on Chastity*, constitutes a summary of Christian doctrine.

Methotrexate, chemical substance (an antimetabolite) that interferes with the synthesis of the vitamin folic acid, an essential component of cells. It is used to treat malignant TUMOURS, LEUKAEMIA and PSORIASIS.

Methuselah, in the Bible, son of ENOCH and eighth in antecedence from ADAM and EVE. He is said to have died at the age of 969 and been the longest-lived person ever. He was the father of many children, including Lamech, the father of NOAH.

Methyl alcohol. See METHANOL.

Methylated spirit, form of ethanol (ethyl alcohol) sold without excise duty because it is not intended to be drunk. It contains 5% methanol (methyl alcohol), which is extremely poisonous, and enough pyridine to give it a foul taste. It is dyed purple and used as a solvent and fuel.

Methyl mercury. See MINIMATA DISEASE.

Metre, in music, rhythmic groupings, each of which contains an equal number of beats or time units. It is usually expressed in the time signature at the beginning of the stave on the first line, after the clef. A 4/4 metre (simple time) has four crotchet beats per bar; a 6/8 metre (compound time) has two dotted crotchet (threequaver) beats per bar.

Metre, in poetry, pattern that occurs when the rhythm of a poem becomes regular enough to be measured. It imposes a regular recurrence of stresses that divides a line into equal units called metrical feet. The most commonly used feet are anapaest, dactyl, iamb and trochee. The metre of a poem is described according to the kind and number of metrical feet per line. For example, iambic pentameters have five iambs per line.

Metric system, decimal system of weights and measures based on a unit of length called the metre (m) and a unit of mass called the kilogramme (kg). The metre was originally defined as one ten-

millionth of the Earth's quadrant (a quarter of the circumference) and is now defined in terms of a wavelength of light. Larger and smaller metric units are related by powers of 10. Devised by the French in 1791, the metric system is used internationally by scientists and has now been adopted for general use by most of the Western world. *See also* WEIGHTS AND MEASURES; SU pp.*26, 27*.

Metro, name of the underground railway network of certain cities, especially Paris. The first line, from the Porte Vincennes to the Porte Maillot, was opened in 1900. In the mid-1970s there were more than 80 stations served by the Metro. The name Metro also applies to the underground networks in Moscow and Montreal. *See also* MM pp.*142–143*.

Metro-Goldwyn-Mayer. See MGM.

Metronome, instrument designed to mark the tempo of music, invented by Nikolaus Wenkel but patented by J. N. MAELZEL in 1816. It consists of a ticking pendulum that swings back and forth at a rate controlled by an adjustable weight that can be slid up and down the pendulum. Some modern metronomes are electric clocks. By setting the metronome at the speed indicated on the music, a performer can be sure he is adopting the tempo intended by the composer.

Metropolitan Opera Company, New York company famous for the high standard of its productions; an appearance with it is coveted by the greatest singers throughout the world. Operas premiered at "the Met" include *Gianni Schicchi* (1918) and *The Girl of the Golden West* (1910), both by PUCCINI. A new Met was opened in the Lincoln Center for the Performing Arts, New York, in 1966.

Metropolitan Police, one of the British police forces. It is responsible for the administration of law and order in the metropolitan area of London, other than the City of London (which is the responsibility of the City of London Police). It is headed by a Commissioner of Police and is under the control of the Home Secretary.

Metsu, Gabriel (1629–67), Dutch painter who worked in Leiden and Amsterdam and was influenced by REMBRANDT. He is perhaps best remembered for his genre pictures of middle-class life; eg *Mother and Sick Child*, which shows an extraordinarily deep feeling for the subject.

Metternich, Prince Klemens Wenzel Nepomuk Lothar von (1773–1859), Austrian statesman who came to loathe revolutionary excesses and who became Foreign Minister in 1809. On taking office he pursued a policy of conciliation towards NAPOLEON but in 1813 formed the QUADRUPLE ALLIANCE with England, Prussia and Russia to defeat him. He reached the zenith of his influence at the Congress of VIENNA (1814–15), which restored Europe to a group of anti-democratic states. He was driven from power by the revolution of 1848. *See also* HC2 pp.82–83, *83*, 108, 170.

Metzinger, Jean (1883–1956), French painter, influenced by NEO-IMPRESSIONISM and later by CUBISM, author of the book *Du Cubisme* (with Albert Gleizes) in 1912. He used violent colours and geometric planes in compositions such as *Tea-time* (1911).

Meuse (Maas), river that rises in NE France, flows across Belgium and The Netherlands to join the Waal River before entering the North Sea. It forms part of the border between Belgium and The Netherlands. There was much heavy fighting in the region of the Meuse during WWI, including the Battle of VERDUN. Length: 934km (580 miles).

MeV, abbreviation for million electron volts, the unit of energy used in particle physics to express the energy of subatomic particles from an ACCELERATOR.

Mexicali, city in NW Mexico, bordering Calexico, California, USA. Its products include cotton and cereals, although today most of its income is derived from tourism. Pop. (1970) 390,411.

Mexican hairless dog, terrier-like toy dog that may derive from a Chinese breed, possibly taken to Mexico in the 16th century. Its only hair is found as tufts on its

head and tail. Weight: to 4.5kg (10lbs). Height: 30–50cm (12–20in) at the shoulder.

Mexican jumping beans, seeds of certain shrubs harbouring larvae of the moth *Carpo-capsa salitarus*. When activated by warmth, the larvae produced "jumping" movements.

Mexican War, (1846–1847), armed conflict between the USA and Mexico, the immediate cause of which was the US annexation of Texas (1845). Mexico claimed that the Nueces River should be the sw boundary of Texas, whereas the Texans insisted that it should be the Rio Grande. Hostilities commenced when US forces, under Gen. Zachary TAYLOR, invaded the disputed area. The Mexicans were overwhelmed and, by the Treaty of Guadalupe Hidalgo (1848), were obliged to relinquish all claims to Texas (the present states of New Mexico, Utah, Nevada, Arizona and California) in return for $15 million. *See also* HC2 p.149.

Mexico, republic in North America. Formerly a Spanish colony, it has been one of the most rapidly developing nations of Latin America during the 20th century. The chief crops include cotton, sugar, fruit, vegetables and coffee. One of the world's leading producers of silver, Mexico also has large deposits of natural gas, coal, copper, sulphur, gold, lead, uranium, zinc and oil. Industries include consumer goods, fishing, timber and tourism. The capital is Mexico City. Area: 1,972,547sq km (761,000sq miles). Pop. (1976 est.) 62,329,000. *See also* MW p.124.

Mexico, Gulf of, arm of the Atlantic Ocean, s of USA and E of Mexico, connected to the Atlantic by the Florida Strait, through which the Gulf Stream flows, and to the Caribbean Sea by the YucatánChannel. It receives the Mississippi River and the Rio Grande. Cuba lies at the entrance to the Gulf. Max. depth: 3,732m (12,245ft). Area: approx. 1,813,000sq km (700,000sq miles). *See also* PE p.83.

Mexico City, capital and largest city of Mexico. The centre of the AZTEC civilization, the city was taken by Hernán CORTÉS in 1521 and it became the seat of the viceroyalty of Spanish colonies from 1521 to 1821, when Mexico gained its independence. Industries include chemicals, tourism, cement, tobacco, petroleum products, textiles and glassware. Pop. (1975 est.) 8,591,750.

Meyer, Deborah ("Debbie") (1952–   ), US freestyle swimmer who, at the 1968 Olympic Games, became the first swimmer to win three gold medals at one Olympics (200m, 400m, 800m).

Meyer, Julius Lothar (1830–95), German chemist who worked on the relationship between atomic volume and atomic weight. He devised the PERIODIC TABLE at the same time as D. I. MENDELEYEV, with whom he received the Davy Medal in 1882.

Meyerbeer, Giacomo (1791–1864), German composer *b.* Jakob Liebmann Beer. He wrote successful Italian operas in the style of ROSSINI after 1816, but his greatest acclaim came in Paris where his works laid the foundations of French grand opera. With libretti by SCRIBE, these included *Robert le Diable* (1831), *Les Huguenots* (1836) and *Le Prophète* (1849).

Meyerhof, Otto (1884–1951), German biochemist. He shared the 1922 Nobel Prize in physiology and medicine with Archibald HILL for work on muscle action and the relationship between the consumption of oxygen and the transformation of lactic acid in muscles. He became professor of biochemistry at the University of Pennsylvania in 1940.

Meyerhold, Vsevolod (1874–1940), Russian actor-director and the first great experimentalist in early 20th-century Russian theatre. He was made a member of the Moscow Art Theatre in 1898, toured Russia with his Society of the New Drama and ran STANISLAVSKY'S experimental theatre studio. The Stalinist regime brought his work to an end.

Mezuzah (Hebrew, "doorpost"), small parchment on which has been inscribed

Metternich was the most influential political figure in Europe from 1815 to 1848.

Meuse: the bridge at Poilvache is one of many that cross the river valley.

Mexican War: American troops storm the fortress of Chapultepec in Sept. 1847.

Meyerbeer was sometimes accused of subordinating his music to theatrical effect.

Miao: these Oriental mountain peoples have become more Westernized since WWII.

Michelangelo's *Pietà* in St Peter's, Rome, is the most important of his earlier works.

Albert Michelson worked for years to determine the exact speed of light.

Michigan: night is illumined by the Ford Company's River Rouge Mills at Detroit.

selected verses from the TORAH. It is encased in a tube and fixed to the doorpost of a Jewish living-room.

Mezzanine, in architecture, a low-ceilinged storey in a building, situated between the levels of two other floors. A mezzanine floor is usually between the ground floor and the first floor.

Mezzo-soprano, literally "middle" soprano, range of the human voice falling between SOPRANO and CONTRALTO. It may roughly be described as occupying the octaves above and below the G above middle C.

Mezzotint, method of ENGRAVING by scraping a design into a copper plate with a rocker, a tool with a serrated edge. The roughened surface produced, called a burr, results in light and dark tones in the print.

MGM (Metro-Goldwyn-Mayer), US film production company, formed in 1924–25. Perhaps the most successful of all Hollywood studios, it prospered during the Depression and WWII, releasing *Gone With the Wind* (1939). The rise of television and the European film industry contributed to MGM's decline since the 1950s.

Miami, resort city in SE Florida, USA, on Biscayne Bay, the second-largest city in the state. Originally a small agricultural community, it developed quickly after 1895, when Henry M. Flagler extended the Florida East Coast Railroad, dredged the harbour and began building recreational facilities. Modern Miami has many sporting facilities, including the Orange Bowl Stadium. Industries: tourism, clothing, concrete, metal products. Pop. (1970) 334,859.

Miao (Meo), people who live in the mountains of China, Vietnam, Laos and Thailand. Predominantly agricultural, they cultivate maize, rice and the opium poppy. They traditionally practise spirit and ancestor worship.

Miaskovsky, Nikolai (1881–1950), Russian post-Romantic composer who became professor of composition at the Moscow Conservatory (1921–50). His work includes 27 symphonies, ten string quartets and many songs.

Mica, group of common rock-forming minerals of the sheet silicate (SiO_4) type. All contain aluminium, potassium and water in the form of OH^- ions; other metals such as iron and magnesium may be present. Micas have perfect basal cleavage; common members are muscovite, biotite, phlogopite and lepidolite.

Micah, Old Testament book and name of the sixth of the 12 minor prophets. A contemporary of AMOS, HOSEA and ISAIAH, he anticipated the destruction of JERUSALEM, but incorporated elements of hope in his writings, including a well-known Messianic prophecy.

Michael, Saint, one of the four archangels mentioned in the Bible, the others being, Gabriel, Raphael and Uriel. In the Old Testament Michael is the guardian of Israel and the highest of the archangels. In Revelation in the New Testament he is said to have thrown down the Dragon (Satan). Feast: 29 Sept. He is given prominence also in Islam.

Michael, name of nine Byzantine emperors. Michael I (*d.*845; *r.*811–13) recognized the claim by CHARLEMAGNE to be ruler of the Western Empire. Michael II (*r.*820–29) founded the Amorian dynasty. Michael III (838–67; *r.*842–67) is regarded by historians as a successful leader in the wars against the Arabs. Michael IV (*r.*1034–41), born a peasant's son, was proclaimed emperor by request of the Empress Zoe whom he married in 1034. Michael V (*r.*1041–42) was deposed after a political uprising. Michael VI (*r.*1056–57) was deposed by Isaac COMNENUS. Michael VII (1059–78; *r.*1071–78) neglected the defence of the empire, losing part of Asia Minor and the remaining Byzantine possessions in southern Italy. Michael VIII (1224–82; *r.*1259–82) was founder of the Palaeologan dynasty and the restorer of the Byzantine Empire. Michael IX (*c.*1277–1320; *r.*1295–1320) was emperor with his father, Andronicus II.

Michael (1596–1695), Tsar of Russia (1613–45), founder of the ROMANOV DYNASTY. His election as tsar ended the chaotic period known as the Time of Troubles that had existed since the death of FEODOR I in 1598. Michael made peace with Sweden in 1617 and Poland in 1618 and, during his reign, some Western influences were introduced but the peasants were further reduced to serfdom.

Michael (1921–), King of Romania, 1927–30, 1940–47. From 1927 to 1930 power was actually in the hands of a council of regents. He aligned with the Allies in 1944, but abdicated in 1947.

Michaelmas, feast for St Michael and all angels celebrated on 29 Sept. Goose is traditionally eaten at Michaelmas in Britain, where it is also the name for a quarterly court term (12 Oct. to 21 Dec.) and academic terms at the universities of Oxford and Cambridge.

Michaelmas daisy, perennial herbaceous plant of the aster family. Introduced to Britain from N America, subsequently hybridization and selection has resulted in a wide variety of colour and habit. Common garden varieties are *Aster amellus* and *Aster novi-belgii.*

Michaux, French family of bicycle pioneers. Pierre (*d.*1883) and his son Ernest together designed the first bicycle with a rotating crankshaft in 1861, and founded a successful factory producing *Vélocipèdes* shortly afterwards. They also designed a steam-driven motorcycle in 1869.

Michelangeli, Arturo Benedetti (1920–), Italian pianist and teacher. He won the International Musical Competition in Geneva in 1939 and since WWII has divided his time between performing, teaching and making recordings, mostly of Romantic music.

Michelangelo Buonarotti (1475–1564), Italian sculptor, painter, architect and poet, an outstanding RENAISSANCE figure and a creator of MANNERISM. He trained in Florence, first in the technique of FRESCOE, under GHIRLANDAIO, and then in the Medici School, where he was influenced by the ideas of the humanists. In 1499 he sculptured the *Pietà* for St Peter's, Rome, which established his reputation. Michelangelo spent most of his productive years in Rome or Florence. He considered himself primarily a sculptor and created numerous monumental and heroic figures. Among these are *David* (1501–04), *Moses* (1513–16), *The Slaves* (*c.*1513) and the symbolic figures *Day, Night, Dawn* and *Evening* for the Medici Chapel, Florence, which he also designed (from 1520). His masterpiece in painting was his fresco cycle (1508–12) and *Last Judgement* (completed 1541) for the Sistine Chapel, Rome.

The Laurentian Library, Florence, which Michelangelo started planning in 1524, is among the earliest important examples of Mannerist architecture. In 1539 he began laying out the new Roman Capitol, and from 1546 to 1564 he was chief architect of St Peter's. During these last years he also wrote some of his finest sonnets. Probably no artist has been more influential. *See also* HC2 pp.256–257, *256, 264, 264.*

Michelozzo, Michelozzi (1396–1472), Florentine sculptor and architect, one of the first generation of RENAISSANCE architects. He received many commissions in Florence from the MEDICI family. With Lorenzo GHIBERTI he worked on such projects as the north doors for the Baptistery in Florence. He worked with the sculptor DONATELLO on many tombs. His most important work was the Medici Palace (now the Rucellai Palace), Florence (*c.*1444–59). *See also* HC1 p.*251.*

Michelsen, Alfonso López (1913–), Colombian politician. He began his career as a lawyer and was elected to the Senate in 1962 as a radical. He served as Foreign Minister (1967–70) and became President (1974–).

Michelson, Albert Abraham (1852–1931), US physicist, *b.* Germany. In 1887 he conducted an experiment with Edward MORLEY to determine the velocity of the Earth through the ETHER, using an INTER-FEROMETER of his own design. The negative result prompted the development of the theory of RELATIVITY. He determined the speed of light, and was awarded the 1907 Nobel Prize in physics. *See also* SU pp.*104, 105,* 105–106.

Michelson-Morley experiment, notable experiment performed in 1887 by the American physicists A. A. MICHELSON and E. W. MORLEY to determine the motion of the Earth through the ETHER. That such a motion was not detected discredited the ether theory and led to a crisis in physics that was resolved by EINSTEIN's theory of RELATIVITY. *See also* SU pp.104–105.

Michener, James (1907–), US author and winner of the 1947 Pulitzer Prize for his first collection of short stories, *Tales of the South Pacific* (1947). His novels include *The Bridges at Toko-ri* (1953), *Sayonara* (1954), *Hawaii* (1959), *The Source* (1965) and *Centennial* (1974).

Michigan, state in N central USA, bordered by four of the GREAT LAKES. Michigan is made up of two peninsulas separated by the Straits of Mackinac, which connect lakes Michigan and Huron. The Upper Peninsula ranges from the swampland of the NE lake shore to the mountains of the W. Copper and iron ore are mined and timber is a valuable resource. The Lower Peninsula is also forested and mineral deposits include oil, gypsum, sandstone and limestone. In the S cereal crops are cultivated and livestock rearing is important. The Lower Peninsula has most of Michigan's population and industries, which include motor vehicles, primary and fabricated metals, chemicals and food products. The state capital is Lansing and the largest city is Detroit.

The French first settled the region, but lost it to Britain after the SEVEN YEARS WAR. The British finally relinquished the area in 1796 and the Territory of Michigan was organized in 1805. The opening of the Erie Canal in 1825 aided Michigan's growth by linking it to the Atlantic through the Great Lakes, but the real industrial boom came with the development of the motor vehicle industry in the second decade of the 20th century. Area: 381,086sq km (147,137sq miles). Pop. (1975 est.) 748,000. *See also* MW p.175.

Michigan, Lake, third-largest of the GREAT LAKES, connected with Lake Huron by the Straits of Mackinac; the only one of the Great Lakes entirely within the USA. It was discovered by the French explorer Jean Nicolet in 1634. The ST LAWRENCE SEAWAY has opened up the lake to international trade. Chicago is on the SW shore. Area: 58,016sq km (22,400sq miles).

Micombero, Michel (1940–), Burundi soldier and statesman. He assumed the presidency after a military coup in 1966, but was deposed in 1976.

Microanalysis, technique in chemical analysis that makes use of extremely small samples – smaller even than those used in the more common semi-microanalysis. *See also* SU p.150, *150.*

Microbe, general name for any microorganism, including BACTERIA, microscopic FUNGI, PROTOZOA, microscopic ALGAE, and various types of animals of microscopic size. VIRUSES are sometimes included in this category but it is debatable whether they can be called living organisms. *See also* MS p.*85;* NW pp.36–37.

Microbiology, study of micro-organisms, their structure, function and significance. Types of micro-organisms include VIRUSES, BACTERIA, PROTOZOANS and microscopic unicellular ALGAE and FUNGI. Aspects of this study involve disease organisms and organisms useful to man, such as bacteria to fight disease or yeasts to promote fermentation. Microbiology began during the 17th century with the invention of the microscope, which enabled scholars to view micro-organisms for the first time. Pioneers in the field include Robert HOOKE, Antonie van LEEUWENHOEK and Louis PASTEUR. *See also* NW pp.26, 36.

Microcephaly, condition of having an extremely small head. It is usually accompanied by severe mental retardation and is thought to be caused by genetic factors, infections during the mother's pregnancy or exposure to radiation during pregnancy.

Microcline, felspar mineral, $KAlSi_3O_8$, with a triclinic crystal structure, hardness of 6–6.5, specific gravity øf 2.56 and a characteristic checkered twinning easily seen through a microscope. It is common in granites.

Microeconomics, one of the two major subdivisions of ECONOMICS, the study of individual components of the economic system. It analyses individual consumers and producers, the market conditions and the law of supply and demand – things that affect individual profitability. *See also* MACROECONOMICS.

Microelectronics, electronic systems designed and produced without wiring or other bulky components. They allow a high packing density that greatly reduces the size of component assemblies. In the years following WWII, the application of such newly developed devices as semiconductor TRANSISTORS and diodes saw the beginnings of the microelectronics industry. This accelerated with the development of PRINTED CIRCUITS, in which metal connections between miniaturized components are etched out in a single piece, on a circuit board. Even further reduction in size, or microminiaturization, was later achieved with INTEGRATED CIRCUITS, in which complex circuits can be made smaller than a fingernail. Molecular electronics is a new development which promises to be the ultimate in size reduction, in which complex circuits of microscopic size are grown inside single crystals. The applications of microelectronics generally have been felt most in the COMPUTER industry, but pocket calculators, miniaturized radios and, more recently, miniaturized television sets are evidence of the growing importance of microelectronics to consumers. *See also* SU pp.128–129.

Microfilm, reduced photographic reproduction of voluminous reading matter such as newspapers, periodicals, books and business records. First used in the 1920s, microcopy storage gained in popularity as a library and commercial information retrieval system in the 1960s and 1970s. Reduction is now so advanced that a microfiche, a single sheet of microfilm, can store up to 1,000 pages of text matter within an area of 75 × 125mm (3 × 5in).

Micrometer, hand-held caliper instrument used in engineering machine shops to check the thickness of metal parts. The caliper of the micrometer is placed over the part to be measured and a screw is turned by the thumb, advancing a spindle until it touches the metal part (which is then just held by the caliper). The thickness is read off the uncovered part of a scale, to an accuracy of about 0.002mm (0.0001in). In astronomy, a filar micrometer can be fitted to the eyepiece of a telescope, and consists of two fine threads, one of which is movable. It is used to measure accurately the apparent distance between stars. *See also* SU p.33.

Microminiaturization. *See* MICROELECTRONICS.

Micronesia, collective term for a large area of island groups in the W Pacific Ocean, generally those s of Japan, w of the INTERNATIONAL DATE LINE, N of Melanesia and E of the Philippines. The larger groups include the Marianas Islands, Marshall Islands, Gilbert and Ellice Islands and Caroline Islands; the native inhabitants are chiefly Polynesian and share related languages and cultures. Micronesia comprises a subdivision of OCEANIA, as do POLYNESIA and MELANESIA.

Micronesian people, the inhabitants of the islands of the W Pacific, one of the three main divisions of OCEANIA. There is considerable cultural diversity in the region, of which complex exchange rituals, feasts and oratorical skills are major features. Micronesians live by fishing and cultivation of crops such as yams and taro. *See also* MS pp.31, 34.

Microphone, device for converting sound into oscillating electric currents of the same frequency. The main types are the carbon microphone, in which the sound pressure causes a variation in the electrical

resistance of carbon granules held between a diaphragm and a carbon block; the crystal microphone, in which the sound impinges upon a crystal, which generates an oscillating current by the PIEZOELECTRIC EFFECT; and the moving coil microphone, in which a coil attached to a diaphragm oscillates in a stationary magnetic field. *See also* MM p.232.

Microscope, optical device for producing an enlarged image of a minute object. The first practical microscope was made by Anton van LEEUWENHOEK in 1668. The modern compound microscope has two converging lens systems, the objective and the eyepiece, both of short focal length. The object to be examined is placed close to the objective lens and illuminated by a strong source of light. The objective produces a magnified image of the object, which in turn is further magnified by the eyepiece to give the image seen by the observer. Most microscopes have three objective lenses on a turret, which may be interchanged to give a choice of low, medium or high magnification. Due to the nature of the visible spectrum of light, an optical microscope can magnify objects only up to 2,000 times, and this magnification is possible only by using lenses immersed in oil.

For extremely small organisms and objects an optical microscope cannot be used, because it is incapable of resolving them. For these purposes the electron microscope was devised. It works in much the same way except that the minute object is "illuminated" with a stream of electrons and the "lenses" consist of magnets that focus the electron beam. Electrons have shorter wavelengths than light and thus provide greater resolution, and so smaller objects may be seen. The image from an electron microscope is obtained by converting the pattern made as the electrons pass through the minute object into a video display, which may be photographed. These microscopes can magnify up to a million times.

A polarizing microscope illuminates mineral specimens with POLARIZED light which refracts into two beams whose velocity difference divided by the specimen's thickness yields an identifying birefringence index. *See also* PE p.95; SU p.157.

Microswitch, electric switch which requires only a small effort and movement to open or close it. The prefix "micro" is derived from its function as a small movement, rather than a small-sized switch. It is used for automatic, or semi-automatic, applications in electro-mechanical machines in which a moving actuator opens or closes the switch to stop or start an operation.

Microtome, precision device for slicing thin sections of specimens for examination under a MICROSCOPE.

Microtone, musical INTERVAL of less than one-quarter of a tone. Most Western instruments cannot play such notes, but the Mexican Julián Carillo invented instruments to play his microtone compositions.

Microwave, ELECTROMAGNETIC RADIATION of wavelength between one millimetre and one metre, and a frequency range of approx. 255 to 300,000MHz. This puts microwave radiation between infra-red and shortwave radio waves on the ELECTROMAGNETIC SPECTRUM. Microwave is being increasingly used today for such purposes as RADAR, TELEVISION broadcasting, moderate distance RADIO communications and in high-speed MICROWAVE HEATING. *See also* MM pp.228, 236.

Microwave heating, use of electromagnetic radiation having a frequency range of about 1,000 to 300,000MHz to heat materials. Microwave ovens are perhaps the major application to date; these offer a very fast method of cooking meat – joints which normally take an hour or more are cooked by microwaves in minutes.

Microwave transmission, accomplished with transmission lines, a system of material boundaries that guide MICROWAVES in a continuous path from one place to another. The structures most commonly used are COAXIAL CABLES and hol-

low WAVEGUIDES. *See also* MM p.228.

Midas, name of several historical Phrygian rulers and one legendary, foolish king. As a reward for rendering a service to the god Dionysus, King Midas asked that anything he touched should become gold. Dionysus agreed and Midas was unable to eat or drink because his food was transmuted. His story was told by OVID.

Mid-Atlantic Ridge, underwater topographic feature along the margin between the American crustal plate on one side and the European and African plates on the other. It follows the mid-line of the Atlantic Ocean, and the plates are growing at a rate of 2cm (0.8in) per year. *See also* PE p.82.

Mid-Bedfordshire, county district in central BEDFORD(SHIRE), England; created in 1974 under the Local Government Act (1972). Area: 504sq km (194sq miles). Pop. (1974 est). 93,900.

Mid-brain, one of the three divisions of the embryonic BRAIN of backboned animals, also called the mesencephalon. In a fully formed human being the mid-brain has become overlaid by the greatly developed cerebral lobes. It is thick-walled and is concerned particularly with sight and hearing. *See also* MS pp.40–41.

Middle age, period of life extending roughly from the 40th to the 60th year. This is the period when the cumulative effects of wear on the body begin to be felt. There is a tendency for vision to be impaired due to hardening of each lens, and a reduction in the ability to hear sounds of high frequency. Heart disease also takes a heavy toll of the middle-aged, particularly of male city-dwellers in high stress occupations. Both sexes are liable to suffer depressive illness, women perhaps more severely than men on the whole, because of the greater physical upset at menopause. However, intellectual functions are not generally impaired; some studies have shown a gradual increase in IQ into the 7th decade of life. In spite of this, middle age is a period of shrinking career opportunities, especially for people in highly developed societies.

Middle Ages, period in Western European history between the disintegration of the West Roman Empire in the 4th and 5th centuries and the period of the RENAISSANCE, beginning in the 15th. Dates for the close of the Middle Ages are imprecise, since the Renaissance developed and flourished at different times in different countries. The term "Middle Ages" was first used by the HUMANISTS of the 15th century who saw themselves as rediscovering the civilization of the Ancients. Classical antiquity was nonetheless a formative influence on the medieval period, as were the Germanic and Scandinavian invaders of w Europe between the 3rd and 12th centuries. But the Middle Ages was marked above all by the rise and dominance of CHRISTIANITY, with the emergence of a strong papacy, the growing importance of monastic orders and the Church's monopoly in education. It is the period not only of liturgical development and doctrinal disputes, but also of the CRUSADES and struggles between a politically powerful Church and increasingly self-conscious state. The decentralization of power in the sub-Roman period gave rise to the FEUDAL SYSTEM, one of the most important political and social developments of the time.

In the East the cultural and political centre was Constantinople, capital of the BYZANTINE EMPIRE. It exerted a considerable influence on the West in spite of increasing ecclesiastical and political differences, culminating in the sack of Constantinople in 1204. Although at one time considered a period void of learning and creativity, the later Middle Ages in particular produced an extension of literacy. The rise of vernacular literatures, the founding of the first universities in the 12th century and the period's own unique form of art and architecture, such as the ROMANESQUE and GOTHIC styles, are among its cultural achievements. Philosophy developed throughout the latter half of the period, reaching its peak in the 13th-century syntheses of theological doctrine

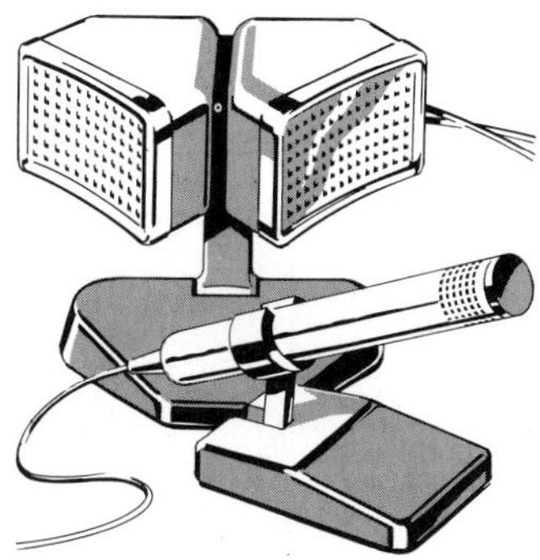

Microphone: the double version is used for recording stereophonic sound.

Microscopes have evolved from glass globes that were filled with clear water.

Midas's golden touch brought him only misery instead of expected happiness.

Middle Ages: the harvest scene dates from about 1340, from the Luttrell Psalter.

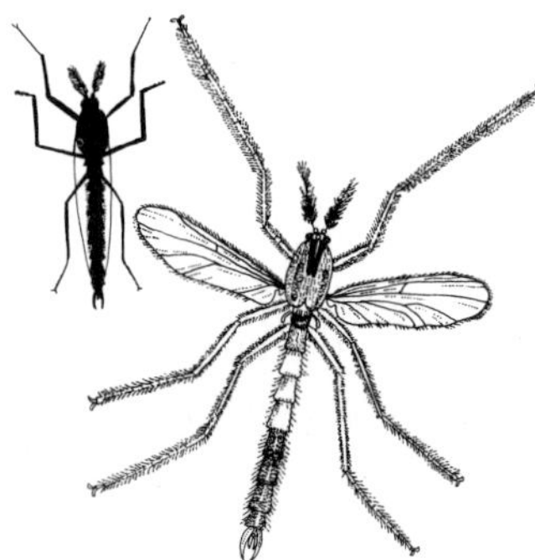

Midges are so small that they can scarcely be discerned by the human eye.

Midsummer Night's Dream: the painting after Landseer shows Titania with Bottom.

Midway: an oil tank burns having been hit by a Japanese bomb in the Battle of Midway.

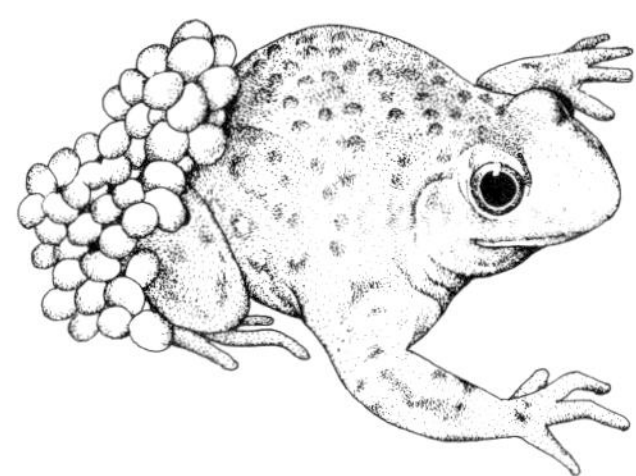

Midwife toads are so named because the male helps the female with her egg-laying.

and Aristotelianism. Much of modern thought, music and political structure derives from the Middle Ages. *See also* DARK AGES; HC1 pp.134–219.

Middle America. *See* MESOAMERICA.

Middle Angles, heathen people of German ANGLE descent who settled an area in England in the 5th century. Their name was used to refer to their geographical position between the kingdoms of East Anglia and Mercia. Although little is known of them, they appear to have retained independence until the 7th century, when they came under the Mercian rule of Peada.

Middle East, general description of the states of NE Africa and SW Asia W of Pakistan. The term was first used in 1902 to describe the area around the Persian Gulf, and was used to describe the larger area during WWII. The justification for such a blanket description lies in the common Islamic culture of most of the countries included; on this basis the N African states, Afghanistan and Pakistan are sometimes included.

Middle English, period in the evolution of the English language from the Norman Conquest until about 1500. This period saw the borrowing of many words from Norman French and the reduction of many inflectional endings. Grammatical gender was superseded by natural gender, and the use of the Anglo-Norman writing system caused radical changes in spellings and shifts in the stress patterns of words.

Middle Kingdom, period in the history of ancient Egypt, from *c.* 2040 BC to 1786 BC. It began with the unification of Egypt by Mentuhotep II. Its capital was at al-Lisht, s of Memphis. It marked a high point of centralized government, particularly under the Eleventh and Twelfth dynasties. *See also* HC1 pp.34–35.

Middlesbrough, county district in S CLEVELAND, England; created in 1974 under the Local Government Act (1972). Area: 47sq km (18sq miles). Pop. (1974 est.) 153,900.

Middlesex, former county of SE England, adjoining London. In 1965 most of the county was absorbed into GREATER LONDON. The Middlesex area of Greater London is considered as a county for the purposes of law.

Middle Stone Age. *See* MESOLITHIC AGE.

Middleton, Osman Edward (1925–), New Zealand author whose short stories observing ordinary life – both European and Maori – appeared in collections such as *A Walk on the Beach* (1964) and *The Loners* (1972).

Middleton, Stanley (1919–), British writer whose novels explored domestic situations and included *A Short Answer* (1958), *Wages of Virtue* (1969) and *Ends and Means* (1977). *Holiday* (1974) won the Booker Prize.

Middleton, Thomas (*c.* 1570–1627), English playwright. He wrote *The Honest Whore* (1604) in collaboration with Thomas DEKKER, *The Changeling* (1621) with William Rowley, and the political satire *A Game at Chesse* (1624).

Middle West (Midwest), region of the USA surrounding the GREAT LAKES and the upper Mississippi River valley. The name is also applied to the area between the Allegheny and Rocky Mts. Part of the interior plains, the area has some of the finest farmland in the USA and is particularly noted for wheat, maize and pigs. It also includes highly industrialized regions and the cities of Chicago, Detroit, St Louis and Minneapolis.

Midgard, in Norse mythology, the Middle Earth or abode of mankind. The body of the giant YMIR, or Aurgelmir, who was killed by the gods, was placed in the centre of the universe and used to make the Midgard, his hair becoming trees and his bones mountains. His skull formed the vault of the heavens.

Midge, small mosquito-like fly of the order Diptera. Most midges have red aquatic larvae called bloodworms, a favourite food of fish. The majority are harmless, but some suck human and animal blood. Families Chironomidae, Caratopogouidae, Cacidomyidae.

Mid-Glamorgan, county in South Wales,

on the Bristol Channel; formed in 1974 from parts of the former counties of GLAMORGAN, BRECKNOCKSHIRE and MONMOUTHSHIRE. The administrative centre is Cardiff. Area: 1,019sq km (393sq miles). Pop. (1975 est.) 540,100.

Midlands, region in central England, generally thought to include the counties of DERBYSHIRE, LEICESTERSHIRE, NORTHAMPTONSHIRE, NOTTINGHAMSHIRE, STAFFORDSHIRE, WARWICKSHIRE, WEST MIDLANDS and the E part of HEREFORD AND WORCESTER. It is a highly industrialized region and includes Birmingham, Coventry, Leicester, Derby, the cities of the POTTERIES, including the area known as the BLACK COUNTRY.

Midlothian, former county in SE Scotland, s of the Firth of Forth; in 1975 it was divided between Lothian and Borders regions. The land rises in the s to the Pentland and Moorfoot Hills, and the main rivers are the Gala Water, Esk, Almond and the Tyne. Sheep are grazed in the hills, and cereal crops and potatoes are grown on the lowlands. Industries include shipbuilding, engineering, brewing, paper and textiles. The county town was Edinburgh. Area: 948sq km (366sq miles). Pop. (1972) 598,500.

Midnight sun, phenomenon observed in regions within the Arctic and Antarctic circles during the summer solstice, when the Sun never sinks totally below the horizon. This persistence of the Sun is a consequence of the tilt of the Earth's axis, which alternatively presents the face of each flattened polar area to the Sun for a period of six months.

Mid-ocean ridge, great median ridge of the sea bottom where new LITHOSPHERE is being formed. The ridges are the spreading edges of the tectonic plates that cover the Earth. They form a world-encircling system that extends, with several side branches, along the MID-ATLANTIC RIDGE up, around, and down through the Indian Ocean (the Mid-Indian Ridge) and across the Pacific (the Pacific-Antarctic Ridge). *See also* SEA-FLOOR SPREADING; PE p.82: 82.

Midrash, in Hebrew literature, rabbinical interpretation and exposition of the text of the Hebrew Scriptures, written from about the 4th century BC to the 11th century AD. These writings are divided into *halakah*, which deals with legal sections of the Bible, and *haggadah*, which deals with biblical legends. A large part of the TALMUD is Midrashic writing.

Midshipman, defunct naval title assigned to young officer recruits below the rank of sub-lieutenant. The name derives from the quarters they were assigned amidships.

Midsummer Day (24 June), in the Christian calendar, the feast of the nativity of St John the Baptist. The day before is known as Midsummer Eve, or St John's Eve. In Celtic times the summer solstice (21 June) was celebrated with much revelry. Bonfires would be lit in honour of the Sun, and it was thought that anyone passing through the embers would enjoy good health. Later the Christian Church took on the tradition of a midsummer celebration. *See also* MS pp.210, 211.

Midsummer Marriage, The (1955), three-act opera by Sir Michael TIPPETT. Its first performance, starring Joan SUTHERLAND at Covent Garden, was not successful but later productions have established its musical reputation.

Midsummer Night's Dream, A (*c.* 1595), five-act comedy by William SHAKESPEARE. The first of his romantic masterpieces, its plot revolves round the romance of Hermia and Lysander and that of the king and queen of the fairies, Oberon and Titania. The work is notable for its idyllic, sylvan setting and its rustics, led by Bottom, who perform their own play, *Pyramus and Thisbe* in the last act.

Mid-Sussex, county district in E WEST SUSSEX, England; created in 1974 under the Local Government Act (1972). Area: 338sq km (130sq miles). Pop. (1974 est.) 105,900.

Midway Islands, coral atoll in the central Pacific Ocean, 2,000km (1,243 miles) WNW of Honolulu. It is US Territory com-

prised of two islands, Eastern and Sand, which are administered by the Department of the Interior. They were annexed to the USA in 1867, came under the authority of the US Navy Department and were made an air base in 1935. They were the scene of the Battle of Midway, an important WWII Allied victory in 1942. Area: 5sq km (2sq miles). Pop. (1975 est.) 2,000.

Midwest. *See* MIDDLE WEST.

Midwife, person who assists in the delivery of a child and the care of the mother. Midwifery gained recognition with the establishment of maternity hospitals and the provision of training for midwives in Europe in the 17th and 18th centuries. Legal recognition of the profession in Britain came with the Midwife's Act of 1902, which regulated the practice of midwifery. Today midwives must undergo formal training and hold a certificate issued by the Central Midwives' Board.

Midwife toad, small European amphibian with dull-grey, warty skin. The male wraps the fertilized eggs, which are laid in strings, around his back legs. After about three weeks the male enters water and the young emerge from the egg cases. Length: to 5cm (2in). Family Discoglossidae; species *Alytes obstetricans*.

Mieris, name of a family of Dutch painters, the most important of whom was Franz Jansz van Mieris (1635–81). A portraitist and genre painter, he was noted for his skilful handling of colour and treatment of silks and fine fabrics.

Mies van der Rohe, Ludwig (1886–1969), German-born US architect who worked in Germany until 1937, when he emigrated to the USA to escape Nazism. He was a member and director (1930–33) of the BAUHAUS. His true greatness as an architect was revealed in his German Pavilion for the Barcelona Exhibition (1929) and in the Tugendhat House, Brno (1930). His concern for quality continued to be central to his work in the USA, where he perfected the elegant, unadorned skyscraper. Typical of his later work are the Farnsworth House in Plano, Illinois (1950), the Lake Shore Drive Apartments in Chicago (1951), and the Seagram Building in New York (1958). *See also* HC2 p.206.

MiG, name of Soviet aircraft originally developed by Col.-Gen. A. Mikoyan and Mikhail I. Gurevich, a mathematician. The MiG-15 jet fighter, built in 1947, was of advanced design, and constant development has produced many successful types. The most recent design is the MiG-25, NATO code-name Foxbat, an all-weather fighter capable of speeds up to Mach 3.2. *See also* MM pp.178–179, *178*.

Mighty five. *See* RUSSIAN FIVE.

Mignard, Pierre (1612–95), French painter, the rival of LEBRUN. After studying under BOUCHER he worked in Rome, painting madonnas which were popularly known as "Mignardes". In 1657 he returned to France, where he began painting court portraits, succeeding LEBRUN as First Painter to Louis XIV in 1690.

Mignonette, any of 70 species of plants that grow in N Africa, Asia and Europe. They have small flowers clustered in a terminal spike, thick stems and coarse, lance-shaped leaves. A common species is the annual *Reseda odorata*, which has strongly scented flowers. Height: to 45cm (18in). Family Resedaceae.

Migraine, splitting, throbbing headache commonly on one side only, often accompanied by nausea, vomiting and visual disturbances such as flashes, and distorted vision. Of no known cause, its sufferers are most frequently women. There is no effective medication except strong ANALGESICS, and it is recommended that sufferers lie in a darkened room during an attack.

Migration, any periodic movement of animals or humans, usually in groups, from one area to another, in order to find food, breeding areas or better conditions. Animal migration, strictly speaking, involves the eventual return of the migrant to its place of departure. Fish migrate between fresh and salt water or from one part of an ocean to another. Birds usually migrate along established routes. Mammals migrate, usually in search of food.

Genuine migration of humans does not include individual exploration or the aimless wanderings of nomads. For thousands of years the deserts of central Asia have widened inexorably and this phenomenon has resulted in the true migration of prehistoric tribes to China, the Fertile Crescent of the Middle East and Europe. Another type of migration occurred in the 1300s AD when the Maoris of New Zealand left their overpopulated homes in the islands of central Polynesia. A classic example of a political migration was provided by the EXODUS of the Israelites from Egypt to seek a new homeland in CANAAN. *See also* MS pp.28–31; NW pp.128–129, 146–147.

Mihajlović, Draža (*c.*1893–1946), Yugoslavian soldier and anti-Communist. He fought with the Serbian guerrillas, the CHETNIKS, in WWI. In 1942 he became Minister of War in the Yugoslav government-in-exile. He clashed with TITO and was executed in 1946.

Mikado, ancient title of the emperor of Japan, meaning "exalted gate". It referred to the imperial palace and took the place of the emperor's personal name, the use of which was forbidden.

Mikado, The (1885), comic opera by William GILBERT and Arthur SULLIVAN, first produced in London. Set in ancient Japan, it is the story of a Japanese emperor (MIKADO) whose daughter loves one man but is betrothed to another, the Lord High Executioner.

Mikoyan, Anastas Ivanovich (1895–), Soviet political leader. He became commissar for foreign trade in 1926 and denounced STALIN in 1953. He supported Nikita KHRUSCHEV and became President in 1964, resigning in 1965.

Milan (Milano), industrial city in N Italy. It was conquered by Rome in 222 BC and became an important city of the Western Roman Empire. It was a free commune by the 12th century and a powerful Italian state under the SFORZA family from 1447 to 1535, when it was taken by the Spanish. Milan fell to NAPOLEON (1796–1814) and became part of Italy in 1860. Its modern manufactures include machinery, textiles and chemicals. Pop. 1,732,500.

Milan, Edict of (313), proclamation by which CONSTANTINE I (the Great) granted religious tolerance throughout the Roman Empire. Thus the long and cruel persecution of the Christians by the emperors of Rome was ended, and prelates were able to assume a position in civil administration.

Milan, School of, school of Italian painting that flourished in the 15th and 16th centuries. Vincenzio FOPPA was its leading artist before the arrival in Milan of LEONARDO DA VINCI who, with his followers, thereafter dominated the school.

Mildenhall Treasure, hoard of 34 separate pieces of mostly 4th-century Roman silverware, including dishes, goblets and spoons; three pieces bear Christian inscriptions. It was discovered in 1942 near Mildenhall, Suffolk. Declared to be treasure trove in July 1946, the best pieces were restored and displayed at the British Museum, London. *See also* WATER NEWTON TREASURE; HC1 p.*109.*

Mildew, external filaments and fruiting structures of numerous mould-like FUNGI, familiarly seen growing on such materials as leather, cloth and wood. More important economically are two groups that attack plants: downy and powdery mildews. Downy mildews (Peronosporales) are parasites of higher plants and cause substantial damage to growing crops. Velvety grey patches of spores form on the leaves, and much damage is done to the internal structure of the plant tissue. The Irish potato famine of 1845–48 was due to *Phytophthora infestans,* which causes blight in potatoes. Powdery mildews (Erysiphales) form a characteristic powdery coating on infected leaves. They cause less damage than downy mildews. *See also* NW pp.40–41; PE pp.*173, 200,* 201, *223.*

Mile End, district of E London where Richard II parleyed with Wat Tyler, leader of the PEASANTS' REVOLT (1381). In the 17th century the urban militia, the train bands, exercised there. Its name is derived either from the fact that Mile End Green was one mile long or from the distance of Mile End from the walls of the city of London. In 1887 a People's Palace was opened by Queen Victoria there providing concerts and exhibitions.

Miles, Antony John ("Tony") (1955–), British chess player. Runner-up in the 1973 world and European junior championships, he won the world junior championship in 1974 and was awarded the title of International Master. He became a Grand Master in 1976.

Miles, Sir Bernard (1907–), British actor-manager and director. His first stage appearance was in *Richard III* in 1930, after which he spent five years in repertory. He formed the Mermaid Theatre Trust which built the riverside Mermaid Theatre at Blackfriars, London. The first play performed there was the musical *Lock Up Your Daughters,* in 1959. He has also appeared in several films.

Milfoil, or yarrow, common weed of the ASTER family, native to Europe and North America. Used since ancient times for medicinal purposes, its scientific name was chosen in honour of Achilles, who was allegedly the first to use it in the treatment of wounds. It bears flattened bunches of small composite flowers. Height: to 90cm (3ft). Species *Anchillea millefolium.*

Milford Haven, Louis Alexander, 1st Marquess of (1854–1921), British admiral from 1904. The son of Prince Alexander of Hesse, he was formerly Prince of Battenburg but became a British subject, relinquishing his German titles and taking the name Mountbatten in 1917. He was Commander-in-Chief of the Atlantic fleet (1908–10) and First Sea Lord in 1912, but resigned his post at the outbreak of WWI in 1914.

Milford Sound, inlet of the Tasman Sea on the SW coast of South Island, New Zealand. A resort area, it is part of the Fjordland National Park. Length: approx. 19km (12 miles). Width: approx. 3km (2 miles).

Milhaud, Darius (1892–1974), leading 20th-century French composer. In the incidental music to Claudel's translation of Aeschylus' *Orestes,* (1913–22), he experimented with polytonality, using several simultaneous themes in different keys. He also included American jazz elements in works such as *La Création du Monde* (1923), which was originally written as a ballet. A prolific, versatile composer, he composed in all the standard forms, his most ambitious work being the opera *Christophe Colombe* (1930). *See also* HC2 pp.270–271, *270.*

Military government, the rule of enemy territory through military occupation. The practices of military government were standardized at the Hague Peace Conferences of 1899 and 1907. WWII saw the most intensive use of this form of government, although the rules of the Hague Conference (1907) were often violated by Germany's use of executions and slave labour. As a result of the experiences and abuses in WWII, a new agreement in military government was signed in 1949 in Geneva. *See also* MARTIAL LAW.

Militia, non-standing reserve military force for use in emergencies, usually organized on a local basis. In Britain it derives from the Anglo-Saxon *fyrd,* raised by the sheriff, in which all freemen were obliged to serve. Under the Tudors it came to be raised by the lords-lieutenant of the counties. It acquired the name militia in the 17th century, when the obligation to provide men at arms for service anywhere in the country was laid upon landowners. In 1757 service was made compulsory and county quotas were filled by ballot. It was then that the militia began to be properly trained. In 1852 service became virtually voluntary. The militia was replaced by the Territorial Force in 1908, which became the TERRITORIAL ARMY in 1921. This was replaced in 1967 by the Territorial and Army Volunteer Reserve.

Milk, liquid food secreted from mammary glands by the females of nearly all mammals to feed their young. The milk of domesticated cattle, sheep, goats, horses, camels and reindeer has been used as food by man since prehistoric times; both directly and indirectly include BUTTER, BUTTERMILK, CHEESE and fermented milks such as yogurt. Milk is essentially a suspension of fat and protein in water, sweetened with lactose sugar. The proportions of these constituents vary with each mammal, those of cows' milk being water 87.1%, protein 3.4%, fat 5.9%. In modern dairying, cows' milk is pasteurized at about 72°C for 16 seconds to kill all harmful microbes, then bottled under aseptic conditions. Nevertheless, milk sours after a day or so because of the action of lactic acid bacteria. *See also* MILK PRODUCTS; PE pp.164, *225, 227–229,* 228–229.

Milkfish, valuable food fish found in tropical waters of the Indian and Pacific oceans. Well adapted for cultivation on fish farms, this silvery fish has a deeply-forked tail and spawns in brackish water. Length: to 1.5m (5ft). Family Chanidae; species *Chanos chanos. See also* PE pp.242–243.

Milking, taking milk from a mammalian mother. Milking dairy cows is nowadays usually done with machines, which are both faster and more hygienic than hand methods. A cow's udder, after being thoroughly cleaned to remove all dirt that could infect the milk, is fitted with teat cups connected by flexible hoses to an electrically driven machine. This produces an intermittent vacuum in the hoses that extracts the milk and conveys it to a tank for temporary storage, before it is removed for pasturization or other processing. *See also* PE pp.226–227.

Milk products, in modern dairy processing, include dried, condensed and evaporated milks, ICE CREAM, bottled and tinned cream, BUTTER, fermented milks, and many kinds of CHEESE. Dried milk, used widely in ice cream and in cheese and meat processing, is made by spraying milk in hot air into an evaporator vessel. Condensed milk is sweetened milk concentrated by warming and placing in a vacuum. Evaporated milk is similarly made but is unsweetened and is HOMOGENIZED to prevent fat separating off. Both these milk concentrates keep well. In the making of fermented milks, milk is deliberately (but selectively) soured with microbes – in YOGURT with a culture of *Lactobacillus* bacteria. *See also* PE pp.226–229.

Milk snake, also called king snake, shiny patterned snake found from Canada to Ecuador. It is usually three-coloured with brown or red, black and yellowish transverse rings or saddles. It was once believed to suck milk from cows. Length: 92cm (3ft). Family Colubridae; species *Lampropeltis triangulum.*

Milkweed, any of numerous species of perennial plants with milky sap in the milkweed family, Asclepiadaceae. They include shrubs, woody vines and some succulent desert plants; they grow in tropical or subtropical regions in Africa and the Americas. Height: to 1.8m (6ft).

Milky Way Galaxy, barred spiral galaxy containing the SOLAR SYSTEM. It is a lens-shaped structure, 100,000 light years across. The broad band of innumerable stars visible from Earth – the Milky Way proper – represents the view along the plane of the galaxy. The centre of the galaxy is never visible, being optically obscured by interstellar dust and gas. Radio astronomy has determined that the centre lies beyond the constellation of Sagittarius. The Sun is situated on one of the spiral arms, about 32,000 light years from the galactic centre. *See also* SU pp.61, 173, *173, 230,* 244–245, *244–245,* 257–260, 263, 266–267.

Mill, James (1773–1836), British philosopher who met Jeremy BENTHAM in 1808 and worked with him towards UTILITARIANISM. He was the father of John Stuart MILL, and wrote an *Analysis of the Phenomena of the Human Mind* in 1829. *See also* HC2 p.33.

Mill, John Stuart (1806–73), British philosopher who was educated by his father, James MILL – recounted in his *Autobiogra-*

The Mikado is perhaps the most popular of all Gilbert and Sullivan's works.

Milan's Gothic cathedral dominates the heart of the city in stately splendour.

Darius Milhaud belonged to "The Six", an influential group of composers.

Milkfish can survive in waters cold enough to kill off most other fish.

Glen Miller's band became the most popular of all dance bands in the 1940s.

Jonathan Miller's talents include those of actor, theatre director and author.

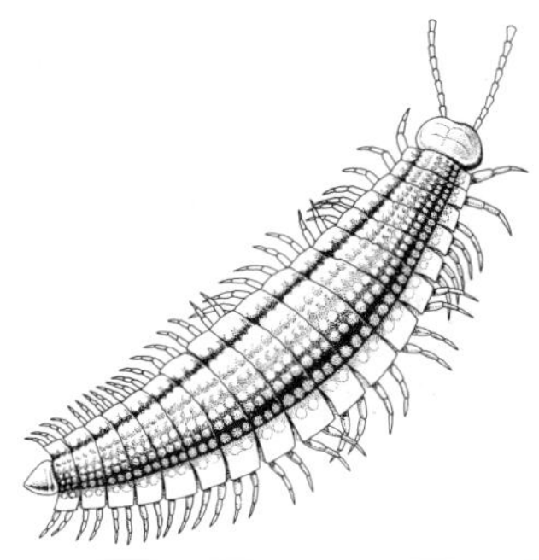

Millipedes, entwined round each other head to head, often mate for hours.

John Milton, hopelessly blind, dictated *Paradise Lost* to his daughters.

phy (1873) – and who advocated UTILITARIANISM. His book *On Liberty* (1859) made him famous as a defender of human rights. *System of Logic* (1843) was an attempt to provide an account of inductive reason. *See also* HC2 p.171; MS pp.228, *232–233, 235,* 273.

Millais, Sir John Everett (1829–96), British painter, a founder member of the PRE-RAPHAELITE BROTHERHOOD. His *Christ in the House of his Parents* (1850) was typical of his attention to realistic detail. From 1870 he painted mainly landscapes, such as *Chill October* (1870), and portraits. *See also* HC2 p.*116.*

Millar, John (1735–1801), British historian. He wrote *Historical View of the English Government from the Settlement of the Saxons in Britain to the Accession of the House of Stewart* (1787), and was a Whig who sympathized with both the American and French revolutions.

Millay, Edna St Vincent (1892–1950), US poet, who wrote *Renascence* (1912), *A Few Figs from Thistles* (1920), *Second April* (1921) and *The Harp-Weaver and Other Poems* (1923), which won a Pulitzer Prize. She was also a political activist.

Millbank, stretch of the Embankment in London on the N side of the River Thames between Lambeth and Vauxhall bridges. Its name derives from a smock mill. Today the area is largely given over to offices. The Tate Gallery is on the site of the notorious Penitentiary (1812–21) where prisoners awaited transportation.

Mille, Cecil Blount de. *See* DE MILLE, CECIL BLOUNT.

Millennium, the second coming of Christ for a 1,000-year reign, during which Satan will be bound in a pit, as described in the Book of Revelation (Ch.20). At the end of 1,000 years, Christ will take the kingdom of saints back to heaven with him and Satan will be freed. This belief was popular until the 4th century and then dormant until the REFORMATION, when it was embraced by ANABAPTISTS, MORAVIAN Brethren, and FIFTH MONARCHY MEN of 17th-century England and later by 19th-century MORMONS and ADVENTISTS.

Miller, Alton Glenn (1904–44), US jazz trombonist and bandleader. He formed what became the most popular dance band of all time, featuring a distinctive reedy sound and a popular theme song, *Moonlight Serenade* (1939), which Miller composed. He created the famous USAAF band in WWII. *See also* HC2 p.272.

Miller, Arthur (1915–), US playwright, who won a Critics' Circle Award for *All my Sons* (1947) and a Pulitzer Prize for *Death of a Salesman* (1947). His other plays include *The Crucible* (1953), *A View From the Bridge* (1955), *After the Fall* (1964) and *The Price* (1968). *See also* HC2 p.289, *289.*

Miller, Henry (1891–), US author. Most of his novels, which were banned until the 1960s in the USA and Britain, were first published in Paris, where he lived between 1930 and 1939. His works include *Tropic of Cancer* (1934), *Tropic of Capricorn* (1939) and *The Air-Conditioned Nightmare* (1945). *See also* HC2 p.289, *289.*

Miller, Jonathan Wolfe (1934–), British director. He wrote parts of, and appeared in, *Beyond the Fringe* (1961). His first production for the NATIONAL THEATRE *The Merchant of Venice* (1970). He was appointed its associate director in 1973.

Miller, Johnny (1947–), US golfer. Tall and blond, he won the 1973 US Open Championship, the 1976 British Open, and the World Cup individual and team awards in 1973 and 1975. In 1974 he equalled the US PGA tour record of eight victories.

Miller, Keith Ross (1919–), Australian cricketer who played for New South Wales, Victoria and Australia. He was a fast bowler and an aggressive batsman; a true all-rounder. In 55 Test matches between 1946 and 1956, Miller took 170 wickets at an average of 22.97 and scored 2,958 runs at an average of 36.97.

Miller, Stanley (1930–), US biochemist. In 1953 he carried out the first deliberate attempt to create life, by passing a mixture of AMMONIA, METHANE, water vapour and

HYDROGEN through water which was continuously sparked by a corona discharge.

Millet, Jean-François (1814–75), French painter who specialized in the portrayal of the everyday life of peasants. His first recognized works, *The Milkmaid* and *The Riding Lesson* (both 1844), were favourably received. He later concentrated on rustic scenes, and *The Gleaners* (1857) and *The Angelus* (1859) are typical of his idealistic approach. His paintings of hard work led critics to regard him as a socialist, *See also* HC2 p.*116.*

Millet, CEREAL grass that produces small, edible seeds, found in most parts of the world. The stalks have flower spikes and the hulled seeds are white. In the USSR, W Africa and Asia it is a staple food. In W Europe it is used mainly for pasture or hay. Pearl millet (*Pennisetum glaucum*) grows in poor soils and is used as food in India and Africa. Height: 1m (39in). Family Gramineae. *See also* PE pp.*156,* 181.

Millikan, Robert Andrews (1868–1953), US physicist who determined the value of the electronic charge. He was also able to verify EINSTEIN's photoelectric equation and he found a precise value for PLANCK's constant. For his work on the electronic charge and the photoelectric effect he was awarded the 1923 Nobel Prize in physics.

Millin, Sarah Gertrude (1889–1968), South African novelist. Her works deal with South African life and include *God's Stepchildren* (1924), *Mary Glenn* (1925), *The South Africans* (1926) and *The Dark Gods* (1931). *General Smuts*, a biography, was published in 1936.

Milling, ore, stage in the treatment of crude ore which mechanically extracts valuable minerals. Primary and secondary crushers grind down large fragments. Fine grinding (comminution) is done with rotating cylindrical mills that contain steel balls, rods or pebbles. The fine powder is then sorted by a variety of methods.

Millionairess, The (1936), four-act play by George Bernard SHAW. An unsuccessful farce, the play was prefaced by an introductory essay about naturally dominant people and the ordering of institutions to thwart them. The play itself deals lightheartedly with the same theme. A film version with Sophia LOREN and Peter SELLERS was directed by Anthony ASQUITH in 1961.

Millipede, any of numerous species of invertebrate arthropod animals found throughout the world. It has a segmented body, one pair of antennae, two pairs of legs per segment and can be orange, brown or black. All species avoid light and feed on plant tissues. Some tropical species squirt a repellant or poisonous secretion from pores in body segments. Length: 2–280mm (0.2–11in). Class Diplopoda. *See also* NW pp.*104, 208, 213.*

Mill on the Floss, The (1860), novel by George ELIOT. The theme of purification through personal trials is nowhere more evident than in this work. Maggie Tulliver, sensitive and intelligent, but dominated by her brother Tom, renounces her own chances of happiness for the general good.

Mills, Frederick ("Freddie") (1919–65), British light-heavyweight boxer. He began to fight as a professional at the age of 16, became British light-heavyweight champion in 1942 and was narrowly defeated by Gus Lesnevitch, the world champion, in 1946. Mills beat Lesnevitch in 1948 to take the world title and was defeated by Joey Maxim in 1950.

Mills, Sir John (1908–), British actor. He became established in the 1930s by such films as *Goodbye Mr Chips* (1939) and played a succession of military roles in films which included *In Which We Serve* (1942) and *This Happy Breed* (1944). In later years he developed character parts such as the idiot in *Ryan's Daughter* (1971), for which he won an Oscar.

Milne, Alan Alexander (1882–1956), British essayist, dramatist and author of children's books. For his young son Christopher Robin he wrote the verses in *When we Were Very Young* (1924) and *Now We Are Six* (1927), and the stories in *Winnie-the-Pooh* (1926) and *The House at Pooh Corner* (1928).

Milne, Edward Arthur (1896–1950), British cosmologist. He postulated kinematic relativity, an alternative to EINSTEIN's GENERAL THEORY OF RELATIVITY, based on time measurement rather than space.

Milne, John (1850–1913), British geologist whose invention of the SEISMOGRAPH in 1880 helped to found the science of SEISMOLOGY. He helped to establish many seismological stations, particularly those for recording earthquakes in Japan. His books include *Earthquakes* (1883) and *Seismology* (1898).

Milner, Alfred Milner, 1st Viscount (1854–1925), British imperial administrator. He was High Commissioner and then Governor of South Africa (1897–1905), and his determination to establish British dominance over the BOERS helped precipitate the SOUTH AFRICAN WAR (1899–1902). As a cabinet minister (1916–21) he formed the unified Allied command in 1918.

Miłosz, Czesaw (1911–), Polish poet and novelist, *b.* Lithuania. His novels *Doliny Issy* (*The Valley of the Issa,* 1955–57) and *Zdobycie Wkadzy* (*The Usurpers,* 1955) demonstrate a critical self-awareness. *Zniewolony umysl* (1953; tr. *The Captive Mind,* 1962) analyses the effects of Communism on contemporary writers. His post-war poetry includes *Światło dzienni* (1953) and is collected in *Selected Poems* (1973).

Milstein, Nathan (1904–), US violinist, *b.* Russia. He studied in Moscow with Leopold AUER, made his first concert tour of Russia in 1923 and appeared in the USA in 1929. His many recordings include the six sonatas for violin composed by J. S. BACH.

Miltiades (*c.*554–489 BC), Athenian nobleman and son of Cimon, credited with the plan to meet the Persians at the Battle of MARATHON in 490 BC.

Milton, John (1608–74), English poet and prose writer. He travelled widely in Europe between 1638 and 1639 and served as Latin secretary to the Commonwealth government (1649–60). In 1652 he became blind. His work is characterized by Latinized language and grandeur of imagery. His theology was unconventional, based on the section ideas prevalent in the 1640's. His greatest works are *L'Allegro* and *Il Penseroso* (both 1632), *Comus* (1634), *Lycidas* (1637), *Areopagitica, a Speech* (1644), *Paradise Lost* (1667), *Paradise Regained* (1671) and *Samson Agonistes* (1671). *See also* HC1 pp.282–283, *290.*

Milton Keynes, town in Buckinghamshire, S central England, 32km (20 miles) s of Northampton; established in 1967 under the New Towns Act. It is the headquarters of the OPEN UNIVERSITY (1961), and light industry is developing. Pop. (1976 est.) 66,000.

Milwaukee, city and port of entry in SE Wisconsin, USA, on the W shore of Lake Michigan. The North West Company established a fur-trading post there in 1795. Milwaukee was founded in 1836 and, during the second half of the 19th century, received many German settlers. There is a college of the University of Wisconsin (1908) in the city. Industries include diesel and gas engines, construction and electrical equipment, brewing. Pop. (1970) 717,372.

Mimas, one of the inner four satellites of the planet Saturn. It is 185,590km (115,270 miles) from Saturn and has a diameter of 500km (310 miles). Its mean sidereal period is less than one a day. The satellite has a low density and is thought to be composed of ice. *See also* SU pp.*210, 213, 213.*

Mime, in drama, the communication of mood, story and ideas by the use of gestures, movements and facial expressions, with no verbal communication. Its antecedents lie in Greek and Roman theatrical traditions and the COMMEDIA DELL'ARTE, although modern mime derives mainly from the performances of the first great French mime artiste, Jean-Baptiste-Gaspard Deburau (1796–1846). The art was further developed by Etienne Decroux (1898–) into an organized language of physical expression. His pupils, who included Jean-Louis BARRAULT and Marcel MARCEAU, consolidated mime's status, both as an art form in its own right and as a factor in dramatic training.

Mimeograph, type of duplicator for making many copies. It consists of a frame in which a stencil is stretched and an inking roller for pressing ink through the porous lines of the stencil on to paper. *See also* MM p.208.

Mimicry, form of animal protection in which the mimic, generally a harmless edible species, imitates the warning shape or coloration of a "model", a poisonous or dangerous species. When coloration increases an animal's chances of survival, it is commonly referred to as PROTECTIVE COLORATION. Batesian mimicry, named after the British naturalist Henry BATES, is exemplified by HOVERFLIES, which mimic inedible wasps; similarly, the VICEROY BUTTERFLY imitates the inedible MONARCH BUTTERFLY. The less common Müllerian mimicry, named after the German naturalist Fritz Müller, involves two or more harmful species that share a similar pattern, thus reinforcing it as one of the warnings to predators. A third form is aggressive mimicry, in which a predatory or parasitic species resembles a harmless one, thus allowing the former to remain undetected by its prey or host.

Mimir, in Norse mythology, wisest of the gods and keeper of wisdom. His tribe, the AESIR, sent him as a hostage to the rival VANIR, who killed him. Mirmir's head was preserved by ODIN, who used it as a source of knowledge. In another story Mimir demanded an eye from Odin in return for a drink from the well of wisdom.

Mimosa, genus of plants, shrubs and small trees native to tropical North and South America. They have showy, feather-like leaves and heads or spikes of white, pink or yellow flowers. Family Mimosaceae. *See also* NW p.45.

Minaret, tower, usually part of a MOSQUE from which MUSLIMS are called to prayer by a MUEZZIN. The earliest minarets were built in Egypt *c.*673 as low square towers; later Persian developments included covered balconies and extensive tiling.

Minas, city in SE Uruguay; capital of Lavalleja department. Founded in 1783, the city was named after its granite and marble quarries. It is noted for its mineral waters and is a resort area for Montevideo, to which it is connected by the Pan-American Highway. Pop. 31,000.

Minas Gerais, state in SE Brazil; the capital is Belo Horizonte. It was declared a province in 1822 and a state of the federal republic in 1891. Located on a plateau, it has a subtropical climate and many resorts and thermal springs. Minas Gerais is also rich in mineral deposits. Its products include coffee, pasture grass, maize, rice and cassava. Area: 587,171sq km (226,707sq miles). Pop. (1970) 11,279,872.

Minch, The (North Minch), arm of the North Atlantic ocean separating the islands of the Outer Hebrides from the NW coast of Scotland. Width: approx. 56km (35 miles).

Mind, hypothetical concept postulated to account for the thoughts and behaviour of conscious beings that cannot be accounted for in terms of purely physical phenomena such as the BRAIN and brain processes. Philosophers such as René DESCARTES distinguish between mind and matter, two totally independent entities. Some hold the world to be a product of the mind, whereas others hold it to be all matter. There are difficulties with all of these views, and Gilbert RYLE has proposed that the separate concepts of mind and matter are due to semantic confusion and that there is no essential difference between the two. *See also* index to *Men and Society.*

Mindanao, second-largest island of the Philippines; Davao is the most important city and port. The island is mountainous and has extensive forests. Products: coconuts, mahogany, iron ore, pineapples, hemp. Area: 94,631sq km (36,537sq miles). Pop. (1970) 7,292,691.

Minden, Battle of (1759), engagement in the SEVEN YEARS WAR (1756–63), fought

near the town of the same name in NW Germany. Anglo-Hanoverian forces under Ferdinand of Brunswick routed a larger French army under the Prince de Broglie. The French were thereafter unable to influence events in Germany during the rest of the war.

Mindszenty, József, Cardinal (1892–1975), Hungarian Roman Catholic prelate. An outspoken opponent of Nazism and Communism, he spent much of his life in jail or self-imposed confinement. He was released briefly by the revolutionaries during the 1956 Hungarian uprising. With the return of Communist power he sought asylum in the US legation in Budapest. He left the legation in 1971 to live in Rome. *See also* HC2 p.*243.*

Mine, excavation from which minerals (mainly coal and metal ores) are extracted. Underground mines are of two main types: shaft mines and drift mines. Shafts are sunk vertically in the Earth's crust until they reach the depth of the seams to be exploited, which are reached by tunnels, or galleries. The gold mines of South Africa, the world's deepest mines, are shaft mines. Drift mines are generally shallower, the seams being reached by a drift, or gradually sloping shaft, which leads on to a gallery system. In opencast mining the seams are near or on the surface and are exposed by giant dragline machines which dig away the topsoil; this is also called strip mining. *See also* COAL-MINING; MM pp.*76–77.*

Mine, concealed or buried explosive device detonated through contact with individuals or vehicles. Underwater mines are used either to protect or to blockade coastal areas. *See also* MM p.243.

Mineralization, FOSSIL-forming process whereby preserved hard parts of plants and animals are altered by the minerals in the soil. Circulating water dissolves certain constituents of bones and shells, which are replaced by silica, iron or other compounds. This replacement can be so exact as to preserve the finest structures. *See also* PE pp.124–130.

Mineralogy, investigation of naturally occurring inorganic substances found on Earth and elsewhere in the Solar System. Major subdivisions are CRYSTALLOGRAPHY, which studies the composition and arrangement of minerals; paramagnetic mineralogy, which deals with the associations and order of crystallization of minerals; descriptive mineralogy, concerned with the physical properties used in identification of minerals; and taxonomic mineralogy, the classification of minerals by chemical and crystal type. *See also* GEOCHEMISTRY; PETROLOGY; PE pp.94–99.

Mineraloid, mineral substance that is non-crystalline or amorphous. Its origin can be that of a gel, ie a colloidal precipitate such as opal; by cooling so rapidly that crystallization could not occur, eg volcanic glass; or its original crystal structure by the disruption of radioactivity. *See also* MINERALS.

Mineral oil, also called liquid paraffin and liquid petrolatum, colourless liquid without taste or smell, distilled from PETROLEUM. It is used as a lubricant, and as a solvent in the manufacture of plastics. It is also used medicinally as a laxative.

Minerals, natural, homogeneous and, with a few exceptions, solid and crystalline materials that form the Earth and make up its ROCKS. Most are formed through inorganic processes, and more than 3,000 minerals have been identified. They are classified on the basis of a chemical make-up, crystal structure and physical properties such as hardness, specific gravity, cleavage, colour and lustre. Some minerals are economically important as ORES from which metals are extracted. *See also* PE pp.94–99.

Mineral waters, originally natural waters valued for their mineral content, now also used to describe carbonated drinks containing salts or various flavourings. "Medicinal waters" from many famous spas were (and still are) bottled and sold. They are classified according to location, use and chemical content. Synthetic mineral waters were developed to imitate natural ones.

Minerva, Roman goddess of the arts, professions and handicrafts, whose cult is believed to have originated in Etruria. She was often identified with the Greek goddess ATHENA, and so became goddess of wisdom and later of war. She was honoured at the festival called the Quinquatria, held annually on 19 March.

Mines Acts, series of acts of the British Parliament regulating conditions and hours of employment in collieries. The first act, passed in 1842, was an early milestone in the growth of state involvement in industry. It was based on the report of the Royal Commission on the Employment of Children in Mines (1840), whose description of children and women suffering underground stirred the public conscience. It was also the work of the Tory reformer Lord Ashley, later Earl of SHAFTESBURY. The act prohibited the employment of women and children underground and the employment of boys under the age of 10. It appointed inspectors to enforce the act and to furnish the Home Office with reports. In 1860 the age limit for boys on enginemen was raised to 18 years. In the mid-19th century there was also a series of acts passed to improve conditions of safety in the mines. The next important act was passed in 1908. It limited the underground shift to eight hours a day. It was the first law to regulate hours of labour for adult men.

Mingan Islands, group of 15 small islands and numerous islets in the St Lawrence River N of Anticosti Island in E Quebec, Canada. They were discovered by Jacques CARTIER in 1535 and purchased by the HUDSON'S BAY COMPANY in 1836.

Ming dynasty (1368–1644), the last of the great Chinese dynasties before the conquest of China by the MANCHUS. The first Ming emperor, CHU YÜAN-CHANG (*r.*1328–98), expelled the Mongol YÜAN dynasty and had unified all of China proper by 1382. The Ming brought a period of cultural and philosophical advance during which China influenced many adjacent areas, including Japan. Great seagoing expeditions were launched to the S and W, reaching the E coast of Africa. Peking was laid out in its present form, and the traditional bureaucracy was reinforced. In the later years of the dynasty, however, a growing population, corrupt officials and weak emperors incited revolts among peasants in the border regions, thereby preparing the way for the Manchu conquest of China. *See also* HC1 pp.127, 222, *222.*

Mingus, Charles (1922–), US jazz bass-player and composer, who started leading his own bands in the 1950s. He is well known for his experiments with atonality and impressionism, and his best-known work is *The Black Saint and the Sinner Lady* (1963). His autobiography, *Beneath the Underdog,* appeared in 1971.

Miniature bull terrier, smaller version of the BULL TERRIER, evolved from small bull terriers and the old toy bull terrier. Height: to 36cm (14in) at the shoulder.

Miniature painting, term which originally meant the art of manuscript ILLUMINATION but was later applied to very small paintings, usually portraits. The earliest miniatures, dating from the late 15th century, were executed in the same materials as illuminated manuscripts – gouache paint on vellum or card – but from the 18th century miniaturists usually painted in watercolour on ivory and sometimes worked in oils on metal. After the mid-19th century the art of miniature painting declined. Famous miniaturists include HOLBEIN THE YOUNGER, Nicholas HILLIARD, Isaac OLIVER and Richard COSWAY in England, Jean and François CLOUET in France and Friedrich Heinrich Füger (1751–1818) in Germany.

Miniature pinscher, small terrier-like dog. It has a small narrow head, a straight, tapering muzzle and erect or drooping ears. The deep-chested body is set on straight, muscular legs and the tail is commonly docked. The smooth, hard coat may be red, black or brown, often with tan markings. Height: to 30cm (12in) at the shoulder.

Miniature poodle, medium-sized breed of POODLE, a dog that is believed to have ori-

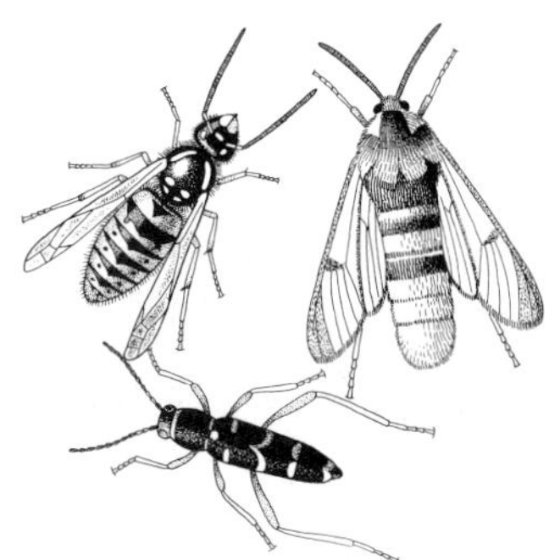

Mimicry: moth (right) and beetle (below) mimic the wasp's markings (left).

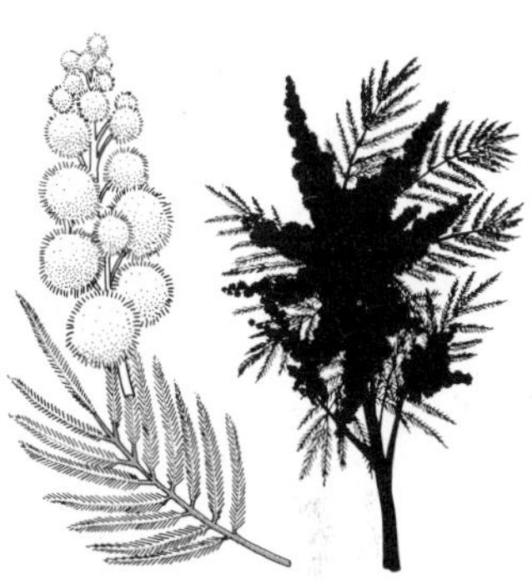

Mimosa: some species are so sensitive that their leaves fold under stimulus.

Minden: in the battle, Anglo-Hanoverian infantry overwhelmed the French cavalry.

Miniature pinschers, from the Rhine Valley, are also called deer terriers.

Miniature schnauzer

Mink hunt on land and in water and often kill more than they need for food.

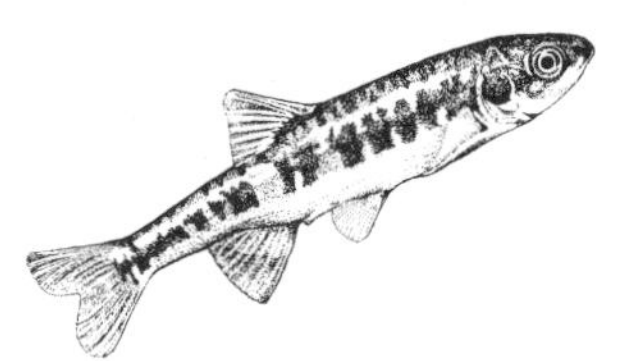

Minnows are scarce in some areas because of over harvesting for live bait.

Minorca: part of the town of Ciudadela huddles round the old harbour.

Minstrels: an early 17th century print shows musicians outside a tavern.

ginated in Germany; the other sizes are standard and toy. It has a small head and a fine muzzle. The slender body is set on delicate legs and the tail is short and erect. The soft, profuse coat, which is curly and fluffy, may be any solid colour. Height: to 38cm (15in) at the shoulder.

Miniature schnauzer, small breed of SCHNAUZER dog. It has a square head, a box-like whiskery muzzle, pricked V-shaped ears and a sturdy neck. The straight body is set on muscular legs and the tail is commonly docked. The wiry coat may be all black or white and grey. Height: to 35cm (14in) at the shoulder.

Minié, Claude-Étienne (1804–79), French army officer who invented the Minié rifle in 1849. His development of the cylindrical bullet solved the problem of rifling a muzzle-loading FIREARM, causing the bullet to spin in flight. The rifle was adopted by the armies of Europe and used in the American Civil War (1861–65).

Minimal art, 20th-century school which claims that anything on a canvas is a painting, any object a work of art. Its sources were MALEVICH's *Black Square on a White Ground* (*c.*1913) and DUCHAMP's exhibition of objects such as the urinal of 1917, called *Fountain*. The school developed an anonymity of line and form, often highly coloured, shown by painters such as Frank STELLA.

Minimata disease, type of poisoning by MERCURY, so-called because it first occurred (1950s) in people living near Minimata Bay, Japan. It affected people who had eaten fish that had absorbed methyl MERCURY, a component of the POLLUTION of the waters of the bay by industrial effluent. Symptoms include anaemia, tremors and paralysis; it can be fatal (in the original outbreak, nearly 50 people died). *See also* PE pp.152, *153*.

Minimum lending rate. See MLR.

Mining, the process of obtaining metallic and non-metallic materials from the Earth's crust. Included are underground, surface, and underwater methods. Mining mostly involves the physical removal of rock and earth. Petroleum, gas, and some sulphur are extracted by techniques which are different from those which would normally be put under the heading "mining". Any mining operation comprises four stages: prospecting, exploration, development and exploitation. Once a valuable deposit has been found and delineated, decisions are made on modes of entry, subsidiary developments and removal techniques. In underground mining, various methods are employed to obtain minerals buried deep in the ground. Access may be through oblique or vertical shafts or horizontal tunnels. Cross cuts are made at various levels and the ore body is divided into blocks by vertical "rises" that connect different levels. The ore is broken up by the use of hand tools, blasting or machinery. Coal, potash and rock salt are generally mined by the room and pillar method in which haulageways open into rooms supported by tenant rock. Underwater mining is done by the use of large conveyor belts which are capable of bringing to the surface many tonnes of material each minute. The parent ships sift the material and deposit the unwanted material overboard. *See also* MM pp.76–77, 98–99; PE pp.132–136.

Ministerial responsibility, British constitutional doctrine which holds that ministers in a government are answerable for their actions and the legislation which they introduce to the House of Commons. The idea evolved gradually in Britain in the 18th and 19th centuries, along with the rise of the CABINET system. It meant that ministers were no longer the servants of the Crown, except in name only, but the servants of the Commons. As a result a government defeated on a major issue in the Commons usually feels compelled to ask for a dissolution of Parliament or to resign. The idea has spread to other countries with a parliamentary system based on the British model.

Mink, small semi-aquatic mammal of the WEASEL family with soft, durable water-repellent hair of high commercial value. Old World minks (*Mustela lutreola*) resemble American minks (*Mustela vison*), but do not have such valuable fur. They have long, slender bodies, short legs, and long, bushy tails. Wild minks have dark brown fur with long black outer hair. Ranch mink have been bred to produce skins of silver, pastels and variations of natural shades. They eat fish, rodents and birds. Length: to 73cm (29in) including the tail; weight: 1.6kg (3.5lb). Family Mustelidae. *See also* NW p.*207*.

Minkowski, Hermann (1864–1909), Russian-German mathematician who contributed to the mathematical foundations of EINSTEIN's work on RELATIVITY and was the discoverer of space-time.

Minneapolis, city and port of entry on the Mississippi River, in SE Minnesota, USA; largest city in Minnesota. Settled in *c.*1839, it was developed as a timber centre. When the plains were planted with wheat and the railways were constructed, the city became and remains a leading milling centre. Minneapolis is the seat of the University of Minnesota (1851). Industries include farm machinery, food processing, electronic equipment, printing and publishing, fabricated metals and textiles. Pop. (1970) 434,400.

Minnelli, Liza (1946–), US singer and actress, the daughter of Judy GARLAND and Vincente MINNELLI. Her films include *Charlie Bubbles* (1967), *Tell Me That You Love Me, Junie Moon* (1970), *Cabaret* (1972) and *New York, New York* (1977).

Minnelli, Vincente (1913–), US film director and producer whose best work has been in musicals, notably *Meet Me in St Louis* (1944), *An American in Paris* (1950), *The Band Wagon* (1953) and *Gigi* (1958). His other films include *Tea and Sympathy* (1956) and *Some Came Running* (1958).

Minnesingers, medieval German poets or singers of courtly love (or Minne), similar in style to the Provençal TROUBADOURS whom they copied originally. An individual German style developed in the 14th century however, and several of the poems are considered the best of Middle High German lyric verse.

Minnesota, state in N central USA, on the Canadian border. The terrain varies from the prairies of the S to the forests of the N. There are mountains in the E. The state is drained chiefly by the Minnesota, St Croix and Mississippi rivers. Wheat and maize are the major crops, grown on the prairies, and there are many farms raising dairy cattle. Since the 1950s manufacturing has replaced agriculture as the main economic activity. Food processing, electronic equipment, machinery, paper products, chemicals, printing and publishing are the chief industries. There are rich deposits of iron ore in the Mesabi Range in the E. The most important cities are St Paul (the state capital), Minneapolis and Duluth.

French fur traders entered the region in the 17th century. The area E of the Mississippi passed to Britain after the SEVEN YEARS WAR, then to the USA after the American War of Independence. The lands W of the Mississippi were acquired from France in the LOUISIANA PURCHASE of 1803. The Minnesota Territory was organized in 1849. Homesteaders from Germany, Norway and Sweden settled the area during the 1880s, after which the state grew rapidly. Area: 217,735sq km (84,067sq miles). Pop. (1975 est.) 3,926,000. *See also* MW p.175.

Minnow, subfamily of freshwater fish found in temperate and tropical regions. The family includes shiners, dace, chub, tench and bream. More specifically, the term includes small fish of the genera *Phoxinus* and *Leuciscus*. Length: 3.8–45.7cm (1.5–18in). Family Cyprinidae. *See also* NW p.*230*.

Minoan civilization (flourished *c.*3000–*c.*1200 BC), ancient Cretan culture, named after the legendary King Minos. Its capital, Knossos, was twice destroyed by earthquakes and rebuilt. Its palaces were among the finest in the world, particularly those at Knossos, Phaestus, Gournia and Cyclonia. The civilization has been divided into three periods: early (*c.*3000–2100 BC), middle (*c.*2100–1550 BC) and late (*c.*1550–*c.*1100 BC). The whole period saw the development from the use of neolithic implements to an advanced system of mercantile trade which, at the height of Minoan influence, extended throughout the Mediterranean area. Two scripts were developed, the "A" and the "B", and a distinctive pottery has since been found throughout the island; the Minoans also became skilled in metalworking. The end of the civilization was probably caused by an earthquake, followed by an invasion by the Myceneans. *See also* MS pp.210, 246, *246*; HC1 pp.38–40, *38–39*.

Minor, in English law, person under 18 years of age. A minor cannot vote at elections, nor hold freehold or leasehold property, nor be made bankrupt. He cannot make a valid will nor a valid marriage without the consent of his parent or guardian. Contracts entered into by a minor may be repudiated by him up to a reasonable time after reaching his majority.

Minor, in music, two kinds of diatonic scale, in distinction to the MAJOR scale. In the harmonic minor scale, the intervals of the third and the sixth above the tonic are minor intervals, a semitone lower than those corresponding in the major scale. The other interval series is the melodic minor scale, which raises the sixth degree of the harmonic minor scale by one semitone in the ascending form and lowers both the seventh degree (leading note) and the sixth degree in the descending form. *See also* KEY.

Minorca, Spanish island, second largest of the Balearic Islands, in the Mediterranean Sea, approx. 40 km (25 miles) NE of Majorca; the capital is Mahón. Occupied at different times by the Carthaginians, Romans and Vandals, it was conquered by James I of Aragón in the 13th century, captured by the British in 1709 and taken by the French in 1756. It was returned to Spain by treaty in 1783. Its products include fruit and vegetables. Tourism is a valuable source of income. Area: 702sq km (271sq miles). Pop. (1970) 50,217.

Minos, in Greek mythology son of EUROPA and ZEUS, King of Crete and a famous law-giver, consigned at his death to HADES to judge human souls. He angered POSEIDON who, in revenge, caused the king's wife Pasiphaë to mate with a bull. She duly gave birth to the monstrous MINOTAUR, half man, half bull, who was kept in the LABYRINTH. *See also* THESEUS.

Minot, George Richards (1885–1950), US physician. He shared the 1934 Nobel Prize in physiology and medicine with George Hoyt Whipple and William Parry Murphy for discovery of a liver therapy diet to combat pernicious ANAEMIA.

Minotaur, in Greek mythology, beast with the head of a bull and the body of a man, the issue of Pasiphaë, wife of MINOS, and a bull. He was confined by Minos in the LABYRINTH built by DAEDALUS. A tribute of seven youths and seven maidens, who were to be fed to the minotaur, was exacted by Minos from many Greek cities. The minotaur was killed by THESEUS.

Minsk, capital city of Belorussia (Belorusskaja SSR), USSR, on the River Svisloč, 644km (400 miles) WSW of Moscow. Founded *c.*1060, it became part of Russia in 1793. Industrial development began in the 1870s and Minsk is now one of the largest manufacturing centres in the country. Products include textiles, machinery, and cement. Pop. (1970) 916,000.

Minster, old name for a monastery church, abbey or priory church. The term is often retained after the disappearance of the monastic foundation; an example is Westminster Abbey. It may also apply to a large church or cathedral, such as York Minster, which has no traceable monastic connections.

Minstrels, itinerant musicians and professional entertainers; more specifically, secular musicians, usually instrumentalists. The term minstrel replaced the earlier Provençal JONGLEUR in the 14th century. Minstrels were popular from the 12th to 17th centuries, and some were attached to courts. Most of their music was based on memory and improvization, and consequently little of their music has survived. In the 14th and 15th centuries, their social importance is reflected in the number of minstrel guilds which were formed throughout Europe. In the 20th century singers and banjoists with blackened faces have been known as minstrels.

Mint, in botany, any of numerous species of aromatic herbs, with a characteristic flavour, of the genus *Mentha*. It is commonly used as a flavouring in cooking, confectionery and medicines. The leaves may be used whole, chopped or dried, or an extract added for its flavour. Most species have oval leaves and spikes of purple or pink flowers. Family Labiateae. *See also* PEPPERMINT; NW p.*59*; PE pp.*212–213.*

Mint, Royal, British government department responsible for providing money for circulation in the United Kingdom. A mint had existed in London since Anglo-Saxon times. In about 1300 it was situated in the Tower of London, and in 1325 its officers were incorporated by Royal Charter. Contractors continued to be employed until 1851, when the Mint became a government department. In 1810–11 it was removed to Tower Hill but since 1968 has been located at Llantrisant, near Cardiff, Wales, manufacture beginning in 1969 – although Tower Hill continued to produce coins until 1975. The Chancellor of the Exchequer is *ex officio* Master of the Mint, which is administered on a daily basis by the Deputy Master and the Royal Mint Advisory Committee.

Minto, Gilbert John Elliot-Murray-Kynynmound, Earl of (1845–1914), British statesman. He was Governor-General of Canada (1898–1904) and Viceroy of India (1905–10). With John MORLEY he began the change to Indian self-rule in their reforms of 1909.

Mintoff, Dominic (1916–), Maltese politician. He helped to reorganize the country's Labour Party in 1944 and was first elected to the legislative assembly in 1947. He became leader of the party in 1949 and was first Prime Minister in 1955–58. He was again elected Prime Minister in 1971.

Minton, Thomas (1765–1836), first of a famous family of potters, born in Shropshire. He worked as an engraver with SPODE in London before settling in Stoke-on-Trent, where in 1793 he founded his own pottery. He was the first to engrave the famous willow pattern.

Minuet, French dance originally from Poitou, fashionable at the court of LOUIS XIV from 1650. It is played in triple time, and consists of graceful and precise glides and steps. It became popular as a dance in the 18th century and was a familiar movement in the SUITES of composers such as HANDEL and MOZART.

Minutemen, colonial militiamen or armed civilians who, in the War of American Independence, agreed to muster at a minute's notice. The term is used especially for those men who were enrolled in 1774 for this purpose by the Massachusetts provincial congress and who fought at LEXINGTON and CONCORD (1775).

Miocene, geological epoch beginning about 26 million and ending about 7 million years ago. It falls in the middle of the Tertiary period and is marked by an increase in grasslands over the globe at the expense of forests and the development of most of the modern mammal groups. No British Miocene deposits are known, although some folding took place caused by the rising of the Alps to the south. *See also* NW p.191.

Miquelon Islands. *See* SAINT PIERRE AND MIQUELON.

Mirabeau, Honoré Gabriel Riqueti, Comte de (1749–91), French political figure who became leader of the Third Estate (bourgeoisie) in 1789. He later tempered his opposition to the court and was President of the National Assembly when he died.

Mira Ceti (Omicron Ceti), red giant star in the constellation of Cetus. It is a long-period variable star with an average period of 331 days, and a range of magnitude from 1.7 to 4 at maximum to 10 at minimum. Mira Ceti is the most celebrated long-period variable. *See also* SU pp.239, *239,* 262–263, *263.*

Miracle, sudden and inexplicable change in the physical circumstances of either a person or an object. Nearly always beneficial to someone, miracles are attributed to the power of God, although in both Christianity and Judaism a human agent is generally involved (eg Jesus or the Old Testament prophets). Many scientific "explanations" of miracles, particularly those of the Old Testament, ignore the fact that for almost every miracle recorded the precise timing of the event, and its significance, was critical for the people concerned.

Miracle in the Gorbals (1944), one-act ballet set in the Gorbals slums of Glasgow. The music is by Sir Arthur Bliss and choreography by Sir Robert Helpmann, and it was first performed in London by the Sadlers Wells Ballet, with sets designed by Edward Burra. The "miracle" occurs when Christ is reincarnated in modern society to experience again His Passion and death.

Miracle plays. *See* MYSTERY PLAYS.

Miraculous Mandarin (1919), pantomine ballet written by Béla BARTÓK; it was first performed in 1926 in Cologne. Stylistically, it owes much to the ballets of STRAVINSKY, but Bartók was already using folk elements in his music.

Mirage, image (often inverted) sometimes seen near the Earth's surface when light is refracted (bent) as it passes between cool dense air to warmer, less dense air. Mirages are most commonly seen as a shimmering on hot, dry roads; the shimmer is a refracted image of the sky. On rare occasions objects such as buildings and trees, situated over the horizon, can be seen as a mirage. *See also* SU p.*104.*

Miranda, Carmen (1904–55), US film actress, dancer and singer. She was famous for her bizarre costumes, particularly her headgear, which was often fashioned from several layers of fruit. Her films include *That Night in Rio* (1941) and *Copacabana* (1947).

Miriam, in the Bible, sister of MOSES and AARON. She was a prophetess, and is said to have been the sister who watched over the baby Moses until he was found in the bullrushes by the pharaoh's daughter. She became leprous for criticizing the marriage of Moses to a Cushite woman, but was later healed.

Miró, Joan (1893–), Spanish painter and graphic artist, a leading exponent of abstract SURREALISM. Many of his works are extremely delicate and simple, free patterns of shapes and lines, influenced by Paul KLEE, and executed with wit and imagination. Famous among his paintings are *Harlequin's Carnival* (1925) and *Women and Bird in the Moonlight* (1949).

Mirror, highly polished surface which, because of the laws of REFLECTION, produces an image of objects in front of it. Most mirrors are made of glass "silvered" on one side with silver, mercury or aluminium. They can be flat (plane) or curved (spherical or paraboloidal). Plane mirrors produce an inverted, virtual image that is unmagnified. Spherical mirrors may be concave (caving inwards) or convex (bulging outwards). The image can be rightway-up or inverted, real or virtual, depending on the position of the object in relation to the focal point of the mirror; it may also be either magnified or reduced in size. A spherical mirror suffers spherical ABERRATION which is absent in a concave paraboloidal mirror, as used in reflecting telescopes. *See also* SU pp.100–101.

Misanthrope, The (1666), psychological comedy by MOLIÈRE. A drawing-room discussion of openness and plain speech turns against its protagonist, because plain speech, without good temper, proves worse than mere politeness. In the original production Molière himself played the part of Alceste, the misanthropist, and his wife Armande took the part of Celimène, the woman he loves.

Miscarriage, common term for a natural (spontaneous) ABORTION, the loss by a woman of a fetus from her womb before it is sufficiently developed to survive.

Miscegenation, interbreeding of human beings of different racial origins. Although modern anthropology has largely discounted myths about its injurious genetic effects, it is still taboo in many societies and in some it is illegal.

Miser, The (1668), prose comedy by MOLIÈRE which was censured by ROUSSEAU as immoral. It demonstrates the follies of avarice through the story of an old miser, Harpagon, who for financial reasons interferes with his children's marriage plans. The plot turns on a scene in which Harpagon is robbed of his money.

Misérables, Les (1862), novel by Victor HUGO. The central character is a convict, Jean Valjean, who has been sentenced to the galleys for stealing bread for his sister's starving family.

Misericord, in church architecture, a wooden bracket attached to the underside of a hinged seat in the choir stalls. When the seat is turned up, the misericord projects outwards providing support for a choir member or priest during a lengthy performance. Common in medieval churches, they were frequently humorously or grotesquely carved.

Mishima, Yukio (1925–70), Japanese author. His novels, short stories, plays and essays reveal his obsession with beauty and violence. An early novel, *Confessions of a Mask* (1949), is a partially autobiographical study of homosexuality. His final work, the four-volume *The Sea of Fertility* (1970), is an epic of modern Japan. He committed ritual suicide in 1970 at Tokyo's military headquarters. *See also* HC2 p.*291.*

Mishnah, collection of Jewish legal traditions and moral precepts that form the basis of the TALMUD. The Mishnah was compiled in about AD 200 under Rabbi Judah ha-Nasi. It is divided into six parts: laws pertaining to agriculture; laws concerning sabbath, fasts and festivals; family laws; civil and criminal laws; laws regarding sacrifices; and laws concerning ceremonial regulations.

Miskolc, city in NE Hungary, on the River Sajó. An important industrial centre, it has a variety of industries including iron-and steel-milling and mining for lignite. Miskolc is also the site of a 13th-century Gothic church. Pop. 193,000.

Missal, Roman Catholic service book, containing the text to be said or sung by priests, choir and congregation during the celebration of MASS throughout the year.

Missiles, unmanned and self-propelled flying weapons, powered by ROCKET, RAMJET or TURBOJET. Ballistic missiles travel in the outer atmosphere and can be powered only by rockets. Cruise missiles travel in the lower atmosphere and can be powered by jet engines. GUIDED MISSILES carry self-contained guidance systems or can be controlled by radio from the ground. Unguided missiles are freeflying, with no control other than initial aim and amount of fuel. *See also* FLYING BOMB; MM pp.169, 175, 178–179.

Missionary Societies, organizations for the promotion of Christianity among non-Christians. The first such society was established in New England in 1649, but of greater importance was the Society for Promoting Christian Knowledge (1698), which was helped in its foreign commitments by the Society for the Propagation of the Gospel in Foreign Parts (1701). In the 18th century both the Baptists and Methodists established societies, and the 19th century saw the emergence of interdenominational and geographically specialized societies. The International Missionary Council was formed in 1921. *See also* HC2 pp.138–139.

Mississippi, state in S central USA, on the Gulf of Mexico. The land slopes down from the hills of the NE to the Delta, a fertile plain between the Mississippi and Yazoo rivers. Pine forests cover most of the S section of the state as far as the coastal plain. Primarily an agricultural state, Mississippi is the leading producer of cotton in the USA; hay and soya beans are also grown. Dairy farming is of great importance. There are valuable reserves of oil and natural gas. The TENNESSEE VALLEY AUTHORITY has helped to introduce new industries to the area, including clothing, wood products, and chemicals. The major cities are Jackson (the state capital), Meri-

Royal Mint: the complex at Llantrisant has a police post at the main entrance.

Miro's ironic outlook can be seen in his *Women and Bird in the Moonlight.*

The Misanthrope featured a self-righteous fool, played by Molière himself.

Missile; an intercontinental ballistic missile rising from its 160-feet-deep silo.

Mississippi: an aerial view of the mighty river between Arkansas and Mississippi.

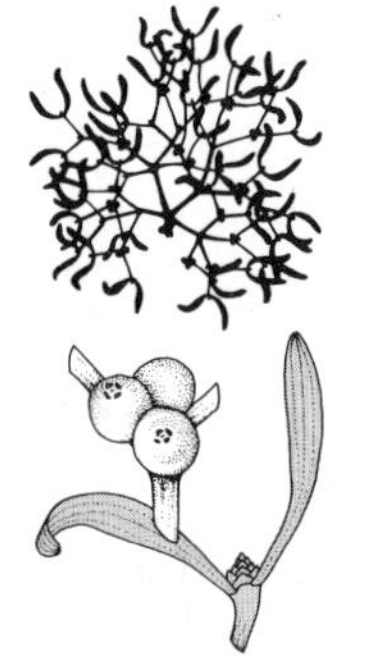

Mistletoe, with its romantic associations, was a holy plant to the Druids.

Frédéric Mistral's *Miréio* was awarded the poet's prize of the French Academy (1859).

Reginald Mitchell's Spitfire forms part of a "Battle of Britain" week display.

dian, Biloxi, Vicksburg and Laurel.

The French claimed the region in 1682, but it passed to Britain after the SEVEN YEARS WAR. The Territory of Mississippi was organized in 1798. The state seceded from the Union in 1861, and Jefferson DAVIS, a Mississippi state senator, became president of the CONFEDERATE STATES. The state was a battleground during the AMERICAN CIVIL WAR and the damage to the economy took decades to repair. Area: 123,584sq km (47,716sq miles). Pop. (1975 est.) 2,346,000. *See also* HC2 pp.150, *151*; MW p.175.

Mississippi, river in central USA. It rises in NW Minnesota and flows SE, forming many state boundaries along its course, to enter the Gulf of Mexico in SE Louisiana. With the MISSOURI, it forms the third-longest river system in the world. Its chief tributaries include the Ohio, Missouri, Arkansas and Tennessee rivers. A major transportation route for raw materials and finished goods, it is connected to the GREAT LAKES and the ST LAWRENCE SEAWAY (N) and the Intracoastal Waterway (E). Discovered by Hernan De SOTO in 1541, it was acquired by the USA in the LOUISIANA PURCHASE of 1803. Since the 1950s improvements have been made to the river's channels, enabling bulkier freight to be transported. The river's upper course freezes over during winter months and springtime flooding is still a major problem. Length: approx. 3,780km (2,348 miles). *See also* PE p.100.

Mississauga, residential and industrial suburb of Toronto in SE Ontario, Canada, at the W end of Lake Ontario. It is also a transport centre, and has port facilities and Toronto's international airport. Industries include aircraft, steel, petroleum products, rubber and chemicals. Pop. (1971) 156,070.

Miss Julie (1888), naturalistic one-act play by August STRINDBERG. It explores with great insight a duel in rank and sex between Fröken Julie and the butler-valet of her country house during one Midsummer Eve's Night. First performed in 1889, it runs continuously for 90 minutes with a balletic interlude as part of the action.

Missolonghi (also called Mesolóngion), town in W central Greece, on the Gulf of Patras. It was an important stronghold in the Greek War of Independence, and eventually fell to Ottoman troops after two sieges (1822–23 and 1825–26). The British poet, BYRON, who sympathized with the Greek fighters, died there of fever in 1824.

Missouri, state in central USA, W of the Mississippi River. To the N of the Missouri River is prairie country, where farmers grow corn and raise livestock; s of the river are the foothills and plateau of the Ozark Mts. In the SW is a small wheat-growing area of the Great Plains, and in the SE are the cotton fields of the Mississippi floodplain. Other farm produce includes soya beans, hay and pigs. Mineral resources include coal, lead, zinc and iron ore. Missouri's economy is based on manufacturing. Major industries are transportation equipment, food processing, chemicals, printing and publishing, fabricated metals and electrical machinery. The state capital is Jefferson City and the largest cities are Saint Louis, Kansas City and Springfield.

The French were the first to settle the area in the mid-18th century. The USA acquired the region in the LOUISIANA PURCHASE of 1803. The Louisiana Territory was organized in 1812 and during the next few decades became a main corridor of westward migration. Missouri was admitted to the Union in 1821 without restrictions on slavery, but when the AMERICAN CIVIL WAR began sympathies were bitterly divided and there was much violence. The state remained in the Union. Area: 180,455sq km (69,674sq miles). Pop. (1975 est.) 4,763,000. *See also* HC2 p.*150*; MW p.175.

Missouri, river in central and NW central USA. It rises at the confluence of the Jefferson, Madison, and Gallatin rivers in s Montana, flows E to central North Dakota, s across South Dakota, and E across Missouri to join the Mississippi River just N of St Louis. It is the chief tributary of the Mississippi, and the longest river in the USA. Sioux City, Iowa, is the head of navigation, and flooding is a serious problem. Explored by the LEWIS AND CLARK EXPEDITION from 1804 to 1806, the river was thereafter used by traders, gold seekers and pioneers as a route to the NW. Length: 4,320km (2,683 miles).

Missouri Compromise (1820), dispute in the US Congress over the accession of Missouri, a slave state, to the Union. To prevent a majority of slave states in the USA, Missouri was only permitted to join at the same time as the free state of Maine. Slavery was prohibited in the rest of the LOUISANA PURCHASE N of Missouri.

Mr Midshipman Easy (1836), novel by Capt. Frederick MARRYAT, describing the adventures of a young MIDSHIPMAN who suffers only the mildest of naval discipline. The work is as a children's classic.

Mistinguett (1875–1956), French singer and dancer, originally named Jeanne-Marie Bourgeois. The star of the Casino de Paris, she became famous for her performances at the Folies Bergère and the songs *J'en ai marre* and *Mon Homme*.

Mistletoe, any of numerous species of evergreen plants that are semi-parasitic on tree branches. It has small, spatulate, yellowish-green leaves and generally forms a large dense ball of foliage. The small yellow flowers produce waxy, white berries, which are poisonous (except to some species of birds). In Britain, Europe and North America, mistletoe retains a ritualistic significance especially at Christmas and midsummer. Families: Loranthaceae and Viscaceae.

Mistral, Frédéric Joseph Etienne (1830–1914), French writer. He was one of the founders of the Félibrige, an organization set up to promote Provençal as a literary language. He published many works in the *langue d'oc*, including novels, memoirs, plays and his poems *Les Isles d'Or* (1875). His poetry is characterized by beautiful language rather than strong thought. He founded the Arles folklore museum in 1899 and shared the Nobel Prize in literature with ECHEGARAY in 1904.

Mistral, Gabriela (1889–1957), Chilean poet, who helped José VASCONCELOS in his revision of the Mexican school system. Her poetry shows the move to simplicity after the exoticism of MODERNISMO. Her three main collections are *Desolación* (1922), *Tala* (1938) and *Lagar* (1954). In 1945 she became the first Latin American to receive the NobelhPrize in literature.

Mistral, N and NW wind prevalent in the NW Mediterranean in the winter. It sweeps from the MASSIF CENTRAL, France, down the Rhône valley reaching the Rhône delta as a strong, dry wind. Orchards and gardens in the area are usually protected against the mistral by cypress hedges.

MIT (Massachusetts Institute of Technology) university in Cambridge, Massachusetts, USA. It is recognized as a leading technical college with facilities which include a large nuclear reactor, a high-energy accelerator and a nuclear science laboratory. There are more than 70 special laboratories in Massachusetts and other states either directly associated or affiliated to MIT.

Mitchel, John (1815–75), Irish revolutionary, lawyer and journalist. He called for an Irish rebellion in his publication the *United Irishman* and was transported to Australia. He escaped to the USA in 1853 and in 1874 returned to Ireland where, in the following year, he was elected to the British Parliament but was declared ineligible and died shortly afterwards.

Mitchell, Arthur. *See* DANCE THEATRE OF HARLEM.

Mitchell, Joni (1943–), Canadian singer and songwriter. She moved from reflecting contemporary California in recordings such as *Clouds* (1969) and *Ladies of the Canyon* (1970) to probing the depths of the problems raised by personal relationships in *Court and Spark* (1974). In *Hejira* (1976) she considered the conflict between the desire for freedom and the yearning for warmth and stability.

Mitchell, Margaret (1900–49), US novelist, famous for her one novel, *Gone with the Wind* (1936; Pulitzer Prize 1937) about the AMERICAN CIVIL WAR period. The film adaptation (1939) was extremely successful. *See also* HC2 pp.268, *269*.

Mitchell, Maria (1818–89), US astronomer. While studying sunspots and nebulae in 1847, she discovered a new comet. She was the first woman to be elected to the American Academy of Arts and Sciences (1848) and the first professor of astronomy at Vassar College (1865–88).

Mitchell, Reginald Joseph (1895–1937), British aircraft designer, best remembered for his work on the Spitfire fighter of WWII. He worked for Vickers-Armstrong (Supermarine) and led a team in designing a number of seaplanes which won the SCHNEIDER TROPHY (1927, 1929 and 1931). *See also* MM pp.148–149, *148–149.*

Mitchell, Sir Thomas Livingstone (1792–1855), Australian explorer, *b.* Scotland. As Surveyor-General of New South Wales (1828–55), he led four expeditions into the interior and in 1836 discovered the fertile Western District of Victoria.

Mitchison, Naomi Margaret (1897–1964), British novelist. She wrote various kinds of fiction, but is best known for her historical novels, such as *The Conquered* (1923), and her science fiction, such as *Memoirs of a Spacewoman* (1962).

Mitchum, Robert (1917–), US film actor. He started his career with small parts in cowboy films and was featured in *The Dancing Masters* (1943), but the 1950s saw his best performances. His relaxed insouciance in *The Lusty Men* (1952) and religious fanaticism in Charles Laughton's *The Night of the Hunter* (1955) reveal Mitchum's range of ability. He was also acclaimed for his performance in *Farewell My Lovely* (1975).

Mite, minute ARTHROPOD found worldwide, many as parasites on plants and animals. The larva has three pairs of legs; the adult has four pairs with claws at the tip, and the head and abdomen are fused. Length: 0.5–3mm (0.02–0.1in). Class Arachnida; order Acarina. *See also* CHIGGER; RED SPIDER; TICK; NW pp.103, *104*, *230.*

Mitford, name of six British sisters, the daughters of Lord and Lady Redesdale (family name Mitford). The most famous, Nancy Mitford (1904–73), was a novelist and biographer. Her first successful novel was *The Pursuit of Love* (1945), which was followed by *Love in a Cold Climate* (1949) and *The Blessing* (1951). Her co-editing of a volume of essays, *Noblesse Oblige* (1956), brought her more fame and led to the practice of labelling speech "U" or "non-U". Her sister Jessica Mitford (1917–) also wrote, protesting about snobbery and privilege in particular and humbug generally, as in *The American Way of Death* (1963) and *Kind and Usual Punishment* (1973). The other two notable sisters, Unity Valkyrie (1914–48) and Diana (1910–) turned to FASCISM; the former travelling to Germany and becoming a disciple of Adolf HITLER and the latter marrying Sir Oswald MOSLEY.

Mitford, Mary Russell (1787–1855), British author. She became popular with the publication of a collection of rural sketches, *Our Village* (five vols, 1824–32). She also wrote a play in blank verse, *Rienzi* (1828).

Mithraism, worship of the Persian god Mithra (Mithras to the Romans) of the sun, justice, contract and war. It became widespread in the Roman Empire during the 2nd century AD, especially among soldiers and slaves. It was a mystical religion with secret rites involving the sacrifice of bulls. Mithraism was suppressed when the Roman emperors adopted Christianity. *See also* MS p.212.

Mithridates VI Eupator, called the Great, King of Pontus (132–63 BC; *r.* 120–63 BC), who attempted to extend his rule southwards but was repeatedly defeated by the Romans. He was overwhelmed by the forces of SULLA in the war of 88–85 and lost his kingdom in a second campaign in 83–82. He reconquered it in 74 but was defeated by POMPEY in 66 and fled to the

Bosphorus. He was planning an invasion of Italy, but his troops revolted, and he committed suicide. *See also* HC1 p.96, *96*.

Mitochondrion, small body found in the cytoplasm of most CELLS, containing enzymes and other agents necessary for energy production.

Mitosis, nuclear division of a CELL resulting in two identical cells. This division includes five phases: interphase (DNA and CHROMOSOMES divide), prophase (chromosomes coil, centrioles start to separate, astral rays and spindles form and the nuclear membrane disappears), metaphase (doubled chromosomes form along the cell's equatorial plate, centrioles move to opposite poles), anaphase (doubled chromosomes separate to opposite poles of the cell and cytoplasmic division starts) and telophase (chromosomes uncoil, nuclear membrane forms, spindle and astral rays disappear). *See also* MEIOSIS.

Mitra, Mithra or Mithras, god who in different forms was worshipped in India, Persia and then the Roman Empire. Originally, in Vedic mythology, Mitra was the spirit of the day, of the rain and of the sun, linked closely with VARUNA, ruler of the night in the administration of justice. The Persian Mithra was a popular deity of the ACHAEMENID Empire, revered as the god of light and power. *See also* MITHRAISM.

Mitre, tall head-dress worn by archbishops and other prelates. It is derived from the papal tiara, and has been worn since about the 11th century.

Mitscherlich, Eilhardt (1794–1863), German chemist who in 1819 discovered that compounds of similar composition tend to have the same crystal structure – the phenomenon known as ISOMORPHISM. This helped in the understanding of the relationship between the molecular composition and crystalline structure of compounds. In 1823 he discovered the monoclinic form of sulphur and in 1827 discovered selenic acid. He named benzene and was the first to synthesize nitrobenzene (1832); he was also among the first to recognize catalysis.

Mitsubishi, Japanese industrial and banking conglomerate set up in 1873. Despite opposition in the 1930s, Mitsubishi has become one of the foremost businesses of Japan, and has electronic, mining, chemical, oil and banking interests. Its head offices are situated in Tokyo.

Mitterand, François Maurice Marie (1916–), French politician. He was active in the French Resistance during WWII and served as Minister for ex-Servicemen (1947–48). He became president of the Federation of the Democratic and Socialist Left in 1965 and leader of the Socialist Party in 1971.

Mix, Tom (1880–1940), US actor and writer-director of silent Western films. Mix and his horse Tony were famed as the heroes of many adventure films, including *Tom Mix in Arabia* (1922), *North of Hudson Bay* (1924) *Destry Rides Again* (1927) and *The Last Trail* (1927).

Mixed economy, system in which part of the gross national product (GNP) is produced by the nationalized industries (the public sector) and the remainder by privately controlled companies (the private sector). *See also* MS pp.*316, 317*.

Mixtec, Central American Indian civilization centred in the province of Oaxaca and in part of Guerrero, Mexico. During the 9th century the Mixtec conquered the neighbouring ZAPOTEC and until the Spanish conquest it was the dominant culture in the Oaxaca and Puebla regions. Due to epidemics and forced migrations, however, their numbers were drastically reduced; today there are about 250,000 descendants.

Miyagi-ken. *See* SENDAI.

MKS units, system of units based on the metre, kilogramme and second. *See* WEIGHTS AND MEASURES.

MLR (minimum lending rate), lowest rate of interest that banks are permitted to charge their borrowing customers. In Britain the rate is fixed by the Treasury in association with the Bank of England, the central bank, and is a means of controlling the flow of currency, supply and demand and the rate of inflation.

Mnemonic device, an aid to MEMORY. Systems designed to improve memory include key words, rhymes and visual imagery. An example of a mnemonic is the phrase "Richard of York gave battle in vain", the initial letters of which are an aid to remembering the colours of a rainbow (red, orange, yellow, green, blue, indigo, violet). *See also* MS pp.44–45.

Mnesicles (5th century BC), Greek architect. He designed the Propylaea, or western entrance to the ACROPOLIS, using pentelic marble and black Eleusian stone. He adjusted the Propylaea to the slope of the Acropolis, and because the Erectheum is similarly handled, it is also sometimes ascribed to him.

Moa, extinct, ostrich-like bird of New Zealand. It is believed that the Maoris hunted it to extinction in the 17th century. It was a flightless bird and had a small head, a long feathered neck and a bulky body; the legs were long and strong. It fed primarily on roots and shoots and was ground-nesting. Height: 3.5m (12ft); species *Dinornis maximus. See also* PE p.*144.*

Moab, nation in hills E of the Dead Sea during biblical times. According to the Bible the Moabites spoke a language similar to Hebrew and were of the same race. Apart from the Bible, the main source of information about them is the Moabite Stone, a set of inscriptions discovered in 1868. Moabite culture lasted from *c.*1400 BC to *c.*580 BC, when they were conquered by the Babylonians.

Moa hunters, pre-Maori culture of South Island, New Zealand. Probably Eastern Polynesian in origin, little is known about them but that they hunted the MOA; the culture was discovered in 1939.

Mobile, form of sculpture developed by Alexander CALDER in the early 1930s and named by Marcel DUCHAMP. A mobile consists of a series of shapes (representational abstract) cut from metal, wood, plastic or other materials and connected by wires so that both the parts and the whole revolve freely when suspended in space. This form breaks with the traditional idea of sculpture as a static object.

Möbius strip, shape or figure formed by giving a strip of paper a half-twist, then joining the ends together. It is of great interest in TOPOLOGY, being a one-sided surface (a line can be drawn along the strip and will cover both sides, returning to the starting-point to meet itself). It was invented by August Ferdinand Möbius (1790–1868). *See also* SU p.53.

Mobutu Sese Seko (1930–), *b.*Joseph(-Désiré) Mobutu, President of Zaire (formerly the Democratic Republic of the Congo). He worked for Congolese independence with Patrice LUMUMBA but deposed him in 1960 and assumed control of the army. In 1965 he led a coup against Joseph KASAVUBU, becoming Prime Minister in 1966 and establishing a presidential government the following year. He changed the Congo's name to Zaire in 1971.

Moby Dick (1851), novel by Herman MELVILLE, considered his masterpiece. Moby Dick is a white whale which is pursued across the Pacific Ocean by Captain Ahab, whose leg the whale had previously bitten off. Melville gives the chase a significance beyond mere externals, the whale becoming a symbol of the awesome forces of the universe.

Mochica, ancient civilization of N Peru, sometimes called "Early Chimu". It dates from the latter part of the Early Intermediate Period (*c.*200 BC–AD 600). Named for their home area in the Moche Valley, these people were famous for their magnificent pottery, large quantities of which have survived.

Mock epic, or mock heroic, literary work that burlesques the grand epic style. Famous examples are *The Rape of the Lock* (1714) by Alexander POPE, in which trivial events are described in highly exalted language; *Dunciad* (1728) also by Pope, a more serious satirical poem; and *Ode on the Death of a Favourite Cat* (1747) by Thomas GRAY.

Mockingbird, any of a group of New World birds, known for imitating other birds. The common mockingbird (*Mimus polyglottos*) of the USA is typical of the group – about 27cm (11in) long, ashy above with brownish wings and tail marked with white. Both sexes build the nest of leaves and twigs. The hen lays pale greenish-blue, brown-spotted eggs. Family Mimidae.

Mockridge, Russell (1928–58), Australian track and road-racing cyclist who won the 1,000m time trial and 2,000m tandem at the 1952 Olympic Games and then won both the amateur and open Grand Prix de Paris. In 1953 he embarked on a successful professional career. He was killed in a collision with a bus during a race.

Model Parliament, English parliament summoned by EDWARD I in 1295. It was called the "Model" parliament by the 19th-century constitutional historian Bishop Stubbs, because at it the knights of the shire and the burgesses (the representatives of the Commons) for the first time treated of the great affairs of the nation with the king and the magnates. This enlargement of the Commons' function was held to be the model for the future. They had previously merely agreed to what the king and the magnates had already decided. It was a change marked in the very language of the writ summoning the parliament, language which, with minor changes, remained the same for the next six centuries.

Moderator, term used in the United Reformed Church and the Scottish Presbyterian Church for various administrative officials. In the United Reformed Church a moderator presides over the General Assembly and is elected annually. It is also a name for one who presides over one of the 12 provinces. This is a permanent position, although renewable every seven years. In a third sense, the term refers to a temporary or interim moderator, who administers local churches without full-time ministers.

Moderator, in nuclear physics, substance consisting of light elements, such as heavy water or graphite, used in fission reactors to slow down the fast neutrons produced by the fission reaction. *See also* MM pp.74, 75.

Modernism, movement in the Roman Catholic Church during the late 19th and early 20th centuries to adapt its beliefs to developments in modern science, philosophy and history. The modernists favoured the application of the critical method to the Bible, acknowledging the limitations of the biblical authors, and objecting to the increasing centralization of the Church authority. The most important of the early modernists was A.F. Loisy, who was dismissed from the *Institut Catholique* in Paris for his views. The movement was initially tolerated by Pope LEO XIII but condemned in 1907 by Pope PIUS X, who later required all suspect clerics to take an anti-modernist oath.

Modernismo, Spanish-American literary movement of the late 19th and early 20th centuries, founded by the Nicaraguan poet Rubén DARÍO. A protest against Europeanized colonial literature and a move towards indigenous South American literature, modernismo was concerned primarily with reforming poetic language and methods of expression.

Modern pentathlon, event of five parts in the OLYMPIC GAMES. The events were originally chosen as being typical of the demands made on an officer bearing a military dispatch through enemy territory in wartime during the 19th century. Thus there is a horse-ride over a 800m cross-country course, a 4,000m cross-country run, a 3,000m swim, foil fencing and pistol shooting (20 shots from 25m at a disappearing silhouette). Rather than trying to win individual events, competitors try to amass points above the set minimum for each test. The competition is both for individuals and teams of three.

Modersohn-Becker, Paula (1876–1907), German painter whose work was concerned with externalizing inner feelings and mood, using simplified colour. She was influenced by such artists as Gaugin and Cézanne. In works such as *Self-portrait with a Camelia* (1907) the overall effect is one of Romanticism.

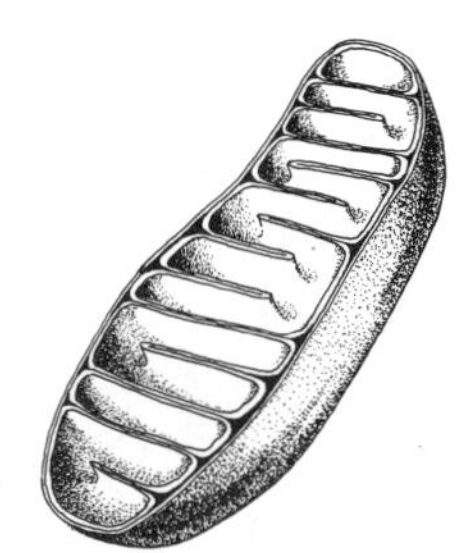

Mitochondrion, present in most cells, is formed of fat, protein and enzymes.

Tom Mix and his horse take part in a pre-war publicity tour of Britain.

Moa: many bones of this extinct New Zealand bird survive in swamps and caves.

Mockingbirds, most versatile songsters, are famed for imitating other birds.

Modes

Modigliani's portrait of Henri Laurens has the usual expressionless features.

Module: the Apollo spacecraft consisted of command, service and lunar modules.

Mogul Empire: the early 17th century drawing shows Akbar helping a farrier.

Mohawk Indian chief Ho Nee Yath No Row was painted by John Verelst in 1710.

Modes, classified scheme, developed during the 4th to 16th centuries AD to systematize the available material of music, and predating harmony. From the scale worked out scientifically by PYTHAGORAS and other Greek thinkers, St Ambrose in the 4th century probably devised four "authentic" modes, the Dorian, Phrygian, Lydian and Mixolydian, all of eight notes within the compass of the octave but different from each other in the arrangement of tones and semitones in between the notes. Pope Gregory (6th century) would seem to have added four "plagal" modes, which were essentially new forms of the Ambrosian modes (Hypodorian, Hypophrygian etc.). Glareanus (16th century) brought the number to twelve, adding the Aeolian and Ionian modes, the respectively, and their "plagal" forms.

Modigliani, Amedeo (1884–1920), Italian painter, sculptor and draughtsman, who studied in Florence and Venice before settling in Paris (1906). He was influenced by the work of TOULOUSE-LAUTREC, CÉZANNE and PICASSO, and by CUBISM and especially African art and his Italian heritage. Modigliani evolved a highly individual style, characterized by its sculptural quality and by the elongation of the head and neck in his portraits and erotic nudes. His superb draughtsmanship (echoing that of BOTTICELLI) and his use of colour (which is close to that of the early SIENNESE masters) make him one of the greatest of 20th-century Italian painters. The legend of his life as a Montparnasse hedonist – drunken and amorous – reinforced by his early death from tuberculosis and drink, overlooks his intense concentration on work in his final years.

Modulation, in physics, the process of varying the characteristics of one wave system in accordance with those of another. It is basic to RADIO broadcasting. In AMPLITUDE MODULATION (AM), the amplitude of a high-frequency radio carrier wave is varied in accordance with the frequency of a current generated by a sound wave. For static-free short-range broadcasting FREQUENCY MODULATION (FM) is used, in which the frequency of the carrier wave is modulated. *See also* MM pp.228–230.

Modulation, in music, the shift from one base key to another by means of a linking passage of chords common to both keys. The term is also applied to natural sounds altered electronically in some contemporary music.

Module, term used in architecture since ancient times. Derived from the Latin *modulus*, it denoted a unit of measure in classical architecture equal to half the width of a column at its base. It was used to achieve good visual proportions. Today it refers to an interchangeable, often mass-produced, unit used in building construction, electronic equipment and spacecraft design.

Modulor, in architecture, system of proportions which LE CORBUSIER advocated in his *Le Modulor* (1949). It is based on the scale of the adult male figure and is used to determine the proportions of building units.

Moe, Jørgen Engebretsen (1813–82), Norwegian poet and folklorist who travelled around Norway collecting tales. The first volume of *Norwegian Folk Tales* appeared in 1841. He also wrote poetry and the children's classic *In the Well and the Pond* (1851).

Mofolo, Thomas Mokopu (1877–1948), African novelist, *b.* Lesotho. He is generally considered to be Africa's first great modern novelist. Praiseworthy among his works are *Pitseng* (1910), *Chaka* (1925) and *The Traveller to the East* (1934), written in the Sesotho language.

Mogadishu, capital city and chief port of Somalia, on the Indian Ocean. Taken in 1871 by the Sultan of Zanzibar, it was leased (1892), then sold (1905) to Italy and occupied by the British in 1941. It is a commercial centre connected to the Gulf of Aden by road. Its main industry is food processing. Pop. 285,000.

Mogok, town in N Burma, 113km (70 miles) NNE of Mandalay. The town was formerly a major ruby and sapphire-mining centre. Pop. (1972) 8,369.

Mogul arts. *See* INDIAN ARTS.

Mogul, or Mughal, Empire (1526–1857), Indian empire founded by BABUR, a Muslim descendant of both TAMERLANE and GENGHIS KHAN, who conquered Delhi and Agra (1526). Under AKBAR (*r.*1556–1605), Babur's grandson, the empire reached its greatest power, extending from Afghanistan to the Bay of Bengal, from Gujarat in the s to northern Deccan. Akbar's grandson, Shah Jahan (*r.*1629–58), built many splendid buildings, including the twin marvels of the Peacock Throne and the TAJ MAHAL. Under AURUNGZEBE (*r.*1658–1707), a Muslim fanatic, Hindu revolts weakened the empire, which began to disintegrate under Muhammed Shah (*r.*1719–48). After 1803 the Mogul emperors were merely puppets of the British. In 1857 the last emperor, BAHADUR SHAH, was forced to abdicate following his part in the INDIAN MUTINY. *See also* HC2 pp.48–51, *48–51*.

Mohacs, Battle of (1526), military engagement in which Louis II of Hungary and Bohemia was defeated by the Ottoman Sultan, SULEIMAN I. The battle marked the beginning of Ottoman domination in Hungary.

Mohair, fibre obtained from the wool of the ANGORA GOAT and used for knitting and weaving yarns. It became important in European textile manufacture in the 19th century, previously being produced only in Asia Minor. Angora goats are now raised in the USA and South Africa.

Mohammad Reza Shah Pahlavi. *See* REZA SHAH.

Mohammed (Muhammad or Mahomet) (*c.*570–632), Arab prophet, leader and the inspiration for ISLAM. He was born in Mecca; his family belonged to the powerful Hashem clan, led by his uncle into whose care he came when his grandfather died, two years after Mohammed had been orphaned. He married a rich older woman, Khadijah, and became a trader. In about 610, Mohammed had a vision commanding him to spread the word of God and he began to preach. He gained followers but also enemies, and in 622 he and many of his followers fled to Medina. Muslims, the followers of Islam, later took this Hegira (*hijrah*, "migration") as initiating the first year in their calendar. Thereafter, Mohammed continued preaching, won more followers, and began to war against his enemies. He conquered Mecca in 630, most of the Arab tribes allied with him, and he became the leading figure in Arabia. In Medina following the death of Khadijah he had married the woman who became his favourite wife, Aishah. After his death, his successors began the rapid conquest of a great empire. Mohammed is revered by Muslims as the Prophet of the one God, Allah, and an ideal man, but he never claimed supernatural powers, and he is not held to be divine. The revelations received by him are recorded in the KORAN, the sacred book of Islam. Orally transmitted sayings of the Prophet are also important guides to the faith and the daily life of Muslims. *See also* HC1 pp.144–145; MS p.220.

Mohammed (name of six Ottoman sultans). *See* MEHMET.

Mohammed V (1909–61), Sultan (1927–57) and King of Morocco (1957–61). A member of the Filali dynasty, he sympathized with the nationalists who wanted freedom from the French. Although deposed and exiled by the French in 1953, he returned in 1955 and was recognized as sultan. He adopted the title of king in 1957.

Mohammedanism. *See* ISLAM.

Mohammed ben Abdullah (1864–1920), Somali leader also known as the "Mad Mullah". In 1899 he proclaimed a JIHAD against the British and Italians, who had annexed large areas of Somalia in the late 1880s. His DERVISH army had considerable success, fighting a protracted war until he died when his base at Taleh was bombed.

Mohave (Mojave), Yuman-speaking tribe of North American Indians who lived in the Mohave Valley of the Colorado River.

In the early 1970s there were about 1,600 members both on and off of the Colorado River Reservation.

Mohawk, Iroquoian-speaking North American Indian tribe of the IROQUOIS CONFEDERACY, formerly inhabiting central New York State, USA. Today there are about 2,000 Mohawk Indians. Most are farmers on two reservations in Ontario, Canada, but some are migrant workers in the US and Canadian steel industries.

Mohegan, tribe of North American Indians of the Algonquian-language family formerly occupying the Thames River area of Connecticut, USA. They constituted the major tribe of the region at the end of the 17th century; their most famous chief was Uncas, although he had been born a PEQUOT Indian. They subsequently sold most of their lands and lived in a small reservation named after them on the Thames River.

Mohenjo-daro, one of the main city centres of the Indus valley civilization (*c.*2300–1750 BC). Remains (in modern Pakistan) show that its inhabitants built large granaries, employed a system of writing, used standard weights and measures, traded extensively, constructed carts and boats and were competent plumbers and metalworkers. *See also* HC1 pp.36, *37.*

Mohican (Mahican), Algonquinian-speaking tribe of North American Indians formerly inhabiting the upper Hudson valley in New York, E of the Housantonic River in Connecticut, USA. They once numbered about 3,000. They joined the Stockbridge Reservation in 1736 and today the surviving 525 Mohicans occupy the Stockbridge-Munsee Reservation in Wisconsin.

Mohl, Hugo von (1805–72), German botanist who did pioneering research into the anatomy and physiology of plant CELLS. He formulated the idea that the cell nucleus is surrounded by a granular, colloidal substance which in 1846 he called PROTOPLASM, although the term had been invented by Jan Purkinje referring to the embryonic material in eggs. Von Mohl was the first to propose that new cells arise from cell division. In 1851 he claimed that the secondary wall of a plant cell is fibrous, and this was subsequently confirmed. He also gave the first distinct account of the function of OSMOSIS and studied the movement of STOMATA.

Moho, also called the Mohorovičić discontinuity, seismic discontinuity layer at the base of the Earth's crust. The layer was discovered by the Yugoslavian geophysicist Andrija Mohorovičić in 1909 when the velocity of seismic waves was measured and found to be different in a number of sites at various distances from an earthquake's epicentre. The Moho is about 10km (6 miles) from the surface in oceanic regions and 35km (20 miles) from the surface under land masses. The Moho is explained by a difference in density of rocks as they change from crust to mantle and therefore act as deflectors of seismic waves. *See also* PE pp.20–21.

Mohole, attempt by US geologists and engineers, begun in the 1950s, to penetrate the Earth's crust to the deeper layers of the mantle, by drilling a borehole at least 10km (6 miles) deep. It was named after the MOHO, the irregular boundary separating the crust and the mantle, which lies between 10 and 35km (6 and 20 miles) deep. Because of rising costs the Mohole was discontinued indefinitely in 1966. *See also* PE pp.20, 21.

Moholy-Nagy, László (1895–1946), Hungarian designer, painter and sculptor. A teacher at the BAUHAUS (1923–28), he created abstract sculptures that cast shadows as they were rotated to show the interplay of light and movement. He often combined transparent materials, such as plastic and Plexiglass, with polished metal discs. *See also* HC2 p.205.

Mohorovičić discontinuity. *See* MOHO.

Mohr, Karl Friedrich (1806–79), German chemist who became the foremost pharmacist of his time in Germany. He improved analytical techniques and invented many items of laboratory equipment. He was also among the first to advance the idea of conservation of energy, in 1837; he became professor of chemistry at the University of Bonn in 1867.

Mohs, Friedrich (1773–1839), German mineralogist. He is best remembered for his scale of the hardness of minerals, the MOHS SCALE. *See also* PE p.99, *99*.

Mohs scale, range of hardness used by geologists to classify the degree of hardness of minerals, devices by the German mineralogist Friedrich Mohs in 1812. The hardest mineral, diamond, has a hardness of 10 on the Mohs scale. It can scratch, or mark, any mineral with a lower Mohs number, including corundum (9); topaz (8); quartz (7); orthoclase (6); apatite (5); fluorite (4); calcite (3); and gypsum (2). Talc (1) can be easily crushed by the fingernail.

Moirae. *See* FATES.

Moiré patterns, series of patterns occurring when a sheet printed with one family of regular curves or lines is superposed on another. The patterns change strikingly as the separate sheets are moved slightly. This effect is used in metal analysis, accurate measurements of flatness, and in OP ART.

Moiseiwitsch, Benno (1890–1963), British pianist, *b.* Russia. In 1904 he studied under Leschetizky in Vienna, and made his London debut in 1909. His tours included the USA, Canada, and Australia as well as many engagements in Europe. He was especially esteemed for his interpretation of Romantic music.

Moiseyev, Igor (1906–), Soviet ballet dancer and choreographer. He danced with the BOLSHOI BALLET in the 1920s and formed his own company in 1936.

Moissan,   Ferdinand-Frédéric-Henri (1852–1907), French chemist. He trained as an apothecary and won international recognition as the first to isolate FLUORINE (1886). He also developed an electric furnace that enabled him to prepare samples of several of the less common metals, such as URANIUM, long before modern metallurgy made them commercially available. He was awarded the Nobel Prize in chemistry in 1906.

Moisture indicator, substance used to detect water vapour, or an instrument (HYGROMETER) used to measure its proportion (ie humidity) in the air. White anhydrous copper sulphate turns blue when it absorbs moisture; paper impregnated with cobolt chloride turns pink. A psychrometer is one type of hygrometer used to measure atmospheric humidity.

Mojarra, any of about 36 species of small, silvery, shallow-water fish found in tropical to temperate seas. They move readily into brackish waters and some species enter fresh water. They have a protrusible mouth that forms a tube when extended, and sheaths of small scales along the bases of the dorsal and anal fins which allow the fins to be depressed into the resulting groove. Family Gerreidae.

Mojave or Mohave, Desert, arid region with low, barren mountains in s California, USA surrounded by mountain ranges on the N and W and the Colorado Desert on the SE. It was formed by volcanic eruptions and deposits from the Colorado River. The desert is the site of the Death Valley National Monument. Area: approx. 38,850sq km (15,000sq miles).

Mokp'o, port in South Chŏlla province, South Korea. The principal industries are fishing and textiles. Pop. (1970) 177,801.

Moksha, "release", term used to denote the Hindu concept of salvation: the release from transmigration by the absorption of the individual self, Atman, into the universal self, BRAHMAN.

Mokuan Reien (*d. c.*1343), Japanese artist. A Zen Buddhist priest, he visited China in 1333 and became the first Japanese to learn the Chinese technique of painting in monochromatic ink. His works include a portrait of Hotei, the god of fortune.

Molar, one of the back teeth of mammals, adapted for grinding and chewing food. Several cusps – the points or ridges on the top surface – occlude with counterparts on the opposing molar. In most herbivores the cusps are fused to form ridges for

grinding plant food. In man there are 12 permanent molars, three on each side of each jaw, a total of 12. *See also* MS pp.130, 131.

Molarity, measure of the concentration of a solution, expressed as the number of moles (gramme-molecules) in a litre of solution. (It is different from molality, which is the number of moles in a litre of solvent.) For example, a solution containing 0.2 moles per litre of solution has a concentration of 0.2 molar (written 0.2M).

Molasses, syrupy by-product of the refining of cane sugar. Canes are cut, crushed, and their juice mixed with water to be boiled and filtered. After some of the sugar has crystallized out, the thick brown syrup remaining is called molasses or treacle. *See also* PE p.197.

Mold, market town in NE Wales, on the River Alyn; administrative centre of Clwyd and of the former county town of Flint(shire). Coal-mining is the chief industry. Pop. (1971) 8,239.

Moldavia (Moldavskaja SSR), constituent republic of the USSR, bounded to the N, E and S by the Ukraine, and to the W by Romania: the capital is Kishinev. It became an ASSR in the Ukraine in 1924 and a constituent republic in 1940. The region is fertile with a mild climate, and grains, fruits and vegetables are cultivated. Industries: food-canning, wine, flour-milling, sugar-refining, tobacco-processing, timber, oil. Area: 33,700sq km (13,010sq miles). Pop. (1976) 3,858,000.

Moldavia, historic region in E Romania, bounded by Wallachia (S), Transylvania (W), the Ukraine (N) and the Moldavian Republic (E), of which most of Moldavia is now part. Suceava, Jassiy, its historic capitals, and Galaţi were the chief cities. Conquered by the Greeks, Romans and Bulgars, and under Mongol rule during the 13th century, the region became a vassal nion of the Ottoman Empire in 1504. MOLDAVIA AND WALLACHIA merged in 1861 to create Romania. Area: approx. 38,047sq km (14,690sq miles). *See also* MW p.147.

Moldavia and Wallachia, historic principalities which united in 1861 to become ROMANIA in the following year. Wallachia was probably founded in the 13th century and Moldavia in the 14th. Although at various times subject to Poland, Hungary, Russia, the Austrian Empire and the Ottoman Empire, both enjoyed comparative independence until, in 1711, they came to be ruled by the Greek governors called the Phanariots. From 1829 they were dominated by Russia, but after the CRIMEAN WAR voted for union and independence.

Mole, any of several species of small, burrowing, mainly insectivorous mammals that live in various habitats throughout the world. The European mole, *Talpa europaea*, has short brown or black fur, a short tail and wide clawed forefeet for digging tunnels. Its eyes are sensitive only to extremes of light and dark; its prey includes mice, worms and small birds. Length: to 18cm (7in). Family Talpidae. *See also* NW pp.*154, 158, 161, 192*.

Mole, basic SI unit of amount of substance. This is the amount of substance which contains as many elementary units as there are atoms in 0.012kg of carbon-12. A mass of one mole of a compound is its gramme molecular weight.

Mole cricket, any of several species of burrowing crickets with a cylindrical body, a pointed head, and forelegs that are modified for shovelling moist earth. The European mole cricket, *Gryllotalpa gryllotalpa*, guards its eggs and young. It feeds on roots and can thus be a pest of growing crops. Length: to 3cm (1.2in). Family Gryllidae. *See also* NW p.*111*.

Molecular biology, biological study of the chemical and physical make-up of molecules comprising living organisms. A major area of study is GENES and their components and functions in HEREDITY, EVOLUTION and the constituents of living organisms. *See also* BIOCHEMISTRY.

Molecular formula, notation for the atoms present in a molecule, written using the

symbols for the atoms with subscripts showing the numbers of atoms present. This is also called the chemical formula of the substance. For instance, C_3H_6O is the molecular formula of acetone. The structural formula gives an indication of the way the atoms are arranged, often by using recognizable groups of atoms, as in CH_3COCH_3 for acetone.

Molecular model, geometrical structure of a molecule made by joining small coloured balls, representing atoms, using stiff metal springs or plastic rods, representing single bonds. They are invaluable aids not only in visualizing shapes of molecules but also in indicating "strain" in bonds and possible conformations that the molecule may adopt in reaction. *See also* SU p.138.

Molecular orbital theory, explanation of how electrons are distributed in stable molecules. In the simpler VALENCY theory of the chemical bond, each atom of a molecule is assumed to retain its own electrons. Molecular orbital theory, however, treats each electron as associated with the molecule as a whole and it describes the configuration of electrons for each molecule. These patterns are unique for each molecule and can be considered as regions surrounding the molecule in which there is a probability of finding an electron. The calculations are very complex and only the simplest molecules can be treated exactly. *See also* SU p.138.

Molecular weight, sum of the atomic weights of all the atoms in a molecule. It is thus the ratio of the average mass per molecule of a specified isotopic composition of a substance to one twelfth of the mass of an atom of carbon-12. The molecular weights of reactants must be known in order to make quantitative calculations about a chemical reaction.

Molecule, smallest particle of substance capable of independent existence and exhibiting the characteristic properties of that substance. Molecules consist of atoms held together by chemical bonds. For example, water molecules consist of two atoms of hydrogen bonded to one atom of oxygen (H_2O). *See also* SU p.138.

Mole fraction, fraction of the number of moles of a given component in a solution or other mixture. If there are n_A moles of component A, n_B of B, etc., the mole fraction (X_A) of A is $n_A/(n_A + n_B ....)$.

Mole rat, any of several species of burrowing rodents that resemble true moles, found in E Europe, Asia and Africa. Its minute eyes can make out only whether it is light or dark. The soft, thick fur may be black, brown or grey. Length: to 30cm (11.8in). Families Spalacidae and Bathyergidae. *See also* NW p.*198*.

Molière (1622–73), French playwright, real name Jean-Baptiste Poquelin. An accurate observer of contemporary manners and types, he is remembered principally for his comedies of character such as *The School for Wives* (1662), *Tartuffe* (1664), *The Misanthrope* (1666) and *The Imaginary Invalid* (1673). He directed his own plays and often played the leading roles himself. *See also* HC1 pp.75, 317.

Molinos, Miguel de (1628–96), Spanish theologian, an important advocate of the doctrine known as QUIETISM. His *Spiritual Guide* (1675) taught that Christian perfection is attainable by contemplation and divine assistance. He was imprisoned for heresy in 1685.

Mollusc, any of more than 80,000 living species of invertebrate animals in the phylum Mollusca. They include the familiar snails, clams and squids, and a host of less well-known forms. Originally marine, members of the group are now found in the oceans, in fresh water and on land. There are six classes: the primitive GASTROPODS, CHITONS, univalves (slugs and snails), BIVALVES (clams, etc.), tusk shells and CEPHALOPODS (squids, etc.). The mollusc body is divided into three: the head, the foot and the visceral mass. Associated with the body is a fold of skin called the mantle which secretes the limy shell typical of most molluscs. The head is well-developed only in snails and in the cephalopods, which have eyes, tentacles and a well-formed mouth. The visceral mass

Henri Moissan's discoveries enabled him to manufacture artificial diamonds.

Mojave Desert supports specialized vegetation such as this Joshua Tree.

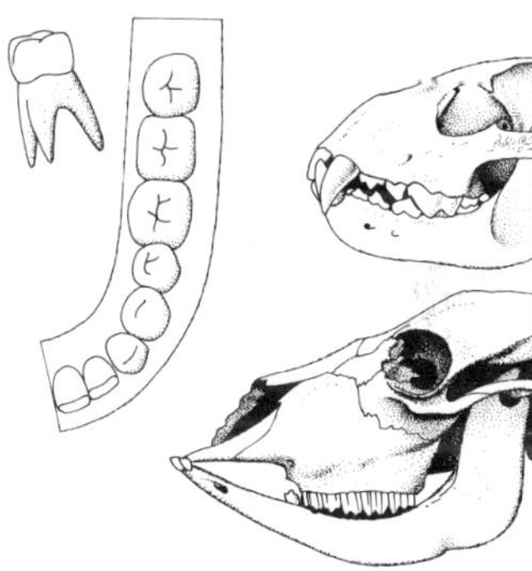

Molar: third molars are called wisdom teeth because they appear in adulthood.

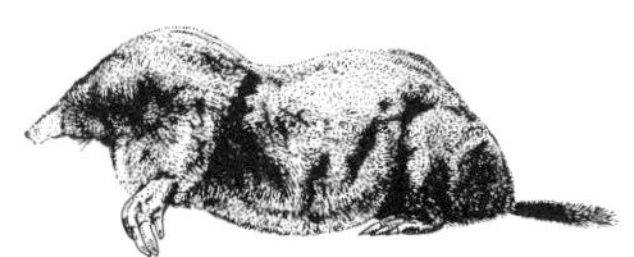

Moles cannot survive more than a few hours without feeding, day and night.

Molnar, Ferenc

Vyacheslav Mikhailovich Molotov: a photograph taken of him in 1953.

Mombasa: these arches were erected to commemorate a visit by Princess Margaret.

Monarch butterflies, when migrating, often gather into sleeping assemblies.

Monastery: St Jovan-Kaneo, Yugoslavia, is isolated in a beautiful setting.

contains the internal organs of circulation (blood vessels and heart), respiration (gills), excretion (kidney), digestion (stomach and intestine) and reproduction (gonads). The sexes are usually separate but there are many hermaphroditic species. *See also* NW pp.72, 90–94, *95*, 96, 178–179.

Molnar, Ferenc (1878–1952), Hungarian author who wrote a large number of light, sophisticated comedies. His reputation was established with the production in 1907 of *The Devil. Liliom* (1909) was a great success in the USA and was made into the musical *Carousel*; *The Guardsman* (1910) was also successful on the New York stage.

Moloch, W Semitic deity commonly thought to have been the object of a cult introduced into Israel after the time of Solomon. Children, usually the first-born, were said to have been offered for sacrifice by fire to Moloch, by being rolled into a furnace from the hands of a huge idol, although many scholars dispute this and even doubt the existence of the cult itself.

Moloch, also known as Thorny Devil, small lizard found in Australia. It is orange and brown and covered in spines; it eats mainly ants. Length: to 20cm (8in). Family Agamidae; species *Moloch horridus*.

Molotov, Vyacheslav Mikhailovich (1890–   ), Soviet political figure. A loyal ally of STALIN, he became a full member of the Politburo in 1926 and was Foreign Minister from 1939 to 1949 and from 1953 to 1956. He was expelled from the Central Committee in 1957. *See also* HC2 pp.*199*, 234, *234*, 240, *307*.

Molotov cocktail, crude hand-grenade made from a bottle filled with petrol or other inflammable liquid, the neck of which is stuffed with a rag as a wick. These "cocktails" were named after the Soviet statesman Vyacheslav MOLOTOV. Such devices were used by resistance movements during WWII; the name was first used by the Finn opponents of Soviet invasion (1940).

Moltke, the name of two German soldiers. Helmuth Karl Bernard von Moltke (1800–91) was chief of the Prussian army staff from 1858. He masterminded victories over Denmark (1864), Austria (1866) and France (1870–71). His organization of the general staff was one of the great innovations in 19th-century warfare. His nephew, Helmuth Johannes Ludwig von Moltke (1848–1916), became chief of the German general staff in 1906. He planned Germany's WWI strategy, modifying the SCHLIEFFEN PLAN, but he was replaced in 1914 when the initial German offensive failed to bring victory. *See also* HC1 pp.*110*, 187.

Moluccas (Maluku), island group and province in E Indonesia; Ambon is the capital. Formerly the Spice Islands, the group includes the larger islands of Halmahera, Ceram, and Buru, and the island groups of Sula, Batjan, Obi, Kai, Aru, Tanimbar, Banda, Babar and Leti. Originally explored by MAGELLAN in the early 16th century, the islands were later settled by the Portuguese. The Dutch took the islands *c.*1605–21 and monopolized the spice trade. Products: spices, copra, timber, sago. Area: 83,675sq km (32,307sq miles). Pop. (1970 est.) 995,000.

Molvig, Jon (1923–70), Australian painter. He studied in Europe from 1949 to 1952 but worked mainly in Brisbane and was noted for his colourful portraits which, in their savagery, often resemble those of Willem DE KOONING. His later paintings were simple, symbolical figures on black backgrounds.

Molybdenite, sulphide mineral, molybdenum sulphide (MoS$_2$). It is a major ore of molybdenum, found in PEGMATITES, IGNEOUS and METAMORPHIC rocks. It has hexagonal system tabular prisms, flakes, and fine granules and is lead-grey in colour with a metallic lustre. Hardness 1–1.5; s.g. 4.7.

Molybdenum, metallic element (symbol Mo) of the second transition series, first identified in 1778 and isolated in 1782. Chief ores are molybdenite (sulphide) and wulfenite (lead molybdate). It is used in alloy steels and molybdenum compounds are used as catalysts and lubricants. It is an essential trace element for plant growth. Properties: at.no. 42; at.wt. 95.94; s.g. 10.22; m.p. 2,610°C (4,730°F); b.p. 5,560°C (10,040°F); most stable isotope Mo^{98} (23.78%). *See also* TRANSITION ELEMENTS.

Molybdenum processing, production of MOLYBDENUM from ores containing MOLYBDENITE (MoS$_2$), usually for molybdenum-iron alloys used in steel-making. The concentrated mineral is roasted to yield molybdenum trioxide, which is mixed with iron in electric furnaces to make ferromolybdenum. The pure metal is produced as a powder and converted to massive metal.

Mombasa, city on Mombasa Island, Kenya, and partly on the mainland, to which it is connected by causeway; the capital of Coast province. From the 11th to the 16th centuries, Mombasa was a centre of the Arab slave and ivory trades. From 1529–1648, it was held by the Portuguese. Taken by Zanzibar in the mid-19th century, the city passed to Britain in 1887, when it was made capital of the British East Africa Protectorate until 1907. An important port, Mombasa exports coffee, fruit and grain. Industries: food-processing, glass, aluminium products. Pop. 255,400.

Moment. *See* TORQUE.

Moment of inertia, in physics, is defined as the sum of the products of elements, dm, of a mass, m, and the squares of their distances, r^2, from a given axis. Thus, $I = \Sigma dmr^2$, where I is the moment of inertia and Σ indicates a SUMMATION.

Momentum, product of the mass and velocity of a body. The principle that the total momentum of any system of bodies is conserved (remains constant) at all times, even during collisions, is one of the fundamental laws of physics, holding an equivalent position to the law of the conservation of mass-energy. *See also* ANGULAR MOMENTUM.

Mommsen, Theodor (1817–1903), German historian. A professor of Roman law, he wrote extensively about Classical Rome and was awarded the 1902 Nobel Prize in literature. His best-known work is a three-volume *History of Rome* (1854–56). An opponent of Otto von BISMARCK, Mommsen served in the Prussian House of Representatives from 1863–79.

Mona. *See* ISLE OF MAN.

Monaco, sovereign state in W Europe, near Italy and bordered by France, which assumed protection of the state in 1860. The chief source of income is tourism: there is some light industry. The capital is Monaco-Ville. Area: 1.5sq km (0.6sq miles). Pop. (1975 est.) 25,000. *See* MW p.125.

Monad, in Greek usage, the number "one", extended in philosophy to apply to any unit. The 17th-century German philosopher Gottfried LEIBNIZ applied the term to autonomous centres of force that, he said, comprise the fundamental metaphysical reality of which the universe is composed. These monads, or souls, do not interact, but are organized to correspond in a pre-established harmony and are therefore eternal.

Monadnock, residual hill rising above a PENEPLAIN, representing the core of the upland area that was eroded to give the plain. The plains of N Australia have many examples. *See also* PE p.107, *107*.

Monaghan, county in NE Republic of Ireland, in Ulster province, on the boundary with Northern Ireland. The S and E are hilly, but the rest of the county is part of the fertile central plain of Ireland. The Blackwater and the Finn are the chief rivers. Primarily an agricultural county, Monaghan's main crops are potatoes, oats and flax. Beef and dairy cattle are raised in large numbers. Industries: linen-milling, footwear, furniture. The county town is Monaghan. Area: 1,290sq km (498sq miles). Pop. (1971) 46,231.

Mona Lisa (*La Gioconda*), half-length oil painting by LEONARDO DA VINCI of the wife of a Florentine businessman, Francesco del Giocondo, probably painted between the years 1503 and 1506. Leonardo probably took the work to France in 1507 and it now hangs in the Louvre, Paris.

Monarch butterfly, or milkweed butterfly, large New World, migratory BUTTERFLY that is believed occasionally to fly across the Atlantic Ocean. It has brownish-orange wings with black veins and borders. Its caterpillars feed on MILKWEED, which gives their bodies a taste disliked by birds. The unrelated VICEROY BUTTERFLY mimics the coloration of the monarch, and so is also avoided by birds. Species *Danaus plexippus. See also* MIMICRY; NW p.*117*.

Monarchy, form of government in which one individual, whose power is usually hereditary, represents the state. There are few present-day monarchs holding absolute power and those states which retain their royal families are governed by constitutional monarchies, with royalty performing only ceremonial functions and retaining few royal prerogatives. *See also* MS pp.264, 273.

Monash, Sir John (1865–1931), Australian soldier. He served at GALLIPOLI (1915), and became commander of the Australian Army Corps in France in 1918. He won two victories, at Hamel (July 1918) and Amiens (Aug. 1918). Monash University, Melbourne, is named after him.

Monastery, communal religious dwelling of monks. In Europe, the houses of BENEDICTINES, the Canons Regular of St AUGUSTINE and the Celtic monasteries of Ireland were the earliest monasteries. Four basic features were essential to them: separation from the world, the autonomy of the government of individual houses, acceptance of the orders, classical rules and commitment to the local community.

Monasticism, ascetic mode of life followed by men and women who have taken religious vows and belong to a recognized religious order. The term generally applies to the strict and often highly ritualized community life of the Enclosed orders (eg the BENEDICTINES and DOMINICANS) who normally have little contact with the outside world. There are, however, other orders that combine asceticism with social welfare work and spiritual guidance to society at large. These orders form by far the larger group, some of whom live in communities but most of whom reside wherever seems required. Monks and nuns of this type are found not only in Christianity but also in HINDUISM, BUDDHISM, JAINISM and TAOISM. The ultimate aim of each order differs according to the religion. Historically in Europe, the purpose of monasticism was to provide spiritual welfare for the deeply religious medieval communities which, because of the hard-working nature of peoples' lives, felt they could not themselves devote sufficient time to earn salvation. *See also* MONASTERY.

Monazite, mineral containing rare-earth metals (LANTHANIDES) such as cerium and lanthanum, and thorium. It occurs widely in granitic and other rocks, and is the major source of those metals.

Mon boddo, James Burnett, Lord (1714–99), Scottish anthropologist who anticipated DARWIN's evolutionary principles. In his main work, *Of the Origin and Progress of Language* (1773–92), he related human beings to the orang-utan and traced the development of society.

Monck, Charles Stanley, 4th Viscount (1819–94), Governor-General of Canada from 1861 to 1868. An energetic and able administrator, he vigorously supported the confederation of Canada and maintained good relations with the USA during the Civil War.

Monck (or Monk), George, 1st Duke of Albemarle (1608–70), English soldier. In the Civil War he fought for CHARLES I (1643–44) but joined the Parliamentarians two years after his capture by them in 1644. He then served in Ireland and under CROMWELL in Scotland. Monck believed in the supremacy of the civil authority over the military and when the protectorate of Richard CROMWELL collapsed he became one of the most influential figures in the RESTORATION of the Stuarts. As Duke of Albemarle, he fought in the Second Anglo-Dutch War (1665–67) and was in charge of London during the great PLAGUE (1665) and the FIRE OF LONDON (1666). *See also* HC1 pp.*291, 299.*

Mond, Ludwig (1839–1909), British chemist and industrialist, *b.*Germany. Mond experimented with alkalis and developed a producer gas. He also discovered nickel carbonyl, a gas formed from carbon monoxide and metallic nickel. Using nickel carbonyl he developed a useful method (the Mond process) for extracting nickel from its ore.

Mondale, Walter (1928–), US politician. He was Attorney-General of Minnesota (1960–64) and DEMOCRATIC Senator from Minnesota (1964–67). In 1976 he was elected Vice-President of the USA.

Monday, second day of the week, the beginning of the working week. It is derived from the Old English for "Moon's day".

Mondrian, Piet (1872–1944), Dutch painter. He studied in Amsterdam and passed through an early naturalistic phase. Later he went to Paris, where CUBISM influenced the development of his geometric, abstract style, which he called NEO-PLASTICISM (an alternative name for De STIJL. Compositions using horizontal and vertical lines to outline rectangular shapes of white and primary colours are typical of his art. *See also* HC2 pp.177, 204, *204,* 280.

Monera, large biological grouping comprising the BACTERIA and blue-green ALGAE. They are sometimes considered distinct enough from other protists (unicellular creatures) to be called a separate kingdom. The main difference lies in the organization of the genetic materials within the CELL. Monerans, also called *procaryotes,* do not have their genetic material packed into a distinct NUCLEUS and lack MITOCHONDRIA and CHLOROPLASTS. Those with a distinct nucleus containing several chromosomes are called *eucaryotes.*

Monet, Claude (1840–1926), French painter; the classic IMPRESSIONIST – and leading member of the movement – who most consistently practised the Impressionist principles of fidelity to visual sensation and of working directly from the subject. Monet was described by Cézanne as "only an eye, but God what an eye!" During the 1860s he studied in Paris where he met RENOIR, SISLEY and most of the future Impressionists. He quickly embarked on his lifelong concern of depicting the endless variations of light and atmosphere in nature, and he succeeded, with hundreds of canvases, in capturing the flickering effects of light by breaking it down into prism-like colour components. In order to observe the transformation of a subject under changing light and atmosphere, he started his series of paintings on the same theme: *The Gare St-Lazare* (1876–78), *Haystacks, Poplars* (both 1890–92) and *Rouen Cathedral* (1892–94).

In 1874 he took part in the first Impressionist exhibition, which was savaged by the critics; they coined the term *Impressionism* after *Impression, Sunrise,* a painting by Monet who, like his fellow Impressionists, continued to work without patronage for many years. In 1883 he settled in Giverny, where he spent most of his remaining years. To this period belong the late *Water-Lily* series (1906–26), painted in his garden, many in his last years when he was almost blind. These huge canvases, bathed in shimmering light and almost formless colour and embodying a cosmic awareness of nature closer to TURNER than Impressionism, were the starting point of much abstract modern art. *See also* HC2 pp.118–119, 119, *118–119.*

Moneta, Ernesto Teodoro (1833–1918), Italian editor and pacifist, winner of the 1907 Nobel Peace Prize. An enthusiastic member of the movement for Italian unification, he served (1859–66) under GARIBALDI. From 1867, as editor of the newspaper *Il Secolo,* he devoted himself to the cause of peace and in 1906 was president of the International Peace Congress in Milan.

Monetarism, economic and monetary

position associated especially with the views of the economist Milton FRIEDMAN. It asserts the primary importance of controlling the MONEY SUPPLY as the means of achieving a non-inflationary, stable economy capable of supporting high employment and economic growth.

Monetary policy, control of the banking system and the MONEY SUPPLY by a government in the pursuit of various ends, such as a stable currency, a favourable balance of exchanges or a desirable level of employment. It is applied by controlling interest rates, the international movement of CAPITAL, the terms of hire-purchase agreements and the lending activities of banks and other financial institutions, such as BUILDING SOCIETIES.

Money, any type of payment that is generally accepted within an economy as a medium of exchange for goods and services; without money, the economy would have to operate on the significantly less efficient barter system. Money serves as a standard of value and facilitates the lending and borrowing of funds. It may take many forms besides currency or cash (coins and banknotes). A large portion of the MONEY SUPPLY, for example, may be in the form of deposits within a banking system. *See also* BANKING; BANK OF ENGLAND; BANK RATE; HC1 pp.45, *69,* 126; HC2 pp.300–301, *300–301.*

Mongolia (Mongol Ard Uls), official name the Mongolian People's Republic, a land-locked nation in central Asia between China and the USSR; also known as Outer Mongolia. Agriculture depends mainly on herd animals. Mineral deposits include coal, tungsten and copper. The capital is Ulan Bator. Area: 1,565,000sq km (604,247sq miles). Pop. (1976 est.) 1,489,000. *See* MW p.125.

Mongolian, group of closely related languages forming a branch of the ALTAIC family. One of its members, Mongol (Khalkha), is the official tongue of the Mongolian People's Republic (Mongolia). Two others, Buryat and Kalmyk, are spoken in parts of the USSR to which they were taken by Mongolian nomads in the 17th century.

Mongolism. *See* DOWN'S SYNDROME.

Mongoloid people, one of the major human racial groupings. Mongoloid physical characteristics include medium skin pigmentation, a rather flat face, EPICANTHIC FOLDS of the eyes, straight black hair and little facial or body hair. Mongoloids make up most of the population of Asia, and in Greenland and North America are represented by Eskimos. *See also* MS pp.32–33, *33–35,* 35.

Mongols, nomadic people of Mongolia, Manchuria and Siberia. They live in felt tents, eat mainly meat and milk, and raise horses, sheep and goats. They were once savage conquerors and rulers of the largest territory in history because of their prowess as archers and horsemen. The Mongols, under GENGHIS KHAN (*c.*1162–*c.*1227), built a huge empire, parts of which survived until the 17th century.

Mongoose, small, agile, carnivorous mammal of the CIVET family, native to Africa, s Europe and Asia. It has a slender, thickly furred body and a long, bushy tail. Mongooses are active by day and night and eat rodents, insects, eggs, birds and snakes. The Indian grey mongoose (*Herpestes edwardsi*) is well known for killing snakes, especially COBRAS. Some species may be domesticated, but most are highly destructive. Length: 46–115cm (18–45in). Family Viverridae. *See also* NW pp.*164,* 165, *200, 232.*

Monism, metaphysical theory that reduces the multiplicity of things to one fundamental reality, either physical or spiritual. *See also* MS p.*233.*

Monitor lizard, any of several species of powerful lizards that live in Africa, s Asia, Indonesia and Australia, including the KOMODO DRAGON (*Varanus komodoensis*). Most species are dull-coloured with yellow markings; many are semi-aquatic. Length: to 3m (10ft). Family Varanidae. *See also* NW pp.*137,* 205, *205, 246.*

Moniz, António de Egas (1874–1955), Portuguese physician and medical researcher. He shared the 1949 Nobel

Prize in physiology and medicine with W.R. HESS for developing the surgical technique known as prefrontal LEUCOTOMY. This treatment of severely mentally ill patients, involving cutting selected brain fibres, was made popular by Moniz in the late 1940s and founded the controversial field of psychosurgery. *See also* MS p.145, *145.*

Monk, George, 1st Duke of Albemarle. *See* MONCK, GEORGE, 1ST DUKE OF ALBEMARLE.

Monk, Thelonius (1920–), US jazz pianist. He helped develop the "bebop", now called "bop", jazz style in the 1940s and displayed a unique style featuring dissonances and distinctive chord structures. His best-known work is probably *Round Midnight. See also* HC2 p.272, *272.*

Monk, member of a community of men living under vows of religious observance such as poverty, chastity and obedience. There are many Christian monastic communities, especially in the Roman Catholic, Orthodox and Anglican churches, but the term "monk" is not confined to Christianity; it also refers to members of Hindu and Buddhist brotherhoods. *See also* MONASTERY.

Monkey, any of a wide variety of mostly tree-dwelling, day-active, omnivorous PRIMATES that live in the tropics and subtropics. Most monkeys have flat, man-like faces, relatively large brains, grasping hands and other characteristics of advanced anthropoid development. They fall into two broad groups – Old World monkeys (family Cercopithecidae) and New World monkeys (Cebidae). The 60 Old World species, sometimes called true monkeys, include MACAQUES, BABOONS, BARBARY APES, LANGUR MONKEYS and COLOBUSES. Some, such as baboons, are mainly ground dwellers, but all are excellent climbers. Most have short or long but always non-prehensile tails. They range in distribution from Japan and N China through s Asia and Africa (except in desert regions). The 70 species of New World monkeys include HOWLER MONKEYS, CAPUCHIN MONKEYS, SPIDER MONKEYS and MARMOSETS. They are all tree-dwellers, and most have grasping (prehensile) tails. The thumbs are less flexible and the nostrils are usually more widely separated than in Old World monkeys. They live in tropical forests of Central and South America. *See also* NW pp.*168,* 172, *173, 201,* 213, 215, *216,* 234–235, *235.*

Monkey-flower, one of a number of annual or perennial plants of the FIGWORT family. Monkey-flowers have two-lipped, tubular flowers resembling those of the snapdragon (antirrhinum), with a spotted pattern suggesting a monkey's face. Family Scrophulariaceae; genus *Mimulus.*

Monkey puzzle, or Chilean pine, evergreen tree native to the South American ANDES mountains. It has tangled branches, with spirally arranged, sharp, flat leaves. The female seeds are edible. It is grown in Europe mainly as an ornamental. Height to 45m (150ft). Family Araucariaceae; species *Araucaria araucana. See also* NW pp.*54,* 210.

Monkfish. *See* ANGLER FISH.

Monkshood. *See* ACONITE.

Monmouth, James Scott, Duke of (1649–85), English nobleman, illegitimate son of CHARLES II. He became a successful general and the champion of the Protestants opposing the succession of the Catholic Duke of York to the English throne. Exiled in 1679, he returned in 1685 to assert his claim to the throne but was defeated at the Battle of SEDGEMOOR by JAMES II and executed. *See also* HC1 p.293, *293.*

Monmouth, municipal borough, former county town of MONMOUTHSHIRE, SE Wales at the junction of the rivers Monnow and Wye. In 1974 it became part of the new non-metropolitan county of GWENT. Remains include an 11th-century castle (in which Henry V was born in 1387), a Norman priory and bridge (over the Monnow), with a gateway built in 1260. The town is a popular tourist attraction.

Monmouthshire, former county in SE Wales; in 1974 it was divided between

Claude Monet's painting *The Cliffs at Dieppe* is a good example of his art.

Mongoose: most species are solitary and form pairs only in the mating season.

Monitor Lizards are included among the largest of living reptiles.

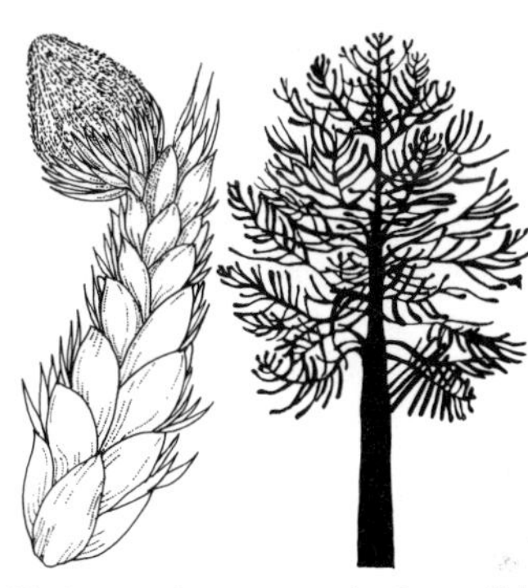

Monkey puzzle: some species have edible seeds and others are good timber trees.

Monmouth's Rebellion

Monmouth's Rebellion ended with the capture and execution of the rebel Duke.

Monotremes lack teeth but males have fangs resembling those of snakes.

Marilyn Monroe's air of fresh innocence contributed to her sex appeal.

Nicholas Monsarrat called his autobiography *Life is a Four-letter Word*.

Gwent and Mid-Glamorgan. The land is generally low-lying and fertile, rising to hills in the NW. Much of the area is pasture, but wheat and fruits are also grown. There are large deposits of coal, and steel is manufactured; iron ore is now imported. Monmouth was the county town. Area: 1,404sq km (542sq miles). Pop. (1971) 461,459.

Monmouth's Rebellion (1685), uprising in the West Country of England led by the Duke of MONMOUTH, illegitimate son of CHARLES II, contesting the accession to the throne of the Roman Catholic, JAMES II. Monmouth was defeated at the Battle of SEDGEMOOR . His mainly peasant followers were tried by Judge JEFFREYS. *See also* HC1 p.293, *293*.

Monnet, Jean (1888–   ), French economist. Active in government since WWI, he was a member of the British Supply Council during WWII, helping to organize the Allied war effort. His Monnet Plan in 1947 envisaged a controlled growth of French industry and he was the first president (1952–55) of the EUROPEAN COAL AND STEEL COMMUNITY. *See also* HC2 p.258, *258*.

Monnoyer, Jean Baptiste (1636–99), Flemish painter and engraver who specialized in floral decorations. He helped decorate the Galerie d'Apollon at the Louvre. In about 1685 he came to London and decorated the town house of the Duke of Montagu. Other stately homes decorated by him include Hampton Court and Kensington Palace.

Monoamine oxidase (MAO), enzyme widely distributed in animals; other names are adrenaline oxidase and tryaminase. Its function in the body is the breakdown of certain biologically active AMINES, three of the most important of which are the tryptamine derivatives, the catechol amines (eg epinephrine and dopamine) and histamine.

Monocline, in geology, sudden downward turn in the direction of a rock strata within a fold, surrounded by horizontal rock.

Monocotyledon, subclass of flowering plants (ANGIOSPERMS) characterized by one seed leaf (COTYLEDON) in the seed embryo; the leaves are usually parallel-veined. Lilies, onions, orchids, palms and grasses are examples of monocotyledons. The larger subclass of plants is DICOTYLEDON. *See also* NW pp.62, 66–69.

Monod, Jacques (1910–76), French biochemist. With François JACOB he developed the idea that messenger RNA carries hereditary information from the nucleus of a CELL to the cellular sites during the synthesis of PROTEINS, and the concept of the operator GENE controlling the activity of other genes. Together with André LWOFF, they were awarded the 1965 Nobel Prize in physiology and medicine.

Monoecious plant, in botany, a plant with both male and female reproductive organs. Flowering species have male flowers, bearing STAMENS, and female flowers bearing one or more carpels.

Monogamy, principle in which it is accepted that MARRIAGE is an exclusive union between two members of opposite sexes. It is commonly supported by legal and/or religious institutions. *See also* POLYGAMY; POLYANDRY; MS pp.168–169.

Monolith, massive freestanding stone used in various Neolithic and Early Bronze Age monuments, alternatively known as MEGALITH. *See also* MENHIRS.

Monomer, chemical compound composed of single molecules, as opposed to a POLYMER (which has molecules built up from repeated monomer units). For example, propylene is the monomer from which polypropylene is made, and methyl methacrylate is the monomer for polymethyl methacrylate, which is sold under the trade names Lucite, Plexiglas and PERSPEX. *See also* MM p.55.

Monophonic sound, sound which is transmitted over a single path or channel to be reproduced by one or more loudspeakers. *See also* STEREOPHONIC SOUND.

Monophysite Controversy, religious dispute centring on the Person of Christ. Monophysitism proposes a single, Divine Nature, whereas Orthodox teaching stresses a double Divine and Human Nature.

EUTYCHES was condemned in 448 for preaching an extreme form of Monophysitism, which was opposed at the Council of Chalcedon (451). Thereafter, many variants of the heresy developed, eventually resulting in the consolidation of three great Monophysite churches: the COPTIC, Syrian Orthodox and Armenian.

Monophysitism, the belief that in Christ after His INCARNATION, there is only one nature which is divine, as distinct from the orthodox Christian doctrine of a double nature, divine and human. Monophysitism was condemned by the Council of CHALCEDON in 451, after which its adherents became separated from the main body of Christendom. The COPTIC, Syrian, Jacobite and Armenian churches are still monophysite. *See also* CHRISTOLOGY; MONOTHELETISM.

Monopolies Commission, British organization set up in 1948 to determine whether monopolistic conditions exist in industries referred to it by the Board of Trade. A monopoly was defined as either restrictive or at least one-third (25% since 1973) of the output of a particular commodity in Britain. Its reports are advisory, and the Commission can also examine any type of service, including professional services. It also deals with business mergers. In 1973 it became re-established as the Monopolies and Merger Commission.

Monopoly, in economics, status of a manufacturer or supplier in which there is imperfect or even no competition, so that the market demand for a product or service is the only demand. A monopolistic industry has complete power over the market for its product and is able to determine levels of output and prices. Such an industry can charge high prices for less output than in a market with free competition. In many countries there are regulations to limit or prevent monopolies. *See also* MS pp.312, *312*.

Monopoly, board game for two or more players. The players, who are allocated an equal sum of money, acquire property according to the throw of the dice, the aim being to take over (or monopolize) their opponents. It was developed in 1933 in the USA by an unemployed salesman, Charles Darrow, who manufactured the game himself after it was initially rejected by games manufacturers. It has been adapted for differing local topographies; for instance the British version, features the streets of London, the German version the streets of Berlin.

Monorail train, passenger train that runs on a single rail; modern versions are gyroscopically stabilized and propelled by electric motors. The passenger cars are unsupported from the bottom or sides; they generally hang from wheeled axles. Bottom-supported cars, such as those used in zoos and amusement parks, are also referred to as monorails. *See also* MM pp.142–144.

Monosodium glutamate (MSG), food additive, a white crystalline powder with a meat-like taste obtained mainly from cereal glutens or from the waste liquor produced in recovering sugar from beet molasses. It is used as a flavour enhancer for meat products.

Monotheism, belief in the existence of a single God in the universe, who is visualized as the creator of all and is both personal and transcendent. Judaism, Christianity and Islam are the three major monotheistic religions.

Monotheletism, or Monothelitism, 7th-century Christian heresy asserting the existence of only one (divine) will in Christ rather than the orthodox belief in the existence of two wills in Him, divine and human. The initial proposing of the heresy was politically motivated, to gain the support of Monophysites and enable the formerly divided Christendom to unite against the Persian and Muslim invasions of the Byzantine Empire. It was formulated in documents drawn up in 638 and 648. Monotheletism was condemned at the LATERAN COUNCIL in 649 and at the Third Council of Constantinople in 680. *See also* CHRISTOLOGY; MONOPHYSITISM.

Monotreme, one of an order of primitive mammals that lay eggs. The only living

monotremes are the platypus and two species of echidna, or spiny anteaters, all native to Australasia. The eggs are temporarily transferred to a pouch beneath the female's abdomen where they eventually hatch and are nourished by rudimentary mammary glands. *See also* NW pp.154, *156*, 158, *159*.

Monotype, in art, method of printmaking in which a design is painted on a glass plate with oil colours and prints are made by pressing paper on to the glass to obtain a mirror image. Etching plates or lithographic stones may be used instead of glass plates.

Monotype, in printing, typesetting machine that casts letters one at a time. *See also* LINOTYPE; PHOTOSETTING; MM pp.206–207, *206–207*.

Monroe, James (1758–1831), fifth President of the United States (1817–25). He served as a member of Congress from 1783–86 and from 1790–94 as a member of the Senate. In 1794 he was made Minister to France, reaching the climax of his diplomatic career when he negotiated the LOUISIANA PURCHASE in 1803. A Jeffersonian, Monroe was chosen to succeed Madison as President and won an easy victory in the election of 1816. His administration was marked by improving relations with Britain and Spain, by the acquisition of Florida, and by the first rumblings of the slavery issue, solved by the MISSOURI COMPROMISE (1820). Monroe is most famous for promulgating the MONROE DOCTRINE (1823).

Monroe, Marilyn (1926–62), US film star, real name Norma Jean Baker. She was a talented comedienne, renowned as a sex symbol until her suicide at the age of 36. Her films include *Gentlemen Prefer Blondes* (1953), *The Seven Year Itch* (1955), *Bus Stop* (1956), *Some Like It Hot* (1959) and *The Misfits* (1961). *See also* HC2 pp.269, *289*.

Monroe Doctrine, foreign policy statement made by President James MONROE to the US Congress on 2 Dec. 1823. The doctrine attempted to prevent European intervention in American countries by stating that any interference in the affairs of any American country would be considered to be an unfriendly act towards the USA. It was established as policy by WWI and continued into the 1960s, when the USA attempted to exercise a proprietary role in the CUBA MISSILE CRISIS of 1962.

Monrovia, port and capital of LIBERIA on the estuary of the St Paul River. It was settled in 1822 by freed US slaves on a site chosen by the American Colonization Society and named in honour of the US President James Monroe. Monrovia exports latex and iron ore; it also has warehouses and facilities for ship repairing. Its own manufactures include bricks and cement. Pop. (1974) 180,000.

Mons, Battle of (24 Aug. 1914) first battle in WWI that involved the British Expeditionary Force (BEF). Two British divisions held up the German advance for 24 hours, and although heavily outnumbered managed to effect an orderly retreat, reaching the River Marne on 6 Sept.

Monsarrat, Nicholas (1910–   ), British author best known for his novel *The Cruel Sea* (1951), which was filmed in 1953. His other works include *The Tribe That Lost Its Head* (1956), *Richer Than All His Tribe* (1968) and *Kappillan of Malta* (1973).

Mons Meg, iron cannon built in *c.*1460. Its calibre is 49.5cm (19.5in), and it weighs 5 tonnes. Mons Meg is on display at Edinburgh Castle in Scotland. *See also* MM p.*167*.

Monsoons, seasonal reversal of winds, and their associated abrupt weather changes, that blow inshore over land masses in summer and offshore over nearby oceans in winter. The monsoon occurs annually in S Africa and E Asia and is centred on the Indian subcontinent. *See also* PE p.74.

Monsoon forest, type of tropical forest found in parts of SE Asia and the Indian sub-continent where rainfall averages between 101.5–203cm (40–80in) annually. It consists of sal and teak trees, both of which are valuable export commodities, with some evergreen undergrowth.

HC1 = History and Culture Vol. 1   HC2 = History and Culture Vol. 2   MS = Man and Society   MM = Man and Machines

Monstera, any of several species of tropical American, climbing or trailing plants with large glossy leaves that are commonly holed or deeply incised. The plants are sometimes confused with species in the genus PHILODENDRON. *Monstera deliciosa* is a popular houseplant; it is often called a cheese plant. Family Araceae.

Monstrance, in the Roman Catholic Church, small decorative dish with a glass cover and a handle to one side, used to display the sacramental wafer or HOST to the congregation during the evening devotional service of Benediction. Usually made of gold, silver or brass in the symbolic form of a sun. It is not used at Mass since the host itself is elevated. The showing of the monstrance at the service itself constitutes a benediction; no final blessing is given thereafter.

Montage, cinematic film-editing technique. It is a style of cutting and splicing a series of shots to obtain a desired narrative, structural or purely aesthetic effect. The Odessa Steps sequence in Sergei EISENSTEIN's *Potemkin* is a classic example of montage.

Montagna, Bartolommeo (*c.*1450–1523), Italian Renaissance painter of stately religious subjects. Many of his works are monumental, and show a mastery of modelling. Among the best known are *Madonna Enthroned between Four Saints* (1499) and the *Pietà* (1500).

Montagnards (The Mountain), nickname of the deputies of the extreme left who occupied the raised seats in the National Convention during the FRENCH REVOLUTION. With the fall of the GIRONDINS (1793), the Montagnards were the dominant political group during the REIGN OF TERROR, when a centralized government was imposed on France. The Montagnards lost power during THERMIDORIAN REACTION in 1794.

Montagu, Lady Mary Wortley (1689–1762), British letter writer. In 1716 she went with her husband to Turkey, where he was briefly ambassador. Returning to England in 1718, she campaigned to educate the public about innoculation against smallpox. Her letters, which contain lively observations of her travels, were published in 1763.

Montague, Charles, Earl of Halifax. *See* HALIFAX, CHARLES MONTAGUE, EARL OF.

Montaigne, Michel Eyquem, Seigneur de (1533–92), French essayist. His *Essays,* begun in 1580, touch on almost every known subject. Written in a style that alternates between high eloquence and racy colloquialism, they constitute an intellectual autobiography that moves from stoicism through SCEPTICISM to a mature acceptance of all that life offered. *See also* HC1 pp.316, *317.*

Montale, Eugenio (1896–   ), Italian poet, journalist, critic and translator who was awarded the 1975 Nobel Prize in literature. In 1922 he helped to found a literary magazine *Primo Tempo* and from 1948 was the literary editor of *Corriere della Sera.* He is considered as one of Italy's major poets. His work is pessimistic, especially in *Ossi di Seppia (Cuttlefish Bones)* (1925), which captures the despondency of postwar Italy. He also translated several works into Italian, including those of T. S. ELIOT, by whom he was influenced, and SHAKESPEARE.

Montana, state in NW USA, on the Canadian border. The W section of the state is dominated by the Rocky Mts. The E is part of the GREAT PLAINS, drained by the Missouri and Yellowstone rivers. Large herds of sheep and cattle are raised on the plains. The principal crops, grown by means of irrigation, are wheat, hay, barley and sugar-beet. The Rockies have large mineral deposits including copper, silver, gold, zinc, lead and manganese. Oil, natural gas and coal are found in the SE. The major industries in the state are timber, food processing and petroleum products; tourism is also an important source of income. The main cities are Helena, the state capital, Billings and Great Falls.

French and Spanish fur traders travelled in the region before 1800, but until the USA acquired the area in the LOUISIANA PURCHASE of 1803, it was little known. LEWIS AND CLARK explored it (1805–06), and this led to the establishment of trading and military posts. The discovery of gold in 1852 brought a rush of immigrants and the Territory of Montana was organized in 1864. The opening of the Northern Pacific Railroad in 1883 provided a great stimulus to the state's growth and development. Area: 381,086sq km (147,137sq miles). Pop. (1975 est.) 748,000. *See also* MW p.175.

Montand, Yves (1921–   ), Italian-born French film actor who began his career as a music hall singer. He made his film debut in the mid-1940s and went to Hollywood in 1961 to make *Let's Make Love,* with Marilyn MONROE. He returned to Europe to make *Z* (1969) and *State of Siege* (1973) with the director COSTA-GAVRAS. *Wages of Fear* (1953) is considered his finest film.

Montanism, 2nd-century apocalyptic movement owing its origin to the "prophet" Montanus of Phrygia. He began preaching about AD 156, describing a heavenly Jerusalem that would descend to Phrygia. He usually spoke while in a state of ecstasy, and his prophecies were often accompanied by behaviour resembling demonic possession. Montanism was condemned by the Asiatic synods which excommunicated the Montanists, probably *c.*177.

Montauban, city in Tarn-et-Garonne dept. S France. It was a centre of the ALBIGENSIANS in the 13th century and the HUGENOTS in the 16th. Industries: aeronautical engineering, textiles, food processing. Pop. 48,600.

Mont Blanc, highest peak in the Alps, and the second-highest peak in Europe, situated on the French-Italian border. The 11km (7 mile) tunnel, cut through the base of Mont Blanc (1958–62), is the longest vehicular tunnel in the world and provides a short route between France and Italy. Height: 4,810m (15,781ft).

Montcalm, Louis-Joseph de, Marquis de Saint-Véran (1712–59), French general in North America. Commander-in-chief of the French army in Canada from 1756–59, he won several victories in the colonial wars against the British, including Fort TICONDEROGA (1758). Lacking support from France, he was defeated by Gen. James WOLFE at Quebec, where he was mortally wounded. *See also* HC2 p.*136.*

Mont Cenis Tunnel, also known as the Fréjus Tunnel, a 13km (8 miles) double-track railway tunnel through the Alps, between Modane (France) and Bardonècchia (Italy). Built between 1857 and 1871, it was the first tunnel through the Alps and the first to use the newly invented pneumatic power drill.

Monte Albán (*c.*300 BC–c.AD 900), ancient ZAPOTEC religious centre located SW of Oaxaca, Mexico. Monte Albán was built on a level hilltop and along the adjacent valleys; stepped platforms supported the religious monuments. *See also* HC1 pp.*46,* 47, 230.

Monte Carlo, town in N Monaco, on the French RIVIERA and Mediterranean Sea. It was founded in 1858 by Prince Charles III of Monaco in agreement with a joint stock company that wished to build and operate the Casino. Today it is a popular resort noted for its scenery and mild climate, and the Casino remains a great tourist attraction. Pop. (1973) 23,850.

Monte Carlo Rally, world-famous motor event held every January, when cars set off from pre-arranged points in Europe to converge on Monaco several days later. Those beating the winter weather to arrive on time then embark of a series of trials around the mountains and in the city. It was first run in 1911 and 1912, then resumed in 1924. Awards are made for outright and class wins. Its importance has decreased in recent years, relative to races such as the RAC and Safari rallies.

Monte Cassino, famous monastery overlooking the town of Cassino in central Italy. The monastery was founded *c.*529 by St BENEDICT OF NURSIA whose rule became the basis of the Benedictine order named after him. Monte Cassino became a medieval centre of Christian learning. The monastery was stormed by the Lombards in 589, the Saracens in 884 and the Normans in 1030. It was reconstructed by Abbot Desiderius in the 11th century and again in the 16th and 17th centuries after an earthquake. It was damaged during WWII but its archives and art treasures were saved.

Montecristo Island, uninhabited Italian island in the Tyrrhenian Sea s of Elba. It is a former penal colony made famous by the Alexandre DUMAS novel *The Count of Monte Cristo.* Area: 16sq km (6sq miles). 43,000.

Montego Bay, port in NW Jamaica, West Indies. It is a commercial centre with a good harbour and extensive rail connections. Products include sugar cane, coffee, ginger, rum and fruit. Tourism is a valuable source of income. Pop. (1970 est.) 43,000.

Montenegro (Crna Gora), constituent republic of SW Yugoslavia, at the s end of the Dinaric Alps, bordered by Albania (SE) and the Adriatic Sea (SW); the capital is Titograd. The region is mountainous and isolated, divided into the Brda region and Montenegro proper to the W of the River Zeta. The population is largely Serbian. The region was an independent principality of Zeta within the Serbian Empire until the 14th century, when the Empire was defeated by the Turks. Formal recognition of independence occurred in 1878 at the Congress of BERLIN. In 1946 Montenegro became one of the six Yugoslavian autonomous republics. Occupations include rearing livestock, mining and farming of cereal crops. Area: 13,812sq km (5,322sq miles). Pop. (1971) 530,361.

Monterrey, city in NE Mexico, 242km (150 miles) s of Laredo, Texas, USA. It is an important commercial centre, having a variety of industries. These include glass blowing, lead smelting and the production of silver, gold and copper. Its equable climate also makes it a popular resort area. Pop. (1970) 830,336.

Montespan, Françoise Athénaïs, Marquise de (1641–1707), mistress of King LOUIS XIV of France. Maid of Honour to Queen Marie Thérèse in 1664, she replaced Mlle de La Vallière as the king's mistress in 1667. She bore him seven children. They were educated by Mme de MAINTENON, who succeeded her in the king's favour. She later retired to a convent.

Montesquieu, Charles-Louis de Secondat, Baron de (1689–1755), French social philosopher. He launched his criticism of contemporary society in the *Persian Letters* (1721). In his *Spirit of Laws* (1748) he attempted to explain the evolution of societies in terms of environmental features: climate, geography and DEMOGRAPHY. *See also* HC2 pp.22, *23.*

Montessori, Maria (1870–1952), Italian educationist who believed that pre-school children, given an environment rich in manipulative materials, would teach themselves. Her method, which she used first with retarded children, was adapted for use in many of the state school systems in Britain and the USA.

Monteux, Pierre (1875–1964), French conductor and founder of the Paris Symphony Orchestra. He toured with the ballet impresario DIAGHILEV (1911–1914), and introduced many important 20th-century works such as STRAVINSKY's *Rite of Spring* and RAVEL's *Daphnis et Chloé.*

Monteverdi, Claudio (1567–1643), Italian composer, the first great opera composer and a pioneer in modern orchestration. He advanced violin technique by trying out new effects in orchestral playing. His many operas include *La Favola d'Orfeo* (1607), *Il Ballo delle Ingrate* (1608) and *L'Incoronazione di Poppea* (1642). He also composed religious music and was the master of the MADRIGAL. *See also* HC1 pp.318–319; HC2 pp.40, 104.

Montevideo, capital of URUGUAY, in the s part of the country on the River Plate 217km (135 miles) E of Buenos Aires, Argentina. Originally a Portuguese fort (1717), it was captured by the Spanish in 1726 and became the capital of Uruguay in 1828. One of South America's major ports, it is the base of a large fishing fleet and handles most of the country's exports.

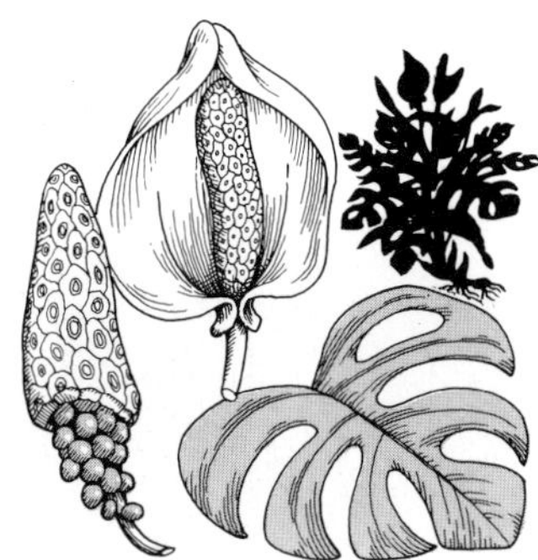

Monstera deliciosa, a popular house-plant, is native to the American tropics.

Montaigne's *essais* were regarded by him as exercises of literary skill.

Monte Cassino, demolished by bombardment during WWII, has now been rebuilt.

Montecristo Island: pirates caused the monastery to be abandoned in 1553.

Montez, Lola

Field Marshal Montgomery, addressing senior NATO officers in 1958.

Montrose, on hearing of the execution of Charles I, led a force to avenge him.

Monument: stairs inside give access to a viewing platform at the top.

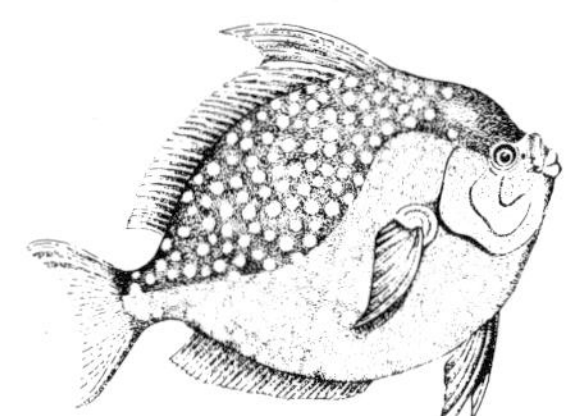

Moonfish are brilliantly coloured and appear irridescent in changing light.

Products include textiles, dairy goods, wine and packaged meat. Pop. 1,230,000.

Montez, Lola (*c.*1818–61), Irish dancer. She became the mistress of Louis I of Bavaria in 1846, but her liberal influence led to his abdication and her expulsion in 1848. She toured the USA in 1851 and Australia in 1855–56.

Montezuma, name of two AZTEC emperors. Montezuma I (*r.*1440–69) increased the size of the empire through a series of wars. Montezuma II (1466–1520) allowed the Spaniards under CORTÉS to enter central Mexico unhindered. Montezuma II was imprisoned by the Spaniards and killed by his own subjects when they rose up against the intruders.

Montfort, Simon de (1208–65), French-born Earl of Leicester. Married to HENRY III's sister, he was the king's adviser until he quarrelled with him and joined the BARONS' WAR. De Montfort defeated Henry at LEWES in 1264 and became virtual ruler of England, but was killed in renewed fighting at EVESHAM. *See also* HC1 pp.*178,* 179, 210, *210.*

Montgolfier, name of two French pioneer balloonists, the brothers Jacques Etienne (1745–99) and Joseph Michel (1740–1810). In 1783 they sent up a large linen bag inflated with hot air. Its flight covered more than a mile (about 2km) and lasted for ten minutes. In the same year they gave their first public demonstration of a manned balloon flight over Paris. *See also* MM pp.146–147.

Montgomery, Lucy Maud (1874–1942), Canadian author. She was a teacher and journalist before writing seven novels dealing with the life of a Canadian orphan girl. *Anne of Green Gables* (1908), the first book, has been widely translated.

Montgomery of Alamein, Bernard Law, Viscount (1887–1976), British soldier. As commander of the British Eighth Army in WWII, he defeated ROMMEL and the AFRIKA KORPS at EL ALAMEIN in Egypt and led the invasion of Sicily and Italy. He helped to plan the Normandy landings in 1944, leading his troops into Germany and commanding the British occupation forces there. He was made a Field Marshal in 1944 and a Viscount in 1946, and was deputy Supreme Allied Commander, Europe, from 1951 to 1958. *See also* HC2 pp.229, 231, *231.*

Montgomery, commercial city in SE central ALABAMA, USA, 137km (85 miles) SSE of Birmingham. It was made the state capital in 1847 and became important as a port and market centre. Today its manufactures include textiles, fertilizers and machinery. Pop. (1970) 133,386.

Montgomeryshire (Trefaldwyn), former county in E central Wales; in 1974 it became part of Powys. The land is mostly mountainous and is drained by the rivers Dyfi, Severn and Vyrnwy. Lead was once mined there, but the region is now devoted almost wholly to sheep-rearing. The county town was Montgomery. Area: 2,064sq km (797sq miles). Pop. (1971) 42,761.

Month, time taken for the Moon to travel completely around the Earth. The sidereal month is the time of one revolution with respect to the stars and is equal to 27.32 days. As the Earth is in motion around the Sun, the synodic month – from full moon to full moon – is longer than the sidereal month and is equal to 29.53 days. The calendar month is not constant in length and is defined for convenience. *See also* MW pp.188–189.

Montherlant, Henri Millon de (1896–1972), French writer. A frequent theme in his novels is the choice between decadence and moral integrity facing the aristocracy. They include *The Bullfighters* (1926), *The Bachelors* (1934) and *Young Girls* (1936). In 1942 he turned to drama, writing *Malatesta* (1946), *Port Royal* (1954) and *La Guerre Civile* (1965).

Montmagny, Charles Jacques Huault de (1583-1653), French colonial governor. He replaced Samuel de CHAMPLAIN as governor of NEW FRANCE (Canada) from 1636 to 1648 and made a peace treaty with the IROQUOIS.

Montmartre, highest point of Paris, on the right bank of the River Seine. In the 19th century writers and artists collected in Montmartre. One of the most impressive landmarks of Paris, the Church of the Sacré Coeur, was built there in 1919. Today Montmartre is a great tourist attraction and centre of night life.

Montmorency, Anne, Duc de (1493–1567), Constable of France who was brought up with the future king, Francis I, whom he followed into Italy (1515). Made Marshal of France in 1522, he defeated Charles V at Susa (1536) and became Constable (1538) but lost favour in 1541. He was restored to court by Henry II (1547). Unpopular with Francis II, he returned once more to favour with Charles IX (1560) and commanded against the Hugenots (1562). He triumphed at St Denis (1567) but was fatally wounded.

Montmorency-Laval, François de (1623–1708), first Bishop of Quebec (1674–88). He was ordained a priest in 1647 and appointed vicar apostolic to the French colonies in 1658, where he organized church schools, directed missionary work and founded the Quebec seminary in 1663.

Montparnasse, quarter of Paris, in the vicinity of the intersection of the Boulevard de Montparnasse and the Boulevard Raspail on the left bank of the River Seine. The Montparnasse cemetery contains the tombs of BAUDELAIRE, MAUPASSANT and César FRANCK, all of whom frequented the area in their youth.

Montpellier, city in S France, 10km (6 miles) N of the Mediterranean coast. It became noticeably larger in the 1960s due to an influx of refugees from Algeria; industry and commerce have been developed accordingly. Manufactures now include textiles, metal goods, printed materials and chemicals. Pop. (1968) 161,910.

Montreal, city on Montreal Island and the N bank of the St Lawrence River, in S Québec province, Canada; largest city in Canada and the country's chief port. Its excellent harbour on the ST LAWRENCE SEAWAY makes it an important entrepôt and it is linked to all the major industrial centres on the GREAT LAKES. Originally the Indian village of Hochelaga, the site was visited by Jacques CARTIER in 1535, and settled by the French in 1642. It was under French control until 1760, when it was taken by the British. The Americans occupied the city during the American War of Independence. The city's growth accelerated with the opening of the Lachine Canal in 1825, connecting it to the Great Lakes. Montreal served as the seat of the Canadian government from 1844 to 1849. A major cultural and educational centre, it has McGill University (1821) and the University of Montreal (1876). It is the focal point of transport routes, commerce and industry in E Canada. Industries include aircraft, electrical equipment, rolling-stock, textiles, metallurgy, chemicals and food processing. Pop. 1,214,355. *See also* MW p.45.

Montrose, James Graham, 5th Earl and 1st Marquess of (1612–50), Scottish military commander. He fought with the COVENANTERS against the English in 1639, but after 1640 took the side of CHARLES I. His wild army was at first successful in Scotland but he was defeated in 1645, went into exile, returned in 1650, and was captured and hanged. *See also* HC1 p.*299.*

Monts, Pierre du Guast, Sieur de (1568–1630), French colonizer of Acadia (Nova Scotia) (1604–48). He was granted a fur-trade monopoly in 1603 and, with Samuel de CHAMPLAIN, founded the first colony at Port Royal in 1605. It proved impossible to enforce the monopoly and they moved (1608) to the St Lawrence and established a fur-trade factory.

Mont Saint Michel, rocky isle 1.6km (1 mile) off NW France, in the Bay of Saint Michel. It is the site of a Benedictine abbey built in 708 by St Aubert. The base of the island is circled with ramparts, towers and bastions rising three stories to support the abbey church. The church spire holds a statue of St Michael, 138m (456ft) above the bay. Built in the GOTHIC style, the abbey attracts many tourists.

Montserrat, island and British colony in the Leeward Islands, in the Lesser Antilles group of the West Indies. Plymouth is the capital and chief port. Discovered in 1493 by Christopher COLUMBUS, it was colonized in 1632 by the British. A former member of the Federation of the West Indies, the island is currently part of the Caribbean Free Trade Association. The shipping of agricultural produce, especially cotton, is the chief economic activity. Area: 98sq km (38sq miles). Pop. (1975 est.) 13,000. *See also* MW p.126.

Monument, The, stone column that stands near Billingsgate in the City of London, England. It was designed by Sir Christopher WREN and completed in 1666. It commemorates the FIRE OF LONDON of 1666 and is thought to stand on the site of the bakery in Pudding Lane where the fire began.

Monza, multi-track autodrome, built in 1922 in a royal park on the outskirts of Milan, Italy. The Italian Grand Prix is raced on the 5.7km (3.5 miles) road-racing circuit, which in places incorporates Monza's famous banked track. It is one of the fastest motor-racing circuits in the world.

Mood, period of stable emotion distinguished by the quality of the feeling – irritability, sadness, optimism, etc – and the inclination to maintain it. The stability of a mood is a measurable factor in human temperament.

Mood, or mode, group of forms in the grammatical conjugation of a verb serving to show the function in which it is used. Indicative states a fact; subjunctive expresses uncertainty; imperative makes a command. Interrogative, asking a question, and conditional, expressing a condition, are sometimes included as moods.

Moody, Helen Wills (1905–), US tennis player. Considered to be the greatest female player of her time, she won eight WIMBLEDON women's singles titles between 1927 and 1938, 19 major singles titles in all, and was ranked first in the world nine times.

Moog synthesizer, computer tool for composing music by translating ideas into electronically produced synthetic sound. Wave forms generated by programmed or randomly manipulated circuitry and modified by altering intensity, frequency, duration, etc are combined to create complex signal patterns. Output is obtained from selected mixtures preserved on magnetic tape. The Moog synthesizer was named after its inventor Robert Moog, a US physicist who developed the instrument in the 1960s.

Moon, natural satellite of a planet, in particular the natural satellite of the planet Earth. The Moon is approx. the same age as the Earth (4,500–5,000 million years old). It emits no light of its own, and reflects less than 10% of the sunlight that falls on it. Man has observed the Moon and recorded its apparent motion across the skies throughout history. It is the Earth's closest neighbour and the only celestial body in the sky, with the exception of the Sun, that is visible to the naked eye as a disc rather than a point of light. With an escape velocity of 2.4km (1.5 miles) per second, the Moon has no atmosphere as such. Mean distance from the Earth: 384,000km (239,000 miles); mass: 0.012 times that of the Earth; volume: 0.02 times that of the Earth; diameter: 3,476km (2,160 miles); rotational and sidereal period; 27.3 days; temperature range: approx. 90°C (195°F) (day) to approx. −130°C (−200°F) (night). *See also* index to *Science and The Universe.*

Moon and Sixpence, The, novel by W. Somerset MAUGHAM, published in 1919, based on the life of the French painter Paul GAUGUIN. Charles Strickland, a conventional stockbroker, becomes interested in painting in middle life and deserts his wife, family and business in order to live and paint in Tahiti.

Moonfish, or opah, sometimes called kingfish, marine fish found in all seas and important as a food fish in Japan. Its oval body is laterally compressed and coloured blue and rose with white spots. Length: to 182.9cm (6ft); weight: to 270kg (600lb).

Family Lampridae; species *Lampris regius*. Other species of fish, notably *Mene maculata* and *Vomer setapinis*, are also called moonfish.

Moon rat, small, rat-like, nocturnal mammal, the largest species of the hairy hedgehogs, found only in SE Asia. Primarily insectivorous, it has bristly black fur and white markings on its head. Length: to 58cm (23in) overall. Family Erinaceidae; species *Echinosorex gymnurus*.

Moonstone, potassium FELSPAR with a milky translucent lustre, valued as a gemstone which is both light-reflecting and hard. Most moonstones of commercial value come from Ceylon. The irridescent quality is produced by crystal planes within the stone.

Moonstone, The (1868), novel by Wilkie COLLINS narrated by several characters and concerning a jewel stolen from an Indian idol and bequeathed to an English girl. Many critics, including T. S. ELIOT, regard *The Moonstone* as the first work of detective fiction in the English language.

Moor, tract of open wasteland, often covered with heather. The soil is acid and commonly marshy or boggy. Found throughout Britain, examples are Dartmoor, Exmoor and the Yorkshire moors.

Moore, Archie (*c.*1915–), US boxer. He turned professional in 1936, and won his first 13 matches by knockouts. In 1952 he won the light-heavyweight championship from Joey Maxim and held the title until 1961. He fought Rocky MARCIANO in 1955 and Floyd Patterson in 1956 for the heavyweight championship, but was beaten on both occasions.

Moore, Brian (1921–), Canadian novelist, *b.* Northern Ireland. His works, humorously written about serious subjects, include *The Lonely Passion of Miss Judith Hearne* (1955), *The Luck of Ginger Coffey* (1960), *Fergus* (1971), *Catholics* (1972) and *The Doctor's Wife* (1976).

Moore, George (1852–1933), Irish novelist, dramatist and short story writer. He studied art in Paris and, inspired by Honoré de BALZAC and Emile ZOLA, turned to writing. Their influence may be detected in the realism of his novels, particularly *Esther Waters* (1894). He was a founder of the Irish National Theatre, later the ABBEY THEATRE.

Moore, George Edward (1873–1958), British philosopher. Concerned with ETHICAL theory, EPISTEMOLOGY and METAPHYSICS, he held that "good in itself" could not be analysed as a concept. His *Principia Ethica* (1903) was a major factor in the declining influence of Hegelianism and Kantianism in British philosophy. Among his other works are *Philosophical Papers* (1959) and *The Common Place Books* (1962), published posthumously. *See also* HEGEL, GEORG; KANT, IMMANUEL.

Moore, Gerald (1899–), British pianist who became widely known as an accompanist to many of the world's leading instrumentalists and singers. His experiences are reported in *The Unashamed Accompanist* (1943, revised 1957) and *Am I Too Loud?* (1962).

Moore, Henry (1898–), British sculptor of international standing. His early work was strongly influenced by pre-Columbian art, but by 1928 his style had become more personal and he had gained an international reputation. His sculpture often incorporates hollows to which he gives as meaningful a shape as solid mass; there are also analogies to landscape (hillsides, caverns, etc) in many of his figures. During WWII he was commissioned by the British government to do his famous drawings of people in the London Underground shelters. The mother and child and the reclining figure are themes central to much of his work. *See also* HC2 pp.205, *205*, 280, *281*.

Moore, Sir John (1761–1809), British soldier. After fighting in the West Indies, Holland and Egypt (1796–1801), and serving as a training officer (1803–06), Moore was sent to fight the French in Spain, where he was mortally wounded at CORUNNA. *See also* HC2 p.79.

Moore, Marianne (1887–1972), US poet. Her poetry covers a wide range of topics and is precise, witty and often satirical.

Volumes of her verse include *Observations* (1924) and *Collected Poems* (1951).

Moore, Patrick (1923–), British astronomer and writer, whose many publications and broadcasts on television and radio established him among the foremost authorities on space-flight, the Moon and the planets. His publications include *Moon Flight Atlas* (1969), *Atlas of the Universe* (1970), *Guide to the Planets* (1971) and *Guide to the Moon* (1976).

Moore, Robert Frederick ("Bobby") (1941–), British footballer who captained England to victory in the 1966 World Cup, won a record 108 caps (1962–73), and led West Ham United to win the 1964 FA Cup and 1965 European Cup-Winners' Cup. A masterly defender, he ended his playing days at Fulham.

Moore, Ronnie (1933–), New Zealand speedway rider who, at 17, was the youngest-ever finalist in the world championships. He won the title in 1954 and 1959, both times with maximum points and despite injuries. He rode for Australia, New Zealand and Britain.

Moore, Stanford (1913–), US biochemist who shared the 1972 Nobel Prize in chemistry with Christian B. ANFINSEN and William H. STEIN for research into the functioning and composition of the ENZYME ribonuclease. Anfinsen had first discovered the composition of ribonuclease in the 1950s; subsequently Moore and Stein explained how it catalyses the digestion of food. They used CHROMATOGRAPHY to analyse AMINO ACIDS and PEPTIDES, obtained from biological samples, and to determine the structure of ribonuclease. Using essentially the same technique, Moore and Stein had analysed the structure of deoxyribonuclease by 1973.

Moore, Thomas (1779–1852), Irish poet and songwriter. His works include *Odes of Anacreon* (1800), *Irish Melodies*, a group of lyrics published between 1808 and 1834, and the famous *Lalla Rookh: An Oriental Romance* (1817).

Moorehead, Alan (1910–), British writer of historical documentaries, *b.* Australia. From 1930 to 1946 he was a journalist and war correspondent. His works include *Mediterranean Front* (1941), *Gallipoli* (1956), *The White Nile* (1960), *The Blue Nile* (1962), *Cooper's Creek* (1963), *Darwin and the Beagle* (1969), and *A Late Education, Episodes in a Life* (1970).

Moorhen, or waterhen, common Old World aquatic bird of the RAIL family, so named because of its liking for rivers and ponds. It has black plumage and a yellow bill, and its long toes lack the webs and lobes typical of other water birds. Length: to 32.5cm (13in). Species *Gallinula chloropus. See also* NW p.230.

Moor macaque, Old World PRIMATE of islands in the Malay archipelago including Celebes and Tonga. The body is baboon-like with a vestigial tail. The omnivorous diet includes shellfish, insects, fruit and plants. Length: to 72cm (28in). Family Cercopithecidae; species *Cynomacaca maurus.*

Moors, nomadic people of N Africa of Berber and Arabic stock. The Moors were early converts to ISLAM and, having crossed over to Spain and Portugal in 711, they quickly conquered most of the Iberian Peninsula. Abd ar-Rahman I, the last survivor of the UMAYYAD dynasty from Damascus, established the emirate (later caliphate) of Córdoba in 756. The Moors were opposed by the Christian rulers of N Spain, who gradually extended their power s, while dissension grew within the Moorish ranks. The puritanical ALMORAVIDS crossed from N Africa in 1085 and conquered the more worldly Spanish Moors. They in turn were replaced by the even more zealous ALMOHADS. Only GRANADA remained Moorish by 1250, although it survived until 1492 when it fell to Ferdinand and Isabella. Many of the Moors who remained suffered under the Spanish INQUISITION after Spain and Portugal were reconverted to Christianity. Today the populations of Algeria, Mauretania, Morocco and Tunisia are of Moorish stock.

Moose. *See* ELK; NW pp.206–207.

Moot, local assembly in medieval England (from the Old English *gemot* for meeting). Moots, which dealt with both legal and administrative matters, were held wherever a community was organized. The most important by the 10th century were the SHIRE, HUNDRED and BOROUGH moots. The greatest was the Witenagemot ("most of the wise men"), the Anglo-Saxon council of the king and his chief nobles.

Moped, simple, lightweight motor cycle which is usually driven by an engine of less than 50cc and may be pedalled if necessary. The moped derives from the practice of attaching an auxiliary engine to bicycles, and was first developed in Germany in the 1950s.

Moraes, Vinicius de (1913–), Brazilian writer. In his poetry his main concern is the conflict between the flesh and the spirit. He also wrote *Orfeu da Conceição* (1956), a folk-tragedy about the annual Rio de Sareiro Carnival on which the Oscar-winning film *Black Orpheus* was based.

Moraine, general term indicating a mound, ridge or other visible accumulation of unsorted glacial drift, predominantly TILL. End moraines are formed when a GLACIER is either advancing or retreating and the rock material is dumped at the glacier's edge. Ground moraines are sheets of debris left after a steady retreat of the glacier. *See also* PE p.116.

Morales Bermudez, Francisco (1921–), Peruvian soldier and politician, grandson of Col. Remiro Morales who was President of Peru from 1890 until 1894. Morales Bermudez became chief of the army general staff in 1974, and in 1975 was made President of Peru.

Morality plays. *See* MYSTERY PLAYS.

Moral Rearmament (MRA), Christian revivalist movement, founded in 1922 by a US Lutheran pastor, Frank N. D. BUCHMAN. The popularity of the movement at Oxford University led it sometimes to be called the Oxford Group. The name Moral Rearmament was adopted in 1938. MRA aims to promote virtuous living and particularly tries to "convert" the rich. It has about 3,000 full-time voluntary workers in many countries.

Morar, Loch, in Highland Region, NW Scotland, 5km (3 miles) SW of Mallaig. With a depth of 310m (1,017ft), it is the deepest Scottish loch and one of the deepest depressions in W Europe. It is said that a monster appears in it before the death of a member of the MacDonald Clan. Length: 20km (12 miles). Width: 0.8–3km ($\frac{1}{2}$–2 miles).

Moravia, Alberto (1907–), pseudonym of the Italian novelist Alberto Pincherle. His early novels, including *The Time of Indifference* (1929) and *The Fancy Dress Party* (1941), were increasingly critical of FASCISM, and Moravia was forced into hiding until 1944. Later works, examining hypocrisy and alienation, include *The Woman of Rome* (1947), *The Conformist* (1951) and *Two Women* (1957).

Moravia, region in central Czechoslovakia, bordered by Bohemia (W), the White and Carpathian Mts (E) and the Sudetic Mts (N). A fertile region, it is famous for its breeding of horses. Coal, oil, iron ore, copper, lignite and graphite are mined. Moravia became part of Czechoslovakia in 1918. It was occupied by the Germans during WWII. Industries: machine tools, beer, motor vehicles, timber, clothing, furniture. Area: 21,287sq km (8,219sq miles).

Moravian Church, Protestant body which originated in Bohemia and Moravia in the 15th century among followers of Jan HUS and developed in the following century under Jan Augusta (1500–72). In the 18th century, Moravians began extensive missionary work. Several groups migrated to North America, where they founded settlements in Bethlehem, Pennsylvania, and Winston-Salem, North Carolina. The Moravian Church has always distrusted doctrinal formulations and its liturgy is based on hymns emphasizing the life and passion of Christ. There are modern Moravian communities in Europe, North

Henry Moore: the abstract grandeur of his work has world-wide appeal.

Moors: interior of the Alhambra, a superb example of Moorish architecture.

Alberto Moravia ridiculed Mussolini in *La mascherata* as an absurd Latin American.

Moravia: an intricately decorated apron, typical of peasant costume of the area.

Sir Thomas More and his daughter watching from his prison as monks go to execution.

Morgans, first used as carriage horses, also make comfortable saddle horses.

Morgan Le Fay: in legend she made many attempts to have King Arthur murdered.

Berthe Morisot's *La Petite Niçoise.* Her later work owes much to Renoir.

and South America, Africa and N India.

Moray, or Murray, James Stewart, 1st Earl of (*c.*1531–70), Scottish nobleman, illegitimate son of JAMES V and half-brother of MARY, QUEEN OF SCOTS. After the return to Scotland of the young Queen Mary (1561), he became her adviser. He opposed her marriage to Lord DARNLEY and, after an unsuccessful rebellion, fled to England. Later, he was reconciled to Mary and, when she was forced to abdicate (1567), he became regent for JAMES VI. He was murdered in 1570 by James HAMILTON.

Moray (Elginshire), former county in NE Scotland, on the Moray Firth; in 1975 it was divided between Highland and Grampian regions. The fertile, low-lying coastal strip rises to hills inland; the chief rivers are the Findhorn, Spey and the Lossie. Sheep are grazed on the hills, and oats and barley are grown. Timber, fishing and the manufacture of whisky and woollen goods are the most important industries. The county town was Elgin. Area: 1,234sq km (467sq miles). Pop. (1971) 51,485.

Moray eel. *See* EEL.

Moray Firth, inlet from the North Sea bounded by the Nairn, Inverness and Ross and Cromarty coastlines. It is being developed for the extraction of oil.

Mordant, in dyeing processes, chemical that reacts with the dye or the fibre being dyed, or both, to "fix" the dye to the fibre, making it less likely to be washed out. Many mordants are metallic HYDROXIDES or salts (such as ALUM).

Mordecai, in the Old Testament, cousin and foster-father of ESTHER. With Esther, he thwarted Haman's efforts to destroy the Jews. His name is a form of MARDUK.

Mordvinia (Mordovskaja USSR), autonomous republic in w central USSR; the capital is Saransk. Annexed by Russia in the 16th century, the region was made an autonomous oblast in 1930, and an autonomous republic in 1934. It is an area of forested steppe and the major products are rye, wheat, oats, beans, potatoes, corn, tobacco and livestock. Industries include beekeeping, timber, motor vehicles, machinery, furniture and food processing. Area: 26,210sq km (10,116sq miles). Pop. (1970) 1,030,000.

More, Hannah (1745–1833), British author and social reformer who was one of the BLUESTOCKINGS. Her ethical tragedies *Percy* and *Fatal Falsehood* were produced by David GARRICK in 1777 and 1779. She was instrumental in the founding of the Religious Tract Society in 1799.

More, Sir Thomas (*c.*1478–1535), English statesman, writer and a Roman Catholic saint. A leading HUMANIST and friend of ERASMUS, his *Utopia* (1516) portrayed an ideal state founded on reason. He became HENRY VIII's adviser and succeeded Cardinal WOLSEY as chancellor, but he enraged the king by refusing to subscribe to the Act of Supremacy which made Henry the head of the Church of England and repudiated papal authority. Henry had him arrested and imprisoned in the Tower of London in 1534 and he was executed on a charge of treason. *See also* HC1 pp.247, *247*, 266, 268, *274*, 282, *282*.

Moréas, Jean (1856–1910), French poet, b. Greece. After going to Paris in 1879, he wrote several collections of SYMBOLIST poetry which show the influence of VERLAINE, including *Les Syrtes* (1884) and *Le Pèlerin Passioné* (1891).

Moreau, Gustave (1826–98), French painter best known for his fantastic and mystical pictures. He was professor at the École des Beaux-Arts, where his pupils included MATISSE and ROUAULT. He refused to sell his paintings and after his death his Paris house (now the Musée Moreau) with his fine art collection, was bequeathed to the nation.

Moreau, Jeanne (1928–   ), French film actress who worked with the NOUVELLE VAGUE (New Wave) directors before venturing into English-speaking cinema. Her films include *Les Liaisons Dangeureuses* (1960), *Jules et Jim* (1961) and *Mr Klein* (1976).

Morel, any of a number of fungi mushrooms genus *Morchella* that have

conical caps resembling pine cones and hollow stems. One species is *M. esculenta,* a highly prized edible mushroom. It grows only in the wild and can be found only in spring. Some "false morels" that outwardly resemble true ones are poisonous. *See also* NW p.*43.*

Morelia, city in SW Mexico; capital of Michoacan state. Founded in 1541 as Valladolid, its name was changed in 1828 in honour of the revolutionary hero Morelos y Pavon. The city has the Colegeo de San Nicolas which, founded in 1540, is the oldest institution of higher learning in Mexico. Industries: timber, handicrafts. Pop. (1975) 209,014.

Moreno, Jacob. *See* PSYCHODRAMA.

Mores, term first used by W. G. SUMNER in 1906 to mean those patterns of shared behaviour common to a particular group that, if violated, elicit shock, horror and moral outrage from the other group members. Mores are often verbal, unlike FOLKWAYS, and violation of them is often met with ostracism.

Moresby, John (1830–1922), British naval explorer. He is remembered for having discovered (1873) the large natural harbour in New Guinea that is now the capital, Port Moresby.

Morgan, Clifford Isaac ("Cliff") (1930–   ), Welsh rugby union player who, as a stocky, elusive flyhalf won 29 caps (1951–58) and toured South Africa with the 1955 British Lions. He played for the Cardiff and Bective Rangers clubs.

Morgan, Conway Lloyd (1852–1936), British psychologist, professor of Zoology at Bristol University (1884–1909). He was one of the founders of modern comparative psychology, which seeks insights into human behaviour through the study of animals. Among his publications are *Animal Behaviour* (1900) and *The Animal Mind* (1930).

Morgan, Daniel (1736–1802), US general who saw service in the French and Indian Wars. He received a gold medal from Congress for his leadership at Cowpens (1781) where he defeated the British during the American War of Independence. In 1794 he helped to suppress the Whiskey Rebellion in Pennsylvania, which tested the newly formed federal government's power to enforce a federal law within a state.

Morgan, Sir Henry (*c.*1635–88), Welsh soldier of fortune who in 1668 became leader of a group of BUCCANEERS commissioned by the British government. Among his conquests were Santa Catalina island, Puerto Príncipe, Puerto Bello and Panama. He was knighted and made Lieutenant-Governor of Jamaica in 1674.

Morgan, Lewis Henry (1818–1881), US anthropologist who did field work among the IROQUOIS and other American Indian groups, Morgan's interest was social organization and he formulated a theory of social evolution from "savagery" to "civilization" in his book *Ancient Society* (1877), but today he is best remembered for his classification of kinship systems.

Morgan, Thomas Hunt (1866–1945), US biologist who was awarded the 1933 Nobel Prize in physiology and medicine for the establishment of the CHROMOSOME theory of HEREDITY. His discovery of the function of chromosomes through experiments with the fruit-fly (*Drosophila*) is related in his book *The Theory of the Gene* (1926, rev. ed. 1928). *See also* NW pp.28–29.

Morgan, William de. *See* DE MORGAN, WILLIAM.

Morgan, breed of light horse developed in New England by Justin Morgan from a versatile horse of the same name foaled in 1793. A superior saddle, driving, racing and farm horse with stamina and docility, it has a high-held head, a round and deep, short-backed body, and wide-set thin legs. Height: to 1.6m (64in) at the shoulder.

Morganatic marriage, marriage between a prince or nobleman and a woman of humble birth or lower rank in which it is agreed that she shall not accede to his rank and that any children of the marriage shall not succeed to their father's hereditary titles, fiefs and entailed properties. Such marriages were first instituted in Germany

but were not adopted by English or French dynasties.

Morgan le Fay, in Arthurian legend, a fairy sorceress. Possibly derived from earlier Celtic goddesses, including the Welsh Morrigan, she appeared in several medieval romances, sometimes as the sister of KING ARTHUR, sometimes as a dangerous enchantress who opposed GUINEVERE. Tutored by MERLIN, she had magical gifts of healing, divination and transformation. Arthur was finally carried to her island, Avalon.

Morgenstern, Christian (1871–1914), German philosopher, poet and translator. His poetical works include love poetry and *Homecoming* (1910), but his fame rests on his two volumes of "nonsense" verse, *Gallow Songs* (1905) and *Palmström* (1910).

Mörike, Eduard Friedrich (1804–75), German poet. After becoming a theology student and later a curate and pastor, he published several volumes of subtle lyric poetry, including *Gedichte* (1938), a collection he added to in 1848, 1856 and 1867. He also wrote some prose works, notably a novel, *Maler Nolten* (1832).

Moriscos, christianized MOORS of Spain. As Christian rulers from the north reconquered Moorish Spain, they generally established a policy of religious tolerance towards the Moors; later it was decreed that all Moors either be converted or be expelled from Spain. Many Moors ostensibly became Christians but were secretly still Muslims, or were suspected of being so. The Spanish kings regarded them as a threat, and they were persecuted by the Spanish INQUISITION. In 1568 they revolted, and were put down two years later. In 1609, PHILIP III expelled all Moriscos.

Morison, Elsie (1924–   ), Australian soprano who studied in Melbourne and London with Clive Carey. Her earliest roles were in oratorio and throughout her career was always in great demand as an oratorio singer. She joined the Sadlers Wells Opera Company in 1948 and the Covent Garden Company in 1953. A highly acclaimed singer, she visited Australia as a guest artist in 1957 and 1960, and in 1964 with her husband Rafael Kubelik.

Morison, Stanley (1889–1967), British typographer whose influence on 20th-century printing was immense. He gained a wide practical experience of design for print and in 1923 joined The Monotype Corporation as typographic adviser. In 1930 Morison was commissioned by *The Times* newspaper to design a typeface especially for use on high-speed newsprint presses, and the resulting Times New Roman face proved to be exceedingly popular. The text of this book is set in Times Roman. Morison's publications included *Four Centuries of Fine Printing* (1924), *First Principles of Typography* (1936) and *Printing The Times* (1953). He was also responsible for many of the volumes of *The History of The Times.*

Morisot, Berthe (1841–95), French IMPRESSIONIST painter. She was Edouard MANET's student and sister-in-law, and each influenced the other's artistic development. Her most notable works, including *Young Woman at the Dance* (1880) and *La Toilette,* are painted in clear, luminous colours. *See also* HC2 pp.118–119.

Morita therapy, Japanese form of psychotherapy in which the patient undergoes a rigid form of conditioning that requires his participation in a number of prescribed social activities. Under a specially trained mentor the patient is discouraged from any introspection, the activity believed to lie at the heart of all neuroses. *See also* MS pp.148–149.

Morland, George (1763–1804), British painter. He was best known for his idealized pictures of rustic life, and his landscapes, such as *Interior of a Stable* (1791).

Morley, Edward William (1838–1923), US chemist. He worked with A. A. MICHELSON on the famous experiment in 1887 that demonstrated the absence in the universe of a stationary hypothetical substance called ETHER, that had been supposed to carry light waves. *See also* SU p.105.

Morley, John, Viscount Morley of Black-burn (1838–1923), British LIBERAL politician and man of letters. He was an MP from 1883 to 1895 and 1896–1908, holding the offices of Chief Secretary for Ireland (1886, 1892–95) and Secretary of State for India (1905–10). He was also editor of the *Fortnightly Review* (1867–83) and the *Pall Mall Gazette* (1880–83). Morley wrote several biographies, among them lives of CROMWELL, BURKE, VOLTAIRE, COBDEN and GLADSTONE.

Morley, Robert (1908–), British actor and playwright. He made his stage debut in 1929 and appeared in many productions in repertory and in London. He made his first film in 1929, and has starred in many comedy roles. His films include *Major Barbara* (1940), *Oscar Wilde* (1960) and *Cromwell* (1970).

Morley, Thomas (1557–*c.*1603), English composer who was organist of St Paul's Cathedral, London. He excelled at MOTETS, compositions for the LUTE and, above all, MADRIGALS. His *Plaine and Easie Introduction to Practicall Music* (1597) is an essay on 16th-century music.

Mormon, Book of, scripture regarded as holy as the Bible by the Church of Jesus Christ of Latter-day Saints, or MORMONS. Joseph SMITH, founder of the Mormon Church, claimed that in 1827 a heavenly messenger called MORONI, delivered a set of inscribed gold plates to him. These plates, which later disappeared, held the record as told by Moroni's father, the prophet named Mormon, of a migration from Jerusalem to the Americas beginning in 600 BC, and of some sayings of Jesus found nowhere else.

Mormons, an ADVENTIST sect, the full name of which is the Church of Jesus Christ of Latter-day Saints. It was established in Manchester, New York, in 1830 by Joseph SMITH who claimed to have received from an angel the Book of MORMON, the source of Mormon doctrine. Believing that they were to found Zion, or a New Jerusalem, Smith and his followers moved west. They tried to settle in Ohio, Missouri, and Illinois, but were driven out. Other settlers resented and attacked Mormons for a number of reasons, one being their belief in polygamy. Joseph Smith was murdered in Illinois in 1844. Brigham Young then rose to leadership and in 1846–47 took the Mormons to Utah Territory, where they established Salt Lake City. Opposition to the Mormons continued, but they prospered through hard work. The Mormon president renounced polygamy in 1890, and Utah became a state of the Union in 1896.

Morning glory, any of several species of twining and trailing vines native to warm climates; the blue, purple, pink or white flowers are funnel-shaped with a flaring disc. Species include common morning glory, *Ipomoea purpurea*, with heart-shaped leaves, and *I. violacea* of tropical America, which bears seeds that contain ALKALOIDS. Family Convolvulaceae.

Morning sickness, nausea and vomiting in pregnancy, usually occurring in the second and third months. *See also* MS p.76.

Morocco, independent nation in NW Africa. The land is dominated by the Atlas ranges; the climate is semi-tropical on the Atlantic side of these mountains and Mediterranean in the N. Agriculture is the main activity, and citrus fruits, vegetables and cereal crops are grown in the W of the country. Morocco is also a leading exporter of phosphates; other mineral deposits include iron, manganese and zinc ores. The capital is Rabat. Area: 458,730sq km (177,116sq miles). Pop. (1975) 17,305,000. *See* MW p.126.

Morón, city in E Argentina, 16km (10 miles) S of and part of Greater Buenos Aires. It was settled in the 16th century as a post on the road between Buenos Aires and Chile. It is now an administrative centre. Pop. (1970) 485,983.

Moroni, angel who appeared to Joseph SMITH, founder of the MORMON Church. Claiming to be the son of Mormon, an earthly prophet and historian, he led Smith to golden plates on which Mormon had recorded the migration of the

Hebrews from Jerusalem to the North American continent. Smith's translation was called *The Book of Mormon*. *See also* MORMON, BOOK OF.

Moroni, Giovanni Battista (*c.*1520–78), Italian painter who worked in the Bergamo and studied under Moretto. Although his religious paintings are dull aed lack originality, his portraits are intimate, penetrating and dignified. Among such works are *Portrait of a Gentleman* (1563), *Antonio Navagero* (1565) and *Portrait of an Old Man* (*c.*1565–70).

Moronobu, Hishikawa (1618–94), Japanese printmaker and master of UKIYO-E (colour print) who is credited with the first edition of single-sheet prints, thereby creating ukiyo-e as an artform.

Moros, in Greek mythology, the supreme power, Doom or Destiny, the son of Night. Even the gods of the Olympic pantheon obeyed his decisions. Although Moros is seldom personified in the myths – being conceived as invisible and shadowy – his presence as a force is pervasive. *See also* FATES.

Morosini, Venetian noble family prominent from the 10th century. Four DOGES of Venice and two cardinals came from the family. Andrea Morosini (1558–1618) was official historian of the Venetian Republic and wrote *Historia Veneta ab anno 1521 usque ad annum 1615*. Francesco (1618–94), an illustrious general and naval commander, was elected doge in 1688.

Morpheme, the smallest unit of speech. It may be an entire word, eg "paint", or a part of one, eg "re" in "repaint". An inflectional morpheme is the added "s, es, ed, ing" of a plural or the endings added to verbs. A derivational morpheme is the "hood" in "motherhood".

Morpheus, in Greek mythology, the god of dreams, supposedly expert in sending the human form into the dreams of men. He was the son of Hypnos, the god of sleep.

Morphine, white crystalline ALKALOID derived from OPIUM. It is a depressant of the central nervous system and is used as an ANALGESIC. As with most strong analgesics, it has addictive properties. Morphine was first isolated in 1806 and named after the Greek god of dreams, MORPHEUS. *See also* HEROIN; METHADONE.

Morpho butterfly, large, tropical American BUTTERFLY noted for the brilliant blue metallic lustre of the upper surfaces of its wings. Family Morphoidae.

Morphology, in grammar, study of the form of words, the branch of PHILOLOGY dealing with inflection and word-formation.

Morphology, biological study of the form and structure of living things and the relation between similar features in different organisms. It ranges from visible characteristics to microscopic structures.

Morris, William (1834–96), British artist, craftsman, writer, social reformer and printer. After attending Oxford University he came into contact with the PRE-RAPHAELITE BROTHERHOOD and began to write poetry and to paint. Both of these activities reflected his love of the Middle Ages and in 1861 he founded a firm of decorators and designers, attempting a return to the craftsmanship of that period. The group produced carvings, wallpapers, carpets and furniture. In the 1880s he became interested in socialism, writing *The Dream of John Ball* (1886–87) and *News from Nowhere* (1890), in which he argued for the replacement of the machine age by skilled labour. His last venture was the KELMSCOTT PRESS, where he was responsible for typography, border design and binding.

Morris, W. R. *See* NUFFIELD, W. R. MORRIS, 1ST VISCOUNT.

Morris dance, English rustic dance probably deriving from pre-Christian rituals, performed as part of the MAY DAY and WHITSUN festivals. It is danced to pipes, tabors, and occasionally bagpipes. The dance was often performed along the roads from one village to another, although its character varied according to locality.

Morrison, Herbert Stanley (1888–1965),

British politician who helped to found the London Labour Party. He became Mayor of Hackney (1920–21) and was an MP three times before 1945. In 1955, after four years as deputy leader of the opposition, he was defeated by Hugh GAITSKELL in the contest for leadership of the LABOUR PARTY.

Morse, Samuel Finley Breese (1791–1872), US inventor and artist. Originally a successful portrait painter, Morse revived his earlier interest in electricity to experiment with a practical system for using electricity to send messages by telegraphy. In 1844 he telegraphed the message "What hath God wrought?" from Washington to Baltimore. He had developed a simple system of dots and dashes, now called the MORSE CODE, by 1837. *See also* MM pp.224–225; MS p.242.

Morse code, code used for sending TELEGRAPH messages, either along wires or by radio telegraphy. It consists of dots and dashes created by the interruption of a continuous electric current or radio signal. The dashes are three times the length of the dots. The codes for letters of the alphabet, or numbers, vary in length from a single dot or dash to six dots or dashes or a combination thereof. *See also* MM p.*224*; MS p.242.

Mortality rate, in population studies, defined as the number of deaths occurring in a year, multiplied by a thousand, divided by the total population at the mid-year. Thus, for a total population of 1 million people and a death toll of 2,000 over a year, the mortality rate for that year is $(2,000 \times 1,000) \div 1,000,000 = 2.0$.

Mortar, in modern military usage, a portable weapon consisting of a short metal tube, baseplate and tripod. With the tube in a near-vertical position, inclined slightly towards the target, shells with explosive charges (often called bombs) are dropped into the open tube, the propulsive charge detonates and the shells shoot out to follow a high curved trajectory to the target. In the earlier history of artillery, the name "mortar" was often given to any large-bore cannon. *See also* MM pp.*167, 168*.

Mortar, in building construction, a material used to bind brick, stone, tile or concrete blocks into a structure. Modern mortar generally consists of a mixture of cement, sand and water; lime may be added to improve its spreading properties. *See also* MM pp.46–47.

Mortar, in chemistry, a bowl of smooth, hard material in which softer substances are ground or beaten with a pestle, a club-shaped tool.

Morte D'Arthur, Le (*c.*1469), Arthurian prose romance by Thomas MALORY, first printed by William CAXTON (1485). Its 21 books deal principally with the quest for the Holy GRAIL, the death of KING ARTHUR and the disbanding of the fellowship of the ROUND TABLE. Malory borrowed much of his material from earlier texts.

Mortgage, transference of the ownership of property pledged as security for a loan of money. Until the loan is fully repaid with all interest due, the property pledged (mortgaged) is technically the possession of the mortgagee who lent the money to the nominal owner (the mortgagor). Often, especially in the buying of land or a house, the item mortgaged is itself the purchase for which the loan is required.

Mortimer, name of a noble and influential Anglo-Norman family which flourished between the 13th and 15th centuries. Roger de Mortimer, 1st Earl of March (1287–1330), secured his position in Ireland, opposed the DESPENSERS (favourites of EDWARD II) in Wales and was imprisoned in 1322. He escaped to France and returned to England in 1326 as the paramour of Isabella, Edward's queen. After Edward's deposition and murder, Mortimer virtually ruled England until seized and hanged by EDWARD III. Edmund de Mortimer, 3rd Earl of March and 1st Earl of Ulster (1351–81), was the son of Roger de Mortimer, 2nd Earl of March. In 1368 he married Edward III's granddaughter, Philippa, and thus gave the house of York its claim to the throne that led ultimately to the Wars of the ROSES. Sir

Morning Glory: its flowers open in early morning but close with the midday sun.

Morpho butterfly: the largest species, *morpho hecuba*, is 17.8cm (7in) across.

Morris dance: the traditional bells and handkerchiefs are essential to the dance.

Samuel Morse: the old Chamber of Deputies, 1822, painted by this artist and inventor.

Mortimer, John Clifford

Mosaic: a detail from the church of San Vitale, Ravenna, of the Emperor Justinian.

Moscow: a view of part of the Kremlin, with the Moscow River in the foreground.

Sir Oswald Mosley, third from right, at a Fascist parade in Rome between the wars.

Mosque: this mosque of Sultan Ahmed in Istanbul dates from the 17th century.

Edmund de Mortimer, son of the 3rd Earl of March, was captured while helping to repress the rebellion of the Welsh prince Owen Glendower in 1402. He subsequently defected to Glendower's side and married his daughter. Roger de Mortimer, 4th Earl of March and 2nd Earl of Ulster (1374–98), son of the 3rd Earl of March, was heir presumptive to RICHARD II and accompanied him on his punitive expedition to Ireland. His subsequent popularity, however, incurred Richard's jealousy. He died fighting the Irish at Kells. Edmund de Mortimer, 5th Earl of March and 3rd Earl of Ulster (1391–1425), was recognized as Richard II's heir but was prevented from reigning by the usurpation of HENRY IV in 1399. He died childless but his claim to the throne passed through his sister Anne to her grandson EDWARD IV.

Mortimer, John Clifford (1923–), British dramatist, novelist and barrister. He published several novels before making his reputation as a dramatist in 1957, when *The Dock Brief*, a play with a legal setting, was produced on radio. His stage plays include *The Wrong Side of the Park* (1960), *The Bells of Hell* (1966) and *A Voyage Round My Father* (1970). He has also made various translations and written film scripts, including *John and Mary* (1969). His first wife, Penelope, is also a novelist. Her works include *The Pumpkin Eater* (1962) and *Long Distance* (1974).

Mortimer's Cross, scene of 1461 battle in the Wars of the ROSES, in which Yorkist forces, under Edward, Duke of York, defeated the Lancastrians. After the battle Edward proceeded to London and proclaimed himself King EDWARD IV.

Mortmain, term (taken from the French for "dead hand") meaning the ownership of land by religious, charitable or other corporations. It had a troubled history throughout the Middle Ages when corporations such as monasteries, exempt from the payment of feudal dues, acquired large properties. The law of mortmain was abolished in Britain in 1960, and such property is restricted to that necessary for charitable use.

Morton, Henry Vollam (1892–), British writer. After the success of his early works, he left journalism and took up writing full-time. His works, mainly travel books, include *In Search of England* (1927), and similar volumes about Ireland (1930), the Middle East (1940) and the Holy Land (1961).

Morton, James Douglas, 4th Earl of (*c*.1516–81), Scottish statesman. As chancellor (appointed 1563) to MARY, QUEEN OF SCOTS he supported, but without enthusiasm, her marriage to Lord DARNLEY. He was subsequently involved in the plot to murder Darnley in 1567. After Mary married Lord BOTHWELL, Morton turned against the queen, who was forced to abdicate. Succeeding as regent in 1572, he administered Scotland efficiently but was criticized for introducing episcopacy. French influence at court finally caused his downfall and execution for complicity in Darnley's death.

Morton, Jelly Roll (1885–1941), US jazz pianist, bandleader and composer, real name Ferdinand Joseph LaMenthe. He played at first in the brothels of Storyville in New Orleans and began recording in 1923 with his group Morton's Red Hot Peppers. His music reflects his Creole origins.

Morton, William Thomas Green (1819–68), US dentist who demonstrated publicly, in 1846, the anaesthetic property of ether by painlessly extracting a tooth. Morton experimented with ether on animals and himself, but first used it on a patient at the suggestion of Charles T. Jackson with whom he later quarrelled for the credit. *See also* LONG, CRAWFORD W.

Morula, embryo cells in the process of splitting before the BLASTULA stage. The morula consists of a number of blastomeres, the cells formed from the fertilized egg as a result of cell division.

Mosaic, technique of surface decoration using tesserae, small pieces of material such as stone, pebbles, tile, glass, shell or metal; they are set closely together in an adhesive substance such as mortar, plaster or resin, to form a decorative design or picture. The design is then grouted; it is caulked to fill the gaps between the tesserae and produce a flat surface. The technique was employed for floor and wall decorations as early as the 4th millennium BC in Mesopotamia. Pebbles were used in the Near East and were replaced in Greece in the 3rd century BC by cut marble tesserae. Roman mosaics commonly featured a central design or a portrait, surrounded by a decorative geometric border. In Byzantine times, the marble was replaced by tesserae of brightly coloured glass, especially golden yellow. The art developed rapidly in early Christian times especially in the 4th to 6th centuries and continued to be used for floors, church and interiors and wall decorations. As a modern technique, used frequently for surface decoration, the tesserae may be mixed with other materials for rich textural effects.

Mosaic disease, any of several VIRUS diseases of plants that result in mottling on leaves, giving rise to light green or yellow blotches. Some mosaics also cause curling and puckering of leaves and stunting of the plants. They are serious diseases affecting many crops, including apples, peaches, beans, cucumbers and tobacco.

Moscow (Moskva), capital and largest city of the USSR, in w central Russian Republic (Rossijskaja SFSR), on the Moscow River; capital of the republic and of Moscow oblast. Archaeological remains indicate that the site has been inhabited since Neolithic times, but Russian records do not mention it until 1147. It became a principality by the end of the 13th century and, in 1367, the first stone walls of the KREMLIN were constructed. In 1341 Moscow captured the principality of Vladimir. By the end of the 14th century the city had emerged as the focus of Russian opposition to the MONGOLS. In 1478, Grand Duke Ivan III led a successful campaign against the principality of Novgorod. Polish troops occupied the city in 1610, but were driven out two years later. Moscow became the capital of the Grand Duchy of Russia from 1547 to 1712, when the capital was moved to St Petersburg. In 1812 Moscow, which was built almost entirely of wood, burned to the ground. NAPOLEON and his troops, who were occupying the city, were forced out by the fire. In 1918 Moscow became the capital of the USSR. The failure of the Germans to take the city during WWII was Germany's first major land defeat of the war. The Kremlin is the centre of the city and the administrative centre of the country. It includes palaces, churches and government buildings within its walls. Adjoining it are Red Square, the Lenin Mausoleum and the 16th-century cathedral of Basil the Beatified (now an anti-religious museum). The city also has the University of Moscow (1755), Tretyakov Gallery, BOLSHOI THEATRE, many institutions of higher education and scientific institutes. Industries: metalworking, oil-refining, motor vehicles, film-making, precision instruments, chemicals, wood and paper products, tourism. Pop. 7,734,000. *See also* HC1 pp.320–321, *321*; HC2 p.*199*.

Moscow Art Theatre, USSR theatre, the most famous of all Russian theatrical organizations. It was founded in 1898 by STANISLAVSKY and NEMIROVICH-DANCHENKO. CHEKHOV's plays were first presented successfully there. The original company was composed of amateur actors from the Society of Art and Literature. It was at the Moscow Art Theatre that Stanislavsky developed his famous "method" of acting, which stressed the importance of the actor's inner identification with the character.

Moscow, Grand Ducy of. *See* MUSCOVY, GRAND DUCHY OF.

Moseley, Henry Gwyn Jeffreys (1887–1915), British physicist. His initial studies involved RADIOACTIVITY; later he discovered a relationship between the X-ray spectra of the elements and their atomic numbers. He discovered that the atomic number of an element, and not its weight, determines its major properties.

Moselle, river in France and Germany which rises in the Vosges Mountains, NE France. It flows NW past Remiremont, Epinal and Toul, turns N to flow past Metz and Thionville and NE to leave France and form the Luxembourg-German boundary. It flows through Germany (where it is called the Mosel), emptying into the River Rhine at Koblenz. The Moselle is connected to the Rhine, Meuse and Seine rivers by canals. Grapes grown along its steep banks between Trier and Koblenz are used to make Moselle wine. Length: 515km (320 miles).

Moselle, dry white wine from the vineyards in the valley of the River Moselle, in France and Germany.

Moser, Annemarie (née Pröll) (1953–), Austrian skier. She was the women's world overall champion in 1971, 1972, 1973, 1974 and 1975, and a runner-up in the downhill and giant slalom events at the 1972 Olympic Games.

Moser, Lukas (*fl. c*.1430–*c*.1450), German painter who worked in Ulm. He is known for the altarpiece of *Mary Magdalene and Martha* (1431) in the church at Tiefenbronn. The detail is realistic and advanced for its period.

Moses (*c*.13th century BC) in the biblical book of Exodus, the spiritual and temporal leader of the ancient Hebrew people. As a male child of the race which was enslaved in Egypt he was, by law, abandoned at birth, but was found by one of the pharaoh's daughters and brought up at court – which accounts for his Egyptian name. Following a call from God, perceived in a burning bush, Moses sought the pharaoh's permission to lead the Hebrews out of Egypt to find a new homeland. Only after several dire natural plagues had infested the land did the EXODUS take place, during which the tablets of stone on which the TEN COMMANDMENTS were inscribed were given to Moses on Mt Sinai. Because of an unrecorded offence against God, Moses was not allowed to set foot in the Promised Land of CANAAN but died within sight of it.

Moses, Grandma (Anna Mary Robertson) (1860–1961), US primitive painter. In her late seventies, when work as a farm wife proved too arduous, she took up painting. Her jolly scenes of country life based on recollections from her youth became world famous through prints and greetings cards. Well-known examples are *Out for the Christmas Trees* and *Thanksgiving Turkey*. She published her autobiography in 1952.

Moshoeshoe II (1938–), King of Lesotho. He was Chief of Basutoland from 1960 to 1966, when he became king of the newly independent Lesotho. He was exiled for a short time, but returned in 1970. *See also* HC2 p.253.

Moskva. *See* MOSCOW.

Moslem League. *See* MUSLIM LEAGUE.

Moslems. *See* MUSLIMS.

Mosley, Sir Oswald Ernald (1896–), British politician and founder of the British Union of Fascists (1932). He was a Conservative Member of Parliament (1918) but later joined the Labour Party, serving as Chancellor of the Duchy of Lancaster (1924). Dissatisfied with both the democratic left and right, he turned to FASCISM coupled with ANTI-SEMITISM. He was interned during WWII (1940–43) but after the war founded the Union Movement (1948), a new right-wing party. *See also* HC2 pp.222–223.

Mosque, Muslim place of worship. Mosques are usually decorated with abstract and geometric designs as the Muslim faith prohibits the imitation of God's creation. The building's parts include a *mihrab,* or prayer niche, which shows the direction of MECCA, a minaret, or tower, from which the MUEZZIN calls prayers, a courtyard for washing before prayer and a maktab, or school. Among the earliest mosques is that of Amr built in Egypt in 641–42. Although in early ISLAM any building could be used as a place of prayer, the Mosque of the Prophet at MEDINA, containing Mohammed's tomb , and the mosque at Mecca, containing the KAABA, have since become the foci of much Muslim pilgrimage and worship. *See also* HC1 pp.144–145.

Mosquito, long-legged, slender-winged insect, found throughout the world. The adult female sucks blood from warm-blooded animals, including man. Some species (such as those of the genus *Anopheles*) carry the parasites of diseases, including MALARIA, YELLOW FEVER, DENGUE, viral ENCEPHALITIS and FILARIASIS. The larvae are aquatic and are found in lakes, rivers, marshes, and other wet places. Adult length: 3–9mm (0.12–0.36in) Family Culicidae. *See also* DIPTERA; NW p.*226.*

Mosquito, British WWII bomber and reconnaissance aircraft, built by De Havilland. Known as the "wooden wonder", it could reach speeds of 640km/h (400mph). Variants of the Mosquito were used as night-fighters.

Mosquito Coast (Mosquitia), E coast of Honduras and Nicaragua, named after the indigenous Miskitto Indians. It was disputed by Britain and the USA in the 19th century, and was made autonomous in 1860. It was divided between Nicaragua and Honduras in 1894; Nicaragua's claim to the whole area was refused by the International Court of Justice in 1960.

Mosquito fish, freshwater fish of the SE USA known for its ability to eat large amounts of mosquito larvae. Its colours are brown and grey with a bluish shimmer. Length: 6.3cm (2.5in). Family Poeciliidae; species *Gambusia affinis.*

Moss, Stirling (1929–), British racing driver who won his first Formula Three race in 1948. He won 16 grand prix events and finished second four times and third three times in the world championship. He retired from the sport after a serious crash at Goodwood in 1962.

Moss, any of about 14,000 species of small, simple non-flowering green plants that typically grow in colonies, often forming dense carpets. Mosses reproduce by ALTERATION OF GENERATION. They do not have specialized tissues for transporting water, food and minerals, although they do have parts resembling the stems, leaves and roots of the higher (flowering) plants. They grow on soil, rocks and tree trunks in a wide variety of land habitats, from the tropics to circumpolar regions, but thrive in shady damp places and some species live in freshwater lakes and streams. Because carpets of moss can absorb large amounts of water, they help to reduce soil erosion. *See also* BRYOPHYTE; NW pp.50–51.

Mossadeq, Mohammad. *See* MUSSADEGH, MUHAMMAD.

Moss animal, any of numerous species of mostly marine, aquatic invertebrates, generally of shallow coastal waters. Some appear moss-like; others assume fan-like or twig-like colonies and are white or pale in colour. Each individual has tentacles to gather food into the mouth. Phylum Bryozoa.

Mössbauer, Rudolf Ludwig, (1929–), German physicist and discoverer of the effect named after him. His doctoral thesis, published in 1958, dealt with the emission of GAMMA RAYS by radioactive nuclei within crystals. The definition of the frequencies of these rays became known as the MÖSSBAUER effect, for which work he shared the 1961 Nobel Prize in physics with Robert HOFSTADTER.

Mössbauer effect, resonance effect observed in certain radioactive substances emitting GAMMA RAYS, is taken up by the atomic nuclei, so that the gamma rays have a sharply defined energy instead of a range of energies. They can then be re-absorbed by a similar nucleus. Motion of the source alters the frequency of the rays, by the DOPPLER EFFECT, and the Mössbauer effect may be used to study nuclear energy levels and molecular structure.

Most-favoured nation status, favourable terms of trade: tariff concessions received by one nation from another. The status is given normally between countries in a trade bloc and which join together under "common market" agreements, eg the EEC. The principle was included in GATT.

Mosul, city in N Iraq, 354km (220 miles) NNW of Baghdad, on the River Tigris. It was the most important city in the area until the 13th century when it was destroyed by the Mongols. It was occupied by the Ottomans (1534–1918) and came under British occupation and mandate in 1918. Mosul was awarded to Iraq by the League of Nations in 1926 after a dispute for its possession between Iraq and Turkey. Industries include oil refining and cement. Pop. 293,100.

Moten, Benjamin (Bennie) (1894–1935), US jazz pianist and band-leader. He was one of the founders of the Kansas City style which emphasized a flowing rhythm. On Moten's death his band was taken over by Count BASIE.

Motet, musical form prominent in all choral church music from *c.* AD 1200–1600. In the 13th and 14th centuries it consisted of three unaccompanied voice parts sung in Latin, the tenor part providing the *canto fermo.* The Renaissance motet of the 15th century, usually for four or five parts, was contrapuntal in style and those of Dunstable, Dufay and Obrecht are typical of this period. The purest examples of the motet are those of PALESTRINA, and after 1600 there were new developments in the form, including occasional obbligato accompaniment and texts in vernacular languages. Motets in English became known as ANTHEMS.

Moth, insect of the order LEPIDOPTERA, found in almost all parts of the world. It is distinguished from a BUTTERFLY mainly by its non-clubbed antennae, although there are a few exceptions. Most moths are nocturnal whereas almost all butterflies are active by day. The largest moths, with a wingspan of 28cm (11in) or more, are found in the tropics. Like a butterfly a moth undergoes complete METAMOPHOSIS and the adult emerges from a pupa formed from its larva (caterpillar). Also like a butterfly, it has a long coiled proboscis for sipping liquid food, particularly the nectar of flowers. *See also* NW pp.*34, 77, 99, 113, 117, 208, 209, 215, 218.*

Moth balls, substance used to protect woollen clothes, carpets, furs, etc against attack by the caterpillar of the clothes moth. Wool can be mothproofed by incorporating insecticides into the dye, by immersing the fabric in a solution of DDT or the salts of aromatic sulphonic acids, or by impregnation with sodium fluorsilicate. Traditional mothproofing agents include camphor, naphthalene, and chlorinated hydrocarbons such as dichlorobenzene and hexachloroethane.

Mother Carey's chicken. *See* STORM PETREL.

Mother Courage and Her Children (1941), play by Bertolt BRECHT, with music by Paul Dessau. An epic chronicle play of the THIRTY YEARS WAR, it is intended as an allegorical indictment of humanity in the form of the profiteer Mother Courage, an itinerant pedlar. Audiences have, however, reacted with sympathy to her sufferings as she follows the war in her wagon, accompanied by her mute daughter and her two sons.

Mother Goose, traditionally in France a figure who told folk-tales. The phrase *Les Contes de ma Mère l'Oye* (Tales of Mother Goose) appeared in Charles PERRAULT'S 1697 collection of tales. The name first appeared in England in a collection of nursery rhymes published by John NEWBERY *c.*1780. He called the collection *Mother Goose's Melody or Sonnets for the Cradle.*

Mother-of-pearl, shiny substance lining many MOLLUSC shells, also called nacre. It is composed of a form of calcium carbonate deposited in layers interspersed with organic material. Diffraction of light causes the lustre and iridescence of mother-of-pearl. It is used in making buttons and jewellery, and for decorative inlay work.

Motherwell, Robert (1915–), US painter and writer. Associated with ACTION PAINTING, he is best known for his ABSTRACT EXPESSIONISTIC works. He was the editor of *The Documents of Modern Art* series (1944–57).

Motion, accelerated, motion in which speed, or VELOCITY, increases. For example, a stone dropped over a cliff accelerates from zero velocity at a rate of 9.81m (32.2ft) per second per second, this acceleration being due to the pull of the Earth's gravity. In general, linear acceleration is equal to the change in velocity divided by the elapsed time. Thus, if the speed of an object increases in 10 seconds from 10m per sec. to 20m per sec. its average acceleration is $(20-10) \div 10 = 1$m per sec. (written 1m sec.$^{-2}$). When an object rotates about an axis, its angular acceleration is given as the change in angular velocity divided by elapsed time, expressed in radians per second per second (written rad. sec.$^{-2}$). A radian is a measure of angle: 2π radians $= 360°$.

Motion, laws of, three laws proposed by Isaac NEWTON in his *Principia* (1687) that form the basis of the classical study of motion and FORCE. According to the first law, a body resists changes in its state of motion–a body at rest tends to remain at rest unless acted on by an external force, and a body in motion tends to remain in motion at the same velocity unless acted on by an external force. The second law states that the change in velocity of a body as a result of a force is directly proportional to the force and inversely proportional to the mass of the body; ie if the change in velocity, or ACCELERATION, is a, the force is F, and the mass is m, then $a = F/m$. According to the third law, to every action there is an equal and opposite reaction. *See also* INERTIA; SU p.70.

Motion, stellar, change in the position of a star including its PROPER MOTION and its RADIAL MOTION. *See also* SU p.165.

Motion picture. *See* CINEMA.

Motivation, psychological process hypothesized to account for observed behaviour. Animals and human beings appear to be motivated to perform certain actions, such as seeking food and water. Motivation is supplied by a hypothetical impulse towards satisfying needs or achieving goals, this force being the drive. Motives may reflect internal, physiological states, eg hunger or thirst; or they may be primarily learned and oriented towards external objects or incentives, eg money, possessions or status.

Motley, joint pseudonym of influential team of theatrical designers comprising Audrey Harris (1902–66), Margaret Harris (1904–) and Elizabeth Montgomery (1902–) who took their title from the name for a jester's diversified and multicoloured dress. In the 1930s, they established a reputation for working on all aspects of stage design from costumes and properties to scenic construction, in close co-operation with a play's director and, often, author. Their first joint designs were for the OUDS production of *Romeo and Juliet* in 1932. After that they worked consistently with major companies in Britain, notably at the OLD VIC and in the US.

Motmot, small bird of Central and South America. The blue diademed motmot (*Motmotus motmota*) has a green body, turquoise head and a serrated bill for crushing insects, nests are tunnelled in earthbanks. Length to 50cm (20in). Family Motmotidae.

Motocross. *See* SCRAMBLING.

Motonobu, Kano (1476–1559), Japanese painter who decorated palace and temple walls with his large works. He combined bright colours with well-defined brushstrokes.

Motor, mechanism that converts energy (eg heat or electricity) into useful work. The term is sometimes applied to the internal combustion ENGINE (which converts heat produced by burning gases into reciprocating or rotary motion), but is more often applied to the ELECTRIC MOTOR (which converts electrical energy into rotary motion). Electric motors vary in size from thumb-sized miniatures used in toys, through those in medium-sized domestic appliances (eg vacuum cleaners, food mixers and hand drills), to large motors such as those used in factories, lifts and diesel-electric locomotives. The linear motor is a new development proposed for high-speed train propulsion. Its principle is similar to that of rotary electric motors, but instead of a number of coils rotating (a rotor) within a fixed ELECTROMAGNET (a stator), the linear motor is

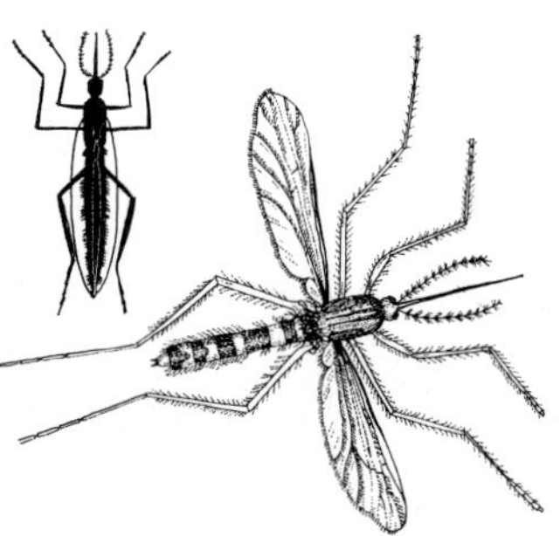

Mosquitoes, found worldwide, spread some of man's most serious diseases.

Mosses may have stem and leaf-like structures although they never flower.

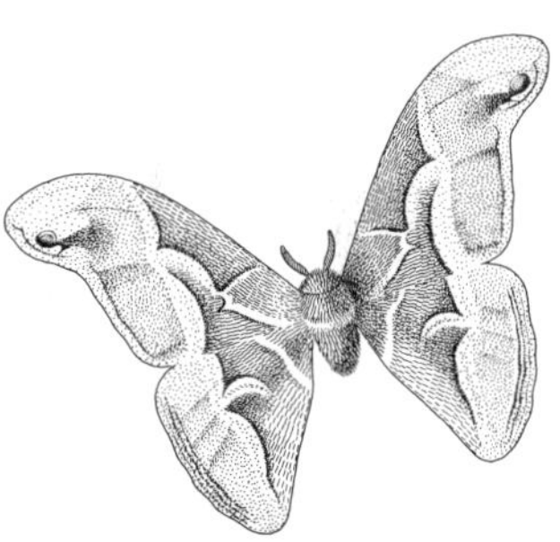

Moths see well in the dark but it is their sense of smell that guides them to flowers.

Motmots are solitary creatures that live on insects, fruit and some small reptiles.

Motor boat

Motor car: C. S. Rolls in a 1900 model with the Duchess of York.

Motorcycle racing: Steve Baker of the USA practising on his 500cc Yamaha.

Mouflon: except at breeding times, males live apart from ewes and their young.

Moulin Rouge: Can-Can dancers depicted by F. de Myrbach at the turn of the century.

"unfolded" so that the "rotor" windings are fixed to the locomotive's path, and the "stator" windings are incorporated into the locomotive, which is both levitated and driven forwards. Rocket engines are motors that can leave the Earth's atmosphere because they carry both fuel and oxidizer. Ion motors are another proposed development, intended for spacecraft propulsion. A stream of ions, possibly from a nuclear reactor, is accelerated in a strong electrostatic field to produce a reaction, similar to that of rocket motors, which drives the spacecraft. *See also* MM pp.140, *141*, 156, *250*–251; SU pp.*69*, *115*, *119*, 122–*123*.

Motor boat, vessel smaller than a ship or motor yacht, powered by an internal combustion engine which usually drives a single screw. Motor boats vary in size from motorized dinghies 2m (6ft) or so in length to large cabin cruisers 20m (60ft) or more long. Engines of smaller craft are often of the easily detachable (outboard) type – which can, however, be extremely powerful. Larger motor boats have inboard marine engines. Fast motor boats include naval chasers, HYDROFOILS, and stepped-hull boats such as power boats, all of which ride high out of the water at speeds of 35km/h (20 knots) upwards.

Motor bus, petrol or diesel powered passenger vehicle. The first motor bus, a petrol-driven coach which could carry eight passengers, was built by Karl Benz in 1895. Motor buses were introduced into Britain at the end of the 19th century and gradually replaced TRAMS because they were more versatile and economic to operate. *See also* MM pp.134–135.

Motor car, road vehicle which first appeared in the 19th century. The first cars were propelled by steam, but such vehicles were not a success and the age of the motor car really dates from the introduction of the petrol-driven horseless carriages of Gottfried DAIMLER and Karl BENZ (1885–86). The internal combustion engine for these cars had been developed earlier by several engineers, most notably by another German, Nikolaus OTTO, in 1876. The main components of a motor car, from then till now, are a body or chassis to which are attached all other parts – including the engine or power plant, the transmission system for transferring the drive to the wheels, and the steering, braking and suspension mechanisms for guiding, stopping and supporting the car. The first cars were assembled by a few experts, but modern mass-production was begun in the early 1900s by Henry FORD and R. E. Olds in the USA. By this means, the cost of a car was drastically reduced, and many more people could afford one. In most modern motor car factories component parts are put together on assembly lines – slow-moving conveyor belts. Each worker usually has a specific task, eg fitting doors or crankshafts. Bodies and engines are constructed on separate assembly lines which converge when the engine is installed. Overhead rail conveyors move heavy components to and along the assembly lines, and lower them into position. At a later stage on the assembly line such items as lamps are fitted, and electrical, braking and control systems are tested. The fully assembled car is road tested before sale. *See also* MM pp.68–69, 94–95, 126–133; MS p.*300*.

Motor cell, in botany, one of several cells which together are able to expand and contract, thus causing movement in a plant.

Motor cycle, two-wheeled automotive vehicle, that combines the principles of a BICYCLE and the INTERNAL COMBUSTION ENGINE. First designed in 1885 by Gottlieb DAIMLER, it did not become popular until about 1910. Motor cycles are classified in terms of engine capacity, which ranges from 50cc to 1200cc. There are two types of engine: four-stroke and two-stroke; transmission is by chain, although some early motor cycles had belt transmission. The clutch, accelerator and front braking controls are usually on the handlebars, and the gear change and rear brake are controlled by foot pedals. *See also* MM p.124.

Motor cycle racing, sport in which motor cyclists compete on road circuits, across country (scrambling and trials), on grass tracks, or on cinders (speedway). Road-racing motor cyclists contest an annual world championship of grands prix in Europe, South and North America, and Japan for 50cc, 125cc, 350cc, 500cc, 750cc and sidecar classes. British, German, and Italian machines were the most successful until the 1960s, when Japanese motor cycles swept aside all but the Italian MV Agustas. Until the 1970s, when the top riders operated a boycott, the most famous event was the Isle of Man Tourist Trophy (TT) held in the first two weeks of June. *See also* MM pp.124–125.

Motor nerve, nerve carrying messages to the muscles from the BRAIN via the SPINAL CORD. The cell bodies of some motor NERVES form part of the spinal cord. The AXONS of these nerves, which are sheathed with MYELIN, pass from the spinal cord to connect with the muscles. These nerves are involved in REFLEX ACTIONS. Motor nerves are, however, also linked to the brain by the descending spinal tract. The cell bodies of this tract lie in the CEREBRAL CORTEX and their axons, which pass down the spinal cord, connect the nerves to the brain, so allowing voluntary muscular control. *See also* MS pp.38–41.

Motor racing, sport in which men and women use cars for competition. The first event of any significance was the 1894 Paris to Rouen motor trial, since when the sport has moved in a number of directions. The Federation Internationale de l'Automobile (FIA) classifies cars as production, prototype or racing, and divides those into specific groups according to the number made, modifications or, with single-seater racing cars, a formula. Within the groups there are further subdivisions into capacity classes. For example, Formula One grand prix cars have since 1966 been limited to an engine capacity of 3 litres (or 1.5 supercharged). Grand prix racing began in France in 1906 but only since 1950 has there been a world championship for drivers and the International Cup for manufacturers. Points are awarded to the first six places in grand prix races held world-wide throughout the year. Sports car drivers, however, have no championship, the International Championship of Makes being solely for constructors. Rallying is in the same position though there is a European championship, as there is for saloon-car racing and hill-climbing. In the USA, popularity is divided between single-seater racing and saloon-car racing on super-fast oval speedways, drag racing and, to a lesser extent, road-racing. The annual Indianapolis 500 is motor sport's richest event. Other branches of the sport, each with an avid following, are autocross, rallycross, stock-car racing, trials, off-road racing, and vintage car racing. Land speed records are kept for the various capacity classes of cars.

Motor scooter, small-wheeled MOTOR CYCLE that originated in Italy soon after WWII; the first was a 125cc model. The wheels are from 20 to 36cm (8–14in) in diameter. Power units are placed low, close to the rear wheel which may be driven by bevel gearing or a chain. *See also* MM pp.124–125.

Motor Sports, set of sports in which competitors use MOTOR CARS, MOTORCYCLES and MOTOR BOATS. Typical motor sports are MOTOR RACING, MOTORCYCLE RACING and SCRAMBLING.

Motorway, in Britain, high-speed roads, often of six lanes, with no direct junctions and on which pedestrians, bicycles and animals are not permitted. Corresponding to freeways in the USA, motorways are much used for both industrial and personal communications. The first motorway, part of the M1 between London and Leeds, was opened in 1959. *See also* MM pp.184–185.

Mott, Sir Frederick Walter (1853–1926), British neurologist. He researched the structure and function of the nervous system and in 1916 postulated the first diagnosis of shell-shock for soldiers who had been fighting in the trenches (WWI).

Mott, John Raleigh (1865–1955), US religious leader who shared the 1946 Nobel Peace Prize with Emily G. Balch for his work in promoting co-operation between the churches. He began his work in 1888 when he became student secretary of the International Committee of the YMCA, a post he held until 1915. From 1926 to 1937 he was president of the World's Alliance of YMCAS. He wrote many books, including *Strategic Points in the World's Conquest* (1897) and *The Larger Evangelism* (1944).

Mott, Lucretia Coffin (1793–1880), US social reformer. A QUAKER, she championed intellectual freedom and opposed slavery. She helped found the American Anti-Slavery Society in 1833. She was also interested in women's position in society and with Elizabeth Cady STANTON, she organized the first women's rights convention at Seneca Falls, New York, in 1848.

Mott, Sir Nevill France (1905–), British scientist. He was professor of experimental physics at Cambridge (1954–71). He made important contributions in the field of QUANTUM MECHANICS, especially in the thoery of atomic scattering, and in 1977 he was awarded the Nobel Prize in physics, jointly with Philip Anderson and John van Vleck, for their work on the electronic structures of magnetic and disordered systems.

Motte and baillie, early medieval style of CASTLE building, most common in England in the late 11th century. It comprised an artificial mound topped by a wooden building (the motte), surrounded by a larger enclosure protected by a wooden palisade (the baillie). Such castles were often later replaced by more permanent stone structures.

Mottelson, Ben Roy (1926–), Danish physicist, *b.* USA, who shared the 1975 Nobel Prize in physics with Aage N. BOHR and L. James RAINWATER for their co-operative research into the structure of atomic nuclei. In the 1950s it was found that atomic nuclei were not spherical, which led Rainwater to postulate that most nuclear particles form an inner nucleus while the remaining particles form an outer nucleus, the shape of each set of particles affecting the other set. He further proposed that enough energy could be created from this interaction to deform an otherwise symmetrical nucleus. From 1950 to 1953 Mottelson and Bohr provided experimental evidence to support this theory, which made Nuclear FUSION a practical possibility.

Mouflon, wild mountain-dwelling sheep of remote parts of Sardinia and Corsica. The coarse coat is mainly reddish brown on the back and white underneath. The male has large, downward-curving horns; the mouflon is smaller than the ARGALI and BIGHORN. Height: up to 70cm (28in) at the shoulder. Family Bovidae; genus *Ovis*. *See also* NW p.*163*.

Mould, any conspicuous mass composed of the MYCELIA (vegetative filaments) and fruiting bodies produced by numerous fungi of the division Mycota. The colour of the mould is due to the SPORES, borne on the mycelia. Many moulds live off fruits, vegetables, cheese, butter, jelly, silage and almost any dead organic material. *Rhixopus nigricans* is commonly seen on stale bread. ROQUEFORT, CAMEMBERT and STILTON cheeses involve the use of mould, eg species of *Penicillium*. Although many species are pathogenic, causing diseases in human beings, animals and plants, PENICILLIN and a few other ANTIBIOTICS are obtained from moulds. *See also* SLIME MOULD; FUNGICIDES; FUNGUS; NW pp.40–41.

Moulin Rouge, Paris dance hall, opened in 1889. Its attractions include a cabaret and a large garden, used for summer dancing and musical entertainment.

Moulmein, port city of S BURMA, at the mouth of the River SALWEEN, on the E shore of the Gulf of Martaban. Its industries include teak milling and shipbuilding. There is also a thriving export trade. Pop. (1974 est.) 202,000.

Moulting, process involving the shedding of the outermost layers of an organism and their replacement. Mammals moult by shedding outer skin layers and hair, often at seasonal intervals – human beings do not moult but lose dead, dry skin continuously as it is replaced from below. Birds moult their feathers, and amphibians and reptiles their skin. In all cases the process is controlled by HORMONES and serves to replace worn out tissues of the skin of a growing animal that has become too small, with new ones. The moulting of insects and other arthropods is a more elaborate affair which is fundamental to growth. The process, also called ecdysis, involves the resorption into the body of calcium and other materials from the hard outer cuticle of the EXOSKELETON, so making the cuticle more fragile. The arthropod then swells its body and bursts free from the old cuticle, and slowly reforms a new one around its swollen body, so increasing in size.

Moulton, Forest (1872–1952), US astronomer who, with Thomas Chamberlain, (1843–1928), put forward a tidal theory of the evolution of the SOLAR SYSTEM. He developed George de Buffon's idea that the planets were formed by the action of a passing star which drew material out of the Sun. *See also* SU pp.174–175.

Mountain, part of the earth's surface that rises conspicuously higher, at least 380m (1250ft) higher than the surrounding area. Mountains have a restricted summit area, comparatively steep sides and considerable bare rock surface. They are identified geologically by their most characteristic features, FOLD, VOLCANICS or fault-block mountains. Mountains may occur as single isolated masses, as ranges and in systems or chains. *See also* PE pp.106–107, *130–131*.

Mountain ash, any of several species of trees of the rose family, ROSACEAE. The European mountain ash, the rowan tree (*Sorbus aucuparia*), has white flowers and, later, orange-red berries. It grows to a height of about 13m (45ft). In Australia, several tall slender trees (eg of the genera *Fraxinus* and *Eucalyptus*) are also known as mountain ash.

Mountain building, geological process by which mountains are formed in one of three ways. First, FOLD mountains such as the HIMALAYAS are formed by a squeezing-in of rock layers, caused by movements of the tectonic PLATES of the Earth's CRUST. Second, block mountains such as the SIERRA NEVADA of North America are formed by vertical movements between geological FAULTS, leading to the tilting of large blocks of STRATA. Third, a typical VOLCANO forms from the molten rock and ash that piles up around its original vent hole. The extent and speed of this process is illustrated by Parácutín, a Mexican volcano that rose out of a cornfield to a height of 450m (1,475ft) from its base between 1943 and 1952.

Mountain cat, or Andean cat, the rarest wild cat in South America. It feeds primarily on rodents. Its coarse fur has striped markings and the tail is prominently ringed. Length 76cm (2.5ft) excluding tail. Family Felidae; species *Felis jacobita.*

Mountain climbing, sport and recreation known to have been enjoyed in the 3rd century BC. It gained popularity in Europe in the 18th and 19th centuries with the ascent of many European mountains including the MATTERHORN in 1865. The first mountaineering club, the Alpine Club, was formed in 1857. One of the most famous ascents was that of EVEREST in 1953 when Sir Edmund HILLARY and Norgay TENZING conquered the world's highest mountain.

Mountain goat, sure-footed, mountain dwelling RUMINANT native to the USA; although goat-like in appearance it is more closely related to the ANTELOPE. Both sexes are alike, with short, pointed horns and woolly underfur with a top coat of coarse, white, shaggy hair. Mountain goats feed primarily on mosses, lichens and scrub foliage. Height: 1m (3.3ft) at the shoulder. Family Bovidae; species *Oreamnos americanus.*

Mountain lion. *See* PUMA.

Mount Athos, mountain peak on the E part of the Acte peninsula, NE Greece. It has been inhabited since the 9th century by an independent community of monks of St BASIL. Mount Athos became a theocratic republic in 1927, using a representative system introduced in 1783 by the patriarch Gabriel IV.

Mountbatten of Burma, Louis Francis Albert Victor Nicholas Mountbatten, 1st Earl (1900–), British admiral. A great grandson of Queen VICTORIA, he had a distinguished naval career during WWII and, as Commander-in-Chief South-East Asia, commanded Allied operations against the Japanese in Burma. In 1947 he became India's last VICEROY and later, as Chief of the Defence Staff (1959–1965), worked to integrate the branches of the armed forces. *See also* HC2 pp.*217, 247.*

Mountbatten-Windsor, name of the British royal family since 1960. The name Windsor was adopted in 1917 (to replace Saxe-Coburg-Gotha) and Mountbatten was added to the name in honour of Prince PHILIP, whose surname it is.

Mount McKinley, park in central ALASKA, with spectacular mountain and forest scenery and a wide variety of wildlife. Mount McKinley, 6,194m (20,320ft), is the highest peak in North America. Area: 785,220 hectares (19,493 acres).

Mount Pleasant, principal sorting office of the Post Office in London and one of the world's largest. It is connected to six other sorting offices and two British rail stations (Liverpool Street and Paddington) by the Post Office underground railway. Completed in 1927, the automatically controlled railway is 10.5km (6.5 miles) long and carries up to 50,000 mailbags a day. The tunnels are 2.7m (9ft) wide and are, on average, 21m (70ft) below street level.

Mount Rushmore Park, memorial in SW SOUTH DAKOTA, USA, in the Black Hills. The busts of four US presidents, WASHINGTON, JEFFERSON, LINCOLN and ROOSEVELT, have been carved out of the granite face of Mt Rushmore and are visible for a distance of 97km (60 miles). Area: 494 hectares (1,220 acres).

Mount Vernon, estate in NE VIRGINIA, USA on the POTOMAC River 24km (15 miles) s of WASHINGTON D.C.; home and burial place of George WASHINGTON. The estate belonged to the Washington family until 1858. Purchased by the Mount Vernon Ladies' Association in 1860, it is now a national monument. The furnishings on the first floor and in Washington's bedroom are original. A museum was founded on the estate in 1853. Area: 3,240 hectares (8,000 acres).

Mount Wilson, observatory in California, USA, administered jointly with Mount PALOMAR as the Hale Observatories. George HALE was responsible for the construction of a 100in (254cm) reflecting telescope there. Completed in 1918, it remained the largest of its kind in the world for over 30 years. *See also* SU pp.170, *221,* 238, 248, 252.

Mourne, mountains of, range in County Down, SE NORTHERN IRELAND, extending 24km (15 miles) NE to SW between Dundrum Bay and Carlingford Lough. They contain the highest peak in Northern Ireland, Slieve Donard, 852m (2,796ft).

Mourning Becomes Electra, (1931), trilogy of plays by Eugene O'NEILL, entitled *Homecoming, The Hunted* and *The Haunted.* They present a modern adaptation of the ELECTRA theme of Classical literature. The trilogy as a whole is an attempt at reproducing, in a modern context, a sense of fate, accessible to the Greeks because of their religious beliefs, but difficult to inculcate into a modern sceptical audience. The plays were made into a film (1947) and into an opera (1967) by Martin David Levy.

Mouse, any of numerous species of small common RODENTS found in a variety of habitats throughout the world; especially the omnivorous, brown-grey house mouse (*Mus musculus*) of the family Muridae, which also includes most of the Old World mice and rats. This prolific nest-builder, often associated with human habitation, is considered a destructive pest and is believed to carry disease-producing organisms. It may grow as long as 20cm (8in) overall, and has been bred for use in laboratories and as a pet. Many species within the family Cricetidae are also called mice, as are pocket mice (Heteromyidae), jumping mice (Zapodidae) and marsupial mice (Dasyuridae). *See also* NW pp.*33, 156, 158, 160, 220, 228.*

Mousebird, also called coly, any of six species of small, crested, long-tailed, fruit-eating birds that live in South Africa; it is considered a pest in fruit-growing areas. Grey with brightly coloured markings, it scampers along branches, sometimes resembling a mouse. This gregarious bird builds a cup-shaped nest. Length: to 33cm (13in). Family Coliidae. *See also* NW p.*141.*

Mouse deer. *See* CHEVROTAIN; ARTIODACTYLA.

Mousetrap, The (1952), mystery MELODRAMA by Agatha CHRISTIE, based on her short story *Three Blind Mice.* Since its London opening in 1952 it has run continuously and has been played in most countries in the world. During a snowstorm a group of strangers takes shelter at a boarding-house. The party is cut off from the outside world and two murders take place before the mystery is solved.

Moussorgsky, Modest Petrovich. *See* MUSSORGSKY, MODEST PETROVICH.

Mousterian culture, Stone Age industry associated with Neanderthal man, inhabiting w Europe and N Africa 80,000–35,000 years ago. It is named after the site at Le Moustier, in the Dordogne region, France, where a Mousterian skeleton was first discovered in 1908. The culture is designated by its type of pointed flaked flint tools. *See also* MS p.27.

Mouth, in animals, the anterior (frontal) end of the ALIMENTARY CANAL. In man and other higher animals it is the buccal cavity within the jaws, containing the teeth and tongue and leading on to the PHARYNX.

Mouth breeders, any of various species of fish, either the male or female of which carries the fertilized eggs in its mouth. After hatching, the young swim with the adult fish, returning to its mouth in times of danger. Included are the CICHLIDS, nat ve to Central and South America, Africa and India. *See also* TILAPIA.

Mouth organ. *See* HARMONICA.

Mowat, Sir Oliver (1820–1903), Canadian statesman. A liberal lawyer, he was Vice-Chancellor of UPPER CANADA from 1864 to 1872. He achieved a certain amount of provincial autonomy and became Premier and Attorney General in 1872. He was knighted in 1892 and made Lieutenant-Governor in 1897.

Moya, John Hidalgo (1920–), British architect, *b.* USA. In collaboration with Philip Powell since 1946, he has won many awards for his works, which include extentions to colleges at Oxford and Cambridge in the 1960s, and the Chichester Festival Theatre (1962).

Mozambique, republic in SE Africa. A former Portuguese colony, it gained independence in 1975. The country has great potential for hydroelectricity, but as yet there is little manufacturing. The chief exports are cashew nuts, copra, cotton, sugar and tea. Invisible earnings come from the transit trade from neighbouring inland nations. The capital is MAPUTO. Area: 783,030sq km (302,328sq miles). Pop. (1976 est.) 9,454,000. *See* MW p.127.

Mozart, Wolfgang Amadeus (1756–91), Austrian composer born in Salzburg. A child prodigy at the piano, he was taken by his father, Leopold (who was also a musician), on performing tours in Europe (1762–64) and England (1764–65) and by 1770 he had written four operas. In the 1780s he devoted himself to full-time composition. To this period belong some of the most sublime works ever composed, such as the operas *Marriage of Figaro* (1786) and *Don Giovanni* (1787) and the last eight of his 41 full-scale symphonies. In 1787, having almost given up hope of an imperial post, he was appointed court composer by the Austrian emperor. In the last year of his life, wracked by ill health, Mozart wrote the operas *The Magic Flute* and *La Clemenza di Tito,* the clarinet concerto and the *Requiem* (completed by a

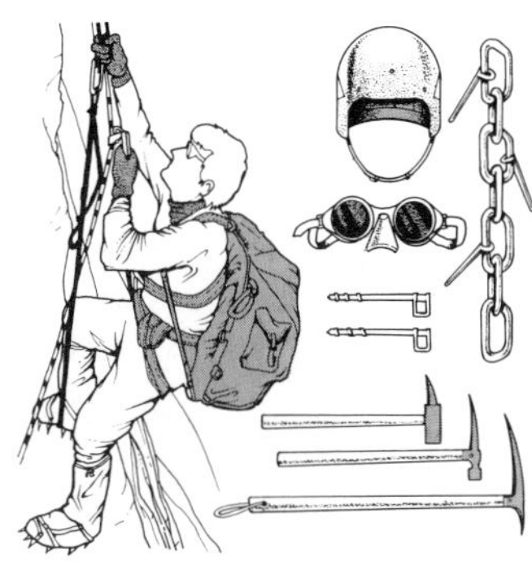

Mountain climbing: ice-axes and hammers are necessary equipment for this sport.

Mountain goats can climb precipices which could be fatal to animals of equal size.

Mouse: its teeth, chisel-sharp for gnawing, continue to grow throughout its life.

Mousebirds run on the back of their legs not on their toes, unlike most birds.

Mudskipper: its eyes, prominent on the head, can move in all directions.

Mulch, applied about 5 to 8cm (2 to 3in) thick, enriches the soil as it decays.

Robert Muldoon visiting a RNZN ship under conversion on the Clyde to a survey vessel.

Mule deer can walk at birth and after less than two years fend for themselves.

pupil). He died in poverty and was buried in an unmarked grave. Together with J.S. BACH, Mozart is today rated as one of the greatest of all composers. In his short life he produced more than 600 compositions and exerted a potent influence on musical development. He perfected the Classical style and looked forward to the early Romanticism of BEETHOVEN and later, BRAHMS, with music that combined elegance of expression with profound emotion. *See also* KÖCHEL; HC1 p.*319*; HC2 pp.41, 120.

Mphahlele, Ezekiel (1919–), South African novelist, dramatist and short-story writer. He went to Nigeria in 1975 and since then has been attacking APARTHEID through his writings. *Down Second Avenue* (1959) is an autobiographical account of a young man's battle with apartheid.

Mrozek, Sławomir (1930–), Polish short-story writer and playwright. His witty and satirical short stories appeared in *Słon* (1957; tr. *The Elephant*, 1962) and *Utwory sceniczne* (1963; tr. *The Ugupu Bird*, 1968). His plays, which belong to the Theatre of the ABSURD, include *Tango* (1965; tr. 1968) and *Policja* (1958; tr. *The Police*, 1967).

Muawiyah (*c.*602–680), early Muslim statesman and first UMAYYAD caliph. He became a scribe of MOHAMMED and later, under OMAR, was appointed the governor of Damascus. He succeeded to the caliphate after refusing to recognize ALI and, although many felt his rule lowered the office of caliph to a mere kingship, he attempted to reconcile the people of his empire and expand his territory in the name of ISLAM.

Mucha, Alphons Marie (1860–1939), Czech designer and painter. He popularized the ART NOUVEAU style of which he was a leading exponent in his posters, eg in the *Poster for Job Cigarette Papers* (1898). Working in Paris, he designed theatre bills for Sarah BERNHARDT's performances, including *Medea* (1898) and *Hamlet* (1899). In his later career he reverted to a more classical mode, and painted decorative historical subjects.

Much Ado About Nothing (*c.*1598), comedy by William SHAKESPEARE. The plot concerns two parallel but contrasting relationships. The real focus of the play, however, is the sexual tension between Beatrice and Benedick, which is manifested in a series of witty verbal battles.

Mucopolysaccharide, class of POLYSACCHARIDE molecules composed of aminosugars linked into repeating units that give a linear unbranched polymeric compound. They are structurally similar to GLYCOGEN and STARCH. CHITIN is a mucopolysaccharide which acts like CELLULOSE as a structural polysaccharide for many phyla of lower plants and insects. The shells of crabs, lobsters, crayfish and many insects mostly contain chitin combined with inorganic salts.

Mucoprotein, type of structural material of the living body, found particularly in the fibres of connective tissue. It is made up of a PROTEIN attached to a MUCOPOLYSACCHARIDE.

Mucor, genus of common MOULDS, found on the surfaces of bread, other foods and dead organic materials. The SPORES are borne in pin-like fruiting bodies which can be seen as dark dots against the white mould. Mucor and related species of FUNGI also reproduce by means of thick-walled spores, which can survive winter cold or drought. *See also* NW p.*41*.

Mucous membrane, lining of all body channels that communicate with the air, such as the mouth and respiratory tract, the digestive and urino-genital tracts and the various glands that secrete MUCUS.

Mucus, slippery, viscous fluid containing mucin, produced by MUCOUS MEMBRANES of the body. It serves for lubrication and protection: nasal mucus traps airborne particles; mucus of the stomach protects the lining from irritation by hydrochloric acid secreted during DIGESTION.

Mud, wet, sticky soil or sediment. The solid matter of mud has a smaller average particle size than that of SAND or SILT, and some of its organic matter is held in colloidal suspension so that it will not settle out upon standing (*see* COLLOID). Marine OOZES are muds formed largely from the shells of microscopic organisms of the PLANKTON. MUDSTONE is a fine-grained rock, a consolidated form of mud. *See also* PE p.102.

Mudfish, name given to a number of fishes that live in muddy water but are otherwise dissimilar. They include the killifish and the bowfin.

Mudie, Charles Edward (1818–90), British library owner and publisher. He started a book-circulating enterprise in 1842 and ten years later this became Mudie's Lending Library. It circulated carefully selected books to the public until it closed in 1937. He published the first English edition of Lowell's *Poems* (1844).

Mudpuppy, aquatic SALAMANDER of streams and rivers in S Canada and the USA. It has dark red gills at the sides of the neck and brown-spotted skin. Length: to 38cm (15in). Family Necturidae; species *Necturus Maculosus*.

Mudskipper, any of several genera of amphibian-like fish of tidal swamps in Africa, Asia and Australia. It can retain moisture in its gill cavities to survive out of water when the tide recedes. With its specialized pectoral fins it can hop on mud, and can even climb trees to cling with a sucker for several hours. Length: to 20cm (8in). Family Periophthalmidae. *See also* NW pp.*126*, *214*, 234, *235*.

Mudstone, rock made of consolidated MUD which, although firmer than clay, lacks the laminated structure and tendency to cleave to SHALE and sometimes decomposes into mud when exposed to the atmosphere.

Mud turtle, any of 12 species of freshwater, bottom-crawling TURTLES of the Americas. It has a short tail, fleshy chin barbels, and hinged, protective flaps at the front and rear of its undershell. Length: 15cm (6in). Family Kinosternidae; genus *Kinosternon*.

Muezzin, one who calls MUSLIMS to prayer. In small MOSQUES, the call is given by the IMAM. In larger mosques, a muezzin is specially appointed for that purpose. If the mosque has a MINARET, or tower, the muezzin calls from the top of it.

Mufti, person who will give an opinion, or *fatwa*, on any point in the sharia (Islamic religious law). The Sharia is thought to be divine revelation, not human creation, and so this interpretation is of great value, although any judgment not concerned with such personal matters as marriage, divorce and inheritance may now be subject to modern legislation. Muftis were given official status in the OTTOMAN EMPIRE and the office of Mufti of Istanbul had considerable political influence. In the British army, mufti also described the civilian dress of one who normally wears a uniform.

Mufulira, city of N central Zambia, 34km (21 miles) NNE of Kitwe on the border with Zaire. Rich local copper deposits make the city the mining centre of the Copperbelt. Pop. (1972 est.) 142,000.

Muggleton, Lodowick (1609–98), English Puritan religious leader. He and his cousin John Reeve, claiming to be the prophetic witnesses in Rev. 11: 3–6, believed in the absence of divine interference in the world after God's revelation to them, denied the usefulness of prayer and the doctrine of the Trinity, and taught that Reason was the Devil's creation. The small sect known as Muggletonians continued until the early 20th century.

Mughals. *See* MOGUL EMPIRE.

Mugwumps, US political faction in the 1880s composed of independent, or liberal, Republicans. In the 1884 presidential election they deserted the Republican candidate, James G. Blaine (whom they considered corrupt) and supported the Democratic candidate, Grover CLEVELAND, who won the election.

Muhammed Ali. *See* ALI, MUHAMMAD.

Muhammed Ali (1769–1849) Egyptian soldier who became Viceroy of Egypt (1804–48). He fought against the French invasion of Egypt (1799) and gradually became independent of his nominal master, the Ottoman sultan. He passed extensive reforms although the cost rather than the benefits fell to his subjects. He attacked the sultan, Mahmud II (1831–34; 1838–41), but an agreement was reached whereby the governorship of Egypt became hereditary in Muhammed Ali's family.

Muir, Edwin (1887–1959), Scottish poet. The contrast between his native Orkney and the harshness of Glasgow, where he moved when young, is a main theme of his *Autobiography* (1954). His poems make powerful use of myth and tradition. With his wife, Willa, he translated the works of many German writers, including Franz KAFKA. He was also known as a literary critic.

Muir, Karen (1952–), South African backstroke swimmer who, in 1965 aged 12 years 10 months 25 days, became sport's youngest ever-world record-holder (68.7 sec for the 110yd event). She went on to set 15 world records over 100m (110yd) and 200m (220yd), but was denied the chance of Olympic Games honours because of South Africa's non-attendance for political reasons.

Mukden. *See* SHENYANG.

Mulattoes, persons having one black (Negro) parent and the other white (European). They occur mostly in Central and South America, and the West Indies. Although some people from these regions are of mixed ancestry, the term is generally applied to those of lighter-coloured skin.

Mulberry, family (Moraceae) of trees and shrubs that grow in tropical and temperate regions and contain a milky latex. Other characteristics are simple leaves, unisexual flowers and the production of fibres or edible fruits. The 1,400 varieties include 12 species of *Morus*, deciduous trees with fleshy, edible fruits. The leaves of some species are used for feeding SILKWORMS. *See also* PE pp.158, 214, *215*.

Mulberry harbour, nickname for two artificial floating harbours used by the Allied forces in WWII to facilitate the D-Day landings of 1944. During the 100 days following the landings, more than 4 million tonnes of stores and 2 million troops passed through the two harbours.

Mulch, layer of loose material placed on top of the soil and around growing plants. Mulches, including leaves, straw, pine-needles, woodchips, stones, newspaper and synthetics (fibreglass, plastic), are used to stabilize soil temperature, keep down weeds, nourish plants and conserve moisture. During winter mulches also protect herbaceous perennials, produce humus (if organic) and keep roots from freezing. Natural mulches include fallen leaves, bracken, grasses and snow.

Muldoon, Robert David (1921–), New Zealand politician and Prime Minister (1975–). He was Finance Minister (1967–72 and 1975–) and became leader of the National Party Opposition in 1974, a year before spearheading a landslide election victory. He became one of the most forceful and controversial leaders in New Zealand political history.

Mule, HYBRID offspring of a female horse and a male ass; it is different from the smaller hinny, which is the result of a cross between a male horse and a female ass. Brown or grey, it has a uniform coat and a similar body to a horse, but has the long ears, heavy head and thin limbs of an ass. Known since ancient times, the hardy mule is commonly used as a draught or pack animal. It is usually sterile. Height: 1.8m (5.8ft). *See also* NW p.*35*.

Mule deer, long-eared deer and a game animal that inhabits W USA from Alaska to Mexico. It is red-brown with a black-tipped white tail; the male bears antlers. It is generally solitary, but often gathers in herds in winter. Height: to 1.1m (3.5ft) at the shoulder. Family Cervidae; species *Odocoileus hemionus*.

Mulgan, name of two New Zealand writers, Alan Edward (1881–1962), father, and John (1911–45), son. Alan wrote the volume of verse *Aldebaran* (1937) and the novel *Spur of Morning* (1940). John's most famous book was the novel *Man Alone* (1939).

HC1 = History and Culture Vol. 1   HC2 = History and Culture Vol. 2   MS = Man and Society   MM = Man and Machines

Mull, island off the w coast of Scotland, in Strathclyde Region, largest of the Inner Hebrides; separated from the mainland by the Sound of Mull and the Firth of Lorn. A mountainous island, its highest point is Ben More, rising to 971m (3,185ft). Sheep and cattle are raised, and the main industries are fishing and tourism. Area: 909sq km (351sq miles).

Mullah, Arabic word meaning "master". It is often given to men versed in the Sharia (Islamic religious law), but can apply to anyone acting in a religious capacity. There are therefore no qualifications that a man must gain to become a mullah, but he will usually have attended a *madrasah*, or religious school.

Mullein, hardy biennial plant of the genus *Verbascum*, including the common mullein (*V. thapsus*) that has 30cm (1ft) leaves and long, dense, yellow flower spikes; height: to 1.8m (6ft). Family Scrophulariaceae.

Müller, Friedrich Max (1823–1900), British linguist and Orientalist, *b*. Germany. Noted for his work on comparative mythology and religion, especially HINDUISM and ZOROASTRIANISM, he edited the *Sacred Books of the East* (51 vols) which include translations of major Oriental, non-Christian scriptures.

Muller, Gerhardt ("Gerd") (1945–), West German footballer whose prolific goalscoring feats helped his country to win the 1974 World Cup and the 1972 European Championship. His club, Bayern Munich, had many successes, including three European Cup wins (1974–76). He was European Footballer of the Year in 1970.

Muller, Hermann Joseph (1890–1967), US geneticist. He found that he could artifically increase the rate of MUTATIONS in fruit-flies (*Drosophila*) by the use of X-rays. He thus highlighted the human risk in exposure to radioactive material. For this work he was awarded the 1946 Nobel Prize in physiology.

Müller, Johannes Peter (1801–58), German physiologist and anatomist. Regarded as one of the founders of modern physiology, he conducted pioneer research into the nervous system, lymph and blood systems, hearing and EMBRYOLOGY.

Müller, Otto (1874–1930), German Expressionist painter and lithographer, known for his paintings of nudes and gypsy women. In 1910 he joined *Die Brücke*, a group of Expressionist painters based in Dresden. He taught at Breslau Academy from 1919 until his death. His works include *The Judgement of Paris* (*c*.1911) and *Self Portrait* (1922).

Müller, Paul Hermann (1899–1965), Swiss chemist who was awarded the 1948 Nobel Prize in physiology and medicine for his discovery of the use of DDT as an insecticide. In 1935 he began the search for a perfect insecticide, one toxic for a wide variety of insects with few poisonous effects on other life forms. He concentrated on chlorine-containing compounds and in 1939 tested DDT. In 1944 DDT was successfully employed against a typhus epidemic in Naples, and for more than 20 years was the most widely used insecticide. In the 1970s, however, it was implicated as a major pollutant and a hazard to animal life, resulting in its use being banned in many countries.

Müller, Wilhelm (1794–1827), German lyric poet, much of whose work was popular for its sympathy with the heritage of Greece in its struggle for independence. He published *Songs of the Greeks* (1821–24) and *Modern Greek Folk Songs* (1825), as well as the cycles *Die Schone Müllerin* and *Die Winterreise*, which were set to music by SCHUBERT.

Müller, William James (1812–45), British landscape painter, of German extraction. Using mainly watercolour, his style was exotic and romantic in the scenes he painted of the Middle East where he travelled extensively. His English paintings show the influence of John CONSTABLE and David COX.

Mullet, or grey mullet, marine food fish found in shoals in shallow tropical and temperate waters throughout the world. Its torpedo-shaped body is green or blue and silver. Size: to 91.4 cm (3ft); weight: 6.8kg (15lb). Among the 100 species is the widely distributed striped *Mugil cephalus*. Family Mugilidae. *See also* PE p.246.

Mulligan, Gerry (1927–), US jazz baritone saxophonist and composer. In the 1940s he was an important arranger, especially for Miles DAVIS; in the 1950s he led his own quartet, playing with a restrained, thoughtful style.

Mulliken, Robert Sanderson (1896–), US chemist. After studying at the Massachusetts Institute of Technology, he became professor at the University of Chicago. In 1966 he was awarded the Nobel Prize in chemistry for his fundamental work on CHEMICAL BONDS and the MOLECULAR ORBITAL THEORY.

Mullion, in architecture, slender, upright member dividing an opening, particularly a division between panes of a window or between windows. Although the mullion occurs in nearly all early architectural styles, it is most characteristic of GOTHIC stone tracery, where it is used to divide large windows into areas suitable for glazing.

Mulready, William (1786–1863), Irish artist trained in classical draughtsmanship at the Royal Academy, but who later specialized in Romantic genre paintings such as *Flight Interrupted* (1816), *The Sonnet* (*c*.1839) and *Shooting a Cherry* (1848). He illustrated many books and in 1840 the illustrated envelope he had designed for Sir Rowland Hill's new postal scheme met with such public ridicule that it was replaced by the Penny Black stamp which could be stuck to any envelope.

Multan, city in E central Pakistan, in the Chenab river valley approx. 322km (200 miles) sw of Lahore. An ancient settlement, the city was taken by Mahmud of Ghazni in 1006 and by TAMERLANE in 1398. The British took the city in 1849. Multan has several ancient Muslim tombs and a Hindu temple. Industries: textiles, iron founding, glassware, food-processing, pottery. Pop. (1972) 542,195.

Multi-party system, democratic political system in which more than two parties compete for office. It is characteristic of most Western democracies, where it is usually accompanied by an electoral system based on some kind of proportional representation.

Multiple birth, bearing of several offspring. Multiple births are common among small mammals but relatively rare in human beings in which the most usual form is TWINS. Identical twins develop from a single OVUM which has divided shortly after FERTILIZATION by a single SPERM; fraternal twins arise from two individually fertilized ova. Multiple egg births are partly genetically determined and partly due to extrinsic factors, such as the mother's age. FERTILITY DRUGS sometimes cause several ova to be released and fertilized simultaneously, resulting in the birth of four or more offspring. *See also* MS pp.76–77.

Multiple myeloma, malignant TUMOUR of the bone marrow, occurring most often in middle age. Mild cases may be treated by surgery or irradiation.

Multiple personality, relatively rare and extreme form of personality dissociation in which two or more distinct and often contradictory personality patterns coexist. They may alternate and through AMNESIA be unknown to one another. Multiple personality should not be confused with the splitting of personality from reality in SCHIZOPHRENIA.

Multiple sclerosis, disease of the nervous system, usually of the white matter (which conducts nerve impulses). Degeneration occurs in the sheath covering the nerve fibres (myelin sheath), resulting in weakness, lack of co-ordination, and speech and visual disturbances. Affected people typically have relapses and remissions over many years. Its cause is unknown, but evidence suggests a possible viral origin. *See also* MS pp.96–97.

Multiple stars, stellar systems consisting of three or more stars orbiting around a common centre of gravity. An example is the Mizar-Alcor system in the constellation of Ursa Major. Mizar, itself a double star, is accompanied by the fainter Alcor. *See also* SU p.237.

Multiplet, in nuclear physics, group of elementary particles, all HADRONS with about the same mass, identical in all other properties except electric charge, usually having two or three members. The NUCLEONS and PIONS form multiplets. In strong interactions, members of a multiplet are all equivalent. A supermultiplet is a larger, more sophisticated and symmetrical grouping of hadrons involving eight QUANTUM NUMBERS, all of whose members have identical spin.

Multiplier, in economics, concept that explains the effect of a change in the level of autonomous consumption, investment or government (public) spending on the GNP (gross national product), through its impact on increasing aggregate demand. Since this increase results in higher spending for goods and services, the incomes of those selling the goods also increase. When the multiplier process has been completed, enough new income has been created for new savings to equal the initial increase in investment or government (public) spending. The multiplier can also apply to changes in tax levels. The theory was first introduced in the USA in the 1930s, later to be propounded by KEYNES.

Mumford, Lewis (1895–), US writer and critic, best known for his essays on town planning and architecture, which include *The City in History* (1961) and *Roots of Contemporary Architecture* (1972). He also published works on a variety of subjects, including a psychological study, *Herman Melville* (1929) and *Renewal of Life* (1934, 1938, 1944, 1951), a tetralogy presenting his philosophy of modern civilization and urging scepticism concerning all systems.

Mumming, traditional British folk-play celebrating the passing of winter into spring and the triumph of good over evil. Such plays, presumably derived from pagan ritual, incorporate a common 17th century source which concerns the killing of a hero, such as St George or Robin Hood, by a villain, often a Turkish knight, and the arrival of the Doctor who revives the hero. Mumming texts were handed down secretly by word of mouth from father to son and were often embellished with references to local events and contemporary heroes such as Nelson. As part of the ritual, mummers (or masqueraders) had to hide their identities behind sooted faces or by wearing masks and beribboned clothes such as those worn at Marshfield, Gloucestershire. Once a common Christmas or Easter entertainment in most British villages, mumming is now rarely performed.

Mummy, body treated with preservatives before burial. Mummification was most commonly practised in ancient Egypt, where the internal organs were first removed, the body soaked in resin and other substances and then wrapped in linen bandages. The Incas of South America, the original inhabitants of the Canary Islands, and several other people also practised mummification. Naturally preserved mummies have been discovered in Scandinavian peat bogs. *See also* MS p.207.

Mumps, contagious disease, most common in children and caused by a myxovirus (one of a large group of VIRUSES that includes the influenza virus). It has an incubation period of about 18 to 22 days, after which fever and painful inflammation of the salivary glands begin, with marked swelling, especially in the parotid glands below and in front of the ears. MENINGITIS occasionally develops in some cases, and the infection may involve other organs such as the pancreas. In males past puberty, inflammation of the testes (orchitis) may occur, with the attendant slight risk of sterility. One attack of mumps generally results in immunity for life against further infection.

Mun, Thomas (1571–1641), English merchant and writer on economics. He became a director of the East India Company in 1615, and wrote *A Discourse of*

Mullein plants, both the stem and the leaves irritate human skin if touched.

Hermann Muller's works include *Out of the Night* and *Genetics, Medicine, and Man.*

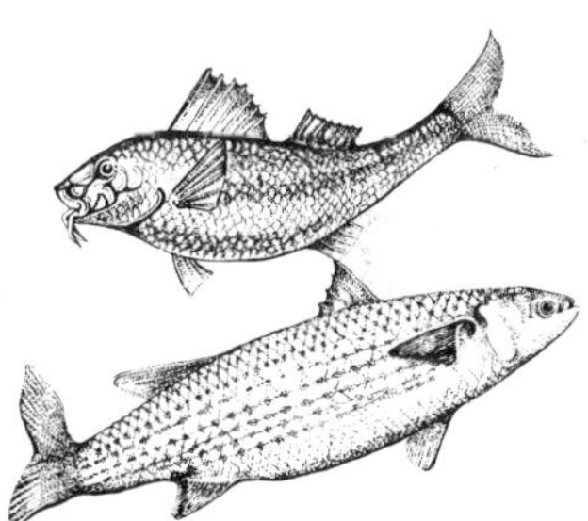

Mullet feed at the surface by sucking in minute items of food.

Gerry Mulligan, in concert with Dave Brubeck, the American jazz pianist.

Münch, Charles

Edvard Munch's *The Sick Child*. Most of his portraits are of women and girls.

Munich Agreement: Chamberlain on his return with the paper signed by Hitler.

Muntjac: the male emits frequent barks to warn other animals of danger.

Murcia is the name of a city and province in SE Spain: a view across the region.

Trade, from England and unto the East Indies (1621), in which he attacked the idea that trade with India weakened England by draining it of bullion. He was an early proponent of the concept of the balance of trade.

Münch, Charles (1891–1968), French conductor. He directed the Paris Conservatoire Orchestra from 1937 to 1949 and succeeded Serge KUSSEVITZKY as conductor of the Boston Symphony Orchestra from 1949 until 1962. He was well known for his interpretations of the works of French composers.

Munch, Edvard (1863–1944), Norwegian painter and printmaker. He made colourful lithographs and woodcuts and painted portraits and murals, but is best known for his paintings showing basic human emotions (such as fear), which were important in the birth of German EXPRESSIONISM. Among his strongest works are *The Shriek* (1893), *The Kiss* (1895) and *Vampire*. *See also* HC2 pp.174, *175*.

München. *See* MUNICH.

Munday, Anthony (1560–1633), English dramatist. After travelling to Europe in 1578 and observing Catholic refugees there, he entered the English College at Rome, masquerading as a Catholic. *The English Roman Life* (1582) is a record of his experiences. He is best remembered for his plays, which include *The Downfall of Robert Earl of Huntingdon* (1601) and *Sir John Oldcastle* (1600).

Mundurucú, large tribe of South American Indians, who speak the Tupí language and live in SW Pará and SE Amazonas in Brazil. They were once one of the largest and most powerful tribes in that area but by about 1800 Brazilian colonists had annexed most of their territory. Today there are few remaining members.

Munich (München), city in s West Germany, on the River Isar; capital of BAVARIA. Founded in 1158 by Henry the Lion, Duke of Saxony and Bavaria, the city became the residence of the Wittelsbach family, the dukes of Bavaria, in 1255. Occupied by the Swedes in 1632 and the French in 1800, Munich developed rapidly in the 19th century, when it became a city of more than 100,000 inhabitants. Bavaria became a state of the German Empire in 1871. In 1923, Munich was the site of the "Beer-hall Putsch", HITLER's unsuccessful attempt to overthrow the Bavarian government. Despite this, Munich became the headquarters of the Nazi Party. The city is world-famous for its annual beer festival, the Oktoberfest. It has the University of Munich (1472), the Renaissance-styled St Michael's Church (1597) and the Propylaen, a neoclassic gate. Industries: chemicals, brewing, pharmaceuticals, motor vehicles, food-processing, tobacco, optical instruments and tourism. Pop. 1,317,000.

Munich Agreement (1938), pact signed by representatives of Britain, France, Germany and Italy. Neville CHAMBERLAIN for Great Britain and Edouard DALADIER for France acceded to Adolf HITLER's demands for immediate German occupation of the Sudetenland area of Czechoslovakia. The pact also provided for plebiscites, but these were never carried out. The agreement averted war for a year but was to become a symbol of the democracies' policy of APPEASEMENT toward Hitler. *See also* HC2 p.218, *218*.

Munich Putsch (1923), attempted coup by Adolf HITLER and his Nazi Party to overthrow the republican government of Bavaria, known as the "Beer-hall Putsch" because it began in a beer-hall. The coup proved abortive and Hitler was arrested and sentenced to five years in the Landsberg fortress, of which he served only nine months.

Municipal Corporations, originally chartered self-perpetuating town councils in England. An act of 1835 abolished those oligarchies in 178 towns and made the councils subject to elections. In the next 50 years 62 more towns were incorporated and by an act of 1882 another 25. That act consolidated the various functions (of which the most important was public health) imposed on the corpora-

tions. The 1888 act made large municipalities county boroughs, with functions similar to those of COUNTY COUNCILS. Municipal corporations of every borough outside Greater London ceased to exist on 1 April 1974. *See also* LOCAL GOVERNMENT.

Munk, Kaj Harald Leininger (1898–1944), Danish playwright and parish priest. Influenced by Shakespeare and Kierkegaard, his works such as *Cant* (1931), *Ordet* (*The Word*, 1932), *De Udvalgte* (*The Elect*, 1933) and *Niels Ebbesen* (1943) are chiefly historical dramas. His play about anti-semitism, *Han sidder ved Smeltediglen* (*He sits at the Melting Pot*, 1938) was banned during the German WWII occupation of Denmark. He became a spiritual leader of the resistance and was finally killed by the Nazis.

Munnings, Sir Alfred James (1878–1959), British artist who specialized in painting sporting scenes and horses. A war artist in WWI, he was president of the Royal Academy (1944–49) and was renowned for his critical views on "modern" art.

Munro, Hector Hugh. *See* "SAKI".

Munro, Sir Leslie (1901–), New Zealand diplomat and politician. He was ambassador to the USA (1952–58) and president of the UN General Assembly (1957–58).

Munrow, David John (1942–76), British musicologist who, without any formal musical education, did much to foster interest in medieval music and the instruments on which it was played. He founded the Early Music Consort of London in 1967 and arranged, composed and played much medieval and Renaissance music for both the theatre and the cinema.

Munster, province in s Republic of Ireland, on the Atlantic coast; largest of Ireland's four provinces. It includes the counties of CLARE, CORK, KERRY, LIMERICK, TIPPERARY (North and South) and WATERFORD. Area: 24,126sq km (9,315sq miles). Pop. (1971) 880,018.

Munsterberg, Hugo (1863–1916), US psychologist, *b.* Germany. He was a pioneer in applied psychology, bringing it into education, law and business. His publications include *Psychology and Industrial Efficiency* (1913) and *Psychology: General and Applied* (1914).

Munthe, Axel Martin Fredrik (1857–1949), Swedish writer. After studying medicine and becoming famous as a psychiatrist, he was appointed physician to the Swedish royal family. His works, mainly reminiscences about international events, include *Letters from a Mourning City* (1897) and *The Story of San Michele* (1929).

Muntjac, small primitive Asian DEER. It is brown with cream markings and has tusk-like canine teeth and short, two-pronged antlers. Muntjacs are generally found alone or in pairs. There are two well known species, the Indian muntjac or barking deer (*Muntiacus muntjak*) and the Chinese muntjac *(M. reevesi)*. Height: to 60cm (24in) at the shoulder; weight: to 18kg (40lb). Family Cervidae. *See also* NW p.*214*.

Münzer, Thomas (*c.*1489–1525), German ANABAPTIST: He became a Protestant in 1518, initially following Martin LUTHER; but by 1522 he was an opponent of Luther's ideas, believing in the authority of the spirit rather than the Scriptures. He preached revolt and attacked the procedure of infant baptism, claiming that God willed the overthrow of the social structure by ordinary men and women and the establishment of a virtuous, simple society with everything owned by the community. When the PEASANTS' WAR broke out, he and Heinrich Pfeiffer established a communistic theocracy at Mühlhausen, but with the defeat of the peasant party Münzer was beheaded.

Muon, negatively charged elementary particle (symbol μ^-), originally thought to be a MESON but now classified as a LEPTON. It has spin $^1/_2$, a mass about 207 times that of the electron, and decays weakly into an ELECTRON, NEUTRINO and antineutrino.

Mural, scene or abstract depicted on a wall and executed in any of a variety of media. The earliest murals were Stone Age hunt-

ing scenes, painted using natural earth colours. The Egyptians, Greeks and Romans used ENCAUSTIC and TEMPERA painting as well as FRESCO which in the works of GIOTTO, MANTEGNA, GENTILE, Fra ANGELICO, MICHELANGELO and RAPHAEL reached the height of its expression in the Renaissance when mural painting was allied with architecture in new efforts to create illusions of space. In the 18th century, TIEPELO made further innovations in representation; PUVIS DE CHAVANNES painted huge canvas murals in oils in a fresco style in the 19th century. The 20th century has accorded more public and social significance to the exterior mural as exemplified by the works of the Mexicans José Clemente OROZCO and Diego RIVERA and the WPA murals. Porcelain and liquid silicate enamels are among the media now used in modern murals.

Murad, name of five Ottoman sultans. Murad (*c.*1326–89, *r.*1360–89) expanded Ottoman rule into Europe. Murad II (1403–51, *r.*1421–51) consolidated these acquisitions and created the JANISSARIES. Murad III (1546–95, *r.*1574–95) continued Ottoman imperialist policy. Murad IV (1612–40, *r.*1623–40), ruled strongly but severely. Murad V (1840–1904), *r.*1876) tried to introduce constitutional government but was deposed after four months because of his insanity. *See also* HC1 p.220; HCs pp.84–85.

Murasaki, Shikibu (*c.*980–*c.*1030), Japanese diarist and novelist. A lady at the court of the Empress Akiko, she kept a diary from 1007 to 1010, showing glimpses of court life in the capital. She is better known for her novel, *The Tale of Genji*, one of the first great works of fiction to be written in Japanese. *See also* HC1 p.*132*.

Murat, Joachim (1767–1815), Marshal of France and King of Naples (1808–15). He helped to bring NAPOLEON to power in the coup of 1799, and as a reward married Napoleon's sister Caroline in 1800. In 1808 he was chosen to succeed Joseph Napoleon as King of Naples. A brilliant and dashing cavalry commander, Murat played an important part in Napoleon's victories, particularly at AUSTERLITZ (1805) and in the Russian campaigns (1812). During the HUNDRED DAYS he rejoined Napoleon but was defeated by the Austrians and fled to Corsica. He failed in an attempt to regain Naples, was arrested and executed.

Murchison, Sir Roderick Impey (1792–1871), Scottish geologist. His work on the rocks underlying the old red sandstone in s Wales led to the definition of the strata known as the SILURIAN System. He worked with Sir Charles LYELL on the Auvergne volcanics and with Adam SEDGWICK on the structure of the Alps. He later collaborated with Sedgwick on a study of the rocks that were to become known as the DEVONIAN System.

Murchison, river in Western Australia that flows w into the Indian Ocean. Once part of a thriving gold-mining region, it is now used mostly for irrigation. Reefs make it unnavigable. Length: 708km (440 miles).

Murchison Falls, waterfall in NW Uganda, E Africa, in the lower Victoria Nile, just above Lake Albert. The falls were discovered by Samuel Baker in the 1860s and are now part of the Murchison Falls National Park. Height: 40m (130ft).

Murcia, city in SE Spain, on the Segura river, 75km (47 miles) SW of Alicante; capital of Murcia province. Founded *c.*225 BC by Carthaginians, it suffered great damage during the SPANISH CIVIL WAR. It has a 14th-century cathedral, an 18th-century episcopal palace and a university (1915). Industries: vegetable canning, textiles, saltpetre, aluminium. Pop. (1970) 243,759.

Murder, unlawful killing of a person, performed with malice aforethought. Committed under sufficient provocation or in self-defence, a killing may not constitute murder; neither is it if the victim of an assault dies after a year and a day have passed. In 1965 the maximum penalty for murder in English law was changed from hanging to life imprisonment. *See also* MANSLAUGHTER.

Murder in the Cathedral (1935), verse play by T.S. ELIOT, written for the 1935 Canterbury Festival, later made into a film, and an opera (1958) by Ildebrando PIZZETTI. It deals with the assassination of St Thomas à BECKET, Archbishop of Canterbury in the reign of HENRY II.

Murdoch, Jean Iris (1919–), British novelist and philosopher, *b.* Dublin. Her novels, complex and puzzling, include *Under the Net* (1954), *The Sandcastle* (1957), *A Severed Head* (1961), *A Fairly Honourable Defeat* (1970), *A Word Child* (1975) and *Henry and Cato* (1976). Apart from her novels, she has written a critique, *Sartre, Romantic Rationalist* (1953).

Murdoch, Keith Rupert (1931–), Australian newspaper publisher. He publishes papers in Australia, the USA and Britain. In 1969 News International Ltd., of which he is chairman, bought the British newspapers, *The Sun* and *The News of the World.*

Murdock, William (1754–1839), Scottish engineer and pioneer of the gas industry. After hearing of the work of Philippe Lebon with wood gas, he began experiments utilizing coal gas, generated in a retort, which proved superior. Murdock devised purifying and washing techniques for coal gas and the first gas-lights were established – in Soho, London – in 1802.

Murillo, Bartolomé Esteban (1617–82), Spanish painter. He was born and lived in Seville, where his series of 11 pictures showing the lives of the Franciscan saints brought him immediate fame. His early works show the influence of ZURBARÁN in the dramatic use of light and shadow. *See also* HC1 p.354.

Murless, Charles Francis Noel (1910–), British horse-racing trainer who saddled more than 1,400 winners and was leading trainer nine times in a long career (1935–76). He trained a record 19 Classics winners: six 1,000 Guineas, two 2,000 Guineas, three Derbys, five Oaks, and three St Legers.

Murmansk, city in NW Russian Republic (Rossijskaya SFSR), USSR, 1,006km (625 miles) N of Leningrad. Capital of Murmansk oblast, it is an ice-free port on the Kola Gulf of the Barents Sea, the largest city N of the Arctic Circle. Developed in 1916 as a supply port, it was occupied by US, British and French forces in 1918. During WWII it was a major port for Anglo-American convoys. There is a polar research station in the city. Exports: fish, lumber, apatite. Industries: fishing, shipbuilding, fish-canning, metal- and wood-working. Pop. (1975) 358,000.

Murngin, area in NE Arnhemland, Northern Territory, Australia after which a group of ABORIGINES is named. Tribal boundaries are usually clearly defined but occasionally there are indeterminate areas that are of no value to either tribe; the Murngin is such an area. The tribe has extremely large kinship groups, to the extent that even distant relatives are regarded as brothers.

Murphy, Alexander John ("Alex") (1939–), British rugby-league player and coach. A tough, dynamic halfback, he won 27 caps (1958–71) and led St Helens, Leigh and Warrington to victory in the Challenge Cup. As player-coach, he guided Warrington to four major trophy wins in the season 1973–74.

Murphy, William Parry (1892–), US physician. He made detailed studies of DIABETES and the liver treatment for pernicious ANAEMIA. For his work on anaemia he shared, with George WHIPPLE and George Richard MINOT, the 1934 Nobel Prize in physiology and medicine.

Murray, Lord George (1694–1760), Scottish Jacobite soldier. Prominent in both JACOBITE REBELLIONS (1715 and 1745), he was the architect of the victory at PRESTONPANS in 1745. He opposed the strategy that led to the defeat at CULLODEN, and afterwards fled to Holland.

Murray, Sir Gilbert (1866–1957), British Classical scholar well known for his translations of Greek drama. The author of several books on politics, he was active in the cause of world peace. He was chairman of the executive council of the LEAGUE OF NATIONS union from 1923 to 1938 and president of the general council of the UNITED NATIONS from 1947 to 1949.

Murray, Sir James Augustus Henry (1837–1915), Scottish philologist and lexicographer. His reputation was secured with the publication of his *Dialect of the Southern Counties of Scotland* (1873) and an article in the *Encyclopaedia Britannica* on the English Language. Planning and editing the Philological Society's *New English Dictionary* (later known as *The Oxford English Dictionary*) occupied him from 1879 to his death. Murray edited about half the work, which was completed in 1928, but was the inspiration for the whole project.

Murray, James Stewart, 1st Earl of. *See* MORAY, JAMES STEWART, 1ST EARL OF

Murray, John (1741–1815), English-American religious leader, founder of the UNIVERSALIST denomination. A convert from METHODISM, he was excommunicated and emigrated to America in 1770 where at Gloucester, Mass., he established the first American Universalist church in 1779.

Murray, Sir John (1841–1914), British oceanographer and marine biologist, noted for his studies of ocean basins, marine deposition and the formation of coral reefs. He was a naturalist on the *Challenger* Expedition (1872–76).

Murray, Leslie (1938–), leading young Australian poet, who has had a great influence on the growth of Australian literary nationalism. His selected poems, *The Vernacular Republic*, were published in 1976.

Murray, river in Australia that rises in the Australian Alps of SE New South Wales, and flows W, forming the border between New South Wales and Victoria. It then flows S across South Australia, through Lake Alexandrina to empty into the Indian Ocean at Encounter Bay. It is navigable for small vessels only in its lower course, and is used for widespread irrigation. With its chief tributary, the Darling, it forms Australia's major river system. Length: 2,590km (1,609 miles).

Murrayfield, rugby union stadium in Edinburgh. The largest rugby union ground in Britain, with a capacity of 78,500, it has been the venue for Scotland's home international matches since 1925.

Murre. *See* GUILLEMOT.

Murrumbidgee, river in SE Australia that rises in the Great Dividing Range and flows W to join the MURRAY River. It is part of the Snowy Mts hydroelectric scheme and the Burrinjuck Dam stores water for irrigation. Length: 1,579km (981 miles).

Murry, John Middleton (1889–1957), British critic and editor. He edited the *Athenaeum* from 1919 to 1921 and founded his own review, the *Adelphi*, in 1923. His literary studies include *Keats and Shakespeare* (1925), *William Blake* (1933) and *Katherine Mansfield and Other Literary Portraits* (1949). He was Katherine MANSFIELD's second husband, and a close friend of D.H. LAWRENCE.

Murry, Kathleen. *See* MANSFIELD, KATHERINE.

Muru, MAORI custom of taking an offender's weapons, clothing, tools or food as a punishment. It served to distribute goods in an economically poor society.

Muscat (Masqat, Maskat), port and capital of OMAN, SE Arabia, on the Gulf of Oman. The city was held by the Portuguese from 1508 to 1648, when it passed to Persia until 1741, after which it became capital of Oman. Exports: fish and dates. Pop. (1970 est.) 18,000.

Muscat and Oman. *See* OMAN.

Muscle, type of body tissue that has the ability to contract. There are three basic types: skeletal muscle, smooth muscle and cardiac muscle. Skeletal muscle, or striated muscle, makes up the largest single tissue part of the human body, comprising about 40% of body weight. It is attached to the bones of the skeleton and is characterized by cross-markings, known as striations; it typically contains many nuclei per cell. Most skeletal muscles require conscious effort for contraction (when they move the limbs and body), and

are therefore also known as voluntary muscles.

Smooth muscle lines the digestive tract, blood vessels and many other organs. It is not striated and typically has only one nucleus per cell; nor is it under conscious control, and is therefore also known as involuntary muscle.

Cardiac muscle is found only in the heart and differs from the other types of muscle in that it beats rhythmically and does not need stimulation by a nerve impulse to contract. Cardiac muscle has some striations (but not as many as in skeletal muscle) and has only one nucleus per cell. *See also* MS p.58.

Muscle tone, continuous state of partial contraction of certain body MUSCLES which helps to maintain erect posture.

Muscone, active component of MUSK, derived from the musk glands of the male MUSK DEER and used as a base for expensive perfumes. It is a ringed compound with a molecular formula of $C_{16}H_{30}O$. *See also* CIVETONE; MM p.*246.*

Muscovite, sheet silicate mineral, hydrous potassium aluminium silicate, $KAl_2(Si_3Al)O_{10}(OH)_2$, and the most common MICA. Found in many kinds of rocks, it crystallizes in hexagonal, tabular forms in the monoclinic system. Its name derives from its use as a glass in MUSCOVY, Russia. It is clear or tinted with varied lustres. Hardness 2–2.5; s.g. 2.9.

Muscovy (Moscow), Grand Duchy of, state that existed in W central Russia, centred on Moscow, from the late 14th to mid-16th centuries. Its slow rise to prominence over adjacent principalities and the Tartars was partially due to the decline in importance of Kiev, the patronage of the rulers of the powerful Vladimir-Suzdal principality and its own strategic location on the Moskva River. It was also a prestigious religious centre and became the seat of the Russian Orthodox Church.

Muscovy Company (founded by royal charter in 1555), company of English merchants involved in trade with Russia. Based in London and working chiefly through the port of ARCHANGEL, it held a monopoly of Anglo-Russian trade until 1698. *See also* HC1 p.278, *278.*

Muscovy duck, tropical American perching DUCK with greenish black plumage and heavy red wattles. It has been domesticated worldwide and is bred for its succulent flesh. Species *Cairina moschata.*

Muscular dystrophy, any of a group of disorders in which the characteristic feature is progressive painless degeneration and ATROPHY of the muscles with no involvement of the nervous system. Of the three main types, the most common is pseudo-hypertrophic muscular dystrophy, in which the symptoms of muscular degeneration begin in childhood and consist of increasing weakness, a peculiar swaying gait and an initial apparent increase in muscle size (pseudohypertrophy), with subsequent atrophy. Those affected rarely reach maturity, since the heart and respiratory muscles become involved. This form of muscular dystrophy is sex-linked and primarily affects males. *See also* MS p.95.

Muses, in Classical mythology, nine daughters of the Titan Mnemosyne (memory) and ZEUS. Each muse presided over a branch of literature, art or science. Calliope was the muse of epic poetry, Clio of history, Erato of love poetry, Euterpe of lyric poetry, Polyhymnia of song, Melpomene of tragedy, Terpsichore of choral dance, Thalia of comedy and Urania of astronomy.

Musgrave, Thea (1928–), British composer, *b.* Scotland. She studied under Nadia BOULANGER and taught at London University (1958–65). Her works include a clarinet concerto (1968) and the operas *The Voice of Ariadne* (1973) and *Mary, Queen of Scots* (1977). *See also* HC2 p.284.

Mushroom, any of numerous relatively large fleshy fungi, many of which are gathered for food. The term is applied especially to stalked fungi with umbrella-shaped caps, such as the common edible field mushroom (*Agaricus campestris*). Inedible mushrooms are

Murillo's *Children with Silver Coins* (1670–80); children were often his subject.

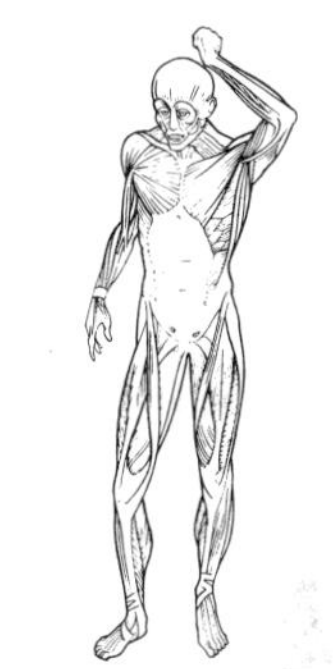

Muscles, tissues that enable animals to move, seen here in a human being.

Muscovy ducks feed mainly on plants, but also small fish, insects and worms.

Muses: a detail from Tintoretto's picture of the daughters of Zeus, at Hampton Court.

Music

Music hall entertainment included comedy, song and dance, and trained animals.

Musk deer: by day this shy creature hides in thickets, emerging at twilight to feed.

Musket of the 17th century. Although at first inaccurate, it revolutionized warfare.

Muskrats have a hairless, flattened tail that acts as a rudder in water.

generally called toadstools. Other fungi commonly called mushrooms include bracket fungi, puff-balls and MORELS. A typical mushroom fungus consists of two parts: an extensive underground cob-webby network of fine filaments – the MYCELIUM – which is the main body of the fungus, and a short-lived fruiting body – the familiar visible mushroom – which may spring up overnight. Since many mushrooms are poisonous, wild mushrooms should be eaten only after they have been unmistakably identified as edible species. Many rough-and-ready tests of edibility, such as indications that a mushroom may be eaten safely by insects, result every year in deaths or severe illness among mushroom gatherers. *See also* FUNGUS; NW pp.38, 40–43.

Music, sound arranged for instruments or voices, in many ways and for many purposes, exhibiting a great variety of forms and styles both from place to place and from age to age. This article treats only of the chief developments in the history of Western music. The roots of Western music lie in the Christian church; medieval church music was based on traditions which came down from the Greeks and the Jews. The Greeks, possessors of a sophisticated mathematics, used scale systems based on the invention of Pythagoras and other thinkers of his time; they named the notes by the letters of the alphabet. The Jews, in their synagogues, developed the art of chanting. In the 4th century St Ambrose systematized the expanding volume of liturgical chants and formulated four MODES which was an early basis of musical organization. By the 11th century the stave (then of four lines only) had been invented. PLAINSONG (plain-chant) music consisted of one vocal line, always set to words from the liturgy. But already in the 10th century there had appeared the anonymous text, *Musica Enchiriadis*, in which was introduced the idea of two parallel voices, moving together at an interval of a fourth or fifth apart. In the early 13th century, Pérotin of France began to write music for three or four voices. With the addition of a religious text to the setting, such works became the first MOTETS.

Early part compositions were based on COUNTERPOINT, the horizontal (linear) aspects of the various parts in combination. The RENAISSANCE was the first great age of Western music, sacred and secular, in which inventiveness was at a peak and the standard musical forms refined. Distinguished composers of the period included GABRIELI, JOSQUIN DES PRÉS, MONTEVERDI and BYRD. Polyphonic writing was a blend of new harmonic thinking and the melodic progression of counterpoint. It reached its highest expression in the age of the BAROQUE which began in Italy, and culminated in Germany with the genius of BACH. Since the time of Bach, Austro-Germany has been the cradle of Western music. The development of harmony, in the classical period of music was the work of HAYDN and MOZART. The late works of BEETHOVEN ushered in the next stylistic period in music, known as ROMANTIC. Famous 19th-century composers included SCHUBERT, MENDELSSOHN, BRAHMS, and WAGNER. By the 20th century the formal principles of composition had broadened and became more flexible. IMPRESSIONISM in music at the turn of the century marked a turn towards freer experimentation. Schoenberg invented a 12-note system which was a radical break from tonal composition and many composers have since written atonal music. Alongside them, however, composers like BRITTEN, SHOSTAKOVICH and Richard STRAUSS continued to compose within established tonal traditions of the preceding centuries. ELECTRONIC MUSIC is a product of this century and its major exponents include STOCKHAUSEN and XENAKIS. *See also* indexes to *History and Culture* volumes.

Music, electronic. *See* ELECTRONIC MUSIC.
Musical, genre of popular dramatic light entertainment exemplified by firm plot, strong songs and vivacious dance numbers developed, at the end of the 19th century, from the best elements of light opera,

revue and BURLESQUE. The earliest example of the form is generally reckoned to be *In Town* (1892), produced in London by George Edwardes, but the most popular musicals after that have originated in the USA with the work of such composers as George GERSHWIN, Jerome KERN, RODGERS and HAMMERSTEIN, Jules Styne and Stephen Sondheim. In Britain in the 1960s, Lionel Bart had considerable successes with *Oliver!* (1960) and *Blitz* (1952). Audiences in the 1970s responded strongly to the work of Tim Rice and Andrew Lloyd Webber's *Jesus Christ Superstar* (1971) which was followed by *Evita* (1978). Successful film musicals such as *West Side Story* (1961), *My Fair Lady* (1964) and *The Sound of Music* (1965) are generally based on stage originals, but original film musicals such as Busby BERKELEY's *Forty-Second Street* (1933), Vincente MINNELLI's *Meet Me in St Louis* (1944), and Stanley Donen's *On the Town* (1949) and *Singin' in the Rain* (1952) exemplified an integrated approach in which neither the cinematic nor the musical aspects were subordinate.

Musical form, the structural scheme which gives shape and artistic unity to a composition. The standard forms are binary, ternary, rondo and sonata. Each consists of a number of musical sections or sub-sections. Binary form copsists of two sections which may be contrasted in idea, key or tempo but which complement each other within the musical entity. Ternary form consists of a re-statement of the first section after a middle section of con-trasted material; an example is the minuet and trio. In rondo form the number of sections varies, but there is at least one re-statement of the first section. Sonata form, as its name suggests, evolved with the SONATA and is used most often for the first movement of a sonata or symphony. The exposition states two subjects which are developed musically in the development, or middle section, before re-stating the subjects of the exposition in the recapitulation. Various types of musical works, eg OPERA, SYMPHONY, SUITE, PRELUDE and FUGUE, and SYMPHONIC POEM, may also be termed forms of music in the broadest sense.

Musical instruments, variety of instruments for producing sound. They may be classified under five headings: stringed, wind, brass, percussion and keyboard. Stringed instruments may be plucked (harp, lute, guitar), sounded by friction with a bow (violin family) or struck (dulcimer). Wind instruments are flue-voiced (recorder, flute) or reed-voiced (single-reeded clarinet, double-reeded oboe and bassoon). Brass instruments are voiced by the vibration of the lips on a metal mouthpiece (horn, trumpet, trombone, tuba). Percussion instruments are struck, either on a stretched membrane (drums), metal (triangle, cymbals) or wood (xylophone). Keyboard instruments combine features of all the others: the piano's strings are struck by hammers; the harpsichord's are plucked; the organ has both flue and reed pipes. There are also various keyboard instruments in ELECTRONIC MUSIC. *See also* articles on individual instruments; BAND; ORCHESTRA; HC2 pp.104–105.

Musical notation, method of writing-down music – the "language" of music. As with any language which must necessarily be understood internationally, a standard form of musical notation is essential. Staff notation, eg using crotchets and quavers to represent notes, defines the absolute and relative pitches of notes and their time values. In the 11th century Guido d'Arezzo (*d.*1050) developed a system which fixed relative pitch by systematically positioning notes on a number of stave lines. Absolute pitch was defined by a coloured line ruled over certain notes (any C was yellow, any F was red). From this system modern clefs have evolved.

Music hall, stage for variety shows, originally tavern annexes, devoted to comic song, acrobatics, magic shows, juggling and dancing. Music halls in London included the Alhambra, The Empire and the London Pavilion. The popularity of the music hall was at its greatest in late

Victorian and Edwardian England, but went into decline with the advent of radio and motion pictures in the 1930s. It continued longer in the USA, where it was often known as vaudeville.

Musick, Master of the Queen's (King's), court position in Britain. Its origins lie in the 15th century, but the official title dates from 1626. The Master was then in charge of the court orchestra. In the 18th century the composition of court odes became an established part of his duties. The post is now virtually honorary, although ceremonial music for state occasions is occasionally composed. Malcolm WILLIAMSON succeeded Sir Arthur BLISS as Master in 1975.

Musicology, academic study of music in most of its aspects such as acoustics, form, rhythm, harmony, melody and mode but usually excluding performances or composition. Its major achievements have perhaps been made in historical research where 19th-century scholars first examined the development of Western music and 20th-century musicologists such as David MUNROW not only rescued the reputations of little-known composers through their researches but also generated wider interest in early music. Musical composition is usually treated as a separate discipline; ethnomusicology – the study of non-Western music – is essentially a branch of ANTHROPOLOGY.

Musique concrète, phrase used by the Frenchmen Pierre Schaeffer and Pierre Henry in 1948 to describe music made from real sounds occurring in the environment. They were recorded on tape as collages of sound and then manipulated by being played faster or slower, backwards or in fragments. A studio for such music was established by Radiodiffusion Française in 1950. A successful early composition was Henry's *The Veil of Orpheus* (1953). Later composers of *musique concrète* have included Iannis XENAKIS.

Musk, strong-smelling, semi-liquid substance obtained from a gland under the belly skin of the male MUSK DEER. It is used in the perfume industry because of its long-lasting and fixative qualities. The name is also given to any penetrating, odoriferous substance secreted by many animals, such as civets, muskrats and musk turtles. *See also* MM *p.246*.

Musk deer, small, timid forest and brush-land DEER of central and NE Asian highlands. Both sexes have long, thick, bristly brown hair. The male has tusks instead of antlers and secretes MUSK, used in perfumes and soap. Height: to 61cm (24in) at the shoulder; weight: to 11kg (24lb). Family Cervidae; species *Moschus moschiferus*. *See also* MM *p.246*.

Musket, smoothbore firearm, fired from the shoulder. The first muskets were handled by two soldiers and fired from a portable rest. Quite inaccurate, they fired lead balls weighing approx. 60gm (2oz) and were about 1.7m (5.5ft) long. They evolved in the early 16th century in Spain from the earlier arquebus (or harquibus) and had an effective range of about 160m (525ft). Later versions were lighter and more accurate. Early muskets were detonated by a glowing fuse (MATCHLOCK) but later versions used the striking of a flint (FLINTLOCK). In the 19th century muskets were replaced by RIFLES, which had spiral grooves inside the barrels, and were capable of far greater accuracy. *See also* MM pp.162–163.

Musk ox, large, wild, shaggy RUMINANT, related to oxen and GOATS, native to N Canada and Greenland. In Europe and Siberia it was exterminated in prehistoric times. Its brown fur reaches almost to the ground, and its down-pointing, recurved horns form a helmet over the forehead. When threatened, the herd forms a defensive circle round the calves. Length: to 2.3m (7.5ft); weight: to 410kg (903lb). Family Bovidae; species *Ovibos moschatus*. *See also* OX; NW *p.226*.

Muskrat, large, aquatic RODENT (a type of VOLE) native to North America. The name comes from the MUSK secreted from two glands at the base of the tail. It is a good swimmer, with partly webbed hind feet and a long, scaly tail. Its commercially valuable fur (musquash) is glossy brown and durable. Muskrats inhabit tunnels or nests by a lake or stream. Length, including tail: to 53.5cm (21in); weight: to 1.8kg (4lb). Family Crisetidae; species *Ondatra obscura* and *O. zibethica. See also* NW p.*160*.

Muslim, one who is a believer in ISLAM and holds MOHAMMED to be the only true prophet. The Arabic word Muslim means "one who submits". There are thought to be 500 million Muslims in the world, mostly in Asia and Africa. *See also* HC1 pp.144–145.

Muslim League, political organization of the Indian subcontinent, founded in 1906 by Aga Khan III to protect and promote the political rights of Muslims in India. At first it co-operated with the Indian National Congress but, fearing Hindu domination, it turned to independent action. Under the leadership of Muhammed Ali JINNAH it called (1940) for the establishment of a separate Muslim state. During WWII the League, in contrast with the Congress Party, supported the British war effort. It became the dominant voice of independent Pakistan (1947), but by 1953 it had to contend with several competing parties. During the martial law imposed by Ayub KHAN (1958–63) it was officially banned. In 1962 it split into three factions: the Convention Muslim League, the Council Muslim League and the Payyum Muslim League. With Ayub's resignation (1969), the Convention faction ceased to exist. The Council faction fared poorly in 1970 elections and ceased to be a major political force in Pakistan.

Muslin, any of several strong, light cotton cloths, woven simply and often sheer; muslin is named after Mosul, a city in Iraq, once known for the manufacture of costly muslin.

Musquash. *See* MUSKRAT.

Mussadegh, or Mossadeq, Muhammad (1880–1967), Iranian statesman. He served in various public offices but retired when REZA SHAH rose to dictatorial power in 1925. Mussadegh resumed government service in 1944 and led the movement to nationalize Iranian oil interests. His movement was successful and he became Prime Minister in 1951. Political and economic crises caused his downfall in 1953. He was imprisoned (1953–56) and then placed under house arrest for the remainder of his life.

Mussel, any of several species of bivalve MOLLUSCS with thin ovoid shells. Marine species of the family Mytilidae are found throughout the world in dense colonies on sea walls and rocky shores, where they attach themselves by means of strands called byssus threads. The edible mussel, *Mytilus edulis,* is sometimes cultivated on ropes hanging from rafts. Freshwater mussels of the family Unionidae, found in northern continents only, produce PEARLS. *See also* NW pp.92–93, 230–231; PE pp.242–243, 248–249.

Musset, Alfred de (1810–57), French writer best remembered for his poems which, after 1834, appeared in the periodical *Revue des Deux Mondes* and reflect the disillusion felt by him and his contemporaries. They were also influenced by his love for the novelist George SAND. He wrote several successful stories and plays, as well as the four famous lyrics *Les Nuits* (1835–37).

Mussolini, Benito (1883–1945), Italian dictator, called Il Duce, founder of Europe's first FASCIST party. An active socialist in his youth he abandoned socialism shortly after WWI and embraced ultra-nationalism, organizing the Fascist Party between 1919 and 1921. His Fascist militia's march on Rome in 1922 ensured his appointment as Prime Minister. Parliamentary government was suspended in 1928 and possible conflict between the Roman Catholic Church and the state was avoided by the LATERAN TREATY (1929). The 1930s saw Mussolini's attack on Ethiopia (1935) and even closer ties with HITLER's Germany (the Rome-Berlin Axis). When France was on the verge of collapse (1940) during WWII, Mussolini entered the war as Germany's ally. Military failure caused his fall from power in

1943, but after his arrest he was freed by the Germans and set up in a puppet government in N Italy. With the German defeat (1945) he was captured and executed by Italian partisans. *See also* HC2 pp.218, 222, *222–223,* 226, *227.*

Mussorgsky, Modest Petrovich (1839–81), Russian composer, one of the "Mighty Five" who promoted nationalism in Russian music. His musical training was sketchy and he composed relatively few works. His finest is the opera *Boris Godunov* (1868–69). Other important works include the piano work *Pictures from an Exhibition* (1874) (later orchestrated by RAVEL) and *A Night on the Bare Mountain* (1867). He was more interested in expression and communication than form, and after his death much of his work was edited and revised, mostly by RIMSKY-KORSAKOV. *See also* HC2 pp.106–107, 120.

Mustafa Kemal. *See* ATATÜRK KEMAL.

Mustang, feral horse of the Great Plains of the USA, descended from horses that were brought from Spain. The mustang has short ears, a low-set tail, round leg bones, and can be any horse colour. During the 17th century there were 2–4 million mustangs, whereas today only about 20,000 survive in SW USA.

Mustard, any of various species of annual and perennial cruciferous plants (the group that also includes radish, cabbage, turnip, alyssum and stock), native to the temperate zone. These plants have pungent-flavoured leaves, cross-shaped, four-petalled flowers and carry pods. They include black mustard (*Brassica nigra*) whose seeds are ground to produce the condiment mustard, and white mustard (*Sinapis alba*) with seeds that produce a hotter mustard. Height: 1.8–4m (6–13ft). Family Cruciferae.

Mustard gas, poisonous gas first used in 1917 during WWI by the Germans. It is a blistering agent – one of the thioethers, a compound of carbon, hydrogen, sulphur and chlorine. By 1918 the Allies were also using this gas. It inflicted many casualties (causing blindness) but relatively few fatalities; its military use was eventually banned. *See also* MM p.*181*.

Mutation, sudden variation in an inherited characteristic of an individual organism that makes it different from the parent organisms. This change occurs in the DNA of the GENES, and can be passed on to the mutant's offspring. Natural mutations are rare, occur randomly, and usually produce an organism unable to survive in its environment. Occasionally the mutant is better adapted and, through natural selection, may become the next evolutionary generation. The mutation rate can be increased by exposing genetic material to X-rays, other ionizing radiation, or mutagenic chemical substance, such as MUSTARD GAS. Natural mutation is one of the means by which organisms evolve. *See also* NATURAL SELECTION; NW p.34.

Mute, person unable to speak. Mutism is a possible consequence of brain injury in which the speech areas are damaged. Most commonly in the past, however, it was a consequence of deafness at birth, particularly that caused by GERMAN MEASLES (rubella). The inability to hear brings with it a correspondingly great difficulty in learning to speak.

Mutiny on the Bounty, mutiny in the South Seas in 1789, led by Fletcher Christian, in which Captain William BLIGH of HMS *Bounty* and most of his ship's crew were cast adrift in an open boat. Bligh survived and three of the leaders were eventually taken back to England from PITCAIRN Island to face trial in 1808.

Mutis, José Celestino Bruno (1732–1808), Spanish naturalist and botanical explorer, who settled in Bogotá, South America in 1761. He cultivated plants for agricultural and medicinal purposes and made a collection of more than 24,000 plants. He wrote numerous botanical papers and was instrumental in founding the first conservatory in South America at Bogotá.

Mutoscope, early form of cinematograph, a peepshow devised by W. K. L. Dickson in 1895 to show "flick" books of photo-

graphs. Dickson made a camera to take the picture and had them printed on to cardboard. The Mutoscope did not need the artificial illumination upon which Thomas EDISON's kinetoscope relied.

Mutton, flesh of sheep slaughtered at about one year old. It is stronger-flavoured, coarser textured and less popular than lamb, the meat of sheep less than one year old. *See also* PE pp.232–233.

Muybridge, Eadweard (1830–1904), US photographer, b. Britain, who emigrated to the USA (1852), where he became a pioneer of motion photography. He invented the Zoopraxiscope (1881) which projected animated pictures on a screen, a forerunner of the motion picture. In his major work, primarily after 1878, he recorded with a series of sequential still cameras, the movements of animals and humans. His *Animals Locomotion* (1887) consisting of 11 volumes containing some 800 plates of men, women, children and animals in action, has been of immense importance to artists. *See also* MM p.*222*.

MVD. *See* KGB.

Myasthenia gravis, disorder found mainly in young people that causes weakness of the muscles. Generally affecting the facial muscles first, it may spread to include those of the neck, trunk and limbs. Its cause is unknown, but symptoms may be repressed by drug treatment.

Mycelium, vegetative body of a FUNGUS, found underground. It is made up of a web of filaments (hyphae), sometimes massed like felt. *See also* NW pp.42, *43*.

Mycenae, ancient city in NE Peloponnesos, Greece, now ruined, approx. 11km (6 miles) N of modern Árgos. Dating from the third millennium BC, Mycenae became the centre of the Mycenean civilization (*c.*1580–1120 BC). The Myceneans entered Greece from the N, bringing with them advanced techniques, particularly in architecture and metallurgy. The city controlled the route from the Peloponnesos to Corinth, being strategically placed to command the Argive Plain. The Myceneans traded with Crete, which also helped develop their culture, and by 1600 BC they dominated the Aegean. After 1200 BC the city began to decline with the invasions of the Dorians. It was destroyed in the 5th century BC but repaired by the Argives in the 3rd. By the 2nd century AD it was in ruins. Archaeological excavations by Heinrich Schliemann, begun in 1876, uncovered such notable remains as the Treasury of Atreus (also known as the tomb of Agamemnon), the Lion Gate which led into the city, beehive tombs, the city walls and many golden ornaments and weapons. *See also* HC1 pp.*22–23,* 39–41, *40–41*.

Mycology, science and study of FUNGUS. *See* NW pp.38–43.

Mycoplasma, any of about 15 species of the smallest known form of cellular life; they are usually considered to be bacteria. Most are parasites on birds and mammals, some live in stagnant water, and some play a part in such diseases as pleuropneumonia in cattle.

Mycorrhiza, or fungus root, association between certain fungi and the root cells of some vascular plants. The fungus may penetrate the root cells or form a mesh around them. Water and minerals enter the roots via these threads. Sometimes the fungus digests organic material for the plant.

Myelin, protective sheath around peripheral and some central nerve fibres which insulates the fibre to prevent loss of electrical impulse and allow rapid conduction. The disorder MULTIPLE SCLEROSIS attacks the myelin. *See also* MS p.97.

Myer Emporium, chain of department stores and supermarkets in Australia. The largest retailer in the country, the company had a turnover of about a \$1,000 million and employed almost 30,000 people in the mid-1970s.

My Fair Lady, musical play with libretto by Alan Jay Lerner and music by Frederick Loewe, originally produced on Broadway in 1956 when it ran for 2,717 performances. It was equally successful when produced at Drury Lane, London.

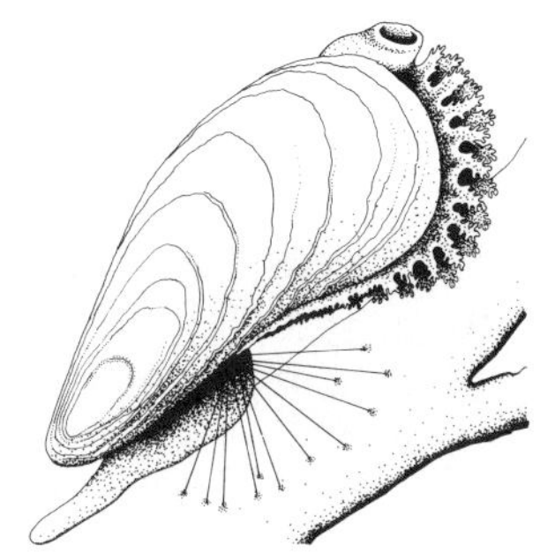

Mussel: its eggs are minute but each female in her life can produce up to 25 million.

Mutiny on the Bounty: the 1935 film starred Clark Gable and Charles Laughton.

Mycenae's hilltop location made it of great strategic importance in the E Peloponnesos.

My Fair Lady: Rex Harrison and Julie Andrews in the Broadway production.

Mynah birds inhabit open country, but people often cage them as pets.

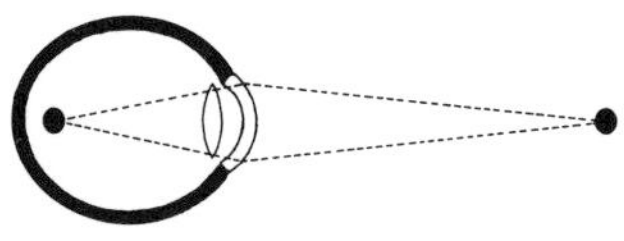

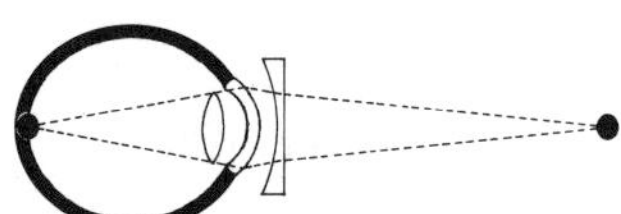

Myopia: the focusing point of a myopic person (top) is corrected by glasses (below).

Mysore; the palace of the Maharaja who was reputed to be the world's richest man.

Nablus was formerly in Palestine but, since 1967, it has been on Israeli-occupied land.

Based on George Bernard SHAW's *Pygmalion*, it is the story of an English professor's successful efforts to educate a street waif into gentility and how he builds a relationship with her in the process. One of the most successful musicals ever produced, it was made into a film in 1964 starring Audrey Hepburn and Rex Harrison.

Mykonos (Mîkonos), mountainous island in SE Greece, in the Aegean Sea. One of the Cyclades (Kikládhes) islands, it has an important tourist trade and fisheries. Area: 83sq km (32sq miles). Pop. (1971) 3,234.

Mylonite, any of several laminated fine-grained rocks formed when layers of parent rock fault, granulate or flow. It is chemically stable but partially melted and reduced to a powder by the movement of rock along a fault line and generally contains fragments of the parent rock. *See also* PE p.103.

Mynah, any of several species of tropical birds of SE Asia, s Africa, Australasia and the Pacific Islands; it is related to the STARLING. A natural mimic, especially the species *Gracula religiosa*, it imitates other birds in the wild and makes a popular pet. It feeds mainly on fruit. Length: to 33 cm (13in). Family Sturnidae.

Myocarditis, inflammation of the muscular tissue of the HEART, commonly caused by a VIRUS. Symptoms are pain, fever and a rapid pulse. Treatment depends on the underlying cause.

Myoglobin, protein found in animals. In vertebrates it is the pigment producing the red colour of muscle tissue. Like HAEMO-GLOBIN, myoglobin combines readily with oxygen for use in rapidly contracting muscles. It has been used extensively in research into the structure of PROTEINS. In 1962 John C. KENDREW shared the Nobel Prize in chemistry for his construction of a three-dimensional crystalline model of sperm whale myoglobin. *See also* SU p.156.

Myopia, near-sightedness, a common disorder of vision in which near objects are seen sharply (in focus) but distant objects are hazy (out of focus). It is caused by the eyeball being either too long or the eye's lens bulging too much, so that light rays entering the eye are brought to a focus in front of the RETINA. It is easily corrected with concave lenses in spectacles or contact lenses. *See also* MS p.49.

Myosin, thick filamentous PROTEIN present in MUSCLE cells, associated with ACTIN in the contractile process. *See also* MS pp.37, 59; SU p.156.

Myrdal, Karl Gunnar (1898–), Swedish economist. A professor of political and international economy at the University of Stockholm (1935–50; 1960–), he also served as executive secretary of the UN Economic Commission for Europe (1947–57). He is well known for *Political Element in the Development of Economic Theory* (1930) and for his massive study of US race relations, *An American Dilemma* (1944). He shared the 1974 Nobel Prize in economics with Friederich von HAYEK.

Myriapoda, class of ARTHROPODS with bodies made up of many similar segments. Each segment has one or more pairs of legs. *See also* CENTIPEDE; MILLIPEDE.

Myron (*fl.c.*480–*c.*440 BC), Greek sculptor who worked mainly in bronze. He was noted for his animals, none of which have survived, and his athletes in action. Most of his work has been lost but is known through descriptions by ancient writers and two of them by copies, the *Discobolus* and *Athena and Marsyas*.

Myrrh, aromatic, resinous, oily gum obtained from thorny, flowering trees such as *Commiphora myrrha*. Known and prized since ancient times, it has commonly been used as an ingredient in incenses, perfumes and medicines. The gum is exuded from ducts when the bark splits or is cut for tapping.

Myrtle, any of numerous species of evergreen shrubs and trees that grow in tropical and subtropical regions; especially the aromatic shrub, *Myrtus communis*, of the Mediterranean region. Its leaves are simple and glossy; the purple-black berries which follow the white flowers were once dried and used like pepper.

Mysore, city in s India, 137km (85 miles) sw of Bangalore, in Karnataka (formerly Mysore) state. The city served as the capital of the Mysore kingdom from 1799 until 1956. It is the seat of Mysore University (1916), and has several palaces. Industries: textiles, chemicals, leather goods, cigarettes, sandalwood oil. Area: (state) 191,756sq km (74,037sq miles). Pop. (1971) (state) 29,224,046; (city) 355,685.

Mystery plays, medieval English dramas based on religious themes. The plays were originally used by the clergy to teach their illiterate congregation the principal stories of the Bible. By the 14th century, however, the plays, although still based on religious topics, had become a form of popular entertainment. Each year, usually in England at the feast of Corpus Christi, when the days were long and there was a chance of good weather, the plays were performed by the various craft guilds in a town. Each guild chose an appropriate biblical episode, eg the carpenters Noah's Ark, the fishmongers Jonah and the Whale and the goldsmiths the Three Wise Men, performing the play on a large open float. The plays were rarely written down, and so varied from place to place and year to year, particularly as the actors used them to make topical jokes and poke fun at unpopular members of the community. In England the mystery plays from four places have survived: CHESTER, YORK, WAKEFIELD and Coventry.

Mysticism, personal religious experience seeking unity of self with God or the Transcendent. It is found in most major religions, and may involve the experience of trances or visions. Christian mystics have included the author of John's Gospel, the author of REVELATION, St PAUL, the author of *The Cloud of Unknowing*, Dame JULIAN OF NORWICH, St TERESA of Avila and St FRANCIS OF ASSISI. Indian mysticism is based on YOGA, while mysticism in Judaism is apparent in HASIDISM. Mystics in the Far East have mostly been followers of TAOISM or BUDDHISM.

Mytens, Daniel (*c.*1590–*c.*1648), Dutch portrait painter born and trained in The Hague. By 1618 he was working in England, and he entered royal service in 1624. As painter to Charles I from 1625 he gained prestige for his portraits which included several of Charles and his queen, Henrietta Maria, and his masterpieces *The Duke of Hamilton* and *The Third Marquess of Hamilton* (both 1629). In about 1635 he returned to Holland.

Mythology, literally, telling of stories, but usually collectively defined as the myths of a particular culture. A myth occurs in a timeless past, contains supernatural elements and seeks to dramatize or explain such themes as the creation of the world and man, the institutions of political power, the cycle of seasons, birth, death and fate. Bronislaw MALINOWSKI held that all myths are validations of the patterns of behaviour and institutions established within a particular society, and so no culture is free from myths. Sir James FRAZER's *The Golden Bough* (first published in 1890) held that all myths are connected with fertility in nature, and showed that birth, death and resurrection are constantly recurring themes. Carl JUNG believed that mankind shares a collective unconscious, which is manifest in certain constant images which he called the archetypes. Robert GRAVES' book *The White Goddess* (1948) sought to base all Greek and Celtic myths on the early supplanting of matriarchical, agricultural cultures by warlike patriarchical peoples. Most mythologies have an established pantheon, or hierarchy, of gods who were more or less anthropomorphic. When one culture has predominated over another it has often taken elements from the defeated culture's mythology. For example, the Romans adopted many aspects of Greek and Etruscan mythology, and later absorbed elements of MITHRAISM, which in turn is believed to have originated in Iran. *See also* MS pp.206–215.

Myxoedema, disease caused by insufficiency of THYROID hormone resulting in fatigue, a tendency towards weight gain

and, in later stages, delusions. Treatment involves the administration of thyroid extracts. It is known as CRETINISM in children.

Myxomatosis, virus disease of rabbits, sometimes spread deliberately by farmers to exterminate them, notably in Australia where early campaigns killed many millions of animals. The disease is generally fatal, although there is evidence that some rabbit populations are becoming resistant. Among rabbit populations the disease is spread by physical contact and by insect vectors such as mosquitoes.

N, the 14th letter of the alphabet, derived from the Semitic letter *nun,* which represented a fish. It was adopted by the Greeks as the letter *nu.* In English the letter is always pronounced in the same way, except in a very few words (such as *condemn* and *hymn*) in which it is silent. *See also* MS pp.244–245.

Naafi, acronym for Navy, Army and Air Force Institutes, a non-profit-making organization that provides canteens and other facilities for British forces both at home and overseas. Its headquarters is at Claygate, Surrey.

Naber, John (1956–), US swimmer who at the 1976 Olympic Games won four gold medals (100m and 200m backstroke, the 4×200m freestyle relay and the 4×100m medley relay) and one silver medal (200m freestyle). His four golds were won in world-record times, and he became the first to swim the 200m backstroke in less than 2min.

Nabis (from Hebrew "the prophets"), group of French painters who exhibited together between *c.*1889 and 1899. Reacting against the principles of IMPRESSIONISM, they based their style on GAUGUIN's pure colours and flat decorative style of painting. BONNARD and VUILLARD are the best known members of the group, which was led by Paul SÉRUSIER and also included ROUSSEL, Vallotton and MAILLOL.

Nablus, city of the w Bank, central Palestine. Excavations have revealed remains dating from *c.*1900 BC. It was the capital of the Samaritans, destroyed in 129 BC by John Hyrcanus I and rebuilt by HADRIAN. Formerly in Jordan, it came under Israeli administration in 1967. Nablus is a market centre for the surrounding region. Products include cereals, olives and soap. Pop. (1971 est.) 44,200.

Nabob, corruption of the title *nawab* which was usually given to a ruling Muslim nobleman in India. During the 18th century the term nabob was applied in Britain to anyone who had amassed great wealth in India, usually while in the service of the British EAST INDIA COMPANY. By extension the word may be applied to any man of wealth or political importance.

Nabokov, Vladimir (1899–1977), Russian-US author. He was brought up in the aristocracy of St Petersburg, but left Russia in 1919 and settled in Germany. He emigrated to the USA in 1940, but left for Switzerland in 1959. As well as being a renowned lepidopterist, he wrote many novels that extended the scope of language by exploring the precise connotations of words, sometimes by means of neologisms and trilingual puns. He wrote first in Russian and later in English. His novels include *Laughter in the Dark* (1933), *Bend Sinister* (1947), *Lolita* (1955) and *Ada* (1969).

Nabonidus (*d.*539 BC), last king of the Chaldean or New Babylonian Empire (*r.*556–539 BC). He was a pious ruler, but his espousal of the god Sin met opposition from the priests of Marduk. The ACHAEMENID Persian king CYRUS THE GREAT took Babylon in 539 BC without a struggle; Nabonidus was captured and exiled.

Nabopolassar (died *c.*604 BC), first king of the Chaldaean or New Babylonian Empire (*r.*625–*c.*604 BC). Babylonia had been a subsidiary state of the ASSYRIAN kingdom but after the death of ASHURBAN-IPAL, Nabopolassar, allied with the MEDES and the Persians, brought about the fall of the empire and established his independence. *See also* HC1 p.57.

Naboth, in the biblical book of 1 Kings, the owner of a vineyard which was coveted by King Ahab in his nearby palace. To possess it, Ahab's queen, Jezebel, had Naboth falsely accused of blasphemy and stoned to death.

Nabu, in Babylonian mythology, the son of MARDUK. Called Nebo in the Old Testament, Nabu was the Babylonian god of wisdom, especially in writing, being the scribe and messenger to the gods, writing man's destiny on the Tablets of Fate, and also being the god of justice. It was his temple which stood beside the huge ziggurat later called the Tower of Babel.

Nabulus. *See* NABLUS.

Nacreous cloud, in meteorology, cloud form resembling cirrus or almond-shaped altocumulus clouds. It shows strong irised (rainbow-tinted) patches. Observations of nacreous clouds are made mainly in Scotland and Scandinavia, and photometric measurements indicate that they occur at an altitude of approx. 20–30km (12–20 miles). *See also* PE p.70.

Nadar (1820–1910), French pioneer photographer, caricaturist and writer, real name Gaspard-Félix Tournachon. Among his superb photographic portraits are those of BAUDELAIRE and DORÉ. The first aerial photograph was taken by him from a balloon *c.*1856 and his studio photographs taken after 1858 were among the first made using electric light. His Parisian studio was a fashionable literary and artistic salon, and the first Impressionist exhibition was held there in 1874.

Nader, Ralph (1934–), US lawyer concerned with the rights of the public as consumers of goods and services. His call for improved car design, *Unsafe at Any Speed* (1965), led to new US legislation on car safety in 1966. He subsequently expanded his range of activities to include hazards in food and drugs, the use of nuclear power, the non-enforcement of job safety standards, and investigations of government organizations. His publications include *You and Your Pensions* (1973) and *Taming the Giant Corporation* (1976).

Nadi, Nedo (1894–1940), Italian fencer expert with foil, sabre and épée. Having won the individual foil at the 1912 Olympic Games, he won the individual and team foil, individual and team sabre, and team épée gold medals at the 1920 Olympic Games. No other fencer has won the foil and sabre titles at one Games, and his record of five gold medals at one Olympics remained unbeaten until 1972.

Nadir, in astronomy, point on the celestial sphere that is diametrically opposite the ZENITH and thus invisible to the observer.

Naevius, Gnaeus (*c.*270–201 BC), Latin epic poet and dramatist. He served in and chronicled the first PUNIC WAR, translated Greek tragedies and comedies into Latin, and was one of the earliest Roman poets. He died in Utica in Tunisia.

Naevus, red, blotchy BIRTHMARK. It consists of many blood vessels tangled together near the surface of the skin. If small enough a naevus can be removed by surgery or ELECTROLYSIS.

Naga, tribe inhabiting the Naga Hills of Assam, NE India, and the upper River Chindwin region of Burma. They were once notorious head-hunters, but today they are farmers, growing mainly rice. The Indian state of Nāgāland was formed in 1961.

Naga, in Hindu mythology derived from a pre-Hindu cult, a divine serpent with a human head. The term later came to mean "cobra". Nagas were able to kill men with their venom but BRAHMA is said to have ordered them to bite only the wicked or those ordained for a premature death. The beautiful females, Nagini, sometimes left their jewelled palaces under the sea to marry mortals.

Nāgāland, state in NE India; the capital is Kohīma. It is an undeveloped region, whose inhabitants, the Nagas, live in a tribal society and derive from the Indo-Mongoloid ethnic group. Nāgāland was established as a separate state in 1961, governed by a chief minister. There is a movement for independence, however, which has led to clashes between state troops and the Naga tribesmen. Area: 16,527sq km (6,381sq miles). Pop. (1971) 516,449.

Nagasaki, seaport in W Kyūshū, Japan. In the 15th century it was the first Japanese port to receive Western ships, although from 1643 to 1854 it was closed to foreigners. It was re-opened in 1859 and again became a centre of European and Christian influence in Japan. In August 1945, during WWII the inner city was destroyed by the second US atomic bomb. Today it is one of the country's leading ports. The largest industry is shipbuilding, and others include fishing and mining. Pop. (1974) 445,655. *See also* MM p.*180.*

Nagoya, city port in central Honshū, Japan, at the head of Ise Bay on the Pacific Ocean. It grew around a daimyo castle built in 1610. Today Nagoya is an industrial centre with iron and steel works, textile mills, fertilizer and chemical plants and aircraft factories. Pop. (1975) 2,080,000.

Nāgpur, city in Mahārāshtra state in W central India, 427km (265 miles) N of Hyderabād. The city was founded in the 18th century as the capital of the kingdom of Nāgpur, Mahratta, and passed to the British in 1853. It became the capital of Berar state (from 1903), and Madhya Pradesh state (1947–56). Nāgpur University (1923) is in the city. Industries: cigarettes, textiles, pottery, glass, leather, pharmaceuticals, brassware. Pop. (1971) 866,076.

Nagy, Imre (1896–1958), Hungarian statesman, Premier 1953–55 and 1956. Expelled from the Communist Party in 1955 for alleged anti-Soviet nationalism, he was recalled to the premiership in 1956 in the wake of the anti-Soviet uprising of late October. He promised free elections, economic reforms and the abolition of the one-party dictatorship. He demanded the withdrawal of Soviet troops and freed Cardinal MINDSZENTY from prison. Although the USSR promised concessions, demonstrations continued and on 4 November Soviet troops and tanks moved in to suppress the insurgents. Nagy was executed in 1958. *See also* HC2 p.*306.*

Naha, seaport city on the W coast of Okinawa island, Japan. Naha is also an important manufacturing centre. Its products include pottery, textiles and sugar. Pop. (1974) 306,446.

Náhua, most important tribal group and language of central Mexico. The best-known representatives are the Aztecs. Today approx. 1 million Indians speak the language and live mainly in the Mexican states of Michoacán, Puebla and Hidalgo. Scattered groups live in Veracruz and parts of Central America.

Nahuatl, American Indian language of the Aztec-Tanoan linguistic stock of North and Central America. It is descended from the ancient Aztec and is spoken today by some 800,000 people, mostly living in Mexico.

Nahum, seventh of the 12 minor prophets in the Old Testament. Perhaps the greatest poet of ancient JUDAH, he predicted the violently destructive fall of the capital of the hated Assyrians, Nineveh (612 BC). His strong nationalism, his hopes for the spiritual salvation of Judah, and his ideals of justice and faith were, however, entirely centred around the events of his own lifetime.

Naiads, in Greek mythology, female figures or NYMPHS identified with streams, rivers and lakes. Their cult was widespread in Greece, where it was believed that mortals could gain inspiration by drinking from springs watched over by particular naiads.

Naidu, Sarojini (1879–1949), Indian poet and politician. She is best known for her books of romantic verse in English about India, the last of which is *The Broken Wing* (1915). She was a social reformer and in 1925 became the first woman to be President of India's National Congress.

Nails, in anatomy, horny growths from the fingers and toes of primates, which grow from a root of cells situated in a fold of skin at the nails' base This fold is easily damaged. Although different in appearance, nails, like FEATHERS are made of the same substance as HAIR, the fibrous protein KERATIN. The nails and claws of animals are worn down naturally; those of human beings are not, and as a result, would grow to extraordinary lengths if not cut. *See also* MS pp.60, 98.

Nails, metal, small spikes used as fasteners since about 1000 BC. Nail manufacturers today use large and heavy machines which are fed with metal wire that passes through dies made of toughened steel. A die grips the wire firmly and a hammer advances to strike a short, exposed length to form the nail head. The nail is then cut and pointed automatically. Spiral-threaded nails such as those used for roofing, are made by passing wire between rollers with spiral grooves or flutes.

Nail varnish, lacquer-based cosmetic applied to the fingernails and toenails. It comprises a solution of nitrocellulose lacquer to which is added a plasticizer – to make the varnish spread and prevent brittleness – and colouring matter. Originally red or pink, nail varnishes are now produced in a wide variety of colours. *See also* MM p.247, *247.*

Naipaul, Vidiadhar Surajprasad (1932–), West Indian novelist and short-story writer. He was educated in Trinidad and at Oxford, and his work clearly reflects these influences. He later settled in London, taking up broadcasting and writing for periodicals. His novels include *A House for Mr Biswas* (1961), *In a Free State* (1971), for which he won the Booker Prize, and *Guerillas* (1975).

Nairn, town in Highland Region, NE Scotland, at the mouth of the River Nairn on the Moray Firth. Former county town of Nairnshire. It is a fishing port and holiday resort. Pop. (1974 est.) 8,900.

Nairne, Baroness Carolina Oliphaunt (1766–1845), Scottish Jacobite songwriter who wrote many well-known "folk" songs perhaps the most popular being *Charlie Is My Darling* and *Will Ye No Come Back Again.*

Nairnshire, former county in NE Scotland, on the S shore of the Moray Firth; since 1975 it has been a district in the Highland Region. On the coastal lowlands oats, potatoes and barley are the chief crops, and cattle are raised. Sheep are grazed on the S uplands. Industries include whisky distilling, granite quarrying, brick making and tourism. The county town was Nairn, where the district authority is now located. Area: 422sq km (163sq miles). Pop. (1974) 8,906.

Nairobi, city in S central Kenya; the capital and largest city of Kenya. Founded in 1899 at the site of a MASAI watering hole, Nairobi replaced Mombasa as the capital of the British East Africa Protectorate in 1905. During the 1950s the city was a centre of the MAU MAU rebellion. The first All Africa Trade Fair was held there in 1972. Nairobi has a National Park (1946), a University (1970), and several institutions of higher education. Industries: cigarettes, textiles, chemicals, food processing, furniture, glass, building materials. Pop. (1975 est.) 700,000.

Naismith, James A. (1861–1939), US sportsman. While working as a physical education instructor at the YMCA college in Springfield, Massachusetts, he originated the game of basketball in 1891. Many of the rules he outlined are still in use today.

Naivasha, lake in SW central Kenya, E Africa, in the Great Rift Valley; it has no outlet. The lake, the volcanic Crescent Island and the town of Naivasha on the NE shore are popular tourist attractions. Area: 280sq km (108sq miles).

Naked and the Dead, The (1948), novel by Norman MAILER. A panoramic work about the brutality and wastefulness of war, it follows one platoon during the US invasion of a small Pacific island held by the Japanese in WWII.

Nakhon Ratchasima (Khorat), city in NE Thailand, on the River Mun; capital of Nakhon Ratchasima province. The city

Ralph Nader, the consumers' champion who successfully challenged big business.

Naga, the serpent-like creature with a human head who bites only the wicked.

Nagasaki: a pre-war street scene before the city's atomic destruction.

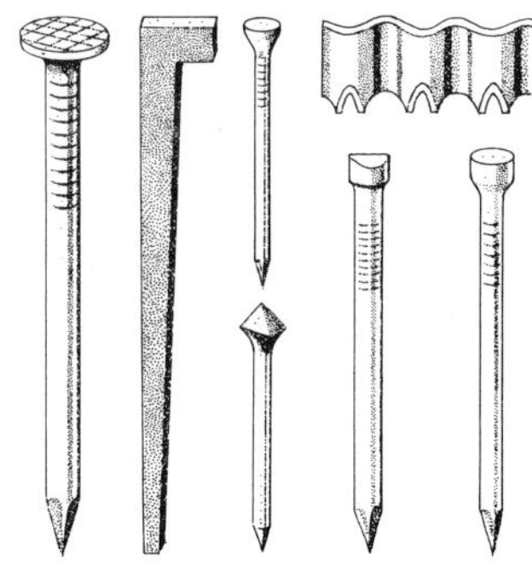

Nails are generally classified either by the shape of the head or by their function.

Namur has bridges spanning the rivers Sambre and Meuse, which meet there.

Fridtjof Nansen achieved fame as an explorer, a scientist, and a statesman.

Edict of Nantes, Henry IV's law which gave freedom of worship to French Huguenots.

John Napier, the inventor of logarithms, also invented many weapons of war.

was founded in the 17th century but did not develop until the construction of the railway to Bangkok in 1890. It is a transport centre, being strategically placed at the entrance to a pass through the Khorat plateau. Copper is mined nearby. Pop. (1970) 102,095.

Nakuru, city in w central Kenya; the capital of Rift Valley province. Founded in 1904 as a European settlement, it is now a developing commercial centre with textiles and food processing industries. Pop. (1969) 47,800.

Nalanda, Buddhist monastic centre N of Rajgir, India. By tradition founded at the time of BUDDHA, the centre was probably built in the 5th century BC, and served as a school for Buddhist philosophers. Manuscripts copied there still exist, as do many religious sculptures in bronze and stone.

Namaqualand (Namaland), region in sw Africa, extending N–s from Windhoek in Namibia (South West Africa) to Cape province in South Africa, and from the Namib Desert w to the Kalahari Desert (E). It is an arid area and pastoral farming is the principal economic activity. Some copper is mined. Area: approx. 388,500sq km (150,000sq miles).

Namath, Joe William ("Broadway Joe") (1943–), US football player. He joined New York as a quarterback in the AMERICAN FOOTBALL League (1965). In 1967, he set the record for the most yards gained – 4,007 (3,664m) – by passing in a single season.

Nam-dinh, town in NE Vietnam, on the delta of the Red River (Hung Ho), 72km (45 miles) SE of Hanoi. It is a trade and transport centre.

Namib Desert, desert region in sw Africa, along the coast of Namibia between the Atlantic Ocean and the interior plateau. The area receives less than 1cm (⅜in) of rain each year and is almost completely devoid of vegetation. Diamonds are mined. Area: 170,000sq km (66,000sq miles).

Namibia, also known as South West Africa, is a territory governed by South Africa, its neighbour to the s. Most of the country consists of high plateau lands with an uninhabited coastal strip. Cattle are reared in the N and sheep in the s but Namibia's economy is based on its mineral resources. Diamonds, lead and uranium are mined. The capital is Windhoek. Area: 823,327sq km (317,887sq miles). Pop. (1975 est.) 883,000. See MW p.127.

Namier, Sir Lewis Bernstein (1888–1960), British historian. He was born in Poland but studied in England and was Professor of history at Manchester (1931–53). His best-known works were *The Structure of Politics at the Accession of George III* (1929) and *England in the Age of American Revolution* (1930).

Namoi, river in SE Australia; it rises in the Liverpool Range, E New South Wales and flows NW to join the Darling River at Walgett. Length: 847km (526 miles).

Namp-o, city in w North Korea, on Korea Bay; formerly known as Chinnampo. Opened to foreign trade in 1897, it has become an important port for Pyongyang, the capital of North Korea. Industries include flour milling. Pop. 130,500.

Namur, town in s Belgium, at the confluence of the Meuse and Sambre rivers, 56km (35 miles) SE of Brussels; it is the capital of Namur province. The city fell to the French in 1692 but it was retaken by WILLIAM OF ORANGE in 1695. It was the scene of many battles during both World Wars. Industries: metallurgy, glass, livestock. Pop. (1970 est.) 32,500.

Nana (1880), novel by Émile ZOLA, written in the Naturalist manner. It is one of the Rougon-Macquart series of novels (1871–1893) tracing the complete social history of a family of the Second Empire. Nana, a beautiful, cruel and misguided Parisian courtesan, captivates and then ruins all the men she meets.

Nanak (1469–1539), founder of SIKHISM. Of Hindu parents and Muslim schooling, Nanak left his family to follow a call from God. In his preaching Nanak had most success in his native Punjab, proclaiming man's oneness under God.

Nance, Willis Raymond ("Ray")

(1913–), US jazz trumpeter, cornetist and violinist. He played with Earl HINES and Horace Henderson in the 1930s and with Duke ELLINGTON after 1940.

Nancy, city in NE France, on the River Meurthe, 287km (178 miles) E of Paris; capital of Meurthe-et-Moselle département. The city developed around the castle of the dukes of Lorraine and became capital of the duchy in the 12th century. The city passed to the French crown in 1766. Industries: iron products, glass, machine tools, textiles. Pop. (1968) 123,428.

Nanda Devi, mountain peak in the Himalayan system, Uttar Pradesh state, N India. It is one of the highest peaks in the country. Nandi Devi was not climbed until 1936, when an Anglo-American team reached the summit. Height: 7,793m (25,662ft).

Nandi, Negroid people of w Kenya. Traditionally a warrior people, they now live by intensive cultivation, especially of millet and corn, breeding cattle for ritual purposes. Their patrilineal society is characteristically divided into age-sets, males passing through seven grades from warriors to elders, the latter holding political and juridical authority.

Nanga Parbat, mountain peak in the Punjab Himalayan system, N India. It is the seventh highest peak in the world. After several unsuccessful attempts, it was scaled in 1953 by a team led by Herman Buhl. Height: 8,131m (26,676ft).

Nanking (Nanjing), city in E China, 242km (150 miles) w of Shanghai, on the River Yangtze. It served as the capital of China until the end of the 14th century, then again in 1928–37 and 1946–49. The treaty of Nanking (1842) ended the OPIUM WAR with Britain and opened five ports to foreign trade; Nanking was declared a treaty port in 1858 but not opened until 1899. It served as the seat of Sun Yat-Sen's provisional presidency in 1912 and fell to Japan in 1937 during the SINO-JAPANESE WAR. The city has an iron and steel complex and oil refining and chemical plants. Pop. (1970 est.) 2,300,000.

Nanking, Treaty of (1842), peace settlement which ended the OPIUM WAR (1839–42) between Britain and China. An indemnity was paid, Hong Kong was ceded to Britain and the ports of Amoy, Foochow, Ninpo and Shanghai opened to British trade. Before 1842 only Canton had been allowed trading rights and Britain had no diplomatic representation. The treaty, which was supplemented by the Treaty of the Bogue (1843), was thus an early milestone in the opening up of China to the West. See also HC2 p.144.

Nansen, Fridtjof (1861–1930), Norwegian explorer and statesman. He crossed the Greenland icecap on foot in 1888, and for his next expedition he constructed a ship, *Fram*, that could be frozen in the Arctic ice. This allowed him to drift to 84°N, and after leaving the *Fram*, he and F.H. Johansen reached 86°14′N in 1895, the farthest north man had yet gone. He was awarded the 1922 Nobel Peace Prize.

Nansen bottle, device, sometimes openended, used for the collection of subsurface water samples.

Nansen International Office for Refugees, founded under the auspices of the League of Nations in 1931 in honour of Fridtjof NANSEN in Geneva. It mainly assisted Armenian refugees from Turkey, "White" Russians and later German Jews.

Nan Shan (Qilianshanmái), mountain range in China, s of the SILK ROUTE through Kansu province. It includes peaks of more than 6,100m (20,000ft).

Nantes, city in NW France, on the River Loire, 172km (107 miles) w of Tours; the capital of Loire-Atlantique département. An ancient Gallic capital prior to the Roman conquest in the 1st century BC, it was captured by the Normans in the 9th century and held by the Dukes of Brittany until 1499 when it became part of France, HENRY IV issued the Edict of NANTES there in 1598. The city has the University of Nantes (1460), and a castle in the Gothic-Renaissance style. Industries: metals, dyes, clothing, bicycles. Pop. (1968) 259,208.

Nantes, Edict of (1598), law granting considerable religious freedom to French Protestants, the HUGUENOTS, promulgated by HENRY IV at Nantes. It guaranteed freedom of conscience, social and political equality, and established a special court, composed of both Catholics and Protestants, to hear disputes arising from the edict. Protestant worship was limited, however, to areas they already held (about 100 fortified towns) and was not permitted within five leagues of Paris. Secret agreements promised the crown's financial support of the armies garrisoned in the 100 towns. The last provision gave the Huguenots a virtual state within the state and was incompatible with the centralizing policies of Cardinal RICHELIEU in the 17th century. LOUIS XIII, acting on Richelieu's advice, withdrew the political and military provisions of the edict in 1629 and LOUIS XIV revoked it entirely in 1685. See also HC1 p.273.

Nantucket, island in SE Massachusetts, USA 40km (25miles) s of Cape Cod, in the Atlantic Ocean. An important whaling centre until the mid-19th century, Nantucket is now a tourist resort and an artists' colony. Area: 148sq km (57sq miles). Pop. (1970) 3,774.

Naomi, in the Bible, wife of Elimelech and mother-in-law of the widow RUTH, whose later marriage to Boaz she arranged.

Napalm, gelatinous petroleum used in wartime to make bombs and as fuel for flame-throwers. It was developed during WWII, and widely used by the American forces in Vietnam. When napalm hits its target, it spreads out, clinging to and burning, everything it touches. See also MM p.181.

Napata, ancient city in NE Sudan and the capital from c. 750 to 590 BC of the kingdom of KUSH which restored Egypt following its destruction by Libyan mercenaries. In 590 Napata was sacked by the 26th Dynasty of Egypt but, although the capital was moved, it remained the religious centre of Kush.

Naphtha, any of several volatile liquid-hydrocarbon mixtures. In the first century AD, "naphtha" was mentioned by Pliny the Elder. Alchemists used the word for various liquids of low boiling-point. Several types of products are now called naphtha, eg coal-tar naphtha, shale naphtha and petroleum naphtha. Petroleum naphtha contains aliphatic hydrocarbons, boils at higher temperatures than petroleum and at lower temperatures than kerosene.

Naphthalene, important HYDROCARBON ($C_{10}H_8$) composed of two benzene rings sharing two adjacent carbon atoms. Naphthalene is soluble in ether and hot alcohol. It is used for making mothballs, dyes and synthetic resins, and is obtained from coal tar and the high-temperature cracking process of petroleum. It crystallizes in white plates; m.p. 80°C (176°F); b.p. 218°C(424°F).

Naphthol, name of two organic compounds derived from NAPHTHALENE. Both are colourless and crystalline and have the formula $C_{10}H_7OH$. 1-Naphthol is produced by heating 1-naptholenesulfonic acid with caustic alkali or by heating 1-naphthylamine with water under pressure. It forms into colourless crystals; m.p. 96°C (205°F); b.p. 288°C (550°F). 2-Naphthol is produced by a similar process to that for 1-naphthol. It, too, forms into colourless crystals; m.p. 124°C (225°F), b.p. 285°C (545°F).

Napier, Sir Charles James (1782–1853), British soldier and colonial administrator. He served in the PENINSULAR WAR in 1810 against the USA in the WAR OF 1812 and kept order in N England during the Chartist agitation (1839–41). He saw service in India from 1841, provoking a war with the Sind in 1842, and governing the conquered province from 1843–53.

Napier, or Neper, John (1550–1617), Scottish mathematician. He treated mathematics as a hobby but in 1614 invented logarithms and the present form of the decimal notation. He developed "Napier's bones", calculating devices for simplifying mathematical calculations. See also SU p.38.

Napier, Robert Cornelis, 1st Baron

Napier of Magdala (1810–90), British army officer, *b*. Ceylon. He was a skilled military engineer and built roads, canals and defences in the Sikh wars (1845, 1848) and during the INDIAN MUTINY. He was Commander-in-Chief of India from 1870–76, became a general in 1874 and a field-marshal in 1883.

Napier, seaport in E North Island, New Zealand, on Hawke Bay. It is a food processing and shipping centre. Its exports include dairy products, fruits, vegetables and meats. Pop. (1970 est.) 38,200.

Naples (Napoli), in s central Italy, 188km (117 miles) SE of Rome, on the Bay of Naples. Founded *c*.600 BC as a Greek colony, Naples came under Roman rule in the 4th century BC, and under Byzantine rule in the 6th century AD. It became capital of the Kingdom of Naples from the 13th to the 19th centuries. Ceded to Austria in 1713, the city joined the Kingdom of Italy in 1860. Notable buildings include the Church of the Holy Apostles, said to have been founded by Constantine, a university (1224) and Virgil's tomb. Industries: textiles, steel, shipbuilding, aircraft. Pop. 1,224,300.

Naples, school of, 17th-century school of painting, which flourished in s Italy and Spain. Based on the example of CARAVAGGIO (who stayed in Naples 1606–07), painters of the school characteristically depicted scenes of violent activity and made much use of heavy CHIAROSCURO. Its most notable painter was RIBERA.

Napo, river of NW central South America, rising in the N central mountains of Ecuador and flowing SE through Ecuador and Peru to join the River Amazon approx. 64km (40 miles) NE of Iquitos, Peru. Length: 1,127km (700 miles).

Napoleon I (1769–1821), general and Emperor of France. He was born Napoleon Bonaparte in Corsica and studied at the military academy in Brienne. He led the French liberation of Toulon from the British in 1793 and was promoted brigadier-general. His future was in doubt during the THERMIDOREAN REACTION, but after saving the NATIONAL CONVENTION from a right-wing uprising in Oct. 1795 he was given command of the army in Italy in 1796. He won important victories there against Austria, but failed in his attempt to drive the British from Egypt in 1798. In 1798–99 France lost most of Italy and Napoleon returned to France in Oct. 1799 and carried out the coup of 18th Brumaire (9 Nov. 1799) against the DIRECTORY. He set up the CONSULATE, and in Dec. 1799 became First Consul. He defeated the Austrians at Marengo in 1800, and at home centralized local government and the tax collection system. He established the Bank of France, reformed the legal system with the CODE NAPOLEON and made peace with the Roman Catholic Church with the Concordat of 1801.

He concluded the Treaty of Amiens with Britain in 1802, and crowned himself Emperor of the French in 1804, with Pope PIUS VII presiding. War broke out again in 1805 and he defeated Russia and Austria at AUSTERLITZ in 1805 and Prussia at JENA in 1806. He reorganized the German states and made members of his family rulers of Spain, Naples and Holland. However, he could not defeat the British at sea and led a disastrous expedition to Russia in 1812. He was forced to abdicate in April 1814 and was confined to the island of Elba. He escaped in March 1815, and seemed triumphant for a HUNDRED DAYS. He was finally defeated at WATERLOO in June 1815, abdicated once more, and was exiled by the British to St Helena, where he died.

In 1796 he married Josephine de Beauharnais. He was divorced from her in 1809 to marry Marie Louise of Austria in 1810. Their son was known as NAPOLEON II. *See also* HC2 pp.74–79.

Napoleon II (1811–1832), only son of NAPOLEON I and Marie-Louise. His full name was François-Charles-Joseph Bonaparte. In 1814 he was taken by his mother to Austria, and in 1818 his grandfather, FRANCIS I of Austria, conferred on him the title of duc de Reichstadt.

Napoleon III (1808–73), Emperor of the French (*r*.1852–70), also known as Louis Napoleon. He was the nephew of NAPOLEON I. He twice attempted a coup in France (1836 and 1840) and won some popularity because of his utopian socialist views. He returned to France from exile in Feb. 1848 and was overwhelmingly elected President in Dec. 1848. He supported Pope PIUS IX against the Italian republicans in 1849, and in reaction to the ensuing opposition staged a coup in Dec. 1851 and was proclaimed emperor in Nov. 1852. During his reign France experienced both economic growth and political authoritarianism. Napoleon undertook expensive wars in Italy, SE Asia and Mexico, and in response to growing opposition instituted a liberal empire in 1870. He was deposed in September 1870 as a result of early French defeats in the Franco-Prussian war, and retired to England where he later died. *See also* HC2 p.111.

Napoleonic Wars, series of wars during which France, under NAPOLEON I fought various coalitions of the other European powers. Britain was constantly at war with France from 1803–14, and then again in 1815; after Napoleon's invasion plans had been frustrated at TRAFALGAR, the main field of conflict between the two opponents was in Spain, during the PENINSULAR WAR (1808–14). Austria and Russia joined with Britain in the War of the Third Coalition in 1805, but Austria was defeated at ULM and AUSTERLITZ and forced to sign the peace of PRESSBURG. In 1806 Prussia entered the war, only to be defeated at Jena and AUERSTADT; and Russia, defeated at Friedland in 1807, signed the Treaty of Tilsit with Napoleon. In 1809, at WAGRAM, Austria was again defeated. Three years later, however, in 1812, Napoleon squandered an army in Russia, and the next year all his opponents united in the Fourth Coalition to defeat him at LEIPZIG. Forced to abdicate in 1814, Napoleon returned to power in 1815, but was finally defeated at WATERLOO. After the wars, Europe was re-organized at the Congress of VIENNA. *See also* HC2 pp.75, 75, 77–79, 77, 79.

Napoli. *See* NAPLES.

Nappe, large-scale fold in rocks, thrown up by mountain-building processes and transported over large distances. The Alps are a system of several such nappes. *See also* PE p.107.

Nappy rash, skin irritation around genitals and anus of babies caused by wet or soiled nappies.

Naqsh-e-Rostam, ACHAEMENID archaeological site in Iran. Four rock tombs, including that of DARIUS THE GREAT, were discovered there together with relief sculpture dating from the 3rd century BC.

Nara, city in s central Honshū, Japan, 42km (26 miles) E of Osaka. Founded in 706, it was the first capital of Japan (710–84). Today it is a cultural and religious centre, having the oldest Buddhist temple in Japan. Pop. (1974) 247,082.

Naram-Sin (*c*.2254–*c*.2218 BC), fourth of the five kings of the s Mesopotamian dynasty of AKKADIA. He spent his reign trying to extend Akkadian power.

Narayan, Rasipuram Krishnaswamy (1906–), Indian novelist and short-story writer. He is famous for his tragicomic novels, written in English set in the fictional town of Malgudi in s India. They include *Waiting for the Mahatma* (1955), *The Man-Eater of Malgudi* (1961) and *A Horse and Two Goats* (1970).

Narayanganj, city in central Bangladesh, at the confluence of the Lakhya and Dhaleswari rivers. It is the region's chief river port and is an important jute trading centre as well as a major industrial city. Pop. (1974) 186,769.

Narbada (Narmada), river in central India. It rises near the MADHYA PRADESH-VINDHYA PRADESH border, flows w between the Vindhya Mountains and the Satpura Range, forms an estuary and enters the Gulf of Cambay. The river forms the traditional border between Hindustan (N) and Deccan (S). Length: 1,290km (801 miles).

Narcissus, in Greek mythology, a beautiful youth who rejected the love of the nymph ECHO, and was induced by her to fall in love with his own reflection in a pond. He pined away because of his unrequited love and was turned into a flower. The term narcissism in psychology is derived from this legend.

Narcissus, any of numerous species of Old World bulbs, now almost universally popular garden flowers. Narcissi bloom in early spring. The long, pointed leaves surround yellow, orange or white trumpet-like flowers. Favourite species are the yellowish DAFFODIL (*Narcissus pseudonarcissus*) and jonquil (*N. jonquilla*). Family Amaryllidaceae.

Narcolepsy, disorder tending towards the chronic, wherein one falls suddenly and uncontrollably asleep during the day. Night sleep is disturbed. This may be caused by hardening of the arteries, syphilis of the central nervous system or brain injury.

Narcotics, depressive drugs used to reduce pain, diminish sensation and induce sleep. They can cause profound stupor, coma or convulsions when given in excessive doses. MORPHINE, CODEINE and meperidine (Demerol) are among the narcotics commonly used in medicine. Narcotics also produce some harmful side-effects, eg constipation and nausea.

Nares, Sir George Strong (1831–1915), British admiral and explorer who commanded the *Challenger* (1872–74) in its investigation of the oceans and ocean beds of the world. *See also* PE pp.82, 90.

Narew, river in NE Poland which rises near the Białowieza Forest, flows w and sw into the Bug River near Warsaw. Length: 477km (296 miles).

Nariño, Antonio (1765–1823), Colombian revolutionary. He published a translation of the French DECLARATION OF THE RIGHTS OF MAN, and was deported to Spain in 1795. He returned to s America in 1797 and following the 1810 Revolution became President of Curdinamarca province. In 1821, after some six years in prison in Spain, he became Vice-President of Columbia. As a political leader, he worked towards a strong, centralist government.

Narlikar, Jayant Vishnu (1938–), Indian physicist who has conducted research in such areas as general relativity and gravitation, quantum theory, cosmology and astrophysics. His many publications include *Action at a Distance in Physics and Cosmology* (1974), written in collaboration with Sir Fred Hoyle.

Narmer. *See* MENES.

Narrenschiff, Das (1494), German satirical poem by Sebastien BRANT. In 115 sections and more than 7,000 lines, the poem castigates all types of human folly by employing the central image of a ship of fools. Evils such as adultery, blasphemy and gluttony are also pilloried, and in the later parts the Church and the Empire are attacked.

Narthex, entrance screen or partition in early Christian and Byzantine churches, found especially in France and Italy. It usually extended across the entire west front of the building creating a vestibule for penitents and converts under instruction who were not admitted to the church proper. As unrestricted entry into the church became normal in the 13th century, the narthex served no further ritual purpose.

Narváez, Pánfilo de (*c*.1480–1528), Spanish conquistador. He was the chief lieutenant to Diego de VELAZQUEZ in the conquest of Cuba in 1514. In 1520 Velázquez sent him to Mexico to arrest Hernán CORTÉS; but instead he was imprisoned by Cortés (1520–22). He returned to Spain but sailed to Florida in 1527, where he searched unsuccessfully for gold. *See also* HC1 p.236.

Narvik, seaport in N Norway, on the Ofot Fjord; it has an ice-free harbour. It was founded in 1887 as a port for the Swedish iron mining towns of Kiruna and Gällivare. The fishing industry is important. Pop. (1971 est.) 13,200.

Narvik Expedition, attempt by British forces in WWII to prevent the Germans, who had occupied the city in April 1940,

Napoleon, as well as being a great general, was a supreme legislator.

Napoleonic Wars; *The Battle of Leipzig,* in which Germany's liberation was completed.

Narcissus; the deeper yellow variety of this flower is better known as daffodil.

Narvik's ice-free port is important for the export of iron ore.

Narwhale

Narwhale; Greenland Eskimos eat its fat and make tools out of the tusks.

John Nash; the west wing of his elegant Park Crescent in London's Regent's Park.

Gamal Nasser's death was mourned by Egyptians throughout the world.

Ilie Nastase, the superb but sometimes temperamental tennis champion.

from using it as a shipping base for Swedish iron ore. A British expeditionary force briefly occupied (28 May–9 June 1940) the port, but were forced to abandon it when allied resistance on the western front crumbled.

Narwhale small, toothed Arctic WHALE. The male has a twisted "horn", half as long as its body, which develops from a tooth and protrudes horizontally through the upper lip. Its function is unknown. Length: up to 5m (16ft). Species *Mondodon monoceros. See also* NW pp.224–225.

NASA, acronym for National Aeronautics and Space Administration, US government body established in 1958 to administrate civilian aeronautical and space research programmes. It has various departments throughout the USA: at Houston, Texas, is the Lyndon B. Johnson Space Center, responsible for manned space flights. Space rockets, both manned and unmanned, are launched from the John F. Kennedy Space Center at Cape Canaveral, Florida. In 1975 it employed 12,000 scientists, engineers and technicians.

Nasby, Petroleum Vesuvius, pseudonym of US journalist David Ross Locke (1833–88). From 1861 onwards he published the "Nasby Letters" in two Ohio newspapers, attacking the racialists of the time by presenting himself as an extreme proponent of their point of view which is damned by his illiterate style and stupidity. Collections of the letters were subsequently published, among them *The Nasby Papers* (1864) and *The Diary of an Office Seeker* (1881).

Nasca. *See* NAZCA.

Naseby, Battle of (14 June 1645), battle of the ENGLISH CIVIL WAR in Leicestershire in which the Royalist troops under Prince RUPERT were outnumbered and defeated by the Parliamentarians under Oliver CROMWELL and Sir Thomas FAIRFAX. It was a decisive battle of the war.

Nash, John (1752–1835), British architect and town planner who was important in the development of the neo-Classic REGENCY STYLE. He is famous for his design of Regent's Park and Regent Street in London. He also started plans for the remodelling of Buckingham Palace and rebuilt the Royal Pavilion Brighton (1815–23). *See also* HC2 p.70.

Nash, John Northcote (1893–1977), British painter and wood engraver. A landscape artist, his style bears affinities with that of his brother Paul NASH, but remains closer to natural forms.

Nash, Ogden (1902–71), US poet. Among his many volumes of humorous and satirical poetry are *Free Wheeling* (1931), *The Face Is Familiar* (1940) and *The Christmass That Almost Wasn't* (1957). He wrote the lyrics for *One Touch of Venus* (1943), a musical comedy in collaboration with S.J. Perelman.

Nash, Paul (1889–1946), British painter and illustrator. The poetic, dreamlike style of his watercolour and oil landscapes was stimulated by SURREALISM, eg *The Menin Road* (1918) and *Landscape from a Dream* (1938). He was appointed an Official War Artist in 1917 and again in WWII.

Nash, Richard ("Beau") (1674–1762), arbiter of English fashion and court behaviour. It was he who set the tone of Bath, the most fashionable spa, in the 18th century by establishing a code of manners and propriety. His disapproval virtually abolished the practice of duelling.

Nash, Sir Walter (1882–1968), New Zealand politician and Prime Minister (1957–60). Born in Britain, he became a New Zealand MP in 1929. As Finance Minister (1935–49) he helped introduce the Labour Party's wide-ranging social security scheme. A doughty and tireless campaigner, he travelled extensively abroad and earned a reputation for passionate humanitarianism. *See also* HC2 p.*250.*

Nashe, Thomas (1567–1601), English poet, pamphleteer and dramatist. His satirical writings include *Pierce Penniless His Supplication to the Divell* (1592) and *The Unfortunate Traveller* (1594), a picaresque prose narrative.

Nashville, state capital and port on N central TENNESSEE, USA, on the Cumberland River. Settled in 1779 as Nashborough, it became state capital in 1843. Nashville was the scene of a decisive victory of Union troops over the Confederate army in December, 1864, during the AMERICAN CIVIL WAR. It is a noted country music centre. There are several universities in the city. Industries: clothing, footwear, food processing, tyres, chemicals, publishing. Pop. (1973) 449,109.

Nasmyth, Alexander (1758–1840), Scottish painter. Influenced by Allan RAMSAY, he became a prominent portrait and landscape painter. Two portraits of *Robert Burns* (1787–1828) are in the Scottish National Portrait Gallery, Edinburgh. His son Patrick (1787–1831) also painted.

Nasmyth, James (1808–90), Scottish inventor and engineer. He developed a foundry for making machine tools and steam-powered machines. In 1842 he patented a steam hammer designed to forge the drive shafts originally intended for the SS *Great Britain.*

Nasrid kingdom, the city of Granada in S Spain during its rule from 1238 to 1492 by the Moorish dynasty named Nasrid. Founded by Muhammad ibn-Yusuf in 1230, it reached its zenith during the reign of Mohammed V (*r.*1356–91), who contributed much to the power and architecture of the city. It was conquered in 1492 by FERDINAND and ISABELLA of Spain and subjected to a brutal policy of Christianization.

Nassau, port of the NE New Providence Island in the British colony of the Bahama Islands. It is now the capital and commercial centre of the Bahamas. Founded in the 1660s by the English and named Charles Towne, it was renamed in 1695. Exports include fruit and vegetables. Pop. (1970) 3,233.

Nassau, House of, European royal family named after a county on the E bank of the Rhine, founded by Walram I. The elder or German branch ruled Nassau until it was annexed by Prussia in 1866. Since 1890, Nassau has been the ruling house of the Duchy of Luxembourg. The younger (Dutch) branch, founded by OTTO I, inherited Orange in 1544. Since the election of WILLIAM THE SILENT (Prince of Orange) in 1579, members of this house have ruled The Netherlands almost continuously under the name of the House of ORANGE.

Nasser, Gamal Abdel (1918–70), Egyptian political figure, first President of the republic of Egypt (1956–70). He graduated from the Royal Military Academy in 1938, was wounded in action in the 1948 Arab-Israeli War and in 1950 was elected Chairman of the Free Officers movement. In 1952 he led the army coup that ousted King FAROUK. As head of the Revolutionary Command Council Nasser controlled Egypt, although Gen. Muhammad Neguib was the nominal premier. Nasser became President in 1956, nationalizing the Suez Canal in the same year. He was President of the United Arab Republic 1958–61. After Egypt's defeat in the 1967 ARAB-ISRAELI WAR he resigned but returned to office amid massive demonstrations in his support. Nasser promoted land reform and economic and social development through a programme called Arab socialism; the completion of the ASWAN Dam (1970), built with Soviet assistance after the USA backed out, was the greatest achievement of this programme. Nasser is remembered for his pan-Arab beliefs and encouragement of THIRD-WORLD unity. *See also* HC2 pp.254–*255.*

Nastase, Ilie (1946–), Romanian tennis player, known for his volatile personality. He was the first player from an Iron Curtain country to win the US Open (1972). In that year also he was beaten by Stan Smith in a classic Wimbledon final. Since then, Nastase's form has fluctuated because of his temperament, although he again reached the Wimbledon final in 1976, to be beaten by Björn BORG.

Nasturtium, annual trailing plant native to central and South America. Cultivated as a garden ornamental, it has round leaves and spurred, trumpet-shaped flowers of yellow, salmon or scarlet. Among 80 species is the common nasturtium, or Indian cress, *Tropaeolum majus.* Family Tropaeolaceae.

Natal, smallest province in the Republic of SOUTH AFRICA, bordered by the Indian Ocean (E), Swaziland (N), and the ORANGE FREE STATE (W); the capital is PIETERMARITZBURG. It was named by VASCO DA GAMA who sighted the entrance to Port Natal (modern DURBAN) on Christmas Day, 1497. The area was then inhabited by Bantu-speaking Zulu people who were defeated in battle by the BOERS in 1838. An Afrikaner republic was established 1838–39. Natal became a British colony in 1843, was granted internal self-government in 1893 and joined the Union of South Africa in 1910. The main industry is sugar refining, others include textiles, tanning and oil refining. Area: 86,967sq km (33,578sq miles). Pop. (1970) 4,236,770.

Nataraja, the Hindu god SHIVA (Siva) in the form of the cosmic dancer or Lord of Dance; he is represented in this guise in most Saiva temples in S India. In the Cola bronzes of the 10th and 11th centuries he is shown dancing within a circle of fire on the figure of a dwarf, a symbol of man's ignorance. Shiva's hair streams out on each side and he has four arms; the posture of his hands and hair are highly significant. As the Nataraja he is said to be the source of motion within the cosmos.

Nathanael, in ST JOHN's Gospel, one of JESUS's disciples. His birthplace was Cana and his call by Jesus is related in John 1: 43–51. However, his name is not included in any of the lists of the twelve APOSTLES in the other gospels, though most biblical scholars regard him as identical with St Bartholomew.

National Aeronautics and Space Administration. *See* NASA.

National anthem, music that at an international or state ceremony represents one specific country. Usually in the form of a song or hymn , the term can refer to the words alone; there are nevertheless examples of national anthems that are without words at all. The oldest is that of Great Britain; *God Save The King/Queen* has been sung on official and ceremonial occasions since 1745. The French national anthem, *La Marseillaise,* was composed as a marching song in 1792; the German *Deutschland, Deutschland über Alles,* authorized in 1922, was sung to the music of HAYDN's *Emperor's Hymn* composed in 1797. The US *The Star-spangled Banner,* although written in 1814, was not officially adopted until 1931. In 1977 the Australians rejected the British national anthem and after a referendum chose *Advance Australia Fair* as the national tune.

National Assembly, name taken by the Third Estate of the French ESTATESGENERAL on 17 June 1789. Ten days later, on the invitation of LOUIS XVI, it was joined by some members of the other two estates. It dissolved itself on 30 September 1791 after the promulgation of the new constitution of 1791. Its most important acts were the abolition of feudal rights and privileges (1789), the DECLARATION OF THE RIGHTS OF MAN (1789) and the Civil Constitution of the Clergy (1790). Various other French parliaments have been given the same name. *See also* HC2 pp.74–75.

National Assistance Board, British government agency set up in 1948 to assist individual cases of poverty. It developed from the Unemployment Assistance Board, which was created in 1934 to aid the unemployed during the Depression and which became known as the Assistance Board in 1940 when it also became responsible for supplementary old age pensions. In 1948 it was named the National Assistance Board. In 1966 it was superseded by the Supplementary Benefits Commission within the Ministry of Social Security.

National Broadcasting Company (NBC), US pioneer radio and television organization, comprising more than 400 independently owned stations. It set up the first radio network in 1926 and introduced a regular television service in 1939.

National Coal Board, British organization that has administered the coal mining and distribution industry since its nationalization in 1946.

National Convention, the governing assembly of France from September 1792 to October 1795. It abolished the monarchy and voted to execute LOUIS XVI and MARIE ANTOINETTE. In 1793 it began the long French wars by declaring war against Britain. It was dominated after June 1793, when it was purged of the GIRONDINS, by the extreme JACOBINS and ROBESPIERRE. Robespierre was overthrown by the Convention in July 1794. It was dissolved and replaced by the Directory, according to the constitution of 1795.

National Covenant (1638), answer by the Scottish Presbyterians to the attempt of CHARLES I and Archbishop LAUD to introduce in 1637 a Scottish Prayer Book. Although it professed loyalty to the Crown and was wi ely signed, it provided a basis for the league of Parliament and Scots Covenanters at the outbreak of the ENGLISH CIVIL WAR.

National Debt. See DEBT, NATIONAL.

National Economic Development Council. See NEDDY.

National Film Theatre, founded in 1951 following the Festival of Britain; it was built on the site of the Telekinema in London. The policy of the NFT is to run seasons of films, normally linked by a common factor, eg director, actor, genre or studio. A second cinema was opened in 1970.

National Front, British political party committed to free enterprise and the repatriation of coloured immigrants. The party was founded in 1967 as a union of the League of Empire Loyalists and the British National Party. Although tainted by a neo-Nazi image, the popularity of the National Front steadily rose in the late 1970s, particularly in inner city areas. See also NATIONAL SOCIALISM.

National Government, coalition government in Britain of CONSERVATIVE, LABOUR and LIBERAL MPS, formed in August 1931 as a result of the economic crisis of 1929–31. Until 1935 it was led by the former Labour Prime Minister Ramsay MACDONALD, thereafter by three Conservatives, Stanley BALDWIN (1935–37), Neville CHAMBERLAIN (1937–40) and Winston CHURCHILL (1940–45). At the elections of 1935 National Government candidates won 432 seats. The coalition was continued to fight WWII. Not until 1945 was there another election, when the three parties dissolved their connection.

National grid (electric and gas), distribution networks that supply consumers in nearly all parts of Britain with electricity and gas. Electricity is fed to the grid by various power sources. On entering the grid, electricity is stepped up in voltage by transformers from thousands of volts to levels as high as hundreds of thousands of volts. This ensures maximum efficiency of transmission across the country along power lines which are mostly carried overhead by pylons, although some travel underground or underwater. Before being used, electricity from the grid is stepped down in voltage in substations, usually to a mains voltage of 240 volts AC. Gas supplies were, until recent years, provided by local coal gasification plants, or gasworks. With the coming of North Sea natural gas, the gas grid is now a national one.

National Health Service (NHS), in Britain, institution founded by the National Health Service Act introduced by a LABOUR government in 1946. The NHS undertook to provide comprehensive coverage for most health services including hospitals, general medical practice and public health facilities. It is administered by the Department of HEALTH AND SOCIAL SECURITY. The NHS is tripartite. General practitioners have registered patients; they may also have private patients and may contract out of the state scheme altogether. They refer patients, when necessary, to specialist consultants in hospitals for outpatient or inpatient treatment. Health visitors such as midwives and district nurses are the third arm of the service. The training of doctors and dentists and supervision of their schools are outside the NHS. Hospitals are administered by regional boards which include governors of teaching hospitals. Public representation in England and Wales is provided by district community health councils which have access to medical information and can visit hospitals.

National Insurance, state scheme in Britain to provide sickness and unemployment benefits and old-age pensions. It dates from the 1911 Act, which provided sickness benefits for all workers and unemployment benefits for some. The scheme was funded, as it still is, by compulsory contributions from employers and employees. During WWI a Ministry of Pensions and a Ministry of Labour (responsible for sickness and unemployment payments) were established. The Ministry of National Insurance was merged with the Ministry of Pensions in 1966 to form the Ministry of Social Security.

Nationalism, ideology according to which all people owe a supreme loyalty to their nation. Nationalist sentiment, drawing upon and extolling a common culture, language and history, can be a powerful unifying force. See also HC2 p.171.

Nationality, legal status denoting membership in a nation or sovereign state. It is determined by birth or residence. The national has obligations to his country such as allegiance, military service and payment of taxes and, in turn, is owed diplomatic protection when abroad. Nationality is defined as an unalienable human right in the UN Universal Declaration of Human Rights.

Nationalization, acquisition for public ownership of business enterprises formerly owned by private individuals or companies. Local authorities have traditionally taken private property for such general purposes as the building of roads or public buildings, but the concept of nationalization has changed in the 20th century in that its purpose is now said to be to enhance social and economic equality; nationalization thus commonly takes place to further socialist principles of society. The Communist states of eastern Europe nationalized all industry and agriculture after WWII. In Great Britain the Labour government (1945–1951) nationalized a number of important industries, including coal, steel and transport. Illegal nationalization of foreign properties is termed expropriation. See also HC2 p.236.

National Labour Party, those Labour MPS in Britain who supported Ramsey MACDONALD in 1931 and joined the NATIONAL GOVERNMENT. The National Labour Party was never an independent political organization; in 1931, 13 MPS were elected under his title, and only eight were elected four years later.

National Liberation Front, *Front de Libération National* (FLN), revolutionary body that directed the Algerian war of independence against France between 1954–62. The war resulted in the Algerian proclamation of independence, and the FLN becoming an established political party.

National parks, protected areas where restraint on the killing of wildlife is enforced and forests, waters and other natural environments are preserved from commercial use. The main purpose of African national parks is game preservation. The USA was the first country to set aside large reserves for such preservation, and also for recreation. In Britain, national parks are not public property, and private land may be enclosed within the park boundaries. There are 10 national parks in England and Wales. There are none in Scotland or Northern Ireland, but in Scotland the government is currently examining the need for a park-system which would provide for both conservation and recreation. See also CONSERVATION; NATURE RESERVE; NW pp.250–251.

National Physical Laboratory, establishment responsible for national systems of measurement and for technical aspects of physical standards. Located at Teddington, England, it carries out research into many fields of physics and engineering. It was set up in 1900, sponsored by the Royal Society.

National Portrait Gallery, art gallery in St Martin's Place, London, housing portraits of outstanding men and women in British history. Built in 1858, the present building exhibits the work of British portrait painters dating from the 1400s.

National Savings, term applied in Britain to money saved by the public, in post office savings banks, trustee savings banks and government bonds and securities. Such money is effectively lent to the government, which uses it for various purposes, but which, like a bank, undertakes to repay the money either on demand or after a certain period, plus interest. The interest is usually tax free, as an incentive to saving. Governments of most countries provide various saving schemes, although the details and names of these vary from country to country. The amount of public saving can be a useful indicator of economic conditions; eg savings tend to increase in periods of economic difficulty.

National Socialism, or Nazism, doctrines and policies of the National Socialist German Workers' Party, which ruled Germany under Adolf HITLER from 1933 to 1945. Members of the party were first called NAZIS as a derisive abbreviation. The party's doctrines were the annulment of the Treaty of VERSAILLES, Aryan supremacy and racial purity, anti-Semitism and anti-Communism, together with German expansionism and the cult of the Führer. See also FASCISM; HC2 pp.193, *193*, 209, *209*, 232.

National Theatre, permanent theatre company usually subsidized by the state and housed in a permanent building where national classics of drama are performed in repertory. Unlike establishments in many other European countries, some more than 200 years old, the National Theatre of Great Britain only became a reality in 1963. Although suggested by David GARRICK in the 18th century, and later by a London publisher, Effingham Wilson in 1848, it was not until 1904 that the critic William ARCHER and the actor-manager Harley GRANVILLE-BARKER, with the support of Bernard SHAW, prepared detailed proposals. In 1913 they purchased a site in Bloomsbury, London, in the name of the Shakespeare Memorial National Theatre Committee. Lack of funds and WWI were the reasons for no further developments before WWII, although in 1938 another site was acquired, in South Kensington. In 1949 Parliament passed the National Theatre Bill, but only under the impetus of a free site and an offer of half the building costs from the (then) London County Council did the National Theatre Board come into existence in 1962. With Laurence OLIVIER as artistic director, the first National Theatre production, *Hamlet*, took place 22 Oct. 1963, at the Old Vic theatre where the company had its home for 13 years. In 1973 Peter HALL succeeded Olivier. New buildings designed by Sir Denys LADSUN were officially opened in Oct. 1976, and comprise The Lyttelton Theatre (proscenium stage), The Olivier Theatre (open stage) and The Cottesloe Theatre (adaptable small-scale arena for experimental and progressive drama).

National Trust, British organization formed by Octavia Hill, Sir Robert Hunter and Canon H.D. Rawnsley in 1895 in order to acquire and own buildings and land worthy of permanent preservation. In 1907 by legislation the trust was given the power to declare its land inalienable. Today it protects about 471,000 acres of which it owns 400,000, the rest being covenanted.

National Youth Theatre, student theatre group formed in 1956 by Michael Croft, after he had organized several school productions of Shakespeare. The group toured Europe and played at the Old Vic in 1965. Their production of Peter Terson's *Zigger-Zagger* at the Jeannetta Cochrane Theatre in 1967 is often considered their biggest success. From 1971 the group was based in London at the Shaw Theatre.

National Government; Ramsay Macdonald was the first leader of the coalition.

Nationalization of British steel was in 1949; it was de-nationalized 1953–67.

National parks; giraffes graze in the safety of the Nairobi National Park.

National Theatre; the foyer of the Lyttleton before a performance.

Nativity

Nativity scene, by Fra Angelico, now in the S. Marco Museum, Florence.

NATO; defence ministers at the start of a committee meeting at the Brussels HQ.

Guilio Natta, the Italian chemist who created the plastic polypropelene.

Natural gas from the North Sea is now supplied to homes and factories in Britain.

Nativity, term for the birth of a biblical figure most commonly applied to the birth of JESUS CHRIST although sometimes referring to that of the Virgin MARY or of JOHN THE BAPTIST. The date now usually given for Christ's nativity is 4 BC; 6 BC is a credible alternative, however. The prophet Micah foretold that Christ's nativity would take place in Bethlehem which, according to the Gospels of St MATTHEW and St LUKE, it did.

NATO, North Atlantic Treaty Organization, Western defensive alliance, which also promotes joint military aid and economic co-operation during peacetime. The original treaty, known as the North Atlantic Pact, was signed in Washington DC, USA in 1949. The original members were Belgium, Britain, Canada, Denmark, France, Iceland, Italy, Luxembourg, Norway, Portugal, The Netherlands and the USA. Greece and Turkey joined in 1952 and West Germany in 1955. NATO's headquarters are in Brussels. Its policy body is called the North Atlantic Council, and its military committee, from which France withdrew in 1966, is divided into three military commands and a regional planning group that aids the defence of North America. *See also* HC2 pp. 235, *235*, 259; MW p.191.

Natrolite, hydrated silicate mineral, hydrous sodium aluminium silicate ($Na_2Al_2Si_3O_{10}.2H_2O$). It has orthorhombic system square, needle crystals, with radiating nodules or compact fibrous masses. It is colourless or white, glassy and brittle. Hardness 5–5.5; s.g. 2.2. *See also* ZEOLITE.

Natta, Giulio (1903–   ), Italian chemist who shared the 1963 Nobel Prize in chemistry with K. ZIEGLER for their work on POLYMERS. In 1953 he began studying very large molecules and succeeded in polymerizing propylene to form polypropylene, a substance of great commercial importance.

Nattier, Jean Marc, the Younger (1685–1766), French painter. He painted historical subjects but was best known as a portrait painter, especially of court ladies. Among his sitters were Peter the Great and the daughters of Louis XV.

Natural childbirth, labour and delivery without analgesics. To achieve this, pregnant women follow a pre-natal course of lectures on physiology and are taught exercises designed to promote relaxation during delivery. This usually helps to minimize pain.

Natural gas, fossil fuel associated geologically with PETROLEUM. The fossil history of the two fuels is, in fact, the same, since they are both formed largely from the decomposition, at remote times in the past, of marine plankton. Among the most chemically stable products of this decomposition are the hydrocarbon compounds, most light of which is methane, CH_4, the main constituent of natural gas. This gas, now a major fuel source for Britain, also contains a little ethane, C_2H_6, and other heavier hydrocarbons.

Natural History and Antiquities of Selborne (1789), collection of letters written by Gilbert White, curate at his birthplace, Selborne, Hampshire, who spent most of his time observing the local countryside and wildlife and passing on notes and observations to men of similar interests, such as Daines Barrington and Thomas Pennant.

Natural History Museum, in South Kensington, London, the principal centre in the UK for research into animal and plant taxonomy. It is also a great tourist attraction, with visitors from all countries coming to see its major collections of animals preserved by taxidermy. These include a blue whale – the largest of all animals. Also exhibited are animal and plant fossils and reconstructions, and ores, rocks and meteorites. The museum is maintained by the Department of Education and Science, and admission is free. *See also* BRITISH MUSEUM; VICTORIA AND ALBERT MUSEUM.

Naturalism, late 19th-century literary movement which began in France and was led by Émile ZOLA. An extension of REALISM, it emphasized the importance of documentation and was a literary adaptation of the principles of DETERMINISM. It emphasized the physiological qualities of man rather than his moral or rational thought processes and writers sought to represent unselective reality in all its emotional and social ramifications. A major exponent of naturalistic drama was August STRINDBERG. Naturalism, which led many writers towards Social REALISM, was over as a movement by the beginning of the 20th century but had a profound influence on the development of the modern US novel. In the visual arts, the term describes an extension of Realism and signifies total fidelity to nature.

Naturalization, acquisition by a person of the NATIONALITY of a country that is not his native one. Requirements for becoming a national vary among countries and also within each country; residence for a minimum period, political asylum, family relationship and marriage are possible grounds for becoming a national. In some countries, naturalized nationals are not necessarily CITIZENS.

Naturalization, in nature, introduction of animals and plants to another country. The process of adaptation to different environmental and climatic conditions is known as acclimatization. It can result in biological changes in plant and animal species; for example, plants when introduced to a warmer climate may become more luxuriant, and in animals acclimatization may lead to changes in the size, colour and density of the coat.

Natural philosophy, term roughly equivalent to the phrase natural science. Natural philosophy embraces both the physical and life sciences as well as mathematics. Isaac NEWTON entitled his great treatise on mathematical physics, published in 1687, *Principia Mathematica Philosophiae Naturalis* (Mathematical Principles of Natural Philosophy).

Natural rights, concept found in the writings of the philosopher John LOCKE, who held that man possessed fundamental rights. Among these were life, property ownership and political equality.

Natural selection, in EVOLUTION, theory that those individuals of a species who possess a better capacity for survival in a particular environment will survive to reproduce and thus increase in numbers. This theory, which was first put forward by Charles DARWIN in his work *On the Origin of Species* (1859), is also expressed as "the survival of the fittest". *See also* NW pp. 22, 31.

Natural theology, teaching concerning God which claims to be derived from the natural light of reason as distinct from revelation. This distinction was first worked out by the medieval Christian Church, which taught that certain religious truths could be perceived by man solely through the use of his natural faculties. In the 17th and 18th centuries the term was also used to denote certain schools of thought, such as DEISM, which were opposed to revealed, dogmatic theological systems requiring the support of a supernatural authority.

Nature reserve, area of land, sometimes including inland waters and estuaries, set aside for the study and conservation of wildlife. In Britain there are many such reserves, some managed by the Nature Conservancy Council, others by local authorities and trusts, including the NATIONAL TRUST and the County Naturalists Trusts. Reserves in Britain are usually looked after by one or more nature wardens. In National Nature Reserves and in others selected for their geological, ecological, zoological or botanical interest, scientific work is carried out by government, university or amateur ecologists. Nature trails are outlined by the British Tourist Authority, and proposals for the protection of hitherto unreserved areas are considered by the Society for the Promotion of Nature Conservation. Great care is taken by the various authorities that no housing or industrial development take place on nature reserves, and considerable public interest has been aroused in the isolated cases in which subjects of special scientific interest have been affected or destroyed. *See also* NW pp.250–251.

Naturopathy, system of medical therapy that relies exclusively on the use of natural substances, such as sunlight and diet. Massage is also a treatment.

Nau, Jacques Jean David (*c.* 1630–71), French pirate who was active against the Spaniards in the Caribbean.

Naucalpan (Naucalpan de Juárez), city in central Mexico 11km (7 miles) NW of Mexico City. The city is noted for its annual religious fiesta, held in September. Now an industrial suburb of Mexico City, the chief industry is textile manufacturing. Pop. (1970) 373,605.

Nauplius, earliest and free-swimming larval form of most CRUSTACEANS and a component of PLANKTON. It has three pairs of appendages, no trunk segmentation and a single median eye. It usually passes through several other larval stages before adult form is attained. *See also* NW p. *100.*

Nauru (Naoero), island republic of the Commonwealth in the W Pacific Ocean, W of the Gilbert Islands. Discovered in 1798 by the British navigator John Hunter, the island gained its independence in 1968. Its economy is based on the mining and export of phosphates. Area: 21sq km (8sq miles). Pop. (1975 est.) 8,000. *See also* MW p.128.

Nausea, expectation of vomiting, which frequently follows; commonly caused by a digestive disorder and accompanied by an excess of saliva. Other causes are pain, shock, or liver or kidney disorders. Accompanied by vertigo, nausea may be a symptom of inner ear or brain disease.

Nausea (1938), early novel by Jean-Paul SARTRE, outlining the precepts of EXISTENTIALISM and attacking middle-class morality. Roquentin, who denies the existence of a divine purpose, feels an overwhelming disgust at the futility of life. Sartre writes with attention primarily to ideas rather than events or characters.

Nautical Almanac, publication of the hydrographic services of several countries that is produced to aid navigators. It includes detailed information on coastal waters, harbour facilities, winds, tides, currents, hazards and other data that cannot be shown on area charts.

Nautical mile, unit used to measure distances at sea. It is defined as the length of one minute of arc of the Earth's circumference at the Equator and is equal to 1,850m (6,069ft), although some countries define it as 1,853m (6,069ft). A speed of one nautical mile per hour is called a KNOT.

Nautilus, or chambered nautilus, CEPHALOPOD found in W Pacific and E Indian oceans at depths down to 200m (660ft). Its large coiled shell is divided into numerous, gas-filled chambers with the body located in the foremost chamber. Its head bear 60–90 retractable, thin tentacles without suckers, and it moves by squirting water from a funnel. Shell size approx. 25cm (10in). Family Nautilidae; genus *Nautilus*. *See also* MOLLUSC; NW pp.22, 94, 96.

Navajo, Athabascan-speaking tribe, the largest Indian group in the USA, famed for fine weaving and silverware. Their tribal reservations in Arizona and New Mexico are the most extensive in the country. In the early 1970s the Navajo numbered about 100,000.

Navarino, Battle of (1827), naval battle in which the British, French and Russian fleets, intervening in the Greek War of Independence, destroyed the Turkish-Egyptian fleet under IBRAHIM PASHA. *See also* HC2 p.182.

Navarre, province and former ancient kingdom in N Spain, on the French border; the present capital is Pamplona. Originally inhabited by Vascones it was conquered by the Romans in the 1st century BC. After the fall of Rome in the 400s, Navarre resisted invasions by the Visigoths, Arabs and Franks for 400 years. The kingdom dominated Christian Spain in the early 11th century but was divided into three kingdoms in 1035. In 1512 the S part was conquered by Ferdinand II of Aragon and incorporated into Castile in 1515. The N part passed to Henry IV,

King of France, in 1589. Its products include timber, cereals, vegetables and sugar beet. Area: 10,502sq km (4,055sq miles). Pop. (1970) 464, 867.

Nave, central and main aisle of a church or cathedral. It extends from the main entrance to the TRANSEPTS, and excludes the secondary north and south aisles. It is traditionally the seating area for the congregation, in contrast to the chancel which is reserved for choir and clergy. The word "nave" derives from the Latin for "ship", perhaps because of the adoption of a ship as a symbol for the Church.

Navel, or umbilicus, depressed scar on the abdomen where the umbilical cord, severed at birth, once joined the fetus to the mother's placenta.

Navigation, science of determining the position of a ship or aircraft, and charting a course for guiding it from one point to another. Four main techniques are used: dead reckoning, piloting, celestial navigation and electronic navigation. Position (the point of the Earth's surface, established by LATITUDE AND LONGITUDE), direction (indicated as angular distance and measured in degrees of arc from true north), speed (rate of travel in nautical miles per hour), and distance may be plotted with special charts and instruments.

Navigation, aircraft, methods of determining the course and location of an aircraft. A pilot can check landmarks to note his progress (a system called pilotage) or he can use dead reckoning to fly a computed heading and speed, timing his arrival at checkpoints. The automatic direction-finder (ADF) can be tuned to low frequency radio stations to determine their relative bearing to his course. The static-free very high frequency omnirange (VOR) and distance measuring equipment (DME) are even more sophisticated radio devices that allow extremely accurate navigation. Inertial Guidance is capable of fixing a point precisely in space without reference to any ground-based equipment. *See also* MM pp.186, 236.

Navigation, celestial, method of determining position on or above the surface of the Earth by comparing computed lines of position with lines established with a sextant. Points of intersection of the plotted lines represent a celestial fix.

Navigation Acts, body of laws imposing protectionist trade restrictions upon foreign commercial shipping, an outgrowth of the economic theory of MERCANTILISM. In Britain, the Navigation Acts were a body of regulations, passed from 1651 onwards, whereby British trade with Asia, Africa and the Americas could be undertaken only by British vessels, and whereby products from the British colonies could not be exported direct to other European nations, but had to pass through a British port first. The Navigation Act of 1651 was one cause of war with the Dutch (1651–54) and later contributed to unrest in the American colonies. The Navigation Acts were not finally repealed until 1854.

Návpaktos, seaport of w central Greece, on the N shore of the Gulf of Corinth. Ancient Navpaktos was the departure point of the Dorians for the Peloponnesos; when it was captured in 456 BC by Athens, it became the chief Athenian naval base on the Gulf of Corinth. Taken by the OTTOMAN TURKS in 1499 and known also as LEPANTO, it was the scene of a major victory by European forces over the Turks in 1571. It has been part of independent Greece since 1832. Pop. (1976) 10,000.

Navvy, labourer who excavates canals, railways, drains and other earthworks. Originally labourers on canals, or navigations, navvies in the canal- and railway-building period of the 1800s moved from one public construction to another. Navvy is also the name for an earth-digging machine, usually called a steam navvy.

Navy. *See* ROYAL NAVY.

Naxalites, Indian nationalist Communist revolutionaries of w Bengal. Their founder was Charu Mazumdar (1915–72). They undertook a guerrilla war against the Indian government in the 1960s, but were severely repressed; several hundred suspected Naxalites and sympathisers were imprisoned in the early 1970s.

Naxos, Grecian island in the Aegean Sea, the largest of the CYCLADES group. It was first colonized by Ionians, became a tributary of Athens in 470 BC and was held by the Ottoman Turks 1566–1830, when it became part of independent Greece. It is famous in myth as the island where THESEUS abandoned ARIADNE. Products include fruit, oil and wine. Area 414sq km (165sq miles) Pop. (1971) 14,201.

Nayler (or Naylor), James (1618–60), prominent English Quaker. In 1656 he rode into Bristol in imitation of Christ's entry into Jerusalem. He was punished and imprisoned for blasphemy.

Naylor, Bernard (1907–), British organist, conductor and composer; he founded the Little Symphony Orchestra in Montreal (1942) which he conducted until 1947. From 1954 he was a lecturer at Reading University. His *King Solomon's Prayer* (1953) for soprano, chorus and orchestra was commissioned by the Canadian Broadcasting Company for the coronation of Queen Elizabeth II.

Nazarenes, group of 19th-century German painters, founded in 1809 by Johann Friedrich OVERBECK and joined the following year by CORNELIUS in the monastery of Isodoru near Rome. Their intention was to paint religious subjects in the style of later medieval and early Renaissance artists, especially DÜRER, PERUGINO and the young RAPHAEL. *See also* HC2 pp.*80,* 81.

Nazareth, town in N Israel, 29km (18 miles) SE of Hai Fu. Mentioned in the New Testament as the home of JESUS CHRIST, it was taken by the Crusaders in 1099 and recaptured by the Turkish sultan Baybars in 1263. It was annexed by the OTTOMAN TURKS in 1517 and during this century lay within the Palestine Mandate (1922–48). Nazareth was taken by Israeli forces in 1948. Today it is a centre of pilgrimage and also a market town for the surrounding area. Manufactures include processed food, textiles and leather goods. Pop. (1970 est.) 34,000.

Nazarite, or Nazirite, in ancient Judaism, one who made a vow for a period of time, usually 30 days, or for life, to follow certain observances, which normally included abstinence from wine. In addition, one did not shave and avoided corpses to maintain ritual cleanliness. The procedure disappeared during the Middle Ages.

Nazca, or Nasca, Indian civilization in the southernmost coastal valleys of Peru; it flourished 200 BC–AD 600. The Nazca, who developed without outside influence, were a relatively small group of farmers notable for their unique and highly stylized ceramics and textiles.

Naze, The (Lindesnes), cape in Vest-Agder county of S Norway, the southernmost point of the Norwegian mainland. It is the site of a 17th-century lighthouse.

Nazerat. *See* NAZARETH.

Nazism. *See* NATIONAL SOCIALISM.

Ndjamena (Fort-Lamy), port and capital city of CHAD on the River Chari. It is an important market for the surrounding region, which produces livestock, dates and cereals. Ndjamena's main industry is meat processing. Pop. 224,000.

Ndola, city in N central Zambia, s central Africa, 274km (170 miles) N of Lusaka. It is the commercial centre of a copper mining region. Other products include cement, chemicals, soap. Pop. (1972 est.) 235,000.

Ndongo, southern African kingdom of the Bantu Mbundu people during the 16th and 17th centuries. It was subject to the depredations of slave traders from the Congo, and gradually conquered by the Portuguese.

Neagh, Lough. *See* LOUGH NEAGH.

Neagle, Dame Anna (1904–), leading British actress, a former chorus girl (1926–30). Her theatre appearances include *As You Like It* (1934), *Emma* (1944), *Charlie Girl* (1965–71), *The First Mrs Fraser* (1976) and *Maggie* (1977). She appeared in the films *Victoria the Great* (1937), *Edith Cavell* (1939), *Odette* (1950), *The Lady With The Lamp* (1951) and *The Lady is a Square* (1958).

Neanderthal man, Middle PALAEOLITHIC man, known from remains in Europe, around the Mediterranean region, and Asia. A Neanderthal man was first discovered in Neander Valley in w Germany in 1856. His body was thick and powerfully built, his face large with small cheek-bones, and his cranial capacity larger than modern man's. *See also* HC1 p.24; MS 26–28, *26–27,* 30–31.

Neap tide. *See* TIDES.

Near-sightedness. *See* MYOPIA.

Nebo, Mount, mountain-top in N Jordan, at the N end of the Dead Sea. According to the Bible, MOSES saw the Promised Land before his death from this point. Height: 806m (2,644ft).

Nebraska, state in w central USA, in the Great Plains. The land rises gradually from the E to the foothills of the Rocky Mts in the w, and is drained chiefly by the Platte River, a tributary of the Missouri. The E half of the state is farmland where farmers grow cereal crops and raise cattle and pigs. Cattle are also grazed on the sand ridges farther w. Nebraska's economy is overwhelmingly agricultural and food processing is the major industry, but since WWII efforts have been made to diversify manufacturing, which now includes electrical machinery and chemicals. There are deposits of oil, and sand, gravel and stone are quarried. Nabraska's state capital is Lincoln; Omaha is the largest city and is also an important insurance centre.

The region was acquired under the LOUISIANA PURCHASE of 1803 and was unknown until the LEWIS AND CLARK EXPEDITION the following year. The territory of Nebraska was created in 1854. After the Homestead Act of 1862, thousands of settlers occupied the free land the act offered. The completion of the Union Pacific Railroad in 1869 brought further development to the region, which had been admitted to the Union in 1867. Area: 200,017sq km (77,227sq miles). Pop. (1975 est.) 1,546,000. *See also* MW p.175.

Nebrija, Elio Antonio de (1444–1522), Spanish scholar. Generally thought to be Spain's greatest Renaissance Humanist, he spent ten years studying in Italy before returning to Spain to write *Introductiones latinae* (1481), a Latin grammar, *Vocabulario latinoespañol* (1492), a lexicon, and his greatest work *Gramática castellana* (1492), the first vernacular scientific grammar in Europe.

Nebuchadnezzar (Nebuchadrezzar), (*d.*562 BC), second king of the Chaldaean or New Babylonian empire (*r.c.*604–562 BC). He defeated the Pharaoh Necho in battle at Carchemish in 605 BC but was himself defeated by Necho in 601 BC. He placed the puppet King Zedekiah on the Judaean throne but was forced to destroy Jerusalem after a revolt. He was responsible for many buildings in BABYLON; for his Median wife he built the hanging gardens there which became one of the seven Wonders of the World. *See also* HC1 pp.33, 54.

Nebula, concentration of gas, chiefly hydrogen, and dust particles in interstellar space. Formerly referred to as galactic nebulae, these objects, from which stars are thought to originate, are assigned to two main classes: emission nebulae and reflection nebulae, both of which are described as bright nebulae. An emission nebula, such as the Trifid Nebula in Sagittarius, is largely gaseous and shines by the absorption and re-emission of energy from stars located within or near it. Reflection nebulae, which contain more dust, shine by reflecting light from nearby stars. An example of this type of nebulosity accompanies the Pleiades cluster. Dark nebulae absorb energy but do not re-emit it as visible light and are only rendered visible when they happen to obscure light from star fields lying beyond them, for example the HORSE'S HEAD NEBULA. *See also* SU pp.228–233, *228–233.*

Neck, in man and other vertebrates, part of the body that connects the head and the torso. It contains the seven cervical vertebrae of the spinal column and, to the front, the throat or pharynx, which leads on to

Naxos; the town and harbour are on the west coast of the island.

Nazareth, although an Arab town, also has many different Christian churches.

Anna Neagle as Florence Nightingale, with the original lamp used in the Crimea.

Nebraska's cultivated fields are protected from erosion by belts of trees.

Needlefish are related to the garb, freshwater fish of North America.

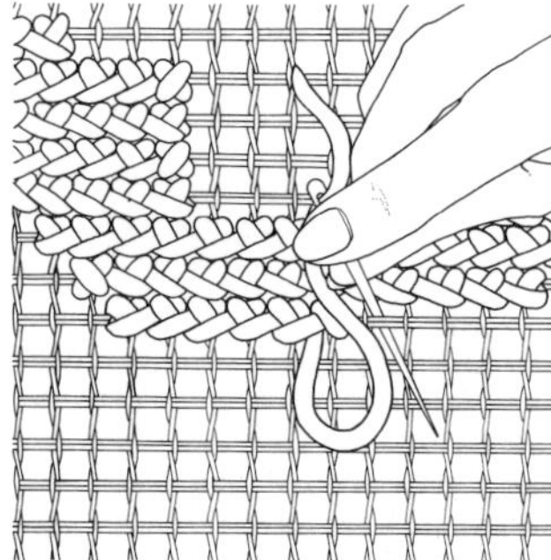

Needlework; this embroidery on an open-mesh canvas is called needlepoint.

Nehru favoured China's UN membership until its forces attacked India in 1962.

Nelson died at the Battle of Trafalgar, regarded as a tactical masterpiece.

the larynx, or voicebox. Farther down, this tube becomes the trachea, which carries air to the lungs. Behind the trachea lies the oesophagus, the tube which carries food to the stomach. Blood vessels of the neck include the carotid arteries and the JUGULAR veins.

Necker, Jacques (1732–1804), French financier and statesman, *b.* Switzerland. Finance minister under Louis XVI from 1776 to 1781, he was a popular minister but had little political acumen and was forced to resign. Recalled in 1788, his programme of reforms earned him the hatred of the court and he was dismissed on 11 July 1789. The storming of the BASTILLE forced Louis to reinstate him, but he finally resigned in 1790. *See also* HC2 p.74.

Necrophilia, sexual attraction towards dead bodies. It is a rare deviation closely related to FETISHISM. Some cases of murder-rape may involve necrophilia.

Necrosis, death of plant or animal tissue caused by an outside agent, such as cuts, burns or diseases. A LESION, a sign of necrosis, is often a valuable guide in diagnosis.

Nectar, sweet liquid secreted by many plants. It consists mainly of a solution of GLUCOSE, FRUCTOSE and SUCROSE in water. The glands that produce it usually lie at the base of the flower petals, but may be found also in parts of the stem. Nectar attracts insects, which help with CROSS-POLLINATION. Bees turn nectar into HONEY.

Nectar, in Greek mythology, drink of the gods. AMBROSIA was their food.

Nectarine, variety of PEACH tree and its sweet, smooth-skinned, fleshy fruit. The tree and the stone within the fruit are indistinguishable from those of the peach. Family Rosaceae; species *Prunus persica nectarina. See also* PE p.192.

Neddy, short for the National Economic Development Council, an advisory body set up by the British Government in 1962 to carry out economic planning, with the aim of increasing the rate of economic growth in Britain. It came under the control of the Department of Economic Affairs in 1964. It is composed of representatives of employers, trade unions and the government, and its other aims are to overcome economic obstacles, stabilize industrial incomes and publicize the importance of new developments in industry. Its function of monitoring prices was taken over by the Prices Commission in 1973–74. Various sub-committees deal with particular industries.

Needlefish, any of several species of primarily marine fish found in tropical and temperate waters. Needlefish have long jaws with fine, needle-sharp teeth and long, slender bodies. Family Belonidae. Length: 1.2m (4ft).

Needles, The, noted geographical feature consisting of three chalk stacks, situated to the w of the Isle of Wight, s England. A lighthouse is situated on the most westerly stack.

Needlework, work done with a needle: sewing and mending or ornamental work such as EMBROIDERY, appliqué, smocking, patchwork and quilting. KNITTING and crochet, which require special eyeless needles, are also included. Evidence survives of quilting and embroidery in Egypt as early as 3400 BC; embroidery was popular in Classical Greece and among the ancient Maoris, and had become an art form in Europe by the 12th century. Appliqué was first used by the Persians in the 9th century BC, and patchwork dates from the 6th century.

Neel, Louis Boyd (1905–), British conductor who founded the Boyd Neel Orchestra (1932), which became known throughout the world as a fine and disciplined string orchestra. After WWII Neel conducted the Sadlers Wells Opera (1944–5), gave many lectures and radio broadcasts and toured extensively. In 1971 he went on a world tour with the Hart House Orchestra of Toronto, which he founded in 1955.

Néel, Louis-Eugène-Félix (1904–), French scientist who shared the 1970 Nobel Prize in physics with H. ALFVÉN for their work in SOLID-STATE PHYSICS. Néel studied the magnetic properties of solids and discovered antiferromagnetism in which the magnetic fields of small groups of atoms are aligned in opposite directions. His discoveries have been widely used in computer memories.

Neerwinden, battles of, two battles fought near the Belgian village of same name. The first, in 1693 during the War of the Grand Alliance, was a victory for a French army against Anglo-Dutch forces. The second, in which Austria defeated France, occurred in 1793 during the FRENCH REVOLUTIONARY WARS.

Nefertiti, or Nefretete (*fl.* 14th century BC), Queen of ancient Egypt; wife of AKHENATON, but overshadowed by her mother-in-law Tiy. She is remembered for her support of her husband's innovatory religious ideas and for her exceptional beauty, seen in the famous bust found at Tell el-Amarna.

Nefertum, in Egyptian mythology, son of PTAH and the lion goddess SEKHMET. Often depicted wearing an open lotus on his head and standing upon a crouching lion, he is sometimes shown with the head, sometimes with the body, of a lion.

Negev, desert region in s Israel that covers approx. 60% of Israel's total land area. The chief city is Beersheba in the NW which has fertile land irrigated by the National Water Carrier Project. Towards the s, the area becomes more desolate. It was the scene of conflict between Egyptian and Israeli forces after the Palestine mandate in 1948. The area has good mineral resources including copper, phosphates, natural gas, gypsum, ceramic clay and magnesium ore. Area: 12,170sq km (4,700sq miles).

Negligence, in law, failure to exercise reasonable care towards material goods, other people or oneself, resulting in unintentional harm. The concept of liability for one's negligence dates from Roman times and is based on the supposed actions of a "prudent and reasonable" man. Most cases involving negligence are decided by a jury. In some circumstances, such as MANSLAUGHTER, negligence can be a criminal offence.

Negombo, city in w Sri Lanka, on the Negombo Lagoon. It is the agricultural trade centre for the produce of the surrounding area, primarily coconuts and cinnamon. The city is most noted for its manufacture of ceramics and brassware. Sri Lanka's international airport is at Negombo. Pop. (1968 est.) 53,000.

Negri Sembilan, state of Malaysia, SE Asia, on the Strait of Malacca; the capital is SEREMBAN. Originally a confederation of nine states, it came under British protection in 1874, became one of the Federated Malay States in 1895 and a state of Malaysia in 1963. Its products include rubber, rice and tin. Area: 6,708sq km (2,590sq miles). Pop. (1970) 479,312.

Negrito, pygmy people of the Philippines. Related to other SE Asian PYGMY groups, they are semi-nomadic forest and grassland dwellers, living by food gathering, hunting and fishing. They possess no distinctive language of their own, but speak the language prevalent in the area in which they live.

Negritude, philosophy first mentioned by Aimé CÉSAIRE in his *Return to My Native Land* (1939). He defined it as the recognition of the fact of being a Negro and the acceptance of this fact and of its cultural and historical consequences.

Negro, in anthropology, one of the major human racial groups, sometimes divided into Congoid and Capoid races, and including PYGMY populations. Generalized Negro physical characteristics include dark skin pigmentation; curly to spiral-tufted hair; full lips; a broad nose; and a high incidence of the Ro blood group. Negro people are indigenous to sub-Saharan Africa, but their migration (at one time forced through the activities of the slave trade) has established them as majority or minority groups in many other parts of the world, particularly in North America and the islands of the West Indies. *See also* MS pp.32–35, *32–35.*

Nehemiah, in the OLD TESTAMENT of the BIBLE, one of the two books with EZRA which tells the history of the JEWS from 538 BC to 432 BC. Many of the documents and lists that make up much of the books are written in Western Aramaic rather than in Hebrew.

Nehru, Jawaharlal Pandit (1889–1964), Indian leader. He joined the Indian independence movement in 1919, and in 1929 was elected President of the Indian National Congress. He became the first Prime Minister and Foreign Minister of India (1947–64). He led the new nation through its troubled beginning, launching programmes to industrialize and socialize India. In 1948 India became involved in a war over Kashmir with Pakistan, although in foreign policy Nehru advocated neutralism and pacificism. *See also* HC2 pp.216, 255.

Neill, Alexander Sutherland (1883–1973), Scottish educator. He founded an International School in Germany (1921), which moved to England (1924) and is now called Summerhill. It continues to be run on his ideas, which included allowing a child maximum freedom and encouraging spontaneity within a structure of self-government. He is the author of *Summerhill* (1962) and *Neill, Neill, Orange Peel* (1973).

Neilson, James Beaumont (1792–1865), Scottish inventor. Working as manager and engineer of Glasgow Gasworks (1817–47), he patented (1828) his invention of the hot-air blast method of smelting iron, which greatly advanced iron production technology.

Neilson, John Shaw (1872–1942), Australian poet who was brought up in the bush and had little formal education. His poems, although simple in structure, are penetrating, mystical insights filled with superb imagery. They include *Heart of Spring* (1919) and *Men of the Fifties* (1948).

Neisse (Nysa, Nysa Łużycka, Nisa), river in central Europe. It rises in NW Czechoslovakia and flows N, forming part of the post-1945 German-Polish border, before joining the Oder River s of Eisenhüttenstadt. Length: 256km (159 miles).

Nejd, plateau region in central Saudi Arabia. The WAHHABI leader IBN SAUD, gradually (1902–25) won it back from Turkish control and, following his capture of the HEJAZ (1925), the two regions became part of modern Saudi Arabia (1932). Area: 1,157,739sq km (447,000sq miles).

Nekrasov, Nikolai Alekseyevich (1821–78), Russian poet and editor of *The Contemporary* (1846–66). He encouraged young talent and published the early work of Ivan TURGENEV, Fyodor DOSTOEVSKY and Leo TOLSTOY. His own poetry expresses compassion for the Russian peasantry, especially in the satirical epic *Who can be Happy and Free in Russia?* (1879).

Nekton, the active swimming organisms at the sea's surface, considered as a whole. It includes the large migrating marine animals such as adult squids, whales and fish. *See also* BENTHOS; PELAGIC; PLANKTON.

Nelson, Horatio Nelson, Viscount (1758–1805), British naval commander. He entered the navy in 1771 and from 1793 served in the Mediterranean under Admiral HOOD; he lost the sight of his right eye in action in Corsica. He fought with great distinction under JERVIS at Cape St Vincent in 1797, and lost his right arm off Tenerife in the same year. He commanded the fleet that won a crushing victory over the French at the Battle of the Nile in 1798 and, promoted to Vice-admiral, he fought in the Battle of Copenhagen in 1801 (which disrupted the ARMED NEUTRALITY). Recalled to the Mediterranean in 1803, he blockaded Toulon for two years until the French fleet finally eluded him. He was killed at the ensuing Battle of TRAFALGAR (1805). *See also* HC2 pp.78, 79.

Nelson, city and port of New Zealand, on Tasman Bay, South Island. It is a centre of light industry and has an impressive Anglican cathedral. It is the home of the noted CAWTHRON INSTITUTE.

Nelson, river in Manitoba province, Canada; it flows NE from Lake Winnipeg into the Hudson Bay at Port Nelson. The mouth of the river was discovered in 1612, and the river became a route for fur traders. Near its mouth the first HUDSON'S BAY COMPANY trading post was founded in 1670. Length: 644km (400 miles).

Nematode. *See* ROUNDWORM.

Nematophyton, large fossil plant, one of the earlier plants to invade the land, dating from about 400 million years ago. It had a stem about 60cm (24in) in diameter and appears to have been a highly developed ALGA, a relative of SEAWEEDS. *See also* NW p.176.

Nemea, city in ancient Greece. It was the site of a temple of ZEUS where and in whose honour the Nemean Games were held. These were established in 573 BC and held biennially. Nemea was also believed to be the place where HERCULES killed the NEMEAN LION.

Nemean Lion, in Greek mythology, beast that terrorized the valley of NEMEA. The slaughter of the lion was the first of HERCULES' twelve labours. Unable to club it to death, Hercules caught it and squeezed the life out of it. He wore its skin as a cape.

Nemertea, ribbon worm, or proboscis worm, also called Nemertinea, phylum or marine flat-worms similar to platyhelminthes and characterized by a protrusile proboscis, an anus and a circulatory system. Reproduction is sexual. Length: 2.54cm (1in) to several metres. The 600 species include the bootlace worm *Lineus geniculatus. See also* NW p.120.

Nemesis, in Greek mythology, personification of the gods' disapproval or jealousy and the retribution for wrong doing or sacrilegious activity. A goddess of fertility also called Nemesis was worshipped in Attica.

Németh, Lázló (1901–), Hungarian dramatist and novelist. After giving up a career in medicine he turned to writing. His novels, mainly autobiographical, include *Gyász* (1935), *Szerdai Fogadónap* (1940) and a social study *Iszony* (1947; tr. *Revulsion* 1965). His plays are mostly historical and include *Galilei.*

Nemirovich-Danchenko, Vladimir (1858–1943), Russian playwright and stage director who co-founded the MOSCOW ART THEATRE with STANISLAVSKY.

Néné, or Hawaiian goose, unusual and rare species of goose found in the Hawaiian Islands. It has a brown-grey barred body and a black face. Length: about 65cm (25in). Family Anatidae; species *Branta sandvicensis.*

Nennius (*fl. c.* 800), Welsh historian to whom is ascribed the set of manuscripts called *Historia Britonum.* It contains much on the early history of Britain and the Anglo-Saxon invasions. It also studies early British legends, especially the Arthurian legend. Most historians believe that the history is a revision by Nennius of an earlier work. *See also* HC1 p.20, 20.

Neo-Classicism, revival of interest in the Classical arts, usually those of Greece and Rome, although in the 20th century the term more specifically refers to the return to classical values in the 18th century. Interest in Greek and Roman art began during the RENAISSANCE, but the self-conscious Neo-Classical movement did not occur until the mid-18th century. Reacting against BAROQUE and ROCOCO styles and inspired by the excavations of HERCULANEUM (in the 1740s), architecture, painting, sculpture and interior design were permeated by Greek and Roman models. Many people throughout Europe began to collect classical artefacts, and painters such as DAVID produced archaeologically exact interpretations of scenes of classical history. PALLADIAN architecture, which used the Greek forms of colonnade and portico, was explored by Robert ADAM, and in Napoleon's France the Empire style explored the possibilities of grandeur contained in Neo-Classicism. Continuing discoveries of classical remains after 1800, such as the ELGIN MARBLES which were procured from the Turks by Lord Elgin in 1803, kept interest in Neo-Classical architecture alive, particularly in the work of SOANE and NASH. In art, Greek line-drawing was popularized by WINCKELMANN and FLAXMAN, and its

austerity, purity of form and elevated emotions reigned supreme until the 1820s. Since 1800, Neo-Classicism had been questioned in all the arts by the advocates of ROMANTICISM, but it was enshrined in academic and official styles until a general widening in artistic tastes brought a move towards eclecticism in the mid 19th-century. In the 20th century, artists such as PICASSO, musicians such as STRAVINSKY and poets such as T. S. ELIOT and Ezra POUND went through a Neo-Classical period in the 1920s and 1930s; in this context the term refers to their return to formalism and reaction to EXPRESSIONISM. *See also* HC2 pp.42, 43, 70, 71.

Neo-classical music, style of composition from *c.* 1920 to 1950. In reaction to the emotionalism and "subjectivity" of Romantic music, the Neo-Classical movement returned to an emphasis on pure sound and the perfection of the form of the composition. Eighteenth-century forms such as the toccata, fugue and concerto grosso were revived, and emotional appeal became more subtle. Composers who explored classicism included STRAVINSKY, PROKOFIEV (eg his Symphony No. 1, the "Classical") and HINDEMITH. Parody and satire is also evident in works by such composers as MILHAUD.

Neo-Darwinism, development of Darwinism in which a distinction is made between genetic and acquired characteristics, the latter not being hereditary.

Neodymium, metallic element (symbol Nd) of the LANTHANIDE SERIES, first isolated in the form of its oxide in 1885. The pure metal was first obtained in 1925. Chief ores are monazite (phosphate) and bastnasite (fluorocarbonate). Properties: at.no. 60; at.wt. 144.24; s.g. 6.8–7.0; m.p. 1,010°C (1,850°F); b.p. 3,127°C (5,661°F); most common isotope Nd^{142} (27.11%).

Neo-Impressionism, late 19th-century painting style, originating in France and involving the use of POINTILLISM (also called Divisionism) and a formal composition far removed from the "snapshot" composition of IMPRESSIONISM. The style is seen at its purest in the works of SEURAT, its chief exponent, whose *Sunday Afternoon on la Grande Jatte* (1885) became the Neo-Impressionist exemplar. Other leading adherents include SIGNAC (the chief theoretician), CROSS and, briefly, GAUGUIN, van GOGH and TOULOUSE-LAUTREC.

Neolithic Age, or New Stone Age, period in man's evolution following the PALAEOLITHIC, in which man first lived in settled villages, domesticated and bred animals, cultivated cereal crops and practised stone-grinding and flint-mining. *See also* HC1 pp.26–27; MS pp.29, 31.

Neomycin, ANTIBIOTIC drug, active against a variety of bacterial diseases. It is produced by a soil actinomycete.

Neon, gaseous non-metallic element (symbol Ne) of the inert gas group, discovered in 1898. It is present in the atmosphere (0.0018% by volume) and is obtained by the fractionation of liquid air. Its main use is in DISCHARGE TUBES for advertising signs. The element forms no compounds. Properties: at.no. 10; at.wt. 20.183; m.p. −248.67°C (−415.6°F); b.p. −246.05°C (−410.89°F); most common isotope Ne^{20} (90.92%).

Neon lamps. *See* DISCHARGE TUBE.

Neon tetra, freshwater tropical characin fish found in the headwaters of the River Amazon; discovered in 1936. It is slender and brilliant blue, red and green, and is a popular aquarium fish. Length: 3.5–4cm (1.3–1.5in). Family Characoidae; species *Paracheirodon innesi.*

Neophyte, in Greek, literally newly planted. The term applied to those newly initiated into the mystery religions such as the ORPHIC or Eleusinian cults during the height of Greek civilization. Later, the newly baptized of the early Christian Church were also called neophytes.

Neoplasm, TUMOUR in human beings and domestic animals caused by various factors, such as viruses, chemicals and injuries. *See also* MS pp.86, 87.

Neo-plasticism, style of painting advocated by the Dutch Group of artists known

as de STIJL in their 1917 periodical of that name. Led by Piet MONDRIAN, this group was important in that they helped to bring about co-operation between artists, sculptors, architects and industrial designers. They used the rectangle as their basic form in design and the primary colours (blue, red and yellow), together with white, black and grey. They believed that to reduce art to abstract, geometrical forms would free it entirely from the artist's personality.

Neo-Platonism, school of philosophy that reached its zenith between about AD 250 and AD 550. It was more than a revival or a new version of Platonic thought. It combined PYTHAGOREAN, STOIC PLATONIC and ARISTOTELIAN ideas with strains from JUDAISM, Oriental religions and CHRISTIANITY. Formative leaders of the movement were two 3rd-century philosophers, PLOTINUS and PORPHYRY, who tried to compete with the rising influence of Christianity. Its end came with the Arab conquest of Alexandria but its influence persisted throughout the Middle Ages and even during the Renaissance.

Neoprene, synthetic rubber very resistant to attack by corrosive chemicals and widely used in the chemical industry. It is a polymer of chloroprene, $CH_2=CCl-CH=CH_2$.

Neoptolemus, in Greek mythology, son of ACHILLES and a noted warrior, sometimes called PYRRHUS (or Pyrrhos). He was taken to the siege of TROY by Odysseus, where he was one of those chosen to be left inside the wooden horse, and where he killed Priam. He died a violent but mysterious death at Delphi. He also appears in *Philoctetes* by SOPHOCLES.

Neo-Pythagoreans, philosophical movement which began in the 1st century BC and continued until it was diluted by the rise of NEO-PLATONISM in the 3rd century AD. It combined Jewish and Hellenistic elements with the more religious and mystical aspects of PYTHAGOREAN thought and influenced Neo-Platonism.

Neorealism, Italian film movement during the period 1945 to 1950, which dealt with the harshness of life and death. Roberto ROSSELLINI directed the first such film, called *Open City* (1945), using non-professionals and real locations. Other directors were able to attack social conditions through specific issues; DE SICA looked at abandoned children in *Shoeshine* (1946) and at unemployment in *Bicycle Thieves* (1948).

Neoteny, persistence in an adult animal of larval characteristics, such as the retention of GILLS, as in some SALAMANDERS. An entire order of TUNICATES, the Larvacea, is permanently larval, never reaching a typical adult form. *See also* MUDPUPPY; NW p.122.

NEP. *See* NEW ECONOMIC POLICY.

Nepal, independent kingdom in central Asia, between India (s) and China (N). The land has three distinct regions: a s lowland area which has some arable farming, a central mountain range, and high mountains in the N extending to the HIMALAYAS. Most of the population is engaged in agriculture, and there is little industrial activity. The capital is Katmandu. Area: 140,797sq km (54,362sq miles). Pop. (1976 est.) 12,904,000. *See* MW p.128.

Nephritis, also called glomerulonephritis or Bright's disease, an inflammatory disease of the KIDNEY. It is frequently caused by streptococcal infection, and may be chronic or acute. Sufferers of acute nephritis are best confined to bed, with a carefully regulated diet; chronic cases may be helped with artificial kidneys or kidney transplants. *See also* MS p.102.

Nephron, microscopic filter of which there are over one million in the human KIDNEY. Each is a cup-like structure called a Bowman's capsule with an attached uriniferous tubule. The capsule surrounds a cluster of blood capillaries, and blood containing waste enters it. The tubule cells extract the water and wastes, and the blood leaves the nephron through the outgoing blood vessel of the capsule. *See also* MS p.69.

Nephrosis, disease of the KIDNEY charac-

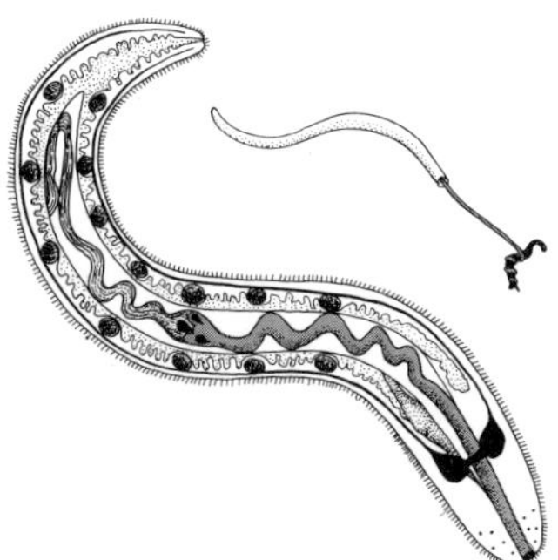

Nemertea, or ribbon worm, shoots out its long proboscis to catch its prey.

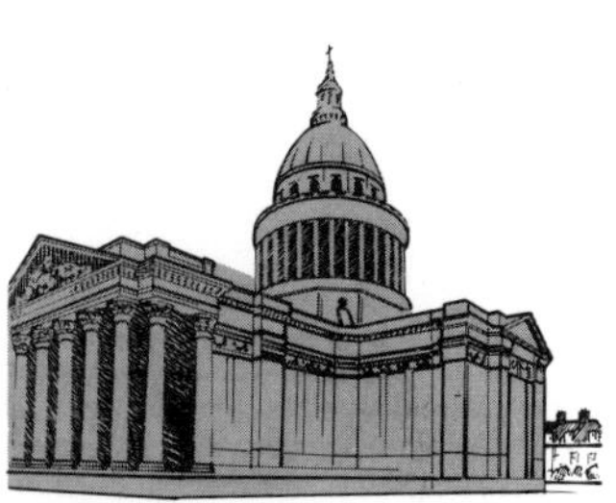

Neo-Classical architecture revived the splendour of Greece and Rome.

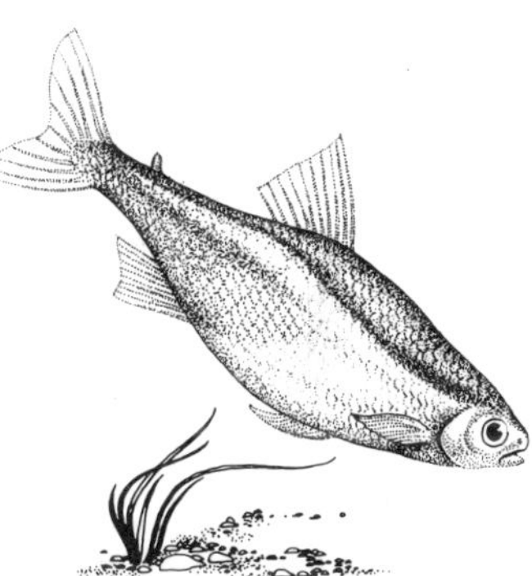

Neon tetra is a species of decorative, highly colourful tropical fish.

Neoptolemus was famed in legend for his courage and also for his cruelty.

Nereis, also called ragworm, is a worm with rows of bristles along its edges.

Nernst's work on chain reactions provided the spur to photochemistry.

Nero, an emperor with an evil reputation, did have talent as an administrator.

Nests vary widely in size, shape and the materials of which they are constructed.

terized by the presence of OEDEMA, large amounts of ALBUMIN in the urine and CHOLESTEROL in the blood. *See also* MS pp.68–69.

Nephthys, also called Maat or Nebhet, in Egyptian mythology, goddess of the dead and mother of ANUBIS by OSIRIS. Although originally probably a mother-goddess in pre-dynastic Egypt, she became a sister and equal of ISIS; both were the deities invoked in the death rites, often represented with outstretched wings protecting the souls of the dead.

Nepia, George (1905–), New Zealand rugby union player. Of Maori birth, he achieved rugby immortality as the 19-year-old full-back of the unbeaten 1924–25 ALL BLACKS, playing in all 30 matches of their tour of Britain and France. He also played in nine internationals for New Zealand (1924–30).

Neptune, Roman god, originally associated with fresh water but later identified with the Greek god POSEIDON and hence the sea. Salacia, a goddess of springwater, became his female counterpart. He was often depicted carrying a trident and riding a dolphin. His festival was in July, when fresh water was most scarce. Offerings were made to Neptune to assuage the water shortage.

Neptune, eighth planet from the Sun, observed in 1846 by Johann Galle and Heinrich d'Arrest. Neptune resembles the other giant planets (Jupiter, Saturn and Uranus) in having an atmosphere of hydrogen, methane and ammonia, a relatively low density and a rapid period of rotation. Neptune has two known satellites, Triton and Nereid. Mean distance from the Sun, 4,497 million km (2,794 million miles); mass, 17.2 times that of Earth; diameter, 49,500km (30,940 miles); rotational period, 15hr 48 min; sidereal period, 164.8 years. *See also* SU pp.176–177, 214–215, 280, 281.

Neptunium, radioactive metallic element (symbol Np) of the ACTINIDE group, first made in 1940 by NEUTRON bombardment of URANIUM. It is found in small amounts in uranium ores. The by-product Np^{237} is obtained in producing plutonium. Properties: at.no. 93; at.wt. 237.0482; s.g. 20.25; m.p. 640°C (1,184°F); b.p. 3,902°C (7,056°F); most stable isotope Np^{237} (2.14×10^6 yr).

Nereid, in Greek mythology, a nymph of the sea or freshwater, a female spirit without malice for mankind. The nereids were said to be the daughters of Nereus, God of the sea, and of Doris, an Oceanid.

Nereid, in astronomy, one of two satellites of the planet Neptune. It has an eccentric orbit and is thought to be a captured asteroid. It comes within 1,400,000km (870,000 miles) of Neptune and has a diameter of less than 300km (200 miles). The sidereal period is almost one Earth year. *See also* SU pp.205, 215, *215*, 280, *281*.

Nereis, or ragworm, polychaete WORM that lives in sand or mud on the seashore. It is usually red or orange with numerous "paddles" on each side of its body. The forepart of its gut can be extended as a proboscis with a pair of jaws. Length: to 91cm (3ft). Family Nereidae. *See also* ANNELID; NW pp.*72*, 88, *89*.

Nergal, in Mesopotamian religion, a secondary god depicted in Assyrian documents as a benefactor who restored the dead to life, heard prayers and protected crops and livestock. He is also depicted as a god of hunger and devastation, and became identified with the king of the underworld.

Neri, Saint Philip (1515–95), Italian priest, mystic and spiritual leader. In c.1533 he travelled to Rome and dedicated himself to God and a life of austerity, becoming a priest in 1551. In the same year he went to San Girolamo della Carità, attracting priests to what later became the Congregation of the Oratory, which received papal approval in 1575. An adviser to several popes and cardinals, he was influential in Henry IV's absolution in 1595 and averted conflict between Rome and France. He was canonized in 1622. Feast day: 26 May.

Nerina, Nadia (1927–), South African prima ballerina, real name Nadine Judd. She studied in England from 1945 under Dame Marie Rambert and became prima ballerina of the Royal Ballet in 1951. In 1963 she created the title role in *Elektra*, a performance which earned her great acclaim.

Nernst, Walther Hermann (1864–1941), German chemist. He became professor at Göttingen and then in Berlin and was awarded the 1920 Nobel Prize in chemistry for his discovery of the third law of THERMODYNAMICS. His other important work was concerned with chain reactions in photochemistry. He wrote *Thermodynamics* (1893) and *The New Heat Theorem* (1918). *See also* SU pp.92–93.

Nero Claudius Caesar (AD 37–68), Roman Emperor (*r.* 54–68), born as Lucius Domitius Ahenobarbus. He was the son of AGRIPPINA II and the stepson of CLAUDIUS I, whom he succeeded. He murdered his brother Britannicus (AD 55), his mother (59) and his wife Octavia (62). He blamed the Christians for a serious fire in Rome in 64 and may have begun the first official persecution of Christians. After discovering a plot against him in 65, he had many distinguished Romans, including SENECA, his former adviser, executed. But he was overthrown in a military revolt and killed himself. *See also* HC1 p.98.

Neruda, Jan (1834–91), Czech lyric poet. He initially adopted ROMANTICISM, but later turned to a more realistic style. Among his works, which include prose and drama, are *Cosmic Songs* (1878) and *Ballads and Romances* (1883).

Neruda, Pablo (1904–73), Chilean poet, real name Neftalí Ricardo Reyes. He occupied a number of diplomatic posts after 1927 and was active in politics; he was elected a Communist Party senator in 1945 and supported the Marxist regime of Salvador ALLENDE (1970–73). Neruda's poetry transcended politics and dealt with the tragedy of the human condition in surreal imagery. His best-known work is *Canto General* (*General Song*; (1950). He was awarded the 1971 Nobel Prize in literature.

Nerva, Marcus Cocceius (*c.* AD 30–98), Roman Emperor from AD 96 to 98. After the despotism of DOMITIAN he reformed land laws in favour of the poor, revised taxation and tolerated the Christians. He adopted TRAJAN and handed over government to him.

Nerval, Gérard de (1808–55), real name Gérard Labrunie, French poet and prose writer. His writings reflect his obsession with mysticism and female dominance, and include *Les Filles du feu* (1854) and the sonnets *Les Chimères* (1854).

Nerve, collection of NEURONS (specifically AXONS) that link the various parts of the vertebrate nervous system with other organs in the body. Afferent or SENSORY NERVES transmit nervous impulses towards the CENTRAL NERVOUS SYSTEM. Efferent or MOTOR NERVES transmit outwards from the central nervous system. Some nerves contain both afferent and efferent fibres, and are thus able to transmit both to and from the CENTRAL NERVOUS SYSTEM. Seven of the 12 pairs of cranial nerves, which link the brain with such sense organs as the eyes and ears, are examples of this. *See also* MS p.38.

Nerve cell. *See* NEURON.

Nerve cell membrane, membrane that surrounds the nerve cell, selectively allowing, or disallowing, the passage of substances in and out of the cell. It differs from other CELL membranes in projecting from the cell body in long structures resembling branches and twigs, called DENDRITES. *See also* MS pp.38–39.

Nerve centre, local concentration of interconnected nerves. In man, these are generally known as plexuses. They include the cervical plexus, supplying neck and diaphragm; the brachial, supplying the arms; the lumbar and sacral plexuses, supplying thighs and legs; and two plexuses of the AUTONOMIC NERVOUS SYSTEM: the solar plexus, supplying organs of the abdomen, and the cardiac plexus serving the heart.

Nerve gas, extremely poisonous form of gas first developed as a weapon by the Germans in WWII but never used. Nerve gases are often organic chemical compounds containing phosphorus. Their effect is to inactivate the enzyme cholinesterase and so prevent nerve messages reaching the muscles. The result is death within minutes by failure of respiratory and other systems. Inhalation of even small amounts of the gas, or any physical contact with the liquid chemical, is enough to cause death.

Nervi, Pier Luigi (1891–), Italian architect, noted for his innovative use of concrete in designs for large structures of great visual impact. His Giovanni Berta stadium at Florence (1932), famous for its cantilevered roof and stairs, and aircraft hangars at Orbetello (1940) established him as a major creative force in modern architecture. In the mid-1940s he invented ferro cemento, a strong, light corrugated concrete that he used in the Turin Exhibition Hall (1948). He also designed the UNESCO building, Paris (1954–58; with Marcel BREUER and Bernard Zehrfuss), the splendid Pirelli skyscraper in Milan (1958; with Gio PONTI), which has floors cantilevered from two tapering concrete pillars, the two domed sports stadiums in Rome (1957; 1959) and the Papal Audience Hall, Vatican (1971).

Nervo, Amado (1870–1919), Mexican poet who was also a journalist and a diplomat. He studied for the priesthood but turned to writing and became a leading figure of MODERNISMO. His best-known collections of verse include *Serenidad* (1914) and *Plenitud* (1918). He worked for several years in his country's diplomatic service, and was Mexico's Minister to Uruguay when he died.

Nervous breakdown, imprecise term commonly used to cover a wide range of conditions that can range from mental exhaustion to a number of mental illnesses. It usually refers to emotional conditions accompanied by such elements as DEPRESSION and ANXIETY. *See also* MENTAL ILLNESS.

Nervous system, mechanism that transmits information from one part of an animal body to another by means of nerve cells or NEURONS. In simple animals such as jellyfish and sea anemones, it consists of a nerve net without a centre, or brain. Stimulation of such animals produces a local muscular response in the region of the stimulus. More advanced animals have a concentration of nerve cells in the head that forms the brain and allows greater co-ordination. The human nervous system has two main parts. The central nervous system controls all voluntary muscles of the body, and is directed by the conscious will. The peripheral nervous system controls the voluntary and involuntary muscles of the body and the transmission of messages from the senses to the CENTRAL NERVOUS SYSTEM. *See also* MS pp.36–41; NW pp.74–75.

Nesbit, Edith (1858–1924), British writer, mainly of books for children. Her best-known stories were written in the period 1899–1913 and included *The Story of the Treasure Seekers* (1899) *Five Children and It* (1902) and *The Railway Children* (1908).

Nesfield, William Andrew (1793–1881), British painter of landscapes and a landscape gardener. He worked on the layout of KEW GARDENS and St James's Park, London.

Ness, Loch, freshwater lake in N Scotland, running SW to NE along the geological fault of Glen More. It forms part of the Caledonian Canal. Since 1933 the Loch has featured in the news several times as the home of the LOCH NESS MONSTER. Length of loch: 38.4km (24 miles); 230m (754ft).

Nest, structure built by a living organism to house itself, its eggs or its young. Nest-builders include some invertebrates, particularly social insects, and members of all the larger groups of vertebrates. The nests of ants, bees, wasps and termites may be highly elaborate and involve tunnels, passages and chambers. The nests of fish may be simple gravel scoops or enclosed structures, sometimes made of bubbles. Birds' nests vary enormously from simple cup-shaped arrangements of

HC1 = History and Culture Vol. 1   HC2 = History and Culture Vol. 2   MS = Man and Society   MM = Man and Machines

twigs and other organic materials, to woven or knotted grass or leaves; some birds scrape a hollow in the ground to make a nest, others make nest-holes in cliffs, earth banks or trees. The most highly evolved animal to make a form of a nest is probably the GORILLA, which builds a new sleeping platform of leafy branches every night.

Nestlé, a Swiss-based industrial corporation that develops and manufactures food products, including canned milk, instant coffee and drinking chocolate. In the late 1970s it employed about 135,000 people throughout the world.

Nestorianism, Christian heresy according to which there are two separate natures in the incarnate Christ, the one divine and the other human, as opposed to the orthodox belief that the incarnate Christ is one person who is at once both God and man. The heresy, which was associated with the name of NESTORIUS (died *c.* 451), bishop of Constantinople, was condemned by the Councils of EPHESUS (431) and of CHALCEDON (451). Nestorius was deposed and banished to the Libyan desert. The supporters of Nestorius, however, gradually organized themselves into a separate Church which called itself the "Church of the East" and had its centre in Persia. They were active in missionary work in Arabia, India and the Far East and have survived as a small community until modern times. *See also* CHRISTOLOGY; MONOPHYSITISM; MONOTHELITISM.

Nestorius (*d. c.*451), patriarch of Constantinople (428–31), born of Persian parents. As patriarch he became involved in a theological dispute about the person of Christ, objecting to the title *Theotokos* (Mother of God) for the Virgin Mary. The Council of EPHESUS (431) condemned his belief in the two divine and human natures of Christ (NESTORIANISM) as a heresy, and Nestorius was deposed as patriarch.

Netball, seven-a-side ball game played by women. It is a variant of BASKETBALL, although the netball player may travel only two steps on receipt of the ball and is restricted to limited areas of the court. Only two players of each team are allowed in the shooting circle in order to take shots at the goal. The goal itself is the same size and height as that in basketball but without a backboard. A match lasts for one hour, divided into four equal parts plus intervals. The game was invented in the final decade of the 19th century, and is played chiefly in the English-speaking countries and the Commonwealth. The All-England Womens' Netball Association was formed in 1926.

Netherlands, The, independent nation in NW Europe. A low-lying country with more than two-fifths of its land below sea-level, it relies on a system of dykes and dams for protection from the sea. Much land has been reclaimed and is used for pasture and arable farming; this forms an important part of the economy. Manufacturing industries have been developed since the end of WWII The capital is AMSTERDAM. Area: 41,160sq km (15,892sq miles). Pop. (1976 est.) 13,796,000. *See* MW p.128.

Netherlands, Revolt of the (1568–1648), also called the Eighty Years War, war by which The Netherlands won independence from Spanish Hapsburg rule. By 1576 the United Provinces (Dutch Republic) was formed. In 1579 the Catholic Walloon provinces defected and the rest of the republic had to struggle against Spain until a truce was called in 1609. The fight resumed in 1621 as part of the THIRTY YEARS WAR. Spain finally recognized Dutch independence in 1648.

Netherlands, The United, union of Belgium, Luxembourg and The Netherlands under one king, William I, son of William V of ORANGE, in 1815. It was arranged by the Congress of VIENNA at the end of the NAPOLEONIC WARS. In 1830 Belgium revolted against the union and declared its independence, which was formally accepted at the London Conference in 1839. In 1848 Luxembourg became an autonomous state with its own constitution and in 1890 it left the dual monarchy.

Netherlands Antilles, autonomous group of five main islands, and part of a sixth, in the West Indies, in the Caribbean Sea. The group includes Aruba, Bonaire, Curaçao (the largest island), Saba, Saint Eustatius, and the s half of Saint Maarten. The capital is Willemstadt, on Curaçao. The islands were discovered in the 1490s by Christopher COLUMBUS, settled by the Spanish in 1527 and captured by the Dutch in 1634. Industries: oil refining, phosphates, tourism. Area: 993sq km (383sq miles). Pop. (1975 est.) 242,000. *See also* MW p.130.

Netherlands East Indies (Dutch East Indies), former overseas territory of The Netherlands which since 1949 has been INDONESIA. The area included Sumatra, Java, Madura, the Celebes, the Moluccas and parts of Borneo and the Lesser Sunda Islands. During WWII the region was occupied by Japan. After the war, international pressure prevented the Dutch from regaining control. Products include rice, rubber, tea, oil, tin, bauxite, coal and manganese.

Neto, Agostinho (1922–), Angolan poet, political leader and doctor. He became a leader of an Angolan cultural and political revival and was often imprisoned for his efforts to achieve Angola's independence. He became President of an independent Angola in 1975. His poetry is included in *Antologia da Poesia Negra de Expressão Portuguesa* (1958).

Netsuke, accessory of traditional Japanese dress, a toggle to secure the cord on which the inrō (a small box, often with several sections, holding medicines or snuff) would hang. In the 18th century it became fashionable to wear decorative wooden, ivory, metal or ceramic netsuke, and a genre of miniature carvings developed, which are now highly prized by collectors. The carvings are usually intricate, finely-worked representations of animals, commonplace objects or mythological subjects, although simple, abstract, or even natural netsuke (such as a weather-worn stone) are also known. Netsuke were also produced in the 19th and 20th centuries for the collectors' market; nevertheless these retained the characteristic pair of holes in the back or base through which a cord could pass.

Nettilling, lake in s Central Baffin Island, Northwest Territories, Canada, which drains w into Fox Basin. The lake is frozen for most of the year. Area: 5,066sq km (1,956sq miles).

Nettle, any of numerous species of flowering plants of the genus *Urtica*, especially the stinging nettle (*U. dioica*), which is typical of the genus in that it has stinging hairs along the leaves and stem. It has heart-shaped serrated leaves, small green flowers and is sometimes used for medicinal or culinary purposes. Family Urticaceae. *See also* NW pp.*59–60*; PE p.*191*.

Nettle rash, or urticaria, also called hives, intensely itchy skin eruption, the raised patches of which resemble nettle stings. It is an allergic reaction, usually to certain foods, but also to insect bites, contact with irritant plants and inoculations. Remedies include CORTISONE injections, the application of ANTIHISTAMINE creams, and avoidance of the particular substances to which the sufferer is hypersensitive. *See also* ALLERGY; MS pp.87, 98.

Network, electric, number of conductors and circuit components, such as resistors, capacitors and inductors, linked together to form a set of interconnected circuits. A passive network includes no source of energy, whereas an active network does. Direct current networks are governed by KIRCHHOFF'S LAWS, one of which states that the algebraic sum of the voltages in a closed circuit is zero and that the sum of all currents entering and leaving any given junction of a closed circuit is zero.

Netzahualcóyotl, city in Mexico state, s central Mexico; a suburb of Mexico City. It is on major transport routes into the capital and is an economically depressed residential area. Pop. (1970) 571,035.

Neue Sachlichkeit (New Objectivity), German art movement and also the title of the exhibition (1925) held at Mannheim. The movement, a return to straightforward realism, rejected EXPRESSIONISM and even IMPRESSIONISM in favour of exact and minute representation of reality. Its principal exponents were Otto DIX and George GROSZ.

Neumann, Franz Ernst (1798–1895), German scientist who, in two papers (1845, 1847), formulated the first mathematical theory of electromagnetic INDUCTION, the process of producing electricity from mechanical energy. In 1831 he extended the law of the heat of elements to molecular heat, postulating that molecular heat is equal to the total heat of each constituent atom. Neumann also did research into optics and crystallography.

Neumann, Johann Balthasar (1687–1753), German Baroque architect influenced by the great geometrically planned buildings of BORROMINI and Guarini. His secular works include the palaces of Bruhl, Bruchsal and Würzburg, famous for their magnificent staircases. His major churches are the Vierzehnheiligen pilgrimage church, the Hofkirche, Würzburg, St Paulinus at Trier and the pilgrimage church at Neresheim. These elegant masterpieces are among the finest of German Baroque churches.

Neuralgia, acute, cyclic pain along a nerve. It may be the result of a VIRUS infection of the nerve, physical injury or VITAMIN DEFICIENCY. Analgesic drugs help relieve neuralgic pain.

Neurasthenia, earlier term for NEUROSIS. The term neurasthenia literally means "weakness of the nerves".

Neuritis, inflammation or degenerative lesion of the nerves accompanied by pain, paralysis and impaired reflexes. Causes are many and treatment is directed toward the cause.

Neurology, branch of medicine dealing with the diagnosis and non-surgical treatment of diseases of the NERVOUS SYSTEM. *See also* MS pp.96–97.

Neuron, nerve cell and basic structural unit of the NERVOUS SYSTEM. It is composed of a cell body and one or more DENDRITES and one axon. The dendrites and axon are nerve cell structures, extending beyond the cell body. Dendrites exchange impulses with neighbouring cells. An axon is usually longer and unbranched; it carries impulses longer distances. *See also* MS pp.38–39, 43.

Neurophysiology, study of the detailed structure and functions of NERVES and the nervous system. This includes the study of how nerve cells transmit their electrical impulses and how these are passed to other nerves and organs.

Neuroptera, order of carnivorous insects that include LACEWINGS and ANT LIONS; it is represented throughout the world. Members undergo complete metamorphosis from an egg, through larval and pupal stages to the adult form. Length: to 7cm (3in). *See also* NW pp.106–107.

Neurosis, personality disturbance marked by continual anxiety, depression, exaggerated fears or feelings of insecurity. It is considered a mild form of illness yet it usually affects significantly the way a person feels and acts towards others. *See also* MS pp.134–135, 140.

Neurosurgery, surgical treatment of diseases and disorders of the NERVOUS SYSTEM. *See also* MS pp.96–97.

Neustria, western section of the kingdom of the FRANKS in the 6th, 7th and 8th centuries, formed when the empire of CLOVIS I was divided among his four sons (511). The ruler of the eastern section of the empire, Pepin of Austrasia, overcame Neustria in 687 and in 911 part of it (the area of present-day Normandy) was ceded to the Scandinavian pirates later known as NORMANS.

Neutrality, legal status of a policy of non-involvement in hostilities existing between nations. It is recognized by international law, mainly in the Declaration of Paris (1856) and the Hague Conventions V and XIII of 1907. The rights and obligations of neutral powers were codified by international treaties and conventions. A nation proclaiming its neutrality must be wholly impartial and refrain from helping or hindering any side. It must also protect its own territory from encroachment by belligerents.

The Netherlands are low-lying and flat and are well-suited to dairy farming.

Neue Sachlichkeit; *A Married Couple* by George Grosz, one of its members.

Johann Neumann designed this staircase at the palace at Wurzburg, Germany.

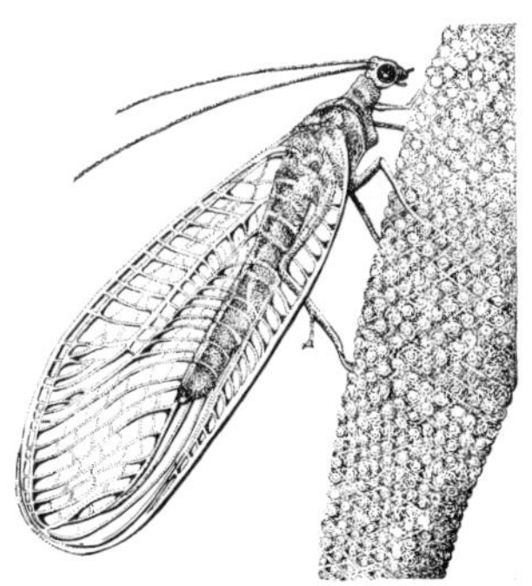

Neuroptera are characterized by transparent wings, netted with tiny veins.

Neutralization

Nevada; Virginia City, now a ghost town, once had a population of 23,000.

New Brunswick was first explored in 1534 by the French navigator Jacques Cartier.

Newcastle upon Tyne's swing bridge, which spans the River Tyne, was built in 1876.

Newcomen's steam engine provided the basic design used for nearly a century.

Neutralization, in chemistry, reaction or TITRATION conducted in an aqueous medium between equivalent weights of an acid and a base to produce water and a salt, which is neither acidic nor basic, but has a pH of approximately 7. This process is often accompanied by the generation of heat. *See also* SU pp.146–147.

Neutrino, uncharged mass-less ELEMENTARY PARTICLE with SPIN $\frac{1}{2}$; it has little reaction with matter and is difficult to detect. There are two kinds: the ELECTRON neutrino (symbol v_e) is closely associated with the electron and is produced when protons and electrons react to form NEUTRONS, as in the Sun. The more common antineutrino occurs when a neutron decays. The muon neutrino (symbol v_μ), associated with MUON, occurs in high-energy reactions. *See also* SU p.66.

Neutron, uncharged ELEMENTARY PARTICLE (symbol n) that occurs in the atomic nuclei of all chemical elements except the lightest isotope of hydrogen. It was first identified by James CHADWICK in 1932. In isolation it is unstable, decaying with a half-life of 11.3 minutes into a PROTON, ELECTRON and antineutrino. Its neutrality allows it to penetrate and be absorbed in nuclei and thus to induce nuclear transmutation and fission. It is a BARYON with SPIN $\frac{1}{2}$ and a mass slightly greater than that of the proton. *See also* SU p.62.

Neutron activation analysis, highly sensitive method of identifying the chemical composition of a substance by bombarding it with high-energy neutrons that are absorbed by the atoms present. The resulting radioactive nuclei emit radiation of an energy and decay rate characteristic of the original atoms, which can thus be identified. The quantity present can also be found with extreme precision.

Neutron flux, speed and rate at which NEUTRONS are released during the fission process in a controlled nuclear chain reaction. In a nuclear reactor, moderators of several kinds (especially DEUTERIUM and HYDROGEN) are used to absorb neutrons, and various spacing devices in the core reduce neutron speeds.

Neutron star. *See* PULSAR.

Neva, Battle of (1240), victory of the Novgorod Russians over the invading Swedes on the banks of the Neva River. Alexander Yaroslovich (later Nevsky) led the Novgorod forces which saved Russia from Swedish conquest.

Nevada, state in W USA. Much of it lies in the Great Basin, but the Sierra Nevada rise steeply from its W edge. Nevada's dry climate and steep slopes have hindered the development of an agricultural economy. Hay and lucerne (alfalfa) are the chief crops, and a little wheat and barley are grown; sheep and cattle grazing are much more important. Most of Nevada's economic wealth comes from its mineral deposits, which include, copper, lead, silver, gold, zinc and tungsten. The principal industrial products are chemicals, timber, electrical machinery, glass products and primary and fabricated metals. It is a tourist area and the main cities are Carson City (the state capital), Las Vegas and Reno, where gambling provides an important source of state revenue. The USA acquired the region in 1848 at the end of the MEXICAN WAR, before which the area had been visited by Spanish and British missionaries and fur trappers. Nevada was situated on the overland trail that took thousands of gold prospectors to California in 1849. When gold and silver were found in Comstock Lode in 1859, settlers flocked to Nevada. The territory of Nevada was organized in 1861 and admitted to the Union in 1864. The ghost towns created when the ores were depleted are today a tourist attraction. Area: 286,297sq km (110,539sq miles). Pop. (1975 est.) 592,000. *See also* MW p.175. **Névé.** *See* FIRN.

Neville, Richard. *See* WARWICK, RICHARD NEVILLE, EARL OF.

Neville's Cross, Battle of (1346), English victory over the Scots. A Scottish army under David II, attempting to profit from Edward III's absence in France, invaded N England, but was shattered by English

forces at Neville's Cross, near Durham. David was captured.

Nevinson, Christopher Richard Wynne (1889–1946), British painter and printmaker whose work as a war artist (1917) is considered his best. He was influenced by CUBISM and FUTURISM and was a founder-member of the LONDON GROUP (1913). He also painted landscapes and portraits in a variety of styles.

Nevis. *See* ST KITTS-NEVIS.

Nevsky, Alexander. *See* ALEXANDER NEVSKY.

Nevus. *See* BIRTHMARK.

Newark, largest city in NE New Jersey USA, on the Passaic River and Newark Bay, connected to nearby New York City by tunnel. Founded in 1666 by the Puritans, Newark's industrial growth began after the American Revolution and with the development of transport routes. It has a college of engineering, dating from 1881. Newark's industries include electrical equipment, paints and chemicals. Pop. (1973) 367,683.

New Atlantis, The (c. 1626), allegorical work of philosophy by Francis BACON. It concerns an imaginary island, Bensalem, where a philosophical commonwealth has been established to study the natural sciences. The pride of Bensalem is the university which, in contrast to English universities of the time is dedicated to serious scientific research.

Newbery, John (1713–67), British publisher, one of the first to produce books exclusively for children. He published a series of books called the *Juvenile Library*, including *Mrs Marjorie Two Shoes* and *Mother Goose's Melody* (c. 1780). The US Newbery Award for children's literature is named after him.

Newberry, John Strong (1822–92), US geologist. After practising medicine in Cleveland (1851–55), he became professor of geology at Columbia University (1866–92). His publications include *Fossil Fishes and Fossil Plants of New Jersey* (1888) and *Paleozoic Fishes of North America* (1889).

Newbolt, Sir Henry John (1862–1938), British poet and writer, best known for his patriotic and naval verse. He practised as a barrister from 1887 to 1889 and from 1900 to 1904 edited *The Monthly Review*. In 1923 he was appointed to write two volumes of the official history of the naval operations of WWI. His works include *Admirals All* (1897), *The Year of Trafalgar* (1905) and *A Perpetual Memory* (1939).

New Britain, largest island of the Bismarck Archipelago, Papua New Guinea, in the SW Pacific Ocean, approx. 90km (55 miles) E of New Guinea; the main towns are Kimbe and Rabaul. Discovered in 1606 by the Dutchman Jacques Lemaire, the island became a German protectorate in 1884; after WWI it was mandated to Australia. The island is mountainous, with several active volcanoes and hot springs. Products: copra, coconuts, cocoa, gold, copper, iron and coal. Area: approx. 36,674sq km (14,160sq miles). Pop. (1970 est.) 154,000.

New Brunswick, maritime province in E Canada, on the USA-Canada border. The land rises gradually from E to W and is drained chiefly by the St John and Miramichi rivers. More than three-quarters of the province is forested. The chief crops are hay, clover, oats, potatoes and fruit; dairy farming is also important. The principal industry in New Brunswick is timber; others include leather goods, pharmaceuticals and machinery, and fishing is important along the coast. There are valuable mineral deposits of zinc, copper, lead, coal and antimony. The principal cities are Fredericton, St John and Moncton.

The region was first explored by Jacques Cartier in 1534. With Nova Scotia, the area formed the French colony of Acadia. The colony was ceded to Britain in 1713, but the British were slow to settle it. Many Loyalists entered the region during the American War of Independence. This influx spurred the establishment of a separate colony of Nova Scotia, and the province of New Brunswick was established in 1784. A boundary dispute with

the USA was settled in 1842, and in 1867 New Brunswick joined NOVA SCOTIA, QUEBEC and ONTARIO to form the Dominion of Canada. Area: 73,437sq km (28,354sq miles). Pop. (1971) 635,000. *See also* MW p.45.

Newby, Percy Howard (1918–), British writer. He was lecturer in English (1942–46) and the controller of the BBC's Third Programme (1958–70). Among his numerous novels are *Something to Answer For* (1968; Booker Prize 1969), *A Lot to Ask* (1973) and *Kith* (1977).

New Caledonia (Nouvelle Calédonie), French overseas territory in the SW Pacific Ocean approx. 1210km (750 miles) E of Australia; the group takes its name from the main island, on which is the capital, NOUMÉA. Discovered in 1774 by Capt. Cook, the island became a French overseas territory in 1946. Products: copra, coffee, cotton, nickel, iron, manganese, cobalt and chromium. Area: 18,342sq km (7,082sq miles). Pop. (1975 est.) 136,000. *See also* MW p.130.

Newcastle, Thomas Pelham-Holles, Duke of (1693–1768), English politician. He was a Whig who supported WALPOLE; he became Secretary of State (1724–54) and was the king's Chief Minister (1754–56; 1757–62). Throughout the period his control of patronage contributed significantly to the Whig ascendancy. *See also* HC1 pp.18, 68.

Newcastle, William Cavendish, Duke of (1592–1676), English politician and soldier. He was appointed governor to the Prince of Wales in 1638. He was commander of royalist forces in the N during the Civil War, but after the Battle of MARSTON MOOR (1644) he lived in exile until 1660. He was created Duke of Newcastle in 1665.

Newcastle, city in New South Wales, SE Australia, 161km (100 miles) NE of Sydney. Founded in 1797 as King's Town, it is situated in the centre of Australia's largest coal-mining region, and is an important port, exporting coal, steel, wool and wheat. The city's university was founded in 1965. Industries: iron and steel, chemicals, textiles, shipbuilding, fertilizers. Pop. (1973) 146,460.

Newcastle upon Tyne, city in Tyne and Wear, NE England, and a major port on the River Tyne. A major wool-exporting port in the 13th century, and then a coal-shipping centre, Newcastle became one of England's chief ship-building centres. Heavy engineering is still important there. The city is the seat of the University of Newcastle (1937), and has an 11th-century castle and the remains of the 13th-century city walls. Pop. (1971) 222,153.

Newcombe, John (1944–), Australian tennis player. The last winner of the WIMBLEDON men's singles title in 1967 before the game became "open", he won it again as a professional in 1970 and 1971. With Tony ROCHE he won all the great doubles titles, including three successive victories at Wimbledon from 1968 to 1970.

New Comedy, type of Greek comedy which developed in the second half of the 4th century BC. Among its innovations were an urbanity of style and a cleverly managed plot. The only remaining complete example of this type of play is *Dyskolos* (in translation *Misanthrope*) by Menander (c. 342–292 BC). *See also* HC1 pp.74–75.

Newcomen, Thomas (1663–1729), British engineer and inventor of the atmospheric steam engine. Assisted by John Calley, a plumber, he spent more than ten years perfecting the invention. A 1698 patent by Thomas Savery covered the principle, and Newcomen went into partnership with him, their first engine being built in 1712. For many years this type of engine was used to power pumps, particularly to drain mines. *See also* MM p.64.

New Deal, in US history, term for the social and economic programme of the administrations of Franklin D. ROOSEVELT, between 1933 and 1939. Elected in 1932 during the worst phase of the Great DEPRESSION (the severe economic crisis supposed to have been precipitated by the stock market crash of 1929), Roosevelt promised a "new deal" to the American people, hence the name that was given to all the domestic reforms of his administration. These included programmes of agricultural and business regulation, inflation and price stabilization, and extensive public works. The second phase of the New Deal in 1935 introduced social security benefits, which included disability pensions, unemployment benefits, old age pensions, central mortgage finance and industrial relations legislation to strengthen the trade unions. Some of these measures were resisted by commercial interests and even the courts, but they eventually gained wide acceptance and became permanent features of US government policy.

New Delhi, capital of INDIA, located in the N of the country on the River Yamuna in Delhi Union Territory. Planned by the British architects Sir Edwin LUTYENS and Herbert Baker, it was constructed in 1912–29 to replace Calcutta as the capital of British India. Whereas the old city of Delhi (to the SW) is primarily a commercial centre, New Delhi has an administrative function. Its industries include textile production. Pop. (1971) 301,800.

New Democratic Party (NDP), Canadian political party founded in 1961 when the CO-OPERATIVE COMMONWEALTH FEDERATION reorganized itself and entered into a close relationship with the Canadian labour unions. The party advocated a moderate socialist programme and, while under the leadership of David Lewis (1971–1974), gained control of provincial governments in Manitoba, Saskatchewan and British Columbia. In 1972 its 31 members in the House of Commons sided with the Liberal Party in a coalition government that lasted until 1974, when the NDP opposed the Liberal government's monetary policy (resulting in the elections of 1974). The NDP lost 15 of its seats in the 1974 election.

New Economic Policy (NEP) (1921–27), Soviet programme to restore the Russian economy and appease the peasantry after the civil war of 1918–20. It relaxed controls on trade and light industry, replaced the seizure of grain from peasants with a fixed amount being taken as tax, allowing the surplus to be sold on the open market, and permitted individual and communal forms of land tenure. The NEP, initiated by LENIN, was seen by some as a denial of Marxist ideals. It ended in 1927 with the forcible collectivization instituted by the first FIVE-YEAR PLAN See also HC2 pp.197, 198; MW p.172.

New England, region in NE USA, made up of the states of MAINE, NEW HAMPSHIRE, VERMONT, CONNECTICUT, MASSACHUSETTS and RHODE ISLAND. The area is geographically cut off from the rest of the country by the Appalachian Mts. The soil is generally poor and agriculture has never been a major economic activity. The coastline provides excellent harbours, and fishing and shipbuilding soon became important. New England was the centre of events leading up to the AMERICAN WAR OF INDEPENDENCE. The region became highly industrialized after the WAR OF 1812 and also developed as a centre of literature and learning, producing such writers as EMERSON, HAWTHORNE and THOREAU. Harvard University was founded as early as 1636. The geographical and political conditions of the region produced a New England "type", the Yankee, characterized by resourcefulness, thrift and self-discipline. The term was later applied to Americans in general. Area: 172,515sq km (66,608sq miles).

New English Art Club, exhibition society founded in 1886 by artists who favoured French IMPRESSIONISM and were out of sympathy with the Classicism of the Royal Academy. Steer, Sicker and Whistler were early members; later members included Augustus and Gwen John, Roger Fry and Paul Nash; Monet, Degas and Bertha Morisot were guest exhibitors. The club played a leading role in British art until the formation of the CAMDEN TOWN GROUP (1911) and the dominant LONDON GROUP (1913).

New Forest, region of forest and heathland in Hampshire, s England. It was established as a royal hunting ground in 1079 by WILLIAM I. His son, William Rufus, was killed by an arrow there in 1100. The Forest includes many species of trees, and pigs, cattle and ponies are reared there. Area: approx. 383sq km (148sq miles).

Newfoundland, province in E Canada, on the Atlantic Ocean; made up of the mainland region of LABRADOR and the island of Newfoundland plus adjacent islands. The two parts are separated by the Strait of BELLE ISLE. The island is a plateau with many lakes and marshes. Labrador has TUNDRA in the N and the cold climate and lack of transport facilities throughout the region have hindered economic development. There are, however, valuable mineral resources that include copper, lead, iron and zinc. Timber is an important industry, and the Grand Banks off the coast is one of the world's best cod-fishing areas. ST JOHN'S is the capital. Norsemen are believed to have landed on the coast of Labrador c. AD 1000, and to have established a settlement on the N tip of Newfoundland. John CABOT reached the island in 1497, and Sir Humphrey GILBERT claimed it for Britain in 1583. The region was settled slowly and did not become a British colony until 1832, remaining apart from the rest of Canada until 1949, when it became that country's tenth province. Area 404,420sq km (156,185sq miles) Pop. (1971) 522,104. See also MW p.45.

Newfoundland, rescue and working dog that was originally bred by fisherman in Newfoundland. It has a massive head with a square, short muzzle and small, triangular ears set close to the head. The full-chested, broad-backed body is set on short, strong legs and the tail is broad and long. The long coat may be black or black and white. Height: 71cm (28in) at the shoulder; weight: 68kg (150lb).

New France, French colony in N America corresponding roughly to Quebec, Ontario and the maritime provinces of Canada. Jacques CARTIER claimed the region for France in 1534. Quebec was founded in 1608. The Company of New France, founded in 1627 to trade in and settle the area, was disbanded in 1663 and the colony placed under a royal governor. In 1713 Acadia (Nova Scotia), Newfoundland and the Hudson Bay area were lost to Britain, and in 1763 the rest of New France was also transferred to Britain. See also HC2 pp.136–137; MW pp.48–49.

Newgate Prison, London prison, demolished in 1902, that used to be located on the site of the OLD BAILEY. It derived its name from its original position, from the beginning of the 13th century at least, at the New Gate to the City of London. It was damaged by fire in the Great Fire of London (1666) and during the Gordon Riots of 1780.

New General Catalogue (NGC), in astronomy, list of stellar clusters, nebulae and galaxies compiled by Danish astronomer Johan Dreyer in 1888, based on the observations of William HERSCHEL and his son John. Dreyer later published two Index Catalogues (IC) bringing the number of stellar objects listed in the catalogue to approx. 13,000. See also SU p.230.

New Georgia, island group in the British Solomon Islands, in the SW Pacific Ocean. The group takes its name from the main island. Area: approx. 5,180sq km (2,000sq miles). Pop. (1970) 22,264.

New Granada, historical administrative area centred on present-day Bogotá, Colombia (founded 1538), and at times including parts of Ecuador, Panama and Venezuela. The viceroyalty of New Granada was founded in 1718 and the name was used to describe independent Colombia during the period 1830–63, after which the name was changed to the United States of Colombia. See also MW p.59.

New Guinea, island in the E Malay Archipelago, in the W Pacific Ocean; second-largest island in the world. It has a tropical climate and is mountainous. The island is divided into Irian Jaya (formerly Irian Barat), a province of INDONESIA, and PAPUA NEW GUINEA, which was administered by Australia until 1975, when it became an independent state. Discovered in the early 16th century, the island was so named because of its resemblance to the Guinea Coast of W Africa. During the next two centuries, New Guinea was colonized by the Dutch (the W half that is now Indonesia), the Germans and the British (the part that is now Papua New Guinea). Products include copra, cocoa, coffee, rubber, coconuts and tobacco. Area: 885,780sq km (342,000sq miles). Pop. (1973 est.) 3,600,000. See also MW p.130.

New Hampshire, state in NE USA, on the Canadian border. Much of the land is mountainous and forested. The principal rivers are the Connecticut and the Merrimack, and there are more than 1,000 lakes, the largest of which is Winnepesaukee. Farming is restricted by the terrain and poor, stony soil, and is mostly concentrated in the Connecticut valley. Dairy and market garden produce, hay, apples and potatoes are the chief products. New Hampshire, however, is highly industrialized. There is abundant hydro-electricity and the main industries are electrical machinery, paper and wood products, printing and publishing, leather goods and textiles. Tourism is also important. The principal cities are Concord, the state capital, Manchester and Nashua. English explorers visited the New Hampshire coast in 1603. The first settlement was founded in 1623. Although no fighting took place on the state's soil during the American Revolution, New Hampshire regiments were active throughout the campaigns. Area: 24,097sq km (9,304sq miles). Pop. (1975 est.) 818,000. See also MW p.175.

New Hebrides (Nouvelles-Hébrides), volcanic island group in the SW Pacific Ocean, approx. 2,300km (1,430 miles) E of Australia; governed jointly by Britain and France. The group consists of approx. 80 islands, which form a chain some 725km (450 miles) in length. The main islands are Espiritu Santo, Efate, which has the capital VILA, Malekula, Pentecost, Malo and Tanna. Discovered in 1606 by Pedro Fernandez de Queiros and explored by Capt. James COOK in 1774, the group was settled by the English and French in the early 1800s. The chief economic activities are fishing, farming and mining. Area: 14,760sq km (5,699sq miles). Pop. (1975 est.) 97,000. See also MW p.131.

New Humanism, movement in literary and social criticism in the early 1900s, led by Irving Babbitt, a Harvard professor, and Paul Elmer More, an editor and writer. Reacting against realism and naturalism in modern literature, the movement sought to avoid the extremes of both religion and science and stressed the importance of human reason and freedom of will in ethical, artistic and intellectual considerations. New Humanist treatises include Babbitt's *Rousseau and Romanticism* (1919) and More's *Shelburne Essays* (1904–21; 1928–36).

Ne Win, U (1911–), Burmese military and political leader born Thakin Shu Maung. At the outbreak of WWII he helped the Japanese invade Burma, but in 1945 his Burmese Independence Army · also turned against the Japanese. He became commander-in-chief of the army upon Burma's independence from Britain in 1948. In 1958 he deposed U NU and had himself appointed Prime Minister. U Nu returned to office in 1960, but Ne Win deposed him again in 1962. He became President in 1974. See also MW pp.43–44.

New Ireland, mountainous island in the Bismarck Archipelago, Papua New Guinea. Sighted by the Dutch in 1616, the island was a German protectorate from 1884 to 1914. The main town and port is Kavieng. Products: coconut, copra, cocoa. Area: 8,651sq km (3.340sq miles). Pop. 50,600.

New Jersey, state in E USA, on the Atlantic coast S of New York City. The N of the state is in the Appalachian highland region; SE of this area are the piedmont plains, and more than half the state is

New Delhi's Parliament House is an elegant, purpose-built administrative building.

New England states, such as New Hampshire, are particularly beautiful in autumn.

Newfoundland waters are among the most productive fishing grounds in the world.

Newfoundland; this North American dog plays a similar role to the St Bernard.

New Kingdom

Cardinal Newman retracted his criticisms of Roman Catholicism in an Oxford journal.

Paul Newman owes much of his success to a careful choice of film roles.

New Orleans: one of a series of evocative scenes by Henri Cartier Breson.

Newspaper presses printing *The Times*, which was established in 1785.

coastal plain. A variety of crops are grown, including potatoes, maize, hay and fruits, and dairy cattle and poultry are also important. New Jersey is, however, overwhelmingly industrial. Principal industries are chemicals, pharmaceuticals, rubber goods, textiles, electronic equipment, missile components, copper smelting and oil refining. The New Jersey coast is a highly developed resort area. The state capital is Trenton and large cities include Newark, Camden and Atlantic City.

The Italian explorer da Verrazano explored the New Jersey coast in 1524, but settlement did not begin until the 1620s, when NE New Jersey was settled at the same time as the Dutch colony of New Netherland. When the British took the colony, in 1664, the land between the Hudson and Delaware rivers was named New Jersey. During the AMERICAN CIVIL WAR and both World Wars, New Jersey's industries were a major supplier of war materials. Area: 20,295sq km (7,836sq miles). Pop. (1975 est.) 7,316,000. *See also* MW p.175.

New Kingdom, or New Empire (1570–332 BC), era of ancient Egyptian history, from the 28th to 30th Dynasties. The 18th Dynasty (to *c.*1342 BC), founded by Ahmose I after a confused century of HYKSOS rule, was the most flourishing period of ancient Egypt. The New Kingdom ended in 332 BC with Alexander the Great's conquest of Egypt.

Newlands, John Alexander Reina (1837–98), British chemist. In 1864 he announced his law of octaves, which arranged the chemical elements in a table of eight columns according to increasing atomic weight. The law was ridiculed until MENDELEYEV included it in his own PERIODIC TABLE five years later.

Newlands, the name of two South African sports grounds, situated near each other in Cape Town. One is for cricket, the other for rugby union. The cricket ground, headquarters of Western Province, was first used for Test matches in 1888, when South Africa played an English touring side. The rugby ground, which is also the home ground for Western Province, saw its first international, against the British Isles, in 1891.

New Look, fashion for women designed in 1947 by Christian DIOR. It featured narrow shoulders, a small waist and a long full skirt, and emphasized the bust.

Newlove, John (1938–), Canadian poet. His poems, many of which deal with alienation and loss, are collected in *Grave Sirs* (1963), *Moving in Alone* (1965), *Black Night Window* (1968), *Lies* (1972) and other volumes.

Newman, Barnett (1905–70), US painter who was closely associated with ABSTRACT EXPRESSIONISM; among the first to abandon the CUBIST idea of spatial composition. In *Who's Afraid of Red, Yellow and Blue* (1969–70) he uses huge expanses of primary colours as the sole source of visual impact. *See also* HC2 p.205.

Newman, Cardinal John Henry (1801–90), British churchman and scholar. A fellow of Oriel College, Oxford (1822), and vicar of St Mary's Oxford, he became the centre of a group of high church Anglican enthusiasts, who made up the Tractarian or Oxford Movement after 1833. Newman joined the Roman Catholic Church in 1845. His autobiography, *Apologia pro vita sua* (1864), was a work of great spiritual revelation, and his *Idea of a University* (1852) argued for a liberal education. He was made cardinal in 1879.

Newman, Paul (1925–), US film actor, director and producer. He is known for his portrayals of cheeky, independent and wryly humorous anti-heroes in such films as *Cat on a Hot Tin Roof* (1958), *The Hustler* (1961), *Hud* (1963), *Cool Hand Luke* (1967), *Butch Cassidy and the Sundance Kid* (1969) and *The Sting* (1973), perhaps his most popular success.

Newmarket, famous English racecourse in Cambridgeshire. It is the headquarters of English racing and the venue of the National Stud. Two Classics, the 1,000 Guineas and 2,000 Guineas, are run there over the straight Rowley Mile course, and other major races include the Cambridge-shire Handicap, the Cesarewitch, and the Champion Stakes.

New Mexico, state in SW USA, on the Mexican border. The Sangre de Cristo Mts in the N flank the Rio Grande, which runs N to S through the state. The terrain includes desert, forested mountains and stark mesa; in the S and SW are semi-arid plains. The S Pecos and Rio Grande rivers are used to irrigate cotton crops; hay, wheat, dairy produce and chili beans are also important. Much of the land is pasture. A large proportion of the state's wealth comes from its mineral deposits, which include uranium, manganese, copper, silver, turquoise, oil, coal and natural gas. Timber and tourism are important industries. Santa Fé is the state capital and Albuquerque is the largest city.

The first permanent Spanish settlement was established in 1598 and Sante Fe was founded in 1610. After 1821 the region became a province of independent Mexico. The Santa Fé trail, a major route westwards opened in 1821, led across New Mexico. The USA acquired the region in 1848 at the end of the MEXICAN WAR, and the Territory of New Mexico was organized two years later. It entered the Union in 1912 as the 47th state. Area: 315,113sq km (121,665sq miles), Pop. (1975 est.) 1,147,000. *See* MW p.175.

New Model Army, national army formed in 1645 that won the ENGLISH CIVIL WAR for Parliament. It was planned to consist of 11 regiments of horse of 600 men each, 12 regiments of foot of 1,200 men each, and 1,000 dragoons (mounted infantrymen). Sir Thomas FAIRFAX was appointed Captain General in 1645. CROMWELL became the leader of horse just before the parliamentary victory at NASEBY on 14 June 1645, and afterwards became its commander. The veterans of the New Model Army gave Cromwell the support he needed to take political control after the Civil War.

New Objectivity. *See* NEUE SACHLICHKEIT.

New Orleans, city in SE Louisiana between Lake Pontchartrain and the Mississippi River; the largest city in Louisiana and a major US port. It was held by the French and the Spanish before it was acquired by the USA under the LOUISIANA PURCHASE of 1803, and was the scene of the British defeat in 1815 by Andrew JACKSON in the WAR OF 1812. The city fell to Union Naval forces during the AMERICAN CIVIL WAR. New Orleans is considered to be the home of JAZZ, centred on Bourbon Street in the French Quarter. It is also the home of the Annual MARDI GRAS and the site of several universities. Industries: food processing, petroleum, natural gas, oil and sugar refining, shipbuilding, tourism, aluminium, petrochemicals, paper products, furniture. Pop. (1973) 513,479.

New Plymouth, city and port in W North Island, New Zealand. Founded in 1841, it is New Zealand's chief dairy centre. Pop. (1971) 34,314.

New Poor Law. *See* POOR LAWS.

Newport, city in SE Wales, on the River Usk, administrative centre of the county of Gwent. It was the scene of CHARTIST riots in 1839. Industries: steel, aircraft, aluminium, metal goods. Pop. (1973) 110,090.

Newport, city in SE Rhode Island, USA, on S Rhode Island in Narragansett Bay. Founded in 1639, it was held by the British during the AMERICAN WAR OF INDEPENDENCE. Newport served as joint state capital with Providence until 1900. The city has the US Naval War College, founded 1885. Tourism is the chief industry. Pop. (1970) 34,562.

New South Wales, state in SE Australia on the Tasman Sea. The terrain ranges from coastal lowlands to the eastern highlands to the western plains. The Murray River and its tributaries are used extensively for irrigation. Wheat, wool, dairy produce and beef are the principal agricultural products. The state has valuable deposits of coal, gold, iron ore, copper, silver, lead and zinc. New South Wales is the most important industrial state. Steel is the chief product. The capital is SYDNEY. Capt. James COOK first visited the area in 1770. He claimed the E coast of Australia for the British. The first Australian settlement (1788) was at Botany Bay, S of Sydney. The territory developed in the first half of the 19th century with the growth of the wool industry. The colony included the territories of Tasmania, Queensland, Victoria, Northern Territory, South Australia and New Zealand until they became colonies in their own right between 1825 and 1863. New South Wales became a state of the Commonwealth of Australia in 1901. Area: 801,430sq km (309,180sq miles). Pop. 4,567,000. *See also* MW p.28.

New Spain, former administrative area within the Spanish colonial empire, centred on Mexico City. New Spain was the first permanent vice-royalty on the American mainland, created in 1535. It covered a vast area: present-day Mexico, Central America, SW USA, Florida, and the West Indies. *See also* MW p.125.

Newspaper, periodical publication, usually daily or weekly, which conveys information and opinion about current events. Julius CAESAR is said to have instituted the first recorded newspaper, the *Acta Diurna*, which was posted daily in public places. But it was the invention and spread of printing after 1430 that primarily influenced the early development of newspapers, and in 1563 the Venetians produced news-sheets which were read aloud in public, and the cost of admission to such readings was a small coin known as a *gazeta*, which word became a common name for such official news-sheets. The *Oxford Gazette* was founded in 1665 and, becoming the *London Gazette* in the following year, is still published. The first English daily newspaper was the *Daily Courrant* of 1702, and then many journals, such as *The Examiner* edited by SWIFT (1710–11), the *Tatler* (1709–11), and ADDISON and STEELE's *The Spectator* (1711–12), were published. The *Daily Universal Register* was founded in 1785 by John WALTER, becoming *The Times* in 1788. The invention of the first practicable telegraph in 1835 allowed news agencies such as REUTER and Associated Press to function, and improvements in high-speed printing have made possible large editions. Modern newspapers in the West rely to a great extent on revenue from advertising. In many countries newspapers carry great influence and governments or politicians have been made or broken by them, as in the US Watergate affair. *See also* MM pp.210–211.

Newsprint, type of paper used for newspapers. It is intended to have a short life, but must be strong enough for high-speed printing presses. Wood pulp, the major constituent of newsprint, is obtained from softwood logs which are crushed and ground. Water is added, along with fibrous materials such as hemp, wood and asbestos as well as glue and fillers. The mixture is spread on to wire screens, pressed, dried and polished by rolling before being sized. *See also* MM pp.60–61.

New Stone Age. *See* NEOLITHIC AGE.

Newt, any of numerous species of tailed AMPHIBIANS of Europe, Asia and North America. The common European newt, *Triturus vulgaris*, is terrestrial except during the breeding season, when it is aquatic and the male develops ornamental fins. Its body is long and slender and the tail is laterally flattened. Length: to 17cm (7in). Family Salamandridae. *See also* NW pp.132–133, *133*.

New Testament, second part of the Bible, consisting of 27 books all originally written in Greek after AD 45 and all concerning the life and teachings of JESUS CHRIST as the Son of God. It begins with three SYNOPTIC GOSPELS (MATTHEW, MARK, LUKE) and a gospel which is more a sermon (JOHN); the ACTS OF THE APOSTLES then precedes 21 letters (EPISTLES), many of which are attributed to St PAUL and addressed to specific early Church communities, while others are to Christians at large; the New Testament ends with the mystic and perhaps prophetic REVELATION (or Apocalypse). The whole represents the new testament (or will) of God, bequeathing salvation to mankind, as opposed to the old covenant (OLD TESTA-

MENT) which Christians regard as having been superseded just as predictions within it prophesied.

New Thought Movement, loose-knit philosophical-religious organization in which the often contradictory beliefs of its adherents are based on an insistence on the power of the mind over the body, within a spirit of optimism. Originating in the USA in the 19th century, it has sometimes been likened to CHRISTIAN SCIENCE.

Newton, Alfred (1829–1907), British zoologist b. Switzerland. He promoted the first laws for the protection of birds, and was the editor of the ornithological journals *Ibis* and *The Zoological Record.* He wrote *A Dictionary of Birds* (1893–96).

Newton, Sir Isaac (1642–1727), English scientist. He studied at Cambridge and became professor of mathematics there (1669–1701). Many of his discoveries were made in the period 1664–66, but not published until later. His main works were *Philosophiae Naturalis Principia Mathematica* (1687) and *Opticks* (1704). In the former he outlined his laws of MOTION and proposed the principle of universal GRAVITATION; in the latter he showed that white light is made up of colours of the SPECTRUM and put forward his particle theory of light. He also created the first system of CALCULUS in the 1660s, but did not publish it until Gottfried LEIBNIZ had published his own system in 1684. He built the first reflecting telescope in c. 1671, and studied alchemy and biblical chronology. He was president of the Royal Society (1703–27), and master of the mint after 1701. In 1705 he became the first man to be knighted for his scientific work. Newton engaged in violent scientific disputes with Robert HOOKE, John FLAMSTEED and Leibniz. *See also* HC1 pp.302–303; index to *Science and the Universe.*

Newton, unit of force in the MKS system of units. One newton is the force that gives a mass of one kilogram an acceleration of one metre per second per second. One pound weight exerts a FORCE of 4.45 newtons.

Newton's Laws, three physical laws that govern the behaviour of FORCES, formulated by the English scientist Isaac NEWTON. The first law is a definition of force: an object remains at rest or moves at constant VELOCITY, unless it is acted upon by a force. The second law enables force to be calculated: force is proportional to the rate of change of MOMENTUM; this statement leads to the result that force is equal to the product of MASS and ACCELERATION. The third law states that every force has associated with it an equal and opposite force (a reaction). *See also* MECHANICS; SU pp.74–75.

New Towns, planned urban communities, designed to rehouse populations from large cities and create local employment. They were established in Britain under the New Towns Acts of 1946 and 1965, and about 30 new towns have been established there since WWII. Each new town is supervised by a Development Corporation.

New Wave. *See* NOUVELLE VAGUE.

New Year's Day, 1 January, one of the oldest of holidays and nowadays celebrated in almost all Western and European lands. In some places (eg Scotland and France) New Year's Day festivities are more elaborate than those for Christmas Day. It is generally celebrated (on the day or its eve) by music-making, sports meetings, general revelry and visiting among friends. It is a paid public holiday (Bank Holiday) in many countries. *See also* HOGMANAY.

New York, state in NE USA, bounded by the Canadian border, the Great Lakes, the Atlantic Ocean and three New England states. Much of it is mountainous, the Adirondacks (NE) and Catskills (SE) being the principal ranges. The W consists of a rolling plateau sloping down to lake Ontario and the St. Lawrence valley. The Hudson and its tributary the Mohawk are the principal rivers. Agricultural produce in New York is varied. The state has orchards and vineyards, cereal crops, hay, potatoes, and market garden and dairy produce are important. New York is the leading manufacturing and commercial state in the USA.

Its industries include clothing, machinery, chemicals, electrical equipment, paper and optical instruments. The state capital is Albany, although New York City is by far the largest city.

Henry HUDSON, an English navigator working for the Dutch government, discovered New York Bay in 1609 and sailed up the river that bears his name, claiming the territory for the Dutch. The New Netherland colony was established in the Hudson valley; in 1664 it was seized by the British and renamed New York. The opening of the Erie Canal in 1825, which created a water route w through the Great Lakes, was an enormous stimulus to New York's economic growth. Throughout US history New York's economic strength and large population have given the state great influence in national affairs. Area: 128,401sq km (49,576sq miles). Pop. (1975 est.) 18,120,000. *See also* MW p.175.

New York City, city in SE New York State, at the mouth of the Hudson River; largest city (by population) in the USA and one of the three largest in the world. It is made up of five boroughs: MANHATTAN, the BRONX, BROOKLYN, QUEENS and Richmond. The greater metropolitan area extends into SW Connecticut, and parts of NE New Jersey and W LONG ISLAND. Manhattan Island was bought from the Indians in 1626 by the Dutch West India Co., and New Amsterdam was founded at the s end of the island. In 1664 the British took the colony and renamed it New York. Revolutionary sentiment in the colony was strong, but the city was unsuccessfully defended by George WASHINGTON in 1776, and the British held the city until 1781. Washington was inaugurated as the first US president in New York's Federal Hall in 1789. The founding of the Bank of New York by Alexander HAMILTON and the opening of the Erie Canal in 1825 made New York the principal US commercial and financial centre. Following the AMERICAN CIVIL WAR the city received a great influx of immigrants. The Greater New York Charter, unifying the five boroughs, was passed in 1897.

New York is a great cultural and educational centre. Its notable areas include WALL STREET, HARLEM, Fifth Avenue and Greenwich Village. Monuments and buildings of interest include the STATUE OF LIBERTY, St Patrick's Cathedral, EMPIRE STATE BUILDING, Rockefeller Centre, the World Trade Centre, United Nations and Pan Am buildings, the Metropolitan Museum of Art, Museum of Modern Art, Guggenheim Museum, Radio City Music Hall, LINCOLN CENTER and CARNEGIE HALL. New York has more than 30 universities and colleges, the METROPOLITAN OPERA COMPANY, the New York City Ballet and the New York Philharmonic Orchestra. One of the world's leading ports and an important financial centre, New York has clothing, chemicals, metal products, food processing, broadcasting, entertainment, tourism and publishing industries. Pop. 7,567,000.

New Zealand, independent Commonwealth nation in the s Pacific, consisting of two main islands, NORTH ISLAND and SOUTH ISLAND, separated by the Cook Strait. Its economy is based on dairy and related products. The most important dairy regions are located in North Island around the fertile plain of the Waikato River. Cereal crops are grown and lambs raised on the CANTERBURY PLAINS in South Island. Industry has been developed since WWII; food processing is important, also the manufacture of machinery and transport equipment. New Zealand's ability to sell meat, wool and dairy produce on the international market has created a prosperous economy. Britain is still the major supplier of imports although the USA and Japan are becoming important trading partners. The capital is WELLINGTON. Area: 268,676sq km (103,736sq miles). Pop. (1977 est.) 3,130,000. *See* MW p.131.

New Zealand spinach, coarse annual plant (*Tetragonia expansa*) native to Australia, New Zealand and Asia, belonging to the family Aizoaceae. It is used in cooking.

Ney, Michel, Duc d'Elchingen, Prince de La Moskowa (1769–1815), French marshal. He fought in Revolutionary and Napoleonic armies from 1792, notably at Friedland (1807) and in the retreat from Moscow (1812). He urged NAPOLEON to abdicate in 1814 and was sent by the Bourbons to stop Napoleon from returning to Paris in 1815. He fought for Napoleon at WATERLOO, was condemned for treason by the Bourbons and executed.

Ngata, Sir Apirana Turupa (1874–1950), New Zealand politician. He became the first Maori barrister and solicitor and represented the Eastern Maori constituency as a Member of Parliament for 39 years. As Minister of Native Affairs (1928–34), he was instrumental in the improvement of the condition of Maoris, especially in education and agriculture.

NGC Catalogue. *See* NEW GENERAL CATALOGUE.

Ngo Dinh Diem. *See* DIEM, NGO DINH.

Ngouabi, Major Marien (1938–77), President of the People's Republic of the Congo from 1969–77. His regime fostered one-party, centralized government. He was assassinated in March 1977.

Ngorongoro Crater, large depression in the GREAT RIFT VALLEY, Tanzania, surrounded by a rim 610m (2,000ft) high. The shallow lakes in its centre support birds and other wildlife. Area: 539sq km (208sq miles).

Nguni, group of Negroid peoples of s Africa, whose languages are closely related. They include the ZULU, SWAZI and XHOSA of Natal and the Transvaal, and a number of dispersed tribes (the Northern Nguni) who were driven north by the rise of the Zulu empire during the 19th century. Nguni society is traditionally polygamous and is based on agriculture and animal husbandry.

Nha-trang, seaport city in the s of the Republic of Vietnam, 80km (50 miles) N of Phan-rang. It was the site of a major US military base during the VIETNAM WAR. It is located on the South China Sea and is a commercial and fishing centre. Pop. (1968 est.) 101,908.

Niagara Falls, waterfalls on the Niagara River on the border of the USA (w New York) and Canada (SE Ontario); divided by Goat Island into the Horseshoe, or Canadian, Falls and the American Falls. Father Louis HENNEPIN discovered the falls in 1678. The Canadian Falls are 48m (158ft) high and 792m (2,600ft), wide; the American Falls are 51m (167ft) high and 305m (1,000ft) wide.

Niamey, capital of Niger, Africa, in sw Niger on the River Niger. Located at the junction of two main roads and with good port facilities, it is the country's largest city and its commercial and administrative centre. Manufactures include bricks, food products and cement. Pop. (1975) 130,299.

Nibelungenlied, The, a Middle High German EPIC poem, written by an anonymous Austrian from the Danube region in the 13th century. The greatest achievement of early German literature, it is founded on old Scandinavian legends contained in the *Völsunga Saga* and the *Poetic Edda*, which deals with some of the same characters (Siegfried, Kriemhild, Brünhilde, Gunther, Gudrun and Hagan), but is primarily concerned with their personal rivalries and adventures rather than with events of national interest performed by idealized heroes. The epic was the inspiration for WAGNER's opera cycle *Der Ring des Nibelungen* (1853–74).

Nicaea, one of the chief cities of ancient Bithynia, Asia Minor. Two ecumenical councils were held there, in AD 325 and 787. It was captured by the SELJUK TURKS in 1078, taken by the Crusaders in 1097 and passed to the OTTOMAN TURKS in 1330. It is the site of the modern city of Iznik.

Nicaea, Councils of, two ecumenical councils held in Nicaea (now Iznik, Turkey). The First Council of Nicaea, held in AD 325, was the first ecumenical council in the history of the Christian Church and was convoked by the Emperor Constantine to resolve the problems caused to the Church and the Empire by the emergence of ARIANISM. It promulgated a creed,

Newton's discoveries have guided the thinking of physicists ever since.

New York City; one of the many bridges that carry traffic across the Hudson River.

New Zealand's capital, Auckland, has a harbour built around 67 extinct volcanoes.

Niagra Falls; about 180,000 tonnes of water plunge 57 metres every minute.

Nicander, or Nicandros

Nice; the pavement near the beach is called the Promenade des Anglais.

Saint Nicholas, seen here in a painting by the Italian artist Piero della Francesca.

Tsar Nicholas II was too late instituting reforms demanded by the people.

Ben Nicholson; *Painting 1937,* a formal abstract much influenced by Mondrian.

affirming belief in the divinity of Christ and condemning Arianism. However, the creed generally known as the NICENE CREED does not seem to have been formulated at this council. The Second Council of Nicaea, held in 787, was the seventh ecumenical council and was summoned by the patriach Tarasius to settle the ICONO-CLASTIC CONTROVERSY.

Nicander, or Nicandros (*fl.* 185–135 BC), Greek poet and grammarian. He was also a priest of Apollo at Claros, Ionia. There remain extant of his works two poems on medical themes.

Nicaragua, republic in Central America, lying between Honduras (N) and Costa Rica (S). More than 50% of the land is covered with forests, which yield fibres, gums and various woods. The main economic activity, however, is agriculture; cotton, coffee beans and rice are grown. Nicaragua's industries include sugar refining and textiles. The capital is Managua. Area: approx. 130,000sq km (50,190sq miles). Pop. (1975 est.) 2,155,000. *See* MW p.134.

Nicaragua, lake in SW Nicaragua; largest lake in Central America. The River Tipitapa carries to it the waters of Lake Managua, and the San Juan River drains it into the Caribbean Sea. It occupies part of an area that was ocean until surrounding land rose up and formed the lake. Saltwater fish such as sharks have adapted to the change from saline to fresh water. Area: 8,030sq km (3,100sq miles).

Niccolite, one of the chief ores of NICKEL, consisting of nickel arsenide (NiAs), often also with some cobalt, iron and sulphur; also called kupfernickel. It has hexagonal system crystals generally in columnar masses, rarely as tabular crystals. It is found in vein deposits and is copper-coloured, with an easily tarnished metallic lustre. Hardness 5–5.5: s.g. 7.8.

Nice, city in SE France, on the Mediterranean coast; capital of Alpes-Maritimes département. Nice was founded by Phocaeans from ancient Massilia. In the 13th and 14th centuries it was part of the House of Savoy. France held it from 1792 to 1814, when it was returned to Savoy until 1860. It is a major centre of the French Riviera, and the Carnival of Nice is an annual event. The city is noted for its villas and such boulevards as the Promenade des Anglais. Industries: tourism, olive oil, perfumes, electronics. Pop. (1968) 322,442.

Nicene Creed, statement of Christian faith named after the First Council of NICAEA, 325. It was once believed that the Creed was enlarged at the First Council of Constantinople in 381. It is, therefore, sometimes called the "Niceno-Constantinopolitan Creed". Its exact origin is, however, uncertain. The Nicene Creed defends the orthodox Christian doctrine of the TRINITY against the ARIAN heresy and affirms that Jesus Christ is "the only-begotten Son of God... of one substance with the Father". Its concluding section affirms belief in the Church, BAPTISM, the RESURRECTION of the dead and eternal life. The Nicene Creed is subscribed to by all the major Christian Churches and is widely used by them in their celebrations of the EUCHARIST. *See also* APOSTLE'S CREED; ATHANASIAN CREED.

Nicephorus, name of three Byzantine emperors. Nicephorus I (r.802–11) deposed Empress Irene in 802, asserted the imperial authority over the Church and imposed heavy taxation on his subjects. He was killed fighting the BULGARS. Nicephorus II Phocas (912–69; r.963–69) undertook military expansion against the ARABS and Bulgars, and was murdered at the command of his successor John I in league with Nicephorus' wife. Nicephorus III Botaneiates was emperor from 1078 to 1081. During his reign the empire disintegrated and internal opposition forced his abdication.

Nicholas, Saint, patron saint of children and sailors, particularly in The Netherlands, Greece and the USSR. Traditionally he was Bishop of Myra in Asia Minor in the 4th century and is the subject of many legends. In one he secretly gave gold to three poor girls as their dowry. From this came the custom of giving presents on his feast day, 6 December, a habit in most countries later transferred to Christmas. His name in one Dutch dialect, Sinte Klaas, became Santa Claus.

Nicholas, name of five popes. Nicholas I (r.858–67) was born c.819. He opposed the divorce of LOTHAIR of Lorraine and supported the claim of IGNATIUS to be Bishop of Constantinople, against Photius, who rejected the NICENE CREED. Nicholas II was pope from 1058 to 1061. Nicholas III (r.1277–80) was born Giovanni Gaetano Orsi c.1225. He reduced the influence in Rome of the German king Rudolf I and the Sicilian King Charles I of Anjou, and reformed the papal states. Nicholas IV was pope from 1288 to 1292. Nicholas V (r.1447–55) was born Tommaso Parentucelli in 1397. He negotiated the Concordat of Vienna in 1448 with the Holy Roman Emperor, reasserting papal authority, and founded the Vatican library.

Nicholas, name of two Russian tsars. Nicholas I (1796–1855), a ruler noted for suppressing all political reform, ascended to the throne in 1825 and immediately crushed the DECEMBRIST revolt led by liberal noblemen who sought constitutional monarchy. He gained strategic territory for Russia in wars against Persia (1826–28) and Turkey (1827–29) and put an end to Polish independence in 1830–31. He expanded the secret police, increased censorship and banned political organizations. He died during the CRIMEAN WAR. Nicholas II (1868–1918) was the last tsar of Russia (1894–1917). During his reign, Russia made economic progress but became militarily involved disastrously in war with Japan (1904–05). Revolution broke out in 1905 and he was forced to introduce representative government in the *Duma.* A new revolution broke out in 1917 following corrupt government and military defeats, and he was forced to abdicate. He and his family were probably executed in July, 1918.

Nicholas Nickleby (1839), novel by Charles DICKENS. Complex and melodramatic, the work criticizes 19th-century educational practices through the graphic descriptions of Dotheboys Hall. It also contains many well-drawn characters.

Nicholas of Cusa (1401–64), German philosopher, theologian, cardinal and mathematician. Writing on philosophy, theology, law and science, his work formed major advances in RENAISSANCE mathematics, astronomy and mysticism.

Nicholas of Verdun (*fl.c.* 1150–1210), French enamellist and goldsmith. He was a pioneer in the development of early Gothic champlevé technique, in which compartments are hollowed from a metal base and filled with enamel. His most famous pieces are the altarpiece at Klosterneuburg (1181) and the reliquary of Our Lady at Tournai Cathedral (1205). *See also* HC1 pp.199, 200.

Nicholas Oresme. *See* ORESME, NICHOLAS.

Nichols, Dudley (1895–1960), US film writer whose impressive credits include *The Lost Patrol* (1934), *Stagecoach* (1939) and *Pinky* (1949). He produced and also directed *Mourning Becomes Electra* (1947). He won an Oscar for the script of John FORD's *The Informer* (1935).

Nichols, Peter Richard (1927–), British playwright who wrote many television scripts before his first stage play, *A Day in the Death of Joe Egg,* was produced in 1967. This was followed by further successes with *The National Health* (1969), *Forget-me-not Lane* (1971), *Chez Nous* (1973) and *Privates on Parade* (1977). He also wrote the screenplays for *Catch Us If You Can* (1965) and *Georgy Girl* (1966).

Nicholson, Ben (1894–), British painter, the son of Sir William NICHOLSON and one-time husband of Barbara HEPWORTH; he became a champion of ABSTRACT ART in Britain. He was influenced by CUBISM and Piet MONDRIAN, and from 1933 he developed a geometric abstract style which he expressed in austere carved and painted reliefs, eg *White Relief* (1935). He later produced a series of freely abstracted still-lifes and landscapes which displayed his outstanding draughtsmanship, and returned in the 1960s to reliefs. Nicholson lived in Cornwall from 1940 to 1956 and took up residence in Switzerland in 1958. Like Henry MOORE he contributed greatly to the international status of post-war British art. *See also* HC2 p.205.

Nicholson, Jack (1937–), US film actor. He made his film debut in 1958 but attracted no attention until 1969 when he took a small part in *Easy Rider.* His other films include *Five Easy Pieces* (1970), *Carnal Knowledge* (1971) and *One Flew Over The Cuckoo's Nest* (1976), for which he won an Oscar.

Nicholson, Sir William Newzam Prior (1872–1949), British painter of portraits and landscapes who, with his brother-in-law James PRYDE, helped to develop the poster as an art form, under the pseudonym of "The Beggarstaff Brothers". He also made woodcuts of well known Victorians.

Nickel, metallic element of the first TRANSITION series, discovered in 1751. Its chief ores are pentlandite and NICCOLITE. The main commercial use is in stainless steels and other special alloys. Nickel is also used in coinage and as a hydrogenation catalyst. The metal is ferromagnetic. Properties: at.no. 28; at.wt. 58.71; s.g. 8.90 (25°C); m.p. 1,453°C (2,647°F); b.p. 2,732°C (4,950°F); most common isotope Ni^{58} (67.84%).

Nickel carbonyl, colourless, inflammable, toxic liquid, $Ni(CO)_4$. It is used in the MOND process for the production of pure nickel. Properties: s.g. 1.32 gm/cm^3; m.p. −25°C (−13°F); b.p. 43°C (109.4°F).

Nickel processing, production of nickel for use in various alloys (mainly with iron, copper and chromium) and in its pure form. The main ores are nickel-iron sulphide (pentlandite) and nickel arsenide (niccolite). Refining is complex and involves the differential separation of the various compounds to obtain concentrated nickel sulphide (or arsenide), which is reduced to the pure nickel. Some nickel is refined using the MOND process.

Niclaus, Jack William (1940–), US golfer. He became a professional golfer in 1961 and since then has won five Masters tournaments (1962, 1965, 1966, 1972, 1975); three US Professional Golf Association championships (1963, 1971, 1973); three US Opens (1962, 1967, 1972) and two British Opens (1966, 1970). By the mid-1970s he was said to have won more money than any other golfer.

Nicobars. *See* ANDAMAN AND NICOBAR ISLANDS.

Nicodemus, Pharisee and member of the SANHEDRIN who is mentioned three times in the Gospel according to St JOHN. Nicodemus came to Jesus by night and evoked a discourse about Christian regeneration (John 3: 1–21). He asked the other Pharisees if they judged any man without a hearing (7: 50–52), and he helped JOSEPH OF ARIMATHAEA bury Jesus (19: 39–42).

Nicol, William (1768–1851), British physicist who in 1828 invented an optical prism named after him. A Nicol prism consists of a rectangular crystal of Iceland spar split along a diagonal and rejoined with Canada balsam. Light passing through it is separated at the join into a reflected (ordinary) ray and a transmitted (extraordinary) ray, both polarized in different planes. *See also* POLARIZED LIGHT.

Nicolai, Otto (1810–49), German composer and conductor who in 1842 helped to found the Vienna Philharmonic Orchestra. His opera, *Il Templario* (1840), was successful at the time but his lasting masterpiece was the comic opera *The Merry Wives of Windsor* (1849).

Nicolet, Jean (1598–1642), French explorer in North America. He travelled with Samuel de CHAMPLAIN, and in 1634 began searching for the NORTHWEST PASSAGE. He was the first white man to explore Michigan, Green Bay and the Fox River.

Nicolle, Charles-Jules-Henri (1866–1936), French bacteriologist. He was awarded the 1928 Nobel Prize in physiology and medicine for his discovery (1909) that TYPHUS is transmitted by the body louse. He later distinguished between classical typhus and murine typhus, which is passed on to man by the rat flea. He was head of the Pasteur Institute in Tunis (1902–32), which became a distinguished centre for bacteriological research under his direction.

Nicol prism, optical device used for the production of POLARIZED LIGHT. It is used in the polarizing microscope and also in GEOLOGY to examine thin sections of rock microscopically.

Nicomachean Ethics, work by ARISTOTLE setting forth part of his scientific and philosophical system based on the view that what is good for man is what man, by his nature, is seeking. In the *Ethics* Aristotle searches for the highest perfection of man and finally concludes that it is thought.

Nicosia (Levkosía), capital city of Cyprus in the central part of the island. For 300 years descendants of Guy Lusignan ruled the island, using Nicosia as their capital. They were defeated by the Venetians in 1489 and many ruins of circular walls and fortifications survive from this period. The Ottoman Turks held the city from 1571 to 1878, when it passed to the British. Nicosia continued to be the capital after Cyprus became independent in 1960. It was the scene of serious fighting prior to 1960 and after the Turkish invasion of the island in 1974. Manufactures include cigarettes, textiles and footwear. Pop. (1971 est.) 117,000.

Nicotiana, genus of more than 100 species of plants of the family Solanaceae. *Nicotiana tabacum,* originally a tropical species, is the source of commercial TOBACCO, although *N. rustica,* a shrubby plant native to eastern USA, has a higher NICOTINE content and was used by American Indians. Other species include JASMINE tobacco, *N. alata,* which has tubular flowers that open at dusk and emit a jasmine odour. *See also* NW p.45.

Nicotine, poisonous ALKALOID obtained from the leaves of TOBACCO, used in agriculture as a pesticide and in veterinary medicine to kill external parasites. Nicotine is the principal addictive agent in smoking tobacco.

Nictitating membrane, inner or third eyelid present in many vertebrates such as birds, reptiles and some mammals. A protective fold of transparent or translucent mucous membrane, it can be drawn over the eye.

Niello, method of decorating silver or gold objects with incised designs filled with a black metallic alloy. The technique was used in Roman and Oriental art and became extremely popular in Italy in the 15th century.

Nielsen, Carl (1865–1931), Danish composer, known internationally for his six symphonies. He also composed concertos for violin, flute and clarinet, two operas, a woodwind quintet, four string quartets and many other choral and piano pieces. Although he did not abandon familiar melody and harmony, his originality lies in the spontaneous use of established procedures in novel contexts, eg melodic shape is built around a particular note which, however, may be only transitory and not suggest true key.

Niemeyer, Oscar (1907–), the best-known Brazilian architect. An early advocate of modern architecture in Latin America, he was influenced by LE CORBUSIER, under whom he worked on the Ministry of Education and Health Building in Rio (1936–45). With Lucio COSTA, he designed the Brazilian Pavilion for the New York World Fair in 1938. With his casino, club and church of S. Francisco at Pampulha and a new holiday resort outside Belo Horizonte (1942–43), he came into his own with a completely new approach to architecture: elegant, lyrical, sub-tropical luxury expressed through curving, sculptural and highly expressive forms. In the late 1950s Niemeyer began designing the main public buildings at Brasília, including the governmental centre, a spectacularly expressive exercise in geometrical forms rising from a flat site.

Niemöller, Friedrich Gustav Emil Martin (1892–), German Protestant minister and theologian who turned to theology after having commanded a submarine in WWI. He opposed Hitler's creation of a "German Christian Church", founded the Confessing Church and was arrested in 1938 because of his opposition to the Nazi Party. He was released in 1945 and in 1961 became a President of the World Council of Churches.

Niepce, Joseph Nicéphore (1765–1833), French chemist, inventor of the heliography process. Using camera and lens, he obtained unstable negatives in 1816, but by 1822 had developed the process sufficiently to achieve permanent images. He worked with Louis DAGUERRE from 1829 until his death, after which Daguerre perfected their process, known as daguerreotypy. Niepce also designed an early bicycle. *See also* MM pp.122, 218.

Nierenberg, William Aaron (1919–), US physicist who worked on the MANHATTAN PROJECT and taught at the University of California where he was appointed Director of the Scripps Institution of Oceanography in 1965. He became senior consultant to the president's science adviser in 1976.

Nietzsche, Friedrich Wilhelm (1844–1900), German philosopher and poet. He studied classical philology at Bonn and taught at Basel in 1869. He met Richard WAGNER in 1878 and his treatise *The Birth of Tragedy* (1872) shows Wagner's influence. In 1879 he abandoned philology for philosophy, and celebrated his new notion of the "superman" in *Also sprach Zarathustra* (*Thus Spake Zarathustra*) (1883–91). His aphoristic style was misunderstood but greatly influenced later German thinkers. His later works include *Beyond Good and Evil* (1886), *On the Genealogy of Morals* (1887) and *Ecce Homo* (1888). *See also* HC2 pp.176, 222, 323; MS p.235.

Nieuwland, Julius Arthur (1878–1936), US chemist and Roman Catholic priest, at one time a leading authority on the chemistry of ACETYLENE. In 1920 he discovered a process for synthesizing a rubbery solid from acetylene and, with the du Pont Chemical Co. in 1929, developed the process and made "neoprene", the first synthetic RUBBER to be marketed. *See also* MM pp.54–55.

Niflheim, in Scandinavian mythology, the misty land to which the dead were sent. It existed in the far north and was eternally dark and cold. From the Hvergelmir well there, 12 rivers flowed. In some accounts it was a place where only evil persons went. It is also known as Hel.

Niger, landlocked, independent nation in W Africa. It consists of flat plateaus and plains, arid in the N with some pasture land in the S and W. The River Niger provides the only drainage. Products include millet, groundnuts and hides. Some uranium and tin ore are mined, although there is little manufacturing industry. The capital is Niamey. Area: 1,267,000sq km (489,189sq miles). Pop. (1975 est.) 4,600,000. *See* MW p.135.

Niger, major river in W Africa, which rises in the Fouta Djallon plateau in the SW Republic of Guinea and flows NE through Guinea into the Mali Republic where it forms an extensive inland delta. It then flows in a great curve NE, E and SE across the border into Nigeria and S into the Gulf of Guinea through another vast delta. The river's course is established by the explorers Mungo PARK (1796–97) and Richard and John LANDER (1830). The inland delta in the Mali Republic is part of an irrigation project that has reclaimed 499,780 hectares (1,235,000 acres) since 1930 and the delta at the Gulf of Guinea is a source of oil and palm oil. The Niger has been dammed at various points to create hydroelectric power. Length: 4,183km (2,600 miles).

Nigeria, independent nation of W Africa. Swamp and tropical rain forests make part of the country inaccessible and the rivers Niger, Benue and Cross provide important transport routes. Most of the people depend on farming and the country is almost self-sufficient in food production. Nigeria's economy is, however, based on oil, which accounts for more than 80% of

all exports. The capital is Lagos. Area: 923,768sq km (356,667sq miles). Pop (1975 est.) 62,925,000. *See* MW p.135.

Night blindness, early symptoms of vitamin A deficiency. The VITAMIN is necessary for the production of RHODOPSIN, a retinal substance vital to vision during times of low illumination. Vitamin A is derived from CAROTENE, a constituent of carrots, and so the old belief that eating carrots helps one see in the dark has some basis in truth.

Night-flowering plant, any of numerous plants with flowers that open at night; the blooms are commonly sweet-scented and tubular in shape. The night-blooming CEREUS (Queen of the Night), with pure white flowers, is commonly cultivated.

Night heron, thick-billed, night-feeding bird of the Mediterranean and the Middle East. It has well developed, ornamental head plumes, especially at breeding time. Length: 71cm (28in). Family Ardeidae; genus: *Nycticorax. See also* HERON.

Nightingale, Florence (1820–1910), British nurse, b. Italy. She founded modern NURSING and is best known for activities in the CRIMEAN WAR. In 1854 she took a unit of 38 nurses to care for wounded British soldiers. Called "the Lady with the Lamp", she believed that nursing should continue night and day. In 1860 she founded the Nightingale School and Home for nurses training at St Thomas's Hospital, London. She was widely honoured and in 1907 became the first woman to be awarded the Order of Merit.

Nightingale, migratory Old World songbird of the THRUSH family (Turdidae). The common nightingale of England and Western Europe (*Luscinia megarhynchos*) is ruddy-brown with light grey underparts. Length: approx. 16.5cm (6.5in).

Nightjar, insect-eating, nocturnal bird found worldwide except for New Zealand and some other islands. It has a huge gape and a whirring cry. Length: 27cm (10.5in). *See also* NW p.141.

Nightshade, any of numerous species of flowering plants of the genus *Solanum* and a common name for DEADLY NIGHTSHADE, (*Atropa belladonna*). Also called bittersweet and woody nightshade, the common nightshade (*S. dulcamara*) yields poisonous red berries. Family Solanaceae.

Nihilism (literally "nothing-ism"), revolutionary theory followed by extremists in imperial (pre–1917) Russia. It called for the total destruction of existing society as a precondition to any new order.

Nihon-shoki, written history of Japan compiled in AD 720. It consists of 30 chapters containing legends of the mythical origins of Japan and historical records of the most powerful clans and the Japanese imperial family, and describes the influence of Chinese civilization and of Buddhism on the old Japanese culture.

Niigata, port on NW Honshū, Japan, at the mouth of the River Shinano; capital of Niigata prefecture. The port was opened to foreign trade in 1868, and exports oil, machinery and textiles. Industries: textiles, fishing, machinery, chemicals, paper, shipbuilding. Pop. (1974) 413,061.

Nijinska, Bronislava (1891–1972), Russian dancer and choreographer. She danced with the Maryinski Company in St Petersburg and the BALLETS RUSSES and for that company choreographed *Les Noces* (1923) and *Les Biches* (1924), the most highly regarded of her choreographies. She became famous for her depiction of the follies of contemporary society. *See also* HC2 p.275.

Nijinsky, Vaslav (1890–1950), Russian dancer, often regarded as the greatest male ballet dancer of the 20th century. Dancing with the BALLETS RUSSES, his most noted roles were in *Petrushka, Daphnis and Chloë, Les Sylphides* and *Scheherazade,* all choreographed by FOKINE. From 1912 he choreographed such ballets as *L'Après-midi d'un faune, Jeux* and *Le Sacre du printemps* for DIAGHELIEV. He retired in 1919. *See also* HC2 pp.39, 274–275, 275.

Nijmegen, city in E Netherlands, on the River Waal near the border with West Germany. One of the oldest Dutch cities,

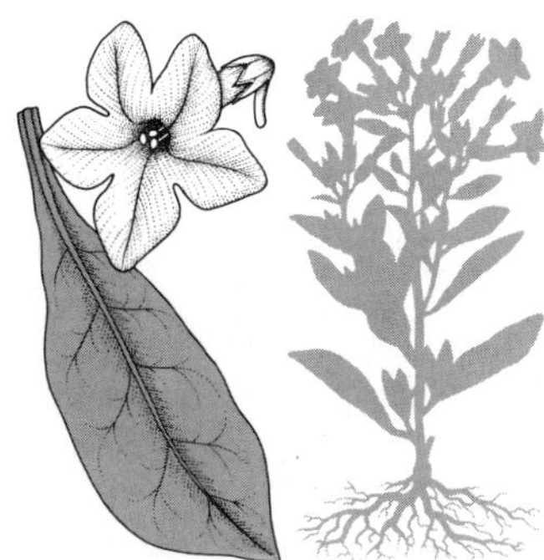

Nicotiana is a family of herbs and shrubs of which tobacco is a member.

Joseph Niepce produced this first photograph in 1826 on a pewter plate.

Nietzche, self-styled immoralist, was opposed to any absolute moral code.

Night herons, silent during the day, commence their squawking cries at sunset.

The Nile is important as a waterway and as a source of irrigation.

Birgit Nilsson performed *Aida* at the New York Metropolitan Opera in 1963.

Chester W. Nimitz, Admiral of the Fleet, is welcomed to New York in 1945.

Marshall Nirenberg illustrates the part of a model of the DNA molecule.

it was founded by the Romans, and became a member of the HANSEATIC LEAGUE. The treaties of Nijmegen (1678–79), ending the DUTCH WARS, were signed there. Industries: machinery, electronic equipment, footwear, paper. Pop. 148,029. *See also* HC1 pp.158, *314–315*.

Nike, in Greek mythology, goddess of victory. Daughter of Pallas, she is associated with the main goddess of Athens, ATHENA; her temple stands on the ACROPOLIS within sight of the PARTHENON.

Nikon (1605–81), Russian priest and political leader, real name Nikita Minin. In 1652 he became head of the Russian Church as Patriarch of Moscow. He reformed Church discipline and introduced a new prayer book in 1855. His reforms created a schism in the Church and he was deposed in 1666.

Nile, river in NE Africa; longest river in the world. The Nile proper is formed at Khartoum by the convergence of the White Nile and the Blue Nile. It flows N from E Africa draining a basin covering 3,348,900sq km (1,293,000sq miles), enters a delta 19km (12 miles) N of Cairo and empties into the Mediterranean Sea. Egypt, Sudan and other African nations depend almost completely on the River Nile as a source of hydroelectric power and irrigation by means of an advanced system of dams, including Aswān High Dam, Gebel Aulia and Sannar. *See also* MM p.*198*.

Nile, Battle of the (1 Aug. 1798), naval action fought between the British fleet, commanded by Horatio NELSON, and the French in the Abukir Bay, N Egypt. NELSON's victory restored British prestige in the Mediterranean and, with the military victory under Sir Ralph ABERCROMBY, cut off NAPOLEON's venture in the Middle East. *See also* HC2 pp.75, 78.

Nilgai, also called bluebuck, large antelope that ranges in wooded areas of the Indian sub-continent; the male is bluegrey and the female red-brown. Both sexes have a short mane and white markings on the belly and throat. The male bears short, curved horns and a tuft of black hair at the throat. Height: to 1.4m (4.6ft). Family Bovidae.

Nilotes, large group of Negroid peoples of the upper River Nile region of the Sudan and Uganda. They are characteristically tall and slender. Nilotic peoples include the Luo of Kenya, the MASAI of Tanzania and Kenya and the Dinka, Shilluk and NUER of the Sudan. There is a wide variety of cultures, ranging from the concept of divine kingship among the Shilluk to the political egalitarianism of the Luo and Nuer. *See also* MS p.33.

Nilotic languages, sub-group of the Sudanic or Nilo-Saharan family of African languages, spoken by peoples of E Africa from Tanzania to the Arab areas of the Sudan. It is divided into three branches – western (Shilluk, NUER), eastern (MASAI, Turkana) and southern (Nandi, Suk).

Nilsson, Birgit (1918–), Swedish soprano who studied in Stockholm with Joseph Hislop. She first came to the public's notice at the Munich Opera from 1954 to 1956, where she sang Brünnhilde in WAGNER's *Die Walküre*; she then became established as the leading Wagnerian soprano of the day. In 1977 she sang the title role in Richard STRAUSS's *Elektra* at Covent Garden, London.

Nimbus, in art, a halo of light, often painted over or behind holy figures. It is usually circular and was transmitted from ancient art to early Christian art in the 4th century.

Nimeiry, Major-General Gaafar. *See* NUMEIRY, MOHAMMED JAAFUR AL.

Nimitz, Chester William (1885–1966), US admiral. He served with the Atlantic fleet's submarine division during WWI. In 1941 he became Commander of the Pacific fleet, a position he held throughout WWII and which gave him a decisive influence on US strategy. He was made chief of naval operations in 1945 and retired from the navy in 1947. *See also* HC2 p.228.

Nimrod, biblical figure, son of CUSH and grandson of NOAH, referred to in Genesis as a "great hunter before the Lord". He founded a kingdom which included Babel,

ERECH (Uruk) and AKKADIA; he probably also built NINEVEH and Calah (NIMRUD).

Nimrud, name of the ancient ASSYRIAN city of Calah, situated in modern Iraq. It was founded in the 13th century BC by Shalmaneser I and was restored and enlarged by Ashurnasirpal II (*c.*883–859 BC). *See also* HC1 pp.*28–29, 33*.

Nin, Anaïs (1903–), US author *b.* France. An early student of JUNG, her novels are all intense psychological studies. They include *The House of aIncest* (1936), *Winter of Artifice* (1939) and *Spy in the House of Love* (1954). She has also published portions of her diaries and correspondence with Henry MILLER.

Nineteen Eighty-Four (1949), novel by George ORWELL about England in a world divided into three constantly warring totalitarian power blocs, where all human rights are suspended. Two lovers seek privacy, but the man is betrayed and broken by Big Brother's Thought Police.

Nineteen Propositions, terms put by parliament to CHARLES I of England on 1 June 1642, requiring that Parliament be given control over the appointment of royal Ministers, the army and the settlement of Church doctrine and organization. Charles' rejection of terms so humiliating to 17th-century kingship led to the ENGLISH CIVIL WAR.

Nineveh, ancient capital of Assyria, on the River Tigris opposite modern Mosul, Iraq. It existed as early as the 18th century BC, but its greatest development occurred during the reigns of SENNACHERIB (704–681 BC) and ASHURBANIPAL (668–*c.*627 BC). When Nineveh fell to Nabopolassar of Babylonia and his allies in 612 BC, the Assyrian Empire fell with it. *See also* HC1 pp.56–57, *56–57*.

Niobe, in Greek mythology, Queen of THEBES, daughter of TANTALUS and wife of Amphion. She boasted of the number of children she had borne (either six or seven of each sex), and jeered at Leto for having had only two children. Leto's offspring, APOLLO and ARTEMIS, sons of Zeus, killed all of Niobe's children. Distraught, Niobe fled from Thebes. ZEUS turned her to stone at Mt Sipylus.

Niobium, once called columbium, metallic element (symbol Nb) of the second transition series, discovered in 1801. Its chief ore is pyrochlore. It is used in special alloy steels. Properties: at.no. 41; at.wt. 92.9064; s.g. 8.47; m.p. 2,468°C (4,474°F); b.p. 4,742°C (8,568°F); most common isotope Nb^{93} (100%).

Nippon, Japanese name for Japan, short for *Nippon-koku*, meaning "Land of the Rising Sun", usually pronounced by the Japanese themselves *Nihon*.

Nippon Steel, international steel and iron industry, based in Japan. In 1976 it employed about 79,000 people throughout the world.

Nirenberg, Marshall Warren (1927–), US biochemist. He was awarded the 1968 Nobel Prize in physiology and medicine, together with Robert W. HOLLEY and Har Gobind KHORANA, for his part in discovering how GENES determine cell function. Nirenberg found the key to the genetic code, and was eventually able to decipher the code triplets for nearly all of the 20 amino acids. *See also* NW pp.28–29; SU pp.152–155.

Nirvana (from the Sanskrit: "blowing out"), conception of salvation in the religions which originated in ancient India: HINDUISM, BUDDHISM and JAINISM. The Hindu conception was expressed in the BHAGAVAD-GITA: "That yogi who is internally happy, internally satisfied and internally illumined, attains extinction in the Supreme Being and becomes that Being". In Buddhism, *nirvana* is the attainment of a transcendent state of enlightenment through the extinction of all desires. In Jainism it means a state of eternal blissful repose. *See also* MOKSHA.

Niš, city in E Yugoslavia, on the River Nišava. The city was taken by the Ottoman Turks in 1386, and passed to Serbia in 1878. It served as the country's capital until 1901. It was occupied by the Germans in WWI and WWII and was taken by Soviet forces in 1944. It is now an important industrial city manufacturing

machinery, leather and tobacco. Pop. (1971) 127,654.

Nīshāpūr, city in NE Iran. It was founded in the 3rd century AD by the SASSANID ruler, Shapur II, and became an important cultural centre with the founding of several colleges by Nizam al-Mulk. Today it is a market centre for the surrounding farming region which produces cotton, fruit and cereals. Local industries include mining for turquoise. Pop. (1972 est.) 40,000.

Nisus, in Greek mythology, name of two characters. Nisus, King of Megara, had an enchanted lock of purple hair which guaranteed his safety. While he was besieged by King MINOS, his daughter Scylla cut off the hair; Nisus was slain and turned into a sea-eagle. In Virgil's *Aeneid*, another Nisus was a Trojan who gave his life for an ARGONAUT, Euryalus.

Niterói, city in SE Brazil on the SE shore of Guanabara Bay opposite Rio de Janeiro. Originally an Indian settlement, it was founded by Portuguese colonists in 1671 and had become a provincial capital by 1835; it became a city in the following year. It is now an important industrial centre, with shipbuilding, metal and textile industries. Pop. (1970) 291,970.

Nitrate, salt or ester of nitric acid, HNO$_3$. Nitrate salts contain the nitrate ion, NO$_3^-$, and some are important naturally occurring compounds, such as saltpetre (potassium nitrate) and Chile saltpetre (sodium nitrate). The salts are used as fertilizers and as a source of nitric acid. Nitric acid esters are covalent organic compounds with the structure R-O-NO$_2$ ("R" represents an organic group such as ethyl in ethyl nitrate).

Nitration, introduction of a nitro group ($-$NO$_2$) to the carbon atom of an organic compound, usually replacing a hydrogen atom. Strong nitric acid or a NITRATE salt with sulphic acid are nitrating agents. The nitro compounds formed are useful chemical intermediates, as in the production of ANILINE from BENZENE, and several are explosives (such as trinitrotoluene, TNT).

Nitrazepam (Mogadon), hypnotic drug with an empirical formula of C$_{15}$H$_{11}$N$_3$O$_3$. A very popular sleeping pill, Mogadon has an advantage over BARBITURATES in that a lethal dose is a high proportion of body weight. It is also used as a minor sedative.

Nitre (salpetre), potassium nitrate, KNO$_3$, known and used since medieval times as an ingredient of GUNPOWDER. It is added to manufactured cigarettes to prevent their going out once lit.

Nitric acid, one of the strongest mineral acids, having the chemical formula HNO$_3$. It is a colourless liquid which fumes slightly in air and soon goes yellow as the result of decomposition into nitrogen dioxide NO$_2$. It attacks most metals resulting in the formation of NITRATES and is a strong oxidising agent. It is mostly made by oxidising ammonia over a CATALYST, and is used for many purposes, including the manufacture of explosives.

Nitriles, organic chemical compounds containing the cyanide group -C≡N, connected to a HYDROCARBON group. They are used in organic synthesis, particularly for the manufacture of acrylic fabrics and synthetic rubber.

Nitro, in chemistry, designates the group $-$NO$_2$.

Nitrobenzene, aromatic nitro compound (C$_6$H$_5$NO$_2$), a pale yellow oily liquid smelling of almonds that is highly toxic. It freezes at 5.7°C (42.3°F) and boils at 211°C (411.6°F). Nitrobenzene is used in the manufacture of ANILINE and other organic chemicals.

Nitrocellulose. *See* CELLULOSE NITRATE.

Nitrogen, common gaseous element (symbol N) of group VA of the periodic table, discovered in 1772. It is the major component of the atmosphere (78% by volume), from which it is extracted by fractionation of liquid air. The main use is in the HABER PROCESS which produces ammonia for fertilizers – nitrogen is essential for plant growth. The element is chemically inert. Properties: at. no. 7; at. wt. 14.0067; m.p. $-$209.86°C ($-$345.75°F); b.p. $-$195.8°C ($-$320.4°F); most common isotope N^{14} (99.76%).

Nitrogen cycle, circulation of nitrogen

through plants and animals in the BIO-SPHERE. Plants obtain nitrogen compounds for producing essential proteins through assimilation. Nitrogen-fixing bacteria in the soil or plant root nodules take free nitrogen from the soil and air to form the nitrogen compounds used by plants to grow. The nitrogen is returned to the soil and air by decay or denitrification, accomplished by denitrifying BACTERIA. *See also* NW p.37; PE pp.170, *170*, 185.

Nitrogen fixation, incorporation of atmospheric NITROGEN into chemicals for use by organisms. Biological nitrogen fixation by micro-organisms provides most of the usable nitrogen. Nitrogen-fixing bacteria live in plant nodules, soil or water. All reduce nitrogen to AMMONIA or ammonium ions that synthesize compounds containing nitrogen or are released into soil or water. *See also* NITROGEN CYCLE.

Nitrogen mustard, any of a group of compounds that are similar in structure to MUSTARD GAS but in which sulphur is replaced by nitrogen. Several of these compounds are used in the treatment of malignant TUMOURS.

Nitroglycerine, oily liquid used in the manufacture of explosives and to relieve the symptoms of ANGINA PECTORIS. It was discovered by Ascanio Sobrero in 1847. *See also* MM pp.242–243.

Nitrous oxide, colourless gas, formula NO, with a pleasant odour, used as an anaesthetic or analgesic during surgical or dental operations. It is also known as "laughing gas" since it produces exhilaration, sometimes accompanied by laughter.

Ni Tsan (1301–74), Chinese landscape painter. Considered one of the Four Great Masters of the Yüan dynasty, he was an influential and widely imitated artist. His landscapes are serene; he used only monochrome ink and popularized the slanting brush stroke technique. *See also* HC1 pp.128–129.

Niue, coral island in the s central Pacific Ocean; a dependency of New Zealand. The port village is Alofi. Exports include copra and bananas. Area: 259sq km (100sq miles) Pop. (1971 est.) 5,100.

Niven, David (1910–), British film actor, *b.* Scotland. In the 1930s he worked in Hollywood, USA, and he was often cast as the typical Englishman. He gave fine performances in *Wuthering Heights* (1939) and *The First of the Few* (1942). His adaptability provided him with a steady succession of roles which included such comedies as *Around the World in Eighty Days* (1956) and *Murder by Death* (1976). He has written two best-selling autobiographies, *The Moon's a Balloon* (1971) and *Bring On The Empty Horses* (1975).

Nixon, Richard Milhous (1913–), US Republican politician and 37th president. He was elected to the House of Representatives in 1946 and 1948 and to the Senate in 1950. He made a name as an anti-Communist on the House Un-American Activities Committee and was nominated as vice-presidential candidate at the Republican convention of 1952. He was Vice-President during EISENHOWER's administrations (1953–60). In 1960 he lost the presidential election to the Democratic candidate John KENNEDY. In 1962 he lost the election for the governorship of California and announced his retirement from politics. In 1968, however, he again won the Republican nomination for president and was elected. He was re-elected in 1972. In 1974 he was forced to resign because of the WATERGATE affair. His administrations were marked by the introduction of prices and incomes controls, the lowering of the voting age to 18 and the ending of military conscription. He was the first US president to pay a state visit to Communist China (1972). In 1973 his administration also ended the US involvement in the VIETNAM WAR.

Niza, Marcos de (c.1495–1558), Franciscan missionary and explorer. He led an expedition into New Mexico in 1539. His belief that the Zuñi civilization contained the fabulous wealth of the Seven Golden Cities of Cibola led to Francisco Vásquez de Coronado's expedition in 1540.

Njal's Saga, long and powerful Icelandic story, written anonymously in the 13th century and relating the violent events of a 50-year blood-feud that had happened 300 years previously. The central character is Njal Thorgeirsson who, at the end, is burned with all his family inside his house; the story has been also translated as *Burnt Njal's Saga*. Pessimistic but with comic episodes, it gives a vivid impression of Icelandic life in the heroic era.

Nkomo, Joshua Rhodes (1917–), Rhodesian politician. From involvement in black union politics he became President of the African National Congress in 1957, living abroad when it was banned in 1959. Returning to Rhodesia in 1960, he became President of the Zimbabwe African Peoples' Union (ZAPU) the following year. He spent much of the 1960s in prison and he was finally released in 1974 to become an executive member of the African National Congress. He led the ANC delegation at the Geneva Conference on Rhodesia (1976). *See also* ZIMBABWE.

Nkrumah, Kwame (1909–72), African political leader. Educated in Africa and the USA, he returned to the Gold Coast and formed the Convention People's Party in 1949. He advocated self-government and was imprisoned by the British in 1950. He was released in 1951, and became Prime Minister in 1952. Under his leadership the Gold Coast became independent in 1957. British Togoland joined it to form the new nation of Ghana and in 1960 Ghana became a republic with Nkrumah as President. Following a build-up of security forces, he made Ghana a one-party state in 1964, but was ousted by the military in 1966 and took refuge in Guinea. *See also* HC2 pp.220–221, *221*.

NKVD. *See* KGB.

No (Noh), form of Japanese symbolic drama during the 12th or 13th to the 15th centuries. It was influenced by ZEN BUDD-HISM in the simplicity of its stage sets and scenery. The plots were taken chiefly from Japanese mythology and poetry; each play portrays but a single incident. With little character or plot development, each No play seeks to convey a moment of experience or insight. The techniques are highly stylized with the use of masks, music, dance and song. A No programme customarily presents a number of plays of contrasting types, and KABUKI drama had its origins partly in No.

Noah, in Genesis, son of Lamech and 10th in descent from ADAM. Chosen by God to be the only man righteous enough to survive the FLOOD, he built the ARK for himself, all his family and a pair of every animal. He and his sons, Ham, Shem and Japheth, and their wives were the ancestors of the human race after the Flood.

Nobel, Alfred Bernhard (1833–96), Swedish chemist, engineer and industrialist. He invented DYNAMITE in 1866, patented a more powerful form of blasting gelatine in 1876 and in 1888 produced ballistite, one of the first NITROGLYCERINE smokeless powders. With the immense fortune he made from the manufacture of explosives and from interests in the BAKU oil fields in Russia, he founded the NOBEL PRIZES, first awarded in 1901.

Nobelium, radioactive metallic element (symbol No) of the actinide group, made in 1957 by bombarding Cm^{244} with C^{13} nuclei. The element has been made only in trace amounts and has not been identified chemically. Properties: at. no. 102, most stable isotope No^{255} (half-life 3 min). *See also* TRANSURANIC ELEMENTS.

Nobel Prizes, awards given each year for outstanding contributions in the fields of physics, chemistry, physiology and medicine, literature and economics, and to world peace. Established in 1901 by the will of Swedish scientist Alfred Bernhard NOBEL, the prizes are awarded annually on 10 December. The economics prize was first awarded in 1969. The winning candidates are selected by four committees: the Swedish Academy of Science for physics, chemistry and economics; the Caroline Institute in Stockholm for physiology and medicine; the Swedish Academy in Stockholm for literature; and the Nobel Committee appointed by the Norwegian parliament, for peace. The Peace Prize is awarded for outstanding work in promoting peace, international brotherhood and disarmament. This prize is presented in Oslo; others are awarded in Stockholm.

Nobili, Leopoldo (1784–1835), Italian physicist who was a pioneer in electrochemistry, a field in which he made important discoveries. He generated electricity using platinum ELECTRODES in an alkaline nitrate ELECTROLYTE and devised the astatic GALVANOMETER to measure the current. He also invented the THERMO-COUPLE and the thermopile (two types of thermometers).

Noble, Dennis (1899–1966), British baritone best known for his performances in operas by Verdi and Puccini. At the suggestion of the conductor Percy Pitt, he auditioned for Covent Garden, London, where in 1924 he made his debut in *Rigoletto*. He appeared subsequently in many of the famous opera houses in Europe and the USA. Although he moved to lighter stage entertainment in the 1940s he returned to the concert world periodically and was widely acclaimed for his performance in Sir William WALTON's *Belshazzar's Feast*.

Noble gases, or inert gases, helium, neon, argon, krypton, xenon and radon–the elements forming the 0 group of the PERIODIC TABLE. They are very unreactive and will combine with only a few other elements (eg krypton, xenon and radon will react with fluorine). This low reactivity is because the noble gases have complete outer electron shells (helium's has two electrons and the rest have eight each), thus offering no VALENCY "hooks". The noble gases were once thought to be completely inert (non-reactive), hence their alternative name. They are also called the "rare gases", although argon and helium are quite common.

Noble metals, unreactive metals found naturally in the native, metallic state and not usually as chemical compounds or ores. They include SILVER, GOLD, PLATINUM, OSMIUM, IRIDIUM, PALLADIUM, RHODIUM and RUTHENIUM. However, COPPER, also sometimes found as the native metal, is not included, whereas MERCURY, found both as the metal and as ores such as cinnabar, often is, so that the noble metals are at best a rather loose assemblage. Gold and platinum are precious metals used for jewellery for their lustre, resistance to tarnishing, ductility and malleability. Osmium, the heaviest of all metals, is alloyed with iridium to make osmiridium, used to tip fountain pen nibs. Platinum and palladium are used in industry as CATALYSTS.

Noctiluca, genus of plant-like flagellate PROTOZOA usually found floating near the surface of the sea. It is phosphorescent and has a round body with one large and one small flagellum at right angles to each other. Width: to 2mm (0.08in). Class Mastigophora.

Noctilucent cloud, cirrus-like cloud visible only at night when directly illuminated by the Sun at between 5°–16° below the horizon. The clouds may be silvery blue or orange-red in colour and occur at heights of 80–85km (50–53 miles). They probably consist of ice particles.

Nocturnal animals, species that hunt and feed by night, and rest or sleep during the day. They include both predators and prey, the balance of nature being maintained as during daylight hours. The term nocturnal is also used for crepuscular species, those which are particularly active at dawn or dusk. Other animals may be irregularly nocturnal in their habits. Truly nocturnal animals usually show marked adaptations to their way of life, large eyes being a common feature. The retina of an OWL's eye contains a great number of rod cells to make use of the smallest amount of light, but contains no cone cells to discriminate colours, which, at low light intensities, become indistinguishable even to the eyes of human beings. BATS generally have poor eyesight, but their special adaptations include a reflected sound ("sonar") navigation system which ensures that they do not collide with obstacles during night flights. In hot deserts nocturnal activity is often an adaptation to climatic conditions.

Nitroglycerine is made by nitrating glycerine with nitric and sulphuric acids.

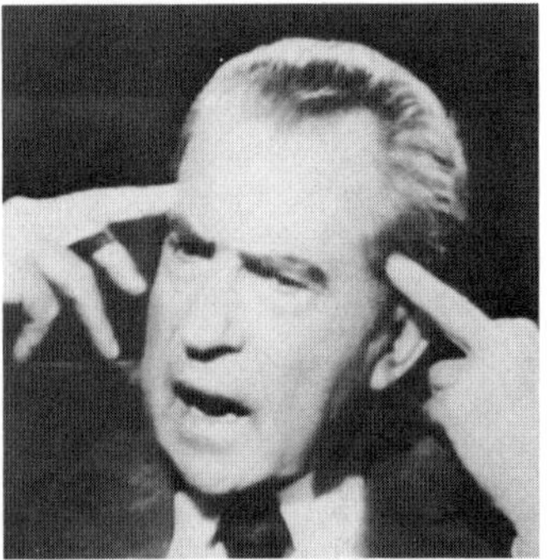

Richard Nixon, photographed during a television interview with David Frost.

Joshua Nkomo is the vociferous black leader in Zimbabwe's national struggle.

Alfred Nobel made a fortune from dynamite and established the prizes named after him.

Nocturne

Noise levels of Concorde being monitored near Kennedy Airport, New York City.

Sydney Nolan in his London studio, photographed with some of his paintings.

Nomads in Mongolia camp outside the feudal stockade of Prince Teh Weng.

Baron Nordenskjöld, in his ship *Vega*, sailed the north east passage in 1881.

Among small rodents and insectivores everywhere, foraging at night offers the advantage of invisibility from predators, except those with well-adapted night vision.

Nocturne, in music, a quiet piece endeavouring to reflect the atmosphere and mood of night-time. First used by John FIELD, the title was later used most by CHOPIN. In painting, the term was given to James McNeill WHISTLER's evocative paintings of the River Thames.

Noddy. See TERN.

Node, in astronomy, either of the two points in the orbit of a planet, comet or satellite etc., at which this orbit intersects the plane of the ECLIPTIC. The ascending node occurs where the body crosses the ecliptic plane from south to north and the descending node from north to south.

Noel-Baker, Philip John (1889–   ), British politician and international statesman. He was a Labour MP (1929–31, 1936–50) and became a junior Minister at the Departments of Air, Commonwealth Relations and Fuel and Power (1946–51). He was also captain of the British Olympic team in 1924. For 40 years' work for international peace he was awarded the Nobel Peace Prize in 1959.

Noguchi, Hideyo (1876–1928), Japanese bacteriologist known for isolating the causative agent of SYPHILIS in the central nervous system. He improved the technique of the WASSERMAN reaction and devised the Wasserman skin test. *See also* MS p.*137*.

Noguchi, Isamu (1904–   ), US abstract sculptor. He was influenced by GIACOMETTI and Alexander CALDER, and his works are marked by great delicacy and employ all sorts of materials. He has constructed much open-air sculpture, including the Japanese garden for the Paris UNESCO building (1956–58).

Noise, any disturbing, often unwanted, sound. Common sources of noise are motor cars and lorries, jet engines, railway trains and factory machinery. Experiments carried out to determine the effects of noise on those workers involved in manufacturing industries show that output seems only to be seriously affected in work which requires mental concentration. Nevertheless, noise is often regarded as a form of acoustic pollution and many noise-abatement pressure groups have been formed to lobby the authorities to produce legislation to reduce noise to a more acceptable level. Noise is usually measured on a decibel scale: conversation is generally about 60dB, and pain starts at about 120dB. *See also* SU pp.81–82, *81*.

Nola, town and episcopal see in Napoli province, Campania region, s Italy. It was an Etruscan city known as Novla before it was taken by the Romans in 313 BC. The town's Gothic cathedral was built between 1395 and 1402. There is a monument to Giordano BRUNO, who was born there, and the Emperor Augustus died in Nola in AD 14. Pop. (1971 prelim.) 26,460.

Nolan, Sidney (1917–   ), Australian painter, the acknowledged leader of the modern Australian movement. His early influences included European painters such as Paul KLEE; later his art was enriched by his admiration for Australian folklore and Aboriginal art and by the untamed grandeur of his native landscape. His famous *Ned Kelly* series (from 1946) expresses with intense power the violent life and death of the legendary outlaw, and the hypnotic atmosphere of the outback. The same almost Surrealist quality pervades all his work on Australian themes, including *Mrs Fraser* and *Glenrowan* (both 1957), as it does on other themes, eg *Leda and the Swan* (1960).

Nolde, Emil (1867–1956), German EXPRESSIONIST painter, a member of Die BRÜCKE, the SEZESSION and the BLAUE REITER. Influenced from *c.*1905 by Van Gogh, Munch and Ensor and after visiting Polynesia and the Far East (1913–14) by primitive art and religion, he painted landscapes and biblical scenes of ferocious impact, which he achieved by the use of intense colour and distortion of drawing. He also worked in etching, wood-cut and lithography.

Noli me tangere ("Do not touch me"), words spoken by JESUS CHRIST after His resurrection to MARY MAGDALENE on her recognition of Him. As quoted the phrase occurs in John 20:17 of the Vulgate (Latin Bible), but it may be used generally as a warning against approach.

Nollekens, Joseph (1737–1823), British sculptor who studied in Rome and returned to London in 1770. His work became fashionable, and among his most famous busts are those of Dr Johnson, the younger Pitt, Charles James Fox and Lawrence Sterne.

Nolo contendere (Latin for "I wish not to contend"), plea in US criminal law which is not itself an admission of guilt but has the same effect on the proceedings as if it were. A defendant pleading it may, however, deny the truth of charges in other proceedings which may arise from the same matter. The phrase has given rise to the popular expression "No contest".

Nomad, member of a wandering group of people who live mainly by hunting or herding and do not have fixed settlements. In contrast to early HUNTERS AND GATHERERS, nomads are often seasonal wanderers who follow the migrations of animals. Nomadism has been regarded as an intermediate state between hunter-gatherer and farming societies, but some peoples have abandoned farming for nomadism. Settled societies have often been invaded by warlike nomads, eg ancient Egypt by the HYKSOS, and, in the Middle Ages, w Asia and Europe by the MONGOLS. Today, nomadic groups survive only in the more remote parts of Africa, Asia and the Arctic. *See also* ARABS; BEDOUINS; ESKIMOS.

Nome, port in w Alaska on Norton Sound, 1,127km (700 miles) w of Fairbanks. It was founded in 1896 when gold was discovered locally. Roald AMUNDSEN arrived there in 1906, having negotiated the NORTHWEST PASSAGE. Nome is a centre for Eskimo handicrafts and has a US Air Force base. Extensive oil reserves exist near by. Industries: tourism and fishing. Pop. (1970) 2,488.

Nominalism, philosophical theory, opposed to realism, that denies the reality of universal concepts. Whereas realists claim that there are universal concepts such as roundness or dog that are referred to by the use of these terms, nominalists argue that such concepts cannot be known, and that the terms refer only to specific qualities common to the particular dogs or circles known. Nominalism was much discussed by the scholastic philosophers of the Middle Ages, especially ROSCELLINUS and William of OCCAM. It was taken up by Thomas HOBBES, by whom it was often seen as a sign of ethical relativism; and again became important in the 20th century, with the work of Nelson Goodman (1906–   ). *See also* MS p.*227*.

Nomura, Kichisaburo (1877–1964), Japanese admiral and diplomat. Commander of the Japanese troops at Shanghai in 1932. He was appointed US ambassador in 1940 and was negotiating at the time of the PEARL HARBOR attack, of which he denied previous knowledge.

Nonalignment. See NEUTRALITY.

Nonconformity, general term for all the sects of British Protestantism which do not conform to the doctrines or discipline of the Church of England. The term arose in the 17th century when the sects, especially the CONGREGATIONALISTS, BAPTISTS and PRESBYTERIANS, but also the QUAKERS, proliferated. Persecuted by the CLARENDON CODE, TEST ACT and Corporation Acts, they were granted freedom of worship in 1689 and civil and political rights in 1828. The METHODISTS and UNITARIANS were added to their ranks in the 18th century.

Non-Euclidean geometry. See GEOMETRY.

Non-figurative art, type of painting, sculpture or decoration that makes no attempt to represent objects from the physical world. ABSTRACT ART is usually equated with non-figurative art.

Non-intervention, term used in international politics to denote a systematic refusal by a state or states to interfere in the affairs of another state.

Nonjurors, clergy of the Church of England and the Scottish Episcopal Church who refused to take the oath of allegiance to King WILLIAM III and Queen MARY II on their accession in 1689, on the grounds that it would break their oath to King JAMES II. They comprised several bishops and about 400 priests, who were all deprived by Act of Parliament. The last nonjuring bishop died in 1805. The Nonjurors believed in the DIVINE RIGHT OF KINGS and NON-RESISTANCE.

Nonnus (*fl.* 5th century AD), Greek poet. His epic *Dionysiaca*, in 48 books, is noted for its inventive style and wealth of mythological lore. Most of it deals with Dionysus's expedition to India.

Nono, Luigi (1924–   ), Italian composer. An early follower of Anton von WEBERN, he gained international recognition with the *Canonic Variations* (1950), an orchestral work based on a note series of Arnold SCHOENBERG. He has composed several political works, including the anti-fascist opera *Intolerance* (1960), which incorporates recorded performance and film sequences and *Song for Vietnam* (1973).

Non-Proliferation Treaty (1968), agreement signed by the USSR, the UK, the USA and many other countries. It requires each participating nation, under the auspices of the UN International Atomic Energy Agency (IAEA), to agree that research and use of nuclear energy for peaceful purposes will not be redirected into military channels. *See also* SALT (STRATEGIC ARMS LIMITATION TALKS).

Non-resistance, doctrine arising in the late 16th century from the theory of the DIVINE RIGHT OF KINGS. It asserted the necessity of passive obedience to civil authority even when the commands of that authority contravened the individual subject's notion of the divine will. In England it was a policy advocated by the Anglican established Church not only for the use of Catholics and Dissenters, but also for Anglicans themselves to justify acquiescence in the Whig settlement of 1689.

Non-specific urethritis. See NSU.

Noot, Jonker Jan van der (*c.*1540–*c.*1595), Dutch poet, the first Dutchman to draw upon French and Italian RENAISSANCE influences. He published his first work, *The Little Wood* (1570), as a political exile in England, and his other writings include Petrarchan sonnets and the incomplete epic *Olympia*.

Nootka, Wakashan-speaking tribe of North American Indians, living along the w coast of Vancouver Island, British Columbia. They were once expert fishermen, and were the only Indians on Canada's Pacific coast to hunt whales. At one time there were approx. 6,000 Nootkas; they now number approx. 1,500.

Noradrenaline, substance secreted by the ADRENAL GLANDS. Chemically it resembles ADRENALINE and has a similar effect in stimulating the AUTONOMIC NERVOUS SYSTEM of the body, although its action is milder. Medically it is used to combat the fall in blood pressure that accompanies shock.

Nordenskjöld, Nils Adolf Erik, Baron (1832–1901), Scandinavian explorer and scientist. A geologist, mineralogist and mapmaker, he made several voyages to SPITSBERGEN, led an exploration of Greenland's inland ice in 1870 and later sailed the NORTH EAST PASSAGE.

Nordic, subdivision of the CAUCASIAN race, characterized by tall stature, fair skin, blond hair, blue eyes and long heads. This physical type predominates in Norway, Sweden and Denmark and occurs throughout N Europe. *See also* MS p.32.

Nordkapp. See NORTH CAPE.

Nördlingen, Battles of, two military engagements during the THIRTY YEARS WAR. The first (1634) was a victory for the Archduke Ferdinand's Hungarian army over the Swedes led by Gustav Horn. In the second (1645), French forces commanded by TURENNE defeated a Bavarian army. *See also* HC1 pp.272–273, *273*, 286–287.

Nore, The, sandbank in the River Thames estuary, off Sheerness, Kent; marked by a lightship established in 1732, the first of its kind to be used in English waters. The Nore is also the name given to the area as a whole, which was much used as an anchorage by naval vessels in the 17th and 18th centuries. In 1797 sailors at the Nore mutinied against their harsh conditions; the mutiny was put down and its leader, Richard Parker, was hanged from the yardarm of his ship.

Norepinephrine. *See* NORADRENALINE.

Norfolk, Dukes of. *See* HOWARD (family).

Norfolk, county of E England. The land is mostly low-lying and is used mainly for agriculture. Its coastline is, for the most part, sandy. The region is drained by the rivers Waveney, Yare, Bure and Ouse, and also by the Broads – a lake and marsh area caused by medieval peat cuttings and a later rise in the sea-level. In the Middle Ages, Norfolk was a centre for the wool industry. Today the county produces cereals and root vegetables, and poultry farming is also important. Norwich, a shoe-manufacturing city, and King's Lynn are the main administrative centres. Area 5,353sq km (2,067sq miles).

Norfolk Broads. *See* BROADS, NORFOLK.

Norfolk Island, territory of Australia in the SW Pacific Ocean, approx. 1,450km (900 miles) E of Australia. Discovered in 1774 by Capt. James Cook, it became a British penal colony from 1788 to 1855. The island is noted for its beautiful pine forests. The chief economic activities are agriculture and tourism. Area: 34sq km (13sq miles). Pop. (1970 est.) 1,380.

Norfolk Island pine, evergreen pine tree (family Araucariaceae) native to the South Pacific, where it grows to 60m (200ft) in height. Branches grow in annual tiers of four to seven and bear bright green needles up to 1.3cm (0.5in) long. The tree is propagated by seeds or cuttings of tip growth. It is widely grown as a house plant in its sapling stage. Species *Araucaria excelsia*. *See also* ARAUCARIA.

Noricum, historical region s of the River Danube, now in Austria and West Germany. It became part of the Roman Empire under an Illyrian proconsul in *c*.15 BC. Its capital was at Vindobona, the modern city of Vienna.

Norma (1831), two-act tragic opera by Vincenzo BELLINI with a libretto by Felice Romani, first performed at LA SCALA, Milan. Set in ancient Gaul, the tragedy involves the attempt of Norma, a Druidic priestess, to regain the love of Pollione, a Roman proconsul, which ends with the death of both. The arias are written for the BEL CANTO style of singing and require considerable technical virtuosity. *See also* HC2 p.120.

Norman, Frank (1931–), British writer whose reputation was established with the musical he wrote for THEATRE WORKSHOP with Lionel BART, *Fings Ain't Wot They Used T'Be* (1959).

Norman architecture, ROMANESQUE architectural style of the Normans in England, N France and S Italy. The main structures were massive castles, churches and abbeys built of stone and free of intricate ornamentation. The rounded arch was widely used. Churches were cruciform in plan and were often topped by one or more square towers.

Norman Conquest, the invasion of England by WILLIAM, Duke of Normandy, who defeated King HAROLD at HASTINGS in 1066 and was crowned King of England. There was considerable resistance to William, but the chief Anglo-Saxon lords had submitted by 1071. William strengthened his position by granting English lands to his Norman barons and establishing Norman feudalism. The Church was also subjected to feudalism and Normanization; William resisted papal supremacy. Norman influence also became dominant in architecture, literature, language, law and the art of warfare. *See also* HC1 pp.167–169, *167–169*.

Normandy, region and former province of NW France, bounded historically by Picardy (NE), Île-de-France (E), Maine (S), Brittany (SW) and the English Channel (W and N). It now includes the départements of Manche, Calvados, Eure, Seine-Maritime and Orne. Part of the Roman province of Gaul, it was absorbed into the Frankish kingdom of NEUSTRIA in the 6th century. Invaded by Norsemen in the mid-9th century, it was the seat of Duke William of Normandy who invaded England in 1066. It was conquered by the French in 1202 and again in 1450. During WWII Normandy was the scene of the Allied invasion (June 1944) of German-occupied France. Forests, flat farmlands, and rolling hills characterize the land; the economy is based on livestock rearing and fishing.

Norman French, dialect of French spoken by the Normans at the time of the conquest of England (1066). Although co-existing with contemporary Anglo-Saxon, it was the language of the Anglo-Norman court and of literature in England for several centuries. It died out around the end of the 14th century.

Normans, warrior people of Scandinavian origin who settled in NW France. In 911, Charles the Simple gave ROLLO, the leader of bands of VIKING pirates, formal control of a coastal area which he and his descendants expanded to form the Duchy of Normandy. The Normans adapted Christianity and feudalism to their needs while still retaining their prowess in war; by 1050, groups of Normans had conquered Sicily, which became an important Mediterranean power under ROGER II, and in 1066 Duke WILLIAM THE CONQUEROR led a Norman army in the conquest of England. Norman knights played an important part in the 1st CRUSADE, and Anglo-Norman families conquered Wales and Ireland for the English crown. The Normans soon became absorbed into the societies they had conquered; Normandy itself was finally taken by the French in the 15th century.

Norms, social, standards of behaviour, either specified or understood, to which members of a social group are expected to conform. Social norms are enforced by the group through both positive and negative sanctions (rewards and punishments). They are in general observed because they are considered functional and moral.

Norns, in Germanic mythology, three maidens who spun or wove the fate of men, akin to the Greek Fates. Their names were Urth (Past), Verthandi (Present), and Skuld (Future). The Norns were depicted as living by Yggdrasill, the world tree, and were linked with both good and evil.

Norodom Sihanouk (1922–), Cambodian political figure; King (1941–55); Prime Minister (1955–60); Head of State (1960–70). Following a right-wing take over of the government in 1970, he established a government in exile in Peking. He returned to Cambodia as head of state with the KHMER ROUGE in 1975 but resigned in April 1976.

Norrie, Lord (1893–1977), British soldier and politician. He served in both World Wars and after governing South Australia (1944–52) he was Governor-General of New Zealand (1952–57).

Norrish, Ronald George Wreyforth (1897–), British chemist who shared the 1967 Nobel Prize in chemistry with Manfred EIGEN and George PORTER for research into very fast chemical reactions. Between 1949 and 1955 Norrish and Porter developed the technique of irradiating a gaseous system at equilibrium with extremely short sparks of high intensity, causing a momentary disturbance of the equilibrium. They measured the time taken for equilibrium to be regained and in this way could study chemical changes occurring in billionths of a second. Although of primarily theoretical importance, this work has potential applications in such areas as the hybridization of large molecules and ultrasonic communications.

Norrköping, port in SE Sweden, at the head of Bråviken, an inlet of the BALTIC SEA. Founded in the 14th century, the city was burned in 1719 by the Russians during the NORTHERN WAR. Industries: furniture, paper, food processing, electrical goods. Pop. 95,851.

Norsemen (Northmen). *See* VIKINGS.

North, Frederick, Lord North, 2nd Earl of Guilford (1732–92), British political figure. Chancellor of the Exchequer from 1767 to 1770 and Prime Minister from 1770 to 1782, he followed the policies of GEORGE III that led ultimately to the loss of the American colonies. He resigned in 1782, but served as Secretary of State the following year with Charles James FOX. *See also* HC2 pp.64, 69.

North, one of the four cardinal points of the compass, opposite SOUTH, the two together defining the axis about which the Earth rotates. The term north is also used to denote one of the poles of a magnet.

North Africa, term applied to the region of the N African continent from the Mediterranean Sea to the Sahara Desert. It includes Morocco, Algeria, Tunisia, Libya and, usually, Egypt. The region is inhabited chiefly by Muslims; most settlements are confined to the coast and the lower valley of the River Nile. Economic and cultural ties with Europe have been pronounced throughout the region's history. During the 19th century much of the area was French or Italian territory. During WWII the Allies defeated German and Italian troops in North Africa. Agriculture in the region depends on irrigation. Mineral deposits include oil and iron ore. Area: approx. 5,570,000sq km (2,140,000sq miles).

North America. *See* AMERICA.

North America Nebula (NGC 7000), bright emission nebula in the constellation Cygnus. Its name derives from its shape, which resembles a map of North America. *See also* SU p.245.

Northampton, town in central England, on the Nene River, 97km (60 miles) NW of London; county town of Northamptonshire. The town has a Norman castle and one of the only four round churches in England. Industries: footwear, leather goods, engineering, motor accessories. Pop. (1973) 128,290.

Northamptonshire, county in central England. The land is undulating and is drained by the Welland and Nene rivers. Much of the land is devoted to pasture, wheat growing and forestry. There are extensive iron ore deposits and the iron and steel industry is important. Northamptonshire is also noted for the manufacture of footwear. NORTHAMPTON is the county town. Area: 2,367sq km (914sq miles). Pop. 505,100.

Northanger Abbey (1817), novel by Jane AUSTEN, completed in 1803. Catherine Morland, an attractive young girl under the influence of Mrs Radcliffe's *Mysteries of Udolpho*, one of the most famous of the so-called "horror school" of English Romanticism, falls in love with Henry Tilney, the son of a general. At the invitation of his father, who mistakenly supposes her to be rich, she visits them at Northanger Abbey where, due to her romantic imagination, she sees mystery and horror everywhere. Jane Austen used this novel to satirize the Radcliffe school of GOTHIC fiction.

North Atlantic Drift, ocean current, also called North Atlantic Current, which extends from SE of the GRAND BANK, Newfoundland, to the NORWEGIAN SEA. It is a warm current which, when mixed with the cold Arctic waters, produces the excellent fishing grounds of the northern European seas.

North Atlantic Treaty Organization. *See* NATO.

Northavon, county district in N and E AVON, England, created in 1974 under the Local Government Act (1972). Area: 461sq km (178sq miles). Pop. (1974 est.) 112,800.

North Borneo, former British protectorate (1888–1963) on the island of Borneo. The region became a state of Malaysia in 1963, when it was called SABAH.

North Briton, The, British newspaper begun in 1762 by John WILKES. It was a stridently anti-government paper, particularly hostile to the court and the Prime Minister, the Earl of BUTE. The seizure of issue no.45, which attacked George III's remarks in the speech from the throne on the PEACE OF PARIS in 1763, led to the prolonged Wilkes affair. *See also* HC2 p.69.

North Cape (Nordkapp), promontory at the N tip of Magerøy island in the far N of Norway, considered to be the most northerly point of Europe. The Cape lies between the BARENTS SEA (E) and the NORWEGIAN SEA (W).

Norfolk Island pines were discovered during James Cook's second voyage.

Norma; set and costume design for the forest of the druids in Bellini's opera.

Norman architecture is characterized by a rounded arch on short heavy columns.

North Cape, at Norway's northernmost tip, is a desolate promontory rising 330m.

North Carolina

North Carolina; the entrance to the Wright Brothers' Memorial at Kill Devil Hill.

Lord Northcliffe believed that large-circulation newspapers wield political power.

Northern Ireland; Lock Veagh, remote from the cities of war-torn Ulster.

North Sea oil should supply Britain's energy needs until the end of the century.

North Carolina, state in E USA, on the Atlantic coast. The coastal plain is swampy near the shore and low-lying. Its W edge rises to rolling hills, and further W are the Blue Ridge and Great Smoky Mts. North Carolina is the leading producer of tobacco in the USA. Other important farm products are corn, soya beans, peanuts, pigs, chickens and dairy produce. Textiles and furniture manufacture are the leading industries. Others include timber, fishing, tourism, electrical machinery and chemicals. Mineral resources include phosphate, felspar, mica and kaolin. The principal cities are Raleigh, the state capital, Charlotte, Greensboro and Winston-Salem. The first English colony in North America was founded on Roanoke Island, just off the coast, in 1585, but this was later abandoned. Permanent settlers moved into the region from Virginia in the 1650s. North Carolina was the first colony to strive for independence from Britain. Before the AMERICAN CIVIL WAR opinions in the state were sharply divided and North Carolina was the last to secede, which it did in May 1861, after fighting had already begun. Area: 136,523sq km (52,712sq miles). Pop. (1975 est.) 5,451,000. *See also* MW p.175.

North Channel, strait between NE Ireland and SW Scotland, separating the Irish Sea from the Atlantic Ocean. Width (max): 23km (14 miles); length: approx. 120km (75 miles).

Northcliffe, Alfred Charles William Harmsworth, Lord (1865–1922), British journalist. He started as a freelance contributor to popular periodicals and gradually formed the world's largest periodical combine, the Amalgamated Press. He bought the London *Evening News* in 1894 and founded the *Daily Mail* (1896) and the *Daily Mirror* (1903). In 1908 he took over and revived the ailing *Times*. He was made Viscount in 1917.

North Dakota, state in N Central USA, on the Canadian border. Situated in the geographical centre of North America, North Dakota experiences very cold winters but only warm summers. The region is generally low-lying and is drained by the Missouri and Red rivers. The state is overwhelmingly rural, and wheat, barley, rye and oats are the chief crops. Cattle rearing is, however, the most important economic activity. Oil is the principal mineral resource. There are also deposits of lignite and natural gas. Food processing is the major industry. The chief cities are Bismarck (the state capital), Fargo, Grand Forks and Minot.

French explorers first visited the region in 1738. The USA acquired the W half of the area from France in the LOUISIANA PURCHASE (1803) and the rest from Britain in 1818 when the boundary with Canada was fixed. Dakota Territory was organized in 1861 and divided into North and South Dakota in 1889, when both were admitted to the Union. The Missouri was a major route for pioneers going W, and the construction of the railways led to further development of the state. Area: 183,022sq km (70,665sq miles). Pop. (1975 est.) 635,000. *See also* MW p.175.

North Downs, range of chalk hills in SE England, stretching W from the coast at Dover and facing the South Downs across the Weald. They are drained by the Wey, Darent, Medway and Stour rivers. The highest point is Leith Hill, 294m (965ft).

North-East Derbyshire, county district in N central DERBYSHIRE, England, created in 1974 under the Local Government Act (1972). Area: 277sq km (107sq miles). Pop. (1974 est.) 91,900.

North East Frontier Agency, Indian territory renamed Arunachal Pradesh ("Dawn's Province") in 1972, in the Himalayas on the India-China border. Once a district of Assam, it is administered separately within the Indian Union. Its population is predominantly Mongoloid. In 1962 it was invaded by the Chinese, who withdrew the next year, returning it to the Indians. Area: 81,426sq km (31,438sq miles).

Northeast Passage, now known as the Bering Strait, route for ships sailing between the Atlantic and Pacific oceans along the N coasts of Europe and Asia. The route was discovered by Vitus J. BERING in 1728 and confirmed as being the Northeast Passage by Capt. James COOK in 1778. A successful navigation of the passage was accomplished in 1878–79 by Baron A. E. NORDENSKJÖLD in the *Vega*. Numerous expeditions had previously sought such a route to the riches and trading opportunities of the East.

Northern Cross (Cygnus or The Swan), northern constellation between Pegasus and Lyra. It contains several bright stars including DENEB (Alpha) of magnitude 1.3 and one of the three components of the Summer Triangle, and ALBIREO (Beta) which is a DOUBLE STAR. Chi is a long-period VARIABLE STAR with a period of 407 days. The MILKY WAY is particularly bright in this region and there are several stellar CLUSTERS. *See also* SU pp.*236, 259, 259.*

Northern Ireland, part of Great Britain, occupying the NE of Ireland. It is composed of fertile plains surrounded by low mountains. There are several large lakes, and of its rivers the principal ones are the Royle, Lagan and Bann. Over 80% of the land is farmed; the chief crops are potatoes and barley. Industry is becoming increasingly important. The country has been disrupted by riots since 1968 and British troops have been stationed in Northern Ireland since 1969. The capital is Belfast. Area: 14,121sq km (5,452sq miles). Pop. (1975 est.) 1,537,000. *See also* MW p.136.

Northern Rebellion (1569–70), revolt against ELIZABETH I of England, led by the Earls of Northumberland and Westmorland. They sought a return to Roman Catholicism and the removal of the Queen's counsellors. They also hoped to release MARY, QUEEN OF SCOTS from imprisonment at Tutbury. Their forces marched S as far as Selby (November 1569), but dispersed in December when opposed by royalist forces under the Earl of Sussex. A second revolt in January 1570 also failed. About 500 rebels were executed.

Northern Rhodesia. *See* ZAMBIA.

Northern Territory, territory in N Australia bounded by the Timor and Arafura seas (N), and the states of Western Australia (W), South Australia (S) and Queensland (E). Darwin is the capital and only seaport. The first settlement in 1824 failed and there was no further attempt at colonization until 1869. It became part of Australia in its own right in 1911, was divided into two regions in 1927 but was reunited in 1931. Today there is little farming but some government-aided stock-breeding. The territory's mineral resources are being exploited; manganese ore, bauxite and iron are mined. Area: 1,347,525sq km (520,280sq miles). Pop. 71,400.

Northern War (1700–21), conflict between Sweden and Russia, with her allies Saxony-Poland and Denmark-Norway, arising chiefly from the desire of the neighbours of Sweden to break Swedish supremacy in the Baltic area. Charles XII of Sweden forced Poland to capitulate in 1706, then attacked Russia (1707). Relying on support from Ivan Mazepa, a Cossack leader, the Swedish forces penetrated the Ukraine but were destroyed at the battle of Poltava (1709) when this support was not forthcoming. The Swedish navy was defeated by Russia in 1714 and Russian armies invaded Sweden. Peace was restored by the treaties of Stockholm between Sweden, Poland and Denmark and of Nystadt, as a result of which Russia gained the Baltic provinces of Estonia, Livonia and part of Karelia, and emerged as a major power.

North Frisian Islands, group of Danish and German islands in the North Sea, off the coast of Schleswig-Holstein, West Germany, and S Jutland, Denmark. The chief islands are Sylt, Föhr, Nordstrand, Pellworm and Amrum, belonging to Germany, and Rømø, Fanø and Manø belonging to Denmark.

North Hertfordshire, county district in Hertfordshire, England, created in 1974 under the Local Government Act (1972). Area: 374sq km (144sq miles). Pop. (1974 est.) 103,700.

North Holland (Noord Holland), province of W Netherlands, between IJsselmeer (E) and the North Sea (W). A dairy-farming region, North Holland is famous for its tulips. Tourism is also important. The capital is Haarlem. Area: 2,911sq km (1,124sq miles). Pop. (1970 est.) 2,244,456.

North Island, smaller of the two islands that comprise New Zealand; separated from SOUTH ISLAND by Cook Strait. Its chief cities are Wellington, Auckland and Hamilton. Most of New Zealand's dairy produce comes from North Island which has four major livestock areas: the narrow northerly peninsula, the Waikato basin, the Taranaki plains and the Bay of Plenty. Other industries include wood pulp and paper at Kaingaroa, mining for iron sand and fishing. Area: 114,729sq km (44,297sq miles). Pop. (1971) 2,051,363. *See* MW p.131.

North Korea. *See* KOREA, NORTH.

Northland, peninsula in North Island, New Zealand, extending NW from Auckland for 320km (200 miles). The area is under-developed and agriculture is the main industry. The only major town is Whangarei. Area:36,713sq km (14,175sq miles). Pop. (1971 est.) 96,000.

North Pacific Current, broad, slow, cyclical current of the N Pacific Ocean. It moves in a clockwise motion from the North Equatorial Current turns N by Japan and E in the N Pacific; it forms the California Current. *See also* PE p.79.

North Point, cape at the N end of W Prince Edward Island, SE Canada, on the Gulf of St Lawrence.

North Pole, northern end of the Earth's axis. Geographic north lies at 90° latitude, 0° longitude; magnetic north is at 76°20′N, 101°ς (1970). The first attempts to reach the North Pole were made in the 17th century but it was not until 1909 that the US explorers Robert E Peary and Matthew Henson were successful.

North Riding, former administrative district of Yorkshire, N England; in 1974 it was divided between the new counties of NORTH YORKSHIRE and CLEVELAND. The administrative centre was Northallerton. Area: 5,571sq km (2,127sq miles). Pop. (1971) 725,658.

Northrop, John Howard (1891–   ), US biochemist who shared the 1946 Nobel Prize in chemistry with James SUMNER and Wendell STANLEY for their work on ENZYMES. In 1930 he crystallized the enzyme PEPSIN and demonstrated that it was a PROTEIN, thus resolving a long-standing dispute concerning the chemical nature of enzymes.

North Sea, part of the Atlantic Ocean, about 960km (600 miles) long and 560km (350 miles) wide, extending between the European continent (S and E) and Britain (W). In the S it is connected to the English Channel by the Strait of Dover. The cod and herring fishing grounds of the North Sea have always been of great value. Another important resource has been tapped recently with the exploitation of oil found under the sea floor in the 1970s. Area: 574,980sq km (222,000sq miles). *See also* PE p.*139.*

North Sea Canal, waterway in The Netherlands connecting Amsterdam with the North Sea. Built during 1865–76 to take ocean-going vessels, it renewed the importance of Amsterdam as a commercial port. Length: 27km (17 miles).

North Star, or Polaris, second magnitude star in the constellation of Ursa Minor, the LITTLE BEAR. It lies within 1° of the north celestial pole and is easily located via the "pointers" in the constellation of Ursa Major, the GREAT BEAR or the Plough. The North Star is of great navigational importance. *See also* SU pp.*226, 256, 256–257, 259, 264, 265.*

Northstead, Manor of. *See* CHILTERN HUNDREDS.

North Tyneside, county district in E central Tyne and Wear, England, created in 1974 under the Local Government Act (1972). Area: 84sq km (32sq miles). Pop. (1974 est.) 205,900.

North Uist, island of the Outer Hebrides off the NW coast of Scotland, in the Western Isles Region. Industries include cattle and sheep rearing, weaving, lobster fishing and seaweed processing. Pop. (1971) 1,807.

Northumberland, John Dudley, Duke of (1502–53), English statesman. He joined the regency council formed during EDWARD VI's minority and on the king's death proclaimed his daughter-in-law, Lady Jane GREY, queen. He was executed by MARY I.

Northumberland, Earls of. See PERCY FAMILY.

Northumberland, county in N England on the border with Scotland. The land slopes down from the Cheviot Hills in the NW, and the region is drained by the Tyne, Tweed, Blythe and Coquet rivers. The county is largely rural, the chief farming activities being cattle and sheep rearing. Coal is mined in the S. Area: 5,033sq km (1,943sq miles). Pop. 287,300.

Northumberland Islands, coral island group off the E coast of Queensland, Australia, between the Great Barrier Reef and Shoalwater Bay. The group is made up of approx. 40 islands extending for 129km (80 miles). The largest is Prudhoe Isle. The islands include several tourist resorts.

Northumberland Strait, arm of the St Lawrence Gulf, separating Prince Edward Island and Nova Scotia and New Brunswick on the Canadian mainland. Length: 290km (180 miles). Width: 19–48km (12–30 miles).

Northumbria, kingdom of, Anglo-Saxon kingdom in N England and S Scotland, formed by the union of BERNICIA and DEIRA in the early 7th century AD. It was the supreme English kingdom in the 7th century. The decision in 664 of its king Oswy to support Roman Christianity rather than Celtic was decisive for the English Church. The arts also flourished at this time; BEDE wrote and the LINDISFARNE GOSPELS were illustrated in Northumbria in the 8th century. The kingdom declined with Danish attacks from 866, and it became an earldom in 944. See also HC1 pp.139, *139*, 154, 164.

North Vietnam. See VIETNAM.

North West Company, Canadian fur trading syndicate, organized by eastern fur traders in 1783 to minimize the disadvantages of competition. It became a rival, however, of the HUDSON'S BAY COMPANY. It sent its traders, called the Nor'Westers, deep into the wilderness, itself assuming the cost of transporting furs to Montreal along the St Lawrence, and extending its trading domain to the Arctic and Pacific Oceans. The trading posts of Fort William (founded 1805) and Astoria (founded 1811) were threatened by the Red River settlement (1812), leading to virtual warfare with the Hudson's Bay Company. Both companies merged in 1821. See also HC2 pp.45–46, *45–46*, 136.

North-West Frontier Province, province in NW Pakistan bounded by Afghanistan (N and W); the capital is Peshawar. Located near the Khyber Pass, it is a region of high mountains divided by fertile valleys. The economy is based on agriculture: wheat, maize, barley and tobacco are grown. There is some exploitation of mineral resources and a limited number of manufacturing industries including food processing and paper making. Area: 103,696sq km (40,037sq miles). Pop. (1970 est.) 7,600,000.

Northwest Passage, sea route around northern Canada and Alaska between the Atlantic and Pacific oceans. Martin FROBISHER first attempted to sail through the Passage in 1576 and was followed by many famous explorers including Sir Humphrey GILBERT who died on his attempt in 1583. Roald AMUNDSEN made the first complete navigation (1903–06), through Lancaster Sound. A Royal Canadian Mounted Police schooner crossed in one season (1944). The route was not regarded as commercially feasible until the giant US ice-breaker the *Manhattan*, equipped with computer-controlled navigational equipment, crashed through the passage in 1969, breaking in excess of 650 miles of surface ice in ten days. It is now an important commercial area because of its huge reserves of oil.

Northwest Territories, province in N Canada, including more than one-third of the country. It is made up of mainland Canada N of latitude 60°N and hundreds of islands in the Arctic Archipelago, the largest of which are Baffin and Victoria islands. The region is divided into three districts: Mackenzie, on the NW mainland; Keewatin, on the NE mainland; and Franklin, which includes the Arctic islands. Much of the N and E of the province is tundra, inhabited by Eskimos and Indians. Most economic development has occurred in Mackenzie district which has large tracts of softwoods and rich mineral deposits that include lead, copper, zinc, gold, uranium and oil. Fishing is also important in the Great Slave Lake. Yellowknife is the capital.

The HUDSON'S BAY COMPANY acquired a large area of NW Canada W of Hudson Bay under charter from Charles II in 1670. In 1869 the Canadian government bought the land from the company. Some of the land was added to the provinces of QUEBEC and ONTARIO. MANITOBA (1870), ALBERTA (1905) and SASKATCHEWAN (1905) were also established in the area. The present boundaries of the Northwest Territories were set in 1912. Area: 3,379,000sq km (1,304,903sq miles). Pop. (1971) 34,807. *See also* MW p.45.

Northwest Territory, region of the Ohio River E of the Mississippi River and S and W of the Great Lakes. It became the first US territory in 1783 by the Treaty of Paris; the states of Indiana, Ohio, Illinois, Michigan, Wisconsin and a part of Minnesota were later created from it. *See also* HC2 p.148.

North Wiltshire, county district in NW WILTSHIRE, England, created in 1974 under the Local Government Act (1972). Area: 778sq km (301sq miles). Pop. (1974 est.) 100,800.

North York, one of the five boroughs of Toronto, S Ontario, Canada, on the Humber River. Since WWII controlled industrial and residential development has taken place, and there are many parks and recreational areas. It includes York University (1959). Pop. 504,150.

North Yorkshire, county in N England, formed in 1974 from parts of the former North, East and West Ridings of YORKSHIRE. The administrative centre is York. Area: 8,309sq km (3,208sq miles). Pop. 768,500.

Norton Sound, inlet of the Bering Sea on the W Alaska coast between Seward Peninsula and the mouth of the Yukon River; it was discovered in 1778 by Capt. James Cook. Length: approx. 322km (200 miles); width: (max.) 150km (90 miles).

Norway, independent nation occupying the western and most northerly parts of the Scandinavian peninsula. The country is generally mountainous; its arable land is restricted to the heads of the fiords and the plains in the E. A highly industrialized country, Norway has abundant natural resources in the way of fast-flowing rivers for hydroelectricity, and offshore oil; the fishing and timber industries are also of great importance. Iron and steel are manufactured. The capital is Oslo. Area: 323,886sq km (125,053sq miles). Pop. (1976 est.) 4,027,000. *See also* MW p.138.

Norwegian current, NE continuation of the NORTH ATLANTIC DRIFT flowing N along the Norwegian coast into the Barents Sea. The warmth of the current running through the Norwegian Sea keeps that sea generally ice-free.

Norwegian elkhound, N European hunting dog that originated in W Norway. It has a wedge-shaped head and small, high-set ears. The square body is set on straight, medium-length legs and the tail is high-set and tightly curled. The dense, smooth coat is usually shades of grey. Height: 52cm (20.5in) at shoulder.

Norwegians, people of Scandinavia who were prominent in the VIKING raids from c.800 to c.1050. The Norwegian and Danish Vikings raided to the West and by the end of the ninth century had settled much of Britain. The Norwegians were prominent in the NW and also reached the

Shetlands, Hebrides, Orkneys and Faroes and Iceland. Norwegian expeditions reached Greenland c.986 and the coast of N America c.1000. Their best-documented settlement was of Iceland which by 1100 had a population of over 70,000. The last attempt to extend power was Harald Hardrada's unsuccessful invasion of England in 1066.

Norwegian Sea, part of the extreme NE Atlantic Ocean, verging on the Arctic Ocean and lying between Norway (E) and Greenland (W). The part along the coast of Greenland is sometimes called the Greenland Sea.

Norwich, city in Norfolk, E England. It has been an important market town since the 11th century and has many fine churches dating from the medieval period; St Giles Hospital dates from the 13th century, and it has a grammar school founded by EDWARD VI. Its industries include textiles, machinery and footwear. Pop. (1971) 121,686.

Norwich School of Artists, English school of landscape painting roughly comparable to the Italian local schools. In 1803 the Norwich Society was founded by John CROME. He developed the school, and annual exhibitions were held from 1805 to 1825. He was succeeded as president by John COTMAN in 1821. The paintings of these artists and others of the group, including James Stark, Robert Ladbrooke and David Hodgson, are well represented in the Norwich museum.

Norwich terrier, small hunting dog bred in England in the 19th century. It has a wide head, a fox-like muzzle and erect ears. The short, compact body is set on short, powerful legs; the tail is commonly docked. Its wiry coat may be red, brown or black. Height: to 28cm (12in) at the shoulder; weight: to 6kg (14lb).

Nose, in human beings, other primates and some vertebrates, the prominent structure between the eyes; it contains the OLFACTORY ORGAN and serves as the opening to the respiratory tract, warming and moistening the air and cleaning it at MUCOUS MEMBRANES. It has two cavities separated by a wall of CARTILAGE; the external openings are the nostrils. The floor of the nose and roof of the mouth are formed by the palatine bone or hard PALATE and the soft palate. During swallowing, the soft palate closes the back of the nose to prevent food from entering.

Nosebleed, haemorrhage from the nose caused by various conditions. Inflammation, uclers, severe head injury, vascular disease, scurvy and haemophilia are among the conditions that may cause nosebleed. It is controlled by rest and application of a cold compress.

Nose cone, rocket, portion of a SPACECRAFT that must be designed to withstand heat caused by friction when re-entering the atmosphere. Blunt shapes heat up less than narrow ones, since they form a bow shock wave that is detached from the cone and heats the atmosphere around it rather than the vehicle itself. Nose cones are often equipped with heat shields to dissipate heat by ABLATION.

Nostradamus (Michel de Nostredame) (1503–66), French seer and astrologer. After practising as a doctor and reputedly curing several people of the plague, he began making astrological predictions in 1547. These were published in rhyming quatrains in *Centuries* (1555) and represented one verse for every year from then till the end of the world (in the 1990s); to avoid prosecution as a magician, however, he completely changed the order of the verses so that no time sequence was discernible. Many individual verses do seem to be of startlingly detailed relevance to historical events from that century to this. *See also* MS pp.196–197.

Notary, or notary public, public official, usually a solicitor, who authenticates certain documents such as deeds, certificates and contracts by witnessing them.

Notation, scientific, representation of numbers, physical quantities, units, operations, atoms and molecules, etc., by symbols in quantifying scientific data. An international notation in science and

The **Duke of Northumberland** hoped to gain his own power through Lady Jane Grey.

Norway; the Leonfjord looking towards the Nordfjord, typical of the coastline.

Norwegian elkhounds are reputed to be able to scent an elk from 5km (3 miles) away.

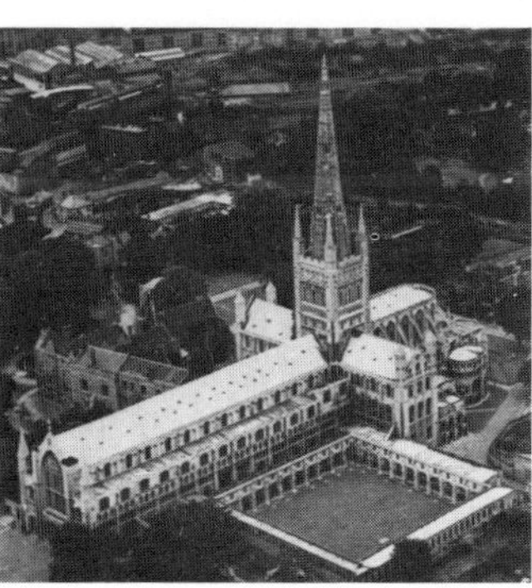

Norwich cathedral has one of the finest Norman towers in England.

Not by Bread Alone

Notre-Dame cathedral is situated on an island in the middle of the River Seine.

Nottingham commemorates the folk-hero Robin Hood with this statue.

Nova Scotia; shipbuilding in Yarmouth is less important now than formerly.

Ivor Novello, at the window of his Aldwych flat, previously the home of his mother.

mathematics is available, which enables scientists to communicate, irrespective of their native languages.

Not by Bread Alone (1956), novel by Vladimir DUDINTSEV, one of the most outspoken in modern Soviet literature. It describes the inventor Lopatkin's fight against military and bureaucratic opposition.

Note, in music, sound of a particular pitch or frequency. The word also refers to the written symbol that represents the sound.

Notes from the Underground (1864), psychological novel by Fyodor DOSTOEVSKY. It deals with a single episode in the life of a man who is alienated from others and unable to form genuine relationships. It is important in Dostoevsky's work because of its exposition of his philosophical viewpoint.

No theatre. *See* NO.

Notochord, in chordates and the early embryonic stages of vertebrates, the flexible, primitive backbone; in mature vertebrates it is replaced by the SPINE of the cushioning intervertebral discs.

Notre-Dame (1163–*c.*1250), early Gothic cathedral in Paris. The design features a wide nave and double ambulatory, and the w façade was imitated in many French churches. Other features include the bold flying BUTTRESSES of the apse, between which were added chapels (1296), and the 13th-century ROSE WINDOWX in the TRANSEPTS.

Nôtre Dame de Paris, novel by Victor HUGO, known in English as *The Hunchback of Nôtre Dame*. First published in 1831, it tells of a hunchbacked bell-ringer who loves Esmeralda, a young gypsy who is condemned to be hanged through the vindictive machinations of a priest whose love she has spurned.

Nottingham, Charles Howard, 1st Earl of (1536–1624), English nobleman. Appointed Lord High Admiral by his cousin ELIZABETH I and serving from 1585 to 1618, he commanded the English fleet against the Spanish ARMADA in 1588, led the Cadiz expedition with the Earl of ESSEX in 1596 and was created earl in 1597. *See also* HC1 p.277.

Nottingham, city in N central England, on the River Trent, county town of Nottinghamshire. It is the traditional birthplace of ROBIN HOOD who is thought to have lived, with his outlaws, in the nearby Sherwood Forest. It is now an important centre of communications and transport. Nottingham is famous for the manufacture of fine lace, cotton and hosiery. Other products include bicycles, electronic equipment, coal and pharmaceuticals. Pop. (1971) 299,758.

Nottinghamshire, county in central England. The land slopes down from the E ridges of the Pennines in the w to the lowlands of the E. The principal river is the Trent. Wheat, barley and sugar-beet are the chief crops; beef and dairy cattle are also important. There are rich deposits of coal in the county. Nottinghamshire has long been noted for its textile industries which include lace, hosiery, cotton and silk. Nottingham is the county town. Area: 2,164sq km (836sq miles). Pop. 977,500.

Notus, in Greek mythology, south wind created from a union between Eos, the goddess of the dawn, and her husband Astraeus. From the same union came the stars, Boreas (the north wind) and Zephyrus (the west wind).

Nouakchott, capital city of Mauritania, in the sw part of the country, approx. 8km (5 miles) from the Atlantic Ocean. Originally a small fishing village, it was chosen as the capital of the republic in 1957 and a development programme was started. Nouakchott now has an international airport, is located on a major road and is the site of modern storage facilities for petroleum. Light industries have also been developed and handicrafts are of significance. Pop.103,500.

Nouméa, seaport city in sw New Caledonia Island, s Pacific Ocean; the capital of the French overseas territory of New Caledonia. The city was used a French penal colony from 1864–94 and served as an Allied airbase during WWII. Pop. (1971 est.) 49,315.

Noun, the part of speech used to name a person, place, object or idea. It can be concrete, abstract, proper, common or collective and generally functions as subject, object or indirect object of the verb in the sentence.

Nous, term used by HOMER to refer to the mind; pre-Socratic philosophers used it to refer specifically to reason, as opposed to perception by the senses. PLATO considered *nous* to be the rational part of the soul.

Nouveau roman (New novel), experimental form pioneered by Alain ROBBE-GRILLET, Samuel BECKETT and Nathalie SARRAUTE in the 1950s, influenced by Franz KAFKA and James JOYCE and by film techniques. Its characteristics are meticulously detailed description, avoidance of value judgments and a consciousness of the artificiality of time sequences.

Nouvelle Vague (New Wave), journalistic term describing the work of young French film directors during the period 1958–61. It embraced the idea that the director of a film was in effect the "author" (auteur) of the piece. Alain RESNAIS, Jean-Luc GODARD and Louis MALLE were amongst those who made personal statements through the art of film. Examples include *Les Quatre Cents Coups* (1959) by TRUFFAUT and *A Bout de Souffle* (1960) by Godard.

Nova (plural novae), faint star that undergoes unpredictable increases in brightness by several magnitudes, apparently due to explosions in their outer regions, and then slowly fades back to normal. These phenomena are relatively common, and some, eg the recurrent novae, show comparatively frequent brightness variations. *See also* VARIABLE STAR; SU pp.235, 240–241, 241.

Novachord, electrophonic musical instrument, a simple electric organ, played on a single manual (keyboard). Tones simulating the organ and other instruments are produced by a series of 12 vacuum tube oscillators with a separate oscillator for each note of the chromatic scale.

Novak, Joseph. *See* KOSINSKI, JERZY.

Novalis (1772–1801), German romantic poet and novelist, real name Friedrich von Hardenberg. His chief work, *Heinrich von Ofterdingen*, was unfinished at the time of his early death from tuberculosis. His grief at the death of his fiancée Sophie von Kühn is beautifully expressed in *Hymns to the Night* (1800).

Nova Lisboa, town in Angola, sw Africa; formerly known as Huambo. It serves as a market centre for the surrounding area and exports cereals, skins and fruit. Pop. 89,000.

Nova Scotia, maritime province in E Canada, made up of mainland peninsula and the adjacent Cape Breton Island and a few smaller islands. The land is generally low-lying, rolling country and there are extensive forests. Dairying is important in the NW, and the principal crops are hay, apples, grain and vegetables. There are valuable coal deposits on Cape Breton Island; gypsum and salt are also mined. Fishing is very important, cod, lobster and haddock being the largest catches. Other industries include shipbuilding, steel-making, fish and food processing. The principal cities are Halifax, the capital, Sydney, Glace Bay and Dartmouth.

The first settlement of Nova Scotia was made by the French at Port Royal in 1605. The French named the area Acadia. During the 17th and 18th centuries the French and British disputed ownership of the region. Thousands of Loyalists fled to the area during the American War of Independence. Nova Scotia joined NEW BRUNSWICK, QUEBEC and ONTARIO to form the Dominion of Canada in 1867. Area: 55,490sq km (21,425sq miles). Pop. (1975 est.) 822,000. *See also* MW p.45.

Novatianus (*c.*200–*c.*258), Roman theologian. Disappointed at Pope Cornelius' lenient treatment of Christians who had compromised with paganism, he joined the rigorist party, becoming antipope (251). Although orthodox in doctrine, his followers (Novatianists) were excommunicated. Novatianus himself was martyred by VALERIAN.

Novaya Zemlya, two islands in the Arctic Ocean off the NE coast of the Russian SFSR. The N island is ice-covered all the year round while the s island has tundra vegetation and a sparse population that exists by trapping and fishing. Area: 81,279sq km (31,382sq miles). Pop. (est.) 400.

Novel, narrative prose fiction that is longer and more detailed than a short story. The word is derived from the Latin *novus* ("new") and the Italian *novella* (short tale with an element of surprise). The novel was established as a literary form in the 18th century and Samuel RICHARDSON's *Pamela* (1740) is generally considered to be the earliest example of serious fiction in English. A novel uses plot and characters to create a picture of life, either past, present or in the future. It may be of almost any length, and among the categories are picaresque, in which a hero lives through certain adventures (eg Daniel DEFOE's *Jonathan Wild*, 1725); psychological, in which the inner life of one or more characters is developed (eg J.D. SALINGER's *Catcher in the Rye*, 1951); and social, which may portray a large number of people (eg Leo TOLSTOY's *War and Peace*, 1863–69).

Novelas ejemplares (1613), 12 tales by Miguel de CERVANTES, including *The Force of Blood, The Little Gypsy* and *Rinconete y Cortadillo*. The last named concerns the adventures of Rincón and Cortado in the Seville underworld. The "exemplary" nature of the tales lies in their moral significance rather than moralizing by the author.

Novella, short, highly structured prose narrative. The form was developed by BOCCACCIO in the *Decameron*, and has been very popular with German authors since the 18th century. In modern times novella refers to a work of prose fiction that is longer than a short story but shorter than a NOVEL.

Novello, Ivor (1893–1951), Welsh composer and playwright of sentimental romantic musicals and plays. The son of the celebrated Welsh soprano, Dame Clara Novello Davies, his first play was *The Rat* (1921). His popular musicals included *Glamorous Night* (1935), *Careless Rapture* (1936) and *The Dancing Years* (1939), in all of which he also starred.

November, eleventh month of the Gregorian calendar, comprising 30 days. It was the ninth month of the ancient Roman calendar (its name derives from Latin *novem*, "nine"). By the Anglo-Saxons it was called "blood-month" (for it was then that cattle were slaughtered) or "wind-month".

Novembergruppe, group of EXPRESSIONIST artists formed in Germany in 1918 by Max PECHSTEIN and César Klein. Other artists associated with it included Walter GROPIUS, Ludwig MIES VAN DER ROHE, Lyonel FEININGER, László MOHOLY-NAGY, Alban BERG, Kurt WEILL and Bertolt BRECHT. The group supported socialism and believed in the unification of the arts; their major exhibition was held in Berlin in 1929.

Noverre, Jean Georges (1727–1810), French choreographer and ballet reformer. His reforms included abolishing the mask and replacing the series of conventional meaningless gestures which then constituted ballet with the *ballet d'action* in which dance and story were united. He published his ideas in *Lettres sur la danse et sur les ballets* in 1760. *See also* HC2 p.39.

Novgorod, city in the NW Russian SFSR, USSR, on the River Volchov, 160km (100 miles) SSE of Leningrad. One of Russia's oldest cities, it was originally a Varangian trading town, conquered by Rurik *c.*862, and governed from Kiev until the 12th century, when it became capital of a vast territory and was called Novgorod the Great. Ruled by Prince ALEXANDER NEVSKI 1238–63, the city fell to Moscow in 1478; its downfall was completed by IVAN the Terrible in 1570 and the Swedes in 1616. During WWII it was held by the Germans (1941–44) and greatly damaged. Its modern industries include distilling, meat packing and flour milling. Pop. (1975) 165,000. *See also* HC1 pp.320, *321*.

Novi Sad, port city in E Yugoslavia, on the River Danube. An important industrial centre, its manufactures include farm machinery, electro-chemical equipment and textiles. Pop. (1971) 214,048.

Novocaine, trade name of a local anaesthetic used by doctors and dentists. It is a synthetic ALKALOID, full name procaine hydrochloride which, unlike the related and more powerful COCAINE, is not habit-forming.

Novosibirsk, city in the S Russian SFSR, USSR, on the River Ob, 628km (390 miles) E of Omsk. Founded in 1896 after the construction of the Trans-Siberian Railway, it grew quickly, surpassing Omsk as Siberia's leading city in the 1930s. During WWII it received complete industrial plants moved from war areas of the W USSR. Its products include agricultural and mining machinery, chemicals and textiles. Pop. (1970) 1,161,000.

Novotny, Antonin (1904–75), Czechoslovak President (1957–68). He was also first secretary of the Czechoslovak Communist Party. He developed a policy of close co-operation with the USSR which became unpopular with the party's nationalistic factions, forcing him to resign in 1968 in favour of DUBČEK. He was reinstated in 1971 following the ascendancy of Soviet factions within the party.

Novum Organum (1620), philosophical work by Francis BACON, named after ARISTOTLE's logical treatises, which Bacon sought to replace by his method of INDUCTIVE LOGIC. Bacon revolted against syllogistic reasoning, which he contended discovered nothing. Man's "natural light" could be extended and his life improved practically by experimental reasoning. Argument should proceed from the specific, through successive stages arranged in tabular form, to general propositions.

Novy Mir (Russian "new world"), influential Soviet literary periodical founded in 1925 and for long edited by the poet Twardovsky. DUDINTSEV's novel *Not by Bread Alone* was first published in *Novy Mir* in 1956. The periodical has also published controversial work by SOLZHENITSYN such as *One Day in the Life of Ivan Denisovich* (1962).

Nowell (Noel), Alexander (c. 1507–1602), English religious scholar. A fellow of Brasenose College, Oxford, he was exiled during the Catholic Queen MARY's reign, but was appointed Dean of St Paul's on his return to England in 1560. He is remembered for his three *Catechisms*, the Small (1549), Middle and Large (1570).

Nowlan, Alden (1933–), Canadian writer whose poems are based on the small New Brunswick town where he lives. His works include *The Rose and the Puritan* (1958), *Wind in a Rocky Country* (1961) and *Playing the Jesus Game* (1970).

Noyes, Alfred (1880–1958), British poet, probably best-known for his poem *The Highwayman*. His works include *Drake* (1906–08), *Tales of the Mermaid Tavern* (1913) and the trilogy *The Torchbearers* (1922–30). Noyes also wrote criticism and fiction.

Noye's Fludde, church pageant opera in one act by Benjamin BRITTEN with libretto based on the text of a Chester MYSTERY PLAY. It was first performed at the Aldeburgh Festival in 1958. Ideally performed in a church, it was written especially for children's voices and boys' bugle classes and recorder groups.

NSU, or nonspecific urethritis, venereal infection, the agent for which is as yet unknown. It is called nonspecific to distinguish it from GONORRHOEA which also causes inflammation of the urethra. *See also* VENEREAL DISEASE.

Nu, or Nun, in Egyptian mythology, the personification of CHAOS or the Ocean from whom all things arose. He is depicted as a man partially immersed in water supporting the beings of his creation.

Nu, U (1907–), Burmese political figure who was active in securing Burma's independence from Britain (1948) and became its first Prime Minister, serving until 1956. He left his post briefly to reorganize the Anti-Fascist People's Freedom League. After the ruling party split, he resigned. Prime Minister again in 1957–

58 and in 1960 he was deposed by Ne Win in 1962 and imprisoned until 1966. He went into exile in Thailand in 1969, returning to Burma in 1970.

Nuba, collective name for a group of several unrelated peoples inhabiting a region of S Sudan. Most Nuba peoples are farmers and many tribes cultivate terraces on rugged granite hillsides. Animal husbandry is also practised. The Nuba peoples are in constant conflict with the Sudanese administration, whose authority many of them refuse to accept. Although ISLAM has made some converts, the predominant religious rituals are closely linked to agricultural fertility rites; in the more remote regions, the men go naked and the women wear lip and nose ornaments.

Nubia, ancient state in NE Africa. It was conquered in the 20th century BC by the Egyptians. Egypt itself came under Nubian rule in the 8th and 7th centuries BC, although this rule was terminated by a series of Assyrian invasions c. 670 BC. The kingdom, with a new capital at Dongola, was converted to Christianity in the 6th century AD but finally collapsed in the 14th century. *See also* HC1 pp.34–35, 48–50.

Nubian Desert, uninhabited region in the NE Republic of the Sudan. It is the W part of the Sahara Desert, between the River Nile and the Red Sea, and consists of a large sandstone plateau. Area: approx. 233,100sq km (90,000sq miles).

Nuclear battery, any packaged source of power that utilizes NUCLEAR ENERGY and provides electricity or heat for long periods. It is usually fuelled by plutonium-238 or spontaneously fissionable elements such as strontium-90.

Nuclear disarmament. *See* DISARMAMENT; SALT (Strategic Arms Limitation Talks).

Nuclear, or atomic, energy, energy released during a nuclear reaction as a result of the conversion of mass into energy, according to Albert EINSTEIN's equation in his theory of RELATIVITY, $E = mc^2$. Nuclear energy is released in two ways – by fission and by fusion. Both methods depend on the release of the binding energy of the atomic NUCLEUS. When fission occurs the nucleus of a heavy atom disintegrates into two smaller ones in which the BINDING ENERGY per NUCLEON is higher than in the original nucleus. The difference in total binding energy is carried away by the two or three NEUTRONS released in the fission. In a fusion reaction two light nuclei combine to form a heavier nucleus, with the release of binding energy. *See also* MM pp.33, 71, 74–75, 84, 180–181; MS pp.182, *182*; SU pp.64–65, 64–65, 68, 111.

Nuclear engineering, branch of engineering concerned with the design, construction and operation of nuclear power plants and particle ACCELERATORS as well as coolant, shielding and moderator devices. *See also* NUCLEAR ENERGY.

Nuclear family, in anthropology, term used to describe a family unit of two adults joined by conjugal link with their children. Ideally the nuclear family provides economic and emotional security to both the parents and children. *See also* KINSHIP; MS pp.250–251, 254.

Nuclear fuel, various chemical and physical forms of URANIUM and PLUTONIUM used in nuclear reactors. Fluid fuels are required in homogeneous reactors; heterogeneous reactors use various forms of fuels – pure metals or alloys, as well as oxides or carbides. The fuel must have a high thermal conductivity, be resistant to radiation damage and be easy to fabricate. *See* MM p.74.

Nuclear fuel enrichment, separation of the fissionable ISOTOPE uranium-235 from the more abundant uranium-238 isotope. Gaseous uranium hexafluoride undergoes diffusion separation utilizing cascades of barriers with microscopically small pores. The difference in mass between the two isotopes is minimal, but sufficient so that the heavier, slower-moving U-238 molecules are concentrated on one side. High-speed centrifugal-force separation methods are also used. *See also* MM p.74.

Nuclear fusion. *See* NUCLEAR ENERGY.

Nuclear ice-breaker, first large floating vessel to be powered by a NUCLEAR REAC-

TOR. The ice-breaker *Lenin* was built in 1957 by the USSR and has been followed by other nuclear ice-breakers.

Nuclear magnetic resonance (NMR), absorption of radio waves by certain nuclei in the presence of a strong magnetic field. In this field the nucleus, as a result of its SPIN, can have slightly different energy values. It can make transitions between these energy values, acquiring the energy by absorbing radio-frequency radiation of the appropriate wavelength. *See also* SU p.*151.*

Nuclear physics, branch of physics concerned with the structure and properties of the atomic NUCLEUS. The principal means of investigating the nucleus is the SCATTERING experiment, carried out in particle ACCELERATORS, in which a nucleus is bombarded with a beam of high-energy ELEMENTARY PARTICLES and the resultant particles analysed. Other information is obtained by studying their magnetic moments and SPIN and any fission and fusion products. Study of the nucleus has led to an understanding of the processes occurring inside stars and has enabled the building of NUCLEAR REACTORS and NUCLEAR WEAPONS. *See also* SU pp.64–65.

Nuclear power, generation of electric power from nuclear fission. The first NUCLEAR REACTOR to generate electric power commercially began operation in 1956 at CALDER HALL, Cumberland. The stimulus to these and later developments was the consideration that, weight for weight, a nuclear fuel such as uranium-235 yields about 3,000,000 times as much energy as the fossil fuels coal and oil. By the 1960s, Britain and the USA had developed large nuclear power industries and had exported nuclear reactors and nuclear technology to many other countries. Other developments included engines powered by nuclear reactors: in 1954 the first nuclear submarine, *Nautilus*, was launched in the USA, and in 1957 the first nuclear-powered ice-breaker, *Lenin*, was launched by the USSR. In the international reactor market, competition was particularly fierce between various British designs involving gas-cooled reactors. By the 1970s, doubts became widespread about the safety and real economics of nuclear power programmes, doubts which have yet to be removed. *See also* MM pp.74–75; SU p.68.

Nuclear radiation, refers to all the types of emissions of waves and particles from the nuclei of atoms as the result of natural or induced RADIOACTIVITY. Natural radioactivity is the spontaneous breakdown of unstable nuclei, such as those of radium or uranium-235, with the release of ALPHA and BETA PARTICLES and GAMMA RAYS. Alpha (α) particles are identical with nuclei of atoms of helium gas: they travel very fast but have little penetrating power and so are relatively harmless. Beta (β) rays or particles are electrically charged and are identical with high energy ELECTRONS or POSITRONS. They have greater penetrating power. Gamma (γ) rays are very high frequency X-rays, and are dangerously penetrating. Other emissions from nuclei include PROTONS, NEUTRONS, and NEUTRINOS. These and yet other particles, called MESONS, result from the interactions of cosmic rays with the upper atmosphere, and from the bombardment of materials with high speed particles in particle accelerators.

Nuclear reactor, device which utilizes the energy of the atomic NUCLEUS to generate useful power. This energy binds together the NUCLEONS, or sub-atomic particles, that make up the nucleus. In a nuclear reaction the fuel is a radioactive heavy metal, usually uranium-235 or plutonium-238. In these metals the nuclei of atoms break down spontaneously by radioactive disintegration, releasing fast NEUTRONS. These sub-atomic particles have great energy but no electric charge and they are able to pass through other atoms without being attracted by positively charged nuclei or negatively charged electrons. In the solid metal fuel however, at least one neutron from each disintegration will strike the nucleus of another atom directly, causing it to break

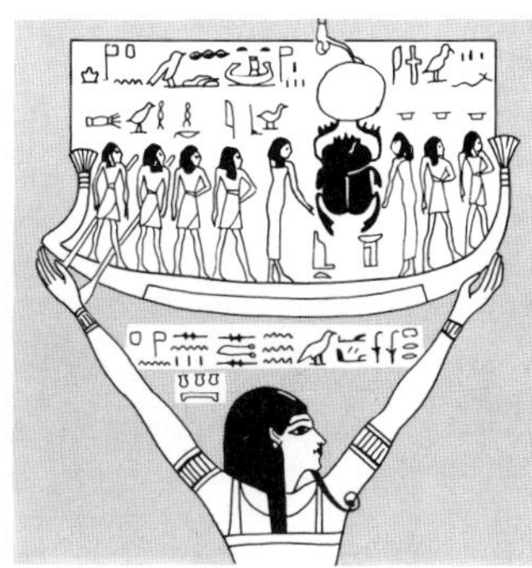

Nu, ocean god from whom all things arose, echoes modern ideas of evolution.

Nubia, rooted in antiquity, has no precise present-day geographical boundaries.

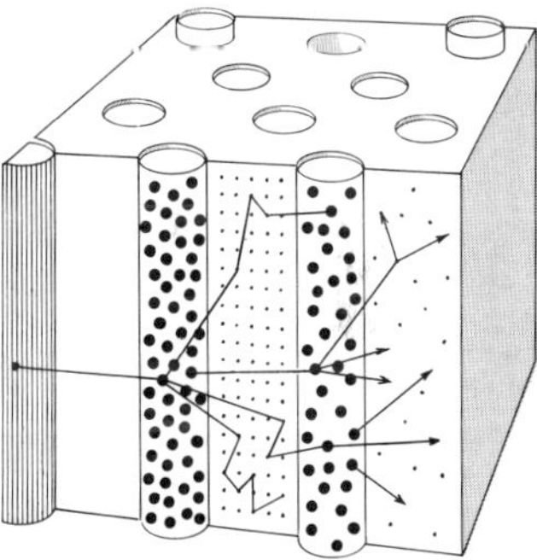

Nuclear energy is released in a reactor by a controlled atomic chain reaction.

Nuclear reactor; a research scientist holds fuel elements at the core of the "pile".

Nuclear submarines

Nuclear submarine; HMS *Revenge*, Britain's fourth Polaris submarine, built in 1969.

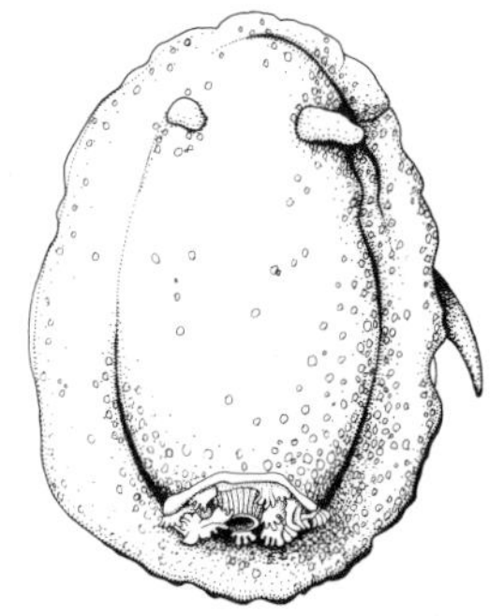

Nudibranch, a shell-less marine gastropod that feeds mainly on sea anemones.

Nuer; a storage gourd used by the farming peoples who live beside the Nile.

Lord Nuffield made his fortune in the motor industry and became a philanthropist.

up and emit more neutrons. In this way a cascade effect, or chain reaction, is set up. In a large enough mass of metal (a CRITICAL MASS), when insufficient neutrons can escape from the surface, the chain reaction will proceed rapidly to a violent explosion, as in the atomic bomb. In a nuclear reactor the chain reaction is controlled by the use of materials such as cadmium which absorb neutrons. These materials take the form of rods which are lowered into the reactor between the fuel elements, also in rod form, so as to limit the chain reaction to provide large amounts of heat. This heat is taken up by a circulating coolant, often liquid sodium, which in turn gives up its heat in a high-pressure boiler to raise steam for the generation of electric power. This description fits fast reactors which utilize the energy of fast neutrons only.

In moderated reactors, neutrons are slowed down by solid materials such as carbon and beryllium, or, in heavy water reactors, by the heavy isotope of hydrogen – deuterium – present in the water (which acts also as a coolant). Fast breeder reactors use enriched fuels blanketed with uranium-238, the commonest uranium isotope. As the fuel becomes used up, neutrons absorbed by the uranium-238 convert, or breed it into plutonium, which is extracted and used as a fuel for other nuclear reactors. Whatever its type, a nuclear reactor generates, not only heat, but also highly lethal amounts of radioactivity, and for this reason is heavily shielded in lead and thick concrete. *See also* MM pp.74–75.

Nuclear submarines, underwater vessels powered by NUCLEAR REACTORS, pioneered by the USA with *Nautilus*, launched in 1954. *Nautilus* travelled 100,000km (62,000 miles) in its first two years of service and in 1958 voyaged under the polar ice cap. A successor travelled more than 57,000km (35,000 miles) submerged. In 1959 the USA launched an improved, more streamlined vessel, with a power plant comprising a nuclear reactor cooled by pressurized water which is fed to a heat exchanger, where it generates steam in a separate water loop; the steam is fed to turbo-generators which drive the submarine. The reactor and pressurized water loop are highly radioactive and are shielded from the rest of the engine. Such submarines have no need to refuel and may remain submerged for years at a time. By 1971 the USA had about 90 of these submarines but the USSR, which began production of nuclear submarines in the early 1960s, overtook US production by the early 1970s. The weaponry in US nuclear submarines is now part of the Trident system, which replaced POLARIS. Each armed nuclear submarine has enormous firepower and is difficult to detect and to destroy. For these reasons, they play a key role in military strategy and are an important deterrent. *See also* MM p.*175*.

Nuclear Test-Ban Treaty (1963), treaty signed in Moscow by the USSR, Britain and the USA. It committed these three major powers to halt all nuclear tests in the atmosphere, under water and in outer space and permitted only underground explosions. It was intended to deter additional countries from acquiring NUCLEAR WEAPONS, and thereby to slow the arms race and ease international tensions. France and China are not signatories to the treaty.

Nuclear weapons, devices whose enormous explosive forces derive from nuclear fusion or fission reactions. The first atomic bombs were dropped by the USA on the Japanese cities Hiroshima and Nagasaki in August 1945. The bombs consisted of two stable subcritical masses of uranium or plutonium which, when brought forcefully together, caused the CRITICAL MASS to be exceeded, thus initiating an uncontrolled nuclear fission reaction. In such detonations huge amounts of energy and harmful radiation are released: the explosive force can be equivalent to 200,000 tonnes of TNT. The much more powerful hydrogen bomb (or thermonuclear bomb), first tested in 1952, consists of a fission bomb that on explosion provides a temperature high enough to cause nuclear fusion in a surrounding solid layer, usually lithium deuteride. The explosive power can be that of several million tonnes (megatonnes) of TNT. Devastation from such bombs covers a wide area: at 20km (12 miles) from a 15 megatonne bomb all flammable material bursts into flame. *See also* NUCLEAR ENERGY; MM pp.180–181.

Nuclease, ENZYME found in animal organs. It can split NUCLEIC ACIDS into NUCLEOTIDES and into nucleosides or their components.

Nucleic acids, giant chemical molecules present in all living cells and in viruses. They are of two types, DNA and RNA, both of which play fundamental roles in heredity. Chemically, they are polymers of NUCLEOTIDES. *See also* GENE; CHROMOSOME.

Nucleon, collective term for any of the constituents of the NUCLEUS of an atom: the NEUTRON and PROTON, the BARYON members of a MULTIPLET of two (doublet). They are considered as manifestations of a single state of matter. *See also* SU p.62.

Nucleosynthesis, production of all the various chemical elements that exist in the universe from one or two simple atomic nuclei. It is believed to have occurred by way of large-scale nuclear reactions during cosmogenesis and still to be in progress in the Sun and other stars. Starting with hydrogen and helium, repeated nuclear fusion reactions can account for most of the elements up to iron. Elements heavier than iron can be explained in terms of repeated reactions involving the capture of NEUTRONS by nuclei.

Nucleotide, complex naturally occurring chemical group that contains a nitrogen base linked with a sugar and an acid phosphate. Nucleotides are the building blocks of NUCLEIC ACIDS, in the molecules of which they are linked together by bonds between the sugar and phosphate groups. Nucleotides also occur freely in a CELL as various COENZYMES, such as ADENOSINE TRIPHOSPHATE (ATP), the principal carrier of chemical energy in the metabolic pathways of the body.

Nucleus, region in most cells which contains the CHROMOSOMES and GENES, the units of inheritance, and is essential for life in most cells. Exceptions include bacteria and blue-green algae in which the chromosomes are scattered in the CYTOPLASM and mature red blood cells. The nucleus, usually spherical in shape, manufactures RIBOSOMES, essential for the synthesis of RNA. Cell division involves the splitting of the nucleus. *See also* NW p.26.

Nudibranch, marine GASTROPOD, usually found on seaweeds. The shell and mantle cavity have been lost and there are finger-like respiratory organs along the sides of the body. Length: 10mm (0.4in) to about 20cm (8in). Species include the sea lemon *Archidoris*.

Nuer, Nilotic-speaking people who live, mostly by raising cattle, in the Sudan. Their population in the 1960s numbered approx. 300,000. There is little unity among the tribes and much feuding is settled by payment in cows, which is made official through the mediation of a priest of their sky-god religion.

Nueva Galicia, former colonial region of Spain in W Mexico; it encompasses the modern states of Jalisco, Nayarit and S Sinaloa. It was the scene of the Mixton War (1541). Known at first as the Presidency of Nueva Galicia, its independence declined as administrative authority was transferred to Mexico City.

Nuevo León, state in NE Mexico; the capital is Monterrey. It has a mountainous terrain, with lowland plains and numerous rivers in the E. The W and N is semi-arid desert country. The area was occupied by US troops during the Mexican War and became a state in 1824. Today the plains area produces cereals, cotton and sugar cane; other products include iron, steel and chemicals. Area: 64,556sq km (24,925sq miles). Pop. (1970) 1,694,689.

Nuffield, William Richard Morris, 1st Viscount (1877–1963), British motor-car manufacturer and philanthropist, who began business in 1892 as a cycle repairer. He opened a car factory at Cowley, near Oxford, in 1912, and the first Morris Oxford car appeared in 1913. He introduced low-price, mass produced cars and revolutionized the British car industry. Morris Motors Ltd was founded in 1919. In 1952 he became chairman of the British Motor Corporation (BMC), which was an amalgamation of the Morris and Austin companies. Equally famous as a public benefactor, he founded Nuffield College, Oxford, in 1937 and the NUFFIELD FOUNDATION in 1943.

Nuffield Foundation, trust founded in 1943 by the motor-car manufacturer Richard Morris (1877–1963), who later became 1st Viscount NUFFIELD. He directed that £10,000,000 of common stock of Morris Motors Ltd be used by the Foundation to improve the health, social welfare and education of the poor.

Nugent, Elliot (1899–), US playwright, actor and director. He worked with his father J. C. Nugent on numerous plays, including *The Poor Nut* (1925) and *Money to Burn* (1930). He wrote *The Male Animal* (1940) with James THURBER.

Nukualofa, capital town of the independent Kingdom of Tonga, on the N coast of Tongatabu Island, in the SW Pacific Ocean. The royal palace and government buildings are located there. Pop. (1966 est.) 20,000.

Nullarbor Plain, large plateau in S Australia, once a sea-bed. It has no surface water and there is little vegetation. The plain is the site of an important rocket research and production centre. Area: approx. 260,000sq km (100,000sq miles).

Numantia, ancient town in Spain, on the River Douro N of modern Soria, in Old Castile. The Celtiberian tribe of Numantia resisted Roman conquest from 195 BC until 133 BC when Scipio Aemilianus took the city after an eight-month siege, completing the Roman conquest of Spain.

Numbat, or banded anteater, squirrel-like Australian marsupial that feeds on TERMITES. The female, unlike most marsupials, has no pouch. The young cling to the teats of her belly until they are a few weeks old. The numbat has a long snout and lateral white bands on its red-brown coat. Length: 46cm (18in). Species *Myrmecobius fasciatus*. *See also* NW p.158.

Number, complex, in mathematics, a number that may be expressed in the form $x + iy$, where x and y are real numbers and i is equal to the square root of minus one ($i = \sqrt{-1}$). The complex numbers were introduced to solve such equations as $x^2 + 1 = 0$ – that is, $x = \pm i$. A number which may be expressed as iy is called imaginary.

Number, prime, positive or negative integer, excluding one and zero, that has no factors other than itself or one. Examples are 2, 3, 5, 7, 11, 13 and 17. The integers 4, 6, 8, ... are not prime numbers since they can be expressed as 2 x 2, 2 x 3, 2 x 2 x 2, ..., that is, as the product of two or more primes. A mathematical function that will generate the series of prime numbers has not been found ; research is presently involved in this problem.

Number, rational, in mathematics, any number that can be written in the form a/b where a and b are INTEGERS. Such numbers include all integers and FRACTIONS. A number that cannot be so expressed is called an IRRATIONAL NUMBER, examples of which include $\sqrt{2}$ and π.

Number base, number of units in a number system that is equivalent to one unit in the next higher counting place. Thus 10 is the base of the decimal system: only the ten digits 0-9 can be used in the units, tens, hundreds, etc. Each number system has a number of symbols equal to its base. In the BINARY SYSTEM (base 2) there are two symbols, 0 and 1. *See also* SU pp.28–30.

Numbers, fourth book of the Bible and of the PENTATEUCH. It includes two census lists of the Israelites (hence its name from the Greek); many laws given by God to MOSES; a history of events during the 40-year march to the Promised Land (hence the Hebrew title *Bemidbar* (In the Wilderness); and the announcement that JOSHUA was to be Moses's successor.

Number systems, counting methods which use various NUMBER BASES. The decimal system, for example, has a base of ten, the duodecimal system a base of twelve. Digital computers base their calculations on the BINARY SYSTEM, of base two.

Number theory, branch of mathematics concerned with the properties of whole numbers or integers and special classes of integers such as PRIME NUMBERS and perfect numbers. The 4th-century BC Greek mathematician EUCLID proved that the number of primes was infinite, and one of the unresolved problems in number theory is the formulas for the generation of the primes. In 1640 Pierre de FERMAT showed that the numbers $F_n = 2^{2^n} + 1$, for $n = 0,1,2,3,$ and 4 were prime. $F_4 = 65,537.$ He thought that F_n was prime for all n, but in the 18th century Leonard EULER showed that $F_5(= 4,294,967,297)$ was a composite number, made up by multiplying together two other primes. Perfect numbers, another facet of number theory, are the sums of their divisors, such as $6 = 3+2+1$ and $28 = 14+7+4+2+1$. It is not known whether there exist any odd perfect numbers.

Numeiry (Nimeiri), Mohammed Jaafar al- (1930–), chief of state in the Sudan. He took control after a military coup in 1969 and was made President in 1971. He was responsible for the end of a 17-year civil war with the southern Sudanese and defeated a Communist revolt in 1971. A Libyan-backed coup in 1976 failed, and led to a strengthening of ties with Egypt.

Numeral, symbol used alone or in a group to denote a number. Arabic numerals are the 10 digits from 0 to 9. Roman numerals consist of seven letters or marks. The formation of numbers from numerals depends on the NUMBER SYSTEM used. *See also* SU p.30.

Numerator, in mathematics, top number in a rational number or fraction. The bottom number is known as the DENOMINATOR. For example, in the fraction $\frac{2}{3}$, 2 is the numerator, 3 the denominator.

Numidia, ancient region of NW Africa roughly corresponding to modern Algeria. It was part of the Carthaginian empire until the PUNIC WARS, when the Numidian king Masinissa allied with Rome. The country prospered independently from 206 BC until the Jugurthine Warr Subjugated by Rome, it survived the VANDAL invasion of the 5th century AD but lost its remaining political importance after the Arab invasion of the 8th century.

Numismatics, study or collection of coins, tokens, medals and similar objects as works of art, for investment, or as a source of historical information. Coins, for example, preserve old forms of writing and bear portraits of eminent people; they also assist in the study of early customs and help to clarify economic and trade relations. The largest coin market in the world is in London.

Nun, female member of a religious order who has taken vows of MONASTICISM. Nuns can belong to either an Enclosed order (one of prayer and meditation within a closed community) or an order which encourages its members to work in the world for the welfare of society at large.

Nun, letter occurring 14th in the Hebrew alphabet and equivalent to the Roman N.

Nunatak, mountain peak or hill that rises above a surrounding glacier. Nunataks become more common towards the edge of ice sheets, and are most common along the coast of Greenland.

Nuneaton, county district in NE WARWICKSHIRE, England, created in 1974 under the Local Government Act (1972). Area: 79sq km (30sq miles). Pop.(1974 est.) 111,900.

Núñez Vela, Blasco (*d.*1548), first Viceroy of Peru (1543–1548). He was sent to Lima to end the civil war and imposed the New Laws of Bartolomé de LAS CASAS. Gonzalo PIZARRO defeated him and he was executed.

Nunivak, island off sw Alaska, USA, in the Bering Sea. It was discovered by the Russians in 1821. Fogbound for most of the year, Nunivak has a small Eskimo population engaged mainly in hunting and fishing. Area: 4,209sq km (1,625sq miles).

Nunn, Trevor (1940–), British stage director, appointed artistic director of the ROYAL SHAKESPEARE COMPANY in 1968. His many notable productions included *As You Like It* (1977) and a widely acclaimed musical version of *The Comedy of Errors.*

Nupe, Kwa-speaking Negroid people of Nigeria who live mainly around the confluence of the Niger and Kaduna rivers. Their society is hierarchical and they live by fishing and agriculture. Their skill as craftsmen is famous throughout West Africa. They are noted particularly for their glass beads, cloth, brass and leatherwork. Most Nupe are Muslims.

Nur-ad-Din (Nureddin) (*c.*1118–74), ruler of Syria. He united Muslim forces in Syria and, making use of the gathering anti-Frankish JIHAD, took Damascus in 1154. While Nur-ad-Din's army fought the Crusaders in Syria SALADIN took control of Egypt; conflict was avoided by the former's death.

Nuremberg (Nürnburg), city in Bayern state, s West Germany, 151km (92 miles) NW of Munich on both sides of River Pegnitz. It became a free imperial city in 1219 and the centre of the German Renaissance during the 15th and 16th centuries. It passed to Bavaria (Bayern) in 1806 and was the location of the annual congress of HITLER's National Socialist Party. After WWII it was the site of the trials of Nazi war criminals (1945–46). Today Nuremberg is an important and commercial centre. Its manufactures include chemicals, textiles and machinery. Pop. (1974) 514,657.

Nuremberg Laws, German laws promulgated in 1935 by the National Socialist government of Adolf HITLER. They deprived Jews, including all persons who were of Jewish descent, of the rights of citizenship and prohibited marriage of Jews and non-Jews.

Nuremberg Trials (1945–46), war crimes trials of WWII German leaders held at Nuremberg, Germany, by the International Military Tribunal established by Britain, France, the USSR and the USA. Under the presidency of Lord Justice Geoffrey Lawrence, Nazis were tried for crimes against humanity. Twelve were sentenced to death, including Hermann GOERING, Joachim von RIBBENTROP and Alfred ROSENBERG; others were imprisoned. German civilians, such as judges, were tried by a US military tribunal at Nuremberg.

Nureyev, Rudolf (1938–), Soviet ballet dancer and choreographer, who became a soloist with the Kirov Ballet, Leningrad in 1958. While on tour in Paris in 1961 he defected from the USSR. He is noted for his spectacular technical virtuosity and sensitive dramatic character portrayal. Major ballets in which he has had leading roles have included *Giselle, Swan Lake* and *Romeo and Juliet* and he has often partnered Margot FONTEYN. In 1977 he choreographed *Romeo and Juliet* (Prokofiev) for the London Festival Ballet and in that year played the title role in the film *Valentino. See also* HC2 p.*275.*

Nurhachi (1559–1626), organizer and creator of the MANCHU state. The Manchus spread s and became all China, ruling as the CH'ING DYNASTY. Nurhachi created the Manchu military banner organization for control and mobilization and developed a writing system for administrative purposes, laying the foundations for almost three centuries of Manchu power.

Nurmi, Paavo Johannes (1897–1973), Finnish middle- and long-distance runner. Known as the "Flying Finn", he set 22 world records between 1920 and 1932 and won 12 Olympic medals, 9 gold and 3 silver. In the 1924 Olympics he won both the 1,500m and the 5,000m within 90 minutes. The favourite for the 1932 Olympic MARATHON, he was disqualified having lost his amateur status.

Nürnberg. *See* NUREMBERG.

Nursery rhymes, verses said or sung to amuse small children, generally brief and usually anonymous. International parallels of many suggest an ancient origin of rhymes such as *Humpty-Dumpty,* while others, including *Twinkle, Twinkle Little Star* and *Mary Had a Little Lamb,* were composed in the 19th century. Many entered the repertoire as popular ballads or songs for adults. Others, like counting and alphabet rhymes, are designed to instruct. Despite many scholarly investigations, there is little to suggest political significance in most of the rhymes, although real figures can sometimes be identified.

Nursery school, school for children between the ages of two and five. Developed in Europe in the 19th and 20th centuries, especially after WWI, such schools aim to prepare children socially and educationally for primary school and help to relieve pressure on working parents.

Nurse shark, carpet SHARK found in shallow tropical and subtropical waters of the Atlantic and E Pacific, particularly in inshore areas. This sluggish fish is yellow to grey-brown above and lighter below. It is recognized by thick, fleshy whiskers near its mouth. Length: 2.5m (8.5ft); weight: 150–170kg (330–370lb). Family: Orectolobidae; species *Ginglymostoma cirratum. See also* CHONDRICHTHYES.

Nursing, profession which has as its general function the care of people who, through ill-health, are unable to care for themselves. Modern nursing, which includes administering prescribed treatment, and patient and family health education, is continually broadening its range of services. Caring for the sick was a concern of ancient civilizations but was particularly emphasised by the early Christian Church; many religious orders and, later, chivalric orders, performed such "acts of mercy". The Sisters of Charity, founded in 1633, have been instrumental in the development, expansion and reform of nursing in a number of countries. In the 18th century the need for reform in nursing was revealed and, by the end of the 19th century, certain principles of Florence NIGHTINGALE's teaching had been adopted in England and the US. These resulted in the general upgrading and secularisation of the profession, with widely improved instruction for nurses by instructor nurses and doctors in independent schools with access to hospitals. Services today include midwifery, care of the mentally ill, medical and surgical nursing in hospital clinics, and public health education programmes.

Nusku, in Mesopotamian mythology, a solar deity, god of fire and light, son of SIN, the moon god. Nusku was the messenger of the gods and sometimes took Sin's place as a dispenser of justice.

Nut, in Egyptian mythology, originally a mother-goddess, later goddess of the arch of the sky. In this role she was depicted touching the Earth with her fingertips and toes while her star-studded belly formed the sky's vault. She also appeared in the form of a cow supported at each of her legs and at her starry underside by Shu, the Egyptian equivalent of Atlas, who later replaced her in this aspect. Nut served also as protector of the dead (whom she provided with fresh air) and can be found on the lids of sarcophagi.

Nut, dry, one-seeded fruit with a hard woody or stony wall. It develops from a flower that has petals attached above the ovary (inferior ovary). Nuts are often formed in association with modified leaves (bracts); the cup of an acorn is formed from fused bracts. A nut does not open at maturity. *See* BEECH; CHESTNUT; HICKORY; WALNUT; PE pp.214–*215.*

Nutation, oscillating movement (period 18.6 years) superimposed on the steady precessional movement of the Earth's axis so that the precessional path of each celestial pole on the CELESTIAL SPHERE follows an irregular rather than a true circle. It results from the varying gravitational attraction of the Sun and Moon on the Earth, due to variations in their distances from Earth and in their relative directions. *See also* PRECESSION.

Nutcracker, widely distributed, crow-like bird of evergreen forests of the Northern Hemisphere. It picks the seeds out of the cones of pine, spruce, cedar and larch, stores them in autumn and eats them in winter. A projection inside the bill turns it into a highly efficient seed cracker or nutcracker. The European thick-billed nutcracker (*Nucifraga caryocatactes*) is a typical species. Family Corvidae. Length: 30cm (12in). *See also* NW p.206.

Numerals; Roman numerals carved on milestones from various provinces.

Nuremberg Trials; the evidence revealed the statistics of Nazi war crimes.

Rudolf Nureyev danced in Paris with the Kirov Ballet before seeking asylum, 1961.

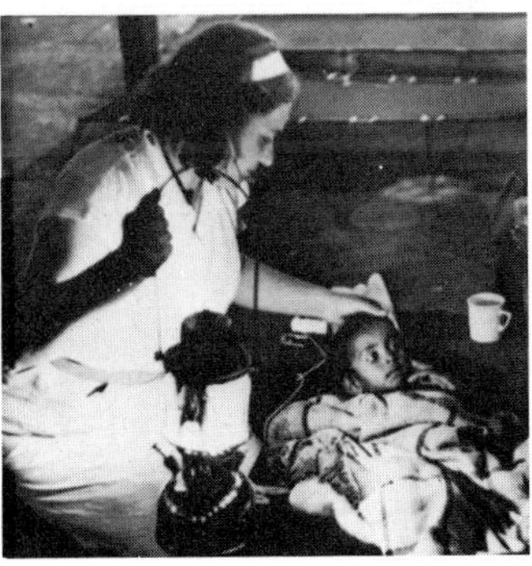

Nursing; Florence Nightingale worked originally in conditions such as these.

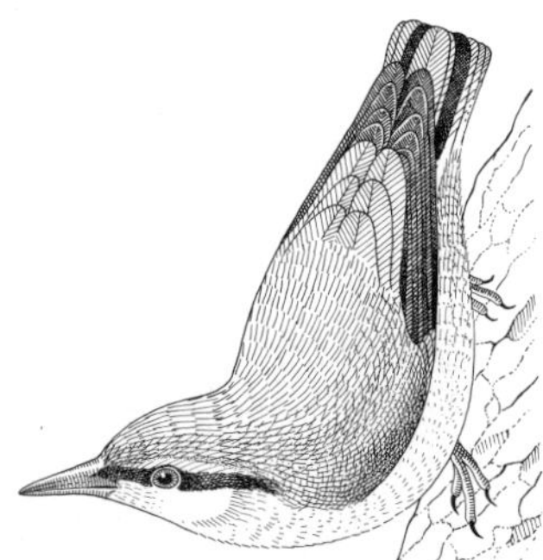

Nuthatches wedge a nut into the bark of a tree before cracking the shell.

Nutmeg trees (*Myristica fragrans*) grow up to 18m (60ft) tall in the Moluccas (Spice Islands).

Julius Nyerere, President of Tanzania, is one of Africa's greatly respected leaders.

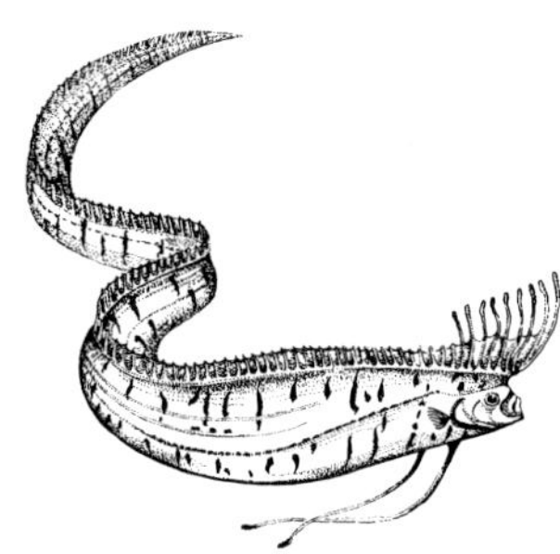

Oarfish swim with a snake-like movement; they are also known as ribbonfish.

Nutcracker, The (1892), ballet in two acts with music by TCHAIKOVSKY, and choreography by Lev Ivanov. The story was adapted by Marius Petipa from *Nutcracker and the Mouse King* by E. T. A. Hoffman. It was first performed in St Petersburg.

Nuthatch, bird found mainly in the Northern Hemisphere and occasionally in Africa and Australia. It is bluish grey above and white, grey or chestnut underneath. It hops in short jerks upwards, downwards and sideways on tree trunks and branches where it wedges a nut into a crevice, opening it with its sharp bill. It also feeds on insects, spiders and seeds, often visiting parks and gardens. Length: 9–19cm (3.5–7.5in). Family Sittidae.

Nutmeg, evergreen tree native to tropical Asia, Africa and America. Cultivated commercially, it has dark brown leaves, pale yellow flowers and yellow, apricot-like fruits. The 500 species include *Myristica fragrans*, native to the Moluccas. Height: up to 18m (60ft). Its seeds yield the spice nutmeg, and its seed covering the spice MACE. Family Myristicaceae. *See also* PE pp.210–211, *211.*

Nutrient solution, either a natural solution, such as seawater, or man-made, that provides most of the substances necessary for an organism to sustain life and growth. Vital nutrients include inorganic phosphates and nitrogen salts, which are needed to produce living tissue, silicates for building shells and trace quantities of vitamins.

Nutrition, all the processes by which whole plants and animals take in and make use of food substances. A study of nutrition involves identifying the kinds and amounts of nutrients necessary for growth and health. Nutrients are generally divided into PROTEINS, CARBOHYDRATES, FATS, MINERALS and VITAMINS. Human nutrition and diet has been made into a special discipline and its study is undertaken by doctors, physiologists, biochemists and agriculturalists. There are between 50 and 60 nutrients essential to life and these are classified according to their chemical structure or function into one of five categories mentioned above. *See also* MS pp.70–71.

Nuvolari, Tazio (1892–1953), Italian motor-racing driver. Small of stature but full of vitality, he is rated among the foremost drivers of the 1930s, and his controlled use of the "four-wheel drift" for cornering gave him great advantages. His many victories include 16 major grands prix, the Le Mans 24-hour race, the Mille Miglia (twice) and the Targa Florio (twice).

Nux vomica, seed of *Strychnos nux-vomica* a tree native to SE Asia; it is one source of the alkaloid STRYCHNINE. Like belladonna and CURARE, strychnine was used as a poison, but is now available in medicine. *See also* MS pp.*86*, 119.

Nyakyusa, Negroid people of SW Tanzania. Their society is traditionally polygamous and their livelihood based on agriculture. Staple crops include bananas, maize and millet; rice and coffee are cash crops. Cattle are important as a symbol of wealth. Traditionally, young people of the same neighbourhood form new villages as they reach puberty, so that a village dies out with its inhabitants.

Nyasa, Lake (Lake Malawi), lake In E central Africa, in the GREAT RIFT VALLEY; bordered by Tanzania (N), Mozambique (E) and Malawi (S and W); third-largest lake in Africa. It is fed chiefly by the Ruhuhu River and drained by the Shire. Discovered by the Portuguese explorer Caspar Boccaro in 1616, the lake was visited and named by LIVINGSTONE in 1859. Area: approx. 29,604sq km (11,430sq miles).

Nye, Robert (1939–), British novelist and poet. After producing several short stories, children's books and a novel, *Doubtfire* (1967), he gained fame with *Falstaff* (1976), a racy novel in the form of fictional memoirs.

Nyenchentanglha Range, mountain range in S Tibet, W China, parallel with the Himalayas and N of the Brahmaputra River. It extends 966km (600 miles) E–W.

The highest point is Nyenchentanglha Peak 7,086m (23,250ft).

Nyerere, Julius Kambarage (1922–), first President of Tanzania, formerly Tanganyika. He served as a legislator and Chief Minister in the Tanganyikan territorial government. An African nationalist leader, he became President of an independent Tanganyika in 1962 and in 1964 manoeuvred the union of Tanganyika and Zanzibar into a republic as Tanzania. As a political theorist, Nyerere's socialist views, published as *Freedom and Unity* (1967) and *Freedom and Socialism* (1968), have had considerable influence in Africa. He is a strong adherent of Pan Africanism and a leading influence in the Organization of African Unity. He has given strong support, both moral and material, to guerrillas fighting in Rhodesia and South Africa.

Nylon, any of numerous synthetic materials consisting of polyamides developed in the US in the 1930s. It is formed into fibres, filaments, bristles or sheets by extrusion through spinnarets and drawing. Nylon is characterized by elasticity and strength and is used chiefly in yarn, cordage and moulded products. Hard and tough or soft and rubbery nylon products can be made by varying the chemical balance. *See also* MM pp.56–57.

Nymph, in Greek mythology, a long-lived female spirit that was said to be a guardian of natural objects. Several forms were said to exist, all identified with a specific location, and most commonly with trees and water. Oceanids, Nereids and Naiads were all water spirits; Oreads inhabited mountains and grottoes; the Napaeae were to be found in groves and glens; Dryads and Hamadryads were always associated with forests and trees.

Nymph, young insect of primitive orders that do not undergo complete METAMORPHOSIS. Nymph is used to designate all immature stages after the egg. The nymph resembles the adult and does so more closely with each successive MOULTING. Some examples are the aquatic nymphs of dragonflies, mayflies and damsel flies.

Nymphomania, insatiable sexual appetite in a female. Frantic and insatiable sexual behaviour is more indicative of this uncommon disorder than heightened sexual desire or strong need for sexual gratification.

Nystagmus, involuntary eye movement, resulting from an early loss of central vision, dizziness, barbiturate intoxication or inner ear or brain disease.

Nzinga Nkuwu, or King John I (died *c.*1506), King (*manikongo*) of the Kongo, a kingdom around the mouth of the River Congo in Africa. He met the first Portuguese explorers of his lands in the 1480s, quickly accepted Christian teachings and helped to establish a Portuguese mission.

O

O, 15th letter of the alphabet, derived in its present form from the Phoenician alphabet, passing then to the Greeks, who called it *omicron*, "short o". It passed unchanged from the Latin alphabet into the various Romance languages, including English. In English *o* is a vowel and may be pronounced with a short sound, as in *not*, or a long sound, as in *note*. An even longer sound is produced by a double *o* (as in *boot*). Combined with *a* the long *o* sound is preserved (*coat*); with other vowels characteristic diphthongs are produced (*hoist* and *rout*). *See also* MS pp. 244–245.

Oahu, mountainous island in Hawaii, between Molokai and Kauai islands. It is the third-largest and economically the most important of the Hawaiian Islands; Honolulu, the state capital, is on the S coast. Oahu has important military installations including Pearl Harbor. Its industries include tourism and fishing. Area: 1,549sq km (598sq miles). Pop. (1970) 629,145.

Oak, common name of almost 300 species of the genus *Quercus*, related to beeches, which are found in abundance throughout temperate areas of the Northern Hemisphere and at high elevations in the tropics. Some species are shrubs, but most are hardwood trees that grow 18 to 30m (60–100ft) tall. Leaves are simple, often lobed and sometimes serrated. Male flowers hang in long yellow catkins; the female ones appear singly or in small clusters and produce the fruit – small nuts called acorns. The texture and colour of the bark vary widely among the many species, and according to the age of each tree; CORK comes from the bark of a Mediterranean species. In medieval England oak timber was prized for shipbuilding, and has long been valued by craftsmen for its durability and the beauty of its intricate grain.

Oakley, Annie (1860–1926), US entertainer, real name Phoebe Anne Oakley Mozee. From childhood she was an expert shot, eventually beating a noted marksman, Frank E. Butler, who married her. She was the star of Buffalo Bill's Wild West Show for 17 years from 1875. Irving BERLIN's musical *Annie Get Your Gun* (1946) is based on her life.

Oaks, The, one of the English Classic horse races, an event for three-year-old fillies, known as "The Ladies Race". It was established in 1779, and run over a 1½-mile (2.4km) course at Epsom. The Oaks was named after the nearby residence of the 12th Earl of Derby, whose horse *Bridget* was the first winner.

Oarfish, any of several deepwater marine RIBBON FISH. Its long thin body has a dorsal fin extending along its entire length. Two long oar-like pelvic fins protrude from beneath the head. Length: to 6m (20ft). Genus *Regalecus*. *See also* NW p.*240.*

OAS, (Organisation de l'Armée Secrete) terrorist group opposed to Algerian independence. It was set up in 1961 by disaffected army officers and French settlers in Algeria when it became clear that President DE GAULLE was preparing to come to an agreement with the FLN (Algerian nationalists) to give the country independence. The terror tactics of the OAS affected both Algeria and metropolitan France, but failed to prevent the declaration of Algerian independence on 3 July 1962.

OAS (Organization of American States), an organization comprising 23 North and South American countries including the USA and Canada, pledged to seek solutions to international disputes affecting America. It is the result of the Pan American Union held in 1948 in Bogatà. The OAS was founded on the principles of the US MONROE DOCTRINE, namely that any attack on an American country by another country would be regarded as an attack on all.

Oasis, area in a desert that is made fertile by the presence of water. It can be formed by the passage of rivers through the desert or by underground waters reaching the surface. Larger oases, eg on the River NILE, support a large agricultural population. *See also* PE pp. 109, *109, 121.*

Oasthouse, or hop kiln, building designed for drying hops used in brewing. A natural or forced draught of heated air is passed through the hops to dry them; the pointed, pivoted cowl of the oasthouse prevents backdraughts. *See also* PE p.*204.*

Oastler, Richard (1789–1861), British radical reformer, nicknamed the "factory king". In the 1830s and 1840s he was a leader of the mass movements of the working class to repeal the Poor Law of 1834 and to gain a maximum ten-hour day in the textile factories, which came into effect as the Ten Hours Act (1847).

Oat, cereal plant native to W Europe and cultivated throughout the world; it grows well even in poor soils. The flower is composed of numerous florets that produce one-seeded fruits. Mainly fed to livestock, oats are also eaten by human beings, particularly as a breakfast food. Family Gramineae; species *Avena sativa*. *See also* PE pp.156–157, 158, 180–181.

Oates, Joyce Carol (1938–), US novelist, short-story writer and poet. Her first book of short stories, *By the North Gate*, appeared in 1963. Her works are a grim assessment of modern American life; they are chronicles of violence and economic and emotional deprivation. Other works are *A Garden of Earthly Delights* (1967), *Them* (1969) and *The Assassins* (1975).

Oates, Captain Lawrence Edward Grace (1880–1912), British explorer who accompanied Capt. Robert SCOTT on the Antarctic expedition of 1910-12. Oates sacrificed his own life by deliberately crawling out into a blizzard because, crippled by frostbite, he was slowing the party on its return journey from the South Pole.

Oates, Titus (1649–1705), English priest who invented the story of the POPISH PLOT. The son of an Anabaptist preacher, he was ordained into the Church of England but was imprisoned for perjury in 1674. He joined the Roman Catholic Church in 1677, but was later expelled from Spanish and Dutch seminaries. In 1678 he and Israel Tonge invented a story of a JESUIT plot to murder CHARLES II. Although Oates was convicted of perjury for this in 1685, he was pardoned in 1688. In 1693 he became a Baptist, but that Church expelled him in 1701.

Oath, appeal to God to witness the truth of something. In a corporal oath, usually taken by a person before he gives evidence in a court of law, the witness swears to the truth of a statement. A promissory oath relates to a promise to do something in the future, eg an Oath of Allegiance. In English courts the witness places his hand on a Bible and repeats after the officer administering the oath "I swear by Almighty God that...". It is also legally permissible to affirm the truth of a statement rather than take an oath (*See* AFFIRMATION).

OAU (Organization of African Unity), association of 42 African states for their co-ordinated development and unity in defence of African independence, founded in 1963 in Ethiopia. It arose out of the Pan-African movement and works towards a united African front in international affairs. Yearly summit conferences are held.

Ob, river in w Siberia, USSR; with its principal tributary, the Irtysh, it is the fifth-longest river in the world, 5,150km (3,200 miles) long. Formed at the confluence of the Diya and Katun' rivers sw of Bisk, the Ob flows NW then NE through the lowlands of w Siberia. It continues N and then E to enter the Gulf of Ob, an arm of the Kara Sea within the Arctic Ocean. The river is approx. 19km (12 miles) wide in its lower course. The Ob is frozen for six months of the year but it remains an important trade route. Length: 3,410km (2,120 miles).

Obadiah, fourth of the 12 minor prophets in the Old Testament; the book of Obadiah has only 21 verses. Sometimes called Abdias, he is thought to have lived in the first half of the 6th century BC because he calls for vengeance upon the Edomites, who refused to help the Israelites at the fall of Jerusalem to the Babylonians (586 BC).

Obbligato, in music, an accompanying vocal or instrumental part that complements the MELODY. It is commonly in double time and continues even when the tune pauses. Originally, however, the term was used to describe the section of the music that had to be played exactly as written in which the composer did not allow the improvisation common in earlier BAROQUE music.

Obelia, genus of plant-like COELENTERATES that are composed of a colony of several sub-individuals or POLYPS. Each species is concerned with the production of free-swimming MEDUSA forms which carry the reproductive organs. Sperm and eggs are released by the medusae and fertilized eggs subdivide, settle, grow and then bud asexually. *See also* NW pp.82–83, *82*.

Obelisk, Egyptian monolith, often carved from red granite, usually a tapering square-based column with a pyramid-shaped point. Pairs of obelisks stood at the entrance to Egyptian temples and their decorative hieroglyphs suggest an association with sun worship. Many obelisks were removed to Italy by the Romans and one of the largest, 32m (105ft) high, weighing *c*.230 tonnes) is still standing in Rome. The oldest extant obelisks date from about 2000 BC, and the two *Cleopatra's needles* in New York's Central Park and on London's Thames Embankment, date from 1500 BC, long before the reign of CLEOPATRA herself.

Oberammergau, small village in upper Bavaria, West Germany, famous for the performances of its PASSION PLAY. This takes place once every 10 years; the practice originated in 1634 in fulfilment of a vow made by the inhabitants during an outbreak of the plague.

Oberon, in medieval folklore, the king of the fairies and husband of TITANIA. He first appeared in English literature in a prose translation (*c*.1534) by Lord Berners of the French romance *Huon de Bordeaux*. Perhaps the most familiar use of the character is in William SHAKESPEARE's *A Midsummer Night's Dream*; here the author is believed to owe much to the earlier French work. Oberon is shown as a magical figure with an impish servant, PUCK. Later Christoph WIELAND, a German poet, wrote the epic *Oberon* (1780), inspired by Shakespeare's character. Carl WEBER also wrote an opera *Oberon* (1826) about him.

Oberon, second-largest of the five satellites of the planet Uranus. It orbits at a distance of 586,230km (364,107 miles) from Uranus and has a diameter of approx. 1,600km (995 miles). It has a sidereal period of 13.46 days. *See also* SU p.214, *214*.

Oberth, Hermann Julius (1894–), German physicist, *b.* Austro-Hungary, who made a fundamental contribution to the development of astronautics. He conducted experiments to simulate and study the effects of weightlessness and designed a long-range, liquid-propellant rocket. During WWII he worked on German rocket development under Wernher VON BRAUN. Oberth later worked on the development of solid-propellant, anti-aircraft rockets.

Obesity, condition in which excessive amounts of fat are stored beneath the skin and within organs. Medically it is defined as an accumulation of body fat sufficient to impair health. People who are overweight risk diseases of the kidney, arteries and heart, and have a shorter life expectancy than those of normal weight. Obesity is usually caused by overeating – the consumption of more calories than the body can use. Hormone imbalance, glandular defects and genetic predisposition may also be factors, although they are far less common. *See* MS pp.70, 101.

Objet trouvé, in art, French term meaning "object found" and usually referring to inanimate items (eg rope and driftwood) which take on aesthetic significance as part of an assemblage. Such objects were often used in DADAIST and SURREALIST art, and have remained as part of the artist's vocabulary, as in works by Robert RAUSCHENBERG.

Oblast, administrative and territorial district in the USSR, a division of one of the constituent republics. If the people in a particular territory are of a nationality different from most others in the republic to which it belongs, that territory usually becomes an autonomous oblast.

Oblomov (1859), Russian satirical novel by Ivan GONCHAROV. Oblomov, an indolent nobleman, succeeds in wasting his life despite the efforts of his friend Stolz to stir him into action. The word "Oblomovism" was coined to describe lethargic aristocrats in Russia *c*.1860.

Oboe, woodwind musical instrument (from the French *hautbois*) with a range of 2¼ octaves. It has a slightly flared bell and, like the BASSOON, is a double-reed instrument. The oboe somewhat resembles the clarinet and has a plaintive tone. The earliest true oboes were used in the French court shortly after the mid-17th century. *See also* HC2 pp.104–105.

Obote, Apollo Milton (1924–), Ugandan Prime Minister and President (1966–71). He created the Uganda People's Congress in 1960, and was instrumental in the country's gaining of independence two years later. He worked for greater national unity, but was ousted by Idi AMIN in 1971.

Obrenović dynasty, Serbian dynasty which ruled 1815–42 and 1858–1903, established at the end of Turkish domination. The first in line, Miloš (1780–1860), led the Serbian revolt against the Turks in 1815 but was forced to abdicate in 1839. His son Milan was briefly reinstated, followed by his brother Michael who (until 1842 and again 1860–68) continued his father's policy of achieving complete Serbian liberation. The line became extinct with the murder of Alexander in 1903. *See also* HC2 p.84.

O'Brien, Edna (1932–), Irish novelist. Her first novels, *The Country Girls* (1960) and *The Lonely Girl* (1962), established her reputation. She has published one play, *A Pagan Place* (1971), based on her novel of the same title.

O'Brien, Michael Vincent (1917–), Irish trainer of racehorses who trained three consecutive Grand National winners and four Derby winners. He trained *Nijinsky*, the first triple crown winner since 1935. He also trained winners of two Irish Derbys, the Prix de l'Arc de Triomphe and the Washington International.

O'Brien, William (1852–1928), Irish nationalist leader, who was an MP from 1883 to 1895. He formed the agrarian movement called the United Irish League in 1898 and took part in the conference which preceded the LAND ACT of 1903. He formed the "All for Ireland" party in 1910, but caused it to withdraw from the 1918 elections, thereby handing power to SINN FEIN.

O'Brien, William Smith (1803–64), Irish nationalist. He entered Parliament as an Irish member in 1828 and supported Roman Catholic emancipation. In 1843 he joined Daniel O'CONNELL's Repeal Association to end Irish union with Britain but left to set up the more militant YOUNG IRELAND Movement. He was arrested in 1848 for insurrection and sentenced to death, but his sentence was commuted to transportation. Released in 1854, he was pardoned in 1856.

Observatory, astronomical, location of telescopes and other equipment for astronomical observations. Large optical TELESCOPES are housed in domed buildings sited usually well away from the smoke of cities, high up in mountain areas. These observatories rotate to face any area of the sky, the telescope being exposed by the retraction of a panel in the dome. Radio observatories are open sites containing one or more large radio telescopes, usually having the shape of a parabolic dish. The largest radio-telescopes have been built in natural mountain hollows: one in Puerto Rico, for example, is 300m (975ft) across. Observatories of the future will include telescopes carried by orbiting satellites. *See also* SU pp.170–171.

Obsidian, rare grey to black glassy volcanic rock. High in silica, it is the uncrystallized equivalent of RHYOLITE and GRANITE. Hardness 5.5; s.g. 2.4. It polishes well and makes an attractive semi-precious stone. *See also* PE p.124.

Obstetrics, branch of medicine that deals with pregnancy, labour, childbirth and the care of women immediately following childbirth. It also treats any abnormalities of pregnancy. *See also* MS p.112.

Ocarina, small keyless musical wind instrument of metal or terracotta, shaped like an elongated egg. It was invented in late 19th-century Italy and is popular for its simplicity and low cost.

O'Casey, Seán (1880–1964), Irish playwright. His first play, *The Shadow of a Gunman* (1923), immediately made him famous. *Juno and the Paycock* (1924), a tragedy of the Dublin slums, was followed by *The Plough and the Stars* (1926), a drama centred on the EASTER RISING of 1916. O'Casey moved to England in 1926 and his later works, such as *The Silver Tassie* (1928), are in an expressionistic style, very different from the realism of his early plays. He never again attained, however, the standard of his first works. Apart from his plays his most important work is a fictional autobiography, written in six volumes, which originally appeared be-

Titus Oates was considered a hero but the Popish Plot was a fabrication by him.

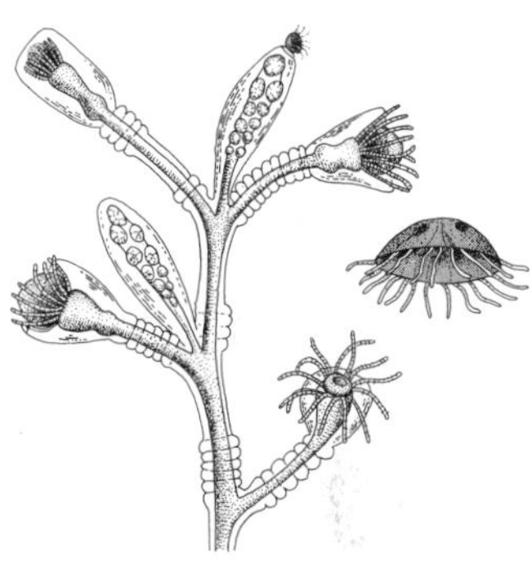

Obelia lives on rocks and around harbours throughout most of the temperate seas.

Oberammergau's quiet streets become busy with visitors during Passion Plays.

Oboe has the highest pitch and is the smallest double reed woodwind instrument.

Occult; a display of some of the macabre ingredients used in the black art.

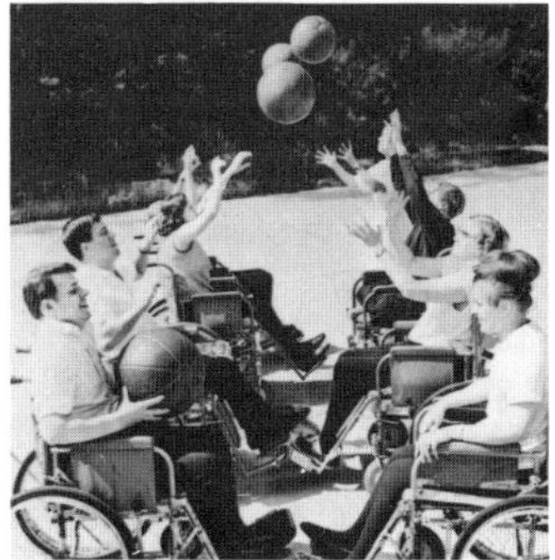

Occupational therapy; ball games are a form of exercise for paraplegics.

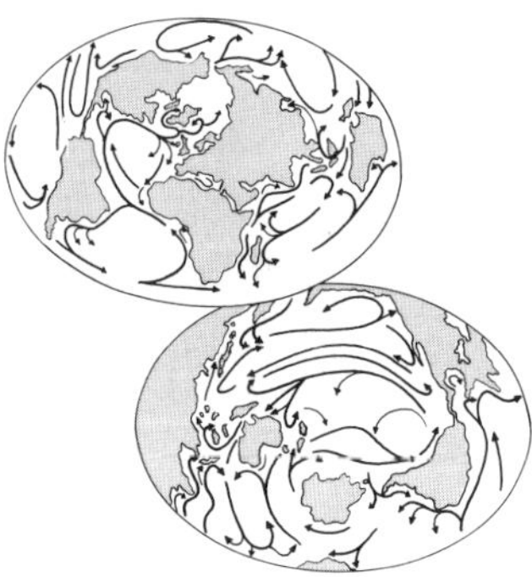

Oceanic currents; some currents flow deep down beneath the surface of the oceans.

Daniel O'Connell, the Liberator, strove to achieve Roman Catholic emancipation.

tween 1939 and 1956 and was re-issued as a collection, *Autobiographies,* in 1963. *See also* HC2 p.283.

Occam (Ockham), William of *c.*1300–*c.*47), English theologian and philosopher. Writing against the Pope, and in favour of Franciscan ideas of poverty and the empire, caused him to be excommunicated. He was a leader of the Nominalist school of philosophy. Contributing to the development of formal logic, he employed the principle of economy known as OCCAM'S RAZOR. *See also* HC1 p.194.

Occam's Razor, principle of economy of explanation named after the 14th-century philosopher William of OCCAM. He said that explanatory principles should not be needlessly multiplied. It is an important scientific principle according to which any process has a minimum number of causal links.

Occident, The, term meaning the WEST. Formerly applied to the countries of Western Europe, it is now generally taken to include also the Americas.

Occiput, also called occipital region, in insects the plate of the EXOSKELETON that forms the back of the head; in vertebrates the area of the head at the joint between the skull and the spinal column.

Occult, the supernatural, especially referring to the forces of darkness. The term is commonly used, however, to describe all man's attempts to predict, forestall or induce events by employing hidden powers, secret knowledge or external forces. In this context the term describes divination by all means other than by mere observation, and includes ASTROLOGY, fortune-telling, the I CHING and the KABBALA. In strong contrast, Satanists and practitioners of black magic in their rites call upon occult forces in an entirely selfish aim: to gain personal power over others. The invoking of such forces, especially by means of such apparatus as OUIJA BOARDS, is extremely dangerous and has led to documented cases of demonic possession. *See also* MS pp.196–197.

Occultation, in astronomy, the temporary concealment of one celestial body by another, especially of a star by the Moon or a planet. Lunar occultation is of particular importance because it has served accurately to pinpoint a number of notable radio sources.

Occupational diseases, disorders to health produced by working conditions. They extend over the whole field of medicine, from the physical injuries so often incurred by miners, tunnellers and divers to the duodenal ulcers and coronary thrombosis suffered widely by people in sedentary, stressful occupations. More specifically, workers in various trades are liable to suffer from the bad effects of prolonged contact with harmful substances: miners' pneumoconiosis, bladder cancer in rubber processors, leukaemia among workers with radioactive materials and mercury and lead poisoning among workers in extractive processes are a few of many examples recorded this century.

Occupational therapy, programme to help physically or mentally disabled and handicapped people develop skills to enable them to lead normal lives and be usefully employed. This type of treatment includes interesting occupations and pastimes to overcome a particular handicap. A person with an artificial leg, for example, might be taught to dance. Occupational therapy is often called "curing by doing", since the patient himself must carry out the activity.

Ocean, continuous body of water that surrounds the continents and fills the Earth's great depressions. There are five main oceans, the ATLANTIC, PACIFIC, INDIAN, Arctic, and Antarctic and they cover 71% of the Earth's surface. The oceans may be divided by distinct region (Littoral, Pelagic and Abyssal) or by depth region (CONTINENTAL SHELF, CONTINENTAL SLOPE, deep sea plain and the deeps). The ocean floor is not flat but has a varied topography like land, with vast mountain chains, valleys and plains. Ocean water consists of about 3.5% dissolved minerals, hence its salty taste. It is constantly moving in currents and waves. *See also* PE pp.76, 82–83.

Oceanarium, large salt-water aquarium for the study and display of sharks and other marine fishes, marine mammals such as dolphins, porpoises and killer whales, and octopus, squid and many lesser invertebrates.

Ocean floor, bottom of the ocean, covered to various depths with marine sediments, both inorganic and organic. Weathered materials, washed and blown off the land, and chemically precipitated from sea water sink to form fine and coarse inorganic sediments. There are always accompanied by organic materials derived from marine organisms, most abundantly the chalky shells of countless tiny organisms of the PLANKTON. Sediments vary in thickness and extent with the topography of the ocean floor, the main zones being the littoral, the continental shelf, submarine mountains and abyssal trenches. *See also* PE pp.82–83.

Oceania, collective name for the lands of the Pacific Ocean. In its widest sense the term includes all islands between Asia and the Americas, excluding the Ryuku and Aleutian Islands and the Japan archipelago, as well as Taiwan, Indonesia and the Philippines (whose people are more closely related to those of the Asian mainland). More generally, however, Oceania refers to the islands of Melanesia, Micronesia and Polynesia and, usually, Australasia (Australia and New Zealand).

Oceanic art, art of the peoples of the islands of the South Pacific (Melanesia, Micronesia and Polynesia). Much of their art involves objects used in religious rites, masks and stylized portrayals of the human face. Among the more notable of these are the giant stone ancestor-cult figures of EASTER ISLAND, MAORI wood carvings and the carved drums, masks, stools and shields of NEW GUINEA. Oceanic artists are greatly admired and respected in their communities.

Oceanic currents, movement of sea water between layers of varying temperature and density, caused by wind friction at the surface. Ocean circulation is produced by CONVECTION, with warm currents travelling away from the equator, cooler water moving from the poles. In the Southern Hemisphere the oceanic currents move in an anti-clockwise system, whereas in the Northern the system is clockwise, an effect of centrifugal force caused by the Earth's rotation. There are about 50 major currents, the most notable of which are: the Gulf Stream of the N Atlantic, the warm Japan current of the N Pacific, the Humboldt (Peru) current off the W coast of South America and the cold Falkland current in the S Atlantic. These currents frequently affect the climates of land masses which they pass. *See also* PE p.74.

Ocean liner, ship used for the conveyance of passengers from one port to another, as distinct from freight or mixed freight cargoes. The modern era of the ocean liner reached its zenith in terms of speed and luxury with the two Cunard liners, *Queen Mary* and *Queen Elizabeth. See also* MM pp.114–115.

Ocean of Storms (Oceanus Procellarum), area on the Moon, a "sea" in the W Northern Hemisphere, separated from the Mare Imbrium by the Carpathian Mts. The area includes the crater Aristarchus, generally considered to be the brightest ray crater on the Moon. *See also* SU p.184, *184.*

Oceanography, science of the marine environment which studies oceans and seas past and present, the shorelines, sediments, rocks, muds, plants, animals, temperatures, tides, winds, currents, formation and erosion of abyssal depths and heights, and the effect of neighbouring land masses. *See also* PE pp.76, 78, 80, 82, 90.

Ocean sunfish, oval fish of the Molidae family, most species of which appear to be tailless. Marine sunfish inhabit warm seas and often swim near the surface of the water. Two species of the genus *Mola* may grow to 2.5m (8ft) in length.

Oceanus, in Greek mythology, a Titan and, with Tethys, parent of 3,000 rivers and numerous daughters, the Oceanids. The Greeks imagined the earth was flat, and that the world was encircled by Oceanus, a river flowing from the Underworld, on whose banks lived the spirits of the dead.

Ocelot, small, spotted cat that lives in scrub and rocky or forested areas of S USA, Central and South America. Its valuable fur is yellowish or rust marked with elongated dark spots. It feeds on small birds, mammals and reptiles. Length: to 1.5m (4.9ft). Family Felidae; species *Felis pardalis. See also* NW pp.*155,* 217.

Ochoa, Severo (1905–), US biochemist *b.* Spain, who emigrated to the USA in 1941. He shared the Nobel Prize in physiology and medicine with Arthur KORNBERG for work on the synthesis of ribonucleic acid (RNA), the hereditary material of some viruses and deoxyribonucleic acid (DNA), the hereditary material of most cells.

Ocho Rios, port in N Jamaica, in the West Indies. It is a popular tourist resort. Pop. (1970 est.) 6,900.

Ochre, yellowish or red pigment made by mixing a type of iron ore (LIMONITE or HAEMATITE) with fine clay and grinding the dried product to a powder.

Ochs, Adolph (1858–1935), US newspaper publisher under whose direction the New York *Times* became one of the world's influential newspapers. He began his career selling papers and then became a printer's apprentice. At the age of 20 he was publisher of the Chattanooga *Times.* He bought the failing New York *Times* in 1896 and gave it the slogan "All the news that's fit to print", which is still printed on its front page. Ochs abhorred sensational journalism and demanded non-partisan reporting from his staff.

Ochtervelt, Jacob (*c.*1635–*c.*1710), Dutch portrait and society painter whose work was much influenced by Pieter de HOOCH and, through him, by his contemporaries VERMEER, TERBORCH and METSU. In his interior scenes, Ochtervelt showed a marked preference for orange-coloured fabrics.

Ockeghem, Johannes (*c.*1425–95), Flemish composer, pupil of DUFAY and teacher of Josquin DES PRÉS. He was composer at the courts of three French kings.

Ockham, William of. *See* OCCAM (OCKHAM), WILLIAM OF.

O'Connell, Daniel (1775–1847), Irish nationalist leader, known as The Liberator. A practising lawyer during the late 1790s, he joined the United Irishmen but refused to take part in the revolt of 1798, preferring peaceful means for the attainment of Catholic emancipation and parliamentary reform. He founded the Catholic Association in 1823 and became an MP at Westminster after the introduction of Catholic emancipation in 1829. He led the small group of Catholic Irish MPs supporting the Whigs in return for reforms which, however, were not forthcoming. O'Connell proceeded to work for the repeal of Anglo-Irish leglislative union and founded the Repeal Association in 1839. In 1844 he was imprisoned for three months after a series of mass meetings throughout Ireland. He then played only a minor political role owing to ill-health. *See also* HC2 p.162, *162.*

O'Connor, Feargus Edward (1794–1855), Irish politician. An MP for Cork (1832–35), he clashed with Daniel o'CONNELL, established the CHARTIST newspaper *Northern Star* in 1837, and gained great popularity in N England. Elected MP for Nottingham in 1847, he presented the Chartists' petition in the following year.

O'Connor, James Arthur (1792–1841), Irish landscape painter, who first exhibited at the Royal Academy in 1822. Trained as an engraver, he turned to painting, modelling his work on that of Richard WILSON. His best landscapes are those he painted in Belgium and Prussia. He was a close associate of Francis DANBY.

O'Connor, Rory (*c.*1116–98), last Irish high king. He became King of Connaught in 1156 and King of all Ireland in 1166. After the Anglo-Norman invasion he was forced to acknowledge HENRY II of England as his liege Lord (1175).

O'Connor, Thomas Power (1848–1929), Irish journalist and political figure. As member for Galway in the British Parliament of 1880, he supported PARNELL and wrote *The Parnell Movement* (1886). He represented Liverpool in the House of Commons (1885–1929), during which time he founded several newspapers, including *T.P.'s Weekly* (1902).

Octane number, indication of the anti-knock properties of a liquid motor fuel. It represents the percentage by volume of iso-octane in a reference fuel consisting of a mixture of iso-octane and normal heptane that matches the knocking properties of the fuel being tested. The higher the number, the less likely the possibility of the fuel detonating. *See also* MM p.81.

Octans, or The Octant, faint constellation in which the south celestial pole is located. Sigma Octantis, the star closest to the pole, is of the fifth magnitude; the brightest star is Nu Octantis. *See also* SU p.261.

Octave, in music, the interval between any given note and another one that is exactly twice (or half) the frequency of the first and thus, acoustically, a perfect consonance. For example, a note of frequency 880Hz sounds an octave higher in pitch than one of 440Hz. This interval is so called because in Western music it encompasses the eight notes of the DIATONIC scale.

Octavian. *See* AUGUSTUS.

October, the tenth month of the year, with 31 days. It was the eighth month in the early Roman calendar and its name is derived from the Latin for eight, *octo*.

October Revolution. *See* RUSSIAN REVOLUTION.

Octopus, predatory cephalopod mollusc with no external shell. Its sac-like body has eight powerful suckered tentacles. It moves by "jet-propulsion" and crawling, and commonly hides in crevices or among debris on the bottom of shallow seas. Many of the 150 species are small, but the common octopus (*Octopus vulgaris*) can grow to a length of 9m (29ft). Family Octopodidae *See also* NW pp.72, 72, 94–95.

O'Dálaigh, Cearbhall (Carroll O'Daly) (1911–), President of the Republic of Ireland (1974–76). He was Irish editor of the *Irish Press* (1931–40), Attorney General (1946–48; 1951–53) and Chief Justice and President of the Supreme Court (1961–73). He was Judge of the Court of Justice of European Communities, Luxembourg in 1973, and became President of the First Chamber of the Court of Justice of European Communities in 1974.

Odds, in betting and games of chance, the financial returns offered to a gambler in proportion to his wager. For example, odds of 3:2 indicate that a bet of two money units will, if successful, win three. Such odds do not indicate true mathematical probabilities, which in horse-racing and similar sports are not precisely calculable. *See also* SU pp.58–59.

Ode, unspecific type of verse, originating in the choral songs of Greek festivals. It is usually lengthy, elaborate in manner and lyrical in mood. The first great writer of odes was PINDAR, who wrote them to a fixed pattern of versification. More simple were the personal lyrical odes of HORACE and CATULLUS. The ode became popular again in Italy during the RENAISSANCE. In 17th-century England it was taken up by JONSON, HERRICK and MARVELL. In the 19th century it was popular with WORDSWORTH, SHELLEY and, especially, KEATS.

Odense, city and port in s central Denmark, 137km (85 miles) wsw of Copenhagen. It was founded in the 10th century and has been an episcopal see since then. Odense has a 14th-century Gothic cathedral and is regarded as a great cultural centre, as well as being an industrial and commercial city. Industries include shipbuilding, metalworking and food processing. Pop. 166,727.

Oder, second-longest river in the catchment basin of the Baltic Sea. It rises in the mountains of NE Czechoslovakia, flows N and w through sw Poland, past Wrocław, turning N to form the Polish–East German border and reaching the Baltic Sea at Szczecin, NW Poland. The river has been an important water route for N Europe

since early times, serving a large area with many navigable tributaries, notably the rivers Neisse and Warta. The main port, Szczecin, is Poland's largest. Length: 886km (550 miles).

Oder-Neisse Line, frontier established between Germany and Poland at the Yalta Conference (Feb. 1945). From the Baltic Sea to the Czechoslovakian border, it followed the rivers Oder and Neisse. The US and British governments originally opposed the line because it would make Poland excessively dependent upon the USSR, and no agreement was achieved at Yalta. After disputed areas, including Danzig, had in effect been arbitrarily incorporated into Poland, the POTSDAM CONFERENCE (Aug. 1945) recognized the line pending a peace treaty with Germany, and allowed large numbers of Germans to be deported to Germany. East Germany and Poland recognized the line in 1950; the West German government did not accept it until 1970.

Odessa, city in the Ukraine (Ukrainskaja SSR), USSR, 40km (25 miles) NE of the mouth of the River Dniester on Odessa Bay. In the 14th century the city was known as Khadzhi-Bei. Under that name, it passed to the Turks in 1764 and came under Russian control in 1791. It was made a port and naval base in 1794 and named Odessa in 1795. It was occupied by Axis forces during WWII. Odessa is now the centre of a fishing and whaling fleet and its other industries include ship-building and repairing, oil-refining and the manufacture of heavy machinery. Pop. (1975) 1,002,000.

Odets, Clifford (1906–63), US playwright and one of the best-known social protest dramatists of the 1930s. Influenced by Anton CHEKHOV and the acting school of STANISLAVSKY, he helped to organize the Group Theatre in 1930. His plays include *Awake and Sing* (1935), his first full-length play, considered to be his best; *Waiting for Lefty* (1935), an expressionist one-act drama; *Golden Boy* (1937); and *The Country Girl* (1950).

Odin, the principal god in Norse mythology. He is usually identified with the Teutonic god Woden and considered to be the god of wisdom, culture, war and death. He lived with the Valkyries in VALHALLA, into which he received the souls of dead warriors. *See also* MS p.213.

Odinga, Oginga (1911–), Kenyan leftist political leader who opposed the British in the Kenyan quest for independence. As the head of the Kenya People's Union he also opposed his former close associate Jomo KENYATTA in the mid-1960s and was imprisoned between 1969 and 1971.

Odo, Bishop of Bayeux (c.1036–97), Norman warrior-priest, half-brother of WILLIAM the Conqueror. Odo became Bishop of Bayeux in 1049 because of his family connections with the ruling house of Normandy. He fought at William's side at the Battle of HASTINGS and was made Earl of Kent in 1067. He then played an important role in Anglo-Norman politics, and may also have commissioned the BAYEUX TAPESTRY.

Odoacer (Odovacar) (c.433–93), chief of the Germanic Heruli people and conqueror of the West Roman Empire. The Heruli were Roman mercenaries until 476 when they declared Odoacer King of Italy. He deposed Romulus Augustulus and established his authority over Italy, proving himself a capable ruler. In 489 the Ostrogoth King Theodoric the Great invaded Italy. Odoacer took refuge in Ravenna which surrendered in 493. Theodoric invited him to a banquet where he was betrayed and murdered.

Odonata, order of primitive winged insects found throughout the world. DAMSELFLIES, in the suborder Zygoptera, have thin bodies with wings held vertically along the body when at rest. The long, slender, aquatic nymphs have three leaf-like gills on the abdomen. DRAGONFLIES, in the suborder Anisoptera, have heavy bodies with wings held horizontally when at rest. The stout nymphs have gills at the anal end. All prey on insects; none attacks man. Length: 18–193mm (0.7–8in). *See also* NW pp.106–107.

Odontology, study of the structure, development and diseases of the teeth. It is closely allied with DENTISTRY, the practice of treating diseases of the mouth.

Oduber Quirós, Daniel (1921–), President of Costa Rica (1974–). He held a number of administrative and diplomatic posts prior to his election as the candidate of the National Liberation Party (1965) and President of the Congress of Costa Rica (1970–74).

O'Duffy, Eoin (1892–1944), Irish politician and soldier. He was Chief of Staff of the IRA (1921–22), in charge of the forces of the Irish Free State (1924–25), and head of its police (1922–33). As a prominent member of the United Ireland Party, his authoritarian ideas led him to organize an Irish Brigade to fight for Franco (1936–37) during the SPANISH CIVIL WAR. This lost him much support.

Odysseus, or Ulysses, Greek hero of HOMER's epic poem, the ODYSSEY. King of the city-state of Ithaca, husband of the faithful PENELOPE and father of Telemachus, he was an astute and brave warrior, although in post-Homeric legend he is depicted as a wily, untruthful man. It was Odysseus who devised the stratagem of the wooden TROJAN HORSE in order to enter Troy. The extent of his wanderings and the problems he encountered have featured throughout literature. *See also* HC1 p.74.

Odyssey, The, epic poem of 24 books attributed to the ancient Greek poet HOMER. The story of ODYSSEUS, King of Ithaca, it tells of his journey home from the Trojan Wars after 10 years of wandering. To win back his wife PENELOPE and his kingdom he enlists the help of his son Telemachus and kills the suitors who have been wooing Penelope in his absence.

OECD, Organization for Economic Co-operation and Development, international consultative body set up in 1961 by the major Western trading nations. Its aims are to stimulate economic growth and world trade by raising the standard of living in member countries and by co-ordinating aid to less developed countries.

Oedema, swelling in the body caused by accumulation of watery liquid, as the result of an abnormal leakage from blood capillories and lymph vessels; formerly called dropsy. It is a symptom particularly of NEPHRITIS (inflammation of the kidney) but also occurs in heart and liver disease as an allergic reaction and in pregnancy.

Oedipus, in Greek mythology and literature, son of Laius, King of Thebes, and of Jocasta, father of ANTIGONE, Ismene, Eteocles and Polynices by his own mother. In the 5th-century BC the playwright SOPHOCLES told how Oedipus was saved from death as an infant and raised in Corinth, how he inadvertently killed his father, solved the riddle of the SPHINX, and became King of Thebes. There he married Queen Jocasta, unaware that she was his own widowed mother.

Oedipus complex, in psychoanalytic theory, incestuous fantasy in which a child desires the parent of the opposite sex. Sigmund FREUD held that children pass through this stage between the ages of three and five. The son desires his mother and subconsciously hopes for his father's death, by which the son may replace him in his mother's attentions. A daughter has analogous feelings; desire for her father and a subliminal desire for her mother's removal. The complex for females is sometimes termed the Electra complex.

Oedipus Rex (c.420 BC), tragedy by SOPHOCLES, one of the greatest in Greek literature. OEDIPUS, King of Thebes, unknowingly kills his father (Laius) and marries his mother, JOCASTA. She hangs herself when the truth is revealed, and Oedipus blinds himself.

Oehlenschläger, Adam Gottlob (1779–1850), Danish poet and dramatist. Influenced by German Romanticism, his poetry includes *The Golden Horns* (1803), *Midsummer Night's Play* (1803) and *Aladdin* (1805). Oehlenschläger used Norse myths in his plays *Hakon Jarl* (1807) and *Baldur the Good* (1807). In 1829 he was given the title of Scandinavian Prince of Poets.

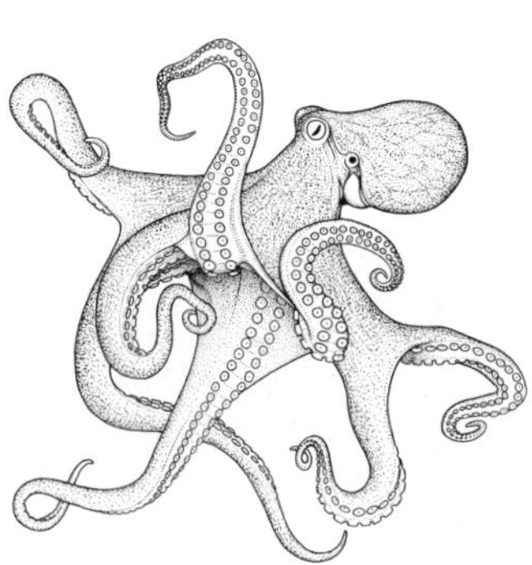

Octopuses; their frightening aspect has led to their also being called devilfish.

Odin: the Scandinavian name of the chief Nordic god the Anglo-Saxons called Woden.

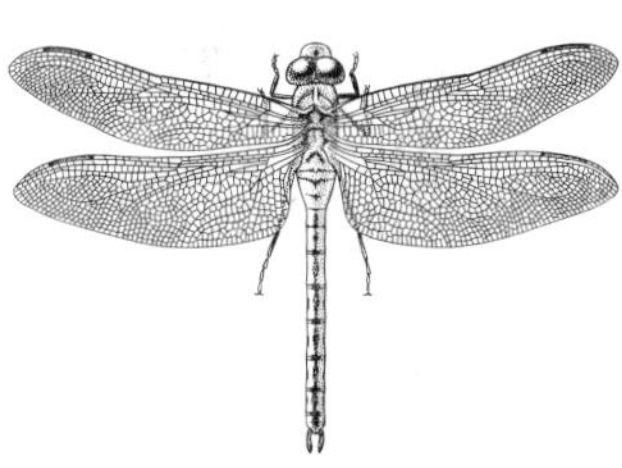

Odonata cannot walk but when flying they are able to catch prey with their legs.

Oedipus solved the Sphinx's riddle, after which she killed herself, freeing Thebes.

Hans Oersted founded the Danish Engineering College and was its president.

Jacques Offenbach is the subject of this caricature by André Gill.

John Ogdon and his wife, Brenda Lucas, are both famed concert pianists.

Georg Ohm formulated the law linking voltage, current and electrical resistance.

Oersted, Hans Christian (1777–1851), Danish physicist and professor at Copenhagen University. He took the first steps in elucidating the relationship between electricity and magnetism, thus founding the science of ELECTROMAGNETISM. The OERSTED unit of magnetic field strength is named after him. He was also the first scientist to isolate pure metallic aluminium in 1825.

Oersted, unit of magnetic field strength in the CGS SYSTEM, equal to the magnetic field that would cause a unit magnetic pole to experience a force of one dyne in a vacuum. It is named after Hans Christian OERSTED.

Oerter, Alfred (1936–), US discus-thrower. He won the Olympic gold medal in the DISCUS event in 1956, 1960, 1964 and 1968. He held the world record for the event from 1962 to 1964.

Oesophagus, muscular tube, part of the alimentary canal (or gut) which carries swallowed food from the pharynx (throat) to the stomach. In adults it is about 30cm (12in) long. *See also* MS p.*68*.

Oestrogen, female SEX HORMONE which is first secreted by a girl at PUBERTY, and leads to the development of the secondary sexual characteristics that turn her body into a woman's: breasts, body hair and redistributed fat. Oestrogen is also a constituent of the contraceptive PILL. *See also* MS pp.65, 74, 75, 80, *101*.

O'Faoláin, Seán (1900–), Irish novelist and short-story writer. His books examine and portray life in Ireland, and that nation's difficulty in reconciling past history with present reality. Collections of stories include *A Purse of Coppers* (1937), *I Remember! I Remember!* (1962), and *The Talking Trees* (1970). Among his novels are *A Nest of Simple Folk* (1933), *Bird Alone* (1936) and *Come Back to Erin* (1940). He also wrote several biographies and studies of Ireland, notably *An Irish Journey* (1940) and *The Irish* (1949). *See also* HC2 p.283.

Offa (d. 796), King of MERCIA (r. 757–96). He made Mercia the foremost Anglo-Saxon kingdom of Britain by gaining territory from WESSEX and the Welsh and controlling the wealthy areas of Kent and Sussex. Offa styled himself *Rex Anglorum* (King of the English). It is probable that he built OFFA'S DYKE. *See also* HC1 p.154.

Offaly, county in Republic of Ireland, Leinster province. It is mainly flat and marshy and drained by the rivers Shannon, Barrow, Brosna and Nore. The chief occupation is agriculture, particularly potatoes, cereals and cattle-raising. There is also a distilling industry. The county town is Tullamore. Area: 1,997sq km (771sq miles). Pop. (1971) 51,834.

Offa's Dyke, earthwork fortification in England and Wales running from the mouth of the River Dee, Clwyd, to Chepstow, Gwent. It was probably built in the reign of OFFA, King of Mercia (r. 757–96), and was designed to mark the Welsh-Mercian border. Its ditch and high rampart were intended to reduce raiding by the Welsh tribes. *See also* HC1 pp.*154, 185.*

Offenbach, Jacques (1819–80), French composer. His great reputation was founded on his brilliance as a composer of operettas: he wrote more than 90, the most famous being *Orpheus in the Underworld* (1858). His only serious opera, *The Tales of Hoffman* (1881), is one of the masterpieces of French opera.

Offertory, in the EUCHARIST of all Christian Churches, the presentation of the elements, bread and wine, upon the altar as a sacrificial offering to God from His people. The whole of this part of the service, with the accompanying prayers and any music, is commonly called the offertory. In Anglican and other Churches following the REFORMATION, the term has been extended to include the money taken at the collection.

Office of the UN High Commissioner for Refugees, organization founded in 1951 to replace the International Refugee Organization (1946–50). Although its administrative costs are financed from the regular budget of the UNITED NATIONS, its fieldwork is carried out with voluntary

pledges because it is believed that governments should take the main responsibility for refugees. Thus the work of the Office, which has branch offices in more than 60 countries, consists mainly of negotiating agreements with governments regarding their acceptance and acknowledgement of refugees' rights. The High Commissioner is responsible to the UN General Assembly and is responsible for the operations undertaken by the Office.

Offset printing, method of printing used widely for magazines and other publications produced in large numbers. It is carried out on presses which have three rollers. The image to be printed – text or illustration – is produced photographically on a printing plate, usually of metal, which is wrapped around one of the rollers. The plate is dampened and the image rejects the moisture, but absorbs the ink which is then applied with rollers. As the press rolls, the image is transferred to a middle, offset, rubber-coated roller, from which it is transferred to paper passing around a third roller. This process, known as offset LITHOGRAPHY, is suitable for long runs because the printing plate with its flat image is not as quickly worn as are directly-printed plates containing raised or sunken images. *See also* LETTERPRESS PRINTING; MM pp.206–207.

Of Human Bondage (1915), novel by W. Somerset MAUGHAM about a lame medical student and his passion for a heartless waitress. The book, partly autobiographical made Maugham's reputation as a serious writer. The film made of it in the USA in 1934 established Bette DAVIS as a major film actress.

O'Flaherty, Liam (1897–), Irish novelist. After working in London and sailing around the world (1918–21) he took up writing, living in Dublin. His novels include *The Informer* (1925), *The Puritan* (1931), *Famine* (1937) and *Insurrection* (1950). He also wrote many volumes of short stories and two autobiographical works, *Two Years* (1930) and *Shame the Devil* (1934).

Of Mice and Men (1937), novel by John STEINBECK. It is about Lennie Small, a huge feeble-minded man, and his friend George Milton, both itinerant labourers whose dreams of a Utopian future together are shattered when Lennie unintentionally kills a woman. Steinbeck successfully dramatized the novel in the same year, and it was filmed in 1939.

Ogaden, region of SE Ethiopia, on the border of the Somali Democratic Republic. It is a desert region inhabited chiefly by Somali nomads. It was conquered by Ethiopia in 1891. It is claimed by Somali nationalists as rightfully theirs.

Ogbomosho, city in SW Nigeria, approx. 80km (50 miles) NNE of Ibadan. It was a HAUSA stronghold against Fulani invasions in the early 19th century. It is now a processing and shipping centre for the surrounding region. Products include tobacco, cattle and fruits. Pop. (1975 est.) 432,000.

Ogdon, John (1937–), British pianist and composer. After studying at the Royal Manchester College of Music, he won considerable recognition in Britain in the late 1950s and established his international reputation in 1962 when he was joint winner of the Tchaikovsky Competition in Moscow with Vladimir ASHKENAZY. His compositions, which are principally for piano, include preludes and a concerto.

Ogeden, city in N Utah, USA, 53km (33 miles) N of Salt Lake City. It was settled by MORMONS in 1846 and is thought to be the oldest continuously settled area in the state. Its industries include food processing and electronic equipment. Pop. (1970) 69,478.

Ogee, double-curved line, half concave and half convex, like the letter S. An ogee arch, introduced into Europe in *c.* 1300, consists of four arcs, so combined to produce the double-curve effect. The word is also used for a double-curved moulding.

Ogham, Ogam or Ogum, alphabetic writing of the Irish (Gaelic) and Pictish languages during and after the 4th century AD. The script took the form of horizontal

or diagonal lines carved across the edges of upright rectangular stones; its usage had dwindled by the late Middle Ages.

Ogilvy, Angus James Bruce. *See* ROYAL FAMILY.

Oglethorpe, James Edward (1696–1785), British general and colonist. After military service in the Austrian army, he returned to England (1722), was elected to Parliament, and became interested in social reform. Taking a group to North America, he founded the colony of Georgia (chartered in 1732) for imprisoned English debtors; he founded Savannah in 1733. He successfully defended the colony in the war between Spain and England.

OGPU. *See* KGB.

Ogwr, county district in SW MID-GLAMORGAN, S Wales, created in 1974 under the Local Government Act (1972). Area: 285sq km (110sq miles). Pop. (1974 est.) 128,100.

O'Hara, John (1905–70), US author, many of whose short stories and novels are sharp observations of American urban life. His novels include *Appointment in Samarra* (1934); *Butterfield 8* (1935); *Pal Joey* (1940), which was also a successful musical comedy; *Ten North Frederick* (1955), for which he won the 1956 National Book Award in fiction; *From The Terrace* (1959); *Ourselves to Know* (1960) and *The Big Laugh* (1962).

O. Henry. *See* HENRY, O.

O'Higgins, Bernardo (1778–1842), South American revolutionary who became commander of Chile's anti-royalist forces in 1813. In 1817, with SAN MARTIN, he defeated the Spanish and was named supreme director of Chile, the independence of which he proclaimed in 1818. His reforms aroused much opposition, however, and in 1823 he was exiled to Peru, where he died.

O'Higgins, Kevin Christopher (1892–1927), Irish politician who joined the SINN FEIN movement during WWI. He became Minister of Justice and Vice-President of the Executive Council in the provisional Irish government of 1922. In 1927 he was appointed Minister of External Affairs, and in the same year was assassinated.

Ohio, state in E central USA, bounded by Lake Erie in the N. Mostly low-lying, rolling country, the state is drained chiefly by the Ohio, Scioto, Miami and Muskingum rivers. Ohio's large farms produce hay, maize, wheat, soya beans and dairy foods, and cattle and pigs are raised. The state is highly industrialized. The leading industries are car and aircraft manufacture, transport equipment, primary and fabricated metals, machinery and plastic and rubber goods. Ohio leads the nation in the production of sandstone. Oil and natural gas, clay, salt, lime and gravel are also produced and there are large deposits of coal. The major cities are Columbus (the state capital), Cincinatti and the lake ports of Toledo and Cleveland, which handle large amounts of iron and copper ore, coal, oil and finished goods.

French and English traders operated competitively in the area in the early 18th century. Britain acquired the land in 1763 at the end of the French and Indian Wars. It was ceded to the USA after the American War of Independence and in 1787 it became part of the Northwest Territory. The first settlement came a year later. Ohio was accepted into the Union in 1803, and the construction of railways and canals accelerated the state's development. Area: 106,764sq km (41,222sq miles). Pop. (1975 est.) 10,759,000. *See also* MW p.175.

Ohio, river in E central USA, formed at the confluence of the Allegheny and Monongahela rivers at Pittsburg, in W Pennsylvania. It flows W and then SW to join the Mississippi River at Cairo, Illinois. The Ohio River valley is a highly industrialized region and large quantities of coal, oil, steel and manufactured articles are shipped along the river. There are various flood-control schemes to prevent spring flooding and to improve navigability. Length: 1,571km (976 miles).

Ohm, Georg Simon (1787–1854), German physicist. He was appointed professor at Munich in honour of his discovery of the relationship (OHM'S LAW) linking current, ELECTROMOTIVE FORCE and resistance in an electric circuit. His name is also given to the unit of electrical resistance.

Ohm, unit of electrical resistance, equal to the resistance between two points on a conductor when a constant potential difference of one VOLT between them produces a current of one AMPERE.

OH-maser emission, stellar, in astronomy, spectral emission of the hydroxyl radical from certain radio sources, characterized by extremely narrow lines of high intensity. One explanation of the effect is that the radio waves are being amplified by stimulated emission in some form of celestial MASER.

Ohm's law, statement that the amount of steady current through a material is proportional to the voltage across the material. Propounded by the German physicist Georg Simon OHM, Ohm's law is expressed mathematically as $V/I=R$ (where V is volts, I is the current in amperes and R is the resistance in ohms). *See also* SU p.126.

Oh, What A Lovely War! (1963), musical satire about WWI leaders and generals, developed by Charles Chilton, Joan LITTLEWOOD and members of THEATRE WORKSHOP within the format of an Edwardian PIERROT show. Richard ATTENBOROUGH directed a film version in 1969.

Oil, general term to describe a variety of substances, whose chief shared characteristics are viscosity at ordinary temperatures, a density less than that of water, inflammability and insolubility in water, and solubility in ether and alcohol. They are in other ways too different to be classified together chemically. Thus there are mineral oils, most notably petroleum, used as fuels, and animal and vegetable oils (fatty oils or fats), used as food, lubricants and as a major ingredient of soap. In addition there are essential oils from plants which, unlike fatty oils, are volatile. The fatty oils, which are fixed, are of two kinds: drying, such as linseed or poppyseed oil, and non-drying, such as olive and castor oil. *See also* SU pp.86–87.

Oil, North Sea, term used for the considerable deposits of the lighter type of crude oil, first discovered off the coast of The Netherlands in 1959. The area has since been divided into sectors, each belonging to a nation whose territorial waters include part of the North Sea. One of the most important is the Ekofisk field off s Norway, which is expected to yield 300,000 barrels of oil per day. It was in this field, however, that a disastrous blowout occurred in 1977, spilling an enormous quantity of oil into the sea. Discoveries have also been made closer inshore, eg in the Moray Firth off NE Scotland. *See also* MM pp.78–79; PE p.*139.*

Oil, vegetable, oil prepared from certain plants and essential to the human diet. Vegetable oils release more energy per unit of weight than any other food. They are commercially extracted from the fruit or seeds of SOYA, GROUNDNUT, MAIZE, COTTON, RAPE, SUNFLOWER, COCONUT and OLIVE. Sunflower oil and soya oil are commonly used in the preparation of margarine or used as cooking oil. *See also* PE pp.184, *184,* 214–215, 216–217.

Oilbird, or guacharo, nocturnal N South American bird, coloured brown and black, with white spots. An ECHOLOCATION system enables it to navigate in the dark. The local Indians used to boil the nestlings to obtain oil for their torches. Length: 33cm (13in); wingspan: 76cm (30in). Family Steatornithidae; species *Steatornis caripensis.*

Oil extraction, process by which crude oil is brought to the Earth's surface and transported to a refinery. A borehole is made into the Earth's crust and the oil reservoir emptied with the help of NATURAL GAS pressure or pumps. The discovery of undersea deposits, as in the North Sea, has meant that many new techniques for extracting and transporting the oil have had to be developed in order to avoid pollution and blowouts. *See also* MM pp.78–79; PE pp.138–139.

Oil painting, method of painting in which

an oil base is used for paint pigments; the term is also applied to pictures painted by this method. Sometimes a diluting material, such as turpentine or a substitute is used to thin the oil-based paint. Jan van EYCK is credited with the innovations that led to the gradual replacement of TEMPERA by oil painting in Europe in the 15th century. The use of canvas as a primary painting surface also began at this time when Venetians such as TITIAN and TINTORETTO used oils successfully with little underpainting. In the 17th century RUBENS used oils in the Flemish tradition, placing transparent colours over a ground primed with white paint. REMBRANDT used a dark ground and thin paint for shadows, and heavier paint for the lighter areas in his pictures. HALS and VELÁZQUEZ also experimented with the use of oils. The traditional use of glazes (layers of thin paint, one on top of the other) was often replaced by direct painting during the 19th century when many schools of painters, eg the IMPRESSIONISTS and NEO-IMPRESSIONISTS, developed different techniques of oil painting, and this experimentation continues today.

Oil palm, tree grown in humid tropical regions of w Africa and Malagasy, source of oil for margarine and soap. The long feather-shaped fronds rise from a short trunk. Height: 9–15m (30–50ft). Family Palmaceae; species *Elaeis guineensis.*

Oil refinery, large industrial chemical complex in which crude oil (petroleum) is refined. This involves many processes whereby the oil is split into its major fractions such as gasoline, kerosene, diesel oil, lubricating oil and bitumen. These are obtained by fractional DISTILLATION, but many refineries carry out further operations on some of these fractions. These include catalytic "cracking", alkylation – increasing the octane rating of petrol – HYDROGENATION and POLYMERIZATION. *See also* MM p.80.

Oil rig, platform, either floating or on legs resting on the sea bottom, used for drilling for offshore oil or natural gas in the NORTH SEA and off the coasts of Australia, California and the N of South America. Rigs are generally sited in comparatively shallow water on the continental shelf. The gas or oil is transported ashore either in tankers or through undersea pipelines.

Oil shale, dark-coloured, soft rock containing hydrocarbon compounds which can be distilled off as shale oil. Oil shales yielding more than a small percentage of oil can be valuable as fuels and as sources of organic chemicals. They are found in many countries, sometimes in vast deposits; that of the Green River shales in the USA alone has been estimated to contain 960,000 million barrels of oil. *See also* PE pp.138–139.

Oil well, shaft through which crude oil (petroleum) is brought to the surface from underground deposits. Wells are usually drilled with a rotating shaft bearing a drill bit which, for cutting through hard rock, may be diamond-tipped. Rock particles are pumped from the hole as a fluid slurry, which also lubricates and cools the drill. Proven wells are cased with steel piping and topped above ground with an oil take-off arrangement, often of the type known as a Christmas tree. *See also* MM pp.78–79; PE pp.138–139.

Oistrakh, David (1908–74), Russian violinist. His interpretation of the violin repertory, both in the concert hall and on records, earned him the reputation of being the greatest violinist of his day. PROKOFIEV and other modern Russian composers wrote works especially for him. His son and pupil, Igor (1931–) is a virtuoso violinist.

Ojibwa, or Chippewa, Indians, group of Algonquinian-speaking North American Indians. In the 17th century they were constantly at war with the SIOUX over possession of the rich fields of wild rice on the shores of Lake Superior but eventually drove them across the Mississippi River. They then continued their expansion westwards into North Dakota. Since then they have lived mainly on reservations in Michigan, Wisconsin, Minnesota and North Dakota. Today they number about

80,000 in both the USA and in Canada.

Okapi, hoofed mammal of the giraffe family. A very shy animal, it lives in the dense rain forests of equatorial Africa and was not discovered until 1900. It is brown with zebra-like stripes on its legs and hindquarters. The tongue is extremely long. Males have small, hair-covered horns. Height: to 1.6m (5.2ft) at the shoulder. Family Giraffidae; species *Okapia johnstoni. See also* NW pp.*212,* 213.

Okara, Gabriel Imomotimi Gbaingbain (1921–), Nigerian author. In 1959 he went to the USA to study journalism, but returned to Africa to write poetry, published in *Black Orpheus* (nos 1,3 and 6) and a novel, *The Voice* (1964). In his writing he draws on African mythology and religion, trying to resolve the conflict between traditional and Western ideas.

Okeechobee, lake in s central Florida, N of the Everglades. The largest fresh water lake in the USA, it is fed by the Kissimmee River and drained by the Caloosahatchee River. It is part of the Okeechobee (Cross-Florida) Waterway System. Area: 1,890sq km (730sq miles).

O'Kelly, Seán Thomas (1883–1966), Irish politician. He was a founder of the SINN FEIN movement and Speaker of the first Dáil Éireann (1919–21). He was Vice-President of the Executive Council and Minister for Local Government and Public Health (1932–39) and Minister of Finance (1939–45). From 1945 to 1959 he was President of Ireland.

Okinawa, largest of the Okinawa Islands, Japan, in the Ryukyu Islands Chain, sw of mainland Japan, in the w Pacific Ocean. It is an island of volcanic origin and is densely forested in the N. The inhabitants, most of whom live in the s, engage in agriculture and fishing. The island was taken by American troops in WWII as it provided the USA with an airfield close to mainland Japan; it was returned in 1971. Area: 1,176sq km (454sq miles). Pop. (1970) 945,111.

Oklahoma, state in central s USA. The w of the state is part of the Great Plains. The E is mountainous. The area is drained chiefly by the Arkansas and Red rivers. Wheat and cotton are the leading cash crops but livestock raising is the most important farming activity. Although many mineral deposits are found in the state, oil and natural gas form the basis of Oklahoma's economic wealth. Most of the state's industries process raw materials. The major cities are Oklahoma City, the state capital, and Tulsa. Much of the area was acquired by the USA from France in the LOUISIANA PURCHASE of 1803. During Andrew JACKSON's presidency (1829–37) Congress created the Indian Territory in the region for AMERICAN INDIANS moved by the Federal Government from states in the E. White cattle farmers who had settled to the w of this area organized the Territory of Oklahoma in 1889. The two territories were merged to form the state of Oklahoma in 1907. Area: 181,089sq km (69,918sq miles). Pop. (1975 est.) 2,712,000. *See also* MW p.175.

Oklahoma!, musical play by Richard RODGERS and Oscar HAMMERSTEIN II, originally produced on the Broadway stage in 1943. The slight story is set in 19th-century Oklahoma Territory. It was also the first of the series of musical successes by the Rogers and Hammerstein partnership.

Oklahoma City, largest city and capital of Oklahoma State, USA, situated on the North Canadian River. The area was opened to homesteaders and settled during the land rush in 1889. The city was made the state capital in 1910, and prospered with the discovery of rich oil deposits in 1928. Industries: oil refining, stockyards, meat packing, grain milling, cotton processing, steel products, aircraft. Pop. (1970) 368,856.

Okra, also called gumbo, annual tropical plant with red-centred yellow flowers. The unripe green fruit pods are eaten as a vegetable. Height: 0.6–1.8m (2–6ft). Family Malvaceae; species *Hibiscus esculentus. See also* PE p.189, *189.*

Olaf, the name of five kings of Norway. Olaf I Tryggvason (*c.*964–*c.*1000;

Oilbird is the only nocturnal bird that depends entirely on fruit for food.

David Oistrakh won the Brussels International Contest for violinists in 1937.

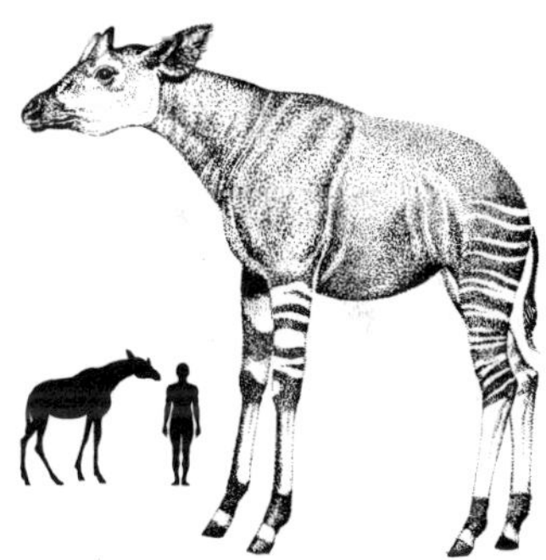

Okapi; it was not until 1958 that an okapi born in captivity survived.

Oklahoma; cereal crops and oil are important factors of the state's economy.

Old Bailey; the facade displays symbols of justice and retribution.

Old English sheepdog's hair needs brushing to keep it clean and untangled.

Old Testament; Adam and Eve are banished from the Garden of Eden.

Old Vic, under Lilian Baylis' management, presented every Shakespeare play.

*r.*995–1000), the great grandson of HAROLD I, (*r.* 995–1000). He spent much of his youth on viking raids but was later converted to Christianity. Olaf II Haraldsson (*c.*995–1030) was also converted to Christianity and continued to gain Norwegian acceptance for Christianity. His reign (1016–28) is remembered for his religious code (1024). He was canonized in 1131 and is the patron saint of Norway. Olaf III Haraldsson (*d.*1093; *r.*1066–93), known as "the quiet" (Kyrri), ruled jointly with MAGNUS II until the latter's death in 1069. Peace and the spread of Christianity characterized his reign. Olaf IV Haakonson (1370–87; *r.*1380–87), became King of Norway in 1380, under a regency, but Norway was governed by his mother as regent, who also succeeded him. Olaf V (1903– ; *r.*1957–) succeeded his father, HAAKON VII. He took an active part in the struggle for liberation after the German occupation (1940) of Norway during WWII. In 1944 he assumed supreme command of the Norwegian forces. He was regent for his father 1955–57.

Olbers, Heinrich Wilhelm Matthias (1758–1840), German astronomer and physician. He discovered five comets and the asteroids, Pallas (1802) and Vesta (1807). He also devised a new method of calculating a comet's orbit. In 1826 Olbers proposed his famous paradox: if there are an infinite number of stars, evenly distributed in space, the sky should remain solidly bright with no darkness on the Earth. His explanation, that interstellar dust obscures the light, was discarded in the 1970s in favour of the theory that expansion of the universe dims the light from distant objects.

Olbrich, Joseph Maria (1867–1908), architect born in Silesia (now Czechoslovakia). He studied under Otto Wagner, one of the founders of the modern movement in European architecture. With Gustav KLIMT, Olbrich was co-founder of the VIENNA SECESSION, and in 1897–98 designed and built the Secession building in Vienna.

Olcott, Col. Henry Steel (1832–1907), American journalist and THEOSOPHIST. He founded the Theosophical Society with Helena Blavatsky in 1875 and went to live and work in India soon afterwards.

Old age, last period of a human being's life, generally after the age of 70. Old age is usually characterized by physical changes, eg loss of hair, failing sight and a general slowing of the body's systems. This physical deterioration is most visible in those body tissues which do not regenerate easily because they are largely composed of non-living materials: the bones, joints, and connective tissue. Brain cells also do not regenerate but their progressive loss is not, apparently, a major factor in ageing. Tissues largely made up of living cells, such as the skin and glands, are more or less in a state of constant repair, although this slows as the body's age increases. *See also* MS pp. 176–179.

Old age pensions. *See* PENSIONS AND NATIONAL INSURANCE, MINISTRY OF.

Old Bailey, common name for the Central Criminal Court, in London, derived from the street in which it is situated. It became a CROWN COURT in 1971. The judges entitled to sit in the court include the Lord Mayor of London and the Aldermen as well as the Recorder of London and the Common Serjeant who are circuit judges. Judges at the Central Criminal Court are still addressed by the title "My Lord".

Oldcastle, Sir John (*c.* 1377–1417), English soldier and LOLLARD leader. Under HENRY IV he performed valuable military service in Wales, where he became a friend of the Prince of Wales (later HENRY V). An enthusiastic supporter of John WYCLIFFE and his teachings, he was condemned for this heresy in 1413. He escaped from the TOWER OF LONDON and was involved in Lollard conspiracies until 1417 when he was recaptured and executed.

Old Catholic Church, religious movement rejecting the dogma of PAPAL INFALLIBILITY which had been announced by the Vatican Council of 1870. The Old Catholics set up churches in German-and

Dutch-speaking Europe, which later united in the Union of Utrecht in 1889, and since then the Archbishop of Utrecht has been head of the International Old Catholic Congress. Old Catholics have much affinity with Anglicans, and achieved communion with the Church of England in 1931. There are more than 450,000 members in Germany, Switzerland, Austria, The Netherlands, Poland and the USA.

Old Curiosity Shop, The (1841), novel by Charles DICKENS. It is famous for its characterizations of Little Nell, the angelic heroine, Quilp, the grotesque villain, and Kit Nubbles, who is devoted to Little Nell.

Oldenbarneveldt, Johan van (1547–1619), Dutch statesman. He fought with WILLIAM THE SILENT for Dutch independence but was executed for treason by MAURICE OF NASSAU.

Oldenburg, Claes (1929–), US painter and sculptor, a leading member of the POP ART school. He has painted grotesquely distorted figures and gigantic sculptures, such as *Lipstick* (1969) at Yale University.

Oldenburg Dynasty, family that ruled Denmark (1448–1863) and Norway (1450–1814). After the death of Christopher III, King of Denmark, Norway and Sweden, Denmark and Norway elected Christian, Count of Oldenburg, as their monarch in 1448 and 1450 respectively. Christian was also King of Sweden from 1450–57. The Treaty of Kiel (1814) separated the two states, and following the death of FREDERICK VII in 1863, the Danish Crown passed to his cousin, the Duke of Glücksburg.

Old English, the English language as spoken and written between *c.*450 and 1100. It is known also as Anglo-Saxon and had a vocabulary of about 50,000 words. It comprised four main dialects, spoken by the Germanic tribes which invaded England in the 5th century: Northumbrian and Mercian, spoken by the Angles; Kentish, spoken by the Jutes; and West Saxon, spoken by the Saxons. The best of Old English literature such as BEOWULF, was written in Northumbrian, but West Saxon became the chief dialect as a result of ALFRED THE GREAT's political influence over much of England in the late 9th century.

Old English sheepdog, working dog, originally bred in England early in the 18th century. It has a square head, square jaws and medium-sized ears carried flat against its head. The compact body is set on sturdy legs. Its tail is commonly docked, which gives rise to its other name – bobtail. The coarse, shaggy coat may be grey, grey-brown or grey-blue with white markings. Height: to 66cm (26in) at the shoulder; weight: to 30kg (66lb).

Oldfield, Barney (1877–1946), US racing driver who in 1903 became the first man to drive at a speed of 60mph (96.6km/h). In 1910 he reached a speed of 131.7mph (212km/h) which set a new world record.

Oldham, county district in Greater Manchester, NW England, created in 1974 under the Local Government Act (1972). Area: 140sq km (54sq miles). Pop. (1976 est.) 227,500.

Old Kingdom, period of Egyptian history from the 3rd to 8th dynasties (*c.*2686–2160 BC), during which the PYRAMIDS were built. Texts of the period tell of trading with Lebanon and of expeditions to Libya, Nubia and Sinai. It is probable that the Old Kingdom ended in a civil war between the royal family and provincial nobles.

Old-man's-beard. *See* CLEMATIS.

Old Moore's Almanac, British astrological magazine. First published in 1699 by Francis Moore, a London physician, to promote his pills, it contained weather predictions based chiefly on astrology. It is still published each year, but now covers more topics than the weather and no longer promotes pills.

Old Norse literature, literature of the Scandinavian Norsemen from the 9th to 12th centuries. It is mainly mythological poetry and sagas. The work was set down in stone and wood, but survived orally and was recorded in the 12th–14th centuries in Eddaic (alliterative) and Skaldic (complex and artificial) Icelandic verse. The SKALDS

became a separate group which only the initiated could comprehend.

Oldowan, type of stone tool dating from the early PLEISTOCENE period (*c.*2.5 million years ago). The name comes from the Olduvai Gorge in Tanzania, where archaeologists found the first tools of this kind. Made from quartz or basalt stones, the edges were chipped away to form tools capable of chopping, scraping or cutting. Oldowan tools were made for about 1.5 million years, but were gradually superseded by better-made ones of the ABBEVILLIAN type. *See also* SU p.21.

Old red sandstone, geological term for fresh-water deposits of the Devonian Period found in Britain, especially Scotland. These strata are noted for their fish fossils among which are jawless fishes (ostracoderms), the first jawed fishes (placoderms) and by the Middle "Old Red" period the first true bony fishes (osteichthyes). Deposits of equivalent age are found in Canada, Greenland, the USA, central Asia and Australia.

Old Stone Age. *See* STONE AGE.

Old Testament, first part of the Bible, of which some dates from the time of the Kingdom of DAVID (*c.*1000 BC), although much more dates from the revision of all religious texts carried out at the time of the BABYLONIAN CAPTIVITY(*c.*580 BC). Originally written for the most part in Hebrew or Aramaic, it comprises the PENTATEUCH (Genesis to Deuteronomy); the Historical Books (Joshua to KINGS); the Wisdom Books (eg JOB, Proverbs); the Major Prophets (eg ISAIAH, Jeremiah); the 12 Minor Prophets (Hosea to Malachi); and the miscellaneous Writings (eg PSALMS, Song of Solomon). There is also a collection of books written in the final three centuries BC, known as the APOCRYPHA (including Tobit, Judith, Esther, Macabbees, Ecclesiasticus, Baruch, some of Daniel, and occasionally the tiny Letter of Menasses), commonly inserted among the Writings and the Prophets. The number, order and names of the books vary slightly between Jewish and Christian traditions; texts for both are based on a 3rd-century BC Greek text, the SEPTUAGINT, a translation into the common language of the Jews of the DIASPORA.

Old-time dances, name for several dances of European and US society in the 19th century, not to be confused with modern ballroom dances of the post-WWI era. The most famous of them, which held sway throughout the 19th century, was the waltz. The basic steps of old-time dancing are the one-step and the two-step.

Old Trafford, name of two sports grounds in Manchester, England. One is the ground of Lancashire County Cricket Club, and the other is that of Manchester United Football Club. The cricket ground was opened in 1857 and TEST MATCHES have been played there regularly since 1884. It has a capacity of about 30,000 people. The football ground was opened in 1908 and holds about 58,000.

Olduvai Gorge, site in Tanzania where remains of primitive man have been found. The gorge is 40km (25 miles) long and 100m (320ft) deep; it runs through the Serengeti Plain. The fossils of ancient man and his forebears are preserved in the layers of lake sediment and volcanic rock which make up the sides of the gorge, enabling man's chronological development to be traced. Louis LEAKEY (1903–72), who spent most of his life working at Olduvai, uncovered four beds, or layers, of remains which date from *c.*2 million years ago to *c.*10,000 years ago. In 1964, he announced the discovery of *Homo habilis*, whom he believed to have been a direct ancestor of modern man.

Old Vic, London theatre. Built in 1818 as the Royal Coburg Theatre, it later became the Royal Victoria Hall and served as a temperance music-hall. It was soon known affectionately as the Old Vic. It became well-known under the management of Lilian BAYLIS for its Shakespearean productions. With the purchase of the Sadler's Wells Theatre in 1931, the Vic-Wells company was formed. Drama and opera and ballet performances were alternated between two theatres. In 1935 the opera and ballet programmes were restricted to Sadler's Wells and the drama to the Old Vic. From 1963 to 1976 it was the home of the NATIONAL THEATRE of Great Britain.

Old Wives' Tales, The (1908), one of the "Five Towns" novels by Arnold BENNETT. It contrasts the lives of sisters Constance and Sophia Baines who were modelled on two undistinguished, elderly women Bennett had seen in a Parisian restaurant.

Oleander, poisonous, Eurasian, evergreen shrubs of the genus *Nerium*. The best-known is probably the rosebay (*N. oleander*). They have milky poisonous sap, clusters of white, pink or purple flowers and smooth leaves. The fruits may be fleshy, berry-like or dry pods that split to release seeds for dispersal by the wind. Family Apocynaceae.

Oleaster, small, spiny-branched, deciduous tree, native of Eurasia. It has narrow leaves with hairy silver undersides and small, fragrant silver-scaled flowers that yield yellow, olive-shaped fruits. Height: to 6m (20ft). Family Elaeagnaceae; species *Elaeagnus angustifolia*.

Olefins, or alkenes, hydrocarbon chemical compounds which have open-chain molecules containing one double bond. The simplest is ethylene, $H_2C{=}CH_2$. The general formula of olefins is C_nH_{2n}. They are important industrially for their ability to form addition compounds: these include the well-known polymers POLYETHYLENE and PVC (polyvinyl chloride).

Oleic acid, organic chemical compound, formula $C_{17}H_{33}COOH$, having a long-chain molecule which is unsaturated (containing a double bond). It is the unsaturated acid found most commonly in edible oils and fats such as mutton fat.

Olfactory nerve, the nerve of SMELL, the name of one of the 12 pairs of cranial nerves present in all vertebrate animals. Its nerve cells (neurons) lie in the brain and their long shafts, or axons, extend to their sense receptors in the upper part of the nasal cavity. *See also* MS pp.54–55.

Olga, Saint (*c*.890–969), first Russian saint. The widow of the assassinated Prince Igor I of Kiev, she was regent for her son from 945 to 964. She was canonized for her unremitting attempts to bring Christianity to the Kievans; she herself was probably baptized in *c* 957 Olga showed little Christian forgiveness however, ordering her husband's murderers to be scalded to death.

Oligarchy, system of government in which power is concentrated in the hands of a few, who rule without the requirement of popular support and without external check on their authority.

Oligocene epoch, extent of geological time from about 38 to 26 million years ago. It is the third of five epochs of the Tertiary Period; during it the climate cooled. Many modern mammals evolved, including early elephants, man-like apes and *Mesohippus*, an ancestor of the modern horse. Only a few archaic mammals, such as titanotheres, survived into the epoch, and they became extinct before it ended. *See also* PE p.*131*.

Oligochaeta, class of ANNELID (segmented) worms, characterized by long, naked bodies bearing a few bristles on each segment. Earthworms are the best-known oligochaetes, but many are fresh-water worms as small as 1mm (0.025in) long. Giant Australian earthworms grow up to several metres long. *See also* NW p.89.

Oliphant, Margaret (1828–97), Scottish novelist, biographer and historical writer. Her best-known novels are *Salem Chapel* (1863) and *Miss Marjoribanks* (1866). Later works exhibit an interest in the OCCULT.

Oliphant, Sir Mark (1901–), Australian nuclear physicist who did much work on the nuclear disintegration of LITHIUM. He was a member of the atomic bomb team at Los Alamos (1943–45) and became Australian representative on the UN Atomic Energy Commission in 1946.

Olivares, Gaspar de Guzman, Condé-Duque de (1587–1645), Chief Minister of PHILIP IV of Spain from 1621–43. The heavy taxes he levied to fight Portuguese and Catalonian rebels brought about his downfall.

Olive, tree, shrub or vine and its fruit. The common olive tree, *Olea europaea*, is native to the Mediterranean region and is cultivated in other warm regions. It has leathery, lance-shaped leaves, a gnarled and twisted trunk and may live for more than 1,000 years. The fruit, which is picked green in an unripe condition or black and ripe, is bitter and inedible before processing; it must be treated with an alkaline solution and brine. Height: to 9m (30ft). Family Oleaceae. *See also* OLIVE OIL; PE pp.*214–215*, 216–217.

Olivenite, rare green mineral, of secondary origin, found in copper deposits. It is a hydrous arsenate (salt of arsenic acid) of copper.

Olive oil, yellowish liquid oil, containing oleic acid, obtained by pressing OLIVES. It is used for cooking, as a salad oil, in the manufacture of soap and in medicine. It contains an extremely low proportion of the fatty acids.

Oliver, Isaac (*c*.1556–1617), French-born painter of miniatures who worked mainly in England. His work includes religious and Classical scenes as well as portraits of Tudor and Stuart gentry. He was a follower of Nicholas HILLIARD but used a more naturalistic style.

Oliver, King (1885–1938), US jazz cornettist, composer and pioneer jazz bandleader, real name Joseph Oliver. He formed King Oliver's Creole Jazz Band (1922), which became the leading exponent of Dixieland jazz and started the careers of many jazz musicians such as Louis ARMSTRONG. Oliver composed many classic blues and jazz songs including *The Canal Street Blues* and *The Chattanooga Stomp*, (with A. Picou).

Oliver Twist (1837–38), novel by Charles DICKENS about Oliver, an orphan who escapes from a workhouse to London and falls in with a gang of juvenile thieves led by Fagin, a master pickpocket. It is occasionally marred by sentimentality but succeeds in its criticism of the Poor Law and the institution of the workhouse.

Olives, Mount of, low range of hills E of Jerusalem, also called Mt Olivet. According to the Bible, Jesus went down from the Mt of Olives to make His triumphal entry into Jerusalem, returning there on the night of His betrayal by JUDAS ISCARIOT in the "garden" of GETHSEMANE. The Mt of Olives is named in Acts 1 as the place from which He ascended into Heaven, and today the Church of the Ascension stands on the spot where the event is believed to have occurred.

Olivier, George Borg (1911–), Maltese politician. He became Prime Minister as leader of the Maltese Nationalist party (1953–55), and, after acting as leader of the opposition (1955–58), became Prime Minister again (1962–71).

Olivier, Laurence Kerr, Baron Olivier of Brighton (1907–), British actor. He made his debut at Stratford-upon-Avon in 1922, as Katharine in *The Taming of the Shrew*. He was a leading actor by 1928, playing romantic roles (*Beau Geste*, 1929) and comic ones (*Private Lives*, 1930). In 1937 he joined the OLD VIC. He played Hamlet, Henry V and Macbeth in his first season, Iago in his second. He was director of the NATIONAL THEATRE (1963–73). He won an Academy Award for the film of *Hamlet* in 1949. He was the first actor to be created a life peer (1970).

Olivine, group of independent tetrahedral silicates $[(Mg,Fe)_2SiO_4]$. The group includes fosterite, tephorite, monticellite fayalite, and peridot. Olivine has orthorhombic system crystals, of usually granular masses. Its colours range among green, brown and grey and it is glassy and brittle. Hardness 6.5–7; s.g. 3.3. *See also* PE p.96.

Olm, amphibian related to the SALAMANDER with degenerate eyes covered by skin. Olms inhabit the underground waters of limestone caves in SE Europe. Because they live in darkness, they lack skin pigmentation. The olm senses its prey by detecting currents or vibrations through the water. Species *Proteus anguineus*. *See also* NW p.*132*.

Olmec, one of the earliest known cultural periods of the pre-Columbian New World, developed by the Olmec Indians (*c*.1100–800 BC). They carved in jade and stone and their artefacts include stone altars and pillars and perfectly ground concave mirrors of polished haematite. Their homeland seems to have been central Mexico, from the Pacific coast of Guerrero to Veracruz, but evidence of their culture has been found as far s as central Guatemala. *See also* HC1 pp.46, 230–231.

Olmedo, José Joaquín (1780–1847), Ecuadorian poet and political figure. He celebrated Latin American independence in his best-known work, *La Victoria de Junín: Canto a Bolívar* (1825). He was elected the first Vice-president of Ecuador in 1830, and sought to promote Latin American culture, particularly that of the indigenous Indians.

Olmstead, Frederick Law (1822–1903), US landscape architect. He designed Central Park in New York City (1857–61) and the Capitol grounds in Washington, DC (1874–95). He helped to acquire the Yosemite Valley, California, for a national park.

Olongapo, city on w Luzon, the Philippines, on the NE coast of Subic Bay. It has excellent port facilities and was formerly the site of a US naval base. The Japanese occupied the city during WWII. Pop. (1975 est.) 134,453.

Olympia, city of ancient Greece, one of its most important religious centres. Games held there (776 BC–AD 393) every fourth year in honour of ZEUS gave rise to the modern idea of the OLYMPIC GAMES. Buildings included the Temple of Zeus, which housed the elaborately adorned statue of the god by PHIDIAS, one of the SEVEN WONDERS OF THE WORLD. Archaeological excavations have unearthed great temples, the celebrated statue of Hermes of Praxiteles, other buildings and the stadium. By repute there was also a hippodrome, which has not been found.

Olympia (1863), oil painting by Édouard MANET, now in the Louvre, Paris. The painting caused a scandal when first exhibited in 1865; its portrayal of a nude woman bypassed all the artistic conventions of the time in showing her as a self-conscious individual, whose nakedness was overtly sexual and in no way idealized.

Olympic Games, the world's major international amateur athletic competition held every four years. The games were first celebrated in 776 BC in OLYMPIA, Greece and were held every four years until 393 AD when they were abolished by the Roman Emperor. They originated from a 200m running race and were gradually extended to include a wide variety of events. The modern summer games were initiated by Baron de Coubertin and were first held in Athens, Greece, in 1896. Women did not compete until 1912. In 1924 a separate Olympics was held for winter sports. The games were cancelled during WWI and WWII. Summer events include archery, basketball, boxing, canoeing, cycling, diving, equestrian sports, fencing, field hockey, handball, judo, gymnastics, polo, rowing, soccer, shooting, swimming, athletics (including the DECATHLON and modern PENTATHLON), volleyball, water polo, weightlifting, wrestling and yachting. Winter events include the biathlon, bobsledding, ice hockey, skating and skiing. Additionally, the host country may name a sport of its choice. The control of the games is vested in the International Olympic Committee, which lays down the rules and chooses venues. Although contestants represent countries, the events are officially competed for individually. A gold medal is awarded for first, a silver for second and a bronze for third. The Modern Olympics are a costly, international event that is broadcast and televised, and held in a different country every four years.

Olympics of the Paralysed. *See* PARAPLEIGIC GAMES.

Olympio, Sylvanus (1902–63), Togolese politician. He was President of the Colonial Assembly (1946–52) and Premier (1958–60). He led the fight for Togo's

Olives live longer than most other trees and bear fruit once every two years.

King Oliver (trumpet) and Louis Armstrong (slide trumpet) in the Creole Jazz Band.

Sir Laurence Olivier in the title role in his 1948 film production of *Hamlet*.

Olympia; excavations reveal the remains of many historic buildings and statues.

Onagers have, in some areas, become rare or even extinct through overhunting.

One Day in the Life of Ivan Denisovich; Tom Courtenay in the film version.

Eugene O'Neill said his task was "to dig at the roots of the sickness of today".

Lars Onsager became a professor at Yale and an American citizen in 1945.

independence from France (1960). After independence he became the first President of the new republic in 1961. He was assassinated in January 1963, in the first successful army coup in Africa s of the Sahara Desert.

Olympus, in Greek mythology, abode of the gods located on the summit of a mountain symbolized by Mt Olympus, the highest mountain in Greece. Beyond a cloud-gate guarded by the Horae (the seasons), dwelt the gods, each in his own palace, repairing to ZEUS' great hall to feast on AMBROSIA and NECTAR.

Omaha, port in E Nebraska, USA, on the Missouri River; the largest city in Nebraska. The area was ceded to the US government in 1854, and it developed rapidly as a supply depot for westward travellers, and later as an industrial centre. It was the capital of Nebraska Territory from 1854 to 1867. A leading livestock market and meat processing centre, it has oil refining, food processing industries, and manufactures farm machinery, ball-bearings, electric signs, paints and railway equipment. It is a major insurance centre. Pop. (1970) 346,929.

Oman, officially named the Sultanate of Oman, an independent Arab nation on the SE Arabian peninsula, on the Arabian Sea and the Gulf of Oman; until 1970 the British dependency of Muscat and Oman. A powerful state in the mid-19th century, it declined after the 1870s, coming under British influence as a dependency of the British Government of India. Oil is the principal export. Fishing is an important industry and sugar-cane, dates, olives and cereals are the main crops. The capital is MUSCAT. Area: 212,457sq km (82,030sq miles). Pop. (1975 est.) 766,000. See MW p.139.

Omar (c.581–644), second CALIPH, or ruler of ISLAM, also called Umar. He was converted to Islam in 618 and became a counsellor of Mohammed. He chose the first caliph, ABU BAKAR, in 632 and succeeded him two years later. Under his rule Islam spread by conquest into Syria, Egypt and Persia and the foundations of an administrative empire were laid. Omar was assassinated by a foreign slave.

Omar Khayyám (c.1048–1122), Persian poet, mathematician and astronomer. He received an extensive education, and early in his life wrote a treatise on algebra which so impressed Sultan Jalāl ad-Dīn that he was asked to reform the calendar. He was a master of philosophy and jurisprudence. His fame in the West is due mainly to a collection of verses attributed to him and which were freely translated by Edward FITZGERALD and published in 1859 as The Rubáiyát of Omar Khayyám.

Omayyads. See UMAYYADS.

Ombudsman, official appointed to safeguard citizens' rights by investigating complaints of injustice made against the government or its employees. The office was created in Sweden in 1809; Denmark adopted it in 1954 and Norway in 1962. New Zealand was the first Commonwealth country to appoint an ombudsman, in 1962; the UK followed in 1967 with a Parliamentary Commissioner for Administration. The first US state to appoint an ombudsman was Hawaii, in 1967.

Omdurman (Umm Durmân), city in NW central Sudan, on the White Nile, opposite Khartoum; chief commercial centre of the Sudan. The city served as the military headquarters of the MAHDI during the uprising of 1884; it was captured by the British in 1898. Places of note include the site of the Mahdi's tomb, the Khalifa museum and the university (1912). Industries include furniture, tanning, pottery, textiles, livestock and gum arabic. Pop. (1971) 258,532.

Omega centauri, GLOBULAR CLUSTER in the southern Constellation of CENTAURUS. The finest example of its type, Omega Centauri is visible to the naked eye as a hazy patch and is resolvable with a small telescope. See also SU pp.243, 261, 266–267.

Omega Workshops, business enterprise established in 1913 by Roger FRY with Duncan GRANT and Vanessa Bell to explore the application of NEO-IMPRESSIONISM and non-representational

art on designs for furniture, textiles, ceramics and other decorative and applied arts. The workshops opened in July 1913 in Fitzroy Square, London, and retained the services of young artists such as Wyndham LEWIS, Edward Wadsworth, William ROBERTS and Henri GAUDIER-BRZESKA. Their work however failed to make any real impact, and the company went into voluntary liquidation in 1919.

Omen, any phenomenon that can be interpreted as a portent of future events, either good or bad. Common omens in folk-beliefs include changes in weather or the behaviour of animals. See also DIVINATION; SUPERSTITION.

Omnibus. See BUS.

Omnivore, any creature that eats both animal and vegetable foods – wo well-known examples are human beings and pigs. Raccoons, badgers and bears are also omnivorous animals, although they are usually classified as CARNIVORES. Omnivorous animals are characterized by having teeth adapted for cutting or tearing and pulping food.

Omsk, city in the Russian Republic (Rossijskaja SFSR) USSR on the Irtysh and Om rivers. Founded as a military town in 1716, it was made the administrative centre of W Siberia in 1824. During the civil war that followed the Russian Revolution of 1917, the city was the headquarters of the counter-revolutionary KOLCHAK government. It is now a major port, and manufactures agricultural machinery and textiles. Pop. (1970) 821,000.

Onager, fast-running animal related to the wild ass, found in semi-desert areas of Iran and India; the "wild ass" mentioned in the Bible (Job 39:5–8). The onager is a dun-coloured animal with a dorsal stripe which reaches the tip of the tail. Height at the shoulder: 0.9–1.5m (2.9–4.9ft). Family Equidae; species Equus hemionus onager. See also NW p.243.

Oncology, in medicine, study of TUMOURS and their effects on the body. See also MS pp.86–87, 96, 125.

One Day in the Life of Ivan Denisovich (1962), novel by Alexander SOLZHENITSYN. Drawn largely from the personal experience of the author in various Soviet prison camps, the novel deals with one day in the life of a convict in such an institution. In addition to its documentary value, it is a work of considerable artistic merit, written with warmth and compassion.

Onega (Onezskoje), lake in Karelia (Karelian ASSR), in the USSR between Lake Ladoga and the White Sea; second largest lake in Europe. It drains SW through the River Svir to Lake Ladoga, and has numerous inlets and islands along its N shore. The chief port is Petrozavodsk. Area: 9,610sq km (3,710sq miles).

O'Neill, Eugene Gladstone (1888–1953), US playwright. His work covers a wide range of subjects and dramatic styles. His first full-length play, Beyond the Horizon (1920), won the Pulitzer Prize and in 1936 he won the Nobel Prize in literature. Much of his work is concerned with human tragedy, and he made use of themes from Classical Greek drama, eg his Mourning Becomes Electra (1931) is based on the Oresteia by AESCHYLUS.

O'Neill of the Maine, Terence Marne, Baron (1914–), ULSTER UNIONIST politician, who entered the Northern Ireland Parliament in 1946. He held office as Minister of Finance (1956–63) and Prime Minister (1963–69). Considered to be one of the most liberal of Unionists, he favoured increased political and civil rights for the Roman Catholic minority; despite this and increasing prosperity during his premiership, he did not appear to satisfy Roman Catholic grievances and failed to win what he saw as a mandate for moderation in the elections of 1969. He resigned and was created a life peer in 1970. See also HC2 p.295.

Onion, hardy, bulbous, biennial plant of the lily family native to central Asia, and cultivated throughout the world for its large, strong-smelling, edible bulb. It has hollow leaves, white or lilac flowers and is most successfully grown in a dryish, well-drained soil, in a sunny location. Height: to 130cm (50in). Family

Liliaceae; common species Allium cepa.

Onnes, Heike Kamerlingh. See KAMERLINGH-ONNES, HEIKE.

Onsager, Lars (1903–76), US chemist, b.Norway, who was awarded the 1968 Nobel Prize in chemistry for his development of a theory of irreversible chemical reactions. In 1931 he presented an explanation of the motion of IONS in solution which, although it received little attention at the time, has since been called the fourth law of thermodynamics. From this theory Onsager developed a technique for producing uranium-235 (the basic fuel of nuclear reactors) from uranium-238. This technique was put into large-scale operation in 1943 with the first gaseous diffraction plant at Oak Ridge, Tennessee, and remained the primary method for obtaining uranium-235 until the 1960s.

Ontario, province in SE Canada, bounded to the s by four of the Great Lakes – Superior, Huron, Erie and Ontario – and the USA. In the N is the forested CANADIAN SHIELD with its lowlands bordering on Hudson and James bays, and to the E and s are the lowlands of the St Lawrence River and the Great Lakes, where most of the people live and agriculture and industry are concentrated. Ontario is the most productive Canadian province in terms of industrial and agricultural output. Cattle, dairy produce and pigs are of importance to the farming economy. The chief crops are tobacco, maize, wheat and vegetables. The major industries are the manufacture of motor cars and transport equipment, food processing, metallurgy, chemicals, paper, machinery and electrical goods. The region of the Canadian Shield has many mineral deposits, including nickel, copper, iron, zinc, gold, silver and uranium. The major cities are Toronto (the capital), Ottawa (the national capital), Hamilton, Windsor and London.

A number of fur-trading posts were established in the region during the 17th century by French explorers. The area became part of New France, but was ceded to Britain in 1763. The population was expanded by British Loyalists fleeing the American colonies during and after the American War of Independence. Ontario was known as UPPER CANADA until 1841, when it joined with Quebec to form Canada. In 1867 The DOMINION OF CANADA was created and the province of Ontario was established. Area: 1,068,587sq km (412,582sq miles). Pop. (1976 est.) 8,300,000. See also MW p.45.

Ontario, Lake, smallest of the GREAT LAKES, bounded by New York State, (s and E) and Ontario Province, Canada (s, w and N). Fed chiefly by the Niagara River, the lake is drained to the NE by the St Lawrence River. Part of the ST LAWRENCE SEAWAY, the lake is a busy shipping route, and pollution is a serious problem. The chief Canadian cities on Lake Ontario are Toronto, Hamilton and Kingston; on the US shore are Rochester and Oswego. Area: 19,684sq km (7,600 sq miles). See also PE p.151, 151.

On the Origin of Species by Means of Natural Selection (1859), book by Charles DARWIN that describes his theory of EVOLUTION. As the naturalist on HMS Beagle (1831–36), Darwin observed that species of plants and animals are not fixed but adapt according to the environment, those adapting to make best use of the environment being the ones which ultimately endure – survival of the fittest. Although Darwin was not the first to suggest an evolutionary theory, he was the first to deny that God caused change by special creation, as proposed by LAMARCK. Because of this, the theory caused great controversy but was so well formulated that eminent scientists, especially T.H. HUXLEY, came forward to defend it and it is now generally accepted. See also NW pp.32–34.

Ontogeny, the total biological development of an organism, including the embryonic stage, birth, growth, body changes and death.

Ontology, in philosophy, branch of METAPHYSICS that studies the basic nature of things – the essence of "being" itself.

Onychophora, class of about 90 species of worm-like terrestrial invertebrates that live mainly in tropical regions. In evolutionary terms they lie between ANNELID worms and arthropods. They have as many as 44 pairs of legs. Length: to 15.2cm (6in).

Onyx, striped variety of the mineral CHALCEDONY. Black and white onyx is often used for making cameos; white and red forms are called carnelian onyx; white and brown, sardonyx. Most are found in India and South America.

Oolite, egg-shaped crystalline deposit, usually a millimetre or less in diameter, found in DOLOMITE or LIMESTONE. Its centre may be a grain of shell or quartz, which suggests to geologists that oolites grow in oceans or streams by accreting layers of calcium carbonate as they roll about in the turbulent water.

Oostende. *See* OSTEND.

Oosterhuis, Peter (1948–), British golfer. He won the British youth championship in 1966, and was chosen for the Walker Cup side in 1967 and for the Ryder Cup side in 1971. In 1968 and 1971 he represented England in the WORLD CUP. He was the leading golfer in the European order of merit from 1971 to 1974. He was accepted as a player on the US professional tour in 1975.

Ooze, deep-sea, fine-grained pelagic deposit containing material of more than 30% organic origin, with the rest comprised of clay derived from colloidal matter. Oozes are divided into two main types according to their chief constituents. Calcareous ooze at depths of 2,000 to 3,900m (6,562–12,792ft) contains the skeletons of animals such as foraminifera and pteropods. Siliceous ooze at depths of more than 3,900m contains skeletons of radiolarians and diatoms. *See also* PE p.83.

Opal, non-crystalline variety of QUARTZ, found in recent volcanoes, deposits from hot springs and sediments. Usually colourless or white with a rainbow play of colour in gem forms, it is the most valuable of quartz gems. Hardness 5.5–6.5; s.g. 2.0. *See also* PE pp.96, 98–99.

Op (Optical) art, US art movement, popular in the mid-1960s, that rejects all signs of representation and insists on non-objectivity. It often incorporates kinetic effects in designs of squares, lines, circles and dots to create optical illusions. Colour, too, is used in a powerful way to give a sense of depth to pictures. In the artist's attempt to generate perceptual response in the viewer, the ultimate effect of the work depends on the viewer's participation. *See also* HC2 pp.278–279, *279.*

OPEC (Organization of Petroleum Exporting Countries), established in 1960 by the world's substantial petroleum producers to determine the best means of safeguarding their interests. Its permanent secretariat, financed by members, provides financial and technical assistance as well as co-ordinating price changes. A biennial conference is held to determine policies, prices and related matters affecting petroleum exporting. The 13 members are Abu Dhabi (United Arab Emirates), Algeria, Ecuador, Gabon, Indonesia, Iran, Iraq, Kuwait, Libya, Nigeria, Qatar, Saudi Arabia and Venezuela.

Opencast mining, stripping surface layers from the Earth's crust to obtain coal, ores and other valuable minerals. Since it does not require the sinking of shafts, tunnelling of galleries or underground working, it is less costly than underground mining. Large dragline excavators are employed to strip away surface layers, bulldozers and other mechanical shovels to distribute minerals and spoil, and lorries, railway wagons and overhead skips to carry minerals away for grading and processing. This form of mining is, however, restricted in many countries because of the damage it does to valuable land. Owners of opencast mines in Britain are obliged to restore the amenity value of land by landscaping and replanting after mining has ceased. *See also* MM p.*98;* PE p.*133.*

Open clusters, stellar CLUSTERS containing between about 20 and 1,000 stars and lying within our galaxy (the Milky Way),

mostly in its spiral arms. They are composed of stars which appear to have condensed from galactic clouds of the Milky Way, the brightest of which are comparatively young blue stars. *See also* SU pp.242–243, *243.*

Open Door policy, arrangement allowing all nations equal commercial access to a particular country. Most often associated with China's early relations with the USA, it was first formally proposed in 1899 by US Secretary of State John Hay. It ended in 1949 following the Communist take-over in China.

Open Golf Championship, British, annual championship held at various golf clubs, open to professionals and amateurs. The first championship is usually considered to be the 36-hole, medal-play event held at Prestwick in 1860, won by Willie Park. It is now a 72-hole, medal-play event.

Open-hearth process, method of producing steel in a reverberatory FURNACE fuelled by natural gas or coal gas. Lime and other slag-forming materials such as scrap iron are added to molten pig iron. The furnace is made of refractory materials containing lime and magnesia for removing impurities. *See also* MM pp.30–31.

Open shop, employment situation in which a worker is able to choose whether or not to join a trade union; membership of the union is not a condition of employment. *See also* CLOSED SHOP.

Open University, The, form of British higher education chartered in 1970 as an alternative to conventional university training. Teaching, which started in 1971, is carried out mainly through television and radio broadcasts in co-ordination with specially prepared course material, as well as some personal tuition. It gives full undergraduate degrees in all its faculties: arts, educational studies, mathematics, science, social science and technology. An honours degree is obtained by completing eight course units which a student may complete over a period of several years. In the mid-1970s there were more than 50,000 students enrolled with the Open University.

Opera, stage drama that is sung. Its combination of several elements – acting, singing, orchestral music, set and costume design – makes for spectacular entertainment. Large opera companies have a permanent staff of singers, designers and stagehands, a musical director, conductor and orchestra. International artists are usually engaged for leading roles, and top producers are often hired for specific productions. For these reasons, opera is extremely expensive to stage. Few, if any, of the major companies run at a profit. The best-known opera houses include La Scala (Milan), the Opéra (Paris), Covent Garden (London), the State Opera (Vienna), the Festspiele (Bayreuth) and the Metropolitan Opera (New York).

Opera began in Italy about 1600. The classical style evolved in about 1750 and its greatest exponent was MOZART. Reforms in favour of more dramatic opera were introduced by GLUCK in the late 18th century, and the 19th century was dominated by the work of Giuseppe VERDI and Richard WAGNER. Twentieth-century opera has been marked by a profusion of styles by composers such as Giacomo PUCCINI, Richard STRAUSS, Alban BERG and Benjamin BRITTEN. *See also* HC1 pp.318–319; HC2 pp.120–121.

Opera, soap, melodramatic serial, originally on radio, now also on television. Soap operas, whose abiding theme is the rocky road to love, began in the USA in the 1930s. In their early days they were sponsored by soap manufacturers.

Opera buffa, the style of Italian comic opera that developed in mid-18th-century Naples from comic intermezzi that in turn derived from interludes between the acts of either a spoken drama or OPERA SERIA. Light and simple in style, *opera buffa* introduced the concerted finale in which the principal singers sang an elaborate finale together. This was soon adopted by *opera seria* and was of the utmost significance for the subsequent development of opera. Fewer artificial male voices were

used than in *opera seria* and bass roles became more prominent. An early example of the style is *La Serva Padrona* (1733) by PERGOLESI.

Opéra comique, style of French opera that began in the late 18th century. A precedent of the style may be seen in the comédie-ballets of the mid-17th century composed by LULLY in collaboration with MOLIÈRE. The hallmarks of *opéra comique* are a witty plot involving some spoken dialogue, romantic rather than heroic subject matter and simple engaging music. André Grétry and Étienne Méhul were early exponents; other composers included AUBER, Adolphe ADAM and Louis Hérold.

Opera seria, the style of Italian opera in the 17th and early 18th centuries. The form of composition was bound by strict convention. The orthodox number roles were for three women and three men, the principal soprano part being invariably for high soprano and the leading and sometimes second male parts for CASTRATI (the third being sung by a tenor). Priority was given to virtuoso vocal display. The formalism and stylization of such operas prompted the development of OPERA BUFFA and once the latter had become established, composers such as CIMAROSA paid equal attention to both styles.

Operation, treatment performed on a person to cure or arrest a disease or to remedy a condition. The most common operations are surgical ones, in which the doctor makes an incision in the patient's body, but strictly speaking any treatment which requires the doctor to use hands (or instruments) on the body is an operation. *See also* MS pp.128–129.

Operator, in mathematics, any symbol indicating that a particular manipulation is to be carried out to convert one function into another. For example, in calculus the operator D signifies differentiation with respect to x, so that Dy means dy/dx. Further, D can itself be manipulated algebraically: eg since $\int (dy/dx)dx = y$, then $I.Dy = y$, where I is the operator signifying integration.

Operetta, type of light OPERA involving songs, dialogue, dancing and an engaging story. Operettas developed from attempts by composers to reach wider audiences. Among the more significant of these were Johann STRAUSS, Arthur SULLIVAN (in association with William S. GILBERT) and Jacques OFFENBACH.

Operon, group of GENES on a CHROMOSOME comprising structural genes and an operator gene. The structural GENES direct the synthesis of ENZYMES involved in the formation of a cell constituent or the utilization of a nutrient. The operator gene responds to a molecule (called a repressor) and can exist open or closed. When the operator gene is open, the genes it controls are functional, producing PROTEINS. When interacting with the repressor, the operator gene is said to be closed.

Ophicleide, now obsolete brass wind instrument used in the early 19th century. It had keys (like a saxophone) and produced low notes, but was superseded by the development of the TUBA.

Ophitism, beliefs of some GNOSTIC sects which arose in Syria in the early Christian era. Ophites worshipped the serpent, which they regarded as the symbol of the Godhead. They accepted the divinity of Christ, but denied the saving efficacy of the Crucifixion.

Ophthalmia, inflammation of the eye, either of the eyeball or of the conjunctiva (CONJUNCTIVITIS). It is usually caused by bacterial infection. In such cases the eye is painful and may appear bloodshot; treatment is usually with an eyewash. Much more serious types of ophthalmia are TRACHOMA, widespread in underdeveloped countries, and ophthalmia neonatorum of the newborn.

Ophthalmology, branch of medicine that specializes in the structure, function and diseases of the eye. It includes surgical and other treatment of eye disorders and the correction of defective vision. *See also* MS pp.48–49.

Ophthalmometer, instrument devised by

Peter Oosterhuis beat Gary Player in this Piccadilly world match-play event.

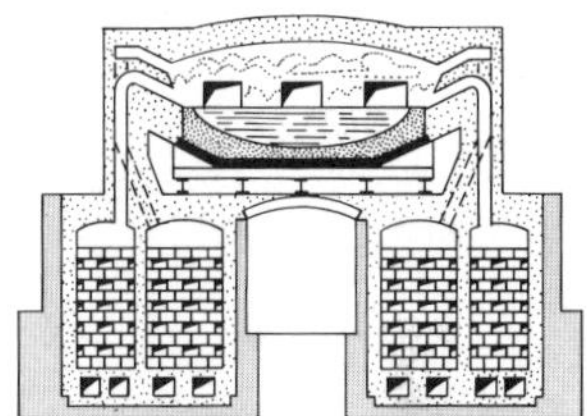

Open-hearth process; the furnace at 1,600°C is lined with fire-resistant bricks.

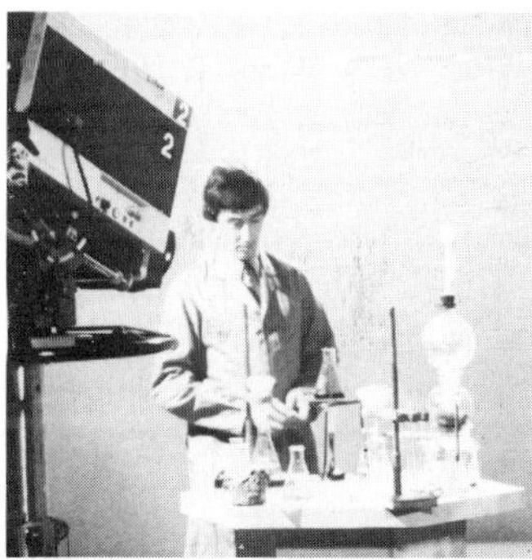

Open University; a lecture being filmed for broadcasting on television.

Opera; a scene from Wagner's *The Flying Dutchman,* written in 1843.

Ophthalmoscope

Martin Opitz; the frontispiece to his influential *Buch von der deutschen Poeterei*.

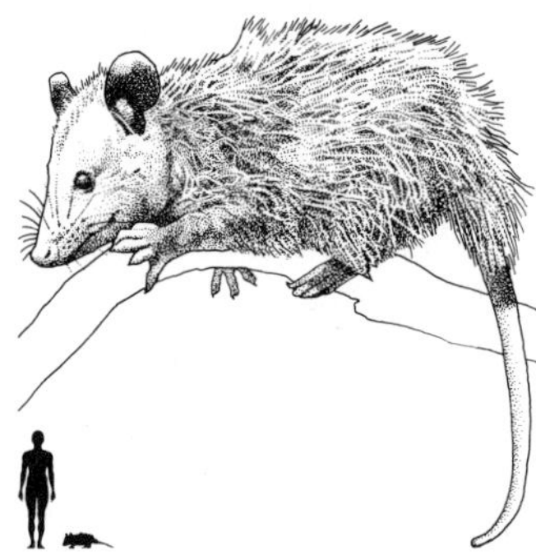

Opossums have opposable toes which enable them to cling to narrow branches.

Julius Oppenheimer is said to have introduced theoretical physics to the USA.

House of Orange; Frederick Henry was the fourth son of William the Silent.

Hermann von HELMOLTZ in the 1850s to measure the curvature of the eye by the images reflected in it. It was later refined to measure the eye's capacity and its refractive irregularities.

Ophthalmoscope, instrument for examining the inner structure of the eye, invented by the German physiologist Hermann von HELMHOLTZ in 1851.

Opie, John (1761–1807), British painter, best known for his historical scenes and portraits. His rather uneven style was influenced in its early stage by the work of REMBRANDT; later works were coarser in style. His paintings include *The Peasant's Family* (1784) and *The Murder of Rizzio* (1787).

Opitz, Martin (1597–1639), German poet who greatly influenced 17th-century German literature. In his verse, written during the THIRTY YEARS WAR, he experimented extensively with styles, and in his *Buch van der deutschen Poeterei* (1624) he theorized about the writing of poetry. His adaptation (1638) of Sir Philip SIDNEY's *Arcadia* and his own *Schäferey von der Nymphen Hercynie* introduced the pastoral form to Germany.

Opium, drug derived from the juice of unripe seed-pods of the opium poppy. Its components and derivatives have been used as NARCOTICS and ANALGESICS for many centuries. It produces drowsiness and euphoria and reduces pain. MORPHINE, CODEINE and papaverine are common opium compounds. British merchants were engaged in smuggling opium into China in the 19th century and opium dens flourished there. Attempts by the Chinese government to curtail the trade resulted in the OPIUM WAR. *See also* HC2 pp.*140*, 144; MS pp.104–105, *105*, 119.

Opium War (1839–42), hostility between Britain and China over Chinese trade restrictions, especially on the illegal import of opium by Western traders, mainly British. By the Treaty of NANKING (1842) the Chinese were required to pay an indemnity for the cost of the war, forced to cede Hong Kong to the British, to open five ports for British trade and to allow British citizens in China to be tried in British courts. This marked the beginning of the so-called unequal treaties forced on China by the West.

Oporto, port city in NW Portugal, on the River Douro, 3km (2 miles) from its mouth on the Atlantic Ocean. An ancient Roman settlement, it was occupied by the Visigoths 540–716, the Moors 716–997 and taken by Alfonso I of Portugal in 1062. Oporto grew as a wine centre in the 17th century. The PORT wine is still exported; other products include fruit, cork and olive oil. Pop. 310,437.

Opossum, New World MARSUPIAL animal. Opossums (known as possums in North America) are the only marsupials found outside Australasia. These tough, cat-sized mammals live mainly in trees and are omnivorous. They have silky grey fur (except on the long prehensile tail), and feign death when in danger – hence the expression "playing possum". The common opossum, *Didelphis marsupialis*, grows up to 50cm (20in) long, including a 30cm (12in) tail. Family Didelphidae. The rat opossum (Family Caenolestidae) is found in South America only. *See also* NW pp.*159, 216*.

Oppenheimer, Sir Ernest (1880–1957), South African mining and industrial magnate, b. Germany. Originally a representative of a diamond company, he helped to organize the conquest of German South West Africa in 1914 on the outbreak of WWI. In 1919 he formed the Consolidated Diamond Mines of South West Africa which soon became the dominant force in the world diamond market. In the 1930s he invested in copper mining in Zambia and gold mining in the Orange Free State. He was the member of parliament for Kimberley (1924–38).

Oppenheimer, Julius Robert (1904–67), US theoretical physicist. After early work in nuclear physics at the University of California, he was appointed director (1943–45) of the Los Alamos laboratory in New Mexico, where he was responsible for the construction of the first atomic bomb. In 1949, however, he strongly opposed the construction of the hydrogen bomb and in 1953 was suspended by the Atomic Energy Commission and declared a security risk. He was subsequently reinstated and presented with the Fermi Award for his contribution to science. *See also* MM p.180.

Opposition, in astronomy, position on the orbit of a superior planet at which the Sun, the Earth and the planet are in line, with the Earth located between the planet and the Sun. Only superior planets, with larger orbits than that of the Earth, can be in opposition to the Sun, and these may be observed most clearly at this time. *See also* SU p.*206*.

Opposition, Leader of the, the leader of the party which is runner-up in a general election and so does not form the government. As a political office in Britain it developed from the emergence of political parties in the House of Commons. The term "HM Opposition" was first used by John Hobhouse in 1826. Official recognition was given to the Leader of the Opposition in 1937, when the holder was granted a salary higher than that of an ordinary MP.

Opsomer, Isidore Edmond Henri, Baron (1878–1967), Belgian artist. He specialized in scenes of his native Antwerp and also achieved a reputation as a portrait painter. He was a leading influence on contemporary Belgian art.

Optical activity, the ability of some chemical compounds in solution and some crystals and transparent substances placed in an intense magnetic field to rotate the plane of POLARIZED LIGHT. Optically active crystals are used in polarimeters, instruments for measuring the degree of optical activity of chemical solutions. Optically active compounds are characterized by having molecules that are asymmetrical. Magnetic rotation of the plane of polarized light was discovered by Michael FARADAY in 1845 and is known as the Faraday effect.

Optical illusion, effect in which visual information is misinterpreted. The commonest optical illusions are those in which the expected visual signals, such as lines of perspective, are distorted or created artificially. *See also* SU p.*46*.

Optic nerve, second cranial nerve, which carries the visual stimuli from the retina of the EYE to the visual centre in the CORTEX of the brain. That part of the retina where the optic nerve enters the eye is known as the blind spot. About 1 million optic nerve fibres comprise the optic nerve, and these fibres are arranged in such a way that impulses from the left side of the visual field travel to the right side of the brain, and vice versa. *See also* MS pp.48–49.

Optics, branch of physics concerned with the study of light and its behaviour. Fundamental aspects are the physical nature of light, both as wave-propagation and as particles (PHOTONS), and the reflection, refraction and polarization of light and its transmission through various media. These studies involve discussion of the properties of mirrors, lenses and lens systems (including that of the eye and those of optical instruments) and of optically active chemicals and crystals which polarize light. More generally, optics also deals with the wider subject of the ELECTROMAGNETIC SPECTRUM ranging from short radio waves to soft X-rays, called the optical spectrum. *See also* SU pp.98–111.

Optometry, science that deals with the examination of eyes and the prescribing of spectacles or contact lenses and exercises to correct defects in vision. It should not be confused with OPHTHALMOLOGY, the medical and surgical treatment of the eyes. *See also* MS p.48.

Opuntia, genus of cactus plants found from Canada to Argentina and characterized by small barbed bristles. They have cylindrical, rounded or, in the PRICKLY PEARS, flattened joints. Family Cactaceae. *See also* NW p.218.

Opus (Latin "work"), in music, term for the numerical classification of a work or group of works in a composer's output. It became widely used in the latter half of the 18th century, and works were ordered according to the date of publication rather than of composition. The works of BEETHOVEN and SCHUBERT are numbered in this way. Later composers have, however, assigned an opus number to a work upon completion, regardless of publication date.

Opus Anglicanum (Latin for "English work"), type of English EMBROIDERY of the period *c.*1250–*c.*1350. Outstanding for the skilful use of silver gilt thread to provide a background to the figures, animals and birds depicted in coloured silks, it was highly valued throughout Europe. It was chiefly used for clerical vestments and many fine examples have survived, including the famous Syon cope.

Opus Dei, name of a Roman Catholic organization of laymen and a few priests in Spain, known particularly for its influence on political and religious opinion since the mid-1950s when Gen. FRANCO appointed several of its members to ministerial posts to further his economic policies. Founded by José María Escrivá de Balaguer y Albás in 1928, it supported the Nationalists during the war and rose to pre-eminence in the subsequent three decades. It aims to promote traditional Christian values. It was approved by the Vatican in 1950, and has several international branches. Its influence appears to have declined in the 1970s.

Oracles, in Greek mythology, communication or advice from the gods in answer to requests from men, usually taking place at the shrine of a god. At Dodona, the shrine of ZEUS, the rustling of oak leaves was interpreted by priests. At the oracle of APOLLO at Delphi, consultation was through the priestess (Pythia) who, in an induced trance, uttered unintelligible sounds. The priests translated these into appropriate although often ambiguous prose or verse. Another way of consulting was by casting lots or by "incubation", in which the enquirer slept in a holy area and received his answer in a dream. *See also* HC1 pp.68, *71*.

Oradea, city in NW Romania, on the River Koros near the Hungarian border. Founded in 1080 by St Ladislas, it was destroyed in 1241 by the Tartars and rebuilt in the 14th century. It was held by the Turks from 1660 to 1692 and ceded to Romania by Hungary in 1919. Today it is the commercial centre for the surrounding agricultural region. Industries include chemicals, mining equipment and canned goods. Pop. (1970 est.) 137,700.

Oral contraception. *See* BIRTH CONTROL.

Oral literature, stories, histories, legends and myths originally conveyed by word of mouth rather than by written records. Their collection and study constitute an important branch of social anthropology. The best-known examples of such oral traditions come from Africa and from the North American Indians, although most preliterate societies preserved their heritage by means of memory and by its repetition in exclusive or public gatherings.

Oran, leading port city and department capital in NW Algeria, on the Gulf of Oran. Founded in the 10th century, it was taken by Spain in 1509, by the Turks in 1708 and re-taken by Spain in 1732. It was captured by the French in 1831 and developed by them as a naval base. Conflict between the French and native Algerians in the 1950s led to an exodus of the European population. It now exports cereals, wine and fruit from the surrounding market gardens. Pop. 327,493.

Orange, House of, royal house of The Netherlands. Orange, a principality in s France from the 11th century, was inherited by the NASSAU house of w Germany. The Prince of Orange, WILLIAM THE SILENT, became Stadtholder of The Netherlands in 1579. Except for brief intervals, the House of Orange has ruled the country ever since, usually with popular support. King WILLIAM III of England was a Prince of Orange.

Orange, evergreen citrus tree and its fruit. There are two basic types. The sweet orange, of which there are many varieties, is native to Asia and widely grown in the USA and Israel. The fruit develops without flower pollination and is often seedless. The sour orange is widely grown in Spain as an ornamental and for the manufacture of marmalade. Height: to 9m (30ft). Family Rutaceae; genus *Citrus. See also* PE p.192, *192.*

Orange, River, principal river of South Africa. It rises in the Drakensberg Mts N Lesotho and runs in an irregular easterly direction, forming the boundary between the Orange Free State and Cape Province, before reaching the Atlantic Ocean at Alexander Bay. The river is used for irrigation and supports many hydro-electric power plants. Length: approx. 2,093km (1,300 miles).

Orange Free State, province in E central South Africa, bounded by Natal (E), Transvaal (N), Lesotho (SE) and Cape Province (S and W). The region was inhabited by the Tswana peoples until the 19th century, when large-scale BOER immigration began. Conflict between the Boers and the British, who had annexed the region in 1848, led to the SOUTH AFRICAN WAR (1899–1902). In 1910 the colony became a province of the Union of South Africa. The economy is based on agriculture; cereals and fruit are produced. Industries include gold and diamond mining. Area: 129,153sq km (48,866sq miles). Pop. (1970) 1,716,350. *See also* HC2 pp.128–129.

Orangemen, members of the Orange Society, founded in Ulster in 1795. It was a political and sectarian society named after the Protestant WILLIAM III, formerly Prince of Orange, who defeated JAMES II at the Battle of the BOYNE on 11 July 1690, a victory celebrated now on 12 July, the date of a further Loyalist victory at the Battle of Aughrim in 1691. The Orangemen sought to maintain the Protestant succession and progressively strengthened their position in Northern Ireland, continuing as proponents of Protestant unionist opinion to the present day.

Orange River Colony (1902–10), name given to the ORANGE FREE STATE in the Treaty of Vereeniging, which recognized British annexation of the province. When it joined the UNION OF SOUTH AFRICA in 1910 it regained its former name.

Orang-utan, stout-bodied, thick-necked great APE native to forests of Sumatra and Borneo. It has a bulging belly and a thin, shaggy, reddish-brown coat. It swings by its arms or walks on branches when travelling through trees, but proceeds on all fours when on the ground. Height: 1.5m (5ft); weight: to 100kg (220 lb); arm span: more than 2.13m (7ft). Species *Pongo pygmaeus. See also* PRIMATES; NW pp.*169, 172, 251.*

Oranjestad, town on the W coast of Aruba Island in the Netherlands Antilles; the principal town of the island. Oil refining is an important industry. Pop. 14,500.

Oratorians, members of the Oratory of St Philip NERI, who began to ordain his followers at a church in Rome in 1564. It is the oldest of several Roman Catholic congregations of secular priests, who live under rules of obedience but without VOWS. PALESTRINA, a penitent of St Philip, composed some of the music for the services. A French Oratory was founded in 1611 by Cardinal de Bérulle and Cardinal NEWMAN founded the English Oratory near Birmingham in 1848.

Oratorio, form of musical composition, usually sacred in character, for solo voices, chorus and orchestra, the first of which were presented in oratories (chapels) in 17th-century Italy. Emilio de Cavalieri, Giovanni Carissimi and Claudio MONTEVERDI were among the first exponents of the genre. Among outstanding examples are HANDEL's *Messiah* (1742) and STRAVINSKY's *Oedipus Rex* (1927).

Oratory, the art of speaking in an inspirational manner. An orator speaks in this way to gain support, eg for a political action or legal opinion. PERICLES, ARISTOTLE, the SOPHISTS and CICERO were early teachers of the art.

Orb, traditional symbol of a monarch's power. It is a sphere mounted by a cross, made of precious metals and stones and originally represented the universe as a harmonious whole. The use of the sphere comes from the Romans; the cross was a Christian addition. An orb was first used at a coronation by the Holy Roman Emperor Henry II in 1014.

Orbigny, Alcide Dessalines d' (1802–57), French naturalist. He studied fossil remains in South America, proposed a theory of fossil stratification, and founded the modern field of micropalaeontology. He wrote *Journey to South America* (1835–47), and was responsible for the first full map of that continent (1842).

Orbit, elliptical path of a celestial body moving round another acting as the former's centre of gravitational attraction. Planetary orbits are described in numerical quantities from which the revolution period and mean orbital velocity are derived. By analogy, in the BOHR model of the atom, the paths ascribed to ELECTRONS are also called orbits. *See also* ORBITALS.

Orbitals, in atomic physics, regions in space around an atomic NUCLEUS in which ELECTRONS can move. There is a high probability of finding an electron in such an orbital, which can accommodate one or two electrons and has a shape and energy characterized by the atom's QUANTUM NUMBERS. In molecules, the bonding electrons move in the combined electric field of all the nuclei. The atomic orbitals then become molecular orbitals – regions encompassing two nuclei, having a characteristic energy and containing two electrons. These molecular orbitals, which can be thought of as formed by the overlap of atomic orbitals, constitute chemical bonds. *See also* SU pp.132, 138.

Orbits, the eye-sockets, cavities in the head containing the eyes.

Orcagna, Andrea (*c.*1308–*c.*1368), Italian painter, sculptor and architect, real name, Andrea di Cione. He was one of the best and most influential Florentine sculptors of his time. His major extant work is a tabernacle for Or San Michele, Florence, of which he was head architect in 1356. *See also* HC1 p.*209.*

Orchardson, Sir William Quiller (1832–1910), Scottish painter. From 1858, Orchardson's first works were white illustrations in the style of the PRE-RAPHAELITES. In 1862 he settled in London and was elected to the Royal Academy in 1877. *Napoleon on Board the Bellerophon* (1880) is his most widely known painting.

Orchestra, group of musicians who play together. The development of the modern symphony orchestra began in the 17th century with the addition of various instruments to the string orchestra: woodwind and later, brass instruments. By the late 18th century the woodwind section had been established, and the brass section by the end of the 19th century. Modern orchestras consist of between 80 and 120 players divided into sections: strings (violin, viola, cello, bass and harp); woodwind (flute, oboe, clarinet and bassoon); brass (trumpet, trombone, French horn and tuba) and percussion.

Orchestration, arrangement of music for various instrumental combinations. It had become an important aid to expression by 1700. BERLIOZ's *Treatise on Orchestration* (1844) was an early attempt to systematize the possibilities within the ORCHESTRA.

Orchid, any plant of the family Orchidaceae. There are about 35,000 species, and they are particularly common in the tropics. All are perennials and grow in soil or as epiphytes – air plants – on other plants. Parasitic and saprophytic species are also known. Colour and shape vary greatly between species, and flowers can be borne singly or in erect or pendant clusters. A club-shaped structure, the columna, results from the fusion of male and female reproductive parts. All orchids have bilaterally symmetrical flower structures, each with three sepals. They range in diameter from approx. 2mm (0.1in) to 38cm (15in). *See also* NW pp.67, 70, 216, *247.*

Orchis, genus of orchids found on rich, wooded slopes of central USA. The purple rose upper petals and the large white lower petal are borne in clusters of 2 to 15 flowers. The two leaves are smooth and long. Height: to 30cm (1ft). Family Or-

chidaceae; species include *O. spectabilis.*

Orczy, Baroness Emmuska (1865–1947), Hungarian novelist who wrote in English. After studying art on the continent and in London, she wrote *The Scarlet Pimpernel,* and the publication of this historical adventure story in 1905 made her famous. She later wrote sequels to this book and detective stories.

Ordeal, ancient manner of criminal trial practised in many societies in Europe during the Middle Ages. The three main types of ordeal were by divination, physical test and combat. In Britain, ordeal by fire, which was administered to freemen, involved grasping very hot metal. If the accused sustained no scars after three days he was pronounced innocent; otherwise he was guilty. Ordeal by cold water, used for the unfree, involved plunging the accused, trussed, into a cold pool; if he floated he was judged to be innocent. The guilt of clergy was ascertained according to whether they choked on a feather placed in food. Trial by battle was introduced by the Normans into England for the trial of knights. Trial by ordeal was abolished in 1215, and was gradually replaced by the jury system.

Order, in the systematic classification of living organisms, a taxonomic category that ranks below a class and above a FAMILY.

Order, in classical architecture, style and decoration of a column, its base, capital and entablature. Of the five orders, the Greeks developed three: the CORINTHIAN, DORIC and IONIC. The Tuscan and COMPOSITE orders were adaptations of these by the Romans.

Orderic, Vitalis (1075–*c.*1143), English monk and chronicler who passed most of his life in Normandy. His *Ecclesiastical History,* which takes events to the year 1143, is a rich source for the history of the NORMANS in Britain and on the continent.

Order of Chivalry, society of distinguished public figures in Britain. There are nine such societies, the oldest of which, the Order of the GARTER, was founded in the mid-14th century as a secular version of the religious orders of knights formed during the CRUSADES. Other orders are the Scottish Order of the Thistle (1687), the Irish Order of St Patrick (1788), and those of St Michael and St George (1818).

Order of Merit, British honour founded by EDWARD VII in 1902. It is limited to 24 holders, people who have distinguished themselves in the armed forces or in the arts and sciences.

Orders in Council, orders issued by the British sovereign on the advice of the Privy Council. During the NAPOLEONIC WARS a series of such orders were used in 1807 to blockade European ports.

Ordinal number, number that indicates the position of a member in an ordered sequence or in a group. Thus "first", "second", and "third", are ordinal numbers; "one", "two", and "three" are cardinal numbers.

Ordination, ceremony of those Churches for whom the ministry is a sacramental office. At ordination a person becomes a member of the clergy, or a member of the clergy is elevated by one rank. Ordination is considered unnecessary by those Churches who see the ordaining of ministers as a personal matter for the individual, and not as a SACRAMENT (a matter for the whole Church).

Ordnance Survey, official mapping of Britain. Undertaken for defence purposes, it was started in 1791 with the creation of the Board of Ordnance. The first sheet, a one-inch map of Kent, was published in 1801. Revision is done by the Ministry of Housing and Local Government.

Ordovician period, the second-oldest division of the PALAEOZOIC era, 500 to 430 million years ago. All animal life was restricted to the sea. Numerous invertebrates flourished and included trilobites, brachiopods, corals, graptolites, molluscs and echinoderms. Remains of bones in coastal deposits mark the first record of the vertebrates. *See also* NW pp.24, 25, 123, 178–179.

Ore, mineral or combination of minerals from which metals and non-metals can be

Orangemen celebrate Orange day with a parade through Belfast.

Orchestra; the Halle Orchestra giving a concert at the Free Trade Hall, Manchester.

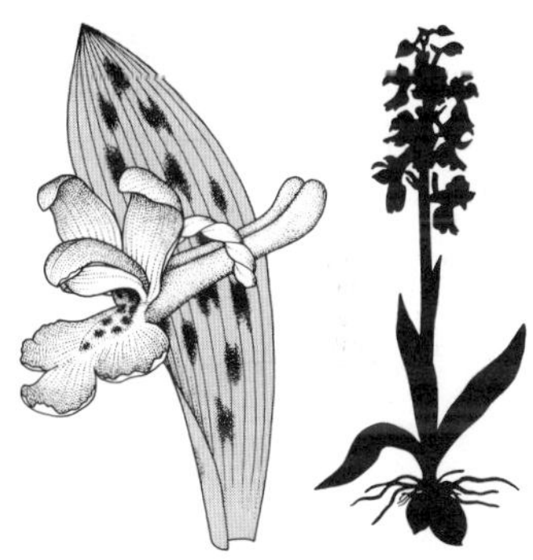

Orchids often have spicate flowers; they grow in the temperate regions of the world.

Ordeals were still used, after they were officially abolished, in witch trials.

Oregon has many areas of great natural beauty, such as the famous Crater Lake.

Carl Orff wrote incidental music for plays after he left Munich Academy.

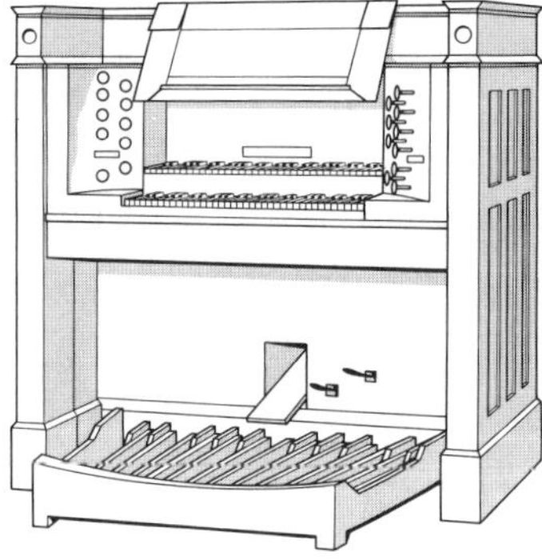

Organs, despite their immense power, can produce extremely delicate music.

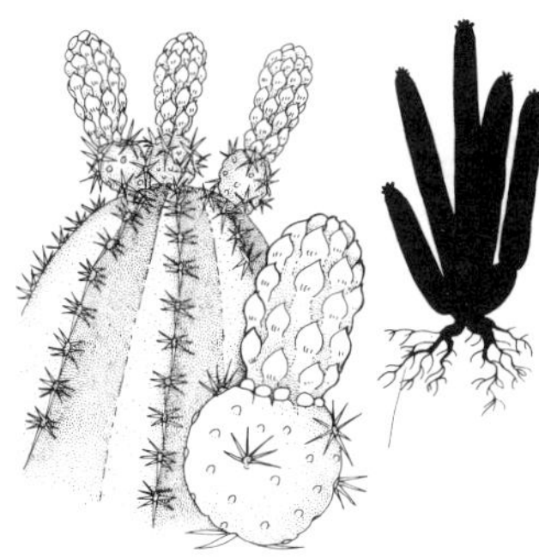

Organ-pipe cactus is closely related to the huge Saguara cactus of Arizona.

extracted. It occurs in veins, beds or seams parallel to the enclosing rock or in irregular masses. Industrial rock deposits in beds, eg gypsum and limestone, are not called ores. *See also* LODE.

Örebro, city in s central Sweden, E of Lake Hjälmaren and capital of Örebro province. It was the site of many diets including the 1529 diet by which the REFORMATION was brought to Sweden. Industries: shoe manufacture, paper and processed foods. Pop. (1970 est.) 90,900.

Oregano, dried leaves and flowers of several perennial herbs native to Mediterranean lands and w Asia. It was introduced into the West as a food seasoning, and has become popular in Italian and Mexican dishes. Family Labiatae.

Oregon, state of NW USA on the Pacific coast. The state is dominated by the forested slopes of the Cascade and the Coast ranges. Between the two lies the fertile Willamette Valley. The Columbia and the Willamette are the major rivers. Of prime importance to the agricultural economy are cattle, dairy produce, wheat and market garden products. Oregon produces more than 20% of the nation's softwood timber and wood processing is a major industry. Other industries include printing and publishing, food processing, fishing, machinery and primary and fabricated metals. Abundant hydro-electric power is produced in the state. Nickel, sand and gravel, and mercury are the main mineral resources. The major towns are Salem, the state capital, Portland and Eugene.

In 1792, Robert Gray sailed up the Columbia River, establishing a US claim to the area drained by the river. Several trading posts were established, mainly by the HUDSON'S BAY COMPANY and the NORTH WEST COMPANY. The OREGON TRAIL brought more settlers from 1842 onwards and the Oregon Territory was formed in 1848. It was admitted to the Union as the 33rd state in 1859. Area 251,180sq km (96,981sq miles). Pop. (1975 est.) 2,288,000. *See also* MW p.175.

Oregon trail, most popular route used by emigrants to the NW of the USA. It ran 3,200km (2,000 miles) from Independence, Missouri, to the Columbia river in Oregon. Fur traders explored parts of the trail from 1811–35 and about 12,000 people used it during the 1840s – so many in fact, that ruts were worn which were visible for generations after. The trail was difficult in places, but was one of the few routes over the Rocky Mountains negotiable by horse or ox-drawn wagons.

O'Reilly, Anthony John Francis ("Tony") (1936–), British rugby union three-quarter who played for Old Belvedere, Leicester, Ireland and the British Lions. He appeared 29 times for Ireland and, playing for the British Lions, he scored a record 16 trics in 1955, and a record 17 in New Zealand in 1959.

Oreopithecus, ancient primate of the Miocene and Pliocene epochs. In a genus by itself, fossil remains of the semi-erect, apparently vegetarian *Oreopithecus* have been found in s Europe and E Africa dating from over 7 million years ago. Height: approx. 1.2m (4ft).

Oresme, Nicholas (*c.* 1325–82), French medieval philosopher who is also called Nicole d'Oresme. He was master of the College of Navarre in Paris and became Bishop of Lisieux in 1377. He wrote on economics, politics and natural science, making notable contributions to geometry which he applied to the laws of motion. He also discussed the possibility that the Earth rotates around other bodies.

Oresteia, trilogy of tragedies by AESCHYLUS, produced in 458 BC and consisting of *Agamemnon, The Libation Bearers* and *Eumenides.* It is the only surviving Greek trilogy and is Aeschylus' last and greatest work. Each play is complete and stands alone; together they explore the themes of crime, revenge and expiation.

Orestes, in Greek mythology, the son of AGAMEMNON and CLYTEMNESTRA, and brother of ELECTRA. He killed his mother and her lover Aegisthus in order to avenge the death of his father whom Clytemnestra and Aegisthus had murdered. He is a

character in Aeschylus' *Oresteia,* Sophocles' *Electra* and Euripides' *Orestes.*

Orff, Carl (1895–), German composer who helped to found the Günter School in Munich in 1925. He wrote opera in a neoprimitive and rhythmic style. His best-known works are the scenic cantatas *Carmina Burana* (1937) and *Catulli Carmina* (1943), and the opera *Trionfo dell'Afrodite* (1953). *See also* HC2 p.270.

Orfila, Matthieu Joseph Bonaventure (1787–1853), French chemist. He published his *Treatise on General Toxicology* in 1813, a work that led him to be considered the founder of toxicology.

Orford, 1st Earl of. *See* WALPOLE, SIR ROBERT, 1ST EARL OF ORFORD.

Orford, 4th Earl of. *See* WALPOLE, HORACE, 4TH EARL OF ORFORD.

Organ, musical keyboard instrument. The player sits at a console and regulates a flow of air to ranks of pipes producing rich and sometimes sombre tones. The early development of the organ is obscure but it was in use in Christian churches in the 8th century. Keyboards appeared in 12th-century Europe and reed pipes – in imitation of other instruments – *c.* 1550. The modern organ dates from the BAROQUE period, when J. S. BACH wrote many highly innovative compositions for it. With the coming of electricity, the electric organ widened the field of possible effects still more. Albert SCHWEITZER and others have revived interest in Baroque organ music.

Organ, in biology, a group of tissues that form a functional and structural unit in a living organism. The kidney, for example, has as its main task the water regulation of the body. Such organs as the liver and the skin have many important tasks. Leaves and roots are examples of plant organs.

Organelle, part of a cell, such as a MITOCHONDRION or a flagellum, with a persistent structure and a specific function. It is to a cell what an organ is to an organism.

Organic analysis, in chemistry, term with two meanings. It is either the determination of the composition and structure of a chemical compound – how the compound is made up from its chemical elements, or it is the determination of the identity and quantity of particular organic compounds in mixtures and in such organic substances as tissues. *See also* SU pp.150–151.

Organic chemistry. *See* CHEMISTRY, ORGANIC.

Organic compounds, COMPOUNDS that contain the element CARBON; they are about hundred times more numerous than inorganic compounds. HYDROCARBONS contain carbon and hydrogen only, and are the basic structures which, when combined with atoms of other elements (eg oxygen, and nitrogen), form a vast range of compounds including those essential to life. *See also* SU pp.152–153.

Organic food, in gardening, food grown with the use of natural fertilizers derived directly from animals and plants, including manures, bone meal, blood, straw and compost. *See also* PE pp.174–175.

Organic functional groups, molecular groups, occurring in organic compounds, which largely determine their chemical properties. For example, alcohols (such as butanol, C_4H_7OH) all contain the ALCOHOL group-OH, and consequently they all form ESTERS and ionize in water. Other organic functional groups include the KETONE group -CO, the ALDEHYDE group -CHO and the CARBOXYLIC ACID group -COOH.

Organisation de l'Armeé Secrète. *See* OAS.

Organization for Economic Co-operation and Development. *See* OECD.

Organization of African Unity. *See* OAU.

Organization of American States. *See* OAS.

Organization of Petroleum Exporting Countries. *See* OPEC.

Organ of Corti, complex structure of the inner EAR of mammals, birds and reptiles, concerned with the final reception of inner ear movements resulting from sound waves striking the eardrum. It rests on a platform of membrane and bone which extends along the COCHLEA, and contains sensory hair cells that detect movements caused by sound waves and connect with nerve fibres that carry messages about these to the brain. *See also* MS p.51.

Organometallic compound, chemical compound in which one or more organic groups or radicals are bonded to an atom of a metal. Metallic carbonates (such as sodium carbonate) and salts of common fatty acids (such as sodium acetate) are usually excluded from this classification. Typical examples are metallic alkyl compounds (such as TETRAETHYL LEAD and triethyl aluminium), GRIGNARD REAGENTS (such as ethylmagnesium iodide) and compounds of transition metals.

Organon, name given to the logical works of ARISTOTLE. These comprise the *Categories,* and *Topics, On Interpretation, Prior Analytics, Posterior Analytics* and *Sophistic Refutations.* In medieval science, the *Organon,* which means "instrument", was thought of as a guide to rigorous scientific investigation. As works of philosophy they deal with the use of language in argument.

Organophosphates, chemicals in which the molecules contain at least one atom of carbon and one of phosphorus. The best-known organophosphates are some insecticides and certain nerve gases, essential nucleic acids and nucleotide enzymes. Organic derivatives of the phosphorus acids are important in the manufacture of fertilizers.

Organ-pipe cactus, CACTUS native to SW USA and South America. It has long columnar stems holding night-blooming flowers. Height: to 10m (33ft). Family Cactaceae; species *Lemairocereus thurberi.*

Orgasm, physiological culmination of sexual stimulation, marked by general release of muscular tension and waves of localized contractions causing EJACULATION in the male and climactic spasms of vaginal muscles in the female.

Orient, The, term meaning the East. Formerly applied to all countries E of the Mediterranean, it is now restricted to Asia, especially the Far East.

Oriental art, painting and the plastic arts of China, Korea and Japan which, although not identical, belong to one essentially Chinese tradition. Their origins lie in the bronze pottery work of the SHANG dynasty of China (*c.* 1600–*c.* 1030 BC). By the 2nd century AD painting was important, culminating in the great landscapes of the 10th to the 14th centuries. The modern flowering of Chinese art occurred in the MING period, when elaborately ornamented ceramics and carved lacquer work achieved their highest expression. *See also* HC1 pp.128–129, 132–133; HC2 pp.54–55.

erienteering, sport of Scandinavian origin. Similar to cross-country running but over rougher and more wooded terrain, it requires both athletic and navigational skills. Runners, each leaving at timed intervals, carry a map and compass with which to locate the control points around the course. The fastest to complete the course wins. Competitors normally choose their own routes between control points. The number of control points varies, there are sometimes over 20. Competitors under 15 years can run on courses under 3km (1.86 miles) long; the senior courses are usually of about 10km (6.21 miles). A variation is score orienteering, in which each control point is allocated a value and competitors score points according to the number of control points visited within a set time.

Orient Express, train that travels between Paris and Bucharest, by way of Vienna. First given the name Orient Express in 1883, its exact route has varied with political circumstances. Its heyday, between 1897 and 1914, ended at the outbreak of WWI. Although it used to travel through Czechoslovakia to Bratislava, after 1951 it ran from Vienna via Hegyashalom, Budapest, Arad and Brashov to Bucharest. The Direct Orient Express, which enabled passengers to travel (without changing trains) from Paris to Istanbul, ran from 1962 until 21 May 1977, when it was discontinued.

Origen (*c.* 185–*c.* 254), leading theologian of the early Christian Church. He was probably born in Alexandria, Egypt, and became famous as a teacher. He studied, wrote and taught in Cappadocia and Palestine, and visited Arabia and Rome. His principal work was *De Principiis* (*On First Principles*). He defended orthodoxy against GNOSTICISM and introduced Greek philosophy into Christian writing.

Original sin, the sin committed by ADAM and EVE that caused their expulsion from the Garden of EDEN. Because Adam and Eve represent mankind, the term also applies to the innate tendency in man's nature to commit sin, and to any mortal condition that is not a state of GRACE.

Orinoco, river in Venezuela, rising in the Sierra Parima Mts in s Venezuela and flowing NW to Colombia, then N forming part of the Venezuela-Colombia border and finally E into the Atlantic Ocean. It was sighted by Christopher COLUMBUS in 1498 and navigated by Diego de Ordaz (1530–31). Length: approx: 2,062km (1,281 miles).

Oriole, name of two unrelated types of medium-sized songbirds. The oriole of the Old World (family Oriolidae) is brightly coloured, arboreal and feeds on insects, and lays two to five speckled white eggs in a cup-shaped nest. The New World oriole (family Icteridae) has similar colouring and builds hanging nests in trees for its two to six speckled white eggs. *See also* HANG-NEST.

Orion, spectacular equatorial constellation situated between Taurus and Lepus. It is visible from every inhabited part of the world. Four young stars, Betelgeuse, Rigel, Saiph and Bellatrix, form a conspicuous quadrilateral containing a row of three other stars representing the belt of Orion, the hunter. This constellation contains the Orion Nebula, and several binary stars. *See also* SU pp.*236, 255, 257, 263–267*.

Orissa, state in NE India on the Bay of Bengal; its capital is Bhubaneswar. It was ruled for centuries by several HINDU dynasties, was taken by Afghans in 1568 but soon afterwards came under the MOGULS. The British occupied it in 1803. The coastal section was made part of Bihar and Orissa province in 1912 and a constituent state of India in 1950. Industries: fishing and mining. Area: 155,782sq km (60,147sq miles). Pop. (1971) 21,944,615.

Orkney Causeway, barriers and road connecting Mainland, Orkney islands, off the N coast of Scotland, with the islands to the s – Burray and South Ronaldsay. The causeway is also called the Churchill Barriers since it was built during WWII to prevent the passage of enemy submarines into the naval base of SCAPA FLOW.

Orkney Islands, group of islands off the N coast of Scotland which constitutes an independent administrative area (Orkney Region). Mainland (Pomona) is the largest in the group of about 70 islands, more than half of which are uninhabited. In recent years it has been one of the centres for the development of the North Sea oilfields but its people are trying to preserve their culture, which is traditionally independent of Scotland. Area: 974sq km (376sq miles). Pop. (1971) 17,075.

Orkneys, South. *See* SOUTH ORKNEYS.

Orlando, Vittorio Emanuele (1860–1952), Italian politician who supported Italy's entry into WWI in 1915 and was elected Prime Minister in 1917. He took part in the VERSAILLES Peace Conference (1919–20) but withdrew in disagreement with Woodrow WILSON and resigned in 1919. In 1920 he was elected President of the Chamber of Deputies. He initially supported the FASCISTS, but resigned in 1925 in protest against a Fascist electoral fraud. In 1946 he returned to the Assembly and became a senator in 1948. *See also* HC2 p.*192*.

Orléans, name of a branch of the French Valois and BOURBON dynasties. It was created in 1344 by Philip VI for his son, who died without male heirs in 1375. In 1498 the duchy was united with the Valois when the Duc d'Orléans became Louis XII. The Bourbon Louis XIV gave the title to his brother Philip, whose descendants have held it since. The only member of the line to become king was Louis-

Philippe (*r.* 1830–48). Because of their relationship with the Bourbons, the house of Orléans was often a source of intrigue against the Crown. After the French Revolution it became identified with constitutional monarchy and 19th-century liberalism. *See also* HC1 p.207.

Orléans, city in N central France, on the River Loire, 113km (70 miles) sw of Paris, and capital of the Loire département. It was besieged by the English during the HUNDRED YEARS WAR and relieved by JOAN OF ARC. After being briefly held by the HUGUENOTS during the Wars of Religion, the city was besieged by Roman Catholics in 1563 and held by them until the Edict of NANTES in 1598. Industries include tobacco, textiles and chemicals. Pop. (1971 est.) 102,200.

Orléans, Siege of (1428–29), defeat for the English during the HUNDRED YEARS WAR. English troops began to besiege the city in Oct. 1428 and French attempts to relieve it failed until JOAN OF ARC took in supplies on 30 April 1429. A week later the Earl of Suffolk raised the siege. This French success was the turning point of the war.

Orley, Bernard van (*c.* 1490–1541), Flemish painter and designer of stained-glass windows and tapestries. He also painted altar-pieces and portraits and became one of the best-known artists of his time.

Ormandy, Eugene (1899–   ), US conductor, *b.* Hungary. He went to the USA in 1921 and has been musical director and permanent conductor of the Philadelphia Orchestra since 1938. He is well known for his interpretations of the Romantic composers and for introducing many new works by 20th-century composers.

Ormolu, elaborate gilded decoration especially popular as ornament on the furniture and clocks of 18th-century France. From a wax or wooden model, a mould is taken and a cast made of bronze alloy. This is then CHASED and gilded and may be treated further for added metallic brilliance. Ormolu was also produced in England by Matthew BOULTON.

Ormonde, James Butler, 12th Earl and 1st Duke of (1610–88), Irish political leader. He supported the Earl of STRAFFORD against the Irish Roman Catholic rebels, defeating them in 1641. He was appointed Lord-Lieutenant of Ireland in 1644 and made peace with the rebels in 1649. Following the defeat of the Stuarts, he joined Charles II in exile in France. After the RESTORATION in 1660, Charles appointed him Lord-Lieutenant of Ireland 1661–69 and again 1677–84.

Ormonde, James Butler, 2nd Duke of (1665–1745), Irish military and political leader. He supported William of Orange at the Battle of the BOYNE in 1690 and in 1711 became commander of British forces during the War of the SPANISH SUCCESSION. In 1715 he was impeached for being a JACOBITE.

Ormuz (Hormuz), island belonging to Iran in the Strait of Ormuz between the Gulf of Oman and the Persian Gulf.

Ormuzd. *See* AHURA MAZDA.

Ornamentation, in music, term used to describe figurative embellishment. It may be written out in full by the composer, or indicated by the appropriate sign of musical notation which the musician then implements in performance. Such signs are those for the trill, turn and mordent. The other form of ornamentation, in widespread practice in Renaissance and Baroque music, was the improvised addition of embellishment by the performer.

Ornithology, study of birds. The science extends to tracing the development of all bird species; the theory currently accepted is that they evolved from the reptiles of the Jurassic period. Included in general ornithological studies are classification, structure, function, evolution, distribution, banding and migration, reproduction, ecology and behaviour. Museums and universities house collections of stuffed birds, skeletons and other preserved specimens. *See also* NW pp.*73, 75, 140–141, 140–141, 142–143, 142–143, 144–145, 144–145, 148–151, 244–245*.

Ornithopoda, suborder of bird-hipped

dinosaurs, including all semi-bipedal herbivorous forms. Early species were *Camptosaurus* and *Iguanodon*, while a later species was the *Hadrosaur* or duck-billed dinosaur. *See also* NW p.185.

Orogeny, in its modern, narrowest usage, the geological process by which a FOLD and FAULT in the upper layers of mountainous areas, and metamorphism in deeper ones, occurs. Each mountain region has its peculiarities, but there are similarities in all the processes of change in this study.

Oroville Dam, dam in N central California on the Feather river. It is the highest dam in the USA and was built between 1957 and 1968 to provide power, and domestic and irrigation water. Height: 235m (770ft). Length: 2,317m (7,600ft).

Orozco, José Clemente (1883–1949), Mexican painter. His wash drawings *Mexico in Revolution* (1911–16) have been likened in thematic treatment to the work of GOYA but Orozco's concern was to demonstrate the futility of war to the illiterate masses. He is renowned for his large murals which are EXPRESSIONIST in style, and often show allegorical subjects with political overtones. *Katharsis* (1934) and *National Allegory* (1947) are both excellent examples of his work.

Orpen, Sir William Newenham Montague (1878–1931), Irish painter. He is best known for his portraits and conversation pieces, which include *Homage to Manet* and *Café Royal* (1912) He was official painter at the Versailles Peace Conference (1918).

Orpheus, in Greek mythology, the son of CALLIOPE by APOLLO, and the finest of all poets and musicians. Orpheus married EURYDICE, who died after being bitten by a snake. He descended into HADES to rescue her and was allowed to regain her if he did not look back at her following him until they emerged together into the sunlight. He could not resist, and Eurydice vanished for ever. A legend has it that later in life he was torn to pieces by the Thracian women, but that his head, when thrown into the River Hebrus, floated, singing, to the island of Lesbos. Orpheus is usually depicted with the lyre he received as a gift from Apollo. *See also* MS p.214.

Orphism, cult in ancient Greece based on the Orphics sacred poems on which the Orphic mysteries were centred. Initiation into the mysteries, an ascetic life and transmigration of the soul after death were the main elements of the cult.

Orphism, or orphic cubism, term invented in 1912 by APOLLINAIRE to describe a new art form combining elements of CUBISM, FUTURISM and FAUVISM. The style was first associated with the work of DELAUNAY and its other exponents exerted considerable influence in Germany through the works of KLEE and KANDINSKY. *See also* BLAUE-REITER.

Orpine, various plant species of the genera *Sedum* and *Supervirum*, hardy perennials with fleshy leaves and stems. Their small flowers are white, yellow or red, and species of both genera have been used medicinally.

Orr, Robin Robert Kemsley (1909–   ), Scottish composer who studied with Alfredo Casella and Nadia BOULANGER and was professor of music at Glasgow University (1956–65) and Cambridge University (1965–76). His works include the operas *Full Circle* (1967) and *Hermiston* (1975) and the choral composition *Te Deum and Jubilate in C* (1953).

Orrery, apparatus designed for Charles Boyle, Earl of Orrery (1676–1731), consisting of a frame which represents the solar system, from which extend spindles ending in balls which can be moved about to illustrate the relative sizes, positions and movements of the bodies in the solar system. *See also* SU p.*167*.

Orrisroot, the rootstock or rhizome of several kinds of S European IRIS, especially the pale-blue Mediterranean species, also known as the "fleur-de-lys". When dug up and dried, orrisroot acquires the distinct scent of violets and is used in medicine and in perfumery.

Ortega y Gasset, José (1883–1955), Spanish philosopher and humanist. He was educated in the classic JESUIT tradition

Orioles eat various types of fruit and, in the breeding season, insects.

Seige of Orléans; Joan of Arc (the Maid of Orléans) making her entry into the city.

Orpheus' music charmed Hades into letting Eurydice leave the underworld.

José Ortega y Gasset left Spain in 1936 to live in exile, not returning until 1945.

Orthoclase

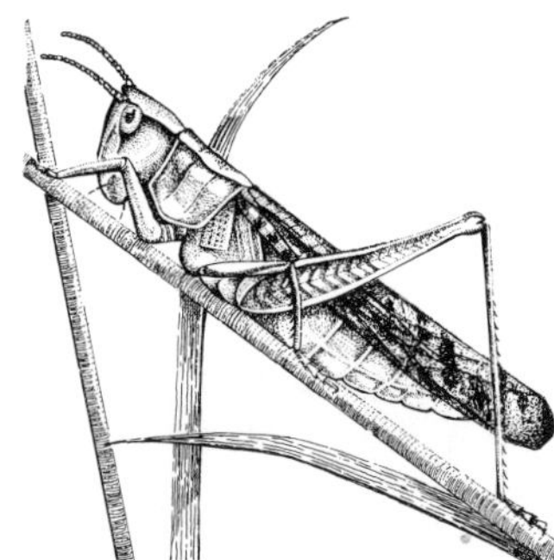

Orthoptera, the order of insects that contains some of the largest known species.

George Orwell's accounts of poverty in his novels were written from experience.

Oryx, or gemsbok, lives in arid areas, feeding at night on moist vegetation.

Oslo; the modern town hall was decorated by artists from throughout Norway.

and influenced by the neo-Kantians at Marburg University but later published an apologia, *Adan en el paraiso* in 1910. His most famous work, *The Revolt of the Masses* (1929), advocated political control by an intellectual élite. His other works include *The Mission of the University* (1930), *Man and People* (1957) and *Man and Crisis* (1956). He founded the Institute of Humanities in Madrid in 1948.

Orthoclase, mineral common in granitic rocks containing mainly potassium feldspar. It is a potassium aluminium silicate, $KAlSi_3O_8$, but also contains varying amounts of sodium felspar, $NaAlSi_3O_8$.

Orthodontics. *See* DENTISTRY.

Orthodox Church, family of Christian national churches mostly of E Europe, also known as the Eastern Orthodox Church. The individual churches are independent but acknowledge the primacy of the Patriarch of Constantinople. They developed from the Church of the BYZANTINE EMPIRE, which separated from that of Rome in 1054 (the GREAT SCHISM) over doctrinal issues and the question of papal authority. After the capture of Constantinople by the Turks in 1453, the Russian Orthodox Church became the most influential, and is still by far the largest in the number of its adherents. The Orthodox Church stresses the role of the first seven ECUMENICAL COUNCILS in the development of its doctrines; it is noted for the splendour of its ritual, its devotion to saints, its restrictions upon married clergy and for the importance that religious art, especially in ICONS, has played in its history.

Orthography, the correct presentation of letters, words, phrases, clauses and sentences in handwriting or in printed type, according to accepted standards of language and society.

Orthopaedics, branch of medicine which deals with the diagnosis and treatment of diseases, disorders and injuries to bones, muscles, tendons and ligaments. The first institute specifically for orthopaedic treatment was founded in 1870 by Jean André Venel in Switzerland. *See also* MS pp.56, 57, 95, 95.

Orthoptera, order of insects represented by more than 20,000 species found throughout the world, especially in tropical regions; it contains KATYDIDS, CRICKETS, GRASSHOPPERS and LOCUSTS in the suborders Ensifera and Caelifera. The term was formerly used to group together the orders Dictyoptera (mantids and cockroaches), Phasmida (leaf and stick insects) and Grylloblattidae (grylloblattids). Most of the species feed on plants, have chewing mouthparts and undergo incomplete METAMORPHOSIS from egg to nymph and adult. EARWIGS, TERMITES and WEBSPINNERS are also considered to be Orthopteroid insects. Length: to 10cm (4in). *See also* NW pp.106–107.

Ortolan, small European BUNTING. It has an olive-green head and chest, a black back streaked with brown, a yellow throat, and pinkish underparts. It lays 3–6 eggs in a cup-shaped nest. Ortolans are regarded as a great delicacy, and thousands are trapped for food each autumn during their migrations to N Africa and the Middle East. Length: 16.4cm (6.5in). species *Emberiza hortulana.*

Orton, Arthur (1834–98), British tradesman who emigrated to Australia in 1852 and settled as a butcher in Wagga-Wagga, New South Wales in 1864. He claimed to be Roger Tichborne, heir to an ancient baronetcy in Hampshire. Orton initiated an ejection suit (1871–72) against Roger's nephew (who had succeeded to the baronetcy) but, in the longest such trial in British history, was found to be an impostor and imprisoned.

Orton, John Kingsley ("Joe") (1933–67), British playwright who specialized in black satirical comedies notable for their energetic dialogue and the outrageous predicaments in which their characters are presented. *Entertaining Mr Sloane* and *Loot* appeared in London in 1964 and 1965. *What the Butler Saw* was produced posthumously in 1969 after Orton's murder by his flatmate.

Oruro, capital and railway centre of Oruro department in W Bolivia. Industries

include mining of silver, tungsten, copper and tin. Pop. (1970) 119,700.

Orvieto, town in central Italy 47km (29 miles) WNW of Terni, on the River Paglia. It became an independent commune in the 12th century and came under papal rule in 1448. There are Etruscan ruins nearby. Industries: ceramics, wine, tourism. Pop. (1971) 25,195.

Orwell, George (1903–50), British journalist, critic and novelist, real name Eric Blair, *b.* India. He served with the Indian imperial police in Burma (1922–27) and was wounded in 1936 while fighting for the Republicans in the SPANISH CIVIL WAR. His books include the autobiographical *Down and Out in Paris and London* (1933), *Homage to Catalonia* (1938), the anti-totalitarian fable *Animal Farm* (1945) and the novel *Nineteen Eighty-Four* (1949). *See also* MM p.212.

Ory, Edward ("Kid") (1886–1973), US jazz trombonist and bandleader. He formed his first band in 1911 and many of the most famous figures of early 20th-century jazz appeared with him, eg Louis ARMSTRONG, King OLIVER, Jimmy Dodds and Sidney BECHET. He made some of the earliest jazz recordings, and composed the well known *Muskrat Ramble,* in 1926.

Oryx, or gemsbok, any of four species of ANTELOPES. The male has a tuft of hair at the throat and both sexes carry long horns ringed at the base. Two species are almost extinct: *Oryx leucoryx* of Arabia and *O. dammah* in the Sahara Desert. *O. gazella beisa* of E Africa and *O. gazella gazella* of the Kalahari Desert in South Africa still survive in considerable numbers. Height: 1.2m (4ft). Family Bovidae. *See also* NW pp. 220–221.

Osage Indians, tribe of North American prairie Indians. Hunters with a strong religious tradition, they settled at different times in Kansas, Missouri and Oklahoma. Numbering about 5,000 today, they have prospered since the discovery of oil on their Oklahoma reservation.

Ōsaka, capital of Ōsaka prefecture in S Honshū, Japan, at the mouth of the River Yodo the second-largest Japanese city and industrial centre. In the 4th century it was called Naniwa and was the capital of the country. The present city grew around the huge 16th-century castle of Hideyoshi. It was the site of the 1970 World's Fair. Industries: textiles, machinery, steel, chemicals, shipbuilding. Pop. (1975) 2,780,000.

Osborne, Dorothy, later Lady Temple (1672–95), English letter-writer. Her letters to Sir William Temple, before and after their marriage, are fine examples of the epistolary art and are also descriptive of 17th-century England.

Osborne, John James (1929–), British dramatist whose *Look Back in Anger* (1956) established his reputation as the "ANGRY YOUNG MAN" of the English theatre and introduced a wave of social realist drama. Other critical and popular successes at the Royal Court Theatre included *The Entertainer* (1957) and *Luther* (1961) but later works by Osborne such as *The Hotel in Amsterdam* (1968), *West of Suez* (1971) and *Watch It Come Down* (1976) provoked consistently critical hostility. His lively screenplay for the film *Tom Jones* (1963) won an OSCAR.

Osborne, Thomas, Earl of Danby. *See* DANBY, THOMAS OSBORNE, EARL OF.

Osborne House, mansion on the Isle of Wight, England, country home of Queen Victoria (who died there). It was built in the style of an Italian Villa between 1845 and 1848 by Thomas CUBITT to the designs of Prince Albert.

Oscar, name of two kings of Sweden and Norway. Oscar I (1799–1859; *r.* 1844–59) freed the press from the restraints his father, Charles XIV, imposed upon it and favoured the union of Scandinavia. Oscar II (1829–1907; *r.* Norway 1872–1905, Sweden 1872–1907) was the son of Oscar I and succeeded his brother, Charles XV. He abdicated the Norwegian throne in 1905, when Norway became independent of Sweden, and was succeeded by his son Gustavus V.

Oscar, properly an Academy Award, the prize awarded annually for services to the

cinema by the US Academy of Motion Picture Arts and Sciences. In a varying number of categories, such as Best Direction, Best Actor, Best Screenplay, five nominees have been voted for by members of the Academy since 1929. *Ben-Hur* (1959) holds the record number of Oscars for a single film (11) while Katharine Hepburn was nominated a record 11 times, receiving the award for Best Actress on three occasions. The gold-plated bronze statuettes stand 25cm (10in) high, weigh 3kg (7lbs) and are reputed to have been nicknamed after the Academy librarian's uncle Oscar.

Oscillator, in the physics of sound, a device for producing sound waves, as in a SONAR or an ultrasonic generator. In electronics, an oscillator circuit converts DC (direct current) electricity into high-frequency AC (alternating current). The many types include harmonic oscillators, which generate sinusoidal wave forms.

Oscilloscope, electronic instrument in which a cathode ray tube (CRT) displays an image of rapidly varying quantities, eg voltage or current. The input signal is usually fed to the vertical plates of the CRT; the electron beam which traces it on the screen being moved by a time-base generator within the oscilloscope. The result is generally a curve or graph on the screen, which can be calibrated and used for measuring the input signal.

O'Shea, William Henry (1840–1905), Irish nationalist. He was a supporter of the IRISH HOME RULE movement in the 1880s and 1890s. He tried to bring PARNELL and the LIBERAL leaders together on the issue in the years 1882 to 1884. By naming Parnell as the co-respondent in his divorce proceedings against his wife in 1890, he ruined his own and Parnell's career.

Oshogbo, city in SW Nigeria 80km (50 miles) NE of Ibadan on the River Oshun. It was part of Ijesha, a YORUBA kingdom, until the 19th century and in 1840 was the scene of a battle in which Ibadan, a Yoruba city state, defeated Ilorin, a Fulani state. Industries: textiles, cigarettes, cotton. Pop. (1975) 282,000.

Osier, any of various willows, especially *Salix viminalis* and *S. purpurea*, the flexible branches and stems of which are used for wickerwork. Osiers have long been cultivated from cuttings, but for strength, cross-fertilized varieties are used.

Osijek, chief city of Slavonia Region, N Yugoslavia, on the River Drava. It was developed around a fortress built on the site of a Roman settlement, and incorporated into Austria-Hungary in 1918. Industries: textiles, distilling, food-processing, shipping, shoe manufacture. Pop. (1971) 93,912.

Osiris, in Egyptian mythology, the god of the dead, son of GEB (the earth) and NUT (the sky). He is generally depicted as a mummified man wearing a feathered crown and bearing the crook and flail of a king. In the myths, Osiris was killed by his bother SETH and dismembered. His sister and wife ISIS retrieved and, in some traditions, reconstructed the corpse, and Osiris' son HORUS avenged his death. *See also* MS pp.207, 214.

Osler, Sir William (1849–1919), Canadian physician who in 1873 was the first to identify platelets in the blood. In 1888 he was appointed professor of medicine at Johns Hopkins University, Baltimore, which he developed into one of the foremost medical schools. He wrote *The Principles and Practice of Medicine* (1892), the leading medical text of the time and which is still published. Probably the most famous physician of that era, he gave his name to several medical terms, such as Osler's nodes and Osler-Vaquez disease.

Oslo, largest city and capital of Norway, in the SE at the head of Oslo Fjord. The city was founded by King Harald III in 1050 and became the capital in 1299. Largely destroyed by fire in 1624 it was rebuilt by Christian IV, who named it Christiania; it was renamed Oslo in 1925. Industries: machinery, wood products, food-processing, textiles, chemicals, shipbuilding and tourism. Pop. (1974) 465,337.

Osman I, or Osman Gazi (1258–*c.*1326), considered the founder of the OTTOMAN

EMPIRE, who conquered NW Asia Minor and proclaimed his independence from the SELJUK TURKS upon the collapse of their empire after 1293. *See also* HC1 p.220.

Osman, Sir Abdul Rahman Muhammed (1902–), Governor-General of Mauritius (1972–). He studied law in Britain, and was appointed senior judge of the Mauritian high court in 1950. He served as acting Governor-General several times between 1970 and 1972.

Osmium, metallic element (symbol Os) of the third transition series, discovered in 1803 by Smithson Tennant. It occurs associated with platinum; the chief source is as a by-product from smelting nickel. Like IRIDIUM, it is used in producing hard alloys. The tetra-oxide (OsO_4) is a powerful oxidizing agent. Properties: at.no.76; at.wt.190.2; s.g.22.57; m.p.3,045°C (5,513°F); b.p.5,027°C (9,081°F); most common isotope $Os^{192}(41.0\%)$.

Osmosis, diffusion of a solvent through a natural or artificial semi-permeable membrane (which blocks the passage of selected dissolved substances) into a more concentrated solution. Osmosis is a vital cellular process. Plant roots absorb water by osmosis; walls of living cells use it selectively to allow the passage of required substances. *See also* SU p.145.

Osmotic pressure, pressure exerted by a dissolved substance by virtue of the motion of its molecules. In dilute solutions it varies with the concentration and temperature as if the solute were a gas occupying the same volume. It can be measured by the pressure which must be applied to counter-balance the process of OSMOSIS into the solution.

Osnabrück, city in N West Germany on the River Haase 93km (58 miles) NE of Dortmund. Negotiations for the PEACE OF WESTPHALIA in 1648 were undertaken there and it was made over to Hanover in 1815. Industries: iron and steel, textiles, papermaking, chemicals, machinery. Pop. (1974) 164,060.

Osprey, hawk that lives beside lakes and on the coast. It dives from the air to seize fish with its talons. It has a short hooked bill, broad ragged wings and a white head; it has brownish-black plumage on its back and a cream breast. Two to three white eggs, spotted with brown, are laid in a stick nest. Length: 51–61cm (20–24in). Family Pandionidae, species *Pandion haliaetus*. *See also* NW p.*251.*

Ossa, Mount, mountain peak in NE Greece. In Greek mythology the Aloadae tried to pile Mt Pelion on top of Ossa, in an attempt to reach heaven and overthrow the gods. Height: 1,978 (6,489ft).

Ossian (Oisin), legendary Gaelic warrior and poet of the 3rd century AD whose work enjoyed a revival of interest in Europe in the 19th century. His name was used by James MACPHERSON as the author of the latter's Ossianic poems in 1765, proven to be fakes after MacPherson's death.

Ossietzky, Carl von (1888–1938), German pacifist and journalist who was awarded the 1935 Nobel Peace Prize. In 1922 he helped to found the *Nie Weider Krieg* (No More War) organization and in 1927 became editor of *Weltbühne*, a liberal newspaper. In this he disclosed secret plans for German re-armament, which caused him to be accused of treason and sentenced to imprisonment in 1931. He was granted an amnesty in 1932 when Adolf HITLER became chancellor, and resumed editorship of the paper. Despite his precarious position he refused to leave Germany and was arrested again in 1933. He was sent to Papenburg concentration camp until 1936 when, suffering from tuberculosis, he was transferred to a hospital in Berlin in which he eventually died.

Ossification, process of BONE formation. Intramembranous bones develop from connective tissue cells which form a network of COLLAGEN fibres. These cells then differentiate into osteoblasts, which secrete bone-forming minerals that combine with the collagen to form the hard bone matrix. Endochondral bones develop from CARTILAGE, when certain cartilage cells change to osteoblasts. *See also*

MS pp.*37*, 56, *56*, 57, *57*, 94–95, *94–95.*

Ostade, Adriaen van (1610–85), Dutch painter. His genre was village life, and he was trained by Frans HALS and later influenced by REMBRANDT. His works include *The Itinerant Fiddler* (1672) and *Carousing Peasants in an Interior* (c.1638). *See also* HC1 p.311.

Osteichthyes, class of fish found in almost every environment. Their characteristics include a bony skeleton; a single flap (operculum) covering the gill openings, and red blood cells with nuclei. Most members of this class have scales. Osteichthyes first appeared during the Devonian period, when they were heavily armoured and lived in fresh water. *See also* NW pp.124–125, 180.

Ostend (Oostende), city and port in W Belgium, West Flanders province. From 1830 until the end of WWI it was a popular social centre; it was occupied by enemy forces during both world wars. Industries: fishing, shipbuilding, tobacco processing. Pop. (1970) 71,227.

Osteoarthritis, degenerative disease in which the CARTILAGE of the joints is destroyed. It may be caused by ageing or by postural or orthopaedic abnormalities, and may be relieved by PHYSIOTHERAPY, ANALGESICS or drugs such as CORTISONE. *See also* MS pp.*86*, 87, 94, *94.*

Osteology, branch of medicine concerned with the study of the growth, structure, functions and diseases of BONES. *See also* MS pp.56–57; 94–95.

Osteomyelitis, inflammation of the BONE or bone marrow, usually as a result of infection. It is accompanied by fever, swelling and pain. The condition may be treated with ANTIBIOTICS or by surgery.

Osteopathy, an approach to medicine based on the theory that body strains interfere with normal body functioning and impair its ability to fight disease. The theory was formulated by Andrew STILL in 1874, and the first school of osteopathy was opened in Kirksville, Missouri in 1892. Treatment is based on manipulative therapy of the musculo-skeletal system. Still's emphasis on treating the whole person remains an ideal of osteopathy.

Ostia, ancient Italian city 26km (16 miles) from Rome at the mouth of the River Tiber. It was Rome's naval base and commercial harbour from the 4th century BC to the 3rd century AD. The town became deserted by the 9th century, and excavation of the site was begun in the 19th century. *See also* HC1 pp.*94–95.*

Ostracism, practice carried out in Athens and some other ancient Greek cities whereby a majority of 6,000 citizens could vote for the banishment of an individual. The device was often used as a means of removing political opponents. Banishment normally lasted for up to ten years. No confiscation of property occurred.

Ostracod, also called seed shrimp, subclass of small CRUSTACEANS common in salt and fresh water. The African species *Mesocypris terrestis* is terrestial, living in damp humus. Each has a rounded or elliptical carapace, resembling two halves of a clam shell. Length: 15–20mm (0.004–0.8in). *See also* NW p.*105.*

Ostrava, city in N central Czechoslovakia, near the confluence of the rivers Oder and Opava. In the Middle Ages, the city guarded the entrance to the Moravian lowlands. Industries: coal, iron, steelmaking, ship and bridge parts. Pop. 274,547.

Ostrich, largest living bird, found in central Africa. It is flightless and has a small, flat, down-covered head and long neck. Plumage is black and white in males, brown and white in females. The female scoops a hole in the sand for 10–12 large yellow eggs, which are incubated by her during the day and by the male at night. Height: to 2.5m (8ft); weight: to 155kg (345lb). Family Struthionidae; species *Struthio camelus*. *See also* NW pp.*140*, 221.

Ostrogoths. *See* GOTHS.

Ostrovsky, Alexsandr Nikolayevich (1823–86), Russian dramatist. Many of his plays deal with the life of the Russian bourgeoisie. *Poverty Is No Crime* (1854), concerns a marriage of convenience. His best-known work is *The Storm* (1860).

Ostwald, Wilhelm (1853–1932), German chemist, b. Russia. He became professor of chemistry at Leipzig University (1887–1906), and remained in Germany for the rest of his life. His work on solutions led him to formulate Ostwald's dilution law. His process for preparing nitric acid and his research on catalysis became important for industry. He was awarded the 1909 Nobel Prize in chemistry.

Oswald, Saint (c.605–41), King of Northumbria (633–41). He was the son of ETHELFRITH but became a Christian and converted his people with the help of St AIDAN from Iona. He defeated the Welsh King Cadwallae in 633 and became King of Northumbria, but was killed in battle by PENDA of Mercia. Feast day: 5 Aug. *See also* HC1 p.154.

Otago, statistical area in the southern part of New Zealand's South Island. The principal town is Dunedin. The chief industry is sheep rearing, and there is also extensive beef and dairy farming and wheat growing.

Othello (1604), play by William Shakespeare. One of Shakespeare's most tragic dramas, perhaps due to its treatment of an accessible theme – sexual jealousy – the play deals with the destruction, through ignorance, to corruption of the purest love. The interplay of Othello (a Moor) and Iago (his malignant tempter) is the crux of the play.

Othman (c.574–656), son-in-law of MOHAMMED. He succeeded Omar, becoming the 3rd caliph (644–56) of the expanding Arab empire, Cyprus and Bactria being conquered during his reign. He compiled an authoritative text of the KORAN, destroying all other versions. Revolts in Egypt and Mesopotamia against his methods of government led to his murder in 656. *See also* HC1 pp.144–145.

Othman I. *See* OSMAN I.

Otis, Elisha Graves (1811–61), US inventor who designed and manufactured lifts and escalators. His first invention (1852) was an automatic safety hoist for a lift in a factory for which he supervised the construction. His fame dates from 1856, when he installed a passenger lift in a New York store. In 1861 he patented a steam-powered lift. *See also* MM pp.96–97.

O'Toole, Peter (1932–), Irish stage and film actor. In the theatre, he gave exciting performances at Stratford-upon-Avon (1960) as Shylock in *The Merchant of Venice* and as Hamlet in the opening presentation of the National Theatre at the Old Vic, London (1963). His first important film role was the title part in *Lawrence of Arabia* (1962), after which he appeared in several other historical films, such as *Becket* (1964), *Night of the Generals* (1966) and *The Lion in Winter* (1968). In 1976 he played the title role in a television adaptation of Geoffrey Household's *Rogue Male*.

Otranto, resort town in S Italy located on the Strait of Otranto. Although it was a large settlement under the Romans, it was devastated in 1480 by the OTTOMAN Turks and declined in importance after that date. Its 15th-century castle is the setting for Horace WALPOLE'S GOTHIC NOVEL *The Castle of Otranto* (1765). Today fishing is the main activity. Pop. (1971) 4,240.

Ottawa, capital of Canada, in SE Ontario on the Ottawa River and the Rideau Canal, which divides the city into upper and lower towns. Queen Victoria chose it as capital of the United Provinces in 1858, and in 1867 it became capital of the Dominion of Canada. At Chaudière Falls hydro-electric power is generated for municipal and domestic use. Rich deposits of silica support a thriving glass-making industry and marble is processed there. Other industries include printing, publishing, sawmilling, pulp-making, food-processing and the manufacture of household appliances. More than 75,000 inhabitants are employed by the federal government. Pop. (1974) 302,345.

Ottawa Agreements (1932), economic agreements negotiated at the Imperial Economic Conference in Ottawa. Commonwealth countries were allowed to export most goods to Britain free of duty,

Ospreys make spectacular dives to scoop up fish in their talons.

Ostriches run at up to 65 km/h (40mph); even one-month-old chicks reach 50 km/h.

Wilhelm Ostwald was famous as a chemist and recognized as a writer and teacher.

Peter O'Toole at the Royal premiere of the film *Lawrence of Arabia* in 1962.

Otter

Otters are usually found in weed-free lakes or small rivers near the sea.

Otterhound is similar to a bloodhound but has a longer coat and shorter ears.

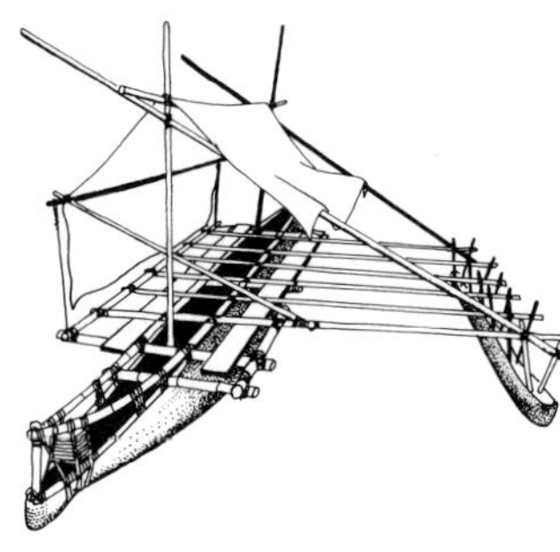

Outrigger; its stability in rough water makes it an excellent sea fishing boat.

Ovenbird; some build nests of clay and grass hidden in the undergrowth.

while Britain imposed new duties on imports from non-Commonwealth countries. This policy, named imperial preference, had been advocated by Chamberlain at the 1897 Colonial Conference. Trade preferences between the colonies and Britain were gradually removed following the Agreement on Tariffs and Trade in 1947.

Otter, semi-aquatic carnivore found everywhere except Australia. Otters have narrow pointed heads with bristly whiskers, sleek furry bodies, short legs with webbed hind feet and long tapering tails. The river otter (genus *Lutra*) of Europe, North and South America, Asia and Africa is small to medium-sized and spends considerable time on land. The giant otter (*Pteronura brasiliensis*) may reach a body length of 1.5m (5ft) with a tail of 60cm (2ft).

Otterhound, DOG used for otter-hunting from the 13th century in England. It has a large, slightly domed, narrow head, and is heavily set with a sickle-shaped tail. Height at shoulder: 61–66cm (24–26in).

Otter shrew, any of three species of aquatic mammals that live in equatorial Africa. They have long, pointed shrew-like faces, long bodies and flattened tails for paddling. They feed mainly on insects. The largest species is up to 64cm (25in) long. Family Tenrecidae; genera *Potamogale* and *Micropotamogale*.

Otto, the name of four Holy Roman emperors. Otto I (Otto the Great) (912–73; *r.*962–73) and German king (*r.*936–73) succeeded his father, Henry I, and after subduing German nobles invaded Italy in 951 to aid Queen Adelaide against Berengar II, who had succeeded her husband. Otto defeated Berengar, assumed the title "King of the Lombards" and married Queen Adelaide. In 955 he routed the Magyars at Lechfeld. On becoming emperor he revived the connection established by CHARLEMAGNE between the Roman empire and the Germans. Otto II (955–83; *r.*973–83), was the son of Otto I. He subdued a revolt by Henry, Duke of Bavaria, and in 978 France invaded Lorraine, which Otto regained in 980. Shortly afterwards he led an expedition to Italy to secure lands in the south but was defeated in 982. Otto III (980–1002; *r.*983–1002) succeeded Otto II while he was still an infant. Regents ruled for him until he assumed the throne in 994. Otto IV (*c.*1175–1218; *r.*1209–15) was elected German anti-king to Philip of Swabia in 1198. After Philip's death in 1208 he was recognized as king and crowned emperor in the following year by Pope Innocent III, but the pope withdrew his support when Otto invaded Italy. After a defeat in 1214 by the French, who were allied with Frederick II, the pope deposed him. *See also* HC1 pp.160, *160, 161, 187.*

Otto I (1815–67), king of the Hellenes (1832–62). His rule of a newly independent Greece was unpopular because of its lack of a constitution. His efforts to overthrow the constitution imposed on him by the revolt of 1843 led to his deposition and exile in 1862.

Otto, Frei Paul (1925–), German architect noted for his lightweight designs, particularly of sports arenas and pavilions. Famous for his suspended roofs and inflated plastic structures, his buildings include the membrane roof for the Olympic Stadium, Munich (1972) and the hotel and conference centre in Mecca, Saudi Arabia (1966–75).

Otto, Nikolaus August (1832–91), German engineer who built a gas-fired engine (1861) which won a gold medal at the Paris Exhibition of 1867. He later built an internal combustion engine, based on the FOUR-STROKE CYCLE (also called the Otto cycle), that was the forerunner of most of today's engines. *See also* MM pp.62, 68, *69, 126.*

Ottoman Empire, former state in Asia Minor, founded by OSMAN I in the 13th century. By the end of the 14th century it had absorbed the BYZANTINE EMPIRE and conquered Egypt and Syria during the 16th century. During the rule of SULEIMAN THE MAGNIFICENT (1520–66) the empire included much of SE Europe, W Asia and N Africa. A series of debilitating wars was fought during the 17th and 18th centuries against Poland, Austria and Russia, bringing about a decline which continued until the dissolution of the empire in 1920 with the Treaty of SÈVRES. The Turkish nationalists finally overthew the sultan in 1922. *See also* HC1 p.221, *221;* HC2 pp.84–85, *84–85.*

Otway, Thomas (1652–85), English dramatist. His best-known plays, *The Orphan* (1680) and *Venice Preserved* (1682) were high points of tragedy during the Restoration period, and notable for their departure from his contemporaries' formal style.

Ouagadougou, largest city and capital of Upper Volta founded in the late 11th century as capital of the Mossi empire. It was regarded as the centre of Mossi power until the French captured it in 1896. Industries: handicrafts, food processing, peanut growing. Pop. (1970 est.) 115,500.

Ouahran. *See* ORAN.

Oud, Jacobus Johannes Pieter (1890–1963), Dutch architect. A member of the de STIJL group of geometric-abstract artists and co-founder of the review *De Stijl* with Theo van DOESBURG in 1917, he designed housing groups in a simplified style using reinforced concrete. He served as Rotterdam's official architect (1918–33), creating housing estates such as those of Tusschendijken (1920), Oud-Mathenesse (1922) and Hook of Holland (1927).

Oudenaarde, Battle of (1708), Allied victory over the French during the War of the SPANISH SUCCESSION. The Duke of MARLBOROUGH and Prince EUGENE of SAVOY led 80,000 men in a march of 80km (50 miles) in 65 hours to relieve Oudenaarde, which was being besieged by 85,000 French troops. The French were forced to retreat, and ultimately to abandon the Spanish Netherlands.

Oudh (Ayodhya), ancient city of N central India in a district of Uttar Pradesh. It is one of the seven holy places of HINDUISM and has many monuments and shrines, including the mosque of RAMA's birthplace. It was a province of the Mogul empire during the 16th century.

Oudry, Jean Baptiste (1686–1755), French painter of animals and still lifes. He was court painter to Louis XV and illustrated La Fontaine's *Fables* in 1755. His *White Duck* (1753) is highly regarded.

Oughtred, William (1575–1660), English mathematician. He invented several trigonometrical and algebraic signs and developed the first slide rule in 1632. He published *Clavis Mathematicae* in 1631 and *Trigonometria* in 1657.

Ouida (1839–1908), British novelist, real name Marie Louise de la Ramée. Her many vigorous, though often factually inaccurate, works include *Under Two Flags* (1867), *A Dog of Flanders* (1872), *Moths* (1880), and *A Village Commune* (1881).

Ouija board, in occultism, a board bearing the letters of the alphabet on top of which is a smaller movable board with a marker which, when pressure is applied to it, may point out letters. This apparatus is used by those wishing to obtain spiritualistic and telepathic messages. The name "ouija" comes from the French and German words for "yes"; *oui* and *ja.*

Oujda, city in NE Morocco, near the Algerian border, founded in the late 10th century by the Berbers and occupied by the French in 1844, 1857 and 1907. It is an important trade and rail centre handling regional agricultural and mineral products. Pop. (1971) 175,532.

Oulu, capital of Oulu province in W central Finland. It began to expand in 1590 and became a commercial centre in the 19th century. Its present-day industries include wood- and leather-processing and metalworking. Pop. (1970 est.) 87,224.

Ounce. *See* LEOPARD, SNOW.

Our Mutual Friend (1864–65), novel by Charles DICKENS. It was his last complete work and suffered adverse criticism at first only to be reinstated later as one of his greatest works. The plot is a powerful exposition of the worth of the individual contrasted with his financial status.

Our Town (1938), three-act drama by Thornton WILDER about the effect of death on an ordinary family. The play is performed without scenery; a few props are arranged by the stage manager, who also introduces, interrupts and comments on the action.

Ouse, English river formed by the confluence of the rivers Ure and Swale, both of which have their source in the Pennines, N Yorkshire. It is navigable to York and is still an important commercial waterway, flowing SE and eventually joining the River Trent to form the River Humber. Length: 72km (45 miles).

Ousel. *See* OUZEL.

Outback, term applied in Australia and New Zealand to the inland areas of the continent. In the early days of settlement the name acquired a mythical quality as a land of adventure and opportunity which lured explorers inland. It remains, however, a rural, largely undeveloped region which is very sparsely populated.

Outboard motor, portable engine used to propel a type of motorboat. It incorporates the drive shaft and propeller and is either clamped to the stern or placed in a "well" specially built into the hull. A boat so equipped is very manoeuvrable, because steering is effected by rotating the whole assembly, including the propeller shaft.

Outbreeding. *See* CROSSBREEDING.

Outcrop, exposure at the surface of the Earth of an edge of rock stratum. This phenomenon may be caused by the erosion of soil by water, wind, ice (especially glaciers) or gravity.

Outer ear. *See* EAR.

Outer Hebrides. *See* HEBRIDES.

Outer Mongolia. *See* MONGOLIA.

Outlawry, in medieval law, act of placing a person who was in contempt of court outside the protection of the law. Conviction of major outlawry, as in a case of alleged treason, involved forfeiture of all property, and implied conviction, rendering the subject liable to be killed on sight. In other cases of civil or criminal contempt of court, outlawry implied the deprivation of civil rights and the forfeiture of chattels. It was abolished in civil proceedings in England in 1879, and in criminal cases in 1938.

Output device, computer, apparatus that transmits or records data from a COMPUTER. It may take the form of a modem for remote receipt of data, a hard-copy printer or recorder (eg a teletypewriter), a line-printer or an alpha-numeric display (a TV-type screen on which letters or numbers are presented). Other forms include digital to analog (D/A) systems that directly control various mechanisms and processes. *See also* MM pp.106–107, *107.*

Outrigger, canoe-type vessel with a float or floats attached outboard of, and parallel to, the main hull. Such vessels are still commonly used in the Pacific and Indian oceans.

Ouzel (ousel), heavy-bodied bird found in the mountains of Asia, Europe and the Western Hemisphere. The ring ouzel (*Turdus torquatus*) has black plumage, with a white chest collar. Family Turdidae.

Oval Cricket Ground, The, home of the Surrey County Cricket Club, in S London. It was first used as a cricket ground in 1845 and was the venue for the first TEST MATCH held in England (1880). It is traditionally the venue of the last match of each Test series played in England. It has a maximum capacity of about 17,000 people. Despite its name, its shape only approximates to an oval.

Ovary, in biology, the part of a multicellular animal or a flowering plant that produces egg cells (ova), or female reproductive cells; in vertebrates it also produces female sex hormones. In female human beings there is an ovary on each side of the uterus (womb). Controlled by the PITUITARY GLAND, each ovary produces two chief female sex hormones; OESTROGEN and PROGESTERONE, which in turn control development and functioning of the female reproductive system. *See also* MENSTRUAL CYCLE; MS pp.65, *65,* 74–75, *74–75,* 77, 102.

Ovenbird, bird found in varying habitats

in South America and named after the oven-shaped nests built by some species, including the shaketail, spinetail and leaf-scraper. They are generally brownish or reddish-brown. Most feed on insects although some feed on seeds. Length: 12–23cm (5–9in). Family Furnariidae. *See also* NW p.*144.*

Overbeck, Johann Friedrich (1789–1869), German religious painter, founder and leader of the NAZARENE group. His importance is chiefly historical since the Nazarene movement was a critical transition from Classicism to ROMANTICISM in German art. *See also* HC2 p.81.

Overbury, Sir Thomas (1581–1613), English writer and courtier who gained a reputation by his personal essarys. For his hostility to the marriage of his friend, Robert Carr, to Frances Howard, he was put in the Tower by JAMES I. There he was slowly poisoned to death by Carr and his wife.

Overfishing, the catching of too many marine creatures to maintain an ecological balance – the cause of the disappearance of stocks of food fishes and whales from the oceans, now a worldwide source of concern. It remains an international problem largely because of the inability of marine law to regulate fishing and WHALING, and to restrain individual nations from reducing once-rich areas to the point of extinction. The great whales, until recently plentiful enough to supply all nations of the world with a sufficiency of their products, have been overfished so badly that some species now face extinction, yet large catches are still being made, particularly by the USSR and Japanese fleets. Many other nations are also restrained only with the greatest difficulty from extinguishing fish stocks, leading to legal proceedings in the Court of INTERNATIONAL JUSTICE. *See also* PE pp. 240, 246, 247, 252.

Overijssel, province in E central Netherlands, bordering West Germany (E), Drenthe province (N), IJsselmeer (W), River IJssel (SW) and Gelderland (S). Industries: machinery, weaving and food-processing. Area: 3,932sq km (1,518sq miles). Pop. (1970 est.) 920,900.

Overpopulation, in economics, situation in which production would be absolutely increased by a reduction in the numbers involved. It is more generally defined as a situation in which the total economic activity of a community is not adequate to supply all its needs. Overpopulation is therefore relative to the economic capacity of the area. It is often considered that most of South-East Asia is overpopulated. The issue of overpopulation, both on a global and a regional scale, has been much discussed since the late 1960s. *See also* MW pp.13–16.

Overtone, usually a harmonic, a constituent of a musical note with a frequency which is a whole-number multiple of the fundamental note. Some instruments (eg cymbals), however, produce non-harmonic overtones. *See also* TIMBRE.

Overture, instrumental prelude to an opera or operetta; the term has now come also to include an orchestral composition in its own right, usually lively in character. Famous operatic overtures were composed by MOZART, eg to *The Marriage of Figaro*; ROSSINI, eg to *William Tell*; and WAGNER, eg to *Tristan und Isolde*. Notable concert overtures include the *1812 Overture* by TCHAIKOVSKY.

Ovid (43 BC–AD 18), Roman poet, full name Publius Ovidius Naso. A popular and prolific poet, he was exiled to an outpost near the Black Sea in AD 8. His works include *Amores*, short love poems (many praising his mistress Corinna), *Ars Amatoria*, a didactic work on how to get and keep a lover, and his masterpiece, the *Metamorphoses*, a collection of skilfully-woven mythological stories. *Tristia* (Sorrows) is an autobiographical work written in exile.

Oviducts, tubes that connect the ovaries with the UTERUS and through which egg cells are released from the OVARY. In mammals they are known as FALLOPIAN TUBES. These are lined with a tissue called the EPITHELIUM.

Oviedo, city of NW Spain, founded *c.*760, 370km (230 miles) NW of Madrid and capital of Oviedo province. It has the University of Oviedo (1604) and a 14th-century cathedral containing the tombs of the Asturian kings. Industries include agriculture, arms manufacture and mining. Area: province 10,564sq km (4,079sq miles). Pop. city (1970) 154,117; province 1,045,635.

Ovulation, the release of a mature OVUM (egg) from the OVARY. In women one egg is released midway through the menstrual cycle, stimulated by FSH (follicle-stimulating hormone) and LH (LUTEINIZING HORMONE) derived from the PITUITARY GLAND. *See also* MENSTRUAL CYCLE; MS pp.74–75.

Ovum, egg or female GAMETE produced in an OVARY. After FERTILIZATION by the sperm it is capable of developing into a new individual. *See also* MS pp.74–75.

Owen, Alun Davies (1925–), British dramatist who used his experiences of Liverpool, where he grew up, in the television play *No Trams to Lime Street* (1959), the musical *Maggie May* (1964), the film *A Hard Day's Night* (1964) for the BEATLES and the stage play *Progress to the Park* (1959). Other plays for the theatre include *A Little Winter Love* (1963) and *The Male of the Species* (1974).

Owen, Daniel (1836–95), Welsh novelist. The son of a miner, he had little formal education, and became a tailor's apprentice at the age of 12. He began writing in his late 20s and vividly portrayed the Wales of his time. His novels include *The Autobiography of Rhys Lewis* (1885) and *The Trials of Enoc Huws* (1891).

Owen, Goronwy (1723–69), Welsh poet, scholar and clergyman. He studied medieval Welsh poetry and revived certain Welsh metrical forms in his work. In 1758 he emigrated to Virginia, where he lived until his death. Two of his more memorable poems are *The Day of Judgement* and *The Gem or The Precious Stone*.

Owen, Robert (1771–1858), Welsh social reformer. He believed that workers would be more productive if their environment were improved. He founded the New Lanark Mills in Scotland, where he paid high wages, provided better living conditions and, in spite of everything his opponents said, still made a profit. He set out his theories in a three-volume work, *New View of Society* (1813–21). *See also* HC2 pp.29, 89, 91, 212.

Owen, Wilfred (1893–1918), British poet. His poems of life at the front line during WWI include *Arms and the Boy* and *Anthem for Doomed Youth*. While recovering from a wound in 1917 he met Siegfried SASSOON who encouraged him to write. After Owen's death in France, one week before the armistice, Sassoon arranged for the publication of his poems. *See also* HC2 p.194.

Owen Falls, waterfall in SE Uganda, on the NILE, just below Lake Victoria. The Falls, 20m (65ft), were submerged when the Owen Falls dam was constructed. The dam supplies hydro-electric power for Uganda and Kenya.

Owen Glendower. *See* GLYN DŶR, OWAIN.

Owens, Jesse (1913–), US black athlete, b. James Cleveland Owens. While studying at Ohio State University he broke several world records for jumping, hurdling and running (1935–36). At the 1936 Olympics in Berlin he won gold medals for the 100m and 200m races, the long jump and his part in the US team's victory in the 400m relay. His 200m and long-jump records remained unbeaten for more than 20 years.

Owl, order of birds occurring throughout the world except at extreme latitudes. Owls are characterized by a round head, hooked bill, large eyes and long, curved talons. They are nocturnal, with keen eyesight and hearing, and feed on small birds, insects and mammals. Owls comprise two families: Strigidae include the tawny, snowy, little and eagle owls, while Tytonidae has the barn owl as its only member. Order Strigiformes. *See also* NW pp.*33, 141, 152, 199, 206,* 207, *209, 217,* 221, *221, 227.*

Ox, two domesticated species of cattle,

Bos taurus and *B. taurus primigenius.* The term is specifically applied to castrated males commonly used as draught animals. Many varieties of wild cattle, such as the African buffalo and the South East Asian gaur are sometimes called wild oxen.

Oxalic acid, poisonous colourless crystalline organic carboxylic acid whose salts occur naturally in some plants, eg sorrel and rhubarb. It is prepared artificially from carbon monoxide and sodium hydroxide and used for metal and textile cleaning and in tanning. Properties: s.g. 1,653; m.p. 101.5°C (214.7°F).

Oxalis, or wood sorrel, genus containing some 800 species of low-growing plants in wooded areas. The common wood sorrel (*Oxalis acetosella*) is a perennial of temperate Europe and North America. Its flowers are white, veined with purple. The plant's name comes from the Latinized Greek *oxalis* (acidic) because of its bitter taste.

Oxbow lake, crescent-shaped lake, often stagnant and filled with silt or marsh vegetation. Such lakes commonly occur in the lower courses of rivers which meander widely until higher pressure of water or mounting silt forces the river to carve a straighter channel. The name derives from the shape of the lake, said to resemble an ox's collar.

Oxenstjerna, Count Axel Gustafsson (1583–1654), Swedish statesman and military strategist. He was appointed Chancellor of Sweden in 1612 under GUSTAVUS ADOLPHUS and was Vice-Regent (1614–16). After Gustavus's death in 1632 he was Regent of Sweden, and so the virtual ruler, until Princess Christina's coronation in 1644. He was a social reformer and, after Queen CHRISTINA's abdication in 1654, served her successor CHARLES X until his death.

OXFAM, formerly the Oxford Committee for Famine Relief, an international charity, formed in WWII and registered as a charity in 1948. It attempts to alleviate suffering due to poverty or natural misfortune in all parts of the world. From its headquarters in Oxford, England, OXFAM promotes research into the causes and prevention of famine and co-ordinates and encourages the public and national response to specific catastrophes, such as earthquakes, floods, droughts or the results of warfare. As a result of its policy to educate people so that they can solve their own problems, nearly three-quarters of OXFAM's overseas budget is devoted to long-term projects which include agriculture, nutrition, family planning and medicine.

Oxford, Edward de Vere, 17th Earl of (1550–1604), English courtier and poet. A favourite of Queen ELIZABETH I, he is sometimes accredited with the works of SHAKESPEARE – against which the best argument is that he died at too early a date. Poetry by him is published in *The Poems of Edward de Vere* (1921).

Oxford, Robert Harley, 1st Earl of. *See* HARLEY, ROBERT.

Oxford, city and county district in N Oxford(shire), S central England. The county town, Oxford, was founded as a trading centre and fort; it was raided by the Danes in the 10th and 11th centuries. Oxford is the seat of OXFORD UNIVERSITY, founded in the 12th century. Other buildings of interest are the Sheldonian Theatre (1662), the Ashmolean Museum (1683), an Observatory (1772) and many old churches and inns. Industries include the manufacture of steel and motor vehicles. Pop. (1971) 108,805. The county district was created in 1974 under the Local Government Act (1972). Area: 36sq km (14sq miles). Pop. (1974 est.) 115,100.

Oxford and Asquith, Earl and Countess of. *See* ASQUITH.

Oxford Committee for Famine Relief. *See* OXFAM.

Oxford Group. *See* MORAL REARMAMENT.

Oxford Movement, also called the Tractarian Movement or the Catholic Revival, the attempt by some members of the Church of England to restore the ideals of the pre-REFORMATION Church, from *c.*1833 to the first decades of the 20th century. It was a reaction to the general

Owl; the barred owl of North America, like most owls, feeds on small rodents.

Oxalis; both the leaves and the flowers of most species close up at night.

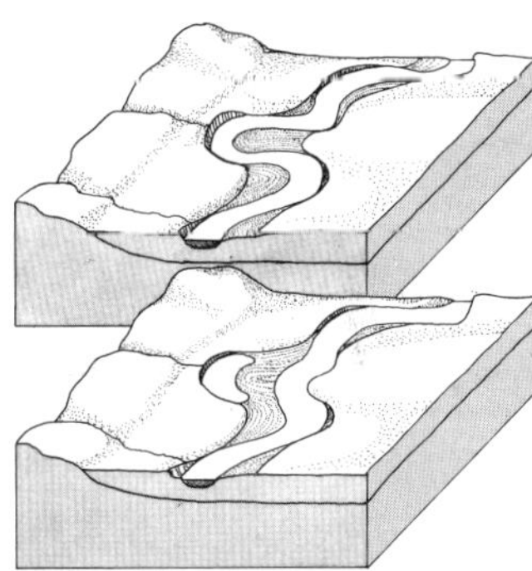

Oxbow lakes are created when a slow-moving river gradually changes course.

Oxford; a view from All Souls and the High Street towards Magdalen Tower.

Oxfordshire

Oxford University; the quadrangle and sundial of Corpus Christi College.

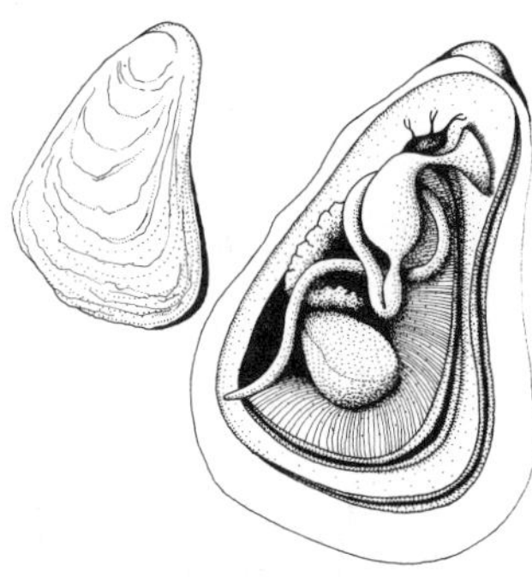

Oysters have been cultivated in undersea oyster beds since pre-Roman times.

Oyster-catchers, despite their name, eat mainly mussels, limpets and worms.

Ozark Mountains; limestone cliffs above the James River in Missouri, USA

decline in religious sentiment and the development of liberal theology; its dominant interest was in the Church's ritual. The main proponents were John KEBLE, Edward PUSEY and John NEWMAN (who eventually became a Roman Catholic cardinal). The legacy of the movement remains as Anglo-Catholicism.

Oxfordshire, county in s central England, bounded in the NW by the Cotswold Hills and in the SE by the Chilterns, and drained by the River Thames. The chief town is Oxford. Area: 2,611sq km (1,008sq miles). Pop. (1974 est.) 504,000.

Oxford Union, The, debating society of Oxford University, founded in 1826 as the Oxford Union Society. For its first three years it had no fixed home but met in private rooms at various colleges. The building in which it is now housed was built in 1856 to 1857 by Benjamin Woodward.

Oxford University, English centre of learning which developed from the group of teachers and students who gathered in Oxford after the exodus of foreign-born scholars from Paris in 1167. The 13th century saw a gradual development of internal organization and a growing hostility between the scholars and the town. There was, however, little sense of community amongst the scholars until the foundation between 1249 and 1264 of the first three colleges, which were to become University, Balliol and Merton. The colleges of Oxford, which quickly increased in number, became almost autonomous. Although the bachelor degree existed in the earliest days of the university, pass and honours examinations were not instituted until 1800 by a statute which was the first of a series of 19th-century reforms, including the admission of women in 1878, which established the university as it is now.

Oxidation, chemical reaction in which oxygen combines with another element or compound to form an oxide. Because oxidation entails a loss of electrons by the molecules with which the oxygen combines, the term is also used to define a loss of electrons even when oxygen is not involved.

Oxidation potential. See ELECTRODE POTENTIAL.

Oxidation-reduction, also called redox, chemical reaction involving simultaneous oxidation and reduction. In general, oxidation and reduction reactions occur together; thus in the reaction $Fe_2O_3 + 3C \rightarrow 2Fe + 3CO$, the iron oxide is reduced by the carbon, and the carbon is oxidized by the iron oxide. Carbon is the reducing agent; iron oxide the oxidizing agent. The term also describes reversible reactions as in the reaction between iron and tin compounds: $2FeCl_2 + SnCl_4 \rightleftarrows 2FeCl_3 + SnCl_2$. Oxidation-reduction reactions are important in many biochemical systems.

Oxidation state, state in which an atom is potentially able to form a compound, depending on the number of electrons available to be transferred. It is quantified on a scale of oxidation numbers which indicate the degree of IONIZATION. Thus in the compound sodium oxide, Na_2O, sodium (Na^+) has an oxidation number of $+1$, and oxygen (O^{2-}) an oxidation number of -2. In CO-VALENT and co-ordination (COMPLEX) compounds, the oxidation number is the electric charge that the atoms would have had if the compound was ionic; eg in the ion $[CuCl_4]^{2-}$, regarded as formed from Cu^{2+} and $4Cl^-$, the copper has an oxidation number of $+2$. Oxidation numbers are often used in the names of chemical compounds, as in iron II chloride ($FeCl_2$) and iron III chloride ($FeCl_3$), formerly called ferrous and ferric chloride.

Oxide, any inorganic chemical compound in which OXYGEN is combined with another element, or any organic compound in which oxygen is combined with one or more RADICALS. Examples of the latter are ethers, which have the general formula R_1-O-R_2, where R_1 and R_2 are carbon radicals. Inorganic oxides are of six kinds, named here with examples: (1) acidic, non-metal oxides generally; (2) basic, metal oxides generally; (3) neutral,

carbon monoxide; (4) amphoteric (capable of being acidic or basic), aluminium oxide; (5) peroxides, barium peroxide; and (6) suboxides, carbon suboxide.

Oxide coatings, coatings of incandescent filaments with the oxide of ALKALI metals or of a CATHODE with the oxide of alkali and ALKALINE-EARTH metals. It allows the production of thermionic emission at low temperatures.

Oxidizing agent, substance that causes oxidation reactions. Thus in the oxidation of carbon, $2C + O_2 \rightarrow 2CO$, oxygen is the oxidizing agent. Other common oxidizing agents include NITRIC ACID, HYDROGEN PEROXIDE, OZONE, potassium dichromate, ferric (iron III) compounds and potassium permanganate. See also OXIDATION-REDUCTION.

Oxley, John Joseph William Molesworth (1781–1828), Australian explorer, b. England, who became surveyor-general of New South Wales. He explored about 1,127km (700 miles) of the E coast between 1823–24, discovering the mouth of the Brisbane river.

Ox-pecker, two species of African birds of the genus *Buphagus*. Both are brown with wide bills and stiff tails and reach 22cm (8.5in) in length. They cling to cattle and big game and remove ticks and maggots. Family Sturnidae.

Oxus, River, in Persian mythology, river dividing the territory of the Persian Prince Rustem from that of his enemies the Turanians. It is mentioned by Matthew ARNOLD in his epic *Sohrab and Rustum* (1853) and is identified with the present day Amu Darya river in Soviet central Asia.

Oxyacetylene welding, the joining of metals by melting and fusing adjacent parts with a hot flame in which acetylene gas burns in pure oxygen. Oxyacetylene torches have two nozzles close together, from which the gases emerge to be ignited. The gases are supplied from pressurized steel "bottles" through flexible high-pressure hoses. See also WELDING; MM p.41.

Oxygen, common gaseous element (symbol O), discovered in 1774 by Joseph PRIESTLEY and independently (c.1772) by C.W. SCHEELE. It is the most abundant element in the earth's crust (49.2% by weight), and is a constituent of water and many rocks. It is also present in the atmosphere (23.14% by weight); it is extracted by fractional DISTILLATION of liquid air. The element is used in steelmaking, welding, the manufacture of industrial chemicals and in breathing and resuscitation apparatus. It is necessary for combustion and for the respiration of plants and animals. It is chemically reactive, and forms compounds with nearly all other elements. Properties: at. no. 8; at.wt15.9994; density 1.429 gm/litre; m.p. $-218.8°C$; ($-361.8°F$); b.p. $-182.96°C$; ($-297.3°F$); most common isotope O^{16} (99.759%). See also OZONE; SU p.134, 134.

Oxygen cycle, interchange of oxygen among agencies such as the atmosphere, the oceans, animal and plant processes and chemical combustion. The main renewable source of the earth's oxygen is the plant process of photosynthesis, wherein oxygen is liberated. Oxygen, dissolved in water, is utilized by aquatic life-forms in the process of respiration, a process essential to most living forms except anaerobic bacteria.

Oxygen process, or L-D process, method of producing steel first developed in Austria in 1952. Unlike the older BESSEMER and OPEN HEARTH PROCESSES which use air, the oxygen process uses jets of pure oxygen blown over the surface of the molten iron ore to convert it to iron.

Oxygen tent, device used in hospitals to provide a patient with an oxygen enriched atmosphere. It is used especially for patients suffering from heart failure or serious respiratory diseases.

Oxyhaemoglobin, the combination of haemoglobin in red blood cells with oxygen from the lungs, in which form oxygen is transported in the blood to all cells of the body. When oxyhaemoglobin gives up its oxygen to cells, a chemical

reaction is promoted which makes carbon dioxide (CO_2) from the tissues more soluble in the blood, for transport back to the lungs and elimination from the body. See also RESPIRATION; MS pp.66–67.

Oxytocin, or pitocin, peptide hormone secreted by the posterior pituitary; its principal effect is on myoepithelial cells of the breast, causing contraction of the ducts and ejection of milk. It also stimulates contraction of smooth muscle of the uterus, and plays a part in the induction of labour. See also MS pp.64, 77.

Oyster, edible BIVALVE mollusc found worldwide in temperate and warm seas. Its shell, usually with valves of unequal size, varies in shape according to environment. The European flat, or edible, oyster *Ostrea edulis* occurs throughout coastal waters. Family Ostreidae. See also NW pp.92–93, 92–93.

Oyster-catcher, sea-shore bird with a strikingly marked black-and-white stocky body and bright orange legs and beak. Oyster-catchers feed on molluscs, such as oysters and mussels, prising them open with their long beaks. Two to four sand-coloured eggs are laid in a depression scraped in the sand. Length: 43cm (17in). Family Haematopodidae; typical genus *Haematopus*. See also NW p.237.

Oyster mushroom, common edible FUNGUS that grows on trees. It has a fleshy oyster- or fan-shaped cap, prominent gills and usually no stalk. Species *Pleurotus ostreatus*.

Ozark Mts (Ozark Plateau), eroded tableland in s central USA, extending from sw Missouri across NW Arkansas into E Oklahoma. Timber is an important industry, as are lead- and zinc-mining and tourism. Area: approx. 129,500sq km (50,000sq miles).

Ozenfant, Amédée (1886–1966), French painter and teacher. He was one of the founders of Purism, which holds that art must be ordered and precise, and in the spirit of the modern machine age. He wrote *Après le Cubisme* (1918) with LE CORBUSIER.

Ozone, unstable gaseous ALLOTROPE of oxygen (O_3). It has a characteristic pungent odour and decomposes into molecular OXYGEN. It is present in the atmosphere mainly in the OZONE LAYER. Ozone is prepared commercially by the irradiation of air, and used as an OXIDIZING AGENT, in the purification of water and for bleaching. See also PE pp.66–67, 66–67.

Ozone layer, general stratum or shell in the atmosphere of the Earth in which OZONE (O_3) is concentrated. It is densest at altitudes of 21 to 26km (13 to 16 miles). Produced by incoming sunlight, the ozone layer absorbs much of the solar ultra-violet radiation, thereby shielding the Earth's surface. Aircraft, nuclear weapons, and some aerosol sprays and refrigerants all yield chemical agents that can decompose the ozone in this layer and could lead to an increase in the destructive ultra-violet radiation reaching the Earth's surface. See also FLUOROCARBONS; PE pp.66–67, 66–67.

P

P, 16th letter of the alphabet, derived from the Semitic letter *pe*, the word for *mouth*. In the Greek alphabet the letter became *pi*, which was taken into Latin and given the shape it has today. In English *p* nearly always has the same sound, as in *pig* and *poke*, although it is silent in a few words, such as *pneumatic*. Combined with *h* it is pronounced like an *f*, as in *phase* and *pheasant*. See also MS pp.244–245.

Pa, MAORI village or group of dwellings. It was usually built on a hill, with elevated stages for ramparts, and stockades built around each terrace.

Paasikivi, Juho Kusti (1870–1956), Finnish statesman who was chairman of the delegation that signed a peace treaty with the USSR in 1920. He was President of Finland from 1946 to 1956.

Pabst, George Wilhelm (1885–1967), Austrian stage actor and film director. Mostly of a pessimistic nature, his works included *Die Freudlose Gasse* (*The Joyless Street*) (1925), *Die Büchse der Pandora* (*Pandora's Box*) (1928), *Komödianten* (*Comedians*) (1941) and *Der Letzte Akt* (The Last Act) (1954).

Paca, also called spotted cavy, shy, nocturnal tailless RODENT of South America; it is brown with rows of white spots and has a relatively large head. A burrow-dweller, it feeds mainly on leaves, roots and fruit. Length: to 76cm (30in). Family Dasyproctidae; species *Cuniculus paca*. *See also* NW p.*216*.

Pacemaker, heart, area of tissue in a vertebrate's heart that sets it beating at about 70 beats a minute. If it fails it can be replaced by a battery-powered electronic unit which stimulates the heart with tiny electrical impulses. The pulse is usually about two volts and lasts two thousandths of a second.

Pacheco, Francisco (1564–1654), Spanish painter and art writer who taught Alonso CANO and Diego VELÁZQUEZ. He became painter to the King of Spain in 1619. He is also the author of *El Arte de la Pintura. . .* (1649), an important document on the history of Spanish painting.

Pacher, Michael (1435–98), Austrian sculptor and painter who studied in Italy and was among the first to encourage the Renaissance in Germany. His masterpiece, the St Wolfgang altarpiece, is in the church of Sankt Wolfgang am Abersee, in the Salzkammergut region of Austria.

Pacific, War of the (1879–84), conflict between Chile and the allied nations, Peru and Bolivia. The threatened take-over of Chilean holdings in the nitrate-rich province of Antofagasta induced Chile to send troops into the area, precipitating the war. Peru and Bolivia were thoroughly vanquished, Peru losing the provinces of Tarapacá and ultimately Arica too, while Bolivia became a land-locked nation with the loss of its only coastal province.

Pacific Islands Trust Territory, US territory including approx. 2,000 islands in the E central Pacific Ocean. Area: 1,860sq km (718sq miles). Pop. (1975 est.) 120,000 *See* MW p.139.

Pacific Ocean, largest and deepest ocean in the world, covering approx. one third of the Earth's surface and containing more than 50% of the Earth's seawater. The Pacific extends from the Arctic Circle to Antarctica, and from North and South America in the w to Asia and Australia in the E. The w Pacific region contains many islands which rise steeply from a deep sea-floor. The E Pacific region is connected with the Cordilleran mountain chain, and there is a narrow CONTINENTAL SHELF. The central region is low-lying. The ocean is ringed by numerous volcanoes, known as the Pacific Ring of Fire. There is also a large number of islands in the Pacific, most of which are in the s and w and are known collectively as OCEANIA. The major islands are New Zealand, the Japanese group and the Malay Archipelago. There are numerous volcanic and coral islands. The principal rivers that drain into the ocean are the Columbia in North America and the Hwang Ho (Yellow) and the Yangtze (Ch'ang Chiang) in Asia. Its average depth is 4,300m (14,000ft). The greatest known depth is that of the Challenger Deep, sw of Guam in the Mariana Trench, which has a depth of 11,033m (36,198ft). The sw and se basins of the Pacific sea bed are separated by the Albatross Cordillera. The current pattern of the Pacific is made up of two gyres: N of the equator are the North Equatorial Current, the Kuroshio Current, the North Pacific Drift and the California Current; s of the Equator are the South Equatorial Current, the East Australian Current and the Peru (Humboldt) Current. The Equatorial Counter Current separates the two gyres. Most fishing in the Pacific Ocean is done on the continental shelves. Crab, herring, cod, sardines and tunny are the principal catch. Area: approx. 166,000,000sq km (64,324,000sq miles). *See also* PE pp.*56–57, 86–87*.

Pacific Scandal, incident in Canadian political history. In 1873 the leader of the Conservative government, Sir John Alexander MACDONALD, was accused of accepting campaign funds from Sir Hugh ALLAN in return for the building contract for the Canadian Pacific Railroad. The government was forced to resign .

Pacific Security Treaty. *See* ANZUS.

Pacific trench. *See* PACDFIC OCEAN.

Pacifism, philosophy opposing the use of war or violence as a means of settling disputes. Elements can be found in ancient Hebrew and early Christian theology and in later Anabaptist and Quaker beliefs. International pacifist groups were organized during the 19th century. The doctrines of Mohandas GANDHI, and in Britain of the Fellowship of Reconciliations, were based on pacifist philosophy.

Pacing. *See* HARNESS RACING.

Packard, Vance Oakley (1914–), US social critic. His works are powerful and uncompromising critical analyses of social evils and have been best-sellers. They include *The Hidden Persuaders* (1957) about advertising, *The Status Seekers* (1959) about social symbols and *The Waste Makers* (1960), which deals with planned obsolescence.

Packer, Sir Douglas Frank (1906–74), Australian magnate, chairman of the Australian Consolidated Press (1936–74).

Packer, Kerry Francis Bullmore (1937–), son of Sir Frank PACKER, whom he succeeded as head of the Australian Consolidated Press (1974). In 1977 he organized international cricketers into professional teams to play in the World Series Cricket tournament in Australia.

Packet ships, early steamships with sails, used to carry mail, passengers and freight at scheduled times, especially in coastal waters. The arrival of British transatlantic steam packets such as the *Sirius* in New York harbour in 1838 spurred US development of scheduled travel and initiated a century of travel by steamship.

Pack rat, or trade rat, any of 22 species of mainly nocturnal, herbivorous wood rats from North and Central USA. They live in deserts, wooded areas or high rocky regions and are generally pale, dark or reddish brown or grey. Length: 20cm (8in). Family Cricetidae; genus *Neotoma*. *See also* NW p.*218*.

Paddlefish, primitive bony fish related to the sturgeon and found in the basins of the Mississippi and Yangtze rivers. Blue, green or grey, it has a cartilaginous skeleton and a long paddle-like snout. Length: 1.8m (6ft). Family Polyodontidae.

Paddle steamer, early steam-powered vessel propelled by one or more paddle-wheels. One of the first paddle steamers, the *Charlotte Dundas*, was constructed by William Symington in Scotland in 1802, and was used for towing shipping on the Forth and Clyde Canal. In 1807, Robert Fulton in New York built the *Clermont*, a passenger steamer. Paddle steamers soon became popular for long river journeys, mostly on the Mississippi river. *See also* MM pp.*111*.

Paddle tennis, bat and ball game invented in the USA after WWI. It is played with a wooden bat and a sponge-rubber ball on a court half the size of a LAWN TENNIS court. The rules are essentially those of lawn tennis, the chief difference being that only one serve is allowed. The net is only 2.5ft (76cm) high. The first official tournament was held at New York in 1924. The United States Paddle Tennis Association was formed in 1936.

Paddy field, field where RICE is cultivated underwater. Unlike most plants, rice thrives when submerged because oxygen is carried from the leaves to the roots. Rice paddies are flooded after the seedlings root, and are drained at harvest time. *See also* PE 181, 183.

Paderewski, Ignacy Jan (1860–1941), Polish pianist, composer and statesman who studied under Theodor LESCHETIZKY in Vienna, and became an internationally famous virtuoso. He wrote a piano concerto (1888) and a symphony (1907), and his opera *Manru* was first produced in Dresden in 1901. He became Prime

Minister and Minister of Foreign Affairs of the newly-created Polish state in 1919. In 1923 he returned to the concert stage. *See also* HC2 p.*270*.

Padua (Padova), city in NE Italy; capital of Padua province, Veneto region. As Patavium, the city prospered under Roman rule. From the 12th to the 14th centuries it was a commune and developed as a centre of the arts. The 13th-century Eremitani church contains what is left of the frescoes by MANTEGNA and the 13th-century basilica of St Anthony has statues and reliefs by DONATELLO . Frescoes by GIOTTO and TITIAN are also in Padua. The University of Padua (1222) is the second-oldest in Italy. Industries: motor vehicles, textiles, machinery. Pop. (1973) 234,203. *See also* HC1 p.*217*.

Padua, School of, Italian school of painting which became important in the 15th century through the works of MANTEGNA, DONATELLO and CRIVELLI.

Paean, in Greek mythology, originally a hymn addressed to APOLLO as healer or deliverer from evil. Later, paeans were addressed also to other gods and were sung to mark military victories.

Paedophilia, adult preference for sexual relations with children. It is severely condemned in many societies and in most countries is a criminal offence. In 1977 a Paedophilia Society was established in England in an attempt to combat the public's prejudice against paedophiles.

Paganism, in modern usage, religious beliefs other than JUDAISM, CHRISTIANITY and ISLAM. First applied in Latin usage to those who did not accept Christianity. The term, which can be derogatory, is often applied to polytheistic religions.

Paganini, Niccolò (1782–1840), Italian violinist, the most famous virtuoso of his day. He enlarged the range of the violin by exploiting harmonics and mastered the art of playing double and triple stops (two or three notes at a time). Many of his compositions were technically demanding pieces for unaccompanied violin.

Page, Sir Earle Christmas Grafton (1880–1961), Australian politician. He was a founder of the Country Party, which he led from 1920 to 1939. He was Treasurer in the Page government of 1923–29, often called the Bruce-Page government. He held successive ministerial posts until 1956, and retired from parliament in 1961.

Page, Sir Frederick Handley (1885–1962), British aeronautical engineer who founded (1909) one of the first aircraft manufacturing companies (Handley Page Ltd) and is best remembered for the aeroplanes that bear his name. His company, the last British aircraft manufacturer to remain independent, went into receivership in 1970. *See also* MM p.*177*.

Page, William (1811–85), US painter of portraits, landscapes and historical subjects. Influenced by Venetian painters during a stay in Italy, he painted many prominent people of his day including John Quincy Adams and Robert and Elizabeth Barrett Browning.

Pageant, dramatic spectacle. The word is thought to derive from the type of cart on which MYSTERY PLAYS were staged in the Middle Ages. In the 20th century a pageant has become a procession of tableaux with a common theme. These were especially common in Britain from 1900 to 1930, and elements survive in London's annual Lord Mayor's Show.

Paget, Sir James (1814–99), British surgeon, a founder of the science of PATHOLOGY. He discovered the parasitic worm *Trichina spiralis* that causes TRICHINOSIS, described the bone disease osteitis deformans (PAGET'S DISEASE), and was among the first to recommend surgical removal of bone-marrow tumours.

Paget's disease, osteitis deformans, a chronic bone disease, most often occurring in middle age, in which the bones soften, causing deformity and the accumulation of calcium in the blood. It is named after Sir James PAGET.

Pagliacci, I (1892), two-act tragic opera written by Ruggiero LEONCAVALLO with libretto by the composer, first produced in Milan. Using the device of a play within a

Paddlefish use their long oar-like snouts for locating food on the river bed.

Paddle steamer *Princess*, a Currier & Ives riverboat that worked the Mississippi River.

Padua: The Church of S. Giustina, one of many beautiful buildings in this city.

Paganini's playing is said to have been such that his audience was moved to tears.

Paint; these three-cylinder refiners are one step in the manufacturing process.

Painted lady; the caterpillars of this butterfly feed mainly on thistle leaves.

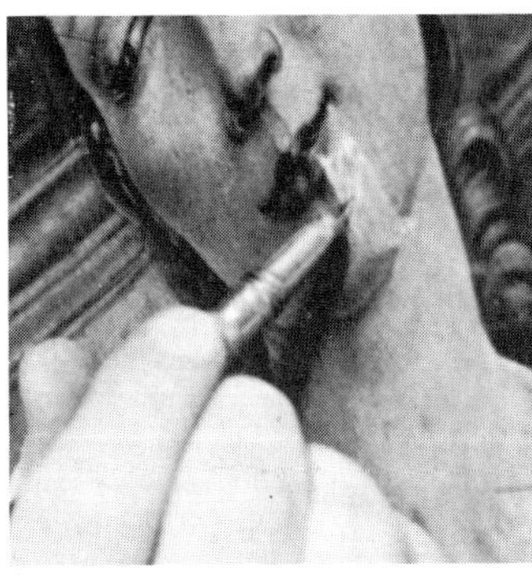

Paintings are restored with modern tools such as an electric retouching pencil.

Ian Paisley is a militant Protestant leader central to Northern Irish politics.

play, the story is based on an actual incident in Calabria which involved the killing of a woman by her actor husband after a theatrical performance.

Pagnol, Marcel (1894–1974), French playwright and film director who wrote *Topaze* (1928), and the trilogy *Marius* (1929), *Fanny* (1931) and *César* (1937), which was later filmed. He directed a number of films, mostly comedies and screen versions of his own plays.

Pagoda, name given to an Eastern building in the form of a tower, which originated in India and spread with BUDDHISM to China, Korea and Japan. Pagodas are polygonal or square in plan and built in superimposed storeys out of stone, brick or wood. They usually form part of a temple and serve as shrines.

Pago Pago, capital of American Samoa, on S Tutuila Island in the SW Pacific Ocean. It has an excellent harbour and an international airport. There is a US naval base in the town. Pop. (1970) 2,451.

Pahang, state in E Malaysia, on the South China Sea. The region is mostly dense jungle and is sparsely populated. Products: rubber, rice, fish, coconuts, tobacco, hemp. Area: 35,931sq km (13,873sq miles). Pop. (1970) 503,131.

Pahoehoe, Hawaiian term for a basaltic lava flow that moves freely and cools with a smooth, rope-like surface. Lava of similar composition but which forms in rough irregular flows, often covered with clinkers, is known as AA.

Pain, intense sensation that warns the brain of injury to, or illness of, any part of the body; it is a sensation that is little understood, and can range from mild discomfort to acute agony. It may arise from the skin or from the deeper tissues of the body, possibly through the overexciting of free nerve endings in those tissues. Individual tolerance of pain varies, perhaps because of genetic differences but more likely because of cultural conditioning, eg members of different ethnic groups tend to have different pain tolerances.

Paine, Thomas (1737–1809), American political and religious writer, *b.*England. His radical and outspoken views greatly influenced colonial opinion during the American War of Independence. In his pamphlet *Common Sense* (1776) he demanded complete independence from England and the setting up of a strong federal union. A series of 16 papers, *The Crisis* was read aloud to encourage the Revolutionary Army. He returned to England as an ardent supporter of the French Revolution, defending its cause so vehemently in *Rights of Man* (1791–92) that the work was suppressed and he was charged with treason and fled to France. There he made enemies by speaking out against Louis XVI's execution. For more than 10 months he was in prison where he worked on *Age of Reason*, an agnostic argument. He sailed to America in 1802, and died on a farm given to him by New York State in appreciation of his writings.

Paint, coating applied to a surface for protective and decorative purposes. Paint is composed of pigment, or colour, and a vehicle (a liquid that suspends the pigment), which adheres to a surface and hardens when dry. Pigments are made of metallic compounds, usually oxides, or synthetic materials. Vehicles may be oils, water mixed with a binding agent, organic compounds, or synthetic RESINS, which may be soluble in water or oil. Water-based paints mix with water, oil-based paints with turpentine or various organic solvents. *See also* MM pp.244–245.

Painted lady, also called thistle butterfly, most widely distributed of all LEPIDO-PTERA. It has brown, black, orange and white wing markings and its 5cm (2in) caterpillars feed on thistles or nettles. Family Nymphalidae; species *Vanessa cardui.*

Painterly, term derived as an attempted translation from the German word *malerisch,* which the art historian Heinrich Wölfflin used to define painters who use form in terms of colour and tone (as do Rembrandt and Titian) rather than in line (as does Botticelli). By extension, a painterly picture may also be one in which

"painterly" values – form, tone, colour, composition – are pre-eminent over other considerations, such as subject matter and literary content.

Painting, the art of using one or more colours, generally mixed with a medium (such as oil or water) and applied to a surface with a finger, brush or some other tool to create pictures. Paintings are among the earliest of historical records. The tradition found in the prehistoric cave art of France and Spain, survives among contemporary primitive tribes of, for example, Australia and Africa.

Painting in early civilizations such as that of Egypt was largely a matter of filling in with colour areas outlined by drawing. Little Greek painting survives (apart from that on pottery), but written records show that the Greeks had discovered the art of shading and of TROMPE-L'OEIL and achieved a standard of painting comparable with their sculpture. The Romans were greatly influenced by Greek art, as the only surviving examples, the fine FRESCO paintings at Pompeii and Herculaneum, demonstrate. In the early Christian and Byzantine periods traditions in mural painting and manuscript ILLUMINATION were established that were to last throughout the Middle Ages, traditions in which a highly spiritual concept was fostered by Church patronage. In Italy the great Byzantine frescoes found an heir in GIOTTO's murals, which were the beginning of ventures into realism and experimentation; various Italian schools of painting arose – that of Florence gained ascendancy in the 14th and 15th centuries.

During the 15th century, as the patronage of the Church was increasingly augmented by that of secular princes, notably the MEDICI, and the humanist ideals of the RENAISSANCE took root, the range of subjects and techniques available to the artist widened enormously. Narrative religious themes now had to vie with pagan mythology, secular figures in landscape settings and pure landscape; and although the greatest masterpieces of MICHELANGELO, RAPHAEL and the Florentine masters were generally murals or altar pieces, this period also saw the first use of oil paint on canvas, the rise of the easel painting and the beginnings of GENRE PAINTING and pure portraiture. It was also the age of the perfection of the study of PERSPECTIVE and of a more natural approach to form and composition. To this creative legacy the great painters of the BAROQUE period added an unrivalled bravura brushwork and drama of vision. North of the Alps, where the Renaissance had spread more gradually than in Italy, the greatest achievements came with the Baroque exuberance of RUBENS and the superb genre, portrait and landscape pictures of Rembrandt and lesser 17th-century Dutch painters, whose choice of intimate, everyday subjects was the antithesis of the grand manner characteristic of the Italian masters and of the Italianate style adopted in France at the court of Versailles.

By the 18th century British painters had become established in portraiture, animal and landscape painting, although overshadowed by the great Venetian masters, TIEPOLO, GUARDI and CANALETTO, whose work was much admired in England. Various schools flourished, divided principally between Baroque and austere Classicism. The 19th century opened with the supremacy of NEO-CLASSICISTS such as Jacques Louis DAVID challenged by the new Romanticism of DELACROIX and GÉRICAULT. But both schools were superseded, first by IMPRESSIONISM and then by the succession of new movements in the late 19th and early 20th centuries. Most of these movements – including Impressionism, POST-IMPRESSIONISM, SYMBOLISM, FAUVISM, CUBISM, DADA and SURREALISM – originated in Paris, although Germany was the cradle of EXPRESSIONISM and Russia contributed SUPREMATISM. More recent major movements include ABSTRACT IMPRESSIONISM, developed fully by US painters, POP ART, OP ART and HARD-EDGE. *See also* articles on individual painters, periods, movements and techniques; and *History and Culture* volumes.

Paisiello, Giovanni (1740–1816), Italian composer, principally of operas of which he wrote more than 100. As court composer to CATHERINE II, THE GREAT of Russia (1776–84), he composed a new setting of Pergolesi's *La Serva Padrona* (1781) and *The Barber of Seville* (1782), which became popular throughout Europe until the version by ROSSINI superseded it.

Paisley, Ian Richard Kyle (1926–), political and religious leader in Northern Ireland. He founded the Free Presbyterian Church of Ulster in 1951, led anti-Catholic demonstrations in the late 1960s, and represented militant Protestant interests in the Northern Ireland parliament in 1970 and in the British House of Commons. In 1973 he was elected to the new assembly of Northern Ireland, but his opposition to the power-sharing executive helped lead to the re-imposition of direct rule from Westminster in 1974. *See also* HC2 p.*295.*

Paisley, town in Strathclyde Region SW Scotland, administrative centre of the former county of Renfrew(shire). Paisley is best known for its textile industry, in particular the Paisley pattern, and thread. Other products include boilers, chemicals and soap. Pop. (1971) 95,344.

Pajou, Augustin (1730–1809), French sculptor known for his decorative work at the Opera Houses in Versailles (1768–70) and for his many portraits. He worked for Mme DU BARRY and then for King LOUIS XVI. His works include *Psyche Abandoned* (1791) and the portrait of his master *Lemoyne* (1758).

Pakistan, officially named the Islamic Republic of Pakistan, nation in Asia NW of the Indian subcontinent. A poor country, Pakistan suffered from the loss of its E province – now the independent state of BAN-GLADESH – in 1971. Two-thirds of the population live off the land; wheat and rice are the chief crops. The manufacture of cotton textiles is the principal industry. Islamabad is the capital. Area: 803,943sq km (310,402sq miles). Pop. (1975 est.) 70,260,000. *See* MW p.140.

Pakokku, port in W central Burma, 121km (75 miles) SW of Mandalay, on the IRRA-WADDY river; capital of Pakokku district. The city was occupied by the British during the 19th century. The Communists held it from 1949 until 1955 when it was liberated by government forces. The Yenangyat oilfields are nearby. Pakokku is a trade and shipping centre. Pop. 30,900.

Palade, George Emil (1912–), US cell biologist, *b.* Romania. He shared the 1974 Nobel Prize in physiology and medicine with Albert CLAUDE and Christian de DUVE for his contributions to an understanding of the structure and function of the CELL. Using sophisticated electron microscope techniques and other methods he discovered RIBOSOMES, which are the sites of PROTEIN synthesis in the cell.

Pala dynasty, Buddhist rulers of Bihar and Bengal from the mid-8th to the mid-12th century, founded by Gopala. His successor, Dharmapala (*r. c.* 770–810), expanded Pala rule as far NW as Kanauj, the capital of the Pratihara dynasty. The Palas carried BUDDHISM into Tibet, but from the mid-9th century their power waned. *See also* HC1 pp.118–119.

Palaeobotany, study of fossil plants, a branch of PALAEONTOLOGY.

Palaeocene, geological epoch that extended from about 65 to 54 million years ago. It is the first epoch of the TERTIARY PERIOD, when the vast majority of the DINOSAURS had disappeared and the small early mammals were flourishing. *See also* PE p.*131;* NW p. *169.*

Palaeography, study of early writing. In its broad sense the term includes all inscriptions; more narrowly it includes only writing done on wax, parchment, papyrus or paper.

Palaeolithic, period in history known also as the Old Stone Age, the earliest stage of man's history, from *c.* 2 million years ago until between 40,000 and 10,000 years ago. It was marked by the use of stone tools and covers the history of the species from Australopithecus through Neanderthal Man to *homo sapiens.*

Palaeologus, Michael. *See* MICHAEL VIII.
Palaeontology, study of geological periods by the examination of fossil remains. The evidence of fossils enables the assessment of the composition, structure and age of rocks. The science, which is divided into palaeozoology and palaeobotany, established its independent status in the early 19th century with the work of CUVIER and LYELL.
Palaeozoic, third major era of geological time, after the Azoic and Proterozoic eras, lasting from 600 million to 225 million years ago. It is considered to have comprised six periods: Cambrian, Ordovician, Silurian, Devonian, Carboniferous and Permian. Invertebrate animals appeared in the Cambrian, fishlike vertebrates in the Ordovician, amphibians in the Devonian and reptiles in the Carboniferous. Plant life declined after its peak in the Carboniferous.
Palamas, Kestis (1859–1943), Greek poet who wrote in vernacular or demotic Greek. His lyrical poetry expressed the modern Greek poetic resurgence. *The Grave* (1898; translated 1930), a collection on the death of his young daughter, has great dignity.
Palate, roof of the mouth, comprised of the bony front part known as the hard palate and the softer fleshy part of the back known as the soft palate, from which the uvula hangs. The palate separates the mouth from the nasal cavities. *See also* MS p.*131.*
Palatinate, two regions of West Germany: the Lower, or Rhenish, Palatinate is on the River Rhine; the Upper Palatinate is on the River Danube and forms part of Bavaria. The regions were politically united in 1329 and were important in the affairs of the HOLY ROMAN EMPIRE after 1356, when the count palatine of the Rhine became an imperial elector. The Palatinate was a centre of the German Reformation and its elector played an essential part in the outbreak of the THIRTY YEARS WAR. In 1946 the Lower Palatinate was incorporated into the state of Rhineland-Palatinate.
Palau Islands, island group in the w Pacific Ocean, in the Caroline Islands, NW of New Guinea and SE of the Philippines. They are made up of approx. 100 islands, many of which are uninhabited; they are administered by the USA. Area: 179sq km (69sq miles). Pop. (1971 est.) 12,500.
Pale, The English, the part of Ireland under the control of the English following the Anglo-Norman invasion of 1171–72. It included Louth, Meath, Trim, Dublin, Kilkenny, Wexford, Waterford, Carlow and Tipperary. It shrank in the 15th century, but expanded again under the Tudors. From it comes the phrase "beyond the pale", meaning outside the bounds of civilized behaviour.
Palembang, port in SE Sumatra, Indonesia, on the Musi River; capital of South Sumatra province. The city served as the capital of the Hindu-Sumatran kingdom of Sri Vijaya in the 8th century. The Dutch took the city and established a trading post there in 1617 and a fortress in 1659. The sultanate of Palembang was abolished by the Dutch in 1825. The city has the Sriwidjaja Universitas Negeri (Sri Vijaya State University) (1959). Exports: oil, rubber, coffee, coal. Industries: food processing, oil refining, rubber, fertilizers. Pop. 474,971.
Palenque, ancient city (*c.* AD 300–900) of the MAYA in Chiapas, s Mexico. Stucco sculpturing and panelling reached their finest expression at Palenque. The Temple of Inscriptions is one of the best preserved Mayan temples; other ruins have hieroglyphic inscriptions, reliefs and a large amount of plasterwork. *See also* HC1 p.*231.*
Palermo, in NW Sicily, on the Tyrrhenian Sea; capital of Sicily and the province of Palermo. The city was founded by the Phoenicians in the 8th century BC. It passed to the Romans in 254 BC and prospered under the Saracens from the 9th to the 11th centuries. It was later ruled by the Normans. The city has many fine buildings, including a 12th-century cathedral and an Arab-Norman Palatine Chapel

(1130–40). Industries: shipbuilding, textiles, food processing, chemicals. Pop. (1975) 662,567.
Palestine, region on the E shore of the Mediterranean Sea; considered as a Holy Land by Jews, Christians and Muslims. The area has been continuously settled since 4000 BC. The Jews moved into Palestine from Egypt *c.* 2000 BC but were subject to the Philistines until 1020 BC, when Saul, David and Solomon established Hebrew kingdoms. The area came under Roman rule in 63 BC. In succeeding centuries Palestine became a focus of Christian pilgrimage. It was conquered by the Muslim Arabs in 640. In 1099, Palestine fell to the Crusaders; in 1291 they in turn were routed by the MAMELUKES. The area was part of the OTTOMAN EMPIRE from 1516 until WWI. It was exposed to Western influences through Russian Jews, immigrating after 1882, who started the ZIONIST movement. The British held a LEAGUE OF NATIONS mandate over Palestine after 1920 and attempted unsuccessfully to divide the country between Arabs and Jews as Arab nationalism developed. The Arabs feared the economic and political results of immigration as WWII and Nazi persecution brought many Jews to Palestine, and in 1947 Britain consigned the problem to the UNITED NATIONS. Most of ancient Palestine became part of the new state of Israel in 1948; the first of several ARAB-ISRAELI WARS occurred in that year. During the 1970s the PLO and other nationalist Arab groups asked for an independent Palestine for the Arabs. *See also* HC2 pp.296–297.
Palestine Liberation Organization. *See* PLO.
Palestine Mandate, authority vested in Britain by the LEAGUE OF NATIONS in 1920 to govern PALESTINE, which had been a Turkish province from 1516 until its conquest by British forces in 1917–18. In 1923 JORDAN was separated from the mandate. Britain withdrew from Palestine on 14 May 1948, when the independent state of ISRAEL was proclaimed. *See also* HC2 pp.296–297.
Palestrina, Giovanni Pierluigi da (*c.* 1525–94), Italian composer who spent most of his life in the service of the Church. His output of church music was immense and included complete masses, magnificats, litanies and about 600 motets in four to eight and twelve parts. The six-part mass *Assumpta est Maria* and the eight-part setting of the motet *Stabat Mater* are profound examples of polyphonic and rhythmic mastery, melodic balance and proportion. He also composed almost 100 secular madrigals.
Palgrave, Francis Turner (1824–97), British poet and anthologist who was professor of poetry at Oxford University from 1885–1895. He is best known for his anthology, *The Golden Treasury of English Songs and Lyrics* (1861).
Pali, ancient Indian language in which the Buddhist scriptures, or canon, were compiled in the first century BC. It is related to classical Sanskrit. It is still the liturgical language of the Theravāda branch of HINAYANA Buddhism, the branch of Buddhism prevailing in Sri Lanka, Burma and Thailand.
Palissy, Bernard (*c.* 1510–89), French Huguenot potter. He experimented in enamel-work for 15 years and after 1556 produced smoothly glazed and richly coloured ware decorated with reptiles, insects and plants.
Palladian, architectural style especially popular in England. It derived from the buildings and theories of the 16th-century Italian architect Andrea PALLADIO, who based his work on Roman classicism, emphasizing symmetrical planning and harmonic proportions. Inigo JONES introduced Palladianism to England after visiting Italy (1613–14). The Queen's House, Greenwich, completed in 1635, is a fine example of this style. The revival of interest in Palladianism in the early 18th century, at the beginning of the GEORGIAN period, was led by Colin CAMPBELL and Lord Burlington. It coincided with the publication, in English translation (1715), of Palladio's *Four Books on Architecture*

and marked a reaction against the grandiose excesses of style during the latter years of Stuart rule. Representative of the style is Holkham Hall, Norfolk, begun in 1734 by William KENT. *See also* HC1 pp.258–259, 296; HC2 pp.70, 71.
Palladio, Andrea (1508–80), Italian RENAISSANCE architect. He studied Roman architecture and published drawings of Roman ruins together with his own designs in *Four Books of Architecture* (1570) and through this work revived symmetrical planning and classical proportioning in architecture. His most notable buildings were villas and palazzos, usually with arch and column façades, such as the Villa Rotunda, Vicenza (*c.* 1550). *See also* HC1 pp.258–259, *258.*
Palladium, in Greek religion, representation of the goddess Pallas Athena. It was also the wooden image kept in the citadel to ensure the safety of Troy. It was supposedly thrown from heaven by ZEUS.
Palladium, precious metallic element (symbol Pd) of the second TRANSITION ELEMENTS, discovered in 1803 by W.H. Wollaston. It is found in nickel ores associated with platinum, to which it is chemically similar. Properties: at.no. 46; at.wt. 106.4; s.g. 12.02; m.p. 1,552°C (2,826°F); b.p. 3,327°C (6,021°F); most common isotope Pd[106](27.3%).
Pallas, Peter Simon (1741–1811), German naturalist and geologist who by the age of 15 had formulated a new taxonomy for certain animal groups. From 1768 to 1774 he travelled throughout Russia and Siberia, finding a wide distribution of fossils and mammoths preserved intact in the tundra; he published the results of the expedition in *Journey Through Various Provinces of the Russian Empire* (3 vols, 1771–76).
Pallas, asteroid (minor planet) discovered in 1802 by Heinrich OLBERS; the second-largest asteroid. Diameter: 450km (280 miles); mean distance from the Sun 414,000,000km (257,000,000 miles); mean sidereal period 4.61yr. *See also* SU pp.204, *205.*
Pallas' cat, rare, small, long-haired cat that lives in burrows or among the rocks of Asian deserts. It has small rounded ears, short legs and a compact body. The coat is silvery grey and yellow brown with dark stripes on the face, back, forelimbs and tail. It feeds on small birds and rodents. Length: to 50cm (20in). Family Felidae; species *Felis manul. See also* NW p.*220.*
Pallava dynasty, rulers of s India from the 4th to the 9th century AD; their capital was at Kanchi. It is noteworthy especially for its DRAVIDIAN architecture, best exemplified in the "Seven Pagodas" of Mahabalipuram. *See also* HC1 pp.118–119.
Palm, family of trees found in tropical and subtropical regions. Palms are ancient flowering plants that date back to the early Triassic (225 million years ago). They have a woody, unbranched, columnar trunk with a crown of large, stiff leaves. The leaves may be either palmate (fan-like) or pinnate (feather-like). Instead of having true bark, palm trunks are covered with fibres which are often derived from the leaf stems. An economically important family, palms are the source of wax, oil, fibres, sugar and other foods. Height: 60m (200ft). Family Palmaceae. *See also* NW pp.*66, 190;* PE p.*195.*
Palma (Palma de Mallorca), port and capital of Majorca (Mallorca) in the Balearic Islands, Spain; capital of Baleares province. The city has a Gothic cathedral (1230–1601), containing the tomb of King James II of Aragon, and a Moorish palace. Industries: tourism, pottery, leather goods, jewellery, wine, food processing. Pop. (1970) 287,038.
Palmas, Las, city on Grand Canary Island, Spain; capital of Las Palmas province. Its port, Puerto de la Luz, is the principal port of the CANARY ISLANDS and one of the busiest in Spain. Industries: fishing, vegetable and fruit production, tourism. Pop. (1975 est.) 328,187.
Palma Vecchio (1480–1528), Venetian painter, real name Jacopo d'Antonio Negretti. He consistently painted women as a fair, full-figured stereotype. His works include *Jacob Greeting Rachel with*

Palermo: the fountain in Piazza Pretorio, by Camilliani and Neccherino (1550).

Palestine: a 16th-century fresco by A. Danti showing an area roughly that of Israel.

Palladian architecture is based on that of Renaissance architect Andrea Palladio.

Palm; dried coconut flesh yields an oil used in soap and margarine.

Samuel Palmer's painting of a garden at Shoreham, where he did much of his work.

Lord Palmerston served longer in office than any other British politician.

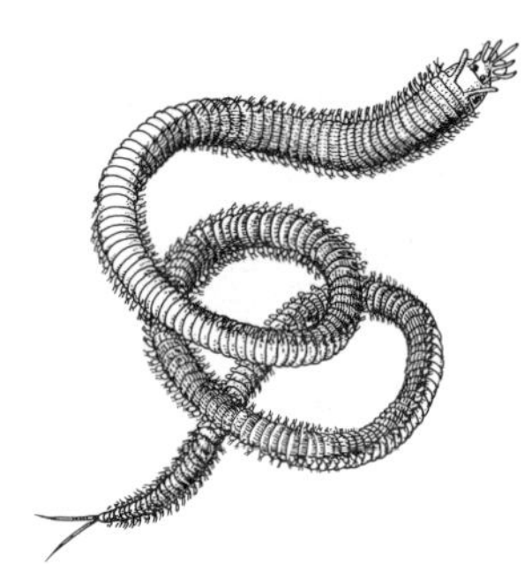

Palolo worms live in the coral reefs of the South Pacific ocean.

Pampas: a gaucho of the Argentine plains hobbles his horse to prevent straying.

a Kiss, The Three Sisters, and his masterpiece, the *St Barbara Altarpiece.*

Palm Beach, town in SE Florida, USA, on the N end of an island separating Lake North, a lagoon, from the Atlantic Ocean. Settled in 1871, it grew rapidly after 1893, when Henry Flagler realized its potential as a resort and financed its planning. Pop. (1970) 9,086.

Palmer, Arnold Daniel (1929–), US golfer who was largely responsible for the growth in popularity of golf. In 1970 he was named Athlete of the Decade. He won the US Open Championship in 1960, the British Open in 1961 and 1962, and the Masters tournament in 1958, 1960, 1962 and 1964.

Palmer, Nathaniel Brown (1799–1877), US sea captain who in 1820 became the first to sight the Antarctic mainland, which he believed to be only an island. He was on a sealing voyage at the time and on the same expedition he also discovered the South Orkney Islands.

Palmer, Samuel (1805–81), British Romantic landscape painter, the most important follower of William BLAKE, whom he met in 1824. By about the age of 15 Palmer was already exhibiting at the Royal Academy. His years in Shoreham, Kent (1826–35) were his most productive. Many works of this period are in the Ashmolean Museum, Oxford. Palmer was rediscovered in the 1920s and 1930s; he influenced modern Romantic painters, including Graham SUTHERLAND. *See also* HC2 p.*103.*

Palmerston, Henry John Temple, 3rd Viscount (1784–1865), British statesman and Prime Minister, who dominated the early and mid-Victorian political scene. He entered the House of Commons in 1807 and between 1809 and 1828 served as Secretary of War in successive Tory governments. In 1830, associating himself with the Whigs, he became Foreign Secretary in Lord GREY's Reform ministry, a position he held until 1834 and again from 1835 to 1841 and 1846 to 1851. During these periods he supported Liberalism in Europe and was a forceful protector of British interests in every part of the world. He was Home Secretary from 1852 and after an inquiry into the government's mishandling of the CRIMEAN WAR, he was Prime Minister from 1855 to 1858. His second administration (1859–65) ended only with his death and is seen in retrospect as a prologue to the Liberal government of 1868–74. *See also* HC2 pp.112–113, 182–183, *182–183.*

Palmerston North, city of s central North Island, New Zealand. It is the seat of Massey University of Manawatu (1964). It is an agricultural trade centre with textiles, pharmaceuticals, electrical equipment and food-processing industries. Pop. (1970 est.) 51,000.

Palmistry, also called chiromancy, character divination and foretelling of the future by interpreting the patterns of lines and fleshy mounds on the human hand. There is no scientific basis for the belief that the structure of the hand has significance, although its condition may give an indication of the subject's occupation.

Palmitic acid, fatty acid which is a major constituent of vegetable oils and animal fats. Together with STEARIC ACID, it is the fatty acid most often present in soaps. Palmitic acid is a saturated compound, containing only single bonds, and has the formula $CH_3(CH_2)_{14}COOH$.

Palmyra, ancient city in central Syria, at an oasis on the N edge of the Syrian Desert. Traditionally founded by Solomon, it developed under the Romans in the 1st century and flourished AD 130–270. After capture by the Emperor Aurelian the city went into decline. It was later taken by the Arabs and the Mongols. *See also* HC1 pp.*98, 115.*

Palolo worm, polychaete ANNELID of the South Pacific that lives in holes among coral reefs, and is a nocturnal swimmer at breeding time. Its rear portion is filled with eggs or sperm; this part develops an eyespot, separates from the rest of the body and swims to the surface to mate. Family Eunicidae; species *Eunice viridis.* *See also* NW p.*89.*

Palomar, Mount, peak in s California, 72km (45 miles) NNE of San Diego. It is the site of the Palomar observatory, which houses the 122cm (48in) Schmidt telescope and a 508cm (200in) reflecting telescope, the second-largest optical telescope in the world. Palomar is administered jointly with MOUNT WILSON OBSERVATORY under the name of the Hale Observatories. *See also* SU pp.168, *169,* 170–171, *171.*

Palomino, light saddle horse which is characteristically yellow gold with a white, silver or ivory mane and tail. Height: to 163cm (64in) at the shoulder.

Palpitation, abnormally rapid beating of the HEART. It may occur after great exertion or strong emotion, or may be due to a disorder.

Palynology, study of spores, seeds and pollen; it is a part of such disciplines as archaeology, geography and palaeontology. Studies of pollen in lake sediments have provided much information about the vegetation, climate and environmental conditions in past ages. Palynology has also helped to deduce the patterns of land use and settlement of primitive man. *See also* NW p.*175.*

Pamir, mountainous region mostly in Tadzhikistan (Tadzikskaja SSR) USSR, partly in NE Afghanistan and China. The region forms a geologic structural knot from which the Tien Shan, Karakoram, Kunlun, and Hindu Kush mountain ranges radiate. The climate is cold during winter with cool summers; the terrain includes grasslands and sparse trees. The main activity is sheep herding although some coal is mined. The highest peaks are Mt Communism, 7,495m (24,590ft) and Lenin Peak 7,134m (23,405ft).

Pampas, large plain in s South America, situated mostly in Argentina. The humid Pampas is extremely fertile; dairy farming is practised and cereals are grown. The larger dry Pampas to the w, which includes the provinces of Buenos Aires, Santa Fe and Cordoba, supports mainly livestock grazing. The Pampas has a complex transport system focused on Buenos Aires.

Pampas grass, species of tall, reed-like GRASS native to South America and widely cultivated in warm parts of the world as a lawn ornamental. Female plants bear flower clusters, 91cm (3ft) tall, which are silvery and plume-like. Family Gramineae; species *Cortaderia selloana.*

Pamplona, city in N Spain, 32km (20 miles) from the French border, on the Arga river; capital of Navarra province. The city was taken in 778 from the Moors by CHARLEMAGNE. It passed to FERDINAND of Aragon in 1512. The French occupied the city during the PENINSULAR WAR. Pamplona is known for its annual fiesta de San Fermin, which includes the running of the bulls through the city streets. There is a 14th-century Gothic cathedral. Industries: sugar refining, brewing, textiles, wine, firearms, furniture, footwear, flour milling. Pop. (1970) 147,168.

Pan, in Greek mythology, the son of HERMES and the god of woods and fields, shepherds and their flocks. He is depicted with the horns, legs and hoofs of a goat. A forest dweller, he pursued and loved the DRYADS and led their dances while playing the syrinx, the pipes of his invention.

Panaetius (*c.* 180–109 BC), Greek philosopher, founder of Roman STOIC philosophy. Although loyal to the fundamental Stoic doctrines, his interpretation was less austere. He wrote less than other Stoics: none of his works has survived complete.

Pan-African Congress, occasional meeting of African leaders in the interest of African unity. It met five times between 1900 and 1927 and worked to bring gradual self-government to African colonial states. Its aims have been continued by the Organization of African Unity (OAU), formed in 1963.

Pan-African Games, major sporting tournament for the nations of Africa, except South Africa and Rhodesia (Zimbabwe). They have been held in 1965 and 1973. As well as track and field events, there are other sports such as basketball, table tennis, boxing, swimming, soccer and handball.

Pan-African Movement, loosely organized effort for the unification and independence of African nations and Black people everywhere. It began officially at the PAN-AFRICAN CONGRESS of 1900 in London, organized by W. E. B. DUBOIS and other Western Blacks. The movement materialized in 1963 as the Organization of African Unity (OAU), and has enjoyed a limited success, both in influence and in internal cohesion.

Panama, republic in Central America on the narrowest part of the isthmus joining North and South America. The economy is almost entirely dependent on the Canal Zone, the area astride the Panama Canal administered by the USA. The principal crops are rubber, coffee, cereals, bananas and cocoa. Deposits of copper were discovered there in 1968. The capital is Panama City. Area: 75,651sq km (29,209sq miles). Pop. (1976 est.) 1,719,000. *See* MW p.141.

Panama, Isthmus of, narrow strip of land in s Central America, which connects South America with Central and North America. *See also* MW p.51.

Panama Canal, waterway built by the USA (1904–14) to connect the Atlantic and Pacific oceans, opening a shorter route for trade to the Far East. The Isthmus of PANAMA, originally owned by Colombia and also the narrowest point of Central America, was chosen as the location. The original agreement between the USA and Columbia of 1903 was superseded by an agreement between the newly independent Republic of Panama, which had declared independence from Columbia and the USA also in 1903. Length: 82km (51 miles). *See also* MM pp.196–197; MW pp.51, 141.

Panama Canal Zone, also called merely Canal Zone, strip of land containing the PANAMA CANAL. The USA bought the strip in 1903 for $10 million and an annuity fixed then at $250,000 from the newly-independent Panama and the canal opened in 1914. From 1974 there were negotiations to give Panama additional revenue and jurisdiction over the canal. Area: 1,432sq km (553sq miles). Pop. (1975 est.) 44,000. *See* MW p.51.

Panama City, capital of PANAMA. It was founded by Pedro Arias de Avila in 1519, and was an important port before its decline in the 17th century. It revived after the construction of the PANAMA CANAL; although no longer a port, it has shoe and textile industries. Pop. (1970) 348,704.

Pan-American Games, athletic competition for all countries in the Americas, similar to the OLYMPIC GAMES. The various events in this competition are modelled after the Olympics and have the same basic rules and regulations. The first games were held in 1951 in Buenos Aires, Argentina, and have since been held every four years.

Pan-American Highway, road system, which will be approx. 25,760m (16,000 miles) long when completed, linking the USA, Mexico, Central and Latin America. From 1923, the USA has financed those parts of the road through the smaller Central American states. The system, which includes the Inter-American Highway, will extend from Nuevo Lavedo, Texas, through Mexico City to Panama City, to Valparaíso, Chile and then w to Buenos Aires, Argentina.

Pan-American Union. See OAS.

Pan-Arab Movement, tendency towards political unification among Arab nations after gaining independence from Ottoman Turkey and Europe. An ARAB LEAGUE was formed in 1945, composed at first of seven states, but growing to 15 countries. It co-ordinated political, economic, and military activities through various treaties. The Arab Bank was formed in 1959. President NASSER of Egypt strengthened the movement, but his UNITED ARAB REPUBLIC (Syria and Egypt) lasted only until 1961. A union between Iraq and Jordan also dissolved in 1958. *See also* OPEC.

Panay, island in the central Philippines in the Visayan group; there are four provinces: Capiz, Antique, Aklan, and Iloilo. The interior is mountainous and drained by the Panay River. The chief occupation is farming, and the island is famous for horse-breeding. Rice and corn are grown in the lowlands. The principal port is Iloilo. Industries: timber, sugar-refining, fishing. Area: 11,515sq km (4,446sq miles) Pop. (1970) 2,144,544.

Panchen Lama, second in the Tibetan Buddhist hierarchy and abbot of the Tashilhumpo monastery. The title was adopted in the 16th or 17th century. The Panchen Lama is believed to be the incarnation of the Buddha of Infinite Light, Amitábha. His role is more limited to spiritual matters than that of the DALAI LAMA, who ranks highest in the hierarchy. When the Dalai Lama fled to India in 1959 following the unsuccessful Tibetan revolt against the Chinese Communists, the Chinese government acknowledged the Panchen Lama as the true Tibetan leader. However, the fate of the last Panchen Lama is unknown. *See also* LAMAISM.

Pancreas, elongated soft gland lying behind the stomach to the left of the mid-line. It is a DIGESTIVE organ, secreting various ENZYMES: trypsin which digests proteins, lipase which acts on lipids, amylase which acts on carbohydrates, and other minor secretions. These pass into the pancreatic duct and then into the common bile duct, which opens to the small intestine. The pancreas also contains the ISLETS OF LANGERHANS, which secrete the hormone INSULIN. *See also* DIABETES MELLITUS; MS pp.68–69, *68–69.*

Panda, two mainly nocturnal mammals of the raccoon family. The lesser panda, *Ailurus fulgens,* ranges from the Himalayas to W China. It has soft, thick, reddish brown fur, a white face and a bushy tail. It feeds mainly on fruit and leaves, but is also a carnivore. Length: 115cm (46in) overall. The giant panda, *Ailuropoda melanoleuca,* is one of the rarest large mammals and inhabits bamboo forests in China (mainly Tibet). It has a short tail and a dense white coat with characteristic black fur on shoulders, limbs, ears and around the eyes. Although classified as a carnivore it eats mainly plant material – particularly bamboo shoots. Length: 1.5m (5ft); weight: 160kg (350lb).

Pandit, Vijaya Lakshmi (1900–), Indian diplomat, daughter of the Motilal Nehru and sister of Jawaharlal NEHRU. She led the Indian UN delegation (1946–48; 1952–53) and was president of the UN General Assembly (1953–54). She was also Soviet (1947–49) and US (1949–51) ambassador. She was High Commissioner to the UK (1954–61) and later became ambassador to Ireland; from 1964 to 1968 she was a member of the Indian Parliament.

Pandora, in Greek mythology, the first woman. She was created on ZEUS's orders as his revenge on PROMETHEUS, who had created man and stolen fire from heaven for man. Pandora was endowed with charm as well as with guile. She was sent to Prometheus's brother, Epimetheus, and brought with her a box that she had been forbidden to open. When she opened it, all the evils of the human race flew out. Hope remained at the bottom of the box.

Pangaea, name for the single supercontinent that is believed to have existed on Earth 200 million years ago. Theorists have reconstructed existing land masses with their continental shelves to form models of this one supercontinent. *See also* PE pp.*26, 27.*

Pangolin (scaly ant-eater), any of several species of toothless insectivorous mammals, covered with horny overlapping plates, that live in Asia and Africa. It has short, powerful forelegs with which it climbs trees and tears open the nests of tree ants, on which it feeds. Length: to 175cm (70in). Family Manidae; genus *Manis. See also* NW pp.*155,* 161, *161.*

Pānīpat, Battles of, series of battles fought on a strategically located plain in NW India. The Sultan of Delhi was defeated by BABUR in 1526, Akbar defeated the Afghans in 1556 and in 1761 Pānīpat was the scene of the Afghan victory over the MAHRATTAS.

Pankhurst, Dame Christabel (1880–1958), British suffragette, eldest daughter of Emmeline PANKHURST and supporter of her mother in the suffrage movement. She was educated for the Bar, but was refused admission because of her sex. She later became a prominent evangelist.

Pankhurst, Emmeline Goulden ("Emily") (1858–1928), British political leader of women's suffrage movement. With her husband Richard Pankhurst she worked to secure married women's property rights. When she set up the Women's Social and Political Union in 1903, the movement for women's suffrage became militant, and she was imprisoned.

Pan Ku (AD 32–92), Chinese historian. He wrote the *History of the Former Han Dynasty,* which was completed by his sister after his death. It dealt with the previous 200 years and consisted of biographies of emperors and other prominent people, descriptions of major events and cultural history. He also wrote prose poems called *fu.*

Panmunjom, village in N South Korea, just S of the 38th parallel that divides North and South Korea. It was in this village in July 1953 that Communist and UN forces signed the armistice ending the KOREAN WAR.

Panna, third and highest level of the Buddhist life. Also written Prajñā, the word means wisdom. The other two levels are *sila* (morality) and *samadhi* (concentration). The wisdom referred to concerns the direct apprehension of transcendent truths. *See also* MS p.*221.*

Pannage, feudal levy paid by VILLEINS for the right to graze their pigs in woodland areas, especially for acorns. The rates ranged from a penny for a yearling pig to a half-penny for a younger pig.

Panning, method used by prospectors for separating silt, sand and earth from coarse pebbles which may contain minerals such as gold and diamonds. Scooped-up sediments are washed in a pan by agitating in a circular motion.

Pannini, Giovanni Paolo (*c.*1692–1765), Italian painter. He is best known for his views of Rome, some of which were imaginary, combining real buildings with ancient ruins. His work anticipated that of Hubert ROBERT. It includes *Roman Ruins with Preaching Apostle* (1753).

Pan-pipes, or syrinx, primitive musical wind instrument, probably from Asia. Several tubes of cane, reed, bamboo, or clay, of different lengths, are joined together side by side. Blown across one end, each pipe produces one note of a scale. Pan pipes are associated with the pastoral Greek god of fertility.

Pan Slavism, attempt to unite the Slavic peoples. In the early 19th century it was a movement interested in publicizing the common culture of the Slavs, but after 1848 it took on more specific political overtones. In the late 19th century the importance of pan-Slavism lay in that the strongest power in E Europe, Russia, was wholly Slav, and often seemed to aspire to the leadership of the Slavs who lived under the declining OTTOMAN and HAPSBURG empires.

Pansy, common name for a cultivated hybrid VIOLET. It is one of the oldest of European cultivated flowering plants. An annual or short-lived perennial, it has velvety flowers, usually in combinations of blue, yellow and white, with five petals. Height: to about 15–30cm (6–12in). Species *Viola tricolor.*

Pantaloon, elderly male character in the English HARLEQUINADE who was invariably the butt of Clown's jokes. In COMMEDIA DELL'ARTE, as Pantalone, he was represented as an elderly Venetian merchant, amorous, gullible and suspicious and always betrayed by his young wife COLUMBINE.

Pantheism, religious system, contrasted with certain forms of DEISM, that stresses that God (or gods) and the universe are identical. According to this philosophy, all life is infused with divinity. No distinction is recognized between the Creator and creatures. The language of mysticism is frequently pantheistic. John Toland, the deist, originated the term in 1705.

Pantheon, in ancient times, a temple for the worship of all the gods in a specific area. By extension, it was a building honouring illustrious public figures. The most famous example is the Pantheon in Rome, originally built by Agrippa (27 BC), rebuilt by Hadrian (125 AD), and converted into a church in the 7th century. Its dome was the largest built until modern times. The Panthéon in Paris was designed as a church (1764) Jacques Germain SOUFFLOT and was dedicated during the French Revolution to the memory of great Frenchmen.

Panther. *See* LEOPARD.

Pantograph, mechanical device sometimes used by artists and draughtsmen to transfer a design from one sheet of paper to another, at the same or at a different scale. It is constructed as a frame of rods, linked together with adjustable pins, pivoted at one of the linkages and with a pointer at another. A pen or pencil at the further end of the frame traces an exact copy as the pointer is made to follow the lines of the original design. The name pantograph is also given to a similar structure on the roof of an electric locomotive which picks up current from an overhead conductor wire. *See also* SU p.44.

Pantomime, originally a dumb-show performance, in which actors conveyed meaning by symbolic movement, facial expression and gesture, sometimes accompanied by music. Popular in ancient Rome and in 17th and 18th-century Italy and France, it survives in Britain today as a Christmas extravaganza for children, with music and comic actors. *See also* MIME.

Pantothenic acid, VITAMIN of the B complex involved in the synthesis of acetyl COENZYME A.

Panufnik, Andrzej (1914–), Polish composer and conductor. He was conductor of the Birmingham Symphony Orchestra (1954-57). His works include the *Sinfonia rustica* (1948) and the *Sinfonia sacra* (1963).

Panzer, German word for armour, commonly used as an abbreviation of *Panzer kampfwagen,* meaning a tank. German Panzer divisions, consisting of tanks and their support units, were very effective in WWII. The main German tanks of WWII oere the *Panzer Mark IV,* the *Panther* and the *Tiger. See also* MM pp.170–171.

Paolozzi, Eduardo (1924–), Scottish-born sculptor who studied in London and worked in Paris (1947–49). Despite several stylistic transformations, the constant theme in his work is his interest in machinery. Paolozzi created boxy chromium-plated sculptures evocative of jazz-age amusement arcades and picture palaces.

Papacy, office of the pope, the head of the Roman Catholic Church. The pope is nominated Bishop of Rome and Christ's spiritual representative on Earth. He is elected by the College of Cardinals as a successor to St PETER, traditionally regarded as the first Bishop of Rome. Until the REFORMATION the pope claimed authority over all Western Christendom. Since then his claim is over all members of the Roman Catholic Church. In Catholic Doctrine his authority is absolute and he is unable to err in a formal statement, such as a PAPAL BULL, concerning matters of faith or morality.

Papal Bull, official letter from the pope, written in solemn style, usually containing decrees relating to doctrine, CANONIZATION, ecclesiastical discipline, promulgation of INDULGENCES or other matters of general importance. The name is derived from the Latin *bulla,* meaning "seal".

Papal infallibility *See* INFALLIBILITY.

Papal legate, personal representative of the pope sent on a mission. Legates can be of three kinds: *a latere,* nuncios or apostolic delegates. The office of legate *a latere* can now be held only by a cardinal. It was especially important during the Middle Ages.

Papal nuncio, diplomatic representative of the VATICAN, accredited to a civil government. Often of ambassadorial rank, he also acts as a liaison between the PAPACY and the Roman Catholic Church of the state to which he is accredited.

Papal States, independent territory in

Panda or wah, a less familiar relative of the giant black and white panda.

Pandora, portrayed by the Pre-Raphaelite painter Dante Gabriel Rossetti in 1869.

Emmeline Pankhurst was a fine orator and fought to secure the vote for women.

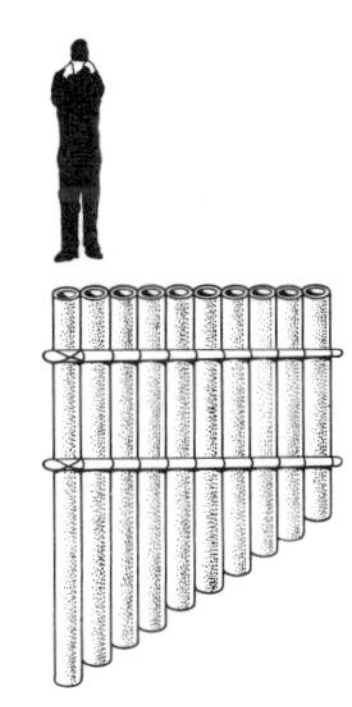

Pan-pipes; the gods turned the nymph Syrinx into the reeds Pan made his pipes from.

Papillon, once called the dwarf spaniel, is named from the French for butterfly.

Paracelsus: he died while under the protection of Archbishop Ernst in Salzburg.

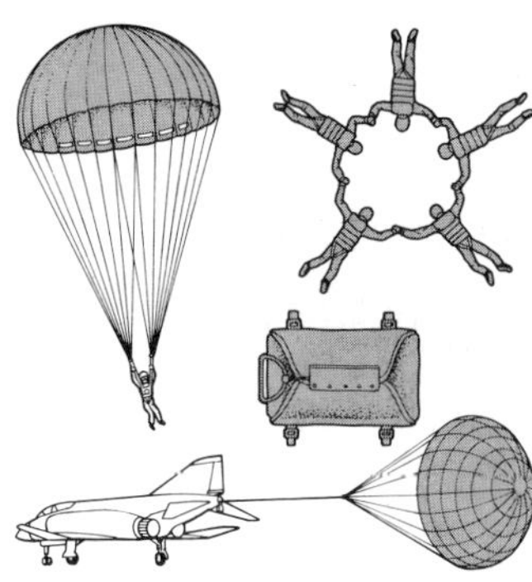

Parachutes enable high-speed aircraft to land in a restricted space.

Paradise fish build nests from bubbles which form a raft on the water.

central Italy under the rule of the popes from 756 to 1870. The territory varied in size at different times but for most of its existence comprised the areas of Italy known today as Lazio, Umbria, Marche and part of Emilia-Romagna. In 756 PEPIN THE SHORT, ruler of the Franks, granted the exarchate of Ravenna to Pope STEPHEN II, thereby establishing the pope's temporal power. Because of the rising power of the communes in Italy, the Babylonian Captivity (1309–77) and the GREAT SCHISM (1377–1417), the control of the PAPACY was weakened in the Middle Ages, never to recover fully. Annexed by France for most of the period between 1798 and 1809, the Papal States were restored in 1815, only to be annexed by Italian forces in the unification movement, Rome falling in 1870. The VATICAN CITY, last bastion of the pope's temporal power, was finally established in its present form by the LATERAN TREATY in 1929. *See also* HC1 pp.*134*, 161, *161*; HC2 p.181.

Papandreou, Georgios (1888–1968), Greek politician. During WWII he led a government-in-exile (1944–45) and after the war he served in several Social Democratic cabinets. He formed the Centre Union Party (1961) and became Prime Minister (1964–65) when his party won the 1964 elections. After the military coup of 1967 he was placed under house arrest.

Papaw. *See* PAPAYA.

Papaya, also sometimes called pawpaw, palm-like tree widely cultivated in tropical America for its fleshy, melon-like, edible fruit. Height: to 6m (20ft). Family Caricaceae; species *Carica papaya*. *See also* PE p.*194*.

Papeete, port town and capital of Tahiti and French Polynesia, on the NW coast of Tahiti in the Society Islands, S Pacific Ocean. It is a trade centre of the islands and the site of an international airport. Its exports include copra, mother-of-pearl and vanilla. Pop. (1971 est.) 24,000.

Papen, Franz von (1879–1969), German politician. He was a member of the Prussian parliament (1921–32) and was Chancellor of Germany in 1932. He lifted the ban on the Nazi militia and in 1933 helped Adolf HITLER to become chancellor, acting briefly as his vice-chancellor. In 1946 he was acquitted by the war crimes tribunal at the NUREMBERG TRIALS. *See also* HC2 p.209.

Paper, sheet or roll of compacted cellulose fibres with a wide range of uses including packaging, writing, wall covering and clothing. The word "paper" derives from papyrus, the plant that the Egyptians used at least 5,500 years ago to make sheets of writing material. The papyrus reed was soaked, slit into strips, laid at right-angles and pounded and pressed into a sheet. The modern process of manufacture originated about 2,000 years ago in China and consists of reducing wood fibre, straw, rags and grasses to a pulp by the action of an alkali, such as caustic soda. The lignin and other non-cellulose material is then extracted and the residue is bleached. After washing and the addition of a filler to provide a smooth and flat surface, the pulp is rolled into thin sheets and dried. Papers of better quality are made from chemical pulps prepared in this way, but newsprint and other cheap papers are mechanical pulps made without chemical treatment and consist of finely-divided wood . *See also* MM pp.60–61.

Paperback, book published in paper covers and usually costing much less than any hardback (stiff-covered) counterpart. Various claims for several 19th-century originals have been made but the so-called Paperback Revolution was started by Allen Lane's Penguin Books in 1935. Developments in printing technology since WWII, especially the growth of large offset presses, have enabled publishers to maintain relatively long print runs and keep down the cost of paperbacks.

Paper money, or bank-note, printed note issued by a bank for a sum of money which it promises to pay to the bearer on demand. Bank-notes originated in England as receipts issued by goldsmiths for money deposited with them, the first being substitutes (or receipts) for money.

Their value depended on their being exchanged for gold. It was not until 1931 that they became money in their own right, ie inconvertible. *See also* HC1 p.*126*; MS pp.308–309.

Paper nautilus. *See* ARGONAUT.

Papier-mâché, 18th-century French art-material made from paper strips soaked in starch of flour and water and pressed into forms which were then painted. The British adopted the technique to produce a thin paper-board to make trays and mouldings which were popular in Victorian times. It is still widely used in the production of decorative objects.

Papilla, small nipple-like protuberance found in mammals and plants. In mammals papillae project into the EPIDERMIS from the DERMIS; they are found in many parts of the body including the mammalian inner EAR and on the surface of the TONGUE. *See also* MS p.54, *54*.

Papillon, small dog developed from a 16th-century dwarf spaniel. It has a small head with large butterfly-like ears and a thin muzzle. The straight-backed body is set on slender legs, and the long, well-arched tail is carried over the back. The long, silky coat is white with darker markings. Height: 28cm (11in) at the shoulder.

Papin, Denis (1647–*c*.1712), French physicist and inventor who helped to develop the steam ENGINE. He invented a steam digester – a pressurized container which demonstrated that increased pressure raises the boiling point of a liquid.

Papineau, Louis Joseph (1786–1871), French-Canadian political figure and insurgent. After a military and legal career, he was elected a member of the Quebec legislature in 1808 and became its Speaker (1815–37). His hostility to the British government in Canada incited his followers to rebellion (1837). With its failure he fled to the USA and then to France (1839). He was granted a full amnesty (1847) and returned to Canadian political life as a member of the House of Commons (1848–54), but he was unable to recapture his former influence.

Papp, Joseph (1921–), US stage director and producer. He founded the New York Shakespeare Festival in 1954 and has given many free performances of Shakespeare's plays. He produced Jason Miller's Broadway play, *That Championship Season*.

Papp, László (1926–), Hungarian boxer. A southpaw, he won three Olympic gold medals (1948 middleweight; 1952, 1956 light-middleweight). In 1962, as the first professional boxer from a Communist country, he won the European middleweight title and retained it until 1965.

Pappus of Alexandria (*fl.* AD 320), mathematician whose commentaries on EUCLID and PTOLEMY provide important sources for their work. He also wrote a handbook to Greek mathematical sciences, *The Collection*, and extended PYTHAGORAS' theorem to apply to any triangle.

Paprika, popular condiment, a red powder ground from the fruit of some species of CHILI pepper that grow in central Europe. It is particularly associated with Hungarian cuisine.

Papua New Guinea, independent nation of the W Pacific made up of the E half of NEW GUINEA and the neighbouring islands of the Bismarck Archipelago and Bougainville. The land is largely mountainous or covered with tropical rain forest. Most farming is at a subsistence level, but coffee, copra and cocoa are produced as cash crops. There are deposits of copper, natural gas, gold and silver. The capital is Port Moresby. Area: 475,369sq km (183,540sq miles). Pop. (1975 est.) 2,756,000. *See* MW p.141.

Papuans, racially and culturally diverse aboriginal inhabitants of Papua New Guinea. The aboriginal highlanders live chiefly by subsistence agriculture; crops include taro and sweet potatoes, with coffee and copra grown as cash crops.

Papyrus, stout perennial water plant, native to S Europe, N Africa and the Middle East, and used by the ancient Egyptians to make a paper-like writing

material. Strips of the stem were arranged in layers, crushed and hammered to form a loosely textured porous kind of paper. Height: to 4.5m (15ft). Family Cyperaceae; species *Cyperus papyrus*. *See also* MM p.*60*.

Parable, short, simple story intended to convey a moral or religious message. Parables differ from allegories in that parables deal with events that might reasonably happen in nature and often feature humans rather than the animals of fables. The best-known parables are those of JESUS CHRIST in the NEW TESTAMENT, including the Good Samaritan, the Prodigal Son, the Sower and the Seed, and the Pearl of Great Price.

Parabola, mathematical curve, a CONIC SECTION traced by a point which moves so that its distance from a fixed point, the focus, is equal to its distance from a fixed straight line, the directrix. It is known as a conic section because it may be formed by cutting a cone parallel to one side. The general equation of a parabola is $y^2 = 4ax$. *See also* SU p.36, *36*.

Paraboloid, mathematical solid figure in which all sections parallel to the axis of symmetry are PARABOLAS, and sections at an angle to this axis are other CONIC SECTIONS: ELLIPSES, HYPERBOLAS or circles.

Paracas, site on the Paracas peninsula of Peru of an Indian culture in three phases: Cavernas, Pinilla and Necropolis (900 BC–AD 400). The Cavernas people produced crude pottery; the Pinilla and Necropolis were noted for their fine pottery and textiles.

Paracelsus (*c*.1493–1541), Swiss physician and alchemist, real name Philippus Aureolus Theophrastus Bombast von Hohenheim. He became professor of medicine at Basel, where his lectures discredited past and contemporary medicine and were preceded by the burning of the works of GALEN and AVICENNA. According to Paracelsus, the body of man is primarily composed of salt, sulphur and mercury, and it is the separation of these elements that causes illness. He introduced mineral baths and made opium, mercury, lead and various minerals part of the pharmacopoeia. According to his enemies, he died in Salzburg after a drunken bout. Others say he was thrown down a hill by hirelings of jealous apothecaries.

Parachute, folding umbrella-shaped device made of a light fabric such as nylon or terylene (dacron), used to reduce the speed of a falling body. It consists of a canopy connected by lines to a harness which holds the jumper or cargo. Before release all but the harness is packed in a small back, seat, or chest pack. The unit opens at the pull of a ripcord. Parachutes are also used as brakes to shorten the landing run of high-speed aircraft and on drag-racing cars.

Parachuting, or sky-diving, sport in which parachutists jump from approx. 3,660m (12,000ft) and free-fall to about 670m (2,200ft) before opening their parachutes. Competition parachuting has events based on accuracy, in which the parachutist "steers" himself as close as possible to a target area, and events based on style, in which a series of manoeuvres are performed during the free-fall. Other events are free-fall relative work by two or more parachutists; para-gliding, using parachutes like wings; and para-ascending, where the parachutist is lifted into the air by a towing vehicle.

Paradise, in mythology and religion, a place of perfect happiness and contentment where all is beautiful and unblemished. In some cultures it is synonymous with HEAVEN. Christianity also associates it with the Garden of EDEN before ADAM and EVE were cast out.

Paradise fish, small, hardy freshwater fish, marked with red and blue bands. Found in S China, it was the first tropical aquarium fish to be introduced to Europe (1869). Family Anabartidae; species *Macropodus opercularis*.

Paradise Lost (1667), epic poem by John MILTON in BLANK VERSE. It originally consisted of 10 books but the revised edition of 1674 has 12. It tells the story of Satan's rebellion against God and of the Fall of

Man and is Milton's attempt to justify the existence of evil in this world. *See also* HC1 pp.282–83, *282–83*.

Paradise Regained (1671), epic poem in BLANK VERSE by John MILTON, a sequel to *Paradise Lost*. Christ resists Satan's temptations in the wilderness and thus creates a new spiritual paradise for man. *See also* HC2 pp.282–283, *282–283*.

Paradox, self-contradictory or absurd statement which conflicts with preconceived notions of what is reasonable or possible, but which is significant when considered from the appropriate viewpoint. A well-known paradox is that of Zeno of Elea, who reasoned that motion is impossible.

Paraesthesia, pricking or tingling sensation on the skin associated with injury or irritation of a sensory nerve.

Paraguay, landlocked republic in E South America. The economy is predominantly agricultural, the rearing of cattle being the most important activity. The principal exports are meat, timber, cotton, coffee and tobacco. Exports are transported through Argentina by river. The capital is Asunción. Area: 406,752sq km (157,047sq miles). Pop. (1976 est.) 2,742,000. *See* MW p.141.

Paraguay, river in S central South America. It rises in SW Brazil, flows S to form part of the Brazil-Paraguay and Paraguay-Argentina border and empties into the River Paraná at the SW corner of Paraguay. Its chief tributaries are the Pilcomayo and Bermejo rivers; Asunción is the chief port. Length: approx. 2,550km (1,584 miles).

Parakeet. *See* PARROT.

Paraldehyde, colourless liquid ($C_6H_{12}O_3$) formed by the POLYMERIZATION OF ACETAL-DEHYDE with sulphuric acid. It regenerates acetaldehyde on heating with dilute acids. It has a pleasant odour but disagreeable taste, and is used mainly as a HYPNOTIC DRUG.

Parallax, apparent change in position of an object, seen against a remote background, when the viewpoint is changed. The parallax of a star (annual parallax) is the angle subtended at the star by the mean radius of the Earth's orbit (one astronomical unit); the smaller the angle, the more distant the star. *See also* PARSEC; SU pp.166 167, *166*.

Parallelogram, quadrilateral (four-sided plane figure) having both pairs of opposite sides parallel. Both the opposite sides and the opposite angles are equal. Its area is the product of one side and its perpendicular distance from the opposite side. A parallelogram with all four sides equal is called a RHOMBUS. *See also* SU p.*45*.

Paralysis, loss of voluntary movement in part of the body. It may result from injury, infection, poisoning, a tumour, damage to nerve cells or hysteria. Temporary paralysis is treated by removing the cause; permanent paralysis may be relieved by physiotherapy or surgery.

Paramagnetism, type of MAGNETISM displayed by such metals as platinum and magnesium, which, when placed in a magnetic field, are magnetized parallel to the field to an extent proportional to the strength of the field. This effect is much weaker than that in ferromagnetic materials such as iron, cobalt and nickel.

Paramaribo, port and capital of SURINAM, on the Surinam River, 21 km (13 miles) from the Atlantic Ocean. It was founded in 1640 as a British colony and was held intermittently by the British and the Dutch until 1816, when the Dutch finally took control. The city has a system of canals similar to those in The Netherlands. Exports include bauxite, sugar cane, rice, rum, coffee and cacao. Pop. (1971) 102,297.

Paramecium, freshwater, ciliated PROTOZOAN characterized by its streamlined "slipper" shape, defined front and rear ends, an oral groove for feeding, food vacuoles for digestion, an anal pore for elimination and two nuclei. Its stiff outer covering is studded with short hair-like cilia. Order Holotricha; species include *Paramecium bursaria* and *Paramecium aurelia*. *See also* NW pp.80, 81.

Paramnesia, in clinical psychology and psychiatry, distortions of memory in which fact and fantasy become merged.

Paramount Pictures Corporation, US film production company formed in 1927. Early films included *The Ten Commandments* (1923) and the "Road" series. Paramount became part of Gulf and Western in 1966 and, although it had expensive failures in the late 1960s, *Love Story* (1970) and *The Godfather* (1971) broke box-office records.

Paraná, river in SE central South America. It rises in SE Brazil, flows s into Argentina, forming the SE and S border of Paraguay and joins the Uruguay River to form the Río de la Plata. It is an important route for inland communications. Length: 2,941 km (1,827 miles).

Paranoia, term in psychology for a disorder characterized by a systematically held, persistent delusion, usually of persecution or grandeur. It was coined by the German psychologist Karl Kahlbaum in 1863. A severe paranoia may be PSYCHOTIC or SCHIZOPHRENIC and it leads its victim to adopt DEFENCE MECHANISMS.

Paranthropus, fossil PRIMATE discovered by the British anthropologists Dr and Mrs Louis LEAKEY in the Olduvai Gorge, Tanzania, and now classified as a type of AUSTRALOPITHECUS. *See also* pp.22–23.

Paraplegia, PARALYSIS of both lower limbs. It may be caused by injury to or disease of the spinal cord or by brain disorders.

Paraplegic Games, annual sports tournament for the semi-paralysed, officially named the International Stoke Mandeville Games, after the English hospital in Buckinghamshire, where they originated in 1948 with an archery competition. When possible, every fourth year they are held in the country staging the OLYMPIC GAMES, and there are also British Commonwealth Paraplegic Games to coincide with the COMMONWEALTH GAMES. Sports contested are archery, swimming, fencing, bowls, table tennis, snooker, weightlifting, throwing the javelin, discus, and hammer and sometimes basketball.

Parapsychology, scientific research into extra-sensory perception, such as TELEPATHY, CLAIRVOYANCE and PRECOGNITION. It developed from PSYCHICAL RESEARCH initiated in England in 1882 by a group of intellectuals who set about sifting and investigating reported paranormal events. In the 1950s sophisticated experiments began, using electronic equipment, and by the end of the 1960s parapsychology had become an accepted scientific study. *See also* MS pp.198–199.

Parasite, any organism that benefits at the expense of another or depends on it totally for its existence; usually the host is not destroyed by this relationship. Parasites occur in many groups of plants and in virtually all major animal groups. A parasite that lives within the host is called an endoparasite; a parasite that subsists on the host's exterior is an ectoparasite. Many parasitic PROTOZOANS, FLEAS, WORMS, insects and other arthropods can carry diseases such as MALARIA, ELEPHANTIASIS and amoebic DYSENTERY, or cause exterior sores or lesions which may then become infected. The CUCKOO and COWBIRD rely on other birds to rear their young, and are parasites. In parasitoidism, the relationship results in the death of the host, eg various flying insects lay their eggs on or in a host which becomes the food for the insect larvae. A hyperparasite is one which parasitizes another parasite. *See also* MS pp.84–85, 93, 106–107; NW p.90.

Parasitology, branch of biology that deals with the study of plant and animal PARASITES, organisms that depend on other organisms for their existence. It has a number of study areas such as veterinary, medical and agricultural parasitology.

Parasol mushroom, any of a family of common terrestrial fungi (Lepiotaceae) that have shaggy umbrella-shaped caps, especially *Lepiota procera*, which is considered one of the best edible species. Some parasols are poisonous and many resemble AMANITAS. *See also* NW p.43.

Parathormone, HORMONE secreted by the PARATHYROID GLAND. It helps to regulate calcium levels in the blood by promoting the release of calcium from the bones.

Parathyroid gland, any of the four small endocrine glands, usually embedded in the back of the THYROID GLAND, that secrete the HORMONE parathormone, which controls the amount of calcium and phosphorus in the blood. Abnormalities of this endocrine gland can usually be treated with parathyroid extract and regulation of calcium and phosphorus intake. *See also* MS pp.64–65.

Paratyphoid fever, infection similar to TYPHOID FEVER but milder and of shorter duration. It is caused by various species of Salmonella BACTERIA (typhoid is caused by *Salmonella typhi*) which are carried in infected food and water.

Parcae, in Roman mythology, the three FATES, Nona, Decuma and Morta, who were sometimes said to be the daughters of Jupiter (Zeus in Greek). In Greek mythology they were the Moirae, named Clotho, Lachesis and Atropos.

Parchment, processed skins of animals such as sheep, goats and calves used as writing material; it was invented in the 2nd century BC, and is named after the ancient city of PERGAMUM. When made from calf or kid skin, parchment is called vellum. Today, "parchment" and "vellum" generally refer to high-quality PAPER made from wood pulp and rags.

Paré, Ambroise (1517–90), French physician regarded by some as the founder of modern surgery. In 1537 he was employed as an army surgeon and in 1552 became surgeon to Henry II, one of four French kings he served during his lifetime. As army surgeon he introduced new methods of treating wounds, described in his book *The Method of Treating Wounds Made by Harquebuses and Other Guns* (1545), and revived the practice of tying arteries during surgery instead of cauterizing them.

Pareja, Juan de (*c*. 1605–*c*. 1670), Spanish painter, sometimes known as El Esclavo (the slave) because he was a mulatto. Pareja studied under VELÁZQUEZ, who painted a greatly admired portrait of him during their visit to Rome (1649–51).

Parenchyma, soft tissue made up of nonspecialized, thin-walled cells. It is one of the chief tissues of plant stems, leaves and fruit pulp; it stores nutrients and water, and helps to support plants. *See also* PHLOEM.

Paresis, tertiary SYPHILIS of the central nervous system which particularly affects the cerebral hemispheres of the brain; the condition is also called GPI or general paralysis of the insane. Untreated it causes gross deterioration of the personality, followed by progressive paralysis and death. It is much less common since the advent of PENICILLIN, which is an effective treatment for syphilis. Paresis can also mean slight or temporary paralysis. *See also* VENEREAL DISEASE.

Pareto, Vilfredo (1848–1923), Italian economist and sociologist. He studied the application of mathematics to economic theory and wrote *Manual of Political Economy* (1906). As a sociological theorist he concentrated on ruling élites and wrote *Mind and Society* (1916). *See also* MS p.*265*.

Pariah, member of a low CASTE in the social order of S India whose touch pollutes those of high caste. In a broader sense, a pariah is any despised person, outside an approved social group.

Parietal bone, large flat bone on each side of the SKULL which forms the vault of the cranium. It joins with the frontal and occipital bones.

Paris, also called Alexandros, in Greek legend, the son of PRIAM and HECUBA. He was abandoned on Mt Ida soon after his birth, rescued by a shepherd and later had a son by Oenone. With APHRODITE's help he abducted HELEN, causing the TROJAN WAR, in which he slew ACHILLES. Oenone at first refused to heal her husband when he was wounded and, relenting too late, committed suicide after his death.

Paris, Matthew (*d*. 1259), also known as Matthew of Paris, English monk and chronicler. He became the historiographer of the monastery of St Albans on the death of Roger of Wendover in 1236. His *Chronica majora* (*Great Chronicle*) is

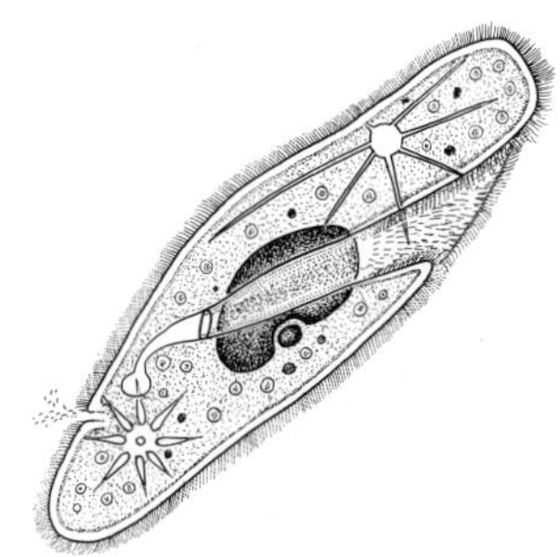

Paramecium beat their cilia to direct food particles into their mouths.

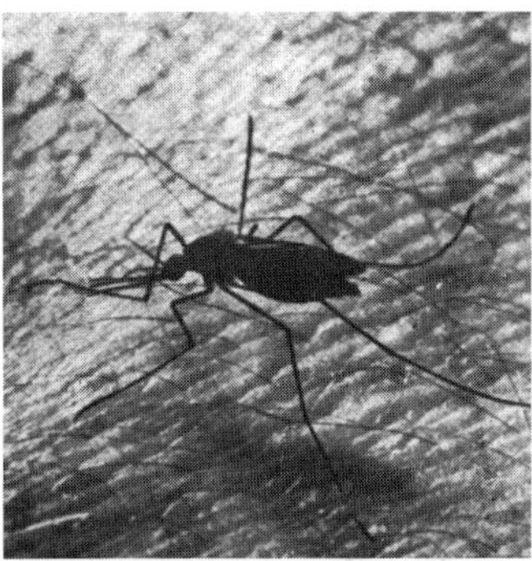

Parasite: the malaria mosquito settles and feeds on a victim's arm.

Vilfredo Pareto examined the way science can be applied to study human behaviour.

Paris: in Rubens' *The Judgement of Paris* he awards the golden apple to Aphrodite.

Paris

Paris: the Arc de Triomphe, begun 1806, stands 50m high in the Place de L'Étoile.

Dorothy Parker was famous for her sharp wit and liberated life-style.

Parliament; a view of the chamber of the House of Commons from the public gallery.

Parmigianino strove for a "grace" beyond that of nature: *Portrait of a Young Woman.*

one of the best sources for the history of Europe from the year 1235 to 1259.

Paris, capital of FRANCE, on the River Seine 375km (233 miles) from its mouth on the English Channel. The city proper consists of the Paris département, Ville-de-Paris, and its suburbs lie in départements of Seine-St-Denis, Val-de-Marne, Hauts-de-Seine, Val-d'Oise, Yvelines and Essone. When the Romans took Paris in 52 BC it was a small village on the Île de la Cité on the Seine. Under the Romans it grew as an administrative centre. The city was the capital of the Merovingian Franks in the 5th century until 584. It was re-established as the French capital by Capetian kings in the 10th century. During the 14th century Paris rebelled against the Crown and declared itself an independent commune. The city suffered from civil disorder during the HUNDRED YEARS WAR. In the reign of LOUIS XIII Cardinal RICHELIEU established Paris as the cultural and political centre of Europe. The outbreak of the FRENCH REVOLUTION focused on the BASTILLE, which was stormed by crowds in 1789. Under the Bonapartes Paris assumed its modern form, especially during the reign of NAPOLEON III when Baron Haussmann was commissioned to plan the boulevards and parks. Although occupied in the Franco-Prussian War (1870–71) and WWII, the city was not badly damaged. Landmarks include the Cathédrale de Notre-Dame, Musée du LOUVRE, Les Invalides, Palais de Justice, Palais de Luxembourg, l'Opéra, Panthéon, Bibliothèque Nationale, Jardin des Plantes, Arc de Triomphe, Palais de l'Élysée, La Madeleine, St-Germain-des-Prés, Palais de la Bourse, La Sorbonne and Tour EIFFEL. Paris still remains the hub of France despite recent attempts at decentralization and retains its importance as a cultural, commercial, and communications centre of Europe. Industries: car, marine and railway engineering, chemicals, textiles, clothing, printing and publishing, luxury goods. Pop. (city proper, 1975) 2,289,800; (metropolitan area) 9,690,000. *See also* HC1 pp.188, 202, 262, *263,* 308–*309;* HC2 pp.38, 42–43, *106, 109,* 123, 171, *173,* 178, *180, 259;* MW p.77.

Paris, Congress of (1856), conference held to determine the peace terms after the CRIMEAN WAR. The Black Sea was made neutral water, open to all merchant marines and closed to warships. Moldavia and Wallachia were given semi-independence from Turkey.

Paris, School of, name given to 20th-century movements of painting which were focused in Paris, irrespective of the nationality of the artists. The school includes most of the modern masters.

Paris, Treaties of, several peace agreements negotiated and signed at Paris. They include three truces in the 13th and early 14th centuries between France and England (1259, 1286 and 1303), two preliminary treaties (1814, 1815) before the Congress of VIENNA and the treaty which ended the CRIMEAN WAR (1856). They also include the treaties made between the Allies and Bulgaria, Finland, Hungary, Italy and Romania in 1946. The three most important such treaties in British history were the PEACE OF PARIS (1763), the PEACE OF PARIS (1783), and the Treaty of VERSAILLES (1919) which resulted from the Paris Peace Conference at the end of WWI. *See also* HC2 pp.85, 192–193, *258.*

Parish, in the Church of England, geographical area under the pastoral care of a clergyman, the "incumbent", usually a vicar or rector. It is a subdivision of a diocese, which is governed by a bishop.

Parity, in physics, term used to denote space-reflection symmetry. The principle of conservation of parity states that physical laws are the same in a left- and right-handed co-ordinate system. This was regarded as inviolable until 1956, when Chen Ning YANG and Tsung-Dao LEE showed that it was transgressed by certain interactions between elementary atomic particles. The term parity is also used in information theory to denote a coding method employed in message transmission to detect errors.

Park, Chung Hee (1917–), President of South Korea (1963–). He seized power in a military coup in 1961, ruled for two years as a general, then resigned from the army. He was elected President in 1963, 1967 and 1971.

Park, Mungo (1771–1806), British explorer. After working as a medical officer on a ship engaged in the East India trade he was asked by the African Association to explore the course of the River Niger. He explored 451km (280 miles) of the river, returning in 1805 to trace the river to its mouth, but was attacked by local people and killed.

Parker, Charles Christopher ("Charlie") (1920–55), US jazz alto saxophonist, nicknamed "Bird". He recorded with Dizzy GILLESPIE in the 1940s and was the most important of the founders of the jazz style called "be-bop" (now called "bop"), which opened up new rhythmic and harmonic possibilities for musicians. *See also* HC2 p.272.

Parker, Dorothy (1893–1967), US author. A drama and book critic, the first of her three collections of satirical verse, *Enough Rope,* appeared in 1926. Her fame rests largely on her short stories satirizing aspects of modern life. Collections include *Laments for the Living* (1930) and *Here Lies* (1939).

Parker, Matthew (1504–75), Archbishop of Canterbury (1559–75). He was chaplain to Anne BOLEYN in 1535, Dean of Lincoln (1552–53), and a moderate in church matters. After ELIZABETH I had appointed him archbishop, he revised the 42 articles to 39 in 1562, and published the "Bishops' Bible" (1572).

Parkes, Sir Henry (1815–96), Australian politician, *b.* England. He emigrated to Australia in 1839. He was five times the Premier of New South Wales between 1872 and 1891. He worked for Australian federation, the ending of transportation of convicts and improvements in education. His latter efforts led to the Public Schools Act of 1866 which, together with legislation in 1880, resulted in compulsory free education.

Parkinson, Cyril Northcote (1909–), British historian and satirist who in 1958 proposed the "law" known as Parkinson's law relating to administration and business practice. It holds that work expands to fill the time available and, consequently, workers increase in numbers even though output remains unchanged.

Parkinson, James (1755–1824), British surgeon and palaeontologist who described PARKINSON'S DISEASE in 1817 and was the first to recognize that perforation of the appendix could be fatal. His many works include *Organic Remains of a Former World* (3 vols, 1804–11).

Parkinson's disease, chronic, progressive nervous disease occurring mostly in older males. Tremors, muscle weakness and fixed facial expression are characteristics of the disorder. The cause is not known, but viruses have been implicated. No cure exists, but physiotherapy and a drug, L-DOPA, may relieve the symptoms. *See also* MS p.97.

Parlements, French courts of the ANCIEN RÉGIME, one at Paris and eventually 12 in the provinces. The provincial parlements were administrative bodies which carried out royal edicts. The Paris parlement, in addition, was the supreme court. They were abolished in 1790.

Parliament, legislative assembly that includes elected members and acts as a debating forum for political affairs. The parliament on which most other parliamentary systems in the world are based is the British Parliament. It emerged in the late 13th century as an extension of the king's council, and has been housed at Westminster since that time. It is the supreme power in the country. Parliament comprises the monarch, for whom members of the government act as servants, and two Houses: the HOUSE OF LORDS, an upper chamber of hereditary and life peers, bishops and law lords, has seen its power effectively limited in the 20th century to being the supreme court in the country and to delaying and altering legislation initiated by the lower house,

the HOUSE OF COMMONS. There are 635 members of the Commons (MPS), elected by constituencies with universal adult suffrage. There is a maximum of five years between elections. Since the 19th century Parliament has been dominated by a two-party system, although there are also members of the other parties and independent members, especially in the Lords. Generally the party that commands the most votes in the Commons forms the government; the second-largest party forms "Her Majesty's Loyal Opposition". The Prime Minister and his Cabinet are almost always MPS (although not necessarily members of the Commons); they are answerable to Parliament for their policies, and the activities and finances of their departments are subject to parliamentary scrutiny. Legislation can be introduced by the government or by private members; after passing three readings in the Commons, a BILL passes to the Lords for three further readings. After it has received the royal assent it becomes an ACT OF PARLIAMENT. *See also* HC1 pp.210–211, 286–293; HC2 pp.68–69, 160–161, 236–239; MS pp.286–287.

Parliament, Houses of. *See* HOUSE OF COMMONS; HOUSE OF LORDS.

Parliament Act (1911), Act passed by the British Parliament to overcome the opposition of the House of Lords to BILLS passed in the Commons. It arose from the Lords' rejection of the Liberal budget of 1909. Money bills were to become law within a month of being sent to the Lords, the peers' opposition notwithstanding. Other public bills passed by the Commons in three successive sessions were also to become law. In 1949 the Lords' delaying power for public bills was reduced to one year.

Parliamentarians, people who sided with Parliament against CHARLES I before and during the ENGLISH CIVIL WAR. In the main they were drawn from the Puritan gentry and squirearchy. They wished to restore Parliament's position after 11 years of non-parliamentary government (1629–40), and to gain control over the king's ministers, the Church and the army. Their most prominent leaders were John PYM and Oliver CROMWELL. *See also* HC1 pp.286–291.

Parliamentary government, various systems of government which have in common an elected assembly of representatives of the community. In Britain the executive, judicial and legislative parts of government are all present in PARLIAMENT. The Crown is the executive (although it delegates its power to its ministers), the House of Lords is the high court of the realm and the House of Commons is the chief legislative organ. In the USA the executive (President), judiciary (Supreme Court) and the legislative (Congress) are separated. *See also* MS pp.276–279.

Parma, Alessandro Farnese, Duke of (1545–92), Spanish general. With JOHN OF AUSTRIA he led the Spanish suppression of the revolt of The Netherlands in 1578, and a year later was appointed governor of The Netherlands.

Parma, School of, Italian school of painting which flourished during the 15th and 16th centuries. Its most influential painters were CORREGGIO and his pupil PARMIGIANINO.

Parmenides (*b.c.* 515 BC), founder of the Eleatic school of Greek philosophy. His main extant work is a poem, *Nature.* Developing the concept of "being" in contrast to Heracleitus' concept of "becoming", he taught that being is eternal, indivisible and immutable. *See also* ELEATICISM; HERACLITUS.

Parmigianino, or Parmigiano, II (1503–40), real name Francesco Mazzola, northern Italian artist, a master of MANNERISM. He worked in Rome, Bologna and Parma. One of the first Italian painters who was also an important etcher, he is also famous for his drawings. His works include *Madonna with St Zachary* (*c.*1530), *Vision of St Jerome* (*c.*1527) and *Madonna with the Long Neck* (*c.*1535). *See also* HC1 p.264.

Parnassians, school of mid-19th-century French poets. They opposed the excessive emotional aspect of Romanticism and sought to achieve meticulous precision in technique. The leader of the school was Leconte de Lisle and members included Charles BAUDELAIRE, François COPPÉE and SULLY-PRUDHOMME. Their poetry appeared in *Le Parnasse contemporain* (1866, 1871 and 1876).

Parnassus, mountain peak in central Greece. In ancient times, it was considered sacred to Apollo, Dionysus and the Muses and was the site of the sacred Castalian spring, which lies just above DELPHI, at the s foot of the mountain. The Corycian Cave, associated with the Bacchic festivals, lies on a plateau between Delphi and the summit. Height: 2,457m (8,061ft).

Parnell, Charles Stewart (1846–91), Irish nationalist. He entered the British Parliament in 1875, vigorously supporting home rule for Ireland and rapidly taking over leadership of the Irish group in the Commons. To gain concessions, he embarked upon a policy of parliamentary obstruction. He was imprisoned (1881–82), but his power reached its peak in 1886 when GLADSTONE introduced the Home Rule Bill. After the bill was defeated, there were attempts to show that Parnell had been implicated in the PHOENIX PARK MURDERS; and although these allegations were shown to be false in 1889, the proof in 1890 of Parnell's adultery with the wife of one of his supporters led to a rapid decline in his influence. *See also* HC2 p.162.

Parnell, Thomas (1679–1718), Irish poet who was a friend of Alexander POPE, Jonathan SWIFT and John GAY, and became a member of the Scriblerus Club. He contributed to the *Spectator*, and assisted Pope in his translation of the *Iliad*. His poems, including *Night Piece on Death* and *Hymn to Contentment*, were published posthumously by Alexander Pope. *See also* HC2 p.72.

Parochial Church Council, administrative body governing a parish of the Church of England. Such councils were established in 1919 and for the first time gave the laity a real chance to participate in parochial administration. At the annual general meeting, presided over by the incumbent, the congregation elects lay council members for the following year (who meet usually once a month), elects one of two churchwardens, and discusses the financing of the church and any other general need.

Parody, literary composition in which another author's language and style are imitated and exaggerated for comic effect. In ancient Greece, ARISTOPHANES wrote parodies of AESCHYLUS and EURIPIDES. In England, Henry FIELDING's *Joseph Andrews* (1742) was a successful parody of Samuel RICHARDSON's novel *Pamela, or Virtue Rewarded*. Among 20th-century writers who have made effective use of parody are Max BEERBOHM, James JOYCE and Stephen LEACOCK.

Parotid gland, largest of the SALIVARY GLANDS which form and secrete saliva. It is located just in front of and a little below the opening of the ear. It is the gland that becomes swollen during MUMPS. *See also* MS p.101.

Parr, Catherine (1512–48), queen consort and sixth wife of HENRY VIII. She married her fourth husband, Thomas SEYMOUR, after the king's death in 1547. As queen, she tried to alleviate the persecution of Catholics and by her intercession the princesses Mary and Elizabeth were reinstated at court.

Parr, Thomas (Old Parr) (*d.*1635), Englishman who is alleged to have lived for 152 years. He died while visiting the court of CHARLES I, where he was introduced by the Earl of Arundel.

Parramatta, city in E New South Wales, Australia, on the Parramatta River. It is a w suburb of Sydney and a manufacturing centre, producing textiles, machinery and vehicle components. It was the first place to be settled in the expansion of the original settlement of Sydney Cove. Pop. (1971) 110,717.

Parrhasius (*fl.*5th century BC), Greek painter who worked in Athens and was famed for his studies of proportion. Particularly notable were two paintings, one of which depicted Theseus and the other the Demos, personifying the people of Athens.

Parrot, common name for many tropical and subtropical birds that are popular as pets. Parrots are brightly coloured and have thick, hooked bills. They include BUDGERIGARS, MACAWS, LORIES, LORIKEETS, PARAKEETS, KEAS, KAKAPOS and others. In the wild all nest in tree holes, rock cracks or on the ground. Pet parrots should be kept in clean, warm, large cages and provided with fresh air, water and proper food. Some can mimic speech. Length: 7.5–90cm (3in–3ft). Family Psittacidae. *See also* NW p.71, 205, 211, 212.

Parrot fever. *See* PSITTACOSIS.

Parrot fish, any of the 80 species of marine fish of tropical Atlantic and Indo-Pacific oceans, identified by fused teeth resembling a parrot's beak. A coral-eating fish that goes through many colour changes, it builds a mucous cocoon for sleeping. Length: 11.3–120cm (4.5in–4ft). Family Scaridae. *See also* NW p.128.

Parry, Sir Charles Hubert Hastings (1848–1918), British composer. His mastery of choral music is best shown in *Blest Pair of Sirens* (1887), a setting of an ode by John MILTON. He is well known for his *Jerusalem* (1916), a unison song based on William BLAKE's poem. He also wrote many songs, five symphonies and an opera. *See also* HC2 p.106.

Parry, John (1710–82), Welsh harpist and music collector, nicknamed Blind Parry. He was harpist to Sir Watkin Williams Wynne of Wynnastay for most of his life. With Evan Williams he edited the earliest published collection of Welsh music, *Ancient British Music* (1742).

Parry, Joseph (1841–1903), Welsh composer. He wrote many hymn tunes, as well as oratorios and cantatas, and in 1896 was honoured by the Llandudno Eisteddfod for his services to Welsh music.

Parry, Sir William Edward (1790–1855), British explorer and navigator. In 1810 he was sent by the Royal Navy to the Arctic to protect the whale population. He conducted five expeditions in the region between 1818 and 1827, on the last of which he attempted unsuccessfully to reach the North Pole.

Parsec, astronomical unit of length (symbol pc) equal to the distance at which the radius of the Earth's orbit subtends an angle of one second. It is thus the distance at which an object would have a PARALLAX of one second, using the Earth-Sun distance as the baseline. One parsec equals 3.2616 light-years or 3.034×10^{13}km (1.916×10^{13} miles).

Parsi, or Parsee, modern descendant of a small number of ancient Persian Zoroastrians who emigrated to Gujerat in India from the 10th century onwards. Modern Parsis follow a mixture of ZOROASTRIANISM and some Indian beliefs and practices. Now concentrated in Bombay, they are a small but active Indian minority.

Parsifal (1882), opera in three acts by Richard WAGNER with libretto by the composer based on a medieval legend. Wagner insisted that the work be performed exclusively at Bayreuth and his wife was successful in ensuring that his wish was honoured after his death. The first performance outside Bayreuth was given at the Metropolitan Opera, New York, in 1903.

Parsley, branching biennial herb, native to the Mediterranean region and widely cultivated for its tender, curled aromatic leaves used for flavouring and as a garnish. It has greenish-yellow flowers and tiny, seed-like fruits. Height: to 0.9cm (3ft). Family Umbelliferae; species *Petroselinum crispum. See also* PE pp.190, 212–213, 213.

Parsnip, biennial vegetable native to Eurasia. It is widely cultivated for its edible white taproot. The plant has many leaves, deeply and finely lobed. Parsnips flourish in deep, rich soil. The roots develop slowly until cool weather sets in, and then they mature quickly. Family Umbelliferae; species *Pastinaca sativa. See also* PE p.188.

Parson bird. *See* TUI.

Parsons, Sir Charles Algernon (1854–1931), British engineer and inventor, owner of an engineering works at Newcastle. He invented the Parsons compound steam turbine, which was introduced in about 1884. He added a condenser in 1891, and adapted it for maritime use in 1897. *See also* MM p.64.

Partch, Harry (1901–74), US avant-garde composer who not only created musical works according to his own system, but also created the instruments on which they were to be played. His works, often borrowing elements from American and Oriental music, includes *The Bewitched* (1955), a dance satire, *Delusion of the Fury* (1966) and music for films.

Parthenogenesis, in biological reproduction, the development of a female GAMETE or sex cell without fertilization (rarely is a male gamete produced), leading to the production of offspring without the involvement of a male gamete. This process occurs naturally among lower plants and some higher ones and invertebrate animals such as APHIDS, BEES, WASPS and ANTS; it can also occur in organisms that are capable of sexual REPRODUCTION. Artificial parthenogenesis may be caused by electric shock, mechanical stimulation, inorganic salts, organic acids or temperature changes. *See also* NW p.100.

Parthenon, temple of the goddess Athena erected (447–432 BC) by Pericles on the ACROPOLIS in Athens. The most perfect example of a DORIC temple, it was created by the architects ICTINUS and CALLICRATES and the master sculptor PHIDIAS. The huge gold and ivory statue of Athena which it once housed was destroyed in ancient times and its inner chambers and porticos were ruined by an explosion in 1687, during the Venetian attack on Athens. Most of the surviving sculptures were removed by Lord ELGIN in 1801–03. The surviving outer structure and sculptural pediments testify to the perfection of its design. *See also* HC1 pp.78–79.

Parthia, region in ancient Persia, generally corresponding to the modern Iranian province of Khurāsān. It was the seat of the Parthian empire (founded *c.*250 BC), which at its peak extended from the Euphrates to the Indus.

Parthians, ancient semi-nomadic people, skilled riders and archers, thought to have been of Scythian origin. In *c.*250 BC they rebelled against their SELEUCID overlords and founded an empire that extended from the River Euphrates to the Indus and reached its height in the 1st century BC. Influenced by Greek and Seleucic culture, they nevertheless created their own art and architecture. They were eventually defeated by the Romans (39–38 BC) and later overrun by the SASSANIDS under Ardashir I (AD 224).

Partial differential equation, type of DIFFERENTIAL EQUATION used when a function depends on two or more independent variables. For example, a wave in three dimensions has an amplitude (U) which depends on time t and also on the three distance measurements x, y and z along mutually perpendicular axes. The differential equation representing the wave is

$$\frac{\partial^2 U}{\partial x^2} + \frac{\partial^2 U}{\partial y^2} + \frac{\partial^2 U}{\partial z^2} = \frac{1}{c^2}\frac{\partial^2 U}{\partial t^2}$$

where c is the wave's velocity. The symbols $\frac{\partial^2 U}{\partial x^2}$ etc., are called partial derivatives and express the rate of change of U in the x direction, etc., only. Partial differential equations are extensively used in physical science.

Partial eclipse. *See* ECLIPSE.

Partial pressure, pressure that each gas in a mixture would exert if it were present alone. The pressure of an ideal-gas mixture is the sum of the partial pressures of its components (Dalton's law of partial pressures). *See also* SU pp.84–85.

Particle accelerator. *See* ACCELERATOR, PARTICLE.

Partita, late 17th- and 18th-century term for a collection of short musical pieces, now generally termed a SUITE; in the early 17th century it meant a variation. J. S.

Parrots can live to a great age. One is reported to have lived 80 years.

Sir Hubert Parry was a British composer, director of the Royal College of Music.

Parsnip, related to the carrot, is rich in vitamins A and C, and sugar.

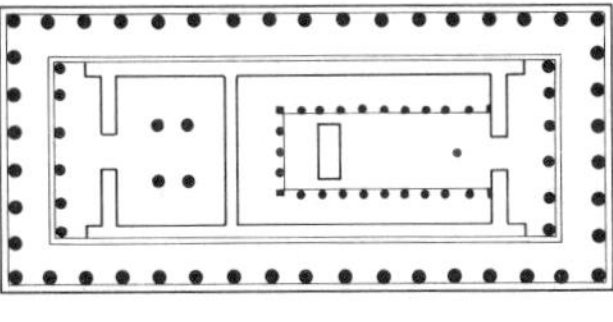

The Parthenon has served as Greek temple, Christian church and Moslem mosque.

Giovanni Pascoli; the birthplace of the poet at San Mauro in northern Italy.

Passenger pigeon, once a common bird in America, has been hunted to extinction.

Passion flowers by tradition symbolize aspects of the passion of Christ.

Boris Pasternak suffered because of the honours he was accorded by the West.

BACH wrote partitas for harpsichord and for violin, and the term has been used by some 20th-century composers, as in WALTON's *Partita* for orchestra (1957).

Partnership, company or business organization owned by two or more people, who may have equal or unequal shares. Usually liability is unlimited, so that each partner is responsible for the debts incurred by the company. *See also* COMPANIES; CORPORATION.

Partridge, Eric Honeywood (1894–), New Zealand philologist and author who lived in Britain after 1921. He is famous for his detailed studies of the English language, among which are *Usage and Abusage* (1947), *A Dictionary of Slang and Unconventional English* (1961) and *Shakespeare's Bawdy* (1968).

Partridge, any of several species of game birds found throughout the world. True partridges of Europe belong to the pheasant family (Phasianidae), and include the common partridge *Perdix perdix*, which has been introduced to North America. It lives on heathland, sandy scrub and farmland and feeds on plants and insects. It makes a ground nest under bushes and lays 9–20 green-brown eggs. *See also* PE pp.*238*, *239*.

Parts of speech, categories that define the ways words can be used according to the grammar of the English language. The main parts of speech are NOUN, adjective (describing a noun), PRONOUN, VERB, adverb (describing a verb), PREPOSITION, conjunction (conjoining two ideas) and interjection. Many of these can be subclassified further. Not all languages, however, can be divided into parts of speech in this way; nouns and verbs are the only universally applicable categories.

Parvati, in Hindu religion, "daughter of the mountain", wife of the god SHIVA; she is the beneficent manifestation of the goddess Shakti. Her children were the elephant-headed Ganesa and the six-headed Skanda. She is always depicted as a beautiful woman.

Parzival, early 13th-century epic poem written in Middle High German by Wolfram von Eschenbach. Based largely on the work by CHRÉTIEN DE TROYES, it deals with the legendary knights of KING ARTHUR and the quest for the Holy GRAIL. It is notable for its lyricism, and the relative seriousness with which it deals with the Anglo-French romances.

Pasay, city in central Luzon in the Philippines, on the E shore of Manila Bay; a residential suburb of Manila. Pasay is a commercial centre and has Araneta University (1946) and Manila's international airport. Many tourists go there for its nightclubs. Pop. (1975) 240,913.

Pascal, Blaise (1623–62), French scientist and mystic. A child prodigy, he had written a book on conic sections by the time he was 17 years old. Later, with Pierre de FERMAT, he laid the foundations of the mathematical theory of probability. He also contributed to calculus and hydrodynamics (devising PASCAL'S LAW) before retiring from science in 1655 to devote himself to religious and philosophical writing, of which *Pensées* is his best-known work.

Pascal's law, concept formulated in 1647 by the French scientist Blaise PASCAL. It states that the pressure applied to an enclosed fluid (liquid or gas) is transmitted equally in all directions and to all parts of the enclosing vessel, if pressure changes due to the weight of the fluid can be neglected. It is the basic principle by which hydraulic machinery works. *See also* SU pp.84–87.

Paschal, the name of two Roman Catholic popes and an ANTIPOPE. Saint Paschal I reigned from 817 to 824. Paschal II (*r.* 1099–1118), a BENEDICTINE, served as a papal legate to Spain under URBAN II. He sponsored the First CRUSADE and opposed lay investiture. The antipope Paschal reigned from 1164 to 1168.

Pascin, Jules (1885–1930), US painter, born in Bulgaria. He is best known for his sensitively-drawn nudes, which include *Nude with Red Sandals* (1927) and *Les Deux Amies* (1924). After making his name in Germany as a cartoonist for

satirical journals, he turned to painting, working in France and the USA and becoming a US citizen in 1920.

Pascoli, Giovanni (1855–1912), Italian poet and scholar. His unhappy youth is reflected in *Myricae* (1891; *Tamarisks*). Later works include the lyrical *Canti di Castelvecchio* (1903; *Songs of Castelvecchio*) and the nationalistic *Poemi del Risorgimento* (1913; *Poems of the Risorgimento*).

Pas-de-Calais, French name for the Strait of DOVER, which connects the English Channel to the North Sea. It is also the name of a French *département*.

Pasha, honorary title given by Ottoman sultans to ministers, governors and military officials within their empire. It was officially abolished by Turkey in 1934 but its use persisted in certain other countries (notably Egypt and Iraq) formerly under Ottoman rule.

Pasht. *See* BAST.

Pashto (Pushtu), one of the two major languages of AFGHANISTAN, the other being Persian. Pashto is spoken by about 12 million people in E Afghanistan and N Pakistan. It is historically the language of the Pathan tribes and is written in an adapted Arabic alphabet. One of the Iranian languages, it thus forms part of the INDO-EUROPEAN family of languages.

Pasig, river in SW Luzon, the Philippines; it drains the Laguna de Bay NW through the city of Manila to empty into Manila Bay. During the Spanish colonial period the river was important to inter-island trade. Its port facilities have now declined. Length: 23km (14 miles).

Pasmore, Victor (1908–), British painter, a founder of the EUSTON ROAD SCHOOL. In the 1950s, possibly due to a period of collaboration with Ben NICHOLSON, his work became more abstract and had an important influence on the development of post-war British painting. His abstract paintings, eg *Abstract in Black, Indian Red and Lilac* (1957), were largely succeeded by three-dimensional constructions in wood or plastic, eg *Projective Painting in White, Black and Ochre* (1963). *See also* HC2 p.*280*.

Pasolini, Pier Paolo (1922–75), Italian poet, novelist and film director. A Marxist, he wrote about urban poverty with great realism in novels such as *Ragazzi di vita* (1955) and *A Violent Life* (1959). Films he directed include *The Gospel According to St Matthew* (1964), *Oedipus Rex* (1967) and *The Decameron* (1970). His last film was *Salo* (1975), a controversial allegory of FASCISM.

Pasque flower. *See* ANEMONE.

Passchendaele, popular name of the WWI battle officially known as the Third Battle of Ypres. It was begun by Britain on 31 July 1917 and lasted for three months. Attempting to advance on a 15-mile front, the appallingly wet weather kept the British army virtually trapped in the Flanders mud. Although German losses were heavy, Britain's were even heavier and the objectives of keeping Russia in the war and preventing Germany from sending reserves to the Italian front were not achieved. *See also* HC2 p.*191*.

Passenger pigeon, extinct species of migratory bird, rather like a large version of the European TURTLE DOVE. Once extremely numerous in North America, the birds were slaughtered by the thousand for meat until the wild flocks were finally destroyed. The last captive specimen died in 1914. Species *Ectopistes migratorius*.

Passer, Ivan (1933–), Czechoslovakian film director. He trained under Milos FORMAN. His first film was *A Boring Afternoon* (1964). *The Legend of Beautiful Julia* (1968) was the last film he made in Czechoslovakia before moving to the USA, where he directed *Born to Win* (1971).

Passeriformes, perching birds, the largest order of birds, containing about half of all known species, in 57 families. It includes the true songbirds, such as robins, blackbirds and thrushes; the crow family; swallows and martins; birds of paradise and many more. Perching birds have grasping feet on which one toe points backwards and the other three point forwards, thus

enabling the birds to grip and remain steady on their perches. *See also* NW p.140.

Passfield, Sidney Webb, Baron. *See* WEBB, SIDNEY.

Passion, The, in Christian theology, the suffering of Christ from the time of His praying in GETHSEMANE till His death on the cross. Passion Sunday is the fifth Sunday in Lent; Palm Sunday then Easter follow.

Passion flower, numerous species of climbing tropical plants which probably originated in tropical America. Flowers are red, yellow, green, or purple; the outer petals ring a fringed centre. The leaves are lobed and some species produce edible fruits, eg granadilla and calabash. Family Passifloraceae. *See also* NW p.*63*.

Passion fruit, fruit of the passion flower. The fruit is a berry, often of delicate flavour, yellow or purple in colour. In many species it is about the size of a hen's egg, but in *Passiflora quadrangularis* it weighs up to 5kg (11lb). Genus *Passiflora*. *See also* NW p.*63*.

Passion play, dramatic presentation of Christ's passion, death and resurrection originally developed in medieval Europe. The best-known example of this tradition is still held every ten years in OBERAMMERGAU, West Germany, to fulfil a vow made by the villagers in 1634 during an outbreak of plague.

Passover (Pesach), also called the Feast of the Unleavened Bread, Jewish festival of eight days commemorating the EXODUS from Egypt and the redemption of the Israelites. Symbolic dishes are prepared, including bitter herbs (maror) and unleavened bread (matzot) to remind the Jews of the haste with which they fled Egypt, and by extension of their heritage. It is a family celebration, at which the Haggadah is read. Christ's Last Supper, at which He instituted the EUCHARIST, was a Passover meal.

Passport, government-issued document facilitating travel to and from different countries. Its use became widespread after WWI. A passport identifies its holder as a citizen or national of a specific country and under that country's protection. Some countries also require visiting passport holders previously to obtain visas (authenticating endorsements).

Passy, Frédéric (1822–1912), French economist and writer. In 1867 he founded the International and Permanent League for Peace and in 1889 was a founder of the Interparliamentary Union. He shared the first Nobel Peace Prize in 1901 with Jean Henri DUNANT. Passy's many works include *The Question of Peace* (1891) and *The Economic Causes of Wars* (1905).

Pasta, starchy food made from semolina, which is derived from durum wheat. Traditionally associated with Italian cooking, it is normally served with a sauce. Pasta comes in a variety of shapes and sizes, including vermicelli, spaghetti, ravioli, macaroni and lasagne. *See also* PE pp.182, 183.

Pastel, painter's medium, used in Western painting at least since the Italian RENAISSANCE of the 15th century and introduced into France in the 18th century. It is a mixture of chalk, pigment and gum water, usually found in the shape of a stick. Its most illustrious modern users have been DEGAS, MANET, WHISTLER and MATISSE.

Pasternak, Boris (1890–1960), Soviet poet, novelist and translator who came to symbolize the struggle between the artist and a hostile political environment. Up to the purges of the 1930s, his published works included *The Twin in the Clouds* (1914), *My Sister, Life* (1922), *Safe Conduct* (1931) and *Second Birth* (1932). He then turned to translating and did not publish any more original work until WWII. After the death of Stalin he began work on *Dr Zhivago* (1957), his novel best known in the West. Although never published in the USSR, its themes, the defence of the individual's right to personal happiness and implicit criticism of the government, offended officials and he was expelled from the Soviet Writers Union. He was also compelled by official pressure to retract his acceptance of the 1958 Nobel

Prize in literature. He lived in virtual exile in an artists' community near Moscow, for the remaining years of his life.

Pasteur, Louis (1822–95), French chemist and one of the founders of the science of microbiology. He made many important contributions to chemistry, bacteriology and medicine. He discovered that micro-organisms can be destroyed by heat, a technique, now known as PASTEURIZATION, used to destroy harmful micro-organisms in food. Pasteur also discovered that he could weaken certain disease-causing micro-organisms – specifically those causing anthrax in animals and rabies in man – and then use the weakened culture to vaccinate individuals against the disease. *See also* HC2 p.156; MS pp.111, 120, 126.

Pasteurella pestis, BACTERIUM responsible for PLAGUE. This highly infectious disease was endemic in Europe until the 17th century, although probably at its worst during the BLACK DEATH in the 14th century. It is now confined to remote areas of Asia. The bacterium is carried by the rat flea and infects man through the wound caused by the flea's bite.

Pasteur Institute, research organization in Paris founded in 1888 as a centre for the prevention and cure of RABIES. It was headed by Louis PASTEUR until his death in 1895. It is still one of the world's foremost biological research centres.

Pasteurization, process of controlled heat treatment to kill bacteria, discovered by Louis PASTEUR in the 1860s. Milk is pasteurized by heating it to 72°C (161°F) and holding it at that temperature for 15 seconds. Alternatively, it may be heated to 62°C (143°F) and held for 30 minutes.

Pastiche, imitation or forgery in which elements from several works by other artists are recombined by the author to give the impression of an original work. It may also mean a work in which the elements are combined in an obvious way to produce a parody. In the past, the word referred only to the visual arts, but it is now applied to any parody in any area of the arts.

Paston letters (1422–1509), correspondence of four generations of the wealthy, landowning Paston family, vividly illuminating contemporary life and underlining Paston support for the Lancastrians during the WARS OF THE ROSES. *See also* HC1 pp.242–243.

Pastoral, in literature, work portraying shepherds or rural life in an idealized manner to contrast this innocence with the corruption of the city or royal court. In Classical times THEOCRITUS and VIRGIL wrote notable pastoral poems, which were known as eclogues. Eclogues were revived during the RENAISSANCE by such poets as DANTE, PETRARCH, BOCCACCIO and SPENSER. John MILTON and Percy Bysshe SHELLEY were noted for their pastoral elegies and certain poets, such as William WORDSWORTH and Robert FROST, have been referred to as pastoral poets because their work has a rural setting.

Pastoralism, a form of subsistence agriculture that involves the herding of domesticated livestock. Societies practising this are small, restricted by the large amount of grazing land needed for each animal. Indigenous pastoralism is widespread in N Africa and central Asia, but in the Americas is confined to the Andes. *See also* NOMAD.

Patagonia, region in Argentina, E of the Andes Mts, extending to the Strait of Magellan. The area was colonized in the 1880s despite protracted border disputes between Argentina and Chile; the present boundaries were set in 1902. Most of Patagonia is located on plateau lands crossed by the Rio Negro, Chubat and the Santa Cruz rivers which flow to the Atlantic Ocean. Until recently sheep rearing was the main source of income. Oil production has now become important and coal and iron ore are mined in the s. Area: 805,490sq km (311,000sq miles).

Patas, large, reddish-grey MONKEY native to grassy woodlands and scrub forests of central Africa. It is ground-foraging and omnivorous, active by day, and lives in groups often led by a large male. Family Cercopithecidae Species *Erythrocebus patas.*

Patchouli, species of East India mint that yields a fragrant brownish yellow essential oil. The leaves yield a heavy perfume used in the Orient and in the manufacture of soaps. Family Labiatae; species *Pogostemon patchouli.*

Patch test, test for hypersensitivity or allergy. Small pieces of linen or paper are impregnated with substances to which the subject may be sensitive. These are placed on the skin for a specific time. The subject's reaction is then observed.

Patella, or kneecap, large flattened, roughly triangular bone just in front of the joint where the FEMUR and TIBIA link. It is surrounded by bursae (sacs of fluid) that cushion the joint. *See also* MS p.57.

Patenier, Joachim (*c.*1485–1524), Dutch painter. His pictures reveal the influences of Antwerp MANNERISM in their use of colour, jagged mountains, rocky shores and tiny figures such as in his *Flight into Egypt* (1520), *Baptism of Christ* (1520) and *St Christopher* (1524).

Patent, privilege granted by LETTERS PATENT to the inventor of a new product or process. A patent excludes others from producing or making use of the invention for a limited period (in Britain, 16 years).

Patent medicine, medicine sold over the counter for which a prescription is unnecessary. Such medicine, labelled only with a registered brandname or trademark, need not list its constituent ingredients; because of this lack of information patent medicine was for long a derogatory term, since its secrecy concerning the contents allowed the sale of many useless concoctions. Now, however, strict legislation controls are enforced.

Pater, Jean Baptiste Joseph (1695–1736), French painter and follower of WATTEAU. His paintings, which include FÊTES GALANTES and military scenes, display sensitive colouring but lack the technical proficiency of Watteau.

Pater, Walter Horatio (1839–94), British author and critic. An admirer of the art of the Italian Renaissance, he believed passionately in the importance of art to life, an attitude which had great influence on Oscar WILDE. Pater put forward his theories in *Studies in the History of the Renaissance* (1873) and his novel *Marius the Epicurean* (1885).

Pater Noster (Our Father), Latin name for the LORD'S PRAYER.

Paterson, Andrew Barton ("Banjo") (1864–1941), Australian poet and novelist. He edited *Town and Country Journal* for many years and was famous for his BUSH BALLADS including *Waltzing Matilda* which appeared in 1917. *The Man from Snowy River* (1895) and *Old Bush Songs* (1905), two books of verse, and the novel *An Outback Marriage* (1906) are among his better-known works.

Paterson, William (1658–1719), Scottish financier. He founded the BANK OF ENGLAND in 1694, but resigned as a director the following year. He then pursued his DARIÉN SCHEME and accompanied the 1698 expedition to Darién. After the failure of this scheme in 1699 he returned to London and became financial adviser to WILLIAM III.

Paterson, city in N New Jersey, USA, on the Passaic River 23km (14 miles) N of Newark. The city was founded in 1791 by Alexander HAMILTON as a planned industrial community. From 1792 to 1794 cotton-spinning mills were set up, and by 1836 Samuel Colt had begun the manufacture of the Colt revolver. A silk industry was established shortly afterwards. Many of the original workers' homes and factories remain on a national historic site. Industries: clothing, chemicals, plastics, paper, food products. Pop. (1973) 143,372.

Pathans (Pashtuns), MUSLIM tribes who constitute the major racial group in SE AFGHANISTAN and NW PAKISTAN. They speak an eastern Iranian language, PASHTO (divided into two dialects), and are composed of 60 tribes practising both PASTORALISM and more settled farming. Disputes among the warlike Pathans commonly result in fierce blood-feuds.

Pathé, Charles (1863–1957), pioneer French film producer. He made a fortune by making and selling phonographs and went on to found Pathé Frères and Pathé Gazette. He produced many short films and is credited with having made one of the first long films, *Les Misérables* (1909).

Pathet Lao, Laotian left-wing political group. It was founded in 1950 and soon joined the Viet Minh in opposing French rule in Indo-China. Although legalized in 1956, the Pathet Lao began armed resistance to the American-backed Laotian government during the 1960s, and by 1975 became the governing party in Laos. *See also* MW p.117.

Pathogen, micro-organism which causes disease in plants or animals. Animal pathogens are most usually BACTERIA or VIRUSES, whereas plant pathogens are most usually FUNGI or viruses.

Pathology, study of diseases, their causes and the changes they produce in the cells, tissues and organs of the body. Of basic importance in medicine, pathology covers three main areas of approach: the symptoms of a disorder; the causative agents, such as bacterium, virus or poison; and the damage to the tissues or organs caused by these agents. While pathologists are not involved in the treatment of disease, their studies are of immense importance to the understanding, and so cure, of diseases.

Patience, any of a multitude of card games played by one person alone. Almost all involve setting out a number of cards in a particular arrangement and either by using the remainder or by exposing some at a time, collecting cards one by one in order in suits.

Patience, or Bunthorne's Bride (1881), light OPERA in two acts by Arthur SULLIVAN, libretto by William GILBERT. It satirizes the aesthetic movement of the 1870s. It was first performed at the Opéra Comique in London, then transferred to the Savoy Theatre. This first run lasted for 408 performances, and it has remained popular ever since.

Patina, green incrustation that forms on the surface of old weathered bronze or copper. Since this has an aesthetic value in sculpture, the term has come to mean any coating wrought by age that gives character to an object or surface.

Patmore, Coventry (Kersey Dighton) (1823–96), British poet. An assistant librarian at the British Museum for many years, his poetry reflects his interest, unusual at the time, in 17th metaphysical poetry. *The Angel in the House* (1854–63) and *The Unknown Eros* (1877), both of which deal with various aspects of love, are among his best-known work. He became a Roman Catholic in 1864 and subsequently also wrote on religious themes.

Patna, city in NE India, on the River Ganges, 464km (290 miles) NW of Calcutta; capital of Bihar state. The city served as the capital of the ancient Mauryan empire (325–185 BC) and the Gupta empire (AD 320–545). Patna was taken by the British in 1763. It has a palace built by Ashoka *c.*270–230 BC and the University of Patna (1917). It is the distribution centre for a rice-growing region, and a commercial centre. Pop. (1971) 473,001.

Pato, Argentinian game in which two teams of four horsemen attempt to put a ball (like a football with six leather handles) into their opponents' goal, a net suspended from a 2.7m (9ft) post. Players punch or throw the ball, ride with it at arm's length, or retrieve it from the ground while at full gallop. A hybrid of polo and basketball, it is played over six eight-minute periods on a rectangular field measuring 210 x 82m (690 × 270ft). Originally a live duck was used.

Paton, Alan Stewart (1903–), South African novelist and reformer. He gave up teaching in 1935 to take charge of a school for delinquent African boys. Strongly opposed to APARTHEID, he helped found and became president of the Liberal Party in 1953. *Cry, the Beloved Country* (1948) and *Too Late the Phalarope* (1953), two of his best-known works, deal with racial oppression.

Pátrai (Patras), port in central Greece, on

Pasteur's research has provided a major contribution to the welfare of mankind.

Pasteur Institute, known for its teaching and research into virulent diseases.

Patas's movement in formation has earnt it the name "hussar monkey".

Pato is the Spanish for "duck", the object used till the football replaced it.

Floyd Patterson was the youngest boxer ever to hold the world heavyweight title.

Paul II: his policies caused conflict with humanists of the Renaissance.

Cesare Pavese's wartime resistance experiences greatly influenced his writing.

Pawnbroker's shop with the traditional three-ball sign on New York's 8th Avenue.

the Gulf of Pátrai; capital of Akhaía department. A leading member of the Second Achaean League, the city led a revolt against the Macedonians in 280 BC. Pátrai was a Roman military colony in the late 1st century BC under Augustus, under whom it developed into a prosperous port. It was held by the Ottoman Turks from 1458 to 1687, and from 1715 to 1828, when it passed to Greece. There is a university (1966). Exports: currants, tobacco, wine, olive oil. Pop. (1971) 111,607.

Patriarch, title given to a few exalted bishops in the Christian Church. In ancient times there were three: of the West, of Antioch and of Alexandria. Constantinople was added in 381, Jerusalem in 451. In the Western Church there are now 11 patriarchs.

Patriarchy, social organization based on the authority of a senior male, usually the father, over a family group. It is based on the absolute authority of the male and purely patriarchal societies are therefore rare. Consequently, the term is now seldom used by anthropologists.

Patricians, aristocrats of the early Roman Republic. The word comes from the Latin *pater* (father), which was used to describe members of the SENATE. The Patrician class controlled the government, the army and the state religion. Until 445 BC a PLEBEIAN could not marry a patrician but by means of a struggle lasting for 200 years the plebeians gradually came to hold more and higher positions. By 287 BC they could hold almost any post and the privileges of the patricians were effectively removed. By the end of the Republic the number of patrician families had declined from the 81 of *c.*367 BC to 30.

Patrick, Saint (*fl.*5th century AD), patron saint of Ireland, *b.*Britain; also called "Apostle of the Irish". Facts about his life are much confused by legend. What is known of him comes almost entirely from his autobiography, *Confessio.* Abducted by marauders at 16, he was carried to Ireland and sold to a local chief. After six years as a herdsman, during which period he became increasingly religious, he escaped. In response to a dream, he returned to Ireland as a missionary, establishing an episcopal see at Armagh. His missionary work was so successful that Christianity was firmly established in Ireland before he died. By tradition he also banished snakes from the country. Feast day: 17 March. *See also* HC1 pp.134–135, *135.*

Patroclus, companion of ACHILLES. Wearing the armour of Achilles during the TROJAN WARS, he was killed by HECTOR. Achilles' bones were buried in the same tomb as those of Patroclus.

Patterson, Floyd (1935–), US boxer. The first man ever to regain the world heavyweight championship, he was also, at 21, the youngest to hold the title. He was the Olympic light heavyweight champion (1952), and beat Archie MOORE (1956) in Chicago for the heavyweight title. He lost the title to Ingemar Johansson (1959) in New York City, regained it from Johansson (1960) in New York City but lost it to Sonny Liston (1962) in Chicago.

Patti, Adelina (1843–1919), Italian soprano, full name Adela Juana Maria Patti; one of the greatest exponents of BEL CANTO. She made her debut in 1859 in New York and sang regularly at Covent Garden from 1861 to 1885 in operas by BELLINI, DONIZETTI, MEYERBEER and others. She was the most popular and highly paid singer of her day. *See also* HC2 p.*120.*

Patton, George Smith, Jr (1885–1945), US general. A graduate (1909), of the US Military Academy at WEST POINT, in WWI he served with the American Expeditionary Force (AEF) in France, where he became familiar with tank warfare. A controversial and highly successful officer, he commanded a tank corps in North Africa and the 7th Army in Sicily in WWII. After the D-DAY invasion in 1944, he commanded the Third Army in its dash across France and into Germany. As military governor of Bavaria after the war he was

criticized for leniency to Nazis and was removed to command the US 15th Army. He was fatally injured in a car accident.

Paul, Saint, apostle of Jesus Christ, missionary and author of many epistles in the NEW TESTAMENT. Named Saul at birth, he was brought up in Tarsus as a Greek-speaking, educated Roman citizen and a Jew, in which religion he became a PHARISEE. Travelling to Damascus to persecute Christians, he received a vision of Jesus and on the spot became a fluent and energetic evangelist and teacher of Christianity. While proclaiming that Jesus was the MESSIAH who sacrificed himself to atone for the sins of mankind, Paul's major theme (described as Hellenistic as opposed to Judaeo-Christian) was that Christianity was not a revised form of Judaism, that the Jewish Law and Priesthood had been superseded, and that Christianity was for the salvation of all. He travelled widely, founding and visiting many small communities before being taken as a prisoner to Rome, where he died in unknown circumstances, *c.*65 AD.

Paul, name of six Roman Catholic popes. Paul I was pope 757–67. Paul II (1417–71; *r.*1464–71) was born Pietro Barbo. A patron of scholars, he established the first printing press in Rome. Paul III (1468–1549; *r.*1534–49) was born Alessandro FARNESE. He inspired the COUNTER-REFORMATION by convening the Council of TRENT in 1545, and patronized the JESUITS. He encouraged Michelangelo to complete his paintings in the SISTINE CHAPEL and was portrayed by Titian. Paul IV (1476–59; *r.*1555–59) was born Gian Pietro Carafa in 1476. An indefatigable opponent of Protestantism and the Jewish faith, he increased the efficiency of the INQUISITION in Rome and established the ghetto in the city. He also tried to abolish corruption in the VATICAN. Paul V (1552–1621; *r.*1605–21) was born Camillo BORGHESE . He forbade English Roman Catholics to take the oath of allegiance to their king, as James I demanded of them. Paul VI (1897– ;*r.*1963–) was born Giovanni Battista Montini. He was ordained in 1920 and entered the Vatican diplomatic service in 1923. He was Archbishop of Milan (1954–63). As pope, he reconvened the Second Vatican Council, and in 1965 became the first pope to visit Asia. He has tried to foster ecumenism. In 1968, he opposed the use of artificial contraception. *See also* HC1 p.272; HC2 pp.304–305.

Paul I (1754–1801), Tsar of Russia (1796–1801). The son and heir of CATHERINE II (the Great), he re-established the principle of hereditary succession and instituted repressive measures to protect Russia from the influence of the FRENCH REVOLUTION. Paul's erratic conduct and his hostility toward his son, Alexander, led to his murder by nobles and military officers, in 1801. *See also* HC2 p.76.

Paul I (1901–64), King of the Hellenes (1947–64), brother and successor to George II. In 1938 he married Princess Frederika of Brunswick. During his reign he followed a pro-Western policy and received US aid to help Greece's economic recovery after WWII. He was succeeded by his son, CONSTANTINE II.

Paul (1893–1976), Prince of Yugoslavia, acted as regent to King PETER II from 1934–41. He failed to solve the problem of increasing pressure from CROATIA and SERBIA for national autonomy and his pro-AXIS policy, culminating in his making Yugoslavia a signatory to the anti-COMINTERN pact in 1941, caused his overthrow in a bloodless coup.

Pauli, Wolfgang (1900–58), US physicist. His work on the QUANTUM THEORY led him to his EXCLUSION PRINCIPLE (the Pauli principle), which relates the quantum theory to properties of atoms. He received the Nobel Prize in physics in 1945 for this work. In 1931 he postulated the existence of the NEUTRINO and lived to see his prediction verified in 1956.

Pauling, Linus Carl (1901–), US chemist. He studied in Europe and at the California Institute of Technology, where he later became a professor. His early

work on the application of WAVE MECHANICS to molecular structure, detailed in his book *The Nature of the Chemical Bond* (1939), led to the Nobel Prize in chemistry in 1954. He also worked on the structure of PROTEINS. A keen protagonist of nuclear disarmament, he was awarded the 1962 Nobel Peace Prize. His work on DNA nearly anticipated the findings of CRICK and WATSON, when he suggested, in the 1950s, that protein molecules were arranged in a helical structure.

Paulinus, Caius Suetonius (*fl.* 1st century AD), Roman governor of Britain from AD 59 to 61. He suppressed the rising of BOADICEA and the Iceni with some brutality in 61. Later he led Otho's troops against Vitellius.

Paullus or Paulus Macedonicus, Lucius Aemilius (*c.*228–160 BC), Roman general and administrator, scion of an illustrious family. As consul (182 BC) he conquered the Ingauni, a Ligurian people. The Third Macedonian war between Rome and King Perseus had been waged indecisively since 171 BC and Paullus accepted a second consulship to fight in Macedonia. He defeated Perseus (168) and organized Macedonia into four republics which had to pay an annual tribute to Rome. Plutarch wrote his life.

Pavage, medieval tax levied in Britain for the repair and maintenance of roads, dating at least from the 12th century and most common in towns. With the growth of towns this duty of road maintenance fell increasingly on parish as well as borough authorities, and from them devolved to their successors, the local councils.

Pavane, stately court dance, fashionable in Italy, Spain, France and England during the 16th and 17th centuries when it was paired with the GALLIARD. The music for the dance was in 2–2 or 4–4 time, and the dance comprised many curtsies, advances and retreats. The pavane as a musical form continued to develop after the dance lost popularity.

Pavese, Cesare (1908–50), Italian poet, novelist and translator. His translations of British and American novels had considerable influence on Italian literature of the time. His imprisonment for anti-FASCIST activities in 1935 and his work with the Resistance between 1943 and 1945 strongly influenced his own creative writing. His works include the novels *The Political Prisoner* (1949) and *The Moon and the Bonfires* (1950).

Pavlov, Ivan Petrovich (1849–1936), Russian neurophysiologist. His early work centred on the physiology and neurology of digestion, for which he received the 1904 Nobel Prize in physiology and medicine. However, he is best known for his classical (Pavlovian) conditioning of behaviour in dogs. This work had a permanent effect on Russian psychology and a profound effect on the development of BEHAVIOURISM in the USA. His major works (in translation) are *Conditioned Reflexes* (1927) and *Lectures on Conditioned Reflexes* (1928).

Pavlova, Anna (1885–1931), Russian ballerina who made her debut at the Maryinsky Theatre, St Petersburg, in 1899. She toured in Europe and the US and left Russia in 1913 to tour with her own company. She was considered the greatest ballerina of her time and excelled in *Giselle, The Dragonfly, Autumn Leaves* and the *Dying Swan,* choreographed for her by Michel FOKINE in 1905. In addition to her conventional repertoire, she often performed ethnic dances of Poland, Russia and Mexico. *See also* HC2 p.274.

Pavo, or the Peacock, southern constellation; one of the four constellations known as the Southern Birds. The only bright star is Alpha, of second magnitude. Kappa is a CEPHEID VARIABLE star with a magnitude range of 4 to 5.5 and a period of 9.1 days. *See also* SU p.262, *262.*

Pawnbroker, moneylender who accepts movable goods as security on loans. Such things as jewellery, furs and household goods are exchanged for a loan of money and a ticket. Within a certain period the article may be redeemed by producing the ticket and paying the interest on the loan.

The rates of interest are high, but regulated by law. The trade existed in China at least 2,000 years ago.

Pawnee, Caddoan-speaking tribe of North American Indians. They are related to the Arikara who once occupied the Central Platte and Republican River areas in Nebraska. In the mid-1970s some 1,000 lived on reservations in Oklahoma.

Pawpaw. *See* PAPAYA.

Pax, in Roman mythology, female personification of peace. She was less honoured in Rome than was VICTORIA (Victory) and became significant only during the reign of the Emperor Augustus. It was not until AD 75, under VESPASIAN, that a temple was finally dedicated to her.

Paxinou, Katina (1900–74), Greek actress. Besides performances in Classical Greek tragedy, she has acted in English films such as *For Whom the Bell Tolls* (1943) and *The Miracle* (1960). Her most famous roles on the stage include Mrs Alving in IBSEN's *Ghosts* (1940) and Bernarda in GARCÍA LORCA's *The House of Bernada Alba* (1954), both in English versions.

Paxton, Sir Joseph (1803–65), British architect and landscape gardener, knighted for his design for the CRYSTAL PALACE to house the Great Exhibition in 1851. As chief gardener to the Duke of Devonshire Paxton became familiar with greenhouse construction and built the Great Conservatory at Chatsworth (1836–40), in which he experimented with a system of glass and metal construction. The Crystal Palace, designed so that all the parts could be factory-made and constructed on site, was epoch-making as the first completely prefabricated building. Paxton, who was elected MP for Coventry in 1854, oversaw the design of many large country houses, including Mentmore. *See also* HC2 p.89.

PAYE, short for Pay As You Earn, the British system of paying income tax by a deduction from current earnings. It was first introduced in 1944 and is now the most common method by which employees pay income tax.

Payer, Julius von (1842–1915), Austrian explorer who accompanied Karl Weyprecht on the expedition (1872–74) which discovered and named Franz Josef Land. In 1874 he turned to the painting of such adventurous subjects as *The Bay of Death, Abandoning the Ship* and *Never to Return*.

Payload, portion of a rocket containing the scientific instruments, warhead, astronauts or other elements involved in the mission objective. Because of the enormous fuel needed to overcome Earth's gravity, only a fraction, about one-thousandth of a rocket's total weight is generally available for payload. In civil aviation the term refers to the revenue-producing part of the aircraft's load, whether it is passengers, freight or both.

Paz, Octavio (1914–), Mexican poet who is considered among the ablest writers of his generation. His work has gone through many phases, from MARXISM to SURREALISM and to oriental philosophies, but its dominant characteristics are a search for harmony and a desire to make language more melodious. His works were published in English in *Selected Poems* (1963). *See also* HC2 p.290.

Pea, climbing annual plant (*Pisum sativum*), probably native to w Asia. It has small oval leaves and white flowers which give rise to pods containing wrinkled or smooth seeds that are a popular vegetable. It grows to 1.8m (6ft). There are many varieties. The early dwarf pea (var. *humile*) is a low-growing plant and has small pods. The snow, or sugar, pea (var. *macrocarpum*) has soft, unlined edible pods. Family Leguminosae. *See also* PE pp.*173*, 184–185, 188.

Peabody, George (1795–1869), US financier and philanthropist. He made an early fortune from a chain of grocery stores, and after settling in London (1837), he helped to support US credit following the panic of 1837. He is chiefly remembered for his philanthropic work, including financing tenement clearance in London and the construction of the

PEABODY BUILDINGS by the Peabody Trust. He also founded the Peabody Institute at Baltimore and the Peabody museums at Yale and Harvard universities.

Peabody Buildings, dwellings erected in London, largely at the instigation of George PEABODY, an American philanthropist. He provided funds for the clearance of slums and built in their place pleasant flats, of small dimensions and with modest rents, to rehouse workers. Today they are administered by the Peabody Trust.

Peace, river in Canada formed at the junction of the Finlay and Parsnip rivers in E central British Columbia. It flows E through the Rocky Mts, then N across the plains to join the Slave River at Lake Athabasca. The river was explored between 1792 and 1793 by Sir Alexander MacKenzie. It was an important route for fur-traders. Length: 1,924km (1,195 miles).

Peace Corps, US government agency which sends young volunteers to developing countries to provide skilled manpower and to promote friendship. The corps was established in 1961 by Pres. John F. KENNEDY and in 1971 it merged with Action, an agency combining several volunteer programmes. Volunteers must be US citizens and at least 18 years old; there is no upper age limit. They are invited to the host country to work on specific projects, often to do with teaching or agriculture.

Peace movement, term for many organizations that have worked to achieve international peace. The first international peace congress met in London in 1843. The most important contemporary societies are the World Council of Peace (1950), the Christian Peace Conference (1958) and the International Peace Research Association (1964).

Peace of Paris (1763), treaty ending the SEVEN YEARS WAR. Britain gained almost all of French Canada, Louisiana E of the Mississippi River, Senegal and the West Indian islands of the Grenadines, Dominica, St Vincent and Tobago. Guadaloupe, Martinique and three smaller West Indian islands occupied by Britain during the war were returned to France. In India, France was forced to recognize the supremacy of the EAST INDIA COMPANY. *See also* HC2 p.64.

Peace of Paris (1783), treaty ending the war between Britain and the 13 American colonies, which were supported by their French and Spanish allies. Britain recognized the independence of the 13 colonies. Their northern boundary was drawn along the line of the St Lawrence River and the Great Lakes. Spain retained most territory, all lands w of the Mississippi river and s of the 31st parallel (w and E Florida). Gibraltar, occupied by Spain during the war, was returned to Britain. *See also* HC2 p.65.

Peace of Westphalia (1648), name for a combination of treaties signed between various states, effectively ending the THIRTY YEARS WAR. For three years the heads of state of all European countries (except the outlying ones of Russia, England, Poland and Denmark), negotiated the treaties at the Westphalian towns of Münster and Osnabrück. It was the first great European diplomatic congress. The chief gainers of territory were Sweden and France, the latter obtaining important influence in Alsace and Lorraine and German lands, at the expense of the Austrian HAPSBURGS. *See also* HC1 p.273.

Peace Prize. *See* NOBEL PRIZES.

Peach, small fruit tree native to China and grown throughout temperate areas. The lance-shaped leaves appear after the pink flowers in spring. The fruit has a thin, downy skin, white or yellow flesh, and is classified as either free-stone or cling-stone. It is eaten fresh or preserved. Height: to 6.5m (20ft). Family Rosaceae; species *Prunus persica*. *See also* PE p.*193*.

Peacock, Thomas Love (1785–1866), British writer of satirical poetry and romances. A friend of SHELLEY, Peacock published several works, including his first volume of poems in 1806. The novels *Headlong Hall* (1816), *Melincourt* (1817)

and *Nightmare Abbey* (1818) were published before he entered the service of the East India Company in 1819.

Peacock. *See* PEAFOWL.

Peacock, any of several butterflies whose wings have a pattern of eyespots, resembling those on the tails of peacocks. In Europe the name is given to *Inachis io*, a vanessid butterfly whose caterpillars feed on nettles. *See also* NW p.112.

Peafowl, any of several species of exotic birds of Asia and Africa. The resplendent male is called a peacock and the female a peahen; peacock has become the common name for both sexes. The male has a 150cm (60in) tail which it can spread vertically as a huge semicircular fan with a pattern of eye-like shapes. The body of the male may be metallic blue, green or bronze, depending on the species. Hens are almost as big as the males; they lack the dramatic tail and head ornaments and are generally brown, red or green. In captivity, white birds may occur. In the wild peafowl inhabit open, lowland forests, flocking and roosting in trees. Eggs are laid in a ground nest. Length of body: 75cm (30in). Family Phasianidae; genera *Pavo* and *Afropavo*. *See also* NW p.*212*.

Peak District (The Peak), plateau area in Derbyshire, N central England, at the s end of the Pennine Chain; the Peak District National Park was established in 1951. The highest point is Kinder Scout, rising to 636m (2,088ft). Area: 1,404sq km (542sq miles).

Peale, Charles Wilson (1741–1827), US miniaturist, portraitist, and museum director. In 1782 he opened a small portrait gallery in Philadelphia and in 1786, his interest in natural history led him to open the Peale Museum, the first major US museum.

Peanut. *See* GROUNDNUT.

Pear, tree and its fruit, native to N Asia and s Europe and grown throughout the world in temperate regions. The tree has a cone shape and clusters of white flowers appear among the glossy, green leaves. The greenish-yellow, brownish or reddish fruit, picked unripe and allowed to mature in storage, is eaten fresh or preserved. Height: 15–23m (50–75ft). Family Rosaceae; species *Pyrus communis. See also* PE pp.192–193, *192*.

Pearl, hard, smooth, iridescent concretion of calcium carbonate produced by certain marine and freshwater bivalve MOLLUSCS. It is composed almost entirely of nacre, or MOTHER-OF-PEARL which forms the inner layer of mollusc shells. A pearl, the only gem of animal origin, results from an abnormal growth of nacre around minute particles of foreign matter, such as a grain of sand. *See also* NW p.93; PE pp.98, 99.

Pearl, The, 14th-century mystical allegorical poem in MIDDLE ENGLISH, and regarded as one of the best poems of its time. A work of great structural complexity, it is possibly by the same author as *Sir Gawain and the Green Knight.*

Pearlfish, any of several species of small, elongated marine fish that live in the bodies of pearl OYSTERS, STARFISH, SEA CUCUMBERS and other invertebrates. Found mainly in shallow tropical waters, they are scaleless and often transparent. Length: to 15cm (6in). Family Carapidae.

Pearl-Fishers, The (Les Pêcheurs de Perles) (1863), three-act opera by Georges BIZET with libretto by Cormon and Carré. It was first produced in Paris. The setting is ancient Ceylon.

Pearl Harbor, virtually landlocked inlet on the island of Oahu, Hawaii; site of a US naval base. On 7 Dec. 1941 much of the US Pacific Fleet was destroyed by the Japanese attack on Pearl Harbor. More than 2,000 soldiers, sailors and civilians were killed, signalling the outbreak of war between the two countries. *See also* HC2 pp.228–229.

Pears, Sir Peter (1910–), British tenor. He studied at the Royal College of Music, London, and later with Elena Gerhardt. From 1939 he toured England, performing with Benjamin BRITTEN and Kathleen FERRIER in many concerts. A lifelong friend of Britten, he created many roles in his operas, including the title roles of *Peter Grimes* (1945) and *Albert Herring* (1949),

Joseph Paxton's famous Crystal Palace at Sydenham, as seen from the south.

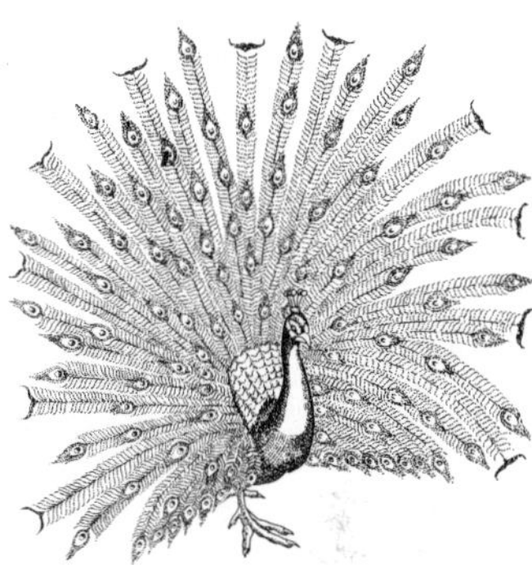

Peafowl, which were sacred to the Greeks, became a fashionable food to the Romans.

The Pearl: the dead girl appears to the poet in a vision.

Pearl Harbor: salvaging one of 19 ships sunk by the Japanese on 7 December 1941.

Robert Peary reached the North Pole with 4 eskimos and a servant, Matthew Henson.

Peccaries have poor sight and hearing but can scent out food which is buried.

Gregory Peck starred in *On the Beach* (1959), a film about atomic war.

Sir Robert Peel, a man of new ideas, did much to improve 19th-century England.

and the Male Chorus in *The Rape of Lucretia* (1946).

Pearse, Patrick Henry (1879–1916), Irish author and political figure. He headed the revival of interest in Gaelic culture, writing poems, short stories and plays. He led the insurgents in the EASTER RISING in 1916, and was court-martialled and executed. *See also* HC2 p.*163.*

Pearson, Sir Cyril Arthur (1866–1921), British publisher and newspaper owner. After forming his own company in 1890, he founded many periodicals, including *Pearson's Weekly.* In 1900 he launched the *Daily Express,* and after acquiring the *Evening Standard* and the *St James Gazette* in 1904, became a pioneer of the new journalism. After becoming blind in 1912, he devoted his energy to the welfare of the blind, founding a home in 1915 and publishing *Victory over Blindness* (1919).

Pearson, Lester Bowles, (1897–1972), Prime Minister of Canada from 1963 to 1968. He was US ambassador (1945–46), chairman of NATO (1951–52) and president of the UN General Assembly (1952–53). A Liberal MP and Secretary of State from 1948, he assumed party leadership in 1958. He was awarded the 1957 Nobel Peace Prize for his work in the Middle East. An energetic leader, he held the office of Prime Minister without a majority, and retired in 1968.

Peary, Robert Edwin, (1856–1920), US Arctic explorer. He made the first of several explorations of Greenland in 1886. Accompanied by his wife in 1891, he found evidence that Greenland was an island. In 1898 he set out on an expedition to the North Pole, and finally reached his goal by sled in 1909 after travelling more than 650km (400 miles) over ice.

Peasant, old name (which has since become somewhat pejorative) for a tenant-farmer or labourer who worked a small plot of land. Under the FEUDAL SYSTEM it was a collective term for villeins and serfs whose rent for the land was paid in labour, produce or money.

Peasants' Revolt, uprisings in England in 1381, especially in the counties of Kent and Essex. The immediate cause of the rebellion was an unpopular poll tax and the uprising took on something of the character of a revolt against the rich. It was led by Wat TYLER and John BALL. Tyler's men beheaded the Archbishop of Canterbury and for a few days held London, until RICHARD II met them and promised them a free pardon and the redress of grievances, in particular the abolition of villein status.

Peasants' War (1524–25), rebellion of German peasants their overlords, the greatest mass uprising in German history. The peasants had long grumbled over ever-increasing dues demanded by the princes and in 1524 there was widespread pillaging in the countryside of s Germany. The peasants hoped for and needed the support of Martin LUTHER but he rejected their charter of liberties. They ignored Luther's call to lay down arms and their rebellion was savagely suppressed in May 1525. *See also* HC1 p.*268.*

Peasant Wedding (*c.*1567), painting by Pieter BRUEGEL the Elder, now housed in Vienna. It depicts a feast following a wedding in Flanders, and is painted with realistic detail. The composition is based on the diagonal trestle tables at which the guests are being served. *See also* HC1 p.*261.*

Peat, dark brown, decayed organic material with a high carbon content, built up in bogs and used as fuel. It is the first stage in the development of COAL. Sphagnum mosses form most peat in the Northern Hemisphere.

Pecan, valuable North American nut tree. Its nut, resembling a small, smooth-shelled WALNUT, is 70% fat. It is used in the USA in pecan pie and ice cream. Family Juglandaceae; species *Carya illinoensis. See also* PE p.215.

Peccary, omnivorous pig-like mammal native to the SW USA and Central and South America. It has coarse, bristly hair, and scent glands on its back. Collared peccaries, or javelinas (*Tayassu tajaçu*), have dark grey hair with a whitish collar.

White-lipped peccaries (*Tayassu pecari*) have brown hair. Weight: 23–30kg (50–66lb). Family Tayassuidae.

Pechora, river in the Russian SFSR, USSR, that rises in the Middle Ural Mountains. It flows N, W and N again to Pechora Bay on the Barents Sea, forming a delta at Naryan-Mar. It receives the Ilych, Shchugor, Usa, Kozhva, Izhma and Tsilma rivers and supports fisheries, farming and livestock raising along its lower reaches. Both the main river and its tributaries are navigable for most of their courses. Length: 1,790km (1,112 miles).

Pechstein, Max (1881–1955), German EXPRESSIONIST painter. He became a member of Die BRÜCKE in Dresden in 1906 and a teacher at the Berlin Academy in 1923. His works include *Green House, The Red Turban* and *Sunrise.*

Peck, (Eldred) Gregory (1916–), US film star, famous for his portrayals of grave and honourable heroes. He began his lengthy film career in *Days of Glory* (1943) and his many appearances include leading roles in *Keys of the Kingdom* (1944), *Spellbound* (1945), *The Gunfighter* (1950), *The Guns of Navarone* (1961), *To Kill a Mockingbird* (1963), *McKenna's Gold* (1968), *The Omen* (1976) and *MacArthur* (1977).

Pecking order, hierarchical system of social organization based on dominance. The leader has power over the entire community, the second-in-command has power over everyone but the leader, and so on. The term derived from the behaviour of hens, which express their dominance by pecking their inferiors, but is now applied to other animals, including human beings. *See also* MS p.256; NW p.172.

Peckinpah, David Samuel ("Sam") (1926–), US director who worked in television, directing *Gunsmoke* and *The Westerner* series before turning to films. The deeper implications of his work are often overlooked because, since *The Wild Bunch* (1969), he has been associated with violent action films. His other films include *Straw Dogs* (1971), *The Getaway* (1973) and *Bring Me the Head of Alfredo Garcia* (1975).

Pecock, Reginald (*c.*1395–*c.*1460), Welsh theologian who became Bishop of St Asaph in 1444 and Bishop of Chichester in 1450. Writing in the vernacular, he was an important influence in the development of English vocabularies. He was the author of *Repressor of Overmuch Blaming of the Clergy* (*c.*1455) against the LOLLARDS. He was accused of heresy and died in confinement.

Pécs, city in S Hungary. Originally the Roman colony of Sopinae, it came under Turkish rule in the 16th century. Todays Pécs is an important industrial centre for the country's chief coal mining region. Its industries include the manufacture of agricultural machinery, leather goods and pottery. Pop. (1970) 145,307.

Pectin, water-soluble SUGAR found in the cell walls and intercellular tissue of certain ripe fruits or vegetables. It is the removal of this substance that causes fruits and vegetables to soften when cooked. It also yields a gel that is the basis of jellies and jams.

Pectoral, relating to the chest or BREAST, as the pectoral muscles, which cover the front of the chest, or the pectoral girdle, the bones of which make up the shoulder.

Pedersen, Christiern (*c.*1480–1554), Danish humanist scholar and translator, leading figure in Denmark at the time of the Reformation. He compiled a Danish-Latin dictionary and translated the New Testament (1521) and the Psalms (1531).

Pediatrics, branch of medicine that studies the growth and development of children. It is also concerned with the diagnosis and treatment of the diseases and disorders of children.

Pediculosis, skin disorder, especially in the scalp and groin, caused by blood-sucking lice. Small red spots marking bites are extremely itchy and may become infected. *See also* LOUSE.

Pedology, scientific study of SOILS. These are divided by pedologists into horizons of various types, according to their physical and chemical composition, including the presence of organic matter and the extent of drainage. Soil types are classified by such names as podzol and CHERNOZEM, reflecting the work done by early Russian pedologists.

Pedometer, surveying instrument used for measuring distance. It consists of a wheel connected to a meter which registers the distance rolled by the wheel.

Pedro, name of five kings of Portugal. Pedro I (1320–67) was a harsh ruler (*r.*1357–67). Pedro II (1648–1706; *r.*1683–1706) consolidated royal absolutism and his kingdom prospered from revenue from Brazilian goldfields. Pedro III reigned (1777–92) jointly with his wife Maria I (*r.*1777–1816). Pedro V (1837–61; *r.*1853–61) abolished slavery in the Portuguese colonies (1857).

Pedro, name of two emperors of Brazil. Pedro I (1798–1834) became Regent of Brazil in 1821 when his father, John VI, returned to Portugal. Although Pedro was recalled by the Portuguese parliament in 1822, he declared Brazil independent and himself emperor. He was forced to abdicate (1831) in favour of his five-year-old son and withdrew to Portugal where, on the death of his father in 1826, he had become titular king as Pedro IV. His son Pedro (1825–91) was under regency until 1840 when he was crowned Pedro II (*r.*1831–89). He abolished slavery in 1888 and was forced to abdicate by the former slave owners in 1889 when Brazil became a republic. *See also* HC2 p.86.

Peebles, market town in Borders Region, s Scotland, at the confluence of Eddleston Water and the River Tweed; county town of the former county of PEEBLES(SHIRE). Parts of the ancient city walls still remain. Industries: tourism, woollens (particularly tweeds). Pop. (1971) 5,881.

Peebles(shire), former county in SE Scotland; since 1975 it has been part of Borders Region. It is a mostly mountainous region, drained by the River Tweed. Beef cattle and sheep are reared. The former county town was Peebles. Industries: tourism, woollen mills, slate and limestone quarrying. Area: 899sq km (347sq miles). Pop. (1971) 13,675.

Peel, Sir Robert (1788–1850), British politician. He was an MP from 1809 to 1850 and chief Secretary (1822–27; 1828–30) in the TORY governments of Lord LIVERPOOL and the Duke of WELLINGTON. He was twice Prime Minister (1834–35; 1841–46). As leader of the Tory party in the House of Commons he played a major part in the reversal of Tory policy and the passing of CATHOLIC EMANCIPATION in 1829. As the first CONSERVATIVE Prime Minister he established the Ecclesiastical Commission (1835) and split his party permanently by the repeal of the CORN LAWS (1846). Those Conservatives who followed him into free trade became known as the PEELITES. *See also* HC2 p.112.

Peele, George (1556–96), English playwright who experimented in a variety of forms including the pageant, comedy and melodrama. One of his best-known works is *The Old Wives Tale* (1595), which may have influenced John MILTON's *Comus.*

Peelites, section of the CONSERVATIVE PARTY, about one-third, who took the side of Robert PEEL when the party split over the repeal of the CORN LAWS (1846). They behaved as a separate party for several years but by 1859 their leaders, of whom the most important was GLADSTONE, had found their way into the LIBERAL PARTY.

Peenemünde, village in East Germany, located on an island at the mouth of the River Peene. It was an important testing site for rockets and missiles during WWII.

Peep-show, miniature theatre contained in an enclosed box and viewed, and usually magnified, through a small hole situated at one end of the box. Peep-shows were invented in the 15th century.

Peerage, British temporal nobility – barons, viscounts, earls, marquesses and dukes. The titles were formerly all hereditary but since 1876 the Lords of Appeal in Ordinary have been included and since 1958 provision has been made for the creation of life peers as a reward for distinction. Peers constitute the Lords Temporal section of the House of Lords, having the right to sit in the House and vote.

Peer group, term used in SOCIOLOGY to mean a group whose members share certain characteristics. A peer group may be defined on the basis of income, mental ability, education, social status, occupation, age or any other such category.

Peer Gynt (1867), five-act poetic drama by Henrik IBSEN, based on Norwegian legend. Peer Gynt, a degenerate but lovable outlaw, leaves the good Solveig and goes travelling. The plot consists of episodes which display Peer Gynt's selfishness and worldliness. He encounters trolls and experiences the death of his mother before returning to be redeemed by the love of Solveig. Edvard GRIEG composed 22 pieces to accompany the drama which he later arranged into the two *Peer Gynt Suites* for orchestra in 1876.

Pegasus, in Greek mythology, winged horse. Born out of the blood of MEDUSA, it carried the thunderbolt of ZEUS. It was tamed by BELLEROPHON and helped him in his battles.

Pegasus, or the Winged Horse, northern constellation between Andromeda and Cygnus. Three of its bright stars form the Giant Square in Pegasus with Alpha Andromedae; the brightest is Epsilon, with a magnitude of 2.30. *See also* SU pp.258, *259,* 265, *265.*

Pegmatite, in geology, any very coarse-grained ROCK, generally light in colour, and often of granitic composition. Pegmatites are the chief sources of gemstones, MICA, and FELSPAR. *See also* PE p.98.

Péguy, Charles (1873–1914), French poet and writer. In 1900 he founded the periodical *Cahiers de la quinzaine,* in which most of his work appeared. He is among the foremost of Catholic writers and his works include *Notre Jeunesse* (1910) and the poem *Le Mystère de la Charité de Jeanne d'Arc* (1910).

Peipus (Čudskoje Ozero), third-largest lake in Europe, situated between Estonia (Estonskaja SSR), USSR and the Russian SFSR, USSR. Its outlet, the River Narva, flows N to the Gulf of Finland, receives the River Velikaya (s) and the River Ema (w) and is connected by a 24km (15 mile) strait to Lake Pskov. Area: 3,600sq km (1,390sq miles).

Peirce, Charles Sanders (1839–1914), US philosopher and logician, a leader of the pragmatists. He explained PRAGMATISM in a series of six papers published between 1877 and 1878. He also helped to develop semiotics, the study of the use of signs and symbols. Although he was a prolific writer, much of his work was published posthumously in eight volumes as *The Peirce Papers: Collected Papers.*

Peking (Bie-jing), capital of The People's Republic of CHINA, on a vast plain between the Pei and Hun Rivers in NE China. There has been a settlement there since c.1000 BC. The city served as China's capital from 1421 to 1911; after the establishment of the Chinese Republic (1911–12) Peking remained the political centre of the country. The seat of government was transferred to Nanking in 1928 and Peking (meaning "northern peace", was known as Pei-p'ing. Occupied by the Japanese in 1937, it was restored to China in 1945 and came under Communist control in 1949. Its name was restored and it was made the capital of the People's Republic.

The city comprises two walled sections: the Inner or Tatar City, which houses the Forbidden City, and the Outer or Chinese city. Peking is the political, cultural, educational, financial and transport centre of China. Heavy industry has been introduced to the area since 1949 and textiles, iron and steel are now produced there. Pop. 7,600,000.

Pekingese, toy dog bred in China from the 8th century. Its soft coat, of any colour, forms ruffs around the collar, legs and tail. The dogs were guarded by the Chinese court, but some were taken to Europe by British soldiers who stole them during the attack on the Summer Palace (1860). Weight: 2.5–6.5kg (6–14lb); height: 15–23cm (6–9in).

Peking Man, *Sinanthropus,* an extinct Pleistocene man known from remains first discovered in 1927 at Choukoutien, China. A hunter and user of stone tools and fire, *Sinanthropus* was more advanced than the related JAVA MAN. *See also* MS pp.*24, 25.*

Pelagianism, Christian heresy associated with the name of the British theologian Pelagius who taught in Rome in the late 4th and early 5th centuries. In opposition to St AUGUSTINE's belief that man could attain salvation only through God's grace, the Pelagians saw man as a creature of inherent spiritual grace and strong will. They, who were free to choose between good and evil, denied original sin and the need of the Church for salvation.

Pelagic, word to describe organisms that live in the sea (but not on the sea-bed) and the area that they inhabit. The organisms are divided into NEKTON (large fish and whales) and PLANKTON (small plants and animals) on which the nekton feed.

Pelagius, name of two popes. Pelagius I (*d.*561) was pope from 556 to 561. He took a prominent part in the Nestorian controversy and established the temporal power of the papacy in Rome. Pelagius II (*d.*590) was pope from 579 to 590. During his reign the Franks were asked to intervene in Italy to help the papacy for the first time.

Pelagius (*c.*354–*c.*418), British monk and theologian who preached the heresy of PELAGIANISM. In *c.*380 he went to Rome and became the spiritual director of many clergy and laymen. His rigorous asceticism – based on the belief that man, through his deeds, is master of his own spiritual salvation – served as a reproach to the spiritual sloth of many Roman Christians. After 410 he preached in Africa and Palestine.

Pelargonium, perennial GERANIUM native to South Africa. The circular or lobed leaves alternate on the stalk and are often aromatic. The five-petalled flowers are red, pink, purple or white. Family Geraniaceae; genus *Pelargonium.*

Pelé (1941–), born Edson Arantes do Nascimento, Brazilian soccer player, often said to be the finest player of all time. He made his first international appearance at the age of 17, led Brazil to three victories in the World Cup (1958, 1962, 1970), and was thought to be one of the highest paid athletes in the world. He made his last international appearance in 1971, and apart from 1975–77, when he played for New York Cosmos, all his club games were played for Santos. During his career he scored more than 1,200 goals.

Pelecaniformes, order of birds which includes the TROPIC BIRDS, FRIGATE BIRDS, and the sub-order Pelecani, of which PELICANS, GANNETS, CORMORANTS and DARTERS are members. *See also* NW p.141.

Pelée Mount, volcanic peak in N Martinique, in the Windward Islands, West Indies. It has erupted many times, most notably in 1902, when it engulfed the town of Saint Pierre killing approx. 30,000 people. It is the highest peak on the island: 1,397m (4,583ft).

Peleus, in Greek mythology, son of Aeacus, King of Phthia in Thessaly. During the expedition by the ARGONAUTS, Peleus, with the help of CHIRON, won the sea nymph Thetis. Their son was ACHILLES, hero of the TROJAN WAR. Peleus also took part in the Calydonian boar hunt, during which he accidentally killed his uncle, King Eurytion.

Pelham, Henry (1696–1754), British politician who was Secretary for War from 1724 to 1730 and Prime Minister from 1743 to 1754. During his administration the War of the AUSTRIAN SUCCESSION ended in 1748 and the national debt was reorganized and reduced.

Pelican, any of several species of stout-bodied inland water birds, with a characteristic distensible pouch under its bill for scooping up fish from shallow water. It is generally white or brown and has a long hooked bill, long wings, short thick legs and webbed feet. Length: to 1.8m (6ft). Family Pelecanidae; genus *Pelecanus. See also* NW p.235.

Pellagra, disease caused by a deficiency of nicotinic acid, a constituent of the B group

of vitamins. Its symptoms are lesions of the skin and mucous membranes, diarrhoea and mental debility.

Pellan, Alfred (1906–), Canadian painter. He worked in Paris from 1926 to 1940 and was influenced by SURREALISM. After his return to Montreal in 1940 he was a guiding force for young Canadian artists and painted complex abstract works, including *Floraison (c.*1945).

Pelléas and Mélisande (1902), opera in five acts by Claude DEBUSSY based on a play by Maurice MAETERLINCK. It was first performed at the Opéra Comique, Paris, where it was reviled as subversive because of its innovatory tonal harmonies, but today it is considered to be a landmark in the history of opera and a masterpiece of musical IMPRESSIONISM.

Pellegrini, Giovanni Antonio (1675–1741), Venetian painter and pupil of Sebastiano RICCI. He travelled widely in Europe and was the first Venetian artist to visit England (1708). His decorative work for buildings such as Kimbolton Castle and Castle Howard influenced the style of English Rococo.

Pelletier, Pierre-Joseph (1788–1842), French chemist, founder of the chemistry of ALKALOIDS. He isolated CHLOROPHYLL in 1817 and then extracted EMETINE. With J. R. Caventou he discovered other alkaloids, such as brucine, cinchonine, quinine and strychnine.

Pellico, Silvio (1789–1854), Italian dramatic poet. His chief work is the play *Francesca da Rimini* (1815). Arrested for joining the CARBONARI, he was imprisoned in 1820. He wrote an account of his eight years in prison in *Le Mie Prigione* (1832).

Peloponnese (Pelopónnisos) or Peloponnesos, peninsula in S Greece, separated from central Greece by the Gulf of Corinth. The region included the ancient cities of Sparta, Corinth, Argos and Megalopolis. The entire peninsula was involved in the Persian Wars (500–449 BC), and it was the site of many battles between Sparta and Athens during the Peloponnesian War (431–404 BC). The region fell to the Romans in 146 BC, who reduced it to a provincial state. Held by the Venetians from 1699 to 1718, then the Ottoman Turks, the peninsula passed to Greece after independence. Products: silk, fish, manganese, chromium, fruits, tobacco, wheat. Tourism is a major industry. Area: 21,756sq km (8,400sq miles). Pop. 986,912.

Peloponnesian War, History of the, THUCYDIDES's masterly account of the war between Athens and Sparta and their allies (431–404 BC). The work, written between *c.*423 BC and *c.*403 BC, provides a detailed and penetrating analysis of the war's causes, and in its balance of detailed scholarship and moral judgement has been influential on all historians since.

Pelops, mythological Greek founder of the Pelopid dynasty of Mycenae. He was the grandson of ZEUS. He was believed to have a shoulder of ivory, a replacement after his father TANTALUS had served him at a banquet for the gods.

Pelorus, one of the Sparti ("sown men") who sprang from the dragon's teeth which Cadmus planted after he had killed the dragon terrorizing Thebes in Greece. Pelorus was a giant. The Theban aristocracy claimed descent from him.

Pelota, generic name given to a variety of games in which a small, hard ball is hit with gloved or bare hand or with a basket-type racket known as a *cesta.* They are played mostly in the Basque regions of Spain and France and in South America. Pelota games can be played across a net or against a wall in a two- or three-sided court known as a *fronton. Jaï-alaï,* the fastest of all pelota games, is played by professionals.

Peltier effect, phenomenon in which heat is either given off or absorbed at a junction where an electric current passes from one kind of metal to another. With a current through a THERMOCOUPLE, the temperature at one junction increases while that at the other decreases, so heat is transferred from one to the other.

Pelvis, bowl-shaped bone structure that supports the soft internal organs of the

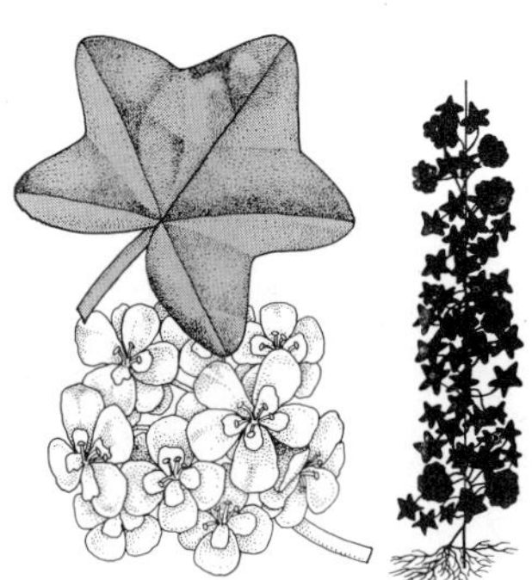

Pelargonium: the large group of plants includes the ornamental geranium.

Pelé: his game has thrilled football fans and inspired players world-wide.

Pelicans occasionally fish collectively, driving fish into shallow water.

Peloponnese: timber and straw huts are typical of rural southern Greece.

Pemba

Pembrokes' castle is in a good state of repair, despite its turbulent history.

Pendulums swing backwards and forwards at a constant rate through any small arc.

Penguin: king penguins are amongst the most popular attractions at the zoo.

Arthur Penn's film *Bonnie and Clyde* starred Faye Dunaway and Warren Beatty.

lower abdomen. The pelvis is formed by the hip bone on each side, joined at the sacro-iliac joint, with the sacrum in the back, and with the pubic bones in front. The pelvis is broader in females than in males and this facilitates childbirth. *See also* MS p.57.

Pemba, coral island in the W Indian Ocean, 64km (40 miles) E of mainland Tanzania; the capital is Wete. Formerly a centre of the slave trade, it is now a major producer of cloves. Area: (including islets) 982sq km (379sq miles). Pop. (1967) 164,321.

Pembroke, Mary Herbert, Countess of (1561–1621), sister of the poet Sir Philip SIDNEY. She was the patron of many poets, including SPENSER, and founded the Pembroke circle, a literary circle devoted to her brother's poetic ideals.

Pembroke, William Marshal, 1st Earl of (c.1146–1219), English nobleman. He was in the service of Prince Henry in 1170 and of both HENRY II and RICHARD I. He upheld John's claim to the throne in 1199 and supported him against the barons. As regent for HENRY III (1216), he concluded the Treaty of Lambeth (1217).

Pembroke, market town in Dyfed, SW Wales, on Milford Haven inlet. The town has an 11th-century priory and castle. HENRY VII was born there in 1457. Industries: tourism, engineering. Pop. (1971) 14,092.

Pembrokeshire, former county in SW Wales; since 1974 it has been part of Dyfed. The area has rolling hills and fertile valleys, and the coastline is deeply indented. The principal rivers are the East and West Cleddau. The economy is overwhelmingly agricultural. Potatoes, wheat and oats are the chief crops, and poultry and dairy cattle are raised. Milford Haven is an important oil-terminal with refineries. The region attracts many tourists. Area: 1,590sq km (614sq miles). Pop. (1971) 97,295.

Pemmican, food of North American Indians, made by drying and pummelling lean meat, then mixing it to a paste with fat and berries. Because it is slow to deteriorate it was a staple of hunters away for long periods in Arctic regions.

P.E.N., abbreviation of the International Association of Poets, Playwrights, Editors, Essayists and Novelists, an international literary organization founded in 1921 by Mrs C.A. Dawson Scott to promote respect for literature and friendship between professional writers. Its presidents have included John GALSWORTHY, H.G. WELLS, Heinrich BÖLL and Maurice MAETERLINCK. Bernard SHAW and Thomas MANN were both active members. Its current membership in 58 countries is more than 8,000. International congresses are held annually in different capitals: London in 1976 and Sidney in 1977.

Penal reform. *See* PRISON REFORM.

Penal settlements, colonies to which convicted criminals were transported for imprisonment and/or hard labour. Areas chosen for such settlements were usually in wilderness or desert and deliberately isolated; occasionally further settlements of freed prisoners grew up nearby. British convicts were sent to North America until the War of AMERICAN INDEPENDENCE, and to Australia until the mid-19th century; the French sent their convicts to their colonies in Africa, to New Caledonia and to French Guiana; and in much the same fashion, some prisoners of the USSR are transported to Siberia.

Penance, the carrying out of a specified act as a mark of sincere regret for sin previously committed. The most common penance, prescribed by a priest after ABSOLUTION, is to say a prayer at a special time, or to repeat it. The term penance is also used to describe the threefold sacrament of CONFESSION, absolution and penance.

Penang, island of Malaysia, off the W coast of the Malay Peninsula; it comprises a state of Malaysia together with Province Wellesley on the Malay Peninsula. Penang, the island's capital, is the principal port of the country. The island was Britain's first possession in Malaya (1786), bought by the British East India Company from the Sultan of Kedah; it was made a Crown colony with Singapore and Malacca. Penang joined the Federation of Malaya in 1957 and became a state of Malaysia in 1963. Its products include rice, rubber and tin. Area (state): 1,040sq km (400sq miles). Pop. (1970) 776,770.

Penates, in Roman mythology, spirits of the household, worshipped privately with images and offerings of food and also regarded as public gods whose protection encompassed Rome itself.

Pen-chi, or Penki, city of S Liaoning province in China, founded in 1915. Its major industries are the mining of iron ore and coal, and it is served by the railway which runs from Shen-yang to North Korea. Pop. (1970 est.) 750,000.

Penda (d.654), Anglo-Saxon King of MERCIA. He was a pagan and killed the Christian Northumbrian kings EDWIN (in c.633) and Oswald (in 642). By defeating Cenwalh of Wessex in 645 he temporarily ruled England. He was slain by Oswiu of Northumbria.

Pendentive, one of the means by which a circular dome is supported above a square or polygonal base or compartment. The pendentive, an inverted concave triangle, springs from a corner of the square or polygonal base, curving upwards and out to meet other pendentives. Their bottom points form a square and the tops form a circle, thus allowing a dome to be placed on a circle from a square base. It was first used on a grand scale by the Byzantines (HAGIA SOPHIA) and is a feature of Romanesque, Renaissance and later architecture. *See also* HC1 p.143.

Penderecki, Krzysztof (1933–), Polish composer. His works, which employ a variety of modern compositional techniques, are original both in style and notation but sometimes retain traditional forms. His reputation was established in 1960 with his *Threnody for the Victims of Hiroshima* for string orchestra. Other pieces include a *Passion According to St Luke* (1963–65), one opera, *The Devils of Loudon* (1968) and *First Symphony* (1973).

Pendlebury, John Devitt Stringfellow (1904–41), British archaeologist who specialized in Minoan civilization. He was curator at KNOSSOS from 1928 to 1934 and excavated Tell et Armarna.

Pendulum, any swinging body supported at a point. A simple pendulum consists of a small heavy mass attached to a string or light rigid rod, and its frequency (rate of swing) is independent of the mass. Small oscillations of such a pendulum have a frequency of $\frac{1}{2}\pi\sqrt{g/L}$, where L is the length and g is the acceleration due to gravity. A compound pendulum has a supporting rod whose mass is not negligible, and its motion cannot be described as a simple relation. *See also* SU p.76.

Penelope, in Greek mythology, wife of ODYSSEUS. She is depicted as a woman of great beauty, fine character and righteous conduct. As described in HOMER's *Odyssey*, she had been married for only a year when her husband left for ten years of war and ten of wandering. She remained faithful, putting off her many suitors with the promise that she would choose one when her weaving was done. By day she wove and by night she undid her work.

Peneplain, in geomorphology, area of land worn down by EROSION almost to a level plain. The sides of a river valley may be eroded by the river, wind, rain and ice, until it becomes a wide, flat plain drained by slow meandering streams. *See* PE pp.107, 114.

Penfield, Wilder Graves (1891–1976), Canadian neurosurgeon, *b.* USA. In treating epilepsy patients, he helped to map the sensory and motor areas of the cortex of the brain by stimulating various areas with tiny electrical probes. His works include *Epilepsy and Functional Anatomy of the Human Brain* (1954) and *Speech and Brain Mechanisms* (1960). *See also* MS p.44.

Penguin, flightless bird that lives in the Southern Hemisphere and ranges from the Antarctic northwards to the Galápagos Islands. The wings of penguins have been adapted to flippers and their webbed feet help to propel their sleek bodies through the water. As a result, although they are awkward on land, they are fast and powerful swimmers, easily able to catch the fish and squid that they feed on. The largest of them are the king and emperor penguins. Height: to 1.22m (4ft). Family Spheniscidae. *See* NW pp.*33, 140*, 144, *147*, 148, *149*, 225, *239*.

Penguin Books, British PAPERBACK publishing company. Founded in 1935 by Sir Allen LANE, a director of The Bodley Head, it was the first British firm to publish paperbacks. The original company has diversified into several imprints, which include Pelican, Puffin and Peregrine.

Penicillin, ANTIBIOTIC agent, one of the greatest discoveries of 20th-century medical science. Discovered by Sir Alexander FLEMING in 1928 and later developed in a soluble form, it is derived from moulds and can now be produced synthetically. It is effective in combating many bacterial disorders. It can produce allergic reactions, ranging from itching to, rarely, fatal shock. *See also* MS pp.*119*, 121.

Penillion, Welsh form of folk-singing in which the singer extemporizes a melody in "counterpoint" to a tune and variations played on the harp. The first written examples date back to the 17th century, although penillion singing is of much older bardic origin.

Peninsular War (1808–14), phase of the NAPOLEONIC WARS, fought between France (under NAPOLEON) and Britain in the Iberian peninsula. In Aug. 1808 British troops under Arthur Wellesley (later Duke of WELLINGTON) landed in Portugal to aid Spanish resistance to French invasion. In Jan. 1809 these troops, then under Sir John MOORE, were re-embarked at La Corunna, but Wellesley returned with another small army in April 1809, and this formed the nucleus of the force which drove the French out of Spain in 1813 (at the Battle of Vitoria) and finally defeated a French army at Toulouse in 1814, just before news of Napoleon's abdication ended hostilities. *See also* HC2 pp.*78–79*.

Penis, male copulatory organ. It contains the URETHRA, and three columns of erectile tissue which, when congested with blood, cause the penis to become erect. *See also* MS pp.74–75, *74–75*, 102–103.

Penitence, deep remorse for SIN committed, often allied with a desire to do PENANCE and avoid further sin. It is not to be confused with repentance.

Penn, Arthur (1922–), US stage and film director. A recurrent theme in his work is that of the individual misplaced in society. His films include *Bonnie and Clyde* (1967), *Alice's Restaurant* (1969), *Little Big Man* (1970) and *Missouri Breaks* (1976).

Penn, Sir William (1621–70), English admiral, father of William PENN. He fought for the PARLIAMENTARIANS in the Civil War and in the First Anglo-Dutch War (1652–54) and captured Jamaica in 1655. Secretly negotiating with the Royalists, he was made Commissioner of the Navy in 1660 and fought in the Second Anglo-Dutch War (1665–67), *See also* HC1 pp.*290*, 293.

Penn, William (1644–1718), founder of the US state of Pennsylvania, *b.* England. He was the son of Admiral William PENN and a leader of the English QUAKERS. Because of his religious views, he was imprisoned four times and wrote *No Cross – No Crown* (1669), explaining Quaker-Puritan morality, while he was in the Tower of London. He persuaded KING CHARLES II to honour an unpaid debt by granting him wilderness land in America to be settled by the Quakers and others seeking religious refuge. In organizing a government for the colony, Penn drew up a frame of government, the first constitution with an amendment clause. *See also* HC2 pp.62–63, *63*.

Penney, William George, Baron (1909–), British atomic physicist who helped to develop the first British atomic bomb. During WWII he was engaged in nuclear research for the British government; in 1944 he went to the Los Alamos Scientific Laboratory and in 1945 was official observer of the atomic bombing of Nagasaki. In 1964 he became chairman of the UK Atomic Energy Authority.

Pennines, range of hills in N England, also known as the Pennine Chain, extending from the Tyne Gap and Eden Valley on the border with Scotland to the valley of the River Trent; commonly referred to as the backbone of England. The hills are a series of highland blocks dissected by rivers such as the Tees, Aire and Ribble. The rearing of sheep is the chief occupation. Tourism and limestone quarrying are also important. The highest peak is Cross Fell, rising to 893m (2,930ft). Length: approx. 260km (160 miles).

Pennsylvania, state in E USA; one of the Middle Atlantic states. Apart from small low-lying areas in the NW and SE, it is composed of a series of mountain ridges and rolling hills with narrow valleys. Farming is concentrated in the SE; the principal crops are cereals, tobacco, potatoes and fruit, and dairy products are important. Pennsylvania has rich deposits of coal and iron ore. The state has long been a leading producer of steel which today accounts for about a quarter of the nation's output. There is much heavy industry, including chemicals, machinery and metal goods. The chief cities are Harrisburg (the state capital), Philadelphia, Pittsburg, Scranton, Bethlehem, Wilkes-Barre and Erie.

Swedish and Dutch settlements were made along the Delaware River in the mid-1600s. By 1664 the area was controlled by the English, and William PENN received a charter from Charles II in 1681. Philadelphia was capital of the colonies during the War of AMERICAN INDEPENDENCE. The DECLARATION OF INDEPENDENCE was signed and the US constitution was ratified in the city. Philadelphia was the national capital from 1790 to 1800. The Union victory at GETTYSBURG in July 1863 was a turning point in the AMERICAN CIVIL WAR. During both world wars, Pennsylvania was a major source of war materials. Area: 117,412sq km (45,333sq miles). Pop. (1975 est.) 11,827,000. *See also* MW p.175.

Penny, coin dating back to Anglo-Saxon times. Originally 240 pennies were cut from one pound weight of silver. Pennies were cast in copper after 1797, and in bronze from 1860. The penny's former symbol (*d*) stems from its equation with the Roman coin the *denarius*. When the British currency was decimalized in 1971, the penny was replaced by the "new penny" (*p*), worth one-hundredth of a pound sterling.

Penny-farthing, bicycle with a large front wheel and a much smaller rear one. It was invented by James STARLEY, and first appeared in England in the early 1870s. *See also* MM p.*123.*

Pennyroyal, common name for a number of plants, including the European *Mentha pulegium,* a sweet herb of the MINT family. It has purple flowers and the leaves are said to discourage mosquitoes.

Penology, branch of CRIMINOLOGY that studies and evaluates the programmes, institutions and organizations that treat criminal offenders. *See also* MS p.296.

Penry, John (1559–93), Welsh Puritan pamphleteer. He was accused of being the author of anti-episcopal pamphlets, published in 1588–89 under the pseudonym of Martin Marprelate. In 1590 Penry was forced to flee to Scotland as searches for his secret press intensified. He returned in 1592 and was arrested, tried and hanged on the charge of writing with intent to incite rebellion.

Pensées, collection of notes on spirituality by the French scientist and JANSENIST, Blaise PASCAL, first published in 1670. It rejects Cartesian Rationalism and argues that God can be discovered only mystically and through Christ. Faith depends, not on rational proof, but on a wager.

Pentagon, the office building of the US Department of Defense in Washington, DC. The complex is made up of five concentric buildings and covers 14 hectares (34 acres). It was completed in 1943. The name Pentagon has come to signify the US military establishment.

Pentameter, verse line of five metrical feet, introduced into English literature by CHAUCER. IAMBIC pentameter, with metrical feet of unaccented followed by accented syllables, is the most durable form of English poetry, of which outstanding examples are William SHAKESPEARE's *Sonnets,* Edmund SPENSER's *The Faerie Queene* and John MILTON's *Paradise Lost.*

Pentateuch, first five books of the Bible, traditionally attributed to MOSES and in Judaism collectively called the Torah. They comprise GENESIS, EXODUS, LEVITICUS, NUMBERS and DEUTERONOMY, the last of which, being also the latest, was written no earlier than 700 BC.

Pentathlon, athletic competition which originated in ancient Greece. It consisted of five events – foot racing, leaping, wrestling, discus- throwing and javelin throwing – all taking place on the same day. It was made an Olympic Games event in 1912. *See also* MODERN PENTATHLON.

Pentatonic scale, in music, scale containing five notes, the octave being reached on the sixth note. Some of the notes are therefore more than a tone apart. It represents an early stage in musical development and is found in primitive and oriental music and some early Gregorian chants. The most common form is without semitones: c-d-f-g-a-c. Some 20th-century composers, including DEBUSSY, have exploited the form for special effects.

Pentecost, in the Jewish calender, the festival seven weeks after the second day of the PASSOVER, commemorating the giving of the Law to MOSES. It is also a celebrated harvest festival, the Day of the First Fruits. In the Christian calender it is known as Whitsunday (seven weeks after Easter) and considered to be the Church's birthday since, according to the ACTS OF THE APOSTLES, it was on this day that Jesus's disciples were in one place together when the HOLY GHOST came down on them like tongues of fire.

Pentecostal Churches, fellowship of Christian believers who stress that its members should seek the Baptism with the Holy Spirit which may be accompanied by a supernatural gift such as the ability to "speak in tongues". The Pentecostal movement, which sought to model itself on instances and practices recorded in the Acts of the Apostles, began in the USA at the beginning of the 20th century and spread rapidly to Britain and other parts of the world. *See also* ELIM FOURSQUARE GOSPEL ALLIANCE.

Penthesileia, in Greek mythology, queen of the AMAZONS. After the death of HECTOR she came to the aid of the Trojans against the Greeks. She was slain by ACHILLES, who mourned her death because of her beauty and courage.

Pentheus, in Greek mythology, king of THEBES. When DIONYSUS came to Thebes, Pentheus refused to allow his people to worship him. They refused to obey. Pentheus went out to spy on the worshippers, was detected, and torn to pieces by women led by his mother Agave. The tragedy is related in *Bacchae,* by EURIPIDES.

Pentimento, in painting, term for a mistake which the artist has painted out, often with an opaque pigment. The original may later become visible through the superimposed layers of paint.

Pentland Firth, channel separating the Orkney Islands from the N Scottish mainland, connecting the Atlantic Ocean to the North Sea. Length 23km (14 miles).

Pentothal, trade name of thiopental, a short-acting BARBITURATE used as an intravenous anaesthetic and for abreaction. Side effects include restlessness, excitement and delirium.

Pentstemon, or beardtongue, genus of North American perennial plants and shrubs. They have terminal clusters of tubular, white or purple flowers with five stamens, one of which is sterile and bearded. Family Scrophulariacea. *See also* FIGWORT.

Penza, city in the central Russian SFSR, USSR, the administrative centre of the Penza region, on the River Sura, 560km (350 miles) SE of Moscow. Founded in 1666 as a fortress, Penza developed in the 18th century as an agricultural centre. Today its industries include machinery, timber, paper and watchmaking. Pop. (1970) 374,000.

Penzance, town in Cornwall, SW England, on Mount's Bay. It is a fishing port and has a sea and air service to the Scilly Isles. The town was sacked by the Spanish in 1595. Industries: tourism, vegetables and flowers. Pop. (1971) 19,352.

Peonage, system of servitude based on the indebtedness of the labourer (peon) to his creditor, prevalent in Spanish America, especially Mexico and Peru. In Mexico a decree against peonage was issued in 1915. The practice persisted but from that date began to go into decline.

Peony, perennial plant native to Eurasia and North America. It has glossy, divided leaves and large white, pink or red flowers, and is frequently cultivated in gardens. Height: to 0.9m (3ft). Tree peonies grow only in hot, dry areas and have brilliant blossoms of many colours. Height: to 1.8m (6ft). Family Ranunculaceae; genus *Paeonia.*

Peoria, city and port of entry in central Illinois, USA, on the Illinois River. Fort Creve Coeur was built on the site in 1680 by Robert La Salle. Abraham LINCOLN condemned slavery in an address there in 1854. It is an agricultural trade centre and shipping point for grain and livestock. Industries: brewing, distilling, farm machinery, bricks, tiles. Pop. (1973) 127,898.

Peperomia, genus of house plants, many of which are SUCCULENTS, native to tropical America with waxy, ridged or smooth, oval or rounded leaves of green, green and white or copper-black. Peperomias grow best in bright indirect light, in soil of equal parts loam, peat moss and sand. They should be kept dry between waterings. Propagation is by stem cuttings. Height: to 25cm (10in). Family Piperaceae.

Pepin, name of two kings of AQUITAINE. Pepin I (*c.*803–38) was the son of Emperor LOUIS I, who made him King of Aquitaine in 817. He led two revolts against his father (830, 833), but later helped him to reassert his authority. Pepin II (*d.c.*870), his son, became king after defeating Charles the Bald in 844. In 852, however, Pepin was captured by Charles. He escaped two years later but was again defeated by Charles after attacking Toulouse in alliance with the Vikings.

Pepin III, the Short (*d.*768), son of CHARLES MARTEL, mayor of Neustria, Burgundy and Provence and King of the Franks (*r.*747–768), the first king of the Carolingian dynasty. He deposed Childeric III in 750, the last MEROVINGIAN ruler, and was anointed by Pope Boniface in 751. A firm supporter of the Roman Church, he was again anointed in 754 by Pope Stephen II and defeated the Lombards on the pope's behalf (754, 756). The new papal lands (the exarchate of Ravenna) thus acquired were known as the Donation of Pepin. This event created the PAPAL STATES, making the papacy a temporal power for the first time. Pepin was the father of CHARLEMAGNE. *See also* HC1 p.152.

Pepper, or capsicum, perennial woody shrub native to tropical America, sometimes cultivated as an ornamental plant. The fruit is a many-seeded, pungent berry whose size depends on the species. Included are bell, red, CAYENNE and CHILI peppers. They all belong to the NIGHTSHADE family, Solanaceae; genus *Capsicum. See also* PE pp.210, *211.*

Pepper, sweet, shrubby perennial plant of the NIGHTSHADE family, sometimes cultivated as an annual, and native to the Americas. The fruits, green or red, are cooked or used raw in salads. An extract of the plant is used in ointments as a heat producer and counter-irritant for rheumatic pains. Height: 91–120cm (3–4ft). Family Solanaceae; species *Capsicum frutescens.*

Peppercorn, dried berry of the black pepper plant, *Piper nigrum,* which is marketed as a spice and is also available ground as black pepper powder. The inner part of the berry yields white pepper.

Peppercorn rent, in law, nominal rent that does not reflect the true rental value of a property or chattel. Its payment does not

Pennsylvania's Golden Triangle, where the Allegheny and Monongahela rivers meet.

Penny-farthing: this 1885 model has the refinement of a separate chain-wheel.

Penzance: sea-urchins for sale on the harbour wall of this Cornish town.

Peony: the common peony blooms in the late spring or early summertime.

Peppermint

Samuel Pepys' diary is a document of great humour and human insight.

Perch are predatory fish which lie in wait behind the stems of water plants.

Père David's deer are named after the French missionary and naturalist.

Giovanni Pergolesi; portrait by Vaccaro of the Italian composer.

normally exempt the tenant or hirer from having to pay the true rental value should it be claimed. The term comes from the early 17th-century usage of the word peppercorn, meaning something small or insignificant.

Peppermint, common name for *Mentha piperita*, a perennial herb of the MINT family cultivated for its essential oil, which is distilled and used in medicine and as a flavouring. *See also* NW p.*59*.

Pepper tree, or Peruvian mastic tree, evergreen tree native to tropical America. Grown as an ornamental, it has feather-like leaves, yellow flowers and reddish berries. Height: to 15m (50ft). Family Anacardiaceae; species *Schinus molle*.

Pepsin, digestive ENZYME secreted by glands of the stomach as part of the gastric juice. In the presence of hydrochloric acid it catalyzes the splitting of PROTEINS in food into smaller PEPTIDE fractions. *See also* MS p.68.

Peptide, compound consisting of two or more AMINO ACIDS linked by bonds between the amino group ($-NH_2$) of one and the carboxyl group ($-COOH$) of the next. This type of linkage is called a peptide bond, and peptides containing three or more amino acids are called polypeptides. PROTEINS consist of polypeptide chains containing up to several hundred amino acids, cross-linked to each other in various ways. *See also* SU pp.154–157.

Pepusch, Johann Christoph (1667–1752), British composer and musical theorist; *b.* Germany. He was a prolific composer of instrumental music and music for the theatre and is remembered for the arrangement of the tunes and the composition of overtures for John GAY's *The Beggar's Opera* (1728).

Pepys, Samuel (1633–1703), English diarist and naval administrator who became Secretary of the Admiralty (1673–89). His *Diary* (1660–69) frankly describes his private life and English society of his time. It includes a vivid account of the RESTORATION, the coronation ceremony in 1661, the PLAGUE in 1663 and the Great FIRE OF LONDON in 1666. Pepys' *Diary* was written in cipher (a system of shorthand) and a deciphered version was not published until 1826, by the Rev. John Smith. *See also* HC1 pp.294–295.

Pequot, tribe of Algonquian-speaking North American Indians who once lived in the Thames River valley region of Connecticut. They were almost entirely wiped out in the Pequot War (1637). In the mid-1970s there were only about 200 people of Pequot ancestry.

Perak, state in Malaysia, on the w coast of the Malay Peninsula, bounded on the w by the Strait of Malacca; IPOH is the capital. It was made a British protectorate in 1875, became a member of the Federated Malay States in 1895 and was made a state of Malaysia in 1963. Its products include rubber, coconuts and rice. Area: 20,798sq km (8,030sq miles). Pop. (1970) 1,562,566.

Percentage, quantity expressed as the number of parts in 100 (considered to be a whole). Thus ¾ as a percentage is obtained by multiplying ¾ by 100, ie *75%*.

Perception, function by which the nervous system transforms energy into an impression of the world. The energy may be external (light, sound waves) or internal (stimulation of muscles and tendons). The end-product of the process is sometimes described as experience and sometimes as behaviour. Perception can be called information processing. In this view, the central nervous system is analogous to a computer with the energy as input and perception as the output of the process. *See also* MS p.48.

Perceval, Spencer (1762–1812), British statesman, son of the 2nd Earl of Egmont. A lawyer, he entered Parliament in 1796 and supported William PITT. He rose from Solicitor General in 1801 to Prime Minister in 1809. He was assassinated by a bankrupt broker.

Perch, freshwater food fish found in lakes, ponds and slow-moving streams of Europe and the USA E of the Rocky Mts. The North American yellow perch (*Pesca*

flavescens) is gold-coloured with six to nine black side-bars. The European perch (*P. fluviatilis*) is found throughout most of Europe. Eggs are laid in long, sticky ribbons. Weight: 1.0–2.7kg (2.2–6lb). Family Percidae. *See also* PE p.244, NW p.*130*.

Percheron, black or grey draught horse bred in the La Perche region of NW France from Flemish and Arabian stock. Introduced to the USA around 1840, it was the most popular draught breed, particularly for farm work. Height: to 170cm (67in) at shoulder. Weight: to 950kg (2,100lb).

Percussion, term for any of several musical instruments which produce their sound when struck with a beater or the hand. They are divided into two groups: ideophones, in which the whole object struck vibrates (such as CYMBALS, GONGS and XYLOPHONES), and membranophones, in which a stretched skin or membrane vibrates a column of air (this group includes all DRUMS). Percussion instruments of either group can be of definite or indefinite pitch. The expressive range of such instruments has long been appreciated by Indian, Japanese and African musical cultures, and in the West began to be exploited during the 20th century. *See also* HC2 p.105; SU p.82.

Percussion cap, device for detonating the charge in a firearm. The principle on which it works was discovered by Alexander J. FORSYTH in 1805, but the actual cap was probably invented in 1815 by Joshua Shaw in Philadelphia, USA. Shaw's cap was a truncated cone of metal containing fulminate of mercury, which ignited when struck by the hammer of the gun.

Percy, Sir Henry (Hotspur). *See* HOTSPUR.

Percy, Thomas (1729–1811), British antiquary, poet and churchman. He is remembered for his collection of traditional ballads published in 1765 as *Reliques of Ancient English Poetry* (also known as *Percy's "Reliques"*). He was chaplain to the Earl of Northumberland and in 1782 was appointed Bishop of Dromore. *See also* HC2 p.72.

Percy family, English noble family of Northumberland. Henry, 1st Earl of Northumberland (1342–1408), fought in France in the HUNDRED YEARS WAR and was Warden of the Scottish Marches. He was created an Earl in 1377. His son, Henry (1364–1403), is generally known as HOTSPUR. Henry, the 4th Earl (1446–89), was a supporter of RICHARD II, but went over to HENRY VII at Bosworth. He was killed by Yorkshire rebels. Thomas, the 7th Earl (1528–72), was Warden of the Scottish marches under MARY I. He took part in the Northern Rebellion against ELIZABETH I and was executed for treason in 1572.

Père David's deer, animal now extinct in its native Chinese marshlands but still surviving in zoological parks throughout the world. The male has long antlers. Height: 1.2m (45in) at shoulder. Family Cervidae; species *Elaphurus davidianus*.

Peregrine falcon, crow-sized, grey, black and white bird of prey. It inhabits craggy open country or rocky coastlines and marshes or estuaries. The largest breeding FALCON in Britain, it flies swiftly with prolonged glides. Length: to 48cm (19in). Family Falconidae; species *Falco peregrinus*.

Peregrinus, Petrus (*fl.* 13th century), French scientist. He described the properties of MAGNETISM and hypothesized that it could be converted to KINETIC ENERGY. He also distinguished between and named the MAGNETIC POLES, and developed the existing rudimentary magnetic compass into a reliable instrument for navigation. *See also* SU p.23.

Pereira, city in w central Colombia; capital of Risaralda department. The city was founded in 1863. It is a market centre for a region producing coffee and cattle, and gold and silver are mined nearby. There is an important textile industry. Pop. (1972 est.) 221,200.

Perennial, plant with a life-cycle of more than two years. It is a common term for flowering herbaceous and woody plants. Perennials are easily cultivated garden plants. They include the LILY, CHRYSAN-

THEMUM, PEONY, DAISY, IRIS, and DELPHINIUM. *See also* ANNUAL; BIENNIAL.

Peretz, Isaac Lieb (1852–1915), Jewish poet, novelist and dramatist. He opposed political ZIONISM and was often accused of radicalism. His early works were in Polish and his later works in YIDDISH. They include Hasidic sketches such as *Stories and Pictures* (1900–01) and the play *The Golden Chain* (1909).

Perez, Rodríguez Carlos Andres (1922–), President of Venezuela (1974–). A member of the Democratic Action Party, he advocated the nationalization of the country's iron, steel and petroleum industries.

Pérez Galdós, Benito (1843–1920), Spanish novelist and dramatist. He wrote a cycle of 46 historical novels, *Episodios nacionales* (1873–1912), describing events in Spain from 1805 to the end of the century. In 1881 he began another cycle of 21 novels examining contemporary society. His work has been described by many as the greatest Spanish literature since Cervantes.

Perfect Fool, The (1923), one-act comic opera by Gustav HOLST with libretto by the composer. It was first produced at Covent Garden, London, and parodies operatic conventions and the works of other composers, particularly Richard WAGNER and Giuseppe VERDI. The title refers to the gullible hero.

Perfume, manufactured substance that produces a pleasing fragrance. The scents of such plants as rose, citrus, lavender and sandalwood are obtained from their essential oils. These are blended and combined with a fixative of animal origin, such as MUSK, ambergris or civet. Fixatives add pungency and prevent the more volatile oils from evaporating too quickly. Liquid perfumes are usually alcoholic solutions containing 10–25% of the perfume concentrate; colognes and toilet waters contain about 2–6% of the concentrate. *See also* MM pp.246–247.

Perga, ancient town of Pamphylia, Asia Minor, 16km (10 miles) NE of modern Antalya, Turkey. ALEXANDER THE GREAT began his invasion of inner Asia Minor from this point in 334 BC, and it was an important town in Roman times. Its ruins include a theatre and a stadium.

Pergamum, ancient city of NW Asia Minor in Mysia (Turkey). It became the capital of an important kingdom under the Attalid dynasty in the 3rd century BC and flourished as a centre of Hellenistic civilization. An agricultural and mining centre, it was also known for its arts and cultural development. It was bequeathed to Rome by Attalus III in 133 BC.

Pergolesi, Giovanni Batlista (1710–1736), Italian composer who, despite his early death, produced the intermezzo *La Serva Padrona* (1733), which became a model for Italian COMIC OPERAS. His *Stabat Mater* (1730) is one of the finest examples of religious music of the BAROQUE period. *See also* HC2 p.41.

Peri, Jacopo (1561–1633), Italian composer. His musical drama *Dafne* was written in collaboration with Jacopo Corsi, with libretto by Ottavio RINUNCCHINI and is generally regarded as the first opera. It received its first private performance in Florence in 1597. Most of this music is now lost, and his next opera *Euridice* (1600), remains the earliest opera which has been preserved in its entirety.

Peri, or Pari, in Persian mythology, fairy or other supernatural being. Peris were considered evil and were held responsible for droughts, crop failures and other natural disasters.

Pericarditis, inflammation of the PERICARDIUM, the membranous sac enclosing the heart. It usually develops during RHEUMATIC FEVER but may also be associated with other disorders. Treatment is directed towards the underlying infection and in acute cases involves surgery.

Pericardium, double membrane that surrounds the HEART, separating it from the rest of the chest cavity and protecting it from mechanical injury. In man, its lower margin is anchored to the DIAPHRAGM, its upper part to the inner surface of the breastbone. The outer tough fibrous pericardial layer is separated from the inner membrane by lubricating pericardial fluid. *See also* MS p.62.

Pericarp, in seed plants, the wall of a ripened carpel, ovary or fruit. The tissues of the pericarp may be fibrous, stony or fleshy.

Perichondrium, in a developing EMBRYO, membrane surrounding the CARTILAGES that eventually become bones. It is well supplied with blood vessels and is the source of the osteoblasts (bone-building cells) that invade the embryonic cartilage and lay down the hard bone matrix. *See also* MS p.56.

Pericles (*c.*490–429 BC), Athenian states-man who was elected to high office every year from 443 to his death. He strengthened Athens' leadership of the Delian League, made a 30-year truce with SPARTA in 445, and attempted to make Athens a great cultural centre. He was responsible for constructing the PARTHENON (447–438 BC) and other notable buildings. His character, oratory (which inspired Athens to war against Sparta in 431) and his rebuilding of the ACROPOLIS caused historians to christen his entire period the "Age of Pericles". *See also* HC1 pp.70, 73, 78, 78.

Pericles, Prince of Tyre (*c.*1608), five-act drama, attributed to William SHAKE-SPEARE, but his authorship is generally regarded as doubtful. Many scholars believe that Shakespeare wrote only Acts III, IV and V.

Peridot, gem variety of transparent green OLIVINE, a silicate mineral. Large crystals are found on St John's Island in the Red Sea and in Burma.

Peridotite, term derived from PERIDOT, the French word for OLIVINE. It is a heavy IGNEOUS ROCK of coarse texture composed of olivine and pyroxene with small flecks of mica or hornblende. It alters readily into SERPENTINE. Rocks that consist mainly of olivine are called dunites. *See also* PE p.98.

Perigee, point in the orbit about the Earth of the Moon or an artificial satellite at which the body is nearest to the Earth. *See also* SU p.224.

Perihelion, point in the orbit of a planet, asteroid, comet or other body (such as a spacecraft) moving around the Sun at which the body is nearest the Sun.

Perimeter, distance around the boundary of a plane closed figure. The perimeter of a circle is its circumference; that of a polygon is the sum of the lengths of its sides.

Period, in astronomy, time taken by one celestial body (such as planet or satellite) to complete one revolution around another. It can be measured in several ways, such as the SIDEREAL PERIOD and the SYNODIC PERIOD.

Periodic law, in chemistry, law first stated by MENDELEYEV in 1869 asserting that the properties of the elements are a periodic function of their ATOMIC WEIGHTS. The groupings of the elements based on this law formed the forerunner of the PERIODIC TABLE. From gaps in these groupings Mendeleyev was able to predict the existence and properties of undiscovered elements. But his table contained anomalies, which were not resolved until H.G.J. Moseley (1887–1915) discovered that periodicity was related to ATOMIC NUMBER (rather than atomic weight) and the later discovery of ISOTOPES. *See also* SU p.134.

Periodic table, arrangement of the chemical elements in order of their ATOMIC NUMBERS in accordance with the PERIODIC LAW stated by MENDELEYEV and later modified by H.G.J. Moseley. In the modern form of the table, the elements are arranged into 18 vertical columns and seven horizontal periods. The vertical columns are numbered 0 to VIII. Columns I to VII are divided into two subgroups, the A subgroup forming the main group and the B subgroup containing the TRANSI-TION ELEMENTS; group VIII contains the iron group of transition elements, and the NOBLE GASES are collected into a ninth group, group O. The elements in each group have the same number of VALENCY electrons and accordingly have similar chemical properties. Elements in the same horizontal period have the same number of electron shells. The elements are

arranged in the periods in order of increasing atomic number from left to right. *See also* SU p.134, *134.*

Periodontics, field of dentistry that deals with the prevention, control and treatment of disorders of the tissues that surround and support teeth, especially the GUMS. PYORRHOEA is the most common gum disorder treated. *See also* MS p.130.

Periodontitis. *See* PYORRHOEA.

Peripatus, genus including any of 65 living species of the subphylum Onychophora, intermediate in structure between ANNE-LIDS and ARTHROPODS. It is confined to humid habitats in tropical and subtropical regions. The worm-like body has about 20 pairs of legs ending in tiny claws. Length: 15cm (6in). *See also* NW pp.98–99.

Peripheral nervous system, 12 pairs of cranial nerves, enclosed in the skull, which control activities in the head and neck, including sight, smell, hearing and taste.

Periscope, optical instrument consisting of a series of mirrors or prisms that reflect or displace the normal direction of the observer's line of sight so that it is possible to see over or around obstacles. *See also* SU p.100.

Perissodactyl, a member of an order of mammals characterized by hoofs with an odd number of toes. The only living members of the order are HORSES, TAPIRS and RHINOCEROSES; there are more than 200 extinct forms known from fossils. *See also* NW pp.155, 162–163.

Peristalsis, series of involuntary muscular contractions that move food along the digestive tract. The route is from the pharynx along the oesophagus to the stomach, then from the stomach through the small and large intestines. In what is known as mass peristalsis (occurring about three or four times every 24 hours), the faeces eventually reach the rectum where nervous impulses lead to defecation. Antiperistalsis (the reverse process) in the oesophagus causes food to be regurgitated. *See also* MS p.68.

Peritoneum, strong membrane of CON-NECTIVE TISSUE that lines the body's abdominal wall and surrounds the abdominal organs. The greater omentum, a fold in the peritoneum, forms an apron over the intestines. Inflammation of the peritoneum is known as PERITONITIS.

Peritonitis, inflammation of the PERI-TONEUM, a membrane that lines the abdominal cavity and the organs within it. It may be general or local and is caused by infection, which causes an accumulation of pus in the abdominal cavity. The principal symptom is severe abdominal pain with vomiting. Treatment is directed at the underlying cause, although in severe cases surgery may be necessary.

Periwinkle, or winkle, any of several marine snails, gastropod molluscs that live in clusters along marine shores. A herbivore, it nestles in cracks among rocks, rather than clinging as does the LIMPET. Many are edible. Length: to 2.5cm (1in). Family Littorinidea; genus *Littorina.*

Periwinkle, any of several species of trailing or erect evergreen plants that are cultivated as ground cover and for hanging baskets. The common creeping periwinkle of W Europe, *Vinca minor,* has small blue, white or pink flowers. Family Apocynaceae.

Perjury, act of bearing false witness under oath. In English law it is an indictable offence, punishable by imprisonment.

Perkin, Sir William Henry (1838–1907), British chemist who, in 1856, accidentally discovered the first artificial dye, aniline purple or mauve, while trying to synthesize quinine. He patented his invention and, a year later, set up a factory for its production, so creating the aniline dye industry.

Perlis, smallest state in Malaysia, bordered by Thailand (N and E), Kedah (SE) and the Andaman Sea (SW). It was made a separate state by the Siamese in 1841, became a British protectorate in 1909, a member of the Federation of Malaya in 1957 and in 1963 a state of Malaysia. Area: 803sq km (310sq miles). Pop. (1970) 121,062.

Perlon, synthetic POLYMER fibre, similar to NYLON, obtained by the POLYMERIZATION

of Caprolactam. Commercially important, it is used for fishing lines, hosiery, tow ropes and in the manufacture of tyres.

Perm, city of the Russian SFSR, USSR, W of the Ural Mountains on the Kama River; capital of Perm oblast. Founded in 1781, it grew rapidly during the 19th century as industry developed. It was known as Molotov from 1942 to 1958. Industries: timber, leather, agricultural machinery, metallurgy, oil extraction. Pop. (1970) 850,000.

Permeability, in geology, ability of rock to transmit water; it is not directly related to porosity. Limestone is a permeable but non-porous rock; water percolates only through the joints and fissures. *See also* PE pp.108–109.

Permeability, in physics, ratio of the magnetic flux density in a body to the external magnetic field inducing it. The permeability of free space is called the magnetic constant and has the value $4\pi \times 10^{-7}$ henry per metre. The relative permeability of a substance is the ratio of its permeability to the magnetic constant.

Permeke, Constant (1886–1952), Belgian painter and sculptor who was the leading Flemish exponent of realistic EXPRESSION-ISM with his monumental depictions of workers in everyday activities.

Permian, geological period and a division of the PALAEOZOIC Era lasting from 280 to 225 million years ago. It was of widespread geologic uplift and mostly cool, dry climates with periods of glaciation in the southern continents. Many groups of marine invertebrate animals became extinct during the period; there was a gradual evolution towards modern types and reptiles diversified. Near the end of the Upper Permian period earth disturbances threw up the Appalachian Mts. *See also* PE pp.*127, 130–131.*

Permittivity, in physics, dielectric property of a material. It is the ratio of the CAPACITANCE of a capacitor (condenser) in which the material is the DIELECTRIC to its capacitance with a vacuum between the capacitor's plates.

Pernicious anaemia, severe ANAEMIA involving progressive decrease in the number of red blood cells, which grow larger in size. It is caused by a deficiency of VITAMIN B_{12}. Treatment is the administration of B_{12} by injection.

Perón, Eva Duarte de (1919–52), Argentine political leader known as "Evita", first wife of President Juan PERÓN. She administered the country's social welfare agencies and was Argentina's chief labour mediator. Her popularity contributed to the longevity of the Peronist regime, and her authority was eclipsed only by that of her husband.

Perón, Juan Domingo (1895–1974), President of Argentina (1946–55, 1973–74), one of the most important political figures of Latin America. Through working-class and union support, and the popularity of his first wife Eva, Perón was elected President in 1946 and again in 1951. The Peronist programme of economic nationalism and social justice gave way to monetary inflation and political violence; Perón was ousted by a military coup in 1955. He returned from exile in Spain and was re-elected president in 1973, but died the following year. *See also* HC2 p.256.

Perón, María Estela (1931–), second wife of Juan PERÓN and first woman chief of state in the Americas, who became President of Argentina when her husband died in office in 1974. She was overthrown on 24 March 1976 in a coup led by the commanders of the armed forces.

Peroxide, chemical compound containing two oxygen atoms united to each other (such as barium peroxide, BaO_2) and yielding a solution of HYDROGEN PEROXIDE (H_2O_2) when treated with an acid. Peroxides are powerful oxidizing agents.

Perpendicular style, in architecture, last period of English Gothic, from the late 14th to the mid-16th centuries. Named after the strong vertical lines of its window tracery and panelling, it is characterized by intricate fan vaulting and flattened arches. Well-known examples of the Perpendicular style include Henry VII's chapel, Westminster Abbey, King's Col-

Periodic law; Mendeleev's discovery speeded the development of chemistry.

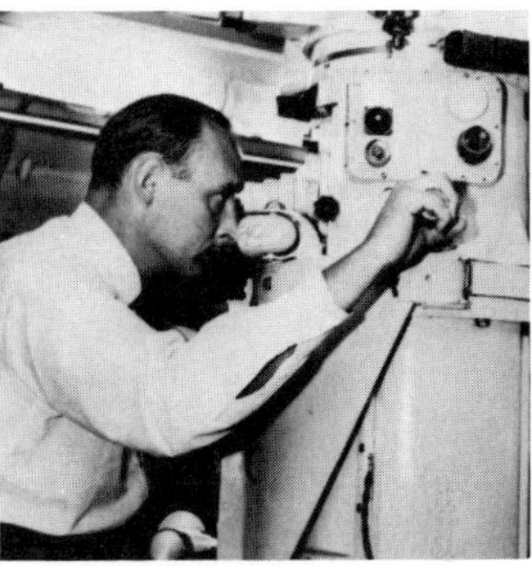

Periscopes are important even in nuclear submarines such as HMS Dreadnaught.

Periwinkle's small blue, white or pink flowers bloom in early spring.

Eva Perón, speaking at the 1951 May Day celebrations in Buenos Aires.

Perpetual motion

Fred Perry was world table tennis champion before turning to lawn tennis.

Persephone's regularly changing abode explained the seasonal variation.

Persepolis; the Persian city was destroyed by Alexander the Great.

Persian cat; pure breeds are rare due to frequent cross-breeding with angoras.

lege Chapel, Cambridge, Sherborne Abbey and Bath Abbey. *See also* HC1 p.*219*.

Perpetual motion, concept of a machine that, once started, would do work continuously without the need for any further input of energy. There are two types of such motion: in the first kind, a machine violates the first law of thermodynamics (conservation of energy); in the second kind, the first law is obeyed but the second law, concerning ENTROPY, is violated. The failure of all the many designs for perpetual motion machines has provided the best verification of the laws of THERMODYNAMICS. *See also* SU p.93, *93*.

Perpignan, city in s France, 155km (96 miles) SE of Toulouse on the River Têt. It was the capital of the Spanish kingdom of Rousillon, passing to France in 1659. It is the site of a 14th-century Gothic cathedral. Today it is a market centre, exporting fruit, vegetables and wine; tourism is also a valuable source of income. Pop. (1968) 102,191.

Perrault, Charles (1628–1703), French critic and poet. He was a member of the Académie Française (1671) and is mainly remembered for his fairy tales. *Tales of Past Time by Mother Goose* was published in 1697. *See also* HC1 p.316.

Perréal, Jean (*c.*1460–1530), French painter. He designed funerary monuments and painted portraits, among them *Charles VIII* and *Anne of Brittany*, both executed *c.*1495, as illuminations.

Perrers, Alice (*d.*1400), mistress of EDWARD III of England. Because of her power and influence (her activities included taking part in lawsuits and sitting on the judges' bench), Parliament passed an act forbidding women to engage in the practice of law. She was banished in 1376, but parliament revoked her banishment three years later.

Perret, Auguste (1874–1954), French architect, a pioneer of REINFORCED CONCRETE as a building material. His apartment house in the Rue Franklin, Paris (1903) was highly influential, as were his garage on the Rue Ponthieu (1905) and Théâtre des Champs Elysées (1910–13), also in Paris. All were built on a reinforced concrete frame, using TRABEATE construction and Neo-Classical features. His Church of Notre Dame at Le Raincy (1922–23) is an early masterpiece in the new material. Perret designed numerous villas and a variety of elegantly monumental industrial buildings. After WWII he contributed designs for the partial rebuilding of Marseilles, Le Havre and Amiens. His research into the standardization and industrialization of building components was influential, notably on his pupil LE CORBUSIER. *See also* HC2 pp.178–179.

Perrin, Jean-Baptiste (1870–1942), French physicist and chemist known for his work on the structure of matter and for his investigation of CATHODE RAYS and BROWNIAN MOVEMENT of particles. Perrin was awarded the 1926 Nobel Prize in physics and wrote several books, including *The Elements of Physics* (1930).

Perronet, Jean Rodolphe (1708–94), French civil engineer. In 1747 he was appointed director of L'Ecole des Ponts et Chaussées (School of Bridges and Roads), the first engineering school in history. He is famous for his stone-arch bridges, such as the Pont de Neuilly and the Pont de la Concorde, in Paris, notable for their flat elliptical arches and narrow piers. *See also* MM p.191.

Perrot, Jules (1810–92), French dancer and choreographer. He trained and partnered Carlotta GRISI, whom he later married. Her title role in *Giselle* (1841) is now generally considered to have been choreographed by him. His other works include *Ondine* (1843), and *Esmeralda* (1844). From 1848 to 1859 he was the leading dancer with the Imperial Theatre, St Petersburg, Russia.

Perry, Fred (1909–), British table tennis and tennis player. World table tennis champion in 1929, he was the first man in the 20th century to win three successive Wimbledon tennis singles titles (1934–36). He also won the US, French and Australian titles and helped Britain to win four Davis Cup challenge rounds (1933–36).

Perry, Matthew Calbraith (1794–1858), US naval officer who established the US Navy's apprentice system in 1837 and organized the first naval engineer corps. From 1833 to 1844 he supervised the application of steam power to warships and in 1837 commanded the first steam vessel in the US Navy, the *Fulton*. He was also responsible for opening-up Japan to the West (1853–54) and negotiating a treaty that secured US trading rights there. *See also* HC2 pp.57, *57*, 146, *154*.

Perry, alcoholic drink, similar to cider, made from fermented pear juice. The name comes from the Old French word *Peré*.

Perse, Saint-John. See LÉGER, ALEXIS ST.-LÉGER.

Perseid meteors, meteor shower visible between 25 July and 18 August, associated with the comet 1862 III and radiating from the constellation of PERSEUS. On a clear night, approx. 70 Perseid meteors can been seen per hour. *See also* SU pp.218–219.

Persephone, in Greek mythology, goddess of spring. She was the daughter of ZEUS and the earth goddess DEMETER. When she was abducted by HADES, king of the underworld, to be his wife, Demeter was grief-stricken. No crops grew and famine spread over the earth. To prevent catastrophe, Zeus commanded Hades to release her. He did so, and thus, each year, spring returns to the earth. Persephone was known as Proserpine to the Romans.

Persepolis, ancient ruined city in Persia, approx. 48km (30 miles) NE of Shiraz, SW central Iran. It served as the capital of the ACHAEMENID Empire 539–330 BC and was the site of the palaces of DARIUS I and Xerxes. Nearby are the ruins of Istakhr which served as the capital of the SASSANIDS *c.* AD 200. Excavations at another nearby site have revealed remains of a village dating from about 4000 BC. *See also* HC1 pp.60–61.

Perseus, in Greek mythology, son of DANAË and ZEUS, who came to her disguised as a shower of gold. Perseus was among the greatest heroes. He beheaded the snake-haired Gorgon MEDUSA, turned ATLAS to stone and rescued the princess ANDROMEDA from being sacrificed to a sea monster.

Perseus, northern constellation between Taurus and Cassiopeia. It contains several notable stellar objects including the eclipsing binary Algol (Beta Perseus) and the Sword Handle, which is made up of two open clusters (h and ξ). The brightest star is Alpha Perseus, or Mirphak, of magnitude 1.8. *See also* SU pp.*237*, 254–255, 258, *258*.

Persia, former name of Iran, in sw Asia. The term was taken from the region of Persis in s Iran and used by the ancient Greeks *c.*540 BC to describe the inhabitants of the whole country. The earliest Persian kingdom was that of the Elamites *c.*2700–600 BC. CYRUS THE GREAT in *c.*550 BC conquered most of Asia Minor but his descendant, Darius, was defeated by the Greeks at MARATHON in 490 BC. They in turn were defeated by the PARTHIANS *c.*250 BC who were conquered by the SASSANIDS in AD 200. ISLAM was introduced to Persia by the Arabs who took Asia Minor *c.*640. Much of the country was destroyed by the MONGOLS in the 13th century. Persia was ruled by the QAJAR DYNASTY from 1780 to 1925, when AHMAD SHAH was deposed and REZA SHAH was elected Shah by the majlis, or parliament. The name of the kingdom was changed to Iran in 1935.

Persian, the principal language of IRAN and one of the two official languages of AFGHANISTAN. It is spoken by about 15 million of Iran's 32 million people, and by about 5 million people in Afghanistan. An Iranian language and thus part of the Indo-European family, it is one of the world's oldest.

Persian cat, long-haired breed of domestic cat that first appeared in Europe in the 16th century. It has a wide head with small, well-covered ears, wide-set blue or copper eyes and a short nose. The compact massive body is set on short, thick

legs and the tail is short and full. The colour of the fine-textured, fluffy coat indicates the variety; it may be black, blue, red tabby, cream, smoke or white.

Persian Empire, empire founded by CYRUS THE GREAT in 549 BC by uniting the kingdom of Persia with the Median empire to the north. It was also called the ACHAEMENID empire, after an ancestor of Cyrus. Its expansion was swift. In 546 BC the LYDIAN empire was annexed, in 538 BC the Chaldean, in 525 BC the EGYPTIAN and over the next 50 years further lands to the N and E. The last of the Persian emperors, DARIUS III, was defeated by ALEXANDER THE GREAT at the Battle of Arbela in 331 BC. Thereafter the Persian empire was absorbed into the empire of Alexander. *See also* HC1 pp.60–61.

Persian Gulf, part of the Arabian Sea, between Arabia and Iran; it is called the Arabian Gulf by the Arabs. In ancient times the gulf was a valuable trade and communication route but it declined in importance after the fall of Mesopotamia. Control of the region was disputed by several Eastern powers but Britain became the major power in the area in 1835. After British withdrawal in the 1960s both the USA and the USSR established bases on the gulf. Since the discovery and development of oil reserves further disputes among neighbouring states have arisen.

Persian lamb. See KARAKUL.

Persian Wars, series of wars fought between the Persian Empire and the Greek city-states from 499 BC to 449 BC. In 449 BC the Ionian city-states of Asia Minor rebelled against the rule of the Persian emperor DARIUS I. Their revolts were suppressed by 494. They had been aided by Athens and Eretria, and in reprisal Darius decided to add Greece to his empire. He conquered Thrace and Macedon in 492, but was defeated by the Athenians at MARATHON in 490. In 480 Darius's successor, XERXES I, captured Athens. His fleet was destroyed by the Athenians at SALAMIS (480) and his army destroyed by the Spartans at PLATAEA (479). The later years of the war saw a growth of Athenian power as the Athenian fleet led the assault on the Persians in Asia Minor. *See also* HC1 pp.69–70.

Persimmon, tropical Asian tree of the genus *Diospyros* which produces reddish-orange fruit which is sour and astringent until completely ripe. Commercially sold persimmons come from the Japanese persimmon (*D.kaki*). Family Ebenaceae.

Personality, the emotional, attitudinal and intellectual characteristics of an individual. Psychologists use the term to refer to the more enduring, long-term characteristics of a person rather than short-lived traits or momentary emotional states. Personality traits are assessed in psychology by means of a variety of devices, including personality tests and projective techniques. Among the most influential theories of personality in psychology have been the Freudian approach (PSYCHOANALYSIS) and theories of BEHAVIOURISM. *See also* MS pp.188–189.

Personality disorder, category of mental illness, differentiated from a NEUROSIS in that the sufferer experiences little or no anxiety. There are several kinds of personality disorders: personality pattern disturbances, including schizoid (shy and seclusive), cyclothymic (alternations of depression and elation) and paranoid (delusions of persecution) behaviour; personality trait disturbances, with a single dominant characteristic such as compulsivity; and sociopathic personality disturbances, marked by asocial or antisocial behaviour, without feelings of guilt. Drug and alcohol ADDICTION are also considered to be sociopathic personality disorders. *See also* MS pp.140–141.

Perspective, in art, way of showing three-dimensional objects and spatial relationships in a two-dimensional image. The linear perspective system is based on the idea of parallel lines and planes converging at a vanishing point as they recede into the distance. Using this principle, an artist can give a sense of perceptual space and volume in drawings and paintings. Instead of using central perspective, with a single vanishing point, an artist can also use angular, or oblique, perspective, with two vanishing points. Another kind of parallel perspective, with the viewpoint from above, is common in Chinese painting. Linear perspective was not used in painting until the late 15th century in Italy, when the architects BRUNELLESCHI and ALBERTI developed mathematical rules for perspective that involved a horizon line or viewer's eye level and a vanishing point. DÜRER used these principles in his woodcuts. The use of linear perspective dominated European painting until the end of the 19th century, when CÉZANNE intentionally "flattened" his canvases and other artists used colour and shading to create depth. *See also* SU p.27.

Perspex, trade-name for polymethyl methacrylate, colourless transparent solid plastic. It is made by POLYMERIZATION of methyl methacrylate, and is widely used as a substitute for glass. *See also* MM pp.36, 54.

Perspiration, clear liquid secreted by the SWEAT GLANDS of mammalian skin as a way of regulating body temperature. The rate of perspiration increases in warm weather, after exercise and for emotional reasons. *See also* MS p.60, 60.

Persuasion (1818), novel by Jane AUSTEN written in 1815–16. It describes the constancy of affection between Anne Elliot and Captain Wentworth, in spite of her having rejected him some years earlier. The tone is softer and the satire less cruel than in some of Austen's other novels.

Perth, city in sw Australia, on the Swan River; capital of Western Australia state. Founded in 1829, the city grew rapidly after the discovery of gold at Coolgardie in the 1890s, the development of the port at Fremantle and the construction of railways to the E in the early 20th century. The University of Western Australia was founded in Perth in 1911. Industries: cement, food processing, motor vehicles. Pop. 739,200.

Perth, city in Tayside Region, central Scotland, on the River Tay. Perth, known as St Johnstoun until the 17th century, used to be an important fortress, being strategically placed between the Highlands and the Lowlands. It was the Scottish capital from the 12th century until the mid-15th century. Today Perth is an important cattle market, and has dye works and textile industries. Pop. (1971) 43,051.

Perth(shire), former county in central Scotland; since 1975 it has been part of Tayside and Central regions. The land is mountainous and drained by the Tay and Forth rivers, and the region has lochs Tay, Earn and Rannock. Its fine scenery attracts many tourists. Wheat, sugar-beet, potatoes and fruits, in particular raspberries, are grown on the fertile lowlands. Sheep and cattle are grazed on the hills. Industries include fishing, timber, textiles and whisky distilling. Area: 6,457sq km (2,493sq miles). Pop. (1971) 127,138.

Pertinax Publius, Helvius (126–93), Roman emperor (*r.* Jan.–March 193). His father the praetorian prefect Laetus, only three months after arranging the murder of Commodus to make Pertinax emperor, had Pertinax assassinated because of his attempts to curb the power of the PRAETORIAN Guard.

Perturbation, disturbance of the motion of a celestial body produced by the gravitational influence of one or more other bodies.

Pertussis. *See* WHOOPING COUGH.

Peru, republic of NW South America, the third-largest nation on the continent, and once the home of the Incas. The Peruvian economy depends upon the country's rich mineral deposits, which include lead, copper, zinc, iron, tin, gold and oil. The principal crops are cotton, rice, coffee and sugar, all of which are exported. The capital is LIMA. Area: 1,285,216sq km (496,222sq miles). Pop. (1975 est.) 16,090,000. *See* MW p.142.

Peru current. *See* HUMBOLDT CURRENT.

Perugia, city of central Italy; capital of Perugia province and Umbria region. A major Etruscan city, Perugia passed to Rome in 310 BC. It came under papal rule in 1540. Perugia was the centre of the Umbrian school of painting in the 15th century. Interesting features include the 13th-century city walls and the Maggiore fountain. Industries: food processing, textiles, machinery. Pop. (1975) 135,011.

Perugino, Pietro Vannucci (*c.*1445–1523), Italian painter of the School of FLORENCE whose works show a fine sense of PERSPECTIVE. In 1481 he was commissioned to paint the frescoes in the SISTINE CHAPEL, Rome. Of these, *Christ Delivering the Keys to St Peter* is famous.

Perutz, Max Ferdinand (1914–), British biochemist, *b.* Austria, who studied the X-ray DIFFRACTION of proteins. In 1953, he discovered that adding a heavy atom, such as gold or mercury, to each molecule of HAEMOGLOBIN produces a slightly different diffraction pattern. By this means he demonstrated the structure of haemoglobin, for which he shared the 1962 Nobel Prize in chemistry with John C. KENDREW, who, by a similar method, discovered the structure of MYOGLOBIN.

Pescadores (P'eng-hu Lieh-tao), group of several small islands controlled by Taiwan, in the Formosa Strait, between Taiwan and the Chinese mainland. The main islands are P'eng-hu, Yüweng and Paisha. Products include sweet potatoes, peanuts and coral. Area: 127sq km (49sq miles).

Pescara, seaport city in central Italy, on the Adriatic coast 153km (95 miles) ENE of Rome. It was badly damaged by bombing in WWII. A seaside resort and tourist centre, it also has local industries which include ship building and textile weaving. Pop. (1975) 132,458.

Peshāwar, city in NW Pakistan, 16km (9 miles) E of the Khyber Pass. The city has always had great strategic importance. It was repeatedly attacked from the 1st century onwards by Afghan, Persian and Mongol invaders. It was held by the Sikhs 1834–49, when it was annexed by Britain, becoming an important military outpost. The modern city is famous for handicrafts, carpets and leather goods. Its industries include the manufacture of textiles, chemicals and paper. Pop. (1972) 268,366.

Pest, any plant, insect or animal regarded by man as a threat or annoyance to himself, to his interests or to his environment. Many pests, such as FLIES, MOTHS, BUTTERFLIES and BEETLES and their LARVAE, as well as APHIDS, have to be controlled systematically using PESTICIDES. Vertebrate pests including RODENTS, foxes, rabbits and birds are difficult to control. *See also* PE pp.148, 149–151, 156, 172–173.

Pestalozzi, Johann Heinrich (1746–1827), Swiss educational reformer whose theories formed the basis of modern elementary education. He wrote many books about new teaching methods, including *How Gertrude Teaches Her Children* (1801). He opposed the then prevailing system of learning by memorizing and believed that education should be based on concrete experience.

Pestalozzi villages, series of villages set up in Europe after WWII to provide homes for refugee children, named after the Swiss educationist Johann PESTALOZZI. The first village was built in 1946 in Switzerland under the direction of Robert Conti.

Pesticides, agents used to protect man or crops from the depredations of harmful plants or animals. An agent which kills weeds is known as a HERBICIDE, one which kills insects, an INSECTICIDE. Pesticides are most often chemicals and an important criterion in their manufacture is that they should decompose after they have performed their function.

Petah Tiqwa, city in w central Israel, 11km (7 miles) ENE of Tel Aviv-Jaffa. It was founded in 1878 as the first modern Jewish agricultural settlement in Palestine. It is now a commercial centre and produces textiles, chemicals, building stone and rubber products. Pop. (1974) 103,000.

Pétain, Henri Philippe (1856–1951), French soldier and chief of state. In WWI he distinguished himself by holding VERDUN against the Germans, was appointed Commander-in-Chief in 1917 and made a marshal in 1918. Between the wars he was the most influential military figure in France and served on the Higher Council of War (1920–30). In 1940, after France had surrendered to Germany at the beginning of WWII, Pétain was recalled from Spain (where he was ambassador) and made head of state of the defeated nation; he led the VICHY GOVERNMENT which collaborated with the Germans. As a great war hero forced to concede German demands, he symbolized the situation of many Frenchmen during WWII. In 1945 he was condemned to death, a sentence which Charles DE GAULLE commuted to life imprisonment.

Petal, one of the four basic parts of a flower (the others are SEPAL, STAMEN and PISTIL). The petals of a flower are together known as the corolla. Surrounded by the sepals, flower petals are usually brightly coloured and often secrete nectar and perfume to attract the insects and birds necessary for CROSS-POLLINATION. Once fertilization occurs, the petals usually drop off. *See also* NW p.71.

Peter, Saint (died *c.*64), APOSTLE of Jesus Christ, always named first in any list of apostles in the gospels and for much of the time leader or spokesman for the others. As Simon son of Zebedee, he was with his brother ANDREW called by Jesus and renamed Cephas (Greek for "rock", translated into Latin as Petrus); Jesus said "On this Rock I will build my Church" (Matthew 16:18). After Jesus's ascension and the descent of the Holy Ghost at PENTECOST, Peter was the first publicly to preach Christianity. It was JAMES, however, who was placed in charge of the community in Jerusalem, and it may be inferred that Peter was either imprisoned or travelling about as a missionary. There is no evidence earlier than the mid-2nd century that Peter ever went to Rome, although according to Roman Catholic theology the POPE is considered Peter's successor.

Peter, name of two kings of MONTENEGRO. Peter I (1747–1830), the Great Vladika (prince-bishop) of Montenegro from 1782, forced the Turkish Sultan Selim III to recognize Montenegro's independence by a treaty in 1799. Peter II (1813–51) became prince-bishop in 1830. While guarding his lands from the constant threat of Turkish invasion, he instituted many far-reaching reforms including the installation of the first printing press.

Peter, name of five kings of Portugal. Peter I (1320–67; *r.*1357–67) succeeded ALFONSO IV after rebelling against him in 1355 because Alfonso had Peter's lover, Ines de Castro, murdered. On his accession, Peter had her murderers executed. Peter II (1648–1706; *r.*1683–1706) became king after acting as regent for his brother Alfonso VI since 1668. He promoted Portugal's prosperity, but was drawn into the War of the SPANISH SUCCESSION. Peter III (1717–86; *r.*1777–86) ruling jointly with his wife and niece, Maria I. (*For* Peter IV, *See* PEDRO I, emperor of Brazil). Peter V (1837–61; *r.*1853–61) succeeded his mother Maria II; his father was regent until 1855. He made many reforms, improving Portugal's transport systems by building railways, and introducing higher education.

Peter, name of three tsars of Russia. Peter I (1672–1725), known as Peter the Great, reigned from 1682 to 1725. He ruled jointly with his half-brother Ivan until 1689. He conquered Azov in 1695, travelled in Europe (1696–97) and took back to Russia many Western ideas and technicians. He began a series of domestic reforms in 1700: he encouraged industry, reorganized the administration, developed a new civil serivc service class, weakened the Church and introduced western fashions. He built ST PETERSBURG as his new capital, and made Russia a European power for the first time. He fought Sweden in the NORTHERN WAR (1700–21) and, although losing Azov in 1711, extended Russia towards Siberia. Peter II (1715–30) became tsar in 1727. Peter III (1728–62) succeeded to the throne in 1762. He withdrew Russia from the SEVEN YEARS WAR, introduced liberal

Perthshire's lochs and hilly countryside attract thousands of visitors each year.

Pescara, capital of Pescara province, is a popular Italian resort.

Pesticides being loaded on to a light plane ready for crop spraying.

Marshal Pétain, a hero in WWI was sentenced as a traitor after WWII.

Peter I

Peter II of Yugoslavia, after his marriage to Princess Alexandra of Greece.

Sándor Petöfi's poetry inspired his compatriots in the revolution of 1848.

Petra, the ancient city in Jordan, is mentioned in the Old Testament.

Petrels sometimes appear to be "walking" on the water as they skim the surface.

reforms and ended conscription for the gentry. He was forced to abdicate and was then killed after a conspiracy in which his wife, who succeeded him as CATHERINE II, took a prominent part. *See also* HC1 pp.322–323.

Peter I (1844–1921; r. 1903–21), King of Serbia. Brought up in exile and educated in France while the OBRENOVICH DYNASTY ruled Serbia, he came to the throne when King Alexander was assassinated. He was a staunch advocate of constitutional government and in 1918 became the first King of the new kingdom of Serbs, Croats and Slovenes (later known as Yugoslavia).

Peter II (1923–70), last king of Yugoslavia. He nominally ascended the throne in 1934 at the age of 11, but the actual ruler was his uncle, Prince Paul, who was deposed in 1941 by a military coup. Peter ruled for a month until the invasion of the AXIS POWERS, when he fled to London. After the monarchy was abolished in 1945 he settled in the USA.

Peter, the Epistles of St, two New Testament letters traditionally ascribed to St PETER. Both content and style, however, make it practically certain that they were not written by him, that they date from 50 to 90 years after his death, and that they were both probably by different authors. The first letter, written c. 117 is addressed to communities of Greek Christians apparently experiencing persecution; the second, written c. 150, encourages faith in the Second Coming of Christ.

Peter, Laurence Johnston (1919–), Canadian-born educator and writer who is well-known for *The Peter Principle: Why Things Always go Wrong* (1969). This states that in a hierarchical system an employee tends to rise to his level of incompetence. The actual work is done by those employees who have yet to reach their level of incompetence. *See also* MS p.300.

Peter and the Wolf (1936), by Serge PROKOFIEV, musical tale for children, first performed by speaker and orchestra in Moscow. Various instruments are used to represent characters in the narrated story – the flute portrays a bird, the clarinet a duck.

Peterborough, Charles Mordaunt, 3rd Earl of (1658–1735), English political figure and military commander. He became First Lord of the Treasury in 1689 after JAMES II was deposed, but was imprisoned in 1697 on suspicion of plotting to assassinate WILLIAM III. Commanding the British fleet in Spain in the War of the SPANISH SUCCESSION, he captured and raised the siege of Barcelona in 1706.

Peterborough, county district in N CAMBRIDGE (SHIRE), England; created in 1974 under the Local Government Act (1972). The city of Peterborough, an important industrial and agricultural centre, is built round St Peter's Cathedral, a renowned example of Romanesque architecture. Area: 346sq km (136sq miles). Pop. (1974 est.) 111,300.

Peter Grimes (1945), three-act opera by Benjamin BRITTEN with libretto by Montagu Slater, based on George CRABBE'S poem *The Borough*. It was first produced at Sadler's Wells Theatre, London, and the title role was created by Peter PEARS. *See also* HC2 p.121.

Peter Lombard (c. 1100–60), Italian religious writer. After teaching in Paris from 1136 he wrote his most famous work, *Libri quattuor sententiarum* (1148–51). This deals with the Trinity, Creation and Sin, the Incarnation and the Virtues, and the Sacraments. It was only superseded by the Summa Universae Theologia of Thomas AQUINAS.

Peterloo Massacre (1819), dispersal of a radical meeting in St Peter's Fields, Manchester. A crowd of some 60,000 had assembled to demonstrate for the repeal of the CORN LAWS and for parliamentary reform. The magistrates ordered the speakers to be arrested by untrained yeomanry who also attacked the crowd. A cavalry charge in aid of the yeomanry resulted in 11 deaths and an estimated 500 injured. The government's endorsement of the magistrates' action caused widespread resentment and intensified the

general demand for parliamentary reform. *See also* HC2 p.94, *94*.

Peter Pan (1904), five-act play by J. M. BARRIE, published in 1928. Peter Pan is a boy who never grows up. He flies in through a nursery window, teaches Wendy and her two brothers to fly, and takes them to Never-Never Land. They encounter Indians, Captain Hook and his pirates, and a crocodile with a ticking clock in its stomach.

Peter principle. *See* PETER, LAURENCE JOHNSTON.

Peterson, Oscar (1925–), Canadian jazz pianist. Gifted with a formidable technique, he led his own trios from the early 1950s, and was much in demand to accompany other musicians and singers such as Ella FITZGERALD.

Peter's Pence, annual payment by the English Crown to the Pope, beginning as a national payment in the reign of ALFRED in the late 9th century. Originally a hearth tax (based on the number of hearths in the country) it was fixed at £201.9s. a year. It was abolished by Parliament in 1534.

Peter the Cruel (1334–69), Spanish King of Castile and León from 1350 to 1369. Constantly at war with his illegitimate brother, Henry of Trastámara (later HENRY II), Peter allied himself with England and EDWARD THE BLACK PRINCE while Henry had the backing of ARAGÓN. Henry defeated Peter in 1366 and had himself crowned. The next year, however, Peter and Edward defeated Henry. In 1369 Henry won the decisive battle of Montiel and killed Peter.

Peter the Great. *See* PETER I (Tsar).

Peter the Hermit (c. 1050–1115), French priest who helped to start the First CRUSADE. After Pope URBAN II proclaimed the Crusade in 1095, Peter began preaching in central France in support of the war. In 1096, having made his way to Cologne, he set out to Constantinople with a motley army. Most were killed by the Turks, but Peter escaped and succeeded in reaching Jerusalem; he returned to Europe in 1100, becoming prior of a monastery in Belgium.

Petesuchos, the sacred crocodile of ancient Egypt. He was worshipped as the incarnation of Sebek, and a live specimen, representing the god, was decorated with gold jewellery and fed delicacies by priests and visitors.

Pethidine, synthetic ANALGESIC drug. It is a strong NARCOTIC and ADDICTION may result from its repeated use.

Petipa, Marius (1819–1910), Russian dancer and choreographer. He rose to fame in 1847 as premier danseur at the Imperial Theatre, ST PETERSBURG, and was choreographer of the Imperial Russian Ballet (1862–1903). His *Don Quixote* (1869), *Sleeping Beauty* (1890) and *Raymonda* (1898) laid the foundations of classical ballet. *See also* HC2 pp.274.

Petition of Right (1628), statement of civil liberties sent by the English Parliament to CHARLES I. The petition was based on earlier statutes and charters and asserted four principles: no taxes may be levied without the consent of Parliament; no subject may be imprisoned without cause shown; no soldiers may be quartered upon the citizenry; and martial law may not be employed in time of peace. The king accepted these rights and in return received subsidies from Parliament, although he continued to collect TONNAGE and POUNDAGE duties without Parliament's authorization. The petition is of great importance as a safeguard of civil liberties. *See also* HC1 p.286.

Petit mal, type of EPILEPSY in which brief episodes of unconsciousness, usually lasting less than 15 seconds, occur several times a day. It is found mainly in children. *See also* MS p.97.

PETN, pentaerythritol tetranitrate, percussion explosive. Its empirical formula is $C_5H_8N_4O_{12}$ and its configuration tetragonal with four CH_2ONO_2 groups about a central carbon atom. More sensitive to shock than TNT, PETN is used to make detonating fuses such as Primacord, which is a waterproof textile with a core of powdered PETN. It is also used in treatment of heart conditions.

Petöfi, Sándor (1823–49), Hungarian poet and revolutionary, whose poems reflected his radical views. His writings include *Versek* (1844) and a novel *A Hóhér Kötele* (1845). His poetry was an inspiration to the patriots of the Hungarian revolution of 1848 during which he was killed.

Petra, ruined ancient city in sw Jordan. It was held by the Nabataeans from the 4th century BC until the Roman occupation of AD 106. It was an early centre of Christianity, captured by the Muslims in the 7th century and recaptured by the Crusaders in the 12th century. Petra was also an important city on the caravan trade routes. The ruins were discovered in 1812 by Johann Burckhardt; they include temples, palaces and a theatre carved from local pink limestone.

Petrarch, Francesco Petrarca (1304–74), Italian lyric poet and scholar, one of the best known poets of the late Middle Ages. His work influenced Geoffrey CHAUCER and was taken as a model by the poets of the English RENAISSANCE, besides exerting an important influence on his Italian successors. Most of his poems have as their subject "Laura", a woman idealized by Petrarch in the style of earlier Italian lyric poets but seen in a more realistic and human light. *See also* HC1 pp.207, 246.

Petrassi, Goffredo (1904–), Italian composer. His early work was neo-tonal and neo-classical in the manner of STRAVINSKY. After WWII he turned to TWELVE-TONE music and, later, to SERIALISM.

Petrel, any of several small oceanic birds related to the ALBATROSS. They are known by a variety of names including FULMAR, gadfly petrel and storm petrel. Most of them nest in colonies and fly over open water, feeding on squids and small fish. They have webbed feet and tubular nostrils. Length: to 42cm (16in). Order Procellariiformes. *See also* NW p.*239*.

Petrie, Sir William Matthew Flinders (1853–1942), British archaeologist and Egyptologist. He helped to develop important excavation methods and techniques, founding the Egypt Research Account in 1894, later known as the British School of Archaeology in Egypt. His most important work was at Memphis but he also excavated Greek trading colonies and in Palestine. *See also* HC1 pp.22–23.

Petrification, fossilizing process in which organic material (eg wood) changes into stone. This is caused by mineral-rich water seeping into the empty spaces of dead, buried trees or animals, which eventually become stone. Although petrified remains can be up to 300 million years old, the stone often reproduces the original living material so clearly that the cell structure is identifiable. The Petrified Forest National Park in eastern Arizona, USA, has many examples of petrified wood.

Petrochemicals, chemical substances derived from PETROLEUM or NATURAL GAS. The refining of petroleum is undertaken on a large scale not only for the fuels obtained (petrol, KEROSENE, fuel oil and natural gas) but also for the wide range of chemicals that can be derived from it. These chemicals include the common alkanes (paraffins) and alkenes (olefins), cyclohexone, BENZENE, TOLUENE and NAPHTHALENE. AMMONIA is sometimes regarded as a petrochemical because the hydrogen used in its manufacture is often derived from petroleum. *See also* MM pp.78–81.

Petrograd. *See* LENINGRAD.

Petrol, also called gasoline, a major HYDROCARBON fuel, a mixture consisting mainly of hexane, octane and heptane. One of the products of oil refining, petrol is extracted from crude oil, which is formed over geological time from the fossilized remnants of early life forms. Frequently other fuels and substances are added to petrol to alter its properties. Most modern cars have high-compression engines, and the mixture of air and petrol vapour supplied to it can explode too quickly, pushing against a rising instead of a descending piston. This effect is known as "knock", and to eliminate it many petrol manufacturers add TETRAETHYL LEAD (an anti-knock agent) to slow the rate of combustion. But this additive has been found to be implicated in atmospheric pollution, and so some manufacturers omit the additive and instead improve the octane rating of the petrol – a measure of its anti-knock properties – by altering the mixture of hydrocarbons in it. *See also* OIL REFINERY; PETROLEUM; MM pp.33, 81; PE pp.138–139.

Petrol engine, commonest type of ENGINE; it uses the expansion of hot gases produced when a mixture of petrol (a light, highly inflammable HYDROCARBON) and air are burnt, to drive a reciprocating (to-and-fro) piston, or rotating cylinder. Unlike the diesel engine, the petrol engine requires an electrical system to ignite the petrol vapour produced in a CARBURETTOR before combustion. *See also* MM pp.62–63.

Petroleum, fossil fuel that is chemically a complex mixture of HYDROCARBONS. It accumulates in underground deposits and is generally a mixture of gaseous, liquid and solid materials. Some deposits consist mainly of gas (NATURAL GAS) whereas others are predominantly liquid (crude oil). The chemical composition of petroleum strongly suggests it originated from the bodies of long-dead organisms, particularly marine plankton. After death these sank to the ocean floor, to be broken down by bacteria into simpler organic materials, including hydrocarbons. Most petroleum is extracted via OIL WELLS from reservoirs in the Earth's crust sealed by upfolds of impermeable rock or by salt domes. *See also* PETROL; PETROLEUM REFINING; MM pp.78–79; PE pp.138–139.

Petroleum refining, process by which crude oil is converted into usable substances. These include PETROL, PETROCHEMICALS and oils for fuel and lubrication. Crude oil is a mixture of many HYDROCARBONS, as well as compounds of sulphur, oxygen, nitrogen and various ASPHALTS. The heavier hydrocarbons, which usually have higher boiling points than lighter ones, are distilled (separated from the original crude substance) as the first stage of refining. The next stage is CRACKING, to break the heavy hydrocarbons down into more economically useful products, eg petrol, KEROSENE. Purification of the various products to remove impurities, such as sulphur and nitrogen compounds, completes the refining process. The most versatile end products are ETHYLENE and PROPYLENE which are widely used in the plastics and chemical industries. *See also* MM pp.80–81.

Petrology, study of rocks, including their origin, chemical composition and occurrence. Formation of the three classes of rocks – IGNEOUS, of volcanic origin; SEDIMENTARY, deposited by water, and METAMORPHIC, which is either of the other two changed by temperature and pressure – are studied. *See also* PE pp.100–103.

Petronius (*d. c.* AD 66) Roman satirist, nicknamed Patronius Arbiter because, according to TACITUS, he was the arbiter of elegance at NERO's court. He committed suicide after he was arrested on a charge of plotting Nero's death. His *Satyricon* (*c.* AD 60), a romance in both prose and verse, survives only in fragments.

Petropavlovsk, city in N Kazkhstan (Kazachskaya SSR), USSR, on the River Isim 274km (170 miles) W of Omsk. Founded in 1752 as a Russian fortress, it was also a caravan trading centre for silk, skins and carpets. Today its products include clothing, canned goods, bricks and machinery. Pop. (1975) 193,000.

Petropolis, city in SE Brazil 43km (27 miles) N of Rio de Janeiro. It was settled by German immigrants under the guardianship of Emperor Dom Pedro II, in whose honour the city was named. It is known as the "summer capital" of Brazil, and its main industry is tourism. Pop. (1970) 116,080.

Petrovic, George. *See* KARAGEORGE.

Petrozavodsk, capital of the Karelian autonomous republic which lies within the Russian Republic (Rossiskaja SFSR), USSR; a port on the W coast of Lake Onega. Its industries include shipbuilding, timber and engineering. Pop. (1975) 210,000.

Petrushka (1911), ballet in four scenes, choreographed by Michael FOKINE with music by Igor STRAVINSKY. First produced by DIAGHILEV'S BALLETS RUSSES in Paris, it is considered one of Fokine's greatest works and is in the repertoire of most US and European ballet companies. *See also* HC2 pp.270, 275.

Petty Sessions, local criminal and civil courts in Britain presided over by two or more JUSTICES OF THE PEACE or a metropolitan or stipendiary magistrate. Developed during Tudor times to deal summarily with minor charges, over the centuries their scope has been enlarged to include many minor criminal and civil matters.

Petunia, genus of flowering plants of the nightshade family that originated in Argentina, and the common name for any of the crossbred varieties that are popular as garden bedding plants. Most varieties derive from the white flowered *P. axillaris* and the red *P. violacea*; they may be erect, shrubby or pendant. The bell-shaped flowers have five petals and may be almost any colour. Family Solanaceae.

Peugeot, international manufacturer of cars, bicycles, tools and steel. Based in France, in the mid-1970s it employed about 97,000 persons. *See also* MM p.126.

Pevensey, village in East Sussex, SE England, once on the English Channel, but now a mile inland. It was the landing place of WILLIAM the Conqueror, in 1066. The modern town is a tourist resort. Pop. 2,151.

Pevsner, Antoine (1886–1962), French sculptor, *b.* Russia, who was initially influenced by CUBISM. Later he became a leading CONSTRUCTIVIST sculptor, creating works in bronze and other materials. In the 1930s he concentrated on non-figurative structures such as *Projections in Space.* A later work is *Monument Symbolizing the Liberation of the Spirit* (1956).

Pevsner, Sir Nikolaus (1902–), British architectural historian, *b.* Germany. He is famous for his monumental survey of the buildings of England and for his work in justifying 20th-century architecture to the English mind.

Pewter, any of several alloys that consist mainly of tin and lead. The most common has four parts of tin and one of lead, combined with small amounts of antimony and copper. Owing to the poisonous properties of lead and antimony, pewter has been replaced by other alloys in the making of domestic utensils.

Peyote, or mescal, either of two species of cactus of the genus *Laphophora* that grow in the USA. The soft-stemmed *L. williamsii* has pink or white flowers in summer and a blue-green stem. *L. diffusa* has white or yellow flowers. Peyote contains many ALKALOIDS, the principal one being MESCALINE, well-known as a hallucinogenic drug. Its use is now prohibited by law in some places. *See also* MS p.105.

Pfeffer, Wilhelm Friedrich Philipp (1845–1920), German botanist. He developed a method for measuring OSMOTIC PRESSURE and showed that such pressure is caused by the accumulation of molecules on one side of a semi-permeable membrane because they are too large to pass through it.

pH, logarithmic measure of the acidity of a solution. It measures the concentration of hydrogen ions in gramme molecules per litre; the symbol for this concentration is $[H^+]$. Pure water at ordinary temperatures dissociates slightly into equal numbers of hydrogen ions and hydroxyl ions: $H_2O \rightleftharpoons H^+ + OH^-$. The concentration of each type of ion is defined as being 10^{-7} gramme molecules per litre. The pH of water is: $pH = \log_{10} 1/10^{-7} = 7$. The concentration of hydroxyl ions is similar: $pOH = \log_{10} 1/10^{-7} = 7$. Thus for water $pH + pOH = 7 + 7 = 14$. If acid is added to water the number of hydrogen ions will increase and the hydroxyl ions decrease such that the sum $pOH + pH = 14$. Thus if ph is 2, pOH is 12, that is $[H^+] = 10^{-2}$ and $[OH^-] = 10^{12}$. If the pH is less than seven the solution is acidic, if greater than seven it is alkaline.

Phaedrus, Latin fabulist (*c.* 15 BC–*c.* AD 50). He was born in Thrace and taken to Rome as a slave in the household of the Emperor AUGUSTUS, who granted him his freedom. Phaedrus was the first Roman to treat the fable as a serious exercise in composition, and his animal fables are based largely on those of AESOP.

Phaestos, ancient city situated on the fertile plain of Mesara in S Crete where early tablets of LINEAR SCRIPT have been found. Excavations have revealed substantial buildings dating from 2000 BC. These appear to have been destroyed *c.* 1700 BC, rebuilt and again destroyed *c.* 1450 BC. Phaestos remained in ruins until colonized by MYCENEANS from mainland Greece *c.* 1200 BC. *See also* HC1 pp.38–39, 40–41.

Phaëton, in Greek mythology, son of HELIOS, the sun god. He drove his father's sun chariot across the sky but lost control of the horses, causing the earth to burn. To save the world Zeus struck him from the reins with a thunderbolt and Phaëton fell to earth at the mouth of the River Eridanus. His sisters, the Heliads, mourned beside his tomb and were transformed into trees; their tears were turned into amber.

Phaeton, light, open, four-wheeled carriage, usually drawn by a pair of horses. It was especially high-perched and the rear wheels were larger than the front. It was a favourite carriage of the fashionable set in Regency England.

Phagocytes, any cells able to engulf other cells, such as bacteria. Two of the LEUCOCYTES, or white blood cells, are phagocytes. These digest what they engulf in the defence of the body against infection. They also act as scavengers by clearing the bloodstream of the remains of the cells that die as part of the body's natural processes. *See also* MS p.67.

Phalanger, any of about 45 species of mainly nocturnal arboreal marsupials of Australasia; they are commonly called possums. It has opposable digits for grasping branches and the tail is long and prehensile. It is mainly a herbivorous, but it also eats insects and small animals. Family Phalargeridae. *See also* NW pp. 159, 210.

Phalanges, bones in the toes and fingers. In human beings there are 14 phalanges in each hand and foot, 2 in each thumb and big toe and 3 in the remaining digits. They are connected to the METACARPALS in the hand and to the METATARSALS in the foot.

Phalanx, tactical formation of ancient Greek armies, consisting of eight or more files of infantrymen armed with long pikes, often 4m (13ft) long. The use of the phalanx reached its height in the Macedonian army of ALEXANDER THE GREAT (*r.* 356–323 BC); thereafter it became obsolete. *See also* HC1 p.68, 71, 80.

Phalarope, any of three species of small, long-necked, web-footed shorebirds; they are wading birds that regularly swim. The female, larger and more brightly coloured than the male, fights for nesting spaces and initiates the courting procedure. Males undertake the nesting duties. Species may be grey, or grey and white, with red. Length: to 25cm (10in). Family Phalaropodidae; genus *Phalaropus*.

Phanerogam, obsolete term for flowering plants now called SPERMATOPHYTES.

Phan-thiet, port in S South Vietnam, on the South China Sea, 180km (112 miles) NE of Saigon; capital of Bin Thuan province. It is now the principal fishing port of South Vietnam and the headquarters of the Vietnamese fishing co-operatives. Industries: fish processing, bricks, tiles. Pop. (1971) 76,652.

Phantom pregnancy. *See* FALSE PREGNANCY.

Pharaoh, title given to the rulers of ancient Egypt after the Eighteenth Dynasty, 1570–1320 BC. The Pharaoh was considered to be the son of the sun god RA and the incarnation of the god HORUS. *See also* HC1 pp.34–35, 48–49.

Pharisees, Jewish religious group, prominent in Israel in the 1st century AD, before the destruction of the second temple in 70. The name of the sect means "separated", indicating their emphasis on purity and freedom from sin. They are seen as the ancestors of the RABBIS of today. Contrary to popular belief, the New Testament condemnation of the

Petroleum; petrochemical plants refine commercial products from crude oil.

Petrushka; Stravinsky (left) and Nijinsky were key figures in Diaghilev's ballet.

Petunias, which thrive in the sun, have large funnel-shaped flowers.

Peugeot; a 1913 racing car made by the famous international motor company.

Pharmaceutical chemistry

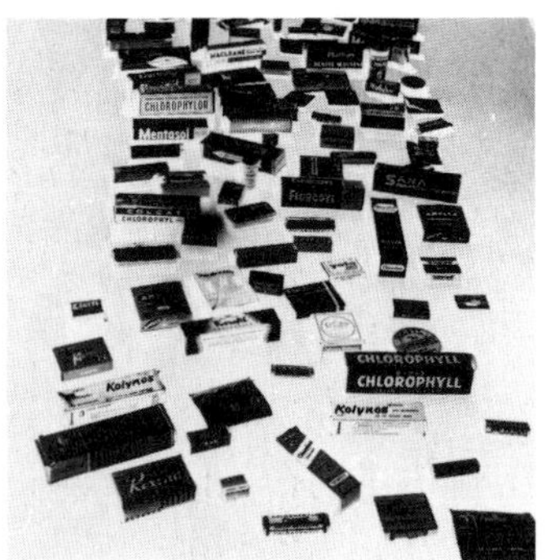
Pharmaceutical industry; new products proliferate in this competitive business.

Pheasants feed and nest on the ground, moving into trees only to roost.

Samuel Phelps, whose 20 years with Sadlers Wells established its prestige.

Philadelphia's town hall with the statue of William Penn on top of the tower.

Pharisees as "hypocrites" was not directed against the entire group, but only against a minority.

Pharmaceutical chemistry, study of the methods of preparation and properties of the inorganic and organic compounds used in medical prophylaxis and therapeutics. Pharmaceutical chemistry, which includes aspects of medicine, biochemistry and preparative organic chemistry, is largely concerned with the synthesis of new types of drug and with testing these for safety and clinical effectiveness.

Pharmaceutical industry, section of industry which manufactures the various remedies prescribed by doctors. These include ANTIBIOTICS, synthetic and natural analgesics, TRANQUILLIZERS, anti-depressants, sleeping pills and contraceptive devices.

Pharmacology, study of medicines, drugs and other substances and their effects on the body. It includes a group of related sciences, such as therapeutics, which looks at the way drugs are used to treat disorders; toxicology, the study of poisons; and pharmacy, the preparation of drugs.

Pharos, island off the coast of N Egypt, in the Mediterranean, connected to the mainland by a stone structure built by ALEXANDER THE GREAT. A lighthouse was completed by Ptolemy II in 280 BC and was considered one of the SEVEN WONDERS OF THE WORLD. It was destroyed by an earthquake in 1346. Pharos is now part of the city of Alexandria.

Pharyngitis, inflammation of the throat (pharynx) usually caused by viral or bacterial infections. The mucous membrane becomes swollen, causing soreness of the throat. Fever, headache and enlargement of the LYMPH nodes are also common symptoms. These persist for about five days; treatment is usually with antibiotics. *See also* MS p.88.

Pharynx, region at the back of the mouth that is the common passageway for food and air. It extends from the nasal cavities (from which it is separated by a flap of tissue called the uvula in the back of the mouth) to the glottis, the opening into the TRACHEA (or windpipe), which is closed off by the epiglottis. The pharynx is continuous with the OESOPHAGUS, which leads to the stomach. *See also* MS pp.66–69.

Phase diagram, graph showing the conditions under which different equilibrium phases of substances exist. For example, a curve of melting point against pressure of a pure solid divides the graph into two regions. Points in one represent temperatures and pressures at which the substance is solid; points in the other represent the liquid conditions. Graphs of composition against temperature are used to indicate such features as solubilities and the ranges of stability of alloy phases. *See also* SU p.85.

Phases, in astronomy, apparent changes in the shape of the Moon as seen from Earth, related to the extent to which the surface of the Moon is illuminated by the Sun, at different positions in the lunar orbit. The phases range from new, when the Moon lies between the Earth and the Sun and is totally invisible, to full, when the Moon reaches opposition and the whole surface is illuminated. Similar phases of the inferior planets, Mercury and Venus, can be seen from Earth. *See also* SU pp.178, *178, 188, 188, 190.*

Phasmida, order of insects, including LEAF INSECTS and STICK INSECTS; some species are wingless or may have tegmina (leathery forewings) shorter than wings, or modified and enlarged hindwings. All have legs developed for walking, and mouthparts for chewing.

Pheasant, game bird of the genus *Phasianus* originally native to Asia, introduced into Europe and naturalized in North America. PEACOCKS and GUINEA FOWL belong to the same family. Males are resplendent with brownish green, red and yellow feathers; females are smaller and brownish. The shooting season in Britain lasts from October to January. Length: up to 89cm (35in). Family Phasianidae; species: *Phasianus colchicus. See also* PE pp.*238,* 239.

Phèdre (1677), tragedy by Jean RACINE, based on *Hippolytus,* by EURIPIDES.

Phèdre, the wife of Thésée, declares her love for her stepson, Hippolyte, who is horrified. Oenone, Phèdre's old nurse and confidante, goes to Thésée and accuses Hippolyte of dishonourable advances to her mistress. Hippolyte dies and Phèdre, who has taken poison, dies confessing Hippolyte's innocence and her own guilt. *See also* HC1 p.*317.*

Phelps, Samuel (1804–78), British actor and theatre manager who became known for his Shakespearean productions. In 1844 he helped to establish the SADLER'S WELLS Theatre in London as a centre for Shakespearean drama.

Phenacetin, or acetophenetidin, white crystalline compound ($C_{10}H_{13}NO_2$) derived from coal tar. It is a mild, non-addictive drug used for reduction of fever and relief of pain. The prolonged use of phenacetin may result in dizziness, kidney damage and haemoglobin changes.

Phenacite, orthosilicate mineral, beryllium silicate (Be_2SiO_4), found in PEGMATITES and high-temperature veins. It has hexagonal system rhombohedral crystals and is either colourless or glassy white, yellow, red or brown in colour. Hardness 7.5–8; s.g. 3. It is sometimes used as a gem.

Phenelzine, anti-depressant DRUG also known as Nardil, Estinerval or Nardelzine. It is a MONOAMINE OXIDASE inhibitor and as such is antagonistic to many foods, especially cheese and dairy products which should therefore not be eaten by people taking the drug. The side-effects of Phenelzine include drowsiness, dizziness and liver damage.

Phenobarbital, barbiturate (whose trade names include Gardenal and Luminal) that acts as a central nervous system depressant frequently used to treat epilepsy and convulsions. Its side-effects include sedation and when its use is discontinued, symptoms of withdrawal may occur.

Phenobarbitone, drug prescribed as a mild sedative in the relief of nervous tension, insomnia and anxiety. As a BARBITURATE it is addictive and highly poisonous if taken with alcohol. It is widely used as an anticonvulsant in the control of epileptic seizures.

Phenol, any of a family of organic compounds that are known by the attachment of a minimum of one hydroxyl group to a carbon atom forming part of the BENZENE ring. Phenol (carbolic acid) is the specific name for monohydroxybenzene (C_6H_5OH) and the generic name for compounds containing one or more hydroxyl groups attached to an aromatic ring. Phenols are colourless liquids or white solids at room temperature with higher melting and boiling points than the parent hydrocarbons from which they are derived. Phenol is used by the chemical and pharmaceutical industries for conversion to such products as asprin, dyes, fungicides, and bactericides, in addition to its application as a starting material for nylon and epoxy resins.

Phenolphthalein, organic compound ($C_{20}H_{14}O_4$), prepared by a reaction between phenol and phthalic anhydride in the presence of sulphuric acid. It is used as a chemical indicator (colourless in acid solutions, red-pink in alkalis), in laxatives and in dyes.

Phenomenalism, philosophical proposition that human knowledge is restricted to what appears either as physical objects or as the content of mental impressions. Phenomenalists either postulate an intrinsic reality behind experience or affirm that appearances alone are sufficient for understanding reality. Both KANT and HUME held to a form of phenomenalism.

Phenomenological movement, school of philosophy at the turn of the century that arose with the work of Edmund HUSSERL. Phenomenology is the study of appearance; Husserl moved away from causal explanations of experience towards direct description of experienced phenomena, to understand their essential structures and relationships.

Phenomenon of Man, The, (1959), translation of *Le Phénomène Humain* (1938–40), by the French palaeontologist and

Catholic priest, Pierre TEILHARD DE CHARDIN. In this book he attempted to reconcile the theory of evolution with theological ideas about the origin of man. These ideas caused great controversy in his Church when the book was first published.

Phenothiazine, widely used worming agent effective against a broad range of animal parasites. It is highly toxic to human beings but its derivatives are used as TRANQUILLIZERS and as antihallucinogenic drugs in the treatment of schizophrenia.

Phenotype, physical characteristic of an organism – anatomical and physiological – resulting from heredity and environment. *See also* GENOTYPE.

Phenthylamine, class of compounds which includes ADRENALINE (epinephrine) and ephedrine. Adrenaline is an important hormone secreted in the body at times of stress and ephedrine is an AMPHETAMINE once used as an antidepressant and for treating allergies such as hay fever.

Phenylalanine, crystalline soluble AMINO ACID found in PROTEINS. It is essential to human nutrition.

Phenylketonuria, or PKU, hereditary condition in which the amino acid phenylalanine cannot be metabolized normally because the ENZYME phenylalanine hydroxylase is absent or inactive. Infants with PKU excrete phenylalanine in the urine. Special diets are used to treat the condition.

Pheromone, substance secreted externally by certain animals that influences the behaviour of members of the same species. Common in mammals, insects and fish, these substances are often sexual attractants and may be a component of body products such as urine, or they may be secreted by specific glands. *See also* NW pp.76–79.

Phidias (*fl.*490–430 BC), Greek sculptor whose greatest achievements were the statue *Athena Parthenos* for the Acropolis and *Zeus* for the temple at OLYMPIA. The *Athena* was the chief treasure of Athens and the *Zeus,* in gold and ivory, one of the SEVEN WONDERS OF THE WORLD. Although his works are lost he is traditionally credited with the design of several sculptures on the PARTHENON. *See also* HC1 p.78.

Philadelphia, city and port in SE Pennsylvania, USA, at the confluence of the Delaware and Schuylkill rivers: fourth largest city in the USA. The city was founded by William PENN in 1681. By 1774 it was a major commercial, cultural and industrial centre of the American colonies. Philadelphia played an important part in the colonies' fight for independence. The CONTINENTAL CONGRESSES were held in the city and the Declaration of Independence was signed there in 1776. The Constitutional Convention met in Philadelphia and adopted the US Constitution in 1787. Philadelphia served as capital of the USA from 1790 to 1800. Benjamin FRANKLIN founded the University of Pennsylvania in 1749. Landmarks include the Liberty Bell and Edgar Allan Poe's house. Industries: shipbuilding, textiles, chemicals, clothing, electrical equipment, metal products, publishing and printing, oil refining, food processing. Pop. 1,862,000.

Philadelphus, or mock orange, or sweet syringa, well-known ornamental deciduous shrub native to the Western Hemisphere and Asia. It has solitary, white or yellowish, fragrant flowers. Family Hydrangeaceae; species: *Philadelphus coronarius.*

Philae, former island of SE Egypt, in the River Nile above the Aswan High Dam, now covered by Lake Nasser. It was the site of many temples built from 300 BC – 600 AD, most of which were removed prior to the submerging of the island.

Philemon (*c.*365–265 BC), poet of ancient Greece. Most of his 97 plays were written in the style of the NEW COMEDY, which used stereotyped characters and dealt with uncontroversial subjects; only fragments of these plays survived. Philemon was considered second only to MENANDER.

Philemon, Christian to whom the apostle PAUL addressed his New Testament epistle on behalf of Onesimus, a slave who had fled from Philemon and been converted to Christianity. In this letter Paul teaches that freedom for all, slave or master, is attained through Jesus Christ.

Philip, Saint, one of the twelve APOSTLES of Jesus Christ. According to the Gospel of St John, Philip was from "Bethsaida, the city of Andrew and Peter" (John 1:44). Philip's name appears in all the lists of Christ's disciples in the gospels and the Acts of the Apostles.

Philip, name of six kings of France. Philip I (1052–1108) reigned from 1060. He supported Robert II of Normandy in order to prevent an Anglo-Norman union. He slightly enlarged the French realm, but the monarchy remained weak. Philip II or Philip Augustus (1165–1223) reigned from 1179. He enlarged France at the expense of Flanders and England. His victory over the Holy Roman Empire at Bouvines (1214) established France as a European power. He also greatly strengthened the monarchy by the creation of a salaried class of royal administrators, the bailiffs. He built the first Louvre. Philip III, the Bold (1245–85), reigned from 1270. He added Poitou, Toulouse, Navarre and Champagne to the French realm. Philip IV, the Fair (1268–1314), ruled from 1285. His taxation of the clergy led to a prolonged quarrel with the papacy and led eventually to the BABYLONIAN CAPTIVITY. He imposed French rule on the Flemish. He convened the first ESTATES-GENERAL in French history (1302). Philip V (c.1294–1322) reigned from 1316. Brother of Louis X, he became king in the absence of a male heir, thereby helping to establish the SALIC LAW in France. Philip VI (1293–1350) reigned from 1328. His reign was dominated by the early years of the HUNDRED YEARS WAR.

Philip, name of kings of Macedon, of whom Philip II (382–336 BC) and Philip V (238–179 BC) are of historical importance. Philip II ruled from 359 to 336 BC. He was appointed regent to Amyntas, but took power for himself. He defeated Athens and Thebes in 338 and established Macedonian control over Greece. He established a federation of Greek states and built the army of his son, ALEXANDER THE GREAT. Philip V ruled from 221 to 179 BC. He extended Macedonian influence in the Balkans and fought the First and Second Macedonian Wars against Rome (215–205 and 200 BC), in the second of which he was defeated.

Philip, name of five kings of Spain. Philip I (1478–1506) became King of Castile in the year of his death. Philip II (1527–98) reigned from 1556. He was also King of Portugal from 1580. In 1554 he married Queen MARY I of England. He supported the INQUISITION and centralized the absolute Spanish monarchy. The second half of his reign was dominated by the revolt of the NETHERLANDS. The southern part (present-day Belgium) was retained by Spain. He sent the ARMADA against The Netherlands' ally, England, in 1588. He founded colonies in America and annexed the Philippines, named after him. Philip III (1578–1621) became king on the death of his father, Philip II. Under him Spain was ruled by his favourite, the Duke of Lerma. Philip IV (1606–65) reigned from 1621, succeeding Philip III. Weak like his father, he left the government in the hands of the Count of Olivares. In 1640 Portugal revolted against the union with the Spanish crown and chose JOHN IV as king. Philip V (1683–1746) reigned as the first Bourbon king of Spain from 1700. He was placed on the throne by his grandfather, LOUIS XIV of France, who thereby precipitated the War of the SPANISH SUCCESSION.

Philip, name of two dukes of Burgundy. Philip the Bold (1342–1404) effectively became Duke of Burgundy in 1364. He married Margaret, heiress of Flanders, and was a regent for his nephew Charles VI during his minority, a position he used to further his own dynastic ambition. Philip the Good (1396–1467), Duke of Burgundy from 1419, maintained a long alliance with HENRY V of England and gave asylum to the dauphin (later LOUIS XI of France). He was a colourful patron of the

arts, and his court was the most splendid in Western Europe. *See also* HC1 p.252.

Philip, Prince, Duke of Edinburgh (1921–), husband of Queen ELIZABETH II of Britain and Prince Consort. He was born in Corfu, the son of Prince Andrew of Greece; he is a grandson of GEORGE I of Greece and a great-great-grandson of Queen VICTORIA. He was brought up and educated in Britain. He became a naturalized British citizen and took the surname Mountbatten. In 1947 he married Elizabeth after becoming the Duke of Edinburgh. He was created a prince in 1957. In 1968 he was awarded the Order of Merit. In 1956 he founded the Duke of Edinburgh awards for creative achievement by young persons.

Philip, Duke of Parma (1720–65; r.1748–65), son of the Bourbon King Philip V of Spain and Elizabeth Parnese (Isabella of Parma) and brother of Charles III, King of Naples and Sicily and of Spain. Although Philip initially brought Spanish culture to the court of Parma, the French influence of his wife Louise Elizabeth, daughter of LOUIS XV of France, soon predominated.

Philip, Arthur (1738–1814), first governor of New South Wales, Australia. He was born in London and in 1786, as a navy captain, led a fleet of convict ships to Botany Bay. In 1788 he founded a settlement at Sydney, the first permanent European colony in Australia. Although successful there was always friction between the settlers and native ABORIGINES. *See also* HC2 p.126, *126.*

Philipe, Gérard (1922–59), French stage and screen actor who worked with most of the leading French film directors, appearing in *La Ronde* (1950), *Le Rouge et le Noir* (1954) and *Les Liaisons Dangereuses* (1959). In the theatre he worked closely with Jean Vilar at the Théâtre National Populaire, and classics in which he excelled included Corneille's *Le Cid* (1951) and Shakespeare's *Richard II* (1955). He was buried in the costume he wore during performances of the former.

Philip of Hesse (1504–67), called the Magnanimous, a German nobleman. He introduced the REFORMATION in Hesse and formed the SCHMALKALDIC LEAGUE against the Catholic Holy Roman Emperor CHARLES V. He was imprisoned by Charles from 1547 to 1552.

Philippi, ancient city of E Macedonia, renamed by Philip II (c.358 BC). It was the site of two battles in which the armies of CASSIUS and BRUTUS were defeated by MARK ANTONY and Octavian (AUGUSTUS) (42 BC). It was the first European city to be visited by a Christian missionary – St Paul twice visited Philippi and founded a church there.

Philippians, New Testament epistle written by the apostle PAUL during his first imprisonment in Rome. In this letter to the Church he had established at Philippi in Macedonia, he stresses his joy in serving Christ, even when threatened with death.

Philippics, orations (351–321 BC) by DEMOSTHENES against King PHILIP II of Macedonia designed to arouse the Athenians against his imperial aspirations. The word is now used to describe any discourse full of invective; CICERO's 14 orations against MARK ANTONY in 44 BC are often described as philippics.

Philippicus (r.711–13), Byzantine emperor. He was exiled by the Emperor TIBERIUS III but in 711 joined the service of Justinian II. Turning against his protector, Philippicus seized the throne and put Justinian to death. He was an advocate of MONOTHELETISM, which led the pope to refuse to recognize his imperial title. Following failures in foreign policy, he was overthrown and blinded.

Philippine Sea, part of the w Pacific Ocean lying E of the Philippines, s of Japan, sw of the Mariana Islands and NW of New Guinea. It contains the Philippine Trench, which follows the line of the E coast of the archipelago. Depth: (max.) 10,539m (34,578ft).

Philippines, island republic in SE Asia, in the sw Pacific Ocean; made up of more than 7,000 islands, the largest of which is Luzon, which includes the capital Quezon

City. Much of the land is mountainous and forested: timber is the chief export. Rice and coconuts are the principal crops. There are rich mineral deposits, many of which remain unexploited. Area: 300,000sq km (115,830sq miles). Pop. (1976 est.) 43,751,000. *See* MW p.143.

Philips' Gloeilampenfabrieken, international industrial corporation with headquarters in Eindhoven, The Netherlands. It makes electrical and electronic goods. In the mid-1970s its total number of employees was about 400,000, about 70% of whom were employed in Europe.

Philistia, ancient coastal region of sw Palestine. Its five chief city-kingdoms formed a confederacy against the frequent invasions by foreign enemies, particularly the Hebrews. King David finally succeeded in conquering it but, after the division of the Kingdom of Israel, Philistia regained its independence.

Philistine, member of a non-Semitic people who lived on the s coast of modern Palestine known as PHILISTIA, from c.1200 BC. They clashed frequently with the Hebrews and are often mentioned in the Bible, particularly in the book of Judges. Today the term philistine may be applied to a person with a material outlook or to one indifferent to culture.

Phillipsite, white or reddish ZEOLITE mineral with a composition $KCa(Al_2Si_6O_{16})6H_2O$. It occurs in vein formations as monoclinic crystals, hardness 4.5–5, s.g. 2.2.

Phillpotts, Eden (1862–1960), British novelist and dramatist who wrote more than 100 novels. Most are set in Devon and centred around Dartmoor. The rural atmosphere and rich dialect is vividly conveyed. He is best known, however, for two plays, *The Farmer's Wife* (1916) and *Yellow Sands*, (1926) in which his daughter Adelaide Eden Phillpotts collaborated.

Philoctetes, in Greek mythology, son of Poeas, archer on the expedition to TROY. He received a snake-bite and was left behind on the uninhabited island of LEMNOS. ODYSSEUS and Diomedes returned for him, however, after discovering that Troy could not be taken without the bow and arrow of HERCULES possessed by Philoctetes. There is a tragedy by SOPHOCLES entitled *Philoctetes*.

Philodendron, genus of house plants native to tropical America. They have shiny, heart-shaped leaves, which are sometimes split; large-leaved, climbing types need support. Philodendrons grow best in bright, indirect light and well-watered soil (equal parts of loam, peat moss and sand). Propagation is by stem cuttings. Height: 10cm–1.8m (4in–6ft). Family Araceae.

Philo Judaeus, or Philo of Alexandria, (b. c.12 BC), Jewish philosopher, b. Egypt. He belonged to a school of thinkers who tried to blend the theology of the Jewish scriptures with Greek philosophy. His view of the nature of God was that He is pure being, present everywhere by His power but nowhere as a substance.

Philology, the study of both language and literature. Besides phonetics, grammar and the structure of language, philology includes textual criticism, ETYMOLOGY, art, archaeology, religion and any system related to ancient or classical languages. Comparative philology, which compares the history, forms and relationships of languages, began in the late 18th century when Sir William Jones, an English orientalist, noticed that SANSKRIT resembled Latin and Greek.

Philosophes, group of 18th-century French philosophers, influenced by DESCARTES and the ideals of the ENLIGHTENMENT. Their ideals included rationalism, toleration, secularization and the practical improvement of the conditions of life. VOLTAIRE, ROUSSEAU and DIDEROT were eminent *philosophes. See also* HC1 pp.324–325; HC2 pp.22,23.

Philosophiae Naturis Principia Mathematica. *See* PRINCIPIA MATHEMATICA.

Philosophy, study of the nature of reality, knowledge, ethics and existence by means of rational enquiry. The term derives from the ancient Greek word meaning "love of knowledge or wisdom". Western philoso-

Philip II of Spain: a detail of Rubens' portrait of him, in the Prado Museum, Madrid.

Philippines: a bamboo and cogon grass hut, typical of the island dwellings.

Eden Phillpotts, seen here at the age of 89, wrote more than 100 novels.

Philodendrons have thick leaves which vary in shape from plant to plant.

Philoxenus

Philosophy; detail of Raphael's painting of Plato and Aristotle, in the Vatican.

Phnom-Penh; rioter attacking a foodstall during the anti-Chinese riots of 1975.

Phoenix; the metropolitan area houses over half the population of Arizona.

Phosphorus is obtained industrially from natural deposits of phosphates.

phy began as a systematic subject among the Greeks in the 6th century BC, and was a general study of all phenomena. Philosophy was a branch of theology in the Middle Ages, but has become narrowed in scope in modern times as science has become more based on observation than on speculation. Today philosophers are mainly concerned with aspects of the questions "How can we know something?" and "What is the real nature of the objects that we perceive?". Questions of the purpose of life and of ethics are only rarely the concern of modern academic philosophers. *See also* MS pp.226–235.

Philoxenus (*fl.* late 4th or early 3rd century BC), Greek painter. According to Pliny he painted a battle scene in which Alexander and Darius are shown and it is assumed that the famous Alexander Mosaic from POMPEII is a copy of this.

Phiz (1815–82), pseudonym of Hablot Knight Browne, British illustrator and cartoonist, best known for his illustrations for Charles DICKENS' novels. He was chosen by Dickens to illustrate *Pickwick Papers* (1836) and subsequently worked with him on several novels.

Phlebitis, inflammation of the wall of a vein. It may be caused by infection, trauma, underlying disease, or by the presence of VARICOSE VEINS. Symptoms include localized swelling and redness. Phlebitis may be long-lasting and complicated wth blood clots. Treatment includes rest and anticoagulant therapy. *See also* MS p.90.

Phlebotomus fever, also called pappataci fever, acute infectious tropical disease of human beings. It is transmitted by the female sandfly, *Phlebotomus papatasii*, when it sucks blood. Symptoms include lassitude, dizziness, severe headache and pains in the joints and abdomen. After a few days of treatment, the symptoms usually subside; convalescence may take several weeks. *See also* DENGUE; LEISHMANIASIS; MS p.106.

Phlebotomy, letting of blood in treatment of a disorder. *See also* MS p.83.

Phloem, vascular tissue for distributing food materials in plants. It consists mainly of elongated sieve-tube cells, companion cells and parenchyma cells. In non-woody species it functions throughout the life of the plant; in woody plants, vascular CAMBIUM is formed. *See also* XYLEM.

Phlogiston, odourless, colourless and weightless material believed by early scientists to be the source of all heat and fire. Combustion was believed to involve the loss of phlogiston. *See also* HC2 p.32; SU pp.24–25.

Phlogopite, also called brown mica, white or brown mica mineral with a composition $KMg_3Fe_3(AlSi_3)O_{10}(OH)_2$. It is found in limestones as monoclinic crystals, hardness 2.5–3, specific gravity 2.8.

Phlox, genus of mostly perennial plants native to North America. The flowers are of various colours and are often cultivated as garden plants. Height: to 1.5m (5ft). Family Polemoniaceae.

Phnom-Penh (Phnum Pénh), capital of Cambodia, in s Cambodia at the confluence of the Mekong and Tonle Sap rivers. Founded in the 14th century, the city was the capital of the Khmers after 1434. It was extensively damaged during the Cambodian civil war before the Communists took over in 1975. At the beginning of the war in 1970 the population was initially swelled by refugees but was subsequently greatly reduced as the new regime evacuated inhabitants. It is a port and transportation centre. Industries: rice-milling, brewing, distilling, fish processing, textiles. Pop. (est.) 470,000.

Phobia, irrational and uncontrollable fear that persists despite reassurance or contradictory evidence. Psychoanalytic theory suggests that phobias such as fear of high places, closed spaces or infection are actually symbolic subconscious fears and impulses. *See also* MS pp.*141–142*, 144, *146*.

Phobos, larger of Mars' two satellites, discovered in 1877 by Asaph Hall. Phobos lies at a mean distance of 9,350km (5,800 miles) from Mars. Its longest diameter is 28km (17 miles) and the shortest is 20km

(12 miles). The sidereal period is 0.32 days. *See also* SU pp.*176*, 202–203, *202–203*.

Phoebe, outermost of the ten satellites of the planet Saturn. It is extremely small, with a diameter of 200km (124 miles) and its retrograde motion suggests that it is a captured ASTEROID. Phoebe lies at a distance of almost 13 million km (8 million miles) from Saturn and has a mean sidereal period of 550.3 days. *See also* SU pp.205, 212–213, *213*.

Phoenicia, ancient region of w Asia, along the E Mediterranean Sea coast; its great city-states included Tyre, Sidon, Tripoli and Byblos. It was founded c.1600 BC. The Phoenicians were traders and colonizers controlling the Mediterranean Sea trade, by the 12th century BC. During the 6th century BC Persia began to absorb Phoenicia, completing the process by Roman times, although the Phoenicians fought Alexander the Great to maintain their autonomy. They originated an alphabet that was later developed by the Greeks and they were famous for their purple Tyrian dye. *See also* HC1 pp.54–55.

Phoenicid meteors, meteor shower visible between 4 and 5 December, and radiating from the constellation of Phoenix. *See also* SU p.219.

Phoenix, state capital in sw central Arizona, USA, on the Salt River; the largest city of Arizona. Founded in 1870, the city grew after agriculture in the area had been made possible by using the water of the Salt for irrigation. Industries: data processing, tourism, aircraft, fabricated metals, machinery, textiles, clothing, food products. Pop. (1973) 1,861,719.

Phoenix, fabulous mythological eagle-like bird linked with sun-worship, especially in ancient Egypt. Of gold and scarlet plumage, only one phoenix could exist at a time, usually with a life-span of about 500 years. When death approached, the phoenix built a nest of aromatic plant material, and was then consumed by fire. From the ashes of the pyre rose a new phoenix.

Phoenix, southern constellation just s of the bright star ACHERNAR in the constellation of Eridanus; one of the four constellations known as the Southern Birds. The only bright star is Alpha (Ankaa) of second magnitude. Beta is a DOUBLE STAR, both components of which are of magnitude 4. Zeta Phoenicis is an ECLIPSING VARIABLE star with a magnitude range of 3.6 to 4.1. *See also* SU pp.262, *263*.

Phoenix Islands, small group of coral islands in the central Pacific Ocean, E of the Gilbert Islands and NW of American Samoa. The islands are associated with the Gilbert and Ellice Islands; two of them are controlled jointly by the USA and Britain. They are an important stopping point on air routes from Hawaii to the islands of the sw Pacific. Area: 28sq km (11sq miles). Pop. 130.

Phoenix Park murders, assassination in Dublin of Lord Frederick CAVENDISH, Secretary for Ireland in the British government, and Thomas Burke, Under-Secretary, on 6 May 1882. Five assassins, connected with the FABIANS, were hanged; three were imprisoned.

Pholidota, small order of mammals, containing only one genus, *Manis*, in the family Manidae; its members are called PANGOLINS. Some species are entirely toothless and feed almost exclusively on ants and termites. *See also* NW p.155, *155*.

Phoneme, minimum unit of significant sound; a speech sound distinguishing meaning. The phoneme "p" distinguishes "tap" from "tab". A phoneme may have two letters such as "th", as long as it is perceived aurally as one sound.

Phonetics, science of speech sounds and the symbols by which they are written down. It is based on a study of the shape of the mouth and position of the tongue when making a particular sound. *See also* MS p.239.

Phonograph. *See* GRAMOPHONE.

Phoronida, phylum of invertebrate marine animals, of which there are only about 15 species. They have worm-like, non-segmented bodies. Adults are seden-

tary and live in membranous tubes, sometimes covered with sand or shell particles, into which they retreat when disturbed. Length: 0.5 to 25cm (0.25–10in). *See also* NW p.121.

Phosgene, colourless, toxic gas, chemical name carbonyl chloride $(COCl_2)$. It was used as a poison gas in WWI, but is now used in the manufacture of various dyestuffs and resins. Properties: b.p. 8.2°C (46.8°F), m.p.–118°C (–180.5°F).

Phosphates, numerous chemical compounds derived from orthophosphoric acid (H_3PO_4). One group is composed of salts with the phosphate ion (PO_4^{3-}), the hydrogen phosphate ion (HPO_4^{2-}), the dihydrogen phosphate ion $(H_2PO_4^{-})$ and the appropriate positively charged ion or ions. For example, sodium forms three phosphates: Na_3PO_4, Na_2HPO_4 and NaH_2PO_4. Another group is composed of esters in which organic combining groups replace one or more of the hydrogen atoms of phosphoric acid.

Phospholipid, type of LIPID which contains a phosphoric acid group or groups and an alcohol base. Phospholipids are found in egg yolk and brain tissue. *See also* SU p.152.

Phosphor, substance capable of luminescence (storing energy and later releasing it as light). There are two main types: the zinc sulphide phosphors, as used on cathode-ray tubes; and the oxygen type, as used on fluorescent light tubes. Zinc sulphide is often mixed with cadmium sulphide and a small quantity of metallic phosphates, silicates or fluorides. *See also* MM p.230, *230*.

Phosphorescence, form of luminescence in which a substance emits ligwt of one wavelength. Unlike fluorescence, it may persist for some time after the initial excitation. In biology, phosphorescence is the production of light by an organism without associated heat, as with a glow-worm.

Phosphoric acid, group of ACIDS, the chief forms of which are orthophosphoric acid (H_3PO_4), metaphosphoric acid (HPO_3) and pyrophosphoric acid $(H_4P_2O_7)$. Orthophosphoric acid is a colourless liquid obtained by the action of sulphuric acid on phosphate rock (calcium phosphate) and used in fertilizers, soaps and detergents. Metaphosphoric acid is obtained by heating orthophosphoric acid and is used as a dehydrating agent. Pyrophosphoric acid is formed by moderately heating orthphosphoric acid or by reacting phosphorus pentoxide (P_2O_5) with water and is used as a catalyst and in metallurgy.

Phosphoros, in Greek mythology, the light-bearer. The name is synonymous with Lucifer and the morning star, VENUS. Generally portrayed as a man bearing a torch, he was the herald and harbinger of the dawn. The name of the element PHOSPHORUS is of the same derivation.

Phosphorus, common nonmetallic element (symbol P) of group V of the periodic table, discovered by Hennig Brand in 1669. It occurs, as PHOSPHATES, in many minerals; APATITE is the chief source. The element is used in making PHOSPHORIC ACID for detergents and fertilizers (phosphorus is an essential element for plant growth). Small amounts are used in insecticides and in matches. Phosphorus has several ALLOTROPES, including the highly reactive white phosphorus, which ignites spontaneously on exposure to air, and the more stable form, red phosphorus. Properties: at.no. 15; at.wt. 30.9738; s.g. 1.82 (white), 2.34 (red); m.p.44.1°C (111.38°F) (white); b.p. 280°C (536°F) (white); most common isotope P^{31} (100%).

Photius (c.820–91), Patriarch of Constantinople. Elected in 858 to succeed St IGNATIUS OF CONSTANTINOPLE, he quarrelled with the Pope over the autonomy of the Eastern Church and was deposed in 867. He became Patriarch again from 877 to 886, when he may have been forced to resign by the Byzantine Emperor LEO VI. *See also* HC1 p.146.

Photocell. *See* PHOTOELECTRIC CELL.

Photochromic glass, form of glass, used in spectacle lenses and electronic devices, which darkens on exposure to ultraviolet or bright visible light. It fades to its original clear state when the light is removed. Silver chloride or silver bromide crystals throughout the glass interacts with light to cause this change.

Photocopying, reproduction of the written word, drawings or photographs by duplicating machines which utilize light, heat, chemicals or electrostatic charge. *See also* MM pp.208–209.

Photoelectric cell (or photocell), type of electric cell whose operation depends upon the extent to which it is exposed to light. Formerly consisting of an electron tube with a photosensitive cathode, nearly all modern photocells are made using light-sensitive semiconductors. They are used as switches (electric eyes), devices to measure light intensity (light meters) or as power sources (solar cells). *See also* SU p.108.

Photoelectric effect, liberation of ELECTRONS from matter by light, ultraviolet radiation, X-rays or gamma rays falling on its surface. The effect can be explained only by the QUANTUM THEORY. Photons in the incident radiation are absorbed by atoms in the substance and enable electrons to escape by transferring the requisite amount of energy to them. A stream of such electrons flowing in a circuit constitute an electric current. *See also* SU p.108.

Photoelectric emission, production of electrons from the surface of a substance (such as silicon) as a result of light falling on it. The mechanism of this PHOTOELECTRIC EFFECT was first described by Albert EINSTEIN using QUANTUM THEORY. *See also* PHOTOELECTRIC CELL.

Photo engraving, process of preparing illustrations for letterpress printing in which the image is transferred by photography upon metal or plastic. It involves two steps: the preparation of a photographic negative copy of the material to be reproduced, and the making of a positive printing plate. The negatives are made on sensitized films with a 303 process camera which can produce an image of any size desired. Plates are made of zinc, copper or magnesium, coated with a photosensitive solution. The light passing through clear sections of the negative affects the coating and makes it insoluble in water. After exposure the plates are washed, leaving an image formed by the insoluble portions. The nonprinting surface is then etched with an acid. *See also* MM pp.206, 208.

Photogrammetry, science of taking measurements from aerial photographs in order to make surface maps. The technique became important during WWI, when it was used by intelligence corps. Since then its use has been extended to civil purposes, eg for mapping otherwise inaccessible areas.

Photographic memory, rare ability to retain and recall information very easily. Although this phenomenon is not fully understood, it appears to involve eidetic images, exceptionally vivid visual pictures. A person with a photographic memory seems to be able to visualize an object at will, such as a page of a book, and "read off" information from this image.

Photography, process of obtaining a permanent image of an object, either in black and white or in colour, on treated paper or film. In black and white photography a CAMERA is used to expose a film to an image of the object to be photographed for a set time. The film is covered on one side with an emulsion containing silver bromide or silver chloride. The effect of the exposure is to make the silver compound easily reduced to metallic silver when treated with a DEVELOPER. The action of the developer is to produce a black deposit of metallic silver particles on those parts of the film that were exposed to light, thus providing a "negative" image. After fixing (in "hypo") and washing, the negative can be printed by placing it over a piece of sensitized paper and exposing it to light so that the silver salts in the paper are affected in the same way as those in the original film. The dark portions of the negative let through the least light and the image on the paper is thus reversed back to a positive. Colour photo-

graphy works on a similar, but more complex process. *See also* MM pp.218–223.

Photometer, instrument used to measure the luminous intensity of a light source by comparing it with a source of known intensity. This is achieved by comparing the positions of the sources when they produce equal illumination on a reference surface.

Photomultiplier, electronic device that converts a light signal or light beam into an equivalent amplified electrical signal, (electric current). ELECTRONS, produced by PHOTOELECTRIC EMISSION from an illuminated CATHODE, are reflected from a series of other electrodes and are increased in number at each bombardment. An amplified signal is obtained from the final ANODE. *See also* SU p.109.

Photon, quantum of electromagnetic radiation, such as light, which can be considered as streams of photons. The energy of a photon equals the frequency of the radiation multiplied by PLANCK'S CONSTANT. Absorption of photons by atoms and molecules can cause excitation or ionization. A photon may be classified as a stable elementary particle of zero rest mass, zero charge and spin 1, travelling at the velocity of light. It is its own antiparticle. Virtual photons are thought to be continuously exchanged between charged particles and thus to be the carriers of the electromagnetic force. *See also* LIGHT; SU pp.108–110.

Photosetting, composition of text to be printed by photographic means, developed in the 1950s as a speedier and more flexible alternative to setting type in metal. The size of type to be set is obtained from a master negative font by photographic reduction or enlargement. This book has been set in Times Roman type on a photosetting machine, and condensed, expanded or *italic* variants of the face can be achieved with great facility. Computer units allied to most photosetters enable the text to be set as fast as the operator can key it in. Once programmed these units automatically control line width, column length, spacing, justification and hyphenation. *See also* LINOTYPE; MONO-TYPE; TYPESETTING; TYPEMETAL; MM pp.*109*, 207, *207.*

Photosphere, surface of the SUN, which radiates the light and heat produced in the solar interior. It has a general temperature of 6,000K and has a granular appearance, which is often disturbed by SUNSPOTS, faculae and associated transient phenomena. *See also* SU p.222, *222.*

Photosynthesis, chemical process occurring in green plants by which minerals, water and carbon dioxide are converted into food and oxygen using energy absorbed from sunlight. The reactions take place almost instantaneously in the CHLOROPLASTS, chlorophyll-containing bodies in the leaf cells. During the first part of the process, light is absorbed by chlorophyll and splits water into hydrogen and oxygen. The hydrogen attaches to a carrier molecule and the oxygen is set free. The hydrogen and light energy build a supply of cellular chemical energy, ADENOSINE TRIPHOSPHATE (ATP). Hydrogen and ATP convert the carbon dioxide into sugars, including glucose, and starch. Photosynthesis is possibly the most important biological process occurring on the Earth. *See also* NW pp.44, 45, 48–49, 62–63, 194; PE pp.*67*, 153.

Phototropism, also called heliotropism, growth of a plant in response to the stimulus of light, which increases cell growth on the shaded side of the plant. Leaves and stems respond positively to light and roots respond negatively or not at all. Indoor plants lean towards windows; leaves usually grow at right-angles to light and are positioned to ensure that overlapping occurs as little as possible.

Phra Nakhon Si Ayutthaya, city and province in central Thailand. The city is a commercial centre for a region producing rice, maize, tobacco and sesame. Founded *c.*1350, the city flourished as the political seat of one of the most powerful states in SE Asia. Tourism is a major industry. Area (province): 2,480sq km (958sq miles). Pop. (province) 421,000; (city) 39,291.

Phrenology, theory that an individual's personality and character traits can be determined by examining the shape and configuration of the skull. It was elaborated in the early 19th century by two Germans, Franz Josef Gall (1758–1828) and Johann Spurzheim (1776–1832). The theory has been discredited in the 20th century. *See also* MS pp.136, 137.

Phrygia, ancient region in Asia Minor, corresponding roughly to modern central Turkey. It was originally settled by Balkan peoples (PHRYGIANS) *c.*13th century BC; it was ruled by Lydia in the 7th century BC and by Persia in the 6th century BC. The Greeks used it as a source of slaves, and it was associated in their legends with the names of MIDAS and Gordius. It came under Roman rule *c.*133 BC.

Phrygians, people who settled in Asia Minor *c.*1400 BC. They dominated w central Anatolia, PHRYGIA, after the end of the HITTITE empire (*c.*1300 BC). Their last king, MIDAS, was overthrown by the Cimmerians *c.*700 BC, and they became a subject people. The Greeks thought highly of them as slaves.

Phthalocyanine dyes, highly stable organic dyes. Most are bright green or blue in colour and they are mainly used in enamels, inks and paints. They are porphyrin complexes like HAEMOGLOBIN and the CHLOROPHYLLS. *See also* MM p.245, 245.

Phthalsulphathiazole, sulphonamide drug used as an intestinal antimicrobial agent. Other names under which it is sold include Enteramida. Taleudron and Thalazole.

Phthiraptera, order of wingless insects divided into the Mallophaga, which are parasitic chewing or biting lice, and the Anoplura, the sucking lice. Mallophaga are parasitic on mammals and birds, whereas Anoplura are parasitic only on mammals. The bites of these insects may transmit disease. *See also* NW pp.106–107, *106.*

Phylloxera, or grape phylloxera, small yellowish insect of the order Homoptera that is a pest on grape plants in Europe and w USA. It attaches itself to the leaves and roots and sucks the plant's fluids, resulting in the the eventual rotting of the plant. The complex life cycle includes parthenogenetic wingless stages. Family Phylloxeridae; species: *Phylloxera vitifoliae.*

Phylum, in the systematic categorization of living organisms, a major group within the animal kingdom. It is comprised of a diverse group of organisms with a common fundamental characteristic. In plant classification the analogous category is sometimes called division. *See also* CLASSIFICATION; NW pp.72–73, 254.

Physical chemistry, study of the physical changes associated with chemical reactions and the relationship between physical properties and chemical composition. The main branches are THERMODYNAMICS, concerned with the changes of energy in physical systems; chemical kinetics, concerned with rates of reaction; and molecular and atomic structure. Other topics included are electrochemistry, thermochemistry and some aspects of nuclear physics, radiation physics and combustion chemistry.

Physics, study of matter and energy. Both matter and energy appear in many forms and physics seeks to see beyond these forms and to connect them in their manifestations. Modern physics recognizes four basic FORCES in nature and these are categorized according to the interactions they induce. The first of these is gravitational force, which was first adequately described by Isaac NEWTON. The second is the weak force, which is responsible for the decay of some sub-atomic particles. The third is the electromagnetic force, which was codified in Maxwell's Equations in the 19th century. The strong force is that which binds together atomic nuclei. This force is some 10^{12} times stronger than the weak force and is the least well understood in physics. Einstein's theory of RELATIVITY showed the basic relation between energy and matter, such that mass in particle physics is expressed in terms of energy. Physics may

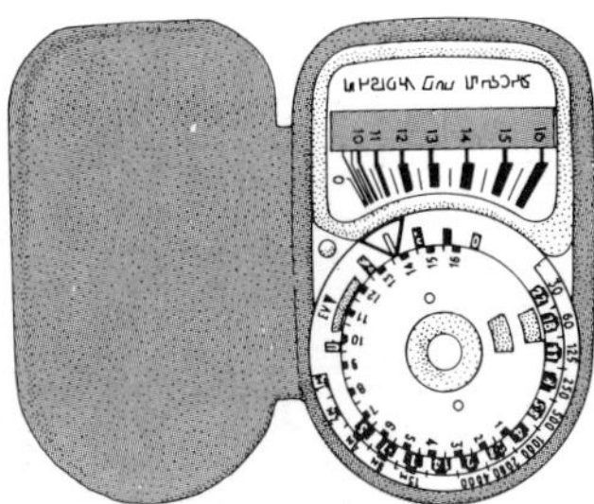

Photoelectric cell in a photographer's light meter shows illumination levels.

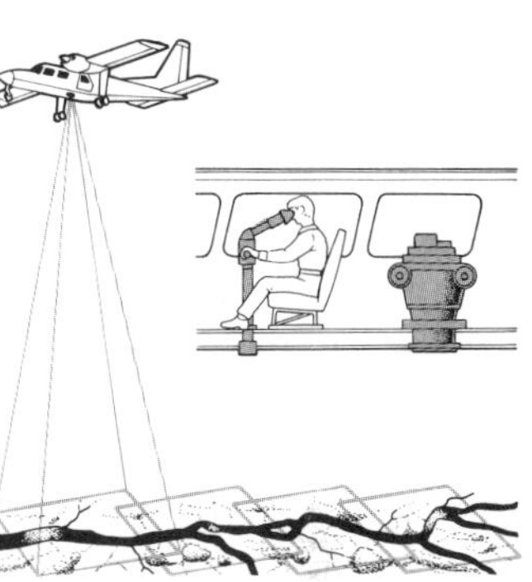

Photogrammetry uses aerial photography for making accurate maps.

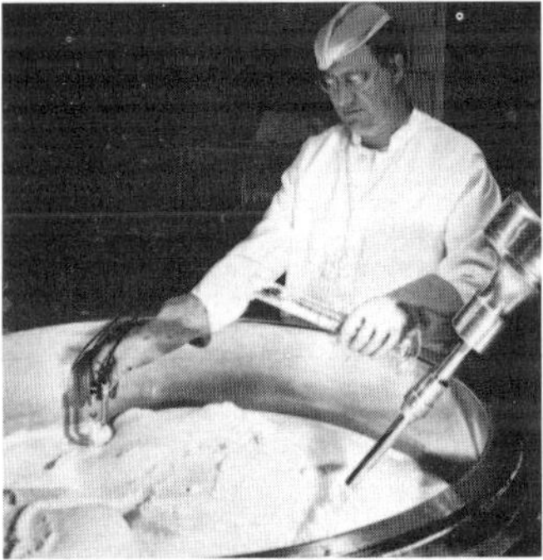

Photography; acidity of an emulsion being measured during manufacture.

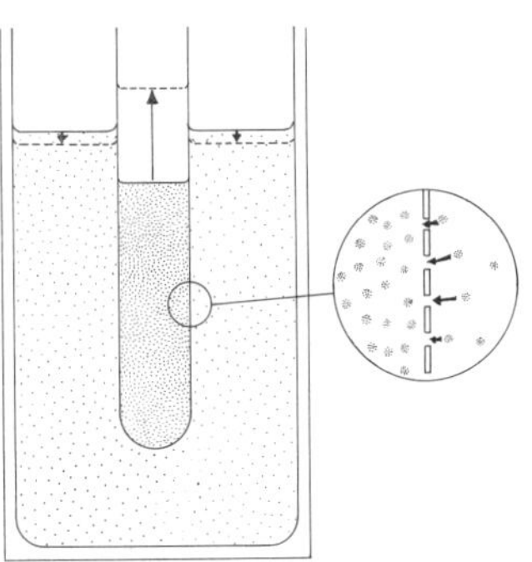

Physical chemistry includes a study of osmosis and semi-permeable membranes.

Pianola rolls were often transcribed directly from virtuoso piano performances.

Piccolo, the smallest of the woodwinds, has a timbre like the military fife.

Picketing occurred during the Grunwick dispute in London in 1977.

Mary Pickford, star of the 20s, was dubbed "the world's sweetheart".

be divided into six fundamental theories: Newtonian mechanics, THERMODYNAMICS, ELECTROMAGNETISM, STATISTICAL MECHANICS, QUANTUM MECHANICS and relativity.

Physics, atomic. See NUCLEAR PHYSICS.

Physiocrats, school of 18th-century French economists. Led by François QUESNAY, they argued that agriculture, rather than industry or commerce, was the basis of all wealth. They had some influence in France under LOUIS XV and in Austria under JOSEPH II; in Britain, Adam SMITH was also strongly influenced by the physiocrats.

Physiognomy, the study of a person's outward appearance to judge the inner character, especially when considering the face as an indicator of individual character. Once a popular theory of human personality, physiognomy has been largely discredited by modern psychology. See also MS p.197.

Physiology, branch of biology concerned with physical and chemical activities and functions of a living organism; these may be necessary to maintain life. Vast in scope, it includes the study of single cells as well as multicellular organisms.

Physiotheraphy, also known as physical therapy, use of various techniques to treat disorders of the musculo-skeletal system. It attempts to relieve pain and restore the use of the affected parts. Its techniques include massage, exercise, heat and water treatment, the use of ultrasonics and diathermy. See also MS pp.88, 90.

Phytoplankton, free-floating oceanic plant life as opposed to ZOOPLANKTON, the free-floating animal life. Most of the organisms are microscopic, such as DIATOMS. See also ALGAE; NW p.238; PE pp.79, 90.

Pi (π), symbol used for the ratio of the circumference of a circle to its diameter. It is an IRRATIONAL NUMBER and an approximation to five decimal places is 3.14159. The ratio 22/7 is often used as a rougher approximation. See also SU pp. 30, 30, 44–45.

Piaf, Edith (1915–63), French cabaret singer, real name Edith Giovanna Gassion. Piaf (Parisian argot for sparrow) began singing at the age of 17 in the cafés and streets of Paris. Her recording career spanned nearly 30 years, and the emotional impact of her voice in songs such as Non, je ne regrette rien won her an international reputation.

Piaget, Jean (1896–   ), Swiss psychologist who developed a comprehensive theory of the intellectual growth of children. His books describe their thought processes from early infancy to adulthood. He became director of the Institut Jean-Jacques Rousseau in Geneva and director of The International Centre of Genetic Epistemology in 1955. With associates he has published more than 30 books on child psychology, the major ones being The Origin of Intelligence in Children (1954) and The Early Growth of Logic in the Child (1964). See also MS p.152.

Piano (pianoforte), musical instrument whose sound is made by strings struck by hammers which are moved from a keyboard. Its invention (c.1709) is attributed to Bartolomeo Cristofori, who combined the hammer-striking action of the CLAVICHORD and the dampers of the HARPSICHORD in one instrument. Its name, from the Italian piano (soft) and forte (strong), was adopted because its range of volume (as of tonal quality) far exceeded that of earlier instruments. The modern grand piano, much larger, louder and more resonant than the 18th-century piano, was developed in the early 19th century.

Pianola, instrument for reproducing piano music mechanically, patented by an American, E. S. Votey, in 1897, but derived from earlier inventions. A roll of perforated paper passes over a "tracker-bar", which activates the piano's keys and hammers. It is generally powered by a foot pedal.

Piatigorsky, Gregor (1903–76), US cellist and teacher, b. Russia. He studied at the Imperial Conservatory, Moscow, and was first cellist of the Berlin Philharmonic from 1924 to 1928. In 1929 he moved to the US and appeared as a soloist in recitals and with major orchestras. He also played in ensembles with HOROWITZ, RACHMANINOV, SCHNABEL and HEIFETZ, and concertos were written for him by several composers including PROKOFIEV.

Piazzetta, Giovanni Battista (1683–1754), Italian painter. He studied in Bologna, but spent most of his life in Venice, where he was influential in the change of style from BAROQUE to ROCOCO. He is noted for his CHIAROSCURO effects, especially in paintings such as Fortune Teller (1740).

Piazzi, Giuseppe (1746–1826), Italian astronomer who discovered the first known asteroid, Ceres in 1801. He also founded the observatory of Palermo and produced a catalogue of 7,646 stars. See also SU p.204.

Pica, abnormal craving for unnatural foods, such as chalk or ashes. It occurs both in humans and animals and is usually a sign of nutritional deficiency.

Picabia, Francis (1879–1953), French painter of Cuban Spanish descent who worked in an IMPRESSIONIST style from 1903 until 1909. He was then influenced by CUBISM and was one of the first exponents of DADA both in Europe and the USA. He was also associated for a time with the Paris SURREALISTS.

Picardy, region and former province of N France, on the English Channel; includes Somme, and parts of Pas-de-Calais, Oise and Aisne départements. It was a French province from 1477 until the French Revolution, when it was replaced by a smaller département. Picardy was the scene of heavy fighting during WWI. The area is made up of the plateau to the N of Paris where wheat and sugar-beet are grown; the valley of the Somme where industrial centres such as Amiens are located; and the coast where fishing is important. Calais and Boulogne-sur-Mer are cross-Channel ferry ports.

Picaresque (from Spanish pícaro, "rogue" or "knave"), term first applied to an early genre of prose fiction in which a roguish hero has a series of adventures, providing the author with a means for satirical comment. Examples include MENDOZA's Lazarillo de Tormes (1554) and LE SAGE's Gil Blas (1715–35). The latter influenced early English novels such as FIELDING's Tom Jones (1749). In a general sense, the term is often used to refer to fiction which is episodic in structure, rather than with its original "roguish" implication. In this wider sense CERVANTES' Don Quixote de la Mancha is often described as a picaresque novel.

Picasso, Pablo (1881–1973), Spanish painter and sculptor who lived in France. His first canvases, those of the "blue period" (1901–04), showed scenes from poverty in predominantly sombre blue tints. There followed in the "rose period" (1905–08) lighter paintings, often of circus life. In 1907 his Les Demoiselles d'Avignon appeared, often regarded by critics as analytical cubism; or the manifestation of the first stage of CUBISM. Thereafter he developed into an artist of great power and variety, producing a steady stream of paintings, sculptures, ceramics, drawings, graphic illustrations and stage designs. See also HC2 pp.176, 177, 203.

Piccard, name of a Swiss family of scientists. Auguste (1884–1962) became professor of physics at Brussels University in 1922. During 1931–32 he made experimental balloon flights into the stratosphere, reaching heights of 15,787m and 16,916m (51,793ft and 55,000ft). He later turned to deep-sea exploration, in bathyscaphes of his own design, and reached a depth of 3,150m (10,330ft) in the Mediterranean Sea (1953). His son and collaborator Jacques (1922–   ) dived to 10,912m (35,800ft) in the Marianas Trench in the Pacific Ocean (1960).

Piccinni, Niccolò (1728–1800), Italian composer of operas. He became a well-known figure in the late 18th century when the opponents to the opera reforms of GLUCK made Piccinni their unwilling champion. Although he wrote more than 120 operas they are only rarely performed today, but his La Cecchina, ossia La buona figliuola (1760), based on Samuel RICHARDSON's Pamela, was probably the most popular OPERA BUFFA of its time.

Piccolo, woodwind musical instrument of the FLUTE family. About half the size of the flute and pitched one octave higher, it is played in the same way – by blowing across the mouth hole and fingering the keys. It has a bright, penetrating tone and is used in symphony orchestras. See also HC2 pp.104–105, 104.

Pic du Midi de Bigorre, observatory in the Pyrenees Mountains, France. Its altitude of approx. 2,877m (9,440ft) means that conditions for observation are excellent. The 24in refracting telescope was used extensively by the French astronomer Audouin DOLLFUS in his planetary studies. See also SU pp.212–213.

Pichergru, Charles (1761–1804), French general in the FRENCH REVOLUTIONARY WARS. He was successful against the Austrians on the River Rhine in 1793, and in the following year invaded The Netherlands. In 1795 he captured the Dutch fleet, ice-bound at Amsterdam. His intrigues with the Austrian royalists led to his being recalled to France, where he was arrested. He escaped to England but returned to France in 1803 and was again arrested. He was later found strangled in his cell before he could be brought to trial.

Piciformes, order of birds characterized by having their first and fourth toes pointing backwards. It includes WOODPECKERS, TOUCANS, BARBETS, HONEY-GUIDES and their relatives; most species nest in holes in trees and eat insects and fruit. Length: to 61cm (24.5in). See also NW p.141.

Pickering, Edward Charles (1846–1919), US physicist and astronomer. He was the first to employ the meridian PHOTOMETER for measuring the magnitude of stars, and in 1884 published Harvard Photometry, the first photometric catalogue. See also SU p.226.

Pickering, William Henry (1858–1938), US astronomer who discovered PHOEBE, the 9th satellite of SATURN in 1899. With his brother Edward. C. PICKERING, he established the Arequipa Observatory in Peru in 1891. He erected the Lowell Observatory in Flagstaff, Arizona, in 1894 for the noted astromer Percival LOWELL. See also SU p.213.

Picketing, the patrolling outside a place of employment by members of its labour force who have stopped their normal work because of a dispute with the employers. The aim of the patrol is to persuade those of their colleagues who are still working also to stop, to dissuade drivers of supply vehicles from making their deliveries, and to gain publicity for their cause. There is no special law for or against pickets; verbal persuasion is naturally allowed, although conduct liable to lead to a breach of the peace may lead to arrest.

Pickford, Mary (originally Gladys Mary Smith) (1893–   ), US film actress famous for her portrayal of innocent heroines in silent films. These include Poor Little Rich Girl (1917), Pollyanna (1919) and Little Lord Fauntleroy (1921). In 1919 she established the United Artists Corporation with Charles CHAPLIN, Douglas FAIRBANKS, whom she married in 1920, and D.W. GRIFFITH. See also HC2 pp.200, 268.

Pick's disease, circumscribed atrophy of the frontal lobes of the BRAIN. It progressively impairs judgement and intellect and may be accompanied by psychotic features. See also ALZHEIMER'S DISEASE; MS p.139.

Pickwick Papers, The (1836–37), first humorous, anecdotal novel by Charles DICKENS. It recounts the adventures of Samuel Pickwick, a benevolent gentleman, as he and his friends travel throughout England. It contains some of Dickens' greatest comic characters and immediately established him as a popular author.

Pico della Mirandola, Giovanni, Conte (1463–94), Italian philospher and scholar, an exponent of Renaissance Platonism. He came from a wealthy family, received a humanist education and studied canon law and Aristotelian philosophy. He wrote The Oration on the Dignity of Man in

1486, but was arrested in 1488 when some of the 900 theses he had written were denounced as heretical.

Picric acid, yellow crystalline solid, chemical name 2,4,6-trinitrophenol, $C_6H_2OH(NO_2)_3$. It was widely used as an explosive before and during WWI, but is now used in the manufacture of dyes and pigments.

Pictogram, or pictograph, symbolic method of visually expressing an object (as opposed to an idea) and used to convey messages before alphabets evolved. Today, pictograms play an important role in the graphic presentation of information. *See also* MS p.244.

Pictorow, Treaty of. *See* YAM ZAPOLSKY, PEACE OF.

Picts, ancient inhabitants of E and N Scotland, first named ("Pict" means "painted") by the Roman writer Eumenius in AD 297 as northern invaders of Roman Britain. By the 8th century they had a united kingdom extending from Caithness to Fife, and had adopted Celtic Christianity. To the W and S of the Picts, invaders from Ireland had established the kingdom of DALRIADA; in 843 its king, KENNETH I, became king also of the Picts, uniting the two kingdoms into what later became the kingdom of Scotland. *See also* HC1 pp.*100*, 101, 182.

Picture of Dorian Gray, The (1891), novel by Oscar WILDE. A fantastic, allegorical work, it describes the degenerate life of a handsome young man who until the end remains youthful while his portrait grows old and corrupt.

Picturesque, the, term used by art historians to describe those elements – ruggedness, random nature, ruins – which lent a "Gothic" aspect to late 18th-century landscapes, especially in England. The picturesque was thus a harbinger of ROMANTICISM.

Picture stage, in theatres, deep acting area with little forestage, which in the 17th century became framed by the PROSCENIUM arch and in which all the illusions of the THEATRE were completely maintained. This concept of a set as an entirely naturalistic entity without a fourth wall was widely employed during the 19th century, but reactions to its inevitable artificiality grew and many new theatres were designed to incorporate the more flexible OPEN STAGE. *See also* STAGE.

Picture tube, type of cathode ray tube used to produce the picture in a television set or the display in a radar set. The picture is "drawn" on the inside face of the tube by a beam or beams of ELECTRONS that scan the tube and stimulate PHOSPHORS to emit light. *See also* MM pp.230, *230*, 236, *236*.

Piculet, any of several species of small WOODPECKERS. They are insectivorous, stubby-tailed birds, most species of which occur in Central and South America, Asia or Africa. Their colouring is generally mottled grey-green to brown above and white below. They dig their nest holes in soft wood. Length: 8–13cm (5–8in). Family Picidae.

Piddock, bivalve distinguished by its ability to bore into rock. This it does by moving its serrated shell back and forth wearing a hole that enlarges to a comma-shape as the animal grows. Species *Pholas dactylus*. *See also* NW p.*92*; PE p.*125*.

Pidgeon, Walter (1897–), Canadian film actor who enjoyed a long Hollywood career. He made his debut in *Mannequin* (1925) and worked steadily throughout the 1930s and 1940s, starring in films such as *How Green Was My Valley* (1941) and *Madame Curie* (1943). Later films included *Funny Girl* (1968) and *Skyjacked* (1972).

Pidgin English, form of language developed by English traders, missionaries and officials abroad for communication with foreign peoples. It has a simplified grammar and a small vocabulary, a mixture of English words with Chinese, Malay and Portuguese. The word supposedly derives from the Cantonese corruption of the word "business".

Pieck, Wilhelm (1876–1960), German political leader. He was founder of the German Communist Party in 1918, and

was in exile from 1933 to 1945. Pieck was President of East Germany (1949–60).

Piedmont (Piemonte), region of NW Italy, made up of the provinces of Allessandria, Asti, Cuneo, Novara, Torino, and Vercelli; bounded to the N, W and S by mountains, and to the E by the Po Valley which has some excellent farmland. The Po River and its tributaries supply water for agriculture, and hydroelectric schemes provide energy for industry, transportation and domestic use. Products: grain, vegetables, fruit, dairy. Industries: wine-making, motor vehicles, textiles, glass, chemicals. Area: 25,400sq km (9,807sq miles). Pop. (1968 est.) 4,316,500.

Pied Piper of Hamelin, in German folklore, a magician. According to the legend, in 1284 a piper rid the town of Hamelin of rats by playing his pipe and leading the rats to the River Weser, where they drowned. Although he was promised money by the townspeople for his deed, they refused to pay. In retaliation the piper mesmerized the children of the town with his playing and led them away to a door in the hills where they vanished. The legend is the basis of works by Robert BROWNING and the brothers GRIMM. The source of the legend may be the emigration of young Germans in the 13th century to colonize the east, or perhaps the ill-fated departure of the German children on the CHILDREN'S CRUSADE of 1212.

Pierce, Franklin (1804–69), 14th US president (1853–57) who served during the turbulent years before the AMERICAN CIVIL WAR. He was known for his pro-slavery views and supported the highly controversial Compromise of 1850 (which attempted to solve the slavery question) and the Kansas-Nebraska Act (1854). The Gadsden Purchase (1853) added land from Mexico. *See also* HC2 p.150.

Pierné, Henri Constant Gabriel (1863–1937), French conductor and composer who studied under MASSENET. He wrote the cantata *La Croisade des Enfants* (1902), several operas and ballets, orchestral works and many songs.

Piero della Francesca, (Piero dei Franceschi), Italian painter. Influenced by many Florentine artists including DONATELLO and DOMENICO VENEZIANO, he combined broad masses of soft colour with subtle light effects. His altarpiece for S. Bernardino, Urbino, is a fine example of his sensitive handling of perspective. Famous among his works are the frescoes of the *Legend of the Holy Cross* (c.1465).

Piero di Cosimo (c.1462–1521), Italian painter who adopted the name of his master, Cosimo Rosselli. As well as conventional religious works which show the influence of LEONARDO DA VINCI, he painted many pictures showing the life of primitive man. These include the *Hunting Scene* and *The Discovery of Honey*. Other well-known works by Piero are the *Death of Procris* and the portrait *Simonetta Vespucci*.

Pierrot, in Britain, pantomine figure and member of a seaside entertainment troupe who traditionally wore a loose white silk shirt and a dunce's cap and had a whitened face. He derived originally from the COMMEDIA DELL'ARTE character Pedrolino and owed his vogue in Britain up to WWII to the London success of the mime play *L'Enfant prodigue* (1891) in which Pierrot, as developed by Deburau, featured.

Piers Plowman (c.1362–92), poem in Middle English attributed to William LANGLAND. There are three extant versions which differ from each other considerably. The basic framework of the story is that of an allegorical vision on the subject of the vices of the times. *See also* HC1 pp.*195*, 215.

Pietà (Italian for "pity"), in art, painting or sculpture showing the suffering Virgin bearing the dead Christ across her knee. The idea developed in 14th-century Germany and emphasizes the emotional rather than the symbolic aspects of Christ's suffering. Famous pietàs include one from the School of Avignon (c.1460, Louvre), and the marble sculpture (1498–1500) by Michelangelo in St Peter's, Rome.

Pietermaritzburg, city in the Republic of South Africa, 64km (40 miles) N of Durban, on the River Umsunduzi. It became the capital of NATAL in 1856. Today it is an industrial and administrative centre, its products including furniture, footwear, motor vehicles and diesel engines. Pop. (1970) 113,747.

Pietism, Christian movement founded in the late 17th century by Philipp Spener, a German, to revitalize evangelical Christianity in opposition to the severe and intellectualized formality of 17th-century Protestantism. In *Pia Desideria* (1675), Spener emphasized the personal nature of faith and salvation. His influence spread throughout Europe.

Pietra dura, work inlaid with hard, semiprecious stones such as agate, quartz or jasper. It was produced mainly in Florence during the 16th and 17th centuries and was used for decorating furniture, particularly cabinets and table-tops. The stones were cut to fit into a geometric or floral pattern and then polished.

Piezoelectric effect, creation of positive electric charge on one side of a nonconducting crystal and a negative charge on the other when the crystal is subjected to mechanical pressure. The pressure polarizes the crystal by separating the centre of the positive charge from that of the negative in a crystal in which there is no centre of structural symmetry. This charge separation results in an electric field that can be detected as voltage between the opposite crystal faces. The effect is used in gramophone pickups, crystal microphones and cigarette lighters. *See also* MM p.93, *249*; SU p.116.

Piezoelectricity. *See* PIEZOELECTRIC EFFECT.

Pig, any of numerous species and varieties of domestic and wild swine of the family Suidae. The male is generally called a boar; the female, a sow. A castrated boar is usually known as a hog. It is generally a massive, short-legged omnivore with a thick skin bearing short bristles. Wild pigs include the warthog, wild BOAR, bush pig and babirusa. *See also* NW p.162; PE pp.224, 230–231, 233–235.

Pigeon, also often called dove, any of a large family of wild and domestic birds found throughout temperate and tropical parts of the world, but concentrated in S Asia and the Australian region. Many have long been domesticated, used for food, for racing and some for carrying messages. Pigeons have small heads, short necks, plump bodies and scaly legs and feet. Plumage is loose but thick and may be brown, grey, white, blue, green, yellow or orange. Length: to 46cm (18in). Family Columbidae; typical genus, *Columba*. See *also* NW pp.*141*, *145*, *151–153*, 211, *217*.

Pigeon racing, sport in which homing pigeons are taken to a common starting point from which they fly back to their own lofts. There the pigeon's race rubber is removed and the time of arrival stamped on by the owner's pigeon clock. Flying speed is calculated in yards or metres per minute, and the fastest bird wins.

Piggott, Lester (1935–), British jockey. Champion apprentice with 52 wins in 1950, Piggott went on to become one of racing's most successful jockeys, winning 17 classics between 1954 and 1970. He is generally regarded as a master tactician, combining great determination with superb judgement.

Pigment, coloured, insoluble substance used to impart colour to an object and incorporated for this purpose into paints, inks and plastics. Unlike DYES, pigments are insoluble in the coating medium. They generally function by absorbing and reflecting light, although some modern luminescent pigments emit coloured light. *See also* MM p.244.

Pigmentation, in biology, the coloration of tissues by PIGMENTS. In human beings, the pigmented areas are skin, hair and the iris of the eye. The pigments MELANIN and CAROTENE, combined with the body's HAEMOGLOBIN, give the colour.

Pignon, Edouard (1905–), French painter. He began painting in 1934, encouraged by PICASSO, and in a style which shows the influence of both Picasso

The Pied Piper of Hamelin punished the city by luring away their children.

Piero di Cosimo painted many portraits and helped decorate the Sistine Chapel.

Pigs are very useful food animals: almost the entire carcase is edible.

Lester Piggott sits astride yet another winner at fashionable Ascot.

Pig-nut

Pike are fierce predatory fish which can devour prey of almost their own size.

Pilâtre de Rozier flew for 25 minutes over Paris in a hot air balloon in 1783.

Pillbox: these concrete gun emplacements can still be seen in southern England.

Pillory; early 17th-century stocks in the village of Great Tew, Oxfordshire.

and MATISSE in his use of space and colour. A miner before he took up painting, his realistic choice of subject matter is seen in such works as *The Dead Miner* (1952).

Pig-nut, also called earth-nut or hog-nut, slender, perennial European plant of the carrot family (Umbelliferae); its name derives from its edible tuber. It has small white flowers and complex divided leaves. Height: to 1m (39in). Species *Conopodium majus. See also* PE p.*191*.

Pigou, Arthur Cecil (1877–1959), British economist who held that the pursuit of self-interest by individuals did not benefit society as a whole which was responsible for and should bear the costs of unemployment, health and housing. He was professor of political economy at Cambridge (1908–43). His works include *The Economics of Welfare* (1920), *The Theory of Unemployment* (1933), *Socialism versus Capitalism* (1937) and *Income: an Introduction to Economics* (1946). *See also* MS p.311.

Pigs, Bay of. *See* CUBAN MISSILE CRISIS.

Pika, any of 12 species of short-haired relatives of the RABBIT. They live in cold regions of Europe, Asia and western USA. Length: to 20cm (8in). Genus *Ochotona*.

Pike, predatory freshwater fish found in E North America and parts of Europe and Asia. A popular fish with anglers, it has a shovel-shaped mouth and a mottled, elongated body. Length: to 137.2cm (54in); weight: 20.9kg (46lb) Family Esocidae; genus *Esox. See also* NW p.*231*.

Pilate, Pontius, Roman governor of JUDAEA when Jesus Christ was crucified. Pilate was made procurator of Judaea in AD 26 and earned a reputation for arrogance and cruelty. He died after AD 36. The New Testament gospels tell of Pilate's role in condemning Jesus. *See also* HC1 p.111.

Pilâtre de Rozier, Jean François (1756–85), French pioneer of ballooning. In Nov. 1783 he and the Marquis d'Arlandes were the first to fly in an untethered balloon. During a flight lasting about 25 minutes, they covered a distance of 8km (5 miles). In 1785, Pilâtre de Rozier was killed while attempting to cross the English Channel by balloon. *See also* MM pp.146–147.

Pilchard, marine, food fish resembling a herring, found in shoals along most coasts except those of Asia. They are caught in millions, and support a huge canning industry. The young are sometimes called SARDINES. Length: less than 45.7cm (18in). Family Clupeidae; species *Sardina pilchardus. See also* PE p.*247*.

Pile-driver, machine for driving post-like foundation members (piles) into the ground. It usually consists of a high frame with attachments that raise and drop a pile-hammer on to the pile, thus forcing it into the ground with repeated blows. *See also* MM p.*49*.

Piles. *See* HAEMORRHOIDS.

Pilgrimage, journey to a shrine or other holy place for religious motives, eg to gain spiritual help or for thanksgiving. Pilgrimages are common to many religions, particularly in the East. A Muslim should make the pilgrimage to MECCA, where devotions last two weeks, at least once in his life. Since the second century, Christians have travelled to Palestine, to the tomb of the Apostles Peter and Paul in Rome, and of James in Santiago de Compostela in Spain. *See also* HC1 pp.172–173, 186–187.

Pilgrimage of Grace (1536), uprising of Roman Catholics in N England in protest against ENCLOSURES and the DISSOLUTION OF THE MONASTERIES following the abolition of papal supremacy in England. After a small rising in Lincolnshire had failed, Robert Aske and his followers occupied York and then marched to Doncaster with about 30,000 armed men. The Duke of NORFOLK, Henry VIII's emissary, promised Aske a pardon from the king and Aske dispersed his men. Further minor outbreaks were suppressed and Aske and more than 200 others were executed in 1537. The repression in N England after the Pilgrimage of Grace put an end to substantial opposition to the government's religious policy. *See also* HC1 p.*267*.

Pilgrim Fathers, English Puritans who left from Plymouth on the *Mayflower* in 1620 and founded a colony in North America at Plymouth, New England. They had originally sought refuge from persecution in England by fleeing to Leiden in 1608, where they decided to found a new religious society in America.

Pilgrim's Progress, The (1678–84), famous narrative by John BUNYAN, in the form of a dream ALLEGORY of the life of Christian from his conversion to his death. The second part deals with Christiana, who follows the same road as Christian, who has gone before. The whole work is remarkable for its simple style and has greatly influenced English prose.

Pilgrims' Way, popular name for an ancient English track running from Hampshire to Kent. It was used in the Middle Ages by pilgrims travelling from the SW to St Thomas à BECKET's shrine at Canterbury.

Pilgrim Trust, charitable concern founded in 1930 by a US philanthropist and railway magnate, Edward Stephen Harkness (1874–1940). Harkness provided £2,000,000 with which the Trust was to promote cultural relationships between Britain and the USA, preserve historical buildings and monuments and provide educational and welfare grants.

Pill, the, popular name for hormonal preparations used as oral contraceptives. They work by preventing ovulation at its usual time at the middle of the MENSTRUAL CYCLE. Two types of hormones, OESTROGENS and PROGESTERONES, are generally used, although the former alone are effective in preventing ovulation; the latter help to regulate the menstrual cycle. From the time of their introduction in the 1950s such contraceptives have been controversial. Controversy has centred around possible side-effects, notably amennorhea, headaches, nausea and weight gain, and around the increased liability of deep-vein thrombosis which is associated with their use. On the other hand their convenience and efficacy make them the most reliable form of BIRTH CONTROL available. *See also* MS pp.80–81.

Pillars of Hercules, in Greek mythology, two promontories in the Mediterranean, on each side of the Strait of Gibraltar. They were called Calpe and Abyla in ancient times. They are usually identified as Gibraltar in Europe and Mt Acha (Mt Hacho) in Africa. According to some legends they were joined together until HERCULES tore them apart in order to reach Cadiz.

Pillbox, military term for a dug-in machine-gun emplacement. It is sometimes reinforced with concrete, steel, logs or filled sandbags and often has overhead protection.

Pilon, Germain (*c.*1535–90), French sculptor hailed as the successor to GOUJAN. His work was different in style, however, and is representative of the transition from MANNERISM towards the BAROQUE. The marble figures of Henry II and Catherine de Medici on their tomb (1563–70) are relaxed flowing forms. Other notable works include the bronze *René de Birague* (1583–85), which shows the influence of MICHELANGELO. *See also* HC1 pp.263, *263*.

Pillory, form of punishment in which offenders had their hands or head locked through the holes of a wooden framework. They were then mocked by passers-by, who often bombarded them with food and debris. (A device for holding the feet of the offender was called the STOCKS.)

Pillow lava, LAVA extruded under water that commonly takes the form of a distorted globular mass, resembling the shape of a pillow. It apparently results from the rapid chilling of the outer skin, thus making a more or less spherical "balloon" that grows and flattens under its own weight. *See also* PE p.82.

Pilocarpine, ALKALOID drug ($C_{11}H_{16}N_2O_2$) obtained from the jaborandi plant as an oily syrup that crystallizes when pure. Pilocarpine is used chiefly in the form of its hydrochloride or nitrate to treat GLAUCOMA or in counteracting mydriasis caused by ATROPINE poisoning.

Pilot fish, marine fish that lives in warm seas, often found swimming close to sharks, ships and other large objects. A blue fish with five to seven dark bar markings on the sides and a white tail, it feeds on smaller fish. Length: to 61cm (2ft). Family Carangidae; species *Naucrates ductor*.

Pilot whale, or blackfish, gregarious small, black, toothed WHALE found in all seas except those around the poles. Specimens have been trained to perform in captivity. Length: to 8.5m (28ft). Genus *Globicephala. See also* DOLPHIN; NW p.*166*.

Pilsen, or Plzen, city in W Czechoslovakia. Before 1918 it was part of the Austro-Hungarian Empire. It is a centre for heavy industry (machine tools, motor cars, armaments), and is famous for its internationally exported beer. Pop. (1974) 154,126.

Pilsudski, Joseph (1867–1935), Polish general and statesman who established the independence of Poland. As a youth he was exiled to Siberia for five years for an alleged attempt to murder Tsar Alexander III. Thereafter he was dedicated to raising a private army. In WWI he fought against the Russians but was interned by the Germans when he refused to fight further because of their interference. He was Minister of Defence and the de facto ruler of Poland from 1926 until his death.

Piltdown man, name given to a skull found in Sussex, England, in 1912–13, which appeared to be the "missing link" in the evolution of man from the apes. In 1953 it was proved to be a fake, perpetrated by Charles Dawson. *See also* MS p.22.

Pimenov, Yuri Ivanovich (1903–), Soviet painter and designer. His painting, mainly in water-colour, concentrated on portraying scenes in towns near Moscow. His early work was influenced by IMPRESSIONISM but he later painted in a more conventional style. He also designed many theatre sets, and became professor at the National Film Institute at Moscow in 1945.

Pimento. *See* ALLSPICE.

Pimpernel, small, trailing annual plant native to Britain and the USA. The single, small, five petalled flowers are scarlet, white or blue. Family Primulaceae; genus *Anagallis*.

Pinafore, HMS. *See* HMS PINAFORE.

Pinchbeck, alloy of COPPER and ZINC, but with much less zinc than is contained in BRASS. Invented by the London watchmaker Christopher Pinchbeck (*c.*1670–1732), it is used to simulate gold. Hence, as an adjective "pinchbeck" is used to mean sham, or in false taste.

Pincus, Gregory Goodwin (1903–67), US biologist who helped to develop the contraceptive PILL using synthetic HORMONES. These hormones simulate the physiological state of women when pregnant, thus preventing fertilization. Pincus wrote several articles on hormones and a book, *The Eggs of Mammals* (1936). *See also* MS p.80.

Pincushion cactus, CACTUS native to W USA, Mexico and Cuba. It has round or oval globular stems, large brilliantly coloured flowers and bright edible berries. Diameter: 1–61cm (0.5in–2ft). Family Cactaceae; genus *Coryphantha*.

Pindar (*c.*522–*c.*440), Greek poet known for his choric lyrics and triumphal ODES. Of his complete works 44 odes written to celebrate victories in athletic games survive. Each was to be sung in procession for the winner on his return home. *See also* HC1 p.74.

Pindling, Lynden Oscar (1930–), chief of state of the Bahamas. A barrister, educated in Britain, he became leader of the Progressive Liberal Party in 1956 and served as Premier from 1957 until the Bahamas achieved independence in 1973.

Pine, any of various evergreen cone-bearing trees, most of which are native to cooler temperate regions of the world. They have scale-like deciduous leaves or evergreen needle-like clusters. The flowers may be catkin-like stalks or pine cones. Many species are valued for soft wood, wood pulp, oils and resins. Family Pinaceae; genus *Pinus. See also* NW pp.56–57, 206–207; PE pp.215, *217–218*, 219, *222–223*.

Pineal body, small organ attached to the lower surface of the brain, the function of which is unkown. In human beings, it becomes calcified by the age of about 60 years. In some lizards it forms a "third eye" and is thought to influence sexual cycles. Evidence suggests that the pineal may be an ENDOCRINE GLAND.

Pineapple, tropical herbaceous perennial plant that is cultivated in the USA, South America, Asia, Africa and Australia, and the fruit of the plant. The fruit is formed from the flowers and bracts of the plant and grows on top of a short stout stem bearing stiff fleshy leaves. The fruit is popularly eaten fresh, tinned or made into juice. Hawaii is the major producer of pineapples. Height: to 1.2cm (4ft). Family Bromeliaceae; species *Ananas comosus.*

Pineapple Poll (1951), one-act ballet choreographed by John CRANKO to music by Arthur SULLIVAN (arranged by Charles Mackerras) with libretto derived from W. S. GILBERT's Bab Ballard, *The Bumboat Woman's Story.* With sets designed by Osbert Lancaster, the ballet was first produced at Sadlers Wells and starred Elaine Fifield, David Blair and Daivid Poole. Its plot bears strong affinities to that of Gilbert and Sullivan's *HMS Pinafore.*

Pinel, Philippe (1745–1826), French physician, a pioneer in the humane treatment of the mentally disturbed. He released patients from imprisonment, developed case histories and attempted to talk and reason with patients. His theories were published in *Traité médico-philosophique sur l'aliénation mentale ou la manie* (1809).

Pine nut, edible seed obtained from any of several species of low-growing PINES called piñon pines (eg, *Pinus cembroides*) of SW North America. The nuts were collected by American Indians for food and they are now used in confectionery.

Pinero, Sir Arthur Wing (1855–1934), British dramatist who wrote carefully constructed plays in realistic style. *The Second Mrs Tanqueray* (1893) raised a protest because of its sympathetic portrayal of a woman with a questionable past. He also wrote farces, such as *The Magistrate* (1885), and sentimental comedies, such as *Sweet Lavender* (1888).

Pinesap, one of a number of saprophytic plants, including the Indian pipe, of Britain and E North America. The plants are whitish or reddish and live on dead vegetable matter or other plant roots. Family Monotropaceae; genus *Monotropa.*

Pingo, large conical mound of soil-covered ice that is raised through the perma-frost of the arctic regions by water pressure from below. Pingos may be up to 90m (330ft) high and 800m (2,600ft) across.

Pink, common name for the several genera of the pink family, especially the genus *Dianthus* of more than 300 species, most of which are native to the Mediterranean region. Primarily short herbaceous perennials, many are hardy evergreens with showy flowers. The stems are often woody at the base, leaves are simple and usually opposite, and the radially symmetrical flowers are usually bisexual. Family Caryophyllaceae. *See also* CARNATION; NW p.*223.*

Pinkerton, Allan (1819–84), US detective, b. Scotland. He moved to the USA in 1842, became a detective on the Chicago police force, resigning in 1850 to establish his own agency, Pinkerton's National Detective Agency. He organized and headed a federal intelligence service (1861).

Pink-eye. See CONJUNCTIVITIS.

Pinkie, Battle of (1547), invasion of Scotland by Protector SOMERSET. It failed to weaken French influence in Scotland or to secure the marriage of EDWARD VI and MARY QUEEN OF SCOTS. The Scots were routed at Pinkie, east of Edinburgh, but the English did not follow up and were forced to leave Scotland by the Treaty of Boulogne (1550).

Pinna, flap of skin and cartilage which, with the auditory canal, makes up the outer EAR in mammals.

Pinnipedia, suborder of meat-eating aquatic mammals formerly classified as carnivores. The order is made up of eared seals (Family Otariidae) including fur seals and SEALIONS; earless, true, or hair seals (Family Phocidae); and the WALRUS (Family Odobenidae). They have fin-like limbs and their digits are united by a membrane. *S e also* NW pp.155, 165.

Pinocchio (1881), children's story by Carlo Collodi. It recounts the adventures of Pinocchio, a wooden puppet who comes to life. The story was made into a cartoon film by Walt Disney in 1939.

Pinochet Ugarte, Augusto (1915–), became President of Chile in 1973, when he led the revolt of the armed forces that overthrew Salvador ALLENDE. As head of a four-man military junta, Pinochet brought in strict policies to eradicate Marxism in Chile and to improve an economy damaged by inflation.

Pinochle, card game for two to four players using a 48-card pack made up of two each of the ace to nine in each suit, the cards ranking ace, ten, king, queen, jack and nine. The most popular form of the game is auction pinochle for three players, each of whom receives 15 cards, the remaining three being turned face down on the table. The players "bid" for the three cards and the winner adds them to his hand, discarding three cards to restore his hand to 15. Points are awarded for combinations of cards displayed by each player and tricks are then collected, the winner of the auction aiming at least to reach the amount of his bid through the points he originally declared and those he wins in play.

Pinocytosis, intake and transport of fluid by living cells. Rather than entering and passing through the cell membrane as individual molecules, a droplet becomes bound or absorbed to the membrane which forms a pocket and pinches off to form a vesicle in the CYTOPLASM. The vesicle may then pass across to the far side of the cell where reverse pinocytosis occurs.

Piñon pine, also called Mexican stone pine, evergreen tree that grows in Mexico and SW USA. It has large, edible seeds which are commonly called PINE NUTS. Height: to 7.6m (25ft). Family Pinaceae; species *Pinus cembroides.*

Pins and needles, term for the numbing sensation caused by the irritation of a nerve. It is experienced when one of the limbs "goes to sleep", from being in an awkward position where its circulation is temporarily restricted.

Pinscher. See DOBERMAN PINSCHER.

Pint, unit of liquid capacity; 1/8 of a gallon, ½ of a quart, 0.833 US pint, or 0.6 litre. Although the United Kingdom has begun to convert to the metric system, the pint will be retained, at least until the 1980s, because of its importance as a traditional measure for beer and milk.

Pintail, surface-feeding DUCK. It has webbed feet, a longish neck, short legs, and a flat, blunt beak. The female is a paler brown than the male. Length: 56cm (22in) plus a 10cm (4in) "pin" tail on the male. Family Anatidae; species *Anas acuta.*

Pinter, Harold (1930–), British playwright who began his career as an actor with provincial companies. His first play, *The Room* (1957), was followed by others such as *The Birthday Party* (1958), *The Caretaker* (1960), *The Homecoming* (1965), *Old Times* (1971) and *No Man's Land* (1975). Most of his plays are so-called "comedies of menace". Ordinary characters and settings are found in an atmosphere of fear, horror and mystery.

Pinto, also called paint, breed of light saddle horse, usually with irregular white markings spreading up from the belly, or down from the back. The name originally implied "a spotted horse"; it may be piebald, skewbald or any other colour, with white markings. Height: to 157cm (62in) at shoulder; weight: to 454kg (1,000lb).

Pinturicchio, or Pintoricchio (c.1454–1513), Italian painter, real name Bernardino di Betto. He decorated the Borgia apartments in the Vatican and the cathedral library at Siena.

Pinza, Ezio (1892–1957), Italian bass who studied in Italy and made his debut in Rome in 1921. He became the principal bass of the METROPOLITAN OPERA in New York, achieving great success in a series of leading roles from 1926 to 1948. In 1949 he starred on BROADWAY in *South Pacific.*

Pinzón, Martín Alonzo (c.1441–1493), Spanish navigator and explorer. He was one of three brothers to take part in Columbus's first voyage to America (1492). As commander of the *Pinta,* he deserted Columbus to look for the legendary land of gold. He again parted from Columbus on their return to Spain in 1493. *See also* HC1 p.234.

Pinzón, Vicente Yáñez (c.1460–c.1523), Spanish explorer and navigator who commanded the *Niña* during Columbus's first voyage to America (1492). When Columbus's own ship, the *Santa María,* was wrecked he completed the journey in the *Niña.* Pinzón later led expeditions to explore the coasts of South and Central America and discovered the mouth of the Amazon. He was made governor of Brazil, which he discovered in 1500.

Pion, elementary particle (symbol π) that is a MESON. There are the three types, forming a nuclear MULTIPLET (triplet). The charged pions, π^+ and π^-, have equal mass, about 280 times the electron mass, and are antiparticles of each other; the neutral π^0, of slightly lower mass, is its own antiparticle. All have zero SPIN. Charged pions decay into MUONS and muon NEUTRINOS; the π^0 decays into PHOTONS. Virtual pions are thought to be exchanged between nucleons bound together by strong interaction. *See also* SU pp.66–67.

Pioneer Space Mission, series of unmanned interplanetary probes launched by the USA. In 1960 Pioneer 5 measured distances within the SOLAR SYSTEM and studied magnetism and the SOLAR WIND. Pioneer 10, launched in March 1972, reached Jupiter in December 1973. It investigated the planet's atmosphere and magnetic field and sent back more than 300 pictures of the planet. Pioneer 11, launched in April 1973, reached Jupiter in December 1974 and is scheduled to reach Saturn in 1979. *See also* SU pp.208–209, 211, 269, *269,* 282, 284, *285.*

Piotrkow Trybunalski, city in central Poland. Of considerable importance from early medieval times until the late 18th century, it was for more than 200 years the seat of the national assembly (1347–1578). Industries: textiles, heavy machinery, bricks. Pop. (1970) 59,700.

Pipefish, any of numerous species of marine fish found in the shallow warm and temperate waters of the Atlantic and Pacific oceans. Closely related to the seahorse, it has a pencil-like body covered with bony rings. Its mouth is at the end of a long snout. Length: to 58.4cm (23in). Family Syngnathidae. *See also* NW p.*130.*

Pipelines, continuous lines of piping, usually sited underground, used for carrying liquids or gases. Pumping stations built at intervals along the length of a pipeline help to maintain the flow of oil, water or gas, and valves inserted at various points help to control the rate of flow. The longest pipeline is in the Soviet Union; it runs 4,444km (2,760 miles). *See also* MM p.79.

Piper, John (1903–), British watercolour painter and art critic who studied at the SLADE SCHOOL (1926–30). Although his early paintings were abstract, his style after the late 1930s became naturalistic, a favourite subject being that of buildings in ruins. Since 1950, he has also worked on stage designs for Covent Garden and Glyndebourne and he also designed stained glass for the cathedrals at Coventry and Llandaff. *See also* HC2 p.281.

Piper, David (1918–), British art historian, critic and novelist. His publications include *The English Face* (1957) and *Enjoying Paintings* (1964); among his novels (written under the pen-name Peter Towry) is *It's Warm Inside* (1953). He became director of the Ashmolean Musuem, Oxford, in 1973.

Pipe Rolls, records of the English EXCHEQUER, kept in rolls in the shape of a pipe. They begin in 1131 and end in 1834 and contain principally the annual accounts of each county sheriff with the

Pinesap; this plant is one of those that feed on decaying organic matter.

Pintails breed in the Northern Hemisphere and migrate south for the winter.

Harold Pinter received two Screen Writers Guild awards in 1964.

Ezio Pinza sang with the Metropolitan Opera Company for over 20 years.

Pipit

Piranha-infested rivers in South America are dangerous to creatures that fall in them.

Pisa's leaning tower increases its tilt by about one millimetre every year.

Camille Pissarro was one of the founders of the French Impressionist movement.

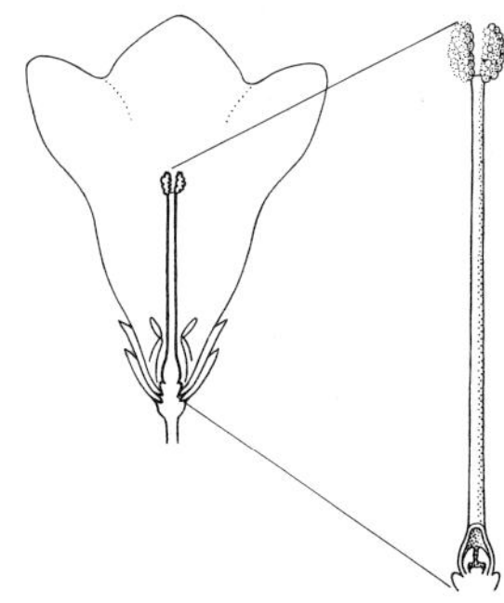

Pistil: the stalk-like style, bulbous ovary at its foot, and the stigma.

Exchequer. They are deposited in the PUBLIC RECORD OFICE in London.

Pipit, also called fieldlark, or titlark, any of more than 50 small, brown inconspicuous birds that resemble LARKS in habits and appearance. They are found throughout the world. The winter and summer plumage of both sexes is a similar streaked brown or greyish colour, with a long, white-edged "wag" tail. Length: 15cm (6in). Family Motacillidae; genus *Anthus*.

Piquet, card game for two players using a 32-card pack, 7 (low) to ace (high) in each suit. Each player receives 12 cards and the remaining eight are left on the table face down. The non-dealer must discard at least one card and can choose to discard up to five, picking up an equal number from the table, while the dealer may exchange any or all of the remaining cards, although this is not obligatory. The players then compare their hands and receive points according to the most cards in a suit, the longest sequence and the highest set of three or four of a kind. Play of cards from the hands follows with points scored for tricks won, the winner being the player with most points after six deals.

Piracy, robbery by force of arms on the high seas. Pirates usually attacked ships of all nations indiscriminately and were therefore distinguished from the crews of PRIVATEERS, who were in the service of a country and held a commission authorizing them to attack the shipping of certain belligerent powers. Piracy has been practised since the earliest times and was prevalent until the 19th century.

Piraeus (Piraiévs), port in E central Greece, on the Saronic Gulf; capital of Piraeus department and the chief port of ATHENS. Piraeus was the naval and maritime headquarters of ancient Athens. It was planned *c.*490 BC by Themistocles, and built *c.*450 BC by Hippodamus of Miletus; the Long Walls connected it to Athens. The modern development of Piraeus began after Greece achieved independence in the 19th century. Industries: shipbuilding, chemicals, textiles. Pop. (1971) 187,362. *See also* MW pp.88–89.

Piran, Saint (*fl.* 5th or 6th century), British holy man, sometimes mistakenly identified as St Kieran. A hermit in Cornwall for much of his life, he is titular of the church of the canons regular in Truro. He is patron saint of miners. Feast: 5 March.

Pirandello, Luigi (1867–1936), Italian novelist, dramatist and short-story writer. His work expresses his desire to distinguish between reality and illusion and was widely translated. Among his plays are *Six Characters in Search of an Author* (1921), *Henry IV* (1922) and *As You Desire Me* (1930). His novels include *The Late Mattia Pascal* (1923). He won the 1934 Nobel Prize in literature.

Piranesi, Giovanni Battista (1720–78), Italian engraver and architect. He lived in Rome where he became famous for his *Vedute*, 137 etchings of the ancient and modern city (1745). His architectural plates are notable for their accuracy and grandeur. The one existing building that he designed is the Church of Sta Maria del Priorato in Rome (1764–65).

Piranha, also called piraya, tropical freshwater bony fish that lives in rivers in South America. It is a notoriously voracious predator, with formidable teeth and an aggressive temperament. It usually travels and attacks in schools and can pose a serious threat to creatures very much larger than itself. Length: to 61cm (24in). Family Characidae; genus *Serrasalmus*. *See also* NW pp.128, *129*.

Pirates of Penzance, The (1879), satirical operetta by Arthur SULLIVAN with libretto by W. S. GILBERT. First produced in Paignton, England, it is a parody of Italian opera and one of the most successful of the many works by Gilbert and Sullivan.

Pire, Dominique Georges (1910–69), Belgian Dominican priest who was awarded the 1958 Nobel Peace Prize for his work in helping refugees from WWII. After being ordained in 1934, he taught philosophy at the Dominican monastery of La Sarte in Belgium and during this period organized various charities to help the poor. In 1949 he founded the Aide aux Personnes Déplacées to help refugees, and from 1950 to 1962 set up homes for them throughout Europe. In 1960 he established the Mahatma Gandhi International Peace Centre at Huy, Belgium, a university of peace. He also wrote *Bâtir la Paix* (*Building Peace*, 1966).

Pirenne, Henri (1862–1935), Belgian historian. Although he wrote the first major history of his country, *History of Belgium* (1900–32), he was better known as a medieval historian. In his *Medieval Cities* (1925) he contradicted accepted views in attributing the revival of cities in the late Middle Ages to the recovery of trade. His *History of Europe* (1936) was written in remarkable circumstances – while interned in Germany during WWI, during which time he had no access to a library.

Pirosmanishvili, Niko (1863–1919), Russian painter, trained as a signwriter. His paintings, in a primitive style similar to that of Douanier ROUSSEAU, depict scenes from the life of the peasantry.

Pisa, city in Tuscany, central Italy, on the River Arno; capital of Pisa province. The city prospered as a Roman colony from *c.*180 BC. From the 10th to the 11th centuries, it was a powerful maritime republic that rivalled Genoa and Venice. It was defeated by Genoa in 1284 and by Florence in 1406. GALILEO, who was born in Pisa (1564), studied and taught at the university which was a centre of Renaissance learning. Landmarks include the Leaning Tower. Industries: textiles, glass, tourism, machine tools. Pop. (1975) 103,412. *See also* SU p.75.

Pisanello (*c.*1395–*c.*1455), real name Antonio Pisano, Italian painter and medalist whose works are in the INTERNATIONAL GOTHIC style and include notable drawings and frescoes. He is best known for his medals of important people of his time, which are of great historic value.

Pisano, name of a family of sculptors, the first of whom was Nicola (*c.*1225–*c.*1284), who founded a new school of Italian sculpture. His first great work was the marble pulpit for the Baptistry in Pisa (1260). Nicola's emphasis on the human figure betrays his use of models from classical antiquity. In his work on the pulpit for the cathedral at Siena (1265–68) he was aided by his son, Giovanni (*c.*1250–*c.*1320), whose taste in elaborate decoration was influenced by the French Gothic style. Giovanni executed two other pulpits – for S Andrea, Pistoia (1298–1301), and for the cathedral at Pisa (1302–10). Giovanni also designed the façade for the cathedral at Siena. *See also* HC1 p.217.

Piscator, Ervin Friedrich Max (1893–1966), German stage director whose EXPRESSIONIST productions used new optical and mechanical resources, such as documentary material, film and slide projections and undisguised stage machinery, to strengthen a play's emotional impact. He developed the concept of EPIC THEATRE which BRECHT incorporated into the work of the Berliner Ensemble and Piscator's work has remained a strong influence on Young German directors, notably Peter Stein.

Pisces, or the Fishes, inconspicuous equatorial constellation situated on the ecliptic between Aquarius and Aries; it is the 12th sign of the Zodiac. No stars are brighter than the fourth magnitude. It contains a distant spiral galaxy, M74 (NGC 628). *See also* MS pp.200–201; SU pp.*254*, 258, 259.

Piscis Austrinis, or The Southern Fish, southern constellation between Capricornus and Sculptor. The brightest star is Alpha, or Formalhaut, of the first magnitude. *See also* SU p.262, *262*.

Pisemsky, Alexey Feofilautovich (1820–81), Russian novelist and playwright who sympathetically portrayed peasants in his realistic description of the countryside. His works include the novel *A Thousand Souls* (1858) and the play *A Bitter Lot* (1859).

Pissarro, Camille (1830–1903), French IMPRESSIONIST painter. He was influenced by Jean Baptiste COROT but later joined the Impressionists and was represented in all eight of their exhibitions from 1874 to 1886. In the 1880s he experimented with the POINTILLIST theories of Georges SEURAT but abandoned them in the 1890s for a freer interpretation of nature. His works include *Louvre from Port Neuf* (1902).

Pissarro, Lucien (1863–1944), French Impressionist painter and engraver and the eldest son of Camille PISSARRO. He commonly employed the POINTILLIST technique. In England he was a member of the CAMDEN TOWN GROUP and, later, the LONDON GROUP.

Pistachio, deciduous tree native to the Mediterranean region and E Asia. It is grown commercially for the edible greenish seed (the pistachio nut) of its wrinkled red fruit. Height: to 6m (20ft). Family Anacardiaceae; species *Pistacia vera*. *See also* PE p.215.

Pistil, female organ located in the centre of a flower. It consists of a slender STYLE, the OVARY and the STIGMA, which receives POLLEN.

Pistol, firearm held and fired in one hand. The first were MATCHLOCKS, in which a glowing fuse ignited the charge; by the end of the 16th century WHEEL-LOCKS and the cheaper FLINTLOCKS were also in use. Of these, only expensive and cumbersome wheel-locks were reliable. The invention of the PERCUSSION CAP in 1815 enabled pistol technology to advance rapidly, and Samuel COLT's REVOLVER of 1835 was the first reliable repeating firearm. Since then pistols have become capable of automatic fire and some, with detachable butts, are practically indistinguishable from SUBMACHINE GUNS. *See also* MM pp.162–165.

Pistol shooting. *See* SHOOTING.

Piston, Walter (1894–   ), US composer. He is known principally for his orchestral music, which includes the popular ballet *The Incredible Flutist* (1938), several concertos and a number of symphonies. He has contributed to the development of 20th-century Neo-Classicism in the USA.

Piston engine. *See* ENGINE.

Pit, in theatre, the ground floor of the auditorium, commonly below ground-level. Although most modern theatres have no pit, theatres were constructed in this way from Elizabethan times until the 19th century.

Pit and the Pendulum, The (1843), suspense story by Edgar Allan POE. One of Poe's most famous tales, it is a condemnation of the tortures and horrors of the INQUISITION.

Pitcairn Island, volcanic island in the central S Pacific Ocean; British possession administered by New Zealand. Pitcairn is inhabited by the descendants of the mutineers from the British ship HMS *Bounty* who landed on the island in 1790. The principal economic activity is the growing and exporting of fruit. Area: 6.5sq km (2.5sq miles). Pop. (1976) 67. *See also* NW p.143.

Pitch, quality of sound that determines its position in a musical scale. It is measured in terms of the frequency of sound waves – the higher the frequency, the higher is the pitch. It also depends to some extent on loudness and timbre: increasing the intensity decreases the pitch of a low note but increases the pitch of a high one. *See also* SU p.82.

Pitch, in aerodynamics, angular displacement of the longitudinal axis of an aircraft relative to the horizon. The elevators control pitch by lowering (or raising) the tail relative to the nose to increase (or decrease) the angle of attack (angle at which air strikes the wing). *See also* MM p.152; SU p.82.

Pitchblende, or uraninite, mineral, the chief ore of uranium which occurs in it in the form of uranium dioxide, UO_2 (with some UO_3). It is found in several parts of the world as a constituent of igneous rocks. It forms cubic crystals, of hardness 5.5, s.g. 7.5–9.7.

Pitcher plant, any of several species of carnivorous bog plant of the tropics and sub-tropics. The vase-shaped leaves, veined with red and green, are lined with bristles. Trapped insects decompose and are absorbed as nutrients by plant cells. The flower is usually red. Height: 20–61cm (8–24in). Family Sarraceniacea;

genera *Sarracenia* and *Nepenthes*. *See also* INSECTIVOROUS PLANTS; NW pp.59, *215*.

Pithecanthropus erectus, name (meaning erect ape-man) originally given to the first specimen of *Homo erectus* when it was discovered in Java in 1891. Pithecanthropines lived about 500,000 years ago, walked erect, made and used crude stone tools and were possibly capable of speech. Their brain capacity was about half way between those of apes and modern man. *See also* MS p.*24*.

Pitman, Sir Isaac (1813–97), schoolmaster who invented the Pitman SHORTHAND system. After teaching for many years he opened his own private school in Bath, England. Influenced by Samuel Taylor's system of shorthand, he worked on a system based on sound. In 1837 he published *Stenographic Sound Hand.* He also set up a phonetic institute and brought out his *Phonetic Journal* in 1842 and moved to London in 1845.

Pitot tube, device for determining pressure in a moving fluid, either liquid or gas, invented by Henri Pitot (1695–1771). For liquids, it is generally a MANOMETER with one open end facing upstream and the other open end out of the stream. The different pressures at the two ends cause a liquid to shift position within the two arms of the tube. For gases, a Pitot tube is generally L-shaped, with one end open and pointing towards the flow of gas and the other end connected to a pressure-measuring device. This type of Pitot tube is commonly used as an air-speed indicator in aircraft.

Pitris, in Hindu mythology, the revered semi-divine ancestral spirits of the departed. They reside in a paradise where they feast with the gods. Their condition, however, is directly related to the rites and sacrifices celebrated by their living descendants.

Pitt, William (The Elder). *See* CHATHAM, WILLIAM PITT, 1ST EARL OF.

Pitt, William (1759–1806), British politician, second son of William Pitt, Earl of CHATHAM. He is known as the younger Pitt. He entered Parliament in 1781, became Chancellor of the Exchequer in 1782 and, at the age of 24, Prime Minister in 1783. He was Prime Minister until 1801, when he resigned in the face of GEORGE III's refusal to consider CATHOLIC EMANCIPATION, and then again from 1804 until his death. His reputation rests on his reorganization of the national finances and his resolute prosecution of the war against France and NAPOLEON. During his administrations were passed the India Act (1784), the Constitutional Act (1791), dividing Canada into French and English provinces, and the Act of UNION with Ireland (1800). He considered himself, if anything, a WHIG and did not attempt to build a party. Yet he is rightly considered as the founder of the second TORY party and, by implication, of the modern CONSERVATIVE PARTY. *See also* HC2 pp. 68–69.

Pitta, bright, colourful bird that lives deep in the jungles of Africa, s Asia, Malaysia, the Philippines, New Guinea and N Australia. It is a plump bird, with a large head, short rounded wings, a short square tail and long legs. Some of the birds are migratory. Most feed on termites, other insects and worms, and build large, globular nests in the fork of a bush or low tree. Length: 15–28cm (6–11in). Family Pittidae; genus *Pitta*.

Pitt-Rivers, Augustus Henry Lane-Fox (1827–1900), soldier and archaeologist. The scientific method of excavation developed by him emphasized sociology rather than art history. His large archaeological and ethnological collection formed the basis of the Pitt-Rivers Museum, Oxford.

Pittsburgh, city and port in sw Pennsylvania, USA, at the confluence of the Allegheny and Monongahela rivers; second-largest city in Pennsylvania and a major steel-producing centre. Fort Duquesne was founded on the site by the French c.1750. It was captured by the British in 1758 and renamed Fort Pitt. Pittsburgh grew rapidly as a steel manufacturing centre during the 19th century. The University of Pittsburgh dates from 1787. Other industries include glass, machinery,

petroleum products, electrical equipment, printing and publishing, railway maintenance, coal mining, and oil and natural gas extraction. Pop. (1973) 479,276.

Pituitary gland, or hypophysis, endocrine gland, in human beings about the size of a small cherry, connected to the lower surface of the brain by a stalk (infundibulum). It is composed of an anterior lobe, and a posterior lobe, each of which has its own functions.

Pit viper, any of 100 species of poisonous SNAKES, including the BUSHMASTER, copperhead, lancehead, moccasin and RATTLESNAKE, found chiefly in the New World, Europe and Asia. Pit vipers get their name from a heat-sensitive pit located on each side of the head, which enables them to detect warm-blooded prey in the dark. Family Crotalidae. *See also* VIPER; NW pp.*138, 214*.

Pityriasis rosea, skin disease of unknown origin characterized by the formation of pimples on the shoulders and abdomen. These gradually spread over the body causing itching; the symptoms last for about six weeks.

Piura, city in NW Peru, on the Piura River, capital of Piura department. Founded in 1523 by PIZARRO, Piura is the oldest Spanish city in Peru. The city is a commercial centre for an area of the Peruvian coastal desert, where rice, corn, cotton and sugar cane are produced by means of irrigation. The city has a technical university founded in 1961. Pop. (1972) 126,010.

Pius, name of 12 popes. Pius I, Saint, was pope *c.*140–155. Pius II (1405–64, *r.*1458–64) was born Aeneas Silvius Piccolomini. He was a Humanist scholar, and as pope tried to organize a crusade against the Turks. Pius III (*c.*1440–1503) was pope briefly in 1503. Pius IV (1499–1565; *r.*1559–65) was born Giovanni Angelo de Medici. He furthered the COUNTER-REFORMATION, reconvening the Council of TRENT for its third and last session (1562–63). He also drafted the Index of Forbidden Books. Pius V (1504–72; *r.*1566–72), born Antonio Ghislieri, was an energetic reformer of the Church and enemy of Protestantism, excommunicating ELIZABETH I of England in 1570 and serving as commissary general of the INQUISITION before being elected pope. Pius VI (1717–99; *r.*1775 99), was born Giannangelo Braschi. He opposed the attempts of the HOLY ROMAN EMPIRE and France to restrict papal authority in their respective lands. Pius VII (1742–1823; *r.*1800–23) was born Barnaba Gregorio Chiaramonti. He secured the Concordat of 1801 with NAPOLEON, and was present at Napoleon's coronation in Paris (1804). After Napoleon took Rome in 1808 Pius excommunicated him and was imprisoned, remaining in exile until 1814. On his restoration he encouraged the reform of religious orders and education. Pius VIII (1761–1830; *r.*1829–30) was a disciple of Pius VII. Pius IX (1792–1878; *r.*1846–78) was born Giovanni Maria Mastai-Ferretti. His pontificate was the longest in history. He was driven from Rome (1848–50), but restored by Napoleon III. The PAPAL STATES were seized by the Italian nationalists in 1860, and Rome itself incorporated into the kingdom of Italy, which in 1870 Pius refused to accept. He also defended Catholics in Germany from persecution by BISMARCK, and in 1869 called the first VATICAN COUNCIL, which proclaimed PAPAL INFALLIBILITY. Pius X, Saint (1835–1914; *r.*1903–14), was born Giuseppe Melchiorre Sarto. He opposed religious MODERNISM, placing several modernist books on the index in 1907, condemned the separation of Church and state in France, and recodified CANON LAW (published 1917). Pius XI (1857–1939; *r.*1922–39) was born Ambrogio Damiano Achille Ratti. He signed the LATERAN TREATY with MUSSOLINI in 1929, but later condemned FASCISM, and denounced COMMUNISM in 1937. He called for unity between the Eastern and Western Churches, and supported scientific research. Pius XII (1876–1958; *r.*1939–58) was born Eugenio Pacelli. He tried to remain neutral in WWII, his conduct during which had given

rise to controversy. In 1946 he denounced Catholic collaboration with Communists. *See also* HC2 pp.*110, 181*, 304.

Pixii, Hippolyte (1808–35), French physicist and instrument maker who devised the first practical electric generator. Pixii's original machine generated alternating current electricity, but with the introduction of the COMMUTATOR it was able to generate direct current electricity. *See also* SU p.*118*.

Pizarro, name of four Spanish brothers, all adventurers and explorers. Francisco (*c.*1471–1541) led the conquest of Peru. Together with his partner and rival, Diego de ALMAGRO, he took control of Peru in 1533 but was later assassinated by Almagro's son. Hernando (*c.*1475–1578), Francisco's half-brother, helped his brothers in the conquest of Peru, and in 1538 defeated Almagro and had him executed. For this he served 20 years imprisonment in Spain. Juan (*c.*1500–36) served as a lieutenant to his brother Hernando and died in battle during the reconquest of Cuzco. Gonzalo (*c.*1506–48), the youngest of the brothers and the most famous after Francisco, was ruthless and tyrannical. After the Peruvian conquest, he explored much of NW South America in search of gold. In defiance of Spain, he set himself up as governor and captain-general of Peru and was eventually beheaded for treason. *See also* HC1 p.236, *236*.

Pizzetti, Ildebrando (1880–1968), Italian composer. His most important works are operas, among them *Fedra* (1905), *Fra Gherardo* (1927) and *Vanna Lupa* (1949).

Pizzicato, note or passage that is played on a bowed stringed instrument by plucking or pinching the string with the finger, rather than by bowing it. Examples of this technique are in all musical periods from the 18th century, eg in HANDEL's *Agrippina* (1709) and the second movement of the string quartet (1903) by RAVEL.

Place, Francis (1771–1854), British tailor and political radical. He campaigned for trade unionism (1814–24) and the passing of the Reform Bill (1832). He helped to draft the People's Charter (1838), the demands of the CHARTISTS. *See also* HC2 p.94.

Placebo, harmless, unmedicated drug substitute given to satisfy a patient's insistence for medicine. It is also used as a control in medical experiments. *See also* HYPOCHONDRIA.

Placenta, organ in mammals (except monotremes and marsupials) that connects the FETUS to the uterus of the mother and serves as an organ of nutrition, respiration and excretion for the fetus. Part of the placenta contains tiny blood vessel branches through which oxygen and food are carried from the mother to the embryo via the umbilical cord and wastes are carried from the embryo to the mother's bloodstream to be excreted. It secretes hormones that maintain pregnancy and is discharged from the mother's body as the afterbirth, immediately after birth of the baby. *See also* MS pp.76–77.

Placental mammals, mammals whose young develop to an advanced stage attached to the placenta – a life-support organ inside the mother's uterus. All mammals except the MONOTREMES and most MARSUPIALS are placentals.

Placer deposits, concentrations of heavy minerals by the action of gravity, usually found in streams. Minerals that occur as placer deposits include gold, copper, rutile, cassiterite and magnetite.

Placoderm, group of primitive jawed fishes that existed in the Devonian period, characterized by armoured plates on the front part of the body. Most were quite small, but some grew to 9m (30ft). *See also* NW p.180, *180*.

Placoid fish, fish, including SHARKS, RAYS, and SAWFISH, with plate-like scales that are actually dermal teeth. The tip of the scale is dentine layered with enamel and the lower part is bone anchored to the skin. Dermal teeth are modified into large spines in rays and into teeth on the snout of the sawfish.

Plagioclase, type of FELSPAR (the most

Pitta: this highly colourful bird grows to about the size of a thrush.

Pius II was a prolific author; his writings include a play called *Chrysis*.

Francisco Pizarro seizing the Inca of Peru, from the painting by Millais.

Placoid fish, relative of the nurse shark (above), has plate-like scales.

Plague

Max Planck's ideas about radiant energy led to the modern quantum theory.

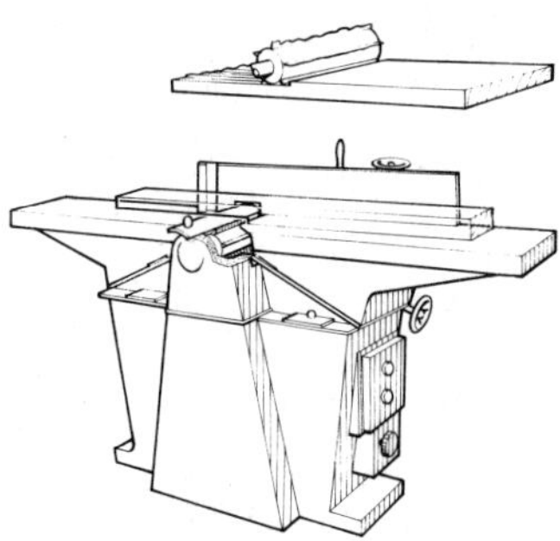

Plane may be a machine tool for the high-speed smoothing of wood.

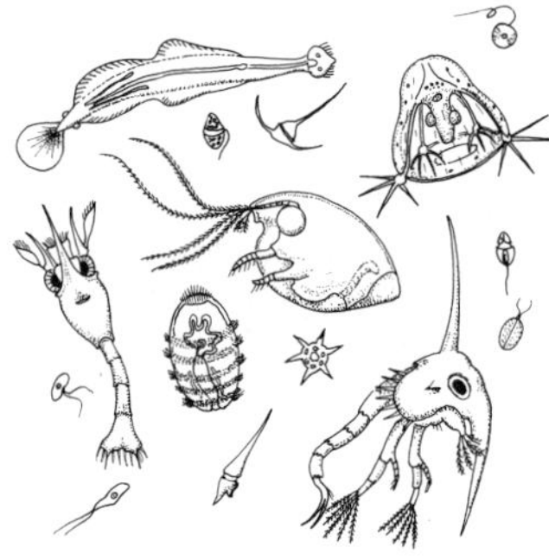

Plankton, mass of tiny plant and animal life which provides food for fish.

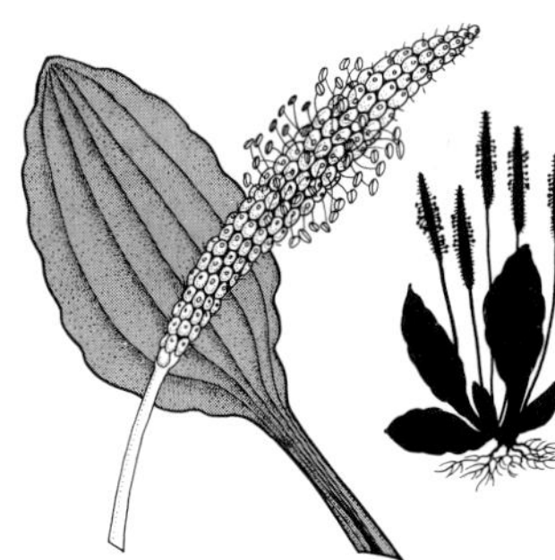

Plantain are broad leaved plant with terminal spikes of small flowers.

abundant group of minerals on Earth). Plagioclases show an oblique cleavage, as opposed to ORTHOCLASE or MICROCLINE, and are composed of varying proportions of the silicates of sodium and calcium with aluminium.

Plague, acute infectious disease of man and rodents caused by the bacillus *Pasteurella pestis.* In man it occurs in three forms: bubonic plague, most common and characterized by swellings called buboes; pneumonic plague, in which the lungs are infected; and septicaemic plague, in which the bloodstream is invaded. Treatment is the administration of vaccines, bed rest, antibiotics and sulpha drugs. *See also* BLACK DEATH.

Plague, The Great (1665–66), last great visitation of the bubonic PLAGUE upon England. It raged in London from April 1665 to the autumn of 1666. The court and parliament moved out of London, but the Lord Mayor remained. It is estimated that between 75,000 and 100,000 people died. It was the subject of Daniel DEFOE's *Journal of the Plague Year* (1722).

Plaice, or European flounder, marine flatfish found along the w European coast. An important food fish, it is brown or grey with orange spots, and has its eyes displaced towards its upper side. Length: to 90cm (3ft); weight: to 11.8kg (26lb). Family Pleuronectidae; species *Pleuronectes platessa.*

Plaid Cymru (Party of Wales), Welsh nationalist political party, founded in 1925 as Plaid Genedlaethol Cymru. It fought its first general election in 1931. Its first MP was elected in May 1966 when its president, Gwynfor Evans, was returned at the Carmarthenshire by-election. In the General Election of October 1974 it returned three MPs.

Plainsong, or plainchant, liturgical music of the Western Church, beginning early in the Christian era. It is unadorned by accompaniment or HARMONY. Its RHYTHM is free so that the natural rhythm of the words may be followed as if they were spoken. In GREGORIAN CHANT its range is not more than five notes. It uses, for its notation, a four-line STAVE. See also MODES.

Plaintiff, party in civil law proceedings that instigates the legal action.

Planché, James Robinson (1796–1880), British dramatist, popular for his PANTOMIMES and BURLESQUES. He introduced historically accurate costumes and accessories in theatre productions, and was the author of *History of British Costume* (1834). He also wrote the libretto for Carl WEBER's last opera, *Oberon.*

Planchon, Roger (1931–), French director, actor and playwright. His productions used all the resources and concepts of Brechtian EPIC THEATRE to bring classics such as Shakespeare's *Henry IV* and Brecht's *The Good Woman of Setzuan* to new regional and working-class audiences. He became internationally recognized with an adaptation of Dumas's *The Three Musketeers* (1957–58), in which he played d'Artagnan.

Planck, Max Karl Ernst Ludwig (1858–1947), German theoretical physicist, professor at Kiel and later in Berlin, where he studied the characteristics of the radiation emitted by black bodies. In 1900 he came to the conclusion that the frequency distribution of the radiation could be accounted for only if the radiation was emitted in separate "packets" called quanta, rather than continuously. An explanation of radiant heat energy distribution given off from a heated surface was proved by Planck's radiation law (1900). PLANCK'S CONSTANT (1900) indicates wave and particle behaviour on the atomic scale. His equation, relating the energy of a quantum to its frequency, is the basis of QUANTUM THEORY. He was awarded the 1918 Nobel Prize in physics for his work. *See also* SU pp.63, *63,* 108–109.

Planck's constant, universal constant (symbol h) of value $6{,}626 \times 10^{-34}$ joule second, equal to the energy of a quantum of ELECTROMAGNETIC RADIATION (a photon) divided by the radiation frequency. It appears in formulas describing physical

quantities that can assume only certain discrete values. *See also* QUANTUM THEORY.

Plane, in mathematics, flat surface such that a straight line joining any two points on it lies entirely within the surface. Its general equation in three-dimensional CARTESIAN CO-ORDINATES is $ax + by + cz + d = 0$ where a,b,c and d are constants. In wood working, a plane is a tool used to smooth wood by cutting away thin strips as it passes over the surface. Any flat shape or surface such as a wing or craft can also be called a plane, hence aeroplane and HYDROPLANE.

Plane geometry, form of GEOMETRY in which the angles of a triangle add up to 180°; all lines, angles and figures are represented in two-dimensional (PLANE) form. In plane geometry EUCLID's axioms are usually applied. THEOREMS are deduced and proved from axioms. *See also* SU pp.44–45.

Planet, celestial body that revolves in an orbit around the Sun or some other star. The nine planets revolving around the Sun are Mercury, Venus, Earth, Mars (the terrestrial planets), and Jupiter, Saturn, Uranus, Neptune (the giant planets), and Pluto. *See also* SU pp.176–177.

Planetarium, domed building in which an instrument is used to project an artificial sky in order to demonstrate the positions and motions of the sun, moon, planets, and stars relative to the Earth. The projector was invented in 1913 by Walter Bauersfeld of the Zeiss Optical Company.

Planetary nebula, type of celestial object consisting of a shell or ring of highly ionized (charged) gas surrounding a hot central star. Such objects appear to be stars departing from the red giant stage, which have thrown off their atmospheres (thus producing the gaseous rings) and are evolving into WHITE DWARFSTARS. *See also* SU pp.232–233.

Planetary probe, spacecraft sent to fly pass, orbit or land on a planet. The era of probes began in 1962 when Mariner 2, the us probe, passed Venus and sent back information that its surface is hot. *See also* MARINER SPACE MISSIONS; SPACE PROBE.

Planetismal hypothesis, theory to explain the solar system's origin, developed by T. C. Chamberlin and F. R. Moulton early in the 20th century. It states that a star that once passed the Sun exerted enough force to tear off bits of matter from it. These cooled and went into orbit round the Sun. Those planetismals were eventually drawn into larger bits of matter, which became planets.

Plane tree, deciduous tree indigenous to most of the Northern Hemisphere. The English plane is a cross of the us sycamore and the Oriental plane. Its use is chiefly ornamental. Family Platanaceae; genus *Platanus.*

Planimeter, instrument invented by J. H. Hermann in 1814 for measuring the area of a plane surface. It consists of a hinged arm which is made to trace the perimeter of the area being measured and turn a graduated wheel which indicates the measured area.

Plankton, all the floating or drifting life of the ocean, especially that near the surface. The organisms are very small or microscopic and move with the currents. There are two main kinds: phytoplankton, floating plants such as DIATOMS and dinoflagellates; and zooplankton, floating animals such as radiolarians, plus larvae and eggs. *See also* NW pp. *238–239.*

Planography, in printing, method of obtaining prints from a smooth surface, most commonly used in LITHOGRAPHY. The most recent of the major graphic techniques, it was discovered in 1798.

Plant, living organism diverse in size, form, activity and habitat, and generally able to manufacture its own food. More than 500,000 plants have been classified. They vary from the short-lived, single-celled bacteria to the slow-growing oaks and redwoods. Sizes vary from a few millimetres to 100m or, higher, forest trees. The fundamental differences between plants and animals are mode of nutrition, scheme of growth, cell wall composition and locomotion. Plants depend on inor-

ganic food materials from soil, water and air to manufacture their own food, particularly by PHOTOSYNTHESIS for which the green pigment CHLOROPHYLL is essential. Some non-green plants, such as fungi, are parasites, existing on other organisms.

Botanists have named 11 divisions of plants and include nine small primitive groups under THALLOPHYTES, the fungi, bacteria, algae and lichens. BRYOPHYTES have a more highly developed reproductive system and include mosses and liverworts. TRACHEOPHYTES, or vascular plants, have strong roots, a water-conducting system and green tissue. They include the club mosses, horsetails, ferns, conifers and flowering plants. *See also* NW pp.*23,* 38–39, 44–45, 58–67, 70–71, 76, 176–177, 232, 246–248.

Plantagenet dynasty, English royal line that ran from HENRY II (*r.* 1154–89) to RICHARD III (*r.* 1483–85); the name has been commonly used since the 17th century. The line descends from Geoffrey, Count of Anjou, and Matilda, daughter of HENRY I, and is properly called ANGEVIN down to the deposition of RICHARD II (1399), and thereafter Lancastrian and Yorkist. The name probably derives from the sprig of broom (planta genista) which Geoffrey of Anjou wore in his cap. It was first used as a nickname, then taken as a surname by Richard Duke of York, the father of EDWARD IV.

Plantain, plant with a rosette of basal leaves and spikes of tiny, greenish white flowers; it grows in temperate regions and was used for medicinal purposes. Family Plantaginaceae; genus *Plantago.* The name plantain is also given to a tropical banana plant believed to be native to SE Asia and now cultivated throughout the tropics. It has fleshy stems, bright green leaves and green fruit that is larger and starchier than a banana and is eaten cooked. Height: to 10m (33ft). Family Musaceae; species *Musa paradisiaca.*

Plantain lily. See HOSTA.

Plantation of Ireland, English policy of permanently settling and subduing Ireland by introducing Protestant settlers to farm areas confiscated from the Irish. The first plantations were set up in Offaly and Leix in 1556 (which were renamed King's and Queen's counties), and plantation continued until the late 17th century. Official plantations were found in most of w and central Ireland; unofficial Scottish plantations were most common in ULSTER. Plantations were a major grievance in the Irish rebellion of 1641. *See also* HC2 pp.24–25, *24.*

Plant classification, system devised to group plants according to certain relationships among them. Plants are known by common names which often vary from area to area, but have only one correct scientific name, which is usually composed of a binomial derived from Latin. The system of classification is made up of seven groups which are, from the largest to the smallest, kingdom, phylum or division, class, order, family, genus and species. All plants belong to the plant kingdom, Plantae. The species is the basic unit; members of a species have common characteristics and identifiable differences from all other species. The present method of classification took many years to develop, but is based on the work of the Swedish naturalist Carolus LINNAEUS.

Plant fibres, thread-like tissues derived from plants. They consist mainly of cellulose (long chains of glucose molecules), and many types are processed by man for spinning and weaving. These include seed fibres (COTTON and KAPOK), leaf fibres (SISAL) and stem fibres (LINEN). *See also* PE p.216.

Plant food. See FERTILIZER.

Plant hormone, or phytohormone, organic chemical produced in plant cells and functioning at various sites to effect plant growth, leaf and fruit drop, healing, cambial growth and possibly flowering. Hormones are transported away from the stem tip. They include abscisic acid (leaf fall), AUXINS (growth), cytokinins (leaf and bud growth and development) and gibberellins (growth).

Plantin, Christophe (1514–89), printer

and typographer, *b.* France. He began working as a bookbinder in Antwerp in 1549. In 1555 he established a production and publishing company and attained high standards of accuracy and appearance in his books; the Polyglot Bible (1568–73) is his most famous publication.

Plant louse. *See* APHIS.

Plant pigment, organic compound present in plant cells and tissues that colours the plant. The most common of these is green chlorophyll, which exists in all higher plants. CAROTENOIDS colour plants yellow to tomato red. Located in CHLOROPLASTS and chromoplasts, there are more than 150 varieties of these durable pigments. Many are essential to PHOTOSYNTHESIS and are a source of vitamin A. Anthocyanins, responsible for pink, red, blue and purple, are found in the cell sap. The shorter days and lower temperatures of autumn cause these pigments to combine with other substances and produce the brilliant foliage colours of DECIDUOUS trees.

Plaque, film of saliva and bacteria that accumulates on teeth. It causes colouration and serves as a breeding place for potentially harmful bacteria. It is usually removed as part of preventive dentistry because it is believed to cause PYORRHOEA. *See also* MS pp.134–135.

Plasma, in biology, liquid portion of the BLOOD. It contains an immense number of ions, inorganic and organic molecules such as immunoglobulins, and hormones and their carriers but not cells. It clots upon standing. *See also* MS pp.62–63.

Plasma, in physics, highly ionized state of matter in which substances consist almost entirely of ELECTRONS (negatively charged) and atomic nuclei or IONS (positively charged). This state, often described as the fourth state of matter, occurs at enormous temperatures, eg in the interiors of stars and in fusion reactors.

Plasmodium, genus of parasitic PROTOZOA that causes MALARIA. It infects the red blood cells of mammals, birds and reptiles throughout the world, being transmitted by the bite of a female *Anopheles* mosquito. Four species cause human malaria, passing from mosquito to man as sporozoites in the mosquito's saliva. Once in the red blood cells, they divide, forming up to 24 daughter parasites, then destroy the red blood cells. Entering the PLASMA, they infect new cells. *See also* MS p.106; NW p.*80.*

Plassey, Battle of (1757), decisive defeat of the Nawab of Bengal, Siraj-ud-Dawlah, by the British under Robert CLIVE. This battle made possible the British acquisition of Bengal.

Plaster, mixture, originally of slaked lime (calcium hydroxide), sand and water, often with hair or fibres added as a binder, that is applied wet to interior walls and ceilings to form a smooth hard surface for papering or painting when dry. Most modern plasters are made from gypsum (calcium carbonate). *See also* PLASTER OF PARIS.

Plaster of Paris, or gypsum cement, powdered form of calcium sulphate hemihydrate, $(CaSO_4)_2H_2O$, obtained by heating GYPSUM to around 150°C (300°F). After the addition of water it sets and hardens and is used as a PLASTER for a wide range of purposes, including the setting of broken limbs and the making of moulds.

Plastics, synthetic materials composed of organic molecules, often in long chains called POLYMERS. The weight and structure of the molecules determine the physical and chemical properties of a given compound. Plastics are synthesized from common materials such as cellulose from cotton or wood pulp; organic acids from coal tar; CASEIN from skimmed milk; and from chemicals derived from petroleum, potatoes, peanuts and soya beans. The plastics industry began in 1869 with the development of celluloid by John Wesley HYATT. *See also* MM pp.54–55.

Plastics fabrication, methods of manufacturing plastic products. There are three basic processes: calendering, casting and extrusion. Calendering is used to make composite material from a THERMOSETTING RESIN such as phenol-formaldehyde (Bakelite) and materials like paper and

cloth. Casting uses the thermosetting plastics such as EPOXY, POLYESTER, phenolic or urea RESINS and enables plastic to be cast around metal parts, as in making screwdrivers. Extrusion forces the softened plastic through a die which shapes it into the desired form, such as a tube, rod or sheet. This is employed with THERMOPLASTIC RESINS such as POLYETHYLENE, POLYSTYRENE and NYLON. *See also* MM p.54.

Plastic surgery, branch of surgery that involves the rebuilding of deformed, damaged or disfigured parts of the body. Tissue can be moulded in the shape of a missing part, such as an ear or nose. Skin grafts are usually taken from one part of the body and attached to another until blood is supplied adequately to the new site. Cosmetic surgery is performed solely to improve appearance. A typical example is face-lifting, which involves stretching the skin until it is taut, and then removing the excess. *See also* MS pp.127, *127, 179.*

Plata, Río de la, anglicized to River Plate, estuary in SE South America formed by the junction of the Paraná and Uruguay rivers. It is 270km (170 miles) long and 190km (120 miles) wide at its mouth.

Plataea, Battle of (479 BC), decisive victory of Greek forces under Pausanias over the Persians under XERXES, which assured Grecian independence. There is a modern village named Plataiai on the site of ancient Plataea. *See also* HC1 p.70.

Plate, Battle of the River, WWII naval engagement that took place in the Río de la PLATA estuary off Montevideo harbour, in Dec. 1939. The British cruisers *Exeter, Ajax* and *Achilles* engaged the German POCKET BATTLESHIP *Admiral Graf Spee,* eventually forcing it to be scuttled.

Plate, crustal, any of six major crustal plates plus many smaller ones that cover the Earth. Their motions are complex and any two plates may be either separating, converging or sliding past each other. *See also* PLATE TECTONICS; PE pp.24, 106.

Plate camera, camera in which the image is exposed on a chemically prepared plate as opposed to a portion of roll film. Before 1888 when George EASTMAN marketed his celluloid roll film, photographers made exposures known as daguerreotypes on polished and silvered copper plates or on heavy glass plates, both of which required coating with complicated chemical solutions until gelatin dry plates came onto the market in the early 1870s. Plate cameras are still used by professional photographers for studio work with technical and stationary subjects. *See also* MM pp.218–219, *218–219.*

Plate glass, optically flat glass generally made by rolling, grinding and polishing, so that its surfaces are plane and parallel. It was first made in France in the 17th century. *See also* MM pp.44–45.

Platelet, colourless, usually round structure found in all mammalian BLOOD. Chemical compounds in platelets, known as factors and cofactors, are essential to the mechanism of clotting. The normal platelet count is about 300,000 per cubic millimetre of blood. *See also* MS pp.62–63.

Plate tectonics, theory proposed in 1962 by H.H. Hess to explain the mechanism of continental drift. Using the evidence of the mid-ocean ridges and deep trenches, he suggested that magma rises by convection from the deep Earth and spreads along the ocean-floor and cools. At the same time a heavier layer under the continental crust, the lithosphere, is also being spread apart by the rising magma. The plates of lithosphere push against each other at a rate of between 1–15cm (0.5–6in) per year and one plate is forced to bend down into the deep mantle where it becomes liquid magma again. When one edge of the top crust pushes against another, it wrinkles, forming new mountain ranges like the Andes. *See also* PE pp.24, 106.

Plath, Sylvia (1932–63), US poet, wife of Ted HUGHES. Her intensely personal verse includes *The Colossus* (1960) and *Ariel* (1965). The latter was written in the last months of her life and was published posthumously, as were *Crossing the Water* (1971) and *Winter Trees* (1972). She

wrote one novel, *The Bell Jar* (1962). *See also* HC2 p.289.

Plating. *See* ELECTROPLATING.

Platinum, precious metallic TRANSITION ELEMENT (symbol Pt), discovered in 1735. Its chief source is certain ores of nickel. It is used in jewellery and in electrical resistance wire, THERMOCOUPLES, electrodes and other laboratory apparatus. It is chemically unreactive; it does not react with oxygen at normal temperatures nor with common acids, ie it resists CORROSION. Properties: at. no. 78; at. wt. 195.09; s. g. 21.45; m. p. 1,769°C (3.217°F); b. p. 3,800°C (6,800°F); most common isotope Pt^{195} (33.8%). *See also* SU pp. *134–135.*

Plato (*c.* 427–347 BC), Greek philosopher. He studied (407–399 BC) under SOCRATES, who appears as a central figure in many of his writings. Plato lived in Athens and set up an Academy there in *c.* 387 BC, where he taught. He visited Syracuse three times in about 388, 367 and 361–36 BC in the hope of setting up an ideal political system there. All of the 36 works of Plato survive, many of which form lasting works of literature and deal with aspects of the relationship between the individual and the state. These dialogues include *Gorgias,* the *Republic* (in which Plato outlined his view of the ideal state), *Phaedo* and the *Symposium* (both aesthetic and mystical works). Plato's work has been continually studied, from the ime of his pupil ARISTOTLE to the 20th century. Perhaps his most influential concept was that of Ideas or Forms, absolute archetypes more real than earthly phenomena that derive from them. *See also* HC1 pp.*75,* 82; MS pp.226–227, *232–233,* 234, 272.

Platonist, one who derives the basis of his thought from PLATO's ideas. Platonists believe in the absolute values of an unchanging reality, distinct from the changing experiences of the senses.

Platyhelminth. *See* FLATWORMS.

Platypus, MONOTREME mammal of Australia and Tasmania. It is amphibious, lays eggs and has webbed feet, a broad tail and a soft duck-like bill. The male has a poison spur on the hind foot. It is 60cm (24in) long and eats small invertebrates. Family Ornithorhynchidae; species *Ornithorhynchus anatinus. See also* NW p.158, *158.*

Platyrrhini, group of monkeys of the Americas (the New World monkeys), as opposed to those from Africa and Asia. The distinction is made because of their broad nostrils, and the group includes the SPIDER MONKEY and MARMOSETS. Many have prehensile tails. *See also* NW pp.168–169.

Plautus, Titus Maccius (*c.* 225–184 BC), Roman comic playwright whose works such as *Miles Gloriosus (The Braggart Soldier, c.* 211 BC) were modelled on Greek originals and combine complicated and farcical plots with amusing, low-life characters and dialogue spiced with puns, topical allusions and repartee. The best-known play based on a Plautus work is Shakespeare's *The Comedy of Errors* (1593) which derives from *Menaechmi (The Two Menaechmuses).* The musical *A Funny Thing Happened on the Way to the Forum* (1962) is based on several of Plautus' plots. *See also* HC1 p.112.

Playbill, poster advertising a play or other theatrical production. It originated in a basic form in the 17th century and gradually grew in size to accommodate as much pertinent information as possible about the production being advertised. It was the direct precursor of the modern theatre programme, which largely superseded it in the mid-19th century.

Playboy of the Western World, The (1907), three-act play by J.M. SYNGE. It was first produced at the ABBEY THEATRE in Dublin where it caused riots among the audience for its portrayal of the Irish peasantry. The central figure is Christy Mahon who, boasting that he has killed his father, becomes a hero. *See also* HC2 p.*283.*

Player, Gary (1935–), South African golfer. He won the Masters Tournament twice (1961, 1974), the US Open Championship (1965), two PGA Championships (1962, 1972) and three British Opens

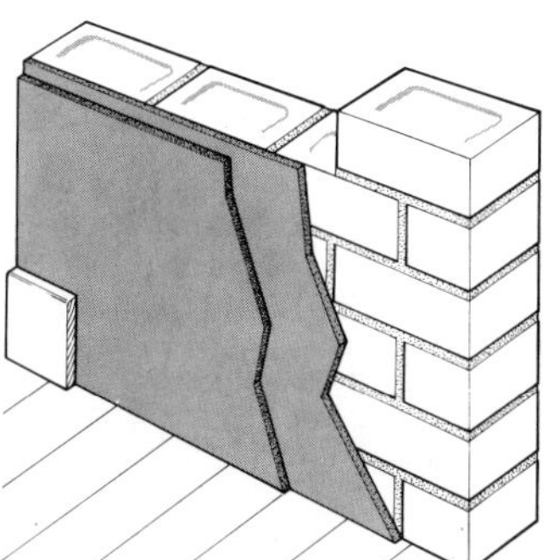

Plaster; brickwork is covered with coarse plaster before a smooth coat is added.

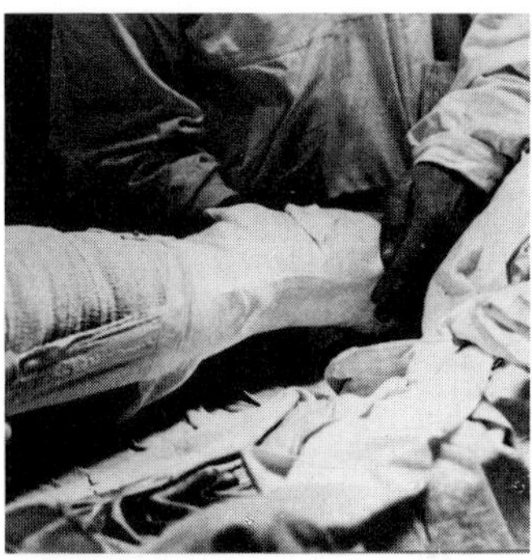

Plastic surgery; after skin has been grafted the limb must be immobilised.

Battle of the River Plate; a scene from the film based on the famous WWII battle.

Gary Player, the South African golfer, on the Wentworth course in England.

Samuel Plimsoll fought to outlaw shipowners who put their crews in danger.

Maya Plisetskaya became the prima ballerina of the Bolshoi Ballet in 1945.

Plovers often use the "broken wing" decoy display to protect their nests.

Plums are eaten fresh, in jams, or dried out to make prunes.

(1959, 1968, 1974). He won more than 60 major tournaments, and was renowned for his dedication to improving his game and to maintaining physical fitness.

Playhouse, permanent THEATRE where dramatic performances of the spoken word, as opposed to opera, are staged. Playhouses in England derived from the tiered courtyards of Elizabethan inns, and playhouse seating developed as unpartitioned seats on several levels grouped around the PICTURE STAGE which with its PROSCENIUM arch was adopted in the 17th century. Restoration playhouses contrasted with the majority of European public theatres, whose higher tiers tended to have boxed accommodation for richer patrons and whose primary acoustic requirements were for opera. *See also* STAGE.

Playing cards, pack of usually 52 cards, used in playing games of chance or skill. They are divided into four categories called suits: diamonds and hearts (which are red) and clubs and spades (which are black). Each suit has 13 ranks from one (ace, theoretically the lowest but paradoxically the highest card in most games) to 10 plus jack (or knave), queen and king. In gambling games such as POKER an extra one or two cards, called jokers are often used to represent whatever is advantageous to the holder. Playing cards probably originated in China between AD 600 and 900 and took their present form in England in about the 15th century.

Playing Fields Association, National, British registered charity whose aims are to encourage and provide recreational and sporting facilities, especially for children and young people in inner-city areas.

Pleasance, Donald (1919–), British actor whose recent work has been in films and television, often cast as a villain. His many films include *Cul de Sac* (1966), *The Greatest Story Ever Told* (1965), *Soldier Blue* (1970) and *The Last Tycoon* (1976).

Pleasure principle, in psychoanalytic theory, need of the ID to be satisfied by bodily pleasures such as food and sex. An individual has to weigh his response to the pleasure principle with the demands of the "reality principle", the need of the EGO to cope with the problems of the everyday world. An imbalance between the two can be a sign of mental disorder.

Plebeians, or Plebs, general body of Roman citizens as distinct from the PATRICIANS, or privileged class. They were originally forbidden to hold any public office except that of military tribune and were prohibited from marrying Patricians. The gulf between Plebs and Patricians gradually closed until 287 BC, when a plebeian dictator was appointed who obtained political equality.

Plebiscite (from the Latin *plebis citum*), a law enacted by the plebs or common people by a direct vote of an entire district or nation. Plebiscites are usually concerned with matters of a national importance such as the election of a leader or choice of government.

Plecoptera. *See* STONEFLY.

Plectrum, small, shaped piece of wood, ivory, horn, tortoiseshell, metal or plastic used to pluck the strings of musical instruments such as the harpsichord, guitar, banjo and zither.

Pléiade, La, group of seven 16th-century French poets. They were Pierre de RONSARD, the leader, Joachim du Bellay, Jean-Antoine de Baïf, Rémy Belleau, Estienne Jodelle, Pontus de Tyard and Jean DORAT. They advocated writing in French instead of Latin and Greek.

Pleiades (M45), stellar cluster in the northern constellation of Taurus. It is a genuine OPEN CLUSTER of approx. 500 stars, of which at least seven are visible to the naked eye. The brightest stars are hot and white. The cluster is surrounded by a large reflection NEBULA. *See also* SU pp.*231, 242, 242,* 258, 265, *265.*

Plein air (French for "open-air"), term used of pictures painted out of doors. It tends to be used most frequently of the works by artists of the BARBIZON SCHOOL and by the IMPRESSIONISTS.

Pleistocene, geological epoch that began about 2 million years ago, during which man and most forms of familiar mammalian life evolved. Episodes of climatic cooling in this epoch led to widespread glaciation in the Northern Hemisphere and the Pleistocene is the best known glacial period or Ice Age in the Earth's history. The present HOLOCENE EPOCH succeeded the Pleistocene around 10,000 BC.

Plekhanov, Georgy Valentinovich (1857–1918), Russian Marxist and revolutionary leader. Originally a populist, he went into exile in 1880 and became an adherent of Marxism. A major theoretician and organizer of the Russian Social Democratic Workers Party, founded in 1898, he greatly influenced V.I. LENIN, with whom he worked until he joined the Mensheviks in 1903. He returned to Russia in 1917 and formed an anti-Bolshevik party, but died shortly after the October revolution. *See also* HC2 p.*168.*

Plesiosaur, any of several extinct marine reptiles of JURASSIC to CRETACEOUS times. It had a stout body propelled by powerful paddle-like limbs. Most plesiosaurs were long-necked with small heads (pliosaurs were short-necked with elongated heads). Length: 4.5–12m (15–40ft). *See also* NW p.*24, 185, 186.*

Pleura, double membrane that lines the space between the lungs and the walls of the chest. It is serous (thin and moist with serum, the fluid containing natural antibodies of the body's immune system). Dry PLEURISY results from inflammation of the pleura, and pleurisy with effusion results from the accumulation of excess serum in the pleural cavity. *See also* MS p.89.

Pleurisy, painful inflammation of the pleura, the membrane lining of the chest cavity, caused by infection in the lung.

Pleydenwurff, Hans (*c.*1420–72), German painter. He became a citizen of Nuremberg in 1457, and was a highly skilled exponent of Netherlandish art in Germany. Famous among his altarpieces is that for the Church of St Elizabeth, Breslau (1462).

Plimsoll, Samuel (1824–98), British reformer who was particularly interested in the dangers faced by merchant seamen. A member of parliament, he secured the passage of the 1875 Merchant Shipping Act. This required every owner to mark his ship with a circular disc with a horizontal line drawn through its centre (the "Plimsoll line"), indicating the maximum depth to which the vessel might be loaded.

Plinlimmon, or Plynlimon, mountain in w Wales, the source of the rivers Wye and Severn. It has three peaks. Height: 752m (2,468ft).

Pliny the Elder (*c.* AD 23–79), Roman naturalist whose full name was Gaius Plinius Secundus. His one surviving work is *Historia Naturalis,* a compilation of all the Greek science then known. He died while investigating the eruption of Vesuvius. *See also* HC1 pp.79, 116.

Pliny the Younger, (*c.*62–114), Roman author and administrator, whose full name was Gaius Plinius Caecilius Secundus Pliny. He began practising law when he was 18 years old, and later become senator, praetor and consul. His literary fame rests on the collection of letters, written probably for publication. *See also* HC2 pp.*94,* 113.

Pliocene, last era of the Tertiary period that lasted from 7 to 2 million years ago and preceded the Pleistocene Ice Age. Animal and plant life was not unlike that of today. *See also* NW pp.*24, 191.*

Pliopithecus, genus of extinct gibbon-like apes from the MIOCENE period of Czechoslovakia. Although a true ape it possessed many features of the PLATYRRHINI. It walked on all fours but the skeleton indicates that it could also swing by its arms.

Plisetskaya, Maya Michailovna (1925–), Soviet ballet dancer who joined the BOLSHOI BALLET, Moscow, as a soloist in 1943 and became its PRIMA BALLERINA in 1945. Her most famous role is Odette-Odile in *Swan Lake*; she has also appeared as an actress in several films, such as *Anna Karenina* (1968).

PLO (Palestine Liberation Organization), agency of guerrilla forces devoted to reclaiming from Israel for Palestine Arabs the w bank of the River Jordan. Founded in 1964, it is officially pledged to the dissolution of Israel. In 1974 the UN recognized the PLO as a government-in-exile. *See also* HC2 pp.297, *297,* 299.

Ploieşti, city in SE central Romania, 56km (35 miles) N of Bucharest. It prospered as the largest oil-producing centre in SE Europe in the 19th century. As a result of Romania's cooperation with the AXIS POWERS in 1940, supplying Germany with oil, the city was heavily bombed by the Allied powers in WWII. Its present industries include oil refining and the manufacture of petrochemicals. Pop. (1974) 175,527.

Plombières, Pact of, (July 1858), secret agreement between Napoleon III of France and Count Cavour, Prime Minister of Piedmont-Sardinia. Napoleon promised to help to expel Austria from Italy in return for the cession of Nice and Savoy to France. *See also* HC2 p.110.

Plomer, William Charles Franklyn (1903–73), British writer. He was born in South Africa, educated at Rugby and returned to South Africa after WWI to become a farmer and trader. He later settled in Britain. His best-known novels are *Turbott Wolfe* (1925) and *At Home* (1958). His poems include *Collected Poems* (1960) and among his short stories are *I speak of Africa* (1927) and *Paper Houses* (1929).

Plotinus (205–270), Roman philosopher, founder of NEO-PLATONISM. Greatly influenced by his teacher, Ammonius Saccas, he opened a school in Rome. His theories were basically Platonic with elements of other Greek philosophies. Although he opposed Christianity, his teachings have affected Christian thought. His pupil, PORPHYRY, compiled the *Enneads* of Plotinus.

Plough, agricultural implement used to cut furrows in soil for aeration and in preparation for sowing or planting. The first ploughs appeared in the NEOLITHIC period and in the BRONZE AGE became metal-tipped wooden wedges fastened to a single handle and a beam, pulled by men or oxen. This form remained virtually unchanged until the 19th century when the mould-board was introduced in the USA. This was a curved board that turned over the slice of earth cut by the plough's blade or share. *See also* HC2 pp.*21, 155:* PE pp. *158, 164.*

Plough. *See* GREAT BEAR.

Plough and the Stars, The (1926), four-act tragedy by Sean O'CASEY. The play presents a picture of the conflict between the ideals and the actuality of the situation of the Irish during the EASTER RISING of 1916. The title is taken from the plough and stars on the flag of the so-called Irish Citizen Army.

Plovdiv, city in central Bulgaria, on the River Maritsa. Seized by the Turks in 1364, it remained under Turkish rule for 500 years, becoming part of Bulgaria in 1885. Today it is a major transport centre that manufactures lead, zinc, textiles and shoes. Pop. 287,700.

Plover, any of several species of wading shorebirds, many of which migrate long distances over open seas from Arctic breeding grounds to Southern Hemisphere wintering areas. It has a large head, a plump grey, brown or golden speckled body and short legs. Length: to 28cm (11in). Family Charadriidae; genera include *Charadrius* and *Pluvialis*.

Plowright, Joan Anne (1929–), British actress whose reputation was established in the 1950s and whose versatile talents were displayed in both classical and contemporary work. The former were mainly at Chichester and The National Theatre with her husband Laurence OLIVIER; the latter included the first performances of John OSBORNE's *The Entertainer* (1957), Arnold WESKER's *Roots* (1959) and the English adaptations by Keith Waterhouse and Willis Hall of Eduardo de FILIPPO's *Il figlio di Pulcinella* (1959) (*Saturday, Sunday, Monday,* 1973) and *Filumena Marturano* (1946) (*Filumena,* 1977).

Plum, fruit tree, mostly native to Asia and naturalized in Europe and North America. The Japanese plum (*Prunus salicina*) is crossed with European varieties to give several cultivated strains. Plum trees have alternate simple leaves, white flowers and edible, smooth-skinned, oval fruits of a purple, red, blue or green colour, according to the variety. Plums are dried to make prunes. Height: to 9m (30ft). Family Rosaceae. *See also* PE pp.*192, 193*.

Plumbago. *See* GRAPHITE.

Plunket, Oliver (1629–81), Irish Roman Catholic prelate. Appointed Archbishop of Armagh and Primate of Ireland in 1669, he was forced into hiding after the TEST ACT (1673). Accused of conspiring in the POPISH PLOT (1678), he was executed, the last man to be martyred for the Roman Catholic faith in England. He was beatified by Pope Benedict XV in 1920.

Plunket, William Conyngham, 1st Baron Plunket (1764–1854), Irish lawyer and politician. He was the WHIG leader in the Irish Parliament before its abolition in 1800. He was the leader of the movement for CATHOLIC EMANCIPATION in the House of Commons in the 1820s, and Lord Chancellor of Ireland from 1830 to 1841.

Plunket Shield, until 1975 the premier cricket trophy in New Zealand. Presented in 1906 by Lord Plunket, it was contested annually in a series of first-class matches between the four major provinces and two district sides. It is now the trophy for matches between North and South Islands.

Plunkett, Sir Horace Curzon (1854–1932), Irish political figure and agricultural reformer. He worked as a cattle rancher in the USA (1879–89). Returning to Ireland, he founded the Irish Agricultural Organization Society (1894), a co-operative. As a Member of Parliament (1892–1900), he campaigned for agricultural reform. He was a senator of the Irish Free State (1922–23).

Pluralism, in politics, theory that power in the state is wielded by a number of groups with conflicting interests, none of which is able to establish absolute authority. In philosophy, pluralism is the name given to the theory that there are many ultimate substances, rather than one, as in MONISM. LEIBNIZ and Bertrand RUSSELL are considered philosophical pluralists. Pluralism can also mean the holding of more than one office at the same time, especially within the Church. *See also* MS p.*275*.

Plutarch (*c.*46–120 AD), Greek biographer and essayist. His best known work is *The Parallel Lives*, biographies of soldiers and statesmen. They are arranged in pairs, first a Greek, then a comparable Roman. The style is pleasant, although the facts are not always accurate. Less known but of great interest are Plutarch's *Moralia*. They consist of essays and dialogues on ethical, literary and historical subjects.

Pluteus, free-swimming larval stage of an echinoderm such as the SEA URCHIN, STARFISH, SEA CUCUMBER and SEA LILY, commonly found in PLANKTON. Most species bear hair-like cilia, sometimes on "arms", to waft food into the gut and to provide a means of larval dispersal. *See also* NW p.*119*.

Pluto, Latin name for the Greek god HADES, sovereign of the lower world and god of the dead. He was brother of ZEUS and husband of PERSEPHONE, and a hard and inexorable character who dwelt beneath the secret places of the earth in a gloomy palace in the barren fields. Pluto was considered also to be the giver of wealth in that he possessed all that sprang from the earth.

Pluto, ninth planet from the Sun, discovered in 1930 by C.W. TOMBAUGH. Mean distance from the Sun, 5,890,000,000km (3,658,000,000 miles); estimated mass, 0.9 times that of the Earth; estimated diameter, approx. 5,998km (3,725 miles); rotation period, 6.4 Earth-days; period of sidereal revolution, 247.7 years; estimated surface temperature, −230°C (−382°F). *See also* SU pp.*176*, 215.

Plutonium, radioactive metallic element (symbol Pu) of the ACTINIDE group, first made in 1940 by deuteron (heavy hydrogen) bombardment of uranium. It is found in small amounts in uranium ores. Pu^{239} (half-life 24,360yr) is made in large quantities in BREEDER REACTORS. It is a fis-

sionable material used in reactors and nuclear weapons. The element is a strong ALPHA PARTICLE emitter and is absorbed by bone, making it a dangerous radiological hazard. Properties: at. no. 94; density 19.84 (25°C); mt. p. 641°C (1,186°F); b. p. 3,327°C (6,021°F); most stable isotope Pu^{244} (half-life 8×10^7yr).

Plymouth, major port and county district in Devon, SW England, on the Tamar estuary; England's most important naval base (Devonport) after Portsmouth. In the 16th century the adventurers Francis DRAKE, John HAWKINS and Walter RALEIGH sailed from Plymouth, and the *Mayflower* sailed for America from there in 1620. Granite, kaolin and manufactured goods are exported. Area: (county district) 79sq km (30sq miles). Pop. (county district, 1974 est.) 251,200; (city, 1971) 239,314.

Plymouth Brethren, strictly Puritan sect of evangelical Christians, founded in Ireland in the late 1820s by J. N. Darby, an ordained Anglican. They are also sometimes known as Darbyites. Their more usual name comes from their having established their first English centre at Plymouth in 1831. In 1849 they split into two groups, the "Open Brethren" and the "Exclusive Brethren", and have since split further.

Plymouth Colony, settlement on the coast of Massachusetts in 1620. About 100 pilgrims and other colonists from the MAYFLOWER landed near Cape Cod in December 1620 after poor weather diverted the ship from Virginia. Government by majority vote was established by the Mayflower Compact, because no royal charter existed. The hardships of the first winter killed nearly half the settlers but, under the leadership of William BRADFORD and with the aid of nearby friendly Indians, the colony survived and in October of the following year the settlers celebrated the first THANKSGIVING DAY. *See also* HC2 p.62, 62.

Plynlimon, Plynlimmon. *See* PLINLIMMON.

Plywood, sheet made of three or more layers of wood glued together with the grain of successive layers at right-angles to the preceding layer. It is usually much stronger and cheaper than most woods of equal thickness.

Plzen. *See* PILSEN.

Pneumatic tool, device powered by compressed air in a rotary or a reciprocating (back and forth) motion to speed up operations such as sawing, grinding, digging, hammering and riveting. The pneumatic drill used in roadworks has a reciprocating, pounding motion at speeds of between 80–500 revs per minute.

Pneumatic tyre. *See* TYRE.

Pneumoconiosis, occupational disease of miners working in confined and dusty conditions; it affects the lungs. Caused by inhaling irritants, often only as minute specks, the disease inflames and can finally destroy lung tissue.

Pneumonia, inflammation of the LUNG tissue, usually caused by bacterial infection. Mycoplasmal pneumonia affects mainly children and young adults; it is caused by the micro-organism *Mycoplasma pneumoniae*, which grows on the mucous membrane lining of the lungs. Pneumococcal pneumonia is more severe and is caused by *Diplococcus pneumoniae*. Both forms produce symptoms including high fever, chest pain, coughing and bloody sputum; treatment for either form is with ANTIBIOTICS. *See also* MS pp. 88–89.

Pneumothorax, air in the space between the membranes covering the lungs and those lining the chest cavity, most commonly caused by injury and more rarely by disease. It causes the lung to collapse for a short time; it is sometimes deliberately induced to allow a lung to heal, as in a treatment for TUBERCULOSIS.

Pnom-Penh. *See* PHNOM-PENH.

Po, longest river in Italy. It rises in the Cottian Alps in NW Italy, and flows E through the Piedmont to empty into the Adriatic Sea. The Po valley is an important industrial and agricultural region, and water from the river is used extensively in irrigation schemes. It is navigable upstream from the mouth to as far as Pavia. Length: 650km (405 miles).

Pocahontas (1595–1617), daughter of American Indian chief Powhatan and heroine of a romantic colonial story which tells how she saved the life of Capt. John Smith, leader of the Jamestown settlement, who was about to be clubbed to death by her father. *See also* HC2 p.*63*.

Pochard, freshwater diving duck of SE England and Europe. It inhabits dense vegetation close to fresh water or coastal lagoons. The drakes have a bright chestnut head, a black neck and breast, and grey underparts. Length: 46cm (18in). Family Anatidae; species *Aythya ferina*. *See also* NW p.*144*.

Po-Chü-i (772–846), Chinese poet who held several high government posts and was Governor of Chung-chou (818) and Mayor of Lo-yang (829). He wrote more than 3,000 short poems in simple, clear language. His most famous is *The Everlasting Wrong* (806).

Pocket battleship, CRUISER armed with large guns. These were built by Germany in the 1930s when, under the terms of the Treaty of Versailles, it was prevented from constructing any battleship of more than 10,000 tonnes. The treaty was violated by the launching of the *Graf Spee* (12,500 tonnes) in 1934. *See also* GRAF SPEE MM pp.*174–175*.

Pocket Borough. *See* ROTTEN BOROUGH.

Podgorny, Nikolai Viktorovitch (1903–), Soviet chief of state from 1965 to 1977. A member of the Communist Party from 1930 and of the Central Committee from 1960, he became Soviet President in 1965. His elevation to the presidency marked his removal from effective power in the confusion following the fall of KHRUSHCHEV. In 1977 he was replaced by Leonid BREZHNEV.

Podicipediformes, *See* GREBE.

Podocarpus, genus of evergreen shrubs and trees native to warm and tropical regions of the Southern Hemisphere. They have narrow, pointed, flat leaves. Family Podocarpaceae. *See also* NW p.*210*.

Podolia, region in the Ukraine (Ukrainskaja SSR), USSR, in the valley of the southern Bug River. The E part of Podolia passed to Russia in 1793; the W part to Austria in 1772, to Poland from 1918 to 1939 and was annexed by the USSR after WWII. Products: sugar beet, tobacco, wheat, sunflowers and dairy produce. Food processing is the principal industry. Kamenec-Podol'skij is the main city.

Podolsk, city in the Russian SFSR, USSR 40km (25 miles) S of Moscow, at the junction of the Moscow-Warsaw and the Moscow-Crimea roads. Industries include machinery plants, oil-refining equipment and the manufacture of cables. Pop. (1975) 187,000.

Poe, Edgar Allan (1809–49), US poet and short story writer, regarded by many as the creator of the modern detective story. Poe's life was brief and tragic. An orphan at three, he had a wealthy godfather with whom he quarrelled while at university. He married his young cousin of 13 who died a few years later of tuberculosis. Much of his finest poetry, such as *The Raven* (1845), deals with fear and horror in the GOTHIC tradition. Other works include the poem *Annabel Lee* (1849) and the stories *The Fall of the House of Usher* (1839), *The Murders in the Rue Morgue* (1841) and *The Pit and the Pendulum* (1843).

Poelzig, Hans (1869–1936), German architect who had an enormous influence on the development of modern architecture in Germany. His office building at Breslau (1911) anticipated the favourite motif of the 1920s – horizontal strip windows. Poelzig designed some highly original industrial architecture, eg the water tower at Posen (1911), a forerunner of his EXPRESSIONIST phase. The apogee of this was the Grosses Schauspielhaus, Berlin (1919). From the late 1920s he began designing buildings of a monumental simplicity, eg the Farben administrative block, Frankfurt (1928–31).

Poetics (335 BC), ARISTOTLE's work of literary criticism which deals with the origins and forms of poetry. In it he argues that poetry is an "imitation" (*mimesis*) of

Pluto became ruler of the underworld by casting lots for his brothers' kingdoms.

Pochards are poor walkers because their legs are adapted for making swift dives.

Pocket battleships complied with the limit imposed on the German navy's size.

Edgar Allan Poe used horror purely as an expression of an inner reality.

Pointers "freeze" when they scent game, pointing out the location of the bird.

Poland; view of Warsaw which was destroyed before the Russian advance.

Roman Polanski's traumatic personal life added intensity to his vision as director.

Polar bears crush sleeping seals with a single powerful blow from one paw.

events and is either EPIC (narrative) or dramatic. He defines TRAGEDY as depicting nobler characters than comedy, and inspiring pity and terror. The second book, on comedy, is lost.

Poet Laureate, title conferred by the British monarch on a distinguished contemporary poet whose duty is then to write commemorative verse on important royal or public occasions. The appointment is a development of the medieval custom of having minstrels in the king's retinue. Ben JONSON is generally regarded as the first poet laureate, although John DRYDEN was the first to receive the official title. The appointment has been held by, among others, Robert SOUTHEY (1813–1843), William WORDSWORTH (1843–1850), Alfred, Lord TENNYSON (1850–1892), John MASEFIELD (1930–1967) and Sir John BETJEMAN (1972–   ).

Poetry, art form in which words are used for their allusions; their sound, rhythm and meaning are all equally important, in contrast to prose, where the meaning usually predominates. Until modern times poetry was usually written in regular lines with carefully structured metres, often with rhymes, and in earlier times with alliteration within the line. With the introduction of free verse by Walt WHITMAN, the boundary between poetry and prose was weakened. Many rigid poetic formats have been developed, such as the SONNET or gnomic forms. *See also* individual styles and poets.

Pogonophora, phylum of 80 known species of sedentary, worm-like, marine invertebrates that inhabit chitinous tubes on the ocean-bed. Its body is divided into three regions: the protosome, bearing tentacles; a short mesosome and a long trunk or metasome. Length: to 50cm (20in). *See also* NW p.*121.*

Pogrom, Russian term meaning "riot", later applied to series of mob attacks on Jews in Russia in the 19th and 20th centuries.

Pohl, Frederik (1919–   ), US science fiction writer and historian, the author or co-author of some 40 science fiction books.

Pohutukawa, flowering tree native to New Zealand. Its red flowers are used by the Maoris for decoration. Species *Metrosideros tomentosa.*

Poi, stuffed flaxen ball swung rhythmically from a cord by Maori women during traditional work songs. Also, a Polynesian food made from TARO roots.

Poincaré, Jules-Henri (1854–1912), French mathematician. Appointed professor at the University of Paris he worked on CELESTIAL MECHANICS, winning an award from the king of Sweden for his contribution to the theory of orbits. In 1906, independently of Albert EINSTEIN, he obtained some of the results of the special theory of RELATIVITY. He developed many new mathematical techniques and attempted to make mathematics accessible to the general public in such works as *The Value of Science* (1905) and *Science and Method* (1908).

Poincaré, Raymond Nicholas Landry (1860–1934), French statesman, premier three times and President of France from 1913 to 1920. He was an ardent nationalist and after WWI, when Germany defaulted on reparation payments to France, he ordered troops into the Ruhr to force the payment. During the 1926–28 financial crisis he stabilized the French economy and was called "saviour of the franc". *See also* HC2 p.193.

Poinciana, genus of evergreen trees and shrubs native to E Africa and Madagascar. They have long pinnate leaves and red, orange or yellow flowers. Royal poinciana (*Poinciana regia*) has orchid-like yellow-striped, scarlet flowers. Height: to 12m (40ft). Family Leguminosae.

Poinsettia, showy house plant native to Mexico. It has tapering leaves and tiny yellow flowers centred in leaf-like red, white or pink bracts. It is a favourite Christmas plant and grows best in bright light and soil of equal parts loam, peat moss and sand. Propagation is by tip cuttings. In its natural environment the tree grows to about 5m (16ft). Height: to 60cm

(2ft) when potted. Family Euphorbiaceae; species *Euphorbia pulcherrima.*

Point Counter Point (1928), novel by Aldous HUXLEY. Containing portraits of Oswald MOSLEY, D.H. LAWRENCE and Middleton MURRY, it debates politics, science, religion and morality in a complex novel of ideas. The title derives from its structure, which imitates that of a fugue.

Pointer, smooth-coated sporting and gun dog that was developed, probably in Britain, in the 17th century for hunting. It can be trained to indicate the direction in which game lies by standing motionless, aligning its muzzle, body and tail. It has a wide head, with a solid muzzle. The strong lean body is set on muscular legs and the tail tapers to a point. The short dense coat can be white with black, or brown markings. Height: to 63cm (25in); weight: to 27kg (60lb).

Pointillism, also called divisionism, technique of oil painting developed by the NEO-IMPRESSIONIST French painter, Georges SEURAT. The method of building up canvases by the application of small dots of pure colour was probably suggested to Seurat by his reading of Charles Blanc. The name comes from the French *pointiller* (to dot, stipple). *Sunday Afternoon on la Grande Jatte,* exhibited in 1886, was taken as a manifesto of the new technique.

Poiseuille's equation, in fluid mechanics, states that the relation between the rate of flow, R, of an incompressible fluid of viscosity η, in a tube of radius a and length l is $R = \pi p a^4 / 8 l \eta$, where p is the difference in pressure between the two ends of the tube.

Poison gas. *See* GAS WARFARE.

Poison ivy, North American shrub that causes a severe, itchy rash on contact with human skin. It has greenish flowers and white berries. Species *Toxicodendron radicans.* Poison sumac (*Rhus vernix*) is a shrub native to the E USA. It has long, featherlike leaves and whitish fruit and grows to 6m (20ft). Family Anarcardiaceae.

Poisson, Siméon-Denis (1781–1840), French scientist who applied mathematics to many areas of physics, including ELECTROMAGNETISM and MECHANICS, and wrote *Treatise on Mechanics* (1811), a standard text for many years. He extended the work of Joseph-Louis LAGRANGE in CELESTIAL MECHANICS and worked on definite INTEGRALS and FOURIER SERIES. In 1837 he published *Researches on the Probability of Opinions,* in which he presented the Poisson distribution, which gives the probability of the occurrence of a specific event among a large number of events. Poisson's ratio, which relates lateral contraction to longitudinal extension in stressed materials, is also named after him.

Poitier, Sidney (1924–   ), US actor who began his career at the American Negro Theater. His first film was *No Way Out* (1950) and he was oustanding in *A Man is Ten Feet Tall* (1956). He won an Oscar for *Lilies of the Field* (1963). His later films include *Guess Who's Coming to Dinner, In the Heat of the Night* (1967) and *For Love of Ivy* (1968).

Poitiers, town in W central France, site of two battles. At the first Battle of Poitiers of AD 732, also called the Battle of Tours, the Franks led by CHARLES MARTEL defeated the Arabs, who were thereby confined to Spain. The Second Battle of Poitiers (1356) was an important encounter in the HUNDRED YEARS WAR. The French King John II was captured by the English under Edward the BLACK PRINCE, and was later ransomed. *See also* HC1 pp.*152, 203.*

Poitou, region in W France, extending from the Atlantic coast to beyond the River Vienne. It includes the départements of Vienne, Deux-Sèvres, and Vendée; the capital is Poitiers. Part of the duchy of Aquitaine, the region was held by England from 1152 to 1204 and from 1356 to 1369. It became part of the French Crown lands in 1416. The region includes some rich farmland; wheat is grown and beef and dairy cattle are raised.

Poker, card game, believed to have originated in Europe in the 16th century. Basi-

cally a gambling game, the object is to win the pot (all the bets that are made after each card is dealt) by holding the best combination of cards or by bluffing the other players into withdrawing. The two main variations of the game are draw poker and stud poker, in both of which a standard pack of 52 PLAYING CARDS is used.

Polack, Hans (*d.*1519), German painter who worked in Bavaria and produced many Gothic frescoes and altarpieces. Among his best work is the high altar of the Church of St Peter in Munich.

Poland, country in central Europe; a member of the Communist Bloc. Throughout history it has been strategically placed in European wars and boundary disputes, and during WWII it suffered great devastation. Poland is one of Europe's leading producers of coal; other mineral deposits include copper, lead and zinc. Agricultural output is high and the principal crops are potatoes and rye. Most industries are state-owned, the chief ones being steel, engineering and chemicals. The capital is WARSAW. Area: 312,674sq km (120,724sq miles). Pop. (1976 est.) 34,481,000. *See* MW p.143.

Polanski, Roman (1933–   ), Polish actor and director, who established his reputation as a director with the short film *Two Men and a Wardrobe* (1958). His first full length film, *Knife in the Water* (1962), was followed by *Repulsion* (1965) and *Cul-de-Sac* (1966), both made in Britain. Polanski's later films included *Rosemary's Baby* (1968), *Macbeth* (1971), *Chinatown* (1974) and *The Tenant* (1976).

Polanyi, Michael (1891–1976), British scientist and philosopher. He was a leading exponent of the view that scientific discoveries owe much to intuition and chance, in contrast to the established opinion that scientific advances proceed logically. His books include *Personal Knowledge* (1958) and *Knowing and Being* (1969).

Polar bear, large white bear that lives on Arctic coasts and ice floes. It spends most of its time at sea on drifting ice floes, often swimming for many miles. It preys chiefly on seals and is hunted for fur and meat. Length: 2.3m (7.5ft); weight: to 405kg (900lb). Species *Thalarctos maritimus. See also* NW pp.224, *224,* 243.

Polarimetry. *See* POLARIZED LIGHT.

Polaris. *See* POLE STAR.

Polaris submarine, nuclear-powered submarine operated by the US and British navies, armed with long-range Polaris missiles. These missiles carry nuclear warheads and can be fired, in rapid succession if necessary, while the craft remains submerged. Such submarines can travel thousands of kilometres without surfacing or refuelling, and so attempt to conceal their location until their missiles are launched. *See also* MM pp.*175, 180.*

Polarized light, light waves whose electric and magnetic components vibrate in a predictable fashion so that the wave-form is no longer symmetrical about the direction in which the wave is travelling. Scientists distinguish between three types: plane-polarized light, circularly-polarized light and elliptical-polarized light, each depending on the net direction of the vibrations, seen in cross-section to the wave. *See also* SU pp.102–103.

Polarography, method of chemical analysis which can measure minute concentrations of ions in solution. Two ELECTRODES are inserted into the solution: one, the reference electrode, is kept at a constant voltage; the other has a variable voltage and consists of a glass capillary tube containing a head of mercury, with tiny drops of mercury at its end becoming the face of the electrode. As these drops form, a current grows to a peak between the two electrodes and then suddenly falls as the drop leaves the capillary tip. By gradually increasing the voltage a polarogram – a graph of the current versus the voltage, which looks like a series of steps – is produced. From these the reduction potentials of the ions may be deduced and, as these are known for each metal ion, the ions can be identified.

Polaroid camera. *See* LAND, EDWIN.

Pole, generally either of the two points of

intersection of the surface of a sphere and its axis of rotation. The Earth has four poles: the north and south geographic poles, where the Earth's imaginary axis meets its surface; and the north and south magnetic poles, where the Earth's magnetic field is most concentrated. A bar magnet has a north pole, where the magnetic flux leaves the magnet, and a south pole, where it enters. A pole is also one of the terminals (positive or negative) of a battery, electric machine, generator or circuit. *See also* SU pp.*73, 116–117.*

Pole, de la, English noble family descended from Sir William de la Pole (*d.*1366), a rich Hull merchant and baron of the exchequer (1339). His son Michael (*c.*1330–89), created 1st Earl of Suffolk (1385), was chancellor to Richard II but died in exile. Michael's grandson William (1396–1450), earl from 1415, was created Duke of Suffolk (1448), but was banished after serving Henry VI and was beheaded on his way to France. William's son John (1442–91) married Elizabeth, sister to Edward IV, and their sons John (*c.*1464–87), Edmund (*c.*1472–1513) and Richard (*d.*1525) unsuccessfully claimed the English throne.

Pole, Reginald (1500–58), English Roman Catholic prelate. His father and HENRY VII were cousins, but Pole would not support Henry VIII against Pope CLEMENT VII when Henry sought a divorce. He went to Italy in 1532, became a cardinal in 1536 and then presided at the Council of TRENT. As Papal legate in England (1553–57) and Archbishop of Canterbury (1556–58) under MARY I, he was extremely influential in the government of the country.

Polecat, any of several species of small, carnivorous, nocturnal mammals that live in wooded areas of Eurasia and N Africa; especially *Mustela putorius,* the common polecat. It has a slender body, long bushy tail, anal scent glands and brown to black fur known as fitch. It eats small animals, birds and eggs. Length: 45.7cm (18in). Family Mustelidae. *See also* NW p.198.

Pole Star, also called Polaris or North Star, important navigational star, nearest to the N celestial star and less than 1° from it. It is in the constellation Ursa Minor and always marks due N.

Pole vault, standard event in field athletics in which contestants use a pole to lift themselves over a horizontal bar. The bar is raised in a fixed progression and vaulters failing to clear each height within three attempts are eliminated. Pole vaulting has been included in the Olympic Games since 1896.

Poliakoff, Serge (1906–69), French painter who worked in London at the SLADE SCHOOL (1935–37). Inspired by Wassily KANDINSKY and Robert DELAUNAY in the late 1930s, he turned to abstract painting and was later influenced by the more mathematical style of Kasimir MALEVICH.

Police forces, legal bodies of people within a state concerned with maintaining civil order and investigating breaches of the law. Their members represent public authority and are pledged to act in the interests of society as a whole. The first police force acting independently of the judiciary was established in Paris in 1667, becoming a uniformed force in 1829. Britain's first effective regular professional force was the Marine Police Establishment in 1800, later absorbed into the Metropolitan Police as the Thames Division. The Metropolitan Police was created by Sir Robert PEEL (hence the nickname "bobbies") in Sept. 1829. The ordinary police forces of today are supplemented by special forces which manage government establishments, and rail, sea and air transport complexes. Private security organizations are also growing rapidly.

Polidoro da Caravaggio (1490–1543), Italian painter. He worked in RAPHAEL's studio until 1520, and in 1521 painted two scenes in the Vatican, *Meeting of Solomon and Sheba* and *Building of the Temple.* His masterpiece is the *Ascent to Calvary,* probably painted before 1534.

Polignac, Auguste-Jules-Armand-Marie, Prince de (1780–1847), French states-man. A firm advocate of royal and papal authority, he was made a prince by the Holy See (1820), and appointed Prime Minister (1829) by Charles X. His autocratic regime brought about the July Revolution (1830) and he was imprisoned and later exiled.

Poliomyelitis, acute VIRUS infection, usually afflicting children (it is also called infantile paralysis), but sometimes adults. It is divided into three types, each immunologically separate. The disease is almost always fatal if it attacks the nerve cells in the brain (bulbar poliomyelitis). Only about 25% of patients whose central nervous system is attacked suffer permanent severe paralysis. It is endemic or epidemic throughout the world, but rare in the West since SALK vaccine was introduced in 1955. *See also* MS pp.*37, 97, 107, 111–112, 143.*

Polish, national language of Poland, spoken by virtually all of the country's 34 million people. It is a Slavonic language and a member of the INDO-EUROPEAN family. Polish is written in the Roman (Latin) alphabet, but with a large number of diacritical marks to represent the various Slavonic vowels and consonants. Documents written in Polish date from the 12th century AD.

Polish Corridor, narrow belt of land along the River Vistula established in 1919 by the Treaty of VERSAILLES to give Poland access to the Baltic Sea. The city of Gdansk (Danzig) was designated a free city, linked to Poland. This settlement was the source of much dispute with Germany and was one of the reasons given for Germany's invasion of Poland in 1939.

Polish Succession, War of the (1733–38), war to settle the Polish throne after the death of Augustus II the Strong in 1733 between the claims of Stanislaw Leszczynski and the Elector Frederick Augustus of Saxony. Russia and the Empire supported Augustus; France, Spain and Sardinia-Savoy supported Leszczynski. It was fought mainly in Italy, but also on the Rhine. It placed Augustus III on the throne, he ruled until 1763.

Politburo (political bureau), administrative and policy-making body of the Soviet Communist Party. Formerly called the Presidium, it consists of 11–12 full members and 6–9 candidate members chosen by the party's central committee. In theory, the government and party structures are separate entities. In practice, they are so interrelated because of overlapping membership that the Politburo virtually runs the country. *See also* HC2 p.*199.*

Politian. *See* POLIZIANO, ANGELO.

Political party, group organized for the purpose of electing candidates to office and for promoting a particular set of political principles. In Britain, political parties originated in the split between the WHIGS and TORIES in the late 17th century, but the two-party parliamentary system did not emerge until the 18th century. *See also* articles on individual parties.

Political science, study of governments at state, local, national and international levels. Political science also looks at the relationship of citizens, with a government and examines how various types of government solve similar problems. *See also* MS pp.*274–275.*

Politics, sphere of action in human society in which power is sought in order to regulate the ways in which people shall live together. For a society to engage in politics, it must conceive of society as being in a state of perpetual change. A static society regulates its social action only by judicial methods, by resolving disputes in accordance with prescriptive rules. A political society accepts the need for perpetually changing the rules in order to make them accord with altered circumstances. *See also* MS pp.*272–281.*

Politics, philosophical tract, of which eight parts remain, by the Greek philosopher ARISTOTLE. It deals with the problems of human activity in communal life, starting from the premise that man is naturally politically inclined.

Poliziano, Angelo (1454–94), also called Ambrogini or Politian, Italian Renais-sance poet and scholar who taught Greek and Latin at the University in Florence. His Latin poetry rivalled that of Giovanni PONTANO. His great work, in the Tuscan vernacular, included the unfinished poem *Stanzas for a Joust* (1475–78) which celebrates the jousting skill of Giuliano de Medici, the brother of his patron, Lorenzo the Magnificent. *See also* HC1 pp.*246.*

Polk, James Knox (1795–1849), 11th US president (1845–49) during whose administration the entire Southwest was secured as a result of the American victory in the Mexican War (1846–48). In his productive term he reduced the tariff (Tariff Act, 1846), restored the independent treasury, settled the Oregon boundary dispute with Britain and acquired California. Polk's poor health prevented his seeking a second term. *See also* HC2 p.*149.*

Polka, lively Bohemian folk dance in 2/4 time. It became fashionable in Paris in about 1830, and thereafter in Europe and the Americas. It is often danced as a ballroom dance, the basic step being three quick steps and a hop.

Pollack, or pollock, marine food fish of the COD family found in large shoals on both sides of the North Atlantic. Coloured green with yellow or grey, it has a jutting lower jaw. Other names for the fish are coalfish and saithe. Length: to 106.7cm (3.5ft); weight: to 15.9kg (35lb). Family Gadidae; species include *Pollachius pollachius* and *Pollachius virens. See also* PE pp.*240, 246.*

Pollaiuolo, Antonio (*c.*1432–98), Italian sculptor and goldsmith who became head of one of the chief Florentine workshops. He was a great draughtsman and probably the first artist to study anatomy by dissection. He collaborated with his brother Piero (*c.*1441–*c.*1496) in paintings which include the *Martyrdom of St Sebastian* (1475). Antonio's major public works were the tombs for Pope Sixtus IV (1493) and Innocent VIII (*c.*1495).

Pollard, Albert Frederick (1869–1948), British historian. He was assistant editor of *The Dictionary of National Biography* from 1893 to 1901, professor of constitutional history at London University from 1903 to 1931, and founder and director of the Institute of Historical Research from 1920 to 1931, then becoming honorary director until 1939. His most outstanding works were in the field of TUDOR history.

Pollen, yellow, powder-like male sex cells in flowering plants. Pollen grains are produced in the anther chambers on the STAMEN, and have thick resistant walls with a pattern of spines, plates or ridges, according to species. These markings are the basis of pollen analysis (PALYNOLOGY) used to identify fossil plants and sediments. *See also* POLLINATION; NW pp.*70–71.*

Pollination, transfer of POLLEN from the STAMEN to the STIGMA of a flower. Self-pollination occurs on one flower and cross-pollination between two flowers. Cross-pollination is more common and results in a great variety of genetic combinations. Pollination occurs mainly by wind (anemophily) and insects (entomophily). Wind-pollinated flowers are usually small and clustered and produce a large quantity of small, light, dry pollen grains. Insect-pollinated flowers are brightly coloured, strongly scented, contain nectar, and produce heavy, sticky pollen. *See also* NW pp.*70–71.*

Pollock, Jackson (1912–56), US painter. Influenced by SURREALISM and PICASSO he moved towards a highly ABSTRACT ART and abandoned the use of brushes in 1947, pouring paint straight on to the canvas. He became one of the chief exponents of ACTION PAINTING and his works include *The Blue Unconscious* (1946) and *Lavendar Mist* (1950). *See also* HC2 pp.*205, 205.*

Pollock, Robert Graeme (1944–), South African cricketer. A forceful left-handed batsman from Eastern Province, he holds the record for the highest score (274) made by a South African in a Test Match. He has only 23 Test caps (1963–70), but this merely reflects his country's isolation from Test cricket.

Poll tax, tax levied on every person in the

Pole vault is the most technically difficult of all the athletic events.

Political party: British labour politicians ending their 1977 conference.

Polka: this dance became a big success when it was introduced in Paris in 1843.

Pollack are attracted by lights at night, when they are easily caught by line.

Pollution

Marco Polo: mosaic by Francesco Salviati; no genuine portrait of him exists.

Polo: Oxford and Cambridge university teams at Kirtlington Park. Oxford won 7–0.

Polynesian fishermen make and repair nets as they have done for centuries.

Polynesian girls wearing fragrant garlands of leis charm visitors to Hawaii.

community, on either a fixed or a sliding scale. The poll tax in Britain became regular during the HUNDRED YEARS WAR in the 14th century; its harshness was a major factor in the PEASANTS' REVOLT. Poll taxes were collected in England until 1698. *See also* HC1 p.215.

Pollution, the spoiling of the natural environment, generally by industrialized society. Pollution is usually a result of an accumulation of waste products, although an excess of noise and heat which has adverse effects on the surrounding ecology is also included in the term. *See also* PE pp.146–153.

Pollux, in Greek mythology, son of ZEUS and LEDA, brother of HELEN and twin brother of Castor, whose fath.r was Tyndareus, King of Sparta. The twins were inseparable adventurers. They fought the boar on the Calydonian hunt and rescued Helen from her kidnapper, Theseus. When Castor was slain in a battle over possession of a herd of oxen, the inconsolable Pollux offered his life instead. Zeus rewarded him by permitting each of the twins alternate days on earth and in the underworld.

Polo, Marco (1254–1324), first European traveller to cross the length of Asia. As a boy he went with his father, Niccolo, and uncle, Maffeo, on a trading mission (1275) that took them to the court of Kublai Khan, the MONGOL ruler of China. Returning to Venice from the Chinese coast in 1295, Marco Polo became involved in a war with Genoa and was taken prisoner, during which time he compiled an account of his travels which remained an important Western source of information on Asia for nearly 600 years. He was soon released and returned to Venice, where he spent the remainder of his life. *See also* HC1 pp.*222, 223,* 225.

Polo, field game played on horseback. Two teams of four players on a field up to 182m (600ft) by 273m (900ft) each try to hit a small ball, about 8.3cm (3.25in) in diameter, into a goal, 7.3m (24ft) wide by 3m (10ft) high, using flexible mallets, 122–132cm (48–52in) long. A game consists of four, six or eight chukkas (periods), each 7.5 minutes long; additional chukkas may be played to decide a game if the scores are tied. Each player may use several ponies during a game. Players are handicapped on a scale of from −2 to 10 goals; this is a device for measuring their worth to their team. Polo originated in Persia in ancient times, and spread throughout Asia. It was revived in India in the 19th century, and was taken up by British army officers there. It was first played in Britain in 1868, and is also played in the USA.

Polonaise, stately graceful dance in 3/4 time. In 1573, it was adapted as a formal march for the coronation of Henry of Anjou as King of Poland, and is now the national dance of that country.

Polonium, rare radioactive metallic element (symbol Po) of group VIA of the periodic table, discovered in 1898 by Marie CURIE. It is found in trace amounts in uranium ores and may also be synthesized. Properties: at. no. 84; density 9.40; m.p. 254°C (489°F); b.p. 962°C (1,764°F); most stable isotope Po^{209} (half-life 103 yrs). *See also* SU pp.*64, 137.*

Poltava, city in the Ukraine (Ukrainskaja SSR), USSR, on the River Vorskla 290km (180 miles) ESE of Kiev. The site has been settled since the 9th century. It was ceded by Lithuania to the Tartars in 1430 and it was a Cossack stronghold in the 17th century. It was the birthplace of Nikolai GOGOL. Today Poltava is an important industrial centre; its products include sugar, cereals, machinery, leather goods and textiles. Pop. (1970) 220,000.

Poltergeist, noisy spirit or ghost, supposedly responsible for rappings, movement of furniture and other unexplained sounds and activity. Such noises often occur during seances, and allegedly attest to the presence of supernatural beings.

Polyamide, man-made POLYMER with the generic name of nylon, formed from the reaction between a dibasic acid (such as adipic acid) and a diamine (such as hexamethylenediamine). Nylon consists of long-chain synthetic polyamides in which the diamine and acid molecules are joined by amide linkages. *See also* SU pp.142–143, 156–157.

Polyandry, marriage in which the woman has several husbands. In Nepal, Tibet and neighbouring countries where polyandry is practised the men are brothers.

Polyanthus, any of a group of spring-flowering, perennial, garden primroses of the genus *Primula*. They occur in many habitats, mainly in the N temperate zone, and may be almost any colour. They have basal leaves and disc-shaped flowers, branching from a common stalk to form a ball-like cluster. Polyanthus hybrids are probably derived from a cross between a PRIMROSE and a COWSLIP. Height: to 15cm (6in). Family Primulaceae.

Polybius (*c.*200–*c.*120 BC), Greek historian and politician. A leader in the Achaean Confederation, he was deported to Rome in 168 where he obtained the patronage of Scipio, probably accompanying him to Spain and Africa. He was ambassador to Achaea in 146. He wrote his *Universal Histories* (40 vols, of which only the first five are wholly exant) describing the rise of Rome in the Mediterranean world from 220 to 146 BC. *See also* HC1 p.75.

Polycarp, Saint (*c.*69–*c.*155), Greek bishop of SMYRNA who, according to tradition, was taught by the APOSTLES. He vigorously opposed contemporary heresies and upheld Christianity in Asia Minor. He went to Rome to discuss the dating of EASTER but on his return was burned to death on the orders of the Roman proconsul.

Polychaete, any of a species of marine ANNELID worms characterized by distinctive segmentation of the body; it includes RAGWORMS and bristleworms. It has a distinct head with sensory projections and most segments have outgrowths or bristles. *See also* NW pp.88–89.

Polycleitus (Polyclitus) (*c.*480–*c.*420 BC), Greek sculptor who chiefly worked in bronze and did a number of statues of athletes. His most famous embodied his ideal of physical perfection and is known to us through a Roman copy, the *Doryphorus* or *Spear-bearer.* No recognized originals by Polycleitus exist.

Polycythaemia, abnormal increase of red blood cells. It may cause clot formation in the circulatory system. Its symptoms are headache, dizziness, enlargement of the spleen and reddening of the face.

Polyester, class of organic substances composed of large molecules arranged in either a chain or a network and formed from many smaller molecules through the establishment of ester linkages. They are usually prepared from equal amounts of glycols and dibasic acids. Polyester fibres are resistant to chemicals and may be washed in alkaline solutions or dry-cleaned and are made into woven and knitted fabrics. *See also* MM pp.54–57.

Polyethylene, POLYMER of ETHYLENE, a partially crystalline lightweight thermoplastic RESIN, with high resistance to chemicals, low moisture absorption and good insulating properties. *See also* MM pp.54–55, 55.

Polygamy, conjugal union where more than one spouse is permitted. More often it is used to denote polygyny (several wives) than POLYANDRY (several husbands).

Polygnotus (*c.*500–*c.*440 BC), Greek painter whose most famous works depicted the *Sack of Troy* and *Odysseus in the Underworld* in the hall of the Cnidians at Delphi. Although none of his paintings have survived, they are known from descriptions left by the Greek traveller Pausanius (*fl.* 143–176 AD).

Polygon, plane geometric figure having three or more sides intersecting at three or more points (vertices). They are named according to the number of sides or vertices: triangle (three-sided), quadrilateral (four-sided), hexagon (six-sided). A regular polygon is equilaterial (has sides equal in length) and equiangular (has equal angles). *See also* pp.48–49.

Polyhedron, in geometry, three dimensional figure (called a solid) whose surface is made up of polygons. These are called the faces of the polyhedron, and the points at which they meet are the vertices. *See also* SU p.48.

Polymer, substance formed by the union of from two to several thousand simple molecules (monomers) to form a large molecular structure. Some polymers, such as cellulose, occur in nature; others form the basis of the PLASTICS and synthetic resin industry. *See also* MM pp.54–55.

Polymerization, chemical combination of several molecules to form straight-chain molecules, cross-linked giant molecules or a combination of both, all called POLYMERS. It is the major industrial process in the manufacture of PLASTICS. In nature, large biochemical compounds (such as PROTEINS and NUCLEIC ACIDS) are also formed by polymerization. *See also* SU pp.142–143, 156–157; NW pp.28–29.

Polymorphism, variation among members of a biological species. Difference in structure or function may be genetically determined or due to differing environments. The most common variations are in colour, physical proportions, behaviour and chemical factors, eg blood group. Polymorphism is common in many plant and animal species and is frequently found to be of adaptive advantage.

Polynesia, one of the three divisions of OCEANIA and the general term for the islands of the central Pacific Ocean. MICRONESIA and MELANESIA lie to the w. The principal islands in Polynesia are the Hawaiian Islands, Phoenix Islands, Tokelau Islands, the Samoa group, Easter Island, Cook Islands, French Polynesia and the two larger islands of New Zealand. The islands are mostly coral or volcanic in origin.

Polynesians, the peoples of POLYNESIA. The inhabitants share common linguistic, racial and cultural roots – the reason for the grouping of the islands under one name. Before European contact in the 18th century the Polynesians, of Pacific stock, were a maritime fishing people whose technology relied on wood, stone, shell and bone, and whose diet consisted of fish, yams, sweet potatoes and coconuts. The major languages are Samoan, Maori, Tahitian and Hawaiian.

Polynomial, sum of terms that are powers of a variable: eg, $5x^4 - 3x^3 + 2x^2 + x + 5$ is a polynomial of the fourth degree (the highest power of four). In general a polynomial has the form $a_0x^n + a_1x^{n-1} + a_2x^{n-2} + ...a_{n-2}x^2 + a_{n-1}x + a_n$, although certain powers of x and the constant term a_n may be missing. The values $a_n, a_{n-1},$ etc., are the coefficients of the polynomial.

Polyp, body type of various species of animals within the phylum Cnidaria. It has a mouth surrounded by extensible tentacles and a lower end which is adapted for attachment to a surface. It is distinct from the free-swimming MEDUSA. It may be solitary, as in the SEA ANEMONE, or more often, an individual of a colonial organism such as CORAL. *See also* NW pp.82–83.

Polyp, in medicine, swollen mass projecting from the wall of a cavity lined with mucous membrane, such as the nose. Although usually benign growths, some can be cancerous.

Polyphemus, in Greek mythology, the CYCLOPS, a giant one-eyed son of POSEIDON. He is described as a shepherd who tended his flocks on the shore of Sicily. Before his eye was put out by ODYSSEUS, he was an unrequited suitor of the NEREID Galatea.

Polyphony, vocal or instrumental part music in which the compositional interest centres on the "horizontal" aspect of each moving part rather than on the "vertical" structure of chords, as in formal harmony. The golden age of polyphonic music was the 16th century, and masters of that time included Giovanni PALESTRINA and William BYRD.

Polyptych, carving or painting that consists of more than three panels hinged together; a similar work with three panels is called a triptych. The polyptych was used in Europe in the Middle Ages, often for altar pieces.

Polysaccharides, POLYMERS (repeating units) of such sugars as GLUCOSE and FRUCTOSE. Many naturally-occurring CARBO-

HYDRATES, including CELLULOSE, GLYCOGEN and STARCH, are polysaccharides.

Polystyrene, synthetic organic POLYMER, composed of long chains of the AROMATIC COMPOUND, styrene. It is a strong THERMOPLASTIC RESIN, acid and alkaline-resistant, non-absorbent and an excellent electrical insulator. *See also* MM pp.54–56.

Polytechnics (from Greek, "many arts"), British term for tertiary education institutions at which many academic and technical subjects are taught. The first polytechnics were founded in the late 19th century and had a technical bias, in contrast to the mainly academic universities. Today, polytechnics have university status.

Polytheism, belief in or worship of many gods and goddesses. Such deities usually have specific attributes or functions. For example, NEPTUNE is the Roman god of the sea; LAKSHMI is the Hindu goddess of good fortune and prosperity. *See also* ANCESTOR WORSHIP; ANIMISM; MONOTHEISM.

Polythene. *See* POLYETHYLENE.

Polyurethane, POLYMER used chiefly in making flexible foams (eg, for upholstery and mattresses), rigid foams (eg, for cores of aeroplane wings), ELASTOMERS and RESINS (for adhesives and coatings). *See also* MM pp.54–55.

Pombal, Sebastião José de Carvalho e Mello, Marquês de (1699–1782), Portuguese statesman. He was Minister for Foreign Affairs (1750–56) and was virtual ruler of the country until the death of King Joseph in 1777. He increased royal power at the expense of the old nobility, the INQUISITION and the JESUITS, whom he expelled in 1759. Influenced by the principles of enlightened despotism, he also reformed the administration, economy, education and the army, and encouraged trade with Brazil. *See also* HC1 p.325.

Pome, fleshy fruit formed from the flower receptacle or base. It is not developed from the carpel and is a false fruit. Familiar examples are the APPLE, PEAR and QUINCE.

Pomegranate, deciduous shrub or small tree native to W Asia. It has shiny, oval leaves and showy, orange-red flowers. The round fruit has a red, leathery rind and numerous seeds coated with edible pulp. Family Punicaceae; species *Punica granatum*. *See also* pp.194–195.

Pomerania (Pomorze), region in N Europe, extending along the Baltic Sea from w of Stralsund, in East Germany, to the River Vistula in Poland. Part of the North German plain, the region is low-lying and fertile. Cereals, sugar-beet and potatoes are the principal crops; stock raising and forestry are also important. Shipbuilding, metallurgy, food processing and paper are the main industries. From 1919 to 1939 the region was divided between Germany (Prussia), Poland and the free city of Danzig. After the POTSDAM CONFERENCE, the part of Prussian Pomerania w of the River Oder was included in Soviet-occupied Mecklenburg, now in East Germany, while the rest of Pomerania came under Polish administration.

Pomeranian, breed of toy dog. Its coat is long and straight, full round the neck, and one of a variety of colours, including white, brown, black and reddish brown. It has been bred down in size from a large sheepdog. Height: 14–18cm (5.5–7in) at the shoulder; weight: 1.5–3kg (3.3–6.6lb).

Pomona, in Roman mythology, the youthful and beautiful goddess of fruit trees. She was accorded a sacred area 19km (12 miles) from Rome. Her male counterpart, VERTUMNUS, overcame her reluctance to marry after courting her in various guises, finally in the shape of a gentle old woman.

Pompadour, Jeanne-Antoinette Poisson le Normant d'Etioles, Marquise de (1721–64), influential mistress and confidante of LOUIS XV after 1745. She rose to prominence in Parisian society and was a great patron of the arts, befriending DIDEROT. Louis had champagne glasses, of the design common today, modelled on the shape of her breasts.

Pomp and Circumstance Marches (1901–30), set of five (planned as six) orchestral military pieces by Edward Elgar. The first four were composed between 1901 and 1907 and their content and title, a quotation from OTHELLO's "Pride, pomp and circumstance of glorious war", reflect the nature of pre-WWI British JINGOISM. A.C. Benson added the words "Land of Hope and Glory" to the trio of *March No. 1*, which was used as Edward VII's Coronation Ode in 1902 and was swiftly adopted as an unofficial national anthem. After WWI Elgar bitterly regretted its nationalist associations.

Pompano, marine food fish found throughout the world in tropical and temperate waters. It has a blunt head and a deep body. Length: to 46cm (18in); weight: to 9kg (2lb). Family Carangidae; species include the common, or Florida, pompano *Trachinotus carolinus* and the African pompano *Alectis crinitis*.

Pompeii, ancient city in SE Italy, at the foot of Mt Vesuvius, 23km (14 miles) SE of Naples. A prosperous port, it came under Roman rule in the 1st century BC but was completely buried by an eruption of Vesuvius in AD 79. Excavations have revealed ruins of the city which show in great detail footpaths, shops, villas and murals. *See also* HC1 pp.94–95, *107*.

Pompey, Gnaeus, also called Pompeius Magnus (106–48 BC), Roman general. He fought for SULLA in 83 BC and campaigned in Sicily, Africa and Spain. He was named consul with CRASSUS in 70, crushed the pirates in the Mediterranean and fought brilliantly against Mithradates VI of Pontus in 66 BC. In Rome in 59 he formed the first triumvirate with Crassus and CAESAR, his fierce rival and father-in-law. He was defeated by Caesar at Pharsalus in 48 in the course of a civil war. After fleeing to Egypt he was stabbed to death by one of his soldiers. *See also* HC1 pp.96–97.

Pompidou, Georges Jean Raymond (1911–74), French Premier (1962–68), and President (1969–74). He was a school teacher until 1944, when he served in the postwar government of Charles DE GAULLE as an adviser. In 1946 he was appointed to the Council of State but he resigned in 1957 to join the Rothschild banking firm and became its director-general in 1959. When De Gaulle became President, he appointed Pompidou Prime Minister. He dealt with the May riots and strikes which shook France in 1968. Upon De Gaulle's resignation he was elected President and, by reversing government policy, made it possible for Britain to enter the EEC.

Ponce, Manuel (1882–1948), Mexican composer who studied in Italy and Germany. His works include a piano concerto (1912) and a violin concerto (1942), in which his famous song *Estrellita* is the theme for the second movement. He also composed *Concierto del Sur* (1940) for guitar and orchestra for Andrés SEGOVIA, the Spanish guitarist.

Ponce, port in s Puerto Rico, 225km (140 miles) SW of the capital San Juan; Puerto Rico's principal port. Founded in 1692, it is one of the oldest cities in the Americas. It has the Catholic University of Puerto Rico (1948) and an 18th-century fort. Sugar is exported, and other industries include cement, oil refining, paper products, rum distilling and tourism. Pop. (1970) 128,233.

Ponce de León, Juan (*c.*1460–1521), Spanish explorer who discovered the Florida peninsula in 1513 while searching for the legendary fountain of youth. He returned in 1521 with a colonizing expedition and was fatally wounded when local Indians attacked his men.

Ponce de León, Luis (1527–91), Spanish monk, author and lyric poet who entered the AUGUSTINIANS and became professor of theology at Salamanca, Spain, in 1561. He was imprisoned by the INQUISITION from 1572 to 1576 for, among other things, his translation into Spanish of the *Song of Solomon*. His prose works include *The Perfect Housewife* (1583) and *Christ's Names* (1583–85).

Ponchielli, Amilcare (1834–86), Italian composer, chiefly of opera, of which he wrote nine. The first, *I Promessi Spose*, was produced at Cremona in 1856. The most famous, and the only one still performed, is *La Gioconda*, first produced at Milan in

1876. He also wrote two ballets, *Le Due Gemelle* and *Clarina* (both 1873).

Pondicherry, seaport city in SE India; the capital of Pondicherry Union Territory and former French India. Its products include textiles, sugar and steel. Pop. (1971) 90,637.

Pondoland, 19th-century name for E part of CAPE PROVINCE between the Kei River and the Natal border, now part of the TRANSKEI. The Xhosa, Tembu and Pondo tribes who lived there were disrupted during the 19th century by the effects of the tribal regrouping, the *mfecane* (The Time of Troubles), and then by the encroachment of white settlers. In 1894 the British government claimed all Pondoland. *See also* HC2 pp.129, 143.

Pondweed, any of numerous species of a large family of aquatic perennial seed plants, found mostly in temperate zone regions in freshwater lakes and ponds, but also in brackish and salt water. Most pondweeds have spike-like flowers that stick out of the water and submerged or floating leaves arranged alternately along the stem. Family Potamogetonaceae. *See also* NW p.231.

Pons, Lily (1904–76), French-born US coloratura soprano who studied at the Paris Conservatoire and made her début in 1928. She first sang at the METROPOLITAN OPERA in New York in 1931 and was principal soprano there until 1959. She appeared in three films, and in 1938 married André Kastelanetz.

Pons, upper segment of the human brain-stem, which contains the nerve fibres connecting the two halves of the CEREBELLUM. The brain-stem is the lower part of the BRAIN and structurally continuous with the SPINAL CORD. Below the pons is the medulla oblongata, which transmits ascending and descending nerve fibres between the spinal cord and the brain. *See also* MS pp.42–43.

Ponta Delgada, seaport city in the AZORES (an island group belonging to Portugal), situated on the SW shore of Sao Miguel island. It contains the best harbour and is the leading commercial centre of the islands. It is also a popular winter resort. Products include sugar, alcoholic drinks, tea and soap. Ponta Delgada exports dairy products and citrus fruits. Pop. (1970) 69,930.

Pontano, Giovanni (1426–1503), also called Jovianus Pontanus, Italian Renaissance poet, historian and statesman. He was head of the Neapolitan humanist academy (1471). From 1447 to 1495 he was military secretary and adviser to the kings of Naples and as Chancellor (1486–95), virtually ruled the kingdom.

Pont Aven School, name associated with a group of painters who were influenced by SYNTHETISM, which GAUGUIN and Émile BERNARD evolved (between 1888 and 1894) in the village of Pont Aven, Brittany. Those influenced included Laval, SÉRUSIER and Seguin. The life, landscape and religion of Brittany inspired Gauguin and his followers. Gauguin's *Jacob Wrestling with the Angel* (1888) was the first masterpiece of Synthetism, which later influenced the NABIS.

Pont du Gard, Roman-built aqueduct across the River Gard at Nîmes, in s France. Its construction is generally attributed to Agrippa in the Augustan period. Built to supply water to Nîmes, it has three tiers, the lowest of which now serves as a bridge. It owes its elegance to the variations in the size of the arches and piers.

Ponti, Giovanni (1891–   ), Italian theatrical and film designer, architect and draughtsman. He was the founder and editor (1928–49) of the magazine *Domus* which influenced styles of interior decoration in Italy during that time. He was also influential in the development of modern architecture. Among his famous buildings is the Pirelli Building, Milan (1955–59), an hexagonal skyscraper with sloping sides which he designed in collaboration with NERVI. *See also* HC2 p.207.

Pontiac (*c.*1720–69), North American Indian chief who united many tribes south of the Great Lakes against the British. In the early years of the FRENCH AND INDIAN WAR (1754–63) Pontiac kept his confede-

Pomegranates are as large as apples and are bright orange-yellow when ripe.

Marquise de Pompadour had an enormous influence over the policies of Louis XV.

Pompeii: view of the last excavation of the ancient ruins carried out in 1876.

Georges Pompidou, Gaullist President of France, lectured before entering politics.

Pontianak

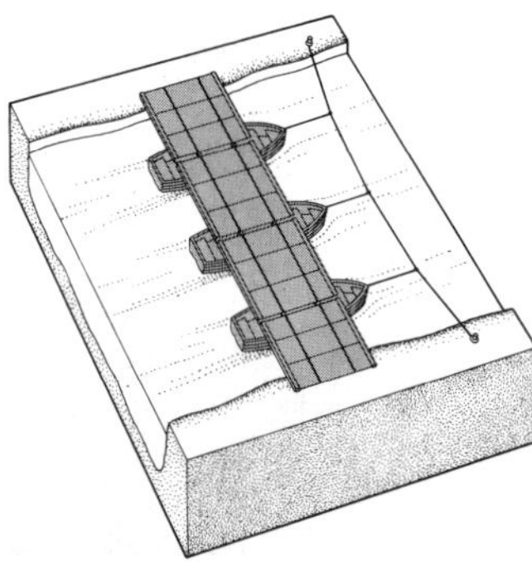

Pontoon bridges are vital in wartime to maintain strategic river crossings.

Ponies are popular with children who are learning to ride.

Pony Express rider on his way from the Missouri River to San Francisco.

Alexander Pope's deformity may have influenced his satirical works.

racy of Ottowa, Potawatomi and Ojibwa neutral, but in 1762 he organized many tribes into a union against the British. From 1763 to 1764 this union, known by the British as Pontiac's Conspiracy, fought a bitter struggle until Pontiac concluded peace.

Pontianak, capital of w Kalimantan Barat province in w Borneo, Indonesia, on the Kapuas-Ketyil River. It is w Borneo's principal city and major port. Exports include rubber, sugar and palm oil, and shipbuilding and food processing are the most important industries. Pop. 150,220.

Pontianus, Saint (*d.c.*236), pope who reigned from 230 to 235. His papacy coincided with the inauguration of the persecution of the Christians by the Emperor Maximinus. Exiled to the mines of Sardinia, Pontianus abdicated to St Antherus and died soon after. Feast: 19 Nov.

Pontifex (Latin, pontiff), Roman priest, a member of the highest college of priests in Rome whose leader, pontifex maximus (supreme pontiff), was the high priest of the Roman religion. Since the 5th century the title pontifex maximus has been in use to designate the POPE.

Pontius Pilate. *See* PILATE, PONTIUS.

Pontoon, gambling card game the object of which is to collect a value-total of as near 21 as possible, but not over. (The name itself is a corruption of the French *vingt-et-un,* 21.) Court cards count as 10, aces as 1 or 11 (whichever is required) and all other cards have their face values. The player chosen as banker deals one card to each player, including himself, who places a bet before receiving a second card. In turn, each player may then, as many times as required, "twist" (receive another card face up), "buy a card" (receive another card face down but add to his stake), or "stick" (decline any further cards – sometimes also called "sitting"). If a player's hand exceeds 21, he is out for the hand and his stake is given to the banker at once. Finally, the banker himself turns to his two cards; having similarly reached as near 21 as he can, but not over, he collects the stakes of all players with hands equal to or less than his own, but pays out to the people with hands better than his own an amount equal to their stakes. If the banker's hand exceeds 21, all players still in the hand receive an amount equivalent to their stakes. "Pontoon" itself, an ace with a court card comprising the first two cards received, is unbeatable unless the banker also draws it; at each pontoon the cards are shuffled, and the pontoon-holder becomes the banker. There are various other scoring combinations.

Pontoon bridge, floating bridge, used mainly for military purposes. Persian engineers built a bridge of boats for King Xerxes' army in 480 BC across the Hellespont. Modern armies use rubber pontoons pressurized by air compressors. *See also* MM p.190.

Pontoppidan, Henrik (1857–1943), Danish novelist. He studied contemporary Denmark in his novel cycles *The Promised Land* (1891–95), *Lucky Peter* (1898–1904) and *The Kingdom of the Dead* (1912–16). His early tales include the volume of short stories *The Apothecary's Daughters* (1886). He shared with Karl GJELLERUP the 1917 Nobel Prize in literature.

Pontormo, Jacopo Carucci (1494–1557), Italian painter who was an early exponent of MANNERISM. In his youth he worked under Andrea del SARTO. In 1518 he painted *The Madonna,* one of the first mannerist pictures. His subjects and themes were usually religious, and his works include *The Visitation* (1516) and *Deposition* (*c.*1527). A major commission was the decoration of the choir of S. Lorenzo, Florence (1545–57).

Pontus, in Greek mythology, the personification of the primeval sea. GAEA, the earth, created him when she made the mountains and the sky. Pontus has no identity other than his name – his character was never developed in the mythology. *See also* OCEANUS.

Pony, any of several breeds of small horses, usually solid and stocky. It is commonly used as a children's saddle horse, for show and for draught. Types include the hardy Shetland pony (the smallest natural breed) of the Shetland Islands, where the climate is harsh and food scarce; the Dartmoor and Exmoor ponies of Cornwall, Somerset and Devon, where they have lived wild for many centuries; the grey Highland pony, a popular saddle animal; the Welsh pony and the Welsh Cob, which is known for its high-stepping gait. Height: from 115 to 145cm (45–57in) at the shoulder.

Pony Express, in US history, relay mail service between St Joseph, Mo. and Sacramento, Calif., that operated only briefly (1860–61). Riders changed horses at least every 16km (10 miles) and took some ten days to carry the 9kg (20lb) mail bags the 2,827km (1,800 miles) between the two cities, a saving of about two weeks over the time taken by a stagecoach. The express was gradually discontinued after the completion of the transcontinental telegraph system in 1861.

Pony trekking, leisure activity in which sure-footed ponies carry their riders over moorlands, mountain sides, forests and glens. A relatively inexpensive way of enjoying riding and the countryside, it is popular as a form of holiday-making.

Poodle, breed of dog believed to have originated in Germany. Bred originally to retrieve from water, its intelligence has made it a popular pet. It has a rounded skull and long straight body, small feet and high-set, docked tail curled up. The thick coat, woolly underneath with a wiry top coat, is a solid colour and is commonly clipped into an ornate style. Three main sizes are bred, standard, miniature and toy. Height: (standard) over 38cm (15in) at shoulder.

Pool, billiards game of American origin but popular today in many countries. It is played with eight single-colour balls and seven striped balls numbered 1 to 15, plus a white cue ball, on a rectangular table with four corner pockets and two side pockets. Rules vary locally. Players must nominate the ball and pocket they are attempting before cueing the ball.

Poole, county district in E DORSET, SW England; created in 1974 under the Local Government Act (1972). Area: 64sq km (24sq miles). Pop. (1974 est.) 112,800.

Poona, city in Maharashtra State, w India, 129km (80 miles) ESE of Bombay. It served as the capital of the MAHRATTA confederacy in the 18th century until 1818, when it came under British control. It was then used as a British military station and has several palaces and temples dating from this period. Today it is a flourishing commercial centre; products include metal goods, chemicals, textiles and paper. Pop. (1971) 853,226.

Poor laws, English legislation designed to alleviate poverty and prevent begging and vagrancy. Introduced in the mid-16th century and consolidated in the Poor Law of 1601, they required individual parishes to provide for the local poor. Because there was everywhere local reluctance to support the poor from other areas, settlement laws were introduced in 1662 to limit migration. From *c.*1700 workhouses were established where the poor were expected to support themselves by work. Poor-law amendments of 1834 sought to provide uniform assistance by placing it under national supervision, but since poverty was assumed to be the consequence of unwillingness to work, relief was maintained at a level lower than that of the poorest labourer. In 1929 the Local Government Act abolished the poor law boards, making relief to the poor the responsibility of local bodies. The Poor Law Act of 1930 placed poor law under the Minister of Health. The poor law was gradually superseded by the introduction of old age pensions (1908) and unemployment and health insurance (1911), which was followed in the 1940s by social security. Workhouses were finally abolished in 1948. *See also* HC1 p.275, *275.*

Poor Richard's Almanack, almanac first issued in 1732, and annually until 1757, by Benjamin FRANKLIN under the pseudonym Richard Saunders. Containing much statistical data, the almanac was also a vehicle for Franklin's famous adages. Its title was adapted from that of the English almanac *Poor Robin* (*c.*1661).

Pop art, movement in art which used popular images and materials in its constructions. The first pieces were produced in Britain in the 1950s. Robert RAUSCHENBERG, an early exponent, used given material – surplus tins of paint, X-ray photographs and newspaper images – in works. Andy WARHOL's painting of a can of Campbell's soup proved to be the image by which pop art was best known. Another series of images were provided by the work of Roy LICHTENSTEIN, who took pictures from popular comics and transferred them to large scale works, such as *Blonde Waiting* (1964). *See also* HC2 p.279.

Pope. *See* PAPACY.

Pope, Alexander (1688–1744), British poet. Self-taught, he was a master of the classical form of poetry. He showed early metrical skills in *Pastorals* (1705), wrote lyric and elegiac poetry and published fine translations of HOMER (1720 and 1726) and *Imitations of Horace* (1733). He is well known also for his own satires, which include *The Rape of the Lock* (1714), *The Dunciad* (1728) and *An Epistle to Dr Arbuthnot* (1735). *See also* HC2 p.22.

Popham, Sir Home Riggs (1762–1820), British admiral. He was naval commander at the capture of the Cape of Good Hope in 1806 and in the same year took Buenos Aires with William Carr BERESFORD, following which he was recalled to England. He served in Jamaica from 1817 to 1820.

Popish Plot, intrigue invented by Titus OATES and Israel Tonge in 1678. The two conspirators claimed that there was a plan to assassinate CHARLES II and replace him with the Catholic James, duke of York. Although there was no evidence to support the claim, many people were killed in a wave of anti-Catholic murders. Oates was later imprisoned. *See also* HC1 pp.292, *293.*

Poplar, any of a number of deciduous, softwood trees native to cool to temperate regions. The oval leaves grow on stalks, and flowers take the form of catkins. Some species are called cottonwoods because of the cotton-like fluff on their seeds. Height: to 60m (197ft). Family Salicaceae; genus *Populus.*

Poplin, fabric made from silk and wool. It has a corded surface and is used for making shirts and dresses. Some kinds are made by substituting other materials eg, cotton or rayon. *See also* MM pp.56–57.

Pop music, songs and tunes which are intended to reach a large audience. Short for popular music, the term was coined in the early 1950s. Pop resulted from the development of radio and gramophones and is a manifestation of MASS CULTURE. *See also* HC2 pp.272–273.

Popocatépetl, snowcapped active volcano in central Mexico, on the Puebla-Mexico state border. The crater contains sulphur deposits but these remain unexploited. Height: 5,452m (17,887ft).

Popova, Liubov (1889–1924), Russian painter. She studied in Paris in 1912, and returning to Russia in 1915 worked in a CUBIST style before producing abstract suprematist works from 1916. After 1917 she abandoned painting and made several constructivist sculptures.

Popper, Sir Karl Raimund (1902–), British philosopher of natural and social sciences, *b.* Austria. Although not a LOGICAL POSITIVIST he was in close contact with the VIENNA CIRCLE and had considerable influence on Rudolf CARNAP. His writings include *The Open Society and its Enemies* (1944) and *The Poverty of Historicism* (5th ed 1966). *See also* MS p.229.

Poppy, common name for plants of the poppy family, especially of the genus *Papaver.* This is made up of about 100 species of annual and perennial erect herbs, most of which are native to central and s Europe but also found in the Far East, North America and Iceland. The plants contain a milky juice and have showy blooms. The seeds are used in cooking. The opium poppy of Greece, Turkey and the Far East is grown for medical purposes and for the manufacture of illicit drugs. Family Papaveraceae. *See also* OPIUM; NW p.*59*; PE pp.*210*, 211.

Popular front, combination of left wing political parties formed to meet the threat of fascism during the 1930s. The basis of most popular fronts was an alliance of Communist parties with less extreme social democrat groups, a union made possible by the Soviet government's decision of 1934 that fascism had to be halted even at the cost of allying with liberal political groups. Popular fronts were formed in many countries but came to power only in France (where Leon BLUM led a ministry in 1936–37) and in Spain, where the election of a popular front government in Feb. 1936 led to the fascist insurrection of June 1936, which began the SPANISH CIVIL WAR. In the 1960s and 1970s various revolutionary groups in Africa and Asia had the words "popular front" as part of their names.

Population, estimated world total (1978), 4,100 million. In 1977 the rate of population growth was about 1.7%, a marked fall from the rate of nearly 2% in 1970. That change makes predictions of future population totals hazardous. Predictions of population in the year 2000 range from 4,500 million to 7,600 million.

Population I and II, in astronomy, classification of regions within a GALAXY according to their stellar nature, devised by Walter BAADE. Population I areas contain much interstellar material; the brightest stars are hot and white. Such regions are found in the spiral arms of galaxies. In Population II areas much of the interstellar material has been used up in star formation; the brightest stars are red giants. Such regions are relatively old and are in the nuclei of galaxies. *See also* SU pp.*230*, 231, 243–244, *243*, 247–248.

Poquelin, Jean-Baptiste. *See* MOLIÈRE.

Porcelain, white, vitreous, non-porous, translucent ceramic. It was first developed by the Chinese (7th or 8th century) and made of kaolin (white china clay) and petuntse (powdered felspar). This true, or hard paste, ceramic ware is fired at 1,400°C. MEISSEN ware (early 18th century) was the first successful European attempt to imitate this hard paste porcelain, referred to as "china". Other kinds are English soft paste, which is composed of clay and powdered glass, fired at a comparatively low temperature, lead glazed and refired; and English bone china, a soft paste modified by the addition of calcined bone. *See also* MM pp.44–45.

Porcupine, short-legged, mostly nocturnal herbivorous rodent with erectile, defensive quills in its back. Old World porcupines of the family Hystricidae are heavy, have brown to black fur with white-banded quills and are terrestrial. New World porcupines of the family Erethizontidae are smaller with yellow to white quills and are arboreal. The largest European and African rodent, the African crested porcupine (*Hystrix cristata*), attains a length of about 80cm (31in).

Porcupine fish, any of several species of tropical marine fish found throughout the world. White and brown, it has sharp spines and can gulp water to inflate its body. Length: to 91.4cm (3ft). Family Diodontidae; genus *Diodon*.

Pordenone (1483–1539), Italian painter, named Giovanni Antonio de Sacchis. His frescoes, such as those for Cremona Cathedral (1521), show him to have had an excellent sense of colour.

Porgy and Bess (1935), three-act Negro folk-opera by George GERSHWIN. The LIBRETTO by Ira Gershwin and Du Bose Heyward is based on Dorothy and Du Bose Heyward's play *Porgy*. The première of this tragedy was held in Boston, Massachusetts. The work is set in a Carolina coastal town.

Pori (Björneborg), city in SW Finland near the mouth of the Kokemäki River. Founded in 1365, Pori was initially overshadowed by the ports in the HANSEATIC LEAGUE, but by 1840 the city had the largest commercial fleet in the country. Industries: textiles, paper, machinery. Pop. (1975) 80,129.

Porifera, or pore-bearers, also called Parazoa, animal phylum made up of the SPONGES. The Porifera are many-celled, stationary animals representing an evolutionary dead end. *See also* NW p.72.

Pornography, visual or aural material that is sexually titillating. The boundary between what is and what is not pornographic is impossible to define, because the response of individuals varies according to age and experience. Consequently, although pornography is subject to legal controls in most countries, interpretation of the law is itself entirely dependent on the subjective judgement of those presiding. In Britain, material is pornographic (obscene) if it is deemed to deprave or corrupt those who see or hear it.

Porosity, in geology, degree to which rock is capable of holding groundwater. This depends on the numbers of AQUIFERS, cracks, fissures and holes present. Porosity is measured and expressed as a percentage of the rock's total volume. *See also* HYDROLOGY; PERMEABILITY.

Porphyria, disease, often genetic, of man and some animals (most commonly cattle) characterized by abnormalities of PORPHYRIN metabolism and subsequent excretion of porphyrins in the urine as well as extreme sensitivity to light. There is no specific treatment, but therapy can alleviate the symptoms.

Porphyrin, class of red-pigmented organic compounds consisting of four pyrrole residues (a ring compound, formula C_4H_4NH) bonded together. The porphyrins form part of the active nucleus of HAEMOGLOBIN, a and b CHLOROPHYLLS and the enzymes catalase and peroxidase. *See also* SU p.*155*.

Porphyry (*c.*234–305), Greek Neo-Platonic philosopher who studied under PLOTINUS. He lectured on the philosophy of Plotinus and was the teacher of Iamblichus. He wrote biographies of PYTHAGORAS and Plotinus but his chief work was *Isagoge*, an introduction to the logic of ARISTOTLE.

Porpoise, small, toothed whale with a blunt beakless snout, that lives in most oceans. The best-known is the common porpoise of the Northern Hemisphere. Its body is black above and white below. Length: to 1.5m (5ft). Family Delphinidae; species *Phocaena phocaena*. *See also* NW pp.*155*, 166.

Porritt, Sir Arthur (1900–), British surgeon, b. New Zealand; consultant surgeon to the British royal family between 1936 and 1967. He was Governor-General of New Zealand (1967–72) and made a life peer in 1973.

Porsche, Ferdinand (1875–1951), German car manufacturer who designed the Volkswagen "Beetle" (1934–38) for Hitler as the "people's car". He also designed the sports car which bears his name.

Port, fortified wine produced in N Portugal in the Douro Valley, E of Oporto after which the wine is named. It may be white or red, and contains 17–20% alcohol; a vintage port is aged for 15–20 years. Port has been popular in Britain since the 18th century.

Portal of Hungerford, Charles Frederick Algernon, 1st Viscount (1893–1971), British Air Chief Marshal. He was Director of Organization in the Air Ministry (1937–38), and Chief of Air Staff (1940–45).

Portal vein, large vein that carries blood from the stomach and surrounding organs to the liver.

Port Arthur, former name of Lüshun, city in SW LIAONING province, NE China, on the Liaotung Peninsula. At the entrance to the Gulf of Po Hai, Port Arthur is strategically placed and is a naval base and a terminus of the South Manchurian railway. The city was held by the Russians from 1898 to 1905, and by the Japanese until 1945, when it came under joint Sino-Soviet control. Soviet forces withdrew in 1955.

Port-au-Prince, capital city of Haiti, on the SE shore of the Gulf of Gonave. It was founded in 1749 and became the capital in 1770. The city manufactures tobacco, textiles and cement, and exports coffee and sugar. Pop. (1971) 419,947.

Port Darwin. *See* DARWIN.

Port Elizabeth, port in Cape Province in S Republic of South Africa, on Algoa Bay in the Indian Ocean. Founded in 1799 by the British, the city grew after the completion of the railway line to Kimberley in 1873. The University of Port Elizabeth was founded in 1964. Exports include diamonds, fruit and wool. Industries: motor vehicles, chemicals, glass, electrical engineering. Pop. (1970) 386,577.

Porteous Riots (1736), incidents that occurred after the Edinburgh city guard fired into a mob gathered at the execution of the smuggler Andrew Wilson, killing several people. John Porteous, captain of the guard, was tried and sentenced to death, but was reprieved. An indignant mob then took him from prison and hanged him. No reprisals were taken against those involved in the hanging.

Porter, Cole (1892–1964), US composer and lyricist, who studied music at Harvard. Most of his many MUSICALs for stage and film were vastly successful. They include *Gay Divorcee* (1932), *Anything Goes* (1934), *Kiss me Kate* (1948) and *High Society* (1956). Among his most popular songs are *Night and Day*, *Let's Do It*, *Begin the Beguine* and *In the Still of the Night*.

Porter, Edwin (1870–1941), US inventor who collaborated with Thomas EDISON in the development of the motion picture camera. In 1903 he made for Edison *The Life of an American Fireman*, the first US documentary film using close-ups and cross-cutting; he also filmed *The Great Train Robbery* (1903) and *The Eternal City* (1915).

Porter, Sir George (1920–), British chemist and director of the Royal Institution of Great Britain since 1966. He shared the 1967 Nobel Prize in chemistry with R. G. W. NORRISH and M. EIGEN for their work on very fast chemical reactions. Porter and Norrish disturbed the equilibrium between atoms and molecules in chlorine gas by irradiating it with very short pulses of energy, and observed it as it returned to equilibrium within as little as 10^{-9} sec.

Porter, Hal (1911–), Australian short-story writer and poet. His first stories were published in 1943, his first poems, *The Hexagon*, in 1956. His autobiographical *The Watcher on the Cast-Iron Balcony* (1956) is much admired.

Porter, Hugh (1940–), British track cyclist whose strength was the individual pursuit. He won a Commonwealth Games gold medal in 1966, and in professional cycling was the first man to win four world pursuit titles (1968, 1970, 1972, 1973).

Porter, Katherine Anne (1890–), US writer noted for the technical accomplishment of her short stories and novellas. She won acclaim with *Flowering Judas* (1930). Subsequent works include *Pale Horse, Pale Rider* (1939) and *The Leaning Tower* (1944). Her long novel *Ship of Fools* (1962) was highly acclaimed.

Porter, Rodney Robert Gerald (1917–), British biochemist. He shared with M. EDELMAN the 1972 Nobel Prize in physiology and medicine for their work on the chemical structure of the GAMMA GLOBULIN molecule.

Port Harcourt, city in S Nigeria, on the River Bonny in the Niger delta system approx. 435km (270 miles) ESE of Lagos. It is a shipping centre and terminus for the railway from the interior. Its exports include coal, tin, palm products and groundnuts. Pop. (1975) 242,000.

Port Hedland, port on the NW coast of Australia, in Western Australia about 1,320km (820 miles) NNE of Perth. It is the largest port in Australia defined in terms of the tonnage of iron handled. Pop. (1971) 7,200.

Port Jackson, in New South Wales, SE Australia, inlet of the Pacific Ocean comprising Sydney Harbour, North Harbour and Middle Harbour. The port is an excellent deep-water anchorage and its ship-repair facilities include one of the world's largest dry docks. Garden Island on the S shore is a base for the Royal Australian Navy. Area: 57sq km (22sq miles).

Port Jackson shark, species of SHARK living in the coastal waters of Australia; it is a

Popular front; Leon Blum led a French socialist ministry in the 1930s.

Porcupine quills consist of long strands of hair fused together into spines.

Ferdinand Porsche left the highly successful "Beetle" as a memorial to his genius.

Cole Porter's success lay in combining witty lyrics and inspired music.

Port Lairge

Port Said has for long been a busy seaport and a major Egyptian naval base.

Portsmouth, known to sailors as "Pompey", has had a naval dockyard since 1496.

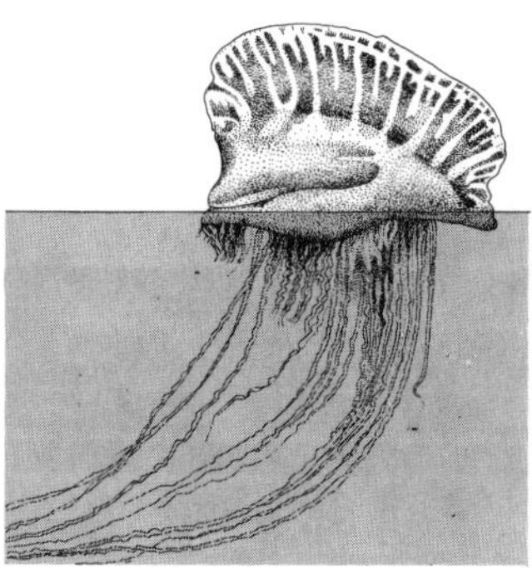

Portugese man-of-war: their stings are dangerous even when dry on a beach.

Emily Post told Americans that they should not eat off their knives.

member of the family Heterodontidae, the most primitive of the shark groups. It has two dorsal fins and five pairs of gill slits. Length: to 1.5m (5ft). Species *Heterodontus philippi*.

Port Lairge. *See* WATERFORD.

Portland, William Bentinck (1649–1709), Anglo-Dutch diplomat. As personal adviser to WILLIAM III he arranged his marriage to MARY II and settled in England in 1688 where he was made Earl of Portland. He fought at the Battle of the BOYNE in 1690 and negotiated the Treaty of RYSWICK in 1697.

Portland, William Henry Cavendish Bentinck (1738–1809), British politician who was the nominal head of the coalition government of Lord NORTH and Charles FOX in 1783. His second ministry from 1807 to 1809 was dominated by Robert CASTLEREAGH and George CANNING.

Portland, port of entry in NW Oregon, USA, on the Willamette River; largest city in Oregon. Settled in 1845, it developed as a major port for exporting timber and grain after 1850. It was a supply station for the California goldfields and the Alaska gold rush (1897–1900). Educational establishments include Clark College (1867), the University of Portland (1901) and Reed College (1904). Industries: shipbuilding, timber, wood products, textiles, metals, machinery. Pop. (1973) 378,134.

Portland, Isle of, peninsula in S Dorset, SW England. It is noted for Portland stone, the quarrying of which is a major industry.

Portland Bill, tip of the Isle of Portland, a limestone peninsula in Dorsetshire, connected to the mainland by a stretch of shingle about 200m (660ft) wide.

Portland cement. *See* CEMENT.

Portland vase, funerary urn discovered in an ancient tomb near Rome in the 17th century. It was made of blue-violet glass overlaid with cameo relief, and probably dates from the 1st century AD. It was acquired by the Duke of Portland in the late 18th century and by the British Museum in 1845.

Port Louis, capital and seaport city in NW Mauritius in the Indian Ocean. It was founded by the French in 1735. The main export is sugar; other industries include the manufacture of cigarettes and food products. Pop. (1969 est.) 138,200.

Port Mahon. *See* MAHON.

Portman, Eric (1903–69), British stage and film actor. He began his acting career in 1924, joining the Old Vic in 1927 and appearing in *Major Barbara* (1929). His films include *The Murder in the Red Barn* (1935), *The Deep Blue Sea* (1955) and *Deadfall* (1968).

Port Moresby, capital of PAPUA NEW GUINEA, on the SE coast of New Guinea. It was discovered in 1873 and occupied by the British ten years later. Its sheltered harbour was the site of an important Allied base in WWII. Port Moresby's exports include gold, copper and rubber. Pop. (1970 est.) 56,200.

Porto. *See* OPORTO.

Pôrto Alegre, city in SE Brazil on the River Guaiba. It was founded by colonists from the Azores in 1742 and later settled by German and Italian immigrants. It is the country's most important river port. Industries include meat-packing, shipbuilding and ore smelting. Pop. (1970) 869,795.

Porto Bello, Battle of, British naval victory over the Spanish at Porto Bello, in Darien, on 22 November 1739. It was a battle in the War of JENKINS' EAR. The British fleet, commanded by Admiral Vernon, captured the town and destroyed its fortifications.

Port of Spain, capital of TRINIDAD AND TOBAGO, on the Gulf of Paria. The city was formerly the interim capital of the Federation of the West Indies (1958–62). It is a major Caribbean shipping centre, and exports agricultural produce and asphalt. The city has attractive gardens and is a popular tourist resort. Pop. (1970) 11,032.

Porto-Novo, capital of BENIN, W Africa. Settled by 16th-century Portuguese traders it later became a shipping point for slaves to America. Today it is a port and market for the surrounding agricultural region; its exports include palm oil, cotton and kapok. Pop. (1975 est.) 104,000.

Portoviejo, town in Ecuador, 145km (90 miles) NW of Guayaquil. Founded in 1534, it was moved to its present site in the 17th century. Its products include baskets, hats and hammocks. Pop. (1970 est.) 49,700.

Portrait, in visual art, the likeness of a person or persons; examples exist in most media. The likeness may be primarily physical, as in portraits by GAINSBOROUGH, or it may be a psychological study, such as the portraits and self-portraits of REMBRANDT and VAN GOGH, or the photographs of Diane Arbus. The subject may be realistically or ideally represented. Examples of portraits include coins, sculpture and miniatures.

Portrait of a Lady, The (1881) novel by Henry JAMES. Set in England and Italy in the late 19th century, it is a brilliant study of the marriage of an unscrupulous English dilettante to a rich but innocent American woman, and is generally regarded as one of James' finest works.

Portrait of the Artist as a Young Dog (1940), series of semi-autobiographical prose sketches by Dylan THOMAS. Concerned with growing up in rural Wales, the lively style, detailed sense of place and vivid characters, such as Grandpa, anticipated his better-known *Under Milk Wood*.

Portrait of the Artist As A Young Man, A (1916), autobiographical novel by James JOYCE. It was Joyce's first experiment in the technique of STREAM OF CONSCIOUSNESS, and is a tirade against his bigoted and poverty-stricken family background, the Roman Catholic religion and Ireland itself. *See also* HC2 pp.*283, 290*.

Port-Royal-des-Champs, former Cistercian covent 11km (7 miles) S of Versailles, founded in 1204. The nuns moved to Paris in 1626 and the abbey at Port-Royal-des-Champs became a centre of JANSENIST teaching and a retreat for men. The buildings were destroyed in 1710 by order of Louis XIV.

Port Said, city in NE Egypt, at the entrance to the Suez Canal, on a narrow peninsula between Lake Manzala and the Mediterranean Sea. Although its harbour was closed to shipping (1967–74), it remained a fuelling station for ships using the canal and also the location of canal maintenance workshops. Fishing is an important source of income. Pop. (1971 est.) 320,000.

Portsmouth, Louise de Kéroualle, Duchess of (1649–1734), French mistress of CHARLES II of England from 1670 until his death in 1685, and mother of the Duke of Richmond.

Portsmouth, port and county district in Hampshire, S England, on the Spithead; the city has been Britain's principal naval base (Portsea) since the reign of HENRY VIII. Nelson's flagship HMS *Victory*, stands there in dry dock. Landmarks include Southsea Castle and Charles DICKENS' birthplace, which is now a museum. Industries include aircraft engineering. Area (county district) 37sq km (14sq miles). Pop. (county district, 1974 est.) 200,000; (city, 1971) 196,973.

Portsmouth, city in SE Virginia, USA, on the Elizabeth River opposite Norfolk, to which it is connected by bridge and tunnel; part of one of the largest naval dockyards and shipyards in the USA. Industries: shipbuilding, chemicals, fertilizers, railway equipment, plastics. Pop. (1973) 109,295.

Portsmouth, Treaty of (1905), agreement ending the RUSSO-JAPANESE WAR, signed at Portsmouth, England. Russia recognized the independence of Korea and Japan's political and economic paramountcy there, returned Manchuria to Chinese sovereignty and ceded to Japan the railway lines which it had built in Manchuria.

Port Sudan, city in NE Sudan, on the Red Sea. It was founded as a harbour during the construction of a railway linking the River Nile with the Red Sea. It is now the chief port of entry for the Sudan; exports include hides, resin and cotton. Pop. (1971) 101,091.

Port Sunlight, town in the outskirts of Birkenhead, Merseyside, N England. It was founded by Lord LEVERHULME in 1888 around the Lever soap factory to provide homes for the factory workers. It includes an art gallery, opened in 1922 in memory of Lady Lever.

Port Talbot, town in West Glamorgan, S Wales, at the mouth of the River Avon on Swansea Bay. Coal is exported and large quantities of iron ore imported. The manufacture of steel is the principal industry; iron and tinplate are also made. Pop. (1971) 50,658.

Portugal, republic in SW Europe on the Iberian Peninsula. Once the centre of a great colonial empire, it is now one of Europe's poorest countries. The economy is largely agricultural, the principal being wheat, maize, oats, rye and barley. Mineral deposits include copper and tungsten. The chief industry is food processing. Lisbon is the capital. Area: (including Madeira and the Azores) 91,640sq km (35,382sq miles). Pop. (1975 est.) 8,762,000. *See* MW p.145.

Portuguese Guinea. *See* GUINEAU-BISSAU.

Portuguese man-of-war, colonial COELENTERATE animal found in marine subtropical and tropical waters, recognized by its bright blue gas float and long, trailing tentacles with strongly poisonous stinging cells. It is not a true jellyfish. The tentacles are actually a cluster of several kinds of modified MEDUSAE and POLYPS. Length: to 18m (60ft). Class Hydrozoa; phylum Coelenterata; genus *Physalia. See also* NW pp.*85, 240*.

Portuguese Timor, former Portuguese colony in the E part of Timor island in the Lesser Sunda Islands of the Malay Archipelago; in 1975 it became part of INDONESIA. The most important product is coffee, accounting for 80% of exports. Others include hides, tea, rubber and copra. Dili is the chief town. Area: 14,925sq km (5,763sq miles). Pop. (1970) 610,541. *See also* MW p.99.

Poseidon, in Greek mythology, god of all waters, earth-shaker and brother of ZEUS and HADES, identified with the Roman god NEPTUNE. Poseidon controlled the monsters of the deep, reputedly created the horse (he was the father of PEGASUS) and sired Orion and POLYPHEMUS. Zeus called upon him to bring the Flood when he was displeased with men.

Positivism, in philosophy, doctrine that asserts that "positive" knowledge can be attained through direct experience. The application of the methods of inductive logic, based on empirical observation, would permit the growth of an ever-increasing corpus of known fact. Positivism was first proposed by Auguste COMTE, and was a dominant system of 19th-century philosophy. LOGICAL POSITIVISM developed in the 20th century, attempting to base "positive knowledge" on the strict application of logic.

Positron, particle (symbol e$^+$) that is identical to the ELECTRON, except it is positively charged, ie it is the antiparticle of the electron. It was postulated as the first antiparticle by P. DIRAC in 1928 and observed in 1932 in COSMIC RAYS; it is also emitted from certain radioactive nuclei. It can exist only for a brief period before ANNIHILATION by combination with an electron. Electron-positron pairs can be produced when GAMMA RAYS interact with matter. *See also* ELEMENTARY PARTICLES; SU pp.*66–67*.

Positronium, unstable, bound state of an ELECTRON and a POSITRON (the electron's antiparticle) held together by electromagnetic attraction, first observed in 1951. It has a half-life of either 1.39×10^{-7} sec or 1.25×10^{-10} sec depending on whether the spin axes of the two particles are parallel or antiparallel.

Possessed, The (1873), novel by Feodor Mihailovich DOSTOEVSKY. It attacks the NIHILISM of the younger generation and is a violent denunciation of the leftists and revolutionaries he previously admired.

Possum, popular name for any of the PHALANGERS of Australasia. Family Phalarperidae. The term is also a misnomer for the OPOSSUM of the USA. *See also* NW p.*71, 159, 210*.

Post, Emily (c. 1872–1960), US authority on etiquette, who began her career as a novelist. Her best-known book is *Etiquette: The Blue Book of Social Usage*

(1922), which soon became the basic practical guide to proper social behaviour. It has been published in several revised editions.

Poster, printed art form in multiple copies for public notice. Originally a striking advertisement, posters developed after the invention of LITHOGRAPHY, and brilliantly coloured posters could be cheaply produced after 1870. TOULOUSE-LAUTREC's posters of the 1890s showed a bravura of execution matched by Alphonse MUCHA's involuted ART NOUVEAU designs. With WWI and the Russian revolution, posters entered the political process as effective and vivid propaganda. POP ART and gramophone record covers in the 1960s continued the development of the poster as an art form in its own right.

Postgate, Raymond William (1896–1971), British historian and novelist, imprisoned for being a conscientious objector in 1916. He worked on the *Daily Herald* and the *Encyclopaedia Britannica,* and founded the *Good Food Guide* in 1951. His political works include *The Bolshevik Theory* (1920) and *History of the British Workers* (1926). Among his novels are *Verdict of Twelve* (1940) and *The Ledger is Kept* (1953). He also wrote, with G. D. H. Cole, *The Common People: 1746–1946* (1946).

Post-Impressionism, term invented by English critic Roger FRY to denote works by artists who exhibited at the Grafton Gallery, London, in 1910. They included MANET, GAUGUIN, van GOGH, CÉZANNE, SEURAT, MATISSE and PICASSO. The term does not refer to one particular style of painting, although all these artists reacted against IMPRESSIONIST and NEO-IMPRESSIONIST preoccupation with visual appearances.

Post-mortem, or autopsy, examination of a body to determine the cause of death. When suspicious, sudden or unexplained human death occurs, a post-mortem is usually performed. Often it entails the DISSECTION of organs for microscopic study and the chemical analysis of certain body liquids and tissues. Such analyses form part of PATHOLOGY.

Post-natal depression, psychotic disorder manifested by expressions of intense sadness which may occur within a day or two of childbirth. It may be engendered by chemical changes in the body, the stresses of giving birth or feelings of inadequacy, but authorities doe not agree on the causes.

Post Office (PO), British public corporation formed in 1969 from the General Post Office (GPO). Mail delivery, its sole function until the 19th century, is still a Post Office monopoly and derives from the Crown's early needs to maintain reliable royal communications. Private post, at rates related to distance, was first delivered in 1635; HILL's penny post of 1840 standardized the rate. The GPO set up a savings bank (1861), a telegraph service (1870) and nationalized existing private telephone companies in 1912.

Postulate, statement or proposition that is to be assumed to be true without proof and that forms a framework for the derivation of THEOREMS. The term as now used is synonymous with "axiom".

Potash, potassium carbonate (K_2CO_3). The term is also used loosely to describe any potassium salt, ore or a mineral with a fairly large proportion of potassium, such as potash alum or potash mica. Potash is mined for use as a fertilizer. *See also* PE pp.*133, 170, 171.*

Potassium, common metallic element (symbol K) first isolated in 1807 by Sir Humphry DAVY. Its chief ores are sylvite (a chloride), carnallite and polyhalite. The metal is used as a heat-transfer medium in nuclear reactors, but has few other commercial uses. Chemically it resembles sodium, being rather more reactive. The natural element contains a radioisotope K^{40} (half-life 1.3×10^9yr), which is used in the radioactive dating of rocks. Properties: at. no. 19; at. wt. 39.102; density 0.86; m.p. 63.65°C (146.6°F); b.p. 774°C (1,425°F); most common isotope K^{39} (93.1%). *See also* ALKALI ELEMENTS.

Potassium carbonate, white solid (K_2CO_3), known as POTASH and usually produced commercially by electrolysis of potassium chloride, followed by treatment of the resulting potassium hydroxide with carbon dioxide. It is used as a fertilizer, to make glass and textile dyes, and in cleaning and ELECTROPLATING metals.

Potassium chloride, colourless or white potassium salt pKCl) extracted from lake brines and from minerals such as sylvite, kainite and carnallite. It is used as a fertilizer and as a raw material in the production of potassium hydroxide and potassium carbonate.

Potassium hydroxide, caustic potash, commercially produced white solid (KOH) prepared by ELECTROLYSIS of potassium chloride. It is a strongly alkaline substance used for making soaps and detergents. *See also* ALKALI.

Potassium nitrate, transparent solid (KNO_3), a naturally occurring mineral called saltpetre or nitre, which is used as a fertilizer for its nitrogen content. It is also used in the manufacture of explosives and as a food preservative.

Potato, plant native to Central and South America and introduced into Europe by the Spaniards in the 16th century. Best grown in a moist, cool climate, it has oval leaves and violet, pink or white flowers. The potato itself is a tuber (an underground modified stem). The leaves and green potatoes contain the alkaloid solanine and are poisonous if eaten raw. Family Solanaceae; species *Solanum tuberosum.* *See also* PE pp.*158, 178, 186–187, 186–187.*

Potato beetle. See COLORADO BEETLE.

Potchefstroom, city in NE Republic of South Africa, in S Transvaal province on the Vaal River. It was the scene of the Transvaal revolt of 1880–81, during which the British garrison was besieged for three months, and finally defeated. The city's growth dates from 1933, when gold was discovered. It is now the centre of one of the world's richest gold-mining regions. Industries: timber, metal goods, malt. Pop. (1970) 60,000.

Potemkin, Grigori Aleksandrovich, Prince (1739–91), Russian statesman. Involved in the coup that brought CATHERINE II to power in 1762, he became her lover for a time and remained until his death the most powerful man in Russia. He played an important part in the annexation of the Crimea in 1783, for which he was created a prince, and was an able administrator of the new province.

Potemkin, The Battleship. See BATTLESHIP POTEMKIN.

Potential difference. See ELECTROMOTIVE FORCE (emf).

Potential energy, energy that an object possesses due to its vertical position in the Earth's gravitational field; also the energy stored in a system such as a spring, or an oscillating system (eg a pendulum) at the point of greatest amplitude. An object on a shelf, for example, has potential energy given by mgh, where m is its mass, g the acceleration due to gravity, and h the height of the shelf. If the object falls to the ground its potential energy is converted into KINETIC ENERGY. *See also* SU pp.74–77, *74–77.*

Potentilla, genus of perennial herbaceous plants of the family Rosaceae; species are commonly called CINQUEFOIL, referring to the five leaflets usually found on the compound leaf. The flowers are solitary and five-petalled. *See also* NW p.*58.*

Potentiometer, instrument used to measure an unknown voltage or potential difference by comparing it with a known one, eg that of a standard cell. It consists of a length of resistance wire fixed taut on a wooden base, and connected across a battery. Both the known and the unknown voltages are in turn used to balance, through a GALVANOMETER and sliding contact, the voltages in a portion of the wire, the ratio of whose lengths equals the ratio of the voltages.

Pothole, in geology, term used for various formations that are pot-shaped. Most commonly it denotes a circular, bowl-shaped hollow formed in a rocky stream-bed by the grinding action of sand and stones whirled around by eddies or the force of the stream. Such potholes are usually found in rapids or at the foot of a waterfall. It is also used to describe the vertical pits worn in limestone rocks by the action of water, which often provide the entrances to cave systems. *See also* POTHOLING; PE pp.110–112, *110–112.*

Potholing, sport devoted to the exploration of underground systems, such as potholes and caves; it is also called caving.

Potidaea, ancient city in NE Greece, at the narrowest point of the Pallene Peninsula (now Kassandra). A Corinthian colony founded in 609 BC, it joined the Delian League but revolted against Athens and was recaptured 429 BC. It was taken and destroyed by Philip II of Macedon in 356 BC, and rebuilt after 316 BC by Cassander and named Kassandreia.

Potomac, river in E USA. It rises in West Virginia at the confluence of the North Branch and South Branch rivers, and flows E and SE to Chesapeake Bay, forming the boundaries of Maryland-West Virginia and Virginia-Maryland. The river is navigable for large ships as far as Washington, DC. Its chief tributary is the Shenandoah. Length: 462km (287 miles).

Potomac Charter, one of a series of grants awarded to the Potomac Company (organized in 1785 with George Washington as its first president) by the state of Maryland and the US Government for the construction of a canal from the Potomac River to the Ohio Valley. A canal connecting the Potomac with Harpers Ferry was completed in 1808, but the project was unprofitable and was finally administered by receivers.

Potosi, city in S Bolivia, approx. 80km (50 miles) S of Sucre. Located at an altitude of 4,200m (13,700ft), Potosi is one of the highest cities in the world. Furniture and electrical appliances are manufactured there. Pop. (1969 est.) 63,600.

Potsdam, city in central East Germany, on the River Havel 27km (17 miles) SW of West Berlin; (17 miles) SW of West Berlin: capital of Potsdam district. It was the scene of the Peace of Potsdam (1805), which strengthened the alliance between Russia and Prussia against France, and the POTSDAM CONFERENCE. It is the centre of the East German film industry; other industries include food processing, textiles, pharmaceuticals and electrical equipment. Pop. (1975) 118,134.

Potsdam Conference (1945), meeting of leaders of principal Allies of WWII to clarify and implement agreements reached at the YALTA CONFERENCE. The participants included President TRUMAN of the USA, Premier STALIN of the USSR, the Prime Minister, CHURCHILL, of Great Britain and, after the latter's defeat in the British elections, ATTLEE. The so-called Potsdam agreement established the Council of Foreign Ministers to prepare draft treaties and to make proposals for settling territorial issues. It also transferred the chief authority in Germany to the US, British, French and Soviet commanders in their respective zones of occupation. German industry was decentralized and an ultimatum calling for unconditional surrender was issued to Japan.

Potsherd, pottery fragment, one of the archaeologist's chief aids in dating historic sites and tracing cultural contacts after the invention of pottery in NEOLITHIC times. Sir Flinders PETRIE developed the system of using pottery for dating during the 1880s.

Potter, Beatrix (1866–1943), British author who created the characters of Peter Rabbit, Jemima Puddleduck, Squirrel Nutkin and others in her animal stories. In the 1890s she sent illustrated animal stories to a sick child and, inspired by their success, published *Peter Rabbit* in 1900 and *The Tailor of Gloucester* in 1902. More than 23 books, enhanced by her delicate drawings, followed. They are now considered classics.

Potter, Paulus (1625–54), Dutch painter, famous for his landscapes with animals. He worked mainly in Amsterdam and the Hague. His works, which are noted for their realism, include the life-size *Young*

Posters and advertising were often satirized in 19th-century magazines.

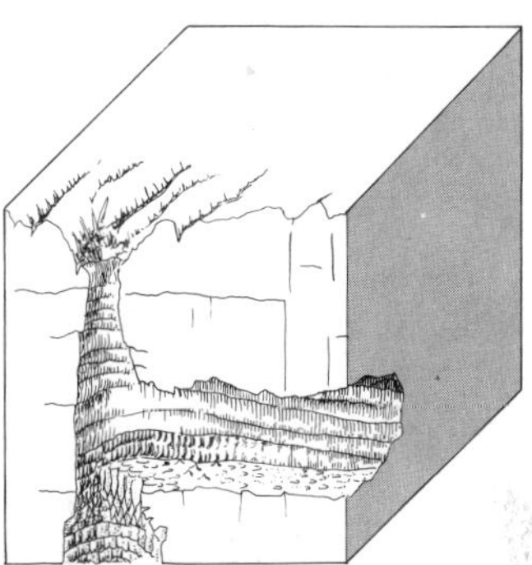

Potholes are worn in the limestone rocks by the slow dissolving action of water.

Beatrix Potter's stories have delighted children throughout the world.

Paulus Potter's painting *The Opening of The Hunt* is typical of his landscapes.

The Potteries: photograph of the unloading of clay from barges at Hanley.

Ezra Pound was famous for his obscure and beautiful poetry.

Nicolas Poussin: detail from the *Gathering of the Ashes of Phocion* (1648).

Enoch Powell's controversial opinions have earned him the mistrust of both parties.

Bull (1647) and *Early Morning* (1651).

Potter, Stephen (1900–69), British critic and humorist. After writing several critical works including *D. H. Lawrence: A First Study* (1930) and *The Muse in Chains* (1937), he turned to satires on the ethics of the English middle classes. These include *The Theory and Practice of Gamesmanship* (1947), *Some Works on Lifemanship* (1950) and *One-Upmanship* (1952).

Potteries, The, district in Staffordshire, w central England, that extends for 14km (9 miles) NW to SE in the upper Trent Valley. The area has been famous since the 16th century for its china and earthenware: it was the home of Josiah WEDGWOOD, Josiah SPODE and Thomas MINTON. Clay for pottery was once mined locally, but is now taken there from SW England. The area includes Burslem, Fenton, Hanley, Longton, Tunstall and Stoke-on-Trent.

Potter's wheel, device used in POTTERY to fashion clay objects. It consists of a horizontal disc revolving on a vertical spindle, powered by a motor or a foot-operated crank. Clay is centred on the rapidly turning disc and is manipulated by the potter into the desired form.

Pottery, in its widest sense, includes all objects shaped of clay and hardened by fire or dried in the sun. The term is derived from the French *poteric*, related to the Latin *potare* (to drink). Pottery is dependent on two properties of clay, its plasticity and its durability after firing. Pottery can be divided into three categories according to the degree of hardness and special constitution: EARTHENWARE, the ordinary pottery dating from primitive times, baked at 700°C (1,292°F) or lower; STONEWARE, fired at up to 1,150°C (2,102°F), more vitrified or close-grained and non-porous and until modern times produced more commonly in the Far East than in Europe; and PORCELAIN, fired at 1,400°C (2,552°F), a Chinese invention and the most refined ceramic material, known as hard paste because of its homogeneous composition. To "throw a pot" the potter uses a special wheel which is rotated by foot or by power, leaving the potter free use of his hands for manipulating the moist clay. After the pot is formed and dried, it is fired in a kiln; GLAZE is then applied, and the pot is refired. Glazes include alkaline, lead, felspathic and salt. *See also* MM pp.44–45, and articles in the *History and Culture* volumes.

Potto, slow-moving, squirrel-sized, primitive African primate with large eyes and a pointed face; it is nocturnal and arboreal. The common potto has sturdy limbs, a short tail and small spines formed by the neck vertebrae. Its woolly fur is grey-red. Length: 37cm (15in), excluding the tail. Species *Perodicticus potto.*

Poulenc, Francis (1899–1963), French composer and a member of Les SIX. His music shows the formative influence of Erik SATIE. Spontaneity and simple melodies, delicate yet free from sentimentality, characterize his works, which include ballets – notably *Les Biches* (1923) – orchestral works, chamber music, piano music and songs. He also composed two operas, *Les Mamelles de Tirésias* (1944) and *Dialogue des Carmélites* (1957). *See also* HC2 pp.270, *270.*

Poultry, collective term for domestic fowl reared as a source of meat and eggs. They include chickens, ducks, geese, guineafoul and turkeys. *See* PE pp.162–163, *224, 236–238, 239.*

Pound, Ezra (1885–1972), US poet and critic, a dominant influence on 20th century Anglo-American literature. He spent much of his life in Europe where his earliest works, *Exultations* (1909) and *Personae* (1909), attracted critical acclaim for their dazzling originality and erudition. He encouraged the work of artists such as T. S. ELIOT and James JOYCE, and was involved with the IMAGISTS. From 1924 he made his home in Italy and at the outbreak of war between the USA and Italy he supported the Italians, making scurrilous broadcasts on Italian radio about the Jews and disparaging the Allied war effort. He was later arrested for treason, but was judged unfit to stand trial and was

confined to a mental hospital (1946–58). His major works include the poem *Homage to Sextus Propertius* (1917), *Hugh Selwyn Mauberley* (1920) and *Cantos. See also* HC2 p.289.

Pound, imperial unit of weight equal to 0.453kg. It became a unit of currency when a pound (lb) weight of silver was divided into 240 PENNY units. The pound STERLING has been the main unit of English currency since the Middle Ages. The first coin of one pound (£) value was coined in 1489; it was, however, known as a SOVEREIGN, and its value was not fixed at 20 SHILLINGS until 1604.

Poundage. *See* TONNAGE.

Pound force. *See* WEIGHTS AND MEASURES.

Pourbus, family of Flemish painters of the 16th and 17th centuries. Pieter (1510–84) worked in Bruges and painted portraits and religious subjects in a naturalistic style. His son Frans the Elder (1545–81) painted similar subjects. His grandson Frans the Younger (1569–1622) was court painter at Mantua (1600–09) and painter to Marie de Médicis in Paris (1609–22).

Pousseur, Henri (1929–), Belgian composer. He studied under BOULEZ and worked with STOCKHAUSEN and BERIO. He is the foremost Belgian exponent of ELECTRONIC MUSIC, as in his *Seismogrammes* (1953) for magnetic tape.

Poussin, Gaspard (1615–75), French painter, born Gaspard Dughet but who changed his name to that of his teacher and brother-in-law Nicolas POUSSIN. His successful paintings were landscapes which blended the Classic and Romantic qualities of Poussin and CLAUDE LORRAIN; they were much admired in 18th-century England. *See also* HC1 pp.308–309.

Poussin, Nicolas (*c.* 1594–1665), French painter of religious and mythological scenes and Classical landscapes. His mature works evoked the spirit of the French Classic ideal as did no other paintings of that period, and after 1650 he enjoyed international prestige. His early works include *The Martyrdom of St Erasmus* (1628) in the BAROQUE style and the Classical *Inspiration of the Poet* (1628–29), which used Venetian colouring. His later works include *The Eucharist* (1644–48), and *The Seven Sacraments* (1648). *See also* HC1 pp.308–309, *308.*

Poverty and Progress (1941), sociological study of working-class urban life by B. S. ROWNTREE. It derived from a survey undertaken in 1936 and updated his *Poverty: A Study of Town Life* (1901).

Powdered milk, form of dried whole milk which contains 26.5% of fat and at least 95% of milk solids and is soluble in water. This type is produced mainly for use in the making of foods, but it is also available as a consumer product. Dried skim milk is non-fat dried milk, and usually contains more than 50% lactose.

Powderhall handicap, colloquial name given to the professional handicap sprint race for men over 120yd (110m) as part of the Powderhall New Year Handicaps, held annually at Powderhall, Edinburgh. It was first raced in 1870 over 130yd at the Powderhall stadium, the distance being reduced to 120yd in 1958. The metric equivalent was adopted in 1970, when the meeting moved to the Meadowbank stadium, also in Edinburgh.

Powder metallurgy, manufacture of metal powders and their use in producing metal parts. Powder particles are compressed to the desired shape and then sintered. The use of powders is more economical than molten metal in making such items as small gears. Melting may also prove impractical, eg when high melting-point metals are used, or when an alloy of unfusable materials is involved. Powder metallurgy is also used to make porous metal parts.

Powell, Anthony Dymoke (1905–), British novelist specializing in social comedies; he was also a scriptwriter and journalist. His work includes *A Dance to The Music of Time*, a series of novels which portray the snobbish, insular world of the English upper classes after WWI, beginning with *A Question of Upbringing* (1951) and including *At Lady Molly's*

(1957), *Temporary Kings* (1973) and *Hearing Secret Harmonies* (1975).

Powell, Cecil Frank (1903–69), British physicist who was awarded the 1950 Nobel Prize in physics for developing a photographic method for studying nuclear processes and for the resultant discovery of the pi MESON (PION). During the 1930s he developed the technique whereby ELEMENTARY PARTICLES were recorded directly onto film, rather than being seen as tracks in a cloud chamber. In 1947 he used this method at high altitude to investigate COSMIC RAYS and discovered a new particle – the pion. This discovery lent support to the theory of nuclear structure proposed by Hideki YUKAWA, who had predicted the existence of such a particle in 1935. Powell subsequently discovered the antiparticle of pion and, in 1949, the decay process of K mesons (KAONS). *See also* SU p.66.

Powell, John Enoch (1912–), British politician. He became a Conservative MP in 1950, and was Minister of Health (1960–63). He was known for his campaign in 1968 against coloured immigration into the UK and for his opposition to Britain's entry into the Common Market. Powell was elected MP for the United Ulster Unionist Coalition in 1974 after breaking with the Conservative Party.

Powell, Sir Philip (1921–), British architect who, in partnership with John MOYA, established himself by winning the competition for the design of the Churchill Gardens flats, Pimlico, London (1948–62). The clarity and precision of the partners' style has won them many prestigious contracts, such as the additions to St John's College, Cambridge (1967), and the Museum of London (1976).

Power, in physics, the rate of doing work, ie a measure of the output of an engine or other power source. James WATT was the first to measure power; he used the unit called HORSE-POWER. The modern unit of power is the watt.

Power, electric. *See* POWER SOURCES, ELECTRIC.

Powerboat racing, sport in which boats powered by outboard or inboard motors race on offshore or inland waters. Powerboats race mainly at speeds of up to 100mph (160km/h), unlike HYDROPLANES, in which power and speed are almost unlimited. The sport was inaugurated internationally in 1903 when Sir Alfred Harmsworth (later Lord Northcliffe) offered the first powerboat racing trophy.

Powers and roots, in mathematics, number of times a number is multiplied or divided by itself. If 2 is raised to the power 3 (written 2^3, called 2 cubed, or 2 to the 3rd power) it is $2 \times 2 \times 2$, or 8. A negative power indicates a fraction: eg 2^{-3} is $1/2^3$, or 1/8. A fractional power indicates a root – the power $\frac{1}{2}$ is a square root, and the power $\frac{1}{3}$ is a cube root: eg 8 to the power $\frac{1}{3}$ (written $8\frac{1}{3}$) means a number (the cube root) which, when multiplied by itself three times, yields 8; the root is 2. Any number to the power 0 equals 1.

Power sources, electric, primary means by which modern civilization obtains electricity. The chief source in Britain relies on the burning of fossil fuels such as coal or oil. In countries such as Sweden and New Zealand, fast-flowing rivers are harnessed to drive hydro-electric plants. Nuclear power relies on the controlled decay of radio-active nuclei and is being increasingly employed. Experiments in the use of tides to power turbines have been made in Brittany, and the use of geothermal power has been tried in Iceland, New Zealand and the USA. At present electricity is distributed by means of a centralized reticulation of power (the NATIONAL GRID). In 1977 the JET project was set up in England to try to develop nuclear fusion which, because its fuel is easily available and its wastes are harmless, would be the ideal power source. *See also* MM pp.70–75, 84–85.

Powles, Sir Guy (1905–), New Zealand diplomat, High Commissioner for Western Samoa (1949–60), ambassador to Nepal (1960–62). He was New Zealand's first ombudsman (1962).

Powys, John Cowper (1872–1963), British writer. He wrote essays and criticism, the best known of which are *The Meaning of Culture* (1929) and *A Philosophy of Solitude* (1933). His novels include *Wolf Solent* (1929), *A Glastonbury Romance* (1932), *Owen Glendower* (1940) and *Atlantis* (1954).

Powys, county in E central Wales, formed in 1974 from the former counties of MONTGOMERY(SHIRE), RADNOR(SHIRE) and most of BRECKNOCK(SHIRE). The administrative centre is Llandrindod Wells. Area: 5,077sq km (1,960sq miles). Pop. (1975 est.) 100,800.

Poynings, Sir Edward (1459–1521), Lord Deputy of Ireland from 1494 to 1495. He gave his name to Poynings' Laws (1494–95), passed by the parliament which he summoned at Drogheda. It subordinated the Irish legislature to the English. It was repealed in 1782.

Poznań, city in w Poland on the Warta River. It is one of the oldest Polish cities, became the seat of the first Polish bishopric in 968 and was the centre of Polish power in the 15th–17th centuries. In 1793 it passed to Prussia. The Grand Duchy of Poznań was created in 1815, as part of Prussia, but the area reverted to Poland in 1919. Industries include metallurgy, chemicals, textiles and food processing. Pop. (1970) 469,000.

Pozzo, Andrea (1642–1709), Italian painter, one of the leading artists of the BAROQUE style. He worked for the Jesuits and painted church ceilings, for example in S. Ignazio, Rome (1691–94), which was a masterpiece of PERSPECTIVE. *See also* HC1 p.*307*.

Prado, in Madrid, the Spanish National Museum of Fine Art. Founded in 1818, it is an example of Spanish NEOCLASSICAL architecture. The major part of its collection derives from the royal collection of the Hapsburg and Bourbon kings of Spain. Numerous important works by VELÁZQUEZ and a full representation of El GRECO and GOYA are contained there. Bosch, Titian, Tintoretto, Veronese, Rubens, van Dyck, and Bruegel are also well represented.

Praesepe, or The Beehive, open cluster containing several hundred stars in the constellation of Cancer. It is visible to the naked eye as a small bright hazy patch. It is 525 light-years from Earth and was first observed by Galileo. *See also* SU p.242, *242*.

Praetor, powerful Roman magistrate. From 242 BC two praetors were appointed, the urban praetor (*praetor urbanus*), who decided cases involving citizens, and the peregrine praetor (*praetor peregrinus*), who decided cases between foreigners. By the 1st century BC there were eight praetors. *See also* HC1 p.92, *92*.

Praetorians, bodyguard of the Roman emperors from AUGUSTUS to CONSTANTINE I. They were first selected by Augustus in 2 BC from the *cohors praetoria*, which had guarded the commanding general in Rome. The number of cohorts (from 500 to 1,000 men each) forming the guard varied from 9 to 16. They had special privileges and, during the period of the emperor's decline, held great authority. Constantine I disbanded them in 312. *See also* HC1 p.*114*.

Praetorius, Michael (1571–1621), German composer and musicologist. He wrote a vast amount of choral music, including more than 1,220 works in the nine-volume *Musae Sioniae* (1605–10). His *Syntagma musicum* (1615–19) included a detailed exposition of the musical instruments and practices of his day.

Pragmatic sanction, term of Byzantine origin meaning a public decree, used later by European monarchs for state documents such as those defining their powers or settling the royal succession. Unless qualified, it usually refers to the pragmatic sanction of 1713 by which Charles VI of Austria settled the succession to his throne upon his daughter MARIA THERESA.

Pragmatism, philosophical school that proclaims that the truth of a proposition has no absolute standing but depends on its practical value or use. Knowledge and thought are products of the existing social situation, and their results relate only to the specific problems that gave rise to them. Pragmatism, primarily supported by US philosophers, was first proposed by C. S. PEIRCE, and was adopted by William JAMES and John DEWEY.

Prague (Praha), capital of CZECHOSLOVAKIA, on the River Vltava 258km (160 miles) NW of Vienna, Austria. Founded in 722, it grew rapidly after Wenceslaus I established a German settlement there in 1232. It remained a great religious and cultural centre, becoming the capital of the new Czechoslovak republic in 1918. It was occupied throughout WWII by the Germans and liberated by Soviet troops in 1945. Prague was the centre of Czech resistance to the Soviet invasion of the country in 1968. It is an important commercial centre, having good communication and transport links and a variety of industries, including engineering, iron and steel, chemicals and printing. Pop. (1974) 1,095,615.

Prague, School of, group of painters who were active in Bohemia during the second half of the 14th century. Painters of the Prague School, such as the Masters of Hohenfurth and Wittingau, are little known as individuals, but they were influenced by contemporary humanism, and in their search for realism used GIOTTO's techniques of perspective.

Praia, on the s shore of São Tiago Island, in the Cape Verde group; capital of the Cape Verde Islands. It is the site of a meteorological station, and is a key centre in the Atlantic telegraph cable network. A trading centre for the Islands, Praia exports castor oil, sugarcane, oranges and coffee. Industries include fishing and the manufacture of straw hats. Pop. (1970) 21,494.

Prairie, region of treeless plain. The prairies of North America extend from Ohio through Indiana, Illinois and Iowa to the Great Plains, and N into Canada. The PAMPAS of s South America, the LLANOS of N South America and the STEPPES of central Europe and Asia correspond to the prairies of North America.

Prairie chicken, chicken-sized, pale brown GROUSE of western USA. It has brown and black, stiffly pointed tail feathers and white neck feathers. During courtship displays the neck feathers of the male stand up when he inflates his orange neck air sacs. Groups of males shuffle about with necks stiff and heads bobbing as part of their displays, which precede mating. Species *Tympanuchus cupido*.

Prairie dog, squirrel-like rodent of North America, named after its barking cry. It has a short tail and its fur is grizzled brown to buff. Active by day, it feeds on plants and insects and lives in communal burrows that are interconnected to form colonies. Length: 30cm (12in). Genus *Cynomys*. *See also* GROUND SQUIRREL.

Praise of Folly, The (*Encomium moriae*) (1509), satire by Desiderius ERASMUS, written in part in the home of Sir Thomas MORE. The narrative, told by Folly herself, shows how all classes of men and women are subject to foolishness. It ridiculed corruption among church dignitaries (and the rest of the world), rather than striking at the church as an institution.

Prajapati, in Vedic mythology, name applied to several gods, such as INDRA, Hiranyaarbha, SAVITRI and SOMA. Prajapati was later identified with BRAHMA as the lord of creation, the chief deity and father of the gods. As the creator, Prajapati spoke the words which brought into being the earth and the sky.

Prakrit, group of languages used in India c.5th century BC to c.12th century AD, after SANSKRIT and before the modern Indic languages (although they overlap). The word Prakrit means "natural" or "vulgar", as opposed to Sanskrit which means "refined" or "cultivated". The finest literature is in PALI and in the canonical and narrative writing of JAINISM.

Prampolini, Enrico (1894–1956), Italian painter and sculptor. He was associated with many aspects of the European avant-garde, being influenced early on by CUBISM, and then keeping in close touch with (among others) the DADA school, VIL-LON, MONDRIAN and the NEO-PLASTICISTS. In 1932 he joined the Paris Abstraction-Création group. His most important commitment was to FUTURISM, and his *Manifesto on Aeropainting* (1929) was an expression of futurist ideas.

Prandtauer, Jakob (1660–1726), Austrian architect who worked primarily on monastic commissions. His most famous buildings are the impressive BAROQUE monastery and church at Melk (1702–38) on a spectacular site overlooking the Danube, and the monastery of St Florian at Linz, on which he worked from 1708.

Prasad, Rajendra (1884–1963), Indian politician and lawyer who renounced his legal practice in 1920 to follow GHANDI. He was a member of the All-India Congress in 1912 and its president four times between 1934 and 1948. Prasad was the first President of the Republic of India from 1950 until 1962. His autobiography, *Atma Katha*, was published in 1958, other works include *India Divided* and *At the Feet of Mahatma Ghandi* (both 1958).

Praseodymium, metallic element (symbol Pr) of the LANTHANIDE SERIES, first isolated in 1885. Its chief ores are monazite (phosphate) and bastnasite (fluorocarbonate). Praseodymium is used in carbonarc lamps and its salts are used in coloured glasses. Properties: at.no.59; at.wt. 140.9077; s.g. 6.77 (α), 6.64 (β) for the α form m.p. 931°C (1,708°F); b.p. 3,212°C (5,814°F); most common isotope Pr^{141} (100%).

Prassinos, Mario (1916–), French artist and graphic designer, *b.* Turkey. Strongly influenced by SURREALISM in the 1930s, he worked in many media; most excitingly in book illustration.

Prat, Jean (1924–), French rugby union player. A goalkicking wing-forward from the Lourdes club, he played 38 times for France against International Board countries (1947–55). France's captain, he later became a selector and coach.

Pratincole, any of about half a dozen species of Old World birds. All are brown with white rumps, forked tails and long, pointed wings. They feed on insects near rivers, and nest in colonies on the ground. The black-winged pratincole (*Glareola pratincola*) is sometimes called the locust bird in its winter home in Africa. Length: to 22cm (9in). Family Glareolidae.

Prato, city in central Italy, 17km (11 miles) NW of Florence. Prato has been famous for its woollen goods since the 13th century; these are still produced. Pop. (1971) 142,913.

Pratt, Edwin John (1883–1964), Canadian poet who broke away from the old romantic tradition to write imaginative narratives of epic events. His works include *Titans* (1926), *Dunkirk* (1941), *Behind the Log* (1947) and *Towards the Last Spike* (1952).

Pratt, John Jeffreys, 2nd Earl and 1st Marquess Camden (1759–1840), British politician. As Lord Lieutenant of Ireland (1795–98) his repressive policies resulted in the Irish rebellion of 1798 and English military intervention in Ireland under Lord CORNWALLIS. *See also* HC2 p.25.

Prawn, any of numerous species of crustaceans in the order Decapoda; it is generally larger than a SHRIMP but smaller than a LOBSTER. Typical genera include *Penaeus, Pandalus, Crangon* and *Nephrops*, which includes the DUBLIN BAY PRAWN or Norway lobster. Many species are regarded as delicacies. *See also* NW pp.100–101; PE pp.242–243.

Praxiteles (*fl.* 370–330 BC), Greek sculptor whose graceful style epitomized the 4th-century Greek ideal. One of the most widely imitated of Greek sculptors, he created *Hermes Carrying the Infant Dionysus* (350 BC), found in Olympia. His most famous work is *Aphrodite from Cnidus* (350 BC), a sensuous work that created a new ideal female. *See also* HC1 p.79.

Prayer, act of thanking, adoring, conferring with or petitioning God and the form of words used for this purpose. Many religions have set forms for praying. Muslims recite prayers while facing Mecca; Judaism has prayers for most occasions of life; and Christians say the LORD'S PRAYER

Prague: the city has 13 bridges which cross the Vltava river.

Prairie dogs identify their family members by a curious "kissing" ritual.

Rajendra Prasad was the first president of the Indian Republic.

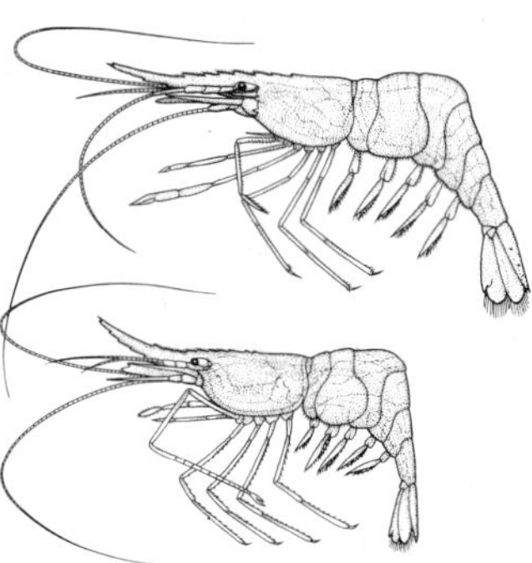

Prawns may be caught by dragging a net along the ocean floor.

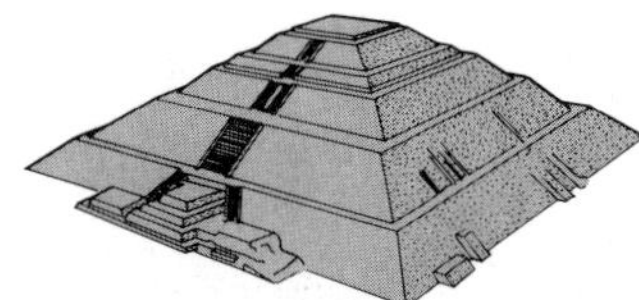

Pre-Columbian architecture: temple design of the classical Mayan period.

Giovanni Ambrogio Preda may have painted this portrait of Lucrezia Crivelli.

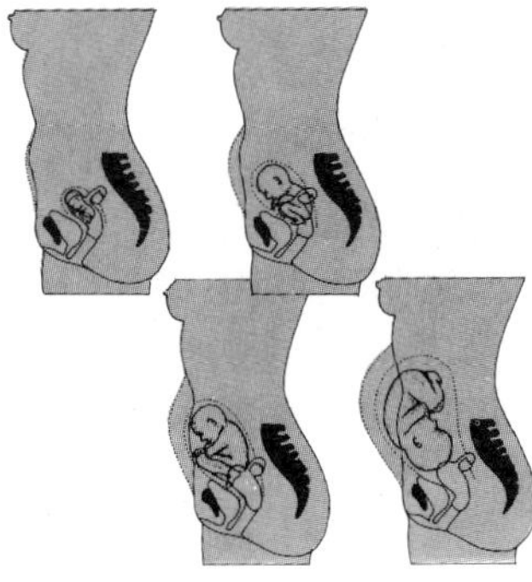

Pregnancy lasts nine months and transforms a single cell to a human being.

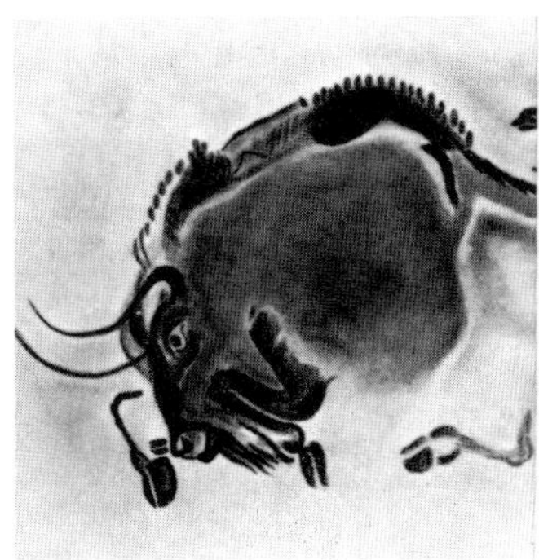

Prehistoric art often featured animals such as this bison from Altamira, Spain.

as taught by Jesus. Prayer often means a form of public worship. The service book of the Anglican Communion is called The Book of COMMON PRAYER. Prayer can also mean the private act of an individual using his or her own words to praise God or petition God for guidance or strength or the fulfilment of his or her special requests.

Prayer Book. *See* COMMON PRAYER, BOOK OF.

Prayer plant, also called banded arrowroot, tropical plant with leaves that fold up in darkness. Often grown as an ornamental, its leaves are marked with narrow white or red streaks. Family Marantaceae; species *Maranta leuconeura.*

Praying mantis. *See* MANTIS.

Precambrian, oldest and longest major division of Earth's history, lasting from the beginning of the Earth more than 4,000 million years ago to the beginning of a good fossil record 570 million years ago. Precambrian fossils are extremely rare because the earliest life forms are presumed not to have had hard parts suitable for preservation. Also, Precambrian rocks have been greatly changed and deformed. Nonetheless, primitive bacteria and blue-green algae have been identified in deposits more than 3,000 million years old. *See also* PE pp.130–131, *131.*

Precedent, in law, a judgement or decision of a court which is binding on a lower court and provides a guide to future judgements. Most of the rules of equity and common law are unwritten and are contained in precedents established by the courts over the years.

Precession, wobble of the axis of a spinning object. If an object like a coin is set spinning about an axis, the axis itself starts slowly to rotate in ever-increasing circles. This effect occurs as a result of the torque on the spin axis, which increases as the angle of precession increases. The Earth precesses about a line through its centre and perpendicular to the plane of the ECLIPTIC extremely slowly (a complete revolution taking 25,800 years) at an angle of 23.5°. Its effects make the equinoxes precess backwards through the ZODIAC, and the positions of the stars change by a small amount each year. The motion of a GYROSCOPE is another consequence of precession, because the entire ring containing the spinning wheel and its axle precesses around the support pivot.

Precious Bane (1924), novel by Mary Webb which won the Prix Fémina for the best English novel of the year. Set in Shropshire, it is written in a rustic and melodramatic mode which has been compared to Thomas HARDY's, although lacking his subtlety. Its great popular success has been attributed to the fact that it was praised by Stanley BALDWIN.

Precipitation, in meteorology, all forms of water particles, whether liquid or solid, that fall from the atmosphere to the ground. Distinguished from cloud, fog, dew and frost, in that it must fall and reach the ground, precipitation includes rain, drizzle, snow (ice pellets and crystals) and hail. Measured by rain and snow gauges, the amount of precipitation is expressed in millimetres or inches of liquid water depth. Precipitation occurs with the condensation of water vapour in clouds into water droplets that coalesce into drops as large as 7mm (0.25in) in diameter or form from melting ice crystals in the clouds. Drizzle consists of fine droplets, and snow of masses of many-sided ice crystals. Sleet is produced by freezing of raindrops into small ice pellets, and hail by the freezing of concentric layers of ice in cumulo-nimbus clouds, with lumps measuring from 0.5 to 10cm (0.2–4in) in diameter. *See also* PE p.70.

Precognition, the ability to foretell future events by extra-sensory means rather than through any known process of rational inference. *See also* CLAIRVOYANCE; ESP; PARAPSYCHOLOGY; MS pp.198, *199.*

Pre-Columbian art and architecture, the arts of Mexico, Central America and the Andean region of South America before the arrival of Europeans. The architecture achieved great artistic complexity in its many phases. In the MAYA Classical period, beginning *c.*200 AD, many cities or ceremonial centres were built in Central America. Their pyramid temples are sculptured monuments paralleled by the TEOTIHUACAN, ZAPOTEC and MIXTEC cultures, and were succeeded by TOLTEC and AZTEC civilizations in the post-classical period (900–1300 AD). Monumental building was achieved without wheels or the use of the arch; surfaces were often decorated with dazzling patterns. In the Andean region the early Chavin sculptures were succeeded by the Mochica, the Tiahuanaco and finally, in the 14th century, by the rich temple architecture and sophisticated engineering of the INCAS. Gold-working, weaving and sculpture were other important Pre-Columbian arts. *See also* HC1 pp.46–47, *46–47,* 230–233.

Preda, Giovanni Ambrogio or Ambrogio da Predis (*c.*1455–*c.*1517), Italian painter, miniaturist and illuminator. He was court painter to the Sforza family of Milan from 1482 and much influenced by LEONARDO DA VINCI. He may have painted the side panels of Leonardo's *The Virgin of the Rocks* (1483), but the only work which can positively be identified as being by Preda is his signed portrait of Maximilian I (1502).

Predator. *See* PREY.

Predestination, the fore-ordaining of a person's spiritual salvation or condemnation by God. According to this doctrine, a person is at birth predestined to the events of life and the fate at death. The question of how this leaves room for the will and spirituality of the individual is at the heart of a controversy that has raged since early Christian times. As possible solutions to the doctrine, which is based mostly on Romans 8:29–30, three propositions have been put forward at different times: to refute the doctrine of predestination altogether (Pelagianism), to state that God never intended to save everybody (Predestinarianism), or to qualify the premise by seeing God's prevision as conditional and subject to possible revision depending on the will and spirituality of the individual, which is the solution to which most Christians adhere. The concept of predestination is also found in ISLAM. *See also* HC1 p.269, MS p.215.

Predis, Ambrogio da. *See* PREDAR, GIOVANNI AMBROGIO.

Prefect, name of the general administrator and representative of the national government in each *département* in France. A prefect is the chief executor for the *conseil général,* the elected departmental government. He is also the head of the local police and supervisor of the *communes,* the units of local government within the *département.* The post was established by Napoleon in 1800.

Pregl, Fritz (1869–1930), Austrian chemist who taught at the university of GRAZ. He was awarded the 1923 Nobel Prize in chemistry for developing techniques in microanalysis of organic compounds. He also worked out a method of making reliable measurements of carbon, hydrogen, nitrogen and sulphur, using only 7–13 milligrammes of material.

Pregnancy, period of time from conception until birth. In humans it is generally divided into three 3-month periods called trimesters. In the first trimester, the fetus grows from a small ball of cells to about 7.6cm (3in) in length, during which time the skeleton, brain and such vital organs as the heart and lungs develop. At the beginning of the second trimester movements are first felt and the fetus grows to approx. 36cm (14in). In the third trimester the fetus attains its full body weight. *See also* MS pp.76–77.

Pregnancy testing, methods of determining whether or not a developing embryo or fetus is present in the uterus. There are several different methods of testing. In the Friedman test a rabbit is injected with urine from the woman suspected to be pregnant; if this causes the animal to ovulate, pregnancy is confirmed. Most modern tests rely on the detection of human chorionic gonadotrophin (CG, or HCG) in the urine. Tests for HCG have a high reliability (about 97%), but are effective only when a woman's period is at least 14 days overdue. Radioactive tracer tests for HCG in the urine are now being developed, however, and these make it possible to detect pregnancy on the first day after a missed period. *See also* MS p.76.

Prehistoric art, earliest known aesthetic expression of modern man dating from the Upper PALAEOLITHIC period (35,000–10,000 years ago). Artefacts are divided into two categories: the mobiliary, objects that can be moved, such as female statuettes called Venuses; and the parietal, such as paintings, sculptures and engravings on cave walls. ALTAMIRA, LASCAUX, Font de Gaume and Les Combarelles are among the most famous caves with depictions of bison, horses, deer, other animals and men. The Abbé Henri BREUIL demonstrated the antiquity of prehistoric art. Scandinavian rock engravings, Siberian amber carvings and Saharan rock paintings also attest to man's creativity. *See also* HC1 pp.24–25; *24–25.*

Prehistoric Man. *See* MAN, PREHISTORIC.

Prelate, religious term restricted in the 20th century to Bishops in the Church of England, and in the ROMAN CATHOLIC CHURCH to officers of the Roman Curia. Originally, a prelate was anyone of either church of high rank. "Prelacy", now usually a derogatory term, signifies a system of government by bishops.

Prelog, Vladimir (1906–   ), Swiss chemist, *b.* Yugoslavia, who shared the 1975 Nobel Prize in chemistry with John W. Cornforth for their work on stereochemistry, the branch of chemistry which deals with the arrangement of atoms in a molecule. Prelog devised a procedure for determining whether a molecule is right- or left-handed. *See also* SU p.44.

Prelude, in music, a complete and separate preliminary movement which serves to introduce a work of which it may or may not formally be a part. An unrestricted musical form, it was often used as the first movement of a SUITE. The popularity of the Chopin *Preludes* led to its association with a short piece of an imaginative nature.

Prelude, The, long autobiographical poem by William WORDSWORTH. Begun in 1759 and continually revised, it showed the growth of a poet's mind by tracing Wordsworth's own life from childhood on. A short version was published after his death in 1805, but the full text did not appear until 1926.

Prem Chand. *See* CHAND, PREM.

Premier, title used by the chief executive in a parliamentary system of government, in contrast with the head of state, who may be a constitutional monarch or an elected official. *See also* PRIME MINISTER.

Preminger, Otto Ludwig (1906–   ), US film director, *b.* Austria, whose predominant interest has been in the evolution and analysis of character. *Laura* (1944) marked the beginning of his career and *The Moon is Blue* (1953) was his first independent production. In 1960 he made the epic *Exodus,* using the wide screen for the first time. His later films include *Hurry Sundown* (1967) and *Skidoo* (1968).

Premium Bonds, properly called Premium Savings Bonds, type of LOTTERY run by the British Post Office. Differing from a true lottery in that stake money may be reclaimed at any time, its prizes range from £25 to £100,000, some prizes being presented weekly, others monthly, and all tax-free. Premium Bonds were first issued in 1956; the present minimum purchasable unit is a £5 Bond. Winning numbers are chosen at random by an electronic machine known as ERNIE.

Premolar, in the dentition of adult human beings and other mammals, the two crushing or cutting TEETH between the CANINES and MOLARS on both sides of the upper and lower jaws; there are usually eight in all.

Premonstratensians, or White Canons, religious order founded by St Norbert in 1119 in the valley of Premontre, France. It is a teaching and preaching order that wears the Benedictine habit and follows a modified Benedictine rule.

Prempeh, name of two kings of ASHANTI, part of the former British colony of the GOLD COAST. Prempeh II (*d.* 1931), the last king (1888–1896) of Ashanti, was deposed and exiled to the Seychelles. He was permitted to return in 1924 and two years later was given chief's rank.

Prendergast, Maurice Brazil (1859–1924), US painter noted for his fine water-colours. A European visit in 1898 fired his admiration for POST-IMPRESSIONISM and he exhibited with the EIGHT in their famous show of 1908. His works, which have the quality of a bright tapestry, include *Umbrellas in the Rain* (1899) and *Central Park* (1901). After 1904 he painted increasingly in oils with a rich IMPASTO technique.

Prendergast, Paddy (1910–), Irish horse-racing trainer who was leading trainer in Britain from 1963 to 1965. He has saddled winners in four of the English Classics, only the Derby eluding him. His successes in the Irish Classics include all but the Oaks.

Pre-Raphaelite Brotherhood (PRB), group of British painters formed in 1848, which included Dante Gabriel ROSSETTI, John Everett MILLAIS, and William Holman HUNT. In reaction to the inflated forms and sentiments that they found in the followers of RAPHAEL, they preferred the sincerity they saw in the Florentine and Sienese who predated him. The group was known simply as PRB until 1850, and by 1853 it had disbanded.

Prerogative, royal, rights, privileges and powers pertaining to the British Crown. The three chief prerogative powers – to appoint and dismiss ministers, to initiate policy and to dissolve Parliament – had all been taken from the Crown by the mid-19th century. The royal VETO on legislation was last used by Queen ANNE. Prerogative courts, which exercised prerogative law through the Privy Council, were widely used by the Tudors, especially the COURT OF STAR CHAMBER and of High Commission. It administered the COMMON LAW mixed with EQUITY and aroused the opposition of common law lawyers, and was finally abolished by the LONG PARLIAMENT in 1641.

Presbyter, historical form of the word "priest" re-introduced by the followers of John CALVIN (1509–64) to mean a non-ordained minister of the Reformed Church. In its original pre-Christian Greek, however, the term denoted an elder of the Jewish synagogue. In the early Christian Church it came to signify an adviser on ecclesiastical matters. The word is rarely used today.

Presbyterianism, major form of Protestant Christianity that became the state Church of Scotland in 1690. One of the Reformed Churches, it was based first on the beliefs of John CALVIN (1509–64) in Switzerland, and brought to Britain by John KNOX (c.1513–72). Its particular form of Church government embodied the desire to return to the organization described in the New Testament. Ministers, occasionally called pastors, are elected by their congregations and affirmed in their office by the Presbytery, the ministers of the area who are responsible for ordaining and installing (and removing) ministers. The minister is assisted by deacons (sometimes known as elders) and trustees. Annually each presbytery sends delegates to a synod (a meeting to discuss Church order) and to a General Assembly (with much wider scope); of both, the chairman is called the Moderator. In 1972 the Presbyterian Church of England united with the CONGREGATIONAL Church of England and Wales. There are Presbyterian Churches all over the world, particularly in North America. *See also* HC1 pp.298–299; HC2 p.304.

Prescott, Alan (1927–), British rugby league player for Halifax, St Helens and Great Britain who captained the latter two from the front row. He played the last of his 28 matches for Britain (1951–58) with a broken arm, guiding his team to victory over Australia in what is known as "Prescott's Match".

Prescott, Samuel (1751–c.1777), US colonial patriot. Along with William DAWES and Paul REVERE he helped to warn the Concord colonists of the impending arrival of British troops (1775). He was cap-

tured while fighting at Fort Ticonderoga (1776) and died in prison in Nova Scotia, probably in 1777.

Prescott, William Hickling (1796–1859), US historian who, despite partial blindness completed authoritative historical surveys. His first book *The History of the Reign of Ferdinand and Isabella the Catholic* (1838) achieved immediate success. He also wrote the *History of the Conquest of Mexico* (1843) and the *History of the Conquest of Peru* (1847), which are considered to be his masterpieces. As history his work is now out of date because of subsequent research but it is still highly regarded as literature.

Prescription, list of drugs or other medication giving correct dosage and means of administration. It is usually given by a doctor to a patient, who takes it to a pharmacist who prepares or dispenses the drugs for the patient. Prescriptions eliminate the need for doctors to dispense drugs and allows the government a measure of control of the dispensing process.

President, the elected head of a republic. In the USA the president is elected by voters through the ELECTORAL COLLEGE for a term of four years and not exceeding two terms. The president is supreme military commander, appoints Supreme Court justices, ambassadors and other high officials, has the authority to make treaties with foreign countries (with the advice and consent of two-thirds of the Senate), grant pardons and veto legislation. Presidential power is limited by the system of checks and balances. *See also* MS p.276.

Presidium, in Soviet government, an influential standing committee. From 1952 to 1966, it replaced the POLITBURO as the chief administrative organ. The chairman of the legislative Presidium of the Supreme Soviet is a member of the Politburo.

Presley, Elvis (1935–77), US pop singer whose first recordings in 1953 led him to dominate rock and roll until 1963. He used the three idioms of blues, country and western music and folk, but it was his pelvic gyrations and sexy delivery that made him a superstar. *Jailhouse Rock, Hound Dog, Heart-break Hotel* and *Love Me Tender* are among his most successful songs. His films, such as *Love Me Tender* (1956) and *Follow That Dream* (1962), failed to make his appeal more solid, but in *Flaming Star* (1960) he turned in a bravura performance as a half-breed Indian. *See also* HC2 p.273.

Press, name of two Soviet sportswomen, sisters who dominated their events in field athletics. Tamara (1937–) won the shot and was second in the discus in the 1960 Olympic Games and won both events at the 1964 Olympics. Irina (1939–) won the 80m hurdles at the 1960 Olympic Games and the pentathlon in 1964. Both held world records in their events.

Press. *See* NEWSPAPERS.

Pressburg, Treaty of (1805), peace treaty between Emperor FRANCIS II and NAPOLEON I. In a swift campaign (1805), Napoleon defeated the Austrians at ULM and the Russians and Austrians at AUSTERLITZ. By the Treaty of Pressburg, Austria ceded Venetia, Istria and Dalmatia to Italy; the Tyrol, Vorarlberg and Augsburg to Bavaria; and certain Hapsburg lands to Württemberg and Baden. Austria was allowed to annex Salzburg but was obliged to recognize Napoleon as king of Italy. France gained Piedmont, Parma and Piacenza. *See also* HC2 p.76.

Press gang, men who went about the streets and taverns of Britain from the 13th century, but especially in the 18th, impressing men into the ROYAL NAVY. They also raided ships of the MERCHANT NAVY. The method ceased to be important after 1815, when improved conditions and higher pay induced sufficient voluntary enlistment. *See also* HC2 p.78.

Pressure, in physics, force exerted on an object divided by the area over which the force is acting. In SI units, pressure is measured in Newtons/sq m, called pascals (pounds/sq inch in the British system). Atmospheric pressure is measured in millimetres of mercury, one standard atmo-

sphere of pressure being 760mm. In meteorology, the unit is the millibar, one standard atmosphere being 1013.25mb.

Pressure gauge, instrument used to measure fluid pressure. Liquid-column gauges, such as MANOMETERS, use the shifts of liquid in U-shaped tubes to measure pressure; PITOT TUBES measure gas pressures between two points. Mechanical gauges, such as Bourdon tubes or bellows-element gauges, utilize the elasticity of metals to measure pressure. *See also* SU p.96, *96*.

Pressurization, reducing the volume of a substance by forcing it into an increasingly smaller space (eg compressing a gas), or restricting expansion of a heated substance (eg in PRESSURE-COOKING). Pressurization is accompanied by increased temperature, but pressure-reduction is accompanied by cooling; this is the principle of the refrigerator. *See also* SU pp.84–85, 94–95.

Prester John, or John the Elder, legendary Christian king, said to live in either Asia or Africa. Belief in the legend originated in Europe in the 12th century, and during the Crusades he was supposed to have inflicted a serious defeat on the Muslims. He may have been identified with several historical characters including GENGHIS KHAN. Several explorers set out in search of him in both Mongolia and Ethiopia. The 15th-century Portuguese who found the route around Africa hoped they would join forces with him against the OTTOMAN Turks.

Preston, county district in central LANCASHIRE, NW England; created in 1974 under the Local Government Act (1972). Area: 142sq km (54sq miles). Pop. (1974 est.) 132,000.

Preston, Battle of (1648), decisive battle of the second ENGLISH CIVIL WAR, in which CROMWELL defeated the Scots led by the Duke of Hamilton. The battle ended CHARLES I's hopes of retaining the throne. *See also* HC1 p.288.

Prestonpans, Battle of (1745), battle in E central Scotland in which the Jacobite forces led by the Young Pretender, Charles Edward Stewart, defeated the English under General Cope. *See also* HC2 p.28, *28*.

Prestressed concrete, concrete that is compressed along its length. The wet concrete is poured over lightly stretched steel reinforcing bars which extend the length of the particular piece. When the concrete is sufficiently dry the steel bars are cut. This releases their tension into the length of concrete, compressing it longitudinally. Thus a piece of prestressed concrete may carry a large direct or bending tension without itself going into tension. This method of multiplying the amount of stress a beam could withstand was greatly used during WWII. After reinforced concrete, prestressed concrete has revolutionized civil engineering and has led to the construction of long bridge spans and large roofed structures. *See also* MM p.49.

Pretender, Young. *See* STUART, CHARLES EDWARD.

Pretender, Old. *See* STUART, JAMES FRANCIS EDWARD.

Pretoria, capital of Transvaal Province and of SOUTH AFRICA. It was founded in 1855 by Marthinus PRETORIUS, the son of Andries Pretorius, the Boer leader after whom the city was named. It became the capital of the Transvaal in 1860, and of the Union of South Africa in 1910. The Peace of VEREENIGING (1902), which ended the Boer War, was signed there. Pretoria is an important communications centre, laid out on a spacious scale. Industries include steel production and diamond mining. Pop. (1970) 561,703. *See also* HC2 p.129; MW p.157.

Pretorius, Marthinus Wessel (1819–1901), Boer political leader. He helped to establish the TRANSVAAL and was president both of the Transvaal and the ORANGE FREE STATE (1859–63), which he tried to unite. He assisted Paul KRUGER in his opposition to British rule in the Transvaal after 1877.

Preventive medicine. *See* IMMUNIZATION; PUBLIC HEALTH; MS pp.112–113.

Prévert, Jacques (1900–77), French film scriptwriter best known for his work with

Maurice Prendergast painted many scenes of crowded public places.

Elvis Presley, the "king" of rock and roll, in a scene from *Girl Happy* (1955).

Press gangs recruited men for service under a law which still applies today.

Pretoria; a monument to the Voortrekkers who were killed in the Boer War.

Previn, André George

André Previn, a notable classical conductor, is also a popular jazz pianist.

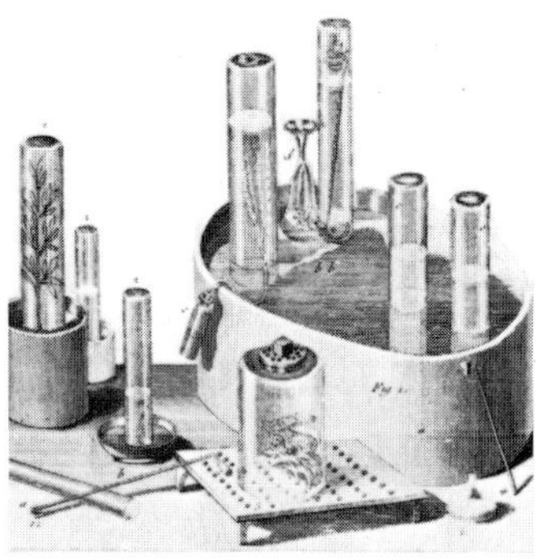

Joseph Priestly's experiments on the respiration of plants and animals.

Primates: the group of mainly tree-dwelling mammals which includes man.

Prime Minister: Robert Walpole may have been the first to use the title.

director Marcel CARNÉ. His films include *Le Jour se Lève* (1939), *Les Visiteurs du Soir* (1942) and *Les Enfants du Paradis* (1945).

Previn, André George (1929–), US conductor, pianist and composer, b. Germany. He studied composition (with Mario CASTELNUOVO-TEDESCO) and conducting (with Pierre MONTEUX) before pursuing a career as a jazz pianist and musical director of Hollywood film scores, for which he won an OSCAR on four occasions. He was appointed principal conductor of the London Symphony Orchestra in 1968. His orchestral compositions include a cello concerto (1968) and a guitar concerto (1970).

Previtali, Andrea (*c.* 1470–1528), Italian painter who was influenced by his Venetian teacher Giovanni BELLINI. His fine colouring and sense of composition can be seen in his *St Catherine* (1504) and *Madonna* (1506).

Prévost, L'abbé (1697–1763), French journalist and novelist who lived an unpredictable life as, alternately, a Jesuit novice, soldier and forger. His most famous work *Manon Lescaut* (1731) is the seventh novel of a series *Mémoires et aventures d'un homme de qualité*.

Prévost, Constant (1787–1856), French geologist who explained mountain formation as a process of gradual retraction of the Earth's crust. He also opposed the theory of successive catastrophes as an explanation of prehistoric fossils.

Priam, in Greek mythology, the King of Troy at the time of the war with Greece. He had been installed as king in his youth by HERCULES, but by the time of the ten years' war had become an old man. He lived to see his sons HECTOR and PARIS killed by the enemy; he himself was killed by Neoptolenus, the son of ACHILLES.

Priapulid, cylindrical marine WORM with a retractable proboscis. Length: to 8cm (3in). *See also* NW p.120.

Priapus, in Greek religion, a god of animal and vegetable fertility. The son of Dionysus and APHRODITE, he was represented in a caricature of the human form, with a grotesquely enlarged phallus. The ass was sacrificed in his honour because it symbolized lechery and was associated with the god's potency.

Pribilof Islands, group of four volcanic islands in the SE Bering Sea, Alaska, 403km (250 miles) NW of the Aleutian Islands. Discovered in 1786 by the Russian navigator Gerasim Pribilof, they were bought by the USA from Russia as part of the Alaska purchase in 1867. St Paul and St George, the larger islands, are breeding places for blue and white foxes and Alaska fur seals. Pop. (1970) 613.

Price, Richard (1723–91), British philosopher who is chiefly known for his *Review of the Principal Questions and Difficulties in Morals* (1758) which assesses the power of reason in making moral judgements. His writings on governmental finance were also well known.

Price, Vincent (1911–), US actor. His career in films began with *Service Deluxe* (1938) but it was not until *House of Wax* (1953) that he first took the macabre roles for which he was best known. His partnership with Roger Corman produced several excellent horror films, including *The Raven* (1963) and *Masque of the Red Death* (1964). Other films included *The Abominable Dr Phibes* (1971) and *Theatre of Blood* (1974).

Price index, measurement of the average price of goods and services at any given time. The RETAIL PRICE INDEX and wholesale price index are most commonly quoted.

Prices and Incomes Board, organization established in Britain in 1965 by the Department of Economic Affairs to administer the government's policy on wages and prices. In 1971 it was replaced by the Pay Board and Prices Commission.

Prichard, Katherine Susannah (1884–1969), Australian writer whose best works are tragedies of social protest in Australia. Her most famous novel is *The Pioneers* (1915). She also published *Brumby Innes* (1927), a play, and *Earth Lover* (1930), a book of poetry.

Prickly heat, itchy skin irritation caused by malfunctioning of the sweat glands in hot, humid weather. It occurs most often in infants and in overweight people. The irritation disappears as the body cools. *See also* MS p.98.

Prickly pear, cactus with flat or cylindrical joints; it grows in the USA and South America and has been introduced into Europe, Africa and Australia. The jointed pads have tufts of bristles and the edible fruit is red and pulpy. Its showy yellow blossoms and interesting shape make it a popular garden plant. Family Cactaceae; genus *Opuntia. See also* NW p.218.

Pride, Thomas (*d.* 1658), military commander on the Parliamentary side in the English Civil War. He commanded regiments at the battles of NASEBY (1645), DUNBAR (1650) and WORCESTER (1651). He also conducted the expulsion of the Presbyterians from the Long Parliament, an event known as PRIDE'S PURGE (1648).

Pride and Prejudice (1813), novel by Jane AUSTEN written in 1796–97. The most widely read of her books, it contains some of her greatest minor creations, such as the toady Mr Collins and the proud Lady Catherine de Bourgh. *See also* HC2 p.98.

Pride's Purge (1648), illegal expulsion of about 140 Presbyterian members from the English House of Commons. It was carried out by Colonel Thomas Pride's regiment, on the orders of the republican army council, who feared that the relatively moderate Presbyterians were royalist sympathizers because they opposed the trial of the king. The remaining RUMP parliament, completely under army control, then arranged the trial and execution of CHARLES I. *See also* HC1 p.288.

Priestley, John Boynton (1894–), British author, a prolific writer in a variety of genres. He started literary criticism as a student and later wrote *The English Novel* (1927) and *Literature and Western Man* (1960). His novels include *The Good Companions* (1929), *Angel Pavement* (1930) and *Bright Day* (1946). Among his plays are *An Inspector Calls* (1946) and *The Glass Cage* (1957). He also wrote many mystery stories.

Priestley, Joseph (1733–1804), British chemist. He studied the properties of carbon dioxide (then called "fixed air") and invented carbonated drinks. Although an advocate of the PHLOGISTON theory, he discovered oxygen (which he called "dephlogisticated air") in 1774 and a number of other gases, including ammonia and the oxides of nitrogen. He emigrated to the USA in 1794, where he renewed a friendship with John ADAMS. He also wrote about history and political theory. *See also* HC2 p.32.

Priest's hole, hidden room or secret hiding place for priests in time of persecution. They were constructed in larger houses in England in the 16th and 17th centuries at the time when the Catholic priesthood was being suppressed.

Prima ballerina, principal female dancer of a ballet company, who is generally one of the best ballerinas of her time. Notable examples from this century include Anna PAVLOVA, Alicia MARKOVA and Margot FONTEYN.

Primal therapy, treatment for neurosis first outlined by Arthur JANOV in *The Primal Scream* (1968). It seeks to take the patient in a group setting to the point of trauma, hoping for a cure by means of "primal" emotional upheaval. *See also* MS p.149.

Primate, in the Church, regional head of an episcopally structured hierarchy. In the Church of England the term refers to the Archbishop of York, who is "Primate of England", and to the Archbishop of Canterbury, "Primate of all England".

Primates, order of mammals that includes MONKEYS, APES, and human beings. Primates, native to most tropical and subtropical regions, are mostly herbivorous, day-active, tree-dwelling animals. Their hands and feet, usually with flat nails instead of claws, are adapted for grasping. Most species have opposable thumbs, and all but man have opposable big toes. They have a poor sense of smell, good hearing and acute binocular vision. The outstanding feature of primates, especially man, is the large complex brain and high intelligence. Primate characteristics are least pronounced in the relatively primitive prosimians (including tree shrews, LEMURS, BUSHBABIES, LORISES and TARSIERS) and are most pronounced in the more numerous and advanced anthropoids (monkeys, apes and human beings). *See also* NW pp. 168–169, 172–173, 189.

Primaticcio, Francesco (1504–*c.* 1570), Italian painter. After assisting Giulio Romano in projects for the Duke of Mantua until 1532, he spent the rest of his life in France at Fontainebleau and Paris as court painter. He is the creator of the FONTAINEBLEAU school of French MANNERISM. At Fontainebleau, his decorative stucco work, a combination of painting and high relief, is partially preserved. *See also* HC1 p.263.

Primavera, or Allegory of Spring (*c.* 1478), tempera painting by BOTTICELLI now in the Uffizi gallery, Florence. The greatness of the work lies in Botticelli's ability to define an ideal style of beauty and to set it in a composition of great restraint; yet he also manages to suggest a wealth of emotional activity between his stylized figures.

Prime meridian, meridian adopted as the zero of longitude on a planet. The prime meridian on Earth passes through Greenwich, England.

Prime Minister, chief executive and head of government in a country with a parliamentary system. He is usually the leader of the largest political party in parliament. The office evolved in Britain in the 18th century along with the CABINET system and the shift of power away from the Crown towards the House of Commons. Some historians consider Robert WALPOLE to be the first Prime Minister; others reserve the title for the younger PITT. Since the time of Pitt, the Prime Minister (the title has no official recognition) has always been the First Lord of the Treasury.

Prime number. *See* NUMBER, PRIME.

Prime of Miss Jean Brodie, The (1961), novel by Muriel Spark. It describes the political and moral dilemmas of a "progressive" Scottish school teacher.

Primitive, in art, a term with three different meanings. (1) It is applied to the art of eg, the Eskimos, the inhabitants of Oceania and of Black Africa – peoples whose culture evolved outside the direct influence of the great Oriental and Western centres of civilization. In the early 20th century primitive art from Africa and Oceania profoundly influenced Western art through the work of PICASSO and others. (2) In Western art the term has been used to describe early phases in the development of painting and sculpture, eg, it has been applied to pre-RENAISSANCE art because it pre-dated the scientific use of PERSPECTIVE and anatomy. (3) The name "primitive" is also given to certain self-taught artists whose work is neither traditional, academic nor *avant-garde* and whose approach and vision is usually naive and often immensely poetic and powerful. Examples of such primitives are Henri ROUSSEAU, Grandma MOSES and Alfred WALLIS.

Primitivism, Russian form of EXPRESSIONISM. It developed *c.* 1905–1920 and was influenced by Russian folk art, FAUVISM and CUBISM. It was characterized by simplified forms and powerful colour, used principally to depict scenes from working-class life. MALEVICH worked in the style early in his career; typical exponents were Larionov and Gontcharova.

Primo de Rivera, José Antonio. *See* RIVERA, MIGUEL PRIMO DE.

Primogeniture, in a FEUDAL SYSTEM, rule of inheritance whereby land descended to the oldest son. In most parts of France and Germany *parage*, involving the partition of properties held in FIEF, was more common, but in England primogeniture became the prevailing custom by the 12th century, so that the father's land was kept intact for the support of the son, who rendered the required military service. When feudalism declined and the payment of a tax replaced military service, the need for primogeniture disappeared. In England, however, the practice is not uncommon to this day.

Primrose, Archibald Philip, Lord Rosebery. See ROSEBERY, ARCHIBALD PHILIP PRIMROSE, 5TH EARL OF.

Primrose, William (1904–), US viola player, *b.* Scotland. One of the world's leading soloists, he toured the USA as a viola player with the London String Quartet (1930–35) and formed his own quartet in 1938.

Primrose, any of numerous species of herbaceous, generally perennial plants that grow in the cooler climates of Europe, Asia, Ethiopia, Java and North America. It has a tuft of leaves rising from the rootstock and clustered flowers of pale yellow to deep crimson. Family Primulaceae; genus *Primula. See also* NW pp.*71, 209.*

Primrose League, unofficial organization or club of the CONSERVATIVE PARTY, formed in 1883 by the small coterie of Conservative MPs, led by Lord Randolph CHURCHILL and other members of what was known as the Fourth Party. Its object was to uphold the CHURCH OF ENGLAND, the monarchy, the aristocracy and Britain's imperial strength. It took its name from the belief that DISRAELI's favourite flower was the primrose. In 1910 the League had two million members, but such popularity did not last.

Primula, genus of flowers of the family Primulaceae. Species include the common PRIMROSE (*Primula vulgaris*), oxlip (*P.elatior*) and COWSLIP (*P.veris*). In common usage the term refers to a cultivated variety of the group. *See also* NW p.*71.*

Prince, Harold (1928–), US theatrical producer. In the 1950s he co-produced several successful musicals, including *Pajama Game* (1954–56), *Damn Yankees* (1955–57) and *West Side Story* (1957). He won an award in 1962 for *A Funny Thing Happened on the Way to the Forum.* His other productions include *Fiddler on the Roof* (1964) and *Cabaret* (1966), which he also directed.

Prince, royal title first used after the break-up of the CAROLINGIAN Empire in the 9th century. From the Latin *princeps* ("chief person"), the heir to the English throne has customarily received the title PRINCE OF WALES since 1301. Other members of the British ROYAL FAMILY have been designated "Prince" by the monarch at various times.

Prince Edward Island, island province in E Canada in the Gulf of St Lawrence off the coast of New Brunswick and Nova Scotia. Most of the island is part of the Prince Edward Island National Park and is a popular resort area. Fishing is the most important industry; food processing is the chief manufacturing industry. Charlottetown, the capital, and Summerside are the chief cities. The island was discovered by Jacques CARTIER in 1534. Area: 5,657sq km (2,184sq miles). Pop. (1976 est.) 118,229. *See also* MW p.45.

Prince Igor (1890), opera in prologue and four acts by Alexander BORODIN, with libretto by the composer based on a sketch by Vladimir Stassov. Unfinished at the time of Borodin's death, the work was completed by Nicolai RIMSKY-KORSAKOV and Alexander GLAZUNOV, and was first produced in St Petersburg (Leningrad).

Prince of Wales, royal title in Britain. It was originally a native Welsh title, but it was taken over by the English monarchy in 1301 when EDWARD I, having crushed Welsh independence, bestowed the title on his son EDWARD II. The first and last Welsh prince to be recognized by the English was LLYWELYN AP GRUFFYDD, slain at Builth in 1282. Since 1343 the title has been given regularly to the eldest son of the English monarch.

Princes in the Tower, EDWARD V and Richard, Duke of York, the sons of EDWARD IV. On their father's death Edward V, then 12 years old, became king with his uncle, Richard, Duke of Gloucester as Protector of the Realm. The princes were housed in the Tower of London following factional disputes between the Woodville group of nobles, the Queen's party, and the Duke of Gloucester. Gloucester summoned the

lords and commons, who proclaimed the children illegitimate and Gloucester king. The princes disappeared; it has been supposed that they were murdered by Gloucester's agents, but the evidence is inconclusive.

Princess Royal, title sometimes bestowed on the eldest daughter of the sovereign. Princess Mary (1897–1965), daughter of GEORGE V, received the title in 1952. She married Viscount Lascelles, later 6th Earl of Harewood, in 1922 and they had two sons.

Princip, Gavrilo (1893–1918), Serbian nationalist. His assassination of the Archduke FRANZ FERDINAND and his wife at Sarajevo in 1914 precipitated WWI. He was sentenced to 20 years imprisonment but died of tuberculosis in 1918. Princip is regarded as a national hero in Yugoslavia. *See also* HC2 pp.*187*, 188.

Principia Ethica (1903), early work by the philosopher G. E. MOORE which marks a movement from his belief in the reducibility of all definitions of universals to the belief, shared with Bishop Butler whom he quotes, that "Everything is what it is and not another thing". Moore concludes that the concept of Good cannot be analysed.

Principia Mathematica (1910–13), three-volume work resulting from a collaboration between A. N. WHITEHEAD and Bertrand RUSSELL. It propounded that mathematics can be deduced from formal logical premises, and notably influenced 20th-century philosophy.

Principle of moments, in mechanics, states that the moments of two bodies balanced about a central pivot or fulcrum are equal. (The moment of a body is the product of its weight and its distance from the pivot.) This principle may be observed in balances and see-saws.

Principles of Political Economy. See MALTHUS, THOMAS ROBERT.

Principles of Political Economy. See MILL, JOHN STUART.

Pringsheim, Nathanael (1823–94), German botanist who investigated the reproductive systems of lower plants. He discovered the occurrence of sexuality in ALGAE and, after further studies, concluded that natural selection plays only a minor part in evolution; he believed that variations are spontaneous and tend towards greater complexity. Pringsheim was the first to demonstrate apospory, the production of a sexual from an asexual generation without the intervention of spores.

Printed circuits, network of electrical conductors chemically etched from a layer of copper foil on insulating plastic, used to interconnect components (eg capacitors, resistors and transistors). The printed circuit board represents one stage in the miniaturization of electronic circuits (MICROELECTRONICS); it has not been replaced by the newly developed INTEGRATED CIRCUIT, but serves to reduce the maze of connecting wires and to compact even more circuitry (eg a complete calculating complex) on a single board or module. *See also* MM p.255.

Printing, reproduction of graphic material by multiple impressions. Although GUTENBURG, CAXTON and others established the use of LETTERPRESS, and TYPEFOUNDING and TYPESETTING progressed steadily, real production improvements came only with the INDUSTRIAL REVOLUTION. Then, printing expanded rapidly with the introduction of STEREOS, PHOTOLITHOGRAPHY, ROTARY PRESSES, HALFTONES and mechanical typesetters. Progress in the 20th century has stressed speed with the application of PHOTOGRAVURE, WEBOFFSET, PHOTOSETTING and computer-aided production. *See also* HC1 pp.242, 246–247, *246–247*; HC2 p.98; MM pp.204–207, *204–207.*

Printmaking, in graphics, process of making printed impressions from wood blocks, metal plates, flat stones or silk. There are four main groups of graphic techniques. The first is the relief method, in which the parts of a wood block or metal plate that are to be printed are left in relief, and the remainder is cut or etched away. Ink is rolled over and adheres to the raised por-

tions, and a print may be made by hand or on a printing press. Some examples are WOODCUTS, WOOD ENGRAVINGS, linocuts and metal cuts. The second is the intaglio method, in which the surface of the block or plate does not print. Instead ink is held in the engraved or etched furrows, damp paper is pressed by a copper plate press into these cuts and the ink is picked up from them. Examples are line ENGRAVINGS, DRYPOINTS, ETCHINGS and AQUATINTS. The third method is the surface or PLANOGRAPHIC method, in which greased areas on a slab of limestone or a metal plate reject printing ink. The ink adheres to the ungreased areas and is transferred to paper. In the stencil method, colour is brushed through holes in a protecting sheet onto the surface below; examples are SILKSCREEN PRINTS, or serigraphs. *See also* MM pp.204–207.

Prior, Matthew (1664–1721), English poet and diplomat who as special envoy, helped to conclude the Peace of UTRECHT in 1713. He was imprisoned by the WHIGS for two years on Queen Anne's death (1714). As a poet he is best remembered for his light verse which includes *Alma, or the Progress of the Mind.*

Prior, in the Christian Church, head or deputy head of a monastery. The term had a vague usage until the 13th century, being applied to various monastic officials and (in Italy) to magistrates. But in the Benedictine order it came to refer to the monk ranking next below the ABBOT, and to the head of a priory of monks or a house of canons regular or friars.

Prism, in mathematics, a solid geometrical figure whose ends are similar polygons (most commonly triangles) with the other faces parallelograms or, in a right prism, rectangles. In physics, a prism (usually triangular) is a piece of transparent material, such as glass, plastic or quartz, in which a light beam is refracted and split into its component colours by dispersion or undergoes internal reflection. *See also* SU p.100.

Prison, place in which people are kept either for safe custody while awaiting trial or as a punishment after conviction for a crime. Prisons developed as penal institutions in England in about the 18th century as an alternative to execution, corporal punishment or transportation, although there were instances of the use of institutional prisons much earlier. The first modern prison was probably Walnut Street Jail in Philadelphia (1790). As a result of pressure to improve conditions inside prisons in Britain, the Prison Act (1839) was passed and in 1877 prisons were made the responsibility of the Home Office rather than local bodies, and the prison system was centralized under the Prison Commission. *See also* MS pp.296–297.

Prisoner of Zenda, The (1894), novel by Anthony HOPE. A popular romance of impersonation and intrigue, it has given to English language the words "Ruritania" and "Ruritanian", which pertain in the book to a fictional kingdom.

Prisoners of war, in international law, military personnel captured by the enemy in an armed conflict between nations. Their treatment is generally expected to be humane. The terms of the first international convention on prisoners of war, signed at the HAGUE PEACE CONFERENCE of 1899, were widened by the 1907 Hague Convention and by the subsequent GENEVA CONVENTION of 1949.

Prison reform, movement aimed at improving conditions inside prisons and exploring alternative systems of rehabilitation. It developed in Britain at the beginning of the 19th century, when the efforts of early prison reformers such as John Howard and Elizabeth FRY led to the setting up of a parliamentary committee to investigate prison conditions. The Prison Act (1839) was the first major legislative attempt to improve conditions. The establishment of government-controlled prisons in 1877 paved the way for major advances, and today prison reform includes such factors as study facilities, home leave and outside work for prisoners. *See also* MS pp.296–297.

Primroses are attractive flowers that thrive well in gardens and greenhouses.

Prince Igor; the score frontispiece of Borodin's opera, unfinished when he died.

Matthew Prior's personal charm and shrewdness made him an excellent negotiator.

Prison Reform; Elizabeth Fry reading to female prisoners in Newgate Prison.

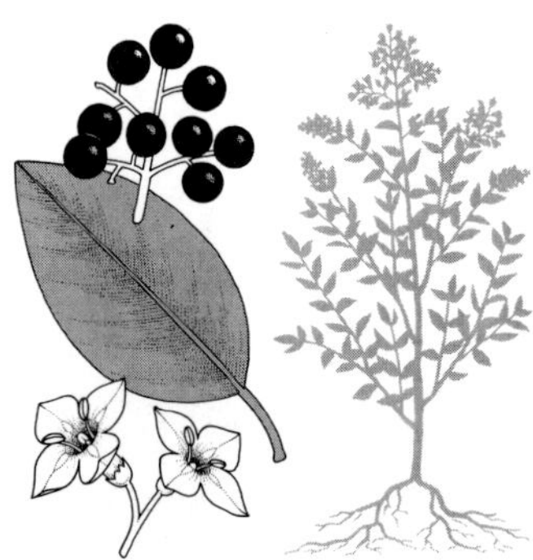

Privet is a member of the olive family and is commonly used for hedges.

Prix de l'Arc de Triomphe, official 1977 photo; Lester Piggott wins on *Alleged.*

Proboscis monkey: its large nose is less developed in the infants and females.

Probus was murdered by troops angered at having to carry out menial public works.

Pritchett, Sir Victor Sawdon (1900–), British journalist, critic and novelist. His works include a novel *Clare Drummond* (1929), many critical works and two volumes of autobiography, *A Cab at the Door* (1968) and *Midnight Oil* (1971).

Private enterprise, economic organization based upon the investment and initiatives of an individual or an independent commercial body as distinct from the state. *See also* MS pp.312–313.

Privateer, armed, privately-owned vessel under commission of a belligerent government to capture enemy shipping. Letters of MARQUE (commission) distinguished it from a pirate craft. Crews were not paid but were allowed to keep most or all of their booty and often acquired great wealth. Privateering was most in evidence from the late 16th to the early 19th century. It was outlawed by the Declaration of Paris (1856) and finally abolished by the Hague Conference of 1907.

Private Lives (1930), sophisticated social comedy by Noël COWARD. It revolves around the chance meeting of a man and his ex-wife, both on honeymoon with their new partners.

Private Member's Bill, BILL introduced in the British parliament by an individual MP without the official support of the government. Only 10 days' time each session is reserved for such bills, and their order of precedence is determined by ballot. Only a few of them reach the statute book.

Private schools, in Britain, independent educational institutions usually run for profit. A few are either Church-run schools or "progressive" schools, as some people feel that the national educational system does not meet adequately the needs of Christians or those interested in unorthodox educational methods. A minority of private schools teach those who fail to get into independent PUBLIC SCHOOLS, and these are generally run on traditional lines. All are inspected by and registered with the Ministry of Education. In the USA, the term private school applies to institutions that would be known as public schools in Britain.

Privet, deciduous shrub native to Australia, Asia, Europe and N Africa, frequently planted to form a hedge. It has smooth, lance-shaped leaves and loose clusters of tiny white flowers that appear in summer; it bears small black berries. Species include the common, or European, privet (*Ligustrum vulgare*) and the hardier California privet (*L. ovalifolium*), which is native to Japan. Height: to 4.6m (15ft). Family Oleaceae.

Privilege, parliamentary, certain rights enjoyed by members of both Houses of the British Parliament. They include the right of FREE SPEECH and immunity from civil arrest, the right to regulate their own proceedings, the right to punish an MP for CONTEMPT OF PARLIAMENT, and the Commons' right to determine its own membership.

Privy Council, group of leading advisers to the British monarch. It developed in the Middle Ages out of the King's Council (Curia Regis). As the CABINET system of government developed, the Privy Council became increasingly restricted in its powers. Its Judicial Committee, established by legislation in 1833, is the final appeal court for most Commonwealth countries. The Privy Council also maintains non-political committees on medical and industrial research.

Privy Purse, Keeper of, British royal officer in charge of the monarch's domestic accounts. The office is derived from that of the medieval lord CHAMBERLAIN. Since the 19th century he has been responsible for the administration of the CIVIL LIST.

Privy Seal, Lord, royal officer in Britain, keeper of the privy seal. The seal is inferior to the GREAT SEAL and is used as proof of royal assent to a document, especially an act of parliament. The office dates from medieval HOUSEHOLD; it now usually carries a CABINET post without portfolio.

Prix de l'Arc de Triomphe, French horse race run over 2.4km (1.5 miles) at Longchamp on the first Sunday in October. A weight-for-age event for three-year-olds and over, it is the richest horse race in Europe. It was first run in 1920.

Prix Goncourt, French literary prize set up in 1903 by the Académie Goncourt, endowed by Edmond de GONCOURT. Awarded annually to the best prose work published in the year, it carries great prestige in the literary world. Usually awarded for novels, it has in the past been won by such authors as Marcel PROUST, André MALRAUX and Simone de BEAUVOIR.

Probability, number representing the likelihood of a given occurrence. The probability of a specified event is the number of ways that event may occur divided by the total possible number of outcomes. For instance, in one throw of a six-sided die there are six possible outcomes, and three of these result in an even number: the probability of throwing an even number is thus $\frac{3}{6}=\frac{1}{2}$. This assumes that each possibility is equally likely. A less circular idea of probability utilizes the concept of a limit. If the die were thrown a large number of times the number of even numbers resulting divided by the total number of throws would tend to the value $\frac{1}{2}$. Probability theory is concerned with the analysis of random events of this type. *See also* SU pp.56–59.

Probate, legal term for the certification by a court of law that a document purporting to be the will of a person who has died is valid. The term also applies to the official copy of a will, with the certificate of its having been proved valid.

Probation, in Britain, sentence of a court of law on a person aged 17 or over which allows the offender to remain at liberty subject to certain conditions, such as good behaviour, and under the supervision of a probation officer. It is not regarded as a conviction. The sentence may last from one to three years, and breach of probation is an offence that invites reconsideration of the original probation order and may lead to a fine or to the issuance of a community service order.

Proboscidea, order of mammals that has lived on Earth from the Eocene to the present day, but now represented only by ELEPHANTS (genera *Elephas* and *Loxodon*). They are characterized by their primitive limbs, but have specalized jaws and teeth. Family Elephantidae. *See also* NW pp.154, *155.*

Proboscis monkey, large monkey, primarily of swamp regions in Borneo, with a protruding nose that is upturned in young monkeys and long and pendulous in older males. These monkeys are gregarious, day-active herbivores that appear to swim and dive freely. Family Ceropithecidae; species *Nasalis larvatus. See also* NW pp.215, 235, *235.*

Probus, Marcus Aurelius (*d.*282), Roman emperor (276–82). A distinguished soldier, he became emperor after a long period of political instability, and was notable for allowing BARBARIANS to settle within the empire. He reformed the administration, and tried to impose strict discipline on the army. His troops mutinied and he was killed. *See also* HC1 p.114.

Procaccini, Ercole the Elder (1515–95), Italian painter who was born and worked in Bologna. Various Bolognese churches contain paintings by him, notably *Descent from the Cross* in the church of San Stefano. He later moved to Milan, where he founded an academy which enjoyed a high reputation in contemporary Italy.

Procellariiformes, order of birds that includes the ALBATROSS, SHEARWATER, STORM PETREL and diving petrel. *See also* NW p.140.

Process engraving. *See* PHOTO ENGRAVING.

Proclamation of 1763, British document designed to restrain encroachment on Indian lands by settlers following the FRENCH AND INDIAN war. Drawn up by Lord SHELBURNE, President of the Board of Trade, and modified by his successor Lord Hillsborough, it formulated a policy that would placate the Indians. The provisions forbade American settlement west of the line formed by the Appalachians and ordered all settlers to vacate the area, which was to be reserved for the Indians.

Four new provinces for the settlers were established – Quebec, East and West Florida, and Grenada.

Proclamations, commands issued in Britain under the royal prerogative. They were first used by the Tudors in place of medieval ordinances to supplement statutes, and their use was approved by the Statute of Proclamations (1539); they were enforced by the Court of STAR CHAMBER. In 1610 the judges ruled, on a Commons' petition, that proclamations could not be used to make new laws or to override the common law. A proclamation is now used chiefly to announce a public holiday or the dissolution of Parliament.

Proclus (*c.*410–85), Greek Neo-platonic philosopher and disciple of the PLATONIST Syrianus. He synthesized Neo-platonic doctrines and gave them their most systematic form. He also wrote commentaries on many of PLATO's dialogues. He opposed Christianity and stressed the Neo-platonic view that material things are appearances, whereas thought constitutes reality.

Proconsul, in ancient Rome, title originally used for a CONSUL whose term of office had been extended beyond the usual 12 months. This extension became increasingly common, especially in Roman territories, and in later Roman times the term came to denote the almost dictatorial governors of Roman provinces. The term has sometimes been applied in modern times to a British colonial governor with far-reaching powers.

Procrustes, in Greek mythology, a brigand in Attica. He tortured travellers whom he captured by placing them on a bed and stretching them to fit if they were too short, or cutting off their limbs if they were too long. THESEUS had him killed by his own methods. He is also known as Damastes.

Procter, Mike (1946–), South African cricketer. A spectacular all-rounder, he played in only seven Test matches (1967–70) before South Africa's exclusion from Test cricket. In England he proved himself an inspirational captain of Gloucestershire; in South Africa he played for Natal, Western Province and Rhodesia.

Proctology, branch of medicine that deals with the diagnosis and treatment of diseases of the RECTUM and lower intestine.

Procurator-fiscal, officer appointed to the sheriff's court in Scotland who acts as public prosecutor and also conducts inquiries into suspected offences. Some of his duties correspond roughly with those of a CORONER in England.

Procyon, or Alpha Canis Minoris, the brightest star in the constellation of Canis Minor and one of the stars nearest to the Sun. The star, known as Procyon A, has a faint white dwarf companion, Procyon B. Characteristics: apparent mag. 0.34 (A), 10.8 (B); absolute mag. 2.6 (A), 13.1 (B); spectral type F5 (A), wF (B); distance 11.4 light-years. *See also* SU pp.244, 257, *257.*

Producer gas, mixture of gases, the inflammable carbon monoxide and hydrogen and the nonflammable nitrogen and carbon dioxide. It is made by the partial combustion of coal or coke in air and steam. It has a lower heating value than other fuels, but can be manufactured with simple equipment. Producer gas is often used as a fuel in large industrial furnaces.

Progesterone, steroid HORMONE secreted mainly by the corpus luteum of the mammalian OVARY, and by the PLACENTA in pregnancy. Its principal function is to prepare and maintain the uterus for pregnancy. Synthetic progesterone is one of the main components of the contraceptive PILL, in which its action is to simulate pregnancy artificially, thus preventing conception.

Prognosis, prediction relating to the expected final outcome in a patient with a particular disorder. The prognosis is based on a doctor's knowledge of the disorder and the probable effects of treatment.

Programmed learning, method of instruction in which a student proceeds through a course of highly structured material at his own pace. The course is designed both to instruct and test the student and to provide an immediate assessment of his progress. *See also* MM p.*109*.

Programme music. *See* ILLUSTRATIVE MUSIC.

Programming, preparation of a COMPUTER so that it can provide the answers to questions put to it. Before being given data, a computer must be given a set of instructions, or a program, telling it how to deal with the data. Each instruction is a single step and all information must be in the form of binary numbers. For computer programmers, languages have been developed, such as FORTRAN for mathematics and science and PL/1 for general use. *See also* MM pp.106, 108, *108*.

Progression. *See* ARITHMETICAL PROGRESSION; GEOMETRIC PROGRESSION.

Prohibition (1919–33), period in US history in which the manufacture, transport, sale and consumption of alcoholic beverages was illegal, enacted through the 18th Amendment to the Constitution. Enforcement of the ban proved extremely difficult. Smuggling could not be prevented and the illegal manufacture of liquor sprang up with such rapidity that the authorities were unable to suppress it. Illegal drinking of inferior and often dangerous alcoholic beverages became commonplace, while corruption of government officials and local police, paralysis of the courts and the growth of organized crime financed by immense BOOTLEGGING profits all followed. In 1933 the US government had no alternative but to repeal the law under the 21st amendment. *See also* HC2 p.208.

Proinsulin, precursor of the hormone INSULIN, a polypeptide formed in the PANCREAS by the beta cells of the ISLETS OF LANGERHANS. Its long-chain molecule splits to form the insulin molecule.

Projection, map, method of representing geographical features of a curved surface, such as those of the Earth, on a flat surface. Map projections can be cylindrical, conic or azimuthal, depending on the purpose of the map. They all produce some distortions of the features. *See also* PE pp.34–35; SU pp.54–55.

Projector, instrument with a lens system, used to cast images onto a screen from an illuminated flat object. An episcope is a projector for opaque objects such as a printed page; it uses light that is reflected from the object. A diascope is a projector for transparent objects such as photographic slides and films; it uses light transmitted through the object. An epidiascope can project images from both transparent and opaque objects. A cine projector produces moving images from the many frames (pictures) on cine film. *See also* MM p.223, *223*.

Prokhorov, Alexander Mikhaylovich (1916–), Soviet physicist who shared the 1964 Nobel Prize in physics with Nikolai G. BASOV and Charles H. TOWNES for research in QUANTUM electronics that resulted in the development of the MASER and LASER. In 1955 Prokhorov and Basov proposed the maser principle, although it was Townes who built the first maser.

Prokofiev, Sergei (1891–1953), Soviet Neo-Classical pianist and composer, whose musical style is characterized by melody, biting dissonances, rich harmony and brilliant orchestration. His most popular works include the ballets *Romeo and Juliet* (1935) and *Cinderella* (1944); the *Classical* (first) *Symphony* (1918); *Peter and the Wolf* (1936), an orchestral suite with narrator for children; and the opera *The Love of Three Oranges* (1921). He also composed piano music, six other operas, six other symphonies, five piano concertos, chamber music and film music, including scores for *Lieutenant Kizhe* (1934), *Alexander Nevsky* (1938) and *Ivan the Terrible* (1942–45).

Prolactin, HORMONE secreted by the pituitary gland in women which stimulates the production of milk after childbirth. The secretions are inhibited by OESTROGEN and PROGESTERONE during pregnancy, but the inhibition ceases with the loss of the placenta at childbirth.

Prolapse, displacement of an internal part of the body, such as the uterus, bladder or rectum, especially after childbirth. It is also the displacement of one of these organs or a structural part, such as an intervertebral disc.

Proletariat, Marxist term for the working class in an industrial society, which has no source of income other than wages. The term derived from the Latin *proletarius,* which described the landless freeman who was a member of the poorer class of ancient Rome. According to Marx, the proletariat is the true creator of the objects produced by industry, and therefore would become an irresistible force when the internal contradictions of CAPITALISM weaken the authority of the factory-owning middle class.

Puget, Pierre (1620–94), French BAROQUE sculptor, painter and architect. In 1667 he began work on the doorway of the City Hall at Toulon. His sculpture, which was influenced by BERNINI, includes *Milo of Crotona* (1683).

Promenade concerts, series of annual summer concerts organized by the BBC in the Royal Albert Hall, London, where cheap standing room is available in the gallery and the stalls. Such concerts originated in Paris in 1833, conducted by Musard, and from 1838 promenade concerts were given at Drury Lane, Covent Garden and elsewhere where the opportunity for strolling to strictly instrumental music was then possible. "Proms" in their modern series began at the Queen's Hall, London in 1895, under the baton of Sir Henry WOOD who established the practice of including new British music in a wide-ranging repertoire. After Wood's death in 1944, a succession of principal conductors took his place, the most popular being Sir Malcolm SARGENT.

Prometheus, in Greek mythology, the fire-giver; his name means "forethought". He created man, provided him with reason, and stole fire from the gods. For this theft, ZEUS had him chained to a rock in the Caucasus where an eagle consumed his liver for eternity. In some myths he was rescued by HERCULES. Prometheus is the subject of many literary works.

Prometheus Bound (*c.*459 BC), tragedy by the Greek dramatist AESCHYLUS. In this, the first part of a trilogy, Prometheus incurs the wrath of ZEUS by giving man the gift of fire. The second and third parts of the trilogy, *Prometheus Unbound* and *Prometheus the Fire-bearer,* are lost.

Prometheus Unbound (1820), allegorical verse drama by Percy Bysshe SHELLEY based on Greek myth. It expounds the triumph of human love.

Promethium, radioactive metallic element (symbol Pm) of the lanthanide group, made in 1941 by particle bombardment of neodymium and praseodymium. The element does not occur naturally on Earth. Properties: at. no. 61; m. p. 1,080°C (1,980°F); b. p. 2,460°C (4,460°F); most stable isotope Pm^{145} (half-life 17.7yr).

Prominence, solar, gaseous jet or cloud of hydrogen or helium gas in the CHROMOSPHERE of the Sun. Prominences have been observed extending more than 50,000km (312,500 miles) above the Sun's surface. They are of two types: eruptive, which are in violent motion, and quiescent, which are much more stable. Prominences are often associated with SUNSPOTS. *See also* SU pp.222, *222–223*.

Promissory note, unconditional promise made in writing to pay the bearer of the note a sum of money either on demand or at a fixed time. It differs from a bill of exchange or a bank CHEQUE in having only two, not three, parties to the contract.

Promoter, in chemistry, a substance used in small quantities in conjunction with a CATALYST to increase its activity. For example, in the Haber-Bosch process an iron catalyst is used to speed up the combination of hydrogen and nitrogen under pressure to form ammonia; this catalyst is usually promoted with small amounts of potassium oxide.

Prompt, in the theatre, to supply an actor with forgotten words. The prompter is stationed within earshot of the players but

out of sight of the audience in the prompt box, usually to the left of the stage.

Pronghorn, only extant member of the family Antilocapridae, related to the ANTELOPE. It is a horned, hoofed herbivorous animal of W USA and N Mexico. The swiftest North American mammal, it is said to be capable of a speed of 80km/h (50mph) over short distances. Height: 3ft (90cm); weight: 45kg (100lb).

Prontosil, trade name for sulphamidochrysoidine, an azo dye which was the first synthetic antimicrobial drug to be widely used. The active part of the molecule was found to be sulphanilamide, and this discovery led to the development of the sulpha drugs.

Propaganda, systematic manipulation of public opinion by use of oratory, publications and photography. Although allusions to propaganda are found in ancient writings, the organized use of propaganda developed only after the Industrial Revolution when modern methods of communication enabled propagandists to reach mass audiences. Modern propaganda differs from other forms of communication in that it is deliberately used to control group attitudes.

Propagation, plant, means of producing several individual plants from one original. This may be in the form of sexual reproduction in which the plant is pollinated and the seed produced in the normal way, or vegetative propagation, in which new plants are grown from parts of the parent, eg leaves, shoots, bulbs or runners. Man has taken all these processes and developed them for commercial use. *See also* PE pp.178– 179.

Propane, colourless inflammable gas (formula C_3H_8), the third member of the alkane series of hydrocarbons. It occurs in natural gas, from which it is obtained; it is also obtained during petroleum refining. It is used (as bottled gas) as a fuel, as a solvent and in the preparation of many chemicals. Properties: m.p.−187.7°C (−305.9°F); b.p. −42.5°C (−44.5°F). *See also* MM p.78.

Propellant, material that undergoes chemical, nuclear or thermoelectric reactions to propel a rocket. Liquid propellants consist of a fuel such as KEROSENE or HYDRAZINE (N_2H_4), which reacts with an oxidizer such as liquid oxygen. Solid propellants contain both fuel and oxidizer in powder form. Nuclear propellants include URANIUM and PLUTONIUM. Ion propellants include metallic CAESIUM, which boils off ions into an electric field that accelerates them to high exhaust velocities. *See also* MM p.156; SU p.268.

Propeller, device for producing thrust, usually mounted on a rotating shaft. The cross-section of an aircraft propeller reveals an aerofoil shape. This allows it to function as a rotating wing; it generates forward thrust by producing aerodynamic lift. A ship's propeller, or screw, "screws" the ship through the water.

Proper motion, in astronomy, transverse movement of a star against the background of more distant stars. The real movement of a star in space is a combination of its proper and radial motions. The largest annual proper motion for a star is approx. 10 seconds of arc (Barnard's Star). *See also* SU p.*165*.

Properties, in the theatre, loose items required during the action of a play which do not count as furniture, costumes or scenery. Properties (usually called merely props) include such items as letters, cigarettes, food, drink, crockery and telephones. Hand props are those that are handled by the actors.

Propertius, Sextus Aurelius (*c.*54–*c.*16 BC), Roman elegiac poet. Although starting as a student of Law, he began writing at an early age, and his first book of poetry, concerned with his mistress Cynthia, was published in 29 BC. His verse is significant for its sensitivity and passion and its frequent abrupt change of mood.

Prophets, in all the major religions, people who believe they have a message from God to deliver or are able to foretell events. For Jews in Old Testament times the prophets were those whose writings had the authority of YAHWEH. They pri-

Prohibition; a US government agent closes down an illegal bar.

Promenade concerts; Sir Malcolm Sargent conducting the BBC Symphony Orchestra.

Prometheus was sentenced to eternal torture for stealing fire from the gods.

Propaganda: William Joyce broadcast as Lord Haw Haw for Nazi Germany.

Proportion

Proscenium arches have been created by set designers to frame a play's action.

Protectorate: a contemporary print shows Cromwell dissolving parliament.

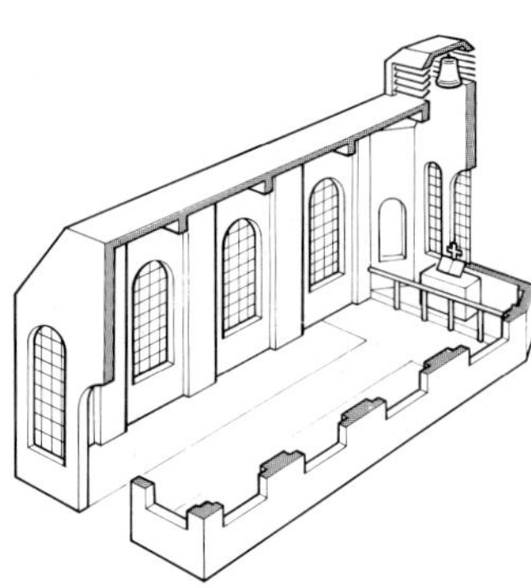

Protestantism has led architects to plan churches that are simple and functional.

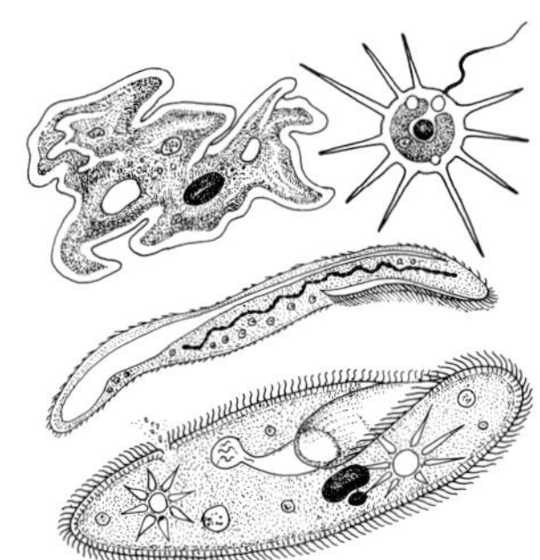

Protista is a term used by scientists to classify such organisms as those above.

marily interpreted God's will to contemporaries rather than foretold the future. The term was also applied to the persons of MOSES, DAVID and others. For Christians, from the time of the early Church onwards, the term signified the writers of the Old Testament (such as ISAIAH) whose sometimes obscure and esoteric passages can be interpreted as predicting the events and significance of the prophet Christ's life. The New Testament also refers to prophets in the early Church such as JOHN THE BAPTIST. Among Muslims, the term is applied exclusively to MOHAMMED, although various individuals proclaimed themselves as his successors. In both Buddhist and Hindu literature, predictions occur and many prophetic reformers have occurred in HINDUISM.

Proportion, mathematical relation of equality between two ratios, having the form $a/b = c/d$. A continued proportion is a group of three or more quantities, each bearing the same ratio to its successor, as in $1:3:9:27:81$. *See also* SU p.28.

Proportional representation (PR), system of electoral representation in which the allocation of seats reflects the percentage of the vote commanded by each party. There are many different systems of proportional representation, but most incorporate a transferable vote principle, by which first-choice votes cast for the least popular candidates (or for the most popular above a certain quota) are transferred to the voter's second choice candidate. Proportional representation is used in many European countries and for elections to the European parliament of the EEC. *See also* MS p.176, *176*.

Proprioception, perception of internal nervous stimuli through which an animal is aware of its own movement, posture and internal condition.

Propyl alcohol, also called propanol, colourless alcohol used as a solvent and in the manufacture of various chemicals. It exists as two ISOMERS. Normal propanol, $CH_3CH_2CH_2OH$, is a by-product of the synthesis of methyl alcohol (methanol). Isopropyl alcohol, $(CH_3)_2CHOH$, is a secondary alcohol which is easily oxidized into acetone.

Propylene (propene), colourless, aliphatic hydrocarbon, C_3H_6, manufactured by the thermal "cracking" of ETHYLENE. It is used in the manufacture of a wide range of chemicals, including vinyl and acrylic resins. Properties: b.p. $-48°C(-54.4°F)$; m.p. $-185°C(-301°F)$.

Proscenium, in the THEATRE, the front part of the stage, especially the arch, first used in the 17th century Italian theatre to create a picture-frame effect.

Prose, in literature, a relatively unstructured form of language. The term derives meaning from the contrast with the metrical discipline of POETRY. The patterning of artistic prose is not prescribed but emerges from the structure of the writer's thoughts.

Proserpine, wife of Pluto and queen of the underworld, the Roman equivalent to PERSEPHONE. She was unable to escape permanently from Pluto because she had eaten a seed of a pomegranate, the fruit of the dead; consequently each year she had to spend six months above ground and six months below.

Prosimians, small to medium-sized nocturnal PRIMATES of the suborder Prosimii. All except the INDRI have long tails. Included are the AYE-AYE, GALAGO, LORIS, LEMUR, POTTO and TARSIER. *See also* NW pp.168–169.

Prospecting, search for exploitable mineral deposits. The earlier methods of chance and divining have been replaced by means based on geology and mineralogy. These involve extensive sampling – the analysis and examination of materials taken from holes drilled at regular intervals – the analysis of the velocity of waves resulting from underground explosions, measurement of variations in the Earth's magnetic field, the detection of gravity anomalies and, for radioactive minerals, the use of Geiger counters. Aerial photography has also proved to be another versatile prospecting method. *See also* PE pp.132–135.

Prostaglandins, series of related fatty acids, with hormone-like action, present in semen, liver, brain and other tissues. Their biological effects include the lowering of blood pressure and the stimulation of contraction in a variety of smooth-muscle tissues, such as the uterus.

Prostate gland, gland in the male reproductive tract surrounding the URETHRA. It secretes specific chemicals that mix with sperm cells and secretions of other accessory glands to make up the sperm-containing fluid, SEMEN, that is released at ejaculation. In many men there is a slow enlargement of the prostate with age. This causes it to press on the neck of the bladder, interfering with urination. The urine builds up in the bladder, gradually weakening it. Prostate enlargement in men past middle age is generally treated by surgical removal of the gland.

Prostatitis, inflammation of the PROSTATE GLAND, caused by bacteria or their toxins. Chronic prostatitis may occur in men over 50, following acute inflammation of the prostate. *See also* MS p.102, *102*.

Prosthesis, man-made substitute for a missing organ or part of the body. Until the 17th century artificial limbs were either wooden or solid metal, but innovations in metallurgy and engineering design have enabled lighter, jointed limbs to be made. Methods of attachment, too, have greatly improved. More recent prosthetic devices include artificial heart valves made of silicone materials. *See also* MS p.127, *127*.

Prostitution, provision of sexual services for reward, usually money. Most prostitutes are women. In Britain prostitution is not strictly illegal, but soliciting, living off the earnings of prostitution and brothel-keeping are all criminal offences. Prostitution has existed in many societies, and has been subject to official regulation.

Protactinium, rare radioactive metallic element (symbol Pa) of the actinide group, first identified in 1913. Its chief source is PITCHBLENDE. Properties: at.no. 91; at.wt. 231.0359; s.g. 15.4; m.p. 1,200°C (2,192°F); b.p. 4,000°C (7,232°F); most stable isotope Pa^{231} (half-life 3.25×10^4 yr).

Protagoras (*c.*485–410 BC), leader of the Greek SOPHISTS. An AGNOSTIC, his belief that we know only our perceptions, not the actual things perceived, led to his observation that " man is the measure of all things".

Protective coloration, natural camouflage or warning colours of organisms that serve to blend it in with the surrounding environment or to ward off predators. TIGERS and some moths have permanent protective colouring. CHAMELEONS and some FLATFISH can change colour to match the background. Warning colours of an animal usually mean it is poisonous, aggressive or distasteful to most predators. Predators learn to recognize and avoid these colorations, which may be mimicked by harmless species. *See also* MIMICRY.

Protectorate (1653–59), in English history, period of absolute rule by CROMWELL, his authority resting on the power of the army. Following the execution of CHARLES I, England was declared a COMMONWEALTH under the rule of the RUMP PARLIAMENT. In 1653, however, Cromwell replaced the Rump with the nominated BAREBONE'S PARLIAMENT, the majority of which was submissive to his will. In December of that year, Cromwell assumed the title of Lord Protector of England, Scotland and Ireland. A virtual dictator, Cromwell divided the country into 11 districts, each under the administration of a major-general. At his death (1658) he was succeeded by his son, Richard CROMWELL, who proved inept. After a period of chaos, the Rump Parliament was recalled and in 1660 General George MONCK brought about the RESTORATION of CHARLES II. *See also* HC1 pp.290–291, *290–291*.

Protein, organic compound containing many AMINO ACIDS linked together. Living cells use about 20 amino acids, but as proteins have thousands of amino acids in each molecule the number of possible proteins is very large. The order of amino acids in proteins is controlled by the genes in the cell's DNA. The most important proteins are the ENZYMES, which determine all the chemical reactions in the cell, and ANTIBODIES, which combat infection. *See also* MS pp.68–71, *102*, 107; NW pp.20, *21*, 26–27; SU p.142.

Protestant Episcopal Church, US Church formally organized in Philadelphia in 1789, and once part of the ANGLICAN COMMUNION. It is self-governing, with the laity having a large role in administrative duties. A presiding bishop is elected by the House of Bishops of the General Convention. The Apostles' Creed and the Nicene Creed are accepted. Membership in the mid-1970s was more than 3,200,000.

Protestantism, branch of Christianity formed in protest against the practices and doctrines of the old Catholic Church. Protestants sought a vernacular Bible to replace the Latin VULGATE, and to express individual elements of nationalism, as in N Germany and England in the mid-16th century. The movement is generally considered to have started when Martin LUTHER nailed his 95 theses to Wittenberg church door. His co-protesters included John WYCLIFFE and Jan HÚS, and later Huldrych ZWINGLI and particularly John CALVIN, whose interpretation of the Bible and concept of PREDESTINATION had great influence. The Protestants held the EUCHARIST to be a symbolic celebration as opposed to the Roman Catholic dogma of TRANSUBSTANTIATION, and claimed that there was no medium between man and God. This led to an emphasis on individual contemplation of the scriptures. *See also* ANGLICANISM.

Proteus, in Greek mythology, a sea god, son of OCEANUS and Tethys. He is depicted as a little old man of the sea. Proteus possessed the gift of prophecy and the ability to alter his form at will – in an instant he could become fire, flood or a wild beast.

Proteus, generic name given to both the cystitis-producing bacterium (*Proteus mirabilis*) and the amphibian the OLM (*Proteus anguinus*). Such duplication in scientific nomenclature is usually resolved by giving the disputed name to the first organism described, but here the matter has not yet been settled. *See also* NW pp.36–37, *36–37*, 132.

Prothrombin, precursor of the blood enzyme thrombin, which is converted into thrombin by thromboplastin. It plays an essential role in the formation of blood clots, which help to stop bleeding and aid in the healing of wounds.

Protista, proposed classification of certain unicellular and simple multicellular organisms, including PROTOZOA, ALGAE, BACTERIA and FUNGI. The term was introduced to overcome the difficulty of distinguishing such organisms from the true plant and animal kingdoms (Plantae and Animalia). This kingdom may include organisms with many nuclei within one cell wall (coenocytes). *See also* NW p.72.

Protoactinium. *See* PROTACTINIUM.

Protoceratops, horned ornithischian DINOSAUR of Cretaceous times that lived in central Asia. It had a frill of bone on the back of its skull but, unlike later forms, its nose was hornless. Length: 1.8m (6ft).

Protochordate, group of prehistoric animals that have certain features in common with the modern CHORDATE group of animals. The protochordates are divided into three subphyla: HEMICHORDATA, UROCHORDATA and CEPHALOCHORDATA.

Protocol, universal code of diplomatic or military behaviour, based on precedence and deference to rank. The term also refers to the record of a diplomatic conference or to a diplomatic document itself, less formal than a treaty.

Protocols of the Learned Elders of Zion, fraudulent document used for the purposes of ANTI-SEMITISM. It reported alleged meetings of Jews in Europe in the late 19th century at which plans were made to overthrow Christianity and control the world. Published in a Russian newspaper in 1903, it was believed by semi-official circles and the public generally, greatly contributing to the anti-semitism of the time.

Proton, stable elementary particle (symbol p) with a positive charge equal in magnitude to that of the ELECTRON. It forms the nucleus of the lightest isotope of hydrogen, and with the NEUTRON is a constituent of the nuclei of all other elements. The number of protons in the nucleus of an element is equal to its ATOMIC NUMBER. Protons also occur in primary cosmic rays. It is a BARYON with spin ½ and a mass 1836.12 times that of the electron. Beams of high-velocity protons, produced by particle accelerators, are used to study nuclear reactions. *See also* SU p.62.

Protoplasm, semi-fluid, essential living matter within all plant and animal CELLS. In modern science, it is considered not as the living material, but as the basic substance of the CYTOPLASM.

Protopopov, name of two Soviet ice figure skaters, Oleg (1932–) and his wife Ludmila (1935–) (née Belousova). With their smooth, ballet-style skating incorporating the breathtaking "death spiral", they won the pairs title at the 1964 and 1968 Winter Olympic Games and were world and European champions 1965–68.

Prototheria, subclass of Mammalia that is represented in living fauna only by the egg-laying ECHIDNA and PLATYPUS.

Protozoa, phylum of one-celled organisms found throughout the world in marine or fresh water, free living and as parasites. First seen in 1674 by Anton van LEEUWENHOEK, these microscopic animals have the ability to move (by cilia or pseudopodia) and have a nucleus, cytoplasm and cell wall; some contain CHLOROPHYLL. They are transparent, green, iridescent, blue, rose or yellow. Reproduction is by fission or encystment. Length: 0.3mm (0.1in). The 30,000 species are divided into four classes – Flagellata, Cnidospora, Ciliophora and Sporozoa. *See also* NW pp.72, 80–81, *80–81*.

Protura, order of about 150 species of primitive, pale, wingless and blind insects. It includes the most primitive hexapods and is represented throughout the world. Most species live in damp soil, feeding on decaying organic matter. Length: to 2mm (0.07in). *See also* NW p.106–107.

Proudhon, Pierre-Joseph (1809–65), French anarchist whose theories of liberty, equality and justice conflicted with those of Karl MARX. In his first book, *Qu'est-ce que la propriété?* (1840), he argued that "property is theft"; his greatest work was *Système des contradictions économiques* (1846). He was exiled to Belgium in 1858 for criticizing NAPOLEON III, but was pardoned in 1862. *See also* HC2 p.*212*.

Proust, Joseph-Louis (1754–1826), French chemist who specialized in analysis, with little interest in theory. As a result of his researches, however, he concluded that compounds were of constant and determinate composition, which was important in the establishment of the atomic theory of chemical composition.

Proust, Marcel (1871–1922), French novelist whose *A la recherche du temps perdu* (*Remembrance of Things Past*; 16 vols, 1913–27) is regarded as one of the great works of literature. Proust was a frequenter of the aristocratic salons of Paris at the turn of the century, and he was actively involved in the DREYFUS AFFAIR (1897–99). He suffered from asthma and his growing ill-health caused him in 1907 to withdraw from society and devote himself to writing. *Remembrance of Things Past* is written in long cascading sentences. It is founded on the effects of involuntary memory, the moments when a chance impression obliterates the present and propels one into a past moment.

Provençal, ROMANCE LANGUAGE spoken in France, in the region of PROVENCE. Most of its users are fluent in French and it has no official status in the country, but it does not appear to be dying out.

Provence, region and former province of SE France, encompassing the present départements of Var, Vaucluse and Bouches du Rhône, and parts of the Basses-Alpes and Aix-en-Provence. The coastal area was settled *c.*600 BC by the Greeks, and the Romans established colonies there in the 2nd century BC. It passed to Louis XI of France in 1481 and was a province of France until the FRENCH REVOLUTION, when France was redivided into départements. Fruit and vegetables are grown on the Riviera and along the Rhône Valley; there is also some stock raising in the Camargue. Tourism is the major industry.

Proverb, short, pithy saying expressing a familiar truth or moral lesson. Proverbs originated in primitive folklore and often exist in similar form in different countries. *The Book of Proverbs,* a section of the OLD TESTAMENT, consists of proverbs written by several authors, formerly thought to include Solomon.

Proverbs (of Solomon and others), book of the Old Testament, probably the oldest existing example of Hebrew "Wisdom" literature, in which much practical counsel, most of it secular, is given.

Providence, port of entry in NE Rhode Island, USA, on Providence Bay; state capital and largest city of Rhode Island. The city was founded in 1636 by Roger Williams as a refuge for religious dissenters from Massachusetts. Brown University (1764) and Bryant College (1863) are in the city. Industries: jewellery, electrical equipment, silverware, machine tools, rubber goods. Pop. (1973) 169,931.

Proxima Centauri, closest star to the Sun, in the southern constellation of Centaurus. Proxima is a red dwarf star and is part of a triple star system with Alfa and Beta Centauri. Characteristics: apparent mag. 10.7; absolute mag. 15.1; spectral type M5; distance 4.3 light-years. *See also* SU pp.227, 236, 242, 261, *261*, 282, *284*.

Prudentius, Aurelius Clemens (348–after 405), Latin Christian poet who, although he held a high place in the Roman court, eventually devoted himself to religious writing. His works include *Apotheosis* (*On the Nature of God*), and *Psychomachia* (*The Battle for the Soul*) – his best-known work, which was especially popular in the Middle Ages.

Prudhoe Bay, inlet on the north Alaskan coast 320km (200 miles) east SE of Point Barrow. Oil discovered on Alaska's North Slope in 1968 made it a focus for investigations to determine whether or not it was to become a significant source of crude oil.

Prud'hon, Pierre Paul (1758–1823), French painter who, in the Neo-Classical age of Jacques-Louis DAVID, painted Romantic, gently modelled portraits. The most delightful of these include portraits of the Empress Joséphine and of his pupil Mlle Constance Mayer. He also painted some large decorative pictures, such as *Justice and Divine Vengeance pursuing Crime,* and designed Neo-Classical decorations and furniture for the empress.

Prufrock, The Lovesong of J. Alfred, poem by T. S. ELIOT, originally published in *Prufrock and other Observations* (1917), his first published work. The poem differed radically from contemporary war poetry, depicting the fantasy in the empty life of an ageing man bound by convention.

Prune, any sweet variety of plum that has been dried and cured to a wrinkled, dark-brown appearance. *Prunus domestica* is the variety most suitable for drying, which is achieved with hot air. Prunes may be eaten raw or cooked; they have a laxative effect.

Pruning, removal of branches, stems or buds from a tree or shrub. It is usually carried out to improve the appearance and flowering of ornamental trees and shrubs and the productivity of fruit trees and berry bushes. Fruit trees are usually pruned in the winter or early spring, as is the HYDRANGEA, ROSE and other flowering shrubs that bloom on new wood. LILAC, FORSYTHIA and other spring-blooming shrubs (all of which bloom on old wood) are usually pruned after flowering to avoid removing flower buds. *See also* PE p.*179*.

Prunus, large genus containing about 200 species of flowering trees and shrubs. It includes the APRICOT, PLUM, CHERRY, ALMOND, NECTARINE and PEACH. Many species are cultivated and are highly important as fruit crops. Family Rosaceae.

Prus, Bolesaw (1845–1912), pen-name of Aleksander Gowacki, a Polish author whose career began in journalism. From 1875 he wrote weekly newspaper articles of witty, detailed social commentary. He published several novels and collections of short stories, and his best work is thought to be *Lalka* (*The Doll*, 1890).

Prussia, former German state which occupied most of N Germany (1918–1945). Ducal Prussia passed to the electors of BRANDENBURG in 1618 who took the title of "King of Prussia" in 1701; previously Royal Prussia had applied only to the region acquired by Poland in the 15th century and later known as West Prussia. Prussia became a strong military power under FREDERICK WILLIAM I and FREDERICK THE GREAT in the 18th century, absorbing Silesia (1740) and parts of Poland, including West Prussia (1772–92). Prussia was defeated by the French in 1795, and took no part in the wars against NAPOLEON until 1806, when it was defeated again at JENA and AUERSTADT. Prussian power was restored and its territory increased by the Congress of Vienna, and in 1834 Prussia encouraged the other German states to unite in the ZOLLVEREIN. Under BISMARK, chancellor from 1862, Prussia assumed leadership of the movement for German unity, and its authority was confirmed in the constitution of the second German Reich. In 1918 what was left of the kingdom of Prussia became a state of the Weimar Republic, but Prussian aristocratic and militaristic traditions greatly influenced German policy in the rise of Nazism. *See also* HC2 pp.*82*, 110, 180.

Prussic acid. *See* HYDROCYANIC ACID.

Pryde, James (1866–1941), Scottish painter who, with William NICHOLSON, created the posters by the "Beggarstaff Brothers" (their pseudonym). His paintings were often architectural compositions, which he clothed in an atmosphere of mystery.

Prynne, William (1600–69), English politician and Puritan pamphleteer. His criticism of the theatre in his book *Histrio Mastix* (1632) was interpreted as an attack on CHARLES I . His ears were cut off and he was imprisoned from 1634 to 1640. After his release he upheld the parliamentary cause in his writing during the ENGLISH CIVIL WAR. He supported the restoration of CHARLES II in 1660, and was made archivist of the Tower of London.

Przewalski's horse, also called Mongolian wild horse, only surviving species of original wild horse. Once common on Eurasian plains, it is now found only in Mongolia and Sinkiang. It is small and stocky with an erect black mane and no forelock. Its red-brown coat is marked with a darker line on the back and shoulders and leg stripes. Height: to 1.5m (4.8ft) at the shoulder. Family Equidae; species *Equus caballus przewalskii*.

Psalm, originally a musical hymn or poem sung to the accompaniment of a stringed instrument called the PSALTERY. The most famous are in the Book of Psalms, in the Old Testament; some are traditionally attributed to David, King of Judah. They are a prescribed part of both Matins and Evensong in the CHURCH OF ENGLAND, and are chanted and sung.

Psalter, the biblical PSALMS in the version found either in the Book of COMMON PRAYER or the Roman Catholic Breviary, often pointed (marked up) for singing by choirs.

Psaltery, medieval stringed musical instrument similar to the ZITHER. Several strings on a frame with a shallow sounding board are plucked with fingers or with a plectrum. It is of Middle Eastern origin and is referred to in the Old Testament. It was depicted in classic art as a variant of the CITTERN or LYRE.

Psephology, statistical and sociological analysis of the results of elections, voting patterns and electoral behaviour. *See also* MS pp.275, *278*.

Pseudepigrapha, Jewish writings of between 200 BC and AD 200, falsely attributed to a biblical author; they follow the style and content of authentic Old Testament works. The term refers more widely to almost all ancient Jewish texts that have not been accepted as canonical by the Christian Church.

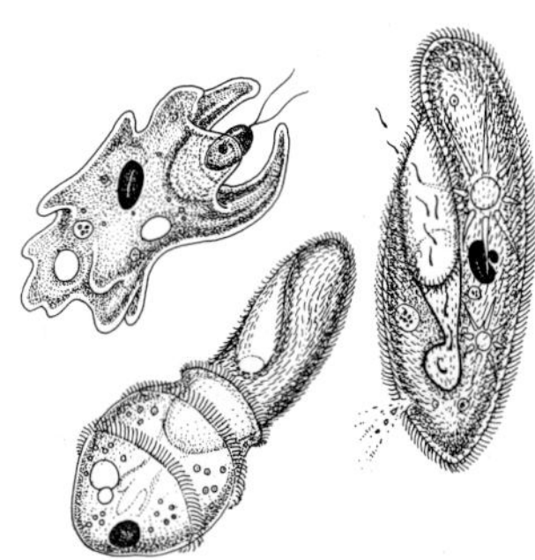

Protozoa are divided into classes according to their means of locomotion.

Marcel Proust compared his major novel to the structure of a cathedral.

Pierre Paul Prud'hon's painting *Jeune zéphyr se balançant au dessus de l'eau.*

Pruning enhances the appearance and improves the growth of trees and shrubs.

Pseudomorphism

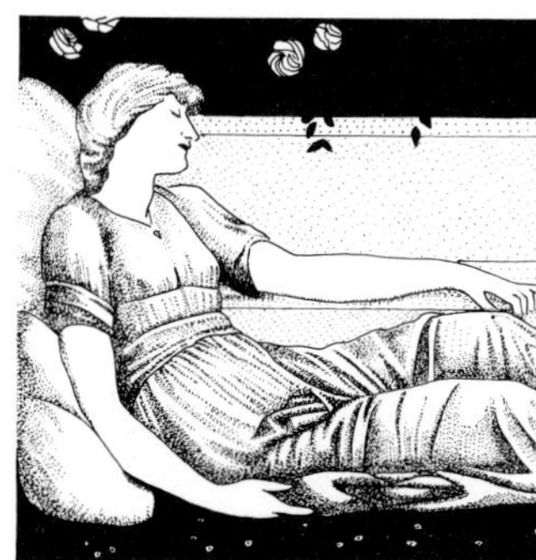

Psyche, a beautiful mortal in Greek mythology, came to represent the human mind.

Psycho; Anthony Perkins in a scene from Hitchcock's thriller made in 1960.

Ptarmigan may lay up to 16 eggs and the chicks can fly when about 10 days olds.

Pterosaurs were well adapted to gliding, but lacked a bird's manoeuvability.

Pseudomorphism, in mineralogy, chemical or structural alteration of a mineral without change in shape. If substitution has occurred, the original substance has been removed and replaced by another. This is exemplified by petrified wood.

Pseudoscorpion, any of numerous species of small ARACHNIDS found throughout the world. It has pedipalps (paired, pincerlike appendages) and resembles a true scorpion, but lacks a tail. Length: to 7.5mm (0.04in). Order Chelonethida; family Arachnida. *See also* NW p.*104.*

Psi, broad term covering the phenomena of PARAPSYCHOLOGY; it generally pertains to any of the four main branches; PSYCHOKINESIS, TELEPATHY, ESP (extra-sensory perception) and CLAIRVOYANCE. *See also* MS pp.198–199.

Psittaciformes, order of strong-flying birds. It contains only one family, Psittacidae, which includes the COCKATOO, LOVEBIRD, BUDGERIGAR, MACAW, parakeet and PARROT. It has been represented since the Miocene epoch. Many species are brightly coloured, with stout hooked bills. They eat seeds, fruit or nectar. Length: to 9cm (38.5in). *See also* NW p.141.

Psittacosis, or parrot fever, disorder usually affecting the respiratory system of birds. Probably caused by bacteria, it can be transmitted to man through the handling of infected birds, in bites from them or from the dust of their feathers. In birds, there is diarrhoea and difficulty in breathing. In man, headache, cough and fever are symptoms. Treatment is generally the administration of ANTIBIOTICS.

Pskov, city in the NW Russian SFSR, USSR, on the River Velikaya near the SE shore of Lake Pskov, 250km (155 miles) sw of Leningrad. Its industries include flax processing, linen, rope, footwear and the manufacture of agricultural machinery. Pop. (1975) 151,000.

Psoriasis, chronic disorder of the skin in which there are patches, placques or papules. The red or brown lesions are slightly elevated and usually covered with white scales. They frequently appear on the chest, knees, elbows and scalp. The cause is unknown and there is no specific treatment, although sunlight sometimes helps.

Psyche, in Greek mythology, beautiful mortal loved by EROS (CUPID). Psyche is the Greek word for the mind and the soul. Apuleius, a writer of the 2nd century AD, tells how the Pythian oracle at Delphi warned her that she would love no mortal, how she wed Cupid, lost him, and how they were reunited. The allegory suggests the freeing of the soul after purification through suffering.

Psychedelic art (1960s) style of painting and poster art. Subjects included the human nude, fantasy and calligraphy, rendered in "day-glo" or neon-bright colours and swirling line; sometimes reminiscent of the sensuality and caricature of ARTNOUVEAU.

Psychedelic drugs. *See* LSD.

Psychiatry and psychotherapy, branches of medicine specializing in the analysis, diagnosis and treatment of mental illnesses and behavioural disorders. Psychiatry includes treatment carried out by means of physiological assistance such as drugs and surgery; psychotherapy on the other hand seeks to restore the patient's mental health by exploring his fundamental emotional difficulties. In its most basic (and often brutal) forms psychiatry has existed since earliest times, but it only became curative in the 20th century. Psychotherapy originated in the late 19th century, and was made popular by Sigmund FREUD and Carl JUNG in the 1900s. The term psychiatrist is often used to describe practitioners of both disciplines. *See also* MS pp.144–145, 148, 149.

Psychical research, or parapsychology, investigation into extrasensory perception (ESP) or the paranormal. There are three areas of study, including telepathy, by means of which one can transmit thoughts to another over a distance. The second area is precognition, whereby one has foreknowledge of an event, and the third is déjà vu, when one feels in a new place that somehow one has been there before.

Psychical research is also carried out into PSYCHOKINESIS, by which physical objects may be moved by psychic power. *See also* MS pp.198–199.

Psycho (1960), classic film thriller directed with typical panache by Alfred HITCHCOCK, in which he skilfully manipulated audiences by a series of elaborate cinematic diversions. The film starred Anthony Perkins and Janet Leigh, was based on a novel by Robert Bloch and concerned grisly murders at a lonely country motel. In a shrewd publicity exercise, its exhibitors by contract refused admission to see the film once it had begun.

Psychoanalysis, method of therapy devised by FREUD and BREUER in the 1890s for treating behaviour disorders, particularly neurosis. It is characterized by its emphasis on unconscious mental processes and on the treatment of mental disorder by the use of FREE ASSOCIATION.

Psychodrama, in group psychotherapy, a form of treatment where a patient acts out, or watches others act out, a personal problem. Patients serve as actors and audience with the help of a therapist who guides the dialogue and discussion. *See also* MM pp.148–149, *149.*

Psychokinesis (PK), in PARAPSYCHOLOGY, change in an object or physical system that is supposedly directly caused by the mind or will rather than by any known physical agency. An example might be influencing the fall of dice by the will alone. *See also* TELEKINESIS; MS pp.198–199.

Psychological warfare, use of psychological techniques calculated to destroy a military enemy's will to fight. It often takes the form of propaganda designed to undermine faith in leaders or causes, or to induce schism within the ranks.

Psychology, study of mental activity and behaviour. It includes the study of perception, thought, problem solving, personality, emotion, mental disorders and the adaptation of the individual to society. It overlaps with many other disciplines, including physiology, philosophy and social anthropology. Probably the first known psychological work is Aristotle's *De anima,* but it was not until the work of John LOCKE and Thomas HOBBES in the 17th century on perception that psychology became a serious study. In the 19th it became a scientific discipline with the work of William JAMES, Hermann EBINGHAUS, Sir Charles SHERRINGTON, Sigmund FREUD and others. In the 20th century psychologists, anxious to give their work a scientific status equal to that of physics, have sought to establish verifiable principles in carefully controlled experiments on human beings and animals, but relatively little indisputable knowledge has been amassed. Perhaps more fruitful have been the developments of clinical PSYCHIATRY, founded on a variety of different conceptual bases, and of anti-psychiatry, which stresses the humanistic side of the discipline and denies the validity of overall theorizing. Applied psychology aims to use the discipline's insights into human behaviour in practical fields such as education and increasingly, in industry. *See also* index to *Man and Society.*

Psychometrics, area of PSYCHOLOGY concerned with the measurement of abilities, capacities and normal behaviour.

Psychopath, person characterized by a personality disorder that leads to antisocial and amoral behaviour. Such people are regarded as impulsive, insensitive to others and unable to think ahead. There is little anxiety, guilt or neurosis in a psychopath and psychoanalytic theory regards his behaviour as stemming from an incompletely developed SUPEREGO. *See also* MS p.*140.*

Psychosis, inability to adjust realistically to one's environment, more severe than NEUROSIS. It is often exhibited in extreme swings of mood between depression and mania, by delusions and distortions of judgement, and by an inappropriateness of response. The organic psychoses may spring from brain damage, advanced SYPHILIS, senile dementia and late stages of advanced EPILEPSY. The functional psychoses, for which there is no organic explanation, include PARANOIA, SCHIZO-

PHRENIA and manic-depressive psychosis. *See also* MS pp.138–139.

Psychosomatic disorder, physiological complaint thought to be caused, at least in part, by unresolved emotional disturbance. Extending the PSYCHIATRIC concept of mind and body (psyche and soma) unity, it follows that emotional conflict can manifest itself in physical symptoms. Organs affected are usually under the control of the INVOLUNTARY or autonomic NERVOUS SYSTEM. Complaints include bronchial asthma, stomach and intestinal ulcers and even circulatory and heart disease.

Ptah, in Egyptian mythology, creator of the universe and patron of craftsmen, depicted as a man standing, often wrapped in a winding sheet. The original centre of his worship was at Memphis.

Ptarmigan, any of three species of northern or alpine grouse, all of which have feathered legs; especially the Eurasian ptarmigan, *Lagopus mutus.* Its plumage changes colour seasonally. The wings and breast are white in colder months, but in the spring it becomes a mottled greybrown. It inhabits high barren regions, feeding on leaves and lichens. Length: to 36cm (14in). Family Tetraonidae.

Pteranodon, any of several species of extinct flying reptiles of the Cretaceous period. It had long toothless jaws and an elongated crest on the back of its head; the crest may have served as a counterbalance for the jaws. It is believed to have been a glider, since it lacked the muscle development and breast keel on its turkey-sized body to flap the wings, the span of which may have been as much as 7.5m (25ft). *See also* PTEROSAUR; PTERODACTYL; NW p.*187.*

Pteridophyte, commonly used name for any of a group of spore-bearing plants. At one time pteridophytes were taken to include CLUB MOSS, HORSETAILS, FERNS and several fossil groups. These plants have similar life cycles but in other respects are quite distinct and are now often classified separately. *See also* TRACHEOPHYTE; NW pp.52–53, 177.

Pteriodidiformes. *See* SAND GROUSE; NW p.141.

Pterodactyl, any of several species of small PTEROSAURS. Almost tail-less, it had a large toothed beak and flimsy membranous wings. Fossil remains show a lack of muscular development and the absence of a breast keel. It is therefore believed that pterodactyls were gliders, incapable of sustained flapping flight. *See also* NW pp.185, *186,* 187.

Pterosaur, extinct flying reptile with wing membranes supported by a single, elongated finger on each side. Distributed throughout the world during Jurassic and Cretaceous times, early forms had teeth and tails; later forms were tailless and had toothless beaks. *See also* NW p.*185.*

Pterygota, subclass of INSECTS, defined as winged in contrast to the Apterygota, or wingless, insects. *See also* NW p.106.

PTFE, polytetrafluoroethylene, a chemically inert solid plastic, known also by the trade name Teflon. It is used as a heat-resistant material for heat-shields on spacecraft (it is stable up to about 300°C; 572°F), and is used as an insulator, lubricant and "non-stick" coating, eg on cooking pans. *See also* MM p.100.

PTH, parathyroid hormone, HORMONE secreted by the PARATHYROID GLANDS. It is important in stabilizing the level of calcium in the blood, too little calcium leading to muscular spasms and convulsions. *See also* MS p.64, *64.*

Ptolemy, name of a 15-member Greek dynasty that ruled Egypt for 300 years (323–30 BC). Ptolemy I Soter, troop commander of ALEXANDER THE GREAT, took over Egypt upon the latter's death in 323 BC. He assumed the title of king in 305 and established the Saropis cult at Memphis. His son Ptolemy II Philadelphus succeeded him, and during his reign (285–246) he made Alexandria a cultural and commercial centre. Under Ptolemy III Euergetes (*fl.*246–221), Egyptian fleets gained control of the eastern Mediterranean. Ptolemy IV Philopator's reign (221–205) was a decadent one and soon after his death in 205 the Seleucid king and the King of Macedonia seized all Egypt's provinces. The ministers of the infant King Ptolemy V Epiphanes appealed to Rome for help and from about 200 BC the power of the dynasty was superseded by the influence of Rome.

Ptolemy, or Claudius Ptolemaeus (*c.* 90–168), Egyptian-born Greek astronomer and geographer. He charted many new stars and his remarkable *Almagest* (*c.* 150) influenced astronomy for the next 1,400 years. His *Geography*, which included Africa and Asia, also had great subsequent influence. He devised new mathematical theorems and proofs and wrote *Optica*, a treatise on optics. *See also* HC1 pp.83, *83*, 188; MS pp.201, *201*, *260*.

Ptomaine, any of several basic organic chemical compounds, some of which are poisonous to human beings. They are derived from the decomposition or putrefaction of animal or vegetable PROTEINS and bear a resemblance to ALKALOIDS. Ptomaine poisoning is a form of food poisoning caused by bacteria or toxins.

Puberty, time in human development when the sex glands first become active and the secondary sexual characteristics start to become evident. In females these include a slight deepening of the voice, the growth of pubic hair, the beginning of menstruation and the development of breasts. In males they entail the enlargement of the testes, a deepening of the voice and the growth of pubic hair. Sexual maturity is attained at about the age of 14 for most males and 12 for most females. The process of puberty is begun by the production of HORMONES by the PITUITARY GLAND. *See also* ADOLESCENCE; MS pp.164–167.

Puberty rites, ceremonies or rituals celebrating the attainment of maturity by an adolescent. With boys they usually occur between the ages of nine and 20; with girls they accompany the MENARCHE. The rites initiate young people into the society of adults and often involve, in the case of boys, an ordeal (such as scarification among the NUER) and, with girls, purification. *See also* RITES OF PASSAGE.

Pubis, in the pelvic girdle, either of a pair of small bones at the base of the abdomen. They are almost U-shaped and meet at the pubic symphysis, which is closed by a pad of cartilage. The symphytic pad spreads during childbirth in mammals. Each pubis joins an ILIUM and an ISCHIUM at a triangular suture in the hip socket.

Public Bill, any bill introduced in the British Parliament of general public concern (as opposed to the private interest of a corporation, company or individual). It may be introduced by the government or it may be a PRIVATE MEMBER'S BILL.

Public health, branch of medicine that attempts to protect and improve the health of people in a community. It is involved with disease control, (such as mass vaccination programmes and screening for TB) and with the establishment and maintenance of health standards for housing, food, water, waste disposal and air quality. *See also* MS p.110.

Public Health Act (1848), act of the British parliament empowering, but not requiring, town corporations to establish local health boards to improve sanitation. By 1854 only 182 boards had been established and only 13 of them had installed waterworks and sewage schemes.

Public opinion polls, unofficial means of assessing and predicting popular responses to events and issues of national importance, especially elections. Established as reliable in the 1930s with the scientific SAMPLING pioneered by GALLUP, the many organizations retained in Britain to conduct such polls include Opinion Research Centre (ORC) and National Opinion Polls (NOP). *See also* MARKET RESEARCH.

Public Record Office, national depository for legal and government records in Britain, established in 1838. Medieval records are kept in Chancery Lane, London; since 1977 modern and departmental records have been kept at Kew.

Public school, in Britain, private fee-paying, usually residential, school independent of state finance. Neither the number of public schools nor their defining traits may be exactly stated. The term came into use in the 18th century, when the best known included Eton, Rugby, Harrow, Winchester and Westminster. They taught a chiefly classical curriculum. In the USA the term applies specifically to non-private schools.

Public trustee, public official who acts as a trustee or executor, in pursuance of the Public Trustee Act (1906), by which his powers and duties are prescribed and restricted. His duties may include the selling of property, investing money or granting of leases.

Publishing, production of multiple copies of printed matter, including books, NEWSPAPERS, magazines, posters and pamphlets. The copying and circulation of manuscripts flourished in the ancient world, but the first printed book to be published in Europe was the Bible, printed by Johann GUTENBERG in 1455 in an edition of fifty copies. The first newspaper was probably *Nieuwe Tijdinghen*, published in Antwerp from 1605. Publishing is complemented by PRINTING and involves the selection and editing of the matter to be printed, and its distribution and marketing. *See also* MM pp. 210–215.

Puccini, Giacomo (1858–1924), Italian composer of operas. The combination of dramatic libretti and magnificent music have ensured Puccini operas a permanent place in the repertory of the world's opera houses. Among his best known works are: *La Bohème* (1896), *Tosca* (1900), *Madame Butterfly* (1904), *The Girl of the Golden West* (1910) and his last work, incomplete at his death, *Turandot* (1926).

Puck, also called Robin Goodfellow or Hobgoblin in English folklore of the Middle Ages, mischievous FAIRY or minor demon. He was one of the key characters in William Shakespeare's *A Midsummer Night's Dream*, portrayed as the messenger of Oberon, king of the fairies.

Puddling process, conversion of pig iron into wrought iron. It involves subjecting the pig iron to heat and frequent stirring in a furnace which can be agitated. Present in the process are oxidizers which free the wrought iron of its carbon and other impurities.

Pudney, John Sleigh (1909–77), British writer. His war poems, influenced by his RAF service and first published in *Ten Summers* (1945) established his reputation for literary versatility.

Pudovkin, Vsevolod Illarionovich (1893–1953), Soviet film director. His first film was *Mechanics of the Brain* (1925–26), a film about Pavlov's theories. In 1927 he was commissioned to make *The End of St Petersburg* for the celebrations of the 10th anniversary of the Russian Revolution. He also published the books *Film Technique* (1933) and *Film Acting* (1935). *See also* HC2 p.201.

Pudu, genus of South American deer that is almost tailless and has short, spike-like antlers. It measures only 38cm (15in) at the shoulder. Adults are a rich brown to grey in colour. Family Cervidae.

Puebla (Puebla de Zaragoza), city in central Mexico, in the foothills of the Sierra Madre Oriental; capital of Puebla state. Founded by the Spanish in 1532 as Puebla de los Angeles, the city has always been of strategic importance, being on the route from Mexico City to the port of Veracruz. US forces occupied the city during the MEXICAN WAR, and the French held the city from 1862 until 1867. There is a university and a 16th-century cathedral in the city. Puebla is an agricultural trade centre and is noted for its onyx working, ceramics and textiles. Pop. (1970) 401,603.

Pueblo Indians, generic name for the several North American Indian tribes inhabiting the MESA and RIO GRANDE regions of Arizona and New Mexico. They belong to several language families including Keresan, Tewa, Hopi and Zuni. Their mud and stone architecture is famous throughout the South-West, and the many-storeyed buildings of the Zuni gave riase to the legendary "Seven Cities of Cíbola" eagerly sought by the Spaniards. *See also* MS p.*254*.

Puerperal fever, infection that can occur in women after childbirth. It is caused by bacteria (streptococci) entering the body because of the lack of aseptic techniques. With modern hospital methods, the disorder has become a rarity.

Puerperium, period of about four weeks after CHILDBIRTH, during which the uterus returns to normal size. After birth there is a discharge from the uterus known as lochia, lasting up to two weeks. The recovery of skin tone and the loss of stretch marks is aided by exercises both before and after childbirth. *See also* MS p.78.

Puerto Armuelles, port in W Panama, on the Pacific coast near the border with Costa Rica. Port facilities were improved in the 1930s to handle the produce from the nearby banana plantations. Pop. (1970) 12,022.

Puerto Plata, city in the N Dominican Republic, formerly San Felipe de Puerto Plata. It is an important seaport; exports include tobacco, sugar, hardwoods, coffee and hides. Pop. (1970) 74,480.

Puerto Rico, official name the Commonwealth of Puerto Rico, a self-governing island in union with the USA, in the West Indies approx. 1,600km (1,000 miles) SE of Florida. Many Puerto Ricans have emigrated to the USA because there is great unemployment on the island. The principal crops are sugar, tobacco, coffee, pineapples and maize. Industries include cement, tourism and food processing. The capital is San Juan. Area: 8,870sq km (3,425sq miles). Pop. (1975 est.) 3,087,000. *See also* MW p.146.

Pufendorf, Samuel (1623–94), German political theorist, author of *On the Laws of Nature and Peoples* (1672), an important contribution to natural law theory. He defined the state as the sum of constituent individual wills. According to his theory state power exceeded that of the Church.

Puff adder, widely distributed African VIPER. Its skin pattern varies, but it usually has yellow, crescentic markings on brown. It hunts large rodents and its poisonous bite is often fatal to man. Up to 80 young are born at one time. Length: to 1.2m (4ft). Family Viperidae; species *Bitis arietans*. *See also* HOG-NOSED SNAKE: NW p.*200*.

Puffball, any of a large family of MUSHROOMS (Lycoperdaceae) whose SPORE masses become powdery at maturity and are expelled in "puffs" when the case is pressed. Puffballs are stemless and some species – but not all – are edible. *See also* NW p.43.

Puffer fish, also called blowfish, or swellfish, marine fish found in warm and temperate seas. When threatened, it inflates itself to nearly twice its body size by a special adaptation of its gullet. Its colours vary widely. Smaller, sharp-nosed varieties are often kept in home aquariums. Length: to 91.4cm (36in). Family Tetraodontidae; there are about 100 species. *See also* NW pp.128, *131*.

Puffin, small, parrot-like diving-bird of the AUK family (Alcidae), found in large colonies in the Northern Hemisphere. The Atlantic puffin (*Fratercula arctica*) has a short neck, triangular bill with red, yellow and blue stripes, and reddish legs and feet; length: about 30cm (12in). The puffin lays a single white, sometimes spotted, egg in a burrow about 1–2m (3.3–6.6ft) deep on a cliff. *See also* NW p.*149*.

Puffing Billy, simple steam locomotive built by William Hedley of England in 1813. Previous locomotives had used cogs and rack-rails but the *Puffing Billy* was an adhesion locomotive, relying on the friction between wheels and rails. It was employed in hauling coal wagons.

Pug, small dog that probably originated in China and was taken West by Dutch traders. It has a large, round head, a blunt, square muzzle and deep facial wrinkles. The wide-chested, short body is set on strong, straight legs and the tail is tightly curled over the hip. The short coat may be grey, light brown or black with a characteristic black face. Height: to 28cm (11in) at the shoulder; weight: to 8kg (18lb).

Pugachev, Emelyan Ivanovich (*c.* 1726–75), Russian COSSACK leader. He

Puccini's operas combine delightful theatre with orchestral brilliance.

Puerto Rico; many people have left the shanty towns to seek a better life in the USA.

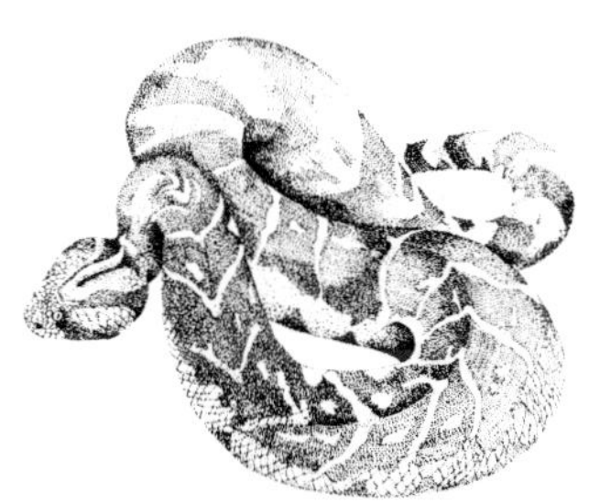

Puff adders make a loud hissing or puffing noise before they strike.

Puffing Billy, the name of William Hedley's steam locomotive of 1813.

Puli has a long matted coat which protects it from the winter weather.

Pumas are powerful animals and have been known to leap more than 12m (40ft).

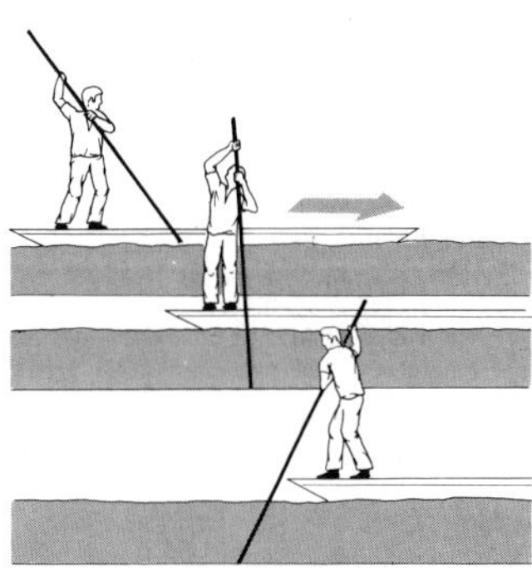

Punting is easier on rivers with shingle beds than on those with muddy bottoms.

Puppet theatres are found worldwide: these are Sicilian string puppets.

led a massive popular revolt against CATHERINE II and in 1773 proclaimed himself Peter III, promising freedom from serfdom, taxes and military service, and the elimination of landlords and other officials. Defeated in late 1774, he was taken to Moscow for trial and executed.

Puget, Pierre (1620–94), French BAROQUE sculptor, painter and architect. In 1667 he began work on the doorway of the City Hall at Toulon. His sculpture includes *Milo of Crotona* (1683).

Pugin, Augustus Welby Northmore (1812–52), British architect who helped to design the Houses of Parliament with Sir Charles BARRY. He also designed many churches, including St Giles, Cheadle, Staffordshire (1841–46), and St Augustine, Ramsgate (1846). By his works and writings, especially *True Principles of Pointed or Christian Architecture* (1841), he was a leading promoter of the GOTHIC REVIVAL.

Puglia. *See* APULIA.

Pulcinella (1920), one-act ballet with music by STRAVINSKY, adapted from PERGOLESI. It was performed by the BALLETS RUSSES in Paris with libretto and choreography by MASSINE and décor by PICASSO. Written in NEO-CLASSICAL style, the story involves COMMEDIA DELL'ARTE characters. *See also* PUNCH AND JUDY.

Puli, Hungarian shepherd dog, a breed that is at least 1,000 years old. It has a slightly domed head with a medium-long muzzle and V-shaped hanging ears. The straight-backed body is set on strong legs and the tail is curved over the back. The unusual long, double coat mats into felt-like cords and may be almost any solid colour. Height: to 48cm (19in); weight: to 13.5kg (30lb).

Pulitzer, Joseph (1847–1911), Hungarian-born US newspaper publisher who founded the PULITZER PRIZES. The pioneer of sensational journalism, he made the *New York World* the USA's largest circulation daily newspaper by crusading powerfully for oppressed workers and against alleged big business and government corruption, using the "yellow-press" techniques that William Randolph Hearst copied successfully.

Pulitzer Prizes, annual American awards presented for outstanding achievement in journalism, letters and music. The cost of the prizes is met by the income from a trust fund left by publisher Joseph PULITZER to the trustees of Columbia University. The first prize was awarded in 1917. Categories include best cartoon, best news photograph and outstanding editorial writing. There are prizes for fiction, drama, US history, biography, poetry and musical composition. Journalism awards are $1,000 each and the letters and music awards are $500 each.

Pulley, simple machine used to multiply force or to change the direction of its application. A simple pulley consists of a wheel, often with a groove, attached to a fixed structure. Compound pulleys consist of two or more such wheels, some movable, that allow a person to raise objects much heavier than he could lift unaided. *See also* MM pp.88, *88*, *96*, *96*.

Pullin, John (1941–), British rugby union player who captained England to victories over South Africa in 1972 and New Zealand in 1973. A hooker from the Bristol club, he became England's most-capped player (42 caps, 1966–76) and toured South Africa (1968) and New Zealand (1971) with the British Lions.

Pullman, George Mortimer (1831–97), US industrialist and inventor of the railway sleeping car. A cabinet maker in his youth, he began converting old railway coaches in 1859 to make long-distance travelling more comfortable and built the first modern sleeping car, known as the Pullman Car, some six years later.

Pulmonary embolism, chronic and sometimes fatal obstruction of a pulmonary blood vessel by a massive travelling clot. It generally leads to symptoms of shock, acute high blood pressure in the pulmonary arteries and eventual failure of the right ventricle of the heart.

Pulsar, rapidly rotating star of extreme density that, although only between about 1.4–3 times as massive as the Sun, has been compressed by gravitational collapse to only a few hundred miles across. Pulsars regularly emit radio pulses and in some cases X-ray and visible pulses; they were first detected in 1967. The magnetic field and gravity are also extreme by terrestrial standards. Pulsars are now identified with neutron stars, postulated in 1939 as dying stars that would be too massive to become stable white dwarfs and which would collapse further with tremendous release of energy. *See also* SU pp.232–235.

Pulse, regular wave of raised pressure in arteries that results from the flow of blood injected into the arterial system at each beat of the HEART. The pulse is usually taken at the wrist, over the radial artery, although it may be observed at any point where an artery is near the surface.

Puma, also called mountain lion or cougar, large New World cat found in mountains, forests, swamps and jungles of the Americas. It has a small round head, erect ears and a heavy tail. The coat is tawny or pale brown with dark brown on the ears, nose and tail; the underparts are white. It preys mainly on deer and small animals, but has been known to take livestock. Length: to 2.3m (7.5ft), including the tail; height: to 75cm (30in) at the shoulder. Family Felidae; species *Felis concolor*. *See also* NW p.217.

Pumice, rhyolitic LAVA blown when it is molten to a low density rock froth by the sudden discharge of gases during a volcanic action. When ground to a powder and pressed into cakes it is used as a light abrasive.

Pump, device for raising, compressing, propelling or transferring fluids (liquids or gases). The lift pump, for raising water from a well, and the bicycle pump, for compressing air into pneumatic tyres, are examples of reciprocating (to-and-fro) pumps. Most modern high efficiency pumps have rotating impellers (blades) which create suction at their input points and force fluid out at their exit points. The centrifugal pump, a type of rotary pump, has an impeller which creates a pressure difference due to centrifugal force. *See also* MM p.*71*; SU pp.*78*, *79*.

Pumpkin, orange hard-rind garden fruit of a trailing annual VINE found in warm regions of the Old World and the USA; a variety of *Cucurbita pepo*. In the USA the pumpkin is also called a squash, especially the winter pumpkin (or squash) *Cucurbita maxima* and *C. moschata*. In Britain, the SQUASH is a member of the GOURD family (Cucurbitaceae). *See also* NW p.*61*; PE pp.*188, 189*.

Punch and Judy, British glove-puppet show. Performed in a tall booth, often out of doors, the show features a comic, grotesque and sadistic protagonist, Mr Punch, and his long-suffering wife, Judy, together with a host of gullible characters ranging from a hangman to a crocodile. The origins of this form of entertainment date from the PULCINELLA of the COMMEDIA DELL'ARTE and it was as an Italian MARIONETTE show that it made its first appearance in London in the 1660s.

Punched card, medium for storing and entering data into computers and accounting machines. The information is typed on the key-board of a machine which encodes it into a pattern of holes on a paper card, which the machine can interpret. *See also* MM pp. 106–109.

Punched tape, medium on which information (eg, data or programs) is converted into machine codes and fed into a computer. The information is typed on the keyboard of a machine which converts it into a pattern of holes on a paper tape. Each letter and digit corresponds to a unique combination of hole positions on the tape. *See also* MM pp. 106–107, *107*.

Punic Wars (265–241, 218–201 and 149–146 BC), three conflicts between Rome and Carthage, during which Rome became the predominant Mediterranean power. The first war arose over a dispute about control of the Straits of Messina. Although the Romans built their first fleet and gained command of the sea, they could not at first take Carthaginian strongholds in Sicily, nor defeat Carthage in N Africa. Eventually a naval victory off the Aegates in 241 led Carthage to sue for peace, and Sicily was surrendered. The second war arose out of Carthage's conquest of Spain (237–219 BC). HANNIBAL marched into Italy from Spain in 218 via the Alps, but failed to take Rome. Roman victories in Spain and N Africa led to Hannibal's recall, and his defeat at ZAMA in 202 BC. Carthage surrendered the next year, and became a Roman vassal. The third war arose out of Roman determination to destroy Carthage and, after a long blockade of the city, the Romans (led by SCIPIO AFRICANUS MINOR) razed Carthage to the ground. *See also* HC1 p.*91*.

Punjab, state in N India; the capital is Chandīgarh. It was formed in 1956 by the merging of East Punjab state, the union territory of Patiala and the East Punjab States Union and further reorganized in 1966 into two states, HARYANA and the Punjab. Apart from the capital, other importan t cities are Amritsar and Jullundur. The Punjab is a largely irrigated agricultural area. Products include textiles, woollens, cereals, cotton and sugar. Area: 50,376sq km (19,450sq miles). Pop. (1971) 13,551,060.

Punjab, province in NE Pakistan. It was formed during the partition of India in 1947 when the original Punjab province of British India was divided into E Punjab (mainly Hindu) which went to India and W Punjab (mainly Muslim) which became the Pakistani province of the Punjab. The area is mainly agricultural with wheat and cotton the major crops. Industries include textiles, machinery and electrical appliances. The capital is Lahore. Area: 206,432sq km (79,703sq miles); pop. (1969 est.) 30,427,000.

Punjabi, INDO-ARYAN language spoken by more than 17 million people in the Punjab in India and Pakistan. Punjabi is divided into two linguistic groups: eastern Punjabi is spoken around Lahore and Amritsar and has a flourishing literature; western Punjabi has no literary tradition. *See also* MS p.*240*.

Punting, sport and recreational pursuit in which a narrow, flat-bottomed boat is propelled by pushing a long pole against the riverbed. As a sport it features in regattas in singles, doubles and mixed doubles forms.

Pupa, quiescent, non-feeding developmental stage of insects that undergo complete metamorphosis, such as the BUTTERFLY and BEETLE. The pupa, often called a chrysalis in butterflies, generally occurs in a four-stage life cycle from the egg, through larva and pupa, to adult. *See also* NW p.*112*, *112*.

Pupil, in the structure of the EYE, the circular aperture through which light passes to the lens; it is located in the centre of the IRIS. Its diameter changes by reflex action of the iris, from about 3 to 8mm (0.12–0.32in), to control the entry of light. *See also* MS pp.*48–49*.

Pupin, Michael Idvorsky (1858–1935), US physicist and inventor, b. Yugoslavia. He devised a means of extending the range of long-distance telephone communications by placing loading coils at intervals along the transmitting wire, and invented a method of taking short-exposure X-ray photographs. His autobiography *From Immigrant to Inventor* (1923) won the Pulitzer Prize.

Puppet theatre, miniature stage for shows of glove puppets, MARIONETTES, rod puppets and flat and shadow figures. Such theatres existed in ancient Egypt and in Greece in the 5th century BC, in China, Java, and elsewhere in Asia, and reached their peak of popularity in 18th-century Europe. They also flourished in the Puritan period in England when the live theatres were closed. The use of dummies by ventriloquists has been developed in the 20th century, and in recent years many puppet shows have been presented on television.

Puppis, or the Poop, southern constellation that with Carina (the Keel) and Vela (the Sails) made up the old constellation of the Ship (Argo). The brightest star is of the second magnitude. *See also* SU pp.260, *260*, 266, *266*.

Purcell, Edward Mills (1912–), US scientist who shared with Felix BLOCH the 1952 Nobel Prize in physics. Purcell's discovery of the nuclear magnetic resonance (NMR) in liquids and solids is instrumental in the measurement of the magnetic moments of nuclei and molecules. *See also* SU p.*151*.

Purcell, Henry (1659–95), English composer and organist of the BAROQUE period. In 1677 he was appointed composer for the king's band and in 1679 became organist at Westminster Abbey. It is his church music for which he is most famous; much of it is still frequently performed. His only opera *Dido and Aeneas* (1689) is regarded as an early masterpiece of the form, and his many other stage works include *The Fairy Queen* (1692). He also excelled at writing music for special occasions, such as his coronation anthem for James II, *My Heart is Inditing* (1685). *See also* HC1 pp.318, *318–319*.

Purdy, Alfred Wellington (1918–), Canadian poet. His experiences while travelling in Canada and abroad form the basis for much of his poetry. His volumes of verse include *The Enchanted Echo* (1944), *The Crafte So Longe to Lerne* (1951), *North of Summer* (1967) and *Storm Warning* (1971).

Purgatorio (*c.*1321), second part of DANTE's *Divine Comedy*. Describing the penance for, and purification of, the seven deadly sins, the work marks the middle stage of the poet's allegorical journey from the study of error and ignorance to the contemplation of perfection. In preparation for his ascent into paradise Dante's guide, hitherto the pagan VIRGIL, becomes, appropriately, the embodiment of human perfection, Beatrice, at the end of this section of the work.

Purgatory, place between heaven and hell where a soul that has died in a state of grace is purged of its sins before entering heaven. In the teachings of the Roman Catholic Church souls which die with unforgiven venial and forgiven mortal sins go to purgatory. The suffering there differs from that in hell, in that the soul in purgatory knows its punishment is only temporary.

Purim, ancient Jewish celebration of thanksgiving, which usually takes place in March, commemorating the deliverance of the Jews of Persia from Haman's plot to destroy them. The story, which appears in the Book of ESTHER in the Old Testament, is read on this festival in all synagogue services. The gift of alms to the poor is obligatory on this occasion, and the festival is associated with an atmosphere of carnival.

Purine, white crystalline organic compound, $C_5H_4N_4$, related to uric acid. ADENINE and GUANINE are both purines and two of the four bases which comprise DNA. Another derivative of purine is ADENOSINE TRIPHOSPHATE (ATP), a nucleotide essential to the transfer of energy within living cells.

Puritans, English Protestant fundamentalists who were particularly influential during the 16th and 17th centuries. They originated in the reign of Elizabeth I; their chief aim was to make a truly Protestant Church out of the CHURCH OF ENGLAND, which was (and still is) technically not one. Following the ideas of Jean CALVIN, their opposition to Anglicanism was initially directed against liturgical vestments, elaborate ritual and other elements of "popish" practice, but they subsequently also demanded the abolition of church government by bishops and the setting up of a policy of PRESBYTERIANISM. According to Calvin's concept of PREDESTINATION, they thought of mankind as divided into the Elect – those destined inevitably for Heaven – and the Reprobate. The Puritans' zenith was reached when Oliver CROMWELL, an Independent (Congregationalist), assumed power and disestablished the Church of England; the Church was re-established in 1660, although Presbyterianism became the state Church in Scotland in 1690.

Purkinje, Jan Evangelista (1787–1869), Czech physiologist who discovered the Purkinje effect, in which red objects are perceived to fade faster than equally bright blue objects as the light intensity decreases. In 1825 he discovered the nucleus of the unripe ovum, which was named after him; he also discovered sweat glands in the skin (1833) and that pancreatic extracts can digest proteins (1836). He is probably best known for the discovery in 1837 of Purkinje cells; large neurons with many branching extensions in the cortex of the CEREBELLUM.

Purple Heart, Order of the, US military decoration. It was first instituted by George WASHINGTON in 1782 during the War of AMERICAN INDEPENDENCE but only three of these original awards were made. The present award, a rectangular mauve ribbon with white at both ends, was recreated in 1932 as an award for US servicemen wounded in battle.

Purpura, leakage of blood from the blood vessels which causes purple spots under the skin. Vitamin deficiencies, some allergies and certain drugs are some of the factors that can cause it. Treatment depends on the underlying cause.

Purse seining, common means of commercial fishing from a boat, in which a vertical net is used to surround a shoal of fish. A rope running around the base of the net is drawn tight, trapping the fish. *See also* PE p.*241*.

Purslane, any of several species of small annual plants with red stems, oval leaves and small red, brown, white or yellow flowers. Common European purslane, *Portulaca oleracea*, is typical in that it grows well in dry soil and can still bloom and produce seeds if uprooted. Family Portulacaceae.

Purvey, John (*c.*1353–*c.*1421), English religious reformer. He was secretary to the LOLLARD leader, John WYCLIFFE, and helped to finish (*c.*1382) Wycliffe's translation of the Bible, the first in English. A more careful and scholarly translation of the Bible (*c.*1396) has been ascribed to him. He was arrested as a heretic in the 1390s; he recanted and was released in 1401. He resumed his Lollard activities in 1403 and was arrested again in 1421, after which nothing more is known of him. *See also* HC1 p.195.

Pus, thick, yellow fluid produced by the body as a defence against bacterial infection. It comprises blood serum, dead PHAGOCYTES (types of white blood cells) and BACTERIA, and damaged tissue. It may contain live bacteria and is therefore a possible source of further infection.

Pusan, port in SE South Korea, 320km (200 miles) SSE of Seoul; capital of S Kyŏngsang province. The city was held by the Japanese from 1910 to 1945. During the KOREAN WAR, Pusan was the only port through which the UN army could be supplied. Industries: shipbuilding, textiles, metallurgy. Pop. (1970), 1,800,710.

Pusey, Edward Bouverie (1800–82), British theologian and leader of the OXFORD MOVEMENT. Appointed Professor of Hebrew at Oxford University in 1828, he opposed theological RATIONALISM and with John KEBLE and the future Cardinal NEWMAN wrote *Tracts for the Times* (1834) on the difference between Anglican and Roman Catholic doctrine and practice. His other writings include *The Doctrine of the Real Presence* (1855–57) and the three books *Eirenicon* (1865–70).

Pushkin, Alexander Sergeievich (1799–1837), Russian poet and novelist. He was exiled in 1820 for his political views and in the same year his first major poem *Ruslan and Lyudmila* was published. *The Prisoner of Caucasus* (1822) and *Boris Godunov* (1831) show his response to the beauty of the CRIMEA and CAUCASUS, and reveal the influence of Lord BYRON. Later, under the protection of tsar NICHOLAS I, he completed his masterpiece *Eugene Onegin* (1833). He also wrote *The Queen of Spades* (1834) and *The Captain's Daughter* (1836). *See also* HC2 p.100, *100*.

Pushtu. See PASHTO.

Puskas, Ferenc (1926–), Hungarian footballer who, having won 84 caps for his country (1945–56), became an exile after the 1956 Uprising. He joined the Spanish club Real Madrid in 1958, for which his dynamic left-foot shooting brought many honours. One of football's great players, he won three caps for Spain (1961–62) and later became a successful manager in Greece.

Putnik, Radomir (1847–1917), Serbian general. He commanded the Serbian army from the beginning of WWI until the occupation of Serbia by the CENTRAL POWERS in 1916.

Putrefaction, decomposition of organic matter, especially PROTEINS, by FUNGI, BACTERIA and OXIDATION. It results in foul-smelling products. Putrefaction of meat, for example, yields hydrogen sulphide, amines and mercaptans.

Putting the weight. *See* SHOT PUT.

Putto (Italian, "little boy"), in art, word to describe the chubby, naked, often winged babe found in many pictures and sculptures. As representations of the god Eros or Cupid, putti have been a common motif in decorative art since classical times.

Putumayo (Içá), river in NW South America. It rises in the SW part of Colombia, on the slope of the Andes; runs SE forming a large part of the Colombia-Peru border; flows into Brazil, where it is known as the Içá, and then into the River Amazon. Length: 1,578km (980 miles).

Puvis de Chavannes, Pierre (1824–98), French mural painter. He gained recognition with his *Return from Hunting* (1859). His huge murals adorned many public buildings and include those for the Hôtel de Ville (1893) and the Panthéon, Paris (begun 1876).

Puy-de-Dôme, département in the Auvergne and part of Bourbonnais, central France. Situated in the MASSIF CENTRAL, it includes an extinct volcano (of the same name), height 1,465m (4,806ft), which has an observatory, built in 1876, on its level summit. Pop. (département) (1972 est.) 573,200.

Pu Yi, Henry (1906–67), Chinese emperor. As Hsuan Tung, he ruled from 1908–12. He became President of Manchukuo, a Japanese puppet state, in 1932 and styled himself Emperor (1934). Captured by Soviet forces (1945), he was delivered to Mao Tse-tung and imprisoned (1949–59). He later served in the Chinese parliament from 1964.

Puzo, Mario (1920–), US novelist. His first novel, *The Fortunate Pilgrim*, was published in 1964. He achieved prominence, however, with the publication of *The Godfather* (1969), a sensational bestseller about a New York MAFIA family. Selling over 500,000 copies in hardback and over 10 million copies in paperback, the book was the basis of two equally successful films. *See also* MM p.*212*.

PVC, polyvinyl chloride, a synthetic THERMOPLASTIC resin made by the POLYMERIZATION of vinyl chloride ($CH_2{=}CHCl$), obtained from natural gas or from crude oil. It is water-repellent and has good insulating properties. Among its many everyday applications are plastic plumbing, guttering, floor covering, waterproof clothing and insulation for electrical cables. *See also* SU p.*143*.

Pyelitis, inflammation of the pelvis of the KIDNEY, where urine collects to drain along the URETER and into the BLADDER. It sometimes occurs in pregnant women. The symptoms include pain in the lower back, chills, fever and frequency of urination. The urine may be bloody, or cloudy with pus. It is usually treated with SULPHA DRUGS or ANTIBIOTICS.

Pyelonephritis, worsening of PYELITIS, in which the initial bacterial infection has spread throughout the KIDNEY. It may persist for many years with concurrent damage.

Pygmalion (1913), five-act comedy by George Bernard SHAW. A modern adaptation of the Pygmalion-Galatea legend, it tells how Henry Higgins, a professor of phonetics, transforms a Cockney flower girl, Eliza Doolittle, into a perfect lady. Shaw added new scenes and speeches for the 1938 film version, and the play was later made into the successful musical *My Fair Lady* (1956).

Pygmy, member of a group of people whose men average less than 1.5m (4.6ft) in height. African Pygmies inhabit parts of central Africa and Zaïre; Asian Pygmies, or Negritos, are found mainly in the Phil-

Henry Purcell's scores were restricted by the instruments of his day.

Puritans destroyed the Cheapside cross as an image of the Church and monarchy.

Pushkin's conflicts with the authorities stemmed from his liberal opinions.

Puy-de-Dôme, a fertile area of central France, also has coal and mineral deposits.

John Pym was an important parliamentarian in the years before the Civil War.

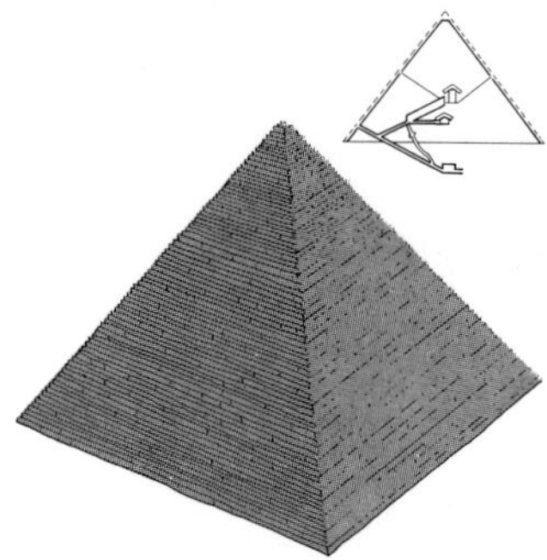

Pyramids of specific sizes have properties of renewal and were used as tombs.

Pyrethrum, which blooms in spring or early summer, resembles the daisy.

Pythagoras: a page from a 16th century book teaching Pythagoras' arithmetic.

lippines and Malaysia. Most pygmies are hunters and gatherers, with few crafts. Their numbers are declining rapidly and they seem likely to become extinct fairly soon. *See also* MS p.33.

Pym, John (1584–1643), English politician. He was a leader of the opposition to the Duke of BUCKINGHAM in the 1620s, and helped to pass the Petition of Right in 1629. He was the leading member of the parliamentary opposition to CHARLES I in 1640–43, attacking STRAFFORD and LAUD. Charles tried but failed to arrest him in Parliament in 1642, and Pym organized the alliance between Parliament and the Scottish COVENANTERS.

Pynas, Jan (c. 1583–1631), Dutch painter of religious and historical subjects. His paintings include *The Raising of Lazarus* (1615) and *The Entombment* (1607).

Pynchon, Thomas (1937–   ), US novelist whose works are noted for their off-beat humour and inventiveness. He has shunned all publicity, so little is known about his life. His works include the science fiction novels *V.* (1963) and *The Crying of Lot 49* (1966), and *Gravity's Rainbow* (1973).

P'yongyang, city in w North Korea, on the Taedong River; capital of North Korea with provincial status. An ancient city, it served as the capital of the Chosor (12th-century BC – 108 BC), Koguryo (c. 37 BC – AD 668) and Koryo (10th–12th centuries) kingdoms. It became the capital of North Korea in 1948. During the KOREAN WAR, P'yŏngyang was held by UN forces but fell to the Communists in 1950. It suffered considerable damage by US air raids during the Korean War and since 1953 has been substantially rebuilt. Parts of the ancient city walls remain and there is a university (1946) in the city. Industries: cement, iron and steel, chemicals, rubber, textiles. Pop. (1966) 1,364,000.

Pyorrhoea, disease of the gums and associated tissues. It starts with bleeding followed by receding of the gums and loosening of the teeth as their roots are exposed. Its causes include malnutrition, poor dental hygiene and dental tartar.

Pyracantha, or fire thorn, genus of thorny evergreen shrubs found from SE Europe to China. They have toothed, leathery leaves, white flowers, and red, orange or yellow berries. Easily grown in subtropical and temperate regions, they are popular garden shrubs. Family Rosaceae.

Pyramid, large monument on a square base with sloping sides rising to a point. Pyramids are found mostly in Egypt and served as burial chambers for the pharaohs. Similar structures, but with stepped sides, called ziggurats, were built in Mesopotamia and Central America. The earliest pyramid was built c. 2700 BC for Zoser; pyramid building reached its peak in the 4th, 5th and 6th Dynasties of ancient Egypt. The largest pyramid was built for KHUFU (Cheops) c. 2500 BC; it was 146m (480ft) high and 231m (758ft) long at the base. The largest New World pyramid was built in TEOTIHUACÁN, Mexico, in the 1st century AD; it was 66m (216ft) high with a surface area of 50,600sq m (547,200sq ft). *See also* HC1 pp.*35, 47.*

Pyramid, in geometry, a solid figure having a polygon as one of its faces (the base), the other faces being triangles with a common vertex. Its volume is one-third of the base area times the vertical height.

Pyramids, Battle of the (July 1798), during the French Revolutionary Wars, a battle fought between French troops under NAPOLEON and the Egyptian MAMELUKES. Napoleon's victory gave the French control over Egypt, but was soon nullified by NELSON's destruction of the French fleet at the Battle of the NILE (1798).

Pyrenean mountain dog, large dog that was originally used in the Pyrenees to guard sheep from wolves and bears. It has a large domed head with a solid muzzle and pendulous ears. The deep-chested, stocky body is set on stout muscular legs and the tail is long. The long, silky, irregular coat may be white or off-white with grey or tan markings. Shoulder height: to 81cm (32in); weight: to 57kg (125lb).

Pyrenees, range of mountains in s France

and N Spain, extending in an almost straight line E to w from the Mediterranean Sea (Cape Creuse) to the Bay of Biscay (Cape Higuer). There are few passes through the mountains, which were formed in the Tertiary era. The Pyrenees contain deposits of marble, gypsum and oil, and there are extensive forests. Sheep and goat grazing is the chief farming activity. The highest point is Pico de Aneto, 3,406km (11,168ft). Length: 435km (270 miles).

Pyrethrum, popular name for the painted daisy (*Chrysanthemum coccineum*) and for several other species of the genus CHRYSANTHEMUM. It is also the name for the insecticide made from the dried flower heads of the Dalmatian chrysanthemum (*C. cinerariaefolium*).

Pyridine, compound (C_5H_5N) of the heterocyclic series characterized by a six-membered ring structure composed of five carbon atoms and one nitrogen atom. It occurs in bone oil and coal tar and is synthetically produced from acetaldehyde and ammonia. Pyridine is used as a solvent and converted to sulphapyridine (a drug used to combat bacterial and viral infection), pyribenzamine and pyrilamine (ANTIHISTAMINES).

Pyridoxine, VITAMIN of the B complex (B_6) involved in the metabolism of AMINO ACIDS in the body.

Pyrimidine, colourless liquid ($C_4H_4N_2$) of the heterocyclic series characterized by a ring structure composed of four carbon atoms and two nitrogen atoms. It may be prepared from uracil, a dihydroxy pyrimidine compound, by chemical reactions that remove two oxygen atoms. It is used in the manufacture of various SULPHA DRUGS and BARBITURATES.

Pyrites, most common and widespread sulphide minerals, iron sulphide (FeS_2), occurring in all types of rocks and veins. Often called "fool's gold" because of its deceptive colour, it is used chiefly to produce sulphuric acid. It crystallizes as cubes, pyritohedra and octahedra, often twinned, and also as granules and globular masses. It is opaque, metallic and brittle, with a hardness of 6.5 and s.g. 5.01.

Pyromania, symptom of a mental disorder characterized by a compulsion to start fires. The pyromaniac enjoys watching fires burn and often lights them following a set, ritualistic pattern.

Pyrometer, thermometer for use at extremely high temperatures, well above the ranges of ordinary thermometers. An optical pyrometer consists essentially of a small telescope, a RHEOSTAT and a filament. When the telescope is aimed at a furnace or other hot object, the filament appears dark against the background. As the electric current flaring through the filament is increased slowly using the rheostat, it grows brighter until it matches the intensity of the furnace. The amount of current is a measure of the temperature. *See also* SU p.91.

Pyrope, magnesium aluminium GARNET, and the most common form of garnet; its composition is $Mg_3Al_2(SiO_4)_3$. It may be purplish or brownish red and is found mainly in South Africa and the USA. Often used in jewellery, it has been given such misleading names as ruby and Bohemian garnet.

Pyrotechnics. *See* FIREWORKS.

Pyroxene, any of a large group of single-chain SILICATES, which include many important rock-forming minerals. Enstatite, augnite, JADEITE and wollastonite are well-known pyroxenes. Their colours range from white and yellow to brown. Hardness 2.3–4; s.g. 5.5–6. *See also* PE pp.96, 101.

Pyrrho (c. 360–270 BC), Greek philosopher, considered the founder of SCEPTICISM. He went with Alexander the Great's expedition to the East, after which he enjoyed distinction in Athens. His doctrine was that nothing can be known because every statement can be plausibly contradicted; therefore wisdom is in reserved judgement.

Pyrrhonism, sceptical philosophy named after the Greek philosopher PYRRHO.

Pyrrhus (319–272 BC), King of Epirus (r. 307–302, 295–272 BC). He greatly

expanded his kingdom into Illyria, Macedon and Thessaly. While aiding Tarentum he defeated the Romans at Heraclea in 280 and Asculum in 279, but sustained grievous losses (hence the term "Pyrrhic victory"). He was killed by a street mob in Argos.

Pyrrole, organic chemical compound consisting of a five-membered ring with four carbon atoms and one nitrogen atom (C_4H_5N). It is a colourless liquid that smells like chloroform, obtained from coal tar and bone oil. It forms part of the molecules of various natural pigments, such as CHLOROPHYLL and HAEMOGLOBIN.

Pyrus, genus of plants, including several species of PEAR, especially *P. communis*, one of the most important fruit trees of the world. Family Rosaceae. *See also* PE p.*192.*

Pyruvic acid, colourless liquid that can be formed by the distillation of TARTARIC ACID; it has a vinegary odour. A derivative, phenylpyruvic acid, occurs in the urine of patients suffering from PHENYLKETONURIA.

Pythagoras, (c. 580–500 BC) Greek philosopher and founder of the Pythagorean school. Little is known of his life except that he was born on Samos and lived in Crotona. The Pythagoreans believed in reincarnation and that numbers constitute the true nature of things. They treated the sexes as equal, were humane to slaves and respected animals; they saw all relations as numerical, made important contributions to geometry, medicine and philosophy, and were probably the first to teach that the Earth is a sphere revolving about the Sun. *See also* HC1 pp.82–83.

Pythian Games, in ancient Greece, games held every four years at DELPHI in honour of APOLLO. They took their name from the Pythia, the medium of the oracle, and included musical and athletic contents.

Pythias, (or, correctly, Phintias), Pythagorean friend of Damon, the subject of a story by CICERO to illustrate true friendship. Condemned to death by Dionysius of Syracuse in the 4th century BC, Pythias asked leave to arrange his affairs and Damon pledged his life against Pythias' failure to return. Such loyalty so impressed Dionysius that he spared the life of Pythias.

Python, name of more than 20 species of non-poisonous snakes of the BOA family (Boidae), found in tropical regions. Like boas, pythons are constrictors which kill their prey (brids and mammals) by squeezing them in their coils. Unlike boas, pythons lay eggs. The reticulated python (*Python reticulatus*) of SE Asia and the Philippines vies with the anaconda as the world's largest snake, reaching up to about 9m (30ft). Subfamily Pythoninae.

Pyx, vessel used to contain consecrated bread; in the Roman Catholic Mass. It may be ornately shaped or a simple cylindrical box with a conical lid. As the liturgy changed, the pyx came to be placed on a stand; this probably led to the use of the MONSTRANCE, in which the host is exposed to view.

Q

Q, 17th letter of the alphabet, derived from the Semitic letter *qoph* (meaning the eye of a needle), which became *qoppa* in the Greek alphabet. It was superseded by *kappa* and retained its *k* sound when incorporated into the Latin alphabet, which gave it the following *u* that accompanied it into all the subsequent Romance languages, and thence into English. Thus the *qu* combination retains a *kw* or *k* pronunciation in such English words as *queen* or *opaque*. In a few words (mostly proper names) of Arabic origin, the *q* has no accompanying *u* and is pronounced as a *k*, as in *Iraq*. *See also* MS pp.244–245.

Qaddafi, Muammar al- (1942–   ), Libyan political leader and army officer who in 1969 led a coup against the monarchy of Idris I, naming himself Libya's commander-in-chief and Chairman of the Revolutionary Command Council. A militant Arab nationalist, he helped to support the Palestinian guerrillas in their actions against Israel and sought to unite Libya with other Arab nations. *See also* HC2 pp.296–297.

Qajar dynasty, ruling house of Persia (Iran) from 1794 to 1925. It was established by Agha Muhammad, a Turkoman eunuch, who gained control of the north, defeated the last Zand ruler in 1794 and was crowned Shah in 1796. He made Teheran his capital. The last Qajar ruler, AHMAD SHAH, was exiled by REZA KHAN in 1923 and deposed in 1925. *See also* MW p.101.

Qandahār. *See* KANDAHAR.

Qantas, international airline of Australia, originally the internal Queensland and Northern Territory Aerial Service. It came under government control in 1947.

Qaraism. *See* KARAISM; KARAITES.

Qatar, independent Arab nation occupying a peninsula extending N into the Persian Gulf from the Arabian mainland. Most of the people live along the coast because the land which rises sharply behind it is a barren plateau. The economy is based on the production of oil and natural gas; the capital is Doha. Area: 11,000sq km (4,247sq miles). Pop. (1975 est.) 180,000. *See* MW p.146.

Qingdao. *See* TSINGTAO.

Quadragesima, the 40 days of LENT, before Easter. The first Sunday in Lent is Quadragesima Sunday.

Quadrant, in plane geometry, a quarter of a circle, bounded by radii at right-angles to each other and the arc of the circle. In analytic geometry it is one of the four sections of a plane divided by an x-axis and a y-axis. A quadrant is also a device for measuring angles, based on a 90° scale.

Quadratic equation, algebraic equation in which the highest exponent of the variable is 2; an equation of the second degree. A quadratic equation has the general form $ax^2 + bx + c = 0$, where a, b and c are constants. It has two solutions (roots), given by the formula $x = [-b \pm \sqrt{(b^2 - 4ac)}]/2a$.

Quadrille, group ballroom dance popular in Europe in the late 18th and 19th centuries. It consisted of four or five short dances, each with prescribed combinations of figures. It is also a North American square dance based on the same form, but often with the different patterns of the shorter dances combined.

Quadruple Alliance, name given to several alliances between European countries. That of 1718 was formed by England, France, The Netherlands and the Holy Roman Emperor against Philip V of Spain. That of 1814 was made by Britain, Austria, Prussia and Russia against Napoleon I. That of 1834 was formed by Britain, France, Spain and Portugal to support the Spanish throne against the CARLISTS.

Quaestor, or questor, magistrate of ancient Rome, with financial rather than judicial authority.

Quagga, extinct ZEBRA once numerous in southern Africa. Excessive hunting, particularly for its skin, led to its extermination by 1883. Its brown coat was marked with black and cream stripes on the head, neck and shoulders only. The legs were pale-coloured and unstriped. A few photographs and preserved museum specimens survive. Family Equidae; species *Equus quagga*.

Quail, any of a group of Old World chicken-like game BIRDS. The European quail (*Coturnix coturnix*), a small short-tailed bird with white throat and mottled brownish plumage, is found throughout Europe, Asia and Africa. The Australian quail, a variety which is known as little quail (*Turnix velox*), is a stocky, sedentary, brownish bird. Mainly ground-living birds, they scrape for fruits and seeds and nest on the ground. *See also* NW p.205.

Quakers, nickname for the Society of Friends, a Christian sect which arose in England in the 1650s, founded by George Fox. The name derived from the injunction given by early Quaker leaders that their followers tremble in the sight of the

Lord. Quakers rejected the episcopal organization of the Church of England, believing in the priesthood of all believers and the direct relationship between a man and the spiritual light of God. Quakers originally worshipped God in silence unless someone was moved by the Holy Spirit to speak or offer up a prayer but have introduced hymns and readings into their meetings since the mid-19th century. The largest Quaker Church is in the USA, where it began with the founding of a settlement by William PENN in 1681.

Qualitative analysis, identification of the chemical elements or ions in a substance or mixture. *See* SU pp.150–151.

Quant, Mary (1934–), British fashion designer whose off-the-peg, youth-oriented clothing – especially the mini-skirt – helped to focus attention on London as a world fashion centre in the 1960s.

Quantitative analysis, identification of the amount of chemical constituents in a substance or mixture. Two methods are employed, chemical and physical. Chemical methods use reactions such as precipitation, neutralization and oxidization, and measure volume or weight (volumetric or gravimetric analysis). Physical methods measure qualities such as density and refractive index. *See* SU pp.150–151.

Quantum. *See* QUANTUM THEORY.

Quantum mechanics, in physics, use of QUANTUM THEORY to explain the behaviour of ELEMENTARY PARTICLES, such as electrons. In essence quantum mechanics makes energy (waves) and particles (matter) interchangeable concepts. In 1924 Louis de BROGLIE suggested that particles have wave properties, the converse having been postulated by Max PLANCK in 1900. In 1926 Erwin SCHRÖDINGER used this hypothesis to predict particle behaviour on the basis of wave properties, and a year earlier Werner HEISENBERG had produced a mathematical equivalent to Schrödinger's theory, without using wave concepts. In 1928, Paul DIRAC unified these approaches while incorporating RELATIVITY into quantum mechanics.

Quantum numbers, in physics, a set of numbers that correspond to the discrete values of a quantized physical property. They are either integers (1, 2, 3, etc) or half integers (1/2, 3/2, 5/2, etc). For example, an electron in a hydrogen atom can have energies E_0/n^2, where n (the principal quantum number) can take values of 1, 2, 3, etc, and E_0 is the lowest energy level (corresponding to $n = 1$). Similarly the charge of any other subatomic particle can be thought of as quantized in units of electron charge, so a proton has a charge of $+1$. An ELEMENTARY PARTICLE is defined uniquely by its mass and its quantum numbers. *See also* QUANTUM THEORY; SU pp.63, *63*, 108–109.

Quantum theory, together with the theory of RELATIVITY, the foundation of 20th-century physics. It is concerned with the relationship between matter and energy at the elementary or sub-atomic level and with the behaviour of ELEMENTARY PARTICLES. According to the theory, all radiant energy is emitted and absorbed in multiples of tiny "packets" or quanta. This breaks down the traditional distinction between energy (as a wave motion) and matter (as occupying a fixed space), because quanta have properties of both. Waves and particles therefore become dual aspects of the same phenomenon, and any instance of either aspect can be quantized – explained in terms of a definite number of quanta. The idea that energy is radiated (and absorbed) in quanta was proposed by Max PLANCK in 1900. He calculated that the energy E of a quantum is related to the frequency ν of radiation according to the formula $E=nh\nu$, where h (Planck's constant) is a constant with the value of 6.63×10^{-34} joule-second, and n (the principal QUANTUM NUMBER) equals 1, 2, 3, 4 ... Using this formula Einstein quantized light radiation and explained the PHOTOELECTRIC EFFECT in 1905. In 1913 Niels BOHR used quantum theory to explain atomic structure and the relationship between the energy levels of an atom's electrons and the frequencies of radiation emitted or absorbed by the

atom. *See also* QUANTUM MECHANICS; SU pp.63, *63*, 108–109.

Quarantine, originally a 40-day waiting period during which ships were forbidden to discharge passengers or freight to prevent transmission of plague or other diseases. Today the term refers to any period of isolation legally imposed on people, animals, plants or goods in order to prevent the spread of disease.

Quark, in physics, any one of three particles and their anti-particles (antiquarks) that have been postulated as a constituent of the HADRON group of ELEMENTARY PARTICLES. In 1964 Murray GELL-MANN and G. Zweig proposed independently that all fundamental particles are composed of quarks and "gluons", which hold the quarks together. Originally three kinds of quarks were proposed: two with a fractional charge of $-\frac{1}{3}$ and one with a charge of $+\frac{2}{3}$, and all with a spin of $\frac{1}{2}$. BARYONS consist of three quarks; MESONS consist of a quark and an anti-quark tightly bound together. Physicists have yet to discover free quarks. *See also* SU pp.*51*, 67.

Quarrying, method of obtaining stone or other types of non-metallic rock from open-pit mines. Dimension (cut) stones for building (such as sandstone, granite and marble) are obtained by separating large blocks from the parent mass using various power-driven drilling or cutting tools and then carving them to size. Ballast (rough, small stones) for road building is obtained by drilling and blasting. It may then be further crushed, screened and sized. *See also* MM p.242; PE p.154.

Quarter Sessions, criminal courts once held in Britain four times a year by justices of the peace in the district to which they were appointed by the Lord Chancellor. They began in 1368 after the heavy expenses involved had ended the king's travelling around the country to hear cases. They were abolished in 1971.

Quartet. *See* CHAMBER MUSIC.

Quartz, rock-forming mineral (silicon dioxide). It may be opaque, translucent or transparent, coloured or uncoloured. It is formed of hexagonal crystals and is classified as crystalline or cryptocrystalline. The most common crystalline varieties are colourless quartz (known as rock crystal), rose, yellow, milky and smoky. The most common cryptocrystalline varieties, whose crystals can be seen only under a microscope, are CHALCEDONY and FLINT. *See also* PE pp. 96–99.

Quasar, also called quasi-stellar object, in astronomy an object that appears to be a massive, highly compressed, extremely powerful source of radio and light waves, characterized by a large red shift. If such red shifts are due to the DOPPLER EFFECT, it can be deduced that quasars are more remote than any other objects previously identified; many are receding at velocities greater than half the speed of light. Their energy may result from the gravitational collapse of a GALAXY or from many SUPERNOVAS exploding in quick succession, although there seems to be no reason why such events should be occurring. *See also* SU pp.172, 235, 250–251.

Quasimodo, Salvatore (1901–68), Italian poet. He published five volumes of verse in the 1930s, the first of which, *Acque e Terra* (1930), established him as the leading Italian symbolist. He was imprisoned for anti-FASCIST conduct in WWII. He was awarded the 1959 Nobel Prize in literature.

Quaternary Period, in the history of the Earth, the most recent period of the Cenozoic era, beginning about 2 million years ago and extending to the present. It marks the age of man. It is divided into the PLEISTOCENE, the glacial epoch, characterized by a periodic succession of four great ice ages, and the HOLOCENE, which started some 10,000 years ago. *See also* NW pp.190–191.

Quaternions, type of higher COMPLEX NUMBER, useful in MECHANICS. An ordinary complex number has the form $a + bi$ (where a and b are real numbers, i an imaginary number, the square root of -1). A quaternion has the form $a + bi = cj + dk$ (in which i, j and k are imaginary,

Quagga: the last specimen of this species died in the London Zoo in 1872.

Quail are common migratory birds in Europe, Asia and N Africa.

Quakers were often the subject of satirical attacks in the 17th century.

Mary Quant was a key figure in the decade known as the "swinging 60s".

Quebec: the Citadel and the Chateau Frontenac overlook the city.

SS Queen Elizabeth: she caught fire while at anchor in Hong Kong harbour.

Queensberry Rules are still followed in modern professional boxing.

Queensland is Australia's warmest state; sugar cane is an important crop.

defined by the equations $i^2 = j^2 = k^2 = ijk = -1$).

Quathlambra. See DRAKENSBERG RANGE.

Quattara, desert basin in N Egypt in the Libyan Desert. Impassable by armies and vehicles, it was the s end of British defence at El-Alamein in 1942 and the point at which Rommel's invasion was halted. It contains the lowest point in Africa, 134m (440ft) below sea-level. Area: approx. 19,425sq km (7,500sq miles).

Quatre Bras and Ligny, twin battles fought on 16 June 1815 during the last campaign of the NAPOLEONIC WARS. Hoping to prevent the concentration of two allied armies, NAPOLEON attacked the Prussians under BLÜCHER at Ligny while Marshal NEY led forces against the British army under WELLINGTON at Quatre Bras. Both allied armies were driven bakc, but they recombined to defeat Napoleon at WATERLOO on 18 June. See also HC2 p.79.

Quattrocento (Italian, "fifteenth century", ie 1400s), in art history, term applied to the 15th-century period of the Italian RENAISSANCE. Venice was its cultural centre and leading figures included the painters Fra ANGELICO, Fra FILIPPO LIPPI, MASACCIO and UCCELLO; the architects BRUNELLESCHI and ALBERTI; and the sculptors DONATELLO and GHIBERTI. See also HC1 pp.250–251, 250–251.

Quebec (Québec) province in E Canada; largest province in area and second-largest in population. Most of the state is on the CANADIAN SHIELD and is relatively uninhabited. The lowlands by the St Lawrence River are the centre of industry and agriculture, the area's small farms yield vegetables, tobacco and dairy produce. Major industries are food processing, brewing, transportation equipment, chemicals, and metal and paper products. The province yields vast quantities of hydro-electric power and timber, and copper, iron, zinc, asbestos and gold are mined. The largest city is Montreal and the capital is the city of Quebec.

Jacques CARTIER found the E coast of Canada in 1534 and later sailed up the St Lawrence River. The port of Québec was the first settlement to be established (1608) and from there the French pushed w founding trading and military posts. The entire area was ceded to Britain by treaty in 1763 and became a British colony. With the establishment of the DOMINION OF CANADA in 1867, Quebec became a province under that name. In the mid-20th century French inhabitants of the province have intensified their demands for recognition of their cultural heritage, including independence of Quebec. Area: 1,540,687sq km (594,860sq miles). Pop. (1974 est.) 6,212,000. See also MW p.45.

Quebec, port in S Quebec province, Canada, on the St Lawrence River; capital of Quebec province. Samuel de CHAMPLAIN established a French trading-post on the site of Quebec (Québec) in 1608. Captured by the British in 1629, the city was returned to France and became the capital of New France from 1663 to 1763. The British defeated the French in 1759 and Québec became a British colony. The city served as the capital of LOWER CANADA (1791–1841) and the United Provinces of Canada (1851–55, 1859–65). The city has become a focal-point for Canada's French-speaking separatists. Quebec is a cultural and tourist centre, and its industries include shipbuilding, paper, leather, textiles, machinery, canned food, tobacco and chemicals. Pop. (1974) 186,090. See also MW p.45.

Quechua, most widely spoken of all South American Indian languages, with about 5 million speakers in Peru, 1.5 million in Bolivia and 500,000 in Ecuador. Originally the language of the great Inca empire, it is related to AYMARÁ, the two forming the Quechumaran family.

Queen, Ellery, pen-name of Frederic Dannay (1905–), US detective fiction writer. He has written more than 60 crime and mystery novels, of which 34 have Ellery Queen as the detective hero. He has also written four novels under the pen-name Barnaby Ross.

Queen Anne Style, in art history, term applied to style of decorative arts (especially furniture) which flourished in England during the reign (1702–14) of Queen Anne – although it developed before and continued after that time. The style is characterized by elegance and simplicity. Curved cabriole legs are a distinctive feature, as are inlay, veneering and lacquerwork.

Queen Charlotte Islands, archipelago off w British Columbia, Canada, separated from the mainland by Hecate Strait. The chief islands are Graham, Moresby, Louise and Lyell and the inhabitants are mostly Haida Indians. Industries: mining, timber, fishing. Area: 9,596sq km (3,705sq miles). Pop. (1966) 2,222.

Queen Elizabeth, SS, British passenger liner. Launched in 1938, the ship was designed as a sister ship for the QUEEN MARY on the weekly service between Southampton and New York. Sailing throughout WWII as a troop-ship, she sailed commercially for the first time in 1946. She was destroyed by fire in 1972 in Hong Kong.

Queen Mary, SS, British passenger liner. Launched in 1934, she served as a troop-ship during WWII. She was withdrawn from service in 1967 and is now a museum and conference centre at Long Beach, California.

Queen Mother. See ELIZABETH, QUEEN MOTHER.

Queen of Spades, The (1834), short story by Alexander PUSHKIN, in which the ambitious and ruthless hero Hermann risks and loses his money and his sanity in a game of cards. The story was used as the basis for an opera by Peter TCHAIKOWSKY (1890).

Queens, largest borough of New York City on the w end of Long Island; connected to the mainland by bridge and tunnel. A residential and industrial area, its manufacturing industries produce consumer goods for New York City. It was the site of New York's World Fairs (1939–40, 1964–65) and has La Guardia and John F. Kennedy International Airports, and St John's University (1870). Area: 280sq km (108sq miles). Pop. (1970) 1,986,473.

Queen's Award to Industry, British scheme, devised in 1965 by a committee chaired by Prince PHILIP, which provides for awards to be made to industry by Queen ELIZABETH in recognition of achievements in exports or technological innovation. Various other aspects of industrial efficiency are also rewarded. Companies receiving an award may display a special emblem for five years. Between 1966 and 1970 nearly 500 awards were made to more than 4,000 applicants.

Queen's (or King's) Bench, highest court of English COMMON LAW. Its name derives from its medieval origin, when the monarch travelled about the country with his court to hear criminal cases. Since 1873 it has been a division of the High Court of Justice. It consists of the Lord Chief Justice and a number of puisne (or minor) judges.

Queensberry, John Sholto Douglas, Marquess of (1844–1900), British nobleman who sponsored the QUEENSBERRY RULES – the basis of modern boxing rules. In 1895 Queensberry publicly insulted Oscar WILDE because of the latter's association with Queensberry's son, Lord Alfred Douglas. Wilde unsuccessfully sued Queensberry for libel, and was convicted for homosexuality.

Queensberry Rules, the basis of modern boxing rules, published under the sponsorship of the Marquess of QUEENSBERRY in 1866. Drafted mainly by John G. Chambers of the British Amateur Athletic Club, they were standardized in 1889 and first used in a championship in the world heavyweight bout of 1892.

Queen's Counsel (QC), member of the Bar, appointed letters patent to be Her Majesty's counsel, in England on the recommendation of the lord CHANCELLOR, in Scotland of the Lord Justice General. He wears a silk robe and is colloquially called a "silk".

Queensland, state in NE Australia. Approx. half the state lies north of the Tropic of Capricorn and there are rain forests in the N. The GREAT DIVIDING RANGE separates the fertile coastal strip from the extensive interior plains. Predominantly a primary producing state, the chief crops are sugar-cane, wheat, cotton and tropical fruits; the main industries are beef cattle and mining. There are valuable mineral deposits, including copper, coal, lead, zinc, bauxite, oil and natural gas. The capital is Brisbane. Queensland was originally part of New South Wales and served as a penal colony from 1824 to 1843; it was separated and became a British colony in 1859. The area became a state of the Commonwealth of Australia in 1901. Area: 1,727,530sq km (667,000sq miles). Pop. 1,799,200. See also MW p.28.

Queen's Medal, short name of two British awards; the Queen's Police Medal and the Queen's Fire Services Medal. Created in 1954 to replace the King's Police and Fire Services Medal, they are awarded, posthumously, for gallantry.

Queen's Men, The, theatrical company which received a royal patent in 1574. After entertaining Elizabeth I at Kenilworth in 1575, the company (under James BURBAGE), acted in a theatre provided by John Braynes until 1583, when another company, the new Queen's Men, was formed. This company was directly in the Queen's service until 1593.

Queen's Prize, British award for service-calibre rifle shooting. Founded by Queen Victoria in 1860, it is the most prized award in British and Commonwealth target shooting. It involves a three-stage contest, usually entered by about 1,000 people, 300 of whom qualify for the second stage. The final stage is contested by 100 marksmen.

Queen's Proctor, name for the solicitor who represents the Crown in the Family Division of the High Court of Justice in England. His duties are assigned to him in the Matrimonial Causes Act of 1973.

Queenston Heights, Battle of (Oct. 1812), decisive battle of the WAR OF 1812. The advance guard of the US army, 1,600 strong, seeking to conquer Canada, was annihilated by the British army, composed chiefly of Canadian volunteers. The British commander, General Brock, was killed.

Queenstown, mountain town in New Zealand's South Island, on Lake Wakatipu, about 282km (175 miles) from DUNEDIN. Formerly a gold-digging town, it is now a leading tourist resort, famed for its scenic beauty.

Quellin (Quellinus), name of two Flemish sculptors, the cousins Artus I (1609–68) and Artus II (1625–1700). The elder did the allegorical figures which decorate the royal palace at Amsterdam, the younger the BAROQUE figure of God in the rood screen of Bruges Cathedral.

Quemoy (Chinmen Tao), island group off SE China, on the Formosa Strait, 242km (150 miles) w of Taiwan. It is comprised of Quemoy and Little Quemoy islands and 12 islets. Quemoy Island has been garrisoned by Taiwan since the Communist take-over in 1949. The main occupation is farming, and crops include sweet potatoes, groundnuts and cereals. Pop. 61,305.

Queneau, Raymond (1903–), French poet and novelist, much influenced by his early association with the SURREALISTS. His poems include Les ziaux (1943) and Exercices de style (1948), a PARODY of literary forms. Among his novels are Le chiendent (1953) and Zazie dans le métro (1959). He became director of the Encyclopédie de la Pléiade in 1955.

Quental, Antero de (1842–91), Portuguese poet and socialist. He was an enemy of the monarchy and of ROMANTICISM, in both its literary and political expressions. He helped to form the Portuguese Socialist Party. His best volumes of verse are the Odes Modernas (1865) and the Sonetos (1881).

Quercia, Jacopo della (c.1374–1438), Italian sculptor whose work linked GOTHIC and RENAISSANCE styles. One of his earliest works was the marble tomb of Ilaria del Caretto (c.1406) in the Cathedral of S Martino at Lucca, and his is known for his imposing figures on the Gaia fountain in Sienna. His stone reliefs outside S Petranio, Bologna, were admired by MICHELANGELO.

Quesnay, François (1694–1774), leader of the 18th-century French PHYSIOCRATS, who devised an economic table for France in 1758 (*tableau économique*), which demonstrated the circular flow of economic activity throughout the society. *See also* HC2 p.*23*.

Quetzal, rare perching forest bird of Central America. The male is bright green above and crimson below with shimmering, iridescent green tail plumes forming a 60cm (2ft) train. The Aztecs and Mayas regarded the quetzal as sacred and wore the tail plumes ceremonially. The duller female nests in a hole, often in a tree, and lays two pale, nearly spherical eggs which are incubated by both parent birds. Family Trogonidae; species *Pharomachrus mocinno*.

Quetzalcóatl, god of wind and fertility, worshipped by the Toltecs and later by the Aztecs of central Mexico and represented as a feathered serpent. Legend also referred to a priest-king named Quetzalcóatl who was banished from his homeland and sailed east into the Gulf of Mexico. The Aztecs believed that CORTÉS was the returning Quetzalcóatl, and put up little resistance, so aiding the Spanish conquest of their empire.

Quezon City, capital and second-largest city of the Republic of the Phillippines, on Luzon island, adjacent to Manila. It was named after Manuel Luis Quezon (1878–1944), the Philippines' first President, and became the capital of the country in 1948. It is mainly a residential area but has a thriving textile industry. Pop. (1973) 896,200.

Quiberon Bay, Battle of (1759), British naval defeat of the French in the SEVEN YEARS WAR (1756–63). During a storm Admiral Edward Hawke's fleet destroyed nine of the 21 opposing French ships, thereby ending French hopes of invading Britain.

Quicksand, sand that is water saturated, so losing its supporting ability. Natural quicksand can trap people and animals. If the victims lie flat on their backs, they cannot drown, because their density is less than the combined density of the sand and water.

Quicksilver. *See* MERCURY.

Quidde, Ludwig (1858–1941), German politician and pacifist. He was at various times a member of the Bavarian diet, the Weimar Assembly and the Reichstag. He was imprisoned for his attack on WILLIAM II in the pamphlet *Caligula: a Study in Roman Caesarean Madness* (1894). In 1927 he shared the Nobel Peace Prize with Ferdinand BUISSON.

Quietism, mystical Christian movement, begun by the Spanish priest Miguel de Molinos in the 17th century. It achieved great influence in 17th-century France and in the WESLEYAN movement of 18th-century Britain. Its adherents believed that only in a state of absolute surrender to God was the mind able to receive the saving infusion of GRACE. Worldly ambition was therefore foresworn.

Quiller-Couch, Sir Arthur Thomas (1863–1944), British writer and critic who published his prose and poetry under the pen-name "Q". A Cornishman, he wrote lively, colourful novels of Cornwall and the sea, including *Dead Man's Rock* (1887) and *The Ship of Stars* (1899). He compiled *The Oxford Book of English Verse* (1900) and *The Oxford Book of Ballads* (1910).

Quillwort, spore-bearing plant native to cool swamps of Eurasia and North America. Most quillworts grow in water but a few are terrestrial. Their grassy, quill-like, hollow leaves are spirally arranged and a large spore capsule is embedded in each leaf base. There are 60 species. Height: 15–50cm (6–20in). Genus *Isoetes. See also* NW p.*53*.

Quilmes, city in E Argentina, 14km (9 miles) SE of Buenos Aires. Founded in 1668, it was the scene (1806) of a British invasion of Buenos Aires. Its modern industries include brewing, oil refining and tourism. Pop. (1970) 355,265.

Quilter, Roger (1877–1953), British composer, a member of the Frankfurt Group. He chiefly wrote songs, many of which were settings of English poetry, especially of SHAKESPEARE. He also wrote three light operas, *Julia*, *Love at the Inn* and *Love and the Countess*.

Quince, shrub or small tree native to the Middle East and central Asia. Its greenish-yellow, hard, pear-shaped fruit is used in preserves. Height: to 6.1m (20ft). Family Rosaceae; species *Cydonia oblonga. See also* PE p.*192*.

Quincey, de. *See* DE QUINCEY, THOMAS.

Quinine, white crystalline ALKALOID, isolated in 1820 fro the bark of the cinchona tree. It was formerly used in the treatment of MALARIA.

Quinn, Anthony Rudolph Oaxaca (1916–), US actor. He won the Venice Film Festival award for *La Strada* (1954) and Academy awards for best supporting actor in *Viva Zapata* (1952) and *Lust for Life* (1956).

Quinone, group of aromatic organic compounds containing a benzene ring in which two hydrogen atoms have been replaced by oxygen atoms. Paraquinone (1,4-benzoquinone, $O{:}C_6H_4{:}O$) is a yellow crystalline solid obtained by the oxidation of aniline with chromic acid; it is used in the manufacture of hydroquinone for photographic developers.

Quinquagesima (from the Latin for "fiftieth"), the Sunday before Ash Wednesday, the beginning of LENT, roughly 50 days before Easter.

Quinsy, suppurative inflammation of the tonsils, caused by an abscess, often a complication of TONSILLITIS. It is generally treated with ANTIBIOTICS.

Quintana, Manuel José (1772–1857), Spanish poet, and tutor to Queen ISABELLA II. The last exponent of 18th-century CLASSICISM in Spanish poetry, he was famous for his patriotic odes, such as *A la batalla de Trafalgar* (1805).

Quintero, name of two brothers, Serafín Álvarez (1871–1938) and Joaquín Álvarez (1873–1944), Spanish playwrights who worked together. Their plays include *El Patio* (1900), *Malvaloca* (1912) and *La Calumniada* (1919).

Quintilian (*c.* AD 35–*c.* 96), Roman scholar from Spain. He is best known for his *Institutes of Oratory*, in 12 books, which covers the range of moral, literary and rhetorical education.

Quiriguá, ancient Mayan city in E Guatemala, in the Motagua Valley. It is the site of temple ruins and carved monoliths. *See also* HC1 pp.*46–47, 47*.

Quirinus, in Roman mythology, major god ranking below JUPITER and MARS, later identified with ROMULUS, one of the founders of ROME. His cult centred on the Quirinal, the traditional settlement site of the SABINES. The Quirinalia festival was celebrated on 17 Feb.

Quisling, Vidkun (1887–1945), Norwegian Fascist leader whose name became synonymous with "traitor" during WWII. He was Minister of Defence (1931–33), but left the Agrarian Party to found the Fascist National Union Party. In 1940 he collaborated with Germany in the invasion of Norway. He was made Premier of the puppet government in 1942 and stayed in power until Norway's liberation, when he was arrested, convicted of high treason and shot. *See also* HC2 p.*228*.

Quito, capital of ECUADOR, in the N central part of the country 182km (114 miles) from the Pacific Coast; it lies almost on the Equator and 2,850m (9,260ft) above sea-level. The site was originally settled by Quito Indians and captured by the INCAS in 1487. It was taken by Spain in 1534, and liberated in 1822 by Antonio José de Sucre. A cultural and political centre, it is the site of Central University of Ecuador (1787). Products include textiles and handicrafts. Pop. (1974) 597,000.

Quit rent, feudal payment in medieval England made by a COPYHOLDER or FREEHOLDER to the lord of the MANOR. It was paid in acquittal of all other dues. It is now only a nominal rent, such as that paid annually by the City of London to the monarch.

Quixote, Don. *See* DON QUIXOTE DE LA MANCHA.

Qumrān, ancient village of NW Jordan, on the NW shore of the Dead Sea. In 1947 the DEAD SEA SCROLLS (writings of a Jewish sect that settled in Qumrān *c.*100 BC–AD 68) were found in nearby caves. The land was taken by the Israelis in 1967. *See also* MS p.*246*.

Quoirez, Françoise. *See* SAGAN, FRANÇOISE.

Quoits, or coits, old British country game. The object is to toss an iron ring (the quoit) on to a pin in the ground, or to get closer to it than one's opponent – thus scoring one point. A version called deck quoits is played with rope rings on passenger and cruise ships.

Quorum, minimum number of members of a legislative or deliberative body that must be present so that business can be carried out.

Quo Vadis? (1896), novel by Henryk SIENKIEWICZ. The work deals with the conflict between paganism and Christianity against the colourful background of Nero's Rome. It was filmed (1951) by Mervyn le Roy, starring Robert Taylor and Deborah KERR.

Quo warranto proceedings (1274), national inquiry conducted by EDWARD I of England to determine by what authority (*quo warranto*) lords of manors held LIBERTIES, and whether they were exceeding their rights. It was thus a grand survey of landholding and the legal system, the first "census" since DOMESDAY BOOK. The results were published in 1274, but quo warranto proceedings continued until the reign of EDWARD III.

Qurān. *See* KORAN.

R

R, 18th letter of the alphabet, derived from the Semitic letter *resh*, the symbol for *head*. It passed almost unchanged into the Greek alphabet as *rho* (which resembled more the English *p*) and thence to the Romans, who gave it its present form. In English an initial *r* is nearly always pronounced in a similar way, as in *rug*. *See also* MS pp 244–45.

Ra. *See* RE.

Rabat, capital of Morocco, in N of the country on the Atlantic coast, approx. 92km (57 miles) NE of Casablanca. Settlement there dates from Phoenician times, but the fortified city was founded in the 12th century, by Abd al-Mumin. It has the 12th-century Hasasane 12th-century Hassane Tower (55m; 180ft tall) and the University of Rabat (1957). Industries: hand-woven rugs, textiles, food processing. Pop. 525,000.

Rabaul, port on the N end of New Britain Island, Papua New Guinea, in the SW Pacific Ocean. Formerly the capital of the Territory of New Guinea, Rabaul was a major Japanese naval and air base during WWII. Copra is the chief export. Pop. (1971) 24,778.

Rabbi, meaning "my master", from *c.* 100 AD an individual responsible for religious education, guidance and services in a SYNAGOGUE. His position is based on his learning and carries no special privileges. His duties include interpreting Jewish Law and guiding the spiritual lives of his congregation.

Rabbit, long-eared, herbivorous, prolific mammal of the family Leporidae, including the European common rabbit and the American COTTON-TAIL. Although usage of the term varies, most so-called rabbits typically are smaller than HARES (*Lepus*), have shorter ears, and run without leaping. The common rabbit is *Oryctolagus cuniculus* and has thick, soft, greyish-brown fur. The wide variety of domesticated rabbits are also of this species. Length: 35–45cm (14–18in); weight: 1.4–2.3kg (3–5lb). *See also* NW pp.*75, 160, 191*; PE pp.*238, 239*.

Rabbit fish. *See* CHIMAERA.

Rabelais, François (*c.* 1494–*c.* 1553), French humanist and satirist. He left a Franciscan monastery in 1530 to become a physician in Lyons. He wrote *Pantagruel* (1532), *Gargantua* (1534), *Tiers Livre*

Quetzalcóatl; a victim goes before the Aztec god formerly known as Kukulkan.

Quince is rarely eaten raw, but is often used in jellies and jams.

Vidkun Quisling imposed Nazism on Norway, including the persecution of Jews.

Quito; this monastery is typical of the Moorish architecture found in the city.

Rabi, Isidor Isaac

Rack railway enables sightseers to admire the view from Mt Snowdon, Wales.

Racoons can eat almost anything and some live as scavengers in urban areas.

RADA students listen to criticism after a performance of *The Venetian* by Clifford Bax.

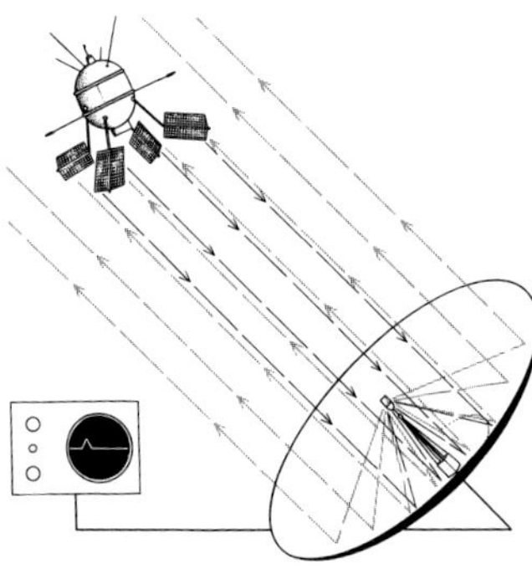

Radar is used in radio astronomy for tracking satellites and calculating distances.

(1546) and *Quart Livre* (1548, 1552). His own views on the education, politics and philosophy of his day underline the broad and ribald humour of his writing. Although condemned as obscene by theologians and the SORBONNE, his books became widely popular and gained him the protection of influential patrons. *See also* GARGANTUA AND PANTAGRUEL; HC1 pp.316–317, *316.*

Rabi, Isidor Isaac (1898–   ), US physicist, known for his work on MAGNETISM, molecular beams and QUANTUM MECHANICS. He was awarded the 1944 Nobel Prize in physics for his discovery and measurement of the radio-frequency spectra of atomic nuclei whose magnetic spin has been disturbed. He also worked on the development of MICROWAVE radar.

Rabies, (hydrophobia), fatal contagious viral infection of the central nervous system. It can occur in all warm-blooded animals and is especially dangerous in domestic dogs as it is transmissible to man through the bite of an infected animal. Its symptoms are severe thirst (but attempting to drink causes painful spasms of the larynx, hence the name hydrophobia – fear of water), foaming at the mouth and wild behaviour. Once the symptoms have appeared, death usually occurs within about four days. *See also* MS pp.*97, 111.*

Rabinowitz, Solomon. *See* ALEICHEM, SHOLOM.

Race, in biology, classification of the human species according to hereditary (genetic) differences. The term "geographical race" denotes large groupings of man that contain many "local races", varying in number from a few hundred to many million. Different racial characteristics arise through genetic mutation and adaptation to a particular environment through many generations. *See also* MS pp.32–33, *32–33,* 34–35, *34–35.*

Race Relations Acts (1965, 1968, 1970), legislation in Britain which made racial discrimination illegal in housing, employment and the provision of services. The 1976 Act set up the Commission for Racial Equality to enforce the Acts.

Racer, any of 20 species of slender, fast-moving SNAKES that live in the Americas, ranging from s Canada to Guatemala. They are broad-headed, large-eyed and varied in colour. Length: to 1.8m (6ft). Family Colubridae; species include *Coluber constrictor.*

Racerunner, active long-tailed LIZARD that is found throughout the Americas. There are 12 species in the USA. They are generally dark, with light, longitudinal stripes. Length: to 40cm (16in). Family Teiidae; genus *Cnemidophorus.*

Rachel, in the biblical Book of Genesis, second daughter of LABAN. Having spent seven years working for Laban in order to marry Rachel, JACOB was tricked into marrying her elder sister, Leah. He then had to serve his father-in-law a further seven years to obtain Rachel's hand. She was the mother of JOSEPH and of BENJAMIN, and died giving birth to the latter.

Rachmanism, charging extortionate rents for inferior premises and using violence against tenants. The term is derived from Peter Rachman (1920–62), a London landlord of the 1950s and early 1960s, who grew rich by this means, and whose dealings were exposed in 1963.

Rachmaninov, Sergei (1873–1943), Russian composer and pianist. As a composer he was greatly influenced by TCHAIKOVSKY, and was one of the greatest pianists of the 20th century. His *Piano Concerto No.2* (1901), *Rhapsody on a Theme of Paganini* (1934) and *Symphony No.2* (1907) are among his most popular works. He composed songs, preludes and études for piano, four piano concertos and various other works. *See also* HC2 p.*106.*

Racine, Jean Baptiste (1639–99), French classical dramatist whose early plays such as *La Thébaïde* (1664) and *Alexandre le Grand* (1665) were influenced by contemporaries such as CORNEILLE, whom he later surpassed as France's leading tragic dramatist. MOLIÈRE, who had staged his first two plays, broke with him when Racine removed *Alexandre*, giving the script to a competing company. His other plays

include *Britannicus* (1669), *Mithridate* (1673) and *Phèdre* (1677), after which Racine retired from the theatre. *See also* HC1 p.317, *317.*

Racing car, MOTOR CAR designed for high performance on specially constructed tracks. Ordinary road cars were used for the first motor races, in the 1890s, but by the 1920s cars built only for racing were in use. Competition took place within different categories, each of which was closely defined; Grand Prix cars (since 1950 called "Formula One") were the most powerful. There are now more than 15 classes of racing car.

Racism, doctrine advocating the superiority of one human race over some or all others. From time to time it has been the avowed policy of certain countries which, as a result, sanctioned slavery and discriminatory practices – eg the USA before the abolition of slavery, Nazi Germany during the 1930s and 1940s, and South Africa today. Racism was defined by UNESCO in 1967 as "anti-social beliefs and acts which are based on a fallacy that discriminatory inter-group relations are justifiable on biological grounds". *See also* MS pp.268–269, *268–269.*

Rack, instrument of torture on which the victim was laid flat and stretched by levers and pulleys. It was used mainly to extract confession in cases of suspected treason, particularly in 16th-century England. *See also* MS p.*296.*

Rack, in engineering, a straight rod on which grooves are cut to mesh with a cylindrical gear (pinion). Rotary motion of the pinion is converted into reciprocating (to-and-fro) motion at the rack. It is used extensively for steering systems in motor vehicles. *See also* MM p.*131.*

Rackham, Arthur (1867–1937), British illustrator and watercolourist, well known for his imaginative pen drawings for children's books. Among these are Charles LAMB's *Tales from Shakespeare* (1899), *Grimm's Fairy Tales* (1900), and *The Arthur Rackham Fairy Book* (1933).

Rack railway, form of mountain transport in which the locomotive powers a cogwheel that engages a cogged rail (rack) set between the track. The first rack railway in Britain came into use in 1812 between Middleton colliery and Leeds in the N of England, a distance of 5.5km (3.5 miles). A rack railway is still used on Mt SNOWDON in Wales.

Rack rent, in law, the amount of payment that can reasonably be charged for the lease of property on the open market; it has come to mean an excessive rent. The term was used particularly in the 19th century by tenant farmers in southern Ireland, who complained of the high rents charged by English landlords.

Racoon, stout-bodied, omnivorous, mostly nocturnal mammal of North and Central American wooded areas near water, related to the PANDA and COATI. Racoons have a black, mask-like marking across their eyes and a long, black-banded tail. They have agile and sensitive front paws and typically dip for food and other objects in water. The seven species include the North American *Procyon lotor.* Length: 40–61cm (16–24in); weight: 10–22kg (22–48lb). Family Procyonidae.

Racquets, game played by two or four people in an enclosed court. The court is 18.3 by 9.1m (60 × 30ft) and is surrounded by three walls 9.1m (30ft) high and a back wall 4.6m (15ft) high. Each player uses a gut-strung racket approx. 68cm (27in) long that has a circular head 17.8 to 20.3cm (7–8in) in diameter and that weighs approx. 255g (9oz). The ball, with a diameter of 2.5cm (1in), is tightly-wound cloth and twine with an adhesive tape cover. A service line is painted on the front wall at a height of 2.9m (9.6ft) and a fixed wooden board (*telltale*), also on the front wall, extends 68.6cm (27in) up from the floor. These are the markers that determine when a ball is in play. The serve must be put in play above the service line and must then land behind a short line, marked 7.3m (24ft) from the back wall. Games are played to 15 points.

Racquetball, game developed in the USA in

the mid-1960s and played by two or four people on a standard four-wall handball court. All the walls, ceiling and floor are in play. Gut-strung tennis-like rackets are used and the ball is of rubber. Points are won (only by the serving side) when the ball hits the floor twice before being returned. A game is won by the side that is the first to score 21 points; the first side to win two games wins the match.

RADA, the Royal Academy of Dramatic Art, leading British school for actors, founded in London by Sir Herbert Beerbohm Tree in 1904 at His Majesty's Theatre. It moved to its present premises in Gower Street, Bloomsbury, in the following year and was granted a royal charter in 1920.

Radar, short for Radio Detecting and Ranging, electronic system for determining the direction and distance of objects. First developed by the British and Germans during WWII, it works by the transmission of pulses of radio-frequency ELECTROMAGNETIC WAVES to an object from the radar transmitting equipment. The object reflects the pulses, which return to the radar station. By measuring the time it takes for the reflected waves to return, the object's distance may be calculated, and its direction ascertained from the alignment of the receiving radar aerial. *See* MM p.236.

Radar astronomy, branch of astronomy in which radar pulses reflected back to Earth from celestial bodies inside the solar system are studied for information concerning the distance from Earth of the bodies, their orbital motion, and large surface features. *See also* SU pp.172–173.

Radar telescope. *See* TELESCOPE.

Radcliffe-Brown, Alfred Reginald (1881–1955), British anthropologist who did field work in the Andaman Islands and Australia. He fostered the development of social anthropology as a science and contributed to the study of social organization. His works include *The Social Organisation of Australian Tribes* (1931). *See also* HC2 p.266.

Radek, Karl (1885–1939), Russian revolutionary who joined the COMMUNIST Party after the BOLSHEVIK revolution. In 1918 and 1919 he helped to organize the German Communist movement and on his return to the USSR became a leading member of the COMMUNIST INTERNATIONAL. In the 1930s he was accused of treason and confessed during his trial in 1937. *See also* HC2 pp.196–199.

Radetzky, Count Joseph (1766–1858), Austrian soldier. He fought against Napoleon, became a field-marshal in 1836 and helped to suppress the Italian revolts against Austrian rule in 1848–49. He was Governor-General of Lombardy-Venetia (1850–57).

Radhakrishnan, Sir Sarvepalli (1888–1975), Indian politician and scholar. He had a distinguished academic career, was elected Vice-President of India in 1952 and served as President from 1962 to 1967. His written works include *Indian Philosophy* (2 vols, 1923–27) and *An Idealist View of Life* (1932).

Radial motion, in astronomy, movement of a celestial body towards or away from an observer on Earth. A star that is receding is said to have positive radial motion; a star that is approaching is said to have negative radial motion. The actual motion of a star through space includes both radial motion and PROPER MOTION. *See also* SU p.*165.*

Radian, angle formed by the intersection of two radii at the centre of a circle, when the length of the arc cut off by the radii is equal to one radius in length. Thus, the radian is a unit of angle equal to approx. 57.295°, and there are 2π radians in 360°.

Radiant heat, form of energy transfer through the atmosphere or space. The Sun's energy reaches the Earth as radiant heat or infra-red RADIATION, as well as from the visible spectrum. Domestic electric fires emit radiant heat, which circulates by CONVECTION.

Radiation, propagation of energy by subatomic particles and ELECTROMAGNETIC WAVES. All electromagnetic radiation can propagate without a medium, which distinguishes it from such effects as conduction and convection and the transmission of sound, all of which require media. In a vacuum, electromagnetic waves travel at the speed of light (3×10^8 m/s; 186,000 miles per second), but this velocity is slightly decreased as the waves pass through a medium, an effect which explains REFRACTION. In addition this type of radiation betrays some of the properties of a stream of particles, which are known as PHOTONS. These particles are said to have no mass but, paradoxically, they may be affected by a gravitational field, (as are massive particles). Similarly streams of particles such as electrons betray properties like those of waves, which allows for electron DIFFRACTION as in an electron MICROSCOPE. *See also* SU pp.101–102, *102*.

Radiation, heat, energy radiated from all solids, liquids or gases as a result of their temperature, however low. The energy derives from the vibrations of atoms in a body and is emitted as ELECTROMAGNETIC WAVES, usually in the form of INFRA-RED RAYS, although in bodies with temperatures of 800°C (1500°F) or more (red hot) visible light is also present. *See also* SU pp.90–91, *90*.

Radiation, cosmic, also called cosmic rays, streams of subatomic particles from space that constantly bombard the Earth at velocities approaching the speed of light. Primary cosmic rays are high-energy RADIATION that comes from the Sun and other, mostly unidentified, sources in outer space. They consist mainly of atomic nuclei (particularly those of hydrogen) and PROTONS. When primary cosmic rays strike gas molecules in the upper atmosphere they yield showers or cascades of secondary cosmic rays (some of which reach the Earth's surface) consisting of energetic protons, NEUTRONS and PIONS. Further collisions yield MUONS, ALPHA RAYS, POSITRONS, ELECTRONS, GAMMA RADIATION and PHOTONS. Secondary rays are harmless near ground level, but pilots flying at high altitudes have "seen" flashes of light, attributed to cosmic rays destroying one or more of their retinal or brain cells.

Radiation cytology, study of the effects of radiation on living tissue including nonionizing radiation such as heat and light and ionizing radiation from radioactive elements, X-ray machines and other high-energy sources. The most frequently seen effects of radioactivity on cells include CHROMOSOME damage which, if it does not kill the cells may give rise to genetic defects in succeeding generations. For example, transformation into cancerous cells can occur.

Radiation, nuclear, particles or ELECTROMAGNETIC WAVES emitted spontaneously and at high energies from atomic nuclei. It can result from RADIOACTIVE DECAY, which yields alpha particles (helium nuclei), beta particles (ELECTRONS), GAMMA-RADIATION and, more rarely, POSITRONS (antielectrons). It can also result from spontaneous fission of a nucleus, with the ejection of NEUTRONS or gamma-rays. *See also* RADIOACTIVITY.

Radiation therapy, use of X-RAYS and gamma-ray RADIATION to treat disease; often used alone or in combination with surgery to eradicate cancer. *See also* MS pp.124–125, *124–125*.

Radiator, space heater in a CENTRAL HEATING system. Commonly it is a slim fluted steel or iron container through which hot water or steam is circulated. Electrically-powered appliances, such as convector heaters and storage heaters, can also function as radiators. In cars, lorries and other vehicles powered by an internal combustion engine, a radiator is a heat exchanger, which dissipates the heat in circulating water (to the surrounding air) to keep the engine cool.

Radical, expression used to indicate the root of a number. It consists of the number of which the root is to be taken (the radicand), the symbol $\sqrt{\ }$ known as the radical sign and the indicated root (the index). Thus the cube root of nine in radical form is $\sqrt[3]{9}$.

Radical, in politics, someone who desires or advocates fundamental solutions to social, economic and political problems. *See also* RADICALISM.

Radicalism, in political theory, doctrines that advocate going to the root (Latin *radix*) of a problem. Historically radicalism has advocated extreme solutions to political problems and is generally associated with the LEFT WING. But it has occasionally been used to describe reformist politics, such as the French Radical Socialist parties of the 1930s and the Radical Republicans who brought about the abolition of slavery in the USA.

Radio, method of communication between people who are "connected" only by means of a transmitter and a receiver of invisible ELECTROMAGNETIC WAVES (radio waves). The term is, however, most commonly applied only to the apparatus for receiving and reproducing a sound broadcast (the receiver). At the transmitter a radio wave of fixed FREQUENCY – the carrier – is continuously generated (the carrier is the "signature" by which different radio stations are distinguishable; the station is selected by tuning to its frequency). The sound to be broadcast is converted by a MICROPHONE into a varying electrical signal which is combined with the carrier by means of MODULATION. The modulated carrier is passed to an AERIAL from which it is transmitted into the atmosphere. At the receiver, another aerial intercepts the signal which is subjected to the reverse of modulation (detection), which retrieves the sound signal from the carrier. The sound signal is amplified so that it can activate a loudspeaker which reproduces the original transmitted sound. *See also* MM pp.228–229.

Radioactive decay, processes by which radioactive ISOTOPES (radioisotopes) progressively and spontaneously change, or transmute, into other isotopes due to disintegration of the nuclei of their atoms. Many chemical elements have radioisotopes. The most stable atomic nuclei are those approximating in size to that of the iron atom, which has the atomic number 26 (ie its nucleus contains 26 protons), but lighter or heavier nuclei may be unstable. A decaying isotope releases one or all of the three natural radioactivities: ALPHA RAYS, BETA RAYS and GAMMA RADIATION. *See also* RADIOACTIVE SERIES; SU pp.64–67; PE p.131, *131*.

Radioactive elements, chemical elements having ISOTOPES which give off RADIATION, ie undergo RADIOACTIVE DECAY. Radioactive isotopes occur in nature along with the stable isotopes of an element, but they can also be made artificially by bombarding other isotopes with high energy subatomic particles in a NUCLEAR REACTOR or a particle ACCELERATOR. Radioactive isotopes are formed naturally in the atmosphere as the result of collisions between atoms in the air and COSMIC RAYS from outer space. Among the elements with radioactive isotopes of economic importance are URANIUM, PLUTONIUM, RADIUM, COBALT, STRONTIUM and HYDROGEN. *See also* SU pp.64–65.

Radioactive series, chain of ISOTOPES, each being a product of the RADIOACTIVE DECAY of its predecessor, starting with a radioactive isotope and ending at one of the stable isotopes of lead. The three naturally occurring series are headed by uranium-238, thorium-232 and uranium-235. The neptunium-237 series does not occur in nature because the parent isotope has a HALF-LIFE (about 2 million years) which is much less than the age of the Earth, and has long decayed; this series is artificially produced. Each step in a radioactive series is called a transmutation. *See also* SU pp.64–65.

Radioactivity, disintegration of the nuclei of radioactive ISOTOPES (radioisotopes), such as uranium-238, usually with the emission of ALPHA RAYS (helium nuclei) or BETA RAYS (ELECTRONS), often accompanied by GAMMA RADIATION. These two processes of RADIOACTIVE DECAY and alpha decay or beta decay cause the radioisotope to be transformed into a chemically different atom. Alpha decay results in the nucleus losing two PROTONS and two NEUTRONS; beta decay occurs when a neutron changes into a proton, an electron (beta particle) being emitted in the process. Thus the ATOMIC NUMBER (number of protons in the nucleus) changes in both types of decay and an isotope of another element is produced, which might also be radioactive. The stability of different isotopes varies widely. It is impossible to predict when a nucleus will disintegrate but in a large collection of atoms there is a characteristic time (the HALF-LIFE) after which one-half of the total number of nuclei would have decayed. This time varies from about ten thousand million years to a ten-millionth of a second, depending on the isotope concerned. The activity of any radioactive sample decreases exponentially with time. *See also* CARBON DATING; SU pp.64–65.

Radioastronomy, detection, measurement and study of radio waves received on Earth from space. Radio signals are known to be emitted by the planets, the Sun, many other stars and interstellar material, galaxies, and particularly by QUASARS. Radioastronomers were the first to determine the shape of our GALAXY by studying the radio frequency of hydrogen gas (21cm). Among the radio sources investigated, the most remarkable are the PULSARS, or neutron stars, which emit regular pulses of radio waves as they rotate rapidly. Radioastronomy has also been used to determine the distances, speeds of rotation, temperatures and surface features of the planets.

Radio telescopes are essentially giant antennae, or radio aerials, usually dish-shaped. Often several dishes are used in conjunction to collect radio waves. The telescopes are moveable, either on tracks, or by rotation of the dish, or both. The 76m (250ft) telescope at JODRELL BANK in Cheshire, England, was for a long time the largest steerable radio telescope in the world. *See also* SU pp.61, 172, *173*, 208, 222–225, 234, 244, 250–251, 285.

Radio beacon, strategically located radio transmitter which emits a regular, non-directional and easily recognizable signal by which mobile receivers can obtain a bearing. Such beacons are used particularly by marine and air navigators who collate the information from two or more such beacons to fix their positions. Directional beacons are used at airports to aid landing aircraft. *See also* MM p.*187*.

Radio carbon dating. *See* CARBON DATING.

Radio control, remote control of any mechanism by radio signals. A familiar example is the hobby of flying miniature aircraft, the ailerons of which are moved by a SERVOMECHANISM in response to radio signals from a hand-held transmitter. Similar systems are used by the armed forces to fly target aircraft and to guide certain missiles. Other guided missiles have their own radio control systems built in and seek and collate their own guidance information. An expanding field of radio control has been created by space research.

Radio frequencies and wavelengths, radio waves categorized into FREQUENCY ranges, called bands, for radio wave communications. In 1959 an international conference decided on the bands, and agreed that they are reserved for specified uses such as television, radio, amateur, scientific and navigational broadcasts. *See also* MM pp.228–229.

Radio GALAXY, galaxy that emits strong ELECTROMAGNETIC RADIATION of radio frequency. These emissions seem to be produced by the high-speed motion (probably caused by explosions occurring in the nuclei of the galaxies concerned) of ELEMENTARY PARTICLES in strong magnetic fields. Some radio galaxies correspond to visual sources, such as Cygnus A, which is a distant double galaxy. Radio galaxies are frequently double and many are of immense dimensions. One, 3C-236, appears to be the largest galaxy known, being 18,000,000 light-years across. *See also* SU pp.172, 250–251.

Radiography, recording as photographs the interiors of opaque bodies using X-RAYS. Industrial X-ray photographs can show assembly faults and defects in the crystal structure of metals; in medicine and dentistry they are invaluable for diag-

Radiator water is cooled by the flow of air through the filaments.

Radical political criticisms of society have been expressed by Germaine Greer.

Radio, transistor: Japanese radios are known for their quality and compactness.

Radioactivity: waste disposal has caused controversy at Windscale, Cumbria.

Radio isotopes

Radiolaria are ocean-dwelling protozoa with complex silica skeletons.

Sir Henry Raeburn's *Boy with Rabbit* is typical of his romantic portraits.

Ragged robin, which grows up to 45cm (18in) tall is a common perennial in Britain.

Lord Raglan used his diplomatic training in dealings with Britain's allies in the Crimea.

nosing bone damage, tooth decay and internal disease. The images show bones predominantly, as well as some tissue structure and air spaces. Using sensitive modern techniques, cross-sectional outlines of the body can be obtained showing organs, blood vessels and diseased parts. *See also* MS pp.124–125.

Radio isotopes. *See* ISOTOPES.

Radiolaria, subclass of unicellular animals found in the topmost layers of the oceans. The silicious skeleton forms a lattice-like structure and is so resistant to decay that great areas of the ocean floor are covered by oozes made of nothing else. Reproduction is ASEXUAL by budding or fission. *See also* NW p.*121*; PE pp.*83*, 90.

Radiology, science and technology of using X-RAYS and other penetrating ms of RADIATION, particularly in medicine, for treating various pathological conditions such as cancer. *See also* RADIOGRAPHY; RADIOTHERAPY; MS pp.113, 124–125.

Radiometer, instrument for detecting and measuring radiant energy. Crooke's radiometer consists of pivoted vanes, painted black on one side, which rotate in a partial vacuum when exposed to light or heat, because of their differential absorption of radiant energy. More sophisticated radiometers can measure radiation in any part of the electromagnetic spectrum. *See also* GEIGER COUNTER; SU pp.*109*, 137.

Radiosonde, instrument package carried by a meteorological balloon which transmits to the ground data on weather conditions in the upper atmosphere. It is recovered when possible for re-use.

Radio-telephone, instrument of communication used on aircraft, ships and in other locations which lack a telephone cable linkage. Transmission and reception are similar to those in other RADIO devices.

Radio-telescope. *See* RADIOASTRONOMY; TELESCOPE.

Radiotherapy, in medicine, use of RADIATION to diagnose and to treat tumours or other pathological structures in the body. It may be done either by implanting a pellet of a radioactive source in the part to be treated, dosing the patient with a radioactive ISOTOPE (eg radioactive iodine for the thyroid gland) or by exposing the patient to precisely focused beams of radiation from a machine such as an X-RAY machine, particle ACCELERATOR or machine containing cobalt-60 which delivers GAMMA-RADIATION and other penetrating radiation. *See also* MS pp.124–125.

Radio waves. *See* ELECTROMAGNETIC SPECTRUM.

Radish, annual garden vegetable developed from a wild plant native to the cooler regions of Asia. Its leaves are long and deeply lobed; the fleshy root, which may be red, white or black, is eaten raw. Red radishes are small and globular; white radishes are long and cylindrical. Family Brassicaceae; species *Raphanus sativus.* *See also* PE pp.188, *189–190*, 191.

Radisson, Pierre Esprit (*c.* 1636–*c.* 1710), French-born Canadian fur trader and explorer. He was instrumental in convincing GEORGE III to charter the HUDSON'S BAY COMPANY but soon fell out with its administrators and led a French expedition against the English forts on Hudson Bay. Rejoining the Company in 1684, he became a British subject in 1688.

Radium, radioactive metallic element (symbol Ra) of the ALKALINE-EARTH ELEMENTS, first isolated as radium salts in 1898 by Pierre and Marie CURIE; the metal was isolated by Mme Curie in 1911. It is present in uranium ores such as pitchblende. The metal is used in research, luminous paints and medical RADIOTHERAPY. It has 13 isotopes, which emit alpha, beta, and gamma radiation; the gas RADON is a decay poduct. Properties: at.no. 88; at.wt. 226.0254; density 5; m.p. 700°C (1,292°F); b.p. 1,140°C (2,084°F); most stable isotope Ra[226] (half-life 1,600yr). *See also* MS pp.124–125; SU pp.64, *64*, *134–135*, *137*.

Radius, in anatomy, one of the two forearm bones, extending from the elbow to the wrist. The radius rotates around the ulna, the other forearm bone, permitting the hand to rotate and be flexible. A projection just above the thumb side of the wrist marks the end of the radius. *See also* MS p.*57*.

Radius, in geometry, the distance or a line from the centre of a circle or sphere to any point on the circumference of the circle or surface of the sphere. *See also* SU p.*35*.

Radnor, former county in E Wales; since 1974 it has been part of Powys. It is a region of moorland, hills and forest, and is sparsely populated. Sheep and cattle rearing is the principal occupation. Presteigne was the county town. Area: 1,220sq km (471sq miles). Pop. (1971) 18,262.

Radon, radioactive non-metallic gaseous element (symbol Rn) of the NOBLE GAS group, first discovered in 1900. Rn[222] (half-life 3.8 days) is a decay product of RADIUM. Thorium decays to Rn[220] (half-life 54.5sec) and actinium gives Rn[219] (half-life 3.92 sec).. The isotopes, which are ALPHA RAY emitters, are present in the Earth's atmosphere in trace amounts. Chemically the element is known to form fluorides. Properties: at.no. 86; density 9.73g g dm^{-3}; m.p. −71°C (−95.8°F); b.p. −61.8°C (−79.24°F); most stable isotope Rn[222]. *See also* SU pp.*134–135*.

Raeburn, Sir Henry (1756–1823), British portrait painter who worked exclusively in Scotland. Well known among his many portraits are *The MacNab* (*c.* 1803–13) and *Lord Newton* (1806–11); Sir Walter SCOTT and James BOSWELL were also among his numerous subjects.

Raedwald (*d.* 625), King of the East Angles, who was converted to Christianity under the influence of ETHELBERT of Kent. In 617 he defeated and slew Ethelfrid, King of Northumbria, in a battle near the River Idle (Nottinghamshire), thus enabling EDWIN to regain the Northumbrian throne. He may have been the king royally interred in the SUTTON HOO ship burial. *See also* HC1 pp.*138*, *164*.

Ra expeditions, two voyages (1969, 1970) undertaken by Thor HEYERDAHL. He was trying to prove the papyrus boats, as used by the ancient Egyptians, could have crossed the Atlantic. On the second expedition he set out from Morocco and reached Barbados.

RAF. *See* ROYAL AIR FORCE.

Rafah, town on the GAZA STRIP; Egyptian town which since 1967 has been in Israeli-occupied territory. The town is 3km (2 miles) from the Mediterranean coast. It was the scene of the defeat of Antiochus the Great by Ptolemy IV in 217 BC. Pop. (1968 est.) 49,800.

Raffia, African palm tree with large pinnate leaves which yield a useful fibre commonly used for matting and baskets. Species *Raphia ruffia.*

Raffles, Sir Thomas Stamford (1781–1826), British colonial administrator. He served as Lieutenant Governor of Java (1811–16) and then of Bengkulu, Sumatra (1818–23), introducing administrative reforms and new crops. In 1819 he established the port of Singapore for the British EAST INDIA COMPANY. *See also* HC2 p.*133*.

Rafflesia, also known as monster flower, parasitic plant native to Sumatra and Java. It grows on the roots of a jungle vine and has no stem or leaves. The foul-smelling, reddish-brown flowers are 1m (3.25ft) in diameter, and are the world's largest flowers. Family Rafflesiaceae; species *Rafflesia arnoldii.* *See also* NW p.*215*.

Raft of the "Medusa", The. *See* GÉRICAULT.

Rāga (Sanskrit for "colour"), in Indian music, a sequence of five to seven notes that is used exclusively for the duration of a performance, usually for about half an hour; its basic structure can be written in the form of a scale. Mood or atmosphere are created by emphasizing certain parts of the scale. There are hundreds of rāgas in use, although a far greater number are possible in theory.

Ragged robin, common name for a flowering plant found in Europe, Asia, Canada and NE USA. It has pink petals divided into four narrow and unequal lobes. Family Caryophyllaceae; species *Lychnis floscuculi.*

"Ragged Schools" Union, organization founded in England in 1844 to bring together various voluntary charity schools for the education of the children of the poor. Ragged schools were begun in about 1818 by John Pounds, a Portsmouth shoemaker. Their chief promoter and benefactor was Lord SHAFTESBURY. The Union provided elementary education, industrial training and Sunday schools. It died out after the introduction of state primary education in 1870.

Raglan, Fitzroy James Henry Somerset, Lord (1788–1855), British military commander who succeeded the Duke of WELLINGTON in 1852 and led the British forces in the CRIMEAN WAR. Promoted to Field-Marshal after the battle of Inkerman he was subsequently criticized for slow progress and failure to capture Sebastopol, where he died.

Ragnarök, in Scandinavian mythology, the doom of the Gods. Heralded by bitter cold and moral decline, it signalled the end of the world, the defeat of the gods and heroes of VALHALLA by the forces of fiery destruction – giants, demons and an all-devouring wolf.

Ragtime, name given to an early style of jazz, particularly associated with piano-playing. Its essential ingredient is the constant syncopation, or "ragging", of a straightforward tune. There was little true jazz improvisation; rags were, rather, a theme and a set of written variations. Scott JOPLIN was the most famous ragtime composer.

Ragworm, any of several species of segmented, polychaete, marine invertebrates, worms of the phylum Annelida. Predominantly marine, it has bristle-covered "paddles" called parapodia on its body. It lives in a burrow in mud or sand on the sea shore and feeds on small marine animals, planktonic organisms and detritus from the mud. Many sea anglers use ragworms as bait. *See also* NW pp.*72*, *88–89*, *89*.

Ragwort, any of several ragged-leaved plants, including golden ragwort or golden groundsel (*Senecio aureus*), a biennial or perennial that bears flat-topped clusters of yellow flower heads. Height: to 1.3m (4ft). Family Compositae. *See also* NW p.*61*.

Rahab, in the Old Testament, prostitute who aided JOSHUA during the siege of Jericho. She sheltered Joshua's spies and in return was spared when Joshua destroyed the city.

Rahere (*d.* 1144), English churchman of Frankish descent. He founded St Bartholomew's Church and St Bartholomew's Hospital at Smithfield, London, in 1123.

Rahman Putra, Tunku Abdul (1903–), Malaysian politician. He became leader of the United Malays' National Organization in 1952–55. He was the first Prime Minister of Malaya after independence (1957) and became Prime Minister of Malaysia when it was established in 1963. He resigned in 1970, after race riots between Malays and Chinese.

Rahu, in HINDU mythology and religion, the mischievous four-armed deity who causes eclipses and the fall of meteors, and who presages trouble and sickness. Rahu is especially worshipped by the Dosadhs and Dhangar of India, who show their devotion by walking on a path of hot coals. *See also* MS p.*212*.

Raikes, Robert (1736–1811), British publisher and Sunday-school founder. He became proprietor of the *Gloucester Journal* in 1757 and founded the first Sunday school in about 1780.

Rail, slender, long-legged, drab-coloured marsh bird that is well camouflaged in its dense swampland home. Rails are shy, generally nocturnal and often emit melodious calls. They lay 8–15 marked white or buff-coloured eggs in a reed-and-grass ground nest. Length: 10–45cm (4–18in). Family Rallidae. Typical genus *Rallus.*

Railways, system of transport in which carriages are guided by flanged wheels along rails. Railways began in the 17th century, when trucks used in mining and other industries were drawn along rail tracks. Rail transport of freight properly began in 1823 with the construction of the Stockton and Darlington Railway, on which George STEPHENSON's steam locomotive *Locomotion* pulled the world's first passenger train in 1825. Thereafter an explosion of railway-building swept the world, and by the 1870s railways were a principal means of transport in Europe, North America, India, Russia and Australia, and were starting up in such countries as Japan, South Africa and New Zealand. Railway tracks are laid by placing sleepers (load-bearing rectangular beams) on firm well-prepared ground, then laying lengths of steel rails on tie plates which distribute their load on the sleepers. Lengths of rail are joined together by joint bars or fishplates, but more often nowadays are welded into continuous lengths. On these rails run passenger coaches and freight wagons drawn by diesel, electric, or diesel-electric locomotives (except in the large Indian system, where steam still reigns supreme). Freight wagons are sorted into trains at marshalling yards, generally large computer-controlled complexes; rail traffic direction and control is now commonly by computer. Modern developments include high speed trains travelling at more than 240km/h (150mph), yet railways are in financial difficulties wherever they are asked to pay their way without large subsidies from the state. *See also* MM pp.*111, 140–145.*

Raimondi, Marcantonio (*c.*1480–*c.*1534), Italian engraver. Chosen by RAPHAEL in 1510 to copy his designs and paintings, Raimondi became the first eminent engraver of reproductions. His works include *Pyramus and Thisbe* (1505) and *Massacre of the Innocents* (*c.*1512).

Rain, water drops that fall from the Earth's atmosphere to its surface, as opposed to fog or dew which drift as suspensions, and snow or hail which fall in the form of ice particles. Clouds are aggregates of minute droplets of moisture and raindrops are formed when these enlarge by further condensation and by the coalescence with other drops as they fall. Warm air passing over the sea absorbs water vapour and rises in thermal currents, or on reaching a mountain range. The water vapour condenses and forms clouds, which account for the usually heavier annual rainfall on windward, compared to leeward, mountain slopes. Rainfall is measured by a RAIN GAUGE. *See also* PF p.70.

Rain, artificial, means of modifying the climate in a limited area by inducing rain to fall. It has been attempted only since the late 1940s and so far with only limited success. The method is used to "seed" a cloud with tiny particles to provide nuclei around which water may condense to form raindrops. Dry ice and silver iodide crystals are the usual substances employed when the cloud is supercooled (at a temperature of below 0°C), but in warmer clouds sodium chloride or a water spray is used. These substances are introduced into the cloud from aeroplanes or by artillery shells or rockets.

Rainaldi, Carlo (1611–91), Italian Baroque architect who worked chiefly on ecclesiastical commissions. Roman churches with which he was associated include Sta Agnese (begun 1652) in the Piazza Navona, Sta Maria in Campitelli (begun 1662) and the twin churches of Sta Maria de Monte Santo and Sta Maria de'Miracoli (1662–67), completed by BERNINI and FONTANA 1671–79) which were designed to given an illusion of visual symmetry in the Piazza del Popolo.

Rainbow, bright multicoloured band, usually seen as a circular arc formed opposite to the sun or other light source. The primary bow is the one usually seen; in it the colours are arranged from red at the top to blue beneath. A secondary bow, in which the order of the colours is reversed, is sometimes seen beyond the primary bow. The colours are caused by reflection of light within the spherical drops of falling rain, which causes the white light (eg, of the sun) to be dispersed into its constituent wavelengths. The colours usually seen are those of the visible SPECTRUM: red, orange, yellow, green, blue, indigo and violet. *See also* SU pp.98–99.

Rain dance, ritualistic and ceremonial dance performed to beg or coerce a deity

to send rain. The dance is often performed in imitation of rain and was once widely practised, notably among American Indians of the SW USA.

Raine, Kathleen (1908–), British poet. Her volumes of poetry, including *Stone and Flower* (1940), *Living in Time* (1946), *Hollow Hill* (1965) and *Defending Ancient Springs* (1967), examine the visionary role of the poet. Influenced by William BLAKE, she has written several studies of his work.

Rainey, Gertrude ("Ma") (1886–1939), US BLUES singer known as the Mother of the Blues. She was the star of the Rabbit Foot Minstrels, with whom she met Bessie SMITH, who became greatly influenced by her. *See also* HC2 pp.272, 273.

Rain gauge, instrument for measuring the amount of rain that falls at a given location. It is usually a cylinder from which rain is funneled into a tube, where its depth is measured. *See also* PE p.72.

Rainier III, Prince (1923–), ruler of Monaco, who succeeded his grandfather Louis II in 1949. He married the American actress Grace KELLY in 1956. They had three children, ensuring the survival of the royal family and hence Monaco's independence from France.

Rainier, Priaulx (1903–), South African composer. She has lived in England since 1920. She was professor of counterpoint and harmony at the Royal Academy of Music (1942–59). Her compositions are chiefly for small chamber groups and the piano.

Rain, Steam and Speed – the Great Western Railway (1844), late work by J. M. W. TURNER. It not only celebrated the coming of the railway age but also, in the firm control of atmosphere and colour and the absence of strong outlines, to some extent anticipated later artistic developments.

Rain tree, or saman, tree native to the West Indies and Central America. It gets its popular name from the fact that cicadas regularly eject juice on it, thus giving the impression under its branches that it is always raining. Height: to 32m (100ft). Family Leguminosae; species *Samanea saman.*

Rainwater, Leo James (1917–), US physicist who in 1950, following the discovery that atomic nuclei are not spherical, proposed a new theory of atomic structure. He postulated that most nuclear particles form an inner NUCLEUS while the remaining particles form an outer nucleus, the shape of each set of particles affecting the other set. He further postulated that if some of the outer particles moved in similar orbits, sufficient energy could be created from their interaction to permanently deform an otherwise symmetrical nucleus. By 1953 Ben R. MOTTELSON and Aage N. BOHR had provided experimental evidence to support this theory. For this research, which made nuclear fusion a practical possibility, Rainwater, Mottelson and Bohr shared the 1975 Nobel Prize in physics. *See also* NUCLEAR ENERGY.

Rais (Retz), Gilles de (1404–40), French lord. He retired after serving at Orléans with JOAN OF ARC, and was later tried in an ecclesiastical court after rumours of vicious doings in his castle. He confessed to kidnapping and murdering over 100 children and was executed.

Raisin, dried, sweet, seedless GRAPE. Special varieties are grown, particularly in Australia and W USA, and are dried either in the sun or artificially. The product may be eaten on its own or used in cooking.

Rājasthān, state in NW India, on the border with Pakistan; Jaipur is the capital. The state was formed in 1947 from several former principalities of Rajputana. The Thar Desert in the w is inhabited by pastoral nomads. The E is part of the Deccan plateau and wheat, millet and cotton are grown with the aid of irrigation. Coal, marble, mica and gypsum are mined. The state is a stronghold of the conservative Hindu Jan Sangh and Swatantra political parties. Industries: handicrafts, cotton milling. Area: 342,266sq km (132,149sq miles). Pop. (1971) 25,724,142.

Rajput states, historic region of NW India,

chiefly corresponding to RĀJASTJĀN. The Rajputs, a warrior class, first became politically powerful in the 9th century AD, and after the Muslim conquests in the 12th century retained their independence in Rajasthan and much of central India. They submitted to the MOGULS in the 16th century, and to the Maratha chiefs in the 18th century. Many of the Rajput states kept their Rajput rulers under British suzerainty.

Rake's Progress, The (*c.*1732–35), sequence of eight anecdotal paintings by William HOGARTH. These illustrations of the decline and fall of a dissolute man exemplify the artist's strong sense of composition and control of contrast and the works were later issued as a set of engravings. *See also* HC2 pp.*70, 71.*

Raleigh, Sir Walter (*c.*1552–1618), English soldier, explorer, courtier and man of letters. A favourite of ELIZABETH I, he organized expeditions to North America and is credited with introducing the potato plant and tobacco to England. After the accession of JAMES I in 1603 he was found guilty of conspiring against the king and imprisoned in the Tower of London, where he wrote his *History of the World* (1614). In 1617 he led an unsuccessful expedition to the Orinoco, Guyana, in search of gold, was arrested on his return and executed under the terms of his original sentence. *See also* HC1 pp.277–9, 279; HC2 p.62.

Rallycross. *See* MOTOR SPORT.

Rallying. *See* MOTOR SPORT.

Ralph Roister Doister (*c.*1553), five-act comedy by Nicholas UDALL. Regarded as the first complete English comedy, the influence of PLAUTUS is obvious but the play is distinguished by its native English humour.

Rama, in the RAMAYANA, Hindu mythology, the perfect Hindu. A devoted and chivalrous husband, obedient to sacred law, courageous, patient and possessed of a great sense of duty, he was considered in the folk legends to be an incarnation of the god VISHNU and his name became synonymous with God.

Ramadan, ninth month of the MUSLIM year, set aside for fasting. The month begins and ends with the sighting of the new moon and for its duration the faithful abstain from food, drink and sexual inter course between sunrise and sunset.

Ramadhin, Sonny (1930–), West Indian cricketer who played for the West Indies, Trinidad and Lancashire. A slow bowler with a deceptive delivery, he took 26 wickets against England in his first Test series and in 43 Test Matches (1950–61) took 158 wickets at an average of 28.96.

Ramakrishna (1836–86), Indian spiritual teacher and Hindu saint. He became a Brahman monk at the Temple of Kali at Dakhineswar, near Calcutta. He taught that all religions were basically united in a common goal of union with God. His chief disciple, Swami Vivekananda, founded the Ramakrishna Mission in India in 1897.

Raman, Sir Chandrasekhara Venkata (1888–1970), Indian physicist who greatly influenced and contributed to the growth of science and research facilities in his country. He was professor of physics at Calcutta University from 1917 to 1933. In Bangalore he directed the Indian Institute of Science and, from 1946, the Raman Institute. He received the 1930 Nobel Prize in physics for his research on the diffusion of light and his discovery of the RAMAN EFFECT.

Raman effect, slight change in the frequency of monochromatic (single-wavelength) light that has been scattered. The effect, discovered by C.V. RAMAN, is seen as secondary spectral lines on either side of the primary spectral line. *See also* SCATTERING.

Ramayana, the romance of Rama, a great epic poem of ancient India, written *c.*300 BC and ranking with the MAHABHARATA. It is ascribed to the poet VALMIKI and comprises 24,000 couplets in seven books. It concerns the adventures of Rama, his wife Sita, his brother Lakshmand and others amid royal intrigues and warfare.

Rambert, Dame Marie (1888–), British

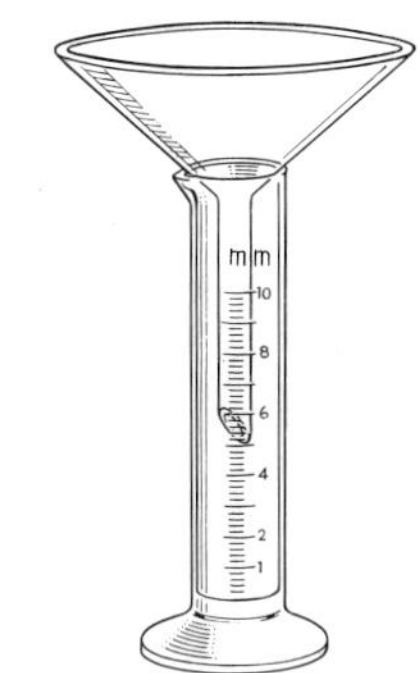

Rain gauges are calibrated containers with a funnel 10 times the tube's area.

Rain, Steam and Speed, by Joseph Turner, shows a train on a bridge at Maidenhead.

Sir Walter Raleigh, a favourite of Elizabeth I, was executed by James I.

Sir Chandrasekhara Raman received the 1930 Nobel Prize for his work on spectra.

Rameau, Jean Philippe

Jean Philippe Rameau did important work on the modern theories of harmony.

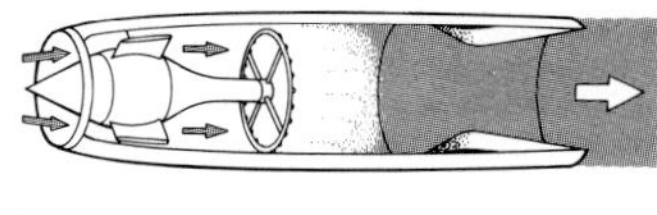

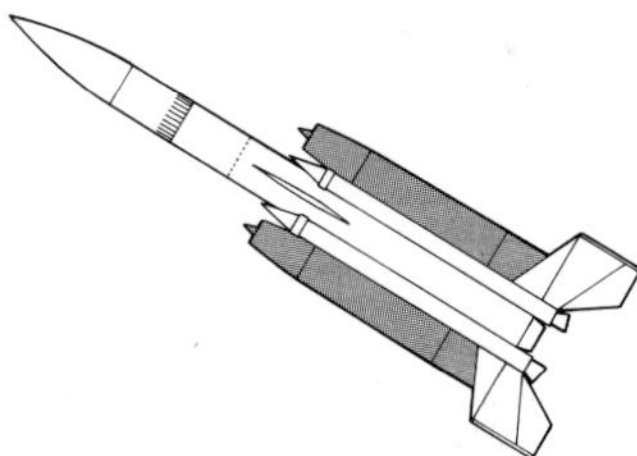

Ramjets have few moving parts and are little more than tube-like containers.

Allan Ramsay, painter to George III, painted this portrait of *Mrs Bruce of Arnot.*

Ramsey, as Archbishop of Canterbury, advocated unity among Christian Churches.

ballet dancer, teacher and choreographer *b.* Poland as Cyvia Rabbam, later changing her name to Miriam Ramberg. She was a member of DIAGHILEV's BALLETS RUSSES (1912–13), was special advisor to NIJINSKY in the first performance of Stravinsky's *The Rite of Spring* (1913) and founded her own school in 1920 which became known as the Ballet Rambert in 1935. Her influence on British ballet, particularly her fostering of new talent, was immense. Her autobiography, *Quicksilver,* appeared in 1972.

Rameau, Jean Philippe (1683–1764), French composer and musical theorist. His most famous opera is *Castor et Pollux* (1737), and among his other stage-works is the dramatic ballet *Les Indes Galantes* (1735).

Ramesses, or Rameses, the name of eleven Egyptian kings of the 19th and 20th Dynasties. Ramesses I (*r. c.* 1320–1318 BC) founded the 19th Dynasty when he succeeded HOREMHEB and re-established the Egyptian Empire in the Near East. His grandson, Ramesses II, (*r.* 1304–1237 BC) brought Egypt a period of unprecedented prosperity and power. His efforts to confirm Egypt's dominant position in Palestine and to regain her possessions and influenc, in Syria led to a major clash with the HITTITES at KADESH in 1299 BC. The war with the Hittites continued until a truce was agreed in 1283, and Ramesses later married a Hittite princess. He built many luxurious monuments, including the temple at ABU SIMBEL, and is believed to be the Pharaoh at the time of the Hebrew Exodus mentioned in the Old Testament. Ramesses III (*r. c.* 1198–1166 BC), the last important Ramesses, was the second king of the 20th Dynasty and the husband of TIY. He defended Egypt from attacks by Libya and the Sea Peoples, but the expenses of defensive wars and ambitious building programmes together with administrative inefficiency led to the dynasty's decline. Egypt then withdrew into political and cultural isolation. The priesthood became the centre of power, and the dynasty ended with the reign of Ramesses XI (*r.* 1113–1085 BC). *See also* HC1 pp.48, 49, *49,* 52.

Ramgoolam, Sir Seewoosagur (1900–   ), Mauritian politician. He became Mayor of Port Louis in 1958, and served as Mauritian Minister of Finance from 1960 to 1972. He was appointed Premier in 1965 and Prime Minister in 1968. He served as President of the Organization of African Unity (OAU) in 1976–77.

Ramie, or ramee, coarse, perennial plant that grows in warm climates. The strong bast fibre of its rod-like stems is used to make fine cloth. Family Urticaceae; species *Boehmeria nivea.*

Ramillies, Battle of (1706), engagement fought near the villages of Ramillies and Offus, Belgium, during the War of the SPANISH SUCCESSION. In one of his most brilliant victories, MARLBOROUGH, commanding British, Dutch and Danish troops, defeated the French under the Duc de Villeroi. The victory enabled the Allies to capture Antwerp, Ghent and Bruges and overrun the Spanish Netherlands. The Allies suffered 6,000 casualties, the French 17,000.

Ramjet engine, aircraft engine, a reaction motor that relies on the speed of the aircraft to ram air (for combustion of a HYDROCARBON fuel) into the intake, thereby compressing it. The only moving part is the variable diameter intake, or diffuser, which varies the volume, and therefore the pressure of the incoming air. The remainder of the engine is an elongated tube which forms the combustion chamber. The engine operates best at speeds above that of sound; it must be first accelerated by other means (such as solid-fuel rockets) until sufficient pressure builds up.

Ramón y Cajal, Santiago (1852–1934), Spanish histologist. He was a professor at Valencia University (1884–87) and was professor of histology at the universities of Barcelona (1887–92) and Madrid (1892–1922). He shared the 1906 Nobel Prize in medicine and physiology with the Italian Camillo GOLGI for his work in

establishing the terminal branching of nerve cells and for research into the structure of the nervous system.

Ramos, Graciliano (1892–1953), Brazilian novelist. Sometimes called the "Brazilian Dostoevsky", in such novels as *São Bernado* (1934) and *Barren Lives* (1938), he shows keen psychological insight into the hard, bitter lives of the people of NE Brazil. His most famous novel is *Anguish* (1936).

Ramsay, Allan (1713–84), Scottish portrait painter, the son of the poet of the same name. The Scottish counterpart of GAINSBOROUGH and REYNOLDS, he trained both at home and in Italy before settling in London where, during 1760, he was appointed painter to George III in preference to his rival Reynolds. His style, graceful and Italianate, lent itself especially well to female portraiture, eg *The Artist's Wife* (1755).

Ramsay, Sir William (1852–1916), British chemist. Working with John RAYLEIGH he discovered ARGON in air. Later he discovered HELIUM, NEON and KRYPTON. He was knighted in 1902 and awarded the 1904 Nobel Prize in chemistry. Ramsay was also interested in RADIOACTIVITY and showed that helium is produced by RADIOACTIVE DECAY.

Ramsden, Jesse (1735–1800), British maker of astronomical instruments. He made significant improvements to the THEODOLITE, BAROMETER, MICROMETER and SEXTANT.

Ramsey, Sir Alfred ("Alf") (1920–   ), British footballer and manager of England (1963–74), who created and guided the national side that won the 1966 WORLD CUP. He began his managerial career with Ipswich Town, taking them to the Football League Championship in 1962. He was a fullback for Southampton and Tottenham Hotspur, and played for England 32 times (1948–53).

Ramsey, Arthur Michael (1904–   ), British churchman, Archbishop of CANTERBURY (1961–1974). A distinguished Anglican theologian, he held the chairs of divinity at the Universities of Durham (1940–1950) and Cambridge (1950–1952) before becoming Bishop of Durham (1952–1956) and Archbishop of York (1956–1961). As Archbishop of Canterbury he did much to further the ECUMENICAL MOVEMENT. Notable events of his primacy included his historic visit to Pope PAUL VI in 1966 and to the patriarchs of the Eastern Orthodox Churches; his support of the scheme of Anglican-Methodist reunion in England; and the inauguration of synodical government in the Church of England.

Rance, river in Brittany, NW France. The Rance estuary is famous for its tidal barrage, which has the world's first operational tidal power station. The Rance was chosen because of its very large tidal range, which at St Malo is approx. 10m (30ft). *See also* MM p.85.

Rand, Mary, née Bignal (1940–   ), British athlete who at the 1964 Olympic Games won a gold medal in the long jump (setting a world record), a silver in the pentathlon and a bronze in the relay. She also won the 1966 Commonwealth Games long jump. In 1969 she married 1968 Olympic decathlon champion Bill Toomey, her second husband.

Rand, The (Witwatersrand), region of S Transvaal, NE South Africa, extending approx. 100km (62 miles) E and W of Johannesburg. The region is one of the world's richest gold-mining areas. Gold was first discovered in 1884 and mining began two years later. Many of the older mines are now almost exhausted but the Rand still produces about a third of the total world output of gold. Silver is recovered by the Wohlwill process as a by-product of gold refining. Coal and manganese are also mined and there are many ancillary industries in the region.

Ranfurly Shield, the premier rugby union trophy in New Zealand, presented in 1901 by the Earl of Ranfurly for the winner of a competition between provincial teams and first awarded to Auckland (in 1902). Since then it has been played on a challenge basis, the holders retaining the "log

of wood" until beaten in an official challenge match.

Rangefinder, optical attachment to guns and cameras, used for measuring distances. Optical rangefinders split the light from the subject into two paths in such a way that the angle between them is proportional to the range of the subject. This angle is measured by the rangefinder.

Ranger, series of unmanned US spacecraft, launched in 1964–65 and aimed to hit the Moon. Ranger 7 (July 1964) obtained close-up television pictures of the Moon's surface, the first ever taken, before it crashed. *See also* SU pp.180, 186.

Rangoon, capital and largest city of Burma, on the Rangoon River approx. 34km (21 miles) N of the Andaman Sea. It is one of SE Asia's largest seaports. The site of an ancient Buddhist shrine, the town grew in the 18th century when its port facilities were developed. It came under British rule in 1852. The city suffered heavy damage in an earthquake in 1930 and was repeatedly bombed in WWII. Its industries include oil refining and shipbuilding. Pop. (1973), 2,055,365.

Ranji Trophy, silver cup awarded annually to the national cricket champions of India. Contested by associations affiliated to the Indian Board of Control, it was played on a knockout basis until 1956, when zonal leagues, leading to knockout final rounds, were introduced. It was instituted in 1934 in memory of the Indian cricketer Kumar RANJITSINHJI.

Ranjitsinhji, Kumar Shri (1872–1933), Indian cricketer who played for England, Cambridge University and Sussex. Known as "Ranji", he thrilled crowds with his majestic batting, twice scoring more than 3,000 runs in a season. He played in 15 Test Matches between 1896 and 1902 (scoring 989 runs, average 44.95).

Rank, Joseph Arthur (1888–1972), British industrialist and film magnate, chairman of many film companies, including GAUMONT BRITISH and Cinema-Television. He promoted the British film industry when HOLLYWOOD and the US film companies had a virtual monopoly.

Rankine, William John Macquorn (1820–72), Scottish engineer and physicist. He became professor of engineering at Glasgow University in 1855, where he made valuable contributions to civil and mechanical engineering and THERMODYNAMICS. He wrote manuals on these subjects and also devised the absolute temperature scale based on the degree Fahrenheit, known as the Rankine scale.

Rank Organization, British film production and distribution company founded in 1946 by J. Arthur RANK. He had previously established General Film Distributors and gained control of the GAUMONT BRITISH and Odeon cinema circuits. Film output declined in the 1950s, but the company expanded into hotels and leisure services.

Ranks Hovis McDougall, international food processing company based in Britain. Its products include flour, bread, cakes and groceries. In the mid-1970s it employed about 58,000 people.

Rank Xerox, international manufacturing company based in Britain, partly owned by the RANK ORGANIZATION. Its products include office copying and duplicating machines, electronic typewriting systems and facsimile transmission equipment. In the mid-1970s its international workforce numbered more than 30,000.

Ransome, Arthur Mitchell (1884–1967), British writer. As a journalist he covered the Russian Revolution of 1917 and its aftermath. Later, he wrote very successful children's novels set in the English Lake District and the Norfolk Broads. These included *Swallows and Amazons* (1930) and *Peter Duck* (1932). His autobiography was published in 1976.

Ransom, money demanded in return for the release of a hostage. In medieval times, the demanding of ransoms was legally sanctioned and the right to have a ransom paid by his tenants was one of the feudal rights of a lord captured in battle, the amount payable varying according to the rank and status of the captive. Heavy taxes had to be levied in England to pay the 150,000 marks ransom demanded by the Emperor HENRY VI in 1193 for the release of RICHARD I. Today, ransom demands are associated with HIJACKING and kidnapping and are a criminal offence. *See also* HC1 p.213.

Ranunculaceae, family of flowering plants, containing about 2,000 species including the BUTTERCUP, ANEMONE and COLUMBINE. *See also* NW pp.*58, 70.*

Raoult's law, statement that the changes in certain properties of a liquid that occur when a substance is dissolved in it are proportional to the number of molecules of solute (dissolved substance) present for a given quantity of solvent molecules. Discovered by François-Marie Raoult in 1886, it has been fundamental to many theories of solution.

Rapacki Plan (1957), proposal put to the United Nations by the Polish Foreign Minister, Adam Rapacki, that a nuclear-free zone be created in East and West Germany, Czechoslovakia and Poland. It was rejected by the Western powers, who were fearful of the USSR's superiority in conventional land forces.

Rape, unlawful sexual intercourse with a female against her will by force, fear or fraud. Legal definitions of rape have often tended to put victims at a disadvantage in trying to prove the crime. In Britain, a husband cannot be accused of raping his wife unless she is legally separated from him. A man cannot be convicted of rape if he had reason to believe that the victim consented.

Rape, plant grown for animal fodder and for its small black seeds, which yield rape oil, used industrially as a lubricant and in the manufacture of various products. It has curly, blue-green leaves, small yellow flowers and slender seed pods. Family Brassicaceae. *See also* PE p.217, *217.*

Rape of Lucrece, The (1594), narrative poem by William SHAKESPEARE, written in Rhyme Royal. It is based on the story of LUCRETIA and was dedicated to Henry Wriothesley, Earl of Southampton.

Rape of Lucretia, The (1946), two-act chamber opera by Benjamin BRITTEN with libretto by Ronald Duncan based on Livy, Shakespeare and *Le Viol de Lucrèce* (1931) by André OBEY.

Rape of the Lock, The (1714), famous poetic satire by Alexander POPE, published in five cantos. It is based on a true incident in which a nobleman cut off a lock of a lady's hair, so causing a bitter feud between the two families.

Rape of the Sabine (1582), marble sculpture by Giambologna, situated in the Loggia dei Lanzi, Florence. The sculpture is a masterpiece of MANNERISM, and demonstrated the sculptor's ability to work equally in large-scale marble as well in small-scale bronze modelling. *See also* HC1 p.*265.*

Raper, Johnny (1939–), Australian rugby league player. One of the game's greatest loose-forwards, he was a key member of the dominant Australian sides of the 1960s, captaining them to victory in the 1968 World Cup. At club level he played mostly for St George of Sydney.

Raphael, biblical archangel who, with MICHAEL, GABRIEL and Uriel, serves as a messenger of God, relieving suffering and healing the sick. Since the 16th century he has been portrayed as the patron of travellers, probably because of his identification in the biblical book of TOBIT.

Raphael, or Raffaelo Sanzio (1483–1520), Italian painter, one of the finest artists of the High RENAISSANCE. Born in Urbino, he went to Perugia and studied under PERUGINO. He moved to Florence in 1504, and there learned his draughtsmanship. In Florence he painted many Madonnas, famous for their delicate portraiture based on the example of LEONARDO DA VINCI. In 1508 he moved to Rome at the request of Pope Julius II, and painted the frescoes of *The School of Athens* and the *Disputa* (1511) in the Stanza della Segnatura in the Vatican. *See also* HC1 pp.256–257, *256–257.*

Raphael, Frederic Michael (1931–), British writer. His work includes the novels *Richard's Things* (1973) and *California Time* (1975), and a film screenplay,

Darling (1965) which won an OSCAR. A television sequence, *The Glittering Prizes* (1976) won wide acclaim.

Rapid eye movement (REM), phenomenon observed in sleeping people. REM has been calculated to take up to 20% of a normal night's SLEEP and has been associated with the time spent dreaming. The sleeper cycles in and out of REM sleep and, although if necessary he can do with very little sleep, suffers if deprived of REM sleep. *See also* MS pp.*43, 135, 198.*

Rare earths. *See* LANTHANIDE SERIES.

Rare gases, or inert gases, also called noble gases, chemical elements, constituting group 0 of the Periodic Table, ie HELIUM, ARGON (which are not particularly rare), NEON, KRYPTON, XENON and RADON. All these gases are unreactive because the outermost shell of electrons in their atoms is filled (they have the stable eight electron configuration). Radon is a radioactive element. *See also* SU pp.*135, 136–137.*

Rarotonga, principal island of the Southern (or Lower) COOK ISLANDS, in the Pacific Ocean, and site of Avarua, capital of the Cook Islands. The island was discovered in 1823 by the British missionary John Williams. Products: tropical fruits, copra. Area: 67sq km (26sq miles). Pop. (1971) 11,437.

Ras al-Khaimah, sheikhdom of E Arabia and, since 1972, a member state of the UNITED ARAB EMIRATES on the Persian Gulf. It was part of the Sharjah sheikhdom until 1921, when it became independent. It is an important oil producing area. Area: 1,684sq km (650sq miles). Pop. (1968) 31,480.

Rasbora, genus of freshwater fish found in SE Asia. Popular with pet fish keepers, they have distinctive black markings that vary according to species. Length: to 5cm (2in). Species include the harlequin (*Rasbora heteromorpha*) and pigmy rasbora (*R. maculata*). Family Cyprinidae.

Rash, pimples, HIVES or weals on the skin. It may be associated with fever or diseases such as MEASLES and SCARLET FEVER. It also occurs as a reaction to some drugs and other substances that give rise to an ALLERGY. The symptoms of heat rashes and hives are caused by the body's production of Histamine. *See also* MS p.98.

Raspberry, bramble fruit grown in polar and temperate regions of Europe, North America and Asia. The black, purple or red fruit is eaten fresh or preserved. Canes, rising from perennial roots, bear fruit the second year. Family Rosaceae; genus *Rubus. See also* PE pp.178, 214.

Rasputin, Grigori Yefimovich (*c.*1872–1916), Russian peasant mystic. He exercised great influence at the court of NICHOLAS II because of his apparent ability to cure the crown prince Alexis' haemophilia. He was influential in Russia during WWI but was assassinated in 1916. *See also* HC2 p.*169.*

Rat, any of numerous kinds of small RODENTS found throughout the world in almost all habitats. Most species are herbivorous and inoffensive to man. The best-known are the black rat (*Rattus rattus*) and brown rat (*R. norvegicus*), both of the family Muridae. They are prolific, aggressive, and eat almost anything. They also carry diseases and destroy or contaminate property and food. Both live everywhere that man lives. Brown rats are slightly heavier and longer, with shorter tails and smaller ears. Most white rats used in laboratory experiments are albino strains of the brown rat. *See also* NW pp.*155, 160.*

Ratchet, mechanism that permits leverage in one direction only. It consists of a wheel or bar with teeth sloping in one direction, which is engaged by a pawl so that the wheel or bar is unable to rotate backwards. Ratchets are commonly parts of machine tools in which non-slip movement is essential. The rear sprocket of a free-wheelable bicycle incorporates a ratchet.

Ratel, or honey badger, solitary, aggressive burrowing mammal of the WEASEL family, native to Africa and S Asia. It has a heavy body with a light back and dark underparts. It is omnivorous, but is parti-

cularly fond of honey. Length: to 76cm (30in). Species *Mellivora capensis.*

Rates, taxes assessed on the owners of property. In Britain the rateable value is determined by the Department of Inland Revenue on the basis of rent that could be charged for a particular property. Rates are assessed by local councils as a percentage in the £ of the rateable value of the property. The collection of rates originated in Elizabethan times. Today rates make up about 25% of local government income, the remainder coming from grants from the central government (49%) and charges.

Ratfish. *See* CHIMAERA.

Rathlin, island in North Channel off the NE coast of Northern Ireland; part of county Antrim. There are ruins of a castle in which Robert the Bruce was supposed to have taken refuge in 1306. Farming and fishing are the chief occupations. Area: 13sq km (5sq miles). Pop. (1971) 109.

Ratio, quotient of two numbers or of two quantities of the same kind, such as two prices or of two lengths, that indicates their relative magnitude. Ratios, as of the numbers 3 and 4, can be written as a fraction (¾ or 3/4) or with a colon (3:4). *See also* SU p.28.

Rationalism, philosophical theory that holds that reason alone, unaided by experience, can arrive at basic truths about the world. Associated with this theory is the doctrine of innate ideas and the method of deducing truths about the world logically from self-evident premises. Eminent rationalists include DESCARTES, LEIBNIZ and SPINOZA. *See also* MS pp.231, *231, 233.*

Rational number. *See* NUMBER, RATIONAL.

Rationing, allocation of goods and services in an economic system. Some form of rationing occurs in most societies because supplies of goods are limited. During wartime governments attempt to ration goods of which there are shortages (particularly food), by issuing ration coupons which must be used in addition to money to make purchases. *See also* HC2 pp.194, 232–233, 236, *236.*

Ratitae, group of large, usually FLIGHTLESS BIRDS with flat breastbones instead of the keel-like prominences found in most flying birds. Ratites include the OSTRICH, RHEA, CASSOWARY, EMU, KIWI and the unusual flying TINAMOU.

Ratsiraka, Didier (1936–), Malagasy naval officer and politician. Educated in France he became Minister of Foreign Affairs in 1972, Prime Minister in 1975 and President in 1976.

Rat snake, any of about 50 species of terrestrial and arboreal SNAKES, distributed widely in North America and Europe. Common in North America are the black and yellow rat snakes, fox snakes and corn snakes. The four-lined and Aesculapian snakes are European. Length: to 1.8m (6ft). Family Colubridae.

Rattan, climbing PALM native to the East Indies and Africa. Its stems grow to 152m (500ft) by climbing over trees and other plants. They are used for making ropes, furniture and mats. Family Palmae; genus *Calamus.*

Rattigan, Sir Terence Mervyn (1911–77), British dramatist. His well-written, carefully constructed plays enjoyed popular, if not critical, success. His works include *French Without Tears* (1936), *The Winslow Boy* (1946), *Separate Tables* (1954), *In Praise of Love* (1973) and *Cause Célèbre* (1977). He wrote many notable film scripts, including *Goodbye Mr Chips* (1968).

Rattlesnake, any of about 30 species of venomous New World PIT VIPERS characterized by a tail rattle of loosely connected, horny segments of unshed skin. It ranges from Canada to South America, usually in arid regions. Most are blotched with dark diamonds, hexagons or spots on a lighter background. They feed mostly on rodents, and bear their young live. Length: 30cm–2.5m (1–8ft). Family Viperidae. *See also* SNAKE; NW pp.*136, 138, 199.*

Rattlesnake fern, conspicuous grape FERN, widely distributed in the Northern Hemisphere. Its clustered sporangia resemble a

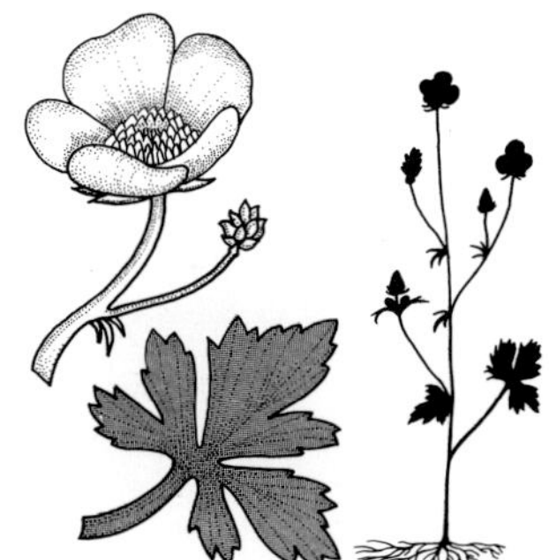

Ranunculaceae; bachelor's button is a member of this large family of plants.

Rape of the Lock; an illustration from an 1828 edition of Pope's famous satire.

Raphael; *The Madonna of the Stool,* one of the many madonnas by this Italian master.

Ratels reverse the usual colours of most animals, with light fur uppermost.

Rauch, Christian Daniel

Ravel's most popular work is his *Bolero*, which drives on to a frenzied climax.

Lord Rayleigh's researches spanned much of physics and chemistry.

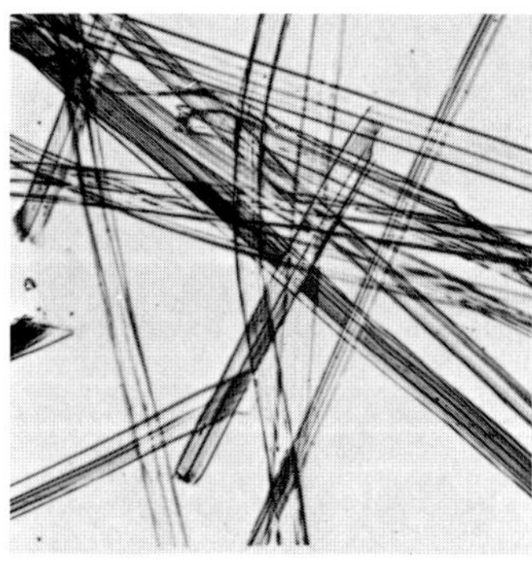

Rayon, patented in 1884 by Hilaire Chardonnet, was originally called artificial silk.

Razorbills, more than any other birds, have suffered from oil pollution at sea.

rattlesnake's tail. It has triangular leaves. Height: to 60cm (24in). Family Ophioglossacea; species *Botrychium virginianum*.

Rauch, Christian Daniel (1777–1857), German sculptor. He is famous for his colossal bronze equestrian monument to Frederick the Great (1839–52) in Berlin.

Rauschenberg, Robert (1925–), US painter and graphic artist, who gained recognition in the 1950s with his collages and assemblages. His works combined various techniques, including dripped paint and collage, as in *The Bed* (1955). From 1955 he composed works that juxtaposed SURREALIST images, eg stuffed animals, with ABSTRACT EXPRESSIONIST painting, as in *Monogram* (1959). Rauschenberg's work is considered a link between Abstract Expressionism and POP ART. *See also* HC2 p.203.

Rauwolfia, drug derived from a genus of shrub of the same name native to India. It is used to treat hypertension, anxiety and psychoses. *See also* RESERPINE; MS p.*118*.

Ravana, in Hindu mythology, one of the Rakshasas, evil spirits in the grip of fate. In the RAMAYANA, Ravana appears as a giant with 10 heads and 20 arms. He was killed by RAMA and a host of monkeys after kidnapping Rama's wife, SITA.

Ravel, Joseph Maurice (1875–1937), French composer. A student of Gabriel FAURÉ, he was a leading exponent of IMPRESSIONISM. Ravel wrote original music within the outlines of classical forms and his piano compositions include *Jeux d'eau* (1901), *Gaspard de la nuit* (1908) and *Le Tombeau de Couperin* (1917), which he also scored for orchestra, and two piano concertos. Among his orchestral works are *Rhapsodie Espagnole* (1907) and *Bolero* (1927).

Raven, Simon Arthur Noel (1927–), British novelist. His first novel was *The Feathers of Death* (1959). Since then he has published almost yearly, including *The Rich Pay Late* (1964) and *The Judas Boy* (1968).

Raven, large bird of the crow family found in deserts, forests and mountainous areas of the Northern Hemisphere. It has a long, conical bill, shaggy throat feathers, a wedge-shaped tail and black plumage with a purple sheen. It eats carrion or any other animal food. Length: to 68cm (27in). Family Corvidae.

Ravenna, city in N Italy 72km (45 miles) ESE of Bologna; capital of Ravenna province. It was a Roman naval base in the 1st century BC, and the capital of the Western Roman Empire in the 5th century AD. Under papal rule from the 16th–19th centuries, it became part of the Kingdom of Italy in 1860. The city is noted for its outstanding Roman and Byzantine remains, particularly the mosaics in the churches of the 5th and 6th centuries AD. Industries: petroleum, furniture, cement, fertilizers, sugar refining. Pop. (1975) 137,303.

Ravensbrück, CONCENTRATION CAMP, situated 81km (50 miles) N of Berlin, built by the Nazis in 1938. Its inmates were mainly women, the majority Russians and Poles. It is estimated that 50,000 people were killed there.

Ravenscroft, George (1618–81), English glassmaker who developed a type of glass stronger and more durable than VENETIAN GLASS. He was commissioned to produce a glass that could be manufactured from English materials; the resulting "flint glass" was granted a patent in 1674.

Rāwalpindi, city in NE Pakistan, approx. 258km (160 miles) NNW of Lahore. It became an important military base after Britain occupied the Punjab in 1849. Today it is an industrial centre with an iron foundry, textile mills, an oil refinery and chemical plants. Pop. (1972) 615,392.

Rawlinson, Sir Henry Creswicke (1810–95), British diplomat, soldier and Assyriologist. He entered the EAST INDIA COMPANY's army in 1827 and helped to reorganize (1833–39) the Persian army. During this time he studied CUNEIFORM inscriptions and deciphered DARIUS's Bisitun inscription. His works include the *Outline of the History of Assyria* (1852).

Rawsthorne, Alan (1905–71), British composer who devoted himself to full-time composition in 1934. He soon established a reputation with *Theme and Variations* for two violins (1938) and *Symphonic Studies* (1938). Other pieces include three symphonies and the fantasy overture *Cortèges* (1945).

Ray, John (1627–1705), English naturalist whose work on plant and animal classification later influenced Carl LINNAEUS and Cuvier. He was the first to distinguish the two main types of flowering plants as MONOCOTYLEDONS and DICOTYLEDONS. *See also* NW p.32.

Ray, Man (1890–1976), US photographer and painter, the founder of the New York DADA movement with Marcel DUCHAMP and Francis PICABIA. In the 1920s he began making his "Rayographs" in which he placed objects on light-sensitive paper and exposed and developed them.

Ray, Satyajit (1921–), Indian film director with a worldwide following. His trilogy of films about life in modern India, *Pather Panchali* (1954), *The Unvanquished* (1956) and *The World of Apu* (1959), is regarded as a classic. His other films include *Kanchenjunga* (1962).

Ray, any of several species of cartilaginous, mostly marine fish related to the SKATE, SHARK and CHIMAERA; it is primitive, having cartilage not bone as its skeletal material. Like a skate, the ray is flattened dorso-ventrally; its body extends sideways into large, wing-like pectoral fins which are "flapped" while swimming, often near the ocean floor. The tail is narrow and may be whip-like or bear poisonous spines. The giant manta ray grows to a span of 6m (20ft), but it is a harmless plankton feeder. *See also* NW pp. 124, *126*.

Ray. *See* ELECTROMAGNETIC RADIATION.

Rayleigh, John William Strutt, Lord (1842–1919), British physicist. His work was chiefly concerned with various forms of wave motion. For his discovery (with William RAMSAY) of the rare gas ARGON he was awarded the 1904 Nobel Prize in physics.

Raymond, name of seven counts of Toulouse, of whom the most important are Raymond IV (c.1038–1105), the leader of the First Crusade; Raymond VI (1156–1222), repeatedly excommunicated for his toleration of the ALBIGENSIANS, and Raymond VII (1197–1247), who surrendered much of S France to LOUIS IX and persecuted the Albigensians.

Raynaud's disease, arterial disorder in which the fingertips or other extremities become pale and then blue, especially when exposed to cold. It is believed to be a result of spasms in certain arterioles, preventing adequate circulation of blood.

Rayon, fine, smooth, man-made textile fibre produced from solutions of the modified CELLULOSE of soft wood pulp or cotton linters. Viscose rayon, the most common, is spun-dried and has a strength approaching that of NYLON. The more costly cuprammonium rayon, known as Bemberg, resembles silk. *See also* MM p.56.

Rayonnism, in art, transitory movement that developed in Russia c.1911 in response to a CUBISM. The painter Mikhail Larionov (1881–1964) propounded that art should exist in a fourth dimension by projecting rays of contrasting colour from objects into space.

Razin, Stenka (d.1671), Cossack leader of the peasant revolt in Russia in 1670. As commander he proclaimed freedom from landlords and officials, but was defeated at Simbirsk (Ulyanovsk) and executed.

Razor-billed auk, or razorbill, chunky penguin-like seabird that lives along coast-lines in the cold parts of the Northern Hemisphere. It is black and white with a white-ringed, narrow bill, and dives for its food. Length: 41cm (16in). Species *Alca torda*.

Razor blades, double- or single-edged shaving blades. The forerunner of the modern safety razor was made in England in 1828. The first double-edged, disposable blade was patented by King Camp Gillette in 1901 and first produced commercially in 1903.

Razor shell, bivalve MOLLUSC related to the CLAM, MUSSEL and SCALLOP. Its long, hinged shells, each shaped like the blade of a cut-throat razor, are common on British beaches. *See also* NW pp.*92, 93*.

RCA, US electronics and communications company, founded in 1919 as the Radio Corporation of America. In the mid-1970s it employed approximately 113,000 people in the manufacture of television cameras and receivers, tape recorders, records and record-players.

R Coronae, in astronomy, type of irregular star whose brightness remains normally at maximum, but suffers sudden, unpredictable drops to minimum. Fewer than 50 stars of this class are known. *See also* SU pp.240–241, 256.

Re, or Ra, in Egyptian mythology, sun god and lord of the dead. In some traditions he is identified with OSIRIS or sometimes as the creator of man. In the myths, Re sailed his sun boat across the sky by day and through the underworld by night. He is depicted as a falcon-headed man with a solar disc on his head. The centre of his worship was at HELIOPOLIS.

Reaction, chemical. *See* CHEMICAL REACTION.

Reaction propulsion, propulsion utilizing NEWTON's third law of motion: for every action there is an equal and opposite reaction. Both ROCKETS and JET ENGINES build up internal pressure from gases or PLASMAS that are allowed to escape in one direction, creating a propulsive force in the opposite direction. *See also* MM p.156; SU p.268.

Reaction rate, in chemistry, the speed at which chemical reactions move to completion or to equilibrium. The rate is accelerated by one or more of the following: the presence of a CATALYST or a promoter, heat, pressure, or reduction of the concentration of the reaction products.

Reaction time, interval between the onset of a stimulus and the conscious execution of a response.

Reactor, fast breeder, NUCLEAR REACTOR that produces ("breeds") more fissionable material than it consumes. Excess neutrons from the fission of a fuel such as uranium-235, instead of being absorbed in control rods, are used to bombard atoms of relatively inactive uranium-238 which transmutes to (or "breeds") the active ISOTOPE plutonium-239. When the original fuel is spent, the plutonium can be used as a nuclear fuel in other reactors or nuclear weapons. *See also* SU pp.64–65; MM pp.74–75, 180–181.

Read, Sir Herbert (1893–1968), British critic and poet who helped to revive an interest in English ROMANTIC poetry and who strongly defended movements in modern art against fierce criticism. His publications include *The Meaning of Art* (1931), *Form in Modern Poetry* (1932), *Art and Society* (1936) and *A Concise History of Modern Painting* (1959).

Read, Phil (1939–), British motorcyclist. He won five world lightweight titles (125cc 1968; 250cc 1964, 1965, 1968, 1971) riding Yamahas and two 500cc titles (1973, 1974) with MV Augusta. Making a comeback at the 1977 Isle of Man TT, he won the Formula I world championship, on a Honda.

Reading, county town of Berkshire, s central England, at the confluence of the Thames and Kennet rivers. It has the ruins of a Benedictine abbey, founded in 1121 by Henry I (who was buried there), and the University of Reading (1926). Industries: ironware, engineering, brewing. Pop. (1973) 132,280.

Reading, assimilation and comprehension of visual symbols representing language. Opinions differ over the precise physical nature of this process, as they do also over whether the language involved – phonetic non-phonetic or ideographic – is influential in the speed of learning and the retention of vocabulary. Visual familiarity with the symbols of any kind of language, however, together with a fair vocabulary, results in the ability to read faster than the words can be spoken. Reading then becomes purely a thought process. It is estimated that just less than 60% of adults in the world can read; the long-term literacy campaign of UNESCO is slowly increasing this figure. *See also* MS p.44, *44*.

Realism, in the arts, a style that attempts to describe life with objective honesty and eschews romanticism and idealization. Classic realist painters include Jean-Baptiste CHARDIN, Gustave COURBET, and REMBRANDT. As a literary school, realism was important in the 19th century. Major realist writers include Gustave FLAUBERT, Honoré de BALZAC, George ELIOT, Henry JAMES and Mark TWAIN. Naturalism is an outgrowth of realism.

Real earnings, earnings calculated by dividing money wages by the RETAIL PRICES INDEX. If prices have risen faster than money wages, then real earnings have decreased.

Realism, medieval doctrine that universal concepts and tangibles exist outside the mind, ie independently of human perception. Some philosophers rejected this view for what may be termed moderate realism, which held that UNIVERSALS exist only in the mind of God. See also IDEALISM; MS pp.227, 226–227.

Realism, socialist, policy of Communist governments that the arts and literature should realistically portray themes of concern to the proletariat, thus furthering the aims of communism.

Reality, perceived and acknowledged concreteness of actuality. PLATO believed that what is common to things described by the same name (eg any common noun) is their Form or Idea. These Forms comprise reality and their various manifestations are but copies.

Reality principle. See PLEASURE PRINCIPLE.

Real number, NUMBER which can be expressed by a single series of digits, such as 241.037... Such numbers include INTEGERS and rational and IRRATIONAL NUMBERS. They are distinguished from complex numbers which are of the form $x + iy$ (where $i=\sqrt{-1}$) and require two series of digits, x and y, to be defined.

Realpolitik, state policy which places national interests above fraternal ones and material considerations above moral ones. It is associated especially with European politics of the second half of the 19th century, above all with the statecraft of BISMARCK.

Real Presence, theological term that describes various doctrines which assert that the body and blood of Christ are actually, not merely symbolically, present in the bread and wine of the EUCHARIST.

Real tennis, racket and ball game played on an indoor court according to rules perhaps more complicated than those for any other ball game. The word "real" is from the old French for "royal". In France, where the game began, it is still called *jeu de paume* (palm-game), from the times when the ball was hit with the hand rather than a racket. In Australia it is called royal tennis; in the USA court tennis. The asymmetric four-walled court, which is surrounded on three sides by structures with sloping roofs, is similar to the monastic cloisters in which the game was first played in the 12th or 13th centuries. See also LAWN TENNIS.

Réamur, René-Antoine Ferchault de (1683–1757), French scientist best remembered for his thermometer scale which designates zero as the freezing point of water and 80 degrees as its boiling point. He wrote widely in natural history and did research into mining, metallurgy, fossils and insects.

Reaper, machine that cuts cereal crops. Early reapers cut the crop and dropped it on the field. COMBINE HARVESTERS incorporate cutting, threshing and grain containment in one operation. See also PE p.164, *164*.

Rearmament, generally the rebuilding of a nation's depleted stock of arms, but more especially the rearmament question in Western European politics in the 1930s. When it became known that HITLER was (in contravention of the Treaty of VERSAILLES) rearming Germany, a section of the Conservative Party in Britain, notably Winston CHURCHILL, strove to persuade the government to rearm. The Prime Minister, Stanley BALDWIN, would not make it an issue at the 1935 elections, but British rearmament began in 1936.

Reasoning, formation of logical conclusions. The two main processes of reasoning are INDUCTIVE LOGIC and DEDUCTIVE

LOGIC. Induction involves putting together observed facts to form a general principle. Deduction is the opposite – a specific fact is reasoned from a general principle. Probably first to systematize reasoning was ARISTOTLE, who emphasized the importance of categories or factors which may be common to all of the things involved in a logical process. See also MS pp.226–229.

Rebab, Moorish form of REBEC.

Rebec, early stringed instrument, played with a bow. It is one of the ancestors of the VIOLIN and closely related to the LYRE. It was made in several sizes to produce notes in the various ranges of the human voice.

Rebecca (Rebekah), in Genesis, of Bethuel, sister of Laban, wife of ISAAC and mother of the twins Esau and JACOB. Jacob was her favourite and she helped him win Isaac's blessing that was rightfully Esau's.

Rebecca Riots (1839, 1842–43), destructive campaign by Welsh tenant farmers, against rates, rents, tithes and particularly turnpike tolls. The rioters took their name from Genesis 24:60 and were suppressed by the military.

Rebel Without a Cause (1955), US film starring James DEAN as a teenager. The anger and frustration he feels with his world lead him to tragedy.

Recall of information, process of retrieving data. The efficiency of this process depends largely on the systems by which the information is classified. Two such systems used in libraries are the Dewey decimal system and the Library of Congress system, but these and others still contain disadvantages. The main problem is the defining of categories. See also MS pp.44–45.

Récamier, Jeanne-Françoise-Julie-Adélaïde ("Juliette") (1777–1849), society figure who kept a fashionable intellectual salon in Paris during the 1830s. She gathered around her many important political and literary figures of the times including CHÂTEAUBRIAND, SAINTE-BEUVE, Madame de STAËL and Benjamin Constant. Jacques-Louis DAVID painted her portrait in 1800.

Recapitulation, biological theory which says that the embryonic development of a human being repeats the evolutionary development of the species. It was called the biogenetic law by HAECKEL.

Receptor, any of several kinds of NEURONS specialized to transform physical energy into the electrochemical impulses of the nervous system. For example, the RODS AND CONES of the RETINA of the eye are receptors that transform light energy into the nerve impulses sent to the brain.

Recession, in economics, phase of the business cycle associated with a declining economy. Its manifestations are rising unemployment, contracting business activity and decreasing purchasing power of consumers. A recession is generally associated with pessimism in industrial and business circles, and a reluctance to invest in factories and machinery for further production. Government policy, such as cuts in government spending or taxes, may be used to stimulate and expand the economy during a recession. If a recession is not checked, it can degenerate into a DEPRESSION.

Recessive, in genetics, describes a GENE which does not express itself in the PHENOTYPE (physical characteristics) of an organism when paired with a dominant gene. Recessive genes were discovered by Gregor MENDEL, who found that a cross between pure bred red and white flowering garden peas always produced red flowers in the offspring. The gene for red coloration was dominant; the gene for white was recessive. In humans the genes for blue eyes are recessive to those of brown eyes; genes for HAEMOPHILIA are also recessive. See also NW pp.30–31.

Recife, Atlantic port and city in NE Brazil; the capital of Pernambuco state. First settled by Portuguese fishermen and sailors in the 1530s, it is located partly on an island and connected to the mainland by numerous bridges. It is now an excellent port and shipping centre, exporting sugar, cotton and coffee. Pop. (1970) 1,046,454.

Reciprocal, number or quantity equal to the number 1 (unity) divided by a speci-

fied number. For example, the reciprocal of 2 is $\frac{1}{2}$ and the reciprocal of $\frac{1}{2}$ is $1 \div \frac{1}{2} = 2$.

Reciprocating engine, any engine in which a piston moves to and fro, such as a steam ENGINE or an INTERNAL COMBUSTION ENGINE, commonly used to power motor vehicles and aircraft. There are two basic designs of reciprocating internal combustion engine. An in-line engine has a row of cylinders or two rows in a V arrangement or horizontally opposed. A radial engine, used only for aircraft, has cylinders arranged in a circle. See also MS p.152.

Recitative, in music, declamatory passages in opera and oratorio, in contrast to the lyrical sections. Recitativo secco, the earliest form of recitative, allows the singer the freedom of normal speech rhythms and its accompaniment, often only a harpsichord or piano, is kept to a minimum. In recitativo accompagnato, the accompaniment is fuller and more rhythmical.

Reclamation of land. See LAND RECLAMATION.

Reconstruction, in US history, the rebuilding of the US South after the AMERICAN CIVIL WAR. President Andrew JOHNSON attempted to preserve in power the former slave-owning rulers of southern society, but the Radical Republicans won enough seats at the 1866 elections to thwart him. The 14th amendment, guaranteeing Blacks political equality, was passed in 1866 and ratified by the states in 1868.

Reconstruction, Ministry of, British government department established in 1917 and lasting until 1919. It was formed out of the committees of reconstruction appointed in 1916 and 1917. Its purpose was to supervise the return to peace-time industrial activity after the dislocation of WWI. It was particularly concerned with finding employment for ex-servicemen. A similar ministry was created in 1944–45.

Record, gramophone, disc physically coded to reproduce sound. The sound to be recorded on a disc is first recorded on tape. The tape, when played back, controls a stylus which cuts a spiral groove in a lacquer-coated aluminium disc. The stylus vibrates horizontally for monophonic sound, vertically and horizontally for stereophonic sound. This master disc is then used to prepare a metal stamp to reproduce the groove on pressings, which may be sold as records. See also MM p.232, *232*.

Recorder, simple woodwind musical instrument, popular in Europe from the 15th to the 18th centuries, and now commonly played by children. It has a whistle-shaped mouthpiece and an end-blown straight tube with eight finger holes. Modern recorders include soprano, descant, tenor and bass instruments.

Recorder, in current English law, part-time judge of a crown or county court. A recorder must be a barrister or solicitor of at least ten years' standing and must take the judicial oath (swearing to uphold legal principles and to be impartial) before commencing office. Before 1972 a recorder was appointed to replace JUSTICES OF THE PEACE as a sole judge of a court of QUARTER SESSIONS.

Recording, storing of signals that represent sound or images, on a medium such as a plastic disc or magnetic tape. In disc recording, the sounds or images being recorded are electronically modified and converted into movements of a stylus which cuts an original lacquer disc, which is then electroplated to make a master negative for pressing out plastic copies. In magnetic recording, sound is recorded on tape with a TAPE RECORDER; a VIDEO RECORDING machine records sound and vision. Optical recording of sound is used for the sound track of motion pictures. The sounds cause variations in thickness or brightness of a strip of film running alongside the motion picture photographs. By a reverse process this is converted back into sound when the picture is run. See also MM pp.*103*, 232–236.

Record player, machine for reproducing sound from gramophone records. It originated with the phonograph invented by Thomas A. EDISON in 1877. In modern record players the stylus of a pick-up head

Madame Récamier; David's portrait shows the elegance of the Napoleonic era.

Records still work using the same basic principle as in clockwork gramophones.

Recorder; a craftsman checking the pitch of a newly manufactured instrument.

Recording; sound mixers equalise and balance signals from many channels.

Rectifier

Red Cross members on their way to help resettle Tibetan refugees in Nepal.

Red deer; one theory suggests the antlers act as radiators to dissipate heat.

Vanessa Redgrave has become increasingly associated with left-wing politics.

Redstarts are migratory birds, the European species wintering north of the equator.

follows the variations in depth or width of grooves on a rotating disc. The variations are converted into electrical signals and amplified before being converted into sound by one or more loudspeakers. *See also* MM pp.232–233.

Rectifier, electronic component that converts an alternating current (AC) into a direct current (DC). It consists of a set of plates, a diode valve or semiconductor that presents a high resistance to a current flowing in one direction and a low resistance to current flowing in the opposite direction. *See also* SU pp.115, 123, 128.

Rector, church official. In the Church of England a rector of a parish is distinguished from the VICAR in that the rector may dispose of the living's property. The heads of Exeter and Lincoln Colleges at Oxford University are termed rectors, and the title is also used in Europe for university heads. JESUITS use the term for heads of their religious and educational institutions.

Rectum, in human beings and many other vertebrates, the terminal part of the large INTESTINE, where the faeces form and collect before being expelled as waste through the anus. Length: approx. 15cm (6in) in human beings.

Recusants, term applied to those (especially Roman Catholics) in England who refused to attend services of the Church of England. The word derives from the Latin *recusantes*, "those refusing". In 1559 the fine for non-attendance was put at 1s per Sunday; it was raised to £20 per month in 1581. Fines were irregularly imposed, and all restrictions were finally removed by the passing of the Catholic Emancipation Act in 1829.

Recycling, natural and man-made processes by which substances are broken down and reconstituted. In nature, elemental cycles include the CARBON CYCLE, NITROGEN CYCLE and HYDROLOGIC CYCLE. Natural cyclic chemical processes include the metabolic cycles, eg the CITRIC ACID CYCLE, in the bodies of living organisms. Man-made recycling includes using bacteria to break down excreta and many kitchen and factory organic wastes to harmless, or even beneficial, substances. Large quantities of inorganic waste such as metal scrap, glass bottles and building spoil, are recycled. Polymer wastes are often not recycled but burned, with consequent atmospheric POLLUTION and loss of material resources. *See also* MM pp.50–51, 200–203, 241; SU p.*155.*

Red admiral, distinctive European butterfly with red bars on the wings and black wing-tips spotted with white. It flies from May to August. The caterpillar, which feeds on nettles, is dark with light sidestripes and branching spikes. Family Nymphalidae; species *Vanessa atalanta.*

Red algae, group of typically reddish mostly marine ALGAE (division Rhodophyta), especially numerous in tropical and subtropical seas. Some, including DULSE and IRISH MOSS, are common on N coasts. Most are many-celled plants growing as strands that form shrub-like masses. *See also* NW pp. 46–47, *47.*

Red Army, army of the USSR. It is characterized by a high degree of political control, institutionalized at all levels with a system of commissars, whose functions are political rather than military. TROTSKY organized it as an army to defeat the Whites and the interventionist powers after the revolution of Oct. 1917. The high command of the Red Army was decimated during the purges of the 1930s. During WWII the Red Army grew to more than 20 million men. It was concentrated in E Europe from 1945, but since the early 1960s it has also been aligned along the Russian border with China.

Red blood cell. *See* ERYTHROCYTE.

Red bug. *See* CHIGGER.

Red Crescent, The, name by which the International RED CROSS is known in Muslim countries. In these countries the red cross on the international flag of the organization is replaced by the red crescent. Iran is an exception to this, however; there a red lion and sun are substituted for the red cross.

Red Cross, founded 1863, international organization that seeks to ameliorate human suffering, particularly through disaster relief, aid to war victims of all nations and services to members of the armed forces. The Red Cross was created at the prompting of Jean Henri DUNANT, a Swiss citizen who had been shocked by the suffering of the wounded after the battle of SOLFERINO in 1859. Today it is composed of more than 100 independent national societies in most countries, with central headquarters in Geneva, Switzerland, and staffed largely by volunteers. The name Red Cross comes from its symbol, a red cross on a white field, which reverses the colours of the Swiss flag. The International Red Cross was awarded the Nobel Peace Prize in 1917 and 1944.

Redcurrant, widely cultivated fruit-bearing shrub and its small, round, red, edible fruit; it is closely related to the BLACKCURRANT. The fruit may be eaten raw, cooked, or as a jelly with meat. Species *Ribes rubrum.* Family Saxifragaceae.

Red deer, largest of the British DEER, with a body length of up to 2.2m (7ft) and a height of up to 1.5 (5ft) at the shoulder. It has a reddish-brown coat in the summer but a thicker darker winter coat. Its preferred habitat is a large deciduous wood. Species *Cervus elaphus. See also* PE p.*253;* NW p.*207.*

Redemption, in Christian theology, the redeeming or reclaiming of mankind from the powers and the effects of sin, by JESUS CHRIST. Christ's death on behalf of humanity was the price of redemption; the RESURRECTION was the sign that His sacrifice of Himself had brought the chance of salvation for all.

Red fescue. *See* FESCUE.

Red Flag, international symbol of SOCIALISM. It was first used by the French government after the storming of the BASTILLE (1789), when its being flown signified that Paris was under martial law. After troops fired on a mob in the *Champ de Mars* in 1791, the flag was adopted by revolutionaries, who inscribed on it the words "Martial law of the people against the Revolt o the Court".

Red Flag, The, British socialist song written by James Connell, secretary of the Workmen's Legal Friendly Society, in 1889. It was originally sung to the tune of *The White Cockade,* now replaced by the tune of *Der Tannenbaum.*

Redford, Robert (1937–), leading US film actor who came into prominence in the late 1960s. His films include *The Chase* (1966), *Barefoot in the Park* (1967), *Butch Cassidy and the Sundance Kid* (1969), *The Candidate* (1972), *The Sting* (1973), *The Great Gatsby* (1974) and *All the President's Men* (1976).

Redgrave, name of a family of British actors and actresses. Sir Michael (1908–), also a director and writer, made major stage appearances in Shakespeare's *Hamlet, Macbeth* and *As You Like It.* He also appeared in many films, including *The Way to the Stars* (1945) and *The Browning Version* (1951). He married the actress Rachel Kempson in 1935 and their children are all actors. Vanessa (1937–) frequently appeared on the London stage and in films such as *Isadora* (1968) and *Julia* (1977). Corin (1939–) has acted on stage and television, and Lynn (1943–) received an Oscar for *Georgy Girl* (1966).

Red Guards, paramilitary groups of Chinese students and young people formed at the suggestion of Chairman MAO TSE-TUNG to further the Chinese CULTURAL REVOLUTION. Named after the red army units of the late 1920s, the Red Guards marched into Peking from outlying China in eight giant units in late 1966. Their greatest influence was during the period 1966–67 as part of the cultural revolution.

Red hake. *See* LING.

Red-hot poker, cultivated garden perennial plant that originated in South Africa. It has an underground rhizome and narrow tubular red or yellow flowers. Height: 1m (3.3ft). Family Liliaceae; species *Kniphofia uvaria.*

Redi, Francesco (1626–97), Italian scientist, physician and poet. Through controlled experiments he showed that certain living organisms, notably maggots in rotting meat, did not arise, as had been alleged, from spontaneous generation. His chief poetic work was *Bacchus in Tuscany* (1685).

Red Indians. *See* AMERICAN INDIANS.

Redon, Odilon (1840–1916), French painter and graphic artist. He created two different types of picture: glowing, semi-Impressionistic landscapes and flower paintings, and highly mysterious fantasies based on his personal inner vision. From the late 1860s he was a stout opponent of naturalism in art, was hailed as the leading SYMBOLIST and acted as a direct stimulus for GAUGUIN. He was admired by the AVANT-GARDE of the era, including MATISSE, The NABIS and BONNARD.

Redouté, Pierre Joseph (1759–1840), Belgian painter and engraver, born near Liège, who studied botany in KEW GARDENS and with Sir Joseph BANKS. Patronized by MARIE ANTOINETTE, JOSÉPHINE Bonaparte and her Bourbon successors, he executed over 6,000 delicate floral watercolours and, apart from his illustrations, is also remembered for his botanical studies such as *Les Roses* (1817–21).

Redpoll, seed-eating bird related to the FINCHES, found in colder parts of the Northern Hemisphere. It has a conical bill, a short forked tail and brownish plumage with a red patch on the crown. Length: 12.5cm (5in). Genus *Acanthis.*

Red River Settlement, colony established in territory granted by the HUDSON'S BAY COMPANY in British North America in 1811–12 to the Earl of Selkirk. In 1815 and 1816 the colony, officially known as Assiniboia, was subject to two attacks by the rival North West Company. It was purchased by the Hudson's Bay Company in 1836. It became part of Manitoba in 1870.

Red Sea, narrow arm of the Indian Ocean between NE Africa and the Arabian Peninsula, connected with the Mediterranean Sea by the Gulf of Suez and the Suez Canal. Known in biblical times as the passageway for the Israelite exodus from Egypt, it was an ancient trade route whose importance declined with the discovery of an all-water route around Africa in 1498. Its importance greatly increased in 1869 with the building of the Suez Canal. But with the closing of the Canal (1967–75) during Arab-Israeli conflicts, the building of vessels too large for the canal and the development of pipelines, the Red Sea's importance as a trade route once again diminished. Area: 438,000sq km (169,000sq miles).

Redshank, Eurasian wading bird of the SANDPIPER family. It has a long slender bill for probing soft sand and mud to find food. It has mottled grey, brown and white plumage and characteristic slender red legs. Length: to 28cm (11in). Family Scolopacidae species *Tringa totanus. See also* NW pp.*235, 237.*

Red shift, shift in the spectral lines of a star, galaxy or other celestial object towards the red end of the visible spectrum, relative to the positions of equivalent lines in a spectrum produced on Earth. Although sometimes a gravitational effect, it is usually due to the DOPPLER EFFECT and results from an object moving rapidly away from the Earth (an approaching star has a blue shift). The red shift of galaxies is evidence for the expansion of the universe. *See also* QUASAR; SU pp.248–249, *249,* 252.

Red spider, or spider mite, red vegetarian MITE found throughout the world. A serious plant pest, it may develop resistance to pesticides, making control difficult. Length: 0.3–0.8mm (0.01–0.03in). Family Tetranychidae.

Redstart, name of two unrelated robin-like birds of the THRUSH family. The male European redstart (*Phoenicurus phoenicurus*) is greyish with rust-red breast and tail. Length: 14cm (5.5in). The American flycatching redstart (*Setophaga ruticilla*), a wood warbler, is slightly smaller.

Red tide, red discoloration of ocean water caused by an unusually large "bloom" of single-celled dinoflagellate organisms. The dinoflagellates release a poison into

the water and make swimming unsafe. The lack of oxygen in red tide waters may also cause the death of fish in the locality.
Reduction. *See* OXIDATION REDUCTION.
Redwing, also called red-winged BLACKBIRD, small well-known bird of the USA. It has striking red and yellow wing markings, feeds primarily on insects, and inhabits marshy areas. Length: to 20cm (8in). Family Icteridae; species *Agelaius phoeniceus*. A small European thrush, *Turdus iliacus*, which has reddish feathers under its wings, is also called a redwing.
Redwood. *See* SEQUOIA.
Reed, Sir Carol (1906–77), British film director whose long career began in the 1930s. He won an Oscar in 1968 for *Oliver*. His other films include *Penny Paradise* (1938), *The Fallen Idol* (1948), *The Third Man* (1949), *Trapeze* (1956), and *The Agony and Ecstasy* (1965).
Reed, Oliver (1938–), British film actor. He has appeared in many films, usually cast in tough roles, including *Beat Girl* (1959), *Women in Love* (1969), *Sitting Target* (1971), *The Three Musketeers* (1974) and *The Sellout* (1976).
Reed, aquatic GRASS native to wetlands throughout the world. The common reed (*Phragmites communis*) has broad leaves, feathery flower clusters and stiff smooth stems. Dry reed stems are used for thatching, construction and musical pipes. Height: to 3m (10ft). Family Gramineae. *See also* NW pp. 68–69.
Reedbuck, light, graceful ANTELOPE native to Africa s of the Sahara, usually near water. Its body is brown to grey, and when disturbed it raises its bushy tail erect. The male has short horns, curved forwards at the tips. Length: to 1.4m (4.5ft). Family Bovidae; genus *Redunca*.
Reed instruments, musical instruments that produce sound when an air current vibrates a fibre or metal tongue. The HARMONIUM, in which wind flows in both directions, uses a free reed. In a CLARINET or reed organ pipe, in which air passes in only one direction, a beating reed vibrates against a hole at the end of the tube. The OBOE and BASSOON have double-reed mouthpieces, the two tongues vibrating against each other when blown. *See also* HC2 p.*105*.
Reed International Ltd, British manufacturer of paper, printing and packaging, employing in the mid-1970s about 86,000 people throughout the world. The company began as a newsprint manufacturer in 1894. It is now also involved in publishing and do-it-yourself materials.
Reedling, also called bearded reedling or reed pheasant, small bird with orange brown plumage, a characteristic tufted beard and a long tail. It lives and nests in areas in Eurasian wetlands, and feeds primarily on seeds and insects. Length: to 15cm (6in). Family Panuridae; species *Panurus biarmicus*.
Reed mace. *See* BULRUSH.
Reef, rocky outcrop lying in shallow water, especially one built up by CORALS or other organisms. *See also* PE p.123.
Re-entry, in aerospace technology, the process of returning to Earth's atmosphere from space. Re-entry involves speeds of up to 24,000km/h (15,000mph) and extremely high temperatures due to atmospheric friction.
Reese, Terence (1913–), British master BRIDGE player, writer and teacher. He has won almost every honour in the game, including the World Championship and the European Championship four times. During the 1965 World Championship he was accused by the Americans of cheating but was eventually exonerated by an independent inquiry.
Reeve, any of several kinds of English medieval officers. Each lord of a MANOR had a local representative to look after the daily work on his farms. He was called a reeve and was usually a VILLEIN, elected by his fellows. There were also shire-reeves in the shires and port-reeves in the towns.
Reeves, Jim (1924–64), US popular singer. After periods as a baseball player and a radio station manager, he began making records in 1949, and had his first success with *Mexican Joe* (1952). He soon moved away from the pure COUNTRY AND WESTERN MUSIC, and his greatest successes *Four Walls, He'll have to go* and *Distant Drums* are pop ballads. He died in a plane crash and his records were subsequently very successful.
Reeves, William Pember (1857–1932), New Zealand politician. He was Minister of Labour in New Zealand's first Liberal government in 1891. He introduced the world's first legislation providing for conciliation and arbitration in industrial disputes (The Industrial Conciliation and Arbitration Act of 1894).
Reference frame, *See* FRAME OF REFERENCE.
Referendum, political process in which legislation or constitutional proposals are put before all voters for approval or rejection. This direct form of voting was known in Greece and other early democracies and is widely used today in certain countries, most notably Switzerland. In Great Britain a referendum was used for the first time in 1975, when the British people voted to remain in the EEC. *See also* PLEBISCITE.
Refinery, any industrial plant based on processes such as DISTILLATION and ELECTROLYSIS. Spirit drinks such as whisky, foods such as SUGAR, and materials such as ores, metals or PETROLEUM are produced, purified or separated into their components in refineries. *See also* MM pp.21, 30, 34, 80–81.
Reflation, economic recovery from a depression before full employment and rising prices are arrived at. It is characterized by the easing of credit restrictions to encourage expansion and other measures designed to increase the amount of money in circulation. This in turn creates a demand for goods and services which contributes to full employment.
Reflecting microscope. *See* MICROSCOPE.
Reflection, change in direction of part of a wave motion when a wave, such as light, encounters a surface separating two different media, such as air and metal, and is "bounced" back into the original medium. The remaining part passes into the second medium. The incident wave, reflected wave, and the line perpendicular to the surface at the point of incidence (the normal) all lie in the same plane; the incident and reflected waves make equal angles with the normal. *See also* SU pp.100–101.
Reflex action, simple type of unlearned and involuntary behaviour, for example, the "knee-jerk" reflex that occurs when a part of the bent knee is struck. At least two NEURONS (receptor and effector) must be involved in the reflex arc, although one or more intervening neurons (interneurons) are often present. Additional nerve impulses travel from the arc to the brain via the CENTRAL NERVOUS SYSTEM. *See also* MS pp.38–39, *38–39*.
Reflex camera, camera which allows the user to view and focus through the lens of the camera. A plane mirror and prism reflects the scene from through the lense on to a ground glass screen. A single-lens reflex (SLR) camera eliminates PARALLAX, which is a problem with viewfinder cameras. When the shutter is pressed the mirror, which is hinged, moves away and light through the lens is able to reach the film. A twin-lens reflex (TLR) camera has two sets of lenses. One set, for viewfinding, involves a mirror and the other, focused in parallel, passes light directly on to the film. *See also* MM pp.218, *218*.
Reflex, conditioned, learned pairing of a response to an artificial stimulus. The connection between a stimulus, such as meat to a dog, and response on the part of the dog, salivation, is known as a stimulus-response (SR) arc. By associating some other stimulus, such as a bell's ringing, with the original stimulus, a conditioned reflex may be established, where the dog will salivate at the presentation of the new stimulus, the sound of the bell. This leads to a new SR arc and this and such-like experiments were made by the originator of classical conditioning, Ivan Petrovich PAVLOV. *See also* MS p.146.
Reformation, The, 16th-century reform of the universal Catholic Church which resulted in the formation of the Anglican and Protestant Churches. More than merely a revolt against the authority, ecclesiastical and doctrinal, of the Church, it also represented a protest by many theologians and scholars against the many questionable activities of the contemporary clergy. The influence of Martin LUTHER from the 1520s was significant, and the effect of the Reformation was felt first in Germany then in Scotland and England, Scandinavia and Switzerland, and finally some of France, but hardly touched Italy or the Iberian peninsula. It took two main forms: Lutheran (emphasizing justification, or salvation, by faith alone) and Calvinist (after John CALVIN, who emphasized predestination); Lutheranism predominated, except in Calvinist Scotland and Switzerland, and in Anglican England. All the REFORMED CHURCHES established Church organizations independent of the pope. *See also* HC1 pp.268–273.
Reform bills, name given to three British measures that liberalized representation in Parliament in the 19th century. The Reform Bill of 1832, enacted under the Whig administration of Lord GREY, redistributed seats in the interest of larger cities, such as Manchester, that had hitherto been unrepresented; it also extended the franchise in the boroughs to those who occupied premises of an annual value of £10 and simplified registration and voting procedure (*see also* HC2 pp.160, *160*, 164). Benjamin DISRAELI's bill of 1867, by lowering property qualifications, enfranchised the urban working class, thereby almost doubling the electorate (*see also* HC2 pp.160, *160*, *161*, 164). In 1884 William GLADSTONE's bill added 2,500,000 new voters to the list by enfranchising most agricultural workers (*see also* HC2 pp.113, 161, 164). The franchise was not extended to women until the REPRESENTATION OF THE PEOPLE ACT (1918) granted suffrage to women over 30. Universal suffrage was not granted until the Representation of the People (Equal Franchise) Act of 1928.
Reformed Church, any Christian denomination that came into being during the REFORMATION by separating, as a congregation, from the old universal Catholic Church. These denominations are generally called PROTESTANT Churches, although the CHURCH OF ENGLAND, the Anglican Church, while certainly Reformed, should not technically be included within that term. More specifically, Reformed Churches are those Churches which adopted the CALVINIST faith and organization in preference to the LUTHERAN. When the Congregational Church of England and Wales and the Presbyterian Church of England combined in 1972 they took the name of the United Reform Church.
Reform Judaism, one of the three divisions in modern JUDAISM. It is also called Progressive or Liberal Judaism, and originated in 19th-century Germany. Israel Jacobson, one of the founders, made changes in rituals and used the vernacular for some prayers. Its main tenet stresses the progressive nature of revelation.
Refraction, change in the direction of a wave, such as a light wave when it crosses the boundary between two transparent media, such as air and glass, and undergoes a change in velocity. The incident wave, refracted wave, and the line perpendicular to the surface at the point of incidence (the normal) all lie in the same plane. The incident and refracted waves make an angle of incidence, i, and an angle of refraction, r, with the normal. The new direction of motion is given by Snell's Law, which states that the ratio of the sines of these angles, $\sin i : \sin r$, always has the same value for two particular media. This constant value is the refractive index for the two media. *See also* SU p.100.
Refractive index, measure of the ability of a transparent medium, such as glass, to refract (bend) light. It is the ratio of the speed of light in a vacuum to its speed in the medium and is calculated by dividing the size of the angle of refraction into the size of the angle of incidence of the incoming light.

Sir Carol Reed was a key figure in the British film industry after WWII.

Reedbucks can swim well and may take to water in order to escape pursuit.

Referendum; in 1975 more than 3 million votes had to be counted in London.

Reformation of the Church in England led to the suppression of the monasteries.

Regency style architecture in Cumberland Terrace near Regents Park, London.

Max Reger composed many works for the organ as well as more than 300 songs.

Reims, in whose cathedral the kings of France used to be crowned.

Reindeer provide for nearly all the material needs of the people of Lapland.

Refractometer, instrument for measuring the REFRACTIVE INDEX of transparent materials, ie the degree to which they bend light that passes through them. Since the refractive index of a chemical solution changes according to its composition, refractometers are frequently employed in chemical analysis.

Refractory material, substance that withstands repeated heating to high temperatures without melting, vaporizing or crumbling. Refractories include FIREBRICK and SILICA cement, used to make industrial FURNACES, and QUARTZ (a natural form of silica), from which smaller crucibles and laboratory furnace linings are often made.

Refrigeration, process by which the temperature in a refrigerator is lowered. In a domestic refrigerator a refrigerant gas such as AMMONIA or FREON is alternately compressed and expanded. The gas is first compressed by a pump and cooled in a condenser where it liquefies. It is then passed into an evaporator where it expands and boils, absorbing heat from its surroundings and thus cooling the refrigerator. It is then passed through the pump again to be compressed.

Refrigerators, insulated containers, maintained at a constant cool temperature, used for storing and preserving perishable foods. Modern REFRIGERATION is powered by electricity or gas; refrigerators were invented by Ferdinand Carré in 1859 and used to ship meat from Australia to Europe. The domestic refrigerator was first built in 1913 in the USA and marketed in Britain in the early 1920s. *See also* MM p.253.

Refugee, one who leaves his native land because of expulsion or to avoid persecution and seeks asylum in another country. Before the 20th century there was little systematic attempt to help refugees, but in 1921 the LEAGUE OF NATIONS appointed Fridtjof NANSEN as High Commissioner for refugee work, a task that was later taken on by the NANSEN INTERNATIONAL OFFICE FOR REFUGEES and later by the UNITED NATIONS High Commission for Refugees. Recent history has provided many refugee problems for such organizations, including the flight of Jews from Germany during Hitler's rule, the flight of refugees from persecution in Russia and Eastern Europe and the problems of Palestinian refugees in Jordan and Lebanon.

Refuse disposal, collection and disposal of domestic waste and other rubbish. Special vehicles collect household refuse from rubbish bins and carry it to the disposal site. There it may be sorted to reclaim recyclable materials such as glass, metal and paper. The remaining refuse is disposed of either by controlled tipping (sanitary landfill), in which it is finally buried under a layer of soil, or it is burned in special incinerators.

Réfusés, Salon des, art exhibition in Paris, France, in 1863, of artists whose works had been rejected by the conservative jury of the Salon (the exhibition of the French Academy). The exhibitors included CÉZANNE, PISSARRO, WHISTLER and MANET.

Regalia. *See* CROWN JEWELS.

Regency style, British style of art and architecture used from 1811 to 1820, when George IV was Prince Regent, and extended to 1830, the end of his reign. Furniture style and decoration were marked by refined elegance and the attenuation of Classical forms.

Regeneration, biological term for the ability of an organism to replace one of its parts if it is lost. *See also* NW p.101.

Regent, person appointed to rule in place of a monarch, either because the monarch is deemed too young to govern or because he is incapacitated.

Reger, Max (1873–1916), German composer who became a church organist at the age of 13 and was professor of composition at the Leipzig Conservatory from 1907 to 1916. His music is characterized by the complexity of its harmonies and its idiosyncratic modulations, and includes *Serenade for Orchestra* and *Fantasie and Fugue in D Minor* for organ.

Reggae, form of West Indian popular music. It can be traced back to CALYPSO and even New Orleans RHYTHM AND BLUES, but did not come into prominence until the mid-1960s, growing out of rock-steady and ska. It is characterized by a hypnotically repetitive beat.

Regina, capital and largest city in Saskatchewan province, Canada, 575km (375 miles) W of Winnipeg. The city was founded in 1882, and was the capital of Northwest Territories from 1882 to 1905, when the province of Saskatchewan was formed. Regina is the N headquarters of the Royal Canadian Mounted Police and the site of the University of Regina. An important transport centre and a distribution centre for the surrounding wheat-growing region, Regina has food-processing, oil refining, motor vehicle and printing industries. Pop. (1971) 139,468.

Registrar-General, British government officer, first appointed under the Registration Act (1836) to keep records of baptisms, marriages and deaths of citizens of every religious denomination. He is also responsible for the national census.

Reich, Wilhelm (1897–1957), Austrian psychoanalyst and clinical assistant to Sigmund FREUD (1922–28) He broke with Freud and went on to introduce the concept of "character armour", investigate sexual repression and expound a controversial theory about orgasm. In the USA from 1939 he claimed to have discovered "orgone" energy, a supposed primal force in the atmosphere. For selling "orgone boxes" he was imprisoned in 1956, and he died in gaol.

Reich, name for the German state or commonwealth. There have been three Reichs, the first being the HOLY ROMAN EMPIRE (962–1806), the second Imperial Germany (1871–1918) and the so-called Third Reich, the period of NAZI rule (1933–45).

Reichenbach, Hans (1891–1953), German-born US philosopher. Closely associated with the development of LOGICAL POSITIVISM, he also contributed to the study of probability and INDUCTIVE LOGIC. He edited the *Journal of Unified Science* with Rudolph CARNAP and his works include *Elements of Symbolic Logic* (1947) and *The Rise of Scientific Philosophy* (1951).

Reichstag fire (1933), conflagration that destroyed part of the German Reichstag building (the lower house of parliament). A Dutchman, Marinus van der Lubbe, was convicted of the crime; HITLER used the incident to eliminate the Communists.

Reichstein, Tadeus (1897–   ), Swiss chemist and pharmacologist, b. Poland. After teaching chemistry at the Institute of Technology in Zurich from 1922 to 1938 he became head of the department of pharmacy at the University of Basel. He shared the 1950 Nobel Prize in physiology and medicine with Edward KENDALL and Philip S. HENCH for his work on the hormones of the cortex of the ADRENAL GLANDS. In 1933 Reichstein became the first to synthesize ascorbic acid (vitamin C).

Reid, Sir George Houston (1845–1918), Australian politician. He led the free trade group in the federal parliament, and was Prime Minister of New South Wales from 1894 to 1899. From 1904 to 1905 he was Prime Minister of Australia.

Reid, John Richard (1928–   ), New Zealand cricketer who led his country to its first-ever Test Match victory (over the West Indies in 1956). He was captain in 34 of his 58 Tests between 1949 and 1965 (scoring 3,431 runs and taking 85 wickets). He played for Wellington and Otago, and in the 1962–63 season hit a world-record 15 sixes in an innings.

Reid, Thomas (1710–96), Scottish philosopher who taught at the Universities of Aberdeen and Glasgow. He was the founder of the common sense school of philosophy, also known as the Scottish school, whose theory is known as COMMON SENSE REALISM. His works include *An Inquiry into the Human Mind on the Principles of Common Sense* (1764).

Reidy, Affonso Eduardo (1909–64), Brazilian architect who worked with LE CORBUSIER on the Ministry of Education building (1937–43) in Rio de Janeiro. His other commissions include the Pedregulho estate (1948–50) in Rio de Janeiro.

Reigate and Banstead, county district in E SURREY, England; created in 1974 under the Local Government Act (1972). Area: 99sq km (38sq miles). Pop. (1974 est.) 115,600.

Reign of Terror, phase of the FRENCH REVOLUTION, from the overthrow of the GIRONDINS in the CONVENTION by the alliance of the JACOBINS and the Parisian SANS CULOTTES in June 1793 to the overthrow of ROBESPIERRE in July 1794. The Terror was so called because of the authoritarian and often ruthless methods the government used to resist invasion and widespread rebellion.

Reims (Rheims), city in NE France, on the River Vesle, 136km (83 miles) ENE of Paris; a port on the Aisne-Marne Canal. Reims Cathedral (built 1211–c.1320) is a noted example of French Gothic architecture. During WWI much of it was destroyed, but it was restored between 1927 and 1938 with funds from the Rockefeller Foundation. Germany's unconditional surrender was signed in Reims on 7 May 1945. The city has a university (1547). Reims is the centre of the champagne industry, and other industries include woollen goods and glass. Pop. (1971 est.) 163,500. *See also* HC1 p.200., *200*

Reincarnation, also called transmigration, rebirth, or metempsychosis, the passage of the soul through successive bodies, the process of existence thereby being infinite. In HINDUISM and BUDDHISM karma (one's earthly conduct) determines the condition into which one is next born. The successive states of rebirth (samsāra) can only end if the soul is released by knowledge or ardous effort (Hinduism) or by union with the universe (Buddhism).

Reindeer, large DEER of northern latitude which ranges from Scandinavia across Siberia to North America, where it is known as the caribou. Its adaptations to the cold climate include thick fur (which is usually grey-brown, becoming lighter underneath) and broad hoofs that help to spread the animal's weight on the snow. It stands up to 1.4m (4.6ft) tall at the shoulders, and feeds on grasses and saplings in the summer and lichens (reindeer moss) it finds beneath the snow in the winter. It is domesticated for meat and as a pack animal by the Lapps. Both sexes carry antlers. Species *Rangifer tarandus. See also* NW pp.*49, 163, 207, 207.*

Reinforced concrete, type of strong concrete widely used in bridge construction. It is set on embedded steel rods, bars or mesh, which and tensile strength. Reinforced concrete can withstand heavy stresses over considerable spans. *See also* PRESTRESSED CONCRETE; MM pp.*49, 49, 183, 192.*

Reinforcement, in psychology, any event following a behavioural response that makes the response more or less likely to happen again. Positive reinforcement is roughly synonymous with "reward", whereas negative reinforcement is "punishment". Reinforcement is often advocated by behaviourist psychologists, notably B. F. SKINNER. *See also* MS p.153.

Reinhardt, Jean-Baptiste ("Django") (1910–1953), Belgian jazz guitarist. Born a gypsy, he blended folk music with JAZZ and SWING styles and he is noted for his guitar improvisations. In 1934 he formed a quintet with the violinist Stéphane Grappelly. In the late 1940s he played in the USA with Duke ELLINGTON.

Reinhardt, Max (1873–1943), Austrian actor and director, real name Max Goldmann. He specialized in spectacular productions in arenas such as Olympia, London, and the Berlin Schauspielhaus, and attempted to involve audiences completely in each performance by using the theatre's flexible mechanical resources to the full and controlling mass crowd scenes with great skill. He was equally capable of directing small-scale presentations.

Reis, Johann Phillip (1834–74), German scientist who c.1861 invented a device which transmitted sound electrically, ie an early telephone. Its capacity to reproduce speech was limited; the first practical tele-

phone was patented by Alexander Graham BELL in 1876. *See also* MM p.*226*.
Reisz, Karel (1926–), British film director, *b.* Czechoslovakia, who was originally a film critic. His films include *Saturday Night and Sunday Morning* (1960), *Morgan: A Suitable Case for Treatment* (1966) and *Isadora* (1968).
Reith, Sir John Charles Walsham, 1st Baron of Stonehaven (1889–1971), general manager of the BBC from 1922 and then its Director-General (1927–38). He was the main architect of the pattern of publicly owned but independent corporations in Britain. He became chairman of British Overseas Airways Corporation (BOAC) in 1939 and was Director of Combined Operations Material at the Admiralty (1943–45). From 1946 to 1950 he was chairman of the Commonwealth Telecommunications Board, leaving to become chairman of the Colonial Development Corporation (1950–59).
Rejection, in the grafting of animal body tissues, the failure of the new tissue to grow as a part of the organism. It occurs when the ANTIGENS of the donor and recipient are incompatible; the LYMPHOCYTES, the main white blood cells that cause immune reactions, migrate to and damage the organ and the muscle areas that surround it. In TRANSPLANT surgery involving whole organs, rejection is the main cause for failure of an operation. In an attempt to combat this, the patient is given IMMUNOSUPPRESSIVE DRUGS to control the reactions of the IMMUNE SYSTEM and to prevent rejection of the graft.
Rejlander, Oscar Gustav (*c.* 1813–1875), Swedish art photographer working chiefly in London. His reputation was established in 1857 with an ambitious pseudo-painting, *Youth and Age (The Two Ways of Life)*, a composite print made with 16 groups of models and 30 separate negatives.
Rejuvenation, being restored to youth. There is no known means of rejuvenation, although the lifespan of certain animals (eg female octopuses) can be extended by severing the glandular links that cause death at the normal completion of the life-cycle. Humans have no such system; all attempts to discover a means of rejuvenation has so far failed.
Relative density, *See* SPECIFIC GRAVITY.
Relative wind, in aviation, air moving past the AEROFOIL with respect to its attitude or position. The relative wind is always parallel and opposite to the flight path and equal in velocity to the speed of the aerofoil through the air.
Relativity, theory which linked the concepts of space and time, proposed in two parts by Albert EINSTEIN. The special theory of relativity, published in 1905, dealt with events as they are measured by observers in a state of uniform relative motion (constant VELOCITY) with respect to each other. Each observer of an event is said to have a frame of reference, a basic concept in relativity. This frame is the base-line from which physicists make measurements of various physical phenomena, such as velocity, acceleration and force. The relations of space and time underlie all these phenomena. NEWTON'S LAWS seemed to establish the absolute nature of space and time, inasmuch as each measurement in a particular frame was independent of any made in another, and that to transfer between the two merely required an allowance for the velocities between the two frames of reference. In 1887 the MICHELSON-MORLEY EXPERIMENT showed that there was no difference between the velocity of light as measured in the direction of the Earth's motion and its velocity at right-angles to this direction. In other words, it did not change with a new frame of reference travelling at a different velocity. To explain this the Lorentz-Fitzgerald transform was devised, which allowed for the experiment's results by postulating that an object's length contracts as its velocity increases (in the direction of motion), an effect appreciable only at velocities approaching that of light. Einstein postulated that not only does an object's length contract, but its sense of elapsed time

expands, with the result that the velocity of light remains constant for that object. Another consequence is that an object's mass increases appreciably as it nears the velocity of light.
In the general theory of relativity, published in 1916, Einstein dealt with the transforms relating non-uniform (accelerating) frames of reference. This showed the relation of space and gravitation. It may be put as the idea that in the presence of a gravitational field space becomes curved. The existence of BLACK HOLES is postulated as a consequence of this. *See also* SU pp.106–107.
Relay, electromagnetic switch that is operated (opened or closed) by variations in an electric current in a coil of wire. Current flowing through the wire coil sets up a magnetic field which moves a metal contact that can open or close another circuit. *See also* MM p.119; SU pp.*191*, *121*.
Relay races, track, field and swimming events involving teams rather than individuals, in which each team member must run or swim an equal portion of the course. In running a baton is passed from one runner to the next at the completion of his part of the relay.
Relief printing, form of printing whereby a print is achieved from a raised, inked surface; the ink-bearing surface stands out above the surrounding non-printing area. The commercial, typographic LETTERPRESS process includes printing from one-piece stereotype plates and from formes which are commonly made up of both text matter and illustrations. In PRINTMAKING, relief processes include linocutting, WOODCUTTING and WOOD ENGRAVING, wherein portions of the surface of the linoleum or wood are cut away if they are not to be inked and printed as part of the design. *See also* LITHOGRAPHY; INTAGLIO; MM pp.204–207.
Relief sculpture, three-dimensional sculpture projecting from a flat background in varying degrees of predominance. The term derives from the Italian noun *rilievo*, mearning relief or projection. In *alto-rilievo* (high relief) the protrusion is great, *basso-rilievo* (bas-relief) protrudes only slightly, and *mezzo-rilievo* is between the two. In a coelanaglyphic relief (*cavo-rilievo*), used in ancient Egypt, the highest surface of the relief is level with the surface of the original stone which surrounds the design.
Religion, code of beliefs and practices formulated in response either to faith in a state of existence after mortal death, or to a desire for union with a supreme and omnipotent spiritual Being, or to a combination of the two. Throughout the history of all cultures some form of religion has been present, generally with some form of sacred literature, and frequently in the worship of many distinct gods or personifications of nature, such as those of ancient Egypt, Greece and Rome. Many such cultures classified their deities into hierarchies – a supreme god or group of gods, with major and minor gods; some modern religions still have pantheons (eg HINDUISM). From within such structures other cults occasionally emerge: one god, even of an originally subsidiary order, may be made the focus of new devotional rites (eg the Dionysiac mysteries of Greece, and the worship of MITHRAS, introduced from Persia during the early Christian era). Other ancient religions, some of which also incorporated belief in a state of existence after death, were more a system of ethical philosophy concentrating on metaphysical contemplation (eg BUDDHISM and TAOISM).
The ancient Israelites were among the first to worship a single omniscient Being; their record of this forms the basis of the OLD TESTAMENT of the Bible. Common to all religions dominated by an omnipotent force (monotheistic religions such as JUDAISM, ISLAM and CHRISTIANITY) is the ascription to it of the two qualities of immanence and transcendence: that the power is all places at once, and that it is beyond the physical plane occupied by humans. Also common to monotheistic religions is the notion of sacrifice, either in

propitiation (appeasement) or to redeem the faithful from what is considered a condition of sin; devotional alms-giving and charity are another consequence of this notion. *See also* MS pp.216–219, 222–225.
Religious orders, term applied to groups of Christian clergy, such as monks, nuns and friars, who have bound themselves together by similar vows and aims. Roman Catholic examples include the BENEDICTINES, CARMELITES and JESUITS. The OXFORD MOVEMENT revived Anglican interest in religious orders; the emphasis of those established was on social service as well as on contemplation and prayer.
Religious toleration, state policy which allows members of all religious sects to practise their religion without hindrance and to enjoy full civil rights and liberties. In Europe it is largely a development of the 18th and 19th centuries. In Britain the last obstruction to it was removed in 1886, when an atheist, Charles BRADLAUGH, was allowed to take his seat in Parliament. *See also* TOLERATION, ACT OF.
REM. *See* RAPID EYE MOVEMENT.
Remak, Robert (1815–65), German physician and physiologist who laid the foundation of modern EMBRYOLOGY. In 1842 he simplified Karl Baer's four germ layers to three, which he named ectoderm, mesoderm and endoderm, and showed their significance in embryological development. In 1838 he became the first to discover that sympathetic NERVE fibres are non-myelinated and that the axons of peripheral nerves arise from the SPINAL CORD. In 1844 he identified neurofibrils in NEURONS, the existence of six distinct cell layers in the CEREBRAL CORTEX and the sympathetic ganglia in the heart – Remak's ganglia. He was also a pioneer in the use of electrotherapy for treating nervous disorders.
Remarque, Erich Maria (1898–1970), German novelist. A wounded veteran of WWI, he settled in Switzerland in the 1930s when his books were banned in Germany. His novels, including *All Quiet on the Western Front* (1929), *The Road Back* (1931) and *Flotsam* (1941), are about war and post-war adjustment.
Rembrandt (Harmensz van Rijn) (1606–69), Dutch painter and graphic artist, prodigious creator of more than 600 paintings, about 300 etchings and nearly 2,000 drawings, many regarded as masterpieces. A miller's son, he was a pupil of Pieter LASTMAN in Amsterdam in 1624. His early, small, realistic works show the influence of CARAVAGGIO. During his Leiden period (1625–31) he began to paint self-portraits which were to number almost 100. He moved to Amsterdam in 1632, where he spent the rest of his life and became highly regarded as a painter of such group portraits as the *Anatomy Lesson of Dr Tulp* (1632). His marriage to Saskia van Uylenburch, the daughter of a wealthy burgomaster, widened him social connections and brought him numerous commissions. With his new prosperity he began to collect works of art, especially ornate costumes which he often used in his portraits. By 1636 he was painting in the richly detailed BAROQUE style typified by *The Blinding of Samson* (1636) and the *Sacrifice of Abraham*. But a few years later a wave of misfortune assailed him. He ran into financial trouble with the purchase of an expensive house, and his wife died (1642) after the birth of their only surviving son, Titus. That year he finished his most famous group portrait *The Corporalship of Captain Frans Banning Cocq's Civic Guards*, better known as *The Night Watch*. When the painting was cleaned in 1946–47 it was shown to have a daylight setting. During the 1640s his interest centred on portraying inner character rather than outward appearances, an unfashionable pursuit at the time. By 1656 he was bankrupt and had withdrawn from society, and he began to paint for himself, choosing religious subjects and the old and downtrodden, for whom he had a life-long sympathy. During these later years he produced some of his greatest works, such as *Jacob Blessing the Sons of Joseph* (1656), *The Jewish Bride*

Lord Reith was a skilful negotiator in overcoming objection to a young BBC.

Relativity, introduced by Albert Einstein, revolutionized concepts of space and time.

Erich Remarque, called the Father of the Best-Seller, became a US citizen in 1947.

Rembrandt's *The Jewish Bride* shows the sensitivity which marks his finest work.

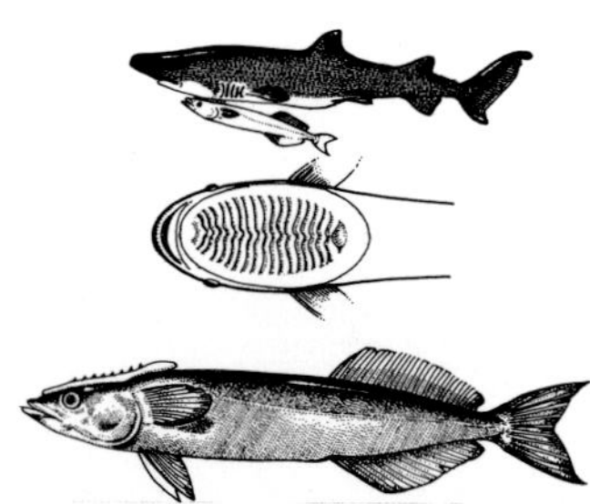

Remoras feed on the scraps left by the host, to which they are attached.

Renaissance arts prospered in Florence under the patronage of Lorenzo de' Medici.

John Rennie built Southwark Bridge, as well as others and docks on the Thames.

Renoir often portrayed young girls and children; this is *The Harvest*.

(*c.* 1669) and *The Return of the Prodigal Son* (*c.* 1669). *See also* HC1 pp.310–311.

Remembrance Sunday, second Sunday of November, the annual commemoration in Britain of the armed forces who died in WWI and WWII. After WWI the dead were remembered on Armistice Day, 11 November, the date of the WWI ceasefire. Remembrance Sunday was introduced in Britain after WWII. In 1956 it was fixed as the second Sunday of November.

Remington, Philo (1816–89), US inventor. He entered his father's small-arms factory and was its president from 1861 to 1865, when he separated the agricultural implement works from the gun factory (which he controlled until 1889). During this time he perfected the Remington breech-loading rifle and began to manufacture typewriters and sewing machines.

Remizov, Alexei Mikhailovich (1877–1957), Russian writer. After his exile for political agitation in 1897, he returned to Moscow (1917–21). In 1923 he settled in Paris. His early stories, *Chasy* (1904; tr. *The Clock*, 1924), dealt with suffering and humiliation, and much of his work returns to the same themes. His works, *Flaming Russia* (1927), which shows the Russian Revolution through the medium of dreams, and *In the Pink Glimmer* (1952), have influenced much of subsequent Soviet writing.

Remonstrants, revisionist Dutch CALVINIST group condemned at the Synod of Dort (1618–19). Followers of ARMINIUS, their position was defined by the Remonstrance (1610) presented to the States General of The Netherlands. Their theological ideas were, however, unacceptable to the Reformed Church of the Netherlands, from which their ministers were expelled. They were persecuted less intensely after the death of MAURICE of Orange in 1625, but did not receive full toleration until 1795. The movement is still strong.

Remora, marine fish found throughout the world in tropical and temperate seas. Grey, reddish, or brown, it has a ridged sucking disc on the top of its head which it uses to cling to sharks, turtles, rays, whales and sometimes boats. Length: 18–91cm (7–36in). Family Echeneidae.

Remsen, Ira (1846–1927), US chemist who, with Constantin Fahlberg, in 1879 discovered saccharin and formulated the law that bears his name concerning the oxidation of methyl groups. At the age of 30 he became the first professor of chemistry at Johns Hopkins University.

Remus. *See* ROMULUS AND REMUS.

Renaissance, The, period of European history lasting roughly from the mid-15th century to the end of the 16th century. The word means rebirth and was used by scholars in late 15th-century Italy to describe the revival of interest in classical learning and classical principles in art and architecture. The Renaissance was immeasurably helped by the fall of Constantinople to the Ottoman Turks in 1453, which resulted in the transport of classical texts to Italy and closed Europe's traditional trading routes to the East; and by the invention, in Germany, of a printing press with moveable type, which greatly assisted the diffusion of the new scholarship. For the Renaissance meant the re-introduction of secular and humanist ideas into a Europe long dominated by the ideology of Christendom united under the Pope and the Holy Roman Emperor. In religion the spirit of questioning led to the Reformation. In politics the Renaissance saw the rise of assertive sovereign states – Spain, Portugal, France and England, and the expansion of Europe beyond its own shores, with the building of trading empires in Africa, the East Indies and America, an expansion led by Portugal and Spain. The growth of a wealthy urban merchant class led to a tremendous flowering of the arts, as city merchants became patrons of artists, jewellers, sculptors and musicians. *See also* HC1 pp.246–284.

Renaissance arts, style in the arts that emerged in Italy in the 15th century, heavily influenced by classical Greek or Roman models and by the new spirit of HUMANISM. While the Renaissance produced its most spectacular artistic achievements in painting, sculpture and architecture, the movement was also apparent in literature, music and dance. Italy was the centre of most of the earliest developments. Many of the greatest Renaissance artists notably MICHELANGELO, the giant of Renaissance art, worked successfully in several different fields.

In painting, the decisive differences between Gothic and Renaissance painting emerged in Florence early in the 15th century, although these had been foreshadowed by the work of GIOTTO almost a century before; these differences included the development of PERSPECTIVE, a new interest in composition and colour harmonies, the increasing use of secular or pagan subject matter as patronage spread beyond the Church, the rise of portraiture, constant experimentation to develop new skills, and a growing concern for the expression of the individual artist. The creators of High Renaissance painting were LEONARDO, Michelangelo and RAPHAEL. The ideas of the Italian artists were taken to France and N Europe and emulated with national variations.

In architecture, Gothic vaulting, pointed arches and spires were replaced by pilasters, arches and columns, and domes. Classical models were adapted for both decoration and composition, and a new concern arose for overall harmony and spaciousness. The dome of Florence cathedral, designed by BRUNELLESCHI in the 1420s, initiated the style. It was developed by ALBERTI, but only reached its fullest form early in the 16th century. The buildings of PALLADIO were particularly highly regarded throughout Europe, although the style did not reach England until the 17th century.

Renaissance literature, which thrived on both scholarship and formal poetry, found an early exponent in PETRARCH. Closely linked with the spread of humanism and with the cult of the courtier, it inspired much poetry and history writing and culminated, in England, in the dramas of Shakespeare.

In music, changes took place in the form of composition as HARMONY took precedence over melody as till then largely exemplified by PLAINSONG. New instruments were developed, increasing the range and variety of musical tones. *See also* HC1 pp.246–265.

Renan, Joseph Ernest (1823–92), French historian, philosopher and philologist. Once a trainee priest, he left the Catholic Church in 1845 and became a rationalist Christian historian and Hebraist who caused controversy by suggesting that Jesus was merely "an incomparable man". His works include *Vie de Jésus* (1863) and *La Réforme intellectuelle et morale* (1871).

Renault, Louis (1843–1918), French jurist who shared the 1907 Nobel Peace Prize with Ernesto T. Moneta. Renault was professor of international law at Paris from 1881 and in 1890 became counsellor to the French ministry of foreign affairs. He wrote many legal works, including *Introduction to the study of International Law* (1897).

Renault, Mary (1905–), pseud. of Mary Challans, British author best known for her fictional works depicting ancient times. *The Last of the Wine* (1956) and *The King Must Die* (1958) have been praised for their authenticity.

Renault, international manufacturing concern owned by the French government since 1945. It manufactures tractors, machine tools and automobiles and in the late 1970s was listed among the 10 leading industrial groups in Europe.

René, France Albert (1935–), politician and lawyer of the Seychelles. He was a founder and the first leader of the People's United Party. In 1976 be became Prime Minister.

René I (1409–80), King of Naples, Duke of Anjou, Bar and Lorraine and Count of Provence. René's court in Provence saw a remarkable flowering of late medieval culture.

Renfrew, county town of the former county of Renfrew(shire), W central Scotland, on the River Clyde; now in Strathclyde Region. Renfrew has been an important port since the 12th century and has shipbuilding, rubber, paint and soap industries. Pop. (1971) 18,589.

Renfrewshire, former county in SW Scotland, on the Firth of Clyde; since 1975 it has been part of Strathclyde Region. It was one of Scotland's smallest but most populous counties, and is an important industrial region. Shipbuilding, engineering, electric cables and food processing are. the principal industries. Renfrew was the county town, but the administrative centre was Paisley. Area: 583 km (225sq miles). Pop (1971) 362,144.

Reni, Guido (1575–1642), Italian painter who worked in Bologna and Rome. His paintings were influenced by Lodovico CARRACCI, with whom he studied. After Carracci died in 1619, Reni became the leading master of Bolognese art. His powerful composition made him much sought after as a painter. His most celebrated works include *Massacre of the Innocents* (1611), *Aurora* (1613) and *Atlanta and Hippoinenes* (*c.* 1625).

Renin, polypeptide HORMONE produced by the KIDNEY in response to lowered blood pressure. It activates angiotensin (a powerful VASOCONSTRICTOR), which reduces venous and arterial volume with a subsequent rise in blood pressure.

Rennell, James (1742–1830), English cartographer, geographer and oceanographer. He was an expert on the geography of W Asia and N Africa, and constructed the most accurate map of India of his time (1783). He pioneered the scientific study of winds and ocean currents.

Rennes, city in NW France, at the confluence of the Ille and Vilaine rivers, 311km (193 miles) WSW of Paris. During the Middle Ages Rennes served as capital of Brittany under the Angevin dukes. The city was destroyed during the HUNDRED YEARS WAR and damaged by fire in 1720. The University of Rennes was founded in 1735. Industries: leather goods, printing, textiles, motor vehicles. Pop. (1968) 180,943.

Rennet, substance used to curdle milk in cheesemaking. It is obtained as an extract from the inner lining of the fourth stomach of calves and other young ruminants, and is rich in rennin, an ENZYME which coagulates the casein (protein) of milk. *See also* PE p.229.

Rennie, John (1761–1821), Scottish civil engineer, builder of bridges and docks on the River Thames in London, including Southwark and Waterloo bridges and the East India and London docks. He also renovated dockyards at Plymouth and Portsmouth.

Reno, city in W Nevada, USA, on the Truckee River. Settled in 1859, the city grew with the construction of the Union Pacific Railroad in 1868. Reno is famous for its free port privileges, legalized gambling and its provision of speedy divorces. It has the University of Nevada (1874) and a rodeo, state fair, and national air race which are held annually. Industries: tourism, mining, meat packing, flour milling, beverages, sheet metals. Pop. (1970) 72,863.

Renoir, Pierre-Auguste (1841–1919), French IMPRESSIONIST painter. His early career in the 1860s was influenced by Gustave COURBET, but by 1868 he and Claude MONET, inspired by MANET's *Le Déjeuner sur l'Herbe* (1863), started painting oat of doors. In 1874 he contributed to the first Impressionist exhibition and masterpieces of this period include *La Loge* (1874) and *Le Moulin de la Galette* (1876). In the early 1880s he became dissatisfied with the Impressionist technique. His interest in the human figure led to an enthusiasm for the Classical style of RAPHAEL with such works as *Bathers* (1884–87) and *After the Bath* (*c.* 1895). After 1900 he was increasingly afflicted by arthritis, but continued to paint with the brush strapped to his hand. *See also* HC2 pp.*118, 119*.

Renoir, Jean (1894–), French film director and actor, son of the painter Auguste RENOIR. His films include *Nana*

(1926), *La Grand Illusion* (1936), *French-Cancan* (1955) and *C'est la Révolution* (1967). *See also* HC2 p.*276*.

Reparations, term used after both world wars for war damage payments, especially those demanded by the victorious ALLIES from the defeated CENTRAL POWERS at the Treaty of Versailles (1919). The USA, however, did not ratify the treaty and waived all reparation claims.

Repeating rifle, firearm developed in the 1880s. Bullets stored in a magazine were introduced into the breech by a simple mechanical action, so that more then one round could be fired without reloading.

Repellent, animal, natural or artificial substance that is obnoxious to animals and causes them to move away. Repellents are produced by many animals themselves, especially insects, usually to warn off predators.

Repertory, theatrical presentation of a variety of plays in turn by a permanent ensemble rather than the more common practice of playing the same work each evening until the end of its run. In Britain that practice has rarely been properly realized except by the large state-subsidized ROYAL SHAKESPEARE COMPANY and the NATIONAL THEATRE. The term "repertory" or "rep" is also used to describe small provincial companies who perform different plays for one or two weeks at a time.

Repin, Ilya (1844–1930), Russian painter whose work was influential in bringing REALISM to Russian painting. *The Volga Boatmen* (1870–73) brought him international fame.

Replacement reaction, chemical reaction in which one atom of a molecule is replaced by another atom. A hydrogen atom in methane, CH_4, for example, may be replaced by a chlorine atom to yield chloromethane, CH_3Cl.

Report on the Sanitary Condition of the Labouring Population (1842), document by Edwin Chadwick which advocated that water supplies and the efficient disposal of sewage should be made available to the whole population. It is considered one of the best formulated and most far-reaching of such reports of its time.

Representation of the People Act (1918), Parliamentary reform act in Britain which gave the vote to women over the age of 30. It redistributed seats, producing constituencies with an average of 70,000 electors and leaving only 10 two-member constituencies. Voting at a general election, hitherto spread over weeks, was to take place on one day.

Repression, process in Freudian psychology whereby the primal and instinctual urges of the id find no outward expression. Unless the repressed desires find an outlet in a sublimated form – such as physical contact – they may emerge in episodes of psychosis.

Reproduction, in art, the copying of a work of art by any of various means. The earliest methods in the 15th century used WOODCUTS and metal ENGRAVING to make reproductions of an original. Prior to this all reproductions were executed by hand. Marcantonio RAIMONDI, in his engravings of paintings by RAPHAEL, perfected that means which was used until about 1800. In the 17th century the invention of MEZZOTINT made possible the fine gradations of tone. The 19th century brought important new developments: the discovery of LITHOGRAPHY combined the steps of drawing and engraving so that multiple copies could be taken from the artist's original design on stone; and the introduction of photography minimized the intervention of the reproductive engraver. The photo-mechanical HALF-TONE PROCESS, developed at the end of the 19th century, is that in most widespread use today, and other processes include photogravure (an INTAGLIO process) and COLLOTYPE, a process which guarantees reproduction of the highest fidelity. *See also* MM pp.205, 205.

Reproduction, the process by which living ORGANISMS – plants and animals (with the exception of certain sterile hybrids such as the mule) – create new organisms similar to themselves. Reproduction falls into two chief classes, sexual and asexual, the first being the fusion of two special reproductive, or sex, cells and the second being the separation of a single organism into two or more organisms. Asexual reproduction is found mostly in PROTOZOA and some lower INVERTEBRATES. Most animals reproduce sexually. They may be (as most are) bisexual, the special having two kinds of individuals – male and female – with different sex functions. The male produces the sperm which fertilizes the female egg (ovum), to produce the ZYGOTE from which a new individual develops. Or they may be HERMAPHRODITIC (like the earthworm and leech), each individual of the species having male and female functions, so that when two of them mate each individual fertilizes the other's eggs. Some animals, such as the TAPEWORM, can fertilize themselves. A few animals fall into a third class, parthenogenetic. Having lost the male sex, they reproduce from unfertilized eggs. Usually, however, such animals have alternating (not always regularly) generations of asexual and bisexual individuals. Examples are the APHID, JELLYFISH and some parasitic worms. Sexually reproducing plants (or generations) are called GAMETOPHYTE; ones which reproduce asexually, SPOROPHYTE. *See also* NW pp.70–71, 76, *109*, 127, 133, 144–145, *156*; MS pp.74–77; ASEXUAL REPRODUCTION; PARTHENOGENESIS; POLLINATION; SEXUAL REPRODUCTION; SPORULATION; REGENERATION.

Reptile, any one of about 6,000 species of VERTEBRATES, including TURTLES, Crocodilians, LIZARDS, SNAKES and the TUATARA, distributed throughout the world. Reptiles are cold-blooded. Most lay shelled, fluid-filled, yolky eggs on land. Some species carry eggs in the body and bear live young. The skin is dry and covered with scales or embedded with bony plates. Reptiles keep their skin clear by rubbing themselves through dry grass or leaves. Their limbs are poorly developed or non-existent. Those with limbs usually have five clawed toes on each foot; this distinguishes them from AMPHIBIANS, which do not have claws. The first reptiles appeared during the Carboniferous Period and flourished during the Mesozoic Era. There are now four living orders: Chelonia (turtles); Rhynchocephalia (tuatara); Squamata (scaly reptiles); and Crocodilia (alligators and crocodiles). *See also* NW pp.73, 134–139, 184–187.

Repton, Humphry (1752–1818), British landscape gardener who initially followed the ideas of Lancelot ("Capability") BROWN but modified his style by re-introducing ordered flower-beds and straight paths close to houses. His 400 or more *Red Books*, containing watercolour sketches for his clients, are a proof of his success. *See also* HC2 p.71.

Republic, state in which sovereignty is vested in the people or their elected or nominated representatives. The word formerly denoted a form of government that was both free from hereditary or monarchial rule and had some type of popular control of the state. Today, in addition to this, a republic may also be understood to be a state in which all segments of society are enfranchised and the power of the state is limited.

Republic, The, philosophical treatise by PLATO in which he describes the working of an imaginary but ideal state. It also serves as an introduction to his thought, containing his theory of Forms, theory of knowledge and views on the proper roles of music and poetry. Divided into 10 books, it is written as a dialogue whose central figure is SOCRATES.

Republican Party, US political party, also known as the Grand Old Party (GOP). It was organized in 1854 as an amalgamation of WHIGS and Free-Soilers with businessmen, workers and professional people who formerly had been known as Independent Democrats, Know-Nothings, Barnburners or Abolitionists. Its first successful presidential candidate was Abraham LINCOLN, who was elected in 1860. After 1932 only two Republicans have been elected president, Dwight EISENHOWER (1952, 1956) and Richard NIXON (1968, 1972), and both had to work with largely Democratic Congresses. *See also* HC2 pp.150, 264, *264–265*.

Requiem, solemn choral service for the dead sung in Christian churches, beginning with the Latin word *requiem* (repose). The chief characteristics are the Introit (or opening section) "Requiem aeternam" and the Dies Irae. There are seven other sections. Many composers have The Dies Irae, a plainsong melody to a 13th-century poem, is a long part of the service and is often divided into sections.

Reredos, in church architecture, an ornamental wall or screen behind the high altar. Originally a moveable hanging tapestry, it later became adorned with elaborate sculpture, tracery or painting. *See also* TRIPTYCH.

Resale price maintenance, trading practice whereby the manufacturer specifies at what price the retailer should sell his goods. Resale price maintenance was abolished in Britain in 1964 since it was seen as an infringement of the shopkeeper's right to compete for custom. The only exceptions are those goods for which the manufacturer can demonstrate that the public interest is served by resale price maintenance (notably books and medicaments) and those goods that are subsidized by government funding.

Research and development, or R & D, systematic scientific investigation and experimentation to make new discoveries, prove or disprove theories, and find practical applications for ideas. Basic R & D refers to the search for new knowledge without specific direction; applied R & D is concerned with the practical laboratory refinement of new processes and products; developmental R & D transfers the knowledge out of the laboratory and into the manufacture of useful products.

Reserpine, trade name Serpasil, derivative of RAUWOLFIA used to treat HYPERTENSION by acting either as a sedative or by reducing cardiac output. Its side-effects include serious depression. *See also* MS pp.118, *118*, 144, *144*.

Reserves, in economics, the portion of deposits that a commercial bank has in its physical possession at any given time to meet the withdrawal demands of its depositors. Commercial banks use only fractional reserves, to limit the amount of capital they must hold.

Reservoir, man-made storage place for water. Reservoirs are usually formed by dams, but sometimes pipelines are used to carry water from rivers to natural depressions in the land-surface. Sedimentation causes most reservoirs to become useless after several decades; few last more than 100 years. *See also* MM pp.198–200.

Reshevsky, Samuel Herman (1911–), US chess player, *b.* Poland. A child prodigy, he toured Europe in a series of exhibitions at the age of six. In 1920 he emigrated to the USA, where he was national champion five times between 1936 and 1946.

Resin, or rosin, viscous substance secreted by various plants. It is impermeable to water and when present in large amounts (eg in pine) makes wood resistant to rot and weather. Oleoresin secreted by conifers is distilled to produce turpentine; rosin – a yellow, brown or black material – remains after the oil of turpentine has been distilled off.

Resistance, in an electric circuit, the ratio of the potential difference (voltage) to the current. It is measured in ohms and represents the opposition to the flow of ELECTRONS in a conductor, the electrical energy being converted into heat. By OHM'S LAW, $R = V/I$, where R is resistance, V is voltage and I is current. In an alternating current circuit, resistance is the real part of the impedance. *See also* SU pp.124–131.

Resistance, psychological, in PSYCHOANALYSIS, the resistance of the patient to the process of bringing unconscious elements into consciousness. This is usually taken to refer to the patient's reluctance to accept the analyst's interpretations. It may also describe the mechanism of the SUPEREGO which prevents repressed material from becoming conscious.

Reproduction; some photocopiers reproduce in colour and can enlarge or reduce.

The Republic was where Plato (here portrayed by Raphael) introduced Socrates.

Republican Party; Lincoln, the party's great leader, at his inauguration in 1865.

Resin is produced by distillation or by extraction using solvents.

Resistance movements

Resistance; the ruins of the Warsaw Ghetto, which the Germans totally destroyed in 1943.

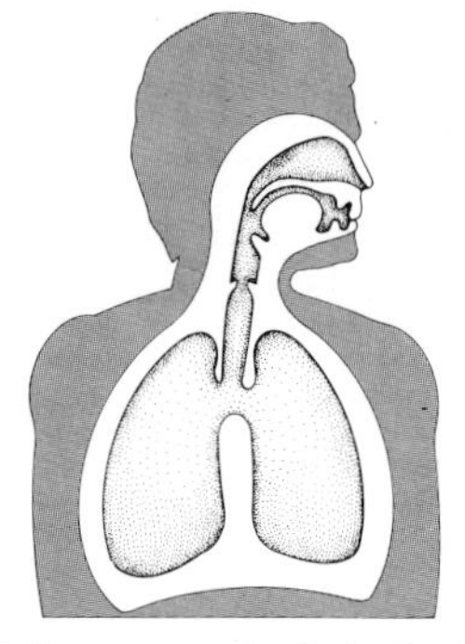

Respiration occurs primarily in the lungs, which receive air via the windpipe.

Restoration of Charles II: a woodcut showing the king's entry into London.

Retrorocket of a space-craft is fired to slow down the vehicle.

Resistance movements, underground organizations which worked against Nazi rule in those countries whose governments had capitulated to German occupation in WWII. They were strongest in Poland, Yugoslavia and France, but they were also helpful to the Allied cause in Belgium, Holland, Denmark and Norway. The most dramatic episodes in European resistance were the two Warsaw risings. The first was the revolt of the Jews in the Warsaw ghetto in April 1943. Before being put down and exterminated, they killed about 1,000 Germans. The second was the equally unsuccessful rising of the "Home Army" in August 1944, to which STALIN refused to lend Soviet support.

In general the resistance movements rendered four great services to the Allies. They carried out acts of sabotage; they acted as sources of information for the Allied intelligence divisions; they helped prisoners of war and crews of crashed aircraft to escape and, especially in France and Yugoslavia, they made the occupying forces maintain large numbers of troops, which they could ill-afford. *See also* MAQUIS.

Resistivity, electrical property (symbol ρ) of materials. Its value is given by $\rho = AR/l$, where A is the cross-sectional area of a conductor, l is its length and R is its RESISTANCE in OHMS. As the temperature of the conductor rises, the resistivity usually also rises. Resistivity is also called specific resistance and is generally expressed in units of ohm-metres.

Resistor, electrical or electronic circuit component that has a specified RESISTANCE. Resistors for electronic circuits usually consist of finely ground carbon particles mixed with a ceramic material and enclosed in an insulated tube. The value of the resistance is denoted by a coded set of coloured rings on the outside of the tube. Resistors for carrying larger currents consist of a coil of insulated wire. *See also* SU pp.124, 126, 130, *130.*

Resnais, Alain (1922–), French film director. He made a number of short films in the late 1940s and 1950s, his first full length feature being *Hiroshima Mon Amour* (1959). Alain ROBBE-GRILLET scripted *Last Year in Marienbad* (1961) which is regarded by many as Resnais' best film. Later work included *Je t'aime, Je t'aime* (1969) and *Stavisky* (1974). *See also* HC2 p.*277.*

Resolutioners, members of the moderate party in the Scottish General Assembly which, in 1650, accepted the resolution rehabilitating those who had not been enemies of the SOLEMN LEAGUE AND COVENANT (1643). The resolution enabled CHARLES II to be crowned king of Scotland.

Resonance, particle, extremely short-lived elementary particle produced in high-energy nuclear reactions occurring in particle accelerators. Such particles decay in a time of the order of 10^{-23} seconds, characteristic of HADRONS. More than 150 have been detected, all since the 1960s. A resonance can be considered as a marked increase in the probability of interaction between two energetic colliding particles, occurring when the combined energy of the particles attains a particular value, the resonance energy.

Resonance, wave, phenomenon in which a mechanical or acoustic system, set into forced vibrations by the application of an external vibration, responds with maximum amplitude. It occurs when the FREQUENCY of the applied force becomes equal to the natural vibrational frequency of the system. The large vibrations can cause damage to the system. *See also* SU pp.77, 83.

Resources, in economics, a country's collective means of support. Economists divide resources into four categories: land, labour, capital and raw materials.

Respighi, Ottorino (1879–1936), Italian composer who studied in Bologna and was later influenced by RIMSKY-KORSAKOV. He is best known for the symphonic poems *Fountains of Rome* (1917) and *Pines of Rome* (1924). He also composed songs, chamber music and operas.

Respiration, exchange of gases between an organism and its environment. The term refers to the overall process by which oxygen is taken from the atmosphere, transported to cells for the oxidation of organic molecules and the return of carbon dioxide and water to the environment. In single-celled organisms the exchange of gases occurs at the cell membrane. In more complex animals specialized systems transfer the gases to and from the internal organs.

In human beings, respiration (or BREATHING) involves the mechanical pumping of air into and out of the lungs. Oxygen in the air diffuses across the capillary membranes of the lung's ALVEOLI into blood capillaries where it combines with the red blood protein HAEMOGLOBIN to form oxyhaemoglobin. This is then distributed throughout the body to the cells. In exhalation the process is reversed. Carbon dioxide diffuses across the cell membrane into the blood and is taken to the lungs, where it passes into the alveoli.

Other types of respiration include AEROBIC and ANAEROBIC. In plants the gas exchange occurs in special organs, the stomata, which are mostly found in the leaves. In fish gaseous exchange occurs across the gill membranes which are supplied with blood vessels. *See also* MS p.66; NW pp.44, *44,* 126, *126.*

Restif, Nicholas-Edme (1734–1806), also called Restif de la Bretonne, French novelist. His familiarity with Paris and its underworld furnished him with vivid material for his prolific output of novels. His best-known works are *Nights of Paris* (1788–90) and *Monsieur Nicholas* (1794).

Rest mass, in RELATIVITY, mass of an object when it is at rest. It is therefore a measure of the INERTIA of the object and is sometimes called the inertial mass. *See also* SU pp.106–107.

Restoration (1660), the return of CHARLES II to the English throne after the COMMONWEALTH and PROTECTORATE had faltered and a strong reaction had set in against the PURITANS and against military control. By the Declaration of BREDA in 1660, Charles II promised religious toleration, a general pardon to all former enemies of the house of Stuart (except regicides) and payment of arrears in salary to the army. He landed in England in May 1660 to general rejoicing. ANGLICANISM was restored but certain civil and religious restrictions were placed upon Roman Catholics and some Protestant sects, such as QUAKERS. *See also* HC1 pp.291–193, *291–293.*

Restoration theatre, plays and performances in the period following CROMWELL and the COMMONWEALTH when, with the RESTORATION of CHARLES II, the theatres were reopened. It was a licentious period and drama reflected the freeness of court morals, by broad satire, farce, wit and bawdy comedy. The most distinguished playwrights of the time were DRYDEN CONGREVE, ETHEREGE and WYCHERLEY.

Restraint of Appeals, Act in (1533), act of the English Parliament that ended the practice of making appeals to the Papacy in disputes over marriages and wills. It made possible HENRY VIII's divorce from CATHERINE OF ARAGON. Its preamble asserted the supremacy of the English Crown in spiritual and temporal affairs.

Restrictive trade practices, in industry and professions, mutual agreements that establish fixed prices or charges. This is usually achieved between companies that sell similar products; professional groups which decide on minimum fees for their services; and manufacturers who "blacklist" retailers for selling goods at lower than the manufacturer's recommended price. The latter practice was declared illegal in Britain by the Restrictive Trade Practices Act (1956), which set up a court to ascertain whether restrictive practices were in the public interest.

Resurrection, the rising of the dead to new life, either in heaven or on Earth. Both Christianity and Judaism hold that at the end of the world there will come the Day of Judgement on which those worthy of eternal joy will be allowed to draw near God, while those unworthy will be condemned to weeping, wailing and gnashing of teeth. Christianity further teaches that all true believers will be raised from the dead with glorified and imperishable bodies.

Resurrection (1889), novel by TOLSTOY. His first full length work after *Anna Karenina* (1877), it is more moralistic than his previous novels. It deals with the repentance of the nobleman, Nekhlyudov, for his seduction of a servant girl.

Resurrection plant, popular name for Rose of Sharon, a plant of the mustard family. It survives dessication by curling up its leaves tightly when dry and unfolding them when moistened. Species *Anastatica hieronochuntica.*

Retail Price Index, governmental measure of changing retail prices in Britain, from which the rate of inflation is calculated. It is based on a constant selection of goods, weighted according to their importance in a family budget. The average cost of a weekly shopping basket is estimated each month from surveys carried out in several parts of the country.

Rethel, Alfred (1816–59), German painter and engraver. Historical paintings such as his frescoes of episodes in the life of CHARLEMAGNE for the Aachen City Hall comprised the bulk of his work. The series of six woodcuts *Another Dance of Death in the Year 1848* (1849) is a masterpiece .

Reticular formation, complex mechanism in the vertebrate CENTRAL NERVOUS SYSTEM, located in the brainstem. It consists of interconnected clusters of nerve cell bodies (grey matter) and is believed to influence many aspects of behaviour, including sleep, blood pressure and coordination. *See also* MS pp.41, *41, 105.*

Retina, inner layer of the wall of the EYE, composed mainly of different kinds of NEURONS (nerve cells), some of which are the visual receptors of the eye. These receptor cells are stimulated by light. Some, known as cones, respond primarily to the spectrum of visible colours; others, known as rods, respond mainly to shades of grey and to movement. The RODS AND CONES connect with sensory neurons that in turn connect with the OPTIC NERVE that carries the visual stimuli to the brain. *See also* MS p.48.

Retinal detachment, separation of the layers of the RETINA in the eye, after which the outermost layer (the pigment epithelium) remains attached to the choroid. It occurs mainly in older people and results in blurred vision. The retina can be restored to its position by draining fluid and applying heat or cold to the wall of the eyeball, or by surgery using a laser beam. *See also* MS p.48; SU p.*110.*

Retirement, period after a person's normal working life. In Britain the usual age of retirement is 65 for men and 60 for women. In some countries, retired people draw pensions from private or public funds. *See also* MS p.*307.*

Retriever, sporting dog originally used only as a water dog to kill or cripple downed game and return it to the hunter; today it is also used to locate game and is a popular pet. It has a water-repelling coat, a keen scent, a soft mouth and swims well. The main breeds include the golden retriever and the Labrador retriever.

Retrorocket, rocket engine used to slow a space vehicle to separate it from another one, place it in a desired orbit, or allow it to make a soft landing.

Reuben, in the biblical book of Genesis, eldest son of JACOB, and the founder of one of the 12 tribes of Israel. He stopped his brothers from murdering JOSEPH by suggesting he should be sold into slavery in Egypt, and later he offered his own two sons as hostages to Jacob as pledges for the safe return from Egypt of Jacob's youngest son, Benjamin.

Reuchlin, Johann (1455–1522), German lawyer and humanist. His *Rudimenta Hebraica* (1506) was the first Hebrew grammar by a Christian. His defence of non-polemical Hebrew classics gained him the fierce opposition of the Roman Catholic Church.

Réunion, island in the Mascarene group in the Indian Ocean, approx. 700km (435 miles) E of Madagascar; St Denis is the capital. Discovered in 1513 by the Portuguese, it was claimed by France in 1638 and used as a penal colony. The island became a French overseas département in 1948. Exports: sugar, rum. Area: 2,510sq km (969sq miles). Pop. (1975 est.) 501,000.

Reuter, Baron Paul Julius von (1816–99), British founder of REUTERS international news agency, *b.* Germany. He went to London in 1851 where he opened a telegraph office. Within ten years, he was supplying British newspapers with overseas information from his worldwide telegraph connections. He was created a German baron in 1871.

Reuters, news agency which transmits international news between its offices in major cities throughout the world. It originated as a service between Britain and the continent, using the telegraph. It is jointly owned by Australian, New Zealand and British newspapers. It was founded by Paul REUTER, in the 1850s.

Revelation, or Apocalypse, last book of the New Testament, written *c.*95 from exile on the island of Patmos by a certain JOHN, sometimes called St John the Divine. He is not generally considered to be identical with John the evangelist although there is controversy over whether he may have been the author of the letters (Epistles) of St John. In highly mystical, allegorical and prophetic terms, he concentrates on depicting the end of Creation, the Day of Judgment.

Revere, Paul (1735–1818), American patriot, silversmith, engraver and printer who is immortalized in a poem by Henry Wadsworth LONGFELLOW for his ride (1775) to warn the Massachusetts colonists that British troops were on the march. He was also a leader of the protest known as the BOSTON TEA PARTY (1773).

Reversible reaction, chemical reaction in which the products can change back into the reactants. Thus nitrogen and hydrogen can be combined to give ammonia, and ammonia may be decomposed to nitrogen and hydrogen. Such processes yield an equilibrium mixture of reactants and products. The equation is written using a double arrow: $N_2 + 3H_2 \rightleftharpoons 2NH_3$. *See also* CHEMICAL EQUILIBRIUM; SU p.146.

Revie, Don (1926–   ), British footballer and manager. An inside-forward, he won FA Cup and England honours (six caps 1954–56) while with Manchester City, where he played as a deep-lying centre forward; he also played for Leicester City, Hull City, Sunderland and Leeds United. As manager of Leeds (1961–74) he took them to domestic and European successes. He was manager of England from 1974 until 1977.

Revolution, agricultural. See AGRICULTURAL REVOLUTION.

Revolution, in a political sense, fundamental and violent change in values, political institutions, social structure and leadership brought about by a large-scale, successful revolt. The totality of change distinguishes it from coups, rebellions and wars of independence, which seek and achieve only particular changes. The term is also used to indicate great economic and technical changes, such as the Industrial Revolution and the Scientific Revolution.

Revolution, orbital motion of a planet or satellite around its primary. A single revolution is the planet's or satellite's "year". *See also* SU p.166.

Revolution, Russian. See RUSSIAN REVOLUTION.

Revolutions of 1848, series of uprisings in western and central European countries, whose object in some places (notably France and Germany) was the establishment of liberal, constitutional regimes; elsewhere (eg, in nations under Austrian rule) national liberation was the aim. Everywhere the ruling authorities – Louis-Philippe in France, Metternich in Austria, Frederick William in Prussia – were forced briefly to give way. But the middle classes, who led the revolutions, were not supported by the peasantry and by the end of 1849 all the revolutions had been crushed. *See also* HC2 pp.108–109.

Revolver, handgun or pistol that can fire several shots without being reloaded, the rounds being fed to the gun barrel from a revolving cylinder. Modern revolvers are essentially similar to the early five-shooters and six-shooters, such as those patented in 1835 by Samuel COLT in the USA, in which pulling the trigger also causes the cylinder to revolve. *See also* MM pp.162–163.

Revolving stage, turntable stage which was invented in Japan in the mid-17th century and introduced into the Western theatre in the late 19th century.

Revue, theatrical entertainment purporting to give a review, usually satirical, of current fashions, events and personalities.

Rex cat, or poodle cat, type of domestic CAT first developed in Britain in 1943. The coat is unusual in that each hair is waved, and guard hairs are short, producing a "permanent wave" effect. There are two main types, Devon Rex and Cornish Rex. The Devon has a shorter coat.

Reyes, Alfonso (1889–1959), Mexican writer. A critic, scholar and poet, he believed that Latin American writers should tackle basic human themes. His writing was collected in *Political Works* (1952) and *Complete Works* (2 vols, 1955–56).

Reyes, N. R. *See* NERUDA, PABLO.

Reykjavík, port and capital of Iceland, on the sw coast of the island. Founded *c.*870, it was the first permanent settlement. The city developed in the 18th century and became capital in 1918. It served as a British and US air base during WWII. Reykjavík has a university (1911), the Althing (parliament) and a National Theatre. Many buildings in the city are heated by means of hot springs nearby. Industries: food processing, fishing, textiles, metallurgy, printing and publishing, shipbuilding. Pop. (1975 est.) 84,900.

Reymont, Władysław Stanisław (1867–1925), Polish novelist whose works were mainly concerned with the detailed lives of workers and peasants. His early novels were written when he was a railway superintendent. He wrote the two-volume *The Promised Land* in 1899 using the novels of Émile ZOLA as models. His best-known work is *The Peasants* (1902–09), a four-volume epic about a peasant's almost mystical attachment to the land. Reymont was awarded the 1924 Nobel Prize in literature.

Reynolds, Sir Joshua (1723–92), English artist and portrait painter. He was the founder and first president of the ROYAL ACADEMY OF ARTS (1768) and one of the leading artistic figures of his time. His fashionable portraits, which include *Nelly O'Brien* (1762), show how the classic manner of the 1760s had gradually evolved into a more relaxed, naturalistic style by the 1780s (*The Duchess of Devonshire and her Daughters* 1786). *See also* HC2 p.71.

Reynolds number, dimensionless quantity characterizing types of fluid flow. For a body of density p and linear dimension l traveling at velocity v in a fluid of viscosity v, the Reynolds number is pvl/v. At low values of the Reynolds number, the fluid flow is laminar, or layered, and is well-understood mathematically; at higher values the flow becomes turbulent and complicated.

Reza'iyeh, city in NW Iran. It was sacked by the SELJUK TURKS in 1184 and later occupied many times by the OTTOMAN Turks. It is also regarded as the traditional birthplace of ZOROASTER. Today it is a market centre for the surrounding agricultural region, which produces fruit and tobacco. Pop. (1971 est.) 120,000.

Reza Shah (1878–1944), King of Iran and founder of the Pahlavi dynasty. He seized power by a military *coup* in 1921, deposed Ahmad Shah in 1925 and was elected Shah in the same year. He abdicated in 1941 after the Anglo-Russian occupation of Iran.

Rhabdomancy, divination by wands or rods, especially the art of discovering ores or water using a divining rod. *See also* DOWSING.

Rhapsody, musical term applied in the 19th and 20th centuries to orchestral works, usually performed in one continuous movement and most often inspired by a nationalist or romantic theme.

Rh disease, also called haemolytic disease of the newborn, anaemic disease of fetuses and of newborn infants due to Rh-factor incompatibility between the mother and the developing embryo. It is marked by destruction of mature red blood cells, presence of many immature red cells, and jaundice. It generally occurs only in second and subsequent pregnancies; severity varies and increases with each successive pregnancy. Exchange transfusions in which compatible blood is substituted in the newborn now save most affected babies. There is also a vaccine that prevents formation of Rh ANTIBODIES in the Rh negative mother. *See also* MS p.76.

Rhea, either of two species of large, brownish, flightless, fast-running South American birds resembling a small OSTRICH. They feed mostly on vegetation and insects. Height: to 1.5m (5ft). Family Rheidae. *See also* NW pp.140, 202, 203.

Rhea Silvia, in Roman mythology, mother of ROMULUS AND REMUS. She was the daughter of Numitor, king of Alba Longa. Forced to become a VESTAL VIRGIN, she nevertheless bore the twin boys by the war god MARS, who later founded Rome.

Rheims. *See* REIMS.

Rhenium, metallic element (symbol Re) of the third transition series, discovered in 1925. It is found in MOLYBDENITE, from which it is obtained as a by-product. It is used in THERMOCOUPLES and various alloys, and is also a useful catalyst. Properties: at. no. 75; at. wt. 186.2; s.g. 21.0; m.p. 3,180°C (5,756°F); b.p. 5,627°C (10,160°F); most common isotope Re^{187} (62.93%). *See also* TRANSITION ELEMENTS.

Rheology, study of the ways in which matter deforms and flows. It includes the investigation of such physical properties as VISCOSITY, ELASTICITY and plasticity (non-elastic deformation). Although fluidity is most widely recognized as a property of liquids and gases, solids also flow to some extent and are included in rheological investigation. Rheological properties are influenced by temperature and pressure, often non-uniformly. Thixotropic liquids decrease in viscosity as they are stirred. These and other rheological phenomena are related to the molecular structure of materials. The science finds its major application in the manufacture of paints and other surface coatings.

Rheostat, device for regulating an electric current by means of variable RESISTANCE. The resistance element may be a metal wire, carbon or a conducting liquid, depending upon the application. Rheostats are used to adjust generator characteristics, to dim lights and to control the speeds of electric motors. *See also* POTENTIOMETER.

Rhesus, medium-sized, yellow-brown MACAQUE MONKEY of India. Short-tailed, it has a large head with a bare face, large ears and closely spaced, deep-set eyes. Used in scientific work, the rhesus gave the first two letters of its name to the Rh blood factor. Height: 60cm (2ft). Species *Macaca mulatta. See also* NW pp.172–173.

Rhesus factor, any of a group of antigens (substances that stimulate ANTIBODY production) found on the surface of red blood cells, so called because first discovered in the rhesus monkey. Rh-negative blood lacks the Rh factor and Rh-positive blood contains it; it is present in about 85% of human beings. Rh incompatibility (an Rh-negative pregnant woman with an Rh-positive fetus) is a major cause of a serious condition in newborn babies. *See also* RH DISEASE, MS pp.62, 76.

Rhetoric, art of discourse and persuasive speaking; language, written or spoken, designed to impress or persuade.

Rheumatic fever, inflammatory disorder caused by streptococcal infection and characterized by fever and swelling and pain in the joints. It is associated with inflammation of the heart and often results in rheumatic heart disease. Treatment is with ANTIBIOTICS, ASPIRIN and CORTISONE. Prevention entails immediate treatment of the streptococcal infections.

Rheumatism, general term for a group of disorders whose symptoms are pain in the joints and bones and their supporting tissues. Rheumatoid ARTHRITIS is a general

Reuter operated a pigeon-post service in Germany before he moved to London.

Paul Revere rode from Charleston to Lexington to warn Massachusetts colonists.

Sir Joshua Reynolds was a dominant figure in British art history (self-portrait).

Rhesus monkeys, used in scientific research, were the first primates in space.

Rhinoceros horn is made up of tubular horny fibres growing from the skin.

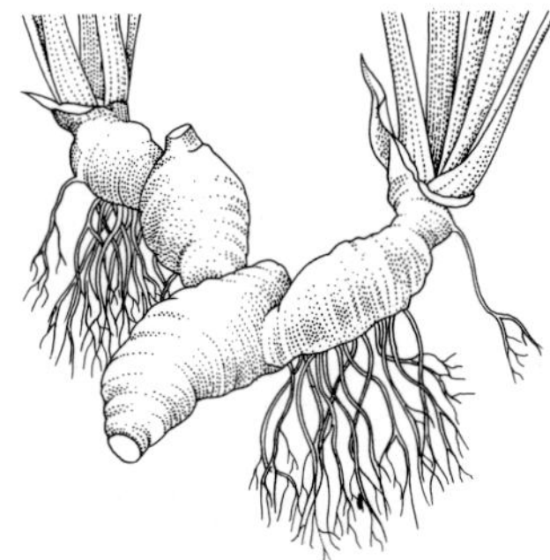

Rhizome is an underground food-storing organ in many flowering plants.

Cecil Rhodes devoted the last years of his life to the development of Rhodesia.

Rhododendron, meaning rose tree, is a flowering plant of the heath family.

constitutional disorder involving mainly the finger joints, although the wrists, knees, ankles, shoulders and sometimes hips may also be affected. Treatment may involve heat treatment or a range of drugs including CORTISONE. RHEUMATIC FEVER attacks the joints, but is due to bacterial infection. *See also* MS p.94.

Rhine, Joseph Banks (1895–), US psychologist. He was a pioneer in the study of extrasensory perception (ESP) and psychic phenomena, bringing these subjects under objective, scientific scrutiny. Among his major publications are *Extra-sensory Perception* (1934) and *Parapsychology Today* (1968). *See also* MS pp.198–199, *198*.

Rhine (Rhein, Rhin or Rijn), major river in w Europe. It rises in SE Switzerland in the Swiss Alps and flows N, bordering on or passing through Switzerland, Austria, Liechtenstein, West Germany, France and The Netherlands to enter the North Sea at Rotterdam. The Rhine is navigable to oceangoing vessels as far as Basel in Switzerland, and is a major transport route for some of w Europe's most industrialized areas. Length: approx. 1,320km (820 miles). *See also* PE p.151.

Rhineland, region in w West Germany along the w bank of the River Rhine. It includes the RHINELAND-PALATINATE, Rhenish Hesse, sw Hesse and NW Baden. The Treaty of Versailles (1919) provided for Allied occupation and demilitarization of the area; the last occupational troops were removed in 1930 (five years early), but by 1936 HITLER had formed the defensive SIEGFRIED LINE and denounced the LOCARNO PACT. The defence system was penetrated by Allied troops in WWII.

Rhineland-Palatinate, industrial state of w West Germany, bordered by France and Saarland (s) and Luxembourg and Belgium (w); the capital is Mainz. The state is drained by the rivers Rhine and Moselle (Mosel), whose valleys are famous for their vineyards. Potatoes, tobacco and fruit are also grown, although most of the people work in industry. Chemicals, machinery and shoes are produced. Area: 19,835sq km (7,658sq miles). Pop. (1974) 3,688,100.

Rhinencephalon, the part of the BRAIN that is concerned with the sense of smell. It is extremely large in sharks (which find their prey by smell) but is small in most higher animals, being completely enveloped in man by the massive growth of the cerebral hemispheres. *See also* MS p.55.

Rhinitis, inflammation of the mucous membrane of the nose. It may be an allergic reaction (ie, HAY FEVER) or a symptom of a disorder such as the common cold.

Rhinoceros, massive, hoofed, herbivorous mammal native to Africa and Asia. Depending on the species, rhinoceroses have either one or two horns on the snout. They have thick skins and poor eyesight, are solitary grazers or browers, and the heat of the day and like to wallow in muddy pools. When alarmed they may charge. Now rare except in protected areas, rhinos were once hunted for their horns (which were believed to have aphrodisiac properties). Weight: 1–3.5 tonnes. Family Rhinocerotidae. *See also* NW pp. 163, *188, 214, 242*.

Rhinoceros beetle, large plant-eating SCARAB BEETLE of tropical and subtropical regions, named after the rhinoceros-like horns of some species. Length: to 150cm (6in), including the horn. Subfamily Dynastinae.

Rhizome, creeping, root-like underground stem of certain plants. It usually grows horizontally, is rich in accumulated starch, and can produce new roots and stems asexually. Rhizomes may be long and slender (as in Solomon's seal), or thick (as in the iris and water lily). They differ from roots in having nodes, buds and scale-like leaves. *See also* TUBER; PE pp.*178*, 186, *187*.

Rhode Island, state in NE USA, on the Atlantic coast in New England; the smallest state in the USA and the second most densely populated. Much of the land is forested, but there is some dairy farming and the state is famous for its Rhode Island Red chickens. Potatoes, hay,

apples, oats and maize are the chief crops. Fishing is of great importance, particularly for shellfish. Mineral deposits include sand, gravel, stone and gemstones. The jewellery industry in Providence is one of the largest in the world. Other industries include textiles, fabricated metals, silverware, machinery, electrical equipment and insurance. The state attracts many tourists and has several leading educational institutions. The principal cities are Providence (the state capital), Warwick, Pawtucket and Cranston.

The region was first settled in 1636 by settlers from Massachusetts seeking religious freedom. Providence was the first settlement. Area: 3,144sq km (1,214sq miles). Pop. (1975 est.) 927,000. *See also* MW p.175.

Rhodes, Cecil John (1853–1902), British financier and statesman. He went to Africa for his health in 1870, where he began mining diamonds and gold. A firm believer in British colonial expansion, he formed the British South Africa Company in 1889 and with it controlled the large areas of SE Africa later called Rhodesia. He was Prime Minister of Cape Colony from 1890 to 1896. Failure to gain control of the Transvaal by means of a conspiracy (the JAMESON RAID, 1895) destroyed his political career. He left his vast fortune to various public works and established scholarships for education. *See also* HC2 pp.133, 143.

Rhodes, Wilfred (1877–1973), British cricketer who played for England and Yorkshire. He took a record 4,187 first-class wickets with his slow left-arm bowling, scored 39,797 runs, and performed the "double" of 100 wickets and 1,000 runs in one season 16 times in his long career (1898–1930). He played in 58 Test matches between 1899 and 1930 (scoring 2,325 runs, at an average of 30.19, and taking 127 wickets, at an average of 26.96).

Rhodes (Ródhos), island in SE Greece, in the Aegean Sea; one of the Dodecanese islands. Colonized by the Dorians c.1000 BC, Rhodes was conquered at different times by Persia, Sparta, Athens, Caria and Alexander the Great. The island was conquered in a special crusade (1308–10) by the KNIGHTS HOSPITALLERS who successfully defended it against the Turks in the Seige of Rhodes. However it was taken by the Turks in 1522, and ceded to Italy in 1923, before being annexed to Greece in 1947. The principal city is Rhodes which is also capital of the Dodecanese department. Products: wheat, tobacco, cotton, olives, fruits, vegetables. Area: 1,400sq km (540sq miles). Pop. (island, 1971) 66,606; (city, 1971) 32,019.

Rhodes, Colossus of, one of seven wonders of the ancient world, a bronze statue of the sun-god overlooking the harbour at Rhodes. It stood more than 100ft (30.5m) high. It was built, at least in part, by Chares of Lindos between c.292 BC and c.280 BC. It was destroyed by an earthquake c.224 BC.

Rhodesia, since 1963 the official name of the British colony which was formerly called Southern Rhodesia. It is now increasingly known by black nations in southern Africa as ZIMBABWE. Northern Rhodesia, which was united with Southern Rhodesia and Nyasaland (now MALAWI) in the Central African Federation of 1953, became the independent country of ZAMBIA in 1964 after the dissolution of the federation.

Rhodesia, Northern, name of ZAMBIA before its independence in 1964. Formed (1911) from the protectorates of Northwest and Northeast Rhodesia, it was taken over by the British Government from Cecil RHODES' British South Africa Company (1924). From 1953–63 it was part of the Federation of Rhodesia and Nyasaland.

Rhodesia and Nyasaland, Federation of, former association founded in 1953 from two British colonies in Africa, also known as the Central African Federation. The common government was primarily white, and the ensuing black protest eventually caused its downfall in 1963. In the following year Northern Rhodesia became

independent as ZAMBIA and most of Nyasaland became MALAWI. Southern Rhodesia, with a change in name to simply RHODESIA, remained officially a British colony, but unilaterally proclaimed independence in 1965. *See also* ZIMBABWE.

Rhodesia man, fossil man whose remains were discovered in a cave at Broken Hill, Zambia (then Northern Rhodesia) in 1921, and dated as late Pleistocene – 40,000 or more years old. The skull is low-browed with brow ridges that are the most massive of all human skulls, but the cranial capacity (brain size) of about 1,300cc is comparable with those of Neanderthal and modern man. Rhodesian man's scientific name is *Homo sapiens rhodesiensis*. *See also* MS pp.*21, 27*.

Rhodes Scholarships, grants for study at Oxford University, England, provided for in the will of the British financier, Cecil RHODES. They are awarded each year to students from the Commonwealth, South Africa and the USA.

Rhodium, metallic element (symbol Rh) of the second transition series, discovered in 1803. It occurs associated with platinum and its chief source is as a by-product of nickel smelting. It is used in hard platinum alloys. Properties: at. no. 45; at. wt. 102.9055; s. g. 12.4; m. p. 1,966°C (3,571°F); b. p. 3,727°C (6,741°F); most common isotope Rh^{103} (100%). *See also* TRANSITION ELEMENTS.

Rhododendron, large genus of shrubs and small trees that grow in the acid soils of cool temperate, often mountainous regions in North America, Europe and Asia. Primarily evergreen, they have leathery leaves and bell-shaped white, pink or purple flowers. Many species are used in landscaping. Family Ericaceae. *See also* AZALEA.

Rhodophyta. *See* RED ALGAE.

Rhodopsin, or visual purple, visual pigment present in the rod cells of the RETINA of the eye. It absorbs light, producing a nerve impulse that is perceived as vision. *See also* MS pp.48–49.

Rhodri Mawr (The Great) (*d*.878), Welsh leader who became King of Gwynedd (844), Powys (855) and two parts of sw Wales, previously united as Seisyllwg (872). Spending much of his career fighting the Danes, he was eventually killed by the Saxons.

Rhombus, PARALLELOGRAM (a four-sided plane figure with opposite sides parallel) with all of its sides equal in length.

Rhondda, former borough in Glamorgan, s Wales, made up of the towns and villages in the valleys of the rivers Rhondda Fawr and Rhondda Fach; since 1974 Rhondda has been a county of MID GLAMORGAN. The region was famous for coal-mining, but many pits have closed in the second half of the 20th century and the population has declined. Light industries are being encouraged. Pop. (1973 est.) 87,710.

Rhône, river in w Europe. It rises in the Rhône Glacier, runs through the Bernese Oberland of Switzerland, flows w to Lake Geneva, and from the sw end of the lake it crosses the French border. It continues s in a valley separating the Massif Central from the French Alps, through Lyons and Avignon to Arles, where it branches into the Grand Rhône and the Petit Rhône. Both rivers flow into the Mediterranean Sea w of Marseilles, to which the river is connected by canal. Length: 813km (505 miles).

Rhubarb, perennial herbaceous plant native to Asia and cultivated in cool climates throughout the world for its edible leaf stalks. It has large poisonous leaves and small white or red flowers. Genus *Rheum*. Height: to 1.2m (4ft). Family Polygonaceae. *See also* MS p.*86*; PE pp.*188*, 189.

Rhuddlan, Statute of (1284), edict issued by EDWARD I of England to provide for the government of Wales. Counties were reorganized, sheriffs appointed and the English criminal law imposed on Wales. It lasted until 1536.

Rhum (Rum), mountainous island of the Inner Hebrides off the w coast of Scotland, in w Highlands Region. The island was made a nature reserve in 1957. The highest point is Askival, rising to 810m

(2,659ft). Area: approx. 182sq km (68sq miles).

Rhyme, similarity of sound in the final syllables of two (or more) words. It has been one of the most important devices, at various times considered almost essential, in poetry. Most common are end rhymes, where two words each at the end of a line of verse rhyme and thereby when read aloud establish the verse structure and reinforce the metre. Internal or LEONINE RHYMES emphasize rhythmic construction. Rhymes are also classified as masculine (when the last and rhyming syllable of a line is stressed) and feminine (when the last of two syllables, both rhyming with those of another line, is unstressed).

Rhymney Valley, county district in E MID GLAMORGAN, S Wales; created in 1974 under the Local Government Act (1972). Area: 176sq km (68sq miles). Pop. (1974 est.) 104,400.

Rhynchocephalia. *See* REPTILE.

Rhyolite, quartz-rich lava of such a fine grain that little or no crystalline structure can be detected in it. The fineness is caused by rapid cooling, giving a glassy texture. *See also* PE p.124.

Rhys, Jean (1894–), British novelist. Among her novels is *Wide Sargasso Sea* (1966), for which she won the W.H. Smith Literary Prize.

Rhythm, in music, duration of the notes, the rapidity with which they follow each other (tempo) and the pattern of sounds formed by changes in duration and tempo. Rhythm in Western music is generally based on metre, ie notes follow one another in a more or less regular pattern at some specified rate. In Oriental and primitive music rhythms may be more free.

Rhythm, in poetry, metrical movement determined by relation of long and short, or stressed and unstressed, syllables.

Rhythm and Blues, form of popular music. It developed as an urban form of the rural negro BLUES. In towns the portable instruments of the original bluesmen were supplemented by pianos, sets of drums and, in the 1950s, electric guitars. Jazz influences, in the form of more insistent rhythms and the use of saxophones, also made themselves felt. This raw, energetic and relatively simple music was the basis of ROCK AND ROLL; it also had a great influence on British popular music of the 1960s, when groups such as the ROLLING STONES imitated American originals.

Rhythm method of contraception. *See* BIRTH CONTROL.

Rib, in architecture, an arch, part of an arch or a projecting band-like member of a vault or ceiling. It is commonly structural, but may be purely decorative, as a separation of the cells of a groined vault. *See also* MS pp.56, 57, *57.*

Rib, curved bones that are arranged in pairs to form part of the front and side support of the chest. There are 12 pairs of ribs in human beings. The first seven pairs, known as true ribs, are joined by costal cartilage to the STERNUM (breastbone); the next three, called false ribs, to the costal cartilage of the seventh pair; and the last two, called floating ribs, are not attached to anything at the front of the body. The head of each rib joins with a thoracic vertebra in the spine. The shafts (main curved part of the bone) of adjacent ribs are joined by intercostal muscles that act to change the capacity of the chest during breathing. *See also* MS pp.56, 57, *57.*

Ribalta, Francisco (c. 1565–1628), Spanish painter. His works include *The Vision of Father Simon* (1612) and *The Vision of St Francis* (c. 1620). He added to the success of his father's studio in Valencia and was one of the first Spanish painters to use TENEBRIST effects in religious paintings.

Ribbentrop, Joachim von (1893–1946), German diplomat and politician. A wealthy merchant, he joined the Nazi party in 1932 and became part of Adolf HITLER's inner circle as a foreign affairs adviser in 1933. From 1936 to 1938 he was ambassador to Great Britain and from 1938 German Foreign Minister. He was instrumental in the conclusion of the Russo-German non-aggression pact of 1939 and in planning the attack in Poland (1939) that precipitated WWII. He was

tried at the Nuremberg Trials (1946), convicted of war crimes and hanged.

Ribbon fish, also called scythe fish or deal fish, marine fish found in cold deep waters. It is identified by its long, thin body and plume-like dorsal fin extending along the entire length of the body. Length: 2.4m (8ft). Family Trachipteridae.

Ribera, José (1591–1652), Spanish BAROQUE painter who after 1616 worked in Naples, then a Spanish possession. Until 1635 he painted religious subjects in sombre colours, influenced by Michelangelo CARAVAGGIO, but between 1635 and 1639 he began to use the brighter colours of TITIAN and Antonio CORREGGIO. One of his early works is *Crucifixion* (c. 1620); his later paintings include *Apollo and Marsyas* (1637) and *The Mystic Marriage of St Catherine* (1648).

Riboflavin, VITAMIN of the B complex, lack of which impairs growth and causes skin disorders. Its formula is $C_{17}H_{20}N_4O_6$, it is soluble in water, and it plays an important part in the health of the skin.

Ribonucleic acid. *See* RNA.

Ribosome, ribonucleo-protein particle in the CYTOPLASM of cells and which appears to be the site of PROTEIN synthesis.

Ricardo, David (1772–1823), British political economist. He was a successful stockbroker, and his book *Principles of Political Economy and Taxation* (1817) made an important contribution to the study of economics.

Ricci, name of two Italian brothers, both of whom were composers, mainly of operas. Federico (1809–77) was involved in composing 19 operas, some of which were written in collaboration with his brother Luigi (1805–59). Their *Crispino e la comare* (1850) enjoyed worldwide success.

Ricci, Matteo (1552–1610), Italian JESUIT who settled in China in 1583. In 1601 he became the first Jesuit to enter Peking. He spent the rest of his life there teaching science, translating many Western works on science into Chinese and writing several of his own, and working as a missionary. He also presented the court with its first maps of the world.

Ricci, Sebastiano (1659–1734), Venetian painter who worked in the Schönbrunn Palace in Vienna before going to Britain in 1712. Some 30 of his paintings are now in the Royal Collection.

Riccio, Andrea Briosco (1470–1532), Italian Renaissance sculptor who was best known for his sculptured bronze miniatures, many of which depict themes from Classical mythology. His great bronze Paschal Candlestick (1507–16) in the Church of St Antony in Padua is his most famous work.

Rice, Elmer (1892–1967), US dramatist whose plays expressed his concern and compassion for the people of the economically depressed 1930s. Among his works are *Street Scene* (1929), of which Kurt WEILL wrote an operatic version (1947), and *Between Two Worlds* (1934). His comedy *Dream Girl* (1945) was a popular success.

Rice, plant native to SE Asia and Indonesia, now cultivated in many warm humid regions and the main grain food for Middle and Far East countries. It provides a staple diet for more than half the population of the world. It is an annual grass, the seed and husk of which is the edible portion. It is usually grown in flooded, terraced paddies with hard subsoil to prevent seepage. Because rice is generally grown underwater, it does not lend itself to regular crop rotation systems and so rice-growing land is often allowed to become fallow after a season or two. Species *Oryza sativa. See also* PE pp.*156*, 158, 169, *181, 183,* 204.

Richard I (1157–99; *r.* 1189–99), King of England, son of HENRY II and ELEANOR of AQUITAINE, also known as Richard Coeur de Lion or Richard the Lion-Heart. He revolted against his father from 1173 to 1174 and again from 1188 to 1189, succeeding to the throne on his father's death in 1189. In 1190 he set out on the Third CRUSADE and took Cyprus and Acre in 1191, but failed to gain Jerusalem. On his journey home he was captured and held for ransom in Austria. He eventually

returned to England in 1194 in time to suppress a revolt raised against him by his brother John. He then went to France where he spent the rest of his life. *See also* HC1 pp.161, *168,* 169, 176, *176, 178.*

Richard II (1367–1400), King of England (*r.* 1377–99), the son of EDWARD THE BLACK PRINCE. He succeeded his grandfather, EDWARD III, to the throne. His uncle, John of Gaunt, controlled the government until the PEASANTS' REVOLT of 1381, and in 1388 the Duke of Gloucester (with four other lords) assumed control. Richard regained authority the following year but Gaunt's heir, Henry, Duke of Lancaster, raised an army against him and was crowned as HENRY IV. Richard was imprisoned and either starved himself to death or was murdered. *See also* HC1 pp.179, *179,* 181, *181,* 211, 214–215, *214,* 242; HC2 p.*309.*

Richard III (1452–85), King of England (*r.*1483–85), brother of Edward IV. When Edward died in 1483 his young son was proclaimed Edward V, but Richard put him and his other young nephew in the Tower of London and assumed the crown himself. His old supporter, the Duke of Buckingham, led an unsuccessful revolt against him shortly after in favour of Henry Tudor. Henry himself landed in Wales in 1485 and killed Richard at the Battle of Bosworth Field. Richard was the last Yorkist king and his death ended the Wars of the ROSES. *See also* HC1 p.242; HC2 p.*309.*

Richard, Cliff (1940–), British popular singer, *b.* India, real name Harold Webb. His first million-selling recording was *Livin' Doll* (1959). After appearing in the film *Expresso Bongo* (1960) he gradually lost his image of a ROCK AND ROLLER and became a family entertainer, with recordings of such songs as *The Young Ones, Bachelor Boy* and *Summer Holiday,* and his series of film musicals, culminating in *Wonderful Life* (1964). He remains popular in the late 1970s.

Richard II (c. 1595), historical play by William SHAKESPEARE, whose chief source was Raphael HOLINSHED's *Chronicles* (1577). Richard II yields his throne to his powerful cousin Henry Bolingbroke, who became Henry IV. Richard is imprisoned, then murdered.

Richard III (c. 1594), historical play by William SHAKESPEARE. Richard, Duke of Gloucester, (a hunchback) murders his brother, the Duke of Clarence; and when his eldest brother, King Edward IV, dies he has Edward's two young sons imprisoned and killed. He siezes the crown but is defeated in battle at Bosworth Field and succeeded by Henry VII. The plot of the play and its characterization bear little relation to historical fact.

Richards, Barry Anderson (1945–), South African cricketer who played for Natal, Hampshire, South Australia and South Africa. A prolific opening batsman, he scored 508 runs (average 72.57) in his four Test matches (1969–70) for South Africa, before his country was barred from international competition.

Richards, Ceri Geraldus (1903–71), British painter who studied at Swansea School of Art and the Royal College of Art, London. Influenced in his early years by PICASSO and Max ERNST, he worked on painted wooden reliefs in the 1930s but returned to his earlier imaginative abstract style after WWII, producing highly coloured works, often based on musical and literary themes. *See also* HC2 p.*287.*

Richards, Dickinson Woodruff (1895–1973), US physiologist. He shared the 1956 Nobel Prize in physiology and medicine with André COURNAND and Werner FORSSMANN for his part in discoveries concerning heart catheterization and pathological changes in the CIRCULATORY SYSTEM, advances that were fundamental to subsequent medical practice.

Richards, Frank (1875–1961), British writer of adventure stories for boys and creator of the character Billy Bunter; real name Charles Hamilton. His early stories appeared in *Gem* (1906–39) and *Magnet* (1908–40). He published his *Autobiography* in 1952.

Richards, Sir Gordon (1904–), British

David Ricardo strongly influenced the newly emerging science of economics.

Richard I, during his entire reign, spent little more than six months in England.

Cliff Richard speaking of the Gospel at a lunchtime concert in London 1970.

Frank Richards' schoolboy character Billy Bunter often appeared in children's comics.

Sir Ralph Richardson played opposite Gina Lollobrigida in *Women of Straw* (1963).

Cardinal Richelieu strengthened the French Monarchy and even ruled France himself.

Manfred von Richthofen (left) talking to an Allied airman whom he shot down.

Leni Riefenstahl used her talents to make films used for Nazi propaganda.

jockey who, in a career lasting from 1921 to 1954, rode an unparalleled 4,870 winners, including a record 269 in 1947 and 14 Classics. He was champion jockey 26 times in 29 seasons (1925–53), another record. He later became a trainer.

Richards, Ivor Armstrong (1893–), British literary critic and theorist, best-known for his innovative approach to the study of literature with its particular stress on semantics. His emphasis on the importance of the texture of literary language may be seen in *The Principles of Literary Criticism* (1924).

Richards, Theodore William (1868–1928), US chemist and professor at Harvard. He made accurate determinations of ATOMIC WEIGHTS and was awarded the 1914 Nobel Prize in chemistry. He provided experimental evidence that supported Frederick Soddy's theory of the existence of ISOTOPES.

Richardson, Dorothy (1873–1957), British novelist. Her most important novel was *Pilgrimage* (12 vols 1915–38). Her use of the "stream of consciousness" technique may have influenced the work of both Virginia WOOLF and James JOYCE.

Richardson, Henry Handel (1870–1946), Australian novelist, real name Ethel Florence Richardson. A naturalistic writer, her work includes a trilogy, *The Fortunes of Richard Mahoney* (1930), which is based on her father's decline after emigrating to Australia, and *The Young Cosima* (1939), her last novel, which reconstructs the love affair of Wagner, Liszt and Cosima.

Richardson, Jonathan (1665–1745), English portrait painter and author. A successful artist, he painted such well-known figures as Alexander POPE and Sir Richard STEELE. He published several theoretical works on art, among them *The Theory of Painting* (1715) and *Essay on the Whole Art of Criticism in Relation to Painting* (1719).

Richardson, Sir Owen Willans (1879–1959), British physicist who was awarded the 1928 Nobel Prize in physics for his work on thermionic emission. In 1911 he showed that ELECTRONS are emitted from hot metals, disproving the theory that they arose from the surrounding air. In the same year he formulated a mathematical equation relating the absolute temperature of a metal to the rate of electron emission, known as Richardson's law, which was of fundamental importance in the development of ELECTRON TUBES (radio valves). He specialized in this field, which he named THERMIONICS, and wrote many works on the subject.

Richardson, Sir Ralph David (1902–), British actor. His distinguished career has included fine Shakespearean performances at both the OLD VIC and STRATFORD-UPON-AVON. He is equally at home, however, in plays such as Harold Pinter's *No man's Land* (1975). Since 1933 he has appeared in a large number of films.

Richardson, Samuel (1689–1761), English novelist and printer. He contributed to the development of epistolary and psychological novels. His novels include *Pamela* (4 vols, 1740–41), and *Sir Charles Grandison* (1753–54).

Richardson, Tony (1928–), British film and stage director. His first stage production was *Look Back in Anger* (1956). He was much admired in the early 1960s for his films: *The Entertainer* (1960), *A Taste of Honey* (1961) and *The Loneliness of the Long Distance Runner* and *Tom Jones* (both 1963). His later films, such as *Joseph Andrews* (1976), were less popular.

Richelieu, Armand-Jean du Plessis de (1585–1642), French cardinal and statesman. He became chief of the royal council in 1624, establishing an ever-increasing royal power that lasted until the FRENCH REVOLUTION. In foreign affairs, he turned the Swedes and Protestant Germans against the Hapsburgs and advanced the French cause in the THIRTY YEARS WAR. A wealthy literary patron, he established the ACADÉMIE FRANÇAISE. *See also* HC1 pp.273, 273, 317.

Richet, Charles Robert (1850–1935), French physiologist. He was awarded the 1913 Nobel Prize in physiology and medicine for his work on immune reactions, which led to research on ALLERGIES.

Richler, Mordecai (1931–), Canadian novelist. Although Richler has lived primarily in England since 1954, his satirical novels are about Canadians and the Jewish ghetto in his native Montreal, and include *The Apprenticeship of Duddy Kravitz* (1959), *Cocksure* (1968) and *St Urbain's Horseman* (1969).

Richmond, state capital and port of entry in E central Virginia, USA, on the James River. Settled in 1637, the city developed as a trading centre. It was made capital of the Commonwealth of Virginia in 1779, and the state capital in 1785. Richmond was the capital of the CONFEDERATE STATES in 1861, and fell to Union forces in 1865. The city includes the University of Richmond (1832). Industries: metal products, tobacco processing, textiles, clothing, chemicals, food processing, printing and publishing. Pop. (1973) 283,087.

Richmond-upon-Thames, borough of SW Greater London, SE England; formed in 1965 by the merging of the former boroughs of Barnes, Richmond and Twickenham. It is a residential area with a number of famous buildings, including HAMPTON COURT PALACE. Pop. (1971) 188,100. *See also* HC2 pp.70–72.

Richter, Adrian Ludwig (1803–84), German painter. He was drawing-master at the MEISSEN porcelain factory. His paintings and woodcut illustrations were a combination of genre and landscape.

Richter, Burton (1931–), US physicist. Working with a very powerful particle accelerator consisting of a pair of electron storage rings which could drive matter and antimatter particles together with extremely high energy, he discovered (1974) a new sub-atomic particle (which he named psi) with the relatively long lifetime of 10^{-20} secs. Samuel TING, working independently of Richter and using a different method, simultaneously discovered the same particle, which he named J. For this work Richter and Ting shared the 1976 Nobel Prize in physics.

Richter, Charles Francis (1900–), US geophysicist and seismologist. With Beno GUTENBERG he developed the RICHTER SCALE to measure earthquake intensity.

Richter, Hans (1843–1916), Austrian conductor, b. Hungary, the personal assistant of Richard WAGNER, whose operas he conducted throughout his career. He conducted at Munich, Bayreuth and Vienna, where he became music director of the Court Opera in 1875. He went to Britain in 1877 and in London conducted what became known as the Richter Concerts until 1897. From 1897 to 1911 he was conductor of the Hallé Orchestra in Manchester, England.

Richter, Sviatoslav (1915–), Soviet pianist. He studied at the Moscow and Odessa Conservatories and won the Lenin Prize in 1961. Acclaimed as a piano virtuoso, he has made extensive concert tours.

Richter scale, classification of earthquake magnitude set up in 1935 by the American geologist Charles RICHTER. The scale is logarithmic and is based on the total energy released by an earthquake as opposed to a scale of intensity that measures the damage done at a particular place. *See also* PE p.28.

Richthofen, Manfred von (1892–1918), German aviator in WWI who shot down 80 enemy aircraft. His fighter wing was called Richthofen's Flying Circus and he was known as the "Red Baron". He was killed in action in April 1918.

Rickets, children's disorder in which the bones fail to harden properly, with a simultaneous excessive growth of cartilage at the ends of the bones. Due to either a dietary lack of VITAMIN D or insufficient sunlight to allow its synthesis in the skin, the disorder results from the inability of the bones to calcify properly, for which vitamin D is essential. A similar disorder, osteomalacia, may occur with pregnant women, who lose calcium to the growing fetus, and with old people.

Rickettsia, group of micro-organisms intermediate between bacteria and viruses, although usually classified with the viruses. They are carried by fleas, lice and ticks and cause many diseases of vertebrates, including TYPHUS, Rocky Mountain spotted fever and Q fever, all of which can be transmitted to humans. Among domestic animals, dogs are afflicted with canine rickettsiosis, of which the symptoms are fever and disorders of the nervous system.

Rickover, Hyman George (1900–), US admiral, b. Russia. From 1947 he managed the US Navy's atomic submarine project and directed the planning and construction of the world's first atomic-powered submarine, USS *Nautilus*, launched in 1954.

Rickshaw, two-wheeled vehicle with a chair-like body and a collapsible hood, pulled by a man between two shafts. Once used widely in the Far East, it was largely superseded by a rickshaw incorporating a bicycle and, later, a motorcycle. *See also* MM p.139.

Riders to the Sea (1903), one-act drama by J. M. SYNGE, first produced in Dublin in 1904. A stark tragedy of man's struggle in the face of the inevitability of death, it tells of a mother whose husband and six sons are drowned at sea.

Ridgeway, The, grassy track in England which dates from prehistoric times and runs along the Berkshire Downs from White Horse Hill to near Streatley. In early SAXON times, when the valleys were wooded, the line of the Ridgeway constituted a buffer zone between the two kingdoms of Mercia and Wessex.

Riding, Laura Jackson (1901–), US poet. While living in Majorca, she collaborated with Robert GRAVES on a critical work *A Survey of Modernist Poetry* (1927) and a novel *No Decency Left* (1932). Collections of her poetry include *Poems: A Joking Word* (1930) and *Poet: A Lying Word* (1933).

Riding, name of the three administrative divisions of the former county of YORKSHIRE, N England. The word means a third.

Riding, equestrian skill required in a variety of sports, including POLO, HORSE RACING and STEEPLECHASING. Riding as a specific skill is judged in SHOWJUMPING and HORSE TRIALS (which include DRESSAGE).

Ridley, Nicholas (c. 1500–55), English religious leader and martyr. He was made Bishop of Rochester in 1547 and Bishop of London in 1550. He helped to compile the Book of COMMON PRAYER, supported Lady Jane GREY and was deprived of his see on Queen MARY's accession. He was martyred in Oxford with LATIMER.

Ridolfi Plot (1571), scheme planned by Roberto Ridolfi, an Italian banker, and the Bishop of Ross to gain the English throne for MARY, QUEEN OF SCOTS. The Duke of Norfolk was to wed Mary after a Catholic uprising had overthrown ELIZABETH I. The plot was discovered when Ridolfi was abroad. Norfolk was executed in 1572.

Riebeeck, Jan Anthoniszoon van (1619–77), Dutch founder of the Cape Town settlement (1652) and its first governor. He was sent to the Cape by the Dutch EAST INDIA COMPANY to establish a provisions station for ships. *See also* HC2 p.128.

Riefenstahl, Leni (1902–), German film director who was employed by Adolf HITLER to make propaganda films, which she did with great talent. Her films include *The Blue Light* (1932), *Triumph Of The Will* (1934) and *Olympische Spiele 1936*. She was also a writer and actress.

Riel, Louis (1844–85), Canadian rebel who led the RED RIVER SETTLEMENT rebellion in MANITOBA from 1869 to 1870. When it collapsed he fled the country but returned and was elected to the Dominion Parliament. He was banished in 1875 and lived in the USA, but in 1885 he led another revolt, after which he was captured and hanged.

Riemann, Georg Friedrich Bernhard (1826–66), German mathematician who, in a life dogged by ill-health, laid the foundations for much of modern mathematics and physics. He worked on integration, functions of complex numbers and differential and non-Euclidean geometry, which was later used in the general theory of relativity. *See also* HC2 p.156.

Riemenschneider, Tilman (c.1460–1531), German sculptor. His work in the late German GOTHIC style is marked by a combination of vigour, emotion and restraint. One of his best-known works is the *Altar of the Virgin* (c.1505) in Creglingen, Württemburg-Baden. *See also* HC1 pp.219, 253.

Rietveld, Gerrit Thomas (1888–1964), one of the leading 20th-century Dutch architects. The Schröder House, Utrecht (1924), his masterpiece, was arguably the most modern European house at the time it was built; it is the classic example of De STIJL architecture. He also designed De Stijl furniture. *See also* HC2 pp.178–179.

Rievaulx Abbey, in North Yorkshire, England, one of the earliest Cistercian buildings in England, founded in 1131, now in ruins. Rievaulx means "Rye Valley", after the river which flows nearby. Built by Cistercian monks from France, it still has some of the finest existing English Gothic architecture.

Rif, Er, range of the Atlas Mts in NE Morocco, NW Africa, that extend along the Mediterranean coast from Ceuta to Melilla. The highest point is Tidiguin (2,457m; 8,060ft).

Rifle, FIREARM with spiral grooves (rifling) along the inside of the barrel to make the bullet spin in flight, thereby greatly increasing range and accuracy over that of a smoothbore weapon. The principle was known in the 15th century, but was impractical for use with contemporary muzzle-loading muskets. Not until the MINIÉ rifle was invented in 1849 were rifled weapons widely used. Since then rifles have been the main small arm used in warfare. During the 19th century, breech-loading and repeating rifles were developed, and since WWII, assault rifles, capable of fully automatic fire, have come into general use. *See also* MM pp.162–165.

Rifleman, one of four types of New Zealand wrens, a brown-yellow perching bird that lives in both North and South Islands. It is a small bird, and derives its name from the "zing" of its call. Length: to 8cm (3in). Species *Acanthisitta chloris.* Family Xenicidae.

Rifle shooting. *See* SHOOTING.

Rift valley, depression formed by the subsidence of land between two parallel faults. Rift valleys are believed to be formed by thermal currents within the Earth's MANTLE that break up the CRUST into large slabs or blocks of rock, which then become fractured. The best example of a rift valley is the GREAT RIFT VALLEY in SW Asia and E Africa. *See also* PE p.105.

Riga, capital city of Latvia (Latvijskaja SSR) USSR, at the extremity of the Gulf of Riga, on the Western River Dvina 14km (9 miles) above its delta. A major Baltic port, it was ceded to Peter the Great of Russia in 1710. The port was closed in 1915 and evacuated by the Russians. Latvia's independence was declared at it in November 1918; from 1918 to 1940, Riga was the capital of Latvia. In WWII, it was taken by the Germans (1941) and retaken by the Soviets (1944). Its manufactures include electrical machinery, telephone and radio equipment, superphosphates and textiles. Pop (1970) 733,000.

Rigaud, Hyacinthe (1659–1743), French painter. He studied at Montpelier and Lyons, and went to Paris in 1681, where he enjoyed immense popularity as a portraitist. His subjects included royalty and the most distinguished people of the day. These included BOSSUET (1699), a celebrated portrait of LOUIS XIV (1701), Cardinal de Bouillon (1708) and Count Zinzendorff (1729). He also explored a more intimate style in unofficial portraits.

Rigel, or Beta Orionis, highly luminous bluish-white super-giant star in the constellation of Orion. Characteristics: apparent mag. 0.08; absolute mag. −7.0; spectral type B8; distance 850 light-years. *See also* SU pp.226–227, *236, 263–264, 263–264, 266, 266.*

Riggs, Robert Larimore ("Bobby") (1918–), US tennis player. He won the US singles championship in 1939 and 1941 and the British singles championship in 1939, and later turned professional. In the early 1970s, he played in exhibition matches with Margaret COURT and Billy Jean KING, beating Court but losing to King.

Right ascension, in astronomy, the angle measured eastwards along the CELESTIAL EQUATOR from the First Point of Aries to the line passing through a CELESTIAL BODY and the CELESTIAL POLES. It is measured in hours, minutes and seconds, the circumference of the celestial equator being 24 hours.

Rights of Man, The, work by Thomas PAINE, in which he defended the FRENCH REVOLUTION in reply to Edmund BURKE's *Reflections on the Revolution in France* (1790). Paine stated that all men held natural rights which could be guaranteed only under a democratic system.

Rights of Man and of the Citizen, Declaration of the, proclamation of human liberties during the FRENCH REVOLUTION (1789). It was inspired by the DECLARATION OF AMERICAN INDEPENDENCE and adopted by the NATIONAL ASSEMBLY to express the new spirit of liberty and equality in France. It was also a more radical document than its US counterpart, guaranteeing rights of "liberty, property, security and resistance to oppression", freedom of speech and equality before the law.

Right whale, any of a group of slow swimming WHALES that are characterized by smooth bodies without grooves, and most lack a dorsal fin. Bouyant when killed, right whales were once the mainstay of the whaling industry and, as a result, are near extinction. Length: 18m (55ft). Family Balaenidae. *See also* NW p.167.

Rigoletto (1851), three-act opera by Giuseppe VERDI, with libretto by Piave, based on Victor HUGO's play *Le Roi s'amuse.* Gilda, the beloved daughter of Rigoletto, a lonely hunchback jester, is mistakenly killed in a plot by her father to murder her lover, the Duke of Mantua.

Rigor mortis, tight contraction of skeletal muscles after death which causes stiffening of the entire body. Onset is gradual from minutes to hours, and disappears within about a day.

Rig Veda (c.1500–1000 BC), one of the four sacred books of Hinduism, consisting of a collection of 1,028 hymns generally directed to the gods of the forces of nature such as ÁGNI, INDRA and VARUNA. *See also* HC1 p.36.

Rijeka (Fiume), city and port in NW Yugoslavia on the Gulf of Quarnero (Kvarner, Rijecki Zaliv); it is Yugoslavia's largest seaport. It was known to the Romans as Tarsatica, and was in turn Frankish, under medieval lords, and within the Austro-Hungarian Empire until 1918. Gabriele D'Annunzio seized it for Italy in 1919, then after being a free state it reverted to Yugoslavia in 1947. Pop. (1971) 132,222.

Riksdag, national parliament of Sweden, formed as a two-chamber assembly in 1966; the second chamber was abolished in 1971. Its 349 members are elected by universal suffrage for a three-year term, elections being held on a system of PROPORTIONAL REPRESENTATION. In 1975, by a new constitution, the Speaker of the Riksdag took over from the monarchy the right to appoint the prime minister.

Riley, Bridget (1931–), British painter, a leading exponent of OP ART. She paints mostly black and white patterns (influenced by VASARELY), creating dazzling optical effects which cover the entire surface of the painting and create an illusion of constant movement and change.

Riley, John (1646–91), English painter. Chiefly a portraitist, he was, with Godfrey KNELLER, official painter at the court of William and Mary.

Riley, Terry (1935–), US musician. Although trained in classical music he has always been popular with ROCK music enthusiasts. In his stage performances he uses tape loops, ostinatos (persistently repeated musical figures) and feedback systems for music that involves repeated patterns or series.

Rilke, Rainer Maria (1875–1926), Austrian lyrical poet and translator. His first volume of poems, *The Book of the Hours* (1905), was inspired by visits to Russia. *Duino Elegies* (1923) and *Sonnets to Orpheus* (1923) praise human existence.

Rill, or rille, on the surface of the Moon, narrow cleft or valley apparently associated with surface collapse. These features are deep, often meandering for several hundred kilometres.

Rimbaud, Arthur (1845–91), French poet who influenced the SYMBOLISTS. After an unhappy childhood he lived with Paul VERLAINE, but they separated in 1873 after an argument. *Une Saison en Enfer,* renouncing his former life, appeared soon after. He then travelled and worked abroad, returning to France in 1891. *Les Illuminations,* containing the major part of his verse, was published in 1886. *See also* HC2 p.100.

Rime of the Ancient Mariner, The (1798), romantic poem by S. T. COLERIDGE first published in *Lyrical Ballads,* published by Coleridge and William WORDSWORTH. This powerful ballad tells of a man alienated from his fellows, who relates his story to a passing stranger.

Rimsky-Korsakov, Nikolai Andreievich (1844–1908), Russian composer and one of the RUSSIAN FIVE. He taught and influenced Igor STRAVINSKY and edited a number of the works of MUSSORGSKY. His operas include *The Snow Maiden* (1881) and *The Golden Cockerel* (1907), and among his most popular orchestral works are *Sheherazade* (1888), *Spanish Capriccio* (1887) and the *Flight of the Bumblebee* from the opera *Tsar Sultan* (1900). *See also* HC2 pp.107, 121, *270.*

Rimu, coniferous tree native to New Zealand, also called the New Zealand red pine. It reaches heights to 45m (150ft).

Rindt, Jochen (1942–70), Austrian motor-racing driver, b. Germany. A leading points-scorer at the time of his death in practice in Monza, he became the first driver to win the world championship posthumously.

Ring of the Nibelungs, The, opera tetralogy by Richard WAGNER composed between 1852 and 1874, with libretto by the composer. Comprising the cycle are *Das Rheingold, Die Walküre, Siegfried* and *Götterdämmerung,* and the first complete performance was given in 1876 at the opening of the festival theatre at Bayreuth (which was built for the purpose). Wagner drew inspiration for the allegoric libretto from the 13th century German epic poem *Nibelungenlied* and the work is a mixture of mythological elements and national feeling.

Ring-tailed cat. *See* CACOMISTLE.

Ring-tailed monkey. *See* CAPUCHIN.

Ringworm, infection of the skin, hair or nails caused by the fungus tinea; so named because it appears as a red, ring-shaped eruption on the skin and spreads to the edges as it heals in the centre. If on the scalp, it is accompanied by burning, itching and loss of hair. *See also* MS p.98.

Rinuccini, Ottavio (1562–1621), Italian poet and author of the first opera libretti. He wrote the texts for *Dafne* (1597) and *Euridice* (1600) by Jacopo PERI, and *Arianna* (1608) by Claudio MONTEVERDI.

Rio de Janeiro, city in SE Brazil, on Guanabara Bay. The capital of the country until 1960, it was originally settled by the Portuguese and later colonized by the French, who were driven out in 1567 by Mem de Sá, governor of the Portuguese colony of Brazil. By the 18th century it had grown as the export centre of all gold mined in the hinterlands. The second-largest city in the country, it is the cultural, commercial and industrial centre of Brazil. Its warm climate makes it a popular tourist resort and the Copacabana beach is world-famous. Pop. 4,296,782.

Rió de la Plata. SEE PLATA RIVER.

Rio Grande, river in North America. It rises in the San Juan Mountains of SW Colorado and flows generally S through New Mexico. It then forms the border between Texas and Mexico, emptying into the Gulf of Mexico just E of Brownsville, Texas, and Matamoros, Mexico. The river is used for irrigation and the generation of hydroelectricity; it is navigable only near the mouth. Length: approx. 3,035km (1,885 miles).

Riga, an industrial city, also retains some of its medieval character.

Bobby Riggs beat Welby van Horn to win the men's national singles trophy.

Arthur Rimbaud coined the phrase "the alchemy of words" to describe poetry.

Rimsky-Korsakov's students who achieved fame included Stravinsky and Prokofiev.

Riopelle, Jean-Paul

Rites of passage; a Balinese bride prepares for her wedding ceremony.

Rivers may have immense energy that can be harnessed to generate electricity.

Primo de Rivera, the Spanish politician, standing by a bust of King Alfonso.

Max Roach (right), the American jazz drummer, seen playing with Sonny Rollins.

Riopelle, Jean-Paul (1922–), French-Canadian painter working mainly in Paris, where he was for a time associated with SURREALISM and TACHISM. His work underwent various phases, by the 1950s showing a pronounced interest in texture and in paint applied with heavy IMPASTO (*Festival*, 1953); in the 1960s he developed a freer, more dramatic painting style, echoed in the sculpture which he also began producing.

Riot Act (1715), British Act of Parliament that charges magistrates with the duty, when 12 or more people are rioting, to command them to disperse in the name of the Queen within one hour. Those who do not disperse are guilty of a felony.

Rio Tinto-Zinc Corporation, international mining, smelting and chemical corporation with headquarters in Britain, where it employed 11,500 people in the mid-1970s, with major holdings in Australia, Canada, South Africa and the USA.

Riouw (Riau) Archipelago, island group in Indonesia, off the SE tip of the Malay Peninsula. Area: 5,903sq km (2,279sq miles). Pop. (1971) 331,136.

Rip current, narrow, swift, short-lived surface current that flows seawards at right-angles to the shoreline. It occurs at the mouth of a bay or along a coast where wind and incoming waves pile up the water until the excess rushes quickly and forcefully back into the sea.

Ripon, Frederick John Robinson. See GODERICH, VISCOUNT.

Rippl-Ronai, Jozsef (1861–1930), Hungarian painter, graphic artist and member of the NABIS in Paris. He produced portraits in pastels, and bold incisive paintings or provincial scenes in Hungary.

Rip Van Winkle (1819), short story by Washington IRVING in which a Dutch colonist falls asleep in the Catskill Mountains of New York State for 20 years and wakes up to find himself a tottering old man.

Rise and Fall of the City of Mahagonny, The (1930), political opera in three acts by Kurt WEILL with libretto by Bertolt BRECHT, first produced in Leipzig. Set in an Alaskan gold-mining community, to illustrate the miseries of CAPITALISM, its score is notable for its jazz-orientated melodies and the social message of Brecht's libretto. See also HC2 pp.121.

Risman, name of two British rugby league players, father and son, who played for Great Britain. Augustus John Ferdinand ("Gus") (1911–) played in 15 Test matches between 1932 and 1946, and Augustus Beverley Walter ("Bev") (1937–) played in five internationals (1968). Bev also played rugby union for England (9 caps, 1959–61).

Risorgimento, ideological and political movement in 19th-century Italy which led to the unification of the country. After NAPOLEON's defeat in 1815, the Austrians regained control of Italy. When the republican revolution of 1848 failed, the Piedmontese House of SAVOY led the movement for unification, expelling the Austrians in 1859 and uniting most of Italy by 1861. The Risorgimento was completed with the annexation of Venetia in 1866 and the occupation of Rome in 1870, with the exception of the area that is today the VATICAN CITY.

Risso's dolphin. See GRAMPUS.

Ritchie-Calder, Peter. See CALDER, PETER RITCHIE-CALDER, BARON RITCHIE.

Rite of Spring (1913), ballet in two scenes, first choreographed by Vaslav NIJINSKY to music by Igor STRAVINSKY. The ballet has been revived many times.

Rites of passage, rites that accompany an individual's transition (passage) from one social status to another. Rites of passage, which include initiation, marriage and funeral, are characterized by special ceremonials and are considered to be a way of reconciling change and continuity in any society. See also MS p.204, *204*.

Ritt, Martin (1919–), US film director who worked originally in television. His films include *Hud* (1963), *Sounder* (1972), *Conrack* (1974) and *The Front* (1977).

Ritter, Johann Wilhelm (1776–1810), Polish physicist who in 1801 discovered

ultra-violet radiation, the previously undiscovered radiation beyond the violet end of the SPECTRUM. He made the discovery while working with silver chloride, which decomposes in the presence of light, especially ultra-violet.

Ritter, Karl (1779–1859), German geographer. His work included detailed descriptions of regions based upon his collections of geographical data. His greatest work was the 20-volume *Earth Science in Relation to Nature and the History of Man*, started in 1817 but never completed.

Rivals, The (1775), comedy by Richard Brinsley SHERIDAN. The rivals are Bob Acres and Ensign Beverly (*alias* Captain Absolute), who contend for the hand of Lydia Languish. Mrs Malaprop, a character in the play noted for her blunders in the use of words, has given literature the word MALAPROPISM to denote such mistakes.

River, large stream. Very few land surfaces are completely flat and rainwater flows down the shortest, steepest slope to collect in depressions creating small streams or rivulets which combine to form rivers. In general the steeper the slope, the faster the river flows; and the greater its volume, the more the underlying earth is eroded, thus deepening the river bed. The river tends to eliminate irregularities in its course, producing a smooth gradient. In its middle, or mature, stage the river's course is less steep, the water flows more slowly and the river begins to meander. In its final stage, or old age, the river approaches its base level, the lowest to which it can erode its bed (usually sea-level). At this stage downwards cutting is replaced by sideways cutting; the river's speed is much reduced and it may cut across the arc of a meander, creating an OXBOW LAKE. In its old age, a river may deposit alluvial material collected in its earlier stages. Deposition is caused when the river's velocity is checked by a less steep gradient, by the river entering a lake or the sea, or by flooding; features include FLOOD-PLAINS, DELTAS and alluvial fans. A river can even build up its bed to a higher level than that of the surrounding flood plain by the deposition of silt. The resultant frequent flooding forms LEVÉES. Throughout history rivers have been important transport and trade routes, and their valleys have provided suitable sites for settlement and agriculture. The power of flowing water in rivers is increasingly being used to generate electricity. See also MM pp.70, *71*, 198, 199; NW pp.230–231, *230–231*; PE pp.112–115, *112–115*, 150–151, *150–151*.

Rivera, Diego (1886–1957), Mexican painter. From 1911 to 1920 he worked in Paris, where he absorbed CUBISM and was also influenced by Henri ROUSSEAU. He also visited Italy, before returning to Mexico in 1921. He is known primarily as a muralist. Characteristic of his style are carefully drawn, simplified, flat geometric forms in expressive colours. He often used symbolism and allegory in his murals to show Mexico's triumph over repression and to express his hope for a Marxist future.

Rivera, José Eustasio (1889–1928), Colombian novelist. His portrayal of the hardship of jungle life in *La vorágine* (1924) was a great advance in the development of the Latin American novel. He also wrote sonnets in classical style about the tropics in *Tierra de promisión* (1921).

Rivera, José Fructuoso (c.1790–1854), Uruguayan general and politician. He was one of the "33 immortals" who liberated Uruguay in 1825. Over the next 30 years he had a tangled political career; he became President twice (1830–35 and 1838–42) but was forced into exile in 1845. See also HC2 p.86.

Rivera, Miguel Primo de (1870–1930), Spanish general and politician. From 1923, when he proclaimed himself dictator after a bloodless coup, Rivera was the most important political figure in Spain. He was forced to resign in 1930 after most other politicians and the king had united against him.

River boat, an essential form of transport in the USA, especially on the MISSISSIPPI,

during the first half of the 19th century. River boats lost their economic role there, with the advent of railways and improved communications, but are still much used in Africa and Asia.

River forecasting, predicting the water level of a river. The forecasting is based on measurements taken from all parts of the river and its tributaries, in areas where floods are frequent and difficult to control. River forecasting is important, too, for efficient use of reservoirs and for people who use river transport.

River herring. See ALEWIFE.

Riverina, district in s New South Wales, SE Australia between the Murumbidgee (N) and Murray (S) rivers, the principal towns are Albury and Wagga Wagga. The area's irrigation system has proved important to agricultural development; products now include early fruits and vegetables and rice. Pop. (1971) 89,968.

Rivers, Larry (1923–), US painter of the New York School, formerly a jazz saxophonist, who began painting in 1944. His *George Washington crossing the Delaware* (1955) was an important influence on the development of POP ART.

Rivers, William Halse (1864–1922), British psychologist and anthropologist who was one of the founders of modern physiological psychology. He studied colour vision, the influence of drugs and mental fatigue, and became director of Britain's first experimental psychology laboratory at London University in 1897. He went on several anthropological expeditions to Australia, India and Melanesia; from his studies of the indigenous populations he wrote *The Todas* (1906), which describes a polyandrous people of s India, and *History of Melanesian Society* (1914).

Riviera, region of SE France and NW Italy, on the Mediterranean Sea, extending 370km (230 miles) from Cannes, France, to La Spezia, Italy. Its spectacular scenery and mild climate make it a leading tourist area of Western Europe; well-known resorts include NICE and CANNES in France, MONTE CARLO in Monaco and San Remo and Alassio in Italy. Products include olives, grapes and citrus fruits, and flowers are grown for export.

Riyadh, capital of SAUDI ARABIA, approx. 378km (235 miles) inland from the Persian Gulf. It is the centre of desert trade and travel; the chief industry is the production of oil. Pop. (1974) 660,800.

Rizal, province on central Luzon island, the Philippines; the capital is Pasig. It includes Quezon city, the capital of the Philippines. The province has an earth satellite station (1968). Products: sugar cane, fruit, rice. Area: 1,898sq km (733sq miles). Pop. (1970) 2,781,081.

Rizzio, David (c.1533–66), secretary and favourite of MARY, QUEEN OF SCOTS, b. Italy. He entered Mary's service as a musician but soon became her secretary. Lord DARNLEY, her husband, and some nobles were jealous of his influence over Mary and had him murdered in 1566.

RKO Radio Pictures Inc., US film production company formed in 1921. Two cinema classics, *King Kong* (1933) and *Citizen Kane* (1940), were made at this studio. The company ceased production in 1953.

RNA, or ribonucleic acid, nucleic acid that controls the synthesis of PROTEINS in a cell and is the genetic material in some viruses. The molecules are copied from DNA and consist of a single strand of NUCLEOTIDES, each containing the sugar ribose, phosphoric acid, and one of four bases: ADENINE, guanine, CYTOSINE or URACIL. Messenger RNA carries the information for protein synthesis from DNA in the NUCLEUS to the RIBOSOMES in the CYTOPLASM. Each AMINO ACID to be formed is specified by a sequence of three bases in messenger RNA. Transfer RNA brings the amino acids to their correct positions in the messenger RNA. See also NW pp.28–29; SU p.156.

Roach, Maxwell Lemuel (1924–), US jazz drummer. In the 1940s he played with many well-known jazz musicians, including Charlie PARKER, Dizzy GILLESPIE and Benny Carter. From the mid-1950s he led his own groups, which included such noted musicians as Clifford Brown and Sonny ROLLINS.

Roach, European freshwater carp found in muddy and, occasionally, in brackish waters. Colours are silver, white and green. Length: to 40cm (16in). Family Cyprinidae; species *Rutilus rutilus*. *See also* PE p.*245*; NW p.231.

Roach. *See* COCKROACH.

Road building, making of roads and motorways, a branch of civil engineering. The course of the road is surveyed and levelled. Then the road bed is prepared: underlying soil is compacted and culverts laid to drain surface water. If the road is to be paved, asphalt may be laid and rolled on to the prepared surfaces. Alternatively, roads are surfaced with concrete, usually reinforced with steel mesh. Road surfaces are cambered, ie curved so that water runs from them. *See also* MM pp.182–183.

Road runner, fast-running desert cuckoo that lives in SW USA. It has a crested head, streaked brownish plumage, long, strong legs and long tail. It feeds on ground animals including snakes, which it kills with a series of quick stabs with its long, pointed beak. Family Cuculidae; species *Geococcyx californianus*. *See also* NW p.*221*.

Roads, paths and tracks made suitable for the passage of wheeled vehicles by being given a levelled, durable surface. The construction of Roman roads was extremely sophisticated, but not until the work of TELFORD and MCADAM in the late 18th century were effective methods of making surfaces permanent rediscovered. The development of the MOTOR CAR led to a great expansion in the scope and size of roads, and in countries such as Britain the road system became more important than the railways for industrial and personal transport. Roads in Britain are classified according to their size and function. Trunk roads are the responsibility of the central government; these roads are now usually MOTORWAYS. Local authorities are responsible for all other roads, although only three classes of roads come under their direct administration: class one are principal roads, signposted as "A" roads, class two are "B" roads, and class three are not normally assigned any further classification. Various unclassified rural and residential roads are also maintained at the public's expense although the local authority commonly shares the cost with other local committees. *See also* MM pp.182–185, *182–185*.

Robbe-Grillet, Alain (1922–), French novelist and theoretician who was one of the originators of the "nouveau roman" (new novel) in the 1950s. He later worked in films, writing the screenplay for *Last Year at Marienbad* (1960) and directing *Trans-Europe Express* (1966). His books include *The Voyeur* (1955) *Towards a New Novel* (1963) and *Project for a Revolution in New York* (1970).

Robber fly, any of numerous species of dull-coloured flies found throughout the world. It preys on other insects, catching them in flight. The largest species is almost 8cm (3in) in length. Family Asilidae.

Robbia, Luca della (*c.*1400–82), Florentine sculptor. He worked in bronze and marble, but his fame resulted chiefly from his numerous, glazed white terra-cotta half-length Madonnas, represented against a blue background.

Robbins, Frederick Chapman (1916–), US physician. He shared the 1954 Nobel Prize in physiology and medicine with John F. ENDERS and Thomas H. WELLER for his part in the discovery of the ability of the poliomyelitis viruses to grow in cultures of various types of tissues. This discovery was vital to the development of a POLIOMYELITIS vaccine.

Robbins, Harold (1916–), US author. His first novel, *The Dream Merchants* (1949), brought him immediate success and his other works include *A Stone for Danny Fisher* (1955), *The Carpetbaggers* (1961), and *The Pirate* (1974).

Robbins, Jerome (1918–), US dancer, choreographer and director. He created roles in FOKINE's *Bluebeard* (1941) and LICHINE's *Helen of Troy* (1942). He also devised ballet sequences for Broadway musicals, including *The King and I* (1951,

and the film version which he directed, 1956), *West Side Story* (1957, filmed 1961) and *Fiddler on the Roof* (1964).

Robbins, Lionel Charles, Baron (1898–), British economist. He was chairman of the government committee that produced the *Robbins Report* (1963), recommending an increase in the number of higher education places.

Robert, name of three kings of Scotland. For Robert I *see* BRUCE, ROBERT. Robert II (1316–90) was the founder of the house of STUART. He acted as regent during the imprisonment of his uncle DAVID II by the English. As king after 1371, he left the details of government to his sons, one of whom, Robert III (*c.*1340–1406), became king in 1390. He was crippled in 1388 and his brother Robert Stewart, Earl of Fife, took over the government.

Robert, name of two dukes of Normandy. Robert I (*r.*1028–35), the father of WILLIAM THE CONQUEROR, assisted EDWARD THE CONFESSOR to become King of England. He died on a pilgrimage to Jerusalem, and passed into legend as Robert the Devil. Robert II (*c.*1054–1134) was known as Robert Curthose. He succeeded to his father's dukedom in 1087. He participated in the First CRUSADE (1096–99) and warred with his brothers WILLIAM II and HENRY I of England. He was imprisoned by Henry (1106–34). *See also* HC1 p.*168*.

Robert, Hubert (1733–1808), French painter and landscape architect. Classical ruins were a favourite subject in his paintings and his decorations for FONTAINEBLEAU are now in the Louvre. He also supervised the layout for part of the gardens at VERSAILLES.

Robert Joffrey Ballet, leading US modern dance company, founded in 1954 by the dancer and choreographer Robert Joffrey. Its repertory includes works of choreographers such as Alvin AILEY and George BALANCHINE.

Robert, Nicholas-Louis (1761–1828), French inventor who launched one of the first hydrogen balloons in Paris in 1783, aided by his brother. In 1799 he patented the first papermaking machine. *See also* MM p.*60*.

Roberti, Ercole d'Antonia de (*c.*1450– 96), Italian painter who in 1486 became court painter in Ferrara. He painted many religious subjects in a decorative style and helped to paint the frescoes *Occupations of the Months* in the Schifanoia Palace.

Roberton, Hugh Stevenson (1874–1952), British choral conductor, arranger and composer, known particularly for establishing and conducting the GLASGOW ORPHEUS CHOIR.

Roberts, Sir Charles George Douglas (1860–1943), Canadian poet and short-story writer. He took up writing after working as a teacher, journalist and professor. His poetry includes *Orion and other Poems* (1880) and *Songs of the Common Day* (1893). He moved to New York in 1897 and wrote animal stories. His *Selected Poems* were published in 1936.

Roberts, Frederick Sleigh, 1st Earl Roberts of Kandahar (1832–1914), British field marshal. He distinguished himself in the INDIAN MUTINY (1857–58) and in the relief of Kandahar in the Second AFGHAN WAR (1878–80). He was Commander-in-Chief in India (1885–93) and Ireland (1895–99). He was made a field-marshal in 1895 and commanded the British forces in the South African (Boer) War (1899–1900).

Roberts, Thomas William ("Tom") (1856–1931), Australian painter known for his portraits and landscapes, who was influential in the development of a purely Australian art. In 1885 he founded the Heidelberg School, a group of IMPRESSIONIST painters in Melbourne. *See also* HC2 p.*127*.

Roberts, William (1895–), British painter who studied at the SLADE SCHOOL. In the early part of his career he was a member of the VORTICIST group, but later modified his abstraction in multi-figure compositions of scenes from everyday life that appear grotesque and mechanical. Among his paintings are *The Playground* (1934–35), and *Vorticists at the Restaurant de la Tour Eiffel, Spring 1915* (1962).

Robertson, Anna M. *See* MOSES, GRANDMA.

Robertson, Thomas William (1829–71), British actor and playwright. He made his stage debut at the age of five, but he is chiefly remembered for his plays, which include *A Night's Adventure* (1851), *David Garrick* (1864), *Ours* (1866), *Caste* (1867) and *M.P.* (1870).

Robeson, Paul (1898–1976), US actor and bass singer who began his career with the Provincetown Players in 1924. He played the title roles in Eugene O'Neill's *Emperor Jones* and Shakespeare's *Othello* in 1930 and 1943. He was also a concert singer, famous for his rendition of negro spirituals.

Robespierre, Maximilian-François-Marie-Isidore (1758–94), French politician, one of the most important figures in the FRENCH REVOLUTION. He was elected to the NATIONAL ASSEMBLY in 1789, where he became one of the main spokesmen of radical democrats, and dominated the JACOBINS, especially after 1791. The outbreak of war in 1792 gave him even more influence. In June 1793 he became a member of the Committee of Public Safety, the ruling executive, which he soon dominated. In spite of seemingly overwhelming odds he and his colleagues, using authoritarian methods known as the REIGN OF TERROR, defeated the internal and external enemies of the republic. Robespierre was overthrown by the coup of 9th Thermidor in 1794 and executed. *See also* HC2 p.75.

Robey, Sir George Edward (1869–1954), British comedian, real name George Edward Wade and billed as the "Prime Minister of Mirth". Noted for his distinctive low bowler hat and large eyebrows, his success was achieved as a performer of music-hall comic songs, but he later progressed to roles in pantomime, musicals and Shakespeare.

Robin, small Eurasian bird with a characteristic red-orange breast and a brown back. It has a highly developed song and feeds on worms and insects. Length: to 14cm (5.5in). Family Turdidae; species *Erithacus rubecula*.

Robin Hood, outlaw in English folk-lore, the subject of medieval ballads. Tradition describes him as a nobleman living in Sherwood Forest during the 12th century. The tales probably originated in the general peasant discontent of the 14th century and became linked to Celtic legends of a green man of the woods.

Robinia, genus of trees of the pea family (Leguminosae) that grow mainly in the USA and Central America; many are known as locust trees. Most familiar in Britain is *Robinia pseudoacacia*, which has long, pinnate leaves and because of this and other resemblances to the acacia is called the false acacia. Height: to 24m (80ft)

Robinson, Edward G. (1893–1973), US film actor, real name Emanuel Goldenberg. His most famous part was that of the snarling gangster in *Little Caesar* (1930). He also played character parts in such films as *Double Indemnity* (1944) and *The Night has a Thousand Eyes* (1948). His last film was *Soylent Green* (1973).

Robinson, Esmé Stuart Lennox (1886–1958), Irish playwright. He had his first play *The Clancy Name* (1908) produced at the ABBEY THEATRE. His plays include *The Cross Roads* (1909), *The Big House* (1926) and *Church Street* (1934). He also wrote a history, *Ireland's Abbey Theatre* (1951).

Robinson, W. Heath (1872–1944), British cartoonist and illustrator of children's books, famed for his drawings of ingenious life-improving contraptions. His autobiography *My Line of Life* was published in 1938.

Robinson, Henry Crabb (1775–1867), British diarist. He was a lawyer, journalist (war correspondent for *The Times*, 1807–1809) and brilliant conversationalist. He knew most of the literary figures of his time including William Blake, Charles Lamb, Samuel Coleridge and William Wordsworth who all appear in his posthumously published *The Diary, Reminiscences and Correspondence of Henry Crabb Robinson* (1869).

Roach sometimes shoal so densely that they make a hiss as they rub together.

Road runners get their name from their habit of running along the highway.

Frederick Robbins shared the 1954 Nobel Prize for his work in preventing polio.

Robespierre eventually fell victim to the Reign of Terror, which he had started.

Robinson, Henry Peach

Robinson Crusoe; a 19th-century print showing him about to leave his island.

Dame Flora Robson in the role of Queen Elizabeth I in *The Sea Hawk.*

Rochdale Pioneers led the way towards the formation of co-operative societies.

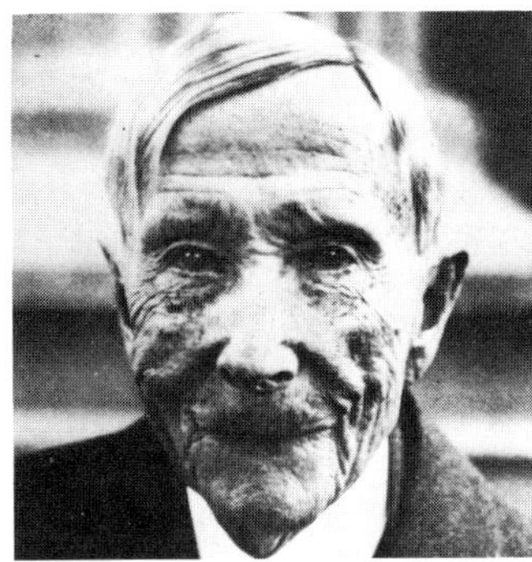

John D. Rockefeller dominated US industry by controlling oil production.

Robinson, Henry Peach (1830–1901), British photographer. After opening a commercial studio in Leamington in 1857 he went on to make what were called "high art photographs" from composite negatives. In 1869 he published *Pictorial Effect in Photography.*

Robinson, John Arthur Thomas (1919–), British theologian who was Bishop of Woolwich 1959–69, and then Fellow and Dean of Trinity College, Cambridge. His writings on theological issues, such as *Honest to God* (1963), caused controversy because of his flexible attitude to conventional morality. *See also* HC2 p.305, *305.*

Robinson, "Perdita" (1758–1800), British actress, real name Mary Darby. Her stage name came from her most successful role in a 1779 production of *The Winter's Tale.* She had a brief affair with the Prince Regent; illness compelled her to abandon her short stage career in 1780.

Robinson, Ray Charles. *See* CHARLES, RAY.

Robinson, Sir Robert (1886–1975), British chemist who was awarded the 1947 Nobel Prize in chemistry for research into ALKALOIDS and other plant chemicals. He succeeded in synthesizing the alkaloid tropinone using a solution of three simple compounds found in plants. This led to his formulating a theory of organic molecular structure in terms of the electronic processes occurring during the formation and disruption of chemical bonds which was to prove important for the understanding of all biosynthetic mechanisms. His other research concerned plant pigments and the genetics of variations in flower colours; he also helped in the synthesis of PENICILLIN and female SEX HORMONES.

Robinson, Stanford (1904–), British composer and conductor noted for his work in radio with the BBC from 1924 to 1966 and for his arrangements of choral and brass band music.

Robinson, Sugar Ray (1920–), US boxer, real name Walker Smith. Undefeated world welterweight champion from 1946 to 1950, he won the middleweight title five times in the 1950s. He fought the last of his 21 world title bouts in 1961.

Robinson Crusoe (1719), famous story by Daniel DEFOE. Robinson Crusoe runs away to sea, is shipwrecked and leads a solitary existence on an uninhabited island. At length he meets a young native, names him "Man Friday" and they become companions. The story is based on the adventures of Alexander Selkirk, a sailing-master who in 1704 was at his own request left for four years on a desolate island off the coast of Chile. *See also* HC2 pp.22, *23,* 34.

Robot, common name for any automated machine used in science or industry to carry out tasks formerly done by human beings. The line separating a robot from other automated machinery is not always distinct. The word was first used in *RUR*, a play (1920) by Karel ČAPEK. Even the most advanced computers are incapable of human qualities such as thought and emotion.

Rob Roy (1671–1734), Scottish bandit, real name Robert Macgregor, immortalized in Sir Walter SCOTT's novel *Rob Roy* (1818). After losing his estates he lived by cattle stealing and plunder. He was sentenced to transportation (1727) but was finally pardoned.

Robsart, Amy (1525–60), maiden name of the wife of Robert Dudley, who was the favourite of Queen ELIZABETH I of England. When Dudley became the Earl of LEICESTER and Amy Robsart was found dead it was rumoured that Dudley had murdered her so that he would be free to marry the Queen. The mystery of her death has never been solved.

Robson, Dame Flora (1902–), British actress. She made her debut in 1921 and soon established a reputation as a strong character actress in plays such as Eugene O'Neill's *Desire Under the Elms* (1931) and J. B. Priestley's *Dangerous Corner* (1932). Her most impressive performances were in such works as *The Innocents* (1952), *Ghosts* (1958) and *John Gabriel Borkman* (1962). A theatre in

Newcastle-on-Tyne named after her was opened in 1962.

Robson, Mark (1913–), US film director b. Canada. After working as a film editor, he became a successful commercial director with such films as *Peyton Place* (1957), *The Inn of the Sixth Happiness* (1958) and *Valley of the Dolls* (1967).

Rochdale, county district in NE Greater Manchester, England; created in 1974 under the Local Government Act (1972). Area: 160sq km (62sq miles). Pop. (1974 est.) 210,600.

Rochdale Pioneers, founders of the first successful CO-OPERATIVE MOVEMENT in Britain. The 28 original subscribers opened a shop at Rochdale, Lancashire, in 1844. Its principle of combining a fixed interest on capital with dividends on purchases was widely followed by other co-operative societies.

Roche, Mazo de. *See* DE LA ROCHE, MAZO.

Rochester, John Wilmot, 2nd Earl of (1647–80), English poet and courtier. The most notorious of the Restoration rakes, he lost the favour of CHARLES II on several occasions. His most famous poem is *Satyr Against Mankind* (1675).

Rochester, Sieges of, three sieges in British history. The first (1088) forced the surrender of Odo of Bayeux, who had led an uprising against WILLIAM II. The second (1215) was an unsuccessful attempt by King JOHN to take the barons' garrison. The third (1264), the first major episode of the BARONS' WARS, was an unsuccessful attempt by Simon de MONTFORT and the Earl of Gloucester to capture the town from HENRY III.

Rock, solid material that comprises the Earth's crust. Although solid, it is not necessarily hard – clay and volcanic ash are also considered to be rocks. Rocks are classified by origin into three major groups. IGNEOUS ROCKS are those formed by the cooling and solidification of molten material from the Earth's interior; volcanic LAVA and GRANITE are igneous rocks. SEDIMENTARY ROCKS are formed from older rock that has been transported from its original position by water, glaciers or the atmosphere, and consolidated again into rock; LIMESTONE and SANDSTONE are examples. METAMORPHIC ROCKS originate from igneous or sedimentary rocks, but have been changed in texture or mineral content or both by extreme pressure and heat deep within the Earth; MARBLE, derived from limestone, is a metamorphic rock. *See also* PE pp.100–103.

Rock, form of popular music characterized by amplified guitars and singing, often repetitive lyrics and driving rhythms. Rock extended from the ROCK AND ROLL and RHYTHM AND BLUES of the fifties and drew on the earlier BLUES and folk music of rural USA to become the cultural expression of the 1960s. Influential British groups were the BEATLES and the ROLLING STONES. Frequently the music relied on gifted guitarists or singers, such as Jimi Hendrix and Janis Joplin, and the lyrics of such writers as Bob DYLAN, Chuck BERRY and Lennon and McCartney Electronic equipment and experiments with lighting and LASERS became features of performances by groups such as Pink Floyd. *See also* HC2 pp.272–273.

Rockall, small rocky island in the Atlantic Ocean, approx. 370km (230 miles) W of the Outer Hebrides, annexed to Britain in 1955. Rockall gives its name to a sea area used in weather forecasting for shipping.

Rock and Roll, form of popular music. Originating in the USA in the early 1950s, it appealed largely to a white audience who found its forerunner, RHYTHM AND BLUES, inaccessible and raw. The term became a household word with the advent of Bill HALEY and the release of the film *Rock Around the Clock* (1956). Its chief proponents were Jerry Lee LEWIS, Elvis PRESLEY and Gene Vincent.

Rockefeller, John Davison (1839–1937), US industrialist and philanthropist. In 1863, when oil drilling began in Pennsylvania, Rockefeller built an oil refinery nearby in Cleveland to exploit its commercial potential. Within two years it had become the largest refinery in Cleveland, becoming incorporated in 1870 into the

Standard Oil Company of Ohio. With mergers, the establishment of a trust and the purchase of other oil refineries, the Standard Oil trust acquired a monopoly in 1882. When he retired he devoted his attention to charitable corporations, to which his donations amounted to about $550 million. *See also* HC2 pp.152, *153.*

Rockefeller, Nelson Aldrich (1908–), US politician. A member of the wealthy Rockefeller family, he held a number of administrative government posts before being elected Governor of New York (1958) as a Republican. He was re-elected three times (1962, 1966, 1970) but resigned in 1973.

Rockefeller Foundation, philanthropic institution established in the USA in 1913 by John D. ROCKEFELLER to promote "the well-being of mankind throughout the world". It finances research into medicine, education, natural and social sciences, humanities and agriculture.

Rocket, slender tapering missile or craft powered by a ROCKET ENGINE. Most of its volume contains fuel; the remainder is the payload (eg, an explosive, scientific instruments or a spacecraft). Liquid-fuelled rockets use a fuel, such as liquid hydrogen, and an oxidizer, usually liquid oxygen, which are burnt together in the engine. Solid-fuelled rockets have both fuel and oxidizer in a solid mixture. A single-stage rocket has one fuel load, or several which are used simultaneously. A multistage rocket has more than one fuel load which are ignited singly in succession as the preceding one burns out. Pioneers of the modern rocket include Konstantin TSIOLKOVSKY (USSR), Hermann OBERTH (Germany) and Robert H. GODDARD (USA). *See also* SU pp.268–269; MM pp.156–159, 168–169.

Rocket engine, engine carrying its own supply of both fuel and oxidizer, as distinguished from a jet engine, which uses oxygen in the surrounding atmosphere as oxidizer. Rocket engines can therefore operate in outer space where there is no atmosphere. They gain thrust from the reaction (referred to in Newton's third law of MOTION) produced by rapid, continuous output of exhaust gases. Chemical rocket engines are powered by solid or liquid propellants that are burned in a combustion chamber and expelled at supersonic velocities from the exhaust nozzle. Nuclear engines heat fuel by RADIATION from reactor cores. Ion engines use thermoelectric power to expel IONS rather than gases. *See also* SU pp.74, *282;* MM pp.156–157.

Rocket warfare, use of self-propelled projectiles in war. The Mongols and Chinese had rockets as early as the 13th century, but they were too inaccurate to be dangerous. The British army experimented with them during the Napoleonic War but even in the mid-19th century when angled vanes (rather like the feather of an arrow) enabled the rocket to be spun in flight, thereby greatly increasing accuracy, conventional artillery was still superior. Rocket warfare came into its own in WWII, when liquid and solid fuel propellants and ballistic principles were better understood. The most important developments were large German missiles, the V1 and V2 rockets, which formed the basis of post-war designs. After WWII, rockets armed with atomic and nuclear warheads, intercontinental ballistic MISSILES (ICBM), became the main strategic armaments of the great powers. *See also* MM pp.169, *169, 178–180.*

Rockingham, Charles Watson-Wentworth, 2nd Marquess of (1730–82), British politician. He led a Whig faction in the 1760s, and was Prime Minister from 1765 to 1766. He repealed the STAMP ACT but failed to effect a reconciliation with the American colonies. He continued to oppose Lord NORTH, and became Prime Minister again in 1782.

Rockling, any of several species of small marine fishes of the COD family, Gadidae. It inhabits coastal waters of the Atlantic. Genus *Gaidropsarus.*

Rockrose, bushy plant or shrub native mainly to the Mediterranean region. It has small, scale-like leaves and five-petalled flowers. Among the 170 species is the common yellow-flowered *Helianthemum chamaecistus*, found in grasslands of Europe and w Asia. Family Cistaceae.

Rockwell hardness test, method of measuring the hardness of a metal or alloy using an apparatus with a diamond-pointed cone. The cone is pressed into the metal with a standard force. Resistance to penetration is automatically indicated by a calibrated dial.

Rocky Mountains, major mountain system in w North America, extending from Mexico to the Bering Strait, N of the Arctic Circle. They form the continental divide, separating the rivers draining to the Atlantic and Arctic oceans from those draining to the Pacific. The mountains, formed in the Mesozoic era, are geologically extremely complex. The highest point is Mt Elbert, Colorado, rising to 4,399m (14,433ft). Length: approx. 4,830km (3,000 miles). *See also* PE p.*107*.

Rocky Mountain sheep. *See* BIGHORN.

Rococo, style of 18th-century decorative art. Colourful, clever and extravagant, it originated at VERSAILLES in the late 17th century and reached its height with the court of Louis XV. From France the style soon spread to the German courts and in some degree to Austria, Italy and to Britain, where its influence appeared in the work of Thomas CHIPPENDALE. A reaction against the pomp and solemnity of the BAROQUE which preceded it, Rococo brought to interior decoration the elegance of swirls, scrolls, shells and arabesques. It was adapted as much to the applied arts of furniture, porcelain and silverware as to the fine arts of painting and sculpture. In painting, its greatest exponents were Antoine WATTEAU and François BOUCHER. *See also* HC2 pp.36–37, *36–37*.

Rodchenko, Alexander (1891–1956), Russian painter, sculptor and designer. Influenced by Kasimir MALEVICH, he turned from the FUTURIST style of his early works to a more ABSTRACT style and became a leading member of the CONSTRUCTIVIST movement. *See also* HC2 p.204.

Rodent, any member of the vast order Rodentia, the most numerous and widespread of all mammals, characterized by a pair of gnawing incisor teeth in both the upper and lower jaws. Numbering close to 2,000 species, rodents live throughout the world. Most rodents are small and light. The distinctive incisor teeth continue to grow as long as the animal lives. Rodent incisors are chisel-shaped and sharpened by use. *See also* NW pp.155, 160–161.

Rodeo, sport with origins in the practical work of a COWBOY. A major entertainment in North America, its seven main events are saddle bronco (unbroken horse) riding, bareback bronco riding, bull-riding, calf-roping, single-steer roping and team-roping, and steer-wrestling. The classic event is saddle bronco riding, in which a competing cowboy must keep his seat on a bucking horse for at least 8 seconds. The rodeo developed in the cattle-raising south-western states of the USA in the mid-19th century.

Rodgers, Richard Charles (1902–), US composer of BROADWAY musicals. He worked first with Lorenz Hart and then with Oscar HAMMERSTEIN II in a series of successful musicals which included *Oklahoma!* (1943), *Carousel* (1945), *South Pacific* (1949), *The King and I* (1951) and *The Sound of Music* (1959).

Rodin, Auguste (1840–1917), French sculptor who brought to a climax the age of ROMANTICISM in sculpture. A visit to Italy (1875–77) familiarized him with the work of MICHELANGELO and DONATELLO and his first major work *The Age of Bronze* was exhibited in 1878. *The Gates of Hell* (unfinished studies for a bronze door for the Musée des arts décoratifs) was a maze of almost 200 tortured figures inspired by GHIBERTI's *Gates of Paradise* (1401–24) and DORÉ's illustrations (1861) for Dante's *Inferno*. These figures provided him with the subjects for further sculpture which included *The Thinker* (1880), *The Kiss* (1886) and *Fugit Amor* (1897). His works exude surface tension

and it was his preoccupation to express emotion and movement. Perhaps his most mature work is the full length-size bronze of Balzac (1897). A commisssion *See also* HC2 p.*99*.

Rodney, George Brydges Rodney, 1st Baron (1718–92), British admiral. He was appointed Commander of the Fleet in the Leeward Islands in 1779. In 1780 he saved Gibraltar from Spanish attack, at Cape St Vincent. His greatest victory was against the French at Les Saintes in the West Indies in 1782.

Rodnina, Irena (1949–), Soviet ice figure skater who won the pairs gold medals at the 1972 and 1976 Olympic Games and eight world championship titles, first with Alexei Ulanov (1969–72) and then with her husband Alexander Zaitsev (1973–76).

Rods and cones, cells in the RETINA of the EYE which are sensitive to light. Located in the pigmented layer, the rod-shaped rhodopsin-secreting cells are the receptors for low-intensity light, and the cone-shaped iodopsin-secreting cells are adapted to distinguish colour. Rods detect only shades of black and white, but are particularly sensitive to movement. *See also* MS pp.48–49.

Roebuck, John (1718–94), British inventor who subsidized James WATT in the development of the first practical condensing steam-engine. In 1746 Roebuck originated a new method for producing sulphuric acid.

Roe deer, small agile Eurasian DEER. It has short, erect antlers and a reddish summer coat which turns brown-grey in winter. The gestation period is almost ten months because of delayed implantation of the fertilized egg. Height: to 86cm (34in) at the shoulder. Family Cervidae; species *Capreolus capreolus. See also* NW p.*209*.

Roeg, Nicholas (1928–), British film director. His films include *Performance* (1968), *Don't Look Now* (1973) and *The Man Who Fell To Earth* (1975).

Roehm, Ernst (1887–1934), leading German Nazi, commander of the SA. After 1930 he made the SA into a powerful army and became HITLER's chief rival. He was murdered on Hitler's orders in June 1934.

Roentgen, Wilhelm Konrad. *See* RÖNTGEN, WILHELM KONRAD.

Roerich, Nikolai Konstantin (1874–1947), Russian painter. He used large areas of flat bright colour and designed sets for Diaghilev's *Ballets Russes* prior to WWI.

Roger II (1095–1154), first Norman King of Sicily who was crowned in 1130. Roger's kingdom was one of the most powerful in the Mediterranean, and his court was noted for its encouragement of scientific thought. He refused to take part in any crusades, because his rule over the mixed Sicilian population was founded on religious tolerance.

Roger of Wendover (*d.c.*1236), English chronicler. His *Flores historiarum* covers the years 1201 to 1235 and contains a derogatory picture of King JOHN.

Rogers, Bruce (1870–1957), US book designer and typographer. He specialized in limited editions and fine printing, and designed the typeface known as Centaur (1915).

Rogers, Claude (1907–), British artist. With Victor PASMORE and Sir William COLDSTREAM he founded the EUSTON ROAD SCHOOL (1937). His paintings are representational and include landscapes, genre and portraits.

Rogers, Derek ("Budge") (1939–), British rugby union player for England, the British Lions and Bedford. A wing-forward, he won a then-record 34 caps (1961–69) for England, and toured South Africa with the Lions in 1962.

Rogers, Ginger (1911–), US actress and dancer. She became famous in the 1930s for several film musicals co-starring Fred ASTAIRE, including *Flying Down To Rio* (1933) and *Top Hat* (1934). She also appeared in the theatre.

Rogers, John (*c.*1500–55), English Protestant martyr. He became a prebendary of St Paul's Cathedral in 1551. He was deprived of his post by MARY I and burned at Smithfield for heresy.

Rogers, William Penn Adair ("Will") (1879–1935), US comedian who appeared in VAUDEVILLE as a cowboy and joined the ZIEGFELD FOLLIES in 1914. He became known through the cinema, radio and a syndicated newspaper column as the "cowboy philosopher".

Roget, Peter Mark (1779–1869), British physician, physiologist and man of letters who helped to establish London University and is especially remembered for his *Thesaurus of English Words and Phrases* (1852).

Rohan, Louis-René-Édouard, Cardinal de (1734–1803), French ecclesiastic and royal servant. He was ambassador to the court of MARIA THERESA of Austria (1772-74). He was implicated in the "affair of the diamond necklace" (1785) and, although acquited, was deprived of his offices and exiled in 1786.

Rohe, Ludwig Mies van der. *See* MIES VAN DER ROHE, LUDWIG.

Róheim, Géza (1891–1953), Hungarian anthropologist. He was one of the earliest thinkers to attempt to apply psychoanalytic theory to the analysis of cultures and national character, in such works as *The Origins and Function of Culture* (1943).

Rohlfs, Christian (1849–1938), German painter. Early in his career he painted in the IMPRESSIONIST manner and after *c.*1905 inclined towards EXPRESSIONISM, abandoning oil paint for watercolour and tempera after 1911.

Rohmer, Eric (1920–), French film director *b.* Jean Maurice Scherer. His works include *My Night with Maude* (1969), *Claire's Knee* (1970) and *Love in the Afternoon* (1972).

Rojas, Fernando de (*c.*1465–1541), Spanish writer. He is noted for his *Celestina* (1499), an extended prose work in 16 acts, about human passion.

Rokeby Venus, The (*c.*1648–49), properly *The Toilet of Venus,* full-length reclining female nude by VELÁZQUEZ. It exemplifies the artist's subtle use of colour and composition and its controlled sensuality made it the object of a suffragette attack in 1914.

Roland, the Song of. *See* CHANSON DE ROLAND.

Rolfe, Frederick William (1860–1913), British writer, pen-name Baron Corvo. A Roman Catholic convert, he twice attempted to become a priest and became a professional writer in 1898. His works include his fantastical autobiography *Hadrian the Seventh* (1904) and *The Desire and Pursuit of the Whole* (1934).

Rolfe, John (1585–1622), English colonist, the first to cultivate tobacco in the JAMESTOWN colony. He married POCAHONTAS in 1614 and returned with her to England. *See also* HC2 p.*63*.

Rolland, Romain (1866–1944), French novelist, playwright and musicologist. Among his best-known works are his 10-volume novel *Jean-Christophe* (1904–12), the pacifist essay *Above the Battle* (1915) and several biographies. He was awarded a Nobel Prize in 1915.

Rolle of Hampole, Richard (*c.*1300–49), English hermit whose writings became popular as manuals of spiritual life. He wrote various mystical tracts and verses, including the *Meditation of the Passion.*

Roller, any of several species of birds of Eurasia which are observed to roll over in flight. An occasional visitor to Britain, *Coracius garrulus,* has blue-green plumage and flies as far north as Sweden; it usually winters in Africa. Family Coraciidae. *See also* NW p.*152*.

Roller bearing, component used in engineering to minimize FRICTION by limiting contact between rotating parts of a machine. It consists of two concentric metal rings separated by, and in contact with, cylindrical, barrel-shaped or tapered rollers made of polished hardened steel. *See also* BEARINGS.

Roller derby, indoor ROLLER SKATING sport popular in the USA since the 1930s. It is played on a banked track by two teams each with five men, who take the rink first, and five women, who follow. Both sexes circle the rink for 12 minutes trying to outlap their opponents; a point is scored for each opposing player passed.

Rococo fountain and railings designed by Jean Lamour at Nancy, France.

Auguste Rodin used modelling techniques to express character and emotion.

Roe deer perform an interesting courtship in which the buck circles with the doe.

Ginger Rogers, seen here with her best-known partner Fred Astaire in *Top Hat.*

Roller skating

Rolling Stones arrive at Milan in 1967 before a concert tour of Italy.

Roman Britain; a doorway in one of the forts which strengthened Hadrian's Wall.

Roman Empire; Barbarian invasions from the north disrupted the empire in Italy.

Romanesque architecture strongly influenced the art of the time.

Roller skating, recreation and sport based on the same principles as ice skating. It consists of gliding on a smooth surface on skates with wheels. The sport first gained popularity in the mid-19th century and in the 20th century, in addition to its recreational use, has taken on several competitive forms. Except for speed skating, which uses oval outdoor tracks or open roads, most skaing is an indoor activity.

Roll film, spool of CELLULOID, coated with photographic emulsion, used to make several exposures which yield printed photographs. *See also* MM pp.218–221.

Rolling, metal, most widely-used method of shaping metal by passing it between revolving rollers. Hot rolling involves reducing the heated metal to thinner and thinner cross-sections (as from ingot to sheet). Cold rolling is used on relatively soft metals to develop in them new mechanical properties. *See also* MM pp.42–43.

Rolling Stones, British rock group of the 1960s and 1970s. Their raucous rhythm-and-blues based music expressed adolescent frustrations and their hit singles included *Satisfaction* (1965) and *Paint it Black* (1966).

Rollins, Sonny (1929–   ), US jazz saxophonist, b. Theodore Walter Rollins. He concei solos based on scales and melodic fragments, rather than on the harmonic approach of his predecessors.

Rollo, Duke of Normandy, (*c.*860–*c.*931), Viking chief who founded NORMANDY. He led an invasion of NW France (*c.*890–910) which was so successful that Charles the Simple bought him off by granting him some land around Rouen in 911. Rollo and his successors increased this area by conquest, to form what became known as Normandy. *See also* HC1 p.156.

Rolls-Royce, British engineering firm famous for the manufacture of motor car and aircraft engines. It was formed in 1904 by the merger of the Derby works of Sir Frederick Henry Royce (1863–1933) and the firm of Charles Stewart Rolls (1877–1910). The Rolls-Royce motor car produced in 1904 was the first fast, luxury motor car. The chassis of the Rolls-Royce has always been designed for large limousine and sedan bodies, but in the 1920s the firm brought out a light, fast Continental under the BENTLEY label. In 1973 part of the company passed under governmental control.

Roma. *See* ROME.

Romagna, historic region of N central Italy, now included in the region of Emilia-Romagna. A centre of Byzantine domination in the 6th century, Romagna resisted effective papal rule in later years until Pope Julius II incorporated it into the PAPAL STATES in the early 16th century. Papal domination continued except for a period of French occupation (1797–1815) until 1860, when Romagna became part of the kingdom of Italy.

Romains, Jules (1885–1972), pseud. of Louis Farigoule, French novelist, poet and playwright. The leading exponent of Unanimism, which posits a collective spirit, his works include *Death of a Nobody* (1911), *Doctor Knock* (1923), and the 27-novel cycle *Men of Good Will* (1933–46) which deals with the early 20th century period of French life.

Roman à clef, literally key-novel, novel containing one or more characters based on actual people (given fictitious names), often in situations that have occurred. The tradition evolved in 17th-century France. Notable modern examples are the political novels of Benjamin DISRAELI, *Point Counter Point* (1928) by Aldous HUXLEY and *The Mandarins* (1954) by Simone de BEAUVOIR.

Roman Britain, period of British history from either CAESAR's invasions of 55–54 BC or CLAUDIUS' occupation of AD 43 until *c.*410. Caesar's invasions failed when unrest in Gaul forced his legions to return to the mainland. The real Roman period began with the arrival of Claudius' army in 43. By 47 lowland Britain had been conquered south of a line from Exeter to the Humber. Under the Flavian emperors the conquest was extended north and west. It was completed by the time of HADRIAN, whose wall (AD 122) marked the northern boundary of Roman Britain, from the Tyne to the Solway. In the south Romanization was fairly thorough, in the north more fitful, although even in the south Britain remained a colony, not an integral part of the Roman world. The Romans brought little change to the "highland" north; in the south they established villas, which were self-contained, unified labour centres as opposed to the isolated pre-Roman farms. The most lasting legacies were the towns, such as Camulodunum (Colchester, established as a *colonia* AD 49), Lindum (Lincoln, AD 71), Glevum (Gloucester, AD 98) and Eboracum (York, AD 237), and the ROMAN ROADS. The Romans exploited Britain's minerals – lead, iron, marble and slate; they also developed the wool and pottery trades. They introduced new plants such as roses, lilies, cabbages, broad beans, apples, walnuts and vines.

Latin was the official language, spoken by some urban artisans, but ignored by the illiterate peasantry, who retained the CELTIC tongue. Unlike Spain and France, Britain never became a Latin-speaking country. Christianity made little progress and even late Roman Britain was essentially pagan. The end of the Roman period began with the withdrawal of part of the Roman garrison by Magnus Maximus in 383 and was completed by Honorius in 409–10. *See also* HC1 pp.100–105.

Roman Catholic Church, Christian denomination which acknowledges the supremacy of the POPE, who in religious matters – particularly in tenets of faith and dogmatic tradition – is regarded as infallible. Before the REFORMATION the term Catholic (from the Greek, meaning "universal") applied to the Western Church as opposed to the Eastern ORTHODOX CHURCH based at Constantinople. Perhaps because of the Reformation, which was in itself a revolt against the Catholic love of elaborate ritual, rich vestments, costly church items and the fervent veneration of saints, the Roman Catholic Church tended to be characterized by rigid adherence to doctrinal tradition from that time until the early 20th century. The desire for a united Christendom has more recently led to a generally more liberal attitude, although the government of the Church is centred solely on the pope's council called the CURIA in Rome, made up of CARDINALS. *See also* HC Indexes.

Romance, literary form, a heroic tale or ballad usually in verse. The form derives from the medieval narratives of TROUBADOURS and the name is due to the fact that they sang their songs in the vernacular, either French or Provençal, rather than Latin (the Old French word *romanz* means "vulgar tongue"). The romance spread through Europe in the 12th century and was used in English by Geoffrey CHAUCER. It continued to be used up to the 16th century, when pastoral romances, such as Sir Philip SIDNEY's prose *Arcadia* were popular. In modern usage romance has come to mean any love story.

Romance languages, languages that evolved from LATIN, the language of the Roman Empire. They include Italian, French, Spanish, Portuguese, Romanian, Catalan, Provençal and Rhaeto-Romanic (spoken in parts of Switzerland). The Romance languages developed as the Vulgar Latin spoken by Roman settlers merged with the language of the region; they all have a grammar and vocabulary based on Latin.

Roman Conquest, conquest of Britain by Rome. The first invasions of Julius CAESAR (55 and 54 BC) failed to establish Roman occupation of the island, which began with the invasion of CLAUDIUS' army in AD 43. Claudius' defeat of CARACTACUS at the Battle of Medway opened the way for his advance to Colchester, capital of the Catuvellauni. By 61 the island south of the river Humber had been conquered. In that year BOADICEA's revolt, although it was partly successful, failed to stem the Roman advance. By 80 Wales had been subdued and by 122 the northern limits of the Roman conquest were established by the building of HADRIAN's WALL. *See also* ROMAN BRITAIN; HC1 pp.100–101.

Roman de la Rose (Romance of the Rose), old French romance of the 13th-century, by Guillaume de Lorris and Jean de Meunq. Guillaume began the poem in *c.*1240 as an allegory of the progress of a courtly lover. A translation of the work into Middle English was one of CHAUCER's earliest literary tasks. *See also* HC1 p.206, *206*.

Roman Empire, that part of the world, centred on the Mediterranean, which was ruled from the Italian city state of ROME from the mid-4th century BC to the Barbarian invasions of Italy and the deposition of the last Roman emperor of the West, ROMULUS AUGUSTULUS, in AD 476. Rome itself was, according to LIVY, founded in 735 BC by ROMULUS AND REMUS and was under ETRUSCAN control until the expulsion of Tarquinius Superbus in *c.*509 BC. His expulsion marks the beginning of the Roman republic, which was governed by two consuls and a senate; the expansion of Rome began soon after. By 340 BC Rome had defeated the cities of the Latin League and established a political and trading control over Italy south of the Po River.

The next phase of expansion took place against the Carthaginians in the PUNIC WARS. The defeat of HANNIBAL and the razing of Carthage in 146 BC left Rome in control of all of Italy, Sicily, Sardinia, Corsica, the southern coast of Iberia and the northern coast of Africa.

Expansion to the east began in the late 3rd century BC. PHILIP II OF MACEDON was defeated in two campaigns (215–205 BC and 200–197 BC) and ANTIOCHUS III of Syria was defeated in 190 BC. By 168 BC Macedonia was a Roman province. Greece was never an imperial province, but it became at the same time subject to Roman hegemony. The next great addition of territory was made by Julius CAESAR's conquest of Gaul between 58 BC and 51 BC. Britain was added to the empire by CLAUDIUS in AD 43. By that time the fact of empire had been recognized by the bestowing of the title emperor upon Octavian, Caesar AUGUSTUS, in 27 BC.

In AD 167 the first Barbarian invasions from the north crossed the Danube and the Alps into Italy. They, and the spread of Christianity, began to disrupt the stability of Roman society and in 285 the empire was divided by DIOCLETIAN into Western and Eastern halves. By the middle of the 5th century the Western empire had disintegrated and its lands were under Barbarian occupation. The eastern, or BYZANTINE EMPIRE, however, survived and was not finally dismembered until 1453. *See also* HC1 pp.90–115.

Romanes, George John (1848–1894), British biologist who wrote popular works on evolution and founded the annual Romanes Lecture at Oxford university, in which speakers deal with a subject in science or literature.

Romanesque, architectural and, by association, artistic style prevailing in W Europe from the late 10th to the 12th century. The term was revised in the 19th century to describe the style derived from Roman models. Romanesque architecture was first explored at Cluny Abbey in France and was common in France, Spain , Italy and in England (where it was known as NORMAN architecture.) Despite many regional variations, its distinctive features included semicircular arches, magnificent west façades to its churches, huge barrel vaulting, complex apses, radiating chapels, heavy, solid walls and rounded pillars, often with their capitals decorated with narrative relief sculptures. In the mid-12th century it was superseded by GOTHIC styles, with higher vaulted roofs and pointed arches. Romanesque arts were associated with the architecture. *See also* HC1 pp.172–175.

Roman-fleuve (River-novel), French term for the type of lengthy novel, dealing with the same character through a number of years, which came to be characteristic of French fiction during the first half of the 20th century. The outstanding example of this form is Marcel PROUST's *Remembrance of Things Past* (1913–1927).

Romania, country in SE Europe, official name the Socialist Republic of Romania, At the end of WWII the USSR aided the Communist Party in taking over the government. Agricultural output is high; most of the farms are collectives and cereals are the principal crops. Romania has rich mineral deposits, in particular oil and natural gas. Industries include machinery, steel, electrical equipment, textiles and chemicals. The capital is Bucharest (București). Area: 237,499sq km (91,698sq miles). Pop. (1975 est.) 21,245,000. *See* MW p.147.

Roman law, system of CIVIL LAW developed between 753 BC and the 5th century AD, which forms the basis of the civil law that applies in many parts of the world. Roman law, based on principles of rights and obligations, was enacted originally by the PATRICIANS then increasingly after 287 BC by the PLEBEIAN assemblies. From 367 BC magistrates, usually PRAETORS, proclaimed the legal principles (*edicta*) which became an important source of law known as *jus honorium*. The emperor could also enact laws and by the mid-2nd century AD became the sole creator of laws.

Roman law can be divided into two parts: *jus civile*, or civil law, which applied only to Roman citizens and which was codified in the TWELVE TABLES of 450 BC; and *jus gentium*, which gradually merged into *jus civile*, originally applying to foreigners in Rome and to others within Roman lands who were not citizens, and taking many principles from Greek commercial law. Roman law was codified by the Emperor JUSTINIAN (*r.*527–64) and was adapted by many of the barbarian invaders of the Empire. *See also* HC1 p.93; MS p.283.

Roman numerals, the letters used by the ancient Romans in their counting system. There were seven: I (one), V (five), X (ten), L (fifty), C (100), D (500) and M (1000). From one to ten they ran: I, II, III, IV (=I less than V), V, VI, VII, VIII, IX (=I before X), X; and the tens: X, XX, XXX, XL (X before L=40), L and so on. Thus 1978 may be expressed as MCMLXXVIII. *See also* SU p.*31*.

Romanov, House of, the Russian imperial family from 1613 until the revolution of 1917. The first tsar, MICHAEL, was chosen in 1613, and his election ended a turbulent period in Russian history. Male inheritance was generally followed until PETER THE GREAT established the principle of a choice of successor by the ruling monarch in 1722. PAUL I however, formulated an order of succession in the Romanov family in 1797. NICHOLAS II, the last tsar, abdicated in 1917 during the Revolution and was apparently executed in 1918 with all the other members of his immediate family by the BOLSHEVIKS.

Roman roads, communications system built by the Romans throughout their empire. They systematized existing trackways and land routes and at the height of the empire about 85,000km (53,000 miles) of efficient roads were in use, based upon 29 military routes centred on Rome itself. The great Roman roads were notable feats of surveying. Renowned for their straightness (still observable on many modern trunk roads which follow their routes), they were equally remarkable in the strength of their construction (in five layers) and because of the attention given to efficient drainage. It has been estimated that messengers on horseback could travel 120km (75 miles) per day on the military roads. After the fall of the Roman Empire, no comparable roads were built in Europe before the 18th century. *See also* HC1 pp.*99, 101*, 104, *104*, 105, *105*; MM pp.182, *182–183*.

Romans, the Epistle to the, letter by St PAUL *c.*57, written probably in Corinth. In it, Paul anxiously tries to counteract the doctrine being propounded by some early Judaeo-Christian teachers in Rome, possibly even St PETER, who were proclaiming Christianity as a new form of Judaism. In his Hellenistic style Paul reveals his concept of salvation through faith for everybody, rather than through the Jewish Law.

Romansh. *See* ROMANCE LANGUAGES.

Romanticism, name of a specific historical movement in the arts. Romantic arts, which can be found in any period, emphasize the emotional perceptions of the artist and his self-expression above the formal requirements of the medium in which he is working. They frequently show an interest in the past or in nature, and stress the importance of the act of creation. The Romantic Movement occurred in the early 19th century in art, architecture, literature, music and political theory. It began as a late 18th-century reaction to the sterile repetition of CLASSICAL formulae and the facile emotions of 18th-century art and ENLIGHTENMENT thought. The Movement's most striking features included the expression of the artist's alienation from society; the depth of his emotions and his spiritual confusion; the attempt to fuse new links between man and the power of nature, and to replace Greek and Roman models with national mythologies. Romantic painting came to stress the terrible and dramatic rather than the lyrical or stylized – even the Romantic landscapes of TURNER are as full of drama as the historical paintings of DELACROIX – and many of the painters used violent brush-strokes to communicate the immediacy of their work. By the mid-19th century the search for immediacy led artists, especially COURBET, to turn to contemporary realistic themes. Romanticism in architecture brought the overthrow of the PALLADIAN style and a new appreciation of the medieval; the Gothic revival of PUGIN in England and VIOLLET-LE-DUC in France and a general carefree eclecticism was typical. Romantic literature developed particular in Germany, with GOETHE and SCHILLER, and in Britain, where BYRON, SHELLEY and WORDSWORTH wrote highly personal poetry and Walter SCOTT created the genre of the historical novel. In France, CHÂTEAUBRIAND and Victor HUGO emulated Byron's Romantic lifestyle and propagated the concept of the Romantic genius, inspired by God and cut off from his own times. Romantic music, epitomized by its freer form and rich orchestration, is first evident in the late work of BEETHOVEN; it was taken up by SCHUBERT, WEBER and WAGNER and held sway until the early 20th century. In political theory, Romanticism was linked with ideas of man's innate goodness, and with movements for national liberation and for liberalism. The European revolutions of 1848 were the last fling of co-ordinated Romantic politics. *See also* HC2 pp.72, 73, 80, 81, 102, 103, 106, 107.

Romanus (*d.*897), pope. He was elected to the papacy after the murder of STEPHEN VI, and his year-long reign (897) was marked chiefly by the feud between his supporters and those of his predecessor, whose acts Romanus nullified.

Romanus, name of four emperors of BYZANTIUM. Romanus I Lecapenus (*r.*920–44) was regent for, and then co-ruler with, his son-in-law, the young Emperor CONSTANTINE VII. His reign marked the high point of Byzantine influence on the Bulgarian kingdom. Romanus II, his grandson, reigned from 959–63. He had no interest in state affairs, and his wife, Theophano, virtually ruled the empire. Romanus III Argyrus (*c.*1028–34) came to the throne when he married the Empress Zöe. A weak ruler, he depleted his treasury and abolished needed taxes. Romanus IV Diogenes (1068–72), a general, became emperor when he married the widow of Constantine X. His defeat by the Seljuk Turks at Manzikert in 1071 marked the end of the BYZANTINE EMPIRE. *See also* HC1 p.143.

Romany, language spoken by the GYPSIES. It is an Indic language, oelated to SANSKRIT and HINDI, and is known to have originated in India. It is spoken by about three million people, but there is little written Romany literature other than translations of the Bible.

Romberg, Sigmund (1887–1951), Hungarian-born US composer whose best-known operettas include *Maytime* (1917), *The Student Prince* (1924) and *The Desert Song* (1926).

Rombouts, Theodor (1597–1637), Flemish painter. His paintings of genre scenes and religious subjects were early influenced by the style of CARAVAGGIO in their use of chiaroscuro and after *c.* 1630, inclined towards the manner of RUBENS.

Rome (Roma), capital of Italy; in the centre of the country on the River Tiber; capital of Lazio region and Roma province. Rome was founded in the 8th century BC from settlements traditionally located on seven hills rising from the marshlands of the Campagna. It was probably an Etruscan city-kingdom in the 6th century.

The Roman Republic was founded *c.*500 BC; by the 3rd century Rome ruled most of Italy and began to expand overseas. The ROMAN EMPIRE was founded in 27 BC. Caesar AUGUSTUS (Octavian) rebuilt the city and encouraged the development of art and education. The great fire of Rome in AD 64 destroyed much of the city. NERO rebuilt it and it remained capital of the Roman Empire until 330 AD. In the Middle Ages Rome was dominated by the papacy. Many great RENAISSANCE and BAROQUE artists and architects worked there. NAPOLEON annexed Rome from 1809 to 1814. Italian troops occupied the city in 1870 and in the following year it became the capital of Italy. The FASCIST march on Rome in 1922 brought MUSSOLINI to power. The city fell to the Allies during WWII in June 1944.

Rome includes VATICAN CITY, which has many fine libraries and museums. Notable buildings and monuments include the COLOSSEUM, the Forum and the Capitoline, the Farnese Palace, many ancient basilicas including ST PETER's and St John Lateran, and the University of Rome (1303). Industries: tourism, printing and publishing, fashion, film, banking and finance. Pop. (1975) 2,868,248. *See also* PE p.*75*; MW p.*107*; and index to *History and Culture 1*.

Rome, School of, Italian school of painting which flourished from the mid-16th to the 19th centuries. MICHELANGELO and RAPHAEL were its most important members; it led the Baroque movement, represented by BERNINI and Pietro da CORTONA; and from the 17th century onwards included POUSSIN, CLAUDE LORRAIN, Piranesi and artists from all over Europe, who came to Rome to study the classical remains.

Rome, Treaties, of, two agreements signed at Rome in 1957. They established the EUROPEAN ATOMIC ENERGY COMMUNITY (Euratom) and the European Economic Community (EEC). The signatory countries were Belgium, France, Italy, Luxembourg, The Netherlands and West Germany. Countries admitted into the EEC in 1973 – Britain, Denmark and Ireland – accepted the terms of the treaty as a requirement of admission. The treaty establishing the EEC provides for the eventual economic, monetary and political union of its member states.

Romeo and Juliet (*c.*1595), tragedy by William SHAKESPEARE. The story is set in Verona, in which two hostile families, the Montagues and the Capulets, reside. Romeo, a Montague, falls in love with Juliet, a Capulet, and an inevitable tragedy ensues. The two lovers die and their families end their feud. *See also* HC1 p.*285*; MS p.*167*.

Rommel, Erwin (1891–1944), German military commander. He proved his ability as a commander of tanks in France in 1940, and later led the AFRIKA KORPS Known as "The Desert Fox", he was defeated by the British at EL ALAMEIN. By 1944, Rommel was disenchanted with Germany's military leadership, and was in communication with plotters who tried to kill HITLER in July; when Rommel's association with the conspirators was discovered, he was forced to commit suicide. *See also* HC2 pp.228, 230–231.

Romney, George (1734–1802), British portrait painter. He visited Italy (1773–75), where he was influenced by the paintings of RAPHAEL, TITIAN and CORREGGIO. He painted more than 50 portraits of Lady Emma HAMILTON, usually in historical roles.

Romney Marsh, region of coastal marsh-

Roman roads formed a network connecting the major towns of Roman Britain.

Rome; the Villa Adriana at Tivoli, a popular summer resort in imperial days.

Romeo and Juliet; an Italian lithograph illustrating the death of Romeo.

George Romney's *A Shepherd's Daughter* contrasts with portraits of society women.

Romulus and Remus; the theme of abandoned children originated in Greece.

Wilhelm Röntgen in his laboratory, where he investigated X-rays and electricity.

Rooks are commonly seen at dusk when they fly back to their nesting colonies.

Franklin Roosevelt addressing Congress during his second term of office.

land in Kent, SE England, mainly between Hythe and Dungeness. The area has been reclaimed and now provides excellent pasture for sheep. Area: approx. 176sq km (68sq miles).

Romulus and Remus, twin brothers and legendary founders of Rome. They were the children of MARS and Rhea Silvia, daughter of the rightful king of Alba Longa. Amulius, the usurper of his throne, fearing they might be a threat to his rule, ordered them to be drowned. They floated safely down the Tiber River, however, where creatures sacred to Mars, a woodpecker and a she-wolf, rescued them. They were brought up by the shepherd Faustulus and founded Rome at the site where they were taken from the river. Later Romulus murdered Remus.

Romulus Augustulus (b.c.461), Western Roman emperor (r.475–76). He was given the throne by his father, Orestes, the master of the soldiers, who had overthrown the Western emperor Julius Nepos. The Germanic mercenaries in Italy then mutinied, killed Orestes and forced Romulus to abdicate; ODOACER, leader of the Barbarian troops, assumed the title of king. The date 476 is therefore the traditional date for the fall of the Western Roman Empire. *See also* HC1 p.115.

Roncalli, Angelo. *See* JOHN XXIII, POPE.

Roncesvalles. *See* CHANSON DE ROLAND.

Rondeau, French medieval verse form, consisting in its full form of four stanzas, with two rhymes only used throughout. The first line of the first stanza acts as a refrain ending the second and third stanzas. The fourth is the same as the first. Rondeaus were originally sung to music composed for several voices; today they are used chiefly as a vehicle for light or witty verse.

Rondo. *See* MUSICAL FORM.

Ronsard, Pierre de (1524–85), French poet and leader of the PLÉIADE. His *Odes* (1550) and *Les Amours* (1552) brought him fame and royal patronage, and he wrote several patriotic poems. His late *Sonnets pour Hélène* (1578) contained some of his finest love poems. *See also* HC1 p.316.

Röntgen, Wilhelm Konrad (1845–1923), German physicist. In 1895, while teaching at Würzburg University, he discovered X-rays, for which he was awarded the first Nobel Prize in physics in 1901. He also did important work on the specific heats of gases, the heat conductivity of crystals, and electricity. *See also* RÖNTGEN.

Röntgen, unit (symbol r) used to measure the X-ray or gamma-ray RADIATION to which a body is exposed. One röntgen causes sufficient ionization to produce a total electric charge of 2.58×10^{-4} coulombs on all the positive (or negative) ions in one kilogram of air.

Rood, old religious term for the cross on which Christ died. It was often set on a beam dividing the nave from the chancel in a church (the ROOD SCREEN), many of which were destroyed in the REFORMATION.

Rood screen, in most medieval and early Renaissance churches, the screen – made of wood or stone and often carved or painted – which separated the chancel from the NAVE. Upon it there was usually a representation of the Crucifixion scene; Christ on the cross and two flanking statues of the Virgin Mary and St John. The name derives from the Old English *rōd*, meaning gallows or cross. The rood and all other objects on many rood screens in England were removed during the REFORMATION, but the screen remained.

Rook, large gregarious European bird of the CROW family. It has glossy black plumage but commonly loses the feathers from about its face. It feeds on grain and insects and has a characteristic raucous cry. Family Corvidae; species *Corvus frugilegus.*

Rooke, Sir George (1650–1709), English admiral. He defeated the French during the War of the Grand Alliance at the Battle of La Hogue in 1692 and in 1702 he triumphed over the combined French and Spanish fleets at Vigo. In 1704, during the War of the Spanish Succession, he captured Gibraltar from Spain.

Room With a View, A (1908), novel by E.M. FORSTER which wittily deflates the snobbery of the upper class in the early 20th century. The heroine falls in love with a young man of lower status and this is seen as a threat by all around her.

Roosevelt, Anna Eleanor (1884–1962), US reformer and humanitarian, wife of Franklin Delano Roosevelt. She visited Britain and the South Pacific in WWII and after the war became US delegate to the UN (1945, 1949–52, 1961) and was chairman of the UN Commission on Human Rights (1946–51). *See also* HC2 p.209.

Roosevelt, Franklin Delano (1882–1945), 32nd US president (1933–45), the only one elected for four consecutive terms. He entered politics as Pres. Woodrow WILSON's assistant secretary of the Navy (1913–20). In 1921 he was stricken by polio which left his legs paralysed. Encouraged by his wife Eleanor ROOSEVELT to resume political life, he was elected governor of New York (1928). In 1932, during the DEPRESSION, he won the presidnetial election as the Democratic Party candidate. His "NEW DEAL", a set of measures designed to restore the economy through direct government intervention antagonized Republicans and conservatives, but Roosevelt's popularity led to a great electoral victory in 1936, and a narrower success in 1940. Early in his third term, he abandoned the traditional US foreign policy of isolationism and aided Britain against Nazi Germany; after the Japanese attack on PEARL HARBOR he led his country into WWII. Re-elected for a fourth time in 1944, he died soon after inauguration. *See also* HC2 pp.209, *209, 218, 235.*

Roosevelt, Theodore (1858–1919), 26th US president (1901–09). After an eventful career, he was elected vice-president in 1900, succeeded to the presidency in 1901 when MCKINLEY was assassinated and was re-elected in 1904. He believed that the presidency should maintain a vigorous involvement in domestic affairs, and initiated much legislation designed to regulate big business; in foreign affairs he extended US power, acquiring the Panama canal zone and reducing European influence in S America. *See also* HC2 pp.*147, 153.*

Root, Elihu (1845–1937), US lawyer and diplomat who advocated US entry into the LEAGUE OF NATIONS and helped to bring the World Court of INTERNATIONAL JUSTICE into existence. He was awarded the 1912 Nobel Peace Prize for his contribution to peace in the Western hemisphere.

Root, John Welborn (1850–91), US architect who became a partner of Daniel BURNHAM in Chicago. Extremely influential in their day, their work includes the Montauk Building (1882) and the Monadnock Building. In these tall buildings Root linked technical innovation and aesthetic design.

Root, underground portion of a VASCULAR PLANT that serves as an anchor and absorbs water and nutrients from the soil. The primary root develops from the lower end of the plant embryo. Some plants, such as the dandelion and carrot, have taproots with smaller lateral branches. Others, such as the grasses (including cereal crops) develop fibrous roots with lateral branches equalling the main root in size. In the mangrove, adventitious roots are produced from the stem above the ground and grow down to provide additional support and means of absorbing water and minerals. Some plants, such as beets and parsnips, have large taproots that store nutrients and are grown as food for men and animals. *See also* NW pp.44–45.

Root, in mathematics, fractional POWER of a number. The square root is written as either $\sqrt{x}$ or $x^{1/2}$, and represents a number that, when multiplied by itself, gives the original number. The fourth root of a number, x, may be written in RADICAL form as $4\sqrt{x}$ or in power form as $x^{1/4}$. The fourth root of 16, for example, is 2 since $2 \times 2 \times 2 \times 2 = 16$.

Root and Branch Petition (1640), petition from London Puritans introduced in the House of Commons, asking for the abolition of the episcopacy in the CHURCH OF ENGLAND "root and branch". The MPs who supported it became known as the Root and Branch party.

Rope, flexible, heavy cord made of twisted natural fibres, synthetic fibres or metal (CABLE). Among natural fibres, HEMP is most commonly used. In rope-making, fibres are first disentangled and spun into yarn, which is then twisted into strands that are laid up into rope. *See also* PE p.216, *216.*

Rops, Félicien (1833–98), Belgian painter and graphic artist. Working mainly in France, he was known for his prolific etchings and book illustrations, many of which were perverse or erotic in nature.

Roque. *See* CROQUET.

Roquefort, softish cheese made from the milk of ewes and goats, named after the town in S France where it originated. It has the veins caused by a mould, *Penicillium roqueforti*, which gives the cheese its individual strong flavour and aroma. *See also* PE pp.228–229, *229.*

Roraima, Mount, flat mountain peak at the junction of the Guyanan, Brazilian and Venezuelan borders; the highest point of the Guyana Highlands. It is the source of tributaries of the rivers Amazon and Orinoco. Height: 2,810m (9,219ft).

Rorqual, any of five species of schooling WHALES, distributed throughout the world; they are the source of baleen, or whalebone. Included are the blue whale, fin whale and humpback whale. Genus *Balaenoptera.*

Rorschach, Hermann (1884–1922), Swiss psychiatrist. He invented the Rorschach Ink Blot Technique, often used to assess personality characteristics. *See also* RORSCHACH TEST.

Rorschach tests, in psychology, familiarly called the "ink blot tests", used to analyse a person's motives and attitudes when these are projected into ambiguous situations. The individual is presented with 10 standardized ink blots and interpretation is based on his description of them. Hermann RORSCHACH devised the tests.

Rosa, Salvator (1615–73), Italian painter and poet who lived mainly in Rome but worked for the MEDICI family in Florence from 1640 to 1649. He is best known for his savage landscapes which were executed boldly, although as painter of religious and historical subjects he was less successful. His poetry satirized his age.

Rosaceae, the rose family, a large group of flowering plants containing about 3,000 species, many of which are important food plants. It includes the APPLE, PEAR, PLUM, CHERRY, APRICOT, PEACH, NECTARINE, BLACKBERRY, RASPBERRY, STRAWBERRY, MEDLAR, QUINCE, LOQUAT, ALMOND and MOUNTAIN ASH. The ROSES themselves, genus *Rosa*, are of considerable variety. *See also* PE pp.178–179, 192–193.

Rosario, port in E central Argentina, on the Paraná River, 306km (190 miles) NW of Buenos Aires; the second largest city in Argentina. Founded in 1725, Rosario grew with the development of the pampas. Today it is a communications and exporting centre, and has food-processing industries. Pop. 810,840.

Rosary, form of meditational prayer in which the worshipper contemplates the life of Jesus and the Blessed Virgin Mary; it is almost entirely restricted to Roman Catholics. The term is also used as a name for the string of beads on which an account may be kept of the number of prayers said.

Rosas, Juan Manuel de (1793–1877), Argentine CAUDILLO and twice Governor of Buenos Aires province. By 1835, his rivals dead or in exile, he had become virtual dictator of Argentina. He ruled with absolute power and survived blockades by the British and French, but was finally defeated by Justo José de URQUIZA in 1852 and died in exile in Britain.

Roscellinus Johannes (c.1050–c.1125), Scholastic philosopher and theologian. Little is known of his life except that he was one of the first defenders of NOMINALISM. His teachings were attacked by his famous pupil, Peter ABELARD.

Roscommon, county in N central Republic of Ireland, in Connacht province. Part of the central plain of Ireland, Roscommon is generally low-lying and the Shannon is the principal river. Sheep and cattle rearing is the chief economic activity; coalmining is important in the NE. Roscommon is the county town. Area: 2,463sq km (951sq miles). Pop. (1971) 53,519.

Rose, Murray (1939–), Australian freestyle swimmer, *b*.Britain. At the 1956 Olympic Games he won three gold medals (400m, 1,500m and 4 × 200m relay), retaining his 400m title in 1960, when he also won a 1,500m silver and a relay bronze. He also won four golds at the 1962 Commonwealth Games and established nine world records.

Rose, wild or cultivated flowering shrub. Most roses are native to Asia, several to America, and a few to Europe and NW Africa. The stems are thorny and erect, climbing or trailing. Flowers appear singly or in loose clusters and colours range from white to yellow, pink, crimson and maroon; many are fragrant. There are about 200 species. Family Rosaceae; genus *Rosa. See also* PE pp.178–179.

Roseau, port and capital of DOMINICA, in the Windward group, on the Roseau River. The port exports tropical vegetables, oils, spices, limes and lime juice. Pop. (1970 est.) 10,200.

Rosebay Willow herb. *See* WILLOW HERB.

Rosebery, Archibald Philip Primrose, 5th Earl of (1847–1929), British Liberal politician. He succeeded GLADSTONE as Prime Minister (1894–95), but resigned following a split in the Liberal Party over imperial policy. He retired from political life in 1905, just before Henry CAMPBELL-BANNERMAN became Liberal Prime Minister.

Rosemary, strongly flavoured, perennial evergreen herb of the mint family. It has small, needle-like leaves and clusters of small pale-blue flowers. Sprigs of rosemary are commonly used as a flavouring when cooking various meats. Family Labiatae; species *Rosmarinus officinalis. See also* PE p.213, *213.*

Rosenberg, Alfred (1893–1946), German Nazi leader, *b*. Estonia. He was an early member of the Nazi Party, and his racial theories contributed to the party's ANTI-SEMITISM. In 1941 he became Minister for the occupied Eastern Regions. After the war he was executed as a war criminal.

Rosenberg, Isaac (1890–1918), British poet. After studying at the Slade School of Art 1911–14 he served as a private in WWI. His best poems were written about his war experiences. He was killed in action.

Rosencrantz and Guildenstern are Dead (1966), two-act comedy by Tom STOPPARD which established his reputation and with great wit and dramatic ingenuity explores the destiny of two minor characters from *Hamlet.*

Rosenfeld, Albert (1885–1970), Australian rugby league player. A fast, elusive wing-threequarter, he toured England in 1908–09 and returned to play for Huddersfield. In the 1911–12 season he scored 80 tries, a record that remained unbeaten in the mid-1970s.

Rosenkavalier, Der (1910), three-act opera by Richard STRAUSS, with libretto by Hugo von HOFMANNSTAL, first performed in Dresden in 1911. The light-hearted story centring on the love of the young rose-bearer Octavian (mezzo-soprano) for the betrothed Sophie (soprano) is set in 18th-century Vienna.

Rose of Sharon, hardy shrub, native to China, which can be grown farther north than many other HIBISCUS plants. The flowers are red, purple, rose or white. Height 4.6m (15ft). Family Malvaceae.

Roseola infantum, mild infectious disease of young children, probably caused by a VIRUS. It is characterized by drowsiness and fever, followed by a mild rash on the trunk and neck. The usual treatment is bed rest and a light diet.

Roses, Wars of the (1455–85), English civil wars fought for the possession of the crown, taking their name from the badges of the House of LANCASTER (red rose) and the House of York (white rose). By 1455 the incompetence of the Lancastrian HENRY VI had led to widespread civil disorder. Henry's interests were promoted by his wife, MARGARET OF ANJOU, who was

determined to protect the inheritance of their son Edward (*b*.1453). In 1455 Richard, Duke of York, and a small group of lords attempted to gain control of the government but were at first defeated. York was killed in 1460, but his son claimed the throne, defeated Henry VI at TOWTON FIELD in 1461 and was crowned as EDWARD IV. In 1470, Edward's closest ally, the Earl of WARWICK, deserted him for the Lancastrians. Briefly deposed, Edward regained his throne and defeated his enemies in 1471. Henry VI and his son were killed. When Edward IV died in 1483, his brother usurped the throne as RICHARD III and Edward's sons were murdered. Their deaths allowed Henry Tudor to rally the Lancastrian faction and Richard III was overcome at Bosworth in 1485 by Henry who, as HENRY VII, married Edward IV's daughter, ELIZABETH OF YORK in 1486, thus reuniting the two warring factions. *See also* HC1 pp.242–243, *243.*

Rosetta stone, slab of black basalt inscribed in HIEROGLYPHIC, demotic and Greek, now in the British Museum. The stone was discovered in 1799 by an officer of Napoleon's army in Egypt. By comparing the three versions of the inscription, first Thomas YOUNG (1818) and later Jean-François CHAMPOLLION (1822) were able to decipher the Egyptian hieroglyphic, and this led to a full understanding of hieroglyphic writing.

Rosewall, Ken (1934–), Australian tennis player. He joined the Australian DAVIS CUP team in 1953 and won the Australian and French titles that year; he remained one of the world's top players for more than 20 years. His major victories after tennis became open to professionals and amateurs include the French Open in 1968, the US Open in 1970, the Australian championship in 1971 and the World Championship of Tennis in 1971 and 1972 (winning both matches against Rod LAVER). He was the beaten men's singles finalist at Wimbledon four times (1954, 1956, 1970 and 1974).

Rose window, decorated wheel-shaped window, popular in GOTHIC and some ROMANESQUE church architecture and usually used at the west end of the nave or in the transepts. It consists of delicate bar TRACERY which radiates symmetrically from a circular carved stone core. The openings are filled with stained glass.

Rosewood, any of several kinds of ornamental hardwoods derived from various tropical trees. The most important are Honduras rosewood, *Dalbergia stevensoni,* and Brazilian rosewood, *D. nigra.* It varies from a deep, ruddy brown to purplish and has a black grain, hard to work because of its high resin content. Once much in demand by cabinetmakers, it is now rare but is occasionally used to make musical instruments. *See also* PE p.218.

Rosh Hashana (Rosh ha-Shanah), the Jewish New Year and first day of the month of Tishri (generally in September). It is also the day on which the ceremonial ram's horn, the *shophar* or *shofar,* is blown to call sinners to repentance; the day called the Day of Judgement or of Remembrance; and the day which begins the Ten Days of Penitence that end with the Day of Atonement, YOM KIPPUR. It is a day for solemn recollection and meditation.

Rosicrucians, secret, worldwide brotherhood using supposedly magical knowledge, founded by Valentine Andrea (1586–1654), who used the pseudonym Christian Rosenkreuz.

Rosin, or colophony, yellowish amorphous resin obtained as a residue from the distillation of turpentine; the chief constituent is abietic acid. It is used in varnishes, soldering fluxes, linoleum and printing inks. Properties: s.g. 1.08; m.p. 100–150°C (212–308°F).

Ross, Sir James Clark (1800–62), British naval officer and explorer. He took part in six Arctic expeditions in search of a NORTH-WEST PASSAGE, and in 1831 discovered the magnetic North Pole. His expedition to the Antarctic in 1839, during which he discovered the Ross Sea in 1841, earned him a knighthood.

Ross, Sir John (1777–1856), British naval

commander and Arctic explorer. His first expedition in 1818 was to explore BAFFIN BAY and attempt a NORTH-WEST PASSAGE. His second, in 1829, produced valuable geographic and oceanographic data. In 1850 he led an unsuccessful search for Sir John FRANKLIN.

Ross, Martin, pen-name of Violet Florence Martin (1862–1915), Irish novelist. With her cousin, Edith SOMERVILLE, she wrote *Some Experiences of an Irish R.M.* (1899).

Ross, Sir Ronald (1857–1932), British bacteriologist, *b*.India, who was awarded the 1902 Nobel Prize in physiology and medicine for his research into MALARIA.

Ross and Cromarty, former county in N Scotland; since 1975 it has been part of Highland Region. It extends from the North Sea to the Atlantic Ocean, and includes some of the Outer Hebrides. The area is largely mountainous and moorland. The principal economic activities are sheep grazing, whisky distilling and fishing. Dingwall was the county town. Area: 8,000sq km (3,089sq miles). Pop. (1971) 58,267.

Ross Dependency, area of Antarctica which includes Ross Island, the coast along the Ross Sea and nearby islands. It has been under the jurisdiction of New Zealand by act of British Parliament since 1923. Area: land mass 427,350sq km (160,000sq miles); ice shell 336,700sq km (130,000sq miles).

Rossellini, Roberto (1906–77), Italian film director and producer. His interest in REALISM is shown in his major films *Open City* (1945), *Paisa* (1946), *L'Amore* (1948), *Stromboli* (1949) and *La Paura* (1954). From 1950 to 1956 he was married to Ingrid BERGMAN. *See also* HC 2 pp.276, *277.*

Rosselli, Cosimo (1439–1507), Italian painter. With BOTTICELLI and GHIRLANDAIO he was commissioned (1481) to paint frescoes in the Sistine Chapel, Rome. These included *The Last Supper* and *The Sermon on the Mount.* Among his other works in *The Annunciation* (1473), and his pupils included Fra BARTOLOMMEO and Piero di Cosimo.

Rossetti, Christina (1830–94), British poet, the sister of Dante Gabriel ROSSETTI. She specialized in religious poetry and verses for children.

Rossetti, Dante Gabriel (1828–82), British poet and painter. He was a founding member of the PRE-RAPHAELITE BROTHERHOOD. Although his poetry was first published when he was 19, he did not achieve recognition as a poet until long after his paintings had won acclaim.

Rossini, Gioacchino Antonio (1792–1868), Italian composer. He was a superb craftsman and his music dominated the world of opera in the early 19th century. Among his comic operas *The Italian Girl in Algiers* (1813) and *The Barber of Seville* (1816) demonstrate the sparkling sense of melody and humour that made him so popular. *See also* HC2 pp. *106–107,* 120, *121.*

Ross Island, name of two islands in Antarctica. One lies in the Weddell Sea and was discovered by NORDENSKJÖLD in 1903. Area: approx. 3,880sq km (1,800sq miles). The other is in the ROSS SEA and was discovered by Sir John ROSS in 1841. Area: approx. 6,500sq km (2,500sq miles).

Rosso, Fiorentino (1495–1540), Italian painter, real name Giovanni Battista Rosso. He worked in Florence, Rome and France, where in 1530 he became painter for FRANCIS I and helped to decorate FONTAINEBLEAU. *See also* HC1 pp.263–265, *264.*

Rosso, Medardo (1858–1928), Italian painter who became a sculptor in 1883. The play of light on surfaces was an absorbing interest of his, and he used it to good effect in his sculpture. He worked mainly in wax and was a friend of RODIN.

Ross Sea, arm of the S Pacific Ocean between Victoria Land and Edward VII Peninsula, Antarctica. It was discovered in 1841 by Capt. James C. Ross.

Rostand, Edmond (1868–1918), French poet and dramatist. His first plays, such as *Les Romanesques* (1894) and *La Princesse Lointaine* (1895), were fanciful and of lit-

Ken **Rosewall** in the men's singles semi-final at Wimbledon in 1955.

Roberto Rossellini with Ingrid Bergman, who starred in several of his films.

Dante Gabriel Rossetti's romantic female quartet *Dallying on the Grass* (1872).

Gioacchino Rossini; Meyerndorff's portrait of the great Italian composer.

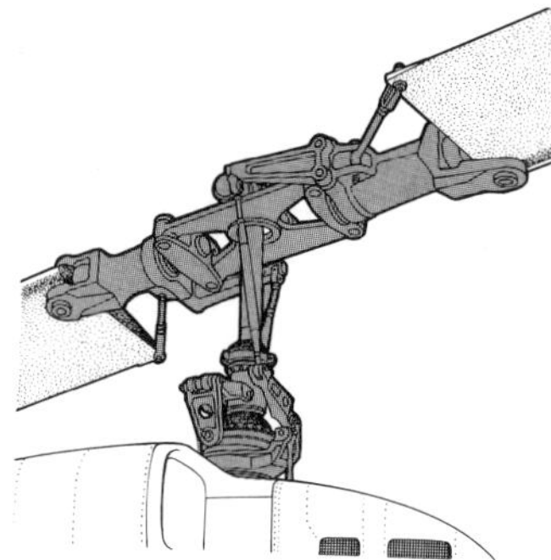

Rotor; when the pitch of the helicopter blades increases, so does lifting power.

Rotterdam was almost totally destroyed by German bombing in autumn 1944.

Rouen, the ancient French town where Joan of Arc was burnt at the stake.

Francis Peyton Rous demonstrated that cancer can be caused by viruses.

tle substance. They were followed by *Cyrano de Bergerac* (1897), *L'Aiglon* (1900) and *Chantecler* (1910). *See also* HC2 p.101.

Rostock, port in N East Germany, on the Baltic Sea; capital of Rostock district. The city received its charter in 1218 and was an important Baltic member of the HAN-SEATIC LEAGUE in the 14th century. Its university, founded in 1419, was a centre of N European learning. The city was badly damaged during WWII. Today it is a fishing and food-processing centre and its industries produce farm machinery, watches, chemicals, diesel engines and furniture. Pop. (1975) 211,723.

Rostov-on-Don (Rostov-na-Donu), city in the Russian Republic (Rossijskaja SFSR), USSR, on the River Don near its mouth in the Sea of Azov; administrative centre of Rostov oblast. It grew around a fortress built between 1761 and 1763. The port was established in 1834 and was a major grain-exporting centre throughout the 19th century. The city suffered great damage during WWII. Industries include agricultural machinery, electrical equipment, barges and chemicals. Pop. (1975) 888,000.

Rostropovich, Mstislav Leopoldovich (1927–), Soviet musician who established a reputation as one of the century's best cellists. Leading composers such as SHOSTAKOVICH, KHACHATURIAN and PROKO-FIEV dedicated works to him, as did Benjamin BRITTEN, with whose ALDEBURGH Festival Rostropovich became closely associated.

Rosyth, town in S Fife Region, central Scotland, on the Firth of Forth; major naval base and dockyard. During WWII the dockyard was expanded and homes were built for workers. Rosyth is the only refitting base for nuclear submarines in Britain. Pop. 6,830.

Rotary Clubs, international organization of clubs for business and professional people. They promote ethical standards in business and professional practice and encourage the concept of public service in commercial dealings. In most countries they are also involved in welfare work.

Rotary printing press, machine which uses a cylinder revolving at high speed to print its image. It may be either a LETTERPRESS or LITHOGRAPHIC machine. In the former, STEREOTYPE plates are curved round the cylinder and print directly on to the paper; in the latter flexible metal plates offset their image to an intermediate rubber roller which then transfers it to the paper. Rotary photogravure presses operate in a similar way to rotary letterpress machines but use etched copper cylinders. *See also* WEB OFFSET; MM pp.204–205, *204–205, 206–207, 206–207.*

Rotation, turning of a celestial body about its axis. In the Solar System the Sun and all the planets, with the exception of Uranus and Venus, rotate from W to E.

Rotation of crops. *See* CROP ROTATION.

Roth, Philip (1933–), US novelist and short-story writer. He established his name with the collection of stories, *Goodbye Columbus* (1959). His best-known novel is *Portnoy's Complaint* (1970).

Rothamsted Experimental Station, agricultural and biological research centre in Harpenden, W Hertfordshire, England, 6.5km (4 miles) N of St Albans. Founded in 1843, it is the oldest such experimental station in the world.

Rothenstein, name of a family of British artists of German stock. Sir William (1872–1945) was noted for his portraiture and was Principal of the Royal College of Art from 1920 to 1935. Sir John Knewstub Maurice (1901–) was an art historian, critic and administrator who was Director of the Tate Gallery (1938–64) and whose numerous publications included *Modern English Painters* (3 vols, 1952, 1956, 1973). Michael (1908–) was an artist particularly noted for his printmaking.

Rotherham, county district in SE South Yorkshire, England; created in 1974 under the Local Government Act (1972). Area: 283sq km (109sq miles). Pop. (1974 est.) 248,100.

Rothermere, Lord. *See* NORTHCLIFFE,

ALFRED CHARLES WILLIAM HARMSWORTH, LORD.

Rothko, Mark (1903–70), US painter, *b.* Russia. He emigrated to the USA in 1913 and became a student of Max WEBER; he later came under the influence of the SURREALISTS. In the 1940s he experimented with ABSTRACT EXPRESSION-ISM. His large canvases generally consist of blurred rectangles of intense colour, such as his work *Green on Blue* (1956). *See also* HC2 p.205.

Rothschild family, commercial and banking dynasty founded by Meyer Amschel Rothschild (1744–1812), son of a money changer in the Jewish ghetto of Frankfurt. His five sons established branches in the financial centres of Europe and the family fortune was established by profitable dealings during the NAPOLEONIC WARS. The Rothschilds were one of the chief financial powers in the 19th century but developments in state financing late in the century greatly reduced their influence.

Rotifer, or "wheel animacule", microscopic METAZOAN found mainly in fresh water. Although it resembles a ciliate PRO-TOZOAN, it is many-celled with a general body structure similar to that of simple WORMS. Rotifers may be elongated or round and are identified by a crown of cilia around the mouth. Class Rotifera. *See also* NW pp. *121, 231.*

Rotor, helicopter, engine-driven rotary wing. Each of the blades is an AEROFOIL that produces lift in the same manner as an aeroplane wing. The hub also moves to permit altering the plane of rotation of the rotors to give sideways or forward and back motion. *See also* MM p.154, *154.*

Rotorua, city in central North Island, New Zealand, founded in the 1870s. Primarily a tourist resort noted for hot springs, geysers and mud pools, it is an important Maori cultural centre. Industries include food processing, cement, timber and engineering. Pop. (1972) 32,600.

Rotten boroughs, constituencies which before the 1832 Reform Act had very few electors and yet were represented in parliament. Such small electorates were entirely the result of historical changes of population but the outcome was that each elector could be approached with offers of bribery.

Rottenhammer, Johann (1564–1625), German painter who painted small mythological subjects on copper and paid particular attention to background landscapes.

Rotterdam, major European and industrial city in W Netherlands, on the New Meuse (Nieuwe Maas) River 24km (15 miles) from the North Sea; second largest city in The Netherlands. Rotterdam is one of the largest and most modern ports in the world. It has a large transit trade with industrial areas of Europe, in particular the Ruhr, West Germany. The city developed with the construction of the New Waterway (1866–72), connecting it to its outer port at Hoek van Holland, and making Rotterdam accessible to ocean-going vessels. During WWII the centre of the city was destroyed by German bombing. Industries: chemicals, textiles, paper, motor vehicle assembly, clothing, food processing, brewing, oil refining, machinery. Pop. 686,600.

Rottweiler, German cattle dog. The short-backed, strong body is set on muscular, medium-length legs and the tail is commonly docked. The short, coarse, flat coat is black with brown markings. Height: to 68.5cm (27in) at the shoulder; 34–41kg (75–90lb).

Rouault, Georges (1871–1958), French painter and printmaker. He painted naturalistic landscapes and pictures with religious and mythological themes. Influenced by FAUVISM, his style became less realistic with his use of dark reds and blues, blots and overpainting to depict the corruption of the world, embodied in his judges, clowns and prostitutes. He turned to printmaking in about 1914 but after 1940 concentrated on religious painting, using heavy black outlines, strong colouring and simplified forms. He destroyed much of his work shortly before his death.

Rouen, port in N France on the River

Seine 114km (71 miles) NW of Paris; capital of Seine-Maritime département. By the 10th century, Rouen was the capital of Normandy and one of Europe's leading cities. It was the scene of Joan of Arc's trial and burning in 1431. Rouen has many fine ecclesiastical buildings including the Cathedral of Notre Dame (12th–15th centuries). The city has a university (1966). Industries: textiles, flour milling, iron foundries, perfumes, leather goods. Pop. (1970 est.) 120,300.

Roulette, game of chance popular at gambling houses. Thought to date from the mid-17th century, it uses a small ivory ball tossed into a wheel which is spun by the gaming-table attendant (the croupier), bets having been placed beforehand on where the ball will come to rest. The wheel is marked off into 37 squares (38 in the USA) with retaining walls. The squares, alternatively red and black, are numbered one to 36; the remaining square, marked 0, is usually green (in the US the extra square is marked 00). Players place a variety of bets on a table which has an arrangement of numbered squares corresponding to the numbers of the wheel.

Round, musical form in which a tune is sung in canon. Performers begin the melody one after another at regular time intervals; the tune is so constructed that the whole is together harmonious.

Round, Theatre in the. *See* THEATRE-IN-THE-ROUND.

Rounders, game resembling BASEBALL. Played mostly by children, it needs two teams, usually of nine players, who alternate as batting and fielding sides for two innings each. The aim is to score rounders by hitting the ball and running around a circuit of four posts or bases. A batsman is out if he is caught, runs inside a post or if a fielder holding the ball touches a post or the batsman while he is between bases. The team scoring the most rounders wins.

Roundheads, name given to Parliamentary party supporters during the ENGLISH CIVIL WAR (1642–51). The name was originally a derogatory term for PURITANS, who cut their hair short in contrast to the long hairstyles worn by CHARLES I and the Royalists. *See also* HC1 p.288.

Round Table, in Arthurian legend, large circular table that seated KING ARTHUR'S knights. A table now in Winchester Cathedral, thought for many years to have been the Arthurian Round Table, was proved in the mid-1970s to be of the 14th century. *See also* HC1 p.21.

Round Table Conferences, two series of talks in British history. The first were held in 1887 on the initiative of Joseph CHAM-BERLAIN in an unsuccessful attempt to reunite the LIBERAL PARTY after the split over IRISH HOME RULE in 1886. The second were the discussions in London (1930–32), attended by British and Indian representatives, which laid the basis for the Government of India Act of 1935.

Roundworm, parasite of the class Nematoda, which inhabits the intestine of mammals. It breeds in the intestine. The larva bores through the intestinal wall, is carried to the lungs in the bloodstream and crawls to the mouth where it is swallowed. Length: 15–30cm (6–12in).

Rous, Francis Peyton (1879–1970), US pathologist who shared the 1966 Nobel Prize in physiology and medicine with Charles B. HUGGINS for the discovery of CANCER-inducing viruses. In 1910 Rous began a series of experiments on sarcomas in hens, injecting healthy birds with a solution made from sarcomas from cancerous ones. He filtered the solution to remove all cells and bacteria but found that the healthy hens developed cancer, which led him to postulate that this cancer could be caused by viruses. The theory was rejected at the time, but later experiments proved him correct.

Rous, Sir Stanley (1895–), British soccer administrator. After 15 years as a referee, he was secretary of the Football Association (1934–62) and president of FIFA (1962–74).

Rousseau, Henri "le Douanier" (1844–1910), French primitive painter who was self taught. He was called le Douanier because before he retired to paint in 1885 he held a post in the Paris customs office. His style is characterized by a strong sense of pattern and colour combined with the use of exotic jungle themes showing tropical plants and beasts.

Rousseau, Jean-Jacques (1712–78), French philosopher born in Geneva, considered to be a founder of ROMANTICISM. He contributed to the ENCYCLOPÉDIE of DIDEROT in the 1740s, and won fame for his *Discourses on Science and the Arts* (1750). In *The Social Contract* (1762) he argued that man had been corrupted by civilization; he envisaged a society in which freedom and equality would be achieved by individuals agreeing to contribute to and be subject to a "General Will", and in which executive power was delegated by the people. His ideas on individual liberation from the constraints of society were developed in *Émile* (1762), in which he outlined his theories of education. He described his early, wandering life in a revealing autobiography, *Confessions*, published posthumously in 1782. *See also* HC2 pp.23, 73; MS pp.*185, 249*, 282.

Rousseau, Théodore (1812–67), French painter, a leading member of the BARBIZON SCHOOL. Strongly influenced by CLAUDE LORRAIN, CONSTABLE and the 17th-century Dutch landscapists, he painted romantic landscapes which nevertheless retained an objective observation of nature.

Roussel, Ker-Xavier (1867–1944), French painter and member of the NABIS who used small IMPRESSIONIST strokes of clear unmixed paint to give a more poetic meaning to landscapes.

Roussillon, region and former province of s France on the Spanish border, bounded by the E Pyrenees (s) and the Gulf of Lion (E). After changing hands many times the region was conquered in 1642 by Louis XIII and annexed to France in 1659 by the Treaty of the Pyrenees.

Rowan. *See* MOUNTAIN ASH.

Rowe, name of two British table tennis players, twin sisters, who became world champions. Diane (1933–) was a left-hander, and her sister Rosalind was a right-hander. They won the world doubles title in 1951 and 1954 and were three times runners-up. Diane ("Di") reached another two doubles and one mixed doubles finals and was twice a European doubles champion.

Rowe, Nicholas (1674–1718), British dramatist. After becoming a lawyer he turned to literature and wrote his first play, *The Ambitious Stepmother*, in 1700. He rose to fame with *The Tragedy of Jane Shore* (1714).

Rowing, way of moving a boat through water using oars, popular as a sport. Modern racing boats hold crews of two, four or eight oarsmen, each crew member using both hands to pull one oar through the water (the use of two oars by a single oarsman is sculling). The oars are attached alternately to opposite sides of the boat, or shell. A COXSWAIN always steers for eights and directs the crew; pairs and fours may or may not have a coxswain. Rowing as a sport has been popular in Britain since 1829, when the first Oxford and Cambridge University crew races were held at Henley-on-Thames. It was unofficially included in the Olympic Games in 1900 and 1904; from 1908 it was an official olympic sport. *See also* BOAT RACE; DIAMOND CHALLENGE SCULLS.

Rowlandson, Thomas (1756–1827), British draughtsman and caricaturist who portrayed life and manners in 18th-century Britain. Among his drawings are *Vauxhall Gardens* (1784) and *The Comforts of Bath* (1798). He used ink and a reed pen, and the fluidity of his line has often been likened to the French ROCOCO style of FRAGONARD. *See also* HC2 pp.*39, 70, 78*.

Rowley, William (*c.*1585–*c.*1642), English dramatist and actor. Specializing in fat clown roles, he was a popular figure in the theatre. He wrote numerous plays, including *A Shoemaker a Gentleman* (1608), *All's Lost by Lust* (1619) and *The Changeling* (1621) with Ruddleton.

Rowling, Wallace Edward (1927–), New Zealand politician and Prime Minister (1974–75). He was Finance Minister (1972–74) and became Prime Minister on the death of Norman Kirk (1974), but in the general election of 1975 his Labour government was heavily defeated.

Rowntree, Benjamin Seebohm (1871–1954), British businessman and sociologist, son of Joseph ROWNTREE. In 1889 he joined the family chocolate firm and introduced employees' pensions (1906), a five-day week (1919) and employee profit-sharing (1923). He published three studies of York, *Poverty: A Study of Town Life* (1901), *Poverty and Progress* (1941) and *Poverty and the Welfare State* (1951).

Rowntree, Joseph (1836–1925), British manufacturer and social reformer. A Quaker, he was aware of the bad industrial conditions endured by the working class, and in the Rowntree cocoa factory he founded at York he strove to establish good working conditions and fair wages. He introduced unemployment relief, set up social and charitable trusts and built a model village in New Earswick (1904) for his workers.

Rowse, Alfred Leslie (1903–), British historian. He was a fellow of All Souls' College, Oxford, from 1925 to 1974. He was principally an Elizabethan and Shakespearean scholar.

Roxburgh, former county in s Scotland on the border with England; since 1975 it has been part of Borders Region. An upland area in the Cheviot Hills, Roxburgh is drained by the Tweed, Liddel Water and Teviot rivers. Sheep rearing is the principal agricultural activity. Textiles is the most important industry. Jedburgh was the county town. Area: 1,722sq km (668sq miles). Pop. (1971) 41,942.

Roy, Gabrielle (1909–), French-Canadian author. Her novels about the Montreal working class include *Bonheur d'occasion* (1945; tr. 1947, *The Tin Flute*) and *Alexander Chenevert caissier* (1954; tr. 1955, *The Cashier*).

Royal Academy of Arts, national academy of the arts in Britain, founded by GEORGE III in 1768 on the advice of William CHAMBERS. Its first president was Sir Joshua REYNOLDS. Since 1867 it has been housed at Burlington House, London. There are 40 academicians at any one time and 30 associates. *See also* HC2 p.71.

Royal Academy of Dramatic Art. *See* RADA.

Royal Academy of Music, teaching institution officially opened in London in 1823 and chartered under royal patronage in 1830. It offers about 70 scholarships and competitive exhibitions to students.

Royal Air Force (RAF), youngest of the armed services of Great Britain. It was formed in 1918 by the amalgamation of the Royal Naval Air Service and the Royal Flying Corps. It was organized on a permanent basis under a scheme drawn up by Lord TRENCHARD in 1919, and controlled by the Air Ministry until this merged with the Ministry of Defence in 1964. In 1939, the RAF had 2,600 operational aircraft and 174,000 personnel. During the war its most famous action was the Battle of Britain. The work of coastal command in protecting convoys against U-boat attack in the Battle of the Atlantic was also important. Towards the end of the war, most resources were devoted to the bombing of Germany by long-range aircraft such as the *Lancaster*. By 1945 the RAF had grown immensely: it then consisted of 9,200 aircraft and 1,079,835 personnel. After WWII there were major technical innovations as jet aircraft came into service. Large manned aircraft, such as the "V" bombers of the 1960s, were gradually superseded as guided missiles became more important. The role of manned aircraft then became one of close support for ground troops, especially using manoeuvrable small planes capable of vertical takeoff. *See also* HC2 pp.*191, 228–230*; MM pp.176–178.

Royal Albert Hall, circular concert hall in Kensington, London, built mainly by public subscription and opened in 1871 as a memorial to Prince ALBERT. It has a seating capacity of about 10,000 and is the venue for the annual Promenade Concerts started by Sir Henry Wood.

Royal and Ancient Golf Club, the leading administrative body, along with the US Golf Association, of world golf. Based at St Andrews in Scotland, it adopted responsibility for the game's rules in 1897.

Royal Ballet. *See* BALLET, ROYAL.

Royal Canadian Mounted Police, federal police of Canada which was originally organized in 1873 as the Northwest Mounted Rifles to bring law and order to the Canadian west. The prefix "Royal" was added in 1904.

Royal College of Music, institution founded to set standards of instruction, award degrees and promote the art of music in Britain generally. It was founded by the Prince of Wales in 1882. Its present premises in Prince Consort Road, Kensington, were opened in 1894.

Royal Court Theatre, London theatre which opened in 1888 and which during a chequered history has had two periods of popular and critical success. From 1904 to 1907 J. E. Vedrenne and Harley Granville-Barker produced seasons of classics and contemporary plays by SHAW, GALSWORTHY, YEATS, Hauptmann, MAETERLINCK and many others. Half a century later, the production by the English Stage Company of John OSBORNE's *Look Back in Anger* (1956) heralded a similarly influential and controversial succession of new plays by BOND, ARDEN, PINTER and WESKER. Their plays revitalized post-WWII British drama and, to encourage further new work, an experimental THEATRE-IN-THE-ROUND, the Theatre Upstairs, was opened at the Royal Court in 1969.

Royal Dutch Shell Oil Group. *See* SHELL.

Royal Exchange, building for the transaction of business by London merchants in Cornhill, London, founded by Sir Thomas Gresham in 1566. It was twice destroyed by fire, in the Great Fire of 1666 and in 1838. In 1928 the third building, opened in 1844, was taken over by the Royal Exchange Assurance Corporation.

Royal Family, the, British monarch and the members of his or her family. The Royal Family is protected and controlled by various statutes, including the Statute of Treasons (1352) and the Act of Settlement (1701). Since the late 17th century, the Royal Family has had its prerogatives severely curtailed as part of the constitutional framework of the nation; the powers and status of each member have been closely defined. The reigning monarch has a great deal of control over the other members of the family, as in such matters as their marriage. A sum known as the CIVIL LIST is paid annually to the sovereign out of the Consolidated Fund. Other members of the Royal Family are also paid from this fund, but separately from the Civil List.

Royal fern, or flowering fern, common, widely distributed bush FERN of wetlands. It has large leaves and cylindrical pinnae on the ends of fronds, which are densely covered with *sporangia* (sacs in which spores are produced) and look like flowers. Height: to 1.8m (6ft). Family Osmundaceae; species *Osmunda regalis*.

Royal Festival Hall, concert hall on the South Bank of the River Thames, opened in 1951. Its design, by Sir Robert Matthew and Sir John Martin, favours excellent acoustics and it originally formed part of the South Bank Exhibition for the Festival of Britain. It holds an audience of about 3,000 people.

Royal Flying Corps, first national organization of military aviation in Britain. It was set up in 1912, and although the Royal Naval Air Service had split off by 1914, the two were merged as the ROYAL AIR FORCE in April 1918. *See also* HC2 p.191, *191*.

Royal Greenwich Observatory, building by Sir Christopher WREN, in Greenwich Park, London. It was formerly the site of Britain's major government observatory, which was gradually moved to Hertsmonceaux Castle in East Sussex after WWI. The prime MERIDIAN, which determines Greenwich Mean Time (GMT), ran through the original observatory in Greenwich Park. *See also* SU p.*171*.

Royal Horticultural Society, body founded in 1804 to encourage all forms of

Jean-Jacques **Rousseau's** theories influenced the leaders of the French Revolution.

Royal Albert Hall auditorium during one of the annual promenade concerts.

Royal Canadian Mounted Policeman in Banff National Park in western Canada.

Royal Greenwich Observatory was built by Sir Christopher Wren in 1675–76.

Royal Institution

Royal Navy during inspection by the Queen aboard the Royal Yacht in 1957.

Royal Pavilion, Brighton, was designed by John Nash in a pseudo-oriental style.

Royal Society's charter is signed by Charles II (the founder) and James (a Fellow).

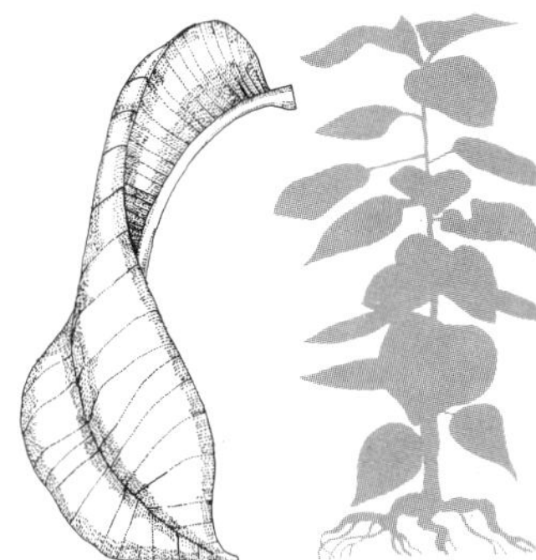

Rubber plants, popular as house plants, are a type of fig tree.

HORTICULTURE and to collect information on plants and trees. The society has scientific gardens at Wisley in Surrey. Its library is probably the finest horticultural library in the world and the society publishes books and holds exhibitions, including the annual flower show at Chelsea.

Royal Institution, organization founded in London in 1799 for the "promotion, diffusion and extension of science and useful knowledge". Distinguished scientists who have used its research laboratories include Sir Humphry DAVY and Michael FARADAY. The "Christmas lectures" for young people, traditional since the 19th century, are now televised.

Royalists, people who supported the Stuarts from 1642 to 1660, during the ENGLISH CIVIL WAR and interregnum. There were few outright republicans in the country, but many people, while desiring to keep a king on the throne, wanted CHARLES I and CHARLES II to alter their policies.

Royal jelly, in HONEYBEE colonies, the special food given to female bee larvae destined to become queens. It is a glandular secretion of young worker bees who themselves are females, but have been fed on honey and pollen and so have not developed into queens.

Royal Marines, soldiers who serve at sea. The names of their ranks correspond with those of the army and their commander is a general, but since their foundation in 1664 they have been intended to operate on or from ships. In 17th- and 18th-century naval warfare, men trained in the use of firearms and able to undertake hand-to-hand combat were an essential part of a ship's complement. The marines were always available, also, as a mobile force which could be put ashore at any time to operate as conventional soldiers. Their first great success in such a role was when they captured Gibraltar in 1704. During WWI they were used in combined operations such as the Zeebrugge raid, and their strength reached 40,000; during WWII the scope for using seaborne troops was even greater, and 74,000 Royal Marines were in service, distinguishing themselves in operations from Iceland to the Irrawaddy River.

Royal Marriage Act (1772), act of the British parliament which established that descendants of GEORGE II, under the age of 25, may marry only with the royal consent.

Royal Military Academy. See SANDHURST.

Royal Navy, fighting force that defends Britain's coastal waters and its merchant shipping. The first naval fleet in Britain was built by ALFRED THE GREAT in 878 to fight off the Viking raids. In the 16th century Henry VII built the first specialist naval ships (previously merchant ships were comandeered when needed) and established the first dockyards, thus paving the way for England's subsequent rise as a naval power. The following centuries saw a struggle for naval supremacy between Britain, France and The Netherlands, culminating in the Battle of TRAFALGAR in 1805 in which the British navy under Lord NELSON defeated the French. This victory heralded the supremacy of the Royal Navy that lasted until well into the 20th century. Since WWII the role of the Royal Navy has diminished because of cuts in the defence budget. The emphasis is now on SUBMARINES and small, fast fighting ships such as FRIGATES and DESTROYERS, many of which are nuclear-powered. *See also* MM pp.118–119, *118–119,* 174–175, *174–175;* HC1 p.277; HC2 pp.78, *78,* 182, *183,* 230.

Royal Opera, The, resident company of the ROYAL OPERA HOUSE, London. It was originally known as the Royal Italian Opera in 1847, and became the Royal Opera in 1892. From this time the custom was established of performing opera in the language in which it was written rather than in Italian. In 1946 it became known as the Covent Garden Opera Company which was formed by a trust founded by John Maynard KEYNES. It was again granted the title "Royal" by a charter of 1968. After WWII, for a time, many foreign operas were sung in English translation. Outstanding musical directors have

included Karl Rankl (1946–51), Rafael KUBELIK (1955–58), Sir Georg SOLTI (1961–71) and the present incumbent Colin DAVIS (1971–   ).

Royal Opera House, originally known as Covent Garden Theatre, home of The ROYAL OPERA and, from 1946, of SADLER'S WELLS Ballet (now the Royal Ballet). The theatre was first opened in 1732 and the present building was completed in 1858 by the architect Edward Barry (1830–80).

Royal palm, ornamental palm native to the Caribbean islands and Central America. It has a tall, light-coloured trunk and feather-shaped leaves. Species include the cabbage palm or palmiste, which grows to 37m (120ft), and the Cuban royal palm, to 21m (70ft). Family Palmaceae; genus *Roystonea.*

Royal Pavilion, the summer palace of George IV, major architectural attraction at the English seaside resort of Brighton. Originally built in 1784–87, it was remodelled by John NASH between 1815 and 1823 at the command of the then Prince Regent.

Royal Shakespeare Company (RSC), state-subsidized British theatrical repertory company based in STRATFORD-UPON-AVON, until 1961 (when it received a Royal charter) known as the Shakespeare Memorial Company. Its chief concern has always been with the plays of Shakespeare, and under the Artistic Directorships of Peter HALL (1960–68) and his successor Trevor NUNN the company became a permanent ensemble. In 1960 it established a second base in London at the Aldwych Theatre. Shakespeare has been presented there alongside other classics and new works by dramatists such as DÜRRENMATT, BECKETT, PINTER, MERCER and STOPPARD, so that plays by Shakespeare can be approached with a contemporary awareness and skilled classical actors can meet the challenge of important modern works. The RSC tours widely and in 1974 it was estimated that its total worldwide audiences had totalled one million for the first time.

Royal Society for the Prevention of Cruelty to Animals. See RSPCA.

Royal Society of London, scientific society founded in 1660 and incorporated two years later. The society's aim was to accumulate experimental evidence on the widest range of scientific subjects, including medicine and botany as well as the physical sciences. In the 20th century, the Royal Society became an independent body of distinguished scientists encouraging a wide variety of research. *See also* HC1 pp.*292,* 303.

Royal tennis. See REAL TENNIS.

Royal Ulster Constabulary, organization established in 1922 by the then Northern Ireland government. Its purpose is primarily that of a civil police force but since 1970 the constabulary has been increasingly involved in security. Its growing membership in the mid-1970s was more than 5,000.

Royal warrant holders, British tradesmen authorized to display the crest, currently of the Queen, the Queen Mother, the Duke of Edinburgh and, from 1980 of Prince Charles. Firms which have satisfactorily supplied royal households for a minimum of three years may apply to the Lord Chamberlain for a warrant.

Royal Worcester. See WORCESTER, ROYAL.

Rozeanu, Angelica (1921–   ), Romanian table tennis player. She won 17 world championship gold medals: six successive singles (1950–55), three doubles (1953, 1955, 1956), three mixed doubles (1951–53) and five team (1950, 1951, 1953, 1955, 1956).

RR Lyrae variable, formerly called a CEPHEID cluster, intrinsic regular short-period variable star of the Cepheid type, usually with a period of less than one day. Stars of this type are of Population II, ie they are old stars commonly found in globular clusters, and appear to be of uniform intrinsic brightness. Like the classic Cepheids, they have been effectively used in distance measurement. *See also* SU pp.*227, 238–239,* 239, 243, *243,* 246.

RSPCA (Royal Society for the Prevention of Cruelty to Animals), organization

established "to prevent cruelty and promote kindness" to animals. Founded in England in 1824, the RSPCA has its headquarters at Horsham, Sussex. In the mid-1970s it had some 225 inspectors in England and Wales, whose duties entail investigating cases of cruelty to animals and, if necessary, bringing offenders to court. There are similar organizations in Northern Ireland, Scotland and the Republic of Ireland.

Ruanda or Rwanda, East African tribe mostly inhabiting the state of RWANDA. Although the land is densely populated the people practise a predominantly agricultural subsistence economy, with families living together in self-contained kraals. The Ruanda language, a member of the BANTU group of languages, is spoken by about five million people.

Ruanda-Urundi. See BURUNDI; RWANDA.

Rubáiyát of Omar Khayyám, The, collection of quatrains said to be by the 11th-century Persian astronomer-poet, OMAR KHAYYÁM. The verses exalt the sensual, pleasure-seeking life and were translated in 1859 by Edward FITZGERALD in a free rather than a literal version. It is doubted whether more than a tenth of the verses can actually be attributed to Omar.

Rub' al-Khali, one of the largest sand deserts in the world, situated in the s Arabian Peninsula. The Arabic name means "empty quarter" and the desert is considered unable to support any life. Area: 650,000sq km (251,000sq miles).

Rubber, elastic solid obtained from the latex of the rubber tree (*Hevea brasiliensis*). Natural rubber consists of a POLYMER of cis-isoprene and is widely used for tyres and other applications, especially after VULCANIZATION. Synthetic rubbers are polymers tailored to excel in certain properties for specific purposes; none have the overall advantages of natural rubber. They include styrene-butadiene rubber (SBR), neoprene, nitrile rubber and the newer stereo-rubbers based on synthetic cis-polyisoprene. *See also* MM p.54.

Rubber plant, evergreen FIG native to India and Malaysia. Tree-sized in the tropics, it is grown as a pot plant in temperate regions. Once cultivated for its white LATEX to make india-rubber, it has large, glossy, leathery leaves and a stout, buttressed trunk. Height: to 30m (100ft). Family Moraceae; species *Ficus elastica.*

Rubber tree, any of several South American trees whose exudations can be made into RUBBER; especially *Hevea brasiliensis* (family Euphorbiaceae), a tall softwood tree native to Brazil but introduced to Malaysia. The milky exudate, called LATEX, is obtained from the inner bark by tapping, and then coagulated by smoking over fires or chemically. *See also* MM p.54.

Rubber tyre, part of a wheel made to make contact with the ground and provide friction for non-slip grip and springiness for a smooth ride. It is made in two parts; the outer rubber tread which holds the road, and the inner body which is a tough fabric impregnated with rubber and bordered by steel wire or nylon threads which hold the tyre to the wheel, and acts as a strengthening framework. Tyres are black in colour because carbon black is added to improve abrasion resistance. The rubber is also hardened by vulcanization or addition of sulphur. The body of early tyres had a cross-ply pattern for the strengthening cords; these are being replaced by radial-ply tyres which offer better road-holding at high speeds. *See also* MM pp.54–55, *130, 133.*

Rubbra, Edmund (1901–   ), British composer, a pupil of HOLST and VAUGHAN WILLIAMS. He has written ten symphonies and concertos for piano, violin and viola. He has also written much choral music, including madrigals and the *Canterbury Mass* for double choir (1945).

Rubella. See GERMAN MEASLES.

Rubénisme, late 17th-century movement in French art. Its protagonists, following the work of TITIAN and RUBEN held that colour was the most important element in painting since it was the most convincing means of achieving reality. This theory was opposed in the French Academy of

Painting by poussinisme. Led by LEBRUN its followers maintained that colour was merely decorative and of less interest than the more formal elements of drawing and design, as shown in the paintings of POUSSIN. The Rubénistes won the debate, and the Academy's acceptance for exhibition of a painting by WATTEAU (1717) consolidated the victory.

Rubens, Peter Paul (1577–1640), Flemish painter and engraver. During two visits to Italy between 1600 and 1608 he painted religious subjects and was greatly influenced in his style by the works of Italian High Renaissance masters. Monumental muscular figures in action are characteristic of his heroic, Mannerist style. When he returned to Antwerp (1608), he established a workshop that became highly successful, producing huge portraits and allegorical series for the wealthy and nobility of western Europe. His best-known works are his crowded Baroque canvases in which he used light and colour to give a dramatic sense of action. His 21 paintings of *The Life of Marie de Medicis* are from this period (1622–25). His later landscapes show a softening of colours. *See also* MANNERISM; HC1 pp.304–305, *305*.

Rubiaceae, large family of flowering plants (the MADDER family) containing about 6,500 species, most of which are tropical. They include COFFEE and CINCHONA; temperate species include madders and GARDENIAS.

Rubicon, small river in N central Italy. It rises in the foothills of the Etruscan Apennines and flows into the Adriatic Sea N of Rimini. The river separated Cisalpine Gaul from Italy in Roman times. Julius CAESAR precipitated civil war against POMPEY after crossing the Rubicon with his army in 49 BC.

Rubidium, metallic element (symbol Rb) of the alkali-metal group, discovered in 1861. The element has few commercial uses; small amounts are used in photoelectric cells. Chemically it resembles SODIUM, but is more reactive. Properties: at. no.37; at. wt.85.4678; s.g.1.53; m.p.38.89°C (102°F); b.p. 688°C (1,270°F); most common isotope Rb^{85} (72.15%). *See also* ALKALI ELEMENTS.

Rubinstein, Artur (1887–   ), US pianist, *b.* Poland. He studied in Warsaw and Berlin, making his debut with the Berlin Symphony Orchestra in 1901. He first played in the USA in 1906 and later became famous for his interpretation of CHOPIN and Spanish composers. He had one of the longest active concert careers of any musician.

Rubinstein, Helena (*c.*1871–1965), Polish-born founder of an international cosmetics business. She opened her first shop in Melbourne, Australia, marketing a Polish cold-cream she had used since childhood. After study with European dermatologists, she opened salons in London (1908), Paris (1912) and New York (1914).

Rublev, Andrei (*c.*1370–1430), Russian fresco and icon painter. Painting in the BYZANTINE tradition, he was Russia's greatest iconographer and famous for his masterpiece *The Old Testament Trinity* (*c.*1411). *See also* HC1 p.*320*.

Rubric, from the Latin *rubrica* ("in red"), originally the heading of chapter or section, but also the part of a manuscript or book, set in a different colour because of special importance. Directions for the conduct of religious services were often distinguished in red ink and are still called rubrics.

Rubus, genus of plants of the ROSE family (Rosaceae), well-known as berry-bearing bushes such as BLACKBERRY, RASPBERRY and LOGANBERRY. Less familiar in Britain are DEWBERRY and cloudberry. *See* PE pp. 178, 214.

Ruby, gem variety of the mineral CORUNDUM (aluminium oxide), whose characteristic red colour is due to impurities of chromium and iron oxides. The traditional source of rubies is Burma. Today synthetic rubies are widely used in industry, especially in LASER technology. *See also* PE pp.98–99, *98–99*; SU p.110,*110*.

Ruda Śląska, town in S central Poland, in the coal-mining region of Upper Silesia. The town dates from the Middle Ages when it was an iron-mining centre. The first coal mine in Poland was opened there in 1751. Pop. (1974) 147,000.

Rudd, or red eye, fish related to the MINNOW. In the USA it is called pearl roach. Found also in Europe N and W Asia, the rudd is a large, full-bodied fish with reddish fins, the dorsal fin being farther back than that of the true ROACH. Length: to 40.6cm (16in); weight: to 2kg (4.5lb). Family Cyprinidae; species *Scardinius erythrophthalmus.*

Rudder, broad surface used to steer aeroplanes or ships. The first rudders, used by the ancient Romans and Greeks, were large oars fitted to one side of the rear of a ship. By the 14th century AD rudders were mounted in the centre of the stern, enabling more control to be exerted. Some modern liners have rudders which are 18m (60ft) long. For a left turn, the rudder is turned to its left, adding a drag to that side of the vessel or aeroplane, causing it to pull round. *See also* MM p.152, *152.*

Ruddigore, or The Witch's Curse (1887), comic opera in two acts by Arthur SULLIVAN, libretto by William GILBERT. It was first performed at the Savoy Theatre, London, and ran for eight months.

Rude, François (1784–1855), French sculptor who suffused CLASSICAL idealism with overt ROMANTICISM in patriotic works such as his famous relief on the ARC DE TRIOMPHE, *The Departure of the Volunteers in 1792*, known also as *La Marseillaise* (1833–36).

Rudolf, the name of two HAPSBURG rulers of Germany. Rudolf I (1218–91), the first Hapsburg monarch, reigned from 1273 to 1291 and laid the foundations of the Hapsburg dominions in 1278 by obtaining Austria, Styria and Carniola from Bohemia. Rudolf II (1552–1612) was Holy Roman emperor (*r.*1576–1612), King of Bohemia (*r.*1575–1611) and of Hungary (*r.*1572–1608). His attempt to force Catholicism on Hungary led to a revolt in 1604. Matthias, his brother, forced Rudolf to cede the rule of Hungary, Austria and Moravia (in 1608) and Bohemia (in 1611) to him.

Rudolf (or Rudolph), Archduke (1858–89), only son of the Austrian Emperor FRANCIS JOSEPH. He was a liberal reformer and committed suicide at Mayerling with his companion Baroness Marie Vetsera in 1889.

Rudolph, Wilma (1940–   ), US track athlete. As world 200m record holder she won two gold medals at the 1960 Olympics for the 200m and 100m sprints and a third gold medal in the relay. She retired after setting a world 100m record of 11.2 seconds in 1961.

Rue, any of about 40 species of evergreen plants or shrubs that grow in warm regions of S Europe and SW Asia. They have aromatic leaves used in medicine or as flavouring. Family Rutaceae; genus *Ruta.*

Ruff, bird of the SANDPIPER family (Scolopacidae). The male is noted for a collar of long feathers about its neck, and for its antic courtship performances. The female is called a reeve. The ruff is polygamous, unlike most sandpipers, and migrates across N Europe and N Asia to Africa and India. Species *Philomachus pugnax. See also* NW pp. *141, 144.*

Rug, loose covering, in general smaller than a CARPET, usually made of wool, now commonly placed on floors. Until the development of power looms in the 19th century, rugs were always manufactured on handlooms.

Rugby fives. *See* FIVES.

Rugby league football. *See* FOOTBALL, RUGBY.

Rugby union football. *See* FOOTBALL, RUGBY.

Ruhr, river in West Germany. It rises in the hills of N central West Germany and flows NW and W to join the River Rhine at Duisburg. The Ruhr valley, an industrial mining district, contains large coal deposits and is one of the most important industrial complexes in the world. The area along the river was occupied by French and Belgian troops between 1923 and 1925 because of a dispute with Germany over reparations. The occupation was greatly resented by German nationalists. Length: 235km (146 miles).

Ruisdael (or Ruysdael), Jacob van (*c.*1628–82), Dutch painter. Greatest of the Dutch landscapists, he studied under his father, Isaak, and was influenced by his uncle, Salomon van Ruisdael. Jacob brought to his paintings of the flat northern landscape and cloud-filled skies an unusual breadth and accuracy. Among his many works are *Wooded Landscape* (*c.*1660) and *Windmill at Wijk* (*c.*1670). *See also* HC1 p.311, *311.*

Ruiz, Juan (*fl.*14th century), Spanish poet. Thought to be the author of what is probably the most remarkable poetic work of the Spanish Middle Ages, the *Libro de buen Amor* (*c.*1343), he employed rhetoric, dialogues, irony and ambiguity to create a vivid picture of a series of love affairs, allegorical and anecdotal.

Ruiz de Alarcón y Mendoza, Juan (*c.*1581–1639), Spanish dramatist. Before abandoning drama to take up a government position in 1626 he wrote 24 plays, many of them moral comedies.

Rum, alcoholic spirit made by the fermentation of MOLASSES and other sugar cane products, which are then distilled. As distilled, rum is colourless, but storage in wooden casks and the addition of caramel give it a brown colour.

Rūm, Muslim term used by Arabs in reference to the BYZANTINE Greeks who considered themlves heirs to the ROMAN EMPIRE and called themselves Romaioi. The term was also used to describe Byzantine territory, thus the sultanate of Rūm was a kingdom in Byzantine Anatolia established by the SELJUK TURKS after the Great Seljuk Sultanate had disintegrated.

Rumania. *See* ROMANIA.

Rumba, popular American ballroom dance with Afro-Cuban origins. The music is in syncopated 4/4 time, and the dance involves two quick side-steps and a slow forward step, the emphasis being on body movements rather than elaborate footwork.

Rumelia, area in S Bulgaria. Historically it was a possession of the Ottoman Turks encompassing modern Yugoslavia, Bulgaria, European Turkey, N Greece, and part of Albania. Bulgaria was made an independent area with Eastern Rumelia an autonomous state under Turkish protection by the Treaty of Berlin in 1878. The Eastern Rumelians voted in 1885 to unite with Bulgaria causing the Serbian War 1885–86. The Serbians were defeated and Prince Alexander of Bulgaria named as governor of Eastern Rumelia. This tacit surrender of the province remained in force until 1908 when Bulgaria and thus Eastern Rumelia became independent.

Rumford, Benjamin Thomson, Count (1753–1814), American-British scientist and administrator. He moved to Britain during the American War of Independence and was made a count of the Holy Roman Empire for service in Bavaria in 1791. In science, he discovered that heat, or KINETIC ENERGY, is a form of motion. In 1799 he founded the ROYAL INSTITUTION of Great Britain.

Rumi (1207–73), Persian poet, full name Jalāl ad-Dīn ar-Rūmī. He is considered the finest exponent of SUFI poetry, his main work being the *Mathnawi* (or *Masnavī*).

Ruminants, cud-chewing, even-toed, hoofed mammals, including the OKAPI, CHEVROTAIN, DEER, GIRAFFE, ANTELOPE, CATTLE, SHEEP and GOATS. All except the chevrotain have four-chambered stomachs. *See also* ARTIODACTYLA.

Rummy, card game for two to six players using a standard pack. In the two-hand game, ten cards are dealt to each player; in a three- or four-hand game, seven cards are dealt to each; in a five- or six-hand game, six cards are dealt to each. The rest of the pack is placed face down on the table and the top card is turned up. Players in turn take either a new card from the pack or the card discarded by the preceding player, before discarding a card themselves. The object is to acquire a hand with

Rubens' *Helen Fourmet with her son Franz* is an example of his portraiture.

Artur Rubinstein, taught by Paderewski, is known for his playing of Chopin.

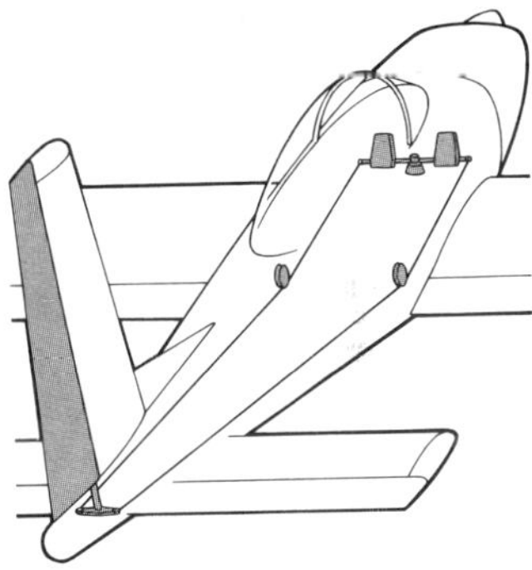

Rudder of a light aircraft is operated by the pilot's foot-controls.

Ruhr, the principal industrial manufacturing region of West Germany.

Rump Parliament

Damon Runyon's stories describe New York and its inhabitants in the 1930s.

Prince Rupert; the original version of this portrait was painted by Van Dyck.

John Ruskin emphasize: the importance of drawing and engraving in art education.

Russian blue cats have a thick coat with two layers of short dense fur.

all of its cards of the same suit in sequence, three or four cards in a sequence or three or four cards of the same rank. The first person to achieve this and set all his cards on the table wins. *See also* GIN RUMMY.

Rump Parliament (1645–53, 1659–60), name given to the Parliament that governed England after the ENGLISH CIVIL WAR had ended with the defeat of CHARLES I. It comprised members who had been elected to the House of Commons in the 1642 election except those hostile to the army, who had been expelled in Pride's Purge in 1648. In 1653 CROMWELL dismissed the Rump Parliament and ruled as virtual dictator of the Commonwealth. The Rump was restored in 1659, and the members expelled by Pride were re-admitted in 1660, clearing the way for a general election and the RESTORATION. *See also* HC1 pp.290, *290–291*.

Rum Rebellion, popular name for the events of 1808 which drove Captain BLIGH from the governorship of New South Wales (1810). Its name derives from the fact that one cause of the revolt was the illegal import of rum stills on the *Parramatta*.

Runeberg, Johan Ludvig (1804–77), Finnish national poet. His simple style helped to check the tendency towards false rhetoric in Scandinavian literature. The first song from his poem on the Russo-Swedish war, *Songs of Ensign Stål* (1848–60), has been adopted as the Finnish national anthem.

Runes, angular letters of the runic alphabet formerly used by Germanic peoples. Also called futhark after its first six sounds (*f, u, th, a, r, k*), runic may have been developed by the GOTHS from the Etruscan alphabet.

Running, activity for sporting and recreational purposes. Many games involve running, and it is a sporting discipline in its own right, as in track athletics, cross-country and marathon running. Jogging, or running within one's limitations, has become internationally popular as a means of keeping fit.

Runnymede, meadow in s England, on the s bank of the River Thames at Egham, Surrey, w of London. It was either there or on nearby Magna Charter Island that King JOHN accepted the MAGNA CARTA in 1215. The meadow was given to the British nation in 1931. Today it also has a memorial to President John F. KENNEDY.

Runyon, Alfred Damon (1880–1946), US author and journalist who wrote humorous stories about colourful New York City characters. *Guys and Dolls* (1932) established his reputation.

Rupert, Prince (1619–82), Bohemian military commander. His uncle CHARLES I made him general of the horse and as the commander of the cavalry in the ENGLISH CIVIL WAR he was undefeated until MARSTON MOOR (1644). He was dismissed after the Royalist defeat at NASEBY (1645) and his surrender to FAIRFAX at Bristol. He led PRIVATEERS against English shipping during the PROTECTORATE and, back in England after the Stuart restoration, served as an admiral in the Dutch wars.

Rupture. *See* HERNIA.

Rural dean, term used in the Church of England to describe the supervisor of the parishes of a certain area. An ancient office, its duties have gradually been absorbed by those of Archdeacons.

Rural Rides (1830), best-known book by the radical British journalist William COBBETT. It is a classic portrayal of rural conditions showing his concern for the rural worker.

Rusalka, in Slavic mythology, lake-dwelling spirit of a drowned virgin or unbaptized child in the form of a beautiful girl robed in mist. She lured passers-by to join her in a dance of death.

Ruse, port city in NE Bulgaria on the River Danube, bordering Romania. Founded in the 2nd century BC, it developed under Turkish rule from the 17th to 19th centuries. It is now an industrial and communications centre. Pop. 145,000.

Rush, tufted, perennial bog plant found in temperate regions. It has long, narrow leaves and small flowers crowded into dense clusters. Among about 700 species is *Juncus effusus* of Europe and N Africa. It has brown flowers and ridged stems. Height: 30–152cm (1–5ft). Family Juncaceae. *See also* NW pp.68, *69*.

Rush-Bagot Convention (1817), British-USA agreement providing for disarmament of the US – Canadian border. It was agreed between Richard Rush, US Acting Secretary of State and Charles BAGOT, British minister in Washington.

Rusk, David Dean (1909–), US politician. In 1950 he was involved in the US decision to enter the KOREAN WAR and as Secretary of State from 1961 to 1969 advocated and defended US intervention in Vietnam.

Ruskin, John (1819–1900), British art theorist and social critic. He attempted to find the solution to many social problems through his approach to art. A strong religious conviction was the basis for his reflection of the Classical tradition in art and for his advocacy of Gothic naturalism as the style through which man could best express his praise for God and creation. These ideas are expressed most forcefully in his books on architecture: *The Seven Lamps of Architecture* (1849) and *The Stones of Venice* (three vols, 1851–53). His notion of the artist was essentially Romantic – as an aesthete, prophet and teacher. His five-volume work *Modern Painters* (1834–60) championed the paintings of J. M. W. Turner and after 1851 he supported the PRE-RAPHAELITE BROTHERHOOD.

Russell, English family, first prominent among the nobility in the 16th century. John Russell (*c.*1486–1555), Lord Keeper of the Privy Seal to HENRY VIII and EDWARD VI, was created 1st Earl of Bedford in 1550. His successors were politically influential and in 1694 the 4th earl was created Duke of Bedford. The two most notable members of the family outside the main line of succession were Lord John Russell later 1st Earl RUSSELL, who played an important part in 19th-century Parliamentary reform and was twice Prime Minister of England, and the philosopher Bertrand RUSSELL.

Russell, Bertrand Arthur William, 3rd Earl (1872–1970), British philosopher, mathematician and social reformer. His philosophy was of empiricism and logical atomism, which he applied to mathematics to demonstrate that mathematics could be explained by the rules of formal logic. His most famous work, Principia Mathematica (1910–13), was written in collaboration with his teacher, A. N. WHITEHEAD. He was a pacifist except during WWII and was a constant campaigner for educational and moral reforms. He was imprisoned twice (1918, 1961) for his activities on behalf of peace. He won the 1950 Nobel Prize in literature and wrote many books, both technical and popular. *See also* PRINCIPIA MATHEMATICA; MS p.228.

Russell, Charles Taze. *See* JEHOVAH'S WITNESSES.

Russell, Countess. *See* ARNIM, MARY ANNETTE VON.

Russell, George William (1867–1935), Irish poet, essayist and journalist who wrote under the pen name A. E. He was a newspaper editor from 1923 to 1930 and published many collections of romantic poetry. Among them are *The Divine Vision* (1904) and *Midsummer Eve* (1928). *See also* HC2 p.282.

Russell, Henry Norris (1877–1957), US astronomer. His work on plotting the stars according to their luminosities and spectral types led to the development of the HERTZSPRUNG-RUSSELL DIAGRAM. *See also* SU p.226.

Russell, John, 1st Earl (1792–1878), British statesman; Prime Minister from 1846–52 and 1865–66. He was the third son of the 6th Duke of Bedford and made an important contribution to the development of British politics in the 19th century, especially in the change from an aristocratic Whig Party to a more broadly based liberalism. He was associated with such major reform issues as Roman Catholic emancipation, parliamentary and municipal reform and the repeal of the CORN LAWS. *See also* HC2 p.*113*.

Russell, John Peter (1858–1931), Australian painter who joined the IMPRESSIONISTS in France and was a friend of MONET and RODIN. Among his works is a portrait of Vincent van GOGH, painted when they were both students together.

Russell, Ken (1927–), British film director. He began work in television and made his first film, *French Dressing*, in 1964. He specialized in biographies, such as *Mahler* (1973), *Lisztomania* (1975), and *Valentino* (1977).

Russell, Morgan (1886–1953), US painter and a founder of synchromism in 1912. This movement involved the arrangement of geometric forms of contrasting colours, the geometric forms being secondary to the colour theory they represented.

Russell, Lord William (1639–83), English politician who first entered parliament in 1660. He was an opponent of Catholicism and the STUART court. For his alleged participation in the RYE HOUSE PLOT he was convicted of treason and executed .

Russell, Sir William Howard (1820–1907), British journalist, b. Ireland. He was one of the first war correspondents, reporting the CRIMEAN WAR for *The Times* where he coined the phrase "thin red streak" which became "thin red line" in popular usage to describe embattled British troops. He also covered the FRANCO-PRUSSIAN WAR (1870) and the ZULU WAR (1879). In 1860 he founded the *Army and Navy Gazette. See also* HC2 p.*182*.

Russell, oldest European settlement in New Zealand, on the Bay of Islands. It was founded in the early 1800s, called Kororareka. It was renamed after Lord John RUSSELL in 1840.

Russia, historical name of the region now forming the Russian SFSR, and the centre of the Russian Empire. The term is popularly used to apply to the entire USSR, of which it is the culturally dominant part. The name derived from the Viking kingdoms of the Rus, or Varangians, around Kiev in the 9th century AD. The Scandinavians were quickly assimilated within the Slavic population. As s Russia was taken over and destroyed by the MONGOLS, the kingdoms' focus shifted to the NE, where the state of MOSCOW won supremacy by the 14th century. Russian colonization E of the Urals began in the 16th century, and early in the 17th century Russia invaded Poland for the first time. PETER the Great expanded Russia to the N into Finland and Livonia after 1700, and also tried to establish Russian influence on the Black Sea coast. By 1800 Russia had annexed much of Poland and taken the CRIMEA. In the mid-19th century, the CAUCASUS and TURKMENIS became Russian, and VLADIVOSTOK was founded in 1860, symbolizing Russian control throughout Asia to the Pacific. By the Treaty of BREST-LITOVSK (1918), areas such as WHITE RUSSIA and the UKRAINE were lost; when they were reincorporated in the Soviet state they were administratively separate from the Russian SFSR. *See also* HC1 pp.320–323, *325*, HC2 pp.168–169, 196–199, 240–241; MW p.169.

Russian blue, short-haired domestic cat of unknown origin. It has a longish face with green eyes and large, pointed ears. The dense double coat is a silvery blue.

Russian Five, group of Russian composers active in St Petersburg during the 1860s and 1870s who hoped to create a truly Russian style of music. They were Mily BALAKIREV, Alexander BORODIN, César CUI, Modest MUSSORGSKY and Nikolai RIMSKY-KORSAKOV.

Russian language, official language of the USSR, the mother tongue of about 142 million of its people (about 6 per cent of the population) and spoken as a second language by about 42 million more. It is the most important of the Slavic languages, which form a subdivision of the INDO-EUROPEAN family. Russian is written in the Cyrillic alphabet.

Russian Orthodox Church. *See* ORTHODOX CHURCH.

Russian Revolution (1905), year of uprisings and protests against Tsarist rule. It began on Bloody Sunday (22 Jan), when a peaceful demonstration of workers petitioning the Tsar at St Petersburg was fired on by troops. Strikes and peasant risings spread, culminating in a general strike in October. The Tsar's October Manifesto promised full civil liberties and a democratically elected DUMA, but apart from the *Potemkin* mutiny the armed forces remained loyal and by the time the first Duma met in 1906 Nicholas had regained control. *See also* HC2 p.169, *169*.

Russian Revolutions (February and October 1917), overthrow of Russian monarchy and BOLSHEVIK accession to power. Incompetent management during WWI, loss of confidence in the regime and riots in the capital occasioned by acute shortage of food forced NICHOLAS II to abdicate in February 1917. A Provisional Government was formed by liberals in the DUMA, with Prince Georgy LVOV as Prime Minister, but had to contend with the growing influence of the soviets (local district councils). The Government's decision to continue the war and its failure to introduce land reform caused popular support in the soviets to shift to the Bolsheviks. In July the Bolsheviks attempted to rally popular support to overthrow the Provisional Government with their call of "All Power to the Soviets" but the Provisional Government was able to retain control. LENIN, who had arrived in July after the provisional government restored order in April, fled to Finland. In September, General KORNILOV unsuccessfully attempted to overthrow KERENSKY who took over as Prime Minister in July from Prince Lvov but was unable to muster support from the army. After the "Kornilov Affair", the Bolsheviks gained a majority in the St Petersburg and Moscow soviets, and increased their popular support with their slogan "Peace, Land and Bread".

Russian Soviet Federated Socialist Republic (Rossiyskaya Sovetskaya Federativnaya), largest constituent republic of the USSR, commonly abbreviated Russian SFSR, bordered by seas of the Arctic Ocean (N) and the seas of the Pacific (E), and occupying more than 75% of the total land area of the USSR. The main cities are Moscow (capital), Leningrad, Gorki, Novosibirsk, and Sverdlovsk. The republic contains the E European lowland, the Urals, the W Siberian Plain and the Siberian Plateau, it is drained by the rivers Volga, Ob, Yenisei, Lena and Amur. It formed a major part of the Tsarist Empire and was the first region to come under Soviet control in 1917. It joined the Ukraine and Belorussia and Transcaucasia to form the USSR in 1922. Industries include the manufacture of machinery, chemicals, textiles and leather goods and mining for coal, iron ore, copper, nickel, lead, manganese, zinc, and platinum. Area: 17,361,400sq km (6,703,235sq miles). Pop. (1976) 139,485,000.

Russlan and Ludmila (1842), five-act opera by Mikhail GLINKA, with libretto by Shirkov and Bakhturin based on a poem by Alexander Pushkin.

Russo-Japanese War (1904–05), conflict originating from rivalry over Manchuria and Korea. In May 1904 Japanese troops crossed the Yalu River from Korea into Manchuria, while naval forces seized PORT ARTHUR, which fell in January 1905. In March 1905 the Japanese captured Mukden and in May destroyed the Russian fleet at TSUSHIMA. The Treaty of Portsmouth (1905) obliged Russia to recognize Japan's "paramount interest" in Korea. Japan was also given the southern half of Sakhalin by the Russians. The war established Japan as a world power and was a primary cause of the Russian Revolution of 1905. *See also* HC2 p.147, *147*.

Russolo, Luigi (1885–1942), Italian artist and musician. In his Futurist manifesto of 1913 he urged composers to "break out of (the) narrow circle of pure musical sounds, and conquer the infinite variety of noise sounds".

Rust, corrosion of iron by a combination of air and water. Carbon dioxide from the air dissolves in water to form an acid solution which attacks the iron to form iron (II) oxide. This is then oxidized by oxygen in the air to reddish-brown iron (III) oxide. Rusting may be prevented by coat-

ing the iron with zinc, a process called GALVANIZING, when the zinc is preferentially attacked by the acid solution.

Rust, in botany, group of fungi that live as parasites on many kinds of higher plants. The rust red colour of their spores gives them their name. Rusts are of particular concern for the damage they do to cereal crops, such as wheat and barley, and to several fruits and vegetables. They have complex life cycles that involve growth on more than one host plant, barberry (BERBERIS) being one of these secondary hosts. *See also* NW p.40.

Ruth, book of the Old Testament that recounts the story of Ruth, a Moabite widow, her fidelity to Naomi, her Hebrew mother-in-law, and her subsequent marriage to Boaz, whereby she becomes the great-grandmother of King DAVID.

Ruthenia, area now constituting an OBLAST within the Ukraine (Ukrainskaja SSR) USSR, its capital is Uzhgorod. Formerly an autonomous region, the name is the Latin for Russia. During the Middle Ages it referred to the Ukrainian population of the NE Carpathians. After 1918, however, the term applied only to the easternmost province of Czechoslovakia. The present day oblast is made up of land ceded to the USSR by Romania, in 1945. The economy is largely dependent on agriculture and much of the area is heavily forested. Industries include coal-mining, lumbering and tobacco. Area: 12,800sq km (4,942sq miles). Pop. (1970) 1,057,000.

Ruthenium, metallic element (symbol Ru) of the second transition series, first isolated in 1844. Ruthenium is alloyed with other platinum metals for electrical contacts and is also used as a catalyst. Properties: at. no. 44; at. wt. 101.07; s.g. 12.30; m.p. 2,310°C (4,190°F); b.p. 3,900°C (7,052°F); most common isotope Ru^{102} (31.61%). *See also* TRANSITION ELEMENTS.

Rutherford, Daniel (1749–1819), Scottish physician who was the first to distinguish between NITROGEN and CARBON DIOXIDE.

Rutherford, Ernest, Lord Rutherford (1871–1937), New Zealand-born British physicist. Working on RADIOACTIVITY and atomic structure, he discovered ALPHA, BETA and GAMMA RADIATION and was the first to show that the atom contains a central nucleus. He carried out the first splitting of the atom, and identified and named the PROTON. Awarded the Nobel Prize in chemistry, for his work on radioactivity, he was made a baron in 1931. *See also* SU pp.62, 64, *64*; HC2 p.*157*.

Rutherford, Dame Margaret (1892–1972), British actress who became famous as a character actress. She appeared as Madame Arcati in *Blithe Spirit* and in 1964 won an Academy Award for her role as a dowager in *The VIPs*.

Rutherfordium, *See* ELEMENT 104.

Rutile, black to red-brown oxide mineral, titanium dioxide (TiO_2), found in igneous and metamorphic rocks and quartz veins. It occurs as long prismatic and needle-like crystals in the tetragonal system and as granular masses. It has a metallic lustre, is brittle and is used as a gemstone. Hardness 6–6.5; s. g. 4.2.

Rutland, former county in central England; since 1974 it has been part of Leicestershire. It is an area of rolling country, drained chiefly by the River Welland. Sheep and cattle grazing is the main farming activity. Oakham was the county town. Industries: iron ore mining, cement, clothing, plastics. Area: 396sq km (153sq miles). Pop. (1971) 27,463.

Ruwenzori, mountain range in central Africa on the Uganda-Zaire border between lakes Albert and Edward. The range was discovered by Henry STANLEY in 1889. The highest peak is Mt Margherita, 5,109m (16,763ft). Length of range: 121km (75 miles).

Ruysdael, Jacob. *See* RUISDAEL, JACOB VAN.

Ružička, Leopold Stephen (1887–1976), Swiss chemist, *b.* Yugoslavia, who shared the 1939 Nobel Prize in chemistry with Adolf BUTENANDT for his research into ring molecules and terpenes, HYDROCARBONS found in the essential oils of many

plants. In 1916 Ružička began investigations into the active compounds in MUSK and CIVET, finding that they were ring structures consisting of 15 and 17 carbon atoms respectively. This discovery had a profound effect on organic chemistry, as until that time rings with more than six atoms were thought to be too unstable to exist. In the 1930s he also synthesized several male sex hormones including TESTOSTERONE, having previously discovered their molecular structure.

Rwanda, landlocked nation in E central Africa. Before 1962 it was part of the Belgian mandate of Ruanda-Urundi. Rwanda is one of the world's poorest nations; most of the people are subsistence farmers. Coffee is the principal commercial crop and tin and tungsten are mined, although there is little manufacturing. Kigali is the capital. Area: 26,338sq km (10,169sq miles). Pop. (1976 est.) 4,321,000. *See also* MW p.148.

Rydberg, Abraham Viktor (1828–95), Swedish writer. His novel *The Last Athenian* (1859) contrasted classical values with contemporary Christian dogmatism. His lyrical but sometimes sombre verse reflected his unhappy childhood.

Ryder, Albert Pinkham (1847–1917), US painter, known for his imaginative works. The sea and the moon predominate in many of his paintings, which have a strange brooding quality.

Ryder Cup, biennial competition in which a team of professional male golfers from the USA pays a team from Britain and Ireland. After an unofficial match at Wentworth, Surrey, in 1926, Samuel Ryder presented a gold cup for the first official match in Massachusetts in 1927. The event consists of eight 18-hole foursomes, eight 18-hole four-balls and 16 18-hole singles. One point is awarded for a win and half a point for a halved game. Britain has won the cup only three times, in 1929, 1933 and 1957.

Rye, hardy cereal grass originating in SW Asia and naturalized throughout the world. It grows in poor soils and colder climates than most other cereals can stand. It has flower spikelets that develop one-seeded grains. It is used for flour, as a forage crop and for making alcoholic drinks. Height: to 0.9m (3ft). Family Gramineae; species *Secale cereale*. *See also* PE pp.158, 180, *181, 182*.

Rye House Plot (1683), Whig conspiracy to kill CHARLES II and his brother James, Duke of York, later JAMES II, as they passed by Rye House on their way from Newmarket to London. The king did not make the journey on the expected day, the plot was revealed and Algernon SIDNEY and Lord William RUSSELL, prominent Whigs, were executed.

Rylands, John (1801–88), textile manufacturer. He contributed to many charities and was a main financier of the Manchester Ship Canal. The John Rylands Library at Manchester University was founded in his memory by his widow.

Ryle, Gilbert (1900–), British philosopher concerned with problems caused by the confusion of grammatical with logical distinctions. He also argued interdependence between body and mind. His works include *Philosophical Arguments* (1945) and *The Concept of Mind* (1949).

Ryle, Sir Martin (1918–), British astronomer. After studying radar during WWII he went to Cambridge University where he pioneered RADIO ASTRONOMY. He also catalogued stellar radio sources, during which he discovered QUASARS. During the 1960s, Ryle developed a one-mile radio telescope with which Antony HEWISH and Jocelyn Bell discovered the first PULSAR in 1968.

Ryswick, Treaty of (1697), agreement that ended the War of the GRAND ALLIANCE, the signatories being France on one side and the Holy Roman Empire, England, Spain and the Netherlands on the other. France surrendered all the territories it had gained since 1679, with the exception of Strasbourg and Alsace. Trading concessions were granted to the Dutch, the independence of Savoy was recognized and France acknowledged WILLIAM III as King of England.

Ernest **Rutherford** with the apparatus he used for counting alpha particles.

Margaret **Rutherford** after her investiture as a Dame Commander of the British Empire.

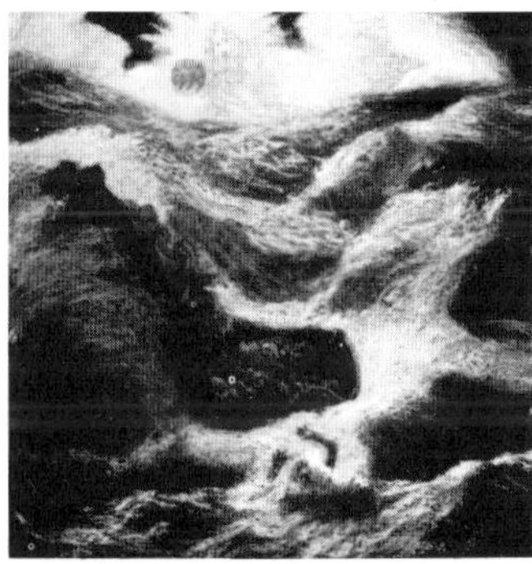

Albert **Ryder's** work is typified in this detail from his painting *Jonah* (1890).

Rye harvest; traditionally the cut rye is gathered into stooks before threshing.

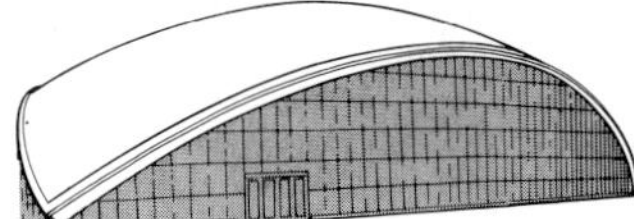

Eero Saarinen was the architect of the Auditorium, MIT, in Cambridge, USA.

Sable martens are bred on farms in parts of the USSR for their valuable pelts.

Nelly Sachs, Jewish poetess, was awarded the 1966 Nobel Prize for literature.

President Sadat was nominated by Parliament for a second term of office in 1976.

Ryukyu Islands (Nansei-skoto), Japanese archipelago in the w Pacific Ocean, extending approx. 965km (600 miles) between Kyushu and Formosa. It separates the East China Sea (w) from the Philippine Sea (E). The island groups, which are volcanic and coral formations, include Amami, Okinawa and Sakishima. Inhabited since early times, they were invaded by China in the 14th century and by Satsuma of Japan in the 17th century, paying tribute to both countries until relinquished by China to Japan in 1879. After wwII they were administered by the USA and restored to Japan in 1972. Agriculture and fishing are the chief occupations. Area: approx. 2,196sq km (848sq miles). Pop. (1974) 171,264.

Ryun, Jim (1947–), US athlete. He first ran the mile in less than 4min at the age of 17. In 1966 he broke the world 1,500m record with a time of 3min 33.1sec, and also set a new world half-mile record with 1min 44.9sec; in 1967 he lowered the world's mile record to 3min 51.1sec.

S, 19th letter of the alphabet, derived originally from the Semitic letter *sin* or *shin*, meaning *tooth*, which had an *sh* sound. The Greeks had various letters for sibilants, but after about 600 BC used mainly *sigma* for *s*. It adopted its present form in the Latin alphabet. The two common pronunciations of *s* in English are voiced and unvoiced – as in *his* and *hiss*. The combination of *s* and *h* gives English virtually another sibilant letter, as in *show* and *wash*, although *s* alone has this sound in such words as *sure* and *mission*, and can have the voiced pronunciation of the combination, as in *division* and *treasure*. In a few words, such as *island* and *aisle*, the *s* is silent. *See also* MS pp.244–245.

SA, the *Sturmabteilung* ("Storm troopers") founded in 1921 in Germany as an armed corps to protect Nazi meetings. Officially recognized by the state as a volunteer force, it was actually a Nazi party organ. Ernst ROEHM took it over in 1931. Roehm and the other SA leaders were murdered by HITLER in the Roehm purge of 30 June 1934.

Saar (Sarre), river in NE France; it rises in the Vosges and flows N into West Germany, then N to join the River Moselle (Mosel) near Trier. The Saar flows through a highly industrialized region and carries much barge traffic. There was heavy fighting in the area during WWII. Length: 240km (150 miles).

Saarinen, Eero (1910–61), US architect and designer, son of Eliel SAARINEN, b. Finland, whose experimental designs using curved concrete were influential in modern architecture. He is famous for the design of the General Motors Technical Center in Warren, Michigan (1948–56), the TWA terminal at Kennedy Airport (1956–62) and the US Embassy in London (1955–61). *See also* HC2 p.329.

Saarinen, Eliel (1873–1950), Finnish-American architect, resident in the USA after 1923. His work had a significant impact on US architecture. In later years he collaborated with his son, Eero SAARINEN, and together they won the highest award of the American Institute of Architects for the design for part of the Smithsonian Institution, Washington DC, in 1947.

Saarland. *See* SAAR.

Saavedra Lamas, Carlos (1878–1959), Argentine jurist who was influential in ending the Chaco War (1932–35) between Bolivia and Paraguay, for which he was awarded the 1936 Nobel Peace Prize. He was President of the League of Nations Assembly in 1936.

Sabah, Sheikh Sabah as-Salim as- (1913–), Amir of Kuwait. He was Minister of Foreign Affairs, deputy Prime Minister (1961–1963) and Prime Minister (1963–1965). He succeeded his brother as head of state in 1965.

Sabah (North Borneo), state of Malaysia, and one of the four political subdivisions of Borneo. It is located on the NE section of the island; Kota Kinabalu is the capital. The terrain is mountainous and forested. Products include coconuts, rice, rubber and timber. Area: 76,522sq km (29,545sq miles). Pop. (1972 est.) 697,000.

Sabatier, Paul (1854–1941), French chemist who discovered that ethane could be obtained from ethylene and hydrogen by heating the two gases with powdered nickel as a CATALYST. This technique of nickel catalysis led to an economical method of manufacturing margarine by the HYDROGENATION of inedible fats. He shared the 1912 Nobel Prize in chemistry with Victor GRIGNARD.

Sabbatarianism, movement which seeks to maintain the Biblical injunction of the 4th commandment to keep the Sabbath as a day of rest. It first became powerful in Britain during the Puritan interregnum (1649–60). After the Sunday Entertainments Act of 1932 empowered local authorities to license Sunday entertainment, Sabbatarianism lost much of its force in England, although it remained strong in parts of Scotland and Wales.

Sabbath, for the Jews the seventh day of the week, set aside specifically as a day of rest (the word is from the Hebrew meaning "rest"). The term is consequently applied by Christians to Sunday, although Saturday is the Jewish Sabbath.

Sabbattini, Nicola (1574–1654), Italian Renaissance stage designer and theoretician. His *Manual for Constructing Theatrical Scenes and Machines* (1638) was a major source of information about 17th century theatricals. *See also* HC1 p.318.

Sabines, or Sabini, ancient people of central Italy who inhabited the Sabine Hills NE of Rome. A significant part of the early Roman population was Sabine in origin. After sporadic fighting the Sabines were conquered in 290 BC and gradually Romanized.

Sabinianus (Sabinian), Pope (r.604–606). A Tuscan who succeeded Gregory the Great, he had previously been Gregory's envoy to the Byzantine Court of Patriarch John IV. His own period of office was troubled by famine and attacks by the LOMBARDS.

Sable, MARTEN native to northern pine forests and Siberia. It has been hunted almost to extinction for its thick, soft, durable fur which is dark brown to black, sometimes flecked with white; but conservation measures have restored the population to a much safer level. Length: to 60cm (24in). Family Mustelidae; species *Martes zibellina*. *See also* NW p.207.

Sabotage, deliberate wrecking of small parts of a system in order to prevent the whole system working properly. The word was first used before WWI to describe the activities of French railway strikers who destroyed the wooden shoes (sabots) which held rails in place. Since WWII sabotage has been associated primarily with military activity.

Sabre, largest weapon used in the sport of FENCING. Fenced only by men, it has a flattened V-shaped blade and a half-circular guard. The target area comprises the head, arms and trunk to the line of the hips.

Sabre-toothed tiger, popular name for a prehistoric member of the CAT family (Felidae) that existed from the OLIGOCENE period to the PLEISTOCENE period. It had extremely long canine teeth adapted to killing large herbivores. Genus *Smilodon*, subfamily Machairodontinae. *See also* NW pp.189, 191.

Saccharin, formula $C_7H_5O_3NS$, man-made substitute for sugar but is about 500 times sweeter. Derived from TOLUENE by a process involving sulphonation, it is used extensively in the food industry.

Sacchi, Andrea (1599–1661), Italian painter who studied in Rome and Bologna. One of the foremost 17th-century Roman painters in the classical tradition, his ideas influenced 18th-century artists through his pupil, MARATTA. Among his works are *The Vision of St Romuald* (c.1638) and the ceiling fresco for the Barberini Palace, *Divine Wisdom* (1629–33).

Sacher, Paul (1906–), Swiss conductor who founded Basel Chamber Orchestra in 1926. Since 1941 he has conducted the Zürich chamber orchestra which is known as the Collegium Musicum.

Sachs, Hans (1494–1576), German playwright and poet, a cobbler by trade. His work is characterized by humour, common sense and moral purpose. Among his most popular works are *Das Schlauraffenland* (1530) and the *Fastnachtspiele*; he also created the Meistersang, a form of poet song based on minstrel tradition, of which he wrote 4,000.

Sachs, Julius von (1832–1897), German botanist, known for his studies of the metabolism of plants. He wrote *Textbook of Botany* (1868).

Sachs, Nelly (1891–1970), German-Jewish poet. She escaped from Nazi Germany in 1940 and her early stories and later poetic works, among them *In the Houses of Death* (1947) and *Later Poems* (1965), bore witness to the sufferings of her people. She shared the 1966 Nobel Prize in literature with S. Y. AGNON.

Sackbut, old English name for the TROMBONE. Its origins are unknown, but it was mentioned in literature of 14th century Spain. Together with the SHAWM and the CORNETT it was an essential part of court bands.

Sackville, Thomas, 1st Earl of Dorset (1536–1608), English poet and statesman. He was the author of *Gorboduc* (1561), the first English tragedy in blank verse. He was Lord Treasurer from 1599 to 1608 and was made an earl in 1604. *See also* p.281.

Sackville-West, Victoria Mary (1892–1962), British poet and novelist, friend of Virginia WOOLF and a member of the BLOOMSBURY GROUP. She was married to the politician and diarist Harry Nicolson. Her best-known novels are *The Edwardians* (1930) and *All Passion Spent* (1931). Her long poem *The Land* (1926) won the 1927 Hawthornden prize.

Sacrament, rite in Christian churches, central to the LITURGY, that is a physical sign or symbol of spiritual reality. Sacraments impart the GRACE of God to men and the Church.

Sacramento, state capital and port in central California, USA, 116km (72 miles) NE of San Francisco. The city grew around the site of a fort built in 1840. It became the state capitol in 1854. The city includes a branch of the University of California (1947). Industries: food processing, soap, machinery, bricks. Pop. (1973) 267,483.

Sacred baboon. *See* HAMADRYAS.

Sacred Heart, devotion to the love of Jesus Christ in his human form, as symbolized by his heart. In its modern form it dates from revelations to St Margaret Mary Alacoque (1647–90), whose visions of Jesus involved His heart being honoured in this way.

Sacrifice, offering or destruction of precious objects – food and drink, flowers and incense, animals and human beings – for religious purposes. Sacrifices are important in many religions. They are made in the hope of winning divine favour, to atone for guilt or for some other mystical purpose. *See also* MS p.205, 205.

Sacrum, or sacral bone, lowest region of the vertebral column, formed by the fusion of five vertebrae. The sacrum articulates with the pelvic girdle. *See also* MS p.56.

Sadat, Mohamed Anwar as- (1918–), Egyptian soldier and politician. He became president of Egypt on the death of NASSER in 1970. He was proclaimed military Governor-General in 1973, and in 1973–74 he also served as Prime Minister. Late in 1977, in an effort to resolve the problems of the Middle East, he went to ISRAEL, the first Arab leader to do so since the formation of the state in 1948.

Sadducees, Jewish sect active in Judaea from c.200 BC until the fall of Jerusalem in AD 70. The Sadducees may have originated, however, as a political party representing an aristocracy that was particularly conservative in religious matters. By the time of Christ the main difference between them and the PHARISEES was their refusal to recognize the oral traditions surrounding the Scriptures as part of the Hebrew Law.

Sade, Donatien Alphonse François, Marquis de (1740–1814), French novelist and playwright. He is one of the founders of the modern French prose style and the licentiousness of his works and the moral freedom which he espoused, especially in sexual relations, have gained him consideration as a forerunner of EXISTENTIAL-ISM. During his 27 years in prison for sexual offences, he wrote many novels, among them *Justine* (1791) and *Juliette* (1797). *See also* SADISM.

S'Adi (Saadi) (*c.*1213–92), Persian poet who is generally regarded as one of the finest SUFI or mystic writers. His masterpiece, *Gulistan* (1258), is a miscellany of prose and verse including anecdotes and maxims. Other works include *Bustan* (1257), a long didactic poem.

Sadism, taking pleasure in hurting or abusing others physically or verbally. Sadism is frequently connected with deviant sexual behaviour in which beating or abuse of another person is a necessary part of sexual arousal and gratification. A person may alternate between these two forms of behaviour. The term is derived from the Marquis de SADE, much of whose writing glorified cruelty and sexual abuse.

Sadko (1898), opera in seven scenes by Nikolai RIMSKY-KORSAKOV, with libretto by the composer and Vladimir Belsky. It was first produced in Moscow.

Sadleir, Michael. *See* FANNY BY GASLIGHT.

Sadler's Wells, theatre in N London. It was first established in 1683 as a music-hall in the garden of a Mr Sadler near a so-called medicinal spring. It was re-built (in 1927–31) to house Lilian BAYLIS's opera and drama company and subsequently, the Vic-Wells Ballet formed by Dame Ninette de VALOIS which, in 1939, became known as the Sadler's Wells Ballet. After the company moved to Covent Garden in 1946, the theatre was the home of Sadler's Wells Opera, until that too moved – to the Coliseum in 1968. *See also* HC2 p.274.

Saenredam, Pieter (1597–1665), Dutch painter who painted carefully planned and accurate pictures of church interiors. His works, typically of the Gothic churches of Utrecht and Haarlem, depict large vaulted spaces in compositions of sensitive colouring and lighting. Among these is the *Interior of the Buurkerk at Utrecht* (1644).

Safavid dynasty, Muslim dynasty which ruled Persia from 1501 to 1736. Its name is taken from Sheikh Safī al-Dīn shāq, the founder of the Muslim order called the Safawiyya. The first Safavid ruler of Persia was Isma'il I. In 1736 the dynasty was overthrown by Nādir Shāh.

Safety glass, form of plate glass, used for windscreens and windows, that does not shatter or splinter when broken. It is laminated with sheets of transparent plastic fused to the glass under pressure. Bullet-proof glass consists of several layers of glass and plastic.

Safety lamp, for coal miners, the best-known example of which was invented by Sir Humphry Davy early in the 19th century. His model had a wire gauze cylinder within which an oil flame burned. Miners also used the lamp as a test for mine gases such as firedamp (methane), in the presence of which the flame would change colour. Modern safety lamps rely on the same principle.

Safety match. *See* MATCHES.

Safeway Stores, American supermarket chain founded in 1926 with headquarters in Oakland, California. In the early 1970s there were more than 2,200 Safeway supermarkets operating in the USA, Canada, Britain, West Germany and Australia, making it the largest chain in the world.

Safflower, annual plant with large, red, orange or white flower heads that are used in making dyestuff. The seeds yield oil which is low in saturated fats and is used in cooking and in the manufacture of MARGARINE, paints and cosmetics. Family Compositae; species *Carthamus tinctorius.*

Saffron, or autumn crocus, perennial CROCUS native to Asia Minor and cultivated in Europe. Its purple or white flowers bloom

in autumn. The golden, dried stigmas of the plant, used as a flavouring or dye, are also called saffron, as is the yellow colour produced by it. Family Iridaceae; species *Crocus sativus.*

Safi, port in central W Morocco, on the Atlantic coast, approx. 200km (125 miles) SW of Casablanca. It was a Portuguese base in the 16th century and is today the centre of Morocco's fishing industry. Phosphates are exported, and industries include chemicals and fertilizers. Pop. (1971) 129,113.

Saga, in OLD NORSE LITERATURE (especially Icelandic), prose narrative which relates the lives of legendary or historical figures. Written between the 7th and 14th century, sagas are generally objective, have skilful characterization and reflect the old Icelandic devotion to honour and family.

Sagan, Françoise (1935–), pen-name of Françoise Quoirez, a French novelist and playwright. Her works portray the disillusionment of the young and innocent, as in *Bonjour Tristesse* (1954), a theme developed in *Un Certain Sourire* (1956) and other works. Among later works are *Aimez-vous Brahms...* (1959) and plays, including *Les Violons, parfois...*(1961).

Sage, common name for a number of plants of the MINT family (Labiatae) native to the Mediterranean region. The best known is *Salvia officinalis,* an aromatic perennial herb used widely for seasoning. Height: 15–38cm (6–15in). *See also* NW p.*60.*

Sagebrush, aromatic shrub common in arid areas of W North America. The common sagebrush has small, silvery-green leaves and bears clusters of tiny white flower heads. Height: to 2m (6.5ft). Family Compositae; species *Artemisia tridentata.*

Sage grouse, or sage hen, largest North American GROUSE. The male fans his long slender tail behind his inflated orange neck air sacs, while bowing, dancing and groaning during courtship displays. Length: to 75cm (30in). Species *Centrocercus urophasianus.*

Sagittarius, or the Archer, southern constellation on the ECLIPTIC between Scorpio and Capricorn. Rich in stellar CLUSTERS, this region of the sky also contains much interstellar matter which obscures the central region of the MILKY WAY GALAXY and is only penetrable with radio telescopes. The brightest star is Epsilon Sagittarii (*Kaus Australis*) of magnitude 1.8. *See also* MS pp.200, *200–201;* SU pp.*254, 262, 262.*

Sagittarius, or the Archer, in astrology the ninth sign of the zodiac. It is represented by a centaur firing a bow and arrow. The Sun is in Sagittarius from 22 November to 21 December.

Sago, starchy substance obtained from the pith of a tropical PALM. A nourishing, highly digestible flour made from sago is used for thickening soups and making puddings. Up to 363kg (800lb) of flour can be made from one tree's yield. Sago palm: family Palmae; species *Metroxylon sagu.*

Sago palm, or fern palm, feather-leaved PALM tree native to swampy areas of Malayasia and Polynesia. Its thick trunk contains SAGO, a starch used in foodstuffs. The sago palm flowers once after growing for 15 years and then dies when the fruit ripens. Height: 1.2–9.1m (4–30ft). Family Palmaceae. Some CYCADS are also called sago palms, but they are not true palms.

Saguaro, large CACTUS native to Arizona, California and Mexico. White, night-blooming flowers appear when the plant is 50 to 75 years old. Its red fruit is edible. Height: to 12m (40ft). Family Cactaceae; species *Carnegiea gigantea.*

Sahara, world's largest desert, in N Africa; extends from the Atlantic Ocean (W) to the Red Sea (E) and from the Atlas Mts Southward to N Sudan. It covers most of Spanish Sahara, Algeria, Niger, Libya, Egypt and Mauritania, the S parts of Morocco and Tunisia and the N parts of Senegal, Mali, Chad and Sudan. The annual rainfall is usually less than 12.7cm (5in). There is very little natural vegetation. Date palms, fruit, vegetables and cereals are grown by means of irrigation in

the oases, where most of the Sahara's estimated 2 million people live. The rest are nomads, herding sheep and goats. Mineral deposits include salt, iron ore, oil and gas. Area: approx. 9,065,000sq km (3,500,000sq miles). *See also* PE pp.120–121, *120–121,* 149, *149.*

Sahel, semi-arid region of Africa, bordered by the Sahara (N) Senegal (W) and Ethiopia (E). It extends through Mauritania, Mali, Upper Volta, Niger, N Nigeria, N Senegal and S central Chad. The area's agricultural economy was devastated by droughts in the late 1960s and early 1970s.

Sa'id, Qabus ibn (1940–), sultan of Oman, who overthrew his father Sa'id ibn Timur in 1970. Previously the country had been isolated and had few diplomatic links, but Qabus ibn Sa'id made efforts to liberalize the government and changed the official name of the country from MUSCAT AND OMAN to the Sultanate of Oman.

Saidpur, town in N Bangladesh, in Rangpur district. Saidpur is a major rail centre and has large railway repair shops. Jute processing and exporting are the chief economic activities. Pop. 60,628.

Saiga, Eurasian ANTELOPE, found only in S Russia and central Asia. It is a sheep-like animal with a "swollen" nose that ends in a pig-like snout. Its coat is brown in summer and white in winter. Males carry short, ridged, slightly curved horns which were prized as hunting trophies. Height: to 80cm (32in) at the shoulder. Family Bovidae; genus *Saiga. See also* NW p.*198, 198.*

Saigon (Ho Chi Minh City), port in S Vietnam, at the mouth of the Saigon River; largest city in Vietnam. The French took it in 1859, and it became the capital of the Union of Indochina. Saigon was chosen as the capital of South Vietnam by the Geneva Convention in 1954. During the VIETNAM WAR it was the military headquarters for US and South Vietnamese forces. The city was severely damaged, and was taken by the North Vietnamese in May 1975. The city is the most important port, commercial and industrial centre in S Vietnam and the focus of road and rail routes and the waterways of the Mekong delta. Industries: textiles, shipbuilding, food processing, building materials, glass, pharmaceuticals, rubber, paper. Pop. 3,500,000.

Sailendra, Buddhist dynasty that ruled central Java *c.*750–850 AD, and the Sumatra-based empire of SRIVIJAYA during the 8th–13th centuries. The Sailendras encouraged sea trade with China and India and were patrons of MAHAYANA BUDDHISM. *See also* HC1 p.*65.*

Sailer, Toni (1935–), Austrian alpine ski racer who, at the 1956 Winter Olympic Games, became the first man to win all three gold medals: downhill, slalom, and giant slalom. The Olympic titles also counted as the world titles; winning all three made him that year's alpine combination world champion. In 1958 he again won the world downhill, giant slalom, and combination titles.

Sailfish, marine bill-fish found throughout the world in tropical seas. A popular sport fish, it is identified by a large, sail-like dorsal fin and sword-shaped upper jaw. Length: to 3.5m (11ft); weight: to 100kg (221lb). Genus *Istiophous.*

Sailing, water sport, either solo or competitive, in a boat propelled through the water by wind upon sail. Sailing for sport began in Holland in the 17th century. The first yacht club (the usual name for a sailing club) was the Water Club of Cork Harbour in Ireland, founded in 1720. The traditional wooden hulls and canvas sailcloth have now given way to fibre-glass hulls and sails of synthetic materials. Dinghy sailing is becoming an increasingly popular sport, helped by these cheaper new materials.

Sailing ships, ships that harness wind power using cloth or canvas sails. From prehistoric times square-rigged sails have been used to drive boats, and in the ancient mediterranean civilizations of Egypt, Crete, Greece and Rome large square sails were fitted to ships to supplement banks of oars. Viking ships of the

Sadler's Wells: the Welsh National Opera Company performs *I Lombardi,* by Verdi.

Saffron: 4,000 flowers are needed to produce 28gm (1oz) of commercial essence.

Saigas were hunted almost to extinction but they are now a protected species.

Toni Sailer, Austrian champion who won all the Olympic skiing gold medals in 1956.

Sail plane

Saint Bernard dogs were originally bred by monks in the monastery of St Bernard.

St Gotthard Pass: tunnel between Faido and Giornico on the Basle to Chiasso route.

St James's Palace: main gate and towers of this Tudor palace, with sentry on duty.

Louis Saint Laurent, in London during the Commonwealth Conference in 1952.

early Middle Ages also used these two methods of propulsion. The use of more sophisticated sets of sails and the development of triangular lateen sails (which enabled craft to make use of even light winds and sail at an angle to the prevailing wind) probably began in China. Sailing ships had reached such a level of sophistication by the late 19th century that CLIPPER ships were faster than many early STEAMSHIPS. By 1900 sail had been outmoded by steam, except for recreation. *See also* MM pp.112–113, *112–113.*

Sail plane, type of GLIDER with particularly long narrow wings. *See also* GLIDING.

Saint, in Christianity, someone who has manifested exceptional love of God and holiness during his or her life. In some Christian writings all believers are called saints, but the title is usually reserved for men and women of the most outstanding merit, regarded as having a special relationship with God. *See also* CANONIZATION.

St Albans, county district in central Hertfordshire, s England; created in 1974 under the Local Government Act (1972). Area: 161sq km (62sq miles). Pop. (1974 est.) 122,700.

St Andrews, golf course in Fife, Scotland, honoured by golfers throughout the world as the headquarters of the ROYAL AND ANCIENT GOLF CLUB. As well as the Old Course, where major championships are played, it has three other courses: the New, the Jubilee and the Eden.

St Anne, town in the Channel Islands, United Kingdom; the only town of any size on the small island of Alderney. Tourism and the rearing of cattle are the chief occupations. Pop. 1,472.

Saint Augustine, city in NE Florida, USA, on the Atlantic Ocean. The oldest city in the USA, it was founded by Pedro Menéndez de Avilés in 1565. It is a port of entry, tourist resort and a fishing centre. Pop. 12,352.

St Basil the Blessed, Cathedral of, 16th-century Russian church in Red Square, Moscow, built between 1554 and 1560. It was commissioned by IVAN IV to commemorate his victories over the Volga Tatar Khanates of KAZAN and ASTRAKHAN. Famous for its ornately carved onion-shaped domes and colourful exterior decoration, it is thoroughly medieval Russian in style, without any trace of Western architectural influence.

St Bernard, two alpine passes in Switzerland. The Great St Bernard links Martigny-Ville, Switzerland, with Aosta, Italy. It was used by the Gauls, Romans, CHARLEMAGNE and NAPOLEON I. There is now a road tunnel linking the two countries. The Little St Bernard connects France with Italy. Altitude: Great St Bernard: 2,469m (8,110ft); Little St Bernard: 2,188m (7,178ft).

Saint Bernard dog, Swiss mountain and rescue dog with excellent scenting abilities; from the 17th century it has been used to track down lost people in deep snow. It has a massive head with characteristic forehead wrinkles, a short deep muzzle and pendulous jaws and ears. The broad straight body is set on long, muscular legs. The dense white and red coat may be either short or long. Height: to approximately 74cm (29in) at the shoulder; weight: to 77kg (170lb).

Sainte-Beuve, Charles Augustin (1804–69), French writer and critic. His major works are a series of *Portraits* (1836–39), his book *Châteaubriand et son groupe* (1849), and collections of biographical and psychological literary criticisms.

St Catharines, city in s Ontario, Canada, on the Welland ship canal. It was incorporated as a city in 1876. The city is the centre of a fruit-growing region and has shipbuilding and repair, machinery, paper and food canning industries. Pop. (1974) 109,725.

St Christopher. *See* Saint KITTS-NEVIS.

Sainte-Claire Deville, Henri Etienne (1818–81), French chemist. He developed a method of producing aluminium commercially in 1854. He was one of the first chemists to produce toluene and the discoverer of nitrogen pentoxide.

St Dunstan's, British organization to assist and retrain servicemen blinded in active service. It was set up in 1915 by Sir Arthur Pearson (1866–1921), who had himself been forced to retire through blindness in 1910.

St-Étienne, city in E central France, 52km (32 miles) sw of Lyons on the Furens River; capital of Loire département. Textile and silk industries were established there in the 15th century, and firearms have been produced there since the 16th century. The manufacture of steel is the other major industry. Pop. (1970 est.) 216,000.

Saint-Exupéry, Antoine de (1900–44), French novelist and aviator. His experiences as a pilot provided the material of his novels, which explore the themes of duty, love and human brotherhood. His novels include *Southern Mail* (1928), *Night Flight* (1931) and *Flight to Arras* (1942).

St Gall, canton in NE Switzerland. It borders on Lake Constance (N), the River Rhine (E) and completely surrounds the canton of Appenzell; the capital is St Gall. The city of St Gall became a free imperial city in 1206 and joined the Swiss confederation in 1454. Products include wine and fruit, and the canton is known for its embroidery and silk. Area: 2,015sq km (778sq miles). Pop. (1970) 384,475.

Saint-Gelais, Mellin de (1487–1558), French poet who was a follower of Clément MAROT. He lived in Italy for many years and the ideas of the RENAISSANCE influenced his work. He was one of the first French poets to use Italian SONNET form.

St George's (St Georges), capital and port on the sw coast of GRENADA, in the West Indies. It was founded in 1650 as a French settlement and later became the capital (1885–1958) of the former British colony of the WINDWARD ISLANDS. Industries include rum distilling and sugar processing. Exports: cacao, nutmeg and mace. Pop. (1975) 8,600.

Saint-Germain, Treaty of (1919), WWI peace agreement between the Allies and Austria at the Paris Peace Conference. It established boundaries for the new republic of Austria, which replaced the AUSTRO-HUNGARIAN EMPIRE and lost a great deal of territory.

St Gotthard Pass, pass approx. 30.6km (19 miles) long and 2,108m (6,916ft) high, over the St Gotthard group of Lepontine ALPS; named after a hospice open since the 14th century. A carriage road built in 1820–30 has been little used since the completion of the St Gotthard Tunnel in 1880.

Saint Helena, island in the s Atlantic, discovered by the Portuguese in 1502. It was captured by the Dutch in 1633, but passed to the British East India Company in 1659. It became a British Crown colony in 1834. It was to St Helena that Napoleon was exiled in 1815 until his death.

St Helens, county district in E MERSEYSIDE, NW England; created in 1974 under the Local Government Act (1972). Area: 133sq km (52sq miles). Pop. (1974 est.) 193,500.

St Helier, chief town of JERSEY, one of the CHANNEL ISLANDS, in the English Channel. Local market-garden produce is sent to St Helier for export. Tourism is another important source of income. Pop. (1971) 28,135.

Sainthill, Loudun (1919–69), Australian theatre designer. He went to England in 1950 and subsequently worked at all the major London theatres, including the Old Vic and Sadler's Wells, as well as at the Shakespeare Memorial Theatre in Stratford-upon-Avon. In 1968 he won an award for his costumes for the New York production of the musical *The Canterbury Tales.*

St James's Palace, royal residence in the centre of London built by HENRY VIII, who surrounded it by St James's Park. It was the London residence of the monarchy from the WHITEHALL fire of 1697 to the accession of Queen VICTORIA (1837).

Saint Joan (1923), play in six scenes and an epilogue by George Bernard SHAW. It depicts Joan of Arc's direct and pure emo-

tions and contrasts them with the impersonal political reasoning of her associates and enemies. Sybil THORNDIKE created the role of Joan.

St John, port in s New Brunswick, Canada, on the N shore of the Bay of Fundy, at the mouth of the St John River. It was founded as a fort and trading post in the early 1630s. A major year-round port, St John has shipping connections with Europe, the West Indies and South America. Industries: shipbuilding and repair, oil refining, paper, food processing. Pop. (1971) 87,910.

St John of Jerusalem, Knights Hospitallers of. *See* KNIGHTS HOSPITALLERS.

St John's, port and capital of Antigua, in the Leeward Islands in the West Indies. During the 18th century St John's was the headquarters of the Royal Navy in the West Indies. Products: rum, sugar, cotton; tourism is an important industry. Pop. (1966) 24,367.

St John's, provincial capital and major port of Newfoundland, Canada, on the SE coast of Newfoundland Island. St John's is one of the oldest settlements in North America: it was founded in 1583 by Sir Humphrey GILBERT. Industries: fishing and fish processing, shipbuilding, textiles, iron. Pop. (1971) 88,102.

St John Ambulance, British organization founded in 1877 to provide voluntary first aid and rescue assistance for any gathering at which such aid might be required. The order developed from the KNIGHTS HOSPITALLERS, and is divided into the Brigade, which provides uniformed volunteers to give first aid to public meetings and events, and the Association, which spreads first-aid education.

St John's wort. *See* HYPERICUM.

Saint-Just, Louis-Antoine-Léon de, (1767–94), politician of the FRENCH REVOLUTION, whose stern and ruthless devotion to the cause of the Republic made him both admired by some and reviled by others. He became deputy in the NATIONAL CONVENTION in 1792, and a member of the COMMITTEE OF PUBLIC SAFETY in May 1793; in this latter capacity he was one of the leaders of France during the REIGN OF TERROR. Saint-Just and his close associate Maximilien ROBESPIERRE were executed after a coup in July 1794.

St Kilda, group of three westernmost islands in the Outer Hebrides approx. 180km (110 miles) w of the Scottish mainland. The island became a bird sanctuary in 1930 and now has the largest gannetry in the world. The islands are entirely uninhabited.

Saint Kitts-Nevis, self-governing state in the Leeward Islands, in the West Indies. The state includes the islands of Saint Kitts (Saint Christopher), Nevis and Sombrero. Basse Terre is the capital, on Saint Kitts. The islands were discovered in 1493 by Christopher COLUMBUS, and settled by the English (1623) and the French (1624). Anglo-French disputes over possession were settled in Britain's favour in 1783 by the Treaty of Paris. The islands achieved self-government in 1967. Tourism is the chief source of income. Sugar, cotton, salt and coconuts are exported. Area: 311sq km (120sq miles). Pop. (1976 est.) 48,000. *See also* MW p.149.

Saint Laurent, Louis Stephen (1882–1973), Canadian statesman. A distinguished lawyer, he was Minister of Justice and Attorney-General (1941–1946) in the Mackenzie KING government. After King's retirement (1948), he became leader of the Liberal Party and served as Prime Minister until 1957. He played an important part in the creation of the UNITED NATIONS. *See also* HC2 p.137.

St Laurent, Yves-Mathieu (1936–), French fashion designer. He popularized trousers for women for all occasions and is associated with the shortened skirts and jackets of the 1960s, the "chic beatnik" look, metallic and transparent fabrics, and geometric shapes and prints. The unique prestige of high couture has been weakened by his ready-to-wear clothes.

St Lawrence, Gulf of, inlet of the Atlantic Ocean between Newfoundland and the Canadian mainland. The Gulf has extensive fishing grounds, but is closed to navigation by ice from Feb. to May. Area: 155,000sq km (60,000sq miles).

St Lawrence, major Canadian river, in SE Ontario and S Quebec provinces; it flows from the NE end of Lake Ontario, to the Gulf of St Lawrence on the SE Canadian shore. Since the completion of the ST LAWRENCE SEAWAY in 1959 the river has been navigable along its entire course to all but the very largest vessels. The river forms the boundary between the USA and Canada for approx. 184km (114 miles). Length: 1,244km (760 miles). *See also* MW p.45; PE p.*151.*

St Lawrence Seaway, international waterway in Canada and the USA, consisting of a series of canals and locks in the St Lawrence River between Montreal and Lake Ontario, and the Welland and SAULT STE MARIE CANALS, connecting the GREAT LAKES with the Atlantic Ocean. The waterway permits the passage of ocean-going vessels to industrial lakeside ports of central North America such as Detroit, Chicago and Toronto. Iron ore, wheat and coal are the chief cargoes carried. Construction of the waterway was authorized by Canada in 1951, and by the USA in 1954; it was built between 1955 and 1959. Length: 3,830km (2,380 miles). *See also* MM p.*196;* MW p.45; PE p.*151.*

Saint Leger, the oldest of the five English classic horse races. It is run at Doncaster in September and is the last leg of the Triple Crown (the other two being the Two Thousand Guineas and the DERBY). The first St Leger took place in 1776. Until 1913 it was run over 2 miles, but since then has been 1 mile, 6 furlongs, 127 yds.

Saint-Léon, Arthur (1821–70), French ballet dancer and choreographer. He won a reputation throughout Europe for his dancing (1838–59), and was ballet master at the Imperial Theatre St Petersburg (1859–69). His choreography included *La Fille de Marbre* (1847) and *Coppélia* (1870).

St Louis, port in E Missouri, USA, on the Mississippi River; largest city in Missouri. It was founded by the French as a fur-trading post in 1763, held by Spain from 1770 to 1800 and then returned briefly to France (1800), before being ceded to the USA in the LOUISIANA PURCHASE of 1803. St Louis was capital of the Territory of Missouri from 1812 to 1821. During the American Civil War the city was a supply base and medical centre. Today it is the seat of three universities, a financial centre and a market for grain and livestock. Industries: motor vehicles, brewing, chemicals, aircraft, food processing. Pop. (1973) 558,006.

St Louis Bridge, bridge over the Mississippi River at St Louis, opened in 1874. It was the world's first major steel bridge, carrying a road above the railway. *See also* MM p.191, *191.*

Saint Lucia, volcanic island in the Windward group in the West Indies; Castries is the capital. The island was discovered by Christopher COLUMBUS in 1502. In 1967 the island became one of the six Associate States of the West Indies and achieved self-government. Bananas are the principal export. Area: 616sq km (238sq miles). Pop. (1975 est.) 112,000. *See also* MW p.149.

St Mark's, Venice, BASILICA renowned for its wealth of mosaics. Begun in 829 to enshrine the remains of the city's patron saint, St Mark, it was restored after a fire in 976 but was later demolished to be rebuilt in the 11th century in the BYZANTINE style. The terrace above its two-tiered façade bears four gilded bronze horses brought from Constantinople in 1204. It contains the Pala d'Oro, a particularly fine altarpiece of jewels, enamel and gold.

Saint-Martin, Yves (1941–), French jockey who dominated French horse racing in the 1960s. He won the national championship nine times riding for the trainer François Mathet. Among his many victories were the PRIX DE L'ARC DE TRIOMPHE, the Washington International, and French and English Classics.

Saint Michael and Saint George, Order of, British honour established in 1818. It is awarded for services abroad and in the

diplomatic service. The ribbon is saxon blue with a scarlet centre; its motto is *Auspicium melioris aevi* ("Token of a better age").

St Michael's Mount, isolated granite outcrop near Penzance, Cornwall, SW England, in Mount's Bay. It is linked to the mainland by causeway passable at low tide and it is surmounted by a castle. Height: 61m (200ft). *See also* PE p.*81.*

St Paul, state capital and port of entry in E Minnesota, USA, on the Mississippi River just E of Minneapolis. The city was founded as Fort Snelling in 1823, and became a fur-trading post. It was made territorial capital in 1849 and state capital in 1858. Industries: iron and steel, motor vehicles, machinery, chemicals, paper, computers, food processing. Pop. (1973) 523,116.

Saintpaulia. *See* AFRICAN VIOLET.

St Paul's Cathedral, (1675–1710), Britain's only cathedral in the CLASSICAL style, built on the site of a medieval cathedral, destroyed in the Great FIRE OF LONDON in 1666. It was designed by Sir Christopher WREN in the shape of a latin cross with a vast circular space at the crossing where eight piers support the great dome. BAROQUE influences are evident in the towers, the main façade, and the sham-perspective window niches and the false walls in the upper storey in the side elevations which serve as buttresses.

St Paul's Cross, canopied pulpit, with a cross on top, in St Paul's Churchyard, London. It was the main place for the preaching of popular sermons in the 17th century. It was destroyed by the FIRE OF LONDON. *See also* HC1 p.*287.*

St Peter Port (St Pierre-Port), chief town of GUERNSEY, one of the CHANNEL ISLANDS. Local market-garden produce is exported from the town. Pop. (1971) 15,245.

Saint Peter's, Rome, world's largest Christian church. The original structure was built upon the traditional site of the grave of St Peter by the Emperor CONSTANTINE. BRAMANTE was appointed as architect in 1506 to rebuild the church, but did not complete it. *See also* HC1 pp.*257, 307, 307.*

St Petersburg, former name of Leningrad, city in the NW Russian Republic (Rossijskaja SFSR), USSR, on the Baltic Sea. The city was renamed Petrograd in 1914 and Leningrad in 1924. Founded in 1703 by PETER the Great with the construction of the Fortress of St Peter and St Paul, the city was the capital of Russia from 1712 to 1918. The cathedral is the burial place of almost all the tsars. St Petersburg was a centre of political unrest that culminated in the overthrow of the Tsars and the RUSSIAN REVOLUTION. *See also* HC1 pp.*322–323.*

Saint Pierre and Miquelon, group of nine small islands SW of Newfoundland, Canada, in the Gulf of St Lawrence; a self-governing French territory. The capital is St Pierre on the island of the same name; Miquelon is the largest island. The group was claimed for France by Jacques CARTIER in 1535. Their proximity to the GRAND BANKS makes fishing by far the most important activity of the main islands. Area: 241sq km (93sq miles). Pop. (1975 est.) 5,000. *See also* MW p.149.

Saint-Quentin, city in N France on the River Somme. During the Middle Ages it was a centre of pilgrimage to the tomb of St Quentin. The city became part of the French crownlands in 1191. Saint-Quentin has been the scene of many battles as a result of its strategic location. It was captured by the Spanish in 1557 during the Wars of Religion. Industries: textiles, furniture, rubber, food processing. Pop. (1968) 64,196.

Saint-Saëns, Charles Camille (1835–1921), French composer, pianist and organist. He composed prolifically in all forms and is best remembered for his opera *Samson et Dalila* (1877), the *Third Symphony* with organ and two pianos (1886), the *Carnival of the Animals* (1886) and the symphonic poem *Danse Macabre* (1874). *See also* HC2 p.107.

Saint-Simon, Claude-Henri de Rouvroy, Comte de (1760–1825), French political reformer and one of the founders of

socialism. As a young man he fought in the American War of Independence; he made a fortune by buying nationalized land during the French Revolution. In *Du Système Industriel* (1820–21) and other works, some written with Auguste COMTE, Saint-Simon proposed a productive industrialized state directed by scientist-businessmen. His writings were largely ignored during his lifetime, but they influenced later socialists. *See also* HC2 p.212.

Saint-Simon, Louis de Rouvroy, Duc de (1675–1755), French courtier and memoir writer. He was an apologist for aristocratic privilege, presenting a biased but colourful view of life in the French court during the latter years of the absolutism of LOUIS XIV. The definitive edition of his diaries in 41 volumes did not appear until 1879–1923, although some parts were published in the 1780s.

St Sophia. *See* HAGIA SOPHIA.

St Valentine's Day, Christian festival, celebrated on 14 Feb. in memory of the martyrdom of St VALENTINE in the 3rd century. "Valentines", cards bearing romantic messages, usually in verse, are traditionally exchanged on that date, often with no signature.

St Valentine's Day Massacre, murder of seven bootleggers of the "Bugs" Moran gang by members of Al CAPONE's gang disguised as policemen in Chicago, USA on 14 Feb. 1929. The murders were part of Capone's attempt to monopolize the illegal traffic in liquor in the era of PROHIBITION. *See also* MS p.290.

Saint Vincent, self-governing island in the Windward group, in the West Indies. It is made up of Saint Vincent Island and the Grenadine Islands to the S. Kingstown, on Saint Vincent Island, is the capital. Saint Vincent achieved self government in 1969. The chief source of income is tourism and the export of arrowroot, bananas, copra and cotton. Area: 389sq km (150sq miles). Pop. (1975 est.) 100,000. *See also* MW p.149.

St Vincent, Cape, SW extremity of Portugal and continental Europe, 97km (60 miles) W of Faro. In 1797 the British, under the command of Admiral John JERVIS and Commodore Horatio NELSON, defeated the Spanish naval forces off the Cape.

St Vincent, Gulf, large inlet of the Indian Ocean, S Australia, E of Yorke Peninsula. The seaport city of Adelaide is on the E shore. Length: 167km (100 miles); width: 72km (45 miles).

St Vitus' Dance, or Sydenham's chorea, disease of the nervous system. Its symptoms are jerky and unco-ordinated movements. The cause is uncertain, but it is commonly associated with rheumatic fever. In the Middle Ages it was believed that St Vitus could heal the mania and its name derives from this.

Sakai, industrial city in W Honshú, Japan, on Osaka Bay approx. 10km (6 miles) S of Osaka, located in the Osaka industrial belt. It was an important port from the 15th to the 17th centuries, but the port business has declined since the silting up of the harbour. Industries include the manufacturing of iron, steel, chemicals and textiles. Pop. (1974) 716,498.

Sake, Japanese rice wine or, more accurately, rice beer. It is brewed by fermenting steamed rice with the mould *Aspergillus oryzae,* a process which takes about five weeks. It contains between 14 and 17% alcohol and is served warm.

Sakhalin, island off the coast of the Far East USSR, between the Sea of Okhotsk and the Sea of Japan. The island is mountainous and forested, with a harsh climate, but grains and potatoes are grown in the S. Timber and fishing are major industries but Sakhalin's chief importance lies in its deposits of coal and iron ore. Oil is extracted in the NE and piped to the Soviet mainland. Area: 76,400sq km (29,500sq miles). Pop. (1970): approx. 600,000.

Sakharov, Andrey Dimitriyevich (1921–), Soviet physicist and social critic. His work in nuclear fusion was instrumental in the development of the Soviet hydrogen bomb. An outspoken defender of civil liberties, he created the Human Rights Committee in 1970 and

Arthur Saint-Léon delighted audiences with both his dancing and choreography.

St Mark's, Venice: the city's cathedral used to be the doge's private chapel.

Yves Saint-Martin with Sassafras, after winning the 1970 Prix de L'Arc de Triomphe.

St Valentine's Day Massacre: a police photograph showing two of the victims.

Saki

"Saki": a photograph by Hoppé of Hector Hugh Monro, who was killed in WW1.

Salamanders are generally small, but the Japanese variety can grow to 1.5m (5ft).

António de Salazar stabilized Portugal's finances but at the expense of freedom.

Salisbury Plain is used by the army for testing tanks and artillery.

received the 1975 Nobel Peace Prize. His books include *Sakharov Speaks* (1974) and *My Country and My World* (1975).
Saki, pen-name of Hector Hugh Munro (1870–1916), British writer, *b.* Burma. He was chiefly a writer of short stories, among them the collections *Reginald* (1904), *Reginald in Russia* (1910), *The Chronicles of Clovis* (1912) and *Beasts and Superbeasts* (1914).
Sakhalin, island off the coast of the Far East USSR, between the Sea of Okhotsk and the Sea of Japan. The island is mountainous and forested, with a harsh climate, but grains and potatoes are grown in the s. Timber and fishing are major industries but Sakhalin's chief importance lies in its deposits of coal and iron ore. Oil is extracted in the NE and piped to the Soviet mainland. The island was settled by Russians and Japanese in the 18th and 19th centuries, and came under Russian control in 1875. Area: 76,400sq km (29,500sq miles). Pop. (1970): approx. 600,000.
Sakkara (Saqqarah), necropolis or burial place of MEMPHIS, on the edge of the Libyan desert. The Step Pyramid, built there by Zoser during Dynasty III, was the first Egyptian tomb to be built entirely in stone. Sakkara also has pyramids dating from the IV and V Dynasties. *See also* HC1 pp.34, *35.*
Saladin (Salah ad-Din) (1137–93), Sultan of Egypt and Syria. As a lieutenant of NUR-AD-DIN, he suppressed the Fatimid dynasty of Egypt, became vizier and then proclaimed himself sultan in 1174. After conquering most of Syria he launched a campaign to drive the Christians from Palestine, gathering around him MUSLIMS of various groups. He won the battle of HATTIN and took Jerusalem in 1187. This resulted in the formation of an army for the third CRUSADE led by RICHARD I of England and PHILIP II of France. *See also* HC1 pp.176–177.
Salamanca, Battle of (1812), decisive battle in the PENINSULAR WAR. The British, led by the Duke of WELLINGTON, defeated the French under Marmont, near Salamanca, w central Spain. The victory allowed Wellington to march unopposed into Madrid. *See also* HC2 p.78.
Salamander, any of 320 species of amphibians found throughout the world, except in Australia and polar regions. It has an elongated body, a long tail and short legs. Most species lay eggs, but some give birth to live young; fertilization is commonly internal. The largest European species, the strikingly coloured fire salamander (*Salamandra salamandra*), lives in hilly wooded areas and may attain a length of 28cm (11in). Order Urodela. *See also* NW pp.73, *132,* 132–133, 208.
Salamis, Battle of, decisive naval engagement (480 BC) during the PERSIAN WARS. THEMISTOCLES lured the Persian fleet into battle and a Greek fleet, largely based on an Athenian contingent, destroyed 200 Persian ships. The Persians, who had expected to conquer all Greece, were forced to retreat northwards. *See also* HC1 pp.54, *60,* 70, *70.*
Sal ammoniac, old name for the salt ammonium chloride, NH$_4$Cl. It does not melt when heated but sublimes (changes from solid to vapour), and was discovered by the ancient Chinese and widely used by alchemists. It has long been used as a flux in soldering and as an electrolyte in DRY CELLS.
Salazar, António de Oliveira (1889–1970), Portuguese politician. He was a professor of economics with right-wing, pro-Roman Catholic beliefs. He became Premier in 1932 and thereafter ruled as virtual dictator until 1968. In the SPANISH CIVIL WAR he supported the Nationalists but, although sympathetic to the Axis powers kept Portugal neutral in WWII. He presided over Portugal's economic revival after the war but fought an unsuccessful battle to retain the Portuguese colonies in Africa.
Salem, Mamdouh Muhammad (1918–), Egyptian politician. He served as Police Commissioner of Alexandria (1946–68), deputy Prime Minister (1971–75) and Prime Minister (1975–).

Salem, state capital of Oregon, USA, 71km (44 miles) s of Portland on the Willamette River. The city was founded in 1840 by Methodist Episcopal missionaries as a mission station and manual training school for Indians. It was made the territorial capital in 1851 and state capital in 1859. Salem is the seat of Willamette University (1842). Industries: timber, paper, textiles, food canning, meat packing. Pop. (1970) 68,296.
Salem, city in NE Massachusetts, USA, on Massachusetts Bay 22km (14 miles) NE of Boston. Fear of witchcraft in colonial America reached its peak in the Salem witch trials of 1692, as a result of which 19 people were hanged. Industries: electrical products and leather goods. Pop. (1970) 40,556.
Salerno, port in s Italy, 47km (29 miles) SE of Naples, on the Gulf of Salerno. Founded by the Romans in 197 BC, it was conquered by the Normans in 1076, sacked by the Swabian Hohenstaufens in 1194 and became part of the Kingdom of Naples in the 15th century. It was the scene of fierce fighting during WWII. The first European medical school was founded at Salerno in the 9th century. Industries include textiles and machinery. Pop. (1975 est.) 159,518.
Salesbury, or Salisbury, William (*c.*1520–*c.*1584), Welsh lexicographer. He was the first translator of the New Testament into Welsh, and edited and published a collection of Welsh proverbs in 1546, probably the first book printed in Welsh. He also issued the first Welsh-English dictionary (1547).
Salford, county district in SW GREATER MANCHESTER, England; created in 1974 under the Local Government Act (1972). Area: 97sq km (37sq miles). Pop. (1974 est.) 273,600.
Salian dynasty, royal and imperial ruling line in Germany from 1024 to 1125. Conrad of Swabia, one of the Salian FRANKS, whs elected King of Germany in 1024 on the extinction of the Saxon dynasty. The last of the Salians, Henry V, was Holy Roman Emperor from 1106 to 1125, when he died heirless.
Salic Law, code promulgated in the 5th century by the Salian Franks, German settlers in Gaul, which included a law that forbade daughters to inherit land. This was cited, incorrectly, after the 14th century as judicial ground for the exclusion of women, and those descended through the female line, from succession to the French throne. It was contravened by EDWARD III's claim through his mother to the French throne, which contributed to the outbreak of the HUNDRED YEARS WAR. It was also invoked in 1593 to prevent the succession of Isabella, the Spanish infanta and Henry II's grand-daughter.
Salicylic acid, colourless crystalline solid, derivatives of which are antirheumatic drugs (including ASPIRIN) and dye-stuff. It is a coal-tar product, being a hydroxy-acid derivative of benzene with the formula $C_6H_4(OH)COOH$.
Salieri, Antonio (1750–1825), Italian composer who lived mainly in Vienna and was the teacher of BEETHOVEN, SCHUBERT and LISZT. The legend that he poisoned MOZART inspired Rimsky-Korsakov's opera *Mozart and Salieri*.
Salina, or playa lake, dried lake bed found in arid areas. It is covered with thick layers of salt and mud deposited when water runs into it from surrounding areas during rainy seasons and forms a lake that evaporates after a few days. *See also* PE p.120.
Salinger, Jerome David (1919–), US novelist whose first book, *Catcher in the Rye* (1951), made an impact in the USA in the 1960s. Salinger was a recluse and shunned publicity. This is perhaps reflected in the tone and style of *Franny and Zooey* (1961) and his two novellas *Raise High the Roof Beam, Carpenters* and *Seymour: An Introduction* (both published in 1963).
Salinity, total amount of dissolved salts in seawater, a concentration which averages 35 parts per thousand. Salinity analyses led to the discovery that the relative proportions of dissolved material in seawaters remain constant regardless of

the degree of salinity, thus proving that ocean water is constantly being mixed over long periods of time. *See also* PE p.76.
Salisbury, Robert Arthur Talbot Gascoyne-Cecil, 3rd Marquess of (1830–1903), British statesman and diplomat. A Conservative, he served in DISRAELI's administration (1874–80), first as Secretary for India and then as Foreign Secretary. On Disraeli's death (1881) he became leader of the Conservative Party and served three terms (1885–86, 1886–92, 1895–1902) as Prime Minister. Notable reforms were introduced at home but his policy abroad of "splendid isolation" left England with few friends during the SOUTH AFRICAN (Boer) WAR (1899–1902). *See also* HC2 p.113, *113.*
Salisbury, Robert Cecil, 1st Earl of. *See* CECIL, ROBERT, EARL OF SALISBURY.
Salisbury, county town and county district in Wiltshire, s England. The town was founded as New Sarum in 1220 when the bishopric was moved from Old Sarum. The city grew around the cathedral, which has the highest spire in England (123m; 404ft) and was added to the original cathedral in the 14th century. Salisbury's library contains one of the original four copies of MAGNA CARTA. Its industries are associated with its function as a market town. Pop. (1971) 35,271. The county district was created in 1974 under the Local Government Act (1972). Area: 1,005sq km (388sq miles). Pop. (1974 est.) 100,800.
Salisbury, city in NE Rhodesia (Zimbabwe), capital of Rhodesia and South Mashonaland province. It was founded in 1890 as a fort from which the Pioneer Column, a mercenary force organized by Cecil RHODES, could seize Mashonaland. The city was named after the British Prime Minister at the time, Lord Salisbury. It was the capital of the Federation of Rhodesia and Nyasaland from 1953 to 1963. It is the seat of the University of Rhodesia (1957). Industries: gold mining, textiles, steel, food processing, clothing, tobacco, chemicals, furniture. Pop. (1970 est.) 435,000.
Salisbury Plain, region of chalk downs in Wiltshire, s England. A number of prehistoric sites have been found on the plain, the most famous of which is STONEHENGE. The army now use much of the area as a military training ground. Area: 777sq km (300sq miles).
Saliva, fluid secreted into the mouth, under the control of the nervous system by the SALIVARY GLANDS. In vertebrates it is composed of about 99% water with dissolved traces of sodium, potassium, calcium and the ENZYME ptyalin. Saliva softens and lubricates food to aid chewing, dissolves some food, starts the digestion of starches, washes the teeth and keeps the mouth moist and flexible. *See also* MS p.68.
Salivary glands, one of three pairs of glands located on each side of the mouth that form and secrete SALIVA. The parotid gland, just below and in front of each ear, is the largest of the salivary glands and the one that becomes enlarged in MUMPS; the submaxillary gland is near the angle of the lower jaw; and the sublingual gland is under the side of the tongue. *See also* MS p.68.
Salk, Jonas Edward (1914–), US medical researcher. He developed the first vaccine against POLIOMYELITIS in 1952, using inactivated poliomyelitis virus as an immunizing agent. Extensive tests of the vaccine began in 1954 and mass immunization programmes followed.
Salk vaccine, preparation developed by Jonas SALK to immunize against POLIOMYELITIS. It is a "killed" virus VACCINE given by means of a series of injections, unlike the Sabin "live" vaccine (which is taken orally).
Sallow, or goat willow, type of WILLOW tree in which large catkins appear in early spring before the unfolding of the leaves. Family Salicaceae; species *Salix caprea.*
Sallust, or Gaius Sallustius Crispus, (86–*c.*34 BC), Roman historian. Having served, often with discredit, as senator, legion commander and governor of Africa

Nova, he retired to devote the remainder of his life to writing. His works include *The Catiline Conspiracy, The Jugurthine War* and the five books of *The History of the Roman Republic. See also* HC1 p.112.

Salmon, marine and freshwater fish of the Northern Hemisphere. A popular sport and commercially farmed fish, most species are silvery and spotted until the spawning season when they turn dark or red. The Pacific salmon (*Oncorhynchus*) hatches, spawns, and dies in freshwater, but spends its adult life in the ocean. The Atlantic salmon (*Salmo salar*) is actually a marine trout which spawns in rivers on each side of the Atlantic Ocean and then returns to the sea. Weight: to 36kg (80lb). Family Salmonidae. *See also* NW pp.129, *130*; PE pp.242–243, *243–245*, 245.

Salmonella, any of several species of rod-shaped bacteria that cause intestinal infections in human beings and animals. *Salmonella typhi* causes TYPHOID FEVER, but the effects of other species may result in only mild GASTROENTERITIS. The bacteria are transmitted by CARRIERS, particularly flies, and in food and water. *See also* MS pp.93, *93*; NW p.36.

Salome (*fl.* 1st century AD), daughter of Herodias, who conspired with her mother to remove JOHN THE BAPTIST. As a reward for her dancing she asked Herod for the head of John on a plate. She is not mentioned by name in the Gospels but the story is recounted by the Jewish historian JOSEPHUS.

Salomé, play (1893) by Oscar WILDE and opera in one act (1905) by Richard STRAUSS. The play was written in French and translated into English by Lord Alfred Douglas in 1894. It had its first performance in Paris in 1896. Strauss began to write the opera in 1904. It was produced at the Court Opera, Dresden in 1905. Both works are treatments of the story of SALOME.

Salon des Indépendants. *See* INDÉPENDANTS, SALON DES.

Salon des Refusés. *See* REFUSÉS, SALON DES.

Salonika. *See* THESSALONIKI.

Salop, county in w England; known as Shropshire until 1974 when its name was changed under the Local Government Act (1972). The county is crossed by the River Severn. To the N of the river the land is generally low-lying, and to the s it rises to the Welsh hills. The economy is primarily agricultural. Mineral deposits include coal, and the principal industries are iron-working, locomotives, farm machinery and electrical goods. Area: 3,490sq km (1,347sq miles). Pop. (1976) 359,000.

Salote, Queen Tupou III (1900–65), Queen of Tonga. She succeeded her father, King George Tupou II, in 1918, and played a major role in reuniting the Tongan Free Church with the Wesleyan Church in 1924. She visited Britain in 1953 for the coronation of Queen Elizabeth II.

Salsify, also called oyster plant, or vegetable oyster, hardy biennial plant with a taproot resembling that of a parsnip. It is grown as a vegetable and prized for its oyster-like flavour. Height: to 1.2m (4ft). Family Compositae; species *Tragopogon porrifolius*.

Salt, Sir Titus (1803–76), British textile manufacturer and inventor. He started wool-spinning in 1834 and was the first to manufacture alpaca fabrics in England. The model village of Saltaire grew up after 1853 round his factories near Bradford, Yorkshire.

SALT agreements, armament control agreements worked out at Strategic Arms Limitations Talks (SALT), from 1969 and signed 26 May, 1972, by the US Pres. Richard NIXON and the Soviet General Secretary Leonid BREZHNEV. The SALT agreements limited anti-ballistic missile systems and offensive missile launchers. *See also* HC2 pp.241, 307.

Salt, chemical compound formed, along with water, when an ACID and a BASE react together. Salts are typically high-melting crystalline compounds that tend to be soluble in water. They are formed of IONS held together by electrostatic forces and in solution they conduct electricity (See ELECTROLYSIS). Sodium chloride – common table salt – is a typical example. *See also* SU p.144.

Saltaire, model village in the town of Shipley, West Yorkshire, England. It was set up in 1853 by Sir Titus Salt for the manufacture of wool.

Saltash, town in s Cornwall, England 6.5km (4 miles) NW of Plymouth (Devon). Situated on the Tamar Estuary, it is linked to Plymouth by a road bridge and a railway bridge, the Royal Albert Bridge (1859), designed by Isambard Brunel.

Saltbush, any of several species of herbs or shrubs of the genus *Atriplex*, found in salt-marshes and desert areas of Australia. To prevent excessive loss of water, the leaves are coated in salts. Family Chenopodiceae. *See also* NW p.204.

Salt flat, level arid salt-encrusted lake floor. The type of shallow basin called a playa is a form of salt flat that results when a desert lake dries up.

Saltglaze, in ceramics, a thin glaze of sodium silicate, one of the traditional glazes used on stoneware. When salt is thrown into the kiln at the maximum temperature sodium from the salt combines with silica in the clay to produce the glaze, also giving the pot a slightly pitted texture.

Saltillo, city in NE Mexico, situated on the plateau of the Sierra Madre Oriental. It was founded in 1575. The chief activities are cereal growing and stock raising but the city also has textile and food processing industries. Pop. (1971) 211,129.

Salt Lake City, state capital in N central Utah, USA 21km (13 miles) E of the Great Salt Lake, on the Jordan River; largest city in Utah. The city was founded in 1847 by the MORMONS. It then became capital of the Territory of Utah in 1856, and of the State of Utah in 1896. Landmarks include the Mormon Temple (1853–93). Salt Lake City is the world headquarters of the Mormon Church. Zinc, gold, silver, lead and copper are mined nearby. Industries: food processing, missiles, rocket engines, tourism, oil-refining, printing and publishing. Pop. (1973) 169,234.

Salto, city in NW Uruguay, situated on the opposite bank of the Uruguay River to Concordia, Argentina. It is the centre of an extensive fruit-growing region. Salto also has boat building and meatpacking industries. Pop. (1963) 57,958.

Salt processing, production of common salt (sodium chloride) from sea water, natural brines and rock salt. Beds of rock salt are mined and quarried and, if sufficiently pure, simply ground and screened. The heat of the sun or artificial heat is used to crystallize sodium chloride from solutions of salt.

Saltwort, or grasswort, annual, weedy herb of temperate seashores in Eurasia and the USA. It has greyish leaves and was once a source of crude soda ash. Height: to 60cm (2ft). Family Chenopodiaceae; species *Salsola kali*.

Saluki, royal coursing dog of Egypt and perhaps the oldest domesticated breed; it was known as long ago as 7000–6000 BC and introduced to England in the 19th century. It may also be called Persian greyhound. A graceful dog, it has a long, narrow head with long pendulous ears. The deep but narrow-chested body has a broad back and a slightly arched loin; it is set on long legs and the tail is long and slender. Height: to 71cm (28in) at shoulder; weight: to 27kg (60lb).

Salvador (Bahia), seaport city in E central Brazil, 1,207km (750 miles) N of Rio de Janeiro. Founded by the Portuguese as Bahia, it was the first formally established colony in Brazil. Salvador was a slave market in the 17th and 18th centuries and the capital of Brazil until 1763. Today the most important industry is food processing; exports include tobacco, sugar, oil and industrial diamonds. Pop. (1970) 998,258.

Salvador, El. *See* EL SALVADOR.

Salvage, compensation paid to someone who saves a ship or its cargo from imminent disaster. In Britain salvage is an ancient maritime right which has been incorporated into statute law, with the courts being authorized to fix the rate of salvage. For salvage to be claimable, the claimant's assistance must have been voluntary and have involved an element of skill and risk.

Salvarsan, first drug that was effective in the treatment of SYPHILIS, introduced by Paul EHRLICH in 1909. It is a yellow powder whose systematic chemical name is arsphenamine and the formula $C_{12}H_{14}As_2Cl_2N_2O_2.2H_2O$. It was also called 606, after the number of previous drugmaking experiments made by Erhlich. Injections of Salvarsan, although a great boon to syphilis sufferers who otherwise might have developed the fatal tertiary forms of the disease, were usually followed by jaundice and other ill-effects. *See also* MS p.120, *120*.

Salvation Army, Christian society devoted to the propagation of the Gospel among the working classes. Its origin was the Christian Revival Association, founded in 1865 in the Whitechapel district of London by William BOOTH. In 1867 it became the East London Christian Mission and in 1878 the Salvation Army, led by "General" Booth. Under the leadership of Booth's son, Bramwell, who was head of the army from 1880 to 1928, its work spread throughout the British Empire and other parts of the world. The Salvation Army also carries out extensive social work among the poor.

Salvi, Nicola. *See* TREVI FOUNTAIN.

Salvia, genus of about 700 species of flowering plants that includes SAGE, *S. officinalis*, which is used as a culinary herb. The genus is represented throughout the world. Family Lamiaceae. *See also* NW p.*60*; PE p.*213*.

Salviati, Francesco de'Rossi (1510–63), Italian painter best known for his decorative frescoes for the Palazzo Farnese in Rome, done in the MANNERIST Style.

Sal volatile, or smelling salts, mixture of ammonium carbonate crystals, alcohol and a perfume, often lavender. The mixture was usually held to the nostrils of fainted persons so that the sharp smell of ammonia could revive them.

Salween (Nu Chiang or Nu Jiang in China; Mae Nam Khong in Burma and Thailand), river in South-East Asia. It rises in the Tibetan Plateau, E Tibet, and flows s through Yünnan province, cutting deep gorges through the terrain almost parallel to the Mekong, Yangtze and Irrawaddy rivers. It empties into Burma's Gulf of Martaban, near Moulmein. It forms many rapids along its course and despite its long total length is navigable for only 120km (75 miles) upstream. Length: approx. 2,400km (1,500 miles).

Salzburg, city in NW Austria on the River Salzach, 114km (71 miles) ESE of Munich, West Germany. It grew around a 7th-century monastery. For more than 1,000 years it was ruled by the archbishops of Salzburg, who were later princes of the HOLY ROMAN EMPIRE. Salzburg became part of Bavaria in 1809, but was later returned to Austria by the Congress of VIENNA. Salzburg University dates from 1622. The city has an annual music festival in honour of MOZART, who was born there. Tourism is an important industry. Pop. 128,845.

Samadhi, in Buddhist and Hindu philosophy, "total collectedness", a state of higher trance which can lead to greater self-knowledge and the awareness of ultimate reality. Death is the final realization of this state of freedom from the body.

Samaria, ancient city of central Palestine, now the site of Sabastiyah in Israel. It was built as the capital of the northern kingdom of Israel in the 9th century BC but was conquered by SHALMANESER in 722–21 BC. Samaria was later destroyed by John Hyrcanus I and rebuilt by HEROD THE GREAT. It gives its name to the religious sect of SAMARITANS.

Samaritans, descendents of Israelites who were not deported from Palestine by the Assyrians in the 8th century BC. The Jews to the south rejected them, which probably forms the basis of Christ's famous parable of the Good Samaritan (Luke 10:23–37). They call themselves "Children of Israel" (Bene-Yisreal) and their sole form of religious observance is the TORAH (the first five books of the Old Testament). *See also* SAMARIA.

Salmon can leap more than 3m (10ft) when swimming upstream to lay their eggs.

Salt Lake City near the Wasatch Range. Most of the population are Mormons.

Salukies are among the swiftest of dogs and were once used to hunt gazelles.

Salzburg; Hohensalzburg castle on top of the Mönchsberg (Mountain of the Monks).

The Samaritans, led by Chad Varah, is a carefully recruited team of volunteers.

Samoyeds are protected from the intense cold by two layers of thick, woolly hair.

Samson destroys his captors and the temple of Dagan, from an engraving by Doré.

San Antonio: a pre-WW11 view, showing the Medical Arts Building (right).

Samaritans, The, organization founded in Britain by Chad Varah in 1953 to prevent suicide by befriending those in despair. Varah set up a telephone service manned 24 hours a day by volunteers of all ages who were able to respond in complete confidence to calls from the suicidal. The organization also has a flying squad of volunteers who can meet those in need of immediate personal contact. By the mid-1970s there were 20,000 Samaritans in 167 branches throughout Britain. Similar organizations also exist in many other countries. *See also* MS p.*191*.

Samarium, metallic element (symbol Sm) of the LANTHANIDE SERIES, first identified spectroscopically in 1879. Its chief ores are monazite (phosphate) and bastänite (fluorocarbonate). The element is used in carbon-arc lamps and as a neutron absorber in NUCLEAR REACTORS. Some samarium alloys are used in making powerful permanent magnets. Properties: at.no. 62; at.wt. 150.35; s.g. 7.5 (α), 7.4 (β); m.p. 1,072°C (1,962°F); b.p. 1,778°C (3,232°F); most common isotope Sm^{152} (26.72%).

Samarkand, city in Uzbekistan (Uzbekskaja SSR), USSR on the W spur of the Alai Mts 270km (180 miles) SW of Tashkent. One of the oldest cities in the USSR, it was conquered by ALEXANDER THE GREAT in 329 BC. Samarkand was an important trading point on the SILK ROUTE from China to the West, flourishing under the Arab UMAYYAD empire of the 8th century AD. It was destroyed in 1220 by GENGHIS KHAN but became the capital of TAMERLANE's empire in 1370. Almost uninhabited by 1700, it was taken by the Russians in 1868 and incorporated into Uzbekistan in 1924. Its products include cotton, silk, leather goods, wine, tea and motor vehicle parts. Pop. (1970) 267,000.

Samarra, town in N central Iraq, on the River Tigris. The existing town was founded as the new ABBASID capital in 836 on the site of an ancient settlement which gave its name to a type of NEOLITHIC pottery. The town contains a gold-domed mosque built in the 17th century and other ruins dating from that period. Pop. (1970 est.) 62,000. *See also* HC1 pp.*26, 148*.

Samaveda, one of the four collections of prayers and hymns that contain the liturgy of the HINDU religion. The Samaveda, or Veda of the Chants, contains the hymns of the RIGVEDA set to music.

Samba, ballroom dance of Brazilian origin which became popular in Europe and the USA in the 1940s. The music is in syncopated 4/4 time and derives from the late 19th century dance called the maxixe.

Samian ware, also known as *terra sigillata*, mass produced bright-red pottery tableware. It was in common use throughout the Roman Empire from the 1st century BC until the third century AD. The moulded reliefs with which it was decorated were known as "sigilla".

Samizdat (Russian: "self-publication") underground political works banned or censored by the Soviet authorities which are distributed usually in typewritten form. *Tamizdat* is the similar distribution of underground literature that has been printed abroad. *Radizdat* is the broadcasting of Samizdat works.

Samnites, ancient warlike people of S central Italy, probably descendants of the SABINES, who formed a Samnite confederation in the S Apennines. After being defeated by the Romans in the Samnite Wars (343–41, 316–04 and 298–90 BC) and again by SULLA in the civil war (82 BC), they were eventually Romanized.

Samoa, group of Pacific islands approximately 2,000km (1,600 miles) N of New Zealand. The group consists of American Samoa, a US dependency, and Western Samoa, an independent country. Pago Pago, the capital of American Samoa, is a tourist centre with an international airport and Western Samoa relies heavily on trade with New Zealand and capital provided by migrant workers in New Zealand. Exports consist of tropical fruits, especially bananas and copra. Pop: American Samoa (1975 est.) 29,000; Western Samoa (1976 est.) 151,275. *See also* MW pp.27, 183.

Sámos, island in SE Greece in the Aegean Sea; one of the Sporades Islands. It was colonized by Ionian Greeks in the 11th century BC. The island was held by the Ottoman Turks from 1500 to 1913, when it was annexed to Greece. Sámos was the home of AESOP, and PYTHAGORAS was born there. Grapes, tobacco and citrus fruits are grown. Industries: boat-building, silk. Area: 477sq km (184sq miles). Pop. (1971) 32,664.

Samothrace (Samothrakí), mountainous island in NE Greece, in the NE Aegean Sea. It was an independent community under Roman protection in the 2nd century BC, and occupied by the Ottoman Turks from 1456 to 1912, when it was annexed to Greece. Excavations in the 19th century unearthed the Winged VICTORY OF SAMOTHRACE. Products: cereals, olive oil, honey, sponges. Area: 179sq km (69sq miles). Pop. 3,830.

Samoyed, ancient sled and guard dog originally bred in N Asia. It has a broad, wedge-shaped head with thick, erect, triangular ears and dark, almond-shaped eyes. The deep-chested, medium-length body is set on moderately long, sturdy legs with large feet, and the long tail is curved over the back. The dense double coat – white, biscuit, or cream – is full and bushy. Height: to 60cm (24in) at the shoulder; weight: to 29.5kg (65lb).

Samoyeds, URALIC-speaking people who inhabit N USSR and consist of a number of groups, including the Nentsy, Yenisey, Tavgis and Selkups. They practise fishing, hunting, PASTORALISM (reindeer and horses) and trapping.

Sampan, small boat used in China, Japan and SE Asia, rigged for sailing but sometimes rowed with one or more oars. Most have wide sterns with the after portion of the deck raised, and are open with a cabin aft. *See also* MM p.*139*.

Samphan, Khieu (1932–), Cambodian politician who succeeded Prince Norodom SIHANOUK as head of state in 1976. He was deputy Prime Minister (1970–76) and became Commander-in-Chief of the KHMER ROUGE High Command in 1973.

Samphire, or glasswort, plants that grow in salt marshes and on seashores. They have fleshy stems with small, scale-like leaves and tiny flowers. Family Chenopodiaceae; genus *Salicornia*.

Sampling, in statistics, the selection of a small sample to be accurately representative of a larger set. Thus in a public opinion poll, a sample of perhaps 1,000 people may be taken to reflect the views of the entire country. Representativeness is sought by choosing the sample randomly, but the randomness itself is carefully protected by seeking the views of a predetermined proportion of people from each region, occupation group, sex, etc.

Samsara. *See* WHEEL OF LIFE.

Samson, Israelite judge and hero renowned for his great strength. When his mistress DELILAH discovered that his strength came from his long hair she had his hair cut off and handed him over to his enemies, the Philistines. Blinded by the Philistines and chained in their temple, Samson regained his strength and destroyed the temple of Dagon in Gaza, killing himself and his captors. *See also* MS p.*214*.

Samson Agonistes (1671), verse tragedy by John MILTON relating the biblical tale of SAMSON in the form of a Greek tragedy.

Samson et Dalila (1877), three-act opera by Camille SAINT-SAËNS, with French libretto by Ferdinand Lemaire. It was first performed in Weimar in a German translation, having been consistently refused by the Paris Opera.

Samuel I and II, two books of the OLD TESTAMENT. Through the stories of the prophet Samuel, and of SAUL and DAVID, Israel's first two kings, they describe the transition of Israel from a collection of tribes under separate chiefs to a single nation under a king.

Samuel Pepys's Diary, private journal by Samuel PEPYS, containing entries from 1660 to 1669. It is a frank, personal diary written in a private shorthand and not intended for publication, which provides an honest view of RESTORATION manners.

A partial edition was published in 1825, a complete edition in 1893–96. *See also* HC1 pp.294–295.

Samuelson, Paul Anthony (1915–), US economist. Noted for his work in MACROECONOMICS and mathematical economics, his *Economics: An Introductory Analysis* is the standard textbook in its field and has been translated into many languages. Other works include *Foundations of Economic Analysis* (1947). He received the 1970 Nobel Memorial Prize in economics for his efforts to "raise the level of scientific analysis in economic theory".

Samurai, warrior class of feudal Japan. Beginning its rise to prominence in the 12th century, the samurai (or *bushi*) class was supreme during the Tokugawa era (1603–1867). Samurai alone were privileged to bear arms, usually wearing two swords, and received a pension from their *daimyo* (feudal lord), to whom they pledged allegiance. Samurai followed BUSHIDO, a special code of conduct. Events such as the Satsuma Rebellion of 1877 had their roots in samurai feudalism.

San'a, capital city of the Yemen Arab Republic, or North Yemen, 145km (90 miles) from Hodeida. The city was occupied by the Ottoman Turks in the 17th century, after which it became the seat of the Rassite dynasty until the second Turkish occupation of 1872–1918. San'a became the capital of the independent Imam of Yemen in 1918 and of the Arab Republic in 1962. Today San'a is an agricultural and trade centre; its products include coffee, leather goods and jewellery. Pop. (1973) 125,000.

San Agustín culture, pre-Columbian culture found in S Colombia, in the region of the town of San Agustín. It made megalithic tombs and large statues of men or human-like deities, which have been dated to the first five centuries AD. *See also* HC1 pp.46, 232.

San Andreas Fault, geological FAULT that forms the boundary between the North American and North Pacific plates and separates SW California from the rest of North America. Over the past 15 million years the total movement along the fault has amounted to more than 320km (200 miles). It is a region in which earthquakes occur (the great San Francisco earthquake of 1906 occurred along it).

San Antonio, city in S central Texas, USA, on the San Antonio River. It was founded in 1793 on the site of the mission San Antonio de Valero, and was the scene of the Mexico-Texas struggles at the Alamo (1836). A military centre, San Antonio has several US air force bases, and it is the seat of two universities. Industries: food processing, aircraft, building materials, chemicals, wood products, tourism. Pop. (1973) 756,226.

Sancho, the name of two kings of Portugal. Sancho I (1154–1211), son and successor of ALFONSO I, ruled from 1185 to 1211 and achieved some success in the Algarve in the struggle against the Moors. His grandson, Sancho II (c. 1209–1248), was faced during his reign (1223–c. 1246) with a powerful and ambitious Roman Catholic Church, and he failed to prevent increasing anarchy. He was deposed at the instigation (1245) of Pope INNOCENT IV and his brother assumed power as Alfonso III.

San Cristóbal, city in W Venezuela, located at the SW end of the Cordillera Mérida in the Andean uplands. It was founded in 1561 and, although severely damaged by an earthquake in 1875, became an important trade centre. Products include coffee, tobacco, asphalt and leather goods. Pop. (1971) 151,717.

Sancroft, William. *See* NONJURORS.

Sanctions, punitive action taken by one or more nations against another stopping short of direct military intervention. Sanctions can include the cessation of trade, the severing of diplomatic relations or the use of a blockade.

Sanctis, Francesco de. *See* SPANISH STEPS.

Sanctuary, holy or reserved part of any religious building. In a Roman temple the sanctuary was called the *adytum* or *cella*; in a Christian church, the chancel or presbytery. The term also applied in the Middle Ages to a church precinct or other sacred place where a fugitive from justice could claim immunity from arrest. This right was abolished in the 17th century.

Sand, George (1804–76), French novelist, real name Amandine-Aurore-Lucie Dupin. After an unhappy marriage she went to Paris with Jules Sandeau, from the first half of whose name she took her pseudonym. Among her later companions were Alfred de MUSSET and Frédéric CHOPIN and her novels from this period, including *Lélia* (1833) and *Mauprat* (1837), examine women's rights to independence. *See also* HC2 pp.98, *107*.

Sand, mineral particles worn away from rocks by EROSION, individually just large enough to be distinguished with the naked eye. Sand is composed mostly of QUARTZ, but black sand (containing volcanic rock) and coral sand also occur.

Sandage, Allan Rex (1926–   ), US astronomer and cosmologist. A leading authority on stellar evolution, he participated in the discovery of the first QUASAR in 1961.

Sandalwood, any of several species of Asian trees, many of which are parasitic on the roots of other plants. They have oval hairy leaves and straw-coloured flowers which turn blood-red. Height: to 10m (33ft). Family Santalaceae.

Sand blasting, industrial method for cleaning the surfaces of materials such as metals and concrete. It is an efficient means of cleaning the outer walls of dirty buildings, but its most frequent use is to remove the oxide scale from metals prior to painting, spraying, electroplating or dip-coating. Articles to be cleaned are subjected to a high pressure blast of sand, where possible inside a closed chamber.

Sandburg, Carl (1878–1967), US poet, biographer and folklorist. His first volume of poetry *Chicago Poems* appeared in 1916. It contains his most famous poem, *Chicago*. Other volumes of poetry include *Cornhuskers* (1918), *Smoke and Steel* (1920), *Good Morning, America* (1928), *The People, Yes* (1936) and *Honey and Salt* (1963). His poetry, inspired by Walt WHITMAN, draws on American history and idiom. His novel *Remembrance Rock* was published in 1948.

Sandby, Paul (1725–1809), British painter who is known for his watercolour landscapes. He reproduced many of his works in AQUATINT, a process which he introduced into Britain.

Sand casting, method of moulding metals and alloys in which the molten metal is poured into a mould made of firmly packed sand. To help the sand grains cohere, they are mixed with clay, water and other binders. The method is used in the casting of steel, brass, bronze and aluminium. *See also* MM p.*42*.

Sand dune, bank of sand found in deserts, formed and kept in motion by the wind. Dunes may be crescent shaped (barchans), linear (seif), or they may be formed in wind shadows. Large numbers merging with one another give rise to a "sand sea" (erg). The term is also used to refer to dunes found together with vegetation, such as those on shorelines. *See also* PE p.121, *121*.

Sand eel, any of about 18 species of small, eel-like, marine fish; it has a long, slender body, a long head and a forked tail. Often found in schools, it lives beneath the sand in shallow water. Length: to 46cm (18in). Families Hypotychidae and Ammodytidae.

Sand fly, any of several species of two-winged flies, not all closely related. They include species of *Phlebotomus*, which, when biting human beings, transmit pappataci fever. Other serious diseases spread in this manner by sand flies include LEISHMANIASIS and DENGUE. Order Diptera. *See also* MS p.106.

Sandgrouse, terrestrial bird of central and s Eurasia and Africa that lives in large flocks in open arid areas, feeding on seeds and vegetable shoots, and sometimes resting in shallow sand depressions. It looks like grouse but is related to the pigeon and, like a pigeon, drinks with its bill held in the water. Length: 22–40cm (9–16in). Family Pteroclidae. *See also* NW pp.*141*, 221.

Sand hopper, also called sand flea, any of several species of small, terrestrial crustaceans. The nocturnal European sand hopper (*Talitrus saltator*) lives on beaches near the high tide mark, emerging to feed on organic debris. Length: to 1.5cm (0.6in). Order Amphipoda; family Talitridae. *See also* NW p.*236*.

Sandhurst, Royal Military Academy, training school for potential officers of the British army. Located near Camberley, Surrey, since 1813, it was first established at Great Marlow in 1802. Cadets generally enter Sandhurst at the age of 19 and follow a course lasting either eight or 16 months. They are then commissioned to the various branches of the army.

San Diego, city in s California, USA; port of entry for s California, Arizona and New Mexico. It was founded in 1769 as a mission site named San Diego de Alcalá. It has a large naval base, is an important centre for scientific research (especially in oceanography) and has a branch of the University of California. Industries: aerospace, electronics, shipbuilding, tourism, fishing, clothing, furniture. Pop. (1973) 757,148.

Sand painting, ritual painting practised by American Indians of the SW USA. Regarded as the vital core of a powerful magic performance designed to heal, the sand picture is believed to absorb the illness from the sick man sitting in its centre.

Sandpiper, wading bird that breeds in cold regions and migrates long distances to winter in warm areas, settling in grass or low bushes near water. It feeds on invertebrates and nests in a grass-lined hole in the ground. Length: 15–60cm (6–24in). Family Scolopacidae.

Sandringham, one of the residences of the British royal family, in Norfolk, England. The estate was bought for the Prince of Wales (later EDWARD VII) in 1861 and the house was built in 1869–71. GEORGE VI died there in 1952.

Sand River and Bloemfontein Conventions (1852 and 1854), recognition by the British of Boer self-government in the Transvaal and the Orange River Sovereignty respectively. The Conventions allowed the Boers but not the Africans to carry arms, thus shifting power to the former. *See also* HC2 p.128.

Sands, David ("Dave") (1926–52), Australian boxer of Aboriginal and Puerto Rican parentage. A devastating puncher with either hand, he held Australian titles at middleweight, light-heavyweight and heavyweight, and was a leading contender for the world middleweight crown when he suffered fatal injuries in a road accident.

Sandstone, SEDIMENTARY ROCK composed of sand grains cemented in such materials as SILICA, iron oxide or carbonate or lime. Its hardness depends on the character of the cementing material. *See also* PE p.102.

Sandstorm, strong wind carrying with it dense clouds of desert sand more coarse-grained than the particles in dust storms. *See also* PE p.70.

Sandwell, county district in SW WEST MIDLANDS, England; created in 1974 under the Local Government Act (1972). Area: 86sq km (33sq miles). Pop. (1974 est.) 320,100.

Sandwich, John Montagu, 4th Earl of (1718–92), British naval administrator. During his second term (1771–82) as First Lord of the Admiralty he presided over British defeats during the War of AMERICAN INDEPENDENCE and became highly unpopular in consequence. In fact he was an able, if corrupt, administrator and was prevented from expanding the navy, as he wished, because of the stringent economies of Lord NORTH.

Sandwich Islands, name given to the islands of HAWAII in the N central Pacific Ocean by Capt. James COOK in 1778. They were named after John Montagu, the Earl of Sandwich.

Sand yachting. *See* LAND YACHTING.

Sanfelice, Ferdinando (1675–1750), Italian architect who was the greatest of the late BAROQUE Neapolitan architects. He is best known for his monumental staircases, as at the Palazzo Serra Cassano, Naples.

San Francisco, port in W California, USA, on a peninsula bounded by the Pacific Ocean (W) and San Francisco Bay (E), which are connected by the Golden Gate Strait. The city was founded by the Spanish as a mission in 1776, and acquired by the USA in 1846. Devastated by an earthquake and fire in 1906, the city was quickly rebuilt and prospered with the opening of the PANAMA CANAL. Landmarks include Chinatown, Fisherman's Wharf, Nob Hill, Telegraph Hill, the Golden Gate Bridge (1937), the city's cable cars, Market Street and the Latin Quarter. There is a branch of the University of California (1855). Industries: tourism, shipbuilding, food processing, oil refining, chemicals, aircraft, fishing, printing and publishing. Pop. (1973) 687,450.

Sangallo family, Italian RENAISSANCE architects. Giuliano da Sangallo (1445–1516), designed the first great Renaissance villa at Poggio a Caiano in Florence. (*See also* HC1 pp.*254*, 255.) His brother, Antonio da Sangallo, the Elder (1455–1534), and their nephew, Antonio da Sangallo, the Younger, (1485–1546) were both influenced by BRAMANTE. Antonio the Younger became the most noted architect of his time. He designed the Farnese Palace, Rome and was a successor to RAPHAEL and Peruzzi on the building work on St Peter's.

Sanger, Frederick (1918–   ), British biochemist who was awarded the 1958 Nobel Prize in chemistry for being the first to determine the exact structure of INSULIN. He discovered a chemical, Sanger's reagent, that will attach itself to only one end of an amino-acid chain (which constitutes a PROTEIN). Then, by breaking up the chain into its component amino-acids, he could tell which acid contained the chemical and therefore which one was at the end of the chain. By repeating this technique he determined the exact order of the 51 amino-acids that make up insulin.

Sanger, Margaret Higgins (1883–1966), US social reformer who established the first birth control clinic in North America. As a public health nurse, she became convinced of the necessity of birth control as a primary means of alleviating urban poverty. Her crusade, begun in 1912, was unpopular at first, and she was jailed for sending birth control information through the post. *See also* MS p.81, *81*.

Sanger's Circus, British circus formed by two brothers, John (1816–89) and George (1825–1911) Sanger. Begun as a small touring entertainment, the circus evolved in 1871 into an equestrian spectacle produced each winter in London and a large travelling circus which toured the provinces in summer.

Sanhedrin, Great, ancient council or court of JUDAISM, prominent in Jerusalem during the period of Roman rule in Palestine until the Second Temple was destroyed in AD 70. Essentially a rabbinical court whose authority was accepted by the Jews, the Great Sanhedrin is believed to have served as a legislative and judicial body on both religious and political issues. JESUS CHRIST appeared before the Sanhedrin.

San Isidro, city in E central Argentina, 21km (13 miles) NW of Buenos Aires. It grew up in the 18th century around a chapel. Modern manufactures include leather, dairy goods and paper; tourism is also an important source of income. Pop. (1970) 250,008.

Sanitation, in civil engineering, provision of public hygiene by various methods, including refuse collection and disposal, drainage maintenance, purification of water and SEWAGE DISPOSAL. *See also* MM pp.200–201; MS p.106.

San José, capital and largest city of Costa Rica, in central Costa Rica, capital of San José province. Founded *c*.1736 as Villa Nueva, San José succeeded Cartago as capital of Costa Rica in 1823. The city soon became the centre of a prosperous coffee trade. It is the seat of the University of Costa Rica (1843). Products: coffee, sugar cane, cacao, vegetables, fruit, tobacco. Industries: food processing, wine, brewing, chocolate, leather goods, textiles, furniture. Pop. (1973) 215,441.

San José, city in W California, USA, 64km

George Sand, photographed in middle age by the pioneer French photographer Nadar.

Sandpipers; most nest on the ground but some use abandoned tree nests.

The 4th Earl of Sandwich invented the "sandwich" to eat at gaming-tables.

San Francisco; statue of Juan Bautista de Anza, the founder of the city.

San Juan

José de San Martín, early in his career, fought in Spain against Napoleon.

San Remo Conference was held in the Municipal Casino of the resort.

Santa Claus distributing presents to the delighted children of a Spanish village.

Santiago; a 19th-century print. Most of the city is today of modern construction.

(40 miles) SE of San Francisco. It was founded in 1777 and was California's capital from 1849 to 1851. San José has a branch of the University of California, and Lick Observatory is nearby. The city is the centre of a rich fruit-growing region. Pop. (1973) 523,116.

San Juan, capital and largest city of Puerto Rico, on the NE coast of the island. It has one of the finest harbours in the West Indies. Founded in 1508, the port prospered during the 18th and 19th centuries, and fell to the USA in 1898. The city has the University of Puerto Rico (1903), a school of tropical medicine, a US air base and an international airport. Exports include coffee, tobacco, fruit and sugar, which go mainly to the USA. Industries: tourism, cigars and cigarettes, clothing, sugar refining, rum distilling, metal products, pharmaceuticals, tourism. Pop. (1970) 452,749.

San Justo, city in E central Argentina, 24km (15 miles) WSW of Buenos Aires. Founded in 1856 as a market for the grain and alfalfa produced in the surrounding region, San Justo is now an important industrial suburb of Buenos Aires. Industries: farm machinery, paper, rubber products, motor vehicle parts. Pop. (1970) 659,193.

Sankara (780–820), Indian philosopher and theologian who founded the Advaita VEDANTA school of philosophy. In his major work, *Brahma-sūtra-bhāsya* (a commentary on *Brahma-sūtra*) he affirmed the one true reality (BRAHMA) as the source of all things and that change is illusory. *See also* MS pp.226, *232–233*.

San Marcos University, university in Lima, Peru. It was founded in 1551 and is consequently the oldest university in South America, although teaching was suspended for many years (1821–61).

San Marino, one of the smallest independent republics in the world, in NE Italy. It was founded by a Christian refugee named Marino, who established a community there in the 4th century AD. Most of the population are farmers. The republic's income is derived from tourism, the export of wine and wool and the sale of postage stamps. The capital is San Marino. Area: 62sq km (24sq miles). Pop. (1976 est.) 20,000. *See also* MW p.149.

San Martín, José de (1778–1850), South American revolutionary, *b.* Argentina. He played a significant part in winning independence for Argentina, Chile and Peru. In 1817 he led an army across the Andes into Chile and in 1821 entered Lima in triumph and became protector of Peru. The following year, however, he resigned in favour of Simon BOLÍVAR and retired to Europe. *See also* HC2 pp.86, *87*.

San Michele, The Story of (1929), book by Axel MUNTHE. Describing his life as a doctor in Paris and Rome, and on the island of Capri, the work won immense popularity chiefly because of its lively characterization and deeply sympathetic portray of human suffering.

Sanmicheli, Michele (*c.*1484–1559), Italian architect and engineer. Trained in Rome, he worked mainly in the Venetian Republic after 1527 designing palaces, villas and military fortifications. Among examples of his work are the Palazzo Bevilacqua in Verona and the Palazzo Grimani in Venice. *See also* HC1 p.259.

San Miguel, city in E central El Salvador, at the foot of the volcanoes of San Miguel and of Chinameca. The city is a trade and processing centre for a region producing cotton, henequen, coffee and sugar cane. The chief industry is textiles. Pop. (1971 est.) 59,500.

Sannazaro, Jacopo (*c.*1456–1530), Italian Renaissance poet. Noted for his highly influential pastoral romance *Arcadia* (1504) and Italian lyrics, he also wrote elegant Latin poetry.

San Pedro de Macorís, city in the Dominican Republic, at the mouth of the Higuamo River. It is the nation's leading sugar-exporting port. Textiles and the distilling of rum are the major industries. Pop. (1970) 70,092.

San Quentin, US prison, first established in 1852 on a peninsula in the San Francisco Bay, NW of San Francisco.

San Remo, Conference of (April 1920), meeting of Belgium, Britain, France, Greece, Italy, and Japan at San Remo, Italy, to discuss the Treaty of VERSAILLES (1919), consider the outlines of the Allied peace with Turkey and award mandates in the Middle East. *See also* HC2 p.296.

San Salvador, capital and largest city of El Salvador, in central El Salvador. Founded in 1524, the city has frequently been damaged by earthquakes. San Salvador served as the capital of the Central American Federation from 1831 to 1839. Products: soap, sugar, beer, textiles, tobacco. Pop. (1971) 337,171.

Sans-culottes, radical supporters of the French Revolution. They received their name because they wore long trousers (*pantalons*) rather than the knee breeches (*culottes*) worn by aristocrats. The term usually refers to the lower classes who demanded that the revolutionary government enact social and economic measures for their benefit.

San Sebastián, seaport city in N Spain, on the Bay of Biscay 77km (48 miles) E of Bilbao. Noted for its scenery and beaches, it is a popular resort area and former summer residence of the Spanish court. Its industries include chemicals, metallurgy and fishing. Pop. (1974) 177,534.

Sanskrit, classical language of India, the literary and sacred language of HINDUISM, and the forerunner of the modern Indo-Aryan languages. Like Latin and Greek it is an INDO-EUROPEAN language and was introduced to India from the north in about 1500 BC. Only about 3,000 Indians are able to speak it today but it has nevertheless been designated one of the national languages of India.

Sansom, William (1912–76), British short story writer and novelist whose writing is precise, detailed and realistic. His stories range over a wide variety of subjects, and collections include *The Ulcerated Milkman* (1966) and *The Vertical Ladder* (1969). His novels include *The Body* (1949) and *Hans Feet in Love* (1971).

Sansovino, Andrea Contucci (1460–1529), Italian religious sculptor of the High Renaissance. His graceful, classical marble carvings were much admired. Among his surviving works are the tombs of Cardinal SFORZA and Cardinal Basso (1505–09) in Sta Maria del Popolo, Rome, and the group *The Virgin and Child with St Anne* (1512) at Agostino.

Sansovino, Jacopo (1486–1570), Italian architect and sculptor. His early work was done in Florence and Rome, but he moved to Venice in 1527 and helped to establish the grand Roman style there. He designed the library of St Mark's (1536) and many other public buildings in Venice. *See also* HC1 pp.258, *259*.

San Stefano, Treaty of (March 1878), treaty concluding the war declared by Russia against the Ottoman Empire in April 1877. According to the treaty Russia won border areas in Asiatic Turkey and southern Bessarabia, and an independent Bulgarian principality was established. *See also* HC2 p.*184*.

Santa Anna, Antonio Lopez de (1794–1876), Mexican President. He identified himself with both liberal and conservative factions and intrigued ceaselessly. While he was President Mexico lost Texas to the USA (1836) and Santa Anna's attempt to recapture the province was crushed. In 1847, after defeat by US forces, he went into exile in Jamaica. He again became President, 1853–55, after which he was exiled again. He returned to Mexico in 1874 and died a pauper.

Santa Claus, variant of the Dutch name Sinte Klaas, itself a version of the name St NICHOLAS (*fl.* 4th century), Bishop of Myra. Santa Claus has become associated with the feast of Christmas and is identified with Father Christmas in Britain and in the USA. Late on Christmas Eve he traditionally brings presents, which he places in the stockings that children have put out as they go to bed.

Santa Cruz, Álvara de Bazán, marqués de (1526–88), Spanish admiral. He distinguished himself against the Turks at the battle of LEPANTO (1571) and against the French off the Azores at Terceira (1583).

He was appointed chief planner of the attempted invasion of England (1588) and commander of the Spanish ARMADA, but died before the expedition could set sail.

Santa Cruz, city in central Bolivia on the Piray River, founded in the 16th century as a Jesuit mission. A railway built in the early 1960s has given the city access to the Atlantic and Pacific oceans, and it is now a trade centre for sugar, coffee, rice and cattle. Pop. (1971 est.) 135,000.

Santa Cruz de Tenerife, city on the NW coast of Tenerife in the Canary Islands, Spain. Founded in 1494, it has a fine harbour and exports fruit, vegetables and sugar cane. Industries include oil refining. Pop. (1974) 157,791.

Santa Fe, state capital in N New Mexico, USA. Founded *c.*1609 by the Spanish, the city was a centre of Spanish-Indian trade for more than 200 years. A seat of government since it was founded, Santa Fe is the oldest capital in the USA. The city became the W terminus of the SANTA FE TRAIL in 1821. Industries: tourism, Indian and Mexican handicrafts, food processing. Pop. (1970) 41,167.

Santa Isabel, former name of Malabo, capital of Equatorial Guinea, on Macías Nguema Biyogo (formerly Fernando Póo island), in the Gulf of Guinea. It is the financial and commercial centre and exports coffee, cacao and timber. Pop. (1976) 19,869.

Sta Maria Maggiore (*c.*432), early Christian BASILICA in Rome. Long perspective is achieved by the combination of a low roof and a row of IONIC marble columns. There are some fine mosaics in the triumphal arch which depict scenes from the childhood of Christ. It is the only basilican church which bears evidence that it was once a pagan place of worship.

Santander, Francisco de Paula (1792–1840), Colombian politician who became the first constitutional President of NEW GRANADA (1833–37).

Santander, city in N Spain in the Bay of Biscay. It was a centre of trade with the New World in the 16th century. In June 1937 it was blockaded and captured by the Nationalists in the SPANISH CIVIL WAR. Much of the town centre was destroyed by fire in 1941. Pop. (1974) 161,947.

Santayana, George, (1863–1952), US philosopher and poet. His works tended towards scepticism but admitted a positive knowledge of the realm of universals or "essences". After WWII he withdrew from the world, which was reflected in the moral detachment of his writing. His works include *The Sense of Beauty* (1896), *The Life of Reason* (1905–06), *Skepticism and Animal Faith* (1923) and a novel, *The Last Puritan* (1935).

Santiago, capital of Chile, in central Chile on the Mapocho River. Founded in 1541 by Pedro de Valdivia, it was destroyed by an earthquake in 1647. It is the economic and cultural centre of the country and also has a number of manufacturing industries, which include textiles and footwear. Pop. (1974) 3,263,000. *See also* HC2 pp.256–257.

Santiago de Compostela, city in N central Spain, 52km (32 miles) S of Lt Coruna. Founded in the 9th century of Alfonso II, who built a chapel on the site of the tomb of St James, the city became a major centre of pilgrimage. It is the site of the annual festival of St James and the cathedral of St James, which was built in the 11th–13th centuries on the site of the original chapel. There are more than 45 churches and a university (1525) in the city. Industries: tourism, handicrafts, wood products, furniture, leather goods. Pop. (1970) 70,893.

Santiago de Cuba, port on the S coast of Cuba; second-largest city in Cuba and capital of Oriente province. Founded in 1514, it was capital of Cuba from 1522 to 1589. Fidel CASTRO's campaign against Fulgencio BATISTA began by an attack on a military garrison there in July 1953. Exports: agricultural produce, timber. Pop. 277,600.

Santillana, Iñigo López de Mendoza, Marqués de (1398–1458), Spanish poet and soldier. His sonnets *Sonetos fechos al* *itálico modo* are imitative of PETRARCH and DANTE, but the lyrical *serranillas* were inspired by folk themes. Santillana's *Proemio e carta* (1449) is the first important work of literary criticism in Spanish.

Santo Domingo, capital of the DOMINICAN REPUBLIC, on the S coast of Hispaniola. Founded in 1496 by Bartholomew Columbus, it is the oldest European-founded community in the Western Hemisphere. The cathedral of Santa Maria (1514) is believed to contain the tomb of Christopher COLUMBUS. The Dominican Republic's leading port, Santo Domingo exports tobacco, coffee and cacao. Industries: textiles, woodworking. Pop. (1970) 673,470.

Santomaso, Giuseppe, (1907–) Italian painter and member of the *Fronte nuova delle arti.* This group set up after WWII and sought to free painting and sculpture from hidebound traditions and assimilate the emerging styles and techniques.

Santos, city in SE Brazil, 72km (45 miles) SSE of Sao Paulo on Sao Vincente Island. Founded by the Portuguese in the 1540s, it is second only to Rio de Janeiro in the tonnage of shipping handled. Santos is the world's largest exporter of coffee and major distributor of other Brazilian products. Pop. (1970) 341,317.

San Vitale (built AD 526–47), Byzantine church in Ravenna, Italy. It has a central octagon which is surrounded by an octagonal ambulatory. The columns of the inner octagon have beautifully carved capitals. The chancel, an apsed extension at the east end, contains famous mosaics of Justinian and Empress Theodora with their court.

Saône, river in E France which rises in the Vosges uplands and flows SW to join the River Rhône at Lyon. The Saône is connected by canal to the rivers Moselle, Marne, Yonne and Loire. Length: 480km (298 miles).

São Paulo, city in SE Brazil, on the Tietê River; capital of São Paulo state and largest city in Brazil. Founded by the Jesuits in 1554, it was the base for expeditions into the interior in search of mineral wealth. São Paulo grew rapidly in the 17th century as a trading centre for a region producing coffee. It now ships large quantities of agricultural produce through its port of Santos. Industries: motor vehicles, printing and publishing, chemicals, electrical equipment, heavy machinery, pharmaceuticals. Pop. 5,869,966.

São Tomé and Príncipe, independent island republic in the Gulf of Guinea, off the W coast of Africa. Formerly a Portuguese colony, the islands became independent in 1975. Cocoa, coffee, bananas and coconuts are grown on plantations and their export provides the republic's only source of income. The capital is São Tomé. Area: 964sq km (372sq miles). Pop. (1975 est.) 80,000. *See also* MW p.149.

Sap, in vascular plants, fluid that circulates water and nutrients through them. Water is absorbed by the roots and carried, along with minerals, through the XYLEM to the leaves. Sap from the leaves is distributed throughout the plant; it travels upwards by OSMOSIS (specialized diffusion), root pressure, and pressure differences created by leaf transpiration coupled with capillary action. The sugar maple (*Acer saccharum*) has sap with a high sugar content that is used to make maple syrup and sugar. *See also* NW pp.44–45.

Sapodilla, fruit-bearing evergreen tree of tropical regions. The fruits are spherical or ovoid, brown, and contain a sweet pulp surrounding shiny black seeds. Species *Achras zapota. See also* PE p.194.

Saponin, GLUCOSIDES present in some plants that produce a lather when agitated with water. Used as a soap substitute, it is found in horse chestnut, soapberry, soapnut and soapwort.

Sapper (1888–1937), British novelist, real name Herman Cyril McNeile. He achieved success with *Bulldog Drummond* (1920) and its numerous sequels involving the same hero, an ex-officer endowed with charm, commonsense, immense chauvinism and a sense of fair play.

Sapper, in the British armed forces, a private in the Royal Engineers, formerly the Royal Sappers and Miners. It was originally his job to dig "saps", which were fortifications, tunnels or trenches, used to approach the enemy unseen.

Sapphire, transparent to transluscent gemstone variety of CORUNDUM. It has various colours, the most valuable being deep blue. They are brilliant when cut and polished, but lack fire. *See also* PE pp.98–99, *98–99*.

Sappho (*fl.* early 6th century BC), Greek poet whose name has come to connote female homosexuality. Her passionate love poetry, written on the island of Lesbos – from which the word lesbian comes – was regarded by PLATO as the expression of "the tenth Muse". Of noble birth, Sappho was the centre of a literary coterie.

Sapporo, city on W Hokkaidō, Japan; capital of Hokkaidō prefecture. A growing industrial and commercial city, it has become the business centre of N Hokkaidō. Hokkaidō University (1918) is in the city. Industries: timber, food processing, printing and publishing, tourism. Pop. (1975) 1,240,000.

Saprophyte, plant that obtains its food from dead or decaying plant or animal tissue. Generally it has no CHLOROPHYLL and grows in humus. Included are fungi such as puff-balls, mushrooms and moulds, and some flowering plants. *See also* PARASITE; NW p.40, *40*.

Saqqara. *See* SAKKARA.

Sarabande, dance of 16th-century Spanish origin. At this time it was accompanied by the castanets and lively music and was regarded with much disfavour by the Church. By the 17th century it had spread to Italy and France where it became popular as a court dance.

Saracens, term applied originally to an Arab tribe in the Sinai Peninsula and then extended to include all Arabs and later all Muslim subjects of the caliph. It was used particularly by medieval christians to denote their Muslim enemies. The Saracens held Sicily from the 9th to the 11th century, leaving a cultural influence that can still be discerned there.

Saragossa, or Zaragoza, capital of Saragossa province and the foremost city of ARAGÓN in NE Spain. It is an ancient city which the Goths captured from the Romans in the 5th century and in turn lost to the Moors in the 8th century. In 1118 it was captured by Alfonso I of Aragon, who made it his capital. Today it is the commercial centre of Aragon and serves the surrounding agricultural region. Industries: timber, food processing. Pop. (1970) 479,845.

Sarah, in the Bible, wife and half-sister of ABRAHAM. Following the custom of the time, and because she was childless until her old age (when she bore ISAAC), she gave Abraham her maid Hagar, who bore Ishmael, as concubine.

Sarajevo, city in SW central Yugoslavia, on the River Bosna, the capital of Bosnia and Herzegovina. The city is first mentioned *c.*1415 as the Vrh-Bosna citadel; it fell to Turks in 1429, passed to Austro-Hungary in 1878 and was incorporated into Yugoslavia in 1918. The city was the scene of the assassination of the Archduke Francis Ferdinand and his wife, in June 1914, an act which precipitated WWI. Its present industries include the manufacture of textiles, steel, beer, carpets, handicrafts and electrical equipment. Pop. (1971) 243,980.

Saratoga, Battle of (17 Oct. 1777), decisive defeat for the British army during the War of AMERICAN INDEPENDENCE. Gen. John BURGOYNE led 8,000 troops from Canada to the Hudson valley, trying to link with British forces in New York, but his advance south was halted in two engagements on the west bank of the Hudson. Burgoyne surrendered to Horatio GATES at Saratoga. The American victory was instrumental in persuading the French to intervene directly on the side of the rebels.

Sarawak, state in Malaysia, on NW Borneo Island; Kuching is the capital. Ruled as an independent state by Britain's James BROOKE after 1841 (when it was ceded from the sultan of Brunei) it was made a British protectorate in 1888 and a Crown colony in 1946 after the Brooke family passed it to Britain. Sarawak became a part of Malaysia in 1963. Products include: coconuts, rice, rubber, oil and sago. Area: 125,205sq km (48,342sq miles). Pop. (1970) 977,013.

Sarcoidosis, systemic disorder of unknown cause, characterized by the formation of scar-like tissue in the lungs, lymph nodes, muscles, liver and elsewhere in the body.

Sarcoma, cancerous growth or TUMOUR made up of tissues from muscles and bones, or cells embedded in connective tissue. It is usually highly malignant. *See also* CANCER.

Sardana, Catalan national dance usually in 6/8 time performed to the folk music accompaniment of pipes and hand-drum. Dancers form a circle and dance with controlled and formal steps, at first slowly, then more rapidly.

Sardine, small marine food fish found throughout the world. It has a laterally compressed body, a large toothless mouth and oily flesh. Length: to 30cm (1ft). Species include the California *Sardinops caerulea*, South American *Sardinops sagax* and the European sardine, or PILCHARD, *Sardina pilchardus.* Family Clupeidae.

Sardinia, mountainous island of Italy, 208km (129 miles) W of the Italian mainland, from which it is separated by the Tyrrhenian Sea. Farming and fishing are the chief occupations; wheat, barley, grapes, olives and tobacco are grown and sheep and goats are reared. Salt extraction is important, and other mineral deposits include gold, silver, coal, zinc, lead and copper. The only large city is Cagliari, the capital. A trade centre for the Phoenicians, Greeks and Carthaginians, Sardinia became a kingdom in 1720. In 1861 it was incorporated into the kingdom of Italy. Area: 23,813sq km (9,194sq miles). Pop. 1,500,008. *See also* MW p.107.

Sardinia, Kingdom of, possessions of the House of Savoy from 1720, when it was ceded Sardinia to compensate for the loss of Sicily. The kingdom included Sardinia, Savoy, Nice, Montferrat and PIEDMONT, and further lands were acquired during the 18th century. Most of the mainland possessions were lost to France between 1792 and 1814. Genoa became part of the kingdom in 1815 and, during the RISORGIMENTO, Lombardy in 1859, most of the PAPAL STATES and the TWO SICILIES in 1860. VICTOR EMMANUEL II of Sardinia became King of Italy in 1861. *See also* HC2 pp.110, *111.*

Sardis, ancient city in Asia Minor, 80km (50 miles) E of Smyrna in the Hermus valley. It was the capital of ancient Lydia *c.*650–519 BC and the site of the first minting of coins in the 7th-century BC. It was taken by the Ionians in 499 BC and the Romans in 133 BC. Sardis was destroyed by TAMERLANE in the 15th century, but its ruins were not discovered until 1958.

Sardou, Victorien (1831–1908), French dramatist. He wrote more than 70 plays, mostly light comedies and historical dramas including *Divorçons!* (1880) and *Dora* (1877). He was particularly well-known for his dramas of a type known as the Well-made Play.

Sargasso Sea, moving region of water in the N Atlantic Ocean which takes its name from the large quantities of floating seaweed (*Sargassum*) covering the surface. Because the sea is choked with seaweed, it resembles an oceanic desert and has the least amount of life of any area of sea water.

Sargent, Sir Harold Malcolm Watts (1895–1967), British conductor. Particularly noted for his choral work, he conducted the Royal Choral Society from 1928 and the BBC Symphony Orchestra from 1950 to 1957. From 1951 he was particularly associated with the Henry WOOD Promenade concerts.

Sargent, John Singer (1856–1925), US painter. He studied in Florence and Paris but most of his work was done in London. He is best known for his numerous portraits, especially those of society ladies which made him the most fashionable portraitist of his generation.

Sappho, Plato's "Tenth Muse", shown in a detail from Paolo's Fidanza's engraving.

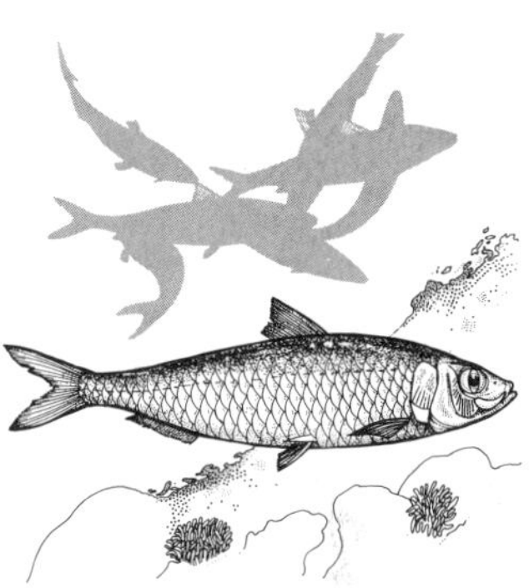

Sardine oil is used in the manufacture of margarine, paint, varnish and cosmetics.

Sardinia; Osilo, a typical hill village near Sassaria in the north of the island.

John Singer Sargent's portrait of Henry James captures his subject's character.

William Saroyan's works are about the poor, whom he applauds for their vitality.

Jean-Paul Sartre has expressed his views in novels, plays and theoretical works.

Sassafras leaves may be crushed to yield a substance used in condiments or tea.

Stefano di Sassetta: a detail from his *Marriage of St Francis and Poverty*.

Sargeson, Frank (1903–), New Zealand novelist and short-story writer who began his career as a lawyer. He achieved recognition with *A Conversation with My Uncle* (1936) and also published a collection of short stories (1964) and a later novel, *The Hangover* (1967).

Sargon I (the Great) (*r. c.*2334–2279 BC), King of Akkad, Mesopotamia. He defeated Lugalzaggisi of Urutz and subdued Sumeria, Syria and part of Asia Minor. The Akkadian Empire was destroyed by the Guti. *See also* HC1 p.31, *31*.

Sargon II (*d.*705 BC), king of Assyria (*r.* 721–705 BC), He conquered Samaria in 721 BC and completely destroyed the northern kingdom of Israel. He also subjugated Babylonia (710). *See also* HC1 pp.33, 59.

Sark, one of the CHANNEL ISLANDS of the United Kingdom, divided into Great Sark, and Little Sark which are connected by a causeway. Sark is part of the Bailiwick of GUERNSEY; its feudal organization dates from the late 17th century. Area: 5sq km (2sq miles). Pop. (1971) 590.

Sarkis, Elias (1924–), Lebanese banker and politician. He was the governor of the Bank of Lebanon from 1968 to 1976. He became President of the Lebanon in 1976.

Sarmatians, ancient nomadic and pastoral peoples who in the 5th century BC lived in central Asia between the Urals and the River Don. By the 2nd century BC they had conquered most of the territory of the SCYTHIANS in S Russia and began to threaten the Romans. They were overpowered by the Goths from the W in the 3rd century AD and the Huns from Asia in the 4th century AD.

Saroyan, William (1908–), US author whose works combine optimism and sentimentality. They include short stories, such as *The Daring Young Man on the Flying Trapeze* (1934), and plays, such as *My Heart's in the Highlands* (1939) and *Time of Your Life* (1939).

Sarraute, Nathalie (1902–), French novelist, *b.* Russia, whose novels are masterpieces of texture rather than of structure. Her preoccupation with psychological detail can be seen in *Tropismes* (1957) and *Les Fruits d'Or* (1963).

Sarsaparilla, wild perennial plant with greenish flowers that grows in the USA. Family Araliaceae; species *Aralia nudicaulis*. It is also a general name for several species of cultivated plants of genus *Smilax*, and for the medicinal preparation and soft drink (root beer) prepared from their dried roots. Family Liliaceae.

Sarto, Andrea del. *See* ANDREA DEL SARTO.

Sartre, Jean-Paul (1905–), French philosopher and writer who was instrumental in the popularization of EXISTENTIALISM. A fighter in the French Resistance during WWII, Sartre's increasing political awareness led him to COMMUNISM, in particular the theories of MARX. He refused the 1964 Nobel Prize in literature. His works include *Nausea* (1938), *Being and Nothingness* (1943), the trilogy *The Roads to Freedom* (1945–49), *Saint Genet* (1952) and *Critique of Dialectical Reason* (1960). *See also* MS pp.231, 232–233, *232–233*.

Saskatchewan, province in W central Canada, in the Great Plains. The cultivation of wheat is the most important agricultural activity; oats, barley, rye and flax are also grown. The province has rich mineral deposits, including uranium, copper, zinc, gold, coal, oil, natural gas and potash. Most industries process raw materials; steel is also manufactured. The principal cities are Regina (the capital and largest city), Prince Albert, Saskatoon and Moose Jaw.

The first permanent settlement was at Cumberland House in 1774. Development was slow until the construction of the transcontinental Canadian Pacific Railroad in 1885. Saskatchewan was admitted to the federation of Canada in 1905. Area: 651,900sq km (251,700sq miles). Pop. (1975 est.) 920,000. *See also* MW p.45.

Saskatchewan, river in S central Canada. Formed at the confluence of the North and South Saskatchewan rivers near Prince Albert, it flows E to empty into Lake Winnipeg. With its tributaries, the Saskatchewan drains most of the Canadian prairie provinces. Length: approx. 550km (340 miles).

Saskatoon, city in S central Saskatchewan, Canada, in the South Saskatchewan River. The city was founded in 1883 and grew with the construction of a railway. The second-largest city in Saskatchewan, it is a major industrial and distribution centre. It is the site of the University of Saskatchewan (1907). Industries: food processing, oil refining, chemicals. Pop. (1974) 126,450.

Sassafras, small E North American tree with furrowed bark, green twigs, yellow flowers and blue berries. An aromatic tea is made from the outer bark of its roots, and oil from the roots is used to flavour root beer. Family Lauraceae; species *Sassafras albidum*.

Sassanids (Sassanians), last native dynasty of Persian kings, founded by ARDASHIR I (*r.* AD 224–241). The Sassanids revived the ACHAEMENID tradition and Zoroastrianism was restored as the state religion. There were about 30 Sassanid rulers, the most important after Ardashir being SHAPUR II (309–79); KHOSHRU I (531–79), who fought the Byzantine empire; and KHOSHRU II (590–628), whose conquest of Egypt marks the height of the dynasty's power. *See also* HC1 pp.60, 114.

Sassetta (Stefano di Giovanni) (*c.*1400–50), Italian painter of the SIENESE SCHOOL. He incorporated elements of the decorative International Gothic style into his imaginative and lyrical style. The best example of his work is the now dismembered altarpiece (1437–44) for S Francesco at Sansepolero.

Sassoon, Siegfried (1886–1967), British writer. His military service in WWI (during which he protested against the conduct of the war) inspired considerable poetry. The autobiographical trilogy *The Complete Memoirs of George Sherston* (1937) includes his most famous novel *Memoirs of a Fox-hunting Man* (1928) an account of his life as a country gentleman.

Sastrugi. *See* ZASTRUGI.

Satan, name for the devil. To the early Jews "Satan" simply meant "opponent", and did not refer to a particular being. Satan appears in the Old Testament by the 6th century BC as an individual angel, subordinate to God. Gradually Satan became considered the source of all evil.

Satellite, celestial body orbiting a planet or star. In the Solar System planets with satellites include the Earth (1), Mars (2), Jupiter (13), Saturn (10), Uranus (5) and Neptune (2). The moons of the Earth and Mars are rocky with little or no atmosphere. The satellites of the other planets are believed to consist of ice or frozen methane and ammonia. Titan, the largest of Saturn's moons, is the only one known to have an appreciable atmosphere.

The Earth's Moon is the largest in relation to its planets: its diameter is 3,476km (2,160 miles). Jupiter's satellite Ganymede is the largest moon in the Solar System: it has a diameter of 5,000km (3,100 miles). *See also* SU pp. 176, *176*, 178–187, 202–203, 212–213, 214–215.

Satellite, artificial, man-made object placed in orbit around the Earth or other celestial body. Sputnik 1 was the first artificial satellite; it was launched on 4 Oct. 1957 by the USSR. Today there are hundreds of satellites of various sizes and descriptions in orbit around the Earth. Satellites that orbit the Earth at heights less than about 160km (100 miles) are slowed by the resistance of the atmosphere and eventually spiral in and burn up. The main uses of satellites are military, for weather studies, communications and scientific research. *See also* SU pp.268–271; MM pp.156–157.

Satie, Erik (1866–1925), French composer. Rebelling against Wagnerian ROMANTICISM, he developed a deceptively simple style in piano pieces such as *Trois Gymnopédies* (1888), which influenced POULENC, MILHAUD and other composers. He also composed the ballets *Relâche* (1924) and *Parade* (1917), and a choral work, *Socrate* (1918). *See also* HC2 p.270.

Satinwood, tree that grows in E India; it has smooth, hard wood that is valued for furniture and veneer. Height to 15m (50ft). Family Rutaceae; species *Chloroxylon swietana*.

Satire, literary work in which human foibles and institutions are mocked and ridiculed. Roman satire was a poem in hexameters on appropriate themes, and was established through the work of LUCILLIUS, HORACE and JUVENAL. In the Middle Ages satire often took the form of the *fabliau*, or bestiary, which used animals to represent typical human failings. Since Thomas MORE's *Utopia*, satire has also found frequent expression in utopian fiction, such as Zemyatin's *We* and SWIFT's *Gulliver's Travels*. Dramatists have often used satire, as demonstrated by the plays of ARISTOPHANES, Ben JONSON, MOLIÈRE, Oscar WILDE and Bertolt BRECHT. Satire may work in two ways: either by inflation, as in DRYDEN's satires which blow figures up to the monumentally ridiculous, or by diminution, as in the works of Alexander POPE, in which heroic pretension is frequently made trivial.

Sato, Eisaku (1901–75), Japanese political figure. In 1948 he became vice-minister of transport. A Liberal-Democratic member of the Diet (parliament), he served as Finance Minister (1958–60) and later as Prime Minister (1964–72). His chief successes in foreign policy were normalizing relations with Asian countries including China, and the signing of a treaty returning the Ryukyu Islands to Japanese jurisdiction (1969). In 1974 he was awarded the Nobel Peace Prize for his part in persuading Japan to sign the treaty on the non-proliferation of nuclear weapons.

Saturated compounds, in organic chemistry, compounds in which the carbon atoms are bonded to one another by single chemical bonds only, never by the more reactive double or triple bonds. For this reason, they tend to be unreactive. A simple example is methane, an inert but inflammable gas in which the molecule has one carbon atom singly bonded to four hydrogen atoms, CH_4.

Saturated fats, organic fatty compounds, the molecules of which contain only saturated FATTY ACIDS combined with GLYCERIN. These acids have long chains of carbon atoms which are bound together by single-bonds only. *See also* SATURATED COMPOUNDS.

Saturated solution, in chemistry, a SOLUTION containing so much of a dissolved compound that, at the same temperature and pressure, no more will dissolve.

Saturation, in physical chemistry, equilibrium condition of a solution when the maximum amount of the dissolved substance (solute) has been taken into solution at a given temperature. The saturation concentration depends on temperature; usually the higher the temperature, the higher the concentration.

Saturday, seventh day of the week and, for Jews and some Christians, the SABBATH. Christians in the early Church worshipped on both the seventh day (prescribed in the Bible as the Sabbath) and the first or Lord's Day, the latter gradually becoming the Sabbath in most Christian churches. From the Latin *Saturnus* (Saturn) and the Old English *daeg* (day), it is the only day named after a Roman god.

Saturn, in Roman mythology, the father of the gods and protector of the harvests. He was the husband of Ops. His festival, the SATURNALIA, held in the third week of December, was a time of popular merry-making that influenced the later festivities of Christmas and New Year.

Saturn, sixth planet from the Sun and second largest in the SOLAR SYSTEM. Saturn has ten satellites and a system of narrow rings composed of icy or ice-covered particles. Mean distance from the Sun, 1,427,000,000km (887,000,000 miles); mass, 95 times that of Earth; volume, 744 times that of Earth; equatorial diameter, 120,000km (75,000 miles); rotation period, 10hr 14min; sidereal period, 29.46 Earth-years. *See also* SU , pp. 176–177, *176–177*, 210–211, *210–211*, 213, *213*, 278–279, *279.* .

Saturnalia, Roman festival commemorating the mythical Golden Age of Rome under SATURN who, identified with the Greek god Cronus, had established a kingdom of infinite happiness in LATIUM. The festival was celebrated by the exchange of gifts and feasting; masters and slaves changed places, business ceased and moral codes were relaxed. It originally took place on 17 Dec. but its duration was later extended. *See also* HC1 p.93.

Satyagraha, political philosophy, developed by Mohandas GANDHI, which was central to the non-violent resistance of the Indian people to British colonial rule. The concept has been translated as "truth force" and it relies upon the satyagrahi's moral certainty in confronting what he considers wrong. *See also* HC2 pp.216–217.

Satyr, in Greek mythology, gods of the woods, fields and mountains. Sensual and lascivious, and associated with Dionysus, they were later depicted by the Romans as goat-legged, goat-bearded men with budding horns. The term is today accordingly used to describe a lecherous person; medically, SATYRIASIS is a morbid condition of overwhelming sexual desire. Satyr is also the common name for any butterfly of the Satyridae family; the name is sometimes applied to a species of orang-utan.

Satyriasis, behaviour pattern in which a man needs continually to seduce women. ORGASM is of secondary importance.

Satyr Play, type of classical Greek drama which was presented as a light contrast to the tragic trilogies performed at festivals. The plays dealt with mythical subjects using a grotesque humour, the chorus always consisted of SATYRS. The plays survive mostly in fragments. *Cyclops* by Euripides is the only complete one.

Saud (1902–69), King of SAUDI ARABIA (r. 1953–64), also known as Ibn Abd al-Aziz Al Faisal Al Saud. The son and successor of IBN SAUD, he became Crown Prince in 1933. His fiscal mismanagement and personal extravagance caused a severe financial crisis in 1958. Soon after his brother FAISAL took over all administrative powers, formally replacing him as king in 1964.

Saudi Arabia, Arab nation occupying much of the Arabian Peninsula in sw Asia. An impoverished desert area before the discovery of oil in the 1930s, the country is now one of the wealthiest lands in the Middle East. The oil revenues are being used to diversify the economy and to improve education. The capital is Riyadh. Area: 2,149,690sq km (829,995sq miles). Pop. (1975 est.) 8,966,000. *See* MW p.149.

Saul (r. c. 1020–1000 BC), anointed by SAMUEL as the first King of the Hebrews. Through much of his reign he waged war against the PHILISTINES, AMALEKITES and other Gentile nations. He was killed in battle and succeeded by DAVID. *See also* HC1 pp.58–59.

Saul (1739), oratorio by George HANDEL, first performed in London. The text is by Charles Jennens and the music includes the famous *Dead March*.

Sault Sainte Marie Canals (Soo Canals), artificial waterway on the USA-Canadian border, between lakes Huron and Superior at the twin cities of Sault Sainte Marie, N Michigan, and Sault Sainte Marie, s central Ontario, Canada; a major link in the GREAT LAKES waterway. There are two canals, which bypass the rapids of the St Marys River. The Canadian canal, 2.2km (1.4 miles) long, was completed in 1895. The US canal, 2.6km (1.6 miles) long, was built between 1881 and 1919.

Saumur, town in the Maine-et-Loire département, w France, noted for its sparkling white wines and its long association with the French cavalry. In the 16th and 17th centuries the town was a centre of the Huguenot movement. Pop. 21,000.

Sauna, wood-lined room in which a heater raises the temperature to between 60° and 135°C. Because the atmosphere in the room is totally dry, the human body can without hardship tolerate a temperature higher than that of boiling water, although perspiration becomes increasingly intense. To increase perspiration still

further, very small amounts of water are thrown on to red-hot stones on top of the heater; the resulting steam rebounds from the walls to give the body a momentary burning sensation. *See also* MS pp.73, *73.*

Saunders, James (1925–), British playwright. The structure of his plays recalls those of PIRANDELLO. His reputation was established with *Next Time I'll Sing to You* (1962), an examination of the life and death of a hermit, and *A Scent of Flowers* (1964), which explored the reasons for a young girl's suicide.

Sausages, cooked or uncooked meat product cased in a thin tubular skin. Sausages that are sold uncooked contain various preparations of assorted fresh meats, including beef, pork, mutton and veal. *See also* PE pp.234–*235.*

Saussure, Horace Bénédict de (1740–99), Swiss geologist. He studied the region of the Alps, and published his study of its weather, geology and botany in *Voyages dans les Alpes* (4 vols, 1779–96). He also developed the first electrometer in 1766 and the first hygrometer for measuring humidity in 1783.

Sauternes, village of the Gironde département, sw France. The sweet, white table wines of the district are named after it. Pop. (1973) 585.

Savage, James (1779–1852), British architect, an early exponent of the Gothic revival. The church of St Luke (1820–24) in Chelsea, London, was the first built in this style and is unusual for its flying buttresses spanning the aisles.

Savage, Michael Joseph (1872–1940), New Zealand politician, *b.* Australia. He moved to New Zealand in 1907, becoming active in the trade union movement. He joined the Labour Party in 1916 and became its leader in 1933. He led Labour to a landslide election victory in 1935 and was Prime Minister of the first Labour government (1935–40).

Savanna, plain with coarse grass and scattered tree growth, particularly the wide plains of tropical and subtropical regions. An extensive example is the savanna of the East African tableland. *See also* NW pp.194, *195.*

Savannah, city and port of entry in E Georgia, USA, on the Savannah River. The oldest city in Georgia, it was chosen as the seat of colonial government and the capital of Georgia Colony in 1754. It was captured by the British in 1778 during the War of AMERICAN INDEPENDENCE and was held until the end of the war. Today it is a major shipping centre, and its exports include tobacco, cotton, sugar, clay and wood pulp. Industries: chemicals, rubber, plastics, timber, petroleum products. Pop. (1973) 105,768. *See also* HC2 p.*151.*

Savannah, first steamship to cross the Atlantic. In 1819 it sailed from Savannah, Georgia, to Liverpool, England, on a voyage lasting 27 days 11 hours.

Savery, Thomas (1650–1715), English engineer. In 1696 he patented a machine for pumping mines, which employed the principle of a vacuum steam-engine. He also designed an early form of paddle-driven boat. *See also* MM pp.64, *65.*

Savings Bank, one of three kinds of bank in Britain, the others being the BANK OF ENGLAND and joint-stock or "commercial" banks. Savings banks arose in the 19th century from the habit of placing money for safe keeping with goldsmiths. They are run on the principle of paying interest to the depositors, or savers, and charging a higher interest from borrowers who are lent money from the accumulated savings.

Savitri, or Savitar, in Hindu mythology, the universal motivating force, personified as a god with golden eyes, tongue and hands, riding a chariot drawn by glittering horses. He is worshipped as the king of heaven who bestows immortality on other gods and remits the sins of those mortals who pray to him.

Savonarola, Girolamo (1452–98), Italian religious reformer. He abandoned medicine to become a Dominican monk, when convinced of the sinfulness of the world. He was a lecturer at San Marco, Florence, from 1482 to 1487, and became a famous preacher. In 1490 he led a crusade against corruption, and after the collapse of

MEDICI rule in 1494 Savonarola became virtual dictator of the city. He was executed by his political enemies following his excommunication.

Savory, any of several hardy aromatic herbs of the MINT family (Labiatae). Its long thin leaves are used, green or dried, for flavouring. Summer savory (*Satureia hortensis*) is an annual; winter savory (*S. montana*) is a perennial. *See also* PE p.*213.*

Savoy, area of SE France, bounded by Lake Geneva (N), the River Rhône (w), the Dauphiné (s) and the Alps of Italy and Switzerland (E); it includes the départements of Haute Savoie and Savoie. It was part of the first Burgundian kingdom, the kingdom of Arles and, in the 11th century, the Holy Roman Empire (as a county). In 1416 it became a duchy and its area was enlarged, incorporating parts of France, Switzerland and Italy. An Italian state in the 16th century, it was part of the kingdom of SARDINIA after 1713. Savoy was annexed by France in 1792, returned to Sardinia in 1815, and finally ceded to France by the Treaty of Turin in 1860, its former rulers then becoming kings of Italy.

Savoy, House of, royal European dynasty and ruling house of Italy from 1861 to 1946. Descended from Humbert the Whitehanded (*d. c.* 1047), the first Count of Savoy, they became dukes in 1416, kings of Sicily in 1713, kings of Sardinia in 1720, and kings of Italy in 1861. Italian unification was led by the house of Savoy in the mid-19th century. *See also* HC2 pp. 110, *111.*

Savoy Operas, the operettas by Arthur SULLIVAN, libretti by William GILBERT. They are so called because they were either first produced or revived at the Savoy Theatre, London, which opened in 1881 with *Patience* (which was transferred from the Opera Comique). The performers in the operas became known as "Savoyards".

Sawchuck, Terrance Gordon (1929–70), Canadian ice hockey player. He was a goalkeeper with the Detroit Red Wings (1949–55, 1957–64), the Boston Bruins (1955–57), the Toronto Maple Leafs (1964–67) and the Los Angeles Kings (1967–70). He won the Vezina Trophy as the National Hockey League's best goalkeeper four times. In 1967 he set an NHL record with his 100th shutout.

Sawfish, any of several species of shark-like, flat bodied RAYS that live in tropical marine and brackish waters. It has a grey or black-brown body with an elongated, saw-toothed snout resembling a flat blade. Length: to 5m (16ft). Family Pristidae; genus *Pristis*.

Sawfly, any of 400 species of primitive, plant-feeding WASPS that lack a narrow waist between thorax and abdomen. Most typical sawflies are included in the family Tenthredinidae. Their LARVAE are caterpillar-like, with eight abdominal legs. The females have a two-bladed, saw-like ovipositor for laying eggs in plant tissue. Length: to 20mm (0.8in). Order Hymenoptera. *See also* NW p.*208;* PE p.*223.*

Sax, Antoine Joseph Adolphe (1814–94), Belgian musician. In the 1840s he invented various instruments such as the saxhorn and SAXOPHONE.

Saxe-Coburg-Gotha, one of four Ernestine Saxon duchies in E central Germany from 1485 to 1918, and a prominent contributor to European royal houses. LEOPOLD I of Belgium was of the house of Saxe-Coburg-Gotha, as were Queen Victoria's husband, Prince ALBERT, and Ferdinand, the prince consort of Maria II of Portugal.

Saxe-Weimar, former duchy in central Germany. It became a separate entity in 1547 when the Ernestine lands were divided into the duchies of Weimar, Gotha, Coburg, Eisenach and Altenburg. In 1741 it became the Saxe-Weimar-Eisenach duchy and developed into a European cultural centre under Grand Duke Carl August (1757–1828), the patron of Johann Wolfgang von GOETHE. In 1920 the duchy was incorporated into the German state of Thuringia.

Saxifrage, perennial plant native to tem-

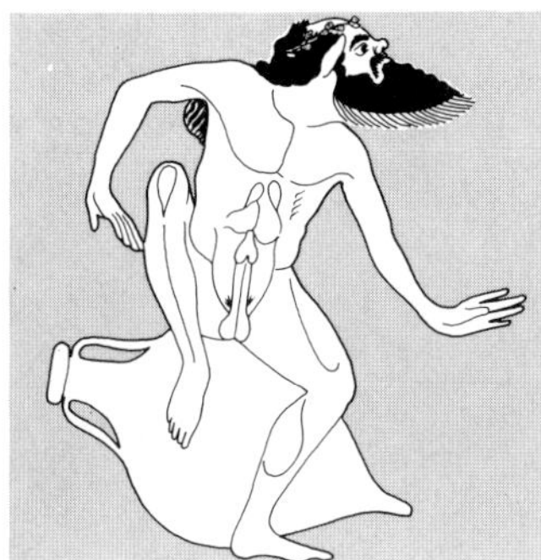

Satyr; young satyrs were reputedly handsome and evil, old ones ugly and idle.

Saumur; the 15th-century castle originally belonged to the dukes of Anjou.

James Savage; one of his most famous works, St Luke's Church, Chelsea.

Sawflies are short lived; their cycle is complete after they have reproduced.

Saxo Grammaticus

Saxony; a typical barn. Many buildings destroyed in WW11 have been rebuilt.

Saxophone; in solo parts, it was used by Debussy, Bizet and Vaughan Williams.

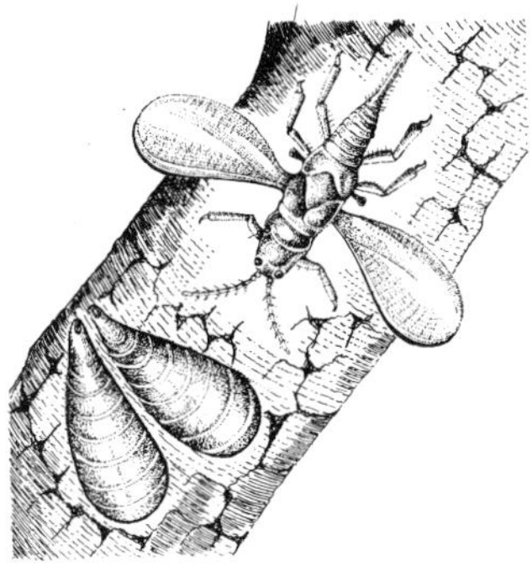

Scale insects produce a secretion that can be made into shellac.

Scapa Flow: some interned German ships before they were scuttled in 1919.

perate and mountainous regions of Europe and North America. The leaves are massed at the base and the small flowers are white, pink, purple or yellow. Height: to 61cm (2ft). Family Saxifragaceae; genus *Saxifraga. See also* NW pp.*58, 223, 226.*

Saxo Grammaticus (*c.*1150–1220), Danish historian who wrote *Danish History*, a history of the Danish kingdom before 1185, at the suggestion of Archbishop Absalon of Lund. The first nine books are mostly based on legend; the remaining seven deal with historical events and his account of those close to his own time, and are a valuable source of reference.

Saxons, ancient Germanic people. Originating in N Germany and S Denmark, they extended S and W, and colonized Britain *c.* AD 450, where they amalgamated with the Angles to form the ANGLO-SAXON civilization. At the same time the continental Saxons, or Old Saxons, expanded into N Germany, which brought them into conflict with the Franks. They were finally subjugated and converted to Christianity by CHARLEMAGNE in the early 9th century. *See also* HC1 pp.*136, 137,* 152.

Saxon shores, name given by the Romans to those shores of the Roman Empire that suffered SAXON raids from the 3rd century AD. They comprised the N coast of Gaul to the Scheldt estuary, and the SE coast of Britain from the Wash to the Solent. *See also* HC1 p.*138.*

Saxony, former German state bordered by Thuringia (W), Bavaria (SW), Czechoslovakia (S) and Brandenburg (N); since WWII it has been part of East Germany. The territory of the Germanic Saxon tribe, it had become by the early 10th century the Duchy of Saxony and by the mid-13th the duke was one of the seven imperial electors of medieval Germany. Saxony included almost all the land between the rivers Elbe (E) and Rhine (W); the ducal and electoral title of Saxony passed to the Margrave of Meissen and the area known as Saxony came to include his territories in Thuringia and Lower Lusatia. In 1815 more than half of Saxony, which had been a kingdom since 1906, was ceded to Prussia and, with several Prussian districts, became the province of Saxony (which included Wittenburg and N Thuringia). In 1945 the state of Anhalt was added to form the Saxony-Anhalt state.

Saxophone, wind musical instrument with single reed, curved conical metal tube and finger keys. It was invented by Adolphe SAX in the 1840s and by 1846 there was a double family of 14, seven in F and C for orchestral music and seven in E and B for bands; the latter are more common today. It is now mostly used in jazz.

Saye and Sele, William Fiennes, 1st Viscount (1582–1662), English politician. He was a Puritan opponent of JAMES I and CHARLES I, although he opposed the execution of Charles. He helped to colonize Connecticut and Providence Island, an island in the Caribbean Sea.

Sayem, Abu Sadat Mohammed (1916–), Bangladesh lawyer and politician. He was Chief Justice of the Supreme Court (1972–75) and President of Bangladesh (1975–77). He was succeeded by Maj.-Gen. Ziaur Rahman.

Sayers, Dorothy Leigh (1893–1957), British novelist and playwright, considered a master of the detective story. Her first novels featured Lord Peter Wimsey, a titled detective who appeared in 10 books including *Whose body?* (1923) and *Strong Poison* (1930). She later wrote religious dramas and several radio plays.

Scabies, skin inflammation caused by a female mite, *Sarcoptes scabiei,* which burrows into the skin to lay eggs. It can be seen as a dark wavy line on the skin and is treated with creams containing either hexachlorophane or benzyl benzoate.

Scabious, garden flowers of the genus *Scabiosa* (the TEASEL family, (Dipsacaceae), including plants with showy heads coloured red, orange, yellow, brown and many shades of blue. The sheep's bit scabious, *Jasione montana,* is a member of the BELLFLOWER family.

Scafell Pike, highest peak in England, in the LAKE DISTRICT, Cumbria, NW England;

part of the Scafell mountain group. Height: 978m (3,210ft).

Scalar, mathematical quantity that has only a magnitude, as contrasted with a VECTOR, which also has direction. Mass, energy and speed are scalars (whereas weight, force and velocity are vectors).

Scald, burn caused by hot liquid or steam. The extent of injury ranges from mere reddening of the skin, to deep damage of underlying tissues, with consequent severe shock and danger to life. Treatment of minor scalds includes immersion of the scalded part in cold water; subsequently they should be kept dry and sterile. Severe scalds should always be attended by a doctor.

Scaldic poetry, courtly verse written in Old Norse and popular in Norway from the 10th to the 13th century. It is characterized by elaborate verse forms combining alliteration, rhyme and other technical devices.

Scale, in biology, small hard plate that forms part of the external skin of an animal; it is usually a development of the dermal or epidermal skin layers. In most fish scales are composed of bone in the dermal skin layer; the scales of some primitive fish contain ganoin, a hard glassy substance. The scales of reptiles, those on the legs of birds and on the tails of scaly-tailed mammals are horny growths of the epidermal skin layer, and are comprised mostly of the fibrous protein KERATIN.

Scale, in music, the ordered arrangement of intervals which forms the basis of musical composition. There is a great variety of scales. In Western music the most important has been the seven-note DIATONIC scale, both in its MAJOR and MINOR forms. The 12-note CHROMATIC scale has a regular progression of semi-tones, represented by all the white and black keys on the piano within the octave. The six-note whole-tone scale was much used by DEBUSSY, but its lack of variety has limited its appeal to composers. The 20th century has seen a number of developments and revivals. There has been a conscious return by some composers to the system of the ecclesiastical MODES; the ancient PENTATONIC SCALE has been assimilated with varying degrees of success; and oriental and experimental scale systems have also been employed. *See also* SERIAL MUSIC.

Scale insect, any of several species of small insects found throughout the world. All species are covered by a waxy, scale-like covering secreted by the insect. It feeds by sucking plant juices. Length: to 25mm (1in). Order Homoptera. *See also* COCHINEAL; NW p.111.

Scaliger, Joseph Justus (1540–1609), French scholar, son of Julius Caesar SCALIGER. One of the foremost classical scholars of his time, he was a historian and linguist. His *Opus de emendatione tempore* (1583) introduced scientific method to chronology.

Scaliger, Julius Caesar (1484–1558), also known as Giulio Cesare Scaligero, Italian scholar and critic. His most famous work is *Poetices* (1561), a manual on the theory of literature. He also wrote commentaries on THEOPHRASTUS, HIPPOCRATES and ARISTOTLE, and it is through his work that Artistotle's theory of the dramatic unities was transmitted to the 17th-century French theatre.

Scallop, edible BIVALVE mollusc. One shell, or valve, is usually convex and the other almost flat. The shell's surface is ribbed (scalloped). Most scallops have a row of eyes that fringe the fleshy mantle. Width: 2.5–20cm (1–8in). Family Pectinidae. *See also* NW pp.*92, 93.*

Scalping, the custom of taking the scalp of an enemy, once practised in Europe, Asia and Africa and among the North American Indians. A circular cut was made around the crown of the head and the skin raised at one side and torn off.

Scampi, any of several types of large PRAWNS familiar as the dish of the same name. They are generally cooked individually by frying in batter. Other foods, such as crayfish and monkfish, are sometimes prepared like and sold as scampi. *See also* PE p.*248.*

Scanderbeg (Skanderbeg) (1405–68),

Albanian hero. A hostage of the Ottoman sultan, Murad II, he was brought up as a Muslim but renounced ISLAM when the Turks were defeated in Serbia. He returned to his country, defeated the Ottoman army and maintained Albanian independence until his death, after which resistance to the Turks collapsed.

Scandinavia, region of N Europe, including SWEDEN, NORWAY, and DENMARK, and from a cultural and historical point of view, FINLAND and ICELAND. Together these are sometimes called the Northern Countries. The climate ranges from subarctic in the N to humid continental in the centre and marine in the W and SW. The terrain is mountainous in the W with swift-flowing streams, and the coastline has many fiords. In the E the land slopes very much more gently and there are thousands of lakes. Some of the region lies within the Arctic Circle where tundra predominates. Denmark and S Sweden have the best farmland. A large proportion of the land is forested and there are rich mineral deposits, particularly of iron ore and copper. Fishing is important off the coast. The largest cities are Stockholm and Gothenburg (Göteborg) in Sweden, Oslo in Norway, Copenhagen in Denmark, and Helsinki in Finland. Area: approx 1,258,000sq km (485,250sq miles).

Scandium, metallic element (symbol Sc) of group IIIB of the periodic table predicted (as ekaboron) by MENDELEYEV and discovered in 1897. It is found in thortveitite and, in small amounts, in other minerals. It is a soft metal with few commercial uses. Chemically it resembles the LANTHANIDES. Properties: at.no. 21; at.wt. 44.9559; s.g. 2.99 (25°C); m.p. 1,539°C (2,802°F); b.p. 2,832°C (5,130°F); most common isotope Sc^{45} (100%).

Scannell, Vernon (1922–), British writer and broadcaster. His first work was a novel, *The Fight* (1953); thereafter he wrote mainly poetry, including *The Masks of Love* (1960) and *The Loving Game* (1975). His autobiography, *The Tiger and the Rose,* appeared in 1971.

Scanning, in medical, diagnostic method in which a patient is given a dose of a radioactive ISOTOPE either orally or by injection, and a RADIATION detector is then used to build up a picture of internal areas by successive sweeps of the detector over the isotope.

Scansion, analysis of the metre of a poem by breaking down each line into "metrical feet". Each foot consists of a group of two or three syllables, stressed (long) or unstressed (short).

Scapa Flow, naval anchorage surrounded by the ORKNEY ISLANDS off NW Scotland. The main entrance is from the Pentland Firth to the S and it was the scene of the scuttling of part of the German fleet in 1919. The British fleet was stationed at Scapa Flow in both World Wars, and HMS *Royal Oak* was sunk there in 1939.

Scapegoat, in ancient Israel, goat used in the ritual on the Day of Atonement. The people touched the goat, confessing their sins, and the high priest then enacted a complex ceremony. The animal was finally driven from the village to die in the wilderness, symbolically bearing the sins of the people. In later times the scapegoat was pushed to its death from a cliff. The term has come to mean anyone blamed unfairly for another's guilt.

Scapolite, or wernerite, group of SILICATE minerals related to the FELSPARS and of variable chemical composition, common in metamorphosed limestones. They are grey and glassy, with a hardness of 5–6 and s.g. 2.7.

Scapula, or shoulder blade, large, roughly triangular flat bone, found on the upper part of the back. It provides for the attachment of muscles that function in arm movement. *See also* MS p.56.

Scapulimancy, ancient method of DIVINATION by heating an animal's shoulder blade (usually that of an ox) on which questions had been scratched and examining the resultant cracks. *See also* HC1 p.*44.*

Scar, mark left on the skin after a wound or sore heals. Scars contain no oil glands, hair follicles or elastic tissue as in normal skin. Severe scars are prone to malignant changes. *See also* MS p.98.

Scarab beetle, any of several different species of broad beetles distributed worldwide. Most, including the June bug, Japanese beetle and RHINOCEROS BEETLE, are leaf chafers. A smaller group, including the DUNG BEETLE, are scavengers. Family Scarabaeidae. *See also* NW p.*107.*

Scarborough, one of the five boroughs of Toronto, SE Ontario, on Lake Ontario, Canada. Originally a farming settlement, Scarborough has grown rapidly since WWII into an industrial and residential area, and includes a college (1964) of the University of Toronto. Pop. (1971) 334,310.

Scarborough, holiday resort and county district in NORTH YORKSHIRE, N England. The town, on the North Sea coast, has fine beaches and a 12th-century castle. The county district was created in 1974 under the Local Government Act (1972). Area: 817sq km (316sq miles). Pop. (1974 est.) 98,700.

Scarborough, port and principal city on Tobago island in Trinidad and Tobago, in the Caribbean Sea. Formerly known as Port Louis, the city replaced Georgetown as the capital of Tobago in 1796. Scarborough is a market for the surrounding region and has saw-milling industries. Pop. 2,029.

Scarfe, Gerald (1936–   ), British caricaturist, famous for his grotesque renderings of public figures. His drawings, made usually with fine-pen lines, are always informed by a sense of moral outrage.

Scarification, also called cicatrization, production of keloids (raised scars) on the human body for decoration. It has generally been done in decorative patterns by cutting the skin and rubbing in charcoal and other substances. It was primarily practised by peoples in much of Africa and by Australian Aborigines, whose skins would not show clearly the effects of tattooing. *See also* MS pp.98, 99 *193.*

Scarlatti, Alessandro (1660–1725), Italian BAROQUE composer who laid the foundations of musical idioms which shaped musical thought to the time of BEETHOVEN. He turned his attention from the principles of contrapuntal writing to harmonic thinking. His genius is displayed in his dramatic music, and in his many operas (only some of which have survived) he developed the use of string accompaniment for the vocal part, in preference to the conventional use of the harpsichord. He is known as the founder of OPERA SERIA.

Scarlatti, Giuseppe Domenico (1685–1757), Italian composer, son of Alessandro SCARLATTI. A harpsichord virtuoso, he settled in Spain and is primarily known for his harpsichord sonatas, of which he composed more than 500. He is considered the founder of modern keyboard technique and the greatest Italian composer of keyboard music of the BAROQUE period.

Scarlet fever, or scarlatina, acute infectious disease, usually affecting children, caused by streptococcal BACTERIA. It is characterized by a bright red body rash, fever and throat infection. It may be transmitted by direct contact or through use of the same utensils. It is not as dangerous as it once was, being easily treated with penicillin or other ANTIBIOTICS.

Scarp, abbreviated form of ESCARPMENT, a line of inland cliffs, of any height, produced by faulting or erosion. The steep margins of a plateau, MESA or terrace are examples. *See also* PE p.105.

Scarron, Paul (1610–60), French burlesque writer who rebelled against the artificiality of the literature of his time. He is best remembered for his novel *Le Roman comique.*

Scattering, in physics, deflection of the path of subatomic particles by atoms. It is the means by which the structure of atoms was discovered. Ernest RUTHERFORD "fired" ALPHA particles through thin metal films, noted their scattering, and postulated the atomic nucleus (positively charged and centrally placed in the atom). Most knowledge on ELEMENTARY PARTICLES and the discovery of new ones has been obtained by scattering experiments carried out in particle ACCELERATORS.

Scattering, light, deflection of light waves from the main direction of a beam by fine particles of solid, liquid or gaseous matter. The effects observed depend on the size of the particles; small particles cause DIFFRACTION and larger particles also produce reflection. *See also* SU p.102.

Scenery, theatrical stage equipment used to represent the place of action, including painted FLATS, curtains and box sets. The setting of earliest drama was the theatre itself and no special scenery was used. With the advent, first of the RENAISSANCE and then OPERA in the 17th century, elaborate scenic effects came into vogue. The 19th century saw a move towards accurate representation in scenery and from stage realism there was a turn towards SYMBOLISM.

Scepticism, philosophical attitude which asserts the uncertainty of all knowledge. The first important sceptic was DEMOCRITUS, who held that sense perceptions are not a reliable means to true knowledge of the material world.

Schaarbeek. *See* SCHÄERBEEK.

Schadow, Johann Gottfried (1764–1850), German sculptor. Schadow was a leading exponent of NEO-CLASSICISM and his works include the Quadriga on the Brandenburg Gate in Berlin and monuments to Frederick the Great (in Szczeczin), and to Blücher von Wahlstatt (in Rostock).

Schäerbeek, suburb of Brussels, in central Belgium. It is an important railway junction and has machinery, clothing and chemical industries. Pop. (1971 est.) 118,947.

Schally, Andrew (1926–   ), US physiologist, *b.* Poland. He studied in Britain and Montreal, and became professor of physiology at the Veterans' Administration Hospital, New Orleans, in 1962. He was awarded a share in the 1977 Nobel Prize in physiology and medicine with Roger Guillemin and Rosalyn YALOW for their work on peptide hormones.

Scharff, Edwin (1887–1955), German sculptor who first studied painting but turned to sculpture while in France (1911–13). He was a founder member of the *Neue Sezession* group in Munich. The monumental quality of his best works after 1920 derives from Aristide MAILLOL.

Scharnhorst, Gerhard Johann David von (1755–1813), Prussian general. After the Prussian defeat by the French in 1807, he became head of the Army Reform Commission. He reorganized the army and circumvented French limitations on its size by calling up small numbers of men at a time who, after quick training, would become reserves.

Scharnhorst, German battle cruiser which was responsible for the sinking of much Allied shipping in WWII. It was launched in 1939 and sunk in 1943 by HMS *Duke of York.*

Scheel, Walter (1919–), West German political leader. As head of the Free Democratic Party, he formed a coalition government in 1969 with the Social Democrats and served as Foreign Minister in the cabinet of Willy BRANDT. He was elected President of West Germany in 1974.

Scheele, Carl Wilhelm (1742–86), Swedish chemist. An apothecary, his interest in chemistry led to an investigation of combustion and discovery of OXYGEN, but publication of his discovery was delayed and the credit went to Joseph PRIESTLEY. He made other important discoveries, including CHLORINE, GLYCERIN and a number of organic acids.

Scheelite, mineral, calcium tungstatemolybdate, $Ca(WO_4,MoO_4)$, an important ore of tungsten. It is found in metamorphic deposits and in pegmatites. It has tetragonal system bipyramidal crystals, also occurring as massive and granular aggregates. It reveals various tints with adamantine lustre, fluoresces under ultraviolet light, and is brittle. Hardness 4.5–5; s.g. 6.

Scheemakers family, Flemish sculptors who did the majority of their work in England. Peter (1691–1781) executed the Shakespeare monument in Westminster Abbey. He made other tombs in the Abbey and at other churches, and his

work was competent but lacking in force. His son Thomas (1740–1808) also worked on monuments, many of which were designed by James STUART.

Schelde (Escaut), river in NW Europe. It rises in N France and flows N and NE through W Belgium to Antwerp, then NW into the North Sea through two estuaries (the East and West Schelde) in the SW Netherlands. The Schelde is navigable for most of its course. Length: 435km (270 miles).

Scheler, Max (1874–1928), German philosopher. Influenced by Franz BRENTANO and some of Edmund HUSSERL's disciples, he made important contributions to the PHENOMENOLOGICAL MOVEMENT. His works include *Formalism in Ethics and Material Value Ethics* (1921), *On the Eternal in Man* (1921) and *The Place of Man in the Universe* (1928).

Schelling, Friedrich Wilhelm Joseph von (1775–1854), German philosopher. Profoundly Romantic, his philosophy of nature linked nature and spirit as real and ideal poles of one essential reality. Ultimately, he asserted, PANTHEISM merges into total MYSTICISM. Clearest accounts of his philosophy appear in *First Sketch of a System of "Naturphilosophie"* (1799) and *System of Transcendental Idealism* (1800).

Scherbius, Arthur. *See* ENIGMA CIPHER MACHINE.

Scherzo, musical piece or, more commonly, movement of a larger work such as a SONATA or SYMPHONY. It is the Italian word for "joke". It may take the place of the MINUET movement. It has the rhythm of the minuet, but is usually quicker (in 6/8 rather than 3/4 time) and lighter.

Schiaparelli, Giovanni Virginio (1835–1910), Italian astronomer, best known for his discovery in 1877 of the markings on the surface of MARS that he called canals. *See also* SU p.194.

Schick, Bela (1877–1967), US paediatrician who devised (1913) the Schick test which determines susceptibility to diphtheria. A small amount of diphtheria toxin is injected into the skin and swelling indicates a low level of antibody production. If the subject has an adequate immunity no skin reaction will occur.

Schick Test, test of immunity to DIPHTHERIA, devised by the Austrian paediatrician Bela SCHICK.

Schiele, Egon (1890–1918), Austrian EXPRESSIONIST painter who was influenced by the IMPRESSIONISTS and then by Gustav KLIMT. Using bright colours he developed a taut, linear style and painted nudes and landscapes.

Schiller, Johann Christoph Friedrich von (1759–1805), German dramatist, historian and philosopher. He wrote blank verse STURM UND DRANG plays: *Don Carlos* (1787), *Wallenstein* (1800), *Mary Stuart* (1801), *Maid of Orleans* (1801) and *William Tell* (1804). His *Ode to Joy* is set to music in the last movement of Beethoven's 9th Symphony.

Schindler, Rudolph (1887–1923), Viennese-born architect who emigrated to the USA in 1913 and worked with Frank Lloyd WRIGHT. With Wright he introduced into the USA such European architectural innovations as cubistic design and austerity of line.

Schinkel, Karl Friedrich (1781–1841), German architect and painter. Working in the classical tradition, he designed the School of Artillery and Engineering, the Royal Theatre and the Old Museum, all in Berlin. *See also* HC2 p.43.

Schipperke, dog originally bred in Flanders several centuries ago. It has a fox-like head with small triangular, erect ears. The short, stocky body is set on medium-length legs and the tail is commonly docked. The harsh short coat is ruffed at the neck and long at the back of the hind legs, and is usually solid black. Height: to 33cm (13in) at the shoulder; weight: to 8kg (18lb).

Schism, in religion, division within a church or a break-away from a church. Before the Protestant Reformation, there were two important schisms within the Christian Church. The first was the schism between East and West, which culminated

Alessandro Scarlatti's overtures consisted of a fast, a slow and a fast movement.

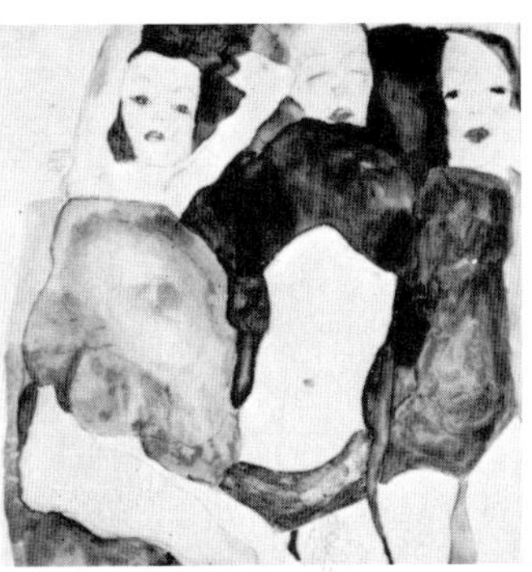

Egon Schiele's *Three Bathers* (1911) is in many ways typical of his paintings.

Johann von Sciller's dramas inspired German liberals in the 1848 revolution.

Schipperkes were used to guard barges and to hurry the horses that pulled them.

Schist

Schleswig-Holstein; Christian IX reigned at a crucial time in the state's history.

Schlieffen Plan; German troops advancing through France in the first weeks of WW1.

Schnauzer; miniature varieties were bred as terriers and as pets.

Martin Schongauer was one of the first artists to use copper for engravings.

in the complete break between the Eastern (ORTHODOX) church and the Western (ROMAN CATHOLIC) church in 1054. The Western GREAT SCHISM involved a split within the Roman Church itself and the election of rival popes between 1378–1417.

Schist, large group of METAMORPHIC ROCKS which have been made cleavable, causing the rocks to split into thin plates leaving a wavy, uneven surface. They are named after their predominant mineral, eg phyllite (or mica) schist. *See also* PE pp.97, 102–103, *103*, 105, *107.*

Schistosomiasis. *See* BILHARZIA.

Schizoid personality, personality pattern disturbance marked by avoidance of close relations with others, inability to express hostility directly, and AUTISTIC thinking. A schizoid person appears shy, aloof and detached, and may be described as introverted and withdrawn.

Schizophrenia, one of a group of psychotic disorders marked primarily by disturbances of cognitive functioning, particularly thinking. Eugen BLEULER, who introduced the term in 1911, saw schizophrenia as a "splitting of the personality" from reality, a tendency of thinking to become wish-fulfilling, idiosyncratic and dominated by fantasy. *See also* MS pp.138, 139.

Schizophyta, division of plants in which some plant taxonomists place single-celled plants that do not have organized nuclei. The Schizophyta contains two classes, the blue-green ALGAE (class Cyanophyceae) and the BACTERIA (class Schizomycetes). In other classifications these subgroups are regarded as divisions of the plant kingdom.

Schlegel, August Wilhelm von (1767–1845), German poet, critic and scholar, noted for his translations of the Bhagavad-Gita and of Shakespeare. In 1798, as a professor in Jena, he led the Romantic movement with his brother Friedrich Johann SCHLEGEL, Ludwig TIECK and Friedrich von Hardenberg, nicknamed NOVALIS.

Schlegel, Friedrich Johann von (1772–1829), German writer and philosopher, brother of August SCHLEGEL. His works include *Lectures on the History of Literature* (1815), *On Language and the Wisdom of India* (1808) and the novel *Lucinde* (1799).

Schleiden, Matthias Jakob (1804–81), German botanist who established the CELL theory for plants in his paper *Contributions to Phytogenesis* (1838), in which he recognized that all parts of a plant are composed of cells.

Schlemmer, Oskar (1888–1943), German painter, sculptor and stage designer who taught at the BAUHAUS from 1920 until 1929. Versatile in a wide range of media, Schlemmer attempted to represent the movements of dance on canvas using unusual perspective. An example of this is his painting *Concentric Group* (1925).

Schlesinger, John Richard (1926–), British film director. His films often explored the influence of social mores on personal relationships. His work included *A Kind of Loving* (1962), *Billy Liar* (1963), *Darling* (1965), *Midnight Cowboy* (1969), *Sunday, Bloody Sunday* (1971), *Day of the Locust* (1975) and *Marathon Man* (1976).

Schleswig-Holstein, state in w Germany, in the s Jutland peninsula. Its capital is KIEL. In the mid-19th century the duchy of Schleswig-Holstein was claimed by both Denmark and the North German Confederation. When CHRISTIAN IX of Denmark tried to incorporate Schleswig in Denmark in 1863, Prussia (at BISMARCK's instigation) and Austria led a campaign to capture the area, which was shared between the two by the Treaty of Gastein. Following the Austro-Prussian war of 1866, the region passed to Prussia. In 1920 N Schleswig became part of Denmark following a plebiscite. *See also* HC2 p.111.

Schlick, Moritz (1882–1936), German philosopher and head of the VIENNA CIRCLE. He studied physics with Max PLANCK and taught the philosophy of inductive science at the University of Vienna. His

works include *General Theory of Knowledge* (1918) and *Problems of Ethics* (1930).

Schlieffen Plan, German war strategy named after Field-Marshall Alfred von Schlieffen (1833–1913). The plan, devised while he was Chief of Staff (1891–1905), called for a lightning attack on France through Belgium by a strong German army. Germany could then concentrate on fighting Russia. The plan was implemented by von MOLTKE in WWI but the attacking force was weakened and the Germans were unable to sweep over France. *See also* HC2 pp.186–187, *186.*

Schliemann, Heinrich (1822–90), German archaeologist. In 1871 he began excavations at his own expense at Hisarlik, Turkey, which he believed to be the site of the Homeric city of TROY, and uncovered four superimposed towns. He also made important discoveries in Greece, especially at Mycenae. *See also* HC1 pp.*22–23, 41.*

Schlüter, Andreas, (1662–1714), German Baroque sculptor and architect whose main work, the Royal Palace, Berlin, (1698–1707), reveals the influence of BERNINI.

Schmalkaldic League, alliance of German Protestant states formed in 1531 to protect Lutheranism against the Holy Roman Emperor CHARLES V. Its leaders were PHILIP OF HESSE and John Frederick of Saxony. War between the Empire and the league broke out in 1546 and the following year the league was decisively defeated at Mühlberg. *See also* HC1 p.268.

Schmidt, Bernhard Voldemar (1879–1935), Estonian optician who, in 1930, developed the telescope that bears his name. The telescope allows large areas of the sky to be photographed with each exposure, and it has a complex correcting plate that reduces optical distortion. *See also* SU p.171, *171.*

Schmidt, Helmut (1918–), West German political leader. Elected to the BUNDESTAG in 1953, he became chairman of the Social Democratic Party in 1967 and Minister of Defence when the Social Democrat-Free Democrat coalition government was formed in 1969. He replaced Willie BRANDT as chancellor of West Germany in 1974.

Schmidt-Rottluff, Karl (1884–1976), German painter and woodcut artist who was a co-founder of die BRÜCKE. A leading German Expressionist, his best works are woodcuts. *See also* HC2 p.174, *174.*

Schnabel, Artur (1882–1951), Austrian pianist and composer, who lived in the USA after 1939. He was best known as an interpreter of the classical repertoire, including Mozart and Beethoven, whose 32 piano sonatas he edited and recorded.

Schnauzer, dog originally bred as a guard dog in Germany. There are now three breeds: standard, giant and MINIATURE SCHNAUZER. The standard breed has a square head with a blunt, heavily whiskered muzzle and folded triangular ears. The strong straight body is set on lean muscular legs and the tail is commonly docked. The coat, which may be black, grey or brown, is harsh and wiry. Height: to 51cm (20in); weight: to 17kg (37lb).

Schneider Trophy, trophy in aviation presented in 1913 by Jacques Schneider to the winner of the international seaplane races. The Frenchman Prevost flying a Deperdussin at 73.7km/h (45.75mph) was the first holder (1913). Such races encouraged improvements in design during the 1920s. The trophy was won outright by Britain after wins in 1927, 1929 and 1931. *See also* MM p.148.

Schnitzler, Arthur (1862–1931), Austrian dramatist, novelist and doctor. His witty plays explore the themes of love and sex. They include *Anatol* and *Merry-go-Round* (1893), *Light-O'Love* (1896) and *The Reckoning* (1897). His famous novels are *None But the Brave* (1901), *The Road to the Open* (1908) and *Flight into Darkness* (1931).

Schockemöhle, Alwin (1937–), West German show jumping rider. In 1976 he was the reigning European champion and won the Olympic Games individual show

jumping gold medal, guiding *Warwick Rex* to the first double clear round in the history of the competition. In 1960 he had won a team gold, and in 1968 a team bronze.

Schöffer, Peter (*c.* 1425–1503), German printer whose name, together with that of Johannes Fust (his father-in-law and associate), appear on a Psalter printed in 1457, the first printed book that shows the name of its printer. *See also* MM p.204.

Scholar-Gipsy, The, poem by Matthew Arnold, published in 1853. A lament for the passing of youth, Arnold called it an elegy, intended to commemorate his wanderings, as an Oxford undergraduate, in the Cumnor Hills. It has 25 stanzas of 10 iambic PENTAMETER lines.

Scholasticism, medieval philosophy that attempted to join faith to reason by synthesizing theology with classical Greek and Roman thought. It was based on the writings of St AUGUSTINE, BOETHIUS, the Pseudo-Dionysius and the Arabic commentaries on ARISTOTLE. Scholasticism was first explored by John Scotus ERIGENA in the 9th century and by ANSELM in the 11th century. Its greatest thinkers were Albertus Magnus and Thomas AQUINAS in the 13th century and DUNS SCOTUS at the turn of the 14th century. The new universities of Paris and Oxford were the centres of scholastic thought. *See also* HC1 pp.188, 189.

Scholes, James Christopher (1852–90), British antiquary. He became a reporter on a local newspaper and was noted for his *Bolton Bibliography and Jottings of Book Lore*, with *Notes on Local Authors and Printers*, published in 1886.

Schollander, Donald Arthur ("Don") (1946–), US freestyle swimmer who, at the 1964 Olympic Games, became the first swimmer to win four gold medals (100m, 400m, 4×100m freestyle relay and 4×200m freestyle relay). He won a fifth gold and a silver in 1968, won three golds at the 1967 Pan American Games, and set 13 individual world records.

Schoenberg, Arnold (1874–1951), Austrian composer who abandoned traditional tonality for an atonal method (without key centre) using "tone-rows" drawn from the 12-note scale. Extending the chromaticism of the later romantics in early works, eg the string sextet *Verklärte Nacht* (1899), he reached strict atonality in his *Suite for Piano* (1924). *See also* SERIAL MUSIC, TWELVE-TONE MUSIC.

Schönbein, Christian Friedrich (1799–1868), German chemist, the discoverer of OZONE. After studying at various universities, he became professor at the University of Basle (1835). He invented guncotton, recognizing its potential as a smokeless substitute for gunpowder. *See also* MM p.242.

Schönberg, Arnold. *See* SCHOENBERG, ARNOLD.

Schönbrunn, palace in the BAROQUE style in Hietzing, a w suburb of Vienna. Begun in 1695 and based on plans by J.B. FISCHER VON ERLACH, it was a royal summer house influenced by VERSAILLES. It was completed during the reign of MARIA THERESA. The gardens were designed by Ferdinand von Hohenberg.

Schongauer, Martin (*c.* 1450–91), German painter and prolific engraver who produced more than 100 signed engravings, mostly religious. His style, rich in detail, was influenced by Rogier van der WEYDEN. *See also* HC1 p.253.

School, term commonly used to describe a place of primary or secondary EDUCATION. In Western countries there are usually two parallel school systems, one provided by the state and the other privately financed (British "public schools" are in fact private). In Communist countries schools are entirely state-run. In almost every country, school attendance is compulsory between certain ages. British state secondary sc schools formerly included academic "grammar schools" and vocational "secondary moderns". They have been largely replaced by comprehensive schools, intended to combine the two functions. In Australia and New Zealand the educational system is broadly similar to Britain's before the introduction of comprehensive schools.

School for Scandal, The (1777), five-act satirical comedy by Richard Brinsley SHERIDAN.

School for Wives (1662), comedy in verse by Molière. A satire on the narrow education given to middle-class girls in 17th-century France, it tells how a man plans to keep his young wife faithful by keeping her uneducated.

Schooner, sailing ship with at least two masts and a bowsprit, and with fore-and-aft sails rather than square sails. Larger schooners appeared in the early 20th century when the small crews needed to operate the fore-and-aft sails allowed them to remain competitive with steamships for several years. *See also* MM pp.112–113.

Schopenhauer, Arthur (1788-1860), German philosopher whose system, described in his main work, *The World as Will and Idea* (1819), establishes the will to endure as an individual's prime motivation and the negation of will and desire as the possibility of escape from pain.

Schouten, Willem Cornelis (c.1580–1625), Dutch navigator. In 1616, on a voyage to find a new trade route to the East, he sailed past a cape at the extreme s of South America. He named it after his birthplace, Hoorn.

Schreiner, Olive Emilie Albertina (1855–1920), South African novelist. In 1881 she moved to Britain and there published a novel, *The Story of an African Farm* (1883), under the pseudonym Ralph Iron. Her next novel, *Trooper Peter Halkett of Mashonaland* (1897), was attacked for its criticism of Rhodesian society.

Schrieffer, John Robert (1931–), US physicist. In 1957, working at the University of Illinois with Leon COOPER and John BARDEEN, he helped to formulate a theory of SUPERCONDUCTIVITY, which is known as the BCS theory after their initials. All three shared the 1972 Nobel Prize in physics.

Schrödinger, Erwin (1887–1961), Austrian physicist who formulated the quantum mechanical wave equation named after him. With Paul DIRAC he shared the 1933 Nobel Prize in physics for the wave equation, which was based on Louis de BROGLIE's suggestion that moving particles have a wave-like nature.

Schröter, Johann Hieronymus (1745–1816), German astronomer. He built an observatory at Lienthal which became, from the excellence of its instruments, an international centre. He was the first man to observe the surface of the moon and the planets systematically over a long period.

Schubert, Franz Peter (1797–1828), Austrian composer whose symphonies represent the final extension of the CLASSICAL sonata form and whose songs are the height of ROMANTIC lyricism. Among his more popular works are symphonies such as the Eighth ("Unfinished") (1822) and the Ninth in C major (1828). He wrote more than 600 songs to the lyrics of such poets as HEINE and SCHILLER and these include such cycles as *Die Schöne Müllerin* (1823) and *Die Winterreise* (1827). In his tragically short lifetime he also composed much chamber music, and his string quintet is regarded as a masterpiece. *See also* HC2 pp. 106, *107*.

Schulberg, Budd Wilson (1914–), US scriptwriter, short story writer and novelist, whose works deal with success and failure in American life. They include the novel *What Makes Sammy Run* (1941) and the script for *On the Waterfront* (1954).

Schultze, Max Johann Sigismund (1825–74), German biologist. He studied the nervous system of vertebrates, the electric organs of fish and the anatomy of worms and molluscs. His most important work was the identification of a cell as an organism containing both nucleus and protoplasm and his recognition that protoplasm is a basic substance of plant and animal cells.

Schumacher, Ernst Friedrich (1911–77), German economist who spent most of his life in Britain. He was economic adviser to the National Coal Board (1950–70) and their director of statistics (1963–70).

Among his publications are *The Roots of Economic Growth* (1962) and *Small is Beautiful: Economics as if People Mattered* (1973).

Schuman, Robert (1886–1963), French politician and Foreign Minister from 1948 to 1952. In 1950 he proposed the creation of a single authority to control the production of coal and steel in Europe. This so-called Schuman Plan became effective in 1952 in the form of the EUROPEAN COAL AND STEEL COMMUNITY, which was the first step towards the creation of the EEC. *See also* HC2 p.258, *258*.

Schuman, William (1910–), US composer. He became president of the Lincoln Centre in 1962. He has written symphonic works and an opera, *The Mighty Casey* (1953). His cantata, *A Free Song* (1942), won the first Pulitzer Prize for music.

Schumann, Clara Josephine Wieck (1819–1896), German pianist, composer and wife of Robert SCHUMANN. She taught at the Frankfurt Conservatory of Music from 1878 to 1892 and was an outstanding interpreter of the works of her husband and of BRAHMS.

Schumann, Elisabeth (1895–1952), German-born US soprano. She made her debut with the Hamburg Opera in 1910, and thereafter excelled in major roles in the operas of MOZART and Richard STRAUSS, with whom she toured the USA in 1921.

Schumann, Robert Alexander (1810–56), German composer and leader of the ROMANTIC movement. He married the pianist Clara Wieck in 1840 and both were friends of Johannes BRAHMS and Frederic CHOPIN. Schumann's piano compositions include *Kinderscenen* (1838), *Carnaval* (1834–5) and *Waldscenen* (1848–9), and in the early 1840s he wrote many songs and orchestral music. He set to music words by such poets as HEINE and GOETHE and among his best song cycles is *Frauenliebe und Leben* (*Women's Love and Life*; 1840). His *Piano Concerto in A Minor* (1841–45) and *Spring Symphony* (1841) represent his best orchestral works. *See also* HC2 p.106.

Schuster, Sir Arthur (1851–1934), British physicist, *b.* Germany, whose work covered a wide variety of subjects, including electrical discharge in gases, calorimetry and SEISMOLOGY and was an authority in the field of SPECTROSCOPY. In 1875, he led the expedition to Siam (now Thailand) to study the solar eclipse.

Schütz, Heinrich (1585–1672), German composer who wrote the earliest German opera, *Dafne* (1627), now lost. Most of his surviving works were written for the Church and they had a great influence on BACH and HANDEL. They include Passions, oratorios, madrigals and *Symphoniae Sacrae* (1619, 1647, 1650). *See also* HC2 p.40.

Schutzenberger, Paul (c.1827–97), French chemist. He worked chiefly in physiological chemistry, especially dyestuffs and fermentation. He wrote *Traité des Matières Colorantes* (1867) and *Les Fermentations* (1875).

Schwann, Theodor (1810–82), German biologist who laid the foundation of modern HISTOLOGY by establishing the CELL theory. In 1836 he became the first to prepare an ENZYME from animal tissue when he isolated and named PEPSIN.

Schwartz, Rudolf (1905–), British conductor, *b.* Austria. He worked for the Jewish Cultural Organization in Berlin (1936–41) and was put in a concentration camp during WWII. He was conductor in chief of the BBC Symphony Orchestra (1957–62).

Schwarz-Bart, André (1928–), French novelist, best-known for his novel, *The Last of the Just* (1959). This work traces the sufferings of Ernie Lévy in a Europe dominated by the Nazis. The novel was awarded the PRIX GONCOURT.

Schwarzenberg, Prince Karl Philipp von (1771–1820), Austrian soldier. In 1812 he led the Austrian forces sent to aid NAPOLEON in the Russian campaign, and the following year commanded the victorious allied armies against Napoleon at Dresden and Leipzig.

Schwarzkopf, Elizabeth (1915–), German soprano known for her versatility in recitals, oratorios and opera. She sang with the Berlin State Opera from 1938 to 1942 and became principal soprano of the Vienna State Opera in 1944. She has made many fine recordings.

Schweitzer, Albert (1875–1965), theologian, musician, medical missionary and philosopher. Born in Alsace, he spent most of his life in Gabon (then French Equatorial Africa) where he founded the Lambaréné Hospital in 1913. Honoured as a scientist and humanitarian, as an organist and an expert on BACH, his ethic was "reverence for life" and he was awarded the 1952 Nobel Peace Prize. His works include *Kant's Philosophy of Religion* (1899) and *Out of My Life and Thought* (1931).

Schwinger, Julian Seymour (1918–), US physicist who shared the 1965 Nobel Prize in physics with R. P. FEYNMANN and S. TOMONAGA for work on quantum ELECTRODYNAMICS. Schwinger's contribution was to combine QUANTUM THEORY with RELATIVITY in treating the interactions between particles and the ELECTROMAGNETIC FIELD.

Schwitters, Kurt (1887–1948), German artist best known for his constructions and collages in which he used refuse items such as old newspapers and tram tickets to create works of remarkable sensitivity. *See also* HC2 p.202, *202*.

Sciatica, NEURALGIA or NEURITIS associated with severe pain in the legs. It is caused by irritation or swelling of the sciatic nerve, which passes from the lower spinal column down the back of the thighs and legs. It can be treated with bedrest, sedation or corrective exercise.

Science fiction (SF), literary genre in which reality is subject to certain transformations in order to explore man's potential and his relation to his environment; usually these transformations are technological and the stories are set in the future or in imaginary worlds. Science fiction became a popular genre only with the comic strip *Amazing Stories* in the USA in 1926. Until the 1960s most science fiction involved adventure stories set in space, with stereotyped middle-American characters. Some writers, such as Isaac ASIMOV, explored the paradoxes contained in purely scientific ideas; others like Ray BRADBURY stressed the moral implications of their stories. A "new wave" of SF writers in the late 1960s, led by Michael Moorcock, stressed the role of SF as a means of liberating the reader from his perceptual confines.

Scientific method, established procedure for scientific study. In the scientific method, a problem is isolated and observations are made (this includes visual analysis as well as making measurements). Next, a theory is postulated and its validity is tested by experiments. If the experiments fit the theory, the theory becomes a law; if not, the theory is rejected or modified.

Scientific Revolution, name given to the great changes which came over the methods and ideas of science, especially cosmology and mathematics, in Europe in the 16th and 17th centuries. In cosmology the movement is traced from COPERNICUS' radical assertion that the Earth revolves around the Sun, through GALILEO and KEPLER, to NEWTON's law of gravity. Along with other aspects of the RENAISSANCE it laid the foundations of modern Western culture. *See also* AGE OF REASON; ENLIGHTENMENT; HC1 pp. 302–303.

Scientific socialism, phrase used by Karl MARX and Friedrich ENGELS to distinguish their political theory from the SOCIALISM of people such as Robert OWEN and SAINT-SIMON. The phrase was meant to convey that Marxist ideas were founded on objective historical analysis of the economic forces which determined social and political development.

Scientology, US movement founded by L. Ron Hubbard in 1952. It advocates confrontation with, and assimilation of, painful experiences from the past to achieve true mental health. The movement has aroused controversy but in the early 1970s it claimed to have 600,000 members.

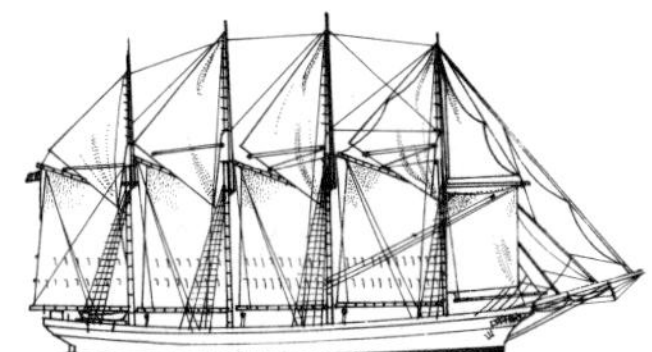

Schooners' rigging allowed them greater speed and less crew than other craft.

Franz Schubert wrote the song *Gretchen at the Spinning Wheel* when he was 18.

Clara Schumann, a brilliant pianist, was taught by her father, Friedrich Wieck.

Robert Schumann, who all his life feared losing his reason, died in an asylum.

Scilly Isles: waves breaking on the rocks of Woolpack Point on St Mary's Isle.

Scipio Africanus, Rome's great military tactician, defeated Hannibal in Africa.

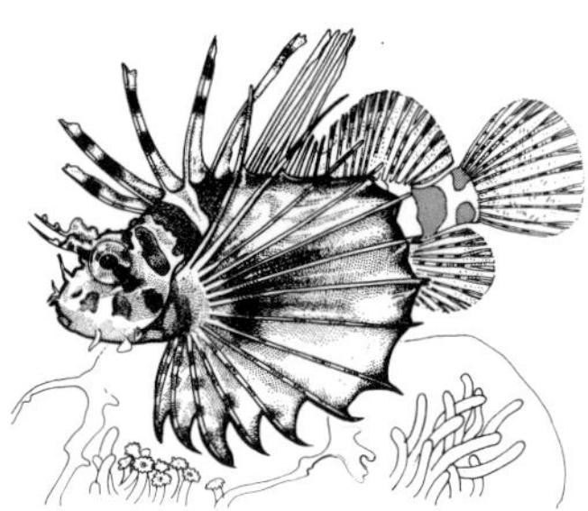

Scorpion fish: the most harmful members of the family live in shallow waters.

Sir Walter Scott's *Heart of Midlothian* is considered one of his best stories.

Scilla, or squill, genus of perennial garden plants of the LILY family native to temperate regions of the Old World. Squat and bulbous, they have white or blue bell-shaped flowers.

Scilly, Isles of, group of more than 140 islands in the Atlantic Ocean off the coast of Cornwall, sw England, 45km (28 miles) sw of Land's End. Five of the islands are inhabited: they are St Mary's (the largest), Tresco, St Martins, St Agnes and Bryher. The climate is mild and flowers and early vegetables are grown; tourism is also important. Pop. (1971) 2,428.

Scintillation, or twinkling, in astronomy, slight variability in the brightness and colour of the stars caused by local disturbances in the Earth's atmosphere that change its REFRACTIVE INDEX.

Scintillation counter, instrument containing a crystal that emits scintillations of light when bombarded by radiation. Each light flash, corresponding to a single particle, is converted into an electric pulse by a PHOTOMULTIPLIER. The number of pulses, counted electronically, indicates the activity of the source of radiation.

Scipio, Publius Cornelius (*d.* 211 BC), Roman general, father of SCIPIO AFRICANUS MAJOR. As consul in 218 his attempt to halt HANNIBAL in N Italy failed when he was forced to put down a revolt of the Gauls. Later that year he lost two-thirds of his army fighting Hannibal at Trebia.

Scipio Africanus Major (Publius Cornelius Scipio) (*c.* 236–183 BC), Roman general, the greatest military genius of the Scipio family. He defeated the Carthaginian forces in Spain in 209 BC and established Roman rule. In 204 BC he captured Tunis with a volunteer army and defeated Hannibal at Zama in 202 BC, earning himself the title of Africanus. *See also* HC1 p.91.

Scipio Africanus Minor (Publius Cornelius Scipio Aemilianus) (*c.* 185–129 BC), Roman general and scholar who was adopted by the son of SCIPIO AFRICANUS MAJOR. He destroyed Carthage in 146 BC and ended the Third PUNIC WAR, and in 134 BC ended a prolonged war in Spain by devastating the city of Numantia.

Scleroderma, skin disorder characterized by thickening and hardening of the skin, which becomes attached to the underlying tissues.

Sclerosis, degenerative hardening of tissue, especially in the arteries and central nervous system. Normal tissue is replaced by connective tissue, as in a SCAR.

Scofield, David Paul (1922–), leading British actor celebrated for his roles in John OSBORNE's *Hotel in Amsterdam* (1968) and Christopher Hampton's *Savages* (1973) at the ROYAL COURT THEATRE. In 1966 his portrayal of Sir Thomas MORE in the film version of Robert BOLT's *A Man for All Seasons* earned him an Oscar.

Scoliosis. *See* CURVATURE OF THE SPINE.

Scone, village in SE Perth, Scotland. Old Scone was the Pictish and Scottish capital where the kings of Scotland resided from 1157 to 1488. The Coronation Stone or Stone of Scone, was removed to Westminster Abbey by EDWARD I of England in 1296. There was an attempt in 1950 by Scottish nationalists to return it to Scone.

Scooter, motor, lightweight, small engine-capacity MOTOR-CYCLE which was popular in many countries during the 1950s and 1960s. *See also* MM p.125, *125.*

Scopas (*fl.* mid-4th century BC), Greek sculptor whose figures are known for their intense facial expression, achieved by carving the eyes deeply into the face and the lips in a slightly parted position. *See also* HC1 p.78.

Scopolamine. *See* HYOSCINE.

Score, written representation of a musical work which includes all the music, not simply that of one part. A full score shows each part separately. The score of a choral work in which each vocal part is shown, but the orchestral parts of which are reduced to a piano version, is called a vocal score.

Scorel, Jan van (1465–1562), Dutch painter who studied in Italy and helped to introduce the Italian High RENAISSANCE to the Netherlands. He was influenced by MICHELANGELO and RAPHAEL, and among his works is *Portrait of Agathe Schoonhaven* (1529).

Scoresby, William (1789–1857), British scientist and Arctic explorer who made several voyages to Greenland between 1799 and 1822. He collected information about the little-known coasts of Greenland and made valuable observations on the Earth's magnetic field.

Scorpio. *See* SCORPIUS.

Scorpio, the eighth astrological sign of the ZODIAC. The ancients identified it with the scorpion which caused the death of the hunter Orion and stung the horses of PHAETHON. The Sun enters the sign on 24 Oct. and leaves it on 21 Nov.

Scorpion, any of numerous species of variously coloured ARACHNIDS that live in warmer regions throughout the world. It has two main body sections, two eyes, a pair of pedipalps (pincers) and a long slender tail ending in a curved, poisonous sting. Length: to 17cm (7in). Class Arachnida; order Scorpionida. *See also* NW pp.*102,* 103, *219.*

Scorpion fish, also called rockfish, any of several species of perch-like marine fish of temperate waters. It can be identified by its venomous fin spines capable of inflicting painful and sometimes fatal injury. Length: to 1m (39in). Family Scorpaenidae. *See also* LIONFISH.

Scorpion fly, any of several species of brown to grey insects found throughout the Northern Hemisphere. Its chewing mouthparts are at the end of a long beaklike structure. Males of some species have an abdomen resembling a scorpion's tail. Length: to 40mm (1.6in). Order Mecoptera. *See also* NW pp.106–107, *107.*

Scorpius (Scorpio) or the Scorpion, southern constellation on the ECLIPTIC between Libra and Sagittarius. It contains several open and globular CLUSTERS. The MILKY WAY also passes through the region. The brightest star is the 1st-magnitude Alpha Scorpii (Antares). *See also* MS pp.200, *200–201;* SU pp. *254,* 255, 261, *261, 266–267,* 267.

Scotland, N part of the United Kingdom of Great Britain, N of England and bounded by the Atlantic Ocean and the North Sea. Most of the people live in the Lowlands but Scotland is noted for its Highlands – its magnificent scenery and the traditions of the Scottish clans. The principal agricultural activity is the rearing of livestock; oats and potatoes are the chief crops. Fishing is important and Scotland's main export is Scotch whisky. Industrial growth has declined since WWI but new industry is being attracted and the development of the North Sea oilfields has stimulated the economy. Edinburgh is the capital. Area: 78,764sq km (30,411 sq miles). Pop. (1975 est.) 5,206,200. *See* MW pp.150, 174.

Scotland, Church of. *See* CHURCH OF SCOTLAND.

Scotland Yard, name given to the headquarters in London of the Metropolitan Police and synonymous with the Criminal Investigation Department (CID). Originally located in Scotland Yard, off Whitehall, when the force was formed in 1829 by Sir Robert PEEL, it was moved to New Scotland Yard on the Embankment in 1890 and to further new premises in Broadway, Victoria, in 1967.

Scots, originally a Celtic people from northern Ireland. They were Goidelic-speaking. Their raids on the w coast of Roman Britain from the 3rd to the 5th century failed to establish independent settlements in Wales or NW England. In the 5th century, however, they were able to establish the kingdom of Dalriada in Pictish territory. From the 11th century the term has been applied to those people living in Scotland.

Scots Guards, British regiment commissioned by CHARLES I. They were originally the Scots Regiment of Guards. The present name dates from 1877. Their motto is *Nemo me impune lacessit* ("No man provokes me with impunity").

Scots Pine. *See* PINE.

Scott, Duncan Campbell (1862–1947), Canadian poet and short story writer. He was a civil servant in the Canadian department of Indian Affairs, and a romantic interest in the Indians can be seen in his poetry. A selection of his poetry was published in 1951.

Scott, George Campbell (1927–), US actor whose brilliant portrayal in the film *Patton* won him the Oscar for best actor in 1970. He was also featured in *The Anatomy of a Murder* (1959) and *The Hustler* (1961).

Scott, Sir George Gilbert (1811–78), British architect, a prominent figure in the Gothic Revival. He achieved a European reputation with his design for the church of St Nicholas, Hamburg. He was involved in much restoration work for churches and cathedrals, such as Bath Abbey and Ripon Cathedral. He designed and built the Albert Memorial (1862–70), the Foreign Office (in Renaissance style) and St Pancras Station – all in London.

Scott, Sir Giles Gilbert (1880–1960), British architect and grandson of Sir George Gilbert SCOTT. He designed the new Anglican Cathedral in Liverpool (1904–24), and other works include the New Bodleian Library (1941), Oxford, and Waterloo Bridge (1945) over the River Thames, London.

Scott, Paul (1920–78), British novelist. He is best known for the four novels set in India before the partition, called *The Raj Quartet: The Jewel in the Crown* (1966), *The Day of the Scorpion* (1968), *The Tombs of Silence* (1971) and *A Division of Spoils* (1975). *Staying On,* set in independent India, won the Booker Prize in 1977.

Scott, Sir Peter Markham (1909–), British ornithologist and artist who was best known for his studies of birds, his *Key to the Wild Fowl of the World* (1949) and his illustrations for GALLICO's *The Snow Goose.* In 1946 he founded the Wildfowl Trust. He was the first president of the WORLD WILDLIFE FUND from 1961 to 1967, president of the Wildlife Youth service from 1969 and chairman of the Survival Service Commission of the International Union for the Conservation of Nature and Natural Resources from 1963.

Scott, Captain Robert Falcon (1868–1912), British Antarctic explorer and naval commander. He reached the South Pole in Jan. 1912, only to find that AMUNDSEN had been there first. Beset by blizzards, Scott and his party perished during their journey across the ice back from the pole. His diaries were published as *Scott's Last Expedition* (1913).

Scott, Robert William Henry "Bob" (1921–), New Zealand rugby player, a fullback. He played for the All Blacks 30 times, the first time in 1946. In 152 first-class matches he scored 830 points.

Scott, Sheila Christine (1927–), British aviator. She took up flying in 1959 and from 1965 undertook a series of record-breaking solo flights, flying around the world, and over the North Pole (1971). She won the light-aircraft London-New York race in 1969.

Scott, Sir Walter (1771–1832), Scottish novelist and poet who started his career as a lawyer. His first published works were translations of two German ballads by Bürger in 1796, followed by a collection of old Scottish ballads, *Minstrelsy of the Scottish Border* (1802–03). His first novel *Waverley* (1814), published anonymously, was an immediate success and was followed by such others as *Rob Roy* (1817), *Ivanhoe* (1819) and *The Talisman* (1825). *See also* HC2 pp.31, *31,* 98, *98.*

Scottish. *See* GAELIC.

Scottish deerhound. *See* DEERHOUND.

Scottish Labour Party, political party formed in 1975 when two Labour MPs resigned the Labour WHIP in order to work for a greater degree of devolution to Scotland than they believed the Labour party would expect. The MPs were James Sillars (Ayrshire South) and John Robertson (Paisley).

Scottish law, separate judicial system from English law, although many STATUTES passed by the British Parliament have force throughout the country. At the Act of UNION (1707) Scotland was allowed to retain its own legal system. Scotland also has its own distinctive structure of courts. The supreme civil court in Scotland is the

COURT OF SESSION, which hears new cases, acts as an Appeal Court and may direct an appeal to the HOUSE OF LORDS. Below it are the sheriff courts, one for each former county. The supreme criminal court is called the High Court of Justiciary. The Scottish jury in criminal cases consists of 15 members, who return a verdict by majority vote. In Scotland there is no CORONER.

Scottish National Party, political party in Britain, founded in 1928. It grew out of the Scottish Home Rule Association, formed in 1886. It elected its first MP at a by-election in 1945. In the late 1960s it entered a period of remarkable growth and at the elections of October 1974 it returned 11 MPs. Its official policy is complete independence from the British parliament. See also HC2 pp.260–261.

Scottish Office, department of state in the British government, responsible for Scottish affairs, in particular agriculture, education, home and health affairs and development. After the union of England and Scotland in 1707, special secretaries of state for Scotland were appointed, down to 1745. Thereafter Scottish matters were charged to one of the two secretaries of state and then to the home secretary. In 1885 a secretary for Scotland was again appointed. The post became that of a secretary of state in 1926.

Scottish terrier, ancient Highland breed of dog that was originally used for hunting. The short, deep-chested body is set on short, heavy legs and the tail is straight. The hard, wiry coat may be grey, brindle, or black. Height: to 25.5cm (10.5in) at the shoulder; weight: to 10kg (22lb).

Scotto, Renata (1934–   ), Italian soprano who studied in Milan and made her debut there in 1953; since that year she has been a resident singer at LA SCALA. She has appeared frequently at Covent Garden since 1962 and made her debut at the Metropolitan Opera, New York, in 1965. Her repertoire includes *La Bohème, La Sonnambula, The Marriage of Figaro* and *La Traviata.*

Scotus, John Duns. See DUNS SCOTUS, JOHN.

Scouts. See BOY SCOUTS.

Scrabble, word game for two to four players. Words are built using individual letters on a board marked out to resemble a crossword puzzle. Each letter has a points value, and certain areas on the board double or triple the value of a letter or word. The player with the most points at the end of a game wins.

Scrambler, electronic device used to render a transmitted message unintelligible to all but its sender and its receiver, used particularly on voice messages sent by radio or telephone. The message is distorted in a prearranged way by the sender's scrambler, for instance by oscillating its frequency or by rearranging its order; the receiver's scrambler reverses this process to make the message intelligible again. See also MS p.243.

Scrambling, or moto-cross, motorcycle sport in which races are held over rough, cross-country circuits featuring natural hazards. It originated in Surrey, England, in the mid-1920s, and was introduced to the Continent in 1946. A team championship, the *Moto-Cross des Nations,* began in 1947, and in 1952 the FIM (the international controlling body) inaugurated a European championship for riders of machines up to 500cc. Since then team and individual championships at 250cc have been introduced. Lightweight bikes of Swedish, Czech and Japanese origin have been the most successful.

Screamer, aquatic South American bird closely related to DUCKS, GEESE and SWANS. It has a hooked, chicken-like bill and short, sharp spurs on its wings. It gets its name for the loud, two-note trumpeting sound it makes. Height: 75cm (30in). Family Anhimidae. *See also* NW p.203.

Scree, or talus, heap of rock waste lying at the bottom of a cliff. It is made up of particles ranging in size from sand grains to boulders that have been loosened from the cliff rock by weathering.

Screech owl. See OWL.

Screw, variant of a simple machine, the inclined plane. It is an inclined plane cut around a cone, usually of metal, in a helical spiral. When force is exerted radially on the screw, eg by a screwdriver or the lever of a screw-jack, the screw advances to an extent determined by its pitch, or the distance between crests of its thread. *See also* MM p.86.

Screwpine, woodly Old World plant distributed throughout tropical Asia, the islands of Polynesia and the Indian Ocean; a few species are found in Africa. Not a true pine, its long, tough, narrow leaves rise from a rosette in a spiral, like the thread of a screw. It bears cone-like fruit. Height: to 12m (40ft). Family Pandanaceae. Genus *Pandarus. See also* NW p.66.

Screw propeller, device that converts the TORQUE of a rotating shaft into propulsion or thrust. It is based on the principle of a helical SCREW, which advances in wood when it is rotated; the propeller advances in air or water when it is rotated by an engine. Simple propellers have a fixed PITCH so that the amount of thrust increases as the speed of rotation increases; variable-pitch propellers are rotated at constant speed, and thrust is varied by adjusting the angle at which the blades meet the air or water. *See also* MM pp.32, 36, *36.*

Screwtape Letters, The (1942), work by C.S. LEWIS. As with his religious allegories for children in the Narnia series, the book deals in an entertaining and instructive way with Christian theological and moral problems.

Screw worm, larva of several species of bluegreen BLUEBOTTLES (blow flies) and parasite of warm-blooded animals. It enters through wounds in the skin and is a serious pest of domestic animals in the USA and South America. Family Calliphoridae; genus *Callitroga.*

Scriabin, Alexander Nicolas (1872–1915), Russian composer and pianist who wrote highly original piano music introducing chords built in fourths instead of the conventional major and minor triads. His most significant works include ten piano sonatas, *The Poem of Ecstasy* (1908), three symphonies and numerous short piano pieces. *See also* HC2 p.270, *270.*

Scribe, Augustin Eugène (1791–1861), French playwright. He wrote more than 300 plays and is generally thought to have originated the Well-made Play. His varied dramatic output includes the historical play *Le verre d'eau (The Glass of Water* 1840) and *Adrienne Lecouvreur* (1849), a tragedy which he wrote with E. Legouvé.

Scribes, in Jewish history, court secretaries who later took on legal functions. Following the return of the Jewish exiles to Palestine led by EZRA, they became teachers and interpreters of the law.

Scriptures, in Christianity, the books of the OLD TESTAMENT and NEW TESTAMENT, often including the APOCRYPHA; also a text or passage from one of these books of the Bible. The term, as used in the New Testament, refers primarily to the sacred Hebraic writings of the Old Testament. Scripture, in the singular, usually refers to the collection of Biblical writings thought of as a whole. The term scriptures is also applied to the sacred writings of other religions, such as the KORAN of Islam.

Scrofula, general name used variously in the history of medicine, especially to describe TUBERCULOSIS of the bones and lymphatic glands. It was also known, in popular tradition, as KING'S EVIL.

Scrope, Richard (c.1350–1405), English prelate. He was appointed Bishop of Coventry and Lichfield in 1386 and Archbishop of York in 1398. For leading a rebellion north of the Tyne with the Earl of Northumberland in 1404 he was executed at York.

Scrotum, in male human beings and most other mammals, a thin external sac of skin, divided into two compartments, each of which contains one of the two TESTES. It is located in the pelvic region behind the PENIS. S¡F¡ee also MS p.74.

Scrub bird, either of two species of small, rare, brown birds of Australia; together they comprise the genus *Atrichornis.* They rarely fly, but run swiftly and have a loud call of various notes. The reddish-hued scrub bird (*A. rufescens*), was discovered in the wet forests of New South Wales in the 1860s. Length: to 18cm (7in). The noisy, or western scrub bird (*A. clamosus*), of Western Australia was believed to have become extinct after 1889, but was rediscovered in 1961. It inhabits dense coastal scrub and bushlands. Length: 22cm (9in). Family Atrichornithidae. *See also* NW p.245.

Scuba diving, water sport. It consists of diving with the use of self-contained underwater breathing apparatus (SCUBA), known as an aqualung. Other equipment includes a wet suit and rubber fins for the diver's feet. The sport has attracted more than a million enthusiasts since the design of the modern aqualung by Jacques Yves COUSTEAU. *See also* PE p.92.

Scud. See AMPHIPOD.

Sculling, water sport which first became competitive in London in 1715. It is similar to ROWING except that the sculler uses two small OARS or sculls (whereas an oarsman rows with one oar only). There are main events for single, double and, for women, quadruple scullers. The most prestigious race is for the DIAMOND CHALLENGE SCULLS at the Henley Royal Regatta, established in 1844.

Sculpin, any of 300 species of bottom-dwelling usually marine fish found in temperate and cold waters of the N Atlantic Ocean. Greyish and mottled with yellow, it has a large bony head covered with prickles. Length: to 60cm (2ft). Family Cottidae.

Sculpture, the art of creating natural or abstract forms in three dimensions, either in the round or in relief. Techniques employed in sculpture include carving (in wood, stone, marble, ivory, etc), modelling (in clay, wax etc.), casting (in bronze and other metals). These materials, and others such as cement and plaster, have long been in use; in modern times artists have used plastics and any other material or object that has appealed to them.

The history of sculpture basically parallels that of PAINTING. The early civilizations of Egypt, Mesopotamia, India and the Far East were rich in sculpture, often of monolithic proportions and of deeply ritualistic significance. The Greeks developed a style of relief and free-standing sculpture, based on the idealized representation of the human form, that reached such perfection during the Classical and Hellenistic periods as to inspire European sculptors for many centuries. Roman sculptors were profoundly influenced by the Greeks, but forsook the Greek ideal in the field of portraiture, which they enriched with individual characterization.

Medieval European sculpture was frequently an integral feature of great ROMANESQUE and GOTHIC churches and cathedrals, many of which were covered externally and internally with thousands of figure and narrative carvings whose function was didactic as well as decorative. Highly stylized in the Romanesque period, this architectural sculpture grew in humanity and realism in the Gothic era, while still retaining perfect unity with its setting.

The sculptor emerged only rarely as an individual (the PISANOS in 13th-century Italy were exceptional), before the Renaissance. Then, however, Florence, its birthplace, was enriched by the works.of such masters as GHIBERTI, DONATELLO, the della ROBBIA and POLLAIUOLO families, CELLINI and MICHELANGELO, who enjoyed Church, civic and private patronage. Important Renaissance sculptors north of the Alps include the elder Peter VISCHER in Germany, and Jean GOUJON and Germain PILON in France, where secular art thrived under royal patronage. High BAROQUE sculpture at its most inventive and technically brilliant is exemplified in the works of the architect-sculptor BERNINI in Rome. PUGET was the movement's leading exponent in France where, in the 18th century, it was superseded by NEO-CLASSICISM. This extended into the 19th century, when it was rivalled and replaced

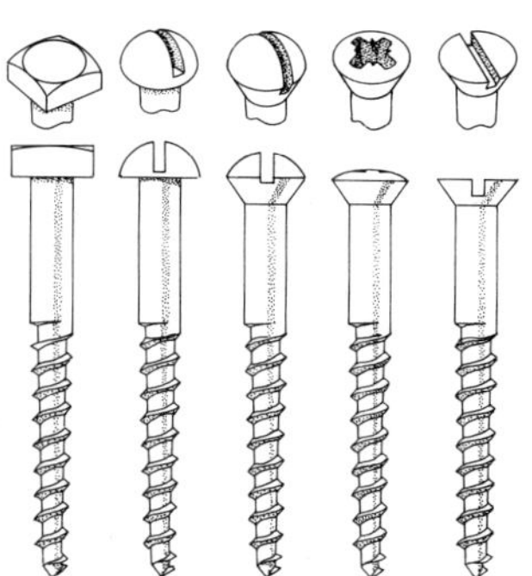
Screws vary in shape and size depending upon their particular purpose.

Screw worms: the larvae of these blowfly maggots often infest animals' wounds.

Alexander Scriabin first won fame as a pianist, touring the USA and Russia.

Scuba diving; named after the self-contained underwater breathing apparatus used.

Sculthorpe, Peter Joshua

Glenn Seaborg with his wife in Stockholm after being awarded the Nobel Prize.

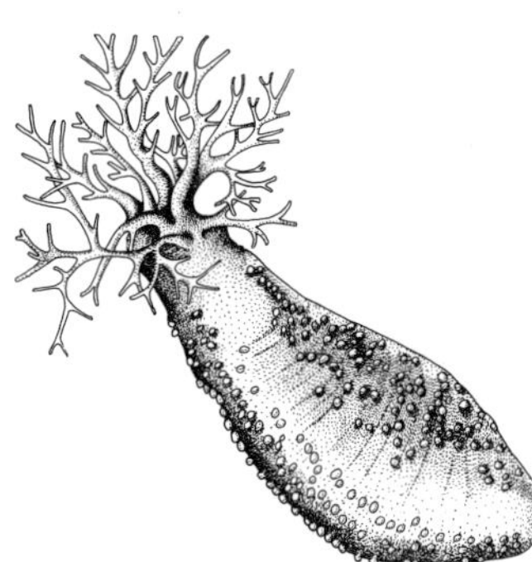

Sea cucumbers either burrow in the sand or lie on one side on the sea bottom.

The Seagull; a scene from a production by the Prose Theatre at Kiev, USSR.

Sea lions are among the few mammals that once lived on land but now inhabit the sea.

by a movement of Realist sculpture.

RODIN above all others led the way in freeing early 20th century sculpture from the dual yoke of Greek domination and slavish subservience to realism. Influences which contributed to this liberation were African, Aztec and other ethnic and ancient sculpture, whose direct and exuberant treatment of form and materials has stimulated great modern sculptors such as PICASSO, MODIGLIANI, Brancusi and Henry MOORE. Movement in sculpture, as used in mobiles and kinetic works, is a relatively recent feature, as is the practice of having large works of sculpture made and assembled in factories to the artist's specifications. *See also* HC1 and HC2.

Sculthorpe, Peter Joshua (1929–), Australian composer. The first public performance of his works was in 1945. His best-known work is the ballet *Sun Music* (1968).

Scumbling, in art, technique used in oil painting. A second layer of opaque paint is applied so that the under layer is not obliterated. The technique is generally associated with such artists as TINTORETTO, TITIAN, VELÁZQUEZ and HALS. In pen-drawing, scumbling refers to loose rambling scribbling or unforming cross-hatching, both of which are used to fill out various areas and to create tones.

Scurvy, disorder caused by a prolonged deficiency of VITAMIN C. It is characterized by weakness, inflamed gums, loose teeth and swollen joints and also absorption by tissues of blood from ruptured vessels, which causes ANAEMIA. *See also* MS pp.70–71.

Scurvy grass, plant of the MUSTARD family native to N Europe and Arctic regions. It has white flowers, and the stems and leaves have an unpleasant taste; these were used on early polar expeditions to treat scurvy. Species *Cochlearia officinalis.*

Scylla and Charybdis, two figures in Greek mythology. Scylla, beloved of Poseidon, had been changed by CIRCE into a long-necked, six-headed beast. Charybdis was changed into a violent whirlpool by the thunderbolt of ZEUS. These two creatures lived on each side of the Straits of Messina and devoured sailors.

Scyphozoa, class of COELENTERATE marine animals that includes the largest of the JELLYFISHES. Their life cycles include a larger jellyfish (MEDUSA) stage and a smaller POLYP, hydra-like stage, although either of these stages may be absent.

Scythia, ancient region, occupied by the Scythians from the 8th to the 3rd centuries BC, considered by the Greeks to be an ill-defined area north of the Black Sea. The Scythians originally came from Siberia.

Scythians, ancient people from central Asia who ruled an empire in S Russia from the 8th–7th centuries BC until they were confined to the Crimea by the SARMATIANS in the 2nd century BC. In the 7th century BC they temporarily extended their power into Mesopotamia, the Balkans and Greece. They were a nomadic tribe with an advanced civilization.

Sea, body of water, containing salt and other mineral deposits, that covers about 71% of the Earth's surface, and represents about 98% of all the water on the face of the Earth. The salt content of seawater, between 3.3% and 3.7% is the result of washout from the land and interchange with the atmosphere over the ages. Water vapour evaporated from the surfaces of the oceans provides the material of clouds. Surface layers of the oceans are normally saturated with atmospheric gases. Light penetrates seawater to a maximum depth of about 300m (1,000ft), below which plant life cannot grow. *See also* SEA-WATER; PE pp.76–77.

Sea anemone, sessile, polyp-type COELENTERATE animal found in marine tidal pools and along rocky shores. It has a cylindrical body with many tentacles around its mouth; the colour varies according to species. Height: 20cm (8in). Class Anthozoa; genera include *Tealia, Anemonia* and *Metridium. See also* NW pp.72, 82–83.

Sea bass, marine food fish found in tropi-

cal and some temperate waters. It includes white and striped BASS in the genus *Roccus,* the groupers, and the black bass *Centropristis striata.* Family Serranidae. *See also* PE p.246.

Sea-bed, floor of the oceans, which extends from the shallow CONTINENTAL SHELF regions to abyssal depths approaching 11km (7 miles) deep in some regions of the PACIFIC OCEAN. The geological composition of the abyssal sea-bed is SIMA, the material that also underlies the SIAL of the continents. *See also* PE pp.82–93.

Seaborg, Glenn Theodore (1912–), US physicist. During WWII he worked on the development of the atomic bomb and was responsible for isolating uranium-233. For his work with the TRANSURANIC ELEMENTS and the discovery of the ACTINIDE ELEMENTS he shared the 1951 Nobel Prize in chemistry with Edwin MCMILLAN.

Sea bream, any of numerous marine, spiny-finned fishes, usually of the family Sparidae. They have deep bodies and massive jaws with heavy crushing teeth. *See also* PE pp.242, 246, 247.

Sea butterfly, any of several species of small, marine gastropod molluscs that swim by undulating a flat elongation of the foot. It lives at or near the surface of the water, feeding on small animals. The shell, if present, is translucent and delicate. Length: to 1cm (0.4in). Order (with shell) Thecosomata; (without shell) Gynanosomata. *See also* NW p.91.

Sea cow. *See* DUGONG.

Sea cucumber, marine ECHINODERM found in rocky areas. It is a cylindrical animal with a soft, fleshy body that can be divided into five parts around a central axis; it has branched tentacles around the mouth. Edible species are smoked and dried; the end product is called bêche-de-mer. Species include the cotton-spinners (*Holothuria* spp). Class Holothuroidea. *See also* NW pp.118–119, 241.

Sea dragon, marine fish of coastal waters of Australia. It is a PIPEFISH, closely related to the SEA-HORSE, and its reddish-brown body is covered with leafy and spine-like projections, making it resemble floating seaweed. Family Syngnathidae.

Sea eagle. *See* EAGLE.

Sea fan, or gorgonian, colonial COELENTERATE animal found in coral reefs in tropical marine waters. The colonies are branching but flat, and eight-tentacled POLYPS live in tiny pits along the horny branches. Class Anthozoa; species *Eunicella verrucosa.*

Seafloor spreading, part of the theory of PLATE TECTONICS that covers the creation of new lithospheric plate material (rocky material of the Earth's crust) at the mid-ocean ridges and its destruction in the deep ocean troughs. The proof of the theory lies in the presence of bands of reversed magnetic polarity in the ocean crust and the increase in age and thickness of sediments away from the ridges. *See also* PE pp.23–25, 24–25, 52, 91.

Seafood, any of numerous edible products of the sea including FISH and SHELLFISH; it includes species of CRAB, LOBSTER, PRAWN, SHRIMP, CLAM, ABALONE, OYSTER, WHELK, WINKLE, COCKLE, SQUID and OCTOPUS. *See also* PE pp.248–249.

Sea gooseberry. *See* COMB JELLY.

Seagram Company, world's largest producer of wine and spirits, founded in Canada in 1924. In 1977 its total sales exceeded $2,000 million. Seagram's head offices in New York, designed by Mies van der Rohe, are a major example of skyscraper building.

Seagull. *See* GULL.

Seagull, The (1896), four-act drama by Anton CHEKHOV. The first of his four plays, it is typical of his tragi-comic depiction of yearning and futility, here involving the love of two men for the same woman. *See also* HC2 p.101.

Sea-horse, marine fish found in shallow tropical and temperate waters. It swims in an upright position and has an outer bony skeleton of platelike rings, a mouth at the end of a long snout, and a curled, prehensile tail with which it clings to seaweed. The male incubates the young in a brood pouch. Length: 3.8–30.5cm (1.5–12in). Family Syngnathidae. *See also* NW p.129.

Sea Island cotton, fine COTTON fabric that was developed in the Sea Islands, a chain of islands off the coasts of Florida, Georgia and South Carolina, USA. The cotton plant, *Gossypium barbadense,* has long staple fibres and is native to tropical America.

Sea kale, any of several species of fleshy, cabbage-like plants that grow in coastal regions of Eurasia. Common sea kale, *Crambe Maritina,* has bluish leaves and small clusters of white flowers. Its young leaves may be cooked and eaten, and have a bitter taste. Family Brassicaceae. *See also* PE p.189, 190.

Seal, any of several species of carnivorous, primarily marine, aquatic mammals. It feeds on fish, crustaceans and other marine animals; various species are hunted by man for meat, hides, oil and fur. Species of true, earless seals such as the leopard seal (*Hydrurga leptonyx*), hooded seal (*Cystophora cristata*) and bearded seal (*Erignathus barbatus*) are included in the family Phocidae. They swim with powerful strokes of their hind flippers and sinuous movements of the whole trunk, but are clumsy on land and move by wriggling. Members of the eared family Otariidae have longer fore flippers, used for propulsion, and use all four limbs when moving on land. They include fur seals (genera *Callorhinus* and *Arctocephalus*) and species of SEA LION. Order Pinnepedia. *See also* NW pp.33, 155, 224–225, 239, 242.

Seal, symbol of authenticity attached to an official or legal document. Seals were originally substitutes for signatures in the days of illiteracy. They have been maintained by custom and by statute. They may be single-sided or double-sided and may be pressed on to the document or left to hang from it. They were originally made of beeswax; they are now commonly plastic. There are three royal seals in Britain, the GREAT SEAL, the PRIVY SEAL and the Signet seal.

Seal, in engineering, liquid or gas-tight union, eg between mating pipe flanges, beneath bolted or clamped hatches and between moving parts of machines. Seals are made from rubber, plastics, fibres such as cotton and flax, and materials made from all these. *See also* MM pp.51, 53, 62, 87, 203.

Sealab, experimental programme conducted by the US navy to determine man's ability to live for long periods at the bottom of the sea. The aquanauts live in habitats placed on the ocean bed and spend some hours outside them working on the ocean floor. *See also* PE p.93.

Sea lavender, widespread plant of coastal areas, particularly in the Near East and parts of Europe. It is a stiff, branching plant with sprays of tiny white, pink, yellow or lavender flowers, and broad basal leaves. Height: 30–61cm (1–2ft). Species include *Limonium vulgare.* Family Plumbaginaceae. *See also* NW p.234.

Sea lettuce, any of various green SEAWEEDS (genus *Ulva*) with edible, crinkly fronds. They resemble the lettuces used in salads. *See also* GREEN ALGAE; NW p.46.

Sea-level, level from which topographic heights are measured. It is usually the mean sea-level and is reckoned on being the average of regular sea-levels taken over a long period of time. It may also be referred to as the Ordnance Datum (OD). *See also* TIDE.

Sea lily, CRINOID ECHINODERM animal found in deep marine waters. Seldom seen, it has many branched arms with ciliated grooves for food-collecting radiating from a tiny body disc. It is spineless and attaches itself to the ocean bottom with a stalk. Class Crinoidea. *See also* NW p.118.

Sea lion, any of five species of SEALS that live in coastal waters of the Pacific and feed primarily on fish and squid. It has a streamlined body and long fore flippers for propulsion. The males of all species, except the California sea lion (*Zalophus californianus*), have manes. The largest species is the Steller sea lion (*Eumetopias jubata*); it may reach 3.3m (11ft) in length. Order Pinnepedia; family Otariidae. *See also* NW pp.77, 165.

HC1 = History and Culture Vol. 1   HC2 = History and Culture Vol. 2   MS = Man and Society   MM = Man and Machines

Sealyham terrier, working dog, with a long head with pronounced whiskers and rounded ears. The deep-chested body is set on short, strong legs and the tail is commonly docked. The hard, wiry top coat is generally white. Height: to 30.5cm (12in) at the shoulder; weight: to 9kg (20lb).

Sea mat. *See* MOSS ANIMALS.

Seami Motokiyo. *See* ZEAMI MOTOKIYO.

Sea mouse, large marine ANNELID WORM with matted bristles on the sides and top of the body. The common sea mouse *Aphrodita aculeata* lives in sandy mud on both sides of the North Atlantic. Length: 15cm (6in). *See also* NW pp.*88, 236*.

Sea otter, marine otter that was once widespread along the Pacific coast of the USA. Now a protected animal, it was almost exterminated for its valuable dark brown fur. Its hind feet are flattened into flippers and it swims and floats on its back in kelp beds, usually close to the shore. It eats fish and invertebrates and often uses stones to open hard shells. Length: to 1.2m (4ft); weight: to 39kg (20lb). Family Mustelidae; species *Enhydra lutris*. *See also* NW p.*242*.

Sea pen, any of several species of marine animals that are related to sea anemones and corals and live in feather-shaped colonies; it resembles a quill pen. Order Pennatulaceae; genera include *Stylatula* and *Funiculina*.

Sea perch, or surf perch, marine fish found in temperate Pacific waters. Often quite colourful, its young are born alive. Length: 45.7cm (18in). Family Embiotocidae; species include *Amphistichus argenteus*.

Seaplane, aeroplane that lands on and takes off from water using floats or pontoons that are shaped to offer minimum water resistance. The first seaplane was built by Glenn H. Curtiss in the USA, and flown in 1911. In the 1930s flying boats with boat-shaped hulls as well as wing pontoons carried more passengers than any other aircraft. They have often been used in air/sea rescue operations. *See also* MM p.*111*.

Search warrant, order authorizing a person to enter premises and seize specified goods, usually stolen ones. In Britain, the order may be given by a JUSTICE OF THE PEACE or by a police officer of at least the rank of superintendent (by the latter, however, only if the person suspected of harbouring stolen goods has been convicted of a crime relating to dishonesty in the previous five years).

Searle, Humphrey (1915–), British composer and pupil of Anton von WEBERN in Vienna (1937–38). His works include *Gold Coast Customs* (1949) and *The Shadow of Cain* (1952) both for speaker, men's chorus and orchestra with texts by Edith Sitwell, as well as a ballet, *Noctambles* (1956), the opera *The Photo of the Colonel* (1964), and *Cello Fantasia* (1972).

Searle, Ronald (1920–), British cartoonist and artist whose cartoons of the outrageously naughty girls of St Trinian's School were the basis for a successful series of films. His work has appeared in satirical magazines such as *Punch* and the *New Yorker*. He attracted attention as a serious artist for his drawings made when a prisoner of the Japanese during WWII.

Sea robin, also known as gurnard, marine fish found throughout the world in tropical and temperate waters. Edible but not popular, it has a bony head, can produce sound, and "walks" on the sea bottom with its large pectoral fins. It is red, grey, yellow or green with dark blotches. Length: to 70cm (28in). Family Triglidae.

Sea serpent, legendary and mythological marine creature that appears in the literature of most ancient cultures. Almost all alleged sightings and carcases have proved to be hoaxes or known marine species. In a very few cases, such as that of the LOCH NESS MONSTER, the results have been inconclusive.

Seasickness, severe form of travel sickness brought on by the rolling and pitching movements of seagoing boats and ships. Many people who are seasick initially in rough weather become accustomed to the ship's motion and lose their discomfort.

Others remain dizzy and nauseated for the duration of a voyage. Anti-seasickness tablets often contain scopolamine or other ALKALOID preparations which desensitize nerves, including those of the Automatic nervous system.

Sea slug, also called nudibranch, any of numerous species of marine gastropod molluscs, related to snails and found throughout the world. They have no shells, quills or mantle cavities, frequent shallow water and feed primarily on sea anemones. Order Nudibranchia. *See also* NW pp.91, *96, 237*.

Sea snail, also called snailfish, any of several species of small marine fish. It has a soft, elongated, scaleless body with a long dorsal fin. Most species are bottom-dwellers, either in shallow coastal waters or deep seas. Length: to 30cm (12in). Family Liparidae. *See also* NW p.241.

Sea snake, any of about 50 species of colourful, highly poisonous marine snakes found in coastal waters of the Indian and Pacific oceans. Length: to 1.2m (4ft). Family Hydrophiidae.

Seasons, four astronomical and climatic periods of the year based on differential solar heating of the Earth as it makes its annual revolution of the Sun. Due to the parallelism of the Earth's axis of rotation, pointed near the Pole Star throughout the year, the Northern Hemisphere receives more solar radiation when its pole is aimed toward the Sun in summer and less in winter when it is aimed away, whereas the opposite holds for the Southern Hemisphere. The seasons are conventionally initiated at the vernal (spring) and autumnal EQUINOXES and the winter and summer SOLSTICES. *See also* PE pp.72–73.

Sea spider, also called Pycnogonid, any of numerous species of small, bottom-dwelling, marine arthropods. It has four to seven pairs of long, slender legs and generally feeds on juices, which it sucks from small marine invertebrates. Length of body: 0.3–50cm (0.1–20in). Class Pycnogonida. *See also* NW p.*104*.

Sea squirt, any of numerous species of small, sac-like, marine animals; it has no true backbone. A sedentary creature, it constantly draws in and discharges water for food and oxygen, and can shoot out water when disturbed. Class Ascidiacea. *See also* NW pp.73, *122, 123*.

Sea star. *See* STARFISH.

SEATO (Southeast Asia Treaty Organization), regional defence agreement signed by Australia, New Zealand, France, Pakistan, the Philippines, Thailand, Britain and the USA in Manila in 1954. It was formed in response to Communist expansion in Southeast Asia. With administrative headquarters in Bangkok, SEATO had no standing forces but combined military exercises were undertaken. The final exercise was held in 1976.

Seaton, John Colborne, Baron of (1778–1863), British general and colonial administrator. Lieutenant-governor of Upper Canada when the Canadian Rebellion of 1839 broke out, he was ordered to assume the office of Governor-general of British North America and the revolt was quickly suppressed. He later served as commander of forces in Ireland.

Sea trout, any of several species of marine fish, related to the Atlantic SALMON (*Salmo salar*); it is commonly confused with the salmon or TROUT. Although it spawns inland, it never leaves the waters around the river mouth. Family Sciaenidae; genus *Cynoscion*.

Seattle, city and port of entry in W Washington, USA between Lake Washington and Puget Sound; the largest city in the Pacific North-West. Settled in 1851, it developed with the construction of the railway and the Alaska Gold Rush (1897). The city has two universities. Seattle is an important shipping point for Alaska and the Far East. Industries: aerospace, shipbuilding, precision instruments, building materials, chemicals, timber, food processing, fishing. Pop. (1973) 503,073.

Sea turtle. *See* TURTLE.

Sea-urchin, spiny ECHINODERM animal found in marine tidal pools along rocky shores. Round with long, radiating (often poisonous) moveable spines, its skeletal

plates fuse to form a perforated shell. Class Echinoidae. *See also* NW pp.118–119; PE p.249.

Sea walnut. *See* COMB JELLY.

Sea wasp. *See* COMB JELLY.

Sea-water, solution of several minerals in the SEA. It has a high concentration of microscopic life and tastes salty. The dissolved salt content varies between 3.3 and 3.7%, most of which is common salt (sodium chloride, NaCl) but nearly all elements are found in seawater, including vast reserves of magnesium, potassium and calcium. Large amounts of nitrogen, oxygen and carbon dioxide are present as dissolved gases. The freezing point of sea water of salt content 3.5% is -1.9°C (28.6°F). *See also* PE pp.76–77.

Seaweed, any of numerous species of brown, green or red ALGAE, found in greatest profusion in shallow waters on rocky coasts. KELP is the largest species, often exceeding 33m (100ft) in length. Many species are commercially important for the manufacture of fertilizers or food, or as a valuable source of chemicals such as iodine. *See also* NW pp.*38, 47*; PE pp.*134*, 191.

Sebaceous gland, skin gland, often occurring along the walls of hair follicles. It secretes SEBUM, a waxy substance containing fats, and waste products from the EPITHELIUM. *See also* MS p.60.

Sebastian, Saint (*d.* AD 288), Roman Christian martyr. Little is known of his life but he was a favourite of the emperor Diocletian who, on learning that Sebastian was a Christian, had him put to death.

Sebastian (1554–78), King of Portugal (1557–78), grandson and successor of John III. Fanatically religious, he was killed during an unsuccessful crusade against the Muslims in Morocco.

Sebastiano del Piombo (Sebastiano Lucian) (*c.*1485–1547), Italian painter of the VENETIAN SCHOOL who went to Rome in 1511 to work at the Villa Farnese, where he was influenced by RAPHAEL.

Sebastopol, port in the SW of the Crimean Peninsula, USSR. A heavily fortified naval base of the Russian Black Sea fleet during the CRIMEAN WAR, it was beseiged in Oct. 1854 by 60,000 British, French and, later, Piedmontese troops. Despite heavy Allied casualties the siege ended with the Russians' evacuation in Sept. 1855.

Seborrhea, disorder of the SEBACEOUS GLANDS, characterized by an increased and altered secretion of fatty matter that results in oily or scaly skin. It may cause baldness and dandruff.

Secant, in TRIGONOMETRY, ratio of the length of the hypotenuse to the length of the side adjacent to an acute angle in a right-angled triangle. The secant of angle A is usually abbreviated to "sec A", and is equal to the reciprocal of its COSINE.

Secchi, Pietro Angelo (1818–78), Italian astronomer who was the first to use photography and SPECTROSCOPY in astronomy. In 1851 he photographed a solar eclipse and between 1864 and 1868 he examined the spectra of 4,000 stars, classifying them into spectral types. His system formed the basis of the classification used today. *See also* SU pp.*218*, 226.

Secession, formal separation from an organized body. In 1860–61 eleven states (South Carolina, Mississippi, Florida, Alabama, Georgia, Louisiana, Texas, Virginia, Arkansas, Tennessee and North Carolina) seceded from the USA and formed the CONFEDERATE STATES of America. *See also* HC2 pp.150–151.

Secondary deviance, term used by sociologists to describe the behaviour of a person who reacts to being labelled deviant for some initial action (primary deviance) by the adoption of even more deviant behaviour.

Secondary emission, in physics, emission of ELEMENTARY PARTICLES by a target, such as a gas or metal surface, that is bombarded with primary (newly generated) radiation or particles. Secondary COSMIC RAYS are emitted when primary cosmic rays bombard the atmosphere. The anode of an ELECTRON TUBE emits secondary electrons when it is bombarded by primary electrons from the cathode. This principle is employed in electron-

Sealyham terriers were originally bred in the 19th century as hunting dogs.

Ronald Searle caricaturing a schoolgirl from the film *The Belles of St Trinians*.

Sea-urchin: the test or skeleton (left) is made up of plates and covered with spines.

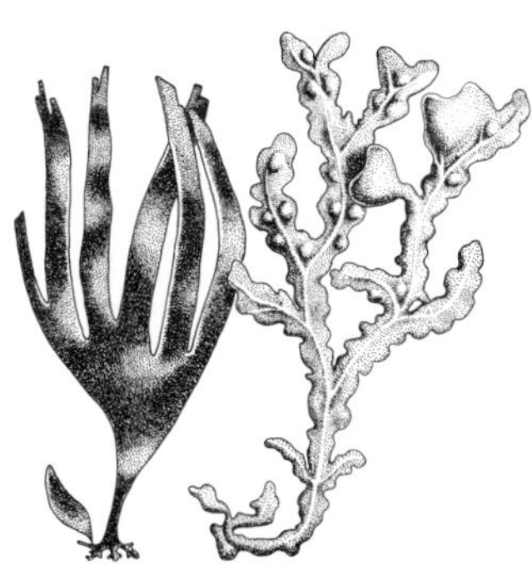

Seaweed; one variety has berry-shaped air bladders, enabling it to float on water.

Second Coming

Secretary birds are long-legged like storks but have characteristic crests.

Security Council; the UN Security Council during one of its first sessions in 1946.

Sedan chairs; a caricature, published in 1796, ridiculing the fashions of that time.

Andrés Segovia, the classical guitarist, with his three-day-old son, Carlos, in 1970.

multiplying devices, eg an image intensifier.

Second Coming, religious term referring to the return in the future of Christ. Christians believe this will end the present world order and involve the judgement of the living and the dead.

Second Empire (1852–70), French government set up by NAPOLEON III after staging a coup against the SECOND REPUBLIC in Dec. 1851. the empire collapsed when France was easily defeated by the Prussians in 1870, and the THIRD REPUBLIC was set up.

Second Mrs Tanqueray, The (1893), four-act drama by Arthur Wing PINERO, the story of a woman's struggle with Victorian ideas of respectability.

Second Republic (1848–52), government of France. After popular insurrection brought the abdication of Louis Philippe, the republic was set up in early 1848. The constitution of 1848 provided for election by male suffrage of a president and assembly; the elections were mainly won by radical or conservative deputies, and Louis Napoleon was elected president of the republic in Dec. 1848. He eventually staged a coup in Dec. 1851 and set up the SECOND EMPIRE in 1852, making himself NAPOLEON III.

Second Vatican Council, (1962–65), 21st Roman Catholic ecumenical council convened by Pope JOHN XXIII. Its purpose was to renew the Church's religious life, bring its discipline, teaching and organization up to date and unify all Christians. The first session (Oct.–Dec. 1962), presided over by Pope John, issued a *Message to Humanity* but reached few definite conclusions. Among the pronouncements of the other three sessions (1963, 1964 and 1965), presided over by Pope PAUL VI, were many concerning the ECUMENICAL MOVEMENT. The Council was closed on 8 Dec. 1965.

Second World War. *See* WORLD WAR II.

Secret Agent, The (1907), novel by Joseph CONRAD. It tells the story of Verloc (the secret agent) among anarchists and in his home life, and gives a particularly nihilistic picture of the political activities it describes. It was filmed by Alfred Hitchcock in 1936.

Secretary bird, bird of prey found in Africa, s of the Sahara. It is pale grey with black markings and has quill-like feathers behind its ears, large wings, and long legs and tail. It feeds on reptiles, birds' eggs, and insects and lays its eggs in a tree nest. Height: 1.2m (4ft). Species *Sagittarius serpentarius*.

Secretary of State, name for several government officers in Britain. The office dates from the 14th century, when the king's secretary was keeper of the SIGNET and a leading adviser on matters of state. From 1540 there were two secretaries. There are now secretaries of state for many departments. In the USA the name is given to the foreign secretary.

Secretin, polypeptide HORMONE secreted by the small intestine. It stimulates the release of pancreatic juice and also stimulates bile secretion by the GALL BLADDER.

Secret Life of Walter Mitty, The (1945), short story by James THURBER. It tells how Mitty escapes from his problems through daydreams, in which he always plays the hero. A film version of the story, starring Danny KAYE was made in 1947.

Secrets Acts, Official, series of acts passed by the British Parliament in the want of a remedy at COMMON LAW to make illegal the breach of official confidence. They were passed in 1889, 1911, 1920 and 1939.

Secret Service, government intelligence organization maintained in most countries to obtain information concerning activities which might endanger the state. In Britain it is divided into two main parts: MI5 (Military Intelligence 5), which is responsible for counter-espionage in Britain and is subject to the HOME OFFICE which in turn reports to the Prime Minister; MI6, responsible for overseas intelligence and answerable to the FOREIGN OFFICE. In the USA the Secret Service is a branch of the Treasury Department. *See also* CIA; FBI; KGB.

Security Council, UNITED NATIONS council

responsible for preserving international peace and security. It consists of five permanent members and ten non-permanent members, elected for two-year terms by the UN General Assembly. The permanent members are China, France, Britain, the USSR and the USA. The Security Council investigates disputes threatening world peace and applies pressure such as SANCTIONS on countries which violate international law or codes of practice. Each permanent member can veto an initiative of the rest of the council. *See also* HC2 pp.298–299.

Sedan, town in NE France, on the River Meuse. It became part of the French Crown lands in 1642 and was a Protestant stronghold in the 16th and 17th centuries. The town was the scene of the defeat of the French in 1870 during the FRANCO-PRUSSIAN WAR and the surrender of Napoleon III. There was heavy fighting there during WWI, and in WWII it was the point of the initial breakthrough of German forces in the invasion of France. Industries: brewing, textiles, metal goods. Pop. 23,000.

Sedan chair, mode of transport common among the upper classes in 17th- and 18th-century Europe and the USA; it was probably introduced into England from Naples. It had poles projecting at each end, by which it was carried by animals or, more usually, two or four servants.

Sedative, drug used to reduce excitement, induce sleep or control convulsions. Sedatives are classified as BARBITURATES and non-barbiturates, the latter including bromides and chloral compounds.

Seddon, Richard John (1845–1906), New Zealand politician known as "King Dick". Emigrating from Britain to Australia and then New Zealand as a gold prospector, he became prominent in local politics and was Prime Minister from 1893 to 1906. He laid down the basis of the social welfare system and in 1893 New Zealand women became the first in the world to have the vote. *See also* HC2 p.*127.*

Sedentary rock, SEDIMENTARY ROCK formed at the place where the rock is found, rather than deposited from material brought from elsewhere. Reef limestone is an example.

Seder, meaning "order" in HEBREW, a religious meal held on the first night of PASSOVER in Jewish homes and returned Jews in Israel, and on the first two nights outside Israel. The Haggadah is recited and the ancient temple service, based on the MISHNA, is commemorated.

Sedge, any of numerous species of grasslike perennial plants widely distributed in temperate, cold and tropical mountain regions, usually on wet ground or in water. Cultivated as ornamental plants, they have flat leaves and spikes of flowers. Family Cyperaceae; genus *Carex*.

Sedgefield, county district in E DURHAM, England; created in 1974 under the Local Government Act (1972). Area: 220sq km (85sq miles). Pop. (1974 est.) 90,400.

Sedgemoor, Battle of (1685), victory of the army of JAMES II of England over the Duke of Monmouth's forces. It took place on 6 July on the plain of Sedgemoor, near Bridgwater, Somerset. The royal army, led by the Earl of Feversham and Lord Churchill, destroyed Monmouth's army and ended his rebellion. *See also* MONMOUTH'S REBELLION.

Sedgwick, Adam, (1785–1873), British geologist. He was appointed professor of geology at Cambridge in 1818 and is famed for his work on the Lower PALAEOZOIC rocks, and the CAMBRIAN PERIOD rocks which he named.

Sediment, in geology, a general term used to describe any material in suspension in air or water; the total load transported by a stream, including materials moved along its bed and those that are in solution as well as sediment in suspension; and any unconsolidated sand and gravel deposit in river valleys and along coastlines. *See also* index to *The Physical Earth.*

Sedimentary rock, rock formed of mineral or organic particles that have been moved by the action of water, wind or glacial ice (or have been chemically precipitated from solution) to a new location. Follow-

ing a process of compaction and cementation the particles form strata of sedimentary rock. *See also* index to *The Physical Earth.*

Sedition, in English law, the misdemeanour of publishing orally ("seditious words") or in print ("seditious libel") anything with the object of exciting disaffection. The main Act is the Act of 1819.

Sedum, genus of succulent perennial plants that grow in dry areas throughout the world. They bear clusters of tiny yellow, pink or blue flowers. Among the numerous species are *Sedum acre*, the yellow STONECROP, with small triangular leaves and yellow star-shaped flowers. Height: to 38cm (15in). Family Crassulaceae.

See, word synonymous with DIOCESE, referring to the geographic area over which a bishop has ecclesiastical jurisdiction. Properly the official seat or throne of a bishop, it normally stands in the cathedral of the diocese.

Seebeck, Thomas Johann (1770–1831), German physicist, *b.* Russia, who discovered the SEEBECK EFFECT in 1821. His discovery was neglected for many years but is widely used today.

Seebeck effect, thermoelectric effect important in THERMOCOUPLES. If wires of two different metals are joined at their ends to form a circuit, a current flows if the junctions are maintained at different temperatures. It is the opposite effect to the PELTIER EFFECT.

Seed, part of a flowering plant that contains the embryo and food store. It is formed in the ovary by fertilization of the female GAMETE by sperm nuclei. The seed coating may be thin as in the GROUNDNUT or hard as in the Brazil nut. The embryo consists of hypocotyl, one or more cotyledons, a growing shoot and root. In some seeds, a nutrient tissue to endosperm, formed by the parent plant, surrounds the seed, as in cereals. In others, such as the bean, food is stored in the embryo itself. *See also* GERMINATION; PE pp.192, 210–211.

Seed drill. *See* TULL, JETHRO.

Seeler, Uwe (1936–), West German footballer. A stocky centre-forward and, later in his career, midfield player, he won a record 72 caps for West Germany (1954–70) and had the distinction of scoring in four World Cups (1958–70).

Seferis, George (1900–71), pseud. of Georgios Seferiadis, Greek poet and diplomat. Educated in Paris, he had a distinguished diplomatic career and was Greek ambassador to Britain from 1957 to 1962. His volumes of lyrical poetry include *The Turning Point* (1931) and *Mythistorema* (1935). In 1963 he became the first Greek to be awarded the Nobel Prize in literature.

Sefton, county district in NW MERSEYSIDE, England; created in 1974 under the Local Government Act (1972). Area: 147sq km (57sq miles). Pop. (1974 est.) 307,200.

Seghers, Hercules Pietersz (c.1589–c.1638), Dutch landscape painter and etcher who usually painted mountain scenes. His conception of landscape was romantic and imaginative, but his rendering realistic and detailed. He had a great influence on early 17th-century Dutch painting.

Segonzac, André Dunoyer de (1884–1974), French engraver and painter. His work is essentially IMPRESSIONIST in style, although it does show some influence of CUBISM. From the 1920s he concentrated on engravings, of which he produced about 1,500, and book illustrations.

Ségou, town and port in s Mali on the River Niger. It was part of the Bambara Kingdom until 1861 when it was captured by a militant Muslim reformer, Al-haji Umar. Ségou was occupied by the French in 1890. It is now the market centre for the surrounding agricultural region producing cotton, cereals and peanuts. Pop. (latest est.) 30,000.

Segovia, Andrés (1893–), Spanish guitarist who successfully established the guitar as a concert instrument. He created many new techniques of GUITAR playing, adapting the instrument to the complex compositions of modern composers. His transcriptions of early contrapuntal music for solo guitar have contributed to the resurgence of interest in the instrument. Composers such as Manuel de FALLA and Heitor VILLA-LOBOS wrote works specially for him.

Segrave, Sir Henry O'Neal de Hane (1896–1930), British motor-racing driver. The first Briton to win a grand prix (the French in 1923), he set a world land speed record of 245 km/h (152.33 mph) in 1926 and went on to raise the land speed record to 372.38 km/h (231.44 mph) in 1929 driving the *Golden Arrow*. He was killed while attempting a record of 193 km/h (120 mph) on Lake Windermere, having just set a new water speed record of 158.91 km/h (98.76 mph).

Segrè, Emilio Gino (1905–), US physicist, *b*. Italy, who shared the 1959 Nobel Prize in physics with Owen CHAMBERLAIN for the discovery of the antiproton. In 1937 Segrè discovered TECHNETIUM, the first element to be artificially produced, and in 1940 he helped to discover another element, ASTATINE. Paul DIRAC had predicted the existence of antiparticles in 1930 and in 1955 Segrè, in collaboration with Chamberlain, produced and identified antiprotons using a bevatron particle ACCELERATOR to bombard copper with high-energy protons. *See also* ANTIMATTER.

Segregation. *See* APARTHEID.

Seguidilla, formal Spanish folkdance performed with light, springing steps; also a verse form that was widely used in Spanish folksongs.

Seiber, Mátyás (1905–1960), British composer, *b*. Hungary. His works include three string quartets, piano music, songs and choral works, including the cantata *Ulysses* (1946–47). His early music shows the influence of Zoltan KODÁLY, with whom he studied. Jazz elements are evident in the inventive rhythms he employed, and he also experimented with the TWELVE-TONE technique of Arnold SCHOENBERG.

Seiche, in oceanography, tide-like rise and fall of water in lakes, semi-enclosed bays and harbours. Sudden changes in barometric pressure, tides or unusual winds can cause the water in such basins to oscillate and tilt back and forth. In harbours, seiches are also known as surges.

Seidler, Harry (1923–), Australian architect, *b*. Austria, who underwent his training at Harvard under Walter GROPIUS. In 1948 he established a firm of architects in Australia. Among his works there which won him international acclaim is the 50-storey circular office block, Australia Square (1970) in Sydney.

Seine, river in N central France. It rises in Langres Plateau and flows NW through Paris to enter the English Channel near Le Havre. It is connected to the rivers Loire, Rhône, Meuse, Schelde, Saône, and Somme by canals. With its tributaries, the Aube, Marne, Oise, Yonne, Loing and Eure, the Seine drains the entire Paris Basin. The most important river of N France, it is navigable for ocean-going vessels as far as Rouen, 560km (350 miles) of its course. Length: 776km (482 miles).

Sei Shonagon (*c*. 967–*c*. 1013), Japanese poet and diarist. Residing in the court of the Empress Sadako from 991, she kept a diary from 991 to 1000 which is noted for its excellent style.

Seismic profile, continuous record of sound waves bounced off sea-bottom sediments. The sounds become SEISMIC WAVES and as such are used to determine the thickness and structure of bottom sediments. *See also* ECHO SOUNDER; PE p.91.

Seismic survey, use of SEISMIC WAVES produced by underground explosions to cause shock waves through the ground so that minerals (eg oil and coal) can be detected and their extent measured. The speed of the shock waves indicates the type of material they are passing through. *See also* GRAVITY SURVEY; PE p.138.

Seismic waves, shock waves produced by EARTHQUAKES. Their velocity (speed and direction) vary according to the material

(type of rock, molten core or oil through which they pass). Primary (P) and secondary (S) waves are transmitted by the solid Earth. P waves vibrate in the direction that they are advancing; S waves vibrate at right-angles to the direction in which they are advancing. Only P waves are transmitted through fluid (liquid or gas) zones. *See also* SEISMOLOGY; PE pp.28–29.

Seismograph, or seismometer, instrument for measuring and recording SEISMIC WAVES caused by movement (earthquake or explosion) in the Earth's crust. *See also* PE pp.*28–29*; SU p.*22*.

Sekhmet, in Egyptian mythology, wife of PTAH and mother of NEFERTUM. She is depicted as a lioness or a woman with the head of a lioness.

Sekondi-Takoradi, seaport city in SW Ghana, on the Gulf of Guinea, 177km (110 miles) WSW of Accra; the capital of the Western Region. The two settlements, Sekondi and Takoradi, were amalgamated in 1946. The city has shipbuilding and railway maintenance industries. Pop. (1970) 89,686.

Selaginella, or spike moss, genus of small-leaved, moss-like plants found throughout the world. Two kinds of spores are borne in cones at the branch tips. Some of the plants grow on trees, others on the ground. Height: 7.6cm (3in). Family Selaginellaceae. There are more than 700 species. *See also* CLUB MOSS; NW p.*52*.

Selangor, state in Malaysia, on the S Malay Peninsula, bounded by the Strait of Malacca (s and w); Shah Alam is the state capital. It was made a British protectorate in 1874, became a member of the Federated States of Malaya in 1895, part of independent Malaya in 1957 and a state of Malaysia in 1963. Products include tin, rubber, rice, pine-apples, coal and oil palms. Area 8,159sq km (3,150sq miles). Pop. (1970) 1,629,386.

Selassie, Haile. *See* HAILE SELASSIE.

Selborne. *See* NATURAL HISTORY AND ANTIQUITIES OF SELBORNE.

Selden, John (1584–1654), English jurist and scholar. Elected to parliament in 1623 he supported parliamentary sovereignty and was co-author of the 1628 PETITION OF RIGHT. One of the most scholarly men of his time, he wrote on a wide range of subjects. A record of his conversations was published posthumously as *Table Talk* (1689).

Selective breeding, process by which stockbreeders and agriculturalists improve the strains of domesticated animals and cultivated plants; it involves selection and pairing of individuals with desirable qualities in the PHENOTYPE. Today a growing knowledge of GENETICS ensures more predictable results.

Selene, in Greek mythology, a moon goddess, daughter of the TITAN, HYPERION, and a more ancient deity than ARTEMIS or DIANA. Each night after bathing she dressed in gleaming robes and drove her chariot across the sky, as her brother, HELIOS, had done during the day.

Selenium, METALLOID element (symbol Se) of group VIA of the periodic table, discovered in 1817 by J. J. BERZELIUS. Its chief source is a by-product in the electrolytic refining of copper. The element is photoactive (its conductivity is proportional to the light shining on it) and is extensively used in PHOTOELECTRIC CELLS, SOLAR CELLS and XEROGRAPHY. Properties: at.no. 34; at.wt. 78.96; sp.gr. 4.79 (grey); m.p. 217°C (423°F); b.p. 684.9°C (1,233°F); most common isotope Se^{80} (49.82%).

Selenology, study of the moon's physical and chemical composition and formative processes. Current research reveals that the Moon greatly resembles the Earth in physico-chemical makeup. Its surface is composed mainly of volcanic basalts and evidence has been found suggesting that the moon was formed at a high temperature along with the other planets of the SOLAR SYSTEM. *See also* SU pp.178–179.

Seleucid Empire, HELLENISTIC kingdom stretching in its heyday from the E Mediterranean to India, set up by the Greek general Seleucus I in 312 BC, and eventually conquered by the Romans in

64 BC. The kingdom, which was centred on Syria, lost most of its Anatolian territory as well as Parthia and Sogdiana in the 3rd century. The BACTRIANS gained their independence in *c*. 250 BC. The kingdom was an important channel for spreading Greek culture through the Middle East.

Seleucus, name of two kings of Syria. Seleucus I (*c*. 355–281 BC) was one of ALEXANDER THE GREAT's most valued generals. By 281 BC he had secured control of Babylonia, Syria and all of Asia Minor, but he was murdered before he could take the vacant throne of Macedonia. Seleucus II (*d*. 226 BC) spent his reign fighting Ptolemy III of Egypt and Antiochus Hierax, eventually losing territory to both.

Self, the part of the person's mind that is sentient as an individual, that allows the person to think individually and not merely react. In Freudian psychology the self is comprised of three parts: id, ego and super-ego. *See also* MS p.266.

Self-denying Ordinance, resolution passed by the English LONG PARLIAMENT in April 1645 which required every member of parliament to relinquish within 40 days any military or civil office he might hold. It was put forward by CROMWELL in order to rid the command of the parliamentary army of peers and was provoked by the indecisive second Battle of Newbury (1644).

Self determination, movement of people of the same race or political or cultural conciousness to establish their own government. It is usually associated with NATIONALISM and was implemented by peace treaties following WWI. It is also recognized by the articles of the UN. *See also* HC2 pp.184–185.

Self-fulfilling prophecy, in social psychology, the unconscious tendencies of individuals to make their behaviour or situation conform to expected or desired outcomes.

Selfridge, Harry Gordon (*c*. 1864–1947) British businessman, born in the USA, the founder of Selfridge's department store in London. He was a partner of Field, Leiter and Co., Chicago, before moving to Britain in 1906. The London store was opened in 1909, the first of its kind in Britain. Selfridge remained in charge of the company until 1939.

Selim, name of three Ottoman sultans. Selim I (1470–1520) also called the Grim or Inexorable came to power in 1512, defeated the SAFAVID dynasty in Persia and destroyed MAMLUK rule in Syria and Egypt, so opening the agriculturally rich Nile region to the Empire. Selim II (1524–74) who ruled from 1566, fought Venice and Spain, losing the battle of LEPANTO in 1571. Although Ottoman rule was later restored in the Mediterranean, the sultan's personal power was curbed by the JANISSARIES. Selim III (1761–1808) who reigned from 1789 to 1807 passed the Nizam-i Cedid (New Order), a series of social and administrative reforms. These and a number of military changes provoked a revolt in which Selim was deposed and later killed.

Seljuk Turks, semi-nomadic tribesmen who conquered Khorasan in the first half of the 11th century. They entered Baghdad in 1055 and their leader, Toghrïl, was proclaimed sultan. His successor, ALP ARSLAN, defeated the Byzantines at Manzikert in 1071, opening the BYZANTINE EMPIRE to Seljuk invasion. This defeat was one of the causes of the CRUSADES. In the early 12th century the Seljuk empire began to disintegrate and the Seljuk states were conquered by MONGOLS in the 13th century. *See also* HC1 pp.143, 187, *176*.

Selkirk, Alexander. *See* ROBINSON CRUSOE.

Selkirk, Thomas Douglas, 5th Earl of (1771–1820), Scottish colonizer. In 1803 he went to Canada and established a settlement on Prince Edward Island for emigrants from the Scottish Highlands. Taking control of the HUDSON'S BAY COMPANY in 1810, he founded the RED RIVER SETTLEMENT in 1812 in what became Manitoba.

Selkirk, former county town of SELKIRK(SHIRE), in SE Scotland, on the Ettrick River 48 km (30 miles) SE of Edinburgh. Since 1975 it has been part of the Borders region. It is a market town and the principal industries are the manufac-

Seine; part of the river valley, showing the tower of the cathedral of Mantes.

Seismic waves; a San Francisco street after the earthquake of 1906.

Seljuk Turks; the Emperor Romanus Diogenes captured by Alp Arslen in 1071.

Selkirk; a bridge at St Mary's Loch, a renowned beauty spot of the area.

David O. Selznick with Jennifer Jones, after filming *A Farewell to Arms* (1957).

Semarang, a general view of the Javanese port, showing warehouses and barges.

Seminole Indians on a reservation at Musa Island, near Green Gables in Florida.

Seneca: an illustrated translation of his work. His plays greatly influenced drama.

ture of woollens. Pop. (1974 est.) 5,600.

Selkirk Mountains, range in SE British Columbia, Canada; extension of the Rocky Mts. They are crossed by the Canadian Pacific Railway; the highest peak is Mt Sir Sanford, rising to 3,535m (11,591ft). Length: 322km (200 miles).

Selkirk(shire), former county in SE Scotland; since 1975 it has been part of Borders region. Part of the southern uplands, the county is mostly hilly, and is drained chiefly by the rivers Ettrick and Yarrow. The rearing of sheep and cattle are the chief occupations. The chief industry is the manufacture of woollens. The region was annexed to Scotland *c.*1020 from the Saxon kingdom of Northumbria. The county town was Selkirk. Area: 694sq km (268sq miles). Pop. (1971 est.) 20,680.

Sellick, Phyllis Doreen (1911–), British pianist who studied at the ROYAL ACADEMY OF MUSIC and was appointed a professor at the ROYAL COLLEGE OF MUSIC in 1964. She gave many performances in Britain and abroad with her husband Cyril SMITH in music for two pianos, including a work specially composed by VAUGHAN WILLIAMS, *Introduction and Fugue "For Phyllis and Cyril"* (1946).

Selwyn, George Augustus (1809–78), British missionary. He was the first bishop of New Zealand (1841) and was famous for his work among the MAORIS. Selwyn College, Cambridge, is named after him.

Selye, Hans (1907–), Canadian physician. Known for his pioneer studies of the effects of stress on human and animal physiology, he outlined the "general adaptation syndrome", a series of stages the body undergoes when subjected to stress. His works include *The Stress of Life* (1956) and *Stress in Health and Disease* (1976).

Selznick, David Oliver (1902–65), US film producer. He founded the independent Selznick International Pictures in 1936. His films included *A Star is Born* (1937), *Gone With The Wind* (1939) and *The Third Man* (1949). He also produced several films starring his wife, Jennifer Jones, including *Duel in The Sun* (1946). His final film was *A Farewell to Arms* (1957).

Semantics, study of words, their meanings and uses, and how they reflect reality. Semantics is a branch of both LINGUISTICS and of PHILOSOPHY in so far as it studies the connection between words and perceptions. *See also* MS p.239.

Semaphore, device or technique that communicates messages visually, a type of optical telegraph. A railway semaphore signal consists of a vertical post on which is mounted a single projecting arm whose angle indicates "all clear" or "danger". In flag semaphore a signaller indicates letters and numerals by the position of his outstretched arms, emphasized with flags. *See also* MS p.242; MM p.224.

Semarang, port in N Java, Indonesia, on the Java Sea; capital of Central Java province. The city was occupied by the Dutch in 1748 and the Japanese during WWII. Exports include kapok, rubber, coffee, sugar and tobacco. Industries: fishing, shipbuilding, textiles. Pop. (1971) 646,590.

Semele, figure in Greek mythology who bore Dionysus to ZEUS. On Semele's request Zeus appeared in divine form, but the fire of his thunderbolts killed her. Zeus seized the unborn Dionysus from her ashes and placed him in his thigh, from which he was later born.

Semele (1744), secular oratorio by George Frederick HANDEL, with text adapted from a libretto by William CONGREVE. The oratorio, a finely proportioned example of Handel's music genius, contains the air *Where'er you walk*.

Semen, liquid ejaculated by the male at his climax of sexual intercourse; it contains sperms and the secretions of various accessory sexual glands. In man the volume of each ejaculate is about 3 to 6ml and contains about 200 to 300 million sperms. *See also* MS pp. 74–75.

Semenov, Nikolai (1896–), Russian chemist. After studying at Leningrad he moved to Moscow State University. He shared the 1956 Nobel Prize in chemistry with C.N. HINSHELWOOD for his work on branched chain reactions in combustion processes.

Semicircular canals, three parts of the inner EAR which function as balance organs. They are tubular ducts which project in different planes and can detect, by the movement of a liquid within them, movement in any direction in space. *See also* MS pp.50–51.

Semi-conductivity, ability of a solid to carry electricity, at room temperature, more efficiently than an INSULATOR, but less efficiently than a CONDUCTOR. At higher temperatures the efficiency increases; at lower, it declines. Semiconductivity is a function of the limited movement of ELECTRONS, determined by the CRYSTAL structure of the solid. There are two types of semi-conductor: N-type, in which the current is carried by negatively-charged electrons, and P-type, in which it is carried by positively-charged holes. *See also* SU pp.128–129.

Seminary, term applied to colleges for the training of priests of the Roman Catholic Church. They were introduced by the Council of TRENT and first established in Italy, largely through the inspiration of S. Charles BORROMEO.

Seminole Indians, North American tribe that separated from the main CREEK INDIAN group in the late 18th century and fled s into Florida under pressure of wars with the white settlers. Bitter conflict with the US Army in the 1800s virtually annihilated them and a treaty in 1842 removed most of the Seminole to territory west of the Mississippi. About 5,000 inhabited Florida and Oklahoma in the mid-1970s.

Semi-Pelagianism, 17th-century term describing doctrines on human nature current in the 5th and 6th centuries. According to these the first steps towards spiritual salvation should ordinarily be taken by man on his own initiative, afterwards aided by divine GRACE. *See also* PELAGIANISM.

Semipermeable membrane, thin sheetlike material that permits the passage of a solvent (such as water) but not larger dissolved substances (such as salt and sugar). The property of permeability depends on the molecular diameter of the dissolved substance and the nature of the membrane. Common membranes include thin palladium foil, pig's bladder, copper ferrocyanide (cyanoferrate) and the walls of plant and animal cells. *See also* OSMOSIS.

Semi-precious stones, minerals and other hard substances worn as jewellery other than the precious stones diamond, emerald, sapphire, pearl and ruby. They include the minerals garnet, turquoise, tourmaline, amethyst, topaz, opal, aquamarine and many others, and the non-minerals, jet, coral and amber.

Semiramis, Greek form of Sammu-ramat, queen of ancient ASSYRIA. Historically she was the wife of Shamshi-Adad V (*r.*823–811 BC) and queen-regent for her son Adad-nirari III (*r.*810–783 BC. In Assyrian mythology she was the daughter of the fish goddess, Ataryatis, and the god of wisdom, Oannes. She is acclaimed as the builder of BABYLON.

Semites, peoples whose native tongue belongs to the SEMITIC LANGUAGES. They originally inhabited an area in Arabia and spoke a common language, Proto-Semitic, from which the Semitic languages descend. The migration of many Semites to Mesopotamia, the Mediterranean coast and the Nile Delta before 2500 BC led to the development of individual states whose inhabitants commonly traced their descent from Shem, Noah's eldest son. Among the modern Semites are Arabs, native Israelis and many Ethiopians.

Semitic languages, group of languages of peoples native to N Africa and the Near East is one of the five branches of the Hamito-Semitic language family. They have quasi-alphabets in which only the consonants are written, with vowels indicated by marks above or below them. Semitic languages are divided into four groups, which contain the modern Hebrew, Arabic, Ethiopic and Assyrian languages.

Semmelweiss, Ignaz Philipp (1818–65), Hungarian physician, probably the first to recognize the importance of ANTISEPTICS in preventing infection. In 1847 he ordered doctors under him at the University of Vienna hospital to wash their hands in strong chemicals before touching a patient – with the immediate result of a dramatic decrease in the incidence of childbed fever. *See also* MS pp.126–127.

Semolina, meal obtained from the grinding of hard WHEAT, the main ingredient in SPAGHETTI, macaroni and other PASTA.

Sempervivum, genus of garden plants grown as ornamental SUCCULENTS, also called houseleeks. Species include perennial herbs and small shrubs having rosettes of fleshy leaves bearing, very occasionally, red flowers. Height: to 30cm (1ft). Family Crassulaceae.

Sénancour, Étienne Pivert de (1770–1846), French writer. Although destined for the Church, he was influenced by the works of Jean-Jacques ROUSSEAU and took up writing. His most famous work was the epistolary novel *Obermann* (1804), which influenced the ROMANTICS.

Senate, Roman, governing body of the Roman republic, first convened during the monarchy as the king's council. By the 2nd century BC it controlled military, religious, financial, domestic and foreign policy. The senators were chosen for life by the CENSORS; they were originally numbered 300, but were 600 under SULLA, 900 under CAESAR and 600 again under AUGUSTUS. Patrician members had privileges of precedence over plebeian members; all enjoyed freedom of speech during the republic. The senate met in Rome to debate legislation; resolutions could be vetoed by the tribunes. After Sulla defeated MARIUS in 83 BC, the Senate was controlled by generals until Caesar defeated Pompey in 48 BC. During the Empire the senate's power was reduced to judicial functions and freedom of debate was lost. *See also* HC1 pp.92–93.

Senate of the United States of America, upper house of the US legislature. It is composed of two senators from each state, who are elected by the people and serve six-year terms. Senators are elected every other year, with about one third of the Senate elected at a time. There are usually 16 standing committees and committee chairmen retain their positions for as long as their party has a majority of the votes. The approval of a simple majority of the Senate is necessary for major presidential appointments and a two-thirds majority for treaties. The Senate can initiate legislation except on fiscal matters.

Sendai, city on NE Honshū, Japan; capital of Miyagi prefecture. It was the seat of the powerful daimyo of Date Masamune from the 17th to the 19th century. Today Sendai is the chief industrial and commercial centre of N Honshū. Tokohu Imperial University (1907) is in the city. Industries: textiles, wood products, brewing. Pop. (1974) 575,603.

Seneca the Elder, Lucius Annaeus (*c.*55 BC–*c.* AD 40), Roman rhetorician and writer. He lived most of his life in his native Spain. His *Controversies* contain model arguments for legal cases and *Persuasions* contains model orations.

Seneca, The Younger, Lucius Annaeus (*c.*4 BC–AD 65), Roman statesman and philosopher, *b.*Spain, the son of SENECA THE ELDER. Tutor to the young NERO, he was extremely influential in the first years of the emperor's reign. A STOIC, he wrote many philosophical essays, including *De Clementia* on the duty of a ruler to be merciful. He also made nine adaptations of Greek tragedies, the best known of which is *Phaedra*. When his influence over Nero waned, he was compelled to commit suicide. *See also* HC1 pp.113, *113*, 282.

Senefelder, Aloys (1771–1834), Bavarian inventor who discovered the process of LITHOGRAPHY (1798) and in 1818 wrote a book *The Complete Course of Lithography.* *See also* MM pp.204, *205.*

Senegal, republic in the extreme w of Africa. Formerly a French colony, it gained independence in 1960. The land is mostly low-lying and 75 per cent of the people are farmers. Ground nuts account for 80 per cent of exports. Fishing is an important industry and phosphates are mined and exported. The capital is Dakar. Area: 197,161sq km (76,124sq miles). Pop. (1976 est.) 5,085,400. *See also* MW p.153.

Sénégal, river in Senegal formed by the confluence of the Bafing Falémé and Bakoye rivers. It rises near the Sierra Leone border, flows N then NW to form the Mauritanian-Senegal border, and empties into the Atlantic Ocean at St Louis. The Sénégal irrigates a large rice-growing region. Length: approx. 1,633km (1,015 miles).

Senghor, Léopold Sédar (1906–), Senegalese politician and poet. He was President of the Federal Assembly in the Mali Federation of Senegal and the Sudan (1959–60) and was elected President in Senegal when his country became a republic in 1960.

Senility, serious decline in mental and physical abilities in old age. Symptoms in elderly people so afflicted include lack of attention, forgetfulness, delusions and incontinence.

Senlac, name given by Ordericus Vitalis in the 12th-century *Historia Ecclesiastica* to the site of the battle commonly known, since DOMESDAY BOOK, as the Battle of HASTINGS.

Senna, plants, shrubs and trees of the pea family native to warm and tropical regions; some species grow in temperate areas throughout the world. They have oblong, feathery leaves and yellow flowers. The dried leaflets yield a cathartic drug of the same name. Family Leguminosae; genus *Cassia*.

Sennacherib, King of Assyria (r.704–681 BC), son and successor of SARGON II. He led expeditions to subdue Phoenicia, Palestine and Philistia in 701, and defeated the Elamite-Chaldean alliance in 691, after which he destroyed Babylon (689). With the peace of his empire thus assured he devoted himself to rebuilding NINEVEH. *See also* HC1 pp.33, 55–56, 57.

Sennett, Mack (1884–1960), US film director, producer and actor whose many short slap-stick comedies enlivened the age of silent films. Films made by his KEYSTONE company brought fame to Charlie CHAPLIN, Mabel Normand, Fatty ARBUCKLE and the Keystone Kops. *See also* HC2 pp.200, *201*.

Sense and Sensibility, novel by Jane AUSTEN written in 1797 and published in 1811. It describes the characters and fortunes in love of two sisters, Elinor and Marianne Dashwood.

Sense receptors, in the body, minute organs situated at nerve endings to detect physical and chemical changes occurring inside the body and on or near its surface. Receptors in the SENSE organs detect external stimuli such as light, sound, touch and taste. Others, eg those in muscles, the ear and visceral organs, tell the brain how the body is positioned and how it moves.

Senses, means by which animals gain information about their environment and physiological condition. The five senses, SIGHT, HEARING, TASTE, SMELL and TOUCH, all rely on specialized RECEPTORS on or near the external surface of the body. Additionally, receptors within the body detect internal physical and chemical changes. *See also* MS pp.40, 46, *47*, 48–55, *61*.

Sensory deprivation, in psychology, depriving someone of variation in sensory stimulation. Such deprivation can be achieved by confining volunteer subjects to small, soundproof rooms and having them wear gloves, goggles and ear muffs. Studies show that normal behaviour is highly dependent on variation in stimulation from the environment. *See also* MS p.*47*.

Sensory nerves, nerves that carry information from outside the body to the brain. Their nerve endings are in the SENSE organs, including the skin, and their fibres carry messages either directly to the brain or via the spinal cord.

Sentence, unit of language, defined grammatically as a string of words consisting of at least a subject noun explicit or understood and a verb. In written language sentences are delimited by capital letters and full stops.

Sentimental Journey, A (1768), episodic novel by Laurence STERNE. Published in two volumes, it is an expanded version of the seventh book of *Tristram Shandy*. Its full title is *A Sentimental Journey Through France and Italy by Mr Yorick* and it tells mainly of the hero's encounters with women. *See also* HC2 p.34.

Seoul (Sŏul), capital of South Korea, on the Han River, 64km (40 miles) E of its port at Inch'on; capital of Kyŏnggi province. Founded in 1392, it became the capital of South Korea (1948) and the headquarters of US occupation forces. The city was occupied by North Korean troops at the beginning of *he* KOREAN WAR and suffered great damage. Industries: textiles, metallurgy, food processing, chemicals, agricultural machinery. Pop. (1970 est.) 5,536,377.

Sepal, modified leaf which makes up the outermost portion, or CALYX, of the flower-parts of a plant. Although usually green and inconspicuous once the flower is open, in some species such as the orchid, the sepals look like the petals, whereas, in others, such as the anemone, they are absent.

Sepik, river in the N of the island of New Guinea, in Papua New Guinea. It rises in the Central Highlands and flows NW and then E to the Bismarck Sea. It drains a large area of central New Guinea. Length: approx. 1,127km (700 miles).

Sepoy, Indian soldier employed in a European army, especially a British one. The term derived from the Urdu word meaning a horseman. The INDIAN MUTINY of 1857–59 is sometimes called the Sepoy Mutiny.

Seppuku. *See* HARA-KIRI.

Sepsis, poisoning of the body by the presence of pathogenic bacteria or their toxins in the bloodstream and tissues. Local or widespread inflammation may occur, possibly followed by NECROSIS, death of the tissues. Treatment includes the use of ANTISEPTICS, ANTIBIOTICS or, in severe cases, surgery.

Sep-Szarzyński, Mikolaj (c.1550–80), Polish poet whose works were collected and published posthumously as *Rytmy abo wiersze polskie* in 1601. His style resembles that of the METAPHYSICAL poets and he wrote some of the best sonnets in early Polish literature.

Septennial Act (1716), act of the British parliament, introduced by the Whigs to put off the election due in 1718. It was meant to protect the Whig government against the danger of riots at elections by raising the maximum term of a parliament's life from three years to seven. It lasted until 1911, when the maximum term was reduced to five years .

Septicaemia, generalized bacterial infection in which BACTERIA, usually STAPHYLOCOCCUS, STREPTOCOCCUS or pneumococcus, multiply in the bloodstream. They may enter from a wound or from an infection within the body.

Septic tank, unit for sewage disposal, usually in rural residences not connected to sewers. It is buried in the ground and is watertight and airtight. ANAEROBIC bacteria in the sewage decompose solids to form a sludge which should be pumped out every 3–6 months, although tanks can function for several years before being emptied. *See also* MM pp.202–*203*.

Septuagesima, in the Christian Church, the third Sunday before LENT, the ninth before EASTER. The Roman Catholic Church suppressed the term in 1969 but the name is still used in the Church of England.

Septuagint (3rd–2nd century BC), Greek translation of the Hebrew OLD TESTAMENT, written for the Greek-speaking Jewish community in Egypt. It is the oldest Greek translation of the BIBLE. The Septuagint contains the entire Jewish CANON plus the APOCRYPHA. It is divided into four sections: the law, history, poetry and prophets. The books of the Apocrypha are inserted throughout where suitable. The Septuagint is still used by the GREEK ORTHODOX CHURCH.

Sequence, numbers or terms in an organized list, eg the set of counting numbers, 1, 2, 3, 4,... Sequences may be finite (having a limited number of terms) or infinite, like the counting numbers. Infinite sequences may be divergent, with the values of the terms increasing indefinitely, or convergent, tending to a limiting value. Sequences are sometimes called progressions. *See also* ARITHMETIC PROGRESSION, GEOMETRIC PROGRESSION, HARMONIC PROGRESSION; SU p.38.

Sequoia, two species of mammoth evergreen trees native to California and S Oregon. They are the giant sequoia or big tree (*Sequoia gigantea*) and the redwood (*Sequoia sempervirens*). The size and scarcity of these trees have made them a natural wonder of the USA. Height: more than 90m (300ft). Family Taxodiaceae. *See also* NW pp.*38, 54*.

Sequoya (c.1760–1843), scholar of the CHEROKEE tribe of American Indians who reduced the CHEROKEE language to writing with an alphabet of 86 characters. He completed his task c.1821 after 12 years work.

Seraglio, The, or The Abduction from the Seraglio, opera in 3 acts by MOZART. The text is by Gottlob Stephanie, from an adapted libretto. First produced in Vienna in 1782, the story tells of the rescue of Constanze by her lover Pedrillo from the clutches of Pasha Osmin.

Seraphim, heavenly beings described in the OLD TESTAMENT as having human form but six wings (Isaiah 6:2–6). Although this is the only mention of them in the BIBLE, the Seraphim became high-ranking angels in Jewish and Christian theology.

Serapias, genus of flowering plants, including about ten species of ORCHIDS of the Mediterranean region. All species have bulb-like flowers formed by the fusion of two sepals and one petal with a long triangular lip. The leaves are lanceshaped. Height: approx. 30cm (12in). Family Orchidaceae.

Serbia, constituent republic in E Yugoslavia; Belgrade is the capital. The region is mountainous in the S and W; the NE is part of the fertile plain drained by the River Danube. Wheat, corn, flax, sugarbeet and fruits are grown, and mining is an important industry. The area was settled by Serbs in the 7th century AD and came under Turkish rule in the mid-15th century. After the Russo-Turkish war (1877–78) Serbia became an independent kingdom but continued to be the focus of expansionist policies of the Great Powers in the Balkans. The assassination of the Austrian Archduke Ferdinand by a Serbian nationalist precipitated WWI. After the war the Kingdom of Serbs, Croats and Slovenes was created; its name was changed to Yugoslavia in 1929. Serbia became a constituent republic in 1940. Area: 88,357sq km (34,115sq miles). Pop. (1971 est.) 8,436,547. *See also* HC2 pp.84, *85*, 185, *184–185*, *186–188*, *187*, 190.

Serbo-Croatian, principal language of YUGOSLAVIA. It is spoken by more than 15 million people in Yugoslavia, is the native tongue in the republics of Croatia, Serbia, Montenegro and Bosnia-Hercegóvina and a second language in Slovenia and Macedonia.

Serbs, Slavic people who settled in the Balkans in the 7th century and who became Christians in the 9th century. They were distinguished from CROATS and SLOVENES by their use of the Cyrillic, not the Roman, alphabet. Most of them now live in Yugoslavia.

Seremban, city in W Malaysia, on the S Malay Peninsula, on the Linggi River; capital of Negeri Sembilan state. Linked by rail to Port Dickson on the Strait of Malacca, Seremban is a trade centre for the rubber-growing region. Tin is mined nearby. Pop. (1970) 79,915.

Serenade, in music, traditionally an "evening song" of love or courtship, sung in the open air. An instrumental serenade, from the late 18th century onwards, is in a number of short movements.

Serfdom, labour system, chiefly of medieval Europe, by which agricultural labourers were bound to their landlords by obligatory ties of service. Such men were distinguished from slaves by their rights to possession of their land and by certain feu-

Senna; both its leaves and pods yield a syrup which may be used as a purgative.

Sense and Sensibility is one of the most gently satirical of Jane Austen's novels.

Serbia; chapel in Belgrade dedicated to the royalty but maintained by the Tito regime.

Serfdom; Russian nobles and serfs in 1861, when serfdom was abolished in Russia.

Sergius IV was nicknamed "Pig-mouth" before he succeeded John XVIII as pope.

Sebastiano Serlio: and Vignola designed the extensions to Fontainebleau.

Sermon on the Mount: is, for Christians, the distillation of Christ's teachings.

Georges Seurat's paintings influenced the Fauves and the Futurists.

dal rights owed to them by the lord. In England serfs were generally called VILLEINS. Serfdom disappeared in England by the end of the 15th century, and in Europe generally by the end of the 18th century. It was abolished in Russia in 1861.

Sergius, name of four Popes. Sergius I (*r*.687–701), elected despite opposition from the archpriest THEODORE, introduced the Agnus Dei into the MASS and improved political relations with the FRANKS. Sergius II (*r*.844–87) was supported by the Roman nobility and became the first Pope to practise nepotism by appointing his brother a bishop. SIMONY was prevalent during his reign. Sergius III (*r*.904–11) was said to have murdered two predecessors and was a supporter of Pope STEPHEN VI. Sergius IV (*r*.1009–12) was the first Pope to call for a crusade.

Sergrè, Emilio Gino (1905–), US physicist, *b*.Italy. He shared the 1959 Nobel Prize in physics with Owen CHAMBERLAIN for discovering the anti-particle of the PROTON, called the anti-proton. They achieved this by bombarding copper atoms with high-energy protons in a particle ACCELERATOR. Segrè also discovered two new elements, TECHNETIUM and ASTATINE, by NEUTRON bombardment of MOLYBDENUM and BISMUTH respectively.

Serial music, technique of musical composition, also known as note-row music, which structures a work on a serial pattern of notes repeated for the duration of the work. The twelve-note system of Arnold SCHOENBERG is the most common form of serial music, although it is only in its most literal application that the notes are heard as a definite repeating entity.

Series, mathematical expression obtained by adding the terms of a SEQUENCE. Thus, the series $1 + 4 + 9 + 16 + \dots$ is formed from the sequence $1, 4, 9, 16\dots$ Series may be finite or infinite and infinite series may converge or diverge. A series formed from increasing powers of a variable is a power series; convergent power series are used for representing many functions. *See also* SU p.38.

Serjeants-at-arms, royal officers in Britain who attend the monarch, two of whom, however, are deputed to attend on the two Houses of Parliament.

Serjeants-at-law, members of the highest order of counsel at the English bar, now an obsolete, although not abolished, order. The office was introduced into England by WILLIAM I. Until the Judicature Act of 1873 judges of the QUEEN'S BENCH and of Common Pleas were serjeants. They had their own inn until their society was dissolved in 1877.

Serkin, Rudolf (1903–), US pianist, *b*.Bohemia, who studied with Arnold SCHOENBERG. In the early 1920s he gave sonata recitals with Adolf BUSCH, and he went to the USA in 1939. He has become famous for his performances of the works of BEETHOVEN and MOZART.

Serlio, Sebastiano (1475–1554), Italian architect who settled in France (1540). His masterpiece, the Grand Ferrare, at Fontainebleau (1544–46), established the standard for the French *hôtel*, or town house, for more than a century. His *L'Architettura* (1537–1551), the first practical rather than theoretical book on architecture, coded the five ORDERS, diffused the style of BRAMANTE and RAPHAEL and gave a vast range of motifs.

Sermon on the Mount, a discourse spoken by JESUS CHRIST to his disciples and others in the hills of GALILEE. Delivered early in His ministry, it is recorded in the Gospel of MATTHEW. Beginning with the BEATITUDES, the Sermon continues with a discussion of social responsibility and sets out the principles of the Christian ethic. *See also* HC1 p.110.

Serotonin, 5-hydroxytryptamine, a chemical found in cells of the gastrointestinal tract, in blood platelets and in brain tissue. In the latter it is concentrated in the midbrain and HYPOTHALAMUS. It is a vasoconstrictor and has an important role in the functioning of the nervous system.

Serov, Valentin (1865–1911), Russian painter and theatrical designer. A well-known portrait and landscape artist, Serov was a member of the WORLD OF ART movement. Among his many portraits is one of Rimsky-Korsakov.

Serowe, town in E central Botswana, S Africa, 258km (160 miles) NNE of the capital Gaborone. It is a trade centre for a fertile farming region. Serowe is the seat of the Bamangwato tribe. Pop. 34,186.

Serpens, or the Serpent, extensive equatorial constellation situated between Libra and Hercules and divided into two parts, Serpens Cauda and Serpens' Caput, by Ophiuchus, the Serpent-bearer. Serpens Caput contains the bright globular cluster M5 (NGC 5904), and Serpens Cauda includes M16 (NGC 6611), an open cluster shining through surrounding nebulosity. The brightest star is Alpha Serpentis (Unukalhai).

Serpent. *See* SNAKE.

Serpent, in music, now obsolete bass wind instrument with a cup mouthpiece resembling that of a trumpet. It was almost always made of wood and had a characteristic undulating serpent-like shape.

Serpent, in religious symbolism, a subtle treacherous and dangerous creature. In the Bible, the tribe of DAN (Gen 49:17) and the SCRIBES and PHARISEES (Matt 23:33) are likened to serpents and the symbol is used throughout both the Old and New Testament to evoke wickedness and slyness. The serpent's most memorable appearance is in Gen 3, where it symbolizes the temptation to disobedience of Adam and Eve. *See also* DRAGON.

Serpentine, group of sheet silicate minerals with a pattern of green mottling like a snake's skin; hydrated magnesium silicate. They are of varied hues and are commonly used in decorative carving; fibrous varieties are used in asbestos cloth. Hardness 2–5; s.g. 2.2.

Serpent star. *See* BRITTLE STAR.

Serra, name of two Spanish painters, the brothers Jaime (*fl. c*.1361–*c*.99) and Pedro (*c*.1343–*c*.1405). Influenced by the SIENNESE SCHOOL, they painted religious subjects in a decorative manner; the altar piece (1394) in Manresa Cathedral, Cetalonia, is a fine example of their work.

Serum, liquid that remains if whole blood is allowed to clot, and the clot removed; essentially the same composition as PLASMA, but without fibrinogen and clotting factors. *See also* MS pp. 62–63.

Sérusier, Paul (1865–1927), French painter of the post-Impressionist style. He was influenced by GAUGUIN and was a founder of the NABIS. His approach to art was highly methodical and he expressed his theories in *ABC de la peinture* (1927).

Servetus, Michael (*c*.1511–53), Spanish physician and theologian. He practised medicine in France and was said to have discovered that blood is circulated from the right chamber of the heart to the lungs. Forced to flee from the Inquisition because of his religious beliefs, he was arrested as a heretic and burnt by Calvinists in Geneva. *See also* HC1 p.269.

Service, Robert William (1874–1958), Canadian writer, *b*.Britain. He went to Canada in 1894 and worked for a bank in British Columbia and The Yukon where his rollicking ballads of the Klondike gold-miners, such as *Songs of A Sourdough* (1907), which included "The Shooting of Dan McGrew", were popular.

Service tree, European tree of the ROSE family, (Rosaceae), native to the Mediterranean region and cultivated for its acid-flavoured fruit. It resembles the MOUNTAIN ASH. Height: 9–15m (30–50ft). Species *Sorbus domestica*.

Servomechanism, device that provides remote control by converting an input signal (eg a radio impulse or mechanical movement) into a mechanical output (eg a lever movement or amplified hydraulic force) to activate a mechanism. It usually forms part of a control system. Some servomechanism, such as the automatic pilot of an aircraft, incorporate a feedback mechanism that makes it independent of human control. *See also* MM pp.104–105.

Servranckx, Victor (1897–1965), Belgian artist. He worked within many of the styles of the early 20th century, including FUTURISM and Purism, and he was one of the first Belgian abstract painters. He was also an influential industrial designer.

Sesame, tropical plant native to Asia and Africa. It is also cultivated in Mexico and SW USA for its oil and seeds, both used in cooking. An annual, it has oval leaves, small pink or white flowers, and seed capsules along the stem. Height: 61cm (2ft). Family Pedaliaceae; species *Sesamum indicum. See also* PE pp.210, 211.

Sesshū (1420–1506), Japanese artist of the Muromachi period who is best known for his intense brushdrawings in ink (Sumi-e). His best works were painted from 1469, after a journey to China, and showed a strong mixture of Chinese motifs executed with Japanese sensitivity.

Session, Court of. *See* COURT OF SESSION.

Sessions, Roger Huntingdon (1896–), US composer whose complex and highly individual works include a *Violin Concerto* (1935), eight symphonies, *Concertino for Chamber Orchestra* (1972) and various piano and organ works.

Seth, in the BIBLE, son of ADAM AND EVE, thought by Eve to be recompense from God for the death of ABEL. According to the NEW TESTAMENT, Seth is an ancestor of JESUS CHRIST.

Seth, Egyptian god. Although a beneficent god in predynastic Egypt, Seth became associated with darkness and was later identified as a god of evil. *See also* HC1 p.328; MS p.*207*.

Sétif, city in NE Algeria; the capital of Sétif province. The city was largely built by the French on the site of the ancient Roman town of Sitifis. Sétif is the trade centre for a region producing cereals and zinc. Textile manufacture is an important industry. Pop. (1966) 98,000.

Seton, Ernest Thompson (1860–1946), US writer and naturalist, *b*.Britain, who spent many years in the prairies of Manitoba. The knowledge he gained there formed the basis of his stories and work as an illustrator of wild animals. Many of his stories were collected in his best-known book *Wild Animals I Have Known* (1898).

Setter, long-haired hunting dog used to locate game and stand on point until the hunter arrives; it also retrieves on command. Setters have good scenting powers and endurance. They were developed in 16th-century England, and modern types include English, Irish and Gordon.

Set theory, branch of mathematics developed by Georg CANTOR in the late 19th century. It is based on George BOOLE's work on mathematical LOGIC, but manipulates sets of abstract or real objects rather than logical propositions. It is concerned with the properties of sets, the relations between them and operations on them, and can be applied to most of the other branches of mathematics. *See also* SU pp.40–41.

Settlement, Act of (1701), statute regulating the succession and defining the rights of the English monarchy. It provided that, if WILLIAM III and the future Queen ANNE died without heirs, the crown would pass to James I's granddaughter, Sophia of Hanover, or her heirs if they were Protestants. It led to Hanoverian succession to the throne in 1714; it also restricted the rights of the Crown to appoint foreigners, and allowed the monarch to leave the country only with the consent of Parliament. *See also* HC2 p.*18*.

Seurat, Georges Pierre (1859–91), French painter and founder of NEOIMPRESSIONISM. During 1876–84 he developed his theories of colour vision based on the juxtaposition of dots which became known as Pointillism. His works include *Bathing at Asnières (1883–84), of iridescent brilliance, and Sunday in Summer on the Island of La Grande-Jatte* (1886), which shows the developed pointillist technique.

Sevastopol. *See* SEBASTOPOL.

Seven, The *See* EFTA.

Seven against Thebes, in Greek mythology, the legend of the battle brought about by the sons of OEDIPUS. When Oedipus was banished, his sons Eteocles and Polyneices agreed to share the throne of Thebes, ruling in alternate years. Eteocles reneged, causing Polyneices to raise an army with seven chieftains against him. The sons met in single combat and killed each other.

HC1 = History and Culture Vol. 1   HC2 = History and Culture Vol. 2   MS = Man and Society   MM = Man and Machines

Seven Bishops, The, the archbishop of Canterbury and the bishops of Bath and Wells, Bristol, Chichester, Ely, Peterborough and St Asaph, who petitioned JAMES II against his order (1688) that the Declaration of Indulgence to Roman Catholics be read in every church. They were imprisoned in the TOWER OF LONDON, tried for sedition and acquitted amid popular rejoicing.

Seven Hills of Rome, site of ROME since its foundation. The hills are the Palatine (central site of the earliest settlement and later of sumptuous houses, palaces and temples); to the NW the Capitoline (settled *c.*6th century BC, it became the ancient city's religious centre and was the site of the great temple of Jupiter); and, in a curve to the E and S, the Quirinal, Viminal, Esquiline, Caelian, and Aventine. *See also* HC1 p.*90.*

Seven Lamps of Architecture, The (1849), treatise by John RUSKIN that deals with the guiding principles of architecture (sacrifice, truth, power, beauty, life, memory, and obedience) and advocating Gothic architecture in particular as displaying these virtues.

Sevenoaks, county district in W KENT, England; created in 1974 under the Local Government Act (1972). Area: 371sq km (143sq miles). Pop. (1974 est.) 101,100.

Seven Pillars of Wisdom (1926), book by T.E. LAWRENCE. It is an account of the campaign against the Turks in the Arabian Peninsula during WWI, in which he became known as Lawrence of Arabia. The book is important as a work of military history and as the self-portrait of a complex personality.

Seven Samurai, The (1954), Japanese film directed by Akira KUROSAWA. It is based on the 16th-century legend of seven SAMURAI warriors who aid a small village under constant attack by bandits.

Seventh-Day Adventists, Christian denomination whose members expect JESUS CHRIST to return to earth in person. They hold the SABBATH on Saturday, the seventh day, and accept the BIBLE literally as their guide for living. The sect was formally organized in 1863 and today, it is the largest ADVENTIST denomination. *See also* HC2 p.*304.*

Seven Wonders of the World, group of sights listed as wonderful by ANTIPATER of Sidon in the 2nd century BC. They were the PYRAMIDS of Egypt, the COLOSSUS OF RHODES, the PHAROS at Alexandria, the temple of ARTEMIS at Ephesus, the MAUSOLEUM at Halicarnassus, the Hanging Gardens of BABYLON and the statue of ZEUS by PHIDIAS at Olympia.

Seven Years War (1756–63), major conflict between Britain and Prussia on the one side and France, Austria, Russia and Sweden on the other. It was fought at sea and on land, in Europe, America, the West Indies and India. It began on 15 May 1756, when Britain declared war on France after France's attack on Minorca. It ended in the signing of the Peace of Paris and the Treaty of Hubertusburg in 1763. The principal results of the war were the supremacy of Britain in India and North America, and the emergence of Prussian power in Europe. *See also* HC2 pp.46, 47, *63.*

Severini, Gino (1883–1966), Italian painter who, influenced by CUBISM, was instrumental in developing the style of FUTURIST painting. His later work became more objective, and he did many murals and mosaics. *See also* HC2 p.*177.*

Severinus, pope (*d.*640). A Roman, he was elected successor of Pope HONORIUS in 638, but Emperor HERACLIUS refused confirmation of his election. He was finally consecrated in 640 and died soon after.

Severn, river in NW Ontario, Canada. It rises in the Finger Lake region and flows NE through the Severn Lake into Hudson Bay. The Hudson's Bay Co. established a trading post at the river's mouth in 1685. Length: 982km (610 miles).

Severn, major river in W England. It rises on Mt Plynlimon, in W Wales, flows NE to Shrewsbury, curves SE and then SW to enter the Bristol Channel through a wide estuary. Major tributaries include the rivers Vyrnwy, Teme, Stour and Wye. The

Severn is connected by canal to the rivers Thames, Mersey and Trent. A railway tunnel passes under the estuary. Length (river): 290km (180 miles).

Severus, Lucius Septimius (AD146–211), Roman emperor. He assumed the throne in 193 and went to Britain in 208. He divided the country into two provinces, with York as the northern capital and London as the southern one. He marched on Scotland, but was forced to retreat and died at York. *See also* HC1 p.114.

Sévigné, Marie, Marquise de (1626–96), French letter-writer. She wrote more than 1,500 letters, mostly to her daughter on personal, social and literary matters. They are distinguished by their elegance of style and acute observation.

Seville, port in SW Spain, on the River Guadalquivir 543km (337 miles) SW of Madrid; capital of Sevilla province. Ruled by the Romans from the 2nd century BC to the 5th century AD, it was taken by the Moors in 712. In 1248 Ferdinand III conquered Seville. The city has a 15th-century cathedral, which contains the tomb of Christopher COLUMBUS and paintings by Goya and El Greco, and the 12th-century Alcazar palace. Industries: agricultural machinery, shipbuilding, chemicals, textiles, shipping. Pop. (1974) 588,784.

Sèvres, in ceramics, high-quality porcelain made from 1765 to the present day at the factory at Sèvres, near Versailles, France. As the MEISSEN factory ceased to hold premier position, Sèvres surpassed it in the production of high-quality European porcelain. Pieces are either of soft-paste (a porcellaneous substance) or hard-paste (true porcelain) and include glazed and unglazed figurines and various vessels.

Sèvres, Treaty of (1920), peace treaty after WWI signed by Turkey and the Allied powers (except the USSR and the USA). It virtually ended the Ottoman empire and Turkish sovereignty in Europe. It was superseded by the Treaty of LAUSANNE (1923).

Sewage disposal, treatment of human waste to render it biologically and ecologically safe before dumping. At a typical sewage plant, incoming sewage is screened free from stones and other hard materials. Heavy sludge is settled off and removed to closed digester vessels in which bacteria break it down to a relatively inoffensive material that can be used as a fertiliser. Liquid sewage passes on to trickle filters or activated sludge tanks, in either of which it is purified by the combined digestive action of bacteria and protozoa; chemical treatment may also be used. Further settling tanks remove the inactive humus materials, after which the remaining water is fully purified and can be reused. *See also* MM pp.202–203.

Sewell, Anna (1820–78), British writer. She wrote prose, verse and children's stories, of which the best-known is *Black Beauty, the Autobiography of a Horse* (1877).

Sewing machine, machine used to join fabrics by interlocking two threads to form stitches. Industrial sewing machines operate at high speeds and are usually specialized, i.e. they produce only one type of stitch. Modern domestic sewing machines can make various types of stitches, including lock stitch, chain stitch and an assortment of zig-zag stitches used for finishing and embroidery, and can also sew buttons and button-holes. *See also* MM p.*252.*

Sex, classification of an organism according to specific qualities and denoting the reproductive function of the organism. There are two divisions of the classification, male and female. The two sexes have fundamental physiological, functional and behavioural differences which distinguish them. In animals the sex glands, OVARIES in the female, TESTES in the male, are primary sexual characteristics. Secondary sexual characteristics such as size, coloration and, in human beings, the development of the breasts and the growth of facial and body hair, are governed by the secretion of SEX HORMONES. *See also* SEX DETERMINATION; SEXUAL REPRODUCTION.

Sexagesima (from the Latin for "sixtieth"), the Sunday of the week in which

the sixtieth day before Easter falls. No longer celebrated as a feast, it is followed by QUINQUAGESIMA and QUADRAGESIMA Sundays.

Sex determination, finding the gender of an organism. Its sex depends on the combination of CHROMOSOMES. There are two types of sex chromosomes, the X type and the Y type. All female eggs has one X chromosome, whereas male sperm has either an X or a Y. If the sperm that fertilizes the egg has an X chromosome, the resulting zygote and the organism that develops from it will be XX, or female. If the egg is fertilized with a Y-carrying sperm, the zygote will be XY, or male. *See also* NW p.30.

Sex hormones, chemical "messengers" secreted by ENDOCRINE glands and carried by the blood, which regulate sexual activity in male and female animals, including the body changes occurring at PUBERTY and during PREGNANCY. In males they include TESTOSTERONE, made by the TESTES, the production of which is itself controlled by a hormone made by the PITUITARY GLAND. In females, the sex hormones OESTROGENS and PROGESTERONE are made by the ovaries, the general development of which is regulated by two further hormones secreted by the pituitary gland. *See also* MS pp.64–65, 164, *164.*

Sexism, discrimination against and subordination of people on the basis of their sex. Sexism may result from prejudice, stereotyping or social pressure. *See also* WOMEN'S LIBERATION ORGANISATION.

Sextant, navigational instrument used to establish latitude (distance north or south of the equator) by determining the angle between the horizon and the Sun or a star. It consists of an arc, marked in degrees, a movable arm with a mirror pivoted at the centre, and a telescope mounted to the framework. *See also* MM p.*90.*

Sexton, Anne (1928–74), US poet. Her poetry relates intimate personal experience, and her works include *To Bedlam and Part Way Back* (1960), a record of her recovery from a nervous breakdown, and *Live or Die* (1966), for which she won a Pulitzer Prize.

Sexual Behavior in the Human Male (1948), pioneering statistical study of sexual behaviour carried out by Alfred KINSEY in the USA. Although its findings were criticized for the methods used it was frequently quoted by other authors in this field. The study was paralleled by *Sexual Behavior in the Human Female* (1953); together these works are often referred to as the Kinsey Reports.

Sexual reproduction, biological process by which living organisms generate others of their own kind. It occurs in different forms throughout the plant and animal kingdoms from PROTOZOA to vertebrates, and involves the exchange of genetic material. This process gives rise to variations of the PHENOTYPE and GENOTYPE within a species. GAMETES (haploid cells), produced by MEIOSIS, are daughter CELLS which contain only half the number of CHROMOSOMES of their parent cells (which are diploid). At fertilization the gametes, generally one from each parent, fuse to form a ZYGOTE with the full complement of chromosomes. The zygote divides repeatedly and the cells differentiate to give rise to an embryo and, finally, a fully formed organism. ASEXUAL REPRODUCTION occurs in many plants and in most lower animals from protozoa to hydra. The type of cell division, called MITOSIS, gives rise to new individuals that are identical to the "parent" organisms. *See also* MS pp.74–75, 191; NW pp.70–71, 76–77, 144–145; PE pp.176–179.

Seychelles, independent island republic in the Indian Ocean, approx. 970km (600 miles) N of Madagascar. It is made up of 85 small islands, the largest of which is Mahé. Formerly a British colony, the Seychelles gained independence in 1976. The main products are copra, cinnamon, tea and fish; tourism is of increasing importance to the economy. The capital is Victoria, on Mahé. Area: 269sq km (105sq miles). Pop. (1976 est.) 60,000 *See also* MW p.153.

Seyfert galaxy, type of SPIRAL GALAXY that

Seven Wonders of the World; an artist's impression of the Colossus of Rhodes.

Severn; a view of the river showing part of the city of Worcester.

Seville, the birthplace of Velázquez and Murillo, lies on the Guadalquivir.

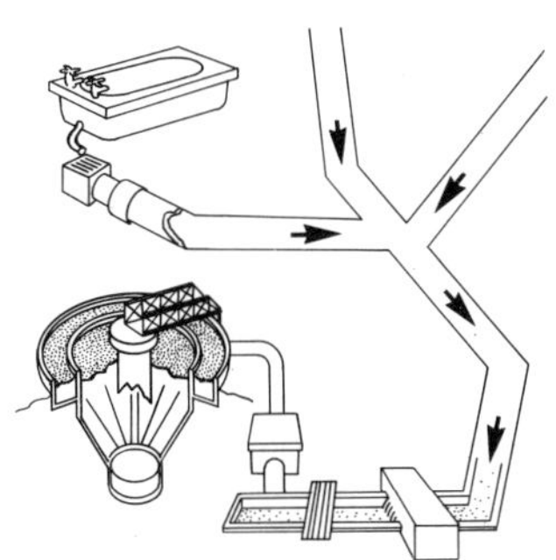

Sewage disposal uses a system of pipes from homes to the treatment plant.

Jane Seymour, Henry VIII's favourite wife, painted by Hans Holbein.

Shackleton's expedition of 1907–09: the snow pillars were used for guidance.

Shad are valued as food fish. Both the flesh and the roe have good flavour.

William Shakespeare: Martin Droeshout's portrait is probably the best likeness.

has a small, extremely bright nucleus and tightly wound arms. The nucleus is almost stellar in appearance and is believed to be made up of clouds of hot gas moving with velocities in the region of 5,000km (3,000 miles) per second. Many of these objects, first discovered by the US astronomer Carl Seyfert in 1943, are radio sources. *See also* SU p.248.

Seymour, James (1702–52), English painter. With John WOOTTON he was a pioneer of sporting art in England and mostly painted hunting and racing pictures and portraits of horses.

Seymour, Jane (*c.*1509–37), queen of England and the third wife of HENRY VIII, whom she married in 1536. She died in 1537 after the birth of her son, the future EDWARD VI.

Seymour, Thomas, 1st Lord Seymour of Sudeley (*c.*1508–1549), brother of Jane SEYMOUR and second husband of HENRY VIII's widow, Catherine PARR. His desire to marry the princess Elizabeth (later ELIZABETH I led to his execution for treason.

Sezession Group. *See* VIENNA SEZESSION.

Sfax (Sàfaqis), seaport city in E Tunisia, on the N shore of the Gulf of Gabès. It was once a Barbary pirate stronghold and is now a trade and industrial centre, exporting phosphates and olive oil. Pop. (1966) 70,472.

Sforza family, Italian dynasty that ruled Milan from 1450 until 1535. Muzio Attendolo (1369–1424) became a *condottiere,* the leader of a strong group of mercenaries that fought for Milan, Florence, Ferrara and Naples. He adopted his nickname "Sforza", which he passed on to his descendants. The first Sforza Duke of Milan was Muzio's son Francesco (1401–1466), who ruled from 1450 until his death. He was one of Italy's most powerful *condottieri.* He subsequently played an important role in Italian politics. Ludovico (1452–1508) was the second son of Francesco and one of the most noted of the Renaissance princes. His ruthless drive for power gave him effective rule in Milan by 1480. He was, however, deposed by Louis XII of France in 1499 and it was not until 1512 that another Sforza – Massimiliano, the son of "il Moro" – was Duke of Milan. He was ousted in 1515 and his brother Francesco Maria was restored by the Emperor Charles V in 1522. On Francesco Maria's death the duchy passed to Charles and the HAPSBURGS.

Sgraffito, in art, the deliberate scratching away of an upper layer of paint, glass or GLAZE to reveal the one beneath. The technique has been used in paintings executed in such mediums as gold leaf, oil paint and TEMPERA; in STAINED GLASS, where a superficial layer of coloured glass may be scratched away; in pottery, before firing, where a layer of coloured or white SLIP (clay or water) is scratched away to reveal a colour that had previously been applied.

's Gravenhage. *See* HAGUE, THE.

Shaba. *See* KATANGA.

Shackleton, Sir Ernest Henry (1874–1922), British antarctic explorer, who was a member of Robert SCOTT's first expedition from 1901–1904. From 1907 to 1909 he led his own expedition, getting within 155km (97 miles) of the south pole. He led another expedition from 1914 to 1916 to the Ross Sea, but his ship was crushed close to Elephant island. From there he and a few companions crossed to South Georgia island where he received aid.

Shad, saltwater food fish of the HERRING family that swims upriver to spawn. Shads are prized for their roe. Deep-bodied, they have a notch in the upper jaw for the tip of the lower jaw. Length: to 75cm (30in). Family Clupeidae.

Shadbolt, Jack Leonard (1909–), Canadian painter *b.*Britain. He studied in England with Victor PASMORE and Sir William COLDSTREAM. His paintings, which show a variety of influences, are inspired by the countryside of British Columbia.

Shadbolt, Maurice (1932–), New Zealand novelist. He has published *The New Zealanders* (1959) and *Strangers and Journeys* (1972), both collections of short stories, and the novels *Among the Cinders*

(1965) and *Danger Zone* (1975).

Shadow, area screened from a light source and thus in relative darkness. In two dimensions it is perceived as an exact representation of the obstruction to the light source, although it can differ very much in size (greater or smaller) depending on the proximity of the light source to the screening object. Sometimes the outline of the shadow is blurred, there is an area of mid-shadow (penumbra) outside the shadow itself (umbra).

Shadow play, type of PUPPET THEATRE originating in E Asia in which flat, jointed puppets, often made of leather, are manipulated by rods or wires between a strong light source and a translucent screen on which their shadows are cast. In 19th-century Britain, the shadow play (or galanty) found popularity in London streets and at fairs.

Shadrach, in the BIBLE, DANIEL's companion who, with Meshach and Abed-nego, was saved by an angel when thrown into the fiery furnace by NEBUCHADNEZZAR for refusing to worship an idol.

Shadwell, Thomas (*c.*1642–92), English dramatist who was influenced by Ben JONSON's comedy of humours. His many plays include *The Sullen Lovers* (1668), *Epsom Wells* (1672) and *Bury Fair* (1689). He is remembered as DRYDEN's great rival and he succeeded Dryden as poet laureate in 1688.

Shaffer, Peter (1926–), British dramatist. After establishing his reputation as a playwright with *Five Finger Exercise* (1960), he wrote a double-bill *The Private Ear and The Public Eye* (1962). *The Royal Hunt of the Sun* (1964), which dealt with the conquest of Peru, *Black Comedy* (1965) and *Equus* (1973) were produced by the National Theatre. His other work includes *The Battle of Shrivings* (1970) and numerous plays for television.

Shafi'i, Abu'Abd Allah ash– (767–820), Muslim theologian who made great contributions to the jurisprudence of Islam. He belonged to the tribe of Quraysh, as did the prophet Mohammed, to whom he was distantly related. He spent much of his time with Bedouins and later travelled widely studying the various legal ideas of Islam, before writing his great legal work, *Risàlah.*

Shaftesbury, Anthony Ashley Cooper, 1st Earl of (1621–83), English statesman, one of the 12 commissioners sent to BREDA to invite CHARLES II to return to England after the collapse of the PROTECTORATE. A member of the CABAL, he was made Earl of Shaftesbury and Lord Chancellor in 1672. He attempted to exclude JAMES II from the succession, supporting the Duke of MONMOUTH. When Monmouth at first held back from open rebellion, Shaftesbury fled to Holland, where he died. *See also* HC1 pp.292, *293.*

Shaftesbury, Anthony Ashley Cooper, 7th earl (1801–85), British social reformer. A Tory MP (1826–51), he led the factory reform movement in the House of Commons, first proposing the ten-hour factory working day in 1833 (passed in 1847). The Coal Mines Act (1842), which prohibited the employment underground of women and children under 13, and the 1850 Reform Act were also fruits of his work. *See also* HC2 p.96.

Shag, large, long-billed seabird of the CORMORANT family, which nests on rocky coasts and is native to most continents. The adult is completely dark except for pale patches at the base of the bill and round the eyes. The feet are webbed. Length: to 100cm (40in). Family Phalacrocoracidae.

Shah Jehan (1592–1666), Mogul emperor of India (*r.*1628–58) responsible for the building of the Taj Mahal, considered one of the most beautiful tombs in the world. During his reign he subjugated the Deccan and took Kandahar from the Persians.

Shah-Nama, Persian epic poem by FIRDAUSI, presented to the Sultan Mahmud of Ghazna in 1010. The title means "book of kings" and it describes the history of Persia in 60,000 verses from mythical times to the 7th century.

Shahn, Ben (1898–1969), US painter, lithographer and photographer, *b.*Lithu-

ania. He emigrated to the USA in 1906. His work consistently reflected his concern with social and political injustices. He painted in the realist and abstract styles.

Shah of Iran. *See* REZA SHAH.

Shaka. *See* CHAKA.

Shakers, millennarian religious organization more properly named the United Society of Believers in Christ's Second Appearing. The practice of shaking, shouting and dancing forms part of its ritual. The Shakers developed in Britain in the mid-18th century and founded their first colony in North America in 1774. They believe that total sexual abstinence is the basis of spiritual salvation.

Shakespeare, William (1564–1616), English poet and dramatist who is generally regarded as the greatest playwright of all time. Despite this, little is known of his life. By 1592 he was established in London, having already written the three parts of *Henry VI.* By 1594 he was a member of the Lord Chamberlain's Men and in 1599 a partner in the Globe Theatre, where many of his plays were presented. He retired to Stratford-upon-Avon in 1614 and died in 1616.

Critics have been tempted into reading personal references in his plays and poetry, especially the 154 *Sonnets,* which were first published in 1609.

The plays are usually divided into four groups: the historical plays, the comedies, the tragedies and the romances. Shakespeare seldom invented the plots of his plays, preferring to take his outlines from earlier works and building characters from hints contained therein and adding others to serve his dramatic intentions. *See also* articles on individual works.

Shakhlin, Boris (1932–), Soviet gymnast. In the Olympic Games of 1956, 1960 and 1964 he won seven gold and four silver medals, and at the World Championships of 1954, 1958, 1962 and 1966 a further six gold and six silver.

Shakti, name for the female partner of a Hindu god. Shakti is the symbol of the creative energy of the divine reality.

Shale, common SEDIMENTARY ROCK made from mud or clay. It may contain various materials such as fossils, carbonaceous matter, and oil. *See also* PE pp.102–103, *103.*

Shallot, perennial plant native to W Asia, and widely cultivated in temperate climates. It has thin, small leaves and clustered bulbs like garlic, but with a milder flavour like onion. It is used in cooking. Family Liliaceae; species *Allium ascalonicum. See also* ONION; PE pp.188, *189.*

Shalmaneser, the name of five kings of Assyria. Shalmaneser I (*r.*1274–1245 BC) liberated the Hittites and took Babylon. Shalmaneser II was king 1019–1008 BC. Shalmaneser III (*r.*858–824 BC), son and successor of ASHURNASIRPAL II, campaigned against Damascus and Urartu and defeated in 842 the alliance of Damascus, Tyre and Sidon, and Jehu of Israel. Shalmaneser IV (*r.*782–772 BC) led expeditions against the Chaldeans. Shalmaneser V (*r.*726–722 BC), successor of TIGLATHPILESER III captured King Hosea of Israel during a revolt but died during the conquest of Samaria. *See also* HC1 pp.33, 56–57, *59.*

Shaman, in primitive cultures, a witch doctor or medicine man believed to have magical powers. Shamanism is found among the ESKIMOS and AMERICAN INDIANS and in Siberia, where the term originated. *See also* ANIMISM; SORCERY; MS pp.122, *217.*

Shamash, in ASSYRO-BABYLONIAN MYTHOLOGY, the sun god, son of the moon god SIN and brother of ISHTAR. Shamash was also a god of DIVINATION who gave messages to men through soothsayers. *See also* HC1 p.*33.*

Shamrock, plant with three-part leaves, usually taken to be *Trifolium repens* or *T. dubium,* the national emblem of Ireland. Legend tells that St Patrick used it to symbolize the Trinity. Other plants also called shamrock include *Oxalis acetosella* and *Medicago lupulina.*

Shang Dynasty, or Yin Dynasty (*c.*1600–*c.*1030 BC), Chinese dynasty, the earliest for which accurate records exist.

Shang rule was based in the valley of the Yellow River. During this period the complex Chinese written language was perfected, as well as techniques of flood control and irrigation. Bronze was also worked for the first time. It was overthrown by the CHOU DYNASTY. *See also* HC1 pp.44–45.

Shanghai, port in SE China, 22km (13 miles) from the Yangtze (Changjiang) delta; largest city in China and the most populous in mainland Asia, and one of the world's major ports. By the Treaty of Nanking in 1842 the city opened to foreign trade, which stimulated tremendous economic growth. The USA, Britain, France and Japan all held their own areas, known as concessions, within the city. The city was restored to China by the end of WWII. It fell to the Communists in 1949. Industries: textiles, steel, chemicals, publishing, food processing, rubber products, farm machinery, shipbuilding, pharmaceuticals. Pop. 10,800,000. *See also* MW p.53.

Shankar, Ravi (1920–), Indian musician. Responsible for popularizing the SITAR and Indian music in general in the West, Shankar founded the National Orchestra of India and was music director of All-India Radio (1948–56). He toured Europe and the USA extensively during the 1960s and 1970s.

Shankly, William ("Bill") (1915–), British footballer and manager. After a distinguished playing career with Preston North End (1933–49) and Scotland, winning five caps (1937–39), he became manager of Liverpool in 1959 and created teams that won League, Cup and European honours. He retired in 1974.

Shannon, longest river in the Republic of Ireland. It rises on Cuilcagh Mountain in NW County Cavan and flows s through loughs Allen, Ree and Derg, s across the central plain of Ireland to Limerick, then w to enter the Atlantic Ocean between Loop Head and Kerry Head. Length: 370km (230 miles).

Shans. *See* TAI.

Shansi (Shanxi), province in N central China, bounded by the Yellow River (Hwang Ho) (w and s), the GREAT WALL OF CHINA (N) and the North China Plain (E); the capital is Taiyuan. Much of the province is a high plateau region. The chief crops, grown by means of irrigation, are winter wheat, cotton, tobacco, grapes and soya beans. Livestock is reared in the N and hides and wool are exported. Forestry is also important. Iron ore, coal and salt are the chief mineral deposits. Industries include fertilizers, chemicals, machinery, cement and locomotives. Area: 157,170 sq km (60,683sq miles). Pop. 18,000,000.

Shantung (Shandong), province in E China; the capital is Tsinan (Tinan). The E part of the province is a peninsula between the Po Hai and the Yellow Sea; the w is part of the Hwang Ho (Yellow River) delta. Products: wheat, cotton, fruits, tobacco, groundnuts, maize, soya beans, timber, silk, fish, salt, oil, coal, iron ore, gold. Area: 153,600sq km (59,304sq miles). Pop. 57,000,000.

Shanty town, city or part of a city in which the common residential building is the shanty, a small house (often single-storied) made of the cheapest materials available – generally wood or corrugated metal sheets. *See also* HC2 p.303.

Shapiro, Karl (1913–), US poet and critic, interested in the role of the poet as a cultural teacher. His volumes of verse include *V-Letter and Other Poems* (1944) and *Poems of a Jew* (1958). He published his essays in such volumes as *In Defense of Ignorance* (1960).

Shapley, Harlow (1885–1972), US astronomer who provided the first accurate model of our GALAXY (The Milky Way). By observing CEPHEID stars in globular clusters, he calculated the distance to each cluster in the galaxy and thereby obtained a picture of its shape and size. *See also* SU pp.243–245.

Shapu, wild mountain sheep of N India and Tibet, living at altitudes of 4,000m (13,100ft) or more in the Himalayas. The Shapu (*Ovis vignei*) is grey-brown, with a white belly and a beard, and lives at higher altitudes than the Red Sheep or Elbur's Urial (*O. occidentalis*), which it resembles. Height: 75cm (2.5ft); weight 50kg (110lbs).

Shapur (Sapor), name of three kings of the Sassamian Empire of Persia. Shapur I (*d.* AD 272) succeeded to the throne in 241. Although defeated by the Roman Emperor Gordian III in 243, he concluded an advantageous treaty with the Emperor Philip in 244 and in the years that followed penetrated deep into Syria. His defeat and capture of the Emperor VALERIAN in 260 was an important factor in the decline of Rome. Shapur II (309–79) was reputedly born king. He defeated the central Asian tribes and fought the Romans over the control of Armenia, which he achieved after the death of the Emperor Julian in battle in 363. Shapur III (*d.*388) came to the throne in 383. He surrendered part of Armenia to the Romans in an attempt to settle a dispute over the region.

Sharaku, Toshusai (*d.*1801), Japanese printmaker. His brilliant UKIYO-E prints, numbering about 145, appear to have been produced in only ten months in 1794–95.

Sharecroppers, tenant farmers in the USA who were provided with land, seed, stock, tools and living facilities in return for their labour and some of their crop. After the AMERICAN CIVIL WAR many former slaves became sharecroppers in the southern states.

Shares, division of the capital of a LIMITED LIABILITY company. Shares are expressed in monetary units, for example in Britain 50p or £1. There are two types of shares: ordinary shares (those that carry a fixed rate of DIVIDEND that is dependent on the net profit of the company) and preference shares (paid before the dividend on ordinary shares). *See also* MS p.309, *309.*

Sharia. *See* ISLAMIC LAW.

Sharjah (Ash-Shāriqah), capital of the Sheikdom of Sharjah, one of the UNITED ARAB EMIRATES, in E Arabia. It is situated on the Persian Gulf. The sheikdom was a British protectorate until 1971. Sharjah is part of a prosperous oil producing area. Pop. (1968) 19,198.

Shark, torpedo-shaped elasmobranch fish found in subpolar to tropical marine waters. They have well developed jaws, bony teeth, a cartilaginous skeleton, skin denticles, usually 5 gill slits on each side of the head and a characteristic lobe-shaped tail with longer top lobe. Lateral lines are used to detect hydrostatic pressure and the location of other things. The sense of smell is well developed. Sharks are carnivorous and at least 10 species are known to attack man. There are about 250 living species. Order Selachii. *See also* CHONDRICHTHYES; DOGFISH; HAMMERHEAD SHARK; MAKO; NURSE SHARK; TIGER SHARK; WHALE SHARK; WHITE SHARK; NW pp.124, *126–127, 130, 240,* 241.

Sharm el Sheikh, bay or inlet at the s end of the Sinai Peninsula, overlooking the Strait of Tiran. Taken from the Egyptians during the 1967 ARAB-ISRAELI WAR, it has since been surrendered by the Israelis as a result of the 1973 war and UN negotiations in 1974.

Sharp, Cecil (1859–1924), British FOLK MUSIC specialist. His life was given to the collection, publication and performance of English folk music, whose preservation owes more to him than to any other man. Cecil Sharp House, London, is the centre of the English Folk Dance and Song Society.

Sharp, in musical notation, an accidental (sign) which is placed before a note to indicate that the note is to be sounded a semitone higher. When used in the key signature at the beginning of a stave of music it indicates a note which is always to be sharpened by a semitone.

Sharpeville, black suburb in Vereeniging, South Africa. It came to international notice on 21 March 1960 when a large gathering of blacks staged a demonstration there against the pass law. In an attempt to disperse the demonstrators, the police fired at the crowd, killing 67 and wounding 186.

Shatt al-Arab, channel in SE Iraq, formed by the confluence of the rivers Tigris and Euphrates. It flows SE to the Persian Gulf through a wide delta and is navigable for ocean going vessels as far as Abadan.

Shaw, George Bernard (1856–1950), British dramatist, critic and member of the FABIAN SOCIETY, *b.*Ireland. He replaced the artificial MELODRAMAS, then dominating the Victorian stage, with vigorous dramas of ideas. His plays, noted for their mastery of prose and brilliant wit, include *Candida* (1895), *Man and Superman* (1903), *Pygmalion* (1913), *Heartbreak House* (1917) and *Saint Joan* (1923). He received the 1925 Nobel Prize in literature. *See also* HC2 pp.101, 210, *212,* 282.

Shaw, Irwin (1913–·), US novelist, dramatist and short story writer. Much of his work deals with the problems facing men in wartime, for instance *Bury the Dead* (1936) an early one-act play, and *The Young Lions* (1948). Other works include the short story collection *Tip on a Dead Jockey* (1957) and a novel *Rich Man, Poor Man* (1970), which was made into a successful television series.

Shaw, Richard Norman (1831–1912), British architect whose domestic architecture recreated Queen Anne and Georgian styles, rather than the more prevalent Tudor and Gothic. An associate of William MORRIS, Shaw emphasized traditional workmanship. In 1875 he created London's first garden suburb at Bedford Park, and designed New Scotland Yard (1888). *See also* HC2 p.122.

Shaw, Robert (1927–), British actor and author. His films include *The Dam Busters* (1955), *A Man for all Seasons* (1966), *The Sting* (1973) and *Jaws* (1975). He wrote the play *The Man in the Glass Booth* (1967) and the novel *The Sun Doctor* (1961).

Shaw, T.E. *See* LAWRENCE, THOMAS EDWARD.

Shawm, double-reed woodwind instrument, the ancestor of the OBOE. Shawms were of various shapes and lengths, between them they encompassed a range from high treble to base. They were much used in European music from the 13th to the 17th century.

Shearer, Norma (1904–), Canadian film actress who made her debut in 1920 and whose film career with MGM was guided by Irving THALBERG, whom she married in 1927. She made notable appearances in *The Barretts of Wimpole Street* (1934), *Romeo and Juliet* (1936) and *The Women* (1939). She received an Oscar for her role in *The Divorcée* (1929), and retired in 1942.

Sheares, Benjamin Henry (1907–), President of Singapore. He was professor of obstetrics and gynaecology at the University of Malaya (1950–60) and then went into private practice. He became President of the Republic of Singapore in 1971.

Shearwater, seabird related to the ALBATROSS and PETREL. Most species are brown or black with pale underparts. Shearwaters live most of their lives on the ocean; they feed on fish and squid and burrow nests in coastal cliffs. Length: 19–56cm (7.5–22in). Family Procellariidae. *See also* NW p.*140.*

Sheba, ancient kingdom of southern Arabia celebrated for its trade in gold, spices and precious stones. Its queen came from the "uttermost parts of the earth" to hear the wisdom of SOLOMON as recorded in the Bible.

Shebelle, river that rises in central Ethiopia. It flows SE into central Somalia, 32km (20 miles) inland from the coast of the Indian Ocean, following the coastline for approx. 320km (200 miles) and ending in swampland N of the Juba River. Length: 2,013km (1,250 miles).

Sheene, Barry (1950–), British motorcycle racer who lived up to the reputation established by superb promotion when, riding Suzukis, he won the 500cc world road racing titles in 1976 and 1977. In 1973 he had won the first-ever international 750cc championship.

Sheep, ruminants of the genus *Ovis,* together with those of the far less numerous genera *Pseudois* and *Ammotragus* (both non-European). Domestic sheep,

Shanghai; rickshaws and bicycles are the most popular forms of transport.

Ravi Shankar gave concerts in 1971 in aid of the Bangladesh fund.

Sharks have graceful, streamlined bodies, enabling them to move at speed.

George Bernard Shaw defended his views in the prefaces to his plays.

Sheepdog

Shelley's cremation: detail of a painting which shows Byron near to the pyre.

Shensi province; the gigantic Buddha and the Yun Kang caves at Tatung.

Sheriffmuir, although indecisive, ended the Jacobite rising of 1715.

Sherlock Holmes; a reconstruction of the detective's revolver, pipe and glass.

Ovis aries, are now bred for several purposes: fine wool, coarse wool, fur (karakul) and mutton. Wild species are found in the mountains of Europe, Asia, Africa and North America. The blue sheep (*Pseudois nayaur*) of W China is the only wild sheep that does not interbreed with domestic sheep. All are of the Family Bovidae. *See also* WOOL; PE pp.*156*, 158, *163*, 224, *225*, 231.

Sheepdog, any of several breeds of dog that were originally bred to herd and guard sheep. The term includes such breeds as the OLD ENGLISH SHEEPDOG, SHETLAND SHEEPDOG, COLLIE, BORDER COLLIE and ALSATIAN. Today, collies are most commonly used for the original purpose.

Sheep liver fluke, parasitic FLATWORM that collects in liver bile passages of cattle and sheep. It develops in a host snail, then a tiny, tailed version of the adult fluke escapes, encysts on grass, and is eaten by livestock. Length: 19mm (0.75in). Phylum Platyhelminthes; class Trematoda; species *Fasciola hepatica*. *See also* NW p.*86*.

Sheerness, seaport and resort town in N Kent, SE England. Strategically located at the confluence of the rivers Thames and Medway, the fortress settlement was captured by the Dutch in 1667.

Sheffield, city and county district in South Yorkshire in N England. The city lies at the confluence of the Don River and its tributaries, the Sheaf, Rivelin and Lordey rivers. It is a major industrial city, noted for its special steels and steel products, in particular cutlery which has been manufactured there since the 14th century. The first BESSEMER steel plant was constructed in Sheffield in 1858. The university college of Sheffield (1897) received its charter as a university in 1905. The county district was created in 1974 under the Local Government Act (1972). Area: 242sq km (150sq miles). Pop. (1976 est.) 558,500.

Sheffield plate, any article of silverware, made of copper coated with a fused layer of silver, not a deposit of silver like that achieved by electroplating. In 1742 Thomas Boulsover discovered that copper coated with a layer of silver remained workable, retaining the pliability of both metals. Sheffield plate was granted an authorized hallmark in 1774 and today it is a collector's item.

Sheffield Shield, Australian cricket trophy awarded to the winners of the annual competition between the five states. First contested in the season 1892–93, it has been dominated by New South Wales. Matches are played over four days with positions determined on a league table system.

Sheik, The (1922), US silent film which starred Rudolph VALENTINO. A number of other films which romanticized the desert and Arab way of life were made in the 1920s. These included *Blood and Sand* (1922) and *Son of the Sheik* (1926), and also starred Valentino.

Shelburne, William Petty FitzMaurice (1737–1805), British statesman. A WHIG, he served in the ministries of GRENVILLE and CHATHAM and as Prime Minister concluded the 1783 peace with the USA. In 1784 he was made Marquis of Lansdowne.

Shelduck, large colourful DUCK that lives in estuaries and on coasts of Europe, North Africa and Asia. The common shelduck has a dark glossy green head and red, brown and black bands about its wings and body. The female is smaller but similarly marked. Length: 59cm (23in). Species *Tadorna tadorna*. *See also* NW pp. *147, 235, 237*.

Shell, in biology, the chalky protective case secreted by various MOLLUSCS. Shells have been important to man as tools, decorations, jewellery and in trade: COWRIE shells are used as money in Africa and as jewellery in the Pacific Islands. WAMPUM was the name given to shell-beads used by North American Indians. *See also* NW pp.90–97.

Shell, in artillery, explosive projectile that ranges in size from a pistol cartridge to a big-gun shell with a weight of tonnes. Types of shell include high explosive, armour-piercing, illuminating and anti-

personnel (designed to maim as many people as possible). *See also* MM pp.242–243.

Shellac, purified RESIN made from the secretions of the lac insect, *Laccifer lacca*. Shellac is soft and fluid (it flows) when heated but hard at room temperature, and is used in the production of adhesives, hair sprays, varnishes and sealants.

Shelley, Mary Wollstonecraft (1797–1851), British novelist, daughter of William GODWIN and Mary Wollstonecraft. She went abroad with Percy Bysshe SHELLEY in 1814 and married him in 1816. Her works include *Frankenstein* (1818), *The Last Man* (1826) and *Lodore* (1835).

Shelley, Percy Bysshe (1792–1822), English Romantic poet whose passions and enthusiasms entered both his life and his writing, making of the two an ambiguous mixture of beauty and sordidness. His first wife committed suicide when he deserted her for Mary SHELLEY. Other poems include *Prometheus Unbound* (1820), *Ode to the West Wind* (1819) and *Adonais* (1821), a poem partly based on the life of John KEATS.

Shellfish, common name for shelled MOLLUSCS and CRUSTACEANS used as food by man. Shelled molluscs include CLAMS, OYSTERS and SCALLOPS; crustaceans include SHRIMPS, LOBSTERS, CRABS and CRAYFISH. *See also* NW pp.92–93; PE pp.242, 248.

Shell Shield, West Indian cricket trophy competed for annually by the major cricketing countries in the Caribbean. It was donated in 1965 by the Shell Oil Company to promote inter-island cricket.

Shell shock, also called battle fatigue, acute hysterical NEUROSIS caused by TRAUMA suffered in warfare as a result of overwhelming stress. Patients suffer from hypersensitivity to noise, movement and light, reacting defensively to such stimuli; they may be irritable and violent and commonly experience sleeplessness and nightmares. Treatment includes rest, food and sedation. *See also* MS p.*135*.

Shem, in GENESIS, eldest son of NOAH. He received the first blessing of Noah after the FLOOD and settled on land between that of his brothers HAM and JAPHETH.

Shen Chou (1427–1509), Chinese ink painter and leader of the WU SCHOOL. Although he copied the style of masters of the YUAN DYNASTY, he is well known for his small hand-scrolls of landscapes. *See also* HC2 p.54, *54*.

Shensi (Shănxī), province in central China; capital Sian (Xi'an). The province can be divided into four geographical regions; in the N is a fertile plateau where crops are raised with the aid of irrigation, and there are rich deposits of coal and iron ore; the Wei River valley which has rich farmlands and where most of the population is concentrated; the Chinling mountain range; and the Han River valley where rice and tea are grown. Textiles and timber are important industries. Area: 195,880sq km (75,630sq miles). Pop. 21,000,000.

Shenyang, capital of Liaoning province, NE China; formerly known as Mukden. One of the largest cities in the country, Shenyang is the centre of a group of communes which grow rice, cereals and vegetables under intensive cultivation. It is also part of an industrial area whose products include aircraft, machine tools, heavy machinery, cables and cement. Pop. 2,800,000.

Shepard, Ernest Howard (1879–1976), British artist whose reputation was established with the definitive illustrations he provided for children's books such as A. A. MILNE's *Winnie-the-Pooh* (1926).

Shepheardes Calender, The (1579), twelve poems by Edmund SPENSER. His earliest major work, they are in the pastoral tradition of THEOCRITUS and VIRGIL.

Shepherd's purse, wild flower growing throughout the world. It has deeply lobed basal leaves and tiny, white flower clusters. The pouch-like seedpods have a peppery flavour. Height: 20–45cm (8–18in). Family Cruciferae; species *Capsella bursa-pastoris*.

Shepstone, Sir Theophilus (1817–93), South African politician. His administra-

tive skill and adroitness in dealings with African peoples led to his appointment as Secretary for Native Affairs in 1856. He carried out the annexation of the Transvaal in 1877.

Sherardizing, industrial method of coating steel parts with zinc for protection from corrosion. The parts, which may be steel castings, nuts and bolts, welded tubes or sections, are tumbled in heated drums containing zinc dust. After some hours the zinc forms a very thin intermetallic surface coating the steel. *See also* GALVANIZING.

Sheraton, Thomas (1751–1806), British furniture designer whose designs became influential through his manuals, such as *Cabinet-Maker and Upholsterer's Drawing-Book* (1791). His style is marked by a delicacy and simplicity with an emphasis on straight, vertical lines.

Sheridan, Richard Brinsley (1751–1816), British dramatist and politician who became part owner and director of the DRURY LANE THEATRE in 1776. He excelled in comedies of manners, such as *The Rivals* (1775) and *The School for Scandal* (1777). Entering parliament as a WHIG member in 1780, he became one of the most brilliant speakers of his time. *See also* HC2 p.27.

Sheriff, royal officer in medieval England; the name is a shortened form of "shire reeve". Originally each sheriff looked after the royal DEMESNE in his shire, but by the 11th century the sheriff had become the chief officer of local administration. After the 13th century the office declined in importance. It is now chiefly ceremonial.

Sheriffmuir, Battle of (1715), bloody but indecisive clash at Sheriffmuir, Scotland, between JACOBITE rebels under the Earl of Mar and Hanoverian forces led by the Duke of Argyll. *See also* HC2 p.28.

Sherlock Holmes, fictional private detective, a character in works by Arthur Conan DOYLE. He first appears in *A Study in Scarlet* (1887). Holmes has many eccentricities and mannerisms, but has amazing powers of observation and detection, and knowledge of obscure subjects. Dr Watson acts as his ponderous foil, and Professor Moriarty is a worthy villain-antagonist.

Sherman, General William Tecumseh (1820–91), US military officer. As a brigadier-general during the AMERICAN CIVIL WAR he took part in the Battle of BULL RUN and in the capture of VICKSBURG. His most noted exploit was his march eastwards through Georgia, including the "march to the Sea" from Atlanta to Savannah in 1864.

Sherpas, Buddhist people who inhabit the Khumba Valley in NE Nepal. Of Tibetan origin, they are renowned for their portering of Himalayan mountaineering expeditions. They cultivate cereals and potatoes and herd sheep and cattle.

Sherriff, Robert Cedric (1896–1975), British dramatist, who wrote the war play *Journey's End* (1929). Other plays include *Windfall* (1933), and *Home at Seven* (1950). He also wrote film scripts, such as *The Invisible Man* (1933), *Goodbye Mr. Chips* (1936) and *Odd Man Out* (1945).

Sherrington, Sir Charles Scott (1857–1952), British physiologist. A pioneer in the study of how the nervous system works, his book *The Integrative Action of the Nervous System* (1906) helped to establish physiological psychology. He shared with Edgar ADRIAN the 1932 Nobel Prize in physiology and medicine. *See also* MS p.37.

Sherry, fortified wine with a characteristic flavour produced by admitting air to the wine during maturation. Sherries range from the light yellow dry (fino) to the sweeter and richer wines of golden brown colour (oloroso). The name sherry derives from Jerez, a town in southern Spain.

Sherwood Forest, ancient royal hunting ground in Nottinghamshire, central England. Little of the original forest remains. The forest is famous as the home of ROBIN HOOD.

Sheshonk (r. c.935–c.914 BC), king of ancient Egypt and founder of the 22nd or Libyan dynasty. After the decline of the dynasty at Tanis, a city in the Nile delta, he assumed the kingship and seized Gaza and Palestine *c.* 930 BC. Excavations at the temple at Karnak revealed records of the tribute paid him in Nubia and Palestine. His tomb was discovered in 1938.

She Stoops to Conquer (1773), five-act comedy by Oliver GOLDSMITH. Originally subtitled *The Mistakes of a Night*, it concerns the relationships of young Marlowe and a friend with the Hardcastle family. *See also* HC2 p.27.

Shetland Islands, group of approx. 100 islands NE of the Orkneys, 209km (130 miles) off the N coast of Scotland, in Shetland Region. The principal islands are Mainland (which has the main town Lerwick), Yell, Unst, Whalsay and Bressay. The islands are generally rocky with thin soils, but oats and barley are grown in places. Fishing and the rearing of livestock are of great importance: the islands are famous for breeding ponies. The region is also noted for its production of Fair Isle woollen goods. Tourism is a major industry. Pop. (1971) 17,298.

Shetland pony, one of the smallest and oldest types of light horses. Originating in the Shetland Islands, it has since been introduced to most parts of the world. A small, hardy horse, it has been used for draught or road work and also makes an ideal children's mount. It can be any horse colour, piebald or skewbald. Height: 71cm (28in) at the shoulder; weight: 169kg (350lb).

Shetland sheepdog, miniature COLLIE used as a guard and farm dog; also called Sheltie. Probably bred from small collies, it has a long blunted, wedge-shaped head; high-set, small ears held three-quarters erect; a moderately long body; medium-length legs and a long tail. The long, straight, double coat is black, blue-grey or sable with white or tan marks; it is short and smooth on the face, feet and ears. Average size: 33–41cm (13–16in) at the shoulder; weight 7kg (15lb).

Shi'a, rite of Islam, whose followers are called Shi'is or SHI'ITES.

Shield, in geology, large, low-relief, exposed mass of PRECAMBRIAN rock, commonly having a gently convex surface and surrounded by belts of younger rock. Shields contain rock that is more than 2,500 million years old, now changed by METAMORPHISM but originally composed of basaltic lava. Shields form the nucleus of continents, such as the one occupying-two-thirds of Canada and known as the Canadian Shield.

Shield, in warfare, form of personal armour that protects the carrier from blows and missiles aimed at his body. Materials for shields ranged from the bronze used by the Ancients, to the remarkably effective hides employed by African and North American tribes, to highly ornate iron or steel in shields of Medieval knights. Shields became obsolete with the advent of gunpowder. *See also* MM pp.160–161.

Shield, radiation, material, such as loaded concrete (concrete containing lead, iron or barium) or lead, surrounding a radioactive source to absorb NEUTRONS and RADIATION. The core of a NUCLEAR REACTOR is surrounded by a massive concrete biological shield to protect personnel, and an inner iron thermal shield to protect the concrete from high temperatures.

Shield, spacecraft, metal or plastic plates or structures that protect the interior of a spacecraft from heat of re-entry, radiation (eg, solar radiation and cosmic rays) or micrometeorites (small objects in space). *See also* SHIELD, RADIATION; ABLATION; PTFE; MM pp.156–157.

Shigella, genus of bacteria. They form no gas as part of their metabolic process, and produce dysenteries (shigellosis) in man and other animals. *See also* DYSENTERY.

Shih Ching, the Classic of Odes, one of the Five Classics in the canonical literature of CONFUCIANISM. The first anthology of Chinese poetry, it was compiled by Confucius (551–479 BC). *See also* HC1 p.125.

Shih Huang Ti (259–210 BC), Emperor of China (r. 221–210 BC). The first king of the CH'IN DYNASTY to rule the whole of China, he organized the country into feudal provinces, reformed the bureaucracy and

completed the GREAT WALL OF CHINA. *See also* HC1 p.122.

Shih tzu, dog bred in the imperial court of China as a pet of the nobility; first brought to the West about 1930. It has a square, short muzzle with a long moustache; a long, broad-chested body; short muscular legs; and a tail that curves up over its back. The long, luxurious coat is slightly wavy and may be any colour. Standard size: to 26.5cm (10.5in) high at the shoulder; weight 5–7kg (9–16lb).

Shi'ites, followers of the MUSLIM Shi'a sect. They are distinguished from the Sunnis by their belief in ALI as the first true CALIPH of Islam and their rejection of the Sunna. The most important subsect is the Isma'ili group. SUFISM also draws on many Shi'a beliefs. The largest Shi'a communities are found in Iran, Iraq and India.

Shikoku, smallest of the four main islands of Japan, S of Honshū and E of Kyūshū. The interior is mountainous and extensively forested and most settlements are on the coast. The principal cities are Matsuyama, Takamatsu and Tokushima. Products: rice, tea, wheat, timber, fish, tobacco, fruit. Area: 18,780sq km (7,245sq miles). Pop. 3,904,014.

Shillelagh, village in County Wicklow, Republic of Ireland. Its forest gave its name to a type of cudgel made from oak or blackthorn.

Shilling, English coin, first made in 1504. It had originally a deeply indented cross and was easily divided into halves and quarters. It was replaced by the new five-pence piece in 1971.

Shiloh, ruins of a Biblical city of Jordan, on Mt Ephraim 24km (15 miles) W of the River Jordan. The modern town is called Khirbat Saylun, and is in the Israeli occupied West Bank area.

Shiloh, Battle of (6–7 Apr. 1862), battle in the AMERICAN CIVIL WAR, also called "Pittsburg Landing". In this engagement in Tennessee, the Confederate Albert S. Johnston surprised a Unionist force led by Ulysses S. GRANT. On the second day, Grant regained the lost ground and eventually won the battle. There were 13,000 Union and 10,000 Confederate casualties. *See also* HC2 p.151.

Shimazaki, Tōson (1872–1943), Japanese novelist and poet who was converted to Christianity in 1888. His work consisted of idealistic, romantic poetry collected in *Wakanashu (Collection of Young Leaves,* 1907) and many novels including *Hakai (The Broken Promise,* 1906) and *Yoake-mae (Before the Dawn,* 1929–35).

Shin Buddhism, sect of Japanese MAHAYANA BUDDHISM known as the Pure Land sect. It was founded by Shinran (1173–1262), and stresses that sinful human nature can only be saved through the grace of Amida, the saviour-BUDDHA. Maintaining the proper attitude of faith is central to Shin's teaching that secular life is no bar to salvation. The sect has more than 13 million followers.

Shingles, or herpes zoster, acute virus infection of the skin and nerves. Groups of small blisters appear along segments of nerves, usually on the side, sometimes preceded by pain in the affected place. *See also* MS p.99.

Shingon ("True Word"), a translation of the Sanskrit *mantra,* the name of a mystical Japanese BUDDHIST sect founded in the 9th century by Kobo Daishi, who was also known as Kukai. Shingon teachings describe the development of spirituality in terms of stages which are reflected in various forms of religious belief and which culminate in the esoteric doctrines of their own school.

Shinto, the "way to the gods", indigenous religion of Japan. Originating as a primitive cult of nature worship, it was shaped by the influence of CONFUCIANISM and BUDDHISM from the 5th century until the 17th century, when a revival of the ancient rites began. Shinto has a hierarchy of gods, which are the forces in mountains, rivers, trees and other forms of nature. It is concerned with ceremonial purity. The followers of Shinto in its present form number more than 60 million. *See also* MS p.216.

Shinty, Scottish version of hockey. Also

known by its Gaelic name *camanachd,* it is a 12-a-side game played on a rectangular pitch up to 182m (600ft) long and 90m (300ft) wide. The object is to hit the ball through the opposing goal (*hail*). The stick (*caman*) is longer than a hockey stock and has a broader blade.

Ship, vessel for conveying passengers and freight by water. Modern ocean-going ships developed from the carracks of the 13th century and the larger early ships called galleons. Fighting ships of the 17th and 18th centuries included frigates of various designs, later models had several rows of guns. Sailing freighters led to the development of the great clippers of the late 19th century, some of which had iron hulls. A century or so earlier, the first steamships had been built. They were powered by wood or coal-burning steam engines that drove large paddle wheels, hence the term PADDLESTEAMER. In 1819 the first Atlantic crossing was made by "steam-assisted sail" and this crossing soon became a regular service. By the mid-19th century steamships driven by propellers, or screws, were in general service and were voyaging on trade missions in all oceans. Marine steam TURBINES were developed at the turn of the 19th century and gradually replaced reciprocating (back-and-forth cranking) steam ENGINES for large vessels, early examples being ocean liners of the 1900s with displacements (weight of water that a ship displaces when fully laden) up to 10,000 tonnes. Oil, rather than coal, was now the favoured fuel for large marine engines. Some of the newest military ships and icebreakers are fitted with nuclear engines in which heat from a NUCLEAR REACTOR raises steam in boilers to drive steam turbines. The larger modern ships are classified as warships, passenger vessels, bulk dry cargo ships, TANKERS for bulk liquids, freighters for mixed cargo and container ships. The largest of these are the supertankers carrying oil, with displacements of about half a million tonnes and lengths of several hundred metres. *See also* BOAT; SAILING SHIPS; MM pp.110–119, 172–175.

Shipbuilding, industry concerned with the design, making and repair of ships. All ships have similar parts. The longitudinal keel, the stempost and the sternpost are the basic components. A series of symmetrically curved ribs, running transversely and fastened to the keel at their centres, give the ship its shape. Longitudinal stringers, or clamps, run the length of the ship, holding the ribs in position. Additional braces extend across the width of the ship; these serve as supports for the deck and superstructure. The skin, which is mounted on the framework of ribs, may be of riveted or welded metal plates, or horizontal wooden planks. Transverse walls, called bulkheads, stiffen the frame and divide the hull into watertight compartments. After the various parts have been prepared, the ship is assembled either on slipways or in a dock. The major shipbuilding nation in the world is Japan, followed by Sweden, West Germany and Great Britain. *See also* MM pp.110, 112–119, 172–175.

Ship money, ancient English tax whereby sea ports provided ships for the Crown in times of war. By the 17th century it had been commuted to a money tax. CHARLES I revived it in 1634 to provide a permanent source of income without calling parliament. When it was extended to inland towns in 1635, John HAMPDEN, a Buckinghamshire squire, created a test case by refusing to pay. The case was the first great blow against Charles' attempt at arbitrary rule. *See also* HC1 p.287.

Ship of Fools. *See* NARRENSCHIFF, DAS.

Ship of the line, large fully-rigged naval sailing ship serving in the major battle formation, "the line", of the 17th and 18th centuries. British warships of the period were rated by the number of cannon carried: first- through to sixth-rate ships carried more than 100, 90, 64, 44, 32 and 20 cannon respectively. Only first-, second- and third-rate ships were considered strong enough to take part in the battle line, and were thus designated ships of the line. *See also* MM p.173, *173.*

Shetland ponies used to pull coal carts in mines because of their strength.

Shilling: an example of one of the first issues, minted in the reign of Henry VII.

Shinto is an unusual religion in that it does not decree a life after death.

Shipbuilding at Malmö, Sweden, an important Shipyard on the Baltic.

Shire horses were bred for use as war horses in the Middle Ages.

Shoebills may stand motionless; they are sluggish birds on land but can fly strongly.

Shooting, as a sport, is organized for shotguns, rifles and pistols.

Shostakovich has been criticised in the USSR for lacking socialist realism.

Shipton, Eric Earle (1907–), British mountaineer. He scaled many of the mountains of E and Central Africa in the 1930s and between 1933 and 1951 was present on five expeditions to Mt EVEREST, paving the way for the successful British ascent in 1953. His published works include *Upon that Mountain* (1948).

Shipworm. See TEREDO.

Shiraz, city in s central Iran, approx. 185km (45 miles) ENE of Bushire. It has been an active trade centre since the 7th century and by the 9th century had also developed into an important Muslim centre. Shiraz was the capital of Persia 1750–94 and many buildings date from that period. The city's products include wines, carpets, metalwork, textiles, cement and sugar. Pop. (1973) 373,000.

Shire, unit of local government in ANGLO-SAXON England that replaced the Roman *provincia* and became the COUNTY of Norman administration. A shire was administered by an EALDORMAN and a shire-reeve, the latter acquiring great power after the NORMAN CONQUEST as the king's representative. *See also* HC1 p.165.

Shire horse, earliest breed of draught horse. It was bred in its present form in England during the 18th century and is a descendent of the great horse of the Middle Ages. Also called war horse and cart horse, this heavy-boned horse has a sluggish temperament. Its colours may be bay, brown or black with white markings. Height: approx. 1.7m (5.7ft) at the shoulder; weight: 1,000kg (2,200lb).

Shirley, James (1596–1666), English dramatist. He wrote many plays in the period before the Civil War, after which he returned to his original profession of schoolmaster. The 36 plays and masques which survive include the comedy *The Gamester* (1633), the masque *The Triumph of Peace* (1634) and a revenge tragedy *The Cardinal* (1641). He is said to have died from fear and exposure in the Great FIRE OF LONDON.

Shirley (1849), novel by Charlotte BRONTË. It is a story of conditions in industry in the mid-19th century, a question of great contemporary concern. The heroine of the title is an heiress who influences workers to accept the changes proposed by the factory owner, Robert Moore.

Shiva (Siva), one of the three great HINDU gods, with BRAHMA and VISHNU. A complex god, he represents both reproduction and destruction, although a combination of contradictory qualities is not uncommon in HINDUISM. His name means "Auspicious One" in SANSKRIT. The cult of Shiva is one of the most popular in modern Hinduism. *See also* HC1 pp.*37, 118*; MS pp.*212, 220*.

Shkodër, town in NW Albania, at the SE end of Lake Scutari. It is the cultural and industrial centre of N Albania. Manufactures include food products, textiles and leather goods; fishing is another important activity. Pop. 55,000.

Shock, acute progressive circulatory failure, due to the heart's inability to pump enough blood through the major organs. It may be caused by injury, burns, haemorrhage, major surgery or poisoning. It is characterized by weakness, shallow breathing, rapid heartbeat and low blood pressure. Treatment involves keeping the patient warm and giving oxygen. *See also* MS p.133.

Shock absorber, elastic mechanism for absorbing shock to prevent damage to a machine. The shock absorbers of motor vehicles typically have a piston, attached to the vehicle body, moving inside a cylinder attached to a wheel. The piston moves against a liquid which escapes from and returns to the cylinder through a spring-loaded valve.

Shockley, William Bradford (1910–), US physicist instrumental in the development of the TRANSISTOR. For this he shared with John BARDEEN and Walter BRATTAIN the 1956 Nobel Prize in physics.

Shock therapy, induction of transient comas or convulsions by electrical, chemical or other artificial means in an attempt to treat severe mental illness. Electroconvulsive therapy (ECT) has been the most common method; another procedure commonly used before the discovery of

tranquillizers was insulin shock, in which hypoglycaemia and shock are induced by injections of INSULIN. *See also* MS pp.144, 145.

Shock wave, in fluids (liquids or gases), region across which sharp discontinuities in pressure and density occur. Shock waves are brought about by objects moving at supersonic velocities. Because the surrounding fluid can propagate disturbances only at the local speed of sound, the moving object "piles up" the disturbances it is causing into a V-shaped "wake" attached to the object. The "sonic boom" of supersonic aircraft is the passage of this shock wave past the eardrum. *See also* EARTHQUAKE; SEISMIC WAVES; SU pp. *79, 81*.

Shoebill stork, tall wading bird found in papyrus marshes of tropical NE Africa. Also known as the whale-headed stork, it has a swollen, shoe-shaped bill with a sharp hook, a short neck, darkish plumage and long legs. It feeds at night on small animals, including lungfish, frogs and turtles. Height: to 1.4m (4.6ft). Family Balaenicipitidae; species *Balaeniceps rex*. *See also* NW p.*232*.

Shofar, ancient Jewish musical instrument, still sounded in synagogues on ROSH HASHANAH and YOM KIPPUR. Made from the curved horn of a ram, it is flattened and bent and a cup mouthpiece is formed at the end.

Shōgun, title of the military ruler of Japan, first conferred upon YORITOMO in 1192, acknowledging him as the supreme military commander in Japan. The term means "Barbarian-quelling general". The Minamoto (1192–1333), ASHIKAGA (1338–1568) and TOKUGAWA shōgunates (1603–1868) provided the *de facto* rulers of feudal Japan, although there was still an emperor. Their authority was in theory purely military, but the bakufu ("tent government") exercised control over all aspects of life. The shōgunate ended with the MEIJI RESTORATION in 1868.

Sholes, Christopher Latham (1819–90), US journalist and politician who developed the first commercially successful typewriter. Development costs forced him to sell his patent rights to the Remington Arms Company who further developed and marketed the machine. *See also* MM p.254.

Sholokhov, Mikhail Alexandrovich (1905–), Soviet novelist who became internationally famous for his epic novel about his native land, *The Silent Don* (1928–40; trans. as *And Quiet Flows the Don*, 1934), and *The Don Flows Home to the Sea* (1940). *Virgin Soil Upturned* (1932–60) deals with the collectivization of agriculture. Several of his short stories are collected in *Tales of the Don* (1926). He was awarded the 1965 Nobel Prize in literature.

Shona, negroid people of E Zimbabwe (Rhodesia). The Shonas' exogamous, patrilineal society is based on subsistence agriculture and is centred on small villages, traditionally abandoned when the local resources are exhausted. Their culture is noted for its pottery, music and dance.

Shooting, competitive sport involving a wide variety of firearms. Its rules are fixed by two bodies: the International Shooting Union (ISU) founded 1907 and the National Rifle Association (NRA) founded in Britain in 1860. Shooting with air weapons (ie guns firing projectiles by the force of compressed air, or in some cases carbon dioxide) takes place at ranges of 10m (32.75ft); the competitor stands without any support. This is not a recognized Olympic sport.

Competition with shotguns consists of firing at clay pigeons, saucer-shaped discs 11cm (4.5in) in diameter which are projected at speed from a springloaded trap. The targets are hit when they are about 36.5m (120ft) from the competitor. There are two forms of competition: "Trench" (an Olympic sport since 1900), which involves firing at 200 clay pigeons from a set position, and "skeet" (an Olympic sport since 1972), in which the competitors move round a set of positions.

Pistol shooting has been an Olympic event since 1896; competition is over ranges from 25–50m (82–164ft). It includes rapid and free fire for automatic, standard and centre-fire pistols. Rifle shooting has also been an Olympic sport since 1896. Ranges may be up to 914m (3,000ft) and the competitors may have to adopt certain stances – such as kneeling or standing – or may be allowed to use the more efficient prone position. There are various classes of rifle shooting, depending on the weapons available to the competitor.

Shop, permanent retail establishment for the sale of goods. Deriving its name from the Teutonic word for a cobbler's stall or booth, it became common usage in about the 14th century. Originally places where craftsmen or merchants sold their own products, shops did not undergo any significant changes until the growth of the manufacturing industries and urbanization in the 19th century. The retail industry flourished during this time, aided by the development of the wholesale industry and the increased purchasing power of people. Department stores and chain stores such as Woolworths developed during this retail boom, and supermarkets and shopping complexes have developed in the 20th century, particularly since WWII. *See also* HC2 p.*159*.

Shops, early closing, in Britain, legal requirement that every shop should close for one full day and one half day each week; the half day should not extend beyond 1 pm. Early closing is regulated by the Shops (Early Closing Days) Act of 1965, but it is also covered by local byelaws, and any shop wishing to stay open for six or more full days a week may do so with local authority permission.

Shop steward, TRADES UNION official elected by his or her colleagues to represent their interests in management/staff negotiations. *See also* MS p.275.

Shorthand, system of codes used to record speech, which date from as early as Roman times. Phonetic shorthand systems first appeared in the 18th century, and the most famous system of this kind, Pitman's shorthand, in 1837. All the sounds of the English language are represented by 49 signs for consonants and 16 signs to indicate vowels. A simplified version was also developed. The Gregg system of phonetic shorthand, widely taught in the USA, has a script based upon the pen movements of ordinary writing. Some recent types of shorthand, often referred to as speedwriting, use no special signs but rely on abbreviations of ordinary written language.

Shorthorn, most widespread breed of BEEF CATTLE. Shorthorns are red, red and white or roan. They were originally developed in Durham, England in the late 1700s. There are two strains, one strictly for beef and one which is also used for milk. Polled shorthorns have no horns. *See also* CATTLE; PE p.*226*.

Short Parliament (1640), English parliament that ended 11 years of personal rule by CHARLES I. It was called because Charles needed money after his defeat in the first of the Scottish Bishops' Wars in 1639. The Short Parliament was dissolved after three weeks, after its members made redress of political grievances the conditions for financial subsidy for the king. *See also* HC1 p.287.

Short wave. See RADIO.

Shoshone Indians, North American Indians of the UTO-AZTECAN language group. They occupy reservations in California, Idaho, Nevada, Utah and Wyoming. They were divided into the COMANCHE, the Northern, the Western and the Wind River Shoshone. In the 1970s, they numbered about 9,600.

Shostakovich, Dmitri Dmitrievich (1906–1975), Soviet composer who worked prolifically in every form and is perhaps the most important modern Russian composer since TCHAIKOVSKY. His works include 15 symphonies, 13 string quartets, ballets, concertos, piano music, film music and vocal works. The opera *Lady Macbeth of Mzensk* (1934) was disapproved of by the Soviet government but he was restored to favour with his *Fifth*

Symphony (1937). *See also* HC2 p.270, *270*.

Shot put, athletics field event in which a heavy spherical shot is projected from a position close to the neck by an extension of the arm. Throughout the put the athlete must remain inside the throwing circle. Men throw a 7.257kg (16lb) shot; women a 4kg (8lb 13oz) shot. The shot put has been an Olympic event since 1909. Modern athletes often put the shot to distances exceeding 21m (68ft).

Shoulder, in human anatomy, joint at the top of the arm. It consists of the ball-and-socket joint between the upper arm (humerus) and the shoulder blade (scapula). It is the most movable joint in the body. The main muscle in the shoulder is the deltoid.

Shoveller, widely distributed river duck that dabbles, or feeds from the water surface. The male engages in a complex courtship routine to attract a female. Family Anatidae. Species *Spatula clypeata.*

Showboat (1927), musical with lyrics by Oscar HAMMERSTEIN II and music by Jerome KERN, originally produced on Broadway and adapted from a novel by Edna FERBER. It was made into a film in 1929.

Showboat, paddleboat or steamboat first used in 1831 to bring entertainment to pioneers settled on western US rivers, especially the Mississippi and the Ohio.

Showjumping, equestrian sport in which horse and rider jump a series of fences arranged in a special order within an arena. As well as being an event in its own right, it is a part of horse trials. Showjumping competitions were first organized in the 1860s. There are now three main types of competition, the most common being that in which penalty points (faults) are incurred for failures at fences and the water jump, for refusals, and for exceeding the time allowed. *Puissance* events test jumping ability alone, and speed competitions test hardiness and obedience.

Shrapnel, Henry (1761–1842), British artillery officer who, in about 1793, invented the shrapnel shell, a device containing a number of "bullets" and a charge of powder fired by a time-fuse that burst the shell and showered its contents over the enemy.

Shrew, the smallest mammal, found throughout the world. It is an active, voracious insectivore that eats more than its own weight daily, feeding on any animal that it can overpower (mainly insects). Some shrews weight only 2g (0.07oz). Family Soricidae. *See also* NW pp.*161, 206, 212.*

Shrewsbury, Battle of (1403), battle in which HENRY IV of England defeated the rebellious Henry "Hotspur", a leader of the PERCY faction that was attempting to place Edmund, Earl of March, on the throne. Hotspur was killed and the rebellion destroyed.

Shrike, or butcherbird, small perching bird found throughout the world, except in South America and Australia. It dives at its prey – insects, small birds, mice – hitting them with its strong hooked bill and then impaling them on a sharp fence post, twig or thorn.

Shrimp, mostly marine, swimming DECAPOD crustacean including true, sand and pistol shrimps. Its compressed body has long antennae, stalked eyes, a beaklike prolongation, segmented abdomen with five pairs of swimming legs and a terminal spine. Large edible shrimps are often called PRAWNS. Length: 5–7.5cm (2–3in); some freshwater shrimps reach 20cm (8in). *See also* NW pp.100, 104, *105, 204, 241.*

Shropshire. *See* SALOP.

Shropshire Lad, A (1896), collection of poetry by A.E. HOUSMAN. It was the first of two of his volumes of verse and was extremely successful. It contains poems in a simple, pastoral style, many of which are ballads.

Shrove Tuesday, in the Roman Catholic Church, the day before ASH WEDNESDAY, which is the beginning of LENT.

Shrub, low, woody perennial plant of limited height, usually less than 10m

(30ft) high. Instead of having a main stem, it branches at or slightly above ground level into several equally strong stems.

Shu, in Egyptian mythology, the god who supports NUT, the sky. In the myth of creation he separated Nut from her brother GEB, the earth, and maintained their separation. Shu is also the god of air, the space between the earth and sky. *See also* MS p.208, *208*.

Shubrā al-Khaymah, suburb of Cairo, N Egypt, on the E bank of the River Nile. Originally a market town for the fertile Nile delta, Shubrā al-Khaymah was developed in the 19th century as an industrial centre. Industries: textiles, glass, ceramics. Pop. (1974) 346,000.

Shubun, Tensho (*fl.* mid-15th century), Japanese painter and sculptor, considered the greatest ink painter of his time. He was the official painter of the SHŌGUN and was a pioneer in the development of monochromatic ink painting.

Shute, Nevil (1899–1960), British novelist, real name Nevil Shute Norway. He wrote more than 20 novels whose fast-moving plots usually have moral themes. They include *Ordeal* (1939), *No Highway* (1948), *A Town Like Alice* (1950), *On the Beach* (1957) and *Trustee from the Toolroom* (1960).

Shuttleworth, Sir Thomas Philips Kay. *See* KAY-SHUTTLEWORTH, SIR THOMAS PHILLIPS.

Si. *See* SI-KIANG (XI-JIANG).

Sial, in geology, uppermost of the two main rock-classes in the Earth's crusts. Sialic rocks are so called because their main constituents are silicon and aluminium. They make up the material of the continents and overlay the SIMA.

Siālkot, city in E Pakistan approx. 110km (70 miles) N of Lahore, situated between the Chenab River and the Indian border. It is the site of an ancient mausoleum and fortress. Industries: surgical instruments, cutlery, rubber products and ceramics. Pop. (1972) 203,779.

Siam. *See* THAILAND.

Siamang, large, black, arboreal, diurnal ape of Sumatra and Malaya. More robust than the related GIBBON, it has an inflatable, hairless throat sac, used to produce a booming call. It swings strongly from branch to branch, and eats fruit, leaves and flowers. Length: to 55cm (21in). Family Pongidae; species *Symphalangus syndactylus. See also* NW p.214.

Siamese cat, exotic short-haired domestic cat breed of disputed origin. An affectionate pet with a distinctive cry, it has a wedge-shaped head, blue almond-shaped eyes, large ears, long slim body and long tapered tail. Small or medium-sized, its albino to fawn coat has darker coloured points on the ears, mask, feet and tail. Recognized varieties: seal point, chocolate point, blue point and lilac point.

Siamese fighting fish. *See* FIGHTING FISH.

Siamese twins, conjoined, identical twins sometimes sharing organs. Usually fusion is along the trunk or at the head. When both are normal, except for fusion, surgical separation is sometimes possible.

Sian (Xi-an), city in NW China 128km (80 miles) W of the confluence of the Wei and Yellow rivers. As Siking, the city was the western capital of the T'ANG DYNASTY (618–906). Since the MING DYNASTY (1368–1644) Sian has been its official name. Products include cotton, textiles, iron, steel, electrical equipment and chemicals. Pop. (1970 est.) 1,900,000.

Sibelius, Jean Julius Christian (1865–1957), Finnish composer who represents the culmination of NATIONALISM in Finnish music. He wrote chamber music, piano music and organ music, but is best known for his orchestral works, including seven symphonies, a violin concerto (1903–5), and the tone poems *En Saga* (1892) and *Finlandia* (1899). *See also* HC2 p.107.

Siberechts, Jan (1627–1703), Flemish landscape painter. His early style is displayed in delicate landscapes. About 1672, he settled in England and his style gained in grandeur, often achieved by the use of powerful blues. He painted simple views of country houses such as Longleat.

Siberia (Sibir'), extensive region of NE Asia, comprising most of Asian USSR, extending from the Ural Mts to the Pacific

Ocean, and from the Arctic Ocean to the S Soviet border with China and Mongolia. The region includes areas of tundra, taiga and steppe. Almost half of the area is forested. Siberia is drained by the Ob, Yenisey and Lena rivers. The region can be divided into four geographical areas: the lowlands of the W, the central uplands, the mountains of the S and the NE mountain systems. Two-thirds of the population live in the SW, where industry is concentrated in cities such as Novosibirsk and Omsk. The KUZNETZ Basin has large deposits of coal and iron ore, and there are rich oil and natural gas fields. E Siberia has deposits of gold, mica, diamonds and aluminium. Forestry is an important industry. The fertile black soils of W Siberia favour dairy farming. Wheat, oats, flax, potatoes, rye and sugar beet are grown. The breeding of reindeer, fishing and seal hunting are the chief occupations in the far N of Siberia. The Russian Cossacks conquered Siberia between 1581 and 1644. The Far Eastern territory was held by the Chinese until 1860. Siberia was administered as a colony of the Russian Empire. Mining was developed in the first half of the 19th century, and the region was used as a penal colony and an exile for political prisoners. Settlement on a large scale came after the construction of the TRANS-SIBERIAN RAILWAY. Area: 13,807,020sq km (5,330,896sq miles). Pop. 35,605,000.

Siberian husky, muscular working dog. It has a broad tapering head with erect triangular ears. Its stout body is set on strong, stocky legs. The dense coat, consisting of a downy underlayer and thick outer coat, may be black, white, grey or tan. Height: to 60cm (24in).

Sibley, Antoinette (1939–), British dancer. She joined Sadler's Wells Ballet in 1956 and became a principal dancer in 1960. She has created many new roles and danced all those of the standard repertory.

Sibyl, prophetess of Greek mythology. The Sibyl of CUMAE offered nine books of her prophecies to Tarquinius Superbus of Rome. He refused her price so she began burning the books until he bought the remaining three for the price she had asked for all nine. The temple of Jupiter on the Capitoline Hill housed these Sybilline books, which were consulted only in times of national emergency.

Sica, Vittorio de (1902–74), Italian film director and actor. He first appeared in films in 1932, establishing his reputation with *Gli uomini…che mascalzoni!* (1932). He appeared in nearly 40 films before co-directing *Rose Scarlatte* (1940). With *Sciuscia* (*Shoeshine*; 1946) he was hailed as a proponent of NEOREALISM; his next film, *Ladri de Biciclette* (*Bicycle Thieves*; 1948) is his best-remembered. His subsequent films include *Umberto D* (1952) and *Il Giardino dei Finzi-contini* (*The Garden of the Finzi-continis*; 1970).

Sicily, Italian island in the Mediterranean Sea off the SW tip of the Italian peninsula; with nearby islands it forms an autonomous region of Italy. The largest island in the Mediterranean Sea, Sicily was once part of the mainland, but is now separated from Italy by the narrow Strait of Messina. Mostly mountainous, it has the volcano Mt Etna 3,323m (10,902ft), the highest in Europe. There are fertile valleys in the central plateau. Fishing and farming are the chief occupations, although tourism is extremely important to the economy. The island's principal cities are Palermo and Messina. Strategically situated between Europe and Africa, Sicily has been invaded many times. Area: 25,706sq km (9,925sq miles). Pop. (1968 est.) 4,867,700 *See also* MW p.107.

Sickert, Walter Richard (1860–1942), British painter; founder member of the CAMDEN TOWN GROUP and later the LONDON GROUP. He was influenced by many artists, including James WHISTLER, Edgar DEGAS and the IMPRESSIONISTS. He painted numerous city scenes in subdued colours.

Sickle-cell anaemia, inheritance of an abnormal HAEMOGLOBIN, occurring mainly in negroes. The haemoglobin is sensitive to a deficiency of oxygen, and red blood cells become rigid and sickle-shaped. If anaemia becomes chronic, there may be

Showjumping; Manfred Kloess on Antea at Hickstead in 1969.

Shrikes are good hunters: Louis XIII of France kept them for hawking.

Siamese cats, born white, develop their true colour in their first year.

Jean Sibelius was enabled, by a state grant, to concentrate on composition.

Siddhartha Gautama

Sidewinders hunt their pray early at night when temperatures are lower.

Siena; the Piazza del Campo, where the annual race on horseback takes place.

Sienese School, 13th-16th centuries: St Bartholomew in S. Michele, Valdelsa.

Sierra Madre, Mexico, has the same name as another mountain range in Spain.

bone and kidney abnormalities. *See also* MS p.35.

Siddhartha Gautama. *See* BUDDHA.

Siddons, Sarah (1755–1831), British actress, née KEMBLE. The most successful member of a noted theatre family, she appeared in her father's repertory company. She achieved success in such classic roles as Ophelia, Desdemona and Lady Macbeth.

Side-drum. *See* DRUM.

Side-necked turtle, TURTLE, found only in southern continents, which can bend its neck sideways under its shell to hide its head. There are two families: hidden-necked turtles (Pelomedusidae) of Africa and South America, and snake-necked turtles (Chelidae) of Australia and South America. Several species of side-necked turtles inhabit fresh water. Length: 15–80cm (6–23in). *See also* MATAMATA.

Sidereal period, length of time taken by a planet to make a complete orbit of the Sun (or a satellite of its planet). Thus the sidereal year of Mars is equal to 686.96 Earth days. The sidereal month, week, day, hour, minute and second are appropriate subdivisions of the sidereal year. *See also* SU p.177.

Siderite, green, brown or white mineral, iron carbonate ($FeCO_3$), found in sedimentary iron ores and as vein deposits with other ores. Its crystals are rhombohedral in the hexagonal system and it occurs as massive deposits or in granular form. Hardness 4; s.g. 3.8. *See also* PE pp.96–97.

Sidewinder, or horned rattlesnake, nocturnal RATTLESNAKE found in deserts of sw USA and Mexico. It has horn-like scales over the eyes and is usually tan with a light pattern. It loops obliquely across the sand, leaving a J-shaped trail. Length: to 75cm (30in). Family Viperidae; species *Crotalus cerastes.*

Sidgwick, Henry (1838–1900), British philosopher and author whose ethical theories were based on UTILITARIANISM; his *Methods of Ethics* (1874) is regarded as the most important ethical work in English. *See also* MS p.198.

Sidi-bel Abbès, city in NW Algeria, on the River Mékerra. Settled by European colonists in the mid-19th century, the modern city developed around an old French town which was the headquarters of the French FOREIGN LEGION until 1962. It is the commercial centre for surrounding vineyards and orchards. Pop. (1966) 86,581.

Sidmouth, Henry Addington, Viscount. *See* ADDINGTON, HENRY.

Sidney, Algernon (c.1622–83), English politician. He declared for Parliament in the Civil War but took no part in the trial of CHARLES I. After CROMWELL's seizure of absolute power he retired from public life and after the RESTORATION, lived on the continent. He was pardoned by CHARLES II in 1677 and returned to England. In 1683, at the time of the RYE HOUSE PLOT (in which he was thought to be implicated), he was arrested and later executed.

Sidney, Sir Philip (1554–86), English poet, diplomat and courtier. Of distinguished family, he was interested in the sciences and the arts and typifies the Renaissance man. His works include the intricate prose allegory *Arcadia* (1590) and the sonnet sequence *Astrophel and Stella* (1591), which was based on the poetry of PETRARCH. He died gallantly at the Battle of Zutphen, an embodiment of gentlemanly virtue. *See also* HC1 p.282.

Sidon, one of the chief Phoenician towns, and bitter rival of TYRE, until conquered by Assyria in 677 BC. Thereafter it was subject to repeated conquests until taken by the Turks in the 16th century.

Siebe, Augustus (1788–1872), German engineer who settled in Britain. In 1837 he introduced the helmet diving suit, the first reliable system of supplying compressed air to divers. *See also* PE p.92.

Siegbahn, Karl Manne Georg (1886–), Swedish physicist who was awarded the 1924 Nobel Prize in physics for research into X-ray SPECTROSCOPY. He improved techniques for measuring the wavelengths of X-rays and used these new methods to discover groups of X-rays both less penetrating and with longer wavelengths than had been thought possible. This discovery lent support to Niels BOHR's shell theory of atomic structure. In 1924 Siegbahn also proved the similarity between X-rays and visible light by showing that X-rays could be diffracted by a prism.

Siegfried, in ancient Germanic literature, hero figure who corresponds with Sigurd in Norse mythology, although accounts vary. In the story of Brunhild he is slain, but elsewhere is generally victorious in his adventures. He plays a major part in the Germanic epic tale of the *Nibelungenlied* and in WAGNER's operatic adaptation of that tale, *The Ring of the Nibelung.*

Siegfried Line, system of fortifications built along Germany's western frontier before WWII. In 1944 it provided a brief respite for the retreating German army and prevented a US breakthrough until the spring of 1945.

Siemens family, four German brothers, electrical engineers and industrialists. Ernst Werner von Siemens (1816–92), developed an ELECTROPLATING process in 1842 and in 1849 developed a telegraph system. With Karl (1829–1906) he set up subsidiaries of the family firm in London, Vienna and Paris. Frederick (1826–1904) and William (1823–1883) developed a regenerative furnace that was used extensively in industry. William introduced the Siemen electroplating process to Britain in 1843 and was head of the British branch of the firm, which became known as Siemens brothers in 1865.

Siemens AG, West German industrial corporation that is a major producer of electrical and electronic equipment. In the mid-1970s it employed 350,000 people.

Siena, city in Tuscany, central Italy; capital of Siena province. It became a commune in the 12th century and reached a peak of prosperity in the 13th century. In the 13th and 14th centuries the city was the centre of the SIENESE SCHOOL, and has many art treasures. The Palazzo Pubblico (1297–1310), which contain frescoes by Simone Martini and Ambrogio LORENZETTI, and the cathedral (11th–14th centuries) are both fine examples of Italian Gothic architecture. Tourism is the major industry. Pop. 69,000. *See also* HC1 p.*217.*

Sienese School, Italian school of painting which flourished between the 13th and 15th centuries. It rivalled the School of Florence although it remained more conservative, combining Byzantine richness of colour and decoration with the grace and elegance of GOTHIC art. Its most important representatives include DUCCIO, Simone MARTINI, the LORENZETTI brothers and SASSETTA.

Sienkiewicz, Henryk (1846–1916), Polish novelist and short-story writer who glorified his country's struggle for national existence in the trilogy *Ogniem i Mieozem* (*With Fire and Sword*; 1884), *Potop* (*The Deluge*; 1886) and *Pan Michael* (1888). He gained international recognition with *Quo Vadis?* (1896), a story of NERO's Rome, and was awarded the 1905 Nobel Prize in literature.

Sierra Leone, republic on the w coast of Africa. A former British protectorate, it gained independence in 1961. Most of the people are subsistence farmers; the principal cash crops are coffee, palm kernels and cacao. Minerals are, however, of greater importance to the economy; diamonds, iron ore and rutile account for 75% of the nation's exports. The capital is Freetown. Area: 71,740sq km (27,699sq miles). Pop. (1975 est.) 2,729,000. *See also* MW p.153.

Sierra Madre, principal mountain range in Mexico, from the US border to SE Mexico and extending s into Guatemala; it is made up of Sierra Madre Occidental, Sierra Madre Oriental, Sierra Madre de Sur, and the sub-range Sierra Madre del Guatemala. The ranges enclose the central Mexican plateau and have long been a barrier to E–W travel. The highest peak is Orizaba (Giltaltepetl) in the Sierra Madre Oriental, rising to 5,700m (18,700ft). Length of main range: 2,400km (1,500 miles); width: approx. 16–480km (10–300 miles).

Sierra Nevada, chief mountain range in s Spain, extending E to w in the provinces of Granada and Almeria. The Sierra Nevada is a source of iron, lead, copper and zinc ores and its southern slopes are intensively cultivated, producing grapes and sugar cane. The highest peak is Mt Mulhacén 3,478m (11,411ft). Length of range: 97km (60 miles).

Sierra Nevada, mountain system in E California, USA, running parallel to the COAST RANGES. In the E it rises steeply from the Great Basin; the w edge slopes more gently down to the Central Valley of California. The snow-fed rivers are used to irrigate the Central Valley and also to provide hydroelectric power. Mount Whitney, 4,418m (14,494ft), is the highest peak. Length: 650km (400 miles).

Siger of Brabant (c.1240–c.1284), Belgian-French philosopher, head of the movement known as Latin Averroism. His belief that something could be true in rational philosophy but false in religious belief was attacked by St Thomas AQUINAS. *See also* HC1 p.188.

Sight, or vision, faculty by which form, colour, size, movement and distance of objects are perceived. The human EYE works in much the same say as a camera. Light rays are focused on to the RETINA, which contains cells known as RODS AND CONES, in which the focused light evokes chemical changes. The cones contain the pigments idopsin and cyanopsin and they provide colour vision. The rods contain visual purple or rhodopsin, and they differentiate the light into shades. The two chemicals are broken down to form a protein and a vitamin A compound (by an unknown mechanism). This biochemical process stimulates an electrical (nerve) impulse to the brain along the optic nerve, which separates into two parts, each going to either the left or right hemisphere of the BRAIN. Here they join the visual cortex in each hemisphere and are combined to be perceived as a single image. *See also* MS pp.48–49.

Sigismund, name of three kings of Poland. Sigismund I (1467–1548), succeeded his brother in 1506. During his reign Polish domains were expanded and LUTHERANISM was introduced. He was succeeded by his son, Sigismund II Augustus (1520–72), last of the JAGELLON kings. He extended Polish rule over Lithuania and Ukraine, tolerated Protestant sects and encouraged the arts. Sigismund III (1566–1632), nephew of Sigismund II, was elected King of Poland in 1587 and was King of Sweden from 1592 to 1604. In his attempt to gain the throne of Russia he took Moscow in 1610 but the Poles were expelled two years later. In spite of such temporary military successes, Polish power declined during his reign. *See also* HC1 p.272.

Signac, Paul (1863–1935), French painter of the NEO-IMPRESSIONIST style who was first influenced by Claude MONET and later associated with Georges SEURAT and POINTILLISM. He was interested in the science of colour and his vigorous, colourful works include *View of the Port of Marseille* (1905). *See also* HC2 p.174.

Signalling, in civil and military communications, sending and receiving of visual and electrical messages. SEMAPHORE is a visual method of hand signalling long used by navies but superseded in WWI by signal codes flashed by lamps. RADAR signalling is used in ground control approach systems for guiding aircraft. Signalling by LASER beams is a new method under development which has many possible applications.

Signet, one-sided seal, smaller than the privy seal, three of which are given to each British secretary of state (four to the home secretary) when he takes office. It was first cast in the reign of EDWARD II and bears the royal coat of arms.

Significant digits, digits of a number that express a magnitude to a desired accuracy, the last digit possibly being rounded up or down. Thus 2.871828 to six significant digits is 2.87183; 2.872 has four significant digits.

Sign languages, non-verbal means of personal communication, especially through hand movements. It is commonly used between people speaking mutually unknown languages. Sign language is extensively used by deaf-mutes as the primary means of communication. Its origins probably lie in the extensions of facial expressions, gestures and other bodily movements which accompany everyday speech.

Signorelli, Luca (*c.* 1441–1523), Italian painter of the Umbrian School and probably a pupil of PIERO DELLA FRANCESCA. He worked in Rome in the 1480s and was probably associated with the Sistine Chapel frescoes with BOTTICELLI and ROSSELLI before starting decorations for the Orvieto Cathedral in 1499.

Signoret, Simone (1921–), French stage and film actress. Her stage performances included the Arthur MILLER production of *The Crucible* in 1956. Her film career began in the 1940s and included *Thérèse Raquin* (1953), *The Witches of Salem* (1957), *Room at the Top* (1959), for which she won an Oscar, and *Ship of Fools* (1965).

Sigurd. *See* SIEGFRIED.

Sihanouk, Norodom (1922–), Cambodian king, head of state and politician. Born into the Cambodian royal family, he succeeded to the throne in 1941 but abdicated in 1955 to become Prime Minister after establishing a socialist government. In 1960 he became head of state and in 1965 broke off diplomatic relations with the USA because of its military involvement in Indo-China. In 1970 he was deposed by Lon Nol in a right-wing military coup but returned from exile when the KHMER ROUGE took over in 1975. *See also* HC2 p.293; MW p.45.

Sikhism, Indian religion founded in the 16th century by NANAK. Combining HINDU and MUSLIM teachings, the Sikh scripture, the ADI GRANTH, stressed the need for a guru's guidance in seeking the One Lord (the word Sikh means "disciple"). Begun by Nanak as a passive way of life, Sikhism became an activist faith in the late 17th century under GOVIND RAI, its tenth guru. Sikhism today has some six million followers, mainly in the PUNJAB.

Sikh Wars (1845–46, 1848–49), military conflicts between the Sikhs and the British in India. After being overrun by the MOGULS, the Sikhs fled to the mountains; they returned to the Punjab to form a powerful state at Lahore with Ranjit Singh (1780–1839) as maharajah. By 1824 they controlled most of northern India. The British took Lahore in the first Sikh War after a victory at Sobrãon (1846) and the Sikhs were finally defeated at Gujcerãt in Feb. 1849, after which the Punjab was annexed. *See also* HC2 p.140.

Si-Kiang (Xi-Jiang), longest river of southern China, rising in the YUNNAN highlands. It flows for 1,957km (1,216 miles) to the South China Sea via CANTON and drains a basin of 329,000sq km (127,000sq miles).

Sikkim, state and protectorate in N India, bounded by Tibet, China (N and NE), Bhutan (SE), India (S) and Nepal (W); the capital is Gangtok. The terrain is generally mountainous, rising to Mt Kanchenjunga, 8,591m (28,185ft), the third-highest peak in the world. The original inhabitants of the area are the Lepchas but after the 17th century Sikkim was ruled by the rajas of Tibet. It had come under British influence by 1816. After British withdrawal from India in 1947, Sikkim became independent but political unrest led to the country becoming an Indian protectorate in 1950. It was made an associate state in 1975. Agriculture is the main source of income. Maize, rice, barley, fruits and cardamon are grown. Area: 7,298sq km (2,818sq miles). Pop. (1971) 208,609.

Sikorski, Władysław (1881–1943), Polish general and statesman. He distinguished himself in the defence of Warsaw against the USSR in 1920. In 1922–23 he was Premier, but fled to Paris in 1928 because of his opposition to Pilsudski, who had taken control by a coup in 1926. When Germany overran Poland (1939), Sikorski became Commander-in-Chief of the Free Polish forces and head of the government in exile. In July 1943 he died in an air crash over Gibraltar.

Sikorsky, Igor Ivan (1889–1972), US aeronautical engineer and designer, *b.* Russia. His first aircraft was a helicopter (1909), which was unsuccessful, and he built and flew the first four-engined aeroplane (1913). He emigrated to the USA in 1919 and four years later established the Sikorsky Aero Engineering Corporation. This was later merged with the United Aircraft Corporation. Sikorsky also designed amphibian planes but from the late 1930s specialized in helicopters; he designed and built the first commercially successful helicopter, which was flown on 14 Sep. 1939. *See also* MM pp.154, 155.

Sila, one of the three fundamentals of BUDDHISM. Sila is morality, and together with SAMADHI (concentration) and PANNA (wisdom) is one basis over which the Buddhist traverses the eightfold path needed to reach NIRVANA. *See also* MS p.221.

Silage, moist, green fodder used to supplement livestock feed. It is made by chopping and storing crops such as hay and alfalfa in airtight SILOS, where controlled fermentation will preserve the silage for months.

Silas, in the NEW TESTAMENT, early Christian disciple and St PAUL's companion on his missionary travels. He is believed to be the Silonus referred to by St Paul when writing to the THESSALONIANS (1 and 2 Thess. 1:1.).

Silbermann, Gottfried (1683–1753), most eminent member of a German family of keyboard instrument manufacturers. He built 47 ORGANS in Saxony. He also made fine CLAVICHORDS and was the first German to build a PIANO (*c.* 1735).

Silchester, village in Hampshire, near Basingstoke and the site of Calleva Atrebatum, the only Roman town in Britain to have been completely excavated. *See also* HC1 pp.102, *102,* 104.

Silence, Theatre of, dramatic theory evolved by Jean-Jacques BERNARD in the 1920s. Based on the ideas of MAETERLINCK, it emphasized the dramatic significance of silences and other devices to express the content of a play, where speech alone is insufficient.

Silent Spring (1962), book by US environmentalist Rachel CARSON. It was one of the first books to draw attention to environmental problems, particularly the depletion of wildlife through industrial pollution. It also describes the dangers of commonly used pesticides and fertilizers.

Silenus, in Greek mythology, a companion of BACCHUS. He was represented as a drunk, fat old man who followed the deity while swaying awkwardly on the back of a donkey. He was also revered because he knew the past and could foretell the future, and had taught Bacchus much of his wisdom.

Silesia, historic region in E Europe, mostly lying in SW Poland; the remainder forms parts of Czechoslovakia and Germany. A former Polish province, it passed from Poles to Bohemians in the 14th century, became part of the Hapsburg empire and was seized by Prussia from Austria in 1742. In WWII it was invaded by the USSR, but in 1945 a greater part of the land was returned to Poland by the terms of the POTSDAM CONFERENCE.

Silhouette, profile of an image in dark colour imposed upon a light ground or the reverse. Named after Etienne de Silhouette (1709–67), a French politician whose hobby was cut-out portraiture, silhouette portraiture as an art form was at its most fashionable during the later 18th and early 19th centuries.

Silica, silicon dioxide, a compound of SILICON and oxygen (SiO_2). Silica is the main constituent of 95% of all rocks and accounts for 59% of the Earth's crust. It is used in the manufacture of glass, ceramics and SILICONES. It is a temperature-resistant material and is frequently of gemstone quality. *See also* PE pp.96, *99,* 101; SU p.143.

Silicate, in chemistry, salt of silicic acid ($HSiO_3$). Silicates form a large part of the material of the Earth's crust. Glass is a mixture of silicates with small amounts of other substances. *See also* SIAL; SIMA.

Silicon, common non-metallic element (symbol Si) of group IVA of the periodic table, discovered by J.J. BERZELIUS in 1824. It is the second most abundant element in the Earth's crust (27.7% by weight), a common constituent of rocks and minerals. It is extensively used in semiconductors (eg in TRANSISTORS). Properties: at. no. 14; at. wt. 28.086; s. g. 2.33 (25°C); m. p. 1,410°C (2,570°F); b. p. 2,355°C (4,271°F); most common isotope Si^{28} (98.21%); *See also* SILICA; SU pp.69, *97, 134–135.*

Silicone, group of polymers based on SILICON and having the general formula $(R_2SiO)_n$, where R is a hydrocarbon radical and n is a whole number. Silicones are inert and stable at high temperatures, and are used in lubricants, adhesives, water repellents and artificial heart valves. *See also* SU p.143.

Silicosis, chronic occupational lung disease, caused by prolonged inhalation of SILICA dust in occupations such as mining and metal grinding.

Silk, natural fibre produced by the spinning glands of the larva (caterpillar) of the silkmoth, *Bombyx mori,* to make its cocoon. A single cocoon can provide between 600–900m (2,000–3,000ft) of unbroken filament. Chemically, silk is a complex mixture of protein, wax and salts.

Silk, artificial, name given at the beginning of the 20th century to RAYON, a fibre made from cotton or wood-pulp and used for the weaving of lightweight fabrics. *See also* MM pp.56–57.

Silk routes, name given to the trade routes leading from China to Asia Minor and Europe in Roman times and particularly in the early Middle Ages. The routes led from Sian, in the Shensi province, China, around the Takla Makan desert to the Pamir mountains then across Afghanistan and on to the Levant. One of the first Europeans known to have reached China by the Silk routes was Marco POLO.

Silk-screen printing, also called serigraphy, in PRINTMAKING, a means of producing a print, generally on paper. A screen composed of a mesh of silk or man-made fibres is stretched over a wooden frame; a design is "stopped out" (painted) on the mesh, using glue, varnish, gelatin or a paper stencil. To make the print, ink is taken, under pressure, across the screen, with a squeegee; the pressure of this pushes the ink through the unstopped areas of the mesh. Several screens may be used to built up a multi-coloured print. Using cylindrical screens with both ink and squeegee inside, it is the most usual method of printing on textiles.

Silkworm, moth CATERPILLAR that feeds chiefly on MULBERRY leaves. The common domesticated *Bombyx mori* is raised commercially for its SILK cocoon. Length: 7.5 cm (3in). Family Bombycidae. *See also* PE p.158.

Silky terrier, small Australian dog originally bred from Australian and Yorkshire terriers. It has a wedge-shaped head and small, V-shaped, erect ears. The long body is set on straight, fine-boned legs and the tail is commonly docked. Height: to 25.5cm (10in) at the shoulder; weight to 4.5kg (10lb).

Sill, sheet-like intrusion of IGNEOUS ROCK that is parallel to the bedding or other structure of the surrounding rock. Sill rock is normally medium-grained; basic sills (DOLERITES) are the most common. *See also* PE pp.30, *101.*

Sillanpää, Frans Eemil (1888–1964), Finnish author whose first novel *Life and Sun* (1916) won him immediate fame. *Meek Heritage* (1919) describes the impact of the Finnish civil war of 1918 with sympathy and realism. *Fallen Asleep While Young* (1931) treats the conflicts of the same period, caused by the disintegration of older values. He also wrote many short stories, and received the 1939 Nobel Prize in literature.

Sillimanite, or fibrolite, mineral, aluminium silicate (Al_2OSiO_4), found in mica SCHISTS and GNEISSES. Its crystals are of the orthorhombic system, usually fibrous masses, and are satin-like or glossy white, brown, green or blue. Pale blue gem variety occurs in Sri Lanka. Hardness 6–7.5; s.g. 3.2.

Luca Signorelli; a detail from *Resurrection of the Flesh*, Orvieto.

General Sikorski, in 1940, walking with Count Raczynski, the Polish ambassador.

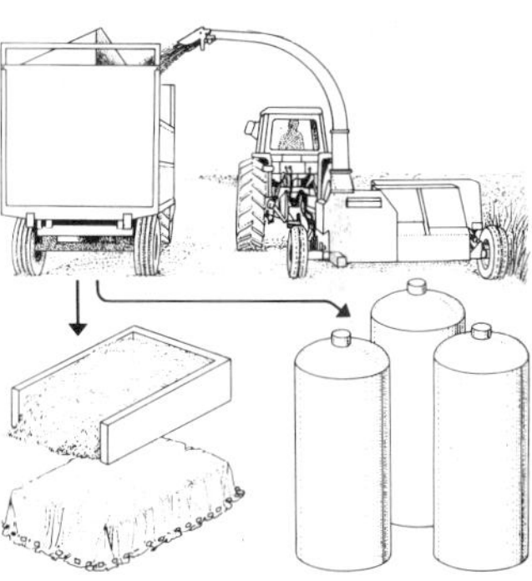

Silage: the processes by which green plants are turned into cattle feed.

Silky terriers were once used on Australian farms for killing rats and snakes.

Silvanus; a mosaic in the Museo Profano, Rome, of the god and his dog.

Lambert Simnel; the imposter as a turnspit in Henry VII's kitchens.

Simon and Garfunkel at the 1971 Hollywood presentation of the Grammy Award.

James Simpson; chloroform anaesthesia was first used in childbirth in 1847.

Sillitoe, Alan (1928–), British novelist and short-story writer whose blunt accounts of the frustrations of working-class life include the novels *Saturday Night and Sunday Morning* (1958) and *The Flame of Life* (1974). He also wrote the collection of short-stories *The Loneliness of the Long-Distance Runner* (1959).

Silo, in agriculture, large container for the storage of SILAGE. Tower silos are built of brick, concrete or steel sheeting, and are usually 10 to 20m (33–65ft) tall and 3 to 6m (10–20ft) in diameter.

Silo, missile, underground bunker reinforced with concrete, in which intercontinental ballistic missiles (ICBMS) are kept aimed and ready for launch. The silo contains all necessary equipment, living quarters and crewmen to service and fire the missile. *See also* MM p.169.

Siloe, Diego de (*c.*1495–1563), Spanish sculptor and architect whose works represent the height of the Spanish Renaissance. After studying sculpture in Florence he returned to Spain and his first masterpiece, the *Escalera Dorada* (Golden Staircase) (1519–23) in the Burgos Cathedral, combines both his sculptural and architectural talents in an exuberant fashion. His major work was the design of the cathedral at Granada.

Silt, mineral particles, produced by the weathering of rock. These particles, varying in size between grains of sand and clay, are carried along in streams and rivers, to be deposited in the gently flowing lower reaches of rivers. When the river changes course or overflows its bank, the silt deposit forms very fertile land. Siltstone (flagstone) is a hard, durable stone which is formed from hardened silt.

Silurian period, third oldest division of the PALAEOZOIC Era, lasting from 430 to 395 million years ago. Marine invertebrates resembled those of Ordovician times, and fragmentary remains show that jawless fishes began to evolve. The earliest land plants (psilopsids) and first land animals (archaic mites and millipedes) developed. Mountains formed in NW Europe. *See also* PE pp.*130–131*.

Silvanus, Roman god associated with the mysterious forces in woodlands; often, like another agricultural god Faunus, represented as a countryman. His character became merged with that of the similarly goatish Greek woodland deity PAN. He was generally worshipped in solitude at a particular tree or group of trees.

Silver, metallic element (symbol Ag) of the second series of TRANSITION ELEMENTS, known from the earliest times. It occurs naturally in argentite (a sulphide) and horn silver (a chloride), and is also obtained as a by-product in the electrolytic refining of copper. It is the most efficient conductor of heat and electricity known and is used in electrical contacts and printed circuits. Other uses include jewellery, decorative ornaments, mirrors and silver salts for photography. The metal does not oxidize in air but quickly tarnishes when exposed to sulphur compounds. Properties: at.no. 47; at.wt. 107.868; s.g. 10.5; m.p. 961.93°C (1,763°F), b.p. 2,212°C (4,104°F); most common isotope Ag^{107} (51.82%). *See also* SU pp.*134*, 136–137, *137*.

Silver chloride, white crystalline compound, insoluble in water, having the formula AgCl; it occurs naturally as the mineral horn silver. It darkens on exposure to light and for this reason is, like silver bromide, used for making photographic emulsions.

Silverfish, or bristle-tail, primitive, grey, wingless insect found throughout the world. It lives in cool, damp places feeding on starchy materials such as food scraps and paper. It gets its name from the silvery scales that cover its body. Length: 13mm (0.5in). Family Lepismatidae; species *Lepisma saccharina*. *See also* NW p.*106*.

Silverius, Saint (*d.*537), Italian pope (*r.*536–37). Defamed by the Byzantine empress THEODORA, he was banished to Palmaria where he died a martyr. Feast day: 20 June.

Silver point, in art, small sharpened rod of silver used for precise drawings on parchment or paper; the result is a permanent drawing, the strokes of which are hard to erase. Lead, copper and gold have been used in this way. The technique was popular with 15th-century Flemish artists such as van EYCK, van der WEYDEN and MEMLING. Albrecht DÜRER used it to great effect in such works as *Self Portrait* (1484).

Silverstone, motor-racing circuit built on a disused airfield in Northamptonshire, England. The grand prix circuit has a lap course of 4.7km (2.932 miles).

Silver sulphide, dark-brown solid having the formula Ag$_2$S; it occurs naturally as the mineral silver glance. The tarnishing of silverware is due to the formation of silver sulphide, which is produced as the silver reacts with atmospheric sulphur compounds such as SULPHUR DIOXIDE (SO$_2$) and HYDROGEN SULPHIDE (H$_2$S).

Silvestri, Constantin (1913–69), Romanian conductor. He was the chief conductor of the Bucharest radio orchestra from 1930, and the Bucharest Opera from 1935. He moved to Britain to become the conductor of the Bournemouth Symphony Orchestra in 1961.

Silviculture, cultivation and propagation of shrubs and trees for ornamentation or shade purposes. It includes planting, pruning and harvesting. *See also* PE pp.218–221.

Sima, in geology, undermost of the two main rock-classes that make up the Earth's crust, so called because its main constituents are silicon and magnesium. It underlies the SIAL of the continents.

Simenon, Georges (1903–), French novelist. A prolific writer of popular fiction, he wrote well over 500 novels and many more short stories. He became famous for his Maigret detective novels, begun in 1931. He later wrote more ambitious novels, including *The Stain on the Snow* (1948).

Simeon Stylites, Saint (*c.*390–459), first of the stylites or pillar ascetics. Born in Sisan, Syria, he spent some years as a monk and hermit before mounting a pillar on which he lived until his death.

Simile, figure of speech comparing two things. It differs from ordinary comparisons in that it compares, for effect, things usually considered dissimilar and sharing only one common characteristic, as "fleece as white as snow". A simile is most often recognized by its use of "like" or "as".

Simla, hill station in the PUNJAB, India, situated in the Himalayas. It came under British rule after the GURKHA War of 1814–16. From 1865 it became the summer capital of the British government of India, because of its cool climate. When India gained independence in 1947, the practice of maintaining a summer capital was abandoned. *See also* HC2 p.*141*.

Simmel, Georg (1858–1918), German sociologist and philosopher. He was influential in establishing sociology in Germany. His works include *Sociology* (1908) and *The Philosophy of Money* (1900). *See also* MS p.263.

Simnel, Lambert (*c.*1475–1535), English impostor. The son of a baker, he was tutored by a priest to impersonate Richard, Duke of YORK, younger son of EDWARD IV. Richard and his elder brother EDWARD V had in fact disappeared and were probably murdered in 1483. The plan was changed and in 1487 Simnel impersonated the Earl of Warwick, the imprisoned nephew of Edward IV. Simnel was accepted as Edward VI by the Irish nobility, and became the focus of a rebellion under Lincoln and Lovell against Henry VII. The rebels invaded England but were defeated at Stoke in 1487. *See also* HC1 p.266.

Simon and Garfunkel, US singers (Paul Simon and Arthur Garfunkel). Both were born in 1942 and they sang together from their early teens. In 1966 their *Sound of Silence* sold more than a million copies and established them as among the most successful recording artists. Their subsequent recordings include *Scarborough Fair* (1966), *Bookends* (1968) and *Bridge over Troubled Water* (1970).

Simonides (*c.*556–*c.*468 BC), Greek poet. He developed the *epinikion* into a choral composition, produced many epigrams of high quality and claimed that poetry is verbal painting. *See also* HC1 p.74.

Simonstown, town in Cape Province, s South Africa, on the w shore of False Bay 32km (20 miles) s of Cape Town. It was founded by the Dutch in 1841 as a naval base; it is now a naval depot for the South African navy. The principal industries are shipbuilding and repair and fishing. Pop. (1970) 12,091.

Simony, trading in sacred offices. The name comes from a NEW TESTAMENT figure. Simon (called a magician) who was rebuked for trying to buy the power of conferring the gifts of the HOLY GHOST (Acts of the Apostles 8:18–24). In the MIDDLE AGES, simony came to mean buying and selling church benefices. The practice was denounced at various times by the papacy and was one of the chief targets of the attacks of the Protestant reformers; it gradually disappeared.

Simple harmonic motion (SHM), periodic motion about a point. A pendulum moves in SHM as does a weight, hung on a spring, which bobs up and down. It is defined as the motion of a point which oscillates about a fixed point such that the restoring FORCE F (the force which maintains the oscillations) is proportional to the distance x of the moving point from the fixed point. Thus $F = -kx$, where k is a constant of proportionality and the minus sign denotes that the restoring force is opposite in direction to the displacement x. As the moving point passes through the fixed point its ACCELERATION is zero. Its VELOCITY, however, is a maximum. At the maximum displacement the velocity is zero and the acceleration is a maximum. *See also* NEWTON'S LAWS; SU pp.76–77, *76*.

Simple multiple proportions, law of, law announced in 1803 by John DALTON. It states that when two elements combine to form more than one compound, the fixed weight of one element that always combines with weights of the other are in ratios of small whole numbers. It was a powerful argument for Dalton's atomic theory.

Simplicius, Saint (*d.*483), pope from 468 to 483. During his pontificate there occurred the first of the schisms over the MONOPHYSITE CONTROVERSY and the last emperor of the w Roman Empire was deposed. Simplicius upheld Chalcedonian orthodoxy against the ideas of the Emperor ZENO. Feast: 2 March.

Simplon Pass, in Switzerland, pass between the Pennine and Leopontine Alps; it reaches an altitude of 2,009m (6,590ft). In 1906, the longest transport tunnel in the world was opened beneath the pass to connect Brig, Switzerland with Iselle, Italy. Surveyed by Alfred Brandt, it runs for 19.8km (12.3 miles) and required a new tunnelling method to overcome problems created by working at depths of as much as 1.6km (1 mile).

Simpson, Sir James Young (1811–70), Scottish physician and obstetrician. He was noted for the discovery of CHLOROFORM as an anaesthetic (1847) and first used it on himself and his assistants to test its efficiency. *See also* MS p.126.

Simpson, Norman Frederick (1919–), British dramatist whose plays such as *A Resounding Tinkle* (1957) and *One Way Pendulum* (1959) are noted for the way they present fantasy within a framework of satire and social comment.

Simpson, Orenthal James (1947–), US football player. He won the 1968 Heisman Trophy at the University of Southern California as a running back. In 1969 he joined Buffalo in the National Football League and set a single season rushing record in 1973 with 1,831m (2,003 yards).

Simpson, Thomas ("Tommy") (1937–67), British road-racing cyclist. In 1962 he became the first Briton to wear the TOUR DE FRANCE yellow jersey. He won the world road-race championship in 1965. He collapsed and died during the 1967 Tour de France, probably from a heart attack caused by a combination of heat, the thin mountain air and drugs he had taken to improve his performance.

Simpson, Mrs Wallis Warfield Spencer, Duchess of Windsor (1896–), wife of Prince Edward, Duke of Windsor, the former King EDWARD VIII of Britain, *b.* USA. In Oct. 1936 she divorced her second husband, Ernest Simpson. Her relationship with Edward VIII caused controversy that resulted in Edward's ABDICATION in Dec. 1936. They married in June 1937 and lived abroad, chiefly in France. *See also* HC2 p.*225.*

Simultaneous equations, two or more equations that can be manipulated to give common solutions. In the simultaneous equations $x + 10y = 25$ and $x + y = 7$, the problem is to find values of x and y. This can be done by substituting the value of x from one equation into the other to give a single equation in y, which can then be solved. Simultaneous equations can be solved only if the number of equations equals the number of unknowns. *See also* SU pp.34–35.

Sin, in ASSYRO-BABYLONIAN MYTHOLOGY, the moon god, father of SHAMASH and ISHTAR. He is depicted as an old man with a long beard who sailed the crescent moon as a boat across the sky. Sin measured time and served as adviser to the gods.

Sin, act in thought, word or deed that is contrary to the ideals of righteousness propounded by Christianity. It is considered to estrange the sinner from God. Reconciliation is attained through PENITENCE, PENANCE and GRACE, although the New Testament states that a sin against the Holy Ghost will never be forgiven. *See also* MS pp.184, *184, 219.*

Sinai Peninsula, peninsula constituting a protectorate of Egypt, bounded by the Gulf of Suez and the Suez Canal (w), the Gulf of Aqaba and the Negev Desert of Israel (E), the Mediterranean Sea (N) and the Red Sea (s); the capital is Al-'Arīsh. It is a barren plateau region sandy in the N, rising to granite ridges in the s, inhabited chiefly by nomads. The peninsula is the site of Jabal Musa or Mt Sinai. The monastery of St Catherine was established there *c.* AD 250, and an early New Testament manuscript was discovered in the monastery in 1844. Ths Sinai Peninsula was the scene of fighting in the ARAB-ISRAELI WARS of 1956, 1967 and 1973. Area: 60,715sq km (23,442sq miles). Pop. 140,000.

Sinān (*c.* 1489–1588), considered one of the greatest Islamic architects. He worked under Suleiman I and other Ottoman sultans, and his huge number of projects included 79 mosques and 34 palaces. His masterpieces include the mosques of Suleiman I at Istanbul (1550–57) and of Selim at Edirne (1568–74).

Sinanthropus pekinensis. *See* PEKING MAN.
Sinatra, Francis Albert ("Frank") (1915–), US popular singer and actor. He sang with the bands of Harry James and Tommy DORSEY and became immensely popular with teenagers of the 1940s for such songs as *Night and Day.* In the 1950s he became a film actor and won an OSCAR for a supporting role in *From Here to Eternity* (1953). *See also* HC2 p.273.

Sinclair, Keith (1922–), New Zealand author and poet. His works include *A Time to Embrace* (1963), a volume of poetry, and *A History of New Zealand* (1959).

Sinclair, Upton (1878–1968), US novelist and social reformer. *The Jungle* (1906), a novel about the Chicago meat-packing industry, led to the reform of food inspection laws. His novels include *The Money Changers* (1908), *King Coal* (1917), about the Colorado miners' strikes, and *Dragon's Teeth* (1942), for which he won the Pulitzer Prize.

Sind, province in s Pakistan, bounded by India (E and s) and the Arabian Sea (sw). It largely consists of the alluvial plain and delta of the River Indus. The region is hot and arid and farming depends on irrigation; the SUKKUR and Kotri barrages are the main schemes. Wheat, rice, millet, cotton, sugarcane, fruits and tobacco are grown. Mineral resources include gypsum and limestone. Hyderābād is the principal industrial city. Fishing is important along the coast. The region was taken by the British in 1843, and held until 1937. Karāchi is the capital. Area: 152,966sq km (58,983sq miles). Pop. 13,965,000.

Sine, ratio of the length of the side oppo-site to an acute angle to the length of the hypotenuse in a right-angled triangle. The sine of angle A is usually abbreviated to "sin A". *See also* SU p.46.

Sine die (without day), in law, adjournment of a case by a court to a date that is to be fixed in the future.

Sine wave. *See* SIMPLE HARMONIC MOTION.
Singapore, island nation in SE Asia at the s end of the Malay Peninsula, made up of Singapore island and 60 smaller islands. Formerly a British colony, it became part of Malaysia in 1963 and an independent republic two years later. Vegetables, tobacco, fruits, rubber and coconuts are the chief agricultural products. Industries include shipbuilding, shipping, steel, tourism and food processing. Singapore city is the capital. Area: 580sq km (224sq miles). Pop. (1976 est.) 2,278,000. *See also* MW p.154.

Singer, Isaac Bashevis (1904–), US novelist and short story writer, *b.* Poland, who emigrated to the USA in 1935. His works, written in Yiddish, include *The Family Moskat* (1950) and *The Slave* (1962).

Singer, Isaac Merrit (1811–1875), US manufacturer, inventor of a rock drill (1839) and in 1852 of a single-thread sewing machine. Singer's machine allowed continuous and curved stitching; it was immediately popular and ensured the financial success of his Singer Manufacturing Company.

Sing Sing, famous US state prison, noted for its stern discipline, in Ossining, New York. Prisoners sentenced to death and awaiting execution are often detained there. In 1969 the prison was renamed the Ossining Correctional Facility.

Singspiel, German vernacular form of opera similar to OPERA BUFFA but with spoken dialogue. Primarily an 18th-century form, several operas by Mozart, including the *Abduction From the Seraglio* (1782) and *The Magic Flute* (1791), strengthened the status of Singspiel.

Sinhalese, people who make up the largest ethnic group of Sri Lanka (Ceylon). They speak an INDO-EUROPEAN language and practise Theravāda Buddhism.

Sink hole, hollow or hole in LIMESTONE formations which extends from the surface all or part of the way down to underground channels and caverns. Such holes are formed by water dissolving the limestone.

Sinkiang-Uigur (Xinjiang Weiwuer Zizhiqu), autonomous region in NW China, bordered by the USSR (NW), Mongolia (NE), Afghanistan and Jammu and Kashmir (sw) and Tibet (s); the capital is Urumchi (Wulumuqi). The region includes the Dzungarian Basin to the E and the Tarim basin to the w. Cotton, wheat, maize, rice, millet, vegetables and fruit are grown; livestock rearing, particularly sheep, is also important. Mineral deposits include oil, copper, zinc, gold and silver. Iron and steel, chemicals, textiles and food processing are the main industries. Area: 1,647,435sq km (636,075sq miles). Pop. 8,000,000.

Sinn Fein, Irish nationalist party which developed out of the Society of Gaels founded in 1902 by Arthur GRIFFITH. Its aim to achieve total Irish independence was not satisfied with the establishment of the Irish Free State in 1921. It has sometimes been regarded as the political wing of the IRA. *See also* HC2 p.163.

Sino-Japanese War (1894–95), military conflict between China and Japan over control of Korea. Japan declared war on 1 Aug. 1894 and her better-equipped forces had an overwhelming advantage. After crippling the Chinese fleet at the mouth of the Yalu River (17 Sept. 1894), Japanese forces captured Port Arthur and Weihuiwei, and the remainder of the Chinese fleet. The war ended with the signing of the Treaty of Shimonoseki (17 April 1895), under the terms of which China ceded territory to Japan and agreed to pay a huge indemnity.

Sintering, process in powder METALLURGY in which compressed particles are heated at temperatures below the melting point of the metal. This forms the particles into a coherent solid body. Glass and ceramics are non-metals that may be sintered.

Sinus, hollow space or cavity, usually in bone. Most often the term refers to the paranasal sinus, any of the four sets of cavities in the skull near the nose. Their function is not fully understood, except that they reduce the weight of bone in the skull.

Sinusitis, acute or chronic disorder involving inflammation of the paranasal SINUSES. Acute sinusitis, causing pain and nasal obstruction, may occur after a cold, a secondary bacterial infection, improper breathing while swimming or sudden changes in barometric pressure. Chronic sinusitis causes obstructed breathing and loss of the sense of smell.

Sioux, or Dakota, group of seven North American Indian tribes inhabiting Minnesota, Nebraska, North and South Dakota and Montana. Their first contact with Europeans occurred in 1640. The tribes concluded several treaties with the US government during the 19th century and finally agreed in 1867 to settle on a reservation in sw Dakota. The discovery of gold in the Black Hills and the subsequent rush of prospectors brought great resistance from Sioux chiefs such as SITTING BULL and CRAZY HORSE. The last confrontation resulted in the massacre of more than 200 Sioux at the battle of Wounded Knee in 1890. The Sioux culture was typical of the GREAT PLAINS Indians. They hunted bison (buffalo) and practised the Sun Dance. All seven tribes speak a language of the Hokan-Siouan family of languages. Today the Sioux live on reservations and number more than 40,000. In 1973 approximately 200 Indians, most of them Sioux, occupied the settlement at Wounded Knee for 69 days and demanded a government investigation of Indian conditions.

Siphon, principle by which a liquid is raised from a container at one level and delivered to a lower level; also any inverted U-shaped tube used for this purpose. The tube is filled with the liquid, and one end of it is placed below the surface of the liquid in the container. When the other end hangs below the level of the surface (outside the container) liquid flows out due to the imbalance of atmospheric pressure. *See also* SU pp.78–79.

Siphonaptera. *See* FLEA.
Siqueiros, David Alfaro (1896–1974), Mexican painter. He was an active socialist, and viewed art as a vehicle for social and political expression. He is best known for his MURALS of revolutionary themes.

Siren, aquatic, tailed AMPHIBIAN of North America. The adult is neotenic (reaches sexual maturity while retaining the larval physical form). These eel-like animals have external gills, tiny forelegs and minute eyes. They have no hind legs. Length: to 72cm (36in). Family Sirenidae. *See also* SALAMANDER.

Sirenia, order of plant-eating aquatic mammals with paddlelike flippers for forelimbs, no hind limbs, and a horizontally flattened tail fin. The four species include the DUGONG (family Dugongidae) and three MANATEES (family Trichechidae). Length: 2–6m (7–20ft); weight: to 356kg (790lb). *See also* NW p.155.

Sirens, The, in Greek mythology, three sea nymphs with women's heads and birds' bodies. They lived on a rocky island near the straits of SCYLLA AND CHARYBDIS and their beautiful singing was believed to attract sailors on to the rocks. ODYSSEUS escaped shipwreck by blocking his sailors' ears with wax and tying himself to the mast so that he might hear their song.

Sir Gawain and the Green Knight, anonymous 14th-century northern English poem written in long, alliterative stanzas each ended by a rhyming quatrain. *See also* HC1 p.206.

Siricius, Saint (*c.* 334–99), Roman Catholic pope elected in 384. He began a new stage in papal authority and ensured the continued separation of the Church in ILLYRIA from Constantinople.

Sirius, or Alpha Canis Majoris, brightest star visible from Earth, in the northern constellation of Canis Major. Sirius is a white MAIN SEQUENCE binary star whose

Frank Sinatra at the end of the concert in the Festival Hall, London, in 1970.

Singapore; a view along Shenton Way, showing the octagonal Telok Ayer market.

Sino-Japanese War; Japanese troops and the Triumphal Arch near Seoul in 1894.

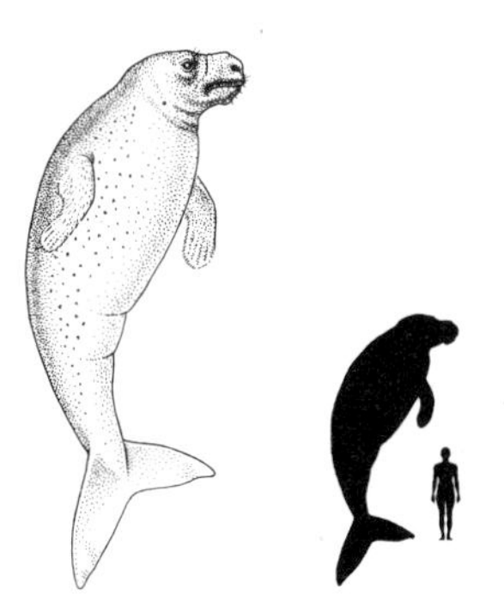

Sirenia are found in swamps, rivers and in other shallow tropical waters.

Alfred Sisley's *Church of Moretal* is in the Musée des Beaux Arts, Rouen.

Sitar; the lower strings vibrate when the main strings are played.

Sitting Bull told Indians to fight to kill, not merely to show bravery.

Skiing: some positions and equipment used in this exciting winter sport.

brightness results more from its relative closeness to the SOLAR SYSTEM than from its power. Characteristics: apparent mag. -1.47; absolute mag. 8.7; spectral type A1; distance 8.7 light-years. *See also* SU pp.*171, 226, 229, 229, 244, 260, 260, 264, 264, 266–267, 266–267.*

Sirocco, wind often arising over the Sahara Desert, picking up moisture from the Mediterranean Sea, and bringing hot, rainy weather to the Mediterranean coast of Europe.

Sisal, or sisal hemp, fibre plant native to Central America and cultivated in Mexico, Java, E Africa and the Bahamas. Fibres from the leaves are used for rope, matting and binder twine. Fibre length: 1–1.5m (3–5ft). Family Agavaceae; species *Agave sisalana. See also* AGAVE; PE p.216.

Sisinnius (*r.*708), pope, a Syrian, who reigned for less than a month.

Siskin, FINCH that nests in evergreen woods in Europe and in cooler climates throughout the world. It has muted green plumage. The male is marked with yellow and black; the female is greyer with white banded underparts. Length 12cm (4.8in). Species *Carduelis spinus.*

Sisley, Alfred (1839–99), Paris-born painter of English parentage. A leading IMPRESSIONIST, he spent most of his life in France. He was almost exclusively a landscape painter, especially of the countryside around Paris. His paintings include *Flood at Port Marly* (1876) and *Effect of Snow* (1874). *See also* HC2 pp.118–119.

Sistine Chapel, private chapel of the popes in the Vatican. It was built between 1473 and 1481 for SIXTUS IV; its ceiling and altar wall were decorated with frescoes of Old Testament scenes by MICHELANGELO. The side walls of the chapel are decorated with frescoes by PERUGINO, Pinturicchio, BOTTICELLI and GHIRLANDAIO. *See also* HC1 p.*256.*

Sistrum, simple ancient percussion instrument, consisting of a clay, metal or wood frame, across which crossbars were set, hung with small metal discs which would jingle when the instrument was shaken.

Sisyphus, in Greek mythology, King of Corinth. He was punished for trying to trick Death by being condemned to work for eternity, pushing a huge rock to the top of a steep hill. The rock always rolled back to the base of the hill when Sisyphus reached the summit.

Sita, in Hindu mythology, the wife of RAMA and the embodiment of purity and devotion. After being kidnapped by RAVANA, her faithfulness was challenged by her husband and she walked into fire, emerging unburned as proof of her purity. *See also* VEDAS.

Sitar, Indian stringed musical instrument with a gourd-like body and long neck. Similar to the lute, it has three to seven gut strings, tuned in fourths or fifths, and a lower course of 12 wire strings.

Sitatunga, or marshbuck, ANTELOPE of central and w Africa, closely related to the BUSHBUCK. It is medium-sized with long hoofs, an adaptation for life in marshes. Height: 82–106cm (32–42in). Species *Tragelaphus spekii. See also* NW p.*232.*

Sitter, Willem de (1872–1934), Dutch astronomer who, by applying the theory of relativity to the universe, deduced that it is expanding. This was confirmed by Edwin P. HUBBLE's observation in 1929 of the DOPPLER Shift in the spectra of distant stars.

Sitting Bull, Chief (*c.*1831–90), American Indian chief of the SIOUX who in 1876 defeated a punitive force led by Gen. George CUSTER's 7th Cavalry detachment at Little Big Horn. He led his men to Canada in 1877, but finally surrendered to the Army in 1881. He appeared in Buffalo Bill CODY's Wild West Show (1885). He was killed during Indian unrest on the Standing Rock Reservation. *See also* HC2 p.*149.*

Sitwell, name of English literary family, Dame Edith (1887–1964) and her brothers Sir Osbert (1892–1969) and Sacheverell (1897–   ). Edith was best known as a poet. Her collected verse was published in 1954. She also wrote a biography of POPE (1930), several works of

criticism and the famous *English Eccentrics* (1933). Osbert wrote verse, stories and novels, but he is famous chiefly for his five volumes of family reminiscences, especially *Left Hand, Right Hand* (1944). Sacheverell made his reputation as an art critic – *Southern Baroque Art* (1924), *German Baroque Art* (1927) and *The Gothick North* (1929). He also wrote biographies of MOZART and LISZT.

SI units, internationally agreed system of units, derived from the Metre Kilogramme Second (MKS) system. Système International (SI) units are now used for many scientific purposes and have replaced the Foot Pound Second (fps) and Centimetre Gramme Second (cgs) systems. The seven basic units are: the metre (m), kilogramme (kg), second (s), ampere (A), kelvin (K), mole (mol) and candela (cd).

Siva. *See* SHIVA.

Six, Les, collective name for six French composers who were organized as a group by Jean COCTEAU in 1917. The members were Georges AURIC, Louis DUREY, Arthur HONEGGER, Darius MILHAUD, Francis POULENC and Germaine TAILLEFERRE. The grouping was somewhat artificial.

Six, The. *See* EEC.

Six Articles (1539), articles imposed on the Church by HENRY VIII to prevent the spread of the practices and doctrines of PROTESTANTISM. TRANSUBSTANTIATION and communion in one kind were maintained, clerical celibacy and monastic vows were upheld and private Masses and auricular confessions were defended.

Six Day War, war between Israel and its Arab neighbours, 5–10 June 1967. It began with surprise Israeli attacks on the airfields of Egypt, Jordan, Syria and Iraq. The immediate cause was Israeli concern at Egypt's closure of the Strait of Tiran and her increasing war hysteria under the leadership of President NASSER. The war resulted in victory for the Israelis, who gained control of the Old City of Jerusalem, Jordanian territory w of the River Jordan, the Golan Heights and the Sinai Peninsula including the Gaza Strip. *See also* HC2 p.297.

Six Dynasties, governments in s China from 220 to 589 with their capital at Nanking, (the Toba TARTARS ruled in the North). In a time of political discord, South China nevertheless developed new forms of literary expression and a sophisticated court life. Gunpowder was invented during this period and the tea plant introduced.

Six Nations, association of North American Indians, after it was joined by the Tuscarora between 1720 and 1727. The original members, known as the Five Nations, were the Cayuga, ONEIDA, Onondaga, MOHAWK and Seneca.

Sixtus, the name of five popes, the most important of whom were the last two. Sixtus IV (1414–84; *r.*1471–84), b. Francesco della Rovere, was a famed patron of the arts and pursued an ambitious territorial policy. He built the SISTINE CHAPEL and opened the VATICAN Library to scholars. His nepotism, however, led to many abuses and the moral authority of the papacy declined during his pontificate. Sixtus V (1521–90; *r.*1585–90), b. Felice Peretti, was an inquisitor under Pope PAUL IV. As Pope, he repressed licence, reformed the administration of the law and improved church finances.

Size, dilute mixture of glue, or resin, used to prepare a canvas for painting. It reduces the absorbency of the canvas and protects it from the corrosive action of oil paints.

Skagerrak, strait between Norway and Denmark. With the Kattegat, the strait connects the North Sea and the Atlantic Ocean. Shallow on the Danish coast, the Skagerrak deepens towards Norway. Length: 204km (130 miles); width: 113km (70 miles).

Skald, Icelandic bard who was an exponent of oral court poetry of the 9th to the 13th century. The style originated in Norway but was developed by skalds, the greatest of whom was probably Egill Skallagrimsson. The metres of the skaldic style are strictly syllabic; the content is subjective and descriptive. Of the 100 verse

forms, the court metre (which uses alliteration and internal rhyme) was most often employed.

Skanda. *See* KARTTIKEYA.

Skate, flattened fish belonging to the RAY family, living mainly in shallow temperate and tropical waters. The pectoral fins are greatly expanded to form wing-like flaps. In Europe, skate is a commercially valuable fish. Length: to about 2.5m (8ft). Family Rajidae. *See also* PE p.246.

Skating. *See* ICE SKATING; ROLLER SKATING.

Skeat, Walter William (1835–1912), British literary scholar and philologist. Although chiefly famous for his editions of early English writers, such as *Complete Works of Geoffrey Chaucer* (7 vols, 1894–97) and *Piers Plowman* (1866), he also published many etymological works, including *The English Etymological Dictionary* (1879–82).

Skeet shooting. *See* CLAY PIGEON SHOOTING.

Skeletal muscle, in human beings and other mammals, one of the three types of muscle; it comprises the bulk of the trunk and limbs, contributing strength and mobility. Mostly attached to bone, either directly or through TENDONS, it includes flexors to bend joints and extensors to straighten them. *See also* INVOLUNTARY MUSCLE; MS pp.58, *58–59.*

Skeleton, bony connective tissue that makes up the general framework of the body. It supports and protects the soft inner organs; serves as a place of attachment for muscles, ligaments, and other structures; provides storage for some minerals; and produces some blood cells. In man the skeleton is divided into two parts. The axial skeleton, or main axis of the body, includes the cranium (skull); the spinal, or vertebral, column; the sternum (breastbone) and the ribs. The appendicular skeleton, serving for the attachment of appendages, includes the shoulder girdle and arm bones and the pelvic, or hip, girdle and leg bones. *See also* MS pp.56–57.

Skelton, John (*c.*1460–1529), English poet who was originally HENRY VIII's tutor and later a parson. He wrote satires on the court, the clergy and Cardinal Wolsey, written in forms he himself devised, called Skeltonics, consisting of short lines and insistent rhymes.

Sketches by Boz, series of articles by Charles DICKENS, published originally in the *Monthly Magazine*, the *Evening Chronicle* and other periodicals and published collectively between 1836 and 1837 with illustrations by CRUIKSHANK. They are satirical essays on English life.

Skibob racing. *See* SKIING.

Skiffle, British musical movement of 1956–58, led by Lonnie Donegan and Johnny Duncan.

Skiing, winter sport in which the participant moves over snow using skis (up to 2.4m (8ft) long) attached to his boots and holding poles to assist balance and movement on the level. Skiing, which also has a utilitarian function, is the national sport of Norway, where it began as a competitive sport in the 19th century. Competition includes four events: Nordic jumping and cross-country, and Alpine downhill and slalom racing. In ski jumping, each participant leaps twice from a specially designed jump slope and scores points on distance and style. Downhill racing is a timed descent, and a slalom course is marked by poles between which the skier must race. The most gruelling event is cross-country, which is over a course 15–50km (9–31 miles) long for men and 5–10km (3.1–6.2 miles) for women. Women compete in all events except jumping. The sport is governed by the Fédération Internationale de Ski, founded in 1924, the same year as the first Winter Olympic Games were held at Chamonix, France.

Skiing, water, sport in which the participant, balancing on one or two skis, is towed by a motor boat and planes on the surface of the water at speeds of up to 80km/h (50mph). The first world championships were held in 1949. Competition takes place in three events: the slalom, in which a course of six diagonally-placed buoys must be negotiated; jumping, over a

6m (20ft) ramp; and figures or tricks in which the skier gains points with a range of freestyle movements.

Ski jumping, winter sport in which skiers gather speed down a ramp and launch themselves through the air to land on the snow below. Points are awarded for distance and style, the latter decided by a panel of judges. World and Olympic Games titles are decided on jumps from heights of 70m and 90m. Only men compete in the Olympic event.

Ski kiting, sport that has developed from water skiing. The participant is towed on water skis while strapped to the framework of a large kite that lifts the skier into the air. Speeds of more than 48 km/h (30mph) can be attained. Ski kiting originated in Australia and the USA in the early 1950s.

Skimmer, any of several species of nocturnal shorebirds that frequent warm seas, rivers and lakes and use their long, knife-like bills to "skim" the water for food. Family Rhynchopidae; genus *Rhynchops*.

Skin, continuous, tough, elastic and sensitive covering of the body, serving many functions. The skin protects the body from mechanical injury, water loss and ultraviolet rays. Nerve endings in the skin provide the sensations of touch, warmth, cold and pain, each perceived at discrete points on the surface. It helps to regulate body temperature through sweating, regulates moisture loss and keeps itself smooth and pliable with an oily secretion from SEBACEOUS GLANDS.

Structurally the skin consists of two main layers: an outer layer (epidermis) and an inner layer (dermis, corium or cutis vera). The epidermis is itself composed of several layers: the stratum corneum, a horny layer made of closely packed dead cells constantly shed as microscopic scales; the stratum lucidum, a layer of flattened cells fixed together to form a kind of membrane; the stratum granulosum in which the cells are changing from soft, living cells to dead, horny cells; the Malpighian layer, living cell layers which contain pigment and nerve fibrils and which divide to replace outer shed layers. The dermis contains dense networks of connective tissue, blood vessels, nerves, glands and hair follicles. *See also* MS p.60.

Skin diving, water sport in which minimum equipment (a mask, snorkel, fins and sometimes a wet suit) is used. As with SCUBA DIVING, one of the main interests in the sport is the exploration of marine life.

Skink, common name for any of more than 600 species of lizards of tropical and temperate regions. They have cylindrical bodies, cone-shaped heads and tapering tails. They live in varying habitats and eat vegetation, insects and small invertebrates. Length: to 66cm (26in). Family Scinidae. *See also* NW p.*137*.

Skinner, Burrhus Fredric (1904–), US psychologist. A key figure in psychological behaviourism, he developed the concept of operant conditioning (the control of behaviour by its consequences or reinforcements. *See also* MS pp.147, *147*, 150, 153–154, 187.

Skinner box, device used in psychological studies of operant conditioning, named after B. F. SKINNER, its inventor. An animal, often a rat, placed in the box learns to press a lever in order to get food or water. The device allows precise control of learning conditions and measurement of changes occurring in behaviour during the learning process. *See also* MS p.*147*.

Skipjack, name of several fishes, not closely related, that swim near or on the surface. Best known is the skipjack TUNA, *Katsuwonus pelamis*. Length: 1m (39in).

Skipjack. *See* CLICK-BEETLE.

Skittles, indoor bowling game, popular in Britain and other countries of Western Europe. It is usually played by two teams on an alley with nine pins (skittles) and a hard flattened ball or cheese-shaped missile. Each player is allowed three tries to knock down the nine pins. A perfect score on each turn (chalk) is 27 points. The highest aggregate score at the end of a predetermined number of chalks wins.

Skoblikova, Lydia (1939–), Soviet ice speed-skater. In 1964, she became the first person to win four gold medals in one Winter Olympic Games. She won two Olympic golds in 1960, and in seven world championships (1959–64 and 1967) was awarded 12 titles, including the overall titles of 1963 and 1964.

Skopje, city in s Yugoslavia on the River Vardar. It was founded in Roman times, became the capital of the Serbian empire in the 14th century, fell to the Ottoman Turks in 1392 and was incorporated into Yugoslavia in 1918. Most of the city was destroyed in an earthquake in 1963; many buildings date from after that time. Products include metals, textiles, chemicals and glassware. Pop. (1971) 312,980.

Skua, any of several species of darkcoloured, swift-flying shorebirds that pirate food from other birds in flight, but also prey on mice, voles and penguin chicks. It nests in colonies on the tundra, but migrates to the oceans in winter. Length: to 56cm (23in). Family Stercorariidae. *See also* NW pp.225, *227*.

Skull, skeleton of the head. It consists of the CRANIUM, or brain case, which in man is made up of eight bones, and the facial bones, 14 irregularly shaped bones that support and protect the eyes, nose and mouth. *See also* MS pp.56–57.

Skunk, nocturnal omnivorous mammal that lives in the USA, Central and South America. It has powerful anal scent glands which eject a foul-smelling liquid, used in defence. It has a small head and a slender, thickly-furred body with short legs and a large bushy tail. The coat is black with bold white warning markings along the back. The most common species is the striped skunk, *Mephitis mephitis*. Length: to 38cm (15in); weight: 4.5kg (10lb). Family Mustelidae.

Skydiving. *See* FREE-FALL PARACHUTING.

Skye, largest island in the Inner Hebrides, off the NW coast of Scotland; the chief town is Portree. It has an irregular coastline, a number of lochs and the Cuillin Hills to the SE; its spectacular scenery makes it a popular tourist resort. Other occupations include livestock raising, weaving and fishing. Area: 1,735sq km (670sq miles). Pop. (1971). 7,372.

Skye terrier, small dog, originally bred in the Isle of Skye, Scotland, in the 16th century. It has a large head with erect or pendulous ears. The body is long and low, with short legs and a long tail. The straight, flat coat is long and parts from head to tail; it may be blue, grey, fawn or cream. Height: to 25.5cm (10in) at the shoulder; weight: 11kg (24lb).

Skyjacking. *See* HIJACKING.

Skylab, US programme of manned orbiting space laboratories. In the mid-1970s three 3-man Skylab crews spent a total of 84 days in space. Further missions using a space shuttle are planned for the 1980s. *See also* SPACE SHUTTLE; SPACE STATION; SU pp.268–271.

Skylark. *See* LARK.

Skyros, largest of the N Sporades group of islands belonging to Greece, situated in the Aegean Sea. Rupert BROOKE, the British poet, is buried there. Products include cereals, figs, citrus fruits, olive oil and cheese. Marble, chromite and iron ore are mined. Area: 210sq km (81sq miles). Pop. (1971) 2,352.

Skyscraper, very tall building, generally constructed on a steel skeleton, for commercial or domestic use. The invention of practical electric lifts and the lack of space in central cities spurred construction of taller buildings in the mid-19th century. The first were of all masonry construction. Next came cage construction – the building's metal framework supported the floors, while the masonry walls supported themselves. True skyscraper construction, in which the metal skeleton supported both floors and walls, was introduced in the Home Insurance Building in Chicago (1885). Many modern skyscrapers have riveted metal skeletons and the walls merely enclose the internal space. The tallest skyscraper existing in 1975 was Chicago's Sears Roebuck Tower, with 110 storeys and 443m (1,454ft) tall (excluding the TV aerial). *See also* HC2 pp.206, 207; MM pp.48, 49.

Slade, Julian (1930–), British composer and dramatist. With Dorothy Reynolds he wrote the book, lyrics and music for *Salad Days* (1954) which ran for 2,329 performances, the third longest run in the British theatre. His other works include a musical *Fresh as Air* (1957), *Follow That Girl* (1960), *Wildest Dreams* (1961) and *Trelawny* (1972).

Slade School, British school of art at University College, London. It was opened in 1871 following the endowment of six scholarships in fine art by the collector Felix Slade, and the school's courses include painting, sculpture, printmaking, theatre design and film studies.

Slag, in the production of iron and steel, liquid layer of ore impurities such as oxides, ash and limestone. It floats on molten steel during refining, protecting the metal from oxidation and removing some unwanted substances. *See also* MM pp.30–31.

Slalom, obstacle race in SKIING or CANOEING. In Alpine skiing, the slalom is raced on a downhill course between a series of gates placed in a zigzag pattern. In the canoe slalom, the racer must negotiate a number of "gates" which are hung just above the water.

Slander, in law, oral defamation of a person's character made in the presence of one or more witnesses. *See also* LIBEL.

Slang, colloquial, vernacular vocabulary not normally accepted as standard written language, and often employing obscure and colourful imagery. Much modern slang is of North American origin, but in Britaiz perhaps the most interesting form of slang is COCKNEY rhyming slang, which employs part of a phrase to stand for an original word which rhymes with the entire phrase. Some of these phrases are ancient and relate to London's history. Thus "Barnet" is short for "Barnet Fair", which rhymes with "hair". A Cockney may therefore say "Barnet" to mean "hair". Slang is not the same as dialect and has nothing to do with a regional accent.

Slash-and-burn agriculture, primitive and wasteful agricultural technique that involves the felling of trees, which are then burned to make land arable. The land is rarely usable after two seasons and the farmer moves on to new woodland to follow the same procedure. *See also* PE pp.158–159.

Slate, grey to blue, fine-grained homogeneous METAMORPHIC ROCK, which splits into smooth, thin layers. It is formed by the metamorphosis of SHALE, and is valuable as a roofing material. *See also* PE pp.102–103, *103*.

Slav, largest ethnic and linguistic group of peoples in Europe. Slavs are generally classified in three main divisions: the East Slavs, the largest division, include the Ukrainians, Russians and Belorussians; the South Slavs include the Serbs, Croats, Macedonians and Slovenes (and frequently also the Bulgarians); the West Slavs comprise chiefly the Poles, Czechs, Slovaks and Wends. *See also* HC1 pp.162, 320; HC2 pp.84–85, 184–185, *185*.

Slave Coast, name given to a part of Africa's Guinea coast that supplied vast numbers of slaves from about 1500–1800. *See also* HC2 pp.44–45, 46, 47, 58–59, *58*, 128.

Slavery, condition in which people are held in involuntary servitude, the slave usually being obliged to perform labour or services as a condition of his servitude. In the New World from the late 16th century, slavery provided a supply of much-needed labour. Although American Indians were the first slaves in the New World, trade in African slaves became highly profitable and was carried on with great brutality. The antislavery movement, which was stimulated by the spread of humanitarian ideals in the 18th century, led to the abolition of slavery in all US States north of Maryland between 1777 and 1804. Eventually growing opposition to the institution in the USA led to the emancipation of the slaves during the AMERICAN CIVIL WAR. Other nations followed suit but in 1962 The International Labour Organisation claimed that slavery still existed in some parts of Asia and Africa. *See also* ABOLITIONISTS; MS pp.190, *190*, 269, *269*; HC1

Skimmers live amicably in flocks, which may include several thousand birds.

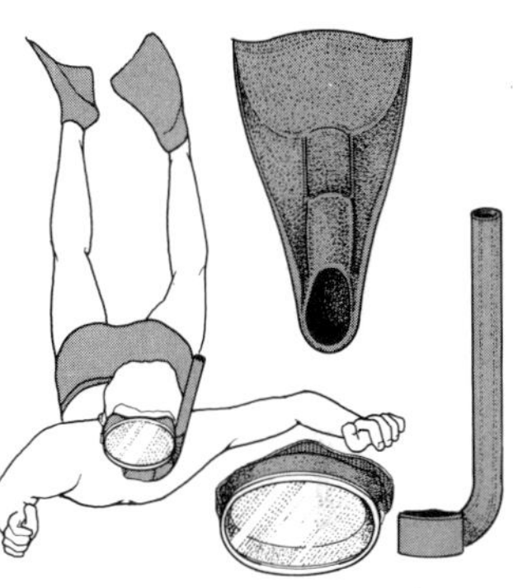

Skin diving; fins, mask and snorkel – the basic equipment this sport requires.

Skunks mature in about a year and can live for 10-12 years in captivity.

Skye fishermen's cottages. All of the island is within five miles of the sea.

Slave trade

Field Marshal Slim was appointed Governor General of Australia in 1953.

Sir Hans Sloane recorded his discoveries in his *Natural History of Jamaica*.

Sloop, the name given today to a vessel with one mast and two sails.

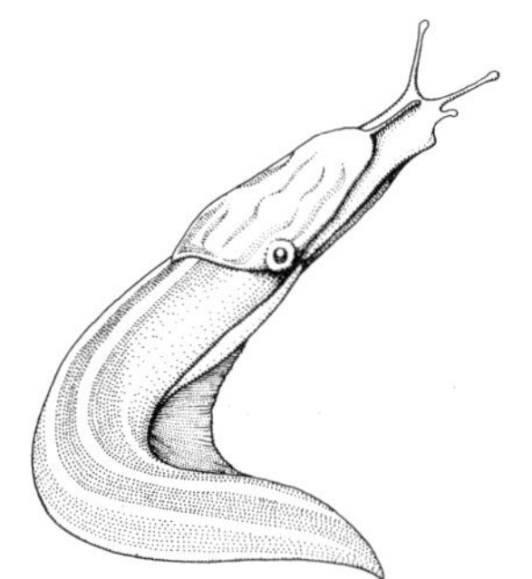

Slugs live underground or in damp places and are usually nocturnal.

pp.72, *72*, 94, 164; HC2 pp.58–59, *58*, 128, 134, 135.

Slave trade, commercial dealing in human beings as an unpaid self-reproducing labour force. The ancient practice of committing captives to a life of servitude was brought to a peak in western Europe – beginning with the Portuguese explorations of the African coast in the 15th century – by the Spanish, British, Dutch and French, all of whose depredations lasted until the 19th century. The British outlawed the slave trade in 1807, having made vast profits by transporting great numbers of Africans to the sugar plantations in the Americas. Slavery continued in British colonies for over 20 more years, however; it was abolished in the USA in 1863. *See also* HC2 pp.44, 46, 58–59, 134–135.

Slaveykov, Petko (1827–95), Bulgarian poet and patriot. He used the vernacular in his poetry so preparing the way for a revival of Bulgarian literature. He campaigned against his country's domination by the Turks and was co-founder of the Democratic Party (1879).

Slavonic languages, group of languages spoken in E Europe and the USSR, constituting a major subdivision of the INDO-EUROPEAN family of languages. Those spoken today are Russian, Ukrainian, Belorussian, Polish, Czech, Slovak, Bulgarian, Serbo-Croatian, Slovenian, Macedonian, and Sorbian or Lusatian (which is spoken in East Germany). Some Slavic languages are written in the Cyrillic alphabet, others in the Roman, depending chiefly on whether the main religion of the region is Eastern Orthodox or Roman Catholic.

Sledge, vehicle for carrying passengers and freight across ice and snow, on runners or skis.

Sleep, periodic relaxation of consciousness. Its precise function is not understood, but its importance is clear. Studies in sleep deprivation have distinguished between two types of sleep: light sleep and rapid eye movement (REM) sleep. Subjects show little to no ill effects when deprived of light sleep, which corresponds to specific brain wave patterns as measured on electroencephalographs (EEGs). During REM sleep a different pattern is registered on an EEG, and the deprivation of this type of sleep leads subjects to become disordered, confused and irritable with possible hallucinations and psychotic interludes. *See also* MS pp.43, 105.

Sleeping Beauty, The (1890), Russian ballet by Marius PETIPA, based on the fairy tale by Charles PERRAULT to music by TCHAIKOVSKY. Its huge popularity in the West dates from the London performances staged by DIAGHILEV in 1921.

Sleeping pill, drug prescribed to induce sleep in cases of restlessness or pain. Depending on the cause of insomnia, a sleeping pill may be a SEDATIVE, a hypnotic drug, an analgesic such as codeine or a PLACEBO.

Sleeping sickness, or trypanosomiasis, disease caused by a parasite transmitted by the TSETSE FLY. It is characterized by fever and may affect the brain and spinal cord, leading to profound lethargy and sometimes death. *See also* MS pp.*85*, 106, *107*.

Sleepwalking, also called somnambulism, neurotic reaction in which a sleeper rises from his bed and walks without waking, sometimes undertaking complex journeys, avoiding furniture and opening doors. Less commonly the sleeper may engage in an activity such as letter-writing, re-enactment of a trauma or past event, or may appear to search for a lost object. Such actions are believed to be the fulfilment of wishes, or the release of tension.

Sleepy sickness, popular name for a form of ENCEPHALITIS; it affects the CEREBRUM and the brain stem, causing swelling, HAEMORRHAGES and destruction of tissue. Other parts of the brain or spinal cord may be affected. Symptoms include drowsiness, restlessness, paralysis or unconsciousness. Treatment includes PHYSIOTHERAPY and long convalescence.

Slide rule, calculating device consisting of two rules engraved with logarithmic scales of numbers, one of which slides next to the other. Most mathematical functions can be accomplished using slide rules *See also* LOGARITHMS; SU p.39, *39*.

Sligo, county in Northern Ireland in Connacht province, bounded by the Atlantic to the N. The land is mountainous, rising in the W to the Ox Mts. The main activity is stock-raising. The county town is Sligo, at the mouth of the River Caravogue on Sligo Bay. It is a fishing centre and seaport. Industries include grain-milling, brewing and meat processing. Area: (county) 1,795sq km (693sq miles). Pop. (1971) (county) 50,236; (town) 14.070.

Slim, William Joseph, 1st Viscount (1891–1970), British general. He achieved international prominence in WWII as commander of the 14th Army in the reconquest of Burma (1943–45). He was British Commander-in-Chief in SE Asia (1944–45) and Chief of the Imperial General Staff (1948–52). He served as Governor-General of Australia from 1953 to 1960. *See also* HC2 p.231, *231*.

Slime eel. *See* HAGFISH.

Slime mould, any of a small group of strange, basically single-celled organisms that are intermediate between the plant and animal kingdoms. During their complex life cycle they pass through several stages. These include a flagellated swimming stage, an amoeba-like stage, a stage consisting of a slimy mass of protoplasm with many nuclei and a flowering sporangium stage.

Slip, in ceramics, cream-like mixture of clay and water, used as an adhesive in pottery for casting ceramic ware, and as a decorative element.

Slipped disc, or herniated disc, intervertebral disc in which the centre has slipped out from between abutting vertebrae, causing pressure against the spinal cord. The patient should stay in bed, and may be treated with analgesic medication, traction, support, physiotherapy and, occasionally, surgical removal of the protruding portion. *See also* MS p.94, *94*.

Sliven, or Slivno, city at the foot of the Balkans Mts. in E central Bulgaria. It holds a strategic position at the entrance to the Balkan passes. Products include textiles, glass, electrical goods and metalware. Pop. (1970) 84,167.

Sloan, John (1871–1951), US painter. A member of The Eight (ASHCAN SCHOOL) he painted realistic city scenes of people engaged in their daily activities.

Sloane, Sir Hans (1660–1753), British physician and naturalist, whose extensive collection of natural history specimens, books and manuscripts formed the nucleus of the British Museum. He was physician to Queen Anne and then to George II. During a brief stay in Jamaica he collected about 800 new species of plants which he catalogued in Latin.

Sloe. *See* BLACKTHORN.

Sloop, small single-masted sailing vessel with fore and aft sails. In the 18th century it was a sail-powered warship with about 20 guns. *See also* MM pp.172–175.

Sloth, any of several species of slow-moving, herbivorous tropical American edentate mammals. It has long limbs with long claws and spends most of its life climbing in trees, where it generally hangs upside down. Length: to 60cm (2ft); weight: to 5.5kg (12lb). Family Brachipodidae. *See also* NW pp.*154*, *191*, *216*.

Slough, county district in NE BERKSHIRE, S England; created in 1974 under the Local Government Act (1972). Area: 28sq km (11sq miles). Pop. (1974 est.) 101,800.

Slovakia, constituent republic of E Czechoslovakia, now divided into three provinces. It is bounded by Poland (N), USSR (E), Hungary (S), and Austria and Moravia (W); the capital is Bratislava. Most of Slovakia lies within the region of the Carpathian Mts. Until 1918 it was primarily under the control of Hungary; it became part of the new country of Czechoslovakia in 1918, but retained broad powers of autonomy. It was occupied by Germany in WWII and by the Soviet Union in 1968. The mountainous parts of the region are rich in minerals; lead, copper and lignite are mined. The soil is fertile and farming, stock rearing and vine growing form the basis of the economy. Area: 49,010sq km (18,923sq miles). Pop. 4,600,000.

Slovenia, constituent republic of Yugoslavia, in the NW part of the country, primarily in the Karst plateau and Julian Alps. It is bounded by Austria (N), Hungary (NE), the Croatian republic (S) and Italy (W); the capital is Ljubljana. It was part of the German Empire until 1918, when it was included in the Kingdom of Serbs, Croats, and Slovenes which was known as Yugoslavia after 1929. In 1946 it was made a constituent republic of Yugoslavia incorporating part of the former Italian region of Venezia Giulia. It is the most industrialized of all Yugoslav republics. Aluminium, steel and iron are produced; other occupations include farming and the raising of livestock. Area: 20,259sq km (7,819sq miles). Pop. (1971). 1,725,088.

Słowacki, Juliusz (1809–49), Polish Romantic poet. His works include the poetical drama, *Kordian* (1834) and two dramatized legends, *Balladina* (1834) and *Lilla Weneda* (1840).

Slow-worm, also called blind-worm, European snake-like, legless lizard of grassy areas and woodlands. It is generally brownish; the female has a black underside. It has pointed teeth and feeds primarily on slugs and snails. Length: to 30cm (12in). Family Anguidae; species *Anguis fragilis. See also* NW p.*134*.

Slug, mostly terrestrial gastropod MOLLUSC identified by the lack of shell and uncoiled viscera. It secretes a protective slime, which is also used to aid locomotion. Length: to 200mm (7.9in). Class Gastropoda; subclass Pulmonata; genera *Arion Limax. See also* SEA SLUG; NW pp.90, 91, *96*.

Slum clearance, removal of buildings in urban areas that have deteriorating and congested housing, poor sanitation and depressed living standards. In Britain the Torrens Act of 1868 provided for the demolition or improvement of slum areas. Slum clearance has become a systematic policy of LOCAL GOVERNMENT and has led to the development of tower blocks and NEW TOWNS. Tower blocks themselves are seen by many as vertical rather than lateral slums, and have given rise to new social problems.

Sluter, Claus (c.1340–1406), Flemish sculptor. Under the patronage of PHILIP The Bold of Burgundy, he worked on the portal sculptures and the *Well of Moses* sculpture for his patron's family mausoleum in Charterhouse at Champmol. *See also* HC1 pp.252, *252*, 262.

Sluys, Battle of (1340), English naval victory over the French in the HUNDRED YEARS WAR. EDWARD III, wishing to invade France through Flanders, attacked and destroyed the French fleet moored at Sluys in the Zwin estuary. It was the first major engagement in the history of the English navy.

Small arms, FIREARMS designed to be carried and fired by one man, as distinguished from ARTILLERY and heavy automatic weapons. They first came into use in the 15th century but were so cumbersome and inefficient that until the early 18th century many infantrymen were armed with pikes instead. Since then, small arms of some kind have been carried by all infantrymen, and the hand-to-hand combat of earlier warfare has now become comparatively rare. *See also* MM pp.162–163, *162–163*.

Small bore shooting. *See* SHOOTING.

Small intestine, part of the DIGESTIVE SYSTEM that extends – about 6m (20ft) coiled and looped – from the STOMACH to the large INTESTINE, or colon. *See also* MS pp.68–69.

Smallpox, highly contagious, often fatal, viral disease (*Variola*). Its symptoms are high fever and a severe rash on the face and extremities, which leaves permanent scars. Death is caused by toxins or when the lungs, heart and brain are infected. The disease has been controlled in most countries by the use of VACCINES. *See also* MS pp. 99, 107–108, *109*, 112.

Smallpox vaccination, inoculation against SMALLPOX. It was first used by Edward JENNER in 1796, when he inoculated an eight-year-old boy with infectious material from the scabs of the milder but related COWPOX, which is caused by the *Vaccinia* virus. *See also* MS pp.*99*, 107–108, *109*.

Smart, Christopher (1722–71), British poet. His lyric poetry, especially *A Song to David* (1763), still attracts praise, as do his classical translations and his *Hilliad* (1753), a satirical work. He was twice committed to an asylum, where he wrote his fine poem *Jubilate Agno.*

Smeaton, John (1724–92), British engineer. He was responsible for building the FORTH AND CLYDE CANAL and rebuilding the EDDYSTONE LIGHTHOUSE.

Smell, or olfaction, one of the five SENSES; stimulus to the brain caused by particles of a substance. The olfactory receptors can detect even a few molecules per million parts of air. The sense of smell in humans is less developed than in some other species (eg dogs and cats). *See also* MS pp.54–55.

Smellie, William (c.1740–95), Scottish printer and natural historian. He co-founded the Newtonian Society (1760) and the Society of Antiquaries of Scotland (1780), and in 1765 became a master printer. He prepared the first edition of the *Encyclopaedia Britannica* (1768–71) for press at the invitation of the engraver Andrew Bell.

Smelt, small, silvery food fish related to SALMON and TROUT. It lives in the N Atlantic and Pacific oceans and in North American inland waters. Family Osmeridae.

Smelting, processes by which metals are separated from their ORES by heating, often involving chemical reduction. The ore (and often other ingredients) is heated in a FURNACE to remove the non-metallic constituents, and the metal is purified by another stage. In the smelting of iron, the ore (iron oxide) is heated in a BLAST FURNACE with COKE and limestone. The oxide combines with carbon (from the coke) to form gaseous oxides of carbon and pig iron (an alloy of carbon, iron and other impurities). *See also* MM pp.30–31.

Smetana, Bedřich (1824–84), Czech Romantic composer who promoted Czech nationalistic style in operas and orchestral music. His masterpiece, *The Bartered Bride* (1866), is one of the greatest folk operas, and is regarded as the national opera of Czechoslovakia. Among other popular works by Smetana is the cycle of symphonic poems, *My Country* (1874–79). *See also* HC2 pp.107, 120.

Smibert, John (1688–1751), Scottish-born colonial American painter. He studied in Florence and Rome (1717–20) and held the first art exhibition in the colonies at Boston in 1730.

Smiles, Samuel (1812–1904), Scottish writer. He wrote a number of biographies of captains of industry (including the three-volume *Lives of the Engineers*; 1861–62) and established his reputation with *Self-Help* (1859) as the great apostle of the mid-Victorian work ethic.

Smilodon. *See* SABRETOOTHED TIGER.

Smith, Adam (1723–90), Scottish economist who laid the foundation of modern economics. Reflecting the emergence in Britain of the industrial system, he argued that the commercial liberty of a competitive market would stimulate production and act in the interests of the public. He wrote his economic treatise *The Wealth of Nations* (1776) after meeting French PHYSIOCRATS. *See also* HC2 pp.23, 31.

Smith, Arthur James Marshall (1902–), Canadian academic and poet. Although he spent little time in Canada, he influenced its poetry, editing *The Book of Canadian Poetry* (1943) and *The Oxford Book of Canadian Verse* (1958).

Smith, Bessie (c.1895–1937), US BLUES singer. She started recording in 1923 and toured with bands as a soloist, her mournful rendering of blues becoming part of JAZZ history. *See also* HC2 p.273.

Smith, Clarence ("Pinetop"), (1904–29), US JAZZ pianist and composer. He originated the term BOOGIE-WOOGIE in his *Pinetop's Boogie Woogie.*

Smith, Cyril (1909–74), British pianist. Disabled in 1956 by a cerebral thrombosis, he began in the following year to perform with the right hand alone and

became well known for his performances (with his wife Phyllis SELLICK) of music for two pianos.

Smith, David (1906–65), leading US sculptor. A pioneer in welded metal sculpture, some of his work has been called "drawing in space" because it seems two-dimensional and linear.

Smith, Dodie (1896–), British novelist and playwright, she achieved success in the 1930s with the plays *Autumn Crocus* (1931), written under the pseudonym C. L. Anthony, and *Dear Octopus* (1938). Her best-known work is *The Hundred and One Dalmatians* (1956), which was made into a film by Walt DISNEY (1960).

Smith, Frederick Edwin, 1st Earl of Birkenhead (1872–1930), British lawyer and politician. He was a Conservative MP (1906–18) and served as Solicitor-General (1915), Attorney-General (1915–19), Lord Chancellor (1919–22) and Secretary of State for India (1924–28). He opposed the inclusion of Ulster in IRISH HOME RULE, and was prominent in the negotiation of the Anglo-Irish Treaty (1921).

Smith, Harvey (1938–), British show jumping rider esteemed by the public as much for his outspokenness as for his skill at training and riding winning horses.

Smith, Henry John Stephen (1826–83), British mathematician who was noted for his work on elliptic functions and the NUMBERS THEORY.

Smith, Ian Douglas (1919–), Rhodesian Prime Minister. He fought in WWII and entered Rhodesian politics in 1948. He became Prime Minister in 1964 and declared RHODESIA's independence (UDI) from Great Britain in 1965. Although he rejected African majority rule of his overwhelmingly black country, he made some efforts to negotiate with black leaders during the 1970s. *See also* HC2 pp.*219*, 253.

Smith, John (c.1580–1631), English soldier and colonist. He took a leading part in establishing JAMESTOWN SETTLEMENT in Virginia, and was instrumental in obtaining maize from the local Indians, thus saving the colony from starvation during the early years of hardship. According to his account, he was rescued from death at hands of Powhatan by the chief's daughter POCAHONTAS.

Smith, Joseph (1805–44), US founder of the MORMON Church of Jesus Christ of the Latter Day Saints (1830). His *Book of Mormon* (1829) was based on sacred writings he claimed to have found on golden tablets. In 1844 he was jailed on a charge of treason with his brother, Hyrum, at Carthage, Illinois. There they were murdered by a mob.

Smith, Joseph Fielding (1838–1918), MORMON leader, nephew of Joseph SMITH, the founder of the Mormon Church. From 1901 to 1918 he served as the 6th president of the Mormon Church.

Smith, Lloyd Logan Pearsall (1865–1946), US critic and essayist, friend of Matthew ARNOLD and WHISTLER, who became a British subject in 1913. His criticism includes *On Reading Shakespeare* (1933) and *Milton and His Modern Critics* (1940).

Smith, name of two brothers who made the first flight from England to Australia; they took off on 12 Nov. and landed at Darwin on 10 Dec. 1919. Sir Keith Macpherson (1890–1955) flew as a pilot in WWI. Sir Ross Macpherson (1892–1922) was a cavalryman who did not learn to fly until 1916. He made his first flight from Cairo to Calcutta in 1918. On their momentous flight the Smiths flew a Vickers Vimy twin engine bi-plane; after landing they were knighted.

Smith, Stevie (1902–71), British poet, real name Florence Margaret Smith, who became famous at public poetry readings during the 1960s. Her poetry is both pathetic and gay and includes the volume *Not Waving But Drowning* (1957).

Smith, Sydney (1771–1845), British clergyman, author and wit, one of the founders of *The Edinburgh Review* (1802).

Smith, William (1769–1839), British geologist. A founder of STRATIGRAPHY, he studied the geological strata of England and Wales, relating his findings to identi-

fied fossils and thus estimating the age of geological formations.

Smith, William Henry (1825–1891), British businessman and politician who entered his father's newsagent business and greatly expanded it by securing the right to sell books and newspapers at railway stations. He entered Parliament in 1868 as a Conservative, serving in DISRAELI's second administration as First Lord of the Admiralty (1877–1880) and in the second SALISBURY administration as First Lord of the Treasury and Leader of the Commons.

Smith, Sir William Sidney (1764–1840), British admiral. He gave distinguished service during the NAPOLEONIC WARS and, while plenipotentiary to Constantinople, hurried to ACRE when that town was threatened by NAPOLEON. He held the town until the siege was raised (1799). In 1807 he destroyed the Turkish fleet off ABYDOS and successfully blockaded the TAGUS. He became an admiral in 1821.

Smithfield, district of the city of London and home of the world's largest meat market. In the 12th century it was the site of a weekly horse fair and various tournaments. Bartholomew Fair, a general livestock and linen fair held as a festival for St Bartholomew and lasting 14 days, was held there from 1133 to 1855.

Smithson, Alison Margaret and Peter Denham, British husband and wife architectural team, exponents of BRUTALISM. Alison (1928–), and Peter (1923–), designed the Hunstanton School, Norfolk (1954) and The Economist Building, London (completed 1964).

Smithson, James (1765–1829), British chemist and mineralogist; the mineral smithsonite (zinc carbonate) was named after him. Angered at the Royal Society's rejection of a paper by him in 1826, he left £105,000 to found an institution, the SMITHSONIAN INSTITUTION in Washington for the "increase and diffusion of knowledge among men".

Smithsonian Institution, US independent trust establishment, in Washington D.C., created (1846) by the bequest of James Smithson of England. Among its activities the Institution funds research, publishes the results of explorations and investigations and preserves for reference over 65,000,000 items of scientific, cultural and historic interest.

Smithsonite, carbonate mineral, zinc carbonate ($ZnCO_3$). It is generally white, but may be other colours; blue specimens from New Mexico were once used as gemstones. Hardness 4–4.5; s.g. 4.4.

Smog, dense atmospheric mixture of smoke and fog or chemical fumes, commonly occurring in urban or industrial areas. It is most dense during temperature inversions. *See also* PE pp.146, *147*.

Smoke signal, column of smoke from a fire used as a means of communication. Smoke signals have been known since the time of Alexander the Great, but the Plains Indians of North America are most famous for employing them. Their use was limited to signalling danger or acting as a prearranged sign.

Smoking. *See* CIGARETTES.

Smollett, Tobias George (1721–71), Scottish novelist and surgeon. His novels include *The Adventures of Roderick Random* (1748), *The Adventures of Peregrine Pickle* (1751) and *The Expedition of Humphrey Clinker* (1771). Smollett also translated VOLTAIRE and edited several periodicals. *See also* HC2 p.34.

Smooth muscle. *See* INVOLUNTARY MUSCLE.

Smuggling, illegal movement of goods or people across national or state boundaries. A corollary of trade restrictions and prohibition, smuggling has flourished in Britain since the 18th century particularly in the importation of highly taxed goods such as silk and spices and prohibited goods such as narcotics. Penalties for smuggling have traditionally been high. In most Western countries it is a criminal offence punishable by a fine or prison sentence. *See also* HC2 p.44.

Smut, group of plant diseases caused by parasitic fungi, also called smuts, that attack many cereals including maize,

Bedřich Smetana, painted by Johan der Södermark. He became deaf in later life.

Frederick Edwin Smith was feared in Parliament for his caustic wit.

William Smith, the geologist, was the first to be awarded the Wollaston medal.

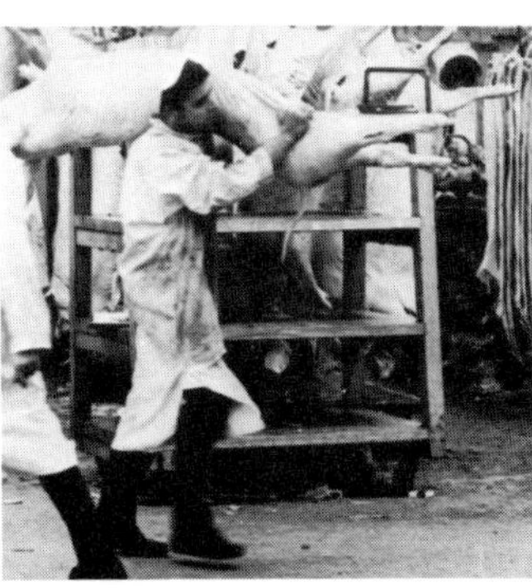

Smithfield market: porters hard at work in London's famous meat market.

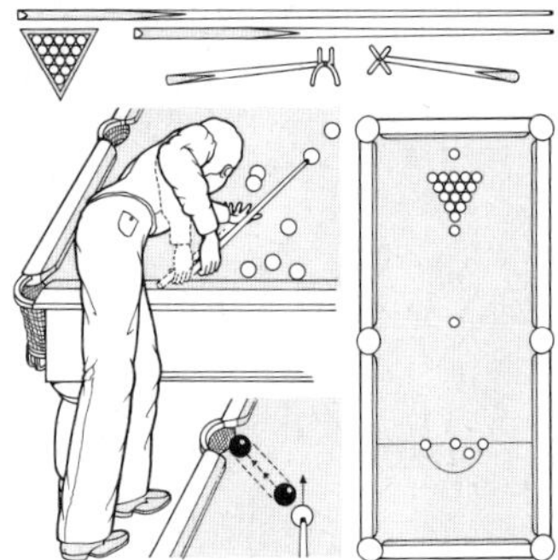

Snooker; rack, cues, bridges and a special table are used in this sport.

Snowy owls inhabit sub-polar regions. The female is larger than the male.

Sir John Soane's library in his home in Lincoln's Inn Fields is now a museum.

Gary Sobers, in London in 1969, was captain of the West Indies Test side.

wheat and other grasses. The diseases are named after the sooty black masses of reproductive spores produced by the fungi.

Smuts, Jan Christiaan (1870–1950), South African political leader. A lawyer, he was alarmed by the JAMESON RAID of 1895 and entered Boer politics. He was a guerrilla leader during the SOUTH AFRICAN WAR (1899–1902) but afterwards worked for the establishment of a unified republic, repressing Boer rebellions during WWI. He served in various ministries under Louis BOTHA and predicted the disastrous consequences of the Treaty of VERSAILLES following WWI. He was Prime Minister from 1919–24 and 1939–48 and helped form the United Nations following WWII. His government was replaced (1948) by the nationalist government of Daniel MALAN. *See also* HC2 p.252, *252*.

Smyrna. *See* IZMIR.

Smyth, Dame Ethel (1858–1944), British composer. She wrote songs, chamber and choral music, and several operas, including *The Wreckers* (1906). She wrote *March of the Women* (1911) for the British SUFFRAGETTE MOVEMENT.

Smythe, Patricia ("Pat") (1928–), British showjumping rider. Women's European champion in 1957, 1961, 1962, 1963 on Flanagan, she was the first woman to ride in Olympic showjumping, winning a team bronze in 1956.

Snail, terrestrial, marine or freshwater gastropod mollusc. It has a coiled protective shell encasing an asymmetric visceral mass, a large fleshy foot and antennae on its head. It may breathe through gills (aquatic species) or through a kind of air-breathing lung (terrestrial species) and has a radula, a rasping organ located within its mouth. Some species such as the Roman snail (*Helix pomatia*), are edible. Length: to 35cm (14in). Class Gastropoda. *See also* NW pp.*22, 34, 90–91, 96–97, 230–232, 241*.

Snake, any of some 2,700 species of legless, elongated REPTILES forming the suborder Serpentes of the order Squamata (which also include LIZARDS). There are 11 families. They range in size from about 10cm (4in) to more than 9m (30ft). There are terrestrial, arboreal, semi-aquatic and aquatic species; one group is entirely marine; many are poisonous. They have no external ear openings, eardrums or middle ears; sound vibrations are detected through the ground. Their eyelids are immovable and their eyes are covered by a transparent protective window. Internal organs are elongated, with the left lung reduced or absent. The long, forked, protractile tongue is used to detect odours. Their bodies are covered with scales. Upper and lower jaws are movable and each half of either jaw can be moved independently of the other, allowing snakes to swallow large prey. *See also* NW p.73, *134*, 135, *137–139, 199, 205, 215, 232, 234*.

Snakebird, or anhinga, bird that lives in or near fresh water of the tropics and subtropics. Snakebirds have dark metallic plumage, straight bills, on long necks, slender bodies, long tails and webbed feet. Length: 90cm (35in). Family Anhingidae.

Snake dance, ritual dance of the HOPI Indians of North America in which men hold live rattlesnakes in their mouths. *See also* RAIN DANCE.

Snapdragon, any of several species of perennial plants with saclike, two-lipped, purple, red, yellow or white flowers. The common snapdragon (*Antirrhinum majus*) is a popular garden plant; height: 15–91cm (0.5–3ft). Family Scrophulariaceae.

Snapper, marine food fish found in tropical waters of the Indo-Pacific and Atlantic oceans. Length: to 91cm (3ft); weight: 50kg (110lb). The 250 species include the red snapper *Lutianus campechanus*, yellowtail *Ocyurus chrysurus* and Atlantic grey *L. griseus*. Family Lutjanidae. *See also* PE p.247.

Snapping beetle. *See* CLICK-BEETLE.

Snare-drum. *See* DRUM.

SNCC (Student Non-Violent Co-ordinating Committee), US organization founded in 1960 to campaign for racial equality in the USA. It worked closely with such integrationist organizations as CORE and the NAACP until 1966, when Stokely CARMICHAEL became its president and it became identified more with the BLACK POWER movement.

Snead, Samuel Jackson (1912–), US golfer. He won more than 100 tournaments, including three US PGA Golf Championships (1942, 1949, 1951), three US Masters (1949, 1952, 1954) and the British Open Championship (1946).

Snell, Peter (1938–), New Zealand athlete. He won three Olympic gold medals, for the 800m in 1960 and 1964, and for the 1,500m in 1964. He set a world mile record of 3 min. 54.1 sec. in 1964.

Snipe, any of several species of migratory long-billed, wading shorebirds found in swampy grasslands and coastal areas over most of the world. It is generally mottled brown and buff. Length: 30cm (12in). Family Scolopacidae; genus *Capella*.

Snoek (snook), any of several species of long, silvery, tropical, marine fish of the Pacific and Atlantic coasts. It has a long head, a projecting lower jaw and two dorsal fins. Length: to 1.5m (5ft). Family Centropomidae; genus *Centropomus*.

Snooker, game usually for two players, played on a BILLIARDS table, using 15 red balls (valued at 1 point each), 6 coloured balls – yellow (2), green (3), brown (4), blue (5), pink (6) black (7) – and 1 white cue-ball. Points are scored by potting (striking the cue-ball with the cue so as to cause to cannon into the pockets) the red balls, which then remain off the table. After potting a red, the player is allowed to attempt to pot any one of the colours. Colours potted thus are returned to their respective spots on the table until all the reds have been potted, when the colours are potted in ascending order of values and this time remain off the table. Penalties are incurred when the cue-ball misses everything, hits a colour without having potted any red remaining on the table, or itself falls into a pocket. When the balls are so arranged that a player cannot strike directly at the ball he is supposed to, he is said to be snookered. The player scoring the highest number of points wins.

Snoring, vibration of the soft PALATE and the narrowed opening between the mouth and throat. It generally occurs during sleep when breathing takes place through the mouth. Mouth breathing may be caused by nasal blockage due to the common cold or CATARRH. Further causes of snoring may be enlargement of the ADENOIDS or TONSILS, or weakening of the soft palate and surrounding tissues due to ill health, smoking or obesity.

Snorri Sturluson (1178–1241), Icelandic historian. His *Prose Edda* is a collection of Norse mythology and a discussion of the art of poetry. *Heimskringla*, sagas of the Norwegian kings to 1184, mingles history and legend.

Snow, Charles Percy, Baron (1905–), British novelist and physicist who held several important positions in the British government. As a novelist he is famous for his sequence of novels, known collectively as *Strangers and Brothers*.

Snow, flakes of frozen water that fall from clouds to the Earth's surface. Snowflakes are symmetric (usually hexagonal) crystalline structures. *See also* SU p.*51*.

Snowberry, shrubby plant native to North America and China. It has pink or white bell-shaped flowers and soft berries. Wild species are coralberry, wolfberry and snow or waxberry. The latter has large, pulpy, white berries. There are 18 species. Family Caprifoliaceae; genus *Symphoricarpos*.

Snowdon, Anthony Armstrong-Jones, Earl of (1930–), British designer and photographer who married, in 1960, HRH Princess MARGARET, the sister of Queen Elizabeth II. *See also* NW p.*252*.

Snowdon, mountain in Gwynedd, NW Wales. Much of the area is included in the Snowdonia National Park which was established in 1951. The summit can be reached by a rack-and-pinion railway from Llanberis. Snowdon has five peaks, one of which is the highest in England and Wales. Height: 1,085m (3,560ft).

Snowdrop, low-growing perennial plant of the Mediterranean region, widely cultivated as a garden ornamental. The green and white, fragrant flowers appear early in spring. The common snowdrop (*Galanthus nivalis*) has narrow leaves; height: to 15cm (6in). Family Amaryllidaceae.

Snow Leopard. *See* LEOPARD, SNOW.

Snowmobile, petrol-powered sled which has ski runners and rubber tracks to propel the vehicle over snow or ice.

Snowshoe rabbit, or varying hare, medium sized HARE of N North America with large furry hind feet that act as snowshoes. It is brownish in summer, white in winter. Species *Lepus americanus*.

Snowy, river in SE Australia. It rises in the Australian Alps and flows through Gippsland into the Bass Strait. It is part of a large hydroelectric project. Length: 448km (278 miles).

Snowy Mountains (Muniong Range), range of the Australian Alps, Australia. They are the location of Australia's major hydroelectric power project, which has been designed to double the country's output of electricity.

Snowy owl, any of several species of white and grey owls, camouflaged to blend with the snowy barren Arctic tundra where they live; especially *Nyctea scandiaca*. It feeds mainly on lemmings and hares, hunting in daylight during the summer. Length: to 60cm (2ft). Family Strigidae. *See also* NW pp.*17, 141, 206, 207, 227*.

Snuff, ground and perfumed tobacco which was inhaled as a stimulant or medicine, most commonly in 17th-century England. The practice was widespread in Europe by the 18th century, when exquisite snuff boxes were made; these are now sought after as collectors' pieces.

Snyder, Gary (1930–), US poet who also worked as a forester, sailor, carpenter and lumberjack. He recorded his experiences in poems, most of which are included in *A Range of Poems* (1966).

Snyders, Frans (1579–1657), Flemish painter. He is known especially for his still-lifes and pictures with animals. He was a pupil of Pieter BRUEGEL the younger and a friend of RUBENS, whom he often helped to paint large, important pictures.

Soane, Sir John (1753–1837), British architect. A leader of the Greek revival, he designed the Bank of England (1795–1827). His own residence at Lincoln's Inn Fields was bequeathed to the nation. *See also* HC2 p.70.

Soap, substance made by the action of ALKALI on FAT, and used to remove dirt and grease. Common soaps are sodium or potassium salts of stearic or oleic acids. Soap consists of long-chain molecules, one end of which attaches to grease and the other dissolves in the water so that the grease loosens and forms a floating scum. *See also* DETERGENT; MM pp.240–241.

Soap Box Derby, event for children in which four-wheeled motorless carts coast downhill, steered on a course, powered only by gravity. The original construction of such a cart was of four wheels, two axles and a soap box. The classic Soap Box Derby is held annually in Akron, Ohio, USA.

Soap opera, dramatic serial programme, either on radio or television, originally sponsored in the USA by companies manufacturing soap and detergents. The story often centres on marital problems and other family situations, frequently involving melodramatic events.

Soapstone, also known as steatite and TALC, ROCK with a soft soapy or greasy texture. There are many types but all contain a large proportion of magnesium silicate, often associated with various amounts of serpentine and carbonates. Food vessels and carvings made from soapstone have been found amongst the remains of prehistoric human cultures.

Sobers, Sir Gary (1936–), West Indian cricketer for West Indies, Barbados, South Australia and Nottinghamshire. A magnificent left-handed all-rounder, he played in 93 Test matches, 1954–74, 39 as captain, scored a world-record aggregate of 8,032 runs, including a world-record Test innings of 365 not out, took 235 wickets, and held 110 catches.

Sobhuza, name of two rulers of Swaziland. Sobhuza I (*r.*1815–36) founded the kingdom and maintained it in the face of Zulu attacks. Sobhuza II (1899–   ) became king of independent Swaziland in 1968, and assumed absolute powers in 1973.

Sobieski, John. *See* JOHN, KINGS OF POLAND.

Sobrero, Ascanio (1812–88), Italian chemist who in 1846 discovered the explosive nitroglycerine (glyceryl trinitrate) by adding glycerine to a mixture of nitric and sulphuric acids. Horrified at the potential of his discovery he made no attempt to develop it.

Soccer. *See* FOOTBALL, ASSOCIATION.

Social contract, concept that society is based on the surrender of natural freedoms by the individual to the organized group or state in exchange for his personal security. The concept can be traced back to the ancient Greeks and was developed by the English philosophers Thomas HOBBES and John LOCKE, and by Jean-Jacques ROUSSEAU in *The Social Contract* (1762).

Social Credit Movement, Canadian political party. In 1935 William ABERHART won Alberta on a programme of redistribution of income through "social dividends", cash payments to consumers, the amount being determined by an assessment of the nation's wealth. Aberhart's attempts to tax banks and control the money supply were declared unconstitutional and the party became divided. It is considered a voice of regional conservatism. In 1971 it lost the election to the Progressive Party.

Social Democrats, name originally taken by MARXIST parties of central and east Europe before 1914, but now used by people who wish to distinguish their socialist beliefs from the dogmas of Marxist parties. In particular, a social democrat in western Europe accepts the pluralism of liberal democracy and wishes to retain a mixed economy.

Socialism, system of social and economic organization in which property is owed not by private individuals but by the community, in order that all may share more fairly in the wealth produced. Many different forms of socialism exist in both theory and practice, differing mainly in their interpetation of how much property the indivudal should be allowed; what definition of "the community" is used, whether it be the factory, the town or the state; and to what degree individual liberty is subordinated to the requirements of the society as a whole.

As a system of social organization, socialism has been practised by many primitive communities. Western society, however, has been founded on the principle of private property since Greek times, and until the 19th century socialism was preached only by radical or millenarian groups, such as the ANABAPTISTS or the DIGGERS of the 16th and 17th centuries. Socialism became seriously considered with the rise of industrialism and the breakdown of paternalist, aristocratic society. The early 19th-century socialists, such as Robert Owen, Henri de SAINT-SIMON and Charles FOURIER, were later dubbed UTOPIAN SOCIALISTS by Karl MARX, who, unlike them, claimed to be "scientific" in that his own vision of the future was based on the economic inadequacies of CAPITALISM, and not simply on the advent of human brotherhood and good sense.

In the 20th century, the term socialism has changed its meaning. Marx made no distinction between socialism and COMMUNISM, defining both as "from each according to his abilities, to each according to his needs", but as the term communism began to apply to those supporters of the Communist Party and therefore of the USSR, socialism was used either to refer to those who adopted reformist rather than revolutionary tactics, or to those who stressed the need for individual liberty within the post-capitalist state. Thus both the British LABOUR PARTY and the Maoists may describe themselves as socialists. In the same way, communities such as Israeli KIBBUTZIM, whose goods are to be sold on the open market within a capitalist economy, may be considered socialist. *See*

also HC2 pp.170–171, 212–213; MS pp. 273, 277.

Socialist parties, political parties in many countries throughout the world, each of which professes a devotion to the general principles of SOCIALISM as they have developed since the 19th century. The first socialist party in Europe was the German Social Democratic Party, founded in 1875. The British LABOUR PARTY was founded in 1906. The Socialist Party of America was founded in 1901. Since 1945 socialist parties have spread to most countries of the Third World, where socialism and NATIONALISM have often developed simultaneously. In Europe socialist parties have been most successful in gaining power in Scandinavia, Germany, Italy, Britain and the Low Countries. Most socialist parties in the Western world call themselves SOCIAL DEMOCRATS, in order to distinguish their policies from those of COMMUNIST PARTIES.

Social realism. *See* REALISM, SOCIAL.

Social security, welfare payments in Britain under two heads, NATIONAL INSURANCE and supplementary benefits, the latter known before 1966 as national assistance. Supplementary benefits ensure that a minimum weekly income is maintained and, unlike national insurance, are not financed by contributions, but are available to the needy as a right. Social security is managed by the Department of HEALTH AND SOCIAL SECURITY.

Social Security, Department of Health and. *See* HEALTH AND SOCIAL SECURITY, DEPARTMENT OF.

Social Security, Ministry of, British government ministry formed in 1966 from the merger of the Ministry of PENSIONS AND NATIONAL INSURANCE and the NATIONAL ASSISTANCE BOARD to provide welfare payments. In 1968 it merged with the Ministry of Health to form the Department of HEALTH AND SOCIAL SECURITY.

Society Islands, two island groups in the s Pacific Ocean, part of w French Polynesia and comprising the WINDWARD ISLANDS and the LEEWARD ISLANDS; the capital is Papeete. They were first sighted in 1607 by the Portuguese navigator Pedro Fernandes de Queiros and were claimed by Great Britain in 1767 and by France in 1786; they were made a French protectorate in 1842. The islands support a tropical fruit industry and export rum, sugar, mother-of-pearl and vanilla. Area: 1,610sq km (620sq miles). Pop. (1971) 100,300.

Society of Friends. *See* QUAKERS.

Society of Jesus. *See* JESUITS.

Sociology, the scientific study of society, its institutions and processes. It examines such areas as social change and mobility and underlying cultural and economic factors. Auguste COMTE invented the term "sociology" in 1843 and from the 19th century numerous complex and sophisticated theories have been expounded by Herbert SPENCER, Karl MARX, Emile DURKHEIM, Max WEBER and others. *See also* HC2 pp.*263,* 266–267, *267,* 270–271, 302.

Socotra, island territory of the People's Democratic Republic of Yemen, in the Indian Ocean, s of the Arabian Peninsula and 256km (160 miles) ENE of Cape Guardafui, Somalia; the capital is Tamridah. The mountainous terrain includes peaks rising to approx. 1,520m (5,000ft). The economy is based on agriculture; farming and herding are important occupations. Exports include tobacco, ghee, dates and pearls. Area: 3,100sq km (1,200sq miles). Pop. 8,000.

Socrates (*c.*469–399 BC), Greek philosopher. Information about his life and philosophy is found in the writings of PLATO and XENOPHON. Believing that the highest meaning of life is attained through self-knowledge, Socrates taught the value of self-analysis. His method of cross-questioning challenged conventional wisdom and his sceptical approach to religion led to his trial for impiety and corrupting the young. Condemned to death, he poisoned himself with hemlock. *See also* MS pp.226, *227,* 232–233.

Soda, sodium carbonate, Na_2CO_3, usually manufactured from common salt (sodium

chloride) and ammonia by the SOLVAY PROCESS. In anhydrous form it is known as soda ash; washing soda is hydrated sodium carbonate, $Na_2CO_3.10H_2O$.

Sodalite, glassy silicate mineral, sodium aluminium silicate with some chloride, found in alkaline IGNEOUS ROCKS. It occurs as small dodecahedral crystals in the cubic system and also as masses. It may be colourless, white, blue or pink; it is sometimes used as gem. Hardness 5–6; s.g. 2.2.

Soddy, Frederick (1877–1956), British chemist who was awarded the 1921 Nobel Prize in chemistry for his studies of radioactive substances and his work on ISOTOPES (he coined the term). In 1920 he revealed the value of isotopes in computing geological age. With Ernest RUTHERFORD, he worked out an explanation of RADIOACTIVE DECAY and later, with William RAMSAY, detected helium as a product of uranium decay.

Söderberg, Hjalmar Erik Fredrik (1869–1941), Swedish novelist and playwright. His novels are about the upper-middle classes are lyrical but disillusioned in tone, and include *Martin Bircks Ungdom* (1901; tr. *Martin Birck's Youth,* 1930) and *Doktor Glas* (1905; tr. *Doctor Glas,* 1963).

Söderblom, Nathan (1866–1931), Swedish theologian who was awarded the 1930 Nobel Peace Prize for his work towards world peace through Church unity. He was ordained in 1893 and spent seven years in Paris as a chaplain to the Swedish legation. In 1914 he was appointed Archbishop of Uppsala, Sweden, and primate of the Lutheran Church of Sweden.

Soderini, Piero (1452–1522), political leader who held the highest office in Florence from 1501 until he was ousted in 1511. During his administration, MACHIAVELLI was commisioned to raise a standing army.

Sodium, common metallic element (symbol Na) of the alkali-metal group, first isolated in 1807 by Sir Humphry DAVY. It occurs in the sea (as salt) and in many minerals. Its chief source is sodium chloride (common salt), from which it is extracted by electrolysis. The metal is used as a heat-transfer medium in nuclear reactors. Chemically, sodium is one of the most reactive elements. Properties: at.no. 11; at.wt. 22.9898; s.g. 0.97; m.p. 97.81°C (208.05°F); b.p. 882°C (1,620°F). *See also* ALKALI ELEMENTS; SU pp.*134,* 136.

Sodium carbonate. *See* SODA.

Sodium chlorate, white, crystalline compound, $NaClO_3$, which is a strong oxidizing agent. It is used as a general herbicide and soil sterilizer and as a laboratory reagent. It is made by the ELECTROLYSIS of a solution of sodium chloride.

Sodium chloride, common salt, NaCl. Found in salt-pans and salt-domes, it is the major mineral component of SEA-WATER. It is also the major ELECTROLYTE of living cells and the loss of too much salt, through evaporation from the skin or through illness, is dangerous. Sodium chloride is also an important chemical in industry, a starting material for various other sodium compounds and for making CHLORINE.

Sodium hydroxide. *See* CAUSTIC SODA.

Sodium oxide, white, crystalline compound, Na_2O, which dissolves violently in water to form sodium hydroxide (caustic soda). It is usually made by heating sodium peroxide, Na_2O_2, with sodium.

Sodium thiosulphate. *See* HYPO, PHOTOGRAPHIC.

Sodom and Gomorrah, in the Bible, two of the five "cities of the plains", probably sited near the Dead Sea, destroyed by God with fire and brimstone for their carnality and vice. The incident is described in Genesis 13–19. The region probably suffered a severe earthquake *c.*1900 BC.

Sodor and Man, the DIOCESE of the ISLE OF MAN. The original diocese of Sodor included the HEBRIDES and other islands w of Scotland. The bishop is a member of the council (upper house) of the Court of Tynwald, the Manx Parliament.

Sofia (Sofija), capital of Bulgaria and Sofia province, in w central Bulgaria. The city was founded in the 2nd century AD and

Socialism; Karl Marx was the chief theorist of modern Socialism.

Society Islands; Cook's Bay on Moorea Island, typical of these South Pacific islands.

Socrates; a 4th-century Roman copy of a Greek original, in the British Museum.

Nathan Söderblom, receiving his Nobel Peace Prize in 1930.

Soissons was a rallying point for the French in 1815 after Waterloo.

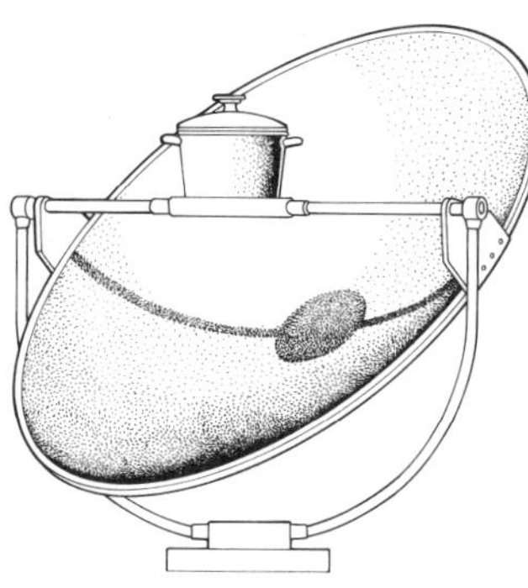

Solar heating for domestic purposes is being developed for commercial use.

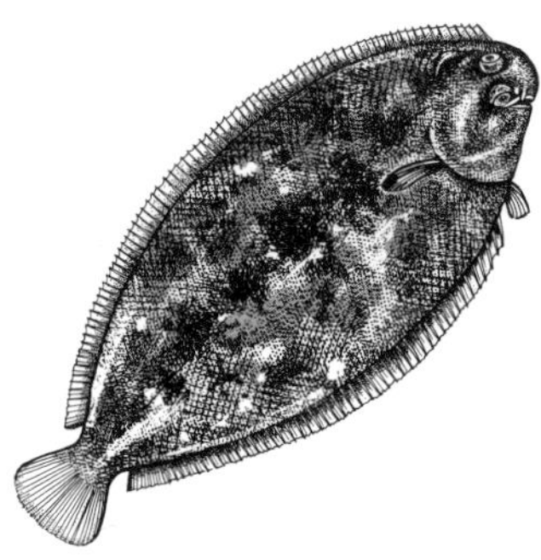

Sole; its close-set eyes are both situated on the same side of its head.

Solenodons dig for worms and insects with their long snouts and sharp claws.

ruled by the Byzantine Empire from 1018 to 1185, and was the residence of Constantine I. Sofia passed to the second Bulgarian empire (1186–1382), and then to the Ottoman Empire (1382–1878). The city was taken by Russia in 1877 and chosen for capital of Bulgaria by the Congress of Berlin. The University in Sofia dates from 1888. Industries: machinery, textiles, rubber, leather goods, food processing. Pop. 946,300.

Softball. *See* BASEBALL.

Soft style, term for the style of art found mainly in Germany in the late 14th and early 15th centuries. Closely related to INTERNATIONAL GOTHIC, it is characterized by gentleness of sentiment and softness in the representation of drapery.

Soho, area of London bounded by Coventry Street, Charing Cross Road, Oxford Street and Regent Street and noted for its busy nightlife and foreign restaurants. Its name is reputed to come from the hunting cry "So-ho" associated with the battle-cry at Sedgemoor of MONMOUTH, who had a house in Soho Square.

Soil, surface layer of earth capable of supporting plant life. It consists of undissolved minerals produced by the weathering and breakdown of surface rocks, organic matter, water and gases. *See also* Pe pp.166–167.

Soil conservation and reclamation, preserving and reconstituting arable soils through programmes to prevent soil EROSION and nutrient depletion and to improve crop yields. Semi-arid soils are being reclaimed by laying moisture barriers below the soil surface to improve water retention. *See also* PE pp.158–177.

Soil profile, vertical view of layers of soil from the surface down to the unaltered parent material. It is used in classifying soils with the ideal being one in which soluble substances from the top layer have reached into the second layer and the third layer is the parent material of the surface soil. *See also* PE p.166.

Soil testing, process of analyzing the chemical and mineral composition and porosity of soil – eg to determine its ability to support plant life.

Soissons, town in N France, on the River Aisne. It was the capital of MEROVINGIAN France, after CLOVIS seized it in 486. It was the scene of fighting in many wars throughout history, and was taken by the Germans in both 1918 and 1940. It has a 13th-century cathedral. *See also* HC1 p.200.

Soke, estate in the DANELAW of England before the Norman Conquest in which groups of peasant proprietors, or sokemen, were given land in return for their service to the lord (socage). They could, however, alienate it as they wished. Sokes were therefore freer than other areas in the early Middle Ages. The modern Soke of Peterborough represents the survival of the ancient system.

Sokolova, Lydia (1896–1974), British ballet dancer, *b.* Hilda Munnings. In 1913 she joined the BALLETS RUSSES of DIAGHILEV, the first British dancer to do so.

Solanaceae, NIGHTSHADE family of plants with about 2,000 species. It includes the POTATO, TOMATO, AUBERGINE and sweet PEPPER (CAPSICUM), as well as TOBACCO, DEADLY NIGHTSHADE (belladonna), HENBANE and other poisonous weeds. Less well-known members include the MANDRAKE and the THORN APPLE (also highly poisonous). The flower is trumpet-shaped or bell-shaped and the fruit is a berry, some of which are particularly rich in poisonous ALKALOIDS.

Solar battery, assembly of SOLAR CELLS, most commonly silicon photocells. A battery of about half-a-dozen photocells will have an output of the order of milliamps at a few volts.

Solar cell, device that converts sunlight directly to electric power. It normally consists of a p-type silicon crystal coated with an n-type silicon crystal. Solar radiation causes charge to polarize in the two crystals and creates a potential difference, so that current flows between electrodes connecting them. All wavelengths shorter than one micrometre can create electrical energy; the cells are about 10% efficient.

Several thousand cells may be deployed in panels to provide power of a few hundred watts. *See also* MM pp.72–73; SU pp.69, 108, 270, 271.

Solar constant, steady rate at which heat from the Sun is received from just outside the Earth's atmosphere, measuring approximately 1.36×10^{-3} joules per square metre (perpendicular to the Sun's rays) per second.

Solar energy, RADIATION consisting of ELECTROMAGNETIC WAVES, including heat (infrared rays), light and radio waves. Approx. 35% of the energy reaching the Earth is absorbed; most is spent evaporating moisture into clouds, and some is converted into organic chemical energy by PHOTOSYNTHESIS of plants. All forms of energy (except NUCLEAR ENERGY) come ultimately from the Sun. SOLAR CELLS are used to power instruments on spacecraft, and experiments are being done to store solar energy in liquids from which electricity can be generated. The effective use of solar energy is hampered by the diurnal cycle, seasonal and climatic variations, and at present there are still cheaper forms of energy available. *See also* MM pp.72–73; PE pp.142–143; SU pp.68–69, 108.

Solar engine, machine that converts the energy of sunlight into mechanical work, particularly for the purpose of providing thrust to a spacecraft. Solar engines now under development have one or more large solar collectors which gather the energy of the Sun's rays to heat up a working fluid such as hydrogen. The heated hydrogen can be used directly to provide thrust or it can be fed to a turbogenerator.

Solar flares, brief but extremely powerful outbursts of radiation from the Sun's surface. They are associated with SUNSPOTS and are seen as rapid local brightenings which may last for only a few seconds. Solar flares are also linked with the appearance of AURORAS at extreme latitudes in the Earth's atmosphere, and with radio disturbances caused by disruptions in the IONOSPHERE. *See also* SU pp.222–223.

Solar heating, domestic heating unit which usually employs roof-mounted heat collectors. These have transparent glass or plastic panels behind which water circulates over black-painted metal surfaces which absorb heat energy from the Sun's rays. The warmed liquid passes on to a tank from which hot water is supplied to domestic radiators. *See also* MM p.72.

Solar physics, study of the Sun's physical behaviour. This includes study of the solar spectrum and of the nature and causes of surface phenomena. From these may be inferred the inner thermal, electrical, magnetic, nuclear and gravitational processes.

Solar plexus, network of ganglia from the autonomic nervous system that controls the functions of the internal organs. *See also* MS p.38.

Solar system, celestial group consisting of the Sun and the assemblage of bodies, gas, and dust particles that revolve around it in closed orbits under the influence of its gravitational attraction. The known constituents include the major planets (Mercury, Venus, Earth, Mars, Jupiter, Saturn, Uranus, Neptune and Pluto), the planetary satellites or moons, thousands of minor planets, or asteroids, many comets, meteorites, dust and gas. *See also* SU pp.174–177.

Solar time, system of time reckoning based upon the interval between successive transits of the Sun across the observer's MERIDIAN (the solar day). Because of variations in the Earth's orbital velocity and changes in the Sun's apparent position as viewed from the orbiting Earth, solar days vary in length throughout the year. A mean solar day has therefore been adopted, giving a mean solar time. As measured on the 0° meridian at the Greenwich Observatory, mean solar time is GMT or Universal time. *See also* MW pp.188–189.

Solar wind, particles accelerated by high temperatures of the solar CORONA to velocities great enough to allow them to escape from the Sun's gravity. The solar

wind deflects the tail of the Earth's MAGNETOSPHERE and the tails of comets away from the Sun. *See also* SU pp.222–223.

Solar year, also called tropical year, interval of time between two successive passages of the Sun through the vernal EQUINOX. It is the calendar year, comprised of 365.2422 mean solar days.

Soldati, Mario (1906–), Italian author and film director. His best-known novel is *Le lettere da Capri* (1954). *OK Nerone* (1951) and *La Provinciale* (1952) are his best-known films.

Solder, low-melting point alloy used to join together metals with a higher melting point. Soft solders consist of alloys of tin and lead in varying proportions; brazing solders are alloys of copper and zinc. *See also* MM pp.40–41.

Soldering, means of joining metals by the use of SOLDERS and the application of heat. Soldering is used to join metal at temperatures less than 430°C (806°F). Firstly, the metal pieces are cleaned, then a soldering flux is applied. This flux may be either ROSIN (extracted from pines) or zinc or ammonium chloride.

Sole, marine flatfish found in the Atlantic Ocean from NW Africa to Norway, especially *Solea solea*. A food fish, which is farmed in some countries, it is green-grey or black-brown with dark spots. Length: to 60cm (24in). Family Soleidae. *See also* PE pp.242, 243, 247, 247.

Solemn League and Covenant (Sept. 1643), agreement between the LONG PARLIAMENT and the Scots that, in return for Parliament's promise to reorganize the established Church on a PRESBYTERIAN basis, the Scots would raise an army in the North of England against CHARLES I, to be paid £30,000 a month. It led directly to the Parliamentary victory at MARSTON MOOR, where Scottish forces gave the Parliamentarians a great advantage.

Solenodon, either of two species of endangered nocturnal INSECTIVORES that resemble large SHREWS; one occurs in Cuba, the other in Hispaniola. They have long, scaly tails and at least one species has poisonous saliva. Length: 30cm (1ft). Genus *Solenodon*. *See also* NW p.243.

Solenoid, ELECTROMAGNET in which a soft iron core moves so as to open or close an electrical circuit, thus acting as a switch or RELAY. *See also* SU p.118.

Solent, The, part of the English Channel between the ISLE OF WIGHT and the Hampshire mainland, S England, giving access to Southampton Water.

Sol-fa, tonic. *See* TONIC SOL-FA.

Solferino, Battle of (1859), indecisive battle between an Austrian army and Franco-Piedmontese forces. Both sides suffered heavy losses (more than 14,000 casualties each) and agreed to an armistice. Horror at the plight of the wounded during this battle led Henri DUNANT to set up the RED CROSS.

Solicitor, legal officer, originally in Britain an officer of the Court of Chancery. By the Judicature Act of 1843 all attorneys, solicitors and proctors were named solicitors of the Supreme Court. To become a solicitor a candidate must be articled to a solicitor for five years and pass examinations set by the Law Society. He is qualified to advise counsel in superior courts and to act as an advocate in inferior courts.

Solicitor-General, second-ranking law officer under the Crown in Britain, appointed on the recommendation of the Prime Minister. He is always a member of the House of Commons and is the deputy of the Attorney-General. It is the custom to grant the holder of the office a knighthood, although in several Labour governments the honour has been refused.

Solid angle, in mathematics, ratio of the surface area of the portion of a sphere enclosed by the conical surface forming the angle, to the square of the radius of the sphere.

Solid circuit, electrical circuit that is present inside a semiconductor or other solid-state device. *See also* SEMICONDUCTIVITY; SU pp.128–129.

Solid geometry, study of figures in three dimensions. It includes the study of planes intersecting with planes, planes intersecting with solids, solids interacting with solids and the mensuration of solid figures. *See also* SU pp.48–49.

Solid-state chemistry, study of the chemical properties of solid chemical compounds. It includes investigations into the structure of crystalline compounds and into the mechanism involved in decomposition, OXIDATION and other reactions of solids. SU pp.88, 89.

Solid-state physics, physics of SOLID materials. From the study of the structure, binding forces, electrical, magnetic and thermal properties of solids has come the development of the SEMICONDUCTOR, MASER, LASER and SOLAR CELL. Solid-state physics is a relatively recent branch of physics, involving far more complicated quantum mechanical calculations than the preceding studies of gases and liquids. *See also* SU pp.88, 89.

Solihull, county district in SE West Midlands, England, created in 1974 under the Local Government Act (1972). Area: 180sq km (70sq miles), Pop. (1974 est.) 199,800.

Solimena, Francesco (1657–1747), Italian Baroque painter. Called "La'Abate Ciccio", he was the most famous and successful Italian painter of his day. Much of his work which was influenced by Gisidaro, was done in churches in Naples and featured dramatic figures and lighting, painted with great energy and style.

Solo man. *See* JAVA MAN.

Solomon (*fl.* 10th century BC), third King of Israel, last-born son of DAVID and traditionally the author of the SONG OF SOLOMON. The first Temple of Jerusalem was built during his reign, but the financial burden of the construction caused the kingdom to divide after his death. *See also* HC1 pp.58, *58–59.*

Solomon (1902–), British pianist, b. Solomon Cutner who used only his first name professionally. He made his debut at the age of eight and after 1916 studied in London and Paris. He toured Europe, the USA and Australia and became internationally famous.

Solomon Islands, volcanic island group in the W Pacific Ocean. Extending for 1,400km (900 miles), the group is made up of the BRITISH SOLOMON ISLANDS, which include Guadalcanal, Malaita, Santa Isabel, New Georgia, Choiseul and the Shortland Islands, and, to the N, Bougainville and Buka which are part of PAPUA NEW GUINEA. Discovered in the 16th century, the islands were not succesuly colonized until the 18th century, when German and British traders and missionaries arrived. The chief exports are copra and timber. Area: approx. 40,150sq km (15,500sq miles). Pop. 250,000 (1972). *See also* MW pp.41, 141.

Solomon's seal, or David's harp, perennial plant native to cool temperate regions of Europe and Asia. It has broad, waxy leaves, white or greenish flowers and red berries. Height: to 0.9m (3ft). Family Liliaceae; species *Polygonatum multiflorum.*

Solon, (*c.* 630–560 BC), Athenian lawgiver, statesman and poet. In *c.* 594 BC he was made archon (annual chief ruler) and in *c.* 574 was empowered to institute reforms. His innovations included the amendment of laws relating to debt, the freeing of slaves, the widening of the basis for political status, and the introduction of a new code of laws. He thereby laid the foundation of the Athenian democracy. *See also* HC1 p.69; MS p.*282.*

Sols, colloidal suspensions of particles in liquid such that the suspensions remain liquid and do not solidify as in a GEL.

Solstice, either of the two days of the year when the Sun is at its greatest angular distance from the celestial equator, leading to the longest day and shortest night (summer solstice) in one hemisphere of the Earth, and the shortest day and longest night (winter solstice) in the other hemisphere. In the northern hemisphere the summer solstice occurs on about 21 June and the winter solstice on about 22 Dec.

Solti, Sir Georg (1912–), Hungarian conductor who studied with KODÁLY, BARTÓK and Ernö DOHNÁNYI. Having directed operas in Munich, Frankfurt and London he became music director of the Chicago Symphony Orchestra in 1968 and of the Orchestre de Paris (1972–75). He has distinguished himself as one of the world's most versatile conductors.

Solubility, mass of a SOLUTE that will saturate 100 grams of SOLVENT under given conditions. Solubility generally rises with temperature, but for a few solutes, such as calcium sulphate, increasing temperature decreases solubility in water.

Solute, gaseous, liquid or solid substance that is dissolved and dispersed homogeneously in a medium (a SOLVENT) to form a SOLUTION. Ionic solids, such as common salt, sugars and starch dissolve in water. Liquids can dissolve in liquids (eg ethanol and water are miscible in all quantities at room temperature) and gases, such as hydrogen chloride (HCl), are soluble in water.

Solution, in chemistry, liquid consisting of two or more chemically distinct compounds, inseparable by filters. A solution consists of a liquid (the SOLVENT) into which the other substance (the SOLUTE) dissolves. Water is the most universal solvent, ie a vast range of substances will dissolve in it to various extents. The amount of a solute dissolved in a solvent is called the concentration of a solution. The ability of one substance to dissolve another depends on the type of chemical BONDING, the temperature, and to a small extent the pressure. Heat can be evolved (an exothermic solution) or absorbed (endothermic) when a solution is formed. *See also* MOLARITY; MIXTURE; SATURATED SOLUTION.

Solution, electrolytic. *See* ELECTROLYTIC SOLUTION.

Solution, molal. *See* MOLARITY.

Solutrean culture, European culture of late Palaeolithic times. Solutrean man, like the Magdalenian cave painters who succeeded him, was a race of *Homo sapiens* or modern man. His artefacts included delicately-flaked tools and weapons and bone needles with eyes.

Solvay, Ernest (1838–1922), Belgian industrial chemist. He discovered a process for preparing sodium bicarbonate (and hence sodium carbonate) from salt and calcium carbonate. He patented the process in 1861 and by 1913 was supplying a large part of the world's sodium carbonate. *See also* SOLVAY PROCESS.

Solvay process, widely used industrial method of making SODA, or sodium carbonate, Na_2CO_3, invented by Ernest SOLVAY in 1863. Brine is saturated with ammonia, and carbon dioxide gas is bubbled into this solution in a Solvay tower. Sodium bicarbonate is formed, which precipitates from the brine, and is then heated to make soda.

Solvent, liquid that dissolves substances without changing their composition. Water is the most universal solvent, and many inorganic compounds dissolve in it. Ethanol, ether, acetone and carbon tetrachloride are common solvents for organic substances. *See also* SOLUTE; SOLUTION; SU pp.86–87.

Solway Firth, inlet of the Irish Sea separating Cumbria, England, from the Dumfries and Galloway region, Scotland. Length: 61km (38 miles).

Solway Moss, Battle of (1542), English victory over the Scots at Solway marshes on the Anglo-Scottish border. The battle stemmed from HENRY VIII's interest in forcing JAMES V to abandon his pro-French, pro-Catholic policy. By the ensuing treaty of Greenwich (1543) the infant Scots queen Mary Stewart (MARY, QUEEN OF SCOTS) was to marry Prince Edward (later EDWARD VI).

Solzhenitsyn, Alexander (1918–), Soviet writer. Sentenced to a forced labour camp in 1945 for criticizing STALIN, he was subsequently exiled to Ryazan but was officially rehabilitated in 1956. A major spokesman for Soviet dissident intellectuals, his novels include *One Day in the Life of Ivan Denisovich* (1962), *The First Circle* (1968) and *Cancer Ward* (1968). Criticism of the Soviet regime in *The Gulag Archipelago* (1974) led to his forcible exile to the West. He was awarded the 1970 Nobel Prize in literature. *See also* HC2 pp.290–291.

Soma, in Vedic mythology, personification of the soma plant or juice and, in the post-Vedic period, the personification of the moon, the waning of which was sometimes attributed to the gods' drinking of soma. Another story is that the god Soma married the 27 daughters of Daksha, one of the Lords of Creation, but allegedly neglected them all except one. Cursed to die by his father-in-law, Soma grew weaker but Daksha relented, making Soma's punishment periodic rather than eternal and thus causing the moon to wax and wane. *See also* VEDAS; SOMA SACRIFICE.

Somalia, republic in NE Africa, on the Indian Ocean. It is a poor, arid country and 80 per cent of the population are nomadic pasturalists. Animals and animal products account for 66 per cent of exports, and the chief crop is bananas. There are a few processing industries. The capital is Mogadisho. Area: 637,657sq km (246,199sq miles). Pop. (1976 est.) 3,258,000. *See also* MW p.154.

Somaliland, British, region on the African coast of the Gulf of Aden, a British protectorate from 1884 to 1960. In 1960 it was joined to Somalia (Italian Somaliland until 1950) to form the independent United Republic of SOMALIA.

Somare, Michael Thomas (1936–), Papua New Guinean politician. A school teacher and then a journalist, he was elected to the House of Assembly in 1968 and became leader of the Pangu party in the same year. He became Prime Minister in September 1975, when his country was granted independence.

Soma sacrifice, one of the most important Vedic rituals. As described in the Rig Veda the juice of the SOMA plant is passed through sheep's wool and mixed with milk and water. It is offered to the gods and the remainder is consumed by the sacrificer and the priests. The plant is supposed to have originated in heaven.

Somatotropin (STH), growth HORMONE made by cells of the anterior, or frontal, PITUITARY GLAND, lying under the brain. It stimulates growth in young animals and human beings, by increasing the amounts of PROTEIN made by their body cells and stimulating fat and carbohydrate metabolism.

Somerset, Edward Seymour, 1st Duke of (*c.* 1500–52), Protector of England, brother of Jane SEYMOUR. Having won the favour of HENRY VIII, he successfully commanded English forces in Scotland in 1544 and at Boulogne in 1545. Appointed protector of EDWARD VI, a minor, on Henry's death in 1547, he enjoyed monarchical powers. His religious reforms, leading to the Act of UNIFORMITY in 1549, consolidated PROTESTANTISM in England. Deposed as protector in 1549 by NORTHUMBERLAND, he was subsequently beheaded. *See also* HC1 pp.270, 280.

Somerset, county in SW England on the Bristol Channel. The land is generally low-lying in the centre, rising to the Mendip Hills in the NE and the Quantock Hills and Exmoor in the W. The region is drained chiefly by the rivers Avon, Exe and Parrett. Dairy farming and fruit-growing are the most important economic activities. Somerset is noted for Cheddar cheese and cider; leather and woollen goods are also manufactured. The county town is Taunton. Area: 3,450sq km (1,332sq miles). Pop. 404,400.

Somerset House, English palace in the Strand, London, first built between 1547–51 for Edward Seymour, Duke of Somerset (then Lord Protector to the boy king, Edward VI). The architect was Sir John Thynne. Somerset House was rebuilt between 1776 and 1786 in the Palladian style and is the work of Sir William CHAMBERS. It houses various government departments. *See also* HC1 p.280; HC2 p.70.

Somerville, Edith Anna Oenone (1858–1949), Irish writer, b. Corfu. She is best known for her literary partnership with her cousin Violet Martin (pseudonym, Martin ROSS). As Somerville and Ross they produced several works, including *Some Experiences of an Irish R.M.* (1899) *Further Experiences of an Irish R.M.* (1908) and *In Mr Knox's Country* (1915).

Somerville (Somervile), William (1675–

Solomon reigned for nearly 40 years and was renowned for his wisdom.

Solway Firth; Sandyhills Bay on the inlet in Dumfries and Galloway, Scotland.

Alexander Solzhenitsyn in Montpelier, Vermont, after expulsion from the USSR.

Somerset House; the main courtyard and facade of this palace by the Thames.

Somes, Michael

Somme; British Mark V howitzers in action in the Fricourt-Mametz valley.

Songs of Innocence was written and illustrated by William Blake.

Sophocles was more than 90 years old when he wrote *Oedipus at Colonus*.

Soul music's best-known performers include the American James Brown.

1742), British poet and writer. After studying law he became a country gentleman, taking up the sports that were later described in his poems. His best-known work is *The Chace* (1735), which documents the history of hunting to 1066.

Somes, Michael (1917–), British dancer and ballet director. He became a principal dancer with the SADLER'S WELLS company in 1938. He was assistant director of the Royal Ballet (1963–70) and since then a principal teacher and répétiteur. He has choreographed the ballet *Summer Interlude* (1950) to music by RESPIGHI.

Somme, Battle of the (1916), WWI battle, the first major predominantly British offensive on the western front. It was a war of attrition, fought in the trenches, begun on 1 July and ending in the mud on 18 Nov. There were 95 German divisions opposed by 55 British and 20 French divisions. The casualties were: British 420,000, French 190,000 and German probably 465,000 (although some authorities have put the German total higher). *See also* HC2 p.*190.*

Sömmering, Samuel Thomas von (1755–1828), German scientist. In 1809, using the voltaic pile invented by Alessandro VOLTA, he developed an early electric telegraph system that sent signals indicated at the receiving end by the appearance of hydrogen bubbles liberated at a gold electrode in a solution of acid.

Somnabulism. *See* SLEEPWALKING.

SONAR, underwater detection and navigation system. The letters stand for Sound Navigation and Ranging. The system emits high-frequency sound that is reflected by underwater objects. *See also* MM pp.236–237.

Sonata, in music, instrumental chamber form of musical composition in several movements. The earliest surviving example is probably the two sonatas included in *Sacrae symphoniae* (1597) by Giovanni GABRIELI. The BAROQUE era sonata contained distinct parts: one or more melodic parts, a bass part and a continuo. In the CLASSICAL period, the sonata became a more clearly defined form for one or two instruments. The movements, usually three or four in number, are based on key relationships. The first movement is arranged with the statement of two subjects in related keys in the Exposition, this is developed in the Development or middle section with varied key changes, and the restatement of the original subjects both in the tonic (principal) key in the Recapitulation, and a coda or conclusion. The second movement of a sonata is generally slow in tempo and the third and perhaps fourth movements faster (allegro or presto).

Sonderbund, defensive pact formed in 1845 by seven Roman Catholic cantons in Switzerland – Lucerne, Uri, Schwyz, Underwalden, Zug, Fribourg and Valais – against a revision of the federal pact of 1815 and a reduction of the sovereignty of the cantons. Sonderbund was declared unconstitutional, and in the civil war which followed the Catholic cantons were defeated.

Son et lumière, form of open-air entertainment which originated in France in the early 1950s. Performed at night, it involves the use of coloured lights, which are played on the façades of buildings to an accompanying soundtrack of music and a pre-recorded commentary which are broadcast by loudspeakers. Effects such as fireworks or smoke bombs may also be included.

Songhay, W African empire, founded *c.* AD 700 which flourished in the 15th and 16th centuries.

Songnim, city in SW North Korea, on the Taedong River. During the Japanese occupation (1910–45), an iron works was established there (1916), since when Songnim has become the largest iron and steel centre in North Korea. The city was destroyed during the KOREAN WAR, but was later rebuilt. Pop. 53,000.

Song of Hiawatha, The (1855), narrative poem by William Wadsworth LONGFELLOW, based on the life of an Ojibway chieftain. Written in a distinctive trochaic metre, it tells of Hiawatha's childhood and

his contest with his father, the West Wind, over his mother Wenonah.

Song of Myself (1855), poem by Walt WHITMAN, considered his most characteristic and possibly his best. It first appeared in his collection *Leaves of Grass* and consists of a sequence of 52 loosely connected sections.

Song of Roland, The. *See* CHANSON DE ROLAND.

Song of Solomon, The (Song of Songs), book of the OLD TESTAMENT. A series of love poems spoken alternately by a man and a woman, it is attributed in the Bible to SOLOMON but was probably edited to its present form in the 3rd century BC.

Songs of Experience (1794), volume of lyric poetry by William BLAKE, including *Earth's Answer, The Tyger, The Garden of Love* and *To Tirzah.* He illustrated the poems himself, which are complementary to those in *Songs of Innocence,* expressing the fallen state of man.

Songs of Innocence (1789), volume of lyric poetry written and illustrated by William BLAKE, including *The Lamb, The Echoing Green* and *Infant Joy.* The poems express a mystical joy in the innocence of youth.

Sonic boom, thunder-like sound produced when the high-pressure SHOCK WAVES formed at the nose and leading edges of wings and tail of aircraft going through the SOUND BARRIER spread out behind the aircraft and strike the ground. *See also* SUPERSONIC FLIGHT; SU pp.78–79.

Sonnet, poem of 14 lines in iambic PENTAMETER, usually employing Petrarchan or Shakespearean rhyme schemes. The Petrarchan consists of an octet and a sextet, usually with an *abbaabbacdecde* rhyme scheme. The Shakespearean, having a final rhyming couplet, is *ababcdcdefefgg.* The form originated in Provence, was widely used in Italy and was introduced into England by Sir Thomas WYATT.

Sons and Lovers (1913), partly autobiographical novel by D. H. LAWRENCE. Attacked on its first publication because of its openness in dealing with sexual matters, it has been praised for its portrayal of life in a mining community.

Soo Canals. *See* SAULT SAINTE MARIE CANALS.

Sophia, (1630–1714), Electress of Hanover, granddaughter of James I of England. In 1658 she married Ernest Augustus, Duke of Brunswick-Lüneberg who became Elector of Hanover. In 1701, by the English Act of Settlement, she was recognized as Protestant heir to the throne. Her son became George I of England. *See also* HC1 p.313.

Sophists, itinerant Greek teachers of the 5th and 4th centuries BC who taught in return for fees. PROTAGORAS was perhaps the first Sophist, and others were Gorgias, Thrasymachus and HIPPIAS. They tended to emphasize rhetoric rather than pure knowledge.

Sophocles (*c.*496–406 BC), Greek playwright, *b.* Colonus. Of his 100 plays only seven tragedies and part of a SATYR PLAY remain. These include *Ajax, Antigone* (441), *Electra, Oedipus Rex* (*c.*428) and *Oedipus at Colonus* (produced 402). His works introduced a third speaking actor, added stage scenery and increased the chorus from 12 to 15 members. *See also* HC1 p.74, *74.*

Soprano, highest range of the human voice, above the mezzo-soprano. The normal range may be given as an octave below and above the B natural above middle C, although exceptional voices may reach notes considerably higher.

Sopwith Camel, British fighter biplane of WWI. It was made of wooden box-girders covered with fabric and was powered, at a top speed of 182km/h (113mph), by a 9-cylinder rotary engine. It was armed with twin synchronized 0.303in Vickers machine guns which fired through the arc of the propeller. *See also* MM p.*176.*

Sorbonne, college of the University of Paris, founded in 1253 by Robert de Sorbon (1201–1274) and located in what is now the Latin Quarter of Paris. Established for the education of students of theology, it was for centuries an intellectual centre of Roman Catholic religious

thought, but towards the end of the 19th century became purely secular in its course. Sorbonne is now often used as a synonym for the whole of Paris University.

Sorcery, manipulation of the natural world by means of magic. Anthropologists sometimes distinguish sorcery from WITCHCRAFT – witches claim inherent spiritual powers whereas sorcerers do not.

Sordello (*c.*1200–*c.*1269), Italian poet who wrote numerous poems in Medieval PROVENÇAL. Although much of his verse is about love, his most famous work is a lament on the death of his patron Blacatz.

Sorghum, tropical cereal grass native to Africa and cultivated throughout the world. Types raised for grain are varieties of *Sorghum vulgare* that have leaves coated with white waxy blooms and panicles that bear up to 3,000 seeds. It yields meal, oil, starch and dextrose (a sugar). Height: 0.5–2.5m (2–8ft). Family Gramineae. *See also* PE pp.*156, 168,* 181.

Sorrel, or dock, herbaceous perennial plant native to temperate regions. It has large leaves which can be cooked as a vegetable and small green or brown flowers. Height: to 2m (6ft). Family Polygonaceae; genus *Rumex acetosa.*

Sorting, in geology, process by which sedimentary particles that are transported by wind or water are segregated according to their physical properties such as weight, shape, durability and density. Flat, thin material is carried farther than round objects of similar weight, weaker materials are more easily fragmented and destroyed, but the most durable, usually quartz objects, survive. *See also* BEDDING PLANE.

SOS, general distress at sea. It is a simple pattern in MORSE CODE consisting of three dots, three dashes and three more dots, which is internationally understood as a call for help.

Sosnowiec. *See* KATOWICE.

Sotatsu, Nonomura (*d.*1643), Japanese painter. An artist of the Edo period, he was one of the founders of the Korin style of Japanese painting, which is based on the older YAMATO-E style.

Soterus, Saint (*d. c.*175), pope from *c.*166, during the persecution of Christians by MARCUS AURELIUS. He is said to have died a martyr.

Soto, Hernando de (*c.*1498–1542), Spanish explorer who led explorations in 1539 to what later became the SE USA. He followed the Mississippi River, hoping to find wealth, but died before the journey was completed.

Soto, Jesús Rafael (1923–), Venezuelan painter and sculptor. In 1950 he went to live in Paris. Using optical illusion and geometric relief he produced a type of KINETIC ART; an example is *Horizontal Movement* (1963).

Soufflot, Jacques-Germain (1713–80), the outstanding French NEO-CLASSICAL architect. He studied architecture in Rome, settling on his return in Lyons, where he made his reputation with his design for the vast Hôtel Dieu (1741). In 1757 he began work on the church of Ste Geneviève (renamed the Panthéon after the Revolution). Regarded as the masterpiece of French Neo-Classicism, its Italianate features and Roman monumentality are reminiscent of St Peter's, Rome.

Soul. *See* SEOUL.

Soulages, Pierre (1919–), French painter well known for his ABSTRACT work, which he began after WWII. His style, similar to that of Franz KLINE is characterized by thick black strokes on an almost neutral background. Among his paintings is *Black, Brown and Grey* (1957).

Soul Music, form of popular music. The term designates black music which developed in the 1960s from rhythm and blues.

Soult, Nicolas-Jean de Dieu (1769–1851), Duke of Dalmatia, created a marshal by Napoleon in 1804. He fought at AUSTERLITZ and JENA, and commanded French troops in Spain from 1809 to 1814. He served three times as Minister of War under Louis-Philippe (*r.*1830–48).

Sound, physiological sensation perceived by the brain via the EAR, caused by an oscillating source, and transmitted through a material medium as a sound wave. The human ear can respond to sounds with frequencies between about 20 and 20,000 hertz (one Hz is one oscillation per second).

The velocity (c) at which a sound wave travels through a medium depends on the ELASTICITY of the medium (K) and its density (ρ): $c=\sqrt{K/\rho}$. If the medium is a gas, the sound wave is longitudinal and the velocity of propagation depends on the gas temperature $c_\theta=c_0\sqrt{1+\alpha\theta}$, where c_0 and c_θ are the velocities at 0°C and θ°C and α is the coefficient of expansion of the gas. The velocity of sound in dry air at Standard Temperature and Pressure (STP) is 331.4m per sec (741mph) and depends on the height above sea level. Every pure sound is characterized by its intensity, PITCH and TIMBRE. The intensity is the rate of flow of sound energy through unit area perpendicular to the direction of flow. *See also* NOISE, SU pp.80–81.

Sound barrier, large and sudden increase in drag experienced by aircraft at near Mach 1 (the speed of sound). This drag is due to the compressibility of air, which is negligible at lower speeds. To overcome this barrier it was necessary to design aircraft with more powerful propulsion systems, less surface area and swept-back wings. At supersonic speeds an aircraft outruns its own sound, but as it passes through the sound barrier it produces a series of shock waves – the SONIC BOOMS. *See also* SU p.79.

Sound distortion, various aural effects. Sounds may be distorted if the organs of hearing are congenitally defective or affected by disease. For example, partially deaf people may be able to hear only the lower sound frequencies. Sound distortion occurs in a concert-hall when its acoustics affect sound waves so that these reach the listener out-of-phase or deadened. Sounds can be distorted intentionally by using electronic devices, as in electronic music.

Sounding, determining the depth of water. The simplest means of sounding is by dropping a measured weighted line till it reaches the bottom. Acoustic SONAR, a sonic depth finder, is often used. *See also* MM p.236.

Sounding machine. *See* ECHO SOUNDER.

Sounding rocket, ROCKET designed for upper-atmosphere studies. It is usually single-staged and follows a vertical trajectory. Data collected may be parachuted or radioed back to a ground tracking station.

Sound of Music, The (1965), US film of the stage musical directed by Robert Wise. The original score was by Richard RODGERS and Oscar HAMMERSTEIN II. The film won five OSCARS.

Sound recording and reproduction, mechanical conversion of sound waves into a form which can be stored and reproduced. Thomas EDISON's phonograph (1877) recorded sound vibrations as indentations made by a stylus on a revolving cylinder wrapped in tinfoil. German-born Emile BERLINER's GRAMOPHONE improved the process by using a zinc disc instead of a cylinder. The volume was amplified by the addition of acoustical horns, which were replaced before WWI by valve amplifiers. Moulded thermoplastic records were introduced in 1901, and subsequently improved plastics allowed finer grooving with reduced surface noise. In 1927 and 1928 patents were issued in the USA and Germany for MAGNETIC TAPE recording processes. Later innovations include HIGH FIDELITY, STEREOPHONIC and quadrophonic reproduction.

Sound track, usually a band of varying width through which a light shines as the film is projected. A photoelectric cell detects the transmitted light signal, converting it into an electric current which is amplified and converted into sound. *See also* MM p.223.

Souphanouvong, Prince (1902–), Laotian nationalist politician. He held several government posts between 1958 and 1962, he became president of the Lao People's Democratic Republic in 1975.

Sousa, John Philip (1854–1932), US composer and bandmaster. He composed about 100 marches, including *Semper*

Fidelis (1888) and *The Stars and Stripes Forever* (1896). He also composed numerous operettas, of which the most famous is *El Capitan* (1896).

Sousaphone, largest brass musical instrument of the TUBA family. It was introduced by John SOUSA to fortify the bass section of US military bands. The tube coils around the player, with the removable flaring bell directed forwards, not upwards, as with orchestral tubas.

Sousse (Susa), coastal town in NE Tunisia, N Africa, on the S shore of the Gulf of Hammamet. Settled by the Phoenicians, it became a Roman port, was destroyed in AD 434 by the Vandals, and was rebuilt by JUSTINIAN. Today it is a trade centre of an agricultural region. Manufactures include textiles and food products. Pop. (1966) 58,161.

South, one of the four cardinal points of the compass, opposite north. Lines of magnetic flux emanate from the Earth's north pole and enter its south pole.

South Africa, republic formed in 1910 from two British territories, Cape Colony and Natal, and two Afrikaaner states, Orange Free State and Transvaal. Since WWI manufacturing has greatly increased in South Africa. The chief industries include engineering, iron and steel, metal working and motor vehicle assembly. There are valuable deposits of gold and South Africa leads the world in the production of diamonds. Maize is the chief crop and dairy cattle and sheep are reared. South Africa has two capitals: Cape Town is the seat of the legislature and Pretoria is the seat of the government. Area: 1,221,042sq km (471,444sq miles). Pop. (1975 est.) 25,471,000. *See* MW p.154.

South African (Boer) Wars, two wars between the BOER and British settlers in southern Africa. The first (1880–81) arose from the British annexation of the Transvaal in 1877. Under KRUGER's leadership the Boers defeated the British at LAING's NEK and MAJUBA Hill and won self-government by the Pretoria Convention (1881). That convention allowed Britain to retain suzerainty over all of southern Africa. The second war (1899–1902) was the outcome of the Boers' desire for entire independence and British desire to take control of the Transvaal gold mines. It began with the Boers' besieging MAFEKING in Oct. 1899 and ended with the signing of the treaty of VEREENIGING in May 1902. *See also* HC pp.128, 129.

South African mackerel. *See* BARRACOUTA.

South America. *See* AMERICA.

Southampton, port and county district in Hampshire, S England. The city, at the head of Southampton Water, is Britain's principal passenger port and a major commercial port. It is the seat of Southampton University (1952). Industries: shipbuilding, engineering, oil refining, food processing. The county district was created in 1974 under the Local Government Act (1972). Area: 49sq km (19sq miles). Pop. (1974 est.) 213,000.

South Arabia, Federation of (1963–67), federation formed by the merger of Aden with the Federation of the Emirates of the South. The latter was formed in 1959. The federation collapsed in 1967, due to the emergence of the National Liberation Front. Aden and South Arabia became the independent state of Southern Yemen. The name People's Democratic Republic of Yemen was adopted in 1970 to distinguish it from the neighbouring North Yemen or Yemen Arab Republic.

South Australia, state in S central Australia, on the Indian Ocean. The area is hilly in the E (Flinders Ranges) and the N (Musgrave Ranges), and in the W are the Great Victoria Desert and the Nullarbor Plain. The Murray River in the SE is the only important river and farming is confined to this area. Barley, oats, wheat, grapes and rye are the chief crops. Livestock are grazed in the N. Important mineral deposits in the state include iron ore, salt, gypsum, coal and natural gas. The principal industries are heavy metals and transport equipment. Whyalla has the largest shipyards in Australia. The main cities are Adelaide (the state capital),

Salisbury and Elizabeth. The S coast of Australia was visited by the Dutch in 1627. The first English colonists arrived in 1836 and the region was federated as a state of the commonwealth in 1901. From 1863 until 1911 South Australia included Northern Territory. Area: 984,380sq km (379,760sq miles). Pop. 1,164,700. *See also* MW p.34.

South Bedfordshire, county district in SW and central BEDFORD(SHIRE), England; created in 1974 under the Local Government Act (1972). Area: 212sq km (82sq miles). Pop. (1974 est.) 94,800.

South Cambridgeshire, county district in S and central CAMBRIDGE(SHIRE), England, created in 1974 under the Local Government Act (1972). Area: 904sq km (349sq miles). Pop. (1974 est.) 97,100.

South Carolina, state in SE USA, on the Atlantic Ocean. The land rises from the coastal plain to the rolling hills of the Piedmont plateau to the Blue Ridge Mts in the NW. The region is drained chiefly by the Pee Dee, Santee and Savannah and Edisto rivers. Tobacco and soya beans have surpassed cotton in importance. Maize, cattle, sweet potatoes and groundnuts are also important. The state's great potential for hydroelectric power has helped to steer South Carolina's traditionally agrarian economy towards manufacturing. The chief industries are chemicals, machinery, textiles, food processing, clothing, timber and fishing. The principal cities are Columbia (the state capital and largest city), Charleston, Greenville and Spartanburg. The Spanish and the French visited the coast in the 16th century but the English were the first permanently to settle the area, from 1663 when Charles II granted a charter of settlement. South Carolina became a royal province in 1729. A plantation society evolved based on rice, indigo and cotton. The state was the first to secede from the union and the first shots of the AMERICAN CIVIL WAR were fired at Fort Sumter in Charleston harbour. The state was devastated in 1865 by Union troops. Area: 80,432sq km (31,055sq miles). Pop. (1975 est.) 2,818,000. *See also* MW p.175.

South China Sea, W part of the Pacific Ocean, surrounded by SE China, Indochina, the Malay Peninsula, Borneo, the Philippines and Taiwan; it is connected to the East China Sea by the Formosa Strait. Its chief arms are the Gulf of Tonkin and Gulf of Thailand, and the Si, Red, Mekong, and Chao Phraya rivers. Area: 2,300,000sq km (848,000sq miles). Depth: (average) 1,140m (3,740ft) but more than 4,920m (15,000ft) in some places.

Southcott, Joanna (1750–1814), British religious fanatic. In about 1792 she proclaimed she was divinely instructed to forewarn the world of the SECOND COMING.

South Dakota, state in N central USA, in the GREAT PLAINS. The land rises gradually from the E to the Black Hills in the W. The Missouri River flows through the centre of the state. One-fifth of the area W of the river is semi-arid plain, inhabited by Indians. The rest is divided into large cattle and sheep ranches. East of the Missouri, livestock rearing is important and wheat, maize, oats, soya beans and flax are also grown. Meat-packing and food processing are by far the most important industries. South Dakota is one of the largest producers of gold in the USA; tin, beryllium, stone, sand and gravel are also important minerals. Pierre is the state capital and Sioux Falls, Aberdeen and Rapid City are the largest cities.

French trappers claimed the region for France in the 1740s. The USA acquired the land in the LOUISIANA PURCHASE of 1803. Lewis and Clark explored the area in 1804–05 but trading and military posts were the only settlements until the late 1850s. Dakota Territory was formed in 1861. The discovery of gold in the Black Hills in 1874 led to an increase in population and the territory was divided into the states of North and South Dakota, both of which joined the Union in 1889. Area: 199,551sq km (77,047sq miles). Pop. (1975 est.) 683,000. *See also* MW p.175.

South Downs, range of chalk hills in SE

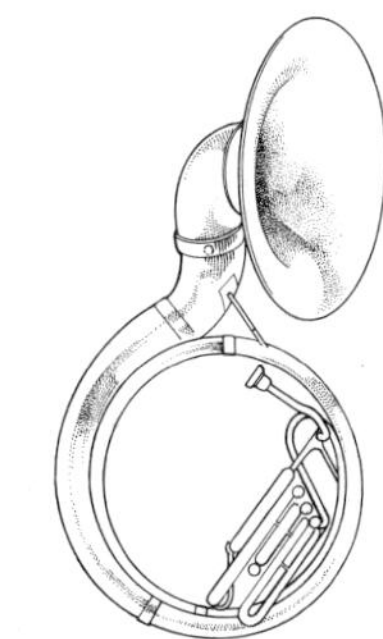

Sousaphones rest on the player's shoulder. They are used in march music.

Southampton, Britain's major passenger and commercial port, on the River Test.

South Carolina: an 18th-century Charleston house where Washington stayed in 1791.

Joanna Southcott's followers believed that she would come to life again.

South-East Asia Treaty Organization

Southend-on-Sea, a popular resort for Londoners, is famous for its pier.

Robert Southey's liberal political opionions changed in later life.

South Pacific; Janet Blair in the 1950 Broadway production of this musical.

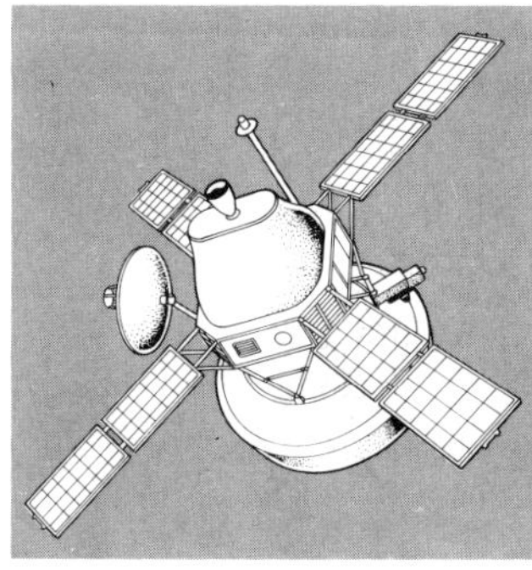

Spacecraft; the solar panels generate electricity to operate the instruments.

England, stretching E–W from E Sussex to W Dorset. They are separated from the North Downs by the WEALD and are drained by the Adur, Arun and Ouse rivers. The highest point is Butser Hill, 271m (889ft).

South-East Asia Treaty Organization. See SEATO.

Southend-on-Sea, county district in SE ESSEX, E England, created in 1974 under the Local Government Act (1972). Area: 42sq km (16sq miles). Pop. (1974 est.) 160,200.

Southern Alps, mountain range on SOUTH ISLAND, New Zealand, extending NE to SW almost the entire length of the island. The most important pass is Arthur's Pass, linking Christchurch (E) with Greymouth (W). The highest peak is Mt Cook, 3,764m (12,349ft).

Southern Cross, or Crux, cross-shaped CONSTELLATION in the Southern Hemisphere, used as a navigational aid because the vertical arm points south. It contains the Coal-sack nebula. See also SU pp.226, 236, 255, 260–261, 260, 266–267.

Southern Rhodesia. See ZIMBABWE.

Southey, Robert (1774–1843), British poet and biographer, related to Samuel Taylor COLERIDGE by marriage. His long epic poems include *Thalaba* (1801), *Madoc* (1805), *The Curse of Kehama* (1810) and *Roderick the Last of the Goths* (1821). He also wrote numerous biographies. He was appointed Poet Laureate in 1813.

South Georgia, island in the S Atlantic Ocean, about 1750km (1100 miles) E of Tierra del Fuego and a dependency of the Falkland Islands. The chief town is Grytviken Harbour.

South Glamorgan, county in S Wales, on the Bristol Channel, formed in 1974 from parts of the former counties of GLAMORGANSHIRE and MONMOUTH(SHIRE). The administrative centre is Cardiff. Area: 416sq km (161sq miles) Pop. (1975 est.) 391,600.

South Island, larger of the two principal islands that comprise New Zealand; separated from North Island by Cook Strait. Its chief cities are CHRISTCHURCH, DUNEDIN and INVERCARGUILL. The SOUTHERN ALPS extend the length of the country and separate the thickly forested W coast from the broad CANTERBURY PLAINS in the E. Cereal growing, sheep and cattle rearing, and dairying are important on the Plains; tourism is a valuable source of income. Area: 150,461sq km (58,093sq miles). Pop. (1971) 811,268. See MW p.131.

South Lakeland, county district in S CUMBRIA, England, created in 1974 under the Local Government Act (1972). Area: 1,549sq km (598sq miles). Pop. (1974 est.) 93,200.

Southland, statistical area of New Zealand, the SW tip of the South Island. It includes Stewart Island. It is chiefly a farming province, important for sheep, cattle and dairy farming.

South Oxfordshire, county district in SE OXFORD(SHIRE), England, created in 1974 under the Local Government Act (1972). Area: 690sq km (266sq miles). Pop. (1974 est.) 138,000.

South Pacific (1949), musical play by Richard RODGERS and Oscar HAMMERSTEIN II, originally produced on Broadway, adapted from stories by James MICHENER. It contains several songs that became famous, including *Some Enchanted Evening* and *Younger than Springtime*. It was filmed in 1958.

South Pole, southernmost geographical point on the earth's surface. The magnetic South Pole is located about 2,400km (1,500 miles) from the geographical South Pole. Roald AMUNDSEN was first to reach the Pole on 14 December 1911.

South Ribble, county district in SW central LANCASHIRE, England, created in 1974 under the Local Government Act (1972). Area: 111sq km (43sq miles). Pop. (1974 est.) 90,500.

South Saskatchewan, river in SW Canada. It rises in the Rocky Mts and forms, with its tributary the River Bow, the longest branch of the River Nelson. Length: 1,940km (1,205 miles).

South Sea Bubble (1720), speculation in the shares of the English South Sea Company, ending in financial collapse. The company, founded by Robert HARLEY in 1711, was formed on the assumption that the terms of the treaty concluding the War of the SPANISH SUCCESSION would give Britain a trading monopoly in the Pacific Ocean and South America. This monopoly did not materialize but the government, to fund the National DEBT, encouraged public confidence in the company, persuading investors to exchange state annuities for its stock. In 1720 the bubble burst; banks failed and many people were ruined. The consequences of the disaster brought Robert WALPOLE to power (1721).

South Seas, name used by early explorers to describe the whole of the Pacific Ocean. It later came to mean the area of the central and S Pacific, but in particular the islands and waters of OCEANIA.

South Tyneside, county district in E central TYNE and WEAR, England, created in 1974 under the Local Government Act (1972). Area: 64sq km (25sq miles). Pop. (1974 est.) 171,700.

South Uist. See UIST.

Southwell, Robert (1561–95), English writer and poet, canonized in 1970. Educated abroad and ordained a JESUIT priest in 1585, he was imprisoned upon returning to England and executed for his faith. His works include *An Epistle of Comfort to the Reverend Priests* (1587) and *Mary Magdalen's Funeral Tears* (1593).

South West Africa. See NAMIBIA.

South Yorkshire, metropolitan county in N central England, formed in 1794 from part of the West Riding of YORKSHIRE. Sheffield is the administrative centre. Area: 1,561sq km (603sq miles). Pop. 1,318,300.

Soutine, Chaïm (1894–1943), Lithuanian-born artist who became associated with the EXPRESSIONIST movement in Paris. His paintings are violent and passionate.

Sovereign, English gold coin first issued in 1489 by Henry VII. It was replaced by the unite in the 17th century but the new sovereign became the standard gold coin in 1817. It ceased to be minted in 1914. Sovereigns are still minted periodically for commemorative purposes.

Sovereignty, ultimate authority, held by a person or institution, against which there is no appeal. In early modern Europe sovereignty came to be ascribed to the absolute monarchs of the new nation states. In Britain (where absolute monarchy never quite established itself) sovereignty, before the country's entry into the EEC, resided in Parliament.

Soviet, Russian word meaning council. Soviets, elected councils of workers, are part of the organizational structure of the Russian government which appeared during the revolutions of 1905 and 1917. After 1917, the Soviets were confirmed as the basic political unit of government. Soviets are organized in a hierarchical system, from the rural council to the Supreme Soviet, the highest legislative body in the USSR. See also HC2 pp.196–197.

Soviet Far East, area in the extreme E of the USSR, coextensive with Primorsky and Khabarovsk krays. As the Far Eastern Republic, it was formed in 1920 with the capital at Chita as a zone between Japan and the USSR territories. The region is almost self-sufficient economically; industries include mining for precious metals, oil refining, and the manufacture of iron and steel. Hunting for pelts and meat and fishing are other important activities. Area: 6,216,000sq km (2,400,000sq miles). Pop. 5,780,000.

Soviet Union. See USSR.

Soweto (South-Western-Township), black township of about 1 million people situated on the outskirts of Johannesburg. A demonstration by students protesting against the compulsory use of Afrikaans in addition to English in Bantu schools in June 1976 gradually culminated in widespread riots against the police, and against unemployment, in which 618 lives were lost. Further sporadic demonstrations and riots occurred in 1977.

Soya bean, annual plant native to China and Japan. It has oval, three-part leaves and small lilac flowers. Grown for food, forage, green manure and oil, its seed is an important source of PROTEIN. Height 60cm (24in). Family Leguminosae; species *Glycine soja*. See also PE pp.184,185, 217.

Soyinka, Wole (1934–), Nigerian writer whose works explore the conflict between the traditional African way of life and corrupting 20th-century influences. His plays include *The Lion and the Jewel* (1966), *Madmen and Specialists* (1970) and *Jero's Metamorphosis* (1972). *The Interpreters* (1965) and *Season of Anomy* (1973) are novels.

Soyuz, series of Soviet manned space missions. Soyuz 1 was launched in April 1967; Soyuz 2 and 3 in 1968 and 4 and 5 in 1969 docked in space. Soyuz 9 established the record for a manned space flight in 1970 – 18 days. The first joint US-Soviet space project took place in 1975, when Apollo and Soyuz spacecraft docked in space. See also SU pp.269, 271.

Spaak, Paul-Henri (1899–1972), Belgian and international statesman. An advocate of European unity, he was Prime Minister of Belgium (1938–39; 1947–50) and Foreign Minister (1936–38; 1939–45; 1945–47; 1954–57; 1961–65). He served as first President of the UN General Assembly (1946) and Secretary-General of NATO (1957–61). See also HC2 p.259.

Space, part of the unlimited four-dimensional continuum in which the physical extensions of matter occur, rather than their temporal extensions, which occur in time. RELATIVITY postulates that space and time are aspects of one thing. More usually space is taken to mean the rest of the Universe which lies beyond the Earth's atmosphere and in which the density of matter is low.

Spacecraft, vehicle capable of flight above the Earth's atmosphere. It is launched from the ground by powerful ROCKETS which fall away once the fuel has been exhausted. Unmanned spacecraft respond to commands from ground stations, but they can function under the control of computers carried aboard. The first unmanned spacecraft, Sputnik 1, was launched by the USSR on 4 Oct. 1957. Manned spacecraft have life-support systems comprising pressure, temperature and oxygen regulators, carbon dioxide and waste management systems, water, food and fuel. The first manned spacecraft, Vostok 1, carried Yuri Gagarin and was launched, again by the USSR, on 12 Apr. 1961. See also MM pp.156–159.

Space exploration, investigation, by spacecraft, of regions of space beyond the Earth's atmosphere. Sputnik 1, launched 4 October 1957 by the Soviet Union, began the research that resulted in the landing of two men on the Moon in July 1969. Major milestones: first space probe (Explorer 1, discoverer of Van Allen radiation belts, launched 31 Jan. 1958); first lunar probe (Luna 2, 12 Sept. 1959); first manned flight (Yuri Gagarin, 12 April 1961); first close-up pictures of Mars (Mariner 4, 1965); first manned lunar landing (Neil Armstrong, 21 July 1969); first pictures from surface of another planet (Venera 9, 22 Oct. 1975). See also SU pp.268–271; MM pp.156–159.

Space heating, in homes, factories and public buildings, heating of the interior of buildings. It is achieved using one or another type of CENTRAL HEATING system. See also CONVECTION; RADIATION.

Space probe, unmanned craft sent to explore regions beyond the Earth's atmosphere. The first US space probe was Explorer 1, launched 31 January 1958; it discovered the inner VAN ALLEN RADIATION BELT. The first lunar probe was Luna 2, launched 12 September 1959. The first PLANETARY PROBE was Mariner 2, launched 27 August 1962, which flew past Venus. Pioneers 10 and 11 photographed Venus (22 Oct. 1975); Viking 1 landed on Mars and collected information to test the Martian surface for life (July 1976). See also SU pp.268–269.

Space research, scientific and technological investigations into new methods for studying and travelling in space. In the last quarter of the 20th century, following the successes of the US Apollo Moon Project, it will probably include (besides exploration of other planets) the establishment of large space stations orbiting the Earth. Through these astronomy and space science generally will enter new phases of advancement. To build, maintain and supply such space stations, the US SPACE SHUTTLE is being developed. New engineering techniques will be evolved and a whole range of new handling equipment and propulsion machinery designed. The medical problems of prolonged stays in space will have to be overcome, and research into the sociology of space communities may be necessary. Among the developments of chemists and biologists will be the design of sophisticated systems of recycling of wastes employing microbes genetically tailored for the purpose. *See also* SU pp.270–283.

Space shuttle, re-usable ROCKET and aerodynamic vehicle carrying men and supplies between Earth and a SPACE STATION. The shuttle is carried on a modified Boeing 747 until a predetermined height is reached. A booster rocket launches the shuttle, which returns to the launch pad to be re-used. The shuttle is planned to have rocket power for manoeuvre into orbit and for deceleration during re-entry, and wings and a tail for flying through the atmosphere. It could land at existing airports.

Space station, orbiting laboratory in space where men can construct and maintain spacecraft and tend scientific and medical experiments. Early stations included the Russian SOYUZ (1971) and the US SKYLAB (1973). *See also* MM pp.268–269.

Spacesuit, pressurized, sealed garment allowing an ASTRONAUT to function in adverse conditions of space, ie extremes of temperature and pressure, RADIATION and meteorite bombardment, and weightlessness. It has a self-contained life-support system, which includes an oxygen supply and a cooling system, and maintains pressures of $0.35–1\text{kg/cm}^2(5–15\text{lb/in}^2)$ with a neoprene coating. Layers of aluminized plastic, fabric, and nylon provide insulation. Plastic visors guard against micrometeorites. *See also* MM pp.156–159.

Space-time, in RELATIVITY theory, concept that unifies the three space dimensions with time to form a four-dimensional frame of reference. In 1907 Hermann MINKOWSKI devised a presentation of relativity theory by extending three-dimensional geometry to four dimensions. A line drawn in this space represents the whole history of a particle as its path both in space and time (its world line). *See also* SU pp.106–107.

Space travel, the launching of manned vehicles to regions in space beyond the influence of the Earth's gravity. Within the next century space travel can be expected to reach the stage when spacecraft are assembled and launched from orbiting space stations, on missions to other stations which orbit sister planets, or probing regions beyond the solar system. For travel within the inner regions of the solar system, spacecraft can employ solar engines (which derive their power from the Sun's radiation). For journeys to outer space nuclear-powered ion engines could provide the thrust needed over longer periods of time. At the end of a space journey, passengers and crew may return to Earth from orbiting terminal in the re-entry vehicle of the SPACE SHUTTLE. *See also* pp.268–283.

Space walks, exciting features of early space flights. Alexei Leonov, in 1965, was the first astronaut to manoeuvre himself outside his spacecraft, Voskhod 2. Astronauts secured themselves to their craft by lifelines and obtained propulsion using jet pistols.

Spadefoot toad, tailless AMPHIBIAN with a horny "spade" on the inside of each hind foot. It is squat and smooth-skinned. Genus *Scaphiopus* is found from Canada to Mexico and genus *Pelobates* is found in s Europe and central Asia. Family Pelobatidae. *See also* NW p.219.

Spadix, in some flowering plants, a spike of small flowers; it is generally enclosed in a SPATHE. A familiar plant, with an inflorescence of this kind, is the cuckoo-pint (*Arum maculatum*). *See also* NW p.70.

Spaghetti, variety of PASTA which takes the form of thin strings which soften when it is cooked in boiling water. Spaghetti is ready to eat after about ten minutes cooking.

Spain, kingdom occupying four-fifths of the Iberian Peninsula in SW Europe. Civil War in the 1930s put Spain under a Fascist dictatorship until the mid-1970s. It is predominantly an agricultural country and three-quarters of the arable land is pasture. The chief crops are cereals, potatoes, sugar-beet and fruits. Mineral deposits include mercury, iron ore, coal, zinc and tin. The principal industries are shipbuilding, motor vehicles, textiles, paper, machinery, iron and steel and cement. Tourism and fishing are also important. The capital is Madrid. Area: 504,750sq km (194,884sq miles). Pop. (1976 est.) 35,972,000. *See* MW p.158.

Spallanzani, Lazzarro (1729–99), Italian biologist who demonstrated the falsity of the theory, then widely believed, that micro-organisms spontaneously generate themselves. The belief was based on the growths of moulds etc. in liquids left standing exposed to the air. Spallanzani, boiling a flask of liquid for an hour, thoroughly sealed it and left it to stand. Even after several days no micro-organisms appeared.

Spandrel, in architecture, triangular space between the curve of an arch, the horizontal drawn from its apex and the vertical drawn from its springing.

Spanghero, Walter (1943–), French rugby union player. From the Narbonne club, he played in 42 matches (1964–73) against International Board countries, captaining France in his final season. Powerfully built, he was equally effective as a lock forward or No.8.

Spaniel, any of several breeds of sporting dogs that may be trained to locate and flush game, to drop for the hunter's shot and sometimes to retrieve on command. Mentioned as early as the 14th century, it was called setting spaniel before the introduction of the gun. The dog's task then was to locate game and sit while the hunter threw a net over it. Land spaniels include the COCKER and toy. Water spaniels are usually RETRIEVERS.

Spanish Armada. *See* ARMADA, SPANISH.

Spanish Civil War (1936–39), conflict between LOYALISTS (often called Republicans) and Nationalists for control of SPAIN. Both sides began as loose alliances; the Loyalists, so called because they were loyal to the government of the Second Republic, consisted of Republicans, Socialists, Communists and BASQUE and CATALONIAN separatists. The Nationalists comprised the right-wing professional army, the more conservative faction of the Roman Catholic Church, monarchists, CARLISTS and the great landowners. In 1931 the Spanish Monarchy was abolished and the Second Republic instituted. Church and state were separated; land was redistributed and military influence curtailed. All these measures were anathema to the traditionalist forces of Spain, especially the Church, the landed interests and the military. War broke out in July 1936, with an uprising of the Spanish army in Morocco. The commander in Morocco, Francisco FRANCO, soon emerged as the leader of the Nationalist side. Nazi Germany and FASCIST Italy gave aid, including troops and supplies, to the Nationalist forces. Franco used this aid unscrupulously and the aerial bombing of civilians was employed for the first time by the Nationalist air-force, notably at GUERNICA in 1937. The democracies, on the other hand, maintained strict neutrality, leaving the USSR to support the Loyalists, which greatly strengthened the COMMUNISTS within the Loyalist ranks. INTERNATIONAL BRIGADES were formed and fought on the side of the Loyalists. The well-organized Nationalists in late 1938 began a major assault on Catalonia. When Barcelona fell in January 1939, the Loyal-

ist cause was doomed. On 28 March 1939 the Nationalists entered Madrid, and the war was over. More than 1,000,000 Spaniards had been killed and more than 250,000 were forced into exile. *See also* HC2 pp.213, 213, 218, 223, 226.

Spanish flu, virulent and contagious form of INFLUENZA which became a world-wide epidemic in 1918–19. An estimated 20 million people died. The young were particularly vulnerable to secondary infection such as bacterial pneumonia for which no cure then existed.

Spanish fly. *See* BLISTER BEETLE; CANTHARIDES.

Spanish Guinea, former name of EQUATORIAL GUINEA, in W central Africa.

Spanish Inquisition. *See* INQUISITION.

Spanish Main, term used to refer to the West Indies and the Caribbean coast of Panama, Colombia and Venezuela during the area's exploration and development. A number of offshore islands provided ideal bases for buccaneers and pirates who, in the 16th century, raided ships taking gold and silver from South America to Europe.

Spanish Sahara, former name of Western Sahara. *See* MW p.183.

Spanish Steps, The (1721–25), stepped passageway in the Piazza di Spagna, Rome with the impressively sited French-built church Trinita dei Monti as a backdrop. A masterpiece of BAROQUE architecture with curvilinear flights of steps, the design was begun by Alessandro Specchi (1668–1729) and completed after 1723 by Francesco de Sanctis (1693–1740). The stairway takes its name from the Spanish Embassy established in the piazza in the 17th century, but the idea was French and the costs were defrayed by a French diplomat.

Spanish Succession, War of the (1701–14), dynastic struggle for the throne of SPAIN. When it became apparent that CHARLES II of Spain would die childless, there were three claimants who had dynastic ties with the Spanish royal family: LOUIS XIV of France, who claimed the throne for his grandson Maximilian II; Emanuel the elector of Bavaria, who claimed the throne for his son Joseph Ferdinand; and the Holy Roman Emperor LEOPOLD I, who claimed the throne for his son. In his first will, Charles II left all his possessions to Joseph Ferdinand but Joseph's sudden death in 1699 reopened the dispute.

While new negotiations were in progress, the dying Charles II named Philip of Anjou, (grandson of Louis XIV) as his heir. When he ascended the throne as PHILIP V in 1701, most of Europe was plunged into war. France was supported by Spain, Bavaria and Cologne, and Mantua and Savoy. The Allies comprised the imperial forces, England, Holland, Prussia and other German states, and Portugal. Despite impressive Allied victories at BLENHEIM, RAMILLIES, OUDENAARDE and MALPLAQUET, the war lasted for more than a decade. In 1713–14, an exhausted France signed the Treaties of UTRECHT and in 1714 those of Rastatt and Baden were concluded. Philip V remained on the Spanish throne but the Allies, and particularly England, gained great territorial and trading concessions. *See also* HC1 pp.312, *312*, 314.

Spanish Town, city in SE central Jamaica, West Indies, on the Cobre River. Founded c.1523, the city was capital of Jamaica from 1692 to 1872. It is a commercial and processing centre for sugar cane, coffee, cocoa, bananas, breadfruit, dyewood and citrus fruits. Pop. (1970) 41,600.

Spanish Tragedy, The (1594), blank verse tragedy by Thomas KYD. One of the best known of 16th-century plays that piled up revenge, blood and horror, it greatly influenced the development of Elizabethan drama. *See also* HC1 p.283.

Spare part surgery, medical treatment that involves replacing the bones or organs of the patient with steel alloy or plastic parts; metal plates may also be used to strengthen existing bones. *See also* MS p.127.

Spark, Muriel (1918–), British novelist who also wrote short stories, plays, poems and criticism. Her novels include

Spain; a statue of Christopher Columbus in Columbus Square, Madrid.

Spaniels are descended from a Spanish breed: the name derives from "Spain".

Spanish Civil War; Nationalist artillery aimed at Government lines at Teruel.

Spanish Steps, one of Rome's attractions, with St Trinita de Monte Church.

Spark plug

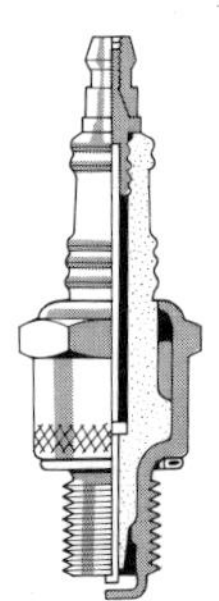

Spark plugs should be kept clean and changed every 16,000km (10,000 miles).

Spartacus: his victory over a Roman force at Mount Vesuvius in 73 BC.

Spearmint grows in temperate zones, and is often cultivated as a garden herb.

Spectacled bears, agile climbers, inhabit the Andes from Ecuador to Chile.

Memento Mori (1959) and *The Prime of Miss Jean Brodie* (1961), which was later adapted for the stage and made into a film in 1969, and *The Abbess of Crewe* (1974).

Spark plug, component in a petrol engine. It has two ELECTRODES separated by an air gap, across which electric current from the ignition discharges to form a spark. The spark plug fits into the cylinder head and the spark ignites the compressed mixture of fuel and air. *See also* MM pp.68–69.

Sparrers Can't Sing (1960), comedy by Stephen Lewis which was first performed by THEATRE WORKSHOP and subsequently filmed under the direction of Joan LITTLE-WOOD in 1962. It is an affectionate portrait of East End Cockney life.

Sparrow, any of a number of small FINCH-like birds that have become almost entirely dependent on man and live on or around human settlements. Typical, and most familiar, is the house sparrow. The male has a chestnut mantle, grey crown and rump and black bib. The female is duller and lacks the bib and grey rump. Highly gregarious, sparrows feed, roost and dust-bathe in noisy, twittering flocks. Basically seed-eaters, with a preference for grain, they can quickly strip a seed crop bare and are widely regarded as pests. They also eat fruit, worms and household scraps. Length: 14.5cm (5.75in). Family Ploceidae; species *Passer domesticus.*

Sparrowhawk, any of several species of birds of prey, especially *Accipter nisus,* the Eurasian sparrowhawk. An inhabitant of wooded areas, it perches and swoops to catch ground-feeding sparrows, other small birds, small mammals and insects or may catch prey in flight. Its plumage is grey and white barred with brown. Length: to 33cm (13in). Family Accipitridae. *See also* NW pp.*209, 217.*

Sparta (Sparti), town in SE Peloponnesos, s Greece, on the River Evrotas; capital of Lakonia department. Just s are the ruins of ancient Sparta, a Greek city-state founded by the Dorians *c.*1000 BC in the province of Lakonia. Spartan society consisted of three classes: the Spartiates (ruling class), perioeci (free inhabitants with no political power) and helots (slaves). After its conquest of Arcadia, Argos and Messinia (*c.*734–716 BC), Sparta flourished as an economic and cultural centre. After a massive helot revolt in late 7th century BC, it changed into an armed camp with all Spartiates trained as soldiers from an early age. It was originally ruled by two kings and later by the *gerousia,* the assembly (*apella*) and *ephors.* In *c.*500 BC the Peloponnesian League was formed with Sparta as the most powerful member. After the Persian Wars (500–449 BC), Athenian power began to rival Sparta's, leading to the Peloponnesian War (431–404 BC), which ended in the defeat of Athens. Sparta dominated Greece until 371 BC when it was defeated by Thebes at Leuctra, and Messinia was freed. Prosperity revived under the Romans but Sparta was destroyed by Visigoths in AD 396. The modern town was built in 1834. Popp. (1971) 13,400. *See also* HC1 p.70–74.

Spartacists, members of the German political party called the Spartacus League, which broke away from the SOCIAL DEMOCRATS during WWI. Led by Karl LIEBKNECHT and Rosa LUXEMBURG, the Spartacists refused to support the war effort and rejected participation in the post-Versailles republican government. They instigated a number of uprisings, including one in Berlin in 1919, after which they were brutally repressed and their leaders murdered. *See also* HC2 pp.*212, 213.*

Spartacus (*d.*71 BC), Thracian gladiator in Rome who led a slave revolt known as the Third Servile (Gladiatorial) War (73–71). His soldiers devastated the land and then moved south towards Sicily, where they were eventually defeated by CRASSUS with POMPEY's aid. Spartacus died in battle.

Spasm, involuntary, abnormal muscle contraction. In clonic spasm, repeated contractions are sudden while frequent relaxations, as in EPILEPSY. Tonic spasm, or cramp, is prolonged contraction, as in TETANUS. Drugs called antispasmodics are used to counteract spasm.

Spassky, Boris Vasilyevich (1937–), Soviet chess player who at the age of 18 became an international grand master. He became the world champion in 1969 but in 1972 lost his title to Bobby FISCHER, a US grand master.

Spastics, people afflicted with a type of CEREBRAL PALSY in which the victim suffers spasms of the muscles. The legs are generally more affected than the arms. Spastic paralysis is caused by damage to the brain's upper motor neurones, generally incurred before or at birth. No other brain functions are implicated and the commonly supposed mental abnormalities of spastics are a myth.

Spathe, broad leaf-like organ that spreads from the base of, or enfolds, the SPADIX of certain plants, such as the GUCKOO-PINT (*Arum maculatum*). *See also* NW p.70.

Speakeasy, bar or nightclub in the USA in which illicit alcohol was sold during the years of PROHIBITION. Personal contacts or a membership card were necessary to obtain entry.

Speaker, radio. *See* LOUDSPEAKER.

Speaker, The, chairman of the House of Commons in the British Parliament and of other assemblies modelled on it. He is an MP elected to the office by his fellows. The speaker does not vote in divisions unless there is a tie.

Spear, weapon consisting of a long wooden shaft to which is attached a pointed head, of various designs. The earliest spears were sticks sharpened to a point which was hardened by fire. The sticks were later tipped with heads of stone, bone, bronze and iron. Spears are of two basic kinds: those used for throwing and those designed to be used at close quarters. *See also* MM pp.160–161.

Spear fishing, fishing underwater with a hand-spear, resembling a trident, or a spear-propelling gun. Divers use either SCUBA or SKIN DIVING equipment, and their guns are usually powered by compressed air, springs or elastic rubber thongs.

Spearman, Charles Edward (1863–1945), British psychologist. A pioneer in the application of statistical and mathematical methods to the analysis of data, he attempted to analyse human abilities into their components by using a method he helped to create called FACTOR ANALYSIS.

Spearmint, common name of *Mentha spicata,* a hardy perennial herb of the MINT family (Labiatae). Its leaves are used for flavouring, especially in sweets. Oil distilled from spearmint is used as a medicine. The plant has pink or lilac flowers that grow in spikes. *See also* NW p.*59.*

Speciation, emergence of new species in EVOLUTION. It results from the separation of parts of a homogeneous population. Over many generations, NATURAL SELECTION operates within the separated groups to produce gradually increasing differences. New species can be said to have evolved when individuals of the separated groups are no longer capable of interbreeding successfully.

Species, in taxonomy, group of physically and genetically similar individuals that interbreed to produce fertile offspring under natural conditions. It ranks below a GENUS and above a variety. More than 300,000 plant species and one million animal species have been identified.

Specific gravity, ratio of the DENSITY of a substance to the density of water. Thus, the specific gravity of gold is 19.3; ie it is about 19 times heavier than an equal volume of water.

Specific heat, heat energy necessary to raise the temperature of a given amount of a substance by one degree centigrade. Substances with a high specific heat, such as water, require much more energy to raise their temperatures than do substances of low specific heat. Thus oceans can exert a moderating influence on the Earth's climate by absorbing excess heat from the land during the day and returning it at night.

Spectacled bear, small black bear that lives in the forests of the Andes. It has light brown, spectacle-like rings around its eyes and is the only native bear of the Southern Hemisphere. Species *Tremarctos ornatus.*

Spectacles, lenses in frames worn in front of the eyes to correct defective vision. They were first made in the late middle ages, and were crude instruments until lens grinding became a fine art in the 17th century. Today most types of visual defect can be prescribed for by an oculist. Short sight is corrected by a diverging, or concave lens; long sight by a converging, or convex lens; and astigmatism by a lens resembling a slice from the side of a cylinder.

Spectator, The (1711–1712, 1714), English periodical edited and largely written by Richard STEELE and Joseph ADDISON, a successor to the *Tatler.* It gave a picture of the social life of the times and had many distinguished contributors, among them Alexander POPE.

Spectrograph, instrument for producing and recording a SPECTRUM. Light from an incandescent source falls on to a prism or DIFFRACTION GRATING, and is split into its component wavelengths. This emission spectrum is then focused on to a photographic plate. To record an absorption spectrum light from a source emitting a continuous spectrum is passed through an absorbing medium before entering the instrument. *See also* SU p.98.

Spectrometer. *See* SPECTROSCOPY.

Spectrophotometer, optical instrument used widely in chemical analysis. Light or other radiation is passed through a filter (monochromator) in which all but a limited range of wavelengths are removed. Emergent radiation of a specific wavelength is then passed through a solution of the specimen to be analysed, and is absorbed by an amount related to its concentration.

Spectroscope, instrument for producing and studying the SPECTRUM of light from some source. Spectroscopes are used in astronomy to study the light from stars, and in chemistry to detect the presence of traces of various elements in samples that are too small to detect by other means. The light entering a spectroscope is collimated into a narrow beam by means of a slit and lens. The beam then passes through either a prism or DIFFRACTION GRATING so that it is dispersed into a spectrum. Combined with the grating or prism is a scale from which the spectral wavelengths may be read directly through the telescope that magnifies the spectrum. *See also* SU pp.*62, 63, 98, 151, 221.*

Spectroscopy, or spectrometry, branch of OPTICS dealing with the measurement of the wavelength and intensity of lines in a SPECTRUM. The main tool in this study is the SPECTROSCOPE. A spectrograph photographs a spectrum and an analysis of the spectrogram can reveal the substances giving rise to the spectrum, because the position of emission and absorption lines and bands characterize each substance. A spectrometer is a calibrated spectroscope, capable of precise measurements. *See also* SU pp.98–99.

Spectrum, distribution obtained when ELECTROMAGNETIC RADIATIONS radiation. frequencies. A well-known example is provided by the rainbow colours produced when white – or visible – light passes through a prism. This effect, which is also produced when visible light passes through a DIFFRACTION GRATING, produces a continuous spectrum, in which all wavelengths (between certain limits) are present. Spectra formed from objects emitting radiations are called emission spectra. These occur when a substance is strongly heated or bombarded by electrons. An absorption spectrum, consisting of dark regions on a bright background, is obtained when white light passes through a semi-transparent medium that absorbs certain frequencies. A line spectrum is one in which only certain wavelengths or "lines" appear. The emission and absorption spectra of a substance are basic characteristics of it and are used as a means of identification. These spectra arise as a result of transitions between different stationary states of the electrons, atoms or molecules of the substance, which give rise to the emission or absorption of electromagnetic radiation. *See also* SU pp.98, 99, 102, *172–173, 221–226.*

Speculum, instrument for widening an aperture or passageway of the body, in order to examine the interior.

Spee, Maximillian Johannes Maria Hubert, Graf von (1861–1914), German admiral. He commanded German forces in the battles of CORONEL and the FALKLAND ISLANDS early in WWI. His flagship, the armoured cruiser *Scharnhorst*, was sunk with all hands in December 1914 by British battle cruisers.

Speech, verbal or vocal expression of ideas and feelings, specifically intended to convey meaning. An infant's babbling is the first stage towards learning to speak; progress thereafter depends on the maturing of speech organs, the ability to hear, intelligence and environment (stimulation, attentive parents, reinforcement of vocalization). *See also* MS pp.238–41, *238, 162–163, 239, 241.*

Speech disorders, speech and language defects resulting from such diverse causes as emotional disorder, organic deficiency, faulty speech mechanisms, brain damage and sensory deficiency. Various remedies for speech disorders have been recorded since ancient times but it was not until the 1900s, and especially since WWII, that scientifically developed techniques were used by trained therapists to treat speech disorders. *See also* MS pp.154–155.

Speed, rate of movement. It is the ratio of the distance covered to the time taken by a moving object. It is a SCALAR quantity. Speed in a specified direction is VELOCITY, a VECTOR quantity.

Speedball, game played on an outdoor field by two teams of 11 players each. The game contains elements of basketball, volleyball, soccer and American football. The object is to advance the ball towards the opponent's goal by throwing or kicking it. Tackling, blocking and running with the ball are forbidden. Field goals (between the goal posts) count three points, a touchdown (catching a forward pass in the end zone) and an end kick (kicking the ball across the end line from within the end zone) each count one point. Single points are also scored by a drop kick (over the crossbar) and a penalty kick.

Speed measurement, of road and track vehicles, is usually made using a SPEEDOMETER. The airspeed indicator of an aeroplane employs a pitot tube, mounted externally and facing into the airstream. Airspeed is then measured automatically as a function of the dynamic pressure of air in the pitot tube and the static air pressure at the aeroplane's particular altitude. A boat's speed is usually calibrated in knots (nautical miles per hour). In the simplest instruments a small propeller mounted on the boat's underside turns a flexible cable similarly to that of a car's speedometer.

Speedometer, apparatus for recording the speed of a vehicle. In a car or lorry, it is driven by a flexible cable connected to the transmission which turns at a speed proportional to the roadspeed. Inside the speedometer the cable turns a magnet which partly rotates a spring-loaded drum to which the indicator needle or coloured strip is attached.

Speed records, for both humans and machines, have been computed since the late 19th century. The fastest speed travelled by a human being is 39,897km/h (24,791mph), achieved by the crew of Apollo 10 on its return from the Moon on 26 May 1969. The world land speed record, achieved by Gary Gabelich of the USA in a jet-powered car on 23 Oct. 1970, is 1,046km/h (650mph); the women's land speed record is held by Kitty O'Neil of the USA, who achieved 843km/h (524mph) on 6 Dec. 1976. The world water speed record, held by Donald Campbell, is 527km/h (328mph), achieved on Coniston Water, England, on 4 Jan. 1967; Campbell was killed during this run. The fastest railway train speed was achieved on 28 March 1974 in Colorado USA, when a Linear Induction Motor Research Vehicle attained 377km/h (234mph).

Speedway, motorcycle sport in which lightweight, specially built machines are raced around oval tracks with cinder, shale, or granite surfaces. It originated in Australia in 1923, and is now popular in Britain, the USA, Scandinavia, Eastern Europe, and Australasia. Riders amass points during a series of races at a meeting, which may be a team match or an individual championship.

Speedwell, common name applied to herbaceous plants of many species of the genus *Veronica*. It is a numerous family, found throughout the world. Germander speedwell, *V. chamaedrys,* is a common British wild flower. Family Scrophulariaceae.

Speenhamland system, form of poor relief in England initiated by the magistrates of Speenhamland, Berkshire, in 1795. Wages were brought up to a minimum level by payments out of the poor rates, according to a schedule based on the price of bread and the size of the recipient's family. It spread throughout much of the south, but was superseded by the Poor Law Amendment Act of 1834.

Speer, Albert (1905–), German architect and NAZI official, a close associate of Adolf HITLER. Speer drew up the plans for Germany's autobahns and for the stadium at Nuremberg. By 1943 his authority over the entire war economy was second only to that of Hermann GOERING. Speer was tried in 1946 by the Nuremberg war crimes tribunal and sentenced to 20 years in Spandau prison.

Speke, John Hanning (1827–1864), British soldier and explorer of Africa. After service in India he made African expeditions with Sir Richard BURTON, during which he alone discovered Lake VICTORIA (1858) and asserted it to be a source of the NILE. *See also* HC2 p.*142.*

Speleology, scientific study of CAVES and cave systems. Included also are the hydrological and geological studies concerned with the rate of formation of stalagmites and stalactites and the influence of groundwater conditions on cave formation. A special aspect is the study of the animals that live in caves. *See also* PE pp.110–111.

Spelthorne, county district in N SURREY, England, created in 1974 under the Local Government Act (1972). Area: 56sq km (22sq miles). Pop. (1974 est.) 95,700.

Spemann, Hans (1869–1941), German zoologist and comparative anatomist. He was awarded the 1935 Nobel Prize in physiology and medicine for his discovery of the "organizer effect" in embryonic development.

Spence, Sir Basil (1907–76), British architect, famous for his design (1951) of Coventry Cathedral, which was consecrated in 1962. Other spectacular works include the Household Cavalry Barracks (1970) at Knightsbridge, London, and the British Embassy in Rome (1971). *See also* HC2 p.*285.*

Spencer, Herbert (1820–1903), British sociologist. He generalized DARWIN's theory of EVOLUTION, developing the idea of a natural progression between primitive and advanced societies based on survival of the fittest. Among his works are *First Principles* (1862) and *The Principles of Ethics* (2 vols 1879–93). *See also* HC2 p.172, *172.*

Spencer, Sir Stanley (1891–1959), British painter. He is famous chiefly for his original and allegorical treatment of religious incidents which are presented in commonplace settings, especially that of Cookham-on-Thames, Berkshire, where he was born and lived for most of his life. *The Resurrection* (1923–27) is set in Cookham churchyard and includes portraits of both himself and his wife, Hilda. A series of his murals (1926–34) decorate the War Memorial Chapel at Burghclere, Hampshire.

Spencer, Sir Walter Baldwin (1860–1929), British ethnologist who, with F. J. GILLEN, explored the interior of Australia and studied the ABORIGINES. Together they wrote books on the aborigines such as *Native Tribes of Central Australia* (1889) and *Across Australia* (1912).

Spencer Gulf (Spencer's Gulf), large inlet of the Indian Ocean off the coast of S Australia, between the Yorke and Eyre peninsulas. The entrance is partly blocked by the islands of Thistle, Gambier and Neptune. Length: 320km (200 miles); width: 145km (90 miles).

Spender, Stephen Harold (1909–), British poet who became well known in the 1930s with works which were inspired by social protest. His autobiography *World Within World* (1951) is a recreation of the political and social atmosphere of the 1930s.

Spengler, Oswald (1880–1936), German philosopher and historian. In *The Decline of the West, Outlines of a Morphology of World History* (2 vols, 1918–22), he expounded a cyclic theory of civilization that established his reputation. He proposed that Western culture was in decline, having passed its creative zenith. Spengler felt that the state needed to have absolute power to reverse this process. Although his theories had some affinity with those of the Nazis, they were critical of his work, and after HITLER came to power in 1933 Spengler went into isolation.

Spenser, Edmund (1552–99), English poet; his *The Faerie Queene* (1589–96) greatly influenced English lyric poetry. His best poetry is to be found in the witty sonnet sequence *Amoretti* (1595). *See also* HC1 p.282.

Spenserian stanza, poetic form devised by Edmund SPENSER. It is a STANZA of nine iambic lines, all of 10 syllables except the last, which is an ALEXANDRINE. Spenser devised it for his *The Faerie Queene* (1589–1596). It was also used in Lord BYRON's *Childe Harold's Pilgrimage,* SHELLEY's *Adonais* and KEATS's *The Eve of St Agnes.*

Speranski, Mikhail Mikhailovich, Count (1772–1839), Russian statesman. The son of a village priest, he rose to become personal assistant of ALEXANDER I in 1807, but his plans to increase the duties of the hereditary nobility and to establish an elected DUMA were unpopular in government circles, and he was dismissed in 1812. In 1819 he was appointed Governor of Siberia, where he instituted many reforms. Under NICHOLAS I he directed the first comprehensive codification of Russian Law.

Sperm, or spermatozoon, male sex cell (GAMETE) in sexually reproducing organisms, corresponding to the female OVUM. Its head contains the genetic material of the male parent, its tail the fertilizing motility.

Spermaceti, white waxy substance from the head of a whale, usually a sperm whale, used in the manufacture of cosmetics and ointments. Its main constituent is cetyl palmitate.

Spermatophyte, seed-bearing plant, including most trees, shrubs and herbaceous plants. It has a stem, leaves, roots and a well-developed VASCULAR system. The dominant generation is the SPOROPHYTE. Spermatophyta was formerly a division in plant classification; now seed plants are classified in the division TRACHEOPHYTA. *See also* NW pp.38–39.

Sperm bank, store of the semen of men or of farm animals, kept alive by carefully controlled freezing methods. Banks of human sperm are used in the treatment of some kinds of infertility. In animal husbandry sperm banks provide the means of propagating favoured strains of animal.

Sperm whale, largest of the toothed whales. It has a squarish head and feeds on squids and cuttlefish. Species *Physeter catodon. See also* WHALE.

Sperry, Elmer Ambrose (1860–1930), US inventor and industrialist who is remembered for the practical development of the gyroscope. He developed a gyroscopic compass and gyroscopic stabilizers for ships and aircraft. He made improvements to equipment in other fields including electricity and illumination.

Speusippus (*fl.* 347–339 BC), Greek philosopher. Nephew and follower of PLATO, he succeeded him as head of the ACADEMY.

Speyer (Spires), port in sw West Germany, on the River Rhine 35km (22 miles) N of Karlsruhe. Notable buildings include an 11th century Romanesque cathedral, which contains the tombs of eight emperors. Industries: paper, chemicals, textiles, oil refining, brewing, wine. Pop. (1970) 42,323.

Albert Speer's stadium at Nuremberg was built for Nazi gatherings in the 1930s.

John Hanning Speke: a studio photograph of the discoverer of the Nile's source.

Stephen Spender is known for his drama and fiction, as well as his poetry.

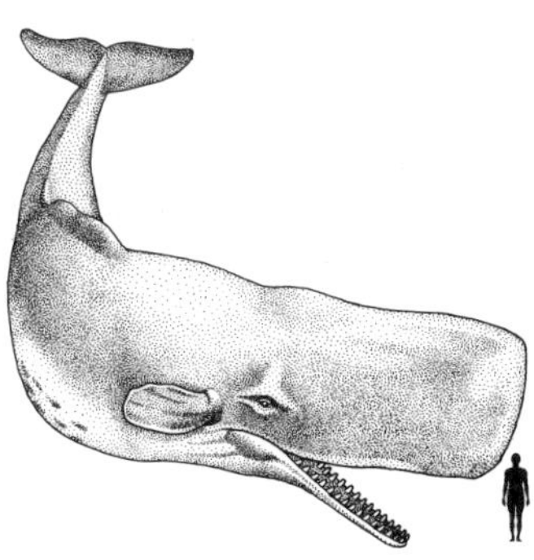

Sperm whale: the greatest recorded length for a male is 26m (84ft).

Sphagnum

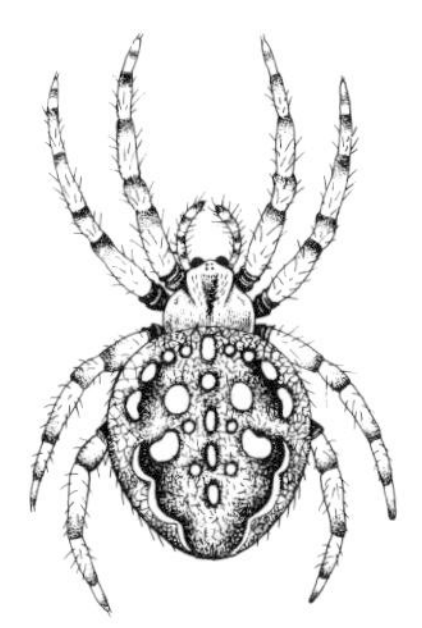

Spider; in many species the female is larger and stronger than the male.

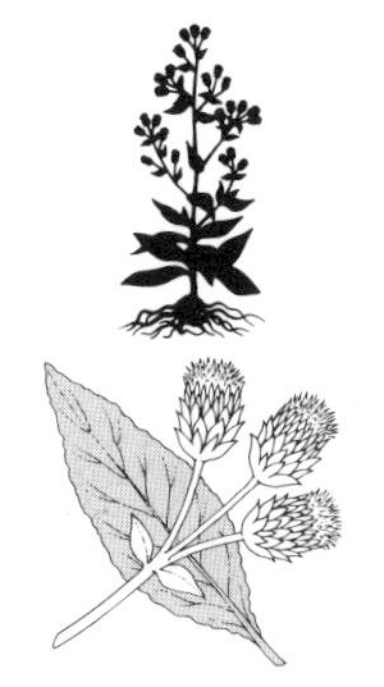

Spikenard; scent is obtained from the cluster of thick stems growing from the root.

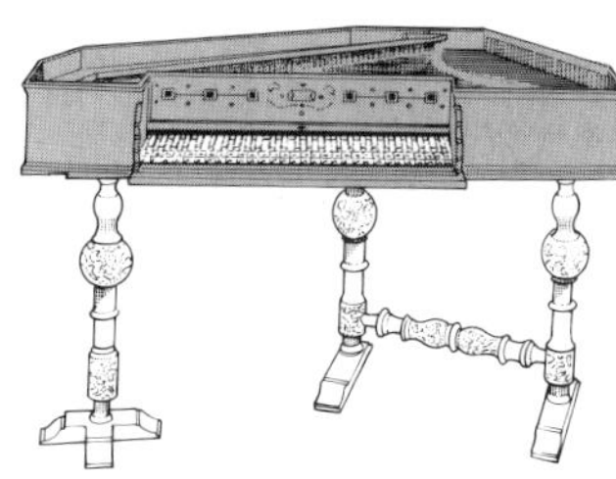

Spinet was probably named after Giovanni Spinetti, who redesigned it.

General Spinola returned to Lisbon in Sept. 1976 after 18 months in exile.

Sphagnum, genus of mosses which grow on boggy soils. They are also known as peat mosses and are gathered for use by horticulturalists, being packed around potted seedlings.

Sphalerite, or blende, sulphide mineral composed of zinc sulphide (ZnS), an important source of zinc. It has cubic system tetrahedral crystals or granular masses. It is white when pure, but impure forms may be yellow, black, red or green with a resinous lustre. Hardness 3.5–4; s.g. 4.

Sphenisciformes. See PENGUIN.

Sphenodon. See TUATARA.

Sphincter, ring of muscles controlling a body passage. Important sphincters include the pyloric sphincter in the stomach, the anal sphincter and the sphincter pupillae in the eye.

Sphinx, mythical beast of the ancient world, usually represented with the head of a man or woman and the body of a lion. Although found throughout the Middle East, sphinxes were especially popular in Egypt, where thousands were built. The most famous was the Great Sphinx at Giza. See also HC1 p.53.

Sphinx, Riddle of the, in Greek mythology, the riddle of the three ages of man which OEDIPUS alone could answer. Those who failed to answer were killed by the SPHINX. When Oedipus solved the riddle the Sphinx killed herself.

Sphygmomanometer, instrument used to measure blood pressure. A rubber cuff connected to the device is wrapped around the upper arm and inflated with a rubber bulb, stopping the flow of blood in a major artery. When this air pressure is slowly released, the systolic pressure of the blood – which corresponds to a contraction in the pumping cycle of the heart – is registered.

Spice, dried part of any of various tropical and sub-tropical plants. Of distinctive aroma and taste, spices are used for enhancing the flavour of food. In the ancient world spices had medicinal and religious, as well as culinary, functions, and were first used in medieval Europe to help in the preservation of food.

Spice Islands. See MOLUCCAS.

Spicule, spear-like column of hot gas ejected from the Sun's CHROMOSPHERE as a jet, often reaching heights of 8,000km (5,000 miles) above the solar surface. Most spicules are less than 500km (300 miles) in diameter. See also SU p.222.

Spider, any of numerous species of terrestrial, invertebrate, arthropod ARACHNIDS found throughout the world in a wide variety of habitats; fossil species have been found that date from the Devonian period (395 million years ago). Species have an unsegmented abdomen attached to a cephalothorax by a slender pedicel. There are no antennae; sensory hairs are found on the appendages (four pairs of walking legs). Most species have pairs of spinnerets on the abdomen for spinning silk to make egg cases, shelters and webs. Many exhibit ingenious methods of trapping or catching prey, which generally consists of insects. See also NW pp.73, 102, 103, 104, 216, 219.

Spider crab, any of 500 species of marine crabs of the Atlantic and Pacific oceans. It has a spiny, sac-shaped carapace that is pointed in front and extremely long, thin legs. The Japanese spider crab (*Macrocheira kaempferi*) may reach 4m (13ft) from claw tip to claw tip, and is the largest species. Family Majidae. See also NW pp.100, 105.

Spider monkey, medium-sized arboreal, MONKEY found from s Mexico to SE Brazil. It has long, spidery legs, a fully prehensile tail, and is an agile climber, using the tail as a fifth hand. It eats mainly fruit and nuts. Genera *Ateles* and *Brachyteles*.

Spiderwort, any of several species of perennial plants with long, keeled, grasslike leaves and blue, rosy-purple or white flowers in flat-topped clusters. Family Commelinaceae; genus *Tradescantia*.

Spikenard, perfume obtained from a herb (*Nardostachys jatamansi*) of the Valerian family, native to India. Substitutes for the perfume are often grown in s Europe and include *Valeriana celtria* and *V. phu*.

Spillane, Mickey (1918–), US writer, real name Frank Morrison Spillane whose tough, often sadistic detective novels include *I, The Jury* (1947), *My Gun Is Quick* (1950), *The Big Kill* (1951) and *The Deep* (1961).

Spin, in quantum mechanics, intrinsic angular momentum possessed by some SUBATOMIC PARTICLES and nuclei which may be regarded by analogy as the spinning of the particle about an axis within itself. It is limited in magnitude to half the whole integral values (eg, 0, $\pm\frac{1}{2}$, ± 1, $\pm 3/2$...), and it is one of the quantum numbers by which a particle is specified.

Spina bifida, congenital disorder in which the bones of the spine do not develop properly to form the channel for the SPINAL CORD. This may deform the cord itself, affecting the nerves controlling the limbs and the bladder. Mild cases can benefit from surgery.

Spinach, herbaceous annual plant widely cultivated in a with cool summers. Its dark green leaves contain substantial amounts of iron and vitamins A and C; they are also high in fibre content. Family Chenopodiaceae; species *Spinacia oleracea*. See also PE p.188.

Spinal anaesthesia, injection of a local ANAESTHETIC into the fluid surrounding the SPINAL CORD, causing loss of sensation in parts of the body served by nerves below the site of the injection.

Spinal column. See SPINE.

Spinal cord, tubular central nerve cord, lying within the vertebral column (spine). With the brain, it makes up the central nervous system. It has 31 pairs of spinal nerves extending from it, each of which has sensory and motor fibres. The spinal cord receives sensory input and conveys messages to all parts of the body and to the brain. See also MS p.38.

Spinal fluid. See CEREBROSPINAL FLUID.

Spinal nerves, nerves branching out from the spinal cord which supply the muscles of the trunk and limbs and carry sensory information to the brain. They are mixed nerves containing both sensory and motor fibres. The motor fibres arise from the anterior (frontal) side of the spinal cord and the sensory fibres from the posterior. See also MS pp.38–39.

Spinal tap (lumbar puncture), puncture of the lumbar (lower back) portion of the spinal cord to remove spinal fluid for examination or to make an injection.

Spine, or spinal column, series of vertebrae running from the skull to the tip of the tail in VERTEBRATES. It encloses the SPINAL CORD.

Spine, curvature of the. See CURVATURE OF THE SPINE.

Spinel, oxide mineral, magnesium aluminium oxide ($MgAl_2O_4$), found in IGNEOUS and METAMORPHIC ROCKS. It has cubic system, frequently twinned, octahedral crystals. In colour it is either glassy black, red, blue, brown or white. Ruby spinel from Sri Lanka is a valuable gemstone. Hardness 7.5–8; s.g. 3.8.

Spinet, early musical instrument of the HARPSICHORD family with one keyboard and one string to each note. The strings, which were plucked with a quill or leather plectrum, were at an angle of 45° or less to the keyboard. Highly decorative in appearance and usually wing shaped, the instrument normally had a range of four to five octaves.

Spinifex, Australian genus of sharp, tufted grasses which live in coastal sandy or desert areas.

Spinning, process of making thread or yarn by twisting together fibres of animal, vegetable or synthetic origin. Various machines were developed for mechanizing the process. See also SPINNING FRAME; SPINNING JENNY; SPINNING MULE; SPINNING WHEEL.

Spinning frame, machine for spinning yarn; driven by horsepower or water power, and invented by Sir Richard ARKWRIGHT in about 1768. The motion of a wheel and belt turned several spindles. See also SPINNING JENNY; SPINNING MULE.

Spinning jenny, in textiles, hand-operated spinning machine invented by James HARGREAVES in about 1764. It used several spindles to spin several strands of yarn

simultaneously, operating upon the same basic principles as the SPINNING WHEEL. See also HC2 p.67; MM pp. 58–59.

Spinning mule, in textiles, spinning machine invented by Samuel CROMPTON in 1779, called a mule because it was a hybrid combining the main features of Hargreave's SPINNING JENNY and Arkwright's SPINNING FRAME. See also SPINNING FRAME; HC2 p.314; MM pp.58–59.

Spinning wheel, simple spinning machine driven by a hand- or foot-cranked wheel attached to a spindle by a belt of heavy twine. See also MM p.58.

Spinola, Ambrogio di, marqués de los Balbases (1569–1630), Italian soldier. He entered the service of Spain and led an army to the Netherlands in 1602. He captured Ostend in 1604, after which he became Commander-in-Chief of the Spanish armies in the Netherlands. He later fought with great distinction in the THIRTY YEARS WAR.

Spinola, Antonio Sebastião Ribeiro de (1910–), Portuguese general and political leader. In April 1974 he led a military revolt that overthrew the dictatorship of Marcello Caetano, and in May he became President. His attempts to form a coalition government of socialists, liberals and military officers were unsuccessful, and he resigned in September 1974.

Spinoza, Baruch (1632–77), Dutch philosopher. A rationalist, he argued that all knowledge could be deduced from the key substance, which he called alternatively God and Nature. In *Ethics* (1667) he held that free will was an illusion which would be dispelled by man's recognition that every event has a cause, and his *Theological-Political Treatise* (1670) contained the first modern historical interpretation of the Bible. See also MS pp.231–232.

Spiny anteater. See ECHIDNA.

Spiracle, external opening for respiration in various animals. In insects and spiders it is the opening to a trachea (air tube). In sharks, rays and some bony fish water passes through a pair of spiracles during gill respiration. In whales the nasal opening is called a spiracle. See also GILLS.

Spiraea, genus of flowing perennial shrubs native to the Northern Hemisphere. They have small flat leaves and clusters of small white, pink or red flowers. Many of the 100 species are grown as ornamentals. Height: 1.5m (5ft). Family Rosaceae.

Spiral galaxy, type of regular GALAXY as classified by Edwin HUBBLE. It is shaped like a flattened disc with a central nucleus from which spiral arms project. Normal spirals have a spherical nucleus. Those galaxies which have spiral arms projecting from a bar-shaped nucleus are known as barred spirals. Spiral galaxies are further classified Sa to Sc, or SBa to SBc in the case of barred spirals, according to the degree to which their arms are tightly or loosely wound. See also SU pp.248–249, 248–249.

Spirit, alcoholic. See ALCOHOL.

Spiritual, religious folk music, especially that associated with US Negro culture of African ancestry. The words are generally adaptions of passages from the Bible and the music is often in four-part harmony with certain specific features like five-note melodies.

Spiritualism, belief that, at death, the soul and spirit of a person are transferred to another plane of existence, wherein communication with those one has left is through ectoplasm. See also MS pp.197, 216–217.

Spirochete, general name applied to a group of protozoa-like BACTERIA that are spiral-rod shaped and capable of flexing and wriggling their bodies as they move about. SYPHILIS is the best known disease caused by a spirochete (*Treponema pallidum*).

Spit, ridge of sand or shingle on a coastline produced by the interaction of long-shore drift and river current. Spits often form across river mouths, forcing them to change course. A British example is Chesil Bank in Dorset. See also PE p.122, 122.

Spithead, E part of the channel between Hampshire, England and the Isle of Wight. A traditional place of anchorage of the Royal Navy for centuries, it was the scene in 1797, of a mutiny. The sailors involved won their claim for better pay and conditions.

Spitsbergen (Svalbard), island group in the Arctic Ocean, approx 930km (580 miles) N of Tromso, Norway, to which it belongs. Although well within the Arctic Circle, the w edge of the islands is ice-free for most of the year as a result of the NORTH ATLANTIC DRIFT. Large coal deposits were discovered on Spitsbergen in the late 19th century, and coal mining is now a major industry. The islands are an important refuge for Arctic wildlife and protective measures have saved certain mammals from extinction. Area: 62,050sq km (29,958sq miles). Pop. (1971) approx. 3,000.

Spitteler, Carl Friedrich Georg (1845–1924), Swiss poet whose reputation rests on his epics, *Prometheus und Epimetheus* (1881) and *Der Olympische Frühling* (1900–05) for which he received the 1919 Nobel Prize.

Spittlebug. *See* FROG HOPPER.

Spitz, Mark Andrew (1950–   ), US swimmer. He set 35 US records and 23 world records. He won two gold medals at the 1968 Olympics and then an unprecedented seven gold medals at the 1972 Olympics. He retired from professional swimming in 1972.

Spleen, important organ of the lymphatic and blood systems. It lies in the abdominal cavity to the left of the stomach. It stores blood and through the action of its macrophages, destroys worn out or damaged red blood cells and produces LYMPHOCYTES.

Spleenwort, any small fern of the genus *Asplenium.* Spleenworts grow from crevices in walls and rocks, and occasionally in hedgerows, and are characterized by the thickened ribs on either side of the veins in their tongue-like leaves.

Splenomegaly, enlargement of the SPLEEN. Usually an indication of disorder elsewhere in the body, such as infectious mononucleosis, LEUKAEMIA or MALARIA. The spleen may be surgically removed in severe cases.

Split, port in Croatia, w Yugoslavia, on the Dalmatian coast. Split was held by Venice from 1420 to 1797, when it passed to Austria. It became part of Yugoslavia in 1918. Industries: shipbuilding, textiles, chemicals, tourism. Pop. (1971) 152,905.

Splitting the atom. *See* NUCLEAR FISSION.

Spock, Benjamin McLane (1903–   ), US paediatrician and writer whose *The Common Sense Book of Baby and Child Care* (1946) reversed the trend in child-rearing by calling for parental warmth and understanding of a child's individual nature, rather than rigid following of feeding and other schedules.

Spode, Josiah (1754–1827), British potter, who learned the trade in the workshops of his father at Stoke-on-Trent. He gave his name to what is known as Spode porcelain, which became the standard English bone china. For this hybrid porcelain he used bone ash and feldspar in the paste as well as china clay and china stone. In 1806 he became potter to George III.

Spode ceramics, porcelain ware produced at the Stoke factory in Staffordshire, England, by Josiah SPODE the younger from c.1800.

Spodumene, silicate mineral, lithium aluminium silicate ($LiAlSi_2O_6$). It displays monoclinic system prismatic crystals or masses and is opaque or transparent in many hues. Spodumene is the chief source of lithium and transparent lilac and green varieties are used as gems. Hardness 6.5–7; s.g. 3.2.

Spohr, Ludwig (1784–1859), German violinist. One of the most famous virtuoso violinists of his day, he was also a prolific composer.

Spondee, metrical foot of two stressed or two long syllables. Spondaic metre is found in Greek poetry in the slow, solemn hymns sung at a "spondee" or drink-offering.

Sponge, primitive, multi-celled aquatic animal. Its extremely simple structure is supported by a skeleton of lime, silica or spongin. There is no mouth, nervous system or cellular co-ordination, nor are there any internal organs. Sponges reproduce sexually and by asexual budding. There are about 5,000 species, including the simple sponge *Leucosolenia*. Length: 1mm–2m (0.4in–6ft). Phylum Porifera. *See also* NW pp.72, 81, *121, 230.*

Spontaneous combustion, outbreak of fire without external application of heat. When combustible material, such as damp hay, paper or rags, slowly oxidized by bacteria or air, the temperature may rise to the ignition point.

Spoonbill, any of several species of wading birds, each with a long bill that is flat and rounded at the tip; species are found in tropical climates throughout the world. It has large wings, long legs, a short tail, and white or pinkish plumage; it feeds on small plant and animal matter. Length: 90cm (3ft). Family: Threskiornithidae. *See also* NW pp.*216, 232.*

Spoonerism (metathesis), slip of the tongue, in which the first syllable or sounds of two or more words are interchanged. The term is named after William Archibald Spooner (1844–1930), warden of New College, Oxford (1903–24), who made many such transpositions including "You have deliberately tasted two worms and you can leave Oxford by the town drain."

Sporades, two island groups in E and SE Greece, in the Aegean Sea. The Northern Sporades, off the coast of Thessaly, include the islands of Skíros, Skíathos, Skópelos and Iliodhrómia. The Southern Sporades, off the coast of Turkey, include islands of the Dodecanese, Ikaría, Sámos and Khíos. Products: olive oil, wine, citrus fruits.

Sporangium, walled structure that produces SPORES in plants and some protozoa. It is found in groups on fern fronds.

Spore, small reproductive body that detaches from the parent plant to produce new plants. Mostly microscopic, spores may consist of one or several cells and are produced in large numbers. Some germinate rapidly, others "rest", surviving unfavourable environmental conditions. Although they occur in all plant groups, spores are particularly characteristic of FUNGI and FERNS.

Sporophyte, asexual stage in the life cycle of plants that produce SPORES and whose nuclei are diploid (having chromosomes in pairs). *See also* ALTERNATION OF GENERATIONS; GAMETOPHYTE.

Sporozoa, class of parasitic PROTOZOA characterized by reproduction by sporulation (asexual multiple fission) and the absence of special locomotive apparatus for much of their lives. The best example is PLASMODIUM, which causes MALARIA. *See also* NW p.80.

Sports, recreational or competitive activities in which physical skill and prowess may be shown. Sports are distinguished from games in that a sport is usually played according to a strict set of specified rules. Individual sports, particularly running, jumping and wrestling, may derive from the psychological need for play; they have been common in most human societies, and have been organized into competitions at least since Greek times. Team sports, on the other hand, are for the most part of 18th- to 20th-century origin. More ancient team games, such as the ball games of the Toltecs, the Mongol game *bushkazi*, American Indian lacrosse and medieval English football, were generally rough and had ill-defined rules. *See also* articles on individual sports and sportsmen.

Sports car racing. *See* MOTOR RACING.

Sprain, in the human body, injury to the ligaments of a joint, usually consisting of the tearing or overstretching of a ligament with associated bleeding. Symptoms include severe pain and visible bruising or swelling of the tissue. Treatment includes bandaging and support, immobilization of the limb and, in severe cases, physiotherapy.

Sprat, also called brisling, small herring-like commercial fish found in the N Atlantic Ocean. It is slender and silvery. Length: to 12.5 cm (5in). Family Clupeidae; species *Clupea sprattus. See also* PE p.247.

Sprengel, Christian (1750–1816), German botanist and teacher whose studies of sex in plants and theories of fertilization form a basis for much knowledge in this field. He discovered that nectaries, organs that produce NECTAR in flowers, have specific colours which attract insects. Insects, he determined, convey POLLEN to the PISTILS of certain flowers from the STAMENS of others.

Spring, Howard (1889–1965), British novelist who produced his first novel *Shabby Tiger* in 1934. His other novels include *O Absalom!* (1938), *Fame is the Spur* (1940) and *I Met a Lady* (1961).

Spring, season of the year between winter and summer. It is calculated astronomically from the vernal EQUINOX (20 or 21 March) to the summer SOLSTICE (21 or 22 June). It is the season of planting and growing, and is celebrated in many countries by festivals based on fertility.

Spring, natural opening for discharge of water from an underground source. Springs are an important part of the water cycle. They may emerge at points on dry land or in beds of streams or ponds. The composition of spring water varies with the surrounding soil or rocks.

Springbok, or springbuck, small horned ANTELOPE native to southern Africa; the national emblem of South Africa. The reddish-brown colour on the back shades into a dark horizontal band just above the white underside. Height: to 80cm (3ft) at the shoulder. Family Bovidae; species *Antidorcas marsupialis. See also* NW p.*201.*

Springboks, name given to South Africa's national rugby union team, but also used of South African teams in other sports. On the playing field, the Springboks have a reputation for strong forward play, but they have also produced many fast-running backs. In the 1970s, political pressure on their traditional rivals prevented several Springbok tours abroad, although the British Lions, New Zealand All Blacks, and French rugby teams visited South Africa.

Springer spaniel, also called English springer spaniel, one of the oldest breeds , originally trained to spring game ; today it is more commonly used as a gun dog. It has a square head with a strong muzzle and long pendulous ears. The straight compact body is set on medium-length feathered legs. The close wavy coat may be brown and white or black and white. Height: to 51cm (20in); weight: to 23kg (50lb).

Springfield, state capital of Illinois, 298km (185 miles) sw of Chicago and founded in 1818. LINCOLN is buried at Oak Ridge cemetery. The centre of a fertile farming area, its industries include machinery, electronics and fertilizers. Pop. (1970) 91,753.

Springing line, in architecture, the point dividing the bottom of an arch from its supporting columns. The first block of the arch above the springing line is known as a springer.

Spring line, geological term describing the emergence of a number of springs above the contact point of a permeable rock strata and an impermeable barrier. In England it is particularly common on the NORTH and SOUTH DOWNS. *See also* PE p.109.

Springtail, tiny, wingless insect found throughout the world. There are about 2,000 species, and they live in damp places such as rotting vegetation and under stones and fallen logs. They vary in colour and can jump 7.5–10cm (3–4in) by forcing downwards and backwards using a lever-like, forked tail under the abdomen. Length: 3–10mm (0.1–0.4in). Order Collembola. *See also* NW p.*106.*

Spring tides. *See* TIDE.

Sprinting. *See* RUNNING.

Sprite, small and elusive supernatural being such as a goblin or an ELF. Often portrayed as being mischievous, it was commonly associated with a specific location such as woodland, marsh or lakeland.

Spruce, various evergreen trees, related to firs, native to mountainous or cooler temperate regions of the Northern Hemisphere. Pyramid-shaped and dense, they have angular rather then flattened needles which fall off easily, and pendulous cones.

Spitsbergen; the islands were discovered by the Vikings, but forgotten until 1596.

Mark Spitz showing his record-breaking five gold medals in the 1972 Munich Olympics.

Spoonbills live together in flocks, often in the company of herons and egrets.

Springboks live in large herds and range over arid lands in search of grass.

Sprue

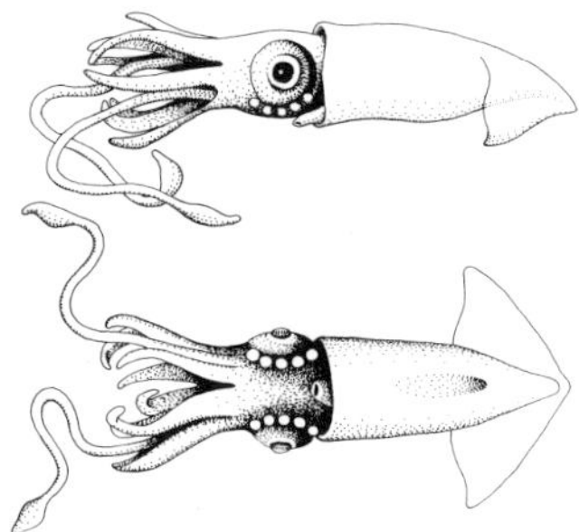

Squids paralyse their prey by squirting them with venom from their salivary glands.

Squirrel monkeys live in large groups, the females leading the members' activities.

Sri Lanka: Kandyan dancers perform before crowds awaiting a religious procession.

Stagecoach; Claire Trevor and John Wayne in the film that made Wayne famous.

The timber is used in cabinet-making, and some species yield TURPENTINE. Height: to 52m (17ft). Family Pinaceae; genus *Picea*. *See also* NW pp.*56–57*.

Sprue, disorder found mainly in E Asia, caused by inadequate diet (insufficient animal PROTEIN, cereal, fruit, vitamins and minerals) and bacterial contamination of the small intestine. Treatment is the administration of FOLIC ACID.

Sprung rhythm, principle of poetic rhythm in which stresses rather than the number of syllables within a line provides the rhythm of the poem. The number of unstressed syllables is not restricted. Gerard Manley HOPKINS is considered to have contributed the principle to English poetry, although sprung rhythm is present in many traditional and popular verses.

Spurge, any of a widely distributed group of herbs, shrubs and trees. Some are SUCCULENTS and others are cactus-like. They exude a milky juice when cut and have small flowers surrounded by large, colourful, flower-like bracts, as in the POINSETTIA. Family Euphorbiaceae; genus *Euphorbia*.

Sputnik, the world's first artificial satellite, launched by the USSR on 4 Oct. 1957. Weighing 83.5kg (184lb) and equipped with a radio transmitter, Sputnik 1 circled the Earth for several months. *See also* SU pp.180, 268, *268*.

Squamata, lizards and snakes, the largest order of reptiles. Although most lizards can be distinguished from snakes by their legs, several species, including the slowworm, are legless. Lizards, however, all have eyelids, which snakes lack.

Square, rectangle with four sides of the same length. Also a square results when a number of quantity is multiplied by itself; the square of 3 is 9, the square of x is x^2. *See also* SU p.28.

Square dance, US folk dance in which an even number of couples, almost always four, form a square, within which framework the dance progresses. The square dance is derived from English country dances and is made up of various patterns called out to the dancers by a "Caller". *See also* QUADRILLE.

Square Deal, phrase used by Theodore ROOSEVELT in 1903 to denote his commitment to fair play between economic interest groups. His policies include new rules that regulated large corporations more closely and limited their political dominance. It became part of the reform programme of the Progressive Party.

Square root, (symbol $\sqrt{}$), number or quantity that must be multiplied by itself to give a specified number or quantity. The square root of 4 is 2, ie $\sqrt{4} = 2; \sqrt{2} = 1.414213$. A negative number has imaginary square roots. *See also* SU p.28.

Squash, any of several species of vine fruits of various shapes, similar in skin texture to pumpkins and marrows, all of which belong to the genus *Cucurbita*. Squashes are native to the Americas, and are cultivated as vegetables. *See also* PE p.*189*.

Squash rackets, ball game played by two or four people on an enclosed, rectangular, four-walled court 9.8m (32ft) long and 6.4m (21ft) wide. The front wall is marked off in three horizontal sections, each divided from the others by a horizontal line. The bottom section, the "telltale", is 47.8cm (19in) high and covered with tin to make a noise when hit by the ball. The next section, extending to 1.8m (6ft) above the floor, is marked by the "cut" line; the service ball must first hit above it. The third line is 4.6m (21ft) above the floor and balls hitting the wall above this are "out". Players use small, round-headed rackets and a small, black or green hard-rubber ball. Balls may bounce off front, side, and back walls but may bounce only once on the floor before being struck. Points are gained when the ball bounces more than once there. Only the server can score points and the first player to score 9 points wins.

Squid, any of numerous species of marine cephalopod MOLLUSCS that have a cylindrical body with an internal horny plate (the pen) which serves as a skeleton. It has eight short suckered tentacles surrounding the mouth, in addition to which there are two longer, arm-like tentacles that can be shot out to seize moving prey. The largest species, the giant squid (*Architeuthis* spp) may reach 20m (65ft) in length. Class Cephalopoda; order Teuthoidea. *See also* NW pp.94–95, *240*.

Squint. *See* STRABISMUS.

Squire. *See* ESQUIRE.

Squirrel, any of numerous species of primary arboreal, diurnal rodents distributed throughout the world. Species of Eurasia, the USA and South America, such as the common grey squirrel, red squirrel and flying squirrels, are the best known. Most species feed on nuts, seeds, fruit, insects, and some eat eggs and young birds. Most have short fur and characteristically bushy tails. Family Sciuridae. *See also* NW pp.*160, 206, 209, 212–215, 249*; PE pp.238–239.

Squirrelfish, any of several species of fish that live throughout the world in tropical seas. Bright red, often with white streaks or spots, they have large eyes and sharp spines on their gill covers and fins. Length: to 61cm (2ft). Family Holocentridae.

Squirrel monkey, either of at least two species of small, diurnal, arboreal MONKEYS of tropical America; it has thick, dark fur and a long, heavy tail. *Saimiri sciureus* has a cap of greyish fur; *S. oerstedi* has a black cap and reddish fur on its back. Both are gregarious and live primarily on fruit, but also eat insects and small animals. Length: to 40cm (16in); tail: 47cm (19in). Family Cebidae.

Sri Lanka, island republic in the Indian Ocean, 32km (20 miles) SE of India; formerly Ceylon. Much of the island is forested or grassland. The economy is largely agricultural, tea, rubber and coconuts being exported. Graphite is the chief mineral export. The capital is Colombo. Area: 65,610sq km (25,332sq miles). Pop. (1975 est.) 13,986,000. *See also* MW p.160.

Srivijaya Empire, BUDDHIST realm in the Malay Archipelago during the 7th–13th centuries. It began on Sumatra but expanded to control the Strait of Malacca. It eventually controlled the international sea trade, establishing commercial relations with China and India, and its capital PALEMBANG was famous as a centre of Buddhist learning and culture.

SS, abbreviation of *Schutzstaffel*, an élite corps of the Nazi Party. Headed by Heinrich HIMMLER, it was originally responsible for domestic security, but its powers after 1936 were enlarged and were absolute throughout German-occupied Europe during WWII.

Ssu-ma Kuang (1019–86), Chinese scholar, statesman and historian of the SUNG DYNASTY who opposed the radical schemes of WANG AN-SHIH. His history of China from 403 BC to AD 959, *Comprehensive Mirror for Aid in Government*, is one of China's greatest historical works.

Stabilator, aircraft, symmetrical aerofoil hinged at its centre for longitudinal control of an aircraft. It is used in place of the stabilizer-elevator system that was found to be ineffective in the transonic speed range. The stabilator also operates efficiently at low speeds.

Stabilizers, in aeronautical engineering, the vertical and horizontal fins usually placed together in the form of a tailplane to prevent erratic rolling and pitching respectively. Like the wings, the horizontal fins have an aerofoil shape and can provide lift at take-off. In flight, however, they act only as stabilizers and in some designs the air thrust on them acts downwards. In marine engineering, fins or tanks are used to keep a submarine or ship steady in rough seas.

Staël, Anne-Louise-Germaine, Madame de (1766–1817), French author whose salon became an intellectual and political centre. She criticized NAPOLEON I and was exiled several times. Her novels *Delphine* (1802) and *Corinne* (1807) are considered among the first FEMINIST novels.

Staël, Nicholas de (1914–55), French painter born in Russia. Best known as an abstract, CUBIST-influenced artist and noted for his use of rich colours, in his later work, though still abstract, he used forms discernibly inspired by landscape and other natural phenomena.

Staffa, small uninhabited island in the Inner Hebrides off the NW coast of Scotland, in Strathclyde Region. It has high cliffs and many caves, among them the famous Fingal's Cave. The island is 1.2km (0.75 miles) long and 0.5km (0.33 miles) wide.

Stafford, county district in W central STAFFORDSHIRE, England; created in 1974 under the Local Government Act (1972). Area: 599sq km (231sq miles). Pop. (1974 est.) 114,300.

Staffordshire, county in W central England. The terrain is composed of rolling hills with moorlands in the N. The region is drained chiefly by the River Trent. Cattle-raising is the chief farming activity, but the county is primarily industrial. It includes the POTTERIES around Stoke-on-Trent, noted for its manufacture of china, and the BLACK COUNTRY, where coal is mined and iron and steel are produced. Stafford is the county town. Area: 2,716sq km (1,049sq miles). Pop. 997,600.

Staffordshire bull terrier, fighting dog that was originally bred in England in the 19th century. It has a short, broad head with small ears and dark eyes. The wide-chested body is set on strong well-boned legs and the tail is short and tapered. Its smooth, short coat may be red, fawn, black or blue, with or without white markings. Height: to 40cm (16in) at the shoulder; weight: to 17kg (38lb).

Staffordshire Moorlands, county district in NE STAFFORDSHIRE, England; created in 1974 under the Local Government Act (1972). Area: 576sq km (222sq miles). Pop. (1974 est.) 93,200.

Stag beetle, large, brown or black BEETLE of Eurasian oak forests; the male bears large antler-like mandibles. The larvae, which take about three years to mature, feed on rotten wood. Length: to 8cm (3in). Family Lucanidae; species *Lucanus cervus*. *See also* NW p.*114*.

Stage, in the theatre, the space in which actors perform before an audience; it generally, but not always, consists of a raised platform. The traditional, PROSCENIUM arch stage has lost some popularity in favour of more OPEN STAGES.

Stagecoach (1939), US film. Directed by John FORD from a screenplay by Dudley NICHOLS, it remains one of the most influential of western films. Made with historical accuracy, it starred John WAYNE, Claire Trevor, Thomas Mitchell and Donald Meek.

Stage-coach, heavy, horse-drawn, four-wheeled vehicle used to transport goods and passengers regularly between two or more points. Such coaches were used in England and France in the 17th century and were more widely used in the first half of the 19th century when the quality of roads improved. The coaches were usually drawn by four or six horses, which were changed at each station, or stage, along the route. *See also* MM pp.120–121.

Stage machinery, in the theatre, devices used to produce rapid changes of scenery, lighting and other effects. Modern stage machinery has developed since the Italian Renaissance and may involve a large revolving stage, or one that is hydraulically controlled to allow it to be raised, lowered, moved sideways or inclined at an angle. Despite this sophistication, however, the 20th century has also witnessed a movement towards simplicity and informality in productions, as epitomized by THEATRE-IN-THE-ROUND and street theatre.

Staggers, disorder of horses, cattle and sheep. Symptoms include unco-ordination, convulsions and paralysis. There are various causes of this disorder. It is often due to a tapeworm in the brain of sheep or to mineral deficiency in horses.

Staghound, breed of English hunting dog related to the St Hubert hound and to the English foxhound. It has a hard dense coat coloured black, white and tan. Height: to 58cm (23in).

Stagnation, in economics, situation in which the country's economy is sluggish, usually characterised by unemployment, inflation and low production.

Stained glass, dyed or superficially coloured glass used for decorative or pictorial effect in windows or other apertures. Most commonly found in church architecture, it was brought to its full flower furing the Gothic period, particularly in France. Of Byzantine origin, stained glass is produced in its purest form by impregnating the components of glass with coloured dyes and arranging shapes cut from the resultant sheets to form patterns or images. These shapes are joined and supported by flexible strips of lead that creat dark emphatic contours. Details are painted on to the glass surfaces in liquid enamel and fused to the surfaces by heat. The earliest complete surviving stained glass is to be found in AUGSBURG Cathedral, Bavaria, (1050–1150) In c.1150 stained-glass windows were created for the Abbey of Saint-Denis, Paris, and shortly after the great west window of Chartres Cathedral was installed. The Île de France soon became the art's great centre, and its influence spread throughout NW Europe, particularly to England. The art of stained glass was revived in the 19th century and in modern times concrete has supplanted the more graceful lead as a support for the glass. *See also* HCl pp.*199, 216*; MM p.*44*.

Stainer, Sir John (1840–1901), British organist, composer and scholar. He was organist of St Paul's Cathedral (1872–88) and wrote much sacred music, including the oratorio *The Crucifixion* (1887).

Stainless steel, alloys of iron that resist tarnishing. Besides carbon, contained in all steels, they contain metals such as nickel and chromium, which are the ingredients that make them rust-proof and resistant to chemical attack. Stainless steel is used to make utensils and cutlery, chemical reactors, and fittings of various kinds including trim for motor cars. Typical stainless steel alloys are known by the descriptions 18/8 and 18/8/2, these being the percentages, respectively, of chromium, nickel and molybdenum. *See also* MM pp. 30–31.

Stair, Sir James Dalrymple, 1st Viscount (1619–95), Scottish statesman, lawyer and writer. He was favoured by CHARLES II, becoming president of the Court of Session (1671–79). *The Institutions of the Law of Scotland,* which he completed in 1681, is a distinguished treatise on Scottish law.

Stair, John Dalrymple, 1st Earl of (1648–1707), Scottish statesman. He was joint Secretary of State for Scotland (1691–95), but was forced to resign for his authorization of the massacre at GLENCOE.

Stakhanovism, Soviet economic policy. It emphasized "socialist competition" in the accelerated industrialization of Josef STALIN's five-year economic plans in the 1930s. It was named after a coal miner, Grigoriyevich Stakhanov who in 1935 was said to have considerably exceeded his quota by mining 102 tonnes of coal.

Stalactite, icicle-like formation of calcite found hanging from the roofs of caves. It is made by the precipitation of LIMESTONE out of water that has seeped into limestone caves.

Stalagmite, deposit of crystalline CALCIUM CARBONATE rising from the floor of cavern, and formed by dripping water.

Stalin, Joseph Vissarionovich Dzhugashvili (1879–1953), Soviet Communist leader and head of the USSR. Educated at the Tiflis theological seminary, from which he was expelled in 1899 as a revolutionary, he became a BOLSHEVIK in 1903. From 1913 he began to use the nickname "Stalin" (man of steel) while editor of the party newspaper, *Pravda.* He played an important role in the Civil War, as Commissar for Nationalities (1919–23). Following LENIN's death in 1924, Stalin was able to manoeuvre for supreme power suppressing his main rival TROTSKY and becoming by 1927 virtual dictator. He introduced forced industrialization and COLLECTIVIZATION, and began the purges of the 1930s to consolidate his power over the state; by 1938 more than 10 million people were believed to have been killed and Stalin controlled the party, the government and the police. In 1941 he became premier and personally supervised the Soviet WWII war effort. His personality cult reached its peak in the postwar era when purges were revived, and his last years were characterized by increasing fear and suspicion. *See also* HC2 pp.197–199, 240.

Stalingrad. *See* VOLGOGRAD.

Stalingrad, Battle of (1942–43), decisive conflict in WWII. In August the Germans attacked the Soviet city of Stalingrad, which stubbornly withstood months of artillery shelling. With great losses on both sides and Stalingrad reduced to rubble, a Soviet counter-offensive surrounded the German 6th army and forced it to surrender in February 1943. The battle marked the turning point in German fortunes on the eastern front, *See also* HC2 pp.*228,* 229, *230.*

Stall, aircraft's abrupt loss of lift. It occurs when the angle of attack increases to such a point that the flow of air over the top of the wing is less that than that below it. Thus the aircraft loses its aerodynamic lift. *See also* MM pp.152–153.

Stamen, pollen-producing male organ of a flower. It consists of an anther and filament. *See also* PISTIL.

Stamford Bridge, Battle of (1066), English victory under HAROLD II over the Norwegians under HARALD HARDRADA and the Northumbrian earl, TOSTIG, at the crossing of the river Derwent north-east of York. Hardrada and Tostig were killed. While Harold was in the north, WILLIAM THE CONQUEROR landed on the south coast. *See also* HC1 166–167.

Stamitz family, family of Bohemian musicians. Johann (1717–57) was concert master of the Court orchestra at Mannheim from 1745. By writing more than 70 symphonies and many concertos – including the first known for clarinet – he was significant in establishing both musical forms. Of his two sons Carl Phillip (1745–1801) was a virtuoso violin and viola player, and a prolific composer too.

Stamp Act (1765), first direct tax levied on the American colonies by the British Parliament. It was devised to raise revenue to provide for troops defending the colonies. The Act required a stamp to be affixed to all legal papers, newspapers, advertisements and other documents. Colonists resisted the tax, attacking the stamp distributors and burning stamped paper. The Act was repealed in 1766. *See also* HC2 p.64.

Stamp duties, tax upon paper used in legal and commercial papers, first levied in Britain in 1670. The most common instance of a stamp duty in Britain today is the one required for a television licence. *See also* HC2 p.64.

Stamping, method of forming metal. A flat blank of metal is stretched over a die and struck with a movable punch into the desired shape. *See also* MM p.42.

Stamp, postage, adhesive token which represents payment for sending a letter or parcel by post. Prepayment of postal charges with stamps started with the British Penny Post (one old pence for each half ounce) in 1840; previously, postal charges had been determined by distance and number of sheets. Postage stamps have long been collector's items.

Stamps, trading, sales scheme whereby a shop offers customers "stamps" in accordance with the amount of money they have spent; the stamps are collected and exchanged for goods. The practice originated in the USA in the late 19th century, and trading stamps became popular in Britain in the 1960s. The British Trading Stamps Act (1964) contains provisions to safeguard the public to whom trading stamps are offered.

Standard, Battle of the (1138), English victory over the Scots at Cowton Moor, Yorkshire. The English were preceded by a mast carrying the banners (standards) of St Peter of York, St John of Beverley and St Wilfred of Ripon.

Standard (reference) electrode, in electrochemistry, a half-cell combined with other half-cells and used as a reference. The basic standard is the hydrogen electrode – a platinum electrode immersed in a molar solution of hydrogen ions and covered with an absorbent layer of platinum black, over which is bubbled molecular hydrogen. This is taken to have a ELECTRODE POTENTIAL of zero.

Standard of living, material well-being of a country measured by the per capita income of the population. It is used to compare the degree of economic development in various countries.

Stanford, Sir Charles Villiers (1852–1924), Irish composer. He composed prolifically in most areas of music and is famous chiefly for his settings of the Anglican services.

Stanford-Binet scales, intelligence tests derived from the first such tests devised by Alfred BINET in 1905. The present Stanford-Binet measures IQ from the age of three and is principally used to predict potential for success in school.

Stanhope, Lady Hester Lucy (1776–1839), British traveller, niece of William PITT. In 1810 she left England, settled among the DRUSES, a Syrian religious sect in Lebanon, and assumed the customs and religion of ISLAM.

Stanislavsky, Konstantin (1863–1938), Russian actor, director and teacher, real name Konstantin Sergeyevich Alekseyev. His teaching, which came to be called the METHOD, stressed ensemble acting with each actor making an exhaustive study of his character's background and psychology, identifying with the inner, emotional life of the character. He was most successful with the works of CHEKHOV and GORKY. *See also* HC2 p.101; MS p.239.

Stanley, Edward, 14th Earl of Derby. *See* DERBY, EDWARD GEORGE GEOFFREY SMITH-STANLEY.

Stanley, Sir Henry Morton (1841–1904), British explorer and journalist, b. John Rowlands. He was sent by the New York *Herald* to central Africa to find David LIVINGSTONE. This he accomplished in 1871. A courageous and tenacious explorer, Stanley made several other expeditions into Africa, making important geographic discoveries and establishing the Congo Free State for Belguim. *See also* HC2 pp.*142,* 143, *154.*

Stanley, Thomas. *See* DERBY, THOMAS STANLEY, 1ST EARL OF.

Stanley, Wendell Meredith (1904–71), US biochemist who shared the 1946 Nobel Prize in chemistry with John NORTHROP and James SUMNER for his research into viruses. In 1935 Stanley crystallized tobacco mosaic VIRUS and showed that it consists of PROTEIN and NUCLEIC ACID MOLECULES in a rod-shaped form. This enalbed the exact molecular structure and means of proliferation of this and other viruses to be determined. He also studied Influenza viruses and developed a VACCINE against the disease.

Stanleyville. *See* KISANGANI.

Stannaries, tin mines and mining towns in Cornwall Stannary Courts regulated trade and all offences, except those concerning life, limb or land, committed by the tin miners. The jurisdiction of these courts was abolished in 1897.

Stansgate, Viscount. *See* BENN, ANTHONY WEDGEWOOD.

Stanton, Elizabeth (1815–1902), US leader of the women's rights movement, which was formed in 1848 to demand suffrage for women in the USA.

Stanza, arrangement of lines to form a verse of a poem. Stanzas are characterized by the number of lines and type of metre, eg SPENSERIAN STANZA. A couplet has two lines; tercet and terza rima, three; quatrain, four, and ottava rima, eight.

Staphylococcus, common genus of bacteria belonging to a group of oval or spherical bacteria called *cocci.* They grow in grape-like clusters and live in air, water and in the bodies of human beings and animals. Pathogenic staphylococci cause a range of disorders, including food poisoning, pneumonia and SEPTICAEMIA. They may be destroyed by ANTIBIOTICS, although some strains have become resistant to such measures.

Staple, in medieval English history, a commodity, trade in which was channelled through one town in order to facilitate its control and the collection of duties. The most important was the wool staple. Nine

Stained glass scenes in medieval churches helped teach Bible stories to the peasants.

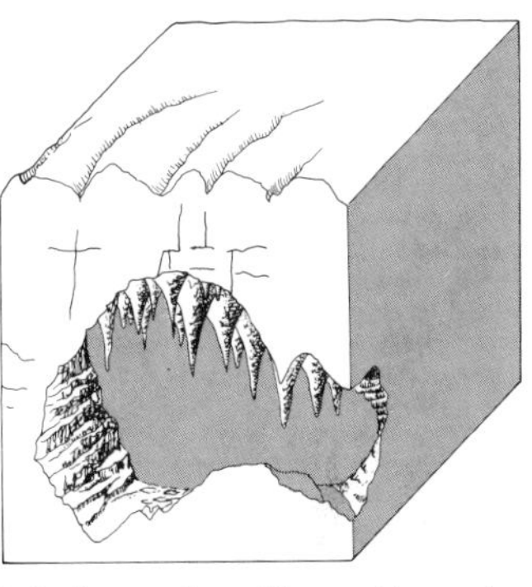

Stalactites are formed from calcium carbonate deposited when water evaporates.

Joseph Stalin allowed no one to oppose him and ruled by terror during his dictatorship.

Henry Morton Stanley, photographed in the clothes he wore when he met Livingstone.

Star

Star Chamber: the court is believed to have originated in medieval times.

Starfishes are commonly yellow, orange, pink and red. There are about 2,000 species.

Starlings were introduced from Europe and Asia to most countries of the world.

Star of David's origins are thought by some to be medieval and cabbalaistic.

English, two Welsh and three Irish towns had been made wool staples in 1326 but the monopolies of these, which included York, Lincoln and Bristol, were abolished in 1328. The wool trade was controlled by the Company of the Merchants of the Staple until exports declined in the 16th century.

Star, hot, self-luminous celestial body composed of gas and held together by its own GRAVITATION. Most stars emit all forms of ELECTROMAGNETIC RADIATION, principally light and heat rays, produced through nuclear reactions occurring in the star's interior. The size of stars varies greatly: their diameters range from several hundred million kilometres for low-density red giant stars to earth-like dimensions for very dense white dwarfs. A neutron star may have a diameter of only a few kilometres. Stellar temperatures also vary enormously, from about 2,000°K to 80,000°K, and stars are grouped according to their luminosity and spectral type. Stars are classed according to their apparent brilliance or magnitude. Their apparent brightness differs because they lie at different distances from the observer or because their intrinsic brightness varies. The spectral classification groups stars on the basis of distinctive features in their spectra. The most commonly used system is the M–K system based on the Harvard Classification. This categorization arranges stars in order of decreasing surface temperature and according to a colour gradient from blue to red. Each class is characterised by distinctive absorption lines in the spectrum representing specific elements and radicals. The six main types are B, A, F, G, K and M-type stars. There are five more groups of rarer type, W, O, R, N and S, which define somewhat different spectral characteristics. When stars are plotted on a graph according to their luminosity and their spectral class (see HERTZSPRUNG-RUSSELL DIAGRAM) most of them lie in a well-defined belt known as the main sequence. The Sun is a typical main sequence star.

Stars are believed to originate in nebulae or concentrations of interstellar material. In an embryo star the material shrinks as a result of gravitational attraction and heat is generated. The course of stellar evolution depends largely on the mass of the star in this initial stage. If the mass is extremely low no nuclear reactions take place and the star radiates feebly until its energy has been dissipated. With a star of a mass similar to that of the Sun, the material shrinks and heat generated at the core is carried to the surface by convection and the star shines brightly. When the core temperature is high enough nuclear reactions begin. Hydrogen nuclei combine to form helium nuclei, resulting in a loss of mass and the release of energy. The star remains stable at this stage (it is now a main-sequence star) for millions of years. The Sun is at this stage in its evolution. When the supply of hydrogen runs low the core heats up again while the surface layers expand and cool. The star is now a red giant. Further reactions follow but eventually there is no nuclear energy left and the star collapses into extremely dense white dwarf. After a very long period, all light and heat has been extinguished and the star becomes a dead black dwarf. A star with greater mass than the Sun evolves much more quickly. Such stars do not collapse into white dwarfs but undergo great changes in structure, producing very high temperatures in the outer layers, and the result is a SUPERNOVA outburst. The stars that we see with the naked eye belong to our GALAXY, the MILKY WAY. They are not evenly distributed but often occur in STAR CLUSTERS. Many stars that appear as a single point of light are in fact DOUBLE STARS or even multiple star systems. VARIABLE STARS do not shine constantly and their luminosity fluctuates either regularly or irregularly. See also SU pp.220–253.

Star, sea. See STARFISH.

Star Carr, archaeological site in N Yorkshire, England. Remains of a lake-side community have been found there, dating back to the 8th millennium BC. See also HC1 p.42.

Starch, form of POLYSACCHARIDE in which CARBOHYDRATE is stored in many plants, becoming an important source of carbohydrate in the diet of animals. Consisting of linked GLUCOSE units it exists in two forms: amylose, in which the glucose chains are unbranched, and amylopectin, in which they are branched. It is made commercially from cereals, potatoes, etc, and used in the manufacture of adhesives and foods. See also SU p.156.

Star Chamber, meeting place of the king's council in Westminster Palace, London. Under the Tudors, it was a royal council constituted as a judicial tribunal and became unpopular for its arbitrary sentences passed without juries and its use of torture. It was especially detested by the opponents of CHARLES I and abolished in 1641. Its name derived from the stars decorating its ceiling.

Star cluster, group of genuinely associated stars moving together through space and having common characteristics. Star clusters are of two types. Globular clusters are symmetrical in shape and contain hundreds of thousands of POPULATION II stars. OPEN CLUSTERS vary greatly in the number of stars they contain and are usually made up of relatively young stars, sometimes surrounded by interstellar material. See also SU pp.242–243.

Starfish, any of numerous species of marine ECHINODERMS with a central disc body and 5 to 25 radiating arms. The mouth is on the underside of the disc and the stomach can be extruded to take in other echinoderms and shellfish. Calcareous spines are embedded in the skin. Starfish move by means of tube feet. Class Asteroidea. See also NW pp.72, 118–119, 237.

Stargazer, any of several species of bottom-dwelling marine fish, the eyes of which are located on top of a large flat head. The electric stargazer (family Uranoscopidae) lives in warm and temperate seas throughout the world; species may have shoulder spines or electric organs. Length: to 55cm (22in). The sand stargazer (family Dactyloscopidae), a much smaller fish, is found in tropical waters. Length: to 10cm (4in).

Stark, Dame Freya Madeline (1893–), British explorer and writer, who travelled widely in the Middle East, especially Arabia. Her explorations in the Hadramaut resulted in her famous book *The Southern Gates of Arabia* (1936). *The Valley of the Assassins* (1934) describes her journeys in Persia (Iran).

Stark, Johannes (1874–1957), German physicist. He received the 1919 Nobel Prize in physics for describing the STARK EFFECT, which he discovered in 1913, and for his work on the DOPPLER EFFECT.

Stark effect, in physics, splitting of the spectral lines of an ATOM or MOLECULE when it is placed in a strong electric field. It was discovered by Johannes STARK.

Starley, James (1831–81), British inventor of sewing machines and bicycles who worked in Coventry from 1857. In 1868 he produced his first bicycle (*The Coventry*). His "ordinary" or "penny farthing" bicycle of 1871 had centre pivot steering. He introduced the *Coventry* tricycle in 1876, adding differential gears in 1877. See also MM pp.122, *123.*

Starling, Ernest Henry (1866–1927), British physiologist who contributed greatly to the knowledge of body functions. He made major discoveries about body tissues, the mechanical controls of the heart and the endocrine system.

Starling, any of several species of small, aggresive birds; exotic colourful species are found throughout the world. The common Eurasian starling, *Sturnus vulgaris*, is mottled black and brown. It feeds on the ground on insects and fruit, often damaging crops, and nests in large noisy colonies; it has been known to imitate the cries of other birds. Length: to 36cm (14in). Family Sturnidae.

Star of Africa, largest cut diamond in the world; it was cut from the Cullinan diamond in 1908. It weighs 530.2 carats (3¾oz) and is now in the British sovereign's scep-tre in the Crown Jewels. See also PE p.99.

Star of Bethlehem, bulbous houseplant native to South Africa. Fragrant, star-shaped, white flowers grow on a central spike rising from slender, arching leaves. It grows best in direct sunlight and slightly dry soil (equal parts of loam, peat moss and sand). Propagation is by cuttings or seeds. Height: to 46cm (18in). Family Liliaceae; genus *Ornithogalum.*

Star of David (Magen David, or Shield of David), six-pointed device formed by opposing two equileral triangles. Used as an emblem by pagans, Christians and Muslims, it has evolved into a symbol of JUDAISM. Its origin is unknown but it is often traced to King DAVID; it was used by Jewish warriors on their shields.

Starr, Ringo (1940–), British musician, real name Richard Starkey. He was the drummer of the BEATLES group, and turned to film acting when the group split up.

Star-spangled Banner, The, US national anthem. The words were written by Francis Scott KEY. The music was taken from "To Anacreon in Heaven", a song by the English composer John Stafford Smith. Although long sung by the US Army and Navy, it did not officially become the national anthem until 1931.

Starvation, prolonged deprivation of food. In human beings it first results in the using-up of glycogen, a carbohydrate stored mainly in the liver. After this reserve has been depleted, fat and protein are progressively broken down to supply energy. As protein is broken down, wasting and weakness of the body develop. Death can be expected when normal body weight is halved. See also ANOREXIA; MALNUTRITION; MS pp.70–71, 108, *109.*

State, Department of, government department in the USA charged with developing and maintaining foreign policy. Originally, the CONTINENTAL CONGRESS conducted foreign affairs. In 1777 a Committee for Foreign Affairs was established but its power was limited. The Department of Foreign Affairs was established in July 1789 but renamed the Department of State in September of that year, Thomas JEFFERSON being appointed its first Secretary in 1790. The department is primarily concerned with the execution of US foreign policy to promote long-term US security and well-being through continuous consultations with other nations. It also negotiates treaties and speaks for the USA in the UN. It is headed by a Secretary of State and is responsible to the President.

Staten Island, island in SE New York State, in New York Bay, 8km (5 miles) SW of Manhattan; separated from Long Island (E) by the Narrows and from New Jersey (W) by the nmrrow Arthur Kill; it is connected with Brooklyn by the Verrazano-Narrows Bridge (1964) and with New York City and Brooklyn by ferries. The island was first visited by Henry HUDSON in 1609 but hostile Indians prevented permanent settlement until 1661. Industries: shipbuilding, printing and publishing, oil refining, soap. Area: 166sq km (64sq miles). Pop. (1970) 295,443.

State of nature, fiction of political theory to describe the condition of man before the advent of any social organization. Thomas HOBBES was an early exponent of the concept as a starting point for his study of society, but where he posited a state of nature that was "nasty, brutish and short", ROUSSEAU claimed that the state of nature was more idyllic than society.

State of the Union Message, annual address given by the US President to CONGRESS, usually before the start of a new session. It is based on the constitutional requirement that the President shall keep Congress informed. The President usually uses the speech to outline his future legislative programme.

State Paper Office, British government department in which official records were kept. It was formed in the reign of ELIZABETH I when a collection of papers of state was installed at the Palace of WHITEHALL. In 1852 state papers were placed in the care of the MASTER OF THE ROLLS and by 1856 the State Papers Office had been absorbed by the PUBLIC RECORD OFFICE.

HC1 = History and Culture Vol. 1   HC2 = History and Culture Vol. 2   MS = Man and Society   MM = Man and Machines

States General. *See* ESTATES GENERAL.

Statham, John Brian (1930–), British cricketer who played for England and Lancashire. A fast bowler of remarkable accuracy, he played in 70 Test matches (1951–65), taking 252 wickets for an average of 24.84 runs. In his first-class career (1950–68) he dismissed 2,260 batsmen at an average of 16.36 runs each.

Static electricity, positive and negative electrical charges which are studied in the branch of physics known as ELECTROSTATICS. Any body with either an excess or a deficit of ELECTRONS is electrically charged. Two objects with a deficit (positive +) will repel each other if brought together, as will two excess (negative −) charges. Two objects with different charges will attract each other. COULOMB's LAW describes the forces that charged bodies have on each other by relating the force to their charge and the distance between them. Static electricity may be produced by friction, eg by combing hair. The hair acquires a negative charge. Similarly, rubbing nylon material will produce static. *See also* SU pp.73, 112–113.

Statics, branch of MECHANICS which deals with the action of forces on objects, the forces being so arranged that the objects are at rest. Its topics include finding the resultant (net) of two or more forces; centres of gravity; moments; and stresses and strains. *See also* DYNAMICS; SU pp.70–71.

Stationers' Company, craft guild of London publishers, printers and booksellers. Formed in 1403 and chartered in 1557, it had a monopoly of publishing until 1694 and acted as the Tudors' and Stuarts' agency for censorship. Compulsory registering of all publications at Stationers' Hall ended in 1923. In 1933 it became the Stationers' and Newspaper Makers' Company.

Stationery Office, HM (HMSO), department of the British CIVIL SERVICE. It supplies the public service with paper, printing, bookbinding and office machinery, and publishes printed matter for sale in its own bookshops and through various agencies. It also publishes HANSARD.

Stations of the Cross, Roman Catholic devotion to commemorate events on the day CHRIST was crucified. Worshippers pray in turn at 14 pictures or sculptures showing the sufferings of Jesus from His sentence by PILATE to His death and burial.

Statistical mechanics, study of large-scale properties of matter based on the underlying laws of QUANTUM MECHANICS in conjunction with the statistical laws of large numbers. Quantum mechanics determines the possible energy states of a substance or system; the large number of molecules in such a system then allows the use of statistics to predict the probability of finding the system in any one of these states. The ENTROPY of the system is related to its number of possible states; a system left to itself will tend to approach the most probable distribution of energy states. *See also* THERMODYNAMICS; SU p.92.

Statistics, science of collecting and classifying numerical data. Statistics can be descriptive (simply summarizing the data obtained) or inferential (leading to conclusions or inferences about larger numbers of which the data obtained is a sample). Inferential statistics are used to give a greater degree of confidence to conclusions, since statistics make it possible to calculate the probability that a conclusion is in error. *See also* CORRELATION; FACTOR ANALYSIS; SAMPLING; VALIDITY; MS p.311; SO pp.56–57.

Statue of Liberty, large copper statue of a woman, standing on Liberty Island in New York Harbour, and a symbol of American democracy. It was built to commemorate the 1876 centenary of US independence and was designed by Frédéric-Auguste BARTHOLDI. The supporting framework was designed and built by Alexandre-Gustave EIFFEL. The statue's correct name is *Liberty Enlightening the World*. It stands 46m (151ft) tall to the top of the torch in the uplifted right hand.

Status, in sociology, a person's social position in a hierarchically arranged society, in which position is based on various factors including lifestyle, prestige, income and education. Status can be inherited or derive from such things as socially enviable possessions (status symbols), valued talents, or friendships with people of high status. A status group is any community of individuals who share the traits of a particular status. *See also* MS p.285.

Statute, law in Britain passed by both Houses of Parliament and given the royal assent. In origin it was simply any written document which altered or confirmed the COMMON LAW. In the 12th century a statute was thus a royal edict. But as Parliament developed in the 13th and 14th centuries it came to be held that for a statute to have force it must receive the consent of Lords and Commons, even though it was rare for it to originate except from the monarch.

Staudinger, Hermann (1881–1965), German chemist, awarded the 1953 Nobel Prize in chemistry for his work on the structure of PLASTICS, which had been thought to be random aggregations of MOLECULES. He showed that such plastics are POLYMERS, similar to CELLULOSE, in which the molecules are bonded together to form a long chain. This discovery contributed to the growth of the modern plastics industry.

Staurolite, silicate mineral, iron magnesium aluminum silicate, found in SCHISTS and GNEISSES. It has orthorhombic system, prismatic and flattened crystals and is glassy brown in colour. Hardness 7–7.5; s.g. 3.6.

Stavanger, port in sw Norway, on the s shore of the Boknafjord. Founded in the 8th century, Stavanger was an episcopal see from the 11th century until 1682 and there is a 12th-century cathedral in the city. Industries: shipbuilding, fishing and fish processing. Pop. (1971 est.) 82,000.

Stave, or staff, in MUSICAL NOTATION, the parallel horizontal lines on or between which notes are written. It was introduced, probably by the monk Guido d'Arezzo (*d.* 1050), who fixed the written representation of notes on four lines. The number of lines in a modern stave is five.

Stawell Gift, Australian professional sprinting race, held annually in the small mining town of Stawell, Victoria. First contested in 1878, it is raced over 130 yards (118.8m) on a handicap basis, a series of heats producing five runners to contest the final.

Stay of proceedings, in law, postponement or cancellation of civil proceedings by a court order.

Stead, Christian Karlson (1932–), New Zealand poet, novelist and critic. His works include *Crossing the Bar* (1972), *The Shed* (1975) and *Tne New Poetics* (1975).

Stead, Christina Ellen (1902–), Australian novelist whose powerful works include *The Man Who Loved Children* (1940), *The People With the Dogs* (1951), *Little Hotel* (1974) and *Miss Herbert* (1976).

Stead, William Thomas (1849–1912), British radical journalist who successfully campaigned against child prostitution in 1885. He died in the *Titanic* disaster.

Steady-state theory, in COSMOLOGY, theory concerning the origin and evolution of the universe. Like the BIG BANG theory, it proposes that the universe is expanding but differs in suggesting that new galaxies are continually created to keep the universe at a constant density. Unlike the big bang theory such a universe would have no beginning. It was first proposed by Sir James JEANS around 1920 and developed after WWII by Sir Fred HOYLE, Thomas Gold and Hermann Bondi. Recent observations in astronomy have produced evidence that is more in favour of the big bang theory.

Steam, gaseous form of water. As commonly seen in the home it contains minute drops of water and is technically called wet steam. The pure gas, dry steam, is obtained by heating steam at temperatures well above the boiling point of water, 100°C (212°F). It is widely used in industry for heating, transporting heat (in heat exchangers) and to drive turbines.

Steam engine. *See* ENGINE.

Steam hammer, heavy hammer powered by steam, invented in 1839 by James NASMYTH to forge the largest metal component to that date, namely the paddle wheel of BRUNEL's steamship *Great Britain*.

Steam pump, apparatus used in factories to circulate liquids such as boiler feedwater and slurries. Steam pumps usually have a steam-powered piston at one end, connected to another piston fitted with a plunger which does the pumping.

Steamroller, heavy road vehicle fitted with a large frontal roller, and propelled by STEAM. Formerly used in roadbuilding, steamrollers have been largely superseded by diesel or petrol driven rollers. *See also* MM p.182.

Steamships, vessels powered by some form of steam engine. They first appeared as French paddle boats of the late 18th century. At the turn of the 19th century larger steamboats of this kind were built in England and the USA to carry river passengers. The first transatlantic crossing by a steamship took place in 1819, although this was assisted by sail. By the 1840s steamships of more than 1,000 tonnes were carrying passengers regularly. They were often built of iron and steel and had coal-fired steam engines; screw propellers began to replace paddles at this time. The next developments were STEAM TURBINE engines (initially for warships) and oil fuel instead of coal. During the second quarter of the 20th century most steamships were superseded by vessels powered by marine diesel engines. *See also* SHIP; MM pp.110–111.

Steam turbine, ENGINE consisting of rotating blades which are turned by steam. TURBINES are commonly used to drive ships and, in tandem with a GENERATOR, to make electricity in power stations. *See also* MM pp.66, 66–67.

Stearate, salt or ester of STEARIC ACID ($C_{17}H_{35}COOH$), one of the most common fatty acids occurring in natural animal and vegetable fats. Stearates, especially the sodium and potassium salts, are used in soaps, lubricants, cosmetics, food additives and pharmaceuticals.

Stearic acid, organic chemical compound commonly found as a glyceride in animal and vegetable fats, together with palmitic acid and oleic acid. It has the formula $CH_3(CH_2)_{16}COOH$ and is used in ointments, creams and soap. Properties: s.g. 0.85; m.p. 70°C (158°F).

Stearin, glyceryl tristearate, $C_{57}H_{110}O_6$. It is present in many animal and vegetable fats, particularly the hard ones such as tallow and cacao butter. Stearin is prepared from STEARIC ACID and glycerol with aluminium oxide as a CATALYST.

Steatite. *See* SOAPSTONE.

Steel, David Martin Scott (1938–), British politician. Liberal MP for Roxburgh, Selkirk and Peebles since 1965, he was elected leader of the Liberal Party in 1976. In March 1975 he announced the formation of the "Lib-Lab Pact" between his party and the Labour government.

Steel, alloy of iron, containing a little carbon, commercially the most important metal. Its malleability distinguishes it from CAST IRON and its freedom of slag from WROUGHT IRON. First made in ancient times, it was not until the development of the BESSEMER and open-hearth processes in the 19th century that steel became available for large-scale uses. Low-alloy steels, with less than 5% of nickel, chromium and molybdenum or other metals added, are exceptionally strong and are used for the structured members of buildings, bridges and machine parts. High-alloy steels, such as stainless steel, with 5% of other metals, are used for tableware and cooking utensils, where appearance and resistance to corrosion are necessary. *See also* MM pp.28–31.

Steele, Sir Richard (1672–1729), British essayist and dramatist, *b.* Dublin. In partnership with Joseph ADDISON he set the civilized tone of the Augustan age in his publications, the *Tatler* (1709–11) and the *Spectator* (1711–12). *See also* HC2 p.34, 34.

Steel engraving, INTAGLIO PRINTING process. Steel plates, unlike copper or

Statue of Liberty; the figure was covered with copper sheeting by designer Bartholdi.

Stavanger is one of the oldest towns in Norway but most buildings are now modern.

Steamships; *L'Elise* arriving in Paris after crossing from London in 1816.

David Steel speaking at the Liberal Party Conference at Scarborough in 1965.

Steel-framed building

Rod Steiger as the gangster Al Capone in the 1958 film *The Al Capone Story*.

Joseph Stella's portrait of Alfred Casella. Many of his works celebrate urban America.

Stendhal expressed the belief that happiness could be achieved through will.

Stephanotis; in winter it develops long branches covered with leaves.

brass, made large numbers of proofs possible, and were popular in the 19th century until superseded by photomechanical processes. *See also* MM pp.204, 207.

Steel-framed building, construction method widely used by architects and developers. PAXTON and EIFFEL pioneered the use of tall cast iron buildings but only with the late 19th-century refinements of mass-produced and standardized steel units could architects such as JENNEY, SULLIVAN and LLOYD WRIGHT develop urban building vertically as large blocks and SKYSCRAPERS. PRE-STRESSED CONCRETE beams further increased the capacity of steel-framed structures to withstand earthquakes and high-altitude winds. *See also* HC2 pp.178–179, 206–207; MM pp.48–49.

Steelmaking, conversion of iron into one of its various alloys known collectively as STEEL. The technique became possible on a large scale only when slag could be removed from the molten iron alloy, and the carbon content carefully controlled. This happened in the mid-19th century with the BESSEMER and open hearth processes, in which excess carbon was air-blasted out from the steel as carbon monoxide, which then combined with oxygen to leave the furnace as carbon dioxide gas, so preventing the formation of oxide slag. This is essentially the blast furnace process still used, although it is being replaced by processes using electric arc furnaces, or by ones in which an oxygen lance is employed to develop heat more quickly. *See also* MM pp.30–31.

Steelyard, weighing balance used since at least 315 BC. It is a metal bar suspended horizontally near one end so that the shorter arm supports the object to be weighed and the longer arm supports a small weight which can be moved on a graduated scale to balance the object being weighed. *See also* SU p.70.

Steelyard, Merchants of the, members of the German merchant guild based at the Steelyard by the River Thames, London. Established *c.* 1250, merchants belonging to this branch of the HANSEATIC LEAGUE enjoyed many privileges, such as exemption from customs dues, that were greatly resented by English merchants. The Steelyard was closed in 1598 by Queen Elizabeth I.

Steen, Jan (1626–79), Dutch painter. He painted everyday scenes including those of tavern life. He used brilliant light and cool colours to emphasize action and expression as in *Skittle Players Outside an Inn* (*c.* 1660).

Steenbok, or steinbok, or grysbok, species of small, slender-legged ANTELOPES native to the grasslands of East and South Africa. Their coats are shades of reddish brown stippled with white, and the males carry short spiked horns. Length: to 85cm (34in); weight: to 14kg (30lb). Family Bovidae; genus *Raphicerus* or *Nototragus*. See also NW p.201.

Steeplechase, athletic test of endurance and hurdling skill. The course, over 3,000m, includes 35 obstacles: 4 hurdles and one water jump in each lap. The steeplechase, which derives from cross-country running, was staged first at the OLYMPIC GAMES in 1900.

Steeplechase, horse-race over a course with obstacles to be jumped. The term originated when church steeples were used as landmarks to guide riders in cross-country horse-races. In modern steeplechasing, horses race over a distance of two miles (3.2km) or more. There must be at least one ditch and six birch fences for every mile, including one water-jump.

Steer, Philip Wilson (1860–1942), British painter of landscapes, portraits and genre pictures in a style influenced by CONSTABLE and TURNER and by IMPRESSIONISM, which he and SICKERT helped to introduce to England.

Steering, in a motor vehicle, mechanism that allows the front wheels to be angled so that the vehicle can go round bends in the road. In rack-and-pinion steering, the most common type, rotation of the steering wheel turns a helical pinion gear against a toothed rack, whose movement actuates the rods that pivot the roads

wheels. In heavy cars and lorries, hydraulic pressure provides power-assisted steering to lessen the effort for the driver. *See also* MM p.*130.*

Stegosaurus, any of several species of ornithischian DINOSAURS of the late Jurassic and early Cretaceous periods of the w USA. It walked on all fours and had a small head, a massive body and a long tapering tail which ended in defensive spikes. Along its spine was a row of vertically-mounted bony plates. Length: to 6m (20ft). *See also* NW pp.*24, 185, 185.*

Steiger, Rod (1925–), US actor whose film career included appearances in *On the Waterfront* (1954), *The Pawnbroker* (1965), *The Illustrated Man* (1969), *Waterloo* (1971) and *W.C. Fields and Me* (1976). He received an OSCAR for his performance in *In The Heat of the Night* (1967).

Stein, Sir Mark Aurel (1852–1943), British archaeologist and scholar, b. Hungary. He was Principal of Oriental College, Lahore, and registrar of Punjab University (1888–99). He transferred to the Archaeological Survey, and between 1906 and 1942 made explorations in Central Asia and Persia.

Stein, Gertrude (1874–1946), US author and critic. She went to Paris in 1903, establishing a cultural salon, where she entertained and influenced many great artists and writers including CÉZANNE, PICASSO, HEMINGWAY and FITZGERALD. Influenced by Cubist theories, her own writing includes *Three Lives* (1909) and *The Autobiography of Alice B. Toklas* (1933). *See also* HC2 p.288.

Stein, John ("Jock") (1923–), British footballer and manager. From 1948 to 1956 he played for Albion Rovers, Llanelli and Glasgow Celtic. In 1965 he rejoined Celtic as manager and under his guidance the club won numerous domestic and international honours, including the EUROPEAN CUP in 1967.

Stein, William Howard (1911–), US biochemist who shared the 1972 Nobel Prize in chemistry with ANFINSEN and MOORE for research into the molecular structure of PROTEINS. Stein developed techniques for the analysis of PEPTIDES and AMINO ACIDS obtained from proteins and applied these methods to determining the molecular structure of the ENZYME ribonuclease which was of great importance for the understanding of processes involved in protein synthesis within CELLS.

Steinbeck, John Ernst (1902–68), US novelist whose works are characterized by realistic dialogue and an understanding for the world's downtrodden. Among his novels are *Of Mice and Men* (1937), *The Grapes of Wrath* (1939), *Cannery Row* (1945), *East of Eden* (1952) and *The Winter of Our Discontent* (1961). He was awarded the Nobel Prize in literature in 1962. *See also* HC2 p.*209.*

Steiner, Jakob (1796–1863), German-Swiss mathematician. He pioneered synthetic geometry, which is concerned with those geometric properties of a figure that remain unchanged when the figure is projected onto a plane surface.

Steiner, Rudolf (1861–1925), Austrian philosopher and educationalist who helped to found the German THEOSOPHY movement. He later developed a philosophy of his own, called anthroposophy, which sought to explain the world in terms of man's spiritual nature or thinking independent of the senses.

Steinheim man, early representative of *Homo sapiens,* the skull of which was discovered in Steinheim, near Stuttgart, Germany, in 1933; it dates from the Mindel-Riss interglacial period, about 250,000 to 200,000 years ago. The brain volume is estimated at 1,150cc. The brow ridges are like those of *Homo erectus*, but the rounded back of the skull more closely resembles that of Neanderthal man, who appeared later. *See also* MS p.26.

Steinway family, German-US family of piano makers. Henry Engelhard (1797–1871) opened a piano business in Germany in 1835, emigrated to the USA in 1849 and, with his sons, founded the New York firm (which still exists) in 1853. It produced the first Steinway grand piano in

1856; the London branch of the firm was opened in 1875.

Stella, Frank (1936–), US painter. His work, in which he reacts against ABSTRACT EXPRESSIONISM, consists of basic geometric shapes set against simple backgrounds.

Stella, Joseph (1880–1946), US painter, b. Italy, who went to New York in about 1902 to study medicine but became an artist instead. He was influenced by CUBISM and FUTURISM, as can be seen in his famous paintings of Brooklyn Bridge.

Stellar parallax, method of using PARALLAX to estimate the distances to the nearer stars. It is based on the apparent changes in star position over periods of six months, during which the Earth has moved from one side of its orbit round the Sun to the other. This known distance is then used as the base of a triangle from which to calculate the star's distance. The apparent movement of the star whose distance is being estimated is judged against that of a fixed or stationary star. For the purpose of the calculation the fixed star is assumed to be at infinity. *See also* SU p.166.

Stellenbosch, town in s South Africa, in Cape Province, 40km (25 miles) E of Cape Town. Founded in 1679, it is one of the oldest European settlements in South Africa. Stellenbosch is situated in a fertile river valley and serves as a centre for the surrounding fruit-growing and wine-making region. There is also a brick and tile manufacturing industry. Pop. (1969 est.) 29,900.

Stem, main, upward-growing part of a plant that bears leaves, buds and flowers. The internal structure is composed of VASCULAR tissue arranged in a ring (in dicotyledons) or scattered (in monocotyledons). Stems are usually erect, but may be climbing (vine) or prostrate (stolon). They may also be SUCCULENT (cactus) or modified into underground structures (rhizomes, tubers, corms, bulbs). Stems vary in size from the thread-like stalks of aquatic plants to tree-trunks.

Stendhal (1783–1842), pseudonym of Marie-Henri Beyle, French novelist. His first novel *Armance* (1823) was scorned by critics, but in 1831 he published the first of his two great novels *Le Rouge et le Noir*. This was followed in 1839 by *La Chartreuse de Parme. See also* HC2 p.98.

Steno, Nicolaus (1638–86), Danish geologist and anatomist. In *The Prodromus* (1669; English trans., 1916) he laid the foundation of the science of CRYSTALLOGRAPHY by reporting equal angles between corresponding faces of quartz crystals. He proposed that fossils are the remains of ancient living organisms and that many rocks arise from SEDIMENTATION. *See also* PE p.94.

Stenosis, abnormal narrowing of a passage, eg that between the stomach and the duodenum. It is usually congenital and may be repaired surgically.

Stephanotis, climbing shrub native to Madagascar, commonly grown as a greenhouse plant elsewhere. Its branches may reach several metres in length and are thickly covered with large, white, strongly scented flowers and thick oval leaves. Family Asclepidaceae; species *Stephanotis floribunda.*

Stephen, Saint (977–1038), Stephen I of Hungary (*r.* 1000–38), the first king of the Árpád dynasty. His chief work was to continue the Christianizing of Hungary begun by his father, by endowing abbeys, inviting foreign prelates to Hungary and suppressing paganism. He was canonized in 1083.

Stephen, name of ten Roman Catholic popes. Stephen I (*r.*254–57), saint and martyr, became involved in disputes with St Cyprian, Bishop of Carthage, of which the most bitter was over the validity, (which Cyprian upheld) of rebaptism of heretics or schismatics. Stephen II (752) died a few days after his election and was never consecrated. He is not usually counted as a pope and the succeeding Stephens are given alternative numbers. Stephen II (or III) (*r.*752–57) sought the aid of the FRANKS against the LOMBARDS who were threatening Rome. He was responsible for the pact with Pepin, King of the Franks, which resulted in the Donation of

Pepin (755) that ceded territories in the exarchate of Ravenna, Venetia and Istria to the papacy, the basis of the PAPAL STATES. *See also* HC1 p.152. Stephen III (or IV) (*r.*768–72) was elected after the expulsion of the Lombard candidate Philip. He tried unsuccessfully to prevent the marriage between CHARLEMAGNE and the daughter of the Lombard king, but later allied himself with the Lombards against the Franks. The other Stephens were Stephen IV (or V) (*r.*816–17); Stephen V (or VI) (*r.*885–91); Stephen VI (or VII) (*r.*896–97); Stephen VII (or VIII) (*r.*929–31); Stephen VIII (or IX) (*r.*939–42); Stephen IX (or X) (*r.*1057–58).

Stephen (*c.*1097–1154), King of England (*r.*1135–54), grandson through his mother of WILLIAM THE CONQUERER. When HENRY I settled the succession on his daughter, MATILDA, Stephen and other barons swore fealty to her. On Henry's death (1135), however, he claimed the throne and was crowned. Harassed by the Scots and Matilda's supporters, Stephen was temporarily deposed in 1141. Although he re-established himself as king, he could no longer control his barons and by a treaty confirmed at Westminster (1153) he conceded the succession to Matilda's son Henry, who on Stephen's death (1154) reigned as HENRY II. *See also* HC1 p.178, *178*.

Stephen I, King of Hungary. *See* STEPHEN, SAINT.

Stephen, Sir James Fitzjames (1829–94), British jurist and historian. He reformed and codified Indian law (1869–72) and subsequently made a major contribution to the codification of English criminal law. He was a high court judge (1879–91), and his standard *History of the Criminal Law of England* was published in 1883.

Stephen, Sir Leslie (1832–1904), British writer and critic. He edited the *Cornhill Magazine* (1871–82), and was the first editor of the *Dictionary of National Biography* (1882–91).

Stephen Harding, Saint (*c.*1048–1134), English monastic reformer. After travelling widely in Europe, he joined a religous community at Molesme, Burgundy, but left because of the lax observance there of the Rule of St BENEDICT. With St Robert de Molesme he established a house at Citeaux, where he became abbot in 1109. The CISTERCIANS date from his abbacy.

Stephens, James (1882–1950), Irish poet and novelist who was a leading figure in the IRISH LITERARY RENAISSANCE. His novels, many of which incorporate Irish legend and folklore include *The Crock of Gold* (1912) and *Deirdre* (1923). Among his volumes of poetry is *The Hill of Vision* (1912). *See also* HC2 p.282.

Stephenson, George (1781–1848), British engineer and inventor, regarded as the father of the locomotive. He built his first locomotive, the *Blucher*, in 1814 in which flanged wheels were incorporated for the first time. His most famous locomotive, the *Rocket*, was built in 1829 and set a record speed of 47km/h (29mph). It later ran on the Liverpool to Manchester line, which he also built. *See also* MM p.140.

Stephenson, Robert (1803–59), British engineer, son of George STEPHENSON. Between 1845 and 1850 he built the tubular Britannia Bridge over the Menai Strait, N Wales. *See also* MM p.191.

Stephenson, Sir William (1896–), Canadian industrialist who worked for British and US intelligence. He was CHURCHILL'S representative in New York and head of British Security Co-ordination in WWII. With J. Edgar HOOVER, he helped to establish the FBI.

Steppe, extensive, semi-arid plains of central Eurasia. The word is a Tartar term that has been adopted by geographers. The steppe includes three zones of vegetation; the forest steppe; the grassland prairie, which is now usually cultivated; and the non-tillable steppe, which is semi-desert and fertile only after irrigation.

Stereochemistry, study of the chemical and physical properties of compounds as affected by the ways in which the atoms of their molecules are arranged in space.

Such arrangements can result in two or more compounds having the same numbers and kinds of atoms but differently shaped molecules, which are called steroisomers. Stereochemistry also deals with optical isomerism, in which the configuration of one molecule is the mirror image of another. These are known as L (*levulo*) and D (*dextro*) forms. An L form may have completely different properties to a D form, and the separation between the two is made from the sense in which each rotates the plane of POLARIZED LIGHT. In the amino acids of higher organisms only the L-amino acid form appears. *See also* SU pp.140–143.

Stereophonic sound, method of reproducing sound so that it gives the illusion of both location and direction; at least two separate channels are required. In stereophonic LONG-PLAYING RECORDS, the groove is modulated in two planes – the lateral and the vertical (groove depth). Depth variations correspond to the difference between the left and right channels of the stereo-recording and lateral modulations correspond to their sum. Two transducers in the pick-up cartridge feed two separate amplifiers and two widely-spaced loudspeakers. *See also* p.232.

Stereoscope, optical binocular device that produces an apparently three-dimensional image by presenting two slightly different plane images, usually photographs, to each eye.

Stereotype, in printing, duplicate printing plate made from the original (which is reserved for making further duplicates as necessary). A traditional method of making stereotypes is to press papier mâché into the original and producing a negative mould from which a duplicate plate is electroplated or cast. *See also* MM p.206.

Stereotyping, tendency to ascribe to anyone characteristics that can be attributed to a group or CLASS of which he is a member. The stereotyped judgement, however, may or may not apply to him, and thus stereotyping is a form of categorization.

Sterility, in men, inability to father children because of a deficiency in the number or quality of SPERM. The cause may be essentially hormonal, or it may be a physical abnormality either inherited, congenital or as a result of disease. In women the condition is called infertility, and is generally a result of a malfunction in OVULATION, sometimes caused by an obstruction of the Fallopian tubes. *See also* HYSTERECTOMY; VASECTOMY; MS pp.102–103.

Sterilization, in medicine and bacteriology, killing of all bacteria, viruses and spores present within a certain environment. This prevents the spread of infection and permits scientists to work in strictly controllable conditions. It is only since the 19th century that sterilization has been widely practised. Joseph LISTER introduced antiseptic methods in surgery, and Louis PASTEUR demonstrated the existence of germs. Sterlization usually involves the application of heat, but chemicals such as ethonol are used when the application of heat is impractiable, for instance in sterilizing the skin.

Sterling, term for British currency thought to derive from the *storling*, the 11th-century penny bearing the design of a star. It is used to distinguish the British pound from pounds of other currencies and can also be used to describe the quality and standard weight of coins. The sterling silver mark on silver (the stamp of a lion *passant*) represents a purity of more than 90%. *See also* HC2 pp.300–301.

Stern, Isaac (1920–), US violinist, b. USSR, who studied at the San Francisco Conservatory and made his debut in 1931. He has made extensive world tours and established a reputation as one of the world's leading virtuosos.

Stern, Otto (1888–1969), German-born US physicist. An outstanding experimental physicist, he was awarded the 1943 Nobel Prize in physics for developing the molecular beam as a tool for studying the characteristics of MOLECULES and for his measurement of the magnetic moment of the PROTON.

Sternberg, Josef von (1894–1969), US film director, b. Austria. He is famed for his sophisticated, sensual films starring Marlene Dietrich, including *The Blue Angel* (1930), *Morocco* (1930) and *The Devil Is a Woman* (1935).

Sterne, Laurence (1713–68), British novelist whose reputation was established with *Tristram Shandy* (1760–67), a richly humourous novel whose techniques anticipated much 20th-century fiction. *A Sentimental Journey Through France and Italy* appeared in 1768. *See also* HC2 p.34.

Sternum, or breast bone, flat, narrow bone in front of the chest between the breasts. It is generally divided into three regions: the manubrium, or upper part; the body; and the xiphoid process, or lower and more flexible cartilaginous part. The top of the manubrium is attached by ligaments to the collarbones and the body is joined to the ribs by seven pairs of costal cartilages. *See also* MS pp.56–57.

Steroid, class of organic compounds characterized by a basic molecular structure of 17 carbon atoms arranged in four rings and bounded by up to 28 hydrogen atoms. Steroids are widely distributed in animals and plants; the most abundant are the sterols (steroid alcohols), such as cholesterol. Another important group are the steroid HORMONES, including the corticosteroids, secreted by the adrenal cortex, and the sex hormones. The sex hormones are the oestrogens (such as oestrone) and progesterone, produced by the ovaries, and androgens (such as testosterone and androsterone) secreted by the testes. Synthetic steroids are widely used in medicine. *See also* SU p.153.

Stesichorus (*c.*630–555 BC), Greek lyric poet, originally called Teisias. Legend holds that he was a master of the choral hymn celebrating the heroes of epic poetry. *See also* HC1 p.74.

Stethoscope, instrument that enables an examiner to hear the sounds of activity in various parts of the body, such as the heart, blood vessels, lungs and intestines. It consists of two earpieces attached to flexible rubber tubes which lead to either a disc or a cone.

Stettin. *See* SZCZECIN.

Stevens, Wallace (1879–1955), US poet and lawyer. His work includes *Harmonium* (1923), *Ideas of Order* (1935), *The Man with the Blue Guitar* (1937) and *Parts of a World* (1942). *See also* HC2 p.289.

Stevenson, Adlai Ewing (1900–65), US politician and ambassador to the UN (1961). He was an adviser at the San Francisco Conference in 1945 that set up the UN. He became governor of Illinois in 1948 and was twice unsuccessful Democratic candidate for the presidency (1952, 1956).

Stevenson, Robert (1772–1850), British civil engineer, famous chiefly for his lighthouses. He invented the intermittent and flashing lights.

Stevenson, Robert Louis (1850–94), Scottish novelist, essayist and poet. His novels include *Treasure Island* (1881), *Dr Jekyll and Mr Hyde* (1886), *Kidnapped* (1886), *The Black Arrow* (1888) and *The Master of Ballantrae* (1889) *A Child's Garden of Verses* appeared in 1885. *See also* HC2 pp.284–284, *285*.

Stevenson, Teophilo (1951–), Cuban amateur boxer. He won the Olympic heavyweight gold medal in 1972 and 1976, became World Amateur Champion in 1974 and won a gold medal in the Pan-American Games of 1975.

Steward, Lord High, office of honorary servant of the British sovereign, originating in the 11th century. HENRY II gave the office to the earls of Leicester, later to be acquired by Simon de Montfort and then passing to members of the Royal Family. Since 1421 appointments have been made only for special occasions.

Steward of the Household, Lord, office of the servant of the British Royal household. Several stewards were appointed in the 11th century, but since the early 14th century the office has been restricted to one person. The office evolved into one of considerable political influence, then reverted to that of personal service to the sovereign, including organization of the Royal household.

Steppes in the USSR. They receive less rain than prairies but more than deserts.

Sternberg's film *Underworld* (1927) led to his long association with Paramount.

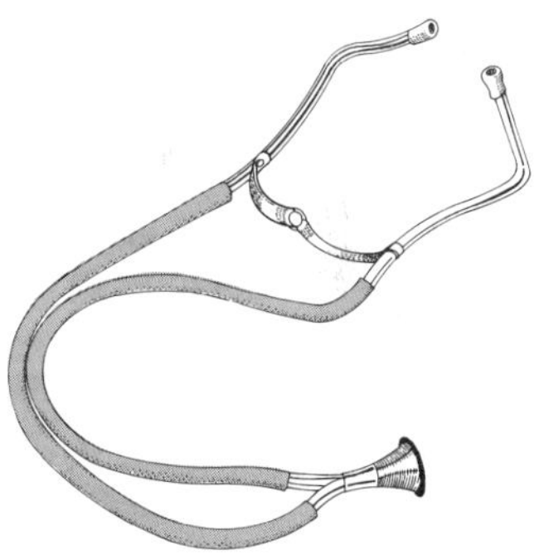

Stethoscope; before its invention a doctor put his ear to the patient's body.

R. L. Stevenson's *Treasure Island* was conceived as an amusement for his stepson.

James Stewart in 1970 with a beard for his part in the film *Fools Parade*.

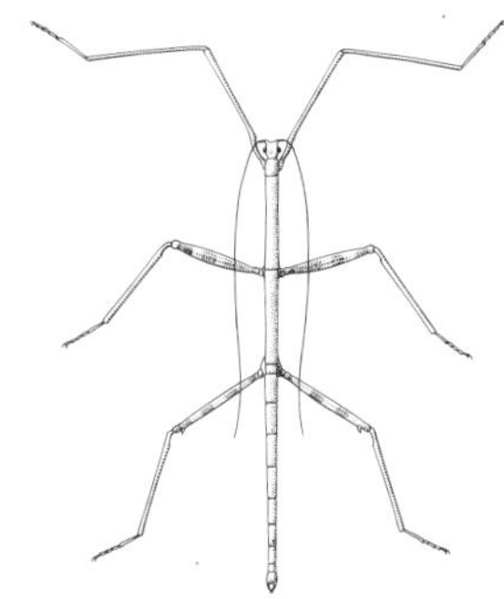

Stick insects feed at night, remaining motionless during the hours of light.

Stirling Castle, which dates from 1460, was a favourite home of the Stuarts.

Stock Exchange; a view of the floor of the London Stock Exchange, taken in 1963.

Stewart, Prince Charles Edward. *See* STUART, CHARLES EDWARD.

Stewart, Douglas Alexander (1913–), New Zealand poet, playwright and critic. He edited the influential literary *Red Page* in the *Sydney Bulletin* from 1940–61 and had a strong interest in Australian folk-literature, editing *Australian Bush Ballads* (1955).

Stewart, Dugald (1753–1828), Scottish philosopher who was one of the main advocates of COMMONSENSE REALISM. Like Thomas REID, Stewart believed that philosophy should employ scientific methods to become a less negulous discipline.

Stewart, Jackie (1939–), Scottish Formula 1 racing driver. He retired in 1973 after winning a record 27 Grands Prix races.

Stewart, James Francis Edward. *See* STUART, JAMES FRANCIS EDWARD.

Stewart, James Innes Mackintosh. *See* INNES, MICHAEL.

Stewart, James Maitland (1908–), US film actor. His notable films include *Mr Smith Goes to Washington* (1939), *The Philadelphia Story* (1940), *Harvey* (1950), *Vertigo* (1958), *Anatomy of a murder* (1959), *The Man who Shot Liberty Valance* (1962), *Shenandoah* (1965) and *The Shootist* (1976).

Stewart, Mary. *See* MARY, QUEEN OF SCOTS.

Stewart, Robert, Viscount Castlereagh. *See* CASTLEREAGH, ROBERT STEWART, VISCOUNT.

Stewart, river in Yukon Territory, Canada. It rises in the Mackenzie Mts and flows w to the Yukon River 81km (50 miles) s of Dawson. It was discovered by Robert Campbell of the Hudson's Bay Co in 1850. Length: 533km (331 miles).

Stewart Island, part of New Zealand, 32km (20 miles) s of South Island. A mountainous and forested region, it has become popular as a tourist resort. Exports include frozen fish. Area: 1,748sq km (675sq miles). Pop. (1975 est.) 410.

Stewarts. *See* STUART.

Steyn, Martinus Theunis (1857–1916), South African statesman. The last president of the ORANGE FREE STATE from 1896 to 1900, he served in the SOUTH AFRICAN WAR and later helped to form the Union of South Africa.

Stibnite, sulphide mineral, ANTIMONY trisulphide (Sb_2S_3) found in low-temperature veins and rock impregnations. It occurs as orthorhombic system aggregates of either prismatic crystals or granular masses and is opaque, sometimes iridescent and grey or black in colour. Hardness 2; s.g. 4.6. Stibnite is an important ore of antimony.

Stick insect, any of numerous species of herbivorous insects of the order Phasmida which strongly resemble the shape and colour of the twigs upon which they rest. Some sway backwards and forwards, mimicking a wind-blown twig, and lay eggs which resemble seeds. Length: to 32cm (11in). *See also* LEAF INSECT; NW pp.*107, 111, 205.*

Stickleback, small fish found in fresh, brackish and salt water. It is usually brown and green, and may be identified by the number of bony plates and spines along its sides and back. Its mating and courtship behaviour have been much studied: the male builds a nest of water plants and, with an elaborate ritual, drives the female into it. He then watches the eggs and cares for the young. Length: 8–11cm (3–4.5in). The 12 or so species include the three-spined *Gasterosteus aculeatus*. Family Gasterosteidae. *See also* NW pp.*35, 125, 230.*

Stigand (*d.*1072), Anglo-Saxon prelate. In 1052 he was appointed Archbishop of Canterbury and Bishop of Winchester. He crowned WILLIAM I King of England but was replaced as Archbishop by LANFRANC in 1070.

Stigma, in botany, the free upper part of the STYLE of the female organs of a flower, to which a pollen grain adheres before FERTILIZATION. *See also* NW p.*71.*

Stijl, De, movement in modern art that originated in The Netherlands with the art periodical of that name founded by Theo van DOESBURG in 1917. It was at first associated with the principles of NEOPLASTICISM, of which MONDRIAN was the major exponent. From the mid-1920s the movement included Dadaists such as ARP and KUPKA and, although lacking in cohesion, De Stijl ideas were representative of the avant-garde movement as a whole. *See also* HC2 pp.178, 204.

Stilbite, hydrated silicate mineral containing calcium, sodium and aluminium; one of the ZEOLITE group. It displays monoclinic-system radiating tabular crystals or aggregates and is white, yellow, red or brown in colour. Hardness 3.5–4; s.g. 2.1.

Stilicho, Flavius (*c.*259–408), Roman general of Vandal origin. He acted as regent for the young Emperor Honorius and in this post was the effective ruler of the Western Empire.

Still. *See* DISTILLING.

Still, Andrew Taylor (1828–1917), US founder of OSTEOPATHY. After three of his children died in an epidemic of spinal MENINGITIS, he devised treatment based on the theory that most physical disorders were caused by "structural derangement" of the body, which could be remedied by manipulations of bones and muscles without the use of drugs.

Still birth, expulsion of a dead foetus after more than 28 weeks within the womb.

Still-life, painting of everyday objects, such as flowers, cooking utensils or books, usually grouped together and viewed in close-up. Before the Renaissance, still-life groupings never formed more than a minor part of the religious and allegorical paintings then in vogue. Still-life became an independent art form in the 16th century, much influenced by the work of CARAVAGGIO, and reached what is often considered its peak in the Netherlands in the 17th century, in the work of such artists as SNYDERS and de HEEM. CHARDIN was the first important French still-life painter. It has remained an art form popular with both artists and the public. Through the works of CÉZANNE and others, still-life has played an important part in the development of abstract art. *See also* HC1 p.*311;* HC2 p.177.

Stilt, long-legged wading marsh bird. It defends its nest with an elaborate display that includes playing dead or distractions in flight. Length: 35–45cm (14–18in). Family Recurvirostridae.

Stilton, blue-veined, crumbly cheese made originally in the English village (in Huntingdonshire) which gave it its name. Penicillium, a blue mould, contributes to the rich aroma and taste of the cheese. *See also* PE p.229.

Stimulant, drug that increases the rate of activity of some part of the body or mind. Many stimulants are used in medicine, such as digitalis, which acts on weakened heart muscles, and oxytocin, which induces contractions of the womb in childbirth. There are a number of stimulants which act on the central nervous system. Amphetamines are experienced as mood elevators and they also increase alertness. Barbiturates are depressants which initially have the effect of stimulants. This paradoxical effect is found in many hypnotic drugs. Many common drinks such as tea, coffee and colas contain the stimulant caffeine in small amounts.

Stimulus, in psychology, any change in energy, either external or internal, that influences the RECEPTORS of the nervous system.

Stinging nettle. *See* NETTLE.

Stingray, any of several species of bottom-dwelling elasmobranch fish that live in marine waters and in some rivers in South America. It is flattened dorso-ventrally, with wing-like fins extending around the head. It has a long, slender tail which can inflict a venomous sting to stun prey, and can cause serious injury to a human being. Width: to 2.1m (7ft). Family Dasyatidae.

Stinkhorn, any of several species of foul-smelling Basidiomycete fungi. At first it resembles a small, whitish "egg", which contains the unripe fruit body (receptacle). When ripe the receptable elongates, rupturing the egg. It carries with it a glutinous spore mass which attracts the flies that disperse the spores on their bodies. Genus *Phallus. See also* NW pp.41–42, *42.*

Stinkwood, tree of the laurel family that grows in southern Africa. It has an unpleasant odour but valuable timber. Species *Ocotea bullata..*

Stirling, James (1820–1909), Scottish philosopher who was responsible for the introduction into Britain of the works of HEGEL. His book, *The Secret of Hegel,* was published in 1865.

Stirling, town in central Scotland, on the River Forth; former county town of Stirlingshire. It is dominated by Stirling Castle, situated 128m (420ft) above the river. Stirling is also the seat of a university, founded in 1967. A market town, it has agricultural engineering, asbestos, carpet and food processing industries, Pop. 29,769.

Stirling Bridge, Battle of (1297), victory by the Scots under William WALLACE over the English under John de Warenne, EDWARD I's guardian of Scotland after his conquest of it in 1296.

Stirling(shire), former county in central Scotland; since 1975 it has been part of Central Region. The county consists of highlands in the NW and lowlands in the SE. The region is drained chiefly by the Forth River. The rearing of livestock, particularly dairy cattle, is a major agricultural activity. Sheep are grazed on the hills. Industries include coalmining, leather goods, brewing and paper. Stirling was the county town. Area: 1,168sq km (451sq miles). Pop. (1971 est.) 209,240.

Stoat, carnivorous mammal of the weasel family. Its slim body is about 30cm (12in) long, including the tail, and like others of the family it has short legs and moves sinuously. It preys upon rabbits and smaller animals in many temperate and northern parts of the world. In the latter regions its fur turns from red-brown and white to white in winter, when it is known as ERMINE. Family Mustelidae; species *Mustela ermina. See also* NW pp.*191, 206.*

Stock, or gilliflower, annual plant native to s Europe, South Africa and parts of Asia, cultivated as a garden flower. It has oblong leaves and pink, purple or white flower clusters. Evening stock has aromatic lilac flowers; height: to 80cm (31in). Family Cruciferae; species *Matthiola bicornis.*

Stock, or share, certificate of ownership of shares in a company. A company may issue shares of stock upon its formation and may issue additional shares when it needs more capital. The ownership of stock implies sharing in the control of the company's operations and majority stockholders may have decisive influence upon the operation of the company.

Stock, mass of plutonic rock (rock heated deep within the Earth) solidified at depth. It is dome-shaped but is regarded as being smaller than a BATHOLITH, which it otherwise resembles. The granites of Cornwall and Devon are examples. *See also* PE p.*101.*

Stockbroker, agent who buys and sells STOCKS or SHARES on behalf of another. He is employed because of his specialized knowledge of the stock market and is paid a commission proportionate to the value of the transaction.

Stock car racing. *See* MOTOR RACING.

Stock exchange, market which deals in STOCK and SHARE transactions. Beginning in coffee houses in the 18th century, the first stock exchange opened in London in 1773, with provincial exchanges quickly being established in other commercial centres. Soon there were stock exchanges in major cities throughout the world. Essentially an auction market, a stock exchange provides a convenient arrangement for both vendors and purchasers because all transactions are conducted by STOCKBROKERS and STOCKJOBBERS. The largest stock exchanges are in London and New York.

Stockhausen, Karlheinz (1928–), German composer and theorist who studied in Paris (1951–53) with MESSIAEN and MILHAUD. One of the foremost and most successful exponents of ELECTRONIC MUSIC. Thwarting the bulk of musical tradition, his interest centres on the fundamental acoustical aspects of music and the relationships between juxtaposed sounds.

Stockholm, port and capital of Sweden, on Lake Mälar's outlet to the Baltic Sea. Founded in the mid-13th century it became an important trade centre and was dominated by the HANSEATIC LEAGUE. GUSTAVUS I made Stockholm the centre of his kingdom, and put an end to the privileges of Hanseatic merchants. The city became the capital of Sweden in 1436, developed as an intellectual centre in the 17th century and attracted such people as DESCARTES. Industrial development dates from the mid-19th century. Built on several islands and peninsulas, modern Stockholm has wide streets, parks and well-planned residential areas. Industries include textiles, clothing, paper and printing, food processing, rubber, chemicals, shipbuilding, beer, metal and machine manufacturing. Pop. (1976 est.) 665,200.

Stockjobber, dealer on the STOCK EXCHANGE who buys and sells stocks and shares at "wholesale" level. He deals in shares on behalf of STOCKBROKERS and does not deal directly with the public. His profit derives from the "jobber's turn", the difference between the buying and selling prices of shares.

Stockport, county district in SE GREATER MANCHESTER, England; created in 1974 under the Local Government Act (1972). Area: 126sq km (49sq miles). Pop. (1974 est.) 294,400.

Stocks, instrument of punishment used in Britain until the 19th century. A wooden device with holes in which the legs or arms were secured, stocks are positioned in the main thoroughfare so that offenders were the object of public ridicule. *See also* PILLORY; MS p.*296*.

Stockton-Darlington Railway, the world's first passenger railway, opened in 1825. It ran from Darlington to Stockton in County Durham, and its steam locomotive *Locomotion* was built by George STEPHENSON. It travelled at 24km/h (15mph) and carried 450 passengers on its first run. *See also* HC2 p.*92*; MM p.140.

Stockton-on-Tees, county region in s central CLEVELAND; created in 1974 under the Local Government Act (1972). Area: 195sq km (75sq miles). Pop. (1974 est.) 162,500.

Stoics, followers of the school of philosophy founded by ZENO in *c.*300 BC. Their name derives from their meeting-place, the *Stoa Poecile* (painted porch) where Zeno lectured in Athens. The ethical aspect of stoicism is the most important. Founded on the premise that virtue is attainable only by living in harmony with nature, it stressed the importance of self-sufficiency and of equanimity in adversity. The philosophy of stoicism was given its first systematic expression by CHRYSIPPUS in the 3rd century BC. It was introduced into Rome in the 2nd century BC by Panaetius of Rhodes. There is found its greatest adherants, SENECA, EPICTETUS and MARCUS AURELIUS.

Stoke-on-Trent, city and county district in NW STAFFORDSHIRE, W central England. The city, on the River Trent, is the centre of the POTTERIES and is noted for its manufacture of china. The county district was created in 1974 under the Local Government Act (1972). Area: 93sq km (36sq miles). Pop. (1974 est.) 258,300.

Stoker, Bram (1847–1912), Irish novelist. His most famous book is *Dracula,* (1897), which has had many stage, film and television adaptations.

Stokes, Sir George Gabriel (1819–1903), British mathematician and physicist. He is remembered for his work with viscous fluids; his law of VISCOSITY, describing the movement of a solid sphere in a fluid (STOKES'S LAW); and for his theorem of VECTOR ANALYSIS. He also worked on FLUORESCENCE (including the origination of the term); the concept of an ETHER; and the wave theory of light. He was a pioneer in GEODESY.

Stokes's law, in fluid mechanics, gives the terminal (final) velocity at which particles fall in a fluid (liquid or gas). For particles of radius r (less than 0.1mm) the terminal velocity is $2(d_1-d_2)gr^2/9\eta$ where d_1 and d_2 are the densities of the particle and the fluid, g is the gravitational constant and η the viscosity of the fluid.

Stokowski, Leopold (1882–1977), US conductor, b. Britain, who was director of the Cincinnati Symphony (1909–12) and conductor for the Philadelphia Orchestra (1912–36). He became known for his individual interpretations and flexibility of approach. He transcribed BACH's organ works for orchestra and was also musical supervisor of Walt DISNEY's *Fantasia* (1939), in which he appeared as conductor.

Stole, ecclesiastical vestment consisting of a long band of silk, coloured according to the liturgical calendar. It is worn by Anglican and Roman Catholic bishops, priests and deacons and by some Protestant clergy, during eucharistic services and when preaching.

Stolon, or runner, modified underground or aerial stem growing from the basal node of a plant. Slender, elongated and prostrate, stolons develop adventitious roots. Aerial stolons are characteristic of strawberries; underground stolons are characteristic of white potato plants. *See also* TUBER.

Stolypin, Peter Arkadievich (1862–1911), Russian statesman, Minister of the Interior and Premier (1906–11). He suppressed the 1905 revolution but introduced reforms, dissolving the Second DUMA in 1907 and changing the electoral laws for the Third Duma. *See also* HC2 p.*169, 169.*

Stomach, J-shaped organ, lying to the left and slightly below the diaphragm in man; one of the organs of the DIGESTIVE SYSTEM. It is connected at its upper end (fundus) to the gullet (oesophagus) at the cardiac orifice. At the lower end (pyloric section) it connects with the small intestine at the pylorus. The stomach itself is lined by three layers of muscle (longitudinal, circular and oblique) and a mucous layer, the gastric mucosa, that lies in folds (rugae) and contains gastric glands. These glands secrete hydrochloric acid that destroys any food bacteria, dissolves salt in the food, and makes possible the action of pepsin, the ENZYME secreted by the glands, that digests PROTEINS alone. Gastric gland secretion is controlled by psychic stimuli – the sight, smell and taste of food – and by hormonal stimuli, chiefly the HORMONE gastrin. As the food is digested, it is churned by muscular action, into a thick liquid state called chyme, at which point it passes the small intestine. *See also* MS p.68, *68.*

Stomach ache, symptom that may be ascribed to systemic disorder, psychosomatic conditions or disorder of other organs as well as to actual stomach disorders. These include ulcers, dyspepsia or gastric indegestion.

Stomata, in botany, pores mostly found on the undersides of leaves through which atmospheric gases pass in and out, for RESPIRATION and PHOTOSYNTHESIS. Surrounding each stoma are two guard cells which swell and close the stoma to prevent excessive loss of water vapour. *See also* TRANSPIRATION.

Stone Age, period in man's evolution during which he used stone tools and weapons. It was during this time that he progressed from making crude flaked pebble tools to finely worked arrowheads and polished and hafted axes. *See also* EOLITHIC AGE; MESOLITHIC AGE; NEOLITHIC AGE; PALAEOLITHIC AGE; HC1 pp.42–43.

Stonechat, small thrush found in open heath and scrubland of Europe, Africa and Asia. Its name comes from its characteristic cry which sounds like stones struck upon one another. The plumage of the male is black, white and rust-brown. Stonechats feed on insects on the ground. Species *Saxicola torquata.*

Stonecrop, any plant of the genus *Sedum* of the family Crassulaceae, especially creeping sedum (*S. acre*), a succulent, low-growing plant of European origin with pungent fleshy leaves and yellow flowers found in rocky areas.

Stonefish, bottom-dwelling marine fish that lives in tropical waters of the Indo-Pacific Ocean. It has a warty, slime-covered body and sharp dorsal spines with which it can inflict a painful, sometimes deadly, sting to human beings. Length: to 33cm (13in). Family Synancejidae; species *Synanceja verrucosa. See also* NW p.*241.*

Stonefly, or salmon fly, soft-bodied insect with long, narrow front wings and chewing mouthparts, found throughout the world. The aquatic nymphs have branched gills, and the adults, used as bait, are brown to black. Length: 5–60mm (0.2–2.5in). Order Plecoptera. *See also* NW p.*107.*

Stone fruit. *See* DRUPE.

Stonehenge, group of prehistoric standing stones on Salisbury Plain, Wiltshire, s England, 13km (8 miles) N of Salisbury. Thought to have been erected in the 2nd millennium BC, its four series of stones are circled by a ditch 91m (300ft) in diameter. The outermost series are made of sandstone and arranged in a circle. They enclose a circle of bluestone MENHIRS. The third series is horseshoe-shaped; the inner ring is ovoid with an Altar Stone within its confines. A huge upright Heelstone is positioned to the N of the circle. Early beliefs that Stonehenge was a Druid temple have been discredited. A later theory proposed that Stonehenge was used as an astronomical instrument. It was claimed, however, that the BRONZE AGE culture in which Stonehenge was created was not sophisticated enough to support this theory. *See also* HC1 pp.*22,* 43, *43,* 84; SU p.*26.*

Stonemasonry. *See* MASONRY.

Stones, Dwight (1953–), US high jumper. World record holder at 2.31m (7ft 7in) before the 1976 Olympic Games, he was only third in the Olympic competition, the wet conditions affecting this technique. At age 18, he had won a bronze medal for coming third in the high jump at the Munich Olympics.

Stoneware, opaque, non-porous, high-fired clay ware, well vitrified and intermediate between PORCELAIN and EARTHENWARE. Stoneware is often salt-glazed for utility pieces and is used for vessels and tiles. *See also* MM p.44.

Stoneworts, largest of the freshwater ALGAE. They are simple plants and are known as stoneworts because they become encrusted with hard calcium salts. Class Charophyceae.

Stoney, George Johnstone (1826–1911), British physicist who named the electron. He worked on the kinetic theory of gases and estimated the number of molecules in a volume of gas at standard temperature and pressure. His estimate of an electron's charge was incorrect because he used an incorrect value for the number of atoms in a gramme of hydrogen.

Stony-iron meteorite, type of METEORITE containing intermixed amounts of both silicate minerals (stone) and metal (nickel and iron). In some, called pallasites, separate stony fragments are contained in a cohesive mass of nickel-iron. Others, called mesosiderites, resemble stony meteorites but with an unusually large amount of interspersed metal. *See also* SU p.218.

Stony meteorite. *See* MASONRY.

Stoolball, bat and ball game of ancient origins. It is played by two teams of 11 on a pitch between two posts surmounted by a small, square board. The bowler or pitcher throws the ball underarm at the board from a point between the posts while the batter defends the board, or wicket, with a short, rounded bat. The batter is out if the ball hits the board or is caught by a fielder. Points are scored by running between the two posts.

Stopes, Marie Charlotte Carmichael, (1880–1958), British pioneer of birth control. After obtaining a doctorate in botany from Munich University in 1904, she began an academic career at Manchester University. She soon became concerned about inadequate contraceptive measures and campaigned for a more rational and open approach to contraception, and established the first birth control clinic in Britain in 1921. She wrote *Wise Parenthood* (1918) and *Contraception: Its Theory, History and Practice* (1923; 1931).

Stoppard, Tom (1937–), British dram-

Stockholm; the State Bank of Sweden, with the parliament building behind.

Stokowski worked to bring music to the greatest number of people possible.

Stoneflies lead short and inactive lives near to water, where they breed.

Stonehenge: the central entrance. People once removed the stones to make bridges.

Storey, David Malcolm

Storks nest in tall trees and occasionally on cliffs and even buildings.

Theodor Storm: an illustration from this German writer's *Immensee*.

Strafford was a critic of Charles I but later opposed Parliamentarian aims.

Stratford-upon-Avon: the river and Trinity Church, where Shakespeare is buried.

atist, *b.* Thomas Straussler in Czechoslovakia, whose work explores the philosophical problems of reality and communication with great wit and dramatic ingenuity. His reputation was established with *Rosencrantz and Guildenstern are Dead* (1966); other plays include *Jumpers* (1972), *Travesties* (1974), a radio play, *Artist Descending a Staircase* (1972), and *Professional Foul* (1977) for television.

Storey, David Malcolm (1933–), British novelist and dramatist whose works are noted for their naturalistic approach to the emotional pressures of working people in everyday situations. His reputation was established in 1960 with the novel *This Sporting Life*. His plays include *In Celebration* (1969) *The Changing Room* (1971) and *Life Class* (1973). *Saville* (1976) won the Booker Prize for Fiction.

Stork, long-legged wading bird that lives along rivers, lakes, and marshes in temperate and tropical regions. Usually black, white and grey, storks have straight bills, long necks, robust bodies, long broad wings and partly webbed toes. They are diurnal birds, feeding on small animals and sometimes carrion. Length: 0.8–1.5m (2.5–5ft). Family Ciconiidae. *See also* NW pp.147, 200, 214, 232.

Storm, Theodor (1817–88), German novelist and lyric poet; his belief that literature should stem from the emotion is reflected in his verse. His best-known work is *The Rider of the White Horse* (1888). *See also* HC2 p.99.

Storm, any strong disturbance of the atmosphere, usually accompanied by high winds and by various forms of precipitation. A storm wind is defined on the BEAUFORT SCALE as a wind with a speed of 103 to 117km/h (64–73mph). Storms may be categorized according to the dominant element in the disturbance. For example, there are snowstorms, thunderstorms, hailstorms and SANDSTORMS. *See also* PE pp.68–70.

Storm and Stress. *See* STURM UND DRANG.

Stormont, suburb in E Belfast, seat and government of NORTHERN IRELAND. The Parliament was established in 1921 with a House of Commons of 52 members and Senate of 25 members. Its legislative responsibility covered domestic matters only. Serious religious disputes led to its collapse in 1974, and Britain assumed direct rule.

Storm petrel. *See* PETREL.

Storm-troopers (Nazi). *See* SA.

Stornoway, largest town in the HEBRIDES, NW Scotland, on the E coast of the Isle of Lewis. It is an administrative centre for the Western Isles region, formed in 1975 by the Local Government Act of 1972. The main activities are fishing and weaving. Pop. (1971) 5,247.

Stoss, Veit (*c.*1440–1533), German GOTHIC sculptor, painter and engraver. He went to Cracow, Poland, in 1477 where he carved his masterpiece, the high altar for the Church of St Mary. Among his other works are the tomb of King Casimir IV in Cracow Cathedral and the painting *Annunciation* (1518).

Stour, name of several rivers in England, of which the chief two are the Suffolk Stour, which rises in Cambridgeshire, flows E forming part of the Suffolk-Essex border, and enters the North Sea at Harwich. Length: 76km (47 miles). The Kentish, or Great, Stour rises in Kent, ESE of Ashford, and flows NE to Pegwell Bay, N of Sandwich. Length: 64km (40 miles).

Stout, Rex (1886–1975), US author, known for his many mystery stories featuring Nero Wolfe, a detective who solved crimes from the safety of his study.

Stout, Sir Robert (1844–1930), New Zealand politician, *b.* Scotland. He became an MP in 1875 as a member of the new Radical Party. He was Prime Minister (1884–87) and Lord Chief Justice (1899–1926).

Stow, John (*c.*1525–1605), British antiquary and historian. He is best known for his *Survey of London* (1598) and *Annales of England* (1592). He also edited the *Works of Geoffrey Chaucer* (1561), and was employed by the Archbishop of Canterbury to edit chronicles dating back to medieval times.

Stow, Julian Randolph (1935–), expatriate Australian poet and novelist. His works, which use a psychological approach to the characters, include *Outrider* (1962), a book of verse, and the novel *Tourmaline* (1963).

Stowe, Harriet Beecher (1811–96), US writer, an advocate of the abolition of slavery. Her novel *Uncle Tom's Cabin*, (1851–52) sold more than 300,000 copies in one year and helped to synthesize the opposition to slavery which led up to the AMERICAN CIVIL WAR. She later wrote a second novel about slavery, *Dred* (1856).

STP (standard temperature and pressure), in physics, reference conditions under which the weights of equal volumes of different gases may be compared. The standard temperature is 0°C (32°F) and standard pressure is 760 mm of mercury.

Strabismus, also called squinting, in human anatomy, condition in which the eyes cannot be aligned to look in the same direction. One eye only may be deviant, resulting in cross-eye (esotropia), walleye (exotropia) or direction of one eye upwards (hypertropia) or downwards (hypotropia). It is generally the result of abnormality in nervous controls; non-alignment due to defects of the eye muscles is rare. The deviant eye may become functionally blind as its unwanted image is suppressed. Severe cases may be treated using surgery.

Strabo (*c.*64 BC–AD 24), Greek historian and geographer. He travelled widely, collecting material for *Historical Sketches*, most of which have been lost. His *Geographical Sketches*, containing geographical and historical information of the ancient world, has survived almost intact.

Stracey, John Henry (1950–), British welterweight boxer. He started boxing professionally in 1969 and in 1973 beat Bobby Arthur to become British Champion. In 1974 he won the European Championship and in 1975 the World Championship, which he lost to Carlos Palamino in 1976.

Strachey, Giles Lytton (1880–1932), British author who was one of the leading members of the BLOOMSBURY GROUP. He revolutionized the art of writing biographies and they include *Eminent Victorians* (1918), *Queen Victoria* (1921) and *Elizabeth and Essex* (1928).

Stradivari, Antonio (*c.*1644–1737), Italian violin maker; his name is often latinized to Stradivarius. Apprenticed to Nicolò AMATI, he was making violins under his own name by 1666. Stradivari perfected violin design and his instruments remain unsurpassed in brilliance of tone. He is believed to have made at least 1,116 violins, violas and cellos.

Strafford, Thomas Wentworth, 1st Earl of (1593–1641), English nobleman, leading adviser of CHARLES I. Wentworth was appointed Lord President of the North in 1628 and a privy councillor in 1629. As Lord Deputy of Ireland (1633–39) he sought to strengthen royal authority, defend PROTESTANTISM and stimulate agriculture and trade. He was made an earl in 1640. Strafford's advocacy of sovereign power led to conflict with Parliament and his IMPEACHMENT in 1640. The charges could not be substantiated and the "inflexibles" in Parliament brought in a bill of ATTAINDER. Wentworth was convicted and executed, the king reluctantly acquiescing in the hope of appeasing Parliament. *See also* HC1 p.287, 287.

Strain, change in dimensions of an object subjected to STRESS. Linear strain is the ratio of the change in length of a bar to its original length while being stretched or compressed. Shearing strain describes the change in shape of a body whose opposite faces are pushed in different directions. HOOK'S LAW for elastic bodies states that strain is proportional to stress up to the body's elastic limit.

Strain gauge, device used to measure mechanical STRAIN such as that caused by the movement of a bridge or glacier. *See also* SU p.96.

Straits Settlements, possessions of the British EAST INDIA COMPANY in SE Asia which in 1867 came under British colonial authority. The Straits Settlements were dissolved in 1946, SINGAPORE becoming a Crown colony and the remainder being incorporated in the MALAYAN Union.

Strange Case of Dr Jekyll and Mr Hyde, The (1886). *See* JEKYLL AND HYDE.

Strangeness, in physics, concept introduced in 1953 by Murray GELL-MANN to explain the unexpected stability of ELEMENTARY PARTICLES such as KAONS and HYPERONS. He assigned an integral strangeness number to other "ordinary" particles, such as electrons and protons.

Strapwork, architectural surface decoration, commonly of ceilings, involving a low relief of braiding, interlaced scrollwork, cross-hatching and shield forms, often perforated with small holes. It may be executed in plaster, wood, stone, metal or papier mâché. Strapwork originated in The Netherlands *c.*1540 and was introduced to Britain in the late 16th century.

Strasberg, Lee (1901–), US theatrical director, *b.* Poland. With Harold Clurman and Cheryl Crawford he founded the Group Theater (1931), where he began teaching the Stanislavsky method of acting, which insists on the actor's inner identification with the character. Then with Elia KAZAN and Crawford he founded the Actor's Studio (1947), later becoming its artistic director.

Strasbourg, city in E France, on the River Ill 4km (2 miles) W of its confluence with the River Rhine; capital of Bas-Rhin département and the commercial capital of Alsace region. Known in Roman times as Argentoratum, the city was destroyed by the Huns in the 5th century. It became part of the HOLY ROMAN EMPIRE in 923 and developed into an important commercial centre. It became a free imperial city in 1262. Strasbourg was a centre of medieval German literature and of 16th-century Protestantism. It was seized by France in 1681, regained by Germany after the FRANCO-PRUSSIAN WAR, but recovered by France at the end of WWI. German troops occupied the city during WWII. The former University of Strasbourg dates from 1621. Industries: metallurgy, oil refining, food processing. Its river port on the Rhine is France's chief grain port. Pop. 260,300.

Strassmann, Fritz (1902–), West German chemist. In 1938 he opened the nuclear era when he discovered, with Otto Hahn, the process of splitting uranium atoms (NUCLEAR FISSION) by bombardment with neutrons. He shared the Fermi Award of the US Atomic Energy Commission with Otto Hahn and Lise Meitner.

Strategic Arms Limitation Talks. *See* SALT AGREEMENTS.

Stratford-on-Avon, county district in S WARWICKSHIRE, England; created in 1974 under the Local Government Act (1972). Area: 977sq km (377sq miles). Pop. (1974 est.) 98,900.

Stratford-upon-Avon, birthplace of William SHAKESPEARE and home of the ROYAL SHAKESPEARE COMPANY. David GARRICK held a small Shakespeare festival in Stratford in 1769 and moves to commemorate the tercentenary of Shakespeare's birth led in 1879 to the opening of a Memorial Theatre. By WWII, the season has lengthened to fill most of the year and the appointment of Peter HALL led to the grant of a royal charter in 1961.

Strathclyde, region in W Scotland, formed in 1975 from most of the former county of ARGYLL and a small part of the former county of STIRLING, and including the city of GLASGOW. Area: 13,727sq km (5,300sq miles). Pop. (1975 est.) 2,504,000.

Strathcona and Mount Royal, Donald Alexander Smith, 1st Baron (1820–1914), Scottish-born Canadian fur trader, financier and statesman. Joining the HUDSON'S BAY COMPANY in 1838, he rose to become its resident governor at Montreal and helped to finance the CANADIAN PACIFIC Railway. He became High Commissioner for Canada in London in 1896 and was made a peer the following year.

Stratification, in sociology, a hierarchical system of social organization in which the extent to which members of groups possess socially esteemed traits determines their position. Usually these traits include wealth, status and power. Basic kinds of stratification are CLASS, CASTE and ESTATE.

Stratification, layering in rocks. It occurs in sedimentary rocks and in those igneous rocks formed from lava flows and volcanic fragmental deposits. Layers range in depth and vary in shape. Strata may occur as thin sheets covering many square kilometres or as thick bodies extending only a few square metres. Separations between individual layers are called stratification planes. They parallel the strata they bound, being horizontal near flat layers and exhibiting inclination on a sloping surface. *See also* STRATUM; STRATIGRAPHY; PE pp.100, 102.

Stratigraphy, branch of geology dealing with stratified or layered rocks. SEDIMENTARY ROCKS are layered in the order of deposition, usually by water. These may then have undergone metamorphosis, folding, faulting or igneous intrusion. Some IGNEOUS ROCKS are also stratified, particularly those associated with volcanic systems that continue to produce material. *See also* PE pp.102, 124, 126.

Stratocirrus, or cirrostratus, cloud, almost transparent cloud that usually lies at a height of more than 7km (4 miles) and constitutes a continuous layer into which wispy cirrus clouds have fused. *See also* PE pp.69–71.

Stratocumulus cloud, low-lying cloud of various forms. In its most distinctive appearance each cloudlet has a puffy shape, arranged in a regular dappled pattern. It is usually at a height of less than 2km (1 mile). *See also* PE pp.69–71.

Stratosphere, section of the Earth's atmosphere between the troposphere and the higher mesosphere. It is about 40km (25 miles) thick and for half this distance the temperature is fairly constant ($-60°$ to $-50°C$ ($-76°$ to $-38°F$) and then rises to a maximum of $-1°C$ (30°F). The stratosphere contains most of the atmosphere's ozone. *See also* PE pp.66–67.

Stratum, in geology, single layer or bed of SEDIMENTARY ROCK. In biology, it is a layer of tissue, as in the stratum basale of the uterus. In archaeology, strata refer to cultural layers.

Stratus cloud, low-lying cloud (below 2km; 1 mile) which is patchy and formless. It is the sort of cloud that drifts raggedly across hilly country, obscuring local features. *See also* PE pp.69–71.

Straub, Johann Baptiste (1704–1784), German ROCOCO sculptor who worked initially in Vienna (1728–34) and later in Munich. His major works include the high altar for the church of St Michael, Berg-am-Laim, Munich, and the altar for the monastery at Ettal.

Strauss, name of a family of Viennese musicians. Johann (1804–49), organized his own orchestra in 1826 and became famous for his waltzes. His son Johann (1825–99) also formed an orchestra, in 1844. It rivalled that of his father, but the two were combined after the elder Johann's death. Johann the younger composed nearly 500 waltzes, including the popular *Blue Danube* (1867) and *Tales from the Vienna Woods* (1868). Two of his brothers, Josef (1827–70) and Eduard (1835–1916), were also successful conductors and composers.

Strauss, Richard (1864–1949), German composer and conductor. He was influenced by Richard WAGNER and was a master of highly evocative richly orchestrated PROGRAMME MUSIC. His best-known works are a number of SYMPHONIC POEMS, which include *Don Juan* (1889), *Till Eulenspiegel* (1895) and *Zarathrustra* (1896). He is also famous for his operas, such as *Salome* (1905), *Elektra* (1908) and *Der Rosenkavalier* (1911). Of the many songs he composed, most famous are those which have become known collectively as the *Four Last Songs* (1948). *See also* HC2 pp.*106*, 107, 121, 270.

Stravinsky, Igor Feodorovich (1882–1971), Russian composer, one of the most influential in the 20th century. After studying with RIMSKY-KORSAKOV (1903–06) he emigrated to Western Europe in 1910 and to the USA in 1939. His early ballets *The Firebird* (1910) and *Petrushka* (1911) earned him popularity and a reputation as a master of ballet music. His violent and primitive orchestral masterpiece *Le Sacre du Printemps* (*The Rite of Spring*, 1913) revolutionized orchestral composition and established him as a major composer. Later works included ballets modelled after TCHAIKOVSKY, such as *The Fairy's Kiss* (1928); Neoclassical instrumental and orchestral works, such as The *Violin Concerto* (1931); and experiments in serialism and 12-tone music, such as the ballet *Agon* (1957). *See also* HC2 pp.*270, 275*.

Strawberry, fruit-bearing plant of the rose family, common in Europe and Asia. It has three-lobed leaves and clusters of white or reddish flowers, and sends out runners that root and produce new plants. There are two types: one that produces fruit in early summer, and one that produces one crop in early summer and another in the autumn. Family Rosaceae; genus *Fragaria. See also* PE pp.178, 214.

Strawflower, Australian annual plant that is widely cultivated for the white, yellow, orange or red flower heads that keep their bright colour when dried. Height: to 91cm (3ft). Family Compositae; species *Helchrysum bracteatum.*

Stream of consciousness, literary style in which the thoughts and feelings of characters are given free rein, as they might come to them, without the usual literary regard for narrative continuity or logical sequence. Its first appearance in literature is generally regarded as Edouard Dujardin's novel *Les Lauriers Sont Coupés* (1888), although the phrase itself comes from William JAMES's *Principles of Psychology,* published two years later. Its foremost exponents in English have been James JOYCE and Virginia WOOLF.

Street, George Edmund (1824–81), British architect and a leader of the GOTHIC REVIVAL. He designed many churches, including St James the Less (1861) in Westminster, London. His best-known building is the Law Courts (1874–82) in London, the last great attempt to apply the ideas of the Gothic revival to public architecture. *See also* HC2 pp.122, *123*.

Street lighting, lamps sited to illuminate roads. Gas street lighting was first introduced in the 19th century to deter crime. Later large cities were lit by electric arc or filament lamps. After WWI street lighting in Britain provided vision by reflection, ie the road was lit strongly to form a bright background against which objects were seen in silhouette. After WWII improved road surfaces were found to be unsuitable for silhouette vision, and the European system of direct vision, whereby objects rather than road are illuminated, was adopted. Today sodium discharge lamps (for traffic routes) and mercury vapour lamps (for city centres) are widely used.

Streeton, Arthur (1867–1943), leading Australian painter and member of the Heidelberg school led by Tom ROBERTS. He is best known for his landscape paintings in a style which was influenced by the IMPRESSIONISTS. He spent many years abroad, and from 1916 to 1918 was the official war artist for the Australian WWI forces in France.

Streisand, Barbra (1942–), US singer and actress. Her performance on Broadway in *Funny Girl* (1964) was highly praised She has sung in concert and made numerous recordings. Her films include *Funny Girl* (1968), *Hello Dolly* (1969), and *A Star is Born* (1976).

Strepsiptera, order of parasitic insects, having about 400 species. The female is wingless and never leaves the HOST, the BEE. The male has back wings and emerges from the bee as an adult.

Streptococcus, genus of BACTERIA characterized by gram-positive spherical cells that grow in bead-like chains. They occur in the mouth, respiratory tract and intestine of man. Some are pathogenic, causing scarlet fever and throat infections, which may be treated with ANTIBIOTICS.

Streptomycin, antibiotic drug used to treat certain bacterial diseases resistant to PENICILLIN. It may produce side effects such as deafness and giddiness. Its discovery in 1944 by Selman A. WAKSMAN led to the development of other related ANTIBIOTICS. Streptomycin is used to treat tuberculosis.

Stresemann, Gustav (1878–1929), Chancellor of Germany (1923) and Foreign Minister (1923–29), the outstanding statesman of the WEIMAR REPUBLIC. He concluded the Pact of LOCARNO in 1925, in one of his many efforts to make a workable post-war settlement with the Allies. He negotiated Germany's entry into the LEAGUE OF NATIONS in 1926 and in that year shared the Nobel Peace Prize with Aristide BRIAND. *See also* MS pp.83, 123, 134, *135*.

Stress, in medicine, mental or physical strain brought on by pressures from the environment. *See also* MS pp.83, 123, 134, *135*.

Stress, in physics, quantity describing the internal pressure of an object being stretched, twisted or squeezed. If any part of the object with cross-sectional area A is subjected to a tensile or compressive force F having components F_n perpendicular to A and F_t tangential to A, then the normal stress is F_n/A and the tangential (or shearing) stress is F_t/A. In a fluid, no shearing stress is possible because the fluid slips sideways. Thus all fluid stresses are normal stresses and are denoted by the term PRESSURE.

Strigiformes, OWLS, an order of birds found worldwide. Owls are birds of prey. Their large eyes contain many more rods (cells that enable night vision) than do the eyes of other birds but no cones (cells that enable colour vision). Owls have soft feathers which allow them to fly silently There are about 120 species. *See also* NW pp.*152, 199, 203, 221, 227*.

Strijdom, Johannes Gerhardus (1893–1958), South African statesman, Prime Minister from 1954 to 1958. He practised law before being elected to Parliament in 1929 and becoming leader of the National Party, advocating APARTHEID.

Strike, employees' concerted withdrawal of their labour, defined in Britain by the Industrial Relations Act (1971) as a furtherance of an industrial dispute in breach of a contract of employment. Historically, the use of strikes developed when workers collectively tried to improve their lot through TRADES UNIONS. *See also* HC2 pp.94, *173, 195*, 210–213, *213, 222, 260*.

Strindberg, Johan August (1849–1912), Swedish dramatist. A master of language, he wrote more than 50 plays in many styles. He is best known outside Sweden for *The Father* (1887) and *Miss Julie* (1888). *See also* HC2 p.107.

Stringed instruments, variety of musical instruments which are sounded by the vibration of strings. They fall into different classes according to the action used to set the strings in motion. There are bowed stringed instruments, chiefly those of the VIOLIN family (the "strings" in an orchestra); plucked stringed instruments, chiefly the HARP, LUTE and GUITAR; and plucked and struck stringed instruments, such as the CITTERN and DULCIMER. KEYBOARD INSTRUMENTS which have strings, such as the HARPSICHORD and the PIANO, are not usually referred to as stringed instruments.

Strip cultivation, method of agriculture prevalent in medieval England. The large open fields of the MANORS were divided into strips and farmed by tenants. The method was gradually superseded by larger holdings during the period of ENCLOSURES. *See also* HC2 pp.20–21, *20–21*; PE pp.*154, 158*.

Strip grazing, method of feeding livestock on pastures that are divided into strips. Up to three or four strips can be used; one is grazed and then left to grow while the stock is moved to the second, and then the third in sequence. *See also* PE pp.*158–159*.

Stroboscope, or strobe, apparatus which emits regular flashes of light, the intervals of which can be varied. Stroboscopes usually have a calibrated scale from which the interval is read. They are used in photography to make multiple exposures of moving subjects, and in engineering to apparently "slow down" moving objects for observation. They are also used to find the speed of rotation of such objects as a wheel by adjusting the rate of flashing until it equals the rotation speed, when the wheel seems to be stationary.

Stroessner, Alfredo (1912–), President

Richard Strauss conducting his *Symphonic Poems,* drawn by Alois Klob in 1926.

Strawberry; the berries are "strewn" among the leaves, from which came its name.

Gustav Stresemann's policies at Locarno were overturned after 1933 by Hitler.

Strindberg opposed female emancipation and much of his work has a misogynist bias.

Stromboli's volcano is constantly active, although usually harmless.

Strophanthus; its seeds give an acid from which cortisone can be derived.

Gilbert Stuart's portrait of Thomas Macdonagh. He was influenced by Gainsborough.

George Stubbs' *Molly Langlegs with Jockey*, showing his mastery of horse anatomy.

of Paraguay (1954–). As commander-in-chief of the armed forces, he led the coup that overthrew the regime of Federico Chavez in 1954. He was re-elected four times (1958, 1963, 1968, 1973). His rule was essentially totalitarian but, using foreign aid, he introduced development programmes, building roads, bridges and schools.

Stroheim, Erich von (1885–1957), US actor and film director, *b.* in Austria as Hans Erich Maria Stroheim von Nordenwall. His films, including *Blind Husbands* (1919), *Foolish Wives* (1921), *Greed* (1923) and *Queen Kelly* (1928), were noted for their originality. He turned to scriptwriting and acting, and appeared in *La Grande Illusion* (1937) and *Sunset Boulevard* (1950). *See also* HC2 p.276, 276.

Stroke, or apoplexy, interruption of the flow of blood to the brain. It is caused by blockage or rupture of an artery and results in temporary or permanent paralysis, difficulty in speaking or loss of muscular co-ordination. It may occur without symptoms or after a period of headaches and irritability. Treatment includes surgery, drugs to reduce clotting and physiotherapy. *See also* MS pp.90, *90*, 96.

Stromboli, volcanic island NE of Sicily that gives its name to a characteristic type of volcanic eruption in which molten rock is blasted out of the crater by large amounts of accumulated gas. *See also* PE p.*30*.

Strong interaction, strongest of the four basic FORCES in physics. Its effects can be observed only in the sub-atomic realm, being of very short range. It is responsible for holding an atomic NUCLEUS together.

Strontianite, carbonate mineral, strontium carbonate ($SrCO_3$), found in limestone veins. It has an orthorhombic system, massive or columnar aggregates or hexagonal twinned crystals, and is pale green, white, grey, yellow or brown. Hardness 3.5–4; s. g. 3.7.

Strontium, metallic element (symbol Sr) of the ALKALINE-EARTH ELEMENTS in group IIa of the periodic table. It occurs naturally in strontianite (the carbonate) and celestite (the sulphate) and is extracted by electrolysis. Strontium salts are used to impart a red colour to flares and fireworks. Strontium-90 is a RADIOACTIVE ELEMENT present in fall-out, from which it is absorbed into milk and bones. At.no. 38, at.wt. 87.62, s.g. 2.554, m.p. 769°C (1,416°F), b.p. 1,384°C (2,523°F).

Strophanthus, genus of tropical Old World trees and shrubs with showy flowers. The seeds and bark of several African species are the source of strophanthins, ALKALOIDS used as cardiac drugs and also as poisons. Species used include *S. kombe* and *S. hispidus*. Family Apocynaceae.

Stroud, county district in SW GLOUCESTERSHIRE, W England; created in 1974 under the Local Government Act (1972). Area: 454sq km (175sq miles). Pop. (1974 est.) 93,400.

Strougal, Lubomir (1924–), Czecholovak politician and Prime Minister. A member of the Czech Communist Party since 1945, he was secretary of the central committee from 1965 to 1968 and was made deputy Prime Minister in 1968. He became Prime Minister in 1970.

Strowger, Almon, (*fl.* 1880s) US inventor of the automatic electromechanical telephone exchange (1889, patented 1891). *See also* MM p.226.

Structural analysis, study of the deformation, faulting, folding and crumpling of rock strata, often reproduced using laboratory models.

Structural functionalism, sociological system proposed by Talcott PARSONS and RADCLIFFE-BROWN. It analyses society with reference to the social and psychological "functions" of the various institutions and status "structures", which include set social relationships and beliefs. The theory concentrates on the stability and integration of the social system.

Structuralism, two related theories in sociology. One involves the study of social status systems and status-related duties in order to discover the society's "social morphology", or structure. The second,

proposed by Claude LÉVI-STRAUSS, is an inter-disciplinary theory that uses anthropology, linguistics, psychology and folklore to analyse all modes of social behaviour – from marriage rites to fashion – to build up a model of the particular society. Structuralists then attempt to construct a general model of society, to particularize the ways in which each society is unique, and to relate social structure to the structure of the human mind.

Strutt, Jedediah (1726–97), British KNITTING-machine pioneer. In 1758 he invented the Derby rib hosiery frame that could be attached to a stocking frame to produce a ribbed weave. He went into partnership with Sir Richard ARKWRIGHT in Nottingham in 1771.

Strutt, John William, Lord Rayleigh. *See* RAYLEIGH, JOHN WILLIAM STRUTT, LORD.

Struve, Friedrich Georg Wilhelm von (1793–1864) German astronomer who catalogued more than 3,000 stars, including many previously undiscovered DOUBLE STARS. In 1838 he became one of the first astronomers to measure the PARALLAX of a star, VEGA.

Struwelpeter (1903), collection of highly moral stories, told in verse by Heinrich Hoffman. The work is fully illustrated and depicts the fate of various children who would not do as they were instructed; Struwelpeter himself would not allow his hair and fingernails to be cut. The children suffer various fates including physical injury and even death.

Strychnine, poisonous ALKALOID obtained from the plant *Strychnos nux-vomica*. In the past it was believed to have a therapeutic value in small doses, as these can stimulate the nervous system. Symptoms of strychnine poisoning include muscular spasms and stiffness of posture.

Strydom, Johannes Gerhardus (1893–1958), South African statesman who became leader of the extremists in the National Party in 1929. He aimed mainly for native apartheid. From 1954 until shortly before his death he was Prime Minister.

Stuart, or Stewart, Scottish family of Breton origin, the senior branch of which inherited the Scottish crown in 1371 (ROBERT II) and the English crown in 1603 (JAMES I). Over a period of two centuries, seven Stuarts ascended the Scottish throne as minors through sudden deaths. The direct male line died out with JAMES V in 1542. His daughter, MARY, QUEEN OF SCOTS, was succeeded by her son, JAMES VI who, on the death of ELIZABETH I, became also JAMES I of England. After the execution of James's son, CHARLES I, in 1649 the Stuarts were excluded from the throne until the RESTORATION of CHARLES II in 1660. With the deposition of JAMES II in 1688, the throne passed to his daughter MARY and her husband, WILLIAM III, who were succeeded by another daughter, Queen ANNE. The throne then passed to the HANOVERIAN dynasty. The last Stuarts of the royal English line were grandsons of James II: Charles Edward STUART (Bonnie Prince Charlie) and Henry Stuart, Cardinal York, known to JACOBITES as Charles III and Henry IX respectively.

Stuart, or Stewart, Charles Edward (1720–88), English prince and claimant to the throne, also called "Bonnie Prince Charlie" and the "Young Pretender". A grandson of the deposed JAMES II and son of James Francis Edward STUART, he landed in Scotland in 1745 to claim the throne. He was initially victorious at PRESTONPANS and reached the city of Derby but was persuaded by his commanders to withdraw. A Hanoverian force under the Duke of CUMBERLAND annihilated his army at CULLODEN (1746) and he went into exile. *See also* HC2 p.28, *28*.

Stuart, Gilbert Charles (1755–1828), US artist who painted portraits in Europe and eventually settled in Boston. His best-known works are three precisely rendered portraits of George WASHINGTON and his many replicas of these.

Stuart, James (1713–1788), British architect who began the revival of Greek architectural forms and motifs in England. He was called "Athenian Stuart" after his *The Antiquities of Athens* (1762). He rebuilt

the interior of the chapel of Greenwich Hospital (1779–88).

Stuart, James Ewell Brown (1833–64), US Civil War Confederate general. At the age of 29, "Jeb" Stuart commanded Robert E. Lee's cavalry, and he became a brilliant leader. *See also* HC2 pp.150–151.

Stuart, or Stewart, James Francis Edward (1688–1766), also called "The Old Pretender", claimant to the British throne, only son of JAMES II. Brought up in France, he was proclaimed King of England in 1701 on the death of his father. In 1715 he made an unsuccessful bid to gain the English throne. He spent most of his remaining years in Rome. *See also* HC2 p.28, *28*.

Stuart, Mary. *See* MARY II.

Stubbs, George (1724–1806), British painter and etcher, famous for his mastery of horse painting. He was interested in anatomy and in 1766 published *Anatomy of the Horse*. A keen observation and a strong sense of form and spatial organization are characteristic of his paintings, which include the highly acclaimed *Phaeton and Pair*.

Stubbs, William (1825–1901), British historian and Bishop of Oxford (1888–1901). He founded the organized study of English medieval constitutional history and wrote *The Constitutional History in Its Origin and Development* (1873–78).

Stucco, coating of rough plaster on walls, generally made of lime, GYPSUM and sand. It can be coloured and variations in texture are easy to obtain.

Stud, National, establishment for the breeding of thoroughbred horses, founded in Kildare, Ireland, in 1915 when Lord Wavertree presented his stud to the British nation. It was resettled at Gillingham, in Dorset, England, in 1943 and, maintained by the Ministry of Agriculture, produced many famous racehorses. It came under the management of the Horserace Betting Levy Board in 1963, and a new National Stud was opened at Newmarket, in Suffolk, in 1967.

Stupa, Buddhist monument, found throughout SE Asia, but mainly in India and Sri Lanka, with a circular dome surrounded by a processional path, a stone railing, and sometimes topped by a mast or pole. Some of the most characteristic works were built *c.*100 BC. *See also* HC1 p.120.

Sture family, Swedish noble family which fought for independence from Danish rule under the Kalmar Union during the 15th and 16th centuries. The family include distinguished regents of Sweden: Sten Sture the Elder (*c.*1440–1503; *r.*1470–1503), his nephew, Svante Nilsson Sture (*c.*1460–1512; *r.*1503–12) and Svante's son, Sten Sture the Younger (*c.*1492–1520; *r.*1512–20).

Sturgeon, William (1783–1850), British electrical engineer who demonstrated (1825) the first electromagnet capable of useful work. It consisted of a soft iron core insulated with varnish and wound with 18 turns of bare copper wire; it was capable of supporting 4kg (9lb) and was powered by a single cell. Sturgeon also built an electric motor, invented the COMMUTATOR, and the suspended coil GALVANOMETER, and studied charges in clouds by flying kites. *See also* SU pp.118–119.

Sturgeon, large, primitive, bony fish found in temperate fresh and marine waters of the Northern Hemisphere. The ovaries of the female are the source of CAVIARE. It has five series of sharp-pointed scales along its sides, fleshy whiskers and a tapering snout-like head. Family Acipenseridae; species Atlantic sturgeon (*Acipenser sturio*) length: to 3.3m (11ft), weight: to 272kg (600lb). The Eurasian freshwater sturgeon is also called BELUGA. *See also* NW p.*124*; PE pp.244–245.

Sturges, John (1911–), US film director. He began his career as a production assistant for David SELZNICK, becoming a director after WWII. He made a number of WESTERNS including *Bad Day at Black Rock* (1954), *Gunfight at the OK Corral* (1956) and *The Magnificent Seven* (1960). He also made *The Great Escape* (1963).

Sturm und Drang ("Storm and Stress"), name of a German literary movement associated principally with the early writings of Johann Gottfried von HERDER, Johann Wolfgang von GOETHE and Friedrich SCHILLER. The term was first employed by Christoph Kaufmann to describe Friedrich Maximilian KLINGER's play *Der Wirrwarr* (1776). Characteristic elements of Sturm und Drang writings include medieval knights, sieges and battles and hostile brothers.

Sturt, Charles (1795–1869), Australian explorer, *b.* India. After military service he went to Australia in 1827. On his first expedition (1828–29), with Hamilton HUME, he discovered the Darling and Murray rivers. Convinced that he would find an inland sea, he led an expedition to the edge of the Simpson Desert (1844–46) but, faced with the hostile conditions, was forced to retreat. Sturt retired to England in 1853.

Stutter, SPEECH DEFECT in which sounds or syllables are continually repeated. Caused by a variety of factors, including a lack of co-ordination of the muscles of the LARYNX, it may also include a psychological component.

Stuttgart, city in sw West Germany, on the River Neckar 61km (38 miles) ESE of Karlsruhe; capital of Baden-Württemberg state. Founded *c.*950, the city became the seat of the counts (later dukes and kings) of Württemberg from 1495 to 1806. Stuttgart was severely damaged during WWII. Buildings of interest include the Solitude Palace (1763–67), New Palace (1746–1807) and the Rosenstein Palace (1824–29); the city is also known for its fine modern architecture. Industries: publishing, textiles, chemicals, paper, wine, iron and steel, machinery, electrical equipment, motor vehicles, musical instruments, tourism. Pop. (1970) 633,158.

Stuyvesant, Peter (*c.*1610–72), Dutch colonial governor. He became Governor of the Caribbean Islands of Curaçao, Bonaire and Aruba in 1643, and in 1647 he was named Director-general of all the Dutch territories, including New Amsterdam (later New York City). As governor he ruled in an autocratic manner, denying religious and political freedom and causing discontent among the colonists. He ended Swedish influence in Delaware (1655). He ruled until the colony was taken over by the English in 1664 and renamed New York.

Stye, infection of an eyelid gland, caused by a STAPHYLOCOCCUS organism. It may occur externally, on the edge of the lid, or internally, under the lining. A small boil or pimple with a central yellow spot erupts, then breaks and discharges its contents.

Style, in botany, the slender tube that forms part of the pistil of a flower, and connects the pollen-receiving STIGMA at its tip to the ovary at its base. *See also* NW p.71.

Styrene, colourless, liquid HYDROCARBON, important as a MONOMER of POLYSTYRENE, a plastics material familiar as hard foams. It is also a monomer of styrene-butadiene rubber, or SBR, the artificial rubber most commonly used in pneumatic tyres. Styrene is a derivative of benzene; its formula is $C_6H_5CH = CH_2$. Properties: m.p. 145.2°C (293.4°F), b.p. −30.6°C (−23.1°F). *See also* MM pp.54–55.

Styx, in Greek mythology, the river across which the souls of the dead were ferried by CHARON on their journey from the world of the living to HADES or the underworld.

Subatomic particles, discrete states of matter characterized by various parameters such as CHARGE, mass, lifetime and SPIN. Matter is fully described in terms of the forces that hold it together of which there are four: GRAVITY; the electromagnetic force, which is millions of times stronger than gravity andsresponsible for the attraction between the nuclei of atoms and orbiting electrons; the STRONG INTERACTION or force, which is the nuclear "cement", some 7,000 times stronger than the electromagnetic force and which acts over extremely short distances (less than 10^{-14}m); and the WEAK INTERACTION, some one-thousandth of the strength of the electromagnetic force, and observed in the transformations of some particles,

such as that in beta decay when a neutron changes to a proton, an electron and an antineutrino. Subatomic particles which have been recognized are classified into two large groups, based on whether they undergo interactions determined by either the electromagnetic or strong interaction. In the first group are the LEPTONS, such as NEUTRINOS, ELECTRONS and POSITRONS; in the second are the HADRONS, which are further split into NUCLEONS, such as PROTONS, NEUTRONS and their anti-particles, HYPERONS such as lambda zero, sigma minus and xi plus, and the third group in the hadrons, the MESONS such as K plus and pi minus. *See also* SU pp.66–67.

Subconscious, any of the mental processes that occur just below the level of awareness. Psychoanalysts define it as a zone between the conscious and the unconscious and believe it is where many important psychological activities occur. Much of psychoanalysis involves bringing into awareness subconscious processes.

Sub-culture, in sociology, any cultural group bound together by shared attitudes or interests. Sub-cultures often develop life-styles that differ from and conflict with those of the larger culture or society within which they exist.

Subjection of Women, The (1869), political treatise by John Stuart MILL. It discusses the political, social and economic reasons for the inferior status of women in British society.

Sublimation, change from solid to gas, without an intervening liquid phase. Dry ice (solid carbon dioxide) sublimes from the solid phase directly to the gaseous phase. Most substances can sublimate at certain pressures, but usually not at atmospheric pressure. *See also* CONDENSATION; EVAPORATION.

Sublimation, in psychology, the substitution of socially acceptable behaviour or motives for needs that are not attainable or acceptable. For example, a hostile person might become a literary critic; an aggressive individual, an athlete. Thus the individual's needs are met by channelling destructive impulses into constructive activities.

Submachine gun, automatic gun smaller than a machine gun, in which the exhaust gases power the reloading mechanism. The earliest example was a Luger, used in WWI, and many submachine guns were used in WWII. *See also* MM p.165.

Submarine, seagoing warship capable of travelling both on and under the water, armed principally with torpedoes. Submarines were planned in the 16th century, but not until Cornelis Brebbel, a Dutch inventor, took a greased leather submarine underwater *c.*1620 were they built. Experimental submarines were used in warfare in the late 18th and 19th centuries, to little effect. Technical advances of the late 19th century led to the general spread of underwater craft in the world's navies. Early submarines were essentially surface ships with a limited ability to remain submerged. Once underwater, they depended on battery-powered electric motors for propulsion and, with a limited air supply, were soon forced to surface. Submerging is accomplished by letting air out of internal ballast tanks; trimming underwater is done by regulating the amount of water in the ballast tanks with pumps; and surfacing is accomplished by pumping ("blowing") the water out of the tanks. The most modern submarines use nuclear power, which eliminates the need to surface while on operations. *See also* NUCLEAR SUBMARINES; MM pp.118, 174.

Submarine cables. *See* CABLE.

Submarine canyon, deep, steep-sided valley that cuts through a continental shelf. Some are aligned with large land rivers, making it likely that they are formed when sediment deposited on the shelf becomes unstable, and gravity causes it to slump. This creates a turbid bottom current that races downslope on to the deep-sea floor, scouring out the canyons. *See also* PE p.82.

Submersible, small mobile underwater craft. It is distinguished from a submarine in that its uses are not military, and from a

bathysphere in that it is self-powered. Submersibles were developed in the late 1950s for engineering and research work; they carry a small crew and often have mechanical tools outside the craft. *See also* MM p.119; PE p.93.

Subparticles, elementary, postulated constituents of the HADRON group of elementary particles. Experimental evidence indicates that PROTONS are made up of tiny point-like bodies, termed partons. There is also evidence that the parton is the theoretical QUARK. Although the eight LEPTONS appear to be point-like particles, the large number of hadrons could be explained by various combinations of two or three quarks. *See also* SU p.66.

Sub-plot, series of actions subsidiary to the main action in a play or novel. It either complements or parallels the main action and many of Shakespeare's plays contain sub-plots. In *King Lear* the Gloucester-Edgar-Edmund sub-plot parallels the development of the tragedy of Lear.

Subpoena, in law, means under penalty, an order that commands a person to appear before a court or judicial officer to give evidence at a specific time and place. Failure to obey a subpoena is a criminal offence.

Subscript, small letter, number or other symbol written or printed below and to the left or right of another symbol, as in a chemical formula (such as C_2H_5) or with different values, x_1, x_2, $x_3,...x_n$, of a mathematical variable.

Subset. *See* SET THEORY.

Subsidiary, in business, a company that is controlled to some extent by another company which holds a majority of the shares.

Subsidies, government assistance to individuals or organizations to benefit the public. Subsidies have been used in Britain since WWI, when they were introduced to keep down the price of food and clothing. They are generally used to protect home industries by enabling them to offer competitive prices in domestic or overseas markets. Other commonly subsidized enterprises include agriculture (the subsidization of farm prices has been widespread in Britain since WWII), business expansion, housing and regional development. Subsidies can be direct (for example, cash payments) or indirect (for example, when the government buys goods at artificially high prices or grants tax concessions, export incentives or rent subsidies).

Subsistence economy, condition in which economic activity provides only the bare necessities for survival; items are produced for consumption by the people who produce them, not for sale in a market. In such primitive economies, money and trade are not yet in evidence and the standard of living is usually very low.

Substitution, chemical reaction in which one atom or group of atoms replaces (usually in the same structural position) another group in a molecule or ion. Chemists differentiate electron-rich incoming groups, such as the hydroxyl and halogen ions, from electron-deficient groups, such as the hydroxonium ion (H_3O^+) and the nitro group ($NO_2^{\ +}$), and unstable uncharged species called radicals.

Succubus. *See* INCUBUS.

Succulent, plant that stores water in its tissues to resist periods of drought. Usually perennial and evergreen, most of the succulent plant body is made up of water storage cells, which give it a fleshy appearance. A well-developed cuticle and low rate of daytime transpiration also conserve water. Succulent plants include CACTUS, MILKWEED, LILY and STONECROP.

Sucker, any of several species of freshwater fish found mainly from N Canada to the Gulf of Mexico. A bottom-grubber similar to minnows, it has a thick-lipped mouth for feeding by suction. Length: to 66cm (26in); weight: to 5.4kg (12lb). Family Catostomidae.

Sucking-fish. *See* REMORA.

Suckling, Sir John (1609–42), English Cavalier lyricist, once described as "the greatest gallant of his time, and the greatest gamester". After plotting unsuccessfully to rescue the Earl of Strafford from the Tower of London he fled to

Stuttgart's ball-bearing and machine-tool factories were heavily bombed in WWII.

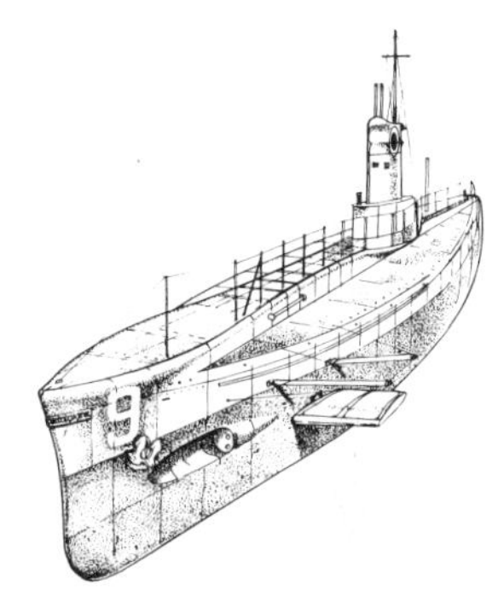

Submarine: tanks fore and aft give buoyancy, those on the side keep it level.

Succulents are found in deserts and low-rainfall areas throughout the world.

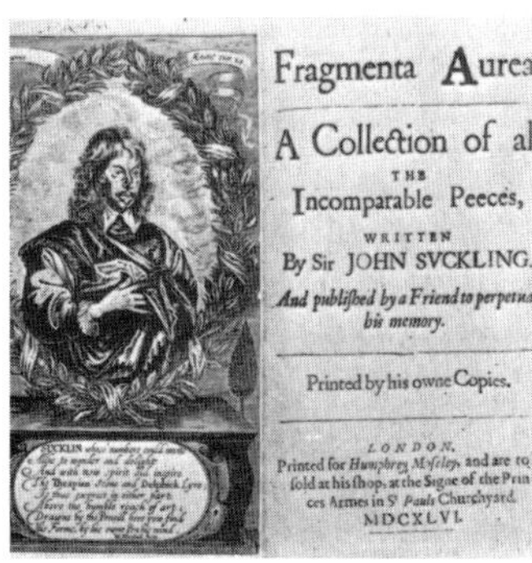

Sir John Suckling: facsimile of his portrait and title-page of Fragmenta Aurea.

Sucrase

Suez Canal; a navy salvage vessel beside a sunken ship blocking the canal in 1956.

Suffolk; a timber-framed house, typical of buildings in the county.

Sugar-beet; the sugar contains up to 98% sucrose, and 99.9% after refining.

Sugar cane; after a crop has been cut, it will send up new crops for many years.

Paris, where he reputedly committed suicide. His works include a collection of poems, *Fragmenta Aurea* (1646) and a play, *Aglaura* (1637).

Sucrase, ENZYME secreted by the small intestine which breaks down SUCROSE (ordinary sugar) into GLUCOSE and FRUCTOSE so they can be absorbed during the digestive process.

Sucre, Antonio José de (1795–1830), South American patriot, BOLÍVAR's chief of staff, key figure in the liberation of Ecuador, Peru and Bolivia. Sucre commanded the rebel forces at the battle of Ayacucho (1824), the decisive engagement in the struggles for independence. He was elected first constitutional President of Bolivia in 1826 and proved an able administrator, but political rivalries forced him to resign in 1828. He took service with Columbia but was assassinated in 1830.

Sucre, city in s central Bolivia; it is the legal capital of Bolivia, the seat of government being La Paz. Known successively as La Plata, Chuguisaca and Charcas, Sucre was renamed in 1839 after the revolutionary leader and first president of Bolivia, Antonio José de SUCRE. The city is the seat of the national university of San Francisco Xavier de Chuquisca (1624). Today it is a commercial and distribution centre for the surrounding farming region, and has cement and oil refining industries. Pop. (1975 est.) 106,590.

Sucrose, disaccharide sugar (formula $C_{12}H_{22}O_{11}$) consisting of linked GLUCOSE and FRUCTOSE units. It occurs in many plants, but its principal commercial sources are sugar cane (*Saccharum officinarum*) and sugar-beet (*Beta vulgaris*). It is also obtained from maple trees, date palms and sorghum, and is widely used for food sweetening and in the manufacture of preserves. *See also* SU p.152, *152.*

Suctorida, order of PROTOZOA having cilia in their immature motile stages before developing into non-motile unciliated adults. They have long hollow protoplasmic extensions which are used to suck in prey. Genera include *Tokophrya.*

Sudan, independent nation in NE Africa. The largest country in the African continent, the Sudan's terrain includes tropical forests, swamplands, arid hills and the Libyan and Sahara deserts. The White Nile flows through the country, providing valuable irrigation in the N. Most of the people are subsistence farmers or nomadic pasturalists. Exports include cotton, groundnuts and sesame. The capital is Khartoum. Area: 2,505,813sq km (967,494sq miles). Pop. (1975 est.) 17,757,000. *See* MW p.160.

Sudd marshes, swampy lowland region in the s central Sudan, fed by the Bahr al-Ghazal and the Bahr-al-Jebel. The area is inhabited by Nilotic NUER tribes. Length: 403km (250 miles). Width: 322km (200 miles).

Sudetenland, name given to a strip of territory along the Sudeten Mts, located in Czechoslovakia but inhabited by German-speaking people. From 1526, under HAPSBURG rule, Germanic influences permeated throughout Bohemia and Moravia, remaining dominant until a Czech nationalist movement began in the 19th century. The new independent country of Czechoslovakia was created after WWI but the German-speaking parts were ceded to HITLER in the MUNICH AGREEMENT of 1938. After WWII Czechoslovakia reclaimed Sudetenland and expelled most of the Germans. *See also* HC2 pp.218, *218, 226,* 227.

Suede, LEATHER which has been given a nap finish by buffing its flesh side with fine sandpaper. After the hide or skin has been cleaned by soaking in water, scraping and soaking again in a mild acidic solution, it is tanned, ie treated with any of various acid solutions which change the skin to leather. The leather is then dyed and given the desired finish, which may be smooth or suede.

Suetonius (Gaius Suetonius Tranquillus) (*c.* AD 69–*c.*140), Roman author. A lawyer and secretary to Emperor HADRIAN, Suetonius is best-remembered as a biographer. Two collections of his work survive: parts of *De viris illustribus* (*On Famous Men*), biographies of literary figures, and *De vita Caesarum* (*The Lives of the Caesars*), biographies of rulers from Julius CAESAR to DOMITIAN.

Suez, city in NE Egypt, at the N end of the Gulf of Suez and the s terminus of the SUEZ CANAL. A naval and trading station in the 16th century, it developed into a leading port in 1869 after the completion of the Suez Canal. The city was occupied by Israelis in the 1973 ARAB-ISRAELI WAR and suffered damage during the fighting. Industries include oil refining and paper making. Pop. (1974 est.) 368,000.

Suez Canal, waterway linking the Mediterranean Sea and the Red Sea. It was planned and built (1859–69) by the Suez Canal Company under the supervision of Ferdinand de LESSEPS. After the introduction of steam vessels, its use increased and DISRAELI purchased (1875) the Egyptian-held stock to make the British government the largest stockholder. In 1955 more than 120 million tonnes of merchandise passed through the canal, much of it oil. Egypt nationalized the canal in 1956, Israel attacked Egypt in the same year and the canal was closed from 1956 to 1957, while repairs were being made. Egypt gained control of the canal and closed it during the ARAB-ISRAELI WAR of 1967; it was not re-opened until 1975. In the intervening period many ships, especially oil tankers, were built for use on alternative sea routes and were too large to pass through the canal. The subsequent loss of revenue forced Egypt to begin to clear and widen the waterway. *See also* HC2 pp.*131, 133,* 155, *219,* 237, 297.

Suez Crisis (1956), Middle East conflict. When Britain and the USA announced that they would not provide financial assistance for Egypt's ASWAN Dam project, the President of Egypt, Gamal Abdel NASSER, nationalized the SUEZ CANAL. Israel, denied use of the canal, invaded Egypt. Britain and France then also invaded the canal area. UN intervention ended the crisis, and treaties gave Egypt control of the canal.

Suffolk, county in E England, on the North Sea coast. The land is mainly low-lying, rising to the East Anglian Heights in the SW. The principal rivers are the Orwell, Stour and Waveney. The economy is mainly agricultural; cereal crops and sugar-beet are grown and sheep, cattle, horses and poultry are reared. Fishing is important along the coast. The principal industries are food processing, farm machinery and fertilizers. The county town is Ipswich. Area: 3,807sq km (1,470sq miles). Pop. 577,600.

Suffolk Coastal, county district in E and SE SUFFOLK, England; created in 1974 under the Local Government Act (1972). Area: 892sq km (344sq miles). Pop. (1974 est.) 94,500.

Suffrage, in politics, the right to VOTE in BALLOTS or ELECTIONS, considered to be one of the primary elements of DEMOCRACY. Although mass suffrage became accepted as desirable only in the 20th century, most countries of the world now profess to accept the right of all adults to vote freely in national elections. In Britain successive REFORM BILLS in the 19th century ensured suffrage for adult males, but not until 1928 were women over 21 finally allowed to vote; Swiss women did not obtain suffrage in federal elections until 1971. *See also* HC2 pp.160–161, *160–161,* 164.

Suffragette movement, campaign of women in Britain in the late 19th and early 20th centuries to win for women the right to vote. It began in the 1860s, and developed until the founding of the National Union of Women's Suffrage Societies in 1897. Emmeline PANKHURST founded the Women's Social and Political Union in 1903. By 1910 the movement had split into several factions, including the Women's Freedom League (1908). In 1913 Sylvia Pankhurst founded the East London Federation, which organized mass marches in London. Women of the age of 30 and over were given the vote in 1918.

Sufism, mystic philosophical movement within ISLAM which developed among the SHI'A communities in the 10th and 11th centuries. Sufis stress the capability of the soul to attain personal union with God. Orders within Sufism include the DERVISHES, who attain this union by ecstatic trances and dancing. The finest examples of classical Persian poetry were written by Sufi devotees, the most famous being OMAR KHAYYÁM, RUMI and HAFIZ.

Sugar, sweet-tasting soluble crystalline monosaccharide or disaccharide CARBOHYDRATE. The common sugar used in food and beverages is the disaccharide SUCROSE. *See also* GLUCOSE; SU p.152.

Sugar Act, or Revenue Act (1764), British legislation imposed on the American colonies to halt smuggling of foreign sugar. The Act reduced the duty on imported molasses and increased duties on refined sugar from England, effectively giving British West Indian sugar producers a monopoly of the American market. It caused much resentment among the Americans and contributed to the War of AMERICAN INDEPENDENCE. *See also* HC2 p.64.

Sugar-beet, variety of BEET grown commercially for its high SUGAR content, which is stored in its thick white roots. Species *Beta vulgaris. See* PE pp.188, 196–197.

Sugar cane, perennial GRASS cultivated in tropical and subtropical regions throughout the world. After harvesting, the stems are processed in factories, becoming the main source of SUGAR. Most cultivated canes are *Saccharum officinarum.* Height: to 4.5m (15ft). Family Gramineae. *See* NW p.68; PE pp.196–197.

Sugar glider, small arboreal marsupial. It can glide up to 55m (180ft) using the flap of skin between its fore and hind legs. Length: to 80cm (32in). Family Petaurus. *See also* NW p.*159.*

Sugar maple. *See* MAPLE.

Suger, Abbot (1081–1151), French cleric who helped to strengthen the French monarchy against the decentralizing forces of feudalism. As adviser to Louis VI and as regent when Louis VII left on the Second Crusade, he devoted his energies to upholding strong central government. He is famous for having commissioned the abbey church of St Denis, built in a new style which became known as Gothic.

Suggestion, in HYPNOSIS, imposition or implanting of notions or ideas in the memory of the hypnotized subject.

Su Han-chen (*fl.* 12th century), Chinese painter who was best known for his pictures of children.

Suharto (1921–), Indonesian military and political figure. A lieutenant-colonel in a guerrilla army that fought for independence from the Dutch (1945–49), Suharto seized power from President SUKARNO in 1966. He was formally elected President in 1968 by the Consultative Congress and was re-elected in 1973.

Suhrawardi, as- (*c.*1155–91), Iranian-born Islamic philosopher and theologian. His thinking is still in evidence in Iran today among mystical sects, including the Nūrbakhshīyah, itinerant holy men who trace the origin of their order to him. His principal work is *Hikmat al-ishrāq* (The Wisdom of Illumination). He was killed by Malikaz-Zâhir, the son of Saladin.

Suicide, intentional act of terminating one's own life; also a person who does so. Although most suicide attempts are committed by females, males are about twice as successful in completing the act. Suicide rates per 100,000 population vary from fewer than 3 in Mexico to about 11 in Great Britain, 15 in the USA, 20 in France and Sweden and 40 in Hungary. *See also* MS pp.104, 262.

Sui dynasty (581–618), Chinese dynasty whose two rulers (Yang Chien and Yang Kuang) re-established strong central rule in China following more than 250 years of division and contention. The Grand Canal was completed, the GREAT WALL OF CHINA refortified and many towns beautified, all at great human cost in taxation and forced labour. After defeat at the hands of Eastern Turks (615), the Sui dynasty was overthrown and succeeded by the T'ANG. *See also* HC1 p.123, *123.*

Suite, musical form comprising a number

of instrumental dances, which differ in metre, tempo, rhythm and mood but are generally all played in the same key. The earliest suites date from the 16th century and involved usually only two dances: the PAVANE and GALLIARD, which were played to open ceremonial balls. By the 18th century the dances included in the form had become standardized: a prelude was followed by an allemande, courante, saraband and gigue. Within this structure, however, there was some flexibility and the minuet, gavotte, bourrée and rondeau were often added. The suite was a popular form of composition in the Baroque era and outstanding composers included LULLY, RAMEAU, CORELLI, PURCELL, COUPERIN and J. S. BACH, who wrote the six *French Suites* and six *English Suites* for harpsichord.

Sukarno, Achmad (1901–70), Indonesian politician, President from 1947 to 1967. A leader of his country's independence movement, he was jailed by the Dutch and exiled (1933–42). When the Japanese invaded Indonesia during WWII, he co-operated with them and at the end of war became the first President of INDONESIA. In 1966 anti-Communist Indonesian military leaders forced a reduction of his powers, and he was deposed in 1967.

Sukkoth, or Sukkat, Jewish festival of tabernacles, which lasts for seven days. It commemorates the wandering of the Jews in the desert and their salvation through God. A sukkat, or simple tent, is raised in the synagogue (formerly in all Jewish homes).

Sukkur, city in W Pakistan, on the River Indus approx. 360km (225 miles) NNE of Karāchi. It is the site of the Sukkur barrage across the Indus which controls one of the largest irrigation schemes in the world, watering more than 2 million hectares (5 million acres). Built between 1923 and 1932, the dam is approx. 1,520m (5,000ft) long. Industries: textiles, foodstuffs. Pop. (1972) 158,876.

Sulawesi. *See* CELEBES.

Suleiman I, or Süleyman the Magnificent (1494–1566), OTTOMAN sultan (*r.*1520–66), son and successor of SELIM I. He extended his father's conquests in the Balkans, the Mediterranean and Asia, captured Belgrade (1521), expelled the KNIGHTS HOSPITALLERS from Rhodes (1522) and defeated the Hungarians under LOUIS II at MOHACS (1526). After the death of JOHN I of Hungary (1540) he annexed much of the country. He made an alliance with FRANCIS I of France against the HAPSBURGS (1536) and his admiral, BARBAROSSA, ravaged the coasts of Spain, Italy and Greece. His Turkish fleet was to control the Mediterranean until 1571. It took Tripoli in 1551 but failed to take Malta. His campaigns against Persia completed the conquest of Iraq and the area around Lake Van. He died during the seige of Szigetvar (Hungary) and his son SELIM II succeeded him. *See also* HC1 pp.220–221, *221.*

Sulla, Lucius Cornelius (138–78 BC), Roman general and statesman who brought the Jugurthine war with NUMIDIA to an end through diplomacy. He was appointed consul in 88 and dictator in 82. He instituted constitutional changes and reformed the courts, but most of his reforms were reversed soon after his retirement (79). *See also* HC1 pp.*92*, 96, *96.*

Sullivan, Sir Arthur Seymour (1842–1900), British composer famous for a series of operettas written with William GILBERT. They included *H.M.S. Pinafore* (1878), *The Pirates of Penzance* (1879) and *The Mikado* (1885). Sullivan also composed one opera *Ivanhoe* (1881), oratorios, cantatas, and church music including many hymns such as *Onward, Christian Soldiers*. His non-vocal music includes overtures, a cello concerto (1866) and piano music.

Sullivan, James ("Jim") (1903–), British rugby league full-back for Great Britain and Wigan. He scored 5,578 points in first-class matches, made three tours of Australia, and played in 25 Tests (1924–34). As a coach he produced outstanding teams for Wigan and St Helens.

Sullivan, John Lawrence (1858–1918), US boxer, who defeated Jake Kilrain in 1889 in Richburg, Missouri, in the last bare-knuckle heavyweight championship bout.

Sullivan, Louis Henry (1856–1924), US architect, a founder of the Chicago School. He believed the exterior forms of buildings should relate to their functions and opposed the Classical revival dominating the later 19th century. Most of his best work was undertaken in partnership with Dankmar Adler (1881–1900). The Wainwright Building in St Louis (1890; which, with its brick-clad steel framework, was the first skyscraper) and the Transportation Building at the 1893 World's Exhibition in Chicago are among his best-known works. *See also* HC2 p.*153.*

Sully, Maximilien de Béthune, Duc de (1560–1641), French Minister and financier under HENRY IV. A HUGUENOT, he narrowly escaped the Massacre of St BARTHOLOMEW'S DAY (1572), and later served the Protestant Henry of Navarre in his struggle for the throne. Once crowned as Henry IV, the King employed Sully as his most trusted counsellor.

Sully-Prudhomme, René-François-Armand (1839–1907), French poet associated with the PARNASSIANS. His early verse is subjective and melancholic. His chief works are two long philosophical poems, *La Justice* (1878) and *Le Bonheur* (1888). In 1901 he was awarded the first Nobel Prize in literature.

Sulpha drugs, class of drugs, often called sulphonamides, extremely effective against bacterial infection. The first sulpha drug, a red dye called PRONTOSIL, was discovered by Gerhard DOMAGK in 1932.

Sulphapyridine, one of the sulphonamides, or SULPHA DRUGS. It was first used in 1937 to treat bacterial infections and showed greater activity than any previous drug of this type. But its poisonous effects, particularly on the kidneys, gradually led to its disuse; it was replaced by other sulpha drugs and by PENICILLIN.

Sulphate. Salt of SULPHURIC ACID, H_2SO_4.

Sulphation, formation of a salt or ESTER of SULPHURIC ACID. The term is frequently used to denote an undesirable side-effect as in the formation of lead sulphate in a car battery.

Sulphide, chemical compound derived from hydrogen sulphide, H_2S, in which another element or group replaces the hydrogen. An inorganic example is lead sulphide, PbS, which occurs as the mineral GALENA; an organic example is ethyl mercoptan, C_2H_5SH.

Sulphonation, introduction of a sulphonic acid group, $-SO_3H$, into a molecule, widely used in the manufacture of dyes, SULPHA DRUGS etc. Many detergents are manufactured by sulphonation of a HYDROCARBON.

Sulphonic acids, organic chemical compounds having the general formula RSO_3H, where R is a HYDROCARBON group. Examples are ethane sulphonic acid, $C_2H_5.SO_3H$, and benzene-sulphonic acid, $C_6H_5.SO_3H$. Sulphonic acids are used as chemical intermediates and their salts, the sulphonates, are widely used in DETERGENTS.

Sulphur, non-metallic ELEMENT in group VIa of the PERIODIC TABLE, known since prehistory. It may occur naturally as a free element, or in sulphide minerals such as GALENA and IRON PYRITES or in sulphate minerals such as GYPSUM. The main commercial source is native (free) sulphur, extracted by the FRASCH PROCESS. Sulphur is used in the VULCANIZATION of rubber and in the manufacture of drugs, dyes, fertilizers, etc. Properties: at.no. 16; at.wt. 32.064; s.g. 2.07 (rhombic), 1.957 (monoclinic); m.p. 112.8°C (235.0°F) (rhombic), 119.0°C (246.2°F) (monoclinic); b.p. 444.6°C (832.3°F). Most common isotope S^{32} (95.1%).

Sulphur dioxide, colourless, poisonous gas, formula SO_2, with a pungent odour. It is used in the manufacture of SULPHURIC ACID and as a refrigerant, bleaching agent and preservative. Properties: m.p. −75.5°C (−103.9°F); b.p. −10.0°C (14.8°F); density 2.2 (air = 1).

Sulphuric acid, or oil of vitriol, colourless, poisonous gas, formula H_2SO_4, one of the strongest acids known. It is used in the manufacture of fertilizers, drugs and a wide range of chemicals. Properties: m.p. 10.3°C (50.5°F); s.g. 1.84.

Sulphur trioxide, inorganic compound, SO_3, the anhydride of SULPHURIC ACID, ie it reacts with water to make the acid. At room temperatures it can exist either as a liquid or as a solid. It further dissolves in sulphuric acid to make fuming sulphuric acid, or oleum, $H_2S_2O_7$.

Sultan, Tuanku Yahya Petra ibni Al Marhum Sultan Ibrahim (1917–), ruler of the state of Kelantan, elected by the rulers of the nine Malay states as the Supreme Head of State of Malaysia in Sept. 1975.

Sumac, any of the shrubs and trees of the genus *Rhus*, widely distributed in temperate regions. They have long feather-like leaves and red, hairy fruit clusters. Some species are poisonous to touch. Height: to 9m (30ft). Family Anacardiaceae. *See also* POISON IVY.

Sumatra (Sumatera), island in W Indonesia; second-largest of the Greater Sunda group. The W coast is rugged and mountainous, the Barisan Mountains rising to 3,810m (12,500ft). Sixty per cent of the lowland area is jungle. By the 7th century, India had established two states in Sumatra – Melayu and Srivijaya. The Portuguese landed on the island in the 16th century and the Dutch followed a century later. Britain held certain parts of the island briefly in the 18th and 19th centuries. Sumatra became part of Indonesia in 1950. The principal cities are Palembang, Medan and Padang. Exports: rubber, tobacco, palm oil, tea, coffee, sisal. Mining and farming are the chief occupations. Area: 473,970sq km (183,000sq miles). Pop. (1971) 20,812,700.

Sumer, the southern region of ancient MESOPOTAMIA, later the southern part of BABYLONIA, now south central Iraq. An agricultural civilization flourished in the area during the 3rd and 4th millennia BC. The Sumerians built canals, established an irrigation system, and were skilled in the use of metals, such as silver, gold and copper, and in the making of pottery, jewellery and weapons. They are credited with the invention of the CUNEIFORM system of writing. Several kings founded dynasties at Kish, Erech and Ur and were able to control large areas of land. SARGON of Akkad brought the region under the SEMITES *c.*2600 BC, and the two cultures merged. The final Sumerian civilization at Ur fell to tribes from ELAM, and the Sumerian nation disappeared completely when it was brought under the control of Hammurabi and Semitic Babylonia. *See also* HC1 pp.28–33.

Sumerians, inhabitants of SUMER, the most southerly part of ancient BABYLONIA and the site of an ancient civilization, the origins of which can be traced back to the 3rd millennium BC. The Sumerian language is the oldest written language. The Sumerians developed a remarkably sophisticated artistic culture and technology, which is represented in surviving artifacts and buildings as well as canals and dams. Their religious practices were centred on a pantheon consisting of many deities, each with special qualities. *See also* HC1 pp.28, *29,* 30–31, *30–31.*

Sumer is icumen in (13th century), English song and the earliest known vocal composition for six voices. It is one of the most important primary source documents of the time for musical scholarship.

Summa Theologica (1266–73), scholastic treatise by St Thomas AQUINAS. In this work Aquinas attempted to reconcile Aristotelianism with faith and revelation, acknowledging both as equal and complementary routes to truth. *See also* HC1 pp.188–189.

Summation, in mathematics, finding the total, as of a column, sequence or array of numbers, or of an infinite series of terms. The sum of all the terms of an infinite series is a finite quantity if the series is convergent (if successive terms are smaller); if the series is divergent (if the terms get larger), the sum will be infinite. The symbol Σ (sigma) is used to indicate a summation.

Sukarno in 1966 at the time of an abortive coup d'etat against his regime.

Louis Sullivan; the Gage building in Chicago, which he built between 1898–99.

De Béthune Sully received 300,000 livres when he resigned on Henry IV's death.

Sumatra; a rubber tapper cutting a strip of bark, from which latex then flows.

Sumo wrestling; two champions during a tournament in Tokyo, in 1956.

Sumter; the fort under attack from Confederate forces in 1861.

Sun bears are adept at climbing and spend most of their day in the tops of trees.

Sunbird; males seldom help females build nests but pairs stay together all year.

Summer, warmest of the four seasons of the year, experienced in all latitudes outside the tropics, during which plant growth and animal populations are generally at their height. In the northern hemisphere the summer SOLSTICE extends from 21 or 22 June to 22 or 23 September. In the southern hemisphere the period is from 22 or 23 December to 20 or 21 March.

Summer time, also called daylight saving scheme, the system by which clocks in Britain are kept one hour ahead of GMT (Greenwich mean time) throughout the summer. Other countries, including Italy, have similar daylight saving schemes. *See also* MW p.188.

Summit meeting, term meaning the meeting of heads of state or their senior representatives to discuss an urgent point at issue between them. Although such meetings are age-old, the term was first applied to the Geneva summit of July 1955, attended by Nikita Khrushchev, Dwight D. Eisenhower, Anthony Eden, Edgar Faure and Nikolai Bulganin.

Summons, in criminal law, notice presented to a person requiring him to attend court to answer a criminal charge. In civil law it is a writ commencing an action and states the nature of the claim.

Sumner, James Batcheller (1887–1955), US biochemist who shared the 1946 Nobel Prize in chemistry with John NORTHROP and Wendell STANLEY for his discovery that ENZYMES could be crystallized. In 1926 Sumner isolated and crystallized the enzyme urease; it was the first enzyme to be crystallized and this achievement finally proved the protein nature of enzymes.

Sumner, William Graham (1840–1910), US sociologist and anthropologist, one of the founders of US sociology. He was ordained an Episcopal clergyman but preferred to teach political and social science at Yale University (1872–1909).

Sumo wrestling, traditional and popular sport of Japan. Wrestlers, who usually weigh more than 159kg (350lb), oppose each other in a match that is quasi-religious in nature and is fought with much ritual. The matches begin from a crouch position on the mat and are generally of brief duration. The object is to force the opponent out of the circular ring. The technique resembles the Western catch-as-catch-can method of wrestling, with a series of holds, trips and falls employed to overcome an opponent. *See also* WRESTLING; MS p.304.

Sumter, Fort, fort in S Carolina, USA, on the S shore of Charleston harbour. The attack on it in Apr. 1861 by the CONFEDERATE STATES marked the beginning of the AMERICAN CIVIL WAR.

Sun, yellow dwarf star at the centre of the SOLAR SYSTEM; one of approx. 100,000 million stars that make up our GALAXY. It is situated in one of the spiral arms, some 32,000 light-years from the galactic centre. Of spectral type G2, the Sun is composed mainly of hydrogen and helium, and has a diameter of 1,392,000km (865,000 miles). Its mass is approx. 333,000 times that of the Earth. The outer surface, or PHOTOSPHERE, of the Sun has a temperature of 5,500°C. The Sun is a source of light, infra-red and ultra-violet radiation as well as radio waves, gamma-rays and X-rays. The temperature at the Sun's core is approx. 10 million degrees centigrade and it is there that the Sun's energy is generated. Characteristics: apparent mag. −26.8; absolute mag. 4.7; distance: 149,596,000km (92,750,000 miles). *See also* MS pp.200, *200–201,* 212, *212;* PE p.81, *81;* SU pp.97, 172, *174,* 175–176, *176–177,* 216, 220–226, *220–226,* 229, 244, *244.*

Sun animal. *See* HELIOZOA.

Sun bear, small omnivorous bear found in SE Asia; it is the smallest species of bear in the world. It has a short black coat with a crescent-shaped yellow mark on its chest and a light grey or orange muzzle. Its diet consists primarily of insects, such as bees and termites, but it also eats honey, fruit and small vertebrates. Length: to 1.2m (4ft); weight: 45kg (100lb). Family Ursidae; species *Helarctos malayanus. See also* NW p.214.

Sunbird, tropical nectar-feeding songbird of the Old World, often considered a counterpart of the New World hummingbird. The males are usually brightly coloured. After the male's courtship display, the female builds a purse-like nest with a porch. Length: 9–15cm (3.5–6in). Family Nectariniidae. *See also* NW p.149.

Sunburn, burning of the skin caused by ultraviolet radiation from the sun. Redness, swelling, blisters and pain are associated with destruction and coagulation of some of the substances in the cells of the skin, along with enlargement of small vessels beneath the EPIDERMIS. In severe cases, large blisters and ulcers may form, accompanied by fever and delirium.

Sunda Islands, group of islands in Indonesia; part of the Malay Archipelago, between the South China Sea and the Indian Ocean. The Greater Sundas includes Sumatra, Java, Borneo and the Celebes (Sulawesi). Pop. (1970) 112,138,700. The Lesser Sundas include Bali, Lombok, Sumbawa, Flores and Timor. Pop. 6,999,000.

Sun dance, religious activity of the North American Plains Indians. The dances took many forms and were usually performed around a TOTEM POLE. They were part of elaborate ceremonies held to reaffirm a tribe's affinity with nature and the universe. Sometimes the dancers would inflict self-torture in the efforts to experience religious visions.

Sunday, the first day of the week, anciently dedicated to the Sun. It is observed by most Christians as the sabbath or day of rest and worship commemorating the resurrection of Christ. In Western countries it is a public holiday.

Sunday schools, institutions run by religious or Church organizations for free instruction in religious and moral matters. The first school was founded by Robert RAIKES in 1780 and the idea quickly spread. In 1803 the London Sunday School Union was founded to promote the movement, which grew rapidly in the Protestant churches of Britain and North America.

Sunderland, county district in SE TYNE AND WEAR, England; created in 1974 under the Local Government Act (1972). Area: 138sq km (53sq miles). Pop. (1974 est.) 292,600.

Sundew, INSECTIVOROUS PLANTS native to temperate swamps and bogs. They have hairy, basal leaves that glisten with a dew-like substance to attract, trap and digest insects. When the hairy projections are touched, the leaves close to trap the insect. Family Droseraceae; genera *Aldrovanda, Dionaea, Drosera* and *Drosophyllum. See also* NW p.59.

Sundial, instrument first used about 5000 years ago in the Near East to measure the time of day. Traditionally a sundial is a short flat-topped pillar on which is mounted the gnomon, or time pointer. This slants upwards away from the dial at an angle determined by the latitude; also it points due north in northern latitudes and south in southern. The position of the shadow of the gnomon indicates the time on a scale marked round the top of the sundial. *See also* MM p.93.

Sun dog, or parhelion, also called mock sun, name for either of two bright coloured spots observed at 22° on each side of the Sun. They are caused by REFRACTION of sunlight by ice crystals in the Earth's atmosphere.

Sunfish, North American freshwater fish. A popular angler's fish, similar in appearance to PERCH, it has a continuous dorsal fin containing spiny and soft rays. It is a nest-building fish and the male guards the nest and young. The 30 species range in size from the blue spotted *Enneacanthus gloriosus* (length: 8.9cm; 3.5in) to the large-mouth bass *Micropterus salmoides* (length: 81.3cm; 32in; weight: 10kg; 22lb). Family Centrarchidae.

Sunflower, coarse annual and perennial plant native to North and South America. The flower heads which turn to face the Sun, resemble huge daisies with yellow ray flowers and a centre disc of yellow, brown or purple. The common sunflower (*Helianthus annuus*) has 30cm (1ft) leaves

and flower heads more than 30cm across; height: to 3.5m (12ft). *See also* PE p.*217.*

Sungari (Songhuajiang), river in NE China. It rises in the Changpai Mts, Kirin province, near the North Korean border, and flows N to the River Amur at the USSR border. It is navigable for most of its 1,850km (1,150 miles).

Sung Dynasty (960–1279), Chinese dynasty. Divided into the Northern (960–1126) and Southern (1127–1279) Sung by the Jurchen conquest and establishment of the Chin Dynasty in the North, it was a period of refinement and cultural flowering in China. Southern ports were opened for trade with South Asia.

Sunglasses, spectacles with dark or tinted "lenses" that protect the eyes of the wearer by absorbing some of the Sun's light and heat radiation. The simplest sunglasses are tinted sheets of glass or plastic, either worn as spectacles or clipped over the wearer's own spectacles. "Polaroid" sunglasses remove glare from sunlight by plane-polarizing it. *See also* POLARIZED LIGHT; SU p.*139.*

Sung Yü (*fl.* 3rd century BC), Chinese poet. Writing towards the end of the CHOU DYNASTY he made major contributions to the development of the *fu,* prose poems especially popular in the HAN DYNASTY. He is reputed to have written the poem *The Summons of the Soul,* which tells of the wandering soul and praises the Chou Dynasty.

Sunni, orthodox rite of ISLAM, whose followers are called Sunnis or Sunnites. It is one of the two main sects of the religion, including 90% of all Muslims and finding authority in the entire Sunna or body of orthodox teachings outside the KORAN. The Sunni differ from the SHI'A sect, who believe in the inspired leadership of a succession of imams, and are divided among four schools of thought: the Hanafites, Malikites, Shafites and Hanbalites.

Sunspots, short-lived dark areas, often several thousand kilometres across, that appear periodically on the Sun's PHOTOSPHERE and travel from one side of the Sun's disc to the other as the Sun rotates. They have a dark centre (umbra) surrounded by a brighter border (penumbra), and are associated with strong magnetic fields. They appear dark because they are about 2,000°C cooler than the surrounding regions. Sunspots occur in groups, and maximum sunspot activity usually occurs every 11 years. *See also* SU pp.220–221, 224.

Sunstroke, disorder caused by overexposure to direct sunlight or to hot air in poorly ventilated places. A victim's temperature will rise quickly and may reach 41°C (106°F) or more; the skin is hot and dry, with no perspiration. The high temperature has a harmful effect on the central nervous system that, in severe cases, may result in convulsions and coma. Treatment includes means of lowering body temperature and complete rest, and saline fluids and sedatives may also be prescribed.

Sun Yat-sen (1866–1925), first President of the Chinese Republic (1911–12) and revolutionary leader of modern China. After China's defeat by Japan in 1895 Sun became convinced that the Manchu dynasty must be overthrown. In exile (1895–1911), he worked through secret societies and with support from overseas Chinese to bring about his aim. His inspiration eventually brought about the revolution of 1911. He joined in reorganizing the KUOMINTANG (1912) and became its leader and ideologist. *See also* HC2 pp.144–145.

Superalloys, name given to ALLOYS of cobalt and nickel used as surface materials for spacecraft to withstand the rapid heating and high temperatures caused by passage through the Earth's atmosphere, particularly on re-entry from space. These alloys oxidize in air above 700°C (1,300°F) but the oxide coating initially formed prevents further oxidation. *See also* ABLATION.

Supercluster, galaxy. *See* GALAXY CLUSTER.

Superconductivity, extraordinary electrical behaviour in metals and alloys that are cooled to temperature approaching ABSOLUTE ZERO. In a supercooled circuit an electric current flows indefinitely, there being no electrical resistance. The highest temperature at which this can occur is about 18°K, or −255°C (−427°F), the maximum superconducting temperature of a niobium-tin alloy. *See also* SU p.95.

Superego, in psychoanalysis, level of personality that sets standards of behaviour. A child starts with the unthinking impulses of the ID. The EGO develops as he learns to deal with reality. Then the superego, or his conscience, develops as he adopts standards determined by society's rewards and punishments. *See also* MS pp.*161, 187.*

Superfluidity, extraordinary physical behaviour shown by the element helium when it is cooled below −271°C (−456°F). At these temperatures, helium (referred to as helium II to distinguish it from helium I, or normal helium) is a liquid which has little or no viscosity and flows into and out from any vessel until the levels inside and outside are the same. *See also* SU p.95.

Supergiant star, largest, most massive and most luminous type of star so far discovered. Supergiants, characterized by low densities, lie at the top of the HERTZSPRUNG-RUSSELL DIAGRAM, above the giants; except for the brightest, they occur throughout the entire surface-temperature range. Red supergiants are often long-period or irregular variables. *See also* SU pp.226–227.

Superheterodyne receiver, most common type of RADIO receiver. An oscillator produces signals of varying frequency such that the difference between them and the frequencies of the incoming modulated carrier waves – the broadcast signal – is a constant, called the intermediate-frequency (i-f). The "i-f" can then be amplified and demodulated to produce the output signal for the loudspeaker.

Superior, Lake, lake in the USA and Canada; largest freshwater lake in the world. The lake is connected to Lake HURON and the ST LAWRENCE SEAWAY by the SAULT STE MARIE CANALS. A centre for commercial and recreational fishing, the lake is also a major commercial transportation route, particularly for grain and iron ore. Superior was discovered in 1622 by Étienne Brulé. Area: 82,362sq km (31,800sq miles).

Supernova, star that undergoes a cataclysmic outburst of energy and matter as the result of internal imbalances caused by the exhaustion of its fuel. These imbalances, which may occur in stars beyond a certain critical mass, take the form of accelerated nuclear reactions in the core, enormously increased internal temperature (up to 5,000 million °K) and gravitational collapse. The catastrophic explosion that follows, in which the star blows itself apart and ejects matter at relativistic speeds, temporarily increases its absolute magnitude to a figure in excess of −16, the brightness of a whole galaxy. These spectacular but also extremely rare phenomena usually leave behind a filamentary remnant, such as the Crab Nebula, and the original core may survive as a neutron star or PULSAR, or even as a BLACK HOLE. *See also* SU pp.228–229.

Superposition, law of, in geology, law that states that in undisturbed strata of sedimentary deposits, younger beds overlie older ones. The law determines the corresponding time equivalence of two geographically distinct formations.

Superscript, small letter, number or other symbol written or printed above and to the left or right of a number, letter or symbol, as indicating a mathematical exponent, such as 3^2 or x^n, or an element's mass number, as U^{238}. *See also* SUBSCRIPT.

Supersonic flight, flight beyond Mach 1, the speed at which sound travels at a particular temperature and pressure. At sea-level 1,223km/h (760mph) is the critical speed. At high altitude the speed might be 1,062km/h (660mph). At these speeds the pressure wave that normally moves away from an aircraft is compressed, creating a cone-shaped SHOCK WAVE. Early attempts to fly through this barrier were unsuccessful. Swept-wing, slender-bodied aeroplanes fly at supersonic speeds by retard-

ing the formation of the wave. *See also* MM pp.151, 178.

Supersonic waves, waves set up in the air by any object moving at speeds exceeding the speed of sound. At sea level, sound travels at about 1,240km/h (770mph), and at an altitude of 10,000m (32,800ft) it is about 1,000km/h (620mph). At these speeds, air waves created by the object's passage pile up at the leading edges of the object into a SHOCK WAVE. An observer on the ground sees a supersonic aeroplane pass overhead before he experiences the shockwave, which spreads out and reaches him as a sonic boom.

Superstition, irrational belief in the inevitable consequences of certain events, which can sometimes be avoided by the performance of counteractive rituals (such as throwing spilt salt over one's left shoulder). Superstition is not specifically linked to religion but is rather connected to ideas of an all-embracing fate and therefore has affinities with astrology.

Suppé, Franz von (1819–95), Austrian composer of Belgian descent, *b.* Francesco Ezechiele Ermenegildo Cavaliere Suppé Demelli. He composed 31 operettas, the best of which include *Fatnitza* (1876) and *Boccaccio* (1879).

Supplementary Benefits. *See* SOCIAL SECURITY.

Supply and demand, law of, economic balance between goods required and produced. The law of supply indicates that, other things being equal, as the price of an item increases suppliers are willing to produce more, and as the price decreases producers are willing to produce less. Thus, price and quantity supplied are directly related. The law of demand states the reverse: as prices increase, consumers demand less, and as prices decrease, consumers demand more. Thus, prices and quantity demanded are inversely related. Through the interaction of supply and demand, an EQUILIBRIUM is reached. *See also* MS pp.*310, 311.*

Suprarenal glands, another name for the ADRENAL GLANDS.

Supremacy, Acts of, two acts passed by the English Parliament in the 16th century. The first (1534) made the monarch the "supreme head" of the Church of England. It thus made the English Church independent of the papacy and also made it subject to temporal authority. The second act (1559) changed the title from supreme head to "supreme governor", thus signifying that authority was vested, not so much in the person of the monarch as in the Crown-in-parliament.

Suprematism, abstract art movement launched in Russia in 1915 by the painter Kasimir MALEVICH. The movement stressed the simplest geometrical forms and is epitomized by Malevich's painting *White on White* (1919).

Supreme Court (Canada), highest court of appeal in Canada, established in 1875. Its judgement became final when appeals to the Judicial Committee of the Privy Council in London were ended – for criminal cases in 1933, for civil cases in 1949. It sits in Ottawa and is composed of a chief justice and eight associated judges. In addition to hearing appeals, it advises the federal government and provincial governments, when asked, on matters requiring constitutional interpretation. Recent decisions in appeal judgements show a tendency to move away from English judicial practice towards that of the USA.

Supreme Court of the United States, US court of final appeal, the highest in the nation, whose duty it is to decide and interpret the constitutionality of state and federal legislation and of executive acts. Once the Supreme Court has arrived at a decision, all lower courts must follow it in similar cases. Cases are decided by majority vote. Created by the Constitution of 1787, the Supreme Court is made up of nine justices appointed for life by the president with the advice and consent of the Senate. A justice may be removed from office only by impeachment. The chief justice presides over all sessions, five judges constituting a quorum.

Supreme Court (USSR), highest court in the

USSR, established in 1922. It not only hears appeals, but acts as the general coordinator of policy in the provinces, issues rulings interpreting the various legal codes of the country and, occasionally, hears cases in the first instance. It is subordinate to the legislature and has no power to declare laws unconstitutional or to check executive action.

Surabaja, port in NE Java, Indonesia, at the w end of the Madura Strait; second-largest city in Indonesia and capital of East Java province. It was an important naval base until WWII, when it was occupied by the Japanese. The city has Airlangga University (1954) and a naval college. Exports: rice, sugar cane, spices, tobacco, maize, tapioca, coffee, cocoa, rubber, copra. Pop. 1,273,000.

Surface tension, molecular forces associated with the boundary layer of a liquid. It makes a liquid behave as if there were a "skin" on the surface. Cohesive forces in the skin tend to resist disruption, so that a needle or razor blade placed carefully on the surface "floats" even though its density is many times that of the liquid. *See also* SU p.87.

Surfactant, substance that lowers the SURFACE TENSION of a SOLVENT in which it is dissolvtd. In water, soaps and detergents act as surfactants. Their molecules concentrate at the surface of oil and water molecules and act as emulsifying or foaming agents, to produce lathering and frothing. *See also* MM pp.240–241, *240–241.*

Surfing, water sport that consists of riding the surf on a specially designed board usually 1.2–1.8m (4–6ft) long. For stronger waves, larger boards (up to 3.7m, 12ft) are used. The technique is to swim out to sea and then wait for a wave before standing on the board and planing back to shore on the wave before it breaks. Surfing originated in Hawaii, and there are now national and international competitions in various parts of the world.

Surfperch, or seaperch, several species of fish which bear their young live and are found mostly off the Pacific Coast of North America. Length: to 45cm (18ins). Order Perciformes; family Embiotocidae.

Surgery, treatment of a disorder by manipulation or operation, usually performed by a surgeon and often using some form of ANAESTHETIC. A doctor's or dentist's consulting room is also known as a surgery.

Surinam, independent country in NE South America, on the Atlantic Ocean. Formerly Dutch Guiana or Netherlands Guiana, it gained independence in 1975. Rice, bananas, coffee, sugar cane, groundnuts, coconuts and citrus fruits are the chief crops. Bauxite is the principal export, and food processing and timber are the main industries. Paramaribo is the capital. Area: 163,265sq km (63,037sq miles). Pop. (1975 est.) 350,000. *See* MW p.161.

Surinam toad, South American aquatic tailless AMPHIBIAN. It is brown with a flat, square body. The female carries fertilized eggs embedded in the skin on her back; the larvae remain there until metamorphosed. Length: 20cm (8in). Family Pipidae; species *Pipa pipa*. *See also* FROG; NW p.133.

Surrealism, influential movement in art and literature, evolved in the mid-1920s from DADAISM. Influenced by Freudian psychology, it represented a reaction against rationalism and advocated creative use of the powers of the unconscious mind. Led by the poet André BRETON, who wrote several Surrealist manifestos, Surrealist writers included Louis ARAGON, Paul ÉLUARD, and Benjamin Peret, while painters included Jean ARP, Max ERNST, René MAGRITTE, Yves TANGUY, Salvador DALI, Joan MIRÓ and Paul KLEE.

Surrey, Henry Howard, Earl of. *See* HOWARD, HENRY, EARL OF SURREY.

Surrey, county in SE England: one of the Home Counties bordering Greater London. The region is traversed from E to W by the North Downs, which slope down to the Thames Valley. The Wey and the Mole are the principal rivers. Much of the land in the w is devoted to farming: dairy and market-garden produce, wheat and oats

Lake Superior; a ship loading grain at Port Arthur, an important outlet.

Supersonic speed; the Anglo-French Concorde, at its first take-off in 1969.

Surfing; many types of surfboard are available but most are made of fibreglass.

Surinam toad females carry fertilized eggs and young on their backs.

Robert Smith Surtees's novels include *Handley Cross* (1843) and *Ask Mamma* (1858).

Suspension bridges are cheaper than most but were not built until 1800.

Sussex: Alfriston, one of the most picturesque villages in the county.

Swallow: the barn swallow will travel up to 10,000 miles on its annual migration.

are the chief products. There is some light industry. Guildford is the county town but the county council is, for the time being, situated in Kingston-upon-Thames, which is no longer in Surrey. Area: 1,679sq km (648sq miles). Pop. 1,002,900.

Surrogate, in psychology, term used to denote a person or thing that is a substitute for another (generally the natural person or object). It may bear no physical relation to the person or object for which it is a substitute. *See also* IMPRINTING.

Surtees, John (1934–), British motorcyclist and motor racing driver, the only man to win world titles at both sports. He was seven times motorcycle champion (350cc 1958–60, 500cc 1956, 1958–60), and won the world racing drivers' championship in 1964. He later became a leading racing-car constructor.

Surtees, Robert Smith (1803–64), British author. His novels of fox-hunting and country life produced one of English literature's memorable characters, the fox-hunting London grocer Mr Jorrocks. From 1831–36 Surtees was editor of *New Sporting Magazine*, which he helped launch and in which his novels were first serialized.

Surveying, accurate measurement of the Earth's surface. It is used in establishing land boundaries, the topography of landforms and for major construction and civil engineering work such as dams, bridges and highways. Measurements are linear or angular, applying principles of geometry and trigonometry. For smaller areas, the land is treated as a horizontal plane. Large areas involve considerations of the Earth's curved shape and are referred to as a geodetic surveys.

Surya, in Hindu mythology, the personification of the Sun. He is depicted as a three-eyed, four-armed man of burnished copper, sometimes seated on a lotus or in a chariot drawn by seven horses. *See also* MS p.212.

Susa, ancient city of ELAM, situated on a plain near the River Karun. It was part of BABYLONIA in the reign of Assurbanipal (*c.*600 BC); ALEXANDER THE GREAT conquered it in 331 BC, obtaining treasures of gold, silver and precious stones. A stele depicting the Code of HAMMURABI was discovered there in 1901–02; it is thought to have been taken from Babylon by an Elamir invader. *See also* HC1 pp.33, 61, 80–81.

Susanna, in The Bible, woman falsely accused of adultery by two elders and who was proved innocent by DANIEL. Christians treat her story as an allegory of salvation.

Susanowo, in Japanese mythology, brother of Amaterasu-o-Mikami, the sun goddess. Susanowo was an agricultural deity, a fertility god associated with snakes and also the god of thunder and rain. His behaviour was influenced by his dual nature, Ara-mi-tama and Nigi-mi-tama, his evil and good souls.

Su Shih (1037–1101), Chinese poet and prose writer, whose penname was Su Tung-p'o. He was one of the outstanding writers of the SUNG DYNASTY. A master of t'zu, the verse form of irregular length which flourished during the Sung period, he freed it from its previous metrical rigidity to write vigorously about heroism and beauty.

Suspension, liquid medium in which small solid particles are uniformly dispersed. If the particles do not settle out on standing but are unable to pass through a semipermeable membrane the suspension is called a COLLOID. The particle size in such colloidal suspensions is usually in the range 10^{-3} to 10^{-6} millimetres.

Suspension bridge, bridge that has its roadway suspended from two or more cables which usually pass over towers and are anchored at the ends. The cables are of wire twisted spirally, and the floor is usually made rigid by stiffening trusses. The Humber Estuary Bridge, planned to be completed in 1979, will have a span of 1,410m (4,626ft) and will be the longest suspension bridge in the world. *See also* MM p.192, *192*.

Sussex, kingdom of Anglo-Saxon England, said to have been settled by the South Saxons under Aelle (*c.* AD 477), and covering roughly the area of the modern county of Sussex. The South Saxons were Christianized (*c.*680) and afterwards Sussex was divided into several kingdoms, to be incorporated into Wessex in the 9th century. *See also* HC1 pp.138–9.

Sussex, former county in SE England, on the English Channel; since 1974 it has been divided into the counties of East Sussex and West Sussex. The area is crossed from E to W by the South Downs. The principal rivers are the Rother, Ouse and the Arun. The county's economy is overwhelmingly agricultural. Dairy and market garden produce, and cereals are the most important products. The Channel coast has many tourist resorts. The county was formerly divided into two administrative districts, West Sussex centred on Winchester, and East Sussex centred on Lewes. Area: 3,773sq km (1,457sq miles). Pop. (1971) 1,241,332.

Süsskind, Walter (1913–), British conductor, *b.* Czechoslovakia. He conducted the Scottish National Orchestra (1946–52) and also held conducting posts with orchestras in Australia (1954–56) and Canada (1956–65) before taking over as chief conductor of the St Louis Symphony Orchestra (1968–75).

Sutherland, Earl Wilbur, Jr (1915–74), US physiologist. He was awarded the 1971 Nobel Prize in physiology and medicine for his work on HORMONES and how they act in the body to control various metabolic processes.

Sutherland, Graham (1903–), British painter, official War Artist in WWII. His work developed under the influence of Samuel PALMER and may be called SURREALIST in its fantasy and suggestion of menace. In later years Sutherland became known for religious subjects, especially for his huge tapestry for Coventry Cathedral (1962). Acknowledged as a modern master of portraiture, his portraits include those of *Maugham* (1949), and *Beaverbrook* (1951).

Sutherland, Joan (1926–), Australian coloratura soprano. She joined the Royal Opera, London, in 1952 and her performance in the title role of DONIZETTI's *Lucia di Lammermoor* at Covent Garden in 1959 earned her world wide acclaim. She is most famous for her BEL CANTO roles, and has sung in all the famous opera houses. In 1954 she married conductor Richard Bonynge.

Sutherland, former county in NW Scotland, on the Atlantic Ocean; since 1975 it has been part of North West Highlands region. The county is noted for its scenery; it is mostly mountainous and there are many lakes. Much of the interior is moorland which is used for grazing sheep. The region has extensive forests which support deer. Fishing, particularly for salmon, tourism, whisky distilling and the manufacture of woollens are the chief industries. The county town was Dornoch. Area: 5,255sq km (2,029sq miles). Pop. (1971) 13,053.

Sutlej (Langchuhe), tributary of the River Indus and the longest of the five rivers of the Punjab; rises in SW Tibet, China, and flows W through the Himalayas and the Punjab, where it is joined by the Beas River. The Sutlej continues into Pakistan where, with the Chenab River, it forms the Panjrad, which collects the waters of all five rivers of the Punjab before joining the Indus. Length: approx 1,370km (850 miles).

Sutra, in Buddhism, a sacred text. It is claimed to have been spoken by BUDDHA himself and recorded by his disciple ANANDA immediately after Buddha's death. Many Sutras, however, were composed centuries later by unknown authors. *See also* HC1 p.131.

Suttee, former Indian custom of a widow throwing herself alive on to her husband's funeral pyre. Originally confined to royalty, it was forbidden under British Rule in 1829. *See also* DEATH RITES; MS p.*181*.

Suttner, Bertha Felice Sophie, Baroness von (1843–1914), Austrian novelist and free-thinking pacifist. Her novels include *Lay Down Your Arms* (1889) and *Das Maschinenzeitalter* (1899). She was active in the peace movement, and largely due to her influence, Alfred NOBEL included a Peace Prize among his awards. She herself received it in 1905.

Sutton, Walter Stanborough (1877–1916), US geneticist and doctor. In 1902, while still a student, he provided the first evidence that CHROMOSOMES occur in pairs and that they are responsible for carrying genetic information.

Suva, seaport on the SE coast of Viti Levu Island, in the SW Pacific Ocean; the capital of the Fiji Islands. It is the manufacturing and trade centre of the islands, with an excellent harbour. Exports include tropical fruits, copra and gold. Pop. (1971 est.) 63,200.

Suwon, city in NW South Korea, 29km (18 miles) S of Seoul; capital of Kyonggi province. The trade centre for a farming region, it has several agricultural research institutes. Pop. (1970 est.) 170,500.

Svedberg, Theodor (1884–1971), Swedish chemist who was awarded the 1926 Nobel Prize in chemistry for his development of the ultra-centrifuge in 1923. He used it to study COLLOIDS and large MOLECULES and it enabled the molecular weight of large PROTEINS to be determined for the first time. He also helped to develop ELECTROPHORESIS techniques for separating proteins on the basis of electrical charge.

Svengali, sinister character in George du Maurier's novel *Trilby* (1894). An Austrian and a brilliant musician, he transforms Trilby into a great singer through hypnosis.

Sverdlovsk. *See* EKATERINBURG.

Svevo, Italo (1861–1928), pseudonym of Ettore Schmitz, Italian novelist and businessman. His writings were unknown until they were discovered by James JOYCE, who taught him English. They are largely psychological and introspective and include *A Life* (1892) and *The Confessions of Zeno* (1923).

Swabia, historic region of SW West Germany, consisting mainly of the S part of the modern states of Baden-Württemberg and SW Bavaria. An important duchy in early medieval Germany, it came under the rule of the house of Hohenstaufen from 1079 to 1268. Swabia was divided in 1268 and most of the lordships became free imperial cities by 1313, and formed powerful Swabian Leagues in 1331–37, 1376–88 and 1488–1534. At the diet of Regensburg (1801–03) many areas were incorporated into Baden, Württemberg and Bavaria.

Swahili, BANTU language of the Niger-Congo family of African languages. A lingua franca in most of E Africa with about one million native speakers, it became the official language of Tanzania in 1967 and of Kenya in 1973. It has a large native literature, using a form of the Arabic alphabet.

Swale, county district in central N KENT, England; created in 1974 under the Local Government Act (1972). Area: 369sq km (143sq miles). Pop. (1974 est.) 105,600.

Swallow, any of 75 species of graceful and agile birds with long tapering wings and a long forked tail. The common European swallow (*Hirundo rustica*) is grey blue with a light brown underside and red throat markings; it feeds primarily on insects, which it catches in flight. If often nests in urban areas, and migrates to Africa for the winter. Length: 20cm (8in). Family Hirundinidae.

Swallowtail, any of numerous species brightly-coloured butterflies found throughout the world; they have characteristic tail-like extensions on their hind wings. Most swallowtails are large and black with bright yellow, blue, green and red markings. The yellow and black giant, *Papilio cresphontes,* has a wingspan of up to 14cm (5.5in). Family Papilionidae. *See also* NW pp.*107, 117, 214*.

Swammerdam, Jan (1637–1680), Dutch naturalist. His researches and his book *General History of Insects* provided a system for classifying insects and laid the foundations of ENTOMOLOGY. He also discovered the existance of red blood corpuscles in 1658. *See also* HC1 p.303.

Swan, Sir Joseph Wilson (1828–1914), British scientist who designed the first electric light bulb. Starting in 1848 he used a carbon filament inside an evacuated bulb and by 1860, 20 years earlier than Thomas Edison, he had produced a workable bulb, though with a short lifetime. His bulbs were used to light the House of Commons in 1881 and the British Museum in 1882.

Swan, any of several species of graceful, white or black waterfowl that nest in N Northern Hemisphere and migrate south for winter. Three species, including the Australian Black swan, live in the Southern Hemisphere. Most have broad, flat bills, long necks, plump bodies, long legs, webbed feet, and dense plumage over down. They dip their heads under water to feed on plant matter. Length: to 1.8m (6ft). Family Anatidae; genus *Cygnus. See also* NW pp.*140*, 148.

Swan Lake, four-act ballet with music by TCHAIKOVSKY and choreography by Wenzel Reisinger. It was first performed in 1877, and revised in 1894–95 by Lev IVANOV and Marius PETIPA. This version remains in the repertoire of many major ballet companies.

Swanscombe man, earliest known form of *Homo sapiens,* the fragmented skull of which was found in gravels of the River Thames at Swanscombe, Kent, SE England in 1935–36. The bones date from the Mindel-Riss interglacial period, approx. 250,000 to 200,000 years ago. *See also* HC1 p.42; MS p.26.

Swansea (Abertawe), city and county district in West Glamorgan, S Wales. The city, on Swansea Bay at the mouth of the River Tawe, is the administrative centre of West Glamorgan and an industrial city, noted for its production of steel, lead and zinc. Pop. (1973) 173,150. The county district was created in 1974 under the Local Government Act (1972). Area: 245sq km (95sq miles). Pop. (1974 est.) 189,800.

Swanson, Gloria (1897–   ), US film actress b. Josephine Swenson, she became famous for bathing-beauty roles in the silent films of Mack SENNETT. Her later films include *Sadie Thompson* (1928), *Indiscreet* (1931), the much-acclaimed *Sunset Boulevard* (1950) and *Airport 75* (1974).

Swarm, gathering of many honey bees around a queen bee, which flies off from a hive to found another colony. This happens when a young queen supplants the old queen, and it is the latter who leads off her retinue in a swarm. *See also* PE p.196.

Swart, Charles Robberts (1894–   ), South African lawyer, journalist and politician. Leader of the National Party of the ORANGE FREE STATE from 1940 to 1959, he was deputy Prime Minister from 1954 to 1959 and State President of the Republic of South Africa from 1961 to 1967.

Swartkrans, archaeological site near Johannesburg, South Africa. Since 1948 remains have been found in a cave there of both *Australopithecines* and of *Homo erectus.* It is the only site where both have been found together. *See also* MS pp.23, 25.

Swastika, emblem of the NAZI Party, originally a mystic sign known as the fylfot or the gammadion (so called because its elaborate cross-shaped design resembles four Greek capital gammas set at right-angles). It was used in Byzantine architecture, and also by North American Indians, as a sign to ward off evil spirits (the word "swastika" comes from the Sanskrit *svastika,* "well-being").

Swaythling Cup, trophy for the men's team event at the world table-tennis championships. First contested in 1927, it was donated by Lady Gladys Swaythling whose son organized the first world event in 1926. The event was dominated by the European nations until the 1950s, but it has since been won almost exclusively by teams from China and Japan.

Swazi, negroid people of southern Africa. Mainly agricultural pastoralists, they have a traditionally polygamous society in which power is shared between a hereditary male ruler and his mother (or recognized substitute). Non-Christian Swazis continue to practise ancestor worship and witchcraft.

Swaziland, small, landlocked and moun-

tainous kingdom in S Africa. Most of the people are farmers; sugar is the principal crop and cattle and sheep are reared. Mineral deposits include iron ore, asbestos and coal, and timber is an important industry. In the second half of the 20th century Swaziland has expanded its economy, reducing its dependence on South Africa. Mbabane is the capital. Area: 17,366sq km (6,705sq miles). Pop. (1976 est.) 482,000. *See also* MW p.161.

Sweat, or perspiration, secretion of the sweat glands (sudoriparous glands) of the skin. Sweat is made up of water, mineral salts and traces of UREA. Sweat glands also help to excrete poisons from the body, but the main function of sweat is to cool the body by evaporation. *See also* MS p.61.

Sweat shops, factories or home industries in which labourers, often women and children, worked long hours for low wages under unsanitary and cramped conditions. In Britain they were prevalent during the late 19th century.

Swede, root vegetable belonging to the mustard family (*Brassicaceae*). The taproot may be eaten cooked as a vegetable or fed to animals as fodder. Height: approx. 30cm (12in). Species: *B.napobrassica. See also* PE p.188.

Sweden, constitutional monarchy occupying the E half of the Scandinavian peninsula. It is a highly industrialized nation and one of the most prosperous in Europe. Sweden is a land of lakes, rivers and forests. The most valuable resources are iron ore, timber and hydroelectric potential. The most important industries are shipbuilding and the manufacture of machinery, motor vehicles and electronic and telecommunications equipment. The principal crops are barley, wheat, oats, potatoes and sugar beet. The capital is Stockholm. Area: 449,964sq km (173,731sq miles). Pop. (1976 est.) 8,226,000. *See* MW p.161.

Swedenborg, Emanuel (1688–1772), Swedish scientist and theologian, whose wide-ranging studies anticipated many later inventions and discoveries. He turned to theological teaching after a spiritual crisis in 1745, when he claimed to have received a revelation of the true meaning of the Scriptures. His writings include *The Economy of the Animal Kingdom* (1740–41) and *True Christian Religion* (1771). Although he did not set out to found a church, a sect, the Swedenborgians, was formed after his death.

Sweeney Agonistes, fragments of an Aristophanic melodrama by T. S. ELIOT, first published in 1932. It is partially a parody of Milton's *Samson Agonistes,* with the protagonists reduced to empty mouthings and sordid assignations in a world too banal for Miltonic heroes.

Sweet, Henry (1845–1912), British philologist and phonetician. He was a pioneer of phonetics research and his writings include *A History of English Sounds* (1874) and *The History of Language* (1900).

Sweetbread, the PANCREAS or sometimes the THYMUS, of a calf or sheep, as food.

Sweetbriar. *See* EGLANTINE.

Sweet corn. *See* MAIZE.

Sweetener, artificial. *See* CYCLAMATES; SACCHARIN.

Sweet pea, climbing, annual plant native to Italy. It has fragrant, butterfly-shaped flowers of white, pink, rose, lavender, purple, red or orange. Height: to 1.8m (6ft). Family Fabaceae; species *Lathyrus odoratus.*

Sweet potato, trailing plant native to South America and cultivated as a vegetable in S Japan, the USSR, USA and the Pacific, including New Zealand (where it is called Kumnia). Its funnel-shaped flowers are pink or violet. The orange or yellow, tuber-like root is edible. Family Convolvulaceae; species *Ipomoea batatas.*

Sweet William, common name for a flowering plant native to Europe. Named after William, Duke of CUMBERLAND, the victor at Culloden, and introduced as a garden plant, it is now a wildflower along roadsides and in waste areas. Its flowers are pink or white. Height: to 60cm (24in). Family Caryophyllaceae; species *Dianthus barbatus.*

Sweyn I Forkbeard (*d.*1014), King of Denmark (987–1014), son of Harald Bluetooth. His attack on England in 994 forced King ETHELRED II to pay tribute and after his invasion in 1013 Ethelred fled to Normandy and Sweyn became king. Sweyn died before his coronation and his son CANUTE the Great eventually succeeded him. *See also* HC1 p.155.

Swift, Jonathan (1667–1745), Irish satirist and novelist. He was ordained (1694) an Anglican priest. His early works include *A Tale of a Tub* (1704) on religious and scholarly corruption, and *The Battle of the Books* (1704), ridiculing the controversy over ancient and modern learning. His *Gulliver's Travels* (1726), a satire of human follies, is now read as a story for children. He wrote numerous works criticizing England's treatment of Ireland, including *A Modest Proposal* (1729), which suggested that the children of the poor be fattened and eaten by the rich. *See also* HC2 pp. 23, *26*, 27, *34*.

Swift, any of several species of fast-flying widely distributed birds; it has a short, sharply-hooked bill, wide mouth, long narrow wings, short weak legs, and darkish plumage. It typically feeds on insects, which it catches in flight, and builds a nest of plant matter held together with saliva, in a chimney or under the eaves of a building. Length: to 23cm (9in). Family Apodidae. *See also* NW pp.214, *217*, *228.*

Swimming, recreational activity and competitive sport. Formal competition was first introduced in 1603 in Japan. The National Swimming Association was formed in England in 1837. By 1908 the Fédération Internationale de Natation Amateur (FINA), the world governing body, was formed. Recognized distances for men and women, established by the federation in 1968, include the 100, 200, 400, 800 and 1,500m freestyle, the 100 and 200m breaststroke, the 100 and 200m butterfly, the 100 and 200m backstroke, the 4 by 100m relay, the 4 by 200m relay (for men), the 200 and 400m individual medley and the 4 by 100m relay medley. For world records, *see* p.848.

Swinburne, Algernon Charles (1837–1909), British poet and critic. *Atalanta in Calydon* (1865), a poetic drama modelled on Greek tragedy, brought him fame and *Poems and Ballads* (1866) were praised for their infusion of new energy into Victorian poetry. His other poems include *The Garden of Proserpine* (1865) and *The Triumph of Time* (1865). *See also* HC2 p.100.

Swine fever, disease of pigs caused by a virus and characterized by high temperature and prostration. The disease is usually fatal but in some cases the animals survive and infect other pigs. Swine fever is widespread in Africa, Europe and the USA; it is controlled mainly with anti-hog-cholera vaccinations.

Swing, form of JAZZ, prevalent in the USA during the 1930s and 1940s. It originated in the music of small groups who played a rhythm of four even beats to the bar, as opposed to the two beats to the bar of the New Orleans style. The groups also made more use of soloists, particularly saxophonists such as Coleman HAWKINS, than had the New Orleans ensembles. Larger groups, such as those of Duke ELLINGTON, Fletcher HENDERSON and especially Count BASIE, made great use of the new possibilities. Their innovations were taken up by white musicians such as Benny GOODMAN, the DORSEY brothers and Glenn MILLER, who created successful popular big bands. *See also* HC2 p.272.

Swinnerton, Frank Arthur (1884–   ), British novelist and critic. He became famous in 1917 with *Nocturne,* his sixth novel, and since then has written more than 30.

Swinton, Sir Ernest Dunlop (1868–1951), British soldier, author and inventor of the TANK. He served in the SOUTH AFRICAN WAR and WWI and rose to the rank of major-general. He held the chair of military history at Oxford (1925–1939) and wrote several books, some under pseudonyms (Backsight-Forethought or O'le Luk-Oie).

Swiss chard, annual plant, related to BEET,

Swans generally feed in the water, but sometimes graze on land in the manner of geese.

Gloria Swanson in *Affairs of Anatol.* She embodied 1920s Hollywood glamour.

Sweden; national costume is a keenly preserved tradition throughout the country.

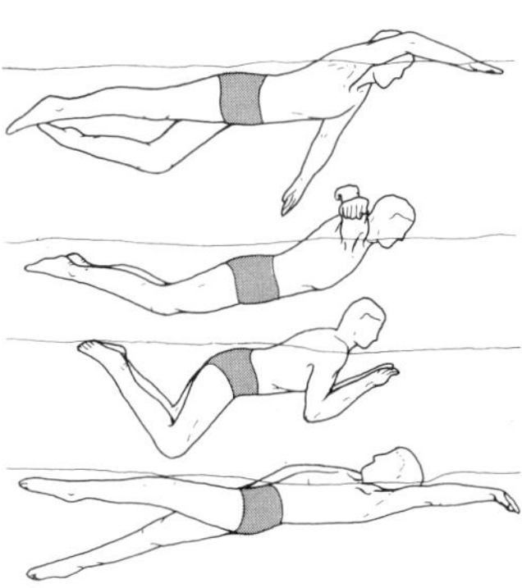

Swimming (top to bottom); the crawl, the butterfly, the breast and back strokes.

Swordfish are solitary and in groups leave some 30–40 feet between individuals.

Sycamore; its leaf, flower and its winged fruit, which is distributed by the wind.

Sydney; yachts and helicopters at the Grand Opening of the Opera House in 1974.

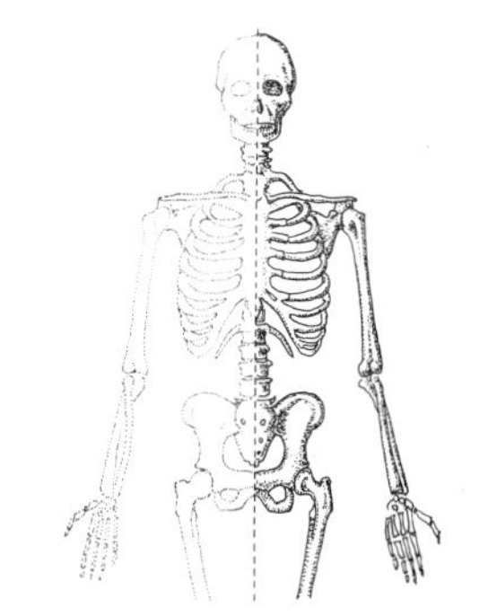

Symmetry; a line drawn down from the nose reveals bilateral symmetry in man.

grown for its spinach-like green or red leaves. The white or red leaf ribs and stems are often cooked separately. Family Chenopodiaceae; species *Beta cicla*.

Swiss Guard, military corps employed in the VATICAN to protect the person of the POPE. It originated during the reign of Pope Julius II (1503–13), who made agreements with the Swiss cantons of Zürich and Lucerne to provide him with 250 guardsmen. They now number about 100 and come from all Swiss cantons.

Swithin (Swithun), Saint (*c.*800–862), ANGLO-SAXON bishop of WINCHESTER from 852, and adviser to the SAXON kings EGBERT and ETHELWULF. His tomb in Winchester cathedral became a famous shrine. According to tradition, whatever the weather is on his feast day (15 July) it will remain the same for the next 40 days.

Switzerland, small, landlocked republic in s Europe. The country is noted for its alpine scenery, its manufacture of clocks and watches and as a centre of finance and banking. Switzerland has one of the highest standards of living in Europe. Dairy farming is the chief agricultural activity. The principal industry is the manufacture of precision instruments; others include heavy engineering, pharmaceuticals, steel and tourism. The capital is Bern. Area: 41,288sq km (15,941sq miles). Pop. (1976 est.) 6,333,000. *See* MW p.163.

Swordfish, or broadbill, marine fish found worldwide in temperate and tropical seas. A popular food fish, it is silvery-black, dark purple or blue. Its long flattened upper jaw, in the shape of a sword, is one-third of its length and used to strike at prey. Length: to 4.57m (15ft); weight: 532kg (1,182lb). Family Xiphiidae; species *Xiphias gladius*. *See also* PE pp.246, 247.

Swordtail, freshwater tropical fish found from s Mexico to Guatemala. The male of this popular aquarium fish can be identified by a long extension of its tail fin. The young are born alive. The many varieties include red-eyed, red wagtail and berlin swordtails. Length: to 12.7cm (5in). Family Poeciliidae; species *Xiphophorus helleri*.

Sycamore, deciduous tree of the MAPLE family, native to central Europe and w Asia but widely naturalized. Also known as the great maple or false plane, it has deeply toothed, five-lobed leaves, greenish yellow flowers and winged brown fruit. Height to 33m (110ft). Species *Acer pseudoplatanus*. In the USA this name is given to *Platanus occidentalis*, an unrelated species. *See also* NW p.65.

Sydenham, Thomas (1624–89), English physician, often called the English Hippocrates. He initiated the cooling method of treating SMALLPOX, made a thorough study of all aspects of epidemics, and wrote descriptions of many diseases including MALARIA, smallpox and GOUT.

Sydney, state capital of New South Wales, SE Australia, on Port Jackson, an inlet on the Pacific Ocean. Sydney is the largest city, the most important financial, industrial and cultural centre and the principal port in Australia. The city was founded in 1788 as the first British penal colony in Australia. Places of interest include the SYDNEY HARBOUR BRIDGE, the Sydney University (1852) SYDNEY OPERA HOUSE and the Australian Museum of natural history. Industries: shipbuilding, textiles, food processing, motor vehicles, oil refining, building materials, chemicals. Exports: wool, wheat, sheepskins, meat. Pop. 2,874,380.

Sydney Cricket Ground, Australian sports ground used for cricket, rugby league and other games. Cricket Test matches were first played there in 1882, and in 1938 the Empire Games were staged there.

Sydney Harbour Bridge, one of the world's largest steel-arch bridges. Opened in 1932, it has a clear 503m (1,650ft) span and carries four urban railway lines and six lanes of road across the great harbour in Sydney, Australia. *See also* MM p.193.

Sydney Opera House, opera house in Sydney, Australia, designed by the Danish architect Joern UTZON. It was completed in 1973, some 18 years after Utzon won the competition for the design in 1956.

Sydow, Max von (1929–), Swedish actor known for his portrayals of madness in Ingmar BERGMAN's films, such as *The Virgin Spring* (1960). Other films include *The Seventh Seal* (1957), *The Greatest Story Ever Told* (1965) and *Exorcist* (1973).

Syllogism, logical argument consisting of at least three propositions: two or more premises and a conclusion. The premises are related in such a way that the conclusion must be valid. *See also* MS p.228.

Sylphides, Les, ballet in one act with choreography by FOKINE and music based on CHOPIN piano pieces orchestrated by GLAZUNOV. It was first produced in St Petersburg, in 1907, and the first performance in its final version after revisions by Fokine was given by the BALLETS RUSSES in Paris in 1909.

Sylvester, name of four Roman Catholic popes and antipopes. St Sylvester I (*r.*314–35) reputedly baptized the Emperor CONSTANTINE I for which he was presented with the Donation of Constantine. He sent two legates to the First Council of NICAEA in 325 which defined the articles of Christianity. Sylvester II (Gerbert) (*r.*999–1003) was the first French pope. He aided the Christianization of Poland and Hungary, and denounced simony, nepotism and the keeping of concubines by clergy. Sylvester III was ANTIPOPE (1045) to Benedict IX; Sylvester IV was antipope (1105–1111) to Paschal II.

Symbiosis, relationship between two or more different organisms that is generally mutually advantageous to each. *See also* COMMENSALISM; PARASITE.

Symbolic logic, also known as formal or mathematical logic, uses symbols to represent the forms of sentences expressing propositions. First developed in the 19th century, it is an extension of previous forms of LOGIC.

Symbolism, European art and literary movement. In art it arose in France in the 1880s in reaction against the pragmatic Realism of COURBET and of IMPRESSIONISM and in favour of forms which would express ideas or abstractions rather than simply imitate the visible world. There were two tendencies in the movement. The first and most important was that which stemmed from GAUGUIN and Émile BERNARD (*c.*1888) and became known as CLOISSONISM. The second, less concerned with formal innovation, took the form of a retreat into the past via religious, historical and mythological themes employed in traditional painting styles. The chief exponents were Gustave MOREAU, Odilon REDON and PUVIS DE CHEVANNES in France and the PRE-RAPHAELITE BROTHERHOOD in England. In literature, the movement included a group of poets active in the 19th century, who were followers of VERLAINE, MALLARMÉ, RIMBAUD in France, all influenced by Baudelaire, and POE and SWINBURNE writing in English. They shared a desire to transcend the limitations of the Realist novel to create poetic truth through suggestion.

Symington, William (1763–1831), British engineer. He applied the principle of the steam engine to a road vehicle in 1786, and to boats in 1788. In 1801 he built the first successful steam-driven paddle boat.

Symmachus, Saint (*d.*514), pope from 498 to 514. Although his election was disputed by the followers of the archpriest Laurentius, Symmachus upheld the Catholic faith against the Henotikon of ZENO and the MANICHEANS, expelling the latter from Rome.

Symmetry, in biology, anatomical description of body form or geometrical pattern of a plant or animal. It is used in the classification of living things (taxonomy), and to clarify relationships.

Symons, Julian Gustav (1912–), British biographer and novelist. After a period as a journalist he took up writing historical studies and biographies, including *Charles Dickens* (1951), *The General Strike* (1957) and *The Thirties* (1960). His detective fiction includes *The Narrowing Circle* (1954) and *The Progress of a Crime* (1960). He also wrote a history of crime fiction, *Bloody Murder* (1972).

Sympathetic nervous system. *See* MS pp.38, *39*, 65.

Symphonic poem, one-movement orchestral work, also called tone poems and often analogous to programme music. It was popular with composers of the 19th century and the term was first used by Franz LISZT, who sought to express the language of the poet in music. Early works by him include *Mazeppa* (composed 1851) and *Hamlet* (1858). The tone poems of Richard STRAUSS are renowned for the mastery of invention and colourful orchestration.

Symphony, large-scale musical work for ORCHESTRA. It has evolved steadily since the 18th century, when it received its first classical definition in the works of HAYDN and MOZART. A symphony usually has four movements and, in the tradition of the CLASSICAL symphony, has its first movement in SONATA form. The classical symphony nearly always included one movement which was either a MINUET or a SCHERZO. The first symphonies were scored almost exclusively for STRINGED INSTRUMENTS of the VIOLIN family, but in the early 19th century the use of large BRASS and WOODWIND sections had become general. *See also* HC2 pp.104–105.

Synaesthesia, sensation experienced at a point different from the point of stimulation. For example, an individual with chromaesthesia, the most common form of synaesthesia, experiences colours (visually) when hearing music (aurally).

Synagogue, place of assembly for Jewish worship, education and cultural development. They serve as communal centres, led by a RABBI; they house the ARK OF THE COVENANT. As actual buildings, synagogues date from the 3rd century BC although homes were used as temporary synagogues after the destruction of Solomon's Temple (Jerusalem) in 586 BC.

Synapse, connection between the nerve ending of one NEURON and the receptive area of the next. In a chain of neurons this gap is bridged by a special "transmitter substance", such as ACETYLCHOLINE, which is produced by structures in the nerve endings. *See also* MS pp.38–39.

Synchrotron, accelerator in which a beam of ELECTRONS or PROTONS is focused and guided around a fixed circular path by changing magnetic fields. Millions of revolutions are made. A high-frequency electric field at one point in the path accelerates the particles. Proton energies can reach hundreds of GeV, electron energies tens of GeV.

Syncline, downward fold in rocks. The bending of rocks is called folding. When rock layers fold down into a trough-like form, it is called a syncline. (An upward arch-shaped fold is called an anticline.) The sides of the syncline are called limbs, and the median line between the limbs along the trough is known as the axis of the fold. *See also* PE pp.104, *105*, 106.

Syncopation, in music, a contradiction or breach of a regular rhythmic pattern. It is normally produced in two ways, by the momentary introduction of an accent off the regular beat of the rhythm or by placing longer notes on weak beats in a bar.

Syndicalism, early 20th-century form of socialism originating in France. It proposed public ownership of the means of production by small worker groups and called for the elimination of central government. Unlike forms of socialism that advocate peaceful change, the movement encouraged revolutionary action by workers to initiate a syndicalist system.

Syndrome, in medicine, a group of symptoms that together form a pattern from which a disorder or disease may be identified.

Synge, John Millington (1871–1909), Irish dramatist and poet who was important in the IRISH LITERARY RENAISSANCE. He was one of the organizers of the ABBEY THEATRE in 1904 and his works, which depict the bleak and tragic lives of Irish peasants and fishermen, include *Riders to the Sea* (1904), *The Well of the Saints* (1905) and *The Playboy of the Western World* (1907). *See also* HC2 pp.282, *283*.

Synge, Richard Laurence Millington (1914–), British biochemist. He shared the 1952 Nobel Prize in chemistry with Archer J.P. MARTIN for developing paper CHROMATOGRAPHY, in which structurally complicated chemicals can be identified by noting the rate at which they diffuse up a column of filter paper. This discovery was used by Frederick SANGER to identify the individual AMINO ACIDS in INSULIN.

Synodic period, interval between two successive conjunctions or oppositions with the Sun of a planet or the Moon as seen from the Earth. For the Moon, the synodic period is the time taken for a single complete cycle of phases, equalling 29.53 days.

Synoptic Gospels, term used to describe the GOSPELS of St MATTHEW, St MARK and St LUKE. These accounts contain similar subject matter, phrasing and interpretation, and their textual interdependence is generally accepted. St Mark's Gospel is generally held to have been the model for St Matthew's and St Luke's, although the latter two have gathered some material from a common non-extant source known as "Q".

Synovitis, inflammation of the synovial membrane of the joints and tendons. It causes pain and may be accompanied by swelling and stiffness. *See also* MS p. 94.

Syntax, in grammar, the construction of sentences, clauses and phrases. The study of syntax examines the arrangement of words and phrases and the relationships between different sentences.

Synthetic diamonds. *See* DIAMOND.

Synthetic dyes, man-made colours for fabrics and plastics. Originating with the discovery of the first aniline dye mauveine by William PERKIN in 1856, the synthetic dye industry rapidly replaced natural dye-stuffs such as INDIGO and MADDER. Chemists have also been able to synthesize these two natural compounds. *See also* MM p.244.

Synthetic fibres. *See* MAN-MADE FIBRES.

Synthetic rubber. *See* RUBBER.

Synthetism, or cloissonism, French style of painting in the 1890s centred around Pont-Aven, Brittany. The major artists involved were Émile BERNARD and Paul GAUGUIN, who attempted to synthesize the colour planes and lines on the canvas with the feeling of the subject.

Syphilis, systemic infectious disorder caused by the SPIROCHETE *Treponema pallidum.* It is usually transmitted through sexual contact, but sometimes occurs congenitally. The first symptom is a small, hard swelling at the site of infection. In the second stage, there are skin lesions or a rash. Tertiary symptoms may be incapacitating or fatal, affecting almost any part of the body. Other forms are endemic and non-venereal. Laboratory tests can detect syphilis in its earliest stage, when treatment with PENICILLIN can achieve a complete cure. *See also* MS p.102.

Syr Darya, shallow river flowing through the USSR, formed by the junction of the Naryn and Kara Darya rivers in the Fergana Valley of Uzbekistan, flowing w past Leninabad in Tadzhikistan and N re-entering Uzbekistan at Begovat. It then flows into Kazakhstan below Chinaz, and past Chardara, forming the E and N limits of the Kyzyl kum desert between Kyzyl-Orda and Kazalinsk, to the Aral Sea. Length: 2,206km (1,370 miles).

Syria, independent nation in the Middle East. Its dominant geographical features are the Anti-Lebanon and Alawite Mts along the coast from Israel to Turkey, the valley of the River Euphrates and the Jebel al-Druse Mts in the s. The economy is based on agriculture and stock raising, and cotton is the major export. The capital is Damascus. Area: 185,123sq km (71,476sq miles). Pop. (1976 est.) 7,585,000. *See* MW p.164.

Syriac, Semitic language belonging to the eastern ARAMAIC group. In ancient times it was spoken in Edessa, now Urfa in SE Turkey. Because of the importance of Edessa as a centre of Christianity in the 2nd century, Syriac was adopted by the neighbouring Aramaic Christians and has been used ever since as a liturgical language by Oriental Christians of the Syrian rite. Syriac literature preserves many translations of Greek Christian texts

which have not survived in the original Greek.

Syrian War, campaigns of 192–189 BC waged between Antiochus III, the SELEUCID king, and Rome. In an attempt to enlarge his kingdom, Antiochus conquered Ptolemaic Syria and Palestine (195 BC) and invaded Greece. He lost two major land battles to the Romans (at Thermopylae and Magnesia), and Seleucid power in the Mediterranean was ended by the peace of Apamea (188).

Syringa, genus of flowering deciduous shrubs and trees native to Europe and NE Asia. Most of the 30 species are cultivated. Family Oleaceae. *See also* LILAC.

Syrinx, vocal organ of a bird, consisting of thin, vibrating muscles at the base of the windpipe.

Systems analysis, method of analysing a process or operation in order to improve efficiency, particularly with the aid of a COMPUTER, which has the capacity to handle the large number of interdependent activities involved in a complex operation. *See also* MS p.275.

Systole, time in the cardiac cycle when the HEART muscle is under contraction, ejecting blood from the heart into the arterial system. *See also* MS pp.62–63.

Szczecin (Stettin), city in NW Poland, on the River Oder near its mouth in Stettiner Haff bay; capital of Szczecin province, a major port and industrial city. Szczecin was the largest city of Pomerania in the 12th century, and an important member of the HANSEATIC LEAGUE. It passed to Sweden in 1648 under the Peace of Westphalia, then to Prussia in 1720. The city was returned to Poland by the Potsdam Conference in 1945. Industries: shipbuilding, iron works, chemicals, food processing, fishing. Pop. (1974) 360,500.

Szechwan (Sihchuan), province in SW China, bounded by the provinces of Tsinghai, Kansu and Shensi (N), Hupei and Hunan (E), Kweichow and Yunnan (s) and completely surrounded by mountains; the capital is Chengtu. The E part of the region corresponds to the large, heavily populated Red Basin, which is the most prosperous area of China. Szechwan is the country's leading producer of rice, corn and sweet potatoes and soya beans, barley and fruit are also grown. Livestock, including cattle, pigs, horses and oxen, are reared, particularly in the w. Salt, coal and iron are mined and other products include rape-oil and silk, for which Szechwan was once world famous. Area: 569,215sq km (219,774sq miles). Pop. 70,000,000. *See* MW p.53.

Szell, George (1897–1970), US conductor, b. Hungary. After studying composition with Max REGER he turned to conducting on the recommendation of Richard STRAUSS and in 1917 became conductor of the Strasbourg Opera. He became conductor at the Metropolitan Opera (1942–46) and then permanent conductor of the Cleveland Orchestra (1946–70)

Szent-Györgyi, Albert von Nagyrapott (1893–), US biochemist, b. Hungary. He was awarded the 1937 Nobel Prize in physiology and medicine for his work on biological oxidation processes and the isolation of vitamin C. He also studied the biochemistry of MUSCLE, discovering the muscle protein ACTIN that is responsible for muscular contraction when combined with the muscle protein MYOSIN.

Szewinska, Irena (1946–), née Kirszenstein, Polish sprinter and long jumper, b. Leningrad. In four Olympic Games she won three gold medals (for the 4×100m relay in 1964, for the 200m in 1968 and for the 400m in 1976), one silver and two bronze medals. She set seven world records for distances of from 100m to 400m.

Szilard, Leo (1898–1964), US physicist, b. Hungary. His early work established the relation between information transfer and ENTROPY. He devised a means of separating ISOTOPES of artificial radioactive elements; developed the chemostat; and proposed aging, recall and memory theories. He was also involved in the creation of the first sustained nuclear chain reaction and in the initiation of the atomic bomb.

Szymanowski, Karol (1882–1937), Polish

post-ROMANTIC composer who began composing in about 1900 and did much to promote the nationalist cause in his country. Early influences were the works of SCRIABIN and CHOPIN and from 1906 to 1908 he studied the works of WAGNER and Richard STRAUSS in Germany. His works include two violin concertos (1917, 1933) and the opera *King Roger* (1926).

T

T, 20th letter of the alphabet, derived from the Semitic letter *taw* (meaning *mark*) and the Greek *tau.* The Roman letter had the same form as the modern *T.* In English *t* nearly always has the sound as in *fat* and *tar,* although it is silent in such words as *listen* and *castle,* and in words derived from the French, eg *depot* and *debut.* The combination with *h* gives rise to two sounds, as in *thin* and *this,* and the combination with *i* sometimes gives the sound *sh,* as in *action* and *militia. See also* MS pp.244–245.

Tabernacle, in Judaism, the portable shrine used by the Jews for worship during their wanderings in Sinai. It is also called Mishkan, or the dwelling of God. It was a rectangular tent covered with skins. It contained the ARK OF THE COVENANT. The Jews believed the tabernacle to embody the presence of God among His people. In Christianity a tabernacle is a receptacle in which the Blessed Sacrament is reserved for the Eucharist, or for spiritual contemplation.

Tables, The, four representative committees which took virtual control of Scotland in 1637 to fight against CHARLES I's attempt to impose the Anglican form of worship on Scotland. They drew up a covenant which denounced the Prayer Book and affirmed PRESBYTERIANISM; and they began to prepare Scotland for war.

Table tennis, table sport (also known as ping pong) played by two or four people. The table is 2.7m (9ft) long, 1.5m (5ft) wide, and 76.2cm (30in) from the floor. The surface is divided by a transverse net 15.2cm (6in) high, and halved longitudinally by a light stripe. A hollow celluloid ball is hit with a wooden bat. On the serve, the ball must bounce once before clearing the net on the near surface and then bounce once on the far surface before it is returned. After the serve, the ball must bounce only once on the far side of the net. If the ball goes foul or fails to be returned properly, a point is scored. Each player in turn has five consecutive serves. The winner is the first player to score 21 points while leading by at least two points; games tied at 20–20 continue until a two-point margin is achieved. In doubles the players must return the ball in a set order. Table tennis is the national sport of China.

Taboo, or tabu, prohibition of a form of behaviour, object, word, etc. In many primitive cultures a thing may be regarded as taboo because it is either unclean or sacred – there is often a taboo on a totem. Breaking a taboo is believed to bring supernatural retribution and often also brings social ostracism or other punishment. The term is of Tongan origin.

Tabriz, city in NW Iran. An ancient city, it served as the capital of Armenia under King Tiridates III in the 3rd century AD. Ghazan Khan developed it as the chief administrative centre for his vast Asian empire in 1295; it was occupied by the Ottoman Turks in 1514, 1585–1603 and 1724–30, and held by the Russians 1827–28. The proximity of Tabriz to Turkey and the Soviet Union has made it an important trading centre; manufactures include carpets, shoes, soap and textiles. Pop. 510,000.

Tabularium, building in ancient Rome of which only the façade remains. It was the state records office and was built in 78 BC. It is one of the earliest buildings to incorporate an arch with the columns of classical architecture and to employ concrete in the vaultings.

Synthetic dyes, produced in the factory, are more colourfast than natural dyes.

Syringa is known as mock orange from its flowers' resemblance to orange blossom.

Szechwan; irrigation wheels driven by the current raising water near Chengtu.

Szymanowski was creating his own folk idiom long before he studied folk music.

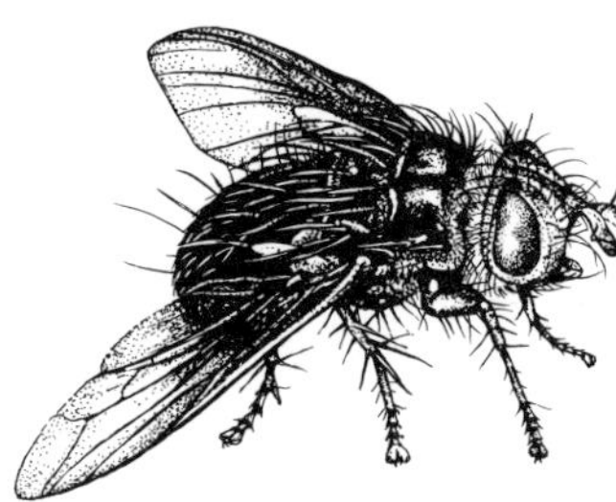

Tachinid flies lay their eggs on other insects; the larvae develop as parasites.

Tacitus's writing reveals much about the period of the early Roman Empire.

Tahrs live in mountains and have thick fur to protect against the cold.

Tailorbirds lay 3 or 4 eggs in nests which are lined with down, grass or hair.

Tachinid fly, any of several species of insects, the large larvae of which are internal parasites of other insects, including the GYPSY MOTH, armyworm and CUTWORM. Many species resemble HOUSE FLIES. Family Tachinidae.

Tachisme (from the French word *tâche* meaning a blot), in art, term used for the method of painting by which irregular or seemingly random dabs of colour are applied to the canvas. The term was coined by Michel Tapié in the early 1950s. Tachiste painters such as Camille Bryen, Georges Mathieu and Jean Atlan reacted against both abstraction and naturalism in art, and claimed supremacy for each individual patch of colour as an element of emotional projection in its own right.

Tachometer, device that measures rate of rotation, as of a wheel or shaft. One type of tachometer is a simple revolution counter, with which average speeds of rotation can be measured with the further aid of a stopwatch. More sophisticated instruments show instantaneous speeds of rotation. These instruments are used widely in high-performance cars and in industry to check machine performance. They may be made on mechanical, electrical or electronic principles.

Tachycardia, rapid heart beat which may occur normally after some form of stress or which may have a pathological origin.

Tacitus, Cornelius (*c.*55–120), Roman historian. The perfection of his style, the appeal of his political ideas and the relative accuracy of his accounts have established him as the greatest Roman historian. His first important works were the *Agricola* (*c.*98), a eulogy of his father-in-law, who was governor of Britain, and the *Germania* (*c.*98), a description of the Rhine frontier. Other works include *Histories* (*c.*109) and *Annals* (*c.*114). *See also* HC1 pp.113, 247.

Tacna-Arica Dispute. *See* ARICA-TACNA DISPUTE.

Taconite, kind of CHERT that contains iron, resulting from replacement processes. It is used as an iron ore in some countries where it is abundant.

Taddeo di Bartolo (*c.*1386–1422), Sienese painter. His work, influenced by Simone MARTINI, is characterized by a profusion of decorative detail. Among his works is the fresco series of Republican heroes in the Palazzo Pubblico, Siena.

Tadema, Lawrence Alma. *See* ALMATADEMA, SIR LAWRENCE.

Tadpole, aquatic larva of a TOAD or FROG; it has a finned tail and gills, and lacks lungs and legs. Unlike adult forms the tadpoles of most species are herbivores, feeding on algae and other aquatic plants. During METAMORPHOSIS limbs are grown, the tail is resorbed, and internal lungs take the place of gills. *See also* NW p.*133, 230.*

Tadzhikistan (Tadzikskaja SSR), constituent republic of the USSR, bounded by Afghanistan (S) and China (E). The region is mountainous and the principal rivers are the Amu Darya, Syr Darya and Zeravshan. Agriculture is the basis of the economy; cotton, fruit and cereals are grown, and livestock raising is also important. Gold, antimony, coal and fluorspar are mined and the manufacture of cotton, silk and leather goods are major industries. The capital is Dushanbe. Area: 143,100sq km (55,250sq miles). Pop. (1976) 3,400,000. *See* MW p.169.

Taegu, city in s central South Korea; capital of North Kyŏngsang province and third largest city in South Korea. The city was successfully defended by UN troops during the KOREAN WAR. Taegu is a trade centre for a farming and mining region. Pop. (1975) 1,309,454.

Taenia. *See* TAPEWORM.

Tafawa Balewa, Alhaji Sir Abubakar (1912–66), Nigerian politician. Leader of the Northern People's Congress, he became Prime Minister of the Federal Republic of Nigeria in 1957 and led Nigeria to independence from British rule in 1960. He was assassinated in a military coup.

Taffeta, thin, tightly woven fabric, finished so that it is rather stiff and has a sheen. Originally woven from spun silk, taffeta may now contain synthetic fibres.

Taff Vale Case (1900), suing of the Amalgamated Society of Railway Servants by the Taff Vale Railway for damages suffered from a strike. Against earlier interpretations of the 1871 Trade Union Act the House of Lords, on appeal, upheld the judgment that the union was liable. The ASRS paid £23,000. The unions gained redress in the Trade Disputes Act of 1906, which made them not liable.

Tafilet, Saharan oasis in modern Morocco which fostered a BERBER trading kingdom from the 8th to the 14th centuries. Tafilet now trades in dates and leather.

Taft, William Howard (1857–1930), 27th President of the USA (1909–13) and 10th Chief Justice of the Supreme Court (1921–30). A practising lawyer, he made his way up the political ladder in the Republican Party. In 1900 he was sent to the Philippines to set up a government and became the first US civil governor in 1901. He was a friend of President Theodore ROOSEVELT, who appointed him Secretary of War (1904), and he became one of Roosevelt's closest advisors and ultimately his successor. Their relationship deteriorated, however, when Roosevelt thought Taft's administration too conservative over anti-trust activities and a high protective tariff. Roosevelt split the Republican Party, with the result that Taft failed to win re-election.

Taft-Hartley Act (1947), US labour legislation, also known as the Labor-Management Relations Act. It outlawed the CLOSED SHOP, required unions to reveal their financial status, limited union political activity, required union leaders to sign a non-Communist affidavit and authorized the government to issue 80-day strike injunctions. Other provisions allowed employers to sue unions and enlarged the National Labor Relations Board.

Tagalog, national language of the Philippines. In 1962 it was made the country's official language and given the new name of Filipino. It is the mother tongue of about 5 million people, but it is estimated that about 37 million more understand the language. Tagalog belongs to the Malayo-Polynesian family of languages.

Tagetes. *See* MARIGOLD.

Taglioni, Filippo (1777–1871), Italian dancer and choreographer; ballet master for the Stockholm Ballet from 1803. His most famous work is *La Sylphide* (1832), created for his daughter, Maria TAGLIONI. This is credited as being the first ROMANTIC ballet. *See also* HC2 p.274, *274.*

Taglioni, Maria (1804–84), Italian ballet dancer who developed a style that became the prototype of Romantic ballet and transformed the use of pointwork from a display of technical virtuosity to an art form of great beauty. Her best performance was considered to be in *La Sylphide*, performed at the Paris Opéra in 1832. *See also* HC2 p.274, *274.*

Tagore, Rabindranath (1861–1941), Indian poet, philosopher and social theorist who received the 1913 Nobel Prize in literature. He wrote mainly in Bengali, to which he introduced several verse forms new to the language, including the ode. He is best known in the west for his collection of poetry *Gitanjali* (1910). *See also* HC2 pp.290, *291.*

Tagus (Tajo, Tejo), river in Spain and Portugal that rises in E central Spain and flows NW then WSW to enter the Atlantic Ocean at Lisbon. It is the longest river on the Iberian Peninsula and forms part of the Spanish-Portuguese border. The river is navigable for 130km (90 miles). Length: approx. 1,000km (620 miles).

Tahiti, island in the S Pacific Ocean, in the Windward group of the Society Islands, French Polynesia; largest of the French Polynesian islands. It was charted in 1767 by the British navigator Samuel Wallis and explored by Capt. James COOK on several visits between 1769 and 1777. The island was colonized by France in 1880. Tahiti is mountainous and fertile and produces tropical fruits, copra, sugar cane and vanilla. Pearl fishing is an important industry. The capital is Papeete. Area: 1,058sq km (408sq miles). Pop. (1971) 79,494.

Tahr, gregarious, goat-like ruminant that lives in the Himalayas, S India and SE Arabia. The long coat varies from light to dark brown. Height: to 1.1m (3.5ft); weight: to 90kg (200lb). Family Bovidae; genus *Hemitragus.*

Tai, people of Chinese origin, who inhabit the mainland of SE Asia. They consist of several Tai-speaking groups: the Siamese, Lao, Shan, Lu, Yunnan Tai and tribal Tai. Agriculturalists, their major crop is rice. Although Theravāda Buddhism is the principal religion, some ANIMISM remains.

Taiaha, wooden MAORI weapon, about 1–2m (3–6ft) long, carved at the head with a tongue protruding from a face. It was the chief two-handed weapon of the Maoris.

Tai Chin (1388–1462), Chinese landscape painter. An important member of the regional Che School, he was known for his clever brushwork.

Taiga, in Asia, extensive northern pine forests that are known in Europe and the USA as BOREAL FOREST. The southern boundary is marked by the increasing number of deciduous trees among the pine trees as the temperature rises. To the north lies the vast frozen wastes of the TUNDRA. *See also* NW pp.206–207, *206–207.*

Taika Reforms (646–702), series of political edicts in Japan. Designed to establish a central Chinese-style government controlled by the emperor, the reforms included the imperial control of land, systematic taxation and the creation of a bureaucracy of officials from which developed the Japanese aristocracy. *See also* HC1 p.130.

Tailings, mine, rock and other waste materials produced by excavation and the washing, crushing and other treatments of ores.

Tailleferre, Germaine (1892–), French composer who studied with Darius MILHAUD. She became prominent as a member of the group of young composers known as Les SIX who reacted against the style of DEBUSSY after WWI. She wrote two operas, *Jeux de Plein Air* (1923) for two pianos, and a concerto for harp and orchestra.

Tailorbird, any of several species of mainly dull-coloured, tropical Asian and African warblers that sew leaves together with plant fibre, cotton, grass or wool, forming a basket, inside which a nest is built. Length: to 12.5cm (5in). Family Sylviidae; genus *Orthotomus. See also* NW p.*144.*

T'ainan, city on the SW coast of Taiwan. One of the oldest cities in Taiwan, it was settled in 1590 and held by the Dutch 1623–62. Fishing is an important activity, and products include textiles, processed food and machinery. Pop. (1970 est.) 475,000.

Táin Bó Cúailgne (*The Cattle Raid of Cooley*), Irish epic-like SAGA of the 7th or 8th century AD. In prose with verse passages, it describes the controversy between Ulster and Connaught over ownership of the Brown Bull of Cooley. It is chiefly read for its descriptions of CÚ CHULAINN. *See also* HC1 p.341.

Taine, Hippolyte-Adolphe (1828–93), French historian whose theories on DETERMINISM made him one of the foremost exponents of POSITIVISM. Among his best-known works is *De l'intelligence* (1870); *Les Origines de la France Contemporaine* (1876–99) deals with the philosophy of history.

T'aipei, capital of Taiwan, at the N end of the island. A major trade centre for tea in the 19th century, the city was enlarged under Japanese rule (1895-1945) and became the seat of the Chinese Nationalist government in 1949. Products include textiles, chemicals, fertilizers, metals and machinery. Pop. (1971) 1,742,626.

Taiping Rebellion (1851–64), revolt in China against the MANCHU led by a Hakka fanatic, Hung Hsiu-ch'uan, who thought himself to be the son of God. The fighting laid waste 17 provinces of China and resulted in more than 20 million casualties. The Manchus never recovered their ability to govern China with their former authority. *See also* HC2 p.144, *144.*

Taira family, leading military group in Japan in the 11th and early 12th centuries

AD. The family was descended from the emperor Kammu and rose to power as servants of the KYŌTO emperors. Under Taira Kiyomori (1118–81) the Taira wielded great power in Japan, but it was heavily defeated by its rival, the Minamoto family, in 1185. *See also* HC1 p.131.

Tairov, Alexander Yakovlevich (1885–1950), Soviet theatrical producer and director. After founding the Kamerny (Chamber) Theatre in Moscow in 1914, he was its producer-director until 1949. A firm opponent of STANISLAVSKY, he subscribed to MEYERHOLD'S opinions, attempting to achieve perfect aesthetic unity in his productions.

T'ai Tsung (597–649), Chinese emperor of the Tang dynasty, son of Li Yüan. He conquered the Eastern Turks in wars from 624 to 630 and controlled the petty states of Turkestan. He also established contacts with India and Persia.

Taiwan. *See* CHINA.

Taiyuan, city in NE China, 427km (265 miles) SW of Peking, on the River Fen; the capital of Shansi province. The region has rich coal and iron ore deposits and Taiyuan is a major industrial city. It has iron and steel, chemical and engineering plants. Pop. (1970 est.) 2,725,000.

Taizé, religious community in SE France, founded in 1940 by Roger Schutz as a home for Jewish and other refugees during WWII. Monastic from 1949 onwards, the community is made up of various Protestant denominations and is dedicated to promoting Christian unity.

Ta'izz, city in S Yemen, 52km (32 miles) E of Mocha. It served as the administrative capital of the Yemen 1948–62. Ta'izz is a market for agricultural goods. Pop. (1970 est.) 84,000.

Taj Mahal, Muslim mausoleum built beside the River Sumna outside Agra, India between 1630 and 1648. It was built for the Mogul emperor Shah Jahan after the death of his favourite wife, and the final design is accredited to Ustād 'Isā. With its bulb-shaped dome, intricate inlays of semiprecious stones and oblong reflecting pool, it is set within a garden and ranks among the world's most beautiful buildings. *See also* HC2 p.*51.*

Takahe, rare, flightless New Zealand bird, related to the RAIL and gallinule; once regarded as extinct, it was rediscovered in 1948. Turkey-sized, it has a heavy curved bill, a reddish shield on the forehead and bright blue-green plumage. Family Rallidae; species *Notornis mantelli. See also* NW p.*153.*

Takin, heavy goat-like mammal of the forests of the Himalayan foothills and the mountains of Central Asia. It has a thick, shaggy brownish coat and backward-pointing horns. It feeds primarily on bamboo shoots and other plants. Height: to 100cm (39in) at shoulder. Family Bovidae, species *Budorcas taxicolor.*

Takis, Vassilakis (1925–   ), self-taught Greek sculptor. After moving to Paris in 1954 he experimented with kinetic sculpture using magnets. Many of his sculptures consist of slender, upright rods which move with a gentle vibrating action.

Takoradi. *See* SEKONDI-TAKORADI.

Talaing, or Mon, Mongoloid people who inhabit the delta region of Burma and W central Thailand. An agricultural people who cultivate rice, they introduced the writing (Pāli) and religion (Buddhism) which is predominant in Burma today.

Talbot, William Henry Fox (1800–77), British scientist who improved on the work of NIEPCE and DAGUERRE by inventing the first practical photographic process capable of producing any number of postive prints from an original paper negative. This calotype (later Talbotype) process was patented in 1841 but lost popularity after the glass negative was introduced in 1851. Talbot's book *The Pencil of Nature* (1844) was the first to include photographs as illustrations. *See also* MM p.218.

Talc, sheet silicate mineral, hydrous magnesium silicate, $Mg_3Si_4O_{10}(OH)_2$; used as base for TALCUM POWDER and in ceramics. It occurs as rare tabulate crystals in a monoclinic system and as masses and is either white, green, blue or brown in colour. Hardness 1; s.g. 2.6.

Talcum powder, powdered form of natural hydrated magnesium silicate. It is used in paints as an extender, as a lubricant and in cosmetic and pharmaceutical preparations.

Talenti, Francesco (*c.*1300–*c.*69), Florentine architect whose principal work includes the completion of the campanile of the cathedral in Florence.

Tale of Genji, Japanese work written in the 11th century by Lady Murasaki Shikibu. The greatest work of the HEIAN period, it relates the amorous adventures of Prince Genji in the Japanese fuedal court with great sensitivity for the subtleties of human emotions, underpinned by a Buddhist philosophy. It is sometimes regarded as the first significant prosework written. It was first translated into English by Arthur WALEY between 1925 and 1935. *See also* HC1 pp.131–132, *131–132.*

Tale of Two Cities, A (1859), historical novel by Charles DICKENS, set in London and Paris against the background of the French Revolution. It tells the melodramatic story of the love of Charles Darnay and Lucie Manette.

Tales of Hoffmann, The (1881), three-act comic opera by Jacques OFFENBACH with libretto by Jules Barbier, derived indirectly from stories by E. T. A. HOFFMANN. The opera occupied Offenbach for many years, but he died before it was produced; it is considered the finest example of his work.

Taliesin, The Book of. *See* BOOK OF TALIESIN.

Tallage, tax in medieval England paid by a VILLEIN to his lord or, more important, by towns and boroughs to the king. First used (by that name) in 1173, it was a special tax, general in its appropriation, beyond regular feudal dues. It was abolished in 1340.

Tallahassee, state capital of Florida, USA, 258km (160 miles) W of Jacksonville. Originally inhabited by Seminole Indians, the settlement was developed as a Spanish mission; it became capital of Florida Territory in 1824. The city is the seat of Florida State University (1857). Industries: timber, food processing, tourism. Pop. (1970) 72,586.

Talleyrand-Périgord, Charles Maurice de (1754–1838), French statesman who was able to follow a successful career throughout the many changes in French politics during his lifetime. Made Foreign Minister under the DIRECTORY in 1797, he also served under NAPOLEON until disagreement over Napoleon's European ambitions led him to help to restore the House of BOURBON in 1814. At the Congress of VIENNA he skilfully exploited the tensions among the victors to the benefit of France, and as a supporter of LOUIS PHILIPPE he became ambassador to London from 1830 to 1834.

Tallin (Talin), capital of Estonia (Estonskaja SSR), USSR, on the Gulf of Finland opposite Helsinki, 320km (200 miles) W of Leningrad. It was founded in 1219 by the Danes and became a member of the HANSEATIC LEAGUE in 1285, but was sold by its ruler to the TEUTONIC KNIGHTS in 1346. It passed to Sweden in 1561 and was ceded to Russia in 1721. It was developed in the 19th century as a port for Russia's Baltic Fleet. It remains a major port and industrial centre, producing processed food, machinery, cables and oilfield equipment. Pop. (1970) 363,000.

Tallis, Thomas (1505–85), English composer, famous for his church music. In 1575 he published with BYRD the *Cantiones Sacrae,* a set of motets. He also wrote a setting of Lamentations, two masses, a number of anthems, and among his motets, *Spem in alium,* in 40 parts.

Tallow tree, decorative evergreen tree that grows in SE China and the USA. The seeds yield a waxy tallow-like substance that is used in the manufacture of soap and candles. Height: to 12m (40ft). Family Euphorbiaceae; species *Sapium sebiferum.*

Tally, piece of wood marked with tranverse notches, used in medieval England as a record of transactions. The notches represented amounts of money or goods. Such a tally stick would be split along its length so that each party to the transaction retained a record of the payment or debt, which could be proved by matching the pieces.

Talmud, meaning "learning", the Oral Law of Judaism as opposed to the Scriptures. It may be divided into two parts: MISHNAH, the traditional text, and Gemara, the interpretation and commentary. Study of the Talmud is central to orthodox Jewish faith.

Talon, Jean-Baptiste (*c.*1625–94), French colonial administrator in Canada. As the first intendant of NEW FRANCE between 1665 and 1672, he encouraged immigration, established industry and developed the political and economic structure of French Canada.

Talus. *See* SCREE.

Tamale, city in N Ghana, 435km (270 miles) N of Accra; capital of the Northern region. It is a centre of education and communications, and a market for agricultural goods. Pop. (1970) 81,612.

Tamandua, small terrestrial or arboreal ANTEATER of tropical America; it has a prehensile tail that it uses for grasping branches. Its coat is thick and generally black, with a white head, neck, limbs and tail. It feeds primarily on ants. Length: to 1m (3.2ft) overall. Species *Tamandua tetradactyla. See also* NW p.216, *216.*

Tamarind, tropical tree native to Asia and Africa. It has pinnate (feather-like) leaves and pale yellow flowers, streaked with red. The fruit pulp is used in beverages, food and medicines. Height: 12–24m (40–80ft). Family Leguminosae; species *Tamarindus indica.*

Tamarisk, any of a group of shrubs usually found in semi-arid areas. They are deciduous and have slender branches covered with blue-green, scale-like leaves and clusters of small, white or pink flowers. Their deep, spreading roots enable them to tolerate drought and the salinity of deserts and coastal regions. Height: to 9.1m (30ft). Family Tamaricaceae.

Tamatave, seaport city on the E coast of Madagascar. It developed around the site of a European trading post in the 17th century and was occupied many times by the French. Modern industries include food processing and meat packing. Exports include spices, rice, coffee and sugar. Pop. (1971 est.) 59,000.

Tambourine, percussion musical instrument, originating in Roman times and much used by wandering musicians in Europe in the middle ages. It comprises a narrow circular frame, made of wood, with a single parchment drumhead and metal jangles attached to the sides.

Tambura, instrument of the LUTE family, found chiefly in the Middle East and India. It has four wire strings and is played with a plectrum. The Caucasian variety, called a *tampur,* has three strings and is played with a bow.

Tamburlaine the Great (*c.*1587), tragedy by Christopher MARLOWE. Rich in imagery and imaginative language, the play describes the rise to power and subsequent moral disintegration of the "scourge of God", Tamburlaine (TAMERLANE). Although imperfect in construction, the work is extremely moving, chiefly because of its great poetic power.

Tamerlane (Timur i Leng) (1336–1405), Mongol conqueror, claiming to be descended from GENGHIS KHAN. He had brought the area around Samarkand under his control by 1369, extending his conquests to the region of the GOLDEN HORDE between the Caspian and Black seas before invading India in 1398. He defeated the Ottoman Sultan BEYAZID I in 1402 and died while planning to invade China. Tamerlane gave his name to the TIMURID DYNASTY.

Tameside, county district in E central GREATER MANCHESTER, England; created in 1974 under the Local Government Act (1972). Area: 103sq km (40sq miles). Pop. (1974 est.) 222,600.

Tamil, language spoken in S India, chiefly in the state of Tamil Nadu on the E coast, by about 31 million people. In addition there are about two million speakers in Sri

Taj Mahal, made of white marble and red sandstone, was built by 20,000 men.

William Henry Fox Talbot had many controversies with other early photographers.

Talleyrand was Foreign Minister under three different regimes during his career.

Tamarisks of Asia Minor exude a honey-like substance eaten by the Bedouins.

Tamil Nadu

Tampere's factories are largely hydroelectric-powered, so reducing pollution.

Tanager, a colourful songbird found throughout the forests of the New World.

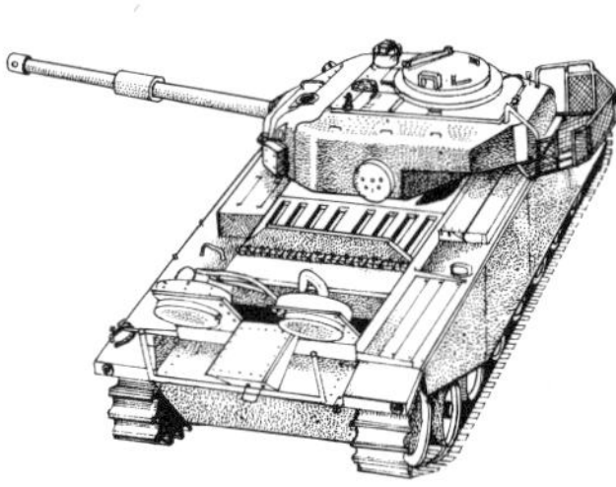

Tanks; named in WWI when they were called "water tanks" to keep their purpose secret.

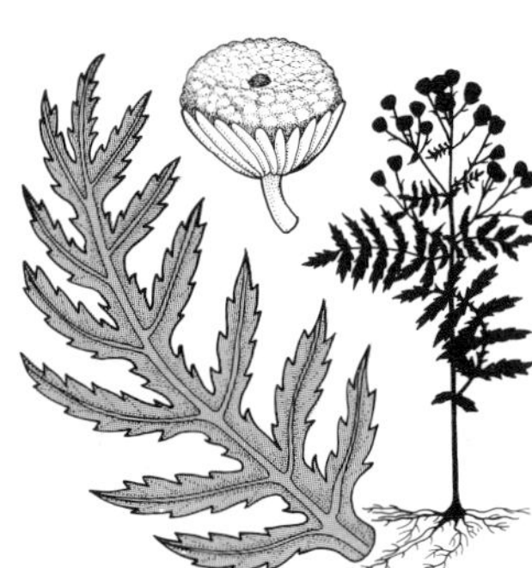

Tansy is related to the thistle and gives off a poisonous oil from its leaves.

Lanka and about one million distributed throughout Malaysia, Singapore, Fiji, Mauritius and Guyana. Tamil belongs to the DRAVIDIAN family of languages.

Tamil Nadu. *See* MADRAS.

Taming of the Shrew, The (*c.* 1594), comedy in five acts by William SHAKESPEARE. Some sources believe it was based on an anonymous play *The Taming of A Shrew* (1594), although the latter may be simply a pirated version of Shakespeare's work. It provided the basis for the Broadway musical *Kiss Me Kate* (1948) and was filmed in 1908, 1929 and 1967.

Tamm, Igor Yevgenyevich (1895–1971), Soviet scientist and noted theoretical physicist. He shared the 1958 Nobel Prize in physics with Pavel A. CERENKOV and Ilya M. Frank for the discovery and interpretation in 1937 of the Cerenkov effect (the radiation emitted by electrons moving faster than light in an optically transparent medium). *See also* MS p.*208.*

Tammany Hall, DEMOCRATIC PARTY organization in New York. It evolved from the fraternal and patriotic order of St Tammany, which was founded in 1789 and rapidly became the focal point of resistance to the Federal Party, which dominated New York politics at the beginning of the 19th century. Politics in the city changed drastically in the next 70 years under the twin influences of manhood suffrage and mass immigration. By 1865 Tammany Hall, under William "Boss" Tweed, had become the most important voice of the Democratic Party in the city, and was synonymous with the organized corruption of US urban politics. Not until the 1930s did its power begin to crumble.

Tammuz, pastoral deity of ancient Mesopotamia, who evolved into a fertility god similar to the Syrian ADONIS. Because he slighted the earth goddess ISHTAR, she condemned him to the underworld. He was also associated with the myth of the cycle of fertility.

Tampere, city in SW Finland, on the lake Näsijärvi; the second largest city in Finland. In 1918 the nationalist White Guards of Finland defeated the Russian-backed Finnish Bolsheviks there. The city has a university (1966). Industries: footwear, paper, textiles, machinery. Pop. (1976) 165,769.

Tamworth Manifesto, election address written by Robert PEEL for his Tamworth constituents at the 1835 elections. It was read by the Cabinet and is thus a prototype of the party manifesto. It stressed the change in Peel's party from die-hard Toryism to moderate reforming Conservatism.

Tanabata, annual Japanese "star festival", held on 7 July to celebrate the "meeting" of two stars, the Star Weaver and Altair. They are believed to represent star-crossed lovers whom the gods allow to meet once a year.

Tanager, small, brightly coloured American forest bird that has a cone-shaped bill. Tanagers feed on insects and fruit. The scarlet tanager (*Piranga olivacea*) of E North America has black on its wings and tail, and after mating loses its red plumage and resembles the olive-green female during its southerly winter migration. Family Emberizidae. *See also* NW p.*217.*

Tanaka, Kakuei (1918–), Prime Minister of Japan from 1972 to 1974. A self-made millionaire, he was elected to the DIET in 1947, served as Minister of Finance from 1962 to 1964 and as Minister of International Trade and Industry from 1971 to 1972. As Prime Minister he established diplomatic relations with China. Allegations of corruption forced his resignation.

Tananarive (Antananarivo), capital and largest city of Malagasy. Founded *c.* 1625, the city became the residence for Imerina rulers in 1794 and the capital of Madagascar. Tananarive was taken by the French in 1895 and became part of a French protectorate. It is the seat of the University of Madagascar (1961). A trade centre for a rice-producing region, it has food processing, textile, tobacco and leather industries. Pop. (1975 est.) 438,000.

Tancred (1847), novel by Benjamin DISRAELI. It takes its name from Tancred

(*d.* 1112), a hero of the first CRUSADE. Disraeli's hero, an early 19th-century heir to a dukedom, goes on a new crusade to the Holy Land.

Tandy, James Napper (1740–1803), Irish revolutionary. He helped Wolfe TONE to found the Society of UNITED IRISHMEN, was charged with sedition in 1793 and fled first to the US and then to France. He landed in Ireland during the unsuccessful rising of 1798. Sentenced to death in 1800, he was released at Napoleon's request.

Tanganyika. *See* TANZANIA.

Tanganyika, Lake, second-largest lake in Africa; second deepest freshwater lake in the world. It lies in E central Africa on the borders of Tanzania, Zaire, Zambia and Burundi, in the GREAT RIFT VALLEY. Area: 32,893sq km (12,700sq miles); depth: 1,437m (4,715ft).

Tangaroa, sea god of Polynesian mythology. In Tongan and Samoan legend, he is said to have created the earth by casting upon the waters a stone which turned into land. *See also* MS p.*208.*

T'ang dynasty (618–907), Chinese dynasty that ruled during imperial China's most vigorous and creative age. The capital Ch'ang-an (modern Sian) was a cosmopolitan centre of international trade and Chinese political and cultural institutions were at their height. It was the time when Neo-Confucianism became the official ideology and was the golden age of Chinese poetry as well as the period when China re-asserted its suzerainty over much of Central Asia and Korea. The dynasty was threatened in 755 by the revolt of the Turkish military leader An Lu-shan, and never recovered its former power. *See also* HC1 pp.123, 126, *126,* 128, *128.*

Tange, Kenzō (1913–), Japanese architect noted for his blending of traditional with modern architecture. He studied at Tokyo University (1935–38 and 1942–45), and became a professor in 1949. He was also a town planner and his work became more dramatic in the 1960s. His buildings include the Peace Centre, Hiroshima (1946–56), St Mary's Cathedral Tokyo (1962–64) and the Communications Centre (1964–67).

Tangent, in TRIGONOMETRY, ratio between the length of the sides opposite and adjacent to an acute angle within a right-angle triangle. (The third side is the hypotenuse.) The expression denoting the tangent of angle A is commonly abbreviated to "tan A". *See also* SU p.*46.*

Tanger. *See* TANGIER.

Tangerine, or mandarin orange, small, edible citrus fruit native to China. It has a thin skin and easily separated segments. Family Rutaceae; species *Citrus reticulata*. *See also* PE p.*192.*

Tangier (Tanger), port in N Morocco, on the Strait of Gibraltar. An ancient Roman port, it was later occupied by Moors and taken by the Portuguese in 1471. Tangier was ceded to England in 1662 as part of the dowry in marriage of Catherine of Braganza to Charles II. The English abandoned the city to the Moors in 1684. It became part of Morocco in 1956. It is the site of a casbah and the old Moorish walled city. Industries: rugs, pottery, shipping, fishing, tourism. Pop. (1973) 187,894.

Tango, ballroom dance that originated in Buenos Aires, Argentina, in the late 19th century. Developed from Argentinian milonga, it quickly became popular in Latin America. By 1915 it was a ballroom favourite in Europe and the USA and is still very popular in Finland. It is characterized by quick long strides and quick reversals of direction on the balls of the feet.

Tangut, former Tibetan-speaking people of a province of NW China; the kingdom, Hsi Hsia, was a part of the Chinese Sung dynasty until 1038, when Li Yüan-hao assumed the title of emperor. The Hsi Hsia (western Hsia) dynasty maintained a troubled existence for the next two centuries, but was conquered by the Mongol troops of Genghis Khan in 1227.

Tanguy, Yves (1900–55), self-taught French painter. Inspired by a painting by CHIRICO, he joined the SURREALISTS in 1925, and his paintings often show pre-

cisely rendered, fantastic objects in a dreamlike setting.

T'ang Yin (1470–1523), Chinese painter considered one of the so-called "Four Great Masters" of the Ming dynasty. He was a prolific artist, and is especially known for his hand scrolls and large hanging scrolls of ink landscapes.

Tanit, chief goddess of ancient Carthage. She was a fertility goddess but was sometimes treated as a deity of the sky. Her cult, dominant from the 5th century BC, probably included child sacrifice.

Tanizaki, Junichiro (1886–1965), Japanese novelist. He was influenced by classical Japanese literature and also by BAUDELAIRE. His works include *Some Prefer Nettles* (1929), *The Makioka Sisters* (1943–48) and *The Key* (1956).

Tank, tracked armoured vehicle mounting a single primary weapon, usually an artillery piece, and one or more machine guns. Modern tanks have an enclosed fully-revolving turret and are heavily armoured; main battle tanks weigh from 35 to 50 tonnes and usually have a crew of four. Developed in great secrecy by the British during WWI, tanks were first employed at the Battle of the Somme in 1916. *See also* MM p.170, *170.*

Tanker, ship that carries bulk liquids as cargo, usually PETROLEUM products in the form of crude oil. Modern tankers can carry up to 500,000 tonnes of liquid cargo. The growth in size of tankers has brought with it great danger from POLLUTION. The sinking of the TORREY CANYON in 1967 was the first such disaster. *See also* MM p.114, *114;* PE p.152–153, *152–153.*

Tannaim, scholars and teachers of the Jewish Oral Law who flourished during the 1st and 2nd centuries AD. Hillel, a *tanna* (scholar), began to spread the MIDRASH (a homiletical interpretation of the Jewish scriptures) of the TALMUD, and the last tanna Juda (died *c.* 220 AD) organized and preserved it.

Tannenberg, Battles of, two battles at Tannenberg, Poland. At the first (1410) Lithuanian and Polish forces defeated the TEUTONIC KNIGHTS. At the second (1914) Germany defeated Russia and advanced into East Prussia.

Tannhäuser (1845), three-act opera by Richard WAGNER with libretto by the composer based on medieval legend. It was first produced at Dresden.

Tannic acid, substance used medically to treat burns and, in styptics, to arrest bleeding from minor wounds. In both instances its function is to precipitate a protective PROTEIN layer over the wound. Chemically, tannic acid is a mixture of gallotannins. *See also* TANNIN.

Tannin, any of a group of complex organic compounds derived from tree bark, unripe fruit and oak galls. They have the property of converting the protein GELATIN into an insoluble compound that will not putrefy, and for this reason are used in tanning to cure hides and make leather.

Tanning, process of converting skins and hides into LEATHER. Traditionally, tanning liquids are based on TANNIC ACID (TANNIN) and, although the process takes weeks, it is still used for heavy leathers. Light leathers are now tanned in a few hours using chromium salts.

Tansy, any of several mostly perennial plants characterized by fern-like aromatic leaves and clusters of yellow, button-like flower heads. *Tanacetum vulgare,* native to Eurasia, is a common weed in North America but is also cultivated as a medicinal herb. Height: to 91cm (3ft). Family Compositae.

Tantā, city in N Egypt, 82km (51 miles), N of Cairo in the delta of the River Nile. It is the scene of three annual festivals for the 13th-century Muslim leader, Ahmad al-Badawi, whose grave is in Tanta. Cotton ginning is the main source of income. Pop. (1974) 278,300.

Tantalum, metallic element (symbol Ta) of the third transition series, first isolated in 1903. Its chief ore is columbite-tantalite. It is used in electrical components in special alloys. Properties: at.no. 73; at.wt. 180.948; s.g. 16.65; m.p. 2,996°C (5,425°F); b.p. 5,425°C

(9,797°F); most common isotope Ta[181] (99.988%).

Tantalus, in Greek mythology, son of ZEUS and a NYMPH. Because he revealed godly secrets to mortals, he was plunged up to his chin in a river of HADES under a fruit-laden tree. When he tried to drink, the waters receded; the fruit was beyond his grasp. From his name comes the verb "tantalize".

Tantrism, term used to describe the various esoteric practices of the Tantra, used by Buddhists, Jains and Hindus to fulfil worldly desires and attain spiritual experiences. These practices include YOGA, MANTRA, MANDALAS and SHAKTI. MM body is purified and the mind controlled by secret practices laid down in a group of post-Vedic Sanskrit writings, so that the believer may attain ultimate spiritual enlightenment. It is practised throughout Tibet, India, Bhutan and Nepal.

Tanzania, independent nation in E Africa, formed in 1961 from the former British territories of Tanganyika and Zanzibar. Most Tanzanians are subsistence farmers. The chief commercial crops are coffee, cotton and sisal; mineral deposits include diamonds. Most manufacturing is concentrated in Dar-es-Salaam, the capital. Tourism is becoming increasingly important to the economy. Areas: 945,087sq km (364,898sq miles). Pop. (1975 est.) 15,312,000. *See* MW p.165.

Tao Ch'ien (365–427), Chinese poet. He developed a simple style that differed from the ornateness of his contemporaries; his verse, which has a predominantly Taoist outlook, is frequently about nature and wine. *See also* TAOISM.

Taoism, Chinese philosophy and religion considered next to CONFUCIANISM in importance. Taoist philosophy is traced to a 6th-century BC classic of LAO-TZE, the *Tao tê Ching.* The recurrent theme is the *Tao,* or path. To follow the *Tao* is to follow the path of the cosmos leading to self-realization. Taoist ethics emphasize patience, simplicity and the harmony of nature. As a religion, Taoism dates from the time of Chang Tao-ling, who organized a group of followers in AD 142. It was a state religion in China for a time, but in modern times membership has dwindled. *See also* HC1 p.125.

Tapa, in the Pacific Islands, fabric made from softened bark, used as a substitute for woven cloth; it is used primarily for clothing, shelter, wall-hangings and mats.

Tap dance, dance in which the toes and heels rapidly tap the floor to emphasize the rhythm of the accompanying music. This type of dance used to be performed in clogs, but is now performed in the lightest of shoes which have metal plates (known as taps) on the toes and the heels.

Tape, magnetic, as used on audio and video tape recorders and computers, thin strip of plastic, coated on one side with a thin layer of iron or chromium oxide. During recording, the oxide layer is magnetized by the recording head in a pattern corresponding to the input signal. During playback the magnetized oxide particles induce an electric current identical to the one that produced them. *See also* MM pp.106–107, 232–235; SU p.120.

Tape recorder, device which records and plays back sound on magnetically treated tape. Sound is transformed into electric current and fed to a TRANSDUCER, which converts it into the magnetic variations that magnetize the particles on the treated tape. *See also* MM p.232.

Tapestry, hand-woven fabric of plain weave, made without shuttle or drawboy, the threads being threaded into the warp with a needle or bodkin. The term now includes many heavy materials that are not true tapestries, such as machine-made carpets and upholstery. Tapestry design is an ancient art and a few examples survive from 15th-century BC Egypt. From the 5th to the 14th century convents and monasteries were the main centres. The first great French woollen tapestry was made at ARRAS in the 14th century. Soon afterwards Brussels became the centre of the craft, and remained important until the rise of the GOBELINS factory at Paris in the 17th century. Tapestry design has now

reached a high degree of perfection and has generally followed parallel movements and styles in art. *See also* HC1 pp.165, 167, *168, 216.*

Tapeworm, or taenia, parasitic flatworm found in the alimentary tract of vertebrate animals, including man. It has a tiny head (Scolex) with hooks and suckers and a long body of segments (proglottides), each having its own reproductive system. Intermediate hosts include cattle, fish and pigs. Human infection occurs when undercooked, infected meat is eaten. Length: 1mm (0.04in) to 15m (50ft). Phylum Platyhelminthes. Class Cestoda. *See also* NW pp.72, 87.

Tapies, Antonio (1923–), Spanish painter whose early works show the influence of Paul KLEE and Torres Garcia. In his later ABSTRACT works he specialized in texture painting, often using sand and plaster. *See also* HC2 pp.278, *278.*

Tapioca. *See* CASSAVA.

Tapir, any of several species of nocturnal, plant-eating, hoofed mammals native to forests of tropical South America and Malaysia. It has a large head, a long, flexible snout, a heavy body, short legs and a tiny tail. The Malayan tapir (*Tapirus indicus*) has a striking half-white, half-black coat, whereas New World species are dark brown. Length: to 2.5m (7.5ft). Family Tapiridae; genus *Tapirus. See also* NW pp.*155,* 163, *163, 214,* 216–217, *216.*

Tapu, MAORI form of TABOO.

Tar, complex mixture of HYDROCARBON compounds, derived from wood, coal and other organic materials by heating them with little or no air, then condensing the tar from the distilled vapours. Coal tar is a major source of hydrocarbons for the synthesis of pharmaceuticals, pesticides and plastics. Cruder tar compounds are used for road surfacing and protecting timber against rot and pests.

Taràbulus. *See* TRIPOLI, LEBANON; TRIPOLI, LIBYA.

Tara Hill, ancient seat of the Ui Neill kings of Ireland until the 6th century, in County Meath about 35km (22 miles) NW of Dublin. Only earthworks remain of Thomas MOORE'S "Tara's Halls", venue of the triennial convention, the Feis of Tara.

Taranaki, New Zealand district in W North Island that includes some of the country's richest dairy land. Chief town: New Plymouth. Area: 9,710sq km (3,750sq miles).

Taranto, city in SE Italy, 80km (50 miles) SSE of Bari, on the Gulf of Taranto. Founded in 8th century BC, it was a leading city of MAGNA GRAECIA, submitting to Rome in 272 BC. It was destroyed by Arabs in AD 927, but rebuilt by the Byzantines in 967. Industries: steel, chemicals, fishing, food processing and shipbuilding. Pop. (1971) 222,826.

Tarantula, large hairy wolf spider of S Europe, once thought to inflict a deadly bite which would cause madness. The bite is not, in fact, dangerous to man. It spins no web, but chases and pounces on its prey. Length of body: to 2.5cm (1in). Family Lycosidae; species *Lycosa tarentula.* The name is also applied to the sluggish, dark, hairy spiders of the SW USA, Mexico and South America. Many species burrow and feed on insects, and occasionally small birds, amphibians or mice. Length of body: to 5cm (2in). Family Theraphosidae; genera *Aphonopelma* and *Eurypelma. See also* NW pp.72, 73.

Tarawa, atoll in the W Pacific Ocean; the capital of Gilbert and Ellice Islands. Located in the N central part of the group, it is the trade centre for the islands. Exports include phosphates, copra and mother-of-pearl. Pop.12,500.

Targum, term meaning interpretation, designating the Aramaic paraphrases of the Old Testament usually supplemented by a commentary, made necessary because of the decline of Hebrew as the Jews' normal language. First produced in Babylon and Palestine several centuries before the Christian era, the targum treats all Old Testament books except Daniel, Nehemiah and Ezra, and is a compilation of explanatory matter added to Scripture in synagogue worship.

Tariff, tax placed on imports either as a

percentage of the value of the item (ad valorem tariff) or per unit (specific duty). Tariffs may be used to discourage the import of certain types of goods or to adjust for price differentials in order to allow the home country's products to be competitive. Tariffs may be levied on particular goods, regardless of the country they came from, or may be specifically aimed at discriminating among goods on the basis of country of origin. General tariff agreements are made between countries every few years in the General Agreement of Tariffs and Trade (GATT).

Tarim, river in W China. It is formed at the confluence of the K'a-shih-ka-erh and Yarkand rivers, flowing E then SE into the Tarim basin in the region of Lop Nor. Length: 2,027km (1,260 miles).

Tariq ibn-Ziyad (died *c.*720), Muslim soldier who led the Moorish invasion of Spain. He captured Gibraltar in 711 with an army of 7,000 and then proceeded to conquer Cordoba, Toledo and other parts of Spain. The name "Gibraltar" derives from the Arabic Jebel-al-Tariq, which means the Mount of Tariq.

Tarkington, Newton Booth (1869–1946), US novelist, best known for his descriptions of middle-class life in small midwestern towns. His works include *The Gentleman from Indiana* (1899); *The Magnificent Ambersons* (1918) and *Alice Adams* (1921), for which he received Pulitzer Prizes.

Tarleton, Sir Banastre (1754–1833), British cavalry soldier who distinguished himself in campaigns in the War of AMERICAN INDEPENDENCE (1776–82). He served under Lord CORNWALLIS, becoming Lieutenant-Colonel of the BRITISH LEGION in 1778. After the surrender at YORKTOWN in 1781, he returned to Britain and became an MP in 1790.

Tarleton, Richard (*d.*1588), English actor and jester, one of the Queen's company of actors, who specialized in yokel roles and was said to be a model for Bottom in Shakespeare's *A Midsummer Night's Dream.* He was a favourite with Queen Elizabeth I and published *Tarleton's Toyes* (1576).

Tarmac, mixture of gravel and tar mainly used to cover roads, drives and paths. It is hard wearing, impervious to water and provides an ideal surface for the rubber tyres of motor vehicles. *See also* ASPHALT; MM pp.182–183.

Taro, large tropical plant native to the Pacific Islands and SE Asia and cultivated in other parts of the world for its edible tuberous root. In Hawaii it is eaten as POI. Family Araceae; species *Colocasia esculenta. See also* LILY; PE p.187.

Tarot, forerunner of modern PLAYING CARDS, believed to have originated from the Middle East in antiquity. In Europe they were first used in Italy in the 14th century. Tarot packs contain 78 cards: the 56-card Minor Arcana, which correspond to the modern pack, and the 22 picture cards of the Major Arcana. They are used for fortune-telling and as meditation symbols by occultists.

Tarpon, tropical marine game fish. Blue and bright silver, it has a long, deeply forked tail. Its eggs develop into eel-like, transparent larvae, and then metamorphose into the juvenile form. Length: to 180cm (6ft); weight: to 150kg (300lb). Main species are the small Pacific *Megalops cyprinoides* and the large Atlantic *M. atlanticus.* Family Elopidae.

Tarquinius Superbus ("Tarquin the Proud") (*r.*534–510 BC), Etruscan, traditionally the seventh and last king of Rome. Largely known only in legend, he is said to have been a despot. *See also* HC1 p.90.

Tarragon, perennial plant with liquorice-flavoured leaves used fresh or dried in salads, pickles, etc. Family Compositae; species *Artemisia dracunculus. See also* PE pp.190, 213, *213.*

Tarsals, one of the three sets of bones in the foot, in human beings the bones that make up the ankle. They are strong compact bones arranged so that the foot can be rotated (to a limited extent) in any direction. In front of the tarsals are the METATARSALS, which form the framework of the foot's two flexible arches.

Tape recorder; the first Magnetophone was demonstrated in Berlin in 1935.

Tapestry; *The Watering Place* (above) woven in Beauvais between 1772 and 1779.

Tapirs are hunted for sport and food and some species are in danger of extinction.

Tarantula; its name is derived from Taranto, a town in southern Italy.

Tarsiers are well suited to an arboreal life with their long hind limbs and tails.

Le Tartuffe, or *The Impostor*, was criticized by the Church when first performed.

Tasmanian devils have strong jaws and teeth and can kill small kangaroos.

Torquato Tasso is considered to be the finest poet of the late Italian Renaissance.

Tar sands, porous, often black, rocks, such as limestones, sands and sandstones found on the surface. They contain deposits of ASPHALT in the spaces between the grains and commonly have a distinct odour of tar. Extensive deposits are found in North America, primarily in Alberta and Texas. Estimates of the amount of asphalt contained in these deposits are extremely high, but its high viscosity makes extraction difficult.

Tarsier, any of several species of nocturnal primates of Indonesia. They are small, squat animals with large eyes, long tails and monkey-like hands and feet; they have silky grey-brown coat fur. Tarsiers live in trees and feed chiefly on insects. Family Tarsiidae; genus *Tarsius. See also* NW pp. 168, *168, 214,* 215.

Tarski, Alfred (1902–), US mathematician and philosopher, *b.* Poland, best known for his work on symbolic logic. After teaching at Warsaw University (1925–39) he moved to the USA to work at Harvard (1939–41) and Princeton (1941–42). From 1942 to 1968 he taught at the University of California. His publications include *Introduction to Logic and the Methodology of Deductive Science* (1936), *Logic, Semantics, Metamathematics* (1956) and *Cylindric Algebras* (Part 1, 1971).

Tarsus, city in s Turkey on the River Tarsus. Conquered by the Romans in 67 BC, it became a leading industrial and cultural centre. It was destroyed by the Arabs in AD 660 and rebuilt by them in 780. The Ottoman Turks took Tarsus in 1515. It is an agricultural market for citrus fruit and cotton grown in the area. Copper, chromium, zinc and coal are mined nearby. Pop. (1970) 78,033.

Tartaglia, Niccolò Fontana (1499–1557), Italian mathematician, the first to obtain a general solution to the cubic equation which lacks the simple term of the form $x^3 + ax^2 + bx + c = 0$. From 1534 he taught in Venice. He studied ballistics and published *Nova Scientia* (1537) and *Trattato di numeri et misure* (1556–60).

Tartan, specific type of plaid cloth, usually woollen. It was worn throughout Scotland and the North of England before becoming adopted by the clans of the Scottish Highlands, each clan having its own patterns. A tartan plaid has cross-bars of different colours and widths on a background usually red or green.

Tartar, unsightly encrustation on and between the teeth consisting of caked salivary secretion, food residue and various calcium and phosphate salts. Tartar that collects between the teeth and along the gums may cause inflammation of the gums, or GINGIVITIS, which can lead to PYORRHOEA and loss of teeth. *See also* MS p.130.

Tartar ASSR, autonomous republic in the Russian SFSR, USSR; the capital is Kazan. The area was conquered by the Mongols of the GOLDEN HORDE in the 13th century. The land is dominated by rolling plains, drained by the Volga, Kama, Sviyaga and Vyatka rivers which are also used for irrigation and the generation of hydroelectricity. Cereals and flax are grown; brown coal, limestone and gypsum are quarried and the republic's manufactures include chemicals and machinery. It is also an important producer of oil and natural gas. Area: 94,000sq km (36,255sq miles). Pop. (1970) 3,131,000.

Tartaric acid, colourless, optically active organic compound, with the formula $(CHOH)_2(COOH)_2$. It is found in fruits and made from maleic anhydride and hydrogen peroxide or by enzyme action on a succinic acid. The acid is widely used in baking powder, cream of tartar, soft drinks and in the textile industry. Properties: s.g. 1.76; m.p. 171°C (340°F).

Tartars. *See* TATARS.

Tartarus, in Greek mythology, the deepest region of the earth, as far below as heaven was above, into whose depths the fallen TITANS were consigned by the victorious Olympian gods after their battle for OLYMPUS.

Tartrate, salt or ester of tartaric acid. Tartrates, such as potassium bitartrate (cream of tartar) and sodium bitartrate, are used in baking powder, in the food industry and in tanning.

Tartu, Treaty of (1920), treaty between Estonia and the USSR. The USSR reaffirmed its recognition of Estonia's independence and granted the territory of Petsamo to Finland.

Tartuffe, Le (1669), three-act comedy by MOLIERE. A satire on hypocrisy and piety, the first version was banned after the performance for LOUIS XIV in 1664, as was the second in 1667. A third and final version was authorized in 1669.

Tarzan, fictional hero created by writer Edgar Rice BURROUGHS in 1912. Tarzan grew up with African apes after surviving an aircraft crash when a child and was able to communicate with and befriend wild beasts. He has been the subject of many films since 1918. The actor Johnny WEISSMÜLLER was probably the most famous Tarzan.

Tasaday, small group of isolated aboriginal people of the rain forests of s Mindanao in the Philippines, first discovered in 1971. They are food-gathering cave dwellers with a STONE-AGE culture.

Tashkent (Taškent), capital of Uzbekistan (Uzbekskaja SSR), USSR, in the Tashkent oasis, watered by the River Chirchik. It was ruled by the Arabs from the 8th until the 11th century and came under the TIMURID Empire in the 13th century. The city was captured by the MONGOLS in 1361. Tashkent was captured by the Russians in 1865. The modern city is a terminus for road, rail and air routes and is the economic centre of the region. Tashkent is a market for cotton and cereals and has manufacturing industries that include textiles, chemicals, food-processing, machinery, paper, porcelain and furniture. Pop. (1975) 1,595,000.

Task force, temporary grouping of armed forces units under one commander for the purpose of carrying out a specific mission or operation.

Tasman, Abel Janszoon (1603–*c.*1659), Dutch explorer. Sent by the Dutch EAST INDIA COMPANY to explore the coasts of Australia and New Guinea, he discovered New Zealand, Tasmania, Tonga and the Fiji Islands in 1642–43, but his voyages were regarded as economic failures by the company. *See also* HC2 p.124, *124.*

Tasmania, island state of Australia, separated from Victoria by the Bass Strait and bounded by the Pacific (E) and Indian (w) oceans. Tasmania is mountainous and forested, with small farms, although sheep rearing and dairy farming are important. The development of hydroelectric power has stimulated the growth of manufacturing; principal industries are metallurgy and textiles. Mineral deposits include copper, tin, zinc, lead and iron ore. The chief cities are Hobart, the state capital, and Launceston.

The island was discovered by Abel TASMAN in 1642 and named Van Diemen's Land. Capt. James COOK visited it in 1777 and claimed it for the British, who established a penal colony there. Tasmania was part of New South Wales until 1825, when it became a separate colony; it was federated as a state of the Commonwealth of Australia in 1901. Area: 68,332sq km (26,383sq miles). Pop. 392,000. *See also* MW p.28.

Tasmanian devil, carnivorous marsupial with a bear-like appearance; it is found only in the forest and scrub of Tasmania. It has a smooth black-and-brown coat with small white areas on the face, forelegs and rear. It feeds mainly on a wide variety of animal food, including carrion. Length: to 80cm (31in). Species *Sarcophilus harrisii. See also* NW pp. *158,* 210–211.

Tasmanian wolf, also called thylacine, largest carnivorous marsupial. It became extinct on the mainland of Australia because of relentless hunting, but a few specimens are believed to have survived in forested areas of Tasmania. It has a wolf-like appearance, but its coat is marked with transverse dark stripes on the back, hindquarters and tail. Species *Thylacinus cynocephalus. See also* NW pp.*158,* 210, *243.*

Tasman Sea, part of the Pacific Ocean between Australia and New Zealand. Sydney is the principal port on the sea, which was named after the Dutch explorer Abel TASMAN. Width: approx. 1,930km (1,200 miles).

Tassili-n-Ajjer, plateau region in s Algeria, the cliff faces of which are covered in prehistoric paintings of scenes depicting people and animals. Discovered in the 19th century, the age and meaning of the paintings remain a mystery, as does the identity of the artists.

Tasso, Torquato (1544–95), Italian Renaissance poet who, after an unhappy childhood, wrote the epic *Rinaldo* (1562) and joined the ESTE court at Ferrara. There he wrote *Aminta* (1573) and his masterpiece *Gerusalemme liberata* (1574). After 1577 he became psychologically disturbed and was confined intermittently in a hospital from 1579 to 1586 where he continued writing. *See also* HC1 p.282.

Taste, gustatory faculty and one of the five SENSES. In human beings the taste buds of the tongue differentiate four qualities: sweetness, saltiness, bitterness and sourness. The finer distinctions in taste are made by the nose. *See also* MS p.54.

TAT. *See* THEMATIC APPERCEPTION TEST.

Tatars, or Tartars, Turkic-speaking people of central Asia. In early medieval times the name Tatar was given to many different Asiatic invaders of Europe. The Tatar tribe proper originated in E Siberia and, mixed with the MONGOLS, invaded Europe in the early 13th century under GENGHIS KHAN. They were converted to Islam in the 14th century and, after the collapse of the Mongol empire, the Tatars living in the Crimea (Little Tartary) came under the Ottoman Empire and those in Tartary (Siberia) came under Muscovite control. By the 16th century they had become settled agriculturalists, and many Tatars achieved eminence within Russia. Crimea became part of Russia in 1783. In the mid-20th century there were more than 5 million Tatars.

Tate, Phyllis (1911–), British composer. She is famous chiefly for chamber music, including a sonata for clarinet and cello (1949) and a string quartet (1952). She also wrote a concerto for saxophone and strings and the operas *The Lodger* (1960) and *Dark Pilgrimage* (1963) for television.

Tate Gallery, art gallery at Millbank, London. It was built in 1897 through the generosity of Sir Henry TATE, and has since been extended. It houses collections of British painting and sculpture from the 16th century and of foreign art from 1800. The gallery is particularly rich in works by TURNER, BLAKE and the modern French painters. Its many temporary retrospective exhibitions are widely admired.

Tati, Jacques (1908–), French film actor and former music-hall artist, *b.* Jacques Tatischeff. After acting in several short comedy films in the 1930s, he directed his own *L'École des facteurs* (1947). His distinctive style of visual humour was developed in *Monsieur Hulot's Holiday* (1953), *Mon Oncle* (1958), *Playtime* (1968) and *Traffic* (1971).

Tatler, The, English periodical launched in 1709 by Richard STEELE, who later collaborated with Joseph ADDISON. It was originally intended to entertain and to provide a news service, but later adopted a moral tone, particularly in its examination of the ideal gentleman.

Tatlin, Vladimir Evgrafovich (1885–1953), Soviet sculptor. Influenced by CUBISM and FUTURISM, in 1913 he instigated the abstract style CONSTRUCTIVISM, in which he used industrial materials such as metal and glass. *See also* HC2 p.*204.*

Tattersall's, British racehorse auctioneers, founded in London in 1766 by Richard Tattersall. These large bloodstock auctions are now held at fixed times annually at Newmarket. It is not to be confused with Tattersall's Committee, which settles questions relating to transactions on horse-racing.

Tattooing, colouring or marking the skin for cosmetic or religious purposes. Pigment is inserted under the skin's surface by means of needles, or scars are made.

Tattooing is an ancient and widespread practice; it was widely employed in Polynesia, especially among the MAORI, and also in Japan. *See also* MS p.*193*.

Tatum, Arthur ("Art") (1910–56), US jazz pianist. Almost blind since birth, he established a standard for jazz piano technique that has never been surpassed. He made his first recording in 1932 and was soon renowned.

Tatum, Edward Lawrie (1900–75), US geneticist and biochemist. He shared the 1958 Nobel Prize in physiology and medicine with George BEADLE and Joshua LEDERBERG for his part in the discovery that genes act by regulating specific chemical processes, a basic principle explaining how genes determine the characteristics of an organism.

Tauber, Richard (1892–1948), British tenor, *b.* Austria. He studied in Germany, made his debut in 1913 and sang in opera and concerts throughout Europe, and first performed at London's Covent Garden in 1931. He was best known for his performances in operettas, especially those of Franz LEHÁR.

Täuber-Arp, Sophie (1889–1943), Swiss painter and sculptor. She was active in the DADA movement from 1916 to 1920. In 1921 she married Jean ARP, with whom she occasionally collaborated on various decoration and design projects.

Tau Ceti, main-sequence yellow star in the constellation of Cetus. It is only slightly smaller and cooler than the Sun and it is relatively close to the Solar System. Characteristics: apparent mag. 3.5; absolute mag. 5.7; spectral type G8; distance 11.9 light-years. *See also* SU pp.226, 285.

Taung skull, remains of a prehistoric hominid identified in 1924 by Raymond Dart in South Africa. Subsequent discoveries of further specimens, mainly by Robert Broom and J. T. Robinson, showed it was an intermediate stage between man and ape, *Australopithecus africanus*, and important in the fossil record of the evolution of man. *See also* MS pp.22–23.

Taunton, county town of Somerset, SW England, on the River Tone. It is a market town and railway junction. Industries: textiles, clothing, precision instruments. Pop. (1971) 37,373.

Taupo, Lake, largest lake in New Zealand, in central North Island, drained by the Waikato River. Area: 606sq km (234sq miles); depth: 159m (522ft).

Taurid meteor showers. *See* METEORITE.

Taurus, or the Bull, in astronomy, northern constellation on the ECLIPTIC between Aries and Gemini; It contains the Pleiades and Hyades stellar clusters and the Crab Nebula. The brightest star is the 1st-magnitude Alpha Tauri (ALDEBARAN). *See also* MS pp.200, *200–201*; SU pp.254, *254*, 258, *258*.

Taurus, the Bull, in astrology, the second sign of the Zodiac. The Sun is in Taurus from about 20 April to 20 May. *See also* MS pp.200–201.

Taurus Mountains (Toros Daǧlari), chain of mountains in S central Turkey, extending 564km (350 miles) roughly parallel to the Mediterranean Coast of S Asia Minor; its NE extension is called the Anti-Taurus. The mountains have important deposits of minerals. The highest peak is Ala Daǧ, 3,724m (12,218ft).

Tautomerism, property exhibited by a compound that exists as a mixture of two ISOMERS in equilibrium. If either isomer is removed from the mixture, the equilibrium is momentarily disturbed but is restored by the spontaneous conversion (due to the migration of an ION or FREE RADICAL) of some of the remaining isomer into the removed type.

Taverner, John (*c.* 1495–1545), English composer. He wrote for only a short period, chiefly during his four years at Cardinal College (now Christ Church), Oxford (1526–30). He composed mostly church music – such as masses, Latin services and motets – but also some secular works for a song book (1530) of Wynkyn de Worde.

Távora, João Franklin da Silveira (1843–88), Brazilian novelist. He contributed to the growth of regionalism in the Brazilian novel. His best-known novel is *Caballeira* (1876).

Tawney, Richard Henry (1880–1962), British social historian, author of *The Acquisitive Society* (1920) and *Religion and the Rise of Capitalism* (1926). He was president of the Workers' Educational Association (1928–44) and professor of economic history at the London School of Economics (1931–49).

Taxes and taxation, compulsory money payments of various kinds made by members of a civil society to supply the expenditure of public authorities. Taxes are of two chief kinds, direct and indirect. Direct taxes are levied on income (wages, salaries, rents, profits, interest, dividends) and capital. Indirect taxes are levied on commodities (goods purchased, imported goods, entertainment) and services (road tax). Local RATES are a form of indirect taxation levied on the assessed value of property. The fundamental purpose of taxation is to defray government expenditure on defence, the social services, administration and the repayment of public debts. But taxation may be used to effect certain social and economic purposes, such as reducing the inequality of wealth in a community, reducing or increasing the purchasing power of consumers according to the requirements of industrial policy, checking the flow of imports or exports in order to alter the balance of trade and influencing the rate and direction of capital investment.

In Britain the origins of taxation lie in the special grants of money occasionally made to the monarch in the Middle Ages. Feudal monarchs were expected to "live of their own", to pay the expenses of their establishments from the income derived from their property and the regular dues owed to them as feudal lords. But for special purposes, especially the cost of defence and foreign wars, limited grants of money, called "grants in aid" or "subsidies", were frequently voted to the king by his council, later by Parliament. Such grants were known as "extraordinary taxation". They might be either direct (the poll tax) or indirect (taxes on movable property). The two most common forms of grants in aid were SCUTAGE and TAILAGE, especially in the 13th and 14th centuries, when extraordinary taxation became a regular feature of royal policy. It was chiefly royal requests for scutage and tallage which established the principle, the very heart of the English constitution, that no subject shall be obliged to pay money to the government except by popular consent. And from that principle flowed its great corollary, that the redress of grievance shall precede the voting of supplies.

In modern British history direct taxation far outweighed indirect taxation until the latter part of the 19th century. Most of the government's revenue in the 17th and 18th centuries came from the large wall of tariff duties which was broken down slowly by the introduction of FREE TRADE budgets in the 19th century. Some came also from the land tax, a direct tax levied on the owners of real property. INCOME TAX was first imposed in 1799 as a temporary expedient for supplying the cost of the Napleonic Wars. It was repealed in 1816 and re-introduced in 1842, since when it has been permanent and is now the central source of government revenue.

Taxi, vehicle available for hire, driven by the owner or his employee. It is named after the taximeter, an automatic instrument that records the fare due. Taxis are sometimes called "cabs" after the cabriolet, a horse-drawn vehicle.

Taxidermy, process of preparing, preserving and mounting skins of animals and birds so that they still appear life-like. It is used mostly to prepare specimens for display in museums.

Taxonomy. *See* CLASSIFICATION.

Tay, river in central Scotland. It rises in the Grampian Mts and flows SE, to enter the North Sea through the Firth of Tay near Dundee. It is the longest river in Scotland and has the largest drainage basin, 6,216sq km (2,400sq miles) in area. Length: 193km (120 miles).

Taylor, Alan John Percivale (1906–), British historian whose interests include diplomatic history and the interpretation of modern wars. Among his books are *Bismarck, The Man and The Statesman* (1955) and *The Origins of the Second World War* (1961). He has reached a wide audience through many appearances on television.

Taylor, Cecil (1933–), US jazz pianist. He was one of the pioneers of free improvisation (ie improvisation related to no prescribed harmonic, melodic or rhythmic guidelines), and his style has been influential, although he has frequently encountered opposition from conservative critics.

Taylor, Elizabeth (1932–), US film actress, *b.* Britain. She began her career as a child star under contract to MGM, attracting attention as the heroine of *National Velvet* (1944). She has remained in demand by producers, making films which include *A Place in the Sun* (1951), *Cat on a Hot Tin Roof* (1958) and *Who's Afraid of Virginia Woolf?* (1966), for which she received an Academy Award.

Taylor, Frederick Winslow (1856–1915), US industrial engineer who was known as the father of scientific management. He developed management methods for many industries, especially steel mills. Among his writings is *The Principles of Scientific Management* (1911).

Taylor, Sir Geoffrey Ingram (1886–1975), British physicist and meteorologist. He specialized in FLUID MECHANICS, and made significant contributions to the understanding of turbulence and diffusion in fluids.

Taylor, Sir Robert (1714–88), British architect. Until about 1753 his main interest was in sculpture, for which he received a number of commissions. As an architect his most distinguished achievement was the surveyorship of the Bank of England. Although much of his work there no longer exists, the building was PALLADIAN in style.

Taylor, Zachary (1784–1850), 12th President of the USA (1849–50), a hero of the Mexican War who died in office after serving only one year of his term. His father was a wealthy planter, but Taylor had little formal education. He received an army commission in 1808 and fought in the War of 1812. In 1845, as commander of US troops on the Texas border, he made a foray into disputed territory and set off the Mexican War. As President, he advocated the early admission of California and New Mexico as states. He was succeeded by Vice-President Millard Fillmore.

Tay-Sach's disease, rare hereditary disorder of fat metabolism. It is characterized by an accumulation of gangliosides in the central nervous system, resulting in severe mental deficiency, blindness and death at about three years of age.

Tayside, region in N central and E Scotland; created in 1975 from the former counties of ANGUS, KINROSS and most of PERTH. Perth and Dundee are the principal cities. Area: 7,665sq km (2,959sq miles). Pop. (1975 est.) 402,000.

Tbilisi (Tiflis), capital of Georgia (Gruzinskaja SSR), USSR, on the upper Kura River. It was founded in AD 455 or 458 as the capital of an ancient Georgian kingdom and was ruled successively by the Persians, Byzantines, Arabs, Mongols and Turks, coming under Russian rule in 1801. Tbilisi's importance lies in its location on the trade route between the Black Sea and Caspian Sea; it is the administrative and economic focus of modern Transcaucasia. Fruit is grown outside the city and industries include beer, wine, spirits, furniture, printing, machine building, and machine tools. Pop. (1975) 1,006,000.

Tchaikovsky, Peter Ilyich (1840–93), Russian late ROMANTIC composer who studied at the St Petersburg Conservatory. He wrote nine operas, four concertos, six symphonies, three ballets, overtures, chamber music and numerous other works. His music includes the orchestral pieces *Romeo and Juliet* (1880) and *Slavonic March* (1876). He also wrote the music for the ballets *Swan Lake* (1876), *The Sleeping Beauty* (1889) and *The Nutcracker* (1892). He died nine days after

Edward Tatum won a Nobel Prize for his innovative work in molecular genetics.

Taurus Mountains; this chain in S Turkey has scattered forests up to 8,000ft.

Taxes; in this cartoon of 1795, William Pitt figures as the remorseless grinder.

Tchaikovsky is still regarded as the master composer for classical ballet.

Teals, or dabbling ducks, nest in grass-lined hollows by the waterside.

Tear gas is seen here being used to disperse a crowd of demonstrators.

Teasels have hooked bracts, used in the textile industry to raise the nap of cloth.

Lord Tedder in the robes of Chancellor of the University of Cambridge.

the premiere of his sixth and most popular symphony, the *Pathétique* (1893). *See also* HC2 pp.*106,* 107, 121, *270,* 274.

Tchelitchew, Pavel (1898–1957), Russian painter who made use of PERSPECTIVE distortion and multiple images. In the 1920s he designed stage sets. After he settled in the USA in 1934, his works became more complex, the best known being *Leaf Children* (1939) and *Hide and Seek* (1940–42).

Tea, family of trees and shrubs with leathery undivided leaves and five-petalled blossoms. Among 500 species is *Camellia sinensis* (*Thea sinensis*), the commercial source of tea. Cultivated in moist tropical regions, tea plants can reach 9.1m (30ft) height, but are kept low by frequent picking of the young shoots for tea leaves; the flowers are white. The leaves are dried immediately to produce green tea and are fermented before drying for black tea. Family Theaceae. *See also* PE pp.208, 209.

Tea Act (1773), legislation by the British Parliament removing export duties on tea. To give relief to the EAST INDIA COMPANY, which was failing financially, the Act withdrew duty on tea exported to the colonies and enabled the company to sell tea directly to the colonies without first going to Britain. Colonial merchants were undersold and the Act led directly to the BOSTON TEA PARTY. *See also* HC2 p.64

Teaching machine, device incorporating an electronic screen on which is relayed programmed instruction. The pupil's progress may be monitored through the use, within the relayed material, of items requiring a response from the pupil via the machine's keyboard. Depending on the complexity of the machine, a negative answer may be supplied, further explanation may be given, or an evaluation of the pupil's capacity may be made from which the machine may choose any of several possible paths by which to continue presentation. *See also* MM p.*109.*

Teak, tree, native to s India, Burma and Indonesia, valued for its hard yellowish-brown wood. Teak wood is water-resistant and takes a high polish and is therefore widely used for furniture and in shipbuilding. Height: 45m (150ft). Family Verbenaceae; species *Tectona grandis. See also* PE pp.218, *219.*

Teal, small, widely distributed river duck that often has bright plumage. The blue-winged teal (*Anas discors*) of North America is a fast, strong flyer. Teal dabble, or feed from the surface of the water. Family Anatidae, genus *Anas. See also* NW p.*148.*

Team handball, indoor or outdoor game. The outdoor version is played by two teams of 11 players on a field and between two goals similar in size to those used in Association FOOTBALL; the indoor version is played on smaller sized courts with smaller goals, by two teams of 7 or 5 players. The object of the game is to score goals by throwing an inflated leather ball, 58–60cm (23–24in) in diameter, over the goal line. Such throws must take place outside a prescribed area near the goal. Players may keep possession of the ball only for three seconds at a time, and little physical contact is allowed.

Teamsters, full name the International Brotherhood of Teamsters, Chauffeurs, Warehousemen and Helpers of America, trade union founded in the USA in 1903. The largest US union, representing lorry drivers and related trades, the Teamsters have been the subject of several government investigations for suspected illegal activities.

Teapot Dome Scandal, in US history, fraudulent leases of the naval oil reserves at Teapot Dome and Elk Hills in South Dakota, granted by Secretary of the Interior Albert B. Fall to oilmen Harry Sinclair and Edward Doheny in 1921–22. Fall collected more than $300,000 from the two leases.

Tear gas, chemical compound known as a lachrymator, a gas or aerosol that causes an excessive flow of tears and blinds and incapacitates temporarily without causing permanent injury. Such gases, as used by military and civil authorities, are generally made of chlorine and bromine derivatives of acetone and acetophenone.

Tea rose, hybrid rose originating from plants imported from China. The fragrant, long-blooming, double flowers range from white to yellow to pink. Family Rosaceae; species *Rosa odorata.*

Tears, fluid that moistens the surface of the CORNEA of the eye secreted by the lachrymal glands. It is antibacterial in nature, and also improves the optical properties of the eye by forming a thin, smooth film on the cornea and compensating for slight surface imperfections.

Teasel, any of several species of plants that grow in Europe, the Middle East and the USA. A prickly, woodland plant, it has hollow leaf bases that trap water, and densely clustered, bristly, egg-shaped flower heads. Species include *Dipsacus fullonum,* a 1.8m (6ft) tall perennial with purple flowers. Family Dipsacaceae.

Technetium, radioactive metallic element (symbol Tc) of the second transition series. It was first made in 1937 by bombarding MOLYBDENUM with DEUTERONS, and was the first element to be synthesized. No technetium has been found in the Earth's CRUST, although it is present in some stars. Properties: at.no. 43; at.wt. 98.9062; s.g. 11.5; m.p. 2,172°C (3,942°F) b.p. 4,877°C (8,811°F); most stable isotope Tc^{97} (half-life 2.6×10^6 yrs). *See also* TRANSITION ELEMENTS.

Technicolor, trade name of the colour film process used in most modern motion pictures. The process has existed in four versions since its development in 1915. A primitive Technicolor was first seen in 1918, and in 1933 Walt DISNEY used three-colour Technicolor for the animated film *Flowers and Trees.*

Technocracy, theory that engineers and scientists should have effective power in economic and social life. The term was coined in the USA in 1919 for the phrase "rule by technicians" and enjoyed brief popularity in the 1930s. The theory gained currency in France in the 1960s when it was identified with ideas of the French socialist thinker Claude Henri de SAINT-SIMON (1760–1825).

Technology, systematic study of the methods and techniques employed in industry, research, agriculture and commerce. More often the term is used to describe any application of scientific discoveries in the production of mechanisms and in the solution of problems which confront man.

Technology, Ministry of (Mintech), former British government department. Set up by the Labour government of 1964, the ministry took over the industrial research functions of the old department of Scientific and Industrial Research. In 1967 the Ministry took over part of the Ministry of Aviation and in 1969 acquired some of the duties of the Board of Trade and Department of Economic Affairs. It also acquired responsibility for atomic energy research. Mintech was abolished by the Conservatives in 1970.

Tecton, architectural team founded in the early 1930s by Berthold Lubetkin, whose members included Denys LASDUN and Michael Dugdale. The team became recognized particularly for its work for London Zoo (1933–38); the style of freely moulded concrete anticipated its use in the 1950s.

Tectonics, study of the Earth's CRUST in terms of the large, moving "plates" which support the continents and float on the liquid rock beneath. Tectonics also studies the effect of this movement on the rims of continents, ocean basins and mountain ranges (which result from crustal movement compression). *See also* PE pp.24, 130, *130.*

Tedder, Arthur William, 1st Baron (1890–1967), British air marshal. He served with the Royal Flying Corps in WWI and from 1941 to 1943 commanded the RAF in the Middle East, where his method of pattern bombing became known as "Tedder's carpet".

Teepee. *See* TEPEE.

Teeside, former county borough in Durham and the North Riding of Yorkshire, NE England; since 1974 it has been part of CLEVELAND. The area included Middlesbrough and Stockton-on-Tees. It is an industrial area producing steel, chemicals and ships. Pop. (1971) 395,477.

Teeth, hard, bone-like protuberances rooted to the jaws of vertebrates, used for chewing food, defence or other purposes. Humans have 16 pairs of teeth when adult, each jaw having four INCISORS, two CANINES, four PREMOLARS and SIX MOLARS. Childrens' teeth lack the premolars and four molars. The incisors and canines are used for cutting and the molars and premolars for crushing and grinding food. The teeth of all vertebrates have a similar structure consisting of three layers. Mammalian teeth have an outer layer of hard ENAMEL. The middle layer consists of DENTINE, a bone-like substance that is capable of regeneration. The core of a tooth contains pulp which is softer and provides nutrition for the dentine. *See also* MS pp.*57,* 130–131, *130–131.*

Teflon. Trade name for the plastic polytetrafluoroethylene. *See* PTFE.

Tegucigalpa, capital of HONDURAS, located in the s central part of the country. Products include sugar, textiles and cigarettes. Pop. (1970 est.) 232,276.

Tehran (Teheran), capital of Iran, 105km (65 miles) s of the Caspian Sea. It was the residence of the QAJAR rulers in the 17th century and became the capital in 1788. Renovation and modernisation of the city began in 1797 under Fath Ali Shah. Tehran is now one of the most important cities in the Middle East. It contains the Gulistan or Rose Garden Palace which houses the Peacock Throne, brought from India by Nadir Shah in 1739. It is the industrial and commercial centre of the country; manufactures include cement and textiles. Tehran is also famous for the exporting of carpets. Pop. (1974 est.) 3,931,000.

Teignbridge, county district in central s DEVON, England, created in 1974 under the Local Government Act (1972). Area: 675sq km (261sq miles). Pop. (1974 est.) 94,300.

Teilhard de Chardin, Pierre (1881–1955), French JESUIT philosopher and palaeontologist. He worked in China between 1923 and 1946 and shared in the discovery of PEKING MAN. His efforts to reconcile scientific views of evolution with Christian faith led to his being asked by his religious superiors not to publish his philosophical works. His best-known work, *The Phenomenon of Man,* was finished in 1938 but only published posthumously.

Teixeira, Pedro (*c.* 1575–1640), Portuguese explorer. Commissioned by the governor of Maranhão in 1637 to undertake an expedition up the AMAZON River, he navigated the Amazon and Napo as far as Quito, Ecuador.

Te Kanawa, Kiri (1948–), New Zealand opera singer, of Maori origin. One of the best-known New Zealand sopranos, she became contracted to the Royal Opera House, Covent Garden, in 1970. *See also* HC2 p.*251.*

Tektites, generally dark, glassy objects, ranging in diameter from 40 microns to 2mm (microtektites) and larger (to 10cm) believed to be of either lunar origin or formed from splashes of liquefied rock during meteorite impact on Earth.

Tel Aviv-Jaffa (Tel Aviv-Yafo), largest city in Israel, on the Mediterranean Sea. It was founded in 1909 as a Jewish suburb by the Jewish population of Arab Jaffa. By 1921 Britain had established it as a separate all-Jewish community. During the British PALESTINE mandate (1923–48), Tel Aviv's immigrant Jewish population grew rapidly. It served as the seat of the transitional government and legislature between 1948 and 1950 (when Isarel's capital was moved to Jerusalem) and it was incorporated with Jaffa in 1950. Products include pharmaceuticals, electrical appliances, textiles, clothing, chemicals, paper, plastics, leather, glass and precision instruments. Pop. 357,600.

Telecine, machine for the direct projection of film on television. It can be used for the insertion of filmed material into live programmes or to transmit complete feature or documentary films. *See also* MM p.231.

Telecommunications, the technology involved in the sending of information over a distance. This information may be in various forms, such as digital signals, sounds, printed words or images. The sending is achieved through TELEGRAPH, TELEPHONE and RADIO and the medium may be wires or electromagnetic (radio) waves, or combinations of the two. Telegraphy was developed in the mid-19th century and radio arrived at the end of the century. TELEVISION made its advent in the mid-20th century. There are two basic types of message: digital, in which case the message to be sent is converted in simple coded pulses and then sent (as in MORSE CODE); and analogue, in which the message – for example, a voice speaking – is converted into a series of electrical pulses which are similar in waveform to the modulations of the original message. *See also* MM pp.224, 226, 228, 230, 236.

Telegraph, any communications system that transmits and receives visible or audible coded signals over a distance. The first, optical, telegraphs were forms of SEMAPHORE. Although most modern devices are electrically operated and connected by wires or cables, telegraph messages can be sent by radio waves, microwaves and communication satellites. Pioneering work was performed by many, but credit for the electric telegraph and its code is generally given to Samuel MORSE who, in 1844, inaugurated the first public line between Washington and Baltimore. In 1866 the first permanently successful telegraph cable was laid across the Atlantic, and in 1875 Thomas A. EDISON invented a method of transmitting several messages simultaneously over the same wire. *See also* MM p.224.

Telekinesis, in parapsychology, the movement of objects, apparently without contact, that cannot be explained by any known physical or natural means. *See also* PSYCHOKINESIS.

Tel-el-Kebir, Battle of (1882), decisive victory, in the British occupation of Egypt, of the British army under Sir Garnet WOLSELEY over the Egyptian nationalists led by Arabi Pasha.

Telemann, Georg Philipp (1681–1767), German composer. He wrote more than 40 operas, 600 overtures and 44 settings of the Passion. His church music, of more historical importance than his operas, shows Telemann's technical mastery.

Telemetry, system of transmitting data, usually measurements, over a distance, such as to the ground from a spacecraft via electromagnetic (radio) waves. In this example, solar cells or chemical batteries provide power for operating high-gain directional antennae on the spacecraft. These beam high-frequency (but low-power) radio waves to large receiving stations on the ground. *See also* MM pp.156–159.

Teleology, type of cosmology explaining final causes. In contrast to mechanism, teleology explains the past and present in terms of the future.

Teleostei, group of fish that have a skeleton composed at least in part of bone, instead of cartilage. There are 40 orders and 20,000 species – most of the existing species of fish. *See also* OSTEICHTHYES.

Telepathy, direct transmission and reception of thoughts. It has never been scientifically proved conclusively that such transference happens, although telepathy has been extensively investigated. Tests have included card-guessing, in which one experimenter (unseen by another) chooses cards at random from a set bearing pictorial symbols; his fellow experimenter attempts to write down or draw the sequence of symbols. As with other tests on ESP, the occasional experimenter produces results which are somewhat outside those that would be expected statistically, but this cannot be said to constitute proof. *See also* MS pp.198, 199.

Telephone, instrument that communicates speech sounds by means of wires or microwaves. In 1876 Alexander Graham BELL invented the prototype, which employed a diaphragm of soft iron which vibrated to sound waves. These vibrations caused disturbances in the magnetic field of a nearby bar magnet, causing an electric current of fluctuating intensity in the thin copper wire wrapped around the magnet. This current could be transmitted along wires to a distant identical device which reversed the process to reproduce audible sound. Later improvements separated the transmitter from the receiver, and replaced the bar magnet with batteries. *See also* MM p.226.

Telephone exchange, central switching system in which incoming telephone calls are routed to their destinations, either by telephone operators or, increasingly, by automatic mechanisms that are called into operation by direct dialling codes. *See also* MM pp.226–227, 226–227.

Telephony, microwave, telecommunications technique which uses microwave radio to transmit telephone messages between stations. The microwaves act as carrier for the telephone signals which are used to modulate them, and enable more than 2,500 simultaneous telephone conversations. *See also* MM p.227.

Telephoto, camera lens with a long focal length. A true telephoto lens has a focal length longer than the physical length of the lens, as opposed to a long-focus lens, in which the focal length is equal to the physical length. There is no theoretical upper limit to the focal length possible but sheer size tends to limit the actual focal lengths practicable. For a 35mm camera, any lens with a focal length of more than about 80mm may be regarded as a telephoto lens. For larger format cameras the focal length may be as much as 1,000mm. Even greater lengths are possible with mirror lenses, which overcome some of the difficulties of very long conventional lenses. *See also* MM p.221.

Teleprinter, machine having a keyboard similar to that of a TYPEWRITER, on which messages can be typed and telegraphed to distant receivers, and which acts also as a receiver, printing out incoming messages.

Telescope, instrument for enlarging a distant object or studying electromagnetic radiation from a distant source. Telescopes are categorized according to the area of the electromagnetic spectrum with which they deal. There are two types of optical telescope: refracting and reflection. The main light-gathering parts or objective, is a lens in a refractor and a mirror in a reflector. The diameter of this lens or mirror is known as the aperture of the telescope. The point at which the objective concentrates the light from the source is its focus, and the distance from the focus to the objective is its focal length. The light-gathering power of a telescope is proportional to the square of the aperture, its ability to resolve, or separate into distinct images two closely spaced objects, is also proportional to the aperture – the larger the aperture, the better the resolution. The earliest invention of the refracting telescope is unknown but there is evidence that it could have been as early as 2000 BC. Extensive use of refracting telescopes, however, followed Hans Luppershay's invention (1608) and GALILEO's telescope (1609). The main disadvantage with refracting telescopes was lens aberration, in which faults in curvature and variations in density caused a single lens to produce a partly distorted image. At first regarded as insurmountable, this problem may be solved by so combining lenses that their aberrations cancel each other out. This initial obstacle was first overcome with the construction of reflecting telescopes and the first astronomical reflector was built by Sir Isaac NEWTON in 1668. There was also a limit to the size of lenses: the world's largest refractor has a diameter of 1m (39in), which is much less than that of one of the largest reflectors (that at Mount Palomar, California, USA, has an aperture of 508cm; 200in). Radio telescopes are complex electronic systems which analyse radio waves from beyond the Earth. Incoming radio waves are collected from a small area of the sky and focused at the centre of a large parabolic, steerable dish aerial. The signals are then amplified and analysed. The first radio telescope was built in 1937 by Grote Reber in the USA. Radio interferometers are arrays of smaller dishes which are tuned in sequence, thus permitting the investigation of even more distant radio sources. X-ray telescopes consist of a pair of mirrors, one a paraboloid surface and the other a hyperparaboloid surface. *See also* SU pp.168, 170, 172.

Telesphorus, Saint (*d. c.* 136), pope (*r. c.* 125 to 136). St Telesphorus, who succeeded St Sixtus I, is said by St IRENAEUS to have observed Easter not on the 14th day of the month Nisan, as was the practice in parts of Asia Minor, but on the following Sunday.

Teletype. *See* TELEPRINTER.

Television, system for the transmission and reception of visual images by radio waves or, sometimes, cable. In a television camera the images are first converted from light rays to electrical signals. These are then amplified and transmitted as very high frequency (VHF) or ultra-high frequency (UHF) radio waves. Typically a television channel has a bandwidth of 5MHz (5 million cycles per second). On their reception these signals are amplified and then converted to light again in a CATHODE RAY TUBE.

The basis of the television camera is an image orthicon tube. Light rays are focused on to thin sheets of photoelectric material, which emits electrons in proportion to the amount of light striking it. Behind these sheets is a positively charged target which emits additional electrons. Between them is a fine mesh screen which collects the secondary emissions from the target. The net result is that the target has a positive charge in proportion to the amount of light in the original scene. At the rear of the orthicon tube is an electron gun emitting a beam of electrons which scans the target from side to side and from top to bottom. Because the target is composed of positively charged areas, the incident electron beam is deflected back to the electron gun, where the electrons are collected by a multiplier, which also amplifies the intensity of the returning beam. The strength of this beam depends on the positive charges on the target and thus the original scene is converted into electrical signals.

The receiver operates in reverse to the camera. In colour television there are three synchronized image orthicon tubes in the camera, one for each of the three primary colours – red, blue and green. The tube of the receiver has three electron guns and the face of the tube is covered with a mosaic of fine phosphors in groups of three, each emitting only red, blue or green light when struck by a beam. These primary colours merge on the face of the screen to reconstitute the originally transmitted image. *See also* MM pp.230–231.

Telex, acronym for Teletype-writer Exchange Service, a telegraphic system using telephone lines through which direct current pulses representing characters typed on a typewriter keyboard are sent. *See also* MM p.225, 225.

Telford, Thomas (1757–1834), Scottish civil engineer; builder of bridges, canals, docks, harbours, waterways and roads. Among his many notable achievements are the 177m (580ft) suspension bridge over the Menai Strait, connecting Anglesey with mainland Wales, begun in 1819: the Ellesmere Canal, connecting the English rivers Dee, Mersey and Severn (1793); and the CALEDONIAN CANAL in Scotland. He was a founder, and first president, of the Institution of Civil Engineers. *See also* HC2 p.31; MM pp.182, 184, 191, 194.

Tell, William, legendary Swiss hero, leader in the 14th-century war of liberation against Albert I of Austria. For refusing to doff his cap to Albert's steward, Gessler, he was made to shoot an arrow through an apple placed on his son's head.

Teller, Edward (1908–), US physicist, b. Hungary. He left Europe and settled in the USA in 1935. During WWII he worked on atomic bomb research with Enrico FERMI and later pioneered the development of the hydrogen bomb. He received the 1926 Enrico Fermi Award.

Tellurium, metalloid element (symbol Te) of group VIA of the periodic table, dis-

Telephone; the end of an undersea telephone cable being brought ashore.

Telescope; the radiotelescope "dish" at Jodrell Bank measures 76m (250ft) across.

Television: colour cameras split pictures into the primary colours red, green and blue.

Thomas Telford; many of his bridges, roads and waterways are still in use today.

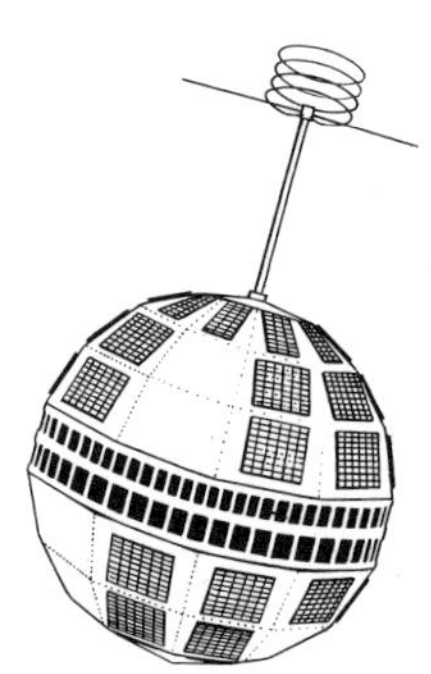

Telstar, the first active communications satellite, re-transmitted signals.

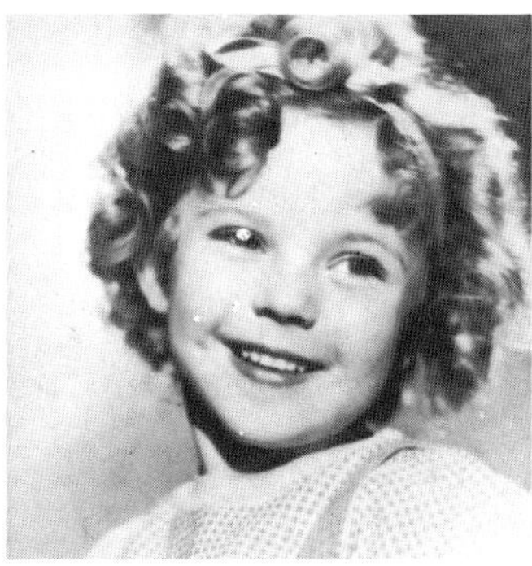

Shirley Temple had already gained an Academy Award at the age of seven.

Ten Commandments; Moses was given the Tablets of Law in the Sinai Wilderness.

Teniers the Younger's scenes of peasant life were often used as tapestry designs.

covered in 1782. It occurs naturally and in sylvanite (Ag, Au)Te$_2$; its chief source is a by-product of the electrolytic refining of copper. It is used in some alloys and semiconductor devices. Properties: at.no. 52; at.wt. 127.6; s.g. 6.24; m.p. 449.5°C (841°F); b.p. 989.8°C (1,814°F); most common isotope Te[130] (34.48%).

Telstar, first active communications satellite, launched on 10 July 1962. It contained a microwave radio receiver, amplifier and transmitter for relaying telephone and television signals. It operated for about 18 weeks, failed for 5 weeks, and worked again for a further 7 weeks before failing for good. *See also* MM pp.156, 228.

Temin, Howard Martin (1934–), US biologist. He shared with D. Baltimore and R. DULBECCO the 1975 Nobel Prize in physiology and medicine for his studies of the interaction between tumour viruses and the genetic material of the cell.

Tempera, painting medium used extensively during the Middle Ages, made of powdered pigments mixed with a base of egg yolk, fig sap or glue. It was used mostly for painting wooden altarpieces and was applied to a smooth surface coated with GESSO. The religious paintings of the Byzantine period are early examples of tempera painting. By the 13th century, Italian painters such as CIMABUE, GIOTTO and DUCCIO had refined the technique. Tempera dries quickly and is applied with a sable brush, one thin layer on another so that the finished effect is semi-opaque and luminous. Some modern painters, including Ben SHAHN and Andrew Wyeth, have revived the technique.

Temperament, general emotional tone or mood of a person. Psychological research has shown that a person's basic temperament is probably partly genetically determined. *See also* MS pp.188, 189.

Temperance movement, organized effort to promote moderation in, or abstinence from, the consumption of alcohol, which was considered to encourage sickness, poverty and crime. It probably began in the USA in the early 19th century and, with assistance from churches, spread to Britain and continental Europe.

Temperate zones, either of two regions of the Earth. The northern temperate zone lies between the Arctic Circle and the Tropic of Cancer, and the southern zone lies between the Tropic of Capricorn and the Antarctic Circle. *See also* PE pp.74–75.

Temperature, measure of the hotness or coldness of an object. Strictly, it is a parameter describing the number of energy states available to a substance or system. Two objects placed in thermal contact exchange heat energy initially but eventually arrive at thermal equilibrium where both are said to have the same temperature, ie each is losing and gaining heat at equal rates so that neither has a net gain or loss of heat. At equilibrium, the most probable distribution of energy states of the atoms and molecules composing the objects has been attained. At high temperatures, the number of energy states available to the atoms and molecules of a system is large; at lower temperatures, fewer states are available (molecules become locked into position and liquids change to solids). At a sufficiently low temperature, all parts of the system are at their lowest energy levels, the ABSOLUTE ZERO of temperature. *See also* TEMPERATURE SCALE; SU pp.90–95.

Temperature, body, internal temperature of an animal. In warm-blooded animals such as birds and mammals, body temperature is maintained within narrow limits regardless of the surrounding air temperature. This is accomplished by muscular activity, operation of cooling mechanisms and normal basal metabolism. In man, the normal body temperature is close to 37°C (98.6°F), but this may vary with degree of activity, reaching 40°C (104°F) during exercise, and falling below normal during sleep. In so-called cold-blooded animals, such as reptiles, body temperature varies between wider limits, depending on the surrounding air temperature.

Temperature measurement, method by which TEMPERATURE is specified. Several

TEMPERATURE SCALES have been established, each of which uses a property (such as electrical resistance or volume expansion) of a thermometric substance (such as a metal wire or a liquid) to define a unit of temperature. The constant-volume gas thermometer is the standard thermometer. Its temperatures are expressed in KELVIN degrees, but it is inconvenient for practical purposes. The International Practical Temperature Scale (IPTS) was adopted in 1968 to provide scientists and technologists with a more convenient scale. The IPTS defines 0°C (32°F) as 273.16°K and defines a number of reference temperatures including the boiling point of oxygen, 90.20°K (−182.96°C, −297.33°F), the melting and boiling points of water and the melting points of zinc, silver and gold. It also specifies the thermometers that are used to span the range. At higher temperatures, eg of furnaces, optical pyrometers which detect variation in colour are used. *See also* SU pp.90–91, 94–95.

Temperature scale, graduated scale of degrees for measuring temperature. The establishment of any temperature scale requires: a thermometric parameter which varies linearly with temperature (eg the volume of a gas at constant pressure, or the expansion of a liquid in a tube); two or more fixed points (readily reproducible reference points such as the boiling and the freezing points of water); and the assignment of arbituary divisions (called degrees) between the fixed points. Gas, alcohol, mercury, electrical resistance and colour have been used as thermometric parameters. Common temperature scales include FAHRENHEIT, CELSIUS (formerly the centigrade) and the KELVIN (or absolute); these are abbreviated to °F, °C and °K (or K). The Fahrenheit scale originally used as fixed points the freezing point of water (taken to be 32 fahrenheit degrees) and the human body temperature (96 degrees, although later found to be 98.6). The interval between these was divided into 64 degrees; by extrapolation, the boiling point of water is 212°F. The Celsius scale uses 0°C and 100°C as the freezing and the boiling point of water, respectively; the interval is divided into 100 degrees. Zero on the Kelvin scale (approx. −273°C, −459°F) coincides with absolute zero, the lower limit of temperature; the degree Kelvin (kelvin) represents the same temperature difference as the degree Celsius. To convert Fahrenheit to Celsius: C = 5(F−32) ÷ 9 to convert Celsius to Fahrenheit, F = (9C÷5) + 32.

Tempering, metallurgical process of hardening and toughening alloys by a cycle of heating and quenching. Tempering is particularly important for the production of tool steels. It works by producing an alloy which is a stable solid solution of small grain size. This can be achieved only by adjusting the crystal structure of an alloy in a calculated way by precise temperature gradients.

Tempest, The, play (1611) by William SHAKESPEARE. The play deals with the theme of reconciliation, and its setting was influenced by contemporary exploration of the Americas.

Templars. *See* KNIGHTS TEMPLAR.

Template, or templet, mould or pattern for making repeated copies of an object. It is usually cut from thin sheets of metal, wood or paper. In architecture, a template is a supporting structure such as a beam placed over a doorway.

Temple, Henry John. *See* PALMERSTON, VISCOUNT.

Temple, Shirley (1928–), US film actress. While still a child she starred in a series of films. These included *The Little Colonel* (1935), *Rebecca of Sunnybrook Farm* (1938) and *The Blue Bird* (1940). She made several films in the late 1940s and then, as Shirley Temple Black, went into politics. She became US Ambassador to Ghana in 1974.

Temple, William (1881–1944), British churchman. He was chaplain to GEORGE V (1915–21), Canon of Westminster (1919–21), Archbishop of York (1929–42) and Archbishop of Canterbury (1942–44). He was a strong and out-

spoken advocate of social reform and also worked for the union of Christian Churches.

Temple, Sir William (1628–99), English statesman, Ambassador to The Hague in 1668. He negotiated the TRIPLE ALLIANCE and in 1677 arranged the marriage of Princess Mary to WILLIAM OF ORANGE.

Temple, in Judaism, the place of worship for Jews in Jerusalem. The first temple, built by King SOLOMON, lasted for more than 400 years until destroyed in 586 BC by the Babylonians. It contained the ARK OF THE COVENANT. It has been replaced by SYNAGOGUES. The term "temple" is also used for Hindu, Buddhist and Sikh places of worship.

Temple, The, the name given to the Middle Temple and the Inner Temple, London, two of the four Inns of Court and originally the seat of the chivalrous order of the Knights Templar. *See also* INNS OF COURT.

Templer, Field Marshal Sir Gerald Walter Robert (1898–), British soldier who was a divisional commander during WWII and later stemmed a Communist terror campaign in Malaya during the early 1950s. From 1955 to 1958 he was Chief of the Imperial General Staff.

Tempo, speed at which a piece of music is performed, usually indicated on a score, in western music, by Italian words, such as *allegro*, fast, and *adagio*, slow.

Temporal lobes, prominent lobes of the cerebral cortex that, in humans, lie directly below the temples. The temporal lobes are directly involved in the interpretation and generation of language. *See also* BRAIN; MS p.40, *40.*

Tenant, person who rents property from another (the landlord) by a tenancy agreement or lease. The concept of tenancy is entrenched in common law, which recognizes different types of tenancy, but most are now defined by the terms of the agreement. Disputes between a tenant and a landlord in Britain may be referred to a Rent Tribunal.

Tench, freshwater food and sport fish of Europe and Asia, belonging to the carp family Cyprinidae. It has a stout, golden yellow body with small scales. It feeds on small animals and plants. Length: to 71cm (28in). Species *Tinca tinca See also* PE p.245.

Ten Commandments, or Decalogue, the code of ethical conduct held in Jewish-Christian tradition to have been revealed by God to MOSES. Representing the moral basis of God's COVENANT with Israel, they appear in both EXODUS 20 and DEUTERONOMY 5, but with different phrasing. Commandments 1–4 exhort obligation and service to the one God, and commandments 5–10 require social responsibility from the individual.

Tender is the Night (1934), novel by F. Scott FITZGERALD. Dick Diver, a psychiatrist, falls in love with Nicole, whom he first met as a patient in a European clinic. After they marry their relationship becomes hollow, his life meaningless and he turns to drink and dissipation. *See also* HC2 p.*289.*

Tendon, strong and flexible band of connective tissue that joins muscle to bone. *See also* MS p.58.

Tendril, part of stem or leaf, a slender, thread-like structure used by climbing plants for support.

Tendring, county district in NE ESSEX, England, created in 1974 under the Local Government Act (1972). Area: 337sq km (130sq miles). Pop. (1974 est.) 107,700.

Tenebrism (from Italian *tenebroso*, murky), style of painting by 17th-century artists in Naples, Spain and The Netherlands, who painted in a very low key and emphasized contrasts of light and shade in imitation of CARAVAGGIO.

Tenerife, largest of the Canary Islands, Spain, in the Atlantic Ocean 64km (40 miles) WNW of Grand Canary Island; It is a mountainous island; its highest peak is Pico de Teide, 3,718m (12,198ft). Products: dates, sugar cane, palms, cotton. Industries: tourism, food processing. Area: 2,059sq km (795sq miles). Pop. 500,381.

Teniers the Elder, David (1582–1649),

Flemish painter who spent many years in Italy, finally settling in Antwerp. Father of David TENIERS THE YOUNGER, he painted religious subjects.

Teniers the Younger, David (1610–90), Flemish GENRE painter who became court painter to the Governor of The Netherlands in 1651. A prolific artist, he painted small precise pictures, his favourite subjects being quiet scenes from peasant life.

Tennant, Charles (1768–1838), British industrial chemist who in 1799 invented a process for manufacturing bleaching powder from chlorine and slaked lime.

Tennessee, state in SE central USA, between the Appalachian Mts and the Mississippi River. In the E are the Great Smoky Mts and the Cumberland Plateau. To the W is the Nashville Basin, a bluegrass region, where livestock are reared – the region is particularly noted for breeding horses. The land is generally not suitable for agriculture but W Tennessee, between the Tennessee and Mississippi rivers, has fertile soil and cotton is grown there. Wheat, hay and tobacco are also grown in the state. Mineral deposits include zinc and coal. Major industries are chemicals, textiles, clothing, food processing and electrical machinery. The chief cities are Nashville (the state capital) and Memphis.

The region was first visited by Spaniards in the 16th century. The French followed a century later but their claim was ceded to Britain in 1763. first permanent settlement was established in 1769. In 1790 the US government created the territory south of the Ohio corresponding to Tennessee, which was admitted to the Union six years later. In the AMERICAN CIVIL WAR opinion in Tennessee was divided, although the Confederacy claimed most supporters. Several major battles were fought in the state, which seceded in 1861. Area: 109,411sq km (42,244sq miles). Pop. (1975 est.) 4,188,000. *See also* MW p.175.

Tennessee Valley Authority (TVA), agency established (1933) during the NEW DEAL as a long-term regional planning project. It was authorized to build dams on the Tennessee River to prevent flooding, produce electricity and enable eroded farmland to be reclaimed.

Tennessee walking horse, light saddle horse, originally bred in the 19th century in Tennessee, USA. The coat may be chestnut, black or golden. Height: to 1.6m (5.2ft) at the shoulder; weight: to 540kg (1,200lb).

Tenniel, Sir John (1820–1914), British painter, illustrator and cartoonist, best known for his illustrations of Lewis CARROLL's *Alice's Adventures in Wonderland* (1865) and *Alice Through the Looking-Glass* (1872).

Tennis. *See* LAWN TENNIS.

Tennis Court, Oath of the (1789), defiant vow, made by members of the Third Estate of the French ESTATES-GENERAL on 20 June 1789, not to separate until a constitution had been established. The name derives from the indoor tennis court in which they met.

Tennis elbow, form of BURSITIS in which there is an inflammation of the lubricating sac over the joint, caused by repeated irritation.

Tennyson, Alfred Lord (1809–92), British poet who was Poet Laureate from 1850 to 1892. His verse, noted for its sensuous use of language, takes its subject matter from (among other things) Arthurian legends and classical mythology. It includes *The Lady of Shalott* (1833), *The Lotus Eaters* (1883), *In Memoriam* (1850) and *Maud* (1855). *See also* HC2 p.100.

Tenochtitlán, AZTEC city founded in AD 1325 on an island in Lake Texcoco, Mexico. Drainage reduced the surrounding swampland to fertile plots surrounded by water, and Tenochtitlán expanded to a metropolis of 500,000 people. It was destroyed by the Spanish, who later built Mexico City on the ruins. *See also* HC1 pp.231, 236, *236.*

Tenor, range of the human voice, falling below CONTRALTO and above BARITONE. It is the highest natural male voice apart from the COUNTER-TENOR.

Tenpin bowling, type of bowling that uses ten pins set on a triangular base, and a heavy ball with holes to enable the participant to hold it with one hand. It originated in England and its rules were standardized in the USA in 1895.

Tenrec, burrowing, insectivorous mammal of Madagascar and the Comoro Islands. The common tenrec (*Tenrec ecaudatus*) is a nocturnal, highly prolific animal with a spiny coat like that of a hedgehog. It is the size of a small domestic cat. Family Tenrecidae.

Tense, in grammar, classification of a verb indicating the time of an action – present, past or future – in relation to the time of speaking.

Tentacle, any slender, flexible organ of an animal, most noticeably those of the OCTOPUS, capable of feeling and grasping.

Ten Thousand Immortals, in ancient Persia, name given to the élite of the army. Skilled in the art of fighting with bows and arrows, they made a decisive contribution to the Persian victory in the campaign against Egypt under the leadership of DARIUS THE GREAT (548–486 BC). *See also* HC1 pp.60–61.

Tenzing Norkey (Norgay) (*c.*1914–), Tibetan mountaineer who went on several expeditions to Mt EVEREST in the mid-1930s. In 1953 he and Edmund HILLARY became the first men to reach the summit of the mountain.

Teotihuacán, ancient city located about 48km (30 miles) N of Mexico City. It dates from *c.*100 BC and flourished until *c.* AD 700. Built by an Olmec-based culture, the city was laid out in a grid pattern centered on the Street of the Dead. The Pyramid of the Sun, the Temple of Quetzalcoatl and the Pyramid of the Moon dominate the E, S and N limits, respectively. Excavations have uncovered sophisticated murals and stylized stone masks. *See also* HC1 pp.47, *47, 230.*

Tepee, or teepee, moveable home of North American Indians, especially those of the plains. It consisted of a hide stretched over poles arranged to make a cone.

Tequila, Mexican alcoholic drink distilled from the fermented juice of the AGAVE plant. It contains 40 to 50% alcohol.

Te Rangi Hiroa. *See* BUCK, SIR PETER.

Te Rauparaha (*c.*1768–1849), New Zealand chief and warrior who is famous for his military exploits during the tribal wars of the 1920s and 1930s.

Terbium, metallic element (symbol Tb) of the lanthanide group, first isolated in 1843, its chief ore is monazite (a phosphate). The element is used in semiconductors. Properties: at.no. 65; at.wt 158.9254; s.g. 8.234; m.p. 1,360°C (2,480°F); b.p. 3,041°C (5,506°F). Most common isotope Tb^{159} (100%).

Terborch, Gerard (1617–81), Dutch painter, best known for his expressive miniature portraits and small interiors with figures, rendered in a NATURALISTIC style. His careful handling of textures of cloth, such as satin, can be seen in *Gallant Conversation* (pre-1655, often known as *Parental Admonition*).

Terbrugghen, Hendrick (1588–1629), Dutch painter. During his travels to Italy, he was influenced by CARAVAGGIO's style. He painted religious subjects, such as the *Calling of St Matthew* (1621) and the *Crucifixion with the Virgin and St John* (1625). He was also a genre painter. Such later works as *Jacob and Laban* (1627) were characterized by a light and delicate colour range.

Teredo, also called shipworm, any of about 15 species of small bivalve MOLLUSCS that bore into wood. It has a worm-like body with the halves of the shell being greatly reduced and modified to act as a drill. It spends its life in submerged wood, doing considerable damage to ships' hulls and wooden pilings. Length: to 30cm (12in). Family Teredidae; genus *Teredo.* *See also* NW p.92.

Terence (*c.*190–159 BC), properly Publius Terentius Afer, Roman author of comedy. He was a slave in Rome where he was educated and freed by his master, the senator Terentius Lucanus. Six of his comedies survive and include *Andria, Phormio* and

Hecyra. He is noted for his deep interest in character and refined style. *See also* HC1 p.112, *112.*

Teresa of Avila, Saint (1515–82), Spanish CARMELITE nun. She played a great part in the COUNTER-REFORMATION. Her literary works, including an autobiography and the meditative *Interior Castle* (1577), led Pope PAUL VI to make her the first woman Doctor of the Church in 1970.

Tereshkova, Valentina Vladimirovna (1937–), Soviet cosmonaut and the first woman in space. She travelled in Vostok 6, launched in June 1963, and spent 71 hours in space. Later in the same year she married Andrian Nikolayev, another Soviet cosmonaut.

Terminator, boundary between the sunlit and dark sides of a planet or satellite. With a body lacking an atmosphere, such as the Moon, the terminator is distinct, although often broken up because of reflections from craters or mountains. Bodies with atmospheres have less well-defined terminators because atmospheric scattering causes twilight. *See also* SU p.178.

Termite, social insect found throughout the world in subterranean nests and above-ground mounds. Termites are popularly but incorrectly called "white ants". They have a caste system, with a king and queen guarded and tended by soldiers, workers and nymphs. They can be distinguished from ants by the lack of a narrow waist. Length: 2–22.5m (0.08–0.9in); queens: to 10cm (4in). Order Isoptera. *See also* NW pp.*107,* 110, *201, 205, 208, 218.*

Terms, University, in Britain, division of the academic year into three terms between October and June. They are traditionally known as Michelmas (at Oxford, Cambridge and others), Trinity (at Oxford and others) and Hilary (at Oxford and Dublin). *See also* LAW TERMS.

Tern, also called sea swallow, any of several species of graceful seabirds that live throughout the world. It is usually white and grey with a darker head, and has a pointed bill, long pointed wings, a forked tail and webbed feet; it dives for fish and crustaceans. After a long migration it lays its eggs in a scraped depression in the ground. Length: to 55cm (22in). Family Laridae; genus *Sterna. See also* NW pp.*142, 237.*

Terpenes, group of unsaturated organic chemicals related to ISOPRENE. They occur in most essential oils and are colourless liquids. Examples are pinene, the chief ingredient of turpentine, and limonene, which is found in the oils of citrus fruits.

Terracette, horizontal ledge formed along the side of a hill by the gradual movement of soil downwards under the influence of weathering and gravity. Terracettes are usually about 1–5m (3–17ft) in width. They form in layers and give a hillside a stepped appearance. *See also* PE p.114.

Terracing, in agriculture, technique involving the building of horizontal or sloping ridges along the contours of hills. This practice traps moisture and prevents the erosion of the soil, and is employed in regions where arable land is scarce.

Terracing, in geology, the process of deposition or erosion which produces step-like formations on a slope. Alluvial terracing is usually caused by periodic reductions in the area of the flood-plain of a river. *See also* PE p.113.

Terracotta, form of porous, unglazed, usually reddish, fired clay used in building, sculpture and pottery. It was first used by ancient civilizations when stone was not available. The term is applied more specifically to figurines made of terracotta.

Terrapin, any of several species of aquatic TURTLES that live in fresh or brackish water in the USA and South America, especially the diamondback terrapin (*Malaclemys terrapin*). Its diet consists mainly of small animals and vegetable matter. Length: to 23cm (9in). Family Emydidae. *See also* NW pp.*134,* 135.

Terrier, any of several breeds of dog. It was originally trained to dig out game. Referred to as early as the 14th century, it has been used to hunt badgers, foxes, weasels and rats. When the quarry is located, the terrier or terriers are sent

Tennyson is regarded by many as the greatest poet of the Victorian era.

Teotihuacán, or "City of the gods" was a commercial and religious centre.

Hendrick Terbrugghen; detail from *A Singer with his Lute,* painted in 1624.

Terns breed in large colonies and are often found on warm oceanic islands.

Terrigenous deposits

Ellen Terry; this detail of a portrait by Sargent shows her as Lady Macbeth.

Test Ban Treaty, signed in 1963, sought to ban nuclear testing in the atmosphere.

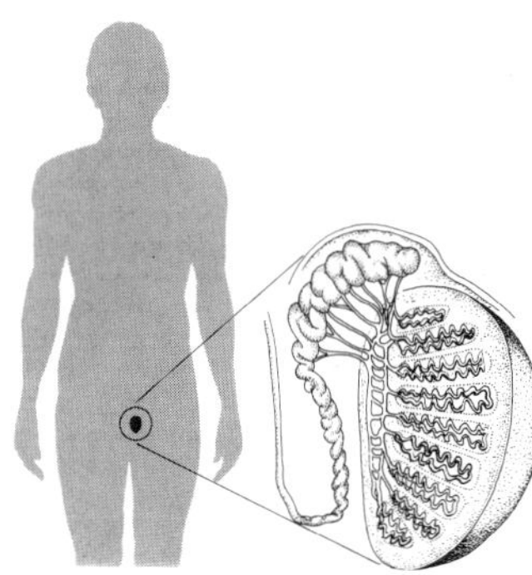

Testis; the numerous tubules in each testis produce reproductive spermatozoa.

Texas has a wealth of industry, much of it located in Houston, the largest city.

down to dig it out of its burrow. Separate breeds emerged in 19th-century England, and include the SEALYHAM TERRIER, FOX TERRIER, MANCHESTER TERRIER, SCOTTISH TERRIER and BEDLINGTON TERRIER. Larger breeds such as the KERRY BLUE TERRIER, AIREDALE TERRIER and IRISH TERRIER are used as guard and police dogs.

Terrigenous deposits, accumulations of sand, silt or mud that form in the sea near land as a result of EROSION.

Territorial Army, former British organization for home defence. Founded in 1908 as the Territorial Force, its name was changed to Territorial Army in 1922. It recruited part-time volunteers who were given training in basic military skills. It merged with the other defence forces 1939–47. In 1967 is was replaced by the Territorial and Army Volunteer Reserve.

Territory, in animal behaviour, the restricted life space of an organism. It is an area selected for mating, nesting, roosting, hunting or feeding and may be occupied by one or more organisms and defended against others of the same, or a different species. The area may be defended or indicated by noise-making, chemical scent, physical displays or aggression. Many invertebrates and most vertebrates display this behaviour.

Terry, Dame Ellen Alice (1847–1928), British actress. She made her debut, aged nine in *The Winter's Tale*, and worked with Sir Henry IRVING for 24 years, touring the USA with him.

Tertiaries, in the Roman Catholic Church, lay people belonging to the third rank of religious orders, who take less severe vows than monks and nuns. Originally, in the 13th century, they were women attracted by the humanitarian rather than the contemplative aspects of religious orders, but they now include people of either sex. Other religions with monastic ideals, such as Buddhism, include comparable groups.

Tertiary Period, earlier division of the CENOZOIC PERIOD, lasting from 65 million to about 2 million years ago. It is divided into four epochs, starting with the Eocene, the earlier part of which is sometimes called the Paleocene, followed by the Oligocene, Miocene and Pliocene. Early Tertiary times were marked by great mountain-building activity (Rockies, Andes, Alps and Himalayas). Both marsupial and placental mammals diversified greatly. Archaic forms of carnivores and herbivores flourished, along with primitive primates, bats, rodents and early whales. *See also* PE pp.130, *131*.

Tertis, Lionel (1876–1975), British viola player. He introduced technical innovations for the making of the viola, and arranged many violin compositions for the instrument.

Tertullian (Quintus Septimius Florens Tertullianus) (*c.* 160–*c.* 230), Roman writer. He developed a systematic approach to theology and Christian APOLOGETICS, and helped to make Latin the official language of Christian theological writing. His works include *Apologeticus De spectaculis* and *De anima*. *See also* HC1 p.113.

Terylene, POLYESTER synthetic fibre made by POLYMERIZATION of a GLYCOL with an organic acid. Its fibres are generally melt-spun through a metal spinneret. Terylene is particularly popular for shirt, dress and stocking fabrics. It is called Dacron in some countries.

Tesla, Nikola (1856–1943), US electrical engineer, *b.* Croatia, who pioneered the applications of high-voltage electricity. He developed arc lighting, generators of high-voltage alternating currents, the Tesla coil and a system of transmitting electric power without wires.

Tesla, SI UNIT of magnetic flux density equal to a density of one WEBER of magnetic flux per square metre.

Tessin, family of Flemish architects who worked in Sweden. Nicodemus the Elder (1615–81) built in the BAROQUE style, and served as royal architect. His son Nicodemus the Younger (1654–1728) studied with BERNINI in Rome.

Test Acts (1672), English laws intended to exclude Protestant dissenters and Roman Catholics from public offices. It required all candidates for such offices to profess the established religion of the Church of England. Not until the 1870s was the legislation finally repealed.

Testator, person who makes a WILL.

Test Ban Treaty (1963), agreement signed in Moscow by the USSR, and USA and Britain not to conduct tests of nuclear weapons in the atmosphere, in space, under water nor in any place where the radioactive debris ("fallout") could spread beyond national borders. All other countries were invited to sign the treaty. Nearly 100 nations signed it, but France and China continued to hold tests above ground.

Testicles. *See* TESTIS.

Testis (plural; testes), male sex gland, found as a pair located in a pouch, the scrotum, external to the body. The testes are made up of seminiferous tubules in which male (SPERM) are formed and mature, after which they drain into ducts, and are stored in the epididymis, from which they are discharged. *See also* MS pp.74–75, *74–75*.

Test match, official international CRICKET match, played under the auspices of the International Cricket Council. The countries which play test cricket are England, Australia, the West Indies, New Zealand, India, Pakistan and South Africa (barred since 1970). The first test match was played between Australia and England at Melbourne in 1877.

Testosterone, STEROID HORMONE produced mainly by the mammalian TESTIS. It is responsible for the growth and development of male sex organs and male secondary sexual characteristics. *See also* MS pp.64–65.

Tetanus, infectious disease of the central nervous system caused by the toxin secreted by the anaerobic bacterium *Clostridium tetani*. The symptoms are extreme stiffness, convulsions and painful muscular spasms, especially of the jaw muscles. This gives the disease its popular name "lockjaw". Tetanus has a high mortality rate but is preventable by immunization. *See also* MS p.96.

Tetany, cramp, symptomatic of metabolic imbalance, in which the muscles of hands and feet contract rhythmically. Spasm may also occur in the larynx, causing difficulty in breathing.

Tethys, outermost of the four inner satellites of the plant Saturn. Tethys lies at a distance of 294,760km (183,080miles) from Saturn and has a diameter of approx. 1,000km (600 miles). It has a mean sidereal period of 1.9 days *See also* SU pp.210, 213, *213*.

Tethys, Sea of, hypothetical sea or trough (geosyncline) which separated the Eurasian part from the African part of the supercontinent Pangaea during the Mesozoic period. The Mediterranean Sea may be regarded as a remnant of this geosyncline. *See also* PE p.26–27.

Tétouan (Tetuan), city in N Morocco. A corsair base in the 14th century, it was rebuilt by Jewish refugees from Spain in 1492. Touan was the former capital of Spanish Morocco (1913–56). Today it is a popular tourist resort. Pop. (1971). 139, 105.

Tetrachloromethane. *See* CARBON TETRACHLORIDE.

Tetracyclines, broad-spectrum antibiotics effective against anaerobic STREPTOCOCCI and certain bacterial and RICKETTSIAL disorders.

Tetraethyl lead, colourless oily poisonous liquid, $Pb(C_2H_5)_4$. It is manufactured either by treating an alloy of lead and sodium with ethyl chloride or by the action of lead chloride on a GRIGNARD REAGENT. It is widely used in petrol as an anti-knock agent. Properties: s.g. 1.65; b.p. 200°C (400°C).

Tetrahedrite, sulphide mineral; composed of varying amounts of copper, iron, zinc, silver, antimony and arsenic sulphides, found in medium- to low-temperature ore veins. It displays cubic system well-formed tetrahedral crystals and also appears as masses. In colour it is metallic grey to black. Hardness 3–5–4; s.g. 4.9. It is an important ore of copper.

Tetrode, ELECTRON TUBE, or valve, that has a fourth electrode, a screen grid, between the anode and the control grid; the cathode is the other electrode. Tetrodes are used in power amplification and oscillator circuits. *See also* SU pp.130–131.

Tetuan. *See* TÉTOUAN.

Teutonic knights, German military and religious order, whose Latin name was *Ordo Domus Sanctae Mariae Teutonicorum* (German order of the Hospital of St Mary). The order was founded *c.* 1190 during the Crusades. It transferred its activities to E Europe in 1211, and in 1233 began the conquest of Prussia. The knights became wealthy through engaging in trade, and their power spread to the lands of the E Baltic and into central Germany. Defeated by the Poles and Lithuanians in 1410, the order lost power and territory until its virtual disintegration in 1525, when its grand master accepted the Reformation and proclaimed Prussia a secular duchy. *See also* HC1 pp.163, *163*, *177*, *177*.

Te Wherowhero (1800–60), first MAORI king. Chief of the Ngati-Mahuta tribe, he was elected King in 1857, at a meeting at Ngaruawahia.

Te Wiata, Inia (1915–), MAORI opera singer. He studied music in London, and took leading roles at Covent Garden.

Tewkesbury, Battle of (1471), Yorkist victory over the Lancastrians which ensured that EDWARD IV retained the English throne. The army raised on HENRY VI's behalf by Queen Margaret and their son Edward was routed.

Texas, state in SW USA, on the Gulf of Mexico; second-largest state in the country. Eastern Texas has pine-covered hills and cypress swamps; cotton and rice are the main crops and the timber industry is important. The rich oil-fields of E Texas are the mainstay of the economy, however. The Gulf coast has popular tourist resorts. Cattle are raised on the plains of the Rio Grande valley, from where the land rises to the Guadalupe Mts of W Texas and the Great Plains area of the Texas Panhandle in the N. Cotton and cereal crops are grown and cattle are raised. Oil has revolutionized the economy in this section of the state also, and industry has developed rapidly. The major Texan industries are oil refining, food processing, aircraft and electronics. Major cities are Austin (the state capital), Houston, Dallas and San Antonio.

The Spaniards explored the region in the first half of the 16th century; it became part of the Spanish colony of Mexico. By the time Mexico attained independence, many Americans had begun to settle in Texas. They revolted against Mexican rule and in 1836, after defeating the Mexican army, established the Republic of Texas, which was recognized by the USA in 1837. Eight years later Texas was admitted to the Union. It joined the Confederate States in 1861 and played an active role in the AMERICAN CIVIL WAR. Area: 692,405sq km (267,338sq miles). Pop. (1975 est.) 12,237,000. *See also* MW p.175.

Texel, one of the West Frisian Islands, NW Netherlands. Largest of the group, it is a popular tourist resort. Area: 184sq km (71sq miles). Pop. 11,394.

Textiles, fabrics produced from fibres by SPINNING and WEAVING. Fibres can be of natural origin – WOOL, COTTON, LINEN or SILK – or synthetic, such as RAYON, NYLON and the POLYESTERS. Fabrics made by hand, as in LACEMAKING, or by machine KNITTING are not strictly textiles (which are woven) but are used in a similar way, especially for clothing and furnishings. The adaptation of machinery to spinning and weaving in the 18th century opened the way to the textile industry and most textiles are now produced in factories. In the 20th century advances in textile manufacturing have included the development of flame-resistant and "easy-care" fabrics that do not crease easily. *See also* MM pp.58–59.

Textured Vegetable Protein, substitute for meat made from vegetables. The protein and fat content of vegetables such as the SOYA BEAN is made into strands, which are knitted toghether to form a product whose texture is similar to that of meat.

Teyte, Dame Maggie (1888–1976), British soprano. She studied at the ROYAL COLLEGE OF MUSIC, London, and made her operatic debut in Monte Carlo in 1907. She was famous for her interpretation of songs by FAURÉ, CHAUSSON and BERLIOZ.

Thackeray, William Makepeace (1811–63), British novelist and satirist. He became famous with *Book of Snobs* and *Vanity Fair* (1847–48), a satirical novel on upper class London society at the start of the 19th century. His other works include *Barry Lyndon* (1844), *Pendennis* (1848–50) and *The Virginians* (1857–59). *See also* HC2 pp.98, *99*.

Thai boxing, spectacular form of boxing in which virtually anything seems allowed. The two gloved fighters in the five-round bouts may punch, knee, elbow, kick or leg throw each other, but use of the hips and shoulders is forbidden. Bouts are preceded by ritual ceremonies.

Thailand (Prathet Thai), kingdom in SE Asia; formerly known as Siam. Although Thailand has a fast-developing economy, it is still based largely on agricultural products. Rice accounts for 20% of exports. Rubber and maize are also important, and Thailand is the world's third largest producer of tin. Tourism is of increasing importance to the economy. The capital is Bangkok. Area: 514,000sq km (198,455sq miles). Pop. (1976 est.) 43,569,000. *See* MW p.166.

Thalamus, area of the forebrain immediately above the HYPOTHALAMUS. Sometimes called the sensory-motor receiving area, it obtains impulses from sensory NEURONS and sends them to other structures in the brain, particularly areas of the CEREBRAL CORTEX. *See also* MS pp.40–41, *41*.

Thalassemia, or Cooley's anaemia, hereditary genetic condition characterized by a deficiency of haemoglobin in the blood. It is prevalent in Italy, Greece, the Middle East, India, Thailand and China, and the mortality rate of victims is high.

Thalben-Ball, George Thomas (1896–   ), Australian organist and composer who spent most of his life in Britain. He was organist at the Temple Church, London (1919–30) and curator organist of the Royal Albert Hall in 1930.

Thalberg, Irving (1899–1936), US film producer. He was production head at UNIVERSAL PICTURES in 1919 and became president of the Mayer Company, later absorbed into MGM, in 1923. At the peak of his career he controlled all MGM productions and supervised the film career of his wife, Norma SHEARER. Although demoted in the early 1930s, Thalberg produced some of MGM's most memorable films, including *The Mutiny on the Bounty* (1935) and *Camille* (1936).

Thales (*c.*640 BC–*c.*546 BC), Greek scientist and philosopher. One of the first known Greek scientists, he founded the Ionian school of natural philosophy which held that water was the substance from which all matter was formed. His studies in geometry include the discovery that the angles at the base of an ISOSCELES TRIANGLE are equal. *See also* HC1 p.82.

Thalidomide, sedative and mild hypnotic drug whose use by women in early pregnancy was associated with the birth of babies with malformed and shortened limbs. It was extensively used in Europe from 1959 until 1961, and the incidence of malformations coincided with a decline in natural miscarriages. It is held that thalidomide caused the womb to retain fetuses that would normally have been rejected.

Thallium, metallic element (symbol Tl) of group IIIA of the periodic table, discovered in 1861. It is usually obtained as a by-product of processing zinc or lead sulphide ores. After roasting these ores thallium is found concentrated in the flue dust. Thallium is an extremely toxic compound. It is used in infra-red detectors and certain specialized glasses. Properties: at.no. 81; at.wt. 204.37; s.g. 11.85; m.p. 303.5°C (578°F); b.p. 1,457°C (2,655°F); most common isotope Tl^{205} (70.5%).

Thallophyte, obsolete term for a sub-kingdom of non-vascular plants containing primitive forms of plant life. They lack clearly differentiated roots, stems or leaves and range in size from one-celled plants to 61m (200ft) SEAWEEDS. Asexual reproduction is by spores and sexual reproduction is by fusion of GAMETES. Chlorophyll-containing thallophytes are ALGAE, Euglenoids, Dinoflagellates and LICHENS; thallophytes lacking chlorphyll include BACTERIA, FUNGI and SLIME MOULDS. *See also* THALLUS.

Thallus, non-vascular plant body of a THALLOPHYTE. Usually flat or ribbon-shaped, it is not differentiated into root, stem or leaves. Examples of plants with a thallus are ALGAE, euglenoids, LICHENS and FUNGI.

Thames, principal river in England. It rises in four headstreams – the Thames, Churn, Coln and Leach – in the Cotswold Hills, E Gloucestershire, then flows E across S England and through London to enter the North Sea at The Nore. The river is navigable by barges below Lechlade, and it is tidal to Teddington. Above London the river is used mainly by pleasure craft and for recreational purposes, such as fishing. The Thames Conservancy Board, established in 1857, controls the freshwater river. The Port of London Authority (1908) administers the river below Teddington, in particular the port of London proper from London Bridge to Blackwall. There the river is navigable for ocean-going vessels below Tilbury. Length: 338km (210 miles) miles).

Thamesdown, county district in NE Wiltshire, England; created in 1974 under the Local Government Act (1972). Area: 222sq km (86sq miles). Pop (1974 est.) 141,100.

Thanet, county district in NE Kent, England, created in 1974 under the Local Government Act (1972). Area: 103sq km (40sq miles). Pop. (1974 est.) 117,600.

Thani, Sheikh Khalifah bin Hamad al- (1934–   ), ruler of Qatar from 1972. He was formerly head of the Security Forces, Minister of Education, Finance Minister and Prime Minister.

Thanksgiving Day, national holiday in the USA. Originating with the PILGRIM FATHERS in 1621, when they celebrated the first harvest of the PLYMOUTH COLONY, it was proclaimed an official holiday in 1863. Since 1941 it has been celebrated on the fourth Thursday in November. Canada celebrates Thanksgiving on the 2nd Monday in October of each year.

Thant, U (1909–74), Burmese diplomat, Secretary-General of the United Nations from 1962 to 1972. He was involved in settling major disputes, including civil war in the Congo (Zaire) in 1963 and in Cyprus in 1964, but was less successful in dealing with the VIETNAM WAR and the Middle East crises. *See also* HC2 pp.246, *298*.

Thar desert (Great Indian Desert), region in NW India and SE Pakistan, between the Aravalli Mountains (E) and the River Indus (W). Area: 200,000sq km (77,000sq miles).

Thatcher, Margaret Hilda (1925–   ), British politician and leader of the Conservative Party. A research chemist and a barrister, she began her parliamentary career as MP for Finchley in 1959. She was chief opposition spokesman for education in 1969 and Secretary of State for Education and Science 1970–74, becoming Party leader in 1975.

Thatching, craft of laying and plaiting vegetable stalks and leaves to make a roof for a hut or a house. English cottages are thatched with straw or with reeds, Norfolk reed being the type particularly favoured. In tropical countries palm leaves are generally used for thatching. Rot-proof man-made fibres can also be used.

Theatre. *See* DRAMA.

Theatre, building where drama is staged. Its architecture has evolved gradually from early times when ritual was most often performed out-of-doors. In medieval Europe, PAGEANTS and churches were utilized as dramatic venues; MYSTERY PLAYS were enacted on open carts. Renaissance architects such as PALLADIO were commissioned to design private theatres with acoustics and perspective in mind. At the same time, popular OPEN STAGES had evolved in Shakespearean England. By the Restoration, however, the PROSCENIUM arch stage had become established as the only viable form of theatre. Pioneers such as William Poel (1852–1934) reacted against the limitations of late-Victorian realism, campaigning for open stages again to ensure that texts were not obscured by presentation. Since WWII, theatrical architecture has again stressed adaptability. The ROYAL SHAKESPEARE COMPANY redesigned their stages along open lines; at the NATIONAL THEATRE, the Cottesloe was designed as an experimental arena suitable for THEATRE-IN-THE-ROUND.

Theatre-in-the-round, form of theatrical presentation derived from the ancient arena stage. The audience is seated on all sides of the players, creating a sense of informality between the actors and the audience. This form of production has become popular in the 20th century.

Theatre of Cruelty, term adopted by Antonin ARTAUD that is often misunderstood. It was intended to embody Artaud's demand for a theatre in which the spectator became so involved with the characters that he left the theatre still sharing that experience.

Theatre of the Absurd. *See* ABSURD, THEATRE OF THE.

Theatre Workshop, drama company founded in 1945 by Joan LITTLEWOOD. It moved to the Theatre Royal, Stratford East, London, in 1953 and created controversy by staging a number of experimental and political plays. Littlewood was director until 1961 and it disbanded in 1964; it reformed in the early 1970s, under the direction of Gerald Raffles. Plays which it produced and transferred to the West End include *The Quare Fellow* (1956), *A Taste of Honey* (1958) and *Oh, What a Lovely War* (1963).

Thebaine, or paramorphine, ALKALOID drug, one of many present in OPIUM. It is a poisonous compound, with the formula $C_{19}H_{21}NO_3$, that may cause convulsions.

Thebes, city of ancient Greece; centre of Mycenaean power, which was destroyed *c.*1200 BC. Rebuilt, it again became powerful, forming alliances with Persia and later with SPARTA in the PELOPONNESIAN WAR. Thebes was victorious (371 BC) in a later clash with Sparta, but was almost destroyed in an uprising against ALEXANDER THE GREAT (336 BC) and fell to the Romans in 197 BC. *See also* HC1 pp.68, 71, 72, 80.

Thebes, ancient capital of Upper Egypt, called Wasi by the Egyptians. It flourished from earliest times, declining in the 10th century BC, and was sacked by the Romans in 30 BC. The well-preserved ruins of temples, obelisks and statues attract many tourists. *See also* HC1 pp.35, 49.

Thecodont, member of the Thecodontia, an order of early Mesozoic reptiles characterized by teeth set in deep sockets. *See also* NW p.184, *184*.

Theft, or larceny, dishonest appropriation of the property of another. A person is guilty of theft if he deprives another of property with the intention (or mens rea) of doing so.

Thegn, in Anglo-Saxon England, a man who held land from his lord in return for service and whose WERGILD was usually at least 1,200 shillings. The status was hereditary, and socially and economically varied – because a thegn's importance depended on that of his lord (who might be the king, an earl or even another thegn). All thegns were bound to serve in the king's forces. *See also* HC1 pp.154, 164.

Theiler, Max (1899–1972), US microbiologist, *b.* South Africa, who was awarded in 1951 the Nobel Prize in physiology and medicine for his research into YELLOW FEVER. In the 1920s he developed a weakened strain of the virus causing yellow fever; he infected monkeys with the disease, transmitted the virus to mice then back to monkeys. This produced only mild symptoms and conferred full immunity on the last monkey in the chain. His later improved vaccine was widely used against yellow fever. *See also* MS p.106.

William Thackeray's novels, especially *Vanity Fair*, are still widely read today.

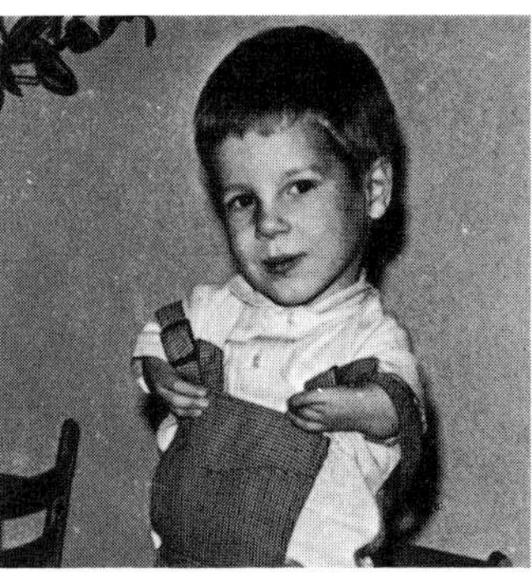

Thalidomide, taken by pregnant women, resulted in the birth of malformed babies.

Thames; a view of England's most important river flowing quietly past Pangbourne.

Margaret Thatcher acknowledges applause at the 1976 Conservative Party Conference.

Theism

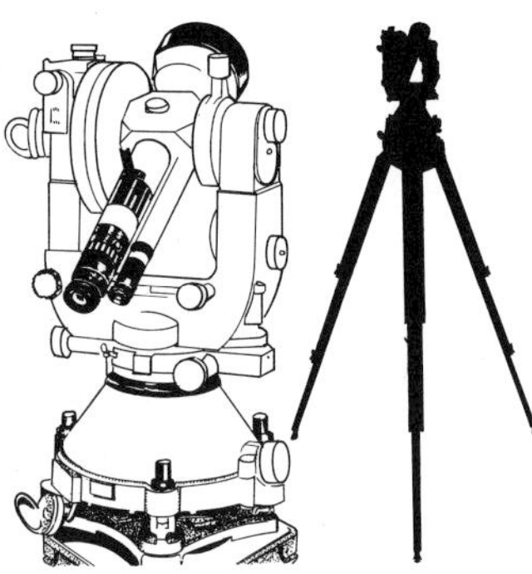

Theodolites are mounted on tripods and give accurate angle measurements.

Theodora: this detail from a Byzantine mosaic shows the first Empress Theodora.

Theodorakis combines traditional folklore and contemporary themes in his songs.

Thermae; remains of the rectangular Roman bathing pool at Bath, Somerset.

Theism, philosophical and theological systems which profess belief in the existence of one supreme being, who is the creator of all; he is perfect and merits man's worship. In most theistic systems man has a FREE WILL and religious doctrines are based on divine Revelation.

Thematic apperception test (TAT), in psychology, technique used to assess personality characteristics, introduced in 1935 by Henry A. Murray. The person to be tested is presented with a series of cards with pictures on them, and he or she must make up a story about the characters in each picture. The method assumes that people reveal their unconscious feelings and desires in their stories. *See also* MS p.*189.*

Themis, in Greek mythology, the goddess of justice, daughter of URANUS and Gaea. Although a TITAN, she was honoured by the gods on OLYMPUS for her wisdom and foresight. She was the mother of the Horae, goddesses of the seasons, and the FATES by ZEUS. Themis is depicted as a stern women bearing a pair of scales.

Themistocles (*c.*528 BC–*c.*460 BC), Athenian statesman. He built up a powerful navy, planned the evacuation of Athens when the Persians invaded and formulated the battle plan for Salamis. *See also* HC1 pp.70, *70–71.*

Thenard, Louis-Jacques (1777–1857), French chemist who discovered HYDROGEN PEROXIDE and the porcelain pigment Thenard's blue. He worked with GAY-LUSSAC and wrote the standard chemical textbook of that time.

Theocracy, government by religious leaders in accordance with divine law. Theocracies were common in primitive societies and existed in ancient Egypt and the Orient. The Puritan colony of New England (1620–60) was predominately theocratic.

Theocritus (*c.*300–260 BC), Greek poet, regarded as the father of pastoral poetry. His work is noted for its humour, vivid expression and perceptive portrayal of contemporary life.

Theodolite, surveying instrument dating back to the 16th century, used to measure horizontal and vertical angles. Its modern form consists of a telescope (with crosshairs in the lens for accurate alignment) mounted to swivel in both directions and levelled with a spirit level.

Theodora, name of three empresses of the BYZANTINE EMPIRE. Theodora (*c.*500–48), wife of JUSTINIAN I, became joint ruler on his accession in 527. Her courage saved the imperial throne for her husband during the Nika political and religious revolt of 532. The second Theodora (*d.*867) ruled for 14 years as regent for her young son MICHAEL III after the death in 842 of her husband, the Emperor Theophilus. In 843 she restored the cult of the ICONS. Another Theodora (980–1056), daughter of CONSTANTINE VIII, became co-Empress after the death of Zoë and Zoë's husband, CONSTANTINE IX Monomachus, in 1055. She was the last ruler of the Macedonian dynasty. *See also* HC1 p.*142.*

Theodorakis, Mikis (1925–), Greek composer. A student of MESSIAEN, his first work, the oratorio *Sinfonia*, was published in 1944. He has composed for films, theatre and ballet and has also written a number of song cycles, including *Ballads* (1975). He was elected to the Greek Parliament in 1964 and imprisoned (1967–70) for political activities.

Theodore, name of two popes. Theodore I (*r.*642–49), a Greek born in Jerusalem, led the opposition to the MONOTHELITE heresy and refused to recognize Patriarch Paul of Constantinople. Theodore II (*d.*897) was pope for less than three weeks in 897.

Theodore of Tarsus (*c.*602–690), Greek prelate and organizer of the Church in much of England. He was appointed Archbishop of Canterbury in 668.

Theodoric I (*d.*451), king of the Visigoths. The son of ALARIC, he ruled from 419 until his death in the Battle of Châlons-sur-Marne, while fighting with the Romans against the Hun army led by ATTILA.

Theodoric of Prague (*fl.* 4th century),
Bohemian artist who painted in a severe, realistic style that had a great influence on later German painting.

Theodoric the Great (*c.*454–526), Ostrogoth king who became ruler of Italy. Encouraged by Zeno, emperor of the Eastern Roman Empire, Theodoric defeated Odoacer, the German king of Italy, and in 493 established a peaceful reign based on Roman law and administration. *See also* HC1 p.137, *137.*

Theodorus of Samos (*fl.*6th century BC), Greek architect. He is reputed to have built a lathe for making columns. The Temple of Hera at Samos (*c.*560 BC) is his most famous building.

Theodulf (AD 750–821), Spanish poet and theologian. He was active at CHARLEMAGNE's court, writing two treatises, *De spiritu sancto* (Concerning the Holy Spirit) and *De ordine Baptismi* (Concerning the Ordinance of Baptism), for the emperor, for which he received the PALLIUM from Pope Stephen IV in 816. He was imprisoned by Louis I (the Pious) in 818 and died in captivity.

Theology, systematic, scientific investigation of the precepts of a religion, usually Christianity. It is intricately related to philosophical and historical studies and strives to achieve an understanding of various beliefs. It is concerned with concepts of a divine being, man, and ethnics.

Theophanes the Greek (*fl.*1378–1405), Byzantine painter. He painted frescoes and icons that combined the late Byzantine with the Russian style. His earliest surviving work is a wall painting for the Church of the Transfiguration, Novgorod.

Theophrastus (*c.*372 BC–*c.*286 BC), Greek philosopher and founder of botanical science. The leading pupil of ARISTOTLE, he succeeded him as head of the LYCEUM. His surviving works include *Historia Plantarum*, *On Metaphysics* and *Characters*. *See also* HC1 p.82.

Theorbo, large double-necked lute, which was commonly used from the 16th–18th centuries. It had up to eight strings over the fingerboard and three or four separate longer bass strings which were played unstopped. Surviving examples of the instrument are 1.2 to 1.8m (4–6ft) long.

Theorell, Axel Hugo Teodor (1903–), Swedish biochemist, Director of the Biochemical Department of the Nobel Medical Institute, Stockholm, from 1937 to 1970. His study of ENZYMES led to a better understanding of the metabolic processes involved in the production of energy, for which he was awarded the 1955 Nobel Prize in physiology and medicine.

Theorem, statement or proposition that is to be proved by logical reasoning from given facts and axioms. In geometry a proposition is considered as a problem (a construction to be effected) or a theorem (a statement to be proved).

Theory of the Earth. *See* UNIFORMITARIAN PRINCIPLE.

Theosophical Society, organization founded in 1875 to spread information about THEOSOPHY.

Theosophy, religious philosophy that originated in the ancient world but was given a new impetus in 1875 when the Theosophical Society was founded by Helena BLAVATSKY and Henry Olcott in New York. They subsequently moved to India (where the movement took root) and after their deaths their work was continued by Annie BESANT. Modern theosophy continues a mystical tradition in Western thought represented by such thinkers as PYTHAGORAS, PLOTINUS and NICHOLAS OF CUSA but is found, most significantly, in Indian thought where it has been transmitted through sacred books such as the UPANISHADS and BHAGAVAD-GITA. The main tenets of the Society lie in its affirmation of a single eternal principle, its advocation of the brotherhood of all men and the identity of each soul with the universal "over-soul" and of the eternity of the universe. Belief in the TRANSMIGRATION OF SOULS also occupies an important place in theosophical doctrine.

Thera, volcanic island in the Aegean Sea to the N of Crete. Although now quiet it is known to have erupted violently in the Bronze Age, spreading thick deposits of
ash over a wide area. *See also* PE p.*31.*

Therapy, in psychology, treatment of personality disorders which seem to be of a psychological origin. Analytic therapy, exemplified by the methods of JUNG, attempts to aid the patient in discovering deep repressed conflicts. Non-directive therapy utilizes the relationship which can grow between the therapist and patient to modify the patient's behaviour and relive guilt and insecurity by frank discussion, often in a group. Behaviour therapy does not seek causes but concentrates on modifying the patient's behaviour by procedures involving rewards and punishments. *See also* MS pp.146–149.

Theresa of Avila. *See* TERESA, SAINT.

Thérèse of Lisieux, Saint (1873–97), known also as Little Flower, or St Thérèse of the Child Jesus. Thérèse entered the Carmelite Convent at Lisieux at the age of 15. She suffered latterly from religious doubts, which she mastered by prayer. She was canonized in 1925. Feast: 3 Oct.

Therm, unit of heat used until recently, particularly, for expressing the calorific value of town gas; it is now obsolete. It is equal to 100,000 British Thermal Units BTU, also obsolete) or 25,200 kilocalories.

Thermae, Roman public baths. They became an important part of Roman civic architecture in the 1st century AD, and remains of these baths can be found in many Roman towns. They were often massive buildings and some included important works of sculpture and design. Typical thermae were the Baths of Carcalla and those at BATH, England. *See also* HC1 p.*94, 104,* 108.

Thermal, or thermal current, small-scale rising current of air produced by local heating of the Earth's surface. Thermals are often used by gliding birds and manned GLIDERS.

Thermal capacity, in physics, capacity of an object to absorb heat. It is the product of the object's mass and its specific heat.

Thermal power, literally, heat power. Any heat engine, for example a steam locomotive, employs thermal power. The heat generated by burning a fuel is converted into useful energy; in this case the energy drives the locomotive's pistons.

Thermic lance, long metal tube tipped with a nozzle, through which oxygen is fed at pressure. Its most important application is in the basic oxygen method of steelmaking, in which the lance, surrounded by a water jacket for cooling, jets oxygen on to a bath of molten steel in a furnace.

Thermidorian reaction, in the FRENCH REVOLUTION, period following the conservative coup of 9 Thermidor Year II of the Revolution (27 July 1794). The coup led to the fall of ROBESPIERRE and the suppression of the JACOBINS. *See also* HC2 pp.74–75.

Thermionic emission, emission of ELECTRONS from the surface of a substance as a result of heating it.

Thermionics, study of the emission of electrons or ions from a heated conductor. This is the principle of operation of the radio valve or electron tube, where the heated conductor is the CATHODE and the emitted electrons are the current carried by the valve. Such valves have now been superseded by TRANSISTORS. A more modern aim for thermionics is the design and construction of thermionic power generators, which will convert heat directly into electricity on a much larger scale than an electron tube.

Thermistor, type of semiconductor whose resistance sharply decreases with increasing temperature. At 20°C the resistance may be of the order of a thousand ohms and at 100°C it may be only 10 ohms. Thermistors are used to measure temperature and to compensate for temperature changes in other parts of a circuit.

Thermite process, smelting process in which a metallic oxide ore is reduced to the metal by heating with finely divided aluminium powder. It is the basis of the Goldschmidt process for extracting such high melting-point metals as chromium, manganese, molybdenum and vanadium. It can also be used for welding cracks in metal, such as large castings and rails.

Thermochemistry, branch of physical chemistry which deals with the heat effects that accompany chemical changes. Examples include the heat given off or absorbed during a chemical reaction, the dissolving of a substance, or changes of state, such as from a liquid to a gas.

Thermocline, middle layers of ocean water between surface and deep waters, which are defined by differing densities and temperatures. It is up to 1,000m (3,300ft) thick with a temperature only a few degrees above freezing. The thermocline is important as a stable boundary that tends to prevent interchange between layers. *See also* PE p.77.

Thermocouple, thermometer consisting of two wires of different alloys joined at one end and with their other two ends maintained at constant temperature. When the joined ends are inserted in the substance to be studied, heat flows at different rates along the two wires because of their different compositions. This difference can then be measured. *See also* SU pp.90–91.

Thermodynamics, the study of the heat content and interactions of systems. Historically, the laws of thermodynamics were developed from observation of large-scale properties of systems, with no understanding of the underlying atomic structure. It is now possible to calculate those laws from statistical and quantum mechanical principles. Thus the historical subject of thermodynamics is now subsumed in the disciplines of STATISTICAL MECHANICS or QUANTUM MECHANICS. There are three laws of thermodynamics. The first law, basically a restatement of the conservation of energy, states that for any substance a quantity called the internal energy can be defined such that the change in the internal energy is the sum of the work done on the system and the heat absorbed by it. The second law states that for any substance a quantity called ENTROPY can be defined such that (a) if the system is left alone, its entropy tends to increase, and (b) if the system absorbs heat, its entropy changes by the ratio of the heat absorbed to the temperature. The third law states that as the temperature approaches absolute zero, the entropy approaches zero. *See also* SU p.92.

Thermoelectricity, various means of converting heat energy into electricity or vice versa. There are three such effects. In the SEEBECK EFFECT the junctions of two metal wires forming a circuit are kept at different temperatures, and an electric current flows; this is the basis of THERMOCOUPLES. The PELTIER EFFECT acts in the reverse manner, and electrical energy is converted to heat energy. Lord KELVIN discovered the third thermoelectric effect (the THOMSON EFFECT): if a temperature gradient is maintained along a wire – a difference in temperature between both ends – it gives rise to a potential difference along the wire. If the current flows from the cooler to the hotter part of the wire, the effect is called the positive Thomson effect; the reverse is the negative Thomson effect.

Thermoelectric propulsion, any of several rocket propulsion systems combining heat and electrical means to accelerate particles to high velocities. An arc jet engine uses an electric arc to heat liquid hydrogen to 50,000°C; the resulting PLASMA of ionized hydrogen is accelerated through a conventional nozzle or, in the plasma engine, through a magnetic field for greater force. Ion rockets accelerate heavy charged particles such as caesium ions. All thermoelectric rockets produce low but long-lasting thrust – a requirement of interplanetary space travel.

Thermogram, photographic or other record of the heat radiated from an object, usually obtained with an infra-red camera. For example, a thermogram can show the different temperature zones of the surface of the human body, which may be useful in the diagnosis of various disorders. Or it may show isothermal contours of the Earth's surface, as recorded by an orbiting spacecraft. *See also* SU p.32.

Thermometer, instrument for measuring temperature. Any substance with physical properties (eg volume) that change with varying temperature may be used as a

measure, provided that the change is correctly monitored and scaled. The mercury thermometer, for example, depends on the expansion of the metal mercury which is held in a glass bulb connected to a narrow, graduated tube. Common scales are FAHRENHEIT, CELSIUS (Centrigrade) and KELVIN. *See also* SU p.90.

Thermonuclear reaction. *See* NUCLEAR ENERGY.

Thermoplastic, type of POLYMER that softens on being heated and can be repeatedly melted or softened by heat without change of properties. Typical examples are POLYETHYLENE, POLYSTYRENE and PVC. *See also* MM p.54.

Thermoplastic resin, synthetic plastic that softens when heated and returns to its original solid state when cooled. Examples include NYLON, PVC, POLYETHYLENE, fluorocarbons and POLYSTYRENE. *See also* MM pp.54–55.

Thermopylae, coastal mountain pass in E central Greece, between Mount Oeta and the s Malian Gulf, SE of Lamia. Its strategic location as an entry point to Greece made it the scene of many ancient battles. *See also* HC1 pp.61, 70, 70.

Thermoreceptors, in anatomy, sensory nerve endings, sited in the skin and in the deep body tissues, which detect changes in temperature and signal them to the brain. There are two types: receptors that are stimulated by cold and receptors responsive to heat.

Thermosetting resin, type of POLYMER that loses its plasticity under heat and pressure. Examples include polyesters, epoxy resins, phenolic resins and silicones. *See also* MM p.54.

Thermosphere, shell of light gases between the MESOSPHERE and the EXOSPHERE, between 100km (60 miles) and 450km (280 miles) above the Earth's surface. The temperature steadily rises with height in the thermosphere.

Thermostat, device for maintaining a constant temperature. A common type contains a strip of two metals, one of which expands and contracts more than the other. Thus the strip bends and breaks an electrical contact at a particular temperature. As it cools, the strip unbends, makes contact and the heating begins again. *See also* SU p.121.

Theropod, any of a group of bipedal carnivorous dinosaurs. They were either large, like TYRANNOSAURUS, or small and lightly built, like *Ornithomimus*. Order Saurischia; sub-order Theropoda. *See also* NW pp.184, 186, 187.

Theroux, Paul (1941–), US novelist, long resident in Britain. His novels, such as *Fong and the Indians* (1968) and *The Family Arsenal* (1975), have been overshadowed by his travel book, *The Great Railway Bazaar* (1975), which has won high praise.

Theseus, in Greek mythology, a great hero of many adventures, the son of Aethra, a princess of Troezen, by Aegeus, King of Athens, and the sea god POSEIDON. Theseus killed many villains (among them PROCRUSTES), but his most famous exploit was the vanquishing of the MINOTAUR of Crete.

Thesiger, Wilfred Patrick (1910–), British writer about travel, adventure and exploration. He explored the Danakil Desert (1933–34). In the Sudan he served in Political Service (1935–40) and during WWII served with the Special Air Service, where he attained the rank of major. His works include *Arabian Sands* (1959) and *The Marsh Arabs* (1964).

Thespis (6th century BC), Greek writer who according to tradition was the inventor of tragedy. He is also said to have introduced a character separate from the chorus, which had previously dominated the stage, who provided dialogue by responding to their comments. *See also* HC1 p.74.

Thessalonians, Epistles to the, two of St PAUL's earliest letters, written to the Church at Thessalonica which, together with neighbouring Christian communities, received encouragement and pastoral guidance from him. He taught them about the RESURRECTION of the dead at the SECOND COMING of Christ.

Thessaloníki (Salonica), port in N Greece, in Macedonia on the Gulf of Thessaloníki; second-largest city in Greece and capital of Thessaloníki prefecture. Founded c.315 BC, it flourished under the Romans after 148 BC as the capital of Macedon province. The city was second only to Constantinople in importance in the BYZANTINE EMPIRE. The city was conquered by Greece in 1913 during the BALKAN WARS. Exports include agricultural products, manganese and hides. An important commercial centre, it has textiles, soap, wine, cement and food processing industries. Pop. (1971) 345,799.

Thessaly (Thessalía), administrative district of N central Greece and region of ancient Greece. The centre of an extensive Neolithic settlement until c.2500 BC, it was cut off from much of the culture and politics of classical Greece. Thessaly was taken by the Macedonians in 352 BC, and became part of the province of Macedonia under Roman rule in 148 BC. It became part of the BYZANTINE EMPIRE with the decline of Roman rule, passed to the Turks in 1393. It was annexed to Greece in 1881.

Thiamine, VITAMIN of the B complex required for carbohydrate metabolism. Its deficiency causes the disease BERIBERI.

Thierry,   Jacques-Nicolas-Augustin (1795–1856), French historian. His reputation rests on two works, *La Conquête de l'Angleterre* (*The Norman Conquest of England*, 1825) and *Lettres sur l'Histoire de France* (*Letters on French History*, 1827).

Thiers, Louis Adolphe (1797–1877), French politician and historian. He was twice Foreign Minister under LOUIS PHILIPPE and lead the Liberal opposition to NAPOLEON III from 1863 to 1870. He was first President of the Third Republic (1871–73). He wrote a 10-volume *History of the French Revolution* (1823–27) and a 20-volume *History of the Consulate and Empire* (1840–55).

Thieu, Nguyen van (1923–), President of the Republic of Vietnam from 1967 to 1975. He participated in the overthrow of President DIEM in 1963 and ruled with KY until his election as president. Criticized for his refusal to accept a political settlement in the VIETNAM WAR, he resigned before the fall of Saigon and fled to Taiwan.

Thieving Magpie, The (1817), two-act opera by Gioacchino ROSSINI with libretto by Giovanni Gherardim. It was first produced at La Scala, Milan.

Thimmonier, Barthélmy (1793–1857), French tailor who patented the first sewing machine that was put to practical use. His patent was taken out in 1830 and it was a wooden machine which used a single thread.

Third man argument, philosophical argument proposed by ARISTOTLE as a criticism of PLATO's theory of FORMS or Ideas. Where Plato had posited the duality of Form and Matter, Aristotle claimed that the qualities that the two have in common can only be expressed in terms of a third concept. *See also* MS p.226.

Third Reich, period in German history from 1933 to 1945, during which Germany was under the totalitarian dictatorship of Adolf HITLER. The term was used to indicate a closeness with other great periods of German history – the First Reich of the Holy Roman Empire and the Second Reich (1871–1918), founded by Otto von BISMARCK.

Third Republic, form of government in France from 1870 to 1940. It was established after the collapse of the SECOND EMPIRE in the FRANCO-PRUSSIAN WAR and its constitution was drawn up as a compromise between conservative republicans and monarchists during the period 1873–75. It was marked by liberalism and a surge of imperial growth until 1914; after 1918 it was beset by economic problems and finally collapsed during WWII with the German invasion in 1940. *See also* HC2 pp.171, 180.

Third World, term of French origin which became fashionable in the 1960s to describe the countries of South America,

Thermoelectricity; this thermoelectric nuclear station is at Berkeley, England.

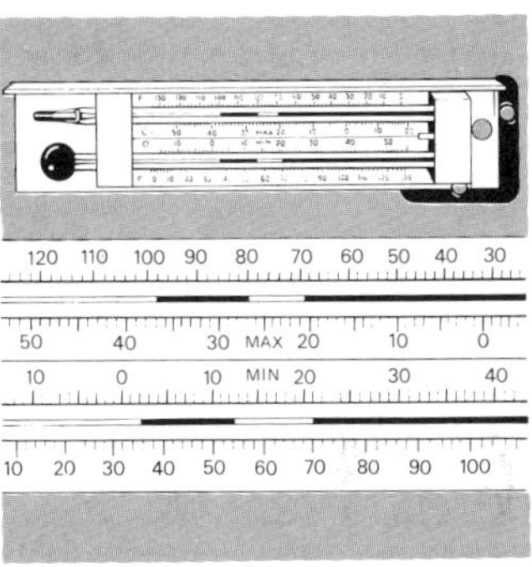

Thermometer; a maximum and minimum version records extremes of temperature.

Thermopylae; Leonidas lead a valiant defence against the Persians in 480 BC.

Thessaloniki's white tower is a well-known landmark on the city's seafront.

Thirteen Colonies

Dylan Thomas, whose native Wales often features in his poetry and prose.

Sir J. J. Thomson's discovery of the electron contributed to atomic research.

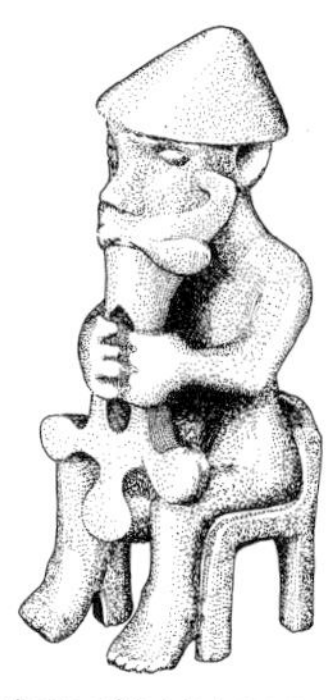

Thor, seen here with his hammer, was worshipped by all early Germanic races.

Thoreau is famous for practising his own form of natural philosophy.

Africa and Asia which are not part of either the Western or the Communist blocs. Many such countries share an under-developed, pre-industrial economy – and a resentment against their former colonial rulers. is felt more for the business interests of the USA than for Most of them are "non-aligned" with any of the recognized world powers. *See also* HC2 pp.247, 254–255, 302–303.

Thirteen Colonies, name given to the 13 territories that, together with Canada, made up British North America prior to the WAR OF AMERICAN INDEPENDENCE. After the war, the colonies declared themselves an independent nation and became the first states in the USA. They are CONNECTICUT, DELAWARE, GEORGIA, MARYLAND, MASSACHUSETTS BAY, NEW HAMPSHIRE, NEW JERSEY, NEW YORK, NORTH CAROLINA, PENNSYLVANIA, RHODE ISLAND, SOUTH CAROLINA and VIRGINIA. *See also* HC2 pp.64–65, 65.

Thirty-nine Articles. *See* ARTICLES OF THE CHURCH OF ENGLAND.

Thirty Years War (1618–1648), European war involving German PROTESTANT princes together with France, Sweden, England and Denmark against the HAPSBURGS and CATHOLIC princes of the HOLY ROMAN EMPIRE. In 1618 Bohemian Protestant princes revolted against the Catholic King Ferdinand (later Emperor FERDINAND II); they were defeated, but the N European Protestant kings of Denmark and Sweden intervened, and the general warfare gave countries such as France the chance to extend their influence. The war left Germany with its lands devastated, its economy in ruins and its population greatly reduced. The Holy Roman Empire fared hardly better, the Hapsburgs losing some of their power. After the war European conflicts arising from religious causes decreased, and the Peace of Westphalia (1648), formally ending the war, led to greater religious tolerance. *See also* HC1 pp.272–273.

Thistle, any of numerous species of plants with thorny leaves and yellow, white, pink or purple flower heads with prickly bracts. The field thistle, *Cirsium discolor,* resembles the heraldic thistle which is the national emblem of Scotland.

Thistle, Most Ancient and Most Noble Order of the, Scottish order of knighthood revived by James II in 1687, consisting of the sovereign and 12 knights (although only eight were nominated). In abeyance from 1689 to 1703, the order was revived again by Queen Anne. The number of companions was raised to 16 from 1827. The order has a chapel in St Giles Cathedral, Edinburgh.

Thistlewood, Arthur (1770–1820), British revolutionary. Fired with radical ideas after visiting Paris and the US, he led the 1820 CATO STREET CONSPIRACY and was hanged for treason.

Tholos, type of monumental stone-built tomb of Mycenaean Greece (1600–1100 BC), sometimes known as "bee-hive tombs" in reference to their circular, dome shape. The finest example, covered with an earth mound and approached by a walled passage, is the "treasury of Atreus" at Mycenae. *See also* HC1 p.41.

Thoma, Hans (1839–1924), German painter. After 1868, his style was influenced by COURBET and the BARBIZON painters. His work includes landscapes, especially of the Black Forest, genre paintings and portraits.

Thomas, Saint, one of the twelve APOSTLES of Jesus Christ. In the Gospel of St John he is called Twin (Greek *Didymus*). He has been called "Doubting Thomas" because, after the RESURRECTION of Christ, he disbelieved the appearance to the other disciples of the risen Lord (John 20: 24-28). The *Gospel of St Thomas* (which is not a part of the CANON of scripture) and three other apocryphal works bearing his name were written well after his time.

Thomas, Dylan Marlais (1914–53), Welsh poet whose powerful verse, although sometimes obscure, is immensely popular. It is at its best in *Deaths and Entrances* (1946). His prose works are collected in such volumes as *Adventures in the Skin Trade* (1955). He wrote the radio play *Under Milk Wood* (1954) and his collected poems were published in 1952. *See also* HC2 p.286, *286.*

Thomas, John. *See* CHRISTADELPHIANS.

Thomas, Philip Edward (1878–1917), British poet who began his career writing essays, travel books and criticisms. Encouraged by Robert FROST, he started writing poetry, mostly about nature, in 1914. His first volume *Poems* was published shortly before he was killed in WWI.

Thomas, Sidney Gilchrist (1850–85), British metallurgist. In 1875 he invented a method, known as the basic OPEN HEARTH PROCESS, for removing phosphorus impurities from iron ore during conversion in the BESSEMER PROCESS.

Thomas à Kempis. *See* KEMPIS, THOMAS À.

Thomas Aquinas, Saint. *See* AQUINAS, SAINT THOMAS.

Thomasites, religious group which later adopted the name CHRISTADELPHIANS, founded in the US in *c.* 1848 by John Thomas. Christadelphians believe in using the Bible as the exclusive source of moral guidance.

Thomism, philosophy of St Thomas AQUINAS one of the major systems in SCHOLASTICISM. Aquinas blended the philosophy of ARISTOTLE with Christian theology. Using Aristotle's concept of matter and form, he conceived a hierarchy in which spirit is higher than matter, soul higher than body, and theology above philosophy.

Thompson, Benjamin, Count Rumford. *See* RUMFORD, BENJAMIN THOMPSON, COUNT.

Thompson, David (1770–1857), British-born Canadian fur trader and explorer. He joined the HUDSON'S BAY COMPANY in 1784, studied surveying and journeyed to Lake Athabasca and the MISSISSIPPI headwaters. In 1807 he established the first trading post on the Columbia River and later charted the US-Canada boundary.

Thompson, Sir Edward Maunde (1840–1929), British palaeographer who was principal librarian and director of the BRITISH MUSEUM (1888–1909). His published works include *An Introduction to Greek and Latin Palaeography* (1912).

Thompson, Francis (1859–1907), British poet. First trained for the Roman Catholic priesthood, he later studied medicine unsuccessfully and became destitute in London until he was found by Alice Meynell. *Poems* (1893) includes his best-known work *The Hound of Heaven.*

Thompson, Sir John Sparrow David (1844–94), Canadian politician. A practising lawyer, he entered politics in 1877. In 1885 he was appointed Minister of Justice; he became Prime Minister in 1892 and died in office.

Thomsen, Christian Jürgensen (1788–1865), Danish museum curator and prehistorian. As head of the National Museum of Denmark (1816–65), he was the first to propose the division of prehistoric remains into the categories STONE AGE, BRONZE AGE and IRON AGE.

Thomson, Sir Charles Wyville (1830–82), Scottish naturalist. He was appointed director of the CHALLENGER voyage (1872–76). Thomson published *The Depths of the Sea* (1873), and an account of his expedition, *The Voyage of the Challenger (1877).*

Thomson, Elihu (1853–1937), US electrical engineer and inventor, b. Britain. With Edwin J. Houston he developed an arc lighting system that led to the founding of the American Electric Company, which later became the Thomson-Houston Electric Company. This merged in 1892 to form the General Electric Company (GEC). Thomson invented the first high-frequency dynamos and transformers.

Thomson, Sir George Paget (1892–1975), British physicist (son of Sir J. J. THOMSON) who shared the 1937 Nobel Prize in physics with Clinton DAVISSON. In 1927 both men, working independently, succeeded in diffracting ELECTRONS, thus confirming their wave nature first predicted in 1923 by Louis DE BROGLIE. Thomson fired fast electrons through thin gold leaf to obtain a diffraction pattern.

Thomson, James (1700–48), Scottish poet. His most famous poem *The Seasons* was published in four parts; *Winter* (1726), *Summer* (1727), *Spring* (1728) and *Autumn* (1730). His sensitivity to nature throughout his work makes him a forerunner of ROMANTICISM.

Thomson, Jeffrey (1944–), Australian cricketer. A fast bowler with a powerful action but sometimes erratic delivery, Thomson made his Test debut in 1974 in the series against England, and in 1977 took his 100th Test wicket in only 22 Test matches.

Thomson, Joseph (1858–95), Scottish explorer. He made his first African journey in 1878 to explore Lake TANGANYIKA. On subsequent expeditions, he advanced British commercial interests in South Africa. His books include *To the Central African Lakes and Back* (1881).

Thomson, Sir Joseph John (1856–1940), British physicist. His work on the conductivity of gases led him to the discovery of the ELECTRON in 1897 and he made the Cavendish laboratory at Cambridge University a world-famous centre of atomic research. He was awarded the 1906 Nobel Prize in physics for his investigations into the electrical conductivity of gases. *See also* SU pp.62, 63.

Thomson, Peter (1929–), Australian golfer. He won the British Open Championship in 1954, 1955, 1956, 1958 and 1965, and the Australian Open in 1951, 1967 and 1972.

Thomson, Virgil (1896–), US critic and composer who was music critic for the New York *Herald Tribune* (1940–54). Much influenced by Erik SATIE, his works include the operas *Four Saints in Three Acts* (1928, first production 1934) and *The Mother of Us All* (1947) with libretti by Gertrude STEIN.

Thomson, William, Lord Kelvin. *See* KELVIN, WILLIAM THOMSON, 1ST BARON.

Thomson effect, potential difference which develops between two points on a metal conductor if the two points are maintained at different temperatures.

Thomsonite, mineral, hydrated aluminium sodium silicate of the zeolite group found in cavities of basaltic rocks. Its crystals are of the orthorhombic system and it is snow-white when pure. Hardness 5–5.5; s.g. 2.2.

Thomson of Fleet, Roy Herbert, Lord (1894–1976), Canadian newspaper proprietor. He established a radio station in 1931 and subsequently took over a weekly newspaper. Soon he had a chain of newspapers and radio stations, becoming a millionaire. In Britain he acquired control of the *Scotsman* and Kemsley newspapers, and from 1966 controlled *The Times.*

Thor, in North European mythology, god of thunder and lightning, corresponding to JUPITER. The eldest and strongest of ODIN's sons, he was represented as a handsome red-bearded warrior, benevolent towards men but a mighty foe of evil. A magic belt doubled his strength.

Thorax, in animal anatomy, the part between the NECK and the ABDOMEN. In mammals it is formed by the RIB cage and contains the LUNGS, the HEART and the OESOPHAGUS. It is separated from the abdomen by the DIAPHRAGM. In insects it consists of several segments to which legs and other appendages are attached.

Thorburn, Archibald (1860–1935), British naturalist, painter, writer and illustrator, who specialized in detailed paintings of British wildlife, particularly of game birds and birds of prey. *A Naturalist's Sketchbook* was published in 1919.

Thoreau, Henry David (1817–62), US writer and naturalist, a follower and friend of Ralph Waldo EMERSON. An ardent individualist, he experimented in living a near-solitary life, rejecting materialism and finding fulfilment in observing plant and animal life. He built himself a cabin at Walden Pond, near Concord in Massachusetts and wrote *Walden,* an account of his life there, in 1854. His essay "Civil Disobedience" (1849) has influenced many passive resistance movements. He was active in the anti-slavery movement in the 1850s. *See also* HC2 p.288.

Thorium, radioactive metallic element (symbol Th) of the ACTINIDE ELEMENTS, first discovered in 1828. The chief ore is monazite (phosphate). The metal is used in photoelectric and thermionic emitters. Chemically reactive, it burns in air but reacts slowly in water. Properties: at.no. 90; at.wt. 232.038; s.g. 11.72; m.p. 1,750°C (3,182°F); b.p. 4,790°C (8,654°F); most stable isotope Th^{232} (1.41 × 10^{10} yrs).

Thorn apple, plant of the genus *Datura*, especially Jimson weed (*D. stramonium*), a poisonous, annual weed of tropical American origin. It has foul-smelling leaves and large white or violet trumpet-shaped flowers that are succeeded by round prickly fruits Family Solanaceae..

Thorndike, Dame Sybil (1882–1976), British actress. She played more than 100 Shakespearean roles and her performances in Euripides's *Medea* and *The Trojan Women* brought her critical acclaim. She also starred in the first production of SHAW's *St Joan* (1924), and in Clemence Darie's *Eighty in the Shade* (1959). She often appeared on stage. In 1969 she performed at the opening of the theatre in Leatherhead named after her.

Thornett, John (1930–), Australian Rugby Union forward who led the WALLABIES during one of their finest eras. As a lock and later in the front row, he won 37 caps (1955–67).

Thornhill, Sir James (1675–1734), British artist and the first British fresco painter in the BAROQUE manner. He decorated the Painted Hall at Greenwich Hospital (1708–27) and the dome of St Paul's Cathedral (1715–17). He also painted portraits and several altar-pieces.

Thóroddsen, Jón (1818–68), Icelandic novelist. He wrote some lyrics and drinking songs, but is chiefly remembered for his novels which include *Lad and Lass* (1850), and the unfinished *Man and Woman* (1876). These works are generally thought to be among the finest fictional Icelandic society.

Thorpe, James Francis (1888–1953), US athlete. Part Sac-fox Indian, he won both the pentathlon and decathlon in the 1912 Olympic Games. A year later, however, he was forced to give up his medals when it was discovered he had played semi-professional baseball. He played baseball in the National League with three different teams (1913–19).

Thothmes. *See* THUTMOSE.

Thousand Islands, group of more than 1,500 islands in the St Lawrence River, extending about 128km (80 miles) along the US-Canada border. Most of the islands belong to Canada, many being forested and popular tourist resorts; others are privately owned. The largest island is Wolfe, with an area of 127sq km (49sq miles).

Thrace, region of SE Europe, now partly in Greece, Turkey and Bulgaria. The Thracians had a flourishing and rich culture. The region came under Macedonian control in 342 BC, and was created a Roman province in AD 46. From the 7th century it was divided between the Byzantine empire and the Bulgar kingdom, eventually falling to the Ottoman Turks in 1453. During the 19th century, Russia tried to extend its influence over Thrace, but the region was divided up amongst its neighbours in the treaties following WWI.

Threadworm, small ROUNDWORM of the phylum Aschelminthes. It is commonest in moist tropical regions and resembles a short length of hair or thread. It may inhabit the intestines of human beings, but can live and breed freely in soil. Species *Oxyurus vermicularis. See also* NW pp.72, 89.

Three-Cornered Hat, The (1919), comedy ballet in one act, choreographed by Leonid MASSINE to music by Manuel de FALLA. The libretto is based on a story (1874) by ALARCÓN. It was first peformed in London, with settings by PICASSO, by Diaghilev's BALLETS RUSSES. The plot concerns a governor's amorous intentions towards a miller's wife.

Three-cushion billiards. *See* BILLIARDS.

Three-day event. *See* HORSE TRIALS.

Three-dimensional films, special-effect films produced in Hollywood which, when viewed through tinted paper glasses, gave

the illusion of depth. One of the more famous productions was *House of Wax* (1953). Their popularity waned when the WIDE SCREEN systems arrived.

Three Estates, classés of citizens in early modern Europe granted a share in the discussion of matters of state. In France the ESTATES-GENERAL, first called in 1302, were the clergy, the nobility and the common people. In England they were the lords spiritual, the lords temporal and the commons (knights of the shires and burgesses).

Three Musketeers, The (1844), historical novel by Alexandre DUMAS (père), containing the characters Athos, Porthos, Aramis and d'Artagnan.

Three Sisters, The (1900), four-act drama by Anton CHEKHOV. The play concerns three sisters who spend a tedious existence in the country and never satisfy their longing to go to Moscow.

Threonine, water-soluble, crystalline essential AMINO ACID found in PROTEINS. It was isolated in 1932.

Threshold, in physiology, theoretical point at which a stimulus can be detected (absolute threshold) or at which it is possible to discriminate between stimuli (difference threshold). Such thresholds are sometimes called limens.

Thrift, any of several species of small, evergreen perennial plants that grow throughout the world in sandy soil. It has blue-green leaves and clusters of white, red or pink flowers. Height: to 30cm (12in). Family Plumbaginaceae; genus *Armeria. See also* NW p.234.

Thrip, any of numerous species of slender, sucking insects found throughout the world. Species vary in colour, but most feed on plants and some carry plant diseases. Often found in large numbers, they have simple fringed wings. Length: to 8mm (0.3in). Order Thysanoptera. *See also* NW pp.106–107, *106–107*, 111.

Throat. *See* PHARYNX.

Thrombin, blood ENZYME that converts FIBRINOGEN to fibrin during the formation of blood clots. *See also* MS p.115.

Thrombocyte. *See* PLATELET.

Thrombophlebitis, condition involving blood clot formation and inflammation of the veins. It may be treated with anticoagulants and hot compresses.

Thromboplastin, substance, present in tissue, which is believed to activate the conversion of prothrombin and calcium to THROMBIN. *See also* BLOOD CLOTTING; MS p.115.

Thrombosis, formation or presence of a blood clot in the circulatory system. It may be due to an injury to the lining of the heart or blood vessels, alterations in normal blood flow and changes in the ability of the blood to coagulate. *See also* MS p.90, *90*.

Thrombus, clot of blood formed within a blood vessel. Unlike an Embolus, it remains attached to its place of origin, and is therefore potentially less dangerous.

Through the Looking Glass (1872), story by Lewis CARROLL, sequel to *Alice's Adventures in Wonderland*. It concerns the same child heroine who dreams that she passes through a mirror into an entirely back-to-front world.

Thrush, any of numerous species of small songbirds of the family Turdidae, which is represented throughout the world. The European song thrush (*Turdus philomelos*) is mottled brown with a lighter, speckled breast. It feeds mainly on insects and fruit and builds a cup-shaped nest in a tree. Length: to 30cm (12in). *See also* NW p.146.

Thrush, or moniliasis, fungal injection of the mucous membranes of the mouth, bronchi, lungs or vagina. Caused by the fungus *Candida albicans* it responds well to treatment, but is rarely serious in otherwise healthy people. *See also* MS p.98, *98*.

Thrust, force needed to drive an Aircraft or ROCKET. For aircraft it is provided by a propeller or jet engine. A propeller imparts thrust from its aerodynamic lift; a jet or rocket derives its thrust from the expulsion, at high velocities, of exhaust gases.

Thucydides (*c.*460–*c.*400 BC), ancient Greek historian. After commanding an

unsuccessful expedition in 424 BC to Amphipolis in the PELOPONNESIAN WAR, he went into exile (423–403) and wrote his *History of the Peloponnesian War.* It marked the beginning of a new style of history in that it viewed events in terms of the general nature and behaviour of man rather than as the result of fate. *See also* HC1 p.75.

Thugs, originally members of a secret society in India who would kill in honour of KALI, the Hindu goddess of destruction; they strangled their victims. Beginning in the 1830s the British stamped out the Thugs through arrests and executions, and by 1848 the menace had ended. The term is now used to denote any brutal robbers or gangsters.

Thulium, metallic element (symbol Tm) of the LANTHANIDE SERIES first discovered in 1879. Its chief ore is monazite (a phosphate). The element has few commercial uses. Properties: at.no. 69; at.wt. 168.9342; s.g. 9.31 (25°C); m.p. 1,545°C (2,813°F); b.p. 1,727°C (3,141°F); most stable isotope Tm^{169} (100%).

Thumbscrew, instrument of torture used to compress and crush the victim's thumbs. It is reputed to have been introduced in the 17th century by James, the Fourth Earl and First Duke of Perth.

Thunder Bay, city in sw Ontario, Canada, on Thunder Bay, an inlet of Lake Superior. It is a major Canadian port and an important shipping point for a rich mining region. The city is the seat of Lakehead University (1965). Industries: shipbuilding, grain milling, paper. Pop. (1974) 108,415.

Thunderstorm, electrical storm which is commonly experienced as lightning and thunder. Thunderstorms are caused by the separation of electrical charges in clouds. Water drops are carried by updraughts to the top of a cloud, where they become ionized and accumulate into positive charges – the base of the cloud being negatively charged. An electrical discharge (a spark) between clouds, or a cloud and the ground, is accompanied by light (seen as a lightning stroke) and heat, which expands the air explosively and causes it to reverberate and produce sounds and echoes called thunder. Thunderstorms are usually accompanied by heavy rain; ozone and the oxides of nitrogen are produced in the air. *See also* PE pp.69.

Thurber, James Grover (1894–1961), US artist and writer. He began his career as a newspaper reporter and in 1927 became a regular contributor of essays, short stories and cartoons to the *New Yorker* magazine. A humorist, his work is ironic and satiric. Collections of his essays and stories include *My Life and Hard Times* (1933), *Fables for our Time* (1940), *My World and Welcome to It* (1942).

Thuringia, historic region of Germany, now in sw East Germany. It became a frontier state in CHARLEMAGNE's empire, and its rulers became powerful princes with the HOLY ROMAN EMPIRE in the 11th century. In the 15th century most of the region passed to the Wettin family. In the 16th century it was one of the first states to declare itself protestant. Thuringia was reconstituted as a state (*Land*) in 1920 under the Weimar republic, but it lost its separate identity in 1952. *See also* HC2 p.*110.*

Thurrock, county district in central s ESSEX, England, created in 1974 under the Local Government Act (1972). Area 163sq km (63sq milles). Pop. (1974 est.) 126,800.

Thursday, fifth day of the week. The name derives from THOR, the Germanic God.

Thus Spake Zarathustra. *See* ALSO SPRACH ZARATHUSTRA.

Thutmose, royal name of four kings of the XVIIIth dynasty of Egypt (*c.*1525–1398 BC). The most significant achievement of Thutmose I (*r.c.*1515 BC) was the subjugation of the Nile Valley up to the 3rd cataract. Thutmose II married his half-sister HATSHEPSUT, who in effect ruled Egypt (*c.*1503–*c.*1482 BC). Thutmose III (*r.c.*1504–1450 BC) ruled at first with Hatshepsut, and extended his empire from the 3rd cataract to the Euphrates. Thutmose

Dame Sybil Thorndike as St Joan in an early production of Shaw's play.

The Three Musketeers is set in 17th-century France at the time of Richelieu.

Thrushes lay three to six eggs in their open, cup-shaped nests.

Thurber's writing and cartoons comment satirically on modern American life.

Tiber flows through Rome. In the background is the Castel Sant'Angelo.

Tiberius; most of his first 20 years of manhood were spent under arms.

Tibetan terriers were considered holy and were given to people for luck.

Tiepolo filled his ceiling paintings with floating, baroque figures.

IV was pharaoh 1425–1416 BC. *See also* HC1 p.48.

Thylacine. *See* TASMANIAN WOLF.

Thyme, aromatic garden herb of the MINT family used as an ornamental plant and in cooking. It yields an oil from which the drug thymol is prepared. It has purplish flowers and curled leaves. Height: 15–20cm (6–8in). Genus *Thymus. See also* NW p.*60*; PE pp.212, 213.

Thymine, in molecular biology, one of the four nitrogen bases in the nucleic acid DNA (the others are ADENINE, GUANINE and CYTOSINE). In RNA thymine is replaced by uracil in the base sequence.

Thymus gland, one of the ENDOCRINE, or ductless, glands, located in the upper chest in mammals; it is large in infancy and shrinks after puberty. It is thought to be a lymphoid organ, secreting a HORMONE that acts on LYMPH tissue and producing ANTIBODIES that function in the body's IMMUNE SYSTEM. Disorder of the thymus is often associated with auto-immune diseases (diseases in which the body produces antibodies that attack its own tissues). *See also* MS pp.64–65.

Thyroid gland, H-shaped ENDOCRINE (ductless) gland lying in the neck region, along the sides of and over the trachea (windpipe), below the Adam's apple. It secretes the hormone THYROXINE, which is essential for the growth and development of the body and brain and for the regulation of metabolism. A lack of thyroxine can produce CRETINISM in children, and MYXOEDEMA in adolescents and adults. Overactivity of the gland, known as Graves disease or thyrotoxicosis, produces weight loss, bulging eyeballs and irritability. *See also* GOITRE; MS pp.64–65.

Thyroid stimulating hormone (TSH), hormone secreted by the frontal (anterior) lobe of the PITUITARY GLAND. It stimulates the THYROID GLAND in the neck to produce the hormone thyroxine, which help control metabolism. When the level of thyroxine in the blood is sufficiently high, it shuts down the production of TSH by the pituitary. *See also* MS pp.64–65.

Thyroxine, hormone secreted by the THYROID GLAND. It contains iodine, and helps regulate the rate of metabolism; it is essential for normal growth and development. *See also* MS pp.64–65.

Thysanoptera, order of 5,000 species of small narrow-bodied insects, which have piercing mouthparts for feeding on plant juices. They are commonly called THRIPS, and may have wings or be wingless. Length: to 15mm (0.6in). *See also* NW pp.106–107.

Thysanura, order of 350 species of primitive wingless insects, commonly called bristletails and including the SILVERFISH. They have elongated scaly bodies, long antennae and bristle-like appendages on the lower abdomen. Subclass Apterygota. *See also* NW pp.106–107.

Tiahuanaco, archaeological site on the s shore of Lake TITICACA in the Bolivian highlands. Tiahuanaco, which flourished AD 600–1000, was a ceremonial centre and possibly the capital of pre-Inca empire. *See also* HC1 pp.233, *233.*

Tianjin, or Tientsin, administratively independent city in the east central part of Hopeh province, China. It is China's third largest city, an international port, and a major centre for heavy industry. Pop. (1970 est.) 4,280,000.

Tibaldi, Pellegrino (1527–96), Italian painter and architect. A follower of MICHELANGELO, he worked mainly in Bologna (his home town), Milan (from 1560) and Spain from 1588. He helped to design Milan Cathedral (1567–76) and was instrumental in the introduction of MANNERISM to Spain.

Tiber, second-longest river in Italy. It rises in the Etruscan Apennines, flows s then sw through Rome and reaches the Tyrrhenian Sea at Ostia, sw of Rome. It is linked to the River Arno by the Paglia River and the Chiana Canal. The upper Tiber is used to generate electricity. Length: 404km (251 miles).

Tiberias, lake port in NE Israel, 48km (30 miles) E of Haifa on the w shore of the Sea of Galilee. The city was founded by the Romans in AD 19 and named after the reigning emperor TIBERIUS. The city was taken by the Crusaders in 1099, and by the Ottoman Empire in the 116th century. It is the site of many ancient synagogues, and the tomb of MAIMONIDES. An agricultural trade centre, Tiberias is one of Israel's major tourist resorts. Pop. (1970 est.) 23,900.

Tiberius (42 BC–AD 37), Roman emperor (AD 14–37). He governed Transalpine Gaul and fought in Germany and Illyricum. He was adopted by AUGUSTUS, whose heirs had died, and became emperor at Augustus' death. He pursued a peaceful foreign policy and accumulated great wealth. He retired from Rome in AD 26, fearing plots upon his life. *See also* HC1 p.98.

Tibesti, mountain range in NW Chad, in the central Sahara. The highest peak is Emi Koussi, 3,415m (11,204ft).

Tibet (Xizang Zizhiqu), high plateau in central Asia bordered by Chinese provinces (N and E), and India (s and w). In 1965 it was declared the Tibet Autonomous Region. Tibet contains the world's highest mountains, the Himalayas, and the average altitude of the country is more than 4,575m (15,000ft). Asia's largest rivers, including the Yantze, Mekong, Indus and Brammaputra, all rise in Tibet, although there is a marked variation in rainfall distribution on either side of the Himalayas. The Tsangpo valley is the only agricultural area; many of the people still engage in nomadic pasturalism and large deposits of gold and copper remain unexploited. The capital is Lhasa. Tibet flourished as a kingdom in the 7th century AD, coming under Mongol rule 13th–18th century. LAMAISM developed in the 11th century under the influence of Indian Buddhist immigrants and the first DALAI LAMA was crowned in 1641. As Mongol power waned China claimed sovereignty of the region; Britain as the dominant power in neighbouring India recognized this in 1907. The Tibetans reasserted their independence in 1912 but the Chinese invaded in 1950. The monastic orders were suppressed and in 1959 the Dalai Lama fled to India. Border areas remain in dispute. Area: 1,222,070sq km (471,841sq miles). Pop. 1,400,000.

Tibetan terrier, terrier-like dog that originated in the Lost Valley of Tibet more than 2,000 years ago. Height: to 38cm (16in) at the shoulder; weight: to 14kg (30lb).

Tibia, larger of the two lower leg bones. It articulates with the femur, or upper leg bone, at the knee and extends to the ankle, where its lower end forms the projecting ankle bone on the inside of the leg. *See also* FIBULA; MS pp.56–57.

Tic, sudden and rapidly repeated muscular contraction, limited to one part of the body, especially the face. Usually involuntary, the condition may also be called a nervous tic.

Tick, any of numerous species of wingless, bloodsucking ARACHNIDS, the most notable of which are ectoparasites of vertebrates and invertebrates. Many species carry diseases (some fatal) in both wild and domesticated animals, and in man. Length: to 3mm (0.1in). Class Arachnida; order Acarina.

Ticonderoga, village in NE New York State, USA, the scene of several incidents in the FRENCH AND INDIAN WAR and the War of AMERICAN INDEPENDENCE. On the route from Canada to New York, and between lakes George and Champlain, the site was defended by the French from 1755, captured by the British in 1759 and taken by the American insurgents in 1775. In spite of the efforts of Benedict ARNOLD, General BURGOYNE took the fort in 1777, on his way to eventual defeat at SARATOGA.

Tidal bore, flow of tidal water from the sea into a funnel-shaped river mouth or estuary which, opposing out-flowing river water, builds into a surface "wall" that accelerates upstream. Notable examples occur in the rivers Amazon and Severn, with bores reaching 5m (15ft) in height and travelling at 15 to 25km/h (9–15mph).

Tidal power, energy harnessed and used by man from tidal movement of the Earth's oceans. This can be considered only where the tidal range is greater than about 4.6m (15ft). Modern schemes involve the use of turbo-generators driven by the passage of water through a tidal barrage. The La Rance power plant in N France has been working successfully since 1966 and in 1969 a small tidal power plant was completed by the USSR on the White Sea. Special turbines have been developed that can be driven by water flowing in either of two directions. *See also* MM p.*85*; PE p.*142.*

Tidal wave, or tsunami, large wave caused by underwater disturbances such as EARTHQUAKES or landslides. It can travel at great speeds in deep water with little surface disturbance, but builds up into a powerful, destructive wave when slowed down by the shallow water around landmasses. *See also* PE pp.28, 80, *81.*

Tide, periodic rise and fall of the surface level of the oceans caused by the gravitational attraction of the Moon and Sun. Tides follow the Moon's cycle of 28 days so they arrive at a given spot 50 minutes later each day. When the Sun and Moon are in conjunction or opposition, the greatest tidal range occurs, called spring tides. When they are in quadrature, when the Moon is half-full, tidal ranges are lowest and are called neap tides. *See also* PE pp.80–81.

Tie-beam, wooden or metal strut used to hold parts of a building or a roof TRUSS together. In the Middle Ages tie-beam roofs were built, particularly in England; St Martin's Church, Leicester, is a typical example.

Tieck, Johann Ludwig (1773–1853), German ROMANTIC poet, novelist and dramatist. The fairytale *Der blonde Eckbert* (1797) ranks among his best works and the novels *William Lovell* (1795–96) and *Franz Sternbald's Travels* (1798) are his most famous.

Tien Shan, mountain system in central Asia, extending from the Pamir Mountains, USSR, through NW China to the border of Mongolia. The highest point is Pobeda Peak 7,439m (24,406ft). Length of range: 2,415km (1,500 miles).

Tientsin. *See* TIANJIN.

Tientsin, Treaties of (1858 and 1885), two of the so-called unequal treaties between China and Western powers. The 1858 treaty with Britain, following the second OPIUM WAR, granted various rights to foreign diplomats, merchants and missionaries, and legalized the opium trade. By the 1885 treaty, which ended the Sino-French War, the Chinese acknowledged the French protectorate of Annam, now in Vietnam.

Tiepolo, Giovanni Battista (1696–1770), Italian painter and decorator, noted for the luminous treatment of his frescoes, superb draftsmanship and the use of steep perspective. He became internationally famous for his frescoes (1749) in the Labia Palace, Venice.

Tierra del Fuego, archipelago of s Argentina and s Chile, forming the southernmost tip of South America, separated from the mainland by the Strait of Magellan. Politically it is divided into two sections: the E belonging to Argentina, the w to Chile. Discovered by Ferdinand MAGELLAN in 1520, the islands were not settled until the 1880s, when the discovery of gold and oil attracted many Europeans, Argentinians and Chileans to the area. The mountainous terrain and harsh climate limit agricultural activity; sheep are reared and there is some fishing. Area: 73,746sq km (28,473sq miles). Pop. 15,658.

Tiffany, Louis Comfort (1848–1933), US painter, designer and a leader of the Art Nouveau style in the USA. In 1878 he formed the interior decorating firm which came to be known as Tiffany Studios in 1900. It specialized in what Tiffany termed "favrile" glass – freely shaped iridescent glasswork sometimes combined with various metals which created a new style in both the USA and Europe.

Tiflis. *See* TBILISI.

Tiger, large powerful cat found throughout Asia, mainly in forested areas. It has a characteristic striped coat of yellow, orange, white and black, with the chin and underparts white. Relying on keen hearing, it hunts for birds, deer, cattle and reptiles, but does not climb trees. The largest tiger is the Siberian variety. Length to 4m (13ft) overall; weight: to 230kg (500lb). Family Felidae; species *Panthera tigris*. *See also* NW pp.*214, 242.*

Tiger lily, tall flowering plant that grows primarily in Japan and China. Its red to orange, black-spotted flowers have looped petals that curl back to the base of the flower. Family Liliaceae; species *Lilium tigrinum.*

Tiger moth, any of numerous species of stout-bodied, nocturnal moths with bright orange, white and black wings. The caterpillars of most species are covered with long hairs, and commonly called woolly bears. Family Arctiidae.

Tiger shark, large shark found in inshore and offshore tropical waters throughout the world. It is grey and has dark vertical bars along its sides and a long, upper lobe on its tail. A noted scavenger, it preys on other sharks, sea turtles and carrion, and is regarded as a maneater. Length: to 6m (18ft). Family Carcharhinidae; species *Galeocerdo cuvieri.*

Tiglath-Pileser, name of three kings of Assyria. Tiglath-Pileser I (*r. c.*1115–*c.*1077 BC) extended the Assyrian empire to the N and restored the internal stability of the empire that had been disturbed in the previous reigns. Tiglath-Pileser II reigned *c.*956–943 BC. Tiglath-Pileser III (*r.*745–727 BC) again restored Assyrian supremacy in Babylonia, where he became king in 729 or 728 BC. He instituted the unpopular Assyrian policy of mass transportation of their defeated enemies. *See also* HC1 pp.56–57.

Tigris, river in SW Asia. It rises in the Taurus Mts of E Turkey and flows SE through Iraq, joining the River Euphrates to form the Shatt al-'Arab. The river is fast-flowing and liable to sudden flooding. There are flood-control schemes and the river's waters are used to irrigate more than 300,000 hectares (750,000 acres). The Tigris is navigable for shallow-draught vessels as far as BAGHDAD. In ancient times it was a major transport route and many great cities of Mesopotamia stood on its banks. Length: approx. 1,900km (1,180 miles).

Tijuana, city in NW Mexico, S of the US border, on the Pacific Coast. It originated around the cattle range Rancho de Tia Juana, in 1862. Tijuana gained popularity during the US prohibition era, because of the availability of alcohol; it was established as a free port in 1933. Manufactures include processed food, electronics and textiles. Pop. 354,805.

Tilapia, genus of freshwater mouth-breeding fish native to the Middle East and Africa. Family Cichlidae. Species include *Tilapia mossambica.*

Tilburg, city in S Netherlands, near the Belgian border, approx. 24km (15 miles) NE of Eindhoven. The city developed as an industrial centre in the 19th century. Industries: textiles, textile machinery, dyes, leather goods. Pop. (1974) 152,318.

Tile, small piece of hard material of regular and simple shape which is used decoratively and structurally in building. Tiles commonly cover roofs, walls or floors. Although tiles are traditionally made of CERAMICS (unglazed, painted or glazed clay), other materials such as stone, glass, rubber, plastic or wood are used where they are better suited to the function required, such as waterproofing, decoration, insulation from sound or heat, or resistance to wear.

Till, in geology, sediment or drift consisting of a heterogeneous mixture of clay, sand, gravel and boulders that is deposited directly by the ice of GLACIERS. It is not sorted into strata or sizes. Tillite is till that has become solid rock.

Tillett, Benjamin (1860–1943), British trade union leader. He helped to found the dockers' union (1887), becoming its general secretary (until 1922) and organizing the 1889 dock strike. He was Labour MP for North Salford, Lancashire (1917–24, 1929–31).

Tilley, Vesta (1864–1952), British music-hall comedienne (from 1878 to 1920) who specialized in impersonating men. She married Walter de Frece (later knighted), who composed many of her songs.

Tillich, Paul Johannes (1886–1965), German theologian and philosopher. He was ordained in the Lutheran Church, and served as chaplain in WWI. He left Germany for the USA in 1933 and taught at Union Theological Seminary and several universities. His many works include *The Courage to Be* (1952), *Dynamics of Faith* (1957), and *Systematic Theology* (three volumes, 1951–63).

Timber. *See* WOOD.

Timbre, characteristic of a musical sound determined by the number and intensity of the overtones (harmonics) in relation to the principal note. Musical instruments of the same type but made of different materials have different sounds because of the different harmonics produced. The skill of the musician can also affect the timbre on most instruments. *See also* SU pp.82–83.

Timbuktu, town in Mali, W Africa. It was founded in the 11th century and later became famous throughout Europe as a market for slaves and gold. It was sacked by the Moroccans in 1591 and seized by the French in 1893. Today its most important trading commodity is salt. Pop. (1970 est.) 10,000.

Time, perception of a sequential order in all experience; also the interval perceived between two events. Time falls within the disciplines of physics, psychology, philosophy and biology. Until Albert EINSTEIN's theory of RELATIVITY, time was conceived of as absolute: a constant one-direction (past to future) flow. Since then the concept of time linked with distance in space ("space-time") has connected time with the relative velocities of those perceiving it. For human beings at velocities approaching that of light, time expands, from the point of view of a stationary observer. Measurement of time, primarily relating to naturally recurring events such as sunrise and sunset, is common in some way to most creatures. Animals and plants have been shown to exhibit a daily (circadian) rhythm in variations in temperature and metabolism. Physiological and psychological experiments have shown the operation of similar cycles in man. For more technical measurements of time expressed in human terms, many ingenious devices have been constructed, from a candle burning at a set rate to an ATOMIC CLOCK. *See also* MW pp.188–189; SU pp.51, *51, 106,* 107, *127.*

Time and motion study, analysis of a job into the separate tasks of a worker doing it and the time for each task. Such analyses innovated by Frederick W. Taylor (1856–1915) and Frank B. Gilbreth (1868–1924), aim to eliminate wasted time and effort and unsafe methods, to increase productivity without creating fatigue. The measured task times can be used as standards in production scheduling. *See also* MS pp.300–301, *300.*

Time of Troubles (1606–13), period of turmoil in Russian history following the death of Boris GODUNOV (1605). The False DMITRI, a Pretender, was made Tsar in place of Boris's son, only to be killed in 1606. The position of his successor, a nobleman Vasily Shuyski, was threatened by the Second False Dmitri who, with Polish aid, besieged Moscow for two years. After his murder in 1610 a militia for defence was organized (1611), leading to the elevation of Michael ROMANOV to the throne in 1613.

Times, The, prestigious British daily newspaper, founded by John WALTER in 1785 as *The Daily Universal Register* and renamed *The Times* in 1788. It became part of the Thompson Organization in Jan. 1967.

Time scale, in geology, system of assigning ages to formations. It is based on the stages of evolution and falls into two grand divisions: the Cryptozoic (before 570 million years ago), which is characterized by a lack of fossils, and the Phanerozoic, divided into eras which are further divided into periods. *See also* PE pp.130–131.

Time-space distortion, in the theory of RELATIVITY, the bending of the SPACE-TIME continuum so that its geometry is no longer Euclidean. This effect is caused by very strong gravitational fields such as that found near a BLACK HOLE where the distortion is so great that light leaving the black hole is bent round upon itself and cannot escape. *See also* EUCLIDEAN GEOMETRY; SU pp.*106,* 107.

Time switch, device used to switch an electric circuit on or off at a pre-arranged time; its mechanism is similar to that of an alarm clock. Time switches are commonly used domestically to control ovens, audio equipment and heating systems.

Time zone, any of 24 longitudinal divisions of the Earth established for the purpose of determining local mean SOLAR TIME. Within each zone the overall local time is standardized and differs from time in a neighbouring zone, usually by one hour. Time zones east of the Greenwich (0°) meridian are various numbers of hours ahead of Greenwich Time; zones west of it are various hours behind. By convention, although zones are meant to be equal, they are usually made to coincide with country or state frontiers for convenience, and are sometimes divided into half-hour intervals. *See also* GMT; MW p.188.

Timişoara, city in W Romania, on the Bega Canal. An ancient Roman settlement, it was ruled by the MAGYARS from 896, annexed to Hungary in 1010 and ruled by the Turks from 1552 to 1716. It was returned by Austria-Hungary in 1716 and passed to Romania in 1920. The city is the site of a 14th–15th century Hunyadi castle (which is now a museum) and a university (1962). Industries: engineering, food processing, tobacco, chemicals, textiles, machinery. Pop. (1974) 210,520.

Timon of Athens (1607–08), tragedy by William SHAKESPEARE, printed in the first Folio of 1623. It has been suggested that it may be a rough draft of a play or the result of collaboration with an unknown author.

Timor, largest of the Lesser Sunda Islands in the Malay archipelago, part of INDONESIA. The Portuguese began trading with the island *c.*1520, and part of the island was taken by the Dutch a century later. Boundaries between Dutch and Portuguese possessions there were set by treaties in 1860 and 1914. The island was occupied by the Japanese during WWII. Dutch Timor passed to Indonesia in 1950, and Portuguese Timor was incorporated in 1975. Industries: fishing, tobacco, coffee, sugar, sandalwood. Area: 33,857sq km (13,074sq miles). Pop. (1970 cst.) 3,000,000.

Timoshenko, Semyon Konstantinovitch (1895–1970), Soviet military commander. He fought for the BOLSHEVIKS during the Russian Civil War (1918–20), and rose to command the Soviet armies that invaded Finland in 1940. He later became Commissar for Defence (1940–41). In 1941 he recaptured Rostov from the Germans and helped to relieve Moscow.

Timothy, Epistles to, two letters in the New Testament attributed to St PAUL, encouraging faith in Christ and demanding orthodox belief. Together with the Epistle to *Titus,* they are called the Pastoral Epistles.

Timpani, or kettledrums, principal percussion instruments in a symphony orchestra. They are hemispherical vessels of copper or brass with single skins, tuned by pedals or screws and struck with sticks with hard felt heads. Military kettledrums were played on horseback by Asian nomads and introduced to Europe by the Crusaders *c.*1100. Modern timpani were introduced in the 17th century.

Timur. *See* TAMERLANE.

Timurid Dynasty (1370–1506), Turkicized Mongol family which ruled Transoxiana, approximating to modern Uzbekistan. It was founded by TAMERLANE (also known as Timur) who became sole ruler in 1369 and made his capital at Samarkand. His campaigns extended his empire, which was largely to die with him, into Iraq, Syria, Caucasia and India. In the following century, the Timurid kingdom was twice

Tiger moth's bright colouring warns birds that it is poisonous.

Tile; part of a factory where tiles are mass-produced with precision.

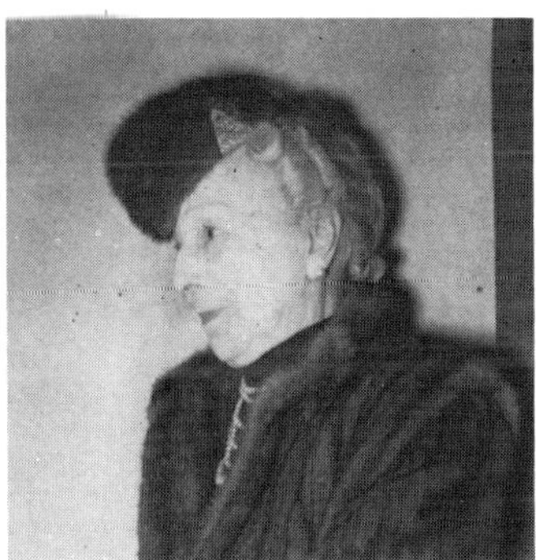

Vesta Tilley was famous for her male impersonations in music hall comedy.

Timpani; the note produced depends on the tautness of the skin of the drum.

Tin

Tinamous remain motionless when in danger, camouflaged by their plumage.

Tintoretto spent most of his life in Venice, where many of his works are found.

Michael Tippett at rehearsals for his 70th birthday concert in London.

Titanic in dock at Southampton. She was regarded as unsinkable in peace time.

divided and came to an end when the Uzbeks expelled BABUR, who went on to found the MOGUL EMPIRE in India.

Tin, METALLOID element (symbol Sn) of group IVA of the PERIODIC TABLE, known from ancient times. Its chief ore is cassiterite (an oxide). It is used as a protective coating for steel in tin plate, and in solder, pewter, type metal, and similar alloys. Two ALLOTROPES exist: the common metallic form (white tin) changes slowly below 13.2°C (55.8°F to a brittle non-metallic form (grey tin). Properties: at.no. 50; at.wt. 118.69; s.g. 5.75 (grey), 7.31 (white); m.p. 232°C(449.6°F); b.p. 2,270°C (4,118°F); most common isotope Sn^{118} (24.03%).

Tinamou, grassland and jungle bird found from Mexico to s South America. It has a heavy body with camouflaging brown or grey plumage, a thin curved bill, small weak wings and a short tail. The male incubates the eggs in a nest on the ground. Length: 23–38cm (9–15in). Family: Tinamidae.

Tinbergen, Jan (1903–), Dutch economist who has developed dynamic models using mathematical methods to analyse economic processes. He shared the first Nobel Prize in economics in 1969, and his publications include *Shaping the World Economy* (1962).

Tinbergen, Nikolaas (1907–), Dutch ethologist. He shared with K. LORENZ and K. von FRISCH the 1973 Nobel Prize in physiology and medicine for his pioneering work in ETHOLOGY. Tinbergen studied how in animals certain stimuli evoke specific responses, and emphasized the importance of observing animals under natural conditions. *See also* MS pp.150, 186.

Tindemans, Léo (1922–), Prime Minister of Belgium from 1974. A Christian Socialist, he was Mayor of Edegem from 1965, Minister of Community Affairs (1968–1971), Minister of Agriculture and Middle Class Affairs (1972–73) and Deputy Prime Minister (1973–74).

Ting, Samuel Chao Chung (1936–), US physicist who researched into ELEMENTARY PARTICLES. He used a SYNCHROTRON to fire PROTONS at a BERYLLIUM target and then observed the resultant decaying particles. From the results he found evidence for the existence of a heavy sub-atomic particle, which he named the J particle. At the same time but working independently, Burton RICHTER discovered the same particle, which he named psi. For this discovery Ting and Richter shared the 1976 Nobel Prize in physics.

Tinguely, Jean (1925–), Swiss sculptor whose work incorporated movement through small motors. He experimented with motorization and produced progressively complicated machines, some of which were self-destructive, such as *Homage to New York* which destroyed itself in 1960 when set into motion. *See also* HC2 p.203.

Tinned food. *See* CANNING.

Tino di Camaino (c. 1285–1337), Sienese sculptor. In charge of the works at Pisa Cathedral from 1315, he was commissioned to make a tomb for the Emperor Henry VII. From 1319 to 1320 he supervised works at Siena Cathedral. Later he worked in Florence and then settled in Naples, where he made monuments for the Angevin rulers.

Tin oxide, either of two oxides: stannous or tin (II) oxide, SnO, a black powder used as a reducing agent; stannic or tin (IV) oxide, SnO_2, a white powder occuring naturally as the mineral CASSITERITE. It is used in ceramics, glass and cosmetics.

Tintoretto (Jacopo Robusti) (1518–94), Italian painter. A major exponent of the Venetian School, he was a leading artist for middle class patrons and Venetian churches. He owed much in style to TITIAN, with whom it is supposed he studied, and his master's influence is particularly evident in the latter part of his career. By 1539 he was a master and established independently. The arresting appeal of his work reflects Tintoretto's sensitivity to lighting and the importance of the angle of portrayal. *The Finding of*

the *Body of St Mark* (1562) and *The Last Supper* (1592–94) are examples of this. His religious paintings are filled with mystery and personal experience and in the series depicting the life of Christ (1565–87) for the Scuola di San Rocco are some of the finest examples of Italian painting.

Tipperary, county in s Republic of Ireland, in Munster province. The region is part of the central plain of Ireland but there are hills in the s; the Suir and Shannon are the principal rivers. The soil is fertile and Tipperary is one of the best farming regions of Ireland. Dairy farming is the most important activity; the principal industry is food processing. Administratively, the county is divided into the North and South Ridings. The county town is Tipperary, a market town. Area of county: 4,255sq km (1,643 (1,643sq miles). Pop. 123,196.

Tippet, Sir Michael Kemp (1905–), British composer. His music includes symphonies, piano and chamber music, the oratorio *A Child of our Time* (1941) and several operas such as *The Midsummer Marriage* (1952), *King Priam* (1962) and *The Ice Break* (1977). In some compositions he used British folk and American Jazz and Negro elements.

Tipu Sahib (1749–99), Indian ruler, Sultan of Mysore (1782–99). He fought the MAHRATTAS and the British in the service of his father, Hyder Ali, but concluded peace with the British in 1784. His negotiations with the French provoked a British invasion of Mysore in 1799 and Tipu was killed while defending his capital at Seringapatam.

Tiradentes (1748–92), real name Joaquim José de Silva Xavier, leader of the first major conspiracy against Portuguese rule in Brazil (1789). Tiradentes ("Toothpuller"), concentrating on the grievances of the inhabitants of the state of Minas Gerais, called for complete independence. The revolt was crushed and Tiradentes was executed.

Tiranë (Tirana), capital of ALBANIA, approx. 27km (17 miles) inland from the Adriatic Sea, in E central Albania. It was founded in the early 17th century by the Ottoman Turks, although most of the present city dates from 1920 when it was chosen as the country's capital. It is the industrial centre of Albania; manufactures include metal goods, textiles and food products. There is also some coal mining. Pop. (1973) 183,500.

Tirpitz, Alfred von (1849–1930), German statesman and admiral. He became Secretary of State in 1897 and Grand Admiral in 1911. He increased the German navy enormously by 1914. His policy of unrestricted submarine warfare against Britain led to the sinking of the *Lusitania* in 1915. He resigned in 1916.

Tirso de Molina (b. 1584–1648), Spanish playwright, b. Gabriel Téllez. After a period as a monk, he was inspired by Lope de VEGA and started writing plays, producing over 300 in his career, 80 of which survive. His play *The Deceiver of Seville* (1630) is the earliest known dramatic portrayal of DON JUAN.

Tischbein, name of a family of 18th-century painters. The best-known, Johann Heinrich Wilhelm (1751–1829), was a friend of GOETHE whose portrait *Goethe in the Campagna*, he painted in 1786. He also supervised the engravings of the Greek vases belonging to Sir William Hamilton. Other Tischbeins of importance include Johann Heinrich the Elder (1722–89), a court painter, and Johann Friedrich (1750–1812) and Anton Wilhelm (1730–1804), both portrait painters.

Tiselius, Arne Wilhelm Kaurin (1902–1971), Swedish biochemist. In 1937, while working with Theodor SVEDBERG on electrophoresis, he observed that since PROTEINS are electrically charged in a colloidal solution a current passing through the solution causes different proteins to move at different speeds and therefore separate. He built a machine that would take advantage of the effect and for this work received the 1948 Nobel Prize in chemistry.

Tissé, Edvard (1897–1961), Russian film

cameraman of Swedish descent. He worked as war photographer in WWI and in 1924 began a long partnership with the Russian director, Sergei EISENSTEIN. Their films included *Strike* (1924), *The Battleship Potemkin* (1925), *Alexander Nevsky* (1939) and *Ivan the Terrible* (1942, 1946).

Tissot, James Joseph Jacques (1836–1902), French painter and etcher. He spent ten years in England after the Franco-Prussian War of 1870, and is remembered mostly for his charming illustrations of Victorian life, eg *My Heart is Poised between the Two* (1877).

Tissot, Janou, née Lefebvre (1945–), French rider, who became the first woman to win two show-jumping world championships, riding *Rocket*, when she won the women's world championship in 1970 and 1974. She was a team silver medallist at the 1964 and 1968 Olympic Games and women's European champion in 1966.

Tissue, in biology, substance of a living body or organ consisting of CELLS. Tissues vary in structure and complexity and may be loosely classified according to function into epithelial, connective, muscular, nervous and glandular tissues, although each of these categories contains more than one different type of cell.

Tissue culture, in biology, artificial cultivation of living TISSUE in sterile conditions in laboratories for research or to help in diagnosis of diseases.

Tit, or titmouse, any of several species of small garden and woodland songbirds, particularly of the family Paridae. European species include the willow tit (*Parus montanus*), the blue tit (*P. caeruleus*) and the great tit (*P. major*). The name is also used for other birds whose appearance or behaviour resembles that of the Paridae. Length: to 17cm (6.5in). *See also* NW pp.*151, 206, 209.*

Titan, largest satellite of the planet SATURN. It is comparable in size with MERCURY and is believed to support a thin atmosphere, chiefly of methane. Diameter 4,800km (3,000 miles); mean distance from planet 1,220,000km (758,000 miles); mean sidereal period 15.97 days. *See also* SU pp.212–213, 278–279.

Titania, in folklore, queen of the fairies and wife of Oberon. In the writing of OVID she represents DIANA at the head of her nymphs. According to Shakespear's *A Midsummer Night's Dream* she quarrelled with her husband over a changeling boy.

Titania, largest of the five satellites of the planet URANUS. It orbits at a distance of 438,370km (272,270 miles) from Uranus and has a diameter of approx. 1,800km (1,120 miles). The satellite has a mean sidereal period of 8.7 days. *See also* SU p.214, *214.*

Titanic, British luxury passenger liner that sank, taking two and a half hours (14–15 April 1912) to do so, on her maiden voyage. The disaster occurred after the liner struck an iceberg in the North Atlantic. Of the 2,224 passengers, many of them US and British celebrities, 1,513 were drowned. The disaster, led to the first International Convention for Safety of Life at Sea (1913).

Titanium, common metallic element (symbol Ti) of the first TRANSITION ELEMENTS, discovered in 1791 by William Gregor. It is found in many minerals; chief sources are ILMENITE (iron titanate) and RUTILE (an oxide). The element is used in steels and other alloys, especially in aircraft and other applications where strength must be combined with lightness. Properties: at.no. 22; at.wt. 47.90; s.g. 4.54; m.p. 1,675°C (3,047°F); b.p. 3,287°C (5,949°F); most common isotope Ti^{48} (73.94%).

Titanium oxide, either of two oxides. Titanium monoxide, TiO, is a white powder formed by reducing the dioxide at 1,500°C (2,732°F). It has no industrial uses. Titanium dioxide, TiO_2, is a white or grey powder, according to its purity. It occurs naturally as the mineral RUTILE from which it is extracted by chlorination or from the mineral ilmenite by treating it with sulphuric acid. It is used as a white pigment in ceramics, glass, cosmetics, paper and paints.

Titans, The, in classical mythology, twelve primeval gods and goddesses who were the sons and daughters of Uranus and Gaea. The Titans were Chronos, Iapetus, Hyperion, Oceanus, Coeus, Creus, Thea, Rhea, Mnemosyne, Phoebe, Tethys and Themis; Atlas, Prometheus and Hecate were descended from them. They were the order of gods which preceded the Olympians who, led by Zeus, overthrew them and condemned them to Tartarus.

Titchener, Edward Bradford (1867–1927), British psychologist. With Wilhelm WUNDT, he was a leader of the STRUCTURALISM school of psychology which encouraged studies in psychology to be more scientific and systematic. His major publication is *Experimental Psychology* (1901–05).

Tithe, levy or tax of one-tenth. It is prescribed in the Old Testament and was adopted by the Christian Church, which required parishioners to give up a tenth of their land, stock, or harvest for the purposes of maintaining churches, the support of priests, and as charity for orphans, widows and the landless. Tithes were obligatory in the Middle Ages, but in the following centuries were gradually abolished in most European countries, being commuted to money rents in England in 1836. *See also* HC1 p.*195*.

Titian (Tiziano Vecellio) (*c.*1487–1576), Venetian painter. He studied under the BELLINIS and GIORGIONE, and his early works include *Votive Picture of Jacopo Pesaro* and *St Peter with Donor* (*c.*1506). From about 1515 Titian's work is characterized by livelier colour, as in the allegorical *Sacred and Profane Love* and *Three Ages of Man* (*c.*1515). In 1518, Titian began a series of mythological pieces for Alfonso d'Este, which include *Worship of Venus* (1518–19), *Bacchanal* (1519–22) and *Bacchus and Ariadne* (1523). After 1530 his style became more restrained; he used related rather than contrasting colours, as in the famous *Venus of Urbino* (1538–39). In 1533 he painted a portrait of the Holy Roman Emperor CHARLES V, and was later patronized by PHILIP II of Spain. His works of the 1550s such as *Rape of Europa* (1559–62), are highly poetic visions, in contrast to the mythological and erotic portrayals of earlier times. After 1560 his works gradually expressed a new warmth, harmony and pathos. Autumnal and sensitive colouring and subtle light effects characterize his later works, such as *Shepherd and Nymph* (*c.*1570) and *Pietà* (1576). *See also* HC1 pp.258–259.

Titicaca, lake on the Peru-Bolivia border, draining s through the River Desaguader into Lake Poopó. It is the highest large navigable lake in the world. Altitude: 3,810m (12,500ft). Length: 193km (120 miles); depth: 280m (920ft).

Tito, (1892–   ), Yugoslav Communist leader, Premier from 1945, and President 1953 b. Josip Broz. As a Croatian soldier in the Austro-Hungarian army in WWI, he was captured by the Tsar's troops (1915) but released by the Communists after the Revolution (1917). Returning to Yugoslavia, he helped organize the Yugoslav Communist Party and, as one of its key members, served a jail term after it was outlawed (1928–34). In WWII he led the partisan resistance forces in Croatia, discrediting his rival Draza Mihajlović, who led the Chetnik guerrillas in Serbia, and became so powerful that he was able to establish a Communist government in 1945 that achieved international recognition. Russian efforts to control Yugoslavia soon led to a split between Tito and STALIN, who expelled the nation from the Communist bloc in 1948. Tito then became an important figure to many THIRD WORLD countries. Internationally, Tito has gained world respect through Yugoslavia's policy of non-alignment. *See also* HC2 pp.243, 254, *255*.

Titration, method used in analytical chemistry to determine the concentration of a SOLUTION by volumetric means. A solution of known concentration is added in measured amounts from a burette (a graduated glass tube) to a liquid of unknown concentration until the reaction is complete (as shown by an INDICATOR). The volume removed from the burette enables the unknown concentration to be calculated. *See also* SU pp.150–151.

Titus (*c.*AD 40–81), Roman emperor (*r.* 79–81), elder son of VESPASIAN. He campaigned in Britain and Germany and captured and destroyed Jerusalem in 70 after a Jewish revolt. He was briefly (79) co-ruler with Vespasian, then sole emperor. As emperor he stopped presecutions for treason, built lavish baths and gained great popularity. He died childless and his brother DOMITIAN erected the Arch of Titus in his memory. *See also* HC1 p.*107*

Titus, Epistle to, letter written by St PAUL to Titus, one of his earliest converts. It is valuable for its elucidations of Christian behaviour and duties: deacons and bishops are warned against false teachers; young men are admonished to be sober-minded and young women to be chaste, discreet and obedient to their husbands. The Epistle to Titus and the two Epistles to TIMOTHY are called the Pastoral Epistles.

Titus Andronicus, (1594), early work attributed to William SHAKESPEARE, but possibly a collaboration. The play, full of bloody horrors, was one of several on the same subject extant in Shakespeare's time.

Tiw, or Tyr, in Germanic mythology, powerful sky god. He was also associated with war, government and justice. The word Tuesday derives from Tiw's day.

Tiwi, semi-nomadic native Autralian people that live on the islands of Melville and Bathurst off Northern Territory. They live by hunting and gathering and practise totemism and patriarchy. Their society is noted for its extreme polygamy, men having up to 29 wives.

Tizard, Sir Henry Thomas (1885–1959), British physical chemist. He is best known for his chairmanship of the Committee for the Scientific Survey of Air Defence (1933–42) and his insistence on the development of radar. In 1942 he became president of Magdalen College, Oxford.

Tlaloc, AZTEC god of rain. He was propitiated by the sacrifice of babies and children, who were then cooked and eaten. If the victims cried, rain was supposed to follow. *See also* MS p.*212*.

Tlatelolco, the commercial centre of the Aztec capital of TENOCHTITLÁN. Most of it was destroyed by the Spanish in 1521, but numerous buildings, shrines and tombs have been excavated.

Tlaxcala, tribe of Mexican Indians who occupy the state of Tlaxcala, Mexico. They supported CORTÉS against the AZTECS, and were largely responsible for the defeat of MONTEZUMA's army in 1521. *See also* HC1 p.*236*.

TNT, (trinitrotoluene), explosive organic compound with the formula $CH_3C_6H_2(NO_2)_3$. It was discovered in 1863 and its stability to shock make it one of the safest high explosives. *See also* MM p.243.

Toad, any of numerous species of tail-less amphibians found throughout the world, except in Australasia. Most species are short and rotund, move with a hopping gait and have no teeth in their upper jaws. Parotid glands behind the eyes secrete an irritating substance to deter predators. Toads are differentiated from frogs by being toothless and having a rougher, bumpier skin and rounder body with shorter legs. Length: 2–25cm (1–10in). Order Anura; family Bufonidae. *See also* TADPOLE; NW pp.73, *73*, 132–133, *204*, *219*, *230*.

Toadfish, any of 40 species of bottom-dwelling fish that live in temperate and tropical seas. It has a large mouth and head, a tapered body and a long ventral fin. Most species are grey, green, yellow or brown with dark speckled markings. The dorsal fin spines of some species are venomous. Length: to 30cm (12in). Family Batrachoididae. *See also* NW p.*131*.

Toadflax, any of several species of annual and perennial plants with leaves resembling those of flax plants. Some species are grown as ornamentals. The common toadflax (*Linaria vulgaris*), or butter-and-eggs, has dense clusters of two-lipped, pale yellow flowers. Family Scrophulariaceae.

Toad of Toad Hall (1929), play by A. A. MILNE, adapted from the novel *The Wind in The Willows* by Kenneth GRAHAME. A popular children's classic, it is revived regularly and tells the story of the animals Ratty, Mole and Toad.

Toadstool, popular name for the fruiting body of a FUNGUS of the class Basidiomycetae. The name usually refers to inedible species and describes the stool-like appearance of the reproductive organ, which is the only visible part of the fungus, the remainder being mycelium hidden in the substratum. It consists of a stem (which may sprout from a cup or vulva) and a head (on which the spores are borne on gills or in tubes). A ring, or annulus, on the stem marks the attachment of the cap before opening. The gills or tubes in which the spores are carried are arranged vertically, so that the spores fall out when ripe and are carried away by the wind. *See also* NW pp.38, *38*, 42, 43, *43*.

Tobacco, herb native to the Americas, but cultivated throughout the world. It has large leaves with no stalk, and white, pink, or red star-shaped flowers. Height: 0.6–2m (2–6ft). *Nicotiana tabacum* is the principal cultivated species. American Indians smoked the leaves of tobacco and used them medicinally well before the arrival of Europeans in the New World. Seeds were brought to Europe in about 1520–30. Settlers in Virginia obtained seeds from the Spanish colonies (1612) and soon tobacco was the major crop of the Virginia colony and America's first export. Leaves are prepared for smoking by curing (drying) and then ageing. Family Solanaceae (NIGHTSHADE family). *See also* PE p.*173*; NW pp.*36*, *45*.

Tobacco mosaic virus (TMV), simple virus used in experiments concerning the transference of the genetic code. It consists of a single helix of RNA containing some 6,400 NUCLEOTIDES. This is coated with some 2,100 molecules of a single PROTEIN, each molecule of which comprises a polypeptide chain of 158 AMINO ACIDS. The sequence of these acids has been determined.

Tobago. *See* TRINIDAD AND TABAGO.

Tobey, Mark (1890–1976), US painter whose work borrowed motifs from Asian calligraphy. A lyrical treatment of light and colour are revealed in his *San Francisco Street* (1941). He used his technique of "white writing" initially to portray city scenes and later to express abstract visions.

Tobias, in the Old Testament, son of TOBIT. Tobias helped the angel RAPHAEL drive away the demon Asmodaeus. Raphael, in turn, cured the blindness of Tobit and induced Tobias to marry SARAH of Ecbatana.

Tobit, Book of, one of the books in the Roman Catholic and Eastern Orthodox Old Testament, regarded by Protestants and Jews as part of the APOCRYPHA. A travel narrative told by Tobit's son, TOBIAS, the story involves Tobias's recovery of his family's wealth and Tobit's recovery of his sight through the efforts of the angel RAPHAEL in disguise.

Toboggan, small sled used to travel down ice or snow-covered slopes.

Tobogganing, winter sport in which the rider either sits on a TOBOGGAN (lugeing), or lies prone on it (Cresta tobogganing). Lugeing originated in Austria and became an Olympic sport in 1964, and there was a Cresta competition at the Olympics in 1924, and regularly after 1948.

Tobruk, port city in NE Libya, on the Mediterranean Sea. Its strategic location on the coastal road made it the objective of both the Germans and the British in WWII; at different times it was taken by both armies in the period 1940–42. Its present source of income is trade. Pop. (1970 est.) 28,000.

Tocanting, river in NE Brazil. It rises in s Goias state and flows N to the River Para. Rapids make the river only partly navigable. Length: 2,700km (1,677 miles).

Toccata, form of instrumental composition from the early 17th century. GABRIELI made early use of the form as a showy

Titian's painting of a woman, *La Bella*, is typical of his portraiture.

President Tito, photographed during a state visit to France in 1956.

Titus Andronicus; Henry Peacham was the first to illustrate a work by Shakespeare.

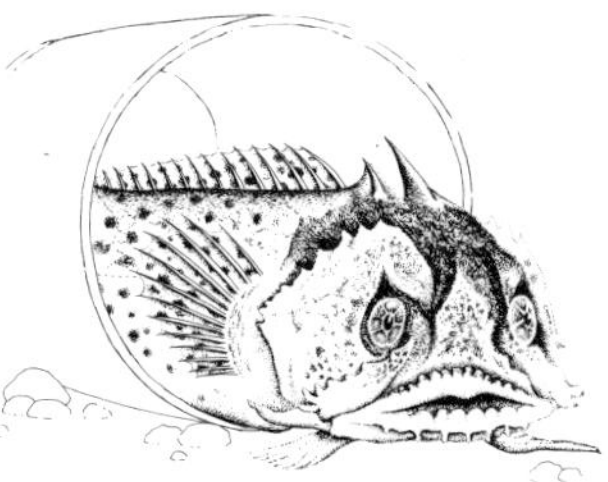

Toadfishes are solitary and usually shelter among rocks on the sea bottom.

Toc H

Alexander Todd taught at Cambridge, Edinburgh and Manchester universities.

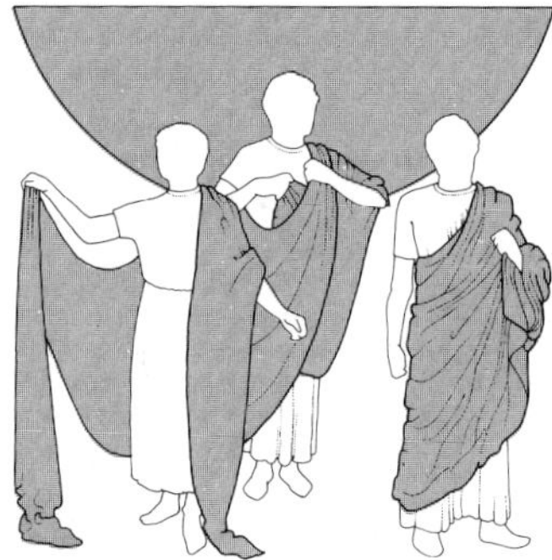

Togas were first worn by both sexes but later women wore the stola instead.

Tokyo; the Imperial Palace moat, with modern factories in the background.

Leo Tolstoy was a young officer when he wrote *Childhood, Boyhood and Youth*.

piece with rapid, even flowing notes and frequent repetition of similar figures. BACH's organ toccatas are well known for their virtuoso display.

Toc H, inter-denominational Christian association devoted to social work. It is open to all men over the age of 16 and now includes a Women's Association. It originated in Talbot House, a soldiers' club opened in 1915 at Poperinghe, Belgium, under an Anglican chaplain, "Tubby" CLAYTON. It was named after Lieut. Gilbert Talbot, who was killed in action at Ypres in 1915. It was re-founded in London in 1920 and expanded rapidly within Britain and other English-speaking countries. It was incorporated by Royal Charter in 1922 and now has many branches and residential houses. Its name is derived from the army signallers' way of pronouncing T H (Talbot House).

Tocqueville, Alexis de (1805–59), French historian. His works include the political study *De la Démocratie en Amérique* (1835) and *L'Ancien Régime et la Révolution* (1856). He was a vigorous opponent of Socialism and Republicanism until Louis Napoleon's coup in 1851.

Todd, Alexander Robertus, Baron (1907–), British biochemist who was awarded the 1957 Nobel Prize in chemistry for his research into the structure and synthesis of NUCLEOTIDES and nucleotide COENZYMES. He synthesized and deduced the structure of all the naturally occurring nucleotide components of the NUCLEIC ACIDS. This work laid the foundation for the determination by James WATSON and Francis CRICK of the structure of DNA and helped towards an understanding of the mechanisms of gene action. Todd synthesized ADP in 1947 and ATP in 1949, compounds essential for the utilization of energy in CELLS.

Todd, Mike (1907–1958), US stage producer. He produced several revues on Broadway, going into film production in 1953 and developing a new WIDE SCREEN process called Todd-AO. His major film is *Around the World in Eighty Days* (1956).

Todd, Reginald Stephen Garfield (1908–), Rhodesian politician, *b.* New Zealand. He was a Southern Rhodesian MP (1943–58) and president of the United Rhodesia Party and Prime Minister (1953–58). In 1961 he founded the New Africa Party. From 1972 to 1976 he was under house arrest. In 1976 he was the adviser to Joshua NKOMO at the Geneva conference on Rhodesia.

Toga, mantle or outer garment worn by Roman citizens, consisting of an oval-shaped piece of cloth. Draped around the body to form a flowing skirt, the remainder of the material was pulled up over the head or shoulders. Its colour and pattern depended upon the status of the wearer.

Togo Heihachiro (1847–1934), Japanese admiral. He was trained at Greenwich naval academy, London (1871–78), and served in the SINO-JAPANESE WAR (1894–95) and the RUSSO-JAPANESE WAR (1904–05). He defeated the Russian fleet at Port Arthur (1904) and destroyed Russia's Baltic fleet at TSUSHIMA (1905). *See also* HC2 p.*147*.

Togo, small republic in W Africa. Formerly mandated to France, Togo became independent in 1960. Most of the population earn a living off the land; the chief cash crops are cocoa and coffee. Phosphates are the principal export. Manufacturing is on a small scale. The capital is Lomé. Area: 56,000sq km (21,622sq miles). Pop. (1975 est.) 2,222,000. *See also* MW p.166.

Togoland, historic region of W Africa. Originally peopled predominantly by the Ewe, the area was raided for slaves until the second half of the 19th century. By the end of the century Germany had established military and economic predominance and it became a German protectorate. After WWI Togoland was divided into a French mandate, which achieved independence in 1960 as the Republic of Togo, and a British mandate, which became part of the independent state of Ghana in 1957.

Tojo, Hideki (1885–1948), Japanese general and political figure. As Prime Minister (1941–44) during WWII he authorized the PEARL HARBOR attack and was overlord of all aspects of the Japanese war effort. He was tried for war crimes by the ALLIES after the defeat of Japan, was found guilty and hanged.

Tokamak, Russian experimental nuclear fusion reactor consisting of a toroidal vessel in which a PLASMA (hot ionized gas) is contained by a magnetic field produced by coils wound over the reaction vessel. Several versions of these devices now exist outside the USSR. *See also* NUCLEAR REACTOR.

Tokelau Islands, group of islands in the W central Pacific Ocean, N of Samoa, it was formerly known as the Union Islands. The group came under British rule in 1877 and became a territory of New Zealand in 1948. The principal product is copra. Area: approx. 10sq km (4sq miles). Pop. 1,687.

Token economy, system of rewards allocated within an institution such as a school or psychiatric hospital. The desired behaviour or work is rewarded with tokens, which may be gold stars or counters, and these tokens may accumulate to be exchanged for either more tangible rewards or different coloured stars. *See also* MS p.*147*.

Tokugawa, Japanese family holding the title of SHŌGUN and maintaining effective control of Japan (1603–1867). Under their shogunate, centralized feudalism was based on an intricate system of allegiances of autonomous daimyō (feudal lords) to the Tokugawa family, who controlled much of the country's wealth and farmland in militarily strategic locations. The period of their rule is notable for its intense isolationism, for improvements in farming, an increase in domestic trade and the growth of literacy and cultural advancement. The regime declined in the 19th century. *See also* HC2 pp.56–57.

Tokugawa Ieyasu. *See* IEYASU TOKUSAWA.

Tokyo, largest city in Japan, at the head of Tokyo Bay. Founded in the 15th century as Edo, it became the capital of the TOKUGAWA Shogunate under IEYASU (1603). The city grew rapidly in the 18th and 19th centuries, and as the Japanese Reformation re-established Imperial power, it was made the capital of Japan in 1868 (replacing Kyoto) and renamed Tokyo ("eastern capital"). Much of Tokyo was destroyed by an earthquake and fire in 1923; the city was reconstructed however, as a modern metropolis. It was further modernized after extensive US bombings caused widespread destruction in 1944 and 1945. Tokyo is now the centre of a massive industrial-commercial belt extending along the W shore of Tokyo Bay, and including Kawasaki and Yokohama, the latter serving as seaport for Tokyo and the vicinity. Modern Tokyo contains more than 200 colleges and universities, including the University of Tokyo (1877). Products include metals, machinery, electronic and transit equipment, chemicals, textiles and consumer goods. Pop. (1975) 8,642,000.

Tolbert, William Richard (1913–), President of Liberia from 1971. He was a member of the House of Representatives (1943–51) and became Vice-President (1951–71).

Toledo, Francisco de (*c.* 1515–84), viceroy of Peru (1569–81), one of the ablest bureaucrats to serve in colonial Spanish America. He maintained order in Peru, drawing up a law code (using many INCA laws) and quelling the revolt (1570–71) of TUPAC AMARU.

Toledo, city in Spain, on the River Tagus 66km (41 miles) SSW of Madrid; capital of Toledo province. The city was captured by the Romans in 192 BC. It flourished as the capital of the Visigoth kingdom in the 6th century AD. Toledo was a centre of Moorish culture from 712 to 1085. It became famous thoughout the world for its sword blades. After the 15th century the city declined as a commercial centre but became the spiritual capital of Catholic Spain; it was the seat of the Grand Inquisitors. Landmarks include the Cathedral (1226), one of the finest Gothic structures in Spain and the Alcázar, a Moorish fortified palace. Many buildings contain the works of El GRECO. Industries: metalwork, silk products, surgical equipment. Pop. 44,382.

Toledo, port in NW Ohio, USA, at the mouth of the Maumee River in Lake Erie; a major GREAT LAKES shipping point. The city grew with the construction of canals and railways, and the development of the coalfields of Ohio. The University of Toledo dates from 1872. Industries: glass, machine tools, oil refining, plastics, machinery. Pop. (1970) 383,818.

Tolerance, in biology, the capacity in an organism to withstand, endure or absorb a substance or its introduction into the body, without ill effect.

Toleration, Act of (1689), English legislation permitting freedom of religious practice to PROTESTANT dissenters, subject to their taking oaths of allegiance to William and Mary. It modifies the TEST ACT, but it did not remove the political and social disabilities of dissenters, nor did it extend to Roman Catholics. *See also* HC1 pp.*312*, 313.

Tolkien, John Ronald Reuel (1892–1973), English philology scholar. He achieved fame with his imaginative epic trilogy, *The Lord of the Rings* (1954–55). Both his prose style and world of fantasy are reminiscent of the Norse sagas and Anglo-Saxon poetry that he taught at Oxford University. The trilogy is introduced by *The Hobbit* (1937).

Toll, originally the right to levy certain dues, later the name for the dues themselves. They were levied in medieval England at town markets for the upkeep of local roads and bridges. As a charge paid at the toll gate at one end of some roads and bridges they survive today.

Toller, Ernst (1893–1939), German EXPRESSIONIST playwright and poet who was imprisoned from 1919 to 1924 for Communist activities. His plays of social protest include *Die Wandlung* (1919, trans. *Transfiguration*, 1935) and *Masse-Mensch* (1920, trans. *Man and the Masses*, 1923).

Tolpuddle Martyrs, name given to six British farm labourers sentenced in 1834 to seven years' transportation to Australia for trade union activities in Tolpuddle, Dorset. The government hoped that punishment of such severity would prevent further working-class unrest but the upsurge of popular support for the six men was so great that their sentences were remitted in 1836. *See also* HC2 pp.95, *95*.

Tolstoy, Aleksei Nikolaievich (1883–1945), Russian novelist, a distant relation of Leo TOLSTOY. Initially opposed to the BOLSHEVIK revolution, he returned to the USSR in 1923 to publish the historical trilogy *Khozhdeniye po Mukam* (1920–41; tr. *The Road to Calvary*, 1946). His unfinished *Pyotr Pervyy* (1929–45; trans. *Peter the Great*, 1956) is considered his best work.

Tolstoy, Count Leo Nikolaievich (1828–1910), Russian novelist and philosopher. While serving in the army he wrote his trilogy *Childhood, Boyhood and Youth* (1852–57). He took part in the defence of Sebastopol in 1854, descriptions of which appeared in the journal *Contemporary* and were noted for their unvarnished picture of the war. He left the army and lived on his estate and in St Petersburg, where he wrote *War and Peace* (1865–69) and *Anna Karenina* (1875–77). In 1879 he underwent a spiritual crisis which culminated in his conversion to the Christian doctrine of love and non-resistance to evil. These beliefs are reflected in his later works, such as *The Death of Ivan Ilyich* (1886) and *Resurrection* (1899). *See also* HC2 pp.99, *99*.

Toltec, major pre-Columbian culture, powerful from AD 900 to 1200, with its centre in the Hidalgo region of Mexico. Toltec is an Aztec derivation from the name of their urban centre at Tollán. A wide-ranging people, their influence extended from the later capital at Tula as far S as Guatemala. *See also* HC1 pp.230, 231.

Toluene, (methylbenzene), aromatic hydrocarbon, formula $C_6H_5CH_3$. Manufactured from the fractional distillation of coal-tar or from petroleum, it is a colourless, inflammable liquid widely used as an industrial solvent and in the manufacture of chemicals. Properties: s.g. 0.87; m.p. −94.5°C (−138.1°F); b.p. 110.7°C (231.3°F).

Tomato, fruit native to the Americas. The plant leaves are deeply toothed or lobed and yellow flowers produce the familiar green fruit which turns reddish-orange as it ripens. The plant was cultivated in Europe as early as 1544 and introduced to the American colonies about 1629. It was not eaten until the 16th century because it was believed to be poisonous and was grown merely as an ornamental. Species *Lycopersicum esculentum.* The small cherry tomato is a variety (*cerasiforme*). *L. pimpenellifolium*, the currant tomato, produces a very small fruit. Family Solanaceae. *See also* PE pp.158, *173*, 188, *189*.

Tombaugh, Clyde William (1906–), US astronomer. Working at the Lowell Observatory, Flagstaff, Arizona, he discovered PLUTO, in 1930. *See also* SU p.*215*.

Tombolo, bar connecting an island with the mainland. It is usually formed by the growth of a sand spit until it reaches the island. *See also* PE p.*123*.

Tomboctou. *See* TIMBUKTU.

Tom Brown's Schooldays (1857), novel by Thomas HUGHES. A description of the author's experiences at Rugby School when Dr Thomas ARNOLD was headmaster, the work was an immediate success. It is the prototype of school stories and is acclaimed not only for its literary merit but also for its documentary value.

Tom Jones, A Foundling (1749), novel by Henry FIELDING. A masterpiece of comic-didactic fiction, the work describes the adventures of the generous, manly but dissipated hero, Tom Jones, from his obscure birth to his ultimate vindication and moral regeneration as Squire Allworthy's heir. *See also* HC2 p.34.

Tommaso da Modena (c. 1325–79), Italian painter who produced decorative frescoes for the Chapter House of S Nicolo, Treviso. This series, depicting Dominican scholars, is reminiscent of earlier manuscript illumination.

Tommy gun, popular name for a machine carbine, the Thompson submachine gun. Patented in the USA in 1920 by a retired army officer, John T. Thompson, it weighed about 4.5kg (10lb). It used 0.45in calibre ammunition from a box magazine holding 20 rounds or a drum holding 50 rounds. It achieved notoriety during the PROHIBITION era as a gangster weapon.

Tomography, technique of X-ray photography in which details of only a single slice or plane of body tissue are shown.

Tomonaga, Shin'ichirō (1906–), Japanese physicist who during WWII derived a formulation of quantum electrodynamics; that branch of QUANTUM THEORY concerned with the interactions of charged elementary particles with each other and with the electromagnetic field. He shared the 1965 Nobel Prize in physics with Richard P. FEYNMAN and Julian SCHWINGER, who worked on the same subject independently of Tomonaga.

Tompion, Thomas (1639–1713), British horologist. He raised the art of clockmaking to a high level, regarding the movement as more important than the exterior casing. There are examples of Tompion's art at Hampton Court and in many British museums and country houses.

Tom Sawyer, The Adventures of (1876), autobiographical novel by Mark TWAIN concerning the adventures of a young boy in a small Missouri town. Its sequel, Huckleberry Finn, proved an equal success. They were followed by *Tom Sawyer Abroad* (1894) and *Tom Sawyer, Detective* (1896).

Tom Thumb (1838–83), US entertainer, real name Charles S. Stratton. He was a midget made famous by the showman P. T. BARNUM, who took him to New York in 1842 and billed him as "General Tom Thomb". Tom Thumb's height never exceeded 102cm (40in).

Tonbridge and Malling, county district in

w central KENT, England, created in 1974 under the Local Government Act (1972) Area: 240sq km (93sq miles). Pop. (1974 est.) 94,800.

Tone, Theobald Wolfe (1768–98), Irish revolutionary leader. Tone, who wanted the Irish to sink religious differences and unite for political independence, founded the United Irishmen in 1791, and promoted the Catholic Relief Act in 1793. After visiting America in 1795, Tone masterminded an abortive French invasion of Ireland in 1796. He committed suicide after being captured in 1798.

Tone languages, languages that use the pitch of the voice as a major device in giving meaning or indicating grammatical relationships. For example, in Chinese a syllable that is transliterated as *ma* means "mother" when spoken on a high tone, "hemp" on a mid-rising tone, "scold" on a high-falling tone and "horse" on a falling-rising tone. *See also* MS p.240.

Tonga, island kingdom in the SW Pacific Ocean, approx. 2,200km (1,370 miles) NE of New Zealand. Formerly the British protectorate of the Friendly Islands, it is made up of approx. 150 islands. Coconuts and bananas are the chief crops and fishing is the most important industry. The capital is Nuku'alofa. Area: 699sq km (270sq miles). Pop. (1975 est.) 102,000. *See also* MW p.166.

Tongking. *See* TONKIN.

Tongue, muscular organ that moves food in the mouth during chewing and to the back of the mouth for swallowing, and functions in speech. The tongue contains the taste buds, groups of cells that distinguish the four basic tastes: bitter, tasted on the back of the tongue; sweet and salty, tasted on the tip and front of the tongue; and sour, tasted mainly on the sides of the tongue. *See also* SENSES; MS pp.68–69.

Tonic sol-fa, system of MUSICAL NOTATION designed to train the ear and voice in correct intonation. The system is the same for every DIATONIC scale, the notes being represented in their relationship to the tonic or keynote (*doh*). The notes are given as syllables (*doh, re, me, fa, soh, la, te*, etc.) and rhythm and accents are indicated by colons and dashes. It was developed in c. 1850 by John CURWEN.

Tonkin (Tongking), large area of N Vietnam. A former French protectorate, it borders China (N), the Gulf of Tonkin (E), the Annam region (S) and Laos (S and W). It was conquered and ruled by the Chinese from 111 BC to AD 939, later becoming an independent kingdom. It was the seat of the Northern rule of Vietnam after 1558, and became part of the union of Indochina in 1887. Tonkin was occupied in WWII by the Japanese and was the scene of a post-war demand for independence from the French, who were finally defeated in 1954. Area: approx. 103,600sq km (40,000sq miles).

Tonkin, Gulf of, shallow arm of the South China Sea bounded by the E coast of North Vietnam, the S coast of China, and Hainan island. The chief ports are Haiphong (Vietnam) and Peihai (China). Length: 500km (300 miles). Width: 240km (150 miles).

Tonnage, in marine engineering, term used to define the size or capacity of a ship. The displacement tonnage is the weight of the volume of water that a ship displaces: it is a measure of the size of the ship. The gross tonnage is the displacement tonnage of a ship and its cargo. The dead-weight tonnage is the difference of the gross and the displacement tonnages. Tonnage is expressed in volumetric tons, 2.83cu m (100cu ft).

Tönnies, Ferdinand Julius (1855–1936), German sociologist and political scientist. He developed the concepts of *Gemeinschaft* (organic community) and *Gesellschaft* (societal associations), which have been useful in analyzing the development of Western society since the INDUSTRIAL REVOLUTION. *See also* MS p.262.

Tønsberg, city in SE Norway, on the Skagerrak. Founded c.870, Tønsberg is one of the oldest cities in Norway. It is a major shipping and whaling centre. Pop. (1971) 11,200.

Tonsillectomy. *See* TONSILLITIS.

Tonsillitis, acute or chronic inflammation of the tonsils, often caused by a STREPTOCOCCUS infection. Tissues surrounding the tonsils frequently form pus during acute attacks, in which white specks or a white exudate may be seen on the surface of the enlarged glands. Chronic tonsillitis is often treated by surgical removal of the TONSILS (tonsillectomy). *See also* MS p.88.

Tonsils, two masses of lymphatic glandular tissue located at the back of the throat, between arch-like folds of membrane (fauces). They have a pitted surface that frequently becomes infected (tonsillitis). *See also* MS p.*88.*

Tonsure, shaving of the head as a sign of pious humility or clerical status. From medieval times in the Christian Churches, the Roman (St Peter's) tonsure left a fringe or a ring of hair to represent the crown of thorns; in the Eastern Churches' (St Paul's) tonsure, the head was entirely shaved.

Tonti (or Tonty), Henri de (1650–1704), Italian explorer of North America. He accompanied LA SALLE to Canada in 1678 and explored the GREAT LAKES. He travelled with La Salle to the mouth of the MISSISSIPPI, where they claimed the Mississippi Basin for France, naming it Louisiana.

Tonton Macoutes, personal police force of Haitian dictator, FRANÇOIS "PAPA DOC" DUVALIER, who empowered them to torture, kill, and terrorize. Their eyes were hidden by dark glasses, presumably to suggest they were zombies, thus invincible. The name is Haitian French for "bogeymen".

Tools, in archaeology, objects used by early man to assist him to shape his environment. The sophistication of the tools is a criterion used by archaeologists to date and evaluate early remains. The first tools were probably unshaped stones, sticks or bones. The techniques used for sharpening flints, in particular, are used to distinguish early cultures. Other important early inventions include fire and the wheel (invented c.3500 BC) but many cultures managed huge engineering works without the wheel. Metal was originally introduced for ornament, with copper the first metal discovered, c.6000 BC. *See also* HC1 pp.23, 27, *42*; MM pp.20–21, 24–27.

Tools, machine. *See* MACHINE TOOLS.

Tooth. *See* TEETH.

Toothache, pain in a tooth or surrounding tissues caused by nerve irritation, infection or decay. Treatment varies but cleaning and filling of cavities is a standard procedure. *See also* MS pp.130–131.

Tooth shell, or scaphophod, any of several hundred species of marine molluscs that have a tubular, toothlike shell open at both ends. It burrows into sand with a foot at the wider end, and has small tentacles to gather tiny organisms for food. The sexes are separated and fertilization is external. Length: 13–130mm (0.4–5in). Class Scaphopoda. *See also* NW p.95.

Topaz, transparent, glassy mineral, aluminium fluosilicate, $Al_2SiO_4(F,OH)$, found in pegmatites. Its crystals are orthorhombic system columnar prisms; it occurs as granular masses. Topaz is colourless, white, blue or yellow; some large crystals are of gem quality. Hardness 8; s.g. 3.5.

Tope, small shark that lives in British waters and is fished for food and sport. It has a grey-brown body and is often found in schools, or near the bottom where it feeds on small fish. Length: to 2m (6.5ft). Family Carcharinidae; species *Galeorhinus galeus*.

Tope, in Buddhist architecture, another name for a dagoba or STUPA. The tope was a burial mound containing relics and usually domed or beehive in shape.

Topeka, state capital of Kansas, USA, on the Kansas River, 90km (55 miles) w of Kansas City. It was founded in 1854 by settlers from New England, and became the state capital in 1861. The Menninger Clinic, world famous for its psychiatric research and treatment of mental illness, is in the city. Topeka is a major shipping centre for cattle and wheat. Industries: food processing, printing and publishing, rubber goods, footwear. Pop. (1970) 125,011.

Tomatoes for table use are harvested when green and allowed to ripen later.

Tom Sawyer's adventures stemmed partly from incidents in Mark Twain's boyhood.

Tønsberg, Norway's oldest town, is noted for the silverware manufactured there.

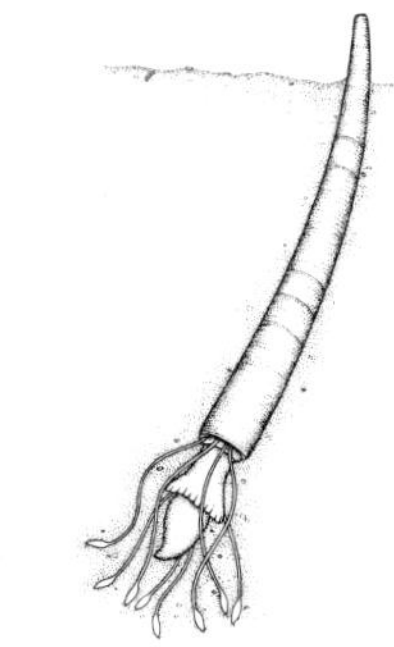

Tooth shells are molluscs whose shells are shaped like an elephant's tusk.

Torah, constituting God's revelation to Israel, is preserved on parchment scrolls.

Toronto, Canada's second largest city, is a financial and commercial centre.

Torricelli, Galileo's assistant, was the first man to create a sustained vacuum.

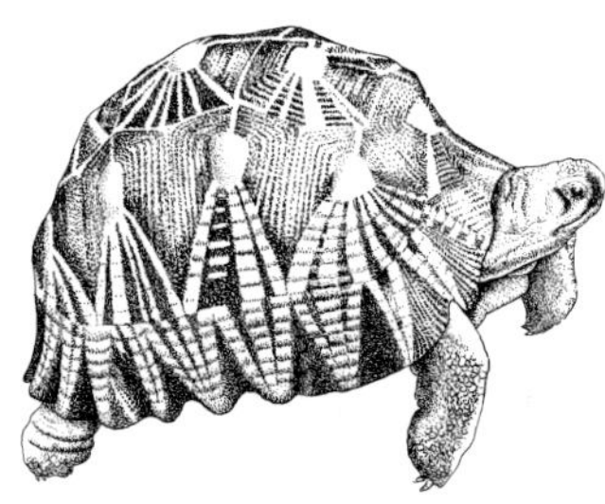

Tortoises carry a bony "shell", which is actually their ribs and backbone.

Topiary, practice of trimming densely-leaved evergreen shrubs and trees into decorative artificial shapes; it was once popular in English and French formal gardens. The best results are obtained with cypress, yew, box and holly.

Topkapi Searayi, or Topkapi Palace, seraglio built by Sultan MEHMET II in Istanbul and completed 1478. Situated on a promontory overlooking the Bosphorus, it was the sultan's residence until the 19th century.

Topography, study of surface features such as hills, valleys, rivers, roads and lakes. It is also the representation of such features on a relief map or a plan for construction. The terrain of a region is explored using surveyors' instruments or aerial PHOTOGRAMMETRY (plotting elevations from photographs). *See also* SURVEYING; PE pp.34–35.

Topology, branch of mathematics concerned with those properties of geometric figures that remain unchanged after a continuous deformation process such as squeezing, stretching, or twisting (but not tearing or breaking). The number of boundaries of a surface is such a property. All plane closed curves (ie any line that eventually comes back to its beginning, all on a single plane) is topologically equivalent to a circle; a cube, a solid cone and a solid cylinder are topologically equivalent to a sphere. These figures can be considered sets of points, each point of one set being transformable into one point in another set. *See also* SU pp.52–53.

Topolski, Feliks (1907–), British draughtsman and illustrator, b. Poland. He settled in England in 1935 and during WWII he was a war artist, both in England and abroad. He was commissioned to paint *Cavalcade of the Commonwealth* for the FESTIVAL OF BRITAIN (1951).

Torah, meaning "teaching", Hebrew name for the PENTATEUCH, the first five books of the Old Testament. By tradition the authorship is ascribed to MOSES, although at least one of the five was written around the time of the BABYLONIAN CAPTIVITY (586 BC). In the synagogue the Torah is kept in the Ark, being removed only to be read during Sabbath services. Torah is also used as a term to describe the complete Jewish Bible and all the laws and customs of Judaism.

Torbay, county district in s Devon England, created in 1974 under the Local Government Act (1972). Area: 63sq km (24sq miles). Pop. (1974 est.) 109,000.

Torch, small portable electric lamp powered by DRY CELL batteries or, rarely, by a portable generator. The first dry-cell torch was made in 1898.

Tordesillas, Treaty of, agreement signed in 1494 by Spain and Portugal in which the newly explored lands were divided into two spheres of influence. The dividing line from pole to pole passed 370 leagues w of the Cape Verde Islands; Spain's portion was all the area to the w of this line. The agreement meant that Brazil, still undiscovered, would be claimed by Portugal. *See also* HC1 p.*237.*

Torelli, Giacomo (1608–78), Italian architect and stage designer who played an important part in the development of stage machinery for the speedy change of heavy sets. In 1640 he designed the Teatro Novissimo in Venice and in 1645 went to Paris, where he was instrumental in the development of spectacle plays.

Torelli, Giuseppe (1658–1709), Italian composer and violinist. He is credited with the origination of the concerto for more than one instrument (his were for two violins), although the attribution to him therefore of the CONCERTO GROSSO form is now considered inaccurate.

Tori Busshi (*fl.* early 7th century), the first great Japanese sculptor. His reputation rests on the gilt bronze triad of Shaka Nyorai (*c.*623).

Torino. *See* TURIN.

Tornado, also called twister, funnel-shaped, violently rotating storm extending downwards from the cumulonimbus cloud from which it forms. At the ground its diameter may be only about 100 metres. Rotational wind speeds range from 160 to 480km/h (100 to 300mph) at the centre.

Tornadoes occur in deep low pressure areas, associated with FRONTS or other instabilities occurring particularly in the Eastern States of the USA. *See also* PE pp.68–69.

Toronto, port in s Ontario, Canada, 580km (360 miles) w of Montreal on the N shore of Lake Ontario; capital of Ontario province, and second largest city in Canada. The site was first visited in 1615 by the French explorer Étienne Brulé. A French fur-trading post, Fort Rouille, was established, followed by a fort in 1749 which was burned in 1759 to prevent capture by the British. In 1787 British Loyalists purchased from the Indians land on which a French fort named Toronto had been sited, and the settlement of York was founded in 1793 as the new capital of UPPER CANADA. During the WAR OF 1812 the city was captured twice by US troops. The city was re-named Toronto in 1834, serving again as the capital of Upper Canada in 1849 and becoming the provincial capital of Ontario in 1867. Today places of interest include Exhibition Park, the Art Gallery of Ontario, the Conservatory of Music and the Royal Ontario Museum. Educational institutions include the University of Toronto (1827) and the University of Trinity College (1852). Exports: grain, livestock, meat. Industries: electrical equipment, brewing, printing, publishing, iron and steel, food processing, chemicals, motor vehicles, aircraft and farm machinery. Pop. (1974) 712,785. *See also* MW p.*46.*

Torpedo, self-propelled underwater missile used by submarines, small surface warships and aircraft to destroy enemy vessels. Modern torpedoes may be launched by rocket boosters and often have internal electronic equipment for guiding the missile to the target. *See also* MM pp.*118.*

Torpedo boat, small, inexpensive, fast coastal naval craft armed with torpedoes and, recently, with antiship guided missiles. Torpedo boats are generally used to defend or attack coastal shipping and installations, and to conduct hit-and-run attacks against larger warships. *See also* MM pp.174–175, *174.*

Torpedo ray, or electric ray, flat-bodied fish found in all tropical and temperate marine waters from shallow depths to 1,000m (3,300ft). It has wing-like fins at the sides of the head and there are electric organs on each side of the head. Length: 1.5m (5ft). Families Torpedinidae, Narkidae and Terneridae. *See also* CHONDRICHTHYES; RAY.

Torquay, popular resort town of TORBAY on the s coast of Devon, sw England.

Torque, in art and archaeology, opened ring of gold or bronze worn as a neck decoration or bracelet. Most examples have a twisted pattern, often with animal heads on the ends. *See also* HC1 p.*86.*

Torque, in physics, VECTOR quantity describing the rotational force about an axis. A force F applied at a distance r from an axis of rotation produces a torque $\tau = rF$. The output of a rotary engine, such as the familiar 4-stroke engine, or an electric motor is rated by the torque which it can develop. *See also* MOMENT.

Torquemada, Tomás de (1420–1498), Spanish churchman and grand inquisitor. A DOMINICAN priest and confessor to FERDINAND II and ISABELLA I, he was appointed (1483) head of the Spanish INQUISITION. He became noted for the severity of his judgements and the harshness of his punishments and it was he who was chiefly responsible for the expulsion of the Jews from Spain in 1492.

Torrens, William Torrens McCullagh (1813–1894), British social reformer. Elected as an independent Liberal MP in 1835, he was influential in the passing of the Artisans' Dwellings Act (1868), relating to slum clearance, and the Education Act (1870), which established the London School Board.

Torres Restrepo, Camilo (1929–66), Roman Catholic priest from Colombia. He adopted Marxist views while studying in Belgium and joined guerillas in Colombia, but was killed by government forces.

Torres Naharro, Bartolomé de (*d.c.*1524), Spanish poet and playwright. After periods as a soldier and a priest, he was captured by the Moors, but survived to write much varied drama, including *Himenea,* a play about honour, and *Serafina,* a comedy.

Torres Strait, channel between Cape York Peninsula, Australia and New Guinea, connecting the Arafura and Coral seas. Numerous reefs make the channel dangerous to shipping. Width: 153km (95 miles).

Torrey Canyon, name of a supertanker carrying crude oil from Kuwait which ran aground off Lands End, Cornwall, in 1967. Nearly 100,000 tonnes of oil escaped, polluting the beaches of sw England, the Channel Islands and France. Thousands of seabirds died and the detergent used to combat the oil destroyed more marine life. *See also* PE p.152, *152.*

Torricelli, Evangelista (1608–47), Italian physicist. Assistant and secretary to GALILEO during the latter's last three months of life, he is credited with the first man-made VACUUM (the Torricellian vacuum) and in 1643 invented the BAROMETER. He also constructed a primitive microscope and improved the telescope. His work in geometry eventually contributed to the development of INTEGRAL CALCULUS.

Torrid zone, or tropics, region of the Earth between the TROPIC OF CANCER and the TROPIC OF CAPRICORN. The Sun is vertical at noon over every point within the tropics on at least one day of the year.

Torrigiano, Pietro (1472–1528), Italian sculptor of the school of FLORENCE. Having incurred the wrath of all Florence for breaking MICHELANGELO's nose, he went to England in 1511; his gilt bronze masterpiece, the tomb of King Henry VII and Elizabeth of York, is preserved in WESTMINSTER ABBEY.

Torsion, in mechanics, strain in material that is subjected to a twisting force. In a rod or shaft, such as an engine drive shaft, the torsion angle of twist is inversely proportional to the fourth power of the rod diameter multiplied by the shear modulus (a constant) of the material. Torsion bars are used in the spring mechanism of some motor car suspensions.

Torsion balance, sensitive device for measuring small forces, whether gravitational, magnetic or electrical. Torsion is the state of strain, set up in a material when it is twisted, which tends to restore it to its original position. A torsion balance has a horizontal arm, suspended at its centre by a fine wire or fibre, which is deflected sideways by the force being measured.

Tort, in British law, wrongful act or omission which can give rise to a civil action at law, other than concerning breach of contract. The law of tort includes negligence, libel, slander, trespass, false imprisonment and nuisance.

Tortelier, Paul (1914–), French cellist and composer. His career as one of the world's leading solo cellists began in 1947. His compositions are chiefly for the cello, but he has also written a piano concerto, a violin concerto and a symphony.

Tortoise, terrestrial or freshwater reptile of the order Chelonia. American usage favours the name TURTLE, reserving "tortoise" for land-living Chelonians of the family Testudinidae. All tortoises are heavily armoured with a high, domed, bony, box-like structure commonly called its "shell". When disturbed, tortoises pull their scaly legs, head and tail into the shelter of the shell. They live in tropical and subtropical regions, and hibernate in temperate countries. They are slow movers, feed almost entirely on plants and live to a great age. Length: usually to 30cm (1ft). A giant species, up to 1.9m (5ft) long, lives in the Galápagos Islands and some Indian Ocean islands. *See also* NW pp.*32–33, 137–138, 219.*

Tortoise-shell butterfly, any of a widespread group of medium sized butterflies native to Europe and temperate regions of Asia. Their wings have tortoise-shell-like markings of orange, black, brown and yellow. Family Nymphalidae.

Tortoise-shell cat, domestic cat with characteristic coloration of black, red and cream in irregular clearly-defined patches; the coat may be short or long. The eyes are generally copper or orange. Most tortoise-shell cats are female; males are usually sterile.

Torture, infliction of pain on a person to extract information, a confession or to indulge sadistic inclinations. It has been practised in many cultures and in some, notably medieval Europe and China, its refinements have reached the level of artistry. Until the 18th century it was considered a legitimate means of extracting a legal confession to a crime. The widespread use of torture for judicial purposes declined in the 18th century under the influence of ENLIGHTENMENT thinking; the 1949 GENEVA CONVENTION included a clause against torture, both of prisoners of war and of civilians of any country. Despite this, torture has continued in the 20th century. There has been increasing recognition of "psychological torture", in which disorientation, fear and loss of sleep and self-respect are used instead of or in addition to the simple application of physical pain. *See also* MS p.*296.*

Tory, Geoffroi (*c.*1480–*c.*1533), French printer and typographer. He is noted for his border design in several *Books of Hours* and for encouraging the departure in France from the use of Gothic to that of Roman letters. He wrote the first known work on type design, *Champfleury* (1529).

Tory Party, former political party in Britain. The term is still used as a nickname for the CONSERVATIVE PARTY, of which it was the forerunner. The word Tory comes from the Irish *toiridhe* (plunderer), which was applied to Irish Catholic bandits and was used as a term of abuse for supporters of JAMES II in the 1680s. During the regrouping which took place in English politics after the GLORIOUS REVOLUTION, a Tory Party based on support for the Anglican Church and opposition to continental entanglements became an important parliamentary force. After 1715, however, traditional Tory support for the STUARTS led the party to be eclipsed by the HANOVERIAN SUCCESSION, and Toryism became a backwater creed. It re-emerged in the 1790s when opponents of the French Revolution rallied round PITT's government. Tory administrations associated with anti-Catholicism, opposition to parliamentary reform and defence of the CORN LAWS dominated British politics until 1830. After the electoral reform in 1832, members of the Tory Party began forming Conservative associations, and Sir Robert PEEL headed the first Conservative administration in 1834. *See also* HC1 pp.292, 293, 312, 313; HC2 pp.68, 69.

Tosca (1900), three-act opera by Giacomo PUCCINI with libretto by Giuseppe Giacosa and Luigi Illica, after Victorien SARDOU's play, first performed in Rome.

Toscanini, Arturo (1867–1957), Italian conductor who became conductor at LA SCALA, Milan in 1898. He conducted the New York Philharmonic Orchestra from 1928 to 1936 and the NBC Symphony, which he formed in New York in 1937. He also worked at the Bayreuth, Salzburg and Lucerne festivals during the 1930s. He instilled his orchestras with remarkable energy and penetrated works with intense emotionalism and musical subtlety.

Tōson, Shimazaki. *See* SHIMAZAKI, TŌSON.

Tostig (*d.*1066), Earl of Northumbria, Anglo-Saxon nobleman, the son of GODWIN. After Tostig provoked a revolt in Northumbria in 1065, his brother, King HAROLD, replaced him with Earl Morcar. Tostig went into exile, but returned in 1066 with the Norwegian King HARALD HARDRADA. Their invasion was defeated and they were slain at the battle of STAMFORD BRIDGE. *See also* HC1 p.166.

Total internal reflection, reflection without refraction of light at a boundary. When light passes from a dense medium, such as glass, to a less dense medium, such as water or air, a range of angles of incidence exists in which no light passes through the boundary; all the light is reflected within the denser medium.

Totalitarianism, form of government in which the state tries to acquire total control of every aspect of social and individual activity or thought, by means of controlling the mass media, suppression of opposition and the often violent use of the police. The term arose in the 1920s to describe FASCIST Italy and has since been applied to NAZI Germany, Stalinist Russia and many other police states. It is not always easy to draw a distinction between totalitarian and "democratic" or "pluralist" societies – totalitarian state systems are rarely as monolithic as they seem and even liberal democratic societies have a monopoly of police and military power, and can have an effective control of the mass media. *See also* MS p.277.

Tote, or totalizator, automatic machine for laying bets, usually on horse or dog races. It divides its takings among the winning ticket-holders, the racing industry and any betting tax.

Totemism, ideas held by certain primitive societies about the relationships of KINSHIP between men and the animals or plants known as totems. Social groups may be distinguished by the totem with which they identify. Members of a totem group are prohibited from marrying others of the same group and from killing or eating their totem. Elaborate, often secret, rituals form an important part of totemistic behaviour. The totem may also be regarded by an individual as his spiritual protector, or as the guardian of his soul.

Totem pole, carved and painted wooden column erected by the Indians of the Pacific Coast of the USA and Canada. They are carved with stylized representations of real and mythical animals and men. Their function is closer to that of heraldic crests than of religious symbols. They are usually erected as roof supports; as doorways; as symbols of greeting; or as mortuary poles or grave-markers. *See also* MS p.*217.*

Totonac, tribe of Mexican Indians who gave their name to a distinct language family. They occupied Puebla and Veracruz as early as AD 100. Their descendants, numbering more than 130,000, still inhabit Veracruz, Puebla and Hidalgo.

Toucan, any of 35 species of colourful, gregarious birds of the forests of tropical America characterized by a large, colourful bill. The plumage is generally red, yellow, blue, black or orange. It lays one to four eggs that are incubated by both parents, and it feeds on fruit and berries. Length: 60cm (2ft). Family Ramphastidae. *See also* NW pp.*149, 216.*

Touch, principal skin sense, occurs when the skin is mechanically deformed (moved). In addition to free nerve endings, the skin of higher vertebrates also has special cell receptors, PACINIAN CORPUSCLES, which are particularly prevalent around muscles and joints. Touch probably interacts with the kinaesthetic sense to produce sensations about body position; the illusion of touch may persist even in a limb that has been amputated. *See also* MS p.52.

Toulon, port in SE France, on the Mediterranean sea, 48km (30 miles) ESE of Marseilles; the principal naval base of France. The city was a Roman naval station in the 3rd century. It passed to France in 1481. Industries: tourism, fishing, shipbuilding, chemicals. Pop. 181,600.

Toulouse, city in s France, on the River Garonne 214km (133 miles) ESE of Bordeaux; capital of Haute-Garonne département. Canals connect the city to the Mediterranean Sea and the Atlantic Ocean. The capital of the Visigoths in the 5th century, Toulouse became part of the French crown lands in 1271. The University of Toulouse dates from 1230. The city is the centre of the French aeronautic industry; other industries include paper, textiles, fertilizers and ammunition. Pop. (1971 est.) 396,800.

Toulouse-Lautrec, Henri Marie Raymond de (1864–1901), French painter and graphic artist. Profoundly influenced by DEGAS and Japanese prints, his paintings and posters depict his fascination for the spontaneity of circus, theatre and cabaret life. He worked mainly in thinned oil paint, and occasionally in watercolour or pastel. *Salon in the Rue des Moulins* (1894) exemplifies his style.

Touraco, or turaco, inquisitive, agile, cuckoo-like bird of SE Africa, known for its special feather pigments. Forest-dwelling touracos, with red wings resulting from the turacin pigment and green plumage due to turacoverdin, are typified by the Hartlaub's touraco (*Tauraco hartlaubi*): length: 35cm (14in). Bushland-dwelling touracos are typified by the grey plantain-eater (*Crinifer africanus*). Family Musophagidae.

Tour de France, the premier professional road cycling race in Europe. Raced over three weeks from the end of June, it travels over flat and mountainous country in a series of stages, beginning at Limoges and circling France until it ends in Paris. Since 1971 it has been limited to 20 stages averaging no more than 200km (124 miles) each. During the race the famous yellow jersey is worn by the overall leader.

Touré, Sékou (1922–), African political leader and first President of Guinea. An active trade union leader and nationalist, he led Guinea's campaign for independence from France and became president in 1958.

Tourel, Jennie (1910–74), French-Canadian mezzo-soprano who studied in Paris and made her début there in Georges BIZET's *Carmen* in 1933. She sang at the Metropolitan Opera, New York in 1937 and from 1943 to 1947.

Tourmaline, silicate mineral, sodium or calcium aluminium borosilicate, found in IGNEOUS and METAMORPHIC ROCKS. Its crystals are hexagonal system and glassy, either opaque or transparent. In colour tourmaline may be either black, red, green, brown or blue. Hardness 7.5; s.g. 3.1. Some crystals are prized as gems.

Tournament, medieval festival at which knights exhibited the martial arts in various sports, chiefly jousting and tilting. It was an element of feudal chivalry and the weapons were usually blunted. A revival, the Eglinton Tournament, was held at Eglinton Castle, Scotland, in 1839. The British armed forces stage a Royal Tournament each summer in London.

Tourneur, Cyril (*c.*1575–1626), English playwright and poet. Little is known about him but his poetical works include *The Transformed Metamorphosis* (1600). Best remembered as a dramatist, his *The Atheist's Tragedy: or The Honest Man's Revenge* was published in 1611.

Tourniquet, device that controls haemorrhage by exerting pressure on an artery usually in the form of a band or strap fastened tightly around a limb, thus preventing the blood from flowing.

Tours, city in w central France on the River Loire, 208km (129 miles) sw of Paris; capital of Indre-et-Loire département. It was the seat of the French government in 1870 during the siege of Paris. A wine market, Tours has tourist, food processing and pharmaceutical industries. Pop. 139,800.

Toussaint l'Ouverture, Pierre Dominique (*c.*1744–1803), Haitian independence leader. He joined the Negro rebellion in 1791 to free the slaves, drove the British and Spaniards from the island and by 1800 controlled all of HISPANIOLA. When NAPOLEON sent troops to re-establish French authority, Toussaint was captured and died in prison in France.

Tovey, Sir Donald Francis (1875–1940), British pianist, composer and music scholar. He is famous chiefly for his musical criticism, including the six-volume *Essays in Musical Analysis* (published 1935–39).

Tower Bridge, cantilever bridge over the River Thames in London, built by Sir Horace Jones between 1886 and 1894. The bridge has twin Gothic towers and a double-leaf bascule mechanism which opens to provide a 76m (250ft) gap. *See also* MM p.192.

Tower of Babel. *See* BABEL, TOWER OF.

Tower of London, fortress on the N bank of the River Thames in London. The original keep, or White Tower, was begun by William the Conqueror in 1078 on the site of British and Roman fortifications. Richard I added concentric CURTAIN

Arturo Toscanini takes the baton to conduct the orchestra of La Scala, Milan.

Toucans, with their huge, colourful bills, are playful and make delightful pets.

Tournament at Eglinton Castle, Scotland, where the medieval festival was revived.

Tower of London was the scene of many macabre incidents throughout history.

Arnold J. Toynbee's scholarship is evident in his history of 26 civilizations.

Tracery was used to create designs which provided attractively adorned openings.

Tractors are among the modern "work horses" for various farming jobs.

Spencer Tracy's rugged features suited his title role in *The Old Man and the Sea*.

WALLS with towers at intervals; later monarchs augmented the building. The Tower, which once housed the Royal Mint was for centuries a royal residence and a prison for state offenders. The Tower is now a barracks and an armoury, and houses the British crown jewels. It is still guarded by "Beefeaters" (Yeomen of the Guard) in traditional Tudor uniform.

Tower of Winds, octagonal tower in Athens built *c.* 100–50 BC by Andronicus of Cyrrhus to measure time. The tower has sundials on its sides and once had a water-clock inside based on a design by Ctesibius. The tower, built of marble, is 12.8m (42ft) high and 7.9m (26ft) in diameter. *See also* HC1 p.*83.*

Town, Harold Barling (1924–), Canadian painter and printmaker. He is known for his abstractions, in which primitivist shapes are elaborately contrasted.

Town, urban community larger than a village and with a significant proportion of inhabitants whose livelihood is not directly derived from the land. Town life is based first on a local market-place, and then on longer distance trade. Towns were first found in ancient SUMER, and they were the focus of civilization until the fall of the Roman Empire. Town life began to revive in the Low Countries in the 9th century, but in England the decisive development that encouraged large towns was the Norman Conquest of 1066. Throughout the Middle Ages towns were the centre of political organization and unrest; many boroughs achieved a degree of independence and their populace, particularly in London, could wield significant political power. The Industrial Revolution caused a major increase in the size and number of towns; and in the 19th century the first attempt was made in Britain to regulate the size and conditions of towns nationally. The Artisans Dwelling Act (1875) gave local authorities the power to clear slums. TOWN PLANNING, a well-known art in the Renaissance, was reintroduced in Britain in the 1880s and the post-WWII period has seen the development of NEW TOWNS and the rebuilding of many old ones. *See also* indexes to *History and Culture* volumes.

Towne, Francis (1739–1816), British watercolour painter. He was esteemed for the spacious simplicity of his designs, and his art was the high point of the 18th-century tinted drawing style of English watercolour painting. The *Source of the Arveiron* (1781) is a fine example of his work.

Townes, Charles Hard (1915–), US physicist. He invented the MASER for which he shared the Nobel Prize in physics in 1964 with Alexander PROKHOROV and Nikolai BASOV. The MICHELSON-MORLEY EXPERIMENT was accurately confirmed with the aid of masers.

Town Planning, arrangement of urban centres to provide healthy living conditions, efficient transport and communications, public facilities and aesthetic surroundings. The ancient cities of Babylon and Nineveh were planned around a central citadel and palace in the plan of a grid-iron. The monumental approach of the Renaissance gave long and wide avenues. In Britain in the 20th century, total town planning has given rise to the GARDEN CITIES.

Townshend, Charles, 2nd Viscount (1674–1738), English politician known as "Turnip Townshend". Townshend, a Whig, was Robert WALPOLE's brother-in-law, and was Secretary of State for the northern department (1714–17, 1721–30). He was responsible for forming the League of Hanover in 1725. Thwarted in his aggressive policy against Austria, he resigned in 1730 and devoted his retirement to agricultural improvements. *See also* HC2 p.20.

Townshend Acts (1767), series of taxes levied on the American colonies by the British Parliament. The Chancellor of the Exchequer, Charles Townshend, introduced them to provide revenue to defray the cost of the colonial government. They taxed numerous imports, including glass, paper, lead, tea and paint. The taxes were so vehemently opposed by the colonists that all were repealed, except that on tea (1770).

Towton Field, near York, site of a battle (1461) in the Wars of the ROSES in which the House of YORK under EDWARD IV defeated the House of LANCASTER.

Toxaemia, condition resulting from the presence of poisons in the blood. It is usually due to bacterial infection at a local site, such as an abscess, but inadequate excretion due to a defective kidney may also cause toxaemia.

Toxicology, the study of poisons and their effects. It has become an important science with the growth of modern industry, many of whose products, although essential for industry, agriculture or medicine, are toxic. Toxicological testing of new products is now indispensable, both within the laboratories of producing industries and in government and other monitoring laboratories.

Toxin, poisonous substance, usually a product of certain plants, animals or BACTERIA. The organisms which cause DIPHTHERIA and TETANUS are examples of toxins. The controlled use of bacterial toxins as immunological agents has been effective in providing protection against those and other disorders. Antitoxins produced against diluted toxins are effective immediately against later infection.

Toxoid, bacterial TOXIN that has lost its virulence but can still stimulate the production of ANTIBODIES in response to infection. This is the basis of VACCINATION.

Toxoplasmosis, infection that occurs in many animals, including man, caused by micro-organisms of the genus *Toxoplasma*. Symptoms are generally mild in adults, but in a pregnant woman the parasite can be damaging to the central nervous system or to the eyes of her child.

Toy dog, any of numerous small breeds of dogs; many have been dwarfed from larger breeds. They range in size from the tiny CHIHUAHUA and MALTESE, to the larger PEKINESE, toy POODLE and PUG.

Toynbee, Arnold Joseph (1889–1975), British historian who worked for the British foreign office during WWI and WWII and was a delegate to the Paris Peace Conference in 1919. He is famous for *A Study of History* (12 vols, 1934–61), in which he emphasized the need to examine whole civilizations rather than individual nations.

Toyonobu, Ishikawa (1711–85), Japanese ukiyo-e painter and printmaker whose earliest works were a series of lacquer prints. His best works were portraits of women and actors, painted with sensitivity and grace.

Trabeate, architectural term for the use of beams in classical Greek building. The beams are horizontal, placed across columns and providing a flat, unvaulted ceiling or doorway.

Trace elements, chemical elements in the body which are essential to life but are needed only in small quantities, supplied in food. They include boron, cobalt, copper, manganese, molybdenum and zinc. They are essential to the reactions of ENZYMES and HORMONES. More generally, trace elements are those found in nature only in small quantities.

Tracer, radioactive, RADIOACTIVE ISOTOPE which, in medicine, is fed to or injected into a patient. Its course is followed in the body by the detection and identification of metabolic compounds "labelled" with the isotope. This technique may help in the diagnosis of disease. Radioactive tracers are used in a similar way to follow chemical reactions in complex organic industrial processes. *See also* MS p.125.

Tracery, in Gothic architecture, ornamental subdivisions of arched openings, especially windows. Beginning with the simple plate tracery of the ROMANESQUE style, designs became increasingly ornate and the window area larger until the culmination of the style in the FLAMBOYANT tracery of the 15th century.

Trachea, or windpipe, tube that extends from the larynx to about the middle of the breastbone. It is lined with hair-like cilia that prevent dirt and other substances from entering the lungs. At its lower end the trachea splits into two branches, the bronchi, which lead to the lungs. *See also* MS p.66, *66.*

Tracheophyte, any of the VASCULAR PLANTS, all of which are members of the phylum Tracheophyta. Within this phylum are the psilopsids, leafless, rootless primitive forms such as whisk ferns; sphenopsids, the HORSETAILS; lycopsids, the CLUB MOSSES; and pteropsids, the FERNS, GYMNOSPERMS and flowering plants.

Tracheotomy, surgical procedure in which an incision is made through the skin into the TRACHEA (windpipe) to facilitate breathing or to give oxygen. The hole cut in the trachea is called a tracheostomy.

Trachoma, chronic, contagious CONJUNCTIVITIS, caused by the micro-organism *Chlamydia trachomatis*, characterized by inflammatory granulations on the conjunctival surfaces of the eye. In early stages it responds to treatment with sulphonamide drugs. It is a major cause of blindness in Africa and Asia.

Tracking, in aerospace technology, following a moving object, usually with radar or radio using aerials that "lock" onto the object. The correct sweep speed of the aerials is calculated by computer from information of the object's position and velocity. Satellite tracking is of two kinds: active tracking uses radar to locate an object; passive tracking uses signals from the object itself.

Tractarianism. *See* OXFORD MOVEMENT.

Traction, in medicine, gradual exertion of a force on a part of the skeleton to overcome muscular contraction and so ensure proper alignment of a fractured bone. The force is exerted gradually over a period of time generally using weights on cords over pulleys.

Traction engine, large, heavy vehicle powered by steam, once used for drawing farm wagons and other heavy loads, but now obsolete. Traction engines were also used at funfairs and circuses to drive dynamos for electricity.

Tractor, four-wheeled or tracked machine designed to pull or push heavy implements over rough terrain. It is commonly used in agriculture, construction and industry. *See also* MM p.*99;* PE p.164, *164.*

Tracy, Spencer (1900–67), US film actor. He appeared in nearly 80 films, of which nine, including *Adam's Rib* (1949), *Pat and Mike* (1957) and his last film, *Guess Who's Coming to Dinner* (1967), were made with his famous screen partner, Katharine HEPBURN.

Trade, Board of, former British government department. It evolved out of a series of committees appointed by the Privy Council in the 17th century to enquire into trade questions. Later committees were called Boards of Trade. By 1970 the Board of Trade was merged with the Ministry of Technology and it became the Department of Trade and Industry. In 1974 it re-emerged as a separate entity, the Department of Trade with its own Secretary of State.

Trade, international, the exchange of goods between countries. Modern international trade developed from the mercantile practices in Europe in the 16th and 17th centuries. Advocated by Adam SMITH as a means whereby nations could specialize in particular products, the growth of such trade paralleled the development of manufacturing and improved sea transport. The development of refrigerated cargo carriers in the late 19th century provided an important additional source of food for Britain from the colonies, thereby enabling it to diversify its manufacturing. Today international trade is complex and diverse with many bilateral and international agreements, such as GATT (General Agreements on Tariffs and Trade), with most countries attempting to sell their produce, obtain needed materials and at the same time protect home industries. *See also* MS pp.314–317.

Trade barriers, methods a country uses to discourage the importing of goods. Trade barriers include TARIFFS, quotas and absolute bans against importing of certain types of goods. Customs regulations may be so complicated as to constitute an informal trade barrier. Trade barriers are used to protect special industrial interests or the national economic interest of a state or a country.

Trade mark, distinguishing mark such as a name, symbol or word, attached to goods by a manufacturer to identify them as made or sold by him. A trade mark must be registered at the PATENT Office, which establishes exclusive right to it. It became important during the INDUSTRIAL REVOLUTION as a means of differentiating between brands of similar products.

Tradescant, name of two British botanists and travellers, father and son, who both became gardeners to Charles I. John (*d.* 1638) travelled to Russia and Algeria gathering plants, among them the "Algier Apricot", and established a garden of exotic plants, at his Lambeth home. His son, also named John (1608–62), travelled to Virginia for plants for the Lambeth garden. He wrote a book about the Lambeth collection, *Musaeum Tradescantium* (1656). Their collection, transferred to Oxford after their deaths, formed the nucleus of the Ashmolean Museum.

Trades Unions Congress. *See* TUC.

Trade Unions, groups of workers organized for the purpose of improving wages and conditions of work. The first trade unions were founded in Britain. Although some craft and agricultural unions developed before industrialization, the growth of trade unionism paralleled the growth of industry, particularly manufacturing. Originally illegal because of their subversive element, trade unions were given restricted legality in Britain in 1825, although their precarious status was illustrated by the conviction of the TOLPUDDLE MARTYRS in 1834. The Trades Union Act (1871) put the unions on a firm legal basis. Legislative provisions affecting unions in Britain are incorporated in the Industrial Relations Act (1971), which provides for the registration of trade unions. *See also* MS pp.275, *275*; HC2 pp.93–95, *170*, 210–213, 238.

Trade winds, steady winds that blow westwards towards the equator from subtropical high pressure zones between 30°N and 30°s.

Trading stamps. *See* STAMPS, TRADING.

Trafalgar, Battle of (1805), naval engagement in the Napoleonic Wars fought off Cape Trafalgar, Spain. NELSON, the British admiral, divided his 27 ships into two squadrons and broke the Franco-Spanish line of 33 ships under Admiral de Villeneuve. Nelson, on board HMS *Victory*, was killed, but the battle established British naval supremacy in the 19th century. *See also* HC2 pp.76, *76*, *78*, *79*.

Traffic control, regulation of the movement of vehicles and pedestrians, especially in cities. At busy crossings traffic lights are used with separate phases for pedestrians and each direction of traffic flow. These lights may be operated by sensitive pads in the road triggered by passing vehicles, pedestrians signalling, by police control of the lights or by computers. Computer control enables traffic volumes along busy roads to be regulated by the automatic allocation of time for each phase of the flow.

Trafford, county district in central s GREATER MANCHESTER, England; created in 1974 under the Local Government Act (1972). Area: 106sq km (41sq miles). Pop. (1974 est.) 227,400.

Tragedy, form of DRAMA in which a noble hero (the protagonist) meets a fate inherent in the drama's action. *Oedipus Rex* by SOPHOCLES is an early example, which was unmatched until the tragedies of Christopher MARLOWE. ARISTOTLE'S *Poetics* systematized tragedy and introduced such ideas as *anagnorisis* (recognition) and *catharsis* (the purging of pity and terror in the spectator).

Traherne, Thomas (1637–74), English writer, chiefly on religious subjects. He wrote a number of mystical poems and the prose works *Roman Forgeries* (1673), *Christian Ethicks* (1675) and, what was to be the most famous of his works, *Centuries of Meditation*, which was not published until 1908.

Traini, Francesco (*fl.* 14th century), Italian painter who was greatly influenced by the work of LORENZETTI and PISANO. Among his works is the altarpiece *Apotheosis of St Thomas Aquinas* (1363) in Sta Caterina, Pisa.

Trains. *See* RAILWAYS.

Trajan (*c.* AD 52–117), Roman emperor (*r.* 98–117) under whom the frontiers of the empire reached their greatest extent. He was born in Spain and was adopted by his predecessor NERVA; he was the first emperor from the provinces. Trajan maintained good relations with traditional Roman institutions such as the Senate. In campaigns of 101–102 and 105–106 he conquered Dacia, N of the Danube, and from 113 to 116 campaigned against the Parthians and added Mesopotamia to the empire. In later years his reign was seen as the summit of Roman power, but some authorities believe that his conquests strained the military resources of the empire. He was succeeded by HADRIAN. *See also* HC1 pp.98, *99*.

Trajectory, path of a projectile. On Earth, and in the absence of air resistance, all trajectories would be portions of an ELLIPSE whose focus was at the centre of the Earth. Since this is 6,440km (4,000 miles) deep, and usually much greater than the height of the trajectory, the mathematically simpler PARABOLA provides a good approximation.

Trakl, Georg (1887–1914), Austrian poet, whose highly expressionist poetry was influenced by BAUDELAIRE and DOSTOEVSKY. *See also* HC2 p.325.

Trampolining, pastime and competitive sport that involves the performance of acrobatic manoeuvres while bouncing on a tightly-sprung canvas springboard.

Trams, passenger carriages that run in streets, fixed to a single course on rails. They were introduced by John Mason in New York in 1832; they were at first pulled by horses, later being powered by steam locomotives or by means of electric overhead cables. In Britain after 1872 when they were first allowed to run by mechanical power, their popularity rose. After WWI they declined, to be replaced by motor buses and the trolleybus. The last tram ran in London in 1952, but many European cities still have a tramway system. *See also* MM pp.134–135.

Tranquillizers, drugs used to quieten mental states characterized by anxiety, tension and mania (or hyperactivity). They are divided into two categories: the antipsychotic drugs, such as RESERPINE and other ALKALOIDS from *Rauwolfia serpentina*, chlorpromazine and its related phenothiazines; and the antineurotics, such as DIAZEPAM, chlordiazepoxide and meprobamate. The antipsychotic drugs, in high doses, can produce serious side-effects. Some of the side-effects of phenothiazines include skin pigmentation, weight gain, sensitivity to light (photosensitivity), lactation and, in a few patients, reactions resembling PARKINSON'S DISEASE. *See also* MS p.137.

Transatlantic cable, any submarine cable laid under the Atlantic Ocean and used to carry electrical signals. The first successful one was the telegraph cable laid in 1866. *See also* MM pp.225–226.

Transatlantic telephone, extension of the European and North American telephone systems to include a link across the Atlantic Ocean via submarine cables or satellite. A major obstacle to the development of long-distance telephone was the weakening of the signal because of the electrical resistance of the cable. Repeaters were developed and incorporated at regular distances into the cable to amplify the signals. The first multi-channel telephone cable was laid in 1956 between Scotland and Newfoundland. *See also* MM pp.226–227.

Transcaucasia (Transcaucasian Federation, or Transcaucasian SR), former federated union of the USSR; now the three Soviet Socialist Republics of Armenia, Azerbaijan and Georgia. It was first formed in 1918 after the Russian Revolution but was soon divided into the three separate republics. In 1919–21 Turkish nationalists fought with the BOLSHEVIKS for control of this region; in 1922 it was reformed and entered the USSR in 1924.

Georgia, Azerbaijan and Armenia were re-established as union republics in 1936.

Transcendentalism, school of philosophy which traced its origin to the idealism of Immanuel KANT. It was concerned not with objects, but with our mode of knowing objects. It spread from Germany to England, where Samuel COLERIDGE and Thomas CARLYLE came under its influence. In the mid-19th century it spread to the USA, where it was propagated by a literary circle based in Concord, Massachusetts, including Ralph EMERSON, Henry THOREAU, Amos Alcott and Margaret Fuller.

Transcendental meditation (TM), type of meditation taught by Maharishi Mahesh Yogi. TM decreases oxygen consumption and heart rate and increases skin resistance and alpha brain waves, yielding a relaxed mental state differing from sleep or hypnosis. *See also* MS pp.195, *224*, *225*.

Transducer, device for converting any nonelectrical signal, such as sound or light, into an electrical signal, and vice versa. Examples are microphones, loudspeakers, gramophone pickups and various measuring instruments used in acoustics.

Transduction, in microbiology, natural process by which genetic material is transferred between one host CELL and another by a VIRUS.

Transept, in church architecture, the transverse section of a church at right-angles to the main axis. The north and south transepts are of the same height as the nave and are positioned immediately next to the choir or chancels.

Transfer orbit (transfer ellipse), path followed by a spacecraft while changing from one orbital path to another. The tangential ellipse is a tangent to the arrival and departure orbits and uses the least energy. Transfer ellipses intersecting the arrival and departure at higher angles use more energy but may be advantageous in other ways.

Transfer printing, in ceramics, a technique of decorating pieces involving an inked copper engraving that is printed onto paper; the print is applied to a glazed pottery surface while wet. The resulting offset monochrome image, which may then be coloured by hand, is fired into the glaze.

Transfer RNA. *See* RNA.

Transfiguration, New Testament occurrence involving the sudden physical transformation of Christ in the presence of St Peter, St James and St John, described, for example, in Matt.17:1–8. His face and garments became radiant, Elijah and Moses appeared, and God spoke from a cloud. This event revealed Christ's true divine nature and earthly mission.

Transfinite number, number denoting the size of an infinite set, its symbol is $\aleph$ (the Hebrew letter aleph) with a subscript. $\aleph_0$ represents the set of all integers; $\aleph_1$ represents the set of all real numbers and is infinitely larger than $\aleph_0$.

Transformer, device for converting alternating current at one voltage to another voltage at the same frequency. It consists of two coils of wire coupled together magnetically. The input current is fed to one coil (the primary), the output being taken from the other coil (the secondary). Large oil-cooled transformers are used in electricity transmission and small ones are employed in domestic appliances and electronic equipment. *See also* MM pp.84–85; SU p.122.

Transform fault, special class of strike-slip fault characteristic of mid-ocean ridges. Because of the transform faults, the Mid-Atlantic Ridge does not run in a straight line but in offset steps. Some geologists think the four major structures of the Earth's crust are mountains, deep-sea trenches, mid-ocean ridges, and strike-slip faults and that they form continuous networks. *See also* PE p.24.

Transfusion, blood. *See* BLOOD TRANSFUSION.

Transhumance, seasonal moving of livestock from one region to another. Transhumance occurs in societies living in zones with extensive climatic changes such as in the mountainous terrain of the Arctic regions or desert of Central Africa, where

Trajan's portrait bust, in the Vatican museum, dates from Roman times.

Trams have been used for public transport in Blackpool, England, since 1885.

Transcendental meditation was established in the West by the Maharishi Mahesh Yogi.

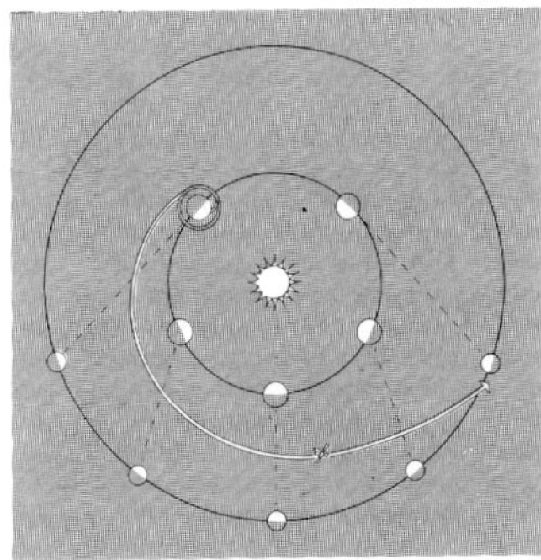

Transfer orbit is the path followed by a spacecraft moving from one orbit to another.

Transient Lunar Phenomena

Transistors have various shapes and sizes, depending on their applications.

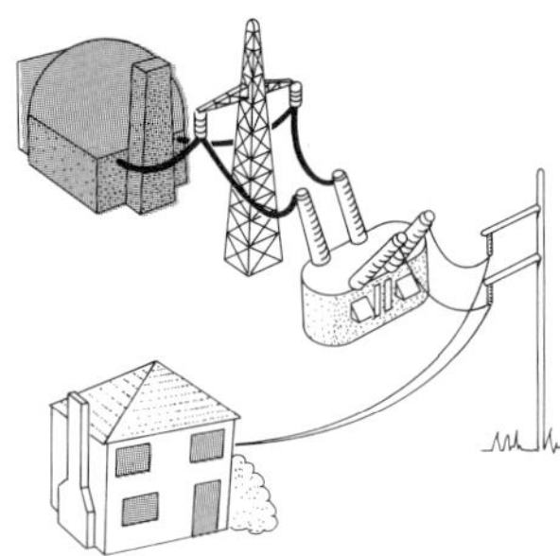

Transmission lines carry electricity from the power station to the consumer.

Transpiration is the process by which plants lose water through their leaves.

Transportation of convicts led to the establishment of penal colonies.

livestock cannot survive for the whole year because of cold, drought or flooding. *See also* PE p.*225.*

Transient Lunar Phenomena (TLP), surface or sub-surface activities on the Moon, possibly volcanic in origin. The sites of TLP are concentrated on the edges of the lunar seas and in regions rich in RIL-LES. *See also* SU pp.183, *183,* 185.

Transients, in music acoustics, the momentary sounds which precede and follow a note played on a musical instrument. These sounds are important, particularly at the onset of the vibration, for the listener's recognition of an instrument. In a certain register, for instance, the sounds of the violin and oboe are very alike but can be distinguished because they have different transients (due to the action of the bow and reed respectively).

Transistor, semiconductor device which can amplify electrical signals. A semiconductor material, such as silicon or germanium, is "doped" with minute amounts of either phosphorus, arsenic or antimony to produce n-type material or with either aluminium, gallium or indium to give a p-type material. Joining together a piece of each gives a DIODE. Sandwiching one type between two of the other produces a transistor. These can thus be of two kinds: a p-n-p or n-p-n transistor. The middle region is known as the base, one of the side regions as the emitter and the other as the collector. In an n-p-n transistor the signal to be amplified is fed across the collector, maintained at a constant voltage with respect to the base. The amplified signal comes out across the base and the emitter. Transistors were first developed by John BARDEEN, Walter Brattain and William SHOCKLEY in 1948 and they have made possible many advances in technology, especially in computers, aerospace, industrial control systems and navigation. *See also* SEMICONDUCTIVITY; SU pp.128–129, *128.*

Transistor radio. *See* RADIO, TRANSISTOR.

Transit, in astronomy, passage of either of the inner planets, Mercury and Venus, across the face of the Sun. Transits of Mercury occur every 13 years, those of Venus every 100 years. The term also signifies the motion of any celestial object across the observer's celestial meridian.

Transitional, in architecture, intermediate phase between about 1145 and 1190 when the transition in style from ROMANESQUE to GOTHIC (or, in Britain, from Norman to EARLY ENGLISH) was taking place.

Transition elements, metallic elements with chemical properties resembling those of their horizontal neighbours in the PERIODIC TABLE. They have incomplete inner electron shells, and are characterized by variable valencies, the formation of coloured ions and a tendency to form stable co-ordination compounds (complexes). They include elements with atomic numbers from 21 (scandium) to 29 (copper), 39 (yttrium) to 48 (cadmium), 57 (lanthanum) to 80 (mercury), and 89 (actinium) to 103 (lawrencium). The lanthanides and actinides are sometimes called inner transition elements. *See also* SU pp.134, 136.

Transjordan, region that today coincides with the Kingdom of Jordan in Asia Minor, bounded by Syria (N), Israel (W), Saudi Arabia (S and W) and Iraq (NW). Called Transjordan from the early 1900s to 1949, the region was part of the British LEAGUE OF NATIONS mandate of PALESTINE in 1920 and became an independent constitutional state in May 1923. By a treaty agreement with Great Britain in 1928, British financial assistance to Transjordan continued, accompanied by the presence of British military forces in the country; the region officially became the Hashemite Kingdom of Jordan in 1949, when national forces took from Palestine the areas W of the River Jordan which had been named Arab territories by the UN.

Transkei, black African "homeland" or Bantustan, E central Cape Province, South Africa. In line with S Africa's APARTHEID policy, Transkei became semi-independent in 1976; this however was not recognized internationally. Transkei is mostly mountainous territory. The prin-

ciple occupation is rearing live-stock. The economy is based on money sent home by migrant labourers working in the Orange Free State and the Transvaal. Area: 39,0008sq km (15,061sq miles). Pop. (1970) 1,733,931. *See* NW p.154.

Translocation, in vascular plants, the movement of food materials in solution through the tissues from one part of the plant to another.

Transmigration of souls, or reincarnation, belief that the soul is reborn in one or more successive mortal bodies. A tenet of Asian religions such as BUDDHISM, it was also taught by the PYTHAGOREANS and ORPHICS in 6th-century BC Greece and is common in tribal religions such as that of the South African Venda.

Transmission, engine, device (a gear or gears and shaft or driving chain) for transmitting power from the engine of a vehicle to the wheels. Speed can usually be varied in discrete steps, with gears or chains providing fixed speed ratios. *See also* MM p.130.

Transmission lines, in electrical engineering, wires or cables used to carry electrical power, or impulses such as telephone and other communications messages. Wires are slung between telegraph or other posts. Heavier cables, carrying electricity at high voltages, are slung on insulators between pylons, or buried underground, or laid underwater. *See also* MM pp.*84, 225.*

Transmitter, radio, electrical apparatus that generates radio waves. It has three main parts: an oscillator which generates high-frequency alternating electric current (a carrier wave); a modulator which encodes the carrier wave with the pattern or sound to be broadcast; and an amplifier which increases the power of the modulated carrier (the signal). *See also* MM pp.228–229.

Transmutation, in physics, formation of one element or ISOTOPE from another by RADIOACTIVE DECAY or by bombardment with energetic particles. The term was formerly applied to alchemists' attempts to convert base metals into gold.

Transpiration, in plants, the release of moisture from leaf surfaces or other plant parts. Of water entering plant roots, most is lost by transpiration. The process is slowed by darkness, extremes of temperature and inadequate water supply.

Transplants, changes or replacements of organs or tissues from one organism to another. Usually these involve a diseased organ such as a kidney being exchanged with a healthy one from a donor. A blood transfusion may be considered to be a transplant, because strictly speaking blood is a tissue. Heart, kidney, eye, testes, bone and skin transplants have worked for various periods. Drugs must be used to prevent the body's immune system from rejecting the organ as "foreign". *See also* MS pp.102, 127, *127.*

Transport, in chemistry and physics, movement of measurable entities such as MOLECULES, IONS, ISOTOPES, electrical charges, mass, momentum or energy through or across a medium, because of the natural or applied non-uniform conditions existing within it. Transport properties include viscosity, diffusivity, and thermal conductivity. Active transport in biochemistry is the movement of a substance against a concentration gradient.

Transport, Ministry of, British government department. It was first formed in 1918 to look after road-building and railways. In 1941 it was joined to the Ministry of Shipping and called the Ministry of War Transport. In 1946 it once more became the Ministry of Transport. In 1953 it became part of the Ministry of Civil Aviation, and in 1969 it became a subordinate department in the department for local government and regional planning. Its duties cover road-building, road licensing and safety and road and rail transport.

Transportation, in geology, movement of particulate matter from a source to an area of deposition. Any movement of material with wind, water, ice or gravity as the transporting medium falls into this category. *See also* PE pp.112, 116.

Transportation of convicts, policy origi-

nating in England in the 17th century of sending criminals to the colonies for life, or shorter periods. They were sent to America until the War of Independence and then in 1788 New South Wales, Australia, was set up as a penal colony. More than 160,000 convicts of both sexes were sent to Australia until 1868, when the policy was finally abolished in all its forms. *See also* HC2 p.126.

Transsexualism, act of permanently changing one's sex or the desire to do so. A sex change may be accomplished partly through hormone treatment or through surgery, and in some countries the change can be accompanied by a legal change of status. Transsexuals may be allowed to marry. For a true hermaphrodite (a person born with both male and female characteristics), choice of sex is literally possible. *See also* MS p.141.

Trans-Siberian Railway, railway that links European USSR to the Pacific. It was begun in 1891 and completed in 1905. It originally started at Chelyabinsk and ended at Vladivostok. The section of it which passed through Marchuria was known as the Chinese Eastern Railway. The present railway has several lines and is linked to the Turkestan-Siberian Railway.

Transubstantiation, Roman Catholic term defining the belief that, during the prayer of consecration at the MASS (the EUCHARIST), the "substance" of the bread and wine are changed into the "substance" of the body and blood of Christ, while the "accidents" (the outward form) remain unchanged. The doctrine was defined at the LATERAN COUNCIL of 1215 and was given classical expression later in the 13th century by St Thomas AQUINAS. The Eastern Orthodox Church holds an essentially similar doctrine, although most of its theologians prefer to avoid the use of the term. The definition involving "substance" and "accidents" was rejected by the reformers of the REFORMATION, but ecumenical discussions in recent years have revealed the underlying similarities in eucharistic belief between the Roman Catholic and other Churches. Many Roman Catholic theologians now regard the use of the term Transubstantiation merely as an attempt to use the Aristotelian philosophical language current in the 13th century to formulate belief in the REAL PRESENCE of Christ in the sacrament. *See also* CONSUBSTANTIATION.

Transuranic elements, those elements with atomic numbers higher than that of URANIUM (at.no. 92), the best known of which are members of the ACTINIDE series in the PERIODIC TABLE. All transuranic elements are radioactive and none occurs naturally. The first to be produced was NEPTUNIUM by E. M. McMillan and P. H. Abelson in 1940, who bomarded uranium with neutrons in a SYNCHROCYCLOTRON. The other elements have since been produced in a similar manner. The only commercially important element in the group is PLUTONIUM which is used in NUCLEAR WEAPONS.

Transvaal, province in the NE Republic of South Africa, between the Limpopo and Vaal Rivers. Inhabited by Bantu-speaking black Africans in the early 19th century, it was taken by the BOERS in the mid-1830s; the SAND RIVER CONVENTION of 1852 recognized the right of Boer self-administration. The South African Republic was formed in 1856, but the land was annexed by Britain in 1877. After a Boer revolt the Transvaal was again proclaimed a republic and granted internal self-government (1881). The discovery of gold in 1886 attracted many foreigners, but the Boers imposed heavy taxation and denied political rights to newcomers. At the close of the Boer War (1902), the Transvaal became a Crown colony and a founding province of the Union of South Africa in 1910. The capital is Pretoria, although the largest city is Johannesburg. Products include livestock, tobacco, cotton, wheat and citrus fruits. There is also mining for gold (it has the richest gold fields in the world), diamonds, uranium, chromium and platinum. Area: 286,065sq km (110,450sq miles). Pop. (1970) 8,717,530. *See* MW p.154.

HC1 = History and Culture Vol. 1   HC2 = History and Culture Vol. 2   MS = Man and Society   MM = Man and Machines

Transverse wave. *See* WAVE.

Transvestism, sexual deviation in which a person identifies with and dresses like the opposite sex.

Transylvania, region in central Romania, surrounded by the Carpathian Mts; the chief city is Cluj. It is located on a plateau drained by the River Mureşul. Originally part of the Dacian Kingdom, Transylvania was conquered by the Romans in AD 106. Colonized by both German and Slavic tribes in the Middle Ages, the area was fiercely disputed by Austria, Hungary and Ottoman Turkey. It became part of Hungary by the AUSGLEICH of 1867 and was ceded to Romania in 1920. Both agriculture and mineral resources are important in modern Transylvania; stock raising and fruit growing are major activities and there is mining for coal, gold, manganese and iron. Area: 55,146sq km (21,292sq miles).

Traore, Moussa (1936–), military commander and politician of MALI. He became president through a coup in 1968 and in 1969 also took the office of premier.

Trapassi, Pietro (1698–1782), Italian poet and librettist, known as Pietro Metastasio. His libretti were intricate and classical.

Trapdoor spider, any of numerous species of brown and black spiders found throughout the world. It digs a tube-like, silk-lined burrow with a hinged lid covering the entrance. When the spider feels vibrations of passing prey, it rushes out and retreats with its captive. Genera include *Actinopus* (family Actinopodidae), of South America, Africa and Australia. Most species belong to the family Ctenizidae. Length: to 30mm (1.2in). *See also* NW p.103.

Trappists, popular name for the CISTERCIANS of the Strict Observance, a religious order of monks and nuns. The order originated in La Trappe Abbey, France, in 1664. They maintain silence, practise VEGETARIANISM and rise for Night Office. Manual labour is compulsory for the monks.

Trauma, injury to a living organism caused by external force or violence; also the emotional shock resulting from the injury. The word derives from the Greek for "wound".

Traumatic neurosis, neurotic disturbance caused by a severe emotional TRAUMA. frustration. The response to emotional shock varies according to the initial traumatic experience, eg an extreme fear of dogs after having been bitten. These reactions can manifest themselves only more generally, with symptoms including hysterical paralysis, irritability and sleep disturbances (frightening dreams).

Travellers' tree, palm-like plant native to Madagascar, so-called because each cup-shaped leaf base holds about 1 litre (1.75 pints) of drinkable water. It has large clusters of white blossoms and blue seeds. Height: to 28m (90ft). Family Strelitziaceae; species *Ravenala madagascariensis*.

Travel sickness, nausea, headache and other bodily upsets caused by the motion of road, air, seagoing and even rail vehicles.

Travers, Ben (1886–), British playwright whose immensely popular farces and comedies were compared with the work of FEYDEAU and PINERO. They include *Rookery Nook* (1926), *Plunder* (1928), revived in 1976 at the National Theatre, London, *Banana Ridge* (1938) and *Corkers End* (1969).

Travers, Morris William (1872–1961), British chemist who, in collaboration with Sir William RAMSAY, discovered the rare gases of the atmosphere. Those found included argon, helium, neon, krypton and xenon. These researches are well described in his book *Discovery of the Rare Gases* (1928).

Travers, Pamela Lyndon (1906–), British writer, *b.* Australia, whose fame rests on the series of children's books about the adventures of a kindly, magical nanny. These were first recounted in *Mary Poppins* (1934), which was the basis of a film of the same name produced by Walt DISNEY.

Traviata, La (1852), three-act opera by Giuseppe VERDI, with Italian libretto by Francesco Piave after *La Dame aux Camélias* by the younger Alexandre DUMAS. One of the most popular of Verdi's operas, it marked a new direction in his career as he broke away from the limiting forms of traditional Italian opera towards a refinement of musical expression. *See also* HC2 p.121.

Travis, William Barrett (1809–36), US soldier and hero of the Battle of the ALAMO (1836). Trained as a lawyer, he moved to Texan in 1831 and soon became a leader in the fight for Texan independence from Mexico. When the Alamo was besieged by Gen. Santa Anna's army, Travis (the commander) was killed in its defence. *See also* HC2 p.149.

Travois, vehicle once used by North American Indians of the Great Plains. It consisted of two poles joined to a man, a dog or a horse which dragged it along the ground. It had a net or platform between the poles at the open end that served as a carrier. *See also* MM p.110.

Trawler, fishing vessel that uses a conical net (or trawl), which is pulled through the water, snaring fish. Giant trawlers with electronic equipment are operated by the USA, Japan and the USSR. Smaller sailing trawlers, such as the Chinese trawling junk, are also still in use.

Trawling, method of catching bottom-dwelling fish, such as shrimps and flat fish. A large net, attached to towboards and pulling cables, is towed by the fishing vessel. This method is generally used only in continental shelf areas. Trawling can also be used in deeper water with an echolocater for finding schools and indicating specific depths. *See also* PE pp.240, *240,* 246.

Trdat (*fl.* late 10th century), Armenian architect who went to Constantinople in 989 to rebuild the dome of HAGIA SOPHIA after earthquake damage. His most famous work is the Cathedral of Ani, which featured pointed arches and clustered piers some 150 years before they appeared in the West.

Treacle, heavy, sweet syrup that is produced as a by-product of SUGAR refining. Brown or almost black in colour, it is generally regarded as an edible grade of MOLASSES. *See also* PE pp.196–197, *197.*

Treadmill, instrument of punishment consisting of a large wheel resembling a water wheel, which a prisoner had to turn by walking up inside it. Treadmills, also known as treadwheels, were invented in the USA in 1818 by William CUBBIT and used in the USA and Britain until the early 20th century. *See also* MS p.297.

Treason, offence against the state or the sovereign. Treason is the most serious criminal offence and is a capital offence in many countries, including Britain. In Britain treason is defined by a series of ACTS OF PARLIAMENT to include the infliction of death or injury on the monarch, violation of defined members of the Royal Family, levying war against the government or giving assistance to the enemy.

Treasure Island (1883), novel by Robert Louis STEVENSON about pirates and buried treasure. Narrated by the boy-hero Jim Hawkins, the story involves such famous characters as Blind Pew and Long John Silver. *See also* HC2 p.284.

Treasure trove, in British law, money, coins, silver, gold, bullion or plate, deliberately buried by an unknown owner and subsequently found. Objects simply lost by the original owner belong to the finder if the owner cannot be traced; but treasure trove belongs to the Crown, and anyone finding it and not reporting the fact to the police is liable for prosecution. *See also* MILDENHALL TREASURE; WATER NEWTON HOARD.

Treasury, British government department responsible for national finance and monetary policy. Dating from the NORMAN CONQUEST, when the chancellor and barons exercised control of royal revenue, the Treasury developed from the office of the Chancellor of the Exchequer. It became a separate ministry in the 19th century. The present department is headed by the Chancellor of the

Exchequer, although the Prime Minister is always the First Lord of the Treasury.

Treasury notes, currency notes first issued by the British TREASURY in 1914 to supplement the Bank of England's issue, in denominations of £1 and 10 shillings. the BANK OF ENGLAND took control of the notes in 1928 and replaced them with Bank of England notes of the same value.

Treasury Solicitor, in Britain, solicitor who advises the TREASURY. The office is generally (but not necessarily) held by a barrister, who is responsible for all legal matters pertaining to the Treasury. The Treasury Solicitor is also HM Procurator General, but is no longer director of public prosecutions.

Trebizond, Empire of, BYZANTINE kingdom on the Black Sea. It was founded by members of the Comnenus family in 1204 after the fall of Constantinople to the 4th Crusade. Despite the onslaughts of the SELJUK TURKS, MONGOLS and the armies of Constantinople, it maintained its autonomy and flourished as a cultural and commercial centre until 1461, when MEHMET II conquered Trebizond and made it part of the OTTOMAN EMPIRE.

Trebuchet. *See* MANGONEL.

Tree, Sir Herbert (Draper) Beerbohm (1853–1917), British actor and theatre manager; half-brother of Max BEERBOHM. He made his debut in 1876 and ran the Haymarket Theatre from 1887 to 1897.

Tree, woody, perennial plant with one main stem or trunk and smaller branches. The trunk increases in diameter each year, and the leaves may be evergreen or deciduous. The largest trees, Redwoods, can grow more than 90m (300ft) tall; the giant SEQUOIAS can live for up to 4,000 years. *See also* NW pp.56–57, 62–64, *65.*

Tree, in mythology, usually part of the initial structure or design of Creation, commonly revered in sacred groves. In Scandinavian myths, the tree YGGDRASIL was the foundation of the world, to be destroyed only at RAGNARÖK, the end of the present age. In the Bible, the Tree of the knowledge of good and evil was, with the serpent, all that stood between innocence and ORIGINAL SIN.

Tree creeper, brownish agile bird that scurries up and down trees in cooler areas of the Northern Hemisphere. It uses its long, slightly down-curved bill to probe for insects under the bark, bracing itself with its long, spine-tipped, stiff tail. Length 13cm (5in). Species *Certhia familiaris. See also* NW p.209.

Tree cricket, slender, whitish or pale green CRICKET found in warmer climates throughout the world. It lays its eggs among the twigs of a tree or shrub. Length: 13mm (0.5in). Family Gryllidae; subfamily Oecanthinae.

Tree fern, tree-like FERN of the family Cyatheaceae. Tree ferns grow in tropical regions, particularly moist mountainous areas. The trunks are crowned with large fronds, bearing spore cases on the lower surfaces. Height: 3–25m (10–80ft). There are 600 species. Genus *Cyathea.* Some unrelated CYCADS are also called tree ferns. *See also* NW p.210, *201.*

Tree frog, FROG found throughout the world, generally living in trees. The tree-living species have long, thin hindlegs and enlarged sucking discs on their toes. Length: to 14cm (2.75in). Family Hylidae; there are 460 species. *See also* AMPHIBIANS; NW p.133.

Tree hopper, any of 2,600 species of mainly tropical, HOMOPTERAN insects that suck plant juices. Most have an enlarged THORAX which extends over the head and forms a spine over the body. Species may be green, blue or bronze and may have spots or stripes. Many are considered to be pests because their feeding habits cause damage to growing crops and help to spread plant diseases. Family Membracidae. *See also* NW p.115.

Tree lungwort, typical "leafy" LICHEN of the north temperate zone; it is possibly the most striking of the British lichens. Its leaf-like structures, which may attain a length of 10cm (4in), are bright green when wet; when dry, yellowish brown. Reproduction is by means of spores, which are produced in cup-shaped struc-

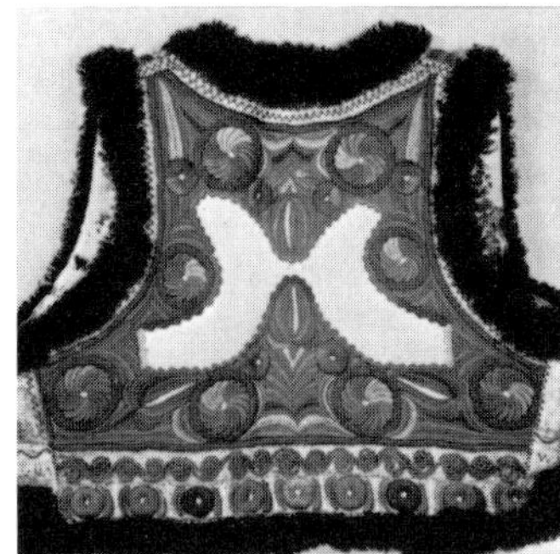

Transylvanian folk arts are shown at their best in examples of costume decoration.

Trapdoor spiders, such as *Cteniza caementaria,* dig complex burrows underground.

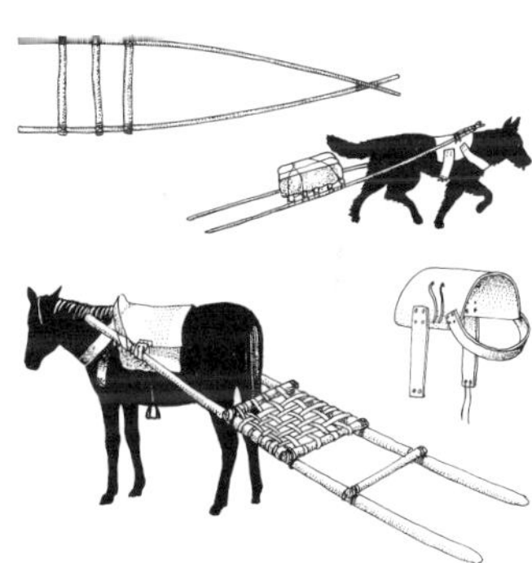

Travois, attached to dog or horse, is a simple device for carrying small loads.

Trawlers may have sterns specially adapted for taking their catch aboard.

Treen

Tree shrews, despite their resemblance to rodents, are more closely related to lemurs.

Trevi Fountain, designed by Nicolo Salvi in 1732, took 30 years to complete.

Richard Trevithick was an important contributor to the Industrial Revolution.

Trial by Jury of Harry Benson for turf frauds at the Old Bailey, London, in 1877.

tures on the surfaces of the "leaves". Species *Lobaria pulmonaria*. *See also* NW p.49.

Treen, small wooden objects in daily use, mostly associated with the kitchen, games, personal adornment and toilet articles. The definition has widened to include utensils made of bone, horn or ivory.

Tree of Jesse, favourite church decoration, sometimes of wood but more usually of glass, showing the genealogy of Jesus from Jesse, the father of DAVID. There are fine examples at St Denis, Chartres, and New College, Oxford.

Tree shrew, small mammal found from India throughout SE Asia to Indonesia. It looks like a long-snouted squirrel without whiskers, but is classified as a primitive PRIMATE. A tree dweller, it is grey-brown in colour and feeds on insects and fruit. Family Tupaiidae; species *Tupaia glis*. *See also* NW pp.168, *168, 214.*

Trefoil, any of numerous plants, such as CLOVER, with leaves divided into three parts. Bird's-foot trefoil is a perennial, used as hay and forage. Family Leguminosae; species *Lotus corniculatus*.

Trelawny of the "Wells" (1893), comedy of stage life by Sir Arthur Wing PINERO. It was first performed in London and has been often revived. In 1972 there was a musical version, *Trelawny*, with music and lyrics by Julian SLADE.

Trematodes, members of a class (Trematoda) of invertebrate animals including nearly 6,000 species of parasitic FLATWORMS and FLUKES. Parasitic trematodes have a thick outer cuticle and one or more suckers for attaching to the tissues of the host. They are bilaterally symmetrical, and commonly found as endoparasites or ectoparasites in all classes of vertebrates, including human beings. Length: to 10cm (4in). *See also* NW pp.86–87.

Tremor, involuntary quivering of a voluntary muscle. It may vary in intensity and duration and often occurs in conjunction with debilitated states or organic disorders.

Trench, deep V-shaped depression of the sea floor. In PLATE TECTONIC theory trenches are places where one plate is being pushed under another. They are the deepest – to 11km (7 miles) – formations on earth and are found primarily along the borders of the Pacific Ocean. *See also* PE p.24.

Trenchard, Hugh Montague, 1st Viscount (1873–1956), British soldier and airman, Marshal of the RAF. He entered the army in 1893 and fought in the SOUTH AFRICAN WAR. As Chief of the Air Staff (1919–29) he organized an offensive air strategy during WWII. He was also Commissioner of the Metropolitan Police (1931–35).

Trench fever, infectious disease caused by a RICKETTSIA transmitted by body lice, characterized by fever and pain in muscles, bones and joints.

Trent, Jesse Boot. *See* BOOT, JESSE.

Trent, river in central England. It rises in Biddulph Moor, Staffordshire, and flows SE and then NE across central England to join the River Ouse and form the Humber estuary. The third-longest river in England, the Trent passes through the POTTERIES. Length: 274km (170 miles).

Trent, Council of (1545–63), the 19th ecumenical council of the Roman Catholic Church convened to state the Church's doctrinal position in the face of PROTESTANTISM and to institute internal reform. With a background of complex diplomatic relations between the Catholic European powers, the Council met at the town of Trent in N Italy during 1545–47 under PAUL III, 1551–52 under JULIUS III and 1562–63 under PIUS IV. Some 16 doctrinal decrees were issued, dealing with original sin, justification, the sacraments, the cult of saints, purgatory and indulgences; the authority of the pope and bishops was reinforced. *See also* HC1 p.272.

Trent Bridge, cricket ground at Nottingham, England, home of the Nottinghamshire County Cricket Club. It was opened in 1838 and has been a test match ground since 1899. It can accommodate about 30,000 spectators.

Trenton, state capital in W New Jersey, USA, on the Delaware River. During the

AMERICAN WAR OF INDEPENDENCE Trenton was the scene of the battle in which Gen. George WASHINGTON crossed the Delaware River to capture Hessian troops in Dec. 1776. The city became the state capital in 1790. Industries: motor vehicle components, plastics, metal products, rubber goods, steel cables, textiles. Pop. (1970) 104,638.

Trepannation, also called trephination, the operation of boring a circular hole in the skull to permit access to the BRAIN. The instrument used is a trepan, which resembles a carpenter's bit with a handle. *See also* MS pp.116, 126, *136, 145.*

Treponema, genus of anaerobic SPIROCHETES. Several species cause diseases in human beings; *T. pallidum* causes SYPHILIS and *T. pertenue* causes yaws, an ulcerative skin infection. They can be treated with PENICILLIN. *See also* BACTERIA.

Trésaguet, Pierre-Marie-Jérôme (1716––1796), French civil engineer whose work formed the basis of the European road system. He introduced lighter and well sorted stones in their construction, saw the importance of good drainage and introduced the idea of regular road maintenance. His methods were later adopted throughout Europe.

Trespass, any breach of the law other than treason. It became a common action in England in the reign of EDWARD I, when there were three chief varieties: trespass *vi et armis* (with force and arms), such as assault and battery; trespass *quare clausum fregit*, the passing on, or allowing animals or objects to pass on, another's property without leave; and trespass *de bonis asportatis*, the stealing of chattels.

Très Riches Heures du duc de Berry, Les, BOOK OF HOURS, attributed to the LIMBURG brothers and unfinished at the time of the death of their patron, Jean de Berry in 1416. A fine example of the French INTERNATIONAL GOTHIC style, the book is a series of calendar illuminations depicting the surroundings and occupations of peasantry and nobles in different months of the year. It is now in the museum at Chantilly, France.

Tressell, Robert (1871–1911), pseud. of Robert Noonan, the Irish author of the influential political novel *The Ragged Trousered Philanthropists*, published posthumously in 1914.

Trevelyan, George Macaulay (1876–1962), British historian, whose great literary gifts found expression in many historical works. He is best known for *British History in the Nineteenth Century* (1922), *History of England* (1926), and *English Social History* (1944).

Trevi Fountain (1732–62), celebrated fountain in Rome designed by Nicola Salvi (1697–1751) and completed by Giuseppe PANNINI. Occupying almost the whole of the small Piazza Barbarini in which it is set, it is a huge marble monument in High BAROQUE style and features Neptune standing on a shell being drawn by horses. According to legend, people who throw coins into it will return to Rome.

Trevino, Lee B. (1939–), US golfer. He twice won both the US Open Championship (1968, 1971) and the British Open (1971, 1972).

Treviranus, Gottfried Reinhold (1776–1837), German naturalist who is best known for his studies, both anatomical and histological, of invertebrates. Because of his six-volume work *Biology, or Philosophy of Living Nature* (1802–22) he has been credited with originating the use of the term "biology".

Trevithick, Richard (1771–1833), British engineer and inventor known for his invention of the high-pressure steam engine. A mining engineer in Cornwall, he experimented with model stationary and locomotive engines from 1797. In 1801 he completed the first steam carriage and in 1803 built in Wales the first steam railway locomotive. Later he adapted his engine to drive an iron-rolling mill (1805), propel a barge (1805) and drive a threshing machine (1812). In 1814 his engines were ordered for use in the silver mines in Peru, and in 1816 he went there himself. He returned penniless in 1827 to find that others were now profiting from his engine.

He died in poverty. *See also*, MM pp.64, *64, 140.*

Trevor-Roper, Hugh Redwald (1914–), British historian. He was a student (or fellow) of Christ Church, Oxford (1946–57) and was appointed regius professor of modern history at Oxford in 1957. His books include *Archbishop Laud* (1940), *The Last Days of Hitler* (1947), *The Gentry, 1540–1640* (1953) and *A Hidden Life, the enigma of Sir Backhouse, Bart.* (1976).

Triad Society, secret society of Chinese in S China opposed to the MANCHU DYNASTY. It existed from the earliest days of the Manchus in the 17th century until the 19th century, when they lent their support to the TAIPING REBELLION. Known in Chinese variously as the Hung League or Heaven and Earth Society, members of the Triad Society have kept the secret organization alive into the 20th century.

Trial, in law, judicial examination or hearing of the facts in a civil or criminal case. A jury may or may not be present.

Trial, The (1925), novel by Franz KAFKA, published posthumously. Joseph K, the solitary central character, wakes to find himself inexplicably under arrest. The work is generally regarded as a symbolic rendering of the hopeless relation between man and bureaucratic authority.

Trial by battle, medieval method of defending oneself against criminal or civil charges, introduced into England by the Normans as an alternative to TRIAL BY ORDEAL. The accused fought the defendant (criminal case) or a named champion (civil case). It was used as late as 1817 and abolished in 1819.

Trial by Jury (1875), one-act operetta by Arthur SULLIVAN and William GILBERT, called by its makers a "novel and original cantata". It was their first major success, produced at the Royalty Theatre, London. It lasts for only about 40 minutes and is the only Gilbert and Sullivan opera to contain no spoken dialogue.

Trial by jury, trial by a number of people (usually 12), who are sworn to deliver a verdict in a court of law upon the evidence presented. As a method of trial it developed from the ANGLO-SAXON judicial custom of acquitting a person of criminal charges if his peers (juratores or compurgatores) swore to his innocence. It is now the main method of trying criminal cases and some civil cases at common law. *See also* JURY; MS pp. 286, *286–270.*

Trial by ordeal, ancient form of trial, practised in England by the Anglo-Saxons and Normans. The theory underlying trial by ordeal was that God would intervene to help the accused. There were three main methods: by hot iron, by cold water, and (for the clergy) by the addition of choking substances to food. *See also* MS p.286.

Triangular Trade, 17th- and 18th-century trading system between the Old and New Worlds, following the prevailing winds. Ships carried slaves from West Africa to the Americas and then brought products such as tobacco back to Europe.

Triangulation, in navigation and land surveying, a method of determining distance. The area under survey is divided into triangles. The baseline of one triangle is measured and the angles that it makes to its vertex are measured with a THEODOLITE. The distances from each end of the baseline to the vertex can be calculated by trigonometry.

Triangulum (The Triangle), faint northern constellation between ANDROMEDA and ARIES. It contains the spiral galaxy M33 (NGC 598), a member of the Local Group located 2,350,000 light years from the Milky Way system.

Triangulum Australe (the Southern Triangle), southern circumpolar constellation. The three brightest stars Alpha, Beta and Gamma (all third magnitude or above) form the conspicuous triangular figure.

Triassic Period, first section of the MESOZOIC PERIOD, lasting from 225 to 190 million years ago. Following upon a wave of extinctions at the close of the Permian, many new kinds of animals developed. On land lived the first DINOSAURS. Mammal-like reptiles were common and by the end of the period the first true MAMMALS existed. In the seas lived the first ichthyosaurs, placodonts and nothosaurs. Also, the first frogs, turtles, crocodilians and lizards appeared. Plant life consisted mainly of primitive gymnosperms, with ferns and conifers predominating. *See also* PE pp.130, *131*.

Tribal society, group which lives in a particular area, speaks the same language, has a unified social organization and is culturally homogeneous.

Tribune, senior military and civil official of ancient Rome. During the Republic there were six tribunes to a legion, each commanding a cohort; during the empire such posts were often filled by the emperor's nominees. In civil politics from 450 BC there were ten tribunes of the plebeian class; their powers of initiating prosecutions and introducing legislation passed to the emperor in 27 BC. *See also* HC1 pp.90, *92*.

Trichina, organism which causes TRICHINOSIS. It is a roundworm, or nematode (full name is *Trichinella spiralis*), the adult worm being a few millimetres long. It lives harmlessly in the human large intestine but the larvae, which cause the symptoms of trichinosis, escape from the body in the faeces and find their way into the bodies of rats, pigs and other animals, where they encyst in muscle.

Trichinosis, disorder of human beings and many other mammals caused by infection with the larvae of a parasitic NEMATODE worm (*Trichinella spiralis*). Most human cases originate from eating raw or undercooked pork. Symptoms include abdominal pain, vomiting, swelling and delirium and possible permanent heart and eye damage may result. Thiabendazole is becoming a widely adopted drug in the treatment of the disorder.

Trichloroethylene, colourless, nonflammable liquid, C_2HCl_3, with a smell of CHLOROFORM. It is manufactured from tetrachloroethane by heating or treating it with lime, and it is used as a degreasing agent and extraction solvent. Properties: s.g. 1.46; b.p. 86.7°C (188.1°F); m.p. −73°C (−99°F).

Trichoptera. *See* CADDISFLY.

Tricyclic compound, organic chemical that has three rings as its primary structure. Examples are ANTHRACENE and phenanthrene.

Tricyclic drugs, drugs belonging to one of the two main groups of antidepressants; they include IMIPRAMINE and amitriptyline, both of which are useful in the treatment of chronic DEPRESSION. The results of treatment by the use of tricyclic drugs compare favourably with those of ECT (electroconvulsive therapy). *See also* MS pp.144, *145*.

Triennial Act (1641), act of the LONG PARLIAMENT in England which compelled the king to summon parliament at least once every three years. It was passed to avoid the possibility of a king ruling without parliament, as CHARLES I did from 1629 to 1640. It was superseded by the SEPTENNIAL ACT of 1716.

Trieste, seaport city in NE Italy, on the Gulf of Trieste 113km (70 miles) ENE of Venice. It was made an imperial free port in 1719 (until 1891) and an Austrian crown land in 1867. It was ceded to Italy in 1919 and occupied by Yugoslavia in 1945. Yugoslavia and Italy agreed to a territorial compromise in 1954 and Trieste was returned to Italy. It is an important industrial and commercial centre with large shipyards and a variety of manufactures, including steel, textiles and petroleum. Pop. (1975) 270,641. *See also* HC2 p.192, *192*.

Triforium, in church architecture, the space in the higher nave between the clerestory windows and the vaulting of the side aisles. In Romanesque times the triforium provided ventilation and light, and with the development of Gothic vaulting it decreased in height and the arcades became more intricate. The decorative effect of the triforium is well displayed in the Angel Choir (1282) in Lincoln Cathedral.

Triggerfish, any of several tropical marine fish found in warm shallow Pacific waters, identified by a dorsal fin spine which can

be erected to lodge the fish in a coral cavity, as a protection against predators. Its laterally compressed body is often beautifully coloured. Length: to 60cm (24in). Family Balistidae; typical genus *Balistoides*. *See also* NW pp.*126, 128, 131*.

Triglyceride. *See* LIPID.

Trigonometry, use of three basic ratios to calculate unknown values in triangles. If three parts (either the length of sides or angles) including one side of a triangle are known, then all the other values may be found. The three basic ratios – TANGENT, COSINE and SINE – are published in tables for any angle. *See also* SU p.46.

Trilby (1894), novel by George DU MAURIER based on his experiences in an artist's studio in Paris. The young model falls into the clutches of SVENGALI, who hypnotized her into becoming a great concert singer.

Trilobite, any of several species within an extinct group of ARTHROPODS found as fossils in marine deposits, ranging in age from Cambrian through Permian times. The body is mostly oval, tapering towards the rear and was covered by a chitinous skeleton. The name refers to the division of the body into three distinct longitudinal segments, consisting of a central axis and two lateral lobes. The cephalon (headshield) carried sensory organs. Transverse divisions show segmentation and bear pairs of jointed limbs. Most species were bottom-crawling, shallow-water forms and range in size from 6mm (0.25in) to 75cm (30in). *See also* NW pp.*22*, 178, *178*.

Trimaran, sailing craft with three hulls, resembling the twin-hulled CATAMARAN.

Trinidad and Tobago, independent nation made up of the two southernmost islands of the West Indies. Formerly a British possession, the islands became an independent member of the Commonwealth in 1962. Major sources of income are sugar production, petroleum and tourism. The capital is Port of Spain on Trinidad. Area: 5,128sq km (1,980sq miles). Pop. (1974 est.) 1,070,000. *See* MW p.166.

Trinitrotoluene. *See* TNT.

Trinity, central doctrine of Christianity, according to which God is three persons: the Father, the Son and the Holy Spirit (or HOLY GHOST). There is only one God, but he exists as "three in one and one in three". The nature of the Trinity is held to be a mystery that cannot be fully comprehended . The doctrine of the Trinity was stated in early Christian creeds to counter heresies such as GNOSTICISM, which denied that Christ was wholly human during His life on earth, and ARIANISM, which denied the true divinity of Christ. Orthodox Christianity holds that God the Son is man but He is also "of one substance with the Father". *See also* APOSTLES' CREED; ATHANASIAN CREED; CHRISTOLOGY; HOLY GHOST; JESUS CHRIST; NICENE CREED.

Trinity House, Corporation of, public corporation of British seamen founded in 1512 when the guild of mariners and lodesmen of Deptford Strand, Kent, requested a charter for pilotage. Its present powers were fixed in 1913. Today it is responsible for pilotage and the administration of lighthouses, lightships and buoys in England and Wales.

Trintignant, Jean-Louis (1930–), French film actor. He appeared in commercially successful films such as *A Man and a Woman* (1966) and took major roles in rather more serious films. In ROHMER's *My Night with Maud* (1965) and BERTOLUCCI's *The Conformist* (1970) his portrayals of characters beset by a conflict of motives won widespread critical acclaim.

Triode. *See* ELECTRON TUBE.

Triolet, poetic form first used by the French in the 13th century. It consists of eight lines with two rhymes. The first two lines are repeated as the seventh and eighth, and the first line is also repeated as the fourth. Triolets were introduced into English verse in the 17th century.

Tripartite Pact (27 Sept. 1940), military and political alliance signed between Germany, Italy and Japan. Known as the Axis powers, the signatories pledged full military and political co-operation.

Triple Alliance, name given to various diplomatic agreements between three countries. The most important was the Triple Alliance formed in 1882 when Italy joined the Dual Alliance of Austria – Hungary and Germany (renewed in 1891) to counterbalance the Alliance of Russia and France. The Alliance intended to bind its members to assist one another in the face of an attack by Russia or an ally of Russia and, in the event of an attack by any other power, to remain at least neutral. Other triple Alliances were those of 1668 between England, Holland and Sweden to oppose the advance of France; of 1717 between England, France and Holland against Spain; and of Brazil, Argentina and Uruguay in the war against Paraguay (1865–70). *See also* HC2 p.180.

Triple Alliance, War of the (1865–70), also known as the Paraguayan War, a conflict between Paraguay on one side and Argentina, Brazil and Uruguay on the other. Long-standing territorial disputes and Brazilian intervention in the Uruguayan civil war prompted Francisco Solano López of Paraguay to declare war on Brazil in March 1865; by May 1865, the conflict involved all four countries. The war devastated Paraguay: 1,000,000 Paraguayan lives and 142,450sq km (55,000sq miles) of Paraguayan territory were lost.

Triple jump, in athletics, an event similar to the long jump with the exception that the contestant makes a hop, a skip (step) and a jump. The jump is a failure if the jumper lands with a wrong foot.

Triple point, temperature and pressure at which all three phases (solid, liquid and gas) of a substance can coexist. For water, the triple point occurs at 0°C and at less than 1% of atmospheric pressure. *See also* SU p.86.

Tripoli, seaport city in Lebanon, thought to have been founded *c.* 700 BC. After capture in 638 by the MUSLIM Arabs, the inhabitants adopted both the Arabic language and the religion of ISLAM. The city was taken in 1109 by Raymond de Saint-Gilles, Count of Toulouse, and sacked in 1289 by the MAMELUKES. It is situated at the terminal of an oil pipeline from Iraq. Once a prosperous town, it suffered severe damage and disruption of trade in the civil war in the Lebanon in the mid-1970s. Industries: cotton goods and soap manufacture, tobacco and fruit processing. Pop. (1970 est.) 158,000. *See also* HC1 p.177, *177*.

Tripoli, capital and chief port of Libya, on the Mediterranean Sea, 644km (400 miles) W of Benghazi. Founded in the 7th century BC by the Phoenicians, it was captured by the Romans in the 1st century BC, by the Vandals in 5th century AD and the Arabs in the 7th century. It developed as an important market centre for trans-Saharan caravan routes. Tripoli was taken by Italy in 1911 and made the capital of the Italian colony of Libya; it is an administrative and commercial centre. Industries include textiles and cigarettes. Pop. 245,000.

Tripolitania, historic region in W Libya. Founded as three Phoenician colonies on the N coast of Africa in the 7th century BC, it was ruled at different times by the Carthaginians, Romans, Vandals, Byzantines, Arabs and the KNIGHTS HOSPITALLERS. In 1557 it fell to the OTTOMAN EMPIRE and in the 18th century engaged in piracy as one of the BARBARY STATES. Tripolitania finally became part of the Italian colony of Libya in 1912, was captured by the British in 1943, and in 1951 became part of Libya. It is now a predominantly agricultural area, producing fruit and vegetables. Manufactures include carpets, metal goods, handicrafts and leather goods.

Triptych, in art, three hinged panels either painted or carved, traditionally used as an altarpiece. The panels may form one picture or the outer panels may be separate and subordinate to the central picture.

Tripura, state in NE India, bordered to the N, W and S by Bangladesh; Agartala is the capital. The region was annexed to Britain in 1808 from the Mogul Empire, and became part of India in 1947. Rodhakishorepar is a Hindu pilgrimage centre.

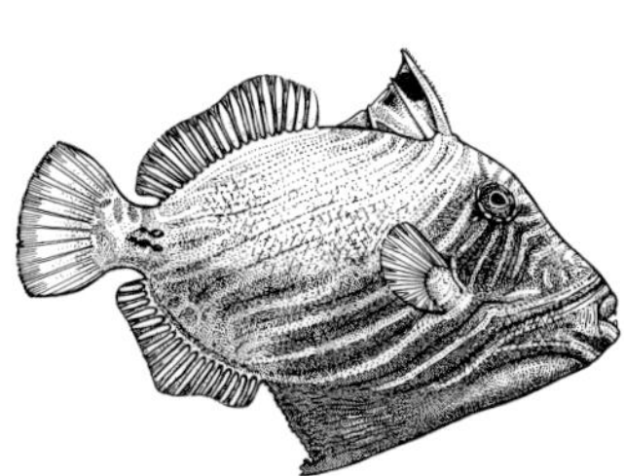

Triggerfish use chisel-like teeth to crack coral and the shells of oysters and crabs.

Trigonometry is the principle used by surveyors to measure angles and distances.

Trilby, staged in London (1895), gave its name to the felt hat worn in the play.

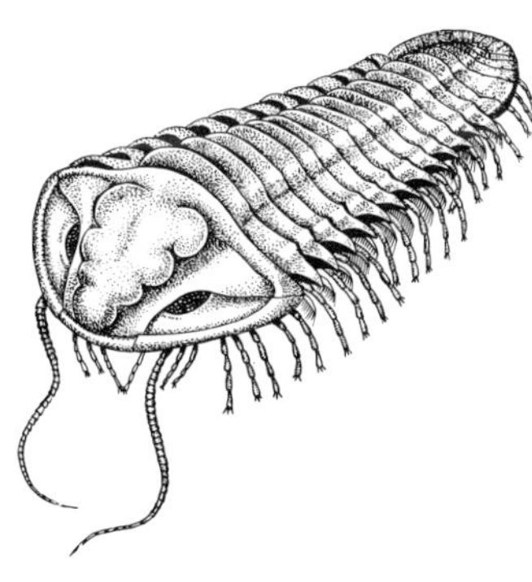

Trilobites are generally considered to be the ancestors of insects, spiders and crabs.

Trireme

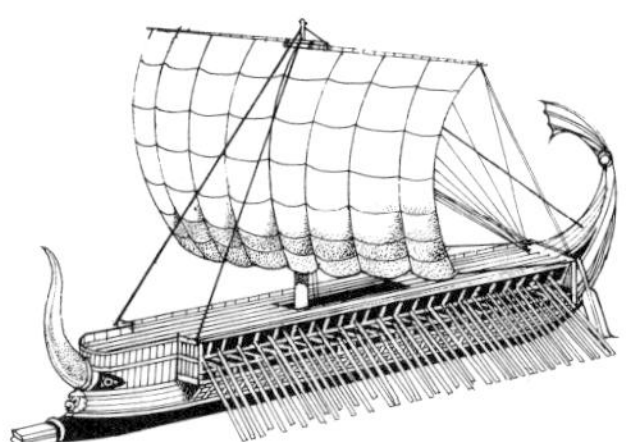

Triremes are believed to have originated with the Phoenicians c. 500 BC.

Trogons have broad bills surrounded by bristles which help in catching insects.

Troll, a figure used by Henrik Ibsen for the mountain king in his *Peer Gynt* (1867).

Leon Trotsky was a forceful and revolutionary Russian Communist leader.

Most of the people work in agriculture; jute is the chief product. Pop. (1971) 1,556,822. Area: 10,477sq km (4,045sq miles).

Trireme, narrow wooden warship used by Mediterranean navies from the 5th century BC. Triremes had three banks or rows of oarsmen on each side, and carried a large ram and a small complement of soldiers. When not in battle, the vessel was powered by a single square sail. *See also* MM p.172.

Tristan, or Tristram, hero of many medieval romances, most commonly as a knight of the Round Table in the Arthurian legend. The tale of his fatal love for the Irish princess Isolde (or Iseult) is the subject of Matthew ARNOLD's poem *Tristram and Iseult* and Richard WAGNER's opera *Tristan and Isolde*.

Tristan and Isolde (c. 1210), courtly romance epic by Gottfried von Strassburg, written in Middle High German. Based on an earlier story by the Anglo-Norman poet Thomas, it recounts one of the most famous and tragic love romances which in c. 1240 was incorporated into Arthurian legend. In this form it has been retold many times by poets and composers, including Thomas MALORY, Alfred TENNYSON and Richard WAGNER.

Tristan and Isolde (1865), three-act opera by Richard WAGNER with libretto by the composer, based on Cornish legend. The expressive potential of the orchestra in opera reached new heights with Wagner's use of LEITMOTIV. This device, and the masterly characterization in the title roles (achieved by penetrating psychological insight) make the work a landmark in operatic history.

Tristan da Cunha, group of three islands in the S Atlantic Ocean located midway between S Africa and South America; the only inhabitable island is Tristan. The group was discovered in 1506 by the Portuguese and annexed by Britain in 1816. In 1961 Tristan suffered a volcanic eruption that caused the complete evacuation of the island, although most of the population returned in 1963. Area of Tristan: 98sq km (38sq miles). Pop. (1970) 280.

Tristram Shandy, The Life and Opinions of (1760–67), novel in nine volumes by Lawrence STERNE. His whimsical imagination and ironic sense made the novel an immense success. *See also* HC2 p.34.

Tritium, radioactive ISOTOPE of HYDROGEN, the nucleus of which consists of one proton and two neutrons. Only one atom in 10^{17} of natural hydrogen is tritium. Tritium compounds are used in radioactive tracing. Properties: at.wt. 3.017; half-life 12.3 yr.

Triton, in Greek mythology, a sea god, son of POSEIDON and AMPHITRITE. He was half man, and half fish, with a scaled body, sharp teeth and claws and a forked fish tail. He had power over the waves and possessed the gift of prophecy. Triton blowing a conch shell seems to have been the personification of the roar of the wild sea.

Triton, large GASTROPOD MOLLUSC found in tropical marine waters near coral reefs. Its shell is used as a religious musical instrument in Shinto temples insJapan. Length: to 40.6cm (16in). Family Cymatiidae; species include the brown-and-pink *Charonia tritonis*.

Triton, nucleus of TRITIUM atom, consisting of one proton and two neutrons.

Trochee, metrical foot consisting of a stressed followed by an unstressed syllable, or a long followed by a short, as in the word "happy". Trochaic metre, is popular in children's verses (*Mary Had a Little Lamb*) and in verses where a regular repetitive rhythm is desired.

Troglodytes, Greek name for various primitive tribes; they were mentioned by a number of Greek and Roman writers, such as Strabo, HERODOTUS and Diodorus. They were thought to live in the N Caucasus, where they lived in caves, and in NW and the interior of N Africa, where they probably lived in wicker huts. They mostly went naked, eating the bones and hides of their cattle as well as the flesh, and drinking a mixture of blood and milk. In

modern biological science, *Troglodytidae* is the name of the family of birds to which the WREN belongs.

Trogon, brilliantly coloured bird of dark tropical forests in America, Africa, and Asia. Trogons nest in holes in trees and feed on fruit and some insect larvae. Both parents incubate the eggs. Length: about 30cm (12in). Family Trogonidae; typical genus *Trogon*.

Trogoniformes. *See* TROGON.

Troilus, son of the Trojan king PRIAM, slain by ACHILLES at the beginning of the TROJAN WARS. His love for CRESSIDA, celebrated by CHAUCER and SHAKESPEARE, has no place in Greek sources, but was first described by Benoît de Sainte-More in the 12th-century romance, *Roman de Troie*.

Troilus and Cressida (1602), five-act play by William SHAKESPEARE. Cressida falls in love with the Trojan prince, Troilus, but when exchanged for a Trojan prisoner of war transfers her affections to the Greek Diomedes.

Troilus and Criseyde (c. 1380), long narrative poem by CHAUCER based on the *Filostrato* of BOCCACCIO. *See also* HC1 pp.214, *215*.

Trois Frères, cave in S France decorated with Upper Palaeolithic paintings and engravings, dating c. 40,000–c. 10,000 BC. Most of the figures are unusual in that they represent half-human beasts; the largest of these is 4m (13ft) high.

Trojan asteroids, minor planets, named after Homeric heroes of the Iliad. They comprise two groups situated in the orbit of Jupiter, one group behind the planet and the other ahead of it.

Trojan horse, in the Homeric epic, *The Iliad*, a colossal hollow wooden horse built by the Greeks in the final days of the siege of TROY. Despite warnings the people of Troy dragged the horse through their gates and in the night those Greeks who had been hiding within its body emerged and opened the gates of Troy to their army.

Trojans, The (composed 1856–58), five-act opera by Hector BERLIOZ with French libretto by the composer based on VIRGIL's *Aeneid*.

Trojan Wars, The, in Greek mythology, a conflict between the Greeks and the Trojans. PARIS, a Trojan prince, abducted HELEN, the wife of MENELAUS, a Spartan. Menelaus' brother AGAMEMNON led the Greeks into war with Troy, and the ten-year siege of Troy culminated in the incident of the TROJAN HORSE. The story of the siege is told by HOMER in the *Iliad*, and the heroes of the battle were ACHILLES, PATROCLUS, ODYSSEUS and Nestor.

Trojan Women (415 BC), tragedy by the Greek playwright EURIPIDES. It deals with the destruction of Troy and with the fate of the wives of the men killed in the War.

Troll, in Icelandic mythology, a malignant one-eyed giant; in Scandinavian mythology, a mischievous dwarf. Trolls were skilled metal workers living in the hills and occasionally making forays to steal from human beings. They were frightened by noise, from a memory of the time when THOR threw his hammer at them.

Trolley bus, public conveyance developed from the TRAM and powered by electricity from overhead cables. A trolley arm reaches from the car to the cable. Trolley buses do not run on metal tracks but on rubber tyres, and so two wires are used, the second being a return circuit. By the early 1900s there were commercial trolley bus systems in Europe; the first British routes began in 1911 in Bradford and Leeds. *See also* MM pp.134–135, *134*.

Trollope, Anthony (1815–82), British novelist who travelled widely while working for the post office. His fame rests chiefly on his series of Barsetshire novels, epitomizing Victorian society. These include *The Warden* (1855), *Barchester Towers* (1857) and *Doctor Thorne* (1858).

Trombone, brass musical instrument with a cylindrical bore, cupped mouthpiece and flaring bell. It is usually played with a slide, except for a variant which has three or four valves. The tenor and bass trombones have a range of three and a half octaves. *See also* HC2 pp.104–105, *105*.

Trompe l'oeil (French "trick the eye"),

painting that attempts to deceive the eye of the beholder by convincing him of the material reality of the objects depicted.

Tromsø, city in NW Norway on Tromsø island; capital of Tromsø county and the chief city of arctic Norway. The German battleship *Tirpitz* was sunk by the British off Tromsø during WWII. There is a university (1968) in the city, which is also the centre of an important herring fishing industry. Other industries include shipbuilding, rope and seal hunting. Pop. (1971 est.) 39,145.

Trondheim, city in central Norway, on the S shore of Trondheim fjord (an inlet of the Atlantic Ocean); the second-largest city in Norway. Founded in 997, it was the political and religious capital of medieval Norway. Trondheim developed as a commercial centre in the 19th century. The city exports wood and metal products, and has brewing, food processing, shipbuilding and hardware industries. Pop 214,000.

Trooping the Colour, British military parade at HORSE GUARDS Parade, London, at which the regimental flag, or colour, is carried between formations of soldiers and presented to the sovereign. The ceremony was probably originally held to familiarize non-British troops with the flag they were to follow. It is now held in early June on the sovereign's official birthday.

Trope, figure of speech in which a phrase or word is used in an unusual context to draw attention to its literal sense. The most common types of trope are METAPHOR and IRONY. Trope, in its musical sense, refers to a phrase or verse introduced as a refinement into some part of the MASS.

Tropical diseases, diseases associated with hot, often wet, climates. They include MALARIA, LEPROSY, BILHARZIA, TRYPANOSOMIASIS (sleeping sickness), LEISHMANIASIS (kala azar, espundia etc), CHOLERA and YELLOW FEVER. The infectious agents of these and other diseases of the tropics range from viruses through bacteria, protozoa, fungi and worms of various kinds. Often more specific to the tropics are the vectors of these disease microbes, eg. the mosquitos which spread the virus of yellow fever and the protozoan parasite of malaria. *See also* MS pp.106–107.

Tropical fish, any of numerous species of small decorative fish, mainly from the Amazon Basin or the waters of SE Asia; they are commonly kept in captivity in domestic aquaria. Species may live in either fresh or saltwater, generally at temperatures at or above 24°C (75°F). Species include the GUPPY and GOURAMI; more exotic types include the TRIGGERFISH and ANGELFISH.

Tropic bird, whitish sea bird with black markings found in Pacfic and Indian Ocean waters where it dives for food. It has a long, bright bill and long, streamer-like central tail feathers. Length: to 50cm (20in). Family Phaethontidae; genus *Phaethon*.

Tropics. *See* CANCER, TROPIC OF; CAPRICORN, TROPIC OF.

Tropism, response in growth, movement or orientation of a plant, a lower animal or a part of either to a directional, external stimulus such as light (PHOTOTROPISM), gravity (GEOTROPISM) or water (HYDROTROPISM). Most responses to such stimuli are orthotropic (they are directed towards the stimulus).

Troposphere, lowest part of the Earth's atmosphere. In it temperature decreases with height, and it is within this region that clouds and other weather phenomena occur. The troposphere extends 16km (10 miles) above the Earth's surface.

Trossachs, The, glen in Central Region central Scotland, between Loch Achray and S end of Loch Katrine. The region was made famous in the poems of William WORDSWORTH and Sir Walter SCOTT. It is a popular tourist centre.

Trotsky, Leon Davidovitch (1879–1940), Russian Bolshevik leader. Chairman of the St Petersburg Soviet, he was prominent in the 1905 revolution, after which he was tried and escaped abroad. He returned to Russia in time for the March 1917 revolution, As head of the Military

Revolutionary Committee of the Petrograd Soviet he was in charge of its defence, resulting in its seizure of power. As commissar for foreign affairs (1917–18) he negotiated for peace at BREST-LITOVSK, but resigned after the treaty was concluded. As commissar for war (1918–25) he created and led the Red Army in the civil war. After LENIN's death he was ousted from power by STALIN. Trotsky was expelled from the Party in 1927 and from the country in 1929 as a leader of the left opposition. He was murdered in Mexico in 1940, probably by an agent of Stalin. Fiercely independent, he was a brilliant polemicist and a fiery speaker. *See also* HC2 pp.196–197, 198, *198*.

Trotskyism, interpretation of MARXISM conceived by Leon TROTSKY. In contrast to STALIN, who tried to create "socialism in one country", Trotsky argued that there must be "permanent revolution" in all the industrialized countries together.

Trotter, Lily. *See* JACANA.

Troubadours, poets in the s of France from the 11th to the 14th century who wrote about love and chivalry. Their poems were sung by wandering minstrels called jongleurs. The name comes from the Provençal *trobar* for "to find" or "to invent". They wrote in the Provençal tongue, the *langue d'oc*.

Troughs, in meteorology, areas of low atmospheric pressure, usually extensions to a DEPRESSION. The opposite are ridges of high pressure.

Trout, any fish of the salmon family (Salmonidae). Trout are good food fish and favourites with anglers. There are three types of the single species of European trout (*S. trutta*), each with a different name. The brown or river trout, found in streams and pools, is small and dark, and does not migrate. The lake trout, of rivers and lakes, is a larger, paler version, and is sometimes migratory. The large silvery sea trout is definitely migratory and is sometimes confused with the salmon. Length: to 1m (3ft); weight: to 13.5kg (30lb). American trout are equally varied and include the rainbow trout (*S. gairdneri*), with a longitudinal red stripe. *See also* NW pp. *123, 131, 246*; PE pp. 242–243, *244, 245*.

Trouvères, aristocratic poet-musicians of N France of the 12th and 13th centuries, similar to TROUBADOURS. Their songs, many of which have survived, deal with chivalry, war, politics and religion. The system of notation used indicates only the pitch and not the metre or note values, and so the songs are difficult to interpret.

Trovatore, Il (1853), four-act opera by Giuseppe VERDI with libretto by Salvatore Cammarano, based on a play by Antonio Gutierrez. Set in 15th-century Spain the plot involves the unravelling of complicated family relationships, and the resultant emotional climaxes are vigorously treated in the music, which is stylistically similar to the earlier opera *Rigoletto*.

Troy, or Ilium, ancient city in NW Asia Minor whose legend is an important theme in Greek literature. The extensive ruins excavated by Heinrich SCHLIEMANN in the 1870s and Wilhelm Dörpfeld in the 1890s have shown nine principal periods when houses were built on the same site, between 3000 and 1100BC (when it was probably abandoned) and between 700 BC and 400 AD. Troy VIIa is probably that of King Priam depicted by HOMER in his *Iliad*. From its early settlement Troy was an important point on the trade route between Europe and Asia.

Troyat, Henri (1911–), French novelist, *b.* Russia as Lev Tarassov. He left the civil service to devote himself to writing and won the Goncourt prize in 1938 for his third novel *L'Araignée*.

Troyes, Chrétien de (*fl.* 1170), French poet, author of the earliest extant Arthurian romances, which were the source of much later similar romances, such as *Le Morte Darthur* by MALORY. Troyes' work includes translations of OVID and the romances *Erec* (after 1155), *Cligès* (*c.* 1176) and the unfinished *Perceval* (*Le Conte du Graal*), which contains the earliest known reference to the legend of the Holy Grail.

Troyes, Treaty of (1420), truce in the HUNDRED YEARS WAR between CHARLES VI of France, HENRY V of England and PHILIP of Burgundy. By its terms Henry married Charles's daughter, Catherine, and was recognized as the heir to the French throne. He died before he could succeed to the throne. *See also* HC1 p.213.

Trubetskoi, Prince Pavel (1866–1938), Russian sculptor who lived also in Paris, the USA and Italy. Strongly influenced by RODIN, his best works date from the early 1900s when he produced IMPRESSIONISTIC portraits and animal statuettes.

Trucial States, former name for the UNITED ARAB EMIRATES, seven emirates on the Persian Gulf: Abu Dhabi, Ajman, Dubai, Fujairah, Ras al-Khaimah, Sharjah and Umm al-Qaiwain. The name "trucial" signifies the truces signed with Britain in 1820 (and a later agreement of 1892), by which the emirates accepted British protection and British control of their defence and foreign policies. They joined together in the UAE when Britain withdrew from the Persian Gulf in 1971.

Trudeau, Pierre Elliott (1919–), Canadian politician. He was elected a Liberal member of Parliament in 1965 and was appointed Minister of Justice and Attorney General in 1967. He succeeded Lester PEARSON as party leader and as Prime Minister in 1968, and during his administration he promoted economic and diplomatic independence for Canada. He recognized the People's Republic of CHINA in 1970 but in that year had temporarily to impose martial law to combat French separatist terrorism. The Liberal Party lost its majority in 1970, but Trudeau continued to govern with the aid of the small NEW DEMOCRATIC PARTY. In 1974 he was re-elected Prime Minister with Liberal support alone. *See also* HC2 p.*137*.

True Intellectual System. *See* CUDWORTH, RALPH.

True Levellers. *See* DIGGERS.

Trueman, Frederick Sewards ("Freddy") (1931–), British cricketer who played for Yorkshire and England. He was a fast bowler, but also a first-class fielder and a fair batsman. He was the first bowler to take more than 300 wickets in Test matches. He had taken 307 at an average of 21.57 when he retired from Test cricket in 1965.

Truffaut, François (1932–), French film director. His first feature film, *The 400 Blows* (1959), won a prize at the Cannes Film Festival, and Truffaut came to be associated with the NOUVELLE VAGUE. His other films include *Shoot the Pianist* (1960), *Jules and Jim* (1961), *Day for Night* (1973) and *Pocket Money* (1976). *See also* HC2 p. *277, 277*.

Truffle, any of several species of ASCOMYCETE fungi that grow underground, mostly among tree roots. Most are edible and are highly prized delicacies. Truffles range in size from that of a pea to that of a potato. Found in Europe, particularly France, and in parts of the USA, they are hunted for by trained pigs and dogs that can scent them out. Family Tuberaceae.

Trujillo, city in NW Peru, approx. 530km (330 miles) NNW of Lima. Located in a fertile coastal region, it is a commercial and export centre, processing sugar cane and producing textiles and cocaine. Pop. (1972 est.) 134,279.

Truman, Harry S. (1884–1972), 33rd President of the USA (1945–53). He took office in April 1945, just before the end of WWII and faced the problems of post-war readjustment, when he succeeded Pres. Franklin D. ROOSEVELT. The decisions he took are still controversial: he ordered the first atomic bombings of Japanese cities and effectively ended the war with Japan; he introduced the TRUMAN DOCTRINE, the MARSHALL PLAN, the NORTH ATLANTIC TREATY ORGANIZATION and organized the NATO Berlin airlift – all intended to challenge the real or imagined threat of Soviet Communism. The COLD WAR pervaded his two terms in office and, at home, Americans watched the witch-hunting trials instigated by Sen. Joseph MCCARTHY on their television sets. Truman's last years in office were occupied with the KOREAN

WAR. *See also* HC2 pp.219, *219*, 234–235, *234–235*, 240, 258, *258*, 264–265, *264–265*, 306.

Truman Doctrine (March 1947), declaration of US foreign policy by Pres. Harry TRUMAN. It stated that the USA would act to prevent the overthrow of democratic institutions by totalitarian governments anywhere in the world. Although the statement was universal in scope, Truman was specifically referring to the protection of Greece and Turkey against Soviet expansion. The doctrine became the primary statement of US policy in the COLD WAR.

Trumbull, John (1756–1843), US painter of portraits and historical subjects. In 1817 he was commissioned to paint a series of scenes from the American War of Independence for the Capitol in Washington, DC, which are severely classical in style. His best talent is displayed in his miniature portrait studies.

Trumpet, brass wind instrument of ancient origin. It has a cylindrical bore in the shape of a flattened loop and three piston valves, added after about 1820 to regulate the pitch. It became an important ceremonial instrument in the 15th century and by the late 17th century had become a standard orchestral instrument. *See also* HC2 pp.104–105, *104–105*.

Trumpeter swan, large black-billed swan with a loud bugle-like call. It is found mainly in national parks in Canada and the NW USA where its numbers continue to increase, having reached near-extinction in the wild. It is one of the heaviest living flying birds. Length: to 1.7m (5.5ft); wingspan: to 3m (10ft); weight: to 17.2kg (38lb). Family Anatidae; species *Cygnus cygnus buccinator*. *See also* NW p.148.

Trunkfish, marine fish that lives in temperate and tropical waters. Its body is almost triangular when seen from the front, with a broad flat ventral region tapering to a narrow dorsal region; it has fused scales which form hexagonal bony plates. Length: to 50cm (20in). Family Ostraciontidae; genus *Lactophrys*.

Truss, in engineering and building, structural member made up of straight pieces of metal or timber formed into a series of triangles lying in a vertical plane. The triangles resist distortion through stress, making the truss capable of sustaining great loads over long spans. It was probably first used almost 5,000 years ago in lake dwellings.

Truss, in medicine, device made of metal, elastic, leather, plastic or canvas and worn to prevent a HERNIA from protruding through the abdominal wall.

Trust, in law, situation in which one person (the TRUSTEE) holds property for the benefit of another (the beneficiary). Trusts are generally created by a legal instrument such as a deed or a will.

Trust. *See* MONOPOLY.

Trustee, in law, the person holding property in TRUST on behalf of the beneficiary.

Trustee, The Public, British official who headed the Public Trust Office and was responsible for the administration of wills and TRUSTS placed under his care. The office was established in 1908, but was taken over by the Treasury in 1972 because of a decline in business.

Trustee Savings Bank, savings banks operated on a non-profit basis, on behalf of investors. First established in Britain in 1810, they now offer a comprehensive range of banking facilities. The largest is the Post Office Savings Bank.

Trust Territories, countries administered by the UNITED NATIONS as a continuation of the MANDATE system of the LEAGUE OF NATIONS. They are former colonial acquisitions of defeated nations which did not have either the inclination or means to become self-governing. The purpose of the trustee system is to prepare them for self-government. After WWII the United Nations established 11 trust territories (Nauru, New Guinea, the Pacific Islands, Ruanda-Urundi, Somaliland and Western Samoa; part of the Cameroon and Togoland was administered by France and the remainder by Britain who also administered Tanganyika). With the exception of

Harry S. Truman at a press conference during his visit to London in 1956.

John Trumbull's *Declaration of Independence* shows his skill at portraiture.

Trunkfish swims by rotary movements of its fins, using its tail as a rudder.

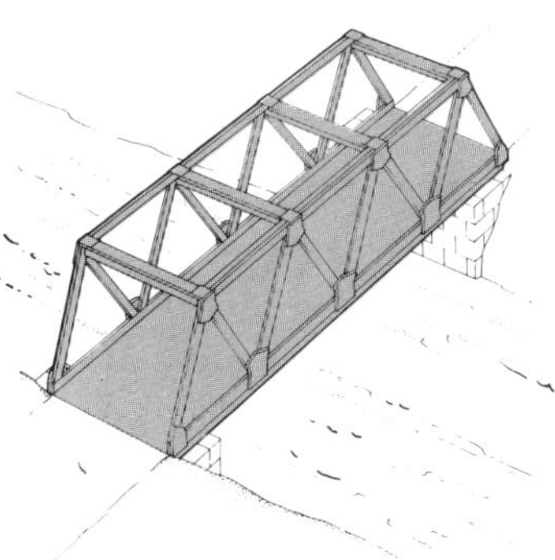

Truss; the principle combines strength with lightness and is used in making bridges.

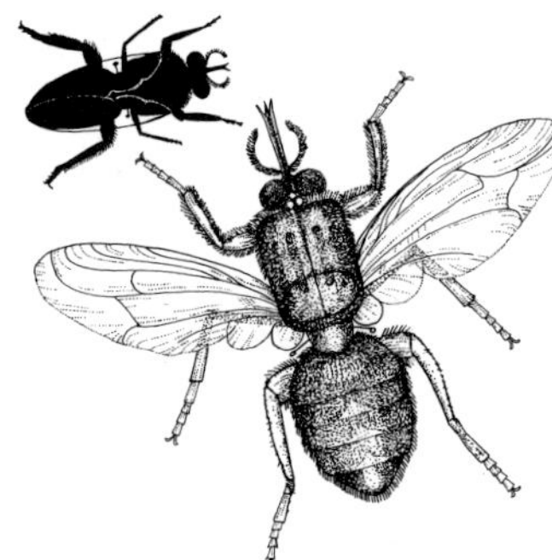

Tsetse flies are controlled by spraying bushes with insecticides, such as DDT.

Tuataras are small burrowing reptiles which feed on insects and small mammals.

Tuberose flowers have a waxy texture and a sweet fragrance.

Mary Tudor; this portrait, by Sir Anthony More, hangs in the Prado, Madrid.

the Pacific Islands, all have since either become independent or passed to the control of other countries.

Truter, Sir Johannes Andreas (1763–1865), South African jurist and administrator. He was appointed secretary to the Court of Justice (1793) and under the Batavian Republic acted as advisor to the Governor, Gen. Janssens. He continued as an administrator under British rule and became President of the Court of Justice in 1812.

Truth, accurate representation of reality, the opposite of falsehood. Truth in various forms has been a major object of philosophical enquiry; ONTOLOGY is the study of the true existence of reality and EPISTEMOLOGY of the correspondence between a concept and reality.

Trypanosome, one of a genus of animal flagellate PROTOZOA that are mainly parasitic in the blood of vertebrates. *Trypanosoma gambiense* is an elongated cell, with long, undulating membrane and flagellum along one side. It is harboured by African game, carried by the TSETSE FLY, and injected into the blood of humans, causing SLEEPING SICKNESS. *See also* NW p.72.

Trypanosomiasis, whole array of debilitating disorders in man and domestic animals, caused by several species of parasitic flagellate protozoans of the genus *Trypanosoma*. The vector is usually a fly or other insect. Examples are human sleeping sickness and Chagas' disease, and nagana and surra in cattle and horses. *See also* MS p.104.

Trypsin, ENZYME that breaks down PROTEINS, particularly those in blood and milk, into PEPTIDES to aid in absorption during the digestive process in the small intestine. *See also* MS p.68.

Tryptamine, organic chemical compound, formula $C_{10}H_2N_2$. Among its many metabolic derivatives is 5-hydroxy tryptamine, or SEROTONIN, which affects (among other things) mental coherence.

Tryptophan, amino acid, first isolated in 1902, which is necessary for the synthesis of the anti-PELLAGRA vitamin nicotinic acid. Tryptophan itself cannot be synthesized by animals and is therefore an essential AMINO ACID that must be supplied in the diet. It is found in most natural PROTEINS, but some CEREALS, notably MAIZE and SORGHUM, contain particularly low levels.

Tsar, name of the rulers of Russia, first adopted by IVAN the Terrible in 1547. It is a corruption of *Caesar*. In 1721 PETER I changed the official title to "emperor", but "tsar" continued to be used in popular language until 1917.

Tsatsos, Konstantinos (1899–), Greek philosopher, lawyer and politician, President of Greece from 1975. After working as professor of law at Athens University (from 1932), he became a member of parliament (1946–50, 1963–67). He held various ministries before becoming President.

Tschaikovsky, Peter Ilyich. *See* TCHAIKOVSKY, PETER ILYICH.

Tsedenbal, Yumjaagiyn (1916–), Mongolian head of state. Secretary-General of the Central Committee of the Mongolian People's Revolutionary Party from 1940 to 1945, he became First Secretary in 1958 and Chairman of the Presidium, Mongolia's governing body, in 1974.

Tsetse fly, any of several species of blood-sucking flies that live in Africa. Larger than a housefly, it has a grey thorax and a yellow to brown abdomen. Females commonly seek large animals such as livestock, transmitting nagana, a disease of cattle; almost 80% of flies that bite human beings are male, carrying SLEEPING SICKNESS. Length: to 16mm (0.6in). Order Diptera; family Muscidae; genus *Glossina*. *See also* NW p.113.

TSH. *See* THYROID STIMULATING HORMONE.

Tshombe, Moise Kapenda (1919–69), Congolese military leader and politician. He became involved in the government of KATANGA (renamed Shaba) and formed a separate nation when the rest of the Congo became independent in 1960. After intervention by the United Nations in 1963 he fled to Europe, but returned as premier of a united Congo (now ZAIRE) in

1964. He was dismissed in 1965 and went into exile. In 1967 he was kidnapped and taken to Algeria, where he was kept under house arrest until his death in 1969.

Tsiolkovsky, Konstantin Eduardovich (1857–1935), Russian scientist who provided the theoretical basis for space travel. The first to stress the importance of liquid propellants in rockets in 1898, he also proposed the technique of multi-stage rockets to overcome gravity and laid down the principles of space stations. *See also* MM p.156; SU pp.268, *268,* 270,*271.*

Tsin Dynasty, Chinese dynasty, also known as the Chin Dynasty. Established in 265 by Chin Wu Ti (Ssu-ma Yen) (the Martial Emperor) and known in history as the West Chin Dynasty, it was destroyed in 316 by the Hsiung-nu or HUNS and its survivors fled s of the Yangtze to establish the East Tsin Dynasty (317–419), one of the Six Dynasties in s China. The migration of the northern aristocracy shifted the centre of Chinese civilization to the s and inaugurated an outstanding period of cultural activity.

Tsinghai (Qinghai), province in N China; the capital is Sining (Xining). It is a mountainous region and the source of some of Asia's greatest rivers, including the HWANG HO (Yellow), YANGTZE (Changjiang) and MEKONG. There is farming of spring wheat, barley and potatoes, and stock rearing; the province is famous for its breeding of horses. The extraction of iron ore, coal, oil, salt and potash is important. It passed to China in 1724 and fell to the Communists in 1949. Area: 721,280sq km (278,486sq miles). Pop. 2,000,000.

Tsingtao, port city in Shantung province, NE China, on the Yellow Sea. The Japanese held Tsingtao from 1914 to 1922 and from 1938 to 1945. It is a major port of China, and has rail links with the rest of the country. Products include textiles, iron and steel, chemicals, tyres and metal goods. Pop. (1975 est.) 1,391,000.

TSR2, British experimental tactical, strike and reconnaissance aircraft designed to penetrate radar defences in high-speed, low-level flight. The first of these aircraft flew in 1964, but the type was scrapped in 1965 for lack of finance.

Tsunami, wave caused by a submarine earthquake, subsidence or volcanic eruption. Tsunamis spread radially from their source in ever-widening circles. In mid-ocean they are shallow, 0.3–6m (1–2ft) high, and are rarely detected. In shallower water they build up in force and height, crashing on shore and causing enormous damage. Tsunami is a Japanese word; such waves are also called seismic sea waves, but tidal wave is an erroneous term for them. *See also* PE pp.28, 80, *81.*

Tsushima, island in s Japan, in the Korea Strait. It was the scene in 1905 of the major naval battle of the RUSSO-JAPANESE WAR. Fishing is the chief occupation. Area: approx. 699sq km (270sq miles).

Tuamotu (Low Archipelago), island group in French Polynesia, in the s Pacific Ocean, ESE of the Society Islands and NE of the Tubuai Islands. Rangiroa is the largest island. The group is made up of approx. 80 small coral islets, which were discovered in 1606 by the Spanish explorer Pedro Fernandes de Queirós. The islands were annexed to France in 1881 and are now administered from Tahiti. Products: coconuts, copra, pearl, shell. Area: 857sq km (331sq miles). Pop. 7,660.

Tuareg, fiercely independent BERBER people of Islamic faith, who inhabit the desert regions of North Africa. Their matrilineal, feudal society is based on nomadic pastoralism, and traditionally maintains a class of black non-Tuareg serfs. Tuareg males wear blue veils, while the women are unveiled. *See also* MS p.*252.*

Tuatara, nocturnal lizard-like reptile of New Zealand; it is the sole surviving member of the primitive order Rhynchocephalia. It is brownish in colour and has an exceptionally well-developed PINEAL BODY, thought to be a vestigial third eye. Length: to 70cm (2.3ft). Species *Sphenodon punctatus*. *See also* NW pp.134, *229.*

Tuba, large brass musical instrument, the lowest instrument of the orchestral brass family. The tuba has a conical bore, a cupped mouthpiece and is oblong in shape. It is held vertically with the bell upward and played with normally four or five valves. *See also* HC2 pp.*104–105.*

Tube, electron. *See* ELECTRON TUBE.

Tube, X-ray. *See* X-RAY TUBE.

Tuber, in plants, the short, swollen, sometimes edible underground stem, modified for the storage of food as in the potato, or swollen root (eg dahlia). In stem tubers new plants develop from the buds, or eyes growing in the axils of the scale leaves. Tubers are propagated by sections containing at least one eye. *See also* PE pp.186–187.

Tuberculosis, infectious disease caused by the bacteria *Mycobacterium tuberculosis* and *M. bovis*. It most often affects the lungs, but may involve the bones and joints, the skin, the lymph nodes, intestines and the kidneys. There are several drugs used to treat this disease, chiefly rifampin, isoniazid and para-aminosalicylic acid (PAS). Preventive measures, which include screening (by chest X-ray and tuberculin testing), pasteurization of milk, and isolation of cases, have controlled the disease successfully in many countries. It remains a major cause of death in countries where such measures are not taken. *See also* MS pp.88, *89.*

Tuberose, Mexican perennial plant grown in the s USA. The white, sweet-scented flowers are used in perfumery. Height: to 91cm (3ft). Family Amaryllidaceae; species *Polianthes tuberosa*.

Tubifex, also called bloodworm or sludge worm, any of several species of freshwater worms that bury themselves in the silt at the bottom of the water. They are commonly used as fishing bait, or to feed fish in aquariums. Family Tubificidae; genus *Tubifex*. *See also* NW pp.89, *230–231.*

Tublidentata. *See* AARDVARK.

Tubman, William Vacanarat Shadrach (1895–1971), Liberian president (1944–71). A descendant of American slaves, he served as a legislator and associate justice of the supreme court (1937–43) before becoming President in 1944.

TUC (Trades Union Congress), permanent association of British trade unions, founded in 1868 to promote trade union principles and gradually acquiring a permanent staff and a limited authority over its member unions. Each year it holds an annual assembly of delegates who discuss common problems. Most of its members are skilled workers. Since 1889 any union can apply for membership. It formed the Labour Representation Committee in 1900, renamed the LABOUR PARTY in 1906, which sponsored candidates for Parliament until 1918, afterwards becoming a national political party. Membership of the TUC was 1,000,000 in 1874, doubling by WWI and numbering 10,000,000 in the mid-1970s. *See also* HC2 pp.95, 164, 210–211, *211, 213.*

Tucson, city in SE Arizona, USA, 165km (103 miles) SE of Phoenix, on the Santa Cruz River. Tucson is the seat of the University of Arizona (1885). Industries: aircraft parts, meat packing, textiles, mining, electronics, tourism. Pop. (1973) 307,551.

Tucu-tucu, any of several species of squirrel-like burrowing rodents of the grasslands of sw South America. Dark brown to a creamy buff, it digs an extensive system of burrows and eats the roots and stems of plants. Length: to 25cm (10in). Family Ctenomyidae; genus *Ctenomys*. *See also* NW p.202.

Tudor, Antony (1908–), British dancer, teacher and choreographer b. William Cook. His ballets such as *Adam and Eve* (1932) and *The Planets* (1934) helped to establish the reputation of Ballet RAMBERT. He was staff choreographer (1940–50) and Associate Director (1974–) of the American Theater Ballet and worked closely with many other national companies.

Tudor, Mary (1516–58), Queen Mary I of England (r. 1553–58), daughter of HENRY

VIII and CATHERINE OF ARAGON. She remained a Roman Catholic and attempted by persecution (which earned her the name "Bloody Mary") to effect a COUNTER-REFORMATION in England. In 1554 she married PHILIP II of Spain. *See also* HC1 pp.270–271.

Tudor, House of, English dynasty that ruled between 1485 and 1603. The family originated in Wales; Owen Tudor (*d*. 1461) married the widow of HENRY V of England and their eldest son, Edmund, married a great-granddaughter of JOHN OF GAUNT. Their son Henry defeated RICHARD III at the Battle of Bosworth Field in 1485 to win the English throne and was proclaimed HENRY VII. By marrying Elizabeth of York he attempted to end the Wars of the ROSES. His second son succeeded him to the throne as HENRY VIII in 1509, while his daughter Margaret married JAMES IV of Scotland, a liaison that led to the STUART succession to the English throne in 1603. Henry VIII was succeeded by his son, EDWARD VI; his sister Mary's granddaughter Lady Jane GREY, and his two daughters MARY and ELIZABETH I. Following the latter's death without an heir in 1603, the throne passed to JAMES VI of Scotland. The House of Tudor produced a series of outstanding monarchs, and its period of rule was decisive for the growth of royal authority and for the emergence of England into a modern nation state. *See also* HC1 pp.266–267, 270–271, 274–281; HC2 p.309.

Tuesday, third day of the week. The name is derived from the Old English Tiwes daeg, the day of TIW (Tyr), the Germanic God of war.

Tufa, deposit of calcium carbonate formed by precipitation from lime-rich water, eg in a limestone cavern. The term is also applied to deposits of silica from hot springs. *See also* PE pp.110–111.

Tuff, SEDIMENTARY ROCK made up of particles of IGNEOUS ROCK from volcanic eruptions. The particles vary in size from fine to coarse but are generally smaller than 4mm (0.16in), and may be either stratified or heterogeneous in their arrangement.

Tu Fu (712–70), Chinese poet of the T'ang Dynasty who wrote about such topics as war, corruption and patriotism. His poetry reflects his troubled personal life and laments the corruption and cruelty that prevailed at court and the sufferings of the poor. *See also* HC1 p.126.

Tug-of-war, athletic contest of strength involving two teams of eight men and a rope. A line or marker separates the teams. The object is to pull the opposing team across the line. Teams are graded according to their combined weight. Tug-of-war was included in the Olympic Games from 1900 to 1920.

Tui, Chinese bronze vessel made during the late Chou dynasty (*c.* 600–221 BC). It is composed of two hemispherical bowls, each of which has handles and three legs. Often decorated with inlay, the bowls fit together to form a closed sphere, which may be used to hold food.

Tui, New Zealand honeyeater, considered among the finest of their forest songbirds. It has two white patches of curly feathers at the throat and is a competent mimic of other birds.

Tuke, William (1732–1822), English Quaker philanthropist, important pioneer in the humane treatment of the mentally disturbed. *See also* MS p.*136.*

Tula, or Tollan, TOLTEC ceremonial centre 73km (45 miles) N of Mexico City, which flourished 900–*c*.1206. It was a relatively large complex containing a central plaza, pyramids and ball-court, and may have been the Toltec capital.

Tulip, hardy bulbous plant native to Europe, Asia and North Africa. Tulips have long, pointed leaves growing from the base and elongated cup-shaped flowers that can be almost any colour or combination of colours. Family Liliaceae; genus *Tulipa*.

Tulip tree, large, broad-leaved tree of the MAGNOLIA family and a misnomer for other species of magnolia. It is native to the USA and has large, TULIP-like, green yellow flowers, each comprised of six petals which are orange at the base, and bears cone-like clusters of winged fruits. Its pale yellow wood is used for furniture and veneers. Height: to 58m (190ft). Family Magnoliaceae; species *Liriodendron tulipifera. See also* NW p.208.

Tull, Jethro (1674–1741), British agriculturist who influenced agricultural methods through his writings which included *The Horse-Hoeing Husbandry* (1733). He invented a mechanical drill for sowing in 1701 and advocated the use of manure and thorough tilling during the growing period. *See also* HC2 pp.20, *20.*

Tumbleweed, plant that characteristically breaks off near the ground in autumn and is rolled along by the wind. The common western US tumbleweed is usually whitestemmed and has pale flowers crowded into the leaf axils. Height: to 51cm (20in). Family Amaranthaceae; genus *Amaranthus.*

Tumour, any uncontrolled, abnormal proliferation of new tissue from pre-existing cells that has no useful function in terms of the body as a whole. Tumours fall into several types and are classified as either benign or malignant. *See also* MS pp.86, *87,* 96.

Tumulus, or barrow, grave mound of prehistoric or early historic peoples. Excavation of barrows has developed into a study of construction and funerary rites. The two most prolific types are the long barrow from the Neolithic period and the round barrow from the Bronze Age. *See also* SU p.20.

Tuna. *See* TUNNY.

Tunbridge Wells, county district in SW KENT, England; created in 1974 under the Local Government Act (1972). Area 331sq km (128sq miles). Pop. (1974 est.) 95,000.

Tundra, Lapp term for the treeless, level or gently undulating plain characteristic of arctic and subarctic regions. The tundra is marshy with dark soil that supports mosses, lichens and low shrubs. It has a permanently frozen subsoil known as permafrost. *See also* NW p.226.

Tung Chung-shu (*c.*179 BC–104 BC), exponent of the YIN-YANG philosophy with which CONFUCIANISM became merged in China in the 2nd century BC.

Tung oil, pungent, yellow, drying oil which is obtained chiefly from the tung tree (*Aleurites fordii*) of China. It is one of the main industrial oils and contains a great amount of free fatty acids and other contaminants such as RESINS and sterols. It is used in the manufacture of paints and varnishes and in the waterproofing of linoleum. *See also* PE p.217.

Tungsten, or wolfram, metallic element (symbol W) of the third transition series, first identified in 1779 and isolated in 1783. Its chief ores are WOLFRAMITE and SCHEELITE. Tungsten has the highest melting point of all metals and is used for filaments in electric-light bulbs and electron tubes (valves). It is also used in high-speed tool steels and other special alloys. The sulphide is used as a lubricant. Chemically, tungsten is fairly unreactive; it oxidizes only at high temperatures. Properties: at.no. 74; at.wt. 183.85; s.g. 19.3; m.p. 3,410°C (6,170°F); b.p. 5,660°C (10,220°F); most common isotope W^{184} (30.64%). *See also* TRANSITION ELEMENTS.

Tungsten carbide, hard, inert, grey powder, formula WC, used for making dies, carbide tools and armour-piercing shells. It is one of the strongest structural materials. It is manufactured by heating tungsten and lamp-black (powdered carbon) in an electric furnace. Properties: s.g. 15.6; m.p. 2,780°C (5.036°F).

Tungsten processing, separation of the metal TUNGSTEN from its chief ores, scheelite and wolframite. The ore is fused with sodium carbonate to yield sodium tungstate, which is treated with acid to produce tungstic acid. The latter is washed and dried to yield tungsten oxide (WO_3), which is reduced by hydrogen to the pure metal powder. This is moulded and pressed into tungsten bars.

Tungus, Ural-Altaic speaking, Mongoloid people who inhabit E Siberia. Consisting of two main groups, the Evenksi and Lamuts, they practise reindeer herding, fishing and agriculture. The Tungus are related to the MANCHUS.

Tunicate. *See* SEA SQUIRT.

Tunis, capital city of Tunisia, N Africa, between the two salt lakes of Tunis and Subkhat as-Sijumi (with canal access to the Gulf of Tunis). It became the capital of the country under the Hafsid dynasty (13th–16th centuries) and in later years was famous as a centre of piracy. Tunisia was occupied by the French from 1881 to 1956. The port facilities of Tunis were greatly improved during the late 1960s. The modern city's products include olive oil, carpets and textiles; tourism is also important. Pop. 642,000.

Tunisia, small republic in N Africa. Formerly a French protectorate, it became independent in 1956. Farming employs approx. 60 per cent of the people; olive oil, wine, fruit and vegetables are the chief farm exports. Mineral deposits include oil, phosphates, iron ore and lead. Manufacturing industries are increasing, and tourism is an important source of income. The capital is Tunis. Area: 164,150sq km (63,378sq miles). Pop. (1975) 5,588,200. *See* MW p.167.

Tunnage and Poundage, subsidies voted by Parliament to the Crown in medieval England, originally separate but after 1350 voted together. Tunnage was a duty on the import of tuns of wine, poundage a charge of 3d. in the pound on the import and export of all goods. From 1415 they were voted for life, until the LONG PARLIAMENT limited the grant to two months in 1641.

Tunnelling, digging of underground passages without removing the overlying rock or soil. Its methods vary with the nature of the material to be cut through. In soft earth it is necessary to shore up the tunnel to prevent its collapsing. Tunnelling through hard rock requires blasting. Compressed air drills were first used on the Mont Cenis Tunnel, 14km (8.5 miles) long, which was opened in 1871. For tunnelling under rivers the shield was developed. This device was composed of metal cylinders which fitted around the walls of the tunnel, the forward end being closed by a diaphragm plate. The shield was pushed forward into the earth, headings being cut through openings in the forward end, and the tunnel walls extended within the cylinder. Modern excavators drill large diameter tunnels rather like the process of using a brace and bit. *See also* MM pp.188–189.

Tunney, James Joseph ("Gene") (1898–), US boxer. Light-heavyweight champion of the USA (1922–1923), he defeated Jack DEMPSEY in 1926 for the world heavyweight championship. He successfully defended his title against Dempsey in 1927 and retired undefeated in 1928. He subsequently became a successful business man.

Tunny, large marine fish related to MACKEREL, found in tropical and temperate seas; often known by its American name, tuna. An important commercial fish, it has a blue-black and silvery streamlined body with a large, deeply divided tail. Length: to 4.3m (14ft); weight: to 810kg (1,800lb). Types of tunny, or tuna, include bluefin, yellowfin, albacore and skipjack. Family Scombridae. *See also* NW pp.*130, 240;* PE pp.*246,* 247.

Tupac Amaru (*c*.1742–1781), b. José Gabriel Condorcanqui, leader of an Indian revolt in colonial Spanish America (1780–81). Tupac's army of Peruvian Indians numbered more than 10,000 and many lives were lost before the revolt was suppressed and its leader brutally executed.

Tupou IV, King Taufa'ahau (1918–), King of Tonga. The eldest son of Queen SALOTE TUPOU III, he was Premier from 1949 to 1965 and succeeded his mother to the throne when she died in 1965. He was crowned in 1967.

Turaco. *See* TOURACO.

Turandot (composed 1924), three-act opera by Giacomo PUCCINI based on a play by Carlo GOZZI, with libretto by Giuseppe Adami and Renato Simoni. The score, unfinished at the time of Puccini's death, was completed by Franco Alfano. It was

Jethro Tull was an agriculturalist and the inventor of early agricultural machines.

Tunis; one of the markets among the Arab quarters in the eastern sector.

Tunnelling; St Bernhard Tunnel, which joins Italy and Switzerland, under construction.

Tunny fish are caught in warmer waters for their strongly flavoured meat.

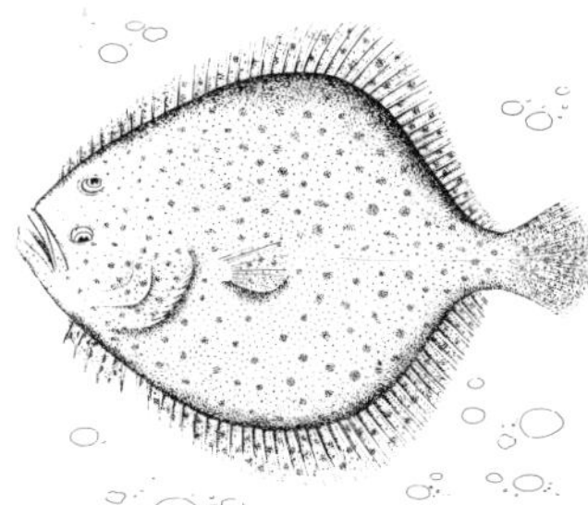

Turbot are one of the most suitable fish for artificial breeding on fish farms.

Turkeys, named for their calls, are bred as poultry, particularly for festive occasions.

Turmeric, related to ginger, is used as a spice, particularly in Indian recipes.

Turner's early paintings are noted for their detail and delicate use of light.

first performed in Milan in 1926. It has an exotic setting in ancient China, fairy-tale elements and a heroic love story.

Turbine, rotary engine that converts the energy of a moving fluid (eg water, steam, or air) into mechanical energy. The basic element in a turbine is a wheel or rotor with blades or buckets arranged on its circumference such that the moving fluid exerts tangential force, which turns the wheel. The rotation of the rotor is transferred through a drive shaft to do work, eg drive a propeller or electricity generator. *See also* MM pp. *62–63*, *66–67*, *69–71*, *153*.

Turbofan engine, TURBINE engine that developed from, and is more efficient than, the TURBOJET ENGINE. It has a compressor, which forces air from the intake into the combustion chamber, and a TURBINE, which is driven by the exhaust gases and drives the compressor. Extra thrust is provided by large blades at the front, driven by the turbine, which push air along a bypass round the combustion chamber and into the tailpipe. *See also* MM p.*63*.

Turbojet engine, aircraft ENGINE (a type of gas TURBINE) that produces power through the reactive force of expanding gases. Air is taken in at the front through a compressor, forced into a combustion chamber, mixed with fuel and burned, producing a rush of expanding gas that propels the aircraft in a direction opposite to that of the rapid outflow. To maintain the cycle the expanding gas also turns a TURBINE which drives the air compressor. *See also* TURBO-FAN ENGINE; MM pp.*69*, *153*.

Turboprop engine, aircraft ENGINE that has a propeller (airscrew) driven by a gas TURBINE through gears. The turbine compresses air which is mixed with fuel and burnt in a combustion chamber, producing exhaust gases which drive the turbine, compressor and propeller. *See also* TURBOFAN ENGINE; TURBOJET ENGINE; MM pp.*63*, *153*.

Turbot, scaleless, bottom-dwelling, European marine FLATFISH; it is farmed in some countries. A typical flatfish, it has a broad flat body with both eyes on its grey-brown, mottled upper surface, which may also be covered in bony knobs; the blind underside is nearly white. Length: to 1m (3.3ft). Family Scophthalmidae; species *Scophthalmus maximus*. *See also* NW p.*130*; PE pp.*242*, *243*, *246*, *247*.

Turenne, Henri de La Tour d'Auvergne (1611–75), French general. He distinguished himself against the Bavarians in the THIRTY YEARS WAR (1618–48) and in the wars of LOUIS XIV. He was involved in the FRONDE (1649–51) but supported the future LOUIS XIV against the forces of the Prince de CONDÉ and Spain.

Turgenev, Ivan Sergeievich (1818–83), Russian novelist, playwright and short story writer who was at his most prolific between 1850 and 1860. The novels of this period opposed social and political evils and received official disapproval. After the appearance of *Fathers and Sons* (1862) he left Russia. *See also* HC2 p.99.

Turgot, Anne Robert Jacques (1721–81), French economist. He was chief administrator for Limoges from 1761 to 1774 and controller general of finance from 1774 to 1776. His attempts to set France's fiscal house in order were unsuccessful and his January edicts of 1776 led to his downfall.

Turin (Torino), city in NW Italy, 125km (78 miles) NW of Genoa, on the River Po; capital of Piedmont (Piemonte) region. Turin was an important Roman town. It became a Lombard duchy from 590 to 636, and recognized the supremacy of Savoy in 1280. From 1720 to 1861 the city was capital of the Kingdom of Sardinia. At the beginning of the 19th century Turin was a centre of the RISORGIMENTO, and from 1861 to 1864 capital of the new Italian Kingdom. The city's university dates from 1405. Turin suffered great damage during WWII. Today it is an important industrial centre and home of the Fiat and Lancia motor companies. Other industries include electronic equipment, machinery, rubber, paper, leather goods, wines, Pop. (1976 est.) 1,202,900.

Turistcheva, Ludmila (1952–), Soviet gymnast whose flawless performances made her the supreme competitor of the early 1970s. She won three Olympic gold medals (combined exercises 1972; team 1968, 1972), was world champion in 1970, and European champion in 1971 and 1973.

Turkey, independent state located between Europe and Asia. The Turkish straits (the Bosphorus, Sea of Marmara and the Dardanelles) link the Black Sea and the Mediterranean Sea. A variety of crops are grown on the coastal plain, wheat is produced on the W Anatolian plateau and the E part of the country is mountainous. Most of the people work in agriculture, and the rest of the economy depends on government-controlled enterprises. The capital is Ankara. Area: 780,574sq km (301,380sq miles). Pop. (1975) 40,198,000. *See* MW p.167.

Turkey, North American GAME BIRD now widely domesticated throughout the world. The common wild turkey (*Meleagris gallopavo*), once abundant in North America, was overhunted and is now protected. The male, or gobbler, is often bearded. Length: 125cm (50in). The ocellated turkey (*Agriocharis ocellata*) of lowland Mexico and adjacent Central American areas has a yellow knob between its eyes, bright metallic plumage and a bare, blue, pimple-covered head and neck. Family Meleagrididae. *See also* PE pp.236, *237–239*, 238.

Turkic languages, group of languages that form a branch of the Altaic family. It is divided into various classes, and its most important member is Turkish; most of the rest are spoken in the USSR.

Turkistan, or Turkestan, extensive geographical region in central Asia, including parts of the USSR (Kazakhstan, Kirgizia, Tadzhikistan, Turkmenistan and Uzbekistan) and China. The region extends from Siberia in the N to Tibet, India, Afghanistan and Iran in the S, and from the Caspian Sea to the Gobi desert. Area: approx. 2.6 million sq km (1 million sq miles).

Turkmenistan, or Turkmen Soviet Socialist Republic, constituent republic of the USSR. It officially entered the USSR in 1925. The capital is Ashkhabad. Turkomans comprise 60% of the population, the other ethnic groups being Russians, Uzbeks, Kazakhs, Tatars, Ukrainians and Armenians. Almost 90% of the land is covered by the Kara-Kum desert, parts of which are irrigated by the Kara-Kum canal. The chief crop is cotton, grown on the sides of the canal and in the Murgab and Tedzhan oases. Area: 488,100sq km (188,455sq miles). Pop. (1976): 2,582,000.

Turks and Caicos Islands, island group of the British West Indies; including more than 30 islands that are, geographically, a SE continuation of the Bahama Islands; the capital is Grand Turk. Discovered in 1512 by Ponce de Léon, the islands were annexed to Jamaica in 1874, and ruled by a British-appointed administrator after August 1962, assisted by an elected legislative assembly. Exports include salt, sponges and shellfish. Area: 430sq km (166sq miles). Pop (1975 est.) 6,000.

Turku (Åbo), port in SW Finland, at the mouth of the Aurajoki on the Baltic Sea; Finland's largest port. Founded in 1229, it is one of Finland's oldest cities; it was the national capital until 1812. Turku has Finnish (1922) and Swedish (1918) universities. Industries: steel, textiles, clothing, shipbuilding. Pop. (1976) 164,380.

Turmeric, herbaceous perennial plant originally found in India, and cultivated in SE Asia. The dried rhizome is powdered for use as seasoning, a yellow dye and in medicines. The plant has thick stalks, long leaves and pale yellow flowers. Family Zingiberaceae; species *Curcuma longa*. *See also* PE pp.210, *211*.

Turner, Dame Eva (1892–), British dramatic soprano. She sang with the Carl Rosa company and in 1924 was engaged by TOSCANINI for La Scala, Milan, where she established herself in operas by VERDI, WAGNER and PUCCINI.

Turner, Joseph Mallord William (1775–1851), British landscape painter, the greatest of the 19th century. His talent was acknowledged early: he was an associate of the Royal Academy at age 24 and was professor of PERSPECTIVE there from 1807 to 1838. He was alway financially successful. He was a withdrawn personality and late in life became virtually a recluse. His paintings were revolutionary in their representation of light, especially of light on water. In his late works, such as *The Slave Ship* (1840) and *Rain, Steam and Speed* (1844), the original subjects are almost obscured, and form comes from the interplay of light and colour. *See also* HC2 p.103, *103*.

Turner's syndrome, hereditary condition in which the sex chromosomes are not XX as in females, nor XY as in males, but X0, the second sex chromosome being absent or dysfunctional. People with the syndrome are characteristically female in general appearance, but short in stature, sterile and with a webbed neck.

Turnip, garden vegetable best grown in cool climates. The edible leaves are large and toothed with thick midribs. A biennial, it has an edible, white, fleshy root. The seeds are used to make margarine. Diameter: 8–15cm (3–6in). Height: to 55cm (20in). Family Cruciferae; species *Brassica rapa*.

Turnpike, method formerly used in Britain to raise money to keep main roads in repair. The first turnpike act (1663) was limited to three counties, but turnpike roads became general in the 18th century, when turnpike trusts were established with power to charge TOLLS from people using the roads. In 1888 responsibility for main roads passed to the county councils and the last turnpike trust was dissolved in 1895, although tolls are still levied on certain British roads and road bridges.

Turnstone, either of two species of migratory shore birds that use their curved bills to turn over pebbles in search of food; they nest on the Arctic tundra. The vividly marked ruddy turnstone (*Arenaria interpres*) ranges widely in winter. The larger black turnstone (*A. melanocephala*) lives mainly on the eastern Pacific coast. Family Scolopacidae. *See also* NW p.*227*.

Turntable, flat circular disc for rotating a gramophone record at a constant speed. *See also* MM p.232.

Turoe Stone, decorated Celtic phallic stone, of about the first century AD, found 6km (4 miles) NNE of Loughrea, in Ireland. Shaped rather like a tall beehive, it is ornamented with three-plane, abstract, curvilinear motifs common to the Celtic style. *See also* HC2 p.26, *26*.

Turpentine, or gum turpentine, sticky liquid obtained from coniferous trees; it contains ROSIN and a volatile oil. Oil of turpentine, often called merely turpentine, is obtained by DISTILLATION of the gum and is used as a paint thinner, solvent and in varnishes and lacquers.

Turpin, Dick (1706–39), British highwayman hanged for horse stealing and murder. His fame is largely derived from William Ainsworth's novel *Rookwood* (1834). His famous ride from London to York on his horse *Black Bess* is fiction.

Turquoise, blue mineral, hydrated basic copper aluminium phosphate, found in aluminium-rich rocks in deserts. It occurs as tiny crystals and dense masses, in crusts and veins, and is a popular gemstone. Hardness 6; s.g. 2.7.

Turret lathe, LATHE in which a number of cutting tools are held in a block, or turret, which can be rotated to present one or more tools to the workpiece. *See also* MM p.*43*.

Turtle, REPTILE found on land or in marine and fresh waters. On the evolutionary scale, turtles have the most ancient lineage of all reptiles, preceding even the dinosaurs. They have a bony, horn-covered, box-like shell that encloses shoulder and hip girdles and all internal organs. The head, neck, limbs and tail project through openings in the shell. Horny jaws, resembling those of birds, replace teeth. All lay eggs on land. Terrestrial turtles are frequently called TORTOISES and some edible species found in brackish waters are called TERRAPINS. Marine turtles usually have smaller, lighter shells. Length: 10cm–2m

(4in–7ft). There are 200–280 species in 22 families. Order Chelonia. *See also* NW pp.73, 134, 136–138, *136–138*, *190, 246, 250.*

Turtle dove, small slender European bird. It has light grey-brown plumage and a white-edged tail. Its voice is soft and purring. Family Columbidae; species *Streptopelia turtur.*

Tuscany, region in central Italy, on the Tyrrhenian Sea, made up of the provinces of Massa Carrara, Lucca, Pistoia, Pisa, Siena, Arezzo, Florence, Grosseto and Leghorn; the capital is Florence. Tuscany is mostly mountainous with fertile valleys particularly in the Arno River Valley and the coastal plains. Agriculture remains the most important activity; products include cereals, olives and grapes. Carrara marble is quarried in the NW and there is mining for lead, zinc, antimony and copper in the SW. Manufactures include textiles, chemicals and artisan industries. Considerable income is also derived from tourism. Area: 22,992sq km (8,877sq miles). Pop (1971 prelim.) 3,292,949.

Tusk, elongated tooth, generally a CANINE, that extends outside the mouth of some vertebrates such as the WALRUS and WILD BOAR. The large tusks of ELEPHANTS are modified upper INCISORS; these have long been sought as a source of valuable ivory, resulting in depletion of the species.

Tussaud, Madame Marie (1761–1850), founder of the "waxworks" exhibition, *b.* France as Marie Grosholtz. She modelled wax figures in Paris, and in 1802 went to London, with her collection. The present site of the exhibition, in Marylebone Road, London, dates from 1884.

Tussock moth, any of several species of moth, the larvae (caterpillars) of which are typically covered with tussocks or tufts of long hairs; many species have stinging hairs. The larvae of most species feed on the leaves of trees and shrubs, often causing much damage. Females are commonly white and brown. *Hemero-campa leucostigma,* the white marked tussock moth, has no wings. Family Lipanidae.

Tutankhamen (*r. c.* 1348–1340 BC), one of the last pharaohs of the 18th dynasty of Egypt. He married the daughter of AKHENATON at the age of 10. During his eight-year reign the god AMON was restored to prominence and the capital was moved from Akhetaten back to THEBES. The discovery of his tomb in 1922, still containing most of its royal burial equipment, has made him one of the best known Egyptian pharaohs. *See also* HC1 p.*48.*

Tutin, Dorothy (1931–), British actress. Trained at RADA, she played many major roles with the Shakespeare Memorial Theatre and the ROYAL SHAKESPEARE COMPANY and toured Russia (1958) with the former and America (1968) with the latter. Throughout her career she has also made film and television appearances.

Tutuola, Amos (1920–), Nigerian writer of great creative energy whose world is a mixture of fantasy and reality. In such books as *The Palm Wine Drinkard* (1952) he conveys a sense of the spiritual validity of the complex beliefs and legends of his countrymen. *See also* HC2 p.291, *291.*

Tuva (Tuvinskaja ASSR), administrative division in the Russian SFSR, USSR, on the Mongolian border. Controlled by the MONGOLS from the 13th to 18th centuries, the region was part of the Chinese Empire from 1857 to 1911. It was made a Russian protectorate in 1914, annexed as an autonomous oblast in 1944 and made an autonomous republic in 1961. Cereals are grown in the lowlands and livestock (including horses, goats, reindeer and camels) is reared on the steppes. Industries include timber-milling, woodworking, leather making and food processing. Area: 170,500sq km (65,800sq miles). Pop. (1975 est.) 251,000.

TVP. *See* TEXTURED VEGETABLE PROTEIN.

Twain, Mark (1835–1910), US writer, journalist and lecturer, real name Samuel Langhorne Clemens. He was among the first to write novels in the American vernacular, such as *The Adventures of Tom Sawyer* (1885) and *The Adventures*

of *Huckleberry Finn* (1876). Both reveal his underlying egalitarian beliefs and a strong desire for social justice. Although he is best known as a humorist his later books such as *The Mysterious Stranger* (1916), and *The Man That Corrupted Hadleyburg and Other Stories and Sketches* (1900) were bitter and pessimistic. *See also* HC2 pp.99, 288, *288.*

Tweed, rough-textured woollen cloth from which warm suits, skirts, etc. are made, particularly in Scotland. Tweed has a twill weave, in which the wool fibres are patterned diagonally. *See also* MM pp.56–59.

Tweed, river in S Scotland. It rises in the Southern uplands and flows E then NE, forming part of the England-Scotland border, to enter the North Sea at Berwick-upon-Tweed. Length: 156km (97 miles).

Tweedsmuir, Lord. *See* BUCHAN, JOHN.

Twelfth Night, the twelfth night after Christmas, ie 6 January. Marking the end of the Christmas season, it is a traditional time for festivals and entertainment. It is the feast of the EPIPHANY in the Christian calendar.

Twelfth Night, or What You Will (1601), romantic comedy by William SHAKESPEARE. The plot was taken through various other sources from an Italian NOVELLA by Matteo Bandillo.

Twelve Tables (*c.* 451–450 BC), laws engraved on wooden tables representing the earliest codification of ROMAN LAW. Written by *decemviri* (a committee of 10) at the probable instigation of the PLEBEIANS, they codified the existing laws and customs of ancient Rome thereby providing a measure of certainty in the administration of the law.

Twelve-tone, or twelve-note, music, post-WWI approach to composition, the introduction of which is accredited to Arnold SCHOENBERG. This new method of composition relies not on the principle of tonality, in which the tonic or keynote is the focal centre, but on the relation of the twelve notes of the CHROMATIC scale one to the other. The composer selects the order in which these notes are to be played and the resultant sequence changes with different compositions. This sequence is not always discernible by ear but is nevertheless the basis of the composition, in the same way as KEY is a fundamental consideration in tonal music. Many composers have experimented with twelve-note music; these include Anton WEBERN, Alban BERG, Ernst KŘENEK, Luigi DALLAPICCOLA and Hans Werner HENZE. *See also* SERIAL MUSIC; HC2 p.271, *271.*

Twelve tribes of Israel, according to the Bible, the groups of Hebrews descended from JACOB and bearing the names of his sons REUBEN, Simeon, JUDAH, ISSACHAR, Zebulun, GAD, Asher, BENJAMIN, DAN and Naphtali. The tribes of Manasseh and EPHRAIM were named after the sons of Jacob's son, JOSEPH. The descendants of Jacob's son Levi, the LEVITES, not counted among the twelve, were devoted to the service of God and acquired no territory in CANAAN, but lived as a priestly class among the others.

Twentieth Century-Fox Film Corporation, US film production and distribution company. Formed in 1935 by the merger of the Twentieth Century Pictures Corporation and the Fox Film Corporation, the company soon became well known for its production of musicals, westerns and epics. These included *The Grapes of Wrath* (1940), *How Green was My Valley* (1941) and *The Gunfighter* (1950). Fox made the first wide-screen feature, *The Robe,* in 1953 but was nearly made bankrupt by *Cleopatra* (1963). Later productions included *The Sound of Music* (1965) and *Star Wars* (1977).

Twenty Thousand Leagues under the Sea (1873), romantic adventure novel by Jules VERNE, remarkable at the time for his prediction of submarines.

Twickenham, rugby ground at Twickenham, London, the headquarters of the Rugby Football Union. It was opened in 1909, since when it has been the home of the England rugby side. It has a capacity for 72,500 spectators.

Twins, two offspring that are born within a short time (usually minutes) of each other. *See also* MS pp.*76, 134,* 191.

Two Gentlemen of Verona (1594–95), comedy by William SHAKESPEARE. The plot is taken from Jorge de Montemayor's pastoral romance *Diana.*

Two New Sciences, Dialogues Concerning (1638), treatise by GALILEO which includes his theory of parabolic BALLISTICS, the proof of the laws of fall in a vacuum, and an outline of the scientific method as well as reiterating the results of some of his earlier experiments. It was the foundation on which Isaac NEWTON based classical MECHANICS.

Two Noble Kinsmen, The (*c.* 1613), five-act play ascribed to William SHAKESPEARE and John FLETCHER. While many scholars attribute to Shakespeare the parts of the play that are not obviously written by Fletcher, others argue that Philip MASSINGER was Fletcher's collaborator. The main plot closely follows CHAUCER'S *Knight's Tale,* while some scenes and situations are borrowed from various plays by Shakespeare.

Two Sicilies, Kingdom of the, state that united S Italy with the island of Sicily from the 15th to the 19th century. In the 11th century the area was united by the NORMANS, but it was divided in 1282 between the French Angevins and the Spanish Aragonese. In 1443 ALFONSO V of Aragon reunited the area and became King of the Two Sicilies. In 1816 FERDINAND IV of Naples (Ferdinand III of Sicily) officially merged the kingdoms to become Ferdinand I of the Two Sicilies. In 1861 the kingdom was incorporated into the new Kingdom of Italy. *See also* HC2 pp.110, *111.*

Two stroke cycle, power cycle of small petrol engines such as those of small motor cycles and lawn mowers. The upward (compression) stroke of the piston compresses a mixture of air and petrol in the upper part of the cylinder, where it is exploded by an electrical spark while more mixture is drawn into the lower part of the cylinder. The downward (power) stroke of the piston uncovers an exhaust port for the escape of the burned gases, and also admits unburned mixture from the lower to the upper cylinder, ready for the next upstroke. *See also* MM p.*63.*

Two-stroke engine. *See* ENGINE.

Tyburn, old name for a part of the Marylebone district of London, named after the river which ran underground from Hampstead to the River Thames at Westminster and supplied London with water. Tyburn gallows, the chief public hanging place in London from the 12th century to 1783, was at the present Marble Arch.

Tyche. *See* FORTUNA.

Tycho, crater in the southern hemisphere of the Moon named after the astronomer Tycho BRAHE. The crater has a system of brilliant rays radiating from it, and is one of the most prominent surface features to be seen from Earth. Tycho is 86km (54 miles) in diameter and has massive walls. *See also* SU pp.*179,* 183, 185, 187, *187.*

Tyler, John (1790–1862), tenth US President (1841–45). In 1840 the Whigs chose Tyler as vice-presidential candidate for William Henry Harrison who died soon after inauguration, and Tyler became the first Vice-President to succeed to the presidency.

Tyler, Wat (*d.* 1381), English leader of the PEASANTS' REVOLT. He was chosen leader of the rebels in Kent at Maidstone, and led their march on London. He was eventually killed by the Lord Mayor of London while parleying with RICHARD II. His death marked the end of the revolt. *See also* HC1 pp.214, 215,*215.*

Tylor, Sir Edward Burnett (1932–1917), British anthropologist, often called the founder of cultural anthropology. *See also* MS pp.249, 250.

Tympanum, in classical architecture, the decorative stonework or masonry found beneath the arch of a doorway or window, or within an arcade on a wall. It usually rests on a lintel and as its shape is defined by the shape of the arch, the tympanum has changed in style and design with different architectural periods.

Turtle doves winter in tropical Africa but migrate to Europe and SW Asia to breed.

Mark Twain first began writing for his brother's newspaper in Hannibal, Missouri.

Twenty Thousand Leagues under the Sea (1873) anticipated 20th-century technology.

Tyburn gallows, with open galleries for the general public, at the time of Charles I.

Tyne, the river at Newcastle – a major ship-building centre since the 19th century.

Typesetting may be carried out on keyboard operated machines such as the Linotype.

Tyrannosaurus, the largest flesh-eating dinosaur, lived in America and Mongolia.

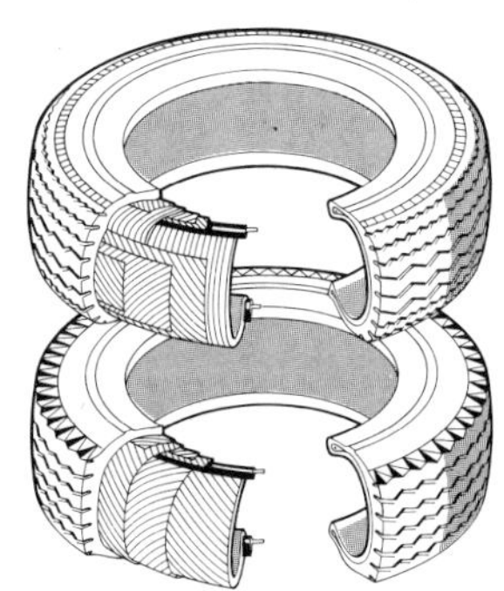

Tyres are of two types; radial-ply (above) and cross-ply should not be used together.

Tyndall, John (1820–1893), Irish physicist who demonstrated that the blue colour of the sky is due to the scattering of light by minute particles of dust. He also studied the magnetic properties of crystals, and light diffusion by dust and large molecules, an effect which bears his name. By 1881 he had helped disprove the theory of spontaneous generation by showing that food does not decay in germ-free air.

Tyndale, William (*c.*1494–1536), British religious reformer and Bible translator. He started printing an English version of the Bible in Cologne in 1525 but, because of harassment, had to complete it in Worms. Despite opposition from Thomas MORE and others, he maintained an output of vernacular scripture and also wrote numerous Protestant works. He was eventually captured and burned at the stake near Brussels.

Tyne, name of two rivers in Britain. One is formed in NE England at the confluence of the North Tyne (which rises in the S Cheviot Hills) and the South Tyne (which rises in Cumbria) and flows E through Newcastle to enter the North Sea near Tynemouth. Length: (from confluence) 48km (30 miles). The second Tyne rises near Borthwick in SE Scotland and flows NE to enter the North Sea near Dunbar. Length: 45km (28 miles).

Tyne and Wear, county in NE England, formed in 1974 from parts of the former counties of NORTHUMBERLAND and DURHAM, and including the former county borough of Newcastle-upon-Tyne, which is the administrative centre of the new county. Area: 540sq km (208sq miles). Pop. (1976) 1,182,900.

Tynwald, court of the Isle of Man. The Government consists of a Lieutenant-governor known as the Lord of Man, a legislative Council (upper house) and the legislative House of Keys (lower house). The two legislative bodies join for some matters to form the Court of Tynwald. All laws passed by the Court are subject to the Royal Assent.

Typesetting, part of the printing process in which metal or other type is set or composed into lines, paragraphs and pages. In early printing, wooden or metal type was set by hand. In modern metal type processes, the LINOTYPE machine casts a complete line of molten type metal into a mould, and the MONOTYPE machine casts individual letters and spaces and arranges these as lines. *See also* PHOTOSETTING.

Type metal, alloy of lead with from 2½–12% of tin and 2½–25% of antimony used for printing type. Both tin and antimony lower the melting temperature, which is usually between 240° and 250°C (460–475°F). Antimony also hardens the alloy and enhances its moulding properties reducing shrinkage during solidification.

Typhoid fever, acute, sometimes epidemic communicable disease marked by fever, chills, prostration, enlargment of the spleen, inflammation of the intestinal tract, and the eruption of pink spots. It is caused by *Salmonella typhi* which is transmitted by contaminated water, milk and food. Inspection of water supplies, pasteurization of milk, typhoid inoculations, and treatment with antibiotics have greatly reduced the incidence of the disease. *See also* MS pp.92–93, *93*.

Typhoon, name given in the Pacific to a HURRICANE, a violent tropical cyclonic storm. *See also* PE pp.68–71, *68–71*.

Typhus, or typhus fever, any of several infectious diseases caused by rickettsial bodies (micro-organisms classified between bacteria and viruses). Epidemic typhus, *Rickettsia prowazeki*, is transmitted in the faeces of human body lice. Vaccinations and antibiotic treatment are now effective against this form. Brill-Zinsser disease, or recrudescent typhus, is a milder form of the same disease. Scrub typhus occurs mainly in tropical and semitropical areas and is transmitted to human beings by various species of mites which carry the infection from rodents.

Typography, the styling of the appearance of printed matter in which the face, size, weight and positioning of the type in which copy is to be produced are specified by a designer, especially where layouts are complicated. The term also refers to the art of fine PRINTING itself. *See also* TYPESETTING.

Typology, system of groupings which aid understanding by establishing relationships between the phenomena being studied. Christian Jorgensen Thomsen, a Danish antiquary, used a typology of materials to establish his division of historical periods into the Stone, Bronze and Iron Ages.

Tyr. *See* TIW.

Tyrannosaurus, any of several species of large, bipedal, carnivorous, THEROPOD dinosaurs that lived during late Cretaceous times. Its head, 1.2m (4ft) long, was armed with a series of dagger-like teeth, some of which were as long as 15cm (6in). The hind legs were stout and well developed, but the forelegs may have been useless except for grasping at close range. The long tail would have served as a counterbalance to the body. The best known species is *T. rex*. Length: 14m (47ft); height: 6.5m (20ft). *See also* NW pp.185, *187*.

Tyre, air-filled rubber and fabric cushion that fits over the wheels of vehicles to grip the road and absorb shock. The pneumatic tyre was invented in 1845 by Robert Thomson but was not commonly used until the end of the century, solid tyres being more popular. The pneumatic tyre consists of a layer of fabric, either rayon, nylon or polyester, surrounded by a thick layer of rubber treated with chemicals to harden it and decrease wear and tear. The angle of the grain of the fabric to the axle defines the kind of tyre: cross-ply tyres have the grain at an angle of about 55°, the more modern radial-ply have the grain parallel to the axle. The latter last longer, and are safer in bad weather; but they are less shock-absorbing and are more expensive.

Tyrian purple, natural dye obtained from the shell of a small purple marine mollusc, *Murex brandaris,* which must be crushed and exposed to light. Originally made from Mediterranean rock whelks by the Phoenicians, it takes its name from the ancient city of Tyre where it was reputedly discovered. *See also* MM p.*244*.

Tyrol (Tirol), federal state in W Austria, bordered by West Germany (N) and Italy (S); Innsbruck is the capital. The region is mountainous and drained by the Inn and Drava rivers. Cattle rearing, forestry and dairy farming are the chief occupations, but tourism is the mainstay of the economy. The province was taken by the HAPSBURGS in 1363. In 1805 Tyrol was awarded to NAPOLEON's ally Bavaria. In 1815 Tyrol was reunited with Austria. Area: 12,647sq km (4,883sq miles). Pop. (1971) 539,000.

Tyrone, largest county of Northern Ireland. Mainly hilly with the Sperrin Mountains in the N and Bessy Bell and Mary Gray in the S; the region is drained by the Blackwater and Mourne rivers. The county town is Omagh. Cereals are grown and dairy cattle are kept. Manufactures include linen, whiskey and processed food. Area: 3,263sq km (1,260sq miles). Pop. (1971) 137,997.

Tyrosine, crystalline AMINO ACID found in PROTEINS. It was isolated in 1849 from casein.

Tyrrhenian Sea, part of the Mediterranean Sea, W of the Italian mainland, E of Corsica and Sardinia, and N of Sicily. It contains several small island groups, including the Lipari and Pontine islands. The chief ports are Naples and Palermo. Width: 97–483km (60–300 miles); length: approx. 765km (475 miles).

Tyus, Wyomia (1945–), US sprinter who won the 100m sprint at the 1964 and 1968 Olympics, establishing a world record in 1968.

Tyutchev, Fyodor Ivanovich (1803–73), Russian lyric poet and essayist. He spent much of his life abroad in the diplomatic service. His work is strongly nationalist.

Tzara, Tristan (1896–1963), French writer, b. Romania. One of the founders of DADAISM in 1916, he tried to write poetry which ignored conventional syntax, as in his *Vingt-cinq poèmes* (1918). He wrote the *Sept manifestes Dada* (1924), but later became more attracted to SURREALISM.

Tzotzil, one of the major Maya-language tribes that live in Mexico. They have retained much of their earlier culture; in the 1970s about 120,000 Tzotzil-speaking people lived in the state of Chiapas.

Tz'u Hsi (1835–1908), Empress Dowager of the CH'ING DYNASTY of China, known as the "Old Buddha". As concubine of the Emperor Hsi'en Feng she learned the politics of the Manchu court and became one of the most powerful and feared women in Chinese history. She was an obstinate opponent of the West and modernization and encouraged the BOXER REBELLION. *See also* HC2 p.144.

U

U, 21st letter of the alphabet, derived from the Semitic letter *vaw* (meaning *hook*), which also gave rise to *v*. It resembled a *Y* and was adopted in similar form by the Greeks, who called it *upsilon*. The Romans gave the letter its present form, and used it both as the vowel *u* and the consonant *v*. In English *u* is a vowel and has many different sounds, chief of which are the short sounds as in the words *bun, bull* and *burr* and the long sounds as in *dune* (diphthong) and *lunar* (in all of these examples the pronunciation of *u* is slightly different). The letter may be silent after *g* or *q* (as in *guilt* and *liquor*) or modify these letters to make the sound *gw* or *kw* (as in *language* and *liquid*). *See also* MS pp.244–245.

Ubac, Raoul (1910–), Belgian painter. He was influenced by the SURREALISTS in the 1930s and then turned to photography and etching.

U-boat, short for *Unterseeboot,* a German SUBMARINE. In WWI Germany was the first nation to employ submarines for large-scale warfare. The U-boat fleets of WWII were initially a great threat to Allied shipping, sinking nearly a million tonnes of British shipping in the first year of the war. After the entry of the USA into the war in 1942, however, more ships and aeroplanes were available for escorting Allied convoys and destroying U-boats, and the struggle to supply Britain (known as the Battle of the ATLANTIC) was won by the Allies. During the war 785 of the 1,162 U-boats launched were sunk. *See also* HC 2 pp.*229*, 230; MM pp.118, *119*.

Uccello, Paolo (1397–1475), Florentine painter, celebrated as an early master of perspective, the invention of which is sometimes wrongly attributed to him. His most creative period was from *c.*1436 to 1460, during which he painted *The Flood* (*c.*1450) and the three panels of *The Rout of San Romano* (1454–57), which were commissioned for the Medici Palace in Florence.

Udall, Nicholas (*c.*1505–56), English dramatist and schoolmaster. He translated several works into English, including the Great Bible (1551); and his play *Ralph Roister Doister* (*c.*1553), is considered the first true comedy written in English. *See also* HC1 p.283.

Udmurt (Udmurtskaja ASSR), autonomous republic in the Russian Republic (Rossijskaja SFSR), USSR, in the W foothills of the Urals. The region's growth was boosted during WWII when many industries were moved there from W USSR. Engineering, steel, machinery, metallurgy and food processing are the principal industries. The region has been under Russian rule since the 16th century. Area: 42,100sq km (16,250sq miles). Pop. (1970) 1,418,000.

Uffizi, Florentine palazzo, built in the 16th century by Giorgio VASARI for the Grand Duke Cosimo I de' Medici as government offices. Vasari's architectural masterpiece, it now contains the Uffizi Gallery, one of the world's greatest art collections. Painters of the Florentine and other Italian schools are represented with works by Piero della Francesca, Botticelli,

Leonardo, Michelangelo, Raphael, Titian and many others, as well as Dutch and Flemish masters.

UFO, unidentified flying object. *See* FLYING SAUCERS.

Uganda, independent nation of E central Africa. There are three main regions, the Lake Victoria lowlands, the N savanna plateau and the Rift Valley zone. Most of the people work in subsistence agriculture; coffee and tea are the most important cash crops. There is some mining for copper and tin, but minerals and chemical manufacturing remain undeveloped. The capital is Kampala. Area: 236,036sq km (91,133sq miles). Pop. (1975 est.) 11,549,000. *See* MW p.168.

Ugarit, ancient city in N Syria, dating from the NEOLITHIC period. It developed as a great commercial power from its trade with Mesopotamia and later from its alliance with Egypt in the 2nd millennium. Its period of greatest prosperity occurred during the 15th and 14th centuries BC. By the 12th century economic change and foreign invasions had caused the city to decline. *See also* HC1 p.55.

Uhland, Johann Ludwig (1787–1862), German poet who held various official posts and taught literature. He became one of the most popular German ROMANTIC poets and his poems, noted for their polished and lucid style, include *The Minstrel's Curse* and *Taillefer.*

Uist, North and South, two islands in the Outer Hebrides, off the NW coast of Scotland, in the Western Isles island authority. Lochmaddy is the chief town on North Uist and Lochboisdale the main town on South Uist. The principal occupations are the rearing of sheep and cattle, fishing, the weaving of tweed and seaweed processing. There is a missile range on South Uist. Area: (North Uist) 572sq km (221sq miles); (South Uist) 362sq km (140sq miles). Pop. (North Uist, 1971) 1,800; (South Uist, 1971) 3,800.

Ujiji, town in Tanzania, now included in the Kigoma Region. In the second half of the 19th century it was a centre for Arab traders in slaves and ivory. It was at Ujiji that Henry STANLEY found David LIVINGSTONE in 1871.

Ujjain, city in W central India, on the River Sipra, 320km (200 miles) E of Ahmadabad. It is one of the seven holy cities of India, and a Hindu pilgrimage centre. The ancient city of Ujjain dates from the 6th century BC and is one of India's oldest cities. It has many notable temples and mosques. Pop. (1971) 203,278.

Ukelele, or ukulele, small guitar developed in Hawaii and taken there by Portuguese seamen. Shaped like a classical guitar with a wooden body, round sound hole and fretted fingerboard, it has nylon strings tuned to A, D, F♯, B or G, C, E, A. It is strummed or plucked to accompany popular songs.

Ukiyo-e, Japanese paintings and woodblock prints that were prevalent in the Edo period (1615–1867). Their subject matter included people engaged in everyday activities as well as Kabuki actors, who were popular in the city of Edo (now Tokyo). Moronobu (c. 1625–95) is generally considered the originator of the true ukiyo-e print and he gained popularity for his woodcut illustrations for popular literature. Other famous ukiyo-e printmakers were Harunobu, HIROSHIGE, HOKUSAI, KIYONAGA, Sharaku and UTAMARO.

Ukraine, or Ukrainian Soviet Socialist Republic, constituent republic of the USSR, in the SW, bordering on the Black Sea. Its capital is Kiev. It contains about 20% of the USSR population and is second in economic important only to Russia. It supplies about 25% of the nation's food, especially wheat, and about 30% of its industrial output: metallurgy, chemicals and machine-building. The Ukrainian language is a Slavic tongue, close to Russian. After the BOLSHEVIK revolution of 1917, the Ukraine proclaimed its independence, but a civil war ended in the defeat of Ukrainian nationalism and the establishment of the Ukraine SSR in Dec. 1919. It joined with the other republics to form the USSR in 1923. From 1920 to 1939 the 6 million inhabitants of the western Ukraine were ruled by Poland. They were rejoined to the Ukraine when Germany and the USSR partitioned Poland in 1939. In 1940 it acquired Northern Bukovina and part of Bessarabia from Romania. In 1945 it gained Ruthenia from Hungary. Area: 445,000sq km (171,815sq miles). Pop. (1976) 49,075,000.

Ulan Bator (Ulaanbaatar), capital of Mongolia, on the River Tola. It was a focus for the Mongolian autonomy movement, and independent Mongolia was proclaimed there in 1911. It was occupied by White Russian forces in 1924 and became the capital of the Republic in the same year. It is the political, cultural and economic centre of the country. Manufactures: textiles, leather, paper, alcohol, food products and glassware. Pop. (1970 est.) 287,000.

Ulanova, Galina (1910–), Russian ballerina who made her debut in 1928. After 1944 she was prima ballerina of the Bolshoi Theatre, Moscow, and excelled in *Swan Lake, Giselle* and the Soviet Première of Prokofiev's *Romeo and Juliet* (1940).

Ulbricht, Walter (1893–1973), Communist leader of East Germany. A founder of the German Communist Party, he fled to Moscow in 1933 but returned with the Soviet army in 1945. He became Secretary-General of the Party in 1950 and leader of East Germany, becoming Chairman of the Council of State in 1960. *See also* HC2 p.243.

Ulcer, sore or lesion on the skin or in the mucosal lining of the gastrointestinal system, where it is frequently caused by abnormally large secretion of gastric juices over a prolonged period. Gastric ulcers may remain superficial, or penetrate the affected organ. This psychophysiological disorder is often precipitated by emotional stress and worry. Symptoms of such ulcers include stomach pains (particularly after eating), nausea and (in untreated cases) haemorrhaging. Treatment varies according to location and type of ulcer, but usually a strict diet is recommended with attention to the underlying cause. Some ulcers may be removed or treated by surgery. *See also* MS p.92, *92.*

Ullswater, lake in Cumbria, NW England, about 8km (5 miles) SW of Penrith. Area: 9sq km (3.5sq miles). Max. depth: 62.5m (205ft).

Ulm, city in S central West Germany, on the River Danube, 72km (45 miles) SE of Stuttgart. One of the most important political and commercial centres of medieval Europe, its industries include textiles, processed food and beer. Pop. (1970) 92,500.

Ulna, long bone of the inner side of the forearm. At its upper end it articulates with the HUMERUS and with the RADIUS, the slightly shorter bone of the outer side of the forearm. At its other end it articulates with the radius only, not directly with the carpal bones of the wrist. *See also* MS p.57.

Ulster, ancient province in NE Ireland, consisting of nine counties. Six of these counties are now in NORTHERN IRELAND. Cavan, Donegal and Monaghan form Ulster province in the Republic of IRELAND. Ulster also refers to Northern Ireland as a political unit of the UK.

Ulster Unionists, political party in Northern Ireland, which arose in the late 19th century to defend the six northern provinces of Ulster from IRISH HOME RULE and to maintain the union with Britain. In 1904 the Ulster Unionist Council was formed to link Orange lodges and constituency associations in the fight against home rule. It is almost exclusively a Protestant party. From 1922 until the impostition of direct rule from Westminster in 1972 it was the ruling party in the Northern Irish assembly.

Ultra high frequency waves (UHF waves), radio waves in the frequency band 300–3,000MHz (megahertz). UHF waves, with a wavelength of about one metre or less, are used for TELEVISION broadcasting.

Ultra-microscope. *See* MICROSCOPE.

Ultramontanism, position taken by Catholics who asserted the supremacy of the papacy over the claims of national churches. The word comes from the Latin *ultra montes* ("beyond the mountains", ie the Alps). It became prominent in the 16th and 17th centuries and was particularly associated with the supporters of papal authority against those of the Gallican church. Following the definition of papal infallibility in 1870, it was opposed by the OLD CATHOLIC CHURCH, in Germany, Austria and Switzerland.

Ultrasonics, study of sound with frequencies beyond the upper limit of human hearing, ie with frequencies in excess of 20,000Hz. Applications of ultrasonics include the agitation of liquids to form emulsions; detection of flaws in metals (the ultrasonic wave passed through a metal specimen is reflected by a hairline crack); cleaning small objects by vibrating them ultrasonically in a solvent; echo sounding in deep water; soldering aluminium; and, in medicine, the location of a tumour, the scanning of a pregnant woman's abdomen to produce a "picture" of the fetus and the treatment of certain neurological disorders. *See also* MM p.236; SU pp.80–81, *81.*

Ultra-violet lamp, lamp which emits radiation of wavelength 2,200 to 4,000 ANGSTROMS. It is used as a germicidal lamp, for fluorescent effects and for producing an artificial "suntan". It uses a mercury vapour discharge lamp, whose walls are made of quartz, which is much more transparent to ultra-violet light than is glass.

Ultra-violet light, ELECTROMAGNETIC RADIATION having wavelengths shorter than those of visible light, from about 40 to about 4,000 angstroms, but larger than those of X-RAYS. Between about 40 and 3,000 angstroms these rays affect living matter in a number of important ways. They kill bacteria and many other parasites and so are useful as a means of sterilizing, whether as sunlight or from ULTRA-VIOLET LAMPS. They tan the skin by their effect on the pigment MELANIN, and help to make VITAMIN D in the body, which plays a part in the prevention of RICKETS. Ultra-violet light also causes certain materials to fluoresce (emit visible light), such as the optical brightness added to some detergents.

Ulysses. *See* ODYSSEUS.

Ummayads, or Ommayads, first great MUSLIM dynasty of Arabian CALIPHS, which ruled from 661 to 750. It was a merchant family of Mecca. The house had two ruling branches, the Sufyānid (r. 661–84) and the Marwānid (r. 684–750).

Umbelliferae, the carrot family of flowering plants, all of which have many small flowers, which are usually white or yellow and are borne in umbrella-like clusters (umbels) at the ends of stalks. Many species are edible, including PARSLEY, CELERY, PARSNIP, FENNEL, DILL, and ANGELICA.

Umberto, name of two kings of Italy. Umberto I (1844–1900; r. 1878–1900). His interests became increasingly imperialistic and he led Italy into the Triple Alliance of 1882 with Germany and Austria. He was assassinated at Monza in 1900. Umberto II (1904–46; r. 1946), the last King of Italy, was the son and successor of VICTOR EMMANUEL III. *See also* HC2 p.180.

Umbilical cord, long, thick cord that connects a developing fetus with the placenta, the hormone-secreting structure through which the fetus receives food and oxygen from the mother's bloodstream and gets rid of waste products. The umbilical cord contains two large arteries and one vein. At birth, the cord is clamped and cut from the placenta; the part of the cord remaining on the baby's abdomen dries and falls off, leaving the scar known as the navel.

Umbrella bird, any of three species of large tropical American birds, each with a retractile, black umbrella-like crest and a long, often tubular-shaped feathered lappet, or tuft, on the throat. The ornate umbrella bird (*Cephalopterus ornatus*) lives high in trees, feeds on fruits and emits loud piping sounds. *See also* NW p.217.

Umbrella pine, also called stone pine, one of the classical Mediterranean landscape trees. It has a characteristic umbrella-like shape which seems to vary remarkably lit-

Ukelele; the instrument was taken to Hawaii from Portugal in the 18th century.

Walter Ulbricht in 1960, after he had become chairman of a new Council of State.

Ullswater; lake's head at Patterdale, with Silver Hows mountain on the right.

Umberto II, while Prince of Piedmont, became a general in the Italian army.

Umbria

Underground railway; some of the workmen constructing London's Victoria Line.

Ungulate; shown here is the hoof formation of a zebra, a member of the order.

Union Movement; Sir Oswald Mosley speaks at a meeting in Trafalgar Square in 1958.

United Arab Emirates; a shelter made of palm fronds at an oasis in Abu Dhabi.

tle from individual to individual. Height: to 7m (20ft). Species *Pinus pinea. See also* NW p.57.

Umbria, region in central Italy made up of the provinces of Perugia and Terni; Perugia is the regional capital. The region is traversed by the APPENNINES and is drained by the River Tiber. Cereal crops, grapes and olives are grown, and cattle and pigs are raised. In the 15th and 16th centuries the city was the centre of the Umbrian school of painting, which included PERUGINO. Industries: iron and steel, chemicals, textiles, food processing. Area: 8,456sq km (3,265sq miles). Pop. 783,274.

Umbriel, satellite of Uranus, at a distance of 267,180km (165,950 miles) from the planet. The moon has a diameter of approx. 1,000km (620 miles). *See also* SU p.214, *214*.

Umm Durman. *See* OMDURMAN.

UN. *See* UNITED NATIONS.

Unamuno y Jugo, Miguel de (1864–1936), Spanish EXISTENTIALIST philosopher and writer, best known for his work *The Tragic Sense of Life in Men and Peoples* (1912) in which he discusses the fundamental dichotomy between reason and faith.

Uncle Remus. *See* HARRIS, JOEL CHANDLER.

Uncle Sam, figure that symbolizes the USA. Samuel Wilson (1766–1854), an army provisioner, stamped the letters "U.S." on barrels of salted meat during the WAR OF 1812. Some people thought it stood for "Uncle Sam". Wilson and those who opposed the war adopted it as an unfriendly nickname for the government.

Unconformity, in geology, break in the time sequence of rocks layered one above the other. This time gap is explained by the erosion of ancient rock layers, which then sank beneath the surface of the sea, and were built upon by later sediments.

Unconscious, term in psychology that part of mental life believed to operate without the individual's immediate awareness or control. In FREUD's system it has two parts, the ID (instinctual drives) and the partly-conscious SUPER-EGO (which represses the *id*). JUNG added that the unconscious is partly inherited concepts or the ability to form them.

Underground arts, movement that flourished in many artistic fields in Britain and the USA in the 1960s among those artists who rejected the conventional structure of artistic success. Underground film-making flourished, as did street theatres and ephemeral magazines that advocated such things as revolution, anarchy and drug-taking.

Underground railway, electrically operated subterranean passenger transport system in urban areas. There are seven cities which run major systems with more than 40km (25 miles) of track: Berlin, Boston, Hamburg, London, Moscow, New York and Paris. More than two dozen cities have smaller systems. The first underground railway was built in London between 1860 and 1863. It was steam-powered and carried passengers from Farringdon to Paddington. The London system, which has approx. 400km (250 miles) of track is the world's second largest. The largest is the New York subway, which has more than 1,130km (700 miles) of track. The first continental European system was opened in Budapest in 1896. *See also* MM pp.142–143.

Under Milk Wood (1953), play by Dylan THOMAS. Often called "A Play for Voices", it was written specifically for radio. Using evocative language and imagery, Thomas depicts a day in the life of a small Welsh coastal town and the memorable characters who live there.

Underpainting, term used to describe (a) what was the initial stage in traditional oil painting, ie the painting of a picture in monochrome before the addition of colour; (b) a layer of colour which is to be glazed or scumbled. *See also* SCUMBLING.

Underwriting, guarantee or cover of a loss either in connection with insurance, particularly marine insurance, or the issue of new STOCKS or SHARES. The technique began at LLOYD's coffeehouse in London in the 17th century. A ship's cargo was recorded on a slip of paper and people willing to share the risk of insuring it would sign their names (underwrite) to it and guarantee what they would pay against loss.

Undset, Sigrid (1882–1949), Norwegian novelist who was awarded the 1928 Nobel Prize in literature. Her masterpiece, *Kristin Lavransdatter* (1920–22), is a trilogy of 14th-century life in Norway.

Undulant fever. *See* BRUCELLOSIS.

Unemployment benefit, financial aid to the unemployed, first introduced into Britain by the unemployment insurance scheme of 1911. Before this the only unemployment benefit was that occasionally paid by small craft unions to its members, although state provision of benefits had been introduced in Germany in the 1880s. Unemployment benefit is now paid by the Department of HEALTH AND SOCIAL SECURITY.

UNESCO, United Nations Educational, Scientific and Cultural Organization, agency founded in 1945 to promote peace by improving the world's standard of education and by bringing together nations in cultural and scientific projects. It gives AID to developing countries. Its head-quarters are in Paris.

U Ne Win. *See* NE WIN, U.

Unfair dismissal, dismissal from a job without just cause, legally distinguished from WRONGFUL DISMISSAL. In Britain the Trades Union and Labour Relations Act (1974) established for employees, subject to certain exceptions, the right not to be dismissed unfairly. The Act enables an employee who thinks he has been unfairly dismissed to seek a remedy by complaining to an Industrial Tribunal. The Act was amended by the Employment Protection Act (1975) which makes substantial changes to these provisions.

Ungava, region of New Quebec, Canada, E of Hudson Bay and N of the Eastmain River. It was transferred from the North-West Territories in 1912 to the province of Quebec; the E portion was assigned to Newfoundland in 1937. Ungava is rich in minerals, especially iron ore. Area: 622,000sq km (240,000sq miles).

Ungulate, MAMMAL with hoofed feet. Most ungulates, including cattle, sheep, pigs and deer, are members of the order ARTIODACTYLA. The order PERISSODACTYLA consists of horses, tapirs and rhinoceroses, while the orders PROBOSCIDEA and Hyracoidea only elephants and hyraxes, respectively. *See also* NW pp.*162–163*.

UNICEF, United Nations International Children's Emergency Fund, agency founded in 1946 to bring AID to children in war-devastated countries and in regions struck by natural disasters. Unlike most UN agencies it depends upon grants from individuals and governments, not on requisitions from UN member states. It is now called the United Nations Children's Fund.

Unicorn, in mythology and heraldry, a magical animal resembling a graceful horse or a young goat with one thin conical or helical horn on its forehead. Traditionally a proud, wild and shy creature, it was believed that a maiden could capture it if she sat very still, for the unicorn would lay its head upon her lap and become docile. *See also* MS p.*206*.

Unidentified flying objects. *See* FLYING SAUCERS.

Unified field Theory, attempt to extend the General theory of RELATIVITY to give a simultaneous representation of both gravitational and electromagnetic fields. A more comprehensive theory would also include the strong and weak interactions. Although some success has been achieved in unifying the electromagnetic and weak interactions, the general problem is still unsolved. *See also* SU pp.106–107.

Uniformity, Act of (1662), act of the English Parliament regulating the form of worship in the CHURCH OF ENGLAND. It required all ordained clergy to follow the Book of COMMON PRAYER, and about 2,000 beneficed clergy were driven out of the Church by it. The act also required the clergy to repudiate the SOLEMN LEAGUE AND COVENANT, to forswear the taking up of arms against the Crown, and to adopt the Liturgy of the Church of England. Those people unable to accept the doctrines and practices of the Church of England were later known as NONCONFORMISTS and the foundation of many such groups dates from 1662.

Union, Act of (1707), act uniting the kingdoms of England and Scotland under one British Parliament at Westminster, with England and Scotland each retaining its own legal system and national church. *See also* HC2 p.28.

Union, Act of (1800; effective 1801), union of Britain and Ireland in one Parliament. The Irish legislature was abolished and Ireland was given four spiritual peers, 28 representative peers and 100 MPs in the British Parliament. The established churches of the two countries were united, and free trade was introduced between the two countries.

Union, acts of (1536–43), series of acts which united England and Wales. The acts abolished the Welsh MARCHES, reorganized the Welsh counties, established English law and administration, in Wales, made English the official language in Wales, and provided for Welsh representation in the English Parliament.

Union Jack, or Union Flag, the flag of the United Kingdom. It first appeared at the accession of JAMES I, with the red cross of England imposed on the Scottish flag, a white cross on a blue field. When Ireland and Britain were united by the Act of 1800, the red saltire of St Patrick was placed on the white saltire of St Andrew. *See also* MW pp.23, 24.

Union Movement, political group in Britain formed in 1931 between Sir Oswald MOSLEY's New Party and some members of the LABOUR PARTY. It was merged in 1932 with the British Union of Fascists, popularly known as the BLACK SHIRTS. The Union movement was revived in 1948 with the slogan "Europe a Nation".

Union of South Africa, the union under one parliament of the four self-governing colonies of Natal, the Transvaal, the Orange Free State and the Cape Colony. It was formed in 1910 and given Dominion status within the British Empire. The name was changed to the Republic of SOUTH AFRICA when the country left the British Commonwealth in 1961.

Union of Soviet Socialist Republics. *See* USSR.

Union of the Two Canadas (1840), act of the British Parliament placing Upper and Lower Canada (present-day Ontario and Quebec) under one Parliament, creating the Province of Canada.

Unions. *See* TRADE UNIONS.

Unitarianism, version of Christianity that denies the doctrine of the Trinity. Originally considered a heresy, it flourished in Poland in the 16th century under the teaching of Faustus Socinus. John Biddle first preached Unitarianism in England in the 1640s, and a Unitarian church established in Philadelphia, USA, in 1796 by Joseph Priestley.

United Arab Emirates (Ittihād al-Imārāt al-'Arabīyah), formerly the Trucial States, a union of seven emirates (Dubai, Ajman, Abu Dhabi, Ras al-Khaimah, Fujairah, Sharja, and Umm al-Qaiwain) bordered by the Persian Gulf (N), Qatar (NW), Saudi Arabia (W and S) and Oman (E). The terrain is flat, consisting mainly of sand and salt flat desert; only in the E does the land rise to any height. Agriculture is limited to the mountain region, oases, and areas accessible to irrigation; the main crops are dates and vegetables. The economy is based on oil which was first produced in 1962 from Abu Dhabi. The temporary capital is Abu Dhabi. Area: 83,600sq km (32,278sq miles). Pop. (1975 est.) 656,000. *See* MW p.173.

United Arab Republic (UAR), political union of Egypt and Syria formed in 1958 with its capital at Cairo and Gamal Abdel NASSER of Egypt as its president. Although Syria withdrew from the Union in 1961, Egypt retained the title "Arab Republic" until Sept. 1971, when it joined the confederation of Arab Republics (Egypt, Syria and Libya).

United Empire Loyalists, Canadian

organization comprising descendants of American Loyalists who fled to Canada during the American War of Independence and after the Treaty of PARIS had failed to make adequate provision for them. Most settled in the Maritime Provinces and Ontario. The organization has been a pro-British conservative force in Canadian politics.

United Irishmen, Society of, underground Irish nationalist society formed in Ulster in 1791 by Theobald Wolfe TONE, James Napper Tandy and Thomas Russell. Its members included both Protestants and Roman Catholics. Its radical objects included the establishment of a representative, independent Irish parliament. Its armed organization was chiefly responsible for the rebellion of 1798.

United Kingdom (UK). *See* GREAT BRITAIN.

United Kingdom Atomic Energy Authority, British public corporation set up by the Atomic Energy Authority Act of 1954. It is now responsible for the control of atomic energy research and development, and is responsible to the Secretary of State for Energy.

United Nations (UN), international organization set up to enable countries to work together for peace and mutual development. It was established by a charter signed in San Francisco in June 1945 by 50 nations; in 1977 the UN had 148 members (*see* MW p. 191). *See also* HC2 pp.191, 219, 246, 296–299.

United Nations International Children's Emergency Fund. *See* UNICEF.

United Nations Relief and Works Agency for Palestine Refugees (UNRWA), organization founded in 1950 to provide relief and welfare services for Palestinian refugees in Lebanon, Syria, E Jordan, the West Bank and the GAZA STRIP. In 1975 the war in Lebanon forced the removal of the UNRWA headquarters from Beirut to Amman, but it returned to Beirut at the end of 1977. The registered refugee population in mid-1977 was 1,706,486.

United Provinces, former state in N India which included the Ganges plain; considered the most historical and religious region of the country. The area was under Muslim rule from the 14th to the 18th century and became the United Provinces of Agra and Oudh in 1902. It is almost coextensive with the modern state of UTTAR PRADESH, which was formed in 1950.

United Provinces of the Netherlands, also known as the Dutch Republic, an historic state that was formed in 1579 out of the Union of UTRECHT and became a world power in the 17th century. *See also* MW p.130.

United States of America. *See* USA.

Unities, theatrical elements of time, place and action, said to derive from the *Poetics* of ARISTOTLE and introduced into French drama by Jean Mairet (1604–86). The convention of a play taking place within the span of a day and comprising one action was used by other French dramatists such as CORNEILLE and RACINE.

Unit One, association formed in 1933 by a number of British artists who, although employing different styles, shared a contemporary approach to the arts. They included Paul NASH, Henry MOORE, Barbara HEPWORTH and Edward BURRA.

Units. *See* CGS SYSTEM; MKS UNITS; WEIGHTS AND MEASURES; SI UNITS.

Unit trust, organization which raises money from small investors and invests in a spread of equity shares in order to reduce the risk of financial loss. Unit trusts were first formed in Britain in the 1860s, but they did not become widespread until the 1950s.

Universal joint, connection between two shafts which allows them to rotate together even when they are at an angle to each other.

Universals, in philosophy, characteristics shared by two or more objects, which are referred to as instances. PLATO considered the utiversals to be distinct from their instances. This theory poses the problem of the relationship between the two. A CONCEPTUALIST view holds that universals are ideas constructed by the mind; this avoids the problem but cannot explain the nature of the external world. For NOMINALISTS, such as William of OCCAM, there are only general words such as "dog" and no universals such as "doghood".

Universal set, in mathematics, collection of all elements that could be included in a particular set. *See also* SU p.40.

Universal time, system of time reckoning based on the mean solar day, the average interval between two successive transits of the Sun across the Greenwich (0°) meridian.

Universe, aggregate of all matter, energy and space, consisting of vast cold, empty regions with a distribution of high-temperature stars and other objects grouped in galaxies. On a large scale the universe is considered uniform: it is identical in every part. It is believed to be expanding at a uniform rate, the galaxies all receding from each other. The origin, evolution and future characteristics of the universe are considered in several cosmological theories. Recent developments in astronomy imply a finite universe, as postulated in the EXPANDING UNIVERSE Theory. *See also* SU pp.252, 282.

Universe (theories of formation). *See* "BIG BANG" THEORY; STEADY STATE THEORY.

University, institution of higher learning, which arose from the *studia generalia* of the 12th century themselves arose out of the need to provide education for priests and monks were attended by students from all parts of Europe. Bologna became an important centre of legal studies in the 11th century. Other great *studia generalia* were founded at Paris, Oxford and Cambridge, where the teaching of the seven liberal arts took on increasing significance. The word *universitas* applied originally to guilds of students or of masters, but by the late 14th century came to designate an institution whose corporate existence had been legally recognized. The first Scottish university was founded at St Andrews in *c.*1412, the first Irish university at Dublin (Trinity College) in 1591. Oxford and Cambridge were the only English universities until the founding of Durham (1832) and London (1836). London was the first university open to Nonconformists; it was also the first British university to award degrees to women undergraduates, in 1847.

University Extension Courses, system of higher education for those unable to attend full-time university because of work commitments. Such courses were first organized in Britain by Cambridge University in 1867.

Unknown warrior, unidentified body of a member of the armed forces that symbolizes all those killed in action who lie in unnamed graves. Many nations accorded a ceremonial burial to such a warrior after WWI. In Britian the tomb of the Unknown Warrior is in the centre of the nave of WESTMINSTER ABBEY, London; in France it is beneath the ARC DE TRIOMPHE in Paris; and in the USA it is in the Arlington National Cemetery, Washington.

Unlawful assembly, in English law, one of three main offences involving public disorder (the others are rout and riot). Its definition in COMMON LAW is a gathering of three or more people with the intention of committing criminal violence or with the effect of causing a breach of the peace. When the people move off to execute their plan it becomes a rout, when the action begins a riot. *See also* RIOT ACT.

Unlearned Parliament, parliament of HENRY IV of England, summoned in 1404 at Coventry. Henry's earlier parliaments had proved intractable and in order to get a parliament more subservient to his wishes he instructed his sheriffs to return no lawyers.

UNRWA. *See* UNITED NATIONS RELIEF AND WORKS AGENCY FOR PALESTINE REFUGEES.

Unsaturated compound, in organic chemistry, compound in the molecule of which two or more carbon atoms are linked, or bonded together, with double or triple bonds. Simple examples are ethylene, $H_2C=CH_2$, and acetylene, $HC\equiv CH$. By breakage of their double or triple bonds, unsaturated compounds can add on more atoms to their molecules.

Unsaturated fats, natural fatty compounds, present in animal and vegetable matter, containing at least one double bond between two carbon atoms. By far the most abundant are oleic acid, $CH_3(CH_2)_{14}CH=CH.COOH$; linoleic acid, $CH_3(CH_2CH=CH)_3(CH_2)_7COOH$. These acids, being unsaturated, confer many of the reactive properties of oils and fats used in industry and for cooking.

Unsaturated solution, chemical solution in which more of the solute (dissolved substance) can be dissolved without causing precipitation. At the same pressure and temperature, it is possible to dissolve more and more of the compound to make first a SATURATED SOLUTION, then a supersaturated solution.

Untouchables, fifth and lowest class of the Indian HINDU CASTE system, so called because it is believed that to touch people of this class is ritually polluting. Although this PARIAH caste was legally abolished in 1949, much discrimination remains. *See also* MS p.270.

U Nu. *See* NU, U.

Upanishads, Hindu texts constituting the final stage of Vedic literature and of uncertain authorship, dating from approximately 650 BC. Of the 108 Upanishads, there are 14 principal ones. In prose and verse, they are in the form of dialogues between teacher and pupil and speculate on reality and man's salvation. *See also* BRAHMAN; HINDUISM.

Upas tree, evergreen tree native to Java; its milky juice is used to poison the tips of arrows. The reddish brown fruit, 2cm (0.8in) in diameter, is pear-shaped. Height: to 30m (100ft), Family Moraceae; species *Antiaris toxicaria*.

Updike, John Hoyer (1932–), US author whose novels, usually about contemporary life, include *The Poorhouse Fair* (1959), *Rabbit Run* (1960), *The Centaur* (1963), *Couples* (1969), *Rabbit Redux* (1972) and *Marry Me* (1976).

Upham, Charles Hazlitt (1908–), New Zealand soldier and farmer, the sole recipient of a VICTORIA CROSS with Bar during WWII. His VC was earned in Crete, its bar at Ruweisat Ridge, north Africa.

Upjohn, Richard (1803–78), US architect, *b.* Britain, the founder and first president (1857–76) of the American Institute of Architects. He is best known for his Neo-Gothic churches.

Upper atmosphere, region of the atmosphere, extending upwards from about 50km (30 miles), which is free of disturbances caused by the weather. It includes the MESOSPHERE, THERMOSPHERE and IONOSPHERE. At this height the air is rarified, at temperatures ranging from −110°C (−170°F) in the lower regions to 250–1,500°C (500–2,700°F) higher up. The behaviour of the upper atmosphere is greatly affected by such extra-terrestrial phenomena as solar radiation and COSMIC RAYS, which cause the gas molecules to produce the ionosphere, and atmospheric tides, which cause turbulence. *See also* PE pp.66–67.

Upper Canada, formed 1791, region that is now the province of Ontario. The name was changed to Canada West in 1841 and to Ontario after the formation of the Confederation of Canada in 1867. *See also* HC2 pp.136–137.

Upper Volta, land-locked republic in W Africa. It is an extremely poor country and 90% of the people are subsistence farmers. Cotton, groundnuts and sesame seeds are the chief crops; the rearing of animals is also important. The capital is Ouagadougou. Area: 274,200sq km (105,869sq miles). Pop. (1975) 6,144,000. *See also* MW p.179.

Uppsala (Upsala), city in E Sweden, 64km (40 miles) NNW of Stockholm, on the Fyrisån River; capital of Uppsala county. The University (1477) is the oldest in N Europe; there is a fine 13th-century Gothic cathedral. Industries: printing, food processing, metal goods. Pop. (1974) 137,543.

Upwelling, the process that brings water of greater density and lower temperature up to the surface of the ocean. It is especially characteristic of the W side of continents where winds blow parallel to the coast and the water carried away by the

United Nations delegates discuss matters of international importance and urgency.

University; Tom Tower and the great quadrangle of Christ Church College, Oxford.

Unknown warrior's tomb between the West Door and the Nave of Westminster Abbey.

Untouchables may not put up their huts in a village and must often sleep by the roadside.

Urban II could not reside in Rome until 1093, when the antipope was expelled.

Uri; the monument to William Tell in Altdorf, capital of the canton.

Urial, or Punjab sheep, are most common in Sind, northern Punjab, Tibet and Iran.

Leon Uris's novel *Mila 18* concerned the Jewish revolt in Warsaw during WWII.

surface current is replaced by the bottom water. *See also* PE pp.78–79.

Ur, ancient city in S Mesopotamia. Settled in the 4th millennium BC, it prospered during its First Dynasty (2,600 BC), and during its Third Dynasty (2,300–2,200 BC) it became the richest city in Mesopotamia, probably in the late 4th century BC. Excavations undertaken in the 1920s and 1930s revealed that Ur was a thriving commercial centre before the First Dynasty. *See also* HC1 pp.31–32.

Uralic, family of languages spoken by about 23,000,000 people in parts of N and E Europe and N Asia. Its two main branches are FINO-UGRIO and SAMOYED.

Urals, range of mountains in the Russian Republic (Rossijskaja SFSR), USSR, traditionally marking the boundary between Europe and Asia. The range extends from the Arctic in the N to the Ural River in the S. Except for the N section of the range, the mountains are extensively forested, and the timber industry is important. The Urals' chief importance lies in their mineral deposits, which include iron ore, oil, coal, copper, nickel, gold, aluminium and many precious stones. These resources have given rise to the Urals industrial region in the central and S Urals and adjacent lowlands. Industrial development was spurred under the first two Soviet five-year plans (1929–39) and during WWII, when many industries were moved there from W USSR. The highest peak is Mt Narodnaya, rising to 1,894m (6,214ft). Length: approx. 2,400km (1,500 miles).

Uraninite, oxide mineral, uranium oxide (UO_2), found in PEGMATITES and medium-temperature veins. Its crystal structure is cubic system as cubes, octahedrons and dodecahedrons; when botryoidal (like a bunch of grapes), with radiating structure, it is called PITCHBLENDE. Greasy or dull black; hardness 5–6; s.g. 7.5–9.7. Pitchblende is a major source of URANIUM.

Uranium, radioactive metallic element (symbol U) of the ACTINIDE group, identified in 1789. It occurs in several minerals, the chief ores being PITCHBLENDE (oxide), autunite and torbernite. The element is important because of its use in fission reactors and bombs. The naturally occurring element contains U^{238}(99.28%), U^{235}(0.71%), and U^{234}(0.0058%). U^{235} (half-life 7.1×10^8 yr) is fissionable and will sustain a neutron chain reaction. Fuels used in reactors are enriched with this ISOTOPE by gaseous diffusion, using the volatile hexafluoride, or by a centrifuge. In breeder reactors the isotope U^{238} is converted into Pu^{239} by neutron capture. Chemically uranium is a reactive metal; it oxidizes in air and reacts with cold water. Properties: at. no. 92; at. wt. 238.029; s.g. 19.05; m.p. 1,132°C (2,070°F); b.p. 3,818°C (6,904°F); most stable isotope U^{238} (half-life 4.51×10^9 yr).

Uranium oxide, one of a series of compounds of which UO_2, U_4O_9, U_3O_8 and UO_3 are the most common, and of which U_3O_8 is the most stable. The latter is green, brown or black with an orthorhombic crystalline structure. URANINITE and PITCHBLENDE (containing UO_2 and UO_3) occur naturally and are used as a source of uranium.

Uranus, in Greek mythology, the starry sky, husband and son of GAEA, the broad-bosomed earth.

Uranus, seventh planet from the Sun discovered in 1781 by William HERSCHEL. It has five satellites. The outer layers of Uranus are gaseous and the surface temperature is extremely low. It has an axial tilt of 98°. Mean distance from the Sun, 2,869,000,000km (1,780,000,000 miles); mass, 14.6 times that of Earth; diameter, 51,800km (32,375 miles); rotation period, 10hr 48min; sidereal period, 84 years. *See also* SU pp.176–177, *176–177*, *214, 214*, 280, *280*.

Urartu, ancient kingdom (*c.*1270–612 BC) situated in present-day E Turkey. In the 8th century BC, when the empire was at its most powerful, it brought most of N Syria under its influence. *See also* HC1 pp.56–57.

Urban, name of eight Roman Catholic popes. Saint Urban I (*r.*222–230) is reputed to have been a MARTYR. Urban II (*c.*1035–99; *r.*1088–99), elected in Terracina, S of Rome, was opposed by ANTIPOPE Clement III, who was supported by the Emperor Henry IV. In 1095, at the Council of Clermont, he launched the idea of the First CRUSADE. His work as a reformer encouraged the development of the CURIA ROMANA and the formation of the College of Cardinals. (*See also* CARDINALS, COLLEGE OF; HC1 p.176, *176*.) Urban III (*d.*1187) was elected in 1185. Urban IV (*c.*1200–64; *r.*1261–64), who was unable to live in Rome because of civil wars there, inherited the conflict between the papacy and the HOHENSTAUFEN family. He was an able administrator and St Thomas AQUINAS worked for the Curia during his papacy. He established the feast of CORPUS CHRISTI. Urban V (*c.*1310–70; *r.*1362–70) crowned at Avignon, attempted in 1367 to return the papacy from Avignon to Rome. Insurrections led him to return to Avignon in 1370. He was beatified in 1870. The election of Urban VI (*c.*1318–89; *r.*1378–89), an Italian, was declared invalid by the 13 French cardinals in the College, who elected the French antipope, Clement VII, at Fondi in 1378, beginning the GREAT SCHISM. The papacy of Urban VI was marked by confusion. Assumed to be insane, he is believed to have murdered at least five cardinals and a bishop. Urban VII (1521–90) died in the year of his election. Urban VIII (*c.*1568–1644; *r.*1623–44), an active and knowledgeable patron of the arts, approved the establishment of new orders, and particularly encouraged missionary activity. *See also* HC1 p.307.

Urbanization, process by which population becomes concentrated in urban areas. Urbanization became widespread in Britain and the USA in the 19th century and the THIRD WORLD in the 20th century. *See also* HC2 pp.*90–91*.

Urbino, town in Pesaro and Urbino province, Marche (The Marches) region in central Italy. Federico da Montefeltro, Duke of Urbino and a great Renaissance patron of arts, transformed the Palace of Urbino into one of the most beautiful early Renaissance palaces. Luciano LAURANA, and PIERO DELLA FRANCESCA both worked for Federico. Today Urbino's industries include tourism and textiles. Pop (1971) 16,296.

Urdu, language belonging to the Indic group of the Indo-Iranian sub-family of the Indo-European languages, one of the official languages of Pakistan. It is the mother tongue of more than 3,000,000 people in Pakistan, but is used as a second language by about as many again. It is also spoken in India by most of the country's Muslims. Urdu is quite similar to HINDI, the chief difference being that it is written in the Arabic script. Its vocabulary contains many Arabic and Persian elements that are absent in Hindi. *See also* MS p.240.

Urea, organic chemical compound, formula $CO(NH_2)_2$. It is the form in which most vertebrates excrete most of their nitrogen wastes; human urine containing about 25 grammes of urea to a litre. Because it is so high in nitrogen, urea is a good fertilizer.

Ureter, in vertebrates, 25cm (10in) long narrow duct that connects the kidney to the urinary BLADDER, transporting URINE from the kidney to the bladder where it is stored until voided along the URETHRA.

Urethra, duct through which URINE is discharged from the bladder in mammals. Urine is produced in the kidney, stored in the bladder until pressure in the bladder triggers specific neural responses that cause urine, under voluntary control, to be released through the urethra. In males the urethra is also the tube through which SEMEN is ejaculated.

Urethritis, inflammation of the URETHRA, most frequently found in males, due to bacterial or viral infection or to mechanical obstruction. Antibiotics or surgical procedures are used for treatment. *See also* MS p.102.

Urey, Harold Clayton (1893–), US chemist who was awarded the 1934 Nobel Prize in chemistry for his isolation of DEUTERIUM. He later isolated ISOTOPES of OXYGEN, NITROGEN, CARBON and SULPHUR. During WWII he helped in the research leading to the production of the atomic bomb. He then turned to GEOPHYSICS and worked on recreating the atmospheric conditions of the primeval earth to elucidate the origin of life. *See also* NW p.21.

Urfé, Honoré d' (1567–1625), French writer. His five-volume *L'Astrée* (1607–10) is regarded by some critics as the first important pastoral novel in French.

Uri, CANTON in central Switzerland; the capital is Altdorf. The terrain is glaciated and forested; there is some agriculture in the river valleys and on the lower mountain slopes. Tourism and forestry are important industries. Area: 1,075sq km (415sq miles). Pop. (1970) 34,091.

Urial, Asian wild sheep. It is reddish-brown in colour; ewes have small horns and rams have massive horns that form a complete arch. Height: to 90cm (36in) at the shoulder. Species *Ovis ammon orientalis*.

Uric acid, end product of PROTEIN metabolism and found in urine, blood and lymph, and also as a salt (urate) in calculi such as KIDNEY STONES An excessive amount of uric acid in the blood causes GOUT, with urate deposits around joints.

Urine, fluid excreted by the KIDNEYS, passing through the URETERS to the bladder and then voided through the URETHRA.

Urinogenital system, major system of the body, containing the urinary and reproductive systems. The urinary system, which is part of the body's excretory system, consists of KIDNEYS, URETERS, the urinary bladder (lying in the pelvic cavity and storing URINE) and the URETHRA. In males, the reproductive system consists of paired TESTES that produce sperm cells and HORMONES and are located in an external pouch known as the SCROTUM; accessory glands; and the PENIS, the organ through which urine is expelled and sperm is ejaculated. In females, the reproductive system consists of paired OVARIES, one on each side of the pelvic cavity; FALLOPIAN TUBES, or oviducts, which provide a passage from the ovaries to the uterus (WOMB); the uterus, or womb, located in the pelvic cavity between the bladder and rectum; and the CERVIX, or lower part of the uterus, which opens into the VAGINA, the opening for intercourse and childbirth. In females, the urethral and vaginal openings are separate but close to each other. *See also* MS pp.69, 74–75.

Uris, Leon Marcus (1924–), US author of historical novels, notably of the bestseller *Exodus* (1958), whose theme was the establishment of the state of Israel.

Urnfield culture, late Bronze Age culture that developed in central Europe and N Italy by the 12th century BC. Characterized by the spread of burial in individual urns, and the increasing use of slashing swords, the culture spread, probably without major migrations, over W Europe, and formed the basis of the Celtic civilization. *See also* HC1 pp.66–67.

Urmia, Lake (Rezā'iyeh, Daryācheh-ye), shallow salt lake in NW Iran, just E of Rezā'iyeh. The largest lake in Iran, it contains Shāhī Island to the N and small islets to the S. It has no outlets, and receives the Talkheh and Zarineh rivers and drainage from nearby mountains. Sharifkhāneh is the chief lake port. Area: approx. 4,701sq km (1,815sq miles). Depth: approx. 15m (49ft).

Urochordata, subphylum of the phylum Chordata; it includes small satentary, marine animals known as Tunicates, or SEA SQUIRTS, which possess few typical chordate features except for paired gill openings. Their free-swimming larvae have a tail and a NOTOCHORD, above which is situated a tubular nerve cord. *See also* NW pp.122–123.

Urodela. See NEWT; SALAMANDER.

Urquiza, Justo José de (1801–70), Argentine general and first constitutional President from 1854 to 1860.

Ursa Major. See GREAT BEAR.

Ursa Minor. See LITTLE BEAR.

Ursinus (*d. c.*385), ANTIPOPE (*r.*366–67). He was elected simultaneously with and in opposition to DAMASUS I, and violence repeatedly broke out between Ursinus' and Damasus' followers. Expelled twice from Rome (in 366 and 368), he was eventually exiled to Cologne after his condemnation by a Roman synod in 378.

Ursula, Saint, legendary virgin and martyr who is especially honoured at Cologne where she is said to have been slain by the HUNS with her 11, or, in some reports 11,000 virgins on the return from a pilgrimage to Rome. She has become the patron of many educational establishments, including the URSULINES. She no longer has a day in the church calendar because her authenticity is dubious. *See also* HC1 p.*135*.

Ursulines, oldest religious order of women in the Roman Catholic Church, named after its patron, St URSULA, and founded by St Angela Merici at Brescia in 1535. The first female teaching order, it soon opened convents in Germany, France and even Canada, where the first congregation of women in North America was established in 1639. It again increased in size during the 19th century and a union of Ursuline convents was created in 1900.

Urticales, order of flowering plants with about 3,100 species ranging from large trees such as elms to herbs such as HOPS and NETTLES. Other members are MULBERRY, Indian hemp (MARIHUANA) and FIG.

Urticaria. *See* HIVES.

Uruguay, official name of the Eastern Republic of Uruguay, nation on the E coast of South America between Brazil (N) and Argentina (S). The rearing of sheep and cattle on the extensive grasslands is the mainstay of the economy. Wool is exported. The capital is Montevideo. Area: 177,508sq km (68,536sq miles). Pop. (1976 est.) 3,101,000. *See* MW p.180.

Uruguay, river in SE South America. Rising in S Brazil, and forming part of the boundary between Rio Grande do Sul and Santa Catarina states, it flows SW to form the boundary between Argentina and S Brazil, and Argentina and Uruguay, providing hydroelectricity for the latter two countries. It empties into the Rio de la Plata. Length: approx. 1,600km (1,000 miles).

Urundi. *See* BURUNDI.

USA, or The United States of America, the world's fourth largest nation and its greatest producer and consumer of industrial goods. The capital is Washington. It consists of 50 states and a few islands, notably Puerto Rico and the Virgin Islands of the United States. It is divided into seven geographical regions: (east to west) the Atlantic-Gulf coastal plain, the Appalachian highlands, the interior plains, the interior highlands, the Rocky Mountain system, the Intermontane region and the Pacific Mountain system. It is the world's largest producer of meat, cheese, soybean and tobacco, the second largest of milk, butter, oats and wheat. It leads the world in producing both electrical and nuclear energy. It is one of the world's largest producers of coal and steel. It is the largest producer of mica, uranium, magnesium, pig iron, motor cars and synthetic rubber. Its major exports are motor cars and parts, aeroplanes and parts, steel and iron goods, chemicals and electronic computers. Its major imports are petroleum and petroleum products, machinery, metal products and paper. Its major trading partners are Canada, West Germany, Japan and Britain. Area: 9,363, 123sq km (3,614,343sq miles). Pop. (1976 est.) 215,135,000. *See* MW pp.175–179.

U.S.A. (1937), novel in three volumes by John DOS PASSOS. Comprising *The 42nd Parallel* (1930), *1919* (1932) and *The Big Money* (1936), the trilogy paints a comprehensive picture of American life, expressed from a radical view point.

Ushant (Ouessant, Île d'), island in the Atlantic Ocean off the NW coast of France. It was the scene of naval battles in 1778 and 1794 between the French and the British. There is a lighthouse on the island. Sheep rearing and fishing are the chief occupations. Area: 15sq km (6sq miles). Pop. 1,814.

US Masters Golf Championship, annual event for professionals and amateurs at the Augusta National Golf Club, Georgia. It is played on a course designed by Bobby JONES and Alister Mackenzie. The first championship was held in 1934. It is officially called the Augusta National Invitation tournament. It is a 72-hole medal-play competition. Until 1977 the only non-US winner was Gary PLAYER of South Africa in 1961.

US Open Golf Championship, annual event held at various golf courses in the USA for professionals and amateurs. The first championship was held over 36 holes at Newport, Rhode Island, in 1895. In 1898 it was made a 72-hole event. There were no championships in 1917–18 and 1942–45. Gary PLAYER of South Africa, who won in 1965, is the only non-US or non-British winner of the championship.

US Professional Golfers' Association Championship, annual event held at various courses in the USA. The association was formed in 1916 and the first event held was in the same year. It was a match-play competition until 1958, when it became a 72-hole medal-play event. Up to 1977 the only non-US or non-British winner was Gary PLAYER of South Africa, who won in 1962 and 1972.

Ussher, James (1581–1656), Irish ANGLICAN clergyman who became Archbishop of Armagh in 1625. He settled in England in 1640 but refused to sit in the WESTMINSTER ASSEMBLY in 1643 and supported CHARLES I. His chronology *Annales Veteris Testamenti* (published 1659), dated the Creation at 4004 BC.

USSR, or Union of Socialist Soviet Republics, world's third largest country (in area the largest), successor to Imperial Russia, organized as a federation of constituent republics in 1923. The capital is Moscow, in the Russian republic, which covers 77% of the area of the country. In 1923 there were four republics in the USSR; there are now 15. It includes more than 100 ethnic groups, the chief ones being the Russians, Ukrainians and Belorussians, who speak Slavic tongues and constitute 77% of the population. There are also the Turkic-speaking Uzbeks, Tatars, Kazakhs and Azerbaijans; and the Armenians, Georgians, Lithuanians and Moldavians. Its argicultural output is about 80% of that of the USA. It is the world's largest producer of wheat, barley, oats, rye and butter and the third largest producer of fish. It is the largest producer of timber and one of the largest of coal and steel. Two-thirds of its trade is with COMECON countries. Area: 22,402,200sq km (8,649,489sq miles). Pop. (1976 est.) 255,413,000. *See* MW p.169.

Ustinov, Peter Alexander (1921–), British author, dramatist and raconteur, of Russian parents. His plays include *Romanoff and Juliet* (1956) and *The Unknown Soldier and his Wife* (1967). He acted in the film *Quo Vadis* (1951), and directed and appeared in the film *Billy Budd* (1962). His autobiography *Dear Me* was published in 1977.

Ust-Kamenogorsk, major centre of the USSR atomic industry. It is situated in Kazakhstan in Central Asian USSR. Pop. (1975) 257,000.

Usury, lending of money at an excessive or unlawful rate of interest. Before the Middle Ages, payment for the use of money was regarded as usury by Christians and was forbidden, particularly by the Roman Catholic Church. In general people regarded interest and usury as synonymous until the late Middle Ages, when a distinction came to be made: reasonable interest on a loan became acceptable when the lender risked capital.

Utagawa, Kuniyoshi (1797–1861), Japanese artist of the UKIYO-E school of painting and print designing. His woodblock prints depicted historical and legendary figures, and his landscapes are particularly fine.

Utah, state in W USA in the Rocky Mts. In the N half of the state the Wasatch Range separates the mountainous E from the Great Basin, which includes the Great Salt Lake, in the W. The Colorado Plateau in the S is dissected by many canyons cut by the Colorado River and its tributaries. The arid climate hinders agriculture. Hay, barley, wheat and sugar-beet are grown by means of irrigation. The chief farming activity, however, is stock raising, especially sheep and cattle. Most inhabitants work in mining. There are rich deposits of copper, petroleum, coal, molybdenum, silver, lead and gold. The main cities are Salt Lake City (the state capital and largest city), Ogden and Provo.

The region was ceded to the USA at the end of the MEXICAN WAR and admitted to the Union in 1896. The influence of the MORMON Church is strong in the state. Area 219,931sq km (84,915sq miles). Pop. (1975 est.) 1,206,000.

Utah War (1857–58), conflict between the MORMON settlers of Utah, USA, and troops of the federal government. The conflict began when President James BUCHANAN removed Brigham YOUNG, the Mormon leader, as territorial governor and replaced him with a non-Mormon. The Mormons rebelled and about 15,000 US troops were sent into the area to subdue them.

Utamaro, Kitagawa (1753–1806), Japanese master of the UKIYO-E woodblock colour print, the first Japanese artist to become well known in the West. He excelled in depicting birds, flowers and feminine beauty. His works were strongly erotic, precise, graceful and immensely popular; women of the court were often his subjects, as in *Catching Butterflies*.

Ute, Shoshonean-speaking tribe of North American Indians, once living in W Colorado, E Utah and part of Nevada. They were fierce, nomadic warriors who, after the introduction of the horse, engaged in warfare with other Indian tribes and hunted bison. Today approximately 3,000 live on reservations in Colorado and Utah.

Uterus. *See* WOMB.

U Thant. *See* THANT, U.

Uthman. *See* OTHMAN.

Utilitarianism, branch of ethical philosophy which holds that actions are to be judged good or bad according to their utility or, in the words of Jeremy BENTHAM, in proportion as they tend towards producing "pleasure, good or happiness" and preventing "mischief, pain, evil or unhappiness". The chief sources for the doctrine are Bentham's *An Introduction to the Principles of Morals and Legislation* (1789) and John Stuart MILL's essay *Utilitarianism* (1862).

Utility clothing, clothing of a uniform, sound standard, made in Britain during WWII. The manufacture, price and rationed sale of such clothing was controlled by the government, through the Consumer Needs department of the Board of Trade, established in 1941.

Uto-Aztecan languages, family of American Indian languages spoken in SW USA and Mexico. It includes Papago, Pima and Hopi of Arizona; Ute of Utah and Colorado; Comanche of Oklahoma and Shoshone, spoken in some western states. In Mexico there are Náhuatl (the language of the Aztecs), Tarahumara and Mayo.

Utopia, book by Sir Thomas MORE, published in Latin in 1516 with the first English edition in 1551. It describes "the Highest State of a Republic" and "the New Island Utopia" ("nowhere"). More sets out an ideal government and economic and social system, with each citizen having equal involvement in all aspects of life, and satirizes the defects he saw in Europe, and particularly in the England of his time. The word Utopia soon became a general name for an imaginary but ideal place. *See also* HC1 pp.*247, 274, 282*; MS p.*298*.

Utopia Limited, or The Flowers of Progress (1893), two-act operetta by Arthur SULLIVAN with libretto by William GILBERT. It was first performed at the Savoy Theatre, London.

Utopian socialism, originally derisory term used by Karl MARX for that form of socialism advocated by Robert OWEN in the early 19th century. While criticizing CAPITALISM, it advocated small-scale cooperative enterprises rather than state ownership.

Utopia Planitia. *See* MARS.

St Ursula and the virgins; their shrine in a church at Bruges, Belgium.

Uruguay; the Palacio Legislativo (Parliament building) in Montevideo.

James Ussher drafted the articles of doctrine of the Irish Protestant church.

Utopia was written as the account of a Portuguese sailor's visit to the island.

Utrecht's cathedral towers above this sunny, tree-lined street and waterway.

Uttar Pradesh; gate of the Kaisr Bagh in Lucknow, the state's capital city.

Uzbek field-hands listening to readings given by a "missionary" denouncing Islam.

Edouard-Marie Vaillant; while an exile in England he worked with Karl Marx.

Utrecht, city in central Netherlands, on the Oude Rijn River, approx. 32km (20 miles) SSE of Amsterdam; capital of Utrecht province. The city has been a trade centre since medieval times. It was the scene of the Union of Utrecht (1579) and the Peace of Utrecht (1713), which ended the War of the SPANISH SUCCESSION. The city has a university (1636) and a 15th-century cathedral. Industries: steel, machinery, textiles, electrical equipment, food processing. Pop. (1974) 259,826.

Utrecht, Peace of (1713–14), series of treaties that ended the War of the SPANISH SUCCESSION and heralded the rise of the British Empire at the expense of French colonial expansion. LOUIS XIV confirmed the renunciation of claims to the French throne by his grandson, PHILIP V of Spain, and surrendered the Spanish Netherlands, Milan, Naples, Sardinia, Mantua and the Tuscan ports, and Breisach, Kehl and Freiburg E of the Rhine to the HOLY ROMAN EMPIRE. France also ceded Newfoundland, Acadia (Nova Scotia), the Hudson Bay Territory and St Kitts to Britain and was obliged to recognize the English succession as being established in the House of HANOVER. Spain ceded Gibraltar and Minorca to Great Britain. Commercial concessions generally favourable to Britain were an important part of the agreements. *See also* HC1 p.312.

Utrecht, Union of (1579), agreement among the seven northern provinces (Holland, Zeeland, Utrecht, Gelderland, Groningen, most of Friesland and Overijssel) and the southern cities of Antwerp, Breda, Ghent, Bruges and Ypres of the Netherlands to continue the fight against Spanish rule. This defence pact ultimately led to a declaration of independence (1581), and formed the constitution of the UNITED PROVINCES until 1795.

Utrecht psalter, 9th-century CAROLINGIAN psalter which is now in the university library at UTRECHT. Each psalm is illuminated with an ink drawing in a freehand style unlike the usual Carolingian style. It was taken to England in the 10th century and influenced the Anglo-Saxon school.

Utrecht School, Dutch school of painters who were influenced (*c.*1610–20) in Rome by the realism of CARAVAGGIO and his follower MANFREDI. They later worked in Utrecht, where they painted religious and genre pictures. The school included Terbruggen, Honhorst and Baburen; Franz Hals, Rembrandt and Vermeer were influenced by it.

Utrillo, Maurice (1883–1955), French artist, son of the painter Suzanne VALADON, best known for his poetic views of Paris (especially of Montmartre, whose popular romantic image his pictures helped to create).

Uttar Pradesh, state in N India; the capital is Lucknow. It is the fourth-largest state in India and the most populous. The region is mountainous in the N and s, enclosing a low-lying plain drained by the Jumara and Ganges rivers. The economy is based on agriculture; cereals, sugar cane, rice and pulses are grown. Industries include the processing of cotton and sugar. Area; 294,413sq km (113,673sq miles). Pop. (1971) 88,341,144.

Utu, Sumero-Akkadian god of sun and of justice, traditionally represented by a sun disc. When depicted as a human being, he has rays issuing from his body.

U-2 Incident (1960), international incident precipitated by the shooting down of a US aircraft, a U-2, while on a photographic reconnaissance mission over the USSR. The pilot, Francis Gary Powers, was captured and confessed to spying. KHRUSHSEV demanded an apology and the incident led to the disruption of the 1960 Summit Conference between the USA, USSR, Britain and France.

Utzon, Jøern (1918–), Danish architect. He is famous for his domestic architecture, but is internationally known as the designer of the recently opened and much admired Sydney Opera House in Australia.

UV Ceti, flare star in the southern constellation of Cetus. An M-type dwarf star, UV Ceti is the prototype of a class of variable stars whose luminosity increases dramatically over a period of minutes as a result of intense flare activity. Its range of magnitudes is from 6.8 to 12. . *See also* SU pp.*238*, 240, *241*.

Uxmal, ancient MAYA ceremonial and administrative centre located on Mexico's Yucatán peninsula. Two pyramids and a grouping of four rectangular buildings called the Nunnery dominated the 60-hectare (150-acre) site. Uxmal flourished AD 600–900 but was abandoned *c.*1450.

Uzbekistan (Uzbekskaya SSR), constituent republic of the USSR, bordered in the s by Afghanistan; the capital is Tashkent. Settled since earliest times, the region was conquered by ALEXANDER THE GREAT and passed in succession to the Arabs, Seljuk Turks, Mongols, Timurids and Uzbek Turks before coming under Russian dominance in the 19th century. Much of NW Uzbekistan lies within the Kyzyl-Kum desert; the SE has fertile soil and both areas are extensively irrigated by the Amu Darya and Syr Darya rivers and a canal system. Cereals, cotton, grapes, tobacco and sesame are grown, livestock is raised in the w and the area is rich in zinc, copper, lead, tungsten and coal. Industries include the manufacturing of textiles, chemicals, machine tools and fertilizers. Area: 447,400sq km (172,740sq miles). Pop. (1976) 14,090,000 *See also* MW p.169.

Uzbeks, Turkic-speaking people, originally of Persian culture, who form two-thirds of the population of UZBEKISTAN and who are the second-largest non-Russian people in the USSR. They took their name from Uzbeg Khan, a chief of the GOLDEN HORDE who died in 1340. By the end of the 16th century the Uzbeks had extended their rule to parts of Persia, Afghanistan and Chinese Turkistan. Their empire was never united and in the 19th century its various states were absorbed by Russia.

V

V, 22nd letter of the alphabet, derived (as were *f*, *u*, and *y*) from the Semitic letter *vaw*, meaning *hook*. It was identical with *u* in the Greek and Roman alphabets, the Romans using it both as a vowel (*u*) and a consonant (*v*), and was not differentiated from *u* in English until the Middle Ages. In modern English *v* has only the one consonant sound, as in *vole* and *wove*. *See also* MS pp.244–245.

Vaccination, injection of a VACCINE in order to produce IMMUNITY against a disease. The term most often refers specifically to the vaccination against SMALLPOX, in which vaccine is scratched into the skin producing a typical "pox" and leaving a small scar. *See also* MS pp.97, 112.

Vaccines, agents used to give immunity against various diseases without producing serious or fatal symptoms of the disease. The agent may be a live attenuated VIRUS, which is low in virulence, or a dead one still able to induce the production of ANTIBODIES in the blood. The first of the vaccines to come into general use was the SMALLPOX vaccine, developed in the late 18th century. Vaccines against POLIOMYELITIS, yellow fever, rabies, measles and other bacterial and viral diseases are available. *See also* IMMUNE SYSTEMS; MS pp.97, 112.

Vaccinia virus, causative agent of cowpox, a minor pustular disease of cattle and human beings. It is related to the virus of SMALLPOX, a much more serious disease, and when cowpox is contracted by human beings the ANTIBODIES it raises in their blood immunizes them to smallpox. This is the basis of protection by VACCINATION, discovered and first carried out by Edward JENNER in 1796.

Vacuole, membrane-bound, fluid-filled cavity within the cytoplasm of a CELL, believed to discharge excess water or wastes. *See also* NW p.26.

Vacuum, region of extremely low pressure. Interstellar space is a high vacuum, with an average density of less than 1 molecule per cubic centimetre; the highest man-made vacuums contain less than 100,000 molecules per cubic centimetre.

Vacuum deposition, method of producing a thin adherent coating, generally of a metal, on plastics, ceramics and other materials. The object to be "plated" is placed in a vacuum chamber that also contains a length of metal wire inside a heavy-duty coil. When a high-voltage electric current is passed through the coil, the metal vaporizes and coats the object. *See also* SU p.96.

Vacuum pump, device that exhausts air from an enclosed space, which becomes a partial VACUUM. The simplest vacuum pump has water or steam ejected through a nozzle into a chamber and then out. As it enters the chamber it forces air out thus creating a partial vacuum. *See also* SU p.96.

Vacuum tube. *See* VALVES, ELECTRONIC.

Vadim, Roger (1927–), French film writer and director, b. Roger Vadim Plemiannikow. His most famous films are *And God Created Woman* (1956), *Les Liaisons Dangereuses* (1959) and *Barbarella* (1968).

Vaduz, capital city of LIECHTENSTEIN, on the right bank of Upper River Rhine, approx. 80km (50 miles) SE of Zurich, Switzerland. Destroyed in a war between the Swiss and the Holy Roman Emperor, it was rebuilt in the early 16th century and became a possession of the Liechtenstein family in 1712. The main source of income is tourism. Pop. (1975 est.) 3,900.

Vagina, female sexual organ, which receives the PENIS during copulation. It is tube-like in shape, and its flexible muscle walls are capable of remarkable dilation during CHILDBIRTH. It opens into the cervix, or mouth of the WOMB. *See also* MS pp.74–75.

Vaginitis, acute infectious inflammation of the VAGINA usually caused by the protozoan *Trichonomas vaginalis* and transmitted during sexual intercourse.

Vagus nerve, tenth CRANIAL NERVE running from the BRAIN to the abdomen. It contains motor, secretory and sensory fibres and supplies branches to the lungs, heart, stomach and other abdominal organs.

Vaihinger, Hans (1852–1933), German philosopher who, in his major book *The Philosophy of "As if"* (1911), extended the philosophy of Immanuel KANT in the direction of PRAGMATISM. He viewed life as a labyrinth of frustrations and in his work proposed a philosophy to make life bearable.

Vaillant, Edouard-Marie (1840–1915), French Socialist politician, influenced by Louis Auguste Blanqui. He was elected to the revolutionary Paris Commune in March 1871, but fled to England after the Commune was crushed, returning in 1880. He was elected to the National Assembly in 1893 and was a Socialist leader from 1898 to 1915.

Vaillant, Wallerant (1623–77), French portrait painter who settled in Amsterdam. He often worked in pastel and was among the first to make MEZZOTINTS, the art of which he learnt from Prince Rupert of the Rhine.

Vaishnavas, followers of one of the three main divisions of HINDUISM, who worship VISHNU and his incarnations Rama and Krishna. A major devotional volume is the *Bhagavad Gita*.

Valadon, Suzanne (1865–1938), French painter, the mother of Maurice UTRILLO. She painted mainly nudes and portraits of working-class people, and a few still-lifes and landscapes. Her style was influenced by DEGAS and GAUGUIN and her treatment, especially of the nude figure, was brutally realistic.

Valdés, Juan de (*c.*1498–1541), Spanish religious writer who, suspected by the Inquisition, went to Naples in 1530. He was influenced by the thought of ERASMUS and his major works include *Diálogo de la doctrina cristiana* (1529; Dialogue on Christian Doctrine) and *Diálogo de la lengua española* (1535; Dialogue on the Spanish Language).

Valdés Leal, Juan de (1622–90), Spanish BAROQUE painter. His early works were monumental renderings of saints. The use of dramatic colour and vigorous brushwork was combined with morbid subject matter. Characteristics of his works are *Triumph of Death* (1672) and *Jesus Disputing with the Doctors* (1686).

Valdivia, Pedro de (*c.* 1498–1554), Spanish CONQUISTADOR and founder of Santiago, capital of Chile, in 1541. He became Governor of Chile in 1548 and in 1550 began to conquer the south, where he founded the city of Concepción. He was killed during a revolt by Araucanian Indians.

Valdivia, province, city and river of Chile. The province stretches from the Andes to the w coast. The city, on the River Valdivia 18km (11 miles) from the ocean, is the capital of the province. Area: 18,473sq km (7,132sq miles). Pop. (prov.) (1970) 274,642; (city) (1970) 90,942.

Valencia, city in E Spain and capital of the province of Valencia. It is situated on the Turia (Guadalaviar) River, 320km (200 miles) SE of Madrid. Originally settled by the Romans, it passed to the Visigoths in the 5th century AD, was conquered by the MOORS in the 8th century and made the capital of the independent Moorish kingdom of Valencia in the 11th century. It was taken in 1328 by JAMES I of Aragon. An agricultural, industrial and communications centre, its manufactures include electrical equipment, chemicals, textiles and machinery. Pop. (1974) 713,026.

Valencia, city in N Venezuela, 120km (75 miles) NE of Caracas. It was the capital of Venezuela in 1812, and in 1830 when the country was proclaimed independent of Greater Columbia. Today it is a major industrial and transport centre. Manufactures include textiles, paper, cement, soap, furniture and vehicle spares. Pop. (1971) 367,171.

Valenciennes, town in N France, on the Escaut (Schelde) River, 46km (28 miles) SE of Lille. Its museum contains the works of French and Flemish artists including Peter Rubens and Jean Antoine Watteau, who was born in Valenciennes. The town's lace industry, which flourished in the 18th century, has been revived. Other industries include coal mining, textiles, machinery and steel tubing. Pop. 47,000.

Valency, or valence, measure of the "combining power" of a particular element, equal to the number of (single) chemical bonds one atom can form. Thus hydrogen has a valency of 1, carbon 4 and sulphur 2, as seen in compounds such as methane (CH_4), carbon disulphide (CS_2) and hydrogen sulphide (H_2S). Covalency refers to formation of covalent bonds; co-ordinate valency to co-ordinate bonds. *See also* SU p.138.

Valency electron, electron in an atom that is transferred or shared in the formation of a chemical bond. The valency electrons of a non-transition metal are those in the outermost shell (farthest from the nucleus), and in forming compounds the atom tends to attain the stable configuration of a NOBLE GAS. TRANSITION ELEMENTS use both the outer shell and the partly filled penultimate shell in chemical bonding. *See also* SU p.138.

Valency orbital, one of the outer atomic orbitals of an atom, containing the ELECTRONS that participate in chemical bonding. In producing IONIC BONDS electrons are lost or gained by valency orbitals. Covalent bonding involves sharing of electrons between atoms, a process often visualized as overlap of valency orbitals with the formation of molecular orbitals. *See also* SU p.138.

Valentin, Moïse, or Le Valentin (*c.* 1591–1632), French painter. He studied in Rome and worked in a style derivative of CARAVAGGIO. He was a friend of Nicholas POUSSIN. His only documented work is the *Martyrdom of SS Processus and Martinian* (1630), but he is better known for his genre paintings of gypsies and soldiers.

Valentine, Saint, name of two legendary saints of Rome: Valentine of Terni and Valentine of Rome. The former, a Roman priest and physician, and the latter, the Bishop of Terni, were both martyred *c.* 3rd century BC. They are commem-

orated on 14 Feb. ST VALENTINE'S DAY is not thought to be related to the saints but is possibly connected with the pagan Roman festival of Lupercalia.

Valentinian, name of three western Roman emperors. Valentinian I (321–75) was proclaimed emperor in 364, and ably defended the frontiers in Gaul and Britain. He was a Christian, and promoted religious toleration. His second son (371–92) became Valentinian II in 375 and ruled (through his mother, Justina) Italy, Africa and Illyricum while his elder brother Gratianus ruled the remainder until 383. Forced to flee from Italy in 382, he was restored to his position in 388. Valentinian III (419–55) was made emperor in 425. He was a weak ruler and lost control of N Africa to the VANDALS in 442, and suffered the invasions of ATTILA in 451. Valentinian killed his general Aetius in 454 and was himself murdered by Maximus the following year.

Valentino, Rudolph (1895–1926), US silent film star, *b.* Italy. He exemplified the mysterious Latin lover, and his films included *The Sheik* (1921), *The Young Rajah* (1922), *The Eagle* (1925) and *Son of the Sheik* (1926). Valentino's early death from a ruptured ulcer caused widespread hysteria among his fans. *See also* HC2 p.268, *268*.

Valentinus (*d.* 827), Pope (*r.* 827). A cardinal deacon under Pope St PASCHAL I, he was respected for his piety and generosity, but held office only for a short period, dying 40 days after his election.

Vale of Glamorgan, county district in S SOUTH GLAMORGAN, Wales; created in 1974 under the Local Government Act (1972). Area: 296sq km (114sq miles). Pop. (1974 est.) 106,400.

Vale of the White Horse, county district in w central OXFORDSHIRE, England; created in 1974 under the Local Government Act (1972). Area: 581sq km (224sq miles). Pop. (1974 est.) 96,900.

Valéra, Eamon de. *See* DE VALÉRA, EAMON.

Valera y Alcalá Galiano, Juan (1824–1905), Spanish novelist and critic. He held various diplomatic posts in Europe and the USA and wrote many letters that formed the basis of his novels, such as *Mariquita y Antonio* (1861).

Valerian, Publius Licinius (*d.* 260), Roman emperor (253–260). In 257 he revived the persecution of the Christians, who were required to perform public acts of worship to the state gods; severe penalties were ordered for those who refused. *See also* HC1 p.114, *114*.

Valerian, or garden heliotrope, plant native to Europe and N Asia and naturalized in the USA. It has pinkish or pale purple flower clusters. Height: to 1.2m (4ft). Family Valerianaceae; species *Valeriana officinalis*.

Valerius Flaccus, Gaius (*fl.* 1st century AD), Latin poet of wealth and position. A contemporary of VESPASIAN and TITUS, he dedicated his unfinished epic, *Argonautica*, to Vespasian.

Vale Royal, county district in central CHESHIRE, England; created in 1974 under the Local Government Act (1972). Area: 384sq km (148sq miles). Pop. (1974 est.) 109,500.

Valéry, Paul (1871–1945), French symbolist poet and critic. Influenced by Stéphane MALLARMÉ, his poems, including *La Jeune Parque* (1917) and *Charmes* (1922), are characterized by lyricism and abstract thought. His critical works include *La Soirée avec Monsieur Teste* (1906) and *Introduction à la Méthode de Léonard de Vinci* (1895). He was elected to the French Academy (1925) and was given a personal chair of poetry at the Collège de France, Paris (1937).

Valhalla, in Scandinavian mythology, the Hall of the Slain, where chosen warriors caroused with the god ODIN. From this glittering palace with golden walls and ceiling of burnished shields, the valiant set out each morning to battle with each other and returned to feast on boar's meat.

Validity, state or condition of a proposition that makes it logically irrefutable or correctly derived from its premises. If an inference has a conclusion that is true, and is false when one or more of its premises

are false, it is valid. *See also* STATISTICS.

Valine, 2-aminoisovaleric acid, a white crystalline essential AMINO ACID found in PROTEINS.

Valium, proprietary name for DIAZEPAN, a mild sedative drug used in the treatment of anxiety and as an anticonvulsant in EPILEPSY.

Valkyries, in Norse mythology, warlike handmaidens of the god ODIN who selected and conducted to VALHALLA those slain heroes who merited a place with him. They were armed and mounted and in some accounts rode to the battlefield, sometimes guarding their favourites from death and condemning others to death.

Valla, Lorenzo (1405–57), Italian scholar who translated HERODOTUS and THUCYDIDES into Latin for Pope NICHOLAS V. His masterpiece, *De elegantia latinae linguae* (1471), is a brilliant defence of classical Latin. *De voluptateac de vero bono* (1431) deals with EPICUREANISM and STOICISM and advises the assimilation of pleasure into the Christian experience. *See also* HC1 p.246.

Valladolid, city in NW central Spain, on the River Pisuerga, 160km (100 miles) NW of Madrid; capital of Valladolid province. The city was liberated from the Moors by Castilian kings in the 10th century. In 1469 it was the wedding-place of FERDINAND and ISABELLA. There is a 12th-century Romanesque church, an art museum, the church of Santa Ana (containing paintings by GOYA), and a monument to Christopher COLUMBUS, who died in the city in 1506. Valladolid's university, founded in 1346, is one of the oldest in Spain. Today its industries produce chemicals, flour, metalwork and textiles. Pop. (prov.) (1970) 413,000; (city) (1974) 275,012.

Valle Inclán, Ramón María del (1869–1936), Spanish novelist, poet and dramatist. His careful use of detail is seen in his vivid description of Galician countryside. Galicia is also used as the setting for several of his plays, notably *Divinas Palabras* (1920), which describes the brutal responses to individuality in a peasant community. Both plot and characters are distorted in *Lights of Bohemia* (1923), a play that heralded the Theatre of the ABSURD.

Vallejo, César (1893–1938), Peruvian poet concerned with the sufferings of the poor and dedicated to the cause of social progress. He went to Europe in self-imposed exile in 1923 and *Poemas humanos*, published in 1939, was written during this time.

Vallès, Jules (1832–85), French novelist and journalist. Vallès was involved in revolutionary activities in Paris, imprisoned (1853) and exiled (1871–80) for his part in the COMMUNE OF PARIS. His most important work is the autobiographical trilogy *Jacques Vingtras*, published between 1879 and 1886. He founded *Le Cri du Peuple* (1871), which became France's leading Socialist newspaper.

Valles Marineris. *See* MARS.

Valletta, port and capital city of Malta, on the NE coast of the island. The city was founded in the 16th century and named after Jean Parisot de La Valette, Grand Master of the Knights of Malta. The city was badly damaged by air raids during WWII. In the city are the Royal University of Malta (1592) and the Cathedral of San Giovanni (1576). Industries: shipbuilding, tourism. Pop. (1971 est.) 15,400.

Valley Forge, camp site 34km (21 miles) NW of Philadelphia where American soldiers under George Washington withstood the winter of 1777–78, living in log huts and with little food and inadequate clothing. The number of ragged, half-starved men dwindled to about 11,000, held together by their loyalty to Washington and the patriotic cause.

Valley of the Kings, narrow valley in w Thebes, Egypt, also called Biban al-Muluk. It contains 60 tombs of almost all the pharaohs of the 18th, 19th and 20th Dynasties. Although hidden, all the tombs except one were plundered in ancient times. TUTANKHAMEN's tomb was found and opened in 1922 by Howard CARTER.

Valley of the Queens, narrow valley in w

St Valentine; legend holds that both saints were beheaded, one in Rome, one in Terni.

Rudolph Valentino won many hearts both on screen and off in scenes such as this.

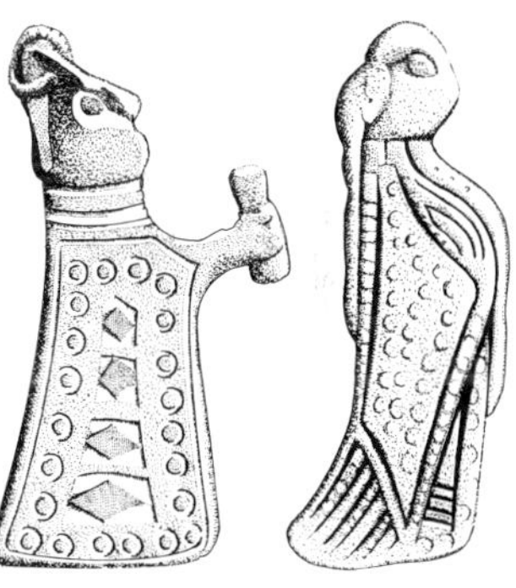

Valkyries; figures such as these were thought to give protection to warriors.

Valladolid's formal flower gardens offer shaded relaxation from the city's bustle.

Valmiki

Valparaíso, the most important port on South America's Pacific coast.

Vampire bats, carriers of rabies, bite into their prey with razor-like teeth.

Sir John Vanbrugh was a notable playwright, as well as an architect.

Martin Van Buren was unpopular due to economic problems during his presidency.

Thebes, Egypt, also called Biban al-Harim. It contains 70 uninscribed tombs of the queens and princes of the 19th and 20th Dynasties.

Valmiki (*fl.* 300 BC), Indian poet who is generally thought to be the author of the epic *Ramayana*. In seven books it deals with the story of RAMA: his courtship, marriage, loss of his wife and her subsequent recapture and the reunion of his family.

Valois, Dame Ninette de. *See* DE VALOIS, DAME NINETTE.

Valois Dynasty, royal house of France (1328–1589). Valois kings survived the HUNDRED YEARS WAR (1337–1453) and challenges by Burgundian and Armegnac rivals and consolidated royal strength over feudal lords. They established the Crown's sole right to tax and to wage war and extended parliaments throughout France. Louis XI (*r*.1461–83) is considered the founder of French royal absolutism. The direct Valois line ended (1498) with Charles VIII, when the dynasty was continued by Louis XII (Valois-Orléans) and, after his death (1515), by Francis I and the Valois-Angouleme line. With the death in 1589 of Henry III, the house of Bourbon, descending from a younger son of Louis IX, ascended the throne in the person of Henry IV. *See also* HC1 pp.212–213, 240, 241.

Valparaíso, port in central Chile; capital of Valparaíso province; founded in 1536. The city was heavily damaged by earthquakes in 1906 and 1971. It is Chile's chief port and is also a cultural centre, with two universities and museums of fine arts and natural history. Industries include chemicals, textiles, sugar refining, vegetable oils and paint. Pop. (1975) 248,972.

Value added tax. *See* VAT.

Valve, in music, piston-like mechanism in the tubes of all modern BRASS instruments except the slide trombone and bugle. Invented about 1815, valves are used to alter the length of the air column, permitting chromatic notes. There are usually three valves, lowering the pitch by a semitone, a whole tone and a minor third.

Valve, in a petrol or diesel ENGINE, device that cyclicly opens and closes the intake and exhaust parts of an engine's combustion chamber or cylinder. It consists of a disc attached to a shank and held against a seat by a spring. A rotating cam pushes on the bottom of the shank, raising the valve from its seat and permitting gas to flow past it. *See also* MM p.68.

Valves, in anatomy, "gates" that prevent the backward flow of blood and lymph in the HEART, veins and lymph vessels. Heart valves separate and connect the two atria and ventricles, the right ventricle and the pulmonary artery, and the left ventricle and the aorta. The various sphincters of the body, as in the stomach, intestine, bladder and anus, are circular muscles which also act as valves, opening and closing periodically to allow materials through. *See also* MS pp.62–63.

Valves, electronic, or vacuum tubes, devices once fitted in radios, now largely replaced by TRANSISTORS, except in high-power circuits such as amplifiers and rectifiers. A valve contains two or more electrodes (current-carrying "plates"). When connected to a source of electricity, the negative electrode (cathode) becomes heated and emits electrons into the partial vacuum inside the valve. The electrons move towards the positive electrode (anode). One or more intermediate electrodes, called grids, can be interposed between the cathode and the anode, to influence the rate of electron emission – ie, the current flowing in the valve. A small increase in grid potential (voltage) causes a large increase in cathode current. In this way, current control and amplification is possible. *See also* SU p.130.

Vampire, in legend, a corpse that lives at night and sucks the blood of the living to sustain itself. The victim in turn becomes a vampire. The monster can be killed by driving a wooden stake through its heart. Bram STOKER's *Dracula* (1897) drew on the legend of the vampire to produce a masterwork of horror.

Vampire bat, small brown bat that lives in tropical and subtropical America. It slices the skin of resting animals and human beings with its sharp teeth then laps up their blood. Length: 7.6cm (3in); wingspan 30cm (12in). Family Desmodontidae; species *Desmodus rotundus. See also* NW pp.*161,* 217.

Vanadium, metallic element (symbol V) of the first transition series, discovered in 1801. The metal is used in special steels, and vanadium pentoxide is an important oxidation catalyst. Chemically vanadium reacts with oxygen and other non-metals at high temperature. Properties: at.no. 23; at.wt. 50.9414; s.g. 6.1 (18.7°C); m.p. 1,890°C (3,288°F); b.p. 3,380°C (6,116°F); most common isotope V^{51} (99.76%). *See also* TRANSITION ELEMENTS.

Van Allen, James Alfred (1914–), US physicist. During WWII he helped to develop the radio proximity fuse. After the war he supervised experiments with captured German V-2 rockets, leading to research with space satellites and his discovery of the two zones of radiation encircling the Earth, commonly known as the VAN ALLEN RADIATION BELTS.

Van Allen radiation belts, two doughnut-shaped regions of radiation trapped by the Earth's magnetic field in the upper atmosphere, named after James A. VAN ALLEN, the US physicist who discovered them in 1958. The belts contain particles carrying energies of from approximately 10,000 electron volts to several million electron volts. The inner belt (of protons) centres about 3,220km (2,000 miles) above the Equator; the outer belt (of electrons) centres at about 18,500km (11,500 miles). *See also* PE p.23.

Vanbrugh, Sir John (1664–1726), outstanding English BAROQUE architect and dramatist, who worked with and was influenced by Sir Christopher WREN. Having taken London by storm with his witty RESTORATION comedies *The Relapse* (1696) and *The Provok'd Wife* (1697), he turned to architecture in 1699, when he designed Castle Howard, Yorkshire. In 1702 he became, with no training, Wren's chief colleague. Blenheim Palace (1705–20), Vanbrugh's masterpiece and the culmination of English Baroque, exemplifies his mature style. *See also* HC1 pp297, *313.*

Van Buren, Martin (1782–1862), 8th US President (1837–41) and one of the founders of the DEMOCRATIC PARTY. During his administration the country suffered its first economic depression (the Panic of 1837), the result of uncontrolled land speculation. Although millions suffered, Van Buren was adamant in denying government aid. He stood for re-election (1840) on Democratic policies but was defeated and in 1848 made another unsuccessful attempt at the presidency.

Vancouver, George (1757–98), British navigator and explorer. He was made commander of his own expedition in 1791 and ordered to survey the NW coast of North America and to search for a passage from the Pacific to Hudson Bay. He made a detailed survey and found no passage to the E. He returned by going around Cape Horn in 1794.

Vancouver, port in S British Columbia, Canada, on the S shore of Burrard Inlet; third-largest city and principal Pacific port of Canada. The area was first explored in 1792 by Capt. George Vancouver, and the first European settlement was established in 1865 and known as Granville; it was renamed Vancouver in 1886. It includes the University of British Columbia (1908) and Simon Fraser University (1963). Exports: wood products, grain, fish. Industries: timber, pulp and paper, tourism, food processing, chemicals, fishing, shipbuilding, meat packing. Pop. 426,260.

Vancouver Island, island off the coast of British Columbia, Canada, in the Pacific Ocean. Capt. James Cook visited the island in 1778; it became a British Crown colony in 1849 and part of British Columbia in 1866. Industries include timber, fishing, tourism and coalmining. Area: 32,137sq km (12,408sq miles). Pop. (1970 est.) 375,000.

Vandals, ancient Germanic people. Their existence in E Europe is recorded during the 1st century AD; early in the 5th century they crossed the Rhine to escape the Huns. They passed through Gaul and Spain, and invaded N Africa in 429. From there they raided the W Mediterranean, briefly capturing Rome in 455. They were Arian Christians, hostile to the Catholic Church. In 533 a Roman army sent by JUSTINIAN destroyed their kingdom. *See also* HC1 pp.115, 136, *136–137.*

Van de Graaff generator, machine that generates high voltages by concentrating electrical charges on the outside of a hollow conductor. It is used as the voltage supply for a type of electrostatic ACCELERATOR (the Van der Graaf accelerator). Positive (or negative) charges, sprayed from a set of high-voltage points on to a moving belt, are conveyed into a large hollow metal sphere to which the charges are carried by a second set of points. A voltage of about 50,000 volts applied to the points can generate up to 10 million volts.

Vanderbilt, Cornelius (1794–1877), US railway owner and financier. He began as a boy running a ferry service in New York and later established a steam boat company, with which he made his fortune. After the California gold rush of 1849, he ran a shipping line between New York and San Francisco. During the American Civil War, he built up a railway empire by gaining control of existing railways and in 1873 he connected New York and Chicago directly by rail.

Vanderlyn, John (1775–1852), US NEO-CLASSICAL painter in historical landscapes and some portraits. He lived mostly in Paris from 1796 to 1815, where he became famous for his *Marius amid the ruins of Carthage* (1807) and *Ariadne Asleep on the Isle of Naxos* (1812).

Van der Post, Laurens Jan (1906–), South African writer. He contributed in Afrikaans to the journal *Voorslag.* On the outbreak of WWII he joined the British Army, and served in the Far East. His novels include *In a Province* (1934), *Flamingo Feather* (1955), *The Hunter and the Whale* (1967) and *A Story Like the Wind* (1972).

Van de Velde, family of 17th-century Dutch artists. Jan (1593–1641) and his cousin Esaias (*c.*1591–1630) painted genre scenes and landscapes. Willem the elder (1611–93), the brother of Esaias, was an important marine painter; he depicted the victories of the Dutch fleet until 1672 when he moved to England. His son Willem the younger (1633–1707), worked with him on many paintings and later became court painter to Charles II. He was the leading marine painter in England of his time.

Van der Velde, Henri (1863–1957), Belgian ART NOUVEAU architect, designer and teacher. His Bloemenwerf House at Uccle, near Brussels (1895), designed for his own family use, was influenced by William MORRIS and the English ARTS AND CRAFTS MOVEMENT; it heralded art nouveau. He moved to Germany in 1899 and in 1905 designed the new School of Applied Arts at Dessau, of which he was the principal.

Van der Waals, Johannes Diderik (1837–1923), Dutch physicist who was awarded the 1910 Nobel Prize in physics for his work on gases and the gas equation which he derived. The ideal gas equation is $PV=RT$ (where P is the pressure, V the volume, T the absolute temperature of the gas and R is the gas constant). The Van der Waals equation is $(P + a/V^2)(V - b) = RT$, where a and b are the Van der Waals constants. This takes into account the volume occupied by the gas molecules and their inter-molecular attraction, which were ignored by the KINETIC THEORY of gases. *See also* IDEAL GAS LAW.

Van der Waals forces, forces of attraction which contribute towards cohesion between neighbouring ATOMS or MOLECULES. They are named after Johannes VAN DER WAALS, who investigated such phenomena during the 19th century. Van der Waals forces are caused by the similar movements of electrons in neighbouring atoms or molecules.

Van Diemen's Land, original name of Tasmania. It was discovered by Abel TASMAN in 1642 and named in honour of Anthony van Diemen, the Governor-General of the Dutch East Indies. It became part of New South Wales in 1803, was made a separate colony in 1825 and given self-governing status in 1850. Van Diemen's Land was renamed Tasmania in 1855. *See also* HC2 p.124.

Van Dongen, Kees. *See* DONGEN, KEES VAN.

Van Dyck, Sir Anthony (1599–1641), Flemish portrait and religious painter who studied in RUBENS' studio before travelling abroad, first to England in 1620. In the following year he went to Italy, where he painted many BAROQUE portraits of the Genoese nobility which show his great gift for aristocratic portraiture. He was invited to England in 1632 by Charles I, who made him court painter and a knight. He was overwhelmed with commissions and his many depictions of English aristocrats greatly influenced future English portrait painting. *See also* HC1 pp.*288, 296–297, 296, 307.*

Vane, the Younger, Sir Henry (1613–62), English politician. He went to Massachusetts in 1635 and became governor (1636–37). As Leader of the House of Commons in 1643 he negotiated the Solemn League and Covenant; he served in the Rump Parliament and opposed its dissolution by Oliver CROMWELL in 1653. He was executed after the Restoration.

Vanern, largest lake in Sweden, in the sw of the country. It is fed by the River Klarälven and drained by the Göta älv sw into the Kattegat. The lake is crossed by the Göta Canal. Pulp and paper mills line the shores. Area: 5,584sq km (2,156sq miles).

Van Eyck. *See* EYCK, VAN.

Van Gogh, Vincent. *See* GOGH, VINCENT VAN.

Vanilla, climbing orchid native to Mexico. The vines bear greenish-yellow flowers that produce seed-pods 20cm (8in) long which are the source of the flavouring vanilla, which is prepared by drying and curing. Family Orchidaceae; species *Vanilla planifolia.*

Vanini, Lucilio (1585–1619), Italian free-thinker, pseudonym Julius Caesar. He studied at Naples and Padua and took holy orders, but preached a modern anti-religious form of pantheism. He was arrested for atheism and witchcraft and later burnt at the stake. His works include *De Admirandis Naturae Arcanis* (1616).

Vanir, in Norse mythology, group of mature gods, the patrons of fertility and wealth who were opposed to the AESIR. They consisted of Nioro, the water god; his wife, Skadi; his son, Frey; and his daughter, Freya.

Vanity Fair, a Novel without a Hero (1848), novel by William Makepeace THACKERAY. It vividly portrays the lives and loves of its main female characters, Becky Sharp and Amelia Sedley, with those of a number of godless, self-satisfied and pompous men.

Van Loo family, French painters of Flemish descent. Jean Baptiste (1684-1745), a portraitist, was acclaimed in Paris and London for his paintings of Louis XV and Sir Robert Walpole. Carle (1705–65), brother of Jean-Baptiste, became principal painter to Louis XV in 1762 and his work was mainly decorative. Louis Michel (1707–71), son of Jean-Baptiste, was court painter to Philip V of Spain (1736–52).

Vansittart, Robert Gilbert (1881–1957), British diplomat. He entered the Foreign Office in 1902 and was Private Secretary to Lord Curzon from 1920 to 1924. As Permanent Under-Secretary for Foreign Affairs between 1930 and 1938, he warned against the growing military power of Germany but his advice went unheeded and in 1938 he was demoted, becoming diplomatic adviser to the government.

Vant' Hoff, Jacobus Henricus. *See* HOFF, JACOBUS HENRICUS, VANT'.

Vantongerloo, Georges (1886–1965), Belgian sculptor, a founder member of the de STIJL movement. After 1938 he became

interested in space and movement, and used wire or clear plastic in curved structures. His *Composition* (1920) is made of concrete.

Van Vollenhoven, Tom (1935–   ), South African rugby wing-threequarter of startling speed. He played seven times for South Africa (1955–56) at rugby union, before changing to rugby league and playing in Britain for St Helens, for whom he scored 381 tries in 393 matches.

Vaporization, or volatilization, the conversion of a liquid or solid into its vapour, eg the heating of water to make steam. Solids such as ammonium chloride, when heated, pass directly into the vapour state, a process known as SUBLIMATION. Vaporization of a solid or liquid can also be achieved by lowering the surrounding pressure or by EVAPORATION, a process which in nature is caused on a large scale by the combined effects of sun and wind.

Vapour, fundamentally the same thing as a gas, but used to mean the gaseous state of a material which is more commonly found at standard temperatures and pressures as a solid or liquid.

Vapour pressure, the pressure exerted by a vapour in equilibrium with its liquid or solid. It is an important property in chemical analysis, because when a solid is dissolved in a liquid, the vapour pressure of the liquid is reduced by an amount proportional to the solid's molecular value.

Vāranasi, also known as Benares and Banoras, city in Uttar Pradesh state, N India, on the River Ganges, about 640km (400 miles) WNW of Calcutta. BUDDHA is reputed to have preached his first sermon near by, and Vāranasi is considered by Hindus to be a holy city. It is a centre of pilgrimage with about 1,500 temples and mosques. Products include silk brocade and brassware. Pop. (1971) 583,856.

Vardon, Harry (1870–1937), British golfer. He won six British Open Championships and the 1900 US Open, and with fellow British professionals James Braid and John Henry Taylor formed the "Great Triumvirate" that dominated golf in Britain for 20 years until 1914.

Varenius, Bernhardus, or Varen, Bernhard (1622–50), German geographer. He is best known for his *Geographia generalis* (1650), in which he attempted to structure geography on a scientific basis and organize and co-ordinate its various branches. He developed the concept of what is now known as regional geography.

Varèse, Edgard (1885–1965), French composer, a leading advocate of 20th-century experimental music. He lived in the USA after 1915 and composed *Hyperprism* (1923) for wind instruments and percussion, *Ionisation* (1931) for two groups of percussion instruments and *Déserts* (1954) for tape-recorded sound. He concentrated on ELECTRONIC MUSIC after the early 1950s. *See also* HC2 p.270.

Vargas, Getúlio Dornelles (1883–1954), twice President of Brazil. He seized the presidency in 1930, establishing dictatorial control and encouraging the country's economic growth, but was deposed in 1945 in favour of a democratic constitution. Re-elected in 1951, he was forced to resign three years later under threat of impeachment, and committed suicide.

Variable, in mathematics, symbol used to represent an unspecified quantity. Variables are used to express a range of possible values, eg in the equation $(x + 1)^2 = x^2 + 2x + 1$, x is the variable.

Variable, in statistics, item that may change or vary unlike a constant, which maintains a given value. The mark given to each student in a class is a variable; it changes from student to student.

Variable star, any STAR that exhibits regular or irregular fluctuations in brightness. Regular variables include eclipsing variables and the pulsating stars of the CEPHEID or RR LYRAE type. Irregular variables include NOVAE and flare stars.

Variation, in biology, difference between members of the same SPECIES due to heredity or environment. The difference may be in type of reproduction, fertility or physical appearance. *See also* ADAPTATION; EVOLUTION.

Variation, in music, a variety of treat-

ments upon a single theme, by such means as simple elaboration, change of KEY, change of TIME SIGNATURE and INVERSION.

Varicella. *See* CHICKEN POX.

Varicose veins, VEINS that are distorted and distended with blood. They can occur anywhere in the body, but are usually found in the legs. The condition arises from the failure of the VALVES in the veins. Treatment includes wearing elastic-type stockings or surgical stripping of the affected veins. *See also* MS p.*91.*

Variola major. *See* SMALLPOX.

Variscite, glassy white to green phosphate mineral, hydrous aluminium phosphate with iron impurities found with aluminium-rich rocks near the Earth's surface. It occurs as crystals in the orthorhombic system. It resembles TURQUOISE, but is greener, and is used in jewellery. Hardness 3.5–4.5; s.g. 2.5.

Varley, Frederick Horsman (1881–1969), Canadian artist, *b.* Britain, who settled in Canada in 1912. He is known chiefly for his penetrating portraits and landscapes which owe much to EXPRESSIONISM.

Varna, port in E Bulgaria, on the Black Sea; founded in the 6th century BC. It came under Turkish rule in 1391. It was ceded to Bulgaria by Turkey in 1878. Industries include textiles, flour-milling, ship building and the manufacture of diesel engines. Pop. (1974) 269,980.

Varnish, solution of a resin in a solvent applied as a film to wood or metal to give a hard glossy transparent protective coating. Pigments may be added for colour. Film formation is the result either of solvent evaporation or of oxidation or polymerization. Shellac in methanol or ethanol (french polish) and cellulose ester in ketones are examples of solvent-evaporation varnishes.

Varro, Marcus Terentius (116–27 BC), Roman scholar and writer. Serving as a deputy of POMPEII in Spain, he was treated kindly by CAESAR when Pompeii was defeated and was made director of the public library in Rome. QUINTILIAN called him "the most erudite Roman of them all". Varro was a prolific writer of historical, biographical, philosophical and geographical treatises.

Varuna, in Vedic mythology, the supreme ruler, guardian of cosmic order, omniscient, omnipresent and possessed of universal power. He is worshipped as the upholder of moral order who presides over the sky, being particularly identified with the moon.

Varve, term applied to a layer of sediment deposited in a single year in a body of still water. Specifically a varve consists of two layers of sediment, a coarse layer deposited in the summer and a fine layer deposited in the winter in glacial melt-water lakes. Varves have been used to date the age of Pleistocene glacial deposits. *See also* PE pp.130–131.

Vasa dynasty (1523–1818), royal dynasty of Sweden. It was established after Gustav Eriksson, a relative of the Stures, a noble Swedish family, revolted against Denmark in 1520 and ended the union. He was elected king in 1523 and moulded Sweden into a nation, stabilized state finances and confiscated property belonging to the Roman Catholic Church, thus precipitating the Reformation in Sweden. The dynasty ended when Jean Bernadotte, one of Napoleon's marshals, was chosen to rule upon the sudden death of the heir to the throne. Bernadotte succeeded to the throne as Charles XIV of Sweden and Norway in 1818.

Vasarély, Victor de (1908–   ), Hungarian painter who worked in Paris after 1930. His early work was influenced by CONSTRUCTIVISM. His paintings, with their emphasis on exciting spatial and optical effects, anticipated OP ART. By the 1960s there was a kinetic effect in his works. Among his paintings is *Ond* (1968). *See also* HC2 p.279.

Vasari, Giorgio (1511–74), Italian painter, architect and art historian. He painted in the mannerist style, mainly in Florence and Rome, but was more well known in his day for his architectural works, which include the Palazzo dei Cavalieri, in Pisa. His fame today rests on

Van Diemen's Land; a drawing by Abel Tasman, showing Dutch ships and native boats.

Vanilla plants live about ten years, producing their first crop after three.

Vanity Fair, like most novels of the time, was published in monthly parts.

Lord Vansittart during the Munich crisis, 1938. His warnings went unheeded.

Vatican; the Sistine Chapel ready for a conclave of cardinals to elect a pope.

Vaughan Williams in 1946, conducting the *London Symphony*, his own composition.

Vaulting; Nelli Kim (USSR) jumping over the horse at the Montreal Olympics.

V-E Day in Trafalgar Square, London, as crowds celebrated victory in Europe.

his history of Italian art, *The Lives of the most excellent Painters, Sculptors and Architects* (1550). *See also* HC1 p.264.

Vasco da Gama. *See* GAMA, VASCO DA.

Vasconcelos, José (1882–1959), Mexican statesman and intellectual. He dedicated himself to raising the cultural level of his country, emphasizing public education and government patronage of the arts. He was also a prolific writer, and is best known for a number of philosophical, sociological and historical works, and for his four autobiographical volumes (1935–39).

Vascular plant, or tracheophyte, plant with vessels or ducts to carry water and food materials within it. All higher plants – FERNS, CONIFERS and FLOWERING PLANTS – have a vascular system (XYLEM and PHLOEM). This system exists in roots, stems and leaves. *See also* NW pp.44–45.

Vas deferens, long narrow tube that carries SPERM from the TESTES, around the bladder, and opens into the seminal vesicles, where the sperm are stored until ejaculation. *See also* VASECTOMY; MS p.74.

Vasectomy, method of sterilizing the male, in which the VAS DEFERENS, through which sperm pass from the testes to the penis, is severed. It is an increasingly acceptable form of BIRTH CONTROL in men.

Vasoconstrictor, any substance, natural or artificial, which causes constriction of blood vessels. An example is the hormone VASOPRESSIN. Vasoconstrictors are administered to reduce the peripheral circulation, so raising BLOOD PRESSURE.

Vasopressin, HORMONE secreted by the posterior lobe of the PITUITARY GLAND situated under the brain. It is also called antidiuretic hormone or ADH, its action being to prevent excessive water loss by reabsorption of water in the KIDNEY tubules during the making of urine.

Vassal, in the Western European FEUDAL SYSTEM, a free man who bound himself to another (the lord) by an oath of FEALTY for mutual protection. In exchange for certain obligations, such as military service, he received direct maintenance in the lord's household, or a benefice or FIEF to work himself.

Vasteras, inland port in E Sweden, at the mouth of the River Svartan on Lake Malaren. Founded in 1100, it was an important medieval city and the scene of several diets, including the diet of Vasteras Recess in 1527 which formed the Lutheran state Church, and the diet of 1544 which made the Swedish throne hereditary. Exports: iron ore, timber. Industries: metal goods, electrical goods, textiles, glass. Pop. 118,044.

Vastitas Borealis. *See* MARS.

VAT, Value Added Tax, indirect tax in Britain, introduced in 1971. It consists of a series of *ad valorem* taxes levied on goods in the various stages of their manufacture at the point of sale: raw material to the manufacturer, wholesale product to the retailer, and retail product to the consumer.

Vatican, The, residence of the Pope in Rome. It is said that a papal residence was first erected on the site near the basilica of St Peter by Pope Symmachus (498–514). Innocent III rebuilt (*c.*1200) an earlier residence and, after the return from the BABYLONIAN CAPTIVITY in 1377, it became the main residence of the head of the Roman Catholic Church. *See also* VATICAN CITY.

Vatican City, independent sovereign state existing as an enclave within Rome. It is the official home of the Pope and the centre of the ROMAN CATHOLIC CHURCH. The area was used as a burial ground until AD 59 when NERO made it a garden. The first papal residence was established there in the 5th century. Vatican City did not achieve full independence until 1929. Area: 44 hectares (109 acres). Pop. (1974 est.) 1,000. *See* MW p.180.

Vatican Council, Second. *See* SECOND VATICAN COUNCIL.

Vatican Councils, series of two ECUMENICAL COUNCILS of the Roman Catholic Church. The First Vatican Council (Vatican I), held in 1869–70, was the twentieth ecumenical council, according to the reckoning of the Church. Convened by Pope

Pius IX to deal with a variety of subjects, such as the effects of rationalism and liberalism, ecclesiastical discipline, canon law, Church-state relations etc the council is mainly associated with the definition of the dogma of papal INFALLIBILITY. During a recess, Piedmontese troops occupied Rome and ended the temporal power of the Pope there; Pius IX then suspended the council. The SECOND VATICAN COUNCIL (Vatican II), held in 1962, dealt with the Church's role in modern society.

Vauban, Sebastien le Prestre, Marquis de (1633–1707), French military engineer, royal engineer (1655–1703) and marshal (1703–07). During the wars of Louis XIV he conducted a long series of sieges and built and improved many fortresses.

Vaudeville, popular US equivalent of the British music hall variety entertainment. Its rise and fall followed the same pattern as its European counterpart, having its heyday in the late 19th century and eventually succumbing to the cinema's popularity.

Vaughan, Henry (*c.*1621–95), English poet, *b.*Wales. After studying law he turned to medicine and became a doctor in Wales. One of the metaphysical poets, his best work was inspired by religious experience and includes *Poems* (1646) and *Silex Scintillans* (1650–55). *See also* METAPHYSICAL POETRY.

Vaughan Williams, Ralph (1872–1958), British composer. His interest in English folk music is apparent in his three *Norfolk Rhapsodies* (1905–07) and his instrumental arrangement *Fantasia on Greensleeves* from the opera *Sir John in Love* (1929). He also used elements of English Tudor music in his *Mass in G Minor* (1923) for unaccompanied chorus. He wrote nine symphonies, the best known of which are the *Sixth Symphony* (1947) and the *Sinfonia Antarctica* (1952), based on his music for the film *Scott of the Antarctic*. *See also* HC2 p.270.

Vault, in architecture, a curved, often ribbed ceiling constructed from bricks, stone blocks or wooden beams, which form a rigid structure. The ancient Egyptians used brick vaults for their drainage systems. During the Middle Ages the technique became complex, with varied constructions. *See also* HC1 pp.198–199, 218–219.

Vaulting, exercise in gymnastics. The apparatus used is called a horse; men vault its length and women the breadth. Vaults are judged on the first flight from take-off (from a small springboard), the positioning of the hands on the horse, the flight from the horse until and including landing, and the difficulty of the vault. The whole performance takes only a few seconds.

Vauxhall, district of the borough of Lambeth, London, s of the River Thames. The name comes from Falkes Hall, from Falkes de Bréauté, the 13th-century Lord of the Manor. Gardens laid out there in 1661 became fashionable in the 18th century but were closed in 1859.

Vavassor, feudal tenant or vassal, indicating, in Normandy, a man who owned less land than a BARON, but who nevertheless owed some military service. The term passed into English usage after 1066 and by the 12th century, as the status of barons rose, so too did the vavassors and the title was replaced by that of KNIGHT.

Vayu, in Vedic mythology, the god of the atmosphere, wind and breath of life. He is gentle and one of the Vedic triad, the other two being Ágni, the fire god, and Surya, the sun. In mythology he is king of the northwest quarter and father of HANUMAN and is symbolized by an antelope.

Vazov, Ivan (1850–1921), Bulgarian poet, novelist and playwright. Known as Bulgaria's national poet, he wrote the revolutionaries' battle hymn *Prvaporets i gusla* (1876) and recorded his country's struggle for independence against the Turks in the novel *Pod Igoto* (trans. *Under the Yoke*, 1894) and the poems *Tâgite na Balgariya* (1877).

Vázquez, Horacio (1860–1936), three times President of the Dominican Republic between 1899 and 1930. He retired from a second term as President during the US occupation of the Republic from

1916 to 1924, but emerged as the dominant figure when US forces withdrew. Rafael Trujillo forced him out of office in 1930.

VD. *See* VENEREAL DISEASE.

Veal, the meat of the calf, which has little fat and a high proportion of muscle, and so tends to be lean, if dry. Larger cuts of veal may be roasted with added fat; small pieces are usually braised to prevent loss of moisture. *See also* PE p.232, *232*.

Veblen, Thorstein Bunde (1857–1929), US economist. He originated the institutionalist school of economics, studying the economy as a whole and institutions and their role in the economic system. His most famous books are *The Theory of the Leisure Class* (1899), in which he criticized conspicuous consumption, and *The Theory of Business Enterprise* (1904).

Vecchietta, Lorenzo di Pietro (*c.*1412–80), Italian sculptor of the Sienese School who also worked as a painter and architect. He was a pupil of Sassetta, and was also greatly influenced by DONATELLO, who visited Siena in 1457. His *Assumption* and *St Catherine* (both *c.* 1462) exemplify his expressive, naturalistic style.

Vector, in mathematics, quantity that has both a magnitude and a direction, as contrasted with a SCALAR which has magnitude only. For example, the VELOCITY of an object is specified by its speed (how fast it is moving) and the direction in which it is moving; similarly, a force has both magnitude and direction. Mass is a scalar quantity, but weight (the force of gravity on a body) is a vector.

Vector, in disease, any agent that carries an infectious agent from one host to another. It may be an insect such as a mosquito or flea, or an inanimate object such as a cup. It also may serve as an intermediate host for the infectious agent.

Vector analysis, study of the mathematical properties of VECTORS. The addition of two vectors is the diagonal of a parallelogram formed with the two vectors as adjacent sides. Multiplication of a vector by a SCALAR a gives a vector with the same direction but its magnitude multiplied by a. Two vectors can also be multiplied; this may occur in two ways: the scalar, or dot product of vectors a and b (written $a. b$) is a scalar given by $ab\cos\theta$, θ being the angle between the vectors and a and b being their magnitudes; the vector, or cross product (written ab or $a \times b$), is a vector given by $ab\sin\theta$, directed at right-angles to the plane of a and b.

Vedanta, in HINDUISM, the *Conclusion* to the sacred literature, the *Vedas*; the best-known and most popular form of Indian philosophy. It was expounded by the 7th–8th century philosopher Sankara, and holds that the natural world is an illusion. There is only one self, Brahman-Atman; ignorance of the oneness of the self with Brahman is the cause of rebirth. Vedanta also believes in the transmigration of souls and the desirability of release from the cycle of rebirth.

Vedas, ancient sacred books of the Hindus. There are four Vedas: the RIG VEDA, containing priestly tradition originally brought to India by the ARYANS; the YAJUR VEDA; the SAMA VEDA; and the *Atharva Veda*. Vedic literature consists of the *Samhitas*, basic hymns contained in the four Vedas, and the elaborations that came to accompany them; the BRAHMANAS, prose explanations of rituals, which are complemented by the *Aranyakas*; and, the final stage of commentary, the UPANISHADS. The Vedas were composed over a long period, probably *c.*1500–1200 BC.

V-E Day (8 May 1945), the day WWII ended in Europe. The Germans surrendered unconditionally to Britain, France, the USA and the USSR.

Vedda, or Veddah, aboriginal people of Sri Lanka, now almost absorbed into the modern Sinhalese population. Physically related to the DRAVIDIANS, they were hunters and gatherers, living in caves and practising an ancestral and SHAMANISTIC religion.

Vega, Garcilasco de la. *See* GARCILASCO DE LA VEGA.

Vega, or Alpha Lyrae, white main-sequence star in the constellation of Lyra; the fifth-brightest star in the sky. It is one of the three brilliant stars making up the Summer Triangle. Characteristics: apparent mag. 0.03; absolute mag. 0.3; spectral type AO; distance 26 light-years. *See also* SU pp.226, *237,* 259, *259,* 264–265, 265, 267, *267.*

Vega Carpio, Lope Félix de (1562–1635), Spanish poet and Spain's first great dramatist. A voluminous writer, he produced epics, pastorals, odes, sonnets and novels and created Spanish Romantic drama. About 300 of his works survive, with a further 100 doubtful attributions, including the plays *Peribanez ye el Comendador de Ocana* (*c.*1610) and *Fuente Ovejuna* (*c.*1613). Other works include the novels *La Arcadia* (1598) and *La Dorotea* (1632), and the poems *La Hermosura de Angélica* (1602).

Vegetable, as opposed to animal, a form of life which builds up its tissues by means of growth using the energy of sunlight, carbon dioxide from the air, and the green pigment CHLOROPHYLL. This process is known as PHOTOSYNTHESIS. Vegetables, or green plants, need also to be supplied with water and mineral salts, which are usually present in the soil. The term vegetable is also used in a more restricted sense for vegetable foods.

Vegetarianism, way of life advocated and practised by those who abstain from eating meat. A minority of vegetarian purists, known as Vegans, also exclude from their diet all animal products such as butter, eggs, milk or cheese. There are also many degrees of vegetarianism between the two categories. Vegetarianism has a religious basis in many cultures, particularly among various sects of Hinduism and Buddhism and the Seventh Day Adventists. It is also advocated on ethical, medical and economic grounds.

Veidt, Conrad (1893–1943), German film actor, often cast as a villain in the many films he made in Britain and the USA. His most famous German film was *Cabinet of Dr Caligari* (1919). Others included *The Thief of Baghdad* (1940), *Nazi Agent* (1942) and *Casablanca* (1942).

Veii, ancient town of Etruria, Italy, 19km (12 miles) NW of Rome. It grew up in the Villanovan epoch, and is therefore much older than Rome. It was the nearest Etruscan city to Rome, which fought many wars against it and finally conquered it *c.* 396 BC after a ten-year siege.

Veil nebula, the remnants of a star which exploded about 60,000 years ago in the constellation Cygnus, the Swan. The gases from this explosion, still moving outwards at a high speed, have a gauzy, veil-like appearance. *See also* SU p.233, *233.*

Veins, in mammals, vessels that, with the exception of the two pulmonary veins, carry deoxygenated blood to the right upper chamber of the heart. The pulmonary veins carry oxygenated blood from the lungs to the left upper chamber. *See also* ARTERY; MS pp.62–63.

Velasco Alvarado, Juan (1910–), President of Peru from 1968 to 1975. A career military officer, he led the 1968 coup against President Fernando Belaúnde over the president's failure to expropriate US-owned oilfields. An advocate of social democracy, Velasco was deposed by his military commanders and replaced by Gen. Francisco MORALES BERMUDEZ.

Velasco Ibarra, José María (1893–), Ecuadorean politician, elected President of Ecuador five times between 1934 and 1972. Deposed or forced to resign before completing any of his terms of office, he became unpopular because of the country's increasing economic and political problems.

Velázquez, Diego (1465–1524), Spanish CONQUISTADOR and first Governor of Cuba (1514–21; 1523–24). He sailed with COLUMBUS to Hispaniola in 1493 and commanded the expedition that conquered and colonized Cuba (1511–14). He commissioned three expeditions to the Mexican coast, placing Hernán CORTÉS in charge of the third venture (1519). Velázquez sent Pánfilo NARVÁEZ (1520) and later Cristóbel de Olid (1524) to compel Cortés to return to Cuba but both were defeated.

Velázquez, Diego Rodriguez de Silva y (1599–1660), Spanish painter, among the greatest masters of realism, hailed in the 19th century as the greatest painter. His early genre works (1617–22) include *The Old Woman Cooking Eggs* (1618) and *The Water Carrier of Seville* (*c.*1619). In 1623 he became court painter to Philip IV, and throughout his career painted many royal portraits. He also executed, with sensitivity and compassion, splendid portraits of dwarfs and jesters. His *Don Juan of Austria* is a typical example. His portrait of Pope Innocent X (*c.*1650) ranks with the greatest of all portrait paintings. Velázquez's later works began to merge reality and illusion and include *The Hilanderas* (*c.*1655) and *Las Meninas* (1656). *See also* HC1 pp.*306,* 307.

Velocipede, early form of BICYCLE or tricycle, also called hobby-horse or dandyhorse. The *Célérifère,* built in 1791 by the Comte de Sivrac, consisted of a wooden frame supported by two in-line wheels. A rider straddled the bench-like frame and by pushing his feet on the ground "walked" the machine along. *See also* MM pp.122–123.

Velocity, rate of change of distance, ie length per unit time in a specified direction. Thus it is a VECTOR, whereas speed, which merely measures length per unit time with specifying direction, is a SCALAR. *See also* SU p.74.

Velvet, soft thick fabric made in a pile weave using silk, cotton or synthetic fibres. It is characterized by a downy pile formed by clipped yarns. *See also* MM p.*59.*

Vena cava, main VEINS of vertebrates that supply the HEART with deoxygenated blood, emptying into its right atrium. The superior vena cava is a large vein formed by the union of the right and left brachiocephalic veins in the lower neck. The inferior vena cava is formed by the union of the right and left iliac veins and also receives venous blood from the renal (kidney), lumbar and hepatic (liver) veins. *See also* MS pp.62–63.

Vendée, Wars of the, counterrevolutionary risings by the peasantry of the Vendée district of France. They were in great part stimulated by ROBESPIERRE'S attack on the Roman Catholic Church. They began in 1793 when Henri de la Rochejaquelein raised a force of 30,000 men to rid the district of revolutionary officials. After initial success, these forces were heavily defeated in 1795 and made a peace which granted them amnesty and freedom of religion. A brief, and again unsuccessful, rising occurred in 1796.

Vendetta, private family blood feud, especially one practised through generations, as in Corsica and parts of Italy.

Vendôme, Louis Joseph, Duc de (1654–1712), French soldier, also known as the Duc de Penthievre. As the commander in Catalonia he captured Barcelona in 1697. He led the French army against Prince Eugene of Savoy at Luzzara in 1702 and defeated the Austrian army at Brihuega and Villaviciosa in 1710.

Veneer, extremely thin sheet of wood, or a thin sheet of a precious material such as ivory or tortoise-shell. Plain, cheaper woods or substitutes (such as plywood, chipboard or blockboard) may be completely covered in a richly coloured or grained veneer to give it the appearance of a decorative hardwood. Veneers may also be used as decorative shapes inlaid or set into the cut patterns in a surface such as wood.

Venera space missions, series of space probes sent to Venus by the USSR after 1965, Venera 3, launched in 1965, was the first spacecraft, a year later, to make impact on another planet. Veneras 9 and 10 made soft landings on Venus in 1975 and transmitted pictures and atmospheric data for an hour. *See also* SU pp.191, *191,* 269, *269,* 276–277, *276.*

Venereal disease, any of the diseases transmitted through sexual contact, chief of which are SYPHILIS, GONORRHOEA and chancroid. Syphilis is the most dangerous and may be passed from an infected mother to her unborn child. Caused by the bacterium *Treponema pallidum,* it follows three stages: firstly a chancre (hard sore) forms on the genital region. Then, about 12 weeks after exposure, a general rash is seen with sores in the mouth and the involvement of the central nervous system and eyes. Syphilis is very infectious at this stage. Thirdly the general symptoms may disappear, but after some years such effects as syphilitic heart disease, blindness, locomotor ataxia and madness may occur. PENICILLIN and its derivatives can still cure syphilis in its early stages.

Gonorrhoea is caused by the gonococcus bacterium and if diagnosed early may be treated with SULPHA DRUGS. Typical symptoms are pain when urinating for a man, and the discharge of pus from the vagina and a burning pain during urination for a woman. Untreated the gonococci reach deeper into the urinary and reproductive systems, possibly attacking the heart valves. All forms of venereal disease respond to the appropriate treatment, but the duration and effectiveness of the treatment depends on an early diagnosis. *See also* MS pp.102–103.

Venetia (Veneto), region in NE Italy, bounded by the Adriatic Sea (E) and Austria (N); Venice is the capital. The N of the region is mountainous and in the S is the fertile Venetian Plain. The chief river is the Po. Most of the population are farmers; the most important crops are cereals, fruits, sugar-beet and hemp. Industries include chemicals, textiles, paper, tourism, fishing, wine and food processing. The region was settled *c.* 900 BC and came under Roman rule in the 2nd century BC. From the 10th century AD cities in the region were powerful under the rule of noble families, and by the 15th century the republic of Venice was dominant. Venetia was later held by Austria, and it finally passed to Italy in 1866. Area: 18,368sq km (7,092sq miles). Pop. (1971 est.) 4,122,000.

Venetian glass, the cristallo soda-glass wares made in Venice, particularly in the 15th and 16th centuries, characterized by its decorativeness, in both colour and structural embellishment. Subsequent technical advances in England and elsewhere developed a glass that was less brittle, but from the 13th to 16th centuries Venetian glass was pre-eminent. In 1291 the island of Murano became the centre of manufacture and there the secret techniques of gilding, enamelling, colouring, millefiori, latticework and engraving were refined and guarded.

Venetian school, school of Italian painting which flourished in the 15th, 16th and 18th centuries. Its centre was Venice. It developed under the influence of ANTONELLO DA MESSINA and was notable for the sumptuousness and radiance of its colour, a legacy from Byzantine art. Its early masters included the BELLINI and VIVARINI families, followed by TITIAN and GIORGIONE, then TINTORETTO and VERONESE. In the 18th century TIEPOLO, CANALETTO and GUARDI revived Venetian painting.

Veneto. *See* VENETIA.

Venezia. *See* VENICE.

Veneziano Domenico. *See* DOMENICO VENEZIANO.

Venezuela, republic in N South America, including 72 islands in the Caribbean Sea. It is the world's fifth-largest producer of oil and one of the wealthiest countries in Latin America – 80% of the country's income comes from oil. Other exports include iron ore, coffee, cocoa, rice, cotton, steel products, sugar and fruit. The capital is Caracas. Area: 912,050sq km (352,143sq miles). Pop. (1976 est.) 12,361,000. *See also* MW p.180.

Venice (Venezia), city in N Italy, 260km (162 miles) E of Milan. It is built on 118 islands, separated by narrow canals, in the Lagoon of Venice, and joined by causeway to the mainland. Settled in the 5th century, it became a vassal of the Byzantine Empire until the 10th century. After defeating Genoa in 1381 Venice became the most important European sea-power, engaging in varied trade in the Mediterranean and Asia. The importance of

Vegetable life derives energy from light by means of photosynthesis.

Conrad Veidt at the height of his fame as a film actor in the early 1940s.

Velázquez painted many royal portraits, such as the *Infanta Margherita* (detail).

Venice; a view of the west façade of St Mark's, one of Italy's finest buildings.

Vening Meinesz, Felix Andries

Eleutherios Venizelos, in 1909, at about the time of the naval mutiny at Salamis.

Verdi's first success was *Nabucco* (1842) and among his best work is *Otello* (1887).

Verlaine's poetry had a lyrical quality, and Debussy and Fauré set it to music.

Vermeer's *Woman with a Water Jug*. A baker later held two paintings against debts.

Venice declined in the 16th century and it was ceded to Austria in 1797. It became part of Italy in 1866. Venice is the site of many churches, palaces and historic buildings. Industries include glass, textiles, tourism and shipbuilding. Pop. (1975) 365,208. *See also* HC1 pp.*159, 190–191, 208, 301.*

Vening Meinesz, Felix Andries (1887–1966), Dutch geophysicist and geodesist who in 1923 developed a method of measuring gravity in ocean basins. He is also noted for research on the effect of solar movements on the Earth's surface and his studies of convection currents within the Earth.

Venison, meat of the deer, commonly the red or fallow deer. It is lean although it tends to be tough and stringy. *See also* PE p.239.

Venizelos, Eleutherios (1864–1936), Greek statesman. In 1910 he became Prime Minister and expanded Greek territories during the Balkan Wars and WWI. He negotiated for Greek interests at the Versailles peace conference but was voted out of office in 1920. He became Prime Minister again in 1928 and 1935 but fought against the monarchy and was forced to flee in 1935. He died in France.

Venn diagram, diagrammatical representation of the relations between mathematical sets or logical statements named after the British logician John Venn (1834–1923). The sets or statements are drawn as geometrical figures, usually circles, which overlap whenever different sets have a common property. *See also* EULER DIAGRAM; SU pp.40–41.

Venom, snake, toxic substance produced in the poison glands of snakes and injected into their victims through ducts in or along their fangs. Many venoms are dangerous and some can be lethal unless counteracted by antisera. The effects of snake venom vary according to the species and the constituents of the poison. Blood coagulation, respiratory effects, neurotoxic effects, haemorrhage and anticoagulant effects are among the most common. *See also* NW p.135.

Venous thrombosis, the formation of a blood clot in a vein, particularly the deep veins of the leg after a surgical operation. The danger of deep vein thrombosis is that part of the clot may break from the original thrombus and lodge elsewhere, such as in the lung or the heart. *See also* MS p.91, 91.

Ventifact, in geology, any stone which has been shaped and sculptured by wind-blown sand. *See also* PE p.121.

Ventris, Michael (1922–56), British architect and archaeologist. In 1952 he deciphered MYCENEAN scripts, particularly Linear B, which was found at Knossos and other sites. His theory (now widely accepted) that Linear B was an archaic form of the Greek language was published, in collaboration with John Chadwick, in *Evidence for Greek Dialect in the Mycenean Archives* (1953).

Venus, Roman goddess of unknown origin originally associated with gardens and cultivation but also with the ideas of charm, grace and beauty. She became identified with the Greek goddess APHRODITE, and hence also personified love and fertility. *See also* MS p.210.

Venus, second planet from the Sun and almost as large as the Earth. It has no satellites. Its dense atmosphere is largely composed of carbon dioxide. Mean distance from the Sun, 108,200,000km (67,200,000 miles); mass and volume, 0.82 and 0.88 times that of Earth respectively; diameter, 12,100km (7,500 miles); rotation period, 243 days; sidereal period, 224.71 days; surface temperature, approx. 485°C (900°F). *See also* SU pp.*176–177, 176–177, 190, 193, 269, 269, 276,277.*

Venus de Milo, Greek statue in marble, generally dated 2nd century BC. Found on the island of Melos in 1820, it was taken by the French ambassador to Turkey and later presented to the LOUVRE by LOUIS XVIII. It is one of the most famous statues in the world and has been generally accepted as the embodiment of the ideal of feminine beauty. *See also* HC1 p.336.

Venus figures, palaeolithic figurines carved from wood, ivory or stone, usually highly stylized representations of a woman. These early sculptures have been found throughout Europe. *See also* HC1 pp.24–25; MS p.210.

Venus of Willendorf. *See* VENUS FIGURES.

Venus's flower basket, SPONGE found in deep marine waters. Its long cylindrical skeleton is formed from separate silica spicules with a latticed framework of silica. Length: 25cm (10in). Phylum Porifera; species *Euplectella aspergillum. See also* NW p.121.

Venus's fly trap, INSECTIVOROUS PLANT, *Dionaea muscipula,* native to the marshlands of North and South Carolina, USA. Its leaves resemble jaws, hinged on the mid-line, with a row of tooth-like hairs and sensitive spines along the edges. When an insect touches one of these spines the halves shut, imprisoning the insect, which is then dissolved by ENZYMES and absorbed by the plant. Family Droseraceae. *See also* NW pp.45, 59.

Veracruz, port in Veracruz State, Mexico, on the Gulf of Mexico. It was the first Spanish colonial post in Mexico, established in 1519; US forces captured it in 1847 during the Mexican War. Today one of Mexico's two chief ports, Veracruz is an important commercial centre for an oil-processing region. Industries: tourism, textiles, footwear, confectionery, brewing. Pop. 266,255.

Verb, in the grammar of most languages, the part of speech (either a word or a phrase) which expresses action, occurrence or existence. The tense of the verb indicates the time of the action, ie past, present or future. A verb is regarded as regular or irregular, according to its series of forms, or "conjugation". A transitive verb requires an object, an intransitive verb does not.

Verbena, genus of annual and perennial trees, shrubs and herbs native to the Western Hemisphere. Some species are popular garden plants and have pink, white or purplish flowers. Family Verbenaceae; there are about 2,600 species.

Vercors, one of the most westerly ranges of the ALPS, which extends between the rivers Isère and Drôme in the French départements of the same name. The MAQUIS were active there during WWII and today the spectacular scenery attracts many tourists. The highest peak is Grand Veymont, 2,346m (7,697ft).

Verdi, Giuseppe (1813–1901), Italian composer, principally of opera. The early operas display an original and lively talent and a promising sense of the dramatic. Up to 1853 his masterpieces were *Rigoletto* (1851), *Il Trovatore* (1853) and *La Traviata* (1853). In *The Sicilian Vespers* (1855) the influence of MEYERBEER is apparent – an influence that led also to *Un ballo in maschera* (1859). *Aïda* (1867), shows a development in style, with richer and more imaginative orchestration. With Verdi's last three operas, *Don Carlos* (1884), *Otello* (1887) and *Falstaff* (1893), Italian opera of the 19th century reached its greatest heights. Among other compositions are songs and several sacred choral works, such as the *Requiem* (1874) and *Stabat Mater* (1898).

Verdigris, green copper carbonate which forms on objects made of copper or bronze which have been exposed to atmospheric corrosion. The word also refers to the blue-green powder (copper acetate) formed from metallic copper and acetic acid. *See also* PATINA.

Verdun, Battle of (Feb.–Sept. 1916), the longest and one of the bloodiest engagements of WWI. On 21 Feb. 1916 the Germans launched a massive offensive against the fortress of Verdun in NE France. A British offensive on the Somme relieved German pressure and by December the French had recovered most of the ground lost. Over 2,000,000 men were engaged at Verdun; and there were more than 600,000 casualties. *See also* HC2 p.188.

Verdun, Treaty of (843), partition of CHARLEMAGNE's empire among his three grandsons, the sons of Emperor LOUIS I. Charles II the Bald received the western part of France, Louis the German territory extending from the Rhine, Germany to the eastern part of the empire and Lothair I the central part of the empire which included the Low Countries, part of Germany and France and much of Italy. Lothair also kept the imperial title. The Treaty of Verdun was significant as marking the end of political unity in Western Europe. *See also* HC1 p.153, 153.

Vereeniging, city in the Transvaal, NE South Africa, on the Vaal River 56km (35 miles) s of Johannesburg. Founded in 1892, the city developed as a coal-mining centre. The Treaty of Vereeniging, ending the South African War, was negotiated there in 1902. Industries include iron and steel and bricks. Pop. 169,553.

Verelst, Simon (1644–c.1720), Dutch still-life painter. He worked in England at the court of CHARLES II, where the Duke of BUCKINGHAM persuaded him to take up the art of court portraiture.

Vereshchagin, Vasili (1842–1904), Russian artist and sailor. Generally his subjects were battles, for which he drew inspiration from campaigns which he himself witnessed, including the Russo-Turkish War of 1877.

Verga, Giovanni (1840–1922), Italian novelist and short-story writer. His works about Sicilian fishermen and peasants include the collection of short stories *Vita del Campi* (1880) and the novel *Maestro Don Gesualdo* (1889). *See also* HC2 p.99.

Vergennes, Charles Gravier, Comte de (1717–87), French Foreign Minister (1774–87). Encouraged by Louis XVI, he actively supported French intervention in the War of AMERICAN INDEPENDENCE. Vergennes advocated secret aid to the American colonies as early as 1775 and made an official alliance with them in 1778. In 1779 he brought Spain into the war against Britain.

Vergil, Polydore (c.1470–1555), Italian historian and humanist scholar. He settled in England in the early 16th century. His chief work was *Anglicae Historiae Libri XXVI* (Twenty-six Books of English History) (1546) which influenced HALL and HOLINSHED and, in 1582, became standard reading in English schools.

Verhaeren, Émile (1855–1916), Belgian poet and critic. He studied law, turned to literature and founded the literary review *La Semaine.* His works include the symbolistic *Les Soirs* (1888), *Les Forces Tumultueuses* (1902) and the war poems *Les Ailes Rouges de la Guerre* (1916).

Veríssimo, Erico Lopes (1905–), Brazilian novelist and critic. In his novels he explores the effects of the changing social structure in Brazil. Among his most acclaimed novels are *Crossroads* (1935), *The Rest is Silence* (1943) and *Time and the Wind* (1950).

Verity, Hedley (1905–43), British cricketer who played for England and Yorkshire. A left-arm spin bowler, he played in 40 Test matches between 1931 and 1939 (144 wickets, average 24.37).

Verlaine, Paul (1844–96), French poet. His early poetry, *Poèmes saturniens* (1866) and *Fêtes galantes* (1869), shows the influence of BAUDELAIRE. An intense relationship with RIMBAUD broke up his marriage and he was imprisoned for shooting Rimbaud in 1873. While in jail (1874–75), he wrote *Romance sans paroles* (1874), an early work of Symbolism. He returned to Catholicism and his later poetry, such as *Sagesse* (1880) and *Jadis et Naguère* (1884), deals with the conflict between the spiritual and the carnal.

Vermeer, Jan (1632–75), Dutch painter of genre subjects, portraits and townscapes. His earliest works (1655–60) are in a style close to early Italian BAROQUE and include *The Procuress, A Girl Asleep, Christ at the House of Mary and Martha* and *Little Street in Delft.* His next group of paintings mark his classic phase, in which he often depicts a single figure, treated in precise detail and bathed in natural lighting. Examples from this period include *Woman with a Red Hat* and *Head of a Young Girl. See also* HC1 pp.310–311.

Vermeyen, Jan Cornelisz (c.1500–59), Dutch painter, tapestry designer and engraver. He may have studied under

MABUSE, but his style is closer to that of his friend Jan van SCOREL. He became court painter to Margaret of Austria, Regent of the Netherlands, in 1525.

Vermiculite, clay mineral. Its flakes are light and are used in plaster and insulation, and as a packing material. The mineral is also used widely for conditioning soil and as a starting medium for seeds.

Vermont, state in NE USA, in New England, on the Canadian border. The N–S ranges of the Green Mountains dominate the terrain. Most of the W border of the state is formed by Lake Champlain. The region is heavily forested and arable land is limited. Dairy farming is by far the most important farming activity and the state is a major producer of maple sugar and syrup. Mineral resources include granite, slate, marble and asbestos. Manufacturing industries include pulp and paper, food processing, computer components and machine tools. Tourism is important to the state's economy. The principal cities are Montpelier, the state capital, and Burlington.

In 1609 Samuel de CHAMPLAIN discovered the lake that bears his name but the region was not settled permanently until 1724. Land grant disputes with the colonies of New Hampshire and New York persisted for many years. In 1777 Vermont declared its independence as the republic of New Connecticut and kept this unrecognized status until it was admitted to the Union in 1791. Area: 24,887sq km (9,609sq miles). Pop. (1975 est.) 471,000. *See also* MW p.175.

Vermouth, white wine-based drink which has been sweetened, infused with aromatic herbs and fortified by adding brandy. Vermouth is traditionally from France or Italy. It is used as an aperitif, and as an ingredient of many cocktails.

Verne, Jules (1828–1905), French writer of adventure stories. His works are characterized by fantastic settings; he is often considered a founder of science-fiction. His novels include *Journey to the Centre of the Earth* (1864), *From the Earth to the Moon* (1865), *Twenty Thousand Leagues Under the Sea* (1870) and *Around the World in Eighty Days* (1873). *See also* SU p.269.

Vernet, name of a notable family of French painters. Among the most important were Claude Joseph (1714–1789), a landscape and marine painter (*see also* HC2 p.118); Antoine Charles Horace (1758–1835), his son, who painted vast battle pieces of MARENGO and AUSTERLITZ, and Emile Jean Horace (1789–1863), one of the great French military painters, who decorated the Gallery of Battles at VERSAILLES.

Verneuil process, method for making gemstones artificially, particularly rubies and sapphires. Both of these gems are essentially forms of aluminium oxide (Al_2O_3) coloured with small amounts of other metal oxides. These are fused together in an oxyhydrogen flame, to form molten droplets of the coloured substance, which builds up gradually into a crystalline boule that is cut to make the gemstone.

Vernier, Pierre (*c.*1580–1637), French mathematician who invented the VERNIER SCALE. He spent the greater part of his life in the service of the King of Spain.

Vernier scale, on instruments for measuring length, small scale that moves alongside the main scale. Its function is to allow more precise measurement by interpolation, and for this purpose it is subdivided more finely than the main scale. Thus, if the main scale extends from 0 to 10 units, the vernier scale may extend from 0 to 1 unit to allow interpolation to one or two places of decimals. On MICROMETERS the vernier scale is on the stem, at right-angles to the spiral main scale.

Vernon, Edward (1684–1757), British admiral. He entered the navy in 1707 and rose steadily in rank. He advocated war with Spain (the War of JENKIN'S EAR) and won great popularity in 1739 by his capture of Portobello. Vernon's nickname, "Grog", was given to the drink – rum diluted with water – that he ordered to be served to his sailors to reduce the amount of alcohol they drank.

Verona, city in NE Italy, 147km (92 miles) E of Milan, on the River Adige; capital of Verona province. The city was captured by Rome in 89 BC. It was a free commune by the 12th century, and prospered under the Della Scala family in the 13th and 14th centuries; it was held by Austria from 1797 to 1866, when it joined Italy. Verona is the site of a Roman amphitheatre dating from the 1st century AD, the church of St Zeno Maggiore (5th–15th centuries), and Gothic tombs of the Scaligeri family. Industries: textiles, chemicals, paper, printing, wine. Pop. 271,079.

Verona, Congress of (1822), conference held in Italy and the last meeting of those powers involved in the Quadruple Alliance of 1814. The Congress decided to use a French army to end a rebellion in Spain against FERDINAND VII, and this decision provoked a rift between the British representative and the other powers. *See also* HC2 p.182.

Veronese, Paolo Caliari (*c.*1528–1588), Venetian painter and decorator, who painted large festive works of pageantry using Biblical, allegorical or historical themes. From 1562 he executed a number of large banquet scenes, including *Marriage at Cana*. His last great commission (from 1577) was the decoration of the duke's Palace in Venice. *See also* HC1 p.258.

Veronica, Saint, traditionally the woman who lent Jesus her head-cloth to wipe his face as he carried the cross. When he returned it, the cloth was supposed to have an imprint of his face and was later thought to perform miracles.

Veronica, or speedwell, widely distributed genus of annual and perennial plants of the figwort family. The small flowers are white, blue or pink. Height: 7.5–153cm (3in–5ft). Family Scrophulariaceae; genus includes about 250 species.

Verrazano, Giovanni da (*c.*1485–*c.*1528), Italian explorer. He explored the Atlantic coast of N America in 1524 in the service of France and discovered New York and Narragansett bays.

Verrio, Antonio (*c.*1639–1707), Italian decorative painter who went to England in 1671. He worked at Windsor and Whitehall, and succeeded Sir Peter LELY as First Painter to the King in 1684. *See also* HC1 p.297, 297.

Verrocchio, Andrea del (*c.*1435–88), Florentine sculptor, painter and goldsmith, whose real name was Andrea di Michele di Francesco Cioni. He is believed to have been a pupil of DONATELLO and put great emphasis on surface finish and elaborate detail. In 1472 he designed the coffins of Piero and Giovanni de' Medici in San Lorenzo, which show skilled metalworking. Among other works are a bronze *David* (pre-1476) and *A Boy with a Dolphin* (*c.*1480), both in Florence and the equestrian statue of Bartolomeo Colconi (*c.*1480), in Venice. *See also* HC1 pp.246, 255, 255.

Verruca, medical term for a wart, commonly on the sole of the foot, which is painful because it is forced to grow inwards. It also refers to tuberculous warts on the hands. Caused by a virus infection of the skin, local epidemics of verrucas can occur, particularly where many people walk barefoot together. Verrucas may be cut out or treated by cauterization or by gradual scraping after they have been softened by formalin.

Versailles, city in N France, 16km (10 miles) WSW of Paris; capital of Yvelines département. Originally just a village, it grew in size and fame in the late 17th century when Louis XIV built the celebrated palace. The palace contains the royal apartments, the famous Hall of Mirrors and a museum; magnificent gardens, a lake, and several outlying buildings, including the Grand and Petit Trianon palaces, complete the estate. The palace was the German headquarters during the Franco-Prussian war (1870–71) and the scene of the signing of the Treaty of Versailles (1919). Pop. (1968) 90,829. *See also* HC1 pp.309, 314.

Versailles, Treaty of (1783). *See* PEACE OF PARIS.

Versailles, Treaty of (1919), chief among the five peace treaties that formally terminated WWI. The leading figures at the conference were Woodrow WILSON for the USA, David LLOYD GEORGE for Britain, Georges CLEMENCEAU for France and Emanuele ORLANDO for Italy. The treaty represented a compromise between Wilson's FOURTEEN POINTS and the demands of Britain, France, Italy and the other Allies for reparations from Germany. It demilitarized Germany and created a triple defence alliance against her. Former German colonies were made mandates of the LEAGUE OF NATIONS, which was created by the treaty; Poland received a corridor to the Baltic; and Yugoslavia was awarded the port of Rijeka (Fiume). The treaty was never ratified by the USA because the high reparations contradicted Wilson's programme. *See also* HC2 pp.192–193.

Vers Libre, or "free verse", term used to describe poetry in which the restrictions of metre and rhyme are relaxed. The term originated in 19th-century France when RIMBAUD and Laforgue attempted to free French verse from the stranglehold of the alexandrine measure. Modern development owes much to Walt WHITMAN, Ezra POUND and T. S. ELIOT.

Vertebra, one of the bones making up the spine (backbone), or vertebral column. Each vertebra is composed of a large solid body from the top of which wing-like processes project to either side; a central opening (vertebral foramen) through which the spinal cord passes below which is often another opening enclosing blood vessels (haemal arch). Further projections extend from the vertebral foramen and haemal arch. *See also* MS p.57.

Vertebral column. *See* SPINE.

Vertebrate, animal with individual discs of bone or cartilage called VERTEBRAE surrounding or replacing the embryonic NOTOCHORD to form a jointed backbone. The head-end of the nerve cord is enlarged into a brain, and the attendant sense organs of smell, taste, sight, hearing and balance are correspondingly complex. The principal division is between the aquatic, fish-like forms (several extinct and living classes) and the partly (AMPHIBIANS) and wholly (REPTILES, BIRDS, MAMMALS) land-adapted forms. Four-legged vertebrates are called tetrapods, and those with fluid-filled embryonic membranes that permit development on land are known as amniotes. Birds and mammals are the only warm-blooded vertebrates with circulatory, respiratory and excretory systems which allow for constant high body temperatures. Phylum Chordata; subphylum Vertebrata. *See also* CHORDATE; FISH; REPTILE; NW pp.74, 75, 122–123.

Vertex (pl. vertices), point at which two sides of a triangle or other polygon intersect or at which three or more sides of a pyramid or other polyhedron intersect. It is also the point of a cone.

Vertigo, disruption of the body's sense of balance or equilibrium. It may be produced by rapid body rotation, some diseases, a brain disorder or reduced flow of blood to the brain due to altitude. It may be accompanied by nausea.

Vertov, Dziga (1896–1954), Soviet film director who specialized in documentaries. A newsreel cameraman, he founded the Kinoki (film-eye) group in 1918 to enhance realism by using the camera as a human eye. He also experimented with crosscutting, camera angles, slow motion and enlarged close ups. His films include *A Sixth of the World* (1927), *Symphony of the Don Basin* (1931) and *Lullaby* (1937). *See also* HC2 p.276.

Vertue, George (1684–1756), British antiquary, collector and engraver. He compiled voluminous notes for a history of the arts which were bought by Horace WALPOLE, and used as the basis for his *Anecdotes of Painting in England* (1761–71).

Vertumnus, Roman god associated with the ripening of fruit and hence the seasons of the year. At his festival on 13 August, timed to coincide with the ripening of fruit, people made offerings of the first fruits of the harvest.

Verona's Roman amphitheatre is only one of many interesting historic buildings.

Paolo Veronese's *Esther and Ahasuerus*, a detail showing this biblical encounter.

Andrea del Verrochio's *Head of a Lady* shows the technique he taught da Vinci.

Versailles; the Chapel and Place d'Armes in the huge palace built for Louis XIV.

Verulamium

Amerigo Vespucci gave his Christian name to America, the coast of which he explored.

Vesuvius; the greatest destruction in recent times was during the 1906 eruption.

Viaducts were first built by the Romans and many fine examples survive.

Vichy; Pétain's government had its headquarters here at the Pavillon de Sévigné.

Verulamium, Romano-British town built to the w of the modern city of St Albans, Hertfordshire. It is thought to have been awarded the status of *municipium* (its inhabitants having the same rights as Roman citizens) in the 1st century AD, soon after the Roman conquest. It was then destroyed by BOADICEA but soon rebuilt. Excavations have revealed a theatre dating from the 2nd century AD, a market hall and a number of mosaics and wall paintings. *See also* HC1 pp.100–101, 104–105.

Vervet monkey. *See* GUENON.

Verviers, city in E Belgium on the River Vesdre. An important weaving centre since the 15th century, it remains the focus of Belguim's woollen trade. Pop. (1971 est.) 33,600.

Verwoerd, Henrik Frensch (1901–66), South African Prime Minister. He held a number of important posts in the Nationalist Party and became Prime Minister in 1958. An ardent segregationist, he led South Africa out of the British Commonwealth in 1961. He was assassinated in 1966. *See also* HC2 p.252.

Vesalius, Andreas (1514–64), Flemish anatomist. His *De Corporis Humani Fabrica,* stating the results of his study and dissection of the human body, marked the beginning of modern anatomy. *See also* HC1 pp.302, 303.

Vespasian (AD 9–79), Roman emperor (*r.*69–79), founder of the Flavian dynasty. He commanded the army in Germany (42) and England (43), served as governor of Judaea under NERO and conducted the war against the Jews from 40 to 68. His soldiers proclaimed him emperor when Nero died. He instituted a period of peace, improved Rome's financial state and started the building and inspired the Romanization of England. *See also* HC1 pp.98, *107.*

Vespers, evening office of the Western Church. It consists of psalms, a reading from the Bible, the Magnificat, a hymn and a collect. Celebrated in the late afternoon, it is the basis of the Anglican evensong service.

Vespucci, Amerigo (1454–1512), Italian navigator and explorer. He was the first to consider America to be a separate continent, which was named after him by Martin Waldseemüller. Vespucci made several expeditions along the coast of Central and South America (1497–1503) in the service of Spain. He developed a method for computing nearly exact longitude. *See also* HC1 p.234, *234.*

Vesta, in Roman religion, the goddess of fire and purity, supreme in the conduct of religious ceremonies. Her priestesses were the VESTAL VIRGINS. Vesta was the guardian of the hearth and patron goddess of bakers. Her festival, the Vestalia, was held on 7–15 June. *See also* HC1 p.93.

Vesta, brightest ASTEROID in the SOLAR SYSTEM, and the only one visible to the naked eye. Its mean distance from the Sun is 352,800,000km (219,130,000 miles) and it has a diameter of approx. 590km (360 miles). *See also* SU pp.204–205.

Vestal virgins, in ancient Rome, the priestesses of the cult of VESTA, who tended the sacred fire in the Temple of Vesta and officiated at ceremonies in the goddess' honour. The Vestals remained in the service of the temple for up to 30 years under vows of absolute chastity, violation of which was punishable by burial alive. They enjoyed great prestige in the social order of Rome. *See also* HESTIA; HC1 p.93, *93.*

Vestigial structure, in biology, an organ or limb that is deformed or degenerate in appearance, and no longer has any recognizable function. Human vestigial structures include the tonsils and the appendix.

Vestris family, family of Italian-French dancers. Gaetan Vestris (1729–1808) studied in Paris and became ballet master to Louis XVI. He introduced the use of facial expressions in mime. His son Auguste (1760–1842) was the leading dancer with the Paris Opéra for many years. *See also* HC2 p.39.

Vesuvianite. *See* IDOCRASE.

Vesuvius, Mount (Vesuvio), active volcano in s Italy, on the Bay of Naples.

There are several villages and numerous vineyards on the fertile lower slopes. The earliest recorded eruption was in AD 79 when Pompeii and Herculaneum were destroyed. It is the site of several seismoloical stations. The height of the volcano changes with each eruption; in 1970 it was 1,280m (4,198ft). *See also* HC1 pp.94–95.

Veterinary medicine, medical science that deals with diseases of animals. It was practised by the Babylonians and Egyptians some 4,000 years ago. Late in the 18th century, schools of veterinary medicine were established in Europe. Today's veterinary surgeons spend most of their time treating household pets and farm animals. Many animal diseases, such as rabies, tuberculosis, psittacosis (parrot fever), tularaemia (rabbit fever) and brucellosis, can be transmitted to human beings, and in these areas a veterinarian's work becomes especially important.

Veto, power to reject. The royal veto in Britain, one of the Crown's PREROGATIVE powers, has not been used to reject parliamentary bills since Queen ANNE vetoed the Scottish militia bill in 1708.

Vevers, Henry Gwynne (1916–), British zoologist. He was appointed Curator of Aquaria and Assistant Director of Science, Zoological Society of London in 1955. His publications include *Nature of Animal Colours* (1960, with H. M. Fox) and *London's Zoo* (1976).

VHF (very high frequency), range or band of radio waves of frequencies 30 to 300MHz and of wavelength of 1–10m. This band is employed in radio broadcasting to provide high-quality reception. *See also* MM pp.228–229.

Viaduct, bridge (usually resting on several narrow, concrete or masonry arches) with high supporting piers, which carries roads or railways. Steel bridges comprising short spans carried on high towers are also referred to as viaducts. *See also* AQUEDUCT; MM p.191.

Vian, Sir Philip Louis (1894–1968), British admiral. In WWII he commanded the 4th destroyer flotilla in the action against the German pocket battleship *Bismarck.* He also took part in the invasion of Sicily, in the Salermo landing and in the Normandy landing in 1944.

Viani, Alberto (1906–), Italian sculptor. He studied at the Academy of Fine Arts, Venice (1944–47), under Arturo Martini and became his assistant. In 1946 he joined the Fronte Nuovo dell' arte group. Influenced by ARP and Brancusi, he later developed his personal, more monumental, abstract form. His works include *Caryatid* (1952) and *Abstract Nude* (1949–53).

Vibraphone, musical instrument with metal bars of different lengths which are struck with sticks or mallets to produce various notes. The notes are amplified by resonance: tubes beneath the bars vibrate at the same frequency as the bar above and magnify the sound. *See also* SU p.83.

Viburnum, genus of flowering shrubs and small trees native to North America and Eurasia. All have small, fleshy fruits containing single flat seeds. The flowers are small, fragrant and clustered. Well-known species are the wayfaring tree, hobblebush, black haw, arrow-wood, highbush, cranberry and snowball. There are about 120 species. Family Caprifoliaceae.

Vicar, in the Church of England, a priest who is the head of a parish. In the Roman Catholic Church "vicar" is used in several senses. The pope is called Vicar of Christ, meaning representative of Christ. A vicar apostolic is a bishop representing the pope in a country that does not have an established branch of the Catholic Church.

Vicente, Gil (*c.*1470–*c.*1536), Portuguese playwright and humanist whose works reveal the influence of Renaissance Italy. His plays vividly portray Portuguese society and include *Auto da Mofina Mendes* (1534), *Comedia de Rubena* (1521) and *Auto da Lusitania* (1532). They were published in 1562 by his son and daughter.

Vicenza, industrial city of NE Italy, 64 km (40 miles) w of Venice. Founded as a Ligurian settlement, it became a member of the Lombard League in the 12th century. It was taken by Venice in 1404 and held by Austria from 1797 until 1866, when it was united with Italy. Industries include the manufacture of machinery and chemicals, and printing. Pop. (1971) 121,056.

Vice-president, deputy for a president. In the USA he may succeed to the presidency in the event of the death, resignation, physical disability or removal from office by impeachment of the president. The American vice-president belongs to the same political party as the president and is elected with the president to serve a four-year term of office.

Viceroy, governor of a country or province, ruling in place of the Sovereign. The office was used in the old Spanish monarchy and by the British in India from 1858 to 1947.

Viceroy butterfly, large BUTTERFLY that has orange-brown wings with black veins and borders. It is noted for its mimicry of the MONARCH BUTTERFLY. The viceroy can be distinguished from the monarch by its smaller size and a transverse black band on its hind wings. Species *Limenitis archippus. See also* NW p.117.

Vichy, town in central France on the River Allier, 320km (200 miles) SSE of Paris. The Romans called it Aquae Calidae ("warm springs") after its hot and cold alkaline springs. This water, now bottled commercially and exported to many parts of the world, is reputed to mitigate stomach and liver complaints. After the fall of France (July 1940) Vichy became the seat of government until the German occupation (Nov. 1942).

Vichy Government, regime set up in the town of Vichy to rule those parts of France and the French overseas empire not occupied by the Germans after the France-German armistice of 22 June 1940. Its prime minister was Marshal PÉTAIN, a hero of WWI, but its most important member was the deputy prime minister, Pierre Laval. From November 1942 German troops occupied all of France, and the Vichy regime lost all but the shadow of power. After the Allied invasions of 1944, the highest officials of the right-wing, authoritarian Vichy regime moved east with the Germans, and the Free French movement of De Gaulle became the dominant political organization. *See also* HC2 p.228, *228.*

Vickers, Jon (1926–), Canadian tenor. He became a principal tenor at COVENT GARDEN in 1957 and made his METROPOLITAN OPERA COMPANY debut in 1959. He is highly regarded as a Wagnerian tenor.

Vicksburg, Siege of (1863), major offensive by Union forces under Gen. Ulysses S. GRANT during the AMERICAN CIVIL WAR. The capture of Vicksburg meant the division of the Confederacy. Confederate troops and civilians suffered greatly to hold the city but it fell on 4 July 1863, after a campaign lasting more than a year.

Vico, Giambattista (1668–1744), Italian historian. In his *New Science* (1725; revised 1730 and 1744) he argued for a study of history as the development of human societies through various historical periods. He is sometimes considered the father of modern historiography.

Victor, name of three popes and two antipopes. Victor I (*r.c.*189–199) was an African. He threatened to excommunicate Roman Catholics of Asia Minor because of the unorthodoxy of the date on which they celebrated Easter. Following the intervention of St Iranaeus and others he relented and avoided a schism. Monarchianism arose during his pontificate. Victor II (*c.*1018–57; *r.*1055–57) executed the duties of his office along the reformed and uncorrupt lines established by his predecessor LEO IX. Victor III (1027–87; *r.*1086–87) campaigned successfully against the Saracens in N Africa and continued the ban on lay investiture, despite continual opposition from the antipope Clement III. Victor IV (*r.*1138) was an antipope chosen in opposition to INNOCENT II by Roger II of Sicily and the Pierleoni family, but was persuaded to submit by St BERNARD OF CLAIRVAUX. Victor V (also known as IV) (*r.*1159–64) was an antipope elected by Frederick I

Barbarossa to oppose ALEXANDER III. Europe was against the imperial control of the papacy and Victor received little support.

Victor Amadeus, name of three dukes of Savoy. Victor Amadeus I (1587–1637) became duke in 1630. Victor Amadeus II (1666–1732) was duke from 1675. He fought in the War of the SPANISH SUCCESSION on the HAPSBURG side after 1703 and was proclaimed King of Sicily in 1713. In 1720 he ceded Sicily to Austria and became King of Sardinia-Piedmont instead, but abdicated in 1730. Victor Amadeus III (1726–96) was Duke of Savoy and King (as Victor Amadeus II) of Sardinia (r. 1773–96). He lost Nice and Savoy to France and was forced to abdicate by NAPOLEON in 1796.

Victor Emmanuel, name of three Italian kings. Victor Emmanuel I (1759–1824) became Duke of Savoy and King of Sardinia in 1802. He joined the First Coalition against France (1792–97), and lost all his territories except Sardinia to NAPOLEON (1802–14). After his restoration, his reactionary policies led to his abdication in 1821. Victor Emmanuel II (1820–78) became King of Sardinia-Piedmont in 1849. He appointed Camillo Cavour to be chief minister in 1852, led the fight to expel Austria from Italy, and encouraged Giuseppe GARIBALDI to conquer Sicily and Naples. He was proclaimed the first King of Italy in 1861. He secCred Venetia as part of Italy in 1866, and the PAPAL STATES in 1870. Victor Emmanuel III (1869–1947) was King of Italy (r. 1900–46). He originally sought neutrality in WWI, and failed to oppose the rise of MUSSOLINI. In 1943 he had Mussolini arrested, but fled to s Italy when the Germans occupied Rome. He abdicated in 1946. *See also* HC2 pp.110, 111.

Victoria, in Roman religion, an ancient agricultural divinity who became identified with Roman military success. As the personification of victory, she was associated with MARS and JUPITER.

Victoria (1819–1901), Queen of Great Britain and Ireland (r. 1837–1901) and Empress of India (r. 1876–1901). She was the daughter of Edward, Duke of Kent (the fourth son of GEORGE III), and of Mary Louisa Victoria of Saxe-Coburg-Gotha, and was the successor to WILLIAM IV. Her reign was the longest in English history. On her accession to the throne she allowed Lord MELBOURNE, the Whig Prime Minister, to become her adviser. She married her first cousin, Prince ALBERT of Saxe-Coburg-Gotha in 1840, and had nine children. He taught her devotion to, and neutrality in, political work. On Albert's death in 1861, she retired to private life at BALMORAL for three years, during which time her popularity declined; but with Benjamin DISRAELI as Prime Minister after 1868 she took a new interest in politics. In her later years her dislike of William GLADSTONE was to influence affairs, especially his policy of Irish Home Rule. Her reign was important at home in winning respect for the monarchy in an increasingly parliamentary age and abroad for extending British influence. *See also* HC2 pp.112–113, *141*.

Victoria, Guadalupe (1789–1843), Mexican soldier and political leader, original name Manuel Félix Fernández. He served as a revolutionary soldier after 1810 and as first President of the Mexican Republic from 1824 to 1829. As president, he organized the economy, established diplomatic relations with other countries and abolished slavery.

Victoria, Tomás Luis de (c. 1548– 1611), Spanish composer, member of PALESTRINA's school in Rome. His works, which include masses, motets, psalm settings and two passions, although influenced by his school, retain a unique Spanish style and made him one of the greatest composers of the 16th century.

Victoria, state in SE Australia bounded by the Indian Ocean, the Bass Strait and the Tasman Sea; the most densely populated state in Australia. The area is crossed by the Australian Alps and other ranges of the Eastern Highlands. Irrigation is used extensively to grow wheat, oats, barley,

fruit and vegetables; sheep and dairy cattle are also important. The principal industries are motor vehicles, textiles and food processing. Coal, natural gas and oil are the chief mineral resources. The main cities are Melbourne (the capital), Geelong, Ballarat and Bendigo.

Sheep farmers from Tasmania began the first permanent settlement in the 1830s. Known as Port Philip District, the region was part of the colony of NEW SOUTH WALES until 1851, when it became a separate colony. The population increased rapidly after 1851, when gold was discovered. Victoria became part of the Commonwealth of Australia in 1901. Area: 227,620sq km (87,813sq miles). Pop. 3,443,800. *See also* MW p.28.

Victoria, city in British Columbia, SW Canada, on SE Vancouver Island; capital, largest city and chief port of British Columbia. Founded in 1843, it developed during the gold rush in 1858 as a supply base for gold prospectors. The city has the University of Victoria (1902). Industries: timber, shipbuilding, fish processing, deep sea fishing, tourism. Pop. (1971) 61,761.

Victoria, capital of the British Crown colony of Hong Kong, situated in the NW of Hong Kong Island. Victoria has one of the finest deep water harbours in the world and is the largest trans-shipment centre in E Asia. The commercial centre of Hong Kong , it has many banking institutions. It is the seat of the University of Hong Kong (1911). Pop. (1971) 520, 932.

Victoria, capital of the Seychelles, a group of islands in the Indian Ocean. Situated on NE Mahé island, it has a deep-water harbour and is the only town of significant size in the group. Pop. (1971 est.) 13,622.

Victoria, or Victoria Nyanza, lake in E central Africa on the border of Tanzania and Uganda. The second-largest freshwater lake in the world, its coastline of more than 3,200km (2,000 miles) provides harbours for coastal towns. It is the chief reservoir of the River Nile. Area: 69,484sq km (26,828sq miles).

Victoria Cross, highest military decoration awarded to the Commonwealth armed services for acts of exceptional bravery. Instituted in 1856 by Queen Victoria, VCs consist of a bronze cross with a crimson ribbon. They were first struck from the metal of cannon captured at Sebastopol but since 1942 have been made from gunmetal.

Victoria de Durango. *See* DURANGO.

Victoria Falls, waterfalls on the Zambezi River at the border of Rhodesia (Zimbabwe) and Zambia. Formed by water erosion along a fracture in the Earth's CRUST, they are divided by several islets into a series of falls. They were discovered by the British explorer David LIVINGSTONE in 1855 who named them after Queen Victoria. Maximum depth: 108m (355ft); width: more than 1,700m (5,580ft).

Victorian Order, Royal, British order of chivalry instituted by Queen Victoria in 1896. It consists of a Grand Master, Chancellor, Secretary, Registrar, Chaplain and five classes: Knight Grand Cross, Knight Commander, Commander, and Member of the fourth and fifth classes.

Victory, HMS, British warship launched in 1765 as a 104-gun ship of the line, 57m (186ft) long and displacing 2,162 tonnes. In 1805 Nelson used it as his flagship. *Victory* is now kept as a museum at Portsmouth. *See* MM p.*173*.

Victory (Nike) of Samothrace (Winged Victory), heroic marble statue of the HELLENISTIC period, depicting a winged female figure swathed in gracefully draped robes. Supposedly carved by Pythecritos of Rhodes (early 2nd century BC), it is in the LOUVRE, Paris.

Victrola, trade mark and name for phonographs (gramophones) manufactured by the Victor Talking Machine Company (1906–29). The name became synonymous with the disc gramophone. *See also* MM p.232.

Vicuña, graceful, even-toed, hoofed South American mammal, humpless and smallest member of the CAMEL family. It lives near the snowline of the Andes Mountains. It resembles the LLAMA and its silky coat is tawny brown with a yellowish

bib under the neck. It can run at 48km/h (30mph). Height: 86cm (34in) at the shoulder; weight: 45kg (100lb). Family Camelidae; species *Vicugna vicugna*. *See also* GUANACO; NW pp.*163*, 251.

Vidal, Gore (1925–), US writer. His first novel, *Williwaw*, appeared in 1946. Other works include *In a Yellow Wood* (1947), *The City and the Pillar* (1948), *Myra Breckinridge* (1968) and its sequel *Myron* (1974). He wrote a trilogy about American political life which began with WASHINGTON DC (1967), continued with BURR (1973) and ended with *1876* (1976). Vidal is also the author of critical essays, plays and screenplays, including *Suddenly Last Summer* (1958).

Videla, Lieut-Gen. Jorge Rafael (1925–), President of Argentina from 1976. He had a military career, becoming Chief of the Army General Staff in 1973 and Commander-in-Chief in 1975. In 1976 he deposed President PERÓN.

Video recording, the principle of recording and reproducing sound and moving pictures, ie television programmes. It is essential in modern television because of the saving in cost and processing and editing time, two disadvantages of films. The video recorder developed from the magnetic tape recorder from which it differs significantly in two respects: video tape is wider to accommodate the picture signals; and the relative speed at which the tape passes the magnetic head is faster to deal with the larger amount of information necessary for recording and reproducing pictures. *See also* MM pp.234–235.

Videotape, magnetic tape that records television signals. The signals are passed through the recording head (an electromagnet) of a videotape recorder, so that they fix, by induction, a magnetic pattern on the tape. These patterns generate the original signals when the tape passes the replay head (also an electromagnet). *See also* MM pp.234–235.

Vienna (Wien), capital of Austria, on the River Danube. Vienna became an important town under the Romans but after their withdrawal in the 5th century it fell to a succession of invaders from E Europe. The first HAPSBURG ruler was instated in 1276 and the city was the seat of the HOLY ROMAN EMPIRE from 1558 to 1806. The city was occupied (1806 and 1809) by the French during the NAPOLEONIC WARS, and was later chosen as the site of the Congress of VIENNA. As the capital of the Austro-Hungarian Empire, it became the cultural and social centre of 19th-century Europe under the Emperor FRANCIS JOSEPH I, but suffered an economic and political collapse following the defeat of the Central Powers in WWI. In 1938 HITLER's armies invaded Vienna after Germany had annexed Austria and after WWII it was occupied (1945–55) by joint Soviet-Western forces. Vienna has many famous historical buildings, including the 12th-century St Stephen's Cathedral, the Schönbrunn (royal summer palace) and the Hofburg (a former residence of the Hapsburgs). Its industries include chemicals, textiles, furniture and clothing. Pop. (1971) 1,614,841. *See also* HC2 pp.82–83, *85*.

Vienna, Congress of (1814–15), assembly of European leaders held to reorganize Europe after the defeat of NAPOLEON. The essential provisions of their treaty lasted for almost a century. The victors – Britain, Russia, Prussia and Austria – together with defeated France, balanced power and territory in Europe with the aim of preventing war, but often at the expense of the national hopes of the peoples affected. The main territorial adjustments were: France lost little, except the Rhineland to Prussia, which also gained some of Saxony; the defunct Holy Roman Empire was superseded by the German Confederation; and Austria received territories in N Italy. Poland was divided between Austria, Russia and Prussia. *See also* HC2 pp.82–83, 108.

Vienna Boys' Choir, music group consisting of 22 boys between the ages of eight and fourteen years. The oldest Viennese musical institution, it was founded in 1498 as the choir of the court chapel.

Victoria's reign saw the height of imperial power and great advances in science.

Victoria's capital city is Melbourne, part of which is shown in this view.

HMS Victory was Nelson's flagship at Trafalgar, a decisive battle in naval history.

Vienna's Opera House was designed to resemble an earlier one destroyed in WWII.

Vienna circle

Viet Cong prisoners after they had appeared at a press conference in 1968.

Marie Vigée-Lebrun; one of her noted portraits was *The Artist and her Daughter.*

De Vigny's *Military Service and Greatness* (1835) portrayed the boredom of service life.

Pancho Villa; the USA denounced Villa when his brutalities became known.

Vienna circle, group of philosophers and mathematicians who met in Vienna from 1922 to 1936 to discuss scientific language and philosophy. The resulting philosophical movement became known as logical positivism. Members of the circle included Moritz Schlick, its leader, and Rudolf Carnan.

Vienna Sezession (1897), radical movement of young Austrian artists who organized their own exhibitions in defiance of the traditional organizations and aligned themselves with their progressive European contemporaries. The first president was Gustav KLIMT and other members were Oskar KOKOSCHKA and Egon SCHIELE. Together with the Sezessionists in Munich and Berlin, these artists were the instigators of the Jugendstil movement.

Vientiane (Viangenan), capital of Laos, in N central Laos on the Mekong River. It was the capital of the Lao kingdom from 1707 to 1828, when it was destroyed by the Siamese. The city passed to the French in 1893 and in 1899 became the capital of the French protectorate. Industries: textiles, hides, wood products. Pop. (1970 est.) 150,000.

Vieta, Franciscus (1540–1603), French mathematician. He devised methods for solving algebraic equations to the fourth degree and his *Isagoge in artem analyticam* (1591) is probably the earliest work on symbolic algebra.

Viet Cong nickname for the guerilla army which fought against the South Vietnamese government and its allies during the 1960s and 1970s. It was formed as an armed opposition to the dictatorial regime of President Diem and from 1960 was directed by the National Liberation Front (NLF). In spite of massive US aid to the South Vietnamese government, the Viet Cong (closely linked to the North Vietnamese government of HO CHI MINH) was remarkably successful during the VIETNAM WAR, and proved its military capacity during the Tet offensive of 1968. American forces were then progressively withdrawn, and a peace was signed at Paris in 1973. In 1975, however, this broke down, and together with North Vietnamese regulars the Viet Cong took over South Vietnam. *See also* HC2 pp.292–293, *292–293.*

Viet Minh, Vietnam League for Independence, founded by HO CHI MINH (then an exile in China) in 1941. From 1943 it conducted operations against the Japanese but in 1945, when the French refused to recognize a Viet Minh provisional government, it began a guerrilla war against the French administration. Brilliantly led by General Giap, Viet Minh forces eventually defeated the French at Dien Bien Phu in 1954. *See also* HC2 pp.292–293.

Vietnam, independent nation in SE Asia. Under foreign influence for centuries, it was divided into two countries in 1954 and reunited in 1975 after the VIETNAM WAR. The northern region is covered by thick jungle and the climate causes frequent flooding. Fertile soil produced good harvests, however, until the escalation of the war in the 1960s, when both agriculture and industry came to a standstill. The capital is Hanoi. Area: 338,392sq km (130,653sq miles). Pop. (1975 est.) 45,211,000. *See* MW p.181.

Vietnamese, the national language of VIETNAM, spoken by about 38 million inhabitants, most of the country's people. It is part of the Muong branch of the Mon-Khmer subfamily of Asiatic languages and derives much of the vocabulary from Chinese.

Vietnam War, military conflict between South Vietnamese government forces, aided by the USA, and the Communists (Viet Cong), supported by North Vietnam. A civil war aimed at reunification of Vietnam following its partition in 1954 into North (Communist) and South (non-Communist) rapidly developed into a major conflict. In the early 1960s the USA became militarily involved and by 1968 there were more than 500,000 US troops in South Vietnam. North Vietnam received aid from the USSR and China. Despite US intervention, the South Vietnamese were unable to defeat the Viet Cong and North

Vietnamese. By the late 1960s pressure developed within the USA for the withdrawal of US troops. Efforts towards a negotiated peace were begun in 1969 and a peace treaty was signed in January 1973 by South Vietnam, the USA, North Vietnam and the National Liberation Front (the Communist provisional revolutionary government in South Vietnam). It provided for an end to hostilities and for the withdrawal of US forces. Fighting between South Vietnamese and Communists continued, however, and in April 1975 the South Vietnamese government fell to the Communists. *See also* HC2 pp.*249, 292–293, 292–293.*

Vigée-Lebrun, Marie Louise Elisabeth (1755–1842), French painter. Her portrait of MARIE ANTOINETTE (1779) led to a lasting friendship with the queen and to her appointment as Painter to the Queen. She also kept a distinguished salon. Her portraits include *Lady Hamilton as Sybil* and *The Artist and her Daughter.*

Vigilius, Roman Catholic pope (*c.* 537–555). He was elected by the persuasive powers of Emperor JUSTINIAN and, with the Empress Theodora, planned the deposition of Pope Silverius, his predecessor. Involved in the MONOPHY SITE controversy, he became a virtual prisoner in Constantinople for eight years and died on the way back to Rome.

Vigneaud, Vincent Du (1901–), US biochemist was was awarded the 1955 Nobel Prize in chemistry for determining the structure and synthesis of VASOPRESSIN and oxytocin, two polypeptide hormones. In the 1930s he discovered the metabolic processes by which a methyl group ($-CH_3$) is moved from one compound to another, called transmethylation. He also deduced the two-ring structure of biotin in 1943 and studied INSULIN, PENICILLIN synthesis and AMINO ACID metabolism.

Vignette, small decorative motif found in books of the 18th and 19th centuries. The term now also applies to vignetted photographs and to any drawn image that fades away towards the edges.

Vignola, Giacomo Barozzi da (1507–73), Italian MANNERIST architect, the leading architect in Rome after the death of MICHELANGELO, whom he succeeded (1567–73) as architect of St Peter's. His Gesú, Rome (1568), mother church of the Jesuits, with its revolutionary design uniting clergy and congregation more closely, has been more widely copied than any other church. His other major works include the Palazzo Farnese, Caprarola (1559), a magnificent pentagonal palace; and the Tempieto di San Andrea (1550) and Sant' Anna dei Palafrenieri (1473), both in Rome.

Vignon, Claude (1593–1670), French painter who worked for LOUIS XIII and RICHELIEU. From 1618 to 1624 he worked in Rome where he was influenced by the followers of CARAVAGGIO, especially Simon VOUET. He later developed a style similar to that of REMBRANDT, who was his friend.

Vigny, Alfred de (1797–1863), French romantic poet, dramatist and novelist. A pessimist, he emphasized the lonely struggle of the individual in a hostile universe. He never expressed personal emotion but presented his ideas through general symbols. His best poems are found in *Poèmes antiques et modernes* (1826), and his novels include *Cinq-Mars* (1826).

Vijayanagar, ancient city of SE India. In the 14th–16th centuries its empire controlled most of S India, and the city itself was an important Hindu cultural centre. Vijayanagar was destroyed by Muslim forces in 1565.

Vikings, Scandinavian marauders and invaders also called Norsemen and in E Europe Varangians. They were skilled shipbuilders and sailors, and raided areas of continental Europe and the British Isles from the 9th to the 11th centuries. In some areas they established colonies, exerting considerable influence on the history of Europe by encouraging the growth of the FEUDAL SYSTEM. *See also* HC1 pp.141, 154, *154,* 156–157, *156.*

Viking space mission, exploration of Mars by two US spacecraft which landed there in the summer of 1976. The mission included

experiments to search for evidence of life. Viking 1 landed on the Chryse basin on 20 July; Viking 2 landed on 3 September on Utopia Planitia. Both spacecraft sent back spectacular colour photographs showing reddish soil strewn with rock. They dug into the soil and collected samples which were analysed in their biology laboratories for signs of life. No organic compounds were detected. Viking detected silicon, iron, calcium, aluminium and titanium as major elements in the Martian soil. *See also* SU pp.196–197.

Vila, capital of the New Hebrides, in the sw Pacific Ocean, on sw Efate island. Vila is the major port and commercial centre of the island group. Pop. (1972) 12,536.

Villa, Pancho (1877–1923), Mexican revolutionary leader. He began as an outlaw and later joined the revolutionary forces of Francisco Madero (1909). He sided with Venustiano Carranza for some time, but later supported Emiliano Zapata. Angered by US recognition of Carranza's government, Villa murdered US citizens in N Mexico and New Mexico. President Woodrow WILSON sent troops to capture him but they failed and in 1920 he was granted a pardon in return for agreeing to retire from politics. He was assassinated three years later. *See also* HC2 pp.256, 257.

Villa, large country house of the Roman and post-Roman period. The private residences of important citizens, villas had spacious reception rooms, often with mosaic floors and sometimes even underfloor heating.

Villafranca, Treaty of (1859), armistice concluded between France and Austria after the French victory at Solferino in the war to drive Austria out of northern Italy. Austria was allowed to retain Venetia, but Lombardy was given up to France so that it might then be ceded by France to Piedmont. The treaty was broken in 1860 when Piedmont annexed the central duchies of Tuscany and Modena. *See also* HC2 p.110.

Village Romeo and Juliet, A, opera in prologue and three acts by Frederick DELIUS; libretto by the composer, based on the story by Gottfried KELLER. It was first produced in Berlin in 1907, and in London in 1910.

Villa-Lobos, Heitor (1887–1959), Brazilian composer and conductor. His work was influenced by South American Indian folk music and the music of DEBUSSY. His range of works includes operas, ballets, symphonies, religious and chamber works. The nine *Bachianas Brasileiras* are a type of Brazilian transcription of the music of BACH; other Brazilian works are fourteen *choros* (serenades) for various instrumental combinations.

Villanueva, Carlos Raúl (1900–), Venezuelan architect. He introduced the European style of modern architecture to his country, particularly in his School of Architecture (1957) and the Auditorium at University City, Caracas (1950–51). His work exploits exposed concrete and is often brightly decorated.

Villard, Paul (1860–1934), French physicist . In 1900, while studying radiation emitted from uranium, he found that some of it was more penetrating than either alpha or beta rays and, unlike these two, was not deflected by a magnetic field. This new form of radiation was similar to X-rays and later came to be called GAMMA RADIATION.

Villard de Honnecourt (*fl.* 13th century), French architect and master mason, known only through his sketchbook, which is now in the Bibliothèque Nationale, Paris. It contains plans of buildings then being constructed, such as REIMS and Leon Cathedrals, projects for his own works, some elementary geometry and notes for younger sculptors.

Villars, Claude-Louis-Hector, duc de (1653–1734), French general. In the War of the SPANISH SUCCESSION (1701–14), he won victories at Friedlingen (1702) and Hochstadt (1703). *See also* HC1 p.*312.*

Villeda Morales, Ramón (1908–71), President of Honduras (1957–63). A civilian and a leader of the Liberal Party, he brought Honduras into the Central

American Common Market in 1960. He was deposed by a military coup.

Villehardouin, Geoffroi de (*c.*1150–*c.*1213), French historian. He was one of the leaders of the fourth crusade, which was responsible for the conquest of Constantinople and the establishment of the Latin Empire there. He received a title and lands in Thrace for his services.

Villein, medieval serf, an unfree peasant in the European FEUDAL SYSTEM. He was the chattel property of his lord and could be bought and sold. The labour shortage caused by the BLACK DEATH resulted in the gradual replacement of villeins by COPY-HOLDERS. Villeinage in England had disappeared by the end of the 15th century.

Villette (1853), novel by Charlotte BRONTË. Taking the common theme of a governess's life, the story reflects Charlotte Brontë's own feelings of desperate love for M. Heger, who is given the character of Paul Emanuel in the book.

Villi, in anatomy, small, finger-like projections of a mucous membrane such as that which lines the inner walls of the small intestine. In digestion intestinal villi absorb most of the products of food broken down in the stomach and duodenum. Each villus contains blood capillaries, which receive PROTEINS and CARBO-HYDRATES, and a lacteal or lymph capillary, which receives FATS. *See also* MS pp.68–69, *69.*

Villiers, Barbara. *See* CLEVELAND, DUCHESS OF.

Villiers, George. *See* BUCKINGHAM, GEORGE VILLIERS, DUKE OF.

Villiers de L'isle-Adam, Philippe Auguste Mathias (1838–89), French poet and dramatist. His short stories are collected in *Contes Cruels* (1883; tr. *Cruel Tales*, 1963), and perhaps his finest achievement is the visionary drama *Axel* (1886).

Villon, François (1430–*c.*1463), French lyric poet now considered the greatest of his day. Born François de Montcorbier or François des Loges, he took the name of his tutor, Guillaume de Villon. He took the degree of master at the Sorbonne in 1452 but led a troubled life after killing a priest in 1455. He wrote his famous *Ballad of a Hanged Man* while awaiting execution in 1462 (the sentence was later commuted to banishment).

Villon, Jacques (1875–1963), real name Gaston Duchamp, French painter and etcher, half-brother of the artists Marcel Duchamp and Raymond Duchamp-Villon. He was first influenced by IMPRES-SIONISM and FAUVISM and later by CUBISM, which is reflected in the style of geometric abstraction which he developed about 1919.

Villot, Cardinal Jean (1905–   ), Roman Catholic churchman who was appointed Secretary of State of the VATICAN CITY and Prefect of the Council for the Public Affairs of the Church in 1959.

Vimy Ridge, Battle of (1917), WWI campaign, fought in France. The Allies launched the attack on the German-held position in order to draw German reserves away from the main action on the river Aisne. On the Allied side the battle was fought almost entirely by the Canadian corps which, in its finest achievement of the war, succeeded in capturing the position in four days (14–17 April).

Vincennes, suburb of Paris, 8km (5 miles) E of Paris, on a wooded plateau in the Val-de-Marne département. Bois de Vincennes in the S, between Bois and the town, is a famous château (begun in the 12th century) and former royal residence. It was a state prison in the reign of Louis XIV. Pop. (1975) 49,143.

Vincent de Paul, Saint (*c.*1580–1660), French priest called the Universal Patron of Works of Charity. As a young man he was captured by Barbary pirates and spent two years as a slave in Tunisia. After escaping back to France, he began a mission to the peasantry, founding the Congregation of the Mission (or Lazarists). Showing great organizing ability, he later helped to found the Daughters of Charity in 1633 to minister to the sick, the old and orphans.

Vincent of Beauvais (*c.*1190–1264), Dominican monk, regarded as a major

Encyclopaedist. Becoming a monk before 1220, he was patronized by Louis IX and compiled three parts of the *Speculum majus,* purportedly a summary of all contemporary knowledge.

Vincent of Lérins, Saint (*d.c.* 450), theologian. Born in Gaul, he became a priest in the Abbey of Lérins where he was noted for his learning. Some scholars take his *Commonitoria* to be in opposition to the theology of St Augustine.

Vinci, Leonardo da. *See* LEONARDO DA VINCI.

Vindication of the Rights of Women (1792), essay by the British writer Mary WOLLSTONECRAFT. Written in the early days of the French Revolution, it argued for the political and social equality of women.

Vine, plant with a long, thin stem that climbs rocks, other plants and supports. Vines develop modifications such as tendrils, disc-like holdfasts, adventitious roots and runners. Examples are tropical LIANAS, wild GRAPES, and MORNING GLORY. *See also* PE pp.*179,* 200–201, *203.*

Vinegar, acid condiment made by fermenting malt to make alcohol, as in BREW-ING, then further fermenting the alcohol to ACETIC ACID. The alcohol fermentation is carried out by brewer's yeast (*Saccharomyces*) and the acid fermentation carried out by bacteria of the genus *Acetobacter.* The French "vinaigre" means sour wine.

Vinegar Hill, Battle of (1798), episode in the Irish Rebellion of 1798. The rebels who had captured Wexford made their headquarters at Vinegar Hill, Enniscorthy. It was attacked on 21 June by a powerful British force led by General Lake. The British victory was followed by the taking of Wexford Town and the collapse of the rebellion.

"Vingt, Les" (The Twenty or Société des Vingt), group of painters who exhibited in Brussels between 1891 and 1893. They were brought together by a common interest in SYMBOLISM. Members of the group included Ensor, SEURAT, GAUGUIN and Van GOGH. Among their significant shows was an exhibition of the works of Henri de TOULOUSE-LAUTREC in 1892.

Vinland (Wineland), area in North America discovered *c* AD 1,000 by Leif ERICSSON. Whether his expedition was planned or accidental is unknown, but the probability is that as he sailed from Norway to Greenland his ship was blown off course to the s and w. What is certain is that he discovered a land containing grapes and self-sown wheat, which he called Vinland. His original landing was probably in Nova Scotia.

Vinland Map, world map showing Vinland. Allegedly drawn *c.*1440, a facsimile of the map was published in 1965 by YALE UNIVERSITY. The existence of the map was thought by some to prove that North America had been discovered before the voyages of Christopher COLUMBUS. The authenticity of the map, based on the script, paper and watermarks, was a subject of controversy until 1973, when it was judged fraudulent after an ink analysis revealed pigments known only after 1920.

Vintage car racing. *See* MOTOR RACING.

Vinyl chloride, gas with an ether-like odour (formula CH_2CHCl) manufactured by the chlorination of ethylene. It polymerizes to form polyvinyl chloride (PVC, the plastic) and is widely used in this form. Properties: m.p. $-153.8°C$ ($-244.8°F$); b.p. $-13.4°C$ ($7.9°F$).

Vinyl plastics, any of several substances composed of large molecules linked by the POLYMERIZATION of vinyl compounds. They are used in plastic film, upholstery, floor tiles, toys, buttons and in fibres. The most common is PVC. *See also* MM p.54.

Viol, fretted stringed musical instrument, played with a bow; it was one of the most important instruments from the 15th to early 18th centuries. It is held on or between the knees and, in its most usual shape, has sloping shoulders and a flat back. The six strings are tuned in fourths, in the same manner as the LUTE. There were four principal sizes, and often a double bass viol, the violone, was added. A possible derivative, the modern DOUBLE BASS, is perhaps the only type of viol to

survive; it shows its ancestry by being tuned in fourths (unlike members of the violin family, which are tuned in fifths). *See also* HC2 p.104.

Viola, stringed musical instrument of the VIOLIN family. It is 5cm (2in) longer than the violin and its four strings are tuned a fifth lower. It is the tenor member of a string quartet. *See also* HC2 pp.104–105.

Viola. *See* PANSY.

Violet, any of about 850 species of herbs, shrubs and small trees of the family Violaceae, found in all parts of the world. All cultivated species are of the genus *Viola.* Violets may be annual or perennial, with 5-petalled flowers that grow singly on stalks; usually blue, violet, lilac, yellow, or white. *See also* NW p.*209.*

Violin, stringed musical instrument. It is thought to have derived from the *lira da braccio,* a Renaissance bowed instrument, and the REBEC. It was perfected in Italy by the AMATI, STRADIVARI and GUARNIERI families between 1650 and 1740. The body is assembled from curved wooden panels, the front pierced by two *f*-shaped sound holes. Four taut strings are played by drawing a horsehair bow across them, or sometimes by plucking them with the fingers (pizzicato). *See also* HC2 pp.104–105; SU pp.*82,* 83.

Violoncello. *See* CELLO.

Viollet-Le-Duc, Eugene (1814–79), French architect who was one of the leaders of the GOTHIC REVIVAL in France and well known for his restoration work on historic French buildings. Among his books on the theory of architecture is *Entretiens sur l'architecture* (1858–72). *See also* HC2 pp.*122, 122,* 178.

Viotti, Giovanni Battista (1755–1824), Italian violinist and composer. In 1784 he became court musician to MARIE ANTOINETTE and after the French Revolution worked in London. He returned to Paris in 1818. As a virtuoso and composer, he made a significant contribution to the violin concerto.

Viper, any of 150 species of poisonous snakes characterized by a pair of long, hollow, venom-injecting fangs in the front of the upper jaw which can be folded back when not in use. The pit-vipers (of which the rattlesnake is an example) have a heat-sensitive organ between the eyes. The common adder (*Vipera berus*) of Europe and E Asia has a dark, zigzag band along its back. Length: to 3m (10ft). Family Viperidae.

Viper's bugloss. *See* BUGLOSS.

Virchow, Rudolf (1821–1902), German pathologist. His discovery that all cells arise from other cells completed the formulation of the cell theory and repudiated the theories of spontaneous generation. He also studied the nature of disease on a cellular level and established the science of cellular pathology.

Viren, Lasse (1949–   ), Finnish athlete. A dominating long distance runner of the early 1970s, he won gold medals for the 5,000m and 10,000m events at the Olympic Games of 1972 and 1976.

Vireo, any of 42 species of American songbirds . They have bristled forehead feathers, slightly downcurved bill and green, grey or yellow plumage. Family Vireonidae.

Virgil, or Vergil (70–19 BC), real name Publius Vergilius Maro, Roman poet. Virgil enjoyed the patronage of MAECE-NAS, who gave him a house near Naples, and was a friend of HORACE. He wrote the *Eclogues* (42–37 BC) and the *Georgics* (36–29 BC), and for the rest of his life wrote the AENEID. *See also* HC1 pp.112, *113.*

Virginal, keyboard musical instrument of the HARPSICHORD family; the strings, a single set running nearly parallel to the keyboard, are plucked by quills. Two instruments, each differing in size and pitch, were sometimes incorporated into the same case; the smaller could be removed and placed over the larger. Virginals were particularly popular in 16th- and 17th-century England.

Virgin Birth, term designating the orthodox Christian doctrine whereby Christ was conceived by the Blessed Virgin Mary through the power of the Holy Spirit. One

Villeins with a churchman and a yeoman in costumes of the time of Henry II.

François Villon, poet and rogue, scoured Paris with the *Coquillarde,* a band of thieves.

Vines, once common in England, are being reintroduced, as in this Surrey garden.

Viper; in the United Kingdom the adder is the only snake to be found in Scotland.

Virginia; a map made in 1624, the year in which the area was declared a royal colony.

Viscachas are of two species, one living on the plains and the other in mountains.

Vistula; boats can sail up as far as Krakow but the river is frozen for part of the year.

Vitamins are bottled either as single vitamins or as a combination of several.

of the basic tenets of Catholicism, most Protestant Churches and of the Muslims, the belief has led to doctrinal argument concerning Mary's perpetual virginity in spite of childbirth.

Virginia, state in E USA, on the Atlantic coast, the most northerly of the southern states. The coastal plain is low lying and drained by the Potomac, Rappahannock, York and James rivers. There are extensive forests. To the w the Piedmont Plateau rises to the Blue Ridge Mts. Virginia's economy is varied. Farming is an important element and the chief crops include tobacco, peanuts, grain, vegetables, apples and peaches. Dairy farming and poultry raising are also important. Chemicals form the state's leading industry. Other industries include shipbuilding, fishing, transport equipment, food processing, textiles and wood products. Coal is the most important mineral deposit; stone, sand and gravel are also quarried. The principal cities are Richmond, the state capital, Norfolk and Newport News.

The first permanent British settlement in North America was made at Jamestown in 1607. Virginia, became a royal colony in 1624 and evolved an aristocratic plantation society based on vast tobacco holdings. Virginia's leaders, who included Thomas JEFFERSON and George WASHINGTON, were in the forefront of the fight for American independence. Richmond became the Confederate capital and Virginia was the chief battleground of the AMERICAN CIVIL WAR. The state was re-admitted to the Union in 1870, but its social and economic recovery after the War was slow. Area: 105,710sq km (40,814sq miles). Pop. (1975 est.) 4,967,000. *See also* MW p.175.

Virginia Companies, two English companies chartered by JAMES I in 1606 to establish colonies in America. The Virginia Company of London was to found a colony 100 miles inland from the coast between latitudes 34°N and 41°N, the Virginia Company of Plymouth one of the same size between latitudes 38°N and 45°N. The London Company founded the first permanent English colony in 1607, but lost its charter in 1624. The Plymouth, which was less successful, was reorganized as the Council for New England in 1620.

Virginia creeper, also called woodbine or American ivy, tendril-climbing vine native to North America, but generally found only in eastern areas. It has leaves divided into five parts, green flower clusters and blue-black inedible berries. Family Vitaceae; species *Parthenocissus quinquefolia.*

Virgin Islands. *See* BRITISH VIRGIN ISLANDS; VIRGIN ISLANDS OF THE UNITED STATES.

Virgin Islands of the United States, group of 68 islands in the Lesser Antilles, in the West Indies, administered by the USA. The chief islands are St Croix and St Thomas, which includes the capital Charlotte Amalie. Livestock raising and the cultivation of sugar cane are the chief farming activities. The islands were Danish until 1917, when they were bought by the USA. Area:344sq km (133sq miles). Pop. (1975 est.) 92,000. *See also* MW p.182.

Virgin Mary. *See* MARY, SAINT OR BLESSED VIRGIN MARY.

Virgo, or the Virgin, equatorial constellation on the ecliptic between Leo and Libra. It lies in a region of the sky that has many galaxies and galaxy clusters. The brightest star is the 1st-magnitude Alpha Virginis, or Spica. *See also* MS pp.200–201, *200–201;* SU pp.*254, 256, 257, 261, 261, 266,* 267.

Virology, study of VIRUSES. The existence of viruses was established in 1892 by D. Ivanovski, a Russian botanist, who found that the causative agent of tobacco mosaic disease could pass through a porcelain filter impermeable to BACTERIA. Bacteriophages (viruses that infect bacteria) were discovered in 1915. The introduction of the electron microscope in the 1940s made it possible to view viruses.

Virtanen, Artturi Ilmari (1895–1973), Finnish biochemist who was awarded the 1945 Nobel Prize in chemistry for his work on fodder preservation. During the 1920s he discovered that fermentation in green fodder, which ruins silage stores, could be prevented by the addition of hydrochloric and sulphuric acids. He also found that this technique, known as the AIV method, had no detrimental effects on the nutritional qualities or edibility of the fodder.

Virus, sub-microscopic infectious organism. Viruses vary in size from 10 to 300 nanometres (1 nanometre = 1 millionth of a millimetre), and contain only genetic material, in the form of deoxy-ribonucleic acid (DNA) or ribonucleic acid (RNA). This is enclosed in a PROTEIN coat known as the capsid.

Viruses can grow and reproduce only when they enter another cell, such as a bacterium or animal cell, because they lack energy-producing and protein-synthesizing functions. When they enter a cell, the host's metabolism is subverted so that viral reproduction is favoured. Pathogenesis is the result of cell death or altered metabolism as the virus multiplies.

Control of viruses is difficult because harsh measures are required to kill them. The animal body has, however, evolved some protective measures, such as production of INTERFERON and of ANTIBODIES directed against specific viruses. One attack of certain viral diseases confers life-time IMMUNITY to the survivors. Where the specific agent can be isolated, VACCINES can be developed, but some viruses (such as influenza) change so rapidly that vaccines become ineffective. *See also* VIRUS INFECTIONS; MS p.84; NW pp.36, 85; SU p.156.

Virus infections, diseases caused by VIRUSES, among the greatest health problems facing man now that ANTIBIOTICS have achieved some measure of control over bacterial infections. Many virus infections can be treated only with specific VACCINES and antisera. Notable amongst viral infections affecting man are SMALLPOX, YELLOW FEVER, RABIES, influenza (FLU), MEASLES and POLIOMYELITIS. Viruses are also responsible for many diseases in plants, in which they are particularly difficult to control. The role of viruses in causing CANCER has been the subject of research since this effect was discovered in 1908. Living cells gain some protection by their own production of INTERFERON, which leads to the formation of another protein virus inhibitor. *See also* MS p.85.

Visayan Islands, also known as Bisayas, island group in the central Philippines, comprised of Cebu, Leyte, Masbate, Negros, Panay, Samar and many small islands. Cebu, Negros and Panay are the commercial heart of the group; the port of Cebu is the second largest in the Philippines, after Manila. Area: approx. 61,077sq km (23,582sq miles). Pop. (1970) 9,273,937.

Viscacha, South American rodent closely related to the CHINCHILLA. Its fur is not commercially valuable. Family Chinchilla; species *Lagostomus maximus* (plains viscacha), *Lagidium peruanum* (mountain viscacha). *See also* NW pp.202, *203.*

Vischer, name of a family of German sculptors and bronze-workers. Peter, the elder (*c.*1460–1529), was one of the greatest bronze casters of all time. He was trained by his father, Hermann (*d.*1488), and influenced to some extent by Italian art. His masterpiece is the tomb of St Sebaldus in Nuremberg (1508–19). Of his five sons who carried on his work, the most celebrated were Hermann, the younger (*c.*1486–1517) and Peter, the younger (1487–1528).

Visconti, Italian family that ruled Milan from the 13th century until 1447. Ottone Visconti (*c.*1207–95), appointed (1262) archbishop of Milan by Pope Urban IV, used his position to become the first Visconti Signore (lord) of Milan. Supporters of the GHIBELLINES, the Visconti established their control over Lombardy in the 14th century, and in 1349 the title of *Signore* became hereditary. Visconti lordship of Milan passed to the Sforza family after the death of Filippe Maria Visconti in 1447.

Visconti, Luchino (1906–76), Italian film director who made his debut in 1942 with *Ossessione.* His other films include *Bellissima* (1951), *Senso* (1953), *Rocco and His Brothers* (1960), *Death in Venice* (1971) and *Conversation Piece* (1976).

Viscose. *See* RAYON.

Viscosity, internal friction of a liquid or gas. In the flow of a liquid through a pipe, the central portion moves the fastest and the layers next to the pipe barely move. Thus layers of the fluid are sliding over one another; the more viscous the liquid, the slower it flows. Viscosity is large for liquids and extremely small for gases. In most liquids, it increases with decreasing temperature. *See also* SU p.87.

Viscount, rank of the nobility in Britain, falling below an earl and above a baron. A viscount is entitled to sit in the House of Lords; he is addressed as the Right Honourable and his coronet has 16 pearls set around its edge.

Vishnu, in Hindu mythology, one of the supreme triad with Brahma and SHIVA. He reigns in heaven with his wife LAKSHMI, the goddess of wealth, as the most important solar deity. Vishnu is depicted as a young man with four hands holding a shell, discus, mace and lotus. As Vishnu he was worshipped as a preserver and restorer; his other incarnations include RAMA and KRISHNA. *See also* MS pp.212–213; HC1 p.*227.*

Visigoths. *See* GOTHS.

Vision. *See* SIGHT.

Vision, in religion, sensory experience of spiritual or earthly events, remote in space and time, by individuals especially chosen by divine election. Biblical examples are numerous and usually involve the transmission to humanity of divine revelation, but visions are part of the experience of many religious groups, notably N American Indians, Hindus and Christian saints and mystics.

Vistula (Wisła), longest river in Poland. It rises in the Carpathian Mts of w Poland and flows NE past Kraków, turns NW through Warsaw and then flows NW through Toruń to enter the Gulf of Danzig at Gdańsk. It is the major waterway of Poland, and serves a large area through its tributary system, which includes the Bug, San, Wisłoka and Pilica rivers. Canals link it with other important rivers. Length: 1,068km (663 miles).

Visual purple. *See* RHODOPSIN.

Vitale da Bologna (*fl.*1330–61), Italian painter and a founder of the Bolognese school of art. His panel and fresco paintings show a strong Sienese influence. The panels entitled *Life of St Anthony Abbot* are highly detailed and original in surface treatment.

Vitalianus, Saint (*d.*672), Pope (*r.*657–72). His election was confirmed by Emperor Constant II, whom Vitalianus ceremonially received into Rome in 663. Constant nonetheless supported the rejection by Maurus, Archbishop of Ravenna, of control from Rome. Vitalianus was interested in the development of the Anglo-Saxon Church and consecrated THEODORE OF TARSUS as Archbishop of Canterbury.

Vitalism, philosophical theory that living matter is essentially different, and not reducible to mechanisms. Vitalists hold that various "powers", eg BERGSON's "élan vital", are operative in the behaviour of living organisms.

Vitamin deficiency, lack of a sufficient amount of a vitamin in the diet, generally resulting in one or another form of deficiency disease. Deficiency of vitamin C causes scurvy; of vitamin D, rickets; of B vitamins, beriberi, pellagra, pernicious anaemia and forms of dermatitis; of vitamin A, dry eyes and night blindness; and deficiency of vitamin E causes sterility in rats, if not in human beings. Most people in the western world get all the vitamins they need from a normal, balanced diet. *See also* MS pp.84–85.

Vitamins, organic compounds that are essential in small amounts to the maintenance and growth of higher animals. The name is a contraction of vital amines. They are usually taken in the diet but some are synthesized in the body: vitamin B_{12} is synthesized in the intestine, and vitamin D is synthesized on the skin when it is exposed to sunlight. Vitamins act as COEN-ZYMES in the regulation of energy transformation and the catalysis of such metabolic processes as the production of HORMONES, the synthesis of cell components and in the detoxification of body wastes. Today most vitamins can be made synthetically, and they are classified in three ways: alphabetically; as either water or fat soluble; or according to their chemical composition. Lack of a particular vitamin can lead to DEFICIENCY DISEASE. *See also* index to *Man and Society*.

Vitrac, Roger (1899–1952), French poet and dramatist who followed in the footsteps of DADAISM and SURREALISM. He was associated with Antonin Artaud and with him founded the Théâtre Alfred Jarry where he produced his own play *Victor ou Les Enfants au Pouvoir* (1928). His plays include *Le Loup-Garou* (1939) and *Le Sabre de mon Père* (1951) .

Vitreous humour, transparent jelly-like medium which fills the eyeball between the lens and the retina. It constitutes the vitreous body, which serves to hold the retina in position and combines with the lens to ensure the clear passage of light to the receptor cells of the retina. *See also* AQUEOUS HUMOUR; MS pp.48–49.

Vitrification, in ceramics, process of producing a glassy phase or close crystallization by firing at a high temperature. Vitrified clay becomes non-porous and loses its plasticity. *See also* MM p.44.

Vitriol. *See* SULPHURIC ACID.

Vitruvius (*fl.* early 1st century AD), Roman architect and engineer, *b.* Marcus Vitruvius Pollio. His ten-volume *De Architectura* (before AD 27), an encyclopaedic work covering almost every aspect of ancient architecture including town planning, types of buildings and materials, was used by Renaissance artists and architects as their chief reference on architectural theory.

Vittoria, Battle of (1813), victory of British, Spanish and Portuguese forces under the Duke of WELLINGTON over a French army under King Joseph Bonaparte. The Battle was part of the PENINSULAR WAR and resulted in the expulsion of the French from Spain.

Vittorio Veneto, town near the foothills of the Alps in the Veneto region of Italy. It is situated on the Piave line, the Italian front in WWI, and was the scene of a series of battles between 24 Oct. and 4 Nov. 1918 which brought about the surrender of the Austrian forces.

Vitus, Saint (*fl.* 4th century), Italian martyr. Secretly raised as a Christian by his nurse, he was put to death during the persecutions of DIOCLETIAN. The patron saint of actors, his intercession is invoked against sudden death, St Vitus' dance and hydrophobia (rabies).

Vivaldi, Antonio (*c.*1675–1741), Italian composer. A master of the concerto and a virtuoso violinist, he helped to standardize the three-movement concerto form and to develop the *concerto grosso* (a concerto for two or more solo instruments. His best-known work is *The Four Seasons* (1725). *See also* HC2 pp.40, 40–41.

Vivarini, Antonio (*c.*1415–*c.*1476), Italian painter. The founding father of a prominent family of Italian painters of the VENETIAN SCHOOL, he worked in the late GOTHIC style. He worked, mostly in collaboration with relatives, on large, elaborately framed altarpieces. His son Alvise (*c.*1445–*c.*1505), also a painter of the Venetian school, was influenced by Giovanni BELLINI and ANTONELLO DA MESSINA. He became an influential teacher, and his pupils included Lorenzo LOTTO and Jacopo de BARBARI.

Vivisection, the dissection of living bodies. Work with laboratory animals in testing drugs, vaccines and pharmaceuticals frequently involves such dissections. The smaller animals such as rats, mice and hamsters are usually killed painlessly after the survey of the test results. The ethical issue of experimenting on living animals is still a matter of controversy, because alternative methods may be available, if not as cheap or convenient to use.

Vizsla, pointer and retriever breed of dog with innate hunting ability; also called Hungarian pointer. The rusty gold or dark sandy yellow coat is short, smooth and dense. Height: to 63.5cm (25in) at the shoulder; weight: to 34kg (75lb).

V-J Day (2 Sept. 1945), the day WWII ended in the Pacific. Japan surrendered to the Allies on board the USS *Missouri* in Tokyo Bay.

Vladimir, name of two Russian rulers. Vladimir I (*c.*956–1015), an Orthodox saint, became Grand Prince of Kiev in 980 after overcoming his half-brother. Converted to Christianity *c.*987, he became the first Christian to rule in Russia and spent much of the rest of his life working to end pagan practices. Vladimir II (Monomakh) (1053–1125), was Grand Prince of Kiev from 1113. A strong, successful ruler, he helped to restore order among his feuding cousins and prevented civil war. *See also* HC1 pp.320–321.

Vladivostock, port in the Russian SFSR (Rossijskaja SFSR), USSR, on the Sea of Japan; chief Soviet Pacific port and a principal city in the Soviet Far East. Founded in 1860 as a military post, the city developed as a naval base after 1872, and it was used as a major supply depot by the Allies in WWI. Vladivostock is the terminus of the Trans-Siberian Railroad which connects the city with Moscow. The harbour is kept open in winter by ice breakers and is a major base for whaling and fishing fleets, and the principal base for the Soviet navy in the Pacific. Industries include ship repairing, metal working and food processing. Pop. (1970) 442,000.

Vlaminck, Maurice (1876–1958), French painter. Largely self-taught, he was an early FAUVIST who was successively influenced by Vincent van GOGH, African sculpture and Paul CÉZANNE.

Vltava, longest river in Czechoslovakia. It rises in the Bohemian Forest in SW Czechoslovakia and flows SE then N to join the River Elbe at Melnik. Length: 435km (270 miles).

Vocal cords. *See* LARYNX; VOICE.

Vocational guidance, advice on the choice of a career. Its extent varies widely, from simple advice often given in schools according to a person's preferences, to that given by psychologists based on the results of aptitude tests, personality tests and interviews.

Vodka, colourless alcoholic spirit distilled from fermented rye and wheat, or sometimes from potatoes. It is commonly made in Russia and Poland. Most natural flavouring constituents of vodka are eliminated in the course of its preparation and so others, such as herbs, berri s, or fruits, are often added after the neutral liquor has been produced. *See also* PE p.207.

Vodun. *See* VOODOO.

Vodyanoy, in Slavic mythology, an evil, destructive freshwater spirit. The Vodyanoy would seize and enslave people imprudent enough to swim in ponds after sunset.

Vogel, Sir Julius (1835–99), New Zealand politician, *b.* London. He moved to New Zealand in 1852, and became colonial treasurer (1869–72) and Prime Minister (1873–75, 1876), during which time he sponsored public works.

Vogt, Alfred Elton van (1912–), Canadian SCIENCE FICTION writer. His first SF work was a short story, *Black Destroyer* (1939), which was later incorporated into the novel *The Voyage of the Space Beagle* (1950). He soon gained international popularity with his many books, which include *The World of Null-A* (1956) and *Battle for Forever* (1971).

Vogt, Hans (1890–), German electrical engineer and inventor. Working with Dr J. Engl and Joseph Massolle between 1918 and 1925, he invented a sound motion pictures film that carried on it the recording of the sound.

Voice, in human beings, sound produced by air from the lungs passing through the vocal cords, causing them to vibrate. The cords are two wedge-shaped bands of fibrous, elastic tissue attached to the cartilage of the the LARYNX (voice-box), which bulges in the front of the neck as the Adam's apple. The respiratory tract above the larynx acts as a resonating chamber, lending the voice its particular timbre. The voice of a bird is produced differently by a vocal apparatus called the SYRINX, located at the base of the windpipe.

Voice, in grammar, mode of inflecting verbs to indicate whether the subject acts (active voice), is acted upon (passive voice) or acts so as to affect itself (middle voice).

Volcanism, or vulcanism, volcanic activity. The term is a general one and includes all aspects of the process: the eruption of molten and gaseous matter, the building up of cones and mountains and the formation of lava flows, geysers and hot springs. *See also* PE p.30.

Volcano, vent from which molten rock or lava, solid rock debris and gases issue. The term is also applied to the pile of rock around the vent. Volcanoes may be of the central vent type, where the material erupts from a single pipe, or of the fissure type, where material is extruded along an extensive fracture, building plains and plateaus. Volcanoes are usually classed as active, dormant or extinct. *See also* MW p.191; PE pp.30–31, 106.

Vole, short-tailed, small-eared, prolific RODENT living wild in the Northern Hemisphere. Most voles are greyish-brown, herbivorous ground-dwellers and quite small. The semi-aquatic water vole is the largest. Length: to 18cm (7in). Family Cricetidae. *See also* NW pp.206, 223.

Volga, river in E European USSR; the longest river in Europe and the principal one in the USSR. The drainage basin of the Volga covers approx. one third of European USSR. The river rises in the Valdai Hills, flows E past Ržev and Kalinin and through the Rybinsk Reservoir to Kazan', where it turns S and continues past Uljanovsk, making a hairpin bend at Kujbyšev. It continues SW to Volgograd (Stalingrad) and SE to enter the Caspian Sea through a wide delta below Astrachan'. The Volga is connected to the Baltic Sea by the Volga-Baltic Waterway, with Moscow by the Moscow Canal, and with the Black Sea and the Sea of Azov by the Volga-Don Canal. Many dams and hydroelectric power stations have been constructed along the river's course, and more than half of all river freight in the USSR is transported on the Volga. Its waters are used to irrigate the steppes of the lower Volga region. The river is navigable for almost all of its course. Length: 3,692km (2,293 miles).

Volgograd, formerly known as Stalingrad, seaport in SW Russian SFSR, USSR, on the River Volga and the E terminus of the Volga-Don Canal. Founded in 1589, the port was taken by COSSACKS in 1670. The city was defended by Soviet troops under STALIN in 1918 and renamed in his honour, but was almost completely destroyed is WWII. Renamed again in 1961, Volgograd is a major rail and heavy industry centre. Industries include oil refining, ship building, flour milling and the production of chemicals and steel. A major hydroelectric power station was completed in 1960. Pop. (1975) 900,000.

Volkswagen, literally "car for the people", name of a German range of cars made by Volkswagenwerk A.G. The best-known, the "Beetle", was planned in 1934 and designed by Ferdinand Porsche. First produced in 1936, the air-cooled rear-engined design proved internationally popular and had changed very little when the model was discontinued in 1977. *See also* MM p.*127.*

Volkswagenwerk A. G., West German car manufacturing corporation. It was founded in 1936, and its mid-1970s workforce numbered about 200,000 people. *See also* MM p.*127.*

Vollard, Ambroise (1865–1939), French art collector, dealer and publisher, who championed unknown artists of the time, including CHAGALL, ROUAULT and VLAMINCK. He opened a gallery in Paris (1893) where the first one-man exhibitions of CÉZANNE (1898), PICASSO (1901) and MATISSE (1904) were held.

Volleyball, sport played indoors or outdoors. A court is divided into two 30-ft (9.1m) squares by an upright net which stands 8ft. (2.4m) from the ground. Two

Vizslas may be descended from dogs taken to central Europe by the Magyars.

Vltava, Czechoslovakia's longest river, has several hydroelectric stations.

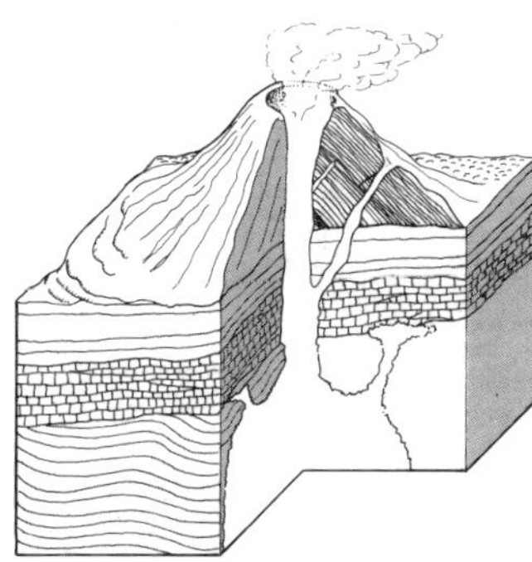

Volcano; molten rock and gases are expelled through the crater.

Vole; species are named after their habitat, an example being the meadow vole.

Alessandro Volta became a count and a senator of the kingdom of Lombardy.

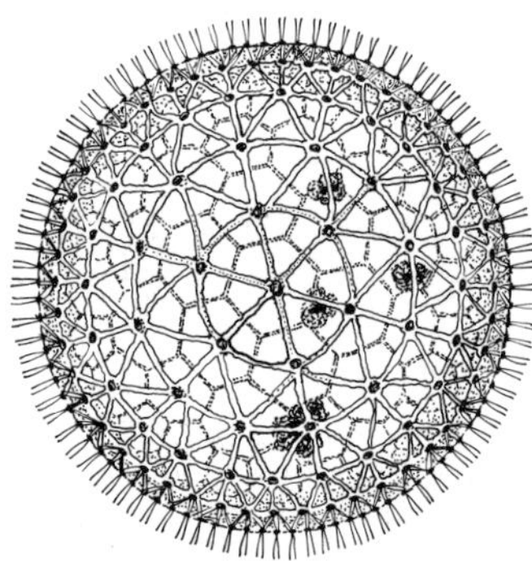

Volvox colonies may contain between 500 and 50,000 individual flagellates.

Vultures will soar effortlessly for long periods on their broad wings.

Andrei Vyshinski, photographed in 1933 when he was deputy state prosecutor.

teams, each of six players, take up positions on either side of the net. The serving team uses an open hand or fist to volley an inflated rubber or leather ball, about 27in (68.6cm) in circumference, over the net and the opponents must return it before it hits the ground. Points are scored only by the serving team and 15 points are needed to win. Volleyball originated in the USA in 1895 and was included in the Olympic Games since 1964.

Völsunga saga, Icelandic SAGA, probably written *c.*1270. A Scandinavian prose version of the German Nibelungenlied, it tells of the mythical Völsungs and how a magic sword, won from ODIN, was broken but reforged by Sigurd (the German SIEGFRIED).

Volt, unit of electric potential difference, or voltage. It is the difference of potential between two points on a conducting wire carrying a constant current of one ampere when the power dissipated between these points is one watt. The volt is also the unit of electromotive force (EMF).

Volta, Count Alessandro Giuseppe Antonio Anastasio (1745–1827), Italian physicist who in 1800 invented the first electric battery, the voltaic pile. He also invented a charge-accumulating machine which incorporated the principle of some modern electrical condensers. His investigations into electricity led him to interpret correctly Luigi GALVANI's experiments with muscles, showing that the metal electrodes and not the tissue generated the current. The volt was named in his honour. *See also* SU p.148, *148*; HC2 pp.32, *33*, 157.

Volta, principal river of Ghana. Formed by the confluence of the Black Volta and White Volta rivers, which rise in Upper Volta, it flows SE to Ada, on the Guinea coast. Length (including Lake Volta, an artificial lake completed in 1967): 483km (300 miles).

Voltage, measure of the electrical potential difference between two points in a circuit. Two points are at a potential difference of one volt if one coulomb of electricity (electric charge) does one joule of work in flowing between them. Voltage is also given by the RESISTANCE multiplied by the CURRENT (using OHM'S LAW).

Voltaire (1694–1778), pseudonym of François Marie Arouet, French philosopher, historian and poet, the outstanding figure of the French Enlightenment. He spent much of his life combating intolerance and injustice. His *Philosophical Letters* were published in 1734. He wrote several tragedies, including *Brutus* (1730) and *Zaire* (1732), the comedy *Mérope* (1743), and the philosophical novel *Candide* (1759). He contributed to the *Encyclopédie* of Diderot, wrote the *Century of Louis XIV* (1751) and outlined his view of morality in *Essay on Morals* (1756). He was a lifelong opponent of the Catholic Church. *See also* HC1 p.325; HC2 pp.22–23, 35.

Volterra, Daniele Ricciarelli da (1509–66), Italian MANNERIST painter and sculptor, a close acquaintance of MICHELANGELO. *Descent from the Cross* (1541) in SS Trinita dei Monti, Rome, is the most famous of his paintings.

Voltmeter, instrument for measuring the potential difference between two points in an electrical circuit. Voltmeters are therefore always connected across (in parallel with) the components whose potential differences (voltages) are being measured. A voltmeter has a high internal resistance (compared with the resistance across which it is connected) so that it takes a negligible proportion of the circuit's current – which affects the voltage. *See also* AMMETER.

Volume, measure of the amount of space taken up by a body. Volume is the mass of a body divided by its density, and is measured in cubic units, eg cm^3, ft^3 or m^3.

Volumetric analysis, in chemistry, widely used method of finding the concentration of solutions. The solution of unknown strength is titrated against one of known strength which reacts with it. The point of complete reaction, the end point, is often shown by a change in colour of an INDICATOR also present. The quantity of known

solution used, together with knowledge of the reaction involved, allows a chemist to calculate the concentration of the unknown solution. End points of titrations may also be detected electrometrically using a voltmeter. *See also* QUANTITATIVE ANALYSIS; SU pp.150, *151*.

Voluntary muscle, also called skeletal, or striped muscle, the muscle that moves the limbs, trunk and head and (unlike muscle of the heart, gut and other internal organs) is under conscious control. Microscopically it is seen to consist of fine fibres having a banded, or striped appearance. These are grouped together in bundles, which are in turn bundled together to make the individual muscles. *See also* INVOLUNTARY MUSCLE; MS pp.58–59.

Volvox, genus of tiny, single-celled green ALGAE. It forms hollow, spherical colonies, barely visible to the naked eye. The lashing of the whip-like flagella of the colony's members makes the whole colony turn slowly about a definite axis. *See also* NW p.46.

Von Braun, Wernher (1912–77), German engineer, known for his role in many aspects of rocketry and space exploration. In 1929 he began experiments with liquid-fuel rockets and he was instrumental in the development of the German v2 ROCKET. In 1945 he went to the USA and directed the Redstone rocket programme which launched the first US satellite, Explorer I (1958).

Vondel, Joost van den (1587–1679), Dutch poet and dramatist. One of the last Dutch writers of Renaissance principles, he struggled against humble birth, limited education and religious persecution to produce outstanding work based on biblical and classical sources. The tragedy *Lucifer* (1654) is his masterpiece.

Von Euler, Ulf. *See* EULER, ULF VON.

Von Frisch, Karl. *See* FRISCH, KARL VON.

Von Haast, Sir Julius (1822–87), New Zealand colonist and scientist, *b.* Germany. He explored the South Island, discovering the coal deposit of the West Coast. He was the first European to see Haast's Pass and the Franz Josef glacier.

Vonnegut, Kurt, Jr (1922–), US writer whose bizarre novels and short stories used black humour to deplore the horrors of the 20th century. His books include *Player Piano* (1952), *Slaughterhouse Five* (1969), *Breakfast of Champions* (1973) and *Slapstick* (1976).

Von Sydow, Max. *See* SYDOW, MAX VON.

Voodoo, religious belief derived from the God Vodun. It is prevalent in parts of Africa, and is the national religion of Haiti. Adherents believe in the reincarnate qualities of Loa, which include deified ancestors, local gods or Roman Catholic saints. Roman Catholicism became incorporated into voodoo during early French colonialism. Loa possesses the believers during dreams or ceremonies, which include dancing and hypnotic trances. In Haiti voodoo is exploited commercially for its tourist potential.

Voroshilov, Klimenty Yefremovich (1881–1969), President of the USSR from Stalin's death (1953) to 1960. He took a military rather than political role during the Russian Revolution (1917) and from 1925 to 1940 was commissar for defence. Removed from this post following initial Soviet defeats in WWII, he continued to serve in responsible positions. He retired in 1960.

Vorster, Balthazar Johannes (1915–), South African political leader. Elected to parliament in 1953 as a member of the Nationalist Party, he was Minister of Justice from 1961 to 1966. He succeeded Hendrik VERWOERD as Prime Minister in 1966. Although he allowed a softening of some APARTHEID policies and met black African leaders, he ruthlessly crushed all internal opposition to his segregationist policies.

Vorticism, British art movement. Derived from CUBISM and Italian FUTURISM, it was originated by Wyndham LEWIS in 1912 and was an attempt to express the zeitgeist (spirit of the time) in harsh angular forms derived from machinery. The artists Edward Wadsworth, David BOMBERG, Henry GAUDIER-BRZESKA and Jacob

Epstein were members of the movement. The Vorticists' only exhibition took place in 1915. *See also* HC2 p.280.

Vos, de, name of two Flemish painters. Cornelis de (*c.*1584–1651) worked for RUBENS. His family portraits are in the style of Rubens and VAN DYCK, and have been mistaken for theirs. His brother Paul (*c.*1596–1678), was a painter of lively hunting scenes and large still-lifes. His art has much in common with that of his brother-in-law, Frans SNYDERS.

Vostok I, first of a series of manned spacecraft placed into Earth orbit by the USSR on 12 April 1961. It was launched by the two-stage Vostok rocket and carried cosmonaut Yuri GAGARIN. Voskhod is an improved version of the Vostok launcher. *See also* SU pp.268, 269, 270.

Vote, right conferred upon members of a group or society to express in an ELECTION, REFERENDUM, parliamentary division or similar formal function their preference, either of people or proposals. The vote may be expressed by voice, by the raising of hands or by Secret BALLOT. *See also* SUFFRAGE; REFORM BILLS.

Voysey, Charles Francis Annesley (1857–1941), British architect and designer. The Pastures (1901), at North Luffenham in Leicestershire, England, is a good example of his style: a long, low building, with rough-cast walls, and low roof.

Vries, Hugo de. *See* DE VRIES, HUGO.

V2 rocket, large liquid-fueled rocket developed by the Germans during WWII; the first ballistic missile used in war. The name was an abbreviation for *Vergeltungswaffen zwei,* German for "Vengeance Weapon Number Two". *See also* MM p.169.

Vuillard, Jean-Edouard (1868–1940), French painter and printmaker, a leading member of the NABIS. He developed a distinctive style of INTIMISM characterized by simplicity and richness of texture and pattern. He specialized in interiors and street scenes.

Vulcan, Roman god of fire and volcanoes, originally called Volcanus. His temples were prudently sited outside city walls. Often invoked to avert fires, he was associated with thunderbolts and the Sun, and was finally interpreted in terms of life-giving warmth. His main festival, the Volcanalia, was held on 23 August.

Vulcanization, chemical process, discovered in 1839, of heating sulphur or its compounds with natural or synthetic rubber to improve its durability and resilience. *See also* MM p.54.

Vulgate, oldest surviving version of the complete Bible, compiled and translated, mostly from Greek, into Latin by St JEROME for Pope DAMASUS I from 382. The text was revised several times and was used universally in the Middle Ages. It was promoted as the official Latin translation in 1546 by the Council of TRENT, reacting to demands for vernacular versions from the leaders of the REFORMATION. It was again revised, by the Benedictines, in 1933.

Vulture, large, keen-sighted, strong-flying bird that feeds on carrion. New World vultures, forming a distinct family, are found throughout the Americas and include the CONDOR, turkey BUZZARD and king vulture; family Cathartidae. Old World vultures, related to eagles, are found in Africa, Europe and Asia and include the Egyptian vulture and the griffon vulture; family Accipitridae. *See also* NW p.148.

Vulva, in females, the external genitalia. A pair of fleshy lips (labia majora) surround the vulvar orifice. Within them two smaller folds of skin (labia minora) surround a small depression called the vestibule, within which are the urethral and vaginal openings. *See also* MS p.74.

Vyshinski, Andrei Yanuarievich (1883–1954), Soviet politician. Originally a Menshevik, he joined the Bolshevik party during the Civil War. With the outbreak of WWII Vyshinski became deputy commissar for foreign affairs (1940–49) and foreign minister (1949–53). In 1953 he was appointed permanent Soviet delegate to the UN.

W

W, 23rd letter of the alphabet which, like *f*, *u*, *v* and *y*, was derived from the Semitic letter *vaw* (meaning *hook*). The Romans wrote it like a *V*, first pronouncing it like a *w* and later as a *v*. Norman-French writers of the 11th century created the modern form of the letter by doubling a *u* or *v* to represent the Anglo-Saxon letter *wyn*, which had no counterpart in their alphabet. In English *w* has a consonant sound, as in *war* and *row*, and is silent in such words as *answer* and *wring*. In some combinations with *h* the *w* is also silent (as in *who* and *whom*) and in others it is the *h* that is not pronounced (as in *which* and *what*). *See also* MS pp.244–245.

Waals, Johannes Diderik van der. *See* VAN DER WAALS, JOHANNES DIDERIK.

Wace, Robert (*c.*1100–75), Anglo-Norman poet and chronicler, *b.* Jersey and educated in France. He wrote several lives of the saints in verse, the *Roman de Rou* (an epic of the exploits of the Dukes of Normandy), and the *Roman de Brut*, based on Geoffrey of Monmouth's *Historia Regum Britanniae.*

Waddell, Helen Jane (1889–1965), British scholar, a specialist in medieval studies. Her books include *The Wandering Scholars* (1927), an account of the poets of the later Middle Ages, and *Medieval Latin Lyrics* (1929).

Wade, Sarah Virginia (1945–), British tennis player. She won the US women's singles title in 1968, the Australian championship in 1972 and Wimbledon in 1977.

Wadi, gorge cut into the solid rock of an arid area. Wadis are water-formed features, but contain streams only during the rainy seasons. *See also* PE p.*121.*

Wad Medani, city in the Sudan on the Blue Nile, 161km (100 miles) SSE of Khartoum and on the railway from Khartoum to Sannar. It is the capital of Blue Nile province. Pop. 82,000.

Wagner, Otto (1841–1918), Austrian architect. He synthesized architecture and technology in an ambitious programme of modern design for public buildings and utilities in Vienna. His book *Moderne Architektur* (1895) had wide influence throughout Europe and particularly on his pupils Joseph OLBRICH and Josef HOFFMAN.

Wagner, Richard (1813–85), German composer. His works consist entirely of opera, for which he provided his own libretti. It was with Wagner's work that German Romantic music received its fullest expression. His early operas include *Rienzi* (1842), *The Flying Dutchman* (1843) and *Tannhäuser* (1845). With *Tristan and Isolde* (1865) and *The Ring of the Nibelungs* (1851–74) the genius of Wagner is fully displayed. The innovations that he introduced to operatic form were vast and served to fuse the elements of music and stage drama as one. Other operas include the *Mastersingers of Nuremberg* (1868) and the sacred stage drama *Parsifal* (1882).

Wagner von Jauregg, Julius (1857–1940), Austrian neurologist and psychiatrist. He was awarded the 1927 Nobel Prize in physiology and medicine for his discovery of a treatment for the general paralysis caused by malaria.

Wagon trains, convoys of large wagons which were the principal method of migration westward in the USA from the late 18th century until the advent of railways. Settlers usually travelled in groups as defence against Indians. The greatest wagon trains were those which crossed the GREAT PLAINS and the ROCKY MOUNTAINS along the OREGON TRAIL and the SANTA FÉ TRAIL.

Wagram, Battle of (1809), military engagement resulting in victory for the French army under NAPOLEON over the Austrian army under the Archduke Charles during the NAPOLEONIC WARS. It took place on the Marchfeld, a plain NE of Vienna, and lasted for two days (5–6 July). The French army numbered about 154,000 men, the Austrian about 158,000. The French casualties were 34,000 men; Austrian losses were 40,000. After their defeat the Austrians were compelled to sign the Treaty of SCHÖNBRUNN.

Wagstaff, Harold (1891–1939), British rugby league centre who made his senior debut for Huddersfield at the age of 15 and later captained their "team of all the talents". One of the game's greatest players, he played 13 times for Great Britain (1911–22), having played for England against the Australians at the age of 17.

Wagtail, any of several species of mainly Old World birds that live near streams; it wags its long tail while foraging for insects. Family Motacillidae.

Wahhabism, puritanical Muslim movement founded in Arabia by Mohammed ibn 'Abd al-Wahhab (1703–92). The movement stresses the absolute oneness of God, literal belief in the KORAN and Hadith (Islamic tradition), the inseparability of belief from ethical action, belief in predestination, rejection of all non-orthodox views (such as SUFISM), and the need to establish the Muslim state on Islamic law alone.

Waikato, longest river in New Zealand, in central and NW North Island; rises from Lake Taupo in the central highlands, and flows NNW into the Tasman Sea. Hydroelectric schemes along the river provide power for most of North Island. The river is navigable for 129km (80 miles). Length: 425km (264 miles). It is also the name of a district.

Wailing Wall, or Western Wall, place in Jerusalem sacred to all Jews. It is a remnant of a wall of the great Temple destroyed by the Romans in AD 70, and is the focus of many pilgrimages. Jews publicly mourn the Temple's destruction at the wall.

Wain, John Barrington (1925–), British poet, novelist and critic who published his poetry in *Mixed Feelings* (1951), *New Lines* (1956) and *Wildtrack* (1965). His novels include *Hurry on Down* (1953) and *The Young Visitors* (1965).

Waitangi, Treaty of (1840), pact between Britain and several New Zealand MAORI tribes. The agreement protected and provided rights for Maoris, guaranteeing them possession of certain tracts of land, while permitting Britain formally to annex the islands and purchase other land areas from the Maoris. *See also* HC2 p.126, *126.*

Waiting for Godot (1952), Theatre of the ABSURD tragi-comedy written by Samuel BECKETT. Illustrating their boredom and disenchantment with life, the two tramp-like characters, Vladimir and Estragon, loiter on a nearly barren stage, awaiting Godot, who never comes. *See also* HC2 p.*283.*

Waitz, Theodor (1821–64), German anthropological psychologist. He wrote extensively on psychology and is the author of *Anthropologie der Naturvölker* (1859–71).

Wajda, Andrzej (1926–), Polish film director whose trilogy *A Generation* (1954), *Canal* (1957) and *Ashes and Diamonds* (1958) was a turning-point in modern Polish cinema. His films examine Poland's historic place in WWII (*Lotna*, 1959), express the unhappiness of adolescent sexuality (*Love at Twenty*, 1962) and include a personal film (*Everything for Sale*, 1968) about the death of his friend, actor Zbigniew Cybulski, the star of his trilogy.

Waka, Japanese courtly poetry, written in the 6th–14th centuries AD. It has various different forms, ranging from long "choka" poetry to the epigrammatic "tanka". The term refers to poetry written in Japanese, or in a Japanese style, rather than in the Chinese language or idiom.

Wakamatsu. *See* KITAKYŪSHŪ.

Wakefield, Edward Gibbon (1796–1862), British colonial reformer. He came to notice with the publication of a *Letter from Sydney* (1829), in which he argued for a rational system of the sale of Crown lands in the colonies, the revenue from which could be used to assist emigration. He organized an association for colonizing New South Wales in 1834 and founded a similar society for New Zealand in 1837.

Wakefield, county district in SE WEST YORKSHIRE, England; created in 1974 under the Local Government Act (1972). Area: 333sq km (129sq miles). Pop. (1974 est.) 305,300.

Wakefield Plays, cycle of MYSTERY plays performed by the workmen of WAKEFIELD and dating from *c.*1450. Apart from the *Second Shepherd's Play,* unique to the Wakefield text, the sequence enacts the traditional biblical events performed in contemporary cycles but employing a more earthy and humorous dialogue. A shortened adaptation of the plays has been frequently performed since 1967.

Waksman, Selman Abraham (1888–1973), US microbiologist, *b.* Russia. He was awarded the 1952 Nobel Prize in physiology and medicine for his discovery of STREPTOMYCIN in 1943. He developed techniques for extracting ANTIBIOTICS from various micro-organisms and discovered several new ones, including NEOMYCIN.

Walachia. *See* WALLACHIA.

Walcheren expedition (1809), British disaster in the NAPOLEONIC WARS. A fleet of about 100 ships with 40,000 men under the Earl of Chatham landed on the island of Walcheren in the Scheldt estuary in July, in an effort to divert French troops. The island was held but the French fleet escaped and the last of the British troops were withdrawn in Dec. 1809 after heavy losses from fever.

Walcott, Derek (1930–), West Indian playwright and poet. His first work was *Twenty-five Poems* (1948). Subsequent works include *In a Green Night* (1962) and *The Castaway* (1966).

Walcott, "Jersey Joe" (1914–), US boxer, *b.* Arnold Raymond Cream. He beat Ezzard Charles in 1951 for the world heavyweight title in Pittsburgh and lost it in 1952 to "Rocky" Marciano.

Wald, George (1906–), US biologist who shared the 1967 Nobel Prize in physiology and medicine with Haldan HARTLINE and Ragnar GRANIT for his research into the chemical processes of vision. He discovered that vitamin A is an essential component of visual pigments and later determined the chemical reactions which occur in the RODS of the retina.

Waldemar, name of four kings of Denmark. Waldemar I (1131–82; *r.*1157–82), known as "the Great", defeated Sweyn III in 1157 and became king, with his foster-brother Bishop Absalon as his adviser. Waldemar II (1170–1241; *r.*1202–41), known as "the Victorious", was forced to relinquish territory by the German princes. He lost Estonia in 1227, but regained it in 1238. His son, Waldemar III, ruled jointly with him from 1218 to 1231. Waldemar IV Atterdag (*c.*1320–75; *r.*1340–75), united the kingdom in 1361.

Waldenses, small Christian sect which had its origin in the "Poor Men of Lyons", as the followers of Peter WALDO of Lyons (*d. c.*1218) were called. Waldo translated the Bible into French and preached without ecclesiastical authorization which led to his excommunication (1184). The Waldenses renounced private property and led an ascetic life. They repudiated many Roman Catholic doctrines and practices such as INDULGENCES, PURGATORY and MASSES for the dead, and denied the validity of SACRAMENTS administered by unworthy priests. The movement flourished briefly in the 13th century, but active persecution extinguished it except in the French and Italian Alps. In the 16th century the remaining Waldenses made contact with the Protestant Reformers, who attended a synod convened by them in 1532 at Cianforan. The synod adopted a confession of faith which included the doctrine of PREDESTINATION. The Waldenses still exist as a small sect, and about 3,000 live in Valdese, North Carolina, founded by members of the sect who emigrated to the USA during the 19th century.

Waldheim, Kurt (1918–), Austrian Foreign Minister between 1968 and 1970, diplomat, and Secretary-General of the United Nations Organization (1972–). He headed Austria's first UN delegation in 1955 and was its representative (1964–

Virginia Wade won the woman's singles championship at Wimbledon in 1977.

Wagner, painted by Renoir; his work greatly influenced contemporary thought.

Wagtail; the yellow wagtail, the only species to reach the New World, breeds in Alaska.

Kurt Waldheim explaining his peace plan for a divided Cyprus to the UN in 1975.

Waldo, Peter

Wales; Caernavon Castle – the site of Prince Charles's investiture in 1969.

Wallaby; *Wallabia elegans* inhabits the coastal areas of Eastern Australia.

Fats Waller began playing the piano as a child; his style influenced many musicians.

Sir Barnes Wallis progressed from airship design to work on the supersonic Concorde.

68; 1970–71). He was elected as successor to U THANT. *See also* HC2 p.*298*.

Waldo, or Valdo, Peter (*c.* 1140–*c.* 1218), French religious reformer after whom the WALDENSES are named. He sent out disciples, known as Poor Men, to read to the common people from the Bible. The authorities forbade them to preach but Waldo persisted and came to oppose many teachings of the Church.

Waldorf, William, Viscount Astor. *See* ASTOR FAMILY.

Wales, Prince of. *See* PRINCE OF WALES.

Wales, principality of Great Britain, occupying a broad peninsula to the w of central Britain. The country is mountainous in the N and s; lowland areas are confined to the borders regions and coastal plains. The principal rivers are the Severn, Dee, Conway and Teifi. Much of the land is used for rough grazing although Wales' major contributions to the economy of the UK come from coal mining and iron and steel manufacture. The capital is Cardiff. Area: 20,761sq km (8,016sq miles). Pop. (1975 est.) 2,765,000. *See* MW p.182.

Waley, Arthur Schloss (1889–1966), British poet and Oriental scholar, best known for his translations of Chinese verse. His collection of *170 Chinese Poems* (1917) influenced such leading 20th-century poets as POUND and YEATS. His other translations include the 16th-century Chinese novel *Monkey* (1942), and Japanese works such as SKIKIBU MURASAKI's *Tale of Genji* (1925–32) and SEI SHONÂGON's *Pillow Book* (1927). His work is noted for its combination of poetic imagination and meticulous scholarship.

Walker, Sir Emery (1851–1933), British process engraver and typographer. He helped William MORRIS to establish the KELMSCOTT PRESS (1890); with T. J. Cobden-Sanderson he founded the Doves Press (1900), which lasted until 1909.

Walker, John (1952–), New Zealand middle-distance runner. He achieved international repute when he came second in Filbert BAYI's record-breaking 1,500m run in the Commonwealth Games in Christchurch in 1974. In May 1975 he broke the world record for the mile, with a time of 3min 49.4sec.

Walker, Robert (*d.c.* 1658), English painter, a follower of VAN DYCK, and best known for his portraits of John EVELYN (1648) and Oliver CROMWELL (1649).

Walker, William (1824–60), US military adventurer. His greatest success was in Nicaragua between 1855 and 1857, where he took advantage of a civil war to make himself president. He was ousted by a coalition of Central American states in 1857 and surrendered to the US navy. After three attempts to return to Nicaragua, he was arrested by the British navy and turned over to the Honduran authorities who executed him.

Walker Cup, biennial competition between teams of amateur male golfers from the USA and Britain. It was founded in 1922 by Cyril Walker, who donated the International Challenge Trophy. Eight 18-hole foursomes and 16 18-hole singles are played over two days.

Walking, in sport, rhythmic action defined by the British Race Walking Association as "a progression by steps so taken that unbroken contact with the ground is maintained". Under the International Amateur Athletic Federation rules, there must be a momentary straightening of the knee in every stride.

Wallabies, name given to Australia's national rugby union team, after the bush kangaroo indigenous to Australia. The Wallabies' first major tour was of Britain in 1908, when they won the rugby union gold medal at the Olympic Games.

Wallaby, any of various medium-sized members of the kangaroo family of marsupial mammals, occurring chiefly in Australia. All species are herbivorous, feeding in open grassland at night, and using the scrubland vegetation for daytime resting. They move fast in a series of leaps, using both strong hind legs simultaneously, balanced by the tail. Length: head and body 45–105cm (18–41in); tail 33–75cm (13–30in). Family Macropodidae. *See also* KANGAROO.

Wallace, Alfred Russel (1823–1913), British naturalist and evolutionist, who developed a theory of NATURAL SELECTION independently of but at the same time as Charles DARWIN. He wrote *Contributions to the Theory of Natural Selection* (1870) which, with Darwin's *Origin of Species*, comprised the fundamental explanation and understanding of the theory of EVOLUTION. He gave his name to WALLACE'S LINE. *See also* NW pp.22–23, 33, *192, 196*.

Wallace, Edgar (1875–1932), English novelist and playwright who produced 170 novels in 25 years. He published his first novel, *The Four Just Men*, in 1905 and went on to become a best-selling writer whose fast-moving and tightly plotted stories remain popular. His first stage success came with *The Ringer* in 1926 and he wrote the screenplay of the classic film *King Kong* (1933).

Wallace, George Corley (1919–), US politician. As Democratic governor of Alabama in 1962, he attempted unsuccessfully to block federal efforts to end racial segregation in Alabama state schools (1962–66). His first wife, Lurleen Wallace, succeeded him as governor in 1966 but he was re-elected in 1970 and 1974. In 1972, while campaigning for the Democratic presidential nomination, he was shot and severely wounded and has remained paralysed from the waist down.

Wallace, Lewis (1827–1905), US writer, diplomat and Civil War general. He is best known for his popular, romantic novel *Ben-Hur* (1880), a story of the Roman Empire and the rise of Christianity. Other works include *The Fair God* (1873) and *The Prince of India* (1893).

Wallace, Sir William (*c.* 1270–1305), Scottish soldier. He led troops to victory over the English at Lanark and STIRLING BRIDGE in 1297 and carried out raids in the N of England. Defeated by EDWARD I's army at Falkirk in 1298, he fled to France to seek foreign aid. He returned to Scotland, was betrayed in 1305 and executed in London. *See also* HC1 p.183, *183*.

Wallace, William Vincent (1812–65), Irish composer who led an adventurous life, visiting Australia, India and South America. He became a violinist and a pianist. His opera *Maritana* (1845) was performed in London with great success. *See also* HC2 p.27.

Wallace's line, imaginary line dividing the islands Borneo and Bali from Celebes and Lombok. It was drawn by Alfred Russel WALLACE to demarcate animal populations which live close to each other and yet are extraordinarily different. Earlier this observation had influenced Wallace to state his theory of natural selection, which was coincident with that of Charles DARWIN. He argued that homogeneous animal populations could, for various reasons, divide and diversify gradually until they had evolved into separate species. *See also* NW p.*196*.

Wallach, Otto (1847–1931), German chemist who was awarded the 1910 Nobel Prize in chemistry for his research into TERPENES. In 1884 he began investigations into the essential oils in plants and, by a method of repeated DISTILLATION, he isolated the components of these complex compounds.

Wallachia, historic region in Romania; formerly the principality between the River Danube and the Transylvanian Alps. It is said to have been established in 1290 by Ralph the Black, vassal of the king of Hungary, from whom the region secured temporary independence in 1330. However, it came gradually under the domination of the Turks, whose suzerainty was acknowledged in 1417. Wallachia and MOLDAVIA became protectorates of Russia under the Treaty of Adrianople (1829) and, by their union in 1859, formed the state of Romania. Wallachia is an important agricultural region, but has been developed industrially in recent years. The Ploeşti oil fields are located near Bucharest; other products include chemicals and heavy machinery. Area: 76,599sq km (29,575sq miles).

Wallenstein, Albrecht Eusebius Wenzel von (1583–1634), German general. He sided with the Holy Roman Empire when

the THIRTY YEARS WAR broke out in 1618, and eventually became the commander of the imperial armies.

Waller, Fats (1904–43), US jazz and blues pianist and composer, real name Thomas Waller. He wrote many successful tunes, including *Honeysuckle Rose* and *Ain't Misbehavin'*, but owed his popular success to his ability to entertain audiences as a comedian as much as to his musical gifts.

Waller, Fred (1886–1954), US film director and inventor in 1928 of the ultra-wide filmscreen process known as CINERAMA. The image was registered by three lenses on three strips of film which were then projected side by side in exact alignment on a curved screen. The process was first used in *This is Cinerama* (1952).

Wall-eye. *See* STRABISMUS.

Wallflower, any of several species of sweet-scented perennial plants that are commonly cultivated in Europe. The European wallflower, *Cheiranthus cheiri*, has lance-shaped leaves and red, orange, yellow, or purple flowers. Height: to 90cm (36in). Family Brassicaceae.

Wallis, Alfred (1855–1942), English PRIMITIVE painter. He was born in Devon and was almost illiterate; he worked as a fisherman in St Ives from 1892. In his seventies he began painting (1928) and continued to do so until his death in a workhouse. His favourite subjects were St Ives and the sea, which he painted with ship's paint, usually on bits of cardboard carton, often allowing the asymmetrical, casually torn shape of the card to dictate the picture's design.

Wallis, Sir Barnes Neville (1887–), British aeronautical engineer and inventor, best-known for his invention of the bouncing bomb during WWII. In the 1920s he designed the airship R100 and after the war the first swing-wing aircraft.

Wallis, Henry (1830–1916), British painter. Among his best works is *The Death of Chatterton* (1856), painted in the manner of the Pre-Raphaelites, for which George MEREDITH was the model.

Wallis, John (1616–1703), English mathematician, who systemized the use of formulas and made a study of the quadrature of circles which he recorded in *Arithmetica Infinitorum* (1655).

Wallis and Futuna, French island territory in the s Pacific Ocean, w of Samoa. The territory is made up of two small groups of volcanic islands, the Wallis Islands and the Hoorn Islands. The principal islands are Uvea, Futuna and Alofi, and the chief town is Matautu. Timber is the main export. The French took the islands in 1842 and they became an overseas territory in 1959. Pop. (1972 est.) 10,000.

Walloons, term for the French-speaking people of southern Belgium as opposed to the Flemish-speaking people of the north. They inhabit chiefly the provinces of Hainaut, Liège (the term originally meant the French dialect spoken there), Namur and southern Brabant. In the early 1970s they numbered about 3,000,000.

Wallpaper, paper wall coverings which first appeared in Europe in the 15th century as replacements for the more expensive cloth wall hangings.

Wall Street, road in lower Manhattan, New York City, which since the mid-19th century has been the centre for the great financial houses of the USA, exchanges, brokerage firms and private banks. It gets its name from the wall built on the site in 1653 to protect the Dutch settlement of New Amsterdam from English attacks.

Walnut, deciduous tree native to North and South America, Europe and Asia. Walnuts have smoother bark than HICKORY trees, to which they are related, and are grown for timber, ornament and nuts. Height: to 50m (165ft). Family Juglandaceae; genus *Juglans. See also* NW p.*65*.

Walpole, Horace, 4th Earl of Orford (1717–97), British man of letters who was an MP from 1741 to 1768. In 1747 he bought Strawberry Hill, near Twickenham, where in 1757 he established a printing press. His literary reputation rests upon his letters which provide invaluable pictures of Georgian Britain. *See also* HC2 pp.34, 70, *71–72*.

Walpole, Sir Hugh Seymour (1884–

1941), British novelist, *b.* New Zealand, who first achieved success with *Fortitude* (1913). His best-known works include *Mr Perrin and Mr Traill* (1911) and the historical Herries tetralogy, comprising *Rogue Herries* (1930), *Judith Paris* (1931), *The Fortress* (1932) and *Vanessa* (1933), which were published in 1942 as *The Herries Chronicle.*

Walpole, Sir Robert, 1st Earl of Orford (1676–1745), British statesman. He was a Whig member of Parliament (1701–42), led parliamentary opposition to the Tories, and became Chancellor of the Exchequer in 1715. Although he resigned in 1717 after developing the first sinking fund, he restored order after the SOUTH SEA BUBBLE in 1720. He returned as Chancellor of the Exchequer and First Lord of the Treasury (1721–42) and in that capacity was England's leading politician. His financial policies served to encourage trade, but he was forced to resign in 1742 through opposition to his pacific foreign policy. During his long tenure of office, the HANOVERIAN SUCCESSION became firmly established. He is considered to have been England's first Prime Minister. *See also* HC2 pp.18, *18*, 312–313.

Walpurgisnacht (1 May), traditional witches' sabbath held in Germany on the eve of one of the feasts of St Walpurgis, or Walburga (*d.*779). The saint was an English missionary who founded a convent at Heidenheim. Her other feast is on 25 Feb.

Walras, Marie-Esprit-Léon (1834–1910), French economist. Much of the work of economic analysts is devoted to studying equilibrium in one particular market (eg the supply and demand conditions for wheat) but Walras was concerned with the interrelations among and between all markets. His method is mathematical in approach and abstract in structure.

Walrus, arctic seal-like mammal; it has a massive body and a large, round head. Its tusks, developed from upper canine teeth, may reach 1m (39in) in length and are used to rake up the sea-bed in search of molluscs. Length: to 3.7m (12ft). Family Otariidae; species *Odobenus rosmarus. See also* NW pp.164, *165, 224.*

Walsall, county district in N West Midlands, England; created in 1974 under the Local Government Act (1972). Area: 106sq km (41sq miles). Pop. (1974 est.) 271,100.

Walsh, Don (1931–), US naval officer. He was the officer in charge of the bathyscaphe *Trieste,* which made a record dive to 10,906m (35,780ft) in Jan. 1960.

Walsingham, Sir Francis (*c.*1532–90), English statesman, Secretary of State (1573–90). He was employed on diplomatic missions and, at his own expense, set up an intelligence network. Through his efficient espionage system he revealed the Throckmorton plot and the BABINGTON CONSPIRACY (1586). *See also* HC1 p.274.

Walter, Bruno (1876–1962), German conductor, real name Bruno Walter Schlesinger. After a series of conducting posts in Europe, he went to the USA in 1939. From 1947 to 1962 he was primarily associated with the Metropolitan Opera, New York, and was conductor of the New York Philharmonic Orchestra (1947–49). He was best known for his interpretations of such German and Austrian composers as MOZART, BRAHMS and MAHLER.

Walter, Hubert (*d.*1205), English statesman and clergyman. After being named Bishop of Salisbury in 1189, he joined Richard I's Third Crusade and was made Archbishop of Canterbury and chief justiciar in 1193. He raised Richard's ransom money, suppressed Prince John's rebellion and virtually ruled England during the king's absence.

Walter, John (1739–1812), British coal merchant and journalist. He was the founder of *The Times* newspaper – originally called *The Daily Universal Register* (1785), but renamed *The Times* in 1788. He took out a patent in 1783 for printing from logotypes (whole words or several letters cast in one piece).

Walter, Lucy (*c.*1630–58), Welsh mistress of the future CHARLES II from 1648 to 1651. Charles was the father of her son James, Duke of Monmouth, born in 1649. She is also known as Mrs Barlow and is frequently (although incorrectly) called Walters or Waters.

Walter, Thomas Ustick (1804–87), US architect. He is best remembered for his extensions to the CAPITOL at Washington, DC. He also designed the interior of the LIBRARY OF CONGRESS. He was president of the American Institute of Architects from 1876 until his death.

Waltherius, epic 9th- or 10th-century poem about the Germanic hero Waltherius, written in Latin, possibly by the monk Ekkehard I the Elder (*d.*973).

Walther von der Vogelweide (*c.*1170–*c.*1230), German poet, considered the greatest MINNESINGER of the Middle Ages. He worked at the court of Duke Frederick I in Vienna until 1198, when Frederick died. *See also* HC1 p.206.

Walton, Ernest Thomas Sinton (1903–), British physicist who shared the 1951 Nobel Prize in physics with Sir John COCKCROFT for the development of the first nuclear particle accelerator. After two unsuccessful attempts they finally succeeded in 1929, using a voltage multiplier to accelerate protons. In 1931, they produced the first artificial nuclear reaction without radioactive isotopes, using high-energy protons to bombard lithium nuclei. This process was of great importance for studying nuclear structure and its modern counterparts are among the most useful tools of NUCLEAR PHYSICS. *See also* SU pp.67, 95.

Walton, Izaak (1593–1683), English author and biographer. He is best known for *The Compleat Angler, or the Contemplative Man's Recreation,* which first appeared in 1653 and was frequently revised with the addition of new material. His biographies include *John Donne* (1640), *Sir Henry Wotton* (1651) and *George Herbert* (1670).

Walton, Sir William Turner (1902–), British composer who, using a largely conservative, post-ROMANTIC style, composed in practically every medium. His orchestral works, noted for their melodic structure, include *Portsmouth Point* (1926) and the comedy overture *Scapino* (1941). He wrote *Crown Imperial* (1937) and *Orb and Sceptre* (1953) as coronation marches for George VI and Elizabeth II. His most widely known works are the jazz-oriented *Façade* (1923), the oratorio *Belshazzar's Feast* (1931) and the opera *Troilus and Cressida* (1954). Later compositions include the *Missa Brevis* (1966) and *Sonata for String Orchestra* (1972). *See also* HC2 p.270.

Waltz, dance performed by couples to music in triple time. This graceful ballroom dance came into fashion in the early 19th century, having developed in the previous century from South German folk dances, such as the Ländler.

Walvis Bay, inlet port and surrounding territory on the W coast of Namibia. Walvis Bay is the home port of several whaling and fishing fleets, and fish canning is an important local industry. Pop. (1970 est.) 23,460.

Walworth, Sir William (*d.*1385), English fishmonger and Lord Mayor of London in 1374 and 1381. He defended London Bridge against Kentish rebels during the PEASANTS' REVOLT of 1381 and killed the peasants' leader, Wat TYLER, at Smithfield. *See also* HC1 p.215.

Wampum, or wampumpeag, beads used by N American Indians as ornaments and as a medium of exchange in early colonial trade. Wampum beads were usually made from shells and were coloured blue and white.

Wanamaker, Sam (1919–), US actor and director of films and theatrical productions. In the cinema, he appeared in *Taras Bulba* (1962) in the *The Voyage of the Damned* (1975) and directed *The Executioner* (1970) and *Sinbad and the Eye of the Tiger* (1975). In the theatre, he was associated with the establishment and direction of the Shakespearean Globe Playhouse on Bankside, London, in 1970.

Wanderers, The, group of thirteen Russian painters who rejected academic formulas in 1870 and attempted to paint realistic pictures of rural and peasant life. They organized their own exhibitions, which they took around Russia themselves. They included Ivan Kramskoy, Ilya REPIN and Vasily VERESHCHAGIN.

Wandering Jew, according to legend and popular literature, a Jew who abused Jesus Christ on His way to Calvary, and was condemned to wander the Earth until the SECOND COMING. Although the story first appeared in 13th-century literature, the wanderer was not identified as a Jew until the 17th century.

Wandering jew, or inch plant, creeping house-plant native to South America. Plants in the genus *Tradescantia* have green-and-white striped oval leaves. Those in the genus *Zebrina* have purplish-green leaves. Family Commeliaceae. *See also* NW p.69.

Wandiwash, Battle of (1760), victory of the British army over the French in the SEVEN YEARS WAR at Wandiwash, India, SW of Madras. The British army was led by Sir Eyre COOTE, the French by the Irishman Count Lally. Together with the capture of PONDICHERRY in 1761 it secured for the British the control of the Carnatic, the SW region of India on the Arabian Sea which was the centre of the Anglo-French struggles for supremacy in India during the 18th century.

Wang An-shih (1021–86), Chinese statesman of the SUNG DYNASTY. To strengthen the central government he introduced administrative reforms such as cheap government credit.

Wang Ching-wei (1883–1944), Chinese political leader. He was a close associate of SUN YAT-SEN and was a leader of the KUOMINTANG. He eventually opposed the rule of CHIANG KAI-SHEK. Wang led the Japanese puppet regime in China from 1940 until his death.

Wangchuk, Jigme Singhe (1955–), King of Bhutan from 1972 (4th sovereign of the Wangchuk dynasty), crowned on 2 June 1974. The youngest reigning sovereign in the world at that time, he continued his father's policy of modernization and represented his country at the nonaligned nations' conference in 1976.

Wang Meng (*c.*1308–85), Chinese landscape painter of the YÜAN DYNASTY. One of the great masters of the 14th century, he was renowned for the virtuosity and variety of his brushwork. *See also* HC1 p.129.

Wang Pi (249–226 BC), one of the creators in China of Neo-TAOISM. A commentator on the Tao te Ching and the I Ching, he conceived Tao as "non-being" which he regarded as a universal and absolute metaphysical reality, and therefore not merely as the antithesis of being.

Wang Yang-ming (1472–1529), Chinese idealist philosopher and government official. He believed that general principles are to be found in the human mind, not through the study of physical objects, and that knowledge and action are identical. His teachings had a great and long-lasting influence on Oriental philosophy. *See also* MS pp.*232–233.*

Wankel rotary engine, most widely-used rotary engine. The rotor is an equal-sided triangular piece that rotates in an orbit within a specially constructed casing. Crescent-shaped combustion chambers, created by the rotor, increase and decrease in size as the rotor turns. At the appropriate time sparks ignite the fuel and air mixture (from a carburettor) in each chamber. *See also* MM p.63.

Wapentake, in Anglo-Saxon England administrative division of the shire equivalent to the HUNDRED, in those parts of the country where Danish influence was strong. Wapentakes were found especially in the Midlands and in Yorkshire.

Wapiti, large deer of North America, closely related to the Old World red deer, second only to the MOOSE in size. It is grey-brown with a whitish rump and dark, brown-black legs, head and neck; its antlers may reach a span of 1.5m (5ft). In the USA it is commonly known as an ELK. Height: to 1.5m (5ft); length: to 2.5m (7.5ft). Family Cervidae; species *Cervus canadensis. See also* NW p.*207.*

Walruses use their tusks to lever clams from rocks and to haul themselves ashore.

Bruno Walter made his debut in the USA in 1923 and became resident there in 1939.

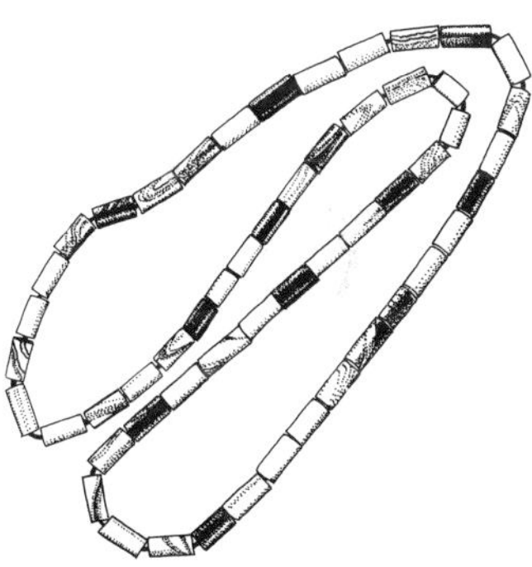

Wampum beads, made from sea shells, were used both as currency and as ornaments.

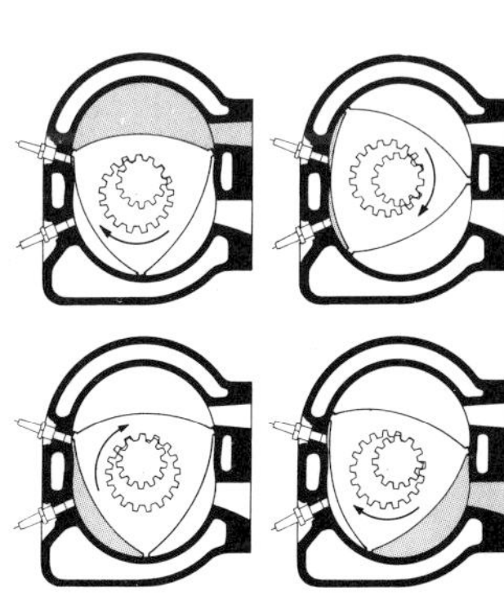

Wankel rotary engines have few moving parts and are quieter than most engines.

War and Peace

War and Peace has over 500 characters and ranks among the greatest of novels.

Warblers are divided into two groups; *Sylvia communis* is an Old World species.

Andy Warhol is sometimes criticised for lowering artistic standards with his art.

Warsaw Pact delegates signing a declaration of member states in Prague.

War and Peace (1865–72), epic novel by Leo TOLSTOY. It is both an historical novel, describing the NAPOLEONIC WARS in Russia, and a family chronicle that examines the lives of several families, centering on the Rostovs.

Waratah, small shrubby flowering plant that grows in acid soils in near-desert conditions in w New South Wales, Australia. It has slender oval alternate leaves and large dense clusters of red blooms up to 10cm (4in) in length. Genus *Telopea. See also* NW pp.*63, 210.*

Warbeck, Perkin (1474–99), Flemish pretender to the English throne. He travelled in Ireland and Europe (1491–93) professing to be Richard, Duke of York, son of Edward IV. He attempted to invade England three times (1495, 1496 and 1497), but despite Scottish support was finally captured and executed.

Warble fly, large, hairy, bee-like fly. It lays its eggs on the feet of animals such as cattle and deer. The larvae work their way through the skin into the animal's body, causing painful swellings, or warbles. Family Hypodermatidae, species *Hypoderma lineatum* and *H.bovis.*

Warbler, name applied to numerous birds of two families, one in the Old World (Sylviidae) and the other the New World WOOD WARBLERS (Parulidae). Old World warblers include the HEDGE SPARROW and TAILORBIRD. Most New World warblers generally have brighter plumage. *See also* NW pp.*33, 148, 211.*

Warburg, Otto Heinrich (1883–1970), German biochemist who was awarded the 1931 Nobel Prize in physiology and medicine for his discovery of respiratory ENZYMES. He made significant contributions to the understanding of the mechanisms of cellular respiration and energy transfer, and of PHOTOSYNTHESIS.

War crimes, violations of international laws of war. The term came into use after WWII when certain German and Japanese leaders were held responsible for violating the laws of war and were tried by Allied military tribunals in Nuremberg and Tokyo (1945–47). In the Vietnam War, Lieut. William Calley was found guilty of war crimes for ordering a massacre of civilians at My Lai in 1968.

Ward, Barbara Mary, Baroness Jackson of Lodsworth (1914–), British economist and author. She taught at Harvard University (1957–68) and at Columbia University (1968–73). Her books include *Policy for the West* (1951), *The Rich Nations and the Poor Nations* (1962), and *Nationalism and Ideology* (1967). She became president of the Conservation Society in 1973, and was made a life peer in 1976.

Ward, Mrs Humphry (1851–1920), British novelist and philanthropist, real name Mary Augusta Arnold, b. Tasmania. She was the granddaughter of Thomas ARNOLD and wife of Thomas Humphry Ward. Her works include *Robert Elsmere* (1888) and *Bessie Costrell* (1895).

Ward, James (1769–1859), British painter and engraver. He was appointed "painter and mezzotinter" to the Prince of Wales in 1794. His large landscape *Gordale Scar* (1815) shows his romantic feeling for grandiose aspects of nature.

Ward, James (1843–1925), British philosopher. He became a fellow of Trinity College Cambridge in 1875 and was professor of mental philosophy at the University from 1897 to 1925.

Ward, Sir Joseph George (1856–1930), New Zealand politician. As leader of the Liberal party he was twice Prime Minister (1906–12 and 1928–30). He introduced the penny postage in 1901 and state fire insurance in 1903. He represented New Zealand at the IMPERIAL WAR CABINETS of 1917–18 and at the Paris peace conference of 1919.

Ward, Lester Frank (1841–1913), US sociologist. He developed a theory of progress for mankind whereby education and development of intellect could direct social evolution. His works include *Dynamic Sociology* (1883).

Warden of the Cinque Ports, English office dating from 1268, when HENRY III gave to the constable of Dover Castle the duty of administering the CINQUE PORTS. The warden was president of Shepway, the royal court established in 1150 as the court of appeal for the five ports. By the 15th century the warden had taken over admiralty jurisdiction over the ports. The office is now purely ceremonial.

Ward of Court, young person who is under the guardianship of a court because of his involvement in a lawsuit concerning property rights. The court's obligations include protection of its ward from financial loss.

Wardrobe, department of the royal HOUSEHOLD in medieval England. It replaced the EXCHEQUER, which had moved "out of court" as the financial department in the reign of HENRY III. It exercised its greatest power in the reign of EDWARD I, when it looked after not merely the household finances but also the expenses of the king's armies.

Wards and Liveries, Court of, English court established by HENRY VII and made into a formal court with its own seal by HENRY VIII. It heard cases arising out of the Crown's enforcement of FEUDAL dues, particularly the right of wardship. The court was abolished by the LONG PARLIAMENT in 1646.

War game, artificial simulation of war, as a pastime or for military planning. As a pastime, war games are played on a table or a map, using model soldiers. In modern military planning, war games are played to assess the possibilities of tactical or strategic situations, and sophisticated equipment, including computers, is used.

Warhol, Andy (1930–), US painter, printmaker, and film-maker. A leader of the POP ART movement, he produced prints of soup cans, packing cartons and faces of film stars. He was the centre of the New York avant-garde in the 1960s, and made a number of films in which the camera impassively observes the everyday life of the actors. *See also* HC2 pp.*277, 279.*

Wari. *See* HUARI.

Warlock, Peter (1894–1930), pseudonym of Philip Heseltine, British composer and musical scholar. He was influenced by DELIUS and Bernard van Dieren and is best known for his song cycle *The Curlew* (1924, based on a poem by W.B. YEATS) and the *Capriol Suite* for string orchestra (1927).

Warlords, regional military leaders in China, particularly in the late 19th and early 20th centuries. From 1912 to 1928 China was torn by rivalry among the warlords, whose struggles prevented growth and consolidation of national power and left China weak and powerless to resist the pressures of imperial Japan. The warlords were separately backed by some European powers and Japan.

Warner, Rex (1905–), British novelist, poet and a classical scholar noted for his translations from Greek and Latin. He was influenced by Franz KAFKA and his early novels, which include *The Wild Goose Chase* (1937) and *The Aerodrome* (1941), are allegories concerned with the problems of power. He also wrote historical novels, among which are *The Young Caesar* (1958) and *Pericles the Athenian* (1963).

Warner Brothers, US film production company formed in 1923 by four brothers, Harry, Jack, Albert and Sam. The first company to introduce sound, they produced *The Jazz Singer* in 1927 and an all-dialogue film, *The Lights of New York*, in 1928. Later films included *My Fair Lady* (1964) and *Bonnie and Clyde* (1967). Warner Brothers were amalgamated with COLUMBIA in 1972. *See also* HC2 pp.*201, 268.*

War of 1812 (1812–1815), conflict between the USA and Britain. Minor points of friction between the two countries, together with the view of some US politicians (known as the "War Hawks") that war with Britain would enable them to seize territory in Canada, led to the outbreak of hostilities. New England, which was economically dependent on trade with Britain, was not enthusiastic about the war. Early attempts by US forces to take Canada were soundly defeated. The small US navy enjoyed some success, but by 1813 the powerful British navy had established an efficient blockade of the US coast. British raids had no long-term success, although one such raid burned Washington, the capital. Peace negotiations were opened at Ghent, Belgium, in 1814. Fifteen days after peace terms were signed, Andrew JACKSON defeated a British army at New Orleans in Jan. 1815.

War Office. *See* DEFENCE, MINISTRY OF.

War of the Worlds, The (1898), science fiction novel by H. G. WELLS. It depicts a Martian invasion of England, and the resultant panic and destruction. The Martians are finally overcome by bacteria in the Earth's atmosphere.

Warrant, legal document of three main kinds. It may be a WRIT conferring some title or authority upon a person, it may be a command delivered to an officer to arrest an offender, or it may be a citation or summons. *See also* SEARCH WARRANT.

Warrant officer, in the Royal Navy, rank between a rating and a commissioned officer, also known as Fleet Chief Petty Officer; in the British Army, rank between non-commissioned and commissioned officer; in the RAF, highest non-commissioned officer in ground trades.

Warren, Robert Penn (1905–), US poet and novelist. In his works he concentrates on Southern themes and characters that are usually related to real incidents. In *All The King's Men* (1946) the central character resembled a ruthless, well-known Southern politician. The book was awarded the 1947 Pulitzer Prize for literature. He won the prize again in 1958 for *Promises*, a book of poetry.

Warren Commission, US commission appointed in November 1963 by President Lyndon B. JOHNSON to investigate the assassination of President John F. KENNEDY. It included US Chief Justice Earl Warren and various senators, congressmen and legal staff. Its conclusion, delivered in September 1964 after taking evidence from 552 witnesses, was that the assassination was a lone act by Lee Harvey OSWALD, and not part of a conspiracy. This, however, has not been universally accepted.

Warrington, county district in central N CHESHIRE, England; created in 1974 under the Local Government Act (1972). Area: 176sq km (68sq miles). Pop. (1974 est.) 163,800.

Warsaw (Warszawa), capital of Poland, on the River Vistula. The largest city in Poland, its first settlements date from the 11th century, after which it developed as a trade centre, and was made the country's capital in 1596. It was taken by Russia in 1813 and occupied by the Germans in both World Wars. The WARSAW PACT, a union of E European Communist nations, was signed in the city in 1955. Warsaw has a fine Gothic cathedral and medieval castle. It is an important transport centre with large concentrations of heavy industry. Products include chemicals, textiles, electrical equipment, iron and steel. Pop. (1975 est.) 1,436,000.

Warsaw, Grand Duchy of, French dependency, created in 1807 by NAPOLEON I out of land Prussia had received in the second and third partitions of Poland, with a constitution modelled on that of France. Polish soldiers who fought with Napoleon's army hoped for independence but after Napoleon's defeat in 1815 about a quarter of the area became the Grand Duchy of POZNAN under Prussian control, and the rest became Congress Poland, an independent nation.

Warsaw Pact, popular name for the Eastern European Mutual Assistance Treaty, signed on 14 May 1955 by Albania, Bulgaria, Czechoslovakia, East Germany, Hungary, Poland, Romania and the USSR. It was intended to give the alliance of Communist bloc countries a more formal organization and to act as a counter to NATO, which in the previous year had admitted West Germany to membership. The member nations pledged themselves to mutual defensive aid and agreed to establish a joint armed command and a political consultative committee. Albania ceased to participate in its activities in 1961 and formally withdrew in 1968 because of the conflict between China and the USSR.

Warsaw Uprising (Aug.–Oct. 1944), action during WWII in which the Polish underground, led by Gen. KOMOROWSKI, attacked the German occupation troops in Warsaw in an attempt to ensure the post-war authority of the government in exile. The breakthrough of Allied armies in France and the rapid approach of Russian forces to the outskirts of Warsaw encouraged the resistance soldiers, but the Soviet advance halted. The Poles fought valiantly for 63 days unaided before they were overwhelmed.

Warship, vessel built for combat at sea. The Egyptians had developed fast strong vessels whose main purpose was warfare by 3000 BC, but the Minoans were probably the first to build warships which were more than a variant on merchant vessels (c. 2000 BC). From about 800 BC the Greeks were using war GALLEYS such as TRIREMES. The efficiency of these ships was improved during the 4th century BC when heavy missile weapons were mounted in them. Galleys continued as the main warships of the Mediterranean until AD 1600. In N Europe, boats such as the Viking longships were used both for war and trading; medieval naval battles were fought by converted merchantmen. During the 15th and 16th centuries the adoption of cannon and new sailing techniques changed naval warfare. By the 18th century, European fleets were composed of specialized vessels, from the heavily armed SHIPS OF THE LINE to the lighter corvettes and FRIGATES. In the 19th century, steam propulsion, rifled artillery and armour plating brought more changes. Fleets of huge BATTLESHIPS (DREADNOUGHTS) were built before WWI. In their turn these were superseded by SUBMARINES and AIRCRAFT CARRIERS, which were the most important ships of WWII. Since then, the destructive power of individual vessels has greatly increased, and many are now armed with nuclear weapons. *See also* MM pp.172–175.

Wars of the Roses. *See* ROSES, WARS OF THE.

Wart, raised and well-defined small growth on the outermost surface of the skin, caused by a VIRUS. It is usually painless unless in a pressure area.

Warthog, wild, tusked PIG native to Africa. It has a brownish-black skin with a crest of thin hair along the back. Height: about 76cm (2.5ft) at shoulder; weight: 90kg (200lb). Family Suidae; species *Phacochoerus aethiopicus. See also* NW p.201.

Warwick, John Dudley, Earl of. *See* NORTHUMBERLAND, JOHN DUDLEY, DUKE OF.

Warwick, Richard de Beauchamp, Earl of (1382–1439), English soldier; son of Thomas, Earl of Warwick. A supporter of Henry IV, he defeated Owain GLYNDWR and fought Sir Henry PERCY at Shrewsbury in 1403. He was appointed Captain of Calais in 1414 and fought for Henry V in France. He was tutor to Henry VI, and in 1437 became Lieutenant of France and Normandy. *See also* HC1 p.243.

Warwick, Richard Neville, Earl of (1428–71), English politican and soldier, known as "the Kingmaker". During the Wars of the ROSES he helped the Yorkists to victory at St Albans in 1455 and captured Henry VI at Northampton in 1460. Warwick was the real ruler of England from 1461 to 1464, in the first part of Edward IV's reign. Warwick gradually lost power, however, and in 1469 turned against Edward, whom he drove into exile in 1470. Edward returned in 1471 and Warwick was killed at the Battle of BARNET. *See also* HC1 pp.242, 243.

Warwick, Sir Robert Rich, 2nd Earl of (1587–1658), English naval commander and colonialist. He was given the patent to administer Massachusetts in 1628 and Connecticut in 1631. He was a leader of the Puritan opposition to CHARLES I and as Lord High Admiral (1643–45) secured the navy for the parliamentary side in the ENGLISH CIVIL WAR.

Warwick, Thomas de Beauchamp, Earl of (*d*.1401), English politican. An important magnate, he was one of the governors of the young RICHARD II, but later opposed Richard and tried to curb his power

(1387). He was imprisoned and banished for alleged treason in 1397, but restored to his estates by Henry IV in 1399.

Warwick, county town and county district in WARWICKSHIRE, central England. The city, on the River Avon, has light industry, but it is best known for its 14th-century castle, which attracts many tourists. The county district was created in 1974 under the Local Government Act (1972). Area: 283sq km (109sq miles). Pop. (county district, 1974 est.) 111,100; (city, 1971 est.) 18,289.

Warwickshire, county in central England. The land is gently rolling, rising to the Cotswold Hills in the S, and is drained chiefly by the River Avon. Cereals are the principal crops and dairy cattle and sheep are raised. Coal is mined in the NE and there is some light industry near Leamington. The county town is Warwick. Area: 1,981sq km (765sq miles). Pop. 471,000.

Wash, The, shallow inlet of the North Sea in E England, between Lincolnshire and Norfolk. It receives the rivers Witham, Welland, Nene and Great Ouse. The Wash is shallow with many sandbars, but dredged channels lead to Boston and King's Lynn. Length: 24km (15 miles). Width: 19km (12 miles).

Washing machine, appliance that washes clothes and linen. The first manually operated machine was sold in 1832, and electrically powered machines appeared in 1914. *See also* MM p.251.

Washing soda. *See* SODIUM CARBONATE.

Washington, Booker Taliaferro (1856–1915), US educator, born a slave in Virginia. After emancipation he educated himself, then taught at a school for black children. In 1881 he founded Tuskegee Institute (Alabama) for the training of blacks as teachers, farmers, mechanics and tradesmen.

Washington, George (1732–99), first US president (1789–97). He was born into a wealthy Virginia family and trained as a surveyor. As a colonel in the Virginia Militia during the French and Indian Wars, he won a reputation as an astute military commander and was chosen to lead the poorly equipped, untrained Continental Army against the British in the War of American INDEPENDENCE. He retired to his estate after the victory but was called back to preside over the Federal Constitutional Convention (1787). When the new country's Constitution was adopted and ratified, Washington was elected President. *See also* HC2 pp.64, 64–65.

Washington, state in NW USA, on the Pacific Ocean and the Canadian border. The W of the state is dominated by Puget Sound, an arm of the Pacific, and the Olympic Mts. The Cascade Range crosses the state from N to S, and to the E is the Columbia Plateau. The Columbia is the principal river. Washington is the leading US producer of apples, and wheat is also important. Among Washington's principal resources are its rivers. The Columbia is one of the world's best sources of hydroelectricity, and is also used for irrigation. Half the state is forested and mineral deposits include magnesium and aluminium ores. The principal industries are food processing, aircraft, missiles and spacecraft, fishing, wood products and tourism. Olympia is the state capital and Seattle is the largest city.

The Spanish discovered the mouth of the Columbia River in 1775. After Captain James COOK's visit in 1778, the fur trade began and traders established a fur trading post at the river's mouth in 1811. The boundary with Canada was fixed by treaty with Britain in 1846, and the Oregon Territory, which included Washington, was established in 1848. Washington Territory was created separately in 1853. Exploitation of its forests and fisheries attracted settlement. Washington was admitted to the Union in 1889. Area: 176,616sq km (68,191sq miles). Pop. (1975 est.) 3,544,000. *See also* MW p.175.

Washington, DC, capital of the USA, on the E bank of the Potomac River, covering the District of Columbia and extending into the neighbouring states of Maryland and

Virginia. In 1790 the site was chosen by President George WASHINGTON as the seat of government, and the city was planned by the French engineer Pierre Charles L'Enfant. Construction of the WHITE HOUSE began in 1793 and of the Capitol the following year. The first Congress was held in Washington in 1800 and Thomas JEFFERSON was the first President to be inaugurated there (1801). During the WAR OF 1812, the city was occupied by the British and many public buildings were burned (1814), including the White House and the Capitol. Other landmarks include the Washington Monument (1848), Lincoln Memorial (1922), the Jefferson Memorial (1943) and the National Archives. Washington is the Legislative, judicial and administrative centre of the USA. Its governmental buildings include the Library of Congress, the PENTAGON, the SUPREME COURT, SENATE and HOUSE OF REPRESENTATIVES and Constitution Hall. Cultural centres include the National Gallery of Art, SMITHSONIAN INSTITUTION and the John F. Kennedy Center for the Performing Arts. There are several universities and institutions of scientific research in the city. There is little industry apart from tourism: the city attracts millions of tourists each year. Pop. (1975 est.) 716,000 *See also* PE p.60.

Washington, Treaty of (1871), agreement between the USA and Britain to submit outstanding disputes to an international arbitration commission. The disputes involved included the ALABAMA CASE, the rights of US fishermen in Canadian waters and the boundary between British Columbia and the state of Washington.

Wasp, any insect of the stinging HYMENOPTERA that is neither a bee nor an ant. The common wasp (*Vespa vulgaris*) has four wings and a yellow body ringed with black. Adults feed on nectar, tree sap and fruit. Length: to 3cm (1.2in). Family Vespidae. *See also* NW pp.79, *79, 107, 113, 116, 206, 208, 209*.

Wasps, The, comedy by ARISTOPHANES produced in 422 BC. It satirizes the Athenian jury system. Philocleon defends it, Bdelycleon attacks it. The wasps of the title are a chorus of jurymen disguised as wasps who rescue Philocleon so that he may attend the court.

Wasserman, August von (1866–1925), German bacteriologist. He made important contributions to immunology, the best-known of which was his development of a diagnostic test for SYPHILIS, the Wasserman test, still used today.

Wasserman, Jakob (1873–1934), German novelist who became famous for *Christian Wahnschaffe* (1919, tr. *The World's Illusion,* 1920), in which he urged a return to the selflessness of early Christianity. His other novels include *Caspar Hauser* (1909).

Waste, radioactive, residues, generally from nuclear reactors, containing radioactive substances. After uranium, plutonium and other useful fission products have been removed, some long-lived RADIOACTIVE ELEMENTS remain, such as caesium-137 and strontium-90. International regulations exist for disposing of such wastes, separate regulations applying to liquids, gases and solids.

Waste disposal. *See* REFUSE DISPOSAL.

Waste Land, The (1922), poem by T. S. ELIOT. In a series of vignettes it attempts to portray the disillusionment and moral disgust of the post-WWI generation with the barrenness and corruption of modern civilization. In it Eliot combined a wealth of literary and mythological allusion with an original and rich use of language. It had immense influence on poets of the day. *See also* HC2 p.289.

Watch, small, portable timepiece usually worn on the wrist. *See also* MM pp92–93.

Watch and ward, system by which a community in England was made responsible for ensuring its own safety. HENRY III issued writs of watch and ward (the first in 1233) requiring that guards be appointed in each township. The system was made permanent by the Statute of WINCHESTER (1285). The statute was repealed in 1827.

Water, odourless, colourless liquid that covers about 70% of the Earth's surface

Warships, such as this Viking longship, were used for fighting as well as trading.

Warthog is docile unless cornered or attacked; the tusks may excede 0.5m (20in).

George Washington is proudly remembered by most Americans as the father of the USA.

Wasps build their nests from paper which they make by pulping wood with saliva.

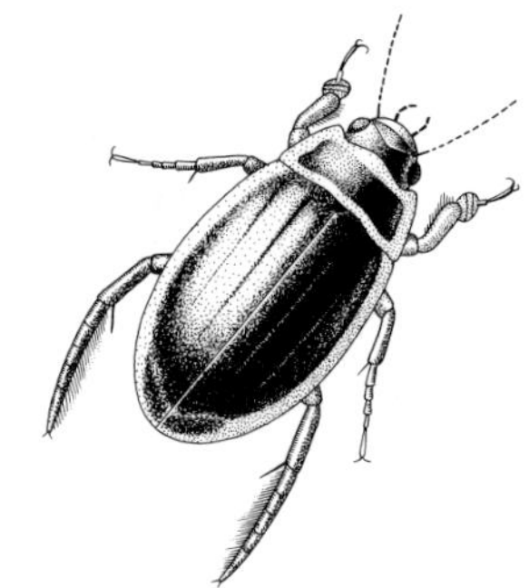

Water beetle paddles itself along with its fringed hind legs, which act as oars.

Waterbucks live near lakes and streams; *Kobus ellipsiprymnus* is a fine swimmer.

Watergate prosecutor Leon Jaworski speaking to reporters during the trials of 1974.

Water skiing relies on the speed of the tow boat and the foil shape of the skis.

and is the most widely used solvent. It is a compound of hydrogen and oxygen (H_2O) with the two H-O links of the molecule forming an angle of 105°. This asymmetry results in polar properties and a force of attraction (hydrogen bond) between opposite ends of neighbouring water molecules. These forces maintain the substance as a liquid, in spite of its low molecular weight, and account for its unusual property of having its maximum density at 4°C. Properties: s.g. (4°C) 1.000; m.p. 0°C (32°F); b.p. 100°C (212°F).

Water, heavy. *See* HEAVY WATER.

Water Babies, The (1863), children's book by Charles KINGSLEY. It is a fantasy account of the underwater adventures of a little chimney sweep who falls into a river and is transformed into a "water baby".

Water beetle, aquatic beetle, especially the predaceous DIVING BEETLE (family Dytiscidae), the water scavenger beetle (family Hydrophilidae) or the whirligig beetle (family Gyrinidae).

Water boatman, large-eyed aquatic insect found worldwide. Its body is grey to black, oval and flat, with fringed, oarlike hind legs. It feeds on ALGAE and micro-organisms in ponds and streams and is an important fish food. Length: about 15mm (0.6in). Order HEMIPTERA.

Waterbuck, or waterbok, large, gregarious, coarse-haired ANTELOPE, native to Africa south of the Sahara, and the Nile Valley. There are six species. Long, ringed horns on males slope backwards, then curve forwards at the tips. Length: 1.4–2.1m (4.5–7ft); height: 1.1–1.5m (3.6–4.9ft) at the shoulder. Family Bovidae; genus *Kobus. See also* NW p.*200.*

Water buffalo, or caraboa, large OX, widely domesticated in much of the tropical world; it is feral in some parts of India. Height: to 1.8m (6ft) at the shoulder. Family Bovidae; species *Bubalus bubalis. See also* NW p.*163.*

Water chestnut, floating, aquatic plant found in Asia, Europe and Africa. It has diamond-shaped leaves and edible, spiny, chestnut-like fruit. Family Trapaceae. Species *Trapa natans.* The water chestnut of Chinese cookery is the succulent corm of a Chinese sedge, *Eleocharis tuberosus.* Family Cyperaceae. *See also* PE p.191.

Watercolour, in art, paint that is made from a colour pigment ground up with a water-soluble gum, such as gum arabic; the term also refers to a painting that is rendered in this medium. The use of transparent paint may be traced to the paintings of ancient Egypt, China and Japan. In Europe, from the 15th century, artists such as DÜRER, and later REMBRANDT and CLAUDE LORRAIN (both of whose watercolour paintings were commonly executed in monochrome) explored the technique. In the late 18th and 19th centuries a freer, bolder use of the medium was made by COZENS, GIRTIN and COTMAN, all of whom helped to establish the English tradition. The work of TURNER was possibly the most adventurous in that he departed from the strictly classical tradition in his use of such techniques as sponging, wiping or scraping the paint to achieve a greater range of highlights.

Watercress, floating or creeping plant found in running or spring waters. The succulent leaves, divided into small, oval leaflets, have a pungent flavour and are used in salads and soups. The clustered flowers are white. Height: 25.4cm (10in). Family Brassicaceae; species *Nasturtium officinale. See also* PE pp.190, *190–191.*

Water cycle. *See* HYDROLOGIC CYCLE.

Waterfall, point in the course of a river at which the water drops perpendicularly. The site of a waterfall usually indicates an outcrop of rock that is particularly resistant to erosion. *See also* PE p.*113.*

Water flea, any of numerous species of small, chiefly freshwater branchiopod CRUSTACEANS, especially those within the genus *Daphnia,* which are common throughout the world. Order Cladocera. *See also* DAPHNIA; NW pp.*100, 105, 230–231.*

Waterford, county in s Republic of Ireland, in Munster province, on the Atlantic Ocean. It is a mountainous region, drained chiefly by the rivers Blackwater

and Suir. The raising of beef and dairy cattle and sheep is the chief agricultural activity. Industries include fishing, food processing, tanning and glassware. Waterford is the county town and an important port serving the whole of s Ireland. Area of county: 1,838sq km (710sq miles). Pop. (1971) 77,315.

Waterford glass, Irish glass first made in the late 18th century. A glass-house existed in Waterford until 1851, making crystal glass with a high lead content (34%). Production recommenced in 1951. The glass is traditionally decorated with deeply cut diamond designs.

Waterfowl, birds, including DUCKS, GEESE and SWANS, found throughout most of the world. Large flocks migrate from cool nesting grounds to warm winter homes. All have short bills, short legs, and dense plumage underlaid by down. Undomesticated species are known as wildfowl in Britain. Order Anseriformes.

Watergate Scandal (1972–74), US political scandal involving the NIXON administration. It arose from an attempted burglary of the DEMOCRATIC PARTY's national offices by persons working for President Nixon's re-election committee. The administration's efforts to hide the connection provoked investigations by the SENATE and Justice Department which ultimately implicated the President. Impeachment proceedings against him were begun in 1973. In August 1974, after being ordered by a SUPREME COURT ruling to relinquish tape recordings that attested to his involvement in the affair, Nixon resigned. Although Nixon was pardoned by his successor, Gerald FORD, other members of the administration were prosecuted and convicted.

Waterhouse, Alfred (1830–1905), British architect. Among his most important works are Manchester Town Hall (1868–77), Balliol College, Oxford (1867–69), and the Natural History Museum, London (1868–80).

Water hyacinth, aquatic herb family, native to tropical America. It has swollen petioles that float in water, and spikes of violet flowers. Family Pontederiaceae; species *Eichhornia crassipes.*

Waterlily, any of about 90 species of freshwater plants widely distributed in temperate and tropical regions. They have thick perennial rootstocks and showy flowers of white, pink, red, blue or yellow. Family Nymphaeceae; genera *Nymphaea, Nuphar, Nelumbo* and *Victoria.*

Waterloo, Battle of, final engagement of the NAPOLEONIC WARS, fought on 18 June 1815. NAPOLEON I led a French army 72,000 strong in an attack on 68,000 men under the Duke of WELLINGTON, who held positions along Mont St Jean near the village of Waterloo, Belgium. French victory would have been certain but for the arrival of Wellington's allies, the Prussians under BLÜCHER, on the French right. There were eventually 45,000 Prussians engaged. From midday until 8 pm, Wellington's army held out until the French were routed. There were 23,000 allied casualties and the French lost 33,000 men. The battle resulted in the final abdication and exile of Napoleon. *See also* HC2 pp.77, *77, 79, 79.*

Waterloo Cup, chief hare COURSING event in Britain. First contested in 1836, it takes place in Lancashire over three days each February.

Waterman, Lewis Edson (1837–1901), US inventor. He is remembered chiefly for his production and manufacture of the first modern fountain pen in 1884. *See also* MM p.248.

Watermark, translucent impression made in otherwise opaque paper, either by machine or by hand. It is formed during the making of paper by a raised design on wires over which the paper is passed while still in a pulpy state. Watermarks have been used since the latter part of the 13th century as a guarantee of authenticity, as in banknotes, and to identify the makers of high-quality papers. *See also* MM pp.*60,* 60–61.

Watermelon, trailing annual VINE native to tropical Africa and Asia and cultivated in warm areas worldwide. Its fruit has a

hard, greenish rind, sweet, red, juicy flesh and many seeds. Family Cucurbitaceae; species *Citrullus lanatus. See also* GOURD.

Water mill, machinery powered by the flow of water past a waterwheel, used in early times to mill food grains. Waterwheel. *See also* MM p.70, *70.*

Water moccasin, or cottonmouth, venomous semi-aquatic SNAKE of SE USA. It is a PIT VIPER, closely related to the COPPERHEADS. It vibrates its tail and holds its white mouth open when excited. Length: to 1.2m (4ft). Family Viperidae; species *Ancistrodon piscivorus.*

Water of crystallization, definite molecular proportion of water that is chemically combined with certain substances in the crystalline state. Much of this water will be lost on heating to about 100°C (212°F) but some, the water of constitution, will be retained to a much higher temperature. Cupric sulphate ($CuSO_4.5H_2O$), loses four molecules of water at 100°C becoming $CUSO_4.H_2O$.

Water Newton Treasure, hoard of Roman Christian silverware discovered in 1975 at Water Newton, Huntingdonshire, within the site of the Roman town Durobrivae. It is the earliest such silver found anywhere in the Roman Empire. It was declared treasure trove and acquired by the British Museum in 1975. *See also* MILDENHALL.

Water plantain, aquatic perennial plant of temperate freshwater swamps and streams. It has large, heart-shaped leaves that float or extend out of the water and small white or pinkish flowers. Family Alismataceae. *See also* ARROWHEAD.

Water pollution. *See* POLLUTION.

Water polo, swimming sport that employs features of BASKETBALL and FOOTBALL. It is played by two teams of seven people in a pool 20–30m (66–99ft) long and a maximum of 20 (66ft) wide. At each end of the pool is a net-enclosed goal, which one of the players defends. The game originated in Britain in 1870 and was introduced into the USA. It has been an Olympic event since 1900.

Water power, energy derived from flowing or falling water. It was first used to power grain mills. Today water is the source of hydroelectric power, in which it is used to turn huge electric turbine generators. Associated with these are large DAMS which store the water in deep reservoirs. *See also* MM pp.70–71, *70–71.*

Waterproofing, rendering materials such as fabrics, leather and wood impervious to water by the application of substances such as rubber, synthetic resins, waxes or metallic compounds.

Water scorpion, any of numerous species of freshwater insects, found throughout the world. It has a long breathing tube at the end of the body and front legs modified for grasping, giving it a scorpion-like appearance. Some species can inflict a painful bite. *See also* NW p.*115.*

Watershed, a term signifying the line, ridge or summit of high ground separating two drainage basins. Watershed lines were often used as boundary lines.

Water skiing, sport in which a skier rides on the surface of the water while being towed by a motor-driven boat; it originated in France in the 1920s. Competition skiing usually includes three events – slalom, jumping and trick riding. In the slalom, skiers are towed through a series of staggered buoys which must be negotiated at high speed. In jumping, the skiers must ski over an inclined wooden ramp 1.8m (6ft) high and 6.1–6.7m (20–22ft) long. Trick riding has no set pattern, the skiers choosing their own intricate routines. Sometimes only one ski is used.

Water snail, also called pond snail, any of numerous species of SNAILS that live in fresh water. Most are more lightly built than terrestrial forms. Families include Lymnaeidae, Planorbidae and Physidae. *See also* NW pp.*91, 230–231.*

Water softener, substance added to water to reduce its hardness, ie its inability to form a lather with soap as a result of the presence of dissolved calcium and magnesium compounds. Some hardness, mostly due to the bicarbonates of these metals, can be removed by boiling. The remaining "permanent" hardness is mostly due to sulphates of the metals and is reduced by adding such compounds as sodium carbonate, sodium phosphate or zeolites (hydrated silicates of calcium and aluminium) to remove or sequester the metallic ions.

Waterspout, column of water formed by a TORNADO travelling over water. The tornado draws the water up inside it, and waterspouts have been known to cause great damage to shipping and coastal installations. They usually occur in tropical regions.

Water supply, method of providing water to homes, farms and factories. It is of two types – treated and untreated. Only water which has been treated to remove all organic matter and bacteria is supplied to domestic consumers, although they can supply themselves with water from a well, butt or other catchment tank. Untreated water pumped straight from rivers, lakes and reservoirs is supplied for industrial processes. On-site water treatment may be necessary, however, as in the water-softening plants which supply boilers. *See also* MM pp.200–201.

Water table, in geology, surface between an upper level, the zone of aeration, and a lower level, the zone of saturation. In the zone of aeration the open spaces are filled mainly with air. In the zone of saturation, a sub-surface level, the openings are filled with water. *See also* PE pp.108–109.

Water vapour, water in its gaseous state. It occurs in the atmosphere and determines humidity and the formation of clouds as it condenses. Most of the atmosphere's water vapour is found in the TROPOSPHERE, mainly below an altitude of 8km (5 miles). Water vapour absorbs infra-red (long-wave) radiation and holds it in the atmosphere, because of the GREENHOUSE EFFECT, and thus plays a vital role in the transfer of energy and the Earth's heat balance. *See also* PE p.66.

Waterways, inland, passages of water that include rivers, interconnected systems of lakes, canals, and inland seaways, all of which are wide enough to carry ships and boats for transport. CANAL systems were much used for this purpose in Britain until the present century, when they have largely fallen into disuse. The ST LAWRENCE SEAWAY is the largest man-made inland waterway, making the Great Lakes of the USA and Canada accessible to large ships. *See also* MM pp.196–197.

Water wheel, mechanical device used since early Roman times to supply power, mainly for grinding cereals. The first water wheels were laid horizontally in flowing water, their shafts pointing upwards and surmounted by a millstone. Later water wheels were vertical, either undershot (in which the water passed under the wheel) or overshot (in which the water was directed by a sluice on to the top of the wheel). The latter had the advantage of being operable almost independently of the level of water in the stream or river that supplied it. *See also* MM p.70.

Watkins, David (1942–), Welsh rugby player. A fly-half with Newport, he won 21 caps for Wales (1963–67) and toured New Zealand with the 1966 British Lions. Playing rugby league for Salford, he won six caps for Great Britain and captained Wales.

Watling Street, Roman road, part of which in the 9th century formed the boundary between English and Danish land. It ran from Dover, via Canterbury, London, St Albans and Wroxeter to Chester. The name is from the Old English Waeclinga-straet, meaning the way to *Waeclin-gaceaster* (St Albans). *See also* HC1 p.101.

Wat's Dyke, great dyke built in England in *c.* 700 to defend Mercia against attacks by Welsh raiders. It stretched almost unbroken from the River Severn at about Shrewsbury to the Dee estuary. It was superseded as the Welsh boundary by OFFA'S DYKE, built about a century later. *See also* HC1 p.*185*.

Watson, James Dewey (1928–), US geneticist and biophysicist. He is known for his role in the discovery of the molecular structure of deoxyribonucleic acid (DNA) and shared the 1962 Nobel Prize in physiology and medicine with Francis CRICK and Maurice WILKINS. Watson later helped to break the genetic code of the DNA base sequences and found the ribonucleic acid (RNA) messenger that transfers the DNA code to the cell's protein-forming structures. *See also* MM p.23.

Watson, John Broadus (1878–1958), US psychologist, the founder of behaviourism in the USA. In the early 1900s he rejected mentalism and introspection and advocated a purely objective psychology that would be concerned solely with observable behaviour. His work did much to make psychological research more objective and rigorous, and his point of view was continued in the work of B. F. SKINNER. *See also* MS pp.158, 187.

Watson-Watt, Sir Robert Alexander (1892–1973), British physicist. As scientific adviser to the Air Ministry in 1940, he was a major influence in the rapid development of RADAR.

Watson-Wentworth, Charles, 2nd Marquess of Rockingham. *See* ROCKINGHAM, CHARLES WATSON-WENTWORTH.

Watt, James (1736–1819), British engineer. He invented the modern condensing steam engine in 1765. In 1782 he invented the double-acting engine. With Matthew BOULTON he founded the Soho Engineering Works at Birmingham in 1775 for the manufacture of steam engines. He and Boulton coined the term "horsepower" and the unit of power called a WATT is named after him. *See also* MM pp.*22*, 65.

Watt, unit of power in the metre-kilogramme-second (MKS) system of units. A machine consuming one JOULE of energy per second has a power output of one watt. One horsepower corresponds to 746 watts. A watt is also a unit of electrical power, equal to the product of voltage and current. For example, an electric fire that uses 12 amps at 250 volts has a power of 3,000 watts (or 3 kilowatts).

Watteau, Jean Antoine (1684–1721), Flemish-born French painter, regarded as one of the major 18th-century French masters. Watteau worked in two styles. The first, influenced by the Flemish genre painter David TENIERS, is represented by the simple realism of his early painting *La Marmotte*. The second is the highly sophisticated style known as FÊTE GALANTE, is characterized in his masterpiece *Departure for the Islands of Cythera* (1717), an early full flowering of the spirit of ROCOCO. He died from consumption in 1721. *See also* HC2 pp.36–37, *36–37*.

Wattle, pendulous, fleshy, often brightly coloured fold of skin that hangs at the neck of certain birds or lizards. The term also refers to any of several Australian trees of the genus *Acacia*.

Wattle and daub, ancient method of building, in which mud or clay (daub) is plastered on to wooden laths or branches (wattle) to form walls. The technique was used by itself, or to fill in spaces such as those of a half-timbered house.

Watts, George Frederick (1817–1904), British painter and sculptor whose works, based on CLASSICAL and RENAISSANCE models, were intended to convey moral precepts through their nobility of line and colour.

Watts, Isaac (1674–1748), British clergyman and author, considered the creator of the modern British hymn. He is remembered for his *Divine and Moral Songs for Children* (1720) and his hymns, notably *When I Survey the Wondrous Cross* and *O God, Our Help in Ages Past*.

Watt's governor. *See* FLY-BALL GOVERNOR.

Waugh, Evelyn Arthur St John (1903–66), British novelist, considered the greatest satirist of his generation. He began his career with a group of humorous novels, satirizing 20th-century life. These include *Decline and Fall* (1928) and *Vile Bodies* (1930). His belief in Roman Catholicism is strongly expressed in *Brideshead Revisited* (1945). He also wrote the Crouchback trilogy concerning WWII and comprising *Men at Arms* (1952), *Officers and Gentlemen* (1955) and *Unconditional Surrender* (1961).

Wave, in oceanography, moving disturbance travelling on or through water but which does not move the water itself. Wind causes waves by frictional drag. Waves not under pressure from strong winds are called swells. Waves begin to break on shore or "feel bottom" when they reach a depth shallower than half the wave's length. When the water depth is about 1.3 times the wave height, the wave front is so steep that the top falls over and the wave breaks. *See also* PE pp.80–81, *80–81*.

Wave, seismic. *See* SEISMIC WAVE.

Wave, tidal. *See* TIDAL WAVE.

Wave amplitude, greatest value of a periodically varying quantity. This peak value may be either positive or negative, as the quantity varies either above or below its zero value (maximum or minimum). Examples include the amplitude of an alternating current and the amplitude of a radio wave.

Wave diffraction. *See* DIFFRACTION.

Wave dispersion, variation of the REFRACTIVE INDEX of a medium with wavelength. It occurs with all electromagnetic waves but is most obvious at visible wavelengths, causing light to be separated into its component colours. This can be achieved by passing a light beam through a refracting medium, such as a glass prism, to form a SPECTRUM. Every spectral colour has a different wavelength, and so the prism refracts each colour present in the light by a different amount, red (long wavelength) being bent less than violet (short wavelength). The effect is present, but unwanted, in lenses, in which dispersion can cause chromatic aberration. *See also* SU p.102.

Wave frequency (symbol ν or f), number of complete oscillations or wave cycles in unit time of a vibrating system, measured in hertz. For a wave it is given by wave velocity divided by wavelength. By QUANTUM THEORY the frequency of any ELECTROMAGNETIC RADIATION (such as light, radio waves and X-rays) is proportional to the energy of the component photons. Many of the characteristics of electromagnetic radiation depend on frequency.

Wave function, in QUANTUM MECHANICS, function which represents the probability of a quantum system being in a particular state s at time t, usually written $\psi(s)$ or $\Psi(s, t)$.

Waveguide, system generally consisting of hollow metal tubes, used to carry and control ultra high frequency radio waves. Many RADAR sets, for instance, employ waveguides to carry their MICROWAVE signals.

Wavelength, distance between successive points of equal phase of a WAVE. For example, the wavelength of water waves could be measured as the distance from crest to crest. It is symbolized by λ (the Greek letter lambda) and equal to the wave VELOCITY divided by the WAVE FREQUENCY.

Wavell, Archibald Percival Wavell, 1st Earl (1883–1950), British field-marshal. As Commander-in-Chief for the Middle East in 1939 he defeated the Italians in N and E Africa (1940–41) but was forced back by the Germans under Field-Marshal Rommel in 1941. He subsequently commanded forces in South-East Asia but lost Malaya, Singapore and Burma to the Japanese (1941–42). He was Viceroy and Governor-General of India between 1943 and 1947.

Wave mechanics, early form of QUANTUM MECHANICS developed in 1926 by Erwin SCHRÖDINGER in which the behaviour of electrons is explained solely in terms of their wave properties. Although quickly superseded by a more abstract formulation by Paul Dirac, it is still widely used in calculations.

Waveney, county district in NE SUFFOLK, England; created in 1974 under the Local Government Act (1972). Area: 371sq km (143sq miles). Pop. (1974 est.) 94,700.

Waverley, county district in SW SURREY, England; created in 1974 under the Local Government Act (1972). Area: 345sq km (133sq miles). Pop. (1974 est.) 108,700.

Waverley, first novel by Sir Walter Scott, published anonymously in 1814. It is also the collective name (the "Waverley novels") for a long series of his works,

James Watson's work on DNA and RNA has advanced the science of genetics.

James Watt, even as a young man, was interested in steam as a motive force.

Watteau's *Love Disarmed* is an example of the influence that Rubens had on his work.

George Frederick Watts's study of George Meredith, an eminent contemporary of his.

Waves

Wax palms yield a resinous wax used in the manufacture of polishes and candles.

John Wayne pausing between scenes during the shooting of *The Longest Day* (1962).

Weaverbirds breed in colonies and are noted for their complex hanging nests.

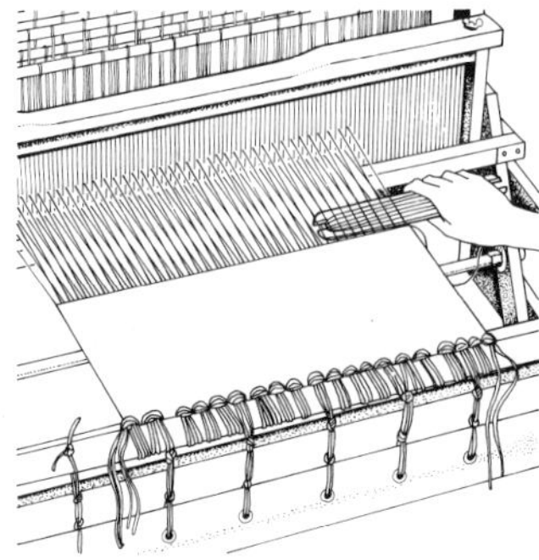

Weaving developed from the simple interlacing of horizontal and vertical threads.

including *Old Mortality* (1816), *Quentin Durward* (1823) and *Redgauntlet* (1824).

Waves, vibrating disturbances, either continuous or transient, by which energy is transmitted through media at velocities dependent on the type of wave and on the medium. ELECTROMAGNETIC WAVES, such as light, consist of varying magnetic and electric fields vibrating at right-angles to each other and to the direction of motion; they are transverse waves. Sound waves are transmitted by the vibrations of the particles of the medium itself, the vibrations being in the direction of wave motion; they are longitudinal waves. Sound waves, unlike electromagnetic waves, cannot travel through a vacuum and cannot undergo POLARIZATION. Both types of waves can be reflected, refracted, and give rise to INTERFERENCE phenomena. A wave is characterized by its WAVELENGTH or frequency, the velocity of wave motion being the product of wavelength and frequency. The velocity of electromagnetic waves greatly exceeds that of sound. *See also* WAVE AMPLITUDE; WAVE FREQUENCY; SU pp.80, 102.

Waves, The (1931), novel by Virginia WOOLF. It has little narrative and no dialogue and is an extension of Woolf's experiments in the "stream of conciousness" technique.

Wax, solid insoluble substance of low melting point. Wax has three forms: animal and vegetable waxes are simple LIPIDS consisting of ESTERS of the higher fatty acids with monohydric alcohols; examples are beeswax and spermaceti (both animal), carnauba and candelilla (both vegetable). Mineral waxes, such as paraffin wax and montan, are esters of the higher HYDROCARBONS. Synthetic waxes are of diverse origins and include POLYETHYLENES and chloronaphthalenes. All types of waxes are used in the manufacture of lubricants, polishes, cosmetics, candles, and so on. Sealing wax is not a wax but is generally made from shellac.

Wax palm, also called carnauba palm, tree that grows in South America. The trunk and the underside of the leaves are covered with a waxy film and waxes prepared from this coating are used in the manufacture of polishes and candles. Family Arecaceae; species *Ceroxylon andicola*.

Waxwing, any of a few species of small, grey-brown birds with distinctive black markings on the head, a small crest and characteristic red or red-and-yellow waxy tips on the secondary wing feathers. They are found in the forests of Eurasia and North America and feed primarily on berries and fruit. Length: to 20cm (8in). Family Bombycillidae; genus *Bombycilla*.

Wayfaring tree. *See* VIBURNUM.

Wayland the Smith, in English folklore, an expert blacksmith and armourer. As Völund in Norse tradition, he exacted a terrible vengeance on the Swedish King Nidud, who seized him from the elves and made him a slave.

Wayne, John (1907–), US film actor, b. Marion Michael Morrison. His first major success was in *Stagecoach*, directed in 1939 by John FORD. He made many films, usually appearing as the archetypal Western hero. These included *She Wore a Yellow Ribbon* (1949), *The Man Who Shot Liberty Valance* (1962), *True Grit* (1969) and *The Shootist* (1976).

Way of All Flesh, The (1903), largely autobiographical satirical novel by Samuel BUTLER about four generations of the Pontifex family. It exposes, with irony and wit, Victorian religious hypocrisy. Ernest rebels against his father, suffers imprisonment and marries a drunkard.

Way of the World, The (1700), comedy by William CONGREVE, a masterpiece of RESTORATION drama. It deals with the relationship between the lovers, Millamant and Mirabell, and Lady Wishfort, Millamant's aunt who opposes her marriage.

Waziristan, mountainous region in NW Pakistan, controlled from the Dera Ismail Khan division as N Waziristan and S Waziristan; it lies between the Kurram and Gumal rivers. The Waziris are a primitive people with a reputation for banditry, and a number of wars were fought with the British in the 19th and 20th centuries. Its industries include weaving and growing rice, wheat and barley. Area: 11,326sq km (4,373sq miles). Pop. (1972 est.) 360,000.

Weak interaction, one of the four basic FORCES in physics, it can be observed only in the sub-atomic realm, being of very short range. It is weaker than the electromagnetic force and the strong interaction (the strongest of the forces) but is stronger than the gravitational force.

Weald, The, wooded area of SE England. It is the tract of land that lies between the South Downs and the North Downs, principally in Kent, but stretching through Sussex and Surrey as far W as Hampshire. The top of the Weald strata is formed of clay, limestone and sandstone and is rich in fossil remains.

Wealden, county district in central EAST SUSSEX, England; created in 1974 under the Local Government Act (1972). Area: 837sq km (323sq miles). Pop. (1974 est.) 113,000.

Wealden, series of strata of the Lower Cretaceous period, about 130 million years old, underlying parts of SE England and NE France. They consist of beds of clay and sandstone laid down when the area was an inland sea bordered by river deltas.

Wealth of Nations, The. *See* INQUIRY INTO THE NATURE AND CAUSES OF THE WEALTH OF NATIONS, AN.

Weapons, instruments of offence or defence. They may be ranged under two heads, shock weapons, which are held in the hands, and missile weapons, which are projected. The first weapons were simply wooden clubs and stones used for hunting. The first stage in their development was the placing of a stone head on a wooden club, then, in the Middle East of *c.*3500 BC, a bronze head. Missile weapons such as SPEARS, JAVELINS, slings and BOWS also originated in prehistoric times, and were tipped with stone before metalworking was invented. Mechanical weapons, such as the BALLISTA, date at least from Roman times. Gunpowder, known to the ancient Chinese, was introduced into Europe by Tatars and Arabs in the 13th century. By the 15th century CANNONS were in general use. Small-arm artillery, operated by one man, first became effective with the 16th-century musket. In the second half of the 19th century, continuous-firing MACHINE GUNS and chemical propellants (nitrocellulose and nitroglycerine) were introduced. The chief 20th-century developments have been in CHEMICAL WARFARE, BIOLOGICAL WARFARE and NUCLEAR WEAPONS. *See also* MM pp.160–181.

Weapons, nuclear. *See* NUCLEAR WEAPONS.

Weasel, any of several species of small, carnivorous, mostly terrestrial mammals of Eurasia, N Africa, the USA and South America. Most species have small heads, long necks, slender bodies, short legs and long tails. Reddish-brown with light coloured underparts, some species turn completely white in winter. Weasels are fierce predators, eating eggs and rodents and often attacking much larger animals and domestic poulty. Length: 50cm (20in) overall. Family Mustelidae; Genus *Mustela*. *See also* NW pp.165, *206*, 207.

Weather, state of the atmosphere in a given locality or over a broad area, particularly as it affects human activities. Weather refers to short-term states (days or weeks) as opposed to the long-term climatic conditions. Weather involves such elements as atmospheric temperature, pressure, humidity, precipitation, cloudiness, brightness, visibility and wind. *See also* CLIMATE; METEOROLOGY.

Weather forecasting, prediction of features and effects of the WEATHER. Factors measured include temperature, precipitation, storms and fair weather, travel and sea conditions, using data from any observations gathered from weather stations and using a variety of calculating and mapping techniques. Because of the intrinsic variability in the complex conditions determining the weather, even short-term, day-to-day weather predictions are fairly limited in accuracy. *See also* PE pp.72–73.

Weathering, in geology, breaking-up and chemical dissolution of rocks by the forces of nature. It is important in the formation of soil and plays a major part in shaping landscapes. In mechanical weathering in cold, wet climates water in rock crevices expands on freezing, so causing the rock to crack further and to crumble. Extreme temperature fluctuations in drier regions, such as deserts, also cause rocks to fragment. Crevices opened by the weather are exploited by plants, the roots of which place further strains upon the rock. Added to these physical effects are those of chemical weathering, which can lead to a weakening of the rock structure, and include oxidation, hydration, silication, desilication and carbonation.

Weather maps, charts made up by meteorologists showing WEATHER elements and conditions over wide areas, such as the whole of a country, to assist them in their forecasts and to inform the public. Compiled for surface or upper-level atmospheric conditions, such maps often show wind speed and direction, isotherms (lines of constant temperature), isobars (lines of constant pressure) depicting high- and low-pressure areas, ridges and troughs, various kinds of fronts and the cyclones and anticyclones associated with them, as well as the types of precipitation expected in various areas. *See also* WEATHER FORECASTING; PE pp.72–73.

Weather modification, the deliberate changing of weather conditions or systems by man, still in the experimental stage. Greater knowledge of precipitation has led to many cloud-seeding experiments to increase or decrease rain or snow in local areas.

Weather vane, or weathercock, device to indicate wind direction which comprises an upright rod with cross-members bearing the cardinal points of the compass and a revolving arm. The arm is arrow-shaped at one end and has a vane at the other whose large surface area is designed to react to wind pressure and point the arrow into the direction from which the wind is blowing. One of the oldest meteorological instruments, vanes are still in wide use, often connected to direction-recording apparatus.

Weaverbird, any of several species of short-billed, often yellow-and-black, finch-like birds that weave complex nests from grass and leaves. The weaverbirds are gregarious insect-eaters of hot, dry areas. The male of some species has several mates and constructs several nests. The African *Ploceus cucullatus* knots strands of grass together. Length: to 22cm (7.5in). Family Ploceidae.

Weaving, process of making fabric by intertwining two sets of threads. A loom is threaded with a set of warp threads. The weft thread is wound round a shuttle and passed between the warp threads, which are separated according to the desired pattern. A beater or reed keeps the woven rows tightly packed. Basic weaves are called plain, twill or satin. An ancient art, weaving is now both a popular craft and a major industry, using natural or man-made fibres. Automatic looms came into widespread use in about 1900. *See also* MM pp.58–59.

Webb, Sir Aston (1849–1930), British architect. His chief buildings in London are the Victoria and Albert Museum, the Royal College of Science, Imperial College, Admiralty Arch and the new eastern facade of Buckingham Palace.

Webb, Beatrice (née Potter) (1858–1943) and **Sidney** (1859–1947), eminent British socialists. Sidney Webb was one of the founders of the FABIAN SOCIETY and one of the chief protagonists of Fabian ideals. The Webbs helped to found the *New Statesman* (1913) and were a source of intellectual leadership for the British Labour Party.

Webb, John (1611–72), English architect. A follower of Inigo JONES, he continued to work in the CLASSICAL style. His best-known work was King Charles's Block (1665), the only part of his design for Greenwich Palace to be built.

Webb, Matthew (1848–83), British swimmer, the first to swim the English Channel.

The feat was not accomplished again until 1923 and Webb's time was not beaten until 1934. Webb died attempting to swim the Niagara Falls.

Webb, Philip (1831–1915), British architect, one of the greatest of the English domestic revival. He specialized in house design and in the Red House at Bexley, built for William MORRIS in 1859, he developed a new, specifically English, style. His most famous work is Clouds, near East Knole, Wiltshire, built in 1876.

Weber, Carl Maria von (1786–1826), German composer, conductor and virtuoso pianist. He influenced WAGNER and helped to establish a German national style in his operas *Der Freischütz* (1821) and *Euryanthe* (1823). He also composed piano and chamber music, concerti and the popular *Invitation to the Dance* (1819). *See also* HC2 pp.*106, 106,* 120, *120.*

Weber, Ernst Heinrich (1795–1878), German physiologist who pioneered the study of sensation and perception. He laid the foundations for the branch of psychology called "psychophysics", influenced other thinkers such as Gustav FECHNER, and encouraged psychologists to be more scientific and methodical.

Weber, Max (1864–1920), German sociologist who advanced the concept of "ideal types", generalized models of social situations as a method of analysis and insisted that the social sciences should be empirical, based on comparative social history and free from value judgements. In his work *The Protestant Ethic and the Spirit of Capitalism* (1904–05), he postulated that CALVINISM was influential in the rise of capitalism. *See also* HC2 pp.*172,* 173, *266*; MS pp.264, 302.

Weber, Max (1881–1961), US painter, *b.* Russia. He studied in Paris and was the first to introduce modern European influences into American art before WWI. Early in his career he embraced FAUVISM and early CUBISM.

Weber, Wilhelm Eduard (1804–91), German physicist who, in 1846, standardized the units used in electricity, relating them to the fundamental dimensions of mass, length, charge and time. He was the first physicist to regard electricity as composed of elementary units each carrying a charge, now known as ELECTRONS. The SI UNIT of magnetic flux is named after him.

Weber, SI UNIT of magnetic FLUX equal to the flux that, linking a wire coil of one turn, will produce in it an ELECTROMOTIVE FORCE of one volt as it reduces to zero at a uniform rate in one second.

Webern, Anton von (1883–1945), Austrian composer, a student and disciple of Arnold SCHOENBERG. His *Passacaglia* was outside the TWELVE-TONE SYSTEM which he adopted soon afterwards. His compositions include *Six Bagatelles* (1913) for string quartet, *Variations* (1940) for orchestra, cantatas and songs. *See also* HC2 pp.270–271, *270–271.*

Weber's Line, imaginary line between the Celebes and Moluccan islands, drawn in 1902 to replace WALLACE's LINE demarcating Oriental from Australasian animal territories.

Web offset, method of printing in which the press utilizes continuous rolls (webs) of paper and lithographic printing plates on rotary printing machines. Web offset is widely used in the newspaper industry, where speed is a vital production factor. *See also* MM pp.205–207, *206–207.*

Webspinner, common name for insects of the order Embioptera. There are 140 species, mostly tropical. They have fragile, yellow or brown bodies, biting mouthparts, short, stout legs and, in the male, two pairs of narrow wings. Larvae and adults have silk-producing glands, but the insects are not of great economic importance. Length: to 20mm (0.8in).

Webster, Daniel (1782–1852), US statesman and orator. As Secretary of State he was influential in preserving the Union of North and South in the debate over slavery. *See also* HC2 p.150.

Webster, John (*c.* 1580–1625), English dramatist whose reputation rests upon his two great tragedies, *The White Devil* (1612) and *The Duchess of Malfi* (1614).

Both plays treat the theme of revenge and generate an atmosphere of gloom and despondency. *See also* HC1 pp.*282,* 285.

Webster, Noah (1758–1843), US lexicographer and writer. He completed his monumental *American Dictionary of the English Language*, containing 70,000 words in 1828. It was the largest English dictionary published up to that time.

Weddell, James (1787–1834), British Antarctic explorer. Aided by unusually open ice conditions, he discovered in 1823 the part of the Atlantic Ocean that projects into Antarctica's coastline, which he called the George IV Sea, later named Weddell Sea. The most sottherly point he reached was 74°15′ s, 34°17′ w, a record unbeaten for almost a century.

Wedekind, Benjamin Franklin ("Frank") (1864–1918), German dramatist whose use of theatre anticipated that of EXPRESSIONISM and influenced BRECHT. His plays include *Spring Awakening* (1891) and *Lulu* which appeared in two parts as *Earth-Spirit* (1895) and *Pandora's Box* (1900).

Wedge, in mechanics, an example of the inclined plane. It is used to multiply an applied force while changing its direction of action. For example, if a metal or wooden wedge is driven into a block of wood then a force is exerted by the wedge at right-angles to the applied force and greater than it. This principle also explains the wood-splitting action of an axe. *See also* MM pp.86–87, *87.*

Wedgwood, Dame Cicely Veronica (1910–), British historian. In 1962 she became an honorary fellow of Lady Margaret Hall, Oxford, and in 1965 an honorary fellow of University College, London. Her books include *The Thirty Years' War* (1938), *The King's Peace* (1955), *The King's War* (1958) and *The Trial of Charles I* (1964).

Wedgwood, Josiah (1730–95), British potter, descended from a family of Staffordshire potters. He pioneered the large-scale production of pottery at his Etruria works near Stoke-on-Trent, and became famous for his creamware. He is perhaps best known for his jasper ware which gave expression to the then current interest in the revival of classical art. *See also* HC2 pp.47, *67.*

Wedgwood, Thomas (1771–1805), physicist, philanthropist and pioneer in photochemistry. In 1802 he succeeded in projecting an image on to paper sensitized with moist silver nitrite but was unable to fix it.

Wednesday, fourth day of the week, named after Woden, the chief god in Teutonic history and the Anglo-Saxon equivalent of MERCURY.

Weed, uncultivated or unwanted plant usually found on wasteland and along roadsides. It produces many seeds, colonizes rapidly and crowds out other plants. Annual or perennial weeds are a threat to commercial crops because they compete for water and sunlight and harbour pests and diseases that can spread to the crop plants. Common weeds include DAISIES, NETTLES and THISTLES. Edible weeds include PURSLANE and DANDELIONS.

Weedkiller. *See* HERBICIDE.

Weekes, Everton de Courcy (1925–), West Indian cricketer for the West Indies and Barbados. A batsman of savage power and a superb range of strokes, he played in 48 Test matches between 1948 and 1958 (scoring 4,455 runs, at an average of 58.61) and established a world record of five consecutive Test centuries.

Weelkes, Thomas (*c.* 1575–1623), English madrigal composer and organist. Almost 100 of his madrigals have survived, the finest being sets of five- and six-part madrigals, many of which are expressively individual and bold in harmony.

Weeping willow. *See* WILLOW.

Weever, any of four species of small fish that commonly bury themselves in sand in European and Mediterranean coastal waters. Poison spines on the dorsal fin and gill covers can inflict a painful sting. Family Trachinidae; genus *Trachinus.*

Weevil, any of numerous species of beetles that are pests to crops, especially

the numerous snout beetles (time weevils), with long, down-curved beaks for boring into plants. Each species is typically specialized for feeding on parts of particular plants. Family Curculionidae. *See also* BOLL WEEVIL; NW pp.*206,* 208–209, *208–209.*

Wegener, Alfred Lothar (1880–1930), German geologist, meteorologist and Arctic explorer. In *The Origin of Continents and Oceans* (1915) he set forth his theory of CONTINENTAL DRIFT. *See also* PE p.26, *26.*

Weidenreich, Franz (1873–1948), US anatomist and physical anthropologist who pioneered the study of Peking man. While associated with the American Museum of Natural History he studied Java man in the 1940s. *See also* MS p.*24.*

Weigela, genus of flowering shrubs native to E Asia and now grown in many parts of the world. There are 12 species. The shrubs grow up to 4m (13ft) tall and have narrow oval leaves. The flowers are white to red, tubular and borne in clusters 3.5cm (1.5in) long. The long narrow seed pods split to disperse the seeds. Family Caprifoliaceae.

Weight, force of attraction on a body due to GRAVITY. A body's weight W is the product of its MASS (m) and the acceleration due to gravity at that point (g): $W = mg$. Mass remains constant, but weight depends on position on the Earth's surface, decreasing with increasing altitude. *See also* MM pp.90–91.

Weightlessness, condition experienced by an object on which the force of GRAVITY is neutralized. Such an object is said to have zero gravity and no weight; it floats and cannot fall. Weightlessness is experienced in space and during a free fall, eg when a car passes at speed over a humpback bridge. Astronauts are trained in the weightless conditions that exist for about 20 seconds in an aircraft that flies in an arc (similar to the car and bridge experience), and the near-weightlessness caused by the buoyancy of water. The adverse effects on the human body of prolonged weightlessness (called hypogravics) include decreased circulation of blood, less water retention in tissues and the bloodstream, and loss of muscle tone. Several protective mechanisms have been developed to combat the effects of weightlessness, and others are being studied. *See also* MM pp.158–159.

Weightlifting, exercise or sport in which weights at the ends of a bar are lifted over the head. As a competitive sport it is popular in Europe, Egypt, Japan, Turkey, the USA and the USSR. Competitions are conducted according to weight classes that range from bantamweight to heavyweight. In a weightlifting competition, each participant uses three standard lifts known as two-hand press, clean-and-jerk, and snatch. The man who lifts the greatest combined total of weights wins. There was a weightlifting competition at the Olympic Games in 1896, but it did not become a regular event until 1920.

Weights and measures, agreed units for expressing the amount of some quantity, such as capacity, length or weight. Early measurements were based on body measurements and on plant grains. The French introduced the metric system (1799), in which the unit of length, the metre, was taken as one ten millionth of the distance from the Equator to the North Pole. A litre was the volume occupied by one kilogramme of water. The modern metre is defined as 1,650,763.73 wavelengths of the red-orange light from an atom of KRYPTON-68 under specified conditions. *See also* CGS SYSTEM; MKS UNITS; SI UNITS; MM pp.90–91; SU pp.26–27, 32–33, 286.

Weight training, method of physical training involving the use of weights in gymnastic exercises. By lifting the weights and working to a prescribed programme of repetitive exercises, weight trainers improve their general fitness, muscular strength and physique. *See also* WEIGHTLIFTING.

Weil, Simone (1909–43), French philosopher who became a teacher and wrote for journals catering for Socialist and

Carl Maria von Weber's music was the forerunner of German Romantic opera.

Josiah Wedgwood's stoneware found an international market in the 1760s.

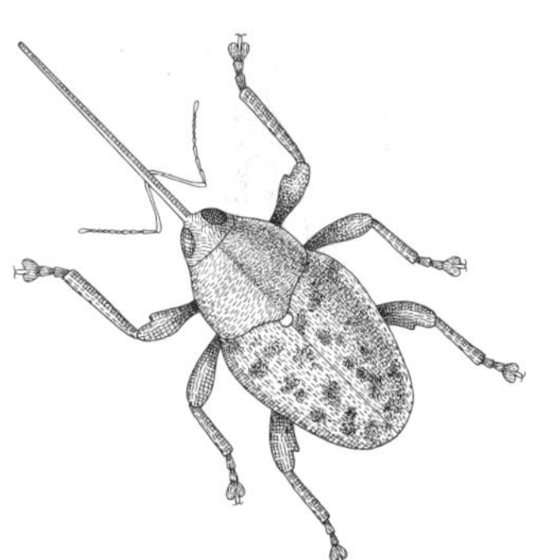

Weevils use their elongated snouts for boring holes in which to lay eggs.

Weightlifting is used as an exercise by sportsmen to develop muscle power.

Chaim Weizmann sums up the Zionist views on the Palestinian Question in 1947.

Duke of Wellington, the "Iron Duke", an honoured soldier and statesman.

Welsh corgis; the two varieties are very similar because of cross-breeding.

Welsh ponies originated in the Welsh mountains and were used as pit ponies.

Communist readerships. Later, and in ill health, she studied biblical and Hindu teachings and was most attracted to Catholicism. Most of her works were published after her death and include *Gravity and Grace* (1947) and *Waiting for God* (1951).

Weill, Kurt (1900–50), German composer who first became known for his satirical opera *Der Protagonist* (1926). His productive collaboration with Bertolt BRECHT began with the *Rise and Fall of the City of Mahgonny* (1927). *The Threepenny Opera* (1928) was a modern version of John GAY's *Beggar's Opera*, again with libretto by Brecht. In 1935 Weill emigrated to the USA, where he wrote scores for Broadway musicals. *See also* HC2 p.121.

Weil's disease. *See* LEPTOSPIROSIS.

Weimaraner, agile hunting dog that was originally bred in Germany in the 19th century. It has a long, tapering head, light amber, grey or blue eyes and folded ears. The medium-length lean body is set on straight muscular legs. Its tail is commonly docked to 15cm (6in), and the short, sleek coat is mouse or silver grey. Height: to 68.5cm (27in) at the shoulder; weight: to 38.5kg (85lb).

Weimar Republic (1919–33), the German federal republic set up after WWI with a constitution that provided for a democratically elected president and a Reichstag of deputies. Forced by the Allies to agree to the humiliating demands of the VERSAILLES Treaty and financially crippled by demands for reparations that could not be met, the Weimar governments were convulsed by a succession of economic and political crises, which facilitated the rise of extremist groups, including the Nazis. The constitution was abolished by the Nazis in 1933. *See also* HC2 pp.222–223.

Weingartner, Paul Felix (1863–1942), Austrian composer and conductor, who became a Swiss citizen in 1937. From 1898 he conducted in Europe and the USA. In 1907 he became conductor of the Court Opera in Vienna and was director of the Vienna State Opera from 1935 to 1936.

Weiskopf, Tom (1942–), US golfer. His best year was 1973, when he won seven major tournaments, including the British Open and the World Series. He was runner-up in the US Masters of 1972 and US Open of 1976.

Weiss, Peter (1916–), Swedish dramatist *b.* Germany. His reputation was established with an exploration of French revolutionary precepts in *The Persecution and Assassination of Jean-Paul Marat as Performed by the Inmates of the Asylum of Charenton Under the Direction of the Marquis de Sade* (1964). Other plays include *The Investigation* (1965) and *Trotsky in Exile* (1970).

Weissmüller, Johnny (1904–), US swimmer and actor. He won gold medals at the 1924 and 1928 Olympics and later played Tarzan in a series of Hollywood films about a noble white man living with apes in darkest Africa.

Weizmann, Chaim (1874–1952), first President of Israel from 1948 to 1952. He was born in Russia and became a naturalized British subject in 1910. A noted chemist, he aided the British munitions industry during WWI by developing a method of making acetone, and played an important role in persuading the British government to agree to the establishment of a Jewish national home in Palestine. He was president of the World Zionist Organization from 1920 to 1931 and from 1935 to 1946, and became president of the Provisional State Council of Israel in 1948. He was elected President of Israel in 1949.

Welding, technique for joining metallic parts by melting some of the metal so that the parts fuse together. A number of processes can be used, depending on the physical properties of the metals to be joined and their intended use. In arc welding, a metal rod forms one of the electrodes in an electric circuit, and the metals being welded form the other. An electric arc of intense heat develops between the rod and the joint when both are close together, which fuses the metals. In OXYACETYLENE WELDING, the heat is provided by burning acetylene gas in oxygen. In resistance or spot welding, the heat is generated by the electrical resistance of the joint; this method is fast and uses low-voltage, high-current electricity. Spot welds are made at regular intervals to join sheet metal. Electron beams and ultrasonics may also be used in welding techniques. *See also* BRAZING; SOLDERING; MM pp.40,*41*.

Weldon, George (1908–63), British conductor. He was permanent conductor of the City of Birmingham Symphony Orchestra (1943–51), conductor for the Sadler's Wells Ballet (1955–56) and from 1952 to 1956 associate conductor of the Hallé Orchestra, Manchester.

Weldon, Walter (1832–85), British chemist. He devised a process for recovering manganese in a useful form (manganese dioxide) from manganese chloride in the manufacture of chlorine, a bleaching agent essential to the textile industry.

Welensky, Sir Roy Roland (1907–), Rhodesian politician. He was a Northern Rhodesian MP (1938–53) and member of the Executive Council (1940–53). He was deputy Prime Minister of the Federation of Rhodesia and Nyasaland (1953–56) and Prime Minister and Minister of External Affairs (1956–63).

Welfare state, description of a state which takes responsibility for the health and subsistence of its citizens. The term is specifically used to describe the system in Britain, where services such as the NATIONAL HEALTH SERVICE were organized by the (1945–51) Labour government. *See also* HC2 pp.97, 296, *296*.

Welhaven, Johan Sebastian Cammermeyer (1807–73), Norwegian lyric poet. In his verse, including the sonnet cycle *The Dawn of Norway* (1834) and *The Spirit of Poetry*, he attacked the nationalism of poets such as WERGELAND.

Well, shaft sunk vertically in the Earth's CRUST through which water, oil, natural gas, brine, sulphur or other mineral substances can be extracted. Artesian wells are sunk into water-bearing rock strata, the so-called AQUIFERS, from which water rises under pressure in the wells to the surface. *See also* FRASCH PROCESS; PETROLEUM.

Well, in architecture, vertical space in a building, such as that occupied by the staircase or the lift shaft.

Welland Canal, Canadian man-made navigable waterway connecting Lake Erie with Lake Ontario, about 16km (10 miles) w of the Niagara River; it was built to avoid the Niagara Falls. The present canal (opened in 1932) replaced an older waterway, partly on a new route. It is 43km (27 miles) long and can take ships up to 210m (700ft) long.

Weller, Thomas Huckle (1915–), US bacteriologist and virologist. He shared the 1954 Nobel Prize in physiology and medicine with John F. ENDERS and Frederick C. ROBBINS for the discovery that the poliomyelitis virus can grow in cultures of various types of tissue (so making it possible to develop a VACCINE for polio).

Welles, Orson (1915–) US actor and director whose first film, *Citizen Kane* (1940), earned him an OSCAR and is regarded as a cinematic milestone. After further success in *The Third Man* (1949) and *Touch of Evil* (1958), Welles directed European productions including *The Trial* (1963), *Chimes at Midnight* (1966) and *The Immortal Story* (1968). *See also* HC2 p.276.

Wellesley, Arthur. *See* WELLINGTON, ARTHUR WELLESLEY, DUKE OF.

Wellington, Arthur Wellesley, Duke of (1769–1852), British general and statesman, commanded allied forces in the PENINSULAR WAR (1808–14) and drove the French back over the Pyrenees in 1814, during the NAPOLEONIC WARS. He was created a duke in 1814 and represented Britain at the Congress of VIENNA (1814–15). Together with the Prussian Gen. von Blücher, he defeated NAPOLEON and the French at the Battle of WATERLOO (1815). He became a Tory cabinet minister in 1818 and Prime Minister (1828–30). He supported the Catholic Emancipation Act (1829), but opposed parliamentary reform, over which he resigned in 1830. *See also* HC2 p.79, *79*.

Wellington, capital of New Zealand, in s North Island, on Port Nicholson, an inlet of Cook Strait. Founded in 1840, it replaced Auckland as capital in 1865. It has an excellent harbour with floating docks, and is an important rail centre. The city has a National Art Gallery (1936) and the Victoria University of Wellington (1962) is a cultural centre. Industries: textiles, clothing, machinery, transport equipment, food processing. Pop. (1976 est.) 139,300.

Wells, Henry (1805–78), US pioneer express postman. In 1852 he formed the Wells Fargo company to serve Californian gold fields and the new west.

Wells, Herbert George (1866–1946), British author whose reputation was established with his science fiction *The Time Machine* (1895), *The Invisible Man* (1897) and *The War of the Worlds* (1898). Novels drawing on his own experiences include *Love and Mr Lewisham* (1900), *Kipps* (1905) and *Tono-Bungay* (1909), while *Ann Veronica* (1909) and *The New Machiavelli* (1911) reflect his strong interest in sociology and politics.

Wells, Horace (1815–48), US dentist, pioneer in the practice of painless dentistry. He used nitrous oxide (laughing gas) in the extraction of teeth. After ether became generally accepted as an ANAESTHETIC, he argued for the use of nitrous oxide as safer and equally effective. *See also* MS p.126.

Welsbach, Carl Auer, Freiherr von (1858–1929), Austrian chemist, inventor of the Welsbach mantle, a shaped gauze which, when heated by gas, produces a bright light. In 1885 he discovered the elements NEODYMIUM and PRASEODYMIUM.

Welsh, language of Wales, spoken by about 25% of the Welsh population (about 600,000 people). Related to GAELIC, and more closely to BRETON, Welsh is one of the Celtic languages and a member of the INDO-EUROPEAN family. *See also* HC2 p.263.

Welsh corgi, small, stocky dog originating in Wales, once widely used to herd cattle and Welsh ponies. There are two varieties: the Pembroke, with a short tail and pointed ears, and the Cardigan, with a long bushy tail and rounded ears. The head is short, pointed and fox-like. Height: 25–30cm (10–12in); weight: 7–10kg (15–24lb).

Welsh Laws, series of medieval laws in Wales, probably first promulgated by Hywel the Good in the first half of the 10th century. The manuscripts of three versions of the laws are extant, named after the lawyers chiefly associated with them, Iorwerth, Blegywryd and Cynferth. The laws partially survived the NORMAN CONQUEST and the Statute of RHUDDLAN, but were superseded by English law at the union of England and Wales in the 1530s.

Welsh Nationalist Party. *See* PLAID CYMRU.

Welsh Office, British government department based in Cardiff and responsible for the administration of Wales. Its responsibilities include social welfare, housing, local government, education, planning and industry. It is under the jurisdiction of the Secretary of State for Wales.

Welsh pony, light saddle horse of a breed known in Wales since Saxon times. Usually a child's mount, it has the physique of a miniature coach horse with good head and neck, short muscular body, and great endurance. The coat may be grey, bay, chestnut or black. Height: to 1.2m (48in) at the shoulder; weight: to 225kg (500lb).

Welsh Presbyterian Church. *See* CALVINISTIC METHODIST CHURCH.

Welsh terrier, dog that was originally bred in Wales to hunt badgers, foxes and otters. Once called old English terrier, it has a long head with chin whiskers, V-shaped ears and small, dark eyes. The short, straight body is set on straight, muscular legs and the tail is carried high up. Its hard, wiry coat may be any combination of black and tan. Height: to 38cm (15in) at the shoulder; weight: to 9kg (20lb).

Welsh trust, charitable society founded by

the London NONCONFORMIST minister, Thomas Gouge, in 1674 to distribute the Bible and other religious literature in Wales. It also established and ran charity schools.

Welwyn Hatfield, county district in central s Hertfordshire, England; created in 1974 under the Local Government Act (1972). Area: 128sq km (49sq miles). Pop. (1974 est.) 93,600.

Wembley, complex of sports stadiums in the suburb of the same name, in NW London. The outdoor stadium with its twin towers, opened in 1923, is used for the finals of the FA Cup, Football League Cup and Rugby League Cup, as well as England's home international football matches. It has also staged speedway and, in 1948, the Olympic Games. In the nearby Empire Pool, a covered arena, ice shows and pop concerts share the facilities with boxing, show jumping, basketball and other sports.

Wenceslas, Saint (c.907–29), Bohemian prince noted for his Christian piety, who was killed by his brother Boleslav. He became recognized as the patron saint of Bohemia; his feast day is 28 Sept. He is famous as the hero of the 19th-century carol, *Good King Wenceslas.*

Wenceslaus (1361–1419), Holy Roman Emperor (1378–1400) and King of Bohemia (1378–1419), as Wenceslaus IV. His reign was characterized by wars, conspiracies and periods of anarchy. He was deposed as emperor in 1400 and his powers as Bohemian king were undermined by the challenges of his brother Sigismund.

Wen Cheng-ming (1470–1559), Chinese painter of the MING DYNASTY and an exponent of *wen-jen* (painting of the *literati* style). He was a pupil of SHEN CHOU and an innovative figure whose combinations of dry brush and wet washes influenced many later painters. *See also* HC2 p.54, 54.

Wends, name given to the Slavs living between the Oder and Elbe rivers in the early Middle Ages. They were conquered by CHARLEMAGNE but not fully Christianized until the 12th century.

Wen Ting-yün (*fl.*9th century), one of the leading Chinese poets of the late period of the T'ang dynasty (618–906). At its best, his verse displays restrained yet vibrant imagery, but at the other extreme its artificiality anticipates the poetic decadence that followed. Li Shang-yin (813–58) wrote similar verse.

Wentworth, William Charles (1793–1872), Australian journalist and politician. In 1824 he founded the *Australian* newspaper, which he used to promote the cause of self-government for the Australian colonies. His agitation was the most important factor leading to the granting of self-government by the British Parliament in 1842.

Wenzel, Carl Friedrich (1740–93), German scientist. He supported the theory of PHLOGISTON, and from his book *Introduction to Higher Chemistry* (1774), he was associated with ALCHEMY. Much of his work was, however, of scientific value, especially that on chemical affinity.

Werewolf, in folklore, a person who metamorphoses into a wolf at night but reverts to human form by day. Some werewolves can change form at will, in others the change occurs involuntarily, under the influence of a full moon.

Wergeland, Henrik Arnold (1808–45), Norwegian poet and patriot who worked hard for cultural independence from Denmark. A champion of democracy, he is considered Norway's national poet.

Wergild, Anglo-Saxon payment made by a murderer or his kin to the kin of a murdered man (*wer*). The payment varied according to the social rank of the victim; in WESSEX, the payment was 200 shillings for a CEORL and 1,200 shillings for a THEGN. It fell into almost entire disuse after the NORMAN CONQUEST.

Werner, Abraham Gottlob (1750–1817), German geologist who was the first to classify minerals systematically. His theory of the Earth's origins, called neptunism, proposed that the Earth was originally a vast ocean from which solid rocks were precipitated to form land.

Werner, Alfred (1866–1919), German-Swiss chemist. After working with Marcelin BERTHELOT in Paris he returned to a professorship at Zurich University. In 1913 he was awarded the Nobel Prize in chemistry for his co-ordination theory of VALENCY.

Werner, Theodor (1886–), German painter who studied at the Stuttgart Academy. He lived in Paris for a time, where he became associated with BRAQUE and MIRÓ. He was strongly influenced by CUBISM, and his paintings became more and more abstract.

Wernerite. *See* SCAPOLITE.

Wertheimer, Max (1880–1943), German psychologist. He was one of the founders of the GESTALT PSYCHOLOGY movement and his early work concerned visual perception. Later he attempted to apply Gestalt principles to thinking and educational problems.

Wesker, Arnold (1932–), British playwright and director whose reputation was established with the trilogy of plays *Chicken Soup with Barley* (1958), *Roots* (1959) and *I'm Talking about Jerusalem* (1960). Other works include *Chips with Everything* (1962), *The Friends* (1970) and *Love Letters on Blue Paper* (1974).

Wesley, John (1703–91), British theologian and evangelist who founded METHODISM. With his brother John he founded the so-called Holy Club at Oxford in 1729. In 1735 they went as missionaries to Georgia, where they established the principles of Methodism which spread quickly after John Wesley's return to England, and many lay preachers were enrolled to help. His *Journal* (1735–90) records the great extent of his itinerant preaching, which often brought Christianity to many who had not known it before.

Wessex, ANGLO-SAXON kingdom, possibly settled by the Saxon Cerdic in 495. Its central core was what later became the counties of Hampshire, Dorset, Wiltshire and Somerset. The invasion by the Danes (865) was resisted by Wessex; ALFRED of Wessex became king (886) of all England not then under Danish rule. *See also* HC1 pp.84, 84, 154, 165.

Wessex culture, name given to the culture of s Britain in the early Bronze Age (shortly after 2000 BC). The culture was marked by more widespread agriculture, and rich burial mounds, probably indicating a more stratified society than hitherto. The Wessex culture was responsible for the last period of building at STONEHENGE, and may have been in contact with Mycenaean Greece. *See also* HC1 p.84, 84.

West, Benjamin (1738–1820), US painter who studied in Italy and then settled in Britain, where he became historical painter to George III and a leader of the NEO-CLASSICAL movement. He succeeded Sir Joshua REYNOLDS as president of the Royal Academy in 1792. His works are mostly of historical, mythological and religious subjects and two of his best-known paintings are *Death of Wolfe* (1771) and *Penn's Treaty with the Indians* (1772).

West, Mae (1893–), US actress and film star. Her first film was *Night After Night* (1932) and others included *She Done Him Wrong* (1933) and *My Little Chickadee* (1940). The overt sexuality and wit of her characterizations made her a legend in showbusiness history.

West, Morris Langlo (1916–), Australian novelist whose first novel, *Gallows on the Sand,* was published in 1955. He also wrote such best-sellers as *The Devil's Advocate* (1959), *Shoes of the Fisherman* (1963), *The Ambassador* (1965) and *The Salamander* (1973).

West, Nathanael (1903–40), US novelist and scriptwriter, *b.* Nathan Weinstein. He wrote only four novels, and they did not gain critical acclaim until after his death. They are *The Dream Life of Balso Snell* (1931); *Miss Lonelyhearts* (1933), a grimly comic story about the writer of a column for the lovelorn; *A Cool Million* (1934); and *The Day of the Locust* (1939), about false dreams and failed lives in Hollywood, where West spent his last years as a scriptwriter.

West, Dame Rebecca (1892–), British novelist and critic, *b.* Cicily Isobel Fair-field. An ardent feminist, she took her pseudonym from the liberated character she had once played in Ibsen's *Rosmersholm.* She worked on many publications after 1911, became the mistress of H.G. WELLS in 1913 and published studies of Henry JAMES (1916) and James JOYCE (1928). Her political works include *The Meaning of Treason* (1949), and she wrote perceptively psychological novels such as *The Thinking Reed* (1936) and *Birds Fall Down* (1966).

West Africa, loose geographical term used to describe the area of the African continent between the Sahara Desert and the Gulf of Guinea, bounded on the E by the E border of Cameroon and on the W by the Atlantic Ocean. Usually included are the political units of Senegal, Guinea Bissau, Gambia, Guinea, Sierra Leone, Liberia, Ivory Coast, Ghana, Togo, Benin, Nigeria, Cameroon, Upper Volta and the parts of Mali and Niger s of the Sahara. *See* articles on individual countries in *The Modern World.*

West Bengal, state in NE India; the capital is Calcutta. It was formed in 1947 after the independence of India and Pakistan and the partitioning of the former British presidency of Bengal into the Hindu or West Bengal region of India, and Muslim or East Pakistan region. In 1950, West Bengal absorbed the state of Cooch Bihar and political instability was caused by Muslim-Hindu disputes, large refugee immigration from Bangladesh and Maoist Naxalite disturbances. Manufactures include chemicals and steel, and there is some mining for coal and manganese; agricultural products include rice, vegetables, fruit and tobacco. Area: 87,853sq km (33,920sq miles). Pop. (1971) 44,312,011.

West Berlin. *See* BERLIN.

West Bromwich, former county borough in Staffordshire, w central England, 8km (5 miles) NW of Birmingham; since 1974 it has been part of the metropolitan county of WEST MIDLANDS. A major industrial city of the Midlands, it has coal mining, metal goods and chemical industries. Pop. (1971 est.) 166,625.

West End, district of London. It is generally acknowledged to cover the area from Trafalgar Square to Oxford Street, and from Charing Cross Road to Hyde Park Corner, and to include SOHO. It is renowned for its big stores, cinemas and its commercial theatres.

Westerlies, air currents or persistent winds from the west. They occur in both hemispheres between 35° and 60° latitude. *See also* EASTERLIES; PE p.68.

Western, type of popular fiction and film, native to the USA. It first appeared in the form of short stories and novels in the "pulp" magazines of the late 19th century. Mark TWAIN's *Roughing It* (1872) became a classic and Zane GREY wrote more than 80 novels on frontier life. Owen Wister's *The Virginian* (1902), about Wyoming cowboys, is still popular. *The Squaw Man* (1914) was one of the first Hollywood westerns and set a trend for a whole new breed of movie cowboy such as Tom MIX, Roy Rogers, Buck Jones and William Boyd, who appeared as "Hopalong Cassidy" in 1935.

Western Australia, largest state in Australia, bordered by the Timor Sea (N), the Indian Ocean (W and S), South Australia and the Northern Territory (E). The climate is tropical. Only the area to the SW, which enjoys a temperate climate, is permanently settled: the Swan River, the state's sole water source, drains this region. The raising of sheep and cattle is the principal agricultural activity but the production of cereals is also important. Western Australia is the country's major gold-producing state; there is also some mining for iron ore, coal and nickel. Industry is still expanding; machinery and transport equipment are manufactured. The capital is Perth. Other major cities include Fremantle, Kalgoorlie and Bunbury. Western Australia was first visited by Dirck Hartog in 1616 although settlement did not begin until 1826 when a penal colony was founded. The first free settlement was in 1829. The state was

Wembley stadium, venue of various sports fixtures including the FA cup finals.

John Wesley met with opposition from the Anglican Church during his early career.

Benjamin West was inspired by classical themes as well as by more recent history.

Mae West; platinum blonde stage and film actress who became a legendary sex-symbol.

Western Isles

West Highland terrier, only all-white Scottish breed; its undercoat is short and soft.

West India Company, Dutch company founded to make use of profits of the slave trade.

Westminster Abbey; Henry VII's chapel, begun in 1503 and completed by Henry VIII.

West Point, site of this civil war monument and the US Military Academy.

governed by New South Wales until 1831 and it became a state of the Commonwealth of Australia in 1901. Area: 2,525,500sq km (975,095sq miles). Pop. 980,000. *See also* MW p.28.

Western Isles, island authority in the Atlantic Ocean off the NW coast of Scotland, formed in 1975 from parts of the former counties of Ross and Cromarty and Inverness. Area: 2,910sq km (1,120sq miles). Pop. (1975 est.) 29,600.

Western Sahara, formerly Spanish Sahara, desert territory on the coast of NW Africa. In 1976 Spain withdrew its forces and the area was divided between Morocco to the N and Mauritania to the s. Industry: phosphate mining. Area: 266,000sq km (102,703sq miles). Pop. (1976 est.) 151,275.

Western Samoa, independent island state in the s Pacific Ocean, made up of the w part of Samoa and nine other main islands including Upolu, which has the capital Apia. Many of the people are farmers, growing yams, bananas, cacao and coconuts and raising pigs and poultry. Industries include food processing and furniture-making; tourism is of increasing importance. Area: 2,841sq km (1,097sq miles). Pop. (1976 est.) 151,275. *See also* MW p.183.

West front, in architecture, main façade of a church or cathedral.

West Germany. *See* GERMANY, WEST.

West Glamorgan, county in s Wales on the Bristol Channel, formed in 1974 from part of the former county of Glamorgan and the county borough of Swansea. The administrative centre is Swansea. Area: 816sq km (315sq miles). Pop. (1975 est.) 371,700.

West Highland terrier, also called West Highland white terrier, hunting dog originally bred in Scotland; the breed is descended from white CAIRN TERRIERS. It has a broad head that tapers to a large nose, small, erect ears and dark eyes under heavy eyebrows. The compact, deep-chested body is set on short legs and the short tail carried up. Bred white to distinguish it from game, it has a double coat. Height: to 30.5cm (12in) at the shoulder; weight: to 11kg (24lb).

West India Company, name of two colonial trading companies, one Dutch, the other French. The Dutch company was founded in 1621 and was given a monopoly of trade with Dutch colonies in America and Africa and a monopoly of ownership of the American colonies. It was the dominant power in the early days of the SLAVE TRADE. It was dissolved in 1794. The French company was incorporated in 1664 and was given a monopoly of trade and ownership in all French colonies in America and West Africa except Newfoundland. The Company was not profitable and its trading rights were revoked in 1674. It was liquidated in 1719.

West Indies, chain of islands encircling the Caribbean Sea and separating it from the Atlantic Ocean, extending from Florida, USA, to Venezuela. Geographically they are divided into three main groups: the Bahamas, the Greater Antilles and the Lesser Antilles, which include the Leeward and Windward islands. Most islands are now independent, but were formerly American, British, Spanish or Dutch possessions. *See* articles on individual islands in *The Modern World.*

West Indies Associated States, association composed of the LEEWARD and WINDWARD ISLANDS, formed in 1967. Full internal power rests with the islands' independent governments; Britain is responsible only for defence and foreign affairs.

Westinghouse, George (1846–1914), US inventor. He received his first patent in 1865 for a rotary engine. He also devised a railway frog (track junction device) and in 1869 the air brake, which made high-speed rail travel safe and which is still standard equipment. He formed the Westinghouse Electric Co., a firm holding more than 400 patents dealing with rail transport.

West Irian. *See* IRIAN BARAT.

West Lancashire, county district in SW LANCASHIRE, England, created in 1974

under the Local Government Act (1972). Area: 327sq km (126sq miles). Pop. (1974 est.) 103,400.

West Lothian, former county in s central Scotland, s of the Firth of Forth; since 1975 it has been divided between LOTHIAN and CENTRAL regions. The land slopes gradually from the Firth to hills in the s. Dairy farming is important in the lowlands and sheep are grazed in the hills. Industries: coal mining, engineering, iron and steel, textiles, distilling, paper, motor vehicles, food processing. The county town was Linlithgow. Area: 311sq km (120sq miles). Pop. (1971) 108,474.

Westmacott, Sir Richard (1775–1856), British sculptor. He worked under CANOVA in Rome, and in London executed the panels for the Marble Arch, the bronze *Achilles* in Hyde Park (1822) and the figures on the pediment of the British Museum portico.

Westmeath, county in Leinster province in the N central Republic of Ireland, bounded by the counties of Meath, Cavan, Roscommon and Offaly. It is mainly lowlying with many lakes or loughs, including Loughs Sheelin and Ree. It is drained by the rivers Shannon, Inny, and Brosna. The main economic activity is cattle raising. The county town is Mullingar. Area: 1,763sq km (681sq miles). Pop. (1971) 53,570.

West Midlands, county in central England, formed in 1974 from parts of the former counties of STAFFORDSHIRE, WORCESTERSHIRE and WARWICKSHIRE, and including the former county boroughs of Birmingham, Warley, Solihull, Coventry, West Bromwich, Dudley, Walsall and Wolverhampton. Wolverhampton is the administrative centre. Area: 899sq km (347sq miles). Pop. 2,743,300.

Westminster, City of, part of the London borough of Westminster since 1965. Westminster was the site of a monastery from 785, and EDWARD THE CONFESSOR built WESTMINSTER ABBEY there, and a palace, which was the principal royal residence until 1512. Parliament met in Westminster Palace until a fire in 1834 led to the building of the Houses of Parliament (1840–67).

Westminster, Statutes of, name given to three medieval legislative promulgations passed by Edward I at Parliament held at Westminster, London, and to one statute passed in 1931. Westminster I (1275) was intended to codify unwritten law. Westminster II (1285) was referred to as *De donis conditionalibus* because of its clause concerning preservation of inheritance of land. Westminster III (1290), referred to as *Quia emptores,* halted subinfeudation, the letting out of land upon feudal terms. Westminster IV (1931) gave legislative independence to the British dominions of Canada, Australia, New Zealand, South Africa, the Irish Republic and Newfoundland. *See also* HC2 pp.248–294.

Westminster Abbey, church in London where most British monarchs have been crowned since William the Conqueror. Edward the Confessor ordered the construction of a cruciform church in 1050 to replace a Benedictine monastery. Henry III began rebuilding the church in 1245. Henry VII's chapel, noted for its fantraceried roof, was begun in 1503. Sir Christopher WREN and Nicholas HAWKSMOOR built the two western towers (1722–40) and Sir George Gilbert SCOTT supervised extensive restoration in the late 19th century. The church as it stands in the 1970s is an impressive conglomeration of Gothic and Perpendicular styles with French influences and Victorian décor. In the south transept, where Poet's Corner is to be found, are the tombs of CHAUCER, BROWNING and TENNYSON. *See also* HC1 pp.165, 195, 200, 219, 266.

Westminster Assembly (1643–52), conference organized by the LONG PARLIAMENT to settle the religion of the CHURCH OF ENGLAND. Most of its 30 lay and 12 clerical members were CALVINISTS, and the Westminster Confession which they drew up has been generally accepted by PRESBYTERIANS. It was rejected in England when the monarchy was restored in 1660.

Westminster Cathedral, in Victoria, Lon-

don, the cathedral of the Roman Catholic Archbishop of Westminster. Begun in 1895, it is built in the Byzantine style. Bands of red brick and stone give the exterior a distinctive pattern. The cathedral was first used for services in 1903.

Westminster Hall, building in London, built by William Rufus between 1097 and 1099. The present roof, completed by RICHARD II in 1399, is one of the largest timber roofs (unsupported by pillars) in the world. The hall, once London's principal law court, survived fires in 1512, 1834 and 1941.

Westmorland, Ralph Neville, 1st Earl of (1364–1425), English nobleman, created an earl in 1397. He married a daughter of JOHN OF GAUNT, supported Henry of Lancaster's revolt in 1399 against RICHARD II and supressed anti-Lancastrian revolts in 1403 and 1405. His daughter, Cecily, became the mother of EDWARD IV and RICHARD III.

Westmorland, former county in NW England; since 1974 it has been part of the new county of CUMBRIA. Largely mountainous, the region includes much of the LAKE DISTRICT and attracts many tourists. Pastoral farming is important to the economy; beef and dairy cattle and sheep are raised. There is also some quarrying for slate, limestone and granite. The county town was Appleby. Area: 2,044sq km (789sq miles). Pop. (1971) 72,724.

West Norfolk, county district in NORFOLK, England, created in 1974 under the Local Government Act (1972). Area: 1,413sq km (546sq miles). Pop. (1974 est.) 114,400.

Weston, Edward (1886–1958), US photographer. He co-founded the *f*/64 group whose sharp images greatly influenced photographic aesthetics.

Weston, Sir Richard (1591–1652), English agricultural writer who introduced the theory of crop rotation. His *Discours of the Husbandrie used in Brabant and Flanders* (1645) spread Flemish agricultural ideas among English farmers. *See also* HC2 p.20.

Westphalia, area of w Germany between the rivers RHINE and Weser. Originally belonging to Saxony, it was separated in 1180 and ruled as a duchy by the Elector of Cologne until 1803. It was conquered by NAPOLEON in 1807, and after the Congress of Vienna in 1815 was given to Prussia. The RUHR region was occupied by the French after WWI and, after heavy destruction in WWII, was in the British zone of occupation until 1946.

Westphalia, Peace of. *See* PEACE OF WESTPHALIA.

West Point, US military post in SE New York State, on the w bank of the Hudson River, approx. 80km (50 miles) NW of New York City; since 1802 it has been the seat of the US Military Academy. Area: 6,470 hectares (16,000 acres).

West Riding, administrative district in the former county of YORKSHIRE, N England; since 1974 the area has been divided into the new counties of NORTH YORKSHIRE and WEST YORKSHIRE.

West Side Story, (1957), musical play with lyrics by Stephen Sondheim, music by Leonard BERNSTEIN, story by Arthur Laurents and choreography by Jerome ROBBINS, first produced on Broadway. It was an adaptation of SHAKESPEARE's *Romeo and Juliet* set in modern New York City. It was made into a successful film in 1961.

West Sussex, county in SE England, formed in 1974 from the administrative district of West Sussex in the former county of SUSSEX and parts of the administrative districts of East Sussex and Surrey. Area: 2,016sq km (778sq miles). Pop. (1976) 627,400.

West Virginia, state in E central USA, in the Appalachian Mt region. Hay, tobacco, corn and apples are the principal crops. West Virginia has rich mineral deposits and is the leading US producer of bituminous coal; oil and natural gas are also extracted. Some 65% of the land is forested, much of it with valuable hardwoods. The principal industries are glass, chemicals, steel, machinery and tourism. Charleston is the state capital and Huntington is the largest city.

HC1 = History and Culture Vol. 1   HC2 = History and Culture Vol. 2   MS = Man and Society   MM = Man and Machines

Settlers from Virginia crossed the Appalachian and Allegheny Mts in the 1700s, first settling the Ohio River valley. The region was part of Virginia, but political and economic disagreements arose between the new settlements and those in the E. When Virginia seceded from the Union in May 1861 there was much opposition in the W, where a loyal government at Wheeling was formed in June. It was admitted to the Union as West Virginia in 1863. Area: 62,629sq km (24,181sq miles). Pop. (1975 est.) 1,803,000. *See also* MW p.175.

West Wiltshire, county district in W central WILTSHIRE, England; created in 1974 under the Local Government Act (1972). Area: 517sq km (200sq miles). Pop. (1974 est.) 92,800.

West Yorkshire, county in central N England, formed in 1974 from parts of the former WEST RIDING of YORKSHIRE and including the former county boroughs of Bradford, Leeds, Huddersfield, Halifax, Dewsbury and Wakefield. Area: 2,039sq km (787sq miles). Pop. (1976) 2,072,500.

Wet, Christiaan Rudolph de. *See* DE WET, CHRISTIAAN RUDOLPH.

Wet and dry thermometer. *See* HYGROMETER.

Wetland, marshy ground in an intertidal zone that has prolific vegetation; coastal wetlands are said to contain a greater concentration of living matter, both flora and fauna, than any other kinds of terrain. *See also* NW p.232.

Wetting agent, substance which, when added to a liquid, lowers the surface tension and so causes the mixture to spread more easily over a surface. *See* SURFACTANT.

Wexford, county in SE Republic of Ireland, in Leinster province. The land is mostly low-lying but rises to the Blackstairs Mts in the W. The chief river is the Slaney. Wexford is primarily an agricultural county; wheat is the chief crop and cattle and pigs are raised. Fishing is an important industry. The county town is Wexford, a fishing port with food-processing industries. Area of county: 2,351sq km (908sq miles). pop. (1971) 86,351.

Weyden, Rogier van der (*C.*1400–64), Flemish painter, among the most influential of the northern European artists of his era. It is believed that he studied at the workshop of Robert Campin in 1427, and left as an accredited master five years later. He was appointed city painter of Brussels in 1436 and pursued much of his career there, but in 1450 travelled to Italy, where his style was influenced by RENAISSANCE art. His painting is characterized by clear, subtly modulated colour. Dirk BOUTS and Hans MEMLING were influenced by him.

Weygand, Maxime (1867–1965), French soldier. He was chief of staff to FOCH (1914–23) and chief of staff to the army (1931–35). He lacked experience as a field commander, however, and handled his forces badly when he was French Commander-in-Chief during May 1940. He advised the French government to capitulate on 12 June and, after a period as a prisoner-of-war, held no more important posts.

Weyl, Herman (1885–1955), US mathematician, *b.* Germany. After holding professorships at Zurich and Göttingen universities he moved to Princeton University, USA, in 1933. His work on geometry and relativity led to a unified field theory in which the electromagnetic and gravitational fields appear as geometric properties of space-time. His *Group Theory and Quantum Mechanics* (1928) helped to form the modern views of QUANTUM THEORY.

Whale, any of several species of large aquatic mammals; it has a fish-like body with paddle-like flippers and the tail is flattened horizontally into flukes for locomotion. It spends its whole life in water, remaining submerged and returning to the surface at intervals of up to two hours to release waste carbon dioxide and to take in air; on the top of its head is a blowhole through which this exchange is made. Two main groups exist: toothed whales and baleen whales. Toothed

whales (Odontoceti) have simple teeth and feed primarily on fish and squid. They include the BOTTLE-NOSED WHALE, KILLER WHALE, SPERM WHALE and BELUGA. Baleen whales (Mysticeti), including the RIGHT WHALE, BLUE WHALE and California grey whale, have no teeth but carry comb-like plates of horny material (whalebone) in the roof of the mouth. These form a sieve, through which the whales strain krill, tiny shrimp-like organisms on which they feed. The blue whale (*Sibbaldus musculus*) is the largest mammal that has ever lived; many specimens reach 30m (100ft) in length and weigh up to 100 tonnes. Species after species have been reduced to near extinction by man's overhunting. Order Cetacea. *See also* WHALING; NW pp.*94*, 166–167, *224*, *239*, 242, *248*.

Whalebone. *See* BALEEN WHALES.

Whale shark, largest species of shark; it lives throughout the world in tropical waters. Brownish to dark grey with white or yellow spots and stripes, this docile, egg-laying fish often travels near the surface. Length: 9m (30ft). Family Rhincodontidae; species *Rhincodon typus.*

Whaling, industry that originated in the Middle Ages, and grew rapidly during the 18th and 19th centuries, of pursuing and catching whales for their oil. The modern whaling era began in the 1850s with the development of harpoons with explosive heads; after 1925 ocean-going factory ships were sent to the Antarctic. Since that time most larger whale species, including the blue whale, have been hunted to near-extinction. *See also* NW p.242.

Wharton, Edith Newbold Jones (1862–1937), US novelist, noted for her polished, ironic stories of New York society at the turn of the century. Her books include *The House of Mirth* (1905), *Ethan Frome* (1911) and *The Age of Innocence* (1920), which won the Pulitzer Prize in 1921.

What Every Woman Knows (1908), comedy by James M. BARRIE about a woman who pushes her husband in his career and corrects his romantic indiscretions, without his realizing it.

Wheat, CEREAL grass originating in the Middle East. Cultivated there since 7000 BC it is now grown worldwide. It has long, slender leaves, hollow stems and flowering heads. *Triticum aestivum* is the large-grained variety used for BREAD, *T. durum* for PASTA and *T. compactum* for cake and pastry flour. Wheat is also used in the preparation of MALT, dextrose and ALCOHOL. Family Gramineae. *See also* PE pp.156, *156*, *176–177*, 180–183, *180–183*, 204.

Wheatear, small bird of the thrush family (Turdidae), common in open spaces in regions of Eurasia and Africa. It is conspicuous by its black wings and white rump. Length: about 15cm (6in). Genus *Oenanthe.*

Wheatley, Dennis Yates (1897–1977), British popular writer. He took up writing in the early 1930s, producing such works as *Forbidden Territory* (1933), *Black August* (1934) and *Traitor's Gate* (1958). Although he often wrote historical novels, he is chiefly remembered for his occult fiction, including *The Devil Rides Out* (1935) and *Strange Conflict* (1941).

Wheatley, Francis (1747–1801), British painter. He studied in London and lived in Dublin (1779–83), painting small portraits similar to those of the French painter GREUZE. He is best known for his series *Cries of London*, engraved in 1795.

Wheatstone, Sir Charles (1802–75), British physicist and inventor. In 1843, with W.F. COOKE, he improved the Wheatstone bridge, a device that accurately measures electrical RESISTANCE, which became widely used in laboratories. They also made major contributions to the electric TELEGRAPH.

Wheatstone bridge, electrical circuit used for measuring RESISTANCE, named after Sir Charles WHEATSTONE. It consists of four resistances arranged to form a square with an ELECTROMOTIVE FORCE (voltage from a battery) across one diagonal, and a GALVANOMETER across the other. When the galvanometer shows no deflection the ratio of the values of one adjacent pair of resistances equals the ratio of the values of

the other two. By adjusting the ratio of one pair (with a sliding contact along a wire which forms an adjacent pair of resistances) the unknown resistance of one of the other two can be calculated.

Wheel, circular structure which revolves evenly round its central axis. The wheel was first used during the BRONZE AGE, and spoked wheels were introduced *c.*2700 BC. First used for locomotion, wheels were being used by potters in the late Bronze Age.

Wheel and axle, machine based on the principle that a small force applied to the rim of a wheel will exert a larger force on an object attached to the axle. The MECHANICAL ADVANTAGE, or multiplication factor for the applied force, is the ratio of the radius of the wheel to that of the axle. *See also* MM pp.88, *88*, 89.

Wheeler, Sir Charles (1892–), British sculptor. He is noted for portrait sculpture and decorative pieces, such as the JELLICOE Memorial Fountain in Trafalgar Square. He was President of the ROYAL ACADEMY OF ARTS (1956–66) and was knighted in 1958.

Wheeler, Sir Robert Eric Mortimer (1890–1976), British archaeologist. He was Keeper and Secretary of the London Museum (1926–44) and became Professor of the Archaeology of the Roman Provinces at the University of London. His works includi *Archaeology from the Earth* (1954) and his autobiography, *Still Digging* (1955). *See also* HC1 p.*23.*

Wheel-lock, mechanism invented in the early 16th century, used to ignite the explosive charge of a firearm. It consists of a steel wheel whose serrated edge protruded through the bottom of a pan. A doghead vice, holding a piece of iron pyrites was lowered into the pan onto the wheel. Pressure on the trigger released the spring-loaded wheel and caused sparks which ignited the priming which ignited the charge. The wheel-lock replaced the MATCHLOCK and was itself replaced by the FLINT-LOCK. *See also* MM pp.162–163.

Wheel of life, Buddhist symbol of the process of reincarnation. The wheel, held by a monster, has at its centre the three basic evils: greed, desire and lust. Between the spokes are the states into which one can be reborn, and around the rim are depicted allegories of the phases of existence. *See also* HC1 p.64.

Whelk, edible marine GASTROPOD MOLLUSC distributed worldwide on seashores. It has a large coiled shell, with a smooth rim and a notch at the lower end. Family Buccinidae. Length: 13–18cm (5–7in). *See also* NW pp.*90, 236*; PE p.249, *249.*

Whig Party, semi-formal parliamentary connection in Britain from the late 17th to the mid-19th century. The word Whig, meaning a Scottish cattle rustler or horse thief, was used by the TORY supporters of JAMES II to describe those politicians who wished to exclude the Duke of York from the throne. The Whig Party thus became those people who promoted the GLORIOUS REVOLUTION of 1688 and who applauded the HANOVERIAN SUCCESSION of 1714. Between 1714 and the accession of GEORGE III in 1760 Toryism was so discredited by its association with Jacobitism that every politician with an eye to office acknowledged himself to be a Hanoverian Whig, even in opposition to a Whig ministry. Then in the reign of George III Toryism gradually reasserted itself and Whiggism again took on a distinctive appearance and even, after 1783, a rudimentary party organization. Under the leadership of the Marquess of ROCKINGHAM, the Duke of PORTLAND and Charles and James FOX it became, broadly speaking, the party of religious toleration, opposition to slavery and parliamentary reform and, in it opposition to Toryism at least, the popular party. From the appointment of PITT as prime minister in 1783 to 1830 the Whigs (except for the brief ministry of the Talents) remained in opposition. They then returned to office under Lord GREY and passed the Great REFORM ACT of 1832. By the mid-19th century they had come to be replaced by, or known as, the Liberal Party. *See also* HC1 pp.312–313; HC2 pp.18–19, 68–69.

Rogier van der Weyden's *Triptych: Seven Sacraments* is typical of the artist's style.

Whaling is now a mechanized operation which gives the whale little chance of escape.

Sir Charles Wheatstone is best remembered for his resistance-measuring bridge.

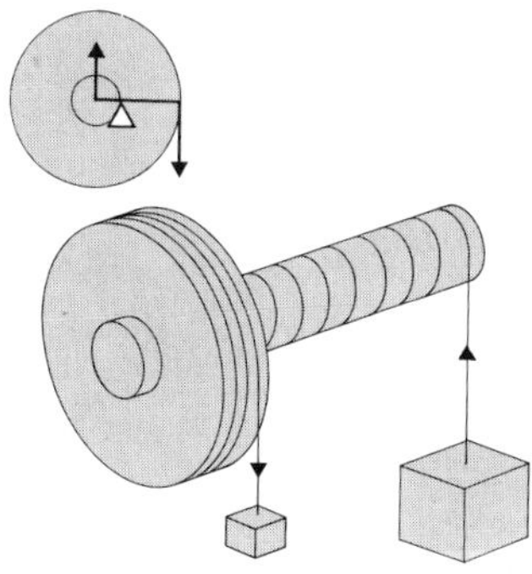

Wheel and axle, simple principle which is still extensively applied in technology.

Whippet, a popular breed descended from greyhounds and English terriers.

Whipsnade zoo, Bedfordshire; home of these North American bisons.

James A. M. Whistler made a disastrous start as a cadet at the US Military Academy.

White-eyes, also called spectacle birds, can cause great damage to fruit crops.

Whigs, one of the two major US political parties from 1834 to 1854. It was formed from the Federalist Party when John Quincy ADAMS and Henry Clay joined in opposition to Andrew JACKSON. It was a coalition party and had the support of eastern capitalists, western farmers and southern plantation owners. Its principal leaders were Henry Clay and Daniel WEBSTER. The party elected two presidents, William Henry Harrison in 1840 and Zachary TAYLOR in 1848. The issue of slavery split the Whigs, however, and the REPUBLICAN PARTY emerged from its disintegration in 1854. *See also* HC2 p.150.

Whin. *See* GORSE.

Whinchat, small Eurasian thrush that commonly inhabits grassy coastal areas in England. It has a brown, mottled back with a white rump and distinctive red breast. The dark head is clearly marked by a white eye-stripe. Length: to 13cm (5in). Species *Saxicola rubetra. See also* NW p.143.

Whineray, Wilson James (1935–   ), New Zealand rugby union front-row forward. One of his country's greatest captains, he led the All Blacks in 30 of his 32 internationals from 1957 to 1965, a period when New Zealand were pre-eminent.

Whinfield, John Rex (1901–66), British chemist who, with John Dickson, invented the synthetic fibre TERYLENE in 1941. His interest was stimulated by the discovery of nylon by W.H. CAROTHERS in 1935. In 1946 terylene was made in the USA and sold as Fibre V and then as Dacron, and in 1947 Imperial Chemical Industries (ICI) began to make and market it to the rest of the world. Whinfield received little recognition for his great invention, although he was awarded the CBE in 1954.

Whip, government officer in Britain whose duty is to see that government supporters attend debates and vote in divisions. It is also the name for the notices which he sends to MPs. Government whips are Lords Commissioners of the TREASURY; the name "whip", is taken from the "whipper-in" at a fox hunt. There are also opposition whips.

Whippet, sporting dog that was originally bred in England for racing and hare coursing. It is capable of running at speeds of 56.5km/h (35mph). It has a long, lean head with a tapered muzzle and small ears thrown back and folded. The long-backed body with arched loin is set on lean, powerful legs and the long, tapering tail is carried low. Its close, smooth coat may be any colour. Height: to 56cm (22in) at the shoulder; weight: to 11kg (24lb).

Whipple, George Hoyt (1878–1976), US pathologist who shared the 1934 Nobel Prize in physiology and medicine with George R. MINOT and William P. MURPHY for his work on anaemia and its treatment with liver extract.

Whippoorwill. *See* NIGHTJAR.

Whipsnade Zoo, British ZOO situated in the Bedfordshire countryside but administered by the Zoological Society of London. It covers about 200 hectares (500 acres) and was opened in 1931. It was the first of the world's major open-range zoos in which animals, often in herds, are housed in large paddocks and fields that to some extent match their natural surroundings. *See also* NW p.252.

Whirling dervish. *See* DERVISH.

Whisky, or whiskey, alcoholic spirit made by distilling fermented cereal grains. These sometimes give their name to the spirit, as in rye whiskey and corn (maize) whiskey or Bourbon, both made in the USA. Scotch whisky and Irish whiskey are both distilled from barley which has been allowed to sprout, then roasted and finally "mashed" and distilled. The unique flavour of Scotch whisky is partly contributed by the fumes of peat burned during the roasting process. *See also* PE p.206, *206*.

Whist, card game for four people playing as two pairs of partners. All 52 cards are dealt, the last card determining the trump suit. The aim is to collect tricks, won by the highest card of four played, one by each player. Every trick won after the first six wins a point. Points can also be won for honours (tricks won using all or three of the ace, king, queen and jack of one suit

dealt to one partnership). A game is won with five points, and a rubber is decided on the best of three games.

Whistler, James Abbott McNeill (1834–1903), US painter and etcher who lived mainly in England. Influenced early in his career by COURBET and VELÁZQUEZ and later by Japanese woodblock prints, he worked in Paris and moved to London in 1859. His painting style became semi-abstract, combining subtle tonal relationships with a judicious arrangement of forms. His pictures, such as *Nocturne in Black and Gold* (1877) are characterized by serenity and aestheticism. He was an accomplished and prolific etcher, and produced some 400 plates. *See also* MM p.205.

Whistler, Reginald John ("Rex") (1905–44), British painter, illustrator and stage designer. He studied at the Slade School of Art, and excelled in a romantic ROCOCO style of painting. His mural for the Tate Gallery restaurant is widely known.

Whistling thorn, species of ACACIA.

Whitby, town in N Yorkshire, England. The Synod of Whitby was held there in the abbey founded in 657, by St HILDA. It was an important fishing and shipbuilding port in the Middle Ages and in the 18th century the ship used by Capt. James COOK was built there. Pop. (1971) 12,717.

White, Edward Higgins II (1930–67), first US astronaut to walk in space. White and James McDivitt were launched in Gemini 4 in June 1965 and during the mission White spent 20 minutes outside the spacecraft. He was one of three astronauts killed when their spacecraft caught fire during a flight simulation of Apollo 1.

White, Gilbert (1720–93), British naturalist. He held curacies at Selbourne in Hampshire, and elsewhere, and devoted himself to the study of natural history around his parish. His famous *The Natural History and Antiquities of Selbourne* (1789) consists of letters to his fellow naturalists Thomas Pennant and Daines Barrington. It combines fresh and entertaining description with accurate observation.

White, John (*fl. c.* 1590), European, probably English, artist and cartographer. While on a pioneering expedition (1585) to VIRGINIA he made numerous drawings and paintings, providing some of the earliest reliable records of the New World.

White, Patrick Victor Martin Sale (1912–   ), Australian novelist whose first book, *Happy Valley*, appeared in 1939. His novels are deeply concerned with the nature of morality, and its relationship to experience. They include *Voss* (1957), *Riders in the Chariot* (1961), *The Vivesector* (1970) *The Eye of the Storm* (1973) and *A Fringe of Leaves* (1976). In 1973 he became the first Australian to win the Nobel Prize in literature. *See also* HC2 p.291.

White, Reginald James (1935–   ), British yachtsman who, with crew John Osborn, completely dominated the Tornado class at the 1976 Olympic Games to win the gold medal. One of the proponents of the Tornado as an international-class catamaran, he was also the Tornado world champion in 1976.

White, Terence Hanbury (1906–64) British author best known for his tetralogy *The Once and Future King* (1939–45), which he adapted into the successful Broadway musical *Camelot* (1960).

White ant. *See* TERMITE.

Whitebait, the young of several types of European HERRING, which is often eaten fried whole. Length: to 5cm (2in). The name is also given to a tropical marine fish found in Australian waters. This fish, *Galaxias attentuatus,* is elongated with its dorsal fin set far back. Family Galaxiidae. Length: to 10cm (5in).

Whitebeam, tree found in S and central Europe, particularly in mountainous regions. The undersides of the leaves have a white felt of hairs, and the scented flowers are dull white in colour. It bears bright red fruits. Species *Sorbus aria.*

White blood cell. *See* LEUCOCYTE.

Whiteboys, members of illegal agrarian groups (usually ROMAN CATHOLIC) in 18th-century Ireland, particularly in

Munster. They were first organized (*c.* 1760) to protest against land enclosure and tithe collecting and their name derives from their white disguises. Protestant counterparts existed in ULSTER. Their early raids were suppressed by 1765, but outbreaks recurred during periods of agricultural hardship until 1800. In the 19th century they were replaced by other secret societies, such as the Molly Maguires.

White currant, small shrub and its round, seedless berry, less acid than the red or black currant. It is now grown in Europe, Asia and North America, and may be made into preserves or dried. Family Grossulariaceae; genus *Ribes. See also* PE p.214.

White dwarf star, faint hot star of planetary dimensions but with a mass comparable to that of the Sun. It results from the gravitational collapse of a star that is not massive enough to have become a SUPERNOVA. White dwarfs represent the last stage in the evolution of stars like the Sun which, having exhausted their fuel, contract under their own WEIGHT to become a million times denser than water, and cool to become cold dark globes. They are found on the bottom left of the HERTZSPRUNG-RUSSELL DIAGRAM. *See also* SU pp.226–229.

White-eye, small greenish bird of Old World tropical forests with a prominent ring of white feathers around the eyes. It has short, rounded wings, a squarish tail, a straight, pointed bill and a brush-tipped, extensible tongue for feeding on nectar. It also eats insects and fruit. Length: to 12.5cm (5in). Family Zosteropidae.

White Fang (1906), novel by Jack LONDON. Offspring of an Indian wolf-dog and a wolf, White Fang is sold to an unscrupulous owner who ill-treats him to make him even more savage. He is rescued and treated with kindness by his new owner. White Fang almost dies while saving his new home from an attack by an escaped convict.

White Fathers, unofficial name for the Society of Missionaries of Africa, a Roman Catholic society founded in 1868 by Cardinal Charles LAVIGERIE, Archbishop of Algiers. Its missionaries wear white tunics.

Whitefield, George (1714–70), British evangelist. He preached in the American colonies 1739–41 and was immensely successful because of his great oratorical powers. In London he emerged as leader of the Calvinistic Methodists in 1741, after breaking with John WESLEY.

Whitefish, also called cisco or lake herring, any of several species of freshwater food fish that live, often in deep waters, in Eurasia and the USA. It is silvery, with large scales and a small mouth. Length: to 150cm (59in); weight: to 29kg (63lb). Family Salmonidae. *See also* PE p.245.

White friars. *See* CARMELITES.

Whitehall, street in London running from Parliament Square to Trafalgar Square. It is the site of the principal government offices including the FOREIGN OFFICE, the HOME OFFICE and the TREASURY. Whitehall Palace, from the time of HENRY VIII to its destruction by fire in 1698, was the London residence of the monarchy. Inigo JONES'S Banqueting House (1622), in front of which CHARLES I was executed, and the HORSE GUARDS Parade are also in Whitehall.

Whitehall farces, series of situation comedies at the Whitehall Theatre, London, in the 1950s and 1960s. The plays, under the direction of the actor Brian Rix, were *Reluctant Heroes* (1950–54), *Dry Rot* (1954–8), *Simple Spymen* (1958–61), *One for the Pot* (1961–4), *Chase Me, Comrade!* (1964–6) and *Come Spy with Me* (1966).

Whitehead, Alfred North (1861–1947), British philosopher and mathematician. He attempted to combine his interests in mathematics, morals and aesthetics into a comprehensive metaphysical system of scientific cosmology. This system is presented in his *Process and Reality* (1929). His *Principia Mathematica* (3 vols, written in collaboration with Bertrand RUSSELL, 1910–13) has had a notable influence on contemporary philosophy. Other works include *Science and the*

Modern World (1925) and *Adventures of Ideas* (1933). *See also* MS pp.*228*, *232–233*.

White hole, hypothetical localized point in space at which matter is suddenly created with explosive violence. It can be considered as being similar to a BLACK HOLE with time reversed. The cosmological BIG BANG is an analogous phenomenon that embraced the whole universe rather than being localized.

White Horse, Vale of the. *See* VALE OF THE WHITE HORSE.

White House, official residence of the President of the US in Washington, DC. It was designed in the NEO-CLASSICAL style by James HOBAN in 1792 and completed in 1800. After being burned down during the British invasion in 1814, it was rebuilt and the porticoes were added in the 1820s. It derives its name from the limestone used, later painted white.

Whiteley, Brett (1939–), Australian painter. After studying in Sydney, he went to Europe on an Italian government scholarship in 1960 and settled in London in 1961. His paintings incorporate some of the devices of POP ART and he has also worked with screenprints in such works as *Drawing about Drawing* (1965).

Whiteman, Paul (1890–1967), US bandleader. He formed an orchestra after WWI which used jazz rhythms and employed jazz soloists such as Bix BEIDERBECKE to found a popular style known as "symphonic jazz". He also commissioned and gave the first performance of various jazz-influenced orchestral pieces, such as George GERSHWIN's *Rhapsody in Blue* (1924).

White Nile, main branch, about 800km (500 miles) long, of the River NILE in NE Africa. It starts at Malakāl, flows N, and continues through the Sudan to Khartoum, where it joins the BLUE NILE to form the River Nile.

White Plains, city in SE New York State, USA. New York's provincial congress met there in 1776 to ratify the Declaration of Independence. During the American War of Independence, the city was the scene of a battle in which the American army was defeated by the British. A largely residential city, White Plains has light industries. Pop. (1970) 50,125.

White Russia. *See* BELORUSSIA.

White Sea, large inlet of the BARENTS SEA on the N coast of the USSR between the Kola Peninsula (W) and the Kanin Peninsula (E). The Mezen, Northern Dvina and Onega rivers drain into it and it is connected to the Baltic Sea by a canal system. Its chief port is ARCHANGEL, which is kept open by icebreakers from Nov. to May. The White Sea is rich in herring and cod and has seal herds. Area: 95,000sq km (36,680sq miles).

White shark, also called great white shark, aggressive shark found throughout the world in tropical and subtropical waters; it is widely regarded as the most dangerous of sharks. It has a heavy solid body, a crescent-shaped tail and saw-edged triangular teeth; it is grey, blue or brown with a white belly. Length: to 11m (36ft); weight: to 2,180kg (7,000lb). Family Isuridae; species *Carcharodon carcharias.*

White slave trade, traffic in women and children to supply enforced prostitution. The movement to suppress it began with the protests against the 1864 Contagious Diseases Prevention Act in Britain (repealed 1886). Josephine BUTLER founded the British, Continental and General Federation for the Abolition of Government Regulation of Prostitution, later known as the International Abolitionist Federation, which held its first world conference in Geneva in 1877. Since then a host of organizations, some sponsored by the League of Nations and the UN, have attempted to eliminate it.

Whitethroat, small bird of the warbler family (Sylviidae). It breeds in undergrowth in W Eurasia and NW Africa and winters in Africa and India. It has drab brown plumage with red-brown wing patches, a long white-edged tail and a white throat. Length:13cm (5in). Genus *Sylvia.* *See also* NW p.*146*.

White whale. *See* BELUGA.

Whitgift, John (*c.*1530–1604), English churchman who became Archbishop of Canterbury. He was ordained in 1560, became professor of divinity at Cambridge in 1567, Bishop of Worcester in 1577 and Archbishop of Canterbury in 1583. He imposed unity on the recently established Anglican Church, although at the cost of alienating the PURITANS. *See also* HC1 p.*271, 271.*

Whiting, John (1917–1963), British actor and playwright. His plays include *A Penny for a Song* (1951) and *Gates of Summer* (1956). The ROYAL SHAKESPEARE COMPANY commissioned him to dramatize Aldous HUXLEY's *The Devils of Loudon* as *The Devils* (1961), which achieved great success.

Whiting, name given to several unrelated food fish. The European whiting (*Merlangus merlangus*) is a haddock-like fish of the COD family, Gadidae. It is found primarily in the North Sea, where it feeds on invertebrates and small fish. It is silver with distinctive black markings at the base of the pectoral fin. Length: to 70cm (28in). Other fish commonly called whitings include the kingfish, *Menticirrhus saxatilis*, and the freshwater WHITEFISH, *Coregonus clupeaformis. See also* NW p.*131*; PE p.*246, 246.*

Whitlam, Edward Gough (1916–), Australian politician. A barrister, Whitlam was elected to the House of Representatives in 1952 and became leader of the parliamentary Labor Party in 1967. He was elected Prime Minister in 1972, but in 1975 his Government was dismissed by the Governor-General, an action unprecedented in Australian political history. In December 1977, after a crushing defeat in the general election, Whitlam announced his resignation from the Labor Party leadership.

Whitman, Walter ("Walt") (1819–92), US poet and essayist. His collection of poems, *Leaves of Grass* (1855), is now considered a classic of American literature, but when it was first published it received little acclaim. His other works include *Drum-Taps* (1865) and *Sequel to Drum-Taps* (1865), both included in a later edition of *Leaves of Grass*, and a collection of essays, *Democratic Vistas* (1871) and *Two Rivulets* (1876). His last book was *November Boughs* (1888).

Whitney, Eli (1765–1825), US inventor of the COTTON GIN (1793), which could clean cotton as quickly as 50 manual pickers. He began producing the machines but demand outstripped production and soon others were manufacturing gins even before he was granted sole right to his patent in 1807. In 1812 he was denied permission to renew it. Meanwhile, he had also set up the first mass production factory in the USA, manufacturing muskets using unskilled workers assembling interchangeable parts. *See also* MM p.*58.*

Whitsun. *See* PENTECOST.

Whitten-Brown, Sir Arthur. *See* BROWN, SIR ARTHUR WHITTEN.

Whittingham, name of two British printers. Charles Whittingham (1767–1840) founded the Chiswick Press in London in 1810 and was one of the first to issue classics in cheap reprints. His nephew, also Charles Whittingham (1795–1876), carried on the business, reviving older type-faces such as Caslon's Old-style type (1844). He later printed several of William MORRIS's works.

Whittington, Richard ("Dick") (*c.*1358–1423), English merchant. The son of a knight, he became wealthy dealing in fine cloths and was Lord Mayor of London (1397–99; 1406–07; 1419–20). He made loans to Henry IV and Henry V and endowed many charitable institutions. He is, however, best known as the subject of a legend and pantomime (probably dating from the early 17th century) about a poor boy who makes his fortune with the aid of his cat.

Whitworth, Sir Joseph (1803–87), British engineer. After learning engineering in Manchester, he went to London, where he invented a technique for making perfectly flat metal surfaces. He returned to Manchester, where he produced extremely accurate machine tools. A system of threads for nuts and bolts is named after him.

Who, The, British rock group. Originally achieving fame in the mid-1960s as a "mod" group, with a recording of Slim Harpo's *Got Love if you want it* (re-titled *I'm the Face*, 1964), The Who maintained their popularity despite the Mods' decline and later made a retrospective record dealing with the period, *Quadrophenia* (1974). Their most famous work is *Tommy* (1968), a rock opera later filmed by Ken RUSSELL.

Wholesaling, marketing activity in which goods are bought and sold by the whole piece or in relatively large quantities. Generally a wholesaler buys directly from a manufacturer and sells to retailers, to other manufacturers, to institutional and commercial users but not normally to the ultimate consumer.

Whooping cough, or pertussis, acute, highly contagious childhood respiratory disease. It is caused by the bacteria *Bordetella pertussis* and marked by spasms of coughing, followed by a long-drawn intake of air, or "whoop", frequently ending with vomiting. Severe attacks can last up to six weeks and serious complications used to be common. Now, immunization VACCINATIONS can be administered to infants.

Whortleberry. *See* BILBERRY.

Whymper, Edward (1840–1911), British mountaineer and wood-engraver. In 1860–69 he conquered many previously unscaled mountains in the ALPS, including the MATTERHORN. He travelled extensively between 1879 and 1880 in the ANDES and made an ascent of CHIMBORAZO. His written works include *Scrambles amongst the Alps* (1871, 1893).

Whyte, Ian (1901–60), Scottish conductor and composer. A pupil of VAUGHAN WILLIAMS, he was chief conductor of the BBC Scottish Symphony Orchestra from 1946. He wrote operas, cantatas and several orchestral works, including the ballet music *Donald of the Burthens* (1951), of which a part is scored for bagpipes, and edited a collection of old Scottish music.

Wiata, Inia Te. *See* TE WIATA, INIA.

Wickham, Sir Henry (1846–1928), British explorer and pioneer of rubberplanting. He established the rubber industry of the Far East by importing seeds from Brazil.

Wicklow, county in E Republic of Ireland, in Leinster province. The terrain is dominated by the Wicklow Mts, but there are fertile lowland areas. The Liffey, Slaney and Avoca are the chief rivers. Sheep and cattle are reared and cereals are grown. Low-grade copper ores are mined and the hill and lake scenery attract many tourists. Wicklow is the county town. Area: 2,025sq km (782sq miles). Pop. (1971) 66,295.

Wicksteed, Philip (1884–1927), British economist who was the first to use Leonard EULER's mathematical theorem to show that, with a particular kind of production function, it is possible to pay each input in accordance with its productivity.

Widerberg, Bo (1930–) Swedish film director and writer. He criticized the influence of Ingmar BERGMAN on Swedish film-making and produced a series of films before *Elvira Madigan* (1967) brought him to the attention of audiences outside Sweden. Other films include *Adalen '31* (1969) and *The Ballad of Joe Hill* (1969).

Wide screen cinema, cinematic process adopted in the 1950s, largely in an attempt to win back audiences from television. Inventions such as Polyvision had appeared as early as the 1920s but the first film made in Cinemascope was *The Robe* (1956). There has been several other developments, including Cinerama, Vistavision and Techmirama; the most successful of which has been Panavision.

Widgeon, river duck with mainly brownish plumage. It feeds on the surface of the water and engages in complex courtship displays. Species include the North American *Mareca americana* and European *Mareca penelope.*

Widmanstätten patterns, lines surrounded by dark bands found in the class of

White House, the official 132-room residence of the US President since 1800.

Paul Whiteman founded the annual Whiteman Award for concert jazz compositions.

Walt Whitman was widely read in England long before achieving fame in the USA.

Dick Whittington, Lord Mayor of London, who became a traditional pantomime hero.

Wiesbaden; many tourists visit the Kurhaus (above) to sample the mineral waters.

William Wilberforce led the first campaign against the slave trade in 1789.

Wild boars are still hunted for sport in many parts of Europe, India and the USA.

Oscar Wilde's wit and creative genius are best expressed in his dramatic works.

METEORITES called siderites (composed entirely of metal). They are formed when the liquefied outer surface of a meteorite cools on entering the Earth's atmosphere, and can be used to determine the meteorite's motions in flight. They appear when the meteorite is treated with nitric acid.

Widor, Charles Marie (1844–1937), French organ virtuoso, composer and teacher. Much of his career was spent as organist at St Sulpice, Paris (1870–1934), and from 1890 he was professor of organ and later of composition at the Paris Conservetoire. He composed many organ works, of which the best known are ten large scale works for the instrument which are termed "symphonies".

Wieland, Christoph Martin (1733–1813), German novelist and poet. His works express his sensuality, and include prose translations of 22 of SHAKESPEARE's plays and the novels *Agathon* (1766), *Peregrinus Proteus* (1791), *Aristipp* (1801), *Der goldene Spiegel* (1772), a verse idyll, *Musarion* (1768), and *Oberon* (1780). He later went to Weimar as a tutor to the duke, where he edited the influential *Der teutsche Merkur. See also* HC2 p.73.

Wieland, Heinrich Otto (1877–1957), German chemist who was awarded the 1927 Nobel Prize in chemistry for his research into BILE acids. In 1912 he discovered that the three bile acids then known were closely related in structure. He showed them to have a STEROID skeleton and thus found that they were also structurally related to CHOLESTEROL. He also did research into oxidation reactions occurring in living tissues and discovered that such oxidation consisted of dehydrogenation (removal of hydrogen atoms), not the addition of oxygen.

Wien, Wilhelm (1864–1928), German physicist who was awarded the 1911 Nobel Prize in physics for his work on radiation. He investigated the radiation from black bodies and, in 1893, calculated that the peak frequency radiated increases with temperature. This is called Wien's displacement law and holds only for high frequencies.

Wien. See VIENNA.

Wiener, Norbert (1894–1964), US mathematician and originator of CYBERNETICS. He contributed much to the study and development of COMPUTERS and feedback mechanisms. His mathematical model of BROWNIAN MOVEMENT established information theory. His book *The Human Use of Human Beings* (1950) was of key importance and his formulation of ENTROPY is still current.

Wiesbaden, city in central West Germany, on the River Rhine, 32km (20 miles) W of Frankfurt am Main; capital of Hessen State. Founded in the 3rd century BC, the city was later a Roman spa town. It is still famous for its mineral springs. Wiesbaden became a free imperial city c.1241 and capital of the duchy of Nassau from 1806 to 1866, when it passed to Prussia. The city was taken by Allies during WWII. Industries: tourism, metal goods, chemicals, plastics, printing. Pop. (1974) 252,457.

Wigan, county district in W GREATER MANCHESTER, England; created in 1974 under the Local Government Act (1972). Area: 199sq km (77sq miles). Pop. (1974 est.) 306,600.

Wiggin, Kate Douglas (1856–1923), US novelist and educationalist, b. Kate Douglas Smith. She opened the first free kindergarten school on the Pacific Coast in San Francisco in 1878. Her writings include kindergarten training literature, but she is best known for her children's books, *Rebecca of Sunnybrook Farm* (1903) and *Mother Carey's Chickens* (1911).

Wight, Isle of. See ISLE OF WIGHT.

Wightman Cup, annual competition between teams of women tennis players from the USA and Britain. The trophy was donated by the US player Hazel Hotchkiss Wightman, and the competition first held in 1923. It consists of five singles and two doubles matches. Up to 1978, the USA had won the Cup 40 times Britain nine times.

Wigner, Eugene Paul (1902–), US physicist who worked during WWII on the Manhattan Project, which developed the first atomic bomb. Wigner was the first physicist to apply GROUP THEORY to QUANTUM MECHANICS. With this technique he discovered the law of conservation of parity. Wigner shared the 1963 Nobel Prize in physics, with Johannes JENSEN and Maria Goeppert MAYER, for his work on the structure of the atomic nucleus.

Wigtown(shire), former county in SW Scotland; since 1975 it has been part of STRATHCLYDE region. Much of the land is hilly moorland. The chief economic activity is the rearing of sheep, cattle and pigs. Most of the population are engaged in farming or the processing of farm produce, and the manufacture of agricultural machinery. The county town was Wigtown. Area: 1,261sq km (487sq miles). Pop. (1971) 27,341.

Wigwam, shelter used by North American Indians in the E woodlands area of the USA. They were made from bark, reed mats or thatch spread over a pole frame, and should not be confused with the conical, skin-covered tepees of the Plains Indians.

Wilberforce, William (1759–1833), British social reformer, and MP for more than 40 years. In 1784 he was converted to Evangelical Christianity. His strong and outspoken opposition to the slave trade led to its abolition in the West Indies in 1807. *See also* HC2 pp.47, 135, 139.

Wilbur, Richard Purdy (1921–), US poet. His collections of poetry include *Things of This World* (1956). His other works include *Advice to a Prophet* (1961) and translation of Molière's *The Misanthrope* and *Tartuffe*.

Wilcox, Ella Wheeler (1850–1919), US poet. She contributed verses to US magazines from the age of seven, but became famous for her *Poems of Passion* (1883).

Wilcox, Sheila (1936–), British horse trials rider who, until breaking her back in 1971, was Britain's finest female rider. She was European champion in 1957, and is the only rider to have won the Badminton trials in three consecutive years (1957–59).

Wild, Jonathan (c.1682–1725), English criminal. He formed a gang of thieves and set up a highly successful organization for the disposal of stolen property. To try and remain on good terms with the authorities, Wild informed on dissident members of his own group as well as on non-members. He was eventually hanged at Tyburn.

Wild boar, tusked, cloven-hoofed mammal of the PIG family that lives wild in forested areas of Eurasia and Africa. Length: to 1.8m (6ft); weight: to 200kg (450lb). Family Suidae; species *Sus scrofa. See also* NW p.206.

Wilde, Jimmy (1892–1969), Welsh flyweight boxer who won the world title in 1916 and held it until 1923. Nicknamed "the ghost with the hammer in his hand", he won 126 of his 140 officially recorded professional fights, many against men heavier than his 7st 10lb and 77 inside the distance.

Wilde, Oscar Fingal O'Flahertie Wills (1854–1900), Irish dramatist and wit. He was the leader of the aesthetic cult of the 1880s. His plays *Lady Windermere's Fan* (1892), *A Woman of No Importance* (1893) and *An Ideal Husband* (1895) were light social comedies. His masterpiece, *The Importance of Being Earnest,* (1895), was a brilliant, epigrammatic comedy. Wilde was convicted of homosexual practices in 1895 and sentenced to two years' hard labour. While in prison he wrote *The Ballad of Reading Gaol* (published in 1898), a rhetorical poem; a prose apologia *De Profundis* (1905) was later published. He wrote one novel, *The Picture of Dorian Gray* (1891). *See also* HC2 p.101.

Wildebeest. See GNU.

Wilder, Billy (1906–), US film director. He began as a scriptwriter in Berlin but left after Hitler's rise to power. He went to HOLLYWOOD in the mid-1930s and worked with Ernst LUBITSCH on *Ninotchka* (1939). As a writer and director he subsequently made a series of films, such as *Seven Year Itch* (1955), *Some Like it Hot* (1959) and *Front Page* (1974). Others dealt realistically with tragic and shocking situations, such as *The Lost Weekend* (1945) and *Sunset Boulevard* (1950).

Wilder, Thornton Niven (1897–1975), US novelist and playwright. He received the Pulitzer Prize for his novel *The Bridge of San Luis Rey* (1927) and for two plays, *Our Town* (1938) and *Skin of Our Teeth* (1942). His *The Matchmaker* became the popular musical *Hello Dolly* (1963).

Wilfrid, or Wilfred, Saint (c.634–c.709), Anglo-Saxon bishop. He devoted his career to establishing close links with Rome, and to rejecting Celtic practices in favour of Roman ones. From 705 he was Bishop of Hexham.

Wilhelm, name of two emperors of Germany and kings of Prussia. Wilhelm I (1797–1888) was King of Prussia from 1861, and Emperor of Germany from 1871 after the FRANCO-PRUSSIAN WAR. In 1862 he appointed BISMARCK Prime Minister, and the latter dominated German politics during the rest of the reign. Wilhelm II (1859–1941) became Emperor of Germany and King of Prussia in 1888. He forced Bismarck to resign in 1890, and took a prominent part in international affairs, contributing to the tension which led to WWI. He was forced to abdicate in 1918 and remained in exile in The Netherlands until his death. *See also* HC2 pp.*110,* 111, *184,* 186, *186–187.*

Wilhelmina (1880–1962), Queen of The Netherlands (r.1890–1948). She succeeded her father, William III, helped to keep the country neutral in WWI and gave the Kaiser asylum after the war ended. When Germany invaded in WWII she took her government into exile to England, returning in 1945. In 1948 she abdicated in favour of her daughter JULIANA.

Wilkes, Charles (1798–1877), US explorer and admiral who is credited with discovering the Antarctic continent in 1838–42. Wilkes Land is named after him.

Wilkes, John (1727–97), British politician. An MP from 1757, he criticized the king and government in his newspaper *North Briton* in 1763. Although arrested and expelled from Parliament for his libels, he was re-elected in 1768 and three times in 1769, despite repeated expulsions. He won radical support and became lord mayor of London in 1774. He was again elected an MP in 1774, but lost popularity for his suppression of the GORDON RIOTS (1780). The controversy surrounding Wilkes was influential in the movement for parliamentary reform and press freedom. *See also* HC2 p.69.

Wilkie, Sir David (1785–1841), Scottish genre painter who worked in London and became popular with his scenes of everyday life. He later turned to portrait and historical painting and became painter-in-ordinary to George IV. Well-known examples of his work include *Pitlessie Fair* (1804) and *The Village Festival* (1811). *See also* HC2 pp.31, 116, *116.*

Wilkie, David (1954–), Scottish breast-stroke swimmer, b. Sri Lanka. Silver medallist in the 1972 Olympic Games 200m breaststroke, he went on to win three world, two European, and two Commonwealth gold medals before crowning his career with a gold in the 200m and a silver in the 100m at the 1976 Olympic Games.

Wilkins, Sir George Hubert (1888–1958), British explorer, photographer, aviator and geographer. He was knighted in 1928 for a flight over the Arctic of a distance of 3,380km (2,100 miles) with a co-pilot, and later did much to develop the use of aeroplanes for polar exploration. He also commanded the submarine HMS *Nautilus* in the Arctic Ocean in 1931 to the latitude 82°15′N.

Wilkins, Maurice Hugh Frederick (1916–), British biophysicist. He shared the 1962 Nobel Prize in physiology and medicine with James D. WATSON and Sir Francis CRICK for his X-ray DIFFRACTION studies which helped to determine the molecular structure of deoxyribonucleic acid (DNA).

Wilkins, William (1778–1839), British architect who spent four years travelling in Italy, Greece and Asia Minor. On his return to Britain he pioneered the Greek revival in architecture with his designs for Downing College (1804), Cambridge, and Haileybury College (1806). His other works include the National Gallery (1832–38) and St George's Hospital (1827), both in London.

Wilkinson, Professor Sir Geoffrey (1921–), British chemist; awarded in 1973, with Ernst FISCHER, the Nobel Prize in chemistry for his work on organometallic compounds. He was appointed professor of inorganic chemistry at Imperial College, London, in 1956.

Will, in law, a clear expression of intent by a person (the testator) concerning the disposal of his effects after death. The testator must be of sound mind and legal age, and the will must be witnessed by two competent people who are not beneficiaries. It may be altered or revoked by the testator at any time, with due legal process. In Britain, most of the laws governing the execution of wills are contained in the Wills Act of 1837.

Willemstad, capital city of the Netherlands Antilles, on Curaçao Island, West Indies. The city is a free port exporting oil from Venezuela; coffee is also exported. Oil refining and tourism are the major industries. Pop. (1960) 43,000.

Willendorf, Venus of. *See* VENUS FIGURES.

William I (the Conqueror) (*c.*1027–87), King of England (1066–87), and Duke of Normandy. He was the illegitimate son of Robert I of Normandy, and was unwillingly accepted as Robert's heir, succeeding to the Dukedom of Normandy in 1035. Supported initially by HENRY I, King of France, he consolidated his position in Normandy against hostile neighbours throughout the 1050s and expanded his territory in 1063. Having apparently been designated EDWARD THE CONFESSOR's successor as King of England, William secured the agreement of Harold, Earl of Wessex, to his accession. Harold's assumption of kingship was the cause of William's invasion of England and the ensuing Battle of HASTINGS in 1066. He crushed all internal resistance but spent most of his reign in France, returning to England only when necessary. He established stable government through astute land distribution and by instituting the FEUDAL SYSTEM in England. The DOMESDAY BOOK (1086) was compiled during his reign. *See also* HC1 pp.166–168.

William II (*c.*1056–1100), King of England (1087–1100), second surviving son of William I. Known as Rufus because of his ruddy complexion, he was of a brutal temperament and his rule in England, although stable, was repressive. *See also* HC1 pp.168, *168*, *196*; HC2 p.308, *308*.

William III, or William of Orange (1650–1702), Prince of Orange, King of England, Scotland and Ireland (1689–1702). Born eight days after the death of his father, WILLIAM II, he was prevented from succeeding him as Stadholder of Holland until 1672. From then he devoted his life to opposing the expansion of France under LOUIS XIV. He married Mary, daughter of the Duke of York (later JAMES II of England), in 1677. Following the GLORIOUS REVOLUTION in 1688, he and Mary replaced James II and ruled jointly until her death in 1694. Using English resources, William carried on his struggle against France, and prepared the alliance which defeated Louis in the War of the SPANISH SUCCESSION. *See also* HC1 pp.293, *293, 297*, 312, *312–313*, 315; HC2 pp.25, *25,* 308, *309*.

William IV (1765–1837), King of Great Britain and Ireland (1830–37), and of Hanover (1830–37). The third son of GEORGE III, he at first pursued a naval career, obtaining the rank of rear admiral in 1790. He became Duke of Clarence and St Andrews in 1789. He took as mistress the actress Dorothea JORDAN, and they had several children before he broke with her, in 1811. He succeeded his brother GEORGE IV and was himself succeeded by his niece VICTORIA. *See also* HC2 p. *309*.

William I and II (of Germany). *See* WILHELM.

William, name of three kings of The Netherlands and Grand Dukes of Luxembourg. William I (1772–1843; *r.*1815–40)

commanded the Dutch forces in the FRENCH REVOLUTIONARY WARS (1793–95). He was first ruler of the Kingdom of The Netherlands. Despite granting a liberalized constitution, he could not truly unite Belgium and Holland and finally signed the Treaty of Separation in 1839. He abdicated in favour of his son William (1792–1849; *r.*1840–49). William II fought France during the NAPOLEONIC WARS and against the Belgian rebels in 1830. He secured a liberal constitution in 1848. William III (1817–90; *r.*1849–90) was a conservative monarch, who accepted the constitution of 1848 unwillingly. The male line of Orange ended with his death, and his daughter WILHELMINA succeeded him.

William, name of two Norman kings of Sicily. William I, "the Bad" (1120–66; *r.*1154–66), was the son and successor of ROGER II. He defeated Byzantine forces in s Italy (1155), concluded peace with the papacy (1156) and repressed baronial conspiracies with great ruthlessness (1160–61), gaining thereby his nickname. William II, "the Good" (1154–89; *r.*1166–89), was under a regency from 1166–71. In contrast to his father, he maintained good relations with the nobility. He failed in attempts to extend his territories into N. Africa and Greece. *See also* HC1 pp.*147, 171.*

William I (1143–1214), King of Scotland (*r.*1165–1214), called "the Lion", successor to his brother Malcolm IV. He rebelled against HENRY II of England but was forced to swear fealty to him, although he bought back his independence from RICHARD I for 10,000 marks in 1189. He established an independent Church responsible only to the see of Rome and not to Canterbury. *See also* HC1 p.182.

William I (the Silent) (1533–84), Prince of Orange and one of the founders of the Dutch Republic. He was one of the great lords of The Netherlands and led a growing opposition to the intolerant Spanish rule, which culminated in the War of Independence from 1568. In the face of immense difficulties, William managed to unite the Calvinist northern provinces and was recognized as the first stadholder in 1579. He was later assassinated by a Catholic fanatic. *See also* HC1 p.*272*.

William II (1626–50), Prince of Orange, Stadholder of The Netherlands (1647–50). He succeeded his father, Frederick Henry, and married Mary, daughter of CHARLES I of England in 1650. He came into conflict with leaders of the states-general and imprisoned some of them in 1650, thus weakening the state-rights movement. He was succeeded by his son WILLIAM III, who became King of England in 1689.

William of Malmesbury (*c.*1095–*c.*1143), English historian who was librarian at Malmesbury Abbey. His most important work is the *Gesta Regum Anglorum*, a history of English kings from 449–1126, and its continuation, the *Historia novella*, which carries the narrative to 1142. He also wrote the *Gesta pontificum*, a source of reference for early ecclesiastical history. *See also* HC1 p.194.

William of Occam. *See* OCCAM, WILLIAM OF.

William of Sens (*d.*1180), French architect. He was master mason of the chancel of CANTERBURY Cathedral, rebuilt after a fire in 1174. In 1175 he started work there but was thrown from a scaffold in 1179 and was forced to abandon it. *See also* HC1 p.*198.*

William of Wykeham (1324–1404), English bishop and politician. He had a varied career, becoming Lord Chancellor in 1367 under Edward III but being deprived of his offices in 1373 because of his opposition to John of Gaunt. In 1389 he was pardoned by Richard III and reinstated as Chancellor until his retirement in 1391. Wykeham was Bishop of Winchester from 1367, and founded New College, Oxford, and Winchester School.

William of Wynford (died *c.*1405–10), English architect. He was master mason of Wells Cathedral in 1365 and protégé of WILLIAM OF WYKEHAM, for whom he

worked at Winchester College and Winchester Cathedral.

Williams, Andy (1930–), US popular singer. He starred in his own television show from 1959 onwards and his many records include *Happy Heart* (1969) and *Can't get used to losing you* (1963).

Williams, Daniel Hale (1858–1931), US surgeon who organized the Provident Hospital, the first for Negroes in the USA. He performed (1893) the first successful closure of a heart wound and from 1899 was professor of clinical surgery at Meharry Medical College at Nashville.

Williams, Elizabeth Mary (1943–), peace organizer from Northern Ireland. She was instrumental in forming the Irish PEACE MOVEMENT (1975), an organization aimed at a non-military solution to the political crisis in Northern Ireland. Together with Mairead Corrigan she organized a women's peace march through Belfast in August 1976. Both women shared the 1976 Nobel Peace Prize.

Williams, Emlyn (1905–), Welsh dramatist and actor. His plays include *Night Must Fall* (1935) and *The Corn is Green* (1938). He appeared in many of his own plays, and in 1951 had a great success playing Charles DICKENS in a series of readings from Dickens' novels.

Williams, Eric Eustace (1911–), Prime Minister of Trinidad and Tobago from 1962. He was a respected scholar and in 1956 became founder-leader of the People's National Movement (PNM), the island's first genuinely modern political party. Six years later he led his country to independence. *See also* HC2 p.*135*.

Williams, Sir George. *See* YMCA.

Williams, Grace (1906–77), Welsh composer who studied with VAUGHAN WILLIAMS and Wellesz. She has written orchestral works, concertos, the opera *The Parlour*, and Welsh folk-songs.

Williams, John (1941–), Australian lutenist and guitarist who settled in Britain. He studied with his father (also a guitarist), with Andrés SEGOVIA, and at the Royal College of Music.

Williams, John Peter Rhys (1949–), Welsh rugby union fullback. A member of the London Welsh club and Bridgend, he became Wales's regular fullback after 1969, winning 41 caps up to 1977; he toured New Zealand (1971) and South Africa (1974) with the British Lions.

Williams, Sir Owen (1890–), British architectural engineer, whose functional, streamlined use of REINFORCED CONCRETE influenced modern British architecture. His works include Boots' Pharmaceutical Factory, Beeston, Nottinghamshire, (1932) and the Empire Pool, Wembley, Middlesex (1934).

Williams, Ralph Vaughan. *See* VAUGHAN WILLIAMS, RALPH.

Williams, Roger (*c.*1603–83), advocate of religious freedom and founder of the US state of RHODE ISLAND, *b.* London. He founded Providence, the earliest settlement on Rhode Island in 1636 on land purchased from Nanagansett Indians. Williams obtained a patent for Rhode Island, allowing full religious freedom (1644) and thereafter the colony served as a refuge from religious persecution.

Williams, Thomas Lanier ("Tennessee") (1911–), US playwright. His first Broadway play *The Glass Menagerie* (1944) was awarded the New York Drama Critic's Circle Award. He received the Pulitzer Prize for *A Streetcar Named Desire* (1947) and *Cat on a Hot Tin Roof* (1955). Many of his plays were set in the South, and some have been made into films.

Williams, William ("Pantycelyn") (1717–91), Welsh Methodist leader. He adopted the Methodist custom of itinerant preaching in unconsecrated places though he considered himself a "minister of the Church of England". He wrote more than 800 hymns, including the one translated from the Welsh as *Guide me, O Thou Great Jehovah.*

Williams, William Carlos (1883–1963), US poet and practising physician. He is regarded as one of the founders of the Objectivist school. His poems are about everyday life. *See also* HC2 p.289.

William the Conqueror, the most powerful feudal lord in England in the Middle Ages.

Betty Williams, Ulster Peace Movement leader, speaking at a UN luncheon.

Emlyn Williams abridges stories and novels by the British author, Charles Dickens.

Tennessee Williams, relaxing in a hotel room between appointments.

Ted Willis, creator of the British television policeman, "Dixon of Dock Green".

Willow pattern shows the story of two lovers turned into birds by the kindly gods.

Richard Wilson's *Foundling Hospital* (1746); he was a founder of the Royal Academy.

Wimbledon's croquet lawns were made into tennis courts as tennis prospered.

Williamson, Alexander William (1824–1904), British chemist who first clarified the relationship between ETHERS and ALCOHOLS in 1850. He also produced the first mixed ether, which contains two different organic radicals.

Williamson, Henry (1895–1977), British novelist. Chiefly famous for his unsentimental animal stories such as *Tarka the Otter* (1927) and *Salar the Salmon* (1935), he also wrote the series of novels *A Chronicle of Ancient Sunlight* (1951–69).

Williamson, Malcolm (1931–), Australian-born composer, pianist and organist. He has composed operas, ballets, film scores, orchestral and keyboard works. Among his operas are *Our Man in Havana* (1963) and *The Winter Star* (1973). He was made Master of the Queen's Musick in 1975, and received the CBE in 1976.

William Tell (1829), opera by Gioacchino ROSSINI. The French libretto, from a play (1804) by SCHILLER, derived from the famous Swiss Legend.

Willibrord, Saint (c. 658–c. 739), Anglo-Saxon missionary. Educated at Ripon in Yorkshire, he preached in Frisia from 690 until 695, when he travelled to Rome to be appointed archbishop of an area centred on Utrecht.

Willis, Sir Ted (1918–), British dramatist whose plays usually have a working-class setting. They include *Woman in a Dressing Gown* (1962) and *Queenie* (1967). He also wrote the television series *Dixon of Dock Green* and *Sergeant Cork*. He was created a life peer in 1963.

Will-o'-the-wisp, or Jack-o'-lantern, mysterious light sometimes seen at night in marshy areas. It is not well understood but it is thought to be due to the spontaneous combustion of marsh gas (methane). Its elusive nature makes it the basis for various superstitious beliefs.

Willow, deciduous shrub and tree native to cool or mountainous temperate regions. It has long, pointed leaves and flowers borne on catkins. Familiar species include the weeping willow (*Salix babylonica*) with drooping branches, and pussy willow (*S. discolor*) with fuzzy catkins. Family Salicaceae. *See also* NW p.64.

Willowherb, also called rosebay willowherb, any of several species of perennial plants with willow-like leaves, especially *Epilobium angustifolium*, the fireweed of the temperate zone, so called because it readily establishes itself on newly burnt ground. It has a long unbranched stem with narrow leaves and purple red flowers. Height: to 1m (3.3ft). Family Onagraceae.

Willow pattern, printed decoration on English blue-and-white earthenware. It derives from Chinese landscape paintings, which it copies in a stylized manner. It became popular in the early 19th century and many of the major manufacturers produced their own designs, some of which have become collectors' pieces. It has retained its popularity as a design to the present day.

Wills, Helen (Newington). *See* MOODY, HELEN WILLS.

Willstätter, Richard (1872–1942), German chemist who revived the technique of chromatography, first discovered by Mikhail Tsvett (1872–1920), and used it to partly deduce the structure of CHLOROPHYLL. He showed that it contained a magnesium atom surrounded by PYRROLE rings and demonstrated its relationship to HAEMOGLOBIN. For this work he was awarded the 1915 Nobel Prize in chemistry.

Wilmington, Spencer Compton, Earl of (c. 1673–1743), British politician. He entered the House of Commons in 1698 and was speaker of the House from 1715 to 1727. He was paymaster-general (1722–30) and lord PRIVY SEAL (1730). He was made First Lord of the Treasury in 1742–43.

Wilmot, Frank Leslie Thompson (1881–1942), Australian poet who was an ardent pacifist, writing under the name of Furnley Maurice, and is best known for his poem *To God: From the Warring Nations* (1917), which is a bitter indictment of war.

Wilson, Angus (1913–), British novelist, real name Angus Frank Johnstone-Wilson. He wrote satirically of English middle-class life, and his novels include *Hemlock and After* (1952), *Anglo-Saxon Attitudes* (1956) and *No Laughing Matter* (1967).

Wilson, Charles Thomson Rees (1869–1959), British physicist who shared the 1927 Nobel Prize in physics with A. H. COMPTON for his invention of the Wilson cloud chamber, a device used in the study of radioactivity, X-rays and cosmic rays. He also devised a way of protecting barrage balloons from lightning.

Wilson, Colin Henry (1931–), British author. He first gained critical recognition with *The Outsider* (1956), a study of an individual who thinks that life is futile and that society is formed to conceal this fact.

Wilson, Edmund (1895–1972), US literary critic and author. He was editor of *Vanity Fair* and literary editor of *New Republic* (1926–31). His work includes *Axel's Castle* (1931), on the Symbolists; *To the Finland Station* (1940), on European revolutionary traditions; and *Patriotic Gore* (1962), on the literature of the American Civil War.

Wilson, Edmund Beecher (1856–1939), US biologist, whose research mainly concerned embryology and cytology. He traced the formation of different kinds of tissues from individual precursor cells, and studied the relationship of CHROMOSOMES to sex-determination and the hereditary function of chromosomes.

Wilson, Sir Henry Hughes (1864–1922), British military commander. He was Chief of the Imperial General Staff between 1918 and 1922 and was made a field-marshal in 1919. After entering politics in 1922, he vigorously opposed Ireland's SINN FEIN and was assassinated by two of its members.

Wilson, Henry Maitland, 1st Baron Wilson (1881–1964), British field-marshal. He served in the SOUTH AFRICAN WAR and in WWI and, at the outbreak of WWII, was appointed Commander-in-Chief in Egypt. In 1943 he became C-in-C Middle East and 1943–44 Supreme Allied Commander Mediterranean. He was called "Jumbo" by his troops because of his large stature.

Wilson, Sir James Harold (1916–), British Labour politician and Prime Minister. He became an MP in 1945 and was appointed President of the Board of Trade in 1947. In 1951 he resigned from ATTLEE's government over the imposition of medical prescription charges. He was elected Labour leader in 1963 and was four times Prime Minister (1964–66, 1966–70, 1974 and 1974–76). He established the OPEN UNIVERSITY (1970) and accelerated the growth of the comprehensive system of education. His government re-nationalized the steel industry, introduced measures for rent control and the control of prices and incomes and provided time for the passing of bills to legalize abortion and homosexual acts. He resigned as Prime Minister in 1976, but retained his seat in the House of Commons. *See also* HC2 pp.238–239.

Wilson, John Tuzo (1908–), Canadian geophysicist and geologist who determined global patterns of faulting and the structure of the continents. His investigations have influenced theories of CONTINENTAL DRIFT, sea-floor spreading and convection currents within the Earth.

Wilson, Richard (1714–82), Welsh landscape painter. He studied in London as a portrait painter, but after visiting Italy (1752–56) turned to landscape in the classical tradition of CLAUDE LORRAIN. He later painted landscapes in England and Wales. *See also* HC2 pp.71, 286.

Wilson, Sandy (1924–), British composer and lyricist. He wrote lyrics for a touring company before writing a pastiche of 1920s musicals, *The Boy Friend* (1953), which ran for five years in London's West End.

Wilson, Thomas Woodrow (1856–1924), 28th US president (1913–21), who administered the country throughout WWI and then worked uncompromisingly for a just international peace settlement. During his administration legislation was launched to give women the vote and to prohibit the sale of alcoholic liquor; anti-trust laws were passed; the tariff was reduced and labour conditions on board ship and for railway workers were reformed. In foreign affairs, WWI overshadowed his first term and as public opinion against Germany mounted during his second term, the USA declared war (1917). In 1918, Wilson formulated his FOURTEEN POINTS, which were to form the basis for the peace; the resulting Treaty of Versailles was not what he had hoped for. Although he played a leading role in establishing the LEAGUE OF NATIONS his own country was strongly opposed to it. *See also* HC2 p.192, *192*.

Wilson, Mount. *See* MOUNT WILSON OBSERVATORY.

Wilson's Promontory, most southerly point of the Australian mainland, in s Victoria, on the Bass Strait between Waratah Bay and Corner Inlet. The area is National Parkland.

Wilton, town in Wiltshire (and after which Wiltshire was named), central s England. The town is famous for its carpets which have been manufactured there for centuries. Wilton was the capital of the ancient kingdom of WESSEX. A noted landmark is Wilton House. Farm machinery is manufactured and the town has an important sheep market. Pop. (1971) 3,815.

Wiltshire, county in SW England. Much of it is part of Salisbury Plain or of the Marlborough Downs to the N. The principal occupation is farming; sheep are grazed on the highland areas and dairy farming is important in the fertile valleys of the Lower Avon, East Avon and Kennet rivers. The chief industries are textiles, farm machinery, food processing and electrical goods. Swindon is the principal manufacturing town, and Salisbury is the county town. Area: 3,481sq km (1,344sq miles). Pop. 512,800.

Wimbledon, popular name for the All England Lawn Tennis Championships played annually at the All England Club, Wimbledon. It is the world's foremost championship played on grass. It was first held in 1877 and was open only to amateurs until 1968. The championships have been held at the present ground since 1922.

Winch, drum mounted vertically, around which is wound rope or cable attached to an object to be hauled. A winch normally has a ratchet wheel and a pawl to prevent slippage of the load, and brakes to regulate the speed of movement. *See also* MM p.96.

Winchester, Oliver Fisher (1810–80), US manufacturer of guns and ammunition who made the Winchester rifle. In 1860 B. T. Henry, who was his chief designer, patented a repeating rifle. This was the forerunner of a long line of Winchester guns; most famous was the Model 73.

Winchester, county town of Hampshire, in s central England, on the River Itchen. Known as Venta Belgarum by the Romans, it was capital of the Anglo-Saxon kingdom of WESSEX from the 6th to the 10th centuries. It became a medieval centre of learning and religion. The city's magnificent cathedral (1070–98) is one of the finest in England, and is the burial place of several Saxon kings and queens. Winchester College, founded in 1387, is the oldest public school in England. Much of the old city remains, including the ruins of a Norman castle. Today Winchester is an important administrative and commercial centre. Pop. (1971 est.) 31,041.

Winchester, city in N Virginia, USA, in the Shenandoah Valley. During the FRENCH AND INDIAN WAR Winchester was a centre for defence against Indian raids. George WASHINGTON, in command of Virginian troops, had his headquarters there. During the AMERICAN CIVIL WAR the city was badly damaged and changed hands several times. Winchester is a trade and distribution centre for a fruit-growing region. Industries: clothing, plastics, furniture. Pop. (1970) 11,463.

Winchester, Statute of (1285), STATUTE issued by EDWARD I of England to provide for the better preservation of peace and order. It laid down penalties for harbouring criminals and concealing crimes. It also made permanent and regular the custom of WATCH AND WARD. Policing was thus made an obligation of citizenship.

Winchester School, painting style of English illuminated manuscripts of the 10th century, produced mainly at Winchester. The greatest example is the *Benedictional of St Aethelwold* (10th century), now in the British Museum. *See also* HC1 p.*165.*

Winckelmann, Johann Joachim (1717–68), German archaeologist and art historian who helped to popularize ancient art, especially that of Greece, and stimulated the Neo-classical revival of the late 18th century.

Wind, air current that moves rapidly parallel to the Earth's surface. (Air currents in vertical motion are called updrafts or downdrafts.) Wind direction is indicated by wind or weather vanes, wind speed by ANEMOMETERS and wind force by the BEAUFORT WIND SCALE. Steady winds in the tropics, eg those in the DOLDRUMS, are called TRADE WINDS. Monsoons are seasonal winds that bring predictable rains in Asia and are caused by high temperature and pressure over continents in winter (for low in summer) compared with those over the oceans. Foehns (föhns) are warm, dry winds produced by adiabatic compression (compression accompanied by temperature rise) as air descends the lee of mountainous areas; found in the Alps, they are called chinooks in the Rockies. Siroccos are hot, humid Mediterranean winds, often bringing rain and fog to continental Europe. *See also PE pp.68–69, 120–121.*

Windaus, Adolf (1876–1959), German chemist. In 1907 he synthesized histamine and in 1928 he was awarded the Nobel Prize in chemistry for his work on the structure of STEROIDS and the photochemistry of VITAMIN D.

Windermere, largest lake in England, in the LAKE DISTRICT in Cumbria, NW England. Windermere town, on the E shore of the lake, is a popular tourist resort. Length: 17km (10 miles); greatest width 1.6km (1 mile).

Wind flower. *See* ANEMONE.

Windhoek, capital and largest city of Namibia. The city originally served as the headquarters of a Nama chief. In 1892 it was made the capital of a German colony but the city was taken by South African troops in WWI. Windhoek is an important world trade market for karakul sheep skins. Pop. 77,000.

Wind instruments, musical instruments which are sounded by blowing, which sets the air inside them vibrating. They may be classified into three types: the first two of which make up the WOODWIND family of instruments, and the third type are BRASS instruments.

Wind in the Willows, The (1908), children's story by Kenneth GRAHAME. It began as a story for his son Alastair, and contains the celebrated animal characters, Mole, Badger, Water Rat, and Toad.

Windlass, hauling device operated by hand, most familiar as that used to raise buckets of water from a well. It consists of a cylinder around which the haulage rope is wound and which is turned by a hand lever or crank, using the principle of the wheel and axle. *See also* MM pp.96–97.

Windmill, mill powered by the wind acting on sails or vanes. The earliest known windmills were in the Middle East in the 7th century. The idea spread to Europe in the Middle Ages and their use was widespread during the early years of the Industrial Revolution, but declined with the development of the steam engine in the 19th century. *See also* WATER MILL; MM pp.70–71.

Windmill Hill culture, name given to one of the earliest NEOLITHIC cultures in Britain, found at Windmill Hill, Wiltshire, and dating from *c.*4000 BC. This culture introduced an agricultural way of life and made the first pottery. *See also* HC1 p.*84.*

Windmill Theatre, playhouse in London's West End which stood on the site of a 17th century windmill. Originally a cinema, the Windmill was best known for its non-stop revues and for the catch phrase "We never closed" which referred to the decision to stay open in the WWII BLITZ. Several comedians including, Tony HANCOCK and Harry

Secombe, began their careers at the Theatre, which closed in 1964 and later reopened as a cinema.

Window tax, British tax levied on the number of windows in a dwelling-place first imposed in 1696 to supply the cost of replacing damaged coinage. It replaced the hearth tax. It was increased in 1782 and 1797. After 1792 houses with fewer than seven windows were exempt from it, after 1825 houses with fewer than eight windows. It was repealed in 1851.

Windpipe. *See* TRACHEA.

Wind power, technology of harnessing the wind to provide ENERGY to drive machines and to generate electricity. Today wind power is being considered as a source of energy, mainly to generate electricity. Initial research, however, shows that windmills would have to be impractically large to match the output of even a small power station. *See also* MM pp.70–71, *84.*

Wind sock, flexible, cone-shaped device that indicates wind direction and force. The sock is designed to swivel so that the small end of the cone points downwind. An idea of windspeed can be judged from the sock's extension. *See also* ANEMOMETER; WEATHERCOCK.

Windsor, Duke of. *See* EDWARD VIII.

Windsor, name by which the British royal family has been known since 1917. Queen VICTORIA's descendants in the male line originally belonged to the German house of Saxe-Coburg Gotha, the family of her husband, Prince Albert. During WWI, however, this connection proved embarrassing. In a proclamation in 1917 GEORGE V decreed that British subjects descended from Victoria in the male line would take the surname "of Windsor". In decrees of 1952 and 1960 Elizabeth II modified this decree to the.effect that her descendants, should take the surname Windsor or Mountbatten-Windsor.

Windsor, Wallis Warfield, Duchess of *See* SIMPSON, MRS WALLIS WARFIELD.

Windsor, town in Berkshire, S central England, on the River Thames. WINDSOR CASTLE has been a residence of the English monarchy since WILLIAM THE CONQUEROR. The castle includes the magnificent St George's Chapel, which contains several royal tombs. Windsor town hall was built by Sir Christopher WREN. Tourism is a major industry. Pop. (1973 est.) 29,700.

Windsor, city in SE Ontario, Canada, on the Detroit River opposite Detroit, Michigan. The city was founded in 1749 by the French. It is a port of entry and, being on the Canadian National and Canadian Pacific railways, an industrial and transportation centre. The university of Windsor was founded in 1963. Industries: motor vehicles, pharmaceuticals, chemicals, salt. Pop. (1974) 203,300.

Windsor and Maidenhead, county district in E BERKSHIRE, England; created in 1974 under the Local Government Act (1972). Area: 198sq km (76sq miles). Pop. (1974 est.) 128,100.

Windsor Castle, originally Norman building 32km (20 miles) W of London, one of the homes of the reigning British monarch. Built on a low hill on the NE edge of Windsor, it covers 5 hectares (13 acres). The first Windsor Castle was on a different site. Begun in 1070, the present castle was added to by Henry II and subsequent monarchs including George IV.

Wind tunnel, chamber in which aircraft components and scale models are tested by being exposed to mechanically produced wind. Modern tunnels can reproduce conditions of temperature and pressure that aircraft may encounter and can generate winds far into the hypersonic range to simulate the most rigorous conditions of flight. *See also* AERONAUTICS SU pp.78–79, *79.*

Windward Islands, group of islands in the SE West Indies. They form part of the Lesser Antilles and extend from the Leeward Islands to the NE coast of Venezuela; the principal islands are MARTINIQUE, GRENADA, DOMINICA, ST LUCIA, ST VINCENT and the Grenadines group. The islands, volcanic in origin, are mountainous. Tropical crops, including bananas, spices, limes and cacao, are grown. Roseau on Dominica and Kingstown on St Vincent are the chief

towns. Tourism is an important industry. Area: approx. 1,810sq km (700sq miles). Pop. 2,149,750.

Windward Passage, strait between Capo Maisi, Cuba (W), and Cap du Môle, Haiti (E), in the West Indies, connecting the Carribean Sea (S) with the Atlantic Ocean (N). It is an important shipping route between the USA and the Panama Canal. Width: approx. 80km (50 miles).

Wine, alcoholic beverage made from the fermented juice of fruits, herbs and flowers, but especially from grapes. Table wines are still, with 7–15% alcohol; sparkling wines are effervescent; fortified wines such as SHERRY contain added BRANDY, giving 16–23% alcoholic content. *See also* PE pp.198–203.

Wine-making, conversion of grapes, or other fruit, to an alcoholic beverage. *Vinis vinifera* is the species of grape most commonly used. It is harvested in late summer or autumn and put into presses to extract the juices. This unfermented juice, known as "must", is left to ferment in vats, casks or bottles. Skins and seeds are included in the vats of must for the production of red wine. Wine yeast (saccharomyces) converts the natural sugar into alcohol and carbon dioxide. During the ageing process it is transferred from cask to cask to get rid of the settled solid matter. Then it is filtered and bottled. The timing and the detailed techniques of these processes determine the type of wine that is made. Leading wine-producing countries include France, Germany, Italy and Spain. *See also* PE pp.198–203.

Wing, aircraft, AEROFOIL whose major function is to provide lift (upward force). They often contain the fuel tanks and support appendages, such as the flaps, ailerons and landing gear. *See also* MM pp.152–153, *152–153,* 178–179, *178–179.*

Wingate, Orde Charles (1903–44), British general famous for his unorthodox ideas and methods. From 1936 to 1939, as an intelligence officer, he organized counter-raids on Arabs attacking Jewish communities. In 1941 he commanded an African force which took Addis Ababa from the Italians. In 1942 he established the Chindits – fighters specially trained in the rigours of jungle warfare – and operated in Burma far behind Japanese lines. Supplied by air, his force caused great disruption to the enemy's supplies and communications. He was killed when his aircraft crashed while attempting to rejoin his command after a staff conference.

Winged Victory. *See* VICTORY (NIKE) OF SAMOTHRACE.

Wings, in biology, specialized organs for flight which are possessed by most birds, many insects and certain mammals and reptiles. The forelimbs of a bird have developed into such structures. Bats have membranous tissue supported by the digits ("fingers") of the forelimbs. Insects may have one or two pairs of veined or membranous wings. *See also* NW pp.108–109, *108–109,* 142–143, *142–143.*

Wings, in the theatre, flats (canvas-covered wooden frames) at either side of the STAGE; they may face the audience directly or obliquely and serve to conceal the side of the set. Wings were first used by Giovan Battista Aleotti (1596–1636) in the 17th century and facilitated rapid and multiple scene changes.

Winkle, or periwinkle, small aquatic gastropod mollusc. The unsegmented body is soft with a distinct head and muscular foot, and is protected by a hard, univalve shell. Winkles feed on rock algae in the intertidal zone. The common marine species *Littorina littorea* is edible. *See also* NW p.72.

Winkler, Hans Gunter (1926–), West German show jumping rider. World champion in 1954 and 1955 riding Halla, he won the individual gold medal at the 1956 Olympic Games and helped the German team to win four golds (1956, 1960, 1964, 1972). He won the first European championship in 1957.

Winner, Michael (1936–), British film director. He began his career in the

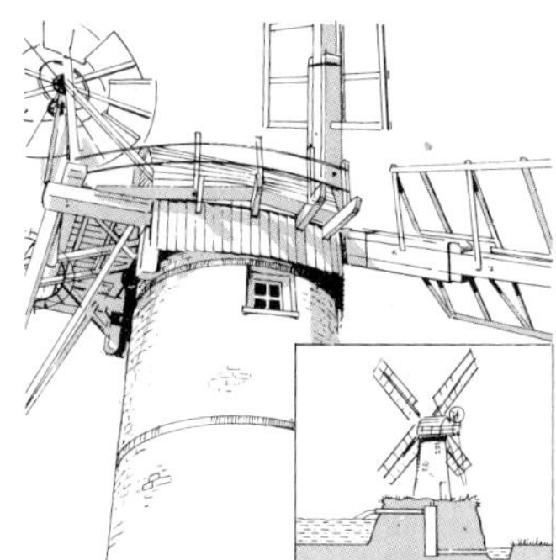

Windmill sails are aligned to the wind by a vane, forcing them to revolve.

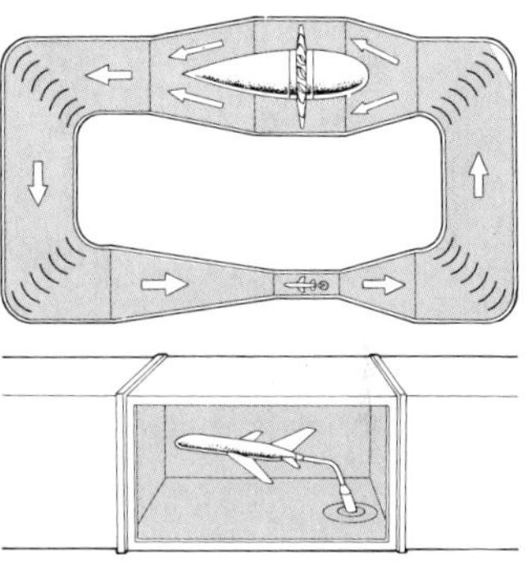

Wind tunnels may be used to study the aerodynamics of aircraft and buildings.

Wine-making; skins are separated from the juice, which is then fermented.

Aircraft wing; flaps are lowered to increase lift and to aid stability.

Winnie-the-Pooh

Wintergreen leaves are used in N America to make the drink known as mountain tea.

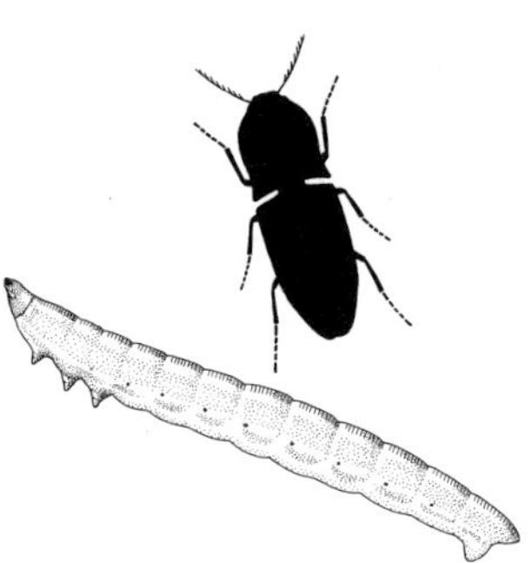

Wireworms live to ten years and are controlled by crop rotation and insecticides.

Wisconsin's largest city is Milwaukee, a major port and industrial centre.

Witch hazel derives its name from the use of its pliant forked twigs as divining rods.

mid-1950s making second features, and progressed to major films in the early 1960s. His films include *I'll Never Forget Whats' Is Name* (1967), *The Nightcomers* (1971) and *The Sentinel* (1976).

Winnie-the-Pooh, popular bear character created in his books by the British novelist A.A. MILNE. The books were written for his six-year-old son, Christopher Robin, and were about his toys. *Winnie-the-Pooh* was written in 1926, and *The House at Pooh Corner* in 1928.

Winnipeg, capital of Manitoba province, and Canada's third-largest city. It stands at the confluence of the Assiniboine and Red rivers. Founded in 1812 as the Red River settlement by the HUDSON'S BAY COMPANY, the town came under the control of the Canadian government in 1870 when the province of Manitoba was established and Winnipeg became the capital. The city grew with the completion of the CANADIAN PACIFIC Railroad. It is now the major city of the Canadian prairies and one of the largest wheat markets in the world. The city is the seat of Manitoba University (1877). The principal industry is food processing. Pop. 246,245.

Winnipeg, Lake, resort lake in S central Manitoba province, third largest lake in Canada. Discovered in 1733, it was used extensively by fur traders and explorers. It is drained by the Nelson River to Hudson Bay and is believed to be a remnant of the glacial Lake Agassiz. Area: 24,514sq km (9,465sq miles).

Winter, Frederick Thomas (1926–), British jockey. He was a flat race rider (1939–42), a National Hunt steeple chaser (1947–64) and won the Grand National twice. In the 1952–53 National Hunt season he rode a total of 121 winners. He became a trainer in 1964.

Winter, coldest season of the year, between autumn and spring, which occurs from December to March in northern latitudes and from June to September in southern latitudes. It is reckoned astronomically from the winter solstice (20 or 21 Dec) to the vernal equinox (20 or 21 March). It is associated with hibernation and dormancy in animals, and there is generally a lull in agricultural production.

Winter cherry. *See* CHINESE LANTERN.

Wintergreen, any of various woody, evergreen plants, especially *Gaultheria procumbens*, the teaberry of the USA, Canada and Britain; the name also refers to the aromatic oil derived from its leaves. Family Ericaceae.

Winterhalter, Franz Xavier (1806–73), German painter and engraver. He painted portraits of various European rulers, including Queen Victoria and her family (1846–47).

Winter Olympics. *See* OLYMPIC GAMES.

Winter's bark, tree, known for its medicinal qualities. The elliptic-shaped leaves are leathery and the fragrant, cream-coloured flowers and are borne in clusters on red-tinged shoots. The bark was used to prevent scurvy. Height: 15m (50ft).

Winter sports, all-embracing term used to describe ice and snow sports, many of which make up the Winter Olympic Games, first held in 1924. There are two distinct ski sports: alpine ski racing (consisting of downhill, slalom and giant slalom) and Nordic skiing. Ice sports include figure skating for single skaters and pairs, ice dancing, ice hockey, speed skating around oval circuits and curling. A third variety of winter sports is tobogganing and bobsledding, in which individuals or teams of two or four (bobsleds only) are timed as they hurtle down specially constructed banked runs of ice. On a non-competitive basis, winter sports are practised wherever there are favourable conditions, and winter sports resorts play a major part in the tourist industries of some countries.

Winter's Tale, The (1610), romance by William SHAKESPEARE. Leontes, King of Sicily, imprisons his wife Hermione for an imagined indiscretion with his friend Polixenes, and orders that her newborn daughter Perdita, be left on a shore to die. Perdita is rescued and brought up by a shepherd. All are reunited when his daughter is wooed by Polixenes's son

Florizel. *See also* HC1 pp.283, *285*.

Winterthur, city in N Switzerland, 19km (12 miles) NE of Zurich. The city was ruled by the counts of Kyburg until 1264, when it passed to the HAPSBURGS. It was a free city of the HOLY ROMAN EMPIRE from 1415 to 1452. It was bought by Zurich in 1467. Winterthur is the site of 15th- and 17th-century castles and has an annual music festival. Industries: railway equipment, textiles, clothes. Pop. (1970) 92,722.

Wire, thin strand of metal, made from rod drawn through metal dies which get progressively smaller until the wire, which is continuously wound onto drums, is of the required diameter. The drawing process toughens steel so that a rope or cable made from steel wire is much stronger than an undrawn steel rod of the same diameter. Copper and aluminium wire are used to make electric CABLES.

Wire-haired terrier. *See* FOX TERRIER.

Wireless. *See* RADIO.

Wirén, Dag Ivar (1905–), Swedish composer. He studied in Stockholm and Paris and has written five symphonies, a cello concerto and several string quartets.

Wireworm, long cylindrical larva of a CLICK- BEETLE of N temperate woodlands. It is generally brown or yellow and is distinctly segmented. Most species live in the soil, and may cause serious damage to the roots of cultivated crops. Family Elateridae. The name also refers to any of the smooth-bodied MILLIPEDES of the family Paraiulidae.

Wiring, electrical, any of several types of metal wires that make up electrical circuits. Homes are supplied with single-phase electricity along a three-core flex consisting of live (brown), neutral (blue) and an earth (green and yellow) wires. *See also* SU pp.122–131.

Wirral, county district in SW MERSEYSIDE, England; created in 1974 under the Local Government Act (1972). Area: 157sq km (61sq miles). Pop. (1974 est.) 349,200.

Wirth, Karl Joseph (1879–1956), German statesman and Chancellor (1921–22) in the WEIMAR REPUBLIC. He advocated fulfilment of Allied demands under the VERSAILLES treaty. In 1922 he negotiated the Treaty of Rapallo with the USSR, which ended Germany's post-war diplomatic isolation.

Wisconsin, state in N central USA, in the region of the GREAT LAKES. The land is rolling plain that slopes gradually down from the N. There are numerous lakes in the state, the largest of which is Lake Winnebago. Wisconsin is the leading US producer of milk and cheese and it has the largest dairy herds. The chief crops are hay, corn, oats, fruit and vegetables. The state's most valuable resource is timber: 45% of the land is forested. Mineral deposits include sand and gravel, and zinc, lead, copper and iron. Food processing is one of the main industries. Others include the manufacture of farm machinery, brewing and tourism. Madison is the state capital and Milwaukee the largest city.

The French claimed the region in 1634, Great Britain took it in 1763 and it was ceded to the USA in 1783. Settlement of the region was slow. Lead deposits were exploited in the SW and more land was obtained after the elimination of hostile Indians. The Territory of Wisconsin was established in 1836 and admitted to the Union in 1848. Area: 145,438sq km (56,154sq miles). Pop. (1975 est.) 4,607,000. *See also* MW p.175.

Wisdom of Solomon, biblical book of the Old Testament APOCRYPHA. The first nine chapters deal with the moral and intellectual aspects of the doctrine of Wisdom, the remaining 10 chapters with the place of the doctrine in history.

Wise, Michael John (1918–), British geographer. He was appointed professor of geography, University of London, in 1958. His books include *A Pictorial Geography of the West Midlands* (1958).

Wise, Thomas James (1859–1937), British bibliographer who between 1890 and 1930 collected the Ashley library of first editions of (mainly British) poets and playwrights.

Wiseman, Nicholas Patrick Stephen (1802–65), British Catholic prelate. He

was rector of the English College at Rome (1828–40) and became a bishop in 1840. His elevation to the archbishopric of Westminster in 1850 provoked the "papal aggression" crisis and the passing of the Ecclesiastical Titles Act (1851), which forbade the assumption of territorial titles by the Catholic hierarchy in England.

Wisent. *See* BISON.

Wishart, George (*c.*1513–46), Scottish religious reformer. Schoolmaster of Montrose, he was charged with heresy in 1538 and fled to England, but after studying in Cambridge, returned to Scotland in 1543. He preached on behalf of Reformed doctrines, and was assisted by John KNOX who became his disciple. He was burnt as a heretic at St Andrews in 1546.

Wisla. *See* VISTULA.

Wissmann, Hermann von (1853–1905), German explorer and colonial administrator in Africa. He explored central and eastern Africa and helped to found the colony of German East Africa, of which he became Imperial Governor (1895–96).

Wisteria, genus of hardy woody vines native to North America, Japan and China. They have showy, fragrant flower clusters of purplish-white, pink or blue. Family Leguminosae.

Witan (witenagemot), national assembly to advise the king in Anglo-Saxon England, from the Old English *witan* (wise man) and *gemote* (a meeting). It consisted of the king, the higher clergy and the great landowners, THEGNS and EALDORMEN. It is considered by some to be the origin of the British Parliament.

Witchcraft, exercise of supernatural occult powers, usually due to some inherent power rather than to an acquired skill such as sorcery. In Europe it originated in pagan cults and in mystical philosophies such as GNOSTICISM, which believed in the potency of both good and evil in the universe. Anthropologists are interested in witchcraft phenomena in some societies where the belief in spirits is associated with attempts to control them for harmful or beneficial ends.

Witch doctor, person regarded as supernaturally powerful, especially in Africa. Witch doctors fight or control evil spirits and may heal the sick, using herbs and rituals. In many societies they wield considerable social power.

Witch hazel, shrubs and small trees of temperate regions, mostly in Asia, that bloom in late autumn or early spring. The common witch hazel (*Hamamelis virginiana*) has yellow flowers. It produces the astringent, also called witch hazel, used for treating sprains. Family Hamamelidaceae.

Wite, fine paid to the king or other public official in Anglo-Saxon England for a number of offences. The principal ones were disturbing the peace and failing to carry out a royal command.

Withering, William (1741–99), British doctor who introduced the use of DIGITALIS to treat cardiac disorders, giving details of his work in *An Account of the Foxglove* (1785). He was the first to establish a connection between dropsy and heart disease; he was also a keen botanist and published *Botanical Arrangement* (1776). *See also* MS p.118.

Witherite, glassy white, yellow or grey mineral, barium carbonate ($BaCO_3$), found in low-temperature lead and barite ore veins. It occurs as twinned, near-hexagonal crystals in the orthorhombic system. Hardness 3–3.5; s.g. 4.5.

Witkiewicz, Stanislaw Ignacy (1885–1939), Polish novelist and literary critic. His novels, full of sexual and political fantasy, include *Pozegnanie jesieni* (1927) and *Nienasycenie* (1930).

Witness, party to an act or transaction whose place in English law derives from Anglo-Saxon times, when COMMON LAW business was done with the aid of two witnesses, one who saw and one who heard. By an act of 1563 plaintiffs could compel witnesses to give SWORN EVIDENCE, but the defence gained the same power only in 1702.

Witt, Johan de (1625–72), Dutch statesman. An opponent of the House of ORANGE, he became Grand Pensionary and virtual Prime Minister of Holland in 1653, playing a leading role in the DUTCH WARS. He helped to form the TRIPLE ALLIANCE with England and Sweden against LOUIS XIV in 1668, but when the French invaded Holland in 1672, popular feeling turned against him and he was assassinated.

Witte, Emanuel de (1617–92), Dutch painter famous for his portraits and genre scenes such as *Adriana van Heusden and her daughter at the fish market* (1672). He also painted architectural views, especially grand-scale church interiors notable for their use of elaborate perspective and control of light and shadow effects. *See also* HC1 pp.*17*, 310–311, *310–311*.

Witte, Count Sergei Yulievich (1849–1915), Russian politician. As Minister of Finance from 1892 to 1903 he reformed Russian finances, stimulated rapid industrialization and encouraged Russian expansion eastwards with the building of the Trans-Siberian Railway. *See also* HC2 p.169, *169*.

Wittekind (Widukind) (*d.*807), SAXON leader. He fought CHARLEMAGNE and the FRANKS in the Rhineland from 778 to 785 and when finally defeated was pardoned by Charlemagne.

Wittelsbach family, German dynasty which ruled Bavaria from 1180 to 1918. Branches of it divided and re-united several times during the Middle Ages. In 1623 the Duchy of Bavaria became an electorate, and in 1806 MAXIMILIAN I became the first King of Bavaria.

Wittenberg, city in central East Germany, on the River Elbe. In 1582 the elector Frederick III the Wise founded a university there which became the centre of the Protestant REFORMATION during the time Martin LUTHER and Philipp MELANCHTHON were teaching there. Martin Luther nailed his 95 theses to the door of the Schlosskirche (castle church) of All Saints in 1517. Today Wittenberg is a mining and industrial centre; products include chemicals, rubber goods and machinery. Pop. (1971) 47,200.

Wittgenstein, Ludwig (1889–1951), Austrian philosopher, one of the most influential of the 20th century. He studied under Bertrand RUSSELL and later taught in Cambridge (1929–47). His work *Tractatus Logico-philosophicus* (1921) influenced the LOGICAL POSITIVISTS, arguing the strict relationships between language and the physical world; he claimed that there were therefore concepts that could not be expressed in language. After 1929 he criticized this hypothesis in highly-acclaimed lectures at Cambridge. These second thoughts were posthumously published in *Philosophical Investigations* (1953), in which he claimed that language was only a conventional "game", in which meaning was affected more by context than by formal relationships to reality.

Wittgenstein, Paul (1887–1961), Austrian pianist who settled in the USA in 1939. He lost his right arm in WWI but continued his concert career, showing great talent with his one hand. He commissioned a number of works to be composed for the left hand, including RAVEL's *Piano Concerto for the Left Hand* (1931).

Witwatersrand. *See* RAND.

Witz, Konrad (*c.*1400–*c.*1445), German painter who worked in Basel and Geneva. His style is characterized by its REALISM, probably derived from Jan VAN EYCK.

Wivallius, Lars (1605–69), Swedish poet and adventurer. Posing as a nobleman, he married a Swedish woman of high birth but was found out and imprisoned. His ballads, written mainly in prison, are notable for their spontaneity and delight in nature.

WMO. *See* WORLD METEOROLOGICAL ORGANIZATION.

Woad, or dyerswoad, biennial or perennial herb once grown as a source of blue dye. It was probably introduced by the early Britons to dye clothing and paint their bodies. A native of Eurasia, it bears small four-petalled, yellow flowers in summer. Height: 90cm (3ft). Family Brassicaceae; species *Isatis tinctoria. See also* MM p.*244*.

Wodehouse, Sir Pelham Grenville

(1881–1975), US novelist, *b.* Britain, famous for his humorous novels set in upper-class society of Edwardian times. His best-known stories are those about Bertie Wooster and his valet Jeeves. He was knighted in 1975.

Woden. *See* ODIN.

Woffington, Margaret ("Peg") (*c.*1714–60), Irish actress who made her English debut in 1740 as Sylvia in *The Recruiting Officer.* She was celebrated for her "breeches" role as Sir Harry Wildair in *The Constant Couple* (1741) and from 1742–48 was GARRICK's leading lady.

Wöhler, Friedrich (1800–82), German chemist who first isolated ALUMINIUM and BERYLLIUM and discovered calcium carbide. His synthesis of UREA (from the inorganic substance ammonium cyanate) in 1828 laid the foundation for modern organic chemistry. *See also* MS p.119.

Wokingham, county district in E central BERKSHIRE, England; created in 1974 under the Local Government Act (1972). Area: 179sq km (69sq miles). Pop. (1974 est.) 108,600.

Wolf, Hugo (1860–1903), Austrian composer. He wrote settings of many poems by GOETHE aHd more than 50 songs based on poems by Eduard MÖRICKE. His only completed large-scale orchestral work is the symphonic poem *Penthesilea* (1885).

Wolf, Max (1863–1932), German astronomer who used photography to discover some 200 new asteroids and various nebulae. *See also* SU p.*204*.

Wolf, wild, dog-like carnivorous mammal, once widespread in the USA and Eurasia, especially the grey wolf (*Canis lupus*), which is now restricted to the USA and Asia. It is powerfully built with a wide head and neck, muscular limbs, large feet and a deep-chested body; the tail is long and bushy. It has earned a reputation for savagery and cunning from occasional attacks on livestock and human beings. Length: to 2m (6.6ft), including the tail. Family Canidae. *See also* NW pp.*191*, *192*, *203*, *206*, *207*.

Wolfe, James (1727–59), British general who fought in North America during the Seven Years War. He took an army down the St Lawrence River, and in a daring manoeuvre his men scaled the Heights of Abraham, taking the city of Quebec from MONTCALM and his French forces. Both Wolfe and Montcalm died in the battle, which ended French ambitions in Canada. *See also* HC2 p.*136*.

Wolfe, Thomas Clayton (1900–38), US writer. His novels were autobiographical, rambling and much edited by his publishers. The first, *Look Homeward, Angel,* was published in 1929. Others include *Of Time and the River* (1935), *The Web and the Rock* (1939) and *You Can't Go Home Again* (1940).

Wolfenden, Lord John Frederick, (1906–), British administrator and academic. He was a fellow and tutor of philosophy at Magdalen College, Oxford (1929–1934), and became Vice-Chancellor of Reading University. He was Chairman of the Committee on Homosexual Offences and Prostitution which produced the "Wolfenden Report" in 1957.

Wolff, Christian von (1679–1754), Polish philosopher. While professor of mathematics at the University of Halle he acquired a reputation for systematizing and popularizing the doctrines of Gottfried LEIBNIZ. The Wolffian system, known for its rigid and narrow rationalism, was satirized by VOLTAIRE in *Candide.*

Wolf-Ferrari, Ermanno (1876–1948), German-Italian operatic composer whose works include a number of comic operas, such as *The School for Fathers* (1906), as well as the tragic opera *The Jewels of the Madonna* (1911).

Wolf fish, any of a number of fishes living in northern areas, of the family Anarhichadidae. They are so-called because of their biting and crushing teeth. They use these to feed on molluscs and crustaceans.

Wolfit, Sir Donald (1902–68), British actor. He retained 19th-century influences and in 1937 formed his own Shakespearean REPERTORY touring company, in the manner of the Victorians.

Wolfram. *See* TUNGSTEN.

Wolframite, black to brown mineral, iron-manganese tungstate, $(Fe,Mn)WO_4$. It is the chief ore of the metal tungsten, and is found extensively in Australia, China and India. It occurs as crystals in the monoclinic system, or as granular masses.

Wolfram von Eschenbach (*c.*1170–1220), German poet who was also a knight and led a restless, roving life. He was one of the great German MINNESINGERS and in 1203 was at the court of Landgrave Herrman of Thuringia. His only complete work is the famous *Parzival.* His other works include the unfinished epic poems *Willehalm* and *Titurel.* He also became a character in WAGNER's opera *Tannhäuser. See also* HC1 p.206.

Wolfsbane, or aconite or monkshood, any of several species of poisonous, perennial plants of the temperate zone. The name wolfsbane is derived from the superstition that the plant repels werewolves. Order Ranunculales.

Wolf spider, any of numerous species of mainly nocturnal, dark or drab spiders found throughout the world. It lives under organic debris or in a tubular nest, although a few species spin webs. Members of the genus *Pirata* often have yellow markings and live near water; those of the genus *Pardosa* have long thin legs with spined feet; species of the genus *Geolycosa* live in burrows and have front legs developed for digging. The large-bodied genus *Lycosa* includes the true tarantula (*L.tarantula*) of S Europe. Body length: to 2.5cm (1in). Family Lycosidae.

Wolgemut, Michael (1434–1519), German painter and designer of woodcuts. A pioneer in the use of woodcuts in book illustration, he was one of the most influential artists of his era and was responsible for the acceptance of the pure, untinted wood cut as a legitimate, self-sufficient medium. Equally important, it was in his workshop that the young Albrecht DÜRER was trained as a painter and graphic artist.

Wollaston, William Hyde (1766–1828), British scientist. He developed a method of making PLATINUM malleable, discovered PALLADIUM and RHODIUM, and invented the reflecting GONIOMETER, the CAMERA LUCIDA, and a lens (named after him) for correcting spherical ABERRATIONS.

Wollongong, city in New South Wales, Australia, 64km (40 miles) S of Sydney. The area was settled in 1815; since that time Wollongong has become a major iron and steel centre. Manufactures include chemicals, textiles and copper. Pop. (1973) 165,240.

Wollstonecraft, Mary (1759–97), British author who was an early proponent of equality between men and women. Her *Vindication of the Rights of Women* (1792) was the first great FEMINIST work. She visited Paris during the FRENCH REVOLUTION and on her return to Britain married William GODWIN. In 1797 she gave birth to a daughter who later became the wife of Percy Bysshe SHELLEY.

Wolof, Negroid people of Senegal, who speak a language belonging to the w Atlantic group of the Niger-Congo family. In the 15th century a Wolof empire dominated West Africa and traded in slaves with the Portuguese. They were converted to ISLAM in the 18th century.

Wols (1913–51), real name Alfred Otto Wolfgang Schulze, German painter. He studied at the Dessau BAUHAUS under MIES VAN DER ROHE and MOHOLY-NAGY before going to Paris in 1932, where he made a living as a photographer until he was interned in 1939. He then began to produce spontaneous, dynamic paintings, TACHISTE in style.

Wolseley, Garnet Joseph Wolseley, 1st Viscount (1833–1913), Commander-in-Chief of the British Army (1895–1901). He saw action during the CRIMEAN WAR, the INDIAN MUTINY, the Anglo-French conflict with China and in Africa. Promoted to Field-Marshal in 1894, he made important reforms in army structure, some of which remain today.

Wolsey, Thomas (*c.*1475–1530), English cardinal and politician. He served as papal legate (1518–30) and Lord Chancellor of England (1515–29). The son of a butcher, he was an ambitious man who rose

Wittenberg; the Luther Memorial stands before the Rathaus in the market square.

P. G. Wodehouse wrote more than 20 film scripts and collaborated on several musicals.

Michael Wolgemut is known for his richly carved altarpieces, eg in Schwabach.

Mary Wollstonecraft worked for the educational, not political, equality of women.

Wolverhampton

Wolverines, or skunk bears, are hunted by no other creature but man.

Sir Henry Wood conducts a lunch-time concert at the National Gallery.

Woodcock, related to the snipe, "drum" on the soil to attract worms.

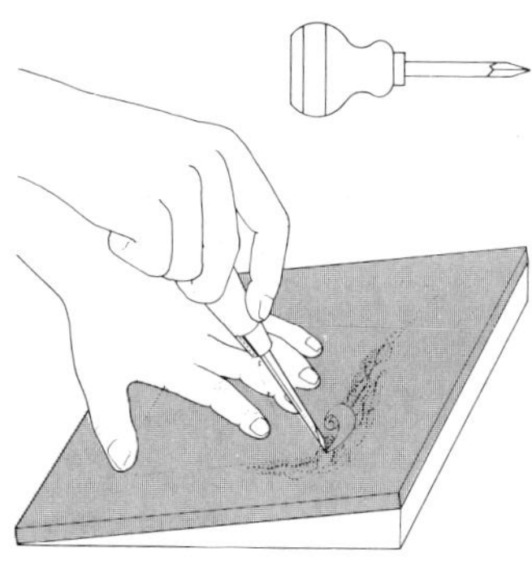

Woodcuts were first used by the ancient Egyptians to imprint designs on bricks.

quickly in Church affairs, becoming HENRY VII's chaplain and then Dean of Lincoln (1509). Appointed a Privy councillor in 1511 by HENRY VIII, he was soon one of the most important men in England. He became Archbishop of York and Bishop of Lincoln in 1514, and a cardinal in 1515. His failure to obtain Henry VIII's divorce from CATHERINE OF ARAGON brought his ruin. *See also* HC1 pp.241, 247, 266, 267.

Wolverhampton, industrial city and county district in West Midlands, England. The city is noted for its metalworking industries. The county district was created in 1974 under the Local Government Act (1972). Area: 69 sq km (27sq miles). Pop. (1974 est.) 268,200.

Wolverine, solitary, ferocious mammal native to pine forests of the USA and Eurasia, the largest member of the WEASEL family. Dark brown, with lighter bands along the sides and neck, it has a bushy tail and large feet. Length: 91cm (36in); weight: 30kg (66lb). Species *Gulo gulo.*

Women in White, The (1860), novel by Wilkie COLLINS, one of the first detective novels in British fiction. The story, told in turn by several of the characters, tells of Sir Percival Glyde's attempt to win his wife's fortune by having her put in a lunatic asylum.

Women Kilde with Kindness, A (acted *c.*1603), play by the English Jacobean dramatist Thomas HEYWOOD. Considered to be his best work, it is a domestic tragedy of middle-class life. *See also* HC1 p.285.

Woman of No Importance, A (1893), play by Oscar WILDE. *See also* HC2 p.101, *101.*

Womb, or uterus, hollow muscular organ in the pelvis of a woman. It protects and nourishes the growing FOETUS until birth. The upper part of it is broad and branches out on each side into the FALLOPIAN TUBES. The lower womb narrows into the CERVIX, which leads to the VAGINA. Its walls are strong, elastic muscle tissue lined with mucous membrane (endometrium), to which the fertilized egg attaches itself after fertilization. *See also* MENSTRUAL CYCLE.

Wombat, either of two species of large, rodent-like marsupial mammals of SE Australia and Tasmania. Both species are herbivorous, primarily nocturnal and live in extensive burrows. The common wombat, *Vombatus ursinus,* has coarse black hair and small ears. The hairy-nosed wombat, *Lasiorhinus latifrons,* has finer, grey fur and large ears. Length: to 1.2m (3.9ft). Family Vombatidae. *See also* NW pp.159, 205.

Women, Married, Property Acts. *See* MARRIED WOMEN'S PROPERTY ACTS.

Women in Love (1920), novel by D. H. LAWRENCE. The author examines the relationships of two couples; two sisters, Ursula and Gudrun, and their lovers Gerald and Rupert. The characters are believed to be based on Lawrence and his wife, Frieda, and John Middleton Murray and his wife, Katherine MANSFIELD.

Women's franchise. *See* SUFFRAGETTE MOVEMENT.

Women's Institutes, organizations for the promotion of women's education and communal activity, particularly in rural areas. Founded in Canada in 1897 by Mrs Adelaide Hoodless, the movement was established in Britain by Mrs Alfred Watt in Anglesey in 1915. In the 1970s there were more than 400,000 members in Britain, in 65 county federations.

Women's liberation, any of various organizations formed in the 1960s and 1970s aimed at obtaining equality and advancement for women. Although there is little international or even national co-ordination, the movement's general aim is to challenge the traditional roles and inferior status assigned to women.

Women's services, military corps of women. In Britain the first corps (apart from nursing) was the Women's Army Auxiliary Corps, formed in 1917; it was followed by corps for the navy and air force and all three were disbanded in 1921. They were revived for WWII as a paid force. They were not trained to fight; women worked chiefly as plotters in the operations rooms, in intelligence and in

the signal corps. In 1949 women's services were incorporated into the permanent establishment of the three forces and their auxiliary and reserve corps were also made regular and permanent.

Wonder, Stevie (1950–), blind US popular singer and composer, real name Stephen Judkins. As a song writer his work spans the gulf between soul music and pop. His records include *Music of my Mind* (1972), *Talking Book* (1972) and *Songs in the Key of Life* (1976).

Wonsan, port in SE North Korea, on the Sea of Japan, capital of Kangwón province and a major naval base. The city was opened to foreign trade in 1883. It served as a Japanese naval base during WWII, and suffered heavy damage during the KOREAN WAR. Industries: engineering, fishing and fish processing, oil refining, food processing, textiles. Pop. 300,000.

Wood, Anthony à (1632–95), English antiquary and historian. He prepared a treatise on the history of the University of Oxford, and also published *Athenae Oxonienses* (1691–92). He was expelled from the University in 1693 for libelling the first Earl of Clarendon.

Wood, Charles (1866–1926), Irish composer and organist. He was a pupil of STANFORD at the Royal College of Music, and became professor of music at Cambridge in 1924. He composed cantatas, church music and many songs.

Wood, Ellen (1814–87), British novelist who wrote under the name of Mrs Henry Wood. Her greatest success was the melodramatic *East Lynne* (1861). Her other novels include *The Channings* (1862) and *Pomeroy Abbey* (1878).

Wood, Garfield Arthur (1880–1971), US speedboat racer. He was a four-time winner of the US Gold Cup (1917, 1919–21) for hydroplane racing and won nine consecutive races for the international Harmsworth Trophy between 1920 and 1933 with his *Miss America* hydroplanes.

Wood, Grant (1892–1942), US painter who was influenced by early German and Flemish painting. In a meticulously detailed style, he depicted aspects of American life and the midwestern landscape. His best-known works are *American Gothic* (1930) and *Daughters of Revolution.*

Wood, Haydn (1882–1959), British composer who wrote more than 200 songs, the most popular of which was *Roses of Picardy* (1916).

Wood, Sir Henry Joseph (1869–1944), British conductor who studied at the Royal Academy of Music and became an opera conductor in 1889. In 1895 he was engaged to conduct the PROMENADE CONCERTS in London and continued to do so until 1944. He introduced many works previously unknown to Britain and taught singing and conducted choral societies.

Wood, Thomas (1892–1950), British composer, a pupil of STANFORD. His compositions, mostly accompanied or unaccompanied choral works, are English in style.

Wood, the name of two English PALLADIAN architects. John Wood (1704–54) was responsible for many of the finest buildings in Bath. He designed Prior Park (1735–38) there and in 1754 began the Circus which was completed by his son, also named John (1728–81). John the younger designed the Royal Crescent and the New Assembly Rooms in Bath and many country houses such as Buckland House, Oxfordshire (1755–57). *See also* HC2 p.70.

Wood, name of a family of Staffordshire potters related by marriage to the WEDGWOOD family and considered among the finest of English potters. Ralph Wood (1715–72) worked under Thomas Whieldon and the Wedgwood family, producing SALTGLAZED figures. His cousin, Aaron Wood (1717–85), was a leading block cutter. Enoch Wood (1759–1840) modelled busts of historical figures, most of which are now museum pieces. The family were the originators of several glazing processes and were credited with the introduction of the Toby mug.

Wood, hard substance that forms the trunks of trees; it is the XYLEM (the

vascular tissue of a woody plant) which comprises the bulk of the stems and roots, supporting the plant. It consists of fine cellular tubes arranged vertically within the trunk, which accounts for the grain found in all wood. The relatively soft light-coloured wood is called sapwood. The non-conducting, older, darker wood is called heartwood, and is generally filled with RESINS, gums, mineral salts and tannin (TANNIC ACID). Easily-worked softwood, generally from a CONIFER such as PINE, is composed of simple tracheids that provide support and conduct water and food. More durable hardwood, generally from a deciduous species such as oak, derives support from woody fibres; water and food are conducted through separate vessels. Wood is commonly used by man as a building material, fuel, to make some types of PAPER, and as a source of CHARCOAL, CELLULOSE, ESSENTIAL OILS, LIGNIN, tannins, dyes and SUGARS. *See also* NW pp.54–56, 62.

Woodbine, name once popularly used in Britain for HONEYSUCKLE, *Lonicera periclymenum,* because of its twining habit.

Wood-carving, art form which includes any kind of sculpture in wood. The pieces may range from bas-relief on a small scale and life-size figures to the ornamentation on furniture and buildings. Wood-carving was one of the earliest art forms but few ancient pieces have survived apart from some Egyptian artefacts. Masks and statuettes were common in pre-colonial Africa, as were the elaborate totem poles of the American Indians. In the Middle Ages wood-carving formed part of the religious art of the age, and later artists, such as Grinling GIBBONS, made realistic carvings for interior decoration.

Woodcock, any of five species of reddish-brown shorebirds that nest in cool parts of the Northern Hemisphere and winter in warm areas. Both the Eurasian *Scolopax rusticola* and American *Philohela minor* insert their long, sensitive, flexible bills into swampy ground to find worms. Length: to 34cm (14in). Family Scolopacidae. *See also* NW pp.143, 209.

Woodcut, in printmaking, one of the oldest means of producing prints. A design is incised on a flat, polished piece of wood, parallel to the grain, using sharp, chisel-like tools. Those parts of the wood that are not to be inked are cut away. In the 16th century, black-line woodcuts reached their highest standards with the works of Albrecht DÜRER and his followers. In the early 19th century, the technique was rivalled in popularity by WOOD ENGRAVING, which eventually lost favour as a means of reproduction to PHOTO ENGRAVING. Many artists throughout the last century, however, such as GAUGUIN and MUNCH, continued to develop the technique. In 16th-century Japan, the UKIYO-E style arose as a result of the work of such artists as Hishikawa MORONOBU, and later by HOKUSAI and HIROSHIGE. The medium generated a popular demand for inexpensive prints, depicting everyday scenes and landscapes.

Wood engraving, in printmaking, technique enabling a print to be made on paper from the flat polished transverse section (through the grain) of a block of hardwood, after a design has been incised with burins (sharp chisel-like tools). Textural and linear effects can be achieved by varying the pressure and direction of the cutting strokes. This technique developed from the WOODCUT in 18th-century England; the first great exponent was Thomas BEWICK who illustrated *A History of British Birds* (1797, 1804). Although wood engravings were used to reproduce works of art, artists such as William BLAKE, Gustav DORÉ and Adolf MENZEL produced drawings specifically intended for this technique.

Wooderson, Sydney (1914–), British middle-distance runner who, in the late 1930s, held world records for the mile, 800m and 880yd, the last standing until 1953. He won European titles at 1,500m (1938) and 5,000m (1946).

Woodfall, British film production company. It was formed in 1958 by playwright John OSBORNE and the director Tony

RICHARDSON for the filming of their stage successes *Look Back in Anger* (1959) and *The Entertainer* (1960). Other productions include *Girl with Green Eyes* (1963), *Tom Jones* (1963) and *The Knack* (1965).

Woodforde, James ("Parson") (1740–1803), British diarist. He wrote the *Diary of a Country Parson* (5 vols, published 1924–31), a charming and vivid picture of the life, interests and activities of a country clergyman of the 18th century. Woodforde held livings in Somerset and Norfolk.

Woodhenge, Neolithic earthwork N of Amesbury, Wiltshire, thought to be older than STONEHENGE. Six concentric rings of holes remain; wooden posts placed in these holes would, like Stonehenge, have indicated the position of the rising sun on Midsummer's Day. *See also* HC1 p.*43*.

Woodland, area of trees, including DECIDUOUS species, EVERGREEN species, or a mixture of both. Such an area is valuable for soil conservation, the safeguarding of water supplies, timber and as a scenic attraction. Modern FORESTRY, for reasons of high yield, concentrates on fast-growing CONIFERS. Woodlands, both natural and man-made, are important habitats for many kinds of wildlife. *See also* NW pp.208–211; PE pp.218–219.

Woodlouse, or sowbug, terrestrial CRUSTACEAN found throughout the world, living under damp logs and stones and in houses. It has an oval, segmented body. Length: 20mm (0.75in). Order Isopoda; Genus *Oniscus*.

Wood-Neumann scale. *See* MERCALLI SCALE, MODIFIED.

Woodpecker, tree-climbing bird found nearly worldwide. Woodpeckers have strong pointed beaks and long protrusible tongues, which in some species often have harpoonlike tips for extracting insect larvae. They have black, red, white, yellow, brown or green plumage and some are crested. Family Picidae. *See also* NW pp.*206, 209, 217, 244*.

Woodruff, any of several species of plants of the genus *Asperula*, especially sweet woodruff (*A. odorata*), which is a herb of European woodlands. It has white flowers and broad leaves. Family Rubiaceae.

Wood's Halfpence, patent granted to William Wood in 1722, allowing him to mint coinage for circulation in Ireland. Another patent, granted in the same year, allowed him to mint coinage for the American colonies. Both patents were withdrawn after public outcry at the large profits he was able to make out of them.

Woodspring, county district in SW AVON, England; created in 1974 under the Local Government Act (1972). Area: 375sq km (145sq miles). Pop. (1974 est.) 147,800.

Woodstock, Oxfordshire borough famous in English history. It was an ancient royal seat (until the ENGLISH CIVIL WAR), described in the DOMESDAY BOOK as a royal forest. The ASSIZE of Woodstock (1184) is the earliest surviving law requiring "the peace of the king's venison". ELIZABETH I was imprisoned there by Mary in 1554. In 1704 the manor of Woodstock was given to the 1st Duke of MARLBOROUGH, and BLENHEIM PALACE was built there for him by the nation.

Woodville, Elizabeth (1437–92), Queen Consort of King EDWARD IV of England. Daughter of Richard Woodville, Earl Rivers, she married Edward secretly in 1464. Her sons, EDWARD V and Richard of YORK, were possibly murdered in the Tower of London, but her daughter, Elizabeth of York (1465–1503), became HENRY VII's Queen and the mother of HENRY VIII.

Woodward, Sir Arthur Smith (1864–1944), British geologist. He was Keeper of Geology at the British Museum (1901–24) but is popularly remembered for his part in the PILTDOWN MAN controversy, his belief that the remains were human contributing to the success of the hoax. *See also* MS p.*22*.

Woodward, Robert Burns (1917–), US chemist. He was awarded the Nobel Prize in chemistry in 1965, in recognition of his synthesis of a number of complex organic substances including QUININE, CHOLESTEROL, CORTISONE, STRYCHNINE, RESERPINE, CHLOROPHYLL and TETRACYCLINE.

Woodwind, family of musical WIND INSTRUMENTS, which are traditionally made of wood but now often metal; they are played by means of a mouthpiece containing one or two reeds. The FLUTE and PICCOLO however are exceptional in that they are played by blowing across a hole. The other woodwind instruments of a symphony orchestra include the CLARINET (single reed) and the OBOE, COR ANGLAIS and BASSOON (all double reed). SAXOPHONES, a group of single-reed instruments always made of metal, are also included in the woodwind family. *See also* HC2 pp.104–5, *104–5*.

Woodworking machinery, fixed power tools used to cut, work and join timber. They include circular bench saws and radial-arm saws for general cutting, such as sawing planks and for ripping, edging and cross-cutting; band saws; surface planers for smoothing and reducing timber; mortising machines for making joints; moulding cutters for making trims and moldings; and sanders for finishing planed surfaces. *See also* CARPENTRY.

Woodworm, larva of various species of beetles that burrow in wood. When present in large numbers woodworms can cause extensive damage. Their presence can be detected by holes in the wood from which the adult beetles have emerged; treatment involves the use of poisons. Genera include *Anobium* and *Lyctus*.

Woodworth, Robert Sessions (1869–1962), US psychologist. He helped to turn US psychology away from the STRUCTURALISM of Wilhelm WUNDT and E. B. TITCHENER towards a consideration of cause and effect in mental processes, which came to be known as the functionalist school of US psychology. *See also* FUNCTIONAL PSYCHOLOGY.

Woody nightshade. *See* NIGHTSHADE.

Wookey Hole, cave system, open to tourists, 5km (3 miles) NW of Wells, on the edge of the Mendip Hills, SW England. Nine chambers (the ninth opened in 1975) follow the course of the River Axe through stalactite grottoes. Human skulls found dating from AD 100–300 suggest early sacrificial rites.

Wool, soft, generally white, brown or black animal fibre that forms the fleece (coat) of sheep. It is also the name of the raw fibres when removed from the animal and the yarns and textiles made from them after spinning, dyeing and weaving. Composed chiefly of KERATIN, the fibres are treated to remove an oily substance (which is purified to make LANOLIN). Selective breeding of sheep has reduced or eliminated a coarse outer coat of long hairs, still present in the MOUFLON, leaving behind the woolly undercoat.

Woolf, Leonard (1880–1969), British writer, husband of Virginia WOOLF, who in 1917 founded the Hogarth Press with his wife. His works include *Socialism and Co-operation* (1921) and *Imperialism and Civilization* (1928).

Woolf, Virginia (1882–1941), British novelist and critic, daughter of Leslie Stephen, wife of Leonard WOOLF and a member of the BLOOMSBURY GROUP. Her works included *Night and Day* (1919), *To the Lighthouse* (1927) and *The Waves* (1931). She suffered from several mental breakdowns and eventually drowned herself.

Woolley, Sir Charles Leonard (1880–1960), British archaeologist. He was director of excavations at the site of the ancient Sumerian city of UR; he also explored CARCHEMISH, a Hittite stronghold, in 1912–14 (with T. E. LAWRENCE), and TELL EL AMARNA, Egypt, in 1921–22. He wrote *Ur of the Chaldees* (1929), and *The Bronze Age* in UNESCO's *History of the Scientific and Cultural Development of Mankind* (1958).

Woolley, Frank Edward (1887–), British cricketer who played for England and Kent. As a left-handed batsman and slow bowler he played in 64 Test matches between 1909 and 1934, scoring 3,283 runs at an average of 36.07 and taking 83 wickets at an average of 33.91. His career aggregate of 58.969 runs in first-class cricket was bettered only by Jack HOBBS and he took more catches in the field, pro-

bably, 1,011 in all, than any other player.

Woolsack, large bag of red wool which is the official seat of the Lord CHANCELLOR as Speaker of the House of Lords. There were originally four woolsacks, probably from the reign of EDWARD III, for the Judges, the Barons of the EXCHEQUER, the SERJEANTS-AT-LAW and the Masters in CHANCERY.

Woolworth, Frank Winfield (1862–1919), US merchant who founded the international chain of stores known as "Woolworth's". In 1879 he opened a five-cent shop in Utica, New York, and eventually developed a large chain of stores in the USA; by 1919 there were more than 1,000 stores. In 1912 the shops were merged into the F. W. Woolworth Company.

Woomera, township in South Australia, 160km (100 miles) NNW of Port Augusta. Since 1945 it has been the site of a missile-testing range (which launched Australia's first earth satellite in 1967) and a satellite-tracking station.

Wootton, John (*c.*1686–1765), British painter who introduced the style of CLAUDE and DUGHET into England. He combined his landscapes with racing scenes and sporting conversation pieces. His *Members of the Beaufort Hunt* (1744) is an important early work of English sporting art.

Wootton of Abinger, Baroness (1897–19), British economist, academic and life peer, *b.* Barbara Frances Wootton. She served on several important Royal Commissions after 1938 and her many publications include *Freedom Under Planning* (1945) and *The Social Foundations of Wage Policy* (1955).

Worcester, Battle of (1651), victory of the English army of CROMWELL over the Scottish army. The Scottish army of about 16,000 was routed by the 20,000-strong English force. The defeat ended the attempt of CHARLES II to regain his throne by Scottish arms.

Worcester, Royal, one of the best-known makes of English PORCELAIN. It is produced by the Royal Worcester Porcelain Co, which was founded in Bristol (1748) and moved to Worcester (1751). Early designs were influenced by English silverwork and Chinese porcelain; MEISSEN, SÈVRES and Japane Imari styles later served as models.

Worcestershire, former county in W central England; since 1974 it has been part of HEREFORD and WORCESTER. The region is hilly and includes the Cotswolds and the Malvern, Clent and Lickey hills. The Severn and the Avon are the chief rivers. The region is noted for its fruit-growing, and vegetables and hops are also important. Sheep, dairy and beef cattle are reared. The N of the county is part of the BLACK COUNTRY and industries include metallurgy and the manufacture of porcelain and carpets. Worcester was the county town. Area: 1,823sq km (704sq miles). Pop. (1971) 692,605.

Wordsworth, William (1770–1850), British poet and a leader of the ROMANTIC movement, who spent much of his youth in Cumberland. He visited France in 1790 and became influenced by the spirit of the FRENCH REVOLUTION. On his return he and his sister Dorothy settled in Dorset where, with Samuel Taylor COLERIDGE, he wrote *Lyrical Ballads* in 1798. He later moved to the Lake District, and explored the influence of nature on the human mind. *See also* HC2 pp.100, *100*.

Work, in physics, energy expended in moving a body against an opposing force. If the opposing force is the body's weight mg, where m is the body's mass and g is the acceleration due to gravity, the work done in raising it a height h is mgh. This work has been transferred to the body in the form of POTENTIAL ENERGY; if the body falls a distance h, the KINETIC ENERGY at the bottom of the fall equals the work done in raising it. *See also* SU p.75.

Work function, in thermodynamics, another name for free energy, the measure of the amount of work that a system can do. It is equal to the total internal energy of a system less the product of its temperature and ENTROPY.

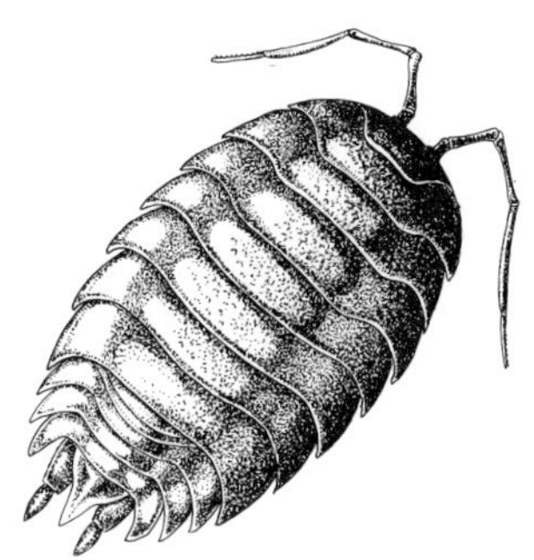

Woodlouse is the only crustacean that is fully adapted to life on land.

Wool; after cleaning the wool is "carded" or combed into loose ropes and twisted.

Worcestershire's county town Worcester, with its 13–14th century cathedral.

William Wordsworth wrote poetry is "the spontaneous overflow of powerful feelings".

Workhouses enforced strict discipline and often had criminal inmates.

World Cup Final in 1966; England beat West Germany by four goals to two.

World War I soldiers on the Western front celebrate Christmas in the trenches.

World War II; a Japanese delegate signs articles of surrender on 2 Sept. 1945.

Workhouse, institution for the unemployed in Britain. It originated in the houses of correction provided for vagabonds by the POOR LAW of 1601 but officially dates from 1696, when a workhouse was established by the Bristol corporation. A general act permitting workhouses to be founded in all parishes was passed in 1723. That act denied relief to those people who refused to enter a workhouse. Workhouses declined under the SPEENHAMLAND system of the late 18th and early 19th centuries, but were revived by the Poor Law of 1834. They had fallen into disuse by the early 20th century.

Working Men's Association, The International, socialist workers' group formed by English and French workers. Although not a founder member, Karl MARX was invited to joint the committee, and became its inspirational leader. It soon became known as the "First International", with branches in most countries.

Working Men's Clubs, social clubs for the working class in Britain, especially in the North. They began in the early 19th century, and were intended chiefly to provide workers with a place to read and to discuss political and social issues.

Works, Ministry of. See ENVIRONMENT, DEPARTMENT OF THE.

Works Progress Administration (WPA), national project in the USA created by Congress in 1935 under Franklin Delano Roosevelt's NEW DEAL policy to stimulate national economic recovery. Billions of dollars were contributed to the scheme in which work programmes provided jobs for the unemployed. An average of about two million people were registered on WPA rolls at any one time between 1935 and 1941.

World Bank (Internationl Bank for Reconstruction and Development), organization with headquarters in Washington DC, established at the BRETTON WOODS CONFERENCE in 1944. Its role is to make long-term loans to member governments to aid their economic development. The major part of the bank's resources are derived from the world's capital markets. See also MS p.316; HC2 p.300.

World Council of Churches, fellowship of Christian Churches formed in AMSTERDAM in 1948 to work for the reunion of all Christian Churches and to establish a united Christian presence in the world. The headquarters of the council are in Geneva. Its membership in the mid-1970s consisted of all the major Christian Churches except the Roman Catholic Church and the Unitarians. The Roman Catholic Church has, however, been sending observers to assemblies of the World Council of Churches since 1961.

World Cup (golf), annual international tournament for two-man teams of professionals. It is played over four rounds with the combined scores of each country's two players determining the placings. A separate award is made for the best individual score.

World Cup (soccer), premier competition in Association football. Held every four years, it is open to member countries of FIFA (Fédération Internationale de Football Associations). They qualify for the final rounds in a series of home and away group matches arranged on a geographical basis. The 14 qualifiers plus the holder and the host nation play in four groups at venues in one country to produce eight quarter-finalists, two from each group. There are then two semifinals, a third-place match between the losing semifinalists, and the final. First played in 1930, the tournament was for the Jules Rimet Trophy until 1970, when Brazil won it for the third time and thus took outright possession. A new trophy, called the FIFA World Cup, was first presented in 1974.

World fairs, large-scale expositions presenting displays of the goods, crafts and customs of participating nations, designed to stimulate trade and reflect cultural achievements. The first, the CRYSTAL PALACE Exposition, was held in London in 1851.

World Health Organization (WHO), UN agency that deals with health problems in the world. It collects and shares medical and scientific information and promotes the establishment of international standards for drugs and vaccines. It was officially established as a UN agency in 1948. Its headquarters are in Geneva, Switzerland. See also HC2 p.298.

World Meteorological Organization (WMO), organization with headquarters in Geneva, founded as the International Meteorological Organization in 1873. It promotes international co-operation in METEOROLOGY through the establishment of a network of meteorological stations throughout the world, and by the mutual exchange of weather information. In 1950 it became part of the United Nations.

World of Art, The, arts group and magazine founded in the 1890s in St Petersburg. Similar to the NABIS, it consisted chiefly of artists but also included musicians and poets. Prominent in the group were BAKST, BENOIS, and DIAGHILEV – the magazine's editor, whose Ballets Russes was the group's most important production.

World War I (1914–18), European conflict of unprecedented extent and ferocity, known also as the Great War. Although it was centred in Europe, hostilities also took place in the Middle East, the Far East and Africa. European territorial and economic rivalries had been intensified by the growing power of the German Empire and the conflict was precipitated by the assassination of Archduke FRANZ FERDINAND of Austria-Hungary (Germany's close ally) by a Serbian nationalist fanatic in 1914. The countries of Europe aligned themselves into two camps: the Allies, comprising Great Britain, France, Russia, Serbia, Montenegro, Japan and later Italy, Romania, Greece and the USA; and the Central Powers, consisting of Germany, Austria-Hungary, Bulgaria and the Ottoman Empire. In France and Belgium (the Western Front) the German schemes for rapid victory (the SCHLIEFFEN PLAN) failed, and the conflict soon degenerated into a war of attrition, both sides digging trenches and erecting fortifications. In the east, the huge manpower of Russia was savaged by the better trained and equipped German army. The outbreak of the RUSSIAN REVOLUTION (1917) effectively removed Russia from the war and peace between Germany and Russia was signed (1918) at BREST-LITOVSK.

Meanwhile, in 1917, the USA had entered the war on the Allied side and began ferrying troops to Europe. Early in 1918 the Germans launched a massive attack on the Western Front. It was frustrated by stubborn Allied resistance and the exhausted German army, and a starving civilian population, were not capable of further effort. An armistice was signed on 11 November 1918 and the ensuing Treaty of VERSAILLES (1919) imposed humiliating terms on Germany. See also 186–191, 186–191, 194–195, 194–195, 218.

World War II (1939–1945), most extensive conflict in human history. After Adolf HITLER had assumed absolute power in Germany, he rebuilt the German armed forces and during the 1930s annexed considerable areas in central Europe. At the same time the FASCIST Italy of Benito MUSSOLINI conquered Ethiopia (1935). Britain and France sought to appease the dictators but when Germany and the USSR (hitherto implacable foes) united to attack Poland (1 Sept 1939) a European conflict could not be avoided.

Poland soon succumbed to the German BLITZKRIEG and was partitioned between Germany and the USSR. In April 1940 the German army rapidly overwhelmed Norway, Holland and France. British forces were evacuated from DUNKIRK and German invasion of England was prevented only by the RAF's Battle of BRITAIN. In June 1941 Hitler invaded the USSR. His troops came within sight of Moscow but were halted in December by the early onset of a particularly harsh winter. In that same month the war became a global conflict. On 7 December the Japanese, determined to secure control of East Asia, bombed PEARL HARBOR, destroying much of the US Pacific fleet. Thereafter they rapidly overran the Philippines, Indonesia, Malaya, Burma and many of the Pacific islands. They were not stemmed until the summer of 1942, when the US navy won victories at CORAL SEA and MIDWAY. In Europe the British drove the Germans and Italians from North Africa and then, together with US forces, invaded Sicily and then Italy, which surrendered in September 1943. Meanwhile, the Russians had inflicted a crushing defeat on the Germans at STALINGRAD and had begun their advance into Europe. On 6 June 1944 Allied forces landed in Normandy. Germany was now fighting on three fronts and was finally completely occupied and forced to capitulate in May 1945. In the Pacific, Japanese forces were gradually driven back, the Allies advancing from island to island. On 6 and 9 August 1945 the USA dropped atomic bombs on the Japanese mainland, which brought about Japanese surrender. Perhaps as many as 45,000,000 people, civilians and armed servicemen together, were killed during WWII which radically altered the balance of power with the USA and the USSR in positions of supreme power. See also HC2 pp.226–233, 248–249.

World Wildlife Fund (WWF), international organization (established in Britain in 1961) which raises voluntary funds for world conservation projects. It is also concerned with the education of mankind on the need to conserve endangered wildlife species. See also NW pp.245.

Worm, any of a large variety of wriggling, limbless creatures with soft bodies. True worms differ from wormlike grubs or insect larvae. They have many cells but most have neither backbone nor notochord. Most worms belong to one or other of four main groups: ANNELIDS, FLATWORMS, NEMATODES, and ribbon worms. See also NW pp.88–89, 88–89.

Worm lizard. See AMPHISBAENID.

Worms, common name for the infestation of animals by parasitic worms. In human beings the most common varieties are the HOOKWORM, the filaria (producing FILARIASIS), the trichinella (producing TRICHINOSIS), the ROUNDWORM and the pinworm.

Worms, port in w West Germany, on the River Rhine, 16km (10 miles) NNW of Mannheim. It was made a free imperial city in 1156, and joined the Rhenish Confederation in 1255. The city was the site of the Diet of Worms in 1521, which was followed by the Edict of Worms in the same year. Worms became part of France in 1801 but passed to Hesse-Darmstadt state in 1815. The French occupied the city from 1918 to 1930 and it was taken by the Allies during WWII. Industries: leather goods, machinery, chemicals. Pop. (1970 est.) 78,000. See also HC1 p.268.

Worms, Concordat of (1122), arrangement between the Holy Roman Emperor Henry V and Pope Calixtus II settling the investiture dispute, a struggle between the empire and the papacy over the control of church offices. After the election of new church officials by the clergy or the emperor in contested elections, the emperor would endow the officials with the Regalia (their due worldly possessions and duties) and the Church would deal with the Spiritualia (religious matters).

Worms, Diet of (1521), conference of the Holy Roman Empire, held in Worms, Germany, and presided over by Emperor CHARLES V. Martin LUTHER was summoned to appear before the Diet to retract his teachings, which had been condemned by the Pope. Luther refused to retract and the Edict of Worms (25 May 1521) declared him an outlaw. The Diet was one of the most important confrontations of the early REFORMATION. See also HC1 p.268.

Wormwood, genus (Artemisia) of aromatic bitter shrubs and herbs, including common wormwood (A. absinthium), a European shrub that yields a bitter, dark green oil used to make ABSINTH. Family Asteraceae. See also NW p.234.

Worrell, Sir Frank Mortimer Maglinne (1924–67), West Indian cricketer who played for the West Indies, Barbados and

Jamaica. A graceful left-handed batsman and left-arm medium-pace bowler, he played in 51 Test matches between 1948 and 1963 scoring 3,860 runs, at an average of 49.48, and taking 69 wickets, at an average of 38.73. From 1960 he won international acclaim as an inspirational captain who drew together the strengths of West Indian cricket.

Worsaae, Jens Jacob Asmussen (1821–85), Danish archaeologist. His extensive travels in Scandinavia, France and Germany resulted in a number of works, two of which appeared in English as *Primeval Antiquities of England and Denmark* (1849) and *The Danes and Norwegians in England* (1852).

Worthington, Henry Rossiter (1817–80), US engineer and inventor. He invented a direct steam pump and set up a factory for its manufacture in New York in 1859. Worthington also developed the duplex steam feed pump, which is much used in oil pipelines and in waterworks.

Wouk, Herman (1915–), US novelist and radio scriptwriter. His novel *The Caine Mutiny* (1951), about the resistance of a junior officer to his tyrannical commanding officer during WWII, won him the 1952 Pulitzer Prize. Other works include *Marjorie Morningstar* (1955), *Youngblood Hawke* (1962), *Don't Stop the Carnival* (1965) and *The Winds of War* (1971).

Wouwerman brothers, Dutch painters based in Haarlem. Philips (1619–68), painted colourful battle scenes, cavalry skirmishes and hunts in which figures and landscape were given equal prominence. His style was imitated by his brothers Pieter (1623–82) and Jan (1629–66).

Woyzeck (1836), play by Georg BÜCHNER. Expressionist in form, it consists of a series of apparently unordered scenes providing a social analysis of the forces that drive a young soldier to murder his wife. In 1920 it was used by Alban BERG as the basis for his opera *Wozzeck*.

Wrangel, Peter Nicholaievich, Baron (1878–1928), Russian General. Commander of a COSSACK division, he joined the anti-BOLSHEVIK "White" army of Denikin and helped to capture Volgograd from the Communists in 1919. He became Commander-in-Chief in 1920 and sought an alliance with Poland, but was defeated in the Ukraine. His army was evacuated to Constantinople from the Crimea and he emigrated to Belgium.

Wrasse, any of about 600 species of brilliantly coloured inshore tropical marine fish; many of the smaller species are popular aquarium species. They have protrusible mouths and sleep on their sides; some remove parasites from larger fish. Length: 7.6cm–3m (3in–10ft). Examples include the rainbow *Coris julis* and the cleaner *Labroides dimidiatus*. Family Labridae.

Wrekin, The, county district in E central SALOP, England; created in 1974 under the Local Government Act (1972). Area: 291sq km (112sq miles). Pop. (1974 est.) 108,600.

Wren, Sir Christopher (1632–1723), English architect, mathematician and astronomer After the FIRE OF LONDON in 1666 he made a plan for the reconstruction of the city, but it was not used. He did, however, design 52 new churches in the city of London, the greatest of which is ST PAUL'S CATHEDRAL. Among his many other works are Chelsea and Greenwich hospitals in London. *See also* HC1 pp.295, *295, 297, 297,* 302.

Wren, Percival Christopher (1885–1941), British novelist. He was a soldier in the British Army and served in the French Foreign Legion, where his experiences inspired his romantic adventure stories, the best known of which are *Beau Geste* (1924) and *Beau Sabreur* (1926).

Wren, small, insect-eating songbird of temperate regions of Europe, Asia and most of the New World. Many species have white facial lines. The typical winter wren (*Troglodytes troglodytes*) has a slender bill, rounded wings, perky tail and dark brownish plumage; length: to 10cm (4in). Family Troglodytidae. *See also* NW pp.*218, 235.*

Wrestling, sport in which two unarmed opponents try to secure a fall by means of body grips, strength and adroitness. The most common styles include Greco-Roman (most popular in continental Europe), which permits no tripping or holds below the waist; and free-style, which permits tackling, leg holds and tripping, but not punching, and is most popular in the USA and Britain. Other forms include SUMO WRESTLING, popular in Japan; yagli, a Turkish form in which the contestants smear themselves with grease to make the holds difficult; sambo, a Russian form of jacket wrestling similar to JUDO; and the English Cumberland-and-Westmorland, an ancient style of wrestling in which the competitors start with their arms clasped behind each other's backs. Amateur wrestling can be conducted in both free-style and Greco-Roman style. It is classified by weight and conducted on a mat. A match consists of three periods of three minutes each, with points awarded for falls (pinning both shoulders to the mat) and other manoeuvres. Professional wrestling uses the free-style method, but is little more than planned entertainment.

Competitive wrestling originated in ancient Greece, where it was regarded as the most important event (after discus throwing) in the Olympic Games. Greco-Roman style wrestling was first included in the modern Olympics in 1896. It was dropped in 1900 but reinstated in 1904, when free-style was also included.

Wrexham Maelor, county district in SE CLWYD, Wales; created in 1974 under the Local Government Act (1972). Area: 367sq km (142sq miles). Pop. (1974 est.) 107,200.

Wright, Frank Lloyd (1869–1959), US architect. His first independent designs were for domestic houses, which he built with low, horizontal lines in his "prairie" style; the Robie house, Chicago (1908) is the most famous. He attempted to combine new mechanical methods and materials with organic architectural design, and to create open planning and free-flowing internal space. Notable buildings include the Larkin Office Building, Buffalo (1904), his winter residence Taliesin West, in Arizona (1938), the Guggenheim Museum, New York (designed 1943), and the Solomon R. Guggenheim Museum, New York City (1957–59). *See also* HC2 pp.*179, 207.*

Wright, Judith (1915–), Australian poet whose lyric poetry is marked by sensitivity and mastery of technique. Collections include *The Moving Image* (1946), *The Gateway* (1953) and *City Sunrise* (1964). She has also published biographies of Australian writers and the historical memoir *The Generations of Men* (1959).

Wright, Richard (1908–60), US short story writer and novelist, the first black American to gain international acclaim as a writer. His works, which deal with racial prejudice in the US, are realistic and brutal. They include *Native Son* (1940), *Black Boy* (1945), an account of the author's boyhood in the South, and *Twelve Million Black Voices* (1941), a folk history of blacks in America.

Wright, William Ambrose ("Billy") (1924–), British footballer who played for England and Wolverhampton Wanderers (Wolves). A stocky halfback, he won 105 caps for England (1946–59), 90 of them as captain. A member of the Wolves' 1949 FA Cup winning team, he led the club to three League titles in the 1950s.

Wright brothers, Wilbur (1867–1912) and Orville (1871–1948), US aviation pioneers. They used their bicycle factory as a laboratory for assembling their first aircraft and experimented with gliders to learn wing control and lateral balancing. The rolling sand dunes and fairly constant winds at Kitty Hawk, North Carolina, allowed them to test gliders (1900–02). Orville made the first piloted flight in a power-driven plane at Kitty Hawk in 1903. *See also* MM pp.148–149.

Wriothesley, Henry. *See* SOUTHAMPTON, HENRY WRIOTHESLEY.

Writ, in English law, a written order issued under seal by an executive officer of the CROWN, such as a judge. It commands the person to whom it is directed to take certain action in relation to a civil or criminal suit, and is the first step in legal proceedings.

Writers to the Signet, ancient society of SOLICITORS in Scotland who alone have the right to prepare Crown writs. Their name comes from the SIGNET Office (abolished in 1851), which kept the Signet seal and whose clerks prepared WARRANTS for the issue of PRIVY SEAL warrants.

Writing, visual record of language. The earliest known writing was developed in Sumeria c.4000 BC. Ancient writing, such as Egyptian hieroglyphs, was at first pictographic, seeking a character for each thing to be named. In the Chinese system there are still thousands of characters, which makes writing extremely difficult to learn and has contributed to the importance of scholars during Chinese history. Most modern alphabets are phonemic (they symbolize only the sounds of the particular language). *See also* MS pp.244–247.

Wrocław, industrial city in SW Poland, on the River Oder, approx. 306km (190 miles) SW of Warsaw and formerly known as Breslau. Originally a Slavic settlement, it was destroyed by the MONGOLS in 1241, rebuilt by the Germans and ceded to Prussia in 1741. It developed as a trade centre in the 19th century and became part of Poland by the Potsdam Conference of 1945. Wroclaw is the site of a 13th-century cathedral and several Gothic churches. Manufactures include heavy machinery, processed food, textiles and chemicals. Pop. (1974) 565,000.

Wrongful dismissal, dismissal of an employee in breach of a contract of employment. It is a ground in common law whereby an employee may sue an employer for damages. It is distinguished from UNFAIR DISMISSAL, which is dismissal on unfair grounds but not necessarily in breach of a contract of employment. In Britain before the Industrial Relations Act of 1971, an employer could dismiss an employee without cause provided it was with adequate notice,

Wrought iron, commercial form of smelted iron (the other is cast iron), containing less than 0.3% carbon with 1 or 2% slag mixed with it. Originally it was made from ore in a forge, and later in a puddling furnace, where it never becomes molten. Wrought iron replaced bronze in Asia Minor (c.2000 BC) at the beginning of the IRON AGE. In the 19th century wrought iron began to be used in building construction, but was replaced by steel after the invention of the BESSEMER and open-hearth processes. Today wrought iron is used principally for decoration, as in ornamental gates and railings. *See also* MM p.28.

Wryneck, bird of the woodpecker family (Picidae) found in deciduous woodlands in Europe. Unlike most of the family, it feeds mainly on the ground. It has brown, grey and white speckles. Length: 16.5cm (6.5in). Species *Jynx torquilla.*

Wuchang. *See* WUHAN.

Wu Chên (1280–c.1354), Chinese scholar-painter of the YUAN DYNASTY. He painted in a traditional but personal style and helped to develop the tradition of landscape painting. He is particularly associated with scenes of fishermen and was also a great painter of bamboo. *See also* HC1 p.129.

Wuhan, city in central China, at the confluence of the Han and Yangtze Rivers; the capital of Hupei province. It was formed in 1950 after the merger of three cities and is now the industrial and commercial centre of central China. Products include textiles, iron, steel, heavy machinery, soap and rice. Wuhan also has the largest cotton mill in China. Pop. (1970 est.) 4,250,000.

Wulfstan, Saint (c.1012–95), English ecclesiastic. He became Bishop of Worcester in 1062, and was the only Anglo-Saxon bishop retained after the Norman conquest. He was a loyal supporter of WILLIAM I THE CONQUEROR and William Rufus in their struggles against the barons and the Welsh. He was canonized by INNOCENT III in 1203.

Wren's typical habitat is dense undergrowth low on the ground; few wrens migrate.

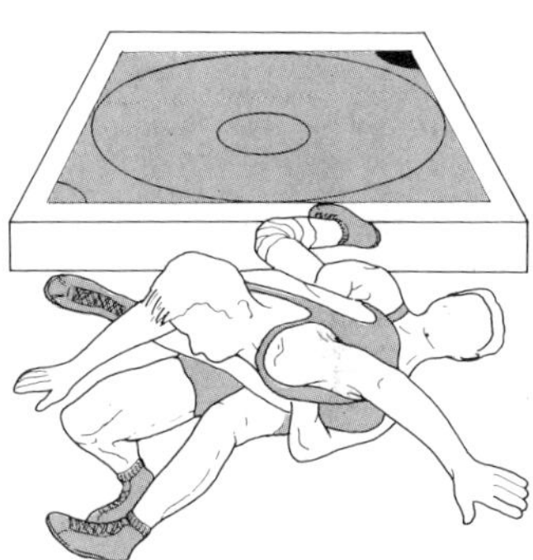

Wrestling is one of the oldest sports, the earliest records dating from before 2000 BC.

Wright brothers: Wilbur in 1903. At first neither thought men would fly at night.

Wrynecks are named after the habit of twisting their necks round when frightened.

Wuppertal, where a monorail railway was built in the early 20th century.

Wuthering Heights filmed by W. Wyler (1939) starred Laurence Olivier and Merle Oberon.

John Wycliffe inspired the first literal translation of the Vulgate Bible into English.

River Wye; Symond's Yat, seen here, lies on a meander between Monmouth and Ross.

Wundt, Wilhelm (1832–1920), German psychologist. He established the first laboratory for experimental psychology in 1879 at Leipzig, and did much to convince early psychologists that the mind could be studied with objective, scientific methods. His major publication is *Principles of Physiological Psychology* (1873–74).

Wuppertal, city in w West Germany, on the River Wupper, 26km (16 miles) ENE of Düsseldorf. It was formed in 1929 by the incorporation of Barmen, Vobwinkel, Elberfeld and several villages, but suffered heavy bomb damage in WWII because of its chemical plants. It remains an industrial centre, producing textiles, pharmaceuticals and rubber. Pop. (1974) 409,715.

Württemberg, former state in sw West Germany, between BAVARIA and BADEN states; its capital was STUTTGART. Württemberg was the scene of much unrest in the early 16th century, after Duke Ulrich was made a vassal of the House of HAPSBURG. Napoleon raised Württemberg to the status of a kingdom, and it enjoyed economic prosperity until the end of the 19th century.

Wurzburg, Konrad von (*c.*1225–87), Middle High German poet, considered among the most important writers in the later period of courtly literature. In addition to his short narrative poems, love lyrics and religious legends, he also wrote an epic history of the TROJAN WAR. The stylized technique and didactic character of his poetry led to the MEISTERSINGERS naming him as one of the 12 great masters of medieval poetry.

Wu Tao-tze (*c.*700–*c.*760), Chinese painter of the T'ANG dynasty. One of the most revered artists China has produced, he is celebrated as the earliest Chinese figure painter and the inventor of various styles and techniques, including a method of printing designs on silk. *See also* HC1 p.128.

Wuthering Heights (1847), novel by Emily BRONTË which, set in the bleak moors of northern England, tells the story of Heathcliff, his benefactors Hindley and Catherine Earnshaw, and their neighbours, Edgar and Isabella Linton. *See also* HC2 p.98.

Wyant, Alexander Helwig (1836–92), US landscape painter whose early work was in the HUDSON RIVER SCHOOL tradition. Influenced by INNES and the BARBIZON School he introduced more atmosphere into his interpretation of nature. Much of his work was done in the CATSKILL MOUNTAINS.

Wyatt, James (1746–1813), British architect. He is famous for his GOTHIC-style country buildings, including Fonthill Abbey, his cathedral restorations, and his design for the Royal Military Academy, Woolwich (1796).

Wyatt, Sir Thomas (1503–42), English poet and statesman who served under HENRY VIII and was a friend, and possibly a lover, of Anne BOLEYN. He introduced the SONNET form into English poetry and 96 of his poems appear in the anthology *Tottel's Miscellany* (1557).

Wyatt, Thomas Henry (1807–80), British architect. He is best known for his Romanesque Wilton Church (1854–55) in Wiltshire – one of the earliest modern British buildings in which mosaic decoration was employed.

Wyatt's Rebellion, Protestant insurrection, led by Sir Thomas WYATT in 1554 to attempt to prevent the marriage of MARY I to PHILIP II of Spain. Wyatt and his followers marched from Maidstone to Ludgate, London, and then gave up the cause as hopeless. The leaders were executed and the Princess ELIZABETH was imprisoned in the TOWER OF LONDON.

Wyatville (Wyatt), Sir Jeffrey (1766–1840), British architect and nephew of James WYATT. He worked with his uncle in the 1790s and later designs included additions to Sidney Sussex College, Cambridge (1821–24). He was architect in charge of remodelling Windsor Castle (1824–28). He changed his name in 1824.

Wych elm, or wild elm, common European and North American tree. Its flowers have no petals, but appear as a bunch of red anthers on the twigs in early spring. The fruit consists of green, winged discs which ripen to brown, fall, and germinate quickly. Species *Ulanus compestries* Height: to 34m (114ft).

Wycherley, William (1640–1716), English dramatist, one of the chief comic playwrights of the RESTORATION. His best known plays are *The Country Wife* (1675), a comedy of sexual manners, and *The Plain Dealer* (1676), a satirical comedy.

Wycliffe, John (*c.*1328–84), English religious reformer. He studied and taught at Oxford University and was vicar of Fillingham (1361) and Ludgershall (1368) and rector of Lutterworth (1374). In the 1370s he acted to protect the interests of the national Church against the PAPACY, and came under the patronage of John of GAUNT. In 1378 he began a series of works attacking Church practices and beliefs, especially TRANSUBSTANTIATION and the secular ambitions of many churchmen. His works were condemned at a synod in 1382. *See also* HC1 p.195.

Wycliffites. *See* LOLLARDS.

Wycombe, county district in sw BUCKINGHAMSHIRE, England; created in 1974 under the Local Government Act (1972). Area: 324sq km (125sq miles). Pop. (1974 est.) 149,700.

Wye, major river in England and Wales. It rises in the hills of central Wales and flows SE into England through Hereford and Worcestershire. It forms the border between Gloucester and Gwent before entering the Severn estuary. Length: approx. 210km (130 miles).

Wyk, Arnold Van (1916–), South African composer, real name Arnoldus Christiaan Vlok van Wyk. He studied in Britain and after working with the British Broadcasting Corporation (1939–44) returned to South Africa where he taught at the University of Capetown. His works include two symphonies (1943, 1952), *Christmas Cantata* (1948) and *Five Elegies* (1941) for string quartet.

Wykeham, William. *See* WILLIAM OF WYKEHAM.

Wyler, William (1902–), US film director. He directed numerous silent films between 1925 and 1929, and then turned to sound production. His many films include *Wuthering Heights* (1939), *The Little Foxes* (1941), *Mrs Miniver* (1942), *The Best Years of Our Lives* (1946), *Ben Hur* (1959) and *Funny Girl* (1968).

Wyndham, John (1903–69), real name John Wyndham Beynon Harris, British science-fiction writer. His novels examine the ethical codes of human behaviour when subjected to the test of unforeseen circumstances. His *Day of the Triffids* (originally serialized *Revolt of the Triffids*) (1951) was later filmed. He also wrote *The Chrysalids* (1955) and *Midwych Cuckoos* (1957), filmed as *The Village of the Damned* (1960).

Wynne, David (1926–), British portrait painter and sculptor who studied at Cambridge University. His paintings include portraits of BEECHAM, GIELGUD, MENUHIN and the BEATLES. Among his sculptures are the marble *Breath of Life* and the bronze *Christ and Mary Magdelen.*

Wynter, Bryan (1915–), British abstract painter who studied at the SLADE SCHOOL. Influenced by the US painters Mark TOBEY and Bradley Tomlin (1899–1953), he turned from early Romantic landscapes to ABSTRACT EXPRESSIONISM in works which are densely and vividly coloured.

Wyoming, state in NW USA. Much of the E lies in the GREAT PLAINS. The Rockies cross the state from NW to SE. The region is the source of many rivers including the North Platte and the Snake. Many of the population are engaged in cattle ranching. Sheep are also reared and sugar-beet, hay, wheat and barley are grown. Oil is the most important mineral deposit: others include uranium, natural gas, gold, iron ore and coal. The manufacture of petroleum products is the principal industry. Food processing and cement products are also important. The magnificent scenery in Wyoming makes tourism a major source of income in the state. The chief cities are Cheyenne, the state capital and largest city, Casper and Laramie.

The Spanish and French may have visited the region in the 1700s but the first known exploration was that of US trappers from the E in the early 19th century. Wyoming Territory was organized in 1869 and admitted to the Union in 1890. Area: 253,596sq km (97,913sq miles). Pop. (1975 est.) 374,000. *See also* MW p.175.

Wyre, county district in w central LANCASHIRE; created in 1974 under the Local Government Act (1972). Area: 280sq km (108sq miles). Pop. (1974 est.) 98,500.

Wyre Forest, county district in NE HEREFORD AND WORCESTER, England; created in 1974 under the Local Government Act (1972). Area: 196sq km (76sq miles). Pop. (1974 est.) 92,200.

Wyspianski, Stanislaw (1869–1907), Polish painter and dramatist who painted many murals and stained-glass windows. He is considered to be the founder of modern Polish drama and his imaginative plays include *Warszawianka* (1898) and *Wesele* (1901).

Wyss, Johann Rudolf (1782–1830), Swiss writer and editor. A professor of philosophy at Bern Academy in 1805, he collected Swiss folklore and published it as *Idyllen, Volkssagen, Legenden und Erzählungen aus der Schweiz* (1815).

X, 24th letter of the alphabet. Its origins are obscure, and the first letter of this form was in the Greek alphabet, in which it had the sound *kh* or *ks*; this was the form adopted by the Romans. In English *x* generally has a *ks* sound, as in *exit* and *exercise*, although in unaccented syllables it is more like *gs* (as in *examine*). With some following vowels *x* is pronounced as *ksh* (as in *luxury*), and as the initial letter in words derived from Greek it is pronounced like a *z* (as in *xylem* and *xanthine*). *See also* MS pp.244–245.

Xanthine, yellow substance found in plants, in most body tissues and fluids, and in urinary tract stones and which can be oxidized to form URIC ACID. It is a muscle stimulant, particularly of cardiac muscle, and its synthetic derivatives are used as diuretics and to dilate blood vessels and bronchi.

Xanthoma, skin condition caused by a disturbance in lipid metabolism. Related to liver disorder, it is marked by flat, raised, yellowish patches or nodules on the skin, especially the eyelids.

Xanthus, ancient city of Lycia, West Asia Minor, now in Turkey. It was captured by the Persians *c.*546 BC and the Romans *c.*42 BC; on both occasions the city was destroyed by the inhabitants before surrender.

Xavier, Francis. *See* FRANCIS XAVIER, SAINT

"X" certificate, classification given to a film in Britain, indicating that it is unsuitable for juvenile audiences. Introduced by the British Board of Film Censors in 1951, the ban applied originally to people under 16 years of age; after 1970 the age limit was raised to 18.

X-chromosome, one of the two kinds of sex CHROMOSOMES in human cell nuclei, the other being the Y-CHROMOSOME. A normal female has two X-chromosomes in her body cells, so that each of her ova has only one X-chromosome. When she passes this on to one of her offspring, it influences the development of female characteristics. Her male offspring will have one X- and one Y-chromosome in his body cells.

Xenakis, Yannis (1922–), Greek composer. He studied music under HONEGGER and MESSAIEN, and was also trained as an architect. His work is 12-tone and serialistic, and he has been a pioneer of the use of computers in writing music. His compositions include *Métastasis* (1955) and *ST/10-1,080262* (1962). *See also* HC2 p.271.

Xenocrates (396–314 BC), Greek philosopher. He was a follower of PLATO and acted as head of the ACADEMY at Athens.

Xenomorphism, formation of a mineral

crystal in a restricted space so that it cannot develop crystal faces. It occurs in the late-forming minerals of IGNEOUS ROCKS. *See also* PE p.*96.*

Xenon, gaseous non-metallic element (symbol Xe) of the NOBLE GAS group, first discovered in 1898. Xenon is present in the Earth's atmosphere (0.000008% by volume) and is obtained by fractionation of liquid air. It is used in discharge lamps. The element forms several compounds including XeF_2, XeF_6, XeF_6, $XePtF_6$ and XeO_3. Properties: at.no. 54; at.wt. 131.30; density 1.88 (air = 1); m.p. $-111.9°C$ ($-169.42°F$); b.p. $-107.1°C$ ($-160.78°F$); most common isotope Xe^{132} (26.89%).

Xenophanes of Colophon (*c.*560–*c.*478 BC), travelling Greek poet and philosopher, often considered to be a precursor of the ELEATIC school. He proposed a version of pantheism, holding that all living creatures have a common natural origin.

Xenophon (*c.*430–*c.*354 BC), Greek historian and soldier. He is best known for the *Anabasis,* an account of the march of a Greek mercenary army across Asia Minor in 401–399 BC in support of a pretender to the Persian throne. An Athenian, Xenophon studied with SOCRATES, about whose teaching he wrote the *Memorabilia. See also* HC1 p.*75.*

Xerography, method of copying documents, known also as photocopying. Photocopying machines contain a drum which is coated with a light-sensitive material. In one widely used process the drum is coated with the element selenium, the electrical resistance of which decreases when it is exposed to light. Where no light falls, as on the areas reflected from the typed areas of a document, the drum remains electrically charged. When the drum is coated with a powdered "ink", this remains stuck, by electrostatic attraction, to the drum in these particular areas, falling away from the other areas. When paper passes over the drum, the "ink" is transferred to its surface, and it is fused on to the paper by a heater before the photocopy leaves the machine. *See also* MM pp.208–209.

Xerophyte, any plant that is adapted to survive in dry conditions, in areas subject to drought or in physiologically dry areas such as saltmarshes and acid bogs. Succulents such as CACTI and AGAVES have thick fleshy leaves and stems for storing water. Other adaptations include the ability to shed leaves during drought and waxy or hairy leaf coatings. *See also* NW p.218,*218.*

Xerxes I (*c.*519–465 BC), King of Persia (*r.*486–465 BC). His first achievement was to bring Egypt under Persian rule (484 BC). He launched a campaign to conquer Greece in 480 BC, marching through Macedonia and burning Athens. His fleet was destroyed by the Athenians at SALAMIS, however, and he returned to Asia; the following year, the Persian army in Greece was defeated at Plataea. Xerxes was later assassinated by his own bodyguard. *See also* HC1 pp.61, *70.*

Xerxes (1738), opera by George Frederick HANDEL, first performed in London. The text, taken from a revised libretto (1654) of Minato, sets the action in ancient Persia. It contains Handel's only purely comic character and the famous aria now known as Handel's *Largo,* originally a satirical *Larghetto.*

Xhosa, or Xosa, group of related BANTU tribes. The Xhosa moved from E Africa to the vicinity of the Great Fish River, s Africa, in the 17th–18th century. They were defeated by the Europeans in 1835, after which they came under European rule. In culture they are closely related to the ZULUS. The 2.5 million Xhosa live in the Transkei and form an important part of South Africa's industrial and mining workforce.

Xi. *See* SI.

Xiamen. *See* AMOY.

Xi'an. *See* SIAN.

X-ray, electromagnetic ray of shorter wavelength, or higher frequency, than visible light, produced when a CATHODE ray impinges upon matter. X-rays were discovered by ROENGTEN in 1895. They are normally produced for scientific use in X-RAY TUBES. Because they are able to penetrate matter which is opaque to light, they are used to investigate inaccesible areas, especially of the body. *See also* RADIOGRAPHY; SU pp.*99,* 235.

X-ray astronomy. *See* ASTRONOMY.

X-ray crystallography, use of X-RAYS to discover the structure of CRYSTALS. *See* PE pp.94–95.

X-ray diffraction. *See* DIFFRACTION.

X-ray tube, evacuated tube used to provide a source of X-rays for medical or other purposes. It consists of an electron gun producing a stream of ELECTRONS that strike an anode, part of which is made of a heavy metal such as tungsten. The tungsten emits X-rays when it is bombarded by the stream of high-energy electrons.

Xylem, woody vascular tissue of a plant. It conducts water and minerals from the roots to the rest of the plant and provides support for the stems, leaves and roots. *See also* PHLOEM.

Xylene, organic chemical compound of formula $C_6H_4(CH_3)_2$ obtained from the distillates of coal tar and petroleum, and important as a solvent and chemical intermediate. Chemically it is dimethyl benzene which exists in three isomeric forms: ortho-, meta-, and para-xylene. The ISOMERS have different physical properties.

Xylophone, tuned percussion instrument, made of hardwood bars arranged as in a piano keyboard and played with mallets of various degrees of hardness. The modern xylophone normally has a range of four octaves, extending from middle C upwards.

XYZ Affair (1797–98), diplomatic incident that strained US relations with France. President John ADAMS sent three representatives, Charles Pinckney, John MARSHALL, and Elbridge Gerry, to negotiate a treaty with Revolutionary France to end France's preying on US commerce. Three French agents, known as X, Y and Z, demanded loans and a $250,000 bribe to be given to TALLEYRAND, the French Foreign Minister, before discussions could begin. The Americans refused The revelation of the attempt at bribery caused an uproar in the USA and its representatives were recalled.

Y

Y, 25th letter of the alphabet, derived (as were *f, u, v* and *w*) from the Semitic letter *vaw,* meaning *hook.* It was adopted by the Greeks, but did not enter the Roman alphabet until the 2nd century AD, when it was used in words of Greek origin. In English *y* may be a consonant (particularly at the beginning of a word such as *yell* or *young*) or a vowel, when it duplicates some of the sounds of the letter *i. See also* MS pp.244–245.

Yachin, Lev (1929–), Soviet footballer, who played for Moscow Dynamos and the Soviet Union. A brilliant goalkeeper with cat-like reflexes, he played a record 74 times for the USSR, including three World Cups, and won a Olympic Games gold medal in 1956.

Yacht, sailing boat or motor-vessel used for pleasure or for racing. Sailing yachts usually have either one or two masts and are built in many sizes, the larger ones being used for ocean racing. Motor-powered yachts are small luxury ships with large crews.

Yacht racing, sailing-boat competition for various classes of YACHT, based on the international rules drawn up by the Yacht Racing Union. The most important, as well as the oldest, international competition is the America's Cup, contested 22 times since 1857. It is named after J. C. Steven's schooner yacht *America,* which won the Royal Yacht Squadron's One Hundred Guineas Cup in 1951. The USA has never failed to defeat challenges for the cup, from Britain, Australia and Canada. Other major races are the Fastnet race, first sailed in 1925, and the Admiral's Cup, first sailed in 1957. Yachting has been an Olympic sport since 1908.

Yahweh, personal name of the God invoked by the ancient Israelites, in the Bible. It was made up of four consonants, YHWH, and may have been a variant of the Hebrew for "I am" – the name revealed to Moses. The lack of vowel symbols has made pronounciation contentious and results in alternatives such as JEHOVAH.

Yajur-Veda, in HINDUISM, a collection of prayers and sacrificial incantations. It is one of four VEDAS or collections of hymns, oblational verses sacred to Hinduism. The others are the RIG-VEDA, SAMA-VEDA and Atharva-Veda.

Yak, large, powerful, long-haired ox, native to Tibet, with domesticated varieties throughout central Asia; it inhabits barren heights up to 6,100m (20,000ft). Domesticated varieties are generally smaller and varied in colour; they breed freely with domestic cattle. Wild yaks have coarse black hair, except on the tail and flanks, where it hangs as a long fringe. The horns curve upward and outward. Height: to 1.8m (6ft) at the shoulder. Family Bovidae; species *Bos grunniens. See also* NW pp.*196,* 222.

Yakuts, Turkic-speaking people who live in Yakut ASSR, USSR. They are noted iron-workers and potters, and formerly lived by semi-nomadic cattle farming and fishing. Horses and blacksmiths were venerated in their shamanistic religion.

Yale, Linus (1821–68), US inventor of various types of locks, including the domestic version that bears his name. *See also* MM p.248, *248.*

Yale University, institute of higher education in New Haven, Connecticut, USA. Founded in 1701 in Branford, Connecticut, it operates several colleges for the study of, for example, divinity, law, medicine, art, drama and architecture. Its current charter dates from 1745.

Yalow, Rosalyn (1921–) US biochemist who shared the 1977 Nobel Prize in physiology and medicine with Roger Guillemin and Andrew Schally for her development of a method of detecting peptide HORMONES in the blood. In the 1950s Yalow found that some people who received regular INSULIN injections developed antibodies against the hormone. She discovered that insulin, labelled with radioactive iodine, combined with the antibodies; from this she developed radio-immunological tests to detect the amount of insulin present.

Yalta Conference (1945), meeting at Yalta in the Crimea between US President Franklin D. ROOSEVELT, Prime Minister Winston CHURCHILL of Britain, and Premier Joseph STALIN of the USSR. They met to plan the final attacks on Germany, its postwar occupation and control and the trial and punishment of war criminals. A conference to establish the UNITED NATIONS was called to meet at San Francisco. In return for concessions in the Far East, the USSR agreed to join the war against Japan, after Germany's defeat. *See also* HC2 pp.218, 234, *235.*

Yam, any of several species of herbaceous vines that grow in warm and tropical regions and the large, tuberous roots of several tropical species, which are edible. They are often confused with the SWEET POTATO. The plant has a long, annual, climbing stem, with lobed or unlobed leaves and small clusters of greenish, bell-shaped flowers. Family Dioscoreaceae; genus *Dioscorea. See also* PE pp.*186,* 187.

Yama, in Indian mythology, lord of death, also worshipped as the first mortal man. He and his sister, Yami, are children of Vivasvat, the sun. Yama is depicted as being green in colour, dressed in red, bearing a club and noose, riding a buffalo. He captures a dying man's spirit with the noose to take it to his kingdom.

Yamasaki, Minoru (1912–), US architect, who designed the Lambert-St Louis Municipal Airport (1953–55) terminal, noted for its concrete vaults. He was also chief designer for the World Trade Centre (1962) in New York City.

Yamato, Japanese clan who gave their

Xenophon's historical narratives are characterized by his love of the truth.

Xerxes's army crossed the Hellespont by bridging it with a double line of boats.

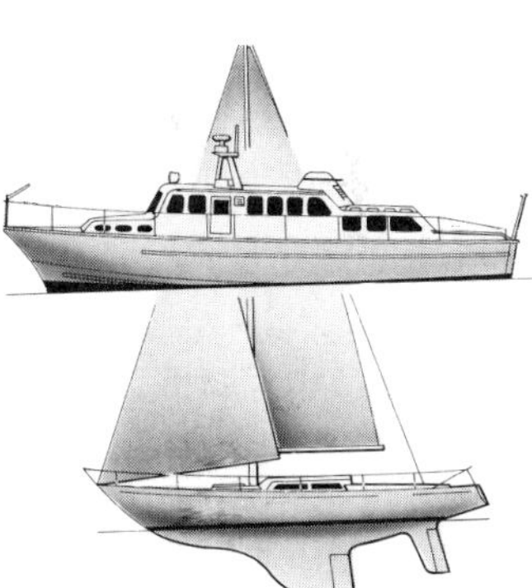

Yachts, generally used for pleasure, may be a sailing craft or motor-driven.

Yaks, despite their size and weight, are agile; if attacked, they are fierce.

Yamato-e

Yangtze-Kiang; boats are pulled up the rapids of its upper reaches.

William Yeats' body was reinterred in Ireland after a year in French soil.

Yellowhammer, whose shrill call sounds like "a-little-bit-of-bread-and-no-cheese".

Yellowstone National Park; a geyser in the salt soil of the Shoshone Geyser Basin.

name to the formative period of the Japanese state (*c.*400–800). Japanese society at this time was based on a hierarchy of feudal and tribal groups. The Yamato clan, claiming descent from the sun goddess, effectively ruled the area around Kyoto and were the ancestors of the present-day imperial line.

Yamato-e, Japanese school of painting which flourished from the 12th to 14th centuries. Mostly found on manuscripts and scrolls, Yamato-e ("the Japanese style") is decorative, secular, anecdotal and flowing; it makes great use of rich and balanced colours. Among the earliest surviving Yamato-e works are the scrolls illustrating the TALE OF GENJI. *See also* HC1 p.*132.*

Yamuna, river in N central India. It rises in the Himalayas and flows S and SE to join the River Ganges at Allāhābād. Navigable for almost its entire length, the Yamuna was once an important trade route. It is now not used for irrigation. Length: approx. 1,385km (860 miles).

Yam Zapolsky, Peace of (1582), settlement between Russia and Lithuania at the end of the Livonian War, conflict fought since 1558 in which Lithuania was supported by Poland and Sweden. Russia renounced all claims to Lithuania and restored Lithuanian territories.

Yang, Chen Ning (1922–), Chinese physicist who shared the 1957 Nobel Prize in physics with Tsung-Dao LEE for their discovery of violations of the conservation of PARITY. They studied the decay of K-mesons, which seemed to break down in two different ways and in 1956 they concluded that in these weak interactions parity need not be conserved.

Yang. *See* YIN-YANG.

Yangchow, city in E China, on the Grand Canal. A thriving city during the SUI and T'ANG dynasties (581–907), it later became a centre of Nestorian Christianity, and was governed by Marco POLO (1282–85). Today it is a market centre. Pop. (1970 est.) 210,000.

Yangtze-Kiang, river in E central China. Rising in the Kunlun Mts in NE Tibet, it flows E through the central Chinese provinces to the East China Sea at Shanghai. Its upper course is known as the Kinsha River; navigation becomes difficult at the Yangtze Gorges, between Ichang and Chungking. The Yangtze and its main tributaries, the Yalung, Min, Chialing and Wu, traverses one of the world's most populated areas, providing water for irrigation and hydroelectricity. It is the longest and most economically important waterway in the country. Length: 5,989km (3,720 miles).

Yankee, nickname for a US citizen, especially from New England or, since the AMERICAN CIVIL WAR, the NE generally. It may derive from *Janke,* the Dutch diminutive for John (from the Dutch settlers in New England).

Yaoundé, capital city of Cameroon, W Africa, 200km (125 miles) E of the Atlantic coast, on the Gulf of Guinea. It was founded in 1888 and became the capital in 1922. Yaoundé is a financial and administrative centre and also serves as a market for the surrounding region. Manufactures include clay and glass products and cigarettes. Pop. (1975 est.) 274,400.

Yardang, elongated ridge formed by wind erosion in arid areas. Rock is eroded into a series of ridges and furrows lying parallel to the prevailing wind direction. *See also* PE p.*120.*

Yarilo, in Slavic mythology, a god of spring and regeneration and of sexual love. He was depicted as a flower-crowned young man on a white horse. Each spring a planting rite was performed in his honour and in summer a ritual was held in which his effigy was ceremonially buried.

Yarmulke, a skullcap worn by Jewish men. Some wear it at all times, while a larger group wear it during prayer and study of the TORAH and TALMUD.

Yaroslav I (980–1054), Grand Duke of Kiev (1019–54), known as "the Wise". He used Viking mercenaries to defeat his brother Sviatopolk the Accursed in 1019. He codified Russian law, rebuilt Kiev on

Byzantine lines and developed close relations with the rest of Europe.

Yarrow. *See* MILFOIL.

Yawata. *See* KITAKYŌSHŌ.

Yawl, two-masted sailing boat with the mainmast higher than the mizzenmast, rigged with one or more jibsails, a mainsail, and a mizzen.

Yaws, or framboesia, contagious skin disease found in the humid tropics. It is caused by a SPIROCHETE (*Treponema pertenue*) indistinguishable from the organism causing syphilis. Yaws, however, is not a VENEREAL DISEASE, but is passed by flies and by direct skin contact with the sores. *See also* MS p.106.

Yayoi culture, Japanese Bronze and Iron age culture (250 BC–AD 300). Introduced from Korea, it brought knowledge of metalworking, agriculture and pottery to Japan. It was succeeded by the Yamato period, in which a single authority began to unify the country. *See also* HC1 p.*130.*

Yazilikaya, ancient Hittite religious centre, located close to Boğhazköy. It consists of natural recesses in rock and is made up of various chambers and galleries. The walls are decorated with processions of Hittite divinities sculpted in relief. It was completed during the 13th century BC. *See also* HC1 p.*52–53.*

Y-chromosome, one of the two kinds of sex CHROMOSOMES in the nuclei of human cells, the other being the X-CHROMOSOME. A normal male has one X and one Y chromosome in his body cells, and his sperm cells contain either an X or a Y chromosome. Female cells contain two X-chromosomes. The Y chromosome is smaller than the X and contains fewer GENES. *See also* SEX DETERMINATION.

Year, length of time taken by the Earth to circle once round the Sun in its orbit. It is defined in various ways, such as the SIDEREAL year. *See also* MW pp.188–189.

Yearling, any animal, especially a horse, that is one year old. A horse becomes a yearling on 1 Jan of the year after which it was foaled, eg a horse foaled in June 1978 would become a yearling on 1 Jan 1979, after which date it is regarded as a yearling for a period of twelve months.

Yeast, any of a group of single-celled microscopic FUNGI (class Ascomycetes) that grow wild in all parts of the world in the soil and in organic matter. Yeasts are also produced commercially for use in baking, brewing and wine-making. *See also* NW pp.40–41; PE pp.*183,* 204–205.

Yeats, Jack Butler (1871–1957), Irish painter, son of the painter John Butler Yeats and brother of William Butler YEATS. He began his career as an illustrator for the family press and painted his first oils *c.*1915. His Irish landscapes are notable for their thick brushwork, which approaches ABSTRACT EXPRESSIONISM. *See also* HC2 p.283.

Yeats, William Butler (1865–1939), Irish poet and dramatist. With the help of Lady Gregory he founded the ABBEY THEATRE as an Irish national theatre and wrote a number of plays for it, including *The Hour Glass* (1904) and *Deirdre* (1907). Among his many memorable poems are *The Tower, The Second Coming* and *The Lake Isle of Innisfree.* He was involved in Irish nationalist politics and served (1922–28) as a senator of the Republic of IRELAND. He was awarded the Nobel Prize for literature in 1923. *See also* HC2 pp.282–283, *282–283.*

Yellow Book, The, yellow-covered quarterly journal published in Britain from 1894 to 1897. It was a leading exponent of the AVANT-GARDE in the 1890s and of ART NOUVEAU. Its contributors included Henry JAMES, Oscar WILDE, Arnold BENNETT and Aubrey BEARDSLEY.

Yellow Emperor (Huang Ti), mythological Chinese figure and earliest ruler recognized by Chinese historians. He supposedly reigned from 2679 BC to 2597 BC and is credited with the defeat of barbarian tribes, introducing government, and inventing coinage and the compass. He is regarded as one of the founders of TAOISM.

Yellow fever, acute infectious disease marked by the sudden onset of headaches,

high fever, jaundice and vomiting. It is caused by a VIRUS transmitted by several species of mosquito, especially in tropical and subtropical regions. It may be prevented by attenuated live-virus vaccines. *See also* MS p.106.

Yellow-green algae, variety of flagellate ALGAE (division Xanthophyta). They differ from GREEN ALGAE by their characteristic motile bodies and unequal flagella.

Yellowhammer, or yellow bunting, small Old World FINCH. It has a brown and yellow body and a yellow head. Length: 16cm (6in).

Yellow River. *See* HWANG HO.

Yellow Sea, branch of the Pacific Ocean, N of the East China Sea between the Chinese mainland and Korea. It is connected to the Chihli and Liaotung gulfs by the Strait of Chihli. The Yellow (Hwango Ho), Liao and Yalu rivers drain into the Yellow Sea. Depth (max.): 76m (250ft). Area: approx. 466,200sq km (180,000sq miles).

Yellowstone National Park, park in Wyoming, Montana, and Idaho, USA. It is the oldest and one of the largest of all US national parks, and it features approx. 10,000 hot springs and 200 geysers, the most famous of which is Old Faithful. It was established in 1872. Area: 8,990sq km (3,470sq miles). *See also* NW p.*250.*

Yemen, People's Democratic Republic of, republic at the SE end of the Arabian Peninsula; it includes the islands of Karaman, Kuria Muria, Perim and Socotra. The highland plateaus behind the coastal plain rise to 1980m (6,500ft), and much of the land is arid. Agriculture is the basis of the economy; cotton, tobacco, coffee, cereals and dates are produced. Petroleum processing at Little Aden is the major industry. The capital is Aden. Area: 287,683sq km (111,046sq miles). Pop. (1975 est.) 1,690,000. *See* MW p.184.

Yemen Arab Republic, independent nation situated at the S end of the Red Sea, also known as North Yemen. One of the poorest countries in the world, the Yemen has a narrow coastal plain backed by arid highlands. The economy is based on agriculture, and the most important crop is coffee. The capital is San'a. Area: 195,000sq km (75,290sq miles). Pop. (1976 est.) 6,668,000. *See* MW p.184.

Yenan, town in N central China, on the River Yen, 258km (160 miles) NNE of Sian. It became the centre of the Chinese Communist Party (1936–48) during its struggle with the Nationalist forces for control of China and it was also the destination of the Red Army on the "Long March". It was captured by Nationalists on 19 March 1947 and fell to the Communists on 22 April 1948. It is now a market town and tourist centre. Pop. (1971 est.) 45,000.

Yenisei, (Jenisej) chief river of central Siberia in the Russian SFSR, USSR. It is formed by the confluence of the Bolshoi Yenisei and the Maly Yenisei at Kysyl. These rivers flow W and N through the Sayan Mountains past Minusinsk into the Yenisei Gulf on the Kara Sea (an arm of the Arctic Ocean). A large hydroelectric station has been built at Krasnoyarsk. The river is a source of sturgeon and salmon and a shipping route for cargoes of cereals and construction materials. Length: 4,030km (2,500 miles).

Yeoman, in medieval England, freeman and minor landowner who was not bound to the lord of the manor for labour or taxation but who owed him military service. Unlike a VILLEIN or SERF, a yeoman was free to sell his extra produce in the local market or to engage in home industry. *See also* MANORIAL SYSTEM.

Yeomanry, class of small freeholders in Britain, historically between the gentry and the labourers, after the 1832 REFORM BILL particularly the enfranchised 40-shilling freeholders. The name also applies to the volunteer cavalry force organized in 1794, which was absorbed into the TERRITORIAL ARMY in 1907.

Yeomen of the Guard, royal bodyguard in Britain, first formed at the coronation of HENRY VII in 1485. They were then a body of 50 men, dressed in scarlet uniforms. In 1669 their number was fixed at 100 by

CHARLES II. They last accompanied a king into battle at DETTINGEN in 1743. They are now a ceremonial force, the Beefeaters (a popular name first used, apparently, in 1645). The name Beefeaters is also given to the Yeomen Warders of the Tower of London who lack the cross-belt worn by the Yeomen of the Guard.

Yeomen of the Guard, The (1888), subtitled *The Merryman and his Maid*, two-act operetta by Arthur SULLIVAN and William GILBERT, first produced at the Savoy Theatre, London. It is the most serious and realistic of their works.

Yeovil, county district in SE SOMERSET, England; created in 1974 under the Local Government Act (1972). Area: 959sq km (370sq miles). Pop. (1974) 121,700.

Yerba maté. *See* MATÉ.

Yerevan, capital of Armenia (Armánskaja SSR) USSR on the River Razdan 176km (110 miles) S of Tbilisi. In the 7th century the city was the capital of Armenia, although under Persian control, and the site of crossroads for caravan routes between India and Transcaucasia. It is the site of a 16th-century Turkish fortress. Products include chemicals, plastics, cables, electrical equipment and wine. Pop. (1970) 767,000.

Yerkes, Robert Mearns (1876–1956), US biologist and psychologist. He was a pioneer in the comparative study of apes and the development of methods to test the abilities of lower animals and humans. His publications include *The Mind of a Gorilla* (1927) and *Chimpanzees: A Laboratory Colony* (1943).

Yersin, Alexandre-Émile-John (1863–1943), Swiss-French bacteriologist. In Hong Kong he discovered (1894) the plague bacillus and developed a serum against it in 1895. Yersin is also reputed to have introduced the rubber tree into Indochina.

Yerushalayim (Al-Quds). *See* JERUSALEM.

Yesenin, Sergey Aleksandrovich (1895–1925), Russian poet, brought up in the country by his peasant grandfather. By 1916 he had been accepted by the literary salons of Moscow and Petersburg, and celebrated the October Revolution in numerous poetic works. In 1922 he married Isadora DUNCAN. Disillusioned, he retired alone to Persia in 1924 and hanged himself the following year in Leningrad. His works include the verse-play *Pugachov* (1921) and the poetry *Chorny chelovek* (1925).

Yeti. *See* ABOMINABLE SNOWMAN.

Yevtushenko, Yevgeny (1933–   ), Soviet poet. He is an outspoken writer frequently critical of Soviet authorities, and so much of his work remains unpublished in the USSR. His verse includes *Precocious Autobiography* (1963) and *Stolen Apples* (1971).

Yew, any of a number of evergreen shrubs and trees native to temperate regions of the Northern Hemisphere. They have stiff, narrow, dark green needles, often with pale undersides, and red, berry-like fruits. Height: to 25m (80ft). Family Taxaceae: genus *Taxus*. *See also* NW pp.57, 210.

Yezidis, or Devil Worshippers, members of a small religious community in and around Kurdistan, whose beliefs contain elements of the CHRISTIAN, ZOROASTRIAN, MANICHAEAN, JEWISH, NESTORIAN and MUSLIM faiths. Yezidis conceive of God as a passive being who has entrusted the affairs of the world to seven angels. The chief of them, Malak Ta'us (peacock angel), fell and became SATAN, but later repented and became good and active. Yezidis worship him.

Yezo. *See* HOKKAIDŌ.

Yggdrasil, in Scandinavian mythology, the world tree, a giant evergreen ash at the centre of the cosmos, linking the underworld with the land of giants and men and extending to ASGARD, the abode of the gods. Its symbolism of renewal may be reflected in the Christmas tree.

Yiddish, language spoken for centuries by Jews living in central and E Europe, and later in other countries of the world (including the USA) to which Jews had migrated. It is basically a variety of German, with many Hebrew and Slavic words added, and is written in the Hebrew alphabet.

Yin dynasty. *See* SHANG DYNASTY.

Yin-yang, in Chinese philosophy, the interaction of complementary forces, two cosmic energy modes comprising the TAO or the eternal dynamic way of the universe. It is a recurring theme in Taoist and CONFUCIAN texts: heaven is yang, or the active, bright, male principle; earth is yin, or the passive, dark, female principle. All things of nature and society are composed of different combinations of these two principles of polarity. The hexagrams of the *I Ching* (see BOOK OF CHANGES) embody yin and yang. *See also* TUNG CHUNG-SHU.

YMCA (Young Men's Christian Association), Christian association for young men established in London in 1844 by George Williams. Its aim is to develop Christian morals and leadership qualities in young people. Clubs were soon formed in the USA and Australia, and the world alliance of the YMCA was formed in Geneva in 1855.

Ymir, in Norse mythology, the first living being, a giant. From him the race of giants was created together with the cow AUDUMULLA, who nourished them all. While licking salt, the cow uncovered Buri, whose son Bor married Bestla, Ymir's daughter and fathered the gods ODIN, Vili and Ve. *See also* MIDGARD.

Yoga (Sanskrit for "union"), term used for a number of Hindu disciplines to aid the union of the soul with God. Based on the Yoga-sutras of Patañjali (written at about the time of Christ), the practice of yoga generally involves moral restraints, meditation and the awakening of physical energy centres through specific postures (*asanas*) or exercises. Devoted to freeing the soul or self from earthly cares, these ancient practices have become recently popular in the West as a means of relaxation, self-control and enlightenment.

Yogācāra, philosophical school of Mahāyāna BUDDHISM. It arose in India in about the 6th century AD and holds that the reality which mankind perceives is actually an illusion; only the consciousness of momentary events exists. This consciousness orders sense perceptions to give the false idea of a consistent reality.

Yokohama, port and major industrial city on SE Honshū, Japan, on the W shore of Tokyo Bay, approx. 30km (18 miles) SSW of Tokyo; capital of Kanagawa prefecture; chief port for Tokyo and third largest city in Japan. The city grew from a small fishing village to a major port after being opened to foreign trade in 1859. There are two universities in the city. Industries: steel, motor vehicles, oil refining, textiles, machinery, chemicals, shipbuilding, electrical equipment. Pop. 2,778,975.

Yom Kippur, the Day of Atonement in JUDAISM. It is the last of the Ten Days of Penitence that begin the new year. On this solemn day set aside for prayer and fasting man is called to account for his sins and reconcile himself with God and man. It is described as the SABBATH of Sabbaths.

Yonge, Charlotte Mary (1823–1901), British novelist. A dedicated Christian, she did much to promote the OXFORD MOVEMENT. Her novels include *The Heir of Redclyffe* (1853), *Heartsease* (1854) and *The Daisy Chain* (1856).

York, Archbishop of, second-highest officer of the Church of England. The acts of the Council of Arles (314) mention a bishop of York, but the Christian community in York was destroyed by Saxon invaders. The see dates without a break from the consecration of Wilfrid in 664. It was raised to archbishopric dignity in 735, when Egbert was given the title Primate of the Northern Province. The Archbishop is now called the Primate of England (the Archbishop of Canterbury is Primate of all England).

York, Richard, Duke of (1411–60), descendant of EDWARD III of England and claimant to the throne. He was the largest landowner in the kingdom, whose attempt to gain the throne ended with the defeat of the Yorkists at Wakefield by the forces of MARGARET OF ANJOU in 1460. York was slain during the battle, but his son, EDWARD IV, became king the next year.

York, city and county district in NORTH YORKSHIRE, N England. The city, at the confluence of the Ouse and Foss rivers, was an important Roman military post named Eboracum. York is the ecclesiastical centre of the North of England: York Minster dates from the 13th century. There is a university (1963) in the city. Tourism is important and other industries include engineering, confectionery and the manufacture of precision instruments. The county district was created in 1974 under the Local Government Act (1972). Area: 29sq km (11sq miles). Pop. (1974 est.) 103,800.

York, one of the five boroughs of Toronto, SE Ontario, Canada. The original town of York was founded in 1793. The borough was formed in 1967 from the amalgamation of York and the town of Weston. Pop. (1971) 147,305.

York, House of, branch of the English PLANTAGENET family, descended from Edmund of LANGLEY, 1st Duke of York (1342–1402), fourth surviving son of EDWARD III. The claim to the English throne of York's grandson, Richard, came also through his mother, who was the great-granddaughter of Lionel, Duke of CLARENCE, third son of Edward III. This claim led to the Wars of the ROSES, a struggle between the Houses of York and LANCASTER. The House of York provided three kings: EDWARD IV, EDWARD V and RICHARD III. With the defeat and death of Richard III at MARKET BOSWORTH (1485) the crown passed to the House of TUDOR.

Yorkists. *See* YORK, HOUSE OF.

York Mystery Plays, The, cycle of 48 medieval MYSTERY PLAYS which starts with the Creation and ends with the Last Judgement. They were performed in York *c.*1300–1569 by craft guilds who each did the particular one relating to their trade. Reformation opposition caused them to cease, but since 1951 annual performances in York have been revived.

Yorkshire, formerly the largest county in England, in N England, on the North Sea between the Humber Estuary (S) and the River Tees (N). Yorkshire was divided into three administrative districts, or ridings: East Riding, with Beverley as its county town; North Riding, with Northallerton as its county town; and West Riding, with Wakefield as the county town. The city of York was a separate administrative area. The Pennine Hills in the W slope down to the fertile Vale of York. Further E are moorland areas: the Yorkshire Wolds, the North York Moors and the Cleveland Hills. The region is drained chiefly by the Ouse, Derwent and Aire rivers. The principal farming activities in the area are the cultivation of cereals and the rearing of cattle and sheep. There are rich coal deposits in the region, particularly in the W. Yorkshire includes some of the great industrial cities of the North of England such as Leeds, Sheffield and Bradford. Industries include iron and steel, chemicals and textiles. Under the Local Government Act (1972) Yorkshire was divided into the new counties of HUMBERSIDE, NORTH YORKSHIRE, SOUTH YORKSHIRE and WEST YORKSHIRE. Part of the North Riding was included in CLEVELAND. Area of former county: 15,859sq km (6,123sq miles). Pop. (1971) 5,047,567.

Yorkshire terrier, small long-haired dog originally bred in Lancashire and Yorkshire, England, in the 19th century. It has a small head with a short muzzle and small V-shaped erect ears. The compact body has a short, straight back and is set on short legs, which are hidden under the coat. The tail is commonly docked. The straight, fine, silky coat is generally blue and tan. Height: to 20cm (8in) at the shoulder; weight to 3 kg (7lb).

Yorktown, Siege of (19 Oct. 1781), last major military campaign of the War of AMERICAN INDEPENDENCE. Trapped on the peninsula at Yorktown, 7,000 British troops under Lord CORNWALLIS surrendered to the allied American and French forces.

Yoruba, people of SW Nigeria of basically Christian or Islamic faith. Most are farmers, growing crops which include yams, maize and cocoa. Many live in towns built

Yevgeny Yevtushenko leads Soviet poets in the cry for greater artistic freedom.

Yin-yang symbol; the two principles, light male and dark female, in balance.

York Minster's great East window is a fine example of late Decorated style.

Yorkshire terriers, bred small in the 1850s, are brave enough to kill rats.

Thomas Young helped to decipher the inscriptions on the Rosetta Stone.

Eugène Ysaye was the main interpreter of the string works of his time.

Yukon Territory's Lake Bennett is frozen over for nearly half the year.

Zagreb's Cathedral, begun in the 11th century, was rebuilt in the 17th.

around the palace of an *oba*, or chief, and commute to their outlying farms. *See also* MS p.*223*.

Young, Brigham (1801–77), US religious leader. An early convert to the Church of Jesus Christ of Latter-Day Saints (MORMONS), Young took over the leadership when Joseph SMITH, the Church founder, was killed by a mob in 1844 in Illinois. Young held the group together through persecutions and led them in their great westward migration (1846–47) to Utah where he directed the settlement that became Salt Lake City. He was later Governor of Utah Territory (1850–57).

Young, Francis Brett (1884–1954), British poet and novelist. His novels include *Deep Sea* (1914), *My Brother Jonathan* (1928) and *The City of Gold* (1939), a novel set in South Africa. His verse included *The Island* (1944).

Young, Thomas (1773–1829), British physicist and physician. He revived the wave theory of light first put forward by Christiaan HUYGENS in the 17th century, and helped to present the Young-HELMHOLTZ three-colour theory. He also detailed the cause of ASTIGMATISM and studied elasticity, giving his name to the tensile modulus. *See also* HC2 p.157, *157*.

Young England, coterie of CONSERVATIVE backbenchers. Its leaders were Benjamin DISRAELI, Lord John Manners and George Smythe, and from 1842 to 1845 they were combined in their aim to re-establish the aristocracy as the agency of social justice. The group provided Disraeli with a platform from which to attack the Conservative Prime Minister, Robert PEEL.

Young Ireland, group of young, chiefly Protestant, Irish nationalists of the 1840s. Its organ was the *Nation* newspaper, and its most important leaders were Charles Gavan Duffy, Thomas Francis Meagher and John Mitchel. It worked for an Irish cultural revival as well as independence from Britain. It declined after an abortive rising in 1848.

Young Italy (*Giovine Italia*), republican and nationalist movement in Italy, founded in 1831 and inspired by the exiled radical, MAZZINI. It sought to provoke an Italian revolution against Austrian rule. It failed to become a mass movement, but fostered the growth of national consciousness in the 1830s and 1840s.

Young Men's Christian Association. *See* YMCA.

Young Pretender. *See* STUART, CHARLES EDWARD.

Young's modulus, ratio of the stress exerted on a body to the longitudinal strain produced.

Young Turks, group of Turks who wished to remodel the OTTOMAN EMPIRE and make it a modern European state with a liberal constitution. Their movement began with unrest in the army and in Universities from the 1880s onward. In 1908 a Young Turk rising deposed Sultan Abdul Hamid II and replaced him with his brother, Mohammed V. From then until the end of WWII, the Young Turks were the dominant party in Turkish politics.

Young Wales, in Welsh *Cymru Fydd* ("Future Wales"), movement founded by Welsh nationalists and Liberal MPs in 1886 in response to the movement for IRISH HOME RULE. It sought home rule for Wales, but it never gained much influence and was a spent force by 1900.

Young Women's Christian Association. *See* YWCA.

Youth hostels, hostels for young people providing inexpensive accommodation for limited lengths of stay. Hostels originated in Germany in the 1900s and spread rapidly throughout Europe after WWI.

Ypres, John Denton Pinkstone French, 1st Earl of (1852–1925), British field-marshal. He served in the SOUTH AFRICAN WAR and (1912–14) was Chief of the Imperial General Staff. From the outbreak of WWI to December 1915 he commanded the BRITISH EXPEDITIONARY FORCE. He was replaced in this post following the heavy casualties at the Battle of LOOS and the first and second battles of YPRES.

Ypres, Battles of, three WWI battles at Ypres, France. The first (12 Oct–22 Nov 1914) stopped the German "race to the sea". There were more than 50,000 British killed. The second (22 April–25 May 1915) failed to gain Germany control of the Ypres salient. There were 59,000 British killed. The third (31 July–4 Nov 1917) was a British offensive which gained the Allies most of the salient. There were 245,000 British killed.

Ysaye, Eugène (1858–1931), Belgian violinist, conductor, teacher and composer. He became professor of music at the Brussels Conservatory of Music in 1886 and later conductor of the Cincinnati Symphony Orchestra (1918–20).

Ysselmeer. *See* IJSSELMEER.

Ytterbium, metallic element (symbol Yb) of the LANTHANIDE SERIES, first isolated in 1878 . Chief ore is monazite (phosphate). The element has few commercial uses. Properties: at.no. 70: at.wt. 173.04; s.g. 6.97 (α); 6.54 (β); m.p. 824°C; (1,515°F), b.p. 1,193°C (2,179°F); most common isotope Yb^{174} (31.84%).

Yttrium, metallic element (symbol Y) of group IIIB of the periodic table, first isolated in 1828. It is found associated with LANTHANIDE elements in monazite (phosphate) and bastnasite (fluorocarbonate) and resembles the lanthanides in its chemistry. Its compounds are used in phosphors and communications devices. Properties: at.no. 39; at.wt. 88.9059; s.g. 4.46; m.p. 1,523°C (2,773°F); b.p. 3,337°C (6,039°F); most common isotope Y^{89} (100%).

Yüan Shih-k'ai (1859–1916), Chinese military commander in the last days of the MANCHU DYNASTY. He became the first President of the Chinese Republic in 1913 when SUN YAT-SEN resigned, but his attempt to start a new dynasty and make himself emperor was thwarted shortly before his death. *See also* HC2 p.145.

Yucatán, state in SE Mexico, in the N part of the Yucatán Peninsula; the capital is Mérida. The terrain is low-lying, covered in places with scrub and cactus thickets. The region is a major producer of henequen. Other products include tobacco, sugar, cotton and tropical fruits. Fishing is important along the coast. The region, once the centre of the MAYA civilization, was conquered by the Spanish in the 1540s. Yucatán became a state in 1824. Area: 38,508sq km (14,868sq miles). Pop. (1970) 758,355.

Yucca, genus of about 40 species of succulent plants native to southern USA, Mexico and the West Indies. Most species are stemless, forming a rosette of leaves, or have a trunk. The flowers grow in clusters and are generally white, but may be tinged with yellow or purple. Height: to 10m (33ft). Family Liliaceae. *See also* NW p.*218*.

Yugoslavia, mountainous country on the E shore of the Adriatic Sea, created at the end of WWI. It has a Communist government but has resisted Soviet domination. There are rich mineral deposits including coal, iron ore, oil and natural gas. Cereals, fruits, tobacco and vegetables are the principal crops and livestock are raised in the mountains. Industries are: engineering, shipbuilding and the manufacture of steel, paper, chemicals and textiles. The capital is Belgrade (Beograd). Area: 255,802sq km (98,756sq miles). Pop. (1976 est.) 21,617.000. *See* MW p.184.

Yukawa, Hideki (1907–), Japanese physicist who was awarded the 1949 Nobel Prize in physics for his prediction of the existence of MESONS. In the 1930s he proposed that there was a nuclear force of very short range (less than 10^{-15}m) which was strong enough to overcome the repulsive force of protons and which diminished rapidly with distance. He postulated that this force manifested itself by the transfer of particles between neutrons and protons. In 1936 Carl ANDERSON discovered the mu-meson, which initially seemed to meet the criteria of Yukawa's theory, but it did not interact sufficiently with the nucleus to fulfil all of the requirements. In 1947 Yukawa's theory was confirmed by the discovery of the pi-meson by Cecil POWELL.

Yukon Territory, region in the extreme NW of Canada, bordered by Alaska (W) and the Arctic Ocean (N). In the N the region consists of desolate arctic waste and is virtually uninhabited. Further s there is spectacular mountain scenery with lakes and forests. The region is drained chiefly by the Yukon and Mackenzie rivers. The climate is harsh; winters are cold and summers short. Farming is extremely limited, but a few cereal crops and vegetables are grown in valleys. The principal activity is mining. Mineral deposits include gold, asbestos, copper, silver, zinc and lead. There is little manufacturing or fishing, but tourism is playing an increasingly important part in the economy. Whitehorse is the territorial capital and the largest town.

The region, then part of Northwest Territories, was first explored by fur traders from the HUDSON'S BAY COMPANY after 1840. The rich gold deposits discovered in the River Klondike in 1896 brought a surge of prospectors to the territory. In ten years gold worth more than $100 million was extracted. The Yukon was made a separate territory in 1898. Area: 536,327sq km (207,076sq miles). Pop. (1976 est.) 21,000. *See also* MW p.45.

Yunnan, province in s central China, bounded by Tibet (NW), Laos and Vietnam (s) and Burma (W); Kunming is the capital. The land is mountainous and allows of only limited farming. The chief crops are rice, maize, wheat and sweet potatoes. There are valuable mineral deposits in the province including tin, iron ore, coal, copper, gold, silver and mercury. Mining and timber are the main industries. Yunnan was independent for centuries but became part of China in 1659. It came under Communist control in 1950. Area: 436,380sq km (168,486sq miles). Pop. 21,000,000.

Yuzovka. *See* DONETSK.

YWCA (Young Women's Christian Association), Christian association for young women, the counterpart of the YMCA. Founded simultaneously in 1855 by two women (Miss Robarts and Lady Kinnaird) in different parts of England, the associations merged in 1877. The YWCA provides accommodation, education, recreation facilities and welfare services to young women and has local branches in more than 80 countries.

Z

Z, 26th and last letter of the alphabet. It is derived from the Semitic letter *zayin*, which passed into Greek as *zeta* and was given its present form. It was an early letter in these alphabets (6th or 7th) and was not needed by the Romans until they began to use Greek words in the 2nd century AD, when they added it as the last letter of their alphabet. In English z is usually pronounced as in the word *zoo*, although it may have a softer *zh* sound, as in *seizure*. *See also* MS pp.244–245.

Zabaleta, Nicanor (1907–), Spanish harpist, probably the greatest of his day. He studied at the Madrid Conservatory and in Paris, and has become famous for his interpretations of harp compositions.

Zacharias (Zachary), Saint, Roman Catholic pope (741–52). The Holy See was much strengthened during his papacy. He achieved a 20-year truce with the LOMBARDS and, with St BONIFACE, established cordial relations with the FRANKS by supporting the accession of PEPIN to the Frankish throne.

Zacharias, in the Gospel according to St LUKE, priest to whom an angel appeared foretelling the birth of his son, JOHN THE BAPTIST.

Zagreb, city in NW Yugoslavia, on the River Sava; second-largest city in Yugoslavia and capital of the Croatian republic. It was probably founded late in the 11th century. During the 19th century, the city was an important centre of the Yugoslav nationalist movement. Zagreb is a major industrial and financial city and a cultural centre. The Yugoslav Academy of Arts and Sciences, founded in 1874, a university (1669) and several museums and art galleries are in the city. Industries: textiles, chemicals, leather goods, machinery. Pop. (1975 est.) 566,000.

Zagros, mountain range in s and sw Iran, extending from the Soviet-Turkish border to the Persian Gulf. The topography varies from rugged peaks in the N to ridges and valleys in the central region and lowland marshes and rock to the s. One of the world's most productive oil fields is located in the w foothills. Sabalān is the highest peak, 4,550m (14,928ft). Length: approx. 900km (560 miles).

Zaire, republic in w central Africa; second-largest country in Africa. Formerly the Belgian Congo, the nation gained independence in 1960. Nearly 70% of the people are subsistence farmers; cattle are reared and the chief cash crops are coffee, cotton, palm products and rubber. Copper is the principal export, and industrial diamonds, cobalt, tin and gold are also mined. Manufacturing is increasing and there is abundant hydroelectric power. The capital is Kinshasa. Area: 2,345,409sq km (905,562sq miles). Pop. (1975 est.) 24,902,000. *See* MW p.185.

Zaire, or Congo, second largest river in Africa, rising in s Zaire and flowing NW to the Atlantic Ocean for 4,380km (2,720 miles). It is Africa's largest potential source of hydroelectric power. The chief ocean port is Matadi; its chief river ports are Kinshasa and Kisangani.

Zambezi, river in s central and SE Africa; it rises in NW Zambia and flows through E Angola and Zambia forming the Zambia-Zimbabwe border, then through Mozambique to empty into the Mozambique Channel of the Indian Ocean. There is great potential for the generation of hydroelectricity along the river's course. The Zambezi was first explored by LIVINGSTONE in 1851. Length: approx. 2,735km (1,700 miles).

Zambia, landlocked republic in s central Africa; formerly the British protectorate of Northern Rhodesia. The Zambian economy depends on copper, which accounted for 94% of exports in 1973; coal is also mined. Approx. 60% of the people are subsistence farmers; tobacco is the chief cash crop. Tourism and manufacturing are of increasing importance. The capital is Lusaka. Area: 752,614sq km (290,584sq miles). Pop. (1975 est.) 4,896,000. *See* MW p.186.

Zamenhof, Ludwik Lejzer (1859–1917), Jewish oculist who practised in Warsaw, chiefly remembered as the inventor of ESPERANTO, an artificial language for universal use.

Zampieri. *See* DOMENICHINO.

Zamyatin, Yevgeny Ivanovich (1884–1937), Russian novelist and dramatist, an early exponent of the anti-utopian novel. His most famous work, *We,* an attack on Soviet society and politics, was written in 1924 and circulated in the USSR, but never published. All of his works were banned and in 1931 he was permitted to emigrate; he died in Paris.

Zantedeschia, genus of mainly tropical, flowering plants of the family Araceae, especially *Z. aethiopica,* the calla, or arum lily. Typical of the order Arales, its has a flower cluster consisting of a spadix (fleshy spike), surrounded or enfolded by a spathe (modified leaf).

Zanzibar, island in the Indian Ocean, off the E coast of Africa. Under Arab rule until the 15th century, it became a Portuguese colony c.1505 and by the end of the 17th century was under the control of the Oman Arabs (when it developed into the main commercial centre of East Africa). In 1890 it became a British protectorate, gaining independence in 1963. The following year the island became part of the United Republic of TANZANIA, but remained self-governing.

Zapata, Emiliano (c.1880–1919), Mexican Indian revolutionary leader. He allied himself with Pancho VILLA and captured Mexico City with him in 1914, then returned to his stronghold in the south where he was led into a trap by forces loyal to President Carranza. Zapata was captured and executed.

Zapotec, central American Indian group that lives in part of the Mexican state of Oaxaca. The Zapotec built great pre-Conquest urban centres at Mitla and MONTE ALBÁN, and fought to preserve their independence from the rival Mixtecs and Aztecs until the arrival of the Spanish. *See also* HC1 pp.46, 47, 230, 230.

Zápotský, Antonin (1884–1957), Czechoslovak Premier from 1948 to 1953 and President from 1953 to 1957. One of the founders of the Communist Party of Czechoslovakia, he wrote numerous articles and documentary novels about the Czech proletarian movement.

Zappa, Francis Vincent ("Frank") (1940–), US rock musician. Classically trained, Zappa's first successes were with film sound-tracks, eg *Run Home Slow* (1963). With his group, The Mothers of Invention, Zappa produced closely edited recordings such as *Freak Out!* (1966) and *We're Only In It For The Money* (1967).

Zarathustra, *See* ZOROASTER.

Zaria, town in N central Nigeria, approx. 144km (90 miles) sw of Kano. Originally one of the seven HAUSA city states, Zaria was taken by British troops in 1901. It is now a trade centre for the surrounding agricultural region. Products include textiles, hides and handicrafts. Pop. (1975) 224,000.

Zastrugi, irregular ridges cut in surface snow by wind erosion. They vary in size according to the force and duration of the wind. Arctic travellers take careful note of them because they make journeying by sled easier and faster.

Zatopek, Emil (1922–), Czechoslovakian athlete. He won a gold medal in the 10,000m race at the 1948 Olympics and three gold medals (5,000m, 10,000m and the marathon) at the 1952 Olympics. From 1948 to 1954 he was undefeated over 10,000m.

Zealots, Jewish sect, active in opposition to Roman rule. They refused to agree that Jews could be ruled by pagans and led resistance to the Roman census of AD 6, pursued a terrorist campaign and played an important role in the rising of AD 66. Their activities continued into the 2nd century AD.

Zeami Motokiyo (1363–1443), Japanese playwright and actor who brought the NŌ drama to its peak of refinement. Using a formal style, music and singing (but no scenery) he wrote more than 100 plays.

Zebra, any of three strikingly patterned, striped, black and white equine mammals of the grasslands of Africa; the stripes are arranged in various patterns, according to species. It has long ears, a tufted tail and narrow hooves, more strongly resembling an ass than a horse in general shape. Height: to 140cm (55in) at the shoulder. Family Equidae; genus *Equus. See also* NW p.163, 163.

Zechariah, biblical author and 11th of the 12 minor prophets. He was HAGGAI's contemporary and shared his concern for rebuilding the TEMPLE in Jerusalem.

Zedekiah, several biblical figures, most notably the last King of JUDAH and JERUSALEM (597–586 BC). The son of JOSIAH, he sided with Egypt and violated his pledge of allegiance to NEBUCHADNEZZAR. He was imprisoned in BABYLON, where he died.

Zeeland, province of the sw Netherlands, on the North Sea, made up of a mainland area bordering Belgium and several islands off the North Sea coast, namely Walcheren, North and South Beveland, Schouwen, Tholen and Sint Philipsland. The islands are connected to the mainland by bridges and dikes as part of the Delta Plan, an extensive flood-control system, which is scheduled to be completed in 1980 to create a rich agricultural region. The capital is Middelburg. Products include cereals, flax and potatoes. Industries include fishing and tourism. Area: 2,700sq km (1,043sq miles). Pop. (1971 est.) 310,320.

Zeeman, Pieter (1865–1943), Dutch physicist. He shared the 1902 Nobel Prize in physics with his teacher Hendrik LORENTZ for their discovery in 1896 of the ZEEMAN EFFECT. Zeeman also detected the magnetic fields at the surface of the Sun.

Zeeman effect, in physics, effect produced by a strong magnetic field on the light emitted by a radiant body, observed as a splitting of its spectral lines. It was first observed by P. ZEEMAN in 1896. The effect has been useful in investigating the charge/mass ratio and magnetic moment of an ELECTRON.

Zeffirelli, Franco (1923–), Italian stage and film director. He worked at Covent Garden on *Cavalleria Rusticana,* at Stratford-upon-Avon on *Othello,* and on Broadway on *The Lady of the Camellias.* His films include *The Taming of the Shrew* (1966), *Romeo and Juliet* (1968) and *Brother Sun and Sister Moon* (1973).

Zeiss, Carl (1816–88), German manufacturer of optical instruments. In 1846 he established a factory at Jena and made various optical components, including lenses, binoculars and microscopes.

Zela, Battles of, two battles of Zela in Asia Minor. At the first (c.67 BC) MITHRIDATES VI, King of Pontus, defeated the Romans. Julius CAESAR defeated Pharances, King of Pontus, at the second battle (47 BC) and recorded the victory in the famous message, "Veni, vidi, vici" ("I came, I saw, I conquered").

Zelenchukskaya Observatory, World's largest optical telescope, built at Zelenchukskaya, USSR. It is a reflecting telescope with a conventional dome and a 6m (20ft) reflector; it started operation in February 1976 and is used to study remote star systems. *See also* SU pp.170, 171.

Zemstvos (1864–1917), locally elected assemblies in Russia which existed on the district and provincial levels. Voters included landed proprietors, urban merchants and peasants. They were active in the 1905 and 1917 revolutions and were abolished by the BOLSHEVIKS.

Zen, Japanese school of BUDDHISM initially developed in China where it is known as Ch'an. Instead of doctrines and scriptures, mind-to-mind instruction from master to disciple is emphasized in order to achieve *satori,* or the awakening of Buddha-nature inherent in everyone. There are two major Zen sects. Rinzai, introduced to Japan from China in 1191, emphasizes sudden shock and meditation on paradoxical statements. The Soto sect, also brought from China (in 1227), advocates the method of quiet meditation. In its secondary emphasis on mental tranquillity, fearlessness, and spontaneity, Zen has had a great influence on Japanese culture. Zen priests inspired art, literature, the tea ceremony and the NŌ play. In recent decades, a number of Zen groups have been formed in Europe and the USA. *See also* MS p.221

Zend-Avesta, alternative name for the sacred book of the ZOROASTRIANS; it is also called AVESTA (probably meaning injunction). Zend means tradition or commentary.

Zenith, in astronomy, point on the celestial sphere which is directly overhead. The zenith distance of a heavenly body is the angle it makes with the zenith. In common use it means the highest point.

Zeno (426–491), E Roman emperor (r.474–491), son-in-law of Leo I. IN 475 he yielded to the usurper Basilicus for 20 months. He sought unsuccessfully to reconcile the heretical Monophysites with the orthodox Christians through his *Henotikon* (482). By persuading Theodoric to invade Italy he finally rid the E empire of the Ostrogoths in 489.

Zenobia (*fl.* 3rd century AD), Queen of PALMYRA and wife of the Syrian nobleman Odenathus. After his murder in 267 she secured power for herself in the name of her young son. Zenobia's ambition led her to expand her territories to Asia Minor and Egypt until the Roman Emperor Aurelian marched against her. She was defeated in AD 272 and taken captive to Rome.

Zeno of Citium (c.334–262 BC), Greek philosopher and founder of the STOIC school. Proceeding from the CYNIC concept of self-sufficiency, he stressed the unity of the universe and the brotherhood of men living in harmony with the cosmos. He claimed virtue to be the only good and wealth, illness and death to be of no human concern.

Frank Zappa's creative imagination ran wild in the film *2000 Motels.*

Emil Zatopek winning the 10,000m event in the 1954 European Championships.

Zebras' stripes help to conceal them from predators such as lions.

Franco Zeffirelli has directed in various media, including opera, theatre and film.

Zeus, "the father of gods and men", was the guardian of law and morality.

Zodiac resting on the back of Atlas; a sculpture in the Villa Albani, Rome.

Johann Zoffany's *The Clandestine Marriage.* He was a founder of the Royal Academy.

Emile Zola's genius was best revealed in his stories, such as *Attaque du Moulin.*

Zeno of Elea (*c.*495–*c.*430 BC), Greek philosopher. A disciple of PARMENIDES, he defended the latter's doctrine of the sole reality of changeless "being". Through his paradoxical arguments, he sought to reveal logical absurdities in theories of motion and change. *See also* ELEATICISM; MS p.*227, 232–233.*

Zeolite, aluminium silicate mineral, similar to FELSPAR from which it is often formed by alteration. There are many zeolites varying in hardness from 3 to 5 and in specific gravity from 2 to 2.4, such as analcite, $Na(AlSi_2O_6).H_2O$, and scolecite, $Ca(Al_2Si_3O_{10}).3H_2O$.

Zephyrus, in Greek mythology, the west wind. He was the brother of Boreas, the north wind, Eurus, the east and NOTUS, the south wind.

Zeppelin, Ferdinand Adolf August Heinrich, Count von (1838–1917), German army officer and inventor. He served in the armies of Württemburg and Prussia and while an observer with the Union army during the US Civil War (1861–65) made his first balloon ascent. In 1900 he invented the first rigid airship, named Zeppelin after him. *See also* MM p.*146.*

Zermelo, Ernst (1871–1956), German mathematician who was one of the founders of axiomatic SET THEORY. The Zermelo-Fraenkel axioms stipulate which collections of objects constitute sets. *See also* SU p.40.

Zernike, Frits (1888–1966), Dutch physicist who in 1934 developed the phase-contrast microscope, in which objects being viewed take on a different colour from their surroundings. He received the 1953 Nobel Prize in physics.

Zero, in mathematics, that number which, when added to any number x, leaves the number unchanged: $x + 0 = x$. Multiplication of a number by zero gives zero. Division by zero is undefined. The concept of zero was unknown to the ancient Greeks and was introduced to the West by the Arabs, who took it from the Hindus. *See also* SU p.30.

Zerqa, or Zarqa, river in N Jordan. It rises W of Amman, and flows N and then W to join the River Jordan. It was known in ancient times as the Jabbok.

Zeta, machine built in Harwell, Britain, in the 1950s by the Atomic Energy Authority to investigate the possibility of harnessing the power of thermonuclear fusion. Zeta is toroidal (doughnut-shaped), its inner space filled with a PLASMA of ionized gas. Surrounding the torus are giant magnets, the fields of which "pinch" the plasma discharge inwards from the walls of the torus, allowing it to be raised even further in temperature.

Zeta Cancri, group of four stars in the constellation of Cancer consisting of two binary systems. Zeta-one has a period of 59.6 years, Zeta-two is a close binary with a period of 17.6 years.

Zetland. *See* SHETLAND ISLANDS.

Zeus, in Greek mythology, the sky god, lord of the wind, clouds, rain and thunder. Zeus was the son of Rhea and the TITAN . His father ate all his offspring in fear of a prophecy that one of his sons would overthrow him, but Rhea hid Zeus and later he duly deposed Cronus. Zeus was the supreme deity.

Zhuangaerpendi (Dzungarian), arid region in NW China, between the Tien Shan and Altai Mts. Ruled by Dzungars, a Mongol tribe, in the 17th century, it was taken by the Chinese in 1873. Some of the population are nomadic, raising cattle, sheep and horses. The sedentary population grow cereals and sugar-beet but there has been a large influx of Chinese to work on industrial schemes. The discovery of oil in 1955 provided the impetus for further development of the area. Area: approx. 777,000sq km (300,000sq miles).

Zhukov, Georgi Konstantinovich (1896–1974), Soviet military commander and political figure. In WWII he led the 1941 defence of Moscow, broke the German sieges of Stalingrad and Leningrad and in 1945 led the final assault on Berlin. He became defence minister in 1955 and was briefly a member of the PRESIDIUM in 1957. He received the Order of Lenin in 1966. *See also* HC2 p.229.

Zick, Januarius Johann (1730–97), German painter. He studied the work of WATTEAU in Paris, and in Germany after 1758 painted many altarpieces in a profuse ROCOCO style.

Ziegfeld Follies, series of theatrical revues staged by the US theatrical producer Florenz Ziegfeld (1869–1932). It ran for 24 years (1907–31) and was famed for its lavish sets, the variety turns and its glamorous showgirls.

Ziegler, Karl (1898–1973), German chemist who shared the 1963 Nobel Prize in chemistry with Giulio Natta for his research into POLYMERS. Ziegler discovered a technique whereby a resin with metal ions attached could be used as a catalyst in the production of POLYETHYLENE. He also did research into aromatic compounds and developed organo-metallic compounds that were more reactive than GRIGNARD REAGENTS.

Ziggurat, Babylonian and Assyrian monumental religious structure, constructed as a truncated pyramid in diminishing tiers, usually square or rectangular. The shrine at the top was reached by a series of ramps. Ziggurats date from 3000 to 600 BC. *See also* HC1 p.*32.*

Zimbabwe, African nationalist name for RHODESIA. *See* MW p.187.

Zimbabwe, ruined city in SE Zimbabwe (Rhodesia), 27km (17 miles) from Port Victoria. The name is believed by some scholars to mean "houses of stone", a reference to the large, well-preserved buildings. The ruins were discovered in 1867 and radiocarbon tests indicate that some are about 1,000 years old. *See also* HC1 p.*229.*

Zimbabwe African People's Union, Rhodesian political party that represents the blacks. It was established in 1961 by Joshua NKOMO as a reorganization of the National Democratic Party, which had been banned the same year for subversion. Banned itself since 1962, it has continued to operate from Dar-es-Salaam and Lusaka. It was a party to the Geneva conference held in late 1976 and early 1977 which aimed to achieve a peaceful settlement to the political crises in Rhodesia (Zimbabwe).

Zinc, metallic element (symbol Zn) of group IIB of the periodic table, known from early times. Chief ores are zinc blende and sphalerite (sulphides), smithsonite (carbonate) and calamine (carbonite). The element is used in many alloys, including brass, bronze, nickel silver and soft solder. It is also used in galvanizing and in producing zinc compounds. Properties: at.no. 30; at.wt. 65.37; s.g. 7.133 (25°C); m.p. 419.58°C (787.2°F); b.p. 907°C (1,665°F); most common isotope Zn^{64} (48.89%). *See also* SU p.148.

Zinc oxide, white powder, formula ZnO, used in ointments as a mild antiseptic, astringent and protection against drying of ECZEMA and other skin conditions.

Zinjanthropus. *See* NUTCRACKER MAN.

Zinnemann, Fred (1907–), US film director, *b.* Austria. His major films include *The Seventh Cross* (1944), *High Noon* (1952), *From Here to Eternity* (1953) *The Nun's Story* (1958) and *Julia* (1978). *See also* HC2 p.*269.*

Zinnia, genus of mainly annual herbs and shrubs, native to the USA and South America. Most garden zinnias are varieties of *Z. elegans* or *Z. angustifolia* and have yellow or brown disc flowers and ray flowers of most colours, except blue. Height: to 91cm (3ft). Family Compositae.

Zinoviev, Grigori Evseyvich (1883–1936), Soviet political leader. A self-educated lawyer, he joined the BOLSHEVIKS in 1903 and was active among the workers in St Petersburg in the RUSSIAN REVOLUTION of 1905. He was a close collaborator of V. I. LENIN in exile (1908–17), but opposed him and supported a coalition government after the 1917 revolution. In the 1920s he fell from power and Joseph STALIN expelled him from the party (1927). He was arrested in 1935 and later executed. *See also* HC2 pp.*197, 198–199, 198, 211.*

Zion, height in Jerusalem, Israel. It is a centre of Jewish spiritual life and symbolic of the Promised Land. *See also* HC1 p.*58.*

Zionism, movement within JUDAISM which advocates the return to the land of Israel or Zion in Palestine. It is based on the conception of the coming of the Messiah connected with the land of the fathers, Israel. In 1897 Theodor HERZL developed the movement into an organized body, and it gained momentum as the Jews gained a sense of nationalism, and a practical, political and cultural sense of Zionism. In 1948, the state of Israel was founded, with Hebrew as the official language. Zionism continues today as an effective movement, combating ANTI-SEMITISM and improving conditions of Jewish life. *See also* HC2 p.296.

Zircon, orthosilicate mineral, zirconium silicate ($ZrSiO_4$) found in IGNEOUS ROCKS, metamorphosed limestones and in sand and gravel. It displays tetragonal system prismatic crystals and is brilliant, either colourless or of varied hues; hardness 7.5; s.g. 4.6. It is used widely as a gemstone because of its hardness and high refractive index. *See also* PE p.*98.*

Zirconium, metallic element (symbol Zr) of the second transition series, first discovered in 1789. Its chief source is the gemstone ZIRCON (silicate) and it has a number of minor commercial uses. Chemically it is similar to titanium. Properties: at.no. 40; at.wt. 91.22; s.g. 6.51; m.p. 1,852°C (3,366°F); b.p. 4,377°C (7,911°F); most common isotope Zr^{90} (51.46%). *See also* TRANSITION ELEMENTS.

Zither, stringed musical instrument. It consists of a resonator in the form of a wooden box, stretched with strings varying in number from 30 to 45. Of these a few are stretched over a fretted board for melody, the rest are used for accompaniment. The melody strings are plucked with the fingers or a plectrum.

Zodiac, region of the heavens in which the Sun, Moon and most of the planets appear to move. It lies 9° on either side of the Sun's apparent path, the ecliptic. The Zodiac was first divided into twelve equal parts by Babylonian astrologers in the 4th century BC, although the constellations of the Zodiac were known at least a thousand years previously. These are now known by the names Aries, Taurus, Gemini, Cancer, Leo, Virgo, Libra, Scorpio, Sagittarius, Capricorn, Aquarius and Pisces. But the constellations now inside the Zodiac do not correspond to those named by the ancients, because PRECESSION of the Earth's axis has meanwhile tilted the Earth in a somewhat different direction. Most of the zodiacal signs represent animals and the name Zodiac derives from a Greek phrase meaning "circle of animals". To modern astronomers the Zodiac has only historical significance. *See also* ASTROLOGY; ASTRONOMY.

Zodiacal dust. *See* DUST, INTERPLANETARY.

Zodiacal light, hazy conical beam of light seen in the region of the ecliptic. Best observed in tropical latitudes, it is generally believed to be caused by the scattering of sunlight by interplanetary dust circling the Sun in the plane of the Earth's orbit.

Zoffany, Johann (1735–1810), German-born painter of portraits and conversation pieces who worked mainly in Britain. He also painted scenes of contemporary theatre, such as *Garrick in "The Farmers Return"* (1762), and his works are valuable records of costume and custom. His domestic scenes filled with detailed animated figures show the influence of HOGARTH. *See also* HC2 pp.*42, 71.*

Zog I, born Ahmed Bey Zogu (1895–1961), King of Albania (1928–39). Son of a Muslim chieftain, he served in the Austrian army during WWI. Zog was Premier (1922–24) and when Albania was proclaimed a republic in 1925, he was elected President. In 1928 he proclaimed himself King. As head of state, he consistently championed modernization and reforms. His rule became autocratic, however, and he was powerless to resist the Italian invasion in 1939.

Zola, Emile Edouard Charles Antoine (1840–1902), French novelist who, after poverty as a journalist, worked as a clerk in the Hachette publishing house, Paris.

He became widely known with the novel *Thérèse Raquin* (1867) and then planned the Rougon-Macquart series, the story of a family during the Second Empire. Twenty volumes of this, over the next 25 years, established him as an exponent of realism and a socialist. In 1898 he wrote a letter in *L'Aurore*, beginning "J'Accuse", which denounced the punishment of Alfred DREYFUS. This led to a brief exile in England and after the vindication of Dreyfus, a hero's return. His works include *L'Assomoir* (1877), *Germinal* (1885) and *La Terre* (1887). *See also* HC2 pp.98, *99*, 119.

Zollverein, German commercial union, formed in 1834 by 18 German states, of which Prussia (where the idea originated in 1818) was dominant. It lowered tariffs among its members, creating economic unification and paving the way for political union. By 1888 all of Germany had joined the Zollverein. *See also* HC2 p.110, *110*.

Zone refining, method of purifying crystals, especially for use in SEMICONDUCTOR devices. The material is placed in a long tube and passed through a furnace in which hot and cold zones alternate. As the rod moves through the furnace, impurities remain in the molten state while melting and recrystallization of the material takes place; the impurities are thus transferred to one end of the tube.

Zoo, abbreviation of zoological gardens, public or private institution in which living animals are kept and exhibited. Wild animals have been kept in captivity since the beginning of recorded history. Organized public zoos, sometimes called menageries or aquariums (for fish), have been operating for more than 500 years in Europe. Today non-profit organizations or zoological societies run most zoos in a scientific manner. Most are organized for public recreation as well as for scientific and educational purposes. Today the emphasis is on conservation of endangered species and exhibiting animals in natural settings. *See also* NW pp.252–253.

Zoogeography, or animal geography, study of the geographic distribution of animals. Formerly, the approach of this science was descriptive; today it utilizes various data, including isotope dating and ocean-bottom core sampling. *See also* NW pp.192–193.

Zoology, study of animals; combined with BOTANY, it comprises the science of BIOLOGY. It is concerned with the structure of the animal and the way in which animals behave, reproduce and function, their evolution and their role in interactions with man and their environment. There are various subdivisions of the discipline, including: TAXONOMY, classification of animals; ZOOGEOGRAPHY, the distribution of animals; ECOLOGY, the relationship of animals to each other and to their environment; PALAEONTOLOGY, the study and interpretation of fossils; ANATOMY, the study of the body and organs. The study of man, ANTHROPOLOGY, is an extension of zoology. *See also* EMBRYOLOGY; GENETICS; MORPHOLOGY.

Zoonosis, infection or infestation of lower animals that can be transmitted to man. Causes include some parasites, such as fleas or canine tapeworms, and BACTERIA (such as *Bacillis anthracis*, a primary cause of disease in sheep and cows).

Zooplankton, animal portion of the PLANKTON. It consists of a wide variety of micro-organisms including COPEPODS and larval forms of higher animals. It is an important constituent of the ocean's food chain. There are few levels or areas of the ocean that have no zooplankton.

Zoroaster, or Zarathustra (*c*.628 BC–*c*.551 BC), Persian religious reformer, founder of ZOROASTRIANISM. Zoroaster is believed to have been born in western Persia. Unable to convert the petty chieftains of his native land, Zoroaster travelled to the East, where in Chorasmia (now in Khorasan province in NE Iran) he converted the royal family, including King Vishtaspa. By the time of Zoroaster's death, his new religion had spread to a large part of Persia. Parts of the AVESTA, the holy scripture of Zoroastrianism, are

believed to have been written by Zoroaster himself.

Zoroastrianism, religion founded by ZOROASTER in the 6th century BC. It became the state religion of Persia in the middle of the 3rd century AD until the mid-7th century when Persia fell to ISLAM. Viewing the world as being divided between the spirits of good and evil. Zoroastrians worship AHURAMAZDA as the supreme deity. They also consider fire sacred. The rise of Islam led to the decline and near disappearance of the religion in Persia. Today the PARSIS of India comprise most of the adherents of Zoroastrianism.

Zorrilla y Moral, José (1817–93), Spanish poet and playwright. Although chiefly remembered for his drama *Don Juan Tenorio* (1844), he also wrote sonorous poetry, including two historical works *Leyendas* and *Granada* (1852).

Zoshchenko, Mikhail Mikhaylovich (1895–*c*.1958), Soviet short story writer. His humorous stories, which deal with the Russian lower middle class and with Soviet petty bureaucracy, brought him into conflict with the Soviet Writers Union, from which he was expelled in the late 1940s. His works include a volume of witty short stories (1922) and *The Woman who Could Not Read* (1940).

Zouaves, French corps of soldiers formed from the BERBER Kabyle tribe of Zwawa. First established in Algeria in 1830, the Zouaves soon lost their native identity, retaining only a semblance of Arab dress.

Zsigmondy, Richard Adolf (1865–1929), Austrian chemist who was awarded the 1925 Nobel Prize in chemistry for his work on COLLOIDS. While employed at a glassworks (1897–1900), he discovered a water suspension of gold and proposed that the shape and size of colloids could be deduced from the way in which the particles scatter light. To aid such studies he developed the ultramicroscope with Heinrich Siedentopf in 1903.

Zuccarelli, Francesco (1702–88), Italian landscape painter of the VENETIAN SCHOOL. His style was influenced by CLAUDE LORRAIN and Marco RICCI. He worked mainly in Venice but visited Britain twice, where he decorated the opera house in London and became well known for his engravings of scenes on the River Thames. He was a founder-member of the ROYAL ACADEMY OF ARTS.

Zuccari, or Zaccaro, Italian artists, two brothers whose late MANNERIST style in Rome was inspired by the works of Raphael and Michelangelo. Taddeo (1529–66) was employed by popes Julius III and Paul IV. Among his many collaborators was his brother Federigo (1543–1609). Together they painted frescoes for the Vatican and scenes from history and mythology for the Caparola Palace of Cardinal Farnese.

Zuckerman, Sir Solly (1904–), British science administrator, *b*. South Africa. He was professor of anatomy at Birmingham (1943–68), and served as chief scientific adviser to the Ministry of Defence (1960–66), as well as working on many other committees dealing with scientific research, including the Royal Commission on Environmental Pollution (1970–74).

Zuckmayer, Carl (1896–1977), US poet and playwright, *b*. Germany. He served in WWI and his first poems appeared in *Die Aktion* (1917); soon afterwards he joined Bertolt BRECHT in Berlin. His first successful play was *Der Fröhliche Weinberg* (1925) and this was followed by many more. After emigrating to the USA because of Nazism in 1939, he wrote his most famous work *Des Teufels General* (1946).

Zugspitze, peak in the Wetterstein Mountains of the Bavarian Alps, West Germany, 87km (54 miles) ssw of Munich. It is the highest peak in Germany, rising to 2,963m (9,722ft).

Zuiderzee, IJsselmeer, former inlet of the North Sea on the N coast of the Netherlands. A dam completed in 1932 divided the area into the Waddenzee in the N, and in the S the IJsselmeer, large areas of which have been reclaimed and are known as polders. Three polders have been formed and two more are planned. *See also* MW p.129; PE p.*44*.

Zulu, BANTU people of South Africa, most of whom live in Natal. They are closely related to the Swazi and the XHOSA. The Zulus have a patriarchal, polygynous society with a strong militaristic tradition, although they are traditionally farmers, concentrating on growing cereals but also possessing large herds of cattle, which they consider to be status symbols. They fiercely resisted the encroachments of White colonialists when unified under their leader Shaka in the 19th century. The predominant religion is now Christianity, although ethnic religions are still common. *See also* NGUNI; HC2 pp.128, *129*, 142–143; MS pp.*202, 255*.

Zulu War (1879), war in southern Africa between the hitherto unconquered ZULUS and the British. Prompted by a fear that the Zulus might attack Natal, the British demanded that the Zulu leader, Cetewaya, disband his army. The Zulus refused, and defeated a British army sent against them, but they were halted at Rorke's Drift and finally defeated at Ulundi, in July 1879. Zululand became a British protectorate in 1887. *See also* HC2 p.*129*.

Zurbarán, Francisco de (1598–1664), Spanish BAROQUE painter. Appointed official painter to Seville in 1628, he worked there for the next three decades, portraying episodes from the lives of the saints.

Zürich, city in N Switzerland on the River Limmat, at the NW end of the Lake of Zürich, in the foothills of the Alps; largest city in Switzerland and capital of Zürich canton. Conquered by the Romans in 58 BC the city later passed to the Alemanni, Franks and to Swabia. It became a free, imperial city in 1218. Zürich joined the Swiss Confederation in 1351, and was an important centre of the Swiss REFORMATION in the 16th century. The city developed as a cultural and scientific centre in the 18th and 19th centuries. It has the largest university (1833) in Switzerland, the Swiss National Museum and many museums and churches. Zürich is the commercial centre of Switzerland and has many banking and financial institutions. Industries: motor vehicles, machinery, paper, textiles, printing and publishing. Pop. (1976 est.) 422,640.

Zwickau, city in S East Germany, on the River Mulde, 64km (40 miles) s of Leipzig. Founded in the 11th century, Zwickau was a free imperial city from 1290 to 1323, when it passed to the Margraves of Meissen. The ANABAPTIST movement of the German REFORMATION started in the city in 1520. Robert SCHUMANN was born in Zwickau in 1810. Places of interest include the Schumann museum, a 12th–15th-century basilica and the 14th-century church of St Catherine. Industries: machinery, textiles, motor vehicles and porcelain. Pop. (1975) 122,604.

Zwingli, Ulrich (or Huldreich) (1484–1531), Swiss theologian. He was a Roman Catholic priest ordained in 1506, but his studies of New Testament in ERASMUS' editions led him to become a Reformer. By 1522 he was preaching reformed doctrine in Zürich and made that city a centre of the REFORMATION. More radical than LUTHER, he saw communion as symbolic and commemorative. He died as a chaplain during a battle against the Catholic cantons at Kappel. *See also* HC1 p.269.

Zworykin, Vladimir Kosma (1889–), US physicist, *b*. Russia, who was a pioneer of television. He left Russia after the October Revolution to settle in the USA. In 1923 he patented the ICONOSCOPE, the forerunner of the modern television camera tube. He also patented a colour television system in 1928 and developed a secondary emission multiplier which was used in the scintillation counter, a sensitive radiation detector.

Zygote, in sexual reproduction, a cell formed by fusion of a male and female GAMETES. A fertilized egg contains a diploid number of CHROMOSOMES, half contributed by the SPERM, half by the OVUM.

Zyrardow, city in Warszawa province, central Poland, 40km (25 miles) wsw of Warsaw. Industries: textiles, leather goods, brewing, food processing. Pop. (1970 est.) 33,200.

Zoroaster, Persian reformer and prophet; this is thought to be a portrait of him.

Zulu chief in tribal costume on a reserve near Durban, Natal, South Africa.

Francisco de Zurbarán's *St Peter Crucified Appears to St Peter Nolasco.*

Zürich is divided by the River Limmat into the "Little City" and the "Great City".

Information tables

Several other volumes in *The Joy of Knowledge Library*, particularly *The Modern World*, contain information in tabular form, and references to the relevant pages are included on these pages.

Animal classification. *See* NW p.254.

Astronomical data. *See* SU pp.288–289.

Athletics records

MEN'S TRACK

Event	Time (min:sec)	Athlete and nationality	Date
100yd	9.0	Ivory Crockett (USA)	11 May 1974
220yd (straight)	19.5	Tommie Smith (USA)	7 May 1966
440yd	44.5	John Smith (USA)	26 June 1971
880yd	1:44.1	Richard Wohlhuter (USA)	8 June 1974
1 mile	3:49.4	John Walker (NZ)	12 Aug. 1975
100m	9.95	James Hines (USA)	14 Oct. 1968
200m	19.83	Tommie Smith (USA)	16 Oct. 1968
400m	43.86	Lee Evans (USA)	18 Oct. 1968
800m	1:43.4	Alberto Juantorena (Cuba)	21 Aug. 1977
1,000m	2:13.9	Richard Wohlhuter (USA)	30 July 1974
1,500m	3:32.2	Filbert Bayi (Tanzania)	2 Feb. 1974
2,000m	4:51.4	John Walker (NZ)	30 June 1976
3,000m	7:35.2	Brendan Foster (UK)	3 Aug. 1974
5,000m	13:12.9	Dick Quax (NZ)	5 July 1977
10,000m	27:30.5	Samson Kimombwa (Kenya)	30 June 1977
20,000m	57:24.2	Jos Hermens (Netherlands)	1 May 1976
25,000m	1hr 14:16.8	Pekka Päivärinta (Finland)	15 May 1975
30,000m	1hr 31:30.4	James Alder (UK)	5 Sept. 1970
Marathon (42.195km)	2hr 08:33.6	Derek Clayton (Australia)	30 May 1969
110m hurdles	13.21	Alexander Casanas (Cuba)	21 Aug. 1977
400m hurdles	47.45	Edwin Moses (USA)	11 June 1977
3,000m steeplechase	8:08.0	Anders Garderud (Sweden)	28 July 1976
4×110yd relay	38.6	Univ. of Southern California	17 June 1967
4×220yd relay	1:21.7	Texas Agricultural and Mechanical College	24 April 1970
4×440yd relay	3:02.4	USA	18 July 1975
4×880yd relay	7:10.4	Univ. Chicago Track Club	12 May 1973
4×100m relay	38.03	USA	3 Sept. 1977
4×200m relay	1:21.4	Arizona State University	30 April 1977
4×400m relay	2:56.1	USA	20 Oct. 1968
4×800m relay	7:08.6	West Germany	13 Aug. 1966
4×1,500m relay	14:38.8	West Germany	17 Aug. 1977

WOMEN'S TRACK

Event	Time (min:sec)	Athlete and nationality	Date
100yd	10.0	Chi Cheng (Taiwan)	13 June 1970
220yd	22.6	Chi Cheng (Taiwan)	3 July 1970
440yd	52.2	Kathleen Hammond (USA)	12 Aug. 1972
880yd	2:02.0	Judith Pollock (Australia)	5 July 1967
1 mile	4:23.8	Natalia Maracescu (Romania)	22 May 1977
100m	10.88	Marlies Oelsner (E Germany)	1 July 1977
200m	22.21	Irena Szewinska (Poland)	13 June 1974
400m	49.29	Irena Szewinska (Poland)	29 July 1976
800m	1:54.9	Tatyana Kazankina (USSR)	26 July 1976
1,500m	3:56.0	Tatyana Kazankina (USSR)	28 July 1976
3,000m	8:27.1	Ludmila Bragina (USSR)	7 Aug. 1977
Marathon (42.195km)	2hr 34:47.5	Christa Vahlensieck (W Germany)	10 Sept. 1977
100m hurdles	12.59	Annelie Ehrhardt (E Germany)	8 Sept. 1972
400m hurdles	55.63	Karin Rossley (E Germany)	13 Aug. 1977
4×110yd relay	44.1	West Germany	18 July 1975
4×220yd relay	1:35.8	Australia	9 Nov. 1969
4×440yd relay	3:29.1	USSR	7 Aug. 1976
4×100m relay	42.5	East Germany	29 May 1976
4×200m relay	1:31.6	UK	20 Aug. 1977
4×400m relay	3:19.2	East Germany	31 July 1976
4×800m relay	7:52.3	USSR	16 Aug. 1976

MEN'S FIELD

Event	Distance	Athlete and nationality	Date
High jump	2.33m (7ft 7¾in)	Vladimir Yashchenko (USSR)	3 July 1977
Pole vault	5.70m (18ft 8¼in)	David Roberts (USA)	22 June 1976
Long jump	8.90m (29ft 2½in)	Robert Beamon (USA)	18 Oct. 1968
Triple jump	17.89m (58ft 8¼in)	João de Oliveira (Brazil)	15 Oct. 1975
Shot	22.00m (72ft 2¼in)	Aleksandr Baryshnikov (USSR)	10 July 1976
Discus	70.86m (232ft 6in)	Maurice Wilkins (USA)	1 May 1976
Hammer	79.30m (260ft 2in)	Walter Schmidt (W Germany)	14 Aug. 1975
Javelin	94.58m (310ft 4in)	Miklós Németh (Hungary)	26 July 1976

WOMEN'S FIELD

Event	Distance	Athlete and nationality	Date
High jump	2.00m (6ft 6¾in)	Rosemarie Ackermann (E Germany)	26 Aug. 1977
Long jump	6.99m (22ft 11½in)	Sigrun Siegl (E Germany)	19 May 1976
Shot	22.30m (73ft 2⅜in)	Helena Fibingerova (Czech.)	20 Aug. 1977
Discus	70.50m (231ft 3in)	Faina Melnik (USSR)	24 April 1976
Javelin	69.26m (227ft 5in)	Catherine Schmidt (USA)	11 Sept. 1977

MEN'S SWIMMING

Event	Time (min:sec)	Athlete and nationality	Date
100m freestyle	49.44	Jonty Skinner (South Africa)	14 Aug. 1976
100m butterfly	54.27	Mark Spitz (USA)	31 Aug. 1972
100m backstroke	55.49	John Naber (USA)	19 July 1976
100m breaststroke	1:03.11	John Henken (USA)	20 July 1976
200m freestyle	1:50.29	Bruce Furniss (USA)	19 July 1976
200m butterfly	1:59.23	Michael Bruner (USA)	18 July 1976
200m backstroke	1:59.19	John Naber (USA)	24 July 1976
200m breaststroke	2:15.11	David Wilkie (UK)	24 July 1976
200m medley	2:05.31	Graham Smith (Canada)	5 Aug. 1977
400m medley	4:23.68	Rod Strachan (USA)	25 July 1976
400m freestyle	3:51.93	Brian Goodell (USA)	20 July 1976
800m freestyle	8:01.54	Bobby Hackett (USA)	21 June 1976
1,500m freestyle	15:02.40	Brian Goodell (USA)	20 July 1976
4×100m freestyle	3:24.85	USA	23 July 1975
4×200m freestyle	7:23.22	USA	21 July 1976
4×100m medley	3:42.22	USA	22 July 1976

WOMEN'S SWIMMING

Event	Time (min:sec)	Athlete and nationality	Date
100m freestyle	55.65	Kornelia Ender (E Germany)	19 July 1976
100m butterfly	1:00.13	Kornelia Ender (E Germany)	4 June 1976
100m backstroke	1:01.51	Ulrike Richter (E Germany)	5 June 1976
100m breaststroke	1:10.86	Hannelore Anke (E Germany)	22 July 1976
200m freestyle	1:59.26	Kornelia Ender (E Germany)	22 July 1976
200m butterfly	2:11.22	Rosemarie Gabriel-Kother (E Germany)	5 June 1976
200m backstroke	2:12.47	Birgit Treiber (E Germany)	4 June 1976
200m breaststroke	2:33.35	Marina Koshevaia (USSR)	21 July 1976
200m medley	2:16.96	Ulrike Tauber (E Germany)	9 July 1977
400m medley	4:42.77	Ulrike Tauber (E Germany)	24 July 1976
400m freestyle	4:09.89	Petra Thumer (E Germany)	20 July 1976
800m freestyle	8:35.04	Petra Thumer (E Germany)	29 July 1977
1,500m freestyle	16:33.94	Jenny Turrall (Australia)	25 Aug. 1974
4×100m freestyle	3:44.94	USA	25 July 1976
4×100m medley	4:07.95	East Germany	18 July 1976

Chemical elements. *See* SU pp.287–288.

Countries of the world. *See* articles on individual countries in *The Modern World*.

MAJOR COUNTRY REGIONS
Australia (States and territories). *See* MW pp.33–34.
Canada (Provinces and territories). *See* MW p.50.
China (Provinces, etc.). *See* MW p.56
England (Counties). *See* MW p.70.
France (Planning regions). *See* MW p.80.
India (States and territories). *See* MW p.99.
Ireland (Provinces and counties). *See* MW p.105.
Italy (Regions). *See* MW p.110.
Scotland (Regions). *See* MW p.153.
South Africa (Provinces and Bantustans). *See* MW pp.157–158.
Union of Soviet Socialist Republics (Republics). *See* MW p.173.
United States (States). *See* MW p.178.
Wales (Counties). *See* MW p.182.
West Germany (Länder). *See* MW p.87.

Engineering achievements

HIGHEST BUILDINGS

Name	m	ft	City
Sears Roebuck Tower	443	1,454	Chicago
World Trade Center	412	1,350	New York
Empire State Building	381	1,250	New York
Standard Oil Building	346	1,136	Chicago
John Hancock Center	337	1,107	Chicago
Chrysler Building	319	1,046	New York

LONGEST BRIDGES

Name	m	ft	Country
Lake Pontchartrain Causeway II	38,422	126,055	USA
Lake Pontchartrain Causeway I	38,353	125,827	USA
Chesapeake Bay Bridge – Tunnel	28,427	93,203	USA
Marathon Key – Bahia Honda Key Bridge	11,869	38,915	USA
San Mateo – Hayward Bridge I	11,341	37,183	USA
San Mateo – Hayward Bridge II	10,891	35,708	USA
Lake Maracaibo Bridge	8,684	28,473	Venezuela
Transbay Bridge	8,322	27,286	USA

LONGEST RAILWAY TUNNELS

Name	m	yd	Country
Simplon No. II	19,815	21,679	Switzerland/Italy
Shin Kanmon	18,588	20,328	Japan
Apennine	18,510	20,252	Italy
Gotthard	14,991	16,402	Switzerland
Hokuriku	13,865	15,169	Japan
Mont Cenis Fréjus	13,651	14,935	France/Italy
Cascade	12,537	13,717	USA
Arlberg	10,246	11,210	Austria

Mathematical symbols. *See* SU p.286.

International organizations. For information on the membership of the world's chief international organizations, *see* MW p.191.

Rulers and heads of state

GOVERNORS-GENERAL

Australia. *See* MW p.33.

Canada:

Name	Term of office		
Lord Monck	1867–68	Duke of Devonshire	1916–21
Sir John Young	1869–72	Lord Byng	1921–26
Lord Dufferin	1872–78	Lord Willingdon	1926–31
Lord Lorne	1878–83	Lord Bessborough	1931–35
Lord Lansdowne	1883–88	Lord Tweedsmuir	1935–40
Lord Stanley	1888–93	Lord Athlone	1940–46
Lord Aberdeen	1893–98	Lord Alexander	1946–52
Lord Minto	1898–1904	Vincent Massey	1952–59
Lord Grey	1904–11	Maj.-Gen. Georges Vanier	1959–67
Duke of Connaught	1911–16	Roland Michener	1967–74
		Jules Léger	1974–

HOLY ROMAN EMPERORS

Name	Reign		
Charlemagne	800–814	Interregnum	1254–73
Louis I the Pious	814–840	Rudolf I of Hapsburg	1273–91
Lothair I	840–855	Adolf of Nassau	1292–98
Louis II	855–875	Albert I of Austria	1298–1308
Charles the Bald	875–877	Henry VII of Luxemburg	1308–13
Interregnum	877–881	Louis IV of Bavaria	1314–47
Charles III the Fat	881–887	Charles IV of Luxemburg	1347–78
Arnulf	887–899	Wenceslaus of Bohemia	1378–1400
Louis III	899–911	Rupert of the Palatinate	1400–10
Conrad I	911–918	Sigismund	1410–37
Henry I the Fowler	919–936	Albert II	1438–39
Otto I the Great	936–973	Frederick III	1440–93
Otto II	973–983	Maximilian I	1493–1519
Otto III	983–1002	Charles V	1519–58
Henry II	1002–24	Ferdinand I	1558–64
Conrad II	1024–39	Maximilian II	1564–76
Henry III	1039–56	Rudolf II	1576–1612
Henry IV	1056–1106	Matthias	1612–19
Henry V	1106–25	Ferdinand II	1619–37
Lothair II of Saxony	1125–37	Ferdinand III	1637–57
Conrad III	1138–52	Leopold I	1658–1705
Frederick I Barbarossa	1152–90	Joseph I	1705–11
Henry VI	1190–97	Charles VI	1711–40
Philip of Swabia	1198–1208	(Interregnum)	1740–42
Otto IV of Brunswick		Charles VII of Bavaria	1742–45
(anti-king)	1198–1215	Francis I	1745–65
Frederick II	1215–50	Joseph II	1765–90
Conrad IV	1250–54	Leopold II	1790–92
		Francis II	1792–1806

KINGS

England (to 1603). *See* HC2 pp.308–309; MW p.74.
Scotland (to 1625). *See* MW p.152.
United Kingdom (from 1603). *See* HC2 p.309; MW p.174.

PRESIDENTS

France

Name	Term of office		
The Third Republic		*Vichy Government*	
Louis Thiers	1871–73	Marshal Pétain, (State)	1940–44
Marshal MacMahon	1873–79	Pierre Laval (Government)	1942–44
Jules Grévy	1879–87	*Provisional Government*	
Sadie Carnot	1887–94	Charles de Gaulle	1944–46
Jean Casimir-Périer	1894–95	Félix Gouin (Jan. 23)	1946
Félix Faure	1895–99	Georges Bidault (June 19)	1946
Emile Loubet	1899–1906	Léon Blum (Dec. 12)	1946
Armand Falliéres	1906–13	*The Fourth Republic*	
Raymond Poincaré	1913–20	Vincent Auriol	1947–54
Paul Deschanel	1920	René Coty	1954–59
Alexandre Millerand	1920–24	*The Fifth Republic*	
Gaston Doumergue	1924–31	Charles de Gaulle	1959–69
Paul Doumer	1931–32	Georges Pompidou	1969–74
Albert Lebrun	1932–40	Giscard D'Estaing	1974–

United States

Name	Term of office		
George Washington	1789–97	James Garfield	1881
John Adams	1797–1801	Chester Arthur	1881–85
Thomas Jefferson	1801–09	Grover Cleveland	1885–89
James Madison	1809–17	Benjamin Harrison	1889–93
James Monroe	1817–25	Grover Cleveland	1893–97
John Adams	1825–29	William McKinley	1897–1901
Andrew Jackson	1829–37	Theodore Roosevelt	1901–09
Martin Van Buren	1837–41	William Taft	1909–13
William Harrison	1841	Woodrow Wilson	1913–21
John Tyler	1841–45	Warren Harding	1921–23
James Knox Polk	1845–49	Calvin Coolidge	1923–29
Zachary Taylor	1849–50	Herbert Hoover	1929–33
Millard Fillmore	1850–53	Franklin Roosevelt	1933–45
Franklin Pierce	1853–57	Harry S. Truman	1945–53
James Buchanan	1857–61	Dwight Eisenhower	1953–61
Abraham Lincoln	1861–65	John Kennedy	1961–63
Andrew Johnson	1865–69	Lyndon Johnson	1963–69
Ulysses Grant	1869–77	Richard Nixon	1969–74
Rutherford Hayes	1877–81	Gerald Ford	1974–76
		James Carter	1976–

PRIME MINISTERS

Australia. *See* MW p.33.
Canada. *See* MW p.51
Ireland. *See* MW p.105.
New Zealand. *See* MW p.134
Northern Ireland. *See* MW p.138.

United Kingdom:

Name	Term of office		
First Lords of the Treasury			
Charles Montagu	1714–15		
Charles Howard	1715		
Robert Walpole	1715–17		
James Stanhope	1717–18		
Charles Spencer	1718–21		
Prime Ministers			
Robert Walpole	1721–42	Edward Stanley	1858–59
Spencer Compton	1742–43	Viscount Palmerston	1859–65
Henry Pelham	1743–54	John Russell	1865–66
Thomas Pelham-Holles	1754–56	Edward Stanley	1866–68
William Cavendish	1756–57	Benjamin Disraeli	1868
Thomas Pelham-Holles	1757–62	William Gladstone	1868–74
John Stuart	1762–63	Benjamin Disraeli	1874–80
George Grenville	1763–65	William Gladstone	1880–85
Charles Watson-Wentworth	1765–66	Robert Cecil	1885–86
William Pitt	1766–68	William Gladstone	1886
Augustus Fitzroy	1768–70	Robert Cecil	1886–92
Frederick North	1770–82	William Gladstone	1892–94
Charles Watson-Wentworth	1782	Archibald Primrose	1894–95
William Petty	1782–83	Robert Cecil	1895–1902
William Bentinck	1783	Arthur Balfour	1902–05
William Pitt the Younger	1783–1801	Henry Campbell-Bannerman	1905–08
Henry Addington	1801–04	Herbert Asquith	1908–16
William Pitt the Younger	1804–06	David Lloyd George	1916–22
William Grenville	1806–07	Andrew Bonar Law	1922–23
William Bentinck	1807–09	Stanley Baldwin	1923–24
Spencer Perceval	1809–12	Ramsay MacDonald	1924
Robert Jenkinson	1812–27	Stanley Baldwin	1924–29
George Canning	1827	Ramsay MacDonald	1929–35
Frederick Robinson	1827–28	Stanley Baldwin	1935–37
Arthur Wellesley	1828–30	Neville Chamberlain	1937–40
Charles Grey	1830–34	Winston Churchill	1940–45
William Lamb	1834	Clement Attlee	1945–51
Robert Peel	1834–35	Winston Churchill	1951–55
William Lamb	1835–41	Anthony Eden	1955–57
Robert Peel	1841–46	Harold Macmillan	1957–63
John Russell	1846–52	Alec Douglas-Home	1963–64
Edward Stanley	1852	Harold Wilson	1964–70
George Gordon	1852–55	Edward Heath	1970–74
Viscount Palmerston	1855–58	Harold Wilson	1974–76
		James Callaghan	1976–

ROMAN EMPERORS

Name	Reign		
Augustus (Octavian)	27 BC–AD 14	Gallienus	260–268
Tiberius	14–37	Claudius II	268–270
Caligula	37–41	Aurelian	270–275
Claudius	41–54	Tacitus	275–276
Nero	54–68	Florian	276
Galba	68–69	Probus	276–282
Otho	69	Carus	282–283
Vitellius	69	Carinus & Numerianus	283–284
Vespasian	69–79	Diocletian	284–305
Titus	79–81	Constantius	305–306
Domitian	81–96	Constantine the Great	306–337
Nerva	96–98	Constantine II	337–340
Trajan	98–117	Constans	337–350
Hadrian	117–138	Constantius II	337–361
Antoninus Pius	138–161	Julian the Apostate	361–363
Marcus Aurelius	161–180	Jovian	363–364
Lucius Aurelius Verus	161–169	Valentinian I (West)	364–375
Commodus	180–192	Valens (East)	364–378
Pertinax	193	Gratian (West)	375–383
Didius Julianus	193	Valentinian II (West)	375–392
Septimius Severus	193–211	Theodosius the Great (East)	379–394
Caracalla	211–217	Maximus (West) (usurper)	383–388
Geta (with Caracalla)	211–212	Eugenius (West) (usurper)	392–394
Macrinus	217–218	Theodosius (sole emperor)	394–395
Heliogabalus or Elagabalus	218–222	Honorius	395–423
Alexander Severus	222–235	Valentinian III	425–455
Maximinus	235–238	Petronius Maximus	455
Gordianus I & II (joint)	238	Avitus	455–456
Balbinus & Pupienus (joint)	238	Majorian	457–461
Gordianus III	238–244	Severus	461–465
Philip the Arabian	244–249	Anthemius	467–472
Decius	249–251	Olybrius	472–473
Gallus	251–253	Glycerius	473–474
Aemilianus	253	Julius Nepos	474–475
Valerian	253–260	Romulus Augustulus	475–476

Swimming records. *See* ATHLETICS RECORDS, p.848.

Weights and measures. *See* SU p.286.

World features. For information on the continents and deserts, islands, lakes, mountains, rivers, volcanoes and waterfalls, *see* MW p.191.

Bibliography

The bibliography relates to the seven thematic volumes in *The Joy of Knowledge Library* and is divided by volume title, beginning with *Science and The Universe*. The numbers down the side of the booklist are an approximate guide to page numbers in those volumes. Thus, the reader wishing to know more about building materials (covered on pages 46–47 of *Man and Machines*) should simply look under that volume heading until he or she comes to the number 46, where a book called *Properties of Building Materials* is cited.

It has not been possible to arrange the bibliography for the two volumes of History and Culture in that way: that part of the bibliography has been divided first into ancient, medieval and modern history and then further divided, in the ancient and medieval sections, by continent and country. The modern section has been divided first into history and arts, then by continent and country, and finally (where appropriate) by artiytic discipline. A work on jazz, for example, will be found in the modern section, in the American subdivision, under music.

Science and The Universe

The Growth of Science
20 Asimov, I.; *Guide to Science*; Penguin, 1975
Lloyd, G. E. R.; *Early Greek Science*; Chatto, 1970
Lloyd, G. E. R.; *Greek Science after Aristotle*; Chatto, 1973
Bronowski, J.; *Ascent of Man*; B.B.C., 1973
Sigerist, H. E.; *History of Medicine*; Oxford U.P., 1951–61 (2 vols)
22 Boas, M.; *The Scientific Renaissance 1450–1630*; Collins, 1962
Debus, A. G. & R. P. Multhauf; *Alchemy and Chemistry in the 17th Century*; Univ. California, 1966
24 Burckhardt, T.; *Alchemy: Science of the Cosmos, Science of the Soul*; Stuart & Watkins, 1967
Burland, C. A.; *Arts of the Alchemists*; Weidenfeld & Nicolson, 1967
Asimov, I.; *Understanding Physics*; New Am. Lib., 1969 (3 vols)
Mathematics
26 Kline, M.; *Mathematics in Western Culture*; Penguin, 1972
Kramer, E. E.; *Main Stream of Mathematics*; Oxford U.P., 1951
Bell, E. T.; *Development of Mathematics*; McGraw-Hill, 1945
Hogben, L.; *Mathematics for the Million*; Pan Books, 1967
28–30 Ore, O.; *Number Theory and History*; McGraw-Hill, 1948
Courant, R. & H. Robbins; *What is Mathematics?* Oxford U.P., 1941
32 Pankhurst, R. C.; *Dimensional Analysis and Scale Factors*; Chapman & Hall, 1964
34 Gardner, K. L.; *Discovering Modern Algebra*; Oxford U.P., 1966
Dubisch, R. & V. E. Howes; *Intermediate Algebra*; Wiley, 1974
36 Lockwood, E. H.; *Book of Curves*; Cambridge U.P., 1961
Thompson, Sir D'A. W.; *On Growth and Form*; Cambridge U.P., 1961 (2 vols)
38 Campbell, H.; *Using Logarithms*; Stillit, 1966
40 Kuratowski, K.; *Introduction to Set Theory and Topology*; Pergamon, 1972
Pinter, C. C.; *Set Theory*; Addison, 1971
Mansfield, D. E. & M. Bruckheimer; *Background to Set and Group Theory: Teaching of Mathematics*; Chatto, 1965
42 Thompson, S. P.; *Calculus Made Easy*; Macmillan, 1965
Kleppner, D. & N. Ramsay; *Quick Calculus*; Wiley, 1972
44 Jacobs, H. R.; *Geometry*; W. H. Freeman, 1974
46–48 Selby, P. H.; *Geometry and Trigonometry for Calculus*; Wiley, 1975
Lunt, W. J.; *Basic Mathematics for Technical College Students*; McGraw-Hill, 1969
Cundy, H. M. & A. P. Rollett; *Mathematical Models*; Oxford U.P., 1973
50 Ledermann, W.; *Introduction to the Theory of Finite Groups*; Oliver, 1961
Schonland, D. S.; *Molecular Symmetry: Introduction to Group Theory and its Uses in Chemistry*; Van Nost. Reinhold, 1965
52 Griffiths, H. B.; *Surfaces*; Cambridge U.P., 1976
54 Wilson, R. J.; *Introduction to Graph Theory*; Longman, 1975
Stephenson, G.; *Introduction to Matrices, Sets and Groups*; Longman, 1965
56 Moroney, M. J.; *Facts from Figures*; Penguin, 1969
58 Weaver, W.; *Lady Luck: Theory of Probability*; Penguin, 1977
Adler, I.; *Probability and Statistics for Everyman*; Dobson, 1963
Atomic Theory
60–68 Asimov, I.; *Realm of Measure*; Houghton, 1960
Matthews, P. T.; *The Nuclear Apple*; Chatto, 1971
Jungk, H.; *Brighter Than a Thousand Suns*; Penguin, 1960
Glasstone, S.; *Source Book on Atomic Energy*; Van Nost. Reinhold, 1968
Henderson, S.; *Chemistry Today*; Macmillan, 1976
Statics and Dynamics
70–78 Jardine, J.; *Physics is Fun*; Heinemann Educ., 1964–67
Jardine, J.; *Nat. Phil. 4*; Heinemann Educ., 1974
Abbott, A. F.; *"O" Level Physics*; Heinemann Educ., 1969
Abbott, A. F. & M. Nelkon; *Elementary Physics*; Heinemann Educ., 1971
Sound
80 Smith, B. J.; *Acoustics*; Longman, 1971
82 Taylor, C. A.; *Physics of Musical Sound*; Eng. U.P., 1965
Matter
84 Mulcahy, M. F. R.; *Gas Kinetics*; Nelson, 1973
86 Walshaw, A. C.; *Mechanics of Fluids*; Longman, 1972
88 Kittel, C.; *Introduction to Solid-State Physics*; Wiley, 1976
Heat
90–96 Starling, S. G. & A. Woodall; *Physics*; Longman, 1957
McKenzie, A. E. E.; *A Second Course of Heat*; Cambridge U.P., 1963
Light
98–104; 108–110 McKenzie, A. E. E.; *A Second Course of Light*; Cambridge U.P., 1956
Sanders, J. H.; *The Velocity of Light*; Pergamon, 1965

Lengyel, B.; *Lasers*; Wiley, 1971
Jenkins, F. A.; *Fundamentals of Optics*; McGraw-Hill, 1976
106 Adler, R.; *Introduction to General Relativity*; McGraw-Hill, 1965
Katz, R.; *Introduction to the Special Theory of Relativity*; Van Nost. Reinhold, 1964
Marks, J.; *Relativity: Elementary Non-Mathematical Introduction (Classical, General and Special)*; Chapman, 1972
Electricity and Magnetism
112–114 Agger, L. T.; *Introduction to Electricity*; Oxford U.P., 1971
Russell, L. A.; *Atoms and Electricity*; Open U.P., 1973
Bleaney, B. I. & B. Bleaney; *Electricity and Magnetism*; Oxford U.P., 1976
Jackson, J. D.; *Classical Electrodynamics*; Wiley, 1975
Page, L. & N. I. Adams; *Electrodynamics*; Dover, 1965
116 Wagner, D.; *Introduction to the Theory of Magnetism*; Pergamon, 1972
118 Foster, K. & R. Anderson; *Electromagnetic Theory*; Butterworth, 1969 (2 vols)
Solymar, L.; *Lectures on Electromagnetic Theory*; Oxford U.P., 1976
McKenzie, A. E. E.; *A Second Course of Electricity*; Cambridge U.P., 1973
122–126 Anderson, J. C. et al.; *Materials Science*; Nelson, 1974
Fitzgerald, A. E. et al.; *Electric Machinery*; McGraw-Hill, 1970
Hindmarsh, J.; *Electrical Machines and their Applications*; Pergamon, 1977
Jurek, S. F.; *Electrical Machines for Technicians and Technician Engineers*; Longman, 1974
Smith, R. J.; *Circuits, Devices and Systems*; Wiley, 1976
Ahmed, H. & P. J. Spreadbury; *Electronics for Engineers*; Cambridge U.P., 1973
128 Wright, D. A.; *Semiconductors*; Methuen, 1966
130 Beck, A. H. W. & H. Ahmed; *Introduction to Physical Electronics*; Arnold, 1968
Turner, J. F.; *Fundamental Electronics*; Merrill, 1972
Chemistry
132 Rossotti, H.; *Introducing Chemistry*; Penguin, 1975
Pauling, L. & P. Pauling; *Chemistry*; W. H. Freeman, 1975
134–136 Puddephat, R. J.; *Periodic Table of the Elements*; Oxford U.P., 1972
Brown, G.; *Introduction to Inorganic Chemistry*; Longman, 1974
Jaffe, B.; *Moseley and the Numbering of the Elements*; Heinemann Educ., 1972
138 Peacocke, T. A. H.; *Atomic and Nuclear Chemistry*; Pergamon, 1967
Holden, A.; *Bonds between Atoms*; Oxford U.P., 1971
Speakman, J. C.; *Valency Primer*; Arnold, 1968
140–146 Parker, D. B. V.; *Polymer Chemistry*; Applied Science, 1974
Perrin, D. D.; *Masking and Demasking of Chemical Reactions*; Wiley, 1970
Thomas, R. W.; *Inorganic Reactions and Volumetric Analysis*; Pitman, 1967
Jackson, R. A.; *Mechanism: Introduction to Organic Reactions*; Oxford U.P., 1972
Whitfield, R. C.; *Guide to Understanding Basic Organic Reactions*; Longman, 1974
148 Denaro, A. R.; *Elementary Electrochemistry*; Butterworth, 1971
Groves, P. D.; *Electrochemistry*; Murray, 1974
Robbins, J.; *Ions in Solution: Introduction to Electrochemistry*; Oxford U.P., 1972
150 Skoog, D. A. & D. M. West; *Fundamentals of Analytical Chemistry*; Holt, 1969
Irving, H. M. N. H.; *Techniques of Analytical Chemistry*; HMSO, 1974
152–154 Stryer, L.; *Biochemistry*; W. H. Freeman, 1975
Rose, S.; *Chemistry of Life*; Penguin, 1970
156 Phillips, D. C. & A. C. T. North; *Protein Structure*; Oxford U.P., 1973

The Universe

General
Calder, N.; *Violent Universe*; Futura Publications, 1975
Hodge, P. W.; *Concepts of Contemporary Astronomy*; McGraw-Hill, 1974
King, I. R.; *Universe Unfolding*; W. H. Freeman, 1976
Abell, G. O.; *Exploration of the Universe*; Holt, 1975
Hinkelbein, A.; *Origins of the Universe*; Collins, 1972
Hoyle, F.; *Ten Faces of the Universe*; Heinemann, 1977
Hoyle, F.; *From Stonehenge to Modern Cosmology*; W. H. Freeman, 1972
Mitton, S. (Ed.); *Cambridge Encyclopaedia of Astronomy*; Cape, 1977
Moore, P.; *New Concise Atlas of the Universe*; Mitchell Beazley, 1978
Roy, A. E. & D. Clarke; *Astronomy: Structure of the Universe*; Hilger, 1977
Roy, A. E. & D. Clarke; *Astronomy: Principles and Practice*; Hilger, 1977
Techniques of Astronomy
164–166 Asimov, I.; *Eyes on the Universe*; Deutsch, 1976
Rackham, T.; *Astronomical Photography at the Telescope*; Faber, 1973
Miczaika, G. R. & W. M. Sinton; *Tools of the Astronomer*; Harvard U.P., 1961
168 Barlow, B. V.; *The Astronomical Telescope*; Wykeham, 1975
Paul, H. E.; *Telescopes for Skygazing*; Amphoto, 1976
170 Thom, A.; *Megalithic Lunar Observatories*; Oxford U.P., 1971
Sidgwick, J. B., *Amateur Astronomer's Handbook*; Faber, 1971
172 Ring, J.; *Infrared Astronomy*; Imperial College, 1968
The Solar System
174–176 Cousins, F. W.; *Solar System*; Baker, 1972
Kopal, Z.; *Solar System*; Oxford U.P., 1973
178–186 Taylor, S. R.; *Lunar Science: Post-Apollo View*; Pergamon, 1975
Guest, P. A. & R. Greeley; *Geology on the Moon*; Wykeham, 1977
Whipple, F. L.; *Earth, Moon and Planets*; Penguin, 1971
Moore, P.; *Guide to the Moon*; Lutterworth, 1976
Kopal, Z.; *Photographic Atlas of the Moon*; Academic Press, 1965
188–214 Abetti, G.; *Stars and Planets*; Faber, 1966
Moore, P.; *Guide to the Planets*; Lutterworth, 1971
Baum, R.; *The Planets*; David & Charles, 1973
188 Antoniadi, E. M.; *Planet Mercury*; Reid, 1974
194 Moore, P.; *Guide to Mars*; Lutterworth, 1977
206 Peek, B. M.; *Planet Jupiter*; Faber, 1958
210 Alexander, A. F.; *Planet Saturn*; Faber, 1962
216–218 Brown, P. L.; *Comets, Meteorites and Men*; Hale, 1973
Moore, P.; *Guide to Comets*; Lutterworth, 1977
Moore, P.; *Comets: Visitors from Space*; Reid, 1973
Wood, J. A.; *Meteorites and the Origins of the Planets*; McGraw-Hill, 1968
The Sun
220–224 Menzel, D. H.; *Our Sun*; Harvard U.P., 1959
Abetti, G.; *Sun*; Faber, 1963
The Stars
226–230 Aller, L. H.; *Atoms, Stars and Nebulae*; Harvard U.P., 1976
Tayler, R. J. & A. S. Everest; *The Stars, Their Structure and Evolution*; Wykeham 1972
232 Smith, F. G.; *Pulsars*; Cambridge U.P., 1977
Gingerich, O. (Ed.); *New Frontiers in Astronomy*; W. H. Freeman, 1975
234 Moore, P. & I. Nicolson; *Black Holes in Space*; Orbach & Chambers, 1974
236–240 Strohmeier, W.; *Variable Stars*; Pergamon, 1972
242 Bok, B. & P. Bok; *Milky Way*; Harvard U.P., 1977

Galaxies
244–248 Brown, P. L.; *What Star is That?*; Thames & Hudson, 1971
Klepesta, J. & A. Rukl; *Constellations*; Hamlyn, 1970
Moore, P.; *Guide to the Stars*; Lutterworth, 1974
Jones, K. G.; *Messier's Nebulae and Star Clusters*; Faber, 1969
250 Bova, B.; *In Quest of Quasars: Introduction to Stars and Starlike Objects*; Cromwell, 1969
Hodge, P. W.; *Galaxies and Cosmology*; McGraw-Hill, 1966
Hey, J. S.; *Radio Universe*; Pergamon, 1971
252 Sciama, D. W.; *Introduction to Modern Cosmology*; Cambridge U.P., 1971
Star Maps
254–256 Norton, A. P.; *Star Atlas and Reference Handbook*; Gall & Inglis, 1973
Klepesta, J. & A. Rukl; *Constellations*; Hamlyn, 1970
Menzel, D. H.; *Field Guide to the Stars and Planets*; Collins, 1966
Man in Space
268–284 Berry, A.; *Next 10,000 Years: Man's future in the Universe*; Cape, 1974
Carter, L. J.; *History of Astronautics*; Visual Publications, 1970
McCall, R. & I. Asimov; *Our World in Space*; P.S.L., 1974
Clarke, A. C.; *Man and Space*; Time-Life, 1970
Moore, P.; *Next 50 Years in Space*; Luscombe, 1976

The Physical Earth

General
New Concise Atlas of the Earth; Mitchell Beazley, 1978
Structure of the Earth
20 Cook, A. H.; *Physics of the Earth and Planets*; Macmillan, 1973
Jacobs, J. A.; *Earth's Core*; Academic Press, 1975
Bott, M. H. P.; *Interior of the Earth*; Arnold, 1971
22–24 McElhinny, M. W.; *Palaeo-Magnetism and Plate Tectonics*; Cambridge U.P., 1975
26 Tarling, D. H. & M. P. Tarling; *Continental Drift*; Penguin, 1972
Calder, N.; *Restless Earth*; B.B.C., 1972
Scientific American; *Continents Adrift and Continents Aground*; W. H. Freeman, 1976
Sullivan, W.; *Continents in Motion: The New Earth Debate*; Macmillan, 1977
28 Fuchs, V. (Ed.); *Forces of Nature*; Thames & Hudson, 1977
30 Francis, P.; *Volcanoes*; Penguin, 1976
Bullard, F. M.; *Volcanoes of the Earth*; Univ. of Texas, 1976
32 Jordan, P.; *Expanding Earth: Dirac's Gravitation Hypothesis*; Pergamon, 1971
Hallam, A. (Ed.); *Planet Earth*; Elsevier-Phaidon, 1977
The Earth in Perspective
34–38 Robinson, A. H. & R. D. Sale; *Elements of Cartography*; Wiley, 1969
Roblin, H. S.; *Map Projections*; Arnold, 1969
40 Dury, G. H.; *British Isles*; Heinemann Educ., 1973
Stamp, L. D.; *Britain's Structure and Scenery*; Collins (New Naturalist Series), 1970
Church, R. J. H. et al.; *Advanced Geography of Northern and Western Europe*; Hulton, 1968
Monkhouse, F. J.; *Regional Geography of Western Europe*; Longman, 1974
42 Clarke, F. I. et al.; *Advanced Geography of Africa*; Hulton, 1975
50 Brice, W. C.; *South West Asia*; Univ. of London Press, 1966
62 Cole, J. P.; *Latin America*; Butterworth, 1975
Robinson, H.; *Latin America*; Macdonald & Evans, 1977
Weather
66–70 Calder, N.; *The Weather Machine*; B.B.C., 1974
Sutton, G.; *Weather*; Teach Yourself Books, 1974
Scorer, P. R.; *Clouds of the World*; David & Charles, 1972
Harvey, J. G.; *Atmosphere and Ocean*; Artemis Press, 1976
72 Watts, A.; *Instant Weather Forecasting*; Adlard-Coles, 1968, Hart-Davis Educ., 1968
74 Griffiths, J. F.; *Climate and the Environment*; Elek, 1976
Seas and Oceans
76–80 Anikouchine, W. A. & R. W. Sternberg; *World Ocean*; Prentice, 1973
Atlas of the Oceans; Mitchell Beazley, 1977
Perry, A. H. & J. M. Walker; *Ocean-Atmosphere System*; Longman, 1977
Bellamy, D.; *Life-giving Sea*; Hamish Hamilton, 1975
82 Shepard, F. P.; *Earth Beneath the Sea*; Johns Hopkins, 1968
Shepard, F. P.; *Submarine Geology*; Harper & Row, 1973
84 Cotter, C. H.; *Atlantic Ocean*; Brown, Son & Ferguson, 1964
86 Menard, H. W.; *Marine Geology of the Pacific*; McGraw-Hill, 1964
88 Toussaint, A.; *History of the Indian Ocean*; Routledge, 1966
90 Maury, M. F.; *Physical Geography of the Sea*; Harvard U.P., 1972
Deacon, G. E. R. (Ed.); *Oceans*; Hamlyn, 1968
92 Sweeney, J. B.; *Pictorial History of Oceanographic Submersibles*; Hale, 1972
Penzias, W. & M. W. Goodman; *Man beneath the Sea*; Wiley, 1973
Geology
94 Buerger, M. J.; *Contemporary Crystallography*; McGraw-Hill, 1970
Buerger, M. J.; *Elementary Crystallography*; Wiley, 1963
Glasser, L. S. D.; *Crystallography and its Applications*; Van Nost. Reinhold, 1977
96 Fraser, R.; *Understanding the Earth*; Penguin, 1967
Robinson, H.; *Physical Geography*; Macdonald & Evans, 1977
100 Bank, H.; *Precious Stones and Minerals*; Warne, 1970
Sparks, B. W.; *Rocks and Relief*; Longman, 1971
Carmichael, I. S. E.; *Igneous Petrology*; McGraw-Hill, 1974
102 Hatch, F. H. & R. H. Rastall; *Petrology of Sedimentary Rocks*; Murby, 1971
104 Miyashiro, A.; *Metamorphism and Metamorphic Belts*; Allen & Unwin, 1973
106 Carson, M. A. & M. J. Kirkby; *Hillslope Form and Process*; Cambridge U.P., 1972
Moore, W. G.; *Mountains and Plateaux*; Hutchinson, 1977
Young, A.; *Slopes*; Oliver & Boyd, 1972
108 Edwards, R. W. & D. J. Garrod; *Conservation and Productivity of Natural Waters*; Academic Press, 1972
110 Waltham, T.; *Caves*; Macmillan, 1974
112 Crickmay, C. H.; *Work of the River*; Macmillan, 1974
114 Dury, G. H.; *Rivers and River Terraces*; Macmillan, 1970
116 Sugden, D. E. & B. S. John; *Glaciers and Landscape*; Arnold, 1976
118 John, B. S.; *Ice Age*; Collins, 1977
Cornwall, I. W.; *Ice Ages: Nature and Effects*; Baker, 1970
120 Cooke, R. U. & A. Warren; *Geomorphology in Deserts*; Batsford, 1973
122 Steers, J. A.; *Coastline of England and Wales*; Cambridge U.P., 1964
Steers, J. A.; *Coastline of Scotland*; Cambridge U.P., 1973
King, C. A. M.; *Beaches and Coasts*; Arnold, 1972
124 Ager, D. V.; *Nature of the Stratigraphical Record*; Macmillan, 1973
126 Seyfert, C. K. & L. A. Sirkin; *Earth History and Plate Tectonics*; Harper & Row

Kummel, B.; *History of the Earth*; W. H. Freeman, 1970
128–130 Tank, R.; *Focus on Environmental Geology*; Oxford U.P., 1976
Ager, D. V.; *Introducing Geology: Earth's Crust as History*; Faber, 1975
Earth's Resources
132–136 Hurlbut, C. S.; *Minerals and Man*; Thames & Hudson, 1969
McDivitt, J. F. & G. Manners; *Minerals and Men*; Johns Hopkins, 1974
Park, C. F. & R. A. Macdiarmid; *Ore Deposits*; W. H. Freeman, 1975
138 Chapman, K.; *North Sea Oil and Gas*; David & Charles, 1976
140 Darmstadter, J.; *Conserving Energy*; Johns Hopkins, 1976
142 Maddox, J.; *Beyond the Energy Crisis*; Hutchinson, 1975
Manners, G.; *Geography of Energy*; Hutchinson, 1971
Odell, P. R.; *Oil and World Power*; Penguin, 1975
Schurr, S. C. (Ed.); *Energy, Economic Growth and the Environment*; Johns Hopkins, 1976
Scott, D.; *World Resources: Energy*; Wheaton, 1976
Arvill, R.; *Man and Environment*; Penguin, 1969
Chisholm, M.; *Rural Settlement and Land Use*; Hutchinson, 1966
Chapman, J.; *Fuel's Paradise (Energy Options for Great Britain)*; Penguin, 1975
144 Foley, G. et al.; *Energy Question*; Penguin, 1976
Hoyle, F.; *Energy or Extinction: The Case for Nuclear Energy*; Heinemann, 1977
Commoner, B.; *Closing Circle*; Cape, 1971
Aldous, T.; *Battle for the Environment*; Collins, 1972
146 Brodine, V.; *Air Pollution*; Harcourt Brace, 1973
148 Barr, J.; *Derelict Britain*; Penguin, 1969
150–152 McCaull, J. & J. Crossland; *Water Pollution*; Harcourt Brace, 1974
Doxat, J.; *Living Thames: Restoration of a Great Tidal River*; Hutchinson, 1977
154 Nicholson, M.; *Environmental Revolution*; Penguin, 1972
Lambert, A. M.; *Making of the Dutch Landscape*; Seminar Press, 1971
156 Birch, G. B.; *Food Science*; Pergamon, 1962
Muller, H. G.; *Introduction to Food Rheology*; Heinemann, 1973
Pirie, N. W.; *Food Resources: Conventional and Novel*; Penguin, 1969
Barrass, R.; *Biology, Food and People*; Eng. U.P. (New Hodder & Stoughton Educ.), 1974
Agriculture
158 Fream, W.; *Elements of Agriculture*; Murray, 1972
160–162 Merriden, J. N.; *Farming*; Eng. U.P., 1974
Lockhart, J. A. R. & A. J. L. Wiseman; *Introduction to Crop Husbandry*; Pergamon, 1975
Purton, R. W.; *Farms and Farming*; Routledge, 1972
Norman, L. & R. B. Coote; *Farm Business*; Longman, 1971
Culpin, C.; *Profitable Farm Mechanization*; Lockwood, 1975
164 Culpin, C.; *Farm Machinery*; Lockwood, 1976
166 Press, F. & R. Seiver; *Earth*; W. H. Freeman, 1974
168 Withers, B. & S. Nipond; *Irrigation: Design and Practice*; Batsford, 1974
Israelson, O. W. & E. H. Vaughan; *Irrigation Principles and Practices*; Wiley, 1962
170 Shewell-Cooper, W. E.; *Soil, Humus and Health*; David & Charles, 1975
Tisdale, S. L. & W. L. Nelson; *Soil Fertility and Fertilizers*; Macmillan, 1966
172 Woods, A.; *Pest control: a survey*; McGraw-Hill, 1974
Irvine, D. E. G. & B. Knights; *Pollution and the Use of Chemicals in Agriculture*; Butterworth, 1974
174 Shewell-Cooper, W. F.; *Compost Gardening*; David & Charles, 1974
Hills, L. D.; *Grow your own Fruit and Vegetables*; Faber, 1971
Furner, B. G.; *Organic Vegetable Growing*; Macdonald, 1976
176 McNair, J. B.; *Studies in Plant Chemistry*; author, 1965
178 Fogg, G. E.; *Growth of Plants*; Penguin, 1963
Cultivated Plants
180 Kent, N. L.; *Technology of Cereals*; Pergamon, 1975
Scade, J.; *Cereals*; Oxford U.P., 1975
182 David, E.; *English Bread and Yeast Cookery*; Allen Lane, 1977
184–188 Loads, F. W.; *Vegetable Growing*; Gifford, 1973
Baker, H. G.; *Plants and Civilization*; Macmillan, 1965
Schery, R. W.; *Plants for Man*; Prentice-Hall, 1972
190 Norman, J.; *Salad Days*; Collins, 1975
192–194 Harrison, S. G. et al.; *Oxford Book of Food Plants*; Oxford U.P., 1969
196 Yudkin, J.; *Sugar*; Butterworth, 1971
Barnes, A. C.; *Sugar Cane*; Hill, 1974
198–206 Johnson, H.; *World Atlas of Wine*; Mitchell Beazley, 1977
Baillie, F.; *Beer Drinker's Companion*; David & Charles, 1975
Lichine, A.; *Encyclopaedia of Wines and Spirits*; Cassell, 1975
Ray, C.; *Wines of France*; Allen Lane, 1976
McDowall, R. J. S.; *Whiskies of Scotland*; Murray, 1971
Read, J.; *Wines of Spain and Portugal*; Faber, 1973
Broadbent, J. M.; *Wine Tasting*; Christie, Manson & Woods, 1974
Jackson, M.; *World Guide to Beer*; Mitchell Beazley, 1977
Jeffs, J.; *Wines of Europe*; Faber, 1971
208 Eden, T.; *Tea*; Longman, 1965
Haarer, A. E.; *Coffee Growing*; Oxford U.P., 1963
Wood, G. A. R.; *Cocoa*; Longman, 1975
210 Humphrey, S. W.; *Spices, Seasonings and Herbs*; Collier, 1973
David, E.; *Spices, Salt and Aromatics in the English Kitchen*; Penguin, 1970
Loewenfeld, C.; *Complete Book of Herbs and Spices*; David & Charles, 1974
212 Dampney, J. & E. Pomeroy; *All about Herbs*; Hamlyn, 1977
214 Eley, G.; *Wild Fruits and Nuts*; E.P., 1976
216 Simmons, N. W. (Ed.); *Evolution of Crop Plants*; Longman, 1976
218–220 Shirley, H. L.; *Forestry and its Career Opportunities*; McGraw-Hill, 1973
222 Mitchell, A.; *Field Guide to the Trees of Britain and N. Europe*; Collins, 1974
Flesh, Fish and Fowl
224 Park, R. D.; *Animal Husbandry*; Oxford U.P., 1970
Alderson, L.; *Observer's Book of Farm Animals*; Warne, 1976
226 Goodwin, D. H.; *Beef Management and Production*; Hutchinson Educ., 1977
Craplet, C.; *Dairy Cow*; Arnold, 1963
228 Porter, J. W. G.; *Milk and Dairy Foods*; Oxford U.P., 1975
Lyon, N. & P. Benton; *Eggs, Milk and Cheese*; Faber, 1971
230 Goodwin, D. H.; *Production and Management of Sheep*; Hutchinson Educ., 1972
Goodwin, D. H.; *Pig Management and Production*; Hutchinson Educ., 1972
232 Comte, M.-C.; *International meat Cookery*; Orbis, 1974
234 Ellis, J.; *Step by Step Guide to Meat Cookery*; Hamlyn, 1973
236 Kay, D.; *Poultry Keeping for Beginners*; David & Charles, 1977
Card, L. E. & M. C. Nesheim; *Poultry Production*; Lea, 1973
238 Allison, S.; *Chicken, Turkey, Duck and Game*; Collins, 1975
Black, M.; *Mrs. Beeton's Poultry and Game Cookery*; Ward Lock, 1974

240–248 Hardy, A. C.; *Open Sea: its Natural History 2. Fish and Fisheries*; Collins, 1971
Grigson, J.; *Fish Cookery*; David & Charles, 1973
Burke, H.; *Good Fish from the Sea*; Deutsch, 1965
Froud, N. & T. Lo; *International Fish Dishes*; Sphere, 1976
250 Cameron-Smith, M.; *Complete Book of Preserving*; Marshall Cavendish, 1976
Ellis, A.; *Complete Book of Home Freezing*; Hamlyn, 1974
Erlandson, K.; *Home Smoking and Curing*; Barrie & Jenkins, 1977
Dubbs, C. & D. Heberle; *Easy Art of Smoking Food*; Black, 1977
252 Pyke, M.; *Technological Eating*; Murray, 1972
Paul, P. C. & H. H. Palmer; *Food Theory and Applications*; Wiley, 1972

The Natural World

How Life Began
20 Orgel, L. E.; *Origins of Life: Molecules and Natural Selection*; Chapman & Hall, 1973
Bernal, J. D. & A. Synge; *Origin of Life*; Oxford U.P., 1972
22 Rhodes, F. H. T.; *Evolution of Life*; Penguin, 1969
Smith, J. M.; *Theory of Evolution*; Penguin, 1969
24 Rugieri, G.; *Earth before Adam*; Collins, 1973
Spinar, Z. V.; *Life before Man*; Thames & Hudson, 1972
Kurten, B.; *Ice Age*; Hart-Davis, 1972
26 Hanawalt & Haynes (Eds); *Chemical Basis of Life*; W. H. Freeman, 1973
Butler, J. A. V.; *Life Process*; Allen & Unwin, 1970
Rose, S.; *Chemistry of Life*; Penguin, 1970
Randle, P. J. & R. M. Denton; *Hormones and Cell Metabolism*; Oxford U.P., 1974
28 Woods, R. A.; *Biochemical Genetics*; Chapman & Hall, 1973
Watson, J. D.; *Double Helix*; Weidenfeld & Nicolson, 1968
Levine, R. P.; *Genetics*; Holt, Rinehart & W. 1968
30 Gurdon, J. B.; *Gene Expression during Cell Differentiation*; Oxford U.P., 1973
32 Moorehead, A.; *Darwin and the Beagle*; Penguin, 1971
Vorzimmer, P. J.; *Charles Darwin: Years of Controversy*; Univ. of London Press, 1972
Darwin, C.; *Origin of the Species*; Dent, 1972
34 Lack, D.; *Ecological Isolation in Birds*; Blackwell, 1971
36 Postgate, J.; *Microbes and Man*; Penguin, 1969
Primrose, S. B.; *Introduction to Modern Virology*; Blackwell, 1974
Fenner, F.; *Biology of Animal Viruses*; Academic Press, 1974

Plants
38 Cronquist, A.; *Introductory Botany*; Harper & Row, 1971
Lowson, J. M.; *Textbook of Botany*; U.T.P., 1973
Heywood, V. H.; *Plant Taxonomy*; Arnold, 1967
40 Duddington, C. L.; *Beginner's Guide to the Fungi*; Pelham, 1972
Burnett, J. H.; *Fundamentals of Mycology*; Arnold, 1968
Webster, J.; *Introduction to Fungi*; Cambridge U.P., 1970
42 Hvass, E. & H. Hvass; *Mushrooms and Toadstools*; Blandford, 1961
Ramsbottom, J.; *Mushrooms and Toadstools*; Collins, 1953
Tribe, I.; *Mushrooms in the Wild*; Orbis, 1977
Brightman, F. H.; *Oxford Book of Flowerless Plants*; Oxford U.P., 1966
44 James, W. O.; *Introduction to Plant Physiology*; Oxford U.P., 1973
Meyer, B. S.; *Introduction to Plant Physiology*; Van Nost. Reinhold, 1973
Hall, D. O. & K. K. Rao; *Photosynthesis*; Arnold, 1972
46 Morris, I.; *Introduction to the Algae*; Hutchinson, 1967
48 Hale, M. E.; *Biology of Lichens*; Arnold, 1974
Richardson, D.; *Vanishing Lichens: History, Biology and Importance*; David & Charles, 1975
50 Watson, E. V.; *British Mosses and Liverworts*; Cambridge U.P., 1968
52 Hyde, H. A. & A. E. Wade; *Welsh Ferns*; Nat. Mus. Wales, 1969
Taylor, P.; *British Ferns and Mosses*; Eyre, 1960
54–56 Sporne, K. R.; *Morphology of Gymnosperms*; Hutchinson, 1967
Johnson, H.; *International Book of Trees*; Mitchell Beazley, 1973
Mitchell, A.; *Field Guide to the Trees of Britain and N. Europe*; Collins, 1974
58–62 Perry, F.; *Flowers of the World*; Hamlyn, 1972
Polunin, O.; *Concise Flowers of Europe*; Oxford U.P., 1972
Polunin, O.; *Flowers of Europe: a Field Guide*; Oxford U.P., 1969
Step, E.; *Wayside and Woodland Blossoms*; Warne, 1963 (3 vols)
Street, H. E. & H. Opik; *Physiology of Flowering Plants*; Arnold, 1976
Stringer, M.; *Wild Flowers*; Bartholomew, 1973
64 Polunin, O.; *Trees and Bushes of Europe*; Oxford U.P., 1976
Pokorny, J.; *Colour Guide to Familiar Trees*; Octopus, 1974
Procter, R.; *Trees of the World*; Hamlyn, 1972
66 Lindley, J.; *Sertum Orchidaceum*; Johnson, 1975
Oplt, J.; *Orchids*; Hamlyn, 1970
68 Hubbard, C. E.; *Grasses: A Guide*; Penguin, 1968
Langer, R. H. M.; *How Grasses Grow*; Arnold, 1972
70 Proctor, M. & P. Yeo; *Pollination of Flowers*; Collins, 1972
Sporne, K. R.; *Morphology of Angiosperms*; Hutchinson, 1975

Animals
72 *Classification of the Animal Kingdom*; Eng. U.P., 1973
74 Alexander, R. M.; *Chordates*; Cambridge U.P., 1975
Buchsbaum, R.; *Animals without Backbones*; Penguin, 1971 (2 vols)
76 Wickler, W.; *Sexual Code*; Weidenfeld & Nicolson, 1974
Austin, C. R. & R. V. Short; *Reproduction in Mammals*; Cambridge U.P., 1972–76 (6 vols)
78 Ardrey, R.; *Social Contract*; Collins, 1970
Tinbergen, N.; *Social Behaviour in Animals*; Chapman & Hall, 1965
80 Vickerman, K. & F. E. G. Cox; *Protozoa*; Murray, 1967
Sleigh, M. A.; *Biology of Protozoa*; Arnold, 1973
82–84 Hardy, A. C.; *Open Sea: Its Natural History Vol 1: World of Plankton*; Collins, 1970
Parks, P. D.; *World You Never See – Underwater Life*; Hamlyn, 1976
86 Erasmus, D. A.; *Biology of Trematodes*; Arnold, 1972
Wright, C. A., *Flukes and Snails*; Allen & Unwin, 1971
88 Dales, R. P.; *Annelids*; Hutchinson, 1967
Edwards, C. A. & J. R. Lofty; *Biology of Earthworms*; Chapman & Hall, 1977
90 Burgess, C. M.; *Living Cowries*; Barnes, 1970
Ellis, A. E.; *British Snails*; Oxford U.P., 1926
Runham, N. W. & P. J. Hunter; *Terrestrial Slugs*; Hutchinson, 1970
92 Morton, J. E.; *Molluscs*; Hutchinson, 1967
Solem, A.; *Shell Makers*; Wiley, 1974
Yonge, C. M. & T. E. Thompson; *Living Marine Molluscs*; Collins, 1976

94 Cousteau, J.-Y.; *Octopus and Squid: Soft Intelligence*; Cassell, 1973
96 Forsyth, W. S.; *Common British Seashells*; Black, 1961
Scase, R. & E. Storey; *World of Shells*; Osprey, 1975
Melvin, A. G.; *Seashells of the World*; Tuttle, 1966
Oliver, A. P. H.; *Hamlyn Guide to Shells of the World*; Hamlyn, 1975
98 Clarke, A. U.; *Biology of the Arthropoda*; Arnold, 1973
100 Warner, G. F.; *Biology of Crabs*; Elek, 1977
102 Bristowe, W. S.; *World of Spiders*; Collins, 1971
104 Schmitt, W. L.; *Crustaceans*; David & Charles, 1973
Kaestner, A.; *Invertebrate Zoology*; Interscience 1967–70 (3 vols)

Insects
106 Farb, P.; *Insects*; Time-Life International, 1969
Imms, A.; *Insect Natural History*; Fontana, 1973
108 Oldroyd, H.; *Insects and their World*; Brit. Mus. (Natural History), 1973
Tweedie, M.; *All Colour Book of Insects*; Octopus, 1973
110 Thomas, J. G.; *Dissection of the Locust*; Witherby, 1963
112 Askew, R. R.; *Parasitic Insects*; Heinemann, 1971
114 Bechyne, J.; *Beetles*; Thames & Hudson, 1956
Evans, G.; *Life of Beetles*; Allen & Unwin, 1975
Lyneborg, L.; *Beetles in Colour*; Blandford, 1977
116 Lewis, H. L.; *Butterflies of the World*; Harrap, 1973
Brian, M. V.; *Ants*; Collins, 1977
Stradbery, J. P.; *Wasps*; Sidgwick, 1973
Butler, C. G.; *World of the Honeybee*; Collins, 1974
118 Clark, A. M.; *Starfishes and their Relations*; British Museum, 1968
120 Barrington, E. J. W.; *Invertebrate Structure and Function*; Nelson, 1967
Buchsbaum, R.; *Animals without Backbones*; Penguin, 1971 (2 vols)
Sutton, S. L.; *Invertebrate Types: Woodlice*; Ginn, 1972
Nichols, D. & J. A. L. Cooke; *Oxford Book of Invertebrates*; Oxford U.P., 1971
Barnes, R. D.; *Invertebrate Zoology*; Saunders, 1974
122 Waterman, A. J.; *Chordate Structure and Function*; Macmillan, 1971
Nixon, M.; *Oxford Book of Vertebrates*; Oxford U.P., 1972
Yapp, W. B.; *Vertebrates: Structure and Life*; Oxford U.P., 1966

Fish
124 Norman, J. R.; *History of Fishes*; Benn, 1975
126–130 Marshall, N. B.; *Life of Fishes*; Weidenfeld & Nicolson, 1965
Hardy, Sir A.; *Open Sea: Its Natural History* Vol. 2 *Fish & Fisheries*; Collins, 1971
Ommaney, F. D.; *Draught of Fishes*; Longman, 1965
Frost, W. E. & M. E. Brown; *Trout*; Collins, 1967
Herald E. S.; *Living Fishes of the World*; Hamish Hamilton, 1961
Sterba, G.; *Freshwater Fishes of the World*; Studio Vista, 1974 (2 vols)
Lindberg, G.; *Fishes of the World: Families and Checklist*; I.P.S.T., 1974
Whitehead, P. J. P.; *How Fishes Live*; Elsevier, 1975

Amphibians and Reptiles
132 Frazer, J. F. D.; *Amphibians*; Wykeham, 1972
134 Bellairs, A. d'A. & J. Attridge; *Reptiles*; Hutchinson Educ., 1975
136–138 Stidworthy, J.; *Snakes of the World*; Hamlyn, 1969
Simms, C.; *Lives of British Lizards*; Goose, 1970
Goode, J.; *Tortoises, Terrapins and Turtles*; Angus, 1971
Goin, C. J. & O. B. Goin; *Introduction to Herpetology*; W. H. Freeman, 1971
Richardson, M.; *Fascination of Reptiles*; Deutsch, 1972

Birds
140 Gooders, J.; *Birds: Illustrated Survey of Bird Families of the World*; Hamlyn, 1975
Peterson, R. et al.; *Field Guide to the Birds of Britain and Europe*; Collins, 1966
Austin, O. L.; *Birds of the World*; Hamlyn, 1968
142 Marshall, A. J.; *Biology and Comparative Physiology of Birds*; Academic Press, 1960
144 Harrison, C.; *Field Guide to the Nests, Eggs and Nestlings of British and European Birds*; Collins, 1975
146 Matthews, G. V. T.; *Bird Navigation*; Cambridge U.P., 1968
Dorst, J.; *Migrations of Birds*; Heinemann, 1962
148 Lorenz, K.; *King Solomon's Ring*; Methuen, 1952
Fisher, J. & J. Flegg; *Watching Birds*; Poyser, 1974
150 Welty, C. J.; *Life of Birds*; Constable, 1964
Nicolai, J.; *Bird Life*; Putnam, 1974
Perrins, C. & A. D. Cameron; *Bird Life: Introduction to the World of Birds*; Elsevier, 1976
Lack, D.; *Life of the Robin*; Witherby, 1965
Tinbergen, N.; *Herring-gull's World*; Collins, 1953
Fry, C. H. & J. Flegg (Eds); *World Atlas of Birds*; Mitchell Beazley, 1974

Mammals
154 Morris, D.; *Mammals*; Hodder & Stoughton, 1965
156 Matthews, L. H.; *Life of Mammals*; Weidenfeld & Nicolson, 1974 (2 vols)
158 Biscoe, H.; *Life of Marsupials*; Arnold, 1973
Ride, W. D. E.; *Guide to the Native Mammals of Australia*; Oxford U.P., 1970
160 Delany, M. J.; *Ecology of Small Mammals*; Arnold, 1974
Sheail, J.; *Rabbits and their History*; David & Charles, 1971
Jewell, P. A. (Ed.); *Island Survivors*; Athlone Press, 1974
162 Douglas-Hamilton, I. & O. Douglas-Hamilton; *Among the Elephants*; Collins, 1976
Geist, V.; *Mountain Sheep: Behaviour and Evolution*; Univ. Chicago Press, 1976
164 Schaller, G. B.; *Serengeti Lion*; Univ. Chicago Press, 1972
Kruuk H.; *Spotted Hyaena*; Univ. Chicago Press, 1972
166 Matthews, L. H. (Ed.); *Whale*; Allen & Unwin, 1969
Gaskin, D. E.; *Whales, Dolphins and Seals*; Heinemann Educ., 1973
168 Napier, J. R. & P. H. Napier; *Handbook of Living Primates*; Academic Press, 1967
170 Moss, C.; *Portraits in the Wild*; Hamish Hamilton, 1976
Bateson, P. P. G.; *Growing Points in Ethology*; Cambridge U.P., 1976
Ewer, R. F.; *Ethology of Mammals*; Elek, 1973
172 Jolly, A.; *Evolution of Primate Behaviour*; Collier-Macmillan, 1972
Goodall, J.; *In the Shadow of Man*; Fontana, 1974

Prehistoric Animals and Plants
174–180 Spoczynaska, J. O.; *Fossils: a Study in Evolution*; Muller, 1971
Hamilton, R.; *Fossils and Fossil-Collecting*; Hamlyn, 1975
Middlemiss, F. A.; *Fossils*; Allen & Unwin, 1969
Rudwick, M. J. S.; *Meaning of Fossils*; Macdonald, 1972
182 Orgel, L. E.; *Origins of Life: Molecules and Natural Selection*; Chapman & Hall, 1973
Butler, J. A. V.; *Life Process*; Allen & Unwin, 1970
Grobstein, C.; *Strategy of Life*; W. H. Freeman, 1974

184–186 Cox, B.; *Prehistoric Animals*; Purnell, 1976
Colbert, E. H.; *Men and Dinosaurs*; Penguin, 1971
Halstead, L. B.; *Evolution and Ecology of the Dinosaurs*; Lowe, 1975
Tweedie, M.; *World of Dinosaurs*; Weidenfeld & Nicolson, 1977
188–190 Kurten, B.; *Age of Mammals*; Weidenfeld & Nicolson, 1971
Animals and their Habitats
192 Illies, J.; *Introduction to Zoogeography*; Macmillan, 1974
Cox, C. B. et al.; *Biogeography*; Blackwell Science, 1976
Elton, C. S.; *Pattern of Animal Communities*; Methuen, 1966
194 Wilson, E. O. (Ed.); *Ecology, Evolution and Population Biology*; W. H. Freeman, 1974
Grzimek, B.; *Encyclopaedia of Ecology*; Van Nost. Reinhold, 1977
Colinvaux, P. A.; *Introduction to Ecology*; Wiley, 1973
Milne, L. & M. Milne; *Nature of Life*; Cassell, 1971
Lack, D.; *Evolution Illustrated by Waterfowl*; Blackwell Science, 1974
196 Lack, D.; *Ecological Isolation in Birds*; Blackwell Science, 1971
Lack, D.; *Island Biology Illustrated by the Land Birds of Jamaica*; Blackwell Science, 1976
198 Spedding, C. R. W.; *Grassland Ecology*; Oxford U.P., 1971
200 Dyer, A.; *Classic African Animals*; Winchester, 1973
Burton, J.; *Animals of the African Year*; Lowe, 1972
Brown, L.; *Africa: A Natural History*; Hamish Hamilton, 1976
Myers, N.; *Long African Day*; Collier-Macmillan, 1973
Grzimek, B.; *Among Animals of Africa*; Collins, 1970
Togawa, Y.; *African Animals*; Ward Lock, 1971
202 Morrison, T.; *Land Above the Clouds*; Deutsch, 1974
204 Bell, A.; *Common Australian Birds*; Oxford U.P., 1969
Morcombe, M. K.; *Illustrated Encyclopaedia of Australian Wildlife*; Macmillan, 1974
Kolar, K.; *Continent of Curiosities*; Souvenir Press, 1969
206–208 Bellamy, D. & M. Boorer; *Green Worlds*; Aldus, 1975
Gohl, H. & E. Krebs; *Living Forests*; Kaye & Ward, 1975
Cousens, J.; *Woodland Ecology*; Oliver & Boyd, 1975
Duffey, E. et al.; *Grassland Ecology and Wildlife Management*; Chapman & Hall, 1974
210 Breeden, S. & K. Breeden; *Living Marsupials*; Collins, 1970
Tyndale-Biscoe, H.; *Life of Marsupials*; Arnold, 1973
212–216 Longman, K. A. & J. Jenik; *Tropical Forest and its Environment*; Longman, 1974
England, M. D.; *Birds of the Tropics*; Hamlyn, 1974
Ripley, S. D.; *Land and Wildlife of Tropical Asia*; Time-Life International, 1966
Stone, R. H.; *Tropical Nature Study*; Cambridge U.P., 1964
Richards, P. W.; *Tropical Rain Forests: Ecological Study*; Cambridge U.P., 1952
220 Cloudsley-Thompson, J.; *Desert*; Orbis, 1977
Leopold, A. S.; *Desert*; Time-Life International, 1969
222 Darling, F. & J. Morton-Boyd; *Highlands and Islands*; Collins, 1969
224–226 Freuchen, P.; *Arctic Year*; Cape, 1959
Herbert, W.; *Polar Deserts*; Collins, 1971
Stonehouse, B.; *Biology of Penguins*; Macmillan, 1975
Stonehouse, B.; *Animals of the Antarctic*; Lowe, 1972
Hicks, J. L.; *Arctic Lands*; Hamish Hamilton, 1976
230 Macan, T. T. & E. B. Worthington; *Life in Lakes and Rivers*; Collins, 1974
Clegg, J.; *Freshwater Life*; Warne, 1974
234 Weber, K. & I. Hoffman; *Camargue*; Harrap, 1971
236 Yonge, C. M.; *Seashore*; Collins, 1963
238–240 Carson, R.; *Sea Around Us*; Panther, 1965
Russell, Sir F. S. & Sir C. M. Yonge; *Seas*; Warne, 1975
Hardy, A. C.; *Great Waters*; Collins, 1967
Marshall, N. & O. Marshall; *Ocean Life*; Blandford, 1971
Saunders, D.; *Seabirds*; Hamlyn, 1971
Scheffer, V. B.; *Year of the Whale*; Penguin, 1973
Conservation
242–252 *International Zoo Year Book*; Zoological Society, 1977
Littlewood, C.; *World's Vanishing Birds*; Foulsham, 1972
Simon, N. & Geroudet, P.; *Last Survivors: Natural History of 48 Animals in Danger of Extinction*; Stephens, 1970
Arvill, R.; *Man and Environment*; Penguin, 1969
International Zoo Year Book; Zoological Society, 1973
Fisher, J.; *Red Book*; Collins, 1969

Man and Society

Evolution of Man
General
Young, J. Z.; *Introduction to the Study of Man*; Oxford U.P., 1974
Bronowski, J.; *Ascent of Man*; B.B.C., 1976
20–34 Pilbeam, D.; *Ascent of Man*; Collier-Macmillan, 1972
Ardrey, R.; *Hunting Hypothesis*; Collins, 1976
Day, M. H.; *Guide to Fossil Man*; Cassell, 1977
Klein, R. G.; *Ice-Age Hunters of the Ukraine*; Univ. Chicago Press, 1973
Clark, G.; *Stone Age Hunters*; Thames & Hudson, 1967
Turnbull, C. M.; *Man in Africa*; David & Charles, 1976
Bender, B.; *Farming in Pre-History*; J. Baker, 1975
Breuil, H. P. E. & R. Lantier; *Men of the Old Stone Age*; Harrap, 1965
Mellersh, H. E. L.; *Story of Man: Human Evolution to end of Stone Age*; Hutchinson, 1959
Leakey, R. & R. Lewin; *Origins*; Macdonald, 1977
Cole, S.; *Races of Man*; British Museum (Natural History), 1965
Harrison, G. A. et al.; *Human Biology*; Oxford U.P., 1977
How Your Body Works
36 Blakelock, C. B.; *Mechanics of the Mind*; Cambridge U.P., 1977
Teyler, T. J.; *Primer of Psychobiology: Brain and Behaviour*; W. H. Freeman, 1975
Eccles, J. C.; *Understanding the Brain*; McGraw-Hill, 1973
Luria, A. R.; *Working Brain: Introduction to Neuropsychology*; Penguin, 1973
38 Bowsher, D.; *Introduction to the Anatomy and Physiology of the Nervous System*; Blackwell, 1970
Noback, C. R. & R. J. Demarest; *Nervous System: Introduction and Review*; McGraw-Hill, 1967
Gibson, J.; *Guide to the Nervous System*; Faber, 1974
Dunkerley, G. B.; *Basic Atlas of the Human Nervous System*; F. A. Davis, 1975
40 Bartlett, Sir F.; *Remembering*; Cambridge U.P., 1967
42 Gregory, R. L.; *Eye and Brain*; Weidenfeld & Nicolson, 1972

Gregory, R. L.; *Intelligent Eye*; Weidenfeld & Nicolson, 1970
Oswald, I.; *Sleep*; Penguin, 1970
Oswald, I.; *Sleeping and Waking*; Elsevier, 1962
44–46 Hunter, I. M. L.; *Memory*; Penguin, 1964
Meredith, P.; *Learning, Remembering and Knowing*; Eng. U.P., 1961
Norman, D. A.; *Memory and Attention: Introduction to Human Information Processing*; Wiley, 1969
48–52 Davson, H. & M. B. Segal (Eds.); *Introduction to Physiology*; Academic Press, 1975–76 (3 vols)
Geldard, F. A.; *Human Senses*; Wiley, 1972
54 CIBA Foundation Symposium; *Taste and Smell in Vertebrates*; Churchill Livingstone, 1970
56 Gray, H.; *Anatomy*; Longman, 1973
Ham, A. W.; *Histology*; Lippincott, 1974
Lumley, J. S. P. et al.; *Essential Anatomy*; Churchill Livingstone, 1975
58 Bourne, G. H. (Ed.); *Structure and Function of Muscle*; Academic Press, 1973–74 (4 vols)
60 Montagna, W.; *Structure and Function of Skin*; Academic Press, 1974
62 Richardson, D. R.; *Basic Circulatory Physiology*; Little, 1976
Navaratnam, V.; *Human Heart and Circulation*; Academic Press, 1975
Gardner, A.; *Heart and Blood*; Hart-Davies, 1974
64 Catt, K. J.; *A. B. C. of Endocrinology*; Lancet, 1971
Malkinson, A. M.; *Hormone Action*; Chapman & Hall, 1975
Marks, J.; *Vitamins in Health and Disease*; Churchill Livingstone, 1968
66 Nagaishi, C.; *Functional Anatomy and Histology of the Lung*; Univ. Park, 1972
68 McMinn, R. M. H. & M. H. Hobdell; *Functional Anatomy of the Digestive System*; Pitman, 1974
70 Bender, A. E.; *Facts of Food*; Oxford U.P., 1975
Fleck, H.; *Introduction to Nutrition*; Macmillan, 1971
Gaman, P. M. & K. B. Sherrington; *Science of Food*; Pergamon, 1977
Pyke, M.; *Man and Food*; Weidenfeld & Nicolson, 1970
72 Royal Canadian Air Force; *Physical Fitness*; Penguin, 1970
74 Page, E. W.; *Human Reproduction*; Saunders, 1976
Comfort, Alex; *Joy of Sex*; Quartet, 1974
Comfort, Alex; *More Joy of Sex*; Mitchell Beazley, 1975
76 Browne, F. J.; *Advice to the Expectant Mother on the Care of her Health and Child*; Churchill Livingstone, 1966
78 Llewellyn-Jones; *Everywoman: Gynaecological Guide for Life*; Faber, 1972
Nilsson, L. et al.; *Everyday Miracle: Child is Born*; A. Lane, 1971
80 Barnes, J.; *Essentials of Family Planning*; Blackwell Science, 1976
Illness and Health
82 Rose, L.; *Health and Hygiene*; Batsford, 1975
Levin, L. S.; *Self-Care: Lay Initiatives in Health*; Croom, 1977
Paul, B. D. & W. B. Miller; *Helath, Culture and Community*; Sage Foundation, 1955
84–86 Fox, J. P. et al.; *Epidemiology: Man and Disease*; Collier-Macmillan, 1970
88 Bouhuys, A.; *Breathing: Physiology, Environment and Lung Disease*; Grune & Stratton, 1974
Geschickter, C. F.; *Lung in Health and Disease*; Lippincott, 1974
Mordrak, M.; *Better Living and Breathing: Manual (Symposium)*; Mosby, 1975
90 Chung, E. K.; *Quick Reference to Cardiovascular Diseases*; Lippincott, 1977
Conn, L. & O. Horwitz; *Cardiac Diseases*; Lea, 1971 (2 vols)
92 Langman, M. J. S.; *Concise Textbook of Gastroenterology*; Churchill, 1973
Truelove, S. C. & P. C. Reynell; *Diseases of the Digestive System*; Blackwell, 1972
94 Rosse, C. & D. K. Clawson; *Introduction to the Musculo-skeletal System*; Harper, 1970
Salter, R. B.; *Textbook of Disorders and Injuries of the Musculo-skeletal System*; Williams & Wilkins, 1970
96 Walton, J. N.; *Essentials of Neurology*; Pitman, 1975
Lyons, J. B.; *Primer of Neurology*; Butterworth, 1974
Brain, W.; *Diseases of the Nervous System*; Oxford U.P., 1977
98 Sauer, G. C.; *Manual of Skin Diseases*; Lippincott, 1973
Coles, R. B. & P. D. C. Kinmont; *Skin Diseases for Beginners*; H. K. Lewis, 1957
100 Dillon, R. S.; *Handbook of Endocrinology*; Lea, 1973
Thomson, J. A.; *Introduction to Clinical Endocrinology*; Churchill, 1976
102 Rous, S. N.; *Understanding Urology*; Karger, 1973
Sturdy, D. E.; *Essentials of Urology*; Wright, 1974
104 Forrest, G. G.; *Diagnosis and Treatment of Alcoholism*; Thomas, 1975
McCord, M. & J. McCord; *Origins of Alcoholism*; Tavistock, 1960
Inglis, B.; *Forbidden Game: Social History of Drugs*; Hodder, 1975
Szasz, T. S.; *Ceremonial Chemistry*; Routledge, 1975
106–108 Brockington, F.; *World Health*; Churchill, 1967
Grant, M. P.; *Biology and World Health*; Abelard, 1970
W.H.O.; *Alternative approaches to meeting basic health needs in developing countries*; W.H.O., 1975
112 Simpson, J.; *Introduction to Preventive Medicine*; Heinemann, 1970
Schwartz, K.; *Preventive Medicine in Medical Care*; Lewis, 1970
114 Fitch, K. L.; *Life, Science and Man: Biological Approach to Health*; Holt, 1973
116 Singer, C. & E. A. Underwood; *Short History of Medicine*; Oxford U.P., 1962
Venzmer, G.; *5,000 Years of Medicine*; Macdonald, 1972
King, L. S.; *History of Medicine: Selected Readings*; Penguin, 1971
118 Benjamin, H.; *Everybody's Guide to Nature Cure*; Thorsons, 1961
Fluck, H.; *Medicinal Plants and their Uses*; Foulsham, 1976
120 Burn, J. H.; *Background of Therapeutics*; Oxford U.P., 1948
Lewis, J. G.; *Therapeutics*; Eng. U.P., 1972
122 Shivananda Sanswati; *Yogic Therapy*; Bramachari Yogeswar, 1969
Maple, E.; *Ancient Art of Occult Healing*; Aquarian, 1974
Manaka, Y. & I. A. Urguhart; *Layman's Guide to Acupuncture*; Weatherhill, 1972
124 John, D. H. O.; *Radiographic Processing in Medicine and Industry*; Focal, 1967
126–128 Cole, W. H.; *Textbook of Surgery*; Butterworth, 1970
Illingworth, C.; *Short Textbook of Surgery*; Churchill, 1972
130 Garfield, S.; *Teeth, Teeth, Teeth*; Arlington, 1972
Charbeneau, G. T. et al.; *Principles and Practices of Operative Dentistry*; Lea, 1975
Pickard, H. M.; *Manual of Operative Dentistry*; Oxford U.P., 1976
132 Docarmo, P. B. & A. T. Patterson; *First Aid Principles and Procedures*; Prentice, 1976
Harrison, R.; *First Aid*; Hodder, 1976
St. John Ambulance; *First Aid Manual*; British Red Cross, 1972
Mental Health
134 Laing, R. D.; *Self and Others*; Penguin, 1971
Szasz, T. S.; *Ideology and Insanity*; Penguin, 1974
Hays, P.; *New Horizons in Psychiatry*; Penguin, 1971

136 Braginsky, B. M.; *Methods of Madness*; Holt, 1969
Copas, J. B.; *Treatment Settings in Psychiatry*; Kimpton, 1974
Szasz, T. S.; *Manufacture of Madness*; Routledge, 1971
Goffman, E.; *Asylums*; Penguin, 1968
138 Bastide, R.; *Sociology of Mental Disorder*; Routledge, 1972
Gordon, R. G.; *Abnormal Behaviour*; Medical, 1946
Hamilton, M.; *Abnormal Psychology*; Penguin, 1967
Maher, B.; *Contemporary Abnormal Psychology*; Penguin, 1973
140 Rycroft, C.; *Anxiety and Neurosis*; Penguin, 1968
Furst, J. B.; *Neurotic: Inner and Outer World*; Citadel, 1954
142 Wolff, S.; *Children under Stress*; Penguin, 1973
Penrose, L. S.; *Biology of Mental Defect*; Sidgwick, 1972
Eastham, R. D. & J. Jancar; *Clinical Pathology in Mental Retardation*; Wright, 1968
Eysenck, H. J. & S. Rachman; *Causes and Cures of Neurosis*; Routledge, 1965
144–146 Rachman, S. & J. Teasdale; *Aversion Therapy and Behaviour Disorders*; Routledge, 1969
Yates, A. J.; *Behaviour Therapy*; Wiley, 1970
148 Evans, C. F.; *Cults of Unreason*; Panther, 1974

Human Development
150 Ausubel, D. P. & E. Sullivan; *Theory and Problems of Child Development*; Grune & Stratton, 1970
152 Vernon, P. E.; *Intelligence and Cultural Environment*; Methuen, 1972
Vernon, M. D.; *Reading and its Difficulties*; Cambridge U.P., 1971
Tanner, J. M.; *Growth at Adolescence*; Blackwell Science, 1969
154 Crystal, D., P. Fletcher & M. Garmon; *Grammatical Analysis and Language Disability*; Arnold, 1976
Brown, R.; *First Language: The Early Stages*; Penguin, 1976
Bruner, J. S. et al.; *Study of Thinking*; Wiley
156–160 Bakwin, H. & R. M. Bakwin; *Behaviour Disorders in Children*; Saunders, 1972
Breckenridge, M. E. & E. L. Vincent; *Child Development*; Saunders, 1965
Carmichael, L. (Ed.); *Manual of Child Psychology*; Wiley, 1970 (2 vols)
Hurlock, E. B.; *Child Development*; McGraw-Hill, 1972
Illingworth, R. S.; *Development of the Infant and Young Child*; Churchill Livingstone, 1975
Illingworth, R. S.; *Normal Child: . . . Problems . . .*; Churchill Livingstone, 1975
Kahn, J. H.; *Human Growth and the Development of Personality*; Pergamon, 1974
Watson, E. H. & G. H. Lowrey; *Growth and Development of Children*; Chicago Year Book Medical Publishers, 1967
Smart, M. S. & R. C. Smart; *Pre-School Children: Development and Relationships*; Collier-Macmillan, 1973
162–168 Coleman, J. C.; *Relationships in Adolescence*; Routledge, 1974
Meyerson, S.; *Adolescence: Crises of Adjustment*; Allen & Unwin, 1975
Mead, M.; *Culture and Commitment: The Generation Gap*; Bodley Head, 1970
168–170 Bernard, J.; *Future of Marriage*; Penguin, 1976
Chesser, E.; *Is Marriage Necessary?*; W. H. Allen, 1974
Levi-Strauss, C.; *Elementary Structures of Kinship*; Eyre, 1969
172–178 Welford, A. T.; *Ageing and Human Skill*; Oxford U.P., 1958
Fogarty, M. P.; *40–60: How We Waste the Middle-Aged*; Bedford Square, 1975
Wallis, J. H.; *Challenge of Middle Age*; Routledge, 1962
Bromley, D. B.; *Psychology of Human Ageing*; Penguin, 1974
Chown, S. M.; *Human Ageing*; Penguin, 1972
Meares, A.; *Why Be Old?*; Collins, 1975
Comfort, A.; *Good Age*; Mitchell Beazley, 1977
180 Hinton, J.; *Dying*; Penguin, 1967
Kastenbaum, R. & R. Aisenberg; *Psychology of Death*; Duckworth, 1974
Matse, J.; *Bereavement: 3 Aspects of Mourning*; Lutterworth, 1971
182 Tribe, D.; *Nucleoethics*; Paladin, 1973
Bresler, F. S.; *Reprieve: Study of a System*; Harrap, 1965
Blom-Cooper, L.; *Hanging Question: Essays*; Duckworth, 1969
184–188 Eysenck, H. J.; *Structure of Human Personality*; Methuen, 1970
Lowe, G. R.; *Growth of Personality from Infancy to Old Age*; Penguin, 1972
Ruddock, R.; *6 Approaches to the Person*; Routledge, 1972
Freud, S.; *Outline of Psychoanalysis*; Hogarth Press, 1969
Morris, D.; *Naked Ape*; Cape, 1967, Mayflower, 1977
Mischel, W.; *Introduction to Personality*; Holt, 1976
190 MacPherson, C. B.; *Political Theory of Possessive Individualism: Hobbes/Locke*; Oxford U.P., 1962
Buchanan, J. M.; *Limits of Liberty: Between Anarchy and Leviathan*; Univ. Chicago Press, 1975
192–194 Humphreys, C.; *Search Within: Course in Meditation*; Sheldon, 1977
Dhiravamsa, V. R.; *Way of Non-Attachment*; Turnstone, 1975
Watts, A. W.; *Meditation*; Celestial Arts, 1974
Thomson, R.; *Psychology of Thinking*; Penguin, 1959

Man and his Gods
196 Devereux, G.; *Psychoanalysis and the Occult*; Souvenir, 1974
Hunt, D.; *Handbook of the Occult*; Barker, 1967
198 Randall, J. L.; *Parapsychology and the Nature of Life*; Souvenir, 1975
Hisey, L.; *Key to Inner Space: Occultism, Metaphysics and the Transcendental*; Julian, 1974
200 Parker, D. & J. Parker; *Compleat Astrologer*; Mitchell Beazley, 1971
Lindsay, J.; *Origins of Astrology*; F. Muller, 1971
Carter, C. E. O.; *Astrological Aspects*; Fowler, 1960
Mayo, J.; *Teach Yourself Astrology*; Eng. U.P., 1964
Gauquelin, M.; *Astrology and Science*; Davies, 1970
202–204 Bocock, R.; *Ritual in Industrial Society*; Allen & Unwin, 1973
Firth, R.; *Symbols: Public and Private*; Allen & Unwin, 1973
206–214 Frazer, J. G.; *Golden Bough*; Macmillan, 1957,
Hooke, S. H.; *Middle Eastern Mythology*; Penguin, 1963
Everyman; *Dictionary of Non-Classical Mythology*; Dent, 1962
Eliade, M.; *Myths, Dreams and Mysteries*; Collins, 1968
Graves, R.; *White Goddess*; Faber, 1952
Graves, R.; *Greek Myths*; Penguin, 1967
Bulfinch, T.; *Mythology*; Spring, 1968
Campbell, J.; *Myths to Live By*; Souvenir, 1972
216 Zaehner, R. C.; *Concise Encyclopaedia of Living Faiths*; Hutchinson, 1971
218 Heaton, E. W.; *Hebrew Kingdoms*; Oxford U.P., 1968
Anderson, B. W.; *Living World of the Old Testament*; Longman, 1971
Vaux, R. de; *Ancient Israel*; Darton, 1973
Eichrodt, W.; *Theology of the Old Testament*; SCM Press, 1961, 1967 (2 vols)
Steinsaltz, A.; *Essential Talmud*; Weidenfeld & Nicolson, 1976

Schweitzer, A.; *Quest of the Historical Jesus*; Black, 1954
Kee, R. & J. Young; *Living World of the New Testament*; Darton, 1972
Kung, H.; *On Being a Christian*; Collins, 1977
Early Christian Writings; Penguin, 1968
Grant, M.; *Jesus*; Weidenfeld & Nicolson, 1977
Copleston, F. C.; *Aquinas*; Penguin, 1970
220 Agency for Cultural Affairs; *Japanese Religion*; Kodansha International (Allen & Unwin), 1973
Ross, N.W.; *Buddhism, Hinduism, Zen: Introduction to Their Meaning and Their Arts*; Faber, 1973
Ameer Ali, S.; *Spirit of Islam*; Chatto, 1923
Klein, F. A.; *Religion of Islam*; Curzon Press, 1972
Palmer, E. H. (Ed.); *Oriental Mysticism*; Octogon, 1974
Stoddart, W.; *Sufism: Mystical Doctrines and Methods of Islam*; Thorsons, 1976
Lienhardt, G.; *Divinity and Experience: Religion of the Dinka*; Oxford U.P., 1961
Nadel, S. F.; *Nupe Religion (Islam in W. Africa)*; Routledge, 1970
222 Parrinder, G.; *Worship in the World's Religions*; Sheldon, 1974
Wach, J.; *Types of Religious Experience*; Univ. Chicago Press, 1951
Goodman, F. D.; *Trance, Healing and Hallucination*; Wiley, 1974
224 Thurian, M.; *Modern Man and Spiritual Life*; Lutterworth, 1968
226 Russell, B.; *History of Western Philosophy*; Allen & Unwin, 1961
Plato; *Republic*; Dent, 1976
Descartes, R.; *Discourse on Method*; Dent, 1975
Hume, D.; *Enquiries (Human Understanding and Principles of Morals)*; Oxford U.P., 1975
Kant, I.; *Prolegomena*; Manchester U.P., 1953
Russell, B.; *Problems of Philosophy*; Oxford U.P., 1967
228 Hodges, W.; *Logic*; Penguin, 1977
Suppes, P.; *Introduction to Logic*; Van Nost. Reinhold, 1957
Nagel, E. & J. R. Newman; *Gödel's Proof*; Routledge, 1971
230–232 Mitchell, B. (Ed.); *Philosophy of Religion*; Oxford U.P., 1971
Ayer, A. J.; *Central Questions of Philosophy*; Penguin, 1976
Passmore, J. A.; *Philosophical Reasoning*; Duckworth, 1970
234 Mill, J. S.; *Utilitarianism*; Dent, 1972
Moore, G. E.; *Principia Ethica*; Cambridge U.P., 1903
Rawls, J.; *Theory of Justice*; Oxford U.P., 1972

Communications
236 Fast, J.; *Body Language*; Souvenir, 1970
Morris, D.; *Manwatching*; Cape, 1977
Critchley, M.; *Silent Language*; Butterworth, 1975
238–240 Aitchison, J.; *Articulate Mammal*; Hutchinson Educ., 1976
Steiner, G.; *After Babel*; Oxford U.P., 1975
Robins, R. H.; *General Linguistics*; Longman, 1976
Crystal, D.; *Linguistics*; Penguin, 1971
242 Higenbottam, F.; *Codes and Cyphers*; Eng. U.P., 1973
Norman, B.; *Secret Warfare*; David & Charles, 1973
Way, P.; *Codes and Cyphers*; Aldus, 1977
244–246 Diringer, D.; *Alphabet*; Hutchinson, 1968
Gelb, I. J.; *Study of Writing*; Univ. of Chicago Press, 1952

Politics
248 Levi-Strauss, C.; *Scope of Anthropology*; Cape, 1967
Coon, C. S.; *Living Races of Man*; Cape, 1966
250 Wendt, H.; *It Began in Babel*; Weidenfeld & Nicolson, 1961
Diop, C. A.; *African Origin of Civilization*; Lawrence Hill, 1974
252–262 Brace, C. L.; *Stages of Human Evolution*; Prentice-Hall, 1967
Sahlins, M. D.; *Tribesmen*; Prentice-Hall, 1968
Lewis, I. M.; *Social Anthropology in Perspective*; Penguin, 1976
Sahlins, M. D. & E. Service; *Evolution and Culture*; Univ. of Michigan Press, 1960
Berger, P. L.; *Invitation to Sociology*; Penguin, 1970
264–270 Fromm, E.; *Fear of Freedom*; Routledge, 1942
Ginsberg, M.; *Psychology of Society*; Methuen, 1963
Ortega y Gasset, J.; *Man and Crisis*; Allen & Unwin, 1959
Parsons, T.; *Social Structure and Personality*; Collier, 1964
Danziger, K.; *Socialization*; Penguin, 1971
Haines, N.; *Freedom and Community: Social Values*; Macmillan, 1966
Lindner, R. M.; *Must You Conform?*; Rinehart, 1956
272–280 Aron, R.; *Democracy and Totalitarianism*; Weidenfeld & Nicolson, 1968
Bronowski, J. & B. Mazlish; *Western Intellectual Tradition*; Penguin, 1960
Duverger, M.; *Idea of Politics*; Methuen, 1966
Fairlie, H.; *Life of Politics*; Methuen, 1968
Finer, S. E.; *Comparative Government*; Penguin, 1974
McKenzie, R. T.; *British Political Parties*; Heinemann Educ., 1964
Mill, J. S.; *On Liberty*; Penguin, 1974
Mills, C. W.; *Power Elite*; Oxford U.P., 1956
Plamenatz, J. P.; *Man and Society*; Longman, 1963 (2 vols)
Popper, K.; *Open Society and its Enemies*; Routledge, 1962 (2 vols)
Schapiro, L.; *Government and Politics of the Soviet Union*; Hutchinson, 1963
Schumpeter, J. A.; *Capitalism, Socialism and Democracy*; Allen & Unwin, 1977

Law
282–284 Friedmann, W.; *Legal Theory*; Stevens, 1967
Allen, Sir C. K.; *Law in the Making*; Oxford U.P., 1967
Plucknett, T. F. T.; *Concise History of the Common Law*; Butterworth, 1966
286 Eldefonso, E.; *Principles of Law Enforcement*; Wiley, 1974
288–290 Radzinowicz, Sir L. & J. King; *Growth of Crime*; Hamish Hamilton, 1977
Feldman, M. P.; *Criminal Behaviour: A Psychological Analysis*; Wiley, 1977
Walker, N.; *Crimes, Courts and Figures*; Penguin, 1971
Hawkins, G.; *Prison: Policy and Practice*; Univ. of Chicago Press, 1976
Johnston, N. et al.; *Sociology of Punishment and Correction*; Wiley, 1970
292–294 Hall, A.; *Crime-busters: FBI, Scotland Yard, etc.*; Verdict, 1976
Kirk, P. L.; *Crime Investigation*; Wiley, 1974
Ullyett, C.; *Criminology*; Watts, 1972
296 Acton, H. B.; *Philosophy of Punishment*; Macmillan, 1969
Thomas, D. A.; *Principles of Sentencing*; Heinemann, 1970

Work and Play
298–302 Weir, D. (Ed.); *Men and Work in Modern Britain*; Fontana, 1973
Weir, M. (Ed.); *Job Satisfaction: Challenge and Response in Modern Britain*; Fontana, 1976
Burns, T. (Ed.); *Industrial Man*; Penguin, 1969
Anthony, P. D.; *Ideology of Work*; Tavistock, 1977
304 Arlott, J. (Ed.); *Oxford Companion to Sports and Games*; Oxford U.P., 1975
Riordan, J. (Ed.); *Sport Under Communism*; Hurst, 1978

Dunning, E. (Ed.); *Readings in the Sociology of Sport*; F. Cass, 1971
Inglis, F.; *Name of the Game: Sport and Society*; Heinemann, 1977
306 Smith, M. (Ed.); *Leisure and Society in Britain*; Allen Lane, 1973
Parker, S.; *Sociology of Leisure*; Allen & Unwin, 1976
Economics
308–316 Atkinson, A. B.; *Unequal Shares: Wealth in Britain*; Penguin, 1974
Culyer, A. J.; *Economics of Social Policy*; Martin Robertson, 1973
Donaldson, P.; *Economics of the Real World*; Penguin, 1973
Donaldson, P.; *Guide to the British Economy*; Penguin, 1970
Friedmann, M.; *Capitalism and Freedom*; Univ. of Chicago Press, 1962
Galbraith, J. K.; *New Industrial State*; Penguin, 1968
Hirsch, F.; *Money International*; Penguin, 1967
Mishan, E. J.; *Costs of Economic Growth*; Penguin, 1969
Myint, H.; *Economics of the Developing Countries*; Hutchinson Educ., 1973
Prest, A. R. & D. J. Coppock (Eds.); *U.K. Economy: Manual of Applied Economics*; Weidenfeld & Nicolson, 1976
Revell, J.; *British Financial System*; Macmillan, 1973
Tew, B.; *Evolution of the International Monetary System*; Hutchinson, 1977
Trevithick, J.; *Inflation: A Guide to the Crisis in Economics*; Penguin, 1977

History and Culture: Ancient (BC-*c*.AD 400)

General
Alexander, J.; *Directing of Archaeological Excavations*; J. Baker, 1970
Heizer, R. F. (Ed.); *Archaeologist at Work*; Greenwood, 1976
Hutchinson, Sir J. et al; *Early History of Agriculture*; British Academy, 1977
Petrie, Sir W. M. F.; *Methods and Aims in Archaeology*; Arno, 1976
Powell, T. G. E.; *Prehistoric Art*; Thames & Hudson, 1966
Singer, C. (Ed.); *Studies in the History and Method of Science*; Arno, 1975
Africa
Freeman-Grenville, G. S.; *Chronology of African History*; Oxford U.P., 1973
Freeman-Grenville, G. S.; *E. African Coast*; R. Collings, 1975
Shinnie, P. L.; *African Iron Age*; Oxford U.P., 1971
Americas
Weaver, M. P.; *Aztecs, Maya and their Predecessors*; Seminar, 1972
Asia
General
Frankfort, H.; *Art and Architecture of the Ancient Orient*; Penguin, 1970
Reischauer, E. O. & J. K. Fairbank; *E. Asia: Great Tradition*; Allen & Unwin, 1961
Schumann, H. W.; *Buddhism*; Rider, 1973
China
Fitzgerald, C. P.; *China: Short Cultural History*; Barrie & Jenkins, 1965
Fung Yu-Lan; *Short History of Chinese Philosophy*; Routledge, 1948
Hay, J.; *Ancient China*; Bodley Head, 1973
Liu Wu-Chi; *Confucius: His Life and Time*; Greenwood, 1972
Loewe, M.; *Everyday Life in Early Imperial China*; Transworld, 1973
Sickman, L. & A. Soper; *Art and Architecture of China*; Penguin, 1971
Watson, B.; *Early Chinese Literature*; Columbia U.P., 1972
Wilhelm, H.; *Confucius and Confucianism*; Routledge, 1972
Yap, Y. & A. B. Cottrell; *Early Civilization of China*; Weidenfeld & Nicolson, 1975
India
Basham, A. L.; *Cultural History of India*; Oxford U.P., 1975
Rowland, B.; *Art and Architecture of India*; Penguin, 1971
Spear, P.; *History of India II*; Penguin, 1970
Thapar, R.; *History of India I*; Penguin, 1969
Watson, F.; *Concise History of India*; Thames & Hudson, 1975
Wheeler, Sir M.; *Indus Civilization*; Cambridge U.P., 1968
Woodruff, P.; *Men Who Ruled India: The Founders*; Cape, 1963
Japan
Morris, I.; *World of the Shining Prince*; Oxford U.P., 1964
Paine, R. T. & A. Soper; *Art and Architecture of Japan*; Penguin, 1975
Sanson, Sir G.; *Japan: A Short Cultural History*; Barrie and Jenkins, 1976
Middle East
General
Bruce, F. F.; *Israel and the Nations*; Paternoster, 1973
Singh, P.; *Neolithic Cultures of Western Asia*; Seminar, 1974
Egypt
Aldred, C.; *Egyptians*; Thames & Hudson, 1961
Aldred, C.; *Egypt to the End of the Old Kingdom*; Thames & Hudson, 1965
Edwards, I. E. S. et al (Eds.); *Cambridge Ancient History I & II*; Cambridge U.P., 1971–75
Gardiner, A. H.; *Egypt of the Pharaohs*; Oxford U.P., 1961
Redford, D. B.; *History and Chronology of the 18th Dynasty of Egypt*; Univ. of Toronto Press, 1968
Hittites
Macqueen, J. G.; *Hittites and their Contemporaries in Asia Minor*; Thames & Hudson, 1975
Sumeria
Kramer, S. N.; *Sumerians: History, Culture and Character*; Univ. of Chicago Press, 1971
Babylon and Assyria
Contenau, G.; *Everyday Life in Babylon and Assyria*; Arnold, 1954
Jastrow, M.; *Civilization of Babylon and Assyria*; Arno, 1976
Laessoe, J.; *People of Ancient Assyria*; Routledge, 1963
Medes and Persia
Olmstead, A. T.; *History of the Persian Empire*; Univ. of Chicago Press, 1958
Phoenicia
Harden, D.; *Phoenicians*; Penguin, 1972
Herm, G.; *Phoenicians*; Gollancz, 1975
Europe
General
Sandars, N. K.; *Prehistoric Art in Europe*; Penguin, 1968
Greek Art
Arnott, P. D.; *Introduction to the Greek Theatre*; Macmillan, 1959
Baldry, H. C.; *Ancient Greek Literature in its Living Context*; Thames & Hudson, 1968
Bowra, C. M.; *The Greek Experience*; Weidenfeld & Nicolson, 1957
Robertson, D. S.; *Greek and Roman Architecture*; Cambridge U.P., 1969
Robinson, M.; *History of Greek Art*; Cambridge U.P., 1976 (2 vols)
Rose, H. J.; *Handbook of Greek Literature*; Methuen, 1965
Greek History
Andrewes, A.; *Greek Society*; Penguin, 1971

Bowra, C. M.; *Periclean Athens*; Weidenfeld & Nicolson, 1971
Chadwick, J.; *Mycenaean World*; Cambridge U.P., 1976
Finley, M. I.; *Ancient Greeks: Life and Thought*; Penguin, 1971
Fox, R. L.; *Alexander the Great*; Futura, 1975
Green, P.; *Alexander the Great*; Weidenfeld & Nicolson, 1970
Hammond, N. G. L.; *History of Greece to 322BC*; Oxford U.P., 1967
Hood, S.; *Minoans*; Thames & Hudson, 1971
Lloyd, G. E. R.; *Early Greek Science*; Chatto, 1970
Lloyd, G. E. R.; *Greek Science after Aristotle*; Chatto, 1973
Meiggs, R.; *Athenian Empire*; Oxford U.P., 1972
Stobart, J. C.; *Glory that was Greece*; Sidgwick & Jackson, 1971
Taylour, W.; *Mycenaeans*; Thames & Hudson, 1964
Roman Art
Duff, J. W. & A. M. Duff; *Literary History of Rome*; Benn, 1962
Richardson, E. H.; *Etruscans: Their Art and Civilization*; Univ. of Chicago Press, 1976
Robertson, D. S.; *Greek and Roman Architecture*; Cambridge U.P., 1969
Strong, E.; *Roman Sculpture from Augustus to Constantine*; Hacker Art, 1975
Wheeler, Sir M.; *Roman Art and Architecture*; Thames & Hudson, 1964
Roman History
Bloch, R.; *Origins of Rome*; Thames & Hudson, 1960
Boak, A. E. R.; *History of Rome to AD 565*; Collier-Macmillan, 1970
Brunt, P. A.; *Social Conflicts in the Roman Republic*; Chatto, 1971
Carcopino, J.; *Daily Life in Ancient Rome*; Penguin, 1970
Cary, M. & H. H. Scullard; *History of Rome to Constantine*; Macmillan, 1960
Charlesworth, M. P.; *Roman Empire*; Oxford U.P., 1968
Homo, L.; *Roman Political Institutions*; Paul, 1929
Jones, A. M. H.; *Constantine and the Conversion of Europe*; Penguin, 1972
Keller, W.; *Etruscans*; Cape, 1975
Liversidge, J.; *Everyday Life in the Roman Empire*; Batsford, 1976
Marsh, F. B.; *Founding of the Roman Empire*; Greenwood, 1976
Parker, H. M. D.; *History of the Roman World 138–337*; Methuen, 1958
Scullard, H. H.; *From the Gracchi to Nero*; Methuen, 1976
Stobart, J. C.; *Grandeur That Was Rome*; Sidgwick & Jackson, 1971
Britain
Chadwick, N. K.; *Celts*; Penguin, 1971
Cunliffe, B.; *Iron Age Communities in Britain*; Routledge, 1974
Renfrew, C. (Ed.); *British Prehistory*; Duckworth, 1974
Richmond, Sir I.; *Roman Britain*; Penguin, 1970
Ross, A.; *Pagan Celtic Britain*; Cardinal, 1974
Wilson, R. J. A.; *Guide to the Roman Remains in Britain*; Constable, 1975

History and Culture: Medieval (*c*.400–*c*.1500)

HISTORY
Africa
Freeman-Grenville, G. S.; *Chronology of African History*; Oxford U.P., 1973
Freeman-Grenville, G. S.; *E. African Coast*; R. Collings, 1975
Americas
Davies, N.; *Aztecs*; Sphere, 1977
Lanning, E. P.; *Peru Before the Incas*; Prentice-Hall, 1968
Thompson, I. F. S.; *Rise and Fall of Maya Civilization*; Gollancz, 1956
Willey, G. R.; *Introduction to American Archaeology*; Prentice-Hall, 1966–72 (2 vols)
Asia
General
Luzzati, E.; *Travels of Marco Polo*; Dent, 1975
Spuler, B.; *History of the Mongols*; Routledge, 1972
Tate, D. J. M.; *Making of Modern S. E. Asia: The European Conquest*; Oxford U.P., 1972
Islamic Empire
Gabrieli, F.; *Muhammed and the Conquests of Islam*; Weidenfeld & Nicolson, 1968
Grunebaum, G. E. V.; *Classical Islam 600–1258*; Allen & Unwin, 1970
Hayes, J. R. (Ed.); *Genius of Arab Civilization*; Phaidon, 1976
Holt, P. M. et al.; *Cambridge History of Islam*; Cambridge U.P., 1971 (2 vols)
Inalcik, H.; *Ottoman Empire 1300–1600*; Weidenfeld & Nicolson, 1973
Watt, W. M.; *Muhammed: Prophet and Statesman*; Oxford U.P., 1961
Europe
General
Barrett, C. K.; *Signs of an Apostle*; Epworth, 1970
Bettenson, H. (Ed.); *Documents of the Christian Church*; Oxford U.P., 1967
Bloch, M.; *Feudal Society*; Routledge, 1965 (2 vols)
Bronsted, J., *Vikings*; Penguin, 1970
Chadwick, H.; *Early Church*; Penguin, 1967
Dodd, C. H.; *Founder of Christianity*; Collins, 1971
Ganshof, F. L.; *Feudalism*; Longman, 1964
Hale, J. R. et al. (Eds.); *Europe in the Late Middle Ages*; Faber, 1970
Huizinga, J. H.; *Waning of the Middle Ages*; Penguin, 1972
Knowles, D.; *Religious Orders in England*; Cambridge U.P., 1948
Knowles, D. & D. Obolensky; *Middle Ages*; Darton, 1969
Lach, D. F.; *Asia in the Making of Europe*; Univ. of Chicago Press, 1965–70 (2 vols)
Leff, G.; *Heresy in the Later Middle Ages*; Manchester U.P., 1967
McArthur, H. K. (Ed.); *In Search of the Historical Jesus*; S.P.C.K., 1970
McNeill, J.; *Celtic Churches: History 200–1200*; Univ. of Chicago Press, 1974
Mascall, E. L.; *Christ, the Christians and the Church*; Longman, 1969
Mayer, H. E.; *Crusades*; Oxford U.P., 1972
Postan, M. M. (Ed.); *Cambridge Economic History of Europe*; Cambridge U.P., 1963–66 (2 vols)
Reumann, J.; *Jesus in the Church's Gospels*; S.P.C.K., 1970
Runciman, Sir S.; *History of the Crusades*; Penguin, 1971
Sawyer, P. H.; *Age of the Vikings*; E. Arnold, 1975
Ullman, W.; *Short History of the Papacy in the Middle Ages*; Methuen, 1972
Wallace-Hadrill, J. M.; *Barbarian West*; Hutchinson, 1967
Wilks, M. J.; *Problem of Sovereignty in the Later Middle Ages*; Cambridge U.P., 1963
Byzantine Empire
Nicol, D. M. (Ed.); *Last Centuries of Byzantium*; Hart-Davis, 1972
Ostrogorsky, G.; *History of the Byzantine State*; Blackwell, 1969
Runciman, Sir S.; *Byzantine Civilization*; Methuen, 1975
Southern, R. W.; *Western Society and the Church in the Later Middle Ages*; Penguin, 1970
Holy Roman Empire
Barraclough, G.; *Medieval Empire: Idea and Reality*; Historical Association, 1964
Heer, F.; *Charlemagne and his World*; Weidenfeld and Nicolson, 1975

France
Evans, J.; *Life in Medieval France*; Phaidon, 1969
   Britain (c.400 – c.1066)
Blair, P. H.; *Introduction to Anglo-Saxon England*; Cambridge U.P., 1956
Kirby, D. P.; *Making of Early England*; Batsford, 1967
Loyn, H. R.; *Anglo-Saxon England and the Norman Conquest*; Longman, 1970
Page, R. I.; *Life in Anglo-Saxon England*; Batsford, 1970
Wallace-Hadrill, J. M.; *Early Germanic Kingship*; Oxford U.P., 1971
Whitelock, D.; *Beginnings of English Society*; Penguin, 1971
Wilson, D. M.; *Anglo-Saxons*; Penguin, 1971
Wilson, D. M.; *Archaeology of Anglo-Saxon England*; Methuen, 1976
   Britain (c.1066 – c.1300)
Brooke, C. N. L.; *Saxon and Norman Kings*; Fontana, 1967
Douglas, D. C.; *William the Conqueror*; Eyre & Spottiswoode, 1964
Holt, J. C.; *Magna Carta*; Cambridge U.P., 1969
Lennard, R.; *Rural England 1086–1135*; Oxford U.P., 1959
Matthew, D.; *Norman Conquest*; Batsford, 1966
Patourel, J. le; *Norman Empire*; Oxford U.P., 1977
   Britain (c.1300 – c.1500)
Barrow, G. W. S.; *Robert Bruce and the Community of the Realm of Scotland*; Edinburgh U.P., 1976
Boulay, F. R. H. du; *Age of Ambition*; Nelson, 1970
Curtis, E.; *History of Medieval Ireland*; Methuen, 1968
Dobson, R. B. (Ed.); *Peasants' Revolt*; Macmillan, 1970
Duncan, A. A. M.; *Scotland: The Making of the Kingdom*; Oliver & Boyd, 1975
Fryde, E. B. & E. Miller; *Historical Studies of the English Parliament*; Cambridge U.P., 1971 (2 vols)
Harriss, G. L.; *King, Parliament and the Public Finance in Medieval England to 1369*; Oxford U.P., 1975
Hilton, R.; *English Peasantry in the Later Middle Ages*; Oxford U.P., 1975
Hoskins, W. G.; *Midland Peasant*; Macmillan, 1957
Knowles, D.; *Monastic Orders in England*; Cambridge U.P., 1963
Knowles, D.; *Religious Orders in England*; Cambridge U.P., 1948
McFarlane, K. B.; *Nobility of Later Medieval England*; Oxford U.P., 1973
Perroy, E.; *100 Years War*; Eyre, 1951
Roderick, A. J. (Ed.); *Wales through the Ages Vol. I*; C. Davies, 1971
Williams, G. A.; *Medieval London: Commune to Capital*; Athlone Press, 1970
ARTS
Islamic
James, D.; *Islamic Art: An Introduction*; Hamlyn, 1974
Rice, D. T.; *Islamic Art*; Thames & Hudson, 1975
European
   General
Conant, K. J.; *Carolingian and Romanesque Architecture*; Penguin, 1974
Fletcher, Sir B.; *History of Architecture*; Athlone, 1975
Frankl, P.; *Gothic Architecture*; Penguin, 1962
Gombrich, E. H.; *Story of Art*; Phaidon, 1972
Janson, H. W.; *History of Art*; Thames & Hudson, 1962
Ker, W. P.; *Epic and Romance*; Dover, 1957
Lund, E. et al.; *History of European Ideas*; Hurst, 1972
Pevsner, N.; *Outline of European Architecture*; Penguin, 1970
Wolff, P.; *Awakening of Europe*; Penguin, 1968
   Byzantine Empire
Krautheimer, R.; *Early Christian and Byzantine Architecture*; Penguin, 1975
Rice, D. T.; *Appreciation of Byzantine Art*; Oxford U.P., 1972
Rice, D. T.; *Byzantine Art and its Influence*; Variorum Reprints, 1973
   France
Evans, J.; *Art in Medieval France*; Oxford U.P., 1948
Lemoisne, P.; *Gothic Painting in France*; Hacker, 1975
   Britain
Kittredge, G. L.; *Chaucer and his Poetry*; Harvard U.P., 1972
Leeds, E. T.; *Early Anglo-Saxon Art and Archaeology*; Oxford U.P., 1970
Lewis, C. S.; *Allegory of Love*; Oxford U.P., 1936
Oxford History of English Art; Oxford U.P., 1952–76 (9 vols)
Woolf, R.; *English Mystery Plays*; Routledge, 1972

History and Culture: Modern (*c.*1500 – the present)

HISTORY
General
Bonhoeffer, D.; *Christology*; Fontana, 1971
Cambridge History of the British Empire; Cambridge U.P., 1940–59 (3 vols)
Grubel, H. G.; *International Monetary System*; Penguin, 1969
Gutteridge, J.; *United Nations in a Changing World*; Manchester U.P., 1970
Jansen, G. H.; *Afro-Asia and Non-Alignment*; Faber, 1966
Kinch, F. W. (Ed.); *Sociology in the World Today*; Addison, 1971
Lijphart, A.; *Trauma of Decolonization*; Yale U.P., 1966
Mar, A. D.; *History of Monetary Systems*; Kelley, 1969
Myrdal, K. G.; *Economic Theory and Underdeveloped Regions*; Duckworth, 1957
Patton, J.; *How the Depression Hit the West*; McClelland, 1974
Pelikan, J. (Ed.); *20th Century Theology in the Making*; Fontana, 1970 (3 vols)
Tetlow, E.; *United Nations: First 25 Years*; P. Owen, 1970
Worsley, P.; *Third World*; Weidenfeld & Nicolson, 1967
Africa
Hallett, R.; *Africa to 1875: A Modern History*; Univ. of Michigan Press, 1970
Hallett, R.; *Africa since 1875: A Modern History*; Univ. of Michigan Press, 1970
Martin, P. M.; *External Trade of the Loango Coast 1576–1870*; Oxford U.P., 1972
Wilson, M. & L. M. Thompson; *Oxford History of S. Africa*; Oxford U.P., 1969–71 (2 vols)
North America
Billington, R. A. & H. E. Huntington; *Westward Expansion*; Collier-Macmillan, 1974
Creighton, D. G.; *Story of Canada*; Faber, 1971
Galbraith, K.; *Affluent Society*; Penguin, 1970
McNaught, K.; *Pelican History of Canada*; Penguin, 1970
Randall, J. G. & D. Donald; *Civil War and Reconstruction*; Heath, 1968
Smith, H. N.; *Virgin Land: American West*; Harvard U.P., 1972
Thistlethwaite, F.; *Great Experiment*; Cambridge U.P., 1955
Wright, E.; *American Revolution*; Leicester U.P., 1967
Central America
Parry, J. H. & P. M. Sherlock; *Short History of the West Indies*; Macmillan, 1968

Prescott, W. H.; *History of the Conquest of Mexico*; Dent, 1970 (2 vols)
South America
Diaz, B.; *Conquest of New Spain*; Penguin, 1969
Furtado, C.; *Economic Development of Latin America*; Cambridge U.P., 1970
Glade, W. P.; *Latin American Economies*; Van Nost. Reinhold, 1969
Hemming, J.; *Conquest of the Incas*; Sphere, 1972
Lambert, J.; *Latin America: Social Structure and Political Institutions*; Univ. of California Press, 1969
Prescott, W. H.; *History of the Conquest of Peru*; Modern Library, 1973
Asia
   General
Crowley, J. B. (Ed.); *Modern E. Asia*; Harcourt Brace, 1970
Fitzgerald, C. P.; *Concise History of E. Asia*; Penguin, 1974
   China
Gittings, J.; *Chinese View of China*; BBC, 1973
Han Suyin; *Morning Deluge: Mao and the Chinese Revolution*; Panther, 1976 (2 vols)
McAleavy, H.; *Modern History of China*; Weidenfeld & Nicolson, 1968
Myrdal, J.; *Report from a Chinese Village*; Pan, 1975
Schram, S.; *Mao Tse-Tung*; Penguin, 1970
Schram, S. (Ed.); *Mao Tse-Tung Unrehearsed*; Penguin, 1974
Schurmann, F. & O. Schell; *Imperial China*; Penguin, 1967
Snow, E.; *China's Long Revolution*; Penguin, 1975
Snow, E.; *Red China Today*; Penguin, 1970
Snow, E.; *Red Star over China*; Penguin, 1972
   India
Brown, J. M.; *Gandhi's Rise to Power*; Cambridge U.P., 1974
Gopal, S.; *British Policy in India 1858–1905*; Cambridge U.P., 1965
Spear, P.; *History of India II*; Penguin, 1970
Woodruff, P.; *Men who Ruled India: The Founders*; Cape, 1963
Woodruff, P.; *Men who Ruled India: The Guardians*; Cape, 1963
   Japan
Beasley, W. G.; *Modern History of Japan*; Weidenfeld & Nicolson, 1973
Sansom, Sir G.; *Western World and Japan*; Barrie & Jenkins, 1966
Storry, R.; *History of Modern Japan*; Penguin, 1968
   Vietnam
Duncanson, D. J.; *Government and Revolution in Vietnam*; Oxford U.P., 1968
Porter, G.; *Peace Denied: US, Vietnam and the Paris Agreement*; Indiana U.P., 1976
Australasia
Marais, J. S.; *Colonization of New Zealand*; Dawsons, 1968
Mills, R. C.; *Colonization of Australia 1829–42*; Dawsons, 1968
Shaw, A. G. L.; *Story of Australia*; Faber, 1973
Sinclair, K.; *History of New Zealand*; Penguin, 1970
Middle East
Rodinson, M.; *Israel and the Arabs*; Penguin, 1973
Europe
   General
Addison, P.; *Road to 1945*; Cape, 1975
Anderson, M. S.; *Eastern Question 1774–1923*; Macmillan, 1966
Andrews, S.; *18th Century Europe*; Longman, 1965
Aster, S.; *Making of the Second World War*; Deutsch, 1974
Bainton, R. H.; *Age of Reformation*; Van Nost. Reinhold, 1956
Bainton, R. H.; *Erasmus of Christendom*; Collins, 1970
Bainton, R. H.; *Here I Stand: Life of Martin Luther*; New American Library, 1950
Barnes, H. E. (Ed.); *Introduction to the History of Sociology*; Univ. Chicago Press, 1966
Becker, C. L.; *Heavenly City of the 18th Century Philosophers*; Yale U.P., 1932
Birnie, A.; *Economic History of Europe 1760–1939*; Methuen, 1967
Calvocoressi, P. & G. Wint; *Total War*; Penguin, 1974
Cameron, R.; *Europe since 1945*; Arnold, 1976
Carsten, F.; *Rise of Fascism*; Methuen, 1970
Clark, Sir G.; *17th Century*; Oxford U.P., 1960
Cole, G. D. H.; *History of Socialist Thought*; Macmillan 1953–60 (5 vols)
Dickens, A. G.; *Counter-Reformation*; Thames & Hudson, 1969
Dickens, A. G.; *Reformation and Society in 16th Century Europe*; Thames & Hudson, 1966
Elton, G. R.; *Reformation Europe 1517–99*; Fontana, 1969
Fejto, F.; *History of the People's Democracies: E. Europe since Stalin*; Penguin, 1974
Fieldhouse, D. K.; *Economics and Empire*; Weidenfeld & Nicolson, 1976
Gollwitzer, H.; *Europe in the Age of Imperialism*; Thames & Hudson, 1969
Green, V. H. H.; *Renaissance and Reformation*; E. Arnold, 1964
Grenville, J. A. S.; *Europe Re-shaped 1848–78*; Fontana, 1976
Hampson, M.; *Enlightenment*; Penguin, 1968
Hartung, F.; *Enlightened Despotism*; Historical Association, 1964
Hayek, F. A. (Ed.); *Capitalism and the Historians*; Univ. Chicago Press, 1954
Hazard, P.; *European Mind 1680–1715*; Penguin, 1973
Keynes, J. M.; *Economic Consequences of the Peace*; Macmillan, 1971
Knapp, W.; *Unity and Nationalism in Europe since 1945*; Pergamon, 1969
Knight, F.; *Exploration*; Benn, 1973
Koebner, R.; *Empire*; Cambridge U.P., 1961
Koenigsberger, N. G. & G. D. Mosse; *Europe in the 16th Century*; Longman, 1971
Laqueur, W. (Ed.); *Fascism: A Reader's Guide*; Wildwood House, 1977
Lichteim, G.; *Short History of Socialism*; Fontana, 1975
Liddell-Hart, B. H.; *History of the First World War*; Pan, 1972
Namier, Sir L.; *1848– Revolution of the Intellectuals*; Oxford U.P., 1971
Nicolson, H.; *Congress of Vienna*; Methuen, 1946
Reid, J.; *Concise Encyclopaedia of the Second World War*; Osprey, 1974
Robertson, E. M. (Ed.); *Origins of the Second World War*; Macmillan, 1971
Sherover, C. M.; *Development of the Democratic Idea*; Mentor, 1974
Smith, D. & D. Newton; *Exploration*; Schofield, 1971
Smith, W. H. C.; *20th Century Europe*; Longman, 1970
Stoye, J.; *Europe Unfolding 1648–88*; Fontana, 1969
Sumner, W. G.; *History of Banking in all the Leading Nations*; Kelley, 1970
Tawney, R. H.; *Religion and the Rise of Capitalism*; Penguin, 1969
Taylor, A. J. P.; *Origins of the Second World War*; Penguin, 1961
Taylor, A. J. P.; *Struggle for Mastery in Europe 1848–1914*; Oxford U.P., 1971
Watkins, F. M.; *Age of Ideology: 1790 – Present*; Prentice-Hall, 1964
Wendel, F.; *Calvin*; Fontana, 1965
   France
Behrens, C. B. A.; *Ancien Regime*; Thames & Hudson, 1967
Goodwin, A.; *French Revolution*; Hutchinson, 1966
Goodwin, A. (Ed.); *American and French Revolutions 1763–93*; Cambridge U.P., 1976

Goubert, P.; *Ancien Regime: French Society 1600–1750*; Weidenfeld & Nicolson, 1973
Hampson, N.; *First European Revolution 1776–1815*; Thames & Hudson, 1969
Hatton, R. M.; *Europe in the Age of Louis XIV*; Thames & Hudson, 1969
Lefebvre, G.; *Napoleon*; Routledge, 1969 (2 vols)
Lefebvre, G.; *French Revolution*; Routledge, 1964 (2 vols)
Ogg, D.; *Louis XIV*; Oxford U.P., 1967
   Germany
Fischer, F.; *Germany's Aims in the First World War*; Chatto, 1967
Taylor, A. J. P.; *Bismarck*; New English Library, 1974
   Italy
Smith, D. M.; *Garibaldi*; Prentice-Hall, 1969
   Spain
Elliott, J. H.; *Imperial Spain 1469–1716*; E. Arnold, 1963
Parry, J. H.; *Spanish Seaborne Empire*; Hutchinson, 1966
   USSR
Auty, R. & D. Obolensky (Eds.); *Introduction to Russian History*; Cambridge U.P., 1976
Carr, E. H.; *Russian Revolution*; Macmillan, 1950–53 (3 vols)
Deutscher, I.; *Stalin: A Political Biography*; Oxford U.P., 1972
Florinsky, M. T.; *Russia: History and Interpretation*; Collier-Macmillan, 1953 (2 vols)
Raeff, M.; *Catherine the Great*; Macmillan, 1972
Riasanovsky, N. V.; *History of Russia*; Oxford U.P., 1969
Schapiro, L. B.; *Communist Party of the Soviet Union*; Methuen, 1970
Seton-Watson, H.; *Russian Empire*; Oxford U.P., 1967
Tatu, M.; *Power in the Kremlin*; Collins, 1968
Trotsky, L.; *Stalin School of Falsification*; New Park, 1974
Ulam, A. B.; *Lenin and the Bolsheviks*; Secker & Warburg, 1966
Vernadsky, G.; *Kievan Russia*; Yale U.P., 1973
Vernadsky, G.; *Mongols and Russia*; Yale U.P., 1953
Vernadsky, G.; *Russia at the Dawn of the Modern Age*; Yale U.P., 1959
Wolfe, B. D.; *Three who made a Revolution*; Penguin, 1966
   Britain (c.1500 – c.1600)
Dickens, A. G.; *English Reformation*; Fontana, 1967
Donaldson, G.; *James V – James VII*; Oliver & Boyd, 1965
Edwards, R.; *Ireland in the Age of the Tudors*; Crown Helm, 1977
Elton, G. R.; *England Under the Tudors*; Methuen, 1974
Elton, G. R.; *Tudor Revolution in Government*; Cambridge U.P., 1953
Hakluyt, R.; *Voyages and Documents*; Oxford U.P., 1958
Lydan, J. F.; *Ireland in the Later Middle Ages*; Gill, 1973
Mattingly, G.; *Defeat of the Spanish Armada*; Cape, 1970
Neale, J. E.; *Elizabeth I and Her Parliaments*; Cape, 1953
Neale, J. E.; *Queen Elizabeth I*; Cape, 1967
Nicholson, R.; *Scotland: The Later Middle Ages*; Oliver & Boyd, 1973
Rowse, A. L.; *England of Elizabeth*; Macmillan, 1950
Scarisbrick, J. J.; *Henry VIII*; Eyre Methuen, 1976
Wernham, R. B.; *Before the Armada*; Cape, 1966
Williamson, J. A.; *Age of Drake*; Black, 1965
   Britain (c.1600 – c.1700)
Bryant, Sir A.; *Samuel Pepys*; Collins, 1967 (3 vols)
Cipolla, C. M.; *Before the Industrial Revolution*; Methuen, 1976
Clark, Sir G.; *Later Stuarts 1660–1714*; Oxford U.P., 1956
Fraser, A.; *Cromwell, Our Chief of Men*; Panther, 1975
Hill, C.; *Century of Revolution 1603–1714*; Cardinal, 1974
Hill, C.; *God's Englishman: Oliver Cromwell and the English Revolution*; Penguin, 1972
Lamont, W. M.; *Godly Rule*; Macmillan, 1969
Mitchison, R.; *History of Scotland*; Methuen, 1970
Ogg, D.; *England in the Reign of Charles II*; Oxford U.P., 1967
Stone, L.; *Crisis of the Aristocracy 1558–1641*; Oxford U.P., 1965
Wedgwood, C. V.; *King's Peace 1637–41*; Collins, 1955
Wedgwood, C. V.; *King's War 1641–47*; Collins, 1958
   Britain (c.1700 – c.1800)
Ashton, T. S.; *Industrial Revolution*; Oxford U.P., 1969
Barnes, D. G.; *George III and William Pitt*; Cass, 1967
Cannon, J.; *Parliamentary Reform 1640–1832*; Cambridge U.P., 1973
Chambers, J. D. & G. E. Mingay; *Agricultural Revolution*; Batsford, 1969
Chandler, D.; *Art of Warfare in the Age of Marlborough*; Batsford, 1976
Churchill, W.; *Marlborough*; Harrap, 1970
Ferguson, W.; *Scotland 1689 to the Present*; Oliver & Boyd, 1968
Hamilton, H.; *Economic History of Scotland in the 18th Century*; Oxford U.P., 1963
James, F. G.; *Ireland in the Empire 1688–1770*; Harvard U.P., 1973
Mathias, P.; *First Industrial Nation*; Methuen, 1969
Pares, R.; *King George III and the Politicians*; Oxford U.P., 1968
Plumb, J. H.; *Sir Robert Walpole*; Allan Lane, 1972 (2 vols)
Seeley, Sir J. R.; *Expansion of England*; Univ. Chicago Press, 1973
Smout, T. C.; *History of the Scottish People 1560–1830*; Fontana, 1972
Williams, E. E.; *Capitalism and Slavery*; Deutsch, 1964
   Britain (c.1800 – c.1900)
Blake, R.; *Disraeli*; Methuen, 1969
Briggs, A.; *Victorian Cities*; Penguin, 1968
Dyos, H. J. & M. Wolff (Eds.); *Victorian City: Images and Realities*; Routledge, 1973
Hayes, P.; *19th Century*; Black, 1975
Hobsbawm, E. J.; *Industry and Empire*; Weidenfeld & Nicolson, 1968
Hunter, J.; *Making of the Crofting Community*; Donald, 1976
Longford, E.; *Victoria RI*; Pan, 1966
Longford, E.; *Wellington*; Panther, 1971 & 1975 (2 vols)
Magnus, Sir P.; *Gladstone*; J. Murray, 1954
Midwinter, E. C.; *Victorian Social Reform*; Longman, 1968
Morton, A. L. & G. Tate; *British Labour Movement 1770–1920*; Lawrence, 1956
Robinson, R. & J. Gallagher; *Africa and the Victorians*; Macmillan, 1966
Roderick, A. J.; *Wales through the Ages Vol II, 1845–1900*; C. Davies, 1971
Samuel, R. (Ed.); *Village Life and Labour*; Routledge, 1975
Seton-Watson, R.; *Britain in Europe 1789–1914*; Cambridge U.P., 1937
Shepherd, T. H. & J. Elmes; *Metropolitan Improvements, or London in the 19th Century*; Arno, 1969
Slaven, A.; *Development of the West of Scotland*; Routledge, 1975
Southey, R.; *Life of Nelson*; Dent
Stewart, R.; *Foundation of the Conservative Party*; Longman, 1978
   Britain (c.1900 – present)
Branson, N. & M. Heinemann; *Britain in the 1930s*; Weidenfeld & Nicolson, 1975

Frankel, J.; *British Foreign Policy 1945–73*; Oxford U.P., 1975
Freeman-Grenville, G. S.; *Chronology of World History*; R. Collings, 1975
Freeman-Grenville, G. S.; *Queen's Lineage*; R. Collings, 1975
Harvie, C.; *Scotland and Nationalism*; Allen & Unwin, 1977
Mansergh, N.; *Commonwealth Experience*; Weidenfeld & Nicolson, 1969
Marwick, A.; *Home Front: British and the Second World War*; Thames & Hudson
Medlicott, W. M.; *Contemporary England 1914–64*; Longman, 1976
Morgan, D.; *Suffragists and Liberals*; Blackwell, 1975
Mowat, C. L.; *Britain between the Wars 1918–40*; Methuen, 1968
Murphy, J. A.; *Ireland in the 20th Century*; Gill, 1975
Orwell, G.; *Road to Wigan Pier*; Penguin, 1970
Richards, D. & A. Quick; *20th Century Britain*; Longman, 1968
Rothstein, A.; *British Foreign Policy and its Critics*; Lawrence, 1969
Rumpf, E. & A. Hepburn; *Nationalism and Socialism in 20th Century Ireland*; Liverpool U.P., 1977
Taylor, A. J. P.; *English History 1914–45*; Penguin, 1970
Williams, D.; *History of Modern Wales*; J. Murray, 1977
ARTS
General
   Architecture
Hatje, G.; *Encyclopaedia of Modern Architecture*; Thames & Hudson, 1963
Sharp, D.; *Sources of Modern Architecture*; Lund Humphries, 1967
Sharp, D.; *Visual History of 20th Century Architecture*; Heinemann, 1972
   Art
Aubin, W.; *Dada and Surrealist Art*; Thames & Hudson, 1969
Lippard, L.; *Pop Art*; Thames & Hudson, 1966
Lucie-Smith, E.; *Movements in Art since 1945*; Thames & Hudson, 1969
Meyer, U.; *Conceptual Art*; E. P. Dutton, 1972
Nadeau, M.; *History of Surrealism*; Penguin, 1973
Richter, H.; *Dada, Art and Anti-Art*; Thames & Hudson, 1965
Seuphor, M.; *Dictionary of Abstract Painting*; Methuen, 1958
Woods, G. (Ed.); *Art without Boundaries*; Thames & Hudson, 1972
   Film
Brownlow, K.; *Parade's Gone By*; Secker & Warburg, 1969
Lindgren, E.; *Art of the Film*; Allen & Unwin, 1970
Mast, G. & M. Cohen (Eds.); *Film Theory and Criticism*; Oxford U.P., 1974
Pratt, G.; *Spellbound in Darkness: History of the Silent Film*; N.Y. Graphic Soc., 1972
   Literature
Fletcher, J.; *New Directions in Literature*; Calder, 1969
Hartnoll, P.; *Concise History of the Theatre*; Thames & Hudson, 1968
   Music
Baines, A. (Ed.); *Musical Instruments through the Ages*; Faber, 1961
Clarke, M. & C. Crisp; *Ballet: An Illustrated History*; Black, 1973
Fox, L. M.; *Instruments of the Orchestra*; Lutterworth, 1971
Africa
Willett, F.; *African Art*; Thames & Hudson, 1971
Americas
   Film
Hampton, B. B.; *History of the American Film Industry*; Dover, 1970
Quigley, M. & R. Gertner; *Films in America 1929–69*; Golden Press, 1970
   Literature
Beaver, H. (Ed.); *American Critical Essays*; Oxford U.P., 1959
Borges, J. L.; *Introduction to American Literature*; Shocken, 1974
Brodhead, R. H.; *Hawthorne, Melville and the Novel*; Univ. Chicago Press, 1976
Cleveland, C. D. (Ed.); *Compendium of American Literature*; Kennikat, 1972
Fiedler, L.; *Love and Death in the American Novel*; Paladin, 1970
   Music
Charters, S.; *Country Blues: Roots of Jazz*; Da Capo, 1976
Gillett, C.; *Sound of the City: Rise of Rock and Roll*; Sphere, 1971
Mingus, C.; *Beneath the Underdog*; Penguin, 1975
Russell, R.; *Bird Lives: High Life and Hard Times of Charlie Parker*; Quartet, 1973
Asia
Fitzgerald, C. P.; *China: Short Cultural History*; Barrie & Jenkins, 1965
Siren, O.; *Chinese Painting: Leading Masters and Principles*; Hacker, 1975 (7 vols)
Swann, P.; *Art of China, Korea and Japan*; Thames & Hudson, 1967
Mukerjee, R.; *Culture of India*; Asia Pub. Ho., 1959
Australasia
Thomas, D.; *Outlines of Australian Art*; Barrie & Jenkins, 1974
Green, H. M.; *Australian Literature 1900–50*; Melbourne U.P., 1963
Hope, A. D.; *Australian Literature 1950– 62*; Melbourne U.P., 1963
Europe (General)
   Architecture & Art
Hitchcock, H. R.; *Architecture: 19th and 20th Centuries*; Penguin, 1971
Appolonio, U. (Ed.); *Futurist Manifesto*; Thames & Hudson, 1973
Clark, T. J.; *Absolute Bourgeois*; Thames & Hudson, 1973
Gaunt, W.; *Impressionists*; Thames & Hudson, 1970
Golding, J.; *Cubism: History and Analysis*; Faber, 1969
Honour, H.; *Neoclassicism*; Penguin, 1968
Kitson, M.; *Age of Baroque*; Hamlyn, 1976
Nochlin, L.; *Realism*; Penguin, 1971
Rewald, J.; *History of Impressionism*; Secker & Warburg, 1973
Romanticism; Heron Books, 1972
Sewter, A. C.; *Baroque and Rococo Art*; Thames & Hudson, 1972
Shearman, J.; *Mannerism*; Penguin, 1967
   Literature
Abrams, M. H.; *Mirror and the Lamp*; Oxford U.P., 1973
Brooks, D.; *Number and Pattern in the 18th Century Novel*; Routledge, 1973
Lukacs, G.; *Studies in European Realism*; Merlin, 1975
Watt, I.; *Rise of the Novel*; Penguin, 1972
Williams, I. (Ed.); *Novel and Romance 1700–1800*; Routledge, 1970
   Music
Blume, F.; *Renaissance and Baroque Music*; Faber, 1975
Bukofzer, M.; *Music in the Baroque Era*; Dent, 1948
Einstein, A.; *Music in the Romantic Era*; Dent, 1947
Harman, A.; *Late Renaissance and Baroque Music*; Barrie & Jenkins, 1969
Munrow, D.; *Instruments of the Middle Ages and Renaissance*; Oxford U.P., 1976
Orrey, L.; *Concise History of Opera*; Thames & Hudson, 1972
Teasdale, M. S.; *Handbook of 20th Century Opera*; Da Capo, 1976
France
   Art
Blunt, A.; *Art and Architecture in France 1500–1700*; Penguin, 1973

Blunt, A.; *Nicholas Poussin*; Pantheon, 1967
Dimier, L.; *French Painting of the 16th Century*; Arno, 1969
Dorival, B.; *School of Paris*; Thames & Hudson, 1962
Elderfield, J.; *Fauvism and its Affinities*; N.Y. Museum of Modern Art, 1976
Levey, M. & W. Kalnein; *Art and Architecture of the 18th Century in France*; Penguin, 1972
Wilenski, R. H.; *Modern French Painters*; Faber, 1963
Zerner, H.; *School of Fontainebleu*; Thames & Hudson, 1969
Literature
Demorest, J.-J. (Ed.); *Studies in 17th Century French Literature*; Cornell U.P., 1962
Graham, V. E. (Ed.); *16th Century French Poetry*; Univ. of Toronto, 1965
Moore, W. G.; *Classical Drama of France*; Oxford U.P., 1971
Italy
Baxendall, M.; *Painting and Experience in 15th Century Italy*; Oxford U.P., 1972
Berenson, B.; *Italian Painters of the Renaissance*; Phaidon, 1952
Burckhardt, J.; *Civilization of the Renaissance in Italy*; Phaidon, 1944
Chastel, A.; *Italian Art*; Faber, 1963
Freedberg, S. J.; *Painting in Italy 1500–1600*; Penguin, 1976
Levey, M.; *High Renaissance*; Penguin, 1975
McCarthy, M.; *Stones of Florence*; Heinemann, 1959
Wilde, J.; *Venetian Art from Bellini to Titian*; Oxford U.P., 1974
Wittowker, R.; *Art and Architecture in Italy 1600–1750*; Penguin, 1973
Spain
Brown, D.; *The World of Velazquez 1599–1660*; Time-Life, 1969
Chabrim, J.-F.; *Goya*; Thames & Hudson, 1965
Guinard, P.; *El Greco*; Skira, 1956
Kubler, G. & M. Soria; *Art and Architecture in Spain and Portugal 1500–1800*; Penguin, 1959
Germany
Art
Vegh, J.; *16th Century German Panel Paintings*; Corvina, 1972
Whitford, F.; *Expressionism*; Hamlyn, 1970
Wolfflin, H.; *Art of Albrecht Durer*; Phaidon, 1971
Literature
Lukacs, G.; *Goethe and his Age*; Merlin, 1968
Prudhoe, J.; *Theatre of Goethe and Schiller*; Blackwell, 1975
Music
Einstein, A.; *Mozart*; Panther, 1971
Hughes, R.; *Haydn*; Dent, 1974
Scott, M. M.; *Beethoven*; Dent, 1971
Low Countries
Clark, Sir K.; *Introduction to Rembrandt*; J. Murray, 1978
Friedlander, M. J.; *Early Netherlandish Painting*; Sitzthoff, 1974
Goldscheider, L.; *Rembrandt: Paintings, Drawings and Etchings*; Phaidon, 1960
Grossman, F.; *Bruegel*; Phaidon, 1955
Rosenberg, J. et al.; *Dutch Art and Architecture 1600–1800*; Penguin, 1972
Britain
Architecture
Maxwell, R.; *New British Architecture*; Thames & Hudson, 1972
Sitwell, O.; *British Architects and Craftsmen*; Pan, 1973
Summerson, Sir J.; *Architecture in Britain 1530–1830*; Penguin, 1969
Summerson, Sir J.; *Georgian London*; Barrie & Jenkins, 1970
Youngson, A. G.; *Making of Classical Edinburgh*; Edinburgh U.P., 1966
Art
Arnold, B.; *Concise History of Irish Art*; Thames & Hudson, 1969
Hammacher, A. M.; *Modern English Sculpture*; Thames & Hudson, 1967
Hardie, W.; *Scottish Painting 1837–1939*; Studio Vista, 1976
Irwin, D.; *English Neoclassical Art*; Faber, 1966
Jones, R. B.; *Anatomy of Wales*; Geverin, 1972
Lindsay, J.; *Hogarth: His Art and World*; MacGibbon, 1977
Lister, R.; *British Romantic Art*; G. Bell, 1973
Rothenstein, J.; *Modern English Painters*; Macdonald, 1956
Shone, R.; *Century of Change*; Phaidon, 1977
Waterhouse, E. K.; *Painting in Britain 1530–1790*; Penguin, 1969
Literature
Chambers, E. K.; *Elizabethan Stage*; Oxford U.P., 1923
Chambers, E. K.; *William Shakespeare*; Oxford U.P., 1930 (2 vols)
Doody, M. A.; *Natural Passion: Novels of Samuel Richardson*; Oxford U.P., 1974
Glen, D.; *Individual and the 20th Century Scottish Literary Tradition*; Akras, 1971
Lewis, C. S.; *English Literature in the 16th Century*; Oxford U.P., 1954
Lindsay, M.; *History of Scottish Literature*; Hale, 1977
Marchand, L. A.; *Lord Byron: Biography*; J. Murray, 1958 (3 vols)
Meisel, M.; *Shaw and the 19th Century Theatre*; Greenwood, 1977
Saul, G. B.; *Traditional Irish Literature and its Backgrounds*; Bucknell, 1975
Sutherland, J.; *Victorian Novelists and Publishers*; Athlone, 1976
Tillyard, E. M. W.; *Elizabethan World Picture*; Penguin, 1972
Willey, B.; *19th Century Studies: Coleridge to Arnold*; Penguin, 1973
Williams, I.; *Realist Novel in England*; Macmillan, 1974
Wilson, F. P.; *Elizabethan and Jacobean*; Oxford U.P., 1945

Man and Machines

The Growth of Technology
20 Kranzberg, M. & C. W. Pursell (Eds.); *Technology in Western Civilization*; Oxford U.P., 1967–68 (2 vols)
Singer, C. et al.; *History of Technology*; Oxford U.P., 1953–58 (5 vols)
Sprague de Camp, L.; *Ancient Engineers*; Tandem, 1977
White, L.; *Mediaeval Technology and Social Change*; Oxford U.P., 1963
Armytage, W. H. G.; *Social History of Engineering*; Faber, 1976
22 Crowther, J. G.; *Discoveries and Inventions of the 20th Century*; Routledge, 1966
24–28 Semenov, S. A.; *Prehistoric Technology*; Cary, Adams & McKay, 1964
Hodges, H.; *Technology in the Ancient World*; Penguin, 1970
30 Tylecote, R. F.; *History of Metallurgy*; Metals Society, 1976
Materials and Techniques
32–42 Alexander, W. O. & A. Street; *Metals in the Service of Man*; Penguin, 1969
Gordon, J. E.; *New Science of Strong Materials*; Penguin, 1968
Scientific American; *Materials*; W. H. Freeman, 1968
Martin, J. W. & R. A. Hull; *Elementary Science of Metals*; Wykeham Publications, 1969
Anderson, J. C. & K. D. Leaver; *Materials Science*; Nelson, 1974
Rollason, E. C.; *Metallurgy for Engineers*; Arnold, 1973
Higgins, R. A.; *Engineering Metallurgy*; Eng. U.P., 1970
Guy, A. G.; *Introduction to Materials Science*; McGraw-Hill, 1978
Vlack, L. van; *Elements of Material Science*; Addison-Wesley, 1975
44 Douglas, R. W. & S. Frank; *History of Glassmaking*; G. J. Foulis, 1972
Haynes, E. B.; *Glass through the Ages*; Penguin, 1969
Hamilton, D.; *Manual of Pottery and Ceramics*; Thames & Hudson, 1974
Godden, G. A.; *Illustrated Encyclopaedia of British Pottery and Porcelain*; Barrie & Jenkins, 1966
Charleston, R. J. (Ed.); *World Ceramics*; Hamlyn, 1968
Honey, W. B.; *English Pottery and Porcelain*; Black, 1969
Stained Glass; Mitchell Beazley, 1976
46 Eldridge, H. J.; *Properties of Building Materials*; M. T. P., 1974
Clifton-Taylor, A.; *The Pattern of English Building*; Faber, 1972
Taylor, G. D.; *Materials of Construction*; Longman, 1974
Komar, A.; *Building Materials and Components*; M. I. R., 1973
48–50 Rogers, P. & M. L. Baltay; *Reinforced Concrete Design for Buildings*; Van Nost. Reinhold, 1973
Abeles, P. W. et al.; *Prestressed Concrete Designer's Handbook*; C. & C. A., 1976
Crawley, S. W.; *Steel Buildings: Analysis and Design*; Wiley, 1970
Dancy, H. K.; *Manual on Building Construction*; I. T. D. G., 1973
52 Boyle, G. & P. Harper (Eds.); *Radical Technology*; Wildwood House, 1976
54 Allen, P. W.; *Natural Rubber and the Synthetics*; Lockwood, 1972
Morton, M.; *Rubber Technology*; Van Nost. Reinhold, 1973
Lubin, G.; *Handbook of Fibreglass and Advanced Plastics Composites*; Van Nost. Reinhold, 1969
56–58 Taylor, M. A.; *Technology of Textile Properties*; Forbes, 1972
Hall, A. J.; *Standard Handbook of Textiles*; Newnes-Butterworth, 1975
Moncrieff, R. W.; *Man-Made Fibres*; Newnes-Butterworth, 1975
60 Britt, K. W.; *Handbook of Pulp and Paper Technology*; Van Nost. Reinhold, 1971
Power
62–68 Rolt, L. T. C. & J. S. Allen; *Steam Engine of Thomas Newcomen*; Moorland, 1977
Rolt, L. T. C.; *George and Robert Stephenson*; Longman, 1960
Dickinson, H. W.; *Short History of the Steam Engine*; F. Cass, 1963
Derry, J. K. and J. I. Williams; *Short History of Technology*; Oxford U.P., 1970
Singer, C. (Ed.); *History of Technology Vols IV and V*; Oxford U.P., 1958
70 Reynolds, J.; *Windmills and Watermills*; Hugh Evelyn, 1970
Beedell, S.; *Windmills*; David & Charles, 1975
Golding, E. W.; *Generation of Electricity by Wind Power*; Span, 1976
Paton, T. A. L. & J. G. Brown; *Power from Water*; L. Hill, 1960
72 Brinkworth, B. J.; *Solar Energy for Man*; Compton Press, 1972
Williams J. R.; *Solar Energy*; Ann Arbor, 1974
Marion, J. B.; *Energy in Perspective*; Academic Press, 1974
Ruedisli, L. C. & M. W. Firebaugh; *Perspectives on Energy*; Oxford U.P., 1975
Wilson, R. & W. R. Jones; *Energy, Ecology and the Environment*; Academic Press, 1974
74 Patterson, W. C.; *Nuclear Power*; Penguin, 1976
Nuclear Power and the Environment; HMSO, 1973
Kenward, H.; *Potential Energy*; Cambridge U.P., 1976
Larsen, E.; *New Sources of Energy and Power*; Muller, 1976
76 *Coal: Technology for Britain's Future*; Macmillan, 1976
Berkovitch, I.; *Coal on the Switchback*; Allen & Unwin, 1977
Griffin, A. R.; *British Coalmining Industry*; Moorland, 1976
78–80 Tugendhat, C. & A. Hamilton; *Oil, the Biggest Business*; Eyre Methuen, 1975
Waddams, A. L.; *Chemicals from Petroleum*; J. Murray, 1973
82 McVeigh, J. C.; *Sun Power: Introduction to Applications of Solar Energy*; Pergamon, 1977
Boyle, G.; *Living on the Sun*; Calder, 1975
84 Starr, A. T.; *Generation, Transmission and Utilization of Electrical Power*; Pitman, 1957
Machines
How Things Work; Paladin, 1972 (2 vols)
86–90 Burstall, A. F.; *History of Mechanical Engineering*; Faber, 1965
Dunsheath, P.; *History of Electrical Engineering*; Faber, 1969
92 Britten, F. J.; *Old Clocks and Watches and their Makers*; Eyre Methuen, 1973
Britten, F. J.; *Watch and Clockmakers Handbook*; Baron, 1976
Gazeley, W. J.; *Watch and Clock Making and Repairing*; Newnes-Butterworth, 1968
Robertson, J. D.; *Evolution of Clockwork*; E. P., 1972
Haswell, J. E.; *Horology*; E. P., 1976
94 Wild, R.; *Work Organization: Study of Manual Work and Mass Production*; Wiley, 1975
Gallagher, C. C. & W. A. Knight; *Group Technology*; Newnes-Butterworth, 1973
96 Hammond, R.; *Mobile and Movable Cranes*; C. R. Books, 1963
Imperial College; *Lifts and Escalators*; I. C. S. T., 1969
Phillips, R. S.; *Electric Lifts: Design, Maintenance, etc.*; Pitman, 1973
98 Means, R. E. & J. V. Parcher; *Physical Properties of Soils*; Constable, 1964
100 Walkinshaw, L.; *Cranes of the World*; Winchester Press, 1974
Dibner, B.; *Moving the Obelisks*; M. I. T. Press, 1970
Barwell, F. T.; *Automation and Control in Transport*; Pergamon, 1973
102 Smith, R. J.; *Electronics: Circuits and Devices*; Wiley, 1974
Crowther, J. G.; *Discoveries and Inventions of the 20th Century*; Routledge, 1966
104 Macmillan, R. H.; *Automation*; Cambridge U.P., 1956
106–108 Maynard, J.; *Computer Programming*; W. H. Allen, 1972
Lodewijk, T.; *Way Things Work: Book of the Computer*; Allen & Unwin, 1975
Overman, M.; *Understanding the Computer*; Lutterworth Press, 1972
Weiss, E. A. (Ed.); *Computer Usage*; McGraw-Hill, 1970
Transport
110 Bagwell, P. S.; *Transport Revolution from 1770*; Batsford, 1974
Brown, B.; *Transport Through the Ages: Drawings*; Barker, 1971
Clarke, D. (Ed.); *Encyclopaedia of Transport*; Marshall Cavendish, 1976
Ridley, A.; *Illustrated History of Transport*; Heinemann, 1969
Gunston, B.; *Transport: Problems and Prospects*; Thames & Hudson, 1972
112–114 Angelucci, E. and A. Culari; *Ships*; Macdonald, 1977
Howarth, D.; *Sovereign of the Seas*; Collins, 1974
116 Larsen, E.; *Hovercraft and Hydrofoils*; Dent, 1970
McLeavy, R.; *Hovercraft and Hydrofoils*; Blandford, 1976
118 Watts, A. J.; *Submarines and Submersibles*; Ward Lock, 1976
Horton, E.; *Illustrated History of the Submarine*; Sidgwick & Jackson, 1974
Roscoe, K.; *Undersea Exploration*; Hamlyn, 1971
120 Bendixon, T.; *Instead of Cars*; Penguin, 1977

Cardy, A. C. C.; *History of Modern Road Transport*; Macmillan, 1970
Tarr, L.; *History of the Carriage*; Vision, 1970
Tristram, W. O.; *Coaching Days and Coaching Ways*; E. P., 1973
122 Ritchie, A.; *Kings of the Road: Illustrated History*; Wildwood, 1975
Woodforde, J.; *Story of the Bicycle*; Routledge, 1970
124 Connolly, H.; *Pioneer Motorcycles (1875–1905)*; Main-Smith, 1974
Hough, R. & L. J. K. Setright; *History of the World's Motorcycles*; Allen & Unwin, 1973
Louis, H. & B. Currie; *Classic Motorcycles*; P. Stephens, 1977
Vanderveen, B. H. (Ed.); *Motorcycles and Scooters from 1945*; Warne, 1976
126–132 Posthumus, C.; *Vintage Cars (1919–30)*; Hamlyn, 1977
Sedgwick, M.; *Cars of the 1930s*; Batsford, 1970
Nicholson, T. R.; *Passenger Cars*; Blandford Press, 1970–72 (3 vols)
Sedgwick, M.; *Passenger Cars (1924–42)*; Blandford Press, 1975
Nicholson, T. R.; *Sports Cars*; Blandford Press, 1969–70 (2 vols)
Day, J.; *Bosch Book of the Motor Car*; Collins, 1975
Young, F.; *Complete Motorist*; E. P., 1973
Plowden, W.; *Motor Car and Politics*; Penguin, 1973
Turner, G.; *Car Makers*; Eyre & Spottiswood, 1963
134 Kaye, D.; *Pocket Encyclopaedia of Buses and Trolleybuses Before 1919*; Blandford, 1972
Joyce, J.; *Trams in Colour (since 1945)*; Blandford, 1970
Kaye, D.; *Buses and Trolleybuses 1919–45*; Blandford, 1970
Booth, G.; *British Motor Bus: Illustrated History*; I. Allan, 1977
136 Newton, K. et al.; *Motor Vehicle*; Newnes-Butterworth, 1972
Champion, R. C. & E. C. Arnold; *Motor Vehicle Calculations and Science*; Arnold, 1970
138 Schumacher, E. F.; *Small is Beautiful*; Sphere, 1974
140–144 Nock, O. S.; *Locomotion*; Routledge, 1975
Reed, B.; *150 Years of Steam (British) Locomotives*; David & Charles, 1975
Ferneyhough, F.; *History of Railways in Britain*; Osprey, 1975
Simmons, F. (Ed.); *Rail 150*; Eyre Methuen, 1975
Nock, O. S.; *Encyclopaedia of Railways*; Octopus, 1978
Nock, O. S.; *Railways Then and Now*; Elek, 1975
Kichenside, G. & A. Williams; *British Railway Signalling*; I. Allan, 1975
Whitehouse, P. B.; *Britain's Main-Line Railways*; New English Library, 1977
Nock, O. S.; *Out the Line*; Elek, 1976
Nock, O. S.; *Railways of the Modern Age since 1963*; Blandford, 1975
146 Brooks, P. W.; *Historic Airships*; Evelyn, 1973
Robinson, D. H.; *Zeppelin in Combat*; G. T. Foulis, 1971
148–152 Taylor, J. W. R. (Ed.); *Guinness Book of Air Facts and Feats*; Guinness, 1973
Taylor, J. W. R. & K. Munson; *History of Aviation*; New English Library 1971–77
Mondey, D.; *Aircraft: A Colour History of Aviation*; Octopus, 1974
Taylor, J. W. R.; *Jets*; Dent, 1976
154 Arkell, B. & J. W. R. Taylor; *Helicopters and VTOL Aircraft Work Like This*; Dent, 1972
156–158 Bono, P.; *Frontiers of Space*; Blandford, 1977
Gatland, K.; *Manned Spacecraft*; Blandford, 1976
Porter, R. W.; *Versatile Satellite*; Oxford U.P., 1977
Gatland, K.; *Missiles and Rockets*; Blandford, 1975
Ryan, P.; *Invasion of the Moon (1957–70)*; Penguin, 1971
Gatland, K.; *Robot Explorers*; Blandford, 1972
Smolders, P. L.; *Soviets in Space*; Lutterworth, 1973

Weapons
160–164 Nickel, A.; *Arms and Armour Through the Ages*; Collins, 1971
Rosa, J. G.; *Colonel Colt, London*; Arms & Armour Press, 1976
Nonte, G. C.; *Firearms Encyclopaedia*; Wolfe, 1974
Wilkinson, F.; *Antique Arms and Armour*; Ward Lock, 1972
Peterson, H. L.; *Book of the Gun*; Hamlyn, 1963
Boothroyd, G.; *Guns Through the Ages*; Sterling, 1962
Wilkinson, F.; *World's Great Guns*; Hamlyn, 1977
Steindler, R.; *Shooting the Muzzle-Loaders*; J. P. O'Hara, 1976
Bailey, D. W.; *Percussion Guns and Rifles*; Arms & Armour Press, 1972
Amber, T. J. (Ed.); *Gun Digest*; Follett, 1976
Smith, W. H. A.; *Small Arms of the World*; Stackpole, 1977
Cormack, A. J. R.; *Small Arms in Profile*; Profile, 1972
166–168 Young, P.; *Machinery of War*; Hart-Davis, 1973
Chant, C.; *How Weapons Work*; Marshall Cavendish, 1976
Johnson, C.; *Modern Military Artillery*; Octopus, 1975
Foss, C. F. & T. J. Gander; *Infantry Weapons of the World*; I. Allan, 1977
Owen, J. I. H.; *Nato Infantry and its Weapons*; Brassey's, 1976
170 Ogorkiewicz, R. M.; *Design and Development of Fighting Vehicles*; Macdonald, 1968
Ogorkiewicz, R. M.; *Armoured Forces*; Arms & Armour Press, 1970
Foss, C.; *Armoured Fighting Vehicles of the World*; I. Allan, 1975
172–174 Watts, A. J.; *Aircraft Carriers and their Aircraft*; Ward Lock, 1977
Macintyre, D.; *Man of War*; Jupiter, 1975
Archibald, E. H. H.; *Wooden Fighting Ships in the R.N.*; Blandford, 1968
Archibald, E. H. H.; *Metal Fighting Ships in the R.N. 1860–1970*; Blandford, 1971
176 Bruce, J. M.; *British Aeroplanes 1914–18*; Putnam & Co., 1969
Bruce, J. M.; *British Military Aircraft of World War I*; Arms & Armour Press, 1976
Turner, J. F.; *British Aircraft of World War II*; Sidgwick & Jackson, 1975
178 Gunston, B.; *Modern Combat Aircraft*; Salamander, 1977
Gunston, B.; *Attack Aircraft of the West*; I. Allan, 1974
Gunston, B.; *Bombers of the West*; I. Allan, 1973
Taylor, J. W. R.; *Modern Combat Aircraft*; Hamlyn, 1976
180 Beaton, R. L. & J. Maddox; *Spread of Nuclear Weapons*; Greenwood, 1977

Engineering
182 O'Flaherty, C. A.; *Highway Engineering*; Arnold, 1974
Overman, M.; *Roads, Bridges and Tunnels*; Aldus, 1968

184 O'Flaherty, C. A.; *Highways and Traffic*; Arnold, 1974
Kefford, C. W.; *Traffic*; Blandford, 1969
186 Henry, B.; *Heathrow Airport, London*; Dent, 1975
188 Beaver, P.; *History of Tunnels*; P. Davies, 1972
190–192 Mare, E. de; *Bridges of Britain*; Batsford, 1975
Metcalfe, L.; *Bridges and Bridge-Building*; Blandford, 1970
Henry, D. & J. A. Jerome; *Modern British Bridges*; Applied Science, 1969
Overman, M.; *Man the Bridge-Builder*; Priory Press, 1975
194 Hartcup, G.; *Code name Mulberry: Harbour Planning, Building and Operation*; David & Charles, 1977
Albion, R. G. & J. B. Pope; *Rise of New York Port 1815–60*; David & Charles, 1970
196 Harris, R.; *Canals and their Architecture*; Evelyn, 1969
198 Smith, N.; *History of Dams*; P. Davies, 1971
200 Overman, M.; *Water*; Aldus, 1968
Tebbutt, T. H. Y.; *Water Science and Technology*; J. Murray, 1973
202 Bolton, R. L. & L. Klein; *Sewage Treatment: Basic Principles and Trends*; Newnes-Butterworth, 1971

Communications
204 Steinberg, S. H.; *500 Years of Printing*; Penguin, 1974
Hutchings, E. A. D.; *Survey of Printing Processes*; Heinemann, 1970
206 Spellman, J. A.; *Printing Works Like This*; Dent, 1964
McMurtrie, D. C.; *Book*; Oxford U.P., 1943
Moran, J. (Ed.); *Printing in the 20th Century*; Northwood, 1974
Moran, J.; *Printing Presses: History and Development from 15th century to 20th century*; Faber, 1973
Durrant, W. R. et al.; *Machine Printing*; Focal Press, 1973
Cimming, R. F. & W. E. Killick; *Single Colour Lithographic Machine Operating*; Pergamon, 1969
Biegeleisen, J. I.; *Complete Book of Silk Screen Printing Production*; Dover, 1963
New, P. G.; *Reprography for Librarians*; Bingley, 1975
208 Chambers, H. T.; *Copying, Duplicating and Microfilm*; Business Books, 1972
Leafe, M.; *Workbook on Duplicating and Photocopying*; Cassell, 1974
210 Grundy, B.; *Press Inside Out*; W. H. Allen, 1976
Looseley, A. E.; *Business of Photojournalism*; Focal Press, 1970
Cudlipp, Lord; *Walking on the Water: Autobiography*; Bodley Head, 1976
Jones, M. W.; *History of the World Newspaper*; David & Charles, 1974
212 Mumby, F. A.; *Publishing and Bookselling*; Cape, 1974
Unwin, Sir S.; *Truth About Publishing*; Allen & Unwin, 1976
Williamson, H.; *Methods of Book Design*; Oxford U.P., 1966
214 Walford, A. J.; *Guide to Reference Material*; Library Association, 1973–77
Kister, K. F.; *Encyclopaedia Buying Guide*; Bowker, 1976
Hulbert, J. R.; *Dictionaries British and American*; Deutsch, 1968
216 Lock, R. N.; *Manual of Library Economy*; Bingley, 1977
Sayers, W. C. B.; *Manual of Classification*; Deutsch, 1975 (Maltby ed.)
Needham, C. D.; *Organizing Knowledge in Libraries*; Deutsch, 1972
218 Hedgecoe, J.; *Book of Photography*; Nat. Mag. Co., 1976
Bowskill, J.; *Photography Made Simple*; W. H. Allen, 1975
220 *Techniques of Photography*; Time-Life, 1975
Feininger, A.; *Successful Photography*; Prentice-Hall, 1975
Lacour, M. & I. T. Lathrop; *Photo Technology*; American Technical Soc., 1972
222 Fellows, M. S.; *Home Movies*; Pelham, 1973
Gilmour, E.; *Photographer's Guide to Movie-Making*; Focal Press, 1969
224–226 Metcalfe, L.; *Post Office and its Services*; Blandford, 1972
Overman, M.; *Understanding Telecommunications*; Lutterworth Press, 1974
228 King, G. J.; *Beginner's Guide to Radio*; Newnes Tech., 1970
Gibson, J.; *Teach Yourself Radio*; Hodder Children's, 1970
230 King, G. J.; *Beginner's Guide to Colour Television*; Newnes Tech., 1973
Cole, H. A.; *Basic Television*; Technical Press, 1972
King, G. J.; *Beginner's Guide to Television*; Newnes Tech., 1970
232 Wasley, J. & R. Hill; *Guide to Hi-fi*; Pelham, 1977
Borwick, J. (Ed.); *Sound Recording Practice*; Oxford U.P., 1976
234 Overman, M.; *Understanding Sound and Video Recording*; Lutterworth, 1977
White, G.; *Video Recording: Record and Replay Systems*; Newnes-Butterworth, 1972
Robinson, J. F.; *Videotape Recording*; Focal Press, 1975
Oliver, W. G.; *Introduction to Video-Recording*; Foulsham, 1971
236 Sonnenberg, G. J.; *Radar and Electronics Navigation*; Newnes-Butterworth, 1970
French, J.; *Small Craft Radar*; Stanford Maritime, 1977
Tucker, D. G.; *Underwater Observation Using Sonar*; Fishing News, 1966
Larsen, E.; *Radar Works Like This*; Dent, 1966

Industrial Chemistry
Derry, T. K. & T. I. Williams; *Short History of Technology*; Oxford U.P., 1970
238 Coulson, J. M. & J. F. Richardson; *Chemical Engineering*; Pergamon, 1968–76 (3 vols)
240 Poucher, W. A.; *Perfumes, Cosmetics and Soap*; Chapman & Hall, 1975 (3 vols)
242 Watkins, T. F. et al.; *Chemical Warfare, Pyrotechnics and the Fireworks Industry*; Pergamon, 1968
244 Nuffield Chemistry; *Colour*; Longman, 1970
246 Balsam, M. S. & E. Sagarin (Eds.); *Cosmetics: Science and Technology*; Wiley, 1972–74 (3 vols)

Domestic Engineering
248–254 Althouse, A. D.; *Modern Refrigeration and Air-Conditioning*; Goodheart, 1975
Wyld, B. A. & D. A. Bell; *Calculator Revolution*; White Lion, 1977
Hall, E.; *Home Plumbing*; Newnes, 1977
Burdett, G. A. T.; *Do Your Own Home Wiring*; Foulsham, 1975
Whalley, J. I.; *Writing Implements and Accessories: Stylus to Typewriter*; David & Charles, 1975
Jewell, F. B.; *Veteran Sewing Machines*; David & Charles, 1975
How Things Work; Paladin, 1972 (2 vols)

Picture and artwork credits

The illustrations in the *Fact Index* are from the Fabbri Picture Library (Milan) and Mitchell Beazley Publishers Ltd, except for the following (illustrations are identified by page number and lettered A to D, top to bottom): Accademia Carrara, Bergamo: 202C; Accademia di Concordi, Ravigo: 466A; Alte Pinakothek, Munich: 32D, 118B, 539A, 683A; A. Meyer Collection, Paris: 197B; Angelo Hornak: 54B, 188A, 219D; Anton Dvorák Museum, Prague: 248D; Architectural Association: 48D, 206B, 249D, 294B, 409A, 452D, 546B, 553C, 601A, 682B, 697C, 710B, 726C, 735A; Archivstelle, Mainz: 582B; Art Institute, Chicago: 367B; Arts Council, London: 151A; Bargello Museum, Florence: 237D; Bayerische Stadtsbibliothek, Munich: 580A; BBC: 68D, 116D, 579C, 659D; Bedrich Smetana Museum, Prague: 242A, 725A; Biblioteca Apostolica Vaticana, Rome: 102B, 214A, 336C, 357D; Biblioteca Trivulziana, Milan: 239A; Bibliothèque de l'Arsenal, Paris: 519C; Bibliothèque de l'Opéra, Paris: 69A, 140A, 563B, 611B; Bibliothèque du Protestantisme, Paris: 131C; Bibliothèque Nationale, Paris: 69B, 97B, 244A, 299A, 317C, 610B, 658A, 681A; Birds Eye Foods Ltd: 96C; Birmingham Museum and Art Gallery: 124B; Bodleian Library, Oxford: 156D; Borghese Gallery, Rome: 194D; Boston Museum of Fine Arts: 134B; British Airways: 102C, 168B, 184C; British Esperanto Association: 269C; British Museum: 83A, 220A, 238C, 314A, 321C, 405C, 484D, 601C, 647B, 730B, 807C, 844A; Camera Press: 356B; Capodimente Museum, Naples: 86B; Casa Carducci, Bologna: 139B; Decca: 55C; Cinema Bookshop: 13D, 73B, 89C, 89D, 102D, 108D, 112A, 119D, 432A, 438C, 495B, 578B, 604D, 644B, 675D, 840B; Civic Museum, Ravenna: 464D; Civico Museo Bibliografico della Musica, Bologna: 656A; Collection Gallini, Milan: 679D; Corocan Gallery of Art, Washington D.C.: 531D; Dahlem Staatliche Museum Cemäldegalerie, Berlin: 633D; Darius Milhaud Collection, Paris: 515C; Deutsches Museum, Munich: 552B; Donald Mill/© Denys Lasdun Redhouse & Softley: 547D; Duncan Schiedt Collection, Indianapolis: 516A, 816C; Earl of Derby Prescott Collection, Lancs.: 634C; École des Beaux Arts, Paris: 91A; Embassy of Israel/© Sidney Harris: 87D; EMI: 75A; Fitzwilliam Museum, Cambridge: 348A; Frederick Gibberd & Partners/David Atkins: 519A; French Government Tourist Office: 60C; Friends of Music Society, Vienna: 353A, 823A; Galleria Accademia, Venice: 776B; Galleria d'Arte Moderna, Turin: 699B; Gallery Saband, Turin: 357A; Garrick Club, London: 846C; General Dental Council: 139D; Giraudon/Louvre: 846D; Glasgow Art Gallery: 143D; Glasgow School of Art: 477D; Glinka Museum, Moscow: 323B, 763D; Greater London Council Photographic Library: 61A; Hammond Organs (UK) Ltd: 347C; Harold Bull: 179D; Harvard University: 821A; Helwan Observatory: 346B; IBM: 665A; Institut de Physique Nucléaire, Paris: 208A; J. Sainsbury Ltd: 188B; Ken Hewis Collection: 105D, 592B, 605B, 620B, 629C; Ken Hewis Collection/© Angus McBean: 347A; Ken Huitson: 263C; Ken Moreman: 257C; Kennedy Gallery, New York: 146A; Kensington Palace, London: 825C; Keystone Press: 302D, 348A, 372D, 413B, 521B, 543A, 560D, 578D, 663A, 672C, 684A, 713B, 768B, 775C, 776C, 818C, 832A; Kunstalle, Hamburg: 448B; Kunsthaus, Zurich: 525A; Kunsthistorisches Museum, Vienna: 712A; London Express: 82C; Louvre, Paris: 150D, 173B, 194C, 216B, 338C, 664D, 815B; Luigi Corbellini Collection, Paris: 522A; Maison Victor Hugo, Paris: 139C; Mme. Scriabin Collection, Paris: 703C; Manchester City Art Gallery: 679C; Mansell Collection: 14D, 22D, 28C, 31C, 32B, 34D, 35A, 35C, 39C, 47B, 47C, 60A, 60B, 61D, 70C, 77B, 77C, 77D, 84C, 85D, 87C, 88A, 88B, 88C, 106B, 106D, 107C, 110A, 114D, 115D, 117B, 119B, 120B, 128D, 147A, 154C, 154D, 155D, 162C, 163A, 164D, 165B, 165D, 170B, 170C, 171A, 171B, 174A, 174D, 177D,

180B, 182A, 185C, 185D, 186D, 187A, 187A, 191B, 195A, 197D, 201A, 205A, 207C, 211A, 214D, 219C, 223A, 225B, 227A, 228C, 229C, 233D, 235C, 239D, 241C, 244B, 263C, 264A, 264D, 265A, 265C, 265D, 267B, 276C, 270C, 271D, 272C, 274A, 274B, 275A, 275D, 277B, 277C, 280C, 281A, 281D, 282B, 283A, 283B, 285C, 287A, 287D, 289A, 292A, 294C, 298A, 298B, 300B, 300C, 300D, 302B, 304C, 305B, 307C, 308D, 312C, 313A, 325A, 328D, 330C, 334B, 335C, 337A, 343C, 343D, 344A, 346A, 346D, 348C, 349A, 349D, 351D, 353B, 353C, 357C, 359B, 359D, 361D, 362C, 362D, 364B, 370C, 371B, 371C, 373C, 374B, 374C, 375B, 381C, 384A, 384C, 385A, 386B, 386C, 386D, 387C, 388C, 388D, 389D, 391D, 392D, 393B, 393C 395B, 396C, 398B, 401A, 401C, 402D, 403A, 403C, 404C, 404D, 405D, 409B, 411A, 412D, 513D, 416A, 417D, 419B, 420D, 429A, 431A, 434D, 437A, 437D, 446B, 447C, 447D, 450B, 452A, 454A, 454B, 456D, 460D, 463B, 468D, 472D, 473A, 473B, 473C, 473D, 476B, 480A, 480D, 482D, 483A, 488B, 490B, 491B, 491D, 492C, 494C, 495C, 496C, 497D, 498D, 499C, 500C, 500D, 501C, 502D, 503A, 503D, 509A, 509D, 512B, 513D, 514B, 517C, 518D, 520C, 522C, 526A, 528B, 528C, 530A, 534A, 534C, 539D, 540C, 544C, 544D, 545A, 545B, 548A, 550D, 551D, 552C, 554C, 554D, 556A, 558B, 558C, 559C, 561D, 565A, 569A, 571A, 572D, 574D, 576C, 579D, 580C, 583B, 583C, 589B, 589D, 590C, 593B, 593C, 594B, 595D, 597B, 607A, 610C, 612B, 613A, 614A, 617A, 618B, 618D, 621C, 624A, 625D, 628A, 629B, 630C, 633A, 636B, 637C, 638D, 639C, 639D, 640D, 642B, 643B, 643C, 645A, 645D, 647A, 647C, 648D, 649C, 651A, 655B, 655C, 658B, 658A, 659B, 661D, 662B, 663B, 664B, 665B, 665C, 666C, 667A, 667B, 668C, 669A, 669B, 670C, 674A, 674C, 674D, 675B, 676B, 676C, 677A, 678B, 681C, 681D, 682C, 684B, 684C, 690A, 692C, 693A, 693C, 694A, 700A, 702B, 706C, 707C, 709B, 709D, 710A, 710C, 712D, 712D, 714C, 715B, 720A, 720B, 720D, 721C, 722C, 724B, 725B, 727C, 729A, 730C, 731D, 732B, 734B, 735B, 739D, 740A, 741C, 747A, 749D, 751C, 752D, 758B, 762B, 763C, 765D, 766D, 767A, 768A, 769A, 770B, 770D, 772D, 774B, 774D, 776D, 780A, 781C, 783A, 784D, 786B, 786C, 786D, 787C, 790D, 791A, 793B, 793D, 794B, 797C, 799A, 799C, 801A, 802C, 802D, 803A, 809C, 810D, 811A, 811B, 819A, 819C, 821B, 825B, 826B, 827C, 828C, 829D, 830B, 830D, 831A, 832B, 835D, 837D, 838A, 839C, 840C, 841B, 846B, 847A; Marlborough Fine Arts, London: 75D; Marquis of Bath, Longleat: 777C; Mary Evans Picture Library: 229B, 267A, 358D, 381D, 481D, 595C, 689A, 754A; Maurice Denis Collection, Saint Germain-en-Laye: 312D; Mauritius, The Hague: 366A; Mendelssohn Archive, Berlin: 260A, 506A; Merion Barnes Foundation, Pa.: 710D; Metropolitan Museum of Art, New York: 191D, 254A, 592A, 806D; MGM: 57B, 247; Mr. & Mrs. Thomas Girtin Collection, London: 199B; M. J. Hendrie: 180D; Mme. de Tinian Collection, Paris: 218D; Mullards Ltd: 784A; Musée, Verailles: 197A; Musée Carnavalet, Paris: 214B; Musée Condé, Chantilly: 527B; 696D; Musée Condé, Paris: 815B; Musée de l'Homme, Paris: 529D, 617B, 785A; Musée de l'Opéra, Paris: 497B; Musée de Peinture et de Sculpture, Grenoble: 279D; Musée des Beaux Arts, Angers: 154B; Musée des Beaux Arts, Besançon: 108B; Musée des Beaux Arts, Budapest: 221A; Musée des Beaux Arts, Grenoble: 405A; Musée des Beaux Arts, Lyons: 317A, 530D; Musée des Beaux Arts, Rouen: 504D, 722A; Musée Fabre, Montpellier: 197C; Musée Pasteur, Paris: 599A; Musée Royale, Anvers: 827A; Musée Royale des Beaux Arts: 766D; Museo Archeologio Nazionale, Naples: 28C; Museo dell'Opera del Duome, Sienna: 244D; Museo del Risorgimento, Udine: 140C; Museo di Storia della Scienza, Florence: 308B; Museo el Greco, Toledo: 335A; Museo Nazionale della Scienza e della Tecnica, Milan: 83B, 252D, 415A, 438D, 655D; Museo Nazionale di Capodimente, Naples:

596D; Museo Nazionale di Palazzo Venezia, Rome: 502B; Museo Teatrale alla Scala, Milan: 151D, 319D, 470D; Museum, Hofuryck: 376D; Museum der Stadt Halberstadt, Gleimhaus: 425C; Museum of Catalan Art, Barcelona: 90A; National Gallery: 152A, 236A, 315B, 316A, 364D, 365D, 383D, 423C, 444B, 670B, 695D, 767C; National Gallery of Art, Washington D.C.: 748C; National Gallery of Scotland, Edinburgh: 196C, 656C; Natural History Museum: 794C; National Maritime Museum: 99B; National Portrait Gallery: 233C, 333A, 357B, 405B, 495A, 516D, 606A, 743D; National Society for Mentally Handicapped Children: 164C; Nelson A. Rockefeller Collection: 306B; New York State Historical Association, Cooperstown: 177C; Österreichische Galerie, Vienna: 425A; Paul Brierley: 220D, 249C, 390D, 481A, 513B, 653A; Pinacoteca di Brera, Milan: 339D; Pinacoteca Moderna, Vatican City: 591B; Pinacoteca Nazionale, Bologna: 810B; Pinacoteca Tosio e Martinengo, Brescia: 468A; Pitti, Florence: 315D; Popperfoto: 13C; 14B; 19A; 19C; 22B; 23A, 23B; 23D; 25B; 27D; 28A; 29A, 30B, 30D, 31A, 32A, 35D, 38C, 41D, 45B, 46B, 49B, 51B, 52D, 54D, 55A, 55B, 57D, 59A, 59B, 61C, 62B, 63A, 63C, 69D, 70B, 73C, 82D, 85A, 87A, 87B, 92C, 92D, 93A, 94B, 95B, 95C, 96A, 98C, 108C, 110B, 110D, 112B, 112D, 113B, 114A, 114C, 115C, 116A, 117C, 118C, 118D, 119A, 120C, 130A, 130B, 130C, 130D, 136C, 136D, 139A, 141C, 143A, 143B, 143C, 147C, 153C, 153D, 155A, 155B, 155C, 160B, 163B, 163D, 166D, 168A, 168D, 169A, 169D, 171C, 174B, 177B, 178C, 179A, 181A, 181C, 181D, 182B, 183C, 185B, 186C, 188C, 188D, 190A, 194B, 198A, 198B, 198C, 198D, 201B, 201C, 204C, 205C, 206D, 207A, 207B, 208C, 208D, 211B, 213A, 213B, 214C, 215C, 217A, 217B, 218A, 219B, 220C, 221C, 221D, 223A, 226B, 226C, 227B, 228A, 229D, 232A, 240B, 242B, 242C, 244C, 246B, 246D, 247A, 247C, 248A, 249A, 249B, 251A, 252A, 253A, 253B, 254D, 263B, 263D, 266C, 266D, 267D, 271B, 271C, 273B, 273C, 273D, 274D, 275B, 276C, 276D, 277D, 279C, 280A, 280B, 283C, 283D, 284B, 284C, 284D, 285D, 289D, 291C, 292D, 294A, 294D, 295A, 295C, 295D, 296A, 296D, 301A, 301B, 302C, 303A, 307B, 307D, 310A, 314B, 314B, 314D, 317B, 318C, 319B, 320B, 324A, 325B, 328A, 333C, 333D, 334C, 336A, 336B, 337C, 339C, 340A, 344B, 344D, 345A, 345B, 348D, 351A, 351B, 352C, 353D, 354B, 354D, 358C, 360B, 361A, 363A, 363B, 363D, 364A, 365A, 366B, 368C, 369C, 372A, 373A, 373D, 375C, 375D, 376A, 376B, 377D, 378C, 379A, 379C, 380A, 384D, 385B, 385D, 386A, 387B, 387D, 389B, 390A, 390C, 390D, 391A, 391B, 392A, 395D, 397B, 398D, 400A, 400D, 402B, 402C, 403B, 403D, 404A, 406B, 407A, 407B, 409C, 410A, 411C, 412A, 412B, 412C, 413A, 413C, 415C, 415D, 416D, 417A, 417B, 417C, 418C, 418D, 419A, 419C, 419D, 420B, 421C, 421D, 422C, 423B, 423D, 424B, 424C, 428C, 429C, 429D, 431C, 431D, 432C, 435D, 436D, 440C, 441B, 442B, 443A, 443C, 443D, 444C, 444D, 446D, 447B, 448A, 448C, 453B, 453C, 453D, 454D, 457B, 460B, 462C, 464C, 465B, 465C, 465D, 467A, 467B, 470A, 470B, 470C, 471A, 471D, 474B, 475A, 475D, 476A, 476D, 477B, 478A, 478B, 479C, 482C, 483C, 483B, 483D, 484B, 487A, 487C, 487D, 488A, 489C, 490A, 490C, 490D, 496D, 497A, 499A, 499B, 499D, 500B, 502A, 504A, 504B, 504C, 505A, 505D, 506B, 507D, 509C, 510B, 511B, 512D, 514A, 518C, 519D, 520A, 520D, 523B, 524A, 524B, 526C, 526D, 528A, 529A, 532D, 534B, 536C, 538B, 542D, 543C, 544B, 546C, 546D, 547A, 547B, 547C, 548C, 548D, 549A, 549B, 549C, 549D, 550C, 553A, 554A, 554C, 555A, 555B, 555C, 556C, 557C, 558A, 560A, 560B, 560C, 561B, 561C, 562A, 562B, 562C, 564A, 564B, 564C, 564D, 565B, 565D, 566A, 566B, 566D, 567D, 568A, 568D, 569B, 569C, 569D, 570C, 572A, 572B, 574B, 574C, 575B, 575D, 576A, 576D, 579A, 580C, 581A, 581B, 582A, 583D, 584B, 585D, 588A, 588D, 589C, 590D, 591A, 592D, 595B, 596A, 596B,

596C, 598D, 599D, 600A, 600D, 601D, 603B, 604A, 604C, 605A, 607B, 607D, 608A, 609A, 609D, 610A, 612A, 612B, 613B, 613C, 614B, 614C, 616A, 616C, 616D, 617D, 619C, 621D, 623B, 623C, 623D, 626B, 626C, 627A, 627B, 628B, 628C, 628D, 629A, 629D, 631A, 631C, 632A, 632A, 632D, 633C, 634A, 634B, 634D, 635A, 635C, 637B, 637D, 638A, 639A, 640B, 641B, 641D, 645A, 649D, 650A, 650B, 650B, 650C, 650D, 651C, 651D, 652A, 652C, 653B, 653D, 656D, 659C, 660A, 661A, 661C, 663C, 664C, 666A, 669C, 670A, 670D, 671B, 672A, 674B, 676A, 678D, 679A, 681B, 682A, 683D, 685B, 685D, 686C, 686D, 687A, 687D, 688B, 688C, 688D, 689C, 690C, 691D, 692A, 692D, 693D, 694C, 695C, 698D, 700B, 700C, 702A, 704A, 705B, 706B, 706D, 707A, 707B, 708A, 708B, 708C, 709C, 711B, 711C, 712B, 713A, 714B, 714D, 716D, 717A, 717D, 719B, 720C, 721A, 721B, 724A, 725D, 726D, 727A, 727B, 727D, 728A, 729B, 729C, 729D, 730A, 730D, 731B, 731C, 733A, 733B, 733C, 733D, 734D, 735C, 736D, 737B, 738C, 739C, 741A, 741D, 742A, 742D, 743A, 744A, 744D, 745A, 745D, 746D, 747C, 750A, 751A, 751D, 752A, 753A, 753B, 754C, 756C, 757D, 759D, 764D, 764D, 765B, 766B, 768D, 769B, 769C, 769D, 770C, 772A, 773A, 773D, 777B, 778A, 778C, 779C, 780B, 781D, 782A, 783B, 783C, 787B, 788D, 789A, 791B, 791C, 794A, 795B, 795C, 795D, 796A, 796C, 796D, 797A, 797B, 797D, 798B, 798C, 798D, 799B, 800A, 800B, 800C, 802A, 803D, 804A, 804B, 804C, 804D, 805B, 806A, 807D, 808D, 809B, 809D, 810A, 811C, 814C, 814D, 815A, 815D, 816A, 816D, 818D, 820C, 822B, 823B, 824A, 825A, 825D, 826D, 828B, 829A, 829B, 830A, 831B, 831C, 831D, 832D, 833C, 834C, 835A, 835B, 836B, 837B, 837C, 838B, 838C, 838D, 840D, 842A, 842B, 842C, 843C, 844C, 844D, 845A, 845B, 845D; Prado, Madrid: 805C, 847C; Press Association: 307A; Rijksmuseum, Amsterdam: 63D, 663D; Rockefeller University, New York: 680D, 763A; Roger-Violett, Paris: 379D, 696B; Royal Academy, London: 654B; Royal Geographical Society: 138B; Royal Greenwich Observatory: 194A; Royal Library, Copenhagen: 231B, 252C; Royal Library, Stockholm: 747D; Royal Netherlands Embassy: 45A; Royal Shakespeare Picture Library, Stratford-on-Avon: 508C; Sanyo Marubeni (UK) Ltd: 653C; Schüler Verlagsgesellschaft, Stuttgart: 507B; Schumann Museum, Zqickau: 701D; Science Museum: 360A, 463D, 602D; Spectrum Colour Library: 14C, 15B, 15C, 16B, 18C, 18D, 19D, 21C, 21D, 25C, 27A, 27C, 29C, 29D, 30A, 30C, 31D, 32C, 34B, 34C, 37A, 37B, 38D, 39D, 42A, 49D, 50C, 52C, 55D, 58A, 58D, 60D, 63B, 65D, 68C, 70D, 85B, 92A, 96D, 98B, 117D, 121A, 157A, 272D, 273A, 481B, 483C, 503C, 605C, 609C, 651B, 689B, 707D, 771C, 771D; Staatliche Kunstsammlungen, Dresden: 362B, 428B; Staatliche Kunstsammlungen, Kassel: 460C; Staatliche Museum preussischer Kulturbesitz, Berlin: 835C; Stadtisches Reiss Museum, Mannheim: 186A; Statens Museum for Kunst, Copenhagen: 700D; Theatre Museum, Monaco: 493B; Theatre Museum, Munich: 181B; The Tate Gallery: 15A, 206A, 281B, 410B, 449A, 519B, 538A, 553B, 558D, 620C, 677D, 729D; The Times: 556D; Turkish Embassy: 58B; Uffizi, Florence: 247D, 657C, 762D, 807B; Victoria and Albert Museum: 187C, 262C, 451B, 809A; Wadsworth Atheneum, Hartford, Mass.: 137C; Walker Art Gallery, Liverpool: 311C, 714A, 714B; Wallace Collection: 316B; Wallraf-Richartz Museum, Cologne: 209C, 427C; Warner Bros: 216C; Washington D.C. National College, Smithsonian Institution: 685C; Western American Picture Library: 52A, 361C, 511C, 522D, 641A, 689D; Whitney Museum of American Art: 637A; World Wildlife Fund: 187B; Yale University Art Gallery: 789B. The Publishers have made every attempt to contact and seek publication permission from the copyright holders of the illustrations, and apologize for any omissions.